Useful Data

M_e	Mass of the earth	5.98×10^{24} kg
R_e	Radius of the earth	6.37×10^6 m
g	Free-fall acceleration on earth	9.80 m/s^2
G	Gravitational constant	6.67×10^{-11} N m^2/kg^2
k_B	Boltzmann's constant	1.38×10^{-23} J/K
R	Gas constant	8.31 J/mol K
N_A	Avogadro's number	6.02×10^{23} particles/mol
T_0	Absolute zero	$-273°C$
σ	Stefan-Boltzmann constant	5.67×10^{-8} W/m^2 K^4
p_{atm}	Standard atmosphere	$101,300$ Pa
v_{sound}	Speed of sound in air at 20°C	343 m/s
m_p	Mass of the proton (and the neutron)	1.67×10^{-27} kg
m_e	Mass of the electron	9.11×10^{-31} kg
K	Coulomb's law constant ($1/4\pi\epsilon_0$)	8.99×10^9 N m^2/C^2
ϵ_0	Permittivity constant	8.85×10^{-12} C^2/N m^2
μ_0	Permeability constant	1.26×10^{-6} T m/A
e	Fundamental unit of charge	1.60×10^{-19} C
c	Speed of light in vacuum	3.00×10^8 m/s
h	Planck's constant	6.63×10^{-34} J s $\quad$ 4.14×10^{-15} eV s
$\hbar$	Planck's constant	1.05×10^{-34} J s $\quad$ 6.58×10^{-16} eV s
a_B	Bohr radius	5.29×10^{-11} m

Common Prefixes

Prefix	Meaning
femto-	10^{-15}
pico-	10^{-12}
nano-	10^{-9}
micro-	10^{-6}
milli-	10^{-3}
centi-	10^{-2}
kilo-	10^3
mega-	10^6
giga-	10^9
terra-	10^{12}

Conversion Factors

Length
1 in = 2.54 cm
1 mi = 1.609 km
1 m = 39.37 in
1 km = 0.621 mi

Velocity
1 mph = 0.447 m/s
1 m/s = 2.24 mph = 3.28 ft/s

Mass and energy
1 u = 1.661×10^{-27} kg
1 cal = 4.19 J
1 eV = 1.60×10^{-19} J

Time
1 day = 86,400 s
1 year = 3.16×10^7 s

Pressure
1 atm = 101.3 kPa = 760 mm of Hg
1 atm = 14.7 lb/in^2

Rotation
1 rad = $180°/\pi$ = 57.3°
1 rev = 360° = 2π rad
1 rev/s = 60 rpm

Mathematical Approximations

Binominal Approximation: $(1 + x)^n \approx 1 + nx$ if $x \ll 1$

Small-Angle Approximation: $\sin\theta \approx \tan\theta \approx \theta$ and $\cos\theta \approx 1$ if $\theta \ll 1$ radian

Greek Letters Used in Physics

Alpha		α	Mu		μ
Beta		β	Pi		π
Gamma	Γ	γ	Rho		ρ
Delta	Δ	δ	Sigma	Σ	σ
Epsilon		ϵ	Tau		τ
Eta		η	Phi	Φ	ϕ
Theta	Θ	θ	Psi		ψ
Lambda		λ	Omega	Ω	ω

Table of Problem-Solving Strategies

Note for users of the five-volume edition:
Volume 1 (pp. 1–443) includes chapters 1–15.
Volume 2 (pp. 444–559) includes chapters 16–19.
Volume 3 (pp. 560–719) includes chapters 20–24.
Volume 4 (pp. 720–1101) includes chapters 25–36.
Volume 5 (pp. 1102–1279) includes chapters 36–42.

Chapters 37–42 are not in the Standard Edition.

Brief Contents

THIRD EDITION

physics

FOR SCIENTISTS AND ENGINEERS

a strategic approach

WITH MODERN PHYSICS

randall d. knight

California Polytechnic State University
San Luis Obispo

PEARSON

Boston   Columbus   Indianapolis   New York   San Francisco   Upper Saddle River
Amsterdam   Cape Town   Dubai   London   Madrid   Milan   Munich   Paris   Montreal   Toronto
Delhi   Mexico City   Sao Paulo   Sydney   Hong Kong   Seoul   Singapore   Taipei   Tokyo

Publisher:	James Smith
Senior Development Editor:	Alice Houston, Ph.D.
Senior Project Editor:	Martha Steele
Assistant Editor:	Peter Alston
Media Producer:	Kelly Reed
Senior Administrative Assistant:	Cathy Glenn
Director of Marketing:	Christy Lesko
Executive Marketing Manager:	Kerry McGinnis
Managing Editor:	Corinne Benson
Production Project Manager:	Beth Collins
Production Management, Composition, and Interior Design:	Cenveo Publisher Services/Nesbitt Graphics, Inc.
Illustrations:	Rolin Graphics
Cover Design:	Jodi Notowitz
Manufacturing Buyer:	Jeff Sargent
Photo Research:	Eric Schrader
Image Lead:	Maya Melenchuk
Cover Printer:	Lehigh-Phoenix
Text Printer and Binder:	R.R. Donnelley/Willard
Photo Credits:	See page C-1

ISBN-13: 978-0-321-82408-0 ISBN 10: 0-321-82408-3

1 2 3 4 5 6 7 8 9 10—DOW—15 14 13 12 11

PEARSON

About the Author

Randy Knight has taught introductory physics for over 30 years at Ohio State University and California Polytechnic University, where he is currently Professor of Physics. Professor Knight received a bachelor's degree in physics from Washington University in St. Louis and a Ph.D. in physics from the University of California, Berkeley. He was a post-doctoral fellow at the Harvard-Smithsonian Center for Astrophysics before joining the faculty at Ohio State University. It was at Ohio State that he began to learn about the research in physics education that, many years later, led to this book.

Professor Knight's research interests are in the field of lasers and spectroscopy, and he has published over 25 research papers. He also directs the environmental studies program at Cal Poly, where, in addition to introductory physics, he teaches classes on energy, oceanography, and environmental issues. When he's not in the classroom or in front of a computer, you can find Randy hiking, sea kayaking, playing the piano, or spending time with his wife Sally and their seven cats.

Builds problem-solving skills and confidence...

... through a carefully structured and research-proven program of problem-solving techniques and practice materials.

At the heart of the problem-solving instruction is the consistent 4-step **MODEL/ VISUALIZE/ SOLVE/ ASSESS** approach, used throughout the book and all supplements. ***Problem-Solving Strategies*** provide detailed guidance for particular topics and categories of problems, often drawing on key skills outlined in the step-by-step procedures of ***Tactics Boxes***. Problem-Solving Strategies and Tactics Boxes are also illustrated in dedicated MasteringPhysics ***Skill-Builder Tutorials***.

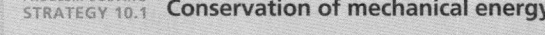
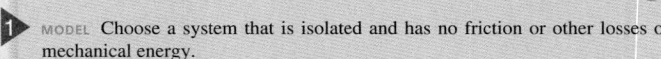

PROBLEM-SOLVING STRATEGY 10.1 **Conservation of mechanical energy** (MP)

1 **MODEL** Choose a system that is isolated and has no friction or other losses of mechanical energy.

2 **VISUALIZE** Draw a before-and-after pictorial representation. Define symbols, list known values, and identify what you're trying to find.

3 **SOLVE** The mathematical representation is based on the law of conservation of mechanical energy:

$$K_f + U_f = K_i + U_i$$

4 **ASSESS** Check that your result has the correct units, is reasonable, and answers the question.

Exercise 8

TACTICS BOX 9.1 **Drawing a before-and-after pictorial representation**

EXAMPLE 4.15 **Analyzing rotational data**

You've been assigned the task of measuring the start-up characteristics of a large industrial motor. After several seconds, when the motor has reached full speed, you know that the angular acceleration will be zero, but you hypothesize that the angular acceleration may be constant during the first couple of seconds as the motor speed increases. To find out, you attach a shaft encoder to the 3.0-cm-diameter axle. A shaft encoder is a device that converts the angular position of a shaft or axle to a signal that can be read by a computer. After setting the computer program to read four values a second, you start the motor and acquire the following data:

Time (s)	Angle(°)
0.00	0
0.25	16
0.50	69
0.75	161
1.00	267
1.25	428
1.50	620

a. Do the data support your hypothesis of a constant angular acceleration? If so, what is the angular acceleration? If not, is the angular acceleration increasing or decreasing with time?
b. A 76-cm-diameter blade is attached to the motor shaft. At what time does the acceleration of the tip of the blade reach 10 m/s²?

1 **MODEL** The axle is rotating with nonuniform circular motion. Model the tip of the blade as a particle.

2 **VISUALIZE** FIGURE 4.38 shows that the blade tip has both a tangential and a radial acceleration.

$\alpha = 2m$. If the graph is not a straight line, our observation of whether it curves upward or downward will tell us whether the angular acceleration us increasing or decreasing.

FIGURE 4.39 is the graph of θ versus t^2, and it confirms our hypothesis that the motor starts up with constant angular acceleration. The best-fit line, found using a spreadsheet, gives a slope of 274.6°/s². The units come not from the spreadsheet but by looking at the units of rise (°) over run (s² because we're graphing t^2 on the x-axis). Thus the angular acceleration is

$$\alpha = 2m = 549.2°/s^2 \times \frac{\pi \text{ rad}}{180°} = 9.6 \text{ rad/s}^2$$

where we used 180° = π rad to convert to SI units of rad/s².

FIGURE 4.39 Graph of θ versus t^2 for the motor shaft.

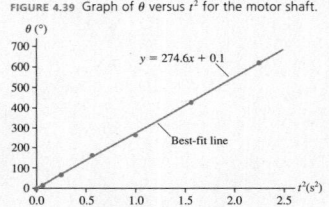

b. The magnitude of the linear acceleration is

$$a = \sqrt{a_t^2 + a_r^2}$$

Worked Examples walk the student carefully through detailed solutions, focusing on underlying reasoning and common pitfalls to avoid.

NEW! Data-based Examples (shown here) help students with the skill of drawing conclusions from laboratory data.

CHALLENGE EXAMPLE 10.10 **A rebounding pendulum**

A 200 g steel ball hangs on a 1.0-m-long string. The ball is pulled sideways so that the string is at a 45° angle, then released. At the very bottom of its swing the ball strikes a 500 g steel paperweight that is resting on a frictionless table. To what angle does the ball rebound?

NEW! *Challenge Examples* illustrate how to integrate multiple concepts and use more sophisticated reasoning.

 NEW! The Mastering Study Area also has **Video Tutor Solutions**, created by Randy Knight's College Physics co-author Brian Jones. These engaging and helpful videos walk students through a representative problem for each main topic, often starting with a qualitative overview in the context of a lab- or real-world demo.

Promotes deeper understanding...

... using powerful techniques from multimedia learning theory that focus and structure student learning, and improve engagement and retention.

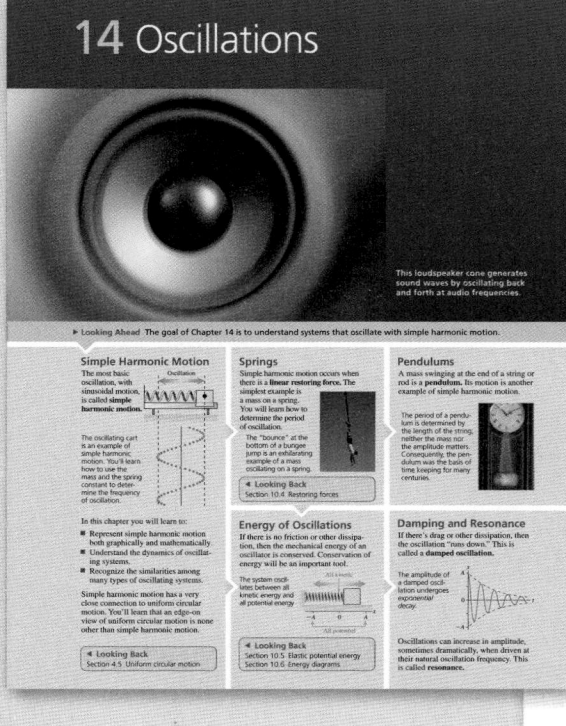

NEW! *Illustrated Chapter Previews* give an overview of the upcoming ideas for each chapter, setting them in context, explaining their utility, and tying them to existing knowledge (through *Looking Back* references).

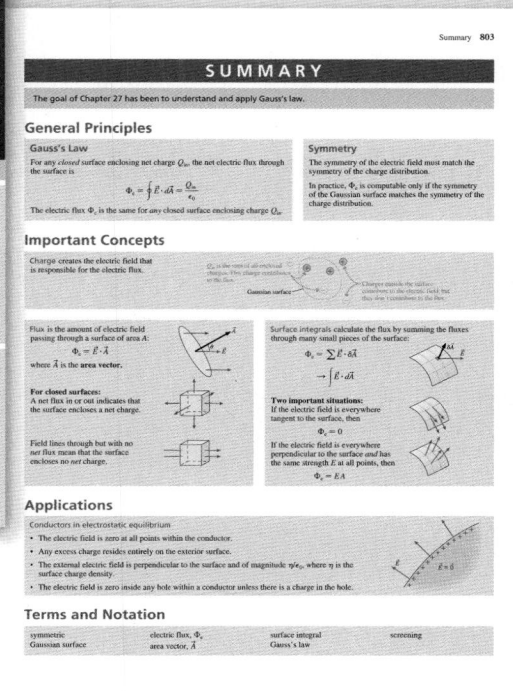

Critically acclaimed *Visual Chapter Summaries* and *Part Knowledge Structures* consolidate understanding by providing key concepts and principles in words, math, and figures and organizing these into a hierarchy.

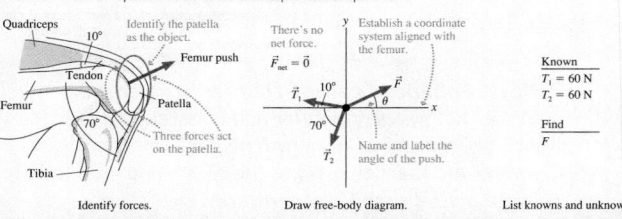

EXAMPLE 6.1 Finding the force on the kneecap

Your kneecap (patella) is attached by a tendon to your quadriceps muscle. This tendon pulls at a 10° angle relative to the femur, the bone of your upper leg. The patella is also attached to your lower leg (tibia) by a tendon that pulls parallel to the leg. To balance these forces, the lower end of your femur pushes outward on the patella. Bending your knee increases the tension in the tendons, and both have a tension of 60 N when the knee is bent to make a 70° angle between the upper and lower leg. What force does the femur exert on the kneecap in this position?

MODEL Model the kneecap as a particle in static equilibrium.

NEW! *Life-science and bioengineering examples* provide general interest, and specific context for biosciences students.

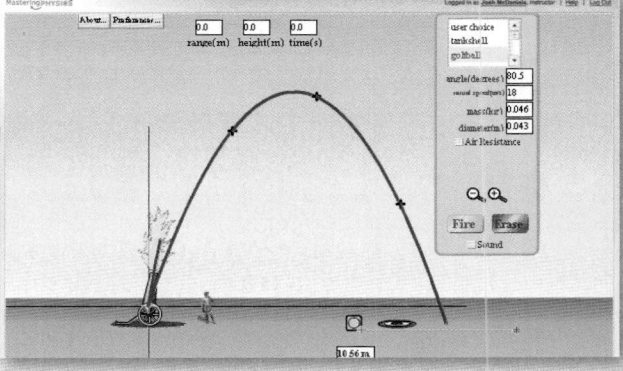

NEW! *PhET Simulations and Tutorials* allow students to explore real-life phenomena and discover the underlying physics. Sixteen tutorials are provided in the MasteringPhysics item library, and 76 PhET simulations are available in the Study Area and Pearson eText, along with the comprehensive library of ActivPhysics applets and applet-based tutorials.

NEW! *Video Tutor Demonstrations* feature "pause-and-predict" demonstrations of key physics concepts and incorporate assessment as the student progresses to actively engage them in understanding the key conceptual ideas underlying deeper physics principles.

Provides research-enhanced problems...

... extensively class-tested and calibrated using MasteringPhysics data.

Data captured by MasteringPhysics® has been thoroughly analyzed by the author to ensure an optimal range of difficulty (indicated in the textbook using a three-bar rating), problem types, and topic coverage are being met.

An *increased emphasis on symbolic answers* encourages students to work algebraically.

56. ‖ A uniform rod of mass M and length L swings as a pendulum on a pivot at distance $L/4$ from one end of the rod. Find an expression for the frequency f of small-angle oscillations.

57. ‖ A solid sphere of mass M and radius R is suspended from a thin rod, as shown in FIGURE P14.57. The sphere can swing back and forth at the bottom of the rod. Find an expression for the frequency f of small-angle oscillations.

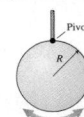

Pivot

R

FIGURE P14.57

58. ‖ A geologist needs to determine the local value of g. Unfortunately, his only tools are a meter stick, a saw, and a stopwatch. He starts by hanging the meter stick from one end and measuring its frequency as it swings. He then saws off 20 cm—using the centimeter markings—and measures the frequency again. After two more cuts, these are his data:

Length (cm)	Frequency (Hz)
100	0.61
80	0.67
60	0.79
40	0.96

Use the best-fit line of an appropriate graph to determine the local value of g.

NEW! *Data-based end-of-chapter problems* allow students to practice drawing conclusions from data (as demonstrated in the new data-based examples in the text).

59. ‖ BIO Interestingly, there have been several studies using cadavers to determine the moments of inertia of human body parts, information that is important in biomechanics. In one study, the center of mass of a 5.0 kg lower leg was found to be 18 cm from the knee. When the leg was allowed to pivot at the knee and swing freely as a pendulum, the oscillation frequency was 1.6 Hz. What

NEW! *BIO problems* are set in life-science, bioengineering, or biomedical contexts.

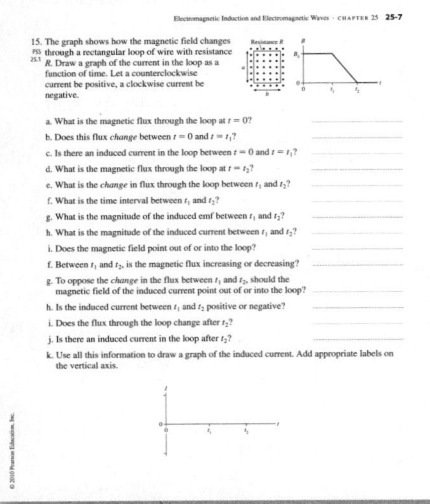

NEW! *Student Workbook exercises* help students work through a full solution symbolically, structured around the relevant textbook Problem-Solving Strategy.

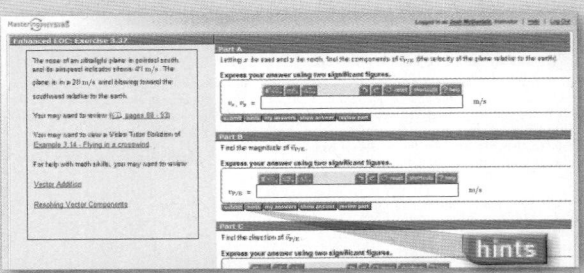

NEW! *Enhanced end-of-chapter problems* in MasteringPhysics now offer additional support such as problem-solving strategy hints, relevant math review and practice, links to the eText, and links to the related *Video Tutor Solution.*

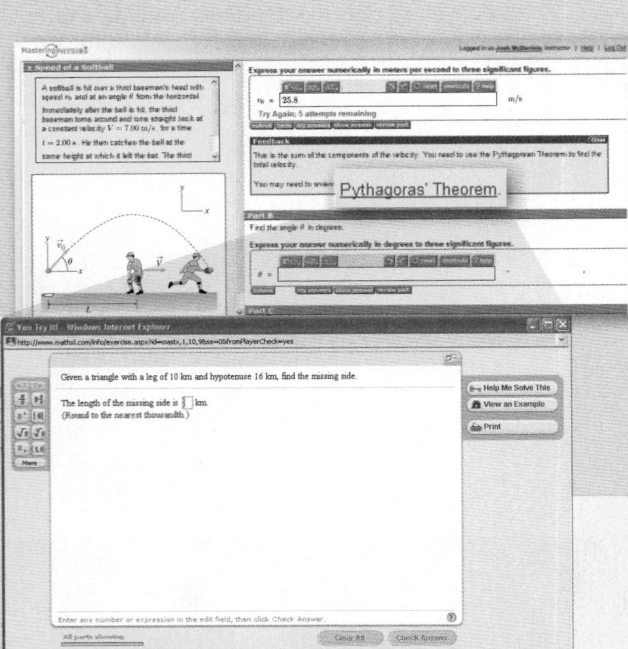

NEW! *Math Remediation* found within selected tutorials provide just-in-time math help and allow students to brush up on the most important mathematical concepts needed to successfully complete assignments. This new feature links students directly to math review and practice helping students make the connection between math and physics.

Make a difference with MasteringPhysics...

... the most effective and widely used online science tutorial, homework, and assessment system available.

MasteringPHYSICS www.masteringphysics.com

Pre-Built Assignments. For every chapter in the book, MasteringPhysics provides pre-built assignments that cover the material with a tested mix of tutorials and end-of-chapter problems of graded difficulty. Professors may use these assignments as-is or take them as a starting point for modification.

NEW! **Quizzing and Testing Enhancements.**
These include options to:
- Hide item titles.
- Add password protection.
- Limit access to completed assignments.
- Randomize question order in an assignment.

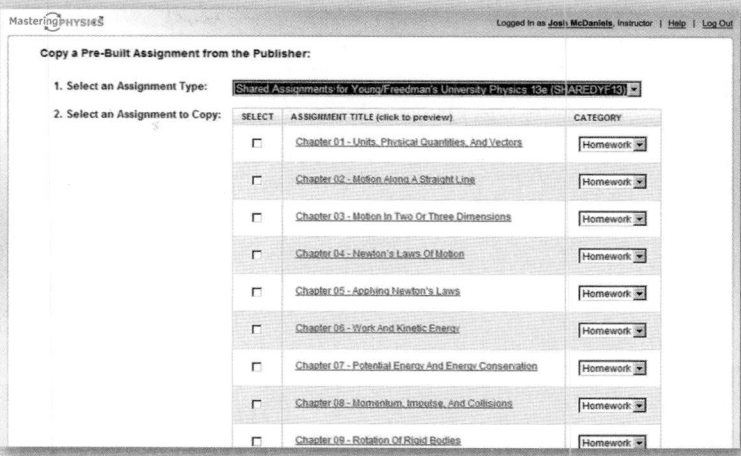

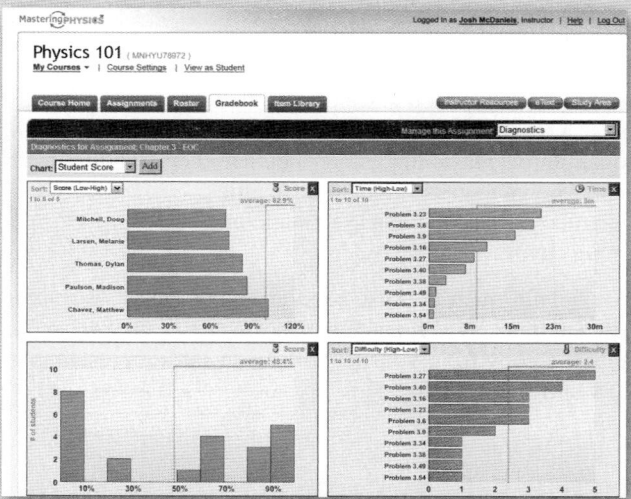

Class Performance on Assignment. Click on a problem to see which step your students struggled with most, and even their most common wrong answers. Compare results at every stage with the national average or with your previous class.

Gradebook

- Every assignment is graded automatically.
- Shades of red highlight vulnerable students and challenging assignments.
- The **Gradebook Diagnostics** screen provides your favorite weekly diagnostics, summarizing grade distribution, improvement in scores over the course, and much more.

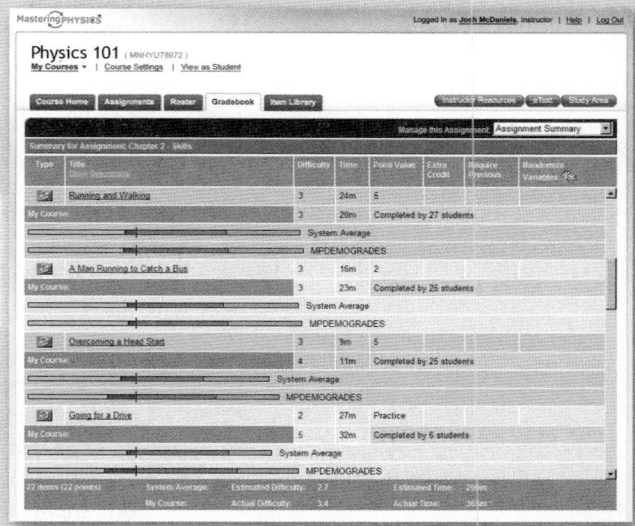

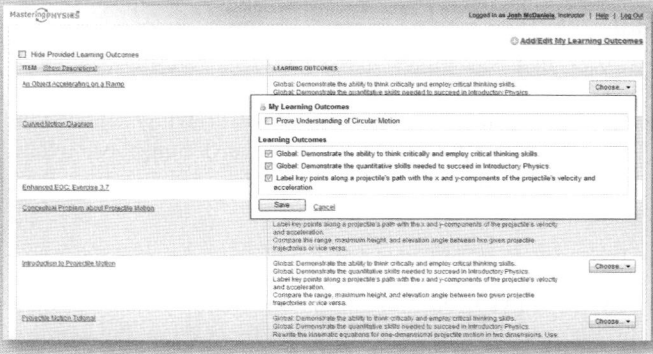

NEW! **Learning Outcomes.** In addition to being able to create your own learning outcomes to associate with questions in an assignment, you can now select content that is tagged to a large number of publisher-provided learning outcomes. You can also print or export student results based on learning outcomes for your own use or to incorporate into reports for your administration.

Preface to the Instructor

In 2003 we published *Physics for Scientists and Engineers: A Strategic Approach.* This was the first comprehensive introductory textbook built from the ground up on research into how students can more effectively learn physics. The development and testing that led to this book had been partially funded by the National Science Foundation. This first edition quickly became the most widely adopted new physics textbook in more than 30 years, meeting widespread critical acclaim from professors and students. For the second edition, and now the third, we have built on the research-proven instructional techniques introduced in the first edition and the extensive feedback from thousands of users to take student learning even further.

Objectives

My primary goals in writing *Physics for Scientists and Engineers: A Strategic Approach* have been:

- To produce a textbook that is more focused and coherent, less encyclopedic.
- To move key results from physics education research into the classroom in a way that allows instructors to use a range of teaching styles.
- To provide a balance of quantitative reasoning and conceptual understanding, with special attention to concepts known to cause student difficulties.
- To develop students' problem-solving skills in a systematic manner.
- To support an active-learning environment.

These goals and the rationale behind them are discussed at length in the *Instructor Guide* and in my small paperback book, *Five Easy Lessons: Strategies for Successful Physics Teaching.* Please request a copy from your local Pearson sales representative if it is of interest to you (ISBN 978-0-8053-8702-5).

FIVE EASY LESSONS

Strategies for Successful Physics Teaching

RANDALL D. KNIGHT

What's New to This Edition

For this third edition, we continue to apply the best results from educational research, and to refine and tailor them for this course and its students. At the same time, the extensive feedback we've received has led to many changes and improvements to the text, the figures, and the end-of-chapter problems. These include:

- New illustrated **Chapter Previews** give a visual overview of the upcoming ideas, set them in context, explain their utility, and tie them to existing knowledge (through **Looking Back** references). These previews build on the cognitive psychology concept of an "advance organizer."
- New **Challenge Examples** illustrate how to integrate multiple concepts and use more sophisticated reasoning in problem-solving, ensuring an optimal range of worked examples for students to study in preparation for homework problems.
- New **Data-based Examples** help students with the skill of drawing conclusions from laboratory data. Designed to supplement lab-based instruction, these examples also help students in general with mathematical reasoning, graphical interpretation, and assessment of results.

End-of-chapter problem enhancements include the following:

- **Data from Mastering Physics® have been thoroughly analyzed** to ensure an optimal range of difficulty, problem types, and topic coverage. In addition, the wording of every problem has been reviewed for clarity. Roughly 20% of the end-of-chapter problems are new or significantly revised.

- **Data-based problems** allow students to practice drawing conclusions from data (as demonstrated in the new data-based examples in the text).
- **An increased emphasis on symbolic answers** encourages students to work algebraically. The *Student Workbook* also contains new exercises to help students work through symbolic solutions.
- **Bio problems** are set in life-science, bioengineering, or biomedical contexts.

Targeted content changes have been carefully implemented throughout the book. These include:

- **Life-science and bioengineering worked examples and applications** focus on the physics of life-science situations in order to serve the needs of life-science students taking a calculus-based physics class.
- **Descriptive text throughout has been streamlined** to focus the presentation and generate a shorter text.
- The chapter on ***Modern Optics and Matter Waves*** has been re-worked into Chapters 38 and 39 to streamline the coverage of this material.

At the front of the book, you'll find an illustrated walkthrough of the new pedagogical features in this third edition. The *Preface to the Student* demonstrates how all the book's features are designed to help your students.

Textbook Organization

The 42-chapter extended edition (ISBN 978-0-321-73608-6/0-321-73608-7) of *Physics for Scientists and Engineers* is intended for a three-semester course. Most of the 36-chapter standard edition (ISBN 978-0-321-75294-9/0-321-75294-5), ending with relativity, can be covered in two semesters, although the judicious omission of a few chapters will avoid rushing through the material and give students more time to develop their knowledge and skills.

There's a growing sentiment that quantum physics is quickly becoming the province of engineers, not just scientists, and that even a two-semester course should include a reasonable introduction to quantum ideas. The *Instructor Guide* outlines a couple of routes through the book that allow most of the quantum physics chapters to be included in a two-semester course. I've written the book with the hope that an increasing number of instructors will choose one of these routes.

The full textbook is divided into seven parts: Part I: *Newton's Laws*, Part II: *Conservation Laws*, Part III: *Applications of Newtonian Mechanics*, Part IV: *Thermodynamics*, Part V: *Waves and Optics*, Part VI: *Electricity and Magnetism*, and Part VII: *Relativity and Quantum Physics*. Although I recommend covering the parts in this order (see below), doing so is by no means essential. Each topic is self-contained, and Parts III–VI can be rearranged to suit an instructor's needs. To facilitate a reordering of topics, the full text is available in the five individual volumes listed in the margin.

Organization Rationale: Thermodynamics is placed before waves because it is a continuation of ideas from mechanics. The key idea in thermodynamics is energy, and moving from mechanics into thermodynamics allows the uninterrupted development of this important idea. Further, waves introduce students to functions of two variables, and the mathematics of waves is more akin to electricity and magnetism than to mechanics. Thus moving from waves to fields to quantum physics provides a gradual transition of ideas and skills.

The purpose of placing optics with waves is to provide a coherent presentation of wave physics, one of the two pillars of classical physics. Optics as it is presented in introductory physics makes no use of the properties of electromagnetic fields. There's little reason other than historical tradition to delay optics until after E&M.

- **Extended edition,** with modern physics (ISBN 978-0-321-73608-6 / 0-321-73608-7): Chapters 1–42.
- **Standard edition** (ISBN 978-0-321-75294-9 / 0-321-75294-5): Chapters 1–36.
- **Volume 1** (ISBN 978-0-321-75291-8 / 0-321-75291-0) covers mechanics: Chapters 1–15.
- **Volume 2** (ISBN 978-0-321-75318-2 / 0-321-75318-6) covers thermodynamics: Chapters 16–19.
- **Volume 3** (ISBN 978-0-321-75317-5 / 0-321-75317-8) covers waves and optics: Chapters 20–24.
- **Volume 4** (ISBN 978-0-321-75316-8 / 0-321-75316-X) covers electricity and magnetism, plus relativity: Chapters 25–36.
- **Volume 5** (ISBN 978-0-321-75315-1 / 0-321-75315-1) covers relativity and quantum physics: Chapters 36–42.
- **Volumes 1–5** boxed set (ISBN 978-0-321-77265-7 / 0-321-77265-2).

The documented difficulties that students have with optics are difficulties with waves, not difficulties with electricity and magnetism. However, the optics chapters are easily deferred until the end of Part VI for instructors who prefer that ordering of topics.

The Student Workbook

A key component of *Physics for Scientists and Engineers: A Strategic Approach* is the accompanying *Student Workbook*. The workbook bridges the gap between textbook and homework problems by providing students the opportunity to learn and practice skills prior to using those skills in quantitative end-of-chapter problems, much as a musician practices technique separately from performance pieces. The workbook exercises, which are keyed to each section of the textbook, focus on developing specific skills, ranging from identifying forces and drawing free-body diagrams to interpreting wave functions.

The workbook exercises, which are generally qualitative and/or graphical, draw heavily upon the physics education research literature. The exercises deal with issues known to cause student difficulties and employ techniques that have proven to be effective at overcoming those difficulties. The workbook exercises can be used in class as part of an active-learning teaching strategy, in recitation sections, or as assigned homework. More information about effective use of the *Student Workbook* can be found in the *Instructor Guide*.

Available versions: Extended (ISBN 978-0-321-75308-3/0-321-75308-9), Standard (ISBN 978-0-321-75309-0/0-321-75309-7), Volume 1 (ISBN 978-0-321-75314-4/0-321-75314-3), Volume 2 (ISBN 978-0-321-75313-7/0-321-75313-5), Volume 3 (ISBN 978-0-321-75312-0/0-321-75310-0), Volume 4 (ISBN 978-0-321-75311-3/0-321-75311-9), and Volume 5 (ISBN 978-0-321-75310-6/0-321-75310-0).

Instructor Supplements

- The **Instructor Guide for *Physics for Scientists and Engineers*** (ISBN 978-0-321-74765-5/0-321-74765-8) offers detailed comments and suggested teaching ideas for every chapter, an extensive review of what has been learned from physics education research, and guidelines for using active-learning techniques in your classroom. This invaluable guide is available on the Instructor Resource DVD, and via download, either from the MasteringPhysics Instructor Area or from the Instructor Resource Center (www.pearsonhighered.com/educator).

- The **Instructor Solutions** (ISBN 978-0-321-76940-4/0-321-76940-6), written by the author, Professor Larry Smith (Snow College), and Brett Kraabel (Ph.D., University of California, Santa Barbara), provide *complete* solutions to all the end-of-chapter problems. The solutions follow the four-step Model/Visualize/Solve/Assess procedure used in the Problem-Solving Strategies and in all worked examples. The solutions are available by chapter as editable Word® documents and as PDFs for your own use or for posting on your password-protected course website. Also provided are PDFs of handwritten solutions to all of the exercises in the *Student Workbook*, written by Professor James Andrews and Brian Garcar (Youngstown State University). All solutions are available

only via download, either from the MasteringPhysics Instructor Area or from the Instructor Resource Center (www.pearsonhighered.com/educator).

- The cross-platform **Instructor Resource DVD** (ISBN 978-0-321-75456-1/0-321-75456-5) provides a comprehensive library of more than 220 applets from **ActivPhysics OnLine** and 76 **PhET simulations**, as well as all figures, photos, tables, summaries, and key equations from the textbook in JPEG format. In addition, all the Problem-Solving Strategies, Tactics Boxes, and Key Equations are provided in editable Word format. PowerPoint® **Lecture Outlines** with embedded **Classroom Response System "Clicker" Questions** (including reading quizzes) are also provided.

- **(MP)** **MasteringPhysics®** (www.masteringphysics.com) is the most advanced, educationally effective, and widely used physics homework and tutorial system in the world. Eight years in development, it provides instructors with a library of extensively pre-tested end-of-chapter problems and rich, multipart, multistep tutorials that incorporate a wide variety of answer types, wrong answer feedback, individualized help (comprising hints or simpler sub-problems upon request), all driven by the largest metadatabase of student problem-solving in the world. NSF-sponsored published research (and subsequent studies) show that MasteringPhysics has dramatic educational results. MasteringPhysics allows instructors to build wide-ranging homework assignments of just the right difficulty and length and provides them with efficient tools to analyze in unprecedented detail both class trends and the work of any student.

MasteringPhysics routinely provides instant and individualized feedback and guidance to more than 100,000 students every day. A wide range of tools and support make MasteringPhysics fast and easy for instructors and students to learn to use. Extensive class tests show that by the end of their course, an unprecedented nine of ten students recommend MasteringPhysics as their preferred way to study physics and do homework.

For the third edition of *Physics for Scientists and Engineers,* MasteringPhysics now has the following functionalities:

- **Learning Outcomes:** In addition to being able to create their own learning outcomes to associate with questions in an assignment, professors can now select content that is tagged to a large number of publisher-provided learning outcomes. They can also print or export student results based on learning outcomes for their own use or to incorporate into reports for their administration.
- **Quizzing and Testing Enhancements:** These include options to hide item titles, add password protection, limit access to completed assignments, and to randomize question order in an assignment.
- **Math Remediation:** Found within selected tutorials, special links provide just-in-time math help and allow students to brush up on the most important mathematical concepts needed to successfully complete assignments. This new feature links students directly to math review and practice helping students make the connection between math and physics.
- **Enhanced End-of-Chapter Problems:** A subset of homework problems now offer additional support such as problem-solving strategy hints, relevant math review and practice, links to the eText, and links to the related Video Tutor Solution.

- **(MP)** **ActivPhysics OnLine™** (accessed through the Self Study area within www.masteringphysics.com) provides a comprehensive library of more than 220 tried and tested ActivPhysics core applets updated for web delivery using the latest online technologies. In addition, it provides a suite of highly regarded applet-based tutorials developed by education pioneers Alan Van Heuvelen and Paul D'Alessandris.

  The online exercises are designed to encourage students to confront misconceptions, reason qualitatively about physical processes, experiment quantitatively, and learn to think critically. The highly acclaimed ActivPhysics OnLine companion workbooks help students work through complex concepts and understand them more clearly. The applets from the ActivPhysics OnLine library are also available on the Instructor Resource DVD for this text.
- The **Test Bank** (ISBN 978-0-321-74766-2/0-321-74766-6) contains more than 2,000 high-quality problems, with a range of multiple-choice, true/false, short-answer, and regular homework-type questions. Test files are provided both in TestGen (an easy-to-use, fully networkable program for creating and editing quizzes and exams) and Word format. They are available only via download, either from the MasteringPhysics Instructor Area or from the Instructor Resource Center (www.pearsonhighered.com/educator).

Student Supplements

- The **Student Solutions Manuals Chapters 1–19** (ISBN 978-0-321-74767-9/0-321-74767-4) and **Chapters 20–42** (ISBN 978-0-321-77269-5/0-321-77269-5), written by the author, Professor Larry Smith (Snow College), and Brett Kraabel (Ph.D., University of California, Santa Barbara), provide *detailed* solutions to more than half of the odd-numbered end-of-chapter problems. The solutions follow the four-step Model/Visualize/Solve/Assess procedure used in the Problem-Solving Strategies and in all worked examples.
- **(MP)** **MasteringPhysics®** (www.masteringphysics.com) is a homework, tutorial, and assessment system based on years of research into how students work physics problems and precisely where they need help. Studies show that students who use MasteringPhysics significantly increase their scores compared to handwritten homework. MasteringPhysics achieves this improvement by providing students with instantaneous feedback specific to their wrong answers, simpler sub-problems upon request when they get stuck, and partial credit for their method(s). This individualized, 24/7 Socratic tutoring is recommended by 9 out of 10 students to their peers as the most effective and time-efficient way to study.
- **Pearson eText** is available through MasteringPhysics, either automatically when MasteringPhysics is packaged with new books, or available as a purchased upgrade online. Allowing students access to the text wherever they have access to the Internet, Pearson eText comprises the full text, including figures that can be enlarged for better viewing. With eText, students are also able to pop up definitions and terms to help with vocabulary and the reading of the material. Students can also take notes in eText using the annotation feature at the top of each page.

- **Pearson Tutor Services** (www.pearsontutorservices.com) Each student's subscription to MasteringPhysics also contains complimentary access to Pearson Tutor Services, powered by Smarthinking, Inc. By logging in with their MasteringPhysics ID and password, they will be connected to highly qualified e-instructors who provide additional interactive online tutoring on the major concepts of physics. Some restrictions apply; offer subject to change.

- **(MP) ActivPhysics OnLine™** (accessed through the Self Study area within www.masteringphysics.com)

provides students with a suite of highly regarded applet-based tutorials (see above). The following workbooks help students work through complex concepts and understand them more clearly:

- **ActivPhysics OnLine Workbook, Volume 1: Mechanics • Thermal Physics • Oscillations & Waves** (ISBN 978-0-8053-9060-5/0-8053-9060-X)
- **ActivPhysics OnLine Workbook, Volume 2: Electricity & Magnetism • Optics • Modern Physics** (ISBN 978-0-8053-9061-2/0-8053-9061-8)

Acknowledgments

I have relied upon conversations with and, especially, the written publications of many members of the physics education research community. Those who may recognize their influence include Arnold Arons, Uri Ganiel, Ibrahim Halloun, Richard Hake, Ken Heller, Paula Heron, David Hestenes, Leonard Jossem, Jill Larkin, Priscilla Laws, John Mallinckrodt, Kandiah Manivannan, Lillian McDermott and members of the Physics Education Research Group at the University of Washington, David Meltzer, Edward "Joe" Redish, Fred Reif, Jeffery Saul, Rachel Scherr, Bruce Sherwood, Josip Slisko, David Sokoloff, Richard Steinberg, Ronald Thornton, Sheila Tobias, Alan Van Heuleven, and Michael Wittmann. John Rigden, founder and director of the Introductory University Physics Project, provided the impetus that got me started down this path. Early development of the materials was supported by the National Science Foundation as the *Physics for the Year 2000* project; their support is gratefully acknowledged.

I especially want to thank my editor Jim Smith, development editor Alice Houston, project editor Martha Steele, and all the other staff at Pearson for their enthusiasm and hard work on this project. Production project manager Beth Collins, Rose Kernan and the team at Nesbitt Graphics, Inc., and photo researcher Eric Schrader get a good deal of the credit for making this complex project all come together. Larry Smith and Brett Kraabel have done an outstanding job of checking the solutions to every end-of-chapter problem and updating the *Instructor Solutions Manual*. Jim Andrews and Brian Garcar must be thanked for so carefully writing out the solutions to *The Student Workbook* exercises, and Jason Harlow for putting together the Lecture Outlines. In addition to the reviewers and classroom testers listed below, who gave invaluable feedback, I am particularly grateful to Charlie Hibbard for his close scrutiny of every word and figure.

Finally, I am endlessly grateful to my wife Sally for her love, encouragement, and patience, and to our many cats, past and present, who understand clearly that their priority is not deadlines but "Pet me, pet me, pet me."

Randy Knight, September 2011
rknight@calpoly.edu

Reviewers and Classroom Testers

Special thanks go to our third edition review panel: Kyle Altman, Taner Edis, Kent Fisher, Marty Gelfand, Elizabeth George, Jason Harlow, Bob Jacobsen, David Lee, Gary Morris, Eric Murray, and Bruce Schumm.

Gary B. Adams, *Arizona State University*
Ed Adelson, *Ohio State University*
Kyle Altmann, *Elon University*
Wayne R. Anderson, *Sacramento City College*
James H. Andrews, *Youngstown State University*
Kevin Ankoviak, *Las Positas College*
David Balogh, *Fresno City College*
Dewayne Beery, *Buffalo State College*
Joseph Bellina, *Saint Mary's College*
James R. Benbrook, *University of Houston*
David Besson, *University of Kansas*

Randy Bohn, *University of Toledo*
Richard A. Bone, *Florida International University*
Gregory Boutis, *York College*
Art Braundmeier, *University of Southern Illinois, Edwardsville*
Carl Bromberg, *Michigan State University*
Meade Brooks, *Collin College*
Douglas Brown, *Cabrillo College*
Ronald Brown, *California Polytechnic State University, San Luis Obispo*
Mike Broyles, *Collin County Community College*
Debra Burris, *University of Central Arkansas*
James Carolan, *University of British Columbia*
Michael Chapman, *Georgia Tech University*
Norbert Chencinski, *College of Staten Island*
Kristi Concannon, *King's College*

Sean Cordry, *Northwestern College of Iowa*
Robert L. Corey, *South Dakota School of Mines*
Michael Crescimanno, *Youngstown State University*
Dennis Crossley, *University of Wisconsin–Sheboygan*
Wei Cui, *Purdue University*
Robert J. Culbertson, *Arizona State University*
Danielle Dalafave, *The College of New Jersey*
Purna C. Das, *Purdue University North Central*
Chad Davies, *Gordon College*
William DeGraffenreid, *California State University–Sacramento*
Dwain Desbien, *Estrella Mountain Community College*
John F. Devlin, *University of Michigan, Dearborn*
John DiBartolo, *Polytechnic University*
Alex Dickison, *Seminole Community College*
Chaden Djalali, *University of South Carolina*
Margaret Dobrowolska, *University of Notre Dame*
Sandra Doty, *Denison University*
Miles J. Dresser, *Washington State University*
Charlotte Elster, *Ohio University*
Robert J. Endorf, *University of Cincinnati*
Tilahun Eneyew, *Embry-Riddle Aeronautical University*
F. Paul Esposito, *University of Cincinnati*
John Evans, *Lee University*
Harold T. Evensen, *University of Wisconsin–Platteville*
Michael R. Falvo, *University of North Carolina*
Abbas Faridi, *Orange Coast College*
Nail Fazleev, *University of Texas–Arlington*
Stuart Field, *Colorado State University*
Daniel Finley, *University of New Mexico*
Jane D. Flood, *Muhlenberg College*
Michael Franklin, *Northwestern Michigan College*
Jonathan Friedman, *Amherst College*
Thomas Furtak, *Colorado School of Mines*
Alina Gabryszewska-Kukawa, *Delta State University*
Lev Gasparov, *University of North Florida*
Richard Gass, *University of Cincinnati*
J. David Gavenda, *University of Texas, Austin*
Stuart Gazes, *University of Chicago*
Katherine M. Gietzen, *Southwest Missouri State University*
Robert Glosser, *University of Texas, Dallas*
William Golightly, *University of California, Berkeley*
Paul Gresser, *University of Maryland*
C. Frank Griffin, *University of Akron*
John B. Gruber, *San Jose State University*
Stephen Haas, *University of Southern California*
John Hamilton, *University of Hawaii at Hilo*
Jason Harlow, *University of Toronto*
Randy Harris, *University of California, Davis*
Nathan Harshman, *American University*
J. E. Hasbun, *University of West Georgia*
Nicole Herbots, *Arizona State University*
Jim Hetrick, *University of Michigan–Dearborn*
Scott Hildreth, *Chabot College*
David Hobbs, *South Plains College*
Laurent Hodges, *Iowa State University*

Mark Hollabaugh, *Normandale Community College*
John L. Hubisz, *North Carolina State University*
Shane Hutson, *Vanderbilt University*
George Igo, *University of California, Los Angeles*
David C. Ingram, *Ohio University*
Bob Jacobsen, *University of California, Berkeley*
Rong-Sheng Jin, *Florida Institute of Technology*
Marty Johnston, *University of St. Thomas*
Stanley T. Jones, *University of Alabama*
Darrell Judge, *University of Southern California*
Pawan Kahol, *Missouri State University*
Teruki Kamon, *Texas A&M University*
Richard Karas, *California State University, San Marcos*
Deborah Katz, *U.S. Naval Academy*
Miron Kaufman, *Cleveland State University*
Katherine Keilty, *Kingwood College*
Roman Kezerashvili, *New York City College of Technology*
Peter Kjeer, *Bethany Lutheran College*
M. Kotlarchyk, *Rochester Institute of Technology*
Fred Krauss, *Delta College*
Cagliyan Kurdak, *University of Michigan*
Fred Kuttner, *University of California, Santa Cruz*
H. Sarma Lakkaraju, *San Jose State University*
Darrell R. Lamm, *Georgia Institute of Technology*
Robert LaMontagne, *Providence College*
Eric T. Lane, *University of Tennessee–Chattanooga*
Alessandra Lanzara, *University of California, Berkeley*
Lee H. LaRue, *Paris Junior College*
Sen-Ben Liao, *Massachusetts Institute of Technology*
Dean Livelybrooks, *University of Oregon*
Chun-Min Lo, *University of South Florida*
Olga Lobban, *Saint Mary's University*
Ramon Lopez, *Florida Institute of Technology*
Vaman M. Naik, *University of Michigan, Dearborn*
Kevin Mackay, *Grove City College*
Carl Maes, *University of Arizona*
Rizwan Mahmood, *Slippery Rock University*
Mani Manivannan, *Missouri State University*
Richard McCorkle, *University of Rhode Island*
James McDonald, *University of Hartford*
James McGuire, *Tulane University*
Stephen R. McNeil, *Brigham Young University–Idaho*
Theresa Moreau, *Amherst College*
Gary Morris, *Rice University*
Michael A. Morrison, *University of Oklahoma*
Richard Mowat, *North Carolina State University*
Eric Murray, *Georgia Institute of Technology*
Taha Mzoughi, *Mississippi State University*
Scott Nutter, *Northern Kentucky University*
Craig Ogilvie, *Iowa State University*
Benedict Y. Oh, *University of Wisconsin*
Martin Okafor, *Georgia Perimeter College*
Halina Opyrchal, *New Jersey Institute of Technology*
Yibin Pan, *University of Wisconsin–Madison*
Georgia Papaefthymiou, *Villanova University*
Peggy Perozzo, *Mary Baldwin College*

Brian K. Pickett, *Purdue University, Calumet*
Joe Pifer, *Rutgers University*
Dale Pleticha, *Gordon College*
Marie Plumb, *Jamestown Community College*
Robert Pompi, *SUNY-Binghamton*
David Potter, *Austin Community College–Rio Grande Campus*
Chandra Prayaga, *University of West Florida*
Didarul Qadir, *Central Michigan University*
Steve Quon, *Ventura College*
Michael Read, *College of the Siskiyous*
Lawrence Rees, *Brigham Young University*
Richard J. Reimann, *Boise State University*
Michael Rodman, *Spokane Falls Community College*
Sharon Rosell, *Central Washington University*
Anthony Russo, *Okaloosa-Walton Community College*
Freddie Salsbury, *Wake Forest University*
Otto F. Sankey, *Arizona State University*
Jeff Sanny, *Loyola Marymount University*
Rachel E. Scherr, *University of Maryland*
Carl Schneider, *U. S. Naval Academy*
Bruce Schumm, *University of California, Santa Cruz*
Bartlett M. Sheinberg, *Houston Community College*
Douglas Sherman, *San Jose State University*
Elizabeth H. Simmons, *Boston University*
Marlina Slamet, *Sacred Heart University*
Alan Slavin, *Trent College*
Larry Smith, *Snow College*

William S. Smith, *Boise State University*
Paul Sokol, *Pennsylvania State University*
LTC Bryndol Sones, *United States Military Academy*
Chris Sorensen, *Kansas State University*
Anna and Ivan Stern, *AW Tutor Center*
Gay B. Stewart, *University of Arkansas*
Michael Strauss, *University of Oklahoma*
Chin-Che Tin, *Auburn University*
Christos Valiotis, *Antelope Valley College*
Andrew Vanture, *Everett Community College*
Arthur Viescas, *Pennsylvania State University*
Ernst D. Von Meerwall, *University of Akron*
Chris Vuille, *Embry-Riddle Aeronautical University*
Jerry Wagner, *Rochester Institute of Technology*
Robert Webb, *Texas A&M University*
Zodiac Webster, *California State University, San Bernardino*
Robert Weidman, *Michigan Technical University*
Fred Weitfeldt, *Tulane University*
Jeff Allen Winger, *Mississippi State University*
Carey Witkov, *Broward Community College*
Ronald Zammit, *California Polytechnic State University, San Luis Obispo*
Darin T. Zimmerman, *Pennsylvania State University, Altoona*
Fredy Zypman, *Yeshiva University*

Preface to the Student

From Me to You

The most incomprehensible thing about the universe is that it is comprehensible.
—Albert Einstein

The day I went into physics class it was death.
—Sylvia Plath, *The Bell Jar*

Let's have a little chat before we start. A rather one-sided chat, admittedly, because you can't respond, but that's OK. I've talked with many of your fellow students over the years, so I have a pretty good idea of what's on your mind.

What's your reaction to taking physics? Fear and loathing? Uncertainty? Excitement? All of the above? Let's face it, physics has a bit of an image problem on campus. You've probably heard that it's difficult, maybe downright impossible unless you're an Einstein. Things that you've heard, your experiences in other science courses, and many other factors all color your *expectations* about what this course is going to be like.

It's true that there are many new ideas to be learned in physics and that the course, like college courses in general, is going to be much faster paced than science courses you had in high school. I think it's fair to say that it will be an *intense* course. But we can avoid many potential problems and difficulties if we can establish, here at the beginning, what this course is about and what is expected of you—and of me!

Just what is physics, anyway? Physics is a way of thinking about the physical aspects of nature. Physics is not better than art or biology or poetry or religion, which are also ways to think about nature; it's simply different. One of the things this course will emphasize is that physics is a human endeavor. The ideas presented in this book were not found in a cave or conveyed to us by aliens; they were discovered and developed by real people engaged in a struggle with real issues. I hope to convey to you something of the history and the process by which we have come to accept the principles that form the foundation of today's science and engineering.

You might be surprised to hear that physics is not about "facts." Oh, not that facts are unimportant, but physics is far more focused on discovering *relationships* that exist between facts and *patterns* that exist in nature than on learning facts for their own sake. As a consequence, there's not a lot of memorization when you study physics. Some—there are still definitions and equations to learn—but less than in many other courses. Our emphasis, instead, will be on thinking and reasoning. This is important to factor into your expectations for the course.

Perhaps most important of all, *physics is not math!* Physics is much broader. We're going to look for patterns and relationships in nature, develop the logic that relates different ideas, and search for the reasons *why* things happen as they do. In doing so, we're going to stress qualitative reasoning, pictorial and graphical reasoning, and reasoning by analogy. And yes, we will use math, but it's just one tool among many.

It will save you much frustration if you're aware of this physics–math distinction up front. Many of you, I know, want to find a formula and plug numbers into it—that is,

(a) X-ray diffraction pattern

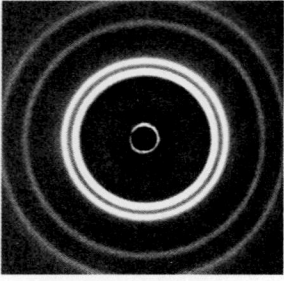

(b) Electron diffraction pattern

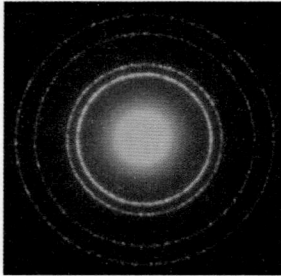

to do a math problem. Maybe that worked in high school science courses, but it is *not* what this course expects of you. We'll certainly do many calculations, but the specific numbers are usually the last and least important step in the analysis.

Physics is about recognizing patterns. For example, the top photograph is an x-ray diffraction pattern showing how a focused beam of x rays spreads out after passing through a crystal. The bottom photograph shows what happens when a focused beam of electrons is shot through the same crystal. What does the obvious similarity in these two photographs tell us about the nature of light and the nature of matter?

As you study, you'll sometimes be baffled, puzzled, and confused. That's perfectly normal and to be expected. Making mistakes is OK too *if* you're willing to learn from the experience. No one is born knowing how to do physics any more than he or she is born knowing how to play the piano or shoot basketballs. The ability to do physics comes from practice, repetition, and struggling with the ideas until you "own" them and can apply them yourself in new situations. There's no way to make learning effortless, at least for anything worth learning, so expect to have some difficult moments ahead. But also expect to have some moments of excitement at the joy of discovery. There will be instants at which the pieces suddenly click into place and you *know* that you understand a powerful idea. There will be times when you'll surprise yourself by successfully working a difficult problem that you didn't think you could solve. My hope, as an author, is that the excitement and sense of adventure will far outweigh the difficulties and frustrations.

Getting the Most Out of Your Course

Many of you, I suspect, would like to know the "best" way to study for this course. There is no best way. People are different, and what works for one student is less effective for another. But I do want to stress that *reading the text* is vitally important. Class time will be used to clarify difficulties and to develop tools for using the knowledge, but your instructor will *not* use class time simply to repeat information in the text. The basic knowledge for this course is written down on these pages, and the *number-one expectation* is that you will read carefully and thoroughly to find and learn that knowledge.

Despite there being no best way to study, I will suggest *one* way that is successful for many students. It consists of the following four steps:

1. **Read each chapter *before* it is discussed in class.** I cannot stress too strongly how important this step is. Class attendance is much more effective if you are prepared. When you first read a chapter, focus on learning new vocabulary, definitions, and notation. There's a list of terms and notations at the end of each chapter. Learn them! You won't understand what's being discussed or how the ideas are being used if you don't know what the terms and symbols mean.
2. **Participate actively in class.** Take notes, ask and answer questions, and participate in discussion groups. There is ample scientific evidence that *active participation* is much more effective for learning science than passive listening.
3. **After class, go back for a careful re-reading of the chapter.** In your second reading, pay closer attention to the details and the worked examples. Look for the *logic* behind each example (I've highlighted this to make it clear), not just at what formula is being used. Do the *Student Workbook* exercises for each section as you finish your reading of it.
4. **Finally, apply what you have learned to the homework problems at the end of each chapter.** I strongly encourage you to form a study group with two or three classmates. There's good evidence that students who study regularly with a group do better than the rugged individualists who try to go it alone.

 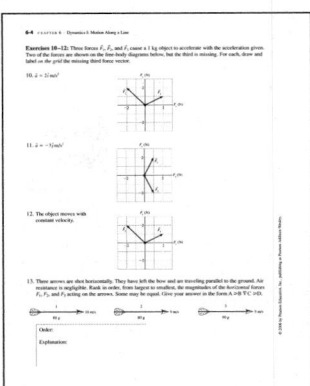 

Did someone mention a workbook? The companion *Student Workbook* is a vital part of the course. Its questions and exercises ask you to reason *qualitatively,* to use graphical information, and to give explanations. It is through these exercises that you will learn what the concepts mean and will practice the reasoning skills appropriate to the chapter. You will then have acquired the baseline knowledge and confidence you need *before* turning to the end-of-chapter homework problems. In sports or in music, you would never think of performing before you practice, so why would you want to do so in physics? The workbook is where you practice and work on basic skills.

Many of you, I know, will be tempted to go straight to the homework problems and then thumb through the text looking for a formula that seems like it will work. That approach will not succeed in this course, and it's guaranteed to make you frustrated and discouraged. Very few homework problems are of the "plug and chug" variety where you simply put numbers into a formula. To work the homework problems successfully, you need a better study strategy—either the one outlined above or your own—that helps you learn the concepts and the relationships between the ideas.

A traditional guideline in college is to study two hours outside of class for every hour spent in class, and this text is designed with that expectation. Of course, two hours is an average. Some chapters are fairly straightforward and will go quickly. Others likely will require much more than two study hours per class hour.

Getting the Most Out of Your Textbook

Your textbook provides many features designed to help you learn the concepts of physics and solve problems more effectively.

- TACTICS BOXES give step-by-step procedures for particular skills, such as interpreting graphs or drawing special diagrams. Tactics Box steps are explicitly illustrated in subsequent worked examples, and these are often the starting point of a full *Problem-Solving Strategy.*

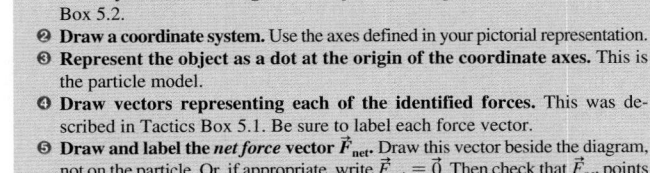

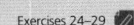

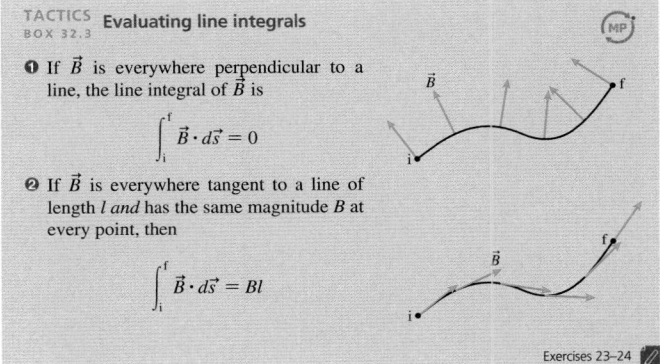

■ PROBLEM-SOLVING STRATEGIES are provided for each broad class of problems—problems characteristic of a chapter or group of chapters. The strategies follow a consistent four-step approach to help you develop confidence and proficient problem-solving skills: MODEL, VISUALIZE, SOLVE, ASSESS.

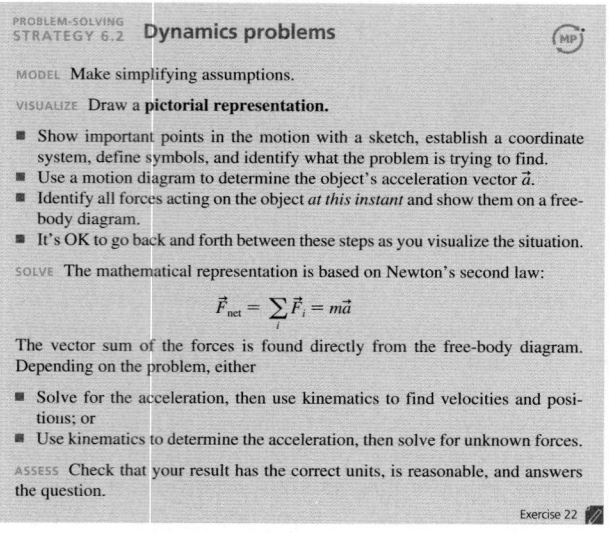

PROBLEM-SOLVING
STRATEGY 6.2 **Dynamics problems**

MODEL Make simplifying assumptions.

VISUALIZE Draw a **pictorial representation.**

■ Show important points in the motion with a sketch, establish a coordinate system, define symbols, and identify what the problem is trying to find.
■ Use a motion diagram to determine the object's acceleration vector $\vec{a}$.
■ Identify all forces acting on the object *at this instant* and show them on a free-body diagram.
■ It's OK to go back and forth between these steps as you visualize the situation.

SOLVE The mathematical representation is based on Newton's second law:

$$\vec{F}_{net} = \sum_i \vec{F}_i = m\vec{a}$$

The vector sum of the forces is found directly from the free-body diagram. Depending on the problem, either

■ Solve for the acceleration, then use kinematics to find velocities and positions; or
■ Use kinematics to determine the acceleration, then solve for unknown forces.

ASSESS Check that your result has the correct units, is reasonable, and answers the question.

Exercise 22

■ Worked EXAMPLES illustrate good problem-solving practices through the consistent use of the four-step problem-solving approach and, where appropriate, the Tactics Box steps. The worked examples are often very detailed and carefully lead you through the *reasoning* behind the solution as well as the numerical calculations. A careful study of the reasoning will help you apply the concepts and techniques to the new and novel problems you will encounter in homework assignments and on exams.
■ NOTE ▶ paragraphs alert you to common mistakes and point out useful tips for tackling problems.
■ STOP TO THINK questions embedded in the chapter allow you to quickly assess whether you've understood the main idea of a section. A correct answer will give you confidence to move on to the next section. An incorrect answer will alert you to re-read the previous section.
■ Blue annotations on figures help you better understand what the figure is showing. They will help you to interpret graphs; translate between graphs, math, and pictures; grasp difficult concepts through a visual analogy; and develop many other important skills.
■ *Pencil sketches* provide practical examples of the figures you should draw yourself when solving a problem.

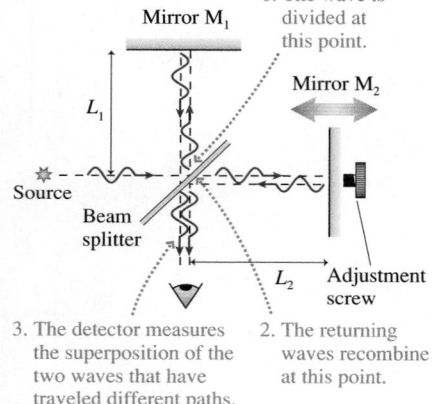

Annotated FIGURE showing the operation of the Michelson interferometer.

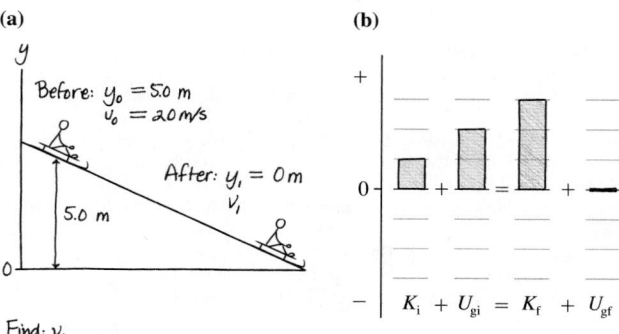

Pencil-sketch FIGURE showing a toboggan going down a hill and its energy bar chart.

- Each chapter begins with a *Chapter Preview*, a visual outline of the chapter ahead with recommendations of important topics you should review from previous chapters. A few minutes spent with the Preview will help you organize your thoughts so as to get the most out of reading the chapter.
- Schematic *Chapter Summaries* help you organize what you have learned into a hierarchy, from general principles (top) to applications (bottom). Side-by-side pictorial, graphical, textual, and mathematical representations are used to help you translate between these key representations.
- *Part Overviews* and *Summaries* provide a global framework for what you are learning. Each part begins with an overview of the chapters ahead and concludes with a broad summary to help you to connect the concepts presented in that set of chapters. KNOWLEDGE STRUCTURE tables in the Part Summaries, similar to the Chapter Summaries, help you to see the forest rather than just the trees.

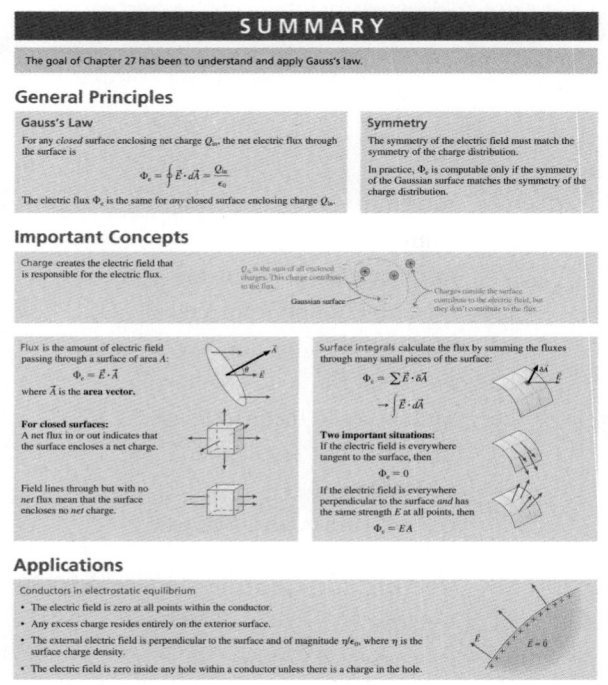

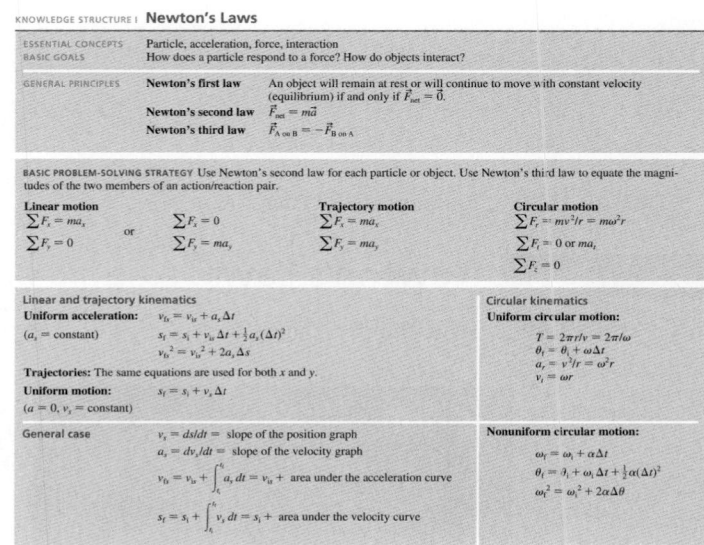

Now that you know more about what is expected of you, what can you expect of me? That's a little trickier because the book is already written! Nonetheless, the book was prepared on the basis of what I think my students throughout the years have expected—and wanted—from their physics textbook. Further, I've listened to the extensive feedback I have received from thousands of students like you, and their instructors, who used the first and second editions of this book.

You should know that these course materials—the text and the workbook—are based on extensive research about how students learn physics and the challenges they face. The effectiveness of many of the exercises has been demonstrated through extensive class testing. I've written the book in an informal style that I hope you will find appealing and that will encourage you to do the reading. And, finally, I have endeavored to make clear not only that physics, as a technical body of knowledge, is relevant to your profession but also that physics is an exciting adventure of the human mind.

I hope you'll enjoy the time we're going to spend together.

Detailed Contents

Part II **Conservation Laws**

Part VI   Electricity and Magnetism

Introduction

Journey into Physics

Said Alice to the Cheshire cat,
"Cheshire-Puss, would you tell me, please, which way I ought to go from here?"
"That depends a good deal on where you want to go," said the Cat.
"I don't much care where—" said Alice.
"Then it doesn't matter which way you go," said the Cat.
  —Lewis Carroll, *Alice in Wonderland*

Have you ever wondered about questions such as

> Why is the sky blue?

> Why is glass an insulator but metal a conductor?

> What, really, is an atom?

These are the questions of which physics is made. Physicists try to understand the universe in which we live by observing the phenomena of nature—such as the sky being blue—and by looking for patterns and principles to explain these phenomena. Many of the discoveries made by physicists, from electromagnetic waves to nuclear energy, have forever altered the ways in which we live and think.

You are about to embark on a journey into the realm of physics. It is a journey in which you will learn about many physical phenomena and find the answers to questions such as the ones posed above. Along the way, you will also learn how to use physics to analyze and solve many practical problems.

As you proceed, you are going to see the methods by which physicists have come to understand the laws of nature. The ideas and theories of physics are not arbitrary; they are firmly grounded in experiments and measurements. By the time you finish this text, you will be able to recognize the *evidence* upon which our present knowledge of the universe is based.

Which Way Should We Go?

We are rather like Alice in Wonderland, here at the start of the journey, in that we must decide which way to go. Physics is an immense body of knowledge, and without specific goals it would not much matter which topics we study. But unlike Alice, we *do* have some particular destinations that we would like to visit.

The physics that provides the foundation for all of modern science and engineering can be divided into three broad categories:

- Particles and energy.
- Fields and waves.
- The atomic structure of matter.

A particle, in the sense that we'll use the term, is an idealization of a physical object. We will use particles to understand how objects move and how they interact with each other. One of the most important properties of a particle or a collection of particles is *energy*. We will study energy both for its value in understanding physical processes and because of its practical importance in a technological society.

A scanning tunneling microscope allows us to "see" the individual atoms on a surface. One of our goals is to understand how an image such as this is made.

Particles are discrete, localized objects. Although many phenomena can be understood in terms of particles and their interactions, the long-range interactions of gravity, electricity, and magnetism are best understood in terms of *fields,* such as the gravitational field and the electric field. Rather than being discrete, fields spread continuously through space. Much of the second half of this book will be focused on understanding fields and the interactions between fields and particles.

Certainly one of the most significant discoveries of the past 500 years is that matter consists of atoms. Atoms and their properties are described by quantum physics, but we cannot leap directly into that subject and expect that it would make any sense. To reach our destination, we are going to have to study many other topics along the way—rather like having to visit the Rocky Mountains if you want to drive from New York to San Francisco. All our knowledge of particles and fields will come into play as we end our journey by studying the atomic structure of matter.

The Route Ahead

Here at the beginning, we can survey the route ahead. Where will our journey take us? What scenic vistas will we view along the way?

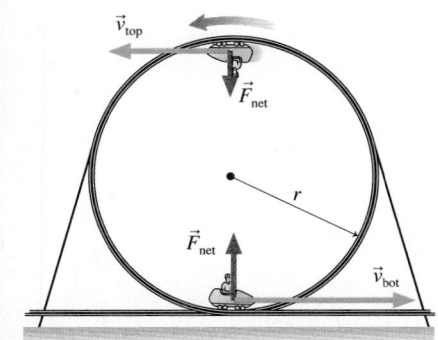

Parts I and II, *Newton's Laws* and *Conservation Laws,* form the basis of what is called *classical mechanics.* Classical mechanics is the study of motion. (It is called *classical* to distinguish it from the modern theory of motion at the atomic level, which is called *quantum mechanics.*) The first two parts of this textbook establish the basic language and concepts of motion. Part I will look at motion in terms of *particles* and *forces.* We will use these concepts to study the motion of everything from accelerating sprinters to orbiting satellites. Then, in Part II, we will introduce the ideas of *momentum* and *energy.* These concepts—especially energy—will give us a new perspective on motion and extend our ability to analyze motion.

Part III, *Applications of Newtonian Mechanics,* will pause to look at four important applications of classical mechanics: Newton's theory of gravity, rotational motion, oscillatory motion, and the motion of fluids. Only oscillatory motion is a prerequisite for later chapters. Your instructor may choose to cover some or all of the other chapters, depending upon the time available, but your study of Parts IV–VII will not be hampered if these chapters are omitted.

Atoms are held close together by weak molecular bonds, but they can slide around each other.

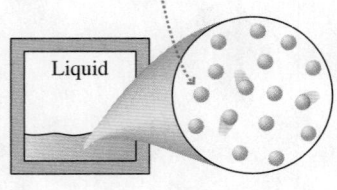

Part IV, *Thermodynamics,* extends the ideas of particles and energy to systems such as liquids and gases that contain vast numbers of particles. Here we will look for connections between the *microscopic* behavior of large numbers of atoms and the *macroscopic* properties of bulk matter. You will find that some of the properties of gases that you know from chemistry, such as the ideal gas law, turn out to be direct consequences of the underlying atomic structure of the gas. We will also expand the concept of energy and study how energy is transferred and utilized.

Waves are ubiquitous in nature, whether they be large-scale oscillations like ocean waves, the less obvious motions of sound waves, or the subtle undulations of light waves and matter waves that go to the heart of the atomic structure of matter. In **Part V,** *Waves and Optics,* we will emphasize the unity of wave physics and find that many diverse wave phenomena can be analyzed with the same concepts and mathematical language. Light waves are of special interest, and we will end this portion of our journey with an exploration of optical instruments, ranging from microscopes and telescopes to that most important of all optical instruments—your eye.

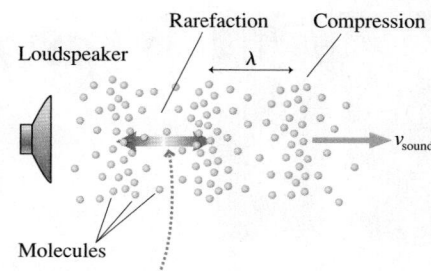

Individual molecules oscillate back and forth with displacement D. As they do so, the compressions propagate forward at speed v_{sound}. Because compressions are regions of higher pressure, a sound wave can be thought of as a pressure wave.

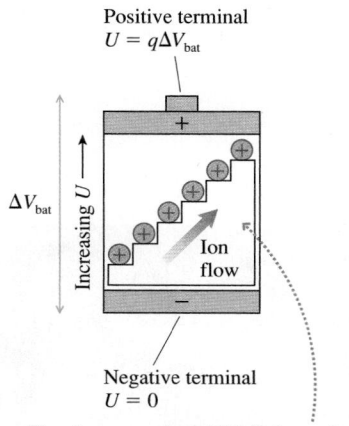

Positive terminal
$U = q\Delta V_{bat}$

ΔV_{bat} Increasing U

Ion flow

Negative terminal
$U = 0$

The charge escalator "lifts" charge from the negative side to the positive side. Charge q gains energy $\Delta U = q\Delta V_{bat}$.

Part VI, *Electricity and Magnetism,* is devoted to the *electromagnetic force,* one of the most important forces in nature. In essence, the electromagnetic force is the "glue" that holds atoms together. It is also the force that makes this the "electronic age." We'll begin this part of the journey with simple observations of static electricity. Bit by bit, we'll be led to the basic ideas behind electrical circuits, to magnetism, and eventually to the discovery of electromagnetic waves.

Part VII is *Relativity and Quantum Physics.* We'll start by exploring the strange world of Einstein's theory of *relativity,* a world in which space and time aren't quite what they appear to be. Then we will enter the microscopic domain of *atoms,* where the behaviors of light and matter are at complete odds with what our common sense tells us is possible. Although the mathematics of quantum theory quickly gets beyond the level of this text, and time will be running out, you will see that the quantum theory of atoms and nuclei explains many of the things that you learned simply as rules in chemistry.

This picture of an atom would need to be 10 m in diameter if it were drawn to the same scale as the dot representing the nucleus.

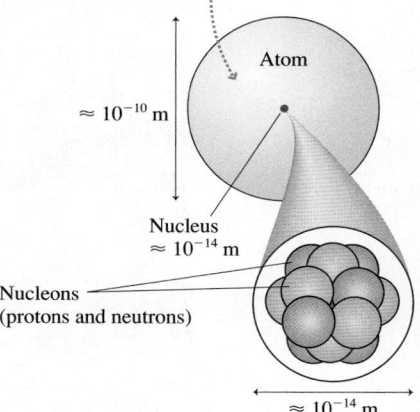

Atom

$\approx 10^{-10}$ m

Nucleus
$\approx 10^{-14}$ m

Nucleons
(protons and neutrons)

$\approx 10^{-14}$ m

We will not have visited all of physics on our travels. There just isn't time. Many exciting topics, ranging from quarks to black holes, will have to remain unexplored. But this particular journey need not be the last. As you finish this text, you will have the background and the experience to explore new topics further in more advanced courses or for yourself.

With that said, let us take the first step.

Newton's Laws

Motion can be exhilarating and beautiful. These sailboats are responding to forces of wind, water, and the weight of the crew as they balance precariously on the edge.

OVERVIEW
Why Things Change

Each of the seven parts of this book opens with an overview to give you a look ahead, a glimpse at where your journey will take you in the next few chapters. It's easy to lose sight of the big picture while you're busy negotiating the terrain of each chapter. In Part I, the big picture, in a word, is *change.*

Simple observations of the world around you show that most things change, few things remain the same. Some changes, such as aging, are biological. Others, such as sugar dissolving in your coffee, are chemical. We're going to study change that involves *motion* of one form or another—the motion of balls, cars, and rockets.

There are two big questions we must tackle:

- **How do we describe motion?** It is easy to say that an object moves, but it's not obvious how we should measure or characterize the motion if we want to analyze it mathematically. The mathematical description of motion is called *kinematics,* and it is the subject matter of Chapters 1 through 4.
- **How do we explain motion?** Why do objects have the particular motion they do? Why, when you toss a ball upward, does it go up and then come back down rather than keep going up? Are there "laws of nature" that allow us to predict an object's motion? The explanation of motion in terms of its causes is called *dynamics,* and it is the topic of Chapters 5 through 8.

Two key ideas for answering these questions are *force* (the "cause") and *acceleration* (the "effect"). A variety of pictorial and graphical tools will be developed in Chapters 1 through 5 to help you develop an *intuition* for the connection between force and acceleration. You'll then put this knowledge to use in Chapters 5 through 8 as you analyze motion of increasing complexity.

Another important tool will be the use of *models.* Reality is extremely complicated. We would never be able to develop a science if we had to keep track of every little detail of every situation. A model is a simplified description of reality—much as a model airplane is a simplified version of a real airplane—used to reduce the complexity of a problem to the point where it can be analyzed and understood. We will introduce several important models of motion, paying close attention, especially in these earlier chapters, to where simplifying assumptions are being made, and why.

The "laws of motion" were discovered by Isaac Newton roughly 350 years ago, so the study of motion is hardly cutting-edge science. Nonetheless, it is still extremely important. Mechanics—the science of motion—is the basis for much of engineering and applied science, and many of the ideas introduced here will be needed later to understand things like the motion of waves and the motion of electrons through circuits. Newton's mechanics is the foundation of much of contemporary science, thus we will start at the beginning.

1 Concepts of Motion

Motion takes many forms. The snowboarder seen here is an example of translational motion.

▶ **Looking Ahead** The goal of Chapter 1 is to introduce the fundamental concepts of motion.

The Chapter Preview

Each chapter will start with an overview of the material to come. You should read these chapter previews carefully to get a sense of the road ahead.

Arrows show the flow of ideas in the chapter.

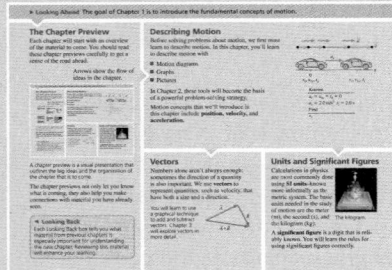

A chapter preview is a visual presentation that outlines the big ideas and the organization of the chapter that is to come.

The chapter previews not only let you know what is coming, they also help you make connections with material you have already seen.

◀ **Looking Back**
Each Looking Back box tells you what material from previous chapters is especially important for understanding the new chapter. Reviewing this material will enhance your learning.

Describing Motion

Before solving problems about motion, we first must learn to describe motion. In this chapter, you'll learn to describe motion with

■ Motion diagrams
■ Graphs
■ Pictures

In Chapter 2, these tools will become the basis of a powerful problem-solving strategy.

Motion concepts that we'll introduce in this chapter include **position, velocity,** and **acceleration.**

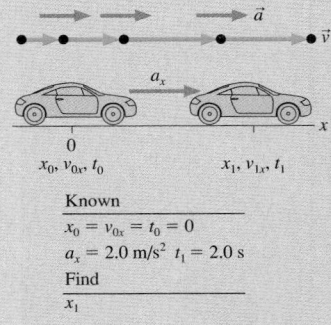

Known
$$x_0 = v_{0x} = t_0 = 0$$
$$a_x = 2.0 \text{ m/s}^2 \quad t_1 = 2.0 \text{ s}$$
Find
$$x_1$$

Vectors

Numbers alone aren't always enough; sometimes the direction of a quantity is also important. We use **vectors** to represent quantities, such as velocity, that have both a size and a direction.

You will learn to use a graphical technique to add and subtract vectors. Chapter 3 will explore vectors in more detail.

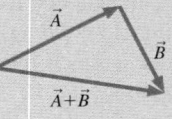

Units and Significant Figures

Calculations in physics are most commonly done using **SI units**—known more informally as the metric system. The basic units needed in the study of motion are the meter (m), the second (s), and the kilogram (kg).

The kilogram.

A **significant figure** is a digit that is reliably known. You will learn the rules for using significant figures correctly.

1.1 Motion Diagrams

Motion is a theme that will appear in one form or another throughout this entire book. Although we all have intuition about motion, based on our experiences, some of the important aspects of motion turn out to be rather subtle. So rather than jumping immediately into a lot of mathematics and calculations, this first chapter focuses on *visualizing* motion and becoming familiar with the *concepts* needed to describe a moving object. Our goal is to lay the foundations for understanding motion.

FIGURE 1.1 Four basic types of motion.

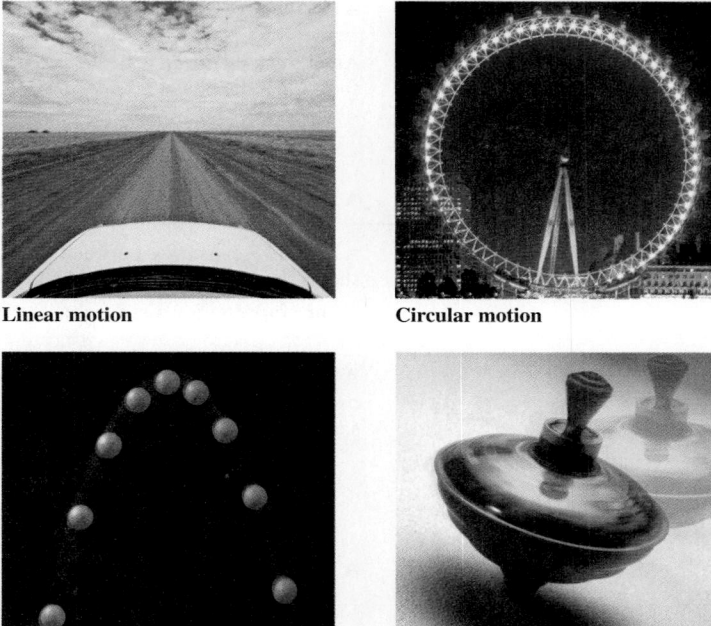

As a starting point, let's define **motion** as the change of an object's position with time. **FIGURE 1.1** shows four basic types of motion that we will study in this book. The first three—linear, circular, and projectile motion—in which the object moves through space are called **translational motion.** The path along which the object moves, whether straight or curved, is called the object's **trajectory.** Rotational motion is somewhat different in that rotation is a change of the object's *angular* position. We'll defer rotational motion until later and, for now, focus on translational motion.

Making a Motion Diagram

An easy way to study motion is to make a movie of a moving object. A movie camera, as you probably know, takes photographs at a fixed rate, typically 30 photographs every second. Each separate photo is called a *frame,* and the frames are all lined up one after the other in a *filmstrip.* As an example, **FIGURE 1.2** shows four frames from the movie of a car going past. Not surprisingly, the car is in a somewhat different position in each frame.

Suppose we cut the individual frames of the filmstrip apart, stack them on top of each other, and project the entire stack at once onto a screen for viewing. The result is shown in **FIGURE 1.3**. This composite photo, showing an object's position at several *equally spaced instants of time,* is called a **motion diagram.** As the example below shows, we can define concepts such as at rest, constant speed, speeding up, and slowing down in terms of how an object appears in a motion diagram.

NOTE ▶ It's important to keep the camera in a *fixed position* as the object moves by. Don't "pan" it to track the moving object. ◀

FIGURE 1.2 Four frames from the movie of a car.

FIGURE 1.3 A motion diagram of the car shows all the frames simultaneously.

The same amount of time elapses between each image and the next.

Examples of motion diagrams

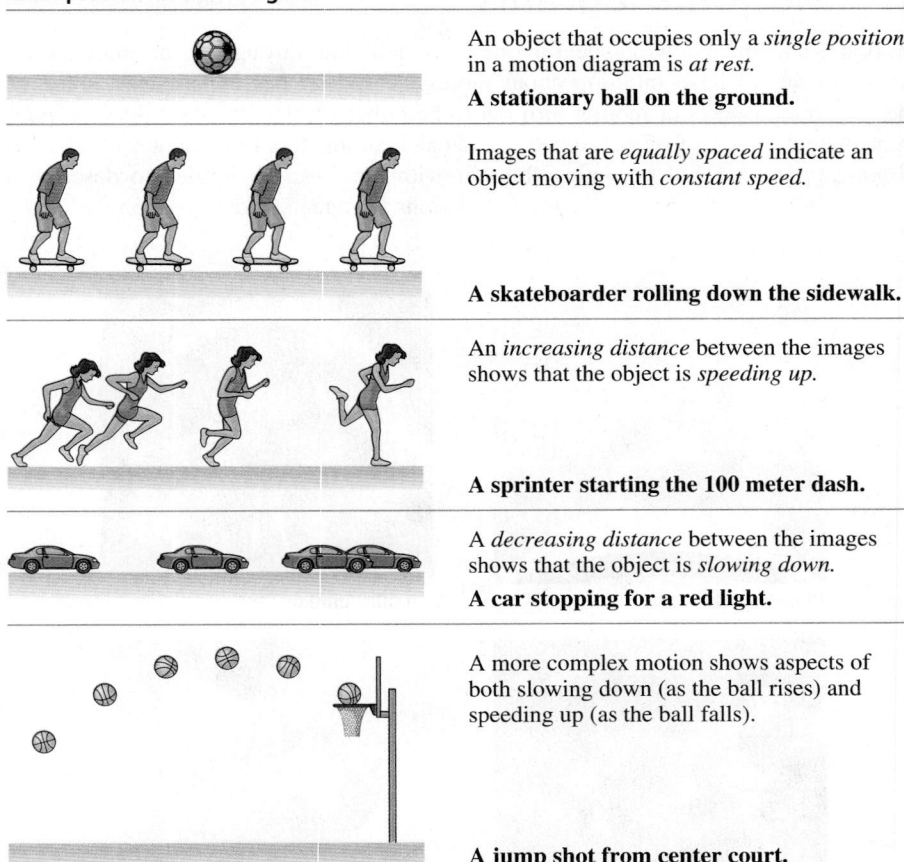

An object that occupies only a *single position* in a motion diagram is *at rest*.

A stationary ball on the ground.

Images that are *equally spaced* indicate an object moving with *constant speed*.

A skateboarder rolling down the sidewalk.

An *increasing distance* between the images shows that the object is *speeding up*.

A sprinter starting the 100 meter dash.

A *decreasing distance* between the images shows that the object is *slowing down*.

A car stopping for a red light.

A more complex motion shows aspects of both slowing down (as the ball rises) and speeding up (as the ball falls).

A jump shot from center court.

STOP TO THINK 1.1   Which car is going faster, A or B? Assume there are equal intervals of time between the frames of both movies.

Car A   Car B

NOTE ▶ Each chapter will have several *Stop to Think* questions. These questions are designed to see if you've understood the basic ideas that have been presented. The answers are given at the end of the chapter, but you should make a serious effort to think about these questions before turning to the answers. If you answer correctly, and are sure of your answer rather than just guessing, you can proceed to the next section with confidence. But if you answer incorrectly, it would be wise to reread the preceding sections before proceeding onward. ◀

1.2 The Particle Model

For many types of motion, such as that of balls, cars, and rockets, the motion of the object *as a whole* is not influenced by the details of the object's size and shape. All we really need to keep track of is the motion of a single point on the object, so we can treat the object *as if* all its mass were concentrated into this single point. An object that can be represented as a mass at a single point in space is called a **particle**. A particle has no size, no shape, and no distinction between top and bottom or between front and back.

If we treat an object as a particle, we can represent the object in each frame of a motion diagram as a simple dot rather than having to draw a full picture. FIGURE 1.4 shows how much simpler motion diagrams appear when the object is represented as a particle. Note that the dots have been numbered 0, 1, 2, . . . to tell the sequence in which the frames were exposed.

Using the Particle Model

Treating an object as a particle is, of course, a simplification of reality. As we noted in the Part I Overview, such a simplification is called a *model*. Models allow us to focus on the important aspects of a phenomenon by excluding those aspects that play only a minor role. The **particle model** of motion is a simplification in which we treat a moving object as if all of its mass were concentrated at a single point. The particle model is an excellent approximation of reality for the translational motion of cars, planes, rockets, and similar objects. In later chapters, we'll find that the motion of more complex objects, which cannot be treated as a single particle, can often be analyzed as if the object were a collection of particles.

Not all motions can be reduced to the motion of a single point. Consider a rotating gear. The center of the gear doesn't move at all, and each tooth on the gear is moving in a different direction. Rotational motion is qualitatively different than translational motion, and we'll need to go beyond the particle model later when we study rotational motion.

FIGURE 1.4 Motion diagrams in which the object is represented as a particle.

(a) Motion diagram of a rocket launch

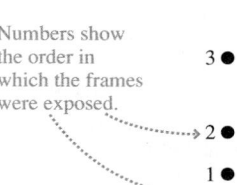

Numbers show the order in which the frames were exposed.

(b) Motion diagram of a car stopping

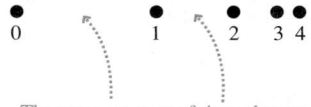

The same amount of time elapses between each image and the next.

STOP TO THINK 1.2　Three motion diagrams are shown. Which is a dust particle settling to the floor at constant speed, which is a ball dropped from the roof of a building, and which is a descending rocket slowing to make a soft landing on Mars?

(a)
0 ●
1 ●
2 ●
3 ●
4 ●
5 ●

(b)
0 ●
1 ●
2 ●
3 ●
4 ●
5 ●

(c)
0 ●
1 ●
2 ●
3 ●
4 ●
5 ●

1.3 Position and Time

As we look at a motion diagram, it would be useful to know *where* the object is (i.e., its *position*) and *when* the object was at that position (i.e., the *time*). Position measurements can be made by laying a coordinate system grid over a motion diagram. You can then measure the (x, y) coordinates of each point in the motion diagram. Of course, the world does not come with a coordinate system attached. A coordinate system is an artificial grid that *you* place over a problem in order to analyze the motion. You place the origin of your coordinate system wherever you wish, and different observers of a moving object might all choose to use different origins. Likewise, you can choose the orientation of the *x*-axis and *y*-axis to be helpful for that particular problem. The conventional choice is for the *x*-axis to point to the right and the *y*-axis to point upward, but there is nothing sacred about this choice. We will soon have many occasions to tilt the axes at an angle.

Time, in a sense, is also a coordinate system, although you may never have thought of time this way. You can pick an arbitrary point in the motion and label it "$t = 0$ seconds."

This is simply the instant you decide to start your clock or stopwatch, so it is the origin of your time coordinate. Different observers might choose to start their clocks at different moments. A movie frame labeled "$t = 4$ seconds" was taken 4 seconds after you started your clock.

We typically choose $t = 0$ to represent the "beginning" of a problem, but the object may have been moving before then. Those earlier instants would be measured as negative times, just as objects on the x-axis to the left of the origin have negative values of position. Negative numbers are not to be avoided; they simply locate an event in space or time *relative to an origin.*

To illustrate, FIGURE 1.5a shows an xy-coordinate system and time information superimposed over the motion diagram of a basketball. You can see that the ball's position is $(x_4, y_4) = (12 \text{ m}, 9 \text{ m})$ at time $t_4 = 2.0 \text{ s}$. Notice how we've used subscripts to indicate the time and the object's position in a specific frame of the motion diagram.

NOTE ▶ The frame at $t = 0$ is frame 0. That is why the fifth frame is labeled 4. ◀

Another way to locate the ball is to draw an arrow from the origin to the point representing the ball. You can then specify the length and direction of the arrow. An arrow drawn from the origin to an object's position is called the **position vector** of the object, and it is given the symbol $\vec{r}$. FIGURE 1.5B shows the position vector $\vec{r}_4 = (15 \text{ m}, 37°)$.

The position vector $\vec{r}$ does not tell us anything different than the coordinates (x, y). It simply provides the information in an alternative form. Although you're more familiar with coordinates than with vectors, you will find that vectors are a useful way to describe many concepts in physics.

A Word About Vectors and Notation

Some physical quantities, such as time, mass, and temperature, can be described completely by a single number with a unit. For example, the mass of an object is 6 kg and its temperature is 30°C. A physical quantity described by a single number (with a unit) is called a **scalar quantity.** A scalar can be positive, negative, or zero.

Many other quantities, however, have a directional quality and cannot be described by a single number. To describe the motion of a car, for example, you must specify not only how fast it is moving, but also the *direction* in which it is moving. A **vector quantity** is a quantity having both a *size* (the "How far?" or "How fast?") and a *direction* (the "Which way?"). The size or length of a vector is called its *magnitude.* The magnitude of a vector can be positive or zero, but it cannot be negative. Vectors will be studied thoroughly in Chapter 3, so all we need for now is a little basic information.

We indicate a vector by drawing an arrow over the letter that represents the quantity. Thus $\vec{r}$ and $\vec{A}$ are symbols for vectors, whereas r and A, without the arrows, are symbols for scalars. In handwritten work you must draw arrows over all symbols that represent vectors. This may seem strange until you get used to it, but it is very important because we will often use both r and $\vec{r}$, or both A and $\vec{A}$, in the same problem, and they mean different things! Without the arrow, you will be using the same symbol with two different meanings and will likely end up making a mistake. Note that the arrow over the symbol always points to the right, regardless of which direction the actual vector points. Thus we write $\vec{r}$ or $\vec{A}$, never $\overleftarrow{r}$ or $\overleftarrow{A}$.

Displacement

Consider the following:

Sam is standing 50 feet (ft) east of the corner of 12th Street and Vine. He then walks northeast for 100 ft to a second point. What is Sam's change of position?

FIGURE 1.5 Position and time measurements made on the motion diagram of a basketball.

(a)

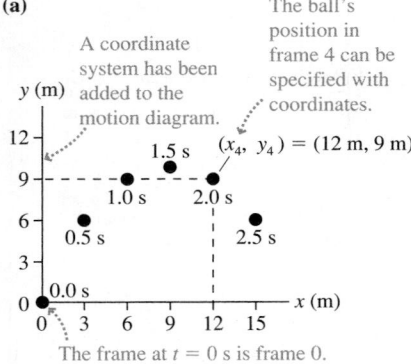

A coordinate system has been added to the motion diagram.

The ball's position in frame 4 can be specified with coordinates.

$(x_4, y_4) = (12 \text{ m}, 9 \text{ m})$

The frame at $t = 0$ s is frame 0.

(b)

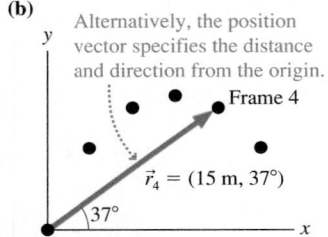

Alternatively, the position vector specifies the distance and direction from the origin.

Frame 4

$\vec{r}_4 = (15 \text{ m}, 37°)$

37°

FIGURE 1.6 shows Sam's motion in terms of position vectors. Sam's initial position is the vector $\vec{r}_0$ drawn from the origin to the point where he starts walking. Vector $\vec{r}_1$ is his position after he finishes walking. You can see that Sam has changed position, and a *change* of position is called a **displacement**. His displacement is the vector labeled $\Delta\vec{r}$. The Greek letter delta (Δ) is used in math and science to indicate the *change* in a quantity. Here it indicates a change in the position $\vec{r}$.

NOTE ▶ $\Delta\vec{r}$ is a *single* symbol. You cannot cancel out or remove the Δ in algebraic operations. ◀

FIGURE 1.6 Sam undergoes a displacement $\Delta\vec{r}$ from position $\vec{r}_0$ to position $\vec{r}_1$.

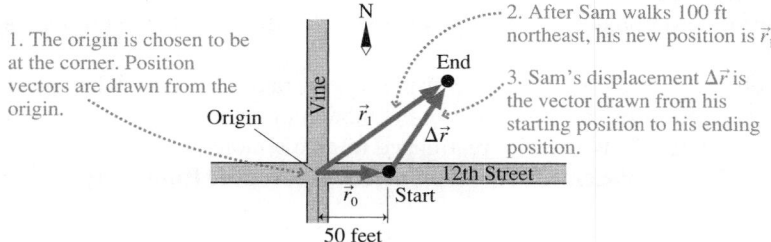

1. The origin is chosen to be at the corner. Position vectors are drawn from the origin.

2. After Sam walks 100 ft northeast, his new position is $\vec{r}_1$.

3. Sam's displacement $\Delta\vec{r}$ is the vector drawn from his starting position to his ending position.

Displacement is a vector quantity; it requires both a length and a direction to describe it. Specifically, the displacement $\Delta\vec{r}$ is a vector drawn *from* a starting position *to* an ending position. Sam's displacement is written

$$\Delta\vec{r} = (100\ \text{ft, northeast})$$

The length, or magnitude, of a displacement vector is simply the straight-line distance between the starting and ending positions.

Sam's final position in Figure 1.6, vector $\vec{r}_1$, can be seen as a combination of where he started, vector $\vec{r}_0$, plus the vector $\Delta\vec{r}$ representing his change of position. In fact, $\vec{r}_1$ is the *vector sum* of vectors $\vec{r}_0$ and $\Delta\vec{r}$. This is written

$$\vec{r}_1 = \vec{r}_0 + \Delta\vec{r} \tag{1.1}$$

Notice, however, that we are adding vector quantities, not numbers. Vector addition is a different process from "regular" addition. We'll explore vector addition more thoroughly in Chapter 3, but for now you can add two vectors $\vec{A}$ and $\vec{B}$ with the three-step procedure shown in Tactics Box 1.1.

TACTICS BOX 1.1   Vector addition

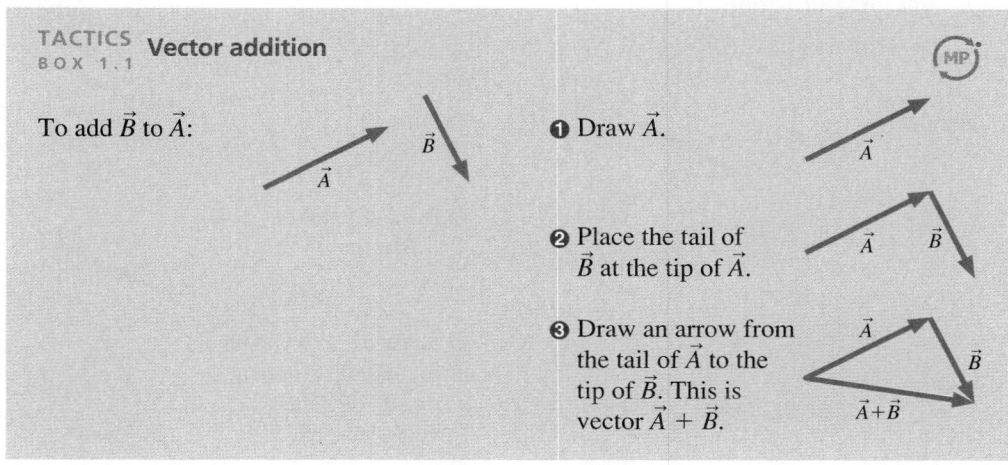

To add $\vec{B}$ to $\vec{A}$:

❶ Draw $\vec{A}$.

❷ Place the tail of $\vec{B}$ at the tip of $\vec{A}$.

❸ Draw an arrow from the tail of $\vec{A}$ to the tip of $\vec{B}$. This is vector $\vec{A} + \vec{B}$.

If you examine Figure 1.6, you'll see that the steps of Tactics Box 1.1 are exactly how $\vec{r}_0$ and $\Delta\vec{r}$ are added to give $\vec{r}_1$.

NOTE ▶ A vector is not tied to a particular location on the page. You can move a vector around as long as you don't change its length or the direction it points. Vector $\vec{B}$ is not changed by sliding it to where its tail is at the tip of $\vec{A}$. ◀

In Figure 1.6, we chose *arbitrarily* to put the origin of the coordinate system at the corner. While this might be convenient, it certainly is not mandatory. FIGURE 1.7 shows a different choice of where to place the origin. Notice something interesting. The initial and final position vectors $\vec{r}_0$ and $\vec{r}_1$ have become new vectors $\vec{r}_2$ and $\vec{r}_3$, but the displacement vector $\Delta\vec{r}$ has not changed! **The displacement is a quantity that is independent of the coordinate system.** In other words, the arrow drawn from one position of an object to the next is the same no matter what coordinate system you choose.

This observation suggests that the displacement, rather than the actual position, is what we want to focus on as we analyze the motion of an object. Equation 1.1 told us that $\vec{r}_1 = \vec{r}_0 + \Delta\vec{r}$. This is easily rearranged to give a more precise definition of displacement: **The displacement $\Delta\vec{r}$ of an object as it moves from an initial position $\vec{r}_i$ to a final position $\vec{r}_f$ is**

$$\Delta\vec{r} = \vec{r}_f - \vec{r}_i \tag{1.2}$$

Graphically, $\Delta\vec{r}$ is a vector arrow drawn from position $\vec{r}_i$ to position $\vec{r}_f$. The displacement vector is independent of the coordinate system.

NOTE ▶ To be more general, we've written Equation 1.2 in terms of an *initial position* and a *final position*, indicated by subscripts i and f. We'll frequently use i and f when writing general equations, then use specific numbers or values, such as 0 and 1, when working a problem. ◀

This definition of $\Delta\vec{r}$ involves *vector subtraction*. With numbers, subtraction is the same as the addition of a negative number. That is, $5 - 3$ is the same as $5 + (-3)$. Similarly, we can use the rules for vector addition to find $\vec{A} - \vec{B} = \vec{A} + (-\vec{B})$ if we first define what we mean by $-\vec{B}$. As FIGURE 1.8 shows, the negative of vector $\vec{B}$ is a vector with the same length but pointing in the opposite direction. This makes sense because $\vec{B} - \vec{B} = \vec{B} + (-\vec{B}) = \vec{0}$, where $\vec{0}$, a vector with zero length, is called the **zero vector.**

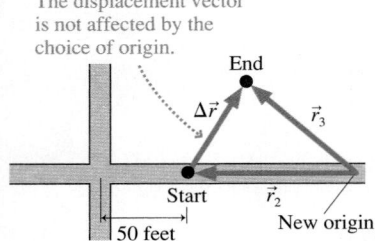

FIGURE 1.7 Sam's displacement $\Delta\vec{r}$ is unchanged by using a different coordinate system.

The displacement vector is not affected by the choice of origin.

End

$\Delta\vec{r}$ $\vec{r}_3$

Start $\vec{r}_2$

New origin

50 feet

FIGURE 1.8 The negative of a vector.

$\vec{B}$

$-\vec{B}$

Vector $-\vec{B}$ has the same length as $\vec{B}$ but points in the opposite direction.

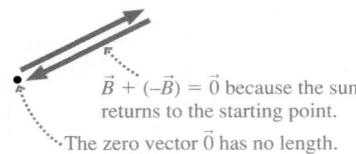

$\vec{B} + (-\vec{B}) = \vec{0}$ because the sum returns to the starting point.

The zero vector $\vec{0}$ has no length.

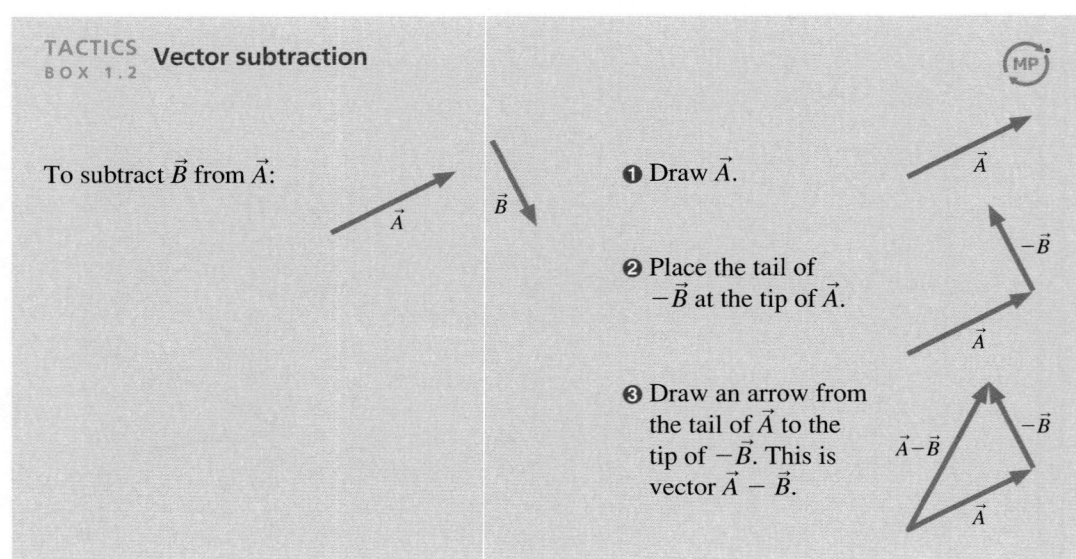

TACTICS BOX 1.2 **Vector subtraction**

To subtract $\vec{B}$ from $\vec{A}$:

$\vec{A}$ $\vec{B}$

❶ Draw $\vec{A}$.

$\vec{A}$

❷ Place the tail of $-\vec{B}$ at the tip of $\vec{A}$.

$-\vec{B}$

$\vec{A}$

❸ Draw an arrow from the tail of $\vec{A}$ to the tip of $-\vec{B}$. This is vector $\vec{A} - \vec{B}$.

$\vec{A} - \vec{B}$ $-\vec{B}$

$\vec{A}$

FIGURE 1.9 uses the vector subtraction rules of Tactics Box 1.2 to prove that the displacement $\Delta\vec{r}$ is simply the vector connecting the dots of a motion diagram.

FIGURE 1.9 Using vector subtraction to find $\Delta\vec{r} = \vec{r}_f - \vec{r}_i$.

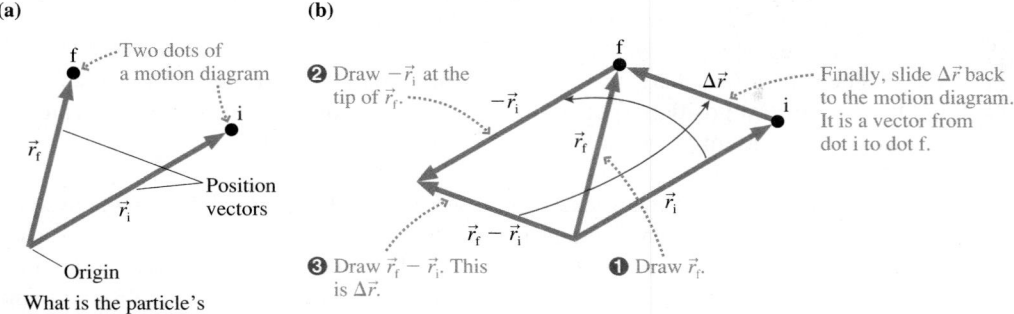

(a)

Two dots of a motion diagram

Position vectors

Origin

What is the particle's displacement vector $\Delta\vec{r}$?

(b)

❷ Draw $-\vec{r}_i$ at the tip of $\vec{r}_f$.

Finally, slide $\Delta\vec{r}$ back to the motion diagram. It is a vector from dot i to dot f.

❸ Draw $\vec{r}_f - \vec{r}_i$. This is $\Delta\vec{r}$.

❶ Draw $\vec{r}_f$.

Application to Motion Diagrams

The first step in analyzing a motion diagram is to determine all of the displacement vectors. As Figure 1.9 shows, the displacement vectors are simply the arrows connecting each dot to the next. Label each arrow with a *vector* symbol $\Delta\vec{r}_n$, starting with $n = 0$. **FIGURE 1.10** shows the motion diagrams of Figure 1.4 redrawn to include the displacement vectors. You do not need to show the position vectors.

NOTE ▶ When an object either starts from rest or ends at rest, the initial or final dots are *as close together* as you can draw the displacement vector arrow connecting them. In addition, just to be clear, you should write "Start" or "Stop" beside the initial or final dot. It is important to distinguish stopping from merely slowing down. ◀

Now we can conclude, more precisely than before, that, as time proceeds:

- An object is speeding up if its displacement vectors are increasing in length.
- An object is slowing down if its displacement vectors are decreasing in length.

FIGURE 1.10 Motion diagrams with the displacement vectors.

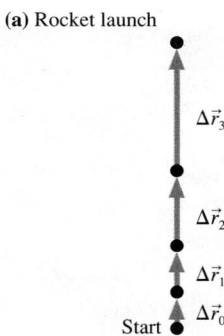

(a) Rocket launch

$\Delta\vec{r}_3$

$\Delta\vec{r}_2$

$\Delta\vec{r}_1$

Start $\Delta\vec{r}_0$

(b) Car stopping

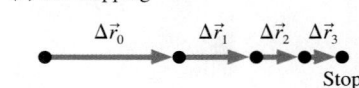

$\Delta\vec{r}_0$ $\Delta\vec{r}_1$ $\Delta\vec{r}_2$ $\Delta\vec{r}_3$

Stop

EXAMPLE 1.1 **Headfirst into the snow**

Alice is sliding along a smooth, icy road on her sled when she suddenly runs headfirst into a large, very soft snowbank that gradually brings her to a halt. Draw a motion diagram for Alice. Show and label all displacement vectors.

MODEL Use the particle model to represent Alice as a dot.

VISUALIZE **FIGURE 1.11** shows Alice's motion diagram. The problem statement suggests that Alice's speed is very nearly constant until she hits the snowbank. Thus her displacement vectors are of equal length as she slides along the icy road. She begins slowing when she hits the snowbank, so the displacement vectors then get shorter until she stops. We're told that her stop is gradual, so we want the vector lengths to get shorter gradually rather than suddenly.

FIGURE 1.11 Alice's motion diagram.

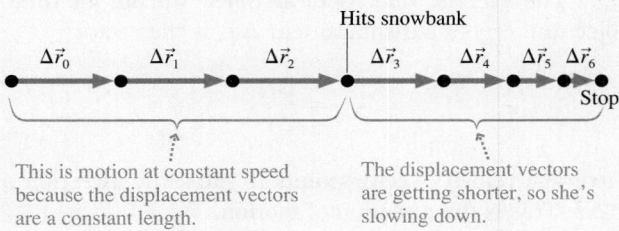

Hits snowbank

$\Delta\vec{r}_0$ $\Delta\vec{r}_1$ $\Delta\vec{r}_2$ $\Delta\vec{r}_3$ $\Delta\vec{r}_4$ $\Delta\vec{r}_5$ $\Delta\vec{r}_6$

Stop

This is motion at constant speed because the displacement vectors are a constant length.

The displacement vectors are getting shorter, so she's slowing down.

A stopwatch is used to measure a time interval.

Time Interval

It's also useful to consider a *change* in time. For example, the clock readings of two frames of film might be t_1 and t_2. The specific values are arbitrary because they are timed relative to an arbitrary instant that you chose to call $t = 0$. But the **time interval** $\Delta t = t_2 - t_1$ is *not* arbitrary. It represents the elapsed time for the object to move from one position to the next. All observers will measure the same value for Δt, regardless of when they choose to start their clocks.

The time interval $\Delta t = t_f - t_i$ measures the elapsed time as an object moves from an initial position $\vec{r}_i$ at time t_i to a final position $\vec{r}_f$ at time t_f. The value of Δt is independent of the specific clock used to measure the times.

To summarize the main idea of this section, we have added coordinate systems and clocks to our motion diagrams in order to measure *when* each frame was exposed and *where* the object was located at that time. Different observers of the motion may choose different coordinate systems and different clocks. However, all observers find the *same* values for the displacements $\Delta\vec{r}$ and the time intervals Δt because these are independent of the specific coordinate system used to measure them.

1.4 Velocity

It's no surprise that, during a given time interval, a speeding bullet travels farther than a speeding snail. To extend our study of motion so that we can compare the bullet to the snail, we need a way to measure how fast or how slowly an object moves.

One quantity that measures an object's fastness or slowness is its **average speed,** defined as the ratio

$$\text{average speed} = \frac{\text{distance traveled}}{\text{time interval spent traveling}} = \frac{d}{\Delta t} \tag{1.3}$$

If you drive 15 miles (mi) in 30 minutes ($\frac{1}{2}$ h), your average speed is

$$\text{average speed} = \frac{15 \text{ mi}}{\frac{1}{2}\text{ h}} = 30 \text{ mph} \tag{1.4}$$

The victory goes to the runner with the highest average speed.

Although the concept of speed is widely used in our day-to-day lives, it is not a sufficient basis for a science of motion. To see why, imagine you're trying to land a jet plane on an aircraft carrier. It matters a great deal to you whether the aircraft carrier is moving at 20 mph (miles per hour) to the north or 20 mph to the east. Simply knowing that the boat's speed is 20 mph is not enough information!

It's the displacement $\Delta\vec{r}$, a vector quantity, that tells us not only the distance traveled by a moving object, but also the *direction* of motion. Consequently, a more useful ratio than $d/\Delta t$ is the ratio $\Delta\vec{r}/\Delta t$. This ratio is a vector because $\Delta\vec{r}$ is a vector, so it has both a magnitude and a direction. The size, or magnitude, of this ratio will be larger for a fast object than for a slow object. But in addition to measuring how fast an object moves, this ratio is a vector that points in the direction of motion.

It is convenient to give this ratio a name. We call it the **average velocity,** and it has the symbol $\vec{v}_{avg}$. **The average velocity of an object during the time interval Δt, in which the object undergoes a displacement $\Delta\vec{r}$, is the vector**

$$\vec{v}_{avg} = \frac{\Delta\vec{r}}{\Delta t} \tag{1.5}$$

An object's average velocity vector points in the same direction as the displacement vector $\Delta\vec{r}$. This is the direction of motion.

NOTE ▶ In everyday language we do not make a distinction between speed and velocity, but in physics *the distinction is very important*. In particular, speed is simply "How fast?" whereas velocity is "How fast, and in which direction?" As we go along we will be giving other words more precise meanings in physics than they have in everyday language. ◀

As an example, **FIGURE 1.12a** shows two ships that move 5 miles in 15 minutes. Using Equation 1.5 with $\Delta t = 0.25$ h, we find

$$\vec{v}_{\text{avg A}} = (20 \text{ mph, north})$$

$$\vec{v}_{\text{avg B}} = (20 \text{ mph, east})$$

(1.6)

Both ships have a speed of 20 mph, but their velocities are different. Notice how the velocity *vectors* in **FIGURE 1.12b** point in the direction of motion.

NOTE ▶ Our goal in this chapter is to *visualize* motion with motion diagrams. Strictly speaking, the vector we have defined in Equation 1.5, and the vector we will show on motion diagrams, is the *average* velocity $\vec{v}_{\text{avg}}$. But to allow the motion diagram to be a useful tool, we will drop the subscript and refer to the average velocity as simply $\vec{v}$. Our definitions and symbols, which somewhat blur the distinction between average and instantaneous quantities, are adequate for visualization purposes, but they're not the final word. We will refine these definitions in Chapter 2, where our goal will be to develop the mathematics of motion. ◀

Motion Diagrams with Velocity Vectors

The velocity vector points in the same direction as the displacement $\Delta\vec{r}$, and the length of $\vec{v}$ is directly proportional to the length of $\Delta\vec{r}$. Consequently, the vectors connecting each dot of a motion diagram to the next, which we previously labeled as displacements, could equally well be identified as velocity vectors.

This idea is illustrated in **FIGURE 1.13**, which shows four frames from the motion diagram of a tortoise racing a hare. The vectors connecting the dots are now labeled as velocity vectors $\vec{v}$. **The length of a velocity vector represents the average speed with which the object moves between the two points.** Longer velocity vectors indicate faster motion. You can see that the hare moves faster than the tortoise.

Notice that the hare's velocity vectors do not change; each has the same length and direction. We say the hare is moving with *constant velocity*. The tortoise is also moving with its own constant velocity.

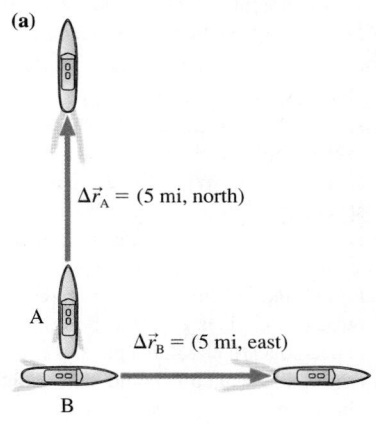

FIGURE 1.12 The displacement vectors and velocities of ships A and B.

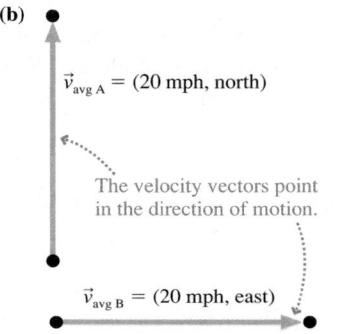

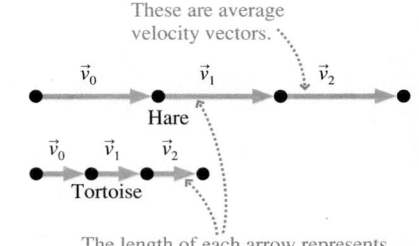

FIGURE 1.13 Motion diagram of the tortoise racing the hare.

EXAMPLE 1.2 Accelerating up a hill

The light turns green and a car accelerates, starting from rest, up a 20° hill. Draw a motion diagram showing the car's velocity.

MODEL Use the particle model to represent the car as a dot.

VISUALIZE The car's motion takes place along a straight line, but the line is neither horizontal nor vertical. Because a motion diagram is made from frames of a movie, it will show the object moving with the correct orientation—in this case, at an angle of 20°. **FIGURE 1.14** shows several frames of the motion diagram, where we see the car speeding up. The car starts from rest, so the first arrow is drawn as short as possible and the first dot is labeled "Start." The displacement vectors have been drawn from each dot to the next, but then they are identified and labeled as average velocity vectors $\vec{v}$.

FIGURE 1.14 Motion diagram of a car accelerating up a hill.

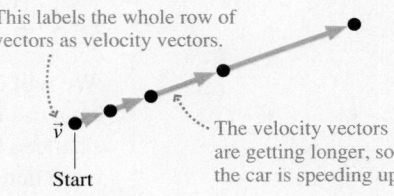

NOTE ▶ Rather than label every single vector, it's easier to give one label to the entire row of velocity vectors. You can see this in Figure 1.14. ◀

EXAMPLE 1.3 **It's a hit!**

Jake hits a ball at a 60° angle above horizontal. It is caught by Jose. Draw a motion diagram of the ball.

MODEL This example is typical of how many problems in science and engineering are worded. The problem does not give a clear statement of where the motion begins or ends. Are we interested in the motion of the ball just during the time it is in the air between Jake and Jose? What about the motion *as* Jake hits it (ball rapidly speeding up) or *as* Jose catches it (ball rapidly slowing down)? The point is that *you* will often be called on to make a *reasonable interpretation* of a problem statement. In this problem, the details of hitting and catching the ball are complex. The motion of the ball through the air is easier to describe, and it's a motion you might expect to learn about in a physics class. So our *interpretation* is that the motion diagram should start as the ball leaves Jake's bat (ball already moving) and should end the instant it touches Jose's hand (ball still moving). We will model the ball as a particle.

VISUALIZE With this interpretation in mind, FIGURE 1.15 shows the motion diagram of the ball. Notice how, in contrast to the car

of Figure 1.14, the ball is already moving as the motion diagram movie begins. As before, the average velocity vectors are found by connecting the dots with *straight* arrows. You can see that the average velocity vectors get shorter (ball slowing down), get longer (ball speeding up), and change direction. Each $\vec{v}$ is different, so this is *not* constant-velocity motion.

FIGURE 1.15 Motion diagram of a ball traveling from Jake to Jose.

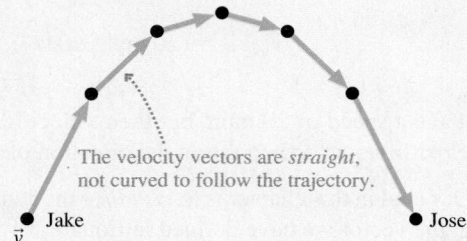

The velocity vectors are *straight*, not curved to follow the trajectory.

Jake $\vec{v}$

Jose

A particle moves from position 1 to position 2 during the interval Δt. Which vector shows the particle's average velocity?

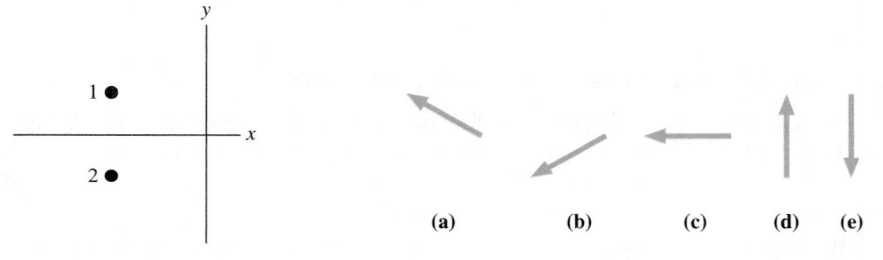

1.5 Linear Acceleration

The goal of this chapter is to find a set of concepts with which to describe motion. Position, time, and velocity are important concepts, and at first glance they might appear to be sufficient. But that is not the case. Sometimes an object's velocity is constant, as it was in Figure 1.13. More often, an object's velocity changes as it moves, as in Figure 1.14 and 1.15. We need one more motion concept, one that will describe a *change* in the velocity.

Because velocity is a vector, it can change in two possible ways:

1. The magnitude can change, indicating a change in speed; or
2. The direction can change, indicating that the object has changed direction.

We will concentrate for now on the first case, a change in speed. The car accelerating up a hill in Figure 1.14 was an example in which the magnitude of the velocity vector changed but not the direction. We'll return to the second case in Chapter 4.

When we wanted to measure changes in position, the ratio $\Delta \vec{r}/\Delta t$ was useful. This ratio is the *rate of change of position*. By analogy, consider an object whose velocity changes from $\vec{v}_1$ to $\vec{v}_2$ during the time interval Δt. Just as $\Delta \vec{r} = \vec{r}_2 - \vec{r}_1$ is the change of position, the quantity $\Delta \vec{v} = \vec{v}_2 - \vec{v}_1$ is the change of velocity. The ratio $\Delta \vec{v}/\Delta t$ is then the *rate of change of velocity*. It has a large magnitude for objects that speed up quickly and a small magnitude for objects that speed up slowly.

The ratio $\Delta \vec{v}/\Delta t$ is called the **average acceleration,** and its symbol is $\vec{a}_{\text{avg}}$. The average acceleration of an object during the time interval Δt, in which the object's velocity changes by $\Delta \vec{v}$, is the vector

$$\vec{a}_{\text{avg}} = \frac{\Delta \vec{v}}{\Delta t} \qquad (1.7)$$

The average acceleration vector points in the same direction as the vector $\Delta \vec{v}$.

Acceleration is a fairly abstract concept. Yet it is essential to develop a good intuition about acceleration because it will be a key concept for understanding why objects move as they do. Motion diagrams will be an important tool for developing that intuition.

NOTE ▶ As we did with velocity, we will drop the subscript and refer to the average acceleration as simply $\vec{a}$. This is adequate for visualization purposes, but not the final word. We will refine the definition of acceleration in Chapter 2. ◀

The Audi TT accelerates from 0 to 60 mph in 6 s.

Finding the Acceleration Vectors on a Motion Diagram

Let's look at how we can determine the average acceleration vector $\vec{a}$ from a motion diagram. From its definition, Equation 1.7, we see that $\vec{a}$ points in the same direction as $\Delta \vec{v}$, the change of velocity. This critical idea is the basis for a technique to find $\vec{a}$.

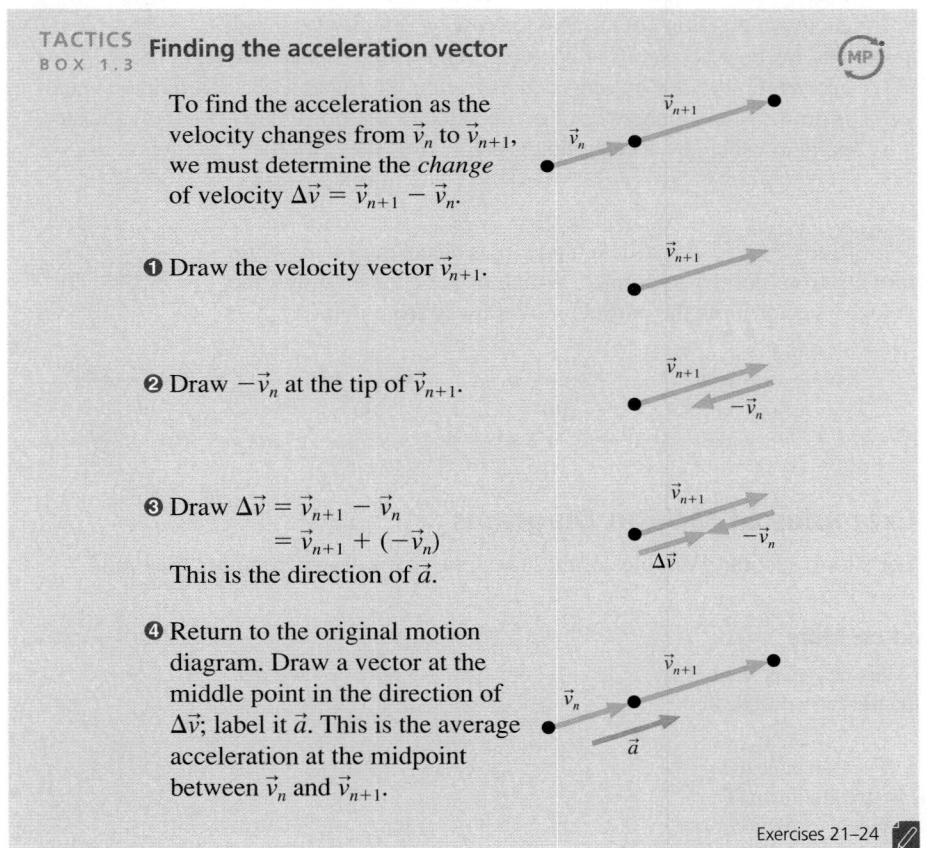

TACTICS
BOX 1.3
Finding the acceleration vector

To find the acceleration as the velocity changes from $\vec{v}_n$ to $\vec{v}_{n+1}$, we must determine the *change* of velocity $\Delta \vec{v} = \vec{v}_{n+1} - \vec{v}_n$.

❶ Draw the velocity vector $\vec{v}_{n+1}$.

❷ Draw $-\vec{v}_n$ at the tip of $\vec{v}_{n+1}$.

❸ Draw $\Delta \vec{v} = \vec{v}_{n+1} - \vec{v}_n$
$= \vec{v}_{n+1} + (-\vec{v}_n)$
This is the direction of $\vec{a}$.

❹ Return to the original motion diagram. Draw a vector at the middle point in the direction of $\Delta \vec{v}$; label it $\vec{a}$. This is the average acceleration at the midpoint between $\vec{v}_n$ and $\vec{v}_{n+1}$.

Exercises 21–24

Many Tactics Boxes will refer you to exercises in the
Student Workbook where you can practice the new skill.

Notice that the acceleration vector goes beside the middle dot, not beside the velocity vectors. This is because each acceleration vector is determined as the *difference* between the *two* velocity vectors on either side of a dot. The length of $\vec{a}$ does not have to be the exact length of $\Delta \vec{v}$; it is the direction of $\vec{a}$ that is most important.

The procedure of Tactics Box 1.3 can be repeated to find $\vec{a}$ at each point in the motion diagram. Note that we cannot determine $\vec{a}$ at the first and last points because we have only one velocity vector and can't find $\Delta\vec{v}$.

The Complete Motion Diagram

You've now seen several *Tactics Boxes* that help you accomplish specific tasks. Tactics Boxes will appear in nearly every chapter in this book. We'll also, where appropriate, provide *Problem-Solving Strategies*.

PROBLEM-SOLVING
STRATEGY 1.1 **Motion diagrams**

MODEL Represent the moving object as a particle. Make simplifying assumptions when interpreting the problem statement.

VISUALIZE A complete motion diagram consists of:

■ The position of the object in each frame of the film, shown as a dot. Use five or six dots to make the motion clear but without overcrowding the picture. More complex motions may need more dots.

■ The average velocity vectors, found by connecting each dot in the motion diagram to the next with a vector arrow. There is *one* velocity vector linking each *two* position dots. Label the row of velocity vectors $\vec{v}$.

■ The average acceleration vectors, found using Tactics Box 1.3. There is *one* acceleration vector linking each *two* velocity vectors. Each acceleration vector is drawn at the dot between the two velocity vectors it links. Use $\vec{0}$ to indicate a point at which the acceleration is zero. Label the row of acceleration vectors $\vec{a}$.

STOP TO THINK 1.4 A particle undergoes acceleration $\vec{a}$ while moving from point 1 to point 2. Which of the choices shows the velocity vector $\vec{v}_2$ as the particle moves away from point 2?

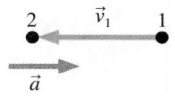

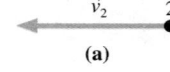   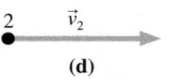

(a)   (b)   (c)   (d)

Examples of Motion Diagrams

Let's look at some examples of the full strategy for drawing motion diagrams.

EXAMPLE 1.4 **The first astronauts land on Mars**

A spaceship carrying the first astronauts to Mars descends safely to the surface. Draw a motion diagram for the last few seconds of the descent.

MODEL Represent the spaceship as a particle. It's reasonable to assume that its motion in the last few seconds is straight down. The problem ends as the spacecraft touches the surface.

VISUALIZE FIGURE 1.16 shows a complete motion diagram as the spaceship descends and slows, using its rockets, until it comes to rest on the surface. Notice how the dots get closer together as it slows. The inset shows how the acceleration vector $\vec{a}$ is determined at one point. All the other acceleration vectors will be similar, because for each pair of velocity vectors the earlier one is longer than the later one.

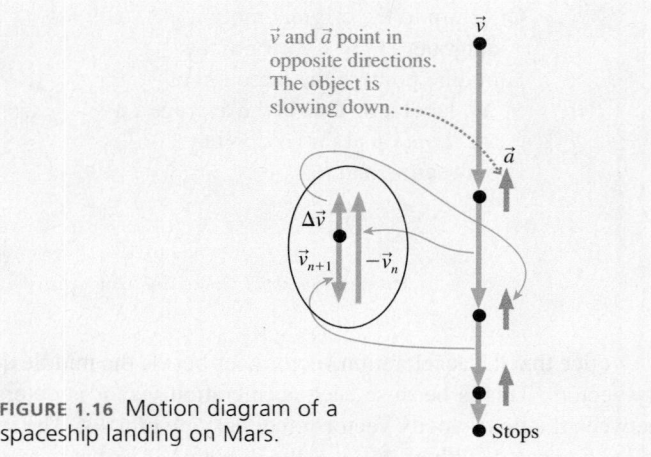

FIGURE 1.16 Motion diagram of a spaceship landing on Mars.

EXAMPLE 1.5 **Skiing through the woods**

A skier glides along smooth, horizontal snow at constant speed, then speeds up going down a hill. Draw the skier's motion diagram.

MODEL Represent the skier as a particle. It's reasonable to assume that the downhill slope is a straight line. Although the motion as a whole is not linear, we can treat the skier's motion as two separate linear motions.

VISUALIZE FIGURE 1.17 shows a complete motion diagram of the skier. The dots are equally spaced for the horizontal motion, indicating constant speed; then the dots get farther apart as the skier speeds up going down the hill. The insets show how the average acceleration vector $\vec{a}$ is determined for the horizontal motion and along the slope. All the other acceleration vectors along the slope will be similar to the one shown because each velocity vector is longer than the preceding one. Notice that we've explicitly written $\vec{0}$ for the acceleration beside the dots where the velocity is constant. The acceleration at the point where the direction changes will be considered in Chapter 4.

FIGURE 1.17 Motion diagram of a skier.

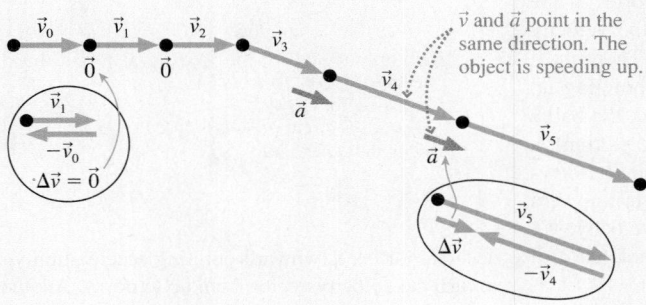

Notice something interesting in Figure 1.16 and 1.17. Where the object is speeding up, the acceleration and velocity vectors point in the *same direction*. Where the object is slowing down, the acceleration and velocity vectors point in *opposite directions*. These results are always true for motion in a straight line. **For motion along a line:**

- **An object is speeding up if and only if $\vec{v}$ and $\vec{a}$ point in the same direction.**
- **An object is slowing down if and only if $\vec{v}$ and $\vec{a}$ point in opposite directions.**
- **An object's velocity is constant if and only if $\vec{a} = \vec{0}$.**

NOTE ▶ In everyday language, we use the word *accelerate* to mean "speed up" and the word *decelerate* to mean "slow down." But speeding up and slowing down are both changes in the velocity and consequently, by our definition, *both* are accelerations. In physics, *acceleration* refers to changing the velocity, no matter what the change is, and not just to speeding up. ◀

EXAMPLE 1.6 **Tossing a ball**

Draw the motion diagram of a ball tossed straight up in the air.

MODEL This problem calls for some interpretation. Should we include the toss itself, or only the motion after the ball is released? Should we include the ball hitting the ground? It appears that this problem is really concerned with the ball's motion through the air. Consequently, we begin the motion diagram at the moment that the tosser releases the ball and end the diagram at the moment the ball hits the ground. We will consider neither the toss nor the impact. And, of course, we will represent the ball as a particle.

VISUALIZE We have a slight difficulty here because the ball retraces its route as it falls. A literal motion diagram would show

Continued

the upward motion and downward motion on top of each other, leading to confusion. We can avoid this difficulty by horizontally separating the upward motion and downward motion diagrams. This will not affect our conclusions because it does not change any of the vectors. **FIGURE 1.18** shows the motion diagram drawn this way. Notice that the very top dot is shown twice—as the end point of the upward motion and the beginning point of the downward motion.

The ball slows down as it rises. You've learned that the acceleration vectors point opposite the velocity vectors for an object that is slowing down along a line, and they are shown accordingly. Similarly, $\vec{a}$ and $\vec{v}$ point in the same direction as the falling ball speeds up. Notice something interesting: The acceleration vectors point downward both while the ball is rising *and* while it is falling. Both "speeding up" and "slowing down" occur with the *same* acceleration vector. This is an important conclusion, one worth pausing to think about.

Now let's look at the top point on the ball's trajectory. The velocity vectors are pointing upward but getting shorter as the ball approaches the top. As the ball starts to fall, the velocity vectors are pointing downward and getting longer. There must be a moment—just an instant as $\vec{v}$ switches from pointing up to pointing down—when the velocity is zero. Indeed, the ball's velocity *is* zero for an instant at the precise top of the motion!

But what about the acceleration at the top? The inset shows how the average acceleration is determined from the last upward velocity before the top point and the first downward velocity. We find that the acceleration at the top is pointing downward, just as it does elsewhere in the motion.

Many people expect the acceleration to be zero at the highest point. But recall that the velocity at the top point *is* changing—from up to down. If the velocity is changing, there *must* be an

FIGURE 1.18 Motion diagram of a ball tossed straight up in the air.

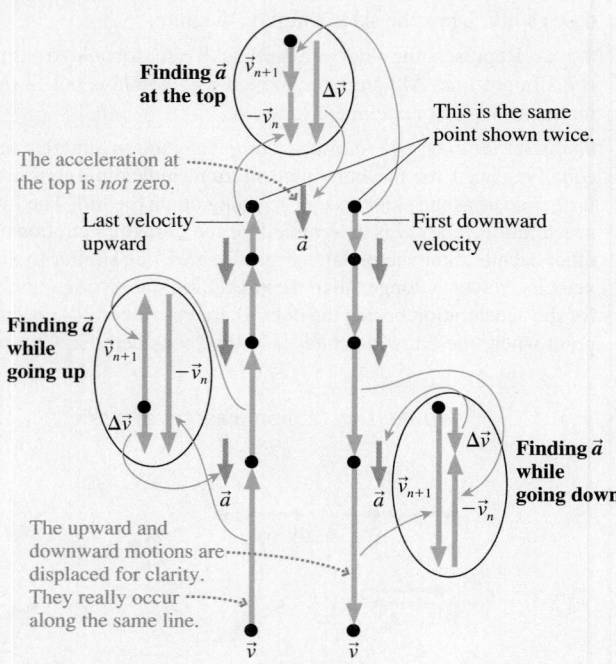

acceleration. A downward-pointing acceleration vector is needed to turn the velocity vector from up to down. Another way to think about this is to note that zero acceleration would mean no change of velocity. When the ball reached zero velocity at the top, it would hang there and not fall if the acceleration were also zero!

1.6 Motion in One Dimension

As you've seen, an object's motion can be described in terms of three fundamental quantities: its position $\vec{r}$, velocity $\vec{v}$, and acceleration $\vec{a}$. These quantities are vectors, having a direction as well as a magnitude. But for motion in one dimension, the vectors are restricted to point only "forward" or "backward." Consequently, we can describe one-dimensional motion with the simpler quantities x, v_x, and a_x (or y, v_y, and a_y). However, we need to give each of these quantities an explicit *sign*, positive or negative, to indicate whether the position, velocity, or acceleration vector points forward or backward.

Determining the Signs of Position, Velocity, and Acceleration

Position, velocity, and acceleration are measured with respect to a coordinate system, a grid or axis that *you* impose on a problem to analyze the motion. We will find it convenient to use an *x*-axis to describe both horizontal motion and motion along an inclined plane. A *y*-axis will be used for vertical motion. A coordinate axis has two essential features:

1. An origin to define zero; and
2. An *x* or *y* label to indicate the positive end of the axis.

We will adopt the convention that **the positive end of an *x*-axis is to the right and the positive end of a *y*-axis is up.** The signs of position, velocity, and acceleration are based on this convention.

Determining the sign of the position, velocity, and acceleration

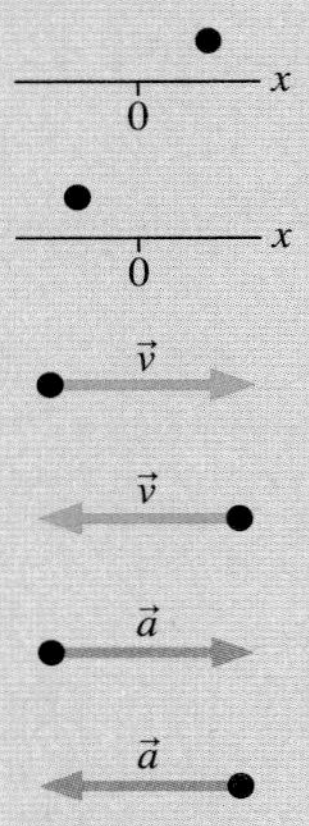

$x > 0$ Position to right of origin.

$x < 0$ Position to left of origin.

$v_x > 0$ Direction of motion is to the right.

$v_x < 0$ Direction of motion is to the left.

$a_x > 0$ Acceleration vector points to the right.

$a_x < 0$ Acceleration vector points to the left.

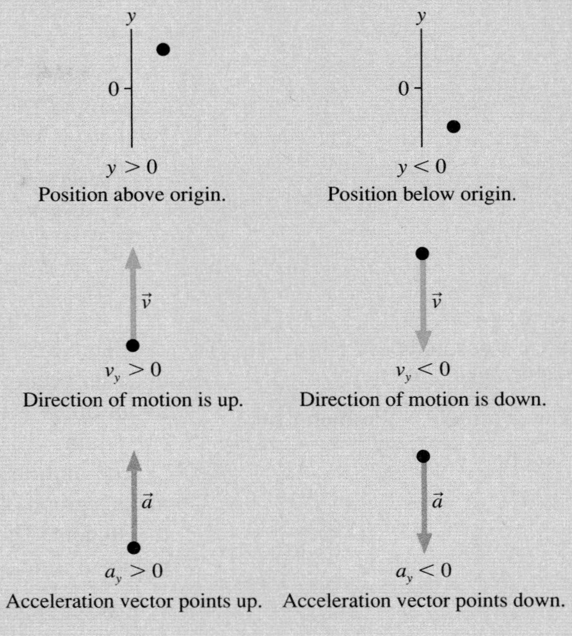

$y > 0$
Position above origin.

$y < 0$
Position below origin.

$v_y > 0$
Direction of motion is up.

$v_y < 0$
Direction of motion is down.

$a_y > 0$
Acceleration vector points up.

$a_y < 0$
Acceleration vector points down.

- The sign of position (x or y) tells us *where* an object is.

- The sign of velocity (v_x or v_y) tells us *which direction* the object is moving.

- The sign of acceleration (a_x or a_y) tells us which way the acceleration vector points, *not* whether the object is speeding up or slowing down.

Exercises 30–31

Acceleration is where things get a bit tricky. A natural tendency is to think that a positive value of a_x or a_y describes an object that is speeding up while a negative value describes an object that is slowing down (decelerating). However, this interpretation *does not work*.

Acceleration was defined as $\vec{a}_{avg} = \Delta\vec{v}/\Delta t$. The direction of $\vec{a}$ can be determined by using a motion diagram to find the direction of $\Delta\vec{v}$. The one-dimensional acceleration a_x (or a_y) is then positive if the vector $\vec{a}$ points to the right (or up), negative if $\vec{a}$ points to the left (or down).

FIGURE 1.19 shows that this method for determining the sign of a does not conform to the simple idea of speeding up and slowing down. The object in Figure 1.19a has a positive acceleration ($a_x > 0$) not because it is speeding up but because the vector $\vec{a}$ points in the positive direction. Compare this with the motion diagram of Figure 1.19b. Here the object is slowing down, but it still has a positive acceleration ($a_x > 0$) because $\vec{a}$ points to the right.

We found that an object is speeding up if $\vec{v}$ and $\vec{a}$ point in the same direction, slowing down if they point in opposite directions. For one-dimensional motion this rule becomes:

- An object is speeding up if and only if v_x and a_x have the same sign.
- An object is slowing down if and only if v_x and a_x have opposite signs.
- An object's velocity is constant if and only if $a_x = 0$.

Notice how the first two of these rules are at work in Figure 1.19.

Position-versus-Time Graphs

FIGURE 1.20 is a motion diagram, made at 1 frame per minute, of a student walking to school. You can see that she leaves home at a time we choose to call $t = 0$ min and makes steady progress for a while. Beginning at $t = 3$ min there is a period where the

FIGURE 1.19 One of these objects is speeding up, the other slowing down, but they both have a positive acceleration a_x.

(a) Speeding up to the right

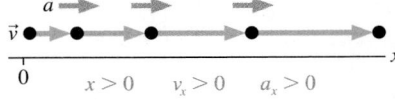

$x > 0$ $v_x > 0$ $a_x > 0$

(b) Slowing down to the left

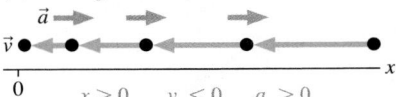

$x > 0$ $v_x < 0$ $a_x > 0$

distance traveled during each time interval becomes less—perhaps she slowed down to speak with a friend. Then she picks up the pace, and the distances within each interval are longer.

FIGURE 1.20 The motion diagram of a student walking to school and a coordinate axis for making measurements.

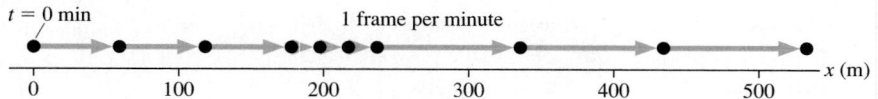

TABLE 1.1 Measured positions of a student walking to school

Time t (min)	Position x (m)
0	0
1	60
2	120
3	180
4	200
5	220
6	240
7	340
8	440
9	540

Figure 1.20 includes a coordinate axis, and you can see that every dot in a motion diagram occurs at a specific position. Table 1.1 shows the student's positions at different times as measured along this axis. For example, she is at position $x = 120$ m at $t = 2$ min.

The motion diagram is one way to represent the student's motion. Another is to make a graph of the measurements in Table 1.1. **FIGURE 1.21a** is a graph of x versus t for the student. The motion diagram tells us only where the student is at a few discrete points of time, so this graph of the data shows only points, no lines.

NOTE ▶ A graph of "a versus b" means that a is graphed on the vertical axis and b on the horizontal axis. Saying "graph a versus b" is really a shorthand way of saying "graph a as a function of b." ◀

FIGURE 1.21 Position graphs of the student's motion.

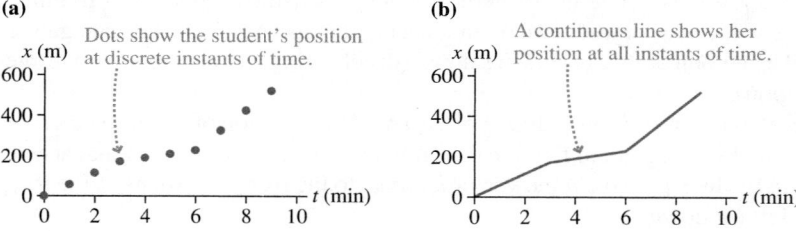

However, common sense tells us the following. First, the student was *somewhere specific* at all times. That is, there was never a time when she failed to have a well-defined position, nor could she occupy two positions at one time. (As reasonable as this belief appears to be, it will be severely questioned and found not entirely accurate when we get to quantum physics!) Second, the student moved *continuously* through all intervening points of space. She could not go from $x = 100$ m to $x = 200$ m without passing through every point in between. It is thus quite reasonable to believe that her motion can be shown as a continuous line passing through the measured points, as shown in **FIGURE 1.21b**. A continuous line or curve showing an object's position as a function of time is called a **position-versus-time graph** or, sometimes, just a *position graph*.

NOTE ▶ A graph is *not* a "picture" of the motion. The student is walking along a straight line, but the graph itself is not a straight line. Further, we've graphed her position on the vertical axis even though her motion is horizontal. Graphs are *abstract representations* of motion. We will place significant emphasis on the process of interpreting graphs, and many of the exercises and problems will give you a chance to practice these skills. ◀

EXAMPLE 1.7 **Interpreting a position graph**

The graph in FIGURE 1.22a represents the motion of a car along a straight road. Describe the motion of the car.

MODEL Represent the car as a particle.

VISUALIZE As FIGURE 1.22b shows, the graph represents a car that travels to the left for 30 minutes, stops for 10 minutes, then travels back to the right for 40 minutes.

FIGURE 1.22 Position-versus-time graph of a car.

(a)

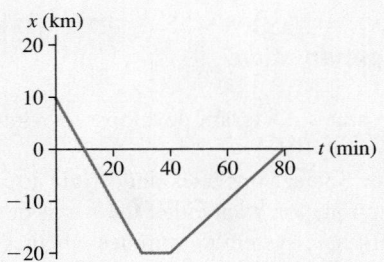

(b)

1. At $t = 0$ min, the car is 10 km to the right of the origin.

2. The value of x decreases for 30 min, indicating that the car is moving to the left.

5. The car reaches the origin at $t = 80$ min.

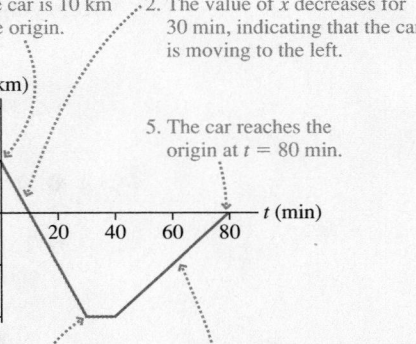

3. The car stops for 10 min at a position 20 km to the left of the origin.

4. The car starts moving back to the right at $t = 40$ min.

1.7 Solving Problems in Physics

Physics is not mathematics. Math problems are clearly stated, such as "What is $2 + 2$?" Physics is about the world around us, and to describe that world we must use language. Now, language is wonderful—we couldn't communicate without it—but language can sometimes be imprecise or ambiguous.

The challenge when reading a physics problem is to translate the words into symbols that can be manipulated, calculated, and graphed. **The translation from words to symbols is the heart of problem solving in physics.** This is the point where ambiguous words and phrases must be clarified, where the imprecise must be made precise, and where you arrive at an understanding of exactly what the question is asking.

Using Symbols

Symbols are a language that allows us to talk with precision about the relationships in a problem. As with any language, we all need to agree to use words or symbols in the same way if we want to communicate with each other. Many of the ways we use symbols in science and engineering are somewhat arbitrary, often reflecting historical roots. Nonetheless, practicing scientists and engineers have come to agree on how to use the language of symbols. Learning this language is part of learning physics.

We will use subscripts on symbols, such as x_3, to designate a particular point in the problem. Scientists usually label the starting point of the problem with the subscript "0," not the subscript "1" that you might expect. When using subscripts, make sure that all symbols referring to the same point in the problem have the *same numerical subscript*. To have the same point in a problem characterized by position x_1 but velocity v_{2x} is guaranteed to lead to confusion!

Drawing Pictures

You may have been told that the first step in solving a physics problem is to "draw a picture," but perhaps you didn't know why, or what to draw. The purpose of drawing a picture is to aid you in the words-to-symbols translation. Complex problems have far more information than you can keep in your head at one time. Think of a picture as a "memory extension," helping you organize and keep track of vital information.

Although any picture is better than none, there really is a *method* for drawing pictures that will help you be a better problem solver. It is called the **pictorial representation** of the problem. We'll add other pictorial representations as we go along, but the following procedure is appropriate for motion problems.

TACTICS
BOX 1.5 Drawing a pictorial representation

❶ **Draw a motion diagram.** The motion diagram develops your intuition for the motion.

❷ **Establish a coordinate system.** Select your axes and origin to match the motion. For one-dimensional motion, you want either the x-axis or the y-axis parallel to the motion. The coordinate system determines whether the signs of v and a are positive or negative.

❸ **Sketch the situation.** Not just any sketch. Show the object at the *beginning* of the motion, at the *end*, and at any point where the character of the motion changes. Show the object, not just a dot, but very simple drawings are adequate.

❹ **Define symbols.** Use the sketch to define symbols representing quantities such as position, velocity, acceleration, and time. *Every* variable used later in the mathematical solution should be defined on the sketch. Some will have known values, others are initially unknown, but all should be given symbolic names.

❺ **List known information.** Make a table of the quantities whose values you can determine from the problem statement or that can be found quickly with simple geometry or unit conversions. Some quantities are implied by the problem, rather than explicitly given. Others are determined by your choice of coordinate system.

❻ **Identify the desired unknowns.** What quantity or quantities will allow you to answer the question? These should have been defined as symbols in step 4. Don't list every unknown, only the one or two needed to answer the question.

It's not an overstatement to say that a well-done pictorial representation of the problem will take you halfway to the solution. The following example illustrates how to construct a pictorial representation for a problem that is typical of problems you will see in the next few chapters.

EXAMPLE 1.8 **Drawing a pictorial representation**

Draw a pictorial representation for the following problem: A rocket sled accelerates horizontally at 50 m/s^2 for 5.0 s, then coasts for 3.0 s. What is the total distance traveled?

VISUALIZE The motion diagram shows an acceleration phase followed by a coasting phase. Because the motion is horizontal, the appropriate coordinate system is an x-axis. We've chosen to place the origin at the starting point. The motion has a beginning, an end, and a point where the nature of the motion changes from accelerating to

coasting. These are the three sled positions sketched in **FIGURE 1.23**. The quantities x, v_x, and t are needed at each of three *points*, so these have been defined on the sketch and distinguished by subscripts. Accelerations are associated with *intervals* between the points, so only two accelerations are defined. Values for three quantities are given in the problem statement, although we need to use the motion diagram, where $\vec{a}$ points to the right, and our choice of coordinate system to know that $a_{0x} = +50 \text{ m/s}^2$ rather than -50 m/s^2.

The values $x_0 = 0$ m and $t_0 = 0$ s are choices we made when setting up the coordinate system. The value $v_{0x} = 0$ m/s is part of our *interpretation* of the problem. Finally, we identify x_2 as the quantity that will answer the question. We now understand quite a bit about the problem and would be ready to start a quantitative analysis.

FIGURE 1.23 A pictorial representation.

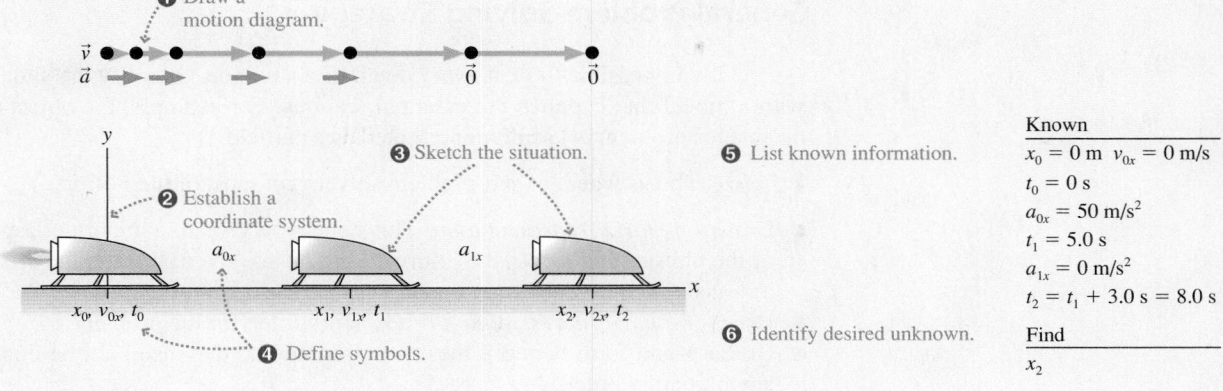

We didn't *solve* the problem; that is not the purpose of the pictorial representation. The pictorial representation is a systematic way to go about interpreting a problem and getting ready for a mathematical solution. Although this is a simple problem, and you probably know how to solve it if you've taken physics before, you will soon be faced with much more challenging problems. Learning good problem-solving skills at the beginning, while the problems are easy, will make them second nature later when you really need them.

Representations

A picture is one way to *represent* your knowledge of a situation. You could also represent your knowledge using words, graphs, or equations. Each **representation of knowledge** gives us a different perspective on the problem. The more tools you have for thinking about a complex problem, the more likely you are to solve it.

There are four representations of knowledge that we will use over and over:

1. The *verbal* representation. A problem statement, in words, is a verbal representation of knowledge. So is an explanation that you write.
2. The *pictorial* representation. The pictorial representation, which we've just presented, is the most literal depiction of the situation.
3. The *graphical* representation. We will make extensive use of graphs.
4. The *mathematical* representation. Equations that can be used to find the numerical values of specific quantities are the mathematical representation.

NOTE ▶ The mathematical representation is only one of many. Much of physics is more about thinking and reasoning than it is about solving equations. ◀

A new building requires careful planning. The architect's visualization and drawings have to be complete before the detailed procedures of construction get under way. The same is true for solving problems in physics.

A Problem-Solving Strategy

One of the goals of this textbook is to help you learn a *strategy* for solving physics problems. The purpose of a strategy is to guide you in the right direction with minimal wasted effort. The four-part problem-solving strategy shown below—**Model, Visualize, Solve, Assess**—is based on using different representations of knowledge. You will see this problem-solving strategy used consistently in the worked examples throughout this textbook, and you should endeavor to apply it to your own problem solving.

Throughout this textbook we will emphasize the first two steps. They are the *physics* of the problem, as opposed to the mathematics of solving the resulting equations. This is not to say that those mathematical operations are always easy—in many cases they are not. But our primary goal is to understand the physics.

General Problem-Solving Strategy

MODEL It's impossible to treat every detail of a situation. Simplify the situation with a model that captures the essential features. For example, the object in a mechanics problem is usually represented as a particle.

VISUALIZE This is where expert problem solvers put most of their effort.

- Draw a *pictorial representation*. This helps you visualize important aspects of the physics and assess the information you are given. It starts the process of translating the problem into symbols.
- Use a *graphical representation* if it is appropriate for the problem.
- Go back and forth between these representations; they need not be done in any particular order.

SOLVE Only after modeling and visualizing are complete is it time to develop a *mathematical representation* with specific equations that must be solved. All symbols used here should have been defined in the pictorial representation.

ASSESS Is your result believable? Does it have proper units? Does it make sense?

Textbook illustrations are obviously more sophisticated than what you would draw on your own paper. To show you a figure very much like what you *should* draw, the final example of this section is in a "pencil sketch" style. We will include one or more pencil-sketch examples in nearly every chapter to illustrate exactly what a good problem solver would draw.

EXAMPLE 1.9 **Launching a weather rocket**

Use the first two steps of the problem-solving strategy to analyze the following problem: A small rocket, such as those used for meteorological measurements of the atmosphere, is launched vertically with an acceleration of 30 m/s². It runs out of fuel after 30 s. What is its maximum altitude?

MODEL We need to do some interpretation. Common sense tells us that the rocket does not stop the instant it runs out of fuel. Instead, it continues upward, while slowing, until it reaches its maximum altitude. This second half of the motion, after running out of fuel, is like the ball that was tossed upward in the first half of Example 1.6. Because the problem does not ask about the rocket's descent, we conclude that the problem ends at the point of maximum altitude. We'll represent the rocket as a particle.

VISUALIZE **FIGURE 1.24** shows the pictorial representation in pencil-sketch style. The rocket is speeding up during the first half of the motion, so $\vec{a}_0$ points upward, in the positive *y*-direction. Thus the initial acceleration is $a_{0y} = 30$ m/s². During the second half, as the rocket slows, $\vec{a}_1$ points downward. Thus a_{1y} is a negative number.

FIGURE 1.24 Pictorial representation for the rocket.

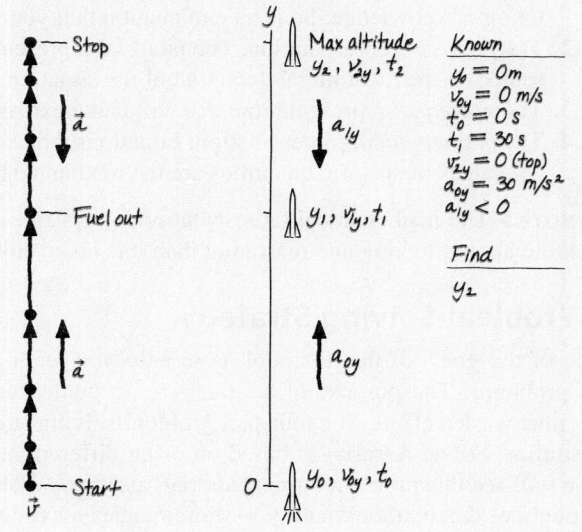

This information is included with the known information. Although the velocity v_{2y} wasn't given in the problem statement, we know it must be zero at the very top of the trajectory. Last, we have identified y_2 as the desired unknown. This, of course, is not the only unknown in the problem, but it is the one we are specifically asked to find.

ASSESS If you've had a previous physics class, you may be tempted to assign a_{1y} the value -9.8 m/s^2, the free-fall acceleration. However, that would be true only if there is no air resistance on the rocket. We will need to consider the *forces* acting on the rocket during the second half of its motion before we can determine a value for a_{1y}. For now, all that we can safely conclude is that a_{1y} is negative.

Our task in this section is not to *solve* problems—all that in due time—but to focus on what is happening in a problem. In other words, to make the translation from words to symbols in preparation for subsequent mathematical analysis. Modeling and the pictorial representation will be our most important tools.

1.8 Units and Significant Figures

Science is based upon experimental measurements, and measurements require *units*. The system of units used in science is called *le Système Internationale d'Unités*. These are commonly referred to as **SI units**. Older books often referred to *mks units,* which stands for "meter-kilogram-second," or *cgs units,* which is "centimeter-gram-second." For practical purposes, SI units are the same as mks units. In casual speaking we often refer to *metric units,* although this could mean either mks or cgs units.

All of the quantities needed to understand motion can be expressed in terms of the three basic SI units shown in Table 1.2. Other quantities can be expressed as a combination of these basic units. Velocity, expressed in meters per second or m/s, is a ratio of the length unit to the time unit.

TABLE 1.2 The basic SI units

Quantity	Unit	Abbreviation
time	second	s
length	meter	m
mass	kilogram	kg

Time

The standard of time prior to 1960 was based on the *mean solar day.* As time-keeping accuracy and astronomical observations improved, it became apparent that the earth's rotation is not perfectly steady. Meanwhile, physicists had been developing a device called an *atomic clock.* This instrument is able to measure, with incredibly high precision, the frequency of radio waves absorbed by atoms as they move between two closely spaced energy levels. This frequency can be reproduced with great accuracy at many laboratories around the world. Consequently, the SI unit of time—the second—was redefined in 1967 as follows:

One *second* is the time required for 9,192,631,770 oscillations of the radio wave absorbed by the cesium-133 atom. The abbreviation for second is the letter s.

Several radio stations around the world broadcast a signal whose frequency is linked directly to the atomic clocks. This signal is the time standard, and any time-measuring equipment you use was calibrated from this time standard.

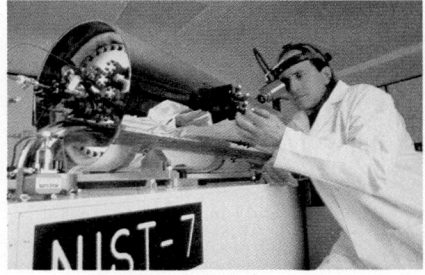

An atomic clock at the National Institute of Standards and Technology is the primary standard of time.

Length

The SI unit of length—the meter—was originally defined as one ten-millionth of the distance from the North Pole to the equator along a line passing through Paris. There are obvious practical difficulties with implementing this definition, and it was later abandoned in favor of the distance between two scratches on a platinum-iridium bar stored in a special vault in Paris. The present definition, agreed to in 1983, is as follows:

One *meter* is the distance traveled by light in vacuum during 1/299,792,458 of a second. The abbreviation for meter is the letter m.

This is equivalent to defining the speed of light to be exactly 299,792,458 m/s. Laser technology is used in various national laboratories to implement this definition and to calibrate secondary standards that are easier to use. These standards

By international agreement, this metal cylinder, stored in Paris, is the definition of the kilogram.

TABLE 1.3 Common prefixes

Prefix	Power of 10	Abbreviation
giga-	10^9	G
mega-	10^6	M
kilo-	10^3	k
centi-	10^{-2}	c
milli-	10^{-3}	m
micro-	10^{-6}	μ
nano-	10^{-9}	n

TABLE 1.4 Useful unit conversions

1 in = 2.54 cm
1 mi = 1.609 km
1 mph = 0.447 m/s
1 m = 39.37 in
1 km = 0.621 mi
1 m/s = 2.24 mph

ultimately make their way to your ruler or to a meter stick. It is worth keeping in mind that any measuring device you use is only as accurate as the care with which it was calibrated.

Mass

The original unit of mass, the gram, was defined as the mass of 1 cubic centimeter of water. That is why you know the density of water as 1 g/cm^3. This definition proved to be impractical when scientists needed to make very accurate measurements. The SI unit of mass—the kilogram—was redefined in 1889 as:

> One *kilogram* is the mass of the international standard kilogram, a polished platinum-iridium cylinder stored in Paris. The abbreviation for kilogram is kg.

The kilogram is the only SI unit still defined by a manufactured object. Despite the prefix *kilo*, it is the kilogram, not the gram, that is the SI unit.

Using Prefixes

We will have many occasions to use lengths, times, and masses that are either much less or much greater than the standards of 1 meter, 1 second, and 1 kilogram. We will do so by using *prefixes* to denote various powers of 10. Table 1.3 lists the common prefixes that will be used frequently throughout this book. Memorize it! Few things in science are learned by rote memory, but this list is one of them. A more extensive list of prefixes is shown inside the cover of the book.

Although prefixes make it easier to talk about quantities, the SI units are meters, seconds, and kilograms. Quantities given with prefixed units must be converted to SI units before any calculations are done. Unit conversions are best done at the very beginning of a problem, as part of the pictorial representation.

Unit Conversions

Although SI units are our standard, we cannot entirely forget that the United States still uses English units. Thus it remains important to be able to convert back and forth between SI units and English units. Table 1.4 shows several frequently used conversions, and these are worth memorizing if you do not already know them. While the English system was originally based on the length of the king's foot, it is interesting to note that today the conversion 1 in = 2.54 cm is the *definition* of the inch. In other words, the English system for lengths is now based on the meter!

There are various techniques for doing unit conversions. One effective method is to write the conversion factor as a ratio equal to one. For example, using information in Table 1.3 and 1.4, we have

$$\frac{10^{-6}\ \text{m}}{1\ \mu\text{m}} = 1 \qquad \text{and} \qquad \frac{2.54\ \text{cm}}{1\ \text{in}} = 1$$

Because multiplying any expression by 1 does not change its value, these ratios are easily used for conversions. To convert 3.5 μm to meters we compute

$$3.5\ \mu\text{m} \times \frac{10^{-6}\ \text{m}}{1\ \mu\text{m}} = 3.5 \times 10^{-6}\ \text{m}$$

Similarly, the conversion of 2 feet to meters is

$$2.00\ \text{ft} \times \frac{12\ \text{in}}{1\ \text{ft}} \times \frac{2.54\ \text{cm}}{1\ \text{in}} \times \frac{10^{-2}\ \text{m}}{1\ \text{cm}} = 0.610\ \text{m}$$

Notice how units in the numerator and in the denominator cancel until only the desired units remain at the end. You can continue this process of multiplying by 1 as many times as necessary to complete all the conversions.

Assessment

As we get further into problem solving, we will need to decide whether or not the answer to a problem "makes sense." To determine this, at least until you have more experience with SI units, you may need to convert from SI units back to the English units in which you think. But this conversion does not need to be very accurate. For example, if you are working a problem about automobile speeds and reach an answer of 35 m/s, all you really want to know is whether or not this is a realistic speed for a car. That requires a "quick and dirty" conversion, not a conversion of great accuracy.

Table 1.5 shows several approximate conversion factors that can be used to assess the answer to a problem. Using 1 m/s ≈ 2 mph, you find that 35 m/s is roughly 70 mph, a reasonable speed for a car. But an answer of 350 m/s, which you might get after making a calculation error, would be an unreasonable 700 mph. Practice with these will allow you to develop intuition for metric units.

> NOTE ▶ These approximate conversion factors are accurate to only one significant figure. This is sufficient to assess the answer to a problem, but do *not* use the conversion factors from Table 1.5 for converting English units to SI units at the start of a problem. Use Table 1.4. ◀

TABLE 1.5 Approximate conversion factors. Use these only for assessment, not in problem solving.

1 cm ≈ $\frac{1}{2}$ in
10 cm ≈ 4 in
1 m ≈ 1 yard
1 m ≈ 3 feet
1 km ≈ 0.6 mile
1 m/s ≈ 2 mph

Significant Figures

It is necessary to say a few words about a perennial source of difficulty: significant figures. Mathematics is a subject where numbers and relationships can be as precise as desired, but physics deals with a real world of ambiguity. It is important in science and engineering to state clearly what you know about a situation—no less and, especially, no more. Numbers provide one way to specify your knowledge.

If you report that a length has a value of 6.2 m, the implication is that the actual value falls between 6.15 m and 6.25 m and thus rounds to 6.2 m. If that is the case, then reporting a value of simply 6 m is saying less than you know; you are withholding information. On the other hand, to report the number as 6.213 m is wrong. Any person reviewing your work—perhaps a client who hired you—would interpret the number 6.213 m as meaning that the actual length falls between 6.2125 m and 6.2135 m, thus rounding to 6.213 m. In this case, you are claiming to have knowledge and information that you do not really possess.

The way to state your knowledge precisely is through the proper use of **significant figures.** You can think of a significant figure as being a digit that is reliably known. A number such as 6.2 m has *two* significant figures because the next decimal place—the one-hundredths—is not reliably known. As FIGURE 1.25 shows, the best way to determine how many significant figures a number has is to write it in scientific notation.

FIGURE 1.25 Determining significant figures.

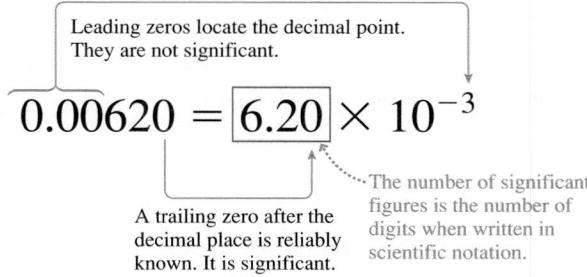

- The number of significant figures ≠ the number of decimal places.
- In whole numbers, trailing zeros are not significant. 320 is 3.2×10^2 and has 2 significant figures, not 3.
- Changing units shifts the decimal point but does not change the number of significant figures.

Calculations with numbers follow the "weakest link" rule. The saying, which you probably know, is that "a chain is only as strong as its weakest link." If nine out of ten links in a chain can support a 1000 pound weight, that strength is meaningless if the tenth link can support only 200 pounds. Nine out of the ten numbers used in a calculation might be known with a precision of 0.01%; but if the tenth number is poorly known, with a precision of only 10%, then the result of the calculation cannot possibly be more precise than 10%.

TACTICS
BOX 1.6   **Using significant figures**

❶ When multiplying or dividing several numbers, or taking roots, the number of significant figures in the answer should match the number of significant figures of the *least* precisely known number used in the calculation.

❷ When adding or subtracting several numbers, the number of decimal places in the answer should match the *smallest* number of decimal places of any number used in the calculation.

❸ It is acceptable to keep one or two extra digits during intermediate steps of a calculation, as long as the final answer is reported with the proper number of significant figures. The goal is to minimize round-off errors in the calculation. But only one or two extra digits, not the seven or eight shown in your calculator display.

Exercises 38–39

EXAMPLE 1.10   **Using significant figures**

An object consists of two pieces. The mass of one piece has been measured to be 6.47 kg. The volume of the second piece, which is made of aluminum, has been measured to be 4.44×10^{-4} m³. A handbook lists the density of aluminum as 2.7×10^3 kg/m³. What is the total mass of the object?

SOLVE   First, calculate the mass of the second piece:

$$m = (4.44 \times 10^{-4} \text{ m}^3)(2.7 \times 10^3 \text{ kg/m}^3)$$

$$= 1.199 \text{ kg} = 1.2 \text{ kg}$$

The number of significant figures of a product must match that of the *least* precisely known number, which is the two-significant-figure density of aluminum. Now add the two masses:

$$\begin{array}{r} 6.47 \text{ kg} \\ + 1.2 \text{ kg} \\ \hline 7.7 \text{ kg} \end{array}$$

The sum is 7.67 kg, but the hundredths place is not reliable because the second mass has no reliable information about this digit. Thus we must round to the one decimal place of the 1.2 kg. The best we can say, with reliability, is that the total mass is 7.7 kg.

Some quantities can be measured very precisely—three or more significant figures. Others are inherently much less precise—only two significant figures. Examples and problems in this textbook will normally provide data to either two or three significant figures, as is appropriate to the situation. **The appropriate number of significant figures for the answer is determined by the data provided.**

NOTE ▶ Be careful! Many calculators have a default setting that shows two decimal places, such as 5.23. This is dangerous. If you need to calculate 5.23/58.5, your calculator will show 0.09 and it is all too easy to write that down as an answer. By doing so, you have reduced a calculation of two numbers having three significant figures to an answer with only one significant figure. The proper result of this division is 0.0894 or 8.94×10^{-2}. You will avoid this error if you keep your calculator set to display numbers in *scientific notation* with two decimal places. ◀

Proper use of significant figures is part of the "culture" of science and engineering. We will frequently emphasize these "cultural issues" because you must learn to speak the same language as the natives if you wish to communicate effectively. Most students know the rules of significant figures, having learned them in high school, but many fail to apply them. It is important to understand the reasons for significant figures and to get in the habit of using them properly.

Orders of Magnitude and Estimating

Precise calculations are appropriate when we have precise data, but there are many times when a very rough estimate is sufficient. Suppose you see a rock fall off a cliff and would like to know how fast it was going when it hit the ground. By doing a mental comparison with the speeds of familiar objects, such as cars and bicycles, you might judge that the rock was traveling at "about" 20 mph.

This is a one-significant-figure estimate. With some luck, you can distinguish 20 mph from either 10 mph or 30 mph, but you certainly cannot distinguish 20 mph from 21 mph. A one-significant-figure estimate or calculation, such as this, is called an **order-of-magnitude estimate**. An order-of-magnitude estimate is indicated by the symbol $\sim$, which indicates even less precision than the "approximately equal" symbol $\approx$. You would say that the speed of the rock is $v \sim 20$ mph.

A useful skill is to make reliable estimates on the basis of known information, simple reasoning, and common sense. This is a skill that is acquired by practice. Many chapters in this book will have homework problems that ask you to make order-of-magnitude estimates. The following example is a typical estimation problem. Table 1.6 and 1.7 have information that will be useful for doing estimates.

EXAMPLE 1.11 **Estimating a sprinter's speed**

Estimate the speed with which an Olympic sprinter crosses the finish line of the 100 m dash.

SOLVE We do need one piece of information, but it is a widely known piece of sports trivia. That is, world-class sprinters run the 100 m dash in about 10 s. Their *average* speed is $v_{\text{avg}} \approx (100 \text{ m})/(10 \text{ s}) \approx 10$ m/s. But that's only average. They go slower than average at the beginning, and they cross the finish line at a speed faster than average. How much faster? Twice as fast, 20 m/s, would be ≈ 40 mph. Sprinters don't seem like they're running as fast as a 40 mph car, so this probably is too fast. Let's *estimate* that their final speed is 50% faster than the average. Thus they cross the finish line at $v \sim 15$ m/s.

STOP TO THINK 1.5   Rank in order, from the most to the least, the number of significant figures in the following numbers. For example, if b has more than c, c has the same number as a, and a has more than d, you could give your answer as b > c = a > d.

a. 82   b. 0.0052   c. 0.430   d. 4.321×10^{-10}

TABLE 1.6 Some approximate lengths

	Length (m)
Circumference of the earth	4×10^7
New York to Los Angeles	5×10^6
Distance you can drive in 1 hour	1×10^5
Altitude of jet planes	1×10^4
Distance across a college campus	1000
Length of a football field	100
Length of a classroom	10
Length of your arm	1
Width of a textbook	0.1
Length of your little fingernail	0.01
Diameter of a pencil lead	1×10^{-3}
Thickness of a sheet of paper	1×10^{-4}
Diameter of a dust particle	1×10^{-5}

TABLE 1.7 Some approximate masses

	Mass (kg)
Large airliner	1×10^5
Small car	1000
Large human	100
Medium-size dog	10
Science textbook	1
Apple	0.1
Pencil	0.01
Raisin	1×10^{-3}
Fly	1×10^{-4}

SUMMARY

The goal of Chapter 1 has been to introduce the fundamental concepts of motion.

General Strategy

Motion Diagrams

- Help visualize motion.
- Provide a tool for finding acceleration vectors.

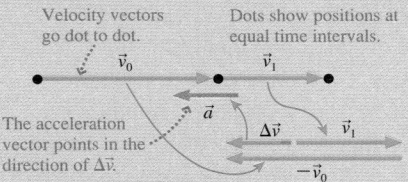

Velocity vectors go dot to dot.   Dots show positions at equal time intervals.

The acceleration vector points in the direction of $\Delta\vec{v}$.

▶ These are the average velocity and the average acceleration vectors.

Problem Solving

MODEL Make simplifying assumptions.

VISUALIZE Use:

- **Pictorial representation**
- **Graphical representation**

SOLVE Use a **mathematical representation** to find numerical answers.

ASSESS Does the answer have the proper units? Does it make sense?

Important Concepts

The particle model represents a moving object as if all its mass were concentrated at a single point.

Position locates an object with respect to a chosen coordinate system. Change in position is called displacement.

Velocity is the rate of change of the position vector $\vec{r}$.

Acceleration is the rate of change of the velocity vector $\vec{v}$.

An object has an acceleration if it

- Changes speed and/or
- Changes direction.

Pictorial Representation

❶ Draw a motion diagram.

❷ Establish coordinates.

❸ Sketch the situation.

❹ Define symbols.

❺ List knowns.

❻ Identify desired unknown.

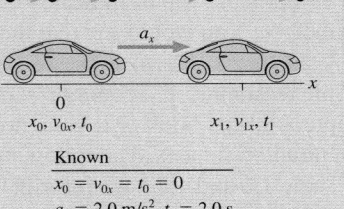

Known
$x_0 = v_{0x} = t_0 = 0$
$a_x = 2.0 \text{ m/s}^2$ $t_1 = 2.0 \text{ s}$

Find
x_1

Applications

For **motion along a line:**

- Speeding up: $\vec{v}$ and $\vec{a}$ point in the same direction, v_x and a_x have the same sign.
- Slowing down: $\vec{v}$ and $\vec{a}$ point in opposite directions, v_x and a_x have opposite signs.
- Constant speed: $\vec{a} = \vec{0}$, $a_x = 0$.

Acceleration a_x is positive if $\vec{a}$ points right, negative if $\vec{a}$ points left. The sign of a_x does *not* imply speeding up or slowing down.

Significant figures are reliably known digits. The number of significant figures for:

- **Multiplication, division, powers** is set by the value with the fewest significant figures.
- **Addition, subtraction** is set by the value with the smallest number of decimal places.

The appropriate number of significant figures in a calculation is determined by the data provided.

Terms and Notation

motion	position vector, $\vec{r}$	average speed	SI units
translational motion	scalar quantity	average velocity, $\vec{v}$	significant figures
trajectory	vector quantity	average acceleration, $\vec{a}$	order-of-magnitude estimate
motion diagram	displacement, $\Delta\vec{r}$	position-versus-time graph	
particle	zero vector, $\vec{0}$	pictorial representation	
particle model	time interval, Δt	representation of knowledge	

CONCEPTUAL QUESTIONS

1. How many significant figures does each of the following numbers have?
   a. 53.2   b. 0.53   c. 5.320   d. 0.0532
2. How many significant figures does each of the following numbers have?
   a. 310   b. 0.00310   c. 1.031   d. 3.10×10^5
3. Is the particle in FIGURE Q1.3 speeding up? Slowing down? Or can you tell? Explain.

   FIGURE Q1.3   • • •   •   •

4. Does the object represented in FIGURE Q1.4 have a positive or negative value of a_x? Explain.
5. Does the object represented in FIGURE Q1.5 have a positive or negative value of a_y? Explain.

   FIGURE Q1.4   FIGURE Q1.5

6. Determine the signs (positive or negative) of the position, velocity, and acceleration for the particle in FIGURE Q1.6.

   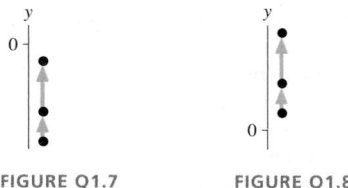

   FIGURE Q1.6   0

7. Determine the signs (positive or negative) of the position, velocity, and acceleration for the particle in FIGURE Q1.7.

   FIGURE Q1.7   FIGURE Q1.8

8. Determine the signs (positive or negative) of the position, velocity, and acceleration for the particle in FIGURE Q1.8.

EXERCISES AND PROBLEMS

Exercises

Section 1.1 Motion Diagrams

1. | A car skids to a halt to avoid hitting an object in the road. Draw a basic motion diagram, using the images from the movie, from the time the skid begins until the car is stopped.
2. | A rocket is launched straight up. Draw a basic motion diagram, using the images from the movie, from the moment of liftoff until the rocket is at an altitude of 500 m.
3. | You're driving along the highway at 60 mph until you enter a town where the speed limit is 30 mph. You slow quickly, but not instantly, to 30 mph. Draw a basic motion diagram of your car, using images from the movie, from 30 s before reaching the city limit until 30 s afterward.

Section 1.2 The Particle Model

4. | a. Write a paragraph describing the particle model. What is it, and why is it important?
   b. Give two examples of situations, different from those described in the text, for which the particle model is appropriate.
   c. Give an example of a situation, different from those described in the text, for which it would be inappropriate.

Section 1.3 Position and Time

Section 1.4 Velocity

5. | You drop a soccer ball from your third-story balcony. Use the particle model to draw a motion diagram showing the ball's position and average velocity vectors from the time you release the ball until the instant it touches the ground.

6. | A softball player hits the ball and starts running toward first base. Use the particle model to draw a motion diagram showing her position and her average velocity vectors during the first few seconds of her run.
7. | A softball player slides into second base. Use the particle model to draw a motion diagram showing his position and his average velocity vectors from the time he begins to slide until he reaches the base.

Section 1.5 Linear Acceleration

8. | a. FIGURE EX1.8 shows the first three points of a motion diagram. Is the object's average speed between points 1 and 2 greater than, less than, or equal to its average speed between points 0 and 1? Explain how you can tell.
   b. Use Tactics Box 1.3 to find the average acceleration vector at point 1. Draw the completed motion diagram, showing the velocity vectors and acceleration vector.

   2 •

   •   •   •   1 •
   2   1   0

   FIGURE EX1.8   FIGURE EX1.9   0 •

9. | a. FIGURE EX1.9 shows the first three points of a motion diagram. Is the object's average speed between points 1 and 2 greater than, less than, or equal to its average speed between points 0 and 1? Explain how you can tell.
   b. Use Tactics Box 1.3 to find the average acceleration vector at point 1. Draw the completed motion diagram, showing the velocity vectors and acceleration vector.

10. ‖ **FIGURE EX1.10** shows two dots of a motion diagram and vector $\vec{v}_1$. Copy this figure and add vector $\vec{v}_2$ and dot 3 if the acceleration vector $\vec{a}$ at dot 2 (a) points up and (b) points down.

FIGURE EX1.10   FIGURE EX 1.11

11. | **FIGURE EX1.11** shows two dots of a motion diagram and vector $\vec{v}_2$. Copy this figure and add vector $\vec{v}_1$ and dot 1 if the acceleration vector $\vec{a}$ at dot 2 (a) points to the right and (b) points to the left.

12. | A car travels to the left at a steady speed for a few seconds, then brakes for a stop sign. Draw a complete motion diagram of the car.

13. | A child is sledding on a smooth, level patch of snow. She encounters a rocky patch and slows to a stop. Draw a complete motion diagram of the child and her sled.

14. | You use a long rubber band to launch a paper wad straight up. Draw a complete motion diagram of the paper wad from the moment you release the stretched rubber band until the paper wad reaches its highest point.

15. | A roof tile falls straight down from a two-story building. It lands in a swimming pool and settles gently to the bottom. Draw a complete motion diagram of the tile.

16. | Your roommate drops a tennis ball from a third-story balcony. It hits the sidewalk and bounces as high as the second story. Draw a complete motion diagram of the tennis ball from the time it is released until it reaches the maximum height on its bounce. Be sure to determine and show the acceleration at the lowest point.

17. | A toy car rolls down a ramp, then across a smooth, horizontal floor. Draw a complete motion diagram of the toy car.

Section 1.6 Motion in One Dimension

18. ‖ **FIGURE EX1.18** shows the motion diagram of a drag racer. The camera took one frame every 2 s.

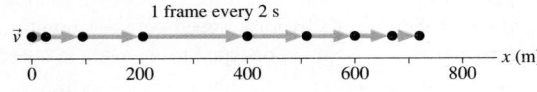

FIGURE EX1.18

   a. Measure the x-value of the racer at each dot. List your data in a table similar to Table 1.1, showing each position and the time at which it occurred.
   b. Make a position-versus-time graph for the drag racer. Because you have data only at certain instants, your graph should consist of dots that are not connected together.

19. | Write a short description of the motion of a real object for which **FIGURE EX1.19** would be a realistic position-versus-time graph.

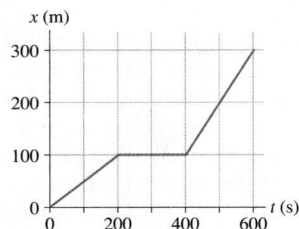

FIGURE EX1.19

20. | Write a short description of the motion of a real object for which **FIGURE EX1.20** would be a realistic position-versus-time graph.

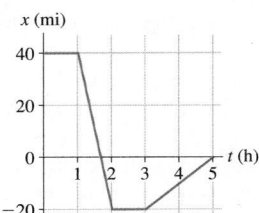

FIGURE EX1.20

Section 1.7 Solving Problems in Physics

21. ‖ Draw a pictorial representation for the following problem. Do *not* solve the problem. The light turns green, and a bicyclist starts forward with an acceleration of 1.5 m/s². How far must she travel to reach a speed of 7.5 m/s?

22. ‖ Draw a pictorial representation for the following problem. Do *not* solve the problem. What acceleration does a rocket need to reach a speed of 200 m/s at a height of 1.0 km?

Section 1.8 Units and Significant Figures

23. | Convert the following to SI units:
   a. 6.15 ms   b. 27.2 km
   c. 112 km/h   d. 72 μm/ms

24. | Convert the following to SI units:
   a. 8.0 in   b. 66 ft/s
   c. 60 mph   d. 14 in²

25. | Convert the following to SI units:
   a. 3 hours   b. 2 days
   c. 1 year   d. 215 ft/s

26. | Using the approximate conversion factors in Table 1.5, convert the following to SI units *without* using your calculator.
   a. 20 ft   b. 60 mi
   c. 60 mph   d. 8 in

27. ‖ Using the approximate conversion factors in Table 1.5, convert the following SI units to English units *without* using your calculator.
   a. 30 cm   b. 25 m/s
   c. 5 km   d. 0.5 cm

28. | Compute the following numbers, applying the significant figure rule adopted in this textbook.
   a. 33.3×25.4 b. $33.3 - 25.4$
   c. $\sqrt{33.3}$ d. $333.3 \div 25.4$

29. | Compute the following numbers, applying the significant figure rule adopted in this textbook.
   a. 12.5^3 b. 12.5×5.21
   c. $\sqrt{12.5} - 1.2$ d. 12.5^{-1}

30. | Estimate (don't measure!) the length of a typical car. Give your answer in both feet and meters. Briefly describe how you arrived at this estimate.

31. | Estimate the height of a telephone pole. Give your answer in both feet and meters. Briefly describe how you arrived at this estimate.

32. | Estimate the average speed with which you go from home to campus via whatever mode of transportation you use most commonly. Give your answer in both mph and m/s. Briefly describe how you arrived at this estimate.

33. | Estimate the average speed with which the hair on your head grows. Give your answer in both m/s and μm/hour. Briefly describe how you arrived at this estimate.

Problems

For Problems 34 through 43, draw a complete pictorial representation. Do *not* solve these problems or do any mathematics.

34. | A Porsche accelerates from a stoplight at 5.0 m/s² for five seconds, then coasts for three more seconds. How far has it traveled?

35. | A jet plane is cruising at 300 m/s when suddenly the pilot turns the engines up to full throttle. After traveling 4.0 km, the jet is moving with a speed of 400 m/s. What is the jet's acceleration as it speeds up?

36. | Sam is recklessly driving 60 mph in a 30 mph speed zone when he suddenly sees the police. He steps on the brakes and slows to 30 mph in three seconds, looking nonchalant as he passes the officer. How far does he travel while braking?

37. | You would like to stick a wet spit wad on the ceiling, so you toss it straight up with a speed of 10 m/s. How long does it take to reach the ceiling, 3.0 m above?

38. | A speed skater moving across frictionless ice at 8.0 m/s hits a 5.0-m-wide patch of rough ice. She slows steadily, then continues on at 6.0 m/s. What is her acceleration on the rough ice?

39. | Santa loses his footing and slides down a frictionless, snowy roof that is tilted at an angle of 30°. If Santa slides 10 m before reaching the edge, what is his speed as he leaves the roof?

40. | A motorist is traveling at 20 m/s. He is 60 m from a stoplight when he sees it turn yellow. His reaction time, before stepping on the brake, is 0.50 s. What steady deceleration while braking will bring him to a stop right at the light?

41. | A car traveling at 30 m/s runs out of gas while traveling up a 10° slope. How far up the hill will the car coast before starting to roll back down?

42. ‖ Ice hockey star Bruce Blades is 5.0 m from the blue line and gliding toward it at a speed of 4.0 m/s. You are 20 m from the blue line, directly behind Bruce. You want to pass the puck to Bruce. With what speed should you shoot the puck down the ice so that it reaches Bruce exactly as he crosses the blue line?

43. ‖ David is driving a steady 30 m/s when he passes Tina, who is sitting in her car at rest. Tina begins to accelerate at a steady 2.0 m/s² at the instant when David passes. How far does Tina drive before passing David?

Problems 44 through 48 show a motion diagram. For each of these problems, write a one or two sentence "story" about a *real object* that has this motion diagram. Your stories should talk about people or objects by name and say what they are doing. Problems 34 through 43 are examples of motion short stories.

44. |

FIGURE P1.44

45. |

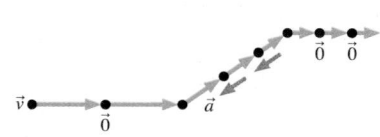

FIGURE P1.45

46. |

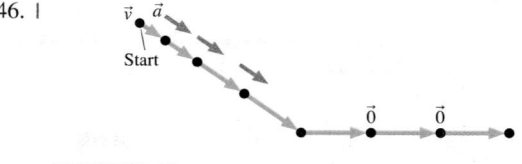

FIGURE P1.46

47. |

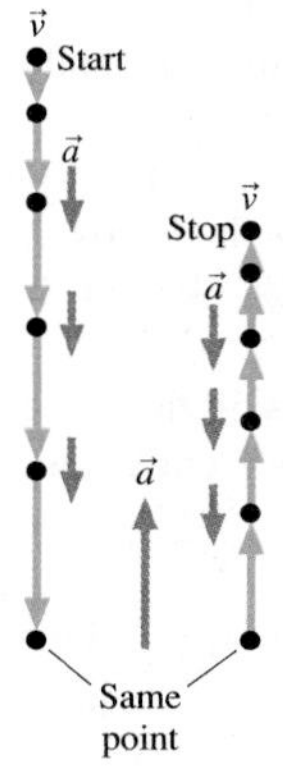

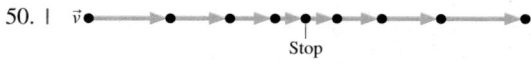

FIGURE P1.47

48. |

Side view of motion in a vertical plane

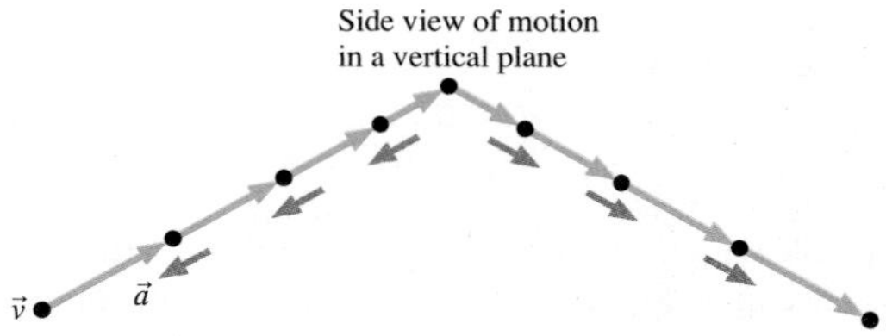

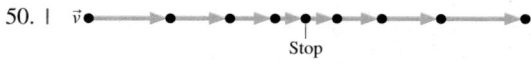

FIGURE P1.48

Problems 49 through 52 show a partial motion diagram. For each:

a. Complete the motion diagram by adding acceleration vectors.

b. Write a physics *problem* for which this is the correct motion diagram. Be imaginative! Don't forget to include enough information to make the problem complete and to state clearly what is to be found.

c. Draw a pictorial representation for your problem.

49. | $\vec{v}$ ●→●→●→●→●→●→●→●

FIGURE P1.49

50. | $\vec{v}$ ●→●→●→●→●→●→●→●→●
Stop

FIGURE P1.50

51. |

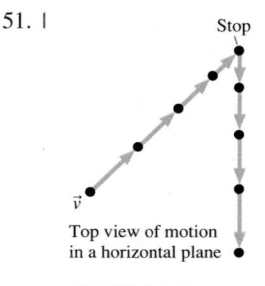

Top view of motion in a horizontal plane

FIGURE P1.51

52. |

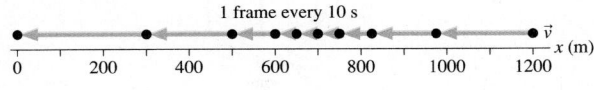

FIGURE P1.52

53. | A regulation soccer field for international play is a rectangle with a length between 100 m and 110 m and a width between 64 m and 75 m. What are the smallest and largest areas that the field could be?

54. ‖ The quantity called *mass density* is the mass per unit volume of a substance. Express the following mass densities in SI units.
   a. Aluminum, 2.7×10^{-3} kg/cm^3
   b. Alcohol, 0.81 g/cm^3

55. ‖ FIGURE P1.55 shows a motion diagram of a car traveling down a street. The camera took one frame every 10 s. A distance scale is provided.

1 frame every 10 s

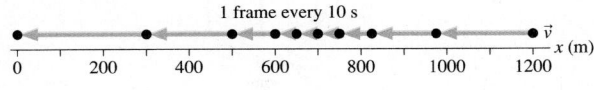

FIGURE P1.55

a. Measure the *x*-value of the car at each dot. Place your data in a table, similar to Table 1.1, showing each position and the instant of time at which it occurred.
   b. Make a position-versus-time graph for the car. Because you have data only at certain instants of time, your graph should consist of dots that are not connected together.

56. | Write a short description of a real object for which FIGURE P1.56 would be a realistic position-versus-time graph.

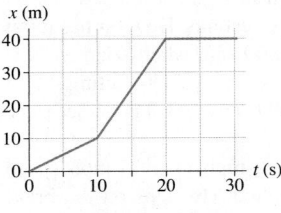

FIGURE P1.56

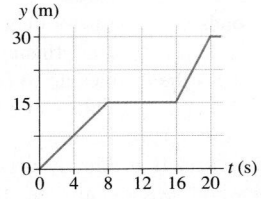

FIGURE P1.57

57. | Write a short description of a real object for which FIGURE P1.57 would be a realistic position-versus-time graph.

STOP TO THINK ANSWERS

Stop to Think 1.1: B. The images of B are farther apart, so it travels a larger distance than does A during the same intervals of time.

Stop to Think 1.2: a. Dropped ball. **b.** Dust particle. **c.** Descending rocket.

Stop to Think 1.3: e. The average velocity vector is found by connecting one dot in the motion diagram to the next.

Stop to Think 1.4: b. $\vec{v}_2 = \vec{v}_1 + \Delta\vec{v}$, and $\Delta\vec{v}$ points in the direction of $\vec{a}$.

Stop to Think 1.5: d > c > b = a.

2 Kinematics in One Dimension

This Japanese "bullet train" accelerates slowly but steadily until reaching a speed of 300 km/h.

▶ **Looking Ahead** The goal of Chapter 2 is to learn how to solve problems about motion in a straight line.

Kinematics

Kinematics is the name for the mathematical description of motion. We begin with motion along a straight line; for example, runners, rockets, and skiers. Kinematics in two dimensions—projectile motion and circular motion—will be taken up in Chapter 4.

The motion of an object is described by its *position*, *velocity*, and *acceleration*. In one dimension, these quantities are represented by x, v_x, and a_x. You learned to show these on motion diagrams in Chapter 1.

Now we will use calculus to give precise meaning to velocity and acceleration.

◀ **Looking Back**
Sections 1.4–1.5 Velocity and acceleration

It is very important to know when velocity and acceleration are positive and when they are negative.

◀ **Looking Back**
Tactics Box 1.4 The signs of position, velocity, and acceleration

Graphical Representations of Motion

Position, velocity, and acceleration are related graphically.

- You will learn how to draw the position-versus-time, velocity-versus-time, and acceleration-versus-time graphs that describe various types of motion.

- You will learn that the slope of the position-versus-time graph is the instantaneous value of velocity.

- Similarly, the slope of the velocity-versus-time graph is the instantaneous value of acceleration.

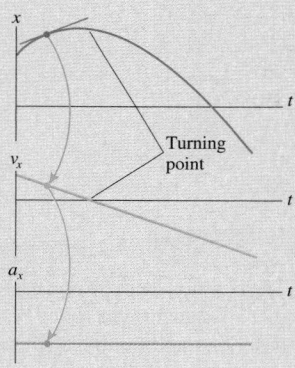

Solving Problems in Kinematics

You will begin learning to solve problems using a four-part *problem-solving strategy*.

MODEL Use the particle model.

VISUALIZE Draw a pictorial representation. Use a graphical representation.

SOLVE Use three kinematic equations that we'll develop in this chapter.

ASSESS Check whether the result makes sense.

◀ **Looking Back**
Tactics Box 1.5 Drawing a pictorial representation

Problems you will learn to solve in this chapter include free fall and motion on an inclined plane.

2.1 Uniform Motion

Riding steadily over level ground is a good example of uniform motion.

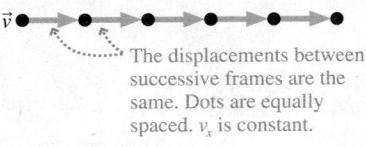

FIGURE 2.1 Motion diagram and position graph for uniform motion.

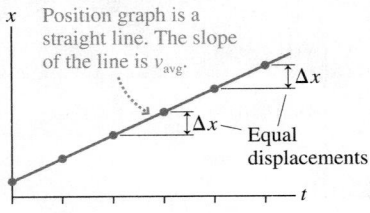

The displacements between successive frames are the same. Dots are equally spaced. v_x is constant.

x Position graph is a straight line. The slope of the line is v_{avg}.
Δx
Δx Equal displacements
t

If you drive your car at a perfectly steady 60 miles per hour (mph), you will cover 60 mi during the first hour, another 60 mi during the second hour, yet another 60 mi during the third hour, and so on. In this case, 60 mi is not your position, but rather the *change* in your position during each hour; that is, your displacement Δx. Similarly, 1 hour is a time interval Δt rather than a specific instant of time. **Straight-line motion in which equal displacements occur during any successive equal-time intervals is called uniform motion.**

FIGURE 2.1 shows how uniform motion appears in motion diagrams and position-versus-time graphs. Because all equal-time intervals have equal displacements, the position graph is a straight line. In fact, an alternative definition of uniform motion is: **An object's motion is uniform if and only if its position-versus-time graph is a straight line.**

The slope of a straight-line graph is defined as "rise over run." Because position is graphed on the vertical axis, the "rise" of a position-versus-time graph is the object's displacement Δx. The "run" is the time interval Δt. Consequently, the slope is $\Delta x / \Delta t$.

Chapter 1 defined the **average velocity** as $\Delta \vec{r}/\Delta t$. For one-dimensional motion this is simply

$$v_{avg} \equiv \frac{\Delta x}{\Delta t} \text{ or } \frac{\Delta y}{\Delta t} = \text{slope of the position-versus-time graph} \qquad (2.1)$$

That is, **the average velocity is the slope of the position-versus-time graph.** Velocity has units of "length per time," such as "miles per hour." The SI units of velocity are meters per second, abbreviated m/s.

NOTE ▶ The symbol $\equiv$ in Equation 2.1 stands for "is defined as" or "is equivalent to." This is a stronger statement than the two sides simply being equal. ◀

In the case of uniform motion, where the slope $\Delta x/\Delta t$ is the same at all times, it appears that the average velocity is constant and unchanging. Consequently, a final definition of uniform motion is: **An object's motion is uniform if and only if its velocity v_x or v_y does not change.** There's no real need to specify "average" for a velocity that doesn't change, so we will drop the subscript and refer to the average velocity as v_x or v_y.

EXAMPLE 2.1 **Skating with constant velocity**

The position-versus-time graph of FIGURE 2.2 represents the motion of two students on roller blades. Determine their velocities and describe their motion.

MODEL Represent the two students as particles.

VISUALIZE Figure 2.2 is a graphical representation of the students' motion. Both graphs are straight lines, telling us that both skaters are moving uniformly with constant velocities.

SOLVE We can determine the students' velocities by measuring the slopes of the graphs. Skater A undergoes a displacement $\Delta x_A = 2.0$ m during the time interval $\Delta t_A = 0.40$ s Thus his velocity is

$$(v_x)_A = \frac{\Delta x_A}{\Delta t_A} = \frac{2.0 \text{ m}}{0.40 \text{ s}}$$

$$= 5.0 \text{ m/s}$$

FIGURE 2.2 Graphical representations of two students on roller blades.

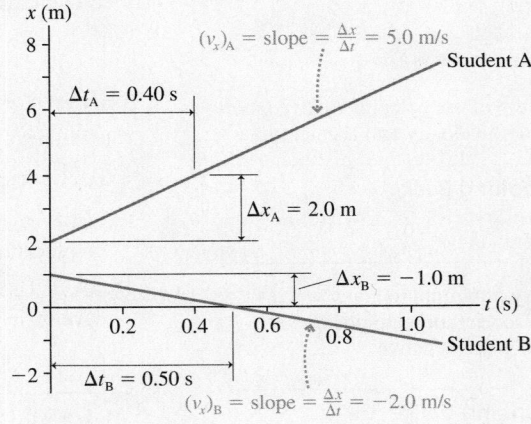

We need to be more careful with skater B. Although he moves a distance of 1.0 m in 0.50 s, his *displacement* is

$$\Delta x_{\text{B}} = x_{\text{at 0.5 s}} - x_{\text{at 0.0 s}} = 0.0 \text{ m} - 1.0 \text{ m} = -1.0 \text{ m}$$

Careful attention to the signs is very important! This leads to

$$(v_x)_{\text{B}} = \frac{\Delta x_{\text{B}}}{\Delta t_{\text{B}}} = \frac{-1.0 \text{ m}}{0.50 \text{ s}} = -2.0 \text{ m/s}$$

ASSESS The minus sign indicates that skater B is moving to the left. Our interpretation of this graph is that two students on roller blades are moving with constant velocities in opposite directions. Skater A starts at $x = 2.0$ m and moves to the right with a velocity of $x = 5.0$ m/s. Skater B starts at $x = 1.0$ m and moves to the left with a velocity of -2.0 m/s. Their speeds, of ≈ 10 mph and ≈ 4 mph, are reasonable for skaters on roller blades.

Example 2.1 brought out several points that are worth emphasizing. These are summarized in Tactics Box 2.1.

TACTICS
BOX 2.1   **Interpreting position-versus-time graphs**

❶ Steeper slopes correspond to faster speeds.
❷ Negative slopes correspond to negative velocities and, hence, to motion to the left (or down).
❸ The slope is a ratio of intervals, $\Delta x / \Delta t$, not a ratio of coordinates. That is, the slope is *not* simply x/t.
❹ We are distinguishing between the *actual* slope and the *physically meaningful* slope. If you were to use a ruler to measure the rise and the run of the graph, you could compute the actual slope of the line as drawn on the page. That is not the slope to which we are referring when we equate the velocity with the slope of the line. Instead, we find the *physically meaningful* slope by measuring the rise and run using the scales along the axes. The "rise" Δx is some number of meters; the "run" Δt is some number of seconds. The physically meaningful rise and run include units, and the ratio of these units gives the units of the slope.

Exercises 1–3 ✎

An object's **speed** v is how fast it's going, independent of direction. This is simply $v = |v_x|$ or $v = |v_y|$ the magnitude or absolute value of the object's velocity. In Example 2.1, for example, skater B's *velocity* is -2.0 m/s but his *speed* is 2.0 m/s. Speed is a scalar quantity, not a vector.

NOTE ▶ Our mathematical analysis of motion is based on velocity, not speed. The subscript in v_x or v_y is an essential part of the notation, reminding us that, even in one dimension, the velocity is a vector. ◀

The Mathematics of Uniform Motion

We need a mathematical analysis of motion that will be valid regardless of whether an object moves along the x-axis, the y-axis, or any other straight line. Consequently, it will be convenient to write equations for a "generic axis" that we will call the s-axis. The position of an object will be represented by the symbol s and its velocity by v_s.

NOTE ▶ In a specific problem you should use either x or y, whichever is appropriate, rather than s. ◀

Consider an object in uniform motion along the s-axis with the linear position-versus-time graph shown in **FIGURE 2.3**. The object's **initial position** is s_i at time t_i. The term *initial position* refers to the starting point of our analysis or the starting point in a

FIGURE 2.3 The velocity is found from the slope of the position-versus-time graph.

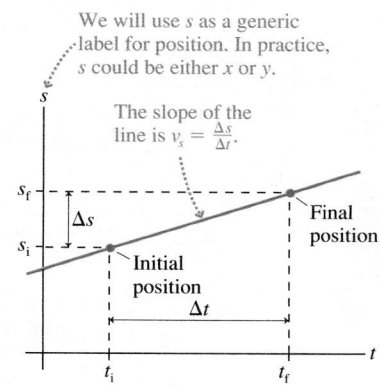

problem; the object may or may not have been in motion prior to t_i. At a later time t_f, the ending point of our analysis, the object's **final position** is s_f.

The object's velocity v_s along the s-axis can be determined by finding the slope of the graph:

$$v_s = \frac{\text{rise}}{\text{run}} = \frac{\Delta s}{\Delta t} = \frac{s_f - s_i}{t_f - t_i} \tag{2.2}$$

Equation 2.2 is easily rearranged to give

$$s_f = s_i + v_s \Delta t \quad \text{(uniform motion)} \tag{2.3}$$

The velocity of a uniformly moving object tells us the amount by which its position changes during each second. A particle with a velocity of 20 m/s *changes* its position by 20 m during every second of motion: by 20 m during the first second of its motion, by another 20 m during the next second, and so on. If the object starts at $s_i = 10$ m, it will be at $s = 30$ m after 1 second of motion and at $s = 50$ m after 2 seconds of motion. Thinking of velocity like this will help you develop an intuitive understanding of the connection between velocity and position.

Relating a velocity graph to a position graph

FIGURE 2.4 is the position-versus-time graph of a car.

a. Draw the car's velocity-versus-time graph.
b. Describe the car's motion.

MODEL Represent the car as a particle, with a well-defined position at each instant of time.

VISUALIZE Figure 2.4 is the graphical representation.

SOLVE

a. The car's position-versus-time graph is a sequence of three straight lines. Each of these straight lines represents uniform motion at a constant velocity. We can determine the car's velocity during each interval of time by measuring the slope of the line. From $t = 0$ s to $t = 2$ s ($\Delta t = 2.0$ s) the car's displacement is $\Delta x = -4.0$ m $- 0.0$ m $= -4.0$ m. The velocity during this interval is

$$v_x = \frac{\Delta x}{\Delta t} = \frac{-4.0 \text{ m}}{2.0 \text{ s}} = -2.0 \text{ m/s}$$

The car's position does not change from $t = 2$ s to $t = 4$ s ($\Delta x = 0$), so $v_x = 0$. Finally, the displacement between $t = 4$ s and $t = 6$ s is $\Delta x = 10.0$ m. Thus the velocity during this interval is

$$v_x = \frac{10.0 \text{ m}}{2.0 \text{ s}} = 5.0 \text{ m/s}$$

These velocities are shown on the velocity-versus-time graph of FIGURE 2.5.

b. The car backs up for 2 s at 2.0 m/s, sits at rest for 2 s, then drives forward at 5.0 m/s for at least 2 s. We can't tell from the graph what happens for $t > 6$ s.

ASSESS The velocity graph and the position graph look completely different. The *value* of the velocity graph at any instant of time equals the *slope* of the position graph.

FIGURE 2.4 Position-versus-time graph.

FIGURE 2.5 The corresponding velocity-versus-time graph.

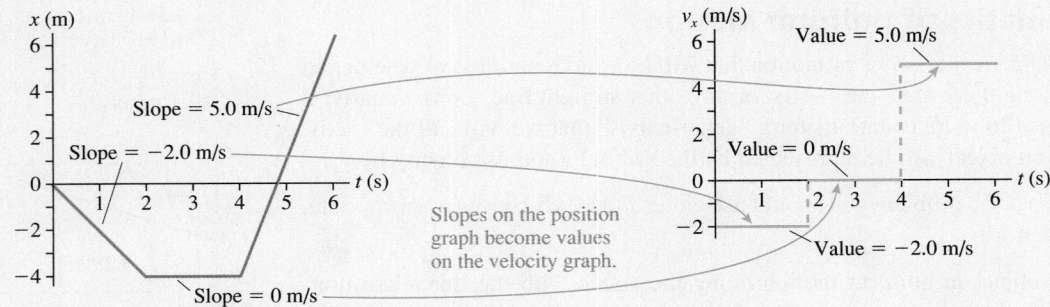

EXAMPLE 2.3 **Lunch in Cleveland?**

Bob leaves home in Chicago at 9:00 A.M. and travels east at a steady 60 mph. Susan, 400 miles to the east in Pittsburgh, leaves at the same time and travels west at a steady 40 mph. Where will they meet for lunch?

MODEL Here is a problem where, for the first time, we can really put all four aspects of our problem-solving strategy into play. To begin, represent Bob and Susan as particles.

VISUALIZE **FIGURE 2.6** shows the pictorial representation. The equal spacings of the dots in the motion diagram indicate that the motion is uniform. In evaluating the given information, we recognize that the starting time of 9:00 A.M. is not relevant to the problem. Consequently, the initial time is chosen as simply $t_0 = 0$ h. Bob and Susan are traveling in opposite directions, hence one of the velocities must be a negative number. We have chosen a coordinate system in which Bob starts at the origin and moves to the right (east) while Susan is moving to the left (west). Thus Susan has the negative velocity. Notice how we've assigned position, velocity, and time symbols to each point in the motion. Pay special attention to how subscripts are used to distinguish different points in the problem and to distinguish Bob's symbols from Susan's.

One purpose of the pictorial representation is to establish what we need to find. Bob and Susan meet when they have the same position at the same time t_1. Thus we want to find $(x_1)_B$ at the time when $(x_1)_B = (x_1)_S$. Notice that $(x_1)_B$ and $(x_1)_S$ are Bob's and Susan's *positions*, which are equal when they meet, not the distances they have traveled.

SOLVE The goal of the mathematical representation is to proceed from the pictorial representation to a mathematical solution of the problem. We can begin by using Equation 2.3 to find Bob's and Susan's positions at time t_1 when they meet:

$$(x_1)_B = (x_0)_B + (v_x)_B (t_1 - t_0) = (v_x)_B t_1$$

$$(x_1)_S = (x_0)_S + (v_x)_S (t_1 - t_0) = (x_0)_S + (v_x)_S t_1$$

Notice two things. First, we started by writing the *full* statement of Equation 2.3. Only then did we simplify by dropping those terms known to be zero. You're less likely to make accidental errors if you follow this procedure. Second, we replaced the generic symbol s with the specific horizontal-position symbol x, and we replaced the generic subscripts i and f with the specific symbols 0 and 1 that we defined in the pictorial representation. This is also good problem-solving technique.

The condition that Bob and Susan meet is

$$(x_1)_B = (x_1)_S$$

By equating the right-hand sides of the above equations, we get

$$(v_x)_B t_1 = (x_0)_S + (v_x)_S t_1$$

Solving for t_1 we find that they meet at time

$$t_1 = \frac{(x_0)_S}{(v_x)_B - (v_x)_S} = \frac{400 \text{ miles}}{60 \text{ mph} - (-40) \text{ mph}} = 4.0 \text{ hours}$$

Finally, inserting this time back into the equation for $(x_1)_B$ gives

$$(x_1)_B = \left(60 \frac{\text{miles}}{\text{hour}}\right) \times (4.0 \text{ hours}) = 240 \text{ miles}$$

While this is a number, it is not yet the answer to the question. The phrase "240 miles" by itself does not say anything meaningful. Because this is the value of Bob's *position*, and Bob was driving east, the answer to the question is, "They meet 240 miles east of Chicago."

ASSESS Before stopping, we should check whether or not this answer seems reasonable. We certainly expected an answer between 0 miles and 400 miles. We also know that Bob is driving faster than Susan, so we expect that their meeting point will be *more* than halfway from Chicago to Pittsburgh. Our assessment tells us that 240 miles is a reasonable answer.

FIGURE 2.6 Pictorial representation for Example 2.3.

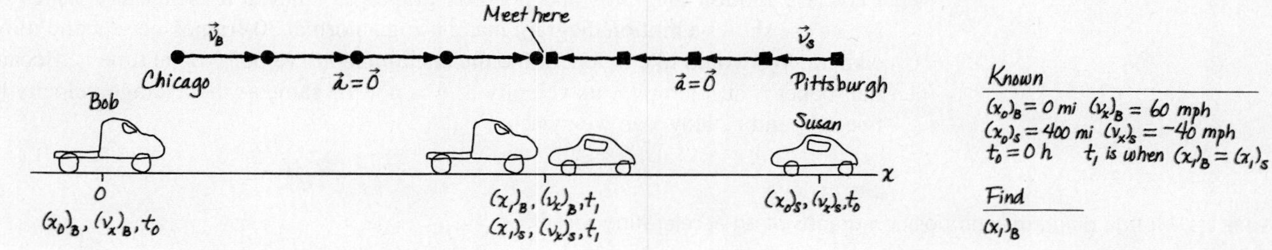

Meet here

$\vec{v}_B$ $\vec{a} = \vec{0}$ $\vec{a} = \vec{0}$ $\vec{v}_S$

Chicago   Pittsburgh

Bob   Susan

O

$(x_0)_B, (v_x)_B, t_0$ $(x_1)_B, (v_x)_B, t_1$ $(x_0)_S, (v_x)_S, t_0$

$(x_1)_S, (v_x)_S, t_1$ x

Known

$(x_0)_B = 0 \text{ mi}$ $(v_x)_B = 60 \text{ mph}$
$(x_0)_S = 400 \text{ mi}$ $(v_x)_S = -40 \text{ mph}$
$t_0 = 0 \text{ h}$ $t_1 \text{ is when } (x_1)_B = (x_1)_S$

Find

$(x_1)_B$

It is instructive to look at this example from a graphical perspective. **FIGURE 2.7** shows position-versus-time graphs for Bob and Susan. Notice the negative slope for Susan's graph, indicating her negative velocity. The point of interest is the intersection of the two lines; this is where Bob and Susan have the same position at the same time. Our method of solution, in which we equated $(x_1)_B$ and $(x_1)_S$, is really just solving the mathematical problem of finding the intersection of two lines.

FIGURE 2.7 Position-versus-time graphs for Bob and Susan.

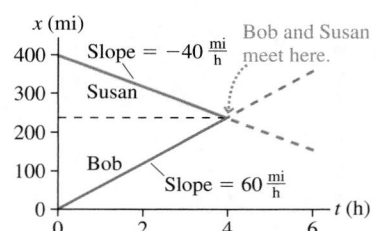

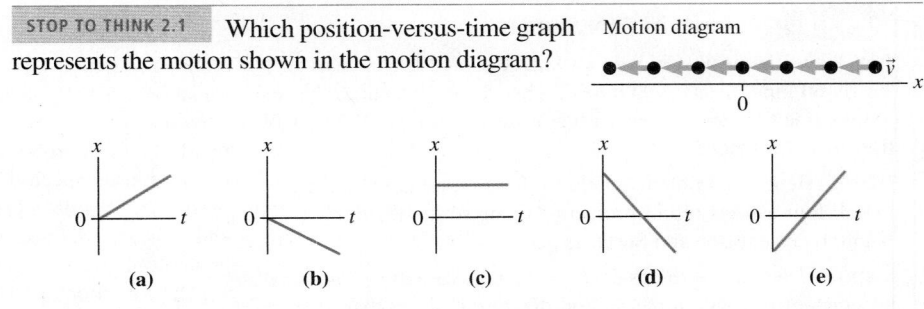

STOP TO THINK 2.1 Which position-versus-time graph represents the motion shown in the motion diagram?

2.2 Instantaneous Velocity

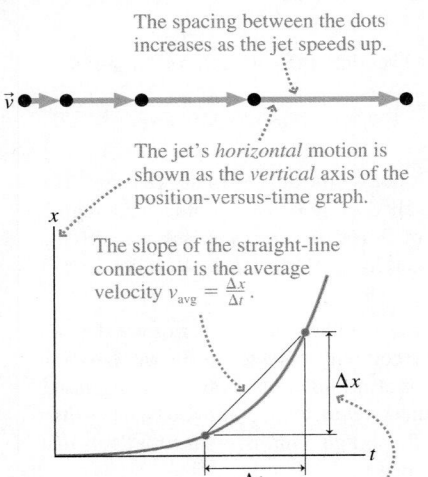

FIGURE 2.8 Motion diagram and position graph of a jet during takeoff.

The spacing between the dots increases as the jet speeds up.

The jet's *horizontal* motion is shown as the *vertical* axis of the position-versus-time graph.

The slope of the straight-line connection is the average velocity $v_{avg} = \frac{\Delta x}{\Delta t}$.

The increasing separation of the dots in the motion diagram means that Δx increases and the graph curves upward.

FIGURE 2.8 shows the motion diagram of a jet as it takes off. The increasing length of the velocity vectors tells us that the jet is speeding up, so this is *not* uniform motion. Consequently, the position-versus-time graph is *not* a straight line.

We can determine the jet's average speed v_{avg} between any two times t_i and t_f by finding the slope of the straight-line connection between the two points. However, average velocity has only limited usefulness for an object whose velocity isn't constant. The jet's average velocity during takeoff might be 30 m/s, but the speedometer in the cockpit would show the jet traveling at less than 30 m/s during the first few seconds. Similarly, the speedometer would read more than 30 m/s just before the wheels leave the ground.

In contrast to a velocity averaged over some interval of time, the speedometer reading tells you how fast you're going *at that instant*. We define an object's **instantaneous velocity** to be its velocity—a speed *and* a direction—at a single *instant* of time t.

As we've seen, velocity is the *rate* at which an object changes its position. Rates tell us how quickly or how slowly things change, and that idea is conveyed by the word "per." An instantaneous velocity of 80 miles *per* hour means that the rate at which your car's position is changing—at that exact instant—is such that it would travel 80 miles in 1 hour *if* it continued at that rate without change. Whether or not it actually does travel at that velocity for another hour, or even for another millisecond, is not relevant.

Using Motion Diagrams and Graphs

Let's use motion diagrams and position graphs to analyze a rocket as it takes off. **FIGURE 2.9a** shows a motion diagram made using a normal 30-frames-per-second movie camera. We would like to determine the *instantaneous* velocity v_{2y} at time t_2. Because the rocket is accelerating, its velocity at t_2 is not the same as the average velocity between t_1 and t_3. How can we measure v_{2y}?

FIGURE 2.9 Motion diagrams and position graphs of an accelerating rocket.

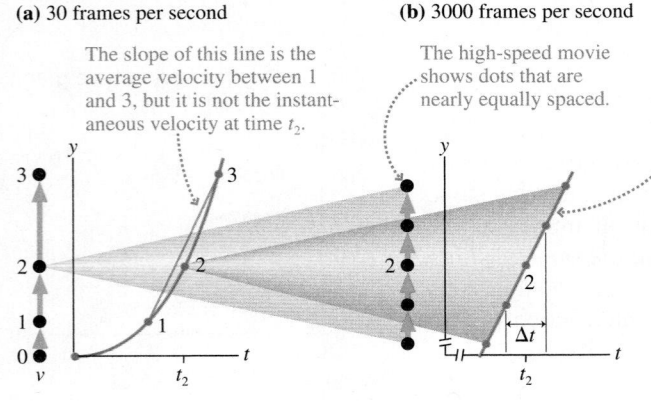

(a) 30 frames per second

The slope of this line is the average velocity between 1 and 3, but it is not the instantaneous velocity at time t_2.

(b) 3000 frames per second

The high-speed movie shows dots that are nearly equally spaced.

The highly magnified section of the graph near point 2 is very nearly a straight line. The slope of this line is a good approximation to the instantaneous velocity at time t_2. The slope *is* the instantaneous velocity in the limit $\Delta t \to 0$.

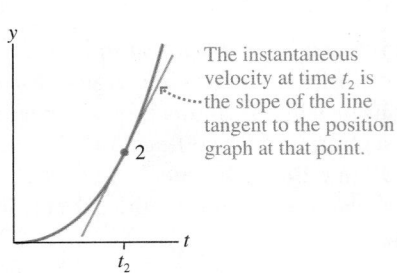

(c) The limiting case

The instantaneous velocity at time t_2 is the slope of the line tangent to the position graph at that point.

Suppose we use a high-speed camera, one that takes 3000 frames per second, to film just the segment of motion right around time t_2. This "magnified" motion diagram is shown in FIGURE 2.9b. At this level of magnification, each velocity vector is *almost* the same length. Further, the greatly magnified section of the curved position graph is *almost* a straight line. That is, the motion appears very nearly uniform on this time scale. If the rocket suddenly changed to *constant*-velocity motion at time t_2, it would continue to move with a velocity given by the slope of the graph in Figure 2.9b.

In other words, the average velocity $v_{avg} = \Delta s/\Delta t$ becomes a better and better approximation to the instantaneous velocity v_s as the time interval Δt over which the average is taken gets smaller and smaller. By magnifying the motion diagram, we are using smaller and smaller time intervals Δt.

We can state this idea mathematically in terms of the limit $\Delta t \rightarrow 0$:

$$v_s \equiv \lim_{\Delta t \to 0} \frac{\Delta s}{\Delta t} = \frac{ds}{dt} \qquad \text{(instantaneous velocity)} \qquad (2.4)$$

As Δt continues to get smaller, the average velocity $v_{avg} = \Delta s/\Delta t$ reaches a constant or *limiting* value. That is, **the instantaneous velocity at time t is the average velocity during a time interval Δt, centered on t, as Δt approaches zero.** In calculus, this limit is called *the derivative of s with respect to t*, and it is denoted ds/dt.

Graphically, $\Delta s/\Delta t$ is the slope of a straight line. As Δt gets smaller (i.e., more and more magnification), the straight line becomes a better and better approximation of the curve *at that one point*. In the limit $\Delta t \rightarrow 0$, the straight line is tangent to the curve. As FIGURE 2.9c shows, **the instantaneous velocity at time t is the slope of the line that is tangent to the position-versus-time graph at time t.** That is,

$$v_s = \text{slope of the position-versus-time graph at time } t \qquad (2.5)$$

EXAMPLE 2.4 **Finding velocity from position graphically**

FIGURE 2.10 shows the position-versus-time graph of an elevator.

a. At which labeled point or points does the elevator have the least speed?
b. At which point or points does the elevator have maximum velocity?
c. Sketch an approximate velocity-versus-time graph for the elevator.

FIGURE 2.10 Position-versus-time graph.

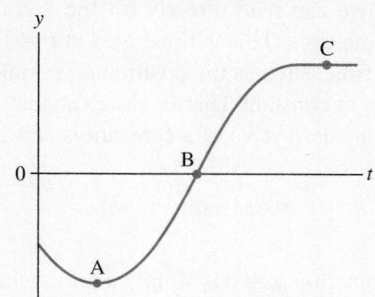

MODEL Represent the elevator as a particle.

VISUALIZE Figure 2.10 is the graphical representation.

SOLVE a. At any instant, the velocity is the slope of the position graph. FIGURE 2.11a shows that the elevator has the least speed—no speed at all!—at points A and C. At point A, the velocity is only instantaneously zero. At point C, the elevator has actually stopped and remains at rest.

FIGURE 2.11 The velocity-versus-time graph is found from the slope of the position graph.

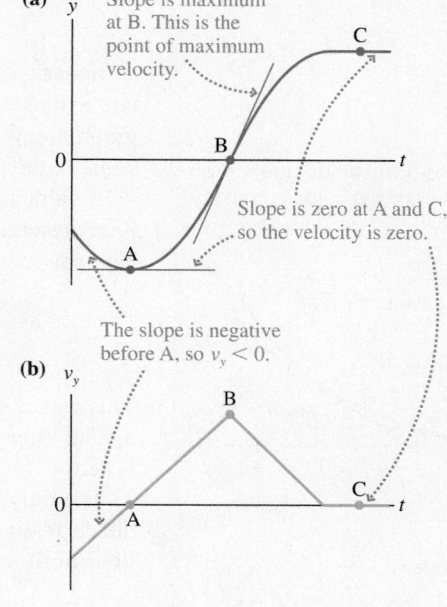

b. The elevator has maximum velocity at point B.

Continued

c. Although we cannot find an exact velocity-versus-time graph, we can see that the slope, and hence v_y, is initially negative, becomes zero at point A, rises to a maximum value at point B, decreases back to zero a little before point C, then remains at zero thereafter. Thus FIGURE 2.11b shows, at least approximately, the elevator's velocity-versus-time graph.

ASSESS Once again, the shape of the velocity graph bears no resemblance to the shape of the position graph. You must transfer *slope* information from the position graph to *value* information on the velocity graph.

A Little Calculus: Derivatives

Calculus–invented simultaneously in England by Newton and in Germany by Leibniz– is designed to deal with instantaneous quantities. In other words, it provides us with the tools for evaluating limits such as the one in Equation 2.4.

The notation ds/dt is called *the derivative of s with respect to t*, and Equation 2.4 defines it as the limiting value of a ratio. As Figure 2.9 showed, ds/dt can be interpreted graphically as the slope of the line that is tangent to the position-versus-time graph at time t.

The only functions we will use in Parts I and II of this book are powers and polynomials. Consider the function $u = ct^n$, where c and n are constants. The following result is proven in calculus:

$$\text{The derivative of } u = ct^n \text{ is } \frac{du}{dt} = nct^{n-1} \tag{2.6}$$

NOTE ▶ The symbol u is a "dummy name." Equation 2.6 can be used to take the derivative of *any* function of the form ct^n. ◀

For example, suppose the position of a particle as a function of time is $s = 2t^2$ m where t is in s. We can find the particle's velocity by using Equation 2.6 with $c = 2$ and $n = 2$ to calculate that the derivative of $s = 2t^2$ with respect to t is

$$v_s = \frac{ds}{dt} = 2 \cdot 2t^{2-1} = 4t$$

FIGURE 2.12 shows the particle's position and velocity graphs. It is critically important to understand the relationship between these two graphs. The *value* of the velocity graph at any instant of time, which we can read directly off the vertical axis, is the *slope* of the position graph at that same time. This is illustrated at $t = 1$ s and $t = 3$ s.

A value that doesn't change with time, such as the position of an object at rest, can be represented by the function $u = c = $ constant That is, the exponent of t^n is $n = 0$. You can see from Equation 2.6 that the derivative of a constant is zero. That is,

$$\frac{du}{dt} = 0 \text{ if } u = c = \text{constant} \tag{2.7}$$

This makes sense. The graph of the function $u = c$ is simply a horizontal line at height c. The slope of a horizontal line—which is what the derivative du/dt measures— is zero.

The only other information we need about derivatives for now is how to evaluate the derivative of the sum of two or more functions. Let u and w be two separate functions of time. You will learn in calculus that

$$\frac{d}{dt}(u + w) = \frac{du}{dt} + \frac{dw}{dt} \tag{2.8}$$

That is, the derivative of a sum is the sum of the derivatives.

Scientists and engineers must use calculus to calculate the trajectories of rockets.

FIGURE 2.12 Position-versus-time graph and the corresponding velocity-versus-time graph.

(a)

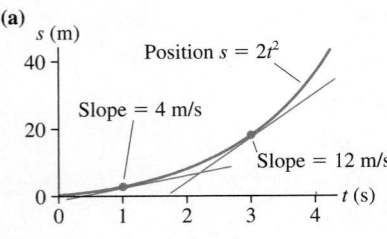

(b)

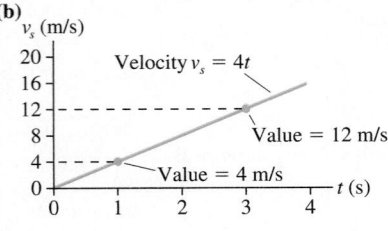

NOTE ▶ You may have learned in calculus to take the derivative dy/dx, where y is a function of x. The derivatives we use in physics are the same; only the notation is different. We're interested in how quantities change with time, so our derivatives are with respect to t instead of x. ◀

EXAMPLE 2.5   **Using calculus to find the velocity**

A particle's position is given by the function $x = (-t^3 + 3t)$ m, where t is in s.

a. What are the particle's position and velocity at $t = 2$ s?
b. Draw graphs of x and v_x during the interval -3 s $\leq t \leq 3$ s.
c. Draw a motion diagram to illustrate this motion.

SOLVE

a. We can compute the position directly from the function x:

$$x(\text{at } t = 2 \text{ s}) = -(2)^3 + (3)(2) = -8 + 6 = -2 \text{ m}$$

The velocity is $v_x = dx/dt$. The function for x is the sum of two polynomials, so

$$v_x = \frac{dx}{dt} = \frac{d}{dt}(-t^3 + 3t) = \frac{d}{dt}(-t^3) + \frac{d}{dt}(3t)$$

The first derivative is a power with $c = -1$ and $n = 3$; the second has $c = 3$ and $n = 1$. Using Equation 2.6, we have

$$v_x = (-3t^2 + 3) \text{ m/s}$$

where t is in s. Evaluating the velocity at $t = 2$ s gives

$$v_x(\text{at } t = 2 \text{ s}) = -3(2)^2 + 3 = -9 \text{ m/s}$$

The negative sign indicates that the particle, at this instant of time, is moving to the *left* at a speed of 9 m/s.

b. **FIGURE 2.13** shows the position graph and the velocity graph. These were created by computing, and then graphing, the values of x and v_x at several points between -3 s and 3 s. The slope of the position-versus-time graph at $t = 2$ s is -9 m/s; this becomes the *value* that is graphed for the velocity at $t = 2$ s. Similar measurements are shown at $t = -1$ s, where the velocity is instantaneously zero.

c. Finally, we can interpret the graphs in Figure 2.13 to draw the motion diagram shown in **FIGURE 2.14**.

■ The particle is initially to the right of the origin ($x > 0$ at $t = -3$ s) but moving to the left ($v_x < 0$). Its *speed* is slowing ($v = |v_x|$ is decreasing), so the velocity vector arrows are getting shorter.

■ The particle passes the origin at $t \approx -1.5$ s, but it is still moving to the left.

■ The position reaches a minimum at $t = -1$ s; the particle is as far left as it is going. The velocity is *instantaneously* $v_x = 0$ m/s as the particle reverses direction.

■ The particle moves back to the right between $t = -1$ s and $t = 1$ s ($v_x > 0$).

■ The particle turns around again at $t = 1$ s and begins moving back to the left ($v_x < 0$). It keeps speeding up, then disappears off to the left.

A point in the motion where a particle reverses direction is called a **turning point.** It is a point where the velocity is instantaneously zero while the position is a maximum or minimum. This particle has two turning points, at $t = -1$ s and again at $t = +1$ s. We will see many other examples of turning points.

ASSESS This example has used three different *representations* of motion: the mathematical equations, the graphs, and the motion diagram. All three describe the motion, but in different ways. Learning to move back and forth among the representations is important for solidifying your understanding of kinematics.

FIGURE 2.13 Position and velocity graphs.

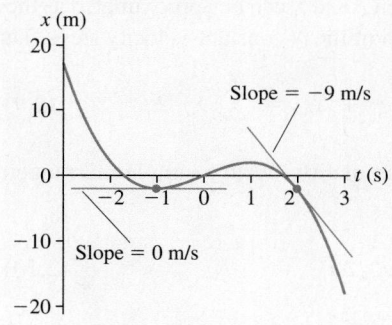

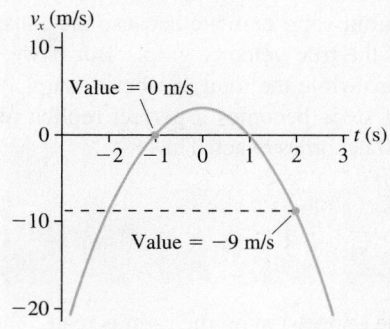

FIGURE 2.14 Motion diagram for Example 2.5.

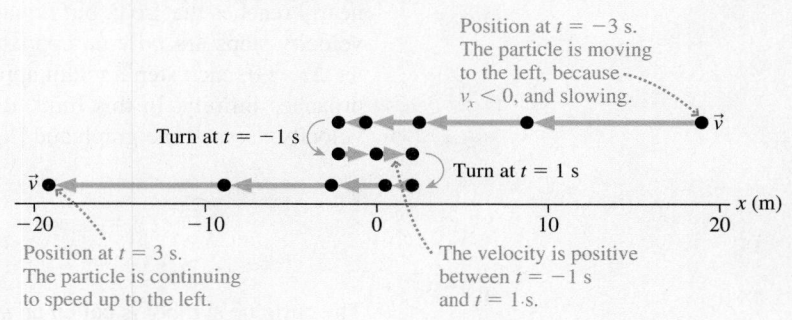

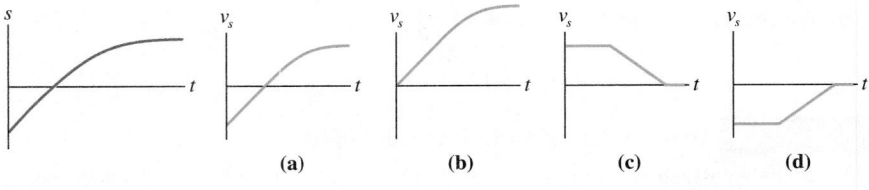

STOP TO THINK 2.2   Which velocity-versus-time graph goes with the position-versus-time graph on the left?

(a)   (b)   (c)   (d)

2.3 Finding Position from Velocity

Equation 2.4 provides a means of finding the instantaneous velocity v_s if we know the position s as a function of time. But what about the reverse problem? Can we use the object's velocity to calculate its position at some future time t? Equation 2.3, $s_f = s_i + v_s \Delta t$, does this for the case of uniform motion with a constant velocity. We need to find a more general expression that is valid when v_s is not constant.

FIGURE 2.15 Approximating a velocity-versus-time graph with a series of constant-velocity steps.

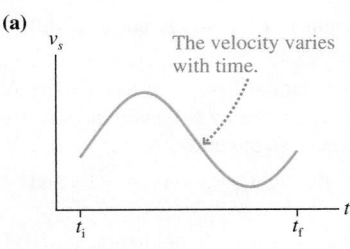

(a)

The velocity varies with time.

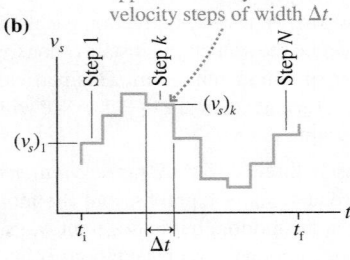

(b)

The velocity curve is approximated by constant-velocity steps of width Δt.

FIGURE 2.15a is a velocity-versus-time graph for an object whose velocity varies with time. Suppose we know the object's position to be s_i at an initial time t_i. Our goal is to find its position s_f at a later time t_f.

Because we know how to handle constant velocities, using Equation 2.3, let's *approximate* the velocity function of Figure 2.15a as a series of constant-velocity steps of width Δt. This is illustrated in FIGURE 2.15b. During the first step, from time t_i to time $t_i + \Delta t$, the velocity has the constant value $(v_s)_1$. The velocity during step k has the constant value $(v_s)_k$. Although the approximation shown in the figure is rather rough, with only nine steps, we can easily imagine that it could be made as accurate as desired by having more and more ever-narrower steps.

The velocity during each step is constant (uniform motion), so we can apply Equation 2.3 to each step. The object's displacement Δs_1 during the first step is simply $\Delta s_1 = (v_s)_1 \Delta t$. The displacement during the second step $\Delta s_2 = (v_s)_2 \Delta t$, and during step k the displacement is $\Delta s_k = (v_s)_k \Delta t$.

The total displacement of the object between t_i and t_f can be approximated as the sum of all the individual displacements during each of the N constant-velocity steps. That is,

$$\Delta s = s_f - s_i \approx \Delta s_1 + \Delta s_2 + \cdots + \Delta s_N = \sum_{k=1}^{N} (v_s)_k \Delta t \qquad (2.9)$$

where Σ (Greek sigma) is the symbol for summation. With a simple rearrangement, the particle's final position is

$$s_f \approx s_i + \sum_{k=1}^{N} (v_s)_k \Delta t \qquad (2.10)$$

Our goal was to use the object's velocity to find its final position s_f. Equation 2.10 nearly reaches that goal, but Equation 2.10 is only approximate because the constant-velocity steps are only an approximation of the true velocity graph. But if we now let $\Delta t \to 0$, each step's width approaches zero while the total number of steps N approaches infinity. In this limit, the series of steps becomes a perfect replica of the velocity-versus-time graph and Equation 2.10 becomes exact. Thus

$$s_f = s_i + \lim_{\Delta t \to 0} \sum_{k=1}^{N} (v_s)_k \Delta t = s_i + \int_{t_i}^{t_f} v_s \, dt \qquad (2.11)$$

The curlicue symbol is called an *integral.* The expression on the right is read, "the integral of $v_s \, dt$ from t_i to t_f." Equation 2.11 is the result that we were seeking. It allows us to predict an object's position s_f at a future time t_f.

We can give Equation 2.11 an important geometric interpretation. **FIGURE 2.16** shows step k in the approximation of the velocity graph as a tall, thin rectangle of height $(v_s)_k$ and width Δt. The product $\Delta s_k = (v_s)_k \Delta t$ is the area (base $\times$ height) of this small rectangle. The sum in Equation 2.11 adds up all of these rectangular areas to give the total area enclosed between the t-axis and the tops of the steps. The limit of this sum as $\Delta t \rightarrow 0$ is the total area enclosed between the t-axis and the velocity curve. This is called the "area under the curve." Thus a graphical interpretation of Equation 2.11 is:

$$s_f = s_i + \text{area under the velocity curve } v_s \text{ between } t_i \text{ and } t_f \qquad (2.12)$$

NOTE ▶ Wait a minute! The displacement $\Delta s = s_f - s_i$ is a length. How can a length equal an area? Recall earlier, when we found that the velocity is the slope of the position graph, we made a distinction between the *actual* slope and the *physically meaningful* slope? The same distinction applies here. The velocity graph does indeed bound a certain area on the page. That is the actual area, but it is *not* the area to which we are referring. Once again, we need to measure the quantities we are using, v_s and Δt, by referring to the scales on the axes. Δt is some number of seconds while v_s is some number of meters per second. When these are multiplied together, the *physically meaningful* area has units of meters. ◀

FIGURE 2.16 The total displacement Δs is the "area under the curve."

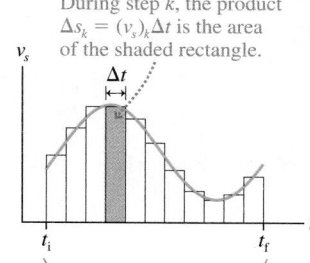

During step k, the product $\Delta s_k = (v_s)_k \Delta t$ is the area of the shaded rectangle.

During the interval t_i to t_f, the total displacement Δs is the "area under the curve."

EXAMPLE 2.6 **The displacement during a drag race**

FIGURE 2.17 shows the velocity-versus-time graph of a drag racer. How far does the racer move during the first 3.0 s?

FIGURE 2.17 Velocity-versus-time graph for Example 2.6.

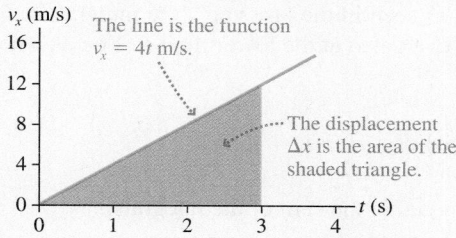

The line is the function $v_x = 4t$ m/s.

The displacement Δx is the area of the shaded triangle.

MODEL Represent the drag racer as a particle with a well-defined position at all times.

VISUALIZE Figure 2.17 is the graphical representation.

SOLVE The question "How far?" indicates that we need to find a displacement Δx rather than a position x. According to Equation 2.12, the car's displacement $\Delta x = x_f - x_i$ between $t = 0$ s and $t = 3$ s is the area under the curve from $t = 0$ s to $t = 3$ s. The curve in this case is an angled line, so the area is that of a triangle:

$$\Delta x = \text{area of triangle between } t = 0 \text{ s and } t = 3 \text{ s}$$
$$= \tfrac{1}{2} \times \text{base} \times \text{height}$$
$$= \tfrac{1}{2} \times 3 \text{ s} \times 12 \text{ m/s} = 18 \text{ m}$$

The drag racer moves 18 m during the first 3 seconds.

ASSESS The "area" is a product of s with m/s, so Δx has the proper units of m.

EXAMPLE 2.7 **Finding the turning point**

FIGURE 2.18 is the velocity graph for a particle that starts at $x_i = 30$ m at time $t_i = 0$ s.

a. Draw a motion diagram for the particle.
b. Where is the particle's turning point?
c. At what time does the particle reach the origin?

FIGURE 2.18 Velocity-versus-time graph for the particle of Example 2.7.

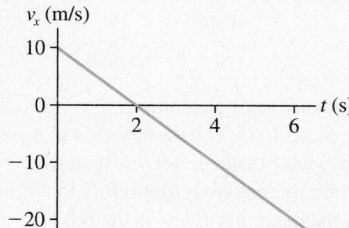

VISUALIZE The particle is initially 30 m to the right of the origin and moving *to the right* ($v_x > 0$) with a speed of 10 m/s. But v_x is

decreasing, so the particle is slowing down. At $t = 2$ s the velocity, just for an instant, is zero before becoming negative. This is the turning point. The velocity is negative for $t > 2$ s, so the particle has reversed direction and moves back toward the origin. At some later time, which we want to find, the particle will pass $x = 0$ m.

SOLVE a. **FIGURE 2.19** shows the motion diagram. The distance scale will be established in parts b and c but is shown here for convenience.

b. The particle reaches the turning point at $t = 2$ s. To learn *where* it is at that time we need to find the displacement during

FIGURE 2.19 Motion diagram for the particle whose velocity graph was shown in Figure 2.18.

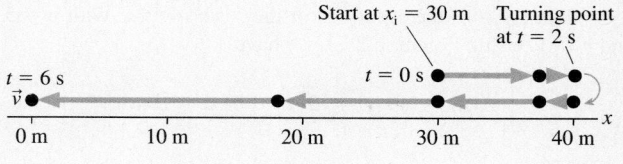

Start at $x_i = 30$ m   Turning point at $t = 2$ s

$t = 6$ s $t = 0$ s

Continued

the first two seconds. We can do this by finding the area under the curve between $t = 0$ s and $t = 2$ s:

x(at $t = 2$ s) $= x_i$ + area under the curve between 0 s and 2 s

$$= 30 \text{ m} + \tfrac{1}{2}(2 \text{ s} - 0 \text{ s})(10 \text{ m/s} - 0 \text{ m/s})$$

$$= 40 \text{ m}$$

The turning point is at $x = 40$ m.

c. The particle needs to move $\Delta x = -40$ m to get from the turning point to the origin. That is, the area under the curve from $t = 2$ s to the desired time t needs to be -40 m. Because the curve is below the axis, with negative values of v_x, the area to the right of $t = 2$ s is a *negative* area. With a bit of geometry, you will find that the triangle with a base extending from $t = 2$ s to $t = 6$ s has an area of -40 m. Thus the particle reaches the origin at $t = 6$ s.

A Little More Calculus: Integrals

Taking the derivative of a function is equivalent to finding the slope of a graph of the function. Similarly, evaluating an integral is equivalent to finding the area under a graph of the function. The graphical method is very important for building intuition about motion but is limited in its practical application. Just as derivatives of standard functions can be evaluated and tabulated, so can integrals.

The integral in Equation 2.11 is called a *definite integral* because there are two definite boundaries to the area we want to find. These boundaries are called the lower (t_i) and upper (t_f) *limits of integration*. For the important function $u = ct^n$, the essential result from calculus is that

$$\int_{t_i}^{t_f} u \, dt = \int_{t_i}^{t_f} ct^n \, dt = \frac{ct^{n+1}}{n+1} \Big|_{t_i}^{t_f} = \frac{ct_f^{n+1}}{n+1} - \frac{ct_i^{n+1}}{n+1} \qquad (n \neq -1) \quad (2.13)$$

The vertical bar in the third step with subscript t_i and superscript t_f is a shorthand notation from calculus that means—as seen in the last step—the integral evaluated at the upper limit t_f *minus* the integral evaluated at the lower limit t_i. You also need to know that for two functions u and w,

$$\int_{t_i}^{t_f} (u + w) \, dt = \int_{t_i}^{t_f} u \, dt + \int_{t_i}^{t_f} w \, dt \qquad (2.14)$$

That is, the integral of a sum is equal to the sum of the integrals.

EXAMPLE 2.8 **Using calculus to find the position**

Use calculus to solve Example 2.7.

SOLVE Figure 2.18 is a linear graph. Its "y-intercept" is seen to be 10 m/s and its slope is -5 (m/s)/s. Thus the velocity can be described by the equation

$$v_x = (10 - 5t) \text{ m/s}$$

where t is in s. We can find the position x at time t by using Equation 2.11:

$$x = x_i + \int_0^t v_x \, dt = 30 \text{ m} + \int_0^t (10 - 5t) \, dt$$

$$= 30 \text{ m} + \int_0^t 10 \, dt - \int_0^t 5t \, dt$$

We used Equation 2.14 for the integral of a sum to get the final expression. The first integral is a function of the form $u = ct^n$ with $c = 10$ and $n = 0$; the second is of the form $u = ct^n$ with $c = 5$ and $n = 1$. Using Equation 2.13, we have

$$\int_0^t 10 \, dt = 10t \Big|_0^t = 10 \cdot t - 10 \cdot 0 = 10t \text{ m}$$

and

$$\int_0^t 5t \, dt = \tfrac{5}{2}t^2 \Big|_0^t = \tfrac{5}{2} \cdot t^2 - \tfrac{5}{2} \cdot 0^2 = \tfrac{5}{2}t^2 \text{ m}$$

Combining the pieces gives

$$x = (30 + 10t - \tfrac{5}{2}t^2) \text{ m}$$

This is a general result for the position at *any* time t.

The particle's turning point occurs at $t = 2$ s, and its position at that time is

$$x(\text{at } t = 2 \text{ s}) = 30 + (10)(2) - \tfrac{5}{2}(2)^2 = 40 \text{ m}$$

The time at which the particle reaches the origin is found by setting $x = 0$ m:

$$30 + 10t - \tfrac{5}{2}t^2 = 0$$

This quadratic equation has two solutions: $t = -2$ s or $t = 6$ s.

When we solve a quadratic equation, we cannot just arbitrarily select the root we want. Instead, we must decide which is the *meaningful* root. Here the negative root refers to a time before the problem began, so the meaningful one is the positive root, $t = 6$ s.

ASSESS The results agree with the answers we found previously from a graphical solution.

Summing Up

As you work on building intuition about motion, you need to be able to move back and forth among four different representations of the motion:

- The motion diagram;
- The position-versus-time graph;
- The velocity-versus-time graph;
- The description in words.

Given a description of a certain motion, you should be able to sketch the motion diagram and the position and velocity graphs. Given one graph, you should be able to generate the other. And given position and velocity graphs, you should be able to "interpret" them by describing the motion in words or in a motion diagram.

STOP TO THINK 2.3   Which position-versus-time graph goes with the velocity-versus-time graph on the left? The particle's position at $t_i = 0$ s is $x_i = -10$ m.

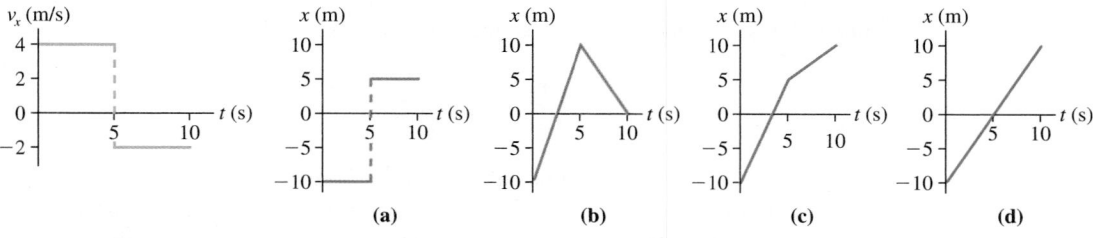

2.4 Motion with Constant Acceleration

We need one more major concept to describe one-dimensional motion: acceleration. Acceleration, as we noted in Chapter 1, is a rather abstract concept. Nonetheless, acceleration is the linchpin of mechanics. We will see very shortly that Newton's laws relate the acceleration of an object to the forces that are exerted on it.

Let's conduct a race between a Volkswagen Beetle and a Porsche to see which can achieve a velocity of 30 m/s (≈ 60 mph) in the shortest time. Both cars are equipped with computers that will record the speedometer reading 10 times each second. This gives a nearly continuous record of the *instantaneous* velocity of each car. Table 2.1 shows some of the data. The velocity-versus-time graphs, based on these data, are shown in FIGURE 2.20.

How can we describe the difference in performance of the two cars? It is not that one has a different velocity from the other; both achieve every velocity between 0 and 30 m/s. The distinction is how long it took each to *change* its velocity from 0 to 30 m/s. The Porsche changed velocity quickly, in 6.0 s, while the VW needed 15 s to make the same velocity change. Because the Porsche had a velocity change $\Delta v_s = 30$ m/s during a time interval $\Delta t = 6.0$ s the *rate* at which its velocity changed was

$$\text{rate of velocity change} = \frac{\Delta v_s}{\Delta t} = \frac{30 \text{ m/s}}{6.0 \text{ s}} = 5.0 \text{ (m/s)/s} \qquad (2.15)$$

Notice the units. They are units of "velocity per second." A rate of velocity change of 5.0 "meters per second per second" means that the velocity increases by 5.0 m/s during the first second, by another 5.0 m/s during the next second, and so on. In fact,

TABLE 2.1 Velocities of a Porsche and a Volkswagen Beetle

t(s)	v_{Porsche}(m/s)	v_{VW} (m/s)
0.0	0.0	0.0
0.1	0.5	0.2
0.2	1.0	0.4
0.3	1.5	0.6
0.4	2.0	0.8
⋮	⋮	⋮

FIGURE 2.20 Velocity-versus-time graphs for the Porsche and the VW Beetle.

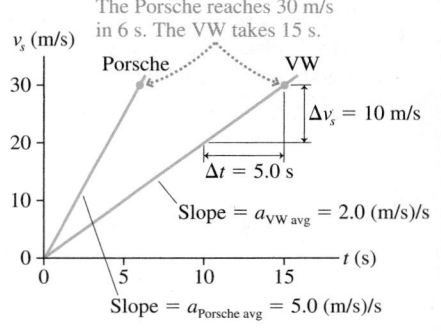

the velocity will increase by 5.0 m/s during any second in which it is changing at the rate of 5.0 (m/s)/s.

Chapter 1 introduced *acceleration* as "the rate of change of velocity." That is, acceleration measures how quickly or slowly an object's velocity changes. In parallel with our treatment of velocity, let's define the **average acceleration** a_{avg} during the time interval Δt to be

$$a_{avg} \equiv \frac{\Delta v_s}{\Delta t} \qquad \text{(average acceleration)} \qquad (2.16)$$

Equations 2.15 and 2.16 show that the Porsche had the rather large acceleration of 5.0 (m/s)/s.

Because Δv_s and Δt are the "rise" and "run" of a velocity-versus-time graph, we see that a_{avg} can be interpreted graphically as the *slope* of a straight-line velocity-versus-time graph. In other words,

$$a_{avg} = \text{slope of the velocity-versus-time graph} \qquad (2.17)$$

Figure 2.20 uses this idea to show that the VW's average acceleration is

$$a_{VW\,avg} = \frac{\Delta v_s}{\Delta t} = \frac{10 \text{ m/s}}{5.0 \text{ s}} = 2.0 \text{ (m/s)/s}$$

This is less than the acceleration of the Porsche, as expected.

An object whose velocity-versus-time graph is a straight-line graph has a steady and unchanging acceleration. There's no need to specify "average" if the acceleration is constant, so we'll use the symbol a_s as we discuss motion along the *s*-axis with constant acceleration.

NOTE ▶ An important aspect of acceleration is its *sign*. Acceleration $\vec{a}$, like position $\vec{r}$ and velocity $\vec{v}$, is a vector. For motion in one dimension, the sign of a_x (or a_y) is positive if the vector $\vec{a}$ points to the right (or up), negative if it points to the left (or down). This was illustrated in Figure 1.19 and the very important Tactics Box 1.4, which you may wish to review. It's particularly important to emphasize that positive and negative values of a_s do *not* correspond to "speeding up" and "slowing down." ◀

EXAMPLE 2.9 **Relating acceleration to velocity**

a. A bicyclist has a velocity of 10 m/s and a constant acceleration of 2 (m/s)/s. What is her velocity 1 s later? 2 s later?
b. A bicyclist has a velocity of −10 m/s and a constant acceleration of 2 (m/s)/s. What is his velocity 1 s later? 2 s later?

SOLVE

a. An acceleration of 2 (m/s)/s *means* that the velocity increases by 2 m/s every 1 s. If the bicyclist's initial velocity is 10 m/s, then 1 s later her velocity will be 12 m/s. After 2 s, which is 1 additional second later, it will increase by another 2 m/s to

14 m/s. After 3 s it will be 16 m/s. Here a positive a_s is causing the bicyclist to speed up.

b. If the bicyclist's initial velocity is a *negative* −10 m/s but the acceleration is a positive +2 (m/s)/s, then 1 s later his velocity will be −8 m/s. After 2 s it will be −6 m/s, and so on. In this case, a positive a_s is causing the object to *slow down* (decreasing speed *v*). This agrees with the rule from Tactics Box 1.4: An object is slowing down if and only if v_s and a_s have opposite signs.

NOTE ▶ It is customary to abbreviate the acceleration units (m/s)/s as m/s^2. For example, the bicyclists in Example 2.9 had an acceleration of 2 m/s^2. We will use this notation, but keep in mind the *meaning* of the notation as "(meters per second) per second." ◀

EXAMPLE 2.10 **Running the court**

A basketball player starts at the left end of the court and moves with the velocity shown in FIGURE 2.21. Draw a motion diagram and an acceleration-versus-time graph for the basketball player.

FIGURE 2.21 Velocity-versus-time graph for the basketball player of Example 2.10.

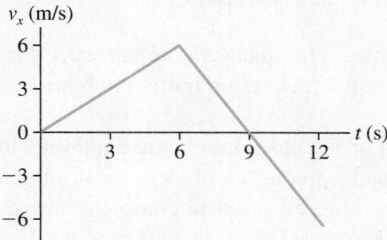

$$a_x = \frac{\Delta v_x}{\Delta t} = \frac{-12 \text{ m/s}}{6.0 \text{ s}} = -2.0 \text{ m/s}^2$$

The acceleration graph for these 12 s is shown in FIGURE 2.22b. Notice that there is no change in the acceleration at $t = 9$ s, the turning point.

FIGURE 2.22 Motion diagram and acceleration graph for Example 2.10.

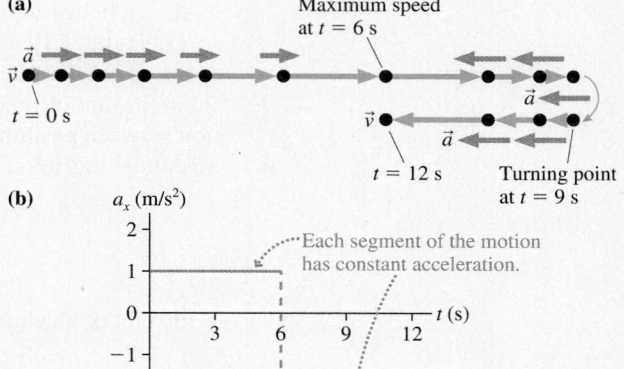

VISUALIZE The velocity is positive (motion to the right) and increasing for the first 6 s, so the velocity arrows in the motion diagram are to the right and getting longer. From $t = 6$ s to 9 s the motion is still to the right (v_x is still positive), but the arrows are getting shorter because v_x is decreasing. There's a turning point at $t = 9$ s, when $v_x = 0$, and after that the motion is to the left (v_x is negative) and getting faster. The motion diagram of FIGURE 2.22a shows the velocity and the acceleration vectors.

SOLVE Acceleration is the slope of the velocity graph. For the first 6 s, the slope has the constant value

$$a_x = \frac{\Delta v_x}{\Delta t} = \frac{6.0 \text{ m/s}}{6.0 \text{ s}} = 1.0 \text{ m/s}^2$$

The velocity then decreases by 12 m/s during the 6 s interval from $t = 6$ s to $t = 12$ s, so

ASSESS The *sign* of a_x does *not* tell us whether the object is speeding up or slowing down. The basketball player is slowing down from $t = 6$ s to $t = 9$ s, then speeding up from $t = 9$ s to $t = 12$ s. Nonetheless, his acceleration is negative during this entire interval because his acceleration vector, as seen in the motion diagram, always points to the left.

The Kinematic Equations of Constant Acceleration

Consider an object whose acceleration a_s remains constant during the time interval $\Delta t = t_f - t_i$. At the beginning of this interval, at time t_i, the object has initial velocity v_{is} and initial position s_i. Note that t_i is often zero, but it does not have to be. We would like to predict the object's final position s_f and final velocity v_{fs} at time t_f.

The object's velocity is changing because the object is accelerating. FIGURE 2.23a shows the acceleration-versus-time graph, a horizontal line between t_i and t_f. It is not hard to find the object's velocity v_{fs} at a later time t_f. By definition,

$$a_s = \frac{\Delta v_s}{\Delta t} = \frac{v_{fs} - v_{is}}{\Delta t} \tag{2.18}$$

which is easily rearranged to give

$$v_{fs} = v_{is} + a_s \Delta t \tag{2.19}$$

The velocity-versus-time graph, shown in FIGURE 2.23b, is a straight line that starts at v_{is} and has slope a_s.

FIGURE 2.23 Acceleration and velocity graphs for constant acceleration.

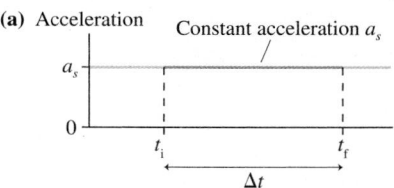

(a) Acceleration

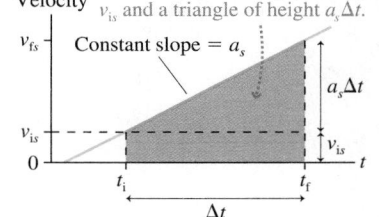

(b) Velocity

As you learned in the last section, the object's final position is

$$s_f = s_i + \text{area under the velocity curve } v_s \text{ between } t_i \text{ and } t_f \qquad (2.20)$$

The shaded area in Figure 2.23b can be subdivided into a rectangle of area $v_{is}\,\Delta t$ and a triangle of area $\frac{1}{2}(a_s\Delta t)(\Delta t) = \frac{1}{2}a_s(\Delta t)^2$. Adding these gives

$$s_f = s_i + v_{is}\,\Delta t + \tfrac{1}{2}a_s(\Delta t)^2 \qquad (2.21)$$

where $\Delta t = t_f - t_i$ is the elapsed time. The quadratic dependence on Δt causes the position-versus-time graph for constant-acceleration motion to have a parabolic shape, as shown below in **FIGURE 2.24**.

Equations 2.19 and 2.21 are two of the basic kinematic equations for motion with *constant* acceleration. They allow us to predict an object's position and velocity at a future instant of time. We need one more equation to complete our set, a direct relation between position and velocity. First use Equation 2.19 to write $\Delta t = (v_{fs} - v_{is})/a_s$. Substitute this into Equation 2.21, giving

$$s_f = s_i + v_{is}\left(\frac{v_{fs} - v_{is}}{a_s}\right) + \tfrac{1}{2}a_s\left(\frac{v_{fs} - v_{is}}{a_s}\right)^2 \qquad (2.22)$$

With a bit of algebra, this is rearranged to read

$$v_{fs}^2 = v_{is}^2 + 2a_s\,\Delta s \qquad (2.23)$$

where $\Delta s = s_f - s_i$ is the *displacement* (not the distance!).

Equations 2.19, 2.21, and 2.23, which are summarized in Table 2.2, are the key results for motion with constant acceleration.

Figure 2.24 is a comparison of motion with constant velocity (uniform motion) and motion with constant acceleration (uniformly accelerated motion). Notice that uniform motion is really a special case of uniformly accelerated motion in which the constant acceleration happens to be zero. The graphs for a negative acceleration are left as an exercise.

TABLE 2.2 The kinematic equations for motion with constant acceleration

$v_{fs} = v_{is} + a_s\,\Delta t$
$s_f = s_i + v_{is}\,\Delta t + \tfrac{1}{2}a_s(\Delta t)^2$
$v_{fs}^2 = v_{is}^2 + 2a_s\,\Delta s$

FIGURE 2.24 Motion with constant velocity and constant acceleration. These graphs assume $s_i = 0$, $v_{is} > 0$, and (for constant acceleration) $a_s > 0$.

(a) Motion at constant velocity

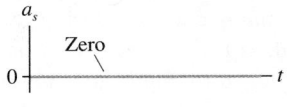

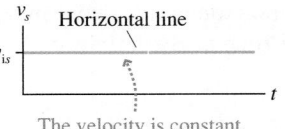

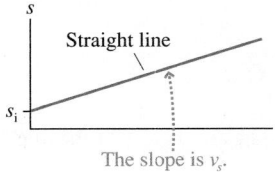

(b) Motion at constant acceleration

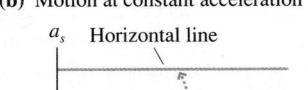

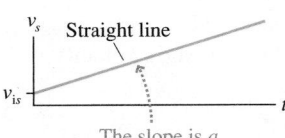

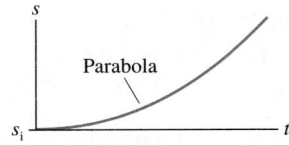

A Problem-Solving Strategy

This information can be assembled into a problem-solving strategy for kinematics with constant acceleration.

PROBLEM-SOLVING STRATEGY 2.1 **Kinematics with constant acceleration**

MODEL Use the particle model. Make simplifying assumptions.

VISUALIZE Use different representations of the information in the problem.

- Draw a *pictorial representation*. This helps you assess the information you are given and starts the process of translating the problem into symbols.
- Use a *graphical representation* if it is appropriate for the problem.
- Go back and forth between these two representations as needed.

SOLVE The mathematical representation is based on the three kinematic equations

$$v_{fs} = v_{is} + a_s \, \Delta t$$

$$s_f = s_i + v_{is} \, \Delta t + \tfrac{1}{2} a_s (\Delta t)^2$$

$$v_{fs}^2 = v_{is}^2 + 2 a_s \, \Delta s$$

- Use x or y, as appropriate to the problem, rather than the generic s.
- Replace i and f with numerical subscripts defined in the pictorial representation.
- Uniform motion with constant velocity has $a_s = 0$.

ASSESS Is your result believable? Does it have proper units? Does it make sense?

NOTE ▶ You are strongly encouraged to solve problems on the Dynamics Worksheets found at the back of the Student Workbook. These worksheets will help you use the Problem-Solving Strategy and develop good problem-solving skills. ◀

EXAMPLE 2.11 **The motion of a rocket sled**

A rocket sled accelerates at 50 m/s^2 for 5.0 s, coasts for 3.0 s, then deploys a braking parachute and decelerates at 3.0 m/s^2 until coming to a halt.

a. What is the maximum velocity of the rocket sled?
b. What is the total distance traveled?

MODEL Represent the rocket sled as a particle.

VISUALIZE **FIGURE 2.25** shows the pictorial representation. Recall that we discussed the first two-thirds of this problem as Example 1.8 in Chapter 1.

SOLVE a. The maximum velocity is identified in the pictorial representation as v_{1x}, the velocity at time t_1 when the acceleration phase ends. The first kinematic equation in Table 2.2 gives

$$v_{1x} = v_{0x} + a_{0x}(t_1 - t_0) = a_{0x}t_1$$

$$= (50 \text{ m/s}^2)(5.0 \text{ s}) = 250 \text{ m/s}$$

We started with the complete equation, then simplified by noting which terms were zero. Also notice that we found an algebraic expression for v_{1x}, *then* substituted numbers. Working algebraically

FIGURE 2.25 Pictorial representation of the rocket sled.

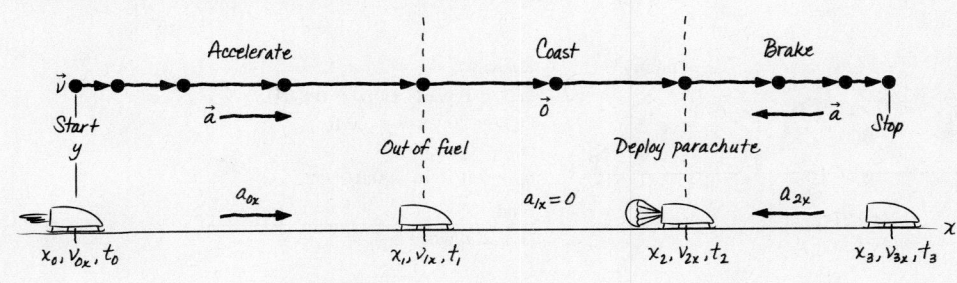

Continued

is a hallmark of good problem-solving technique, and many homework problems will ask you to do so.

b. Finding the total distance requires several steps. First, the sled's position when the acceleration ends at t_1 is found from the second equation in Table 2.2:

$$x_1 = x_0 + v_{0x}(t_1 - t_0) + \tfrac{1}{2}a_{0x}(t_1 - t_0)^2 = \tfrac{1}{2}a_{0x}t_1^2$$

$$= \tfrac{1}{2}(50\,\text{m/s}^2)(5.0\,\text{s})^2 = 625\,\text{m}$$

During the coasting phase, which is uniform motion with no acceleration ($a_{1x} = 0$),

$$x_2 = x_1 + v_{1x}\,\Delta t = x_1 + v_{1x}(t_2 - t_1)$$

$$= 625\,\text{m} + (250\,\text{m/s})(3.0\,\text{s}) = 1375\,\text{m}$$

Notice that, in this case, Δt is not simply t. The braking phase is a little different because we don't know how long it lasts. But

we do know that the sled ends with $v_{3x} = 0$ m/s, so we can use the third equation in Table 2.2:

$$v_{3x}^2 = v_{2x}^2 + 2a_{2x}\,\Delta x = v_{2x}^2 + 2a_{2x}(x_3 - x_2)$$

This can be solved for x_3:

$$x_3 = x_2 + \frac{v_{3x}^2 - v_{2x}^2}{2a_{2x}}$$

$$= 1375\,\text{m} + \frac{0 - (250\,\text{m/s})^2}{2(-3.0\,\text{m/s}^2)} = 12{,}000\,\text{m}$$

ASSESS Using the approximate conversion factor 1 m/s ≈ 2 mph from Table 1.5, we see that the top speed is ≈ 500 mph. The total distance traveled is ≈ 12 km ≈ 7 mi. This is reasonable because it takes a very long distance to stop from a top speed of 500 mph!

NOTE ▶ We used explicit numerical subscripts throughout the mathematical representation, each referring to a symbol that was defined in the pictorial representation. The subscripts i and f in the Table 2.2 equations are just generic "place holders" that don't have unique values. During the acceleration phase we had i = 0 and f = 1. Later, during the coasting phase, these became i = 1 and f = 2. The numerical subscripts have a clear meaning and are less likely to lead to confusion. ◀

EXAMPLE 2.12   **Friday night football**

Fred catches the football while standing directly on the goal line. He immediately starts running forward with an acceleration of 6 ft/s^2. At the moment the catch is made, Tommy is 20 yards away and heading directly toward Fred with a steady speed of 15 ft/s. If neither deviates from a straight-ahead path, where will Tommy tackle Fred?

MODEL   Represent Fred and Tommy as particles.

VISUALIZE   The pictorial representation is shown in FIGURE 2.26. With two moving objects we need the additional subscripts F and T to distinguish Fred's symbols and Tommy's symbols. The axes have been chosen so that Fred starts at $(x_0)_F = 0$ ft and moves to the right while Tommy starts at $(x_0)_T = 60$ ft and runs to the left with a *negative* velocity.

SOLVE   We want to find *where* Fred and Tommy have the same position. The pictorial representation designates time t_1 as *when* they meet. The second equation of Table 2.2 allows us to find their positions at time t_1. These are:

$$(x_1)_F = (x_0)_F + (v_{0x})_F(t_1 - t_0) + \tfrac{1}{2}(a_x)_F(t_1 - t_0)^2$$

$$= \tfrac{1}{2}(a_x)_F t_1^2$$

$$(x_1)_T = (x_0)_T + (v_{0x})_T(t_1 - t_0) + \tfrac{1}{2}(a_x)_T(t_1 - t_0)^2$$

$$= (x_0)_T + (v_{0x})_T t_1$$

FIGURE 2.26 Pictorial representation for Example 2.12.

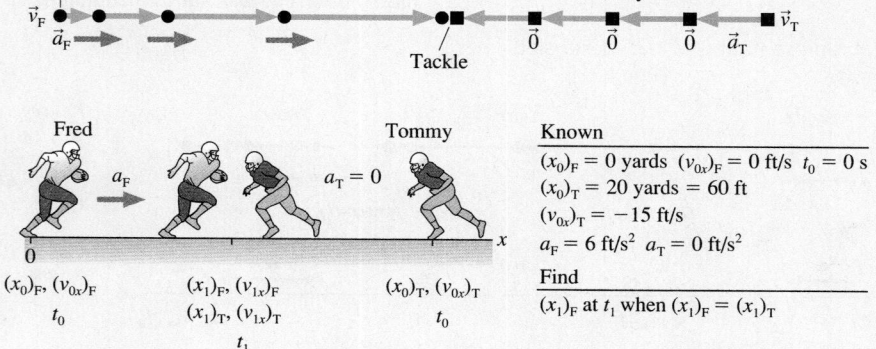

Notice that Tommy's position equation contains the term $(v_{0x})_T t_1$, not $-(v_{0x})_T t_1$. The fact that he is moving to the left has already been considered in assigning a *negative value* to $(v_{0x})_T$, hence we don't want to add any additional negative signs in the equation. If we now set $(x_1)_F$ and $(x_1)_T$ equal to each other, indicating the point of the tackle, we can solve for t_1:

$$\tfrac{1}{2}(a_x)_F t_1^2 = (x_0)_T + (v_{0x})_T t_1$$

$$\tfrac{1}{2}(a_x)_F t_1^2 - (v_{0x})_T t_1 - (x_0)_T = 0$$

$$3t_1^2 + 15t_1 - 60 = 0$$

The solutions of this quadratic equation for t_1 are $t_1 = (-7.62 \text{ s}, +2.62 \text{ s})$. The negative time is not meaningful in this problem, so the time of the tackle is $t_1 = 2.62$ s. We've kept an extra significant figure in the solution to minimize round-off error in the next step. Using this value to compute $(x_1)_F$ gives

$$(x_1)_F = \tfrac{1}{2}(a_x)_F t_1^2 = 20.6 \text{ feet} = 6.9 \text{ yards}$$

Tommy makes the tackle at just about the 7-yard line!

ASSESS The answer had to be between 0 yards and 20 yards. Because Tommy was already running, whereas Fred started from rest, it is reasonable that Fred will cover less than half the 20-yard separation before meeting Tommy. Thus 6.9 yards is a reasonable answer.

NOTE ▶ The purpose of the Assess step is not to prove that an answer must be right but to rule out answers that, with a little thought, are clearly wrong. ◀

It is worth exploring Example 2.12 graphically. **FIGURE 2.27** shows position-versus-time graphs for Fred and Tommy. The curves intersect at $t = 2.62$ s, and that is where the tackle occurs. You should compare this problem to Example 2.3 and Figure 2.7 for Bob and Susan to notice the similarities and the differences.

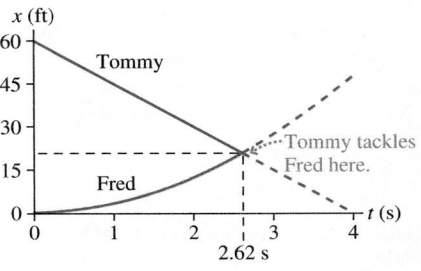

FIGURE 2.27 Position-versus-time graphs for Fred and Tommy.

STOP TO THINK 2.4 Which velocity-versus-time graph or graphs go with the acceleration-versus-time graph? The particle is initially moving to the right.

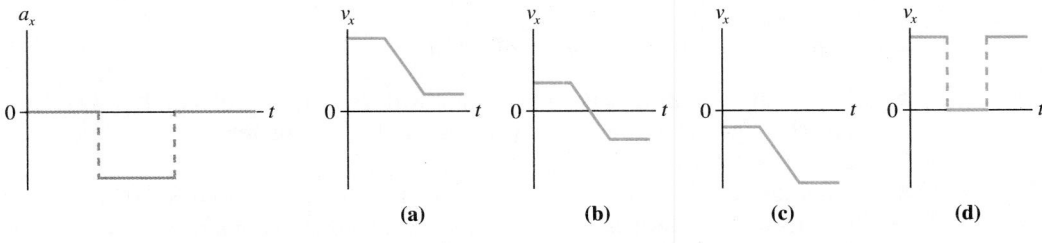

2.5 Free Fall

The motion of an object moving under the influence of gravity only, and no other forces, is called **free fall.** Strictly speaking, free fall occurs only in a vacuum, where there is no air resistance. Fortunately, the effect of air resistance is small for "heavy objects," so we'll make only a very slight error in treating these objects *as if* they were in free fall. For very light objects, such as a feather, or for objects that fall through very large distances and gain very high speeds, the effect of air resistance is *not* negligible. Motion with air resistance is a problem we will study in Chapter 6. Until then, we will restrict our attention to "heavy objects" and will make the reasonable assumption that falling objects are in free fall.

Galileo, in the 17th century, was the first to make detailed measurements of falling objects. The story of Galileo dropping different weights from the leaning bell

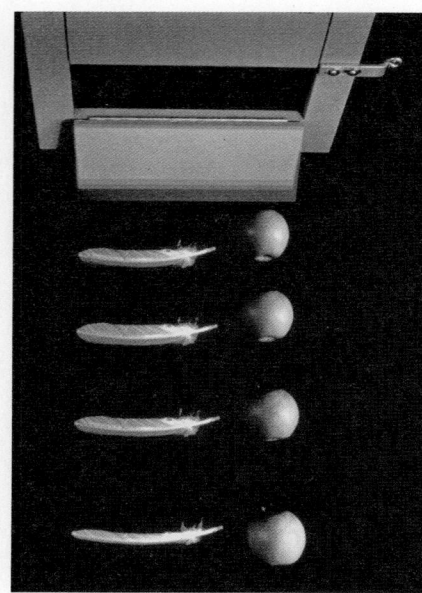

In the absence of air resistance, any two objects fall at the same rate and hit the ground at the same time. The apple and feather seen here are falling in a vacuum.

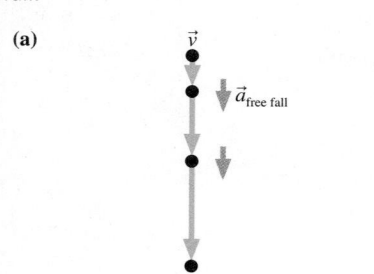

FIGURE 2.28 Motion of an object in free fall.

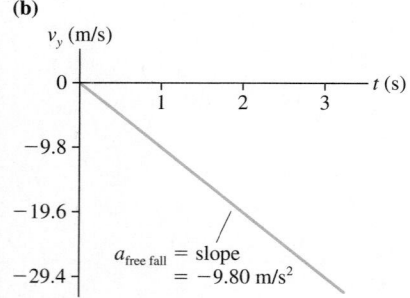

tower at the cathedral in Pisa is well known, although historians cannot confirm its truth. But bell towers were common in the Italy of Galileo's day, so he had ample opportunity to make the measurements and observations that he describes in his writings.

Galileo developed a *model* of motion—motion in the absence of air resistance—that could only be approximated by any real object. His discovery can be summarized as follows:

- Two objects dropped from the same height will, if air resistance can be neglected, hit the ground at the same time and with the same speed.
- Consequently, **any two objects in free fall, regardless of their mass, have the same acceleration** $\vec{a}_{\text{free fall}}$. This is an especially important conclusion.

FIGURE 2.28a shows the motion diagram of an object that was released from rest and falls freely. FIGURE 2.28b shows the object's velocity graph. The motion diagram and graph are identical for a falling pebble and a falling boulder. The fact that the velocity graph is a straight line tells us the motion is one of constant acceleration, and $a_{\text{free fall}}$ is easily found from the slope of the graph. Careful measurements show that the value of $\vec{a}_{\text{free fall}}$ varies ever so slightly at different places on the earth, due to the slightly nonspherical shape of the earth and to the fact that the earth is rotating. A global average, at sea level, is

$$\vec{a}_{\text{free fall}} = (9.80 \text{ m/s}^2, \text{ vertically downward}) \qquad (2.24)$$

For practical purposes, *vertically downward* means along a line toward the center of the earth. However, we'll learn in Chapter 13 that the rotation of the earth has a small effect on both the size and direction of $\vec{a}_{\text{free fall}}$.

The length, or magnitude, of $\vec{a}_{\text{free fall}}$ is known as the **free-fall acceleration,** and it has the special symbol g:

$$g = 9.80 \text{ m/s}^2 \text{ (free-fall acceleration)}$$

Several points about free fall are worthy of note:

- g, by definition, is *always* positive. **There will never be a problem that will use a negative value for g.** But, you say, objects fall when you release them rather than rise, so how can g be positive?
- g is *not* the acceleration $a_{\text{free fall}}$, but simply its magnitude. Because we've chosen the y-axis to point vertically upward, the downward acceleration vector $\vec{a}_{\text{free fall}}$ has the one-dimensional acceleration

$$a_y = a_{\text{free fall}} = -g \qquad (2.25)$$

It is a_y that is negative, not g.
- Because free fall is motion with constant acceleration, we can use the kinematic equations of Table 2.2 with the acceleration being that of free fall, $a_y = -g$.
- g is not called "gravity." Gravity is a force, not an acceleration. The symbol g recognizes the influence of gravity, but g is *the free-fall acceleration.*
- $g = 9.80 \text{ m/s}^2$ only on earth. Other planets have different values of g. You will learn in Chapter 13 how to determine g for other planets.

NOTE ▶ Despite the name, free fall is not restricted to objects that are literally falling. Any object moving under the influence of gravity only, and no other forces, is in free fall. This includes objects falling straight down, objects that have been tossed or shot straight up, and projectile motion. ◀

EXAMPLE 2.13   **A falling rock**

A rock is released from rest at the top of a 100-m-tall building. How long does the rock take to fall to the ground, and what is its impact velocity?

MODEL Represent the rock as a particle. Assume air resistance is negligible.

VISUALIZE FIGURE 2.29 shows the pictorial representation. We have placed the origin at the ground, which makes $y_0 = 100$ m. Although the rock falls 100 m, it is important to notice that the *displacement* is $\Delta y = y_1 - y_0 = -100$ m.

FIGURE 2.29 Pictorial representation of a falling rock.

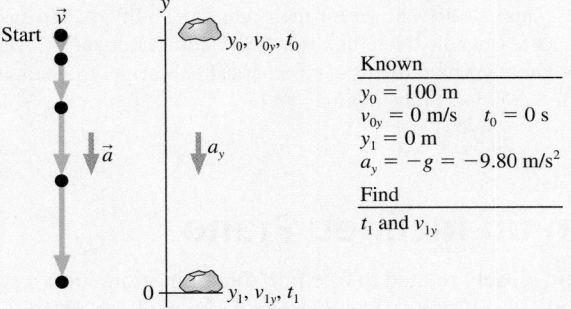

Known
$y_0 = 100$ m
$v_{0y} = 0$ m/s $t_0 = 0$ s
$y_1 = 0$ m
$a_y = -g = -9.80$ m/s^2

Find
t_1 and v_{1y}

SOLVE Free fall is motion with the specific constant acceleration $a_y = -g$. The first question involves a relation between time and distance, so only the second equation in Table 2.2 is relevant. Using $v_{0y} = 0$ m/s and $t_0 = 0$ s, we find

$$y_1 = y_0 + v_{0y}\,\Delta t + \tfrac{1}{2}a_y\,\Delta t^2 = y_0 + v_{0y}\,\Delta t - \tfrac{1}{2}g\,\Delta t^2 = y_0 - \tfrac{1}{2}g t_1^2$$

We can now solve for t_1, finding

$$t_1 = \sqrt{\frac{2(y_0 - y_1)}{g}} = \sqrt{\frac{2(100\text{ m} - 0\text{ m})}{9.80\text{ m/s}^2}} = \pm\,4.52\text{ s}$$

The $\pm$ sign indicates that there are two mathematical solutions; therefore we have to use physical reasoning to choose between them. A negative t_1 would refer to a time before we dropped the rock, so we select the positive root: $t_1 = 4.52$ s.

Now that we know the fall time, we can use the first kinematic equation to find v_{1y}:

$$v_{1y} = v_{0y} - g\,\Delta t = -g t_1 = -(9.80\text{ m/s}^2)(4.52\text{ s})$$
$$= -44.3\text{ m/s}$$

Alternatively, we could work directly from the third kinematic equation:

$$v_{1y} = \sqrt{v_{0y}{}^2 - 2g\,\Delta y} = \sqrt{-2(9.80\text{ m/s}^2)(-100\text{ m})} = \pm\,44.3\text{ m/s}$$

This method is useful if you don't know Δt. However, we must again choose the correct sign of the square root. Because the velocity vector points downward, the sign of v_y has to be negative. Thus $v_{1y} = -44.3$ m/s. The importance of careful attention to the signs cannot be overemphasized!

A common error would be to say "The rock fell 100 m, so $\Delta y = 100$ m." This would have you trying to take the square root of a negative number. As noted above, Δy is not a distance. It is a *displacement*, with a carefully defined meaning of $y_f - y_i$. In this case, $\Delta y = y_1 - y_0 = -100$ m.

ASSESS Are the answers reasonable? Well, 100 m is about 300 feet, which is about the height of a 30-floor building. How long does it take something to fall 30 floors? Four or five seconds seems pretty reasonable. How fast would it be going at the bottom? Using 1 m/s $\approx$ 2 mph, we find that 44.3 m/s $\approx$ 90 mph. That also seems pretty reasonable after falling 30 floors. Had we misplaced a decimal point, though, and found 443 m/s, we would be suspicious when we converted this to $\approx$ 900 mph! The answers all seem reasonable.

EXAMPLE 2.14   **Finding the height of a leap**

The springbok, an antelope found in Africa, gets its name from its remarkable jumping ability. When startled, a springbok will leap straight up into the air—a maneuver called a "pronk." A springbok goes into a crouch to perform a pronk. It then extends its legs forcefully, accelerating at 35 m/s^2 for 0.70 m as its legs straighten. Legs fully extended, it leaves the ground and rises into the air. How high does it go?

MODEL Represent the springbok as a particle.

VISUALIZE FIGURE 2.30 shows the pictorial representation. This is a problem with a beginning point, an end point, and a point in

FIGURE 2.30 Pictorial representation of a startled springbok.

Known
$y_0 = -0.70$ m $t_0 = 0$ s
$v_{0y} = 0$ m/s $a_{0y} = 35$ m/s^2
$y_1 = 0$ m $v_{2y} = 0$ m/s
$a_{1y} = -g = -9.80$ m/s^2

Find
y_2

between where the nature of the motion changes. We've identified these points with subscripts 0, 1, and 2. The motion from 0 to 1 is a rapid upward acceleration until the springbok's feet leave the ground at 1. Even though the springbok is moving upward from 1

Continued

to 2, this is free-fall motion because the springbok is now moving under the influence of gravity only.

How do we put "How high?" into symbols? The clue is that the very top point of the trajectory is a *turning point*, and we've seen that the instantaneous velocity at a turning point is $v_{2y} = 0$. This was not explicitly stated but is part of our interpretation of the problem.

SOLVE For the first part of the motion, pushing off, we know a displacement but not a time interval. The third equation in Table 2.2 is perfect for this situation:

$$v_{1y}^2 = v_{0y}^2 + 2a_{0y}\Delta y = 2(35 \text{ m/s}^2)(0.70 \text{ m}) = 49 \text{ m}^2/\text{s}^2$$

$$v_{1y} = \sqrt{49 \text{ m}^2/\text{s}^2} = 7.0 \text{ m/s}$$

The springbok leaves the ground with a velocity of 7.0 m/s. This is the starting point for the problem of a projectile launched straight up from the ground. One possible solution is to use the velocity

equation to find how long it takes to reach maximum height, then the position equation to calculate the maximum height. But that takes two separate calculations. It is easier to make another use of the velocity-displacement equation:

$$v_{2y}^2 = 0 = v_{1y}^2 + 2a_{1y}\Delta y = v_{1y}^2 - 2g(y_2 - y_1)$$

where now the acceleration is $a_{1y} = -g$. Using $y_1 = 0$, we can solve for y_2, the height of the leap:

$$y_2 = \frac{v_{1y}^2}{2g} = \frac{(7.0 \text{ m/s})^2}{2(9.80 \text{ m/s}^2)} = 2.5 \text{ m}$$

ASSESS 2.5 m is a bit over 8 feet, a remarkable vertical jump. But these animals are known for their jumping ability, so the answer seems reasonable. Note that it is especially important in a multi-part problem like this to use numerical subscripts to distinguish different points in the motion.

2.6 Motion on an Inclined Plane

FIGURE 2.31 Acceleration on an inclined plane.

(a)

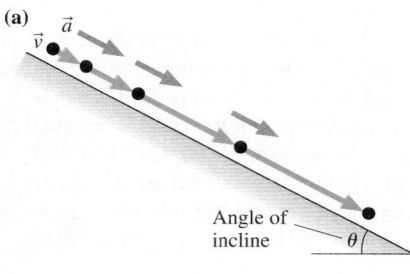

Angle of incline θ

(b)

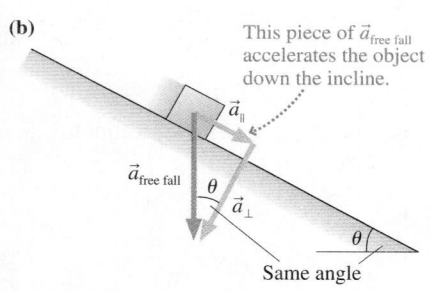

This piece of $\vec{a}_{\text{free fall}}$ accelerates the object down the incline.

$\vec{a}_\parallel$

$\vec{a}_{\text{free fall}}$ θ $\vec{a}_\perp$

θ

Same angle

FIGURE 2.31a shows a problem closely related to free fall: that of motion down a straight, but frictionless, inclined plane, such as a skier going down a slope on frictionless snow. What is the object's acceleration? Although we're not yet prepared to give a rigorous derivation, we can deduce the acceleration with a plausibility argument.

FIGURE 2.31b shows the free-fall acceleration $\vec{a}_{\text{free fall}}$ the object would have if the incline suddenly vanished. The free-fall acceleration points straight down. This vector can be broken into two pieces: a vector $\vec{a}_\parallel$ that is parallel to the incline and a vector $\vec{a}_\perp$ that is perpendicular to the incline. The surface of the incline somehow "blocks" $\vec{a}_\perp$, through a process we will examine in Chapter 6, but $\vec{a}_\parallel$ is unhindered. It is this piece of $\vec{a}_{\text{free fall}}$, parallel to the incline, that accelerates the object.

By definition, the length, or magnitude, of $\vec{a}_{\text{free fall}}$ is g. Vector $\vec{a}_\parallel$ is opposite angle θ (Greek *theta*), so the length, or magnitude, of $\vec{a}_\parallel$ must be $g \sin\theta$. Consequently, the one-dimensional acceleration along the incline is

$$a_s = \pm g \sin\theta \qquad (2.26)$$

The correct sign depends on the direction in which the ramp is tilted. Examples will illustrate.

Equation 2.26 makes sense. Suppose the plane is perfectly horizontal. If you place an object on a horizontal surface, you expect it to stay at rest with no acceleration. Equation 2.26 gives $a_s = 0$ when $\theta = 0°$, in agreement with our expectations. Now suppose you tilt the plane until it becomes vertical, at $\theta = 90°$. Without friction, an object would simply fall, in free fall, parallel to the vertical surface. Equation 2.26 gives $a_s = -g = a_{\text{free fall}}$ when $\theta = 90°$, again in agreement with our expectations. Equation 2.26 gives the correct result in these *limiting cases*.

EXAMPLE 2.15 **Measuring acceleration**

In the laboratory, a 2.00-m-long track has been inclined as shown in FIGURE 2.32. Your task is to measure the acceleration of a cart on the ramp and to compare your result with what you might have expected. You have available five "photogates" that measure the cart's speed as it passes through. You place a gate every 30 cm from a line you mark near the top of the track as the starting line. One run generates the data shown in the table.

FIGURE 2.32 The experimental setup.

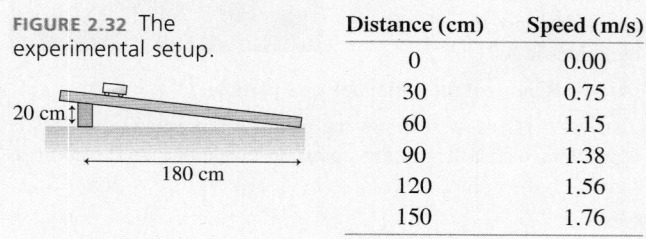

20 cm

180 cm

Distance (cm)	Speed (m/s)
0	0.00
30	0.75
60	1.15
90	1.38
120	1.56
150	1.76

The first entry isn't a photogate measurement, but it is a valid data point because you know the cart's speed is zero at the point where you release it.

NOTE ▶ Physics is an experimental science. Our knowledge of the universe is grounded in observations and measurements. Consequently, some examples and homework problems throughout this book will be based on data. These won't replace an actual laboratory experience, but they will provide you with an opportunity for thinking about how we make sense of the underlying theory. Data-based homework problems require the use of a spreadsheet, graphing software, or a graphing calculator in which you can "fit" data with a straight line. ◀

MODEL Represent the cart as a particle.

VISUALIZE FIGURE 2.33 shows the pictorial representation. We've chosen the x-axis to be parallel to the track, which is tilted at angle $\theta = \tan^{-1}(20 \text{ cm}/180 \text{ cm}) = 6.34°$. This is motion on an inclined plane, so you might expect the cart's acceleration to be $a_x = g \sin \theta = 1.08 \text{ m/s}^2$. In any laboratory situation, it's good to have an idea what to expect.

FIGURE 2.33 The pictorial representation of the cart on the track.

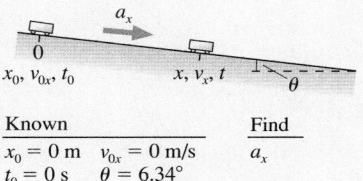

Known		Find
$x_0 = 0$ m	$v_{0x} = 0$ m/s	a_x
$t_0 = 0$ s	$\theta = 6.34°$	

SOLVE In analyzing data, we want to use *all* the data, not just pick out one or two measurements. Further, we almost always want to use graphs when we have a series of measurements. We might start by graphing speed versus distance traveled. This is shown in FIGURE 2.34a, where—recognizing that our data table has inconsistent units—we converted distances to meters. As expected, speed increases with distance, but the graph isn't linear and that makes it hard to analyze.

FIGURE 2.34 Graphs of velocity and of velocity squared.

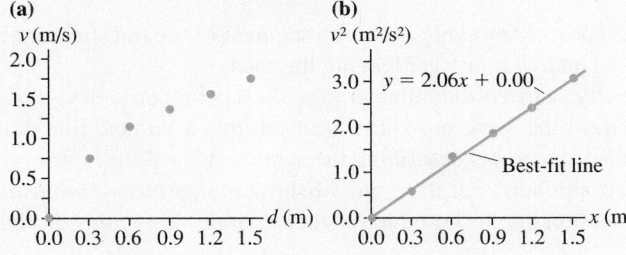

Rather than proceeding by trial and error, let's be guided by theory. We have information about speed and distance, but not about how long it took the cart to reach each photogate. *If* the cart has constant acceleration—which we don't yet know and need to confirm—the third equation of Table 2.2 tells us that velocity and displacement should be related by

$$v_x^2 = v_{0x}^2 + 2a_x \Delta x = 2a_x x$$

The last step was based on starting from rest ($v_{0x} = 0$) at the origin ($\Delta x = x - x_0 = x$). Although we measured speed, the cart is moving

in the $+x$-direction, so we can interpret the speeds as velocities. And we've measured distance from the origin, so the distance values are x.

Rather than graphing v_x versus x, suppose we graphed v_x^2 versus x. If we let $y = v_x^2$, the kinematic equation reads

$$y = 2a_x x$$

This is in the form of a linear equation: $y = mx + b$, where m is the slope and b is the y-intercept. In this case, $m = 2a_x$ and $b = 0$. So if the cart really does have constant acceleration, a graph of v_x^2 versus x should be linear with a y-intercept of zero. This is a prediction that we can test.

Thus our analysis has three steps:

1. Graph v_x^2 versus x. If the graph is a straight line with a y-intercept of zero (or very close to zero), then we can conclude that the cart has constant acceleration on the ramp. If not, the acceleration is *not* constant and we cannot use the kinematic equations for constant acceleration.
2. If the graph has the correct shape, we can determine its slope m.
3. Because kinematics predicts $m = 2a_x$, the acceleration is $a_x = m/2$.

FIGURE 2.34b is the graph of v_x^2 versus x. It does turn out to be a straight line with a y-intercept of zero, and this is the evidence we need that the cart has a constant acceleration on the ramp. To proceed, we want to determine the slope by finding the straight line that is the "best fit" to the data. This is a statistical technique, justified in a statistics class, but one that is implemented in spreadsheets and graphing calculators. The solid line in Figure 2.34b is the best-fit line for this data, and its equation is shown. We see that the slope is $m = 2.06 \text{ m/s}^2$. **Slopes have units**, and the units come not from the fitting procedure but by looking at the axes of the graph. Here the vertical axis is velocity squared, with units of $(\text{m/s})^2$, while the horizontal axis is position, measured in m. Thus the slope, rise over run, has units of m/s^2.

Finally, we can determine that the cart's acceleration was

$$a_x = \frac{m}{2} = 1.03 \text{ m/s}^2$$

This is about 5% less than the 1.08 m/s^2 we expected. Two possibilities come to mind. Perhaps the distances used to find the tilt angle weren't measured accurately. Or, more likely, the cart rolls with a small bit of friction. The predicted acceleration $a_x = g \sin \theta$ is for a *frictionless* inclined plane; any friction would decrease the acceleration.

ASSESS How did we know to graph v_x^2 versus x rather than v_x versus x? We were guided by theory! The analysis of data requires linking measurements with theory, and we had a theoretical prediction, from kinematics, that v_x^2 is proportional to x for constant-acceleration motion that starts from rest at the origin. Thus the shape of a v_x^2-versus-x graph both tests the assertion that the acceleration is constant and, if the assertion is true, allows us to find the acceleration from the slope of the graph. We'll see this procedure over and over: Use theory to suggest a graph that should be linear if the assumptions of the theory are true, graph it, then—if the graph really is linear—match the experimentally determined slope and/or intercept with their theoretical predictions to extract useful results.

In this case, the graph was linear and we could use the slope to determine the cart's acceleration. The value was just slightly less than would be predicted for a frictionless incline, so the result is reasonable.

Thinking Graphically

Kinematics is the language of motion. The concepts we have developed in this chapter will be used extensively throughout the rest of this textbook. One of the most important ideas, summarized in Tactics Box 2.2, has been that the relationships among position, velocity, and acceleration can be expressed graphically.

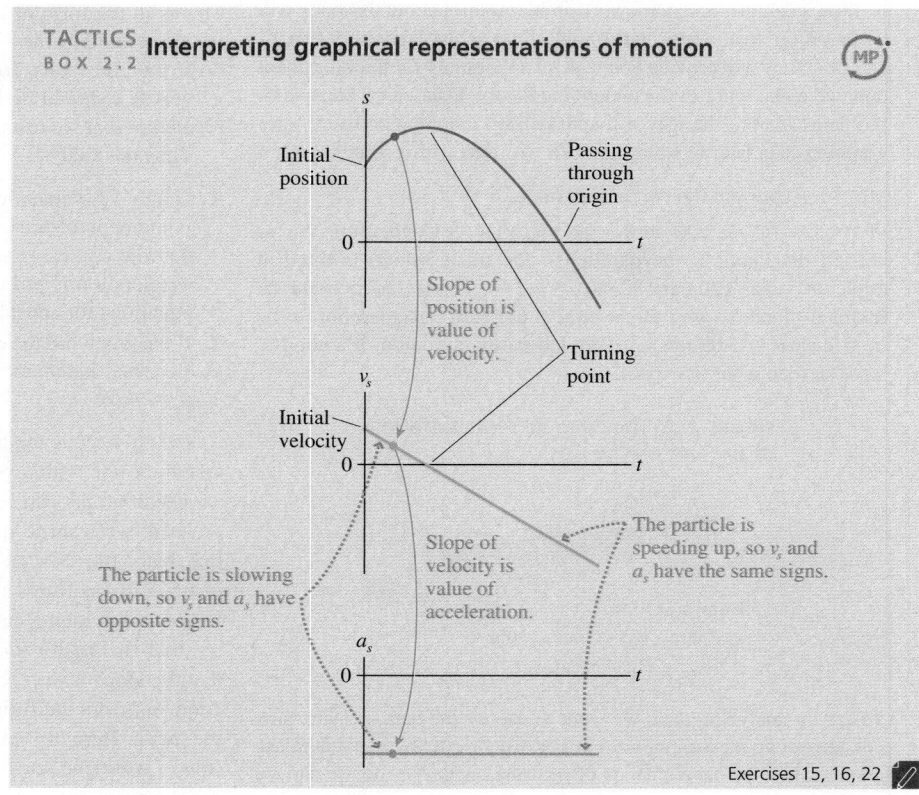

TACTICS BOX 2.2 Interpreting graphical representations of motion

Initial position
Passing through origin
Slope of position is value of velocity.
Turning point
Initial velocity
Slope of velocity is value of acceleration.
The particle is slowing down, so v_s and a_s have opposite signs.
The particle is speeding up, so v_s and a_s have the same signs.

Exercises 15, 16, 22

A good way to solidify your understanding of motion graphs is to consider the problem of a hard, smooth ball rolling on a smooth track. The track is made up of several straight segments connected together. Each segment may be either horizontal or inclined. Your task is to analyze the ball's motion graphically.

There are a small number of rules to follow:

1. Assume that the ball passes smoothly from one segment of the track to the next, with no loss of speed and without ever leaving the track.
2. The position, velocity, and acceleration graphs should be stacked vertically. They should each have the same horizontal scale so that a vertical line drawn through all three connects points describing the same instant of time.
3. The graphs have no numbers, but they should show the correct *relationships*. For example, the position graph should have steeper slopes in regions of higher speed.
4. The position s is the position measured *along* the track. Similarly, v_s and a_s are the velocity and acceleration parallel to the track.

EXAMPLE 2.16 **From track to graphs**

Draw position, velocity, and acceleration graphs for the ball on the frictionless track of **FIGURE 2.35**.

FIGURE 2.35 A ball rolling along a track.

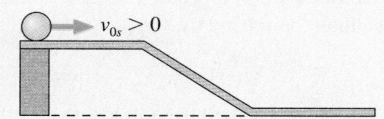

$v_{0s} > 0$

VISUALIZE It is often easiest to begin with the velocity. Here the ball starts with an initial velocity v_{0s}. There is no acceleration on the horizontal surface ($a_s = 0$ if $\theta = 0°$), so the velocity remains constant until the ball reaches the slope. The slope is an inclined plane that, as we have learned, has constant acceleration. The velocity increases linearly with time during constant-acceleration motion. The ball returns to constant-velocity motion after reaching the bottom horizontal segment. The middle graph of **FIGURE 2.36** shows the velocity.

We have enough information to draw the acceleration graph. We noted that the acceleration is zero while the ball is on the horizontal segments, and a_s has a constant positive value on the slope. These accelerations are consistent with the slope of the velocity graph: zero slope, then positive slope, then a return to zero slope. The acceleration cannot *really* change instantly from zero to a nonzero value, but the change can be so quick that we do not see it on the time scale of the graph. That is what the vertical dotted lines imply.

Finally, we need to find the position-versus-time graph. The position increases linearly with time during the first segment at constant velocity. It also does so during the third segment of

FIGURE 2.36 Motion graphs for the ball in Example 2.16.

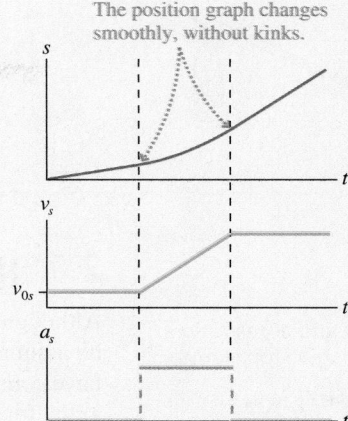

motion, but with a steeper slope to indicate a faster velocity. In between, while the acceleration is nonzero but constant, the position graph has a *parabolic* shape.

Two points are worth noting:

1. The dotted vertical lines through the graphs show the instants when the ball moves from one segment of the track to the next. Because of rule 1, the speed does not change abruptly at these points; it changes gradually.
2. The parabolic section of the position-versus-time graph blends *smoothly* into the straight lines on either side. This is a consequence of rule 1. An abrupt change of slope (a "kink") would indicate an abrupt change in velocity and would violate rule 1.

EXAMPLE 2.17 **From graphs to track**

FIGURE 2.37 shows a set of motion graphs for a ball moving on a track. Draw a picture of the track and describe the ball's initial condition. Each segment of the track is *straight,* but the segments may be tilted.

FIGURE 2.37 Motion graphs of a ball rolling on a track of unknown shape.

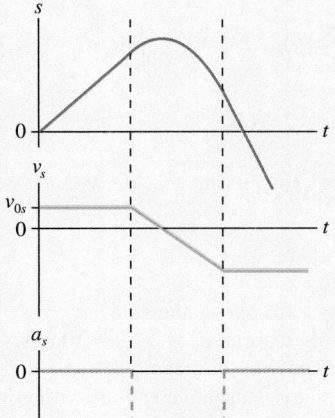

VISUALIZE Let's begin by examining the velocity graph. The ball starts with initial velocity $v_{0s} > 0$ and maintains this velocity for

awhile; there's no acceleration. Thus the ball must start out rolling to the right on a horizontal track. At the end of the motion, the ball is again rolling on a horizontal track (no acceleration, constant velocity), but it's rolling to the *left* because v_s is negative. Further, the final speed ($|v_s|$) is greater than the initial speed. The middle section of the graph shows us what happens. The ball starts slowing with constant acceleration (rolling uphill), reaches a turning point (s is maximum, $v_s = 0$), then speeds up in the opposite direction (rolling downhill). This is still a negative acceleration because the ball is speeding up in the negative s-direction. It must roll farther downhill than it had rolled uphill before reaching a horizontal section of track. **FIGURE 2.38** shows the track and the initial conditions that are responsible for the graphs of Figure 2.37.

FIGURE 2.38 Track responsible for the motion graphs of Figure 2.37.

This track has a "switch." A ball moving to the right passes through and heads up the incline, but a ball rolling downhill goes straight through.

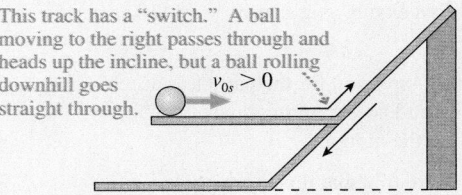

The ball rolls up the ramp, then back down. Which is the correct acceleration graph?

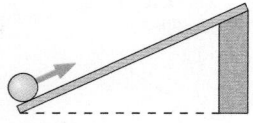

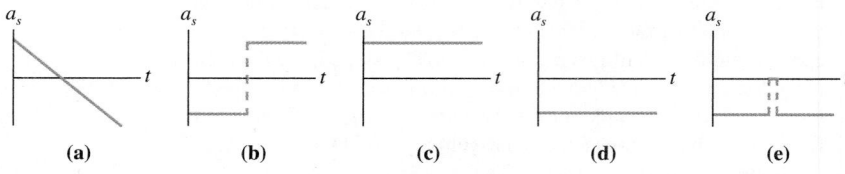

(a)   (b)   (c)   (d)   (e)

2.7 Instantaneous Acceleration

FIGURE 2.39 Velocity and acceleration graphs of a car leaving a stop sign.

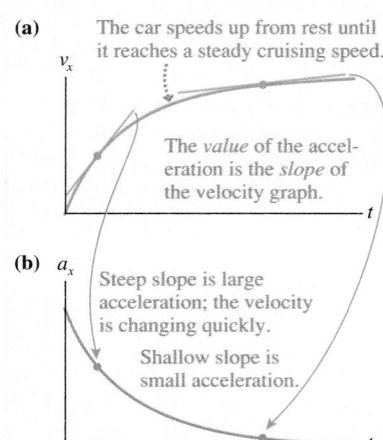

(a)   The car speeds up from rest until it reaches a steady cruising speed.

v_x

The *value* of the acceleration is the *slope* of the velocity graph.

t

(b) a_x

Steep slope is large acceleration; the velocity is changing quickly.

Shallow slope is small acceleration.

t

Although constant acceleration makes for straightforward problems and will often be assumed as part of a simplified model of motion, real moving objects only rarely have constant acceleration. For example, **FIGURE 2.39a** is a realistic velocity-versus-time graph for a car leaving a stop sign. The graph is not a straight line, so this is *not* motion with constant acceleration.

We can define an instantaneous acceleration in much the same way that we defined the instantaneous velocity. The instantaneous velocity at time t is the slope of the position-versus-time graph at that time or, mathematically, the derivative of the position with respect to time. By analogy: **The instantaneous acceleration a_s is the slope of the line that is tangent to the velocity-versus-time curve at time t.** Mathematically, this is

$$a_s = \frac{dv_s}{dt} = \text{slope of the velocity-versus-time graph at time } t \qquad (2.27)$$

The instantaneous acceleration is the rate of change of the velocity. **FIGURE 2.39b** applies this idea by showing the car's acceleration graph. At each instant of time, the *value* of the car's acceleration is the *slope* of its velocity graph. The initially steep slope indicates a large initial acceleration. The acceleration decreases to zero as the car reaches cruising speed.

The reverse problem—to find the velocity v_s if we know the acceleration a_s at all instants of time—is also important. Again with analogy to velocity and position, an acceleration curve can be divided into N very narrow steps so that during each step the acceleration is essentially constant. During step k, the velocity changes by $\Delta(v_s)_k = (a_s)_k \Delta t$. This is the area of the small rectangle under the step. The total velocity change between t_i and t_f is found by adding all the small $\Delta(v_s)_k$. In the limit $\Delta t \rightarrow 0$, we have

$$v_{fs} = v_{is} + \lim_{\Delta t \to 0} \sum_{k=1}^{N} (a_s)_k \Delta t = v_{is} + \int_{t_i}^{t_f} a_s \, dt \qquad (2.28)$$

The graphical interpretation of Equation 2.28 is

$$v_{fs} = v_{is} + \text{area under the acceleration curve } a_s \text{ between } t_i \text{ and } t_f \qquad (2.29)$$

EXAMPLE 2.18   **A nonuniform acceleration**

A particle's velocity is given by $v_s = [10 - (t - 5)^2]$ m/s, where t is in s.

a. Find an expression for the particle's acceleration a_s, then draw velocity and acceleration graphs.
b. Describe the motion.

MODEL We're told that this is a particle.

VISUALIZE FIGURE 2.40a shows the velocity graph. It is a parabola centered at $t = 5$ s with an apex $v_{max} = 10$ m/s. The slope of v_s is positive but decreasing in magnitude for $t < 5$ s. The slope is zero at $t = 5$ s, and it is negative and increasing in magnitude for $t > 5$ s. Thus the acceleration graph should start positive, decrease steadily, pass through zero at $t = 5$ s, then become increasingly negative.

SOLVE a. We can find an expression for a_s by taking the derivative of v_s. First, expand the square to give

$$v_s = (-t^2 + 10t - 15) \text{ m/s}$$

Then use the derivative rule (Equation 2.6) to find

$$a_s = \frac{dv_s}{dt} = (-2t + 10) \text{ m/s}^2$$

where t is in s. This is a linear equation that is graphed in **FIGURE 2.40b**. The graph meets our expectations.

b. This is a complex motion. The particle starts out moving to the left ($v_s < 0$) at 15 m/s. The positive acceleration causes the speed to decrease (slowing down because v_s and a_s have opposite signs) until the particle reaches a turning point ($v_s = 0$) just before $t = 2$ s. The particle then moves to the right ($v_s > 0$) and speeds up until reaching maximum speed at $t = 5$ s. From $t = 5$ s to just after $t = 8$ s, the particle is still moving to the right ($v_s > 0$) but slowing down. Another turning point occurs just after $t = 8$ s. Then the particle moves back to the left and gains speed as the negative a_s makes the velocity ever more negative.

FIGURE 2.40 Velocity and acceleration graphs for Example 2.18.

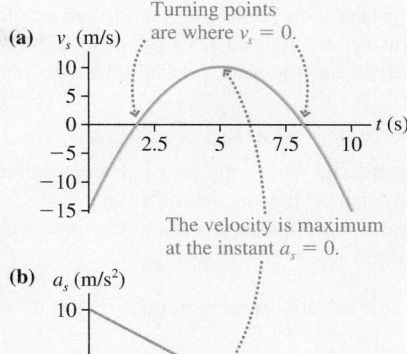

EXAMPLE 2.19   **Finding velocity from acceleration**

FIGURE 2.41 shows the acceleration graph for a particle with an initial velocity of 10 m/s. What is the particle's velocity at $t = 8$ s?

FIGURE 2.41 Acceleration graph for Example 2.19.

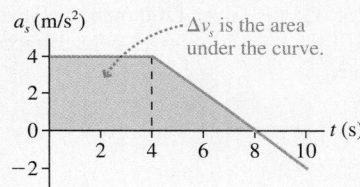

MODEL We're told this is the motion of a particle.

VISUALIZE Figure 2.41 is a graphical representation of the motion.

SOLVE The change in velocity is found as the area under the acceleration curve:

$$v_{fs} = v_{is} + \text{area under the acceleration curve } a_s \text{ between } t_i \text{ and } t_f$$

The area under the curve between $t_i = 0$ s and $t_f = 8$ s can be subdivided into a rectangle ($0 \text{ s} \leq t \leq 4$ s) and a triangle ($4 \text{ s} \leq t \leq 8$ s). These areas are easily computed. Thus

$$v_s(\text{at } t = 8 \text{ s}) = 10 \text{ m/s} + (4 \text{ (m/s)/s})(4 \text{ s})$$

$$+ \tfrac{1}{2}(4 \text{ (m/s)/s})(4 \text{ s})$$

$$= 34 \text{ m/s}$$

STOP TO THINK 2.6   Rank in order, from most positive to least positive, the accelerations at points A to C.

a. $a_A > a_B > a_C$
b. $a_C > a_A > a_B$
c. $a_C > a_B > a_A$
d. $a_B > a_A > a_C$

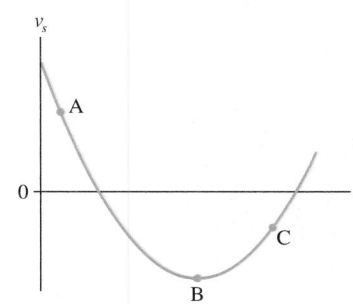

CHALLENGE EXAMPLE 2.20 **Rocketing along**

A rocket sled accelerates along a long, horizontal rail. Starting from rest, two rockets burn for 10 s, providing a constant acceleration. One rocket then burns out, halving the acceleration, but the other burns for an additional 5 s to boost the sled's speed to 625 m/s. How far has the sled traveled when the second rocket burns out?

MODEL Represent the rocket sled as a particle.

VISUALIZE **FIGURE 2.42** shows the pictorial representation. This is a two-part problem with a beginning, an end (the second rocket burns out), and a point in between where the motion changes (the first rocket burns out).

FIGURE 2.42 The pictorial representation of the rocket sled.

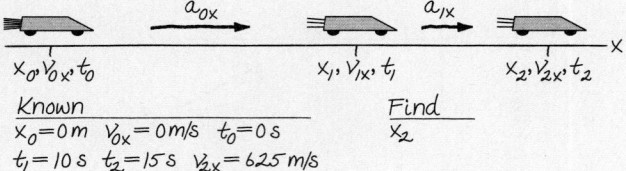

Known
$x_0 = 0 \text{ m}$ $v_{0x} = 0 \text{ m/s}$ $t_0 = 0 \text{ s}$
$t_1 = 10 \text{ s}$ $t_2 = 15 \text{ s}$ $v_{2x} = 625 \text{ m/s}$

Find
x_2

SOLVE The difficulty with this problem is that there's not enough information to completely analyze either the first or the second part of the motion. A successful solution will require combining information about both parts of the motion, and that can be done only by working algebraically, not worrying about numbers until the end of the problem. A well-drawn pictorial representation and clearly defined symbols are essential.

The first part of the motion, with both rockets firing, has acceleration a_{0x}. The sled's position and velocity when the first rocket burns out are

$$x_1 = x_0 + v_{0x}\,\Delta t + \tfrac{1}{2}a_{0x}(\Delta t)^2 = \tfrac{1}{2}a_{0x}t_1^2$$

$$v_{1x} = v_{0x} + a_{0x}\,\Delta t = a_{0x}t_1$$

where we simplified as much as possible by knowing that the sled started from rest at the origin at $t_0 = 0$ s. We can't compute numerical values, but these are valid algebraic expressions that we can carry over to the second part of the motion.

From t_1 to t_2, the acceleration is a smaller a_{1x}. The velocity when the second rocket burns out is

$$v_{2x} = v_{1x} + a_{1x}\,\Delta t = a_{0x}t_1 + a_{1x}(t_2 - t_1)$$

where for v_{1x} we used the algebraic result from the first part of the motion. Now we have enough information to complete the solution. We know that the acceleration is halved when the first rocket burns out, so $a_{1x} = \tfrac{1}{2}a_{0x}$. Thus

$$v_{2x} = 625 \text{ m/s} = a_{0x} \cdot 10 \text{ s} + \tfrac{1}{2}a_{0x} \cdot 5 \text{ s} = (12.5 \text{ s}) \cdot a_{0x}$$

Solving, we find $a_{0x} = 50 \text{ m/s}^2$.

With the acceleration now known, we can calculate the position and velocity when the first rocket burns out:

$$x_1 = \tfrac{1}{2}a_{0x}t_1^2 = \tfrac{1}{2}(50 \text{ m/s}^2)(10 \text{ s})^2 = 2500 \text{ m}$$

$$v_{1x} = a_{0x}t_1 = (50 \text{ m/s}^2)(10 \text{ s}) = 500 \text{ m/s}$$

Finally, the position when the second rocket burns out is

$$x_2 = x_1 + v_{1x}\,\Delta t + \tfrac{1}{2}a_{1x}(\Delta t)^2$$

$$= 2500 \text{ m} + (500 \text{ m/s})(5 \text{ s}) + \tfrac{1}{2}(25 \text{ m/s}^2)(5 \text{ s})^2 = 5300 \text{ m}$$

The sled has traveled 5300 m when it reaches 625 m/s at the burnout of the second rocket.

ASSESS 5300 m is 5.3 km, or roughly 3 miles. That's a long way to travel in 15 s! But the sled reaches incredibly high speeds. At the final speed of 625 m/s, over 1200 mph, the sled would travel nearly 10 km in 15 s. So 5.3 km in 15 s for the accelerating sled seems reasonable.

SUMMARY

The goal of Chapter 2 has been to learn how to solve problems about motion in a straight line.

General Principles

Kinematics describes motion in terms of position, velocity, and acceleration. General kinematic relationships are given **mathematically** by:

Instantaneous velocity $\quad v_s = ds/dt = $ slope of position graph

Instantaneous acceleration $\quad a_s = dv_s/dt = $ slope of velocity graph

Final position $\quad s_f = s_i + \displaystyle\int_{t_i}^{t_f} v_s\, dt = s_i + \begin{cases} \text{area under the velocity} \\ \text{curve from } t_i \text{ to } t_f \end{cases}$

Final velocity $\quad v_{fs} = v_{is} + \displaystyle\int_{t_i}^{t_f} a_s\, dt = v_{is} + \begin{cases} \text{area under the acceleration} \\ \text{curve from } t_i \text{ to } t_f \end{cases}$

The kinematic equations for motion with constant acceleration are:

$$v_{fs} = v_{is} + a_s\, \Delta t$$
$$s_f = s_i + v_{is}\, \Delta t + \tfrac{1}{2} a_s (\Delta t)^2$$
$$v_{fs}^2 = v_{is}^2 + 2a_s\, \Delta s$$

Uniform motion is motion with constant velocity and zero acceleration:

$$s_f = s_i + v_s\, \Delta t$$

Important Concepts

Position, velocity, and acceleration are related **graphically.**

- The slope of the position-versus-time graph is the value on the velocity graph.
- The slope of the velocity graph is the value on the acceleration graph.
- s is a maximum or minimum at a turning point, and $v_s = 0$.

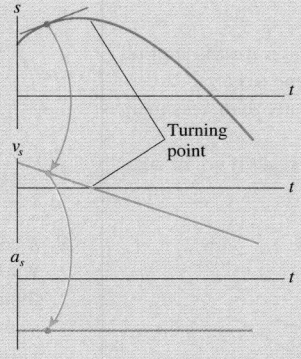

- Displacement is the area under the velocity curve.

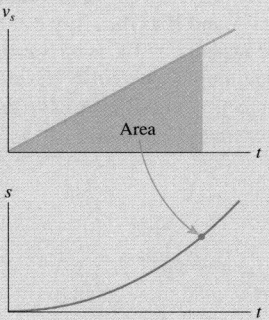

Applications

The **sign of v_s** indicates the direction of motion.

- $v_s > 0$ is motion to the right or up.
- $v_s < 0$ is motion to the left or down.

The **sign of a_s** indicates which way $\vec{a}$ points, *not* whether the object is speeding up or slowing down.

- $a_s > 0$ if $\vec{a}$ points to the right or up.
- $a_s < 0$ if $\vec{a}$ points to the left or down.
- The direction of $\vec{a}$ is found with a motion diagram.

An object is **speeding up** if and only if v_s and a_s have the same sign. An object is **slowing down** if and only if v_s and a_s have opposite signs.

Free fall is constant-acceleration motion with

$$a_y = -g = -9.80 \text{ m/s}^2$$

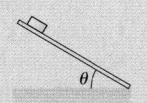

Motion on an inclined plane has $a_s = \pm g \sin \theta$. The sign depends on the direction of the tilt.

Terms and Notation

kinematics	speed, v	instantaneous velocity, v_s	free fall
average velocity, v_{avg}	initial position, s_i	turning point	free-fall acceleration, g
uniform motion	final position, s_f	average acceleration, a_{avg}	instantaneous acceleration, a_s

CONCEPTUAL QUESTIONS

For Questions 1 through 3, interpret the position graph given in each figure by writing a very short "story" of what is happening. Be creative! Have characters and situations! Simply saying that "a car moves 100 meters to the right" doesn't qualify as a story. Your stories should make *specific reference* to information you obtain from the graph, such as distance moved or time elapsed.

1.
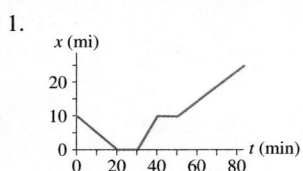
FIGURE Q2.1

2.
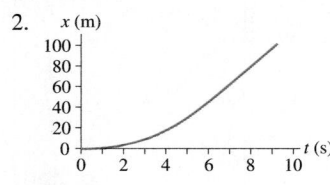
FIGURE Q2.2

3.
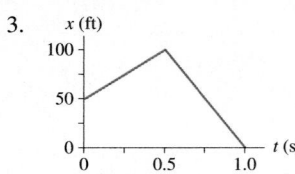
FIGURE Q2.3

4. **FIGURE Q2.4** shows a position-versus-time graph for the motion of objects A and B as they move along the same axis.
   a. At the instant $t = 1$ s, is the speed of A greater than, less than, or equal to the speed of B? Explain.
   b. Do objects A and B ever have the *same* speed? If so, at what time or times? Explain.

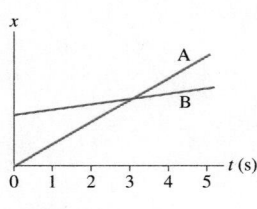
FIGURE Q2.4

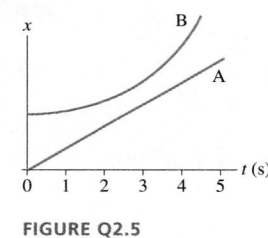
FIGURE Q2.5

5. **FIGURE Q2.5** shows a position-versus-time graph for the motion of objects A and B as they move along the same axis.
   a. At the instant $t = 1$ s, is the speed of A greater than, less than, or equal to the speed of B? Explain.
   b. Do objects A and B ever have the *same* speed? If so, at what time or times? Explain.

6. **FIGURE Q2.6** shows the position-versus-time graph for a moving object. At which lettered point or points:
   a. Is the object *moving* the slowest?
   b. Is the object moving the fastest?
   c. Is the object at rest?
   d. Is the object moving to the left?

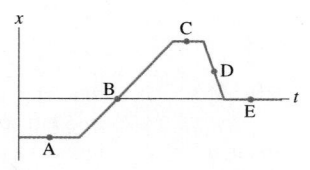
FIGURE Q2.6

7. **FIGURE Q2.7** shows the position-versus-time graph for a moving object. At which lettered point or points:
   a. Is the object moving the fastest?
   b. Is the object moving to the left?
   c. Is the object speeding up?
   d. Is the object turning around?

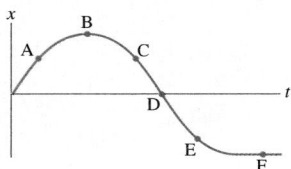
FIGURE Q2.7

8. **FIGURE Q2.8** shows six frames from the motion diagrams of two moving cars, A and B.
   a. Do the two cars ever have the same position at one instant of time? If so, in which frame number (or numbers)?
   b. Do the two cars ever have the same velocity at one instant of time? If so, between which two frames?

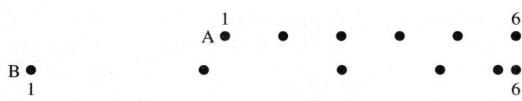
FIGURE Q2.8

9. You're driving along the highway at a steady speed of 60 mph when another driver decides to pass you. At the moment when the front of his car is exactly even with the front of your car, and you turn your head to smile at him, do the two cars have equal velocities? Explain.

10. A bicycle is traveling east. Can its acceleration vector ever point west? Explain.

11. (a) Give an example of a vertical motion with a positive velocity and a negative acceleration. (b) Give an example of a vertical motion with a negative velocity and a negative acceleration.

12. A ball is thrown straight up into the air. At each of the following instants, is the magnitude of the ball's acceleration greater than g, equal to g, less than g, or 0? Explain.
    a. Just after leaving your hand.
    b. At the very top (maximum height).
    c. Just before hitting the ground.

13. A rock is *thrown* (not dropped) straight down from a bridge into the river below. At each of the following instants, is the magnitude of the rock's acceleration greater than g, equal to g, less than g, or 0? Explain.
    a. Immediately after being released.
    b. Just before hitting the water.

14. A rubber ball dropped from a height of 2 m bounces back to a height of 1 m. Draw the ball's position, velocity, and acceleration graphs, stacked vertically, from the instant you release it until it returns to its maximum bounce height. Pay close attention to the time the ball is in contact with the ground; this is a short interval of time, but it's not zero.

EXERCISES AND PROBLEMS

Exercises

Section 2.1 Uniform Motion

1. | Alan leaves Los Angeles at 8:00 a.m. to drive to San Francisco, 400 mi away. He travels at a steady 50 mph. Beth leaves Los Angeles at 9:00 a.m. and drives a steady 60 mph.
   a. Who gets to San Francisco first?
   b. How long does the first to arrive have to wait for the second?

2. ‖ Larry leaves home at 9:05 and runs at constant speed to the lamppost seen in FIGURE EX2.2. He reaches the lamppost at 9:07, immediately turns, and runs to the tree. Larry arrives at the tree at 9:10.
   a. What is Larry's average velocity, in m/min, during each of these two intervals.
   b. What is Larry's average velocity for the entire run?

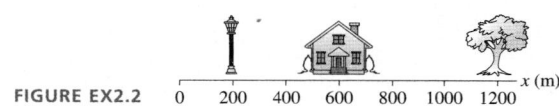
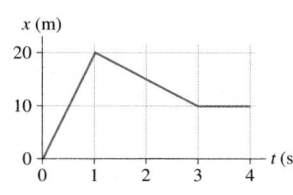
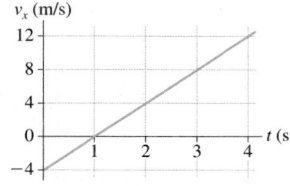

FIGURE EX2.2   0   200   400   600   800   1000   1200 x (m)

3. ‖ Julie drives 100 mi to Grandmother's house. On the way to Grandmother's, Julie drives half the distance at 40 mph and half the distance at 60 mph. On her return trip, she drives half the time at 40 mph and half the time at 60 mph.
   a. What is Julie's average speed on the way to Grandmother's house?
   b. What is her average speed on the return trip?

4. | FIGURE EX2.4 is the position-versus-time graph of a jogger. What is the jogger's velocity at $t = 10$ s, at $t = 25$ s, and at $t = 35$ s?

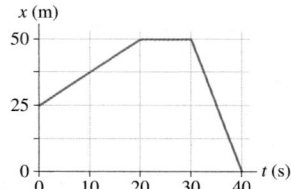

FIGURE EX2.4

Section 2.2 Instantaneous Velocity

Section 2.3 Finding Position from Velocity

5. | FIGURE EX2.5 shows the position graph of a particle.
   a. Draw the particle's velocity graph for the interval $0 \text{ s} \leq t \leq 4 \text{ s}$.
   b. Does this particle have a turning point or points? If so, at what time or times?

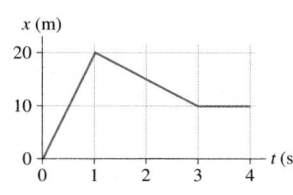

FIGURE EX2.5

FIGURE EX2.6

6. | A particle starts from $x_0 = 10$ m at $t_0 = 0$ s and moves with the velocity graph shown in FIGURE EX2.6.
   a. Does this particle have a turning point? If so, at what time?
   b. What is the object's position at $t = 2$ s, 3 s, and 4 s?

7. ‖ FIGURE EX 2.7 is a somewhat idealized graph of the velocity of
   BIO blood in the ascending aorta during one beat of the heart. Approximately how far, in cm, does the blood move during one beat?

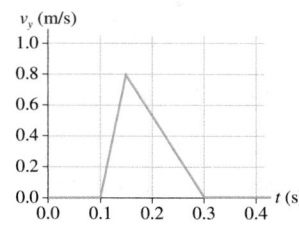
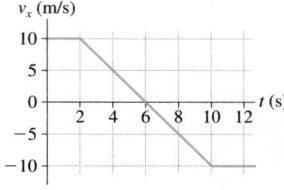

FIGURE EX2.7   FIGURE EX2.8

8. | FIGURE EX2.8 shows the velocity graph for a particle having initial position $x_0 = 0$ m at $t_0 = 0$ s.
   a. At what time or times is the particle found at $x = 35$ m? Work with the geometry of the graph, not with kinematic equations.
   b. Draw a motion diagram for the particle.

Section 2.4 Motion with Constant Acceleration

9. | FIGURE EX2.9 shows the velocity graph of a particle. Draw the particle's acceleration graph for the interval $0 \text{ s} \leq t \leq 4 \text{ s}$. Give both axes an appropriate numerical scale.

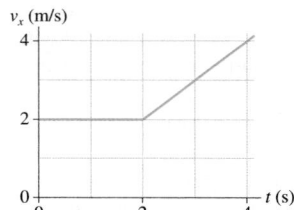

FIGURE EX2.9

10. | FIGURE EX2.7 showed the velocity graph of blood in the aorta.
    BIO Estimate the blood's acceleration during each phase of the motion, speeding up and slowing down.

11. | FIGURE EX2.11 shows the velocity graph of a particle moving along the x-axis. Its initial position is $x_0 = 2.0$ m at $t_0 = 0$ s. At $t = 2.0$ s, what are the particle's (a) position, (b) velocity, and (c) acceleration?

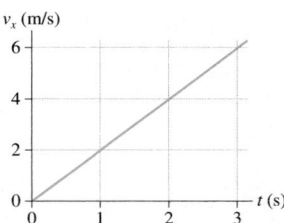

FIGURE EX2.11

12. | FIGURE EX2.12 shows the velocity-versus-time graph for a particle moving along the x-axis. Its initial position is $x_0 = 2.0$ m at $t_0 = 0$ s.
    a. What are the particle's position, velocity, and acceleration at $t = 1.0$ s?
    b. What are the particle's position, velocity, and acceleration at $t = 3.0$ s?

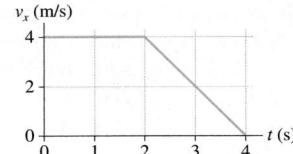

FIGURE EX2.12

13. ‖ A jet plane is cruising at 300 m/s when suddenly the pilot turns the engines up to full throttle. After traveling 4.0 km, the jet is moving with a speed of 400 m/s. What is the jet's acceleration, assuming it to be a constant acceleration?

14. | When you sneeze, the air in your lungs accelerates from rest to BIO 150 km/h in approximately 0.50 s. What is the acceleration of the air in m/s^2?

15. ‖ A speed skater moving across frictionless ice at 8.0 m/s hits a 5.0-m-wide patch of rough ice. She slows steadily, then continues on at 6.0 m/s. What is her acceleration on the rough ice?

16. | A Porsche challenges a Honda to a 400 m race. Because the Porsche's acceleration of 3.5 m/s^2 is larger than the Honda's 3.0 m/s^2, the Honda gets a 1.0 s head start. Who wins?

Section 2.5 Free Fall

17. | Ball bearings are made by letting spherical drops of molten metal fall inside a tall tower—called a *shot tower*—and solidify as they fall.
    a. If a bearing needs 4.0 s to solidify enough for impact, how high must the tower be?
    b. What is the bearing's impact velocity?

18. | A ball is thrown vertically upward with a speed of 19.6 m/s.
    a. What is the ball's velocity and its height after 1.0, 2.0, 3.0, and 4.0 s?
    b. Draw the ball's velocity-versus-time graph. Give both axes an appropriate numerical scale.

19. ‖ A student standing on the ground throws a ball straight up. The ball leaves the student's hand with a speed of 15 m/s when the hand is 2.0 m above the ground. How long is the ball in the air before it hits the ground? (The student moves her hand out of the way.)

20. ‖ A rock is tossed straight up with a speed of 20 m/s. When it returns, it falls into a hole 10 m deep.
    a. What is the rock's velocity as it hits the bottom of the hole?
    b. How long is the rock in the air, from the instant it is released until it hits the bottom of the hole?

Section 2.6 Motion on an Inclined Plane

21. ‖ A skier is gliding along at 3.0 m/s on horizontal, frictionless snow. He suddenly starts down a 10° incline. His speed at the bottom is 15 m/s.
    a. What is the length of the incline?
    b. How long does it take him to reach the bottom?

22. ‖ A car traveling at 30 m/s runs out of gas while traveling up a 10° slope. How far up the hill will it coast before starting to roll back down?

Section 2.7 Instantaneous Acceleration

23. | A particle moving along the x-axis has its position described by the function $x = (2t^2 - t + 1)$ m, where t is in s. At $t = 2$ s what are the particle's (a) position, (b) velocity, and (c) acceleration?

24. ‖ A particle moving along the x-axis has its velocity described by the function $v_x = 2t^2$ m/s, where t is in s. Its initial position is $x_0 = 1$ m at $t_0 = 0$ s. At $t = 1$ s what are the particle's (a) position, (b) velocity, and (c) acceleration?

25. ‖ FIGURE EX2.25 shows the acceleration-versus-time graph of a particle moving along the x-axis. Its initial velocity is $v_{0x} = 8.0$ m/s at $t_0 = 0$ s. What is the particle's velocity at $t = 4.0$ s?

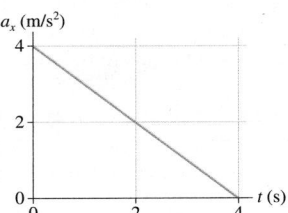

FIGURE EX2.25

Problems

26. ‖ A particle's position on the x-axis is given by the function $x = (t^2 - 4t + 2)$ m, where t is in s.
    a. Make a position-versus-time graph for the interval 0 s $\leq t \leq 5$ s. Do this by calculating and plotting x every 0.5 s from 0 s to 5 s, then drawing a smooth curve through the points.
    b. Determine the particle's velocity at $t = 1.0$ s by drawing the tangent line on your graph and measuring its slope.
    c. Determine the particle's velocity at $t = 1.0$ s by evaluating the derivative at that instant. Compare this to your result from part b.
    d. Are there any turning points in the particle's motion? If so, at what position or positions?
    e. Where is the particle when $v_x = 4.0$ m/s?
    f. Draw a motion diagram for the particle.

27. ‖ Three particles move along the x-axis, each starting with $v_{0x} = 10$ m/s at $t_0 = 0$ s. In FIGURE P2.27, the graph for A is a position-versus-time graph; the graph for B is a velocity-versus-time graph; the graph for C is an acceleration-versus-time graph. Find each particle's velocity at $t = 7.0$ s. Work with the geometry of the graphs, not with kinematic equations.

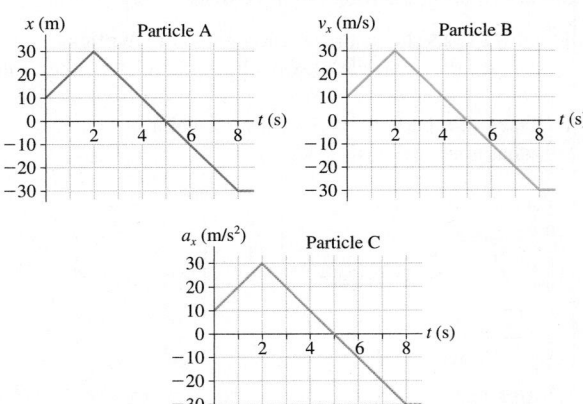

FIGURE P2.27

28. ‖ **FIGURE P2.28** shows the acceleration graph for a particle that starts from rest at $t = 0$ s. Determine the object's velocity at times $t = 0$ s, 2 s, 4 s, 6 s, and 8 s.

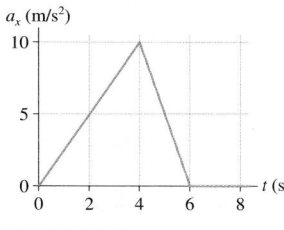

FIGURE P2.28

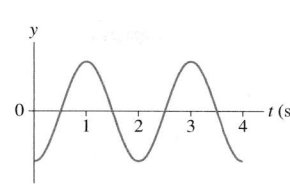

FIGURE P2.29

29. ‖ A block is suspended from a spring, pulled down, and released. The block's position-versus-time graph is shown in **FIGURE P2.29**.
    a. At what times is the velocity zero? At what times is the velocity most positive? Most negative?
    b. Draw a reasonable velocity-versus-time graph.

30. ‖ A particle's velocity is described by the function $v_x = t^2 - 7t + 10$ m/s, where t is in s.
    a. At what times does the particle reach its turning points?
    b. What is the particle's acceleration at each of the turning points?

31. ‖ The position of a particle is given by the function $x = (2t^3 - 9t^2 + 12)$ m, where t is in s.
    a. At what time or times is $v_x = 0$ m/s?
    b. What are the particle's position and its acceleration at this time(s)?

32. ‖ An object starts from rest at $x = 0$ m at time $t = 0$ s. Five seconds later, at $t = 5.0$ s, the object is observed to be at $x = 40$ m and to have velocity $v_x = 11$ m/s.
    a. Was the object's acceleration uniform or nonuniform? Explain your reasoning.
    b. Sketch the velocity-versus-time graph implied by these data. Is the graph a straight line or curved? If curved, is it concave upward or downward?

33. ‖ A particle's velocity is described by the function $v_x = kt^2$ m/s, where k is a constant and t is in s. The particle's position at $t_0 = 0$ s is $x_0 = -9.0$ m. At $t_1 = 3.0$ s, the particle is at $x_1 = 9.0$ m. Determine the value of the constant k. Be sure to include the proper units.

34. ‖ A particle's acceleration is described by the function $a_x = (10 - t)$ m/s², where t is in s. Its initial conditions are $x_0 = 0$ m and $v_{0x} = 0$ m/s at $t = 0$ s.
    a. At what time is the velocity again zero?
    b. What is the particle's position at that time?

35. ‖ A ball rolls along the frictionless track shown in **FIGURE P2.35**. Each segment of the track is straight, and the ball passes smoothly from one segment to the next without changing speed or leaving the track. Draw three vertically stacked graphs showing position, velocity, and acceleration versus time. Each graph should have the same time axis, and the proportions of the graph should be qualitatively correct. Assume that the ball has enough speed to reach the top.

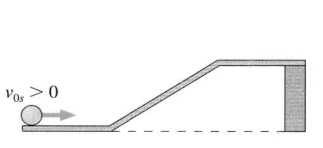

FIGURE P2.35

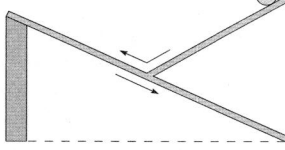

FIGURE P2.36

36. ‖ Draw position, velocity, and acceleration graphs for the ball shown in **FIGURE P2.36**. See Problem 35 for more information.

37. ‖ Draw position, velocity, and acceleration graphs for the ball shown in **FIGURE P2.37**. See Problem 35 for more information. The ball changes direction but not speed as it bounces from the reflecting wall.

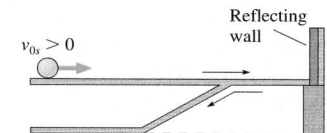

FIGURE P2.37

38. ‖ **FIGURE P2.38** shows a set of kinematic graphs for a ball rolling on a track. All segments of the track are straight lines, but some may be tilted. Draw a picture of the track and also indicate the ball's initial condition.

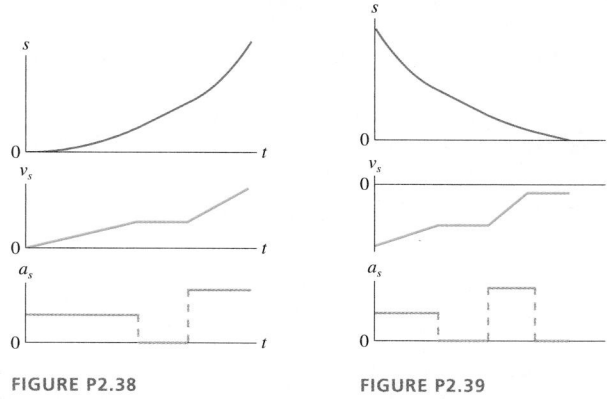

FIGURE P2.38   **FIGURE P2.39**

39. ‖ **FIGURE P2.39** shows a set of kinematic graphs for a ball rolling on a track. All segments of the track are straight lines, but some may be tilted. Draw a picture of the track and also indicate the ball's initial condition.

40. ‖ The takeoff speed for an Airbus A320 jetliner is 80 m/s. Velocity data measured during takeoff are as shown.
    a. What is the takeoff speed in miles per hour?
    b. Is the jetliner's acceleration constant during takeoff? Explain.
    c. At what time do the wheels leave the ground?
    d. For safety reasons, in case of an aborted takeoff, the runway must be three times the takeoff distance. Can an A320 take off safely on a 2.5-mi-long runway?

t (s)	v_x (m/s)
0	0
10	23
20	46
30	69

41. ‖ a. What constant acceleration, in SI units, must a car have to go from zero to 60 mph in 10 s?
    b. What fraction of g is this?
    c. How far has the car traveled when it reaches 60 mph? Give your answer both in SI units and in feet.

42. ‖ a. How many days will it take a spaceship to accelerate to the speed of light (3.0×10^8 m/s) with the acceleration g?
    b. How far will it travel during this interval?
    c. What fraction of a light year is your answer to part b? A light year is the distance light travels in one year.

    NOTE ▶ We know, from Einstein's theory of relativity, that no object can travel at the speed of light. So this problem, while interesting and instructive, is not realistic. ◀

43. | You are driving to the grocery store at 20 m/s. You are 110 m from an intersection when the traffic light turns red. Assume that your reaction time is 0.50 s and that your car brakes with constant acceleration.
    a. How far are you from the intersection when you begin to apply the brakes?
    b. What acceleration will bring you to rest right at the intersection?
    c. How long does it take you to stop after the light turns red?

44. ‖ a. Suppose you are driving at speed v_0 when a sudden obstacle in the road forces you to make a quick stop. If your reaction time before applying the brakes is t_R, what constant deceleration (absolute value of a_x) do you need to stop in distance d? Assume that d is larger than the car travels during your reaction time.
    b. Suppose you are driving at 21 m/s when you suddenly see an obstacle 50 m ahead. If your reaction time is 0.50 s and if your car's maximum deceleration is 6.0 m/s², can you stop in time to avoid a collision?

45. ‖ You're driving down the highway late one night at 20 m/s when a deer steps onto the road 35 m in front of you. Your reaction time before stepping on the brakes is 0.50 s, and the maximum deceleration of your car is 10 m/s².
    a. How much distance is between you and the deer when you come to a stop?
    b. What is the maximum speed you could have and still not hit the deer?

46. ‖‖ The minimum stopping distance for a car traveling at a speed of 30 m/s is 60 m, including the distance traveled during the driver's reaction time of 0.50 s.
    a. What is the minimum stopping distance for the same car traveling at a speed of 40 m/s?
    b. Draw a position-versus-time graph for the motion of the car in part a. Assume the car is at $x_0 = 0$ m when the driver first sees the emergency situation ahead that calls for a rapid halt.

47. ‖ When jumping, a flea accelerates at an astounding 1000 m/s²,
BIO but over only the very short distance of 0.50 mm. If a flea jumps straight up, and if air resistance is neglected (a rather poor approximation in this situation), how high does the flea go?

48. ‖ A cheetah spots a Thomson's gazelle, its preferred prey, and
BIO leaps into action, quickly accelerating to its top speed of 30 m/s, the highest of any land animal. However, a cheetah can maintain this extreme speed for only 15 s before having to let up. The cheetah is 170 m from the gazelle as it reaches top speed, and the gazelle sees the cheetah at just this instant. With negligible reaction time, the gazelle heads directly away from the cheetah, accelerating at 4.6 m/s² for 5.0 s, then running at constant speed. Does the gazelle escape?

49. ‖‖ A 200 kg weather rocket is loaded with 100 kg of fuel and fired straight up. It accelerates upward at 30 m/s² for 30 s, then runs out of fuel. Ignore any air resistance effects.
    a. What is the rocket's maximum altitude?
    b. How long is the rocket in the air before hitting the ground?
    c. Draw a velocity-versus-time graph for the rocket from liftoff until it hits the ground.

50. ‖ A 1000 kg weather rocket is launched straight up. The rocket motor provides a constant acceleration for 16 s, then the motor stops. The rocket altitude 20 s after launch is 5100 m. You can ignore any effects of air resistance.
    a. What was the rocket's acceleration during the first 16 s?
    b. What is the rocket's speed as it passes through a cloud 5100 m above the ground?

51. ‖‖‖ A lead ball is dropped into a lake from a diving board 5.0 m above the water. After entering the water, it sinks to the bottom with a constant velocity equal to the velocity with which it hit the water. The ball reaches the bottom 3.0 s after it is released. How deep is the lake?

52. ‖ A hotel elevator ascends 200 m with a maximum speed of 5.0 m/s. Its acceleration and deceleration both have a magnitude of 1.0 m/s².
    a. How far does the elevator move while accelerating to full speed from rest?
    b. How long does it take to make the complete trip from bottom to top?

53. ‖ A car starts from rest at a stop sign. It accelerates at 4.0 m/s² for 6.0 s, coasts for 2.0 s, and then slows down at a rate of 3.0 m/s² for the next stop sign. How far apart are the stop signs?

54. ‖ A car accelerates at 2.0 m/s² along a straight road. It passes two marks that are 30 m apart at times $t = 4.0$ s and $t = 5.0$ s. What was the car's velocity at $t = 0$ s?

55. ‖ Santa loses his footing and slides down a frictionless, snowy roof that is tilted at an angle of 30°. If Santa slides 10 m before reaching the edge, what is his speed as he leaves the roof?

56. ‖ Ann and Carol are driving their cars along the same straight road. Carol is located at $x = 2.4$ mi at $t = 0$ h and drives at a steady 36 mph. Ann, who is traveling in the same direction, is located at $x = 0.0$ mi at $t = 0.50$ h and drives at a steady 50 mph.
    a. At what time does Ann overtake Carol?
    b. What is their position at this instant?
    c. Draw a position-versus-time graph showing the motion of both Ann and Carol.

57. ‖ a. A very slippery block of ice slides down a smooth ramp tilted at angle θ. The ice is released from rest at vertical height h above the bottom of the ramp. Find an expression for the speed of the ice at the bottom.
    b. Evaluate your answer to part a for ice released at a height of 30 cm on ramps tilted at 20° and 40°.

58. ‖ A toy train is pushed forward and released at $x_0 = 2.0$ m with a speed of 2.0 m/s. It rolls at a steady speed for 2.0 s, then one wheel begins to stick. The train comes to a stop 6.0 m from the point at which it was released. What is the magnitude of the train's acceleration after its wheel begins to stick?

59. ‖ Bob is driving the getaway car after the big bank robbery. He's going 50 m/s when his headlights suddenly reveal a nail strip that the cops have placed across the road 150 m in front of him. If Bob can stop in time, he can throw the car into reverse and escape. But if he crosses the nail strip, all his tires will go flat and he will be caught. Bob's reaction time before he can hit the brakes is 0.60 s, and his car's maximum deceleration is 10 m/s². Is Bob in jail?

60. ‖ One game at the amusement park has you push a puck up a long, frictionless ramp. You win a stuffed animal if the puck, at its highest point, comes to within 10 cm of the end of the ramp without going off. You give the puck a push, releasing it with a speed of 5.0 m/s when it is 8.5 m from the end of the ramp. The puck's speed after traveling 3.0 m is 4.0 m/s. Are you a winner?

61. ‖ a. Your goal in laboratory is to launch a ball of mass m straight up so that it reaches exactly height h above the top of the launching tube. You and your lab partners will earn fewer points if the ball goes too high or too low. The launch tube uses compressed air to accelerate the ball over a distance d, and you have a table of data telling you how to set the

air compressor to achieve a desired acceleration. Find an expression for the acceleration that will earn you maximum points.

    b. Evaluate your answer to part a to achieve a height of 3.2 m using a 45-cm-long launch tube.

62. ‖ Nicole throws a ball straight up. Chad watches the ball from a window 5.0 m above the point where Nicole released it. The ball passes Chad on the way up, and it has a speed of 10 m/s as it passes him on the way back down. How fast did Nicole throw the ball?

63. ‖ A motorist is driving at 20 m/s when she sees that a traffic light 200 m ahead has just turned red. She knows that this light stays red for 15 s, and she wants to reach the light just as it turns green again. It takes her 1.0 s to step on the brakes and begin slowing. What is her speed as she reaches the light at the instant it turns green?

64. ‖ When a 1984 Alfa Romeo Spider sports car accelerates at the maximum possible rate, its motion during the first 20 s is extremely well modeled by the simple equation

$$v_x^2 = \frac{2P}{m}t$$

where $P = 3.6 \times 10^4$ watts is the car's power output, $m = 1200$ kg is its mass, and v_x is in m/s. That is, the square of the car's velocity increases linearly with time.

    a. What is the car's speed at $t = 10$ s and at $t = 20$ s?

    b. Find an algebraic expression in terms of P, m, and t, for the car's acceleration at time t.

    c. Evaluate the acceleration at $t = 1$ s and $t = 10$ s.

    d. This simple model fails for t less than about 0.5 s. Explain how you can recognize the failure.

65. ‖ David is driving a steady 30 m/s when he passes Tina, who is sitting in her car at rest. Tina begins to accelerate at a steady 2.0 m/s² at the instant when David passes.

    a. How far does Tina drive before passing David?

    b. What is her speed as she passes him?

66. ‖ A cat is sleeping on the floor in the middle of a 3.0-m-wide room when a barking dog enters with a speed of 1.50 m/s. As the dog enters, the cat (as only cats can do) immediately accelerates at 0.85 m/s² toward an open window on the opposite side of the room. The dog (all bark and no bite) is a bit startled by the cat and begins to slow down at 0.10 m/s² as soon as it enters the room. Does the dog catch the cat before the cat is able to leap through the window?

67. ‖ Jill has just gotten out of her car in the grocery store parking lot. The parking lot is on a hill and is tilted 3°. Twenty meters downhill from Jill, a little old lady lets go of a fully loaded shopping cart. The cart, with frictionless wheels, starts to roll straight downhill. Jill immediately starts to sprint after the cart with her top acceleration of 2.0 m/s². How far has the cart rolled before Jill catches it?

68. ‖ As a science project, you drop a watermelon off the top of the Empire State Building, 320 m above the sidewalk. It so happens that Superman flies by at the instant you release the watermelon. Superman is headed straight down with a speed of 35 m/s. How fast is the watermelon going when it passes Superman?

69. ‖‖ I was driving along at 20 m/s, trying to change a CD and not watching where I was going. When I looked up, I found myself 45 m from a railroad crossing. And wouldn't you know it, a train moving at 30 m/s was only 60 m from the crossing. In

a split second, I realized that the train was going to beat me to the crossing and that I didn't have enough distance to stop. My only hope was to accelerate enough to cross the tracks before the train arrived. If my reaction time before starting to accelerate was 0.50 s, what minimum acceleration did my car need for me to be here today writing these words?

70. ‖ As an astronaut visiting Planet X, you're assigned to measure the free-fall acceleration. Getting out your meter stick and stop watch, you time the fall of a heavy ball from several heights. Your data are as follows:

Height (m)	Fall time (s)
0.0	0.00
1.0	0.54
2.0	0.72
3.0	0.91
4.0	1.01
5.0	1.17

Analyze these data to determine the free-fall acceleration on Planet X. Your analysis method should involve fitting a straight line to an appropriate graph, similar to the analysis in Example 2.15.

71. ‖ Your engineering firm has been asked to determine the deceleration of a car during hard braking. To do so, you decide to measure the lengths of the skid marks when stopping from various initial speeds. Your data are as follows:

Speed (m/s)	Skid length (m)
10	7
15	14
20	27
25	37
30	58

    a. Do the data support an assertion that the deceleration is constant, independent of speed? Explain.

    b. Determine an experimental value for the car's deceleration—that is, the absolute value of the acceleration. Your analysis method should involve fitting a straight line to an appropriate graph, similar to the analysis in Example 2.15.

In Problems 72 through 75, you are given the kinematic equation or equations that are used to solve a problem. For each of these, you are to:

    a. Write a *realistic* problem for which this is the correct equation(s). Be sure that the answer your problem requests is consistent with the equation(s) given.

    b. Draw the pictorial representation for your problem.

    c. Finish the solution of the problem.

72. $64 \text{ m} = 0 \text{ m} + (32 \text{ m/s})(4 \text{ s} - 0 \text{ s}) + \frac{1}{2}a_x(4 \text{ s} - 0 \text{ s})^2$

73. $(10 \text{ m/s})^2 = v_{0y}^2 - 2(9.8 \text{ m/s}^2)(10 \text{ m} - 0 \text{ m})$

74. $(0 \text{ m/s})^2 = (5 \text{ m/s})^2 - 2(9.8 \text{ m/s}^2)(\sin 10°)(x_1 - 0 \text{ m})$

75. $v_{1x} = 0 \text{ m/s} + (20 \text{ m/s}^2)(5 \text{ s} - 0 \text{ s})$
    $x_1 = 0 \text{ m} + (0 \text{ m/s})(5 \text{ s} - 0 \text{ s}) + \frac{1}{2}(20 \text{ m/s}^2)(5 \text{ s} - 0 \text{ s})^2$
    $x_2 = x_1 + v_{1x}(10 \text{ s} - 5 \text{ s})$

Challenge Problems

76. The two masses in FIGURE CP2.76 slide on frictionless wires. They are connected by a pivoting rigid rod of length L. Prove that $v_{2x} = -v_{1y}\tan\theta$.

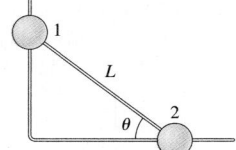

FIGURE CP2.76

77. A rocket is launched straight up with constant acceleration. Four seconds after liftoff, a bolt falls off the side of the rocket. The bolt hits the ground 6.0 s later. What was the rocket's acceleration?

78. Your school science club has devised a special event for homecoming. You've attached a rocket to the rear of a small car that has been decorated in the blue-and-gold school colors. The rocket provides a constant acceleration for 9.0 s. As the rocket shuts off, a parachute opens and slows the car at a rate of 5.0 m/s². The car passes the judges' box in the center of the grandstand, 990 m from the starting line, exactly 12 s after you fire the rocket. What is the car's speed as it passes the judges?

79. Careful measurements have been made of Olympic sprinters in the 100-meter dash. A simple but reasonably accurate model is that a sprinter accelerates at 3.6 m/s² for $3\frac{1}{3}$ s, then runs at constant velocity to the finish line.
    a. What is the race time for a sprinter who follows this model?
    b. A sprinter could run a faster race by accelerating faster at the beginning, thus reaching top speed sooner. If a sprinter's top speed is the same as in part a, what acceleration would he need to run the 100-meter dash in 9.9 s?
    c. By what percent did the sprinter need to increase his acceleration in order to decrease his time by 1%?

80. Careful measurements have been made of Olympic sprinters in the 100-meter dash. A quite realistic model is that the sprinter's velocity is given by

    $$v_x = a(1 - e^{-bt})$$

    where t is in s, v_x is in m/s, and the constants a and b are characteristic of the sprinter. Sprinter Carl Lewis's run at the 1987 World Championships is modeled with $a = 11.81$ m/s and $b = 0.6887$ s⁻¹.

    a. What was Lewis's acceleration at $t = 0$ s, 2.00 s, and 4.00 s?
    b. Find an expression for the distance traveled at time t.
    c. Your expression from part b is a transcendental equation, meaning that you can't solve it for t. However, it's not hard to use trial and error to find the time needed to travel a specific distance. To the nearest 0.01 s, find the time Lewis needed to sprint 100.0 m. His official time was 0.01 s more than your answer, showing that this model is very good, but not perfect.

81. A sprinter can accelerate with constant acceleration for 4.0 s before reaching top speed. He can run the 100-meter dash in 10.0 s. What is his speed as he crosses the finish line?

82. A rubber ball is shot straight up from the ground with speed v_0. Simultaneously, a second rubber ball at height h directly above the first ball is dropped from rest.
    a. At what height above the ground do the balls collide? Your answer will be an *algebraic expression* in terms of h, v_0, and g.
    b. What is the maximum value of h for which a collision occurs before the first ball falls back to the ground?
    c. For what value of h does the collision occur at the instant when the first ball is at its highest point?

83. The Starship Enterprise returns from warp drive to ordinary space with a forward speed of 50 km/s. To the crew's great surprise, a Klingon ship is 100 km directly ahead, traveling in the same direction at a mere 20 km/s. Without evasive action, the Enterprise will overtake and collide with the Klingons in just slightly over 3.0 s. The Enterprise's computers react instantly to brake the ship. What magnitude acceleration does the Enterprise need to just barely avoid a collision with the Klingon ship? Assume the acceleration is constant.
    Hint: Draw a position-versus-time graph showing the motions of both the Enterprise and the Klingon ship. Let $x_0 = 0$ km be the location of the Enterprise as it returns from warp drive. How do you show graphically the situation in which the collision is "barely avoided"? Once you decide what it looks like graphically, express that situation mathematically.

STOP TO THINK ANSWERS

Stop to Think 2.1: d. The particle starts with positive x and moves to negative x.

Stop to Think 2.2: c. The velocity is the slope of the position graph. The slope is positive and constant until the position graph crosses the axis, then positive but decreasing, and finally zero when the position graph is horizontal.

Stop to Think 2.3: b. A constant positive v_x corresponds to a linearly increasing x, starting from $x_i = -10$ m. The constant negative v_x then corresponds to a linearly decreasing x.

Stop to Think 2.4: a and b. The velocity is constant while $a = 0$, it decreases linearly while a is negative. Graphs a, b, and c all have the same acceleration, but only graphs a and b have a positive initial velocity that represents a particle moving to the right.

Stop to Think 2.5: d. The acceleration vector points downhill (negative s-direction) and has the constant value $-g\sin\theta$ throughout the motion.

Stop to Think 2.6: c. Acceleration is the slope of the graph. The slope is zero at B. Although the graph is steepest at A, the slope at that point is negative, and so $a_A < a_B$. Only C has a positive slope, so $a_C > a_B$.

3 Vectors and Coordinate Systems

Wind has both a speed and a direction, hence the motion of the wind is described by a vector.

▶ **Looking Ahead** The goal of Chapter 3 is to learn how vectors are represented and used.

Vectors

A **vector** is a quantity with both a size—the *magnitude*—and a direction.

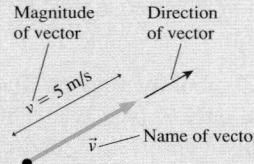

Examples of vectors that you will meet in coming chapters are:

Position	Velocity
Displacement	Acceleration
Force	Momentum

The two most basic vector operations—addition and subtraction—were introduced in Chapter 1.

> ◀ **Looking Back**
> Tactics Box 1.1 Vector addition
> Tactics Box 1.2 Vector subtraction

You may have learned in a math class to think of vectors as pairs or triplets of numbers, such as $(4, -2, 5)$. If so, you already know a lot about vectors even though we will use a different notation in physics.

Graphical Addition and Subtraction of Vectors

You will learn to add vectors $\vec{A}$ and $\vec{B}$:

Tip-to-tail addition

Vector subtraction is addition

$$\vec{A} - \vec{B} = \vec{A} + (-\vec{B})$$

with $-\vec{B}$ defined to point opposite $\vec{B}$:

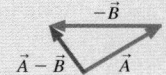

Tip-to-tail subtraction

The net displacement is the vector sum of two individual displacements.

Unit Vectors

Unit vectors define what we mean by the $+x$- and $+y$-directions in space.

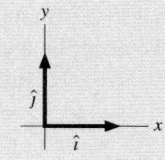

Unit vectors will be very valuable when we later use a tilted coordinate system to analyze motion on an inclined plane.

Components

You will learn how to find the *components* of vectors that are parallel to the coordinate axes. We write this as

$$\vec{E} = E_x \hat{i} + E_y \hat{j}$$

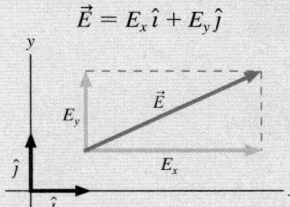

Components will simplify vector math.

3.1 Vectors

A quantity that is fully described by a single number (with units) is called a **scalar quantity.** Mass, temperature, and volume are all scalars. Other scalar quantities include pressure, density, energy, charge, and voltage. We will often use an algebraic symbol to represent a scalar quantity. Thus m will represent mass, T temperature, V volume, E energy, and so on. Notice that scalars, in printed text, are shown in italics.

Our universe has three dimensions, so some quantities also need a direction for a full description. If you ask someone for directions to the post office, the reply "Go three blocks" will not be very helpful. A full description might be, "Go three blocks south." A quantity having both a size and a direction is called a **vector quantity.**

The mathematical term for the length, or size, of a vector is **magnitude,** so we can also say that **a vector is a quantity having a magnitude and a direction.**

FIGURE 3.1 shows that the *geometric representation* of a vector is an arrow, with the tail of the arrow (not its tip!) placed at the point where the measurement is made. The vector then seems to radiate outward from the point to which it is attached. An arrow makes a natural representation of a vector because it inherently has both a length and a direction. As you've already seen, we label vectors by drawing a small arrow over the letter that represents the vector: $\vec{r}$ for position, $\vec{v}$ for velocity, $\vec{a}$ for acceleration, and so on.

NOTE ▶ Although the vector arrow is drawn across the page, from its tail to its tip, this does *not* indicate that the vector "stretches" across this distance. Instead, the vector arrow tells us the value of the vector quantity only at the one point where the tail of the vector is placed. ◀

The *magnitude* of a vector is sometimes shown using absolute value signs, but more frequently indicated by the letter without the arrow. For example, the magnitude of the velocity vector in Figure 3.1 is $v = |\vec{v}| = 5$ m/s. This is the object's *speed.* The magnitude of the acceleration vector $\vec{a}$ is written a. **The magnitude of a vector is a scalar quantity.**

NOTE ▶ The magnitude of a vector cannot be a negative number; it must be positive or zero, with appropriate units. ◀

It is important to get in the habit of using the arrow symbol for vectors. If you omit the vector arrow from the velocity vector $\vec{v}$ and write only v, then you're referring only to the object's speed, not its velocity. The symbols $\vec{r}$ and r, or $\vec{v}$ and v, do *not* represent the same thing, so if you omit the vector arrow from vector symbols you will soon have confusion and mistakes.

FIGURE 3.1 The velocity vector $\vec{v}$ has both a magnitude and a direction.

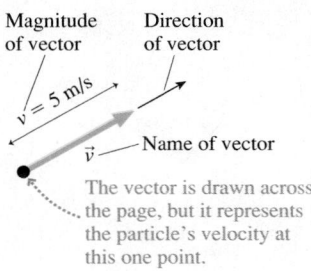

Magnitude of vector · Direction of vector

$v = 5$ m/s

$\vec{v}$ — Name of vector

The vector is drawn across the page, but it represents the particle's velocity at this one point.

3.2 Properties of Vectors

Suppose Sam starts from his front door, walks across the street, and ends up 200 ft to the northeast of where he started. Sam's displacement, which we will label $\vec{S}$, is shown in **FIGURE 3.2a**. The displacement vector is a *straight-line connection* from his initial to his final position, not necessarily his actual path.

To describe a vector we must specify both its magnitude and its direction. We can write Sam's displacement as

$$\vec{S} = (200 \text{ ft, northeast})$$

where the first piece of information specifies the magnitude and the second is the direction. The magnitude of Sam's displacement is $S = |\vec{S}| = 200$ ft, the distance between his initial and final points.

FIGURE 3.2 Displacement vectors.

(a)

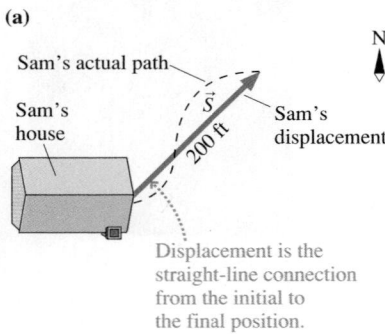

Sam's actual path

Sam's house

N

$\vec{S}$

200 ft

Sam's displacement

Displacement is the straight-line connection from the initial to the final position.

(b)

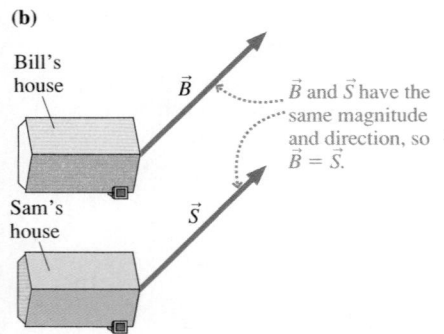

Bill's house

$\vec{B}$

$\vec{B}$ and $\vec{S}$ have the same magnitude and direction, so $\vec{B} = \vec{S}$.

Sam's house

$\vec{S}$

Sam's next-door neighbor Bill also walks 200 ft to the northeast, starting from his own front door. Bill's displacement $\vec{B} = (200 \text{ ft}, \text{northeast})$ has the same magnitude and direction as Sam's displacement $\vec{S}$. Because vectors are defined by their magnitude and direction, **two vectors are equal if they have the same magnitude and direction.** This is true regardless of the starting points of the vectors. Thus the two displacements in FIGURE 3.2b are equal to each other, and we can write $\vec{B} = \vec{S}$.

NOTE ▶ A vector is unchanged if you move it to a different point on the page as long as you don't change its length or the direction it points. We used this idea in Chapter 1 when we moved velocity vectors around in order to find the average acceleration vector $\vec{a}$. ◀

Vector Addition

If you earn $50 on Saturday and $60 on Sunday, your *net* income for the weekend is the sum of $50 and $60. With numbers, the word *net* implies addition. The same is true with vectors. For example, FIGURE 3.3 shows the displacement of a hiker who first hikes 4 miles to the east, then 3 miles to the north. The first leg of the hike is described by the displacement $\vec{A} = (4 \text{ mi}, \text{east})$. The second leg of the hike has displacement $\vec{B} = (3 \text{ mi}, \text{north})$. Vector $\vec{C}$ is the *net displacement* because it describes the net result of the hiker's first having displacement $\vec{A}$, then displacement $\vec{B}$.

The net displacement $\vec{C}$ is an initial displacement $\vec{A}$ *plus* a second displacement $\vec{B}$, or

$$\vec{C} = \vec{A} + \vec{B} \tag{3.1}$$

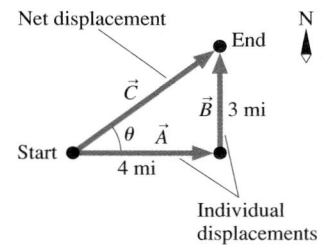

FIGURE 3.3 The net displacement $\vec{C}$ resulting from two displacements $\vec{A}$ and $\vec{B}$.

The sum of two vectors is called the **resultant vector.** It's not hard to show that vector addition is commutative: $\vec{A} + \vec{B} = \vec{B} + \vec{A}$. That is, you can add vectors in any order you wish.

Look back at Tactics Box 1.1 on page 7 to see the three-step procedure for adding two vectors. This tip-to-tail method for adding vectors, which is used to find $\vec{C} = \vec{A} + \vec{B}$ in Figure 3.3, is called **graphical addition.** Any two vectors of the same type—two velocity vectors or two force vectors—can be added in exactly the same way.

The graphical method for adding vectors is straightforward, but we need to do a little geometry to come up with a complete description of the resultant vector $\vec{C}$. Vector $\vec{C}$ of Figure 3.3 is defined by its magnitude C and by its direction. Because the three vectors $\vec{A}$, $\vec{B}$, and $\vec{C}$ form a right triangle, the magnitude, or length, of $\vec{C}$ is given by the Pythagorean theorem:

$$C = \sqrt{A^2 + B^2} = \sqrt{(4 \text{ mi})^2 + (3 \text{ mi})^2} = 5 \text{ mi} \tag{3.2}$$

Notice that Equation 3.2 uses the magnitudes A and B of the vectors $\vec{A}$ and $\vec{B}$. The angle θ, which is used in Figure 3.3 to describe the direction of $\vec{C}$, is easily found for a right triangle:

$$\theta = \tan^{-1}\left(\frac{B}{A}\right) = \tan^{-1}\left(\frac{3 \text{ mi}}{4 \text{ mi}}\right) = 37° \tag{3.3}$$

Altogether, the hiker's net displacement is

$$\vec{C} = \vec{A} + \vec{B} = (5 \text{ mi}, 37° \text{ north of east}) \tag{3.4}$$

NOTE ▶ Vector mathematics makes extensive use of geometry and trigonometry. Appendix A, at the end of this book, contains a brief review of these topics. ◀

EXAMPLE 3.1 **Using graphical addition to find a displacement**

A bird flies 100 m due east from a tree, then 200 m northwest (that is, 45° north of west). What is the bird's net displacement?

VISUALIZE **FIGURE 3.4** shows the two individual displacements, which we've called $\vec{A}$ and $\vec{B}$. The net displacement is the vector sum $\vec{C} = \vec{A} + \vec{B}$, which is found graphically.

FIGURE 3.4 The bird's net displacement is $\vec{C} = \vec{A} + \vec{B}$.

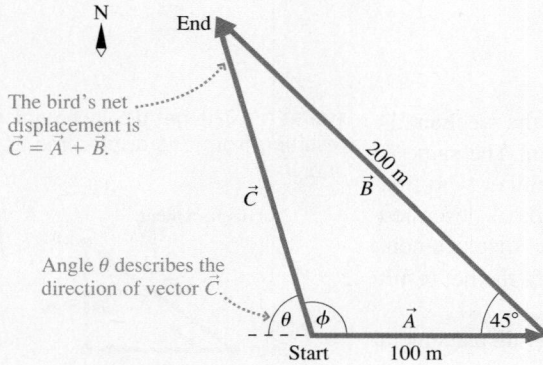

SOLVE The two displacements are $\vec{A} = (100 \text{ m, east})$ and $\vec{B} = (200 \text{ m, northwest})$. The net displacement $\vec{C} = \vec{A} + \vec{B}$ is found by drawing a vector from the initial to the final position. But describing $\vec{C}$ is a bit trickier than the example of the hiker because $\vec{A}$ and $\vec{B}$ are not at right angles. First, we can find the magnitude of $\vec{C}$ by using the law of cosines from trigonometry:

$$C^2 = A^2 + B^2 - 2AB\cos 45°$$
$$= (100 \text{ m})^2 + (200 \text{ m})^2 - 2(100 \text{ m})(200 \text{ m})\cos 45°$$
$$= 21{,}720 \text{ m}^2$$

Thus $C = \sqrt{21{,}720 \text{ m}^2} = 147$ m. Then a second use of the law of cosines can determine angle ϕ (the Greek letter phi):

$$B^2 = A^2 + C^2 - 2AC\cos\phi$$
$$\phi = \cos^{-1}\left[\frac{A^2 + C^2 - B^2}{2AC}\right] = 106°$$

It is easier to describe $\vec{C}$ with the angle $\theta = 180° - \phi = 74°$. The bird's net displacement is

$$\vec{C} = (147 \text{ m, } 74° \text{ north of west})$$

It is often convenient to draw two vectors with their tails together, as shown in **FIGURE 3.5a**. To evaluate $\vec{D} + \vec{E}$, you could move vector $\vec{E}$ over to where its tail is on the tip of $\vec{D}$, then use the tip-to-tail rule of graphical addition. That gives vector $\vec{F} = \vec{D} + \vec{E}$ in **FIGURE 3.5b**. Alternatively, **FIGURE 3.5c** shows that the vector sum $\vec{D} + \vec{E}$ can be found as the diagonal of the parallelogram defined by $\vec{D}$ and $\vec{E}$. This method for vector addition is called the *parallelogram rule* of vector addition.

FIGURE 3.5 Two vectors can be added using the tip-to-tail rule or the parallelogram rule.

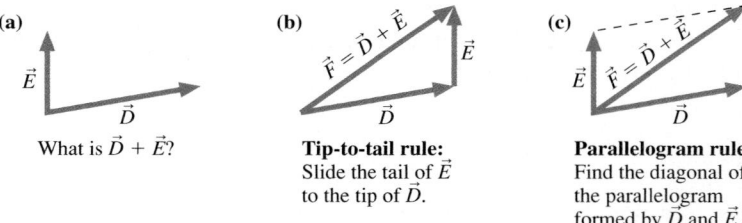

(a) What is $\vec{D} + \vec{E}$?

(b) **Tip-to-tail rule:** Slide the tail of $\vec{E}$ to the tip of $\vec{D}$.

(c) **Parallelogram rule:** Find the diagonal of the parallelogram formed by $\vec{D}$ and $\vec{E}$.

FIGURE 3.6 The net displacement after four individual displacements.

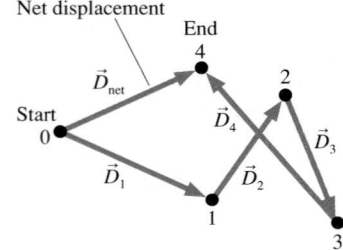

Vector addition is easily extended to more than two vectors. **FIGURE 3.6** shows the path of a hiker moving from initial position 0 to position 1, then position 2, then position 3, and finally arriving at position 4. These four segments are described by displacement vectors $\vec{D}_1$, $\vec{D}_2$, $\vec{D}_3$, and $\vec{D}_4$. The hiker's *net* displacement, an arrow from position 0 to position 4, is the vector $\vec{D}_{\text{net}}$. In this case,

$$\vec{D}_{\text{net}} = \vec{D}_1 + \vec{D}_2 + \vec{D}_3 + \vec{D}_4 \tag{3.5}$$

The vector sum is found by using the tip-to-tail method three times in succession.

STOP TO THINK 3.1 Which figure shows $\vec{A}_1 + \vec{A}_2 + \vec{A}_3$?

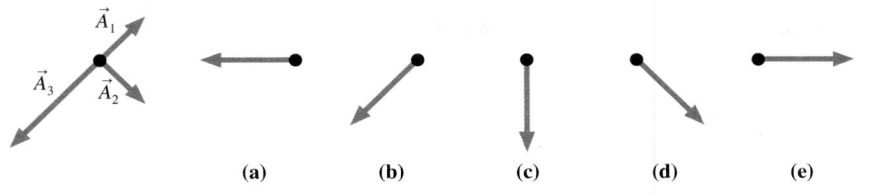

(a)   (b)   (c)   (d)   (e)

More Vector Mathematics

In addition to adding vectors, we will need to subtract vectors, multiply vectors by scalars, and understand how to interpret the negative of a vector. These operations are illustrated in FIGURE 3.7.

FIGURE 3.7 Working with vectors.

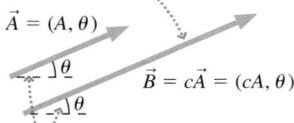

The length of $\vec{B}$ is "stretched" by the factor c. That is, $B = cA$.

$\vec{A} = (A, \theta)$

$\vec{B} = c\vec{A} = (cA, \theta)$

$\vec{B}$ points in the same direction as $\vec{A}$.

Multiplication by a scalar

$\vec{A} + (-\vec{A}) = \vec{0}$. The tip of $-\vec{A}$ returns to the starting point.

Vector $-\vec{A}$ is equal in magnitude but opposite in direction to $\vec{A}$.

The **zero vector** $\vec{0}$ has zero length

The negative of a vector

$-2\vec{A}$

Multiplication by a negative scalar

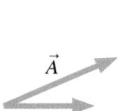

$\vec{A}$

$\vec{C}$

Vector subtraction: What is $\vec{A} - \vec{C}$?
Write it as $\vec{A} + (-\vec{C})$ and add!

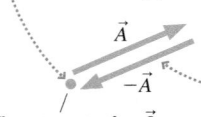

$-\vec{C}$

$\vec{A} - \vec{C}$

$\vec{A}$

Tip-to-tail method using $-\vec{C}$

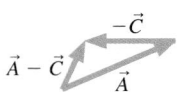

$\vec{A} - \vec{C}$

$\vec{A}$

$-\vec{C}$

Parallelogram method using $-\vec{C}$

EXAMPLE 3.2   Velocity and displacement

Carolyn drives her car north at 30 km/h for 1 hour, east at 60 km/h for 2 hours, then north at 50 km/h for 1 hour. What is Carolyn's net displacement?

SOLVE Chapter 1 defined velocity as

$$\vec{v} = \frac{\Delta\vec{r}}{\Delta t}$$

so the displacement $\Delta\vec{r}$ during the time interval Δt is $\Delta\vec{r} = (\Delta t)\vec{v}$. This is multiplication of the vector $\vec{v}$ by the scalar Δt. Carolyn's velocity during the first hour is $\vec{v}_1 = (30 \text{ km/h, north})$, so her displacement during this interval is

$$\Delta\vec{r}_1 = (1 \text{ hour})(30 \text{ km/h, north}) = (30 \text{ km, north})$$

Similarly,

$$\Delta\vec{r}_2 = (2 \text{ hours})(60 \text{ km/h, east}) = (120 \text{ km, east})$$

$$\Delta\vec{r}_3 = (1 \text{ hour})(50 \text{ km/h, north}) = (50 \text{ km, north})$$

In this case, multiplication by a scalar changes not only the length of the vector but also its units, from km/h to km. The direction, however, is unchanged. Carolyn's net displacement is

$$\Delta\vec{r}_{\text{net}} = \Delta\vec{r}_1 + \Delta\vec{r}_2 + \Delta\vec{r}_3$$

This addition of the three vectors is shown in FIGURE 3.8, using the tip-to-tail method. $\Delta\vec{r}_{\text{net}}$ stretches from Carolyn's initial position

FIGURE 3.8 The net displacement is the vector sum $\Delta\vec{r}_{\text{net}} = \Delta\vec{r}_1 + \Delta\vec{r}_2 + \Delta\vec{r}_3$.

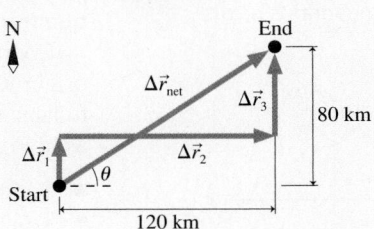

Continued

to her final position. The magnitude of her net displacement is found using the Pythagorean theorem:

$$r_{net} = \sqrt{(120 \text{ km})^2 + (80 \text{ km})^2} = 144 \text{ km}$$

The direction of $\Delta\vec{r}_{net}$ is described by angle θ, which is

$$\theta = \tan^{-1}\left(\frac{80 \text{ km}}{120 \text{ km}}\right) = 34°$$

Thus Carolyn's net displacement is $\Delta\vec{r}_{net} = (144 \text{ km}, 34°$ north of east).

STOP TO THINK 3.2 Which figure shows $2\vec{A} - \vec{B}$?

$\vec{A}$

$\vec{B}$

(a) (b) (c) (d) (e)

3.3 Coordinate Systems and Vector Components

Vectors do not require a coordinate system. We can add and subtract vectors graphically, and we will do so frequently to clarify our understanding of a situation. But the graphical addition of vectors is not an especially good way to find quantitative results. In this section we will introduce a *coordinate representation* of vectors that will be the basis of an easier method for doing vector calculations.

Coordinate Systems

The world does not come with a coordinate system attached to it. A coordinate system is an artificially imposed grid that you place on a problem in order to make quantitative measurements. You are free to choose:

- Where to place the origin, and
- How to orient the axes.

Different problem solvers may choose to use different coordinate systems; that is perfectly acceptable. However, some coordinate systems will make a problem easier to solve. Part of our goal is to learn how to choose an appropriate coordinate system for each problem.

FIGURE 3.9 shows the *xy*-coordinate system we will use in this book. The placement of the axes is not entirely arbitrary. By convention, the positive *y*-axis is located 90° *counterclockwise* (ccw) from the positive *x*-axis. Figure 3.9 also identifies the four **quadrants** of the coordinate system, I through IV.

Coordinate axes have a positive end and a negative end, separated by zero at the origin where the two axes cross. When you draw a coordinate system, it is important to label the axes. This is done by placing *x* and *y* labels at the *positive* ends of the axes, as in Figure 3.9. The purpose of the labels is twofold:

- To identify which axis is which, and
- To identify the positive ends of the axes.

This will be important when you need to determine whether the quantities in a problem should be assigned positive or negative values.

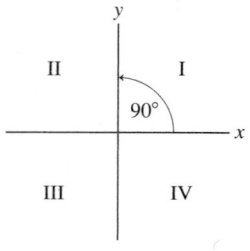

The navigator had better know which way to go, and how far, if she and the crew are to make landfall at the expected location.

FIGURE 3.9 A conventional *xy*-coordinate system and the quadrants of the *xy*-plane.

Component Vectors

FIGURE 3.10 shows a vector $\vec{A}$ and an xy-coordinate system that we've chosen. Once the directions of the axes are known, we can define two new vectors *parallel to the axes* that we call the **component vectors** of $\vec{A}$. You can see, using the parallelogram rule, that $\vec{A}$ is the vector sum of the two component vectors:

$$\vec{A} = \vec{A}_x + \vec{A}_y \qquad (3.6)$$

In essence, we have broken vector $\vec{A}$ into two perpendicular vectors that are parallel to the coordinate axes. This process is called the **decomposition** of vector $\vec{A}$ into its component vectors.

> NOTE ▶ It is not necessary for the tail of $\vec{A}$ to be at the origin. All we need to know is the *orientation* of the coordinate system so that we can draw $\vec{A}_x$ and $\vec{A}_y$ parallel to the axes. ◀

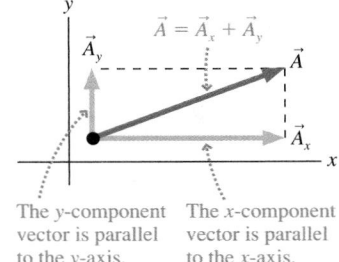

FIGURE 3.10 Component vectors $\vec{A}_x$ and $\vec{A}_y$ are drawn parallel to the coordinate axes such that $\vec{A} = \vec{A}_x + \vec{A}_y$.

The y-component vector is parallel to the y-axis.

The x-component vector is parallel to the x-axis.

Components

You learned in Chapters 1 and 2 to give the kinematic variable v_x a positive sign if the velocity vector $\vec{v}$ points toward the positive end of the x-axis, a negative sign if $\vec{v}$ points in the negative x-direction. The basis of that rule is that v_x is what we call the x-component of the velocity vector. We need to extend this idea to vectors in general.

Suppose vector $\vec{A}$ has been decomposed into component vectors $\vec{A}_x$ and $\vec{A}_y$ parallel to the coordinate axes. We can describe each component vector with a single number called the **component**. The x-component and y-component of vector $\vec{A}$, denoted A_x and A_y, are determined as follows:

TACTICS
BOX 3.1 **Determining the components of a vector** (MP)

❶ The absolute value $|A_x|$ of the x-component A_x is the magnitude of the component vector $\vec{A}_x$.
❷ The *sign* of A_x is positive if $\vec{A}_x$ points in the positive x-direction, negative if $\vec{A}_x$ points in the negative x-direction.
❸ The y-component A_y is determined similarly.

Exercises 10–18

In other words, the component A_x tells us two things: how big $\vec{A}_x$ is and, with its sign, which end of the axis $\vec{A}_x$ points toward. FIGURE 3.11 shows three examples of determining the components of a vector.

FIGURE 3.11 Determining the components of a vector.

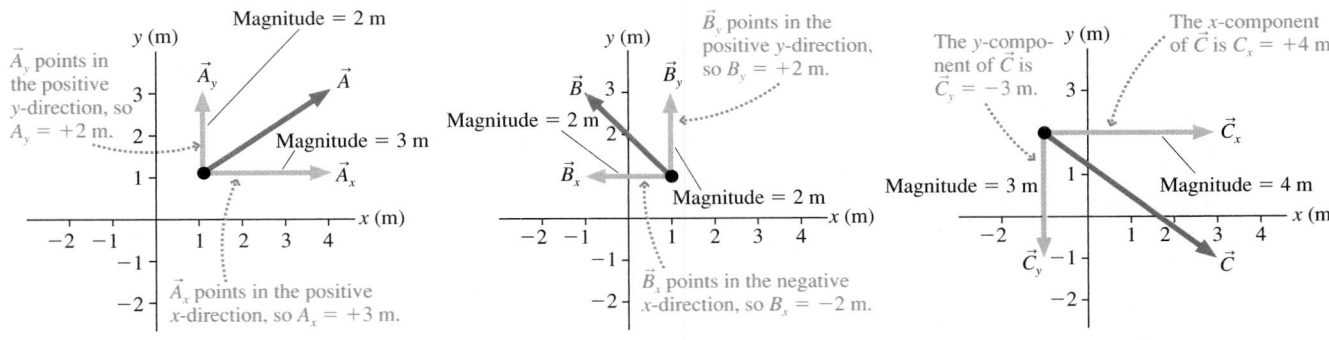

NOTE ▶ Beware of the somewhat confusing terminology. $\vec{A}_x$ and $\vec{A}_y$ are called *component vectors,* whereas A_x and A_y are simply called *components.* The components A_x and A_y are just numbers (with units), so make sure you do *not* put arrow symbols over the components. ◀

We will frequently need to decompose a vector into its components. We will also need to "reassemble" a vector from its components. In other words, we need to move back and forth between the geometric and the component representations of a vector. FIGURE 3.12 shows how this is done.

FIGURE 3.12 Moving between the geometric representation and the component representation.

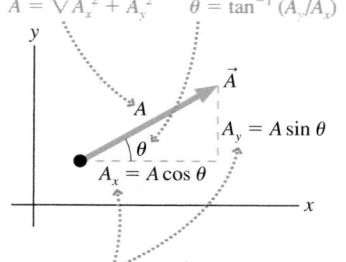

The magnitude and direction of $\vec{A}$ are found from the components. In this example,

$$A = \sqrt{A_x^2 + A_y^2} \qquad \theta = \tan^{-1}(A_y/A_x)$$

The components of $\vec{A}$ are found from the magnitude and direction. In this example, $A_x = A\cos\theta$ and $A_y = A\sin\theta$.

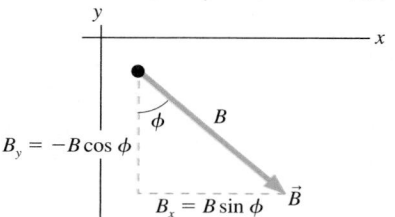

The angle is defined differently. In this example, the magnitude and direction are

$$B = \sqrt{B_x^2 + B_y^2} \qquad \phi = \tan^{-1}(B_x/|B_y|)$$

Here the components are $B_x = B\sin\phi$ and $B_y = -B\cos\phi$. Minus signs must be inserted manually, depending on the vector's direction.

Each decomposition requires that you pay close attention to the direction in which the vector points and the angles that are defined.

- If a component vector points left (or down), you must *manually* insert a minus sign in front of the component, as was done for B_y in Figure 3.12.
- The role of sines and cosines can be reversed, depending upon which angle is used to define the direction. Compare A_x and B_x.
- The angle used to define direction is almost always between 0° and 90°, so you must take the inverse tangent of a positive number. Use absolute values of the components, as was done to find angle ϕ (Greek phi) in Figure 3.12.

EXAMPLE 3.3 **Finding the components of an acceleration vector**

Find the *x*- and *y*-components of the acceleration vector $\vec{a}$ shown in FIGURE 3.13.

FIGURE 3.13 The acceleration vector $\vec{a}$ of Example 3.3.

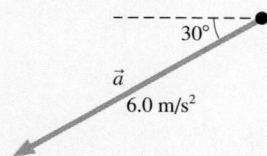

VISUALIZE It's important to *draw* vectors. FIGURE 3.14 shows the original vector $\vec{a}$ decomposed into components parallel to the axes. Notice that the axes are "acceleration axes," not *xy*-axes, because we're measuring an acceleration vector.

SOLVE The acceleration vector $\vec{a} = (6.0\ \text{m/s}^2,\ 30°$ below the negative *x*-axis) points to the left (negative *x*-direction) and down (negative *y*-direction), so the components a_x and a_y are both negative:

$$a_x = -a\cos 30° = -(6.0\ \text{m/s}^2)\cos 30° = -5.2\ \text{m/s}^2$$
$$a_y = -a\sin 30° = -(6.0\ \text{m/s}^2)\sin 30° = -3.0\ \text{m/s}^2$$

FIGURE 3.14 Decomposition of $\vec{a}$.

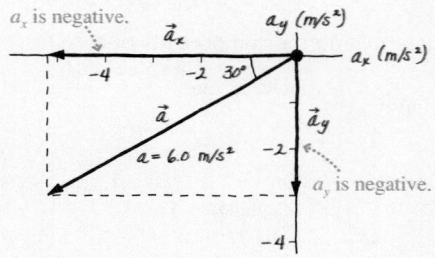

ASSESS The units of a_x and a_y are the same as the units of vector $\vec{a}$. Notice that we had to insert the minus signs manually by observing that the vector points left and down.

EXAMPLE 3.4 **Finding the direction of motion**

FIGURE 3.15 shows a car's velocity vector $\vec{v}$. Determine the car's speed and direction of motion.

FIGURE 3.15 The velocity vector $\vec{v}$ of Example 3.4.

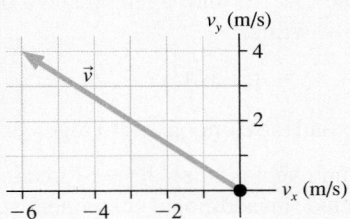

FIGURE 3.16 Decomposition of $\vec{v}$.

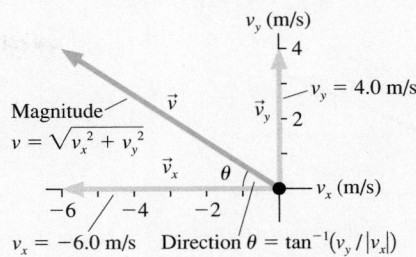

$v_x = -6.0$ m/s   Direction $\theta = \tan^{-1}(v_y / |v_x|)$

VISUALIZE **FIGURE 3.16** shows the components v_x and v_y and defines an angle θ with which we can specify the direction of motion.

SOLVE We can read the components of $\vec{v}$ directly from the axes: $v_x = -6.0$ m/s and $v_y = 4.0$ m/s. Notice that v_x is negative. This is enough information to find the car's speed v, which is the magnitude of $\vec{v}$:

$$v = \sqrt{v_x^2 + v_y^2} = \sqrt{(-6.0 \text{ m/s})^2 + (4.0 \text{ m/s})^2} = 7.2 \text{ m/s}$$

From trigonometry, angle θ is

$$\theta = \tan^{-1}\left(\frac{v_y}{|v_x|}\right) = \tan^{-1}\left(\frac{4.0 \text{ m/s}}{6.0 \text{ m/s}}\right) = 34°$$

The absolute value signs are necessary because v_x is a negative number. The velocity vector $\vec{v}$ can be written in terms of the speed and the direction of motion as

$$\vec{v} = (7.2 \text{ m/s}, 34° \text{ above the negative } x\text{-axis})$$

or, if the axes are aligned to north,

$$\vec{v} = (7.2 \text{ m/s}, 34° \text{ north of west})$$

STOP TO THINK 3.3   What are the x- and y-components C_x and C_y of vector $\vec{C}$?

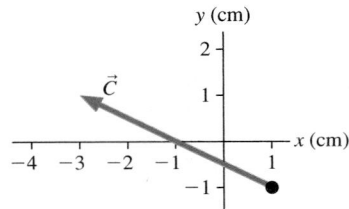

3.4 Vector Algebra

The vectors (1, $+x$-direction and (1, $+y$-direction), shown in **FIGURE 3.17**, have some interesting and useful properties. Each has a magnitude of 1, no units, and is parallel to a coordinate axis. A vector with these properties is called a **unit vector.** These unit vectors have the special symbols

$$\hat{i} \equiv (1, \text{ positive } x\text{-direction})$$

$$\hat{j} \equiv (1, \text{ positive } y\text{-direction})$$

The notation $\hat{i}$ (read "i hat") and $\hat{j}$ (read "j hat") indicates a unit vector with a magnitude of 1. Recall that the symbol $\equiv$ means "is defined as."

Unit vectors establish the directions of the positive axes of the coordinate system. Our choice of a coordinate system may be arbitrary, but once we decide to place a coordinate system on a problem we need something to tell us "That direction is the positive x-direction." This is what the unit vectors do.

The unit vectors provide a useful way to write component vectors. The component vector $\vec{A}_x$ is the piece of vector $\vec{A}$ that is parallel to the x-axis. Similarly, $\vec{A}_y$ is parallel

FIGURE 3.17 The unit vectors $\hat{i}$ and $\hat{j}$.

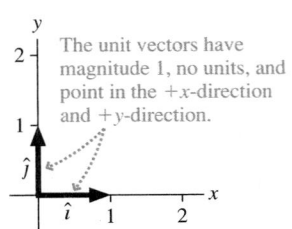

FIGURE 3.18 The decomposition of vector $\vec{A}$ is $A_x\hat{\imath} + A_y\hat{\jmath}$.

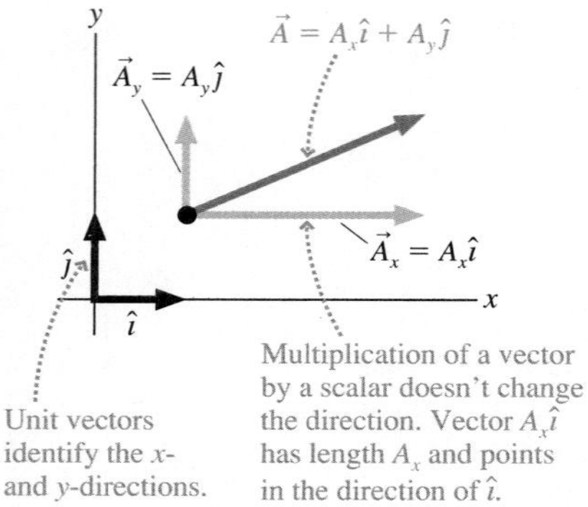

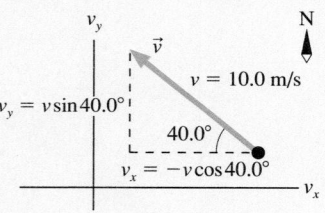

Unit vectors identify the x- and y-directions.

Multiplication of a vector by a scalar doesn't change the direction. Vector $A_x\hat{\imath}$ has length A_x and points in the direction of $\hat{\imath}$.

to the y-axis. Because, by definition, the vector $\hat{\imath}$ points along the x-axis and $\hat{\jmath}$ points along the y-axis, we can write

$$\vec{A}_x = A_x\hat{\imath}$$
$$\vec{A}_y = A_y\hat{\jmath} \tag{3.7}$$

Equations 3.7 separate each component vector into a length and a direction. The full decomposition of vector $\vec{A}$ can then be written

$$\vec{A} = \vec{A}_x + \vec{A}_y = A_x\hat{\imath} + A_y\hat{\jmath} \tag{3.8}$$

FIGURE 3.18 shows how the unit vectors and the components fit together to form vector $\vec{A}$.

NOTE ▶ In three dimensions, the unit vector along the $+z$-direction is called $\hat{k}$, and to describe vector $\vec{A}$ we would include an additional component vector $\vec{A}_z = A_z\hat{k}$. ◀

EXAMPLE 3.5 **Run rabbit run!**

A rabbit, escaping a fox, runs 40.0° north of west at 10.0 m/s. A coordinate system is established with the positive x-axis to the east and the positive y-axis to the north. Write the rabbit's velocity in terms of components and unit vectors.

VISUALIZE **FIGURE 3.19** shows the rabbit's velocity vector and the coordinate axes. We're showing a velocity vector, so the axes are labeled v_x and v_y rather than x and y.

FIGURE 3.19 The velocity vector $\vec{v}$ is decomposed into components v_x and v_y.

v_y

N

$v_y = v\sin 40.0°$

$\vec{v}$

$v = 10.0$ m/s

40.0°

$v_x = -v\cos 40.0°$

v_x

SOLVE 10.0 m/s is the rabbit's *speed,* not its velocity. The velocity, which includes directional information, is

$$\vec{v} = (10.0 \text{ m/s}, 40.0° \text{ north of west})$$

Vector $\vec{v}$ points to the left and up, so the components v_x and v_y are negative and positive, respectively. The components are

$$v_x = -(10.0 \text{ m/s})\cos 40.0° = -7.66 \text{ m/s}$$
$$v_y = +(10.0 \text{ m/s})\sin 40.0° = 6.43 \text{ m/s}$$

With v_x and v_y now known, the rabbit's velocity vector is

$$\vec{v} = v_x\hat{\imath} + v_y\hat{\jmath} = (-7.66\hat{\imath} + 6.43\hat{\jmath}) \text{ m/s}$$

Notice that we've pulled the units to the end, rather than writing them with each component.

ASSESS Notice that the minus sign for v_x was inserted manually. **Signs don't occur automatically; you have to set them after checking the vector's direction.**

Working with Vectors

You learned in Section 3.2 how to add vectors graphically, but it is a tedious problem in geometry and trigonometry to find precise values for the magnitude and direction of the resultant. The addition and subtraction of vectors become much easier if we use components and unit vectors.

To see this, let's evaluate the vector sum $\vec{D} = \vec{A} + \vec{B} + \vec{C}$. To begin, write this sum in terms of the components of each vector:

$$\vec{D} = D_x\hat{\imath} + D_y\hat{\jmath} = \vec{A} + \vec{B} + \vec{C}$$
$$= (A_x\hat{\imath} + A_y\hat{\jmath}) + (B_x\hat{\imath} + B_y\hat{\jmath}) + (C_x\hat{\imath} + C_y\hat{\jmath}) \tag{3.9}$$

We can group together all the x-components and all the y-components on the right side, in which case Equation 3.9 is

$$(D_x)\hat{\imath} + (D_y)\hat{\jmath} = (A_x + B_x + C_x)\hat{\imath} + (A_y + B_y + C_y)\hat{\jmath} \tag{3.10}$$

Comparing the x- and y-components on the left and right sides of Equation 3.10, we find:

$$D_x = A_x + B_x + C_x$$
$$D_y = A_y + B_y + C_y \tag{3.11}$$

Stated in words, Equation 3.11 says that we can perform vector addition by adding the *x*-components of the individual vectors to give the *x*-component of the resultant and by adding the *y*-components of the individual vectors to give the *y*-component of the resultant. This method of vector addition is called **algebraic addition**.

EXAMPLE 3.6 **Using algebraic addition to find a displacement**

Example 3.1 was about a bird that flew 100 m to the east, then 200 m to the northwest. Use the algebraic addition of vectors to find the bird's net displacement.

VISUALIZE FIGURE 3.20 shows displacement vectors $\vec{A} = (100 \text{ m},$ east) and $\vec{B} = (200 \text{ m, northwest})$. We draw vectors tip-to-tail to add them graphically, but it's usually easier to draw them all from the origin if we are going to use algebraic addition.

FIGURE 3.20 The net displacement is $\vec{C} = \vec{A} + \vec{B}$.

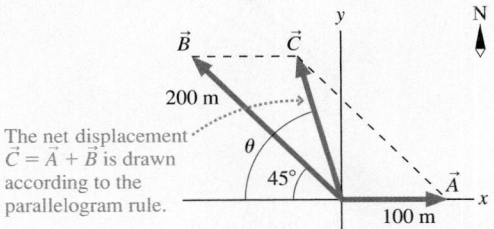

The net displacement $\vec{C} = \vec{A} + \vec{B}$ is drawn according to the parallelogram rule.

SOLVE To add the vectors algebraically we must know their components. From the figure these are seen to be

$$\vec{A} = 100\,\hat{\imath}\,\text{m}$$
$$\vec{B} = (-200\cos 45°\,\hat{\imath} + 200\sin 45°\hat{\jmath})\,\text{m}$$
$$= (-141\hat{\imath} + 141\hat{\jmath})\,\text{m}$$

Notice that vector quantities must include units. Also notice, as you would expect from the figure, that $\vec{B}$ has a negative *x*-component. Adding $\vec{A}$ and $\vec{B}$ by components gives

$$\vec{C} = \vec{A} + \vec{B} = 100\hat{\imath}\,\text{m} + (-141\hat{\imath} + 141\hat{\jmath})\,\text{m}$$
$$= (100\text{ m} - 141\text{ m})\hat{\imath} + (141\text{ m})\hat{\jmath} = (-41\hat{\imath} + 141\hat{\jmath})\,\text{m}$$

This would be a perfectly acceptable answer for many purposes. However, we need to calculate the magnitude and direction of $\vec{C}$ if we want to compare this result to our earlier answer. The magnitude of $\vec{C}$ is

$$C = \sqrt{C_x^2 + C_y^2} = \sqrt{(-41\text{ m})^2 + (141\text{ m})^2} = 147\text{ m}$$

The angle θ, as defined in Figure 3.20, is

$$\theta = \tan^{-1}\left(\frac{C_y}{|C_x|}\right) = \tan^{-1}\left(\frac{141\text{ m}}{41\text{ m}}\right) = 74°$$

Thus $\vec{C} = (147\text{ m}, 74°$ north of west), in perfect agreement with Example 3.1.

Vector subtraction and the multiplication of a vector by a scalar, using components, are very much like vector addition. To find $\vec{R} = \vec{P} - \vec{Q}$ we would compute

$$R_x = P_x - Q_x$$
$$R_y = P_y - Q_y$$ (3.12)

Similarly, $\vec{T} = c\vec{S}$ would be

$$T_x = cS_x$$
$$T_y = cS_y$$ (3.13)

In other words, a vector equation is interpreted as meaning: Equate the *x*-components on both sides of the equals sign, then equate the *y*-components, and then the *z*-components. Vector notation allows us to write these three equations in a much more compact form.

Tilted Axes and Arbitrary Directions

As we've noted, the coordinate system is entirely your choice. It is a grid that you impose on the problem in a manner that will make the problem easiest to solve. As you've already seen in Chapter 2, it is often convenient to tilt the axes of the coordinate system, such as those shown in **FIGURE 3.21**. The axes are perpendicular, and the *y*-axis is oriented correctly with respect to the *x*-axis, so this is a legitimate coordinate system. There is no requirement that the *x*-axis has to be horizontal.

Finding components with tilted axes is no harder than what we have done so far. Vector $\vec{C}$ in Figure 3.21 can be decomposed into $\vec{C} = C_x\,\hat{\imath} + C_y\hat{\jmath}$, where $C_x = C\cos\theta$ and $C_y = C\sin\theta$. Note that the unit vectors $\hat{\imath}$ and $\hat{\jmath}$ correspond to the *axes*, not to "horizontal" and "vertical," so they are also tilted.

Tilted axes are useful if you need to determine component vectors "parallel to" and "perpendicular to" an arbitrary line or surface. This is illustrated in the following example.

FIGURE 3.21 A coordinate system with tilted axes.

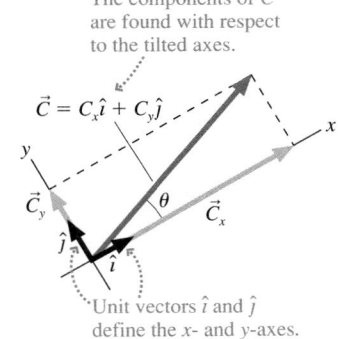

The components of $\vec{C}$ are found with respect to the tilted axes.

$$\vec{C} = C_x\hat{\imath} + C_y\hat{\jmath}$$

Unit vectors $\hat{\imath}$ and $\hat{\jmath}$ define the *x*- and *y*-axes.

EXAMPLE 3.7   **Muscle and bone**

The deltoid—the rounded muscle across the top of your upper arm—allows you to lift your arm away from your side. It does so by pulling on an attachment point on the humerus, the upper arm bone, at an angle of 15° with respect to the humerus. If you hold your arm at an angle 30° below horizontal, the deltoid must pull with a force of 720 N to support the weight of your arm, as shown in **FIGURE 3.22a**. (You'll learn in Chapter 5 that force is a vector quantity measured in units of *newtons*, abbreviated N.) What are the components of the muscle force parallel to and perpendicular to the bone?

VISUALIZE   **FIGURE 3.22b** shows a tilted coordinate system with the x-axis parallel to the humerus. The force $\vec{F}$ is shown 15° from the x-axis. The component of force parallel to the bone, which we can denote $F_\parallel$, is equivalent to the x-component: $F_\parallel = F_x$. Similarly, the component of force perpendicular to the bone is $F_\perp = F_y$.

SOLVE   From the geometry of Figure 3.22b, we see that

$$F_\parallel = F\cos 15° = (720\ \text{N})\cos 15° = 695\ \text{N}$$

$$F_\perp = F\sin 15° = (720\ \text{N})\sin 15° = 186\ \text{N}$$

ASSESS   The muscle pulls nearly parallel to the bone, so we expected $F_\parallel \approx 720$ N and $F_\perp \ll F_\parallel$. Thus our results seem reasonable.

FIGURE 3.22 Finding the components of force parallel and perpendicular to the humerus.

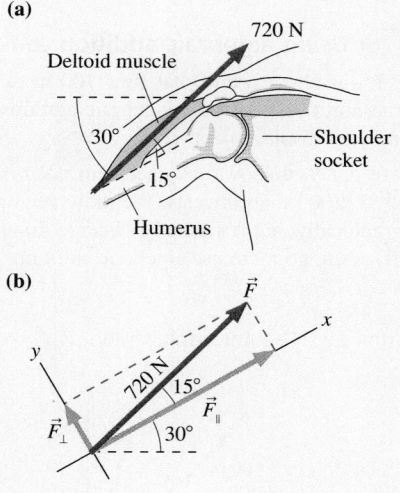

(a)

(b)

STOP TO THINK 3.4   Angle ϕ that specifies the direction of $\vec{C}$ is given by

a. $\tan^{-1}(|C_x|/C_y)$ d. $\tan^{-1}(|C_y|/C_x)$
b. $\tan^{-1}(C_x/|C_y|)$ e. $\tan^{-1}(C_y/|C_x|)$
c. $\tan^{-1}(|C_x|/|C_y|)$ f. $\tan^{-1}(|C_y|/|C_x|)$

CHALLENGE EXAMPLE 3.8   **Finding the net force**

FIGURE 3.23 shows three forces acting at one point. What is the net force $\vec{F}_{\text{net}} = \vec{F}_1 + \vec{F}_2 + \vec{F}_3$?

VISUALIZE   Figure 3.23 show the forces and establishes a tilted coordinate system.

SOLVE   The vector equation $\vec{F}_{\text{net}} = \vec{F}_1 + \vec{F}_2 + \vec{F}_3$ is really two simultaneous equations:

$$(F_{\text{net}})_x = F_{1x} + F_{2x} + F_{3x}$$

$$(F_{\text{net}})_y = F_{1y} + F_{2y} + F_{3y}$$

The components of the forces are determined with respect to the axes. Thus

$$F_{1x} = F_1\cos 45° = (50\ \text{N})\cos 45° = 35\ \text{N}$$

$$F_{1y} = F_1\sin 45° = (50\ \text{N})\sin 45° = 35\ \text{N}$$

$\vec{F}_2$ is easier. It is pointing along the y-axis, so $F_{2x} = 0$ N and $F_{2y} = 20$ N. To find the components of $\vec{F}_3$, we need to recognize—because $\vec{F}_3$ points straight down—that the angle between $\vec{F}_3$ and the x-axis is 75°. Thus

$$F_{3x} = F_3\cos 75° = (57\ \text{N})\cos 75° = 15\ \text{N}$$

$$F_{3y} = -F_3\sin 75° = -(57\ \text{N})\sin 75° = -55\ \text{N}$$

The minus sign in F_{3y} is critical, and it appears not from some formula but because we recognized—from the figure—that the

FIGURE 3.23 Three forces.

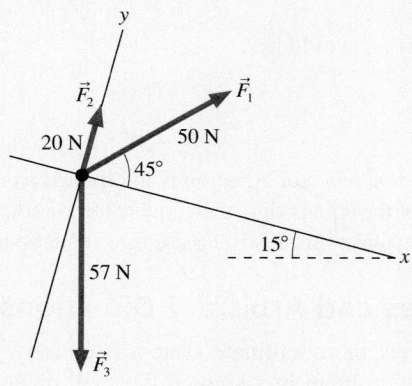

y-component of $\vec{F}_3$ points in the −y-direction. Combining the pieces, we have

$$(F_{\text{net}})_x = 35\ \text{N} + 0\ \text{N} + 15\ \text{N} = 50\ \text{N}$$

$$(F_{\text{net}})_y = 35\ \text{N} + 20\ \text{N} + (-55\ \text{N}) = 0\ \text{N}$$

Thus the net force is $\vec{F}_{\text{net}} = 50\hat{\imath}$ N. It points along the x-axis of the tilted coordinate system.

ASSESS   Notice that all work was done with reference to the axes of the coordinate system, not with respect to vertical or horizontal.

SUMMARY

The goals of Chapter 3 have been to learn how vectors are represented and used.

Important Concepts

A vector is a quantity described by both a magnitude and a direction.

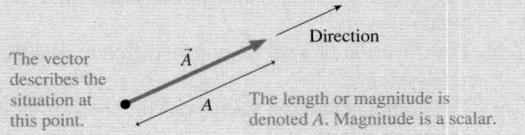

The vector describes the situation at this point.

Direction

$\vec{A}$

A

The length or magnitude is denoted A. Magnitude is a scalar.

Unit Vectors

Unit vectors have magnitude 1 and no units. Unit vectors $\hat{\imath}$ and $\hat{\jmath}$ define the directions of the x- and y-axes.

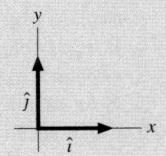

Using Vectors

Components

The component vectors are parallel to the x- and y-axes:

$$\vec{A} = \vec{A}_x + \vec{A}_y = A_x\hat{\imath} + A_y\hat{\jmath}$$

In the figure at the right, for example:

$$A_x = A\cos\theta \qquad A = \sqrt{A_x^2 + A_y^2}$$

$$A_y = A\sin\theta \qquad \theta = \tan^{-1}(A_y/A_x)$$

▶ Minus signs need to be included if the vector points down or left.

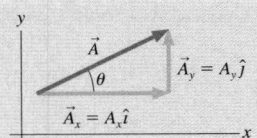

The components A_x and A_y are the magnitudes of the component vectors $\vec{A}_x$ and $\vec{A}_y$ *and* a plus or minus sign to show whether the component vector points toward the positive end or the negative end of the axis.

$A_x < 0$	$A_x > 0$
$A_y > 0$	$A_y > 0$
$A_x < 0$	$A_x > 0$
$A_y < 0$	$A_y < 0$

Working Graphically

Addition $\vec{A} + \vec{B}$

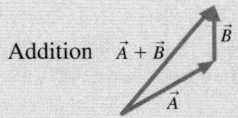

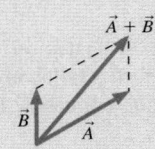

Negative

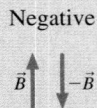

Subtraction

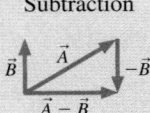

Multiplication

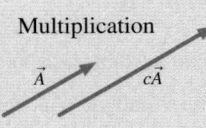

Working Algebraically

Vector calculations are done component by component:

$$\vec{C} = 2\vec{A} + \vec{B} \qquad \text{means} \qquad \begin{cases} C_x = 2A_x + B_x \\ C_y = 2A_y + B_y \end{cases}$$

The magnitude of $\vec{C}$ is then $C = \sqrt{C_x^2 + C_y^2}$ and its direction is found using $\tan^{-1}$.

Terms and Notation

scalar quantity	resultant vector	quadrants	component
vector quantity	graphical addition	component vector	unit vector, $\hat{\imath}$ or $\hat{\jmath}$
magnitude	zero vector, $\vec{0}$	decomposition	algebraic addition

CONCEPTUAL QUESTIONS

1. Can the magnitude of the displacement vector be more than the distance traveled? Less than the distance traveled? Explain.
2. If $\vec{C} = \vec{A} + \vec{B}$, can $C = A + B$? Can $C > A + B$? For each, show how or explain why not.
3. If $\vec{C} = \vec{A} + \vec{B}$, can $C = 0$? Can $C < 0$? For each, show how or explain why not.
4. Is it possible to add a scalar to a vector? If so, demonstrate. If not, explain why not.
5. How would you define the *zero vector* $\vec{0}$?
6. Can a vector have a component equal to zero and still have non-zero magnitude? Explain.

7. Can a vector have zero magnitude if one of its components is nonzero? Explain.
8. Suppose two vectors have unequal magnitudes. Can their sum be zero? Explain.
9. Are the following statements true or false? Explain your answer.
   a. The magnitude of a vector can be different in different coordinate systems.
   b. The direction of a vector can be different in different coordinate systems.
   c. The components of a vector can be different in different coordinate systems.

EXERCISES AND PROBLEMS

Exercises

Section 3.1 Vectors

Section 3.2 Properties of Vectors

1. | Trace the vectors in FIGURE EX3.1 onto your paper. Then find (a) $\vec{A} + \vec{B}$ and (b) $\vec{A} - \vec{B}$.

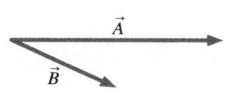

FIGURE EX3.1

2. | Trace the vectors in FIGURE EX3.2 onto your paper. Then find (a) $\vec{A} + \vec{B}$ and (b) $\vec{A} - \vec{B}$.

FIGURE EX3.2

Section 3.3 Coordinate Systems and Vector Components

3. | a. What are the x- and y-components of vector $\vec{E}$ shown in FIGURE EX3.3 in terms of the angle θ and the magnitude E?
   b. For the same vector, what are the x- and y-components in terms of the angle ϕ and the magnitude E?

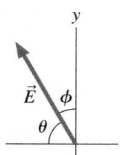

FIGURE EX3.3

4. ‖ A velocity vector 40° below the positive x-axis has a y-component of -10 m/s. What is the value of its x-component?
5. | A position vector in the first quadrant has an x-component of 8 m and a magnitude of 10 m. What is the value of its y-component?
6. ‖ Draw each of the following vectors, then find its x- and y-components.
   a. $\vec{r} = (100$ m, $45°$ below positive x-axis)
   b. $\vec{v} = (300$ m/s, $20°$ above positive x-axis)
   c. $\vec{a} = (5.0$ m/s^2, negative y-direction)
7. ‖ Draw each of the following vectors, then find its x- and y-components.
   a. $\vec{v} = (10$ m/s, negative y-direction)
   b. $\vec{a} = (20$ m/s^2, $30°$ below positive x-axis)
   c. $\vec{F} = (100$ N, $36.9°$ counterclockwise from positive y-axis)

8. | Let $\vec{C} = (3.15$ m, $15°$ above the negative x-axis) and $\vec{D} = (25.6$ m, $30°$ to the right of the negative y-axis). Find the magnitude, the x-component, and the y-component of each vector.
9. | The *magnetic field* inside an instrument is $\vec{B} = (2.0\hat{\imath} - 1.0\hat{\jmath})$ T where $\vec{B}$ represents the magnetic field vector and T stands for *tesla*, the unit of the magnetic field. What are the magnitude and direction of the magnetic field?

Section 3.4 Vector Algebra

10. | Draw each of the following vectors, label an angle that specifies the vector's direction, then find its magnitude and direction.
    a. $\vec{B} = -4\hat{\imath} + 4\hat{\jmath}$ b. $\vec{r} = (-2.0\hat{\imath} - 1.0\hat{\jmath})$ cm
    c. $\vec{v} = (-10\hat{\imath} - 100\hat{\jmath})$ m/s   d. $\vec{a} = (20\hat{\imath} + 10\hat{\jmath})$ m/s^2
11. | Draw each of the following vectors, label an angle that specifies the vector's direction, then find the vector's magnitude and direction.
    a. $\vec{A} = 4\hat{\imath} - 6\hat{\jmath}$ b. $\vec{r} = (50\hat{\imath} + 80\hat{\jmath})$ m
    c. $\vec{v} = (-20\hat{\imath} + 40\hat{\jmath})$ m/s   d. $\vec{a} = (2.0\hat{\imath} - 6.0\hat{\jmath})$ m/s^2
12. | Let $\vec{A} = 2\hat{\imath} + 3\hat{\jmath}$ and $\vec{B} = 4\hat{\imath} - 2\hat{\jmath}$.
    a. Draw a coordinate system and on it show vectors $\vec{A}$ and $\vec{B}$.
    b. Use graphical vector subtraction to find $\vec{C} = \vec{A} - \vec{B}$.
13. | Let $\vec{A} = 4\hat{\imath} - 2\hat{\jmath}$, $\vec{B} = -3\hat{\imath} + 5\hat{\jmath}$, and $\vec{C} = \vec{A} + \vec{B}$.
    a. Write vector $\vec{C}$ in component form.
    b. Draw a coordinate system and on it show vectors $\vec{A}$, $\vec{B}$, and $\vec{C}$.
    c. What are the magnitude and direction of vector $\vec{C}$?
14. | Let $\vec{A} = 4\hat{\imath} - 2\hat{\jmath}$, $\vec{B} = -3\hat{\imath} + 5\hat{\jmath}$, and $\vec{D} = \vec{A} - \vec{B}$.
    a. Write vector $\vec{D}$ in component form.
    b. Draw a coordinate system and on it show vectors $\vec{A}$, $\vec{B}$, and $\vec{D}$.
    c. What are the magnitude and direction of vector $\vec{D}$?
15. | Let $\vec{A} = 4\hat{\imath} - 2\hat{\jmath}$, $\vec{B} = -3\hat{\imath} + 5\hat{\jmath}$, and $\vec{E} = 4\vec{A} + 2\vec{B}$.
    a. Write vector $\vec{E}$ in component form.
    b. Draw a coordinate system and on it show vectors $\vec{A}$, $\vec{B}$, and $\vec{E}$.
    c. What are the magnitude and direction of vector $\vec{E}$?
16. | Let $\vec{A} = 4\hat{\imath} - 2\hat{\jmath}$, $\vec{B} = -3\hat{\imath} + 5\hat{\jmath}$, and $\vec{F} = \vec{A} - 4\vec{B}$.
    a. Write vector $\vec{F}$ in component form.
    b. Draw a coordinate system and on it show vectors $\vec{A}$, $\vec{B}$, and $\vec{F}$.
    c. What are the magnitude and direction of vector $\vec{F}$?

17. ‖ Let $\vec{B} = (5.0 \text{ m}, 60° \text{ counterclockwise from vertical})$. Find the x- and y-components of $\vec{B}$ in each of the two coordinate systems shown in **FIGURE EX3.17**.

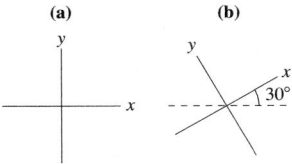

(a)   (b)

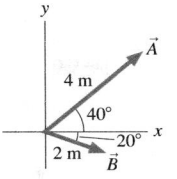

$\vec{v} = (100 \text{ m/s, south})$

FIGURE EX3.17   **FIGURE EX3.18**

18. | What are the x- and y-components of the velocity vector shown in **FIGURE EX3.18**?

Problems

19. ‖ Let $\vec{A} = (3.0 \text{ m}, 20° \text{ south of east})$, $\vec{B} = (2.0 \text{ m, north})$, and $\vec{C} = (5.0 \text{ m}, 70° \text{ south of west})$.
   a. Draw and label $\vec{A}$, $\vec{B}$, and $\vec{C}$ with their tails at the origin. Use a coordinate system with the x-axis to the east.
   b. Write $\vec{A}$, $\vec{B}$, and $\vec{C}$ in component form, using unit vectors.
   c. Find the magnitude and the direction of $\vec{D} = \vec{A} + \vec{B} + \vec{C}$.
20. | Let $\vec{E} = 2\hat{i} + 3\hat{j}$ and $\vec{F} = 2\hat{i} - 2\hat{j}$. Find the magnitude of
   a. $\vec{E}$ and $\vec{F}$ b. $\vec{E} + \vec{F}$ c. $-\vec{E} - 2\vec{F}$
21. | The position of a particle as a function of time is given by $\vec{r} = (5.0\hat{i} + 4.0\hat{j})t^2$ m, where t is in seconds.
   a. What is the particle's distance from the origin at $t = 0, 2$, and 5 s?
   b. Find an expression for the particle's velocity $\vec{v}$ as a function of time.
   c. What is the particle's speed at $t = 0, 2$, and 5 s?
22. ‖ **FIGURE P3.22** shows vectors $\vec{A}$ and $\vec{B}$. Let $\vec{C} = \vec{A} + \vec{B}$.

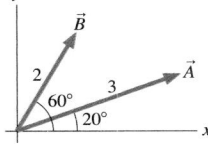

FIGURE P3.22

   a. Reproduce the figure on your page as accurately as possible, using a ruler and protractor. Draw vector $\vec{C}$ on your figure, using the graphical addition of $\vec{A}$ and $\vec{B}$. Then determine the magnitude and direction of $\vec{C}$ by *measuring* it with a ruler and protractor.
   b. Based on your figure of part a, use geometry and trigonometry to *calculate* the magnitude and direction of $\vec{C}$.
   c. Decompose vectors $\vec{A}$ and $\vec{B}$ into components, then use these to calculate algebraically the magnitude and direction of $\vec{C}$.
23. ‖ For the three vectors shown in **FIGURE P3.23**, $\vec{A} + \vec{B} + \vec{C} = 1\hat{j}$. What is vector $\vec{B}$?
   a. Write $\vec{B}$ in component form.
   b. Write $\vec{B}$ as a magnitude and a direction.

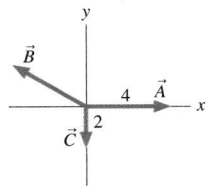

FIGURE P3.23

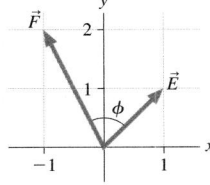

FIGURE P3.24

24. | a. What is the angle ϕ between vectors $\vec{E}$ and $\vec{F}$ in **FIGURE P3.24**?
   b. Use geometry and trigonometry to determine the magnitude and direction of $\vec{G} = \vec{E} + \vec{F}$.
   c. Use components to determine the magnitude and direction of $\vec{G} = \vec{E} + \vec{F}$.

25. ‖ **FIGURE P3.25** shows vectors $\vec{A}$ and $\vec{B}$. Find vector $\vec{C}$ such that $\vec{A} + \vec{B} + \vec{C} = \vec{0}$. Write your answer in component form.

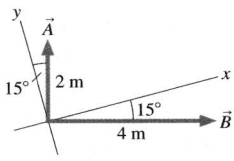

FIGURE P3.25   **FIGURE P3.26**

26. ‖‖ **FIGURE P3.26** shows vectors $\vec{A}$ and $\vec{B}$. Find $\vec{D} = 2\vec{A} + \vec{B}$. Write your answer in component form.
27. ‖ Find a vector that points in the same direction as the vector $(\hat{i} + \hat{j})$ and whose magnitude is 1.
28. ‖ Carlos runs with velocity $\vec{v} = (5.0 \text{ m/s}, 25° \text{ north of east})$ for 10 minutes. How far to the north of his starting position does Carlos end up?
29. ‖ While vacationing in the mountains you do some hiking. In the morning, your displacement is $\vec{S}_{\text{morning}} = (2000 \text{ m, east}) + (3000 \text{ m, north}) + (200 \text{ m, vertical})$. After lunch, your displacement is $\vec{S}_{\text{afternoon}} = (1500 \text{ m, west}) + (2000 \text{ m, north}) - (300 \text{ m, vertical})$.
   a. At the end of the hike, how much higher or lower are you compared to your starting point?
   b. What is the magnitude of your net displacement for the day?
30. ‖ The minute hand on a watch is 2.0 cm in length. What is the displacement vector of the tip of the minute hand
   a. From 8:00 to 8:20 A.M.?
   b. From 8:00 to 9:00 A.M.?
31. ‖ Bob walks 200 m south, then jogs 400 m southwest, then walks 200 m in a direction 30° east of north.
   a. Draw an accurate graphical representation of Bob's motion. Use a ruler and a protractor!
   b. Use either trigonometry or components to find the displacement that will return Bob to his starting point by the most direct route. Give your answer as a distance and a direction.
   c. Does your answer to part b agree with what you can measure on your diagram of part a?
32. ‖ Jim's dog Sparky runs 50 m northeast to a tree, then 70 m west to a second tree, and finally 20 m south to a third tree.
   a. Draw a picture and establish a coordinate system.
   b. Calculate Sparky's net displacement in component form.
   c. Calculate Sparky's net displacement as a magnitude and an angle.
33. ‖ A field mouse trying to escape a hawk runs east for 5.0 m, darts southeast for 3.0 m, then drops 1.0 m straight down a hole into its burrow. What is the magnitude of its net displacement?
34. | A cannon tilted upward at 30° fires a cannonball with a speed of 100 m/s. What is the component of the cannonball's velocity parallel to the ground?
35. | Jack and Jill ran up the hill at 3.0 m/s. The horizontal component of Jill's velocity vector was 2.5 m/s.
   a. What was the angle of the hill?
   b. What was the vertical component of Jill's velocity?
36. | A pine cone falls straight down from a pine tree growing on a 20° slope. The pine cone hits the ground with a speed of 10 m/s. What is the component of the pine cone's impact velocity (a) parallel to the ground and (b) perpendicular to the ground?

37. ‖ Mary needs to row her boat across a 100-m-wide river that is flowing to the east at a speed of 1.0 m/s. Mary can row the boat with a speed of 2.0 m/s relative to the water.
    a. If Mary rows straight north, how far downstream will she land?
    b. Draw a picture showing Mary's displacement due to rowing, her displacement due to the river's motion, and her net displacement.

38. ‖ The treasure map in FIG-URE P3.38 gives the following directions to the buried treasure: "Start at the old oak tree, walk due north for 500 paces, then due east for 100 paces. Dig." But when you arrive, you find an angry dragon just north of the tree. To avoid the dragon, you set off along the yellow brick road at an angle 60° east of north. After walking 300 paces you see an opening through the woods. Which direction should you go, and how far, to reach the treasure?

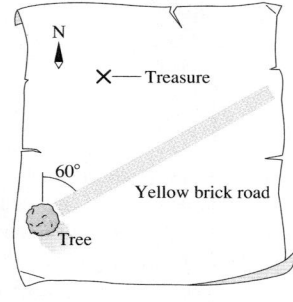

FIGURE P3.38

39. ‖ A jet plane is flying horizontally with a speed of 500 m/s over a hill that slopes upward with a 3% grade (i.e., the "rise" is 3% of the "run"). What is the component of the plane's velocity perpendicular to the ground?

40. ‖
BIO The bacterium *E. coli* is a single-cell organism that lives in the gut of healthy animals, including humans. When grown in a uniform medium in the laboratory, these bacteria swim along zigzag paths at a constant speed of 20 μm/s.

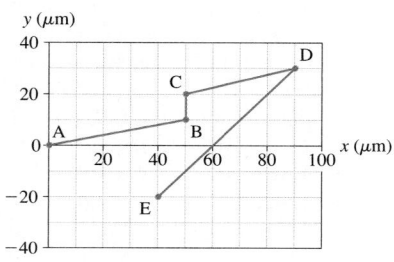

FIGURE P3.40

FIGURE P3.40 shows the trajectory of an *E. coli* as it moves from point A to point E. What are the magnitude and direction of the bacterium's average velocity for the entire trip?

41. ‖ A flock of ducks is trying to migrate south for the winter, but they keep being blown off course by a wind blowing from the west at 6.0 m/s. A wise elder duck finally realizes that the solution is to fly at an angle to the wind. If the ducks can fly at 8.0 m/s relative to the air, what direction should they head in order to move directly south?

42. ‖ FIGURE P3.42 shows three ropes tied together in a knot. One of your friends pulls on a rope with 3.0 units of force and another pulls on a second rope with 5.0 units of force. How hard and in what direction must you pull on the third rope to keep the knot from moving?

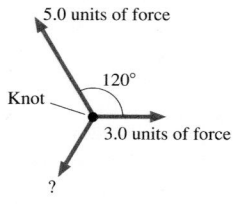

FIGURE P3.42

43. ‖ Three forces are exerted on an object placed on a tilted floor in FIG-URE P3.43. The forces are measured in newtons (N). Assuming that forces are vectors,
    a. What is the component of the *net* force $\vec{F}_{net} = \vec{F}_1 + \vec{F}_2 + \vec{F}_3$ parallel to the floor?
    b. What is the component of $\vec{F}_{net}$ perpendicular to the floor?
    c. What are the magnitude and direction of $\vec{F}_{net}$?

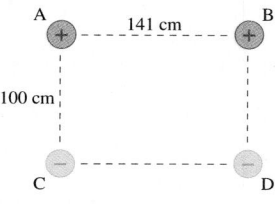

FIGURE P3.43

44. ‖ FIGURE P3.44 shows four electric charges located at the corners of a rectangle. Like charges, you will recall, repel each other while opposite charges attract. Charge B exerts a repulsive force (directly *away from* B) on charge A of 3.0 N. Charge C exerts an attractive force (directly *toward* C) on charge A of 6.0 N. Finally, charge D exerts an attractive force of 2.0 N on charge A. Assuming that forces are vectors, what are the magnitude and direction of the net force $\vec{F}_{net}$ exerted on charge A?

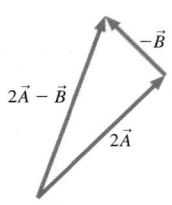

FIGURE P3.44

STOP TO THINK ANSWERS

Stop to Think 3.1: C. The graphical construction of $\vec{A}_1 + \vec{A}_2 + \vec{A}_3$ is shown at right.

Stop to Think 3.2: a. The graphical construction of $2\vec{A} - \vec{B}$ is shown at right.

Stop to Think 3.3: $C_x = -4$ cm, $C_y = 2$ cm.

Stop to Think 3.4: c. Vector $\vec{C}$ points to the left and down, so both C_x and C_y are negative. C_x is in the numerator because it is the side opposite ϕ.

STOP TO THINK 3.1

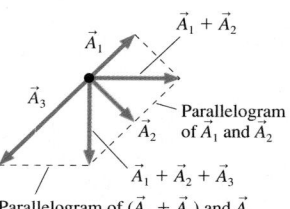

STOP TO THINK 3.2

4 Kinematics in Two Dimensions

The water droplets are following the parabolic trajectories of projectile motion.

▶ **Looking Ahead** The goal of Chapter 4 is to learn how to solve problems about motion in a plane.

Two-Dimensional Motion

An object moving in two dimensions follows a **trajectory**. The object's acceleration has a component associated with changing speed *and* a component associated with changing direction. The latter is perpendicular to the direction of motion.

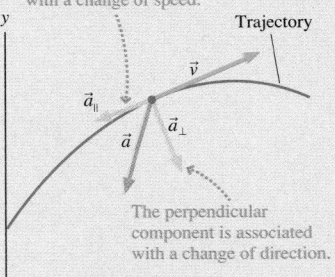

The parallel component is associated with a change of speed.

The perpendicular component is associated with a change of direction.

You will learn to extend the motion diagrams of Chapter 1 and the kinematics of Chapter 2 to motion in two dimensions.

> ◀ **Looking Back**
> Section 1.5 Finding acceleration vectors on a motion diagram

> ◀ **Looking Back**
> Sections 2.5–2.6 Constant acceleration kinematics and free fall

Projectile Motion

Projectile motion is two-dimensional motion under the influence of only gravity.

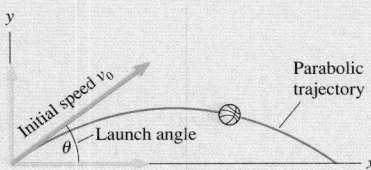

Projectile motion follows a *parabolic trajectory* characterized by the initial speed and the launch angle.

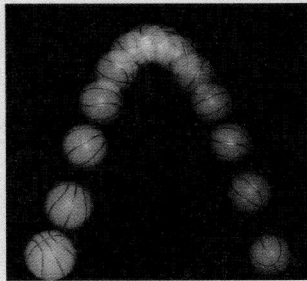

You'll learn to calculate how high and how far a projectile travels.

Circular Motion

Uniform circular motion is circular motion with constant speed. Because the direction is changing, there is a *centripetal acceleration* pointing toward the center of the circle.

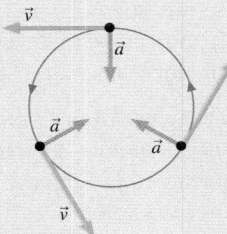

You'll learn that circular motion can be described by angular position θ, angular velocity ω, and angular acceleration α. These are analogous to the familiar linear position x, velocity v, and acceleration a.

The London Eye is a stately example of circular motion.

4.1 Acceleration

In Chapter 1 we defined the *average acceleration* $\vec{a}_{avg}$ of a moving object to be the vector

$$\vec{a}_{avg} = \frac{\Delta\vec{v}}{\Delta t} \tag{4.1}$$

From its definition, we see that **$\vec{a}$ points in the same direction as $\Delta\vec{v}$**, the change of velocity. As an object moves, its velocity vector can change in two possible ways:

1. The magnitude of $\vec{v}$ can change, indicating a change in speed, or
2. The direction of $\vec{v}$ can change, indicating that the object has changed direction.

The kinematics of Chapter 2 considered only the acceleration of changing speed. Now it's time to look at the acceleration associated with changing direction. Tactics Box 4.1 shows how we can use the velocity vectors on a motion diagram to determine the direction of the average acceleration vector. This is an extension of Tactics Box 1.3, which showed how to find $\vec{a}$ for one-dimensional motion.

FIGURE 4.1 Using Tactics Box 4.1 to find Maria's acceleration on the Ferris wheel.

(a)

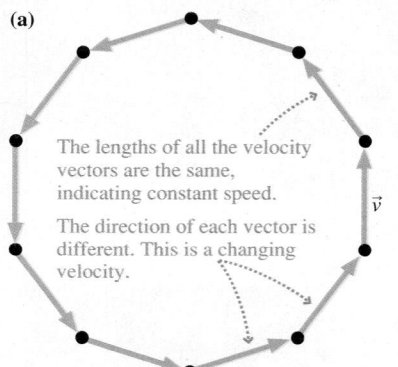

The lengths of all the velocity vectors are the same, indicating constant speed.

The direction of each vector is different. This is a changing velocity.

Maria moves at constant speed but *not* at constant velocity. Thus she is accelerating.

(b)

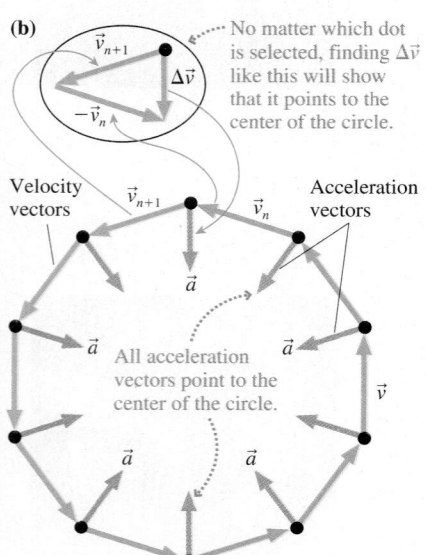

No matter which dot is selected, finding $\Delta\vec{v}$ like this will show that it points to the center of the circle.

Velocity vectors

Acceleration vectors

All acceleration vectors point to the center of the circle.

Maria's acceleration is an acceleration of changing direction, not of changing speed.

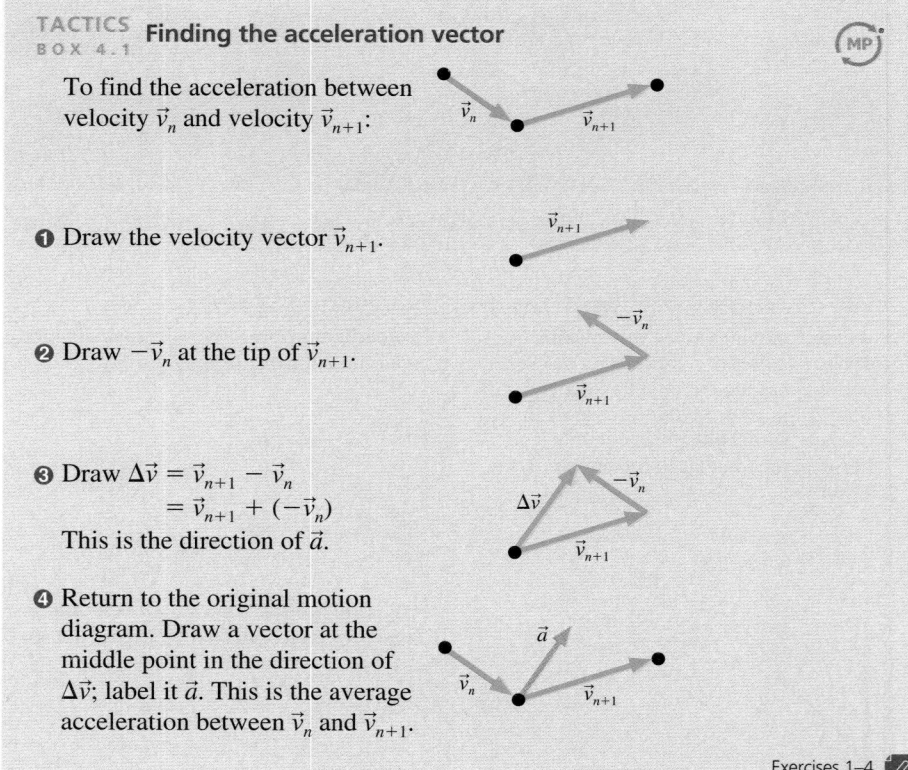

TACTICS BOX 4.1 Finding the acceleration vector

To find the acceleration between velocity $\vec{v}_n$ and velocity $\vec{v}_{n+1}$:

❶ Draw the velocity vector $\vec{v}_{n+1}$.

❷ Draw $-\vec{v}_n$ at the tip of $\vec{v}_{n+1}$.

❸ Draw $\Delta\vec{v} = \vec{v}_{n+1} - \vec{v}_n$
$= \vec{v}_{n+1} + (-\vec{v}_n)$
This is the direction of $\vec{a}$.

❹ Return to the original motion diagram. Draw a vector at the middle point in the direction of $\Delta\vec{v}$; label it $\vec{a}$. This is the average acceleration between $\vec{v}_n$ and $\vec{v}_{n+1}$.

Exercises 1–4

To illustrate, **FIGURE 4.1a** shows a motion diagram of Maria riding a Ferris wheel at the amusement park. Maria has constant speed but *not* constant velocity, so she is accelerating. **FIGURE 4.1b** applies the rules of Tactics Box 4.1 to find that—at every point—Maria's acceleration points toward the center of the circle. This is an acceleration due to changing direction, not to changing speed.

NOTE ▶ Our everyday use of the word "accelerate" means "speed up." The technical definition of acceleration—the rate of change of velocity—also includes slowing down, as you learned in Chapter 2, as well as changing direction. All these are motions that change the velocity. ◀

EXAMPLE 4.1 **Through the valley**

A ball rolls down a long hill, through the valley, and back up the other side. Draw a complete motion diagram of the ball, showing velocity and acceleration vectors.

MODEL Model the ball as a particle.

VISUALIZE **FIGURE 4.2** is the motion diagram. Where the particle moves along a *straight line*, it speeds up if $\vec{a}$ and $\vec{v}$ point in the same direction and slows down if $\vec{a}$ and $\vec{v}$ point in opposite directions.

This idea was the basis for the one-dimensional kinematics we developed in Chapter 2. For linear motion, acceleration is a change of speed. When the direction of $\vec{v}$ changes, as it does when the ball goes through the valley, we need to use vector subtraction to find the direction of $\Delta\vec{v}$ and thus of $\vec{a}$. The procedure is shown at two points in the motion diagram. Notice that the point at the bottom of the valley is much like the bottom point of Maria's motion diagram in Figure 4.1b.

FIGURE 4.2 The motion diagram of the ball of Example 4.1.

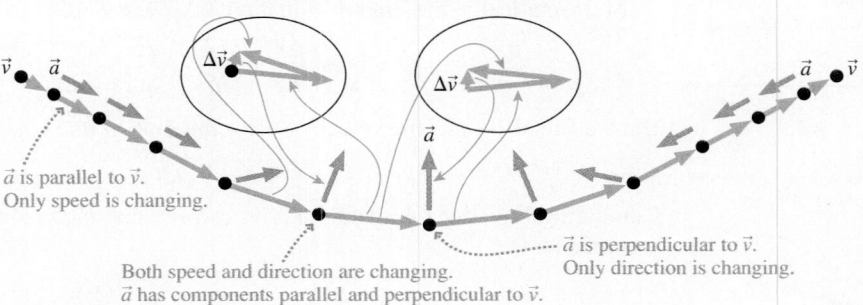

$\vec{a}$ is parallel to $\vec{v}$.
Only speed is changing.

Both speed and direction are changing.
$\vec{a}$ has components parallel and perpendicular to $\vec{v}$.

$\vec{a}$ is perpendicular to $\vec{v}$.
Only direction is changing.

FIGURE 4.3 shows that an object's acceleration vector can be decomposed into a component parallel to the velocity—that is, parallel to the direction of motion—and a component perpendicular to the velocity. $\vec{a}_\parallel$ **is the piece of the acceleration that causes the object to change speed,** speeding up if $\vec{a}_\parallel$ points in the same direction as $\vec{v}$, slowing down if they point in opposite directions. $\vec{a}_\perp$ **is the piece of the acceleration that causes the object to change direction.** An object changing direction *always* has a component of acceleration perpendicular to the direction of motion.

Looking back at Example 4.1, we see that $\vec{a}$ is parallel to $\vec{v}$ on the straight portions of the hill where only speed is changing. At the very bottom, where the ball's direction is changing but not its speed, $\vec{a}$ is perpendicular to $\vec{v}$. The acceleration is angled with respect to velocity—having both parallel and perpendicular components—at those points where both speed and direction are changing.

FIGURE 4.3 Analyzing the acceleration vector.

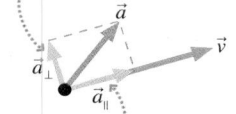

This component of $\vec{a}$ is changing the direction of motion.

This component of $\vec{a}$ is changing the speed of the motion.

STOP TO THINK 4.1 This acceleration will cause the particle to

a. Speed up and curve upward.
c. Slow down and curve upward.
e. Move to the right and down.

b. Speed up and curve downward.
d. Slow down and curve downward.
f. Reverse direction.

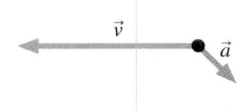

4.2 Two-Dimensional Kinematics

Motion diagrams are an important tool for visualizing motion, but we also need to develop a mathematical description of motion in two dimensions. We're going to begin with motion in which the horizontal and vertical components of acceleration are independent of each other. For convenience, we'll say that the motion is in the *xy*-plane regardless of whether the plane of motion is horizontal or vertical.

FIGURE 4.4 shows a particle moving along a curved path—its *trajectory*—in the *xy*-plane. We can locate the particle in terms of its position vector $\vec{r} = x\,\hat{\imath} + y\,\hat{\jmath}$.

NOTE ▶ In Chapter 2 we made extensive use of position-versus-time graphs, either *x* versus *t* or *y* versus *t*. Figure 4.4, like many of the graphs we'll use in this chapter, is a graph of *y* versus *x*. In other words, it's an actual *picture* of the trajectory, not an abstract representation of the motion. ◀

FIGURE 4.4 A particle moving along a trajectory in the *xy*-plane.

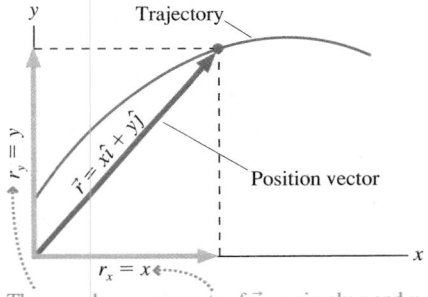

The *x*- and *y*-components of $\vec{r}$ are simply *x* and *y*.

FIGURE 4.5 The instantaneous velocity vector is tangent to the trajectory.

(a)

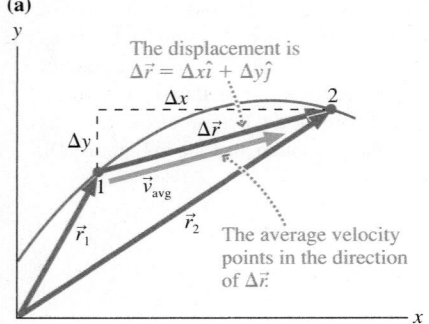

(b)

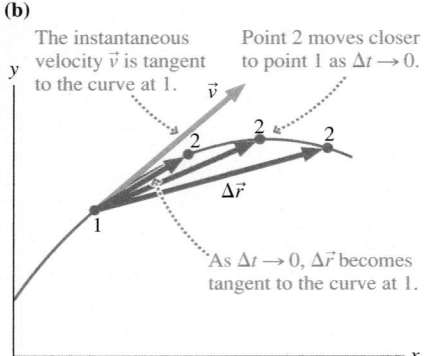

FIGURE 4.6 Relating the components of $\vec{v}$ to the speed and direction.

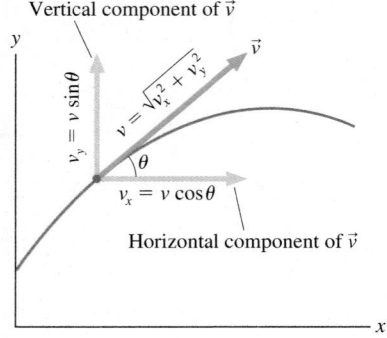

FIGURE 4.5a shows the particle moving from position $\vec{r}_1$ at time t_1 to position $\vec{r}_2$ at a later time t_2. The average velocity—pointing in the direction of the displacement $\Delta \vec{r}$—is

$$\vec{v}_{avg} = \frac{\Delta \vec{r}}{\Delta t} = \frac{\Delta x}{\Delta t}\hat{i} + \frac{\Delta y}{\Delta t}\hat{j} \tag{4.2}$$

You learned in Chapter 2 that the instantaneous velocity is the limit of $\vec{v}_{avg}$ as $\Delta t \rightarrow 0$. As Δt decreases, point 2 moves closer to point 1 until, as **FIGURE 4.5b** shows, the displacement vector becomes tangent to the curve. Consequently, **the instantaneous velocity vector $\vec{v}$ is tangent to the trajectory.**

Mathematically, the limit of Equation 4.2 gives

$$\vec{v} = \lim_{\Delta t \rightarrow 0} \frac{\Delta \vec{r}}{\Delta t} = \frac{d\vec{r}}{dt} = \frac{dx}{dt}\hat{i} + \frac{dy}{dt}\hat{j} \tag{4.3}$$

But we can also write the velocity vector in terms of its x- and y-components as

$$\vec{v} = v_x\hat{i} + v_y\hat{j} \tag{4.4}$$

Comparing Equations 4.3 and 4.4, you can see that the velocity vector $\vec{v}$ has x- and y-components

$$v_x = \frac{dx}{dt} \quad \text{and} \quad v_y = \frac{dy}{dt} \tag{4.5}$$

That is, the x-component v_x of the velocity vector is the rate dx/dt at which the particle's x-coordinate is changing. The y-component is similar.

FIGURE 4.6 illustrates another important feature of the velocity vector. If the vector's angle θ is measured from the positive x-direction, the velocity vector components are

$$v_x = v\cos\theta$$
$$v_y = v\sin\theta \tag{4.6}$$

where

$$v = \sqrt{v_x^2 + v_y^2} \tag{4.7}$$

is the particle's *speed* at that point. Speed is always a positive number (or zero), whereas the components are *signed* quantities (i.e., they can be positive or negative) to convey information about the direction of the velocity vector. Conversely, we can use the two velocity components to determine the direction of motion:

$$\tan\theta = \frac{v_y}{v_x} \tag{4.8}$$

NOTE ▶ In Chapter 2, you learned that the *value* of the velocity component v_s at time t is given by the *slope* of the position-versus-time graph at time t. Now we see that the *direction* of the velocity vector $\vec{v}$ is given by the *tangent* to the y-versus-x graph of the trajectory. **FIGURE 4.7** reminds you that these two graphs use different interpretations of the tangent lines. The tangent to the trajectory does not tell us anything about how fast the particle is moving, only its direction. ◀

FIGURE 4.7 Two different uses of tangent lines.

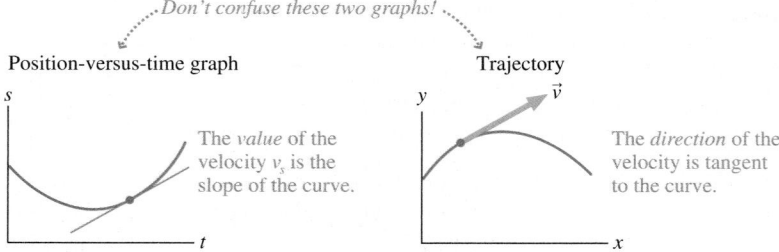

EXAMPLE 4.2 **Describing the motion with graphs**

A particle's motion is described by the two equations

$$x = 2t^2 \text{ m}$$
$$y = (5t + 5) \text{ m}$$

where the time t is in s.

  a. Draw a graph of the particle's trajectory.
  b. Draw a graph of the particle's speed as a function of time.

MODEL These are *parametric equations* that give the particle's co-ordinates x and y separately in terms of the parameter t.

SOLVE a. The trajectory is a curve in the xy-plane. The easiest way to proceed is to calculate x and y at several instants of time.

t (s)	x (m)	y (m)	v (m/s)
0	0	5	5.0
1	2	10	6.4
2	8	15	9.4
3	18	20	13.0
4	32	25	16.8

These points are plotted in **FIGURE 4.8a**; then a smooth curve is drawn through them to show the trajectory.

  b. The particle's speed is given by Equation 4.7. We first need to use Equation 4.5 to find the components of the velocity vector:

$$v_x = \frac{dx}{dt} = 4t \text{ m/s} \quad \text{and} \quad v_y = \frac{dy}{dt} = 5 \text{ m/s}$$

Using these gives the particle's speed at time t:

$$v = \sqrt{v_x^2 + v_y^2} = \sqrt{16t^2 + 25} \text{ m/s}$$

FIGURE 4.8 Two motion graphs for the particle of Example 4.2.

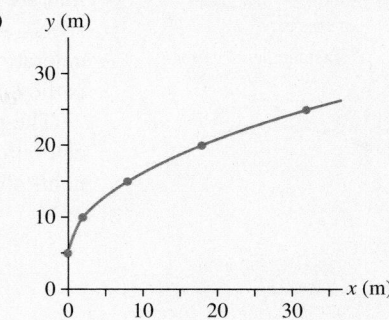

(a)

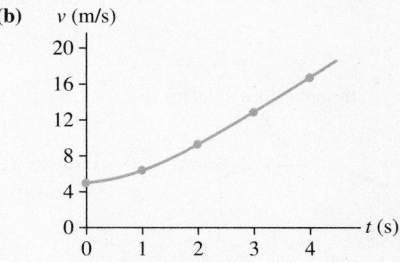

(b)

The speed was computed in the table and is graphed in **FIGURE 4.8b**.

ASSESS The y-versus-x graph of Figure 4.8a is a trajectory, not a position-versus-time graph. Thus the slope is *not* the particle's speed. The particle is speeding up, as you can see in the second graph, even though the slope of the trajectory is decreasing.

Acceleration

Let's return to the particle moving along a trajectory in the xy-plane. **FIGURE 4.9a** shows the instantaneous velocity $\vec{v}_1$ at point 1 and, a short time later, velocity $\vec{v}_2$ at point 2. These two vectors are tangent to the trajectory. We can use the vector-subtraction technique, shown in the inset, to find $\vec{a}_{\text{avg}}$ on this segment of the trajectory.

If we now take the limit $\Delta t \rightarrow 0$, the *instantaneous acceleration* is

$$\vec{a} = \lim_{\Delta t \to 0} \frac{\Delta \vec{v}}{\Delta t} = \frac{d\vec{v}}{dt} \qquad (4.9)$$

As $\Delta t \rightarrow 0$, points 1 and 2 in Figure 4.9a merge, and the instantaneous acceleration $\vec{a}$ is found at the same point on the trajectory (and the same instant of time) as the instantaneous velocity $\vec{v}$. This is shown in **FIGURE 4.9b**.

FIGURE 4.9 The average and instantaneous acceleration vectors on a curved trajectory.

(a) Average acceleration

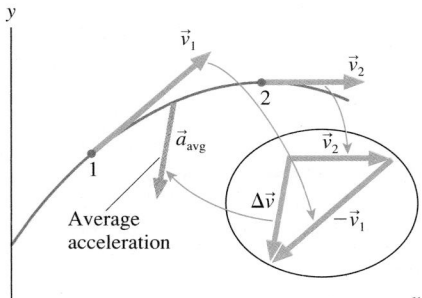

(b) Instantaneous acceleration

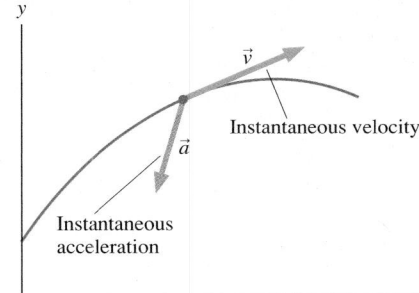

FIGURE 4.10 Decomposition of the instantaneous acceleration $\vec{a}$.

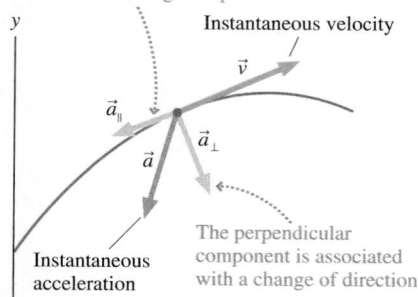

(a) The parallel component is associated with a change of speed.

Instantaneous velocity

The perpendicular component is associated with a change of direction.

Instantaneous acceleration

(b)

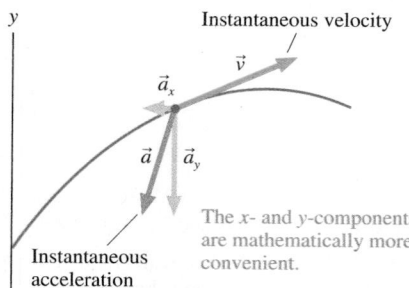

Instantaneous velocity

The x- and y-components are mathematically more convenient.

Instantaneous acceleration

By definition, the acceleration vector $\vec{a}$ is the rate at which the velocity $\vec{v}$ is changing at that instant. To show this, FIGURE 4.10a decomposes $\vec{a}$ into components $\vec{a}_{\parallel}$ and $\vec{a}_{\perp}$ that are parallel and perpendicular to the trajectory. $\vec{a}_{\parallel}$ is associated with a change of speed, and $\vec{a}_{\perp}$ is associated with a change of direction. Both kinds of changes are accelerations. Notice that $\vec{a}_{\perp}$ always points toward the "inside" of the curve because that is the direction in which $\vec{v}$ is changing.

The parallel and perpendicular components of $\vec{a}$ convey important ideas about acceleration, but it's usually more practical to write $\vec{a}$ in terms of the x- and y-components shown in FIGURE 4.10b. Because $\vec{v} = v_x \hat{i} + v_y \hat{j}$, we find

$$\vec{a} = a_x \hat{i} + a_y \hat{j} = \frac{d\vec{v}}{dt} = \frac{dv_x}{dt} \hat{i} + \frac{dv_y}{dt} \hat{j} \tag{4.10}$$

from which we see that

$$a_x = \frac{dv_x}{dt} \quad \text{and} \quad a_y = \frac{dv_y}{dt} \tag{4.11}$$

That is, the x-component of $\vec{a}$ is the rate dv_x/dt at which the x-component of velocity is changing.

Constant Acceleration

If the acceleration $\vec{a} = a_x \hat{i} + a_y \hat{j}$ is constant, then the two components a_x and a_y are both constant. In this case, everything you learned about constant-acceleration kinematics in Chapter 2 carries over to two-dimensional motion.

Consider a particle that moves with constant acceleration from an initial position $\vec{r}_i = x_i \hat{i} + y_i \hat{j}$, starting with initial velocity $\vec{v}_i = v_{ix} \hat{i} + v_{iy} \hat{j}$. Its position and velocity at a final point f are

$$x_f = x_i + v_{ix}\,\Delta t + \tfrac{1}{2}a_x(\Delta t)^2 \qquad y_f = y_i + v_{iy}\,\Delta t + \tfrac{1}{2}a_y(\Delta t)^2$$

$$v_{fx} = v_{ix} + a_x\,\Delta t \qquad\qquad v_{fy} = v_{iy} + a_y\,\Delta t \tag{4.12}$$

There are *many* quantities to keep track of in two-dimensional kinematics, making the pictorial representation all the more important as a problem-solving tool.

NOTE ▶ For constant acceleration, the x-component of the motion and the y-component of the motion are independent of each other. However, they remain connected through the fact that Δt must be the same for both. ◀

EXAMPLE 4.3 **Plotting the trajectory of the shuttlecraft**

The up thrusters on the shuttlecraft of the starship *Enterprise* give it an upward acceleration of 5.0 m/s². Its forward thrusters provide a forward acceleration of 20 m/s². As it leaves the *Enterprise*, the shuttlecraft turns on only the up thrusters. After clearing the flight deck, 3.0 s later, it adds the forward thrusters. Plot a trajectory of the shuttlecraft for its first 6 s.

MODEL Represent the shuttlecraft as a particle. There are two segments of constant-acceleration motion.

VISUALIZE FIGURE 4.11 shows a pictorial representation. The coordinate system has been chosen so that the shuttlecraft starts at the origin and initially moves along the y-axis. The craft moves vertically for 3.0 s, then begins to acquire a forward motion. There are three points in the motion: the beginning, the end, and the point at the which forward thrusters are turned on. These points are labeled (x_0, y_0), (x_1, y_1), and (x_2, y_2). The velocities are (v_{0x}, v_{0y}), (v_{1x}, v_{1y}), and (v_{2x}, v_{2y}). This will be our standard labeling scheme for trajectories, where it is essential to keep the x-components and y-components separate.

FIGURE 4.11 Pictorial representation of the motion of the shuttlecraft.

FIGURE 4.12 The shuttlecraft trajectory.

Pictorial representation

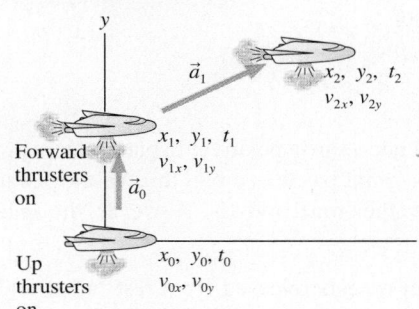

Known		
$x_0 = y_0 = 0$ m	$v_{0x} = v_{0y} = 0$ m/s	$t_0 = 0$
$a_{0x} = 0$ m/s^2	$a_{0y} = 5.0$ m/s^2	$t_1 = 3.0$ s
$a_{1x} = 20$ m/s^2	$a_{1y} = 5.0$ m/s^2	$t_2 = 6.0$ s

Find

x and y at time t

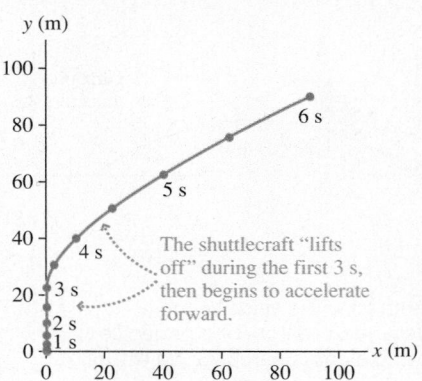

The shuttlecraft "lifts off" during the first 3 s, then begins to accelerate forward.

SOLVE During the first phase of the acceleration, when $a_{0x} = 0$ m/s^2 and $a_{0y} = 5.0$ m/s^2, the motion is described by

$$y = y_0 + v_{0y}(t - t_0) + \tfrac{1}{2}a_{0y}(t - t_0)^2 = 2.5t^2 \text{ m}$$

$$v_y = v_{0y} + a_{0y}(t - t_0) = 5.0t \text{ m/s}$$

where the time t is in s. These equations allow us to calculate the position and velocity at any time t. At $t_1 = 3.0$ s, when the first phase of the motion ends, we find that

$$x_1 = 0 \text{ m} \qquad v_{1x} = 0 \text{ m/s}$$

$$y_1 = 22.5 \text{ m} \qquad v_{1y} = 15 \text{ m/s}$$

During the next 3 s, when $a_{1x} = 20$ m/s^2 and $a_{1y} = 5.0$ m/s^2, the x- and y-coordinates are

$$x = x_1 + v_{1x}(t - t_1) + \tfrac{1}{2}a_{1x}(t - t_1)^2$$

$$= 10(t - 3.0)^2 \text{ m}$$

$$y = y_1 + v_{1y}(t - t_1) + \tfrac{1}{2}a_{1y}(t - t_1)^2$$

$$= \left(22.5 + 15(t - 3.0) + 2.5(t - 3.0)^2 \right) \text{ m}$$

where, again, t is in s. To show the trajectory, we've calculated x and y every 0.5 s, plotted the points in **FIGURE 4.12**, and drawn a smooth curve through the points.

STOP TO THINK 4.2 During which time interval or intervals is the particle described by these position graphs at rest? More than one may be correct.

a. 0–1 s
b. 1–2 s
c. 2–3 s
d. 3–4 s

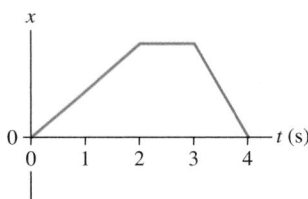

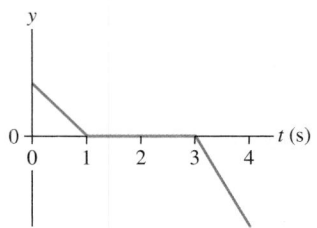

4.3 Projectile Motion

Baseballs and tennis balls flying through the air, Olympic divers, and daredevils shot from cannons all exhibit what we call *projectile motion*. A **projectile** is an object that moves in two dimensions under the influence of only gravity. Projectile motion is an extension of the free-fall motion we studied in Chapter 2. We will continue to neglect the influence of air resistance, leading to results that are a good approximation of reality for relatively heavy objects moving relatively slowly over relatively short distances. As we'll see, projectiles in two dimensions follow a *parabolic trajectory* like the one seen in **FIGURE 4.13**.

The start of a projectile's motion, be it thrown by hand or shot from a gun, is called the *launch,* and the angle θ of the initial velocity $\vec{v}_0$ above the horizontal (i.e., above

FIGURE 4.13 The parabolic trajectory of a bouncing ball.

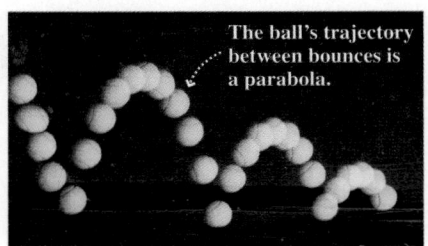

The ball's trajectory between bounces is a parabola.

FIGURE 4.14 A projectile launched with initial velocity $\vec{v}_0$.

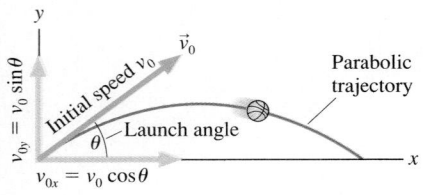

the x-axis) is called the **launch angle**. FIGURE 4.14 illustrates the relationship between the initial velocity vector $\vec{v}_0$ and the initial values of the components v_{0x} and v_{0y}. You can see that

$$v_{0x} = v_0 \cos \theta$$
$$v_{0y} = v_0 \sin \theta \qquad (4.13)$$

where v_0 is the initial speed.

NOTE ▶ The components v_{0x} and v_{0y} are not necessarily positive. In particular, a projectile launched at an angle *below* the horizontal (such as a ball thrown downward from the roof of a building) has *negative* values for θ and v_{0y}. However, the *speed* v_0 is always positive. ◀

Gravity acts downward, and we know that objects released from rest fall straight down, not sideways. Hence a projectile has no horizontal acceleration, while its vertical acceleration is simply that of free fall. Thus

$$a_x = 0$$
$$\qquad \text{(projectile motion)} \qquad (4.14)$$
$$a_y = -g$$

FIGURE 4.15 The velocity and acceleration vectors of a projectile moving along a parabolic trajectory.

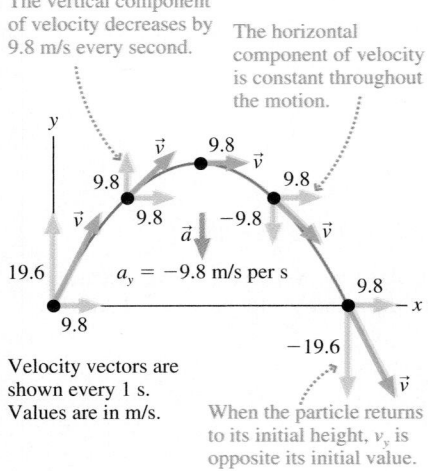

The vertical component of velocity decreases by 9.8 m/s every second.

The horizontal component of velocity is constant throughout the motion.

$a_y = -9.8$ m/s per s

Velocity vectors are shown every 1 s. Values are in m/s.

When the particle returns to its initial height, v_y is opposite its initial value.

In other words, **the vertical component of acceleration a_y is just the familiar $-g$ of free fall, while the horizontal component a_x is zero. Projectiles are in free fall.**

To see how these conditions influence the motion, FIGURE 4.15 shows a projectile launched from $(x_0, y_0) = (0 \text{ m}, 0 \text{ m})$ with an initial velocity $\vec{v}_0 = (9.8\hat{\imath} + 19.6\hat{\jmath})$ m/s. The value of v_x never changes because there's no horizontal acceleration, but v_y decreases by 9.8 m/s every second. This is what it *means* to accelerate at $a_y = -9.8 \text{ m/s}^2 = (-9.8 \text{ m/s})$ per second.

You can see from Figure 4.15 that **projectile motion is made up of two independent motions:** uniform motion at constant velocity in the horizontal direction and free-fall motion in the vertical direction. The kinematic equations that describe these two motions are simply Equations 4.12 with $a_x = 0$ and $a_y = -g$.

EXAMPLE 4.4 **Don't try this at home!**

A stunt man drives a car off a 10.0-m-high cliff at a speed of 20.0 m/s. How far does the car land from the base of the cliff?

MODEL Represent the car as a particle in free fall. Assume that the car is moving horizontally as it leaves the cliff.

VISUALIZE The pictorial representation, shown in FIGURE 4.16, is *very* important because the number of quantities to keep track of is quite large. We have chosen to put the origin at the base of the cliff. The assumption that the car is moving horizontally as it leaves the cliff leads to $v_{0x} = v_0$ and $v_{0y} = 0$ m/s.

FIGURE 4.16 Pictorial representation for the car of Example 4.4.

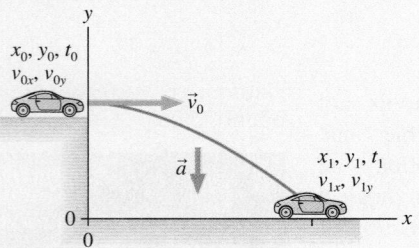

Known	Find
$x_0 = 0$ m $\quad v_{0y} = 0$ m/s $\quad t_0 = 0$ s	x_1
$y_0 = 10.0$ m $\quad v_{0x} = v_0 = 20.0$ m/s	
$a_x = 0$ m/s^2 $\quad a_y = -g$ $\quad y_1 = 0$ m	

SOLVE Each point on the trajectory has x- and y-components of position, velocity, and acceleration but only *one* value of time. The time needed to move horizontally to x_1 is the *same* time needed to fall vertically through distance y_0. **Although the horizontal and vertical motions are independent, they are connected through the time t.** This is a critical observation for solving projectile motion problems. The kinematics equations with $a_x = 0$ and $a_y = -g$ are

$$x_1 = x_0 + v_{0x}(t_1 - t_0) = v_0 t_1$$
$$y_1 = 0 = y_0 + v_{0y}(t_1 - t_0) - \tfrac{1}{2}g(t_1 - t_0)^2 = y_0 - \tfrac{1}{2}g t_1^2$$

We can use the vertical equation to determine the time t_1 needed to fall distance y_0:

$$t_1 = \sqrt{\frac{2y_0}{g}} = \sqrt{\frac{2(10.0 \text{ m})}{9.80 \text{ m/s}^2}} = 1.43 \text{ s}$$

We then insert this expression for t into the horizontal equation to find the distance traveled:

$$x_1 = v_0 t_1 = (20.0 \text{ m/s})(1.43 \text{ s}) = 28.6 \text{ m}$$

ASSESS The cliff height is ≈ 33 ft and the initial speed is $v_0 \approx 40$ mph. Traveling $x_1 = 29$ m ≈ 95 ft before hitting the ground seems reasonable.

The x- and y-equations of Example 4.4 are parametric equations. It's not hard to eliminate t and write an expression for y as a function of x. From the x_1 equation, $t_1 = x_1/v_0$. Substituting this into the y_1 equation, we find

$$y = y_0 - \frac{g}{2v_0^2}x^2 \qquad (4.15)$$

The graph of $y = ax^2$ is a parabola, so Equation 4.15 represents an inverted parabola that starts from height y_0. This proves, as we asserted above, that a projectile follows a parabolic trajectory.

Reasoning About Projectile Motion

Think about the following question:

> A heavy ball is launched exactly horizontally at height h above a horizontal field. At the exact instant that the ball is launched, a second ball is simply dropped from height h. Which ball hits the ground first?

It may seem hard to believe, but—if air resistance is neglected—the balls hit the ground *simultaneously*. They do so because the horizontal and vertical components of projectile motion are independent of each other. The initial horizontal velocity of the first ball has *no* influence over its vertical motion. Neither ball has any initial motion in the vertical direction, so both fall distance h in the same amount of time. You can see this in FIGURE 4.17.

FIGURE 4.18a shows a useful way to think about the trajectory of a projectile. Without gravity, a projectile would follow a straight line. Because of gravity, the particle at time t has "fallen" a distance $\frac{1}{2}gt^2$ below this line. The separation grows as $\frac{1}{2}gt^2$, giving the trajectory its parabolic shape.

Use this idea to think about the following "classic" problem in physics:

> A hungry bow-and-arrow hunter in the jungle wants to shoot down a coconut that is hanging from the branch of a tree. He points his arrow directly at the coconut, but as luck would have it, the coconut falls from the branch at the *exact* instant the hunter releases the string. Does the arrow hit the coconut?

You might think that the arrow will miss the falling coconut, but it doesn't. Although the arrow travels very fast, it follows a slightly curved parabolic trajectory, not a straight line. Had the coconut stayed on the tree, the arrow would have curved under its target as gravity causes it to fall a distance $\frac{1}{2}gt^2$ below the straight line. But $\frac{1}{2}gt^2$ is also the distance the coconut falls while the arrow is in flight. Thus, as FIGURE 4.18b shows, the arrow and the coconut fall the same distance and meet at the same point!

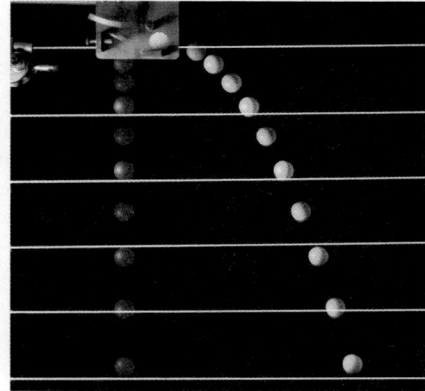

FIGURE 4.17 A projectile launched horizontally falls in the same time as a projectile that is released from rest.

FIGURE 4.18 A projectile follows a parabolic trajectory because it "falls" a distance $\frac{1}{2}gt^2$ below a straight-line trajectory.

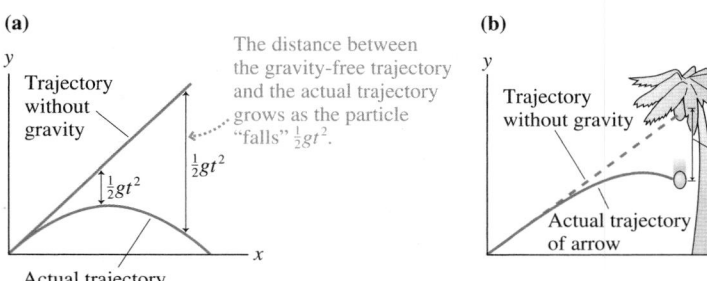

(a)

Trajectory without gravity

The distance between the gravity-free trajectory and the actual trajectory grows as the particle "falls" $\frac{1}{2}gt^2$.

$\frac{1}{2}gt^2$

$\frac{1}{2}gt^2$

Actual trajectory

(b)

Trajectory without gravity

$\frac{1}{2}gt^2$

Actual trajectory of arrow

Solving Projectile Motion Problems

PROBLEM-SOLVING
STRATEGY 4.1 **Projectile motion problems**

MODEL Make simplifying assumptions, such as treating the object as a particle. Is it reasonable to ignore air resistance?

VISUALIZE Use a pictorial representation. Establish a coordinate system with the x-axis horizontal and the y-axis vertical. Show important points in the motion on a sketch. Define symbols and identify what the problem is trying to find.

SOLVE The acceleration is known: $a_x = 0$ and $a_y = -g$. Thus the problem is one of two-dimensional kinematics. The kinematic equations are

$$x_f = x_i + v_{ix}\,\Delta t \qquad y_f = y_i + v_{iy}\,\Delta t - \tfrac{1}{2}g(\Delta t)^2$$

$$v_{fx} = v_{ix} = \text{constant} \qquad v_{fy} = v_{iy} - g\,\Delta t$$

Δt is the same for the horizontal and vertical components of the motion. Find Δt from one component, then use that value for the other component.

ASSESS Check that your result has the correct units, is reasonable, and answers the question.

EXAMPLE 4.5 **Jumping frog contest**

Frogs, with their long, strong legs, are excellent jumpers. And thanks to the good folks of Calaveras County, California, who have a jumping frog contest every year in honor of a Mark Twain story, we have very good data on how far a determined frog can jump.

High-speed cameras show that a good jumper goes into a crouch, then rapidly extends his legs by typically 15 cm during a 42 ms push off, leaving the ground at a 30° angle. How far does this frog leap?

MODEL Represent the frog as a particle. Model the push off as linear motion with constant acceleration. A bullfrog is fairly heavy and dense, so ignore air resistance and consider the leap to be projectile motion.

VISUALIZE This is a two-part problem: linear acceleration followed by projectile motion. A key observation is that **the final velocity for pushing off the ground becomes the initial velocity of the projectile motion.** FIGURE 4.19 shows a separate pictorial representation for each part. Notice that we've used different coordinate systems for the two parts; coordinate systems are our

choice, and for each part of the motion we've chosen the coordinate system that makes the problem easiest to solve.

SOLVE While pushing off, the frog travels 15 cm = 0.15 m in 42 ms = 0.042 s. We could find his speed at the end of pushing off if we knew the acceleration. Because the initial velocity is zero, we can find the acceleration from the position-acceleration-time kinematic equation:

$$x_1 = x_0 + v_{0x}\,\Delta t + \tfrac{1}{2}a_x(\Delta t)^2 = \tfrac{1}{2}a_x(\Delta t)^2$$

$$a_x = \frac{2x_1}{(\Delta t)^2} = \frac{2(0.15\text{ m})}{(0.042\text{ s})^2} = 170\text{ m/s}^2$$

This is a substantial acceleration, but it doesn't last long. At the end of the 42 ms push off, the frog's velocity is

$$v_{1x} = v_{0x} + a_x\Delta t = (170\text{ m/s}^2)(0.042\text{ s}) = 7.14\text{ m/s}$$

We'll keep an extra significant figure here to avoid round-off error in the second half of the problem.

FIGURE 4.19 Pictorial representations of the jumping frog.

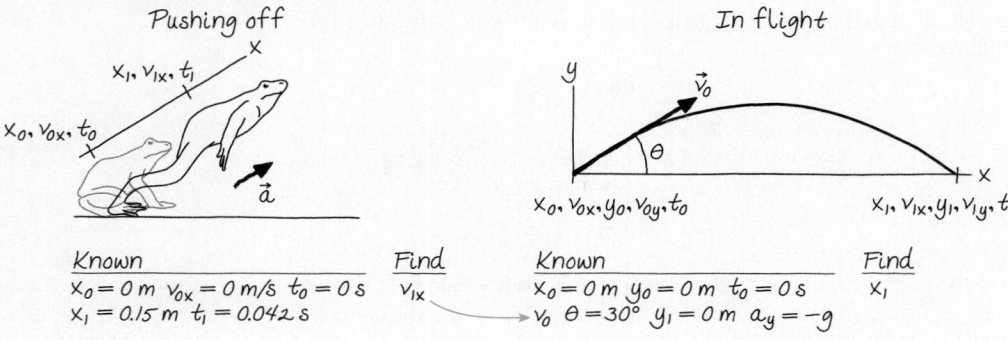

The end of the push off is the beginning of the projectile motion, so the second part of the problem is to find the distance of a projectile launched with velocity $\vec{v}_0 = (7.14 \text{ m/s}, 30°)$. The initial x- and y-components of the launch velocity are

$$v_{0x} = v_0 \cos\theta \qquad v_{0y} = v_0 \sin\theta$$

The kinematic equations of projectile motion, with $a_x = 0$ and $a_y = -g$, are

$$
\begin{aligned}
x_1 &= x_0 + v_{0x}\Delta t \\
&= (v_0 \cos\theta)\Delta t \\
y_1 &= y_0 + v_{0y}\Delta t - \tfrac{1}{2}g(\Delta t)^2 \\
&= (v_0 \sin\theta)\Delta t - \tfrac{1}{2}g(\Delta t)^2
\end{aligned}
$$

We can find the time of flight from the vertical equation by setting $y_1 = 0$:

$$0 = (v_0 \sin\theta)\Delta t - \tfrac{1}{2}g(\Delta t)^2 = (v_0 \sin\theta - \tfrac{1}{2}g\,\Delta t)\Delta t$$

and thus

$$\Delta t = 0 \qquad \text{or} \qquad \Delta t = \frac{2v_0 \sin\theta}{g}$$

Both are legitimate solutions. The first corresponds to the instant when $y = 0$ at the launch, the second to when $y = 0$ as the frog hits the ground. Clearly, we want the second solution. Substituting this expression for Δt into the equation for x_1 gives

$$x_1 = (v_0 \cos\theta)\frac{2v_0 \sin\theta}{g} = \frac{2v_0^2 \sin\theta \cos\theta}{g}$$

We can simplify this result with the trigonometric identity $2\sin\theta \cos\theta = \sin(2\theta)$. Thus the distance traveled by the frog is

$$x_1 = \frac{v_0^2 \sin(2\theta)}{g}$$

Using $v_0 = 7.14$ m/s and $\theta = 30°$, we find that the frog leaps a distance of 4.5 m.

ASSESS 4.5 m is about 15 feet. This is much farther than a human can jump from a standing start, but it seems believable. In fact, the current record holder, Rosie the Ribeter, made a leap of 6.5 m!

As Example 4.5 found, a projectile that lands at the same elevation from which it was launched travels distance

$$\text{distance} = \frac{v_0^2 \sin(2\theta)}{g} \qquad (4.16)$$

The maximum distance occurs for $\theta = 45°$, where $\sin(2\theta) = 1$. But there's more that we can learn from this equation. Because $\sin(180° - x) = \sin x$, it follows that $\sin(2(90° - \theta)) = \sin(2\theta)$. Consequently, a projectile launched either at angle θ or at angle $(90° - \theta)$ will travel the same distance *over level ground*. FIGURE 4.20 shows the trajectories of projectiles launched with the same initial speed in 15° increments of angle.

NOTE ▶ Equation 4.16 is *not* a general result. It applies *only* in situations where the projectile lands at the same elevation from which it was fired. ◀

FIGURE 4.20 Trajectories of a projectile launched at different angles with a speed of 99 m/s.

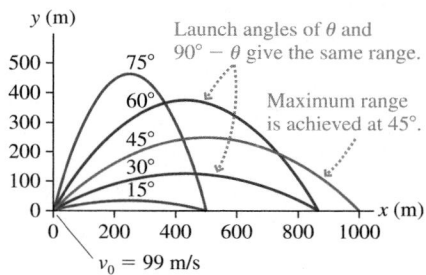

STOP TO THINK 4.3   A 50 g marble rolls off a table and lands 2 m from the base of the table. A 100 g marble rolls off the same table with the same speed. It lands at distance

a. Less than 1 m.   b. 1 m.   c. Between 1 m and 2 m.
d. 2 m.   e. Between 2 m and 4 m.   f. 4 m.

4.4 Relative Motion

FIGURE 4.21 shows Amy and Bill watching Carlos on his bicycle. According to Amy, Carlos's velocity is $v_x = 5$ m/s. Bill sees the bicycle receding in his rearview mirror, in the *negative* x-direction, getting 10 m farther away from him every second. According to Bill, Carlos's velocity is $v_x = -10$ m/s. Which is Carlos's *true* velocity?

Velocity is not a concept that can be true or false. Carlos's velocity *relative to Amy* is $(v_x)_{\text{CA}} = 5$ m/s, where the subscript notation means "C relative to A." Similarly, Carlos's velocity *relative to Bill* is $(v_x)_{\text{CB}} = -10$ m/s. These are both valid descriptions of Carlos's motion.

FIGURE 4.21 Amy and Bill each measure the velocity of Carlos on his bicycle. The velocities shown are in Amy's reference frame.

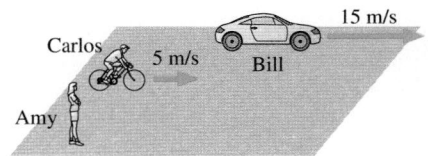

It's not hard to see how to combine the velocities for one-dimensional motion:

The first subscript is the same on both sides.

The last subscript is the same on both sides.

$$(v_x)_{CB} = (v_x)_{CA} + (v_x)_{AB} \quad (4.17)$$

The inner subscripts "cancel."

We'll justify this relationship later in this section and then extend it to two-dimensional motion.

Equation 4.17 tells us that the velocity of C relative to B is the velocity of C relative to A *plus* the velocity of A relative to B. Note that

$$(v_x)_{AB} = -(v_x)_{BA} \quad (4.18)$$

because if B is moving to the right relative to A, then A is moving to the left relative to B. In Figure 4.21, Bill is moving to the right relative to Amy with $(v_x)_{BA} = 15$ m/s, so $(v_x)_{AB} = -15$ m/s. Knowing that Carlos's velocity relative to Amy is 5 m/s, we find that Carlos's velocity relative to Bill is, as expected, $(v_x)_{CB} = (v_x)_{CA} + (v_x)_{AB} = 5$ m/s $+ (-15)$ m/s $= -10$ m/s.

EXAMPLE 4.6 **A speeding bullet**

The police are chasing a bank robber. While driving at 50 m/s, they fire a bullet to shoot out a tire of his car. The police gun shoots bullets at 300 m/s. What is the bullet's speed as measured by a TV camera crew parked beside the road?

MODEL Assume that all motion is in the positive x-direction. The bullet is the object that is observed from both the police car and the ground.

SOLVE The bullet B's velocity relative to the gun G is $(v_x)_{BG} = 300$ m/s. The gun, inside the car, is traveling relative to the TV crew C at $(v_x)_{GC} = 50$ m/s. We can combine these values to find that the bullet's velocity relative to the TV crew on the ground is

$$(v_x)_{BC} = (v_x)_{BG} + (v_x)_{GC} = 300 \text{ m/s} + 50 \text{ m/s} = 350 \text{ m/s}$$

ASSESS It should be no surprise in this simple situation that we simply add the velocities.

Reference Frames

A coordinate system in which an experimenter (possibly with the assistance of helpers) makes position and time measurements of physical events is called a **reference frame**. In Figure 4.21, Amy and Bill each had their own reference frame (where they were at rest) in which they measured Carlos's velocity.

More generally, FIGURE 4.22 shows two reference frames, A and B, and an object C. It is assumed that the reference frames are moving with respect to each other. At this instant of time, the position vector of C in reference frame A is $\vec{r}_{CA}$, meaning "the position of C relative to the origin of frame A." Similarly, $\vec{r}_{CB}$ is the position vector of C in reference frame B. Using vector addition, you can see that

$$\vec{r}_{CB} = \vec{r}_{CA} + \vec{r}_{AB} \quad (4.19)$$

where $\vec{r}_{AB}$ locates the origin of A relative to the origin of B.

In general, object C is moving relative to both reference frames. To find its velocity in each reference frame, take the time derivative of Equation 4.19:

$$\frac{d\vec{r}_{CB}}{dt} = \frac{d\vec{r}_{CA}}{dt} + \frac{d\vec{r}_{AB}}{dt} \quad (4.20)$$

By definition, $d\vec{r}/dt$ is a velocity. The first derivative is $\vec{v}_{CB}$, the velocity of C relative to B. Similarly, the second derivative is the velocity of C relative to A, $\vec{v}_{CA}$. The last derivative is slightly different because it doesn't refer to object C. Instead, this is the velocity $\vec{v}_{AB}$ of reference frame A relative to reference frame B. As we noted in one dimension, $\vec{v}_{AB} = -\vec{v}_{BA}$.

Writing Equation 4.20 in terms of velocities, we have

$$\vec{v}_{CB} = \vec{v}_{CA} + \vec{v}_{AB} \quad (4.21)$$

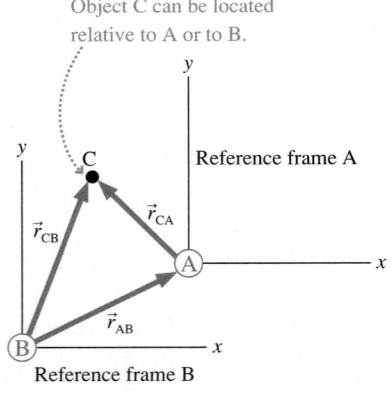

FIGURE 4.22 Object C is measured from two different reference frames.

Object C can be located relative to A or to B.

Reference frame A

Reference frame B

This relationship between velocities in different reference frames was recognized by Galileo in his pioneering studies of motion, hence it is known as the **Galilean transformation of velocity.** If you know an object's velocity in one reference frame, you can *transform* it into the velocity that would be measured in a different reference frame. Just as in one dimension, the velocity of C relative to B is the velocity of C relative to A plus the velocity of A relative to B, *but* you must add the velocities as vectors for two-dimensional motion.

As we've seen, the Galilean velocity transformation is pretty much common sense for one-dimensional motion. The real usefulness appears when an object travels in a *medium* moving with respect to the earth. For example, a boat moves relative to the water. What is the boat's net motion if the water is a flowing river? Airplanes fly relative to the air, but the air at high altitudes often flows at high speed. Navigation of boats and planes requires knowing both the motion of the vessel in the medium and the motion of the medium relative to the earth.

EXAMPLE 4.7 Flying to Cleveland I

Cleveland is 300 miles east of Chicago. A plane leaves Chicago flying due east at 500 mph. The pilot forgot to check the weather and doesn't know that the wind is blowing to the south at 50 mph. What is the plane's ground speed? Where is the plane 0.60 h later, when the pilot expects to land in Cleveland?

MODEL Establish a coordinate system with the x-axis pointing east and the y-axis north. The plane P flies in the air, so its velocity relative to the air A is $\vec{v}_{PA} = 500\,\hat{i}$ mph. Meanwhile, the air is moving relative to the ground G at $\vec{v}_{AG} = -50\,\hat{j}$ mph.

FIGURE 4.23 The wind causes a plane flying due east in the air to move to the southeast relative to the ground.

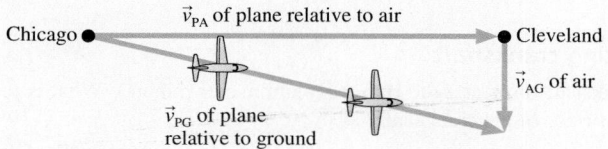

$\vec{v}_{PA}$ of plane relative to air
Chicago ● ————————————————→ ● Cleveland
$\vec{v}_{AG}$ of air
$\vec{v}_{PG}$ of plane relative to ground

SOLVE The velocity equation $\vec{v}_{PG} = \vec{v}_{PA} + \vec{v}_{AG}$ is a vector-addition equation. **FIGURE 4.23** shows graphically what happens. Although the nose of the plane points east, the wind carries the plane in a direction somewhat south of east. The plane's velocity relative to the ground is

$$\vec{v}_{PG} = \vec{v}_{PA} + \vec{v}_{AG} = (500\,\hat{i} - 50\,\hat{j})\ \text{mph}$$

The plane's ground speed is

$$v = \sqrt{(v_x)_{PG}^2 + (v_y)_{PG}^2} = 502\ \text{mph}$$

After flying for 0.60 h at this velocity, the plane's location (relative to Chicago) is

$$x = (v_x)_{PG}\,t = (500\ \text{mph})(0.60\ \text{h}) = 300\ \text{mi}$$

$$y = (v_y)_{PG}\,t = (-50\ \text{mph})(0.60\ \text{h}) = -30\ \text{mi}$$

The plane is 30 mi due south of Cleveland! Although the pilot thought he was flying to the east, his actual heading has been $\tan^{-1}(50\ \text{mph}/500\ \text{mph}) = \tan^{-1}(0.10) = 5.71°$ south of east.

EXAMPLE 4.8 Flying to Cleveland II

A wiser pilot flying from Chicago to Cleveland on the same day plots a course that will take her directly to Cleveland. In which direction does she fly the plane? How long does it take to reach Cleveland?

MODEL Establish a coordinate system with the x-axis pointing east and the y-axis north. The air is moving relative to the ground at $\vec{v}_{AG} = -50\,\hat{j}$ mph.

SOLVE The objective of navigation is to move between two points on the earth's surface. The wiser pilot, who knows that the wind will affect her plane, draws the vector picture of **FIGURE 4.24**. She sees that she'll need $(v_y)_{PG} = 0$, in order to fly due east to Cleveland. This will require turning the nose of the plane at an angle θ north of east, making $\vec{v}_{PA} = (500\cos\theta\,\hat{i} + 500\sin\theta\,\hat{j})$ mph.

The velocity equation is $\vec{v}_{PG} = \vec{v}_{PA} + \vec{v}_{AG}$. The desired heading is found from setting the y-component of this equation to zero:

$$(v_y)_{PG} = (v_y)_{PA} + (v_y)_{AG} = (500\sin\theta - 50)\ \text{mph} = 0\ \text{mph}$$

$$\theta = \sin^{-1}\left(\frac{50\ \text{mph}}{500\ \text{mph}}\right) = 5.74°$$

FIGURE 4.24 To travel due east in a south wind, a pilot has to point the plane somewhat to the northeast.

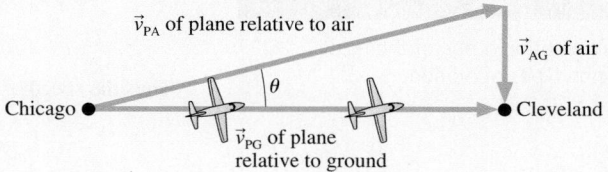

$\vec{v}_{PA}$ of plane relative to air
$\vec{v}_{AG}$ of air
Chicago ● θ ● Cleveland
$\vec{v}_{PG}$ of plane relative to ground

The plane's velocity relative to the ground is then $\vec{v}_{PG} = (500\ \text{mph}) \times \cos 5.74°\,\hat{i} = 497\,\hat{i}$ mph. This is slightly slower than the speed relative to the air. The time needed to fly to Cleveland at this speed is

$$t = \frac{300\ \text{mi}}{497\ \text{mph}} = 0.604\ \text{h}$$

It takes 0.004 h = 14 s longer to reach Cleveland than it would on a day without wind.

ASSESS A boat crossing a river or an ocean current faces the same difficulties. These are exactly the kinds of calculations performed by pilots of boats and planes as part of navigation.

A plane traveling horizontally to the right at 100 m/s flies past a helicopter that is going straight up at 20 m/s. From the helicopter's perspective, the plane's direction and speed are

a. Right and up, less than 100 m/s.

b. Right and up, 100 m/s.

c. Right and up, more than 100 m/s.

d. Right and down, less than 100 m/s.

e. Right and down, 100 m/s.

f. Right and down, more than 100 m/s.

4.5 Uniform Circular Motion

FIGURE 4.25 A particle in uniform circular motion.

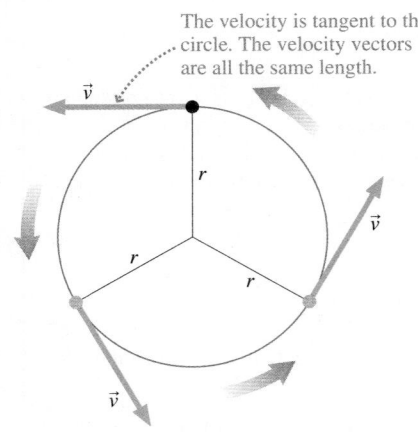

The velocity is tangent to the circle. The velocity vectors are all the same length.

FIGURE 4.25 shows a particle moving around a circle of radius r. The particle might be a satellite in an orbit, a ball on the end of a string, or even just a dot painted on the side of a rotating wheel. Circular motion is another example of motion in a plane, but it is quite different from projectile motion.

To begin the study of circular motion, consider a particle that moves at *constant speed* around a circle of radius r. This is called **uniform circular motion**. Regardless of what the particle represents, its velocity vector $\vec{v}$ is always tangent to the circle. The particle's speed v is constant, so vector $\vec{v}$ is always the same length.

The time interval it takes the particle to go around the circle once, completing one revolution (abbreviated rev), is called the **period** of the motion. Period is represented by the symbol T. It's easy to relate the particle's period T to its speed v. For a particle moving with constant speed, speed is simply distance/time. In one period, the particle moves once around a circle of radius r and travels the circumference $2\pi r$. Thus

$$v = \frac{1 \text{ circumference}}{1 \text{ period}} = \frac{2\pi r}{T} \tag{4.22}$$

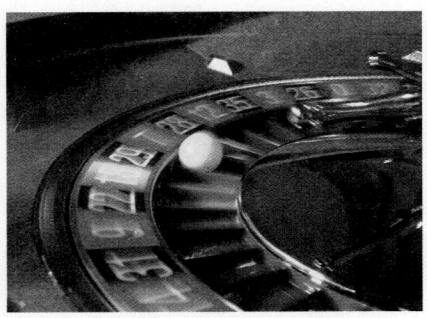

Circular motion is one of the most common types of motion.

EXAMPLE 4.9   **A rotating crankshaft**

A 4.0-cm-diameter crankshaft turns at 2400 rpm (revolutions per minute). What is the speed of a point on the surface of the crankshaft?

SOLVE We need to determine the time it takes the crankshaft to make 1 rev. First, we convert 2400 rpm to revolutions per second:

$$\frac{2400 \text{ rev}}{1 \text{ min}} \times \frac{1 \text{ min}}{60 \text{ s}} = 40 \text{ rev/s}$$

If the crankshaft turns 40 times in 1 s, the time for 1 rev is

$$T = \frac{1}{40} \text{ s} = 0.025 \text{ s}$$

Thus the speed of a point on the surface, where $r = 2.0 \text{ cm} = 0.020 \text{ m}$, is

$$v = \frac{2\pi r}{T} = \frac{2\pi(0.020 \text{ m})}{0.025 \text{ s}} = 5.0 \text{ m/s}$$

FIGURE 4.26 A particle's position is described by distance r and angle θ.

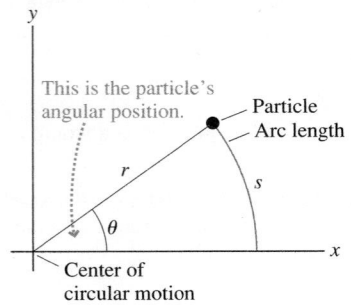

This is the particle's angular position.

Particle

Arc length

Center of circular motion

Angular Position

Rather than using xy-coordinates, it will be more convenient to describe the position of a particle in circular motion by its distance r from the center of the circle and its angle θ from the positive x-axis. This is shown in **FIGURE 4.26**. The angle θ is the **angular position** of the particle.

We can distinguish a position above the x-axis from a position that is an equal angle below the x-axis by *defining* θ to be positive when measured *counterclockwise* (ccw) from the positive x-axis. An angle measured clockwise (cw) from the positive x-axis has a negative value. "Clockwise" and "counterclockwise" in circular motion are analogous, respectively, to "left of the origin" and "right of the origin" in linear motion, which we associated with negative and positive values of x. A particle $30°$ below the positive x-axis is equally well described by either $\theta = -30°$ or $\theta = +330°$. We could also describe this particle by $\theta = \frac{11}{12}$ rev, where *revolutions* are another way to measure the angle.

Although degrees and revolutions are widely used measures of angle, mathematicians and scientists usually find it more useful to measure the angle θ in Figure 4.26 by using the **arc length** s that the particle travels along the edge of a circle of radius r. We define the angular unit of **radians** such that

$$\theta(\text{radians}) \equiv \frac{s}{r} \tag{4.23}$$

The radian, which is abbreviated rad, is the SI unit of an angle. An angle of 1 rad has an arc length s exactly equal to the radius r.

The arc length completely around a circle is the circle's circumference $2\pi r$. Thus the angle of a full circle is

$$\theta_{\text{full circle}} = \frac{2\pi r}{r} = 2\pi \text{ rad}$$

This relationship is the basis for the well-known conversion factors

$$1 \text{ rev} = 360° = 2\pi \text{ rad}$$

As a simple example of converting between radians and degrees, let's convert an angle of 1 rad to degrees:

$$1 \text{ rad} = 1 \text{ rad} \times \frac{360°}{2\pi \text{ rad}} = 57.3°$$

Thus a rough approximation is $1 \text{ rad} \approx 60°$. We will often specify angles in degrees, but keep in mind that the SI unit is the radian.

An important consequence of Equation 4.23 is that the arc length spanning angle θ is

$$s = r\theta \qquad (\text{with } \theta \text{ in rad}) \tag{4.24}$$

This is a result that we will use often, but it is valid *only* if θ is measured in radians and not in degrees. This very simple relationship between angle and arc length is one of the primary motivations for using radians.

NOTE ▶ Units of angle are often troublesome. Unlike the kilogram or the second, for which we have standards, the radian is a *defined* unit. Further, its definition as a ratio of two lengths makes it a *pure number* without dimensions. Thus the unit of angle, be it radians or degrees or revolutions, is really just a *name* to remind us that we're dealing with an angle. Consequently, the radian unit sometimes appears or disappears without warning. This seems rather mysterious until you get used to it. This textbook will call your attention to such behavior the first few times it occurs. With a little practice, you'll soon learn when the rad unit is needed and when it's not. ◀

Angular Velocity

FIGURE 4.27 shows a particle moving in a circle from an initial angular position θ_i at time t_i to a final angular position θ_f at a later time t_f. The change $\Delta\theta = \theta_f - \theta_i$ is called the **angular displacement.** We can measure the particle's circular motion in terms of the rate of change of θ, just as we measured the particle's linear motion in terms of the rate of change of its position s.

In analogy with linear motion, let's define the *average angular velocity* to be

$$\text{average angular velocity} \equiv \frac{\Delta\theta}{\Delta t} \tag{4.25}$$

As the time interval Δt becomes very small, $\Delta t \to 0$, we arrive at the definition of the instantaneous **angular velocity**

$$\omega \equiv \lim_{\Delta t \to 0} \frac{\Delta\theta}{\Delta t} = \frac{d\theta}{dt} \qquad (\text{angular velocity}) \tag{4.26}$$

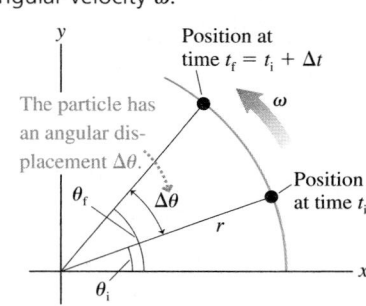

FIGURE 4.27 A particle moves with angular velocity ω.

The symbol ω is a lowercase Greek omega, *not* an ordinary w. The SI unit of angular velocity is rad/s, but °/s, rev/s, and rev/min are also common units. Revolutions per minute is abbreviated rpm.

Angular velocity is the *rate* at which a particle's angular position is changing as it moves around a circle. A particle that starts from $\theta = 0$ rad with an angular velocity of 0.5 rad/s will be at angle $\theta = 0.5$ rad after 1 s, at $\theta = 1.0$ rad after 2 s, at $\theta = 1.5$ rad after 3 s, and so on. Its angular position is increasing at the *rate* of 0.5 radian per second. **A particle moves with uniform circular motion if and only if its angular velocity ω is constant and unchanging.**

Angular velocity, like the velocity v_s of one-dimensional motion, can be positive or negative. The signs shown in **FIGURE 4.28** are based on the fact that θ was defined to be positive for a counterclockwise rotation. Because the definition $\omega = d\theta/dt$ for circular motion parallels the definition $v_s = ds/dt$ for linear motion, the graphical relationships we found between v_s and s in Chapter 2 apply equally well to ω and θ:

FIGURE 4.28 Positive and negative angular velocities.

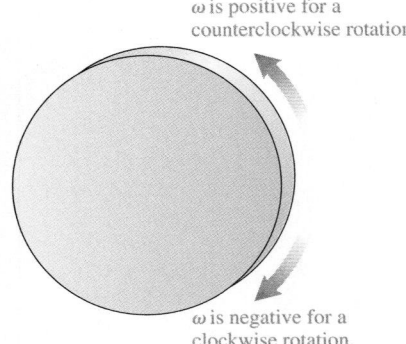

ω is positive for a counterclockwise rotation.

ω is negative for a clockwise rotation.

$$\omega = \text{slope of the } \theta\text{-versus-}t \text{ graph at time } t$$
$$\theta_f = \theta_i + \text{area under the } \omega\text{-versus-}t \text{ curve between } t_i \text{ and } t_f \qquad (4.27)$$
$$= \theta_i + \omega \Delta t$$

You will see many more instances where circular motion is analogous to linear motion with angular variables replacing linear variables. Thus much of what you learned about linear kinematics carries over to circular motion.

EXAMPLE 4.10 **A graphical representation of circular motion**

FIGURE 4.29 shows the angular position of a painted dot on the edge of a rotating wheel. Describe the wheel's motion and draw an ω-versus-t graph.

FIGURE 4.29 Angular position graph for the wheel of Example 4.10.

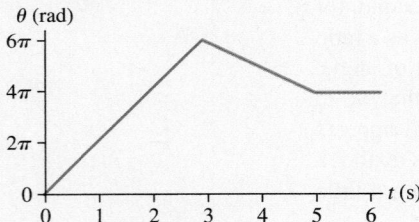

SOLVE Although circular motion seems to "start over" every revolution (every 2π rad), the angular position θ continues to increase. $\theta = 6\pi$ rad corresponds to three revolutions. This wheel makes 3 ccw rev (because θ is getting more positive) in 3 s, immediately reverses direction and makes 1 cw rev in 2 s, then stops at $t = 5$ s

and holds the position $\theta = 4\pi$ rad. The angular velocity is found by measuring the slope of the graph:

$t = 0–3$ s   slope $= \Delta\theta/\Delta t = 6\pi$ rad/3 s $= 2\pi$ rad/s

$t = 3–5$ s   slope $= \Delta\theta/\Delta t = -2\pi$ rad/2 s $= -\pi$ rad/s

$t > 5$ s   slope $= \Delta\theta/\Delta t = 0$ rad/s

These results are shown as an ω-versus-t graph in **FIGURE 4.30**. For the first 3 s, the motion is uniform circular motion with $\omega = 2\pi$ rad/s. The wheel then changes to a different uniform circular motion with $\omega = -\pi$ rad/s for 2 s, then stops.

FIGURE 4.30 ω-versus-t graph for the wheel of Example 4.10.

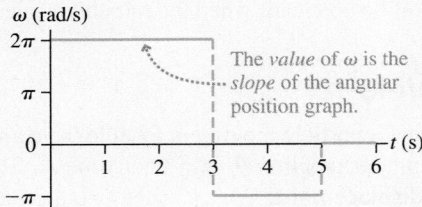

The *value* of ω is the *slope* of the angular position graph.

NOTE ▶ In physics, we nearly always want to give results as numerical values. Example 4.9 had a π in the equation, but we used its numerical value to compute $v = 5.0$ m/s. However, angles in radians are an exception to this rule. It's okay to leave a π in the value of θ or ω, and we have done so in Example 4.10. ◀

Not surprisingly, the angular velocity ω is closely related to the *period T* of the motion. As a particle goes around a circle one time, its angular displacement is

$\Delta\theta = 2\pi$ rad during the interval $\Delta t = T$. Thus, using the definition of angular velocity, we find

$$|\omega| = \frac{2\pi \text{ rad}}{T} \qquad \text{or} \qquad T = \frac{2\pi \text{ rad}}{|\omega|} \qquad (4.28)$$

The period alone gives only the absolute value of $|\omega|$. You need to know the direction of motion to determine the sign of ω.

EXAMPLE 4.11 **At the roulette wheel**

A small steel roulette ball rolls ccw around the inside of a 30-cm-diameter roulette wheel. The ball completes 2.0 rev in 1.20 s.

a. What is the ball's angular velocity?
b. What is the ball's position at $t = 2.0$ s? Assume $\theta_i = 0$.

MODEL Model the ball as a particle in uniform circular motion.

SOLVE a. The period of the ball's motion, the time for 1 rev, is $T = 0.60$ s. Angular velocity is positive for ccw motion, so

$$\omega = \frac{2\pi \text{ rad}}{T} = \frac{2\pi \text{ rad}}{0.60 \text{ s}} = 10.47 \text{ rad/s}$$

b. The ball starts at $\theta_i = 0$ rad. After $\Delta t = 2.0$ s, its position is

$$\theta_f = 0 \text{ rad} + (10.47 \text{ rad/s})(2.0 \text{ s}) = 20.94 \text{ rad}$$

where we've kept an extra significant figure to avoid round-off error. Although this is a mathematically acceptable answer, an observer would say that the ball is always located somewhere between 0° and 360°. Thus it is common practice to subtract an integer number of 2π rad, representing the completed revolutions. Because $20.94/2\pi = 3.333$, we can write

$$\theta_f = 20.94 \text{ rad} = 3.333 \times 2\pi \text{ rad}$$
$$= 3 \times 2\pi \text{ rad} + 0.333 \times 2\pi \text{ rad}$$
$$= 3 \times 2\pi \text{ rad} + 2.09 \text{ rad}$$

In other words, at $t = 2.0$ s the ball has completed 3 rev and is 2.09 rad = 120° into its fourth revolution. An observer would say that the ball's position is $\theta_f = 120°$.

STOP TO THINK 4.5   A particle moves cw around a circle at constant speed for 2.0 s. It then reverses direction and moves ccw at half the original speed until it has traveled through the same angle. Which is the particle's angle-versus-time graph?

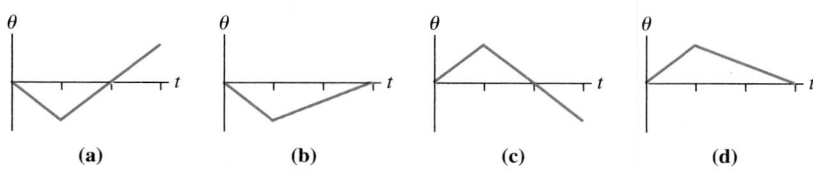

4.6 Velocity and Acceleration in Uniform Circular Motion

For a particle in circular motion, such as the one in FIGURE 4.31, the velocity vector $\vec{v}$ is always tangent to the circle. In other words, the velocity vector has only a *tangential component*, which we will designate v_t.

The tangential velocity component v_t is the rate ds/dt at which the particle moves *around* the circle, where s is the arc length measured from the positive x-axis. From Equation 4.24, the arc length is $s = r\theta$. Taking the derivative, we find

$$v_t = \frac{ds}{dt} = r\frac{d\theta}{dt}$$

But $d\theta/dt$ is the angular velocity ω. Thus the tangential velocity and the angular velocity are related by

$$v_t = \omega r \qquad \text{(with } \omega \text{ in rad/s)} \qquad (4.29)$$

NOTE ▶ ω is restricted to rad/s because the relationship $s = r\theta$ is the definition of radians. While it may be convenient in some problems to measure ω in rev/s or rpm, you must convert to SI units of rad/s before using Equation 4.29. ◀

FIGURE 4.31 Velocity and acceleration of uniform circular motion.

The instantaneous velocity $\vec{v}$ is tangent to the circle at all points.

For uniform circular motion, the acceleration $\vec{a}$ points to the center of the circle.

The angular velocity is constant.

The tangential velocity v_t is positive for ccw motion, negative for cw motion. Because v_t is the only nonzero component of $\vec{v}$, the particle's speed is $v = |v_t| = |\omega|r$. We'll sometimes write this as $v = \omega r$ if there's no ambiguity about the sign of ω.

As a simple example, a particle moving cw at 2.0 m/s in a circle of radius 40 cm has angular velocity

$$\omega = \frac{v_t}{r} = \frac{-2.0 \text{ m/s}}{0.40 \text{ m}} = -5.0 \text{ rad/s}$$

where v_t and ω are negative because the motion is clockwise. Notice the units. Velocity divided by distance has units of s^{-1}. But because the division, in this case, gives us an angular quantity, we've inserted the *dimensionless* unit rad to give ω the appropriate units of rad/s.

Acceleration

Figure 4.1 at the beginning of this chapter looked at the uniform circular motion of a Ferris wheel. You are strongly encouraged to review that figure. There we found that a particle in uniform circular motion, although moving with constant speed, has an acceleration because the *direction* of the velocity vector $\vec{v}$ is always changing. The motion-diagram analysis showed that the **acceleration $\vec{a}$ points toward the center of the circle.** The instantaneous velocity is tangent to the circle, so $\vec{v}$ and $\vec{a}$ are perpendicular to each other at all points on the circle, as Figure 4.31 shows.

The acceleration of uniform circular motion is called **centripetal acceleration,** a term from a Greek root meaning "center seeking." Centripetal acceleration is not a new type of acceleration; all we are doing is *naming* an acceleration that corresponds to a particular type of motion. The magnitude of the centripetal acceleration is constant because each successive $\Delta\vec{v}$ in the motion diagram has the same length.

The motion diagram tells us the direction of $\vec{a}$, but it doesn't give us a value for a. To complete our description of uniform circular motion, we need to find a quantitative relationship between a and the particle's speed v. FIGURE 4.32 shows the velocity $\vec{v}_i$ at one instant of motion and the velocity $\vec{v}_f$ an infinitesimal amount of time dt later. During this small interval of time, the particle has moved through the infinitesimal angle $d\theta$ and traveled distance $ds = r\,d\theta$.

By definition, the acceleration is $\vec{a} = d\vec{v}/dt$. We can see from the inset to Figure 4.32 that $d\vec{v}$ points toward the center of the circle—that is, $\vec{a}$ is a centripetal acceleration. To find the magnitude of $\vec{a}$, we can see from the isosceles triangle of velocity vectors that, if $d\theta$ is in radians,

$$dv = |d\vec{v}| = v\,d\theta \qquad (4.30)$$

For uniform circular motion at constant speed, $v = ds/dt = r\,d\theta/dt$ and thus the time to rotate through angle $d\theta$ is

$$dt = \frac{r\,d\theta}{v} \qquad (4.31)$$

Combining Equations 4.30 and 4.31, we see that the acceleration has magnitude

$$a = |\vec{a}| = \frac{|d\vec{v}|}{dt} = \frac{v\,d\theta}{r\,d\theta/v} = \frac{v^2}{r}$$

In vector notation, we can write

$$\vec{a} = \left(\frac{v^2}{r}, \text{ toward center of circle}\right) \qquad \text{(centripetal acceleration)} \quad (4.32)$$

Using Equation 4.29, $v = \omega r$, we can also express the magnitude of the centripetal acceleration in terms of the angular velocity ω as

$$a = \omega^2 r \qquad (4.33)$$

The centripetal acceleration is enormous in a high-speed centrifuge.

FIGURE 4.32 Finding the acceleration of circular motion.

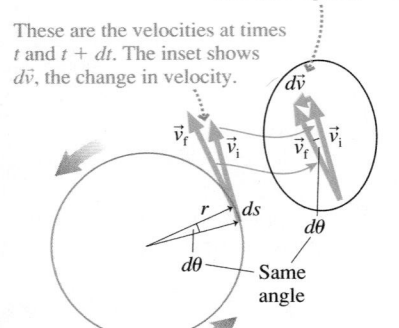

$d\vec{v}$ is the arc of a circle with arc length $dv = v\,d\theta$.

These are the velocities at times t and $t + dt$. The inset shows $d\vec{v}$, the change in velocity.

NOTE ▶ Centripetal acceleration is not a constant acceleration. The magnitude of the centripetal acceleration is constant during uniform circular motion, but the direction of $\vec{a}$ is constantly changing. **Thus the constant-acceleration kinematics equations of Chapter 2 do *not* apply to circular motion.** ◀

EXAMPLE 4.12 **The acceleration of a Ferris wheel**

A typical carnival Ferris wheel has a radius of 9.0 m and rotates 4.0 times per minute. What magnitude acceleration do the riders experience?

MODEL Model the rider as a particle in uniform circular motion.

SOLVE The period is $T = \frac{1}{4}$ min = 15 s. From Equation 4.22, a rider's speed is

$$v = \frac{2\pi r}{T} = \frac{2\pi(9.0\text{ m})}{15\text{ s}} = 3.77\text{ m/s}$$

Consequently, the centripetal acceleration is

$$a = \frac{v^2}{r} = \frac{(3.77\text{ m/s})^2}{9.0\text{ m}} = 1.6\text{ m/s}^2$$

ASSESS This was not intended to be a profound problem, merely to illustrate how centripetal acceleration is computed. The acceleration is enough to be noticed and make the ride interesting, but not enough to be scary.

STOP TO THINK 4.6 Rank in order, from largest to smallest, the centripetal accelerations a_a to a_e of particles a to e.

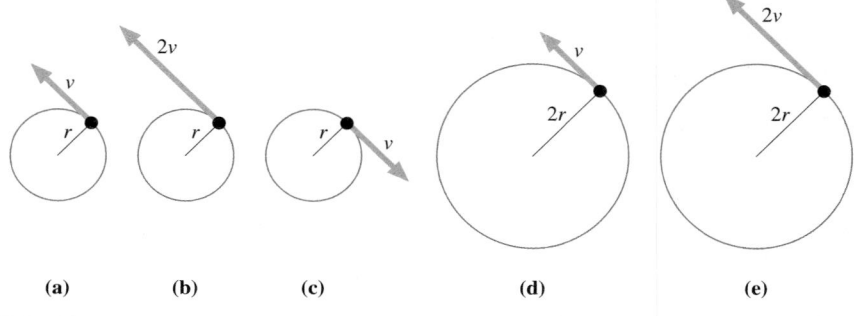

(a)   (b)   (c)   (d)   (e)

4.7 Nonuniform Circular Motion and Angular Acceleration

A roller coaster car doing a loop-the-loop slows down as it goes up one side, speeds up as it comes back down the other. The ball in a roulette wheel gradually slows until it stops. Circular motion with a changing speed is called **nonuniform circular motion.**

To begin our analysis of nonuniform circular motion, **FIGURE 4.33** shows a wheel rotating on an axle. Notice that two points on the wheel, marked with dots, turn through the *same angle* as the wheel rotates, even though their radii may differ. That is, $\Delta\theta_1 = \Delta\theta_2$ during some time interval Δt. As a consequence, any two points on a rotating object have equal angular velocities, $\omega_1 = \omega_2$, and we can refer to ω as the angular velocity *of the wheel.*

Suppose the wheel's rotation is speeding up or slowing down—that is, points on the wheel have nonuniform circular motion. For linear motion, we defined acceleration as $a_x = dv_x/dt$. By analogy, let's define the **angular acceleration** α (Greek alpha) of a rotating object, or a point on the object, to be

$$\alpha \equiv \frac{d\omega}{dt} \quad \text{(angular acceleration)} \tag{4.34}$$

The units of angular acceleration are rad/s^2. Angular acceleration is the *rate* at which the angular velocity ω changes, just as linear acceleration is the rate at which the linear velocity v_x changes. **FIGURE 4.34** on the next page illustrates this idea.

FIGURE 4.33 All points on the wheel rotate with the same angular velocity.

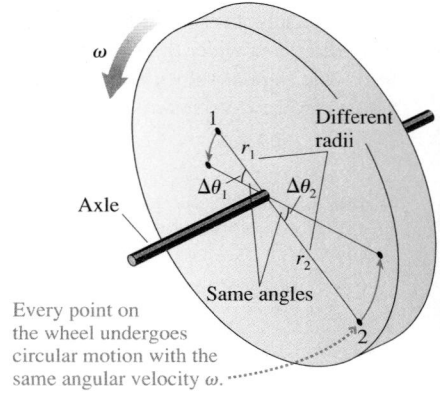

Every point on the wheel undergoes circular motion with the same angular velocity ω.

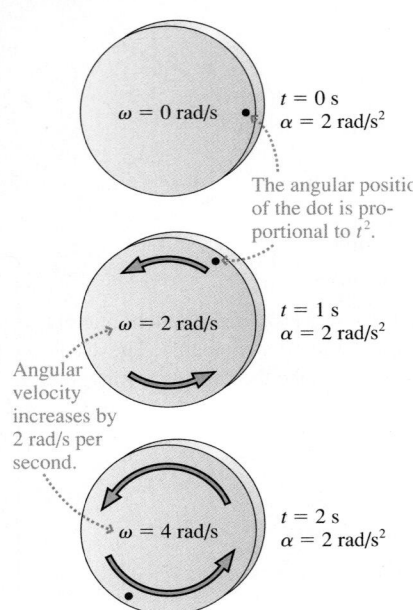

The angular position of the dot is proportional to t^2.

Angular velocity increases by 2 rad/s per second.

For linear acceleration, positive a_x means that v_x is increasing to the right or decreasing to the left; negative a_x means that v_x is increasing to the left or decreasing to the right. For rotational motion, α is positive if ω is increasing ccw (the direction of positive angle) or decreasing cw, negative if ω is increasing cw or decreasing ccw. These ideas are illustrated in FIGURE 4.35.

FIGURE 4.35 The signs of angular velocity and acceleration. The rotation is speeding up if ω and α have the same sign, slowing down if they have opposite signs.

Initial angular velocity

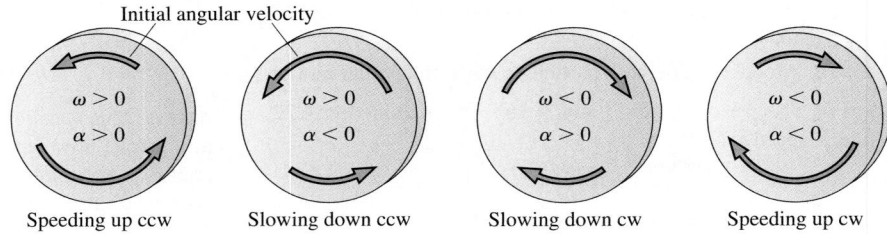

Speeding up ccw   Slowing down ccw   Slowing down cw   Speeding up cw

NOTE ▶ Be careful with the sign of α. You learned in Chapter 2 that positive and negative values of the acceleration can't be interpreted as simply "speeding up" and "slowing down." Similarly, positive and negative values of angular acceleration can't be interpreted as a rotation that is speeding up or slowing down. ◀

Because α is the time derivative of ω, we can use exactly the same graphical relationships that we found for linear motion:

$$\alpha = \text{slope of the } \omega\text{-versus-}t \text{ graph at time } t$$

$$\omega_f = \omega_i + \text{area under the } \alpha\text{-versus-}t \text{ curve between } t_i \text{ and } t_f \qquad (4.35)$$

These relationships involving slopes and areas are illustrated in the following example.

EXAMPLE 4.13 **A rotating wheel**

FIGURE 4.36a is a graph of angular velocity versus time for a rotating wheel. Describe the motion and draw a graph of angular acceleration versus time.

SOLVE This is a wheel that starts from rest, gradually speeds up *counterclockwise* until reaching top speed at t_1, maintains a constant angular velocity until t_2, then gradually slows down until stopping at t_3. The motion is always ccw because ω is always positive. The angular acceleration graph of FIGURE 4.36b is based on the fact that α is the slope of the ω-versus-t graph.

Conversely, the initial linear increase of ω can be seen as the increasing area under the α-versus-t graph as t increases from 0 to t_1. The angular velocity doesn't change from t_1 to t_2 when the area under the α-versus-t is zero.

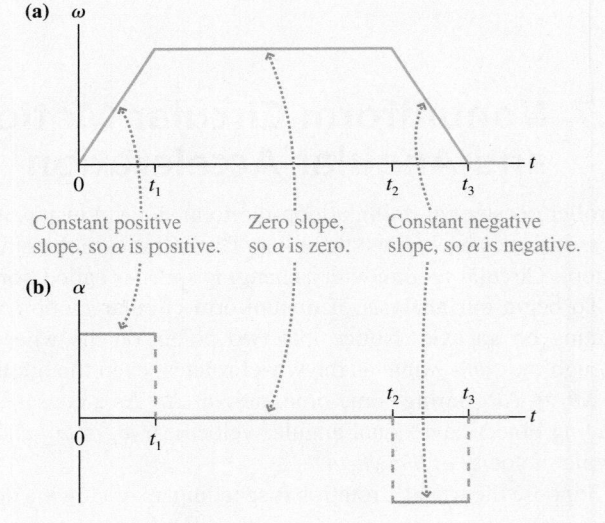

(a) ω

Constant positive slope, so α is positive.   Zero slope, so α is zero.   Constant negative slope, so α is negative.

(b) α

FIGURE 4.36 ω-versus-t graph and the corresponding α-versus-t graph for a rotating wheel.

Table 4.1 shows the kinematic equations for rotational motion with constant angular acceleration. These equations apply to a particle in circular motion or to the rotation of a rigid object. **The rotational kinematic equations are exactly analogous to the linear kinematic equations,** as they must be since the mathematical relationships among θ, ω, and α are identical to the relationships among x, v_x, and a_x. Thus all the problem-solving techniques you learned in Chapter 2 for linear motion carry over to circular and rotational motion.

TABLE 4.1 Rotational and linear kinematics for constant acceleration

Rotational kinematics	Linear kinematics
$\omega_f = \omega_i + \alpha \, \Delta t$	$v_{fs} = v_{is} + a_s \, \Delta t$
$\theta_f = \theta_i + \omega_i \, \Delta t + \frac{1}{2}\alpha(\Delta t)^2$	$s_f = s_i + v_{is} \, \Delta t + \frac{1}{2}a_s(\Delta t)^2$
$\omega_f^2 = \omega_i^2 + 2\alpha \, \Delta\theta$	$v_{fs}^2 = v_{is}^2 + 2a_s \, \Delta s$

EXAMPLE 4.14 **Back to the roulette wheel**

A small steel roulette ball rolls around the inside of a 30-cm-diameter roulette wheel. It is spun at 150 rpm, but it slows to 60 rpm over an interval of 5.0 s. How many revolutions does the ball make during these 5.0 s?

MODEL The ball is a particle in nonuniform circular motion. Assume constant angular acceleration as it slows.

SOLVE During these 5.0 s the ball rotates through angle

$$\Delta\theta = \theta_f - \theta_i = \omega_i \, \Delta t + \frac{1}{2}\alpha(\Delta t)^2$$

where $\Delta t = 5.0$ s. We can find the angular acceleration from the initial and final angular velocities, but first they must be converted to SI units:

$$\omega_i = 150 \, \frac{\text{rev}}{\text{min}} \times \frac{1 \, \text{min}}{60 \, \text{s}} \times \frac{2\pi \, \text{rad}}{1 \, \text{rev}} = 15.71 \, \text{rad/s}$$

$$\omega_f = 60 \, \frac{\text{rev}}{\text{min}} = 0.40\omega_i = 6.28 \, \text{rad/s}$$

The angular acceleration α is

$$\alpha = \frac{\Delta\omega}{\Delta t} = \frac{6.28 \, \text{rad/s} - 15.71 \, \text{rad/s}}{5.0 \, \text{s}} = -1.89 \, \text{rad/s}^2$$

Thus the ball rotates through angle

$$\Delta\theta = (15.71 \, \text{rad/s})(5.0 \, \text{s}) + \frac{1}{2}(-1.89 \, \text{rad/s}^2)(5.0 \, \text{s})^2 = 54.9 \, \text{rad}$$

Because $54.9/2\pi = 8.75$, the ball completes $8\frac{3}{4}$ revolutions as it slows to 60 rpm.

ASSESS This problem is solved just like the linear kinematics problems you learned to solve in Chapter 2.

Tangential Acceleration

FIGURE 4.37 shows a particle in nonuniform circular motion. *Any* circular motion, whether uniform or nonuniform, has a centripetal acceleration because the particle is changing direction; this was the acceleration component $\vec{a}_\perp$ of Figure 4.10. The centripetal acceleration, which points radially toward the center of the circle, will now be called the **radial acceleration** a_r. The centripetal expression $a_r = v_t^2/r = \omega^2 r$ is still valid in nonuniform circular motion.

For a particle to speed up or slow down as it moves around a circle, it needs—in addition to the centripetal acceleration—an acceleration parallel to the trajectory or, equivalently, parallel to $\vec{v}$. This is the acceleration component $\vec{a}_\parallel$ associated with changing speed. We'll call this the **tangential acceleration** a_t because, like the velocity v_t, it is always tangent to the circle. Because of the tangential acceleration, **the acceleration vector $\vec{a}$ of a particle in nonuniform circular motion does not point toward the center of the circle.** It points "ahead" of center for a particle that is speeding up, as in Figure 4.37, but it would point "behind" center for a particle slowing down. You can see from Figure 4.37 that the magnitude of the acceleration is

$$a = \sqrt{a_r^2 + a_t^2} \tag{4.36}$$

If a_t is constant, then the arc length s traveled by the particle around the circle and the tangential velocity v_t are found from constant-acceleration kinematics:

$$s_f = s_i + v_{it} \, \Delta t + \frac{1}{2}a_t(\Delta t)^2$$
$$v_{ft} = v_{it} + a_t \, \Delta t \tag{4.37}$$

Because tangential acceleration is the rate at which the tangential velocity changes, $a_t = dv_t/dt$, and we already know that the tangential velocity is related to the angular velocity by $v_t = \omega r$, it follows that

$$a_t = \frac{dv_t}{dt} = \frac{d(\omega r)}{dt} = \frac{d\omega}{dt}r = \alpha r \tag{4.38}$$

FIGURE 4.37 Acceleration in nonuniform circular motion.

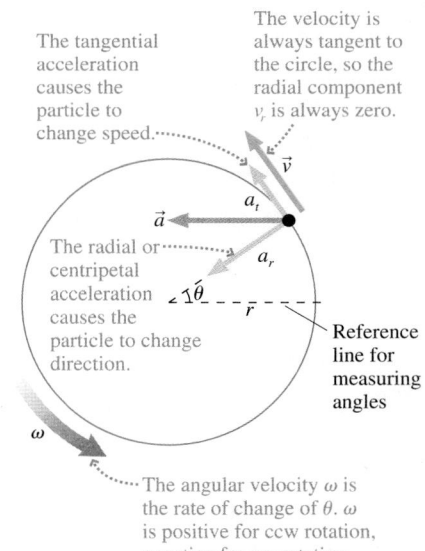

The tangential acceleration causes the particle to change speed.

The velocity is always tangent to the circle, so the radial component v_r is always zero.

The radial or centripetal acceleration causes the particle to change direction.

Reference line for measuring angles

The angular velocity ω is the rate of change of θ. ω is positive for ccw rotation, negative for cw rotation.

Thus $v_t = \omega r$ and $a_t = \alpha r$ are analogous equations for the tangential velocity and acceleration. In Example 4.14, where we found the roulette ball to have angular acceleration $\alpha = -1.89$ rad/s^2, its tangential acceleration was

$$a_t = \alpha r = (-1.89 \text{ rad/s}^2)(0.15 \text{ m}) = -0.28 \text{ m/s}^2$$

EXAMPLE 4.15 Analyzing rotational data

You've been assigned the task of measuring the start-up characteristics of a large industrial motor. After several seconds, when the motor has reached full speed, you know that the angular acceleration will be zero, but you hypothesize that the angular acceleration may be constant during the first couple of seconds as the motor speed increases. To find out, you attach a shaft encoder to the 3.0-cm-diameter axle. A shaft encoder is a device that converts the angular position of a shaft or axle to a signal that can be read by a computer. After setting the computer program to read four values a second, you start the motor and acquire the following data:

Time (s)	Angle(°)
0.00	0
0.25	16
0.50	69
0.75	161
1.00	267
1.25	428
1.50	620

a. Do the data support your hypothesis of a constant angular acceleration? If so, what is the angular acceleration? If not, is the angular acceleration increasing or decreasing with time?

b. A 76-cm-diameter blade is attached to the motor shaft. At what time does the acceleration of the tip of the blade reach 10 m/s^2?

MODEL The axle is rotating with nonuniform circular motion. Model the tip of the blade as a particle.

VISUALIZE FIGURE 4.38 shows that the blade tip has both a tangential and a radial acceleration.

FIGURE 4.38 Pictorial representation of the axle and blade.

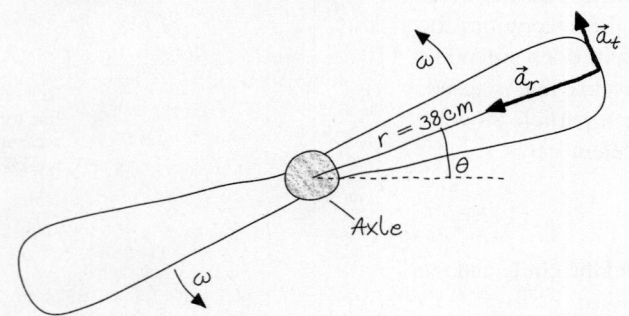

SOLVE

a. *If* the motor starts up with constant angular acceleration, with $\theta_i = 0$ and $\omega_i = 0$ rad/s, the angle-time equation of rotational kinematics is $\theta = \frac{1}{2}\alpha t^2$. This can be written as a linear equation $y = mx + b$ if we let $\theta = y$ and $t^2 = x$. That is, constant angular acceleration predicts that a graph of θ versus t^2 should be a straight line with slope $m = \frac{1}{2}\alpha$ and y-intercept $b = 0$. We can test this. If the graph turns out to be a straight line with zero y-intercept, it will confirm the hypothesis of constant angular acceleration and we can then use its slope to determine the angular acceleration:

$\alpha = 2m$. If the graph is not a straight line, our observation of whether it curves upward or downward will tell us whether the angular acceleration us increasing or decreasing.

FIGURE 4.39 is the graph of θ versus t^2, and it confirms our hypothesis that the motor starts up with constant angular acceleration. The best-fit line, found using a spreadsheet, gives a slope of 274.6°/s^2. The units come not from the spreadsheet but by looking at the units of rise (°) over run (s^2 because we're graphing t^2 on the x-axis). Thus the angular acceleration is

$$\alpha = 2m = 549.2°/s^2 \times \frac{\pi \text{ rad}}{180°} = 9.6 \text{ rad/s}^2$$

where we used $180° = \pi$ rad to convert to SI units of rad/s^2.

FIGURE 4.39 Graph of θ versus t^2 for the motor shaft.

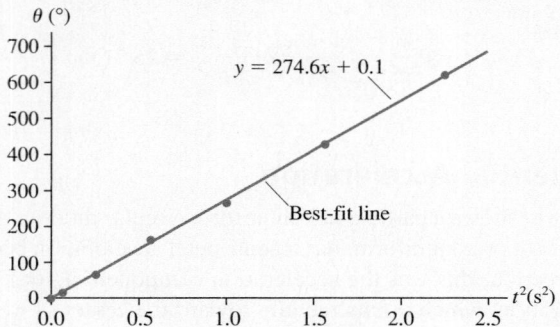

b. The magnitude of the linear acceleration is

$$a = \sqrt{a_r^2 + a_t^2}$$

Constant angular acceleration implies constant tangential acceleration, and the tangential acceleration of the blade tip is

$$a_t = \alpha r = (9.6 \text{ rad/s}^2)(0.38 \text{ m}) = 3.65 \text{ m/s}^2$$

We were careful to use the blade's radius, not its diameter, and we kept an extra significant figure to avoid round-off error. The radial (centripetal) acceleration increases as the rotation speed increases, and the total acceleration reaches 10 m/s^2 when

$$a_r = \sqrt{a^2 - a_t^2} = \sqrt{(10 \text{ m/s}^2)^2 - (3.65 \text{ m/s}^2)^2} = 9.31 \text{ m/s}^2$$

Radial acceleration is $a_r = \omega^2 r$, so the corresponding angular velocity is

$$\omega = \sqrt{\frac{a_r}{r}} = \sqrt{\frac{9.31 \text{ m/s}^2}{0.38 \text{ m}}} = 4.95 \text{ rad/s}$$

For constant angular acceleration, $\omega = \alpha t$, so this angular velocity is achieved at

$$t = \frac{\omega}{\alpha} = \frac{4.95 \text{ rad/s}}{9.6 \text{ rad/s}^2} = 0.52 \text{ s}$$

Thus it takes 0.52 s for the acceleration of the blade tip to reach 10 m/s^2.

ASSESS The motor has not completed 2 full revolutions in 1.5 s, so it has a slow start and modest accelerations. A tangential acceleration of 3.65 m/s^2 seems reasonable, so we have confidence in our final answer of 0.52 s.

STOP TO THINK 4.7   The fan blade is slowing down. What are the signs of ω and α?

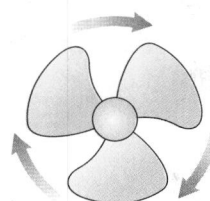

a. ω is positive and α is positive.
b. ω is positive and α is negative.
c. ω is negative and α is positive.
d. ω is negative and α is negative.

CHALLENGE EXAMPLE 4.16   **Hit the target!**

One day when you come into lab, you see a spring-loaded wheel that can launch a ball straight up. To do so, you place the ball in a cup on the rim of the wheel, turn the wheel to stretch the spring, then release. The wheel rotates through an angle $\Delta\theta$, then hits a stop when the cup is level with the axle and pointing straight up. The cup stops, but the ball flies out and keeps going. You're told that the wheel has been designed to have constant angular acceleration as it rotates through $\Delta\theta$. The lab assignment is first to measure the wheel's angular acceleration. Then the lab instructor is going to place a target at height h above the point where the ball is launched. Your task will be to launch the ball so that it just barely hits the target. You'll lose points if the ball doesn't reach the target or if it slams into the target.

a. Find an expression in terms of quantities that you can measure for the angle $\Delta\theta$ that launches the ball at the correct speed.
b. Evaluate $\Delta\theta$ if you've determined that the wheel's diameter is 62 cm, its angular acceleration is 200 rad/s^2, the mass of the ball is 25 g, and the instructor places the target 190 cm above the launch point.

MODEL   Model the ball as a particle. It first undergoes nonuniform circular motion. We'll then ignore air resistance and treat the vertical motion as free fall.

VISUALIZE   FIGURE 4.40 is a pictorial representation. This is a two-part problem, with the speed at the end of the angular acceleration being the launch speed for the vertical motion. We've chosen to call the wheel radius R and the target height h. These and the angular acceleration α are considered "known" because we will measure them, but we don't have numerical values at this time.

FIGURE 4.40 Pictorial representation of the ball launcher.

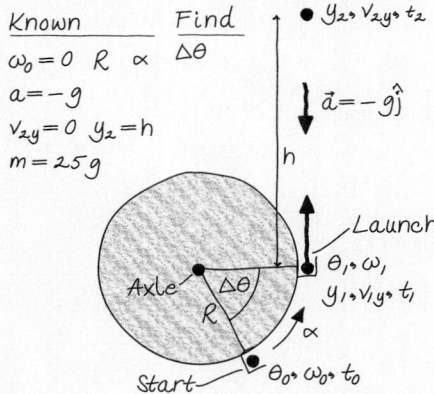

SOLVE

a. The circular motion problem and the vertical motion problem are connected through the ball's speed: The final speed of the angular acceleration is the launch speed of the vertical motion. We don't know anything about time intervals, which suggests using the kinematic equations that relate distance and acceleration (for the vertical motion) and angle and angular acceleration (for the circular motion). For the angular acceleration, with $\omega_0 = 0$ rad/s,

$$\omega_1^2 = \omega_0^2 + 2\alpha\,\Delta\theta = 2\alpha\,\Delta\theta$$

The final speed of the ball and cup, when the wheel hits the stop, is

$$v_1 = \omega_1 R = R\sqrt{2\alpha\,\Delta\theta}$$

Thus the vertical-motion problem begins with the ball being shot upward with velocity $v_{1y} = R\sqrt{2\alpha\,\Delta\theta}$. How high does it go? The highest point is the point where $v_{2y} = 0$, so the free-fall equation is

$$v_{2y}^2 = 0 = v_{1y}^2 - 2g\Delta y = R^2 \cdot 2\alpha\,\Delta\theta - 2gh$$

Rather than solve for height h, we need to solve for the angle that produces a given height. This is

$$\Delta\theta = \frac{gh}{\alpha R^2}$$

Once we've determined the properties of the wheel and then measured the height at which our instructor places the target, we'll quickly be able to calculate the angle through which we should pull back the wheel to launch the ball.

b. For the values given in the problem statement, $\Delta\theta = 0.969$ rad $= 56°$. Don't forget that equations involving angles need values in radians and return values in radians.

ASSESS   The angle needed to be less than 90° or else the ball would fall out of the cup before launch. And an angle of only a few degrees would seem suspiciously small. Thus 56° seems to be reasonable. Notice that the mass was not needed in this problem. Part of becoming a better problem solver is evaluating the information you have to see what is relevant. Some homework problems will help you develop this skill by providing information that isn't necessary.

SUMMARY

The goal of Chapter 4 has been to learn how to solve problems about motion in a plane.

General Principles

The instantaneous velocity

$$\vec{v} = d\vec{r}/dt$$

is a vector tangent to the trajectory.

The instantaneous acceleration is

$$\vec{a} = d\vec{v}/dt$$

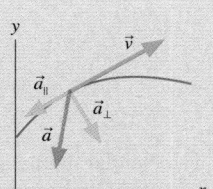

$\vec{a}_\parallel$, the component of $\vec{a}$ parallel to $\vec{v}$, is responsible for change of *speed*. $\vec{a}_\perp$, the component of $\vec{a}$ perpendicular to $\vec{v}$, is responsible for change of *direction*.

Relative motion

If object C moves relative to reference frame A with velocity $\vec{v}_{CA}$, then it moves relative to a different reference frame B with velocity

$$\vec{v}_{CB} = \vec{v}_{CA} + \vec{v}_{AB}$$

where $\vec{v}_{AB}$ is the velocity of A relative to B. This is the Galilean transformation of velocity.

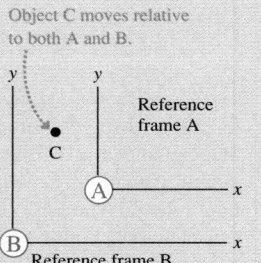

Object C moves relative to both A and B.

Important Concepts

Uniform Circular Motion

Angular velocity $\omega = d\theta/dt$.
v_t and ω are constant:

$$v_t = \omega r$$

The centripetal acceleration points toward the center of the circle:

$$a = \frac{v^2}{r} = \omega^2 r$$

It changes the particle's direction but not its speed.

Nonuniform Circular Motion

Angular acceleration $\alpha = d\omega/dt$.
The radial acceleration

$$a_r = \frac{v^2}{r} = \omega^2 r$$

changes the particle's direction. The tangential component

$$a_t = \alpha r$$

changes the particle's speed.

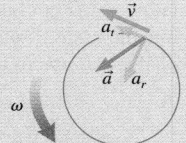

Applications

Kinematics in two dimensions

If $\vec{a}$ is constant, then the x- and y-components of motion are independent of each other.

$$x_f = x_i + v_{ix}\Delta t + \tfrac{1}{2}a_x(\Delta t)^2$$

$$y_f = y_i + v_{iy}\Delta t + \tfrac{1}{2}a_y(\Delta t)^2$$

$$v_{fx} = v_{ix} + a_x\Delta t$$

$$v_{fy} = v_{iy} + a_y\Delta t$$

Projectile motion occurs if the object moves under the influence of only gravity. The motion is a parabola.

- Uniform motion in the horizontal direction with $v_{0x} = v_0\cos\theta$.
- Free-fall motion in the vertical direction with $a_y = -g$ and $v_{0y} = v_0\sin\theta$.
- The x and y kinematic equations have the *same* value for Δt.

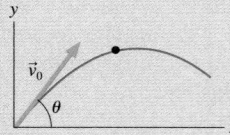

Circular motion kinematics

Period $T = \dfrac{2\pi r}{v} = \dfrac{2\pi}{\omega}$

Angular position $\theta = \dfrac{s}{r}$

$$\omega_f = \omega_i + \alpha\,\Delta t$$

$$\theta_f = \theta_i + \omega_i\,\Delta t + \tfrac{1}{2}\alpha(\Delta t)^2$$

$$\omega_f^2 = \omega_i^2 + 2\alpha\,\Delta\theta$$

Angle, angular velocity, and angular acceleration are related graphically.

- The angular velocity is the slope of the angular position graph.
- The angular acceleration is the slope of the angular velocity graph.

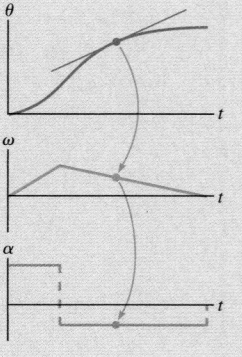

Terms and Notation

projectile	uniform circular motion	angular displacement, $\Delta\theta$	radial acceleration, a_r
launch angle, θ	period, T	angular velocity, ω	tangential acceleration, a_t
reference frame	angular position, θ	centripetal acceleration, a	
Galilean transformation of velocity	arc length, s	nonuniform circular motion	
	radians	angular acceleration, α	

CONCEPTUAL QUESTIONS

1. a. At this instant, is the particle in FIGURE Q4.1 speeding up, slowing down, or traveling at constant speed?
   b. Is this particle curving to the right, curving to the left, or traveling straight?

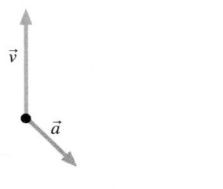

FIGURE Q4.1

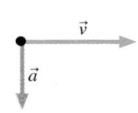

FIGURE Q4.2

2. a. At this instant, is the particle in FIGURE Q4.2 speeding up, slowing down, or traveling at constant speed?
   b. Is this particle curving upward, curving downward, or traveling straight?

3. Tarzan swings through the jungle by hanging from a vine.
   a. Immediately after stepping off a branch to swing over to another tree, is Tarzan's acceleration $\vec{a}$ zero or not zero? If not zero, which way does it point? Explain.
   b. Answer the same question at the lowest point in Tarzan's swing.

4. A projectile is launched at an angle of 30°.
   a. Is there any point on the trajectory where $\vec{v}$ and $\vec{a}$ are parallel to each other? If so, where?
   b. Is there any point where $\vec{v}$ and $\vec{a}$ are perpendicular to each other? If so, where?

5. For a projectile, which of the following quantities are constant during the flight: x, y, r, v_x, v_y, v, a_x, a_y? Which of these quantities are zero throughout the flight?

6. A cart that is rolling at constant velocity on a level table fires a ball straight up.
   a. When the ball comes back down, will it land in front of the launching tube, behind the launching tube, or directly in the tube? Explain.
   b. Will your answer change if the cart is accelerating in the forward direction? If so, how?

7. A rock is thrown from a bridge at an angle 30° below horizontal. Immediately after the rock is released, is the magnitude of its acceleration greater than, less than, or equal to g? Explain.

8. Anita is running to the right at 5 m/s in FIGURE Q4.8. Balls 1 and 2 are thrown toward her by friends standing on the ground. According to Anita, both balls are approaching her at 10 m/s.

Which ball was thrown at a faster speed? Or were they thrown with the same speed? Explain.

FIGURE Q4.8

9. An electromagnet on the ceiling of an airplane holds a steel ball. When a button is pushed, the magnet releases the ball. The experiment is first done while the plane is parked on the ground, and the point where the ball hits the floor is marked with an X. Then the experiment is repeated while the plane is flying level at a steady 500 mph. Does the ball land slightly in front of the X (toward the nose of the plane), on the X, or slightly behind the X (toward the tail of the plane)? Explain.

10. Zack is driving past his house in FIGURE Q4.10. He wants to toss his physics book out the window and have it land in his driveway. If he lets go of the book exactly as he passes the end of the driveway, should he direct his throw outward and toward the front of the car (throw 1), straight outward (throw 2), or outward and toward the back of the car (throw 3)? Explain.

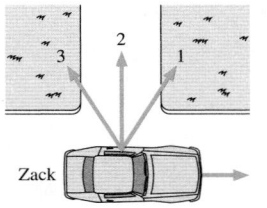

FIGURE Q4.10

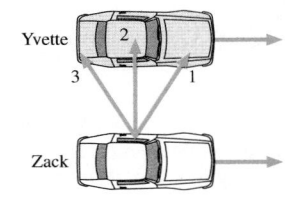

FIGURE Q4.11

11. In FIGURE Q4.11, Yvette and Zack are driving down the freeway side by side with their windows down. Zack wants to toss his physics book out the window and have it land in Yvette's front seat. Ignoring air resistance, should he direct his throw outward and toward the front of the car (throw 1), straight outward (throw 2), or outward and toward the back of the car (throw 3)? Explain.

12. In uniform circular motion, which of the following quantities are constant: speed, instantaneous velocity, tangential velocity, radial acceleration, tangential acceleration? Which of these quantities are zero throughout the motion?

13. **FIGURE Q4.13** shows three points on a steadily rotating wheel.

   a. Rank in order, from largest to smallest, the angular velocities ω_1, ω_2, and ω_3 of these points. Explain.

   b. Rank in order, from largest to smallest, the speeds v_1, v_2, and v_3 of these points. Explain.

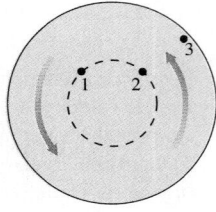

FIGURE Q4.13

14. **FIGURE Q4.14** shows four rotating wheels. For each, determine the signs (+ or −) of ω and α.

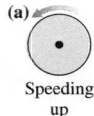 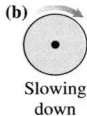 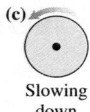 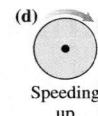

(a) Speeding up (b) Slowing down (c) Slowing down (d) Speeding up

FIGURE Q4.14

15. **FIGURE Q4.15** shows a pendulum at one end point of its arc.

   a. At this point, is ω positive, negative, or zero? Explain.

   b. At this point, is α positive, negative, or zero? Explain.

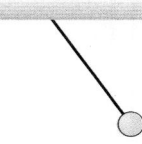

FIGURE Q4.15

EXERCISES AND PROBLEMS

Exercises

Section 4.1 Acceleration

Problems 1 and 2 show a partial motion diagram. For each:

a. Complete the motion diagram by adding acceleration vectors.

b. Write a physics *problem* for which this is the correct motion diagram. Be imaginative! Don't forget to include enough information to make the problem complete and to state clearly what is to be found.

1. |

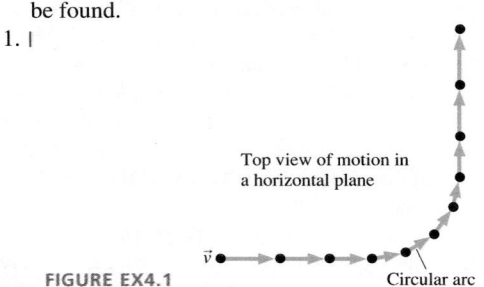

Top view of motion in a horizontal plane

FIGURE EX4.1

Circular arc

2. |

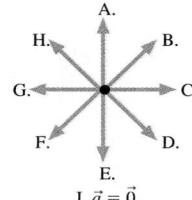

FIGURE EX4.2

Answer Problems 3 through 5 by choosing one of the eight labeled acceleration vectors or selecting option I: $\vec{a} = \vec{0}$.

3. | At this instant, the particle is slowing and curving upward. What is the direction of its acceleration?

FIGURE EX4.3

4. | At this instant, the particle has steady speed and is curving to the right. What is the direction of its acceleration?

FIGURE EX4.4

5. | At this instant, the particle is speeding up and curving downward. What is the direction of its acceleration?

FIGURE EX4.5

Section 4.2 Two-Dimensional Kinematics

6. ‖ A sailboat is traveling east at 5.0 m/s. A sudden gust of wind gives the boat an acceleration $\vec{a} = (0.80 \text{ m/s}^2, 40° \text{ north of east})$. What are the boat's speed and direction 6.0 s later when the gust subsides?

7. ‖ A model rocket is launched from rest with an upward acceleration of 6.00 m/s^2 and, due to a strong wind, a horizontal acceleration of 1.50 m/s^2 How far is the rocket from the launch pad 6.00 s later when the rocket engine runs out of fuel?

8. ‖ A particle's trajectory is described by $x = \left(\frac{1}{2}t^3 - 2t^2\right)$ m and $y = \left(\frac{1}{2}t^2 - 2t\right)$ m, where t is in s.

   a. What are the particle's position and speed at $t = 0$ s and $t = 4$ s?

   b. What is the particle's direction of motion, measured as an angle from the x-axis, at $t = 0$ s and $t = 4$ s?

9. ‖ A rocket-powered hockey puck moves on a horizontal frictionless table. **FIGURE EX4.9** shows graphs of v_x and v_y, the x- and y-components of the puck's velocity. The puck starts at the origin.

   a. In which direction is the puck moving at $t = 2$ s? Give your answer as an angle from the x-axis.

   b. How far from the origin is the puck at $t = 5$ s?

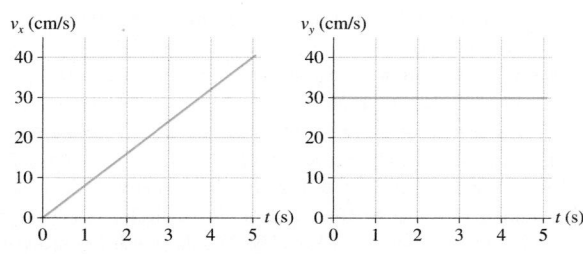

FIGURE EX4.9

10. ‖ A rocket-powered hockey puck moves on a horizontal frictionless table. FIGURE EX4.10 shows graphs of v_x and v_y, the x- and y-components of the puck's velocity. The puck starts at the origin. What is the magnitude of the puck's acceleration at $t = 5$ s?

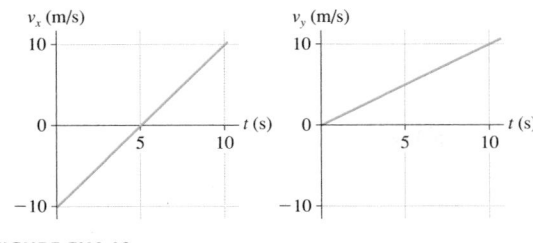

FIGURE EX4.10

Section 4.3 Projectile Motion

11. ‖ A physics student on Planet Exidor throws a ball, and it follows the parabolic trajectory shown in FIGURE EX4.11. The ball's position is shown at 1 s intervals until $t = 3$ s. At $t = 1$ s, the ball's velocity is $\vec{v} = (2.0\,\hat{i} + 2.0\,\hat{j})$ m/s.
   a. Determine the ball's velocity at $t = 0$ s, 2 s, and 3 s.
   b. What is the value of g on Planet Exidor?
   c. What was the ball's launch angle?

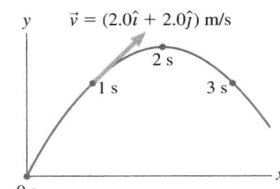

FIGURE EX4.11

12. ‖ A ball thrown horizontally at 25 m/s travels a horizontal distance of 50 m before hitting the ground. From what height was the ball thrown?

13. ‖ A rifle is aimed horizontally at a target 50 m away. The bullet hits the target 2.0 cm below the aim point.
   a. What was the bullet's flight time?
   b. What was the bullet's speed as it left the barrel?

14. ‖ A supply plane needs to drop a package of food to scientists working on a glacier in Greenland. The plane flies 100 m above the glacier at a speed of 150 m/s. How far short of the target should it drop the package?

Section 4.4 Relative Motion

15. ‖ A boat takes 3.0 hours to travel 30 km down a river, then 5.0 hours to return. How fast is the river flowing?

16. ‖ When the moving sidewalk at the airport is broken, as it often seems to be, it takes you 50 s to walk from your gate to baggage claim. When it is working and you stand on the moving sidewalk the entire way, without walking, it takes 75 s to travel the same distance. How long will it take you to travel from the gate to baggage claim if you walk while riding on the moving sidewalk?

17. ‖ Mary needs to row her boat across a 100-m-wide river that is flowing to the east at a speed of 1.0 m/s. Mary can row with a speed of 2.0 m/s.
   a. If Mary points her boat due north, how far from her intended landing spot will she be when she reaches the opposite shore?
   b. What is her speed with respect to the shore?

18. ‖ Susan, driving north at 60 mph, and Trent, driving east at 45 mph, are approaching an intersection. What is Trent's speed relative to Susan's reference frame?

Section 4.5 Uniform Circular Motion

19. ‖ FIGURE EX4.19 shows the angular-position-versus-time graph for a particle moving in a circle. What is the particle's angular velocity at (a) $t = 1$ s, (b) $t = 4$ s, and (c) $t = 7$ s?

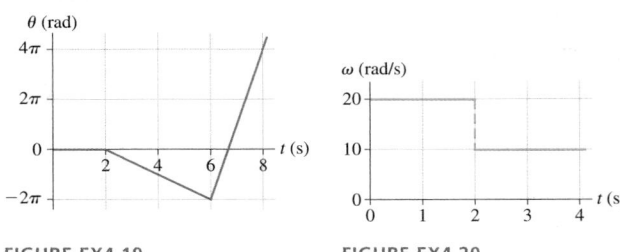

FIGURE EX4.19   FIGURE EX4.20

20. ‖ FIGURE EX4.20 shows the angular-velocity-versus-time graph for a particle moving in a circle. How many revolutions does the object make during the first 4 s?

21. ‖ FIGURE EX4.21 shows the angular-velocity-versus-time graph for a particle moving in a circle, starting from $\theta_0 = 0$ rad at $t = 0$ s. Draw the angular-position-versus-time graph. Include an appropriate scale on both axes.

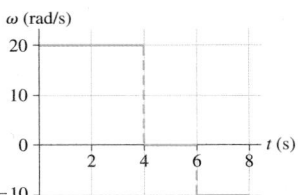

FIGURE EX4.21

22. ‖ An old-fashioned single-play vinyl record rotates on a turntable at 45 rpm. What are (a) the angular velocity in rad/s and (b) the period of the motion?

23. ‖ The earth's radius is about 4000 miles. Kampala, the capital of Uganda, and Singapore are both nearly on the equator. The distance between them is 5000 miles. The flight from Kampala to Singapore takes 9.0 hours. What is the plane's angular velocity with respect to the earth's surface? Give your answer in °/h.

Section 4.6 Velocity and Acceleration in Uniform Circular Motion

24. ‖ A 3000-m-high mountain is located on the equator. How much faster does a climber on top of the mountain move than a surfer at a nearby beach? The earth's radius is 6400 km.

25. ‖ How fast must a plane fly along the earth's equator so that the sun stands still relative to the passengers? In which direction must the plane fly, east to west or west to east? Give your answer in both km/h and mph. The earth's radius is 6400 km.

26. ‖ To withstand "g-forces" of up to 10 g's, caused by suddenly pulling out of a steep dive, fighter jet pilots train on a "human centrifuge." 10 g's is an acceleration of 98 m/s^2. If the length of the centrifuge arm is 12 m, at what speed is the rider moving when she experiences 10 g's?

27. | The radius of the earth's very nearly circular orbit around the sun is 1.5×10^{11} m. Find the magnitude of the earth's (a) velocity, (b) angular velocity, and (c) centripetal acceleration as it travels around the sun. Assume a year of 365 days.

28. ‖ Your roommate is working on his bicycle and has the bike upside down. He spins the 60-cm-diameter wheel, and you notice that a pebble stuck in the tread goes by three times every second. What are the pebble's speed and acceleration?

Section 4.7 Nonuniform Circular Motion and Angular Acceleration

29. | FIGURE EX4.29 shows the angular velocity graph of the crankshaft in a car. What is the crankshaft's angular acceleration at (a) $t = 1$ s, (b) $t = 3$ s, and (c) $t = 5$ s?

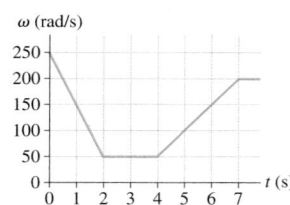

FIGURE EX4.29

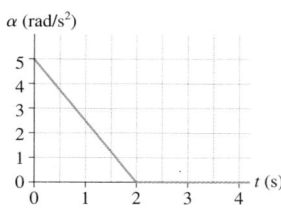

FIGURE EX4.30

30. ‖ FIGURE EX4.30 shows the angular acceleration graph of a turntable that starts from rest. What is the turntable's angular velocity at (a) $t = 1$ s, (b) $t = 2$ s, and (c) $t = 3$ s?

31. ‖ FIGURE EX4.31 shows the angular-velocity-versus-time graph for a particle moving in a circle. How many revolutions does the object make during the first 4 s?

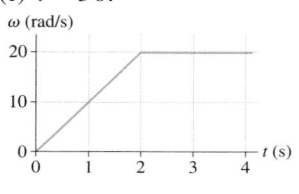

FIGURE EX4.31

32. ‖ A 5.0-m-diameter merry-go-round is initially turning with a 4.0 s period. It slows down and stops in 20 s.
   a. Before slowing, what is the speed of a child on the rim?
   b. How many revolutions does the merry-go-round make as it stops?

33. ‖ An electric fan goes from rest to 1800 rpm in 4.0 s. What is its angular acceleration?

34. ‖‖ A bicycle wheel is rotating at 50 rpm when the cyclist begins to pedal harder, giving the wheel a constant angular acceleration of 0.50 rad/s².
   a. What is the wheel's angular velocity, in rpm, 10 s later?
   b. How many revolutions does the wheel make during this time?

35. ‖‖ A 3.0-cm-diameter crankshaft that is rotating at 2500 rpm comes to a halt in 1.5 s.
   a. What is the tangential acceleration of a point on the surface?
   b. How many revolutions does the crankshaft make as it stops?

Problems

36. ‖ A particle starts from rest at $\vec{r}_0 = 9.0\hat{j}$ m and moves in the xy-plane with the velocity shown in FIGURE P4.36. The particle passes through a wire hoop located at $\vec{r}_1 = 20\hat{i}$ m, then continues onward.

   a. At what time does the particle pass through the hoop?
   b. What is the value of v_{4y}, the y-component of the particle's velocity at $t = 4$ s?

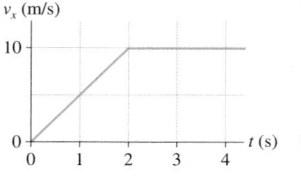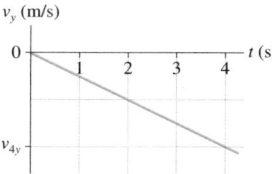

FIGURE EX4.36

37. ‖ A spaceship maneuvering near Planet Zeta is located at $\vec{r} = (600\hat{i} - 400\hat{j} + 200\hat{k}) \times 10^3$ km, relative to the planet, and traveling at $\vec{v} = 9500\hat{i}$ m/s. It turns on its thruster engine and accelerates with $\vec{a} = (40\hat{i} - 20\hat{k})$ m/s² for 35 min. Where is the spaceship located when the engine shuts off? Give your answer as a vector measured in km.

38. ‖ A projectile's horizontal range on level ground is $R = v_0^2 \sin 2\theta / g$. At what launch angle or angles will the projectile land at half of its maximum possible range?

39. ‖ a. A projectile is launched with speed v_0 and angle θ. Derive an expression for the projectile's maximum height h.
   b. A baseball is hit with a speed of 33.6 m/s. Calculate its height and the distance traveled if it is hit at angles of 30.0°, 45.0°, and 60.0°.

40. ‖ A gray kangaroo can bound across level ground with each
BIO jump carrying it 10 m from the takeoff point. Typically the kangaroo leaves the ground at a 20° angle. If this is so:
   a. What is its takeoff speed?
   b. What is its maximum height above the ground?

41. ‖ A projectile is fired with an initial speed of 30 m/s at an angle of 60° above the horizontal. The object hits the ground 7.5 s later.
   a. How much higher or lower is the launch point relative to the point where the projectile hits the ground?
   b. To what maximum height above the launch point does the projectile rise?

42. ‖ In the Olympic shotput event, an athlete throws the shot with an initial speed of 12.0 m/s at a 40.0° angle from the horizontal. The shot leaves her hand at a height of 1.80 m above the ground.
   a. How far does the shot travel?
   b. Repeat the calculation of part (a) for angles 42.5°, 45.0°, and 47.5°. Put all your results, including 40.0°, in a table. At what angle of release does she throw the farthest?

43. ‖ On the Apollo 14 mission to the moon, astronaut Alan Shepard hit a golf ball with a 6 iron. The free-fall acceleration on the moon is 1/6 of its value on earth. Suppose he hit the ball with a speed of 25 m/s at an angle 30° above the horizontal.
   a. How much farther did the ball travel on the moon than it would have on earth?
   b. For how much more time was the ball in flight?

44. ‖ A ball is thrown toward a cliff of height h with a speed of 30 m/s and an angle of 60° above horizontal. It lands on the edge of the cliff 4.0 s later.
   a. How high is the cliff?
   b. What was the maximum height of the ball?
   c. What is the ball's impact speed?

45. ‖ A tennis player hits a ball 2.0 m above the ground. The ball leaves his racquet with a speed of 20.0 m/s at an angle 5.0° above the horizontal. The horizontal distance to the net is 7.0 m, and the net is 1.0 m high. Does the ball clear the net? If so, by how much? If not, by how much does it miss?

46. ‖‖ A baseball player friend of yours wants to determine his pitching speed. You have him stand on a ledge and throw the ball horizontally from an elevation 4.0 m above the ground. The ball lands 25 m away.
   a. What is his pitching speed?
   b. As you think about it, you're not sure he threw the ball exactly horizontally. As you watch him throw, the pitches seem to vary from 5° below horizontal to 5° above horizontal. What are the lowest and highest speeds with which the ball might have left his hand?

47. ‖ You are playing right field for the baseball team. Your team is up by one run in the bottom of the last inning of the game when a ground ball slips through the infield and comes straight toward you. As you pick up the ball 65 m from home plate, you see a runner rounding third base and heading for home with the tying run. You throw the ball at an angle of 30° above the horizontal with just the right speed so that the ball is caught by the catcher, standing on home plate, at the same height as you threw it. As you release the ball, the runner is 20 m from home plate and running full speed at 8.0 m/s. Will the ball arrive in time for your team's catcher to make the tag and win the game?

48. ‖ You're 6.0 m from one wall of the house seen in FIGURE P4.48. You want to toss a ball to your friend who is 6.0 m from the opposite wall. The throw and catch each occur 1.0 m above the ground.
   a. What minimum speed will allow the ball to clear the roof?
   b. At what angle should you toss the ball?

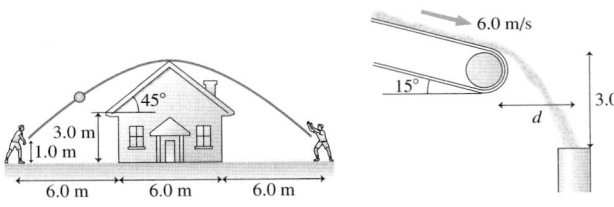

FIGURE P4.48   FIGURE P4.49

49. ‖ Sand moves without slipping at 6.0 m/s down a conveyer that is tilted at 15°. The sand enters a pipe 3.0 m below the end of the conveyer belt, as shown in FIGURE P4.49. What is the horizontal distance d between the conveyer belt and the pipe?

50. ‖ A stunt man drives a car at a speed of 20 m/s off a 30-m-high cliff. The road leading to the cliff is inclined upward at an angle of 20°.
   a. How far from the base of the cliff does the car land?
   b. What is the car's impact speed?

51. ‖ A javelin thrower standing at rest holds the center of the
BIO javelin behind her head, then accelerates it through a distance of 70 cm as she throws. She releases the javelin 2.0 m above the ground traveling at an angle of 30° above the horizontal. Top-rated javelin throwers do throw at about a 30° angle, not the 45° you might have expected, because the biomechanics of the arm allow them to throw the javelin much faster at 30° than they would be able to at 45°. In this throw, the javelin hits the ground 62 m away. What was the acceleration of the javelin during the throw? Assume that it has a constant acceleration.

52. ‖ Ships A and B leave port together. For the next two hours, ship A travels at 20 mph in a direction 30° west of north while the ship B travels 20° east of north at 25 mph.

   a. What is the distance between the two ships two hours after they depart?
   b. What is the speed of ship A as seen by ship B?

53. ‖ A kayaker needs to paddle north across a 100-m-wide harbor. The tide is going out, creating a tidal current that flows to the east at 2.0 m/s. The kayaker can paddle with a speed of 3.0 m/s.
   a. In which direction should he paddle in order to travel straight across the harbor?
   b. How long will it take him to cross?

54. ‖ Mike throws a ball upward and toward the east at a 63° angle with a speed of 22 m/s. Nancy drives east past Mike at 30 m/s at the instant he releases the ball.
   a. What is the ball's initial angle in Nancy's reference frame?
   b. Find and graph the ball's trajectory as seen by Nancy.

55. ‖ While driving north at 25 m/s during a rainstorm you notice that the rain makes an angle of 38° with the vertical. While driving back home moments later at the same speed but in the opposite direction, you see that the rain is falling straight down. From these observations, determine the speed and angle of the raindrops relative to the ground.

56. ‖ You've been assigned the task of using a shaft encoder—a device that measures the angle of a shaft or axle and provides a signal to a computer—to analyze the rotation of an engine crankshaft under certain conditions. The table lists the crankshaft's angles over a 0.6 s interval.

Time (s)	Angle (rad)
0.0	0.0
0.1	2.0
0.2	3.2
0.3	4.3
0.4	5.3
0.5	6.1
0.6	7.0

   Is the crankshaft rotating with uniform circular motion? If so, what is its angular velocity in rpm? If not, is the angular acceleration positive or negative?

57. ‖ A speck of dust on a spinning DVD has a centripetal acceleration of 20 m/s².
   a. What is the acceleration of a different speck of dust that is twice as far from the center of the disk?
   b. What would be the acceleration of the first speck of dust if the disk's angular velocity was doubled?

58. ‖ A typical laboratory centrifuge rotates at 4000 rpm. Test tubes have to be placed into a centrifuge very carefully because of the very large accelerations.
   a. What is the acceleration at the end of a test tube that is 10 cm from the axis of rotation?
   b. For comparison, what is the magnitude of the acceleration a test tube would experience if dropped from a height of 1.0 m and stopped in a 1.0-ms-long encounter with a hard floor?

59. ‖ Astronauts use a centrifuge to simulate the acceleration of a
BIO rocket launch. The centrifuge takes 30 s to speed up from rest to its top speed of 1 rotation every 1.3 s. The astronaut is strapped into a seat 6.0 m from the axis.
   a. What is the astronaut's tangential acceleration during the first 30 s?
   b. How many g's of acceleration does the astronaut experience when the device is rotating at top speed? Each 9.8 m/s² of acceleration is 1 g.

60. ‖ Peregrine falcons are known for their maneuvering ability. In a tight circular turn, a falcon can attain a centripetal acceleration 1.5 times the free-fall acceleration. What is the radius of the turn if the falcon is flying at 25 m/s?

    BIO

61. ‖ As the earth rotates, what is the speed of (a) a physics student in Miami, Florida, at latitude 26°, and (b) a physics student in Fairbanks, Alaska, at latitude 65°? Ignore the revolution of the earth around the sun. The radius of the earth is 6400 km.

62. ‖ Communications satellites are placed in a circular orbit where they stay directly over a fixed point on the equator as the earth rotates. These are called *geosynchronous orbits*. The radius of the earth is 6.37×10^6 m, and the altitude of a geosynchronous orbit is 3.58×10^7 m ($\approx 22{,}000$ miles). What are (a) the speed and (b) the magnitude of the acceleration of a satellite in a geosynchronous orbit?

63. ‖ A computer hard disk 8.0 cm in diameter is initially at rest. A small dot is painted on the edge of the disk. The disk accelerates at 600 rad/s² for $\frac{1}{2}$ s, then coasts at a steady angular velocity for another $\frac{1}{2}$ s.
    a. What is the speed of the dot at $t = 1.0$ s?
    b. Through how many revolutions has the disk turned?

64. ‖ a. A turbine spinning with angular velocity ω_0 rad/s comes to a halt in T seconds. Find an expression for the angle $\Delta\theta$ through which the turbine turns while stopping.
    b. A turbine is spinning at 3800 rpm. Friction in the bearings is so low that it takes 10 min to coast to a stop. How many revolutions does the turbine make while stopping?

65. ‖ A high-speed drill rotating ccw at 2400 rpm comes to a halt in 2.5 s.
    a. What is the drill's angular acceleration?
    b. How many revolutions does it make as it stops?

66. ‖ A wheel initially rotating at 60 rpm experiences the angular acceleration shown in FIGURE P4.66. What is the wheel's angular velocity, in rpm, at $t = 3.0$ s?

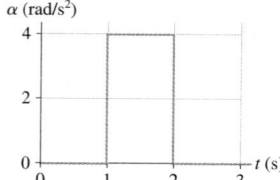

FIGURE P4.66

67. ‖ Your car tire is rotating at 3.5 rev/s when suddenly you press down hard on the accelerator. After traveling 200 m, the tire's rotation has increased to 6.0 rev/s. What was the tire's angular acceleration? Give your answer in rad/s².

68. ‖ The angular velocity of a process control motor is $\omega = (20 - \frac{1}{2}t^2)$ rad/s, where t is in seconds.
    a. At what time does the motor reverse direction?
    b. Through what angle does the motor turn between $t = 0$ s and the instant at which it reverses direction?

69. ‖ A Ferris wheel of radius R speeds up with angular acceleration α starting from rest. Find an expression for the (a) velocity and (b) centripetal acceleration of a rider after the Ferris wheel has rotated through angle $\Delta\theta$.

70. ‖ A 6.0-cm-diameter gear rotates with angular velocity $\omega = (2.0 + \frac{1}{2}t^2)$ rad/s, where t is in seconds. At $t = 4.0$ s, what are:
    a. The gear's angular acceleration?
    b. The tangential acceleration of a tooth on the gear?

71. ‖ On a lonely highway, with no other cars in sight, you decide to measure the angular acceleration of your engine's crankshaft while braking gently. Having excellent memory, you are able to read the tachometer every 1.0 s and remember seven values long enough to later write them down. The table shows your data:

Time (s)	rpm
0.0	3010
1.0	2810
2.0	2450
3.0	2250
4.0	1940
5.0	1810
6.0	1510

What is the *magnitude* of the crankshaft's angular acceleration? Give your result in rad/s².

72. ‖‖ A car starts from rest on a curve with a radius of 120 m and accelerates at 1.0 m/s². Through what angle will the car have traveled when the magnitude of its total acceleration is 2.0 m/s²?

73. ‖‖ A long string is wrapped around a 6.0-cm-diameter cylinder, initially at rest, that is free to rotate on an axle. The string is then pulled with a constant acceleration of 1.5 m/s² until 1.0 m of string has been unwound. If the string unwinds without slipping, what is the cylinder's angular speed, in rpm, at this time?

In Problems 74 through 76 you are given the equations that are used to solve a problem. For each of these, you are to
   a. Write a realistic problem for which these are the correct equations. Be sure that the answer your problem requests is consistent with the equations given.
   b. Finish the solution of the problem, including a pictorial representation.

74. $100 \text{ m} = 0 \text{ m} + (50\cos\theta \text{ m/s})t_1$
    $0 \text{ m} = 0 \text{ m} + (50\sin\theta \text{ m/s})t_1 - \frac{1}{2}(9.80 \text{ m/s}^2)t_1^2$

75. $v_x = -(6.0\cos 45°) \text{ m/s} + 3.0 \text{ m/s}$
    $v_y = (6.0\sin 45°) \text{ m/s} + 0 \text{ m/s}$
    $100 \text{ m} = v_y t_1, \ x_1 = v_x t_1$

76. $2.5 \text{ rad} = 0 \text{ rad} + \omega_i (10 \text{ s}) + \left((1.5 \text{ m/s}^2)/2(50 \text{ m})\right)(10 \text{ s})^2$
    $\omega_f = \omega_i + \left((1.5 \text{ m/s}^2)/(50 \text{ m})\right)(10 \text{ s})$

Challenge Problems

77. You are asked to consult for the city's research hospital, where a group of doctors is investigating the bombardment of cancer tumors with high-energy ions. The ions are fired directly toward the center of the tumor at speeds of 5.0×10^6 m/s. To cover the entire tumor area, the ions are deflected sideways by passing them between two charged metal plates that accelerate the ions perpendicular to the direction of their initial motion. The acceleration region is 5.0 cm long, and the ends of the acceleration plates are 1.5 m from the patient. What sideways acceleration is required to deflect an ion 2.0 cm to one side?

    BIO

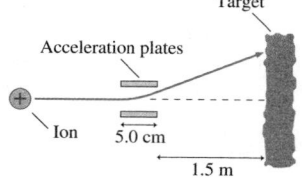

FIGURE CP4.77

78. In one contest at the county fair, seen in FIGURE CP4.78, a spring-loaded plunger launches a ball at a speed of 3.0 m/s from one corner of a smooth, flat board that is tilted up at a 20° angle. To win, you must make the ball hit a small target at the adjacent corner, 2.50 m away. At what angle θ should you tilt the ball launcher?

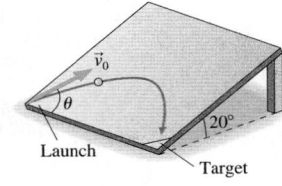

FIGURE CP4.78

79. You are watching an archery tournament when you start wondering how fast an arrow is shot from the bow. Remembering your physics, you ask one of the archers to shoot an arrow parallel to the ground. You find the arrow stuck in the ground 60 m away, making a 3.0° angle with the ground. How fast was the arrow shot?

80. An archer standing on a 15° slope shoots an arrow 20° above the horizontal, as shown in FIGURE CP4.80. How far down the slope does the arrow hit if it is shot with a speed of 50 m/s from 1.75 m above the ground?

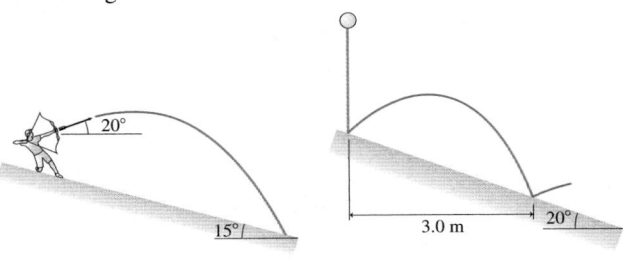

FIGURE CP4.80   FIGURE CP4.81

81. A rubber ball is dropped onto a ramp that is tilted at 20°, as shown in FIGURE CP4.81. A bouncing ball obeys the "law of reflection," which says that the ball leaves the surface at the same angle it approached the surface. The ball's next bounce is 3.0 m to the right of its first bounce. What is the ball's rebound speed on its first bounce?

82. A skateboarder starts up a 1.0-m-high, 30° ramp at a speed of 7.0 m/s. The skateboard wheels roll without friction. How far from the end of the ramp does the skateboarder touch down?

83. A motorcycle daredevil wants to set a record for jumping over burning school buses. He has hired you to help with the design.

He intends to ride off a horizontal platform at 40 m/s, cross the burning buses in a pit below him, then land on a ramp sloping down at 20°. It's very important that he not bounce when he hits the landing ramp because that could cause him to lose control and crash. You immediately recognize that he won't bounce if his velocity is parallel to the ramp as he touches down. This can be accomplished if the ramp is tangent to his trajectory *and* if he lands right on the front edge of the ramp. There's no room for error! Your task is to determine where to place the landing ramp. That is, how far from the edge of the launching platform should the front edge of the landing ramp be horizontally and how far below it? There's a clause in your contract that requires you to test your design before the hero goes on national television to set the record.

84. A cannon on a train car fires a projectile to the right with speed v_0, relative to the train, from a barrel elevated at angle θ. The cannon fires just as the train, which had been cruising to the right along a level track with speed v_{train}, begins to accelerate with acceleration a. Find an expression for the angle at which the projectile should be fired so that it lands as far as possible from the cannon. You can ignore the small height of the cannon above the track.

85. A child in danger of drowning in a river is being carried downstream by a current that flows uniformly with a speed of 2.0 m/s. The child is 200 m from the shore and 1500 m upstream of the boat dock from which the rescue team sets out. If their boat speed is 8.0 m/s with respect to the water, at what angle from the shore should the pilot leave the shore to go directly to the child?

86. An amusement park game, shown in FIGURE CP4.86, launches a marble toward a small cup. The marble is placed directly on top of a spring-loaded wheel and held with a clamp. When released, the wheel spins around clockwise at constant angular acceleration, opening the clamp and releasing the marble after making $\frac{11}{12}$ revolution. What angular acceleration is needed for the ball to land in the cup? The top of the cup is level with the center of the wheel.

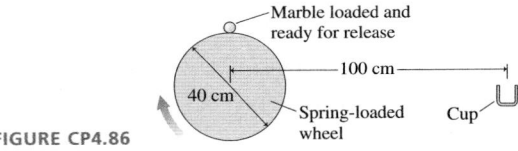

FIGURE CP4.86

STOP TO THINK ANSWERS

Stop to Think 4.1: d. The parallel component of $\vec{a}$ is opposite $\vec{v}$ and will cause the particle to slow down. The perpendicular component of $\vec{a}$ will cause the particle to change direction downward.

Stop to Think 4.2: c. $v = 0$ requires both $v_x = 0$ and $v_y = 0$. Neither x nor y can be changing.

Stop to Think 4.3: d. A projectile's acceleration $\vec{a} = -g\hat{j}$ does not depend on its mass. The second marble has the same initial velocity and the same acceleration, so it follows the same trajectory and lands at the same position.

Stop to Think 4.4: f. The plane's velocity relative to the helicopter is $\vec{v}_{PH} = \vec{v}_{PG} + \vec{v}_{GH} = \vec{v}_{PG} - \vec{v}_{HG}$, where G is the ground. The vector addition shows that $\vec{v}_{PH}$ is to the right and down with a magnitude greater than the 100 m/s of $\vec{v}_{PG}$.

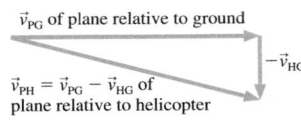

Stop to Think 4.5: b. An initial cw rotation causes the particle's angular position to become increasingly negative. The speed drops to half after reversing direction, so the slope becomes positive and is half as steep as the initial slope. Turning through the same angle returns the particle to $\theta = 0°$.

Stop to Think 4.6: $a_b > a_e > a_a = a_c > a_d$. Centripetal acceleration is v^2/r. Doubling r decreases a_r by a factor of 2. Doubling v increases a_r by a factor of 4. Reversing direction doesn't change a_r.

Stop to Think 4.7: c. ω is negative because the rotation is cw. Because ω is negative but becoming *less* negative, the change $\Delta\omega$ is *positive*. So α is positive.

5 Force and Motion

These ice boats are a memorable example of the connection between force and motion.

▶ **Looking Ahead** The goal of Chapter 5 is to establish a connection between force and motion.

What Causes Motion?

Kinematics describes *how* an object moves. For the more fundamental issue of understanding *why* an object moves, we now turn our attention to **dynamics**.

Dynamics joins with kinematics to form **mechanics**, the science of motion.

◀ **Looking Back**
Section 1.5 Acceleration
Section 3.2 Vector addition

Force

The fundamental concept of dynamics is that of *force*.

- A force is a push or a pull.
- A force acts on an object.
- A force is a vector.
- A force can be a contact force or a long-range force.

Some important forces that we'll study in this chapter are

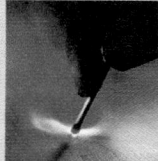

Gravity   Tension   Friction   Drag

Force and Motion

Force causes an object to *accelerate!*

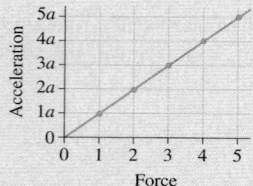

You'll learn that the acceleration of an object is directly proportional to the force exerted on it.

Newton's Laws

You've likely seen Newton's second law, the famous equation $F = ma$. This is the first of several chapters in which you'll learn to use Newton's *three* laws of motion to solve dynamics problems.

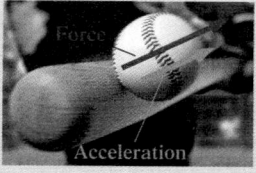

An object accelerates in the same direction as the net force on the object.

Identifying Forces

In this chapter you will learn to *identify* forces and then to represent them on a **free-body diagram.**

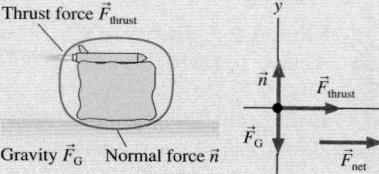

Thrust force $\vec{F}_{thrust}$

Gravity $\vec{F}_G$ Normal force $\vec{n}$

Except for the long-range force of gravity, forces act at points of contact.

5.1 Force

The two major issues that this chapter will examine are:

- What is a force?
- What is the connection between force and motion?

We begin with the first of these questions in the table below.

What is a force?

A force is a push or a pull.

Our commonsense idea of a **force** is that it is a *push* or a *pull.* We will refine this idea as we go along, but it is an adequate starting point. Notice our careful choice of words: We refer to "a force," rather than simply "force." We want to think of a force as a very specific *action,* so that we can talk about a single force or perhaps about two or three individual forces that we can clearly distinguish. Hence the concrete idea of "a force" acting on an object.

A force acts on an object.

Implicit in our concept of force is that a **force acts on an object.** In other words, pushes and pulls are applied *to* something—an object. From the object's perspective, it has a force *exerted* on it. Forces do not exist in isolation from the object that experiences them.

A force requires an agent.

Every force has an **agent,** something that acts or exerts power. That is, a force has a specific, identifiable *cause.* As you throw a ball, it is your hand, while in contact with the ball, that is the agent or the cause of the force exerted on the ball. *If* a force is being exerted on an object, you must be able to identify a specific cause (i.e., the agent) of that force. Conversely, a force is not exerted on an object *unless* you can identify a specific cause or agent. Although this idea may seem to be stating the obvious, you will find it to be a powerful tool for avoiding some common misconceptions about what is and is not a force.

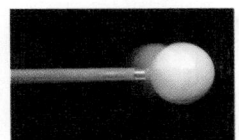

A force is a vector.

If you push an object, you can push either gently or very hard. Similarly, you can push either left or right, up or down. To quantify a push, we need to specify both a magnitude *and* a direction. It should thus come as no surprise that a force is a vector quantity. The general symbol for a force is the vector symbol $\vec{F}$. The size or strength of a force is its magnitude F.

A force can be either a contact force . . .

There are two basic classes of forces, depending on whether the agent touches the object or not. **Contact forces** are forces that act on an object by touching it at a point of contact. The bat must touch the ball to hit it. A string must be tied to an object to pull it. The majority of forces that we will examine are contact forces.

. . . or a long-range force.

Long-range forces are forces that act on an object without physical contact. Magnetism is an example of a long-range force. You have undoubtedly held a magnet over a paper clip and seen the paper clip leap up to the magnet. A coffee cup released from your hand is pulled to the earth by the long-range force of gravity.

NOTE ▶ In the particle model, objects cannot exert forces on themselves. A force on an object will always have an agent or cause external to the object. Now, there are certainly objects that have internal forces (think of all the forces inside the engine of your car!), but the particle model is not valid if you need to consider those internal forces. If you are going to treat your car as a particle and look only at the overall motion of the car as a whole, that motion will be a consequence of external forces acting on the car. ◀

Force Vectors

We can use a simple diagram to visualize how forces are exerted on objects.

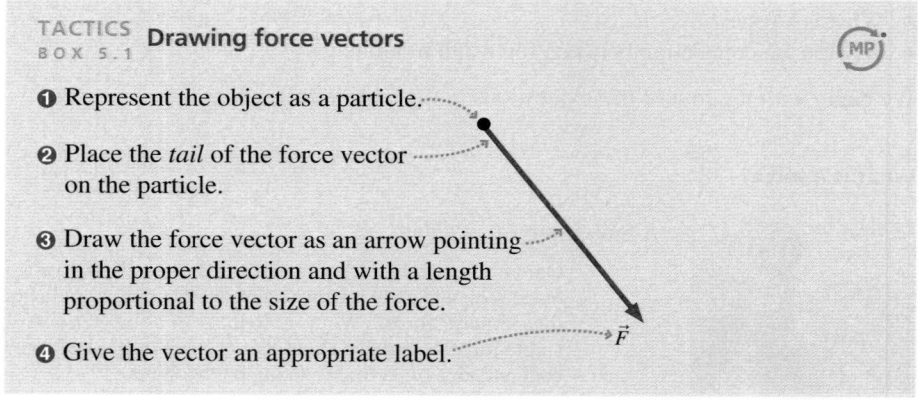

TACTICS
BOX 5.1   **Drawing force vectors**

❶ Represent the object as a particle.

❷ Place the *tail* of the force vector on the particle.

❸ Draw the force vector as an arrow pointing in the proper direction and with a length proportional to the size of the force.

❹ Give the vector an appropriate label.

Step 2 may seem contrary to what a "push" should do, but recall that moving a vector does not change it as long as the length and angle do not change. The vector $\vec{F}$ is the same regardless of whether the tail or the tip is placed on the particle. FIGURE 5.1 shows three examples of force vectors.

FIGURE 5.1 Three examples of forces and their vector representations.

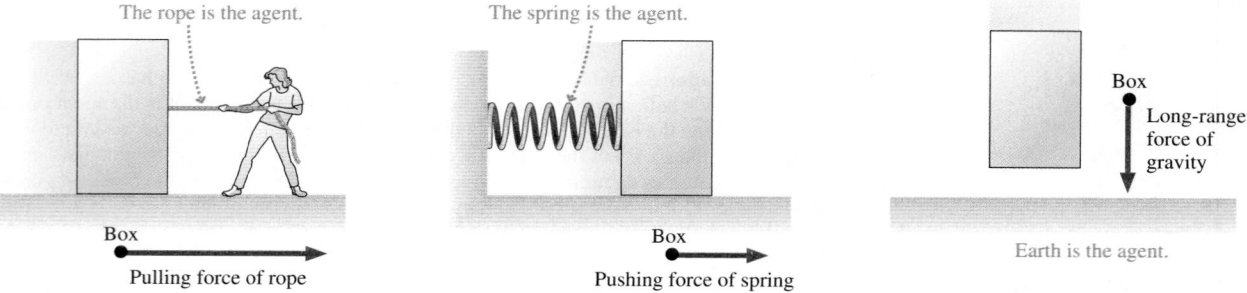

The rope is the agent.

Box
Pulling force of rope

The spring is the agent.

Box
Pushing force of spring

Box
Long-range force of gravity

Earth is the agent.

Combining Forces

FIGURE 5.2a shows a box being pulled by two ropes, each exerting a force on the box. How will the box respond? Experimentally, we find that when several forces $\vec{F}_1$, $\vec{F}_2$, $\vec{F}_3$, . . . are exerted on an object, they combine to form a **net force** given by the *vector sum* of *all* the forces:

$$\vec{F}_{net} \equiv \sum_{i=1}^{N} \vec{F}_i = \vec{F}_1 + \vec{F}_2 + \cdots + \vec{F}_N \tag{5.1}$$

Recall that $\equiv$ is the symbol meaning "is defined as." Mathematically, this summation is called a **superposition of forces**. FIGURE 5.2b shows the net force on the box.

FIGURE 5.2 Two forces applied to a box.

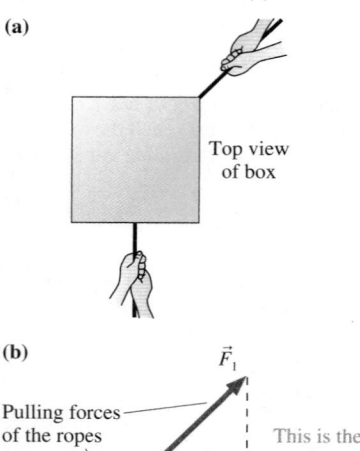

(a)

Top view of box

(b)

$\vec{F}_1$

Pulling forces of the ropes

This is the net force on the box.

$\vec{F}_{net} = \vec{F}_1 + \vec{F}_2$

The box is represented as a particle.

$\vec{F}_2$

STOP TO THINK 5.1   Two of the three forces exerted on an object are shown. The net force points to the left. Which is the missing third force?

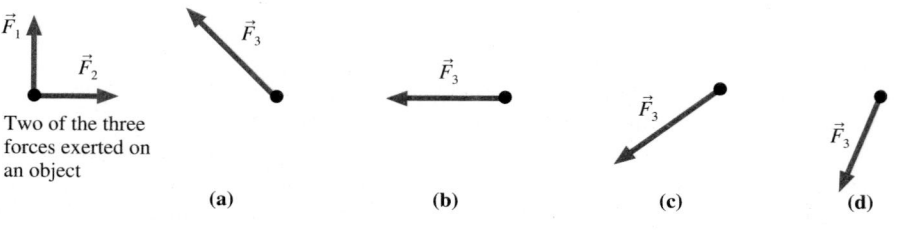

$\vec{F}_1$

$\vec{F}_2$

Two of the three forces exerted on an object

$\vec{F}_3$

(a)

$\vec{F}_3$

(b)

$\vec{F}_3$

(c)

$\vec{F}_3$

(d)

5.2 A Short Catalog of Forces

There are many forces we will deal with over and over. This section will introduce you to some of them. Many of these forces have special symbols. As you learn the major forces, be sure to learn the symbol for each.

Gravity

Gravity—the only long-range force we will encounter in the next few chapters—keeps you in your chair, and the planets in their orbits around the sun. We'll have a thorough look at gravity in Chapter 13. For now we'll concentrate on objects on or near the surface of the earth (or other planet).

The pull of a planet on an object on or near the surface is called the **gravitational force.** The agent for the gravitational force is the *entire planet*. Gravity acts on *all* objects, whether moving or at rest. The symbol for gravitational force is $\vec{F}_\text{G}$. The gravitational force vector always points vertically downward, as shown in FIGURE 5.3.

> NOTE ▶ We often refer to "the weight" of an object. For an object at rest on the surface of a planet, its weight is simply the magnitude F_G of the gravitational force. However, weight and gravitational force are not the same thing, nor is weight the same as mass. We will briefly examine mass later in the chapter, and we'll explore the rather subtle connections among gravity, weight, and mass in Chapter 6. ◀

FIGURE 5.3 Gravity.

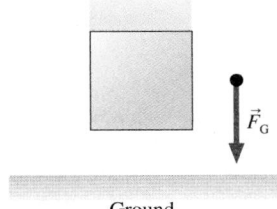

The gravitational force pulls the box down.

$\vec{F}_\text{G}$

Ground

Spring Force

Springs exert one of the most common contact forces. A spring can either push (when compressed) or pull (when stretched). FIGURE 5.4 shows the **spring force,** for which we use the symbol $\vec{F}_\text{sp}$. In both cases, pushing and pulling, the tail of the force vector is placed on the particle in the force diagram.

FIGURE 5.4 The spring force.

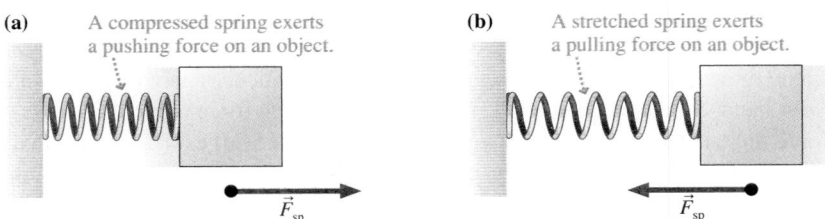

(a) A compressed spring exerts a pushing force on an object.

$\vec{F}_\text{sp}$

(b) A stretched spring exerts a pulling force on an object.

$\vec{F}_\text{sp}$

Although you may think of a spring as a metal coil that can be stretched or compressed, this is only one type of spring. Hold a ruler, or any other thin piece of wood or metal, by the ends and bend it slightly. It flexes. When you let go, it "springs" back to its original shape. This is just as much a spring as is a metal coil.

Tension Force

When a string or rope or wire pulls on an object, it exerts a contact force that we call the **tension force,** represented by a capital $\vec{T}$. The direction of the tension force is always in the direction of the string or rope, as you can see in FIGURE 5.5. The commonplace reference to "the tension" in a string is an informal expression for T, the size or magnitude of the tension force.

> NOTE ▶ Tension is represented by the symbol T. This is logical, but there's a risk of confusing the tension T with the identical symbol T for the period of a particle in circular motion. The number of symbols used in science and engineering is so large that some letters are used several times to represent different quantities. The use of T is the first time we've run into this problem, but it won't be the last. You must be alert to the *context* of a symbol's use to deduce its meaning. ◀

FIGURE 5.5 Tension.

The rope exerts a tension force on the sled.

$\vec{T}$

FIGURE 5.6 An atomic model of tension.

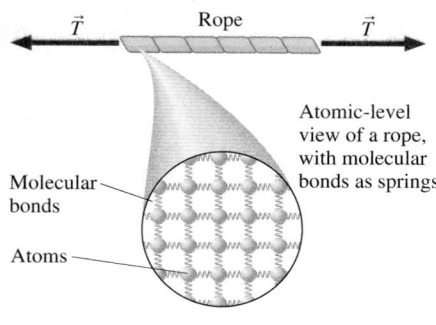

If you were to use a very powerful microscope to look inside a rope, you would "see" that it is made of *atoms* joined together by *molecular bonds*. Molecular bonds are not rigid connections between the atoms. They are more accurately thought of as tiny *springs* holding the atoms together, as in FIGURE 5.6. Pulling on the ends of a string or rope stretches the molecular springs ever so slightly. The tension within a rope and the tension force experienced by an object at the end of the rope are really the net spring force being exerted by billions and billions of microscopic springs.

This atomic-level view of tension introduces a new idea: a microscopic **atomic model** for understanding the behavior and properties of macroscopic objects. It is a *model* because atoms and molecular bonds aren't really little balls and springs. We're using macroscopic concepts—balls and springs—to understand atomic-scale phenomena that we cannot directly see or sense. This is a good model for explaining the elastic properties of materials, but it would not necessarily be a good model for explaining other phenomena. We will frequently use atomic models to obtain a deeper understanding of our observations.

FIGURE 5.7 An atomic model of the force exerted by a table.

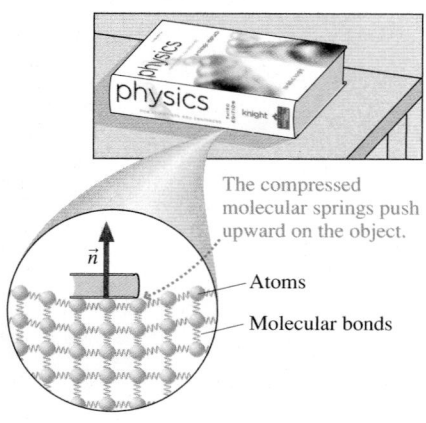

Normal Force

If you sit on a bed, the springs in the mattress compress and, as a consequence of the compression, exert an upward force on you. Stiffer springs would show less compression but still exert an upward force. The compression of extremely stiff springs might be measurable only by sensitive instruments. Nonetheless, the springs would compress ever so slightly and exert an upward spring force on you.

FIGURE 5.7 shows an object resting on top of a sturdy table. The table may not visibly flex or sag, but—just as you do to the bed—the object compresses the molecular springs in the table. The size of the compression is very small, but it is not zero. As a consequence, the compressed molecular springs *push upward* on the object. We say that "the table" exerts the upward force, but it is important to understand that the pushing is *really* done by molecular springs. Similarly, an object resting on the ground compresses the molecular springs holding the ground together and, as a consequence, the ground pushes up on the object.

We can extend this idea. Suppose you place your hand on a wall and lean against it, as shown in FIGURE 5.8. Does the wall exert a force on your hand? As you lean, you compress the molecular springs in the wall and, as a consequence, they push outward against your hand. So the answer is yes, the wall does exert a force on you.

The force the table surface exerts is vertical; the force the wall exerts is horizontal. In all cases, the force exerted on an object that is pressing against a surface is in a direction *perpendicular* to the surface. Mathematicians refer to a line that is perpendicular to a surface as being *normal* to the surface. In keeping with this terminology, we define the **normal force** as the force exerted by a surface (the agent) against an object that is pressing against the surface. The symbol for the normal force is $\vec{n}$.

We're not using the word *normal* to imply that the force is an "ordinary" force or to distinguish it from an "abnormal force." A surface exerts a force *perpendicular* (i.e., normal) to itself as the molecular springs press *outward*. FIGURE 5.9 shows an object on an inclined surface, a common situation.

In essence, the normal force is just a spring force, but one exerted by a vast number of microscopic springs acting at once. The normal force is responsible for the "solidness" of solids. It is what prevents you from passing right through the chair you are sitting in and what causes the pain and the lump if you bang your head into a door.

FIGURE 5.8 The wall pushes outward.

Friction

Friction, like the normal force, is exerted by a surface. But whereas the normal force is perpendicular to the surface, the friction force is always *tangent* to the surface. It is useful to distinguish between two kinds of friction:

■ *Kinetic friction,* denoted $\vec{f}_k$, appears as an object slides across a surface. This is a force that "opposes the motion," meaning that the friction force vector $\vec{f}_k$ points in a direction opposite the velocity vector $\vec{v}$ (i.e., "the motion").

FIGURE 5.9 The normal force.

■ *Static friction,* denoted $\vec{f}_s$, is the force that keeps an object "stuck" on a surface and prevents its motion. Finding the direction of $\vec{f}_s$ is a little trickier than finding it for $\vec{f}_k$. Static friction points opposite the direction in which the object *would* move if there were no friction. That is, it points in the direction necessary to *prevent* motion.

FIGURE 5.10 shows examples of kinetic and static friction.

FIGURE 5.10 Kinetic and static friction.

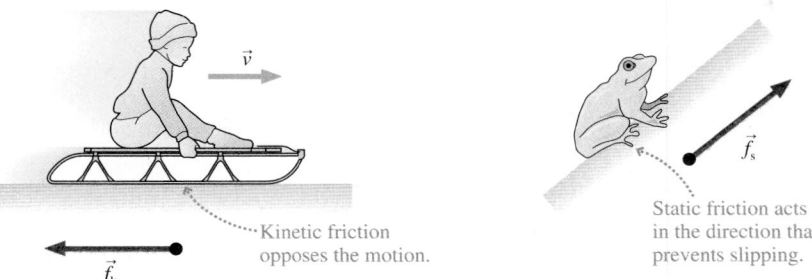

Kinetic friction opposes the motion.

Static friction acts in the direction that prevents slipping.

NOTE ▶ A surface exerts a kinetic friction force when an object moves *relative to* the surface. A package on a conveyor belt is in motion, but it does not experience a kinetic friction force because it is not moving relative to the belt. So to be precise, we should say that the kinetic friction force points opposite to an object's motion *relative to* a surface. ◀

Drag

Friction at a surface is one example of a *resistive force,* a force that opposes or resists motion. Resistive forces are also experienced by objects moving through fluids—gases and liquids. The resistive force of a fluid is called **drag,** with symbol $\vec{D}$. Drag, like kinetic friction, points opposite the direction of motion. **FIGURE 5.11** shows an example.

Drag can be a significant force for objects moving at high speeds or in dense fluids. Hold your arm out the window as you ride in a car and feel how the air resistance against it increases rapidly as the car's speed increases. Drop a lightweight object into a beaker of water and watch how slowly it settles to the bottom.

For objects that are heavy and compact, that move in air, and whose speed is not too great, the drag force of air resistance is fairly small. To keep things as simple as possible, **you can neglect air resistance in all problems unless a problem explicitly asks you to include it.**

Thrust

A jet airplane obviously has a force that propels it forward during takeoff. Likewise for the rocket being launched in **FIGURE 5.12**. This force, called **thrust,** occurs when a jet or rocket engine expels gas molecules at high speed. Thrust is a contact force, with the exhaust gas being the agent that pushes on the engine. The process by which thrust is generated is rather subtle, and we will postpone a full discussion until we study Newton's third law in Chapter 7. For now, we will treat thrust as a force opposite the direction in which the exhaust gas is expelled. There's no special symbol for thrust, so we will call it $\vec{F}_{thrust}$.

Electric and Magnetic Forces

Electricity and magnetism, like gravity, exert long-range forces. We will study electric and magnetic forces in detail in Part VI. For now, it is worth noting that the forces holding molecules together—the molecular bonds—are not actually tiny springs. Atoms and molecules are made of charged particles—electrons and protons—and what we call a molecular bond is really an electric force between these particles. So when

FIGURE 5.11 Air resistance is an example of drag.

Air resistance is a significant force on falling leaves. It points opposite the direction of motion.

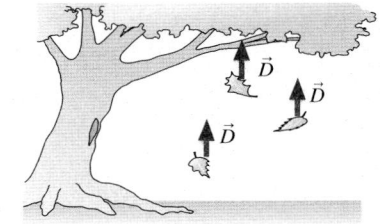

FIGURE 5.12 Thrust force on a rocket.

Thrust force is exerted on a rocket by exhaust gases.

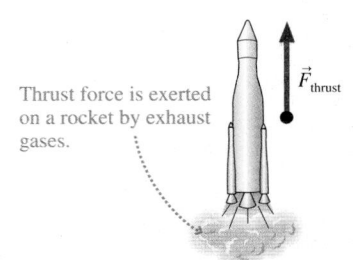

we say that the normal force and the tension force are due to "molecular springs," or that friction is due to atoms running into each other, what we're really saying is that these forces, at the most fundamental level, are actually electric forces between the charged particles in the atoms.

5.3 Identifying Forces

Force	Notation
General force	$\vec{F}$
Gravitational force	$\vec{F}_G$
Spring force	$\vec{F}_{sp}$
Tension	$\vec{T}$
Normal force	$\vec{n}$
Static friction	$\vec{f}_s$
Kinetic friction	$\vec{f}_k$
Drag	$\vec{D}$
Thrust	$\vec{F}_{thrust}$

Force and motion problems generally have two basic steps:

1. Identify all of the forces acting on an object.
2. Use Newton's laws and kinematics to determine the motion.

Understanding the first step is the primary goal of this chapter. We'll turn our attention to step 2 in the next chapter.

A typical physics problem describes an object that is being pushed and pulled in various directions. Some forces are given explicitly; others are only implied. In order to proceed, it is necessary to determine all the forces that act on the object. The procedure for identifying forces will become part of the *pictorial representation* of the problem.

TACTICS
BOX 5.2 **Identifying forces**

❶ **Identify the object of interest.** This is the object whose motion you wish to study.
❷ **Draw a picture of the situation.** Show the object of interest and all other objects—such as ropes, springs, or surfaces—that touch it.
❸ **Draw a closed curve around the object.** Only the object of interest is inside the curve; everything else is outside.
❹ **Locate every point on the boundary of this curve where other objects touch the object of interest.** These are the points where *contact forces* are exerted on the object.
❺ **Name and label each contact force acting on the object.** There is at least one force at each point of contact; there may be more than one. When necessary, use subscripts to distinguish forces of the same type.
❻ **Name and label each long-range force acting on the object.** For now, the only long-range force is the gravitational force.

Exercises 3–8

EXAMPLE 5.1 **Forces on a bungee jumper**

A bungee jumper has leapt off a bridge and is nearing the bottom of her fall. What forces are being exerted on the jumper?

VISUALIZE

FIGURE 5.13 Forces on a bungee jumper.

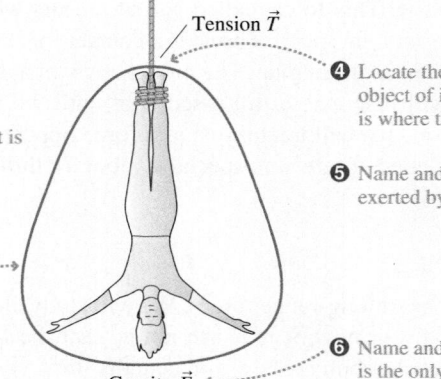

❶ Identify the object of interest. Here the object is the bungee jumper.

❷ Draw a picture of the situation.

❸ Draw a closed curve around the object.

Tension $\vec{T}$

❹ Locate the points where other objects touch the object of interest. Here the only point of contact is where the cord attaches to her ankles.

❺ Name and label each contact force. The force exerted by the cord is a tension force.

❻ Name and label long-range forces. Gravity is the only one.

Gravity $\vec{F}_G$

EXAMPLE 5.2 **Forces on a skier**

A skier is being towed up a snow-covered hill by a tow rope. What forces are being exerted on the skier?

VISUALIZE

FIGURE 5.14 Forces on a skier.

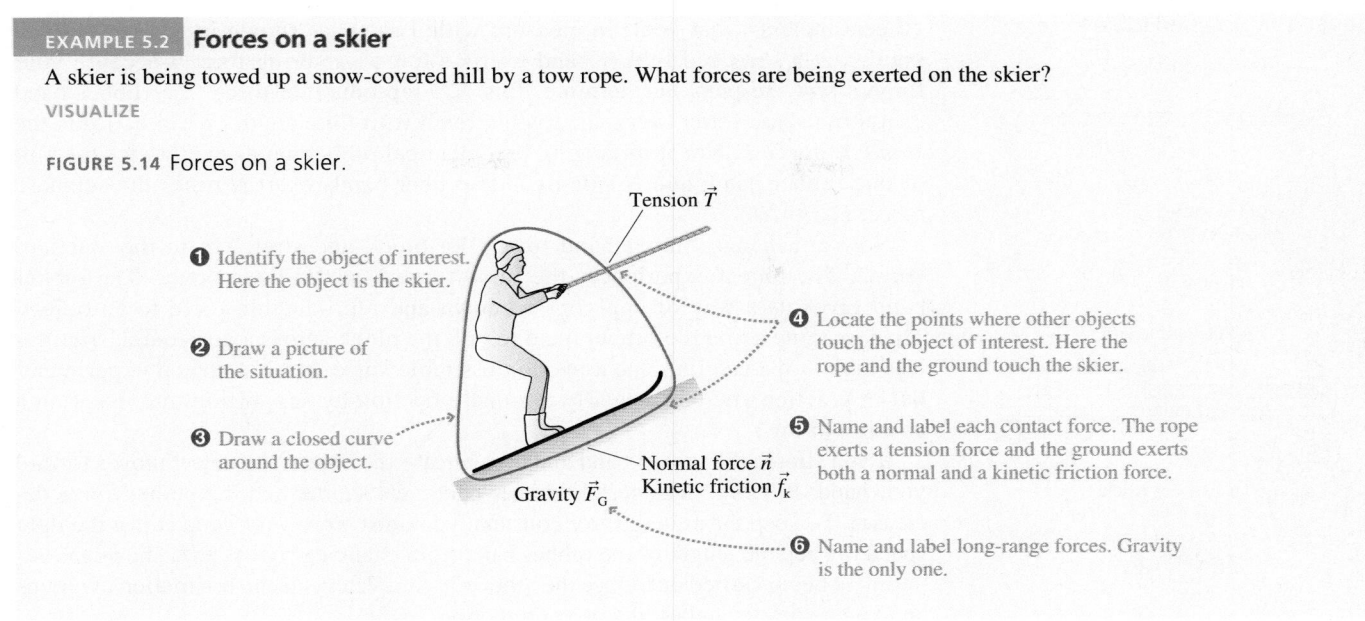

NOTE ▶ You might have expected two friction forces and two normal forces in Example 5.2, one on each ski. Keep in mind, however, that we're working within the particle model, which represents the skier by a single point. A particle has only one contact with the ground, so there is one normal force and one friction force. ◀

EXAMPLE 5.3 **Forces on a rocket**

A rocket is being launched to place a new satellite in orbit. Air resistance is not negligible. What forces are being exerted on the rocket?

VISUALIZE This drawing is much more like the sketch you would make when identifying forces as part of solving a problem.

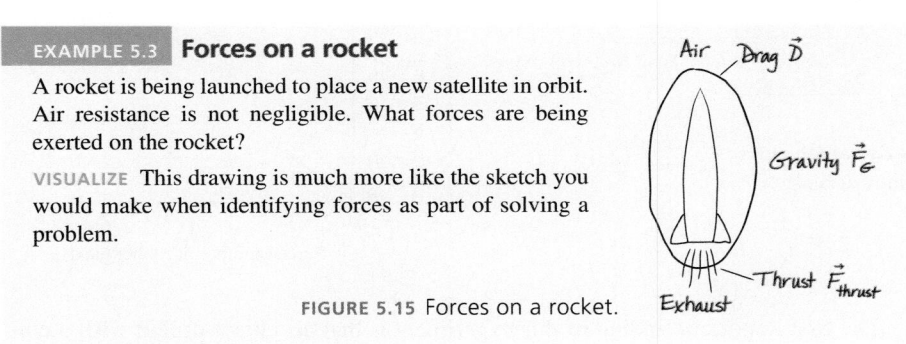

FIGURE 5.15 Forces on a rocket.

STOP TO THINK 5.2   You've just kicked a rock, and it is now sliding across the ground about 2 meters in front of you. Which of these forces act on the rock? List all that apply.

a. Gravity, acting downward.
b. The normal force, acting upward.
c. The force of the kick, acting in the direction of motion.
d. Friction, acting opposite the direction of motion.

5.4 What Do Forces Do? A Virtual Experiment

Having learned to identify forces, we ask the next question: How does an object move when a force is exerted on it? The only way to answer this question is to do experiments. Let's conduct a "virtual experiment," one you can easily visualize. Imagine using your fingers to stretch a rubber band to a certain length—say

FIGURE 5.16 A reproducible force.

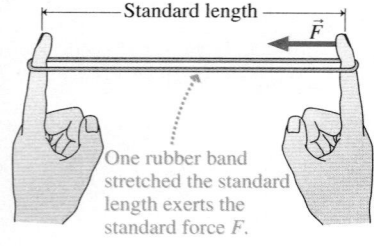

One rubber band stretched the standard length exerts the standard force *F*.

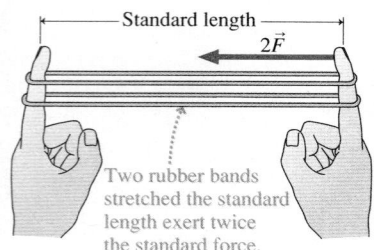

Two rubber bands stretched the standard length exert twice the standard force.

10 centimeters—that you can measure with a ruler, as shown in FIGURE 5.16. You know that a stretched rubber band exerts a force—a spring force—because your fingers *feel* the pull. Furthermore, this is a reproducible force; the rubber band exerts the same force every time you stretch it to this length. We'll call this the *standard force F*. Not surprisingly, two identical rubber bands exert twice the pull of one rubber band, and *N* side-by-side rubber bands exert *N* times the standard force: $F_{net} = NF$.

Now attach one rubber band to a 1 kg block and stretch it to the standard length. The object experiences the same force *F* as did your finger. The rubber band gives us a way of applying a known and reproducible force to an object. Then imagine using the rubber band to pull the block across a horizontal, frictionless table. (We can imagine a frictionless table since this is a virtual experiment, but in practice you could nearly eliminate friction by supporting the object on a cushion of air.)

If you stretch the rubber band and then release the object, the object moves toward your hand. But as it does so, the rubber band gets shorter and the pulling force decreases. To keep the pulling force constant, you must *move your hand* at just the right speed to keep the length of the rubber band from changing! FIGURE 5.17a shows the experiment being carried out. Once the motion is complete, you can use motion diagrams and kinematics to analyze the object's motion.

FIGURE 5.17 Measuring the motion of an object that is pulled with a constant force.

(a)

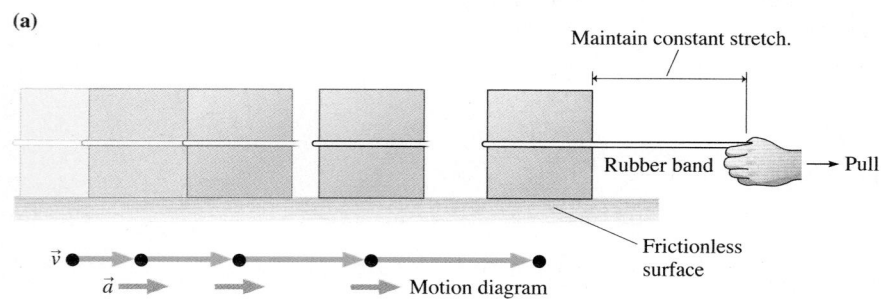

Maintain constant stretch.

Rubber band → Pull

Frictionless surface

$\vec{v}$

$\vec{a}$ Motion diagram

(b)

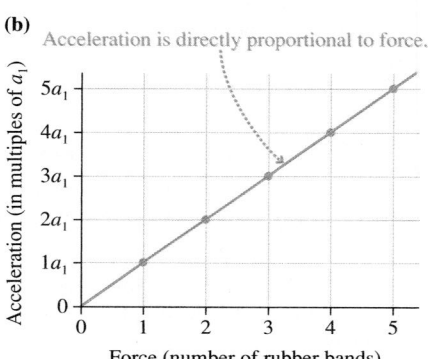

Acceleration is directly proportional to force.

Acceleration (in multiples of a_1)

Force (number of rubber bands)

The first important finding of this experiment is that **an object pulled with a constant force moves with a constant acceleration.** That is, the answer to the question What does a force do? is: A force causes an object to accelerate, and a constant force produces a constant acceleration. This finding could not have been anticipated in advance. It's conceivable that the object would speed up for a while, then move with a steady speed. Or that it would speed up, but that the *rate* of increase, the acceleration, would steadily decline. These are conceivable motions, but they're not what happens. Instead, the object accelerates *with a constant acceleration a_1* for as long as you pull it with a constant force *F*.

What happens if you increase the force by using several rubber bands? To find out, use two rubber bands, then three rubber bands, then four, and so on. With *N* rubber bands, the force on the block is *NF*. FIGURE 5.17b shows the results of this experiment. You can see that doubling the force causes twice the acceleration, tripling the force causes three times the acceleration, and so on. The graph reveals our second important finding: **The acceleration is directly proportional to the force.** This result can be written as

$$a = cF \tag{5.2}$$

where *c*, called the *proportionality constant,* is the slope of the graph.

MATHEMATICAL ASIDE **Proportionality and proportional reasoning**

The concept of **proportionality** arises frequently in physics. A quantity symbolized by u is *proportional* to another quantity symbolized by v if

$$u = cv$$

where c (which might have units) is called the **proportionality constant.** This relationship between u and v is often written

$$u \propto v$$

where the symbol $\propto$ means "is proportional to."

If v is doubled to $2v$, then u is doubled to $c(2v) = 2(cv) = 2u$. In general, if v is changed by any factor f, then u changes by the same factor. This is the essence of what we *mean* by proportionality.

A graph of u versus v is a straight line *passing through the origin* (i.e., the y-intercept is zero) with slope $= c$. Notice that proportionality is a much more specific relationship between u and v than mere linearity. The linear equation $u = cv + b$ has a straight-line graph, but it doesn't pass through the origin (unless b happens to be zero) and doubling v does not double u.

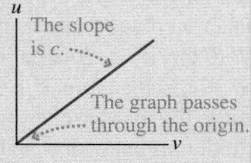
u is proportional to v.

If $u \propto v$, then $u_1 = cv_1$ and $u_2 = cv_2$. Dividing the second equation by the first, we find

$$\frac{u_2}{u_1} = \frac{v_2}{v_1}$$

By working with *ratios*, we can deduce information about u without needing to know the value of c. (This would not be true if the relationship were merely linear.) This is called **proportional reasoning.**

Proportionality is not limited to being linearly proportional. The graph on the left below shows that u is clearly not proportional to w. But a graph of u versus $1/w^2$ *is* a straight line passing through the origin, thus, in this case, u is proportional to $1/w^2$, or $u \propto 1/w^2$. We would say that "u is proportional to the inverse square of w."

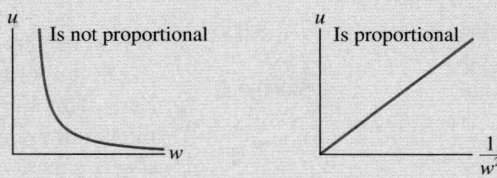

u is proportional to the inverse square of w.

EXAMPLE u is proportional to the inverse square of w. By what factor does u change if w is tripled?

SOLUTION This is an opportunity for proportional reasoning; we don't need to know the proportionality constant. If u is proportional to $1/w^2$, then

$$\frac{u_2}{u_1} = \frac{1/w_2^2}{1/w_1^2} = \frac{w_1^2}{w_2^2} = \left(\frac{w_1}{w_2}\right)^2$$

Tripling w, with $w_2/w_1 = 3$, and thus $w_1/w_2 = \frac{1}{3}$, changes u to

$$u_2 = \left(\frac{w_1}{w_2}\right)^2 u_1 = \left(\frac{1}{3}\right)^2 u_1 = \frac{1}{9} u_1$$

Tripling w causes u to become $\frac{1}{9}$ of its original value.

Many *Student Workbook* and end-of-chapter homework questions will require proportional reasoning. It's an important skill to learn.

The final question for our virtual experiment is: How does the acceleration depend on the mass of the object being pulled? To find out, apply the *same force*—for example, the standard force of one rubber band—to a 2 kg block, then a 3 kg block, and so on, and for each measure the acceleration. Doing so gives you the results shown in FIGURE 5.18. An object with twice the mass of the original block has only half the acceleration when both are subjected to the same force.

Mathematically, the graph of Figure 5.18 is one of *inverse proportionality*. That is, **the acceleration is inversely proportional to the object's mass,** which we can write as

$$a = \frac{c'}{m} \tag{5.3}$$

where c' is another proportionality constant.

Force causes an object to *accelerate!* The results of our experiment are that the acceleration is directly proportional to the force applied and inversely proportional to the object's mass. We can combine these into the single statement

$$a = \frac{F}{m} \tag{5.4}$$

if we define the basic unit of force as the force that causes a 1 kg mass to accelerate at 1 m/s². That is,

$$1 \text{ basic unit of force} \equiv 1 \text{ kg} \times 1 \frac{\text{m}}{\text{s}^2} = 1 \frac{\text{kg m}}{\text{s}^2}$$

FIGURE 5.18 Acceleration is inversely proportional to mass.

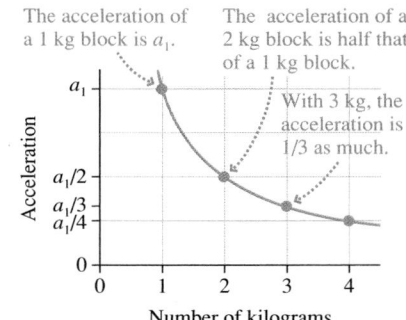

TABLE 5.1 Approximate magnitude of some typical forces

Force	Approximate magnitude (newtons)
Weight of a U.S. quarter	0.05
Weight of a 1 pound object	5
Weight of a 110 pound person	500
Propulsion force of a car	5,000
Thrust force of a rocket motor	5,000,000

This basic unit of force is called a newton:

One **newton** is the force that causes a 1 kg mass to accelerate at 1 m/s². The abbreviation for newton is N. Mathematically, $1 \text{ N} = 1 \text{ kg m/s}^2$.

Table 5.1 lists some typical forces. As you can see, "typical" forces on "typical" objects are likely to be in the range 0.01–10,000 N.

Mass

We've been using the term *mass* without a clear definition. As we learned in Chapter 1, the SI unit of mass, the kilogram, is based on a particular metal block kept in a vault in Paris. This suggests that *mass* is the amount of matter an object contains, and that is certainly our everyday concept of mass. Now we see that a more precise way of defining an object's mass is in terms of its acceleration in response to a force. Figure 5.18 shows that an object with twice the amount of matter accelerates only half as much in response to the same force. The more matter an object has, the more it *resists* accelerating in response to a force. You're familiar with this idea: Your car is much harder to push than your bicycle. The tendency of an object to resist a *change* in its velocity (i.e., to resist acceleration) is called **inertia.** Consequently, the mass used in Equation 5.4, a measure of an object's resistance to changing its motion, is called **inertial mass.** We'll meet a different concept of mass, *gravitational mass,* when we study Newton's law of gravity in Chapter 13.

STOP TO THINK 5.3   Two rubber bands stretched to the standard length cause an object to accelerate at 2 m/s². Suppose another object with twice the mass is pulled by four rubber bands stretched to the standard length. The acceleration of this second object is

   a. 1 m/s²   b. 2 m/s²   c. 4 m/s²   d. 8 m/s²   e. 16 m/s²

Hint: Use proportional reasoning.

5.5 Newton's Second Law

Equation 5.4 is an important finding, but our experiment was limited to looking at an object's response to a single applied force. Realistically, an object is likely to be subjected to several distinct forces $\vec{F}_1, \vec{F}_2, \vec{F}_3, \dots$ that may point in different directions. What happens then? In that case, it is found experimentally that the acceleration is determined by the *net* force.

Newton was the first to recognize the connection between force and motion. This relationship is known today as Newton's second law.

Newton's second law An object of mass m subjected to forces $\vec{F}_1, \vec{F}_2, \vec{F}_3, \dots$ will undergo an acceleration $\vec{a}$ given by

$$\vec{a} = \frac{\vec{F}_{net}}{m} \tag{5.5}$$

where the net force $\vec{F}_{net} = \vec{F}_1 + \vec{F}_2 + \vec{F}_3 + \cdots$ is the vector sum of all forces acting on the object. The acceleration vector $\vec{a}$ points in the same direction as the net force vector $\vec{F}_{net}$.

The significance of Newton's second law cannot be overstated. There was no reason to suspect that there should be any simple relationship between force and acceleration. Yet there it is, a simple but exceedingly powerful equation relating the two. The critical idea is that **an object accelerates in the direction of the net force vector $\vec{F}_{net}$.**

We can rewrite Newton's second law in the form

$$\vec{F}_{net} = m\vec{a} \tag{5.6}$$

which is how you'll see it presented in many textbooks. Equations 5.5 and 5.6 are mathematically equivalent, but Equation 5.5 better describes the central idea of Newtonian mechanics: A force applied to an object causes the object to accelerate.

It's also worth noting that **the object responds only to the forces acting on it *at this instant*.** The object has no memory of forces that may have been exerted at earlier times. This idea is sometimes called **Newton's zeroth law.**

NOTE ▶ Be careful not to think that one force "overcomes" the others to determine the motion. Forces are not in competition with each other! It is $\vec{F}_{net}$, the sum of *all* the forces, that determines the acceleration $\vec{a}$. ◀

As an example, FIGURE 5.19a shows a box being pulled by two ropes. The ropes exert tension forces $\vec{T}_1$ and $\vec{T}_2$ on the box. FIGURE 5.19b represents the box as a particle, shows the forces acting on the box, and adds them graphically to find the net force $\vec{F}_{net}$. The box will accelerate in the direction of $\vec{F}_{net}$ with acceleration

$$\vec{a} = \frac{\vec{F}_{net}}{m} = \frac{\vec{T}_1 + \vec{T}_2}{m}$$

NOTE ▶ The acceleration is *not* $(T_1 + T_2)/m$. You must add the forces as *vectors*, not merely add their magnitudes as scalars. ◀

Forces Are Interactions

There's one more important aspect of forces. If you push against a door (the object) to close it, the door pushes back against your hand (the agent). If a tow rope pulls on a car (the object), the car pulls back on the rope (the agent). In general, if an agent exerts a force on an object, the object exerts a force on the agent. We really need to think of a force as an *interaction* between two objects. This idea is captured in Newton's third law—that for every action there is an equal but opposite reaction.

Although the interaction perspective is a more exact way to view forces, it adds complications that we would like to avoid for now. Our approach will be to start by focusing on how a single object responds to forces exerted on it. Then, in Chapter 7, we'll return to Newton's third law and the larger issue of how two or more objects interact with each other.

FIGURE 5.19 Acceleration of a pulled box.

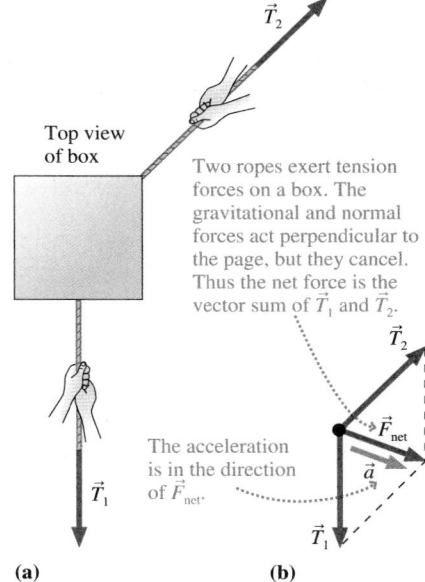

Two ropes exert tension forces on a box. The gravitational and normal forces act perpendicular to the page, but they cancel. Thus the net force is the vector sum of $\vec{T}_1$ and $\vec{T}_2$.

The acceleration is in the direction of $\vec{F}_{net}$.

STOP TO THINK 5.4 Three forces act on an object. In which direction does the object accelerate?

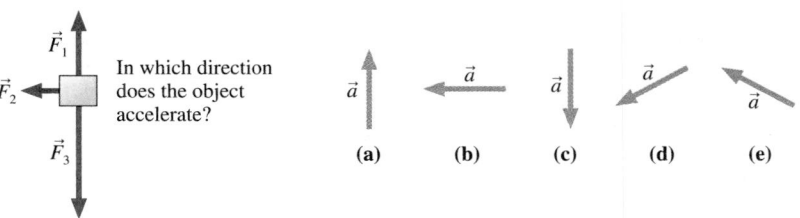

In which direction does the object accelerate?

(a)   (b)   (c)   (d)   (e)

5.6 Newton's First Law

Aristotle and his contemporaries in the world of ancient Greece were very interested in motion. One question they asked was: What is the "natural state" of an object if left to itself? It is easy to see that every moving object on earth, if left to itself, eventually comes to rest. Aristotle concluded that the natural state of an earthly object is to be at rest. An object at rest requires no explanation. A moving object, though, is not in its natural state and thus requires an explanation: Why is this object moving? What keeps it going and prevents it from being in its natural state?

Galileo reopened the question of the "natural state" of objects. He suggested focusing on the *limiting case* in which resistance to the motion (e.g., friction or air resistance) is zero. Many careful experiments in which he minimized the influence of friction led Galileo to a conclusion that was in sharp contrast to Aristotle's belief that rest is an object's natural state.

Galileo found that an external influence (i.e., a force) is needed to make an object accelerate—to *change* its velocity. In particular, a force is needed to put an object in motion. In the absence of friction or air resistance, a moving object would continue to move along a straight line forever with no loss of speed. In other words, the natural state of an object—its behavior if free of external influences—is *uniform motion* with constant velocity! This does not happen in practice because friction or air resistance prevents the object from being left alone. "At rest" has no special significance in Galileo's view of motion; it is simply uniform motion that happens to have $\vec{v} = \vec{0}$.

It was left to Newton to generalize this result, and today we call it Newton's first law of motion.

> **Newton's first law** An object that is at rest will remain at rest, or an object that is moving will continue to move in a straight line with constant velocity, if and only if the net force acting on the object is zero.

Newton's first law is also known as the *law of inertia.* If an object is at rest, it has a tendency to stay at rest. If it is moving, it has a tendency to continue moving with the *same velocity.*

NOTE ▶ The first law refers to *net* force. An object can remain at rest, or can move in a straight line with constant velocity, even though forces are exerted on it as long as the *net* force is zero. ◀

Notice the "if and only if" aspect of Newton's first law. If an object is at rest or moves with constant velocity, then we can conclude that there is no net force acting on it. Conversely, if no net force is acting on it, we can conclude that the object will have constant velocity, not just constant speed. The direction remains constant, too!

An object on which the net force is zero, $\vec{F}_{net} = \vec{0}$, is said to be in **mechanical equilibrium.** There are two distinct forms of mechanical equilibrium:

1. The object is at rest. This is **static equilibrium.**
2. The object is moving in a straight line with constant velocity. This is **dynamic equilibrium.**

Two examples of mechanical equilibrium are shown in FIGURE 5.20. Both share the common feature that the acceleration is zero: $\vec{a} = \vec{0}$.

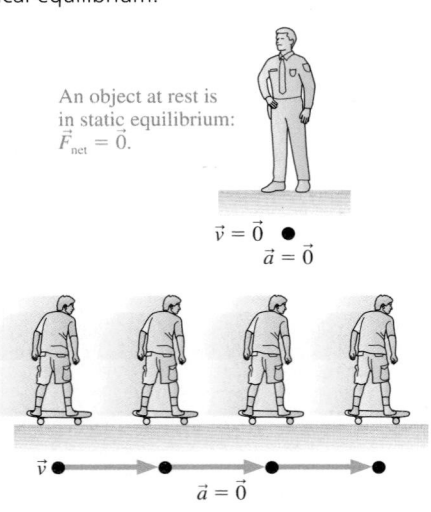

FIGURE 5.20 Two examples of mechanical equilibrium.

An object at rest is in static equilibrium: $\vec{F}_{net} = \vec{0}$.

$\vec{v} = \vec{0}$
$\vec{a} = \vec{0}$

$\vec{v}$
$\vec{a} = \vec{0}$

An object moving in a straight line at constant velocity is in dynamic equilibrium: $\vec{F}_{net} = \vec{0}$.

What Good Is Newton's First Law?

The first law completes our definition of force. It answers the question: What is a force? If an "influence" on an object disturbs a state of equilibrium by causing the object's velocity to change, the influence is a force.

Newton's first law changes the question the ancient Greeks were trying to answer: What causes an object to move? Newton's first law says **no cause is needed for an object to move!** Uniform motion is the object's natural state. Nothing at all is required for it to remain in that state. The proper question, according to Newton, is: What causes an object to *change* its velocity? Newton, with Galileo's help, also gave us the answer. **A *force* is what causes an object to change its velocity.**

The preceding paragraph contains the essence of Newtonian mechanics. This new perspective on motion, however, is often contrary to our common experience. We all know perfectly well that you must keep pushing an object—exerting a force on it—to keep it moving. Newton is asking us to change our point of view and to consider motion *from the object's perspective* rather than from our personal perspective. As far as the object is concerned, our push is just one of several forces acting on it. Others might include friction, air resistance, or gravity. Only by knowing the *net* force can we determine the object's motion.

Newton's first law may seem to be merely a special case of Newton's second law. After all, the equation $\vec{F}_{net} = m\vec{a}$ tells us that an object moving with constant velocity ($\vec{a} = \vec{0}$) has $\vec{F}_{net} = \vec{0}$. The difficulty is that the second law assumes that we already know what force is. The purpose of the first law is to *identify* a force as something that disturbs a state of equilibrium. The second law then describes how the object responds to this force. Thus from a *logical* perspective, the first law really is a separate statement that must precede the second law. But this is a rather formal distinction. From a pedagogical perspective it is better—as we have done—to use a commonsense understanding of force and start with Newton's second law.

Inertial Reference Frames

If a car stops suddenly, you may be "thrown" into the windshield if you're not wearing your seat belt. You have a very real forward acceleration *relative to the car,* but is there a force pushing you forward? A force is a push or a pull caused by an identifiable agent in contact with the object. Although you *seem* to be pushed forward, there's no agent to do the pushing.

The difficulty—an acceleration without an apparent force—comes from using an inappropriate reference frame. Your acceleration measured in a reference frame attached to the car is not the same as your acceleration measured in a reference frame attached to the ground. Newton's second law says $\vec{F}_{net} = m\vec{a}$. But which $\vec{a}$? Measured in which reference frame?

We define an **inertial reference frame** as a reference frame in which Newton's laws are valid. The first law provides a convenient way to test whether a reference frame is inertial. If $\vec{a} = \vec{0}$ (an object is at rest or moving with constant velocity) only when $\vec{F}_{net} = \vec{0}$, then the reference frame in which $\vec{a}$ is measured is an inertial reference frame.

Not all reference frames are inertial reference frames. FIGURE 5.21a shows a physics student cruising at constant velocity in an airplane. If the student places a ball on the floor, it stays there. There are no horizontal forces, and the ball remains at rest relative to the airplane. That is, $\vec{a} = \vec{0}$ in the airplane's reference frame when $\vec{F}_{net} = \vec{0}$. Newton's first law is satisfied, so this airplane is an inertial reference frame.

The physics student in FIGURE 5.21b conducts the same experiment during takeoff. He carefully places the ball on the floor just as the airplane starts to accelerate down the runway. You can imagine what happens. The ball rolls to the back of the plane as the passengers are being pressed back into their seats. Nothing exerts a horizontal contact force on the ball, yet the ball accelerates *in the plane's reference frame.* This violates Newton's first law, so the plane is *not* an inertial reference frame during takeoff.

In the first example, the plane is traveling with constant velocity. In the second, the plane is accelerating. **Accelerating reference frames are not inertial reference frames.** Consequently, Newton's laws are not valid in an accelerating reference frame.

The earth is not exactly an inertial reference frame because the earth rotates on its axis and orbits the sun. However, the earth's acceleration is so small that violations of Newton's laws can be measured only in high-precision experiments. We will treat the earth and laboratories attached to the earth as inertial reference frames, an approximation that is exceedingly well justified.

To understand the motion of the passengers in a braking car, you need to measure velocities and accelerations *relative to the ground.* From the perspective of an observer on the ground, the body of a passenger in a braking car tries to continue moving forward with constant velocity, exactly as we would expect on the basis of Newton's first law, while his immediate surroundings are decelerating. The passenger is not "thrown" into the windshield. Instead, the windshield runs into the passenger!

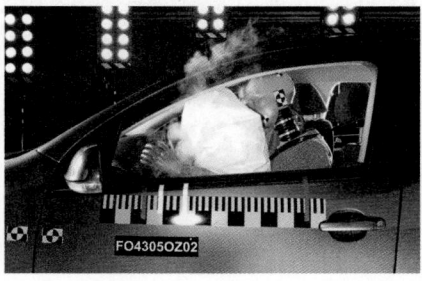

This guy thinks there's a force hurling him into the windshield. What a dummy!

FIGURE 5.21 Reference frames.

(a)

$\vec{a} = \vec{0}$

The ball stays in place.

A ball with no horizontal forces stays at rest in an airplane cruising at constant velocity. The airplane is an inertial reference frame.

(b)

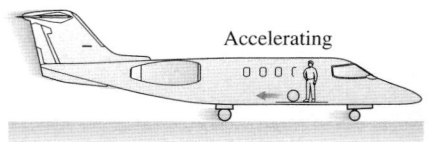

Accelerating

The ball rolls to the back.

The ball rolls to the back of the plane during takeoff. An accelerating plane is not an inertial reference frame.

Thinking About Force

It is important to identify correctly all the forces acting on an object. It is equally important not to include forces that do not really exist. We have established a number of criteria for identifying forces; the three critical ones are:

- A force has an agent. Something tangible and identifiable causes the force.
- Forces exist at the point of contact between the agent and the object experiencing the force (except for the few special cases of long-range forces).
- Forces exist due to interactions happening *now*, not due to what happened in the past.

We all have had many experiences suggesting that a force is necessary to keep something moving. Consider a bowling ball rolling along on a smooth floor. It is very tempting to think that a horizontal "force of motion" keeps it moving in the forward direction. But *nothing contacts the ball* except the floor. No agent is giving the ball a forward push. According to our definition, then, there is *no* forward "force of motion" acting on the ball. So what keeps it going? Recall our discussion of the first law: *No cause is needed to keep an object moving at constant velocity.* It continues to move forward simply because of its inertia.

One reason for wanting to include a "force of motion" is that we tend to view the problem from our perspective as one of the agents of force. You certainly have to keep pushing to move a box across the floor at constant velocity. If you stop, it stops. Newton's laws, though, require that we adopt the object's perspective. The box experiences your pushing force in one direction *and* a friction force in the opposite direction. The box moves at constant velocity if the *net* force is zero. This will be true as long as your pushing force exactly balances the friction force. When you stop pushing, the friction force causes an acceleration that slows and stops the box.

A related problem occurs if you throw a ball. A pushing force was indeed required to accelerate the ball *as it was thrown.* But that force disappears the instant the ball loses contact with your hand. The force does not stick with the ball as the ball travels through the air. Once the ball has acquired a velocity, *nothing* is needed to keep it moving with that velocity.

There's no "force of motion" or any other forward force on this arrow. It continues to move because of inertia.

5.7 Free-Body Diagrams

Having discussed at length what is and is not a force, we are ready to assemble our knowledge about force and motion into a single diagram called a *free-body diagram.* You will learn in the next chapter how to write the equations of motion directly from the free-body diagram. Solution of the equations is a mathematical exercise—possibly a difficult one, but nonetheless an exercise that could be done by a computer. The *physics* of the problem, as distinct from the purely calculational aspects, are the steps that lead to the free-body diagram.

A **free-body diagram,** part of the *pictorial representation* of a problem, represents the object as a particle and shows *all* of the forces acting on the object.

TACTICS **Drawing a free-body diagram**
BOX 5.3

❶ **Identify all forces acting on the object.** This step was described in Tactics Box 5.2.
❷ **Draw a coordinate system.** Use the axes defined in your pictorial representation.
❸ **Represent the object as a dot at the origin of the coordinate axes.** This is the particle model.
❹ **Draw vectors representing each of the identified forces.** This was described in Tactics Box 5.1. Be sure to label each force vector.
❺ **Draw and label the *net force* vector $\vec{F}_{net}$.** Draw this vector beside the diagram, not on the particle. Or, if appropriate, write $\vec{F}_{net} = \vec{0}$. Then check that $\vec{F}_{net}$ points in the same direction as the acceleration vector $\vec{a}$ on your motion diagram.

Exercises 24–29

An elevator accelerates upward

An elevator, suspended by a cable, speeds up as it moves upward from the ground floor. Identify the forces and draw a free-body diagram of the elevator.

MODEL Treat the elevator as a particle.

VISUALIZE

FIGURE 5.22 Free-body diagram of an elevator accelerating upward.

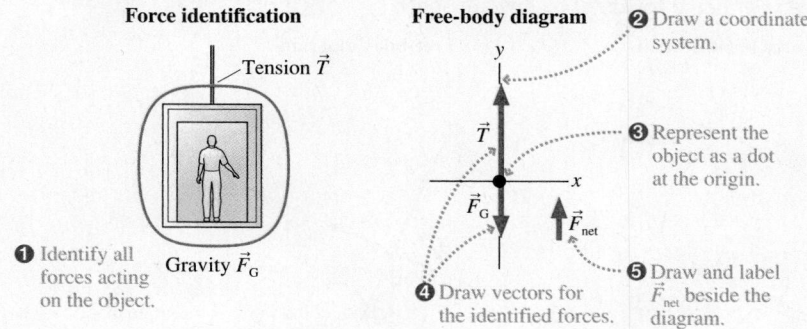

ASSESS The coordinate axes, with a vertical y-axis, are the ones we would use in a pictorial representation of the motion. The elevator is accelerating upward, so $\vec{F}_{net}$ must point upward. For this to be true, the magnitude of $\vec{T}$ must be larger than the magnitude of $\vec{F}_G$. The diagram has been drawn accordingly.

An ice block shoots across a frozen lake

Bobby straps a small model rocket to a block of ice and shoots it across the smooth surface of a frozen lake. Friction is negligible. Draw a pictorial representation of the block of ice.

MODEL Treat the block of ice as a particle. The pictorial representation consists of a motion diagram to determine $\vec{a}$, a force-identification picture, and a free-body diagram. The statement of the situation implies that friction is negligible.

VISUALIZE

FIGURE 5.23 Pictorial representation for a block of ice shooting across a frictionless frozen lake.

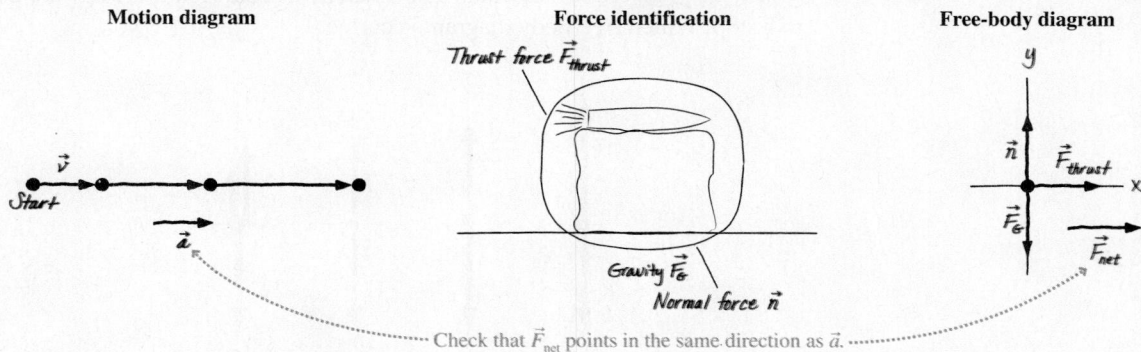

ASSESS The motion diagram tells us that the acceleration is in the positive x-direction. According to the rules of vector addition, this can be true only if the upward-pointing $\vec{n}$ and the downward-pointing $\vec{F}_G$ are equal in magnitude and thus cancel each other $((F_G)_y = -n_y)$. The vectors have been drawn accordingly, and this leaves the net force vector pointing toward the right, in agreement with $\vec{a}$ from the motion diagram.

EXAMPLE 5.6 **A skier is pulled up a hill**

A tow rope pulls a skier up a snow-covered hill at a constant speed. Draw a pictorial representation of the skier.

MODEL This is Example 5.2 again with the additional information that the skier is moving at constant speed. The skier will be treated as a particle in *dynamic equilibrium*. If we were doing a kinematics problem, the pictorial representation would use a tilted coordinate system with the *x*-axis parallel to the slope, so we use these same tilted coordinate axes for the free-body diagram.

VISUALIZE

FIGURE 5.24 Pictorial representation for a skier being towed at a constant speed.

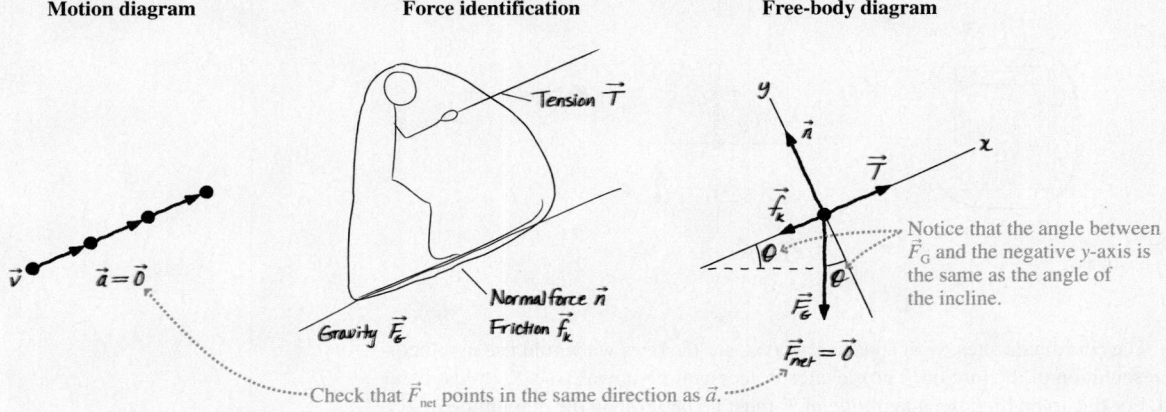

ASSESS We have shown $\vec{T}$ pulling parallel to the slope and $\vec{f}_k$, which opposes the direction of motion, pointing down the slope. $\vec{n}$ is perpendicular to the surface and thus along the *y*-axis. Finally, and this is important, the gravitational force $\vec{F}_G$ is *vertically* downward, *not* along the negative *y*-axis. In fact, you should convince yourself from the geometry that the angle θ between the $\vec{F}_G$ vector and the negative *y*-axis is the same as the angle θ of the incline above the horizontal. The skier moves in a straight line with constant speed, so $\vec{a} = \vec{0}$ and, from Newton's first law, $\vec{F}_{net} = \vec{0}$. Thus we have drawn the vectors such that the *y*-component of $\vec{F}_G$ is equal in magnitude to $\vec{n}$. Similarly, $\vec{T}$ must be large enough to match the negative *x*-components of both $\vec{f}_k$ and $\vec{F}_G$.

Free-body diagrams will be our major tool for the next several chapters. Careful practice with the workbook exercises and homework in this chapter will pay immediate benefits in the next chapter. Indeed, it is not too much to assert that a problem is half solved, or even more, when you complete the free-body diagram.

STOP TO THINK 5.5   An elevator suspended by a cable is moving upward and slowing to a stop. Which free-body diagram is correct?

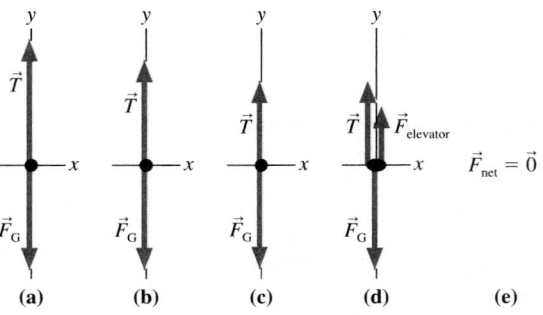

SUMMARY

The goal of Chapter 5 has been to establish a connection between force and motion.

General Principles

Newton's First Law

An object at rest will remain at rest, or an object that is moving will continue to move in a straight line with constant velocity, if and only if the net force on the object is zero.

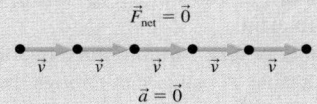

$$\vec{F}_{net} = \vec{0}$$

$$\vec{a} = \vec{0}$$

The first law tells us that no "cause" is needed for motion. Uniform motion is the "natural state" of an object.

Newton's laws are valid only in inertial reference frames.

Newton's Second Law

An object with mass m will undergo acceleration

$$\vec{a} = \frac{1}{m}\vec{F}_{net}$$

where $\vec{F}_{net} = \vec{F}_1 + \vec{F}_2 + \vec{F}_3 + \cdots$ is the vector sum of all the individual forces acting on the object.

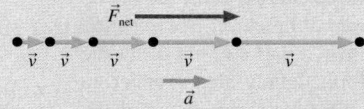

$$\vec{F}_{net}$$

$$\vec{a}$$

The second law tells us that a net force causes an object to accelerate. This is the connection between force and motion that we are seeking.

Important Concepts

Acceleration is the link to kinematics.

From $\vec{F}_{net}$, find $\vec{a}$.
From a, find v and x.

$\vec{a} = \vec{0}$ is the condition for equilibrium.

Static equilibrium if $\vec{v} = \vec{0}$.
Dynamic equilibrium if $\vec{v}$ = constant.

Equilibrium occurs if and only if $\vec{F}_{net} = \vec{0}$.

Mass is the resistance of an object to acceleration. It is an intrinsic property of an object.

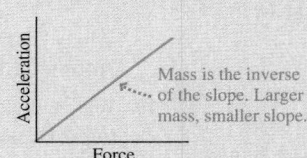

Mass is the inverse of the slope. Larger mass, smaller slope.

Force is a push or a pull on an object.

- Force is a vector, with a magnitude and a direction.
- Force requires an agent.
- Force is either a contact force or a long-range force.

Key Skills

Identifying Forces

Forces are identified by locating the points where other objects touch the object of interest. These are points where contact forces are exerted. In addition, objects with mass feel a long-range gravitational force.

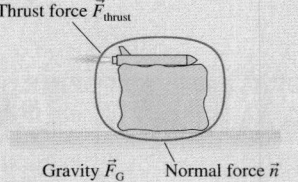

Thrust force $\vec{F}_{thrust}$

Gravity $\vec{F}_G$ Normal force $\vec{n}$

Free-Body Diagrams

A free-body diagram represents the object as a particle at the origin of a coordinate system. Force vectors are drawn with their tails on the particle. The net force vector is drawn beside the diagram.

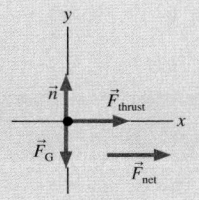

Terms and Notation

dynamics	gravitational force, $\vec{F}_G$	proportionality	Newton's first law
mechanics	spring force, $\vec{F}_{sp}$	proportionality constant	mechanical equilibrium
force, $\vec{F}$	tension force, $\vec{T}$	proportional reasoning	static equilibrium
agent	atomic model	newton, N	dynamic equilibrium
contact force	normal force, $\vec{n}$	inertia	inertial reference frame
long-range force	friction, $\vec{f}_k$ or $\vec{f}_s$	inertial mass, m	free-body diagram
net force, $\vec{F}_{net}$	drag, $\vec{D}$	Newton's second law	
superposition of forces	thrust, $\vec{F}_{thrust}$	Newton's zeroth law	

CONCEPTUAL QUESTIONS

1. An elevator suspended by a cable is descending at constant velocity. How many force vectors would be shown on a free-body diagram? Name them.

2. A compressed spring is pushing a block across a rough horizontal table. How many force vectors would be shown on a free-body diagram? Name them.

3. A brick is falling from the roof of a three-story building. How many force vectors would be shown on a free-body diagram? Name them.

4. In FIGURE Q5.4, block B is falling and dragging block A across a table. How many force vectors would be shown on a free-body diagram of block A? Name them.

5. You toss a ball straight up in the air. Immediately after you let go of it, what forces are acting on the ball? For each force you name, (a) state whether it is a contact force or a long-range force and (b) identify the agent of the force.

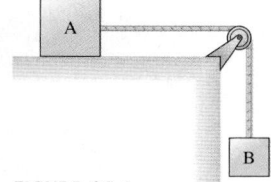

FIGURE Q5.4

6. A constant force applied to A causes A to accelerate at 5 m/s^2. The same force applied to B causes an acceleration of 3 m/s^2. Applied to C, it causes an acceleration of 8 m/s^2.
   a. Which object has the largest mass? Explain.
   b. Which object has the smallest mass?
   c. What is the ratio m_A/m_B of the mass of A to the mass of B?

7. An object experiencing a constant force accelerates at 10 m/s^2. What will the acceleration of this object be if
   a. The force is doubled? Explain.
   b. The mass is doubled?
   c. The force is doubled *and* the mass is doubled?

8. An object experiencing a constant force accelerates at 8 m/s^2. What will the acceleration of this object be if
   a. The force is halved? Explain.
   b. The mass is halved?
   c. The force is halved *and* the mass is halved?

9. If an object is at rest, can you conclude that there are no forces acting on it? Explain.

10. If a force is exerted on an object, is it possible for that object to be moving with constant velocity? Explain.

11. Is the statement "An object always moves in the direction of the net force acting on it" true or false? Explain.

12. Newton's second law says $\vec{F}_{net} = m\vec{a}$. So is $m\vec{a}$ a force? Explain.

13. Is it possible for the friction force on an object to be in the direction of motion? If so, give an example. If not, why not?

14. Suppose you press your physics book against a wall hard enough to keep it from moving. Does the friction force on the book point (a) into the wall, (b) out of the wall, (c) up, (d) down, or (e) is there no friction force? Explain.

15. FIGURE Q5.15 shows a hollow tube forming three-quarters of a circle. It is lying flat on a table. A ball is shot through the tube at high speed. As the ball emerges from the other end, does it follow path A, path B, or path C? Explain.

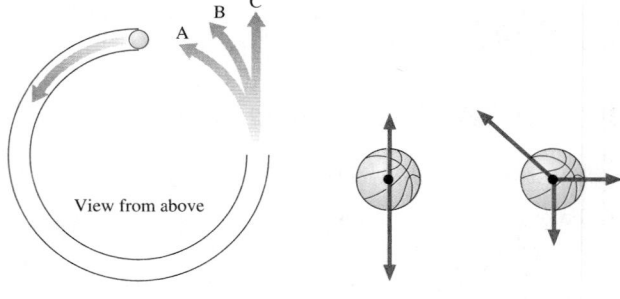

FIGURE Q5.15   FIGURE Q5.16

16. Which, if either, of the basketballs in FIGURE Q5.16 are in equilibrium? Explain.

17. Which of the following are inertial reference frames? Explain.
    a. A car driving at steady speed on a straight and level road.
    b. A car driving at steady speed up a 10° incline.
    c. A car speeding up after leaving a stop sign.
    d. A car driving at steady speed around a curve.

EXERCISES AND PROBLEMS

Exercises

Section 5.3 Identifying Forces

1. | A chandelier hangs from a chain in the middle of a dining room. Identify the forces on the chandelier.

2. | A car is parked on a steep hill. Identify the forces on the car.

3. ‖ A jet plane is speeding down the runway during takeoff. Air resistance is not negligible. Identify the forces on the jet.

4. | A baseball player is sliding into second base. Identify the forces on the baseball player.

5. ‖ A bullet has just been shot from a gun and is now traveling horizontally. Air resistance is not negligible. Identify the forces on the bullet.

Section 5.4 What Do Forces Do? A Virtual Experiment

6. | Two rubber bands cause an object to accelerate with acceleration *a*. How many rubber bands are needed to cause an object with half the mass to accelerate three times as quickly?

7. | Two rubber bands pulling on an object cause it to accelerate at 1.2 m/s^2.
   a. What will be the object's acceleration if it is pulled by four rubber bands?
   b. What will be the acceleration of two of these objects glued together if they are pulled by two rubber bands?

8. ‖ FIGURE EX5.8 shows an acceleration-versus-force graph for three objects pulled by rubber bands. The mass of object 2 is 0.20 kg. What are the masses of objects 1 and 3? Explain your reasoning.

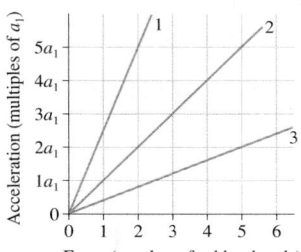
FIGURE EX5.8

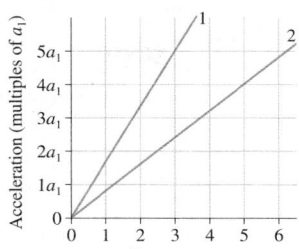
FIGURE EX5.9

9. ‖ FIGURE EX5.9 shows acceleration-versus-force graphs for two objects pulled by rubber bands. What is the mass ratio m_1/m_2?

10. ‖ For an object starting from rest and accelerating with constant acceleration, distance traveled is proportional to the square of the time. If an object travels 2.0 furlongs in the first 2.0 s, how far will it travel in the first 4.0 s?

11. ‖ The period of a pendulum is proportional to the square root of its length. A 2.0-m-long pendulum has a period of 3.0 s. What is the period of a 3.0-m-long pendulum?

Section 5.5 Newton's Second Law

12. | FIGURE EX5.12 shows an acceleration-versus-force graph for a 500 g object. What acceleration values go in the blanks on the vertical scale?

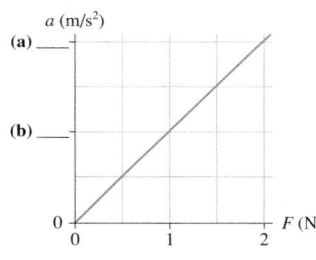
FIGURE EX5.12

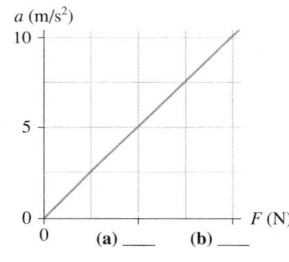
FIGURE EX5.13

13. | FIGURE EX5.13 shows an acceleration-versus-force graph for a 200 g object. What force values go in the blanks on the horizontal scale?

14. | FIGURE EX5.14 shows an object's acceleration-versus-force graph. What is the object's mass?

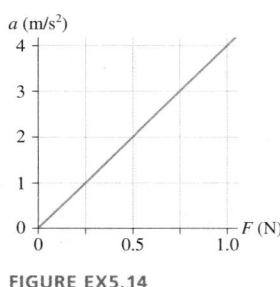
FIGURE EX5.14

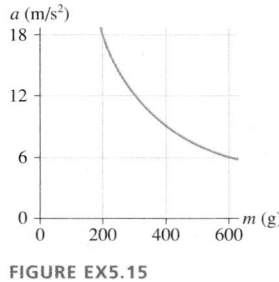
FIGURE EX5.15

15. | FIGURE EX5.15 shows the acceleration of objects of different mass that experience the same force. What is the magnitude of the force?

16. | Based on the information in Table 5.1, *estimate*
   a. The weight of a laptop computer.
   b. The propulsion force of a bicycle.

17. | Based on the information in Table 5.1, *estimate*
   a. The weight of a pencil.
   b. The propulsion force of a sprinter.

Section 5.6 Newton's First Law

Exercises 18 through 20 show two of the three forces acting on an object in equilibrium. Redraw the diagram, showing all three forces. Label the third force $\vec{F}_3$.

18. ‖   19. ‖   20. ‖

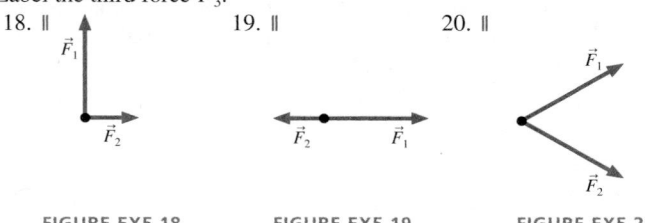

FIGURE EX5.18   FIGURE EX5.19   FIGURE EX5.20

Section 5.7 Free-Body Diagrams

Exercises 21 through 23 show a free-body diagram. For each:
   a. Redraw the free-body diagram.
   b. Write a short description of a real object for which this is the correct free-body diagram. Use Examples 5.4, 5.5, and 5.6 as models of what a description should be like.

21. | 22. | 23. |

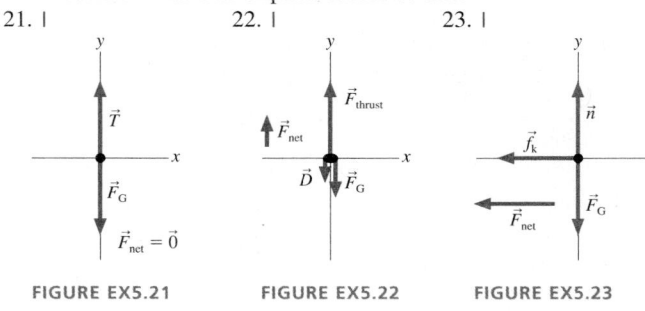

FIGURE EX5.21   FIGURE EX5.22   FIGURE EX5.23

Exercises 24 through 27 describe a situation. For each, identify all forces acting on the object and draw a free-body diagram of the object.

24. | A cat is sitting on a window sill.

25. | An ice hockey puck glides across frictionless ice.

26. | Your physics textbook is sliding across the table.

27. | A steel beam is being lowered at steady speed by a crane.

Problems

28. | Redraw the two motion diagrams shown in FIGURE P5.28, then draw a vector beside each one to show the direction of the net force acting on the object. Explain your reasoning.

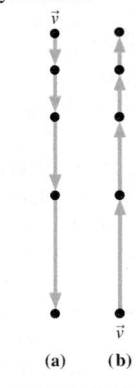

(a)   (b)

FIGURE P5.28

29. | Redraw the two motion diagrams shown in FIGURE P5.29, then draw a vector beside each one to show the direction of the net force acting on the object. Explain your reasoning.

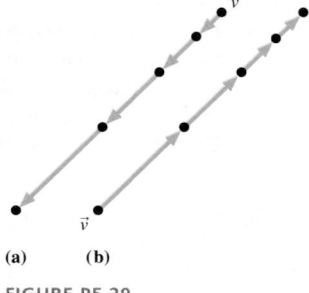

FIGURE P5.29

30. | A single force with x-component F_x acts on a 2.0 kg object as it moves along the x-axis. The object's acceleration graph (a_x versus t) is shown in FIGURE P5.30. Draw a graph of F_x versus t.

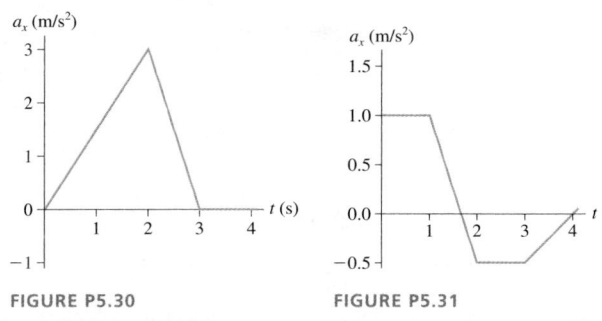

FIGURE P5.30

FIGURE P5.31

31. | A single force with x-component F_x acts on a 500 g object as it moves along the x-axis. The object's acceleration graph (a_x versus t) is shown in FIGURE P5.31. Draw a graph of F_x versus t.

32. | A single force with x-component F_x acts on a 2.0 kg object as it moves along the x-axis. A graph of F_x versus t is shown in FIGURE P5.32. Draw an acceleration graph (a_x versus t) for this object.

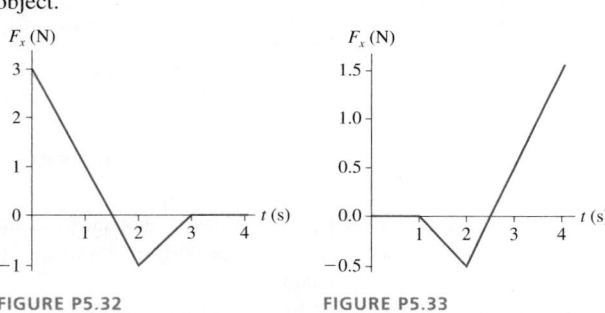

FIGURE P5.32

FIGURE P5.33

33. | A single force with x-component F_x acts on a 500 g object as it moves along the x-axis. A graph of F_x versus t is shown in FIGURE P5.33. Draw an acceleration graph (a_x versus t) for this object.

34. | A constant force is applied to an object, causing the object to accelerate at 8.0 m/s². What will the acceleration be if
   a. The force is doubled?
   b. The object's mass is doubled?
   c. The force and the object's mass are both doubled?
   d. The force is doubled and the object's mass is halved?

35. | A constant force is applied to an object, causing the object to accelerate at 10 m/s². What will the acceleration be if
   a. The force is halved?
   b. The object's mass is halved?
   c. The force and the object's mass are both halved?
   d. The force is halved and the object's mass is doubled?

Problems 36 through 41 show a free-body diagram. For each:
   a. Redraw the diagram.
   b. Identify the direction of the acceleration vector $\vec{a}$ and show it as a vector next to your diagram. Or, if appropriate, write $\vec{a} = \vec{0}$.
   c. If possible, identify the direction of the velocity vector $\vec{v}$ and show it as a labeled vector.
   d. Write a short description of a real object for which this is the correct free-body diagram. Use Examples 5.4, 5.5, and 5.6 as models of what a description should be like.

36. |

37. |

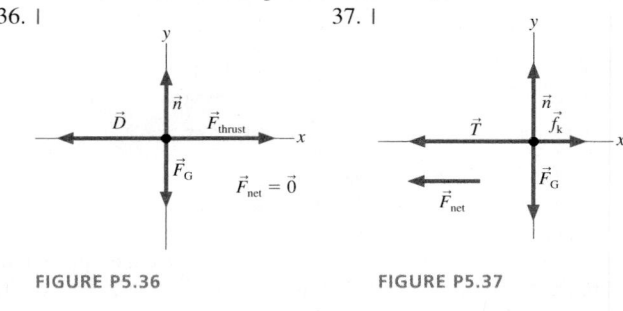

FIGURE P5.36

FIGURE P5.37

38. |

39. |

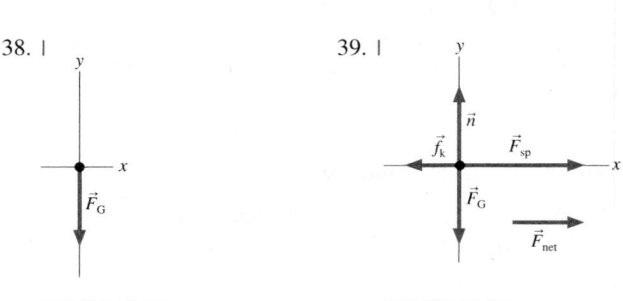

FIGURE P5.38

FIGURE P5.39

40. ‖

41. ‖

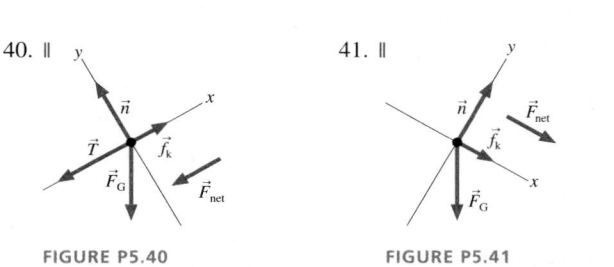

FIGURE P5.40

FIGURE P5.41

42. ‖ In lab, you propel a cart with four known forces while using an ultrasonic motion detector to measure the cart's acceleration. Your data are as follows:

Force (N)	Acceleration (m/s²)
0.25	0.5
0.50	0.8
0.75	1.3
1.00	1.8

   a. How should you graph these data so as to determine the mass of the cart from the slope of the line? That is, what values should you graph on the horizontal axis and what on the vertical axis?
   b. Is there another data point that would be reasonable to add, even though you made no measurements? If so, what is it?
   c. What is your best determination of the cart's mass?

Problems 43 through 52 describe a situation. For each, draw a motion diagram, a force-identification diagram, and a free-body diagram.

43. | An elevator, suspended by a single cable, has just left the tenth floor and is speeding up as it descends toward the ground floor.

44. ‖ A rocket is being launched straight up. Air resistance is not negligible.

45. | A jet plane is speeding down the runway during takeoff. Air resistance is not negligible.

46. | You've slammed on the brakes and your car is skidding to a stop while going down a 20° hill.

47. ‖ A skier is going down a 20° slope. A *horizontal* headwind is blowing in the skier's face. Friction is small, but not zero.

48. ‖ You've just kicked a rock on the sidewalk and it is now sliding along the concrete.

49. | A Styrofoam ball has just been shot straight up. Air resistance is not negligible.

50. | A spring-loaded gun shoots a plastic ball. The trigger has just been pulled and the ball is starting to move down the barrel. The barrel is horizontal.

51. ‖ A person on a bridge throws a rock straight down toward the water. The rock has just been released.

52. | A gymnast has just landed on a trampoline. She's still moving downward as the trampoline stretches.

53. ‖ The leaf hopper, champion jumper of the insect world, can
BIO jump straight up at 4 m/s². The jump itself lasts a mere 1 ms before the insect is clear of the ground.

   a. Draw a free-body diagram of this mighty leaper while the jump is taking place.

   b. While the jump is taking place, is the force of the ground on the leaf hopper greater than, less than, or equal to the force of gravity on the leaf hopper? Explain.

Challenge Problems

54. A heavy box is in the back of a truck. The truck is accelerating to the right. Draw a motion diagram, a force-identification diagram, and a free-body diagram for the box.

55. A bag of groceries is on the seat of your car as you stop for a stop light. The bag does not slide. Draw a motion diagram, a force-identification diagram, and a free-body diagram for the bag.

56. A rubber ball bounces. We'd like to understand *how* the ball bounces.

   a. A rubber ball has been dropped and is bouncing off the floor. Draw a motion diagram of the ball during the brief time interval that it is in contact with the floor. Show 4 or 5 frames as the ball compresses, then another 4 or 5 frames as it expands. What is the direction of $\vec{a}$ during each of these parts of the motion?

   b. Draw a picture of the ball in contact with the floor and identify all forces acting on the ball.

   c. Draw a free-body diagram of the ball during its contact with the ground. Is there a net force acting on the ball? If so, in which direction?

   d. Write a paragraph in which you describe what you learned from parts a to c and in which you answer the question: How does a ball bounce?

57. If a car stops suddenly, you feel "thrown forward." We'd like to understand what happens to the passengers as a car stops. Imagine yourself sitting on a *very* slippery bench inside a car. This bench has no friction, no seat back, and there's nothing for you to hold onto.

   a. Draw a picture and identify all of the forces acting on you as the car travels at a perfectly steady speed on level ground.

   b. Draw your free-body diagram. Is there a net force on you? If so, in which direction?

   c. Repeat parts a and b with the car slowing down.

   d. Describe what happens to you as the car slows down.

   e. Use Newton's laws to explain why you seem to be "thrown forward" as the car stops. Is there really a force pushing you forward?

   f. Suppose now that the bench is not slippery. As the car slows down, you stay on the bench and don't slide off. What force is responsible for your deceleration? In which direction does this force point? Include a free-body diagram as part of your answer.

STOP TO THINK ANSWERS

Stop to Think 5.1: c.

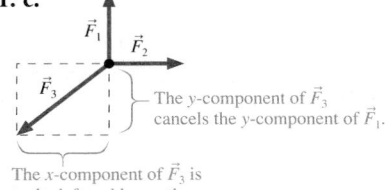

The y-component of $\vec{F}_3$ cancels the y-component of $\vec{F}_1$.

The x-component of $\vec{F}_3$ is to the left and larger than the x-component of $\vec{F}_2$.

Stop to Think 5.2: a, b, and d. Friction and the normal force are the only contact forces. Nothing is touching the rock to provide a "force of the kick."

Stop to Think 5.3: b. Acceleration is proportional to force, so doubling the number of rubber bands doubles the acceleration of the original object from 2 m/s² to 4 m/s². But acceleration is also inversely proportional to mass. Doubling the mass cuts the acceleration in half, back to 2 m/s².

Stop to Think 5.4: d.

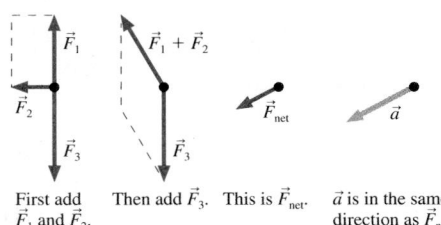

First add $\vec{F}_1$ and $\vec{F}_2$. Then add $\vec{F}_3$. This is $\vec{F}_{net}$. $\vec{a}$ is in the same direction as $\vec{F}_{net}$.

Stop to Think 5.5: c. The acceleration vector points downward as the elevator slows. $\vec{F}_{net}$ points in the same direction as $\vec{a}$, so $\vec{F}_{net}$ also points down. This will be true if the tension is less than the gravitational force: $T < F_G$.

6 Dynamics I: Motion Along a Line

The powerful thrust of the jet engines accelerates this enormous plane to a speed of over 150 mph in less than a mile.

▶ **Looking Ahead** The goal of Chapter 6 is to learn how to solve linear force-and-motion problems.

Forces

In Chapter 5 you learned what a force is and how force and motion are related through Newton's second law.

In this chapter, you'll study in more detail some of the forces introduced in Chapter 5. You'll also learn to solve **equilibrium** problems (with zero net force) and then **dynamics** problems.

The problem-solving procedures developed in this chapter will be used throughout the remainder of the book.

◀ **Looking Back**
Sections 5.3 and 5.7 Identifying forces, drawing free-body diagrams

Equilibrium

An object at rest or moving in a straight line with constant velocity is in equilibrium. The net force is zero.

What tension is needed to tow the car at constant velocity?

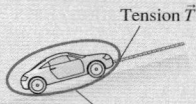

Tension $\vec{T}$

Gravity $\vec{F}_G$ Normal force $\vec{n}$

To solve equilibrium problems, you must be able to identify and work with forces.

◀ **Looking Back**
Section 5.2 A catalog of forces

Dynamics

A net force on an object causes the object to accelerate. This is Newton's second law.

What is the acceleration of the file cabinet?

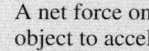

θ

Normal $\vec{n}$
Friction $\vec{f}_s$ Gravity $\vec{F}_G$

To solve dynamics problems, you will need to use constant-acceleration kinematics.

◀ **Looking Back**
Sections 2.4–2.6 Constant-acceleration kinematics

Mass and Weight

You'll learn how mass and weight are different.

- **Mass** is the amount of matter in an object. It is the same everywhere.
- **Weight** is the result of weighing an object on a scale. It depends on gravity and acceleration.

This astronaut on the moon weighs only 1/6 of what he does on earth, but his mass is the same.

Friction and Drag

We'll expand our understanding of friction and drag by developing a model of each.

- Static and kinetic friction depend on the **coefficient of friction**, but not on the object's speed.
- Drag depends on the *square* of the speed and also on the object's cross-section area.

$\vec{D}$

$\vec{F}_G$

A falling object reaches terminal speed when the drag force balances the gravitational force.

Problem Solving

We'll develop a *strategy* for solving force and motion problems, one based on a set of *procedures* rather than a memorized set of equations.

This chapter focuses on motion in a straight line, the motion of bicycles, cars, planes, and rockets.

6.1 Equilibrium

An object on which the net force is zero is in *equilibrium.* The object might be at rest in *static equilibrium,* or it might be moving along a straight line with constant velocity in *dynamic equilibrium.* Both are identical from a Newtonian perspective because $\vec{F}_{net} = \vec{0}$ and $\vec{a} = \vec{0}$.

Newton's first law is the basis for a four-step *strategy* for solving equilibrium problems.

PROBLEM-SOLVING
STRATEGY 6.1 **Equilibrium problems**

MODEL Make simplifying assumptions. When appropriate, represent the object as a particle.

VISUALIZE

- Establish a coordinate system, define symbols, and identify what the problem is asking you to find. This is the process of translating words into symbols.
- Identify *all* forces acting on the object and show them on a free-body diagram.
- These elements form the **pictorial representation** of the problem.

SOLVE The mathematical representation is based on Newton's first law:

$$\vec{F}_{net} = \sum_i \vec{F}_i = \vec{0}$$

The vector sum of the forces is found directly from the free-body diagram.

ASSESS Check that your result has the correct units, is reasonable, and answers the question.

Newton's laws are *vector equations.* The requirement for equilibrium, $\vec{F}_{net} = \vec{0}$, is a shorthand way of writing two simultaneous equations:

$$(F_{net})_x = \sum_i (F_i)_x = 0$$

$$(F_{net})_y = \sum_i (F_i)_y = 0$$

(6.1)

The concept of equilibrium is essential for the engineering analysis of stationary objects such as bridges.

In other words, each component of $\vec{F}_{net}$ must simultaneously be zero. Although real-world situations often have forces pointing in three dimensions, thus requiring a third equation for the z-component of $\vec{F}_{net}$, we will restrict ourselves for now to problems that can be analyzed in two dimensions.

NOTE ▶ The equilibrium condition of Equations 6.1 applies only to particles, which cannot rotate. Equilibrium of an extended object, which can rotate, requires an additional condition that we will study in Chapter 12. ◀

Equilibrium problems occur frequently, especially in engineering applications. Let's look at a couple of examples.

Static Equilibrium

EXAMPLE 6.1 **Finding the force on the kneecap**

Your kneecap (patella) is attached by a tendon to your quadriceps muscle. This tendon pulls at a 10° angle relative to the femur, the bone of your upper leg. The patella is also attached to your lower leg (tibia) by a tendon that pulls parallel to the leg. To balance these forces, the lower end of your femur pushes outward on the patella. Bending your knee increases the tension in the tendons, and both have a tension of 60 N when the knee is bent to make a 70° angle between the upper and lower leg. What force does the femur exert on the kneecap in this position?

MODEL Model the kneecap as a particle in static equilibrium.

FIGURE 6.1 Pictorial representation of the kneecap in static equilibrium.

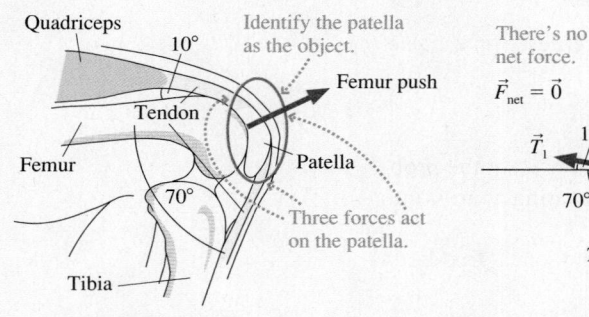
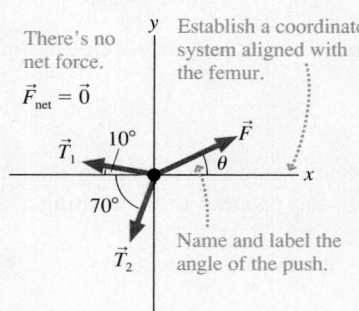
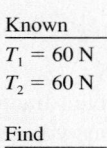

Identify forces.   Draw free-body diagram.   List knowns and unknowns.

VISUALIZE **FIGURE 6.1** shows how to draw a pictorial representation. We've chosen to align the x-axis with the femur. The three forces—shown on the free-body diagram—are labeled $\vec{T}_1$ and $\vec{T}_2$ for the tensions and $\vec{F}$ for the femur's push. Notice that we've *defined* angle θ to indicate the direction of the femur's force on the kneecap.

SOLVE This is a static-equilibrium problem, with three forces on the kneecap that must sum to zero. Newton's first law, written in component form, is

$$(F_{net})_x = \sum_i (F_i)_x = T_{1x} + T_{2x} + F_x = 0$$

$$(F_{net})_y = \sum_i (F_i)_y = T_{1y} + T_{2y} + F_y = 0$$

NOTE ▶ You might have been tempted to write $-T_{1x}$ in the equation since $\vec{T}_1$ points to the left. But the net force, by definition, is the *sum* of all the individual forces. That fact that $\vec{T}_1$ points to the left will be taken into account when we *evaluate* the components. ◀

The components of the force vectors can be evaluated directly from the free-body diagram:

$$T_{1x} = -T_1 \cos 10° \qquad T_{1y} = T_1 \sin 10°$$
$$T_{2x} = -T_2 \cos 70° \qquad T_{2y} = -T_2 \sin 70°$$
$$F_x = F \cos \theta \qquad F_y = F \sin \theta$$

This is where signs enter, with T_{1x} being assigned a negative value because $\vec{T}_1$ points to the left. Similarly, $\vec{T}_2$ points both to the left and down, so both T_{2x} and T_{2y} are negative. With these components, Newton's first law becomes

$$-T_1 \cos 10° - T_2 \cos 70° + F \cos \theta = 0$$
$$T_1 \sin 10° - T_2 \sin 70° + F \sin \theta = 0$$

These are two simultaneous equations for the two unknowns F and θ. We will encounter equations of this form on many occasions, so make a note of the method of solution. First, rewrite the two equations as

$$F \cos \theta = T_1 \cos 10° + T_2 \cos 70°$$
$$F \sin \theta = -T_1 \sin 10° + T_2 \sin 70°$$

Next, divide the second equation by the first to eliminate F:

$$\frac{F \sin \theta}{F \cos \theta} = \tan \theta = \frac{-T_1 \sin 10° + T_2 \sin 70°}{T_1 \cos 10° + T_2 \cos 70°}$$

Then solve for θ:

$$\theta = \tan^{-1}\left(\frac{-T_1 \sin 10° + T_2 \sin 70°}{T_1 \cos 10° + T_2 \cos 70°}\right)$$

$$= \tan^{-1}\left(\frac{-(60\text{ N}) \sin 10° + (60\text{ N}) \sin 70°}{(60\text{ N}) \cos 10° + (60\text{ N}) \cos 70°}\right) = 30°$$

Finally, use θ to find F:

$$F = \frac{T_1 \cos 10° + T_2 \cos 70°}{\cos \theta}$$

$$= \frac{(60\text{ N}) \cos 10° + (60\text{ N}) \cos 70°}{\cos 30°} = 92\text{ N}$$

The question asked What force? and force is a vector, so we must specify both the magnitude and the direction. With the knee in this position, the femur exerts a force $\vec{F} = (92\text{ N}, 30° \text{ above horizontal})$ on the kneecap.

ASSESS The magnitude of the force would be 0 N if the leg were straight, 120 N if the knee could be bent 180° so that the two tendons pull in parallel. The knee is closer to fully bent than to straight, so we would expect a femur force between 60 N and 120 N. Thus the calculated magnitude of 92 N seems reasonable.

Dynamic Equilibrium

EXAMPLE 6.2   **Towing a car up a hill**

A car with a weight of 15,000 N is being towed up a 20° slope at constant velocity. Friction is negligible. The tow rope is rated at 6000 N maximum tension. Will it break?

MODEL We'll treat the car as a particle in dynamic equilibrium.

VISUALIZE This problem asks for a yes or no answer, not a number, but we still need a quantitative analysis. Part of our analysis of the problem statement is to determine which quantity or quantities

allow us to answer the question. In this case the answer is clear: We need to calculate the tension in the rope. **FIGURE 6.2** shows the pictorial representation. Note the similarities to Examples 5.2 and 5.6 in Chapter 5, which you may want to review.

We noted in Chapter 5 that the weight of an object at rest is the magnitude F_G of the gravitational force acting on it, and that information has been listed as known.

FIGURE 6.2 Pictorial representation of a car being towed up a hill.

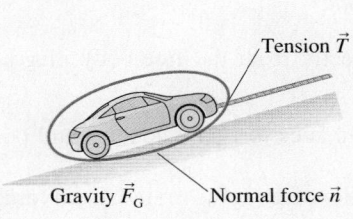

Tension $\vec{T}$

Gravity $\vec{F}_G$ Normal force $\vec{n}$

The coordinate system is chosen with one axis parallel to the motion so that the acceleration vector has only one nonzero component.

The normal force is perpendicular to the surface.

$\vec{n}$

$\vec{T}$ x

y

θ $\vec{F}_{net} = \vec{0}$

Same angle $\vec{F}_G$ θ

Known
$\theta = 20°$
$F_G = 15,000$ N

Find
T

SOLVE The free-body diagram shows forces $\vec{T}$, $\vec{n}$, and $\vec{F}_G$ acting on the car. Newton's first law is

$$(F_{net})_x = \sum F_x = T_x + n_x + (F_G)_x = 0$$

$$(F_{net})_y = \sum F_y = T_y + n_y + (F_G)_y = 0$$

From here on, we'll use $\sum F_x$ and $\sum F_y$, without the label i, as a simple shorthand notation to indicate that we're adding all the x-components and all the y-components of the forces.

We can deduce the components directly from the free-body diagram:

$$T_x = T \qquad\qquad T_y = 0$$

$$n_x = 0 \qquad\qquad n_y = n$$

$$(F_G)_x = -F_G \sin\theta \qquad (F_G)_y = -F_G \cos\theta$$

NOTE ▶ The gravitational force has both x- and y-components in this coordinate system, both of which are negative due to the direction of the vector $\vec{F}_G$. You'll see this situation often, so be sure you understand where $(F_G)_x$ and $(F_G)_y$ come from. ◀

With these components, the first law becomes

$$T - F_G \sin\theta = 0$$

$$n - F_G \cos\theta = 0$$

The first of these can be rewritten as

$$T = F_G \sin\theta = (15,000 \text{ N}) \sin 20° = 5100 \text{ N}$$

Because $T < 6000$ N, we conclude that the rope will *not* break. It turned out that we did not need the y-component equation in this problem.

ASSESS Because there's no friction, it would not take *any* tension force to keep the car rolling along a horizontal surface ($\theta = 0°$). At the other extreme, $\theta = 90°$, the tension force would need to equal the car's weight ($T = 15,000$ N) to lift the car straight up at constant velocity. The tension force for a 20° slope should be somewhere in between, and 5100 N is a little less than half the weight of the car. That our result is reasonable doesn't prove it's right, but we have at least ruled out careless errors that give unreasonable results.

6.2 Using Newton's Second Law

The essence of Newtonian mechanics can be expressed in two steps:

- The forces acting on an object determine its acceleration $\vec{a} = \vec{F}_{net}/m$.
- The object's trajectory can be determined by using $\vec{a}$ in the equations of kinematics.

These two ideas are the basis of a strategy for solving dynamics problems.

PROBLEM-SOLVING
STRATEGY 6.2   **Dynamics problems**

MODEL Make simplifying assumptions.

VISUALIZE Draw a **pictorial representation.**

- Show important points in the motion with a sketch, establish a coordinate system, define symbols, and identify what the problem is trying to find.
- Use a motion diagram to determine the object's acceleration vector $\vec{a}$.
- Identify all forces acting on the object *at this instant* and show them on a free-body diagram.
- It's OK to go back and forth between these steps as you visualize the situation.

SOLVE The mathematical representation is based on Newton's second law:

$$\vec{F}_{net} = \sum_i \vec{F}_i = m\vec{a}$$

The vector sum of the forces is found directly from the free-body diagram. Depending on the problem, either

- Solve for the acceleration, then use kinematics to find velocities and positions; or
- Use kinematics to determine the acceleration, then solve for unknown forces.

ASSESS Check that your result has the correct units, is reasonable, and answers the question.

Exercise 22

Newton's second law is a vector equation. To apply the step labeled Solve, you must write the second law as two simultaneous equations:

$$(F_{net})_x = \sum F_x = ma_x$$

$$(F_{net})_y = \sum F_y = ma_y$$

(6.2)

The primary goal of this chapter is to illustrate the use of this strategy.

EXAMPLE 6.3   **Speed of a towed car**

A 1500 kg car is pulled by a tow truck. The tension in the tow rope is 2500 N, and a 200 N friction force opposes the motion. If the car starts from rest, what is its speed after 5.0 seconds?

MODEL We'll treat the car as an accelerating particle. We'll assume, as part of our *interpretation* of the problem, that the road is horizontal and that the direction of motion is to the right.

VISUALIZE FIGURE 6.3 on the next page shows the pictorial representation. We've established a coordinate system and defined symbols to represent kinematic quantities. We've identified the speed v_1, rather than the velocity v_{1x}, as what we're trying to find.

SOLVE We begin with Newton's second law:

$$(F_{net})_x = \sum F_x = T_x + f_x + n_x + (F_G)_x = ma_x$$
$$(F_{net})_y = \sum F_y = T_y + f_y + n_y + (F_G)_y = ma_y$$

All four forces acting on the car have been included in the vector sum. The equations are perfectly general, with + signs everywhere, because the four vectors are *added* to give $\vec{F}_{net}$. We can now "read" the vector components from the free-body diagram:

$$T_x = +T \qquad T_y = 0 \qquad n_x = 0 \qquad n_y = +n$$
$$f_x = -f \qquad f_y = 0 \qquad (F_G)_x = 0 \qquad (F_G)_y = -F_G$$

The signs depend on which way the vectors point. Substituting these into the second-law equations and dividing by m give

$$a_x = \frac{1}{m}(T - f)$$

$$= \frac{1}{1500 \text{ kg}}(2500 \text{ N} - 200 \text{ N}) = 1.53 \text{ m/s}^2$$

$$a_y = \frac{1}{m}(n - F_G)$$

NOTE ▶ Newton's second law has allowed us to determine a_x exactly but has given only an algebraic expression for a_y. However, we know *from the motion diagram* that $a_y = 0$! That is, the motion is purely along the x-axis, so there is *no* acceleration along the y-axis. The requirement $a_y = 0$ allows us to conclude that $n = F_G$. Although we do not need n for this problem, it will be important in many future problems. ◀

FIGURE 6.3 Pictorial representation of a car being towed.

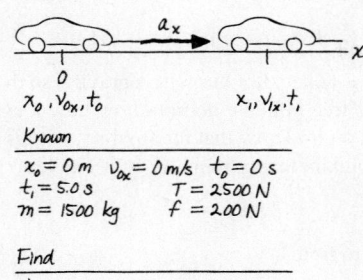

Sketch

Motion diagram and forces

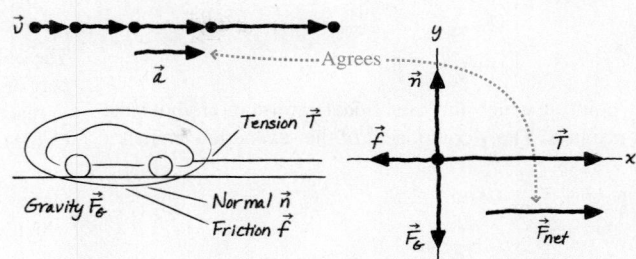

Because a_x is a constant 1.53 m/s², we can finish by using constant-acceleration kinematics to find the velocity:

$$v_{1x} = v_{0x} + a_x \, \Delta t$$
$$= 0 + (1.53 \text{ m/s}^2)(5.0 \text{ s}) = 7.7 \text{ m/s}$$

The problem asked for the *speed* after 5.0 s, which is $v_1 =$ 7.7 m/s.

ASSESS 7.7 m/s ≈ 15 mph, a quite reasonable speed after 5 s of acceleration.

EXAMPLE 6.4   **Altitude of a rocket**

A 500 g model rocket with a weight of 4.90 N is launched straight up. The small rocket motor burns for 5.00 s and has a steady thrust of 20.0 N. What maximum altitude does the rocket reach? Ignore the mass loss of the burned fuel.

MODEL We'll treat the rocket as an accelerating particle. Air resistance will be neglected.

VISUALIZE The pictorial representation of **FIGURE 6.4** finds that this is a two-part problem. First, the rocket accelerates straight up. Second, the rocket continues going up as it slows down, a free-fall situation. The maximum altitude is at the end of the second part of the motion.

SOLVE We now know what the problem is asking, have established relevant symbols and coordinates, and know what the forces are. We begin the mathematical representation by writing Newton's second law, in component form, as the rocket accelerates upward. The free-body diagram shows two forces, so

$$(F_{\text{net}})_x = \sum F_x = (F_{\text{thrust}})_x + (F_G)_x = ma_{0x}$$
$$(F_{\text{net}})_y = \sum F_y = (F_{\text{thrust}})_y + (F_G)_y = ma_{0y}$$

The fact that vector $\vec{F}_G$ points downward—and which might have tempted you to use a minus sign in the y-equation—will be taken into account when we *evaluate* the components. None of

FIGURE 6.4 Pictorial representation of a rocket launch.

Sketch

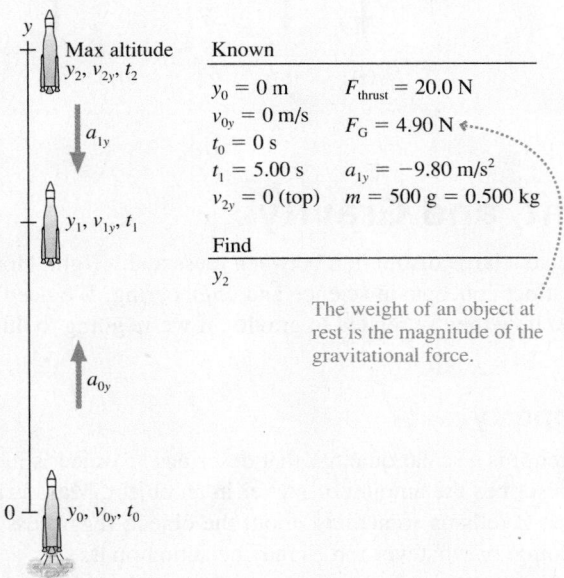

Known

$y_0 = 0$ m	$F_{\text{thrust}} = 20.0$ N
$v_{0y} = 0$ m/s	$F_G = 4.90$ N
$t_0 = 0$ s	
$t_1 = 5.00$ s	$a_{1y} = -9.80$ m/s²
$v_{2y} = 0$ (top)	$m = 500$ g $= 0.500$ kg

Find

y_2

The weight of an object at rest is the magnitude of the gravitational force.

Motion diagram and forces

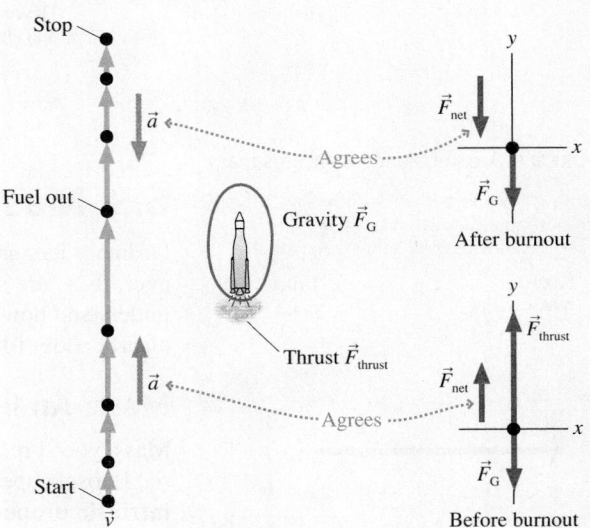

Continued

the vectors in this problem has an x-component, so only the y-component of the second law is needed. We can use the free-body diagram to see that

$$(F_{\text{thrust}})_y = +F_{\text{thrust}}$$

$$(F_G)_y = -F_G$$

This is the point at which the directional information about the force vectors enters. The y-component of the second law is then

$$a_{0y} = \frac{1}{m}(F_{\text{thrust}} - F_G)$$

$$= \frac{20.0 \text{ N} - 4.90 \text{ N}}{0.500 \text{ kg}} = 30.2 \text{ m/s}^2$$

Notice that we converted the mass to SI units of kilograms before doing any calculations and that, because of the definition of the newton, the division of newtons by kilograms automatically gives the correct SI units of acceleration.

The acceleration of the rocket is constant until it runs out of fuel, so we can use constant-acceleration kinematics to find the altitude and velocity at burnout ($\Delta t = t_1 = 5.00 \text{ s}$):

$$y_1 = y_0 + v_{0y}\,\Delta t + \tfrac{1}{2}a_{0y}(\Delta t)^2$$

$$= \tfrac{1}{2}a_{0y}(\Delta t)^2 = 377 \text{ m}$$

$$v_{1y} = v_{0y} + a_{0y}\,\Delta t = a_{0y}\,\Delta t = 151 \text{ m/s}$$

The only force on the rocket after burnout is gravity, so the second part of the motion is free fall. We do not know how long it takes to reach the top, but we do know that the final velocity is $v_{2y} = 0$. Constant-acceleration kinematics with $a_{1y} = -g$ gives

$$v_{2y}^2 = 0 = v_{1y}^2 - 2g\,\Delta y = v_{1y}^2 - 2g(y_2 - y_1)$$

which we can solve to find

$$y_2 = y_1 + \frac{v_{1y}^2}{2g} = 377 \text{ m} + \frac{(151 \text{ m/s})^2}{2(9.80 \text{ m/s}^2)}$$

$$= 1540 \text{ m} = 1.54 \text{ km}$$

ASSESS The maximum altitude reached by this rocket is 1.54 km, or just slightly under one mile. While this does not seem unreasonable for a high-acceleration rocket, the neglect of air resistance was probably not a terribly realistic assumption.

The solutions to these first few examples have been quite detailed. Our purpose has been to show how the problem-solving strategy is put into practice. Future examples will be briefer, but the basic *procedure* will remain the same.

STOP TO THINK 6.1 A Martian lander is approaching the surface. It is slowing its descent by firing its rocket motor. Which is the correct free-body diagram?

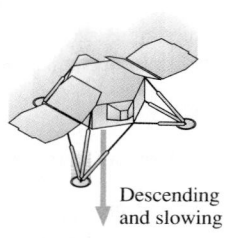

Descending and slowing

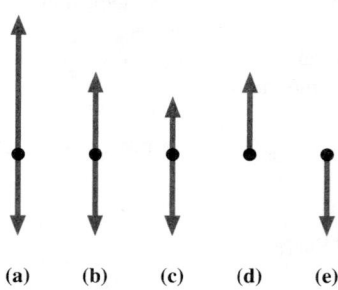

(a) (b) (c) (d) (e)

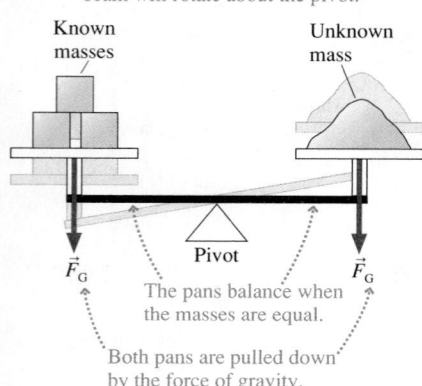

FIGURE 6.5 A pan balance measures mass.

If the unknown mass differs from the known masses, the beam will rotate about the pivot.

Known masses

Unknown mass

$\vec{F}_G$ $\vec{F}_G$

Pivot

The pans balance when the masses are equal.

Both pans are pulled down by the force of gravity.

6.3 Mass, Weight, and Gravity

Ordinary language does not make a large distinction between mass and weight. However, these are separate and distinct concepts in science and engineering. We need to understand how they differ, and how they're related to gravity, if we're going to think clearly about force and motion.

Mass: An Intrinsic Property

Mass, you'll recall from Chapter 5, is a scalar quantity that describes an object's inertia. Loosely speaking, it also describes the amount of matter in an object. **Mass is an intrinsic property of an object.** It tells us something about the object, regardless of where the object is, what it's doing, or whatever forces may be acting on it.

A *pan balance,* shown in FIGURE 6.5, is a device for measuring mass. Although a pan balance requires gravity to function, it does not depend on the strength of gravity. Consequently, the pan balance would give the same result on another planet.

Gravity: A Force

The idea of gravity has a long and interesting history intertwined with our evolving ideas about the solar system. It was Newton who—along with discovering his three laws of motion—first recognized that **gravity is an attractive, long-range force between *any* two objects.**

FIGURE 6.6 shows two objects with masses m_1 and m_2 separated by distance r. Each object pulls on the other with a force given by *Newton's law of gravity:*

$$F_{1\,\text{on}\,2} = F_{2\,\text{on}\,1} = \frac{Gm_1 m_2}{r^2} \qquad \text{(Newton's law of gravity)} \qquad (6.3)$$

where $G = 6.67 \times 10^{-11}\ \text{N m}^2/\text{kg}^2$, called the *gravitational constant,* is one of the basic constants of nature. Notice that the force gets weaker as the distance between the objects increases.

The gravitational force between two human-sized objects is minuscule, completely insignificant in comparison with other forces. That's why you're not aware of being tugged toward everything around you. Only when one or both objects is planet-sized or larger does gravity become an important force. Indeed, Chapter 13 will explore in detail the application of Newton's law of gravity to the orbits of satellites and planets.

For objects moving near the surface of the earth (or other planet), things like balls and cars and planes that we'll be studying in the next few chapters, we can make the **flat-earth approximation** shown in FIGURE 6.7. That is, if the height above the surface is very small in comparison with the size of the planet, then the curvature of the surface is not noticeable and there's virtually no difference between r and the planet's radius R. Consequently, a very good approximation for the gravitational force of the planet on mass m is simply

$$\vec{F}_G = \vec{F}_{\text{planet on }m} = \left(\frac{GMm}{R^2}, \text{straight down} \right) = (mg, \text{straight down}) \qquad (6.4)$$

The magnitude or size of the gravitational force is $F_G = mg$, where the quantity g—a property of the planet—is defined to be

$$g = \frac{GM}{R^2} \qquad (6.5)$$

In addition, the direction of the gravitational force defines what we *mean* by "straight down."

But why did we choose to call it g, a symbol we've already used for free-fall acceleration? To see the connection, recall that free fall is motion under the influence of gravity only. FIGURE 6.8 shows the free-body diagram of an object in free fall near the surface of a planet. With $\vec{F}_{\text{net}} = \vec{F}_G$, Newton's second law predicts the acceleration to be

$$\vec{a}_{\text{free fall}} = \frac{\vec{F}_{\text{net}}}{m} = \frac{\vec{F}_G}{m} = (g, \text{straight down}) \qquad (6.6)$$

Because g is a property of the planet, independent of the object, **all objects on the same planet, regardless of mass, have the same free-fall acceleration.** We introduced this idea in Chapter 2 as an experimental discovery of Galileo, but now we see that the mass independence of $\vec{a}_{\text{free fall}}$ is a prediction of Newton's law of gravity.

But does Newton's law predict the correct value, which we know from experiment to be $g = |a_{\text{free fall}}| = 9.80\ \text{m/s}^2$? We can use the average radius ($R_{\text{earth}} = 6.37 \times 10^6\ \text{m}$) and mass ($M_{\text{earth}} = 5.98 \times 10^{24}\ \text{kg}$) of the earth to calculate

$$g_{\text{earth}} = \frac{GM_{\text{earth}}}{(R_{\text{earth}})^2} = \frac{(6.67 \times 10^{-11}\ \text{N m}^2/\text{kg}^2)(5.98 \times 10^{24}\ \text{kg})}{(6.37 \times 10^6\ \text{m})^2} = 9.83\ \text{N/kg}$$

You should convince yourself that N/kg is equivalent to m/s², so $g_{\text{earth}} = 9.83\ \text{m/s}^2$. (Data for other astronomical objects, which you may need for homework, are provided inside the back cover of the book.)

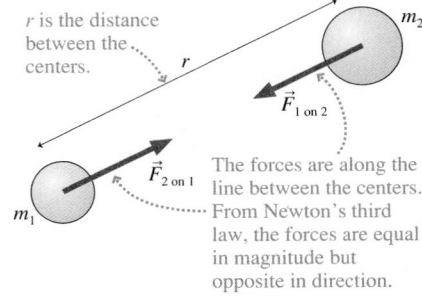

FIGURE 6.6 Newton's law of gravity.

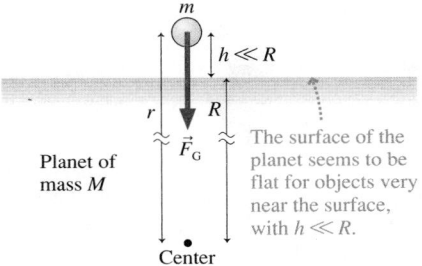

FIGURE 6.7 Gravity near the surface of a planet.

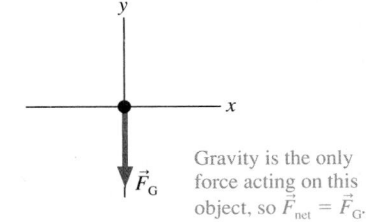

FIGURE 6.8 The free-body diagram of an object in free fall.

Newton's prediction is very close, but it's not quite right. The free-fall acceleration *would* be 9.83 m/s² on a stationary earth, but, in reality, the earth is rotating on its axis. The "missing" 0.03 m/s² is due to the earth's rotation, a claim we'll justify when we study circular motion in Chapter 8. Because we're on the outside of a rotating sphere, rather like being on the outside edge of a merry-go-round, the effect of rotation is to "weaken" gravity.

Our goal is to analyze motion from within our own reference frame, a reference frame attached to the earth. Strictly speaking, Newton's laws of motion are not valid in our reference frame because it is rotating and thus is not an inertial reference frame. Fortunately, we can use Newton's laws to analyze motion near the earth's surface, and we can use $F_G = mg$ for the gravitational force *if* we use $g = |a_{\text{free fall}}| = 9.80$ m/s² rather than $g = g_{\text{earth}}$. (This assertion is proved in more advanced classes.) In our rotating reference frame, $\vec{F}_G$ is the *effective gravitational force*, the true gravitational force given by Newton's law of gravity plus a small correction due to our rotation. This is the force to show on free-body diagrams and use in calculations.

Weight: A Measurement

When you weigh yourself, you stand on a *spring scale* and compress a spring. You weigh apples in the grocery store by placing them in a spring scale and stretching a spring. The reading of a spring scale, such as the two shown in FIGURE 6.9, is F_{sp}, the magnitude of the force the spring is exerting.

With that in mind, let's define the **weight** of an object as the reading F_{sp} of a calibrated spring scale on which the object is stationary. That is, **weight is a measurement, the result of "weighing" an object.** Because F_{sp} is a force, weight is measured in newtons.

Suppose the scales in Figure 6.9 are at rest relative to the earth. Then the object being weighed is in static equilibrium, with $\vec{F}_{net} = \vec{0}$. The stretched spring *pulls* up, the compressed spring *pushes* up, but in both cases $\vec{F}_{net} = \vec{0}$ only if the upward spring force exactly balances the downward gravitational force of magnitude mg:

$$F_{sp} = F_G = mg \tag{6.7}$$

Because we defined weight as the reading F_{sp} of a spring scale, the weight of a stationary object is

$$w = mg \quad \text{(weight of a stationary object)} \tag{6.8}$$

The scale does not "know" the weight of the object. All it can do is to measure how much its spring is stretched or compressed. On earth, a student with a mass of 70 kg has weight $w = (70 \text{ kg})(9.80 \text{ m/s}^2) = 686$ N *because* he compresses a spring until the spring pushes upward with 686 N. On a different planet, with a different value for g, the expansion or compression of the spring would be different and the student's weight would be different.

> NOTE ▶ **Mass and weight are not the same thing.** Mass, in kg, is an intrinsic property of an object; its value is unique and always the same. Weight, in N, depends on the object's mass, but it also depends on the situation—the strength of gravity and, as we will see, whether or not the object is accelerating. Weight is *not* a property of the object, and thus weight does not have a unique value. ◀

Surprisingly, you cannot directly feel or sense gravity. Your *sensation*—how heavy you feel—is due to contact forces pressing against you, forces that touch you and activate nerve endings in your skin. As you read this, your sensation of weight is due to the normal force exerted on you by the chair in which you are sitting. When you stand, you feel the contact force of the floor pushing against your feet.

But recall the sensations you feel while accelerating. You feel "heavy" when an elevator suddenly accelerates upward, but this sensation vanishes as soon as the elevator reaches a steady speed. Your stomach seems to rise a little and you feel lighter than normal as the upward-moving elevator brakes to a halt or a roller coaster goes over the top. Has your weight actually changed?

FIGURE 6.9 A spring scale measures weight.

(a)

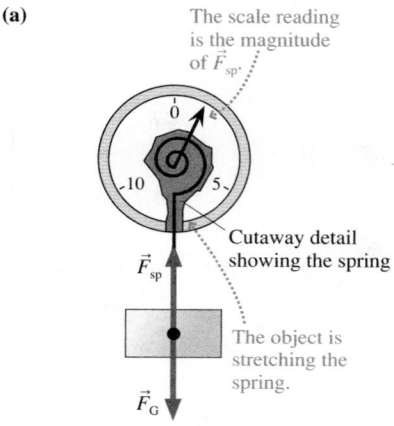

The scale reading is the magnitude of $\vec{F}_{sp}$.

Cutaway detail showing the spring

$\vec{F}_{sp}$

The object is stretching the spring.

$\vec{F}_G$

(b)

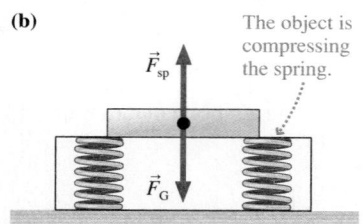

The object is compressing the spring.

$\vec{F}_{sp}$

$\vec{F}_G$

To answer this question, **FIGURE 6.10** shows a man weighing himself on a spring scale in an accelerating elevator. The only forces acting on the man are the upward spring force of the scale and the downward gravitational force. This seems to be the same situation as Figure 6.9b, but there's one big difference: The man is accelerating, hence there must be a net force on the man in the direction of $\vec{a}$.

For the net force $\vec{F}_{net}$ to point upward, the magnitude of the spring force must be *greater* than the magnitude of the gravitational force. That is, $F_{sp} > mg$. Looking at the free-body diagram in Figure 6.10, we see that the y-component of Newton's second law is

$$(F_{net})_y = (F_{sp})_y + (F_G)_y = F_{sp} - mg = ma_y \qquad (6.9)$$

where m is the man's mass.

We defined weight as the reading F_{sp} of a calibrated spring scale *on which the object is stationary*. That is the case here as the scale and man accelerate upward together. Thus the man's weight as he accelerates vertically is

$$w = \text{scale reading } F_{sp} = mg + ma_y = mg\left(1 + \frac{a_y}{g}\right) \qquad (6.10)$$

If an object is either at rest or moving with constant velocity, then $a_y = 0$ and $w = mg$. That is, the weight of an object at rest is the magnitude of the (effective) gravitational force acting on it. But its weight differs if it has a vertical acceleration.

You *do* weigh more as an elevator accelerates upward ($a_y > 0$) because the reading of a scale—a weighing—increases. Similarly, your weight is less when the acceleration vector $\vec{a}$ points downward ($a_y < 0$) because the scale reading goes down. Weight, as we've defined it, corresponds to your sensation of heaviness or lightness.*

We found Equation 6.10 by considering a person in an accelerating elevator, but it applies to any object with a vertical acceleration. Further, an object doesn't really have to be on a scale to have a weight; an object's weight is the magnitude of the contact force supporting it. It makes no difference whether this is the spring force of the scale or simply the normal force of the floor.

NOTE ▶ Informally, we sometimes say "This object weighs such and such" or "The weight of this object is. . . ." We'll interpret these expressions as meaning mg, the weight of an object of mass m at rest ($a_y = 0$) on the surface of the earth or some other astronomical body. ◀

Weightlessness

Suppose the elevator cable breaks and the elevator, along with the man and his scale, plunges straight down in free fall! What will the scale read? When the free-fall acceleration $a_y = -g$ is used in Equation 6.10, we find $w = 0$. In other words, *the man has no weight!*

Suppose, as the elevator falls, the man inside releases a ball from his hand. In the absence of air resistance, as Galileo discovered, both the man and the ball would fall at the same rate. From the man's perspective, the ball would appear to "float" beside him. Similarly, the scale would float beneath him and not press against his feet. He is what we call *weightless*. Gravity is still pulling down on him—that's why he's falling—but he has no *sensation* of weight as everything floats around him in free fall.

But isn't this exactly what happens to astronauts orbiting the earth? If an astronaut tries to stand on a scale, it does not exert any force against her feet and reads zero. She is said to be weightless. But if the criterion to be weightless is to be in free fall, and if astronauts orbiting the earth are weightless, does this mean that they are in free fall? This is a very interesting question to which we shall return in Chapter 8.

*Surprisingly, there is no universally agreed-upon definition of *weight*. Some textbooks define weight as the gravitational force on an object, $\vec{w} = (mg, \text{down})$. In that case, the scale reading of an accelerating object, and your sensation of weight, is often called *apparent weight*. This textbook prefers the definition of *weight* as being what a scale reads, the result of a weighing measurement.

FIGURE 6.10 A man weighing himself in an accelerating elevator.

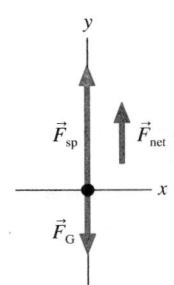

The man feels heavier than normal while accelerating upward.

Spring scale

Astronauts are weightless as they orbit the earth.

FIGURE 6.11 Static friction keeps an object from slipping.

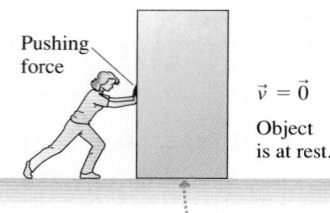

Pushing force

$\vec{v} = \vec{0}$

Object is at rest.

Static friction is opposite the push to prevent motion.

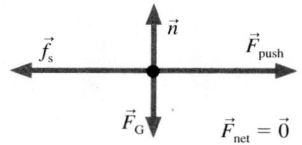

FIGURE 6.12 Static friction acts in *response* to an applied force.

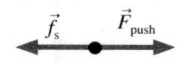

$\vec{F}_{push}$ is balanced by $\vec{f}_s$ and the box does not move.

As $\vec{F}_{push}$ increases, f_s grows . . .

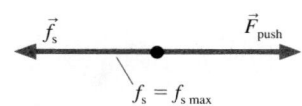

$f_s = f_{s\,max}$

. . . until f_s reaches $f_{s\,max}$. Now, if $\vec{F}_{push}$ gets any bigger, the object will start to move.

FIGURE 6.13 The kinetic friction force is opposite the direction of motion.

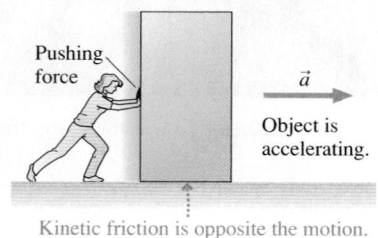

Pushing force

$\vec{a}$

Object is accelerating.

Kinetic friction is opposite the motion.

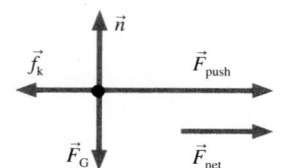

An elevator that has descended from the 50th floor is coming to a halt at the 1st floor. As it does, your weight is

a. More than *mg*. b. Less than *mg*. c. Equal to *mg*. d. Zero.

6.4 Friction

Friction is absolutely essential for many things we do. Without friction you could not walk, drive, or even sit down (you would slide right off the chair!). Although friction is a complicated force, many aspects of friction can be described with a simple model.

Static Friction

Chapter 5 defined *static friction* $\vec{f}_s$ as the force on an object that keeps it from slipping. **FIGURE 6.11** shows a person pushing on a box that, due to static friction, isn't moving. The box is in static equilibrium, so the static friction force must exactly balance the pushing force:

$$f_s = F_{push} \tag{6.11}$$

To determine the direction of $\vec{f}_s$, decide which way the object would move if there were no friction. The static friction force $\vec{f}_s$ points in the *opposite* direction to prevent the motion.

Unlike the gravitational force, which has the precise and unambiguous magnitude $F_G = mg$, the size of the static friction force depends on how hard you push. The harder the person in Figure 6.11 pushes, the harder the floor pushes back. Reduce the pushing force, and the static friction force will automatically be reduced to match. Static friction acts in *response* to an applied force. **FIGURE 6.12** illustrates this idea.

But there's clearly a limit to how big f_s can get. If you push hard enough, the object slips and starts to move. In other words, the static friction force has a *maximum* possible size $f_{s\,max}$.

- An object remains at rest as long as $f_s < f_{s\,max}$.
- The object slips when $f_s = f_{s\,max}$.
- A static friction force $f_s > f_{s\,max}$ is not physically possible.

Experiments with friction show that $f_{s\,max}$ is proportional to the magnitude of the normal force. That is,

$$f_{s\,max} = \mu_s n \tag{6.12}$$

where the proportionality constant μ_s is called the **coefficient of static friction**. The coefficient is a dimensionless number that depends on the materials of which the object and the surface are made. Table 6.1 on the next page shows some typical coefficients of friction. It is to be emphasized that these are only approximate. The exact value of the coefficient depends on the roughness, cleanliness, and dryness of the surfaces.

Kinetic Friction

Once the box starts to slide, as in **FIGURE 6.13**, the static friction force is replaced by a kinetic friction force $\vec{f}_k$. Experiments show that kinetic friction, unlike static friction, has a nearly *constant* magnitude. Furthermore, the size of the kinetic friction force is *less* than the maximum static friction, $f_k < f_{s\,max}$, which explains why it is easier to keep the box moving than it was to start it moving. The direction of $\vec{f}_k$ is always opposite to the direction in which an object slides across the surface.

The kinetic friction force is also proportional to the magnitude of the normal force:

$$f_k = \mu_k n \tag{6.13}$$

where μ_k is called the **coefficient of kinetic friction.** Table 6.1 includes typical values of μ_k. You can see that $\mu_k < \mu_s$, causing the kinetic friction to be less than the maximum static friction.

Rolling Friction

If you slam on the brakes hard enough, your car tires slide against the road surface and leave skid marks. This is kinetic friction. A wheel *rolling* on a surface also experiences friction, but not kinetic friction. The portion of the wheel that contacts the surface is stationary with respect to the surface, not sliding. To see this, roll a wheel slowly and watch how it touches the ground.

No wheel is perfectly round and thus, as **FIGURE 6.14** shows, a wheel has an area of contact with the ground. Molecular bonds are quickly established where the wheel presses against the surface. These bonds have to be broken as the wheel rolls forward, and the effort needed to break them causes **rolling friction.** (Think how it is to walk with a wad of chewing gum stuck to the sole of your shoe!) The force of rolling friction can be calculated in terms of a **coefficient of rolling friction** μ_r:

$$f_r = \mu_r n \qquad (6.14)$$

Rolling friction acts very much like kinetic friction, but values of μ_r (see Table 6.1) are much lower than values of μ_k. This is why it is easier to roll an object on wheels than to slide it.

A Model of Friction

These ideas can be summarized in a *model* of friction:

> Static: $\vec{f}_s \leq (\mu_s n$, direction as necessary to prevent motion)
> Kinetic: $\vec{f}_k = (\mu_k n$, direction opposite the motion) $\qquad (6.15)$
> Rolling: $\vec{f}_r = (\mu_r n$, direction opposite the motion)

Here "motion" means "motion relative to the surface." The maximum value of static friction $f_{s\,max} = \mu_s n$ occurs at the point where the object slips and begins to move.

NOTE ▶ Equations 6.15 are a "model" of friction, not a "law" of friction. These equations—a simplification of reality—provide a reasonably accurate, but not perfect, description of how friction forces act. They are not a "law of nature" on a level with Newton's laws. ◀

FIGURE 6.15 summarizes these ideas graphically by showing how the friction force changes as the magnitude of an applied force $\vec{F}_{push}$ increases.

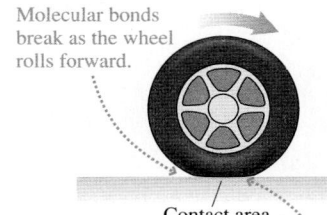

FIGURE 6.14 Rolling friction is due to the contact area between a wheel and the surface.

Molecular bonds break as the wheel rolls forward.

Contact area

The wheel flattens where it touches the surface, giving a contact area rather than a point of contact.

TABLE 6.1 Coefficients of friction

Materials	Static μ_s	Kinetic μ_k	Rolling μ_r
Rubber on concrete	1.00	0.80	0.02
Steel on steel (dry)	0.80	0.60	0.002
Steel on steel (lubricated)	0.10	0.05	
Wood on wood	0.50	0.20	
Wood on snow	0.12	0.06	
Ice on ice	0.10	0.03	

FIGURE 6.15 The friction force response to an increasing applied force.

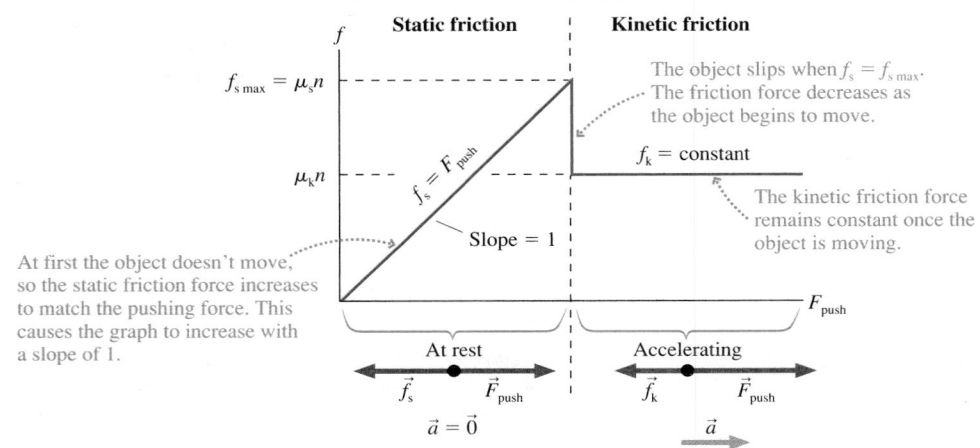

Static friction | Kinetic friction

$f_{s\,max} = \mu_s n$

The object slips when $f_s = f_{s\,max}$. The friction force decreases as the object begins to move.

$f_k = $ constant

$\mu_k n$

$f_s = F_{push}$

The kinetic friction force remains constant once the object is moving.

Slope = 1

At first the object doesn't move, so the static friction force increases to match the pushing force. This causes the graph to increase with a slope of 1.

F_{push}

At rest

$\vec{f}_s \quad \vec{F}_{push}$

$\vec{a} = \vec{0}$

Accelerating

$\vec{f}_k \quad \vec{F}_{push}$

$\vec{a}$

STOP TO THINK 6.3 Rank in order, from largest to smallest, the sizes of the friction forces $\vec{f}_a$ to $\vec{f}_e$ in these 5 different situations. The box and the floor are made of the same materials in all situations.

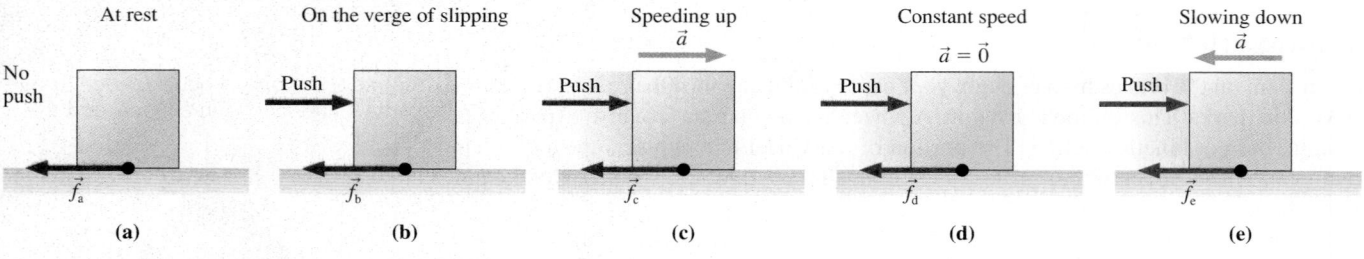

| At rest | On the verge of slipping | Speeding up | Constant speed | Slowing down |

(a)   (b)   (c)   (d)   (e)

EXAMPLE 6.5 How far does a box slide?

Carol pushes a 50 kg wood box across a wood floor at a steady speed of 2.0 m/s. How much force does Carol exert on the box? If she stops pushing, how far will the box slide before coming to rest?

MODEL We model the box as a particle and we describe the friction forces with the model of static and kinetic friction. This is a two-part problem: first while Carol is pushing the box, then as it slides after she releases it.

VISUALIZE This is a fairly complex situation, one that calls for careful visualization. **FIGURE 6.16** shows the pictorial representation both while Carol pushes, when $\vec{a} = \vec{0}$, and after she stops. We've placed $x = 0$ at the point where she stops pushing because this is the point where the kinematics calculation for "How far?" will begin. Notice that each part of the motion needs its own free-body diagram. The box is moving until the very instant that the problem ends, so only kinetic friction is relevant.

SOLVE We'll start by finding how hard Carol has to push to keep the box moving at a steady speed. The box is in dynamic equilibrium ($\vec{a} = \vec{0}$), and Newton's first law is

$$\sum F_x = F_{push} - f_k = 0$$

$$\sum F_y = n - F_G = n - mg = 0$$

where we've used $F_G = mg$ for the gravitational force. The negative sign occurs in the first equation because $\vec{f}_k$ points to the left and thus the *component* is negative: $(f_k)_x = -f_k$. Similarly, $(F_G)_y = -F_G$ because the gravitational force vector—with magnitude mg—points down. In addition to Newton's laws, we also have our model of kinetic friction:

$$f_k = \mu_k n$$

Altogether we have three simultaneous equations in the three unknowns F_{push}, f_k, and n. Fortunately, these equations are easy to solve. The y-component of Newton's law tells us that $n = mg$. We can then find the friction force to be

$$f_k = \mu_k mg$$

We substitute this into the x-component of the first law, giving

$$F_{push} = f_k = \mu_k mg$$
$$= (0.20)(50 \text{ kg})(9.80 \text{ m/s}^2) = 98 \text{ N}$$

This is how hard Carol pushes to keep the box moving at a steady speed.

The box is not in equilibrium after Carol stops pushing it. Our strategy for the second half of the problem is to use Newton's second law to find the acceleration, then use constant-acceleration kinematics to find how far the box moves before stopping. We

FIGURE 6.16 Pictorial representation of a box sliding across a floor.

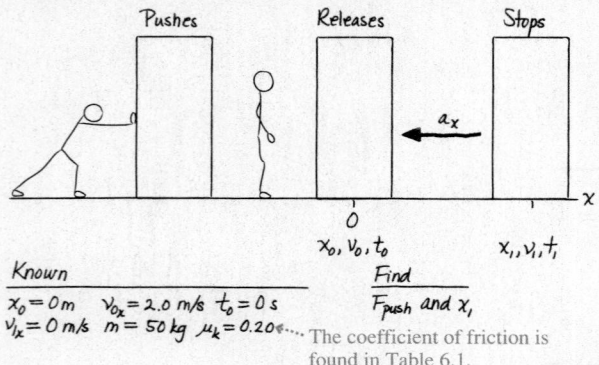

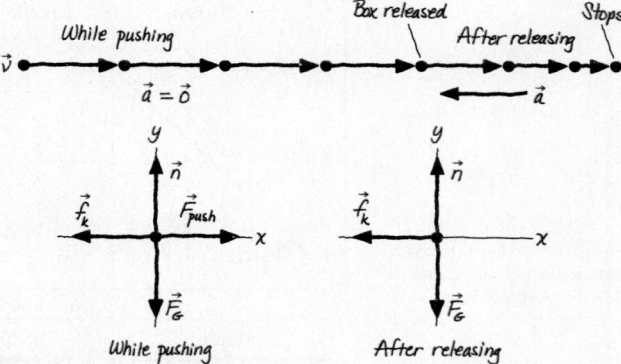

know from the motion diagram that $a_y = 0$. Newton's second law, applied to the second free-body diagram of Figure 6.16, is

$$\sum F_x = -f_k = ma_x$$

$$\sum F_y = n - mg = ma_y = 0$$

We also have our model of friction,

$$f_k = \mu_k n$$

We see from the y-component equation that $n = mg$, and thus $f_k = \mu_k mg$. Using this in the x-component equation gives

$$ma_x = -f_k = -\mu_k mg$$

This is easily solved to find the box's acceleration:

$$a_x = -\mu_k g = -(0.20)(9.80 \text{ m/s}^2) = -1.96 \text{ m/s}^2$$

The acceleration component a_x is negative because the acceleration vector $\vec{a}$ points to the left, as we see from the motion diagram.

Now we are left with a problem of constant-acceleration kinematics. We are interested in a distance, rather than a time interval, so the easiest way to proceed is

$$v_{1x}^2 = 0 = v_{0x}^2 + 2a_x \Delta x = v_{0x}^2 + 2a_x x_1$$

from which the distance that the box slides is

$$x_1 = \frac{-v_{0x}^2}{2a_x} = \frac{-(2.0 \text{ m/s})^2}{2(-1.96 \text{ m/s}^2)} = 1.0 \text{ m}$$

ASSESS Carol was pushing at 2 m/s $\approx$ 4 mph, which is fairly fast. The box slides 1.0 m, which is slightly over 3 feet. That sounds reasonable.

NOTE ▶ We needed both the horizontal and the vertical components of the second law even though the motion was entirely horizontal. This need is typical when friction is involved because we must find the normal force before we can evaluate the friction force. ◀

EXAMPLE 6.6 **Dumping a file cabinet**

A 50 kg steel file cabinet is in the back of a dump truck. The truck's bed, also made of steel, is slowly tilted. What is the size of the static friction force on the cabinet when the bed is tilted 20°? At what angle will the file cabinet begin to slide?

MODEL We'll model the file cabinet as a particle. We'll also use the model of static friction. The file cabinet will slip when the static friction force reaches its maximum value $f_{s\,max}$.

VISUALIZE **FIGURE 6.17** shows the pictorial representation when the truck bed is tilted at angle θ. We can make the analysis easier if we tilt the coordinate system to match the bed of the truck. To prevent the file cabinet from slipping, the static friction force must point *up* the slope.

SOLVE The file cabinet is in static equilibrium. Newton's first law is

$$(F_{net})_x = \sum F_x = n_x + (F_G)_x + (f_s)_x = 0$$

$$(F_{net})_y = \sum F_y = n_y + (F_G)_y + (f_s)_y = 0$$

From the free-body diagram we see that f_s has only a *negative* x-component and that n has only a positive y-component. The gravitational force vector can be written $\vec{F}_G = +F_G \sin\theta\,\hat{\imath} - F_G \cos\theta\,\hat{\jmath}$,

so $\vec{F}_G$ has both x- and y-components in this coordinate system. Thus the first law becomes

$$\sum F_x = F_G \sin\theta - f_s = mg \sin\theta - f_s = 0$$

$$\sum F_y = n - F_G \cos\theta = n - mg \cos\theta = 0$$

where we've used $F_G = mg$.

You might be tempted to solve the y-component equation for n, then to use Equation 6.12 to calculate the static friction force as $\mu_s n$. **But Equation 6.12 does *not* say $f_s = \mu_s n$.** Equation 6.12 gives only the maximum possible static friction force $f_{s\,max}$, the point at which the object slips. In nearly all situations, the actual static friction force is less than $f_{s\,max}$. In this problem, we can use the x-component equation—which tells us that static friction has to exactly balance the component of the gravitational force along the incline—to find the size of the static friction force when $\theta = 20°$:

$$f_s = mg \sin\theta = (50 \text{ kg})(9.80 \text{ m/s}^2) \sin 20°$$

$$= 170 \text{ N}$$

FIGURE 6.17 The pictorial representation of a file cabinet in a tilted dump truck.

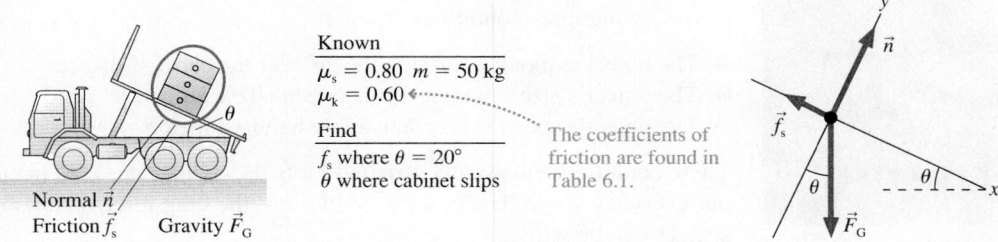

Known
$\mu_s = 0.80$ $m = 50$ kg
$\mu_k = 0.60$

Find
f_s where $\theta = 20°$
θ where cabinet slips

The coefficients of friction are found in Table 6.1.

Normal $\vec{n}$
Friction $\vec{f}_s$ Gravity $\vec{F}_G$

Continued

Slipping occurs when the static friction reaches its maximum value

$$f_s = f_{s\,max} = \mu_s n$$

From the y-component of Newton's law we see that $n = mg\cos\theta$. Consequently,

$$f_{s\,max} = \mu_s mg\cos\theta$$

NOTE ▶ A common error is to use simply $n = mg$. Be sure to evaluate the normal force within the context of each specific problem. In this example, $n = mg\cos\theta$. ◀

Substituting this into the x-component of the first law gives

$$mg\sin\theta - \mu_s mg\cos\theta = 0$$

The mg in both terms cancels, and we find

$$\frac{\sin\theta}{\cos\theta} = \tan\theta = \mu_s$$

$$\theta = \tan^{-1}\mu_s = \tan^{-1}(0.80) = 39°$$

ASSESS Steel doesn't slide all that well on unlubricated steel, so a fairly large angle is not surprising. The answer seems reasonable.

FIGURE 6.18 An atomic-level view of friction.

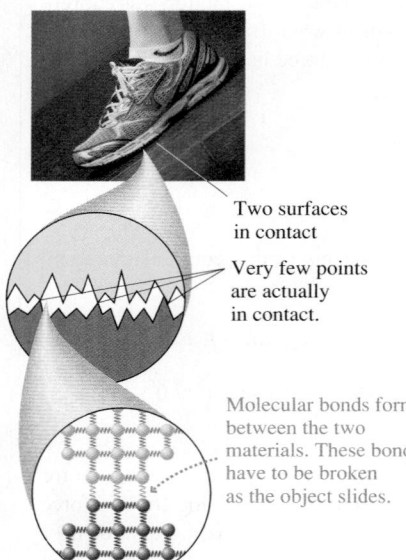

Two surfaces in contact

Very few points are actually in contact.

Molecular bonds form between the two materials. These bonds have to be broken as the object slides.

Causes of Friction

It is worth a brief pause to look at the *causes* of friction. All surfaces, even those quite smooth to the touch, are very rough on a microscopic scale. When two objects are placed in contact, they do not make a smooth fit. Instead, as FIGURE 6.18 shows, the high points on one surface become jammed against the high points on the other surface, while the low points are not in contact at all. The amount of contact depends on how hard the surfaces are pushed together, which is why friction forces are proportional to n.

At the points of actual contact, the atoms in the two materials are pressed closely together and molecular bonds are established between them. These bonds are the "cause" of the static friction force. For an object to slip, you must push it hard enough to break these molecular bonds between the surfaces. Once they are broken, and the two surfaces are sliding against each other, there are still attractive forces between the atoms on the opposing surfaces as the high points of the materials push past each other. However, the atoms move past each other so quickly that they do not have time to establish the tight bonds of static friction. That is why the kinetic friction force is smaller. Friction can be minimized with lubrication, a very thin film of liquid between the surfaces that allows them to "float" past each other with many fewer points in actual contact.

6.5 Drag

The air exerts a drag force on objects as they move through the air. You experience drag forces every day as you jog, bicycle, ski, or drive your car. The drag force $\vec{D}$

- Is opposite in direction to $\vec{v}$.
- Increases in magnitude as the object's speed increases.

FIGURE 6.19 The drag force on a high-speed motorcyclist is significant.

FIGURE 6.19 illustrates the drag force.

Drag is a more complex force than ordinary friction because drag depends on the object's speed. Drag also depends on the object's shape and on the density of the medium through which it moves. Fortunately, we can use a fairly simple *model* of drag if the following three conditions are met:

- The object is moving through the air near the earth's surface.
- The object's size (diameter) is between a few millimeters and a few meters.
- The object's speed is less than a few hundred meters per second.

These conditions are usually satisfied for balls, people, cars, and many other objects in our everyday world. Under these conditions, the drag force on an object moving with speed v can be written

$$\vec{D} = (\tfrac{1}{2}C\rho A v^2, \text{ direction opposite the motion}) \tag{6.16}$$

Notice that the drag force is proportional to the *square* of the object's speed. The symbols in Equation 6.16 are:

- *A* is the *cross-section area* of the object as it "faces into the wind," as illustrated in FIGURE 6.20.
- ρ is the density of the air, which is 1.2 kg/m³ at atmospheric pressure and room temperature.
- *C* is the **drag coefficient.** It is smaller for aerodynamically shaped objects, larger for objects presenting a flat face to the wind. Figure 6.20 gives approximate values for a sphere and two cylinders.

This model of drag fails for objects that are very small (such as dust particles), very fast (such as bullets), or that move in liquids (such as water). We'll leave those situations to more advanced textbooks.

FIGURE 6.20 Cross-section areas for objects of different shape.

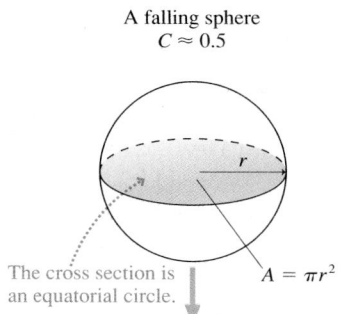

A falling sphere
$C \approx 0.5$

The cross section is an equatorial circle.

$A = \pi r^2$

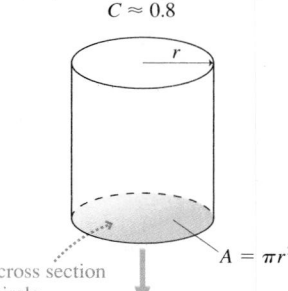

A cylinder falling end down
$C \approx 0.8$

The cross section is a circle.

$A = \pi r^2$

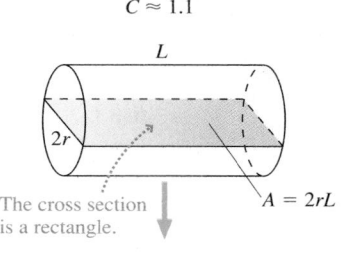

A cylinder falling side down
$C \approx 1.1$

The cross section is a rectangle.

$A = 2rL$

EXAMPLE 6.7   **Air resistance compared to rolling friction**

The profile of a typical 1500 kg passenger car, as seen from the front, is 1.6 m wide and 1.4 m high. Aerodynamic body shaping gives a drag coefficient of 0.35. At what speed does the magnitude of the drag equal the magnitude of the rolling friction?

MODEL Treat the car as a particle. Use the models of rolling friction and drag.

VISUALIZE FIGURE 6.21 shows the car and a free-body diagram. A full pictorial representation is not needed because we won't be doing any kinematics calculations.

FIGURE 6.21 A car experiences both rolling friction and drag.

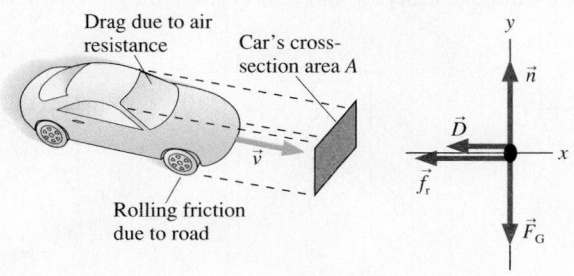

Drag due to air resistance

Car's cross-section area *A*

Rolling friction due to road

SOLVE Drag is less than friction at low speeds, where air resistance is negligible. But drag increases as *v* increases, so there will be a speed at which the two forces are equal in size. Above this speed, drag is more important than rolling friction.

There's no motion and no acceleration in the vertical direction, so we can see from the free-body diagram that $n = F_G = mg$. Thus $f_r = \mu_r mg$. Equating friction and drag, we have

$$\tfrac{1}{2}C\rho A v^2 = \mu_r mg$$

Solving for *v*, we find

$$v = \sqrt{\frac{2\mu_r mg}{C\rho A}} = \sqrt{\frac{2(0.02)(1500 \text{ kg})(9.80 \text{ m/s}^2)}{(0.35)(1.2 \text{ kg/m}^3)\,(1.4 \text{ m} \times 1.6 \text{ m})}} = 25 \text{ m/s}$$

where the value of μ_r for rubber on concrete was taken from Table 6.1.

ASSESS 25 m/s is approximately 50 mph, a reasonable result. This calculation shows that we can reasonably ignore air resistance for car speeds less than 30 or 40 mph. Calculations that neglect drag will be increasingly inaccurate as speeds go above 50 mph.

Terminal Speed

FIGURE 6.22 An object falling at terminal speed.

Terminal speed is reached when the drag force exactly balances the gravitational force: $\vec{a} = \vec{0}$.

The drag force increases as an object falls and gains speed. If the object falls far enough, it will eventually reach a speed, shown in FIGURE 6.22, at which $D = F_G$. That is, the drag force will be equal and opposite to the gravitational force. The net force at this speed is $\vec{F}_{net} = \vec{0}$, so there is no further acceleration and the object falls with a *constant* speed. The speed at which the exact balance between the upward drag force and the downward gravitational force causes an object to fall without acceleration is called the **terminal speed** v_{term}. Once an object has reached terminal speed, it will continue falling at that speed until it hits the ground.

It's not hard to compute the terminal speed. It is the speed, by definition, at which $D = F_G$ or, equivalently, $\frac{1}{2}C\rho Av^2 = mg$. This speed is

$$v_{term} = \sqrt{\frac{2mg}{C\rho A}} \tag{6.17}$$

A more massive object has a larger terminal speed than a less massive object of equal size and shape. A 10-cm-diameter lead ball, with a mass of 6 kg, has a terminal speed of 160 m/s, while a 10-cm-diameter Styrofoam ball, with a mass of 50 g, has a terminal speed of only 15 m/s.

A popular use of Equation 6.17 is to find the terminal speed of a skydiver. A skydiver is rather like the cylinder of Figure 6.20 falling "side down," for which we see that $C \approx 1.1$. A typical skydiver is 1.8 m long and 0.40 m wide ($A = 0.72$ m^2) and has a mass of 75 kg. His terminal speed is

$$v_{term} = \sqrt{\frac{2mg}{C\rho A}} = \sqrt{\frac{2(75 \text{ kg})(9.8 \text{ m/s}^2)}{(1.1)(1.2 \text{ kg/m}^3)(0.72 \text{ m}^2)}} = 39 \text{ m/s}$$

This is roughly 90 mph. A higher speed can be reached by falling feet first or head first, which reduces the area A and the drag coefficient.

FIGURE 6.23 The velocity-versus-time graph of a falling object with and without drag.

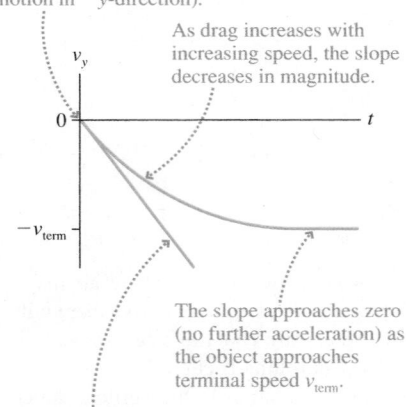

FIGURE 6.23 shows the results of a more detailed calculation for a falling object. Without drag, the velocity graph is a straight line with slope $= a_y = -g$. When drag is included, the slope steadily decreases in magnitude and approaches zero (no further acceleration) as the object reaches terminal speed.

Although we've focused our analysis on objects moving vertically, the same ideas apply to objects moving horizontally. If an object is thrown or shot horizontally, $\vec{D}$ causes the object to slow down. An airplane reaches its maximum speed, which is analogous to the terminal speed, when the drag is equal and opposite to the thrust: $D = F_{thrust}$. The net force is then zero and the plane cannot go any faster. The maximum speed of a passenger jet is about 550 mph.

STOP TO THINK 6.4 The terminal speed of a Styrofoam ball is 15 m/s. Suppose a Styrofoam ball is shot straight down with an initial speed of 30 m/s. Which velocity graph is correct?

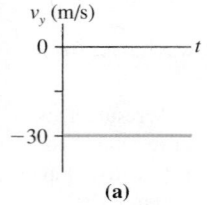 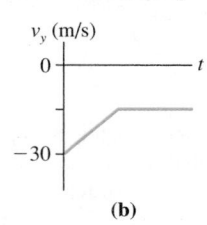 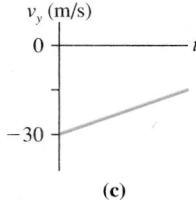 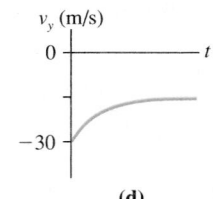 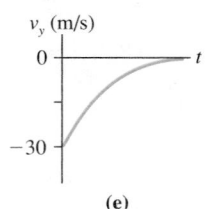

6.6 More Examples of Newton's Second Law

We will finish this chapter with four additional examples in which we use the problem-solving strategy in more complex scenarios.

EXAMPLE 6.8 **Stopping distances**

A 1500 kg car is traveling at a speed of 30 m/s when the driver slams on the brakes and skids to a halt. Determine the stopping distance if the car is traveling up a 10° slope, down a 10° slope, or on a level road.

MODEL We'll represent the car as a particle and we'll use the model of kinetic friction. We want to solve the problem only once, not three separate times, so we'll leave the slope angle θ unspecified until the end.

VISUALIZE FIGURE 6.24 shows the pictorial representation. We've shown the car sliding uphill, but these representations work equally well for a level or downhill slide if we let θ be zero or negative, respectively. We've used a tilted coordinate system so that the motion is along one of the axes. We've *assumed* that the car is traveling to the right, although the problem didn't state this. You could equally well make the opposite assumption, but you would have to be careful with negative values of x and v_x. The car *skids* to a halt, so we've taken the coefficient of *kinetic* friction for rubber on concrete from Table 6.1.

SOLVE Newton's second law and the model of kinetic friction are

$$\sum F_x = n_x + (F_G)_x + (f_k)_x$$
$$= -mg\sin\theta - f_k = ma_x$$
$$\sum F_y = n_y + (F_G)_y + (f_k)_y$$
$$= n - mg\cos\theta = ma_y = 0$$
$$f_k = \mu_k n$$

We've written these equations by "reading" the motion diagram and the free-body diagram. Notice that both components of the gravitational force vector $\vec{F}_G$ are negative. $a_y = 0$ because the motion is entirely along the x-axis.

The second equation gives $n = mg\cos\theta$. Using this in the friction model, we find $f_k = \mu_k mg\cos\theta$. Inserting this result back into the first equation then gives

$$ma_x = -mg\sin\theta - \mu_k mg\cos\theta$$
$$= -mg(\sin\theta + \mu_k\cos\theta)$$
$$a_x = -g(\sin\theta + \mu_k\cos\theta)$$

This is a constant acceleration. Constant-acceleration kinematics gives

$$v_{1x}^2 = 0 = v_{0x}^2 + 2a_x(x_1 - x_0) = v_{0x}^2 + 2a_x x_1$$

which we can solve for the stopping distance x_1:

$$x_1 = -\frac{v_{0x}^2}{2a_x} = \frac{v_{0x}^2}{2g(\sin\theta + \mu_k\cos\theta)}$$

Notice how the minus sign in the expression for a_x canceled the minus sign in the expression for x_1. Evaluating our result at the three different angles gives the stopping distances:

$$x_1 = \begin{cases} 48\text{ m} & \theta = 10° & \text{uphill} \\ 57\text{ m} & \theta = 0° & \text{level} \\ 75\text{ m} & \theta = -10° & \text{downhill} \end{cases}$$

The implications are clear about the danger of driving downhill too fast!

ASSESS 30 m/s ≈ 60 mph and 57 m ≈ 180 feet on a level surface. This is similar to the stopping distances you learned when you got your driver's license, so the results seem reasonable. Additional confirmation comes from noting that the expression for a_x becomes $-g\sin\theta$ if $\mu_k = 0$. This is what you learned in Chapter 2 for the acceleration on a frictionless inclined plane.

FIGURE 6.24 Pictorial representation of a skidding car.

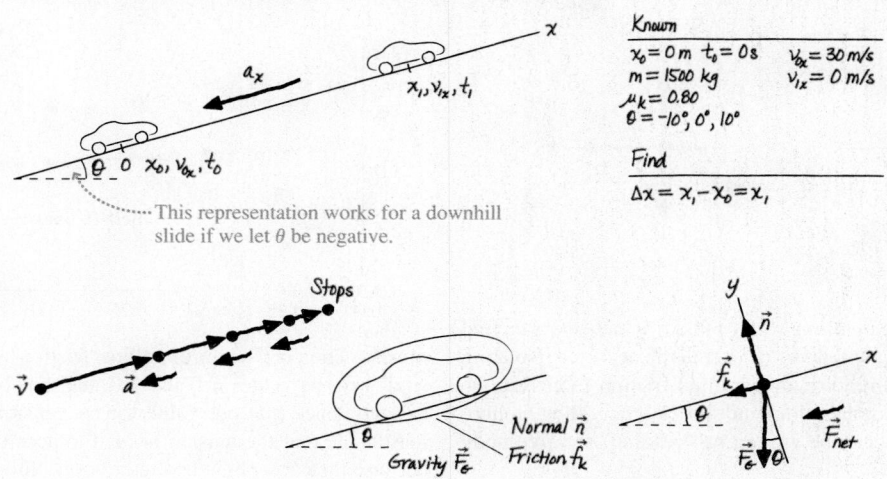

EXAMPLE 6.9 **Measuring the tension pulling a cart**

Your instructor has set up a lecture demonstration in which a 250 g cart can roll along a level, 2.00-m-long track while its velocity is measured with a motion detector. First, the instructor simply gives the cart a push and measures its velocity as it rolls down the track. The data below show that the cart slows slightly before reaching the end of the track. Then, as FIGURE 6.25 shows, the instructor attaches a string to the cart and uses a falling weight to pull the cart. She then asks you to determine the tension in the string. For extra credit, find the coefficient of rolling friction.

Time (s)	Rolled velocity (m/s)	Pulled velocity (m/s)
0.00	1.20	0.00
0.25	1.17	0.36
0.50	1.15	0.80
0.75	1.12	1.21
1.00	1.08	1.52
1.25	1.04	1.93
1.50	1.02	2.33

FIGURE 6.25 The experimental arrangement.

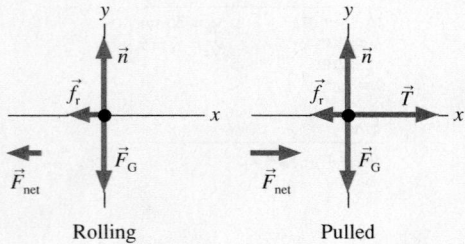

MODEL Model the cart as a particle.

VISUALIZE The cart changes velocity—it accelerates—when both pulled and rolled. Consequently, there must be a net force for both motions. For rolling, force identification finds that the only horizontal force is rolling friction, a force that opposes the motion and slows the cart. There is no "force of motion" or "force of the hand" because the hand is no longer in contact with the cart. (Recall Newton's "zeroth law": The cart responds only to forces applied *at this instant.*) Pulling adds a tension force in the direction of motion. The two free-body diagrams are shown in FIGURE 6.26.

FIGURE 6.26 Pictorial representations of the cart.

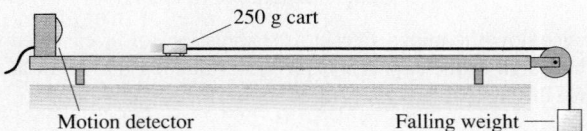

SOLVE The cart's acceleration when pulled, which we can find from the velocity data, will allow us to find the net force. Isolating the tension force will require knowing the friction force, but we can determine that from the rolling motion. For the rolling motion, Newton's second law can be written by "reading" the free-body diagram on the left:

$$\sum F_x = (f_r)_x = -f_r = ma_x = ma_{roll}$$
$$\sum F_y = n_y + (F_G)_y = n - mg = 0$$

Make sure you understand where the signs come from and how we used our knowledge that $\vec{a}$ has only an x-component, which we called a_{roll}. The magnitude of the friction force, which is all we'll need to determine the tension, is found from the x-component equation:

$$f_r = -ma_{roll} = -m \times \text{slope of the rolling-velocity graph}$$

But we'll need to do a bit more analysis to get the coefficient of rolling friction. The y-component equation tells us that $n = mg$. Using this in the model of rolling friction, $f_r = \mu_r n = \mu_r mg$, we see that the coefficient of rolling friction is

$$\mu_r = \frac{f_r}{mg}$$

The x-component equation of Newton's second law when the cart is pulled is

$$\sum F_x = T + (f_r)_x = T - f_r = ma_x = ma_{pulled}$$

Thus the tension that we seek is

$$T = f_r + ma_{pulled} = f_r + m \times \text{slope of the pulled-velocity graph}$$

FIGURE 6.27 shows the graphs of the velocity data. The accelerations are the slopes of these lines, and from the equations of the best-fit lines we find $a_{roll} = -0.124$ m/s^2 and $a_{pulled} = 1.55$ m/s^2. Thus the friction force is

$$f_r = -ma_{roll} = -(0.25 \text{ kg})(-0.124 \text{ m/s}^2) = 0.031 \text{ N}$$

Knowing this, we find that the string tension pulling the cart is

$$T = f_r + ma_{pulled} = 0.031 \text{ N} + (0.25 \text{ kg})(1.55 \text{ m/s}^2) = 0.42 \text{ N}$$

and the coefficient of rolling friction is

$$\mu_r = \frac{f_r}{mg} = \frac{0.031 \text{ N}}{(0.25 \text{ kg})(9.80 \text{ m/s}^2)} = 0.013$$

FIGURE 6.27 The velocity graphs of the rolling and pulled motion. The slopes of these graphs are the cart's acceleration.

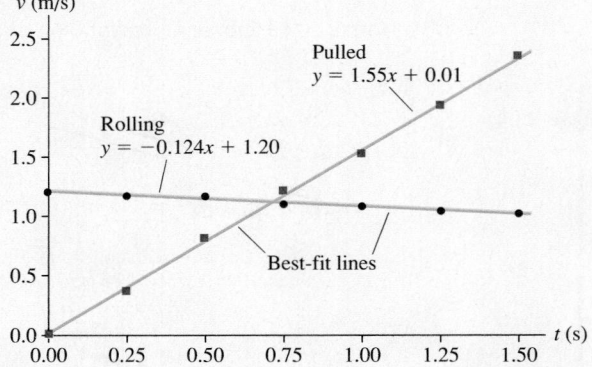

ASSESS The coefficient of rolling friction is very small, but it's similar to the values in Table 6.1 and thus believable. That gives us confidence that our value for the tension is also correct. It's reasonable that the tension needed to accelerate the cart is small because the cart is light and there's very little friction.

EXAMPLE 6.10 **Make sure the cargo doesn't slide**

A 100 kg box of dimensions 50 cm × 50 cm × 50 cm is in the back of a flatbed truck. The coefficients of friction between the box and the bed of the truck are $\mu_s = 0.40$ and $\mu_k = 0.20$. What is the maximum acceleration the truck can have without the box slipping?

MODEL This is a somewhat different problem from any we have looked at thus far. Let the box, which we'll model as a particle, be the object of interest. It contacts other objects only where it touches the truck bed, so only the truck can exert contact forces on the box. If the box does *not* slip, then there is no motion of the box *relative to the truck* and the box must accelerate *with the truck:* $a_{box} = a_{truck}$. As the box accelerates, it must, according to Newton's second law, have a net force acting on it. But from what?

Imagine, for a moment, that the truck bed is frictionless. The box would slide backward (as seen in the truck's reference frame) as the truck accelerates. The force that prevents sliding is *static friction,* so the truck must exert a static friction force on the box to "pull" the box along with it and prevent the box from sliding *relative to the truck.*

VISUALIZE This situation is shown in **FIGURE 6.28**. There is only one horizontal force on the box, $\vec{f}_s$, and it points in the *forward* direction to accelerate the box. Notice that we're solving the problem with the ground as our reference frame. Newton's laws are not valid in the accelerating truck because it is not an inertial reference frame.

SOLVE Newton's second law, which we can "read" from the free-body diagram, is

$$\sum F_x = f_s = ma_x$$
$$\sum F_y = n - F_G = n - mg = ma_y = 0$$

Now, static friction, you will recall, can be *any* value between 0 and $f_{s\,max}$. If the truck accelerates slowly, so that the box doesn't slip, then $f_s < f_{s\,max}$. However, we're interested in the acceleration a_{max} at which the box begins to slip. This is the acceleration at which f_s reaches its maximum possible value

$$f_s = f_{s\,max} = \mu_s n$$

The y-equation of the second law and the friction model combine to give $f_{s\,max} = \mu_s mg$. Substituting this into the x-equation, and noting that a_x is now a_{max}, we find

$$a_{max} = \frac{f_{s\,max}}{m} = \mu_s g = 3.9 \text{ m/s}^2$$

The truck must keep its acceleration less than 3.9 m/s² if slipping is to be avoided.

ASSESS 3.9 m/s² is about one-third of g. You may have noticed that items in a car or truck are likely to *tip over* when you start or stop, but they slide only if you really floor it and accelerate very quickly. So this answer seems reasonable. Notice that the dimensions of the crate were not needed. Real-world situations rarely have exactly the information you need, no more and no less. Many problems in this textbook will require you to assess the information in the problem statement in order to learn which is relevant to the solution.

FIGURE 6.28 Pictorial representation for the box in a flatbed truck.

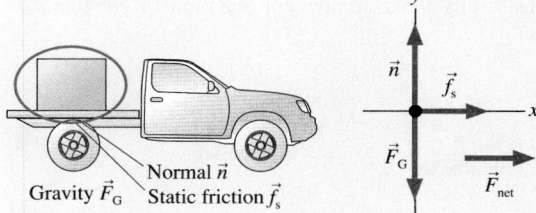

Known
$m = 100$ kg
Box dimensions 50 cm × 50 cm × 50 cm
$\mu_s = 0.40$ $\mu_k = 0.20$

Find
Acceleration at which box slips

Gravity $\vec{F}_G$ Normal $\vec{n}$ Static friction $\vec{f}_s$

The mathematical representation of this last example was quite straightforward. The challenge was in the analysis that preceded the mathematics—that is, in the *physics* of the problem rather than the mathematics. It is here that our analysis tools—motion diagrams, force identification, and free-body diagrams—prove their value.

Acceleration from a variable force

Force $F_x = c \sin(\pi t/T)$, where c and T are constants, is applied to an object of mass m that moves on a horizontal, frictionless surface. The object is at rest at the origin at $t = 0$.

a. Find an expression for the object's velocity. Graph your result for $0 \le t \le T$.

b. What is the maximum velocity of a 500 g object if $c = 2.5$ N and $T = 1.0$ s?

MODEL Model the object as a particle.

VISUALIZE The sine function is 0 at $t = 0$ and again at $t = T$, when the value of the argument is π rad. Over the interval $0 \le t \le T$, the force grows from 0 to c and back to 0, always pointing in the positive x-direction. **FIGURE 6.29** on the next page shows a graph of the force and a pictorial representation.

Continued

FIGURE 6.29 Pictorial representation for a variable force.

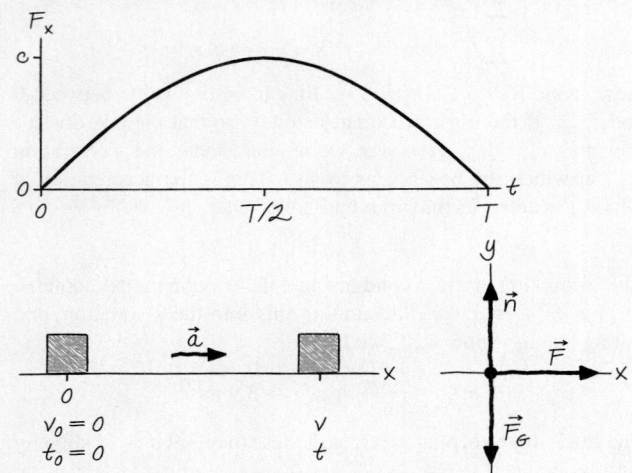

SOLVE The object's acceleration increases between 0 and $T/2$ as the force increases. You might expect the object to slow down between $T/2$ and T as the force decreases. However, *there's still a net force in the positive x-direction, so there must be an acceleration in the positive x-direction.* The object continues to speed up, only more slowly as the acceleration decreases. Maximum velocity is reached at $t = T$.

a. This is not constant-acceleration motion, so we cannot use the familiar equations of constant-acceleration kinematics. Instead, we must use the definition of acceleration as the rate of change—the time derivative—of velocity. With no friction, we need only the x-component equation of Newton's second law:

$$a_x = \frac{dv_x}{dt} = \frac{F_{net}}{m} = \frac{c}{m} \sin\left(\frac{\pi t}{T}\right)$$

First we rewrite this as

$$dv_x = \frac{c}{m} \sin\left(\frac{\pi t}{T}\right) dt$$

Then we integrate both sides from the initial conditions ($v_x = v_{0x} = 0$ at $t = t_0 = 0$) to the final conditions (v_x at the later time t):

$$\int_0^{v_x} dv_x = \frac{c}{m} \int_0^t \sin\left(\frac{\pi t}{T}\right) dt$$

The fraction c/m is a constant that we could take outside the integral. The integral on the right side is of the form

$$\int \sin(ax)\,dx = -\frac{1}{a} \cos(ax)$$

Using this, and integrating both sides of the equation, we find

$$v_x \Big|_0^{v_x} = v_x - 0 = -\frac{cT}{\pi m} \cos\left(\frac{\pi t}{T}\right)\Big|_0^t = -\frac{cT}{\pi m}\left(\cos\left(\frac{\pi t}{T}\right) - 1\right)$$

Simplifying, we find the object's velocity at time t is

$$v_x = \frac{cT}{\pi m}\left(1 - \cos\left(\frac{\pi t}{T}\right)\right)$$

This expression is graphed in FIGURE 6.30, where we see that, as predicted, maximum velocity is reached at $t = T$.

FIGURE 6.30 The object's velocity as a function of time.

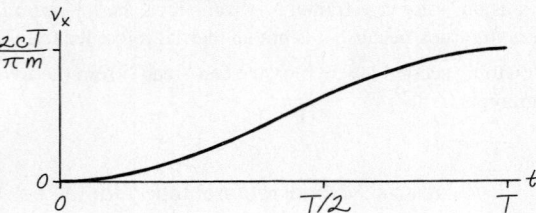

b. Maximum velocity, at $t = T$, is

$$v_{max} = \frac{cT}{\pi m}(1 - \cos \pi) = \frac{2cT}{\pi m} = \frac{2(2.5 \text{ N})(1.0 \text{ s})}{\pi(0.50 \text{ kg})} = 3.2 \text{ m/s}$$

ASSESS A steady 2.5 N force would cause a 0.5 kg object to accelerate at 5 m/s² and reach a speed of 5 m/s in 1 s. A variable force with a maximum of 2.5 N will produce less acceleration, so a top speed of 3.2 m/s seems reasonable.

SUMMARY

The goal of Chapter 6 has been to learn how to solve linear force-and-motion problems.

General Strategy

All examples in this chapter follow a four-part strategy. You'll become a better problem solver if you adhere to it as you do the homework problems. The *Dynamics Worksheets* in the *Student Workbook* will help you structure your work in this way.

Equilibrium Problems

Object at rest or moving with constant velocity.

MODEL Make simplifying assumptions.

VISUALIZE

- Translate words into symbols.
- Identify forces.
- Draw a free-body diagram.

SOLVE Use Newton's first law:

$$\vec{F}_{net} = \sum_i \vec{F}_i = \vec{0}$$

"Read" the vectors from the free-body diagram.

ASSESS Is the result reasonable?

Go back and forth between these steps as needed.

Dynamics Problems

Object accelerating.

MODEL Make simplifying assumptions.

VISUALIZE

- Translate words into symbols.
- Draw a sketch to define the situation.
- Draw a motion diagram.
- Identify forces.
- Draw a free-body diagram.

SOLVE Use Newton's second law:

$$\vec{F}_{net} = \sum_i \vec{F}_i = m\vec{a}$$

"Read" the vectors from the free-body diagram. Use kinematics to find velocities and positions.

ASSESS Is the result reasonable?

Important Concepts

Specific information about three important forces:

Gravity $\quad \vec{F}_G = (mg, \text{downward})$

Friction $\quad \vec{f}_s = (0 \text{ to } \mu_s n, \text{direction as necessary to prevent motion})$

$\qquad\quad \vec{f}_k = (\mu_k n, \text{direction opposite the motion})$

$\qquad\quad \vec{f}_r = (\mu_r n, \text{direction opposite the motion})$

Drag $\quad \vec{D} = (\frac{1}{2}C\rho A v^2, \text{direction opposite the motion})$

Newton's laws are vector expressions. You must write them out by **components**:

$$(F_{net})_x = \sum F_x = ma_x$$

$$(F_{net})_y = \sum F_y = ma_y$$

The acceleration is zero in equilibrium and also along an axis perpendicular to the motion.

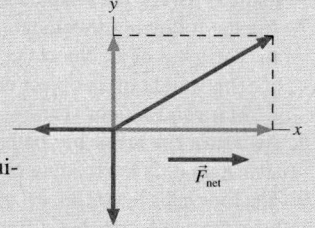

Applications

Mass is an intrinsic property of an object that describes the object's inertia and, loosely speaking, its quantity of matter.

The weight of an object is the reading of a calibrated spring scale on which the object is stationary. Weight is the result of weighing. An object's weight depends on its mass, its acceleration, and the strength of gravity. An object in free fall is weightless.

A falling object reaches terminal speed

$$v_{term} = \sqrt{\frac{2mg}{C\rho A}}$$

Terminal speed is reached when the drag force exactly balances the gravitational force: $\vec{a} = 0$.

Terms and Notation

flat-earth approximation	coefficient of kinetic friction, μ_k	drag coefficient, C
weight	rolling friction	terminal speed, v_{term}
coefficient of static friction, μ_s	coefficient of rolling friction, μ_r	

CONCEPTUAL QUESTIONS

1. Are the objects described here in static equilibrium, dynamic equilibrium, or not in equilibrium at all? Explain.
   a. A 200 pound barbell is held over your head.
   b. A girder is lifted at constant speed by a crane.
   c. A girder is being lowered into place. It is slowing down.
   d. A jet plane has reached its cruising speed and altitude.
   e. A box in the back of a truck doesn't slide as the truck stops.

2. A ball tossed straight up has $v = 0$ at its highest point. Is it in equilibrium? Explain.

3. Kat, Matt, and Nat are arguing about why a physics book on a table doesn't fall. According to Kat, "Gravity pulls down on it, but the table is in the way so it can't fall." "Nonsense," says Matt. "There are all kinds of forces acting on the book, but the upward forces overcome the downward forces to prevent it from falling." "But what about Newton's first law?" counters Nat. "It's not moving, so there can't be any forces acting on it." None of the statements is exactly correct. Who comes closest, and how would you change his or her statement to make it correct?

4. If you know all of the forces acting on a moving object, can you tell the direction the object is moving? If yes, explain how. If no, give an example.

5. An elevator, hanging from a single cable, moves upward at constant speed. Friction and air resistance are negligible. Is the tension in the cable greater than, less than, or equal to the gravitational force on the elevator? Explain. Include a free-body diagram as part of your explanation.

6. An elevator, hanging from a single cable, moves downward and is slowing. Friction and air resistance are negligible. Is the tension in the cable greater than, less than, or equal to the gravitational force on the elevator? Explain. Include a free-body diagram as part of your explanation.

7. Are the following statements true or false? Explain.
   a. The mass of an object depends on its location.
   b. The weight of an object depends on its location.
   c. Mass and weight describe the same thing in different units.

8. An astronaut takes his bathroom scale to the moon and then stands on it. Is the reading of the scale his weight? Explain.

9. The four balls in FIGURE Q6.9 have been thrown straight up. They have the same size, but different masses. Air resistance is negligible. Rank in order, from largest to smallest, the magnitude of the net force acting on each ball. Some may be equal. Give your answer in the form a > b = c > d and explain your ranking.

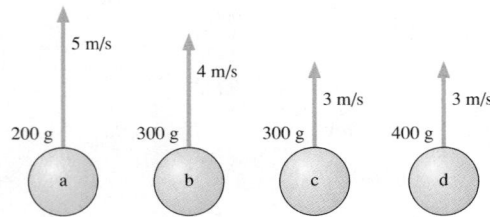

FIGURE Q6.9

10. Suppose you attempt to pour out 100 g of salt, using a pan balance for measurements, while in a rocket accelerating upward. Will the quantity of salt be too much, too little, or the correct amount? Explain.

11. A box with a passenger inside is launched straight up into the air by a giant rubber band. Before launch, the passenger stood on a scale and weighed 750 N. Once the box has left the rubber band but is still moving upward, is the passenger's weight more than 750 N, 750 N, less than 750 N but not zero, or zero? Explain.

12. An astronaut orbiting the earth is handed two balls that have identical outward appearances. However, one is hollow while the other is filled with lead. How can the astronaut determine which is which? Cutting or altering the balls is not allowed.

13. A hand presses down on the book in FIGURE Q6.13. Is the normal force of the table on the book larger than, smaller than, or equal to mg?

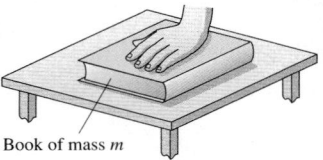

Book of mass m

FIGURE Q6.13

14. Suppose you push a hockey puck of mass m across frictionless ice for a time Δt, starting from rest, giving the puck speed v after traveling distance d. If you repeat the experiment with a puck of mass $2m$,
    a. How long will you have to push for the puck to reach the same speed v?
    b. How long will you have to push for the puck to travel the same distance d?

15. A block pushed along the floor with velocity v_{0x} slides a distance d after the pushing force is removed.
    a. If the mass of the block is doubled but its initial velocity is not changed, what distance does the block slide before stopping?
    b. If the initial velocity is doubled to $2v_{0x}$ but the mass is not changed, what distance does the block slide before stopping?

16. Can the friction force on an object ever point in the direction of the object's motion? If yes, give an example. If no, why not?

17. A crate of fragile dishes is in the back of a pickup truck. The truck accelerates north from a stop sign, and the crate moves without slipping. Does the friction force on the crate point north or south? Or is the friction force zero? Explain.

18. Five balls move through the air as shown in FIGURE Q6.18. All five have the same size and shape. Air resistance is not negligible. Rank in order, from largest to smallest, the magnitudes of the accelerations a_a to a_e. Some may be equal. Give your answer in the form a > b = c > d > e and explain your ranking.

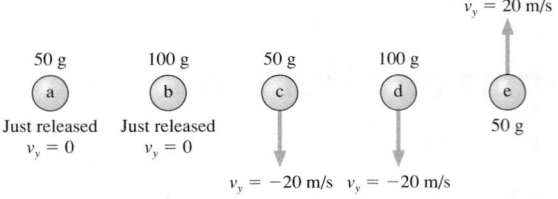

FIGURE Q6.18

EXERCISES AND PROBLEMS

Exercises

Section 6.1 Equilibrium

1. | The three ropes in FIGURE EX6.1 are tied to a small, very light ring. Two of the ropes are anchored to walls at right angles, and the third rope pulls as shown. What are T_1 and T_2, the magnitudes of the tension forces in the first two ropes?

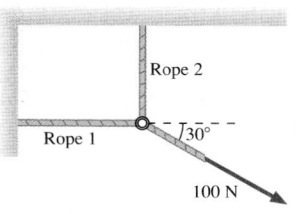

FIGURE EX6.1

FIGURE EX6.2

2. | The three ropes in FIGURE EX6.2 are tied to a small, very light ring. Two of these ropes are anchored to walls at right angles with the tensions shown in the figure. What are the magnitude and direction of the tension $\vec{T}_3$ in the third rope?

3. ‖ A 20 kg loudspeaker is suspended 2.0 m below the ceiling by two 3.0-m-long cables that angle outward at equal angles. What is the tension in the cables?

4. ‖ A football coach sits on a sled while two of his players build their strength by dragging the sled across the field with ropes. The friction force on the sled is 1000 N and the angle between the two ropes is 20°. How hard must each player pull to drag the coach at a steady 2.0 m/s?

5. | A construction worker with a weight of 850 N stands on a roof that is sloped at 20°. What is the magnitude of the normal force of the roof on the worker?

Section 6.2 Using Newton's Second Law

6. | In each of the two free-body diagrams in FIGURE EX6.6, the forces are acting on a 2.0 kg object. For each diagram, find the values of a_x and a_y, the x- and y-components of the acceleration.

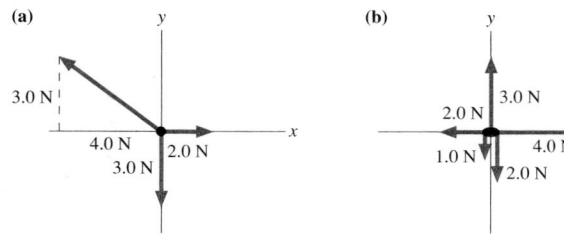

FIGURE EX6.6

7. ‖ In each of the two free-body diagrams in FIGURE EX6.7, the forces are acting on a 2.0 kg object. For each diagram, find the values of a_x and a_y, the x- and y-components of the acceleration.

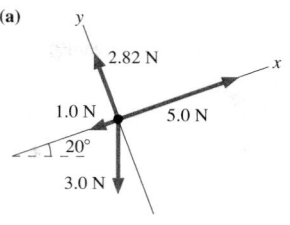 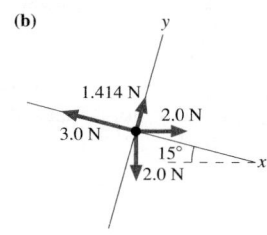

FIGURE EX6.7

8. | FIGURE EX6.8 shows the velocity graph of a 2.0 kg object as it moves along the x-axis. What is the net force acting on this object at $t = 1$ s? At 4 s? At 7 s?

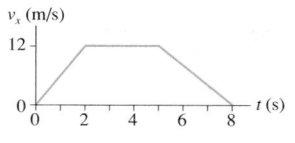 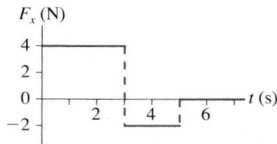

FIGURE EX6.8

FIGURE EX6.9

9. | FIGURE EX6.9 shows the force acting on a 2.0 kg object as it moves along the x-axis. The object is at rest at the origin at $t = 0$ s. What are its acceleration and velocity at $t = 6$ s?

10. | A horizontal rope is tied to a 50 kg box on frictionless ice. What is the tension in the rope if:
   a. The box is at rest?
   b. The box moves at a steady 5.0 m/s?
   c. The box has $v_x = 5.0$ m/s and $a_x = 5.0$ m/s²?

11. | A 50 kg box hangs from a rope. What is the tension in the rope if:
   a. The box is at rest?
   b. The box moves up at a steady 5.0 m/s?
   c. The box has $v_y = 5.0$ m/s and is speeding up at 5.0 m/s²?
   d. The box has $v_y = 5.0$ m/s and is slowing down at 5.0 m/s²?

12. ‖ a. The block in FIGURE EX6.12 floats on a cushion of air. It is pushed to the right with a force that remains constant as the block moves from 0 to 1. The block
      A. Speeds up from 0 to 1.
      B. Speeds up at first, then has constant speed.
      C. Moves with constant speed from 0 to 1.
   b. From 1 to 2, the size of the force steadily decreases until it reaches half of its initial value. The block
      A. Continues to speed up from 1 to 2.
      B. Moves with constant speed from 1 to 2.
      C. Slows down.

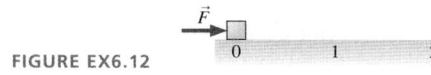

FIGURE EX6.12

Section 6.3 Mass, Weight, and Gravity

13. | A woman has a mass of 55 kg.
   a. What is her weight while standing on earth?
   b. What are her mass and her weight on Mars, where $g = 3.76$ m/s²?

14. | It takes the elevator in a skyscraper 4.0 s to reach its cruising speed of 10 m/s. A 60 kg passenger gets aboard on the ground floor. What is the passenger's weight
    a. Before the elevator starts moving?
    b. While the elevator is speeding up?
    c. After the elevator reaches its cruising speed?
15. ‖ FIGURE EX6.15 shows the velocity graph of a 75 kg passenger in an elevator. What is the passenger's weight at $t = 1$ s? At 5 s? At 9 s?

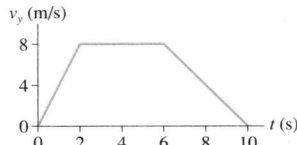

FIGURE EX6.15

16. ‖ What thrust does a 200 g model rocket need in order to have a vertical acceleration of 10 m/s^2
    a. On earth?
    b. On the moon, where $g = 1.62$ m/s^2?

Section 6.4 Friction

17. ‖ Bonnie and Clyde are sliding a 300 kg bank safe across the floor to their getaway car. The safe slides with a constant speed if Clyde pushes from behind with 385 N of force while Bonnie pulls forward on a rope with 350 N of force. What is the safe's coefficient of kinetic friction on the bank floor?
18. | A stubborn, 120 kg mule sits down and refuses to move. To drag the mule to the barn, the exasperated farmer ties a rope around the mule and pulls with his maximum force of 800 N. The coefficients of friction between the mule and the ground are $\mu_s = 0.8$ and $\mu_k = 0.5$. Is the farmer able to move the mule?
19. ‖ A 10 kg crate is placed on a horizontal conveyor belt. The materials are such that $\mu_s = 0.5$ and $\mu_k = 0.3$.
    a. Draw a free-body diagram showing all the forces on the crate if the conveyer belt runs at constant speed.
    b. Draw a free-body diagram showing all the forces on the crate if the conveyer belt is speeding up.
    c. What is the maximum acceleration the belt can have without the crate slipping?
20. | Bob is pulling a 30 kg filing cabinet with a force of 200 N, but the filing cabinet refuses to move. The coefficient of static friction between the filing cabinet and the floor is 0.80. What is the magnitude of the friction force on the filing cabinet?
21. ‖ A 4000 kg truck is parked on a 15° slope. How big is the friction force on the truck? The coefficient of static friction between the tires and the road is 0.90.
22. | A 1500 kg car skids to a halt on a wet road where $\mu_k = 0.50$. How fast was the car traveling if it leaves 65-m-long skid marks?
23. ‖ A 50,000 kg locomotive is traveling at 10 m/s when its engine and brakes both fail. How far will the locomotive roll before it comes to a stop? Assume the track is level.

Section 6.5 Drag

24. ‖ A 75 kg skydiver can be modeled as a rectangular "box" with dimensions 20 cm × 40 cm × 180 cm. What is his terminal speed if he falls feet first? Use 0.8 for the drag coefficient.
25. ‖ A 6.5-cm-diameter tennis ball has a terminal speed of 26 m/s. What is the ball's mass?

Problems

26. ‖ A 5.0 kg object initially at rest at the origin is subjected to the time-varying force shown in FIGURE P6.26. What is the object's velocity at $t = 6$ s?

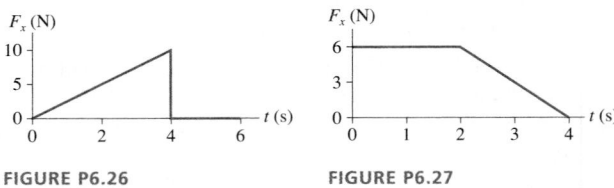

FIGURE P6.26   FIGURE P6.27

27. ‖ A 2.0 kg object initially at rest at the origin is subjected to the time-varying force shown in FIGURE P6.27. What is the object's velocity at $t = 4$ s?
28. ‖ The 1000 kg steel beam in FIGURE P6.28 is supported by two ropes. What is the tension in each?

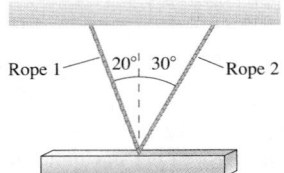

FIGURE P6.28

29. ‖ In an electricity experiment, a 1.0 g plastic ball is suspended on a 60-cm-long string and given an electric charge. A charged rod brought near the ball exerts a horizontal electrical force $\vec{F}_{elec}$ on it, causing the ball to swing out to a 20° angle and remain there.
    a. What is the magnitude of $\vec{F}_{elec}$?
    b. What is the tension in the string?
30. | A 500 kg piano is being lowered into position by a crane while two people steady it with ropes pulling to the sides. Bob's rope pulls to the left, 15° below horizontal, with 500 N of tension. Ellen's rope pulls toward the right, 25° below horizontal.
    a. What tension must Ellen maintain in her rope to keep the piano descending at a steady speed?
    b. What is the tension in the main cable supporting the piano?
31. ‖ Henry gets into an elevator on the 50th floor of a building and it begins moving at $t = 0$ s. FIGURE P6.31 shows his weight over the next 12 s.
    a. Is the elevator's initial direction up or down? Explain how you can tell.
    b. What is Henry's mass?
    c. How far has Henry traveled at $t = 12$ s?

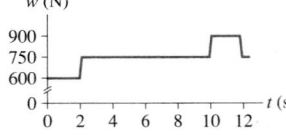

FIGURE P6.31

32. | Zach, whose mass is 80 kg, is in an elevator descending at 10 m/s. The elevator takes 3.0 s to brake to a stop at the first floor.
    a. What is Zach's weight before the elevator starts braking?
    b. What is Zach's weight while the elevator is braking?
33. ‖ An accident victim with a broken leg is being placed in trac-
    BIO tion. The patient wears a special boot with a pulley attached to the sole. The foot and boot together have a mass of 4.0 kg, and the

doctor has decided to hang a 6.0 kg mass from the rope. The boot is held suspended by the ropes, as shown in FIGURE P6.33, and does not touch the bed.

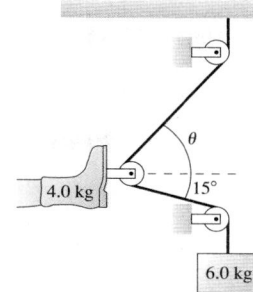

a. Determine the amount of tension in the rope by using Newton's laws to analyze the hanging mass.

b. The net traction force needs to pull straight out on the leg. What is the proper angle θ for the upper rope?

c. What is the net traction force pulling on the leg?

FIGURE P6.33

Hint: If the pulleys are frictionless, which we will assume, the tension in the rope is constant from one end to the other.

34. ‖ Seat belts and air bags save lives by reducing the forces ex-
BIO erted on the driver and passengers in an automobile collision. Cars are designed with a "crumple zone" in the front of the car. In the event of an impact, the passenger compartment decelerates over a distance of about 1 m as the front of the car crumples. An occupant restrained by seat belts and air bags decelerates with the car. By contrast, an unrestrained occupant keeps moving forward with no loss of speed (Newton's first law!) until hitting the dashboard or windshield. These are unyielding surfaces, and the unfortunate occupant then decelerates over a distance of only about 5 mm.

a. A 60 kg person is in a head-on collision. The car's speed at impact is 15 m/s. Estimate the net force on the person if he or she is wearing a seat belt and if the air bag deploys.

b. Estimate the net force that ultimately stops the person if he or she is not restrained by a seat belt or air bag.

c. How do these two forces compare to the person's weight?

35. ‖ The position of a 2.0 kg mass is given by $x = (2t^3 - 3t^2)$ m, where t is in seconds. What is the net horizontal force on the mass at (a) $t = 0$ s and (b) $t = 1$ s?

36. ‖ The piston of a machine exerts a constant force on a ball as it moves horizontally through a distance of 15 cm. You use a motion detector to measure the speed of five different balls as they come off the piston; the data are shown below. Use theory to find two quantities that, when graphed, should give a straight line. Then use the graph to find the size of the piston's force.

Mass (g)	Speed (m/s)
200	9.4
400	6.3
600	5.2
800	4.9
1000	4.0

37. ‖ Compressed air is used to fire a 50 g ball vertically upward from a 1.0-m-tall tube. The air exerts an upward force of 2.0 N on the ball as long as it is in the tube. How high does the ball go above the top of the tube?

38. ‖ a. A rocket of mass m is launched straight up with thrust $\vec{F}_{\text{thrust}}$. Find an expression for the rocket's speed at height h if air resistance is neglected.

b. The motor of a 350 g model rocket generates 9.5 N thrust. If air resistance can be neglected, what will be the rocket's speed as it reaches a height of 85 m?

39. ‖ A rifle with a barrel length of 60 cm fires a 10 g bullet with a horizontal speed of 400 m/s. The bullet strikes a block of wood and penetrates to a depth of 12 cm.

a. What resistive force (assumed to be constant) does the wood exert on the bullet?

b. How long does it take the bullet to come to rest?

c. Draw a velocity-versus-time graph for the bullet in the wood.

40. ‖ A 20,000 kg rocket has a rocket motor that generates 3.0×10^5 N of thrust.

a. What is the rocket's initial upward acceleration?

b. At an altitude of 5000 m the rocket's acceleration has increased to 6.0 m/s². What mass of fuel has it burned?

41. ‖ a. An object of mass m is at rest at the top of a smooth slope of height h and length L. The coefficient of kinetic friction between the object and the surface, μ_k, is small enough that the object will slide down the slope if given a very small push to get it started. Find an expression for the object's speed at the bottom of the slope.

b. Sam, whose mass is 75 kg, stands at the top of a 12-m-high, 100-m-long snow-covered slope. His skis have a coefficient of kinetic friction on snow of 0.07. If he uses his poles to get started, then glides down, what is his speed at the bottom?

42. ‖ Sam, whose mass is 75 kg, takes off across level snow on his jet-powered skis. The skis have a thrust of 200 N and a coefficient of kinetic friction on snow of 0.10. Unfortunately, the skis run out of fuel after only 10 s.

a. What is Sam's top speed?

b. How far has Sam traveled when he finally coasts to a stop?

43. ‖‖ Sam, whose mass is 75 kg, takes off down a 50-m-high, 10° slope on his jet-powered skis. The skis have a thrust of 200 N. Sam's speed at the bottom is 40 m/s. What is the coefficient of kinetic friction of his skis on snow?

44. ‖ A baggage handler drops your 10 kg suitcase onto a conveyor belt running at 2.0 m/s. The materials are such that $\mu_s = 0.50$ and $\mu_k = 0.30$. How far is your suitcase dragged before it is riding smoothly on the belt?

45. ‖‖ You and your friend Peter are putting new shingles on a roof pitched at 25°. You're sitting on the very top of the roof when Peter, who is at the edge of the roof directly below you, 5.0 m away, asks you for the box of nails. Rather than carry the 2.5 kg box of nails down to Peter, you decide to give the box a push and have it slide down to him. If the coefficient of kinetic friction between the box and the roof is 0.55, with what speed should you push the box to have it gently come to rest right at the edge of the roof?

46. ‖ It's moving day, and you need to push a 100 kg box up a 20° ramp into the truck. The coefficients of friction for the box on the ramp are $\mu_s = 0.90$ and $\mu_k = 0.60$. Your largest pushing force is 1000 N. Can you get the box into the truck without assistance if you get a running start at the ramp? If you stop on the ramp, will you be able to get the box moving again?

47. ‖ An Airbus A320 jetliner has a takeoff mass of 75,000 kg. It reaches its takeoff speed of 82 m/s (180 mph) in 35 s. What is the thrust of the engines? You can neglect air resistance but not rolling friction.

48. ‖ A 2.0 kg wood block is launched up a wooden ramp that is inclined at a 30° angle. The block's initial speed is 10 m/s.

a. What vertical height does the block reach above its starting point?

b. What speed does it have when it slides back down to its starting point?

49. ‖ It's a snowy day and you're pulling a friend along a level road on a sled. You've both been taking physics, so she asks what you think the coefficient of friction between the sled and the snow is. You've been walking at a steady 1.5 m/s, and the rope pulls up on the sled at a 30° angle. You estimate that the mass of the sled, with your friend on it, is 60 kg and that you're pulling with a force of 75 N. What answer will you give?

50. ‖ a. A large box of mass M is pulled across a horizontal, frictionless surface by a horizontal rope with tension T. A small box of mass m sits on top of the large box. The coefficients of static and kinetic friction between the two boxes are μ_s and μ_k, respectively. Find an expression for the maximum tension T_{max} for which the small box rides on top of the large box without slipping.

    b. A horizontal rope pulls a 10 kg wood sled across frictionless snow. A 5.0 kg wood box rides on the sled. What is the largest tension force for which the box doesn't slip?

51. ‖ a. A large box of mass M is moving on a horizontal surface at speed v_0. A small box of mass m sits on top of the large box. The coefficients of static and kinetic friction between the two boxes are μ_s and μ_k, respectively. Find an expression for the shortest distance d_{min} in which the large box can stop without the small box slipping.

    b. A pickup truck with a steel bed is carrying a steel file cabinet. If the truck's speed is 15 m/s, what is the shortest distance in which it can stop without the file cabinet sliding?

52. ‖ Your assignment in lab is to measure the coefficient of kinetic friction between a 350 g block and a smooth metal table. To do so, you decide to launch the block at various speeds and measure how far it slides; your data are listed in the table. Use a graph to determine the value of μ_k.

Speed (m/s)	Distance (cm)
0.5	5
1.0	24
1.5	41
2.0	83
2.5	130

53. ‖ You're driving along at 25 m/s with your aunt's valuable antiques in the back of your pickup truck when suddenly you see a giant hole in the road 55 m ahead of you. Fortunately, your foot is right beside the brake and your reaction time is zero! Will the antiques be as fortunate?

    a. Can you stop the truck before it falls into the hole?

    b. If your answer to part a is yes, can you stop without the antiques sliding and being damaged? Their coefficients of friction are $\mu_s = 0.60$ and $\mu_k = 0.30$.

    Hint: You're not trying to stop in the shortest possible distance. What's your best strategy for avoiding damage to the antiques?

54. ‖ The 2.0 kg wood box in FIGURE P6.54 slides down a vertical wood wall while you push on it at a 45° angle. What magnitude of force should you apply to cause the box to slide down at a constant speed?

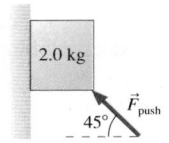

FIGURE P6.54

55. ‖ A 1.0 kg wood block is pressed against a vertical wood wall by the 12 N force shown in FIGURE P6.55. If the block is initially at rest, will it move upward, move downward, or stay at rest?

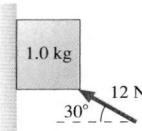

FIGURE P6.55

56. ‖ A person with compromised pinch strength in his fingers can exert a force of only 6.0 N to either side of a pinch-held object, such as the book shown in FIGURE P6.56. What is the heaviest book he can hold vertically before it slips out of his fingers? The coefficient of static friction between his fingers and the book cover is 0.80.

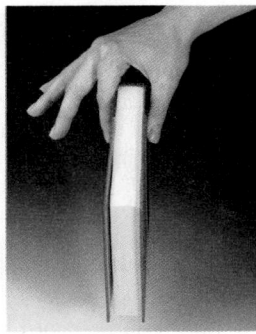

FIGURE P6.56

57. ‖ What is the terminal speed for an 80 kg skier going down a 40° snow-covered slope on wooden skis? Assume that the skier is 1.8 m tall and 0.40 m wide.

58. ‖ A ball is shot from a compressed-air gun at twice its terminal speed.

    a. What is the ball's initial acceleration, as a multiple of g, if it is shot straight up?

    b. What is the ball's initial acceleration, as a multiple of g, if it is shot straight down?

59. ‖ An artist friend of yours needs help hanging a 500 lb sculpture from the ceiling. For artistic reasons, she wants to use just two ropes. One will be 30° from vertical, the other 60°. She needs you to determine the smallest diameter rope that can safely support this expensive piece of art. On a visit to the hardware store you find that rope is sold in increments of $\frac{1}{8}$-inch diameter and that the safety rating is 4000 pounds per square inch of cross section. What diameter rope should you buy?

60. ‖ You've entered a "slow ski race" where the winner is the skier who takes the *longest* time to go down a 15° slope without ever stopping. You need to choose the best wax to apply to your skis. Red wax has a coefficient of kinetic friction 0.25, yellow is 0.20, green is 0.15, and blue is 0.10. Having just finished taking physics, you realize that a wax too slippery will cause you to accelerate down the slope and lose the race. But a wax that's too sticky will cause you to stop and be disqualified. You know that a strong headwind will apply a 50 N horizontal force against you as you ski, and you know that your mass is 82 kg. Which wax do you choose?

61. ‖ Astronauts in space "weigh" themselves by oscillating on a spring. Suppose the position of an oscillating 75 kg astronaut is given by $x = (0.30 \text{ m}) \sin((\pi \text{ rad/s}) \cdot t)$, where t is in s. What force does the spring exert on the astronaut at (a) $t = 1.0$ s and (b) 1.5 s? Note that the angle of the sine function is in radians.

62. ‖ A particle of mass m moving along the x-axis experiences the net force $F_x = ct$, where c is a constant. The particle has velocity v_{0x} at $t = 0$. Find an algebraic expression for the particle's velocity v_x at a later time t.

63. ‖ At $t = 0$, an object of mass m is at rest at $x = 0$ on a horizontal, frictionless surface. A horizontal force $F_x = F_0(1 - t/T)$, which decreases from F_0 at $t = 0$ to zero at $t = T$, is exerted on the object. Find an expression for the object's (a) velocity and (b) position at time T.

64. ‖‖ At $t = 0$, an object of mass m is at rest at $x = 0$ on a horizontal, frictionless surface. Starting at $t = 0$, a horizontal force $F_x = F_0 e^{-t/T}$ is exerted on the object.
    a. Find and graph an expression for the object's velocity at an arbitrary lateer time t.
    b. What is the object's velocity after a very long time has elapsed?

65. ‖‖ Large objects have inertia and tend to keep moving—
BIO Newton's first law. Life is very different for small microorganisms that swim through water. For them, drag forces are so large that they instantly stop, without coasting, if they cease their swimming motion. To swim at constant speed, they must exert a constant propulsion force by rotating corkscrew-like flagella or beating hair-like cilia. The quadratic model of drag of Equation 6.16 fails for very small particles. Instead, a small object moving in a liquid experiences a *linear* drag force, $\vec{D} = (bv,$ direction opposite the motion), where b is a constant. For a sphere of radius R, the drag constant can be shown to be $b = 6\pi\eta R$, where η is the *viscosity* of the liquid. Water at 20°C has viscosity 1.0×10^{-3} N s/m^2.
    a. A *paramecium* is about 100 μm long. If it's modeled as a sphere, how much propulsion force must it exert to swim at a typical speed of 1.0 mm/s? How about the propulsion force of a 2.0-μm-diameter *E. coli* bacterium swimming at 30 μm/s?
    b. The propulsion forces are very small, but so are the organisms. To judge whether the propulsion force is large or small *relative to the organism*, compute the acceleration that the propulsion force could give each organism if there were no drag. The density of both organisms is the same as that of water, 1000 kg/m^3.

66. ‖‖ Very small objects, such as dust particles, experience a *linear* drag force, $\vec{D} = (bv,$ direction opposite the motion), where b is a constant. That is, the quadratic model of drag of Equation 6.16 fails for very small particles. For a sphere of radius R, the drag constant can be shown to be $b = 6\pi\eta R$, where η is the *viscosity* of the gas.
    a. Find an expression for the terminal speed v_{term} of a spherical particle of radius R and mass m falling through a gas of viscosity η.
    b. Suppose a gust of wind has carried a 50-μm-diameter dust particle to a height of 300 m. If the wind suddenly stops, how long will it take the dust particle to settle back to the ground? Dust has a density of 2700 kg/m^3, the viscosity of 25°C air is 2.0×10^{-5} N s/m^2, and you can assume that the falling dust particle reaches terminal speed almost instantly.

Problems 67 and 68 show a free-body diagram. For each:
    a. Write a realistic dynamics problem for which this is the correct free-body diagram. Your problem should ask a question that can be answered with a value of position or velocity (such as "How far?" or "How fast?"), and should give sufficient information to allow a solution.
    b. Solve your problem!

67.

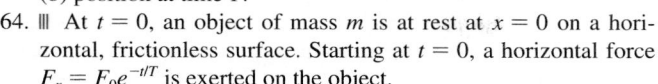

FIGURE P6.67

68.

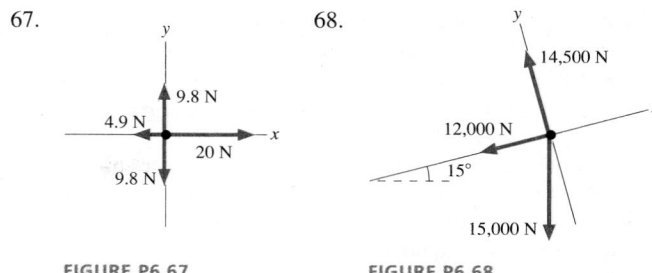

FIGURE P6.68

In Problems 69 through 71 you are given the dynamics equations that are used to solve a problem. For each of these, you are to
    a. Write a realistic problem for which these are the correct equations.
    b. Draw the free-body diagram and the pictorial representation for your problem.
    c. Finish the solution of the problem.

69. $-0.80n = (1500 \text{ kg})a_x$
    $n - (1500 \text{ kg})(9.80 \text{ m/s}^2) = 0$

70. $T - 0.20n - (20 \text{ kg})(9.80 \text{ m/s}^2) \sin 20°$
    $= (20 \text{ kg})(2.0 \text{ m/s}^2)$
    $n - (20 \text{ kg})(9.80 \text{ m/s}^2) \cos 20° = 0$

71. $(100 \text{ N}) \cos 30° - f_k = (20 \text{ kg})a_x$
    $n + (100 \text{ N}) \sin 30° - (20 \text{ kg})(9.80 \text{ m/s}^2) = 0$
    $f_k = 0.20n$

Challenge Problems

72. A block of mass m is at rest at the origin at $t = 0$. It is pushed with constant force F_0 from $x = 0$ to $x = L$ across a horizontal surface whose coefficient of kinetic friction is $\mu_k = \mu_0(1 - x/L)$. That is, the coefficient of friction decreases from μ_0 at $x = 0$ to zero at $x = L$.
    a. Use what you've learned in calculus to prove that

$$a_x = v_x \frac{dv_x}{dx}$$

    b. Find an expression for the block's speed as it reaches position L.

73. The machine in **FIGURE CP6.73** has an 800 g steel shuttle that is pulled along a square steel rail by an elastic cord. The shuttle is released when the elastic cord has 20 N tension at a 45° angle. What is the initial acceleration of the shuttle?

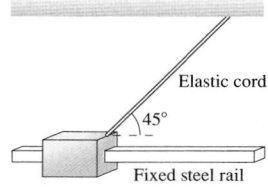

FIGURE CP6.73

74. **FIGURE CP6.74** shows an *accelerometer*, a device for measuring the horizontal acceleration of cars and airplanes. A ball is free to roll on a parabolic track described by the equation $y = x^2$, where both x and y are in meters. A scale along the bottom is used to measure the ball's horizontal position x.

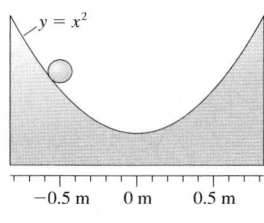

FIGURE CP6.74

    a. Find an expression that allows you to use a measured position x (in m) to compute the acceleration a_x (in m/s^2). (For example, $a_x = 3x$ is a possible expression.)
    b. What is the acceleration if $x = 20$ cm?

75. An object moving in a liquid experiences a *linear* drag force: $\vec{D} = (bv$, direction opposite the motion), where b is a constant called the *drag coefficient*. For a sphere of radius R, the drag constant can be computed as $b = 6\pi\eta R$, where η is the *viscosity* of the liquid.

    a. Find an algebraic expression for $v_x(t)$, the x-component of velocity as a function of time, for a spherical particle of radius R and mass m that is shot horizontally with initial speed v_0 through a liquid of viscosity η.

    b. Water at 20°C has viscosity $\eta = 1.0 \times 10^{-3}$ N s/m². Suppose a 4.0-cm-diameter, 33 g ball is shot horizontally into a tank of 20°C water. How long will it take for the horizontal speed to decrease to 50% of its initial value?

76. An object moving in a liquid experiences a *linear* drag force: $\vec{D} = (bv$, direction opposite the motion), where b is a constant called the *drag coefficient*. For a sphere of radius R, the drag constant can be computed as $b = 6\pi\eta R$, where η is the *viscosity* of the liquid.

    a. Use what you've learned in calculus to prove that

    $$a_x = v_x \frac{dv_x}{dx}$$

b. Find an algebraic expression for $v_x(x)$, the x-component of velocity as a function of distance traveled, for a spherical particle of radius R and mass m that is shot horizontally with initial speed v_0 through a liquid of viscosity η.

c. Water at 20°C has viscosity $\eta = 1.0 \times 10^{-3}$ N s/m². Suppose a 1.0-cm-diameter, 1.0 g marble is shot horizontally into a tank of 20°C water at 10 cm/s. How far will it travel before stopping?

77. An object with cross section A is shot horizontally across frictionless ice. Its initial velocity is v_{0x} at $t_0 = 0$ s. Air resistance is not negligible.

    a. Show that the velocity at time t is given by the expression

    $$v_x = \frac{v_{0x}}{1 + C\rho A v_{0x} t/2m}$$

    b. A 1.6-m-wide, 1.4-m-high, 1500 kg car with a drag coefficient of 0.35 hits a very slick patch of ice while going 20 m/s. If friction is neglected, how long will it take until the car's speed drops to 10 m/s? To 5 m/s?

    c. Assess whether or not it is reasonable to neglect kinetic friction.

STOP TO THINK ANSWERS

Stop to Think 6.1: a. The lander is descending and slowing. The acceleration vector points upward, and so $\vec{F}_{net}$ points upward. This can be true only if the thrust has a larger magnitude than the weight.

Stop to Think 6.2: a. You are descending and slowing, so your acceleration vector points upward and there is a net upward force on you. The floor pushes up against your feet harder than gravity pulls down.

Stop to Think 6.3: $f_b > f_c = f_d = f_e > f_a$. Situations c, d, and e are all kinetic friction, which does not depend on either velocity or acceleration. Kinetic friction is smaller than the maximum static friction that is exerted in b. $f_a = 0$ because no friction is needed to keep the object at rest.

Stop to Think 6.4: d. The ball is shot *down* at 30 m/s, so $v_{0y} = -30$ m/s. This exceeds the terminal speed, so the upward drag force is *larger* than the downward weight force. Thus the ball *slows down* even though it is "falling." It will slow until $v_y = -15$ m/s, the terminal velocity, then maintain that velocity.

7 Newton's Third Law

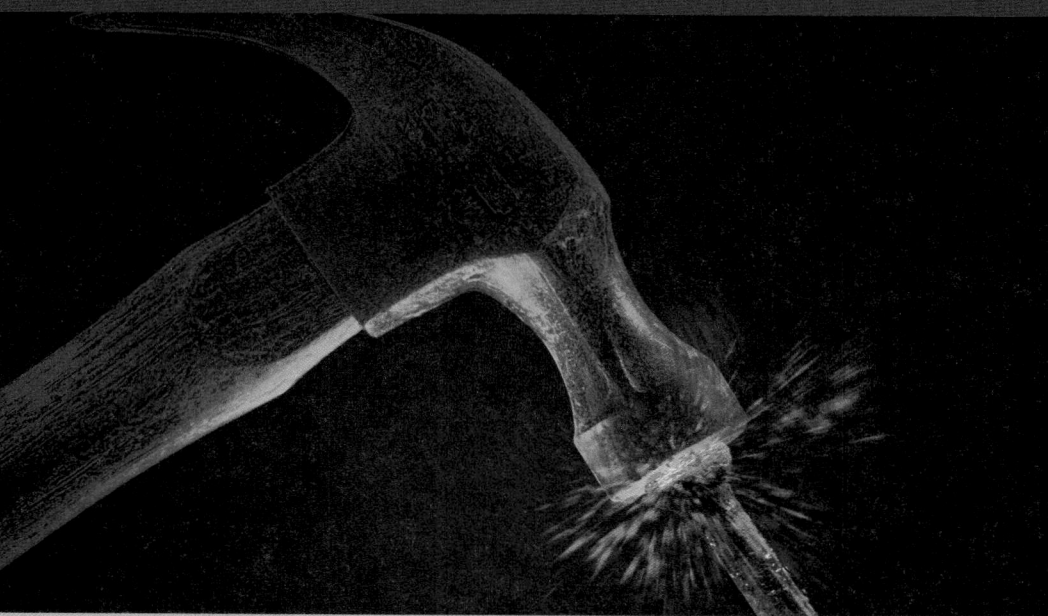

The hammer and nail are interacting. The forces of the hammer on the nail and the nail on the hammer are an action/reaction pair of forces.

▶ **Looking Ahead** The goal of Chapter 7 is to use Newton's third law to understand how objects interact.

Interactions

Newton's second law treats an object as an isolated entity acted upon by external forces.

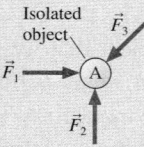

For example, we've looked at the normal force of a table on a book. But what about the book's effect on the table?

Whenever two or more objects exert forces on each other, by touching, being tied together, or via long-range forces, we say that they *interact*.

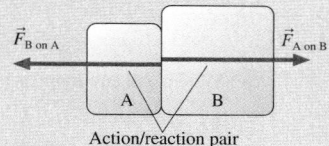

Action/reaction pair

You'll learn that if object A exerts a force on object B, then object B exerts a force on object A. These two forces form what is called an **action/reaction pair.**

◀ **Looking Back**
Sections 5.1–5.3 Basic concepts of force and the atomic model of tension

Newton's Third Law

Interactions are described by **Newton's third law:**

- *Every* force occurs as one member of an action/reaction pair.
- The two members of a pair act on two *different* objects.
- The two members of a pair are equal in magnitude but opposite in direction.

Thrust and propulsion are two important applications of Newton's third law.

Ropes and Pulleys

A common way that two objects interact is via ropes or cables or strings. Pulleys can be used to change the direction of the tension force.

You'll learn that:

- Objects that are connected together must have the same acceleration.
- *Tension is constant* throughout a rope or string if we can model it as being massless and pulleys as frictionless.

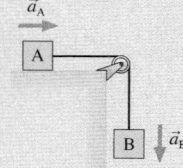

Interaction Diagrams

You'll learn how to draw **interaction diagrams** to show the action/reaction pairs of forces between interacting objects.

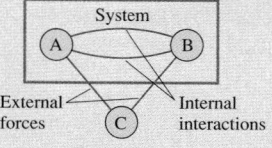

The **system** consists of those objects whose motion we wish to analyze. Objects that exert forces but whose motion is not of interest—such as the earth—form the **environment**.

Problem-Solving Strategy

We will expand the problem-solving strategy that we began in Chapter 6.

- Draw an interaction diagram.
- Identify the system.
- Draw a separate free-body diagram for each object in the system.
- Write Newton's second law for each object.
- Use Newton's third law to relate action/reaction pairs of forces.

◀ **Looking Back**
Sections 6.1–6.2 Problem-solving strategies for force and motion

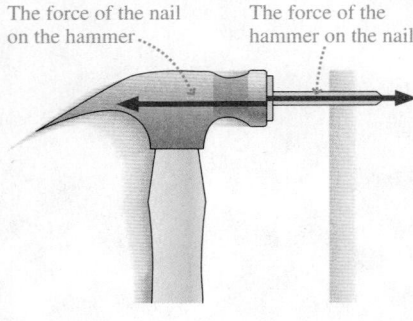

FIGURE 7.1 The hammer and nail are interacting with each other.

The force of the nail on the hammer...

The force of the hammer on the nail

7.1 Interacting Objects

FIGURE 7.1 shows a hammer hitting a nail. The hammer exerts a force on the nail as it drives the nail forward. At the same time, the nail exerts a force on the hammer. If you're not sure that it does, imagine hitting the nail with a glass hammer. It's the force of the nail on the hammer that would cause the glass to shatter.

In fact, any time an object A pushes or pulls on another object B, B pushes or pulls back on A. When you pull someone with a rope in a tug-of-war, that person pulls back on you. Your chair pushes up on you (the normal force) as you push down on the chair. These are examples of an **interaction,** the mutual influence of two objects on each other.

To be more specific, if object A exerts a force $\vec{F}_{\text{A on B}}$ on object B, then object B exerts a force $\vec{F}_{\text{B on A}}$ on object A. This pair of forces, shown in FIGURE 7.2, is called an **action/reaction pair.** Two objects interact by exerting an action/reaction pair of forces on each other. Notice the very explicit subscripts on the force vectors. The first letter is the *agent,* the second letter is the object on which the force acts. $\vec{F}_{\text{A on B}}$ is a force exerted *by* A *on* B.

> NOTE ▶ The name "action/reaction pair" is somewhat misleading. The forces occur simultaneously, and we cannot say which is the "action" and which the "reaction." **An action/reaction pair of forces exists as a pair, or not at all.** In identifying action/reaction pairs, the labels are the key. Force $\vec{F}_{\text{A on B}}$ is paired with force $\vec{F}_{\text{B on A}}$. ◀

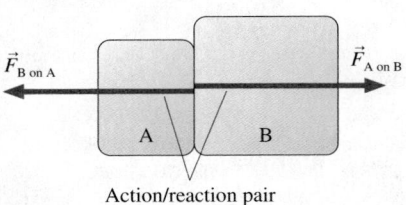

FIGURE 7.2 An action/reaction pair of forces.

$\vec{F}_{\text{B on A}}$

$\vec{F}_{\text{A on B}}$

A   B

Action/reaction pair

The hammer and nail interact through contact forces. The same idea holds true for long-range forces such as gravity. If you release a ball, it falls because the earth's gravity exerts a downward force $\vec{F}_{\text{earth on ball}}$. But does the ball really pull upward on the earth with a force $\vec{F}_{\text{ball on earth}}$?

Newton was the first to realize that, indeed, the ball *does* pull upward on the earth. His evidence was the tides. Astronomers had known since antiquity that the tides depend on the phase of the moon, but Newton was the first to understand that tides are the ocean's response to the gravitational pull of the moon on the earth. As FIGURE 7.3 shows, the flexible water bulges toward the moon while the relatively inflexible crust remains behind.

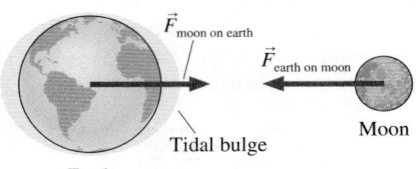

FIGURE 7.3 The ocean tides are an indication of the long-range gravitational interaction of the earth and the moon.

$\vec{F}_{\text{moon on earth}}$

$\vec{F}_{\text{earth on moon}}$

Moon

Tidal bulge

Earth

Objects, Systems, and the Environment

Chapters 5 and 6 considered forces acting on a single object that we modeled as a particle. FIGURE 7.4a shows a diagrammatic representation of single-particle dynamics. We can use Newton's second law, $\vec{a} = \vec{F}_{\text{net}}/m$, to determine the particle's acceleration.

We now want to extend the particle model to situations in which two or more objects, each represented as a particle, interact with each other. For example, FIGURE 7.4b shows three objects interacting via action/reaction pairs of forces. The forces can be given labels such as $\vec{F}_{\text{A on B}}$ and $\vec{F}_{\text{B on A}}$. How do these particles move?

We will often be interested in the motion of some of the objects, say objects A and B, but not of others. For example, objects A and B might be the hammer and the nail, while object C is the earth. The earth interacts with both the hammer and the nail via

FIGURE 7.4 Single-particle dynamics and a model of interacting objects.

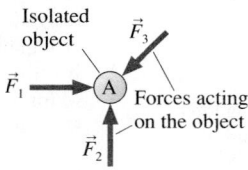

(a) Single-particle dynamics

Isolated object

$\vec{F}_3$

$\vec{F}_1$

A

Forces acting on the object

$\vec{F}_2$

This is a force diagram.

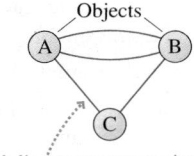

(b) Interacting objects

Objects

A   B

C

Each line represents an interaction and an action/reaction pair of forces. Some pairs of objects, such as A and B, can have more than one interaction.

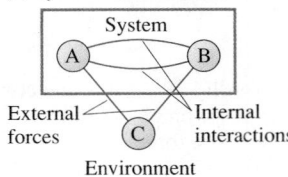

(c) System and environment

System

A   B

C

External forces

Internal interactions

Environment

This is an *interaction diagram.*

gravity, but in a practical sense the earth remains "at rest" while the hammer and nail move. Let's define the **system** as those objects whose motion we want to analyze and the **environment** as objects external to the system.

FIGURE 7.4c is a new kind of diagram, an **interaction diagram,** in which we've enclosed the objects of the system in a box and represented interactions as lines connecting objects. This is a rather abstract, schematic diagram, but it captures the essence of the interactions. Notice that interactions with objects in the environment are called **external forces.** For the hammer and nail, the gravitational force on each—an interaction with the earth—is an external force.

NOTE ▶ The system–environment distinction is a practical matter, not a fundamental distinction. If object A pushes or pulls on object B, then B pushes or pulls on A. *Every* force is one member of an action/reaction pair, and there is no such thing as a true "external force." What we call an external force is an interaction between an object of interest, one we've chosen to place inside the system, and an object whose motion is not of interest. ◀

The bat and the ball are interacting with each other.

7.2 Analyzing Interacting Objects

> **TACTICS**
> **BOX 7.1**   Analyzing interacting objects   (MP)
>
> ❶ **Represent each object as a circle.** Place each in the correct position relative to other objects.
>
>   ■ Give each a name and a label.
>   ■ The surface of the earth (contact forces) and the entire earth (long-range forces) should be considered separate objects. Label the **entire earth EE.**
>   ■ Ropes and pulleys often need to be considered objects.
>
> ❷ **Identify interactions.** Draw connecting lines between the circles to represent interactions.
>
>   ■ Draw *one* line for each interaction. Label it with the type of force.
>   ■ Every interaction line connects two and only two objects.
>   ■ There can be at most two interactions at a surface: a force parallel to the surface (e.g., friction) and a force perpendicular to the surface (e.g., a normal force).
>   ■ The entire earth interacts only by the long-range gravitational force.
>
> ❸ **Identify the system.** Identify the objects of interest; draw and label a box enclosing them. This completes the interaction diagram.
>
> ❹ **Draw a free-body diagram for each object in the system.** Include only the forces acting *on* each object, not forces exerted by the object.
>
>   ■ Every interaction line crossing the system boundary is one external force acting on the object. The usual force symbols, such as $\vec{n}$ and $\vec{T}$, can be used.
>   ■ Every interaction line within the system represents an action/reaction pair of forces. There is one force vector on *each* of the objects, and these forces always point in opposite directions. Use labels like $\vec{F}_{A \text{ on } B}$ and $\vec{F}_{B \text{ on } A}$.
>   ■ Connect the two action/reaction forces—which must be on *different* free-body diagrams—with a dashed line.
>
> Exercises 1–7 ✐

We'll illustrate these ideas with two concrete examples. The first example will be much longer than usual because we'll go carefully through all the steps in the reasoning.

EXAMPLE 7.1 **Pushing a crate**

FIGURE 7.5 shows a person pushing a large crate across a rough surface. Identify all interactions, show them on an interaction diagram, then draw free-body diagrams of the person and the crate.

FIGURE 7.5 A person pushes a crate across a rough floor.

VISUALIZE The interaction diagram of **FIGURE 7.6** starts by representing every object as a circle in the correct position but separated from all other objects. The person and the crate are obvious objects. The earth is also an object that both exerts and experiences forces, but it's necessary to distinguish between the surface, which exerts contact forces, and the entire earth, which exerts the long-range gravitational force.

FIGURE 7.6 The interaction diagram.

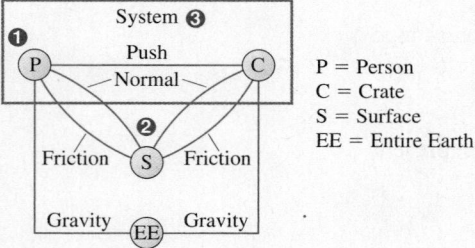

Figure 7.6 also identifies the various interactions. Some, like the pushing interaction between the person and the crate, are fairly obvious. The interactions with the earth are a little trickier. Gravity, a long-range force, is an interaction between each object and the earth as a whole. Friction forces and normal forces are contact interactions between each object and the earth's surface. These are two different interactions, so two interaction lines connect the crate to the surface and the person to the surface. Altogether, there are seven interactions. Finally, we've enclosed the person and crate in a box labeled System. These are the objects whose motion we wish to analyze.

NOTE ▶ Interactions are between two *different* objects. None of the interactions are between an object and itself. ◀

We can now draw free-body diagrams for the objects in the system, the crate and the person. **FIGURE 7.7** correctly locates the crate's free-body diagram to the right of the person's free-body diagram. For each, three interaction lines cross the system boundary and thus represent external forces. These are the gravitational force from the entire earth, the upward normal force

FIGURE 7.7 Free-body diagrams of the person and the crate.

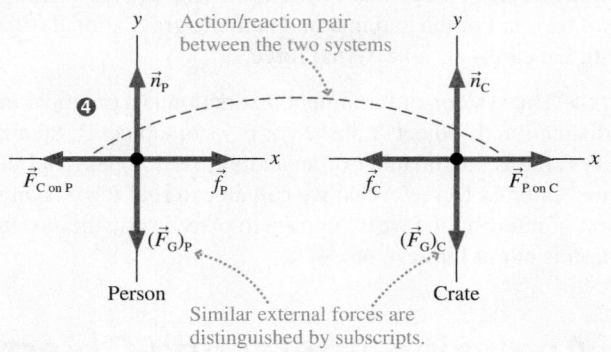

from the surface, and a friction force from the surface. We can use familiar labels such as $\vec{n}_P$ and $\vec{f}_C$, but **it's very important to distinguish different forces with subscripts.** There's now more than one normal force. If you call both simply $\vec{n}$, you're almost certain to make mistakes when you start writing out the second-law equations.

The directions of the normal forces and the gravitational forces are clear, but we have to be careful with friction. Friction force $\vec{f}_C$ is kinetic friction of the crate sliding across the surface, so it points left, opposite the motion. But what about friction between the person and the surface? It is tempting to draw force $\vec{f}_P$ pointing to the left. After all, friction forces are supposed to be in the direction opposite the motion. But if we did so, the person would have two forces to the left, $\vec{F}_{C \text{ on } P}$ and $\vec{f}_P$, and none to the right, causing the person to accelerate *backward!* That is clearly not what happens, so what is wrong?

Imagine pushing a crate to the right across loose sand. Each time you take a step, you tend to kick the sand to the *left,* behind you. Thus friction force $\vec{f}_{P \text{ on } S}$, the force of the person pushing against the earth's surface, is to the *left.* In reaction, the force of the earth's surface against the person is a friction force to the *right.* It is force $\vec{f}_{S \text{ on } P}$, which we've shortened to $\vec{f}_P$, that causes the person to accelerate in the forward direction. Further, as we'll discuss more below, this is a *static* friction force; your foot is planted on the ground, not sliding across the surface.

Finally, we have one internal interaction. The crate is pushed with force $\vec{F}_{P \text{ on } C}$. If A pushes or pulls on B, then B pushes or pulls back on A. The reaction to force $\vec{F}_{P \text{ on } C}$ is $\vec{F}_{C \text{ on } P}$, the crate pushing back against the person's hands. Force $\vec{F}_{P \text{ on } C}$ is a force exerted on the crate, so it's shown on the crate's free-body diagram. Force $\vec{F}_{C \text{ on } P}$ is exerted on the person, so it is drawn on the person's free-body diagram. **The two forces of an action/ reaction pair never occur on the same object.** Notice that forces $\vec{F}_{P \text{ on } C}$ and $\vec{F}_{C \text{ on } P}$ are pointing in opposite directions. We've connected them with a dashed line to show that they are an action/ reaction pair.

ASSESS The completed free-body diagrams of Figure 7.7 could now be the basis for a quantitative analysis.

Propulsion

The friction force $\vec{f}_P$ (force of surface on person) is an example of **propulsion.** It is the force that a system with an internal source of energy uses to drive itself forward. Propulsion is an important feature not only of walking or running but also of the forward motion of cars, jets, and rockets. Propulsion is somewhat counterintuitive, so it is worth a closer look.

If you try to walk across a frictionless floor, your foot slips and slides *backward.* In order for you to walk, the floor needs to have friction so that your foot *sticks* to the floor as you straighten your leg, moving your body forward. The friction that prevents slipping is *static* friction. Static friction, you will recall, acts in the direction that prevents slipping. The static friction force $\vec{f}_P$ has to point in the *forward* direction to prevent your foot from slipping backward. It is this forward-directed static friction force that propels you forward! The force of your foot on the floor, the other half of the action/reaction pair, is in the opposite direction.

The distinction between you and the crate is that you have an *internal source of energy* that allows you to straighten your leg by pushing backward against the surface. In essence, you walk by pushing the earth away from you. The earth's surface responds by pushing you forward. These are static friction forces. In contrast, all the crate can do is slide, so *kinetic* friction opposes the motion of the crate.

FIGURE 7.8 shows how propulsion works. A car uses its motor to spin the tires, causing the tires to push backward against the ground. This is why dirt and gravel are kicked backward, not forward. The earth's surface responds by pushing the car forward. These are also *static* friction forces. The tire is rolling, but the bottom of the tire, where it contacts the road, is instantaneously at rest. If it weren't, you would leave one giant skid mark as you drove and would burn off the tread within a few miles.

What force causes this sprinter to accelerate?

FIGURE 7.8 Examples of propulsion.

The person pushes backward against the earth. The earth pushes forward on the person. Static friction.

The car pushes backward against the earth. The earth pushes forward on the car. Static friction.

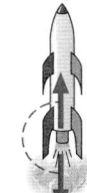

The rocket pushes the hot gases backward. The gases push the rocket forward. Thrust force.

EXAMPLE 7.2 | **Towing a car**

A tow truck uses a rope to pull a car along a horizontal road, as shown in **FIGURE 7.9**. Identify all interactions, show them on an interaction diagram, then draw free-body diagrams of each object in the system.

FIGURE 7.9 A truck towing a car.

VISUALIZE The interaction diagram of **FIGURE 7.10** represents the objects as separate circles, but with the correct relative positions. The rope is shown as a separate object. Many of the interactions are identical to those in Example 7.1. The system—the objects in motion—consists of the truck, the rope, and the car.

FIGURE 7.10 The interaction diagram.

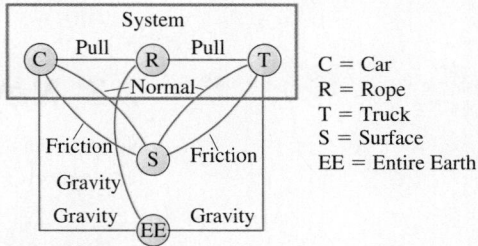

C = Car
R = Rope
T = Truck
S = Surface
EE = Entire Earth

The three objects in the system require three free-body diagrams, shown in **FIGURE 7.11** on the next page. Gravity, friction, and normal forces at the surface are all interactions that cross the system boundary and are shown as external forces. The car is an inert object rolling along. It would slow and stop if the rope were cut, so the surface must exert a rolling friction force $\vec{f}_C$ to the left. The truck, however, has an internal source of energy. The truck's drive wheels

Continued

FIGURE 7.11 Free-body diagrams of Example 7.2.

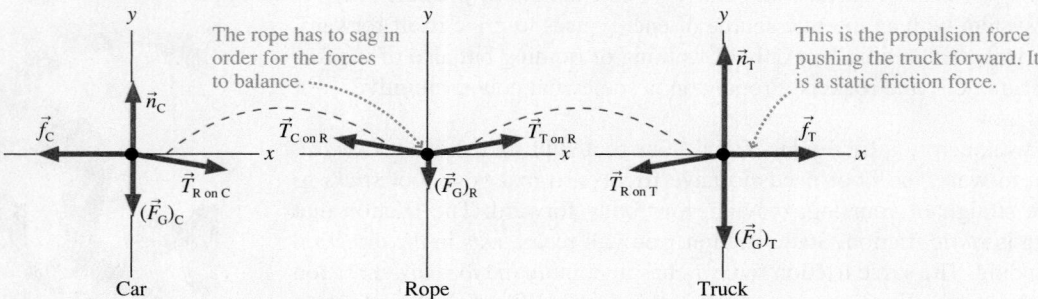

push the ground to the left with force $\vec{f}_{\text{T on S}}$. In reaction, the ground propels the truck forward, to the right, with force $\vec{f}_{\text{T}}$.

We next need to identify the horizontal forces between the car, the truck, and the rope. The rope pulls on the car with a tension force $\vec{T}_{\text{R on C}}$. You might be tempted to put the reaction force on the truck because we say that "the truck pulls the car," but the truck is not in contact with the car. The truck pulls on the rope, then the rope pulls on the car. Thus the reaction to $\vec{T}_{\text{R on C}}$ is a force on the *rope:* $\vec{T}_{\text{C on R}}$. These are an action/reaction pair. At the other end, $\vec{T}_{\text{T on R}}$ and $\vec{T}_{\text{R on T}}$ are also an action/reaction pair.

NOTE ▶ Drawing an interaction diagram helps you avoid mistakes because it shows very clearly what is interacting with what. ◀

Notice that the tension forces of the rope *cannot* be horizontal. If they were, the rope's free-body diagram would show a net downward force, because of its weight, and the rope would accelerate downward. The tension forces $\vec{T}_{\text{T on R}}$ and $\vec{T}_{\text{C on R}}$ have to angle slightly upward to balance the gravitational force, so any real rope has to sag at least a little in the center.

ASSESS Make sure you avoid the common error of considering $\vec{n}$ and $\vec{F}_{\text{G}}$ to be an action/reaction pair. These are both forces on the *same* object, whereas the two forces of an action/reaction pair are always on two *different* objects that are interacting with each other. The normal and gravitational forces are often equal in magnitude, as they are in this example, but that doesn't make them an action/reaction pair of forces.

STOP TO THINK 7.1   A rope of negligible mass pulls a crate across the floor. The rope and crate are the system; the hand is part of the environment. What, if anything, is wrong with the free-body diagrams?

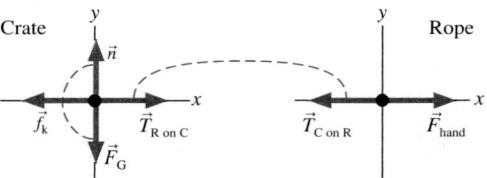

7.3 Newton's Third Law

Newton was the first to recognize how the two members of an action/reaction pair of forces are related to each other. Today we know this as Newton's third law:

Newton's third law Every force occurs as one member of an action/reaction pair of forces.

- The two members of an action/reaction pair act on two *different* objects.
- The two members of an action/reaction pair are equal in magnitude but opposite in direction: $\vec{F}_{\text{A on B}} = -\vec{F}_{\text{B on A}}$.

We deduced most of the third law in Section 7.2. There we found that the two members of an action/reaction pair are always opposite in direction (see Figures 7.7 and 7.11). According to the third law, this will always be true. But the most significant portion of the third law, which is by no means obvious, is that the two members of an action/reaction pair have *equal* magnitudes. That is, $F_{A \text{ on } B} = F_{B \text{ on } A}$. This is the quantitative relationship that will allow you to solve problems of interacting objects.

Newton's third law is frequently stated as "For every action there is an equal but opposite reaction." While this is indeed a catchy phrase, it lacks the preciseness of our preferred version. In particular, it fails to capture an essential feature of action/reaction pairs—that they each act on a *different* object.

NOTE ▶ Newton's third law extends and completes our concept of *force*. We can now recognize force as an *interaction* between objects rather than as some "thing" with an independent existence of its own. The concept of an interaction will become increasingly important as we begin to study the laws of momentum and energy. ◀

Reasoning with Newton's Third Law

Newton's third law is easy to state but harder to grasp. For example, consider what happens when you release a ball. Not surprisingly, it falls down. But if the ball and the earth exert equal and opposite forces on each other, as Newton's third law alleges, why doesn't the earth "fall up" to meet the ball?

The key to understanding this and many similar puzzles is that **the forces are equal but the accelerations are not.** Equal causes can produce very unequal effects. FIGURE 7.12 shows equal-magnitude forces on the ball and the earth. The force on ball B is simply the gravitational force of Chapter 6:

$$\vec{F}_{\text{earth on ball}} = (\vec{F}_{\text{G}})_{\text{B}} = -m_{\text{B}}\,g\,\hat{j} \qquad (7.1)$$

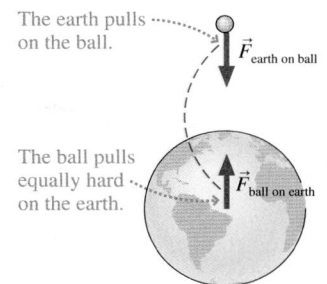

FIGURE 7.12 The action/reaction forces of a ball and the earth are equal in magnitude.

The earth pulls on the ball.

$\vec{F}_{\text{earth on ball}}$

The ball pulls equally hard on the earth.

$\vec{F}_{\text{ball on earth}}$

where m_{B} is the mass of the ball. According to Newton's second law, this force gives the ball an acceleration

$$\vec{a}_{\text{B}} = \frac{(\vec{F}_{\text{G}})_{\text{B}}}{m_{\text{B}}} = -g\,\hat{j} \qquad (7.2)$$

This is just the familiar free-fall acceleration.

According to Newton's third law, the ball pulls up on the earth with force $\vec{F}_{\text{ball on earth}}$. Because $\vec{F}_{\text{earth on ball}}$ and $\vec{F}_{\text{ball on earth}}$ are an action/reaction pair, $\vec{F}_{\text{ball on earth}}$ must be equal in magnitude and opposite in direction to $\vec{F}_{\text{earth on ball}}$. That is,

$$\vec{F}_{\text{ball on earth}} = -\vec{F}_{\text{earth on ball}} = -(\vec{F}_{\text{G}})_{\text{B}} = +m_{\text{B}}\,g\,\hat{j} \qquad (7.3)$$

Using this result in Newton's second law, we find the upward acceleration of the earth as a whole is

$$\vec{a}_{\text{E}} = \frac{\vec{F}_{\text{ball on earth}}}{m_{\text{E}}} = \frac{m_{\text{B}}\,g\,\hat{j}}{m_{\text{E}}} = \left(\frac{m_{\text{B}}}{m_{\text{E}}}\right)g\,\hat{j} \qquad (7.4)$$

The upward acceleration of the earth is less than the downward acceleration of the ball by the factor $m_{\text{B}}/m_{\text{E}}$. If we assume a 1 kg ball, we can estimate the magnitude of $\vec{a}_{\text{E}}$:

$$a_{\text{E}} = \frac{1 \text{ kg}}{6 \times 10^{24} \text{ kg}}\,g \approx 2 \times 10^{-24} \text{ m/s}^2$$

With this incredibly small acceleration, it would take the earth 8×10^{15} years, approximately 500,000 times the age of the universe, to reach a speed of 1 mph! So we certainly would not expect to see or feel the earth "fall up" after we drop a ball.

NOTE ▶ Newton's third law equates the size of two forces, not two accelerations. The acceleration continues to depend on the mass, as Newton's second law states. **In an interaction between two objects of different mass, the lighter mass will do essentially all of the accelerating even though the forces exerted on the two objects are equal.** ◀

The forces on accelerating boxes

The hand shown in FIGURE 7.13 pushes boxes A and B to the right across a frictionless table. The mass of B is larger than the mass of A.

a. Draw free-body diagrams of A, B, and the hand H, showing only the *horizontal* forces. Connect action/reaction pairs with dashed lines.

b. Rank in order, from largest to smallest, the horizontal forces shown on your free-body diagrams.

FIGURE 7.13 Hand H pushes boxes A and B.

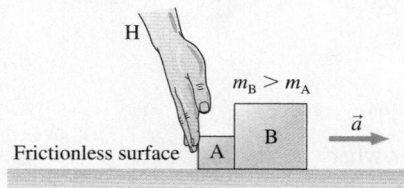

VISUALIZE a. The hand H pushes on box A, and A pushes back on H. Thus $\vec{F}_{\text{H on A}}$ and $\vec{F}_{\text{A on H}}$ are an action/reaction pair.

Similarly, A pushes on B and B pushes back on A. **The hand H does not touch box B, so there is no interaction between them.** There is no friction. FIGURE 7.14 shows the four horizontal forces and identifies two action/reaction pairs. Notice that each force is shown on the free-body diagram of the object that it acts *on*.

b. According to Newton's third law, $F_{\text{A on H}} = F_{\text{H on A}}$ and $F_{\text{A on B}} = F_{\text{B on A}}$. But the third law is not our only tool. The boxes are *accelerating* to the right, because there's no friction, so Newton's *second* law tells us that box A must have a net force to the right. Consequently, $F_{\text{H on A}} > F_{\text{B on A}}$. Thus

$$F_{\text{A on H}} = F_{\text{H on A}} > F_{\text{B on A}} = F_{\text{A on B}}$$

ASSESS You might have expected $F_{\text{A on B}}$ to be larger than $F_{\text{H on A}}$ because $m_B > m_A$. It's true that the *net* force on B is larger than the *net* force on A, but we have to reason more closely to judge the individual forces. Notice how we used both the second and the third laws to answer this question.

FIGURE 7.14 The free-body diagrams, showing only the horizontal forces.

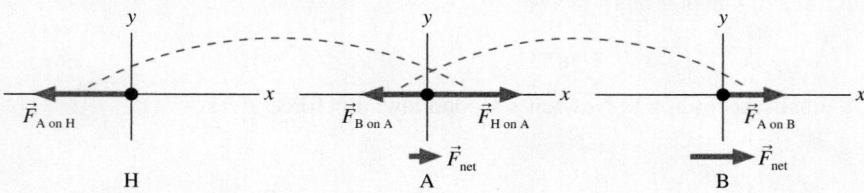

Car B is stopped for a red light. Car A, which has the same mass as car B, doesn't see the red light and runs into the back of B. Which of the following statements is true?

a. B exerts a force on A, but A doesn't exert a force on B.
b. B exerts a larger force on A than A exerts on B.
c. B exerts the same amount of force on A as A exerts on B.
d. A exerts a larger force on B than B exerts on A.
e. A exerts a force on B, but B doesn't exert a force on A.

Acceleration Constraints

Newton's third law is one quantitative relationship you can use to solve problems of interacting objects. In addition, we frequently have other information about the motion in a problem. For example, if two objects A and B move together, their accelerations are *constrained* to be equal: $\vec{a}_A = \vec{a}_B$. A well-defined relationship between the accelerations of two or more objects is called an **acceleration constraint.** It is an independent piece of information that can help solve a problem.

FIGURE 7.15 The car and the truck have the same acceleration.

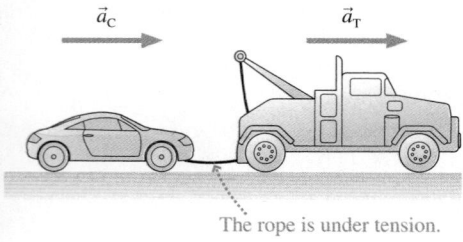

The rope is under tension.

In practice, we'll express acceleration constraints in terms of the *x*- and *y*-components of $\vec{a}$. Consider the car being towed in FIGURE 7.15. This is one-dimensional motion, so we can write the acceleration constraint as

$$a_{Cx} = a_{Tx} = a_x$$

Because the accelerations of both objects are equal, we can drop the subscripts C and T and call both of them a_x.

Don't assume the accelerations of A and B will always have the same sign. Consider blocks A and B in FIGURE 7.16. The blocks are connected by a string, so they are constrained to move together and their accelerations have equal magnitudes. But A has a positive acceleration (to the right) in the *x*-direction while B has a negative acceleration (downward) in the *y*-direction. Thus the acceleration constraint is

$$a_{Ax} = -a_{By}$$

This relationship does *not* say that a_{Ax} is a negative number. It is simply a relational statement, saying that a_{Ax} is (-1) times whatever a_{By} happens to be. The acceleration a_{By} in Figure 7.16 is a negative number, so a_{Ax} is positive. In some problems, the signs of a_{Ax} and a_{By} may not be known until the problem is solved, but the *relationship* is known from the beginning.

A Revised Strategy for Interacting-Objects Problems

Problems of interacting objects can be solved with a few modifications to the basic problem-solving strategy we developed in Chapter 6. A revised problem-solving strategy follows.

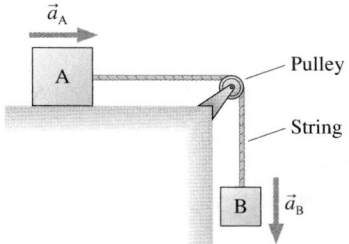

FIGURE 7.16 The string constrains the two objects to accelerate together.

PROBLEM-SOLVING STRATEGY 7.1 Interacting-objects problems

MODEL Identify which objects are part of the system and which are part of the environment. Make simplifying assumptions.

VISUALIZE Draw a pictorial representation.

- Show important points in the motion with a sketch. You may want to give each object a separate coordinate system. Define symbols and identify what the problem is trying to find.
- Identify acceleration constraints.
- Draw an interaction diagram to identify the forces on each object and all action/reaction pairs.
- Draw a *separate* free-body diagram for each object. Each shows only the forces acting *on* that object, not forces exerted by the object.
- Connect the force vectors of action/reaction pairs with dashed lines. Use subscript labels to distinguish forces that act independently on more than one object.

SOLVE Use Newton's second and third laws.

- Write the equations of Newton's second law for *each* object, using the force information from the free-body diagrams.
- Equate the magnitudes of action/reaction pairs.
- Include the acceleration constraints, the friction model, and other quantitative information relevant to the problem.
- Solve for the acceleration, then use kinematics to find velocities and positions.

ASSESS Check that your result has the correct units, is reasonable, and answers the question.

You might be puzzled that the Solve step calls for the use of the third law to equate just the *magnitudes* of action/reaction forces. What about the "opposite in direction" part of the third law? You have already used it! Your free-body diagrams should show the two members of an action/reaction pair to be opposite in direction, and that information will have been utilized in writing the second-law equations. Because the directional information has already been used, all that is left is the magnitude information.

NOTE ▶ Two steps are especially important when drawing the free-body diagrams. First, draw a *separate* diagram for each object. The diagrams need not have the same coordinate system. Second, show only the forces acting *on* that object. The force $\vec{F}_{A\text{ on B}}$ goes on the free-body diagram of object B, but $\vec{F}_{B\text{ on A}}$ goes on the diagram of object A. The two members of an action/reaction pair *always* appear on two different free-body diagrams—*never* on the same diagram. ◀

EXAMPLE 7.4 **Keep the crate from sliding**

You and a friend have just loaded a 200 kg crate filled with priceless art objects into the back of a 2000 kg truck. As you press down on the accelerator, force $\vec{F}_{\text{surface on truck}}$ propels the truck forward. To keep things simple, call this just $\vec{F}_T$. What is the maximum magnitude $\vec{F}_T$ can have without the crate sliding? The static and kinetic coefficients of friction between the crate and the bed of the truck are 0.80 and 0.30. Rolling friction of the truck is negligible.

MODEL The crate and the truck are separate objects that form the system. We'll model them as particles. The earth and the road surface are part of the environment.

VISUALIZE The sketch in **FIGURE 7.17** establishes a coordinate system, lists the known information, and—new to problems of interacting objects—identifies the acceleration constraint. As long as the crate doesn't slip, it must accelerate *with* the truck. Both accelerations are in the positive *x*-direction, so the acceleration constraint in this problem is

$$a_{Cx} = a_{Tx} = a_x$$

The interaction diagram of Figure 7.17 shows the crate interacting twice with the truck—a friction force parallel to the surface of the truck bed and a normal force perpendicular to this surface. The truck interacts similarly with the road surface, but notice that the crate does not interact with the ground; there's no contact between them. The two interactions within the system are each an action/reaction pair, so this is a total of four forces. You can also see four external forces crossing the system boundary, so the free-body diagrams should show a total of eight forces.

Finally, the interaction information is transferred to the free-body diagrams, where we see friction between the crate and truck as an action/reaction pair and the normal forces (the truck pushes up on the crate, the crate pushes down on the truck) as another action/reaction pair. It's easy to overlook forces such as $\vec{f}_{C\text{ on T}}$, but you won't make this mistake if you first identify action/reaction pairs on an interaction diagram. Note that $\vec{f}_{C\text{ on T}}$ and $\vec{f}_{T\text{ on C}}$ are *static* friction forces because they are forces that prevent slipping; force $\vec{f}_{T\text{ on C}}$ must point forward to prevent the crate from sliding out the back of the truck.

SOLVE Now we're ready to write Newton's second law. For the crate:

$$\sum (F_{\text{on crate}})_x = f_{T\text{ on C}} = m_C a_{Cx} = m_C a_x$$

$$\sum (F_{\text{on crate}})_y = n_{T\text{ on C}} - (F_G)_C = n_{T\text{ on C}} - m_C g = 0$$

For the truck:

$$\sum (F_{\text{on truck}})_x = F_T - f_{C\text{ on T}} = m_T a_{Tx} = m_T a_x$$

$$\sum (F_{\text{on truck}})_y = n_T - (F_G)_T - n_{C\text{ on T}}$$
$$= n_T - m_T g - n_{C\text{ on T}} = 0$$

Be sure you agree with all the signs, which are based on the free-body diagrams. The net force in the *y*-direction is zero because there's no motion in the *y*-direction. It may seem like a lot of effort to write all the subscripts, but it is very important in problems with more than one object.

Notice that we've already used the acceleration constraint $a_{Cx} = a_{Tx} = a_x$. Another important piece of information is Newton's third law, which tells us that $f_{C\text{ on T}} = f_{T\text{ on C}}$ and $n_{C\text{ on T}} = n_{T\text{ on C}}$. Finally, we know that the maximum value of F_T will occur when the static friction on the crate reaches its maximum value:

$$f_{T\text{ on C}} = f_{s\text{ max}} = \mu_s n_{T\text{ on C}}$$

FIGURE 7.17 Pictorial representation of the crate and truck in Example 7.4.

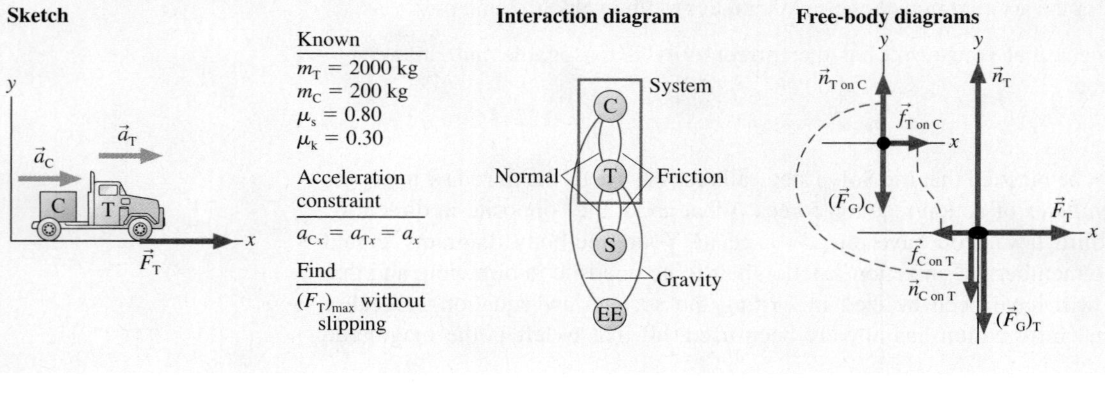

| Sketch | Known | Interaction diagram | Free-body diagrams |

Known
$m_T = 2000$ kg
$m_C = 200$ kg
$\mu_s = 0.80$
$\mu_k = 0.30$

Acceleration constraint
$a_{Cx} = a_{Tx} = a_x$

Find
$(F_T)_{\text{max}}$ without slipping

The friction depends on the normal force on the crate, not the normal force on the truck.

Now we can assemble all the pieces. From the y-equation of the crate, $n_{\text{T on C}} = m_C g$. Thus

$$f_{\text{T on C}} = \mu_s n_{\text{T on C}} = \mu_s m_C g$$

Using this in the x-equation of the crate, we find that the acceleration is

$$a_x = \frac{f_{\text{T on C}}}{m_C} = \mu_s g$$

This is the crate's maximum acceleration without slipping. Now use this acceleration *and* the fact that $f_{\text{C on T}} = f_{\text{T on C}} = \mu_s m_C g$ in the x-equation of the truck to find

$$F_T - f_{\text{C on T}} = F_T - \mu_s m_C g = m_T a_x = m_T \mu_s g$$

Solving for F_T, we find the maximum propulsion without the crate sliding is

$$(F_T)_{\text{max}} = \mu_s (m_T + m_C)g$$

$$= (0.80)(2200 \text{ kg})(9.80 \text{ m/s}^2) = 17{,}000 \text{ N}$$

ASSESS This is a hard result to assess. Few of us have any intuition about the size of forces that propel cars and trucks. Even so, the fact that the forward force on the truck is a significant fraction (80%) of the combined weight of the truck and the crate seems plausible. We might have been suspicious if F_T had been only a tiny fraction of the weight or much greater than the weight.

As you can see, there are many equations and many pieces of information to keep track of when solving a problem of interacting objects. These problems are not inherently harder than the problems you learned to solve in Chapter 6, but they do require a high level of organization. Using the systematic approach of the problem-solving strategy will help you solve similar problems successfully.

STOP TO THINK 7.3 Boxes A and B are sliding to the right across a frictionless table. The hand H is slowing them down. The mass of A is larger than the mass of B. Rank in order, from largest to smallest, the *horizontal* forces on A, B, and H.

a. $F_{\text{B on H}} = F_{\text{H on B}} = F_{\text{A on B}} = F_{\text{B on A}}$

b. $F_{\text{B on H}} = F_{\text{H on B}} > F_{\text{A on B}} = F_{\text{B on A}}$

c. $F_{\text{B on H}} = F_{\text{H on B}} < F_{\text{A on B}} = F_{\text{B on A}}$

d. $F_{\text{H on B}} = F_{\text{H on A}} > F_{\text{A on B}}$

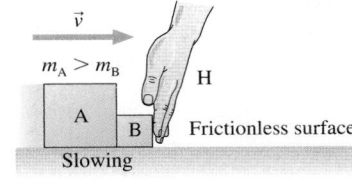

7.4 Ropes and Pulleys

Many objects are connected by strings, ropes, cables, and so on. In single-particle dynamics, we defined *tension* as the force exerted on an object by a rope or string. Now we need to think more carefully about the string itself. Just what do we mean when we talk about the tension "in" a string?

Tension Revisited

FIGURE 7.18a shows a heavy safe hanging from a rope, placing the rope under tension. If you cut the rope, the safe and the lower portion of the rope will fall. Thus there must be a force *within* the rope by which the upper portion of the rope pulls upward on the lower portion to prevent it from falling.

Chapter 5 introduced an atomic-level model in which tension is due to the stretching of spring-like molecular bonds within the rope. Stretched springs exert pulling forces, and the combined pulling force of billions of stretched molecular springs in a string or rope is what we call *tension*.

An important aspect of tension is that it pulls equally *in both directions*. **FIGURE 7.18b** is a very thin cross section through the rope. This small piece of rope is in equilibrium, so it must be pulled equally from both sides. To gain a mental picture, imagine holding your arms outstretched and having two friends pull on them. You'll remain at rest—but "in tension"—as long as they pull with equal strength in opposite directions. But if one lets go, analogous to the breaking of molecular bonds if a rope breaks or is cut, you'll fly off in the other direction!

FIGURE 7.18 Tension forces within the rope are due to stretching the spring-like molecular bonds.

(a)

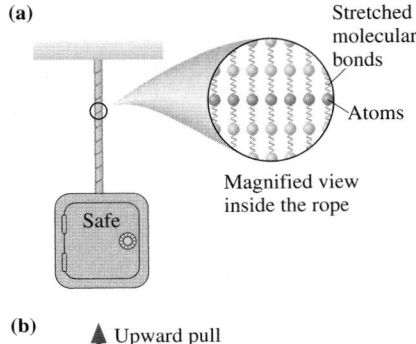

Stretched molecular bonds

Atoms

Magnified view inside the rope

(b)

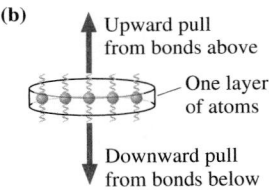

Upward pull from bonds above

One layer of atoms

Downward pull from bonds below

EXAMPLE 7.5 **Pulling a rope**

FIGURE 7.19a shows a student pulling horizontally with a 100 N force on a rope that is attached to a wall. In **FIGURE 7.19b**, two students in a tug-of-war pull on opposite ends of a rope with 100 N each. Is the tension in the second rope larger than, smaller than, or the same as that in the first rope?

FIGURE 7.19 Pulling on a rope. Which produces a larger tension?

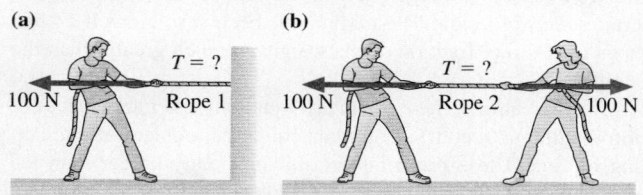

(a)

$T = ?$

100 N   Rope 1

(b)

$T = ?$

100 N   Rope 2   100 N

SOLVE Surely pulling on a rope from both ends causes more tension than pulling on one end. Right? Before jumping to conclusions, let's analyze the situation carefully.

FIGURE 7.20a shows the first student, the rope, and the wall as separate, interacting objects. Force $\vec{F}_{\text{S on R}}$ is the student pulling on the rope, so it has magnitude 100 N. Forces $\vec{F}_{\text{S on R}}$ and $\vec{F}_{\text{R on S}}$ are an action/reaction pair and must have equal magnitudes. Similarly for forces $\vec{F}_{\text{W on R}}$ and $\vec{F}_{\text{R on W}}$. Finally, because the rope is in static equilibrium, force $\vec{F}_{\text{W on R}}$ has to balance force $\vec{F}_{\text{S on R}}$. Thus

$$F_{\text{R on W}} = F_{\text{W on R}} = F_{\text{S on R}} = F_{\text{R on S}} = 100 \text{ N}$$

The first and third equalities are Newton's third law; the second equality is Newton's first law for the rope.

Forces $\vec{F}_{\text{R on S}}$ and $\vec{F}_{\text{R on W}}$ are the pulling forces exerted by the rope and are what we *mean* by "the tension in the rope." Thus the tension in the first rope is 100 N.

FIGURE 7.20b repeats the analysis for the rope pulled by two students. Each student pulls with 100 N, so $F_{\text{S1 on R}} = 100 \text{ N}$ and $F_{\text{S2 on R}} = 100 \text{ N}$. Just as before, there are two action/reaction pairs and the rope is in static equilibrium. Thus

$$F_{\text{R on S2}} = F_{\text{S2 on R}} = F_{\text{S1 on R}} = F_{\text{R on S1}} = 100 \text{ N}$$

The tension in the rope—the pulling forces $\vec{F}_{\text{R on S1}}$ and $\vec{F}_{\text{R on S2}}$—is still 100 N!

You may have *assumed* that the student on the right in Figure 7.19b is doing something to the rope that the wall in Figure 7.19a does not. But our analysis finds that the wall, just like the student, pulls to the right with 100 N. The rope doesn't care whether it's pulled by a wall or a hand. It experiences the same forces in both cases, so the rope's tension is the same in both.

ASSESS Ropes and strings exert forces at *both* ends. The force with which they pull—and thus the force pulling on them at each end—*is* the tension in the rope. Tension is not the sum of the pulling forces.

FIGURE 7.20 Analysis of tension forces.

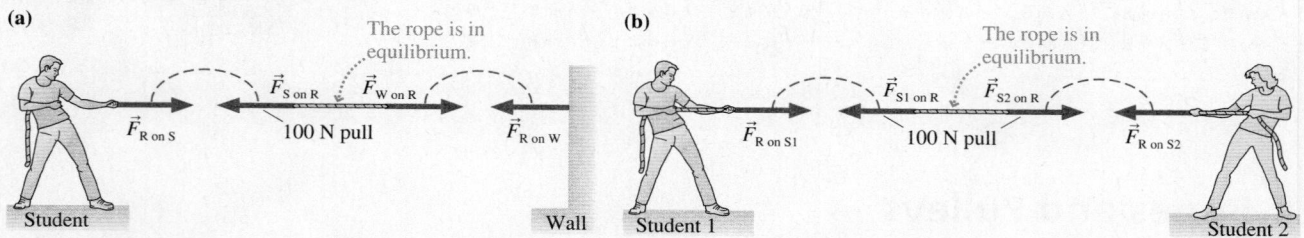

(a)

$\vec{F}_{\text{S on R}}$ $\vec{F}_{\text{W on R}}$

The rope is in equilibrium.

$\vec{F}_{\text{R on S}}$ 100 N pull $\vec{F}_{\text{R on W}}$

Student   Wall

(b)

$\vec{F}_{\text{S1 on R}}$ $\vec{F}_{\text{S2 on R}}$

The rope is in equilibrium.

$\vec{F}_{\text{R on S1}}$ 100 N pull $\vec{F}_{\text{R on S2}}$

Student 1   Student 2

STOP TO THINK 7.4 All three 50 kg blocks are at rest. Is the tension in rope 2 greater than, less than, or equal to the tension in rope 1?

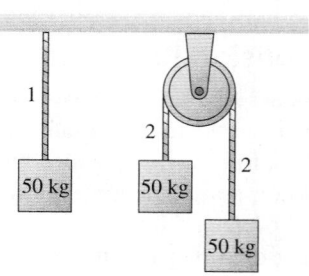

1

2

2

50 kg   50 kg

50 kg

The Massless String Approximation

The tension is constant throughout a rope that is in equilibrium, but what happens if the rope is accelerating? For example, **FIGURE 7.21a** shows two connected blocks being pulled by force $\vec{F}$. Is the string's tension at the right end, where it pulls back on B, the same as the tension at the left end, where it pulls on A?

FIGURE 7.21b shows the horizontal forces acting on the blocks and the string. The only horizontal forces acting on the string are $\vec{T}_{\text{A on S}}$ and $\vec{T}_{\text{B on S}}$, so Newton's second law *for the string* is

$$(F_{\text{net}})_x = T_{\text{B on S}} - T_{\text{A on S}} = m_s a_x \qquad (7.5)$$

where m_s is the mass of the string. If the string is accelerating, then the tensions at the two ends can *not* be the same. The tension at the "front" of the string must be higher than the tension at the "back" in order to accelerate the string!

Often in physics and engineering problems the mass of the string or rope is much less than the masses of the objects that it connects. In such cases, we can adopt the **massless string approximation**. In the limit $m_s \rightarrow 0$, Equation 7.5 becomes

$$T_{\text{B on S}} = T_{\text{A on S}} \qquad \text{(massless string approximation)} \qquad (7.6)$$

In other words, **the tension in a massless string is constant.** This is nice, but it isn't the primary justification for the massless string approximation.

Look again at Figure 7.21b. If $T_{\text{B on S}} = T_{\text{A on S}}$, then

$$\vec{T}_{\text{S on A}} = -\vec{T}_{\text{S on B}} \qquad (7.7)$$

That is, the force on block A is equal and opposite to the force on block B. Forces $\vec{T}_{\text{S on A}}$ and $\vec{T}_{\text{S on B}}$ act *as if* they are an action/reaction pair of forces. Thus we can draw the simplified diagram of FIGURE 7.22 in which the string is missing and blocks A and B interact directly with each other through forces that we can call $\vec{T}_{\text{A on B}}$ and $\vec{T}_{\text{B on A}}$.

In other words, **if objects A and B interact with each other through a massless string, we can omit the string and treat forces $\vec{F}_{\text{A on B}}$ and $\vec{F}_{\text{B on A}}$ as if they are an action/reaction pair.** This is not literally true because A and B are not in contact. Nonetheless, all a massless string does is transmit a force from A to B without changing the magnitude of that force. This is the real significance of the massless string approximation.

NOTE ▶ For problems in this book, you can assume that any strings or ropes are massless unless the problem explicitly states otherwise. The simplified view of Figure 7.22 is appropriate under these conditions. But if the string has a mass, it must be treated as a separate object. ◀

FIGURE 7.21 The string's tension pulls forward on block A, backward on block B.

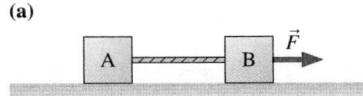

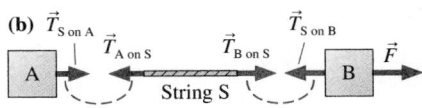

FIGURE 7.22 The massless string approximation allows objects A and B to act *as if* they are directly interacting.

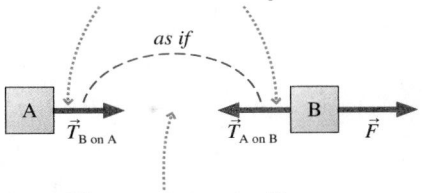

EXAMPLE 7.6 **Comparing two tensions**

Blocks A and B in FIGURE 7.23 are connected by massless string 2 and pulled across a frictionless table by massless string 1. B has a larger mass than A. Is the tension in string 2 larger than, smaller than, or equal to the tension in string 1?

FIGURE 7.23 Blocks A and B are pulled across a frictionless table by massless strings.

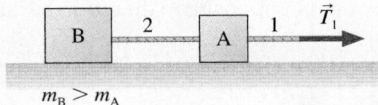

MODEL The massless string approximation allows us to treat A and B *as if* they interact directly with each other. The blocks are accelerating because there's a force to the right and no friction.

SOLVE B has a larger mass, so it may be tempting to conclude that string 2, which pulls B, has a greater tension than string 1, which pulls A. The flaw in this reasoning is that Newton's second law tells us only about the *net* force. The net force on B *is* larger than

the net force on A, but the net force on A is *not* just the tension $\vec{T}_1$ in the forward direction. The tension in string 2 also pulls *backward* on A!

FIGURE 7.24 shows the horizontal forces in this frictionless situation. Forces $\vec{T}_{\text{A on B}}$ and $\vec{T}_{\text{B on A}}$ act *as if* they are an action/reaction pair.

From Newton's third law,

$$T_{\text{A on B}} = T_{\text{B on A}} = T_2$$

where T_2 is the tension in string 2. From Newton's second law, the net force on A is

$$(F_{\text{A net}})_x = T_1 - T_{\text{B on A}} = T_1 - T_2 = m_A a_{\text{Ax}}$$

FIGURE 7.24 The horizontal forces on blocks A and B.

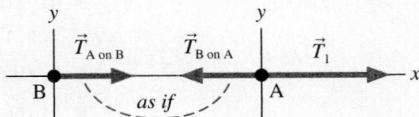

Continued

The net force on A is the *difference* in tensions. The blocks are accelerating to the right, making $a_{Ax} > 0$, so

$$T_1 > T_2$$

The tension in string 2 is *smaller* than the tension in string 1.

ASSESS This is not an intuitively obvious result. A careful study of the reasoning in this example is worthwhile. An alternative analysis would note that $\vec{T}_1$ accelerates *both* blocks, of combined mass $(m_A + m_B)$, whereas $\vec{T}_2$ accelerates only block B. Thus string 1 must have the larger tension.

Pulleys

Strings and ropes often pass over pulleys. The application might be as simple as lifting a heavy weight or as complex as the internal cable-and-pulley arrangement that precisely moves a robot arm.

FIGURE 7.25a shows a simple situation in which block B drags block A across a frictionless table as it falls. FIGURE 7.25b shows the objects separately as well as the forces. As the string moves, static friction between the string and pulley causes the pulley to turn. If we assume that

- The string *and* the pulley are both massless, and
- There is no friction where the pulley turns on its axle,

then no net force is needed to accelerate the string or turn the pulley. In this case,

$$T_{\text{A on S}} = T_{\text{B on S}}$$

In other words, **the tension in a massless string remains constant as it passes over a massless, frictionless pulley.**

FIGURE 7.25 Blocks A and B are connected by a string that passes over a pulley.

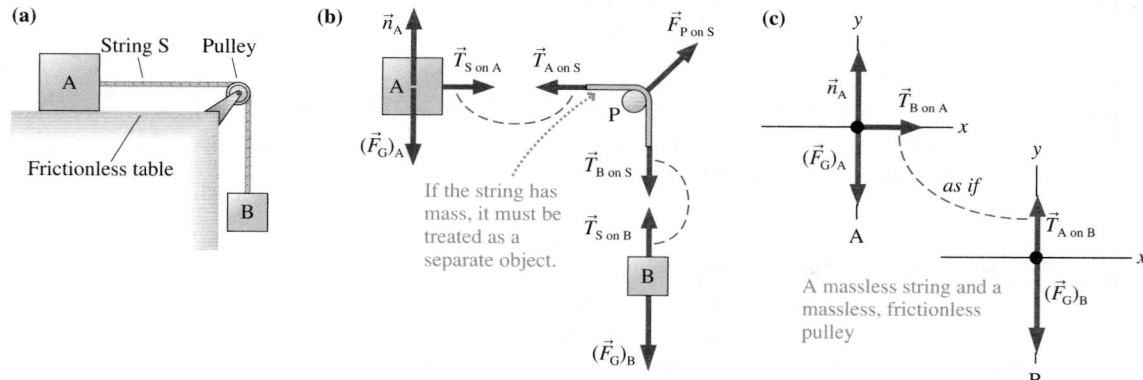

Because of this, we can draw the simplified free-body diagram of FIGURE 7.25c, in which the string and pulley are omitted. Forces $\vec{T}_{\text{A on B}}$ and $\vec{T}_{\text{B on A}}$ act *as if* they are an action/reaction pair, even though they are not opposite in direction because the tension force gets "turned" by the pulley.

TACTICS
BOX 7.2 **Working with ropes and pulleys**

For massless ropes or strings and massless, frictionless pulleys:

- If a force pulls on one end of a rope, the tension in the rope equals the magnitude of the pulling force.
- If two objects are connected by a rope, the tension is the same at both ends.
- If the rope passes over a pulley, the tension in the rope is unaffected.

Exercises 17–22

STOP TO THINK 7.5   In Figure 7.25, is the tension in the string greater than, less than, or equal to the gravitational force acting on block B?

7.5 Examples of Interacting-Objects Problems

We will conclude this chapter with three extended examples. Although the mathematics will be more involved than in any of our work up to this point, we will continue to emphasize the *reasoning* one uses in approaching problems such as these. The solutions will be based on Problem-Solving Strategy 7.1. In fact, these problems are now reaching such a level of complexity that, for all practical purposes, it becomes impossible to work them unless you are following a well-planned strategy. Our earlier emphasis on identifying forces and using free-body diagrams will now really begin to pay off!

EXAMPLE 7.7   **Placing a leg in traction**

Serious fractures of the leg often need a stretching force to keep contracting leg muscles from forcing the broken bones together too hard. This is done using *traction*, an arrangement of a rope, a weight, and pulleys as shown in **FIGURE 7.26**. The rope must make the same angle on both sides of the pulley so that the net force on the leg is horizontal, but the angle can be adjusted to control the amount of traction. The doctor has specified 50 N of traction for this patient with a 4.2 kg hanging mass. What is the proper angle?

FIGURE 7.26 A leg in traction.

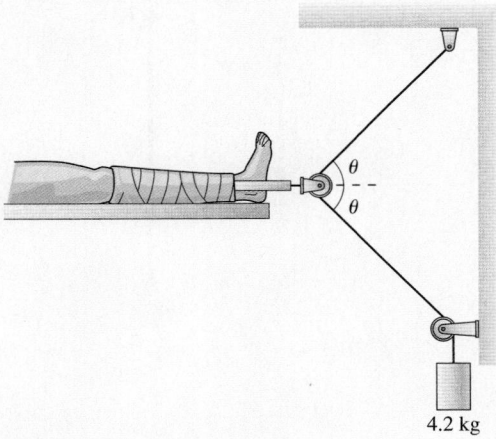

4.2 kg

MODEL Model the leg and the weight as particles. The other point where forces are applied is the pulley attached to the patient's foot, which we'll treat as a separate object. We'll assume massless ropes and a massless, frictionless pulley.

VISUALIZE FIGURE 7.27 shows three free-body diagrams. Forces $\vec{T}_{P1}$ and $\vec{T}_{P2}$ are the tension forces of the rope as it pulls on the pulley. The pulley is in static equilibrium, so these forces are balanced by $\vec{F}_{L\,on\,P}$, which forms an action/reaction pair with the 50 N traction force $\vec{F}_{P\,on\,L}$. Our model of the rope and pulley makes the tension force constant, $T_{P1} = T_{P2} = T_W$, so we'll call it simply T.

FIGURE 7.27 The free-body diagrams.

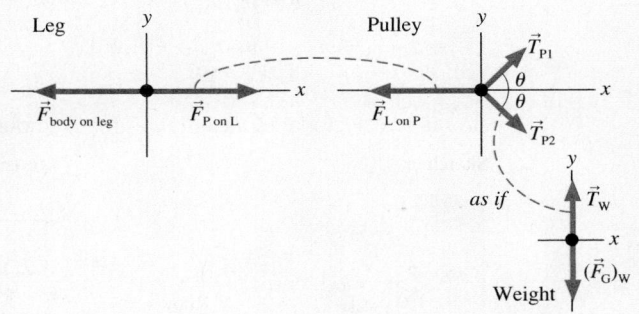

SOLVE The x-component equation of Newton's first law for the pulley is

$$\sum (F_{on\,P})_x = T_{P1}\cos\theta + T_{P2}\cos\theta - F_{L\,on\,P}$$

$$= 2T\cos\theta - F_{L\,on\,P} = 0$$

Thus the correct angle for the ropes is

$$\theta = \cos^{-1}\left(\frac{F_{L\,on\,P}}{2T}\right)$$

We know, from Newton's third law, that $F_{L\,on\,P} = F_{P\,on\,L} = 50$ N. We can determine the tension force by analyzing the weight. It also is in static equilibrium, so the upward tension force exactly balances the downward gravitational force:

$$T = (F_G)_W = m_W g = (4.2\ \text{kg})(9.80\ \text{m/s}^2) = 41\ \text{N}$$

Thus the proper angle is

$$\theta = \cos^{-1}\left(\frac{50\ \text{N}}{2(41\ \text{N})}\right) = 52°$$

ASSESS The traction force would approach 82 N if angle θ approached zero because the two ropes would pull in parallel. Conversely, the traction would approach 0 N if θ approached 90°. The desired traction is roughly midway between these two extremes, so an angle near 45° seems reasonable.

EXAMPLE 7.8 **The show must go on!**

A 200 kg set used in a play is stored in the loft above the stage. The rope holding the set passes up and over a pulley, then is tied backstage. The director tells a 100 kg stagehand to lower the set. When he unties the rope, the set falls and the unfortunate man is hoisted into the loft. What is the stagehand's acceleration?

MODEL The system is the stagehand M and the set S, which we will model as particles. Assume a massless rope and a massless, frictionless pulley.

VISUALIZE FIGURE 7.28 shows the pictorial representation. The man's acceleration a_{My} is positive, while the set's acceleration a_{Sy} is negative. These two accelerations have the same magnitude because the two objects are connected by a rope, but they have opposite signs. Thus the acceleration constraint is $a_{Sy} = -a_{My}$. Forces $\vec{T}_{M\,on\,S}$ and $\vec{T}_{S\,on\,M}$ are not literally an action/reaction pair, but they act *as if* they are because the rope is massless and the pulley is massless and frictionless. Notice that the pulley has "turned" the tension force so that $\vec{T}_{M\,on\,S}$ and $\vec{T}_{S\,on\,M}$ are *parallel* to each other rather than opposite, as members of a true action/reaction pair would have to be.

SOLVE Newton's second law for the man and the set are

$$\sum (F_{on\,M})_y = T_{S\,on\,M} - m_M g = m_M a_{My}$$

$$\sum (F_{on\,S})_y = T_{M\,on\,S} - m_S g = m_S a_{Sy} = -m_S a_{My}$$

Only the y-equations are needed. Notice that we used the acceleration constraint in the last step. Newton's third law is

$$T_{M\,on\,S} = T_{S\,on\,M} = T$$

where we can drop the subscripts and call the tension simply T. With this substitution, the two second-law equations can be written

$$T - m_M g = m_M a_{My}$$

$$T - m_S g = -m_S a_{My}$$

These are simultaneous equations in the two unknowns T and a_{My}. We can eliminate T by subtracting the second equation from the first to give

$$(m_S - m_M)g = (m_S + m_M)a_{My}$$

Finally, we can solve for the hapless stagehand's acceleration:

$$a_{My} = \frac{m_S - m_M}{m_S + m_M}g = \frac{100\ \text{kg}}{300\ \text{kg}}9.80\ \text{m/s}^2 = 3.27\ \text{m/s}^2$$

This is also the acceleration with which the set falls. If the rope's tension was needed, we could now find it from $T = m_M a_{My} + m_M g$.

ASSESS If the stagehand weren't holding on, the set would fall with free-fall acceleration g. The stagehand acts as a *counterweight* to reduce the acceleration.

FIGURE 7.28 Pictorial representation for Example 7.8.

Sketch

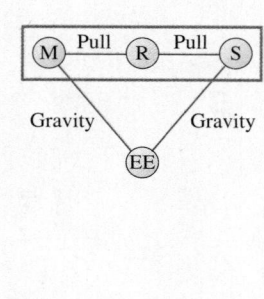

Interaction diagram

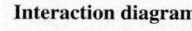

Known
$m_M = 100\ \text{kg}$
$m_S = 200\ \text{kg}$

Acceleration constraint
$a_{Sy} = -a_{My}$

Find
a_{My}

Free-body diagrams

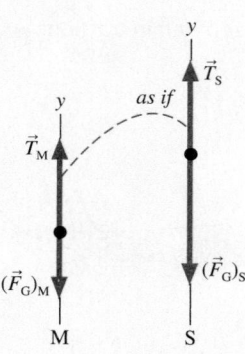

STOP TO THINK 7.6 A small car is pushing a larger truck that has a dead battery. The mass of the truck is larger than the mass of the car. Which of the following statements is true?

a. The car exerts a force on the truck, but the truck doesn't exert a force on the car.
b. The car exerts a larger force on the truck than the truck exerts on the car.
c. The car exerts the same amount of force on the truck as the truck exerts on the car.
d. The truck exerts a larger force on the car than the car exerts on the truck.
e. The truck exerts a force on the car, but the car doesn't exert a force on the truck.

CHALLENGE EXAMPLE 7.9   **A not-so-clever bank robbery**

Bank robbers have pushed a 1000 kg safe to a second-story floor-to-ceiling window. They plan to break the window, then lower the safe 3.0 m to their truck. Not being too clever, they stack up 500 kg of furniture, tie a rope between the safe and the furniture, and place the rope over a pulley. Then they push the safe out the window. What is the safe's speed when it hits the truck? The coefficient of kinetic friction between the furniture and the floor is 0.50.

MODEL   This is a continuation of the situation that we analyzed in Figures 7.16 and 7.25, which are worth reviewing. The system is the safe S and the furniture F, which we will model as particles. We will assume a massless rope and a massless, frictionless pulley.

VISUALIZE   The safe and the furniture are tied together, so their accelerations have the same magnitude. The safe has a y-component of acceleration a_{Sy} that is negative because the safe accelerates in the negative y-direction. The furniture has an x-component a_F that is positive. Thus the acceleration constraint is

$$a_{Fx} = -a_{Sy}$$

The free-body diagrams shown in **FIGURE 7.29** are modeled after Figure 7.25 but now include a kinetic friction force on the furniture. Forces $\vec{T}_{\text{F on S}}$ and $\vec{T}_{\text{S on F}}$ act *as if* they are an action/reaction pair, so they have been connected with a dashed line.

SOLVE   We can write Newton's second law directly from the free-body diagrams. For the furniture,

$$\sum (F_{\text{on F}})_x = T_{\text{S on F}} - f_k = T - f_k = m_F a_{Fx} = -m_F a_{Sy}$$

$$\sum (F_{\text{on F}})_y = n - m_F g = 0$$

And for the safe,

$$\sum (F_{\text{on S}})_y = T - m_S g = m_S a_{Sy}$$

Notice how we used the acceleration constraint in the first equation. We also went ahead and made use of Newton's third law:

$T_{\text{F on S}} = T_{\text{S on F}} = T$. We have one additional piece of information, the model of kinetic friction:

$$f_k = \mu_k n = \mu_k m_F g$$

where we used the y-equation of the furniture to deduce that $n = m_F g$. Substitute this result for f_k into the x-equation of the furniture, then rewrite the furniture's x-equation and the safe's y-equation:

$$T - \mu_k m_F g = -m_F a_{Sy}$$

$$T - m_S g = m_S a_{Sy}$$

We have succeeded in reducing our knowledge to two simultaneous equations in the two unknowns a_{Sy} and T. Subtract the second equation from the first to eliminate T:

$$(m_S - \mu_k m_F)g = -(m_S + m_F)a_{Sy}$$

Finally, solve for the safe's acceleration:

$$a_{Sy} = -\left(\frac{m_S - \mu_k m_F}{m_S + m_F}\right)g$$

$$= -\frac{1000 \text{ kg} - (0.50)(500 \text{ kg})}{1000 \text{ kg} + 500 \text{ kg}}9.80 \text{ m/s}^2 = -4.9 \text{ m/s}^2$$

Now we need to calculate the kinematics of the falling safe. Because the time of the fall is not known or needed, we can use

$$v_{1y}^2 = v_{0y}^2 + 2a_{Sy}\,\Delta y = 0 + 2a_{Sy}(y_1 - y_0) = -2a_{Sy}y_0$$

$$v_1 = \sqrt{-2a_{Sy}y_0} = \sqrt{-2(-4.9 \text{ m/s}^2)(3.0 \text{ m})} = 5.4 \text{ m/s}$$

The value of v_{1y} is negative, but we only needed to find the speed so we took the absolute value. This is about 12 mph, so it seems unlikely that the truck will survive the impact of the 1000 kg safe!

FIGURE 7.29 Pictorial representation for Challenge Example 7.9.

Sketch

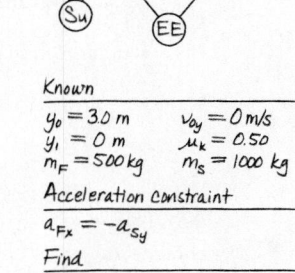

Interaction diagram

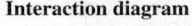

Known
$y_0 = 3.0$ m $v_{0y} = 0$ m/s
$y_1 = 0$ m $\mu_k = 0.50$
$m_F = 500$ kg $m_S = 1000$ kg

Acceleration constraint
$a_{Fx} = -a_{Sy}$

Find
v_1

Free-body diagrams

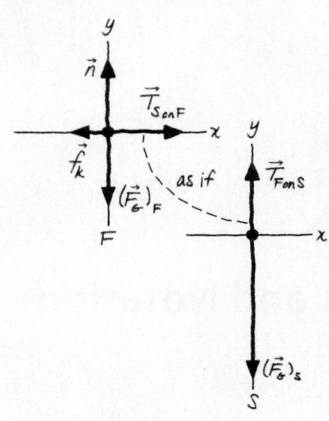

SUMMARY

The goal of Chapter 7 has been to use Newton's third law to understand how objects interact.

General Principles

Newton's Third Law

Every force occurs as one member of an **action/reaction pair** of forces. The two members of an action/reaction pair:

- Act on two *different* objects.
- Are equal in magnitude but opposite in direction:

$$\vec{F}_{\text{A on B}} = -\vec{F}_{\text{B on A}}$$

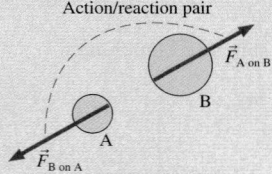

Action/reaction pair

Solving Interacting-Objects Problems

MODEL Choose the objects of interest.

VISUALIZE
Draw a pictorial representation.
  Sketch and define coordinates.
  Identify acceleration constraints.
  Draw an interaction diagram.
  Draw a separate free-body diagram for each object.
  Connect action/reaction pairs with dashed lines.

SOLVE Write Newton's second law for each object.
  Include *all* forces acting *on* each object.
  Use Newton's third law to equate the magnitudes of action/
    reaction pairs.
  Include acceleration constraints and friction.

ASSESS Is the result reasonable?

Important Concepts

Objects, systems, and the environment

Objects whose motion is of interest are the system.
Objects whose motion is not of interest form the environment.
The objects of interest interact with the environment, but those interactions can be considered external forces.

Interaction diagram

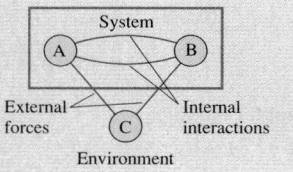

Applications

Acceleration constraints

Objects that are constrained to move together must have accelerations of equal magnitude: $a_\text{A} = a_\text{B}$.
This must be expressed in terms of components, such as $a_{\text{A}x} = -a_{\text{B}y}$.

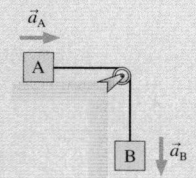

Strings and pulleys

The tension in a string or rope pulls in both directions. The tension is constant in a string if the string is:

- Massless, or
- In equilibrium

Objects connected by massless strings passing over massless, frictionless pulleys act *as if* they interact via an action/reaction pair of forces.

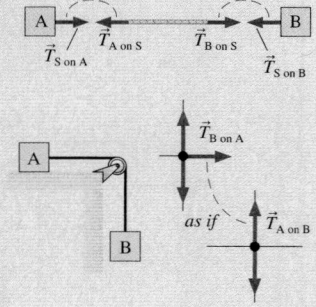

Terms and Notation

interaction	environment	propulsion	acceleration constraint
action/reaction pair	interaction diagram	Newton's third law	massless string approximation
system	external force		

CONCEPTUAL QUESTIONS

1. You find yourself in the middle of a frozen lake with a surface so slippery ($\mu_s = \mu_k = 0$) you cannot walk. However, you happen to have several rocks in your pocket. The ice is extremely hard. It cannot be chipped, and the rocks slip on it just as much as your feet do. Can you think of a way to get to shore? Use pictures, forces, and Newton's laws to explain your reasoning.

2. How do you paddle a canoe in the forward direction? Explain. Your explanation should include diagrams showing forces on the water and forces on the paddle.

3. How does a rocket take off? What is the upward force on it? Your explanation should include diagrams showing forces on the rocket and forces on the parcel of hot gas that was just expelled from the rocket's exhaust.

4. How do basketball players jump straight up into the air? Your explanation should include pictures showing forces on the player and forces on the ground.

5. A mosquito collides head-on with a car traveling 60 mph. Is the force of the mosquito on the car larger than, smaller than, or equal to the force of the car on the mosquito? Explain.

6. A mosquito collides head-on with a car traveling 60 mph. Is the magnitude of the mosquito's acceleration larger than, smaller than, or equal to the magnitude of the car's acceleration? Explain.

7. A small car is pushing a large truck. They are speeding up. Is the force of the truck on the car larger than, smaller than, or equal to the force of the car on the truck?

8. A very smart 3-year-old child is given a wagon for her birthday. She refuses to use it. "After all," she says, "Newton's third law says that no matter how hard I pull, the wagon will exert an equal but opposite force on me. So I will never be able to get it to move forward." What would you say to her in reply?

9. Teams red and blue are having a tug-of-war. According to Newton's third law, the force with which the red team pulls on the blue team exactly equals the force with which the blue team pulls on the red team. How can one team ever win? Explain.

10. Will hanging a magnet in front of the iron cart in FIGURE Q7.10 make it go? Explain.

FIGURE Q7.10

11. FIGURE Q7.11 shows two masses at rest. The string is massless and the pulley is frictionless. The spring scale reads in kg. What is the reading of the scale?

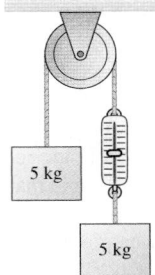

FIGURE Q7.11

12. FIGURE Q7.12 shows two masses at rest. The string is massless and the pulley is frictionless. The spring scale reads in kg. What is the reading of the scale?

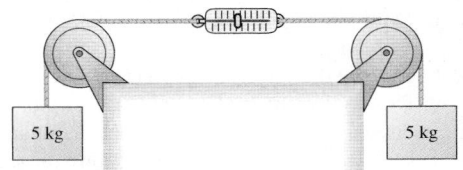

FIGURE Q7.12

13. The hand in FIGURE Q7.13 is pushing on the back of block A. Blocks A and B, with $m_B > m_A$, are connected by a massless string and slide on a frictionless surface. Is the force of the string on B larger than, smaller than, or equal to the force of the hand on A? Explain.

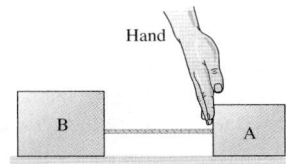

FIGURE Q7.13

14. Blocks A and B in FIGURE Q7.14 are connected by a massless string over a massless, frictionless pulley. The blocks have just been released from rest. Will the pulley rotate clockwise, counterclockwise, or not at all? Explain.

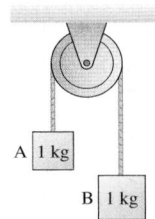

FIGURE Q7.14

15. In case a in FIGURE Q7.15, block A is accelerated across a frictionless table by a hanging 10 N weight (1.02 kg). In case b, block A is accelerated across a frictionless table by a steady 10 N tension in the string. The string is massless, and the pulley is massless and frictionless. Is A's acceleration in case b greater than, less than, or equal to its acceleration in case a? Explain.

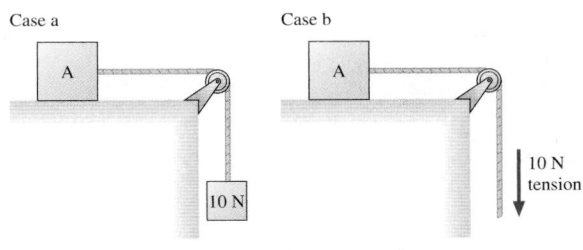

FIGURE Q7.15

EXERCISES AND PROBLEMS

Exercises

Section 7.2 Analyzing Interacting Objects

Exercises 1 through 6 describe a situation. For each:
a. Draw an interaction diagram, following the steps of Tactics Box 7.1.
b. Identify the "system" on your interaction diagram.
c. Draw a free-body diagram for each object in the system. Use dashed lines to connect the members of an action/reaction pair.

1. | A weightlifter stands up at constant speed from a squatting position while holding a heavy barbell across his shoulders.

2. | A soccer ball and a bowling ball have a head-on collision at this instant. Rolling friction is negligible.

3. | A mountain climber is using a rope to pull a bag of supplies up a 45° slope. The rope is not massless.

4. ‖ A battery-powered toy car pushes a stuffed rabbit across the floor.

5. ‖ Block A in **FIGURE EX7.5** is heavier than block B and is sliding down the incline. All surfaces have friction. The rope is massless, and the massless pulley turns on frictionless bearings. The rope and the pulley are among the interacting objects, but you'll have to decide if they're part of the system.

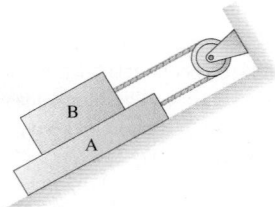

FIGURE EX7.5

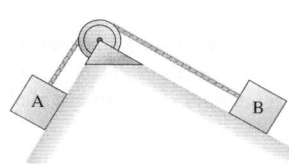

FIGURE EX7.6

6. ‖ Block A in **FIGURE EX7.6** is sliding down the incline. The rope is massless, and the massless pulley turns on frictionless bearings, but the surface is not frictionless. The rope and the pulley are among the interacting objects, but you'll have to decide if they're part of the system.

Section 7.3 Newton's Third Law

7. | a. How much force does an 80 kg astronaut exert on his chair while sitting at rest on the launch pad?
   b. How much force does the astronaut exert on his chair while accelerating straight up at 10 m/s²?

8. ‖ Block B in **FIGURE EX7.8** rests on a surface for which the static and kinetic coefficients of friction are 0.60 and 0.40, respectively. The ropes are massless. What is the maximum mass of block A for which the system is in equilibrium?

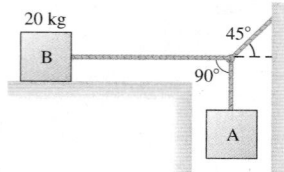

FIGURE EX7.8

9. ‖ A 1000 kg car pushes a 2000 kg truck that has a dead battery. When the driver steps on the accelerator, the drive wheels of the car push against the ground with a force of 4500 N. Rolling friction can be neglected.
   a. What is the magnitude of the force of the car on the truck?
   b. What is the magnitude of the force of the truck on the car?

10. ‖ Blocks with masses of 1 kg, 2 kg, and 3 kg are lined up in a row on a frictionless table. All three are pushed forward by a 12 N force applied to the 1 kg block.
   a. How much force does the 2 kg block exert on the 3 kg block?
   b. How much force does the 2 kg block exert on the 1 kg block?

11. ‖ A massive steel cable drags a 20 kg block across a horizontal, frictionless surface. A 100 N force applied to the cable causes the block to reach a speed of 4.0 m/s in a distance of 2.0 m. What is the mass of the cable?

Section 7.4 Ropes and Pulleys

12. ‖ What is the tension in the rope of **FIGURE EX7.12**?

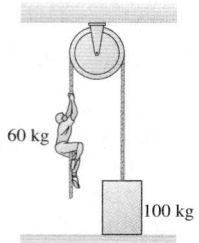

60 kg

100 kg

FIGURE EX7.12

13. ‖ **FIGURE EX7.13** shows two 1.0 kg blocks connected by a rope. A second rope hangs beneath the lower block. Both ropes have a mass of 250 g. The entire assembly is accelerated upward at 3.0 m/s² by force $\vec{F}$.
   a. What is F?
   b. What is the tension at the top end of rope 1?
   c. What is the tension at the bottom end of rope 1?
   d. What is the tension at the top end of rope 2?

$\vec{F}$

A

Rope 1

B

Rope 2

FIGURE EX7.13

14. ‖ Jimmy has caught two fish in Yellow Creek. He has tied the line holding the 3.0 kg steelhead trout to the tail of the 1.5 kg carp. To show the fish to a friend, he lifts upward on the carp with a force of 60 N.
   a. Draw separate free-body diagrams for the trout and the carp. Label all forces, then use dashed lines to connect action/reaction pairs or forces that act as if they are a pair.
   b. Rank in order, from largest to smallest, the magnitudes of all the forces shown on your free-body diagrams. Explain your reasoning.

15. ‖ A 2.0-m-long, 500 g rope pulls a 10 kg block of ice across a horizontal, frictionless surface. The block accelerates at 2.0 m/s². How much force pulls forward on (a) the ice, (b) the rope?

16. ‖ The cable cars in San Francisco are pulled along their tracks by an underground steel cable that moves along at 9.5 mph. The cable is driven by large motors at a central power station and extends, via an intricate pulley arrangement, for several miles beneath the city streets. The length of a cable stretches by up

to 100 ft during its lifetime. To keep the tension constant, the cable passes around a 1.5-m-diameter "tensioning pulley" that rolls back and forth on rails, as shown in FIGURE EX7.16. A 2000 kg block is attached to the tensioning pulley's cart, via a rope and pulley, and is suspended in a deep hole. What is the tension in the cable car's cable?

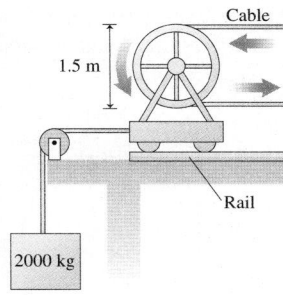

FIGURE EX7.16

17. ‖ A 2.0 kg rope hangs from the ceiling. What is the tension at the midpoint of the rope?

18. ‖ A mobile at the art museum has a 2.0 kg steel cat and a 4.0 kg steel dog suspended from a lightweight cable, as shown in FIGURE EX7.18. It is found that $\theta_1 = 20°$ when the center rope is adjusted to be perfectly horizontal. What are the tension and the angle of rope 3?

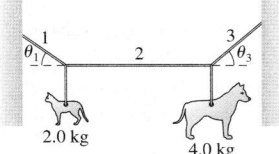

FIGURE EX7.18

Problems

19. ‖‖‖ FIGURE P7.19 shows two strong magnets on opposite sides of a small table. The long-range attractive force between the magnets keeps the lower magnet in place.

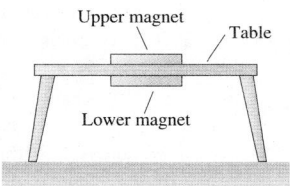

FIGURE P7.19

   a. Draw an interaction diagram and draw free-body diagrams for both magnets and the table. Use dashed lines to connect the members of an action/reaction pair.
   b. Suppose the weight of the table is 20 N, the weight of each magnet is 2.0 N, and the magnetic force on the lower magnet is three times its weight. Find the magnitude of each of the forces shown on your free-body diagrams.

20. ‖ An 80 kg spacewalking astronaut pushes off a 640 kg satellite, exerting a 100 N force for the 0.50 s it takes him to straighten his arms. How far apart are the astronaut and the satellite after 1.0 min?

21. ‖ A massive steel cable drags a 20 kg block across a horizontal, frictionless surface. A 100 N force applied to the cable causes the block to reach a speed of 4.0 m/s in 2.0 s. What is the difference in tension between the two ends of the cable?

22. ‖ FIGURE P7.22 shows a 6.0 N force pushing two gliders along an air track. The 200 g spring between the gliders is compressed. How much force does the spring exert on (a) glider A and (b) glider B?

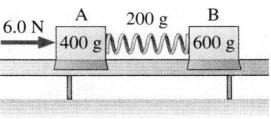

FIGURE P7.22

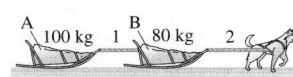

FIGURE P7.23

23. ‖ The sled dog in FIGURE P7.23 drags sleds A and B across the snow. The coefficient of friction between the sleds and the snow is 0.10. If the tension in rope 1 is 150 N, what is the tension in rope 2?

24. ‖ A rope of length L and mass m is suspended from the ceiling. Find an expression for the tension in the rope at position y, measured upward from the free end of the rope.

25. ‖ While driving to work last year, I was holding my coffee mug in my left hand while changing the CD with my right hand. Then the cell phone rang, so I placed the mug on the flat part of my dashboard. Then, believe it or not, a deer ran out of the woods and on to the road right in front of me. Fortunately, my reaction time was zero, and I was able to stop from a speed of 20 m/s in a mere 50 m, just barely avoiding the deer. Later tests revealed that the static and kinetic coefficients of friction of the coffee mug on the dash are 0.50 and 0.30, respectively; the coffee and mug had a mass of 0.50 kg; and the mass of the deer was 120 kg. Did my coffee mug slide?

26. ‖ Two-thirds of the weight of a 1500 kg car rests on the drive wheels. What is the maximum acceleration of this car on a concrete surface?

27. ‖‖‖ A Federation starship (2.0×10^6 kg) uses its tractor beam to pull a shuttlecraft (2.0×10^4 kg) aboard from a distance of 10 km away. The tractor beam exerts a constant force of 4.0×10^4 N on the shuttlecraft. Both spacecraft are initially at rest. How far does the starship move as it pulls the shuttlecraft aboard?

28. ‖ Your forehead can withstand a force of about 6.0 kN before
BIO fracturing, while your cheekbone can withstand only about 1.3 kN. Suppose a 140 g baseball traveling at 30 m/s strikes your head and stops in 1.5 ms.
   a. What is the magnitude of the force that stops the baseball?
   b. What force does the baseball exert on your head? Explain.
   c. Are you in danger of a fracture if the ball hits you in the forehead? On the cheek?

29. ‖ Bob, who has a mass of 75 kg, can throw a 500 g rock with a speed of 30 m/s. The distance through which his hand moves as he accelerates the rock from rest until he releases it is 1.0 m.
   a. What constant force must Bob exert on the rock to throw it with this speed?
   b. If Bob is standing on frictionless ice, what is his recoil speed after releasing the rock?

30. ‖ You see the boy next door trying to push a crate down the sidewalk. He can barely keep it moving, and his feet occasionally slip. You start to wonder how heavy the crate is. You call to ask the boy his mass, and he replies "50 kg." From your recent physics class you estimate that the static and kinetic coefficients of friction are 0.8 and 0.4 for the boy's shoes, and 0.5 and 0.2 for the crate. Estimate the mass of the crate.

31. ‖‖‖ Two packages at UPS start sliding down the 20° ramp shown in FIGURE P7.31. Package A has a mass of 5.0 kg and a coefficient of friction of 0.20. Package B has a mass of 10 kg and a coefficient of friction of 0.15. How long does it take package A to reach the bottom?

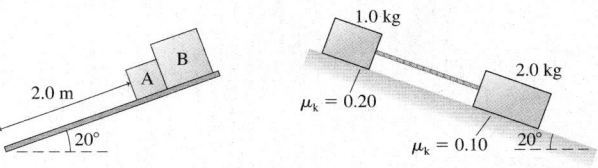

FIGURE P7.31   FIGURE P7.32

32. ‖‖‖ The two blocks in FIGURE P7.32 are sliding down the incline. What is the tension in the massless string?

33. ‖ The 1.0 kg block in FIGURE P7.33 is tied to the wall with a rope. It sits on top of the 2.0 kg block. The lower block is pulled to the right with a tension force of 20 N. The coefficient of kinetic friction at both the lower and upper surfaces of the 2.0 kg block is $\mu_k = 0.40$.
    a. What is the tension in the rope holding the 1.0 kg block to the wall?
    b. What is the acceleration of the 2.0 kg block?

FIGURE P7.33   FIGURE P7.34

34. ‖ The coefficient of static friction is 0.60 between the two blocks in FIGURE P7.34. The coefficient of kinetic friction between the lower block and the floor is 0.20. Force $\vec{F}$ causes both blocks to cross a distance of 5.0 m, starting from rest. What is the least amount of time in which this motion can be completed without the top block sliding on the lower block?

35. ‖‖ The lower block in FIGURE P7.35 is pulled on by a rope with a tension force of 20 N. The coefficient of kinetic friction between the lower block and the surface is 0.30. The coefficient of kinetic friction between the lower block and the upper block is also 0.30. What is the acceleration of the 2.0 kg block?

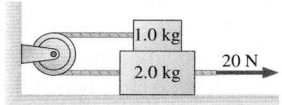

FIGURE P7.35

36. ‖ The block of mass M in FIGURE P7.36 slides on a frictionless surface. Find an expression for the tension in the string.

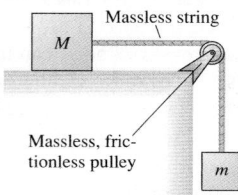

Massless string

Massless, frictionless pulley

FIGURE P7.36

37. ‖‖ A rope attached to a 20 kg wood sled pulls the sled up a 20° snow-covered hill. A 10 kg wood box rides on top of the sled. If the tension in the rope steadily increases, at what value of the tension does the box slip?

38. ‖ The 100 kg block in FIGURE P7.38 takes 6.0 s to reach the floor after being released from rest. What is the mass of the block on the left? The pulley is massless and frictionless.

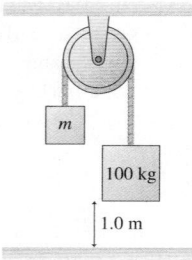   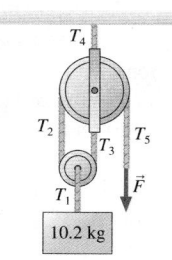

FIGURE P7.38   FIGURE P7.39

39. ‖‖ The 10.2 kg block in FIGURE P7.39 is held in place by a force applied to a rope passing over two massless, frictionless pulleys. Find the tensions T_1 to T_5 and the magnitude of force $\vec{F}$.

40. ‖‖ The coefficient of kinetic friction between the 2.0 kg block in FIGURE P7.40 and the table is 0.30. What is the acceleration of the 2.0 kg block?

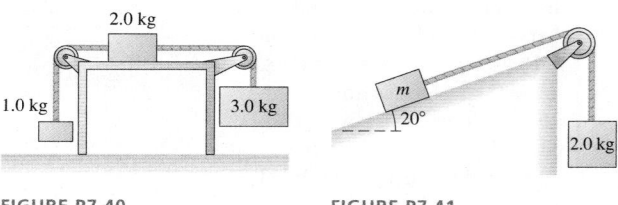

FIGURE P7.40   FIGURE P7.41

41. ‖ FIGURE P7.41 shows a block of mass m resting on a 20° slope. The block has coefficients of friction $\mu_s = 0.80$ and $\mu_k = 0.50$ with the surface. It is connected via a massless string over a massless, frictionless pulley to a hanging block of mass 2.0 kg.
    a. What is the minimum mass m that will stick and not slip?
    b. If this minimum mass is nudged ever so slightly, it will start being pulled up the incline. What acceleration will it have?

42. ‖ A 4.0 kg box is on a frictionless 35° slope and is connected via a massless string over a massless, frictionless pulley to a hanging 2.0 kg weight. The picture for this situation is similar to FIGURE P7.41.
    a. What is the tension in the string if the 4.0 kg box is *held* in place, so that it cannot move?
    b. If the box is then released, which way will it move on the slope?
    c. What is the tension in the string once the box begins to move?

43. ‖ The 1.0 kg physics book in FIGURE P7.43 is connected by a string to a 500 g coffee cup. The book is given a push up the slope and released with a speed of 3.0 m/s. The coefficients of friction are $\mu_s = 0.50$ and $\mu_k = 0.20$.
    a. How far does the book slide?
    b. At the highest point, does the book stick to the slope, or does it slide back down?

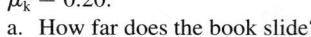

FIGURE P7.43

44. ‖ The 2000 kg cable car shown in FIGURE P7.44 descends a 200-m-high hill. In addition to its brakes, the cable car controls its speed by pulling an 1800 kg

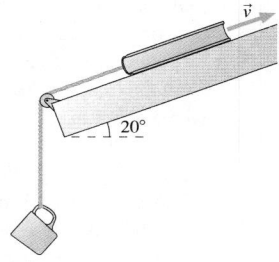

FIGURE P7.44

counterweight up the other side of the hill. The rolling friction of both the cable car and the counterweight are negligible.
    a. How much braking force does the cable car need to descend at constant speed?
    b. One day the brakes fail just as the cable car leaves the top on its downward journey. What is the runaway car's speed at the bottom of the hill?

45. ‖‖ The century-old *ascensores* in Valparaiso, Chile, are small cable cars that go up and down the steep hillsides. As FIGURE P7.45 shows, one car ascends as the other descends. The cars use a two-cable arrangement to compensate for friction; one cable passing around a large pulley connects the cars, the second is pulled by a small motor. Suppose the mass of both cars (with passengers) is 1500 kg, the coefficient of rolling friction is 0.020, and the cars move at constant speed. What is the tension in (a) the connecting cable and (b) the cable to the motor?

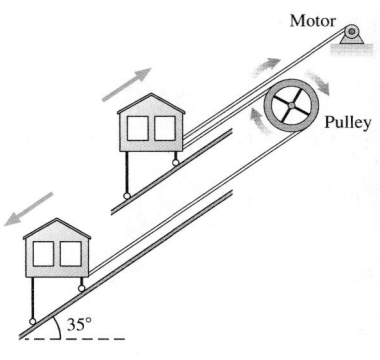

FIGURE P7.45

46. ‖ A house painter uses the chair-and-pulley arrangement of **FIGURE P7.46** to lift himself up the side of a house. The painter's mass is 70 kg and the chair's mass is 10 kg. With what force must he pull down on the rope in order to accelerate upward at 0.20 m/s²?

FIGURE P7.46

FIGURE P7.47

47. ‖‖ Jorge, with mass m, is wearing roller skates whose coefficient of friction with the floor is μ_r. He ties a massless rope around his waist, passes it around a frictionless pulley, and grabs hold of the other end, as shown in **FIGURE P7.47**. Jorge then pulls hand over hand on the rope with a constant force F. Find an expression for Jorge's acceleration toward the wall.

48. ‖‖ A 70 kg tightrope walker stands at the center of a rope. The rope supports are 10 m apart and the rope sags 10° at each end. The tightrope walker crouches down, then leaps straight up with an acceleration of 8.0 m/s² to catch a passing trapeze. What is the tension in the rope as he jumps?

49. ‖ Find an expression for the magnitude of the horizontal force F in **FIGURE P7.49** for which m_1 does not slip either up or down along the wedge. All surfaces are frictionless.

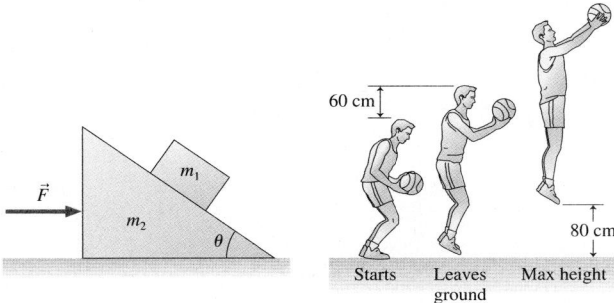

FIGURE P7.49   **FIGURE P7.50**

50. ‖ A 100 kg basketball player can leap straight up in the air to a height of 80 cm, as shown in **FIGURE P7.50**. You can understand how by analyzing the situation as follows:

a. The player bends his legs until the upper part of his body has dropped by 60 cm, then he begins his jump. Draw separate free-body diagrams for the player and for the floor *as* he is jumping, but before his feet leave the ground.
b. Is there a net force on the player as he jumps (before his feet leave the ground)? How can that be? Explain.
c. With what speed must the player leave the ground to reach a height of 80 cm?
d. What was his acceleration, assumed to be constant, as he jumped?
e. Suppose the player jumps while standing on a bathroom scale that reads in newtons. What does the scale read before he jumps, as he is jumping, and after his feet leave the ground?

Problems 51 and 52 show the free-body diagrams of two interacting systems. For each of these, you are to
a. Write a realistic problem for which these are the correct free-body diagrams. Be sure that the answer your problem requests is consistent with the diagrams shown.
b. Finish the solution of the problem.

51.   52.

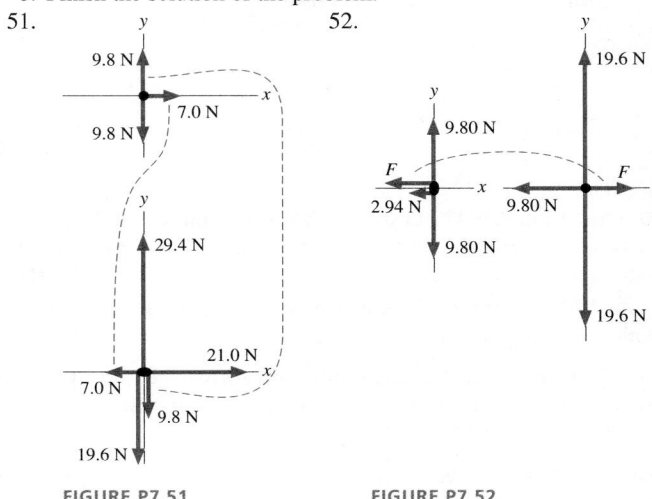

FIGURE P7.51   **FIGURE P7.52**

Challenge Problems

53. A 100 g ball of clay is thrown horizontally with a speed of 10 m/s toward a 900 g block resting on a frictionless surface. It hits the block and sticks. The clay exerts a constant force on the block during the 10 ms it takes the clay to come to rest relative to the block. After 10 ms, the block and the clay are sliding along the surface as a single system.
a. What is their speed after the collision?
b. What is the force of the clay on the block during the collision?
c. What is the force of the block on the clay?

NOTE ▶ This problem can be worked using the conservation laws you will be learning in the next few chapters. However, here you're asked to solve the problem using Newton's laws. ◀

54. In **FIGURE CP7.54**, find an expression for the acceleration of m_1. The pulleys are massless and frictionless.
Hint: Think carefully about the acceleration constraint.

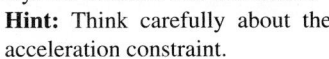

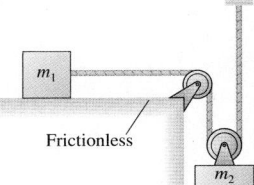

FIGURE CP7.54

55. What is the acceleration of the 2.0 kg block in FIGURE CP7.55 across the frictionless table?

    Hint: Think carefully about the acceleration constraint.

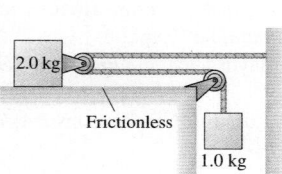

FIGURE CP7.55

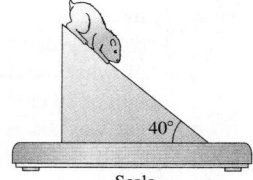

FIGURE CP7.56

56. FIGURE CP7.56 shows a 200 g hamster sitting on an 800 g wedge-shaped block. The block, in turn, rests on a spring scale. An extra-fine lubricating oil having $\mu_s = \mu_k = 0$ is sprayed on the top surface of the block, causing the hamster to slide down. Friction between the block and the scale is large enough that the block does *not* slip on the scale. What does the scale read, in grams, as the hamster slides down?

57. FIGURE CP7.57 shows three hanging masses connected by massless strings over two massless, frictionless pulleys.

    a. Find the acceleration constraint for this system. It is a single equation relating a_{1y}, a_{2y}, and a_{3y}.

    Hint: y_A isn't constant.

    b. Find an expression for the tension in string A.

    Hint: You should be able to write four second-law equations. These, plus the acceleration constraint, are five equations in five unknowns.

    c. Suppose: $m_1 = 2.5$ kg, $m_2 = 1.5$ kg, and $m_3 = 4.0$ kg. Find the acceleration of each.

    d. The 4.0 kg mass would appear to be in equilibrium. Explain why it accelerates.

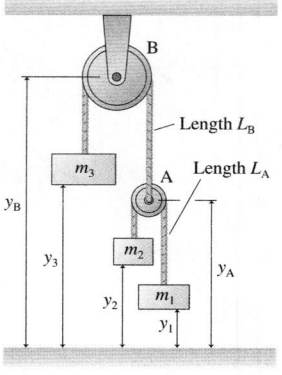

FIGURE CP7.57

STOP TO THINK ANSWERS

Stop to Think 7.1: The crate's gravitational force and the normal force are incorrectly identified as an action/reaction pair. The normal force should be paired with a downward force of the crate on the ground. Gravity is the pull of the entire earth, so $\vec{F}_G$ should be paired with a force pulling up on the entire earth.

Stop to Think 7.2: c. Newton's third law says that the force of A on B is *equal* and opposite to the force of B on A. This is always true. The speed of the objects isn't relevant.

Stop to Think 7.3: b. $F_{B\,on\,H} = F_{H\,on\,B}$ and $F_{A\,on\,B} = F_{B\,on\,A}$ because these are action/reaction pairs. Box B is slowing down and therefore must have a net force to the left. So from Newton's second law we also know that $F_{H\,on\,B} > F_{A\,on\,B}$.

Stop to Think 7.4: Equal to. Each block is hanging in equilibrium, with no net force, so the upward tension force is *mg*.

Stop to Think 7.5: Less than. Block B is *accelerating* downward, so the net force on B must point down. The only forces acting on B are the tension and gravity, so $T_{S\,on\,B} < (F_G)_B$.

Stop to Think 7.6: c. Newton's third law says that the force of A on B is *equal* and opposite to the force of B on A. This is always true. The mass of the objects isn't relevant.

8 Dynamics II: Motion in a Plane

Why doesn't the roller coaster fall off the track at the top of the loop?

▶ **Looking Ahead** The goal of Chapter 8 is to learn how to solve problems about motion in a plane.

Newton's Laws in 2D

This chapter extends Newton's laws to two-dimensional motion in a plane.

One important application is circular motion. You studied the kinematics in Chapter 4; now we want to look at the forces of circular motion.

 You'll learn that the net force on this turning plane is directed toward the center of the circle.

You'll learn to analyze circular motion using a coordinate system with *radial* and *tangential* components—what we'll call the *rtz*-coordinate system.

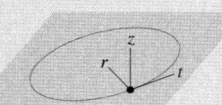

The kinematics of projectile motion was another important topic of Chapter 4. We'll justify those equations and learn how to handle situations where there are forces in addition to gravity.

◀ **Looking Back**
Chapter 4 Kinematics of planar and circular motion

◀ **Looking Back**
Section 6.2 Solving dynamics problems with Newton's second law

Dynamics of Circular Motion

For uniform circular motion, there must be a net force toward the center of the circle to create the centripetal acceleration of changing direction.

Acceleration points toward the center for uniform circular motion at constant speed. You'll learn that non-uniform circular motion has a tangential component of acceleration.

Dynamics in a Plane

Two-dimensional motion with acceleration along both axes often can be analyzed writing Newton's second law in terms of its *x*- and *y*-components.

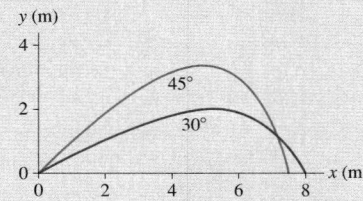

An example is projectile motion with air resistance, for which maximum range no longer occurs for a 45° launch.

Gravity and Orbits

You'll see that an **orbit** can be thought of as projectile motion that never gets any closer to the ground because the ground curves away as fast as the object falls.

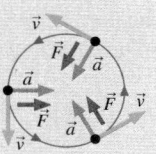 An orbiting projectile is in free fall!

◀ **Looking Back**
Section 6.3 Gravity and weight

Reasoning About Circular Motion

Water in a bucket swung over your head has a downward gravitational pull, but it doesn't fall out.

 These riders at the carnival feel pressed against the wall, yet the force on them points inward, toward the center.

You will learn to understand and explain the physics of these seemingly odd effects.

8.1 Dynamics in Two Dimensions

Newton's second law, $\vec{a} = \vec{F}_{net}/m$, determines an object's acceleration. It makes no distinction between linear motion and two-dimensional motion in a plane. In general, the x- and y-components of the acceleration vector are given by

$$a_x = \frac{(F_{net})_x}{m} \quad \text{and} \quad a_y = \frac{(F_{net})_y}{m} \tag{8.1}$$

Suppose the x- and y-components of acceleration are *independent* of each other. That is, a_x does not depend on either y or v_y, and similarly a_y does not depend on x or v_x. Then Problem-Solving Strategy 6.2 for dynamics problems, on page 142, is still valid. As a quick review, you should

1. Draw a pictorial representation—a sketch and a free-body diagram.
2. Use Newton's second law in component form:

$$(F_{net})_x = \sum F_x = ma_x \quad \text{and} \quad (F_{net})_y = \sum F_y = ma_y$$

The force components (including proper signs) are found from the free-body diagram.
3. Solve for the acceleration. If the acceleration is constant, use the two-dimensional kinematic equations of Chapter 4 to find velocities and positions.

EXAMPLE 8.1 **Rocketing in the wind**

A small rocket for gathering weather data has a mass of 30 kg and generates 1500 N of thrust. On a windy day, the wind exerts a 20 N horizontal force on the rocket. If the rocket is launched straight up, what is the shape of its trajectory, and by how much has it been deflected sideways when it reaches a height of 1.0 km? Because the rocket goes much higher than this, assume there's no significant mass loss during the first 1.0 km of flight.

MODEL Model the rocket as a particle. We need to find the *function* $y(x)$ describing the curve the rocket follows. Because rockets have pointy, aerodynamic shapes, we'll assume no vertical air resistance.

VISUALIZE **FIGURE 8.1** shows a pictorial representation. We've chosen a coordinate system with a vertical y-axis. Three forces act on the rocket: two vertical and one horizontal. The wind force is essentially drag (the rocket is moving sideways relative to the wind), so we've labeled it $\vec{D}$.

SOLVE The vertical and horizontal forces are independent of each other, so we can follow the problem-solving strategy summarized

above. Newton's second law is

$$a_x = \frac{(F_{net})_x}{m} = \frac{D}{m}$$

$$a_y = \frac{(F_{net})_y}{m} = \frac{F_{thrust} - mg}{m}$$

Both accelerations are constant, so we can use kinematics to find

$$x = \tfrac{1}{2}a_x(\Delta t)^2 = \frac{D}{2m}(\Delta t)^2$$

$$y = \tfrac{1}{2}a_y(\Delta t)^2 = \frac{F_{thrust} - mg}{2m}(\Delta t)^2$$

where we used the fact that all initial positions and velocities are zero. From the x-equation, $(\Delta t)^2 = 2mx/D$. Substituting this into the y-equation, we find

$$y(x) = \left[\frac{F_{thrust} - mg}{D}\right]x$$

This is the equation of the rocket's trajectory. It is a linear equation. Somewhat surprisingly, given that the rocket has both vertical and horizontal accelerations, its trajectory is a *straight line*. We can rearrange this result to find the deflection at height y:

$$x = \left[\frac{D}{F_{thrust} - mg}\right]y$$

From the data provided, we can calculate a deflection of 17 m at a height of 1000 m.

ASSESS The solution depended on the fact that the time parameter Δt is the *same* for both components of the motion.

FIGURE 8.1 Pictorial representation of the rocket launch.

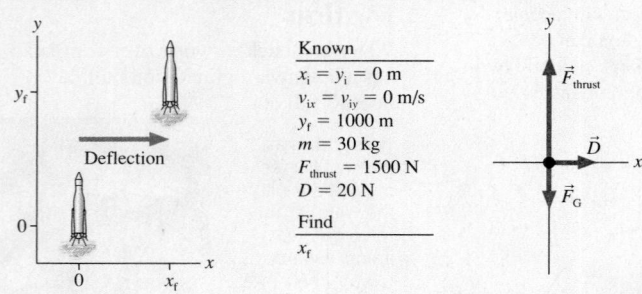

Known
$x_i = y_i = 0$ m
$v_{ix} = v_{iy} = 0$ m/s
$y_f = 1000$ m
$m = 30$ kg
$F_{thrust} = 1500$ N
$D = 20$ N

Find
x_f

Projectile Motion

We found in Chapter 6 that the gravitational force on an object near the surface of a planet is $\vec{F}_G = (mg, \text{down})$. If we choose a coordinate system with a vertical y-axis, then

$$\vec{F}_G = -mg\,\hat{j} \qquad (8.2)$$

Consequently, from Newton's second law, the acceleration is

$$a_x = \frac{(F_G)_x}{m} = 0$$

$$a_y = \frac{(F_G)_y}{m} = -g \qquad (8.3)$$

Equations 8.3 justify the accelerations of Chapter 4—a downward acceleration $a_y = -g$ with no horizontal acceleration—that led to the parabolic motion of a drag-free projectile. The vertical motion is free fall, while the horizontal motion is one of constant velocity.

However, the situation is quite different for a low-mass projectile, where the effects of drag are too large to ignore. We'll leave it as a homework problem for you to show that the acceleration of a projectile subject to drag is

$$a_x = -\frac{\rho CA}{2m} v_x \sqrt{v_x^2 + v_y^2}$$

$$a_y = -g - \frac{\rho CA}{2m} v_y \sqrt{v_x^2 + v_y^2} \qquad (8.4)$$

Here the components of acceleration are *not* independent of each other because a_x depends on v_y and vice versa. It turns out that these two equations cannot be solved exactly for the trajectory, but they can be solved numerically. **FIGURE 8.2** shows the numerical solution for the motion of a 5 g plastic ball that's been hit with an initial speed of 25 m/s. It doesn't travel very far (the maximum distance would be more than 60 m in a vacuum), and the maximum range is no longer reached for a launch angle of 45°. In this case, maximum distance is achieved by hitting the ball at a 30° angle. A 60° launch angle, which gives the same distance as 30° in vacuum, travels only $\approx 75\%$ as far. Notice that the trajectories are not at all parabolic.

STOP TO THINK 8.1   This acceleration will cause the particle to

a. Speed up and curve upward.   b. Speed up and curve downward.
c. Slow down and curve upward.   d. Slow down and curve downward.
e. Move to the right and down.   f. Reverse direction.

When drag is included, the angle for maximum range of a projectile depends both on its size and mass. The optimum angle is roughly 35° for baseballs. The flight of a golf ball is even more complex because the dimples and the high rate of spin greatly affect its aerodynamics. Professional golfers achieve their maximum distance at launch angles of barely 15°.

FIGURE 8.2 A projectile is affected by drag. This example shows trajectories of a plastic ball launched at different angles.

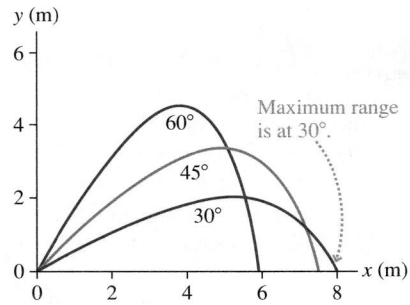

8.2 Uniform Circular Motion

We studied the mathematics of circular motion in Chapter 4, and a review is *highly* recommended. Recall that a particle in uniform circular motion with angular velocity ω has speed $v = \omega r$ and centripetal acceleration

$$\vec{a} = \left(\frac{v^2}{r}, \text{toward center of circle}\right) = (\omega^2 r, \text{toward center of circle}) \qquad (8.5)$$

Now we're ready to study *dynamics*—how forces *cause* circular motion.

The xy-coordinate system we've been using for linear motion and projectile motion is not the best coordinate system for circular dynamics. **FIGURE 8.3** shows a circular

FIGURE 8.3 The rtz-coordinate system.

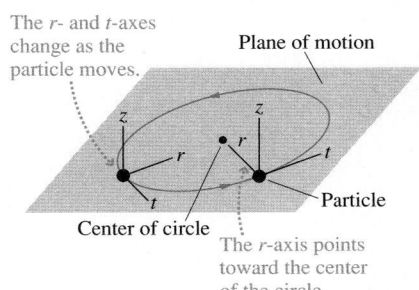

The r- and t-axes change as the particle moves.

Plane of motion

Particle

Center of circle

The r-axis points toward the center of the circle.

trajectory and the plane in which the circle lies. Let's establish a coordinate system with its origin at the point where the particle is located. The axes are defined as follows:

- The *r*-axis (radial axis) points *from* the particle *toward* the center of the circle.
- The *t*-axis (tangential axis) is tangent to the circle, pointing in the ccw direction.
- The *z*-axis is perpendicular to the plane of motion.

The three axes of this *rtz*-coordinate system are mutually perpendicular, just like the axes of the familiar *xyz*-coordinate system. Notice how the axes move with the particle so that the *r*-axis always points to the center of the circle. It will take a little getting used to, but you will soon see that circular-motion problems are most easily described in these coordinates.

FIGURE 8.4, from Chapter 4, reminds you that a particle in uniform circular motion has a velocity tangential to the circle and an acceleration—the centripetal acceleration—pointing toward the center of the circle. Thus the *rtz*-components of $\vec{v}$ and $\vec{a}$ are

$$
\begin{array}{ll}
v_r = 0 & a_r = \dfrac{v^2}{r} = \omega^2 r \\[2mm]
v_t = \omega r & a_t = 0 \qquad\qquad (8.6) \\[2mm]
v_z = 0 & a_z = 0
\end{array}
$$

where $\omega = d\theta/dt$, the angular velocity, must be in rad/s. In other words, **the velocity vector has only a tangential component, the acceleration vector has only a radial component.** Now you can begin to see the advantages of the *rtz*-coordinate system. For convenience, we'll generally refer to a_r as "the centripetal acceleration" rather than "the radial acceleration."

FIGURE 8.4 The velocity and acceleration vectors in the *rtz*-coordinate system.

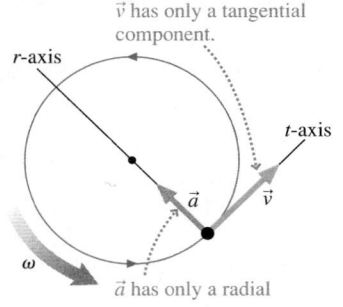

$\vec{v}$ has only a tangential component.

r-axis

t-axis

$\vec{a}$　$\vec{v}$

ω

$\vec{a}$ has only a radial component.

NOTE ▶ Recall that ω is positive for a counterclockwise (ccw) rotation, negative for a clockwise (cw) rotation. Hence the tangential velocity v_t is positive/negative for ccw/cw rotations. Because v_t is the only nonzero component of velocity, the particle's speed is $v = |v_t| = |\omega| r$. We'll sometimes write this as $v = \omega r$ if there's no ambiguity about the signs. ◀

EXAMPLE 8.2　**The ultracentrifuge**

A 17-cm-diameter ultracentrifuge produces an extraordinarily large acceleration of 600,000g, where g is the free-fall acceleration. What is the rotational frequency in rpm? What is the speed of a bacterium at the bottom of the centrifuge tube?

SOLVE The radius of the circular motion is 8.5 cm, or 0.085 m. From Equations 8.6, we see that the angular velocity is

$$
\omega = \sqrt{\frac{a_r}{r}} = \sqrt{\frac{(600,000)(9.8 \text{ m/s}^2)}{0.085 \text{ m}}} = 8320 \text{ rad/s}
$$

Converting to rpm, we find

$$
\omega = 8320 \,\frac{\text{rad}}{\text{s}} \times \frac{1 \text{ rev}}{2\pi \text{ rad}} \times \frac{60 \text{ s}}{1 \text{ min}} = 80,000 \text{ rpm}
$$

This incredibly fast rotation rate is why it's called an *ultra*centrifuge. The tubes pivot outward as the centrifuge spins, so a bacterium at the "bottom" of the tube is rotating at the end of an arm of radius 8.5 cm. Its speed is

$$
v = \omega r = (8320 \text{ rad/s})(0.085 \text{ m}) = 710 \text{ m/s}
$$

This is more than twice the speed of sound!

STOP TO THINK 8.2　Rank in order, from largest to smallest, the centripetal accelerations $(a_r)_a$ to $(a_r)_e$ of particles a to e.

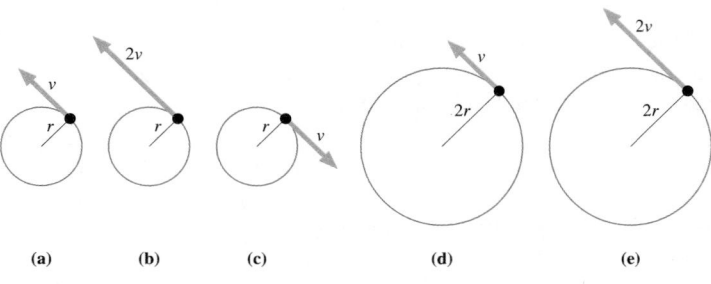

(a)　(b)　(c)　(d)　(e)

Dynamics of Uniform Circular Motion

A particle in uniform circular motion is clearly not traveling at constant velocity in a straight line. Consequently, according to Newton's first law, the particle *must* have a net force acting on it. We've already determined the acceleration of a particle in uniform circular motion—the centripetal acceleration of Equation 8.5. Newton's second law tells us exactly how much net force is needed to cause this acceleration:

$$\vec{F}_{net} = m\vec{a} = \left(\frac{mv^2}{r}, \text{toward center of circle}\right) \qquad (8.7)$$

Highway and racetrack curves are banked to allow the normal force of the road to provide the centripetal acceleration of the turn.

In other words, a particle of mass m moving at constant speed v around a circle of radius r *must* have a net force of magnitude mv^2/r pointing toward the center of the circle. Without such a force, the particle would move off in a straight line tangent to the circle.

FIGURE 8.5 shows the net force $\vec{F}_{net}$ acting on a particle as it undergoes uniform circular motion. You can see that $\vec{F}_{net}$ **points along the radial axis of the *rtz*-coordinate system, toward the center of the circle.** The tangential and perpendicular components of $\vec{F}_{net}$ are zero.

FIGURE 8.5 The net force points in the radial direction, toward the center of the circle.

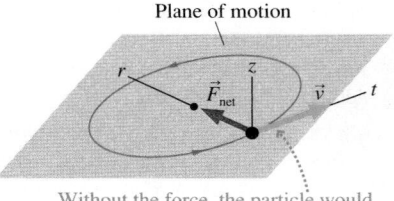

NOTE ▶ The force described by Equation 8.7 is not a *new* force. Our rules for identifying forces have not changed. What we are saying is that a particle moves with uniform circular motion *if and only if* a net force always points toward the center of the circle. The force itself must have an identifiable agent and will be one of our familiar forces, such as tension, friction, or the normal force. Equation 8.7 simply tells us how the force needs to act—how strongly and in which direction—to cause the particle to move with speed v in a circle of radius r. ◀

The usefulness of the *rtz*-coordinate system becomes apparent when we write Newton's second law, Equation 8.7, in terms of the r-, t-, and z-components:

$$(F_{net})_r = \sum F_r = ma_r = \frac{mv^2}{r} = m\omega^2 r$$

$$(F_{net})_t = \sum F_t = ma_t = 0 \qquad (8.8)$$

$$(F_{net})_z = \sum F_z = ma_z = 0$$

Notice that we've used our explicit knowledge of the acceleration, as given in Equations 8.6, to write the right-hand sides of these equations. **For uniform circular motion, the sum of the forces along the *t*-axis and along the *z*-axis *must* equal zero, and the sum of the forces along the *r*-axis must equal ma_r, where a_r is the centripetal acceleration.**

A few examples will clarify these ideas and show how some of the forces you've come to know can be involved in circular motion.

EXAMPLE 8.3   **Spinning in a circle**

An energetic father places his 20 kg child on a 5.0 kg cart to which a 2.0-m-long rope is attached. He then holds the end of the rope and spins the cart and child around in a circle, keeping the rope parallel to the ground. If the tension in the rope is 100 N, how many revolutions per minute (rpm) does the cart make? Rolling friction between the cart's wheels and the ground is negligible.

MODEL Model the child in the cart as a particle in uniform circular motion.

VISUALIZE FIGURE 8.6 on the next page shows the pictorial representation. A circular-motion problem usually does not have starting and ending points like a projectile problem, so numerical subscripts such as x_1 or y_2 are usually not needed. Here we need to define the cart's speed v and the radius r of the circle. Further, a motion diagram is not needed for uniform circular motion because we already know the acceleration $\vec{a}$ points to the center of the circle.

Continued

FIGURE 8.6 Pictorial representation of a cart spinning in a circle.

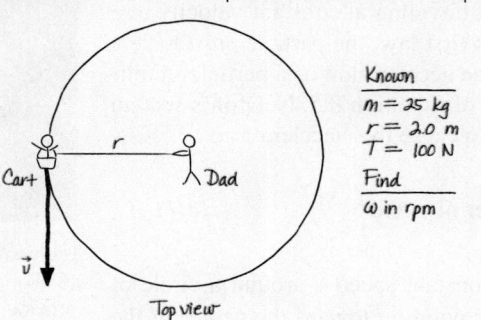

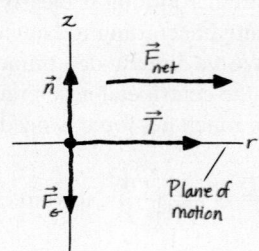

The essential part of the pictorial representation is the free-body diagram. **For uniform circular motion we'll draw the free-body diagram in the rz-plane, looking at the edge of the circle, because this is the plane of the forces.** The contact forces acting on the cart are the normal force of the ground and the tension force of the rope. The normal force is perpendicular to the plane of the motion and thus in the z-direction. The direction of $\vec{T}$ is determined by the statement that the rope is parallel to the ground. In addition, there is the long-range gravitational force $\vec{F}_G$.

SOLVE We defined the r-axis to point toward the center of the circle, so $\vec{T}$ points in the positive r-direction and has r-component $T_r = T$. Newton's second law, using the rtz-components of Equations 8.8, is

$$\sum F_r = T = \frac{mv^2}{r}$$

$$\sum F_z = n - mg = 0$$

We've taken the r- and z-components of the forces directly from the free-body diagram, as you learned to do in Chapter 6. Then

we've *explicitly* equated the sums to $a_r = v^2/r$ and $a_z = 0$. This is the basic strategy for all uniform circular-motion problems. From the z-equation we can find that $n = mg$. This would be useful if we needed to determine a friction force, but it's not needed in this problem. From the r-equation, the speed of the cart is

$$v = \sqrt{\frac{rT}{m}} = \sqrt{\frac{(2.0 \text{ m})(100 \text{ N})}{25 \text{ kg}}} = 2.83 \text{ m/s}$$

The cart's angular velocity is then found from Equations 8.6:

$$\omega = \frac{v_t}{r} = \frac{v}{r} = \frac{2.83 \text{ m/s}}{2.0 \text{ m}} = 1.41 \text{ rad/s}$$

This is another case where we inserted the radian unit because ω is specifically an *angular* velocity. Finally, we need to convert ω to rpm:

$$\omega = \frac{1.41 \text{ rad}}{1 \text{ s}} \times \frac{1 \text{ rev}}{2\pi \text{ rad}} \times \frac{60 \text{ s}}{1 \text{ min}} = 14 \text{ rpm}$$

ASSESS 14 rpm corresponds to a period $T = 4.3$ s. This result is reasonable.

EXAMPLE 8.4 **Turning the corner I**

What is the maximum speed with which a 1500 kg car can make a left turn around a curve of radius 50 m on a level (unbanked) road without sliding?

MODEL Although the car turns only a quarter of a circle, we can model the car as a particle in uniform circular motion as it goes around the turn. Assume that rolling friction is negligible.

VISUALIZE FIGURE 8.7 shows the pictorial representation. The issue we must address is *how* a car turns a corner. What force or forces cause the direction of the velocity vector to change? Imagine driving on a completely frictionless road, such as a very icy road. You would not be able to turn a corner. Turning the steering wheel would be of no use; the car would slide straight ahead, in accordance with both Newton's first law and the experience of anyone

who has ever driven on ice! So it must be *friction* that somehow allows the car to turn.

Figure 8.7 shows the top view of a tire as it turns a corner. If the road surface were frictionless, the tire would slide straight ahead. The force that prevents an object from sliding across a surface is *static friction*. Static friction $\vec{f}_s$ pushes *sideways* on the tire, toward the center of the circle. How do we know the direction is sideways? If $\vec{f}_s$ had a component either parallel to $\vec{v}$ or opposite to $\vec{v}$, it would cause the car to speed up or slow down. Because the car changes direction but not speed, static friction must be perpendicular to $\vec{v}$. $\vec{f}_s$ causes the centripetal acceleration of circular motion around the curve, and thus the free-body diagram, drawn from behind the car, shows the static friction force pointing toward the center of the circle.

FIGURE 8.7 Pictorial representation of a car turning a corner.

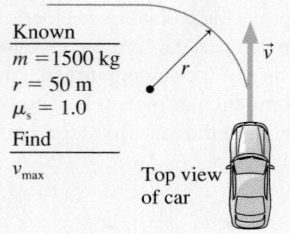

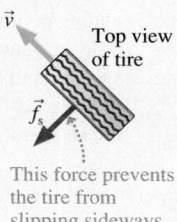

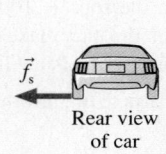

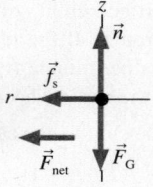

Known
$m = 1500$ kg
$r = 50$ m
$\mu_s = 1.0$
Find
v_{max}

Top view of car

Top view of tire

This force prevents the tire from slipping sideways.

Rear view of car

SOLVE The maximum turning speed is reached when the static friction force reaches its maximum $f_{s\,max} = \mu_s n$. If the car enters the curve at a speed higher than the maximum, static friction will not be large enough to provide the necessary centripetal acceleration and the car will slide.

The static friction force points in the positive r-direction, so its radial component is simply the magnitude of the vector: $(f_s)_r = f_s$. Newton's second law in the rtz-coordinate system is

$$\sum F_r = f_s = \frac{mv^2}{r}$$

$$\sum F_z = n - mg = 0$$

The only difference from Example 8.3 is that the tension force toward the center has been replaced by a static friction force toward the center. From the radial equation, the speed is

$$v = \sqrt{\frac{rf_s}{m}}$$

The speed will be a maximum when f_s reaches its maximum value:

$$f_s = f_{s\,max} = \mu_s n = \mu_s mg$$

where we used $n = mg$ from the z-equation. At that point,

$$v_{max} = \sqrt{\frac{rf_{s\,max}}{m}} = \sqrt{\mu_s rg}$$

$$= \sqrt{(1.0)(50\ \text{m})(9.80\ \text{m/s}^2)} = 22\ \text{m/s}$$

where the coefficient of static friction was taken from Table 6.1.

ASSESS 22 m/s ≈ 45 mph, a reasonable answer for how fast a car can take an unbanked curve. Notice that the car's mass canceled out and that the final equation for v_{max} is quite simple. This is another example of why it pays to work algebraically until the very end.

Because μ_s depends on road conditions, the maximum safe speed through turns can vary dramatically. Wet roads, in particular, lower the value of μ_s and thus lower the speed of turns. Icy conditions are even worse. The corner you turn every day at 45 mph will require a speed of no more than 15 mph if the coefficient of static friction drops to 0.1.

EXAMPLE 8.5 | **Turning the corner II**

A highway curve of radius 70 m is banked at a 15° angle. At what speed v_0 can a car take this curve without assistance from friction?

MODEL The car is a particle in uniform circular motion.

VISUALIZE Having just discussed the role of friction in turning corners, it is perhaps surprising to suggest that the same turn can also be accomplished without friction. Example 8.4 considered a level roadway, but real highway curves are *banked* by being tilted up at the outside edge of the curve. The angle is modest on ordinary highways, but it can be quite large on high-speed racetracks. The purpose of banking becomes clear if you look at the free-body diagram in FIGURE 8.8. The normal force $\vec{n}$ is perpendicular to the road, so tilting the road causes $\vec{n}$ to have a component toward the center of the circle. **The radial component n_r is the inward force that causes the centripetal acceleration needed to turn the car.** Notice that we are *not* using a tilted coordinate system, although

this looks rather like an inclined-plane problem. The center of the circle is in the same horizontal plane as the car, and for circular-motion problems we need the r-axis to pass through the center. Tilted axes are for *linear* motion along an incline.

SOLVE Without friction, $n_r = n \sin\theta$ is the only component of force in the radial direction. It is this inward component of the normal force on the car that causes it to turn the corner. Newton's second law is

$$\sum F_r = n \sin\theta = \frac{mv_0^2}{r}$$

$$\sum F_z = n \cos\theta - mg = 0$$

where θ is the angle at which the road is banked and we've assumed that the car is traveling at the correct speed v_0. From the z-equation,

$$n = \frac{mg}{\cos\theta}$$

Substituting this into the r-equation and solving for v_0 give

$$\frac{mg}{\cos\theta} \sin\theta = mg \tan\theta = \frac{mv_0^2}{r}$$

$$v_0 = \sqrt{rg \tan\theta} = 14\ \text{m/s}$$

ASSESS This is ≈ 28 mph, a reasonable speed. Only at this very specific speed can the turn be negotiated without reliance on friction forces.

FIGURE 8.8 Pictorial representation of a car on a banked curve.

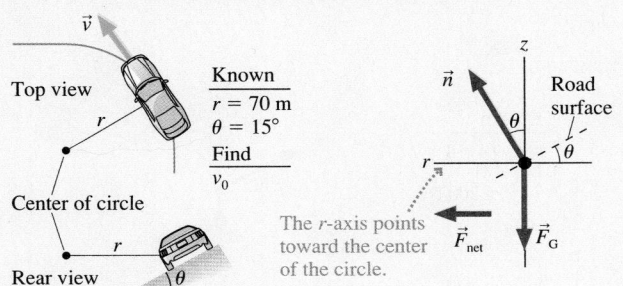

It's interesting to explore what happens at other speeds on a banked curve. FIGURE 8.9 shows that the car will need to rely on both the banking *and* friction if it takes the curve at a speed higher or lower than v_0.

FIGURE 8.9 Free-body diagrams for a car going around a banked curve at speeds higher and lower than the friction-free speed v_0.

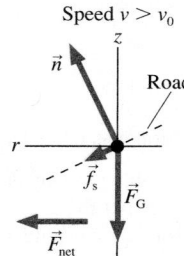

Speed $v > v_0$

Road surface

Static friction must point downhill: A faster speed requires a larger net force toward the center of the circle. The radial component of static friction adds to n_r to allow the car to make the turn. Maximum speed occurs when the static friction force reaches its maximum.

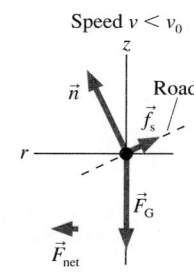

Speed $v < v_0$

Road surface

Static friction must point uphill: Without a static friction force *up* the slope, a slow-moving car would slide *down* the incline! Further, n_r is too much radial force for circular motion at $v < v_0$. Here the radial component of static friction reduces the net radial force.

EXAMPLE 8.6 A rock in a sling

A Stone Age hunter places a 1.0 kg rock in a sling and swings it in a horizontal circle around his head on a 1.0-m-long vine. If the vine breaks at a tension of 200 N, what is the maximum angular speed, in rpm, with which he can swing the rock?

MODEL Model the rock as a particle in uniform circular motion.

VISUALIZE This problem appears, at first, to be essentially the same as Example 8.3, where the father spun his child around on a rope. However, the lack of a normal force from a supporting surface makes a *big* difference. In this case, the *only* contact force on the rock is the tension in the vine. Because the rock moves in a horizontal circle, you may be tempted to draw a free-body diagram like FIGURE 8.10a, where $\vec{T}$ is directed along the r-axis. You will quickly run into trouble, however, because this diagram has a net force in the z-direction and it is impossible to satisfy $\sum F_z = 0$. The gravitational force $\vec{F}_G$ certainly points vertically downward, so the difficulty must be with $\vec{T}$.

As an experiment, tie a small weight to a string, swing it over your head, and check the *angle* of the string. You will quickly discover that the string is *not* horizontal but, instead, is angled downward. The sketch of FIGURE 8.10b labels the angle θ. Notice that the rock moves in a *horizontal* circle, so the center of the circle is *not* at his hand. The r-axis points to the center of the circle, but the tension force is directed along the vine. Thus the correct free-body diagram is the one in Figure 8.10b.

SOLVE The free-body diagram shows that the downward gravitational force is balanced by an upward component of the tension, leaving the radial component of the tension to cause the centripetal acceleration. Newton's second law is

$$\sum F_r = T\cos\theta = \frac{mv^2}{r}$$

$$\sum F_z = T\sin\theta - mg = 0$$

where θ is the angle of the vine below horizontal. From the z-equation we find

$$\sin\theta = \frac{mg}{T}$$

$$\theta = \sin^{-1}\left(\frac{(1.0 \text{ kg})(9.8 \text{ m/s}^2)}{200 \text{ N}}\right) = 2.81°$$

where we've evaluated the angle at the maximum tension of 200 N. The vine's angle of inclination is small but not zero.

Turning now to the r-equation, we find the rock's speed is

$$v = \sqrt{\frac{rT\cos\theta}{m}}$$

Careful! The radius r of the circle is *not* the length L of the vine. You can see in Figure 8.10b that $r = L\cos\theta$. Thus

$$v = \sqrt{\frac{LT\cos^2\theta}{m}} = \sqrt{\frac{(1.0 \text{ m})(200 \text{ N})(\cos 2.81°)^2}{1.0 \text{ kg}}} = 14.1 \text{ m/s}$$

We can now find the maximum angular speed, the value of ω that brings the tension to the breaking point:

$$\omega_{max} = \frac{v}{r} = \frac{v}{L\cos\theta} = \frac{14.1 \text{ rad}}{1 \text{ s}} \times \frac{1 \text{ rev}}{2\pi \text{ rad}} \times \frac{60 \text{ s}}{1 \text{ min}} = 135 \text{ rpm}$$

FIGURE 8.10 Pictorial representation of a rock in a sling.

(a)

Wrong diagram!

(b)

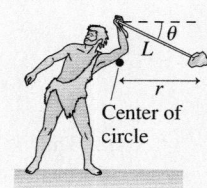

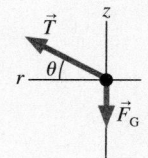

Known
$m = 1.0 \text{ kg}$
$L = 1.0 \text{ m}$
$T_{max} = 200 \text{ N}$

Find
ω_{max}

STOP TO THINK 8.3   A block on a string spins in a horizontal circle on a frictionless table. Rank order, from largest to smallest, the tensions T_a to T_e acting on blocks a to e.

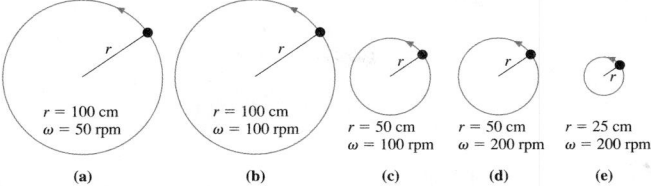

(a) $r = 100$ cm $\omega = 50$ rpm

(b) $r = 100$ cm $\omega = 100$ rpm

(c) $r = 50$ cm $\omega = 100$ rpm

(d) $r = 50$ cm $\omega = 200$ rpm

(e) $r = 25$ cm $\omega = 200$ rpm

8.3 Circular Orbits

Satellites orbit the earth, the earth orbits the sun, and our entire solar system orbits the center of the Milky Way galaxy. Not all orbits are circular, but in this section we'll limit our analysis to circular orbits. We'll look at the elliptical orbits of satellites and planets in Chapter 13.

How does a satellite orbit the earth? What forces act on it? Why does it move in a circle? To answer these important questions, let's return, for a moment, to projectile motion. Projectile motion occurs when the only force on an object is gravity. Our analysis of projectiles assumed that the earth is flat and that the acceleration due to gravity is everywhere straight down. This is an acceptable approximation for projectiles of limited range, such as baseballs or cannon balls, but there comes a point where we can no longer ignore the curvature of the earth.

FIGURE 8.11 shows a perfectly smooth, spherical, airless planet with one tower of height h. A projectile is launched from this tower parallel to the ground ($\theta = 0°$) with speed v_0. If v_0 is very small, as in trajectory A, the "flat-earth approximation" is valid and the problem is identical to Example 4.4 in which a car drove off a cliff. The projectile simply falls to the ground along a parabolic trajectory.

As the initial speed v_0 is increased, the projectile begins to notice that the ground is curving out from beneath it. It is falling the entire time, always getting closer to the ground, but the distance that the projectile travels before finally reaching the ground—that is, its range—increases because the projectile must "catch up" with the ground that is curving away from it. Trajectories B and C are of this type. The actual calculation of these trajectories is beyond the scope of this textbook, but you should be able to understand the factors that influence the trajectory.

If the launch speed v_0 is sufficiently large, there comes a point where the curve of the trajectory and the curve of the earth are parallel. In this case, the projectile "falls" but it never gets any closer to the ground! This is the situation for trajectory D. A closed trajectory around a planet or star, such as trajectory D, is called an **orbit**.

The most important point of this qualitative analysis is that **an orbiting projectile is in free fall**. This is, admittedly, a strange idea, but one worth careful thought. An orbiting projectile is really no different from a thrown baseball or a car driving off a cliff. The only force acting on it is gravity, but its tangential velocity is so large that the curvature of its trajectory matches the curvature of the earth. When this happens, the projectile "falls" under the influence of gravity but never gets any closer to the surface, which curves away beneath it.

In the flat-earth approximation, shown in **FIGURE 8.12a**, the gravitational force acting on an object of mass m is

$$\vec{F}_G = (mg, \text{vertically downward}) \quad \text{(flat-earth approximation)} \quad (8.9)$$

But since stars and planets are actually spherical (or very close to it), the "real" force of gravity acting on an object is directed toward the *center* of the planet, as shown in **FIGURE 8.12b**. In this case the gravitational force is

$$\vec{F}_G = (mg, \text{toward center}) \quad \text{(spherical planet)} \quad (8.10)$$

FIGURE 8.11 Projectiles being launched at increasing speeds from height h on a smooth, airless planet.

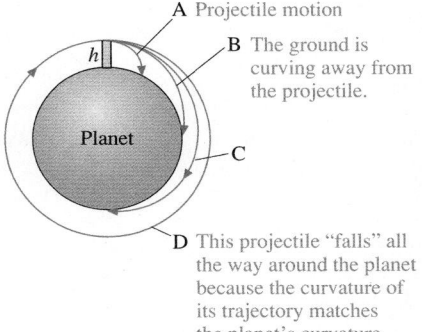

A Projectile motion

B The ground is curving away from the projectile.

C

D This projectile "falls" all the way around the planet because the curvature of its trajectory matches the planet's curvature.

FIGURE 8.12 The "real" gravitational force is always directed toward the center of the planet.

(a)

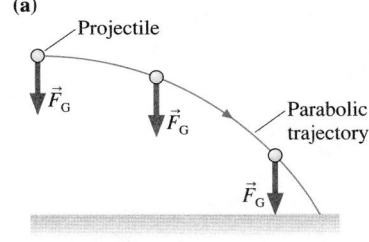

Projectile

$\vec{F}_G$

$\vec{F}_G$

Parabolic trajectory

$\vec{F}_G$

Flat-earth approximation

(b)

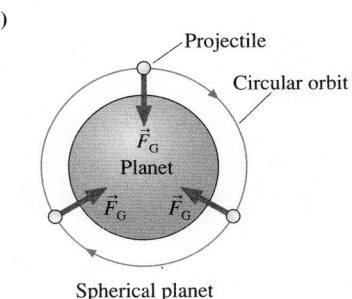

Projectile

Circular orbit

$\vec{F}_G$

Planet

$\vec{F}_G$ $\vec{F}_G$

Spherical planet

The orbiting space shuttle is in free fall.

As you have learned, a force of constant magnitude that always points toward the center of a circle causes the centripetal acceleration of uniform circular motion. Thus the gravitational force of Equation 8.10 on the object in Figure 8.12b causes it to have acceleration

$$\vec{a} = \frac{\vec{F}_{net}}{m} = (g, \text{ toward center}) \tag{8.11}$$

An object moving in a circle of radius r at speed v_{orbit} will have this centripetal acceleration if

$$a_r = \frac{(v_{orbit})^2}{r} = g \tag{8.12}$$

That is, if an object moves parallel to the surface with the speed

$$v_{orbit} = \sqrt{rg} \tag{8.13}$$

then the free-fall acceleration provides exactly the centripetal acceleration needed for a circular orbit of radius r. An object with any other speed will not follow a circular orbit.

The earth's radius is $r = R_e = 6.37 \times 10^6$ m. (A table of useful astronomical data is inside the back cover of this book.) The orbital speed of a projectile just skimming the surface of an airless, bald earth is

$$v_{orbit} = \sqrt{rg} = \sqrt{(6.37 \times 10^6 \text{ m})(9.80 \text{ m/s}^2)} = 7900 \text{ m/s} \approx 16{,}000 \text{ mph}$$

Even if there were no trees and mountains, a real projectile moving at this speed would burn up from the friction of air resistance.

Suppose, however, that we launched the projectile from a tower of height $h = 200$ mi $\approx 3.2 \times 10^5$ m, just above the earth's atmosphere. This is approximately the height of low-earth-orbit satellites, such as the space shuttle. Note that $h \ll R_e$, so the radius of the orbit $r = R_e + h = 6.69 \times 10^6$ m is only 5% greater than the earth's radius. Many people have a mental image that satellites orbit far above the earth, but in fact many satellites come pretty close to skimming the surface. Our calculation of v_{orbit} thus turns out to be quite a good estimate of the speed of a satellite in low earth orbit.

We can use v_{orbit} to calculate the period of a satellite orbit:

$$T = \frac{2\pi r}{v_{orbit}} = 2\pi \sqrt{\frac{r}{g}} \tag{8.14}$$

For a low earth orbit, with $r = R_e + 200$ miles, we find $T = 5190$ s $= 87$ min. The period of the space shuttle at an altitude of 200 mi is, indeed, close to 87 minutes. (The actual period of the shuttle at this elevation is 91 min. The difference, you'll learn in Chapter 13, arises because g is slightly less at a satellite's altitude.)

When we discussed *weightlessness* in Chapter 6, we discovered that it occurs during free fall. We asked the question, at the end of Section 6.3, whether astronauts and their spacecraft were in free fall. We can now give an affirmative answer: They are, indeed, in free fall. They are falling continuously around the earth, under the influence of only the gravitational force, but never getting any closer to the ground because the earth's surface curves beneath them. Weightlessness in space is no different from the weightlessness in a free-falling elevator. It does *not* occur from an absence of gravity. Instead, the astronaut, the spacecraft, and everything in it are weightless because they are all falling together.

Gravity

We can leave this section with a glance ahead, where we will look at the gravitational force more closely. If a satellite is simply "falling" around the earth, with the gravitational force causing a centripetal acceleration, then what about the moon? Is it obeying the same laws of physics? Or do celestial objects obey laws that we cannot discover by experiments here on earth?

The radius of the moon's orbit around the earth is $r = R_m = 3.84 \times 10^8$ m. If we use Equation 8.14 to calculate the period of the moon's orbit, the time it takes the moon to circle the earth once, we get

$$T = 2\pi \sqrt{\frac{r}{g}} = 2\pi \sqrt{\frac{3.84 \times 10^8 \text{ m}}{9.80 \text{ m/s}^2}} = 655 \text{ min} \approx 11 \text{ h}$$

This is clearly wrong. As you probably know, the full moon occurs roughly once a month. More exactly, we know from astronomical measurements that the period of the moon's orbit is $T = 27.3$ days $= 2.36 \times 10^6$ s, a factor of 60 longer than we calculated it to be.

Newton believed that the laws of motion he had discovered were *universal*. That is, they should apply to the motion of the moon as well as to the motion of objects in the laboratory. But why should we assume that the free-fall acceleration g is the same at the distance of the moon as it is on or near the earth's surface? If gravity is the force of the earth pulling on an object, it seems plausible that the size of that force, and thus the size of g, should diminish with increasing distance from the earth.

If the moon orbits the earth because of the earth's gravitational pull, what value of g would be needed to explain the moon's period? We can calculate $g_{\text{at moon}}$ from Equation 8.14 and the observed value of the moon's period:

$$g_{\text{at moon}} = \frac{4\pi^2 R_m}{T_{\text{moon}}^2} = 0.00272 \text{ m/s}^2$$

This is much less than the earth-bound value of 9.80 m/s^2.

As you learned in Chapter 6, Newton proposed the idea that the earth's force of gravity decreases inversely with the square of the distance from the earth. In Chapter 13, we'll use Newton's law of gravity, the mass of the earth, and the distance to the moon to *predict* that $g_{\text{at moon}} = 0.00272 \text{ m/s}^2$, exactly as expected. The moon, just like the space shuttle, is simply "falling" around the earth!

8.4 Fictitious Forces

If you are riding in a car that makes a sudden stop, you may feel as if a force "throws" you forward toward the windshield. But there really is no such force. You cannot identify any agent that does the throwing. An observer watching from beside the road would simply see you continuing forward as the car stops.

The decelerating car is not an inertial reference frame. You learned in Chapter 5 that Newton's laws are valid only in inertial reference frames. The roadside observer is in the earth's inertial reference frame. His observations of the car decelerating relative to the earth while you continue forward with constant velocity are in accord with Newton's laws.

Nonetheless, the fact that you *seem* to be hurled forward relative to the car is a very real experience. You can describe your experience in terms of what are called **fictitious forces**. These are not real forces because no agent is exerting them, but they describe your motion *relative to a noninertial reference frame*. FIGURE 8.13 shows the situation from both reference frames.

Saturn's beautiful rings consist of dust particles and small rocks orbiting the planet.

FIGURE 8.13 The forces are properly identified only in an inertial reference frame.

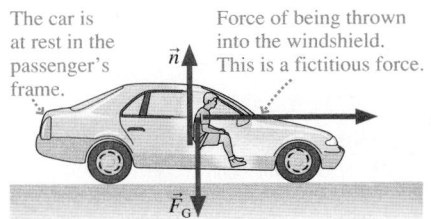

Noninertial reference frame of passenger

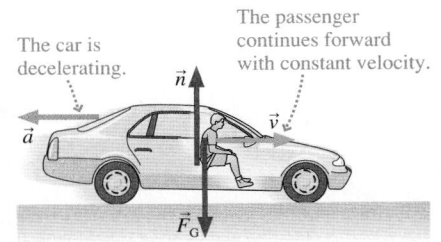

Inertial reference frame of the ground

FIGURE 8.14 Bird's-eye view of a passenger as a car turns a corner.

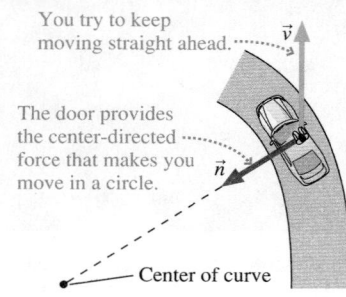

You try to keep moving straight ahead. $\vec{v}$

The door provides the center-directed force that makes you move in a circle. $\vec{n}$

Center of curve

Centrifugal Force?

If the car turns a corner quickly, you feel "thrown" against the door. But is there really such a force? FIGURE 8.14 shows a bird's-eye view of you riding in a car as it makes a left turn. You try to continue moving in a straight line, obeying Newton's first law, when—without having been provoked—the door suddenly turns in front of you and runs into you! You do, indeed, then feel the force of the door because it is now the normal force of the door, pointing *inward* toward the center of the curve, causing you to turn the corner. But you were not "thrown" into the door; the door ran into you. The bird's-eye view, from an inertial reference frame, gives the proper perspective of what happens.

The "force" that seems to push an object to the outside of a circle is called the *centrifugal force.* Despite having a name, the centrifugal force is a fictitious force. It describes your experience *relative to a noninertial reference frame,* but there really is no such force. **You must always use Newton's laws in an inertial reference frame.** There are no centrifugal forces in an inertial reference frame.

> NOTE ▶ You might wonder if the *rtz*-coordinate system is an inertial reference frame. It is, and Newton's laws apply, although the reason is rather subtle. We're using the *rtz*-coordinates to establish directions for decomposing vectors, but we're not making measurements in the *rtz*-system. That is, velocities and accelerations are measured in the laboratory reference frame. The particle would always be at rest ($\vec{v} = \vec{0}$) if we measured velocities in a reference frame attached to the particle. Thus the analysis of this chapter really is in an inertial reference frame. ◀

Gravity on a Rotating Earth

There is one small problem with the admonition that you must use Newton's laws in an inertial reference frame: A reference frame attached to the ground isn't truly inertial because of the earth's rotation. Fortunately, we can make a simple correction that allows us to continue using Newton's laws on the earth's surface.

FIGURE 8.15 shows an object being weighed by a spring scale on the earth's equator. An observer hovering in an inertial reference frame above the north pole sees two forces on the object: the gravitational force $\vec{F}_{M \text{ on } m}$, given by Newton's law of gravity, and the outward spring force $\vec{F}_{sp}$. The object moves in a circle as the earth rotates—it's accelerating—and circular motion *requires* a net force directed toward the center of the circle. The gravitational force points toward the center of the circle, the spring force points away, so Newton's second law is

$$\sum F_r = F_{M \text{ on } m} - F_{sp} = m\omega^2 R$$

where ω is the angular speed of the rotating earth. The spring-scale reading $F_{sp} = F_{M \text{ on } m} - m\omega^2 R$ is *less* than it would be on a nonrotating earth.

The blow-up in Figure 8.15 shows how we see things in a noninertial, flat-earth reference frame. Here the object is at rest, in static equilibrium. If we insist on using Newton's laws, we have to conclude that $\vec{F}_{net} = \vec{0}$ and hence the upward spring force must be exactly balanced by a downward gravitational force $\vec{F}_G$. Thus the spring-scale reading is $F_{sp} = F_G$.

Now both we and the hovering, inertial observer measure the same spring compression and read the same number on the scale. If F_{sp} is the same for both of us, then

$$F_G = F_{M \text{ on } m} - m\omega^2 R \tag{8.15}$$

In other words, force $\vec{F}_G$—what we called the *effective* gravitational force in Chapter 6—is slightly less than the true gravitational force $\vec{F}_{M \text{ on } m}$ because of the earth's rotation. In essence, $m\omega^2 R$ is the centrifugal force, a fictitious force trying—from our perspective in a noninertial reference frame—to "throw" us off the rotating platform.

FIGURE 8.15 The earth's rotation affects the measured value of *g*.

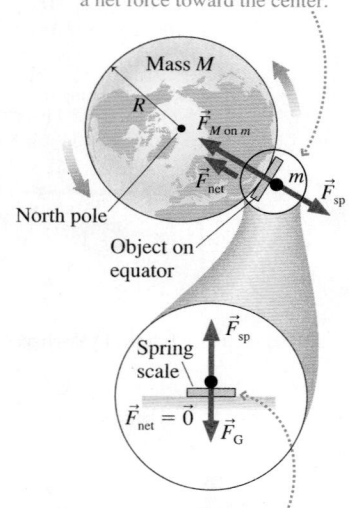

The object is in circular motion on a rotating earth, so there is a net force toward the center.

Mass M

R

$\vec{F}_{M \text{ on } m}$

$\vec{F}_{net}$

North pole

m

$\vec{F}_{sp}$

Object on equator

$\vec{F}_{sp}$

Spring scale

$\vec{F}_{net} = \vec{0}$

$\vec{F}_G$

The object is in static equilibrium in our reference frame on the rotating earth.

This has the effect of "weakening" gravity. There really is no such force, but—this is the important point—**we can continue to use Newton's laws in our rotating reference frame if we pretend there is.**

Because $F_G = mg$ for an object at rest, the effect of the centrifugal term in Equation 8.15 is to make g a little smaller than it would be on a nonrotating earth:

$$g = \frac{F_G}{m} = \frac{F_{M \, \text{on} \, m} - m\omega^2 R}{m} = \frac{GM}{R^2} - \omega^2 R = g_{\text{earth}} - \omega^2 R \qquad (8.16)$$

We calculated $g_{\text{earth}} = 9.83 \text{ m/s}^2$ in Chapter 6. Using $\omega = 1$ rev/day (which must be converted to SI units) and $R = 6370$ km, we find $\omega^2 R = 0.033 \text{ m/s}^2$ at the equator. Thus the free-fall acceleration—what we actually measure in our rotating reference frame—is about 9.80 m/s^2. The purely gravitational acceleration g_{earth} has been reduced by the centripetal acceleration of our rotation.

Things are a little more complicated at other latitudes, but the bottom line is that we can safely use Newton's laws in our rotating, noninertial reference frame on the earth's surface if we calculate the gravitational force—as we've been doing—as $F_G = mg$ with g the measured free-fall value, a value that compensates for our rotation, rather than the purely gravitational g_{earth}.

Why Does the Water Stay in the Bucket?

If you swing a bucket of water over your head quickly, the water stays in, but you'll get a shower if you swing too slowly. Why? We'll answer this question by starting with an equivalent situation, a roller coaster doing a loop-the-loop.

FIGURE 8.16a shows a roller-coaster car going around a vertical loop-the-loop of radius r. Why doesn't the car fall off at the top of the circle? FIGURE 8.16b shows the car's free-body diagrams at the top and bottom of the loop. Now, motion in a vertical circle is *not* uniform circular motion; the car slows down as it goes up one side and speeds up as it comes back down the other. But at the very top and very bottom points, only the car's direction is changing, not its speed, so at those points the acceleration is purely centripetal.

Because the car is moving in a circle, there must be a net force toward the center of the circle. First consider the very bottom of the loop. The only forces acting on the car are the gravitational force $\vec{F}_G$ and the normal force $\vec{n}$ of the track pushing up on it, so a net force toward the center—upward at this point—requires $n > F_G$. The normal force has to *exceed* the gravitational force to provide the net force needed to "turn the corner" at the bottom of the circle. This is why you "feel heavy" at the bottom of the circle or at the bottom of a valley on a roller coaster.

We can analyze the situation quantitatively by writing the r-component of Newton's second law. At the bottom of the circle, with the r-axis pointing upward, we have

$$\sum F_r = n_r + (F_G)_r = n - mg = ma_r = \frac{m(v_{\text{bot}})^2}{r} \qquad (8.17)$$

From Equation 8.17 we find

$$n = mg + \frac{m(v_{\text{bot}})^2}{r} \qquad (8.18)$$

The normal force at the bottom is *larger* than mg.

Things are a little trickier as the roller-coaster car crosses the top of the loop. Whereas the normal force of the track pushes up when the car is at the bottom of the circle, it *presses down* when the car is at the top and the track is above the car. Think about the free-body diagram to make sure you agree.

The car is still moving in a circle, so there *must* be a net force toward the center of the circle to provide the centripetal acceleration. The r-axis, which points toward the

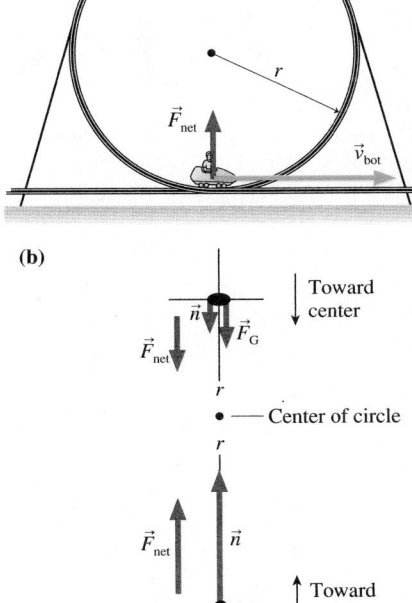

FIGURE 8.16 A roller-coaster car going around a loop-the-loop.

center of the circle, now points *downward*. Consequently, both forces have *positive* components. Newton's second law at the top of the circle is

$$\sum F_r = n_r + (F_G)_r = n + mg = \frac{m(v_{top})^2}{r} \tag{8.19}$$

Thus at the top the normal force of the track on the car is

$$n = \frac{m(v_{top})^2}{r} - mg \tag{8.20}$$

The normal force at the top can exceed *mg* if v_{top} is large enough. Our interest, however, is in what happens as the car goes slower and slower. As v_{top} decreases, there comes a point when *n* reaches zero. "No normal force" means "no contact," so at that speed, the track is *not* pushing against the car. Instead, the car is able to complete the circle because gravity alone provides sufficient centripetal acceleration.

The speed at which $n = 0$ is called the *critical speed* v_c:

$$v_c = \sqrt{\frac{rmg}{m}} = \sqrt{rg} \tag{8.21}$$

The critical speed is the slowest speed at which the car can complete the circle. Equation 8.20 would give a negative value for *n* if $v < v_c$, but that is physically impossible. The track can push against the wheels of the car ($n > 0$), but it can't pull on them. If $v < v_c$, the car cannot turn the full loop but, instead, comes off the track and becomes a projectile! **FIGURE 8.17** summarizes this reasoning for the car on the loop-the-loop.

Water stays in a bucket swung over your head for the same reason: Circular motion requires a net force toward the center of the circle. At the top of the circle—if you swing the bucket fast enough—the bucket adds to the force of gravity by pushing *down* on the water, just like the downward normal force of the track on the roller-coaster car. As long as the bucket is pushing against the water, the bucket and the water are in contact and thus the water is "in" the bucket. As you swing slower and slower, requiring the water to have less and less centripetal acceleration, the bucket-on-water normal force decreases until it becomes zero at the critical speed. At the critical speed, gravity alone provides sufficient centripetal acceleration. Below the critical speed, gravity provides *too much* downward force for circular motion, so the water leaves the bucket and becomes a projectile following a parabolic trajectory toward your head!

Notice the similarity to the car making the left turn in Figure 8.14. The passenger feels like he's being "hurled" into the door by a centrifugal force, but it's actually the pushing force from the door, toward the center of the circle, that causes the passenger to turn the corner instead of moving straight ahead. Here, while it seems like the water is being "pinned" against the bottom of the bucket by a centrifugal force, it's really the pushing force from the bottom of the bucket causing the water to move in a circle instead of following a free-fall parabola.

FIGURE 8.17 A roller-coaster car at the top of the loop.

The normal force adds to gravity to make a large enough force for the car to turn the circle.

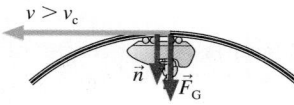

$v > v_c$

At v_c, gravity alone is enough force for the car to turn the circle. $\vec{n} = \vec{0}$ at the top point.

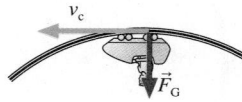

v_c

The gravitational force is too large for the car to stay in the circle!

Normal force became zero here.

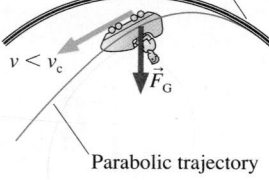

$v < v_c$

Parabolic trajectory

STOP TO THINK 8.4 An out-of-gas car is rolling over the top of a hill at speed *v*. At this instant,

a. $n > F_G$
b. $n < F_G$
c. $n = F_G$
d. We can't tell about *n* without knowing *v*.

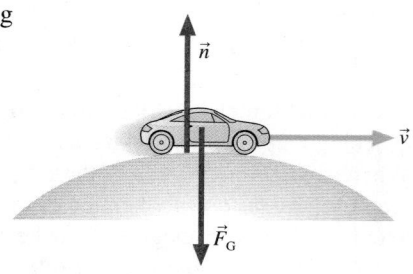

8.5 Nonuniform Circular Motion

Many interesting examples of circular motion involve objects whose speed changes. As we've already noted, a roller-coaster car doing a loop-the-loop slows down as it goes up one side, speeds up as it comes back down the other side. Circular motion with a changing speed is called *nonuniform circular motion.*

FIGURE 8.18, which is borrowed from Chapter 4, reminds you of the key ideas. Here the particle is speeding up or slowing down as it moves around the circle. In addition to centripetal acceleration, a particle in nonuniform circular motion has a tangential acceleration a_t. Tangential acceleration, parallel to $\vec{v}$, is the acceleration of changing speed. Mathematically, the tangential acceleration is simply the rate at which the tangential velocity changes:

$$a_t = \frac{dv_t}{dt} \tag{8.22}$$

It is usually most convenient to write the kinematic equations for circular motion in terms of the angular velocity ω and the angular acceleration $\alpha = d\omega/dt$. In Chapter 4, we found the connection between the tangential and angular accelerations to be

$$a_t = r\alpha \tag{8.23}$$

This is analogous to the similar equation $v_t = r\omega$ for tangential and angular velocity. In terms of angular quantities, the equations of constant-acceleration kinematics are

$$\theta_f = \theta_i + \omega_i\,\Delta t + \tfrac{1}{2}\alpha(\Delta t)^2$$
$$\omega_f = \omega_i + \alpha\,\Delta t \tag{8.24}$$
$$\omega_f^2 = \omega_i^2 + 2\alpha\,\Delta\theta$$

In addition, the centripetal acceleration equation $a_r = v^2/r = \omega^2 r$ is still valid.

Dynamics of Nonuniform Circular Motion

FIGURE 8.19 shows a net force $\vec{F}_{net}$ acting on a particle as it moves around a circle of radius r. $\vec{F}_{net}$ is likely to be a superposition of several forces, such as a tension force in a string, a thrust force, a friction force, and so on.

We can decompose the force vector $\vec{F}_{net}$ into a *tangential* component $(F_{net})_t$ and a radial component $(F_{net})_r$. The component $(F_{net})_t$ is positive for a tangential force in the ccw direction, negative for a tangential force in the cw direction. Because of our definition of the r-axis, the component $(F_{net})_r$ is positive for a radial force *toward* the center, negative for a radial force away from the center. For example, the particular force illustrated in Figure 8.19 has positive values for both $(F_{net})_t$ and $(F_{net})_r$.

The force component $(F_{net})_r$ perpendicular to the trajectory creates a centripetal acceleration and causes the particle to change directions. It is the component $(F_{net})_t$ parallel to the trajectory that creates a tangential acceleration and causes the particle to change speed. Force and acceleration are related to each other through Newton's second law:

$$(F_{net})_r = \sum F_r = ma_r = \frac{mv^2}{r} = m\omega^2 r$$
$$(F_{net})_t = \sum F_t = ma_t \tag{8.25}$$
$$(F_{net})_z = \sum F_z = 0$$

NOTE ▶ Equations 8.25 differ from Equations 8.8 for uniform circular motion only in the fact that a_t is no longer constrained to be zero. ◀

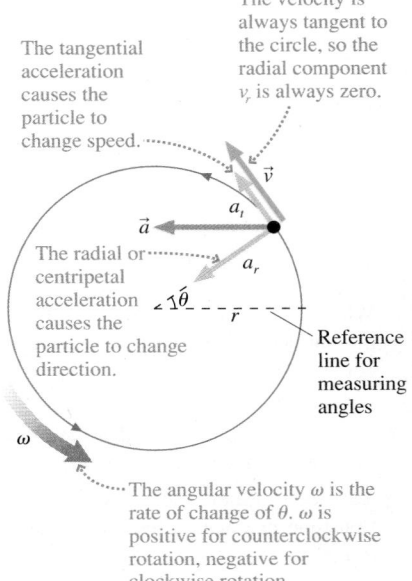

FIGURE 8.18 Nonuniform circular motion.

The velocity is always tangent to the circle, so the radial component v_r is always zero.

The tangential acceleration causes the particle to change speed.

The radial or centripetal acceleration causes the particle to change direction.

Reference line for measuring angles

The angular velocity ω is the rate of change of θ. ω is positive for counterclockwise rotation, negative for clockwise rotation.

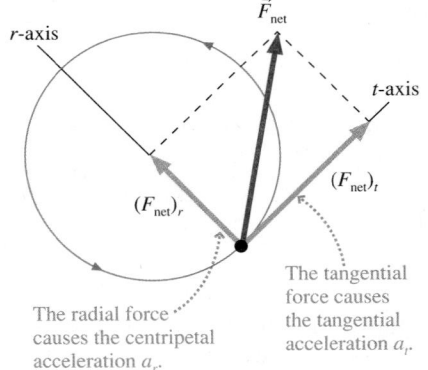

FIGURE 8.19 Net force $\vec{F}_{net}$ is applied to a particle moving in a circle.

r-axis

$\vec{F}_{net}$

t-axis

$(F_{net})_t$

$(F_{net})_r$

The radial force causes the centripetal acceleration a_r.

The tangential force causes the tangential acceleration a_t.

EXAMPLE 8.7 Circular motion of a grinding machine

A machine for grinding small samples down to thin slabs of uniform thickness consists of a 2.0 kg steel block that spins around on a table at the end of an 80-cm-long arm. Samples are glued to the bottom of the heavy block, then dragged across the sandpaper-like surface of the table as the block spins. A tachometer attached to the motor shaft gives the following readings after the motor is disengaged at $t = 0$ s:

Time (s)	rpm
0.0	156
0.5	114
1.0	88
1.5	52
2.0	17

What is the coefficient of kinetic friction between the sample and the surface? How many revolutions does the block make as it comes to a halt?

MODEL Model the block and sample as a particle in nonuniform circular motion. Assume that the mass of the sample is negligible compared to the mass of the block and that the axle is frictionless.

VISUALIZE FIGURE 8.20 shows a pictorial representation. For the first time, we need a free-body diagram showing forces in three dimensions.

FIGURE 8.20 Pictorial representation of the circular motion.

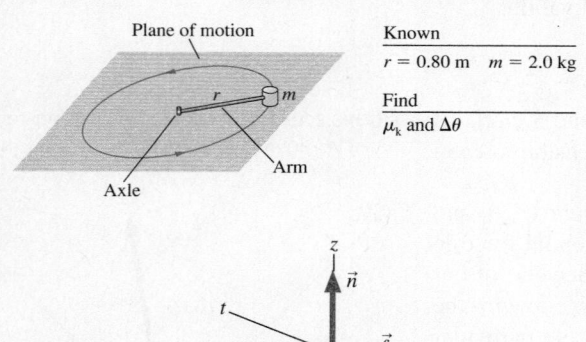

SOLVE The block slows down because kinetic friction between the sample and the table's surface exerts a retarding force $\vec{f}_k$. Kinetic friction is always opposite the direction of motion, so $\vec{f}_k$ is *tangent* to the circle and has magnitude $f_k = \mu_k n$.

There's no net force in the vertical direction, so the z-component of the second law is

$$\sum F_z = n - F_G = 0$$

from which we can conclude that $n = F_G = mg$ and thus $f_k = \mu_k mg$. The friction force is the only tangential component of

force, so we can use the t-component of Newton's second law to find the tangential acceleration:

$$\sum F_t = (f_k)_t = -f_k = ma_t$$

$$a_t = \frac{-f_k}{m} = \frac{-\mu_k mg}{m} = -\mu_k g$$

Thus the angular acceleration is $\alpha = a_t/r = -\mu_k g/r$.

We can find α experimentally as the slope of the ω-versus-t graph. FIGURE 8.21 shows a graph of the data (after conversion of rpm to rad/s) and a best-fit line. We see that the angular acceleration is $\alpha = -7.12$ rad/s², and with this value we can calculate the coefficient of kinetic friction:

$$\mu_k = -\frac{\alpha r}{g} = -\frac{(-7.12 \text{ rad/s}^2)(0.80 \text{ m})}{9.80 \text{ m/s}^2} = 0.58$$

We can now use the kinematic equation $\omega_f{}^2 = 0 = \omega_i{}^2 + 2\alpha \Delta\theta$ to find how many revolutions the block makes as it comes to rest. But what is ω_i? The first data entry is 156 rpm = 16.3 rad/s, but clearly—from the graph—the data have uncertainties. The first entry has no more claim to being "perfect" than any other entry. It's better to use the y-intercept of our best-fit line, 16.1 rad/s. That is, a statistical analysis of *all* the data tells us that 16.1 rad/s (154 rpm) is our best estimate of the angular velocity ω_i when the motor was disengaged at $t = 0$ s. With this:

$$\Delta\theta = -\frac{\omega_i{}^2}{2\alpha} = -\frac{(16.1 \text{ rad/s})^2}{2(-7.12 \text{ rad/s}^2)} = 18.2 \text{ rad} = 2.9 \text{ rev}$$

The block makes 2.9 revolutions while stopping.

FIGURE 8.21 Graph of angular velocity versus time. Angular acceleration is the slope of the best-fit line.

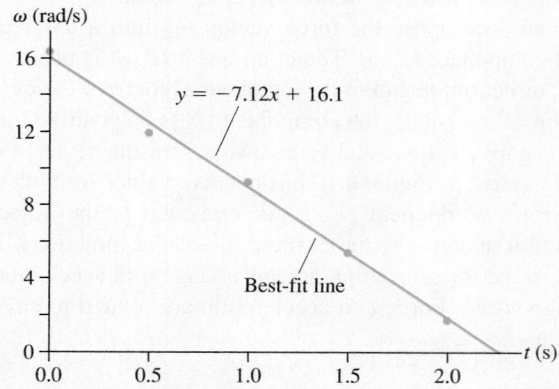

ASSESS A coefficient of kinetic friction of 0.58 is reasonable for a sandpaper-like surface. And even though the friction is fairly large, it's reasonable that the block would make several revolutions before stopping. The purpose of the Assess step, as always, is not to prove that the answer is right but to rule out obviously unreasonable answers that have been reached by mistake.

We've come a long way since our first dynamics problems in Chapter 6, but our basic strategy has not changed.

PROBLEM-SOLVING
STRATEGY 8.1 **Circular-motion problems**

MODEL Make simplifying assumptions.

VISUALIZE Draw a pictorial representation. Use *rtz*-coordinates.

- Establish a coordinate system with the *r*-axis pointing toward the center of the circle.
- Show important points in the motion on a sketch. Define symbols and identify what the problem is trying to find.
- Identify the forces and show them on a free-body diagram.

SOLVE Newton's second law is

$$(F_{\text{net}})_r = \sum F_r = ma_r = \frac{mv^2}{r} = m\omega^2 r$$

$$(F_{\text{net}})_t = \sum F_t = ma_t$$

$$(F_{\text{net}})_z = \sum F_z = 0$$

- Determine the force components from the free-body diagram. Be careful with signs.
- The tangential acceleration for uniform circular motion is $a_t = 0$.
- Solve for the acceleration, then use kinematics to find velocities and positions.

ASSESS Check that your result has the correct units, is reasonable, and answers the question.

Exercise 17

STOP TO THINK 8.5   A ball on a string is swung in a vertical circle. The string happens to break when it is parallel to the ground and the ball is moving up. Which trajectory does the ball follow?

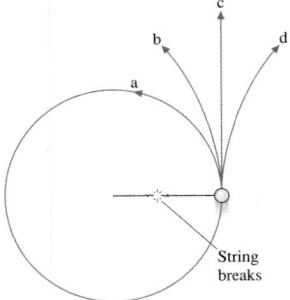

CHALLENGE EXAMPLE 8.8   **Circular motion inside a cone**

A small ball of mass m rolls in a horizontal circle around the inside of the inverted cone shown in FIGURE 8.22. The walls of the cone make an angle θ with a horizontal plane. The coefficient of static friction between the ball and the cone is μ_s; rolling friction is negligible. What minimum speed v_{min} must the ball maintain to remain at a constant height h?

MODEL Model the ball as a particle in uniform circular motion.

FIGURE 8.22 Pictorial representation of the circular motion.

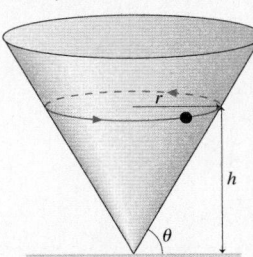

Continued

VISUALIZE The forces on the ball are a normal force, perpendicular to the surface; a static friction force parallel to the surface that keeps the ball from sliding up or down; and gravity. FIGURE 8.23 shows a free-body diagram with the r-axis pointing toward the center of the circle. Notice that the situation is very similar to that of a car on a banked curve. Figure 8.9 showed that the static friction force must point up the slope to keep a slow-moving car from sliding down the slope, and that information was used in drawing Figure 8.23.

FIGURE 8.23 Free-body diagram of the ball.

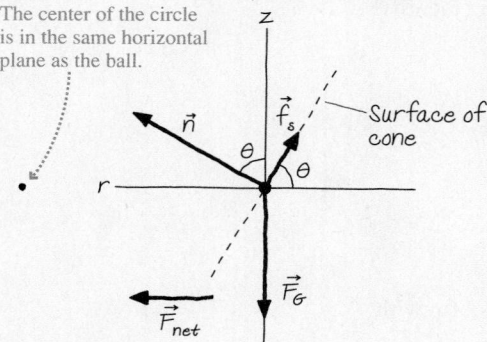

The center of the circle is in the same horizontal plane as the ball.

SOLVE This is uniform circular motion, so we need to consider only the r- and z-component equations of Newton's second law. All the information is on the free-body diagram, but considerable care is needed to write down all the components correctly. The two equations are

$$\sum F_r = n_r + (f_s)_r + (F_G)_r = n \sin\theta - f_s \cos\theta = \frac{mv^2}{r}$$

$$\sum F_z = n_z + (f_s)_z + (F_G)_z = n \cos\theta + f_s \sin\theta - mg = 0$$

The r-axis points toward the center of the circle, here on the left, so $\vec{n}$ has a positive r-component while $\vec{f}_s$ has a negative r-component.

There is one specific speed at which the ball would roll around the inside without friction, just like the car in Example 8.5. For slower speeds, some amount of static friction is needed to keep the ball from sliding down. Our task is to find the *minimum* speed for maintaining motion in a horizontal plane, and that occurs when the static friction force reaches its *maximum* value: $f_{s\,max} = \mu_s n$. Then the r-component equation becomes

$$\frac{mv_{min}^2}{r} = n \sin\theta - \mu_s n \cos\theta = n(\sin\theta - \mu_s \cos\theta)$$

To find n, we use the z-component equation with $f_{s\,max} = \mu_s n$:

$$n \cos\theta + \mu_s n \sin\theta = n(\cos\theta + \mu_s \sin\theta) = mg$$

$$n = \frac{mg}{\cos\theta + \mu_s \sin\theta}$$

Substituting this into the equation for v_{min} and taking the square root, we find

$$v_{min} = \sqrt{rg\left(\frac{\sin\theta - \mu_s \cos\theta}{\cos\theta + \mu_s \sin\theta}\right)}$$

This is a complicated answer, but we can check it because without friction the ball should roll around at the same speed as the car turning a banked curve without friction. If we set $\mu_s = 0$, we find

$$v_{frictionless} = \sqrt{rg\tan\theta}$$

which, indeed, was the answer to Example 8.5.

The information we have is the ball's height h, not the radius of the circle, so the final step, which we can see from Figure 8.22, is to substitute $r = h/\tan\theta$. Thus the minimum speed for the ball to circle at height h is

$$v_{min} = \sqrt{\frac{hg}{\tan\theta}\left(\frac{\sin\theta - \mu_s \cos\theta}{\cos\theta + \mu_s \sin\theta}\right)}$$

ASSESS An important problem-solving skill to learn is checking new results by comparing them to previously known results. In this case, we recognized that the ball rolling around the inside of the cone without the aid of friction is equivalent to a car turning a banked curve without friction. The fact that we could reproduce that earlier result gives us confidence in our answer.

SUMMARY

The goal of Chapter 8 has been to learn how to solve problems about motion in a plane.

General Principles

Newton's Second Law

Expressed in x- and y-component form:

$$(F_{net})_x = \sum F_x = ma_x$$

$$(F_{net})_y = \sum F_y = ma_y$$

Expressed in rtz-component form:

$$(F_{net})_r = \sum F_r = ma_r = \frac{mv^2}{r} = m\omega^2 r$$

$$(F_{net})_t = \sum F_t = \begin{cases} 0 & \text{uniform circular motion} \\ ma_t & \text{nonuniform circular motion} \end{cases}$$

$$(F_{net})_z = \sum F_z = 0$$

Uniform Circular Motion

- v is constant.
- $\vec{F}_{net}$ points toward the center of the circle.
- The **centripetal acceleration** $\vec{a}$ points toward the center of the circle. It changes the particle's direction but not its speed.

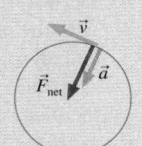

Nonuniform Circular Motion

- v changes.
- $\vec{a}$ is parallel to $\vec{F}_{net}$.
- The radial component a_r changes the particle's direction.
- The tangential component a_t changes the particle's speed.

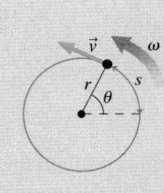

Important Concepts

rtz-coordinates

- The r-axis points toward the center of the circle.
- The t-axis is tangent, pointing counterclockwise.

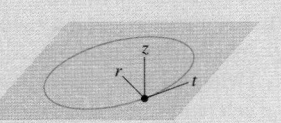

Angular velocity

$$\omega = d\theta/dt$$

$$v_t = \omega r$$

Angular acceleration

$$\alpha = d\omega/dt$$

$$a_t = \alpha r$$

Applications

Orbits

A circular orbit has radius r if

$$v = \sqrt{rg}$$

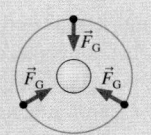

Hills and valleys

Circular motion requires a net force pointing to the center. n must be > 0 for the object to be in contact with a surface.

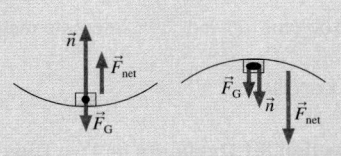

Terms and Notation

orbit
fictitious force

CONCEPTUAL QUESTIONS

1. In uniform circular motion, which of the following are constant: speed, velocity, angular velocity, centripetal acceleration, magnitude of the net force?

2. A car runs out of gas while driving down a hill. It rolls through the valley and starts up the other side. At the very bottom of the valley, which of the free-body diagrams in FIGURE Q8.2 is correct? The car is moving to the right, and drag and rolling friction are negligible.

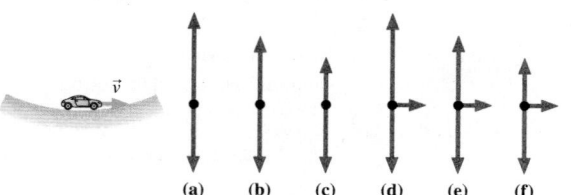

FIGURE Q8.2

3. FIGURE Q8.3 is a bird's-eye view of particles moving in horizontal circles on a tabletop. All are moving at the same speed. Rank in order, from largest to smallest, the tensions T_a to T_d. Give your answer in the form a > b = c > d and explain your ranking.

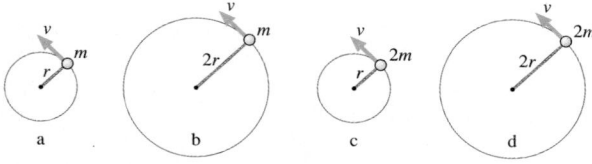

FIGURE Q8.3

4. Tarzan swings through the jungle on a vine. At the lowest point of his swing, is the tension in the vine greater than, less than, or equal to the gravitational force on Tarzan? Explain.

5. FIGURE Q8.5 shows two balls of equal mass moving in vertical circles. Is the tension in string A greater than, less than, or equal to the tension in string B if the balls travel over the top of the circle (a) with equal speed and (b) with equal angular velocity?

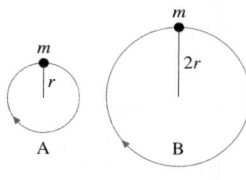

FIGURE Q8.5

6. Ramon and Sally are observing a toy car speed up as it goes around a circular track. Ramon says, "The car's speeding up, so there must be a net force parallel to the track." "I don't think so," replies Sally. "It's moving in a circle, and that requires centripetal acceleration. The net force has to point to the center of the circle." Do you agree with Ramon, Sally, or neither? Explain.

7. A jet plane is flying on a level course at constant speed. The engines are at full throttle.
   a. What is the net force on the plane? Explain.
   b. Draw a free-body diagram of the plane as seen from the side with the plane flying to the right. Name (don't just label) any and all forces shown on your diagram.
   c. Airplanes bank when they turn. Draw a free-body diagram of the plane as seen from behind as it makes a right turn.
   d. *Why* do planes bank as they turn? Explain.

8. A small projectile is launched parallel to the ground at height $h = 1$ m with sufficient speed to orbit a completely smooth, airless planet. A bug rides inside a small hole inside the projectile. Is the bug weightless? Explain.

9. You can swing a ball on a string in a vertical circle if you swing it fast enough. But if you swing too slowly, the string goes slack as the ball nears the top. Explain *why* there's a minimum speed to keep the ball moving in a circle.

10. A golfer starts with the club over her head and swings it to reach maximum speed as it contacts the ball. Halfway through her swing, when the golf club is parallel to the ground, does the acceleration vector of the club head point (a) straight down, (b) parallel to the ground, approximately toward the golfer's shoulders, (c) approximately toward the golfer's feet, or (d) toward a point above the golfer's head? Explain.

EXERCISES AND PROBLEMS

Problems labeled ▨ integrate material from earlier chapters.

Exercises

Section 8.1 Dynamics in Two Dimensions

1. ‖ As a science fair project, you want to launch an 800 g model rocket straight up and hit a horizontally moving target as it passes 30 m above the launch point. The rocket engine provides a constant thrust of 15.0 N. The target is approaching at a speed of 15 m/s. At what horizontal distance between the target and the rocket should you launch?

2. ‖ A 500 g model rocket is on a cart that is rolling to the right at a speed of 3.0 m/s. The rocket engine, when it is fired, exerts an 8.0 N thrust on the rocket. Your goal is to have the rocket pass through a small horizontal hoop that is 20 m above the launch point. At what horizontal distance left of the hoop should you launch?

3. ‖ A 4.0×10^{10} kg asteroid is heading directly toward the center of the earth at a steady 20 km/s. To save the planet, astronauts strap a giant rocket to the asteroid perpendicular to its direction of travel. The rocket generates 5.0×10^9 N of thrust. The rocket is fired when the asteroid is 4.0×10^6 km away from earth. You can ignore the earth's gravitational force on the asteroid and their rotation about the sun.
   a. If the mission fails, how many hours is it until the asteroid impacts the earth?
   b. The radius of the earth is 6400 km. By what minimum angle must the asteroid be deflected to just miss the earth?
   c. The rocket fires at full thrust for 300 s before running out of fuel. Is the earth saved?

Section 8.2 Uniform Circular Motion

4. | A 1500 kg car drives around a flat 200-m-diameter circular track at 25 m/s. What are the magnitude and direction of the net force on the car? What causes this force?

5. | A 1500 kg car takes a 50-m-radius unbanked curve at 15 m/s. What is the size of the friction force on the car?

6. ‖ A 200 g block on a 50-cm-long string swings in a circle on a horizontal, frictionless table at 75 rpm.
   a. What is the speed of the block?
   b. What is the tension in the string?

7. ‖ In the Bohr model of the hydrogen atom, an electron (mass $m = 9.1 \times 10^{-31}$ kg) orbits a proton at a distance of 5.3×10^{-11} m. The proton pulls on the electron with an electric force of 8.2×10^{-8} N. How many revolutions per second does the electron make?

8. ‖ A highway curve of radius 500 m is designed for traffic moving at a speed of 90 km/h. What is the correct banking angle of the road?

9. ‖ Suppose the moon were held in its orbit not by gravity but by a massless cable attached to the center of the earth. What would be the tension in the cable? Use the table of astronomical data inside the back cover of the book.

10. | It is proposed that future space stations create an artificial gravity by rotating. Suppose a space station is constructed as a 1000-m-diameter cylinder that rotates about its axis. The inside surface is the deck of the space station. What rotation period will provide "normal" gravity?

Section 8.3 Circular Orbits

11. | A satellite orbiting the moon very near the surface has a period of 110 min. What is free-fall acceleration on the surface of the moon?

12. ‖ What is free-fall acceleration toward the sun at the distance of the earth's orbit? Astronomical data are inside the back cover of the book.

Section 8.4 Fictitious Forces

13. | A car drives over the top of a hill that has a radius of 50 m. What maximum speed can the car have at the top without flying off the road?

14. ‖ The weight of passengers on a roller coaster increases by 50% as the car goes through a dip with a 30 m radius of curvature. What is the car's speed at the bottom of the dip?

15. ‖ A roller coaster car crosses the top of a circular loop-the-loop at twice the critical speed. What is the ratio of the normal force to the gravitational force?

16. ‖ The normal force equals the magnitude of the gravitational force as a roller coaster car crosses the top of a 40-m-diameter loop-the-loop. What is the car's speed at the top?

17. ‖ A student has 65-cm-long arms. What is the minimum angular velocity (in rpm) for swinging a bucket of water in a vertical circle without spilling any? The distance from the handle to the bottom of the bucket is 35 cm.

18. | While at the county fair, you decide to ride the Ferris wheel. Having eaten too many candy apples and elephant ears, you find the motion somewhat unpleasant. To take your mind off your stomach, you wonder about the motion of the ride. You estimate the radius of the big wheel to be 15 m, and you use your watch to find that each loop around takes 25 s.

a. What are your speed and the magnitude of your acceleration?
b. What is the ratio of your weight at the top of the ride to your weight while standing on the ground?
c. What is the ratio of your weight at the bottom of the ride to your weight while standing on the ground?

Section 8.5 Nonuniform Circular Motion

19. ‖ A new car is tested on a 200-m-diameter track. If the car speeds up at a steady 1.5 m/s², how long after starting is the magnitude of its centripetal acceleration equal to the tangential acceleration?

20. ‖ A toy train rolls around a horizontal 1.0-m-diameter track. The coefficient of rolling friction is 0.10.
   a. What is the magnitude of the train's angular acceleration after it is released?
   b. How long does it take the train to stop if it's released with an angular speed of 30 rpm?

Problems

21. ‖ A popular pastime is to see who can push an object closest to the edge of a table without its going off. You push the 100 g object and release it 2.0 m from the table edge. Unfortunately, you push a little too hard. The object slides across, sails off the edge, falls 1.0 m to the floor, and lands 30 cm from the edge of the table. If the coefficient of kinetic friction is 0.50, what was the object's speed as you released it?

22. ‖ A motorcycle daredevil plans to ride up a 2.0-m-high, 20° ramp, sail across a 10-m-wide pool filled with hungry crocodiles, and land at ground level on the other side. He has done this stunt many times and approaches it with confidence. Unfortunately, the motorcycle engine dies just as he starts up the ramp. He is going 11 m/s at that instant, and the rolling friction of his rubber tires (coefficient 0.02) is not negligible. Does he survive, or does he become crocodile food?

23. ‖‖ Sam (75 kg) takes off up a 50-m-high, 10° frictionless slope on his jet-powered skis. The skis have a thrust of 200 N. He keeps his skis tilted at 10° after becoming airborne, as shown in FIGURE P8.23. How far does Sam land from the base of the cliff?

FIGURE P8.23

24. ‖ Derive Equations 8.4 for the acceleration of a projectile subject to drag.

25. ‖ A 5000 kg interceptor rocket is launched at an angle of 44.7°. The thrust of the rocket motor is 140,700 N.
   a. Find an equation $y(x)$ that describes the rocket's trajectory.
   b. What is the shape of the trajectory?
   c. At what elevation does the rocket reach the speed of sound, 330 m/s?

26. ‖‖ A rocket-powered hockey puck has a thrust of 2.0 N and a total mass of 1.0 kg. It is released from rest on a frictionless table, 4.0 m from the edge of a 2.0 m drop. The front of the rocket is pointed directly toward the edge. How far does the puck land from the base of the table?

27. ‖ A 500 g model rocket is resting horizontally at the top edge of a 40-m-high wall when it is accidentally bumped. The bump pushes it off the edge with a horizontal speed of 0.5 m/s and at the same time causes the engine to ignite. When the engine fires, it exerts a constant 20 N horizontal thrust away from the wall.
    a. How far from the base of the wall does the rocket land?
    b. Describe the trajectory of the rocket while it travels to the ground.

28. ‖ An experimental aircraft begins its takeoff at $t = 0$ s. Every second, an onboard GPS measures and records the plane's distances east and north of a reference marker. The following data are downloaded to your computer:

Time (s)	East (m)	North (m)
0.0	91	0
1.0	86	4
2.0	77	18
3.0	65	39
4.0	39	63
5.0	19	101

Analyze these data to determine the magnitude of the aircraft's takeoff acceleration.

29. ‖ Communications satellites are placed in circular orbits where they stay directly over a fixed point on the equator as the earth rotates. These are called *geosynchronous orbits*. The altitude of a geosynchronous orbit is 3.58×10^7 m ($\approx 22{,}000$ miles).
    a. What is the period of a satellite in a geosynchronous orbit?
    b. Find the value of g at this altitude.
    c. What is the weight of a 2000 kg satellite in a geosynchronous orbit?

30. ‖ A 75 kg man weighs himself at the north pole and at the equator. Which scale reading is higher? By how much?

31. ‖ A 500 g ball swings in a vertical circle at the end of a 1.5-m-long string. When the ball is at the bottom of the circle, the tension in the string is 15 N. What is the speed of the ball at that point?

32. ‖‖ A concrete highway curve of radius 70 m is banked at a 15° angle. What is the maximum speed with which a 1500 kg rubber-tired car can take this curve without sliding?

33. ‖ a. An object of mass m swings in a horizontal circle on a string of length L that tilts downward at angle θ. Find an expression for the angular velocity ω.
    b. A student ties a 500 g rock to a 1.0-m-long string and swings it around her head in a horizontal circle. At what angular speed, in rpm, does the string tilt down at a 10° angle?

34. ‖ A 5.0 g coin is placed 15 cm from the center of a turntable. The coin has static and kinetic coefficients of friction with the turntable surface of $\mu_s = 0.80$ and $\mu_k = 0.50$. The turntable very slowly speeds up to 60 rpm. Does the coin slide off?

35. ‖ You've taken your neighbor's young child to the carnival to ride the rides. She wants to ride The Rocket. Eight rocket-shaped cars hang by chains from the outside edge of a large steel disk. A vertical axle through the center of the ride turns the disk, causing the cars to revolve in a circle. You've just finished taking physics, so you decide to figure out the speed of the cars while you wait. You estimate that the disk is 5 m in diameter and the chains are 6 m long. The ride takes 10 s to reach full speed, then the cars swing out until the chains are 20° from vertical. What is the cars' speed?

36. ‖ A conical pendulum is formed by attaching a ball of mass m to a string of length L, then allowing the ball to move in a horizontal circle of radius r. FIGURE P8.36 shows that the string traces out the surface of a cone, hence the name.
    a. Find an expression for the tension T in the string.
    b. Find an expression for the ball's angular speed ω.
    c. What are the tension and angular speed (in rpm) for a 500 g ball swinging in a 20-cm-radius circle at the end of a 1.0-m-long string?

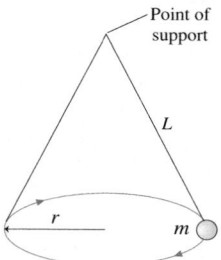

FIGURE P8.36

37. ‖‖‖ Two wires are tied to the 2.0 kg sphere shown in FIGURE P8.37. The sphere revolves in a horizontal circle at constant speed.
    a. For what speed is the tension the same in both wires?
    b. What is the tension?

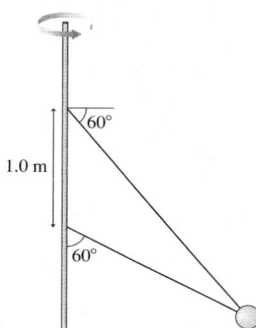

FIGURE P8.37

38. ‖‖‖ In an old-fashioned amusement park ride, passengers stand inside a 5.0-m-diameter hollow steel cylinder with their backs against the wall. The cylinder begins to rotate about a vertical axis. Then the floor on which the passengers are standing suddenly drops away! If all goes well, the passengers will "stick" to the wall and not slide. Clothing has a static coefficient of friction against steel in the range 0.60 to 1.0 and a kinetic coefficient in the range 0.40 to 0.70. A sign next to the entrance says "No children under 30 kg allowed." What is the minimum angular speed, in rpm, for which the ride is safe?

39. ‖ A 10 g steel marble is spun so that it rolls at 150 rpm around the *inside* of a vertically oriented steel tube. The tube, shown in FIGURE P8.39, is 12 cm in diameter. Assume that the rolling resistance is small enough for the marble to maintain 150 rpm for several seconds. During this time, will the marble spin in a horizontal circle, at constant height, or will it spiral down the inside of the tube?

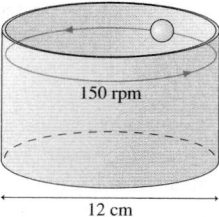

FIGURE P8.39

Exercises and Problems 213

40. ‖ The ultracentrifuge is an important tool for separating and
BIO analyzing proteins. Because of the enormous centripetal accelerations, the centrifuge must be carefully balanced, with each
sample matched by a sample of identical mass on the opposite
side. Any difference in the masses of opposing samples creates a
net force on the shaft of the rotor, potentially leading to a catastrophic failure of the apparatus. Suppose a scientist makes a
slight error in sample preparation and one sample has a mass
10 mg larger than the opposing sample. If the samples are 12 cm
from the axis of the rotor and the ultracentrifuge spins at
70,000 rpm, what is the magnitude of the net force on the rotor
due to the unbalanced samples?

41. ‖ Three cars are driving at 25 m/s along the road shown in
FIGURE P8.41. Car B is at the bottom of a hill and car C is at the
top. Both hills have a 200 m radius of curvature. Suppose each
car suddenly brakes hard and starts to skid. What is the tangential acceleration (i.e., the acceleration parallel to the road) of
each car? Assume $\mu_k = 1.0$.

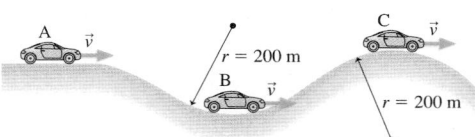

FIGURE P8.41

42. ‖ A 500 g ball moves in a vertical circle on a 102-cm-long string.
If the speed at the top is 4.0 m/s, then the speed at the bottom will
be 7.5 m/s. (You'll learn how to show this in Chapter 10.)
   a. What is the gravitational force acting on the ball?
   b. What is the tension in the string when the ball is at the top?
   c. What is the tension in the string when the ball is at the
      bottom?

43. ‖ In an amusement park ride
called The Roundup, passengers
stand inside a 16-m-diameter
rotating ring. After the ring has
acquired sufficient speed, it tilts
into a vertical plane, as shown in
FIGURE P8.43.
   a. Suppose the ring rotates once
      every 4.5 s. If a rider's mass
      is 55 kg, with how much
      force does the ring push on
      her at the top of the ride? At
      the bottom?
   b. What is the longest rotation period of the wheel that will pre-
      vent the riders from falling off at the top?

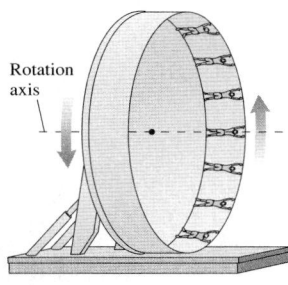

FIGURE P8.43

44. ‖ You have a new job designing rides for an amusement park.
In one ride, the rider's chair is attached by a 9.0-m-long chain
to the top of a tall rotating tower. The tower spins the chair and
rider around at the rate of 1.0 rev every 4.0 s. In your design,
you've assumed that the maximum possible combined weight
of the chair and rider is 150 kg. You've found a great price for
chain at the local discount store, but your supervisor wonders if
the chain is strong enough. You contact the manufacturer and
learn that the chain is rated to withstand a tension of 3000 N.
Will this chain be strong enough for the ride?

45. ‖ Suppose you swing a ball of mass m in a vertical circle on a
string of length L. As you probably know from experience, there
is a minimum angular velocity ω_{min} you must maintain if you

want the ball to complete the full circle without the string going
slack at the top.
   a. Find an expression for ω_{min}.
   b. Evaluate ω_{min} in rpm for a 65 g ball tied to a 1.0-m-long
      string.

46. ‖ A heavy ball with a weight of 100 N ($m = 10.2$ kg) is hung
from the ceiling of a lecture hall on a 4.5-m-long rope. The ball is
pulled to one side and released to swing as a pendulum, reaching
a speed of 5.5 m/s as it passes through the lowest point. What is
the tension in the rope at that point?

47. ‖ A 30 g ball rolls around a 40-cm-
diameter L-shaped track, shown in
FIGURE P8.47, at 60 rpm. What is the
magnitude of the net force that the
track exerts on the ball? Rolling friction can be neglected.
   Hint: The track exerts more than one force on the ball.

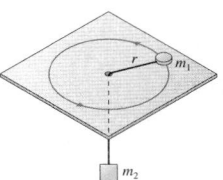

FIGURE P8.47

48. ‖ Mass m_1 on the frictionless table of FIGURE P8.48 is connected
by a string through a hole in the table to a hanging mass m_2.
With what speed must m_1 rotate in a circle of radius r if m_2 is to
remain hanging at rest?

FIGURE P8.48

49. ‖ The physics of circular motion sets an upper limit to the
BIO speed of human walking. (If you need to go faster, your gait
changes from a walk to a run.) If you take a few steps and
watch what's happening, you'll see that your body pivots in
circular motion over your forward foot as you bring your rear
foot forward for the next step. As you do so, the normal force
of the ground on your foot decreases and your body tries to "lift
off" from the ground.
   a. A person's center of mass is very near the hips, at the top of
      the legs. Model a person as a particle of mass m at the top of a
      leg of length L. Find an expression for the person's maximum
      walking speed v_{max}.
   b. Evaluate your expression for the maximum walking speed
      of a 70 kg person with a typical leg length of 70 cm. Give
      your answer in both m/s and mph, then comment, based on
      your experience, as to whether this is a reasonable result. A
      "normal" walking speed is about 3 mph.

50. ‖ A 100 g ball on a 60-cm-long string is swung in a vertical
circle about a point 200 cm above the floor. The tension in the
string when the ball is at the very bottom of the circle is 5.0 N. A
very sharp knife is suddenly inserted, as shown in FIGURE P8.50,
to cut the string directly below the point of support. How far to
the right of where the string was cut does the ball hit the floor?

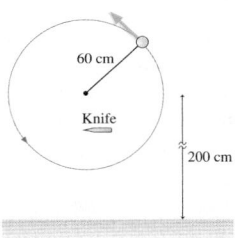

FIGURE P8.50

51. ‖ A 60 g ball is tied to the end of a 50-cm-long string and swung in a vertical circle. The center of the circle, as shown in FIGURE P8.51, is 150 cm above the floor. The ball is swung at the minimum speed necessary to make it over the top without the string going slack. If the string is released at the instant the ball is at the top of the loop, how far to the right does the ball hit the ground?

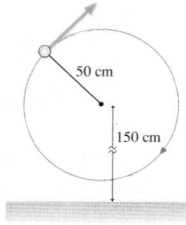

FIGURE P8.51

52. ‖ Elm Street has a pronounced dip at the bottom of a steep hill before going back uphill on the other side. Your science teacher has asked everyone in the class to measure the radius of curvature of the dip. Some of your classmates are using surveying equipment, but you decide to base your measurement on what you've learned in physics. To do so, you sit on a spring scale, drive through the dip at different speeds, and for each speed record the scale's reading as you pass through the bottom of the dip. Your data are as follows:

Speed (m/s)	Scale reading (N)
5	599
10	625
15	674
20	756
25	834

Sitting on the scale while the car is parked gives a reading of 588 N. Analyze your data, using a graph, to determine the dip's radius of curvature.

53. ‖ A 100 g ball on a 60-cm-long string is swung in a vertical circle about a point 200 cm above the floor. The string suddenly breaks when it is parallel to the ground and the ball is moving upward. The ball reaches a height 600 cm above the floor. What was the tension in the string an instant before it broke?

54. ‖‖ A 1500 kg car starts from rest and drives around a flat 50-m-diameter circular track. The forward force provided by the car's drive wheels is a constant 1000 N.
   a. What are the magnitude and direction of the car's acceleration at $t = 10$ s? Give the direction as an angle from the r-axis.
   b. If the car has rubber tires and the track is concrete, at what time does the car begin to slide out of the circle?

55. ‖ A 500 g steel block rotates on a steel table while attached to a 2.0-m-long massless rod. Compressed air fed through the rod is ejected from a nozzle on the back of the block, exerting a thrust force of 3.5 N. The nozzle is 70° from the radial line, as shown in FIGURE P8.55. The block starts from rest.
   a. What is the block's angular velocity after 10 rev?
   b. What is the tension in the rod after 10 rev?

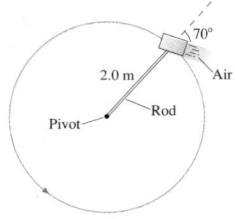

FIGURE P8.55

56. ‖ A 2.0 kg ball swings in a vertical circle on the end of an 80-cm-long string. The tension in the string is 20 N when its angle from the highest point on the circle is $\theta = 30°$.
   a. What is the ball's speed when $\theta = 30°$?
   b. What are the magnitude and direction of the ball's acceleration when $\theta = 30°$?

In Problems 57 and 58 you are given the equation used to solve a problem. For each of these, you are to
   a. Write a realistic problem for which this is the correct equation. Be sure that the answer your problem requests is consistent with the equation given.
   b. Finish the solution of the problem.

57. $60 \text{ N} = (0.30 \text{ kg})\omega^2(0.50 \text{ m})$

58. $(1500 \text{ kg})(9.8 \text{ m/s}^2) - 11{,}760 \text{ N} = (1500 \text{ kg}) v^2/(200 \text{ m})$

Challenge Problems

59. In the absence of air resistance, a projectile that lands at the elevation from which it was launched achieves maximum range when launched at a 45° angle. Suppose a projectile of mass m is launched with speed v_0 into a headwind that exerts a constant, horizontal retarding force $\vec{F}_{wind} = -F_{wind}\hat{\imath}$.
   a. Find an expression for the angle at which the range is maximum.
   b. By what percentage is the maximum range of a 0.50 kg ball reduced if $F_{wind} = 0.60$ N?

60. The father of Example 8.3 stands at the summit of a conical hill as he spins his 20 kg child around on a 5.0 kg cart with a 2.0-m-long rope. The sides of the hill are inclined at 20°. He again keeps the rope parallel to the ground, and friction is negligible. What rope tension will allow the cart to spin with the same 14 rpm it had in the example?

61. A small ball rolls around a horizontal circle at height y inside the cone shown in FIGURE CP8.61. Find an expression for the ball's speed in terms of a, h, y, and g.

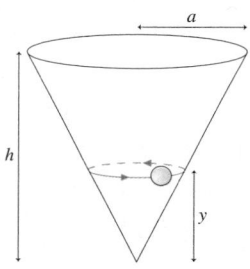

FIGURE CP8.61

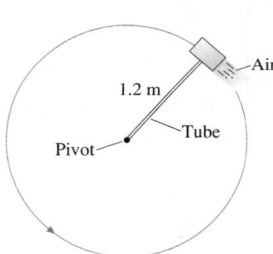

FIGURE CP8.62

62. A 500 g steel block rotates on a steel table while attached to a 1.2-m-long hollow tube as shown in FIGURE CP8.62. Compressed air fed through the tube and ejected from a nozzle on the back of the block exerts a thrust force of 4.0 N perpendicular to the tube. The maximum tension the tube can withstand without breaking is 50 N. If the block starts from rest, how many revolutions does it make before the tube breaks?

63. Two wires are tied to the 300 g sphere shown in FIGURE CP8.63. The sphere revolves in a horizontal circle at a constant speed of 7.5 m/s. What is the tension in each of the wires?

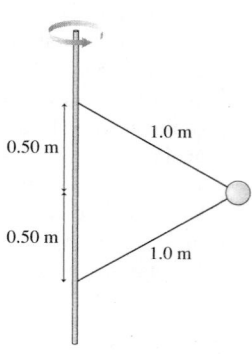

FIGURE CP8.63

64. A small ball rolls around a horizontal circle at height y inside a frictionless hemispherical bowl of radius R, as shown in FIGURE CP8.64.

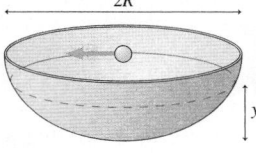

2R

y

FIGURE CP8.64

   a. Find an expression for the ball's angular velocity in terms of R, y, and g.
   b. What is the minimum value of ω for which the ball can move in a circle?
   c. What is ω in rpm if $R = 20$ cm and the ball is halfway up?

65. You are flying to New York. You've been reading the in-flight magazine, which has an article about the physics of flying. You learned that the airflow over the wings creates a *lift force* that is always perpendicular to the wings. In level flight, the upward lift force exactly balances the downward gravitational force. The pilot comes on to say that, because of heavy traffic, the plane is going to circle the airport for a while. She says that you'll maintain a speed of 400 mph at an altitude of 20,000 ft. You start to wonder what the diameter of the plane's circle around the airport is. You notice that the pilot has banked the plane so that the wings are 10° from horizontal. The safety card in the seatback pocket informs you that the plane's wing span is 250 ft. What can you learn about the diameter?

66. If a vertical cylinder of water (or any other liquid) rotates about its axis, as shown in FIGURE CP8.66, the surface forms a smooth curve. Assuming that the water rotates as a unit (i.e., all the water rotates with the same angular velocity), show that the shape of the surface is a parabola described by the equation $z = (\omega^2/2g)r^2$. **Hint:** Each particle of water on the surface is subject to only two forces: gravity and the normal force due to the water underneath it. The normal force, as always, acts perpendicular to the surface.

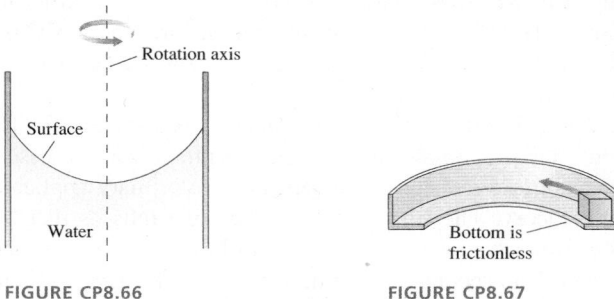

Rotation axis

Surface

Water

Bottom is frictionless

FIGURE CP8.66   **FIGURE CP8.67**

67. FIGURE CP8.67 shows a small block of mass m sliding around the inside of an L-shaped track of radius r. The bottom of the track is frictionless; the coefficient of kinetic friction between the block and the wall of the track is μ_k. The block's speed is v_0 at $t_0 = 0$. Find an expression for the block's speed at a later time t.

STOP TO THINK ANSWERS

Stop to Think 8.1: d. The parallel component of $\vec{a}$ is opposite $\vec{v}$ and will cause the particle to slow down. The perpendicular component of $\vec{a}$ will cause the particle to change directions in a downward direction.

Stop to Think 8.2: $(a_r)_b > (a_r)_e > (a_r)_a = (a_r)_c > (a_r)_d$. Centripetal acceleration is v^2/r. Doubling r decreases a_r by a factor of 2. Doubling v increases a_r by a factor of 4. Reversing direction doesn't change a_r.

Stop to Think 8.3: $T_d > T_b = T_e > T_c > T_a$. The center-directed force is $m\omega^2 r$. Changing r by a factor of 2 changes the tension by a factor of 2, but changing ω by a factor of 2 changes the tension by a factor of 4.

Stop to Think 8.4: b. The car is moving in a circle, so there must be a net force toward the center of the circle. The circle is below the car, so the net force must point downward. This can be true only if $F_G > n$.

Stop to Think 8.5: c. The ball does not have a "memory" of its previous motion. The velocity $\vec{v}$ is straight up at the instant the string breaks. The only force on the ball after the string breaks is the gravitational force, straight down. This is just like tossing a ball straight up.

SUMMARY
Newton's Laws

The goal of Part I has been to discover the connection between force and motion. We started with *kinematics,* which is the mathematical description of motion; then we proceeded to *dynamics,* which is the explanation of motion in terms of forces. Newton's three laws of motion form the basis of our explanation. All of the examples we have studied so far are applications of Newton's laws.

The table below is called a *knowledge structure* for Newton's laws. A knowledge structure summarizes the essential concepts, the general principles, and the primary applications of a theory. The first section of the table tells us that Newtonian mechanics is concerned with how *particles* respond to *forces.* The second section indicates that we have introduced only three general principles, Newton's three laws of motion.

You use this knowledge structure by working your way through it, from top to bottom. Once you recognize a problem

as a dynamics problem, you immediately know to start with Newton's laws. You can then determine the category of motion and apply Newton's second law in the appropriate form. Newton's third law will help you identify the forces acting on particles as they interact. Finally, the kinematic equations for that category of motion allow you to reach the solution you seek.

The knowledge structure provides the *procedural knowledge* for solving dynamics problems, but it does not represent the total knowledge required. You must add to it knowledge about what position and velocity are, about how forces are identified, about action/reaction pairs, about drawing and using free-body diagrams, and so on. These are specific *tools* for problem solving. The problem-solving strategies of Chapters 5 through 8 combine the procedures and the tools into a powerful method for thinking about and solving problems.

KNOWLEDGE STRUCTURE I Newton's Laws

ESSENTIAL CONCEPTS	Particle, acceleration, force, interaction
BASIC GOALS	How does a particle respond to a force? How do objects interact?

GENERAL PRINCIPLES		
	Newton's first law	An object will remain at rest or will continue to move with constant velocity (equilibrium) if and only if $\vec{F}_{\text{net}} = \vec{0}$.
	Newton's second law	$\vec{F}_{\text{net}} = m\vec{a}$
	Newton's third law	$\vec{F}_{\text{A on B}} = -\vec{F}_{\text{B on A}}$

BASIC PROBLEM-SOLVING STRATEGY Use Newton's second law for each particle or object. Use Newton's third law to equate the magnitudes of the two members of an action/reaction pair.

Linear motion

$$\sum F_x = ma_x$$
$$\sum F_y = 0$$

or

$$\sum F_x = 0$$
$$\sum F_y = ma_y$$

Trajectory motion

$$\sum F_x = ma_x$$
$$\sum F_y = ma_y$$

Circular motion

$$\sum F_r = mv^2/r = m\omega^2 r$$
$$\sum F_t = 0 \text{ or } ma_t$$
$$\sum F_z = 0$$

Linear and trajectory kinematics

Uniform acceleration:
(a_s = constant)

$$v_{fs} = v_{is} + a_s \Delta t$$
$$s_f = s_i + v_{is}\Delta t + \tfrac{1}{2}a_s(\Delta t)^2$$
$$v_{fs}^2 = v_{is}^2 + 2a_s\Delta s$$

Trajectories: The same equations are used for both x and y.

Uniform motion:
($a = 0$, v_s = constant)

$$s_f = s_i + v_s \Delta t$$

Circular kinematics

Uniform circular motion:

$$T = 2\pi r/v = 2\pi/\omega$$
$$\theta_f = \theta_i + \omega\Delta t$$
$$a_r = v^2/r = \omega^2 r$$
$$v_t = \omega r$$

General case

$$v_s = ds/dt = \text{ slope of the position graph}$$
$$a_s = dv_s/dt = \text{ slope of the velocity graph}$$
$$v_{fs} = v_{is} + \int_{t_i}^{t_f} a_s\, dt = v_{is} + \text{ area under the acceleration curve}$$
$$s_f = s_i + \int_{t_i}^{t_f} v_s\, dt = s_i + \text{ area under the velocity curve}$$

Nonuniform circular motion:

$$\omega_f = \omega_i + \alpha\Delta t$$
$$\theta_f = \theta_i + \omega_i \Delta t + \tfrac{1}{2}\alpha(\Delta t)^2$$
$$\omega_f^2 = \omega_i^2 + 2\alpha\Delta\theta$$

The Forces of Nature

What are the fundamental forces of nature? That is, what set of distinct, irreducible forces can explain everything we know about nature? This is a question that has long intrigued physicists. For example, friction is not a fundamental force because it can be reduced to electric forces between atoms. What about other forces?

Physicists have long recognized three basic forces: the gravitational force, the electric force, and the magnetic force. The gravitational force is an inherent attraction between two masses. The electric force is a force between charges. The magnetic force, which is a bit more mysterious, causes compass needles to point north and holds your shopping list on the refrigerator door.

In the 1860s, the Scottish physicist James Clerk Maxwell developed a theory that *unified* the electric and magnetic forces into a single *electromagnetic force.* Where there had appeared to be two separate forces, Maxwell found there to be a single force that, under appropriate conditions, exhibits "electric behavior" or "magnetic behavior." Maxwell used his theory to predict the existence of *electromagnetic waves,* including light. Our entire telecommunications industry is testimony to Maxwell's genius.

Maxwell's electromagnetic force was soon found to be the "glue" holding atoms, molecules, and solids together. With the exception of gravity, *every* force we have considered so far can be traced to electromagnetic forces between atoms.

The discovery of the atomic nucleus, about 1910, presented difficulties that could not be explained by either gravitational or electromagnetic forces. The atomic nucleus is an unimaginably dense ball of protons and neutrons. But what holds it together against the repulsive electric forces between the protons? There must be an attractive force inside the nucleus that is stronger than the repulsive electric force. This force, called the *strong force,* is the force that holds atomic nuclei together. The strong force is a *short-range* force, extending only about 10^{-14} m. It is completely negligible outside the nucleus. The subatomic particles called *quarks,* of which you have likely heard, are part of our understanding of how the strong force works.

In the 1930s, physicists found that the nuclear radioactivity called *beta decay* could not be explained by either the electromagnetic or the strong force. Careful experiments established that the decay is due to a previously undiscovered force within the nucleus. The strength of this force is less than either the strong force or the electromagnetic force, so this new force was named the *weak force.* Although discovered in conjunction with radioactivity, it is now known to play an important role in the fusion reactions that power the stars.

By 1940, the recognized forces of nature were four: the gravitational force, the electromagnetic force, the strong force, and the weak force. Physicists were understandably curious whether all four of these were truly fundamental or if some of them could be further unified. Indeed, innovative work in the 1960s and 1970s produced a theory that unified the electromagnetic force and the weak force.

Predictions of this new theory were confirmed during the 1980s at some of the world's largest particle accelerators, and we now speak of the *electroweak force.* Under appropriate conditions, the electroweak force exhibits either "electromagnetic behavior" or "weak behavior." But under other conditions, new phenomena appear that are consequences of the full electroweak force. These conditions appear on earth only in the largest and most energetic particle accelerators, which is why we were not previously aware of the unified nature of these two forces. However, the earliest moments of the Big Bang provided the right conditions for the electroweak force to play a significant role. Thus a theory developed to help us understand the workings of nature on the smallest subatomic scale has unexpectedly given us powerful new insights into the origin of the universe.

The success of the electroweak theory has prompted efforts to unify the electroweak force and the strong force into a *grand unified theory.* Only time will tell if the strong force and the electroweak force are really just two different aspects of a single force, or if they are truly distinct. Some physicists even envision a day when all the forces of nature will be unified in a single theory, the so-called *Theory of Everything!* For today, however, our understanding of the forces of nature is in terms of three fundamental forces: the gravitational force, the electroweak force, and the strong force.

FIGURE I.1 A historical progression of our understanding of the fundamental forces of nature.

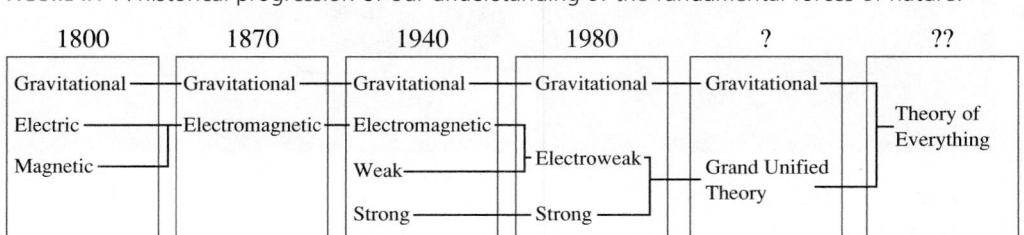

PART II

Conservation Laws

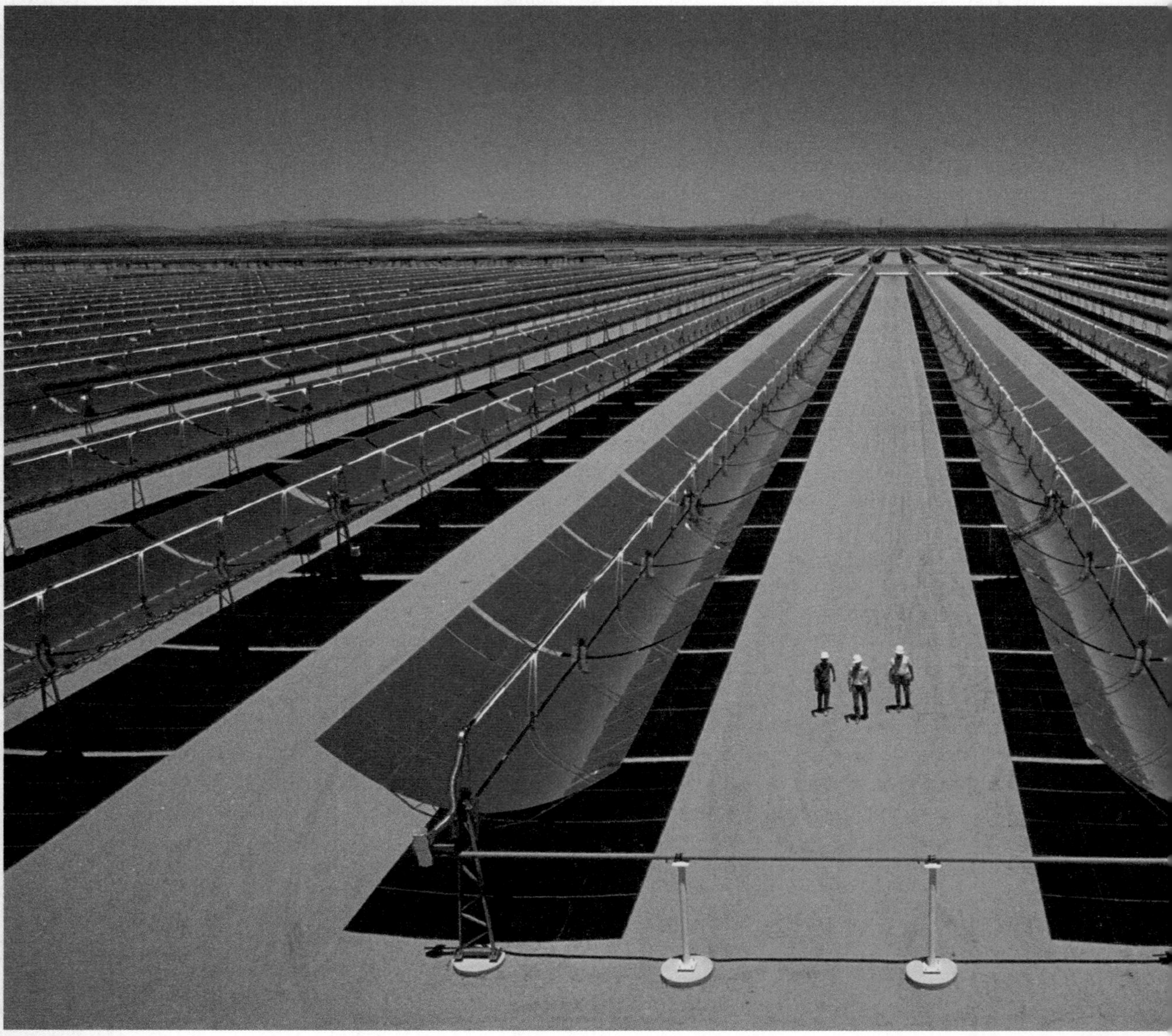

Energy is the lifeblood of modern society. This power plant in the Mojave Desert transforms solar energy into electrical energy and, unavoidably, increased thermal energy.

OVERVIEW
Why Some Things Don't Change

Part I of this textbook was about *change*. One particular type of change—motion—is governed by Newton's second law. Although Newton's second law is a very powerful statement, it isn't the whole story. Part II will now focus on things that *stay the same* as other things around them change.

Consider, for example, an explosive chemical reaction taking place inside a closed, sealed box. No matter how violent the explosion, the total mass of the products—the final mass M_f—is the same as the initial mass M_i of the reactants. In other words, matter cannot be created or destroyed, only rearranged. This is an important and powerful statement about nature.

A quantity that *stays the same* throughout an interaction is said to be *conserved*. Our knowledge about mass can be stated as a *conservation law*:

Law of conservation of mass The total mass in a closed system is constant. Mathematically, $M_f = M_i$.*

The qualification "in a closed system" is important. Mass certainly won't be conserved if you open the box halfway through and remove some of the matter. Other conservation laws we'll discover also have qualifications stating the circumstances under which they apply.

A system of interacting objects has another curious property. Each system is characterized by a certain number, and no matter how complex the interactions, the value of this number never changes. This number is called the *energy* of the system, and the fact that it never changes is called the *law of conservation of energy*. It is, perhaps, the single most important physical law ever discovered.

But what is energy? How do you determine the energy number for a system? These are not easy questions. Energy is an abstract idea, not as tangible or easy to picture as mass or force. Our modern concept of energy wasn't fully formulated until the middle of the 19th century, two hundred years after Newton, when the relationship between *energy* and *heat* was finally understood. That is a topic we will take up in Part IV, where the concept of energy will be found to be the basis of thermodynamics. But all that in due time. In Part II we will be content to introduce the concept of energy and show how energy can be a useful problem-solving tool. We'll also meet another quantity—*momentum*—that is conserved under the proper circumstances.

Conservation laws give us a new and different perspective on motion. This is not insignificant. You've seen optical illusions where a figure appears first one way, then another, even though the information has not changed. Likewise with motion. Some situations are most easily analyzed from the perspective of Newton's laws; others make more sense from a conservation-law perspective. An important goal of Part II is to learn which is better for a given problem.

*Surprisingly, Einstein's 1905 theory of relativity showed that there are circumstances in which mass is *not* conserved but can be converted to energy in accordance with his famous formula $E = mc^2$. Nonetheless, conservation of mass is an exceedingly good approximation in nearly all applications of science and engineering.

9 Impulse and Momentum

An exploding firework is a dramatic event. Nonetheless, the explosion obeys some simple laws of physics.

▶ **Looking Ahead** The goals of Chapter 9 are to understand and apply the new concepts of impulse and momentum.

Momentum

An object's **momentum** is the product of its mass and velocity: $\vec{p} = m\vec{v}$.

An object can have a large momentum by having a large mass or a large velocity.

Momentum is a vector. Paying attention to the *signs* of the components of momentum will be especially important.

You'll learn to write Newton's second law in terms of momentum.

Impulse

A force of short duration is an *impulsive force*. The **impulse** J_x is the area under the force-versus-time curve.

F_x

Impulse = J_x = area

t_i t_f t

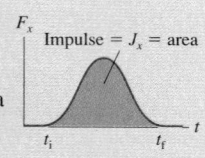

We say that the bat delivers an impulse to the ball.

The **impulse-momentum theorem** says that an impulse changes a particle's momentum: $\Delta p_x = J_x$.

Conservation Laws

Part I of this textbook was about how interactions cause things to change. Part II will explore how some things are *not* changed by the interactions. We say they are *conserved*.

The particles of an **isolated system** interact with each other—perhaps very intensely—but not with the external environment.

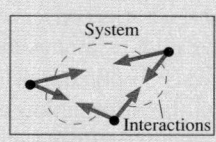

System

Interactions

Environment

The mass, the momentum, and the energy of an isolated system are conserved. Conservation laws will be the basis of a new and powerful problem solving strategy:

final value = initial value

Conservation of momentum for an isolated system is a consequence of Newton's third law.

◀ **Looking Back**
Sections 7.1–7.3 Action/reaction force pairs and Newton's third law

Representations

Conservation laws require new visual tools. You will learn to draw and use:

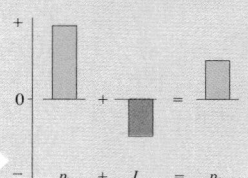

p_{ix} + J_x = p_{fx}

Momentum bar charts to show how impulse changes momentum.

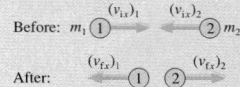

Before: m_1 ① $\xrightarrow{(v_{ix})_1}$ $\xleftarrow{(v_{ix})_2}$ ② m_2

After: $\xleftarrow{(v_{fx})_1}$ ①   ② $\xrightarrow{(v_{fx})_2}$

Before-and-after pictorial representations to compare quantities before and after an interaction.

Collisions and Explosions

You will learn to apply conservation of momentum to the analysis of *collisions* and *explosions*.

A **collision** is when two or more particles come together for a short but intense interaction.

① $\xrightarrow{(v_{ix})_1}$ $\xleftarrow{(v_{ix})_2}$ ②

An **explosion** is when a short but intense interaction causes two or more particles to move apart.

$\xleftarrow{(v_{fx})_1}$ ①   ② $\xrightarrow{(v_{fx})_2}$

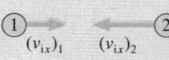

9.1 Momentum and Impulse

A **collision** is a short-duration interaction between two objects. The collision between a tennis ball and a racket, or a baseball and a bat, may seem instantaneous to your eye, but that is a limitation of your perception. A careful look at the photograph reveals that the right side of the ball is flattened and pressed up against the strings of the racket. It takes time to compress the ball, and more time for the ball to re-expand as it leaves the racket.

The duration of a collision depends on the materials from which the objects are made, but 1 to 10 ms (0.001 to 0.010 s) is fairly typical. This is the time during which the two objects are in contact with each other. The harder the objects, the shorter the contact time. A collision between two steel balls lasts less than 1 ms.

FIGURE 9.1 shows a microscopic view of a collision in which object A bounces off object B. The spring-like molecular bonds—the same bonds that cause normal forces and tension forces—compress during the collision, then re-expand as A bounces back. The forces $\vec{F}_{A\ on\ B}$ and $\vec{F}_{B\ on\ A}$ are an action/reaction pair and, according to Newton's third law, have equal magnitudes: $F_{A\ on\ B} = F_{B\ on\ A}$. The force increases rapidly as the bonds compress, reaches a maximum at the instant A is at rest (point of maximum compression), then decreases as the bonds re-expand.

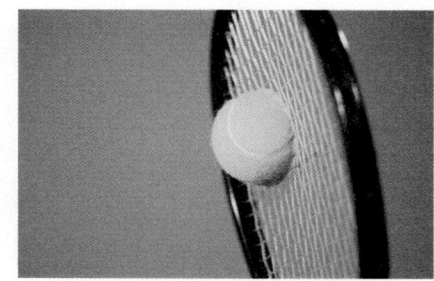

A tennis ball collides with a racket. Notice that the right side of the ball is flattened.

FIGURE 9.1 Atomic model of a collision.

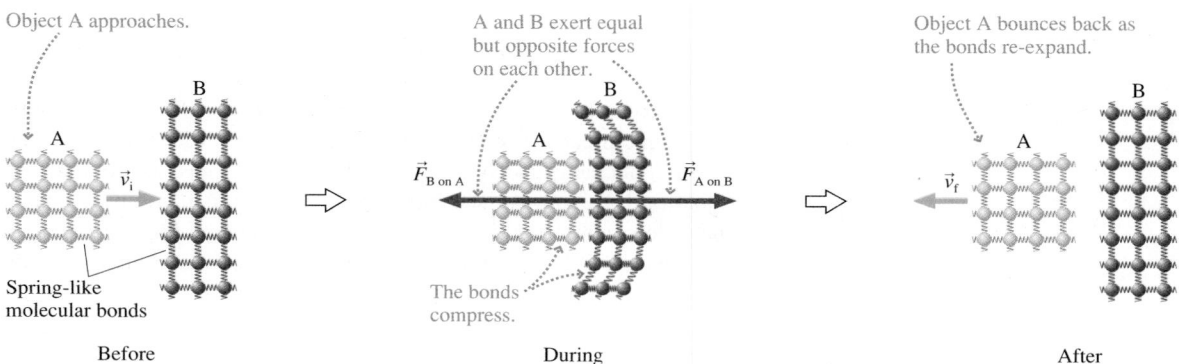

Object A approaches.

A and B exert equal but opposite forces on each other.

Object A bounces back as the bonds re-expand.

Spring-like molecular bonds

The bonds compress.

Before   During   After

A large force exerted for a small interval of time is called an **impulsive force**. FIGURE 9.2 shows that a particle undergoing a collision enters with initial velocity $\vec{v}_i$, experiences an impulsive force of short duration Δt, then leaves with final velocity $\vec{v}_f$. The graph shows how a typical impulsive force behaves, growing to a maximum and then decreasing back to zero. Because an impulsive force is a function of time, we will write it as $F_x(t)$.

NOTE ▶ Both v_x and F_x are components of vectors and thus have *signs* indicating which way the vectors point. ◀

We can use Newton's second law to find the final velocity. Acceleration in one dimension is $a_x = dv_x/dt$, so the second law is

$$ma_x = m\frac{dv_x}{dt} = F_x(t)$$

After multiplying both sides by dt, we can write the second law as

$$m\,dv_x = F_x(t)\,dt \tag{9.1}$$

The force is nonzero only during the interval of time from t_i to t_f, so let's integrate Equation 9.1 over this interval. The velocity changes from v_{ix} to v_{fx} during the collision; thus

$$m\int_{v_i}^{v_f} dv_x = mv_{fx} - mv_{ix} = \int_{t_i}^{t_f} F_x(t)\,dt \tag{9.2}$$

We need some new tools to help us make sense of Equation 9.2.

FIGURE 9.2 A particle undergoes a collision.

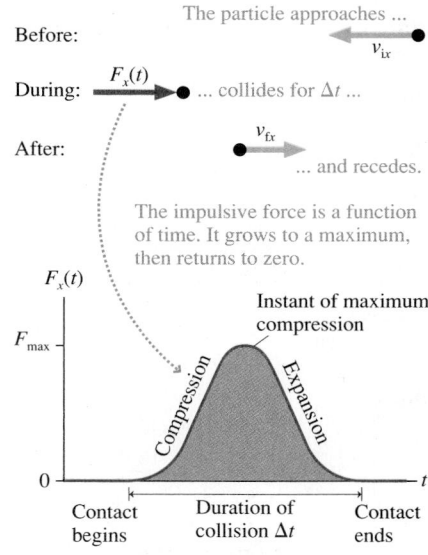

The particle approaches ...

Before:
v_{ix}

During: $F_x(t)$ ... collides for Δt ...

After: v_{fx}
... and recedes.

The impulsive force is a function of time. It grows to a maximum, then returns to zero.

$F_x(t)$

F_{max}

Instant of maximum compression

Compression   Expansion

0

Contact begins | Duration of collision Δt | Contact ends

Momentum

The product of a particle's mass and velocity is called the *momentum* of the particle:

$$\text{momentum} = \vec{p} \equiv m\vec{v} \tag{9.3}$$

Momentum, like velocity, is a vector. The units of momentum are kg m/s. The plural of "momentum" is "momenta," from its Latin origin.

The momentum vector $\vec{p}$ is parallel to the velocity vector $\vec{v}$. FIGURE 9.3 shows that $\vec{p}$, like any vector, can be decomposed into *x*- and *y*-components. Equation 9.3, which is a vector equation, is a shorthand way to write the simultaneous equations

$$p_x = mv_x$$
$$p_y = mv_y$$

An object can have a large momentum by having either a small mass but a large velocity (a bullet fired from a rifle) or a small velocity but a large mass (a large truck rolling at a slow 1 mph).

NOTE ▶ One of the most common errors in momentum problems is a failure to use the appropriate signs. The momentum component p_x has the same sign as v_x. Momentum is *negative* for a particle moving to the left (on the *x*-axis) or down (on the *y*-axis). ◀

Newton actually formulated his second law in terms of momentum rather than acceleration:

$$\vec{F} = m\vec{a} = m\frac{d\vec{v}}{dt} = \frac{d(m\vec{v})}{dt} = \frac{d\vec{p}}{dt} \tag{9.4}$$

This statement of the second law, saying that **force is the rate of change of momentum,** is more general than our earlier version $\vec{F} = m\vec{a}$. It allows for the possibility that the mass of the object might change, such as a rocket that is losing mass as it burns fuel.

Returning to Equation 9.2, you can see that mv_{ix} and mv_{fx} are p_{ix} and p_{fx}, the *x*-component of the particle's momentum before and after the collision. Further, $p_{fx} - p_{ix}$ is Δp_x, the *change* in the particle's momentum. In terms of momentum, Equation 9.2 is

$$\Delta p_x = p_{fx} - p_{ix} = \int_{t_i}^{t_f} F_x(t)\, dt \tag{9.5}$$

Now we need to examine the right-hand side of Equation 9.5.

Impulse

Equation 9.5 tells us that the particle's change in momentum is related to the time integral of the force. Let's define a quantity J_x called the *impulse* to be

$$\text{impulse} = J_x \equiv \int_{t_i}^{t_f} F_x(t)\, dt \tag{9.6}$$
$$= \text{area under the } F_x(t) \text{ curve between } t_i \text{ and } t_f$$

Strictly speaking, impulse has units of N s, but you should be able to show that N s are equivalent to kg m/s, the units of momentum.

The interpretation of the integral in Equation 9.6 as an area under a curve is especially important. FIGURE 9.4a portrays the impulse graphically. Because the force changes in a complicated way during a collision, it is often useful to describe the collision in terms of an *average* force F_{avg}. As FIGURE 9.4b shows, F_{avg} is the height of a rectangle that has the same area, and thus the same impulse, as the real force curve. The impulse exerted during the collision is

$$J_x = F_{avg}\, \Delta t \tag{9.7}$$

Equation 9.2, which we found by integrating Newton's second law, can now be rewritten in terms of impulse and momentum as

$$\Delta p_x = J_x \quad \text{(impulse-momentum theorem)} \tag{9.8}$$

FIGURE 9.3 A particle's momentum vector $\vec{p}$ can be decomposed into *x*- and *y*-components.

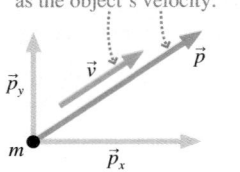

FIGURE 9.4 Looking at the impulse graphically.

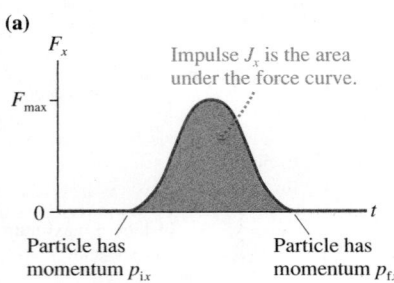

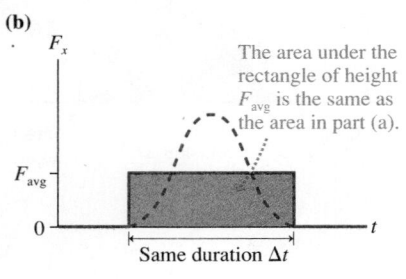

This result is called the **impulse-momentum theorem.** The name is rather unusual, but it's not the name that is important. The important new *idea* is that **an impulse delivered to a particle changes the particle's momentum.** The momentum p_{fx} "after" an interaction, such as a collision or an explosion, is equal to the momentum p_{ix} "before" the interaction *plus* the impulse that arises from the interaction:

$$p_{fx} = p_{ix} + J_x \qquad (9.9)$$

FIGURE 9.5 illustrates the impulse-momentum theorem for a rubber ball bouncing off a wall. Notice the signs; they are very important. The ball is initially traveling toward the right, so v_{ix} and p_{ix} are positive. After the bounce, v_{fx} and p_{fx} are negative. The force *on the ball* is toward the left, so F_x is also negative. The graphs show how the force and the momentum change with time.

Although the interaction is very complex, the impulse—the area under the force graph—is all we need to know to find the ball's velocity as it rebounds from the wall. The final momentum is

$$p_{fx} = p_{ix} + J_x = p_{ix} + \text{area under the force curve}$$

and the final velocity is $v_{fx} = p_{fx}/m$. In this example, the area has a negative value.

Momentum Bar Charts

The impulse-momentum theorem tells us that **impulse transfers momentum to an object.** If an object has 2 kg m/s of momentum, a 1 kg m/s impulse exerted on the object increases its momentum to 3 kg m/s. That is, $p_{fx} = p_{ix} + J_x$.

We can represent this "momentum accounting" with a **momentum bar chart.** **FIGURE 9.6a** shows a bar chart in which one unit of impulse adds to an initial two units of momentum to give three units of momentum. The bar chart of **FIGURE 9.6b** represents the ball colliding with a wall in Figure 9.5. Momentum bar charts are a tool for visualizing an interaction.

> **NOTE** ▶ The vertical scale of a momentum bar chart has no numbers; it can be adjusted to match any problem. However, be sure that all bars in a given problem use a consistent scale. ◀

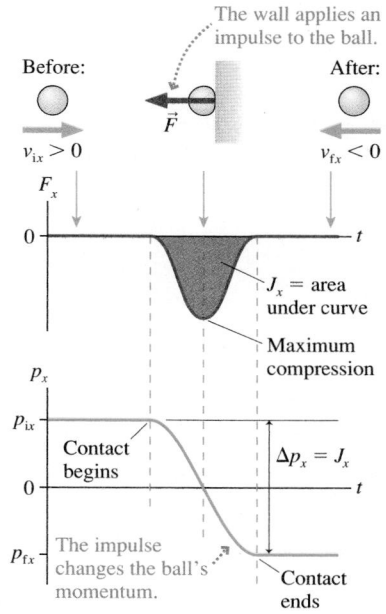

FIGURE 9.5 The impulse-momentum theorem helps us understand a rubber ball bouncing off a wall.

FIGURE 9.6 Two examples of momentum bar charts.

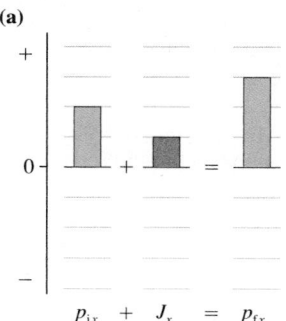

(a)

$$p_{ix} + J_x = p_{fx}$$

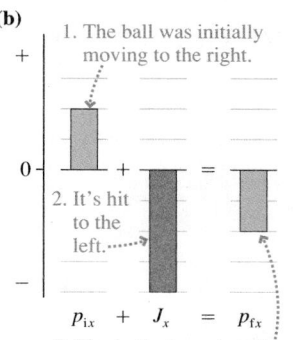

(b)

1. The ball was initially moving to the right.

2. It's hit to the left.

$$p_{ix} + J_x = p_{fx}$$

3. The ball rebounds to the left with no loss of speed.

STOP TO THINK 9.1   The cart's change of momentum is

a. -30 kg m/s
b. -20 kg m/s
c. 0 kg m/s
d. 10 kg m/s
e. 20 kg m/s
f. 30 kg m/s

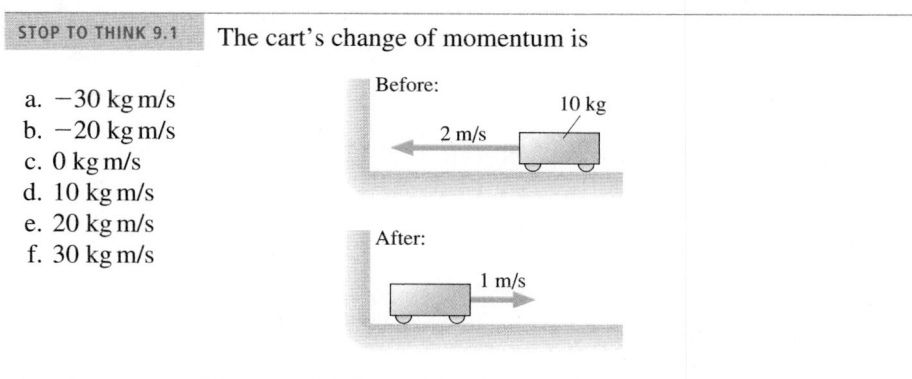

9.2 Solving Impulse and Momentum Problems

Pictorial representations have become an important problem-solving tool. The pictorial representations you learned to draw in Part I were oriented toward the use of Newton's laws and a subsequent kinematic analysis. For conservation-law problems we need a new representation, the **before-and-after pictorial representation.**

TACTICS
BOX 9.1 **Drawing a before-and-after pictorial representation**

❶ **Sketch the situation.** Use two drawings, labeled "Before" and "After," to show the objects *before* they interact and again *after* they interact.

❷ **Establish a coordinate system.** Select your axes to match the motion.

❸ **Define symbols.** Define symbols for the masses and for the velocities before and after the interaction. Position and time are not needed.

❹ **List known information.** Give the values of quantities that are known from the problem statement or that can be found quickly with simple geometry or unit conversions. Before-and-after pictures are simpler than the pictures for dynamics problems, so listing known information on the sketch is adequate.

❺ **Identify the desired unknowns.** What quantity or quantities will allow you to answer the question? These should have been defined in step 3.

❻ If appropriate, **draw a momentum bar chart** to clarify the situation and establish appropriate signs.

Exercises 17–19

NOTE ▶ The generic subscripts i and f, for *initial* and *final*, are adequate in equations for a simple problem, but using numerical subscripts, such as v_{1x} and v_{2x}, will help keep all the symbols straight in more complex problems. ◀

EXAMPLE 9.1 **Hitting a baseball**

A 150 g baseball is thrown with a speed of 20 m/s. It is hit straight back toward the pitcher at a speed of 40 m/s. The interaction force between the ball and the bat is shown in FIGURE 9.7. What *maximum* force F_{max} does the bat exert on the ball? What is the *average* force of the bat on the ball?

FIGURE 9.7 The interaction force between the baseball and the bat.

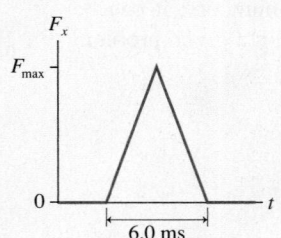

MODEL Model the baseball as a particle and the interaction as a collision.

VISUALIZE FIGURE 9.8 is a before-and-after pictorial representation. The steps from Tactics Box 9.1 are explicitly noted. Because F_x is positive (a force to the right), we know the ball was initially moving toward the left and is hit back toward the right. Thus we converted the statements about *speeds* into information about *velocities*, with v_{ix} negative.

SOLVE Until now we've consistently started the mathematical representation with Newton's second law. Now we want to use the impulse-momentum theorem:

$$\Delta p_x = J_x = \text{area under the force curve}$$

We know the velocities before and after the collision, so we can calculate the ball's momenta:

$$p_{ix} = mv_{ix} = (0.15 \text{ kg})(-20 \text{ m/s}) = -3.0 \text{ kg m/s}$$
$$p_{fx} = mv_{fx} = (0.15 \text{ kg})(40 \text{ m/s}) = 6.0 \text{ kg m/s}$$

FIGURE 9.8 A before-and-after pictorial representation.

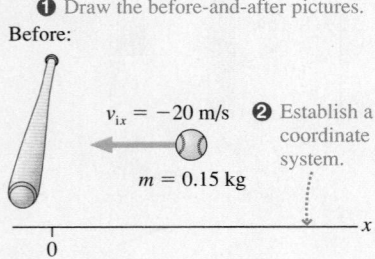

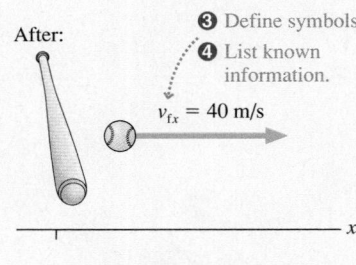

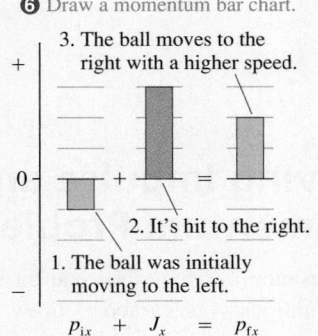

Thus the *change* in momentum is

$$\Delta p_x = p_{fx} - p_{ix} = 9.0 \text{ kg m/s}$$

The force curve is a triangle with height F_{max} and width 6.0 ms. The area under the curve is

$$J_x = \text{area} = \frac{1}{2} \times F_{max} \times (0.0060 \text{ s}) = (F_{max})(0.0030 \text{ s})$$

According to the impulse-momentum theorem,

$$9.0 \text{ kg m/s} = (F_{max})(0.0030 \text{ s})$$

Thus the *maximum* force is

$$F_{max} = \frac{9.0 \text{ kg m/s}}{0.0030 \text{ s}} = 3000 \text{ N}$$

The *average* force, which depends on the collision duration $\Delta t = 0.0060$ s, has the smaller value:

$$F_{avg} = \frac{J_x}{\Delta t} = \frac{\Delta p_x}{\Delta t} = \frac{9.0 \text{ kg m/s}}{0.0060 \text{ s}} = 1500 \text{ N}$$

ASSESS F_{max} is a large force, but quite typical of the impulsive forces during collisions. The main thing to focus on is our new perspective: An impulse changes the momentum of an object.

Other forces often act on an object during a collision or other brief interaction. In Example 9.1, for instance, the baseball is also acted on by gravity. Usually these other forces are *much* smaller than the interaction forces. The 1.5 N weight of the ball is vastly less than the 3000 N force of the bat on the ball. We can reasonably neglect these small forces *during* the brief time of the impulsive force by using what is called the **impulse approximation.**

When we use the impulse approximation, p_{ix} and p_{fx} (and v_{ix} and v_{fx}) are the momenta (and velocities) *immediately* before and *immediately* after the collision. For example, the velocities in Example 9.1 are those of the ball just before and after it collides with the bat. We could then do a follow-up problem, including gravity and drag, to find the ball's speed a second later as the second baseman catches it. We'll look at some two-part examples later in the chapter.

EXAMPLE 9.2 **A bouncing ball**

A 100 g rubber ball is dropped from a height of 2.00 m onto a hard floor. FIGURE 9.9 shows the force that the floor exerts on the ball. How high does the ball bounce?

MODEL Model the ball as a particle subjected to an impulsive force while in contact with the floor. Using the impulse approximation, we'll neglect gravity during these 8.00 ms. The fall and subsequent rise are free-fall motion.

FIGURE 9.9 The force of the floor on a bouncing rubber ball.

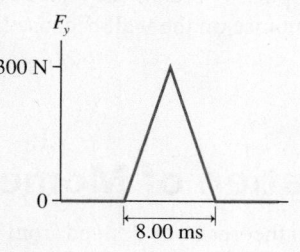

VISUALIZE FIGURE 9.10 is a pictorial representation. Here we have a three-part problem (downward free fall, impulsive collision, upward free fall), so the pictorial motion includes both the before and after of the collision (v_{1y} changing to v_{2y}) and the beginning and end of the free-fall motion. The bar chart shows the momentum change during the brief collision. Note that p is negative for downward motion.

FIGURE 9.10 Pictorial representation of the ball and a momentum bar chart of the collision with the floor.

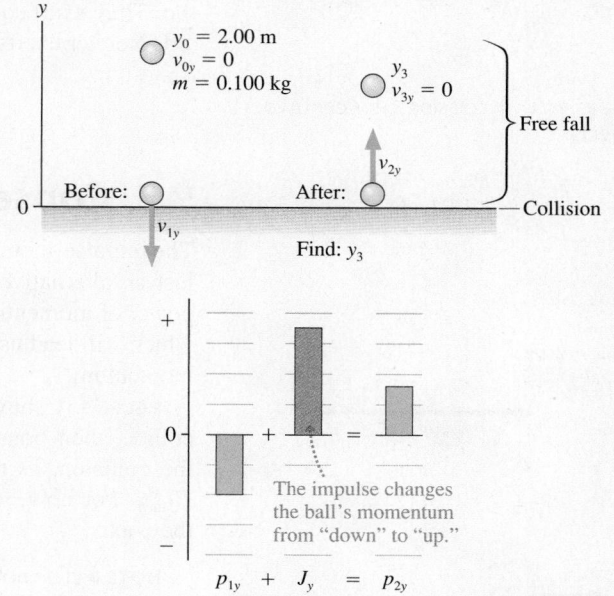

Continued

SOLVE Velocity v_{1y}, the ball's velocity *immediately* before the collision, is found using free-fall kinematics with $\Delta y = -2.0$ m:

$$v_{1y}^2 = v_{0y}^2 - 2g\Delta y = 0 - 2g\Delta y$$

$$v_{1y} = \sqrt{-2g\Delta y} = \sqrt{-2(9.80 \text{ m/s}^2)(-2.00 \text{ m})} = -6.26 \text{ m/s}$$

We've chosen the negative root because the ball is moving in the negative y-direction.

The impulse-momentum theorem is $p_{2y} = p_{1y} + J_y$. The initial momentum, just before the collision, is $p_{1y} = mv_{1y} = -0.626$ kg m/s. The force of the floor is upward, so J_y is positive. From Figure 9.9, the impulse J_y is

$$J_y = \text{area under the force curve} = \frac{1}{2} \times (300 \text{ N}) \times (0.00800 \text{ s})$$

$$= 1.200 \text{ N s}$$

Thus

$$p_{2y} = p_{1y} + J_y = (-0.626 \text{ kg m/s}) + 1.200 \text{ N s} = 0.574 \text{ kg m/s}$$

and the post-collision velocity is

$$v_{2y} = \frac{p_{2y}}{m} = \frac{0.574 \text{ kg m/s}}{0.100 \text{ kg}} = 5.74 \text{ m/s}$$

The rebound speed is less than the impact speed, as expected. Finally a second use of free-fall kinematics yields

$$v_{3y}^2 = 0 = v_{2y}^2 - 2g\Delta y = v_{2y}^2 - 2gy_3$$

$$y_3 = \frac{v_{2y}^2}{2g} = \frac{(5.74 \text{ m/s})^2}{2(9.80 \text{ m/s}^2)} = 1.68 \text{ m}$$

The ball bounces back to a height of 1.68 m.

ASSESS The ball bounces back to less than its initial height, which is realistic.

NOTE ▶ Example 9.2 illustrates an important point: The impulse-momentum theorem applies *only* during the brief interval in which an impulsive force is applied. Many problems will have segments of the motion that must be analyzed with kinematics or Newton's laws. The impulse-momentum theorem is a new and useful tool, but it doesn't replace all that you've learned up until now. ◀

STOP TO THINK 9.2   A 10 g rubber ball and a 10 g clay ball are thrown at a wall with equal speeds. The rubber ball bounces, the clay ball sticks. Which ball exerts a larger impulse on the wall?

a. The clay ball exerts a larger impulse because it sticks.
b. The rubber ball exerts a larger impulse because it bounces.
c. They exert equal impulses because they have equal momenta.
d. Neither exerts an impulse on the wall because the wall doesn't move.

FIGURE 9.11 A collision between two objects.

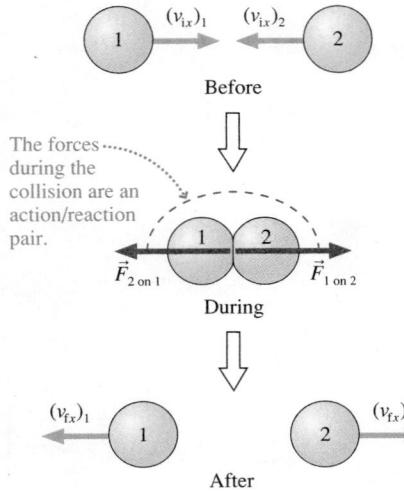

9.3 Conservation of Momentum

The impulse-momentum theorem was derived from Newton's second law and is really just an alternative way of looking at single-particle dynamics. To discover the real power of momentum for problem solving, we need also to invoke Newton's third law, which will lead us to one of the most important principles in physics: conservation of momentum.

FIGURE 9.11 shows two objects with initial velocities $(v_{ix})_1$ and $(v_{ix})_2$. The objects collide, then bounce apart with final velocities $(v_{fx})_1$ and $(v_{fx})_2$. The forces during the collision, as the objects are interacting, are the action/reaction pair $\vec{F}_{1 \text{ on } 2}$ and $\vec{F}_{2 \text{ on } 1}$. For now, we'll continue to assume that the motion is one dimensional along the x-axis.

NOTE ▶ The notation, with all the subscripts, may seem excessive. But there are two objects, and each has an initial and a final velocity, so we need to distinguish among four different velocities. ◀

Newton's second law for each object *during* the collision is

$$\frac{d(p_x)_1}{dt} = (F_x)_{2 \text{ on } 1}$$

$$\frac{d(p_x)_2}{dt} = (F_x)_{1 \text{ on } 2} = -(F_x)_{2 \text{ on } 1} \tag{9.10}$$

We made explicit use of Newton's third law in the second equation.

Although Equations 9.10 are for two different objects, suppose—just to see what happens—we were to *add* these two equations. If we do, we find that

$$\frac{d(p_x)_1}{dt} + \frac{d(p_x)_2}{dt} = \frac{d}{dt}\left[(p_x)_1 + (p_x)_2\right] = (F_x)_{2 \text{ on } 1} + (-(F_x)_{2 \text{ on } 1}) = 0 \quad (9.11)$$

If the time derivative of the quantity $(p_x)_1 + (p_x)_2$ is zero, it must be the case that

$$(p_x)_1 + (p_x)_2 = \text{constant} \tag{9.12}$$

Equation 9.12 is a conservation law! If $(p_x)_1 + (p_x)_2$ is a constant, then the sum of the momenta *after* the collision equals the sum of the momenta *before* the collision. That is,

$$(p_{fx})_1 + (p_{fx})_2 = (p_{ix})_1 + (p_{ix})_2 \tag{9.13}$$

Furthermore, this equality is independent of the interaction force. We don't need to know anything about $\vec{F}_{1 \text{ on } 2}$ and $\vec{F}_{2 \text{ on } 1}$ to make use of Equation 9.13.

As an example, **FIGURE 9.12** is a before-and-after pictorial representation of two equal-mass train cars colliding and coupling. Equation 9.13 relates the momenta of the cars after the collision to their momenta before the collision:

$$m_1(v_{fx})_1 + m_2(v_{fx})_2 = m_1(v_{ix})_1 + m_2(v_{ix})_2$$

Initially, car 1 is moving with velocity $(v_{ix})_1 = v_i$ while car 2 is at rest. Afterward, they roll together with the common final velocity v_f. Furthermore, $m_1 = m_2 = m$. With this information, the sum of the momenta is

$$mv_f + mv_f = 2mv_f = mv_i + 0$$

The mass cancels, and we find that the train cars' final velocity is $v_f = \frac{1}{2}v_i$. That is, we can make the very simple prediction that the final speed is exactly half the initial speed of car 1 without knowing anything at all about the very complex interaction between the two cars as they collide.

Law of Conservation of Momentum

Equation 9.13 illustrates the idea of a conservation law for momentum, but it was derived for the specific case of two particles colliding in one dimension. Our goal is to develop a more general law of conservation of momentum, a law that will be valid in three dimensions and that will work for any type of interaction. The next few paragraphs are fairly mathematical, so you might want to begin by looking ahead to Equations 9.21 and the statement of the law of conservation of momentum to see where we're heading.

Consider a system consisting of N particles. **FIGURE 9.13** shows a simple case where $N = 3$. The particles might be large entities (cars, baseballs, etc.), or they might be the microscopic atoms in a gas. We can identify each particle by an identification number k. Every particle in the system *interacts* with every other particle via action/reaction pairs of forces $\vec{F}_{j \text{ on } k}$ and $\vec{F}_{k \text{ on } j}$. In addition, every particle is subjected to possible *external forces* $\vec{F}_{\text{ext on } k}$ from agents outside the system.

If particle k has velocity $\vec{v}_k$, its momentum is $\vec{p}_k = m_k\vec{v}_k$. We define the **total momentum** $\vec{P}$ of the system as the vector sum

$$\vec{P} = \text{total momentum} = \vec{p}_1 + \vec{p}_2 + \vec{p}_3 + \cdots + \vec{p}_N = \sum_{k=1}^{N}\vec{p}_k \tag{9.14}$$

FIGURE 9.12 Two colliding train cars.

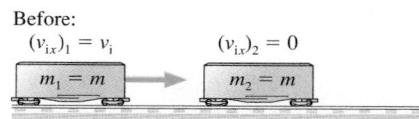
Before:
$(v_{ix})_1 = v_i$ $(v_{ix})_2 = 0$
$m_1 = m$ $m_2 = m$

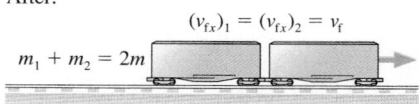
After:
$(v_{fx})_1 = (v_{fx})_2 = v_f$
$m_1 + m_2 = 2m$

FIGURE 9.13 A system of particles.

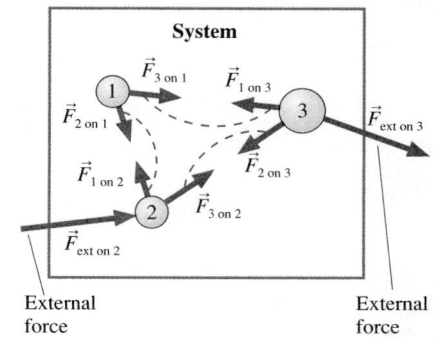
External force   External force

The total momentum of the rocket + gases system is conserved, so the rocket accelerates forward as the gases are expelled backward.

In other words, the total momentum *of the system* is the vector sum of all the individual momenta.

The time derivative of $\vec{P}$ tells us how the total momentum of the system changes with time:

$$\frac{d\vec{P}}{dt} = \sum_k \frac{d\vec{p}_k}{dt} = \sum_k \vec{F}_k \qquad (9.15)$$

where we used Newton's second law for each particle in the form $\vec{F}_k = d\vec{p}_k/dt$, which was Equation 9.4.

The net force acting on particle k can be divided into *external forces,* from outside the system, and *interaction forces* due to all the other particles in the system:

$$\vec{F}_k = \sum_{j \neq k} \vec{F}_{j \text{ on } k} + \vec{F}_{\text{ext on } k} \qquad (9.16)$$

The restriction $j \neq k$ expresses the fact that particle k does not exert a force on itself. Using this in Equation 9.15 gives the rate of change of the total momentum $\vec{P}$ of the system:

$$\frac{d\vec{P}}{dt} = \sum_k \sum_{j \neq k} \vec{F}_{j \text{ on } k} + \sum_k \vec{F}_{\text{ext on } k} \qquad (9.17)$$

The double sum on $\vec{F}_{j \text{ on } k}$ adds *every* interaction force within the system. But the interaction forces come in action/reaction pairs, with $\vec{F}_{k \text{ on } j} = -\vec{F}_{j \text{ on } k}$, so $\vec{F}_{k \text{ on } j} + \vec{F}_{j \text{ on } k} = \vec{0}$. Consequently, **the sum of all the interaction forces is zero.** As a result, Equation 9.17 becomes

$$\frac{d\vec{P}}{dt} = \sum_k \vec{F}_{\text{ext on } k} = \vec{F}_{\text{net}} \qquad (9.18)$$

where $\vec{F}_{\text{net}}$ is the net force exerted on the system by agents outside the system. But this is just Newton's second law written for the system as a whole! That is, **the rate of change of the total momentum of the system is equal to the net force applied to the system.**

Equation 9.18 has two very important implications. First, we can analyze the motion of the system as a whole without needing to consider interaction forces between the particles that make up the system. In fact, we have been using this idea all along as an *assumption* of the particle model. When we treat cars and rocks and baseballs as particles, we assume that the internal forces between the atoms—the forces that hold the object together—do not affect the motion of the object as a whole. Now we have *justified* that assumption.

The second implication of Equation 9.18, and the more important one from the perspective of this chapter, applies to what we call an *isolated system.* An **isolated system** is a system for which the *net* external force is zero: $\vec{F}_{\text{net}} = \vec{0}$. That is, an isolated system is one on which there are *no* external forces or for which the external forces are balanced and add to zero.

For an isolated system, Equation 9.18 is simply

$$\frac{d\vec{P}}{dt} = \vec{0} \qquad \text{(isolated system)} \qquad (9.19)$$

In other words, the *total* momentum of an isolated system does not change. The total momentum $\vec{P}$ remains constant, *regardless* of whatever interactions are going on *inside* the system. The importance of this result is sufficient to elevate it to a law of nature, alongside Newton's laws.

Law of conservation of momentum The total momentum $\vec{P}$ of an isolated system is a constant. Interactions within the system do not change the system's total momentum.

Mathematically, the law of conservation of momentum for an isolated system is

$$\vec{P}_f = \vec{P}_i \qquad (9.20)$$

The total momentum after an interaction is equal to the total momentum before the interaction. Because Equation 9.20 is a vector equation, the equality is true for each of the components of the momentum vector. That is,

$$(p_{fx})_1 + (p_{fx})_2 + (p_{fx})_3 + \cdots = (p_{ix})_1 + (p_{ix})_2 + (p_{ix})_3 + \cdots$$
$$(p_{fy})_1 + (p_{fy})_2 + (p_{fy})_3 + \cdots = (p_{iy})_1 + (p_{iy})_2 + (p_{iy})_3 + \cdots \qquad (9.21)$$

The *x*-equation is an extension of Equation 9.13 to *N* interacting particles.

NOTE ▶ It is worth emphasizing the critical role of Newton's third law. The law of conservation of momentum is a direct consequence of the fact that interactions within an isolated system are action/reaction pairs. ◀

EXAMPLE 9.3 A glider collision

A 250 g air-track glider is pushed across a level track toward a 500 g glider that is at rest. FIGURE 9.14 shows a position-versus-time graph of the 250 g glider as recorded by a motion detector. Best-fit lines have been found. What is the speed of the 500 g glider after the collision?

FIGURE 9.14 Position graph of the 250 g glider.

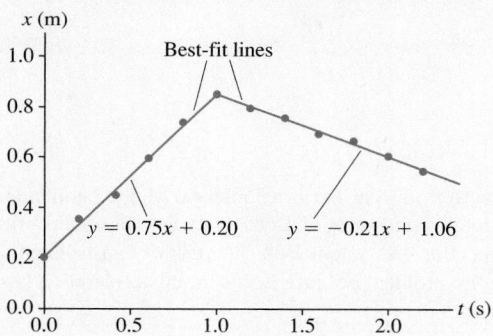

MODEL The two gliders, modeled as particles, are the system. The gliders interact with each other, but the external forces (normal force and gravity) balance to make $\vec{F}_{net} = \vec{0}$. Thus the gliders form an isolated system and their total momentum is conserved.

VISUALIZE FIGURE 9.15 is a before-and-after pictorial representation. The graph of Figure 9.14 tells us that the 250 g glider initially moves to the right, collides at $t = 1.0$ s, then rebounds to the left (decreasing x).

SOLVE Conservation of momentum for this one-dimensional problem requires that the final momentum equal the initial momentum: $P_{fx} = P_{ix}$. In terms of the individual components, conservation of momentum is

$$(p_{fx})_1 + (p_{fx})_2 = (p_{ix})_1 + (p_{ix})_2$$

Each momentum is mv_x, so conservation of momentum in terms of velocities is

$$m_1(v_{fx})_1 + m_2(v_{fx})_2 = m_1(v_{ix})_1 + m_2(v_{ix})_2 = m_1(v_{ix})_1$$

FIGURE 9.15 Before-and-after pictorial representation of a glider collision.

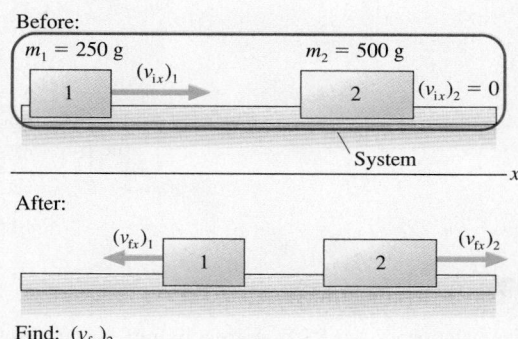

where, in the last step, we used $(v_{ix})_2 = 0$ for the 500 g glider. Solving for the heavier glider's final velocity gives

$$(v_{fx})_2 = \frac{m_1}{m_2}\left[(v_{ix})_1 - (v_{fx})_1\right]$$

From Chapter 2 kinematics, the velocities of the 250 g glider before and after the collision are the slopes of the position-versus-time graph. Referring to Figure 9.14, we see that $(v_{ix})_1 = 0.75$ m/s and $(v_{fx})_1 = -0.21$ m/s. The latter is negative because the rebound motion is to the left. Thus

$$(v_{fx})_2 = \frac{250\ g}{500\ g}\left[0.75\ \text{m/s} - (-0.21\ \text{m/s})\right] = 0.48\ \text{m/s}$$

The 500 g glider moves away from the collision at 0.48 m/s.

ASSESS The 500 g glider has twice the mass of the glider that was pushed, so a somewhat smaller speed seems reasonable. Paying attention to the *signs*—which are positive and which negative—was very important for reaching a correct answer. We didn't convert the masses to kilograms because only the mass *ratio* of 0.50 was needed.

A Strategy for Conservation of Momentum Problems

PROBLEM-SOLVING
STRATEGY 9.1 **Conservation of momentum**

MODEL Clearly define *the system.*

■ If possible, choose a system that is isolated ($\vec{F}_{net} = \vec{0}$) or within which the interactions are sufficiently short and intense that you can ignore external forces for the duration of the interaction (the impulse approximation). Momentum is conserved.

■ If it's not possible to choose an isolated system, try to divide the problem into parts such that momentum is conserved during one segment of the motion. Other segments of the motion can be analyzed using Newton's laws or, as you'll learn in Chapters 10 and 11, conservation of energy.

VISUALIZE Draw a before-and-after pictorial representation. Define symbols that will be used in the problem, list known values, and identify what you're trying to find.

SOLVE The mathematical representation is based on the law of conservation of momentum: $\vec{P}_f = \vec{P}_i$. In component form, this is

$$(p_{fx})_1 + (p_{fx})_2 + (p_{fx})_3 + \cdots = (p_{ix})_1 + (p_{ix})_2 + (p_{ix})_3 + \cdots$$

$$(p_{fy})_1 + (p_{fy})_2 + (p_{fy})_3 + \cdots = (p_{iy})_1 + (p_{iy})_2 + (p_{iy})_3 + \cdots$$

ASSESS Check that your result has the correct units, is reasonable, and answers the question.

Exercise 16 ▨

EXAMPLE 9.4 **Rolling away**

Bob sees a stationary cart 8.0 m in front of him. He decides to run to the cart as fast as he can, jump on, and roll down the street. Bob has a mass of 75 kg and the cart's mass is 25 kg. If Bob accelerates at a steady 1.0 m/s², what is the cart's speed just after Bob jumps on?

MODEL This is a two-part problem. First Bob accelerates across the ground. Then Bob lands on and sticks to the cart, a "collision" between Bob and the cart. The interaction forces between Bob and the cart (i.e., friction) act only over the fraction of a second it takes Bob's feet to become stuck to the cart. Using the impulse approximation allows the system Bob + cart to be treated as an

isolated system during the brief interval of the "collision," and thus the total momentum of Bob + cart is conserved during this interaction. But the system Bob + cart is *not* an isolated system for the entire problem because Bob's initial acceleration has nothing to do with the cart.

VISUALIZE Our strategy is to divide the problem into an *acceleration* part, which we can analyze using kinematics, and a *collision* part, which we can analyze with momentum conservation. The pictorial representation of FIGURE 9.16 includes information about both parts. Notice that Bob's velocity $(v_{1x})_B$ at the end of his run is his "before" velocity for the collision.

FIGURE 9.16 Pictorial representation of Bob and the cart.

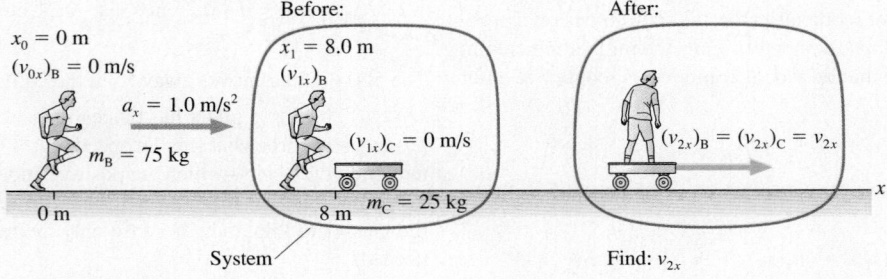

SOLVE The first part of the mathematical representation is kinematics. We don't know how long Bob accelerates, but we do know his acceleration and the distance. Thus

$$(v_{1x})_B^2 = (v_{0x})_B^2 + 2a_x \Delta x = 2a_x x_1$$

His velocity after accelerating for 8.0 m is

$$(v_{1x})_B = \sqrt{2a_x x_1} = 4.0 \text{ m/s}$$

The second part of the problem, the collision, uses conservation of momentum: $P_{2x} = P_{1x}$. Equation 9.21 is

$$m_B (v_{2x})_B + m_C (v_{2x})_C = m_B (v_{1x})_B + m_C (v_{1x})_C = m_B (v_{1x})_B$$

where we've used $(v_{1x})_C = 0$ m/s because the cart starts at rest. In this problem, Bob and the cart move together at the end with a common velocity, so we can replace both $(v_{2x})_B$ and $(v_{2x})_C$ with simply v_{2x}. Solving for v_{2x}, we find

$$v_{2x} = \frac{m_B}{m_B + m_C}(v_{1x})_B = \frac{75 \text{ kg}}{100 \text{ kg}} \times 4.0 \text{ m/s} = 3.0 \text{ m/s}$$

The cart's speed is 3.0 m/s immediately after Bob jumps on.

Notice how easy this was! No forces, no acceleration constraints, no simultaneous equations. Why didn't we think of this before? Conservation laws are indeed powerful, but they can answer only certain questions. Had we wanted to know how far Bob slid across the cart before sticking to it, how long the slide took, or what the cart's acceleration was during the collision, we would not have been able to answer such questions on the basis of the conservation law. There is a price to pay for finding a simple connection between before and after, and that price is the loss of information about the details of the interaction. If we are satisfied with knowing only about before and after, then conservation laws are a simple and straightforward way to proceed. But many problems *do* require us to understand the interaction, and for these there is no avoiding Newton's laws.

It Depends on the System

The first step in the problem-solving strategy asks you to clearly define *the system*. This is worth emphasizing because many problem-solving errors arise from trying to apply momentum conservation to an inappropriate system. **The goal is to choose a system whose momentum will be conserved.** Even then, it is the *total* momentum of the system that is conserved, not the momenta of the individual particles within the system.

As an example, consider what happens if you drop a rubber ball and let it bounce off a hard floor. Is momentum conserved during the collision of the ball with the floor? You might be tempted to answer yes because the ball's rebound speed is very nearly equal to its impact speed. But there are two errors in this reasoning.

First, momentum depends on *velocity*, not speed. The ball's velocity and momentum just before the collision are negative. They are positive after the collision. Even if their magnitudes are equal, the ball's momentum after the collision is *not* equal to its momentum before the collision.

But more important, we haven't defined the system. The momentum of what? Whether or not momentum is conserved depends on the system. FIGURE 9.17 shows two different choices of systems. In FIGURE 9.17a, where the ball itself is chosen as the system, the gravitational force of the earth on the ball is an external force. This force causes the ball to accelerate toward the earth, changing the ball's momentum. The force of the floor on the ball is also an external force. The impulse of $\vec{F}_{\text{floor on ball}}$ changes the ball's momentum from "down" to "up" as the ball bounces. The momentum of this system is most definitely *not* conserved.

FIGURE 9.17b shows a different choice. Here the system is ball + earth. Now the gravitational forces and the impulsive forces of the collision are interactions *within* the system. This is an isolated system, so the *total* momentum $\vec{P} = \vec{p}_{\text{ball}} + \vec{p}_{\text{earth}}$ is conserved.

In fact, the total momentum is $\vec{P} = \vec{0}$. Before you release the ball, both the ball and the earth are at rest (in the earth's reference frame). The total momentum is zero before

FIGURE 9.17 Whether or not momentum is conserved as a ball falls to earth depends on your choice of the system.

(a)

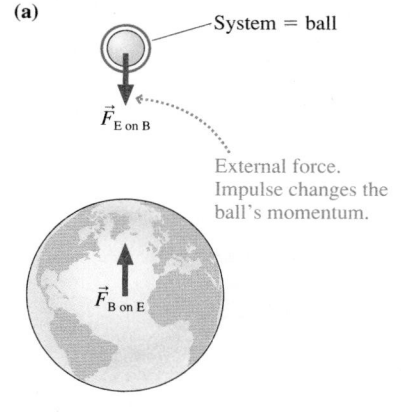

System = ball

$\vec{F}_{\text{E on B}}$

$\vec{F}_{\text{B on E}}$

External force. Impulse changes the ball's momentum.

(b)

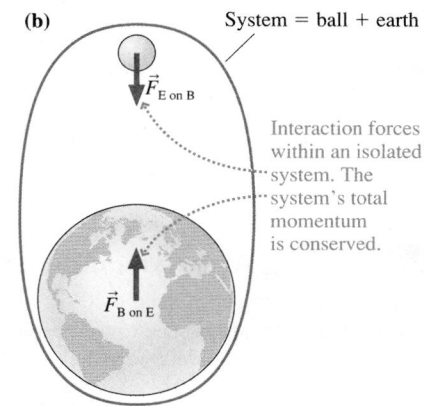

System = ball + earth

$\vec{F}_{\text{E on B}}$

$\vec{F}_{\text{B on E}}$

Interaction forces within an isolated system. The system's total momentum is conserved.

you release the ball, so it will *always* be zero. Just before the ball hits the floor with velocity v_{By}, it must be the case that $m_B v_{By} + m_E v_{Ey} = 0$ and thus

$$v_{Ey} = -\frac{m_B}{m_E} v_{By}$$

In other words, as the ball is pulled down toward the earth, the ball pulls up on the earth (action/reaction pair of forces) until the entire earth reaches velocity v_{Ey}. The earth's momentum is equal and opposite to the ball's momentum.

Why don't we notice the earth "leaping up" toward us each time we drop something? Because of the earth's enormous mass relative to everyday objects. A typical rubber ball has a mass of 60 g and hits the ground with a velocity of about −5 m/s. The earth's upward velocity is thus

$$v_{Ey} \approx -\frac{6 \times 10^{-2} \text{ kg}}{6 \times 10^{24} \text{ kg}}(-5 \text{ m/s}) = 5 \times 10^{-26} \text{ m/s}$$

The earth does, indeed, have a momentum equal and opposite to that of the ball, but the earth is so massive that it needs only an infinitesimal velocity to match the ball's momentum. At this speed, it would take the earth 300 million years to move the diameter of an atom!

STOP TO THINK 9.3 Objects A and C are made of different materials, with different "springiness," but they have the same mass and are initially at rest. When ball B collides with object A, the ball ends up at rest. When ball B is thrown with the same speed and collides with object C, the ball rebounds to the left. Compare the velocities of A and C after the collisions. Is v_A greater than, equal to, or less than v_C?

Before: After:

B ⟶ A   ○   A ⟶ v_A
         $v = 0$
$m_A = m_C$

B ⟶ C   ⟵ ○   C ⟶ v_C

9.4 Inelastic Collisions

Collisions can have different possible outcomes. A rubber ball dropped on the floor bounces, but a ball of clay sticks to the floor without bouncing. A golf club hitting a golf ball causes the ball to rebound away from the club, but a bullet striking a block of wood embeds itself in the block.

A collision in which the two objects stick together and move with a common final velocity is called a **perfectly inelastic collision.** The clay hitting the floor and the bullet embedding itself in the wood are examples of perfectly inelastic collisions. Other examples include railroad cars coupling together upon impact and darts hitting a dart board. FIGURE 9.18 emphasizes the fact that the two objects have a common final velocity after they collide.

In an *elastic collision,* by contrast, the two objects bounce apart. We've looked at some examples of elastic collisions, but a full analysis requires ideas about energy. We will return to elastic collisions in Chapter 10.

FIGURE 9.18 An inelastic collision.

Two objects approach and collide.

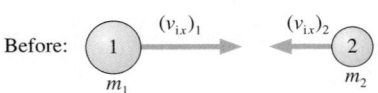

Before:

They stick and move together.

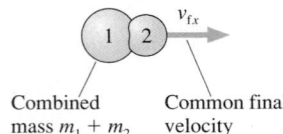

After:

Combined mass $m_1 + m_2$ Common final velocity

EXAMPLE 9.5 **An inelastic glider collision**

In a laboratory experiment, a 200 g air-track glider and a 400 g air-track glider are pushed toward each other from opposite ends of the track. The gliders have Velcro tabs on the front and will stick together when they collide. The 200 g glider is pushed with an initial speed of 3.0 m/s. The collision causes it to reverse direction at 0.40 m/s. What was the initial speed of the 400 g glider?

MODEL Model the gliders as particles. Define the two gliders together as the system. This is an isolated system, so its total momentum is conserved in the collision. The gliders stick together, so this is a perfectly inelastic collision.

VISUALIZE **FIGURE 9.19** shows a pictorial representation. We've chosen to let the 200 g glider (glider 1) start out moving to the right, so $(v_{ix})_1$ is a positive 3.0 m/s. The gliders move to the left after the collision, so their common final velocity is $v_{fx} = -0.40$ m/s.

SOLVE The law of conservation of momentum, $P_{fx} = P_{ix}$, is

$$(m_1 + m_2)v_{fx} = m_1(v_{ix})_1 + m_2(v_{ix})_2$$

where we made use of the fact that the combined mass $m_1 + m_2$ moves together after the collision. We can easily solve for the initial velocity of the 400 g glider:

$$
\begin{aligned}
(v_{ix})_2 &= \frac{(m_1 + m_2)v_{fx} - m_1(v_{ix})_1}{m_2} \\
&= \frac{(0.60 \text{ kg})(-0.40 \text{ m/s}) - (0.20 \text{ kg})(3.0 \text{ m/s})}{0.40 \text{ kg}} \\
&= -2.1 \text{ m/s}
\end{aligned}
$$

FIGURE 9.19 The before-and-after pictorial representation of an inelastic collision.

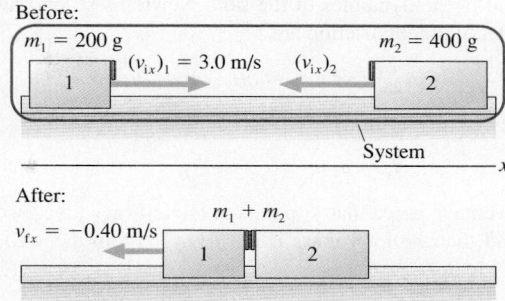

Find: $(v_{ix})_2$

The negative sign indicates that the 400 g glider started out moving to the left. The initial *speed* of the glider, which we were asked to find, is 2.1 m/s.

EXAMPLE 9.6   **Momentum in a car crash**

A 2000 kg Cadillac had just started forward from a stop sign when it was struck from behind by a 1000 kg Volkswagen. The bumpers became entangled, and the two cars skidded forward together until they came to rest. Officer Tom, responding to the accident, measured the skid marks to be 3.0 m long. He also took testimony from the driver that the Cadillac's speed just before the impact was 5.0 m/s. Officer Tom charged the Volkswagen driver with reckless driving. Should the Volkswagen driver also be charged with exceeding the 50 km/h speed limit? The judge calls you as an "expert witness" to analyze the evidence. What is your conclusion?

MODEL This is really *two* problems. First, there is an inelastic collision. The two cars are not a perfectly isolated system because of external friction forces, but during the brief collision the external impulse delivered by friction will be negligible. Within the impulse approximation, the momentum of the Volkswagen + Cadillac system will be conserved in the collision. Then we have a second problem, a dynamics problem of the two cars sliding.

VISUALIZE **FIGURE 9.20a** is a pictorial representation showing both the before and after of the collision and the more familiar picture for the dynamics of the skidding. We do not need to consider forces during the collision because we will use the law of conservation of momentum, but we do need a free-body diagram of the cars during the subsequent skid. This is shown in **FIGURE 9.20b**.

The cars have a common velocity v_{1x} just after the collision. This is the *initial* velocity for the dynamics problem. Our goal is to find $(v_{0x})_{VW}$, the Volkswagen's velocity at the moment of impact. The 50 km/h speed limit has been converted to 14 m/s.

SOLVE First, the inelastic collision. The law of conservation of momentum is

$$(m_{VW} + m_C)v_{1x} = m_{VW}(v_{0x})_{VW} + m_C(v_{0x})_C$$

Solving for the initial velocity of the Volkswagen, we find

$$(v_{0x})_{VW} = \frac{(m_{VW} + m_C)v_{1x} - m_C(v_{0x})_C}{m_{VW}}$$

FIGURE 9.20 Pictorial representation and a free-body diagram of the cars as they skid.

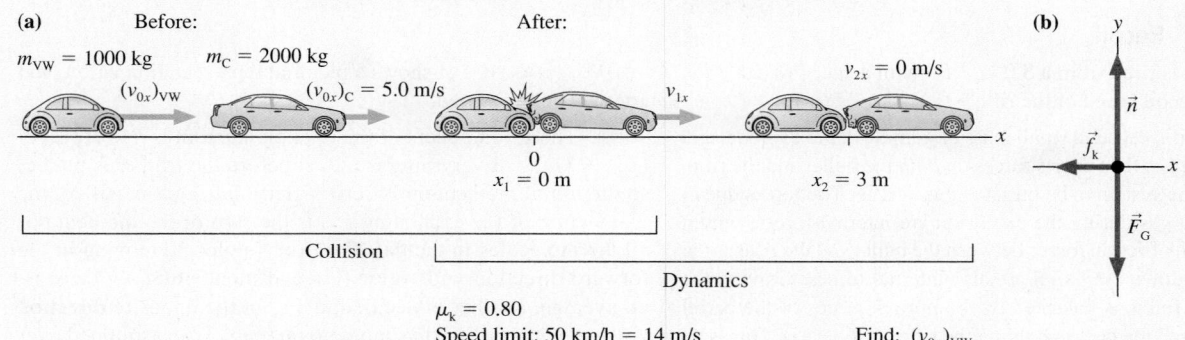

Continued

To evaluate $(v_{0x})_{VW}$, we need to know v_{1x}, the velocity *immediately* after the collision as the cars begin to skid. This information will come out of the dynamics of the skid. Newton's second law and the model of kinetic friction are

$$\sum F_x = -f_k = (m_{VW} + m_C)a_x$$

$$\sum F_y = n - (m_{VW} + m_C)g = 0$$

$$f_k = \mu_k n$$

where we have noted that $\vec{f}_k$ points to the left (negative x-component) and that the total mass is $m_{VW} + m_C$. From the y-equation and the friction equation,

$$f_k = \mu_k(m_{VW} + m_C)g$$

Using this in the x-equation gives

$$a_x = \frac{-f_k}{m_{VW} + m_C} = -\mu_k g = -7.84 \text{ m/s}^2$$

where the coefficient of kinetic friction for rubber on concrete is taken from Table 6.1. With the acceleration determined, we can move on to the kinematics. This is constant acceleration, so

$$v_{2x}^2 = 0 = v_{1x}^2 + 2a_x(\Delta x) = v_{1x}^2 + 2a_x x_2$$

Hence the skid starts with velocity

$$v_{1x} = \sqrt{-2a_x x_2} = \sqrt{-2(-7.84 \text{ m/s}^2)(3.0 \text{ m})} = 6.9 \text{ m/s}$$

As we have noted, this is the final velocity of the collision. Inserting v_{1x} back into the momentum conservation equation, we finally determine that

$$(v_{0x})_{VW} = \frac{(3000 \text{ kg})(6.9 \text{ m/s}) - (2000 \text{ kg})(5.0 \text{ m/s})}{1000 \text{ kg}}$$

$$= 11 \text{ m/s}$$

On the basis of your testimony, the Volkswagen driver is *not* charged with speeding!

> **NOTE** ▶ Momentum is conserved only for an isolated system. In this example, momentum was conserved during the collision (isolated system) but *not* during the skid (not an isolated system). In practice, it is not unusual for momentum to be conserved in one part or one aspect of a problem but not in others. ◀

STOP TO THINK 9.4 The two particles are both moving to the right. Particle 1 catches up with particle 2 and collides with it. The particles stick together and continue on with velocity v_f. Which of these statements is true?

Before   After

a. v_f is greater than v_1.   b. $v_f = v_1$ c. v_f is greater than v_2 but less than v_1.
d. $v_f = v_2$ e. v_f is less than v_2.   f. Can't tell without knowing the masses.

9.5 Explosions

An **explosion,** where the particles of the system move apart from each other after a brief, intense interaction, is the opposite of a collision. The explosive forces, which could be from an expanding spring or from expanding hot gases, are *internal* forces. If the system is isolated, its total momentum during the explosion will be conserved.

EXAMPLE 9.7 | **Recoil**

A 10 g bullet is fired from a 3.0 kg rifle with a speed of 500 m/s. What is the recoil speed of the rifle?

MODEL The rifle causes a small mass of gunpowder to explode, and the expanding gas then exerts forces on *both* the bullet and the rifle. Let's define the system to be bullet + gas + rifle. The forces due to the expanding gas during the explosion are internal forces, within the system. Any friction forces between the bullet and the rifle as the bullet travels down the barrel are also internal forces. Gravity, the only external force, is balanced by the normal forces of the barrel on the bullet and the person holding the rifle, so $\vec{F}_{net} = \vec{0}$. This is an isolated system and the law of conservation of momentum applies.

VISUALIZE **FIGURE 9.21** shows a pictorial representation before and after the bullet is fired.

SOLVE The x-component of the total momentum is $P_x = (p_x)_B + (p_x)_R + (p_x)_{gas}$. Everything is at rest before the trigger is pulled, so the initial momentum is zero. After the trigger is pulled, the momentum of the expanding gas is the sum of the momenta of all the molecules in the gas. For every molecule moving in the forward direction with velocity v and momentum mv there is, on average, another molecule moving in the opposite direction with velocity $-v$ and thus momentum $-mv$. When summed over the enormous number of molecules in the gas, we will be left

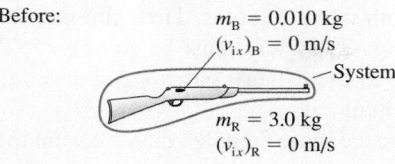

FIGURE 9.21 Before-and-after pictorial representation of a rifle firing a bullet.

Before:
$m_B = 0.010$ kg
$(v_{ix})_B = 0$ m/s

System

$m_R = 3.0$ kg
$(v_{ix})_R = 0$ m/s

After:

$(v_{fx})_R$ $(v_{fx})_B = 500$ m/s

Find: $(v_{fx})_R$

with $p_{gas} \approx 0$. In addition, the mass of the gas is much less than that of the rifle or bullet. For both reasons, we can reasonably neglect the momentum of the gas. The law of conservation of momentum is thus

$$P_{fx} = m_B (v_{fx})_B + m_R (v_{fx})_R = P_{ix} = 0$$

Solving for the rifle's velocity, we find

$$(v_{fx})_R = -\frac{m_B}{m_R}(v_{fx})_B = -\frac{0.010 \text{ kg}}{3.0 \text{ kg}} \times 500 \text{ m/s} = -1.7 \text{ m/s}$$

The minus sign indicates that the rifle's recoil is to the left. The recoil *speed* is 1.7 m/s.

We would not know where to begin to solve a problem such as this using Newton's laws. But Example 9.7 is a simple problem when approached from the before-and-after perspective of a conservation law. The selection of bullet + gas + rifle as "the system" was the critical step. For momentum conservation to be a useful principle, we had to select a system in which the complicated forces due to expanding gas and friction were all internal forces. The rifle by itself is *not* an isolated system, so its momentum is *not* conserved.

EXAMPLE 9.8 Radioactivity

A ^{238}U uranium nucleus is radioactive. It spontaneously disintegrates into a small fragment that is ejected with a measured speed of 1.50×10^7 m/s and a "daughter nucleus" that recoils with a measured speed of 2.56×10^5 m/s. What are the atomic masses of the ejected fragment and the daughter nucleus?

MODEL The notation ^{238}U indicates the isotope of uranium with an atomic mass of 238 u, where u is the abbreviation for the *atomic mass unit*. The nucleus contains 92 protons (uranium is atomic number 92) and 146 neutrons. The disintegration of a nucleus is, in essence, an explosion. Only *internal* nuclear forces are involved, so the total momentum is conserved in the decay.

VISUALIZE FIGURE 9.22 shows the pictorial representation. The mass of the daughter nucleus is m_1 and that of the ejected fragment is m_2. Notice that we converted the speed information to velocity information, giving $(v_{fx})_1$ and $(v_{fx})_2$ opposite signs.

FIGURE 9.22 Before-and-after pictorial representation of the decay of a ^{238}U nucleus.

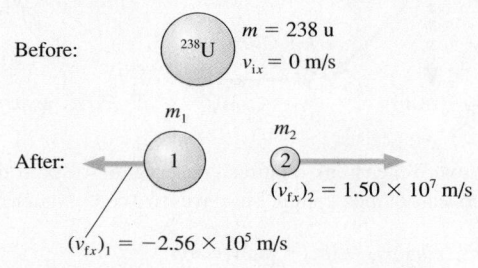

Before: ^{238}U $m = 238$ u
                           $v_{ix} = 0$ m/s

m_1 m_2

After: 1   (2)
                       $(v_{fx})_2 = 1.50 \times 10^7$ m/s

$(v_{fx})_1 = -2.56 \times 10^5$ m/s

Find: m_1 and m_2

SOLVE The nucleus was initially at rest, hence the total momentum is zero. The momentum after the decay is still zero if the two pieces fly apart in opposite directions with momenta equal in magnitude but opposite in sign. That is,

$$P_{fx} = m_1 (v_{fx})_1 + m_2 (v_{fx})_2 = P_{ix} = 0$$

Although we know both final velocities, this is not enough information to find the two unknown masses. However, we also have another conservation law, conservation of mass, that requires

$$m_1 + m_2 = 238 \text{ u}$$

Combining these two conservation laws gives

$$m_1 (v_{fx})_1 + (238 \text{ u} - m_1)(v_{fx})_2 = 0$$

The mass of the daughter nucleus is

$$m_1 = \frac{(v_{fx})_2}{(v_{fx})_2 - (v_{fx})_1} \times 238 \text{ u}$$

$$= \frac{1.50 \times 10^7 \text{ m/s}}{(1.50 \times 10^7 - (-2.56 \times 10^5)) \text{ m/s}} \times 238 \text{ u} = 234 \text{ u}$$

With m_1 known, the mass of the ejected fragment is $m_2 = 238 - m_1 = 4$ u.

ASSESS All we learn from a momentum analysis is the masses. Chemical analysis shows that the daughter nucleus is the element thorium, atomic number 90, with two fewer protons than uranium. The ejected fragment carried away two protons as part of its mass of 4 u, so it must be a particle with two protons and two neutrons. This is the nucleus of a helium atom, ^{4}He, which in nuclear physics is called an *alpha particle* α. Thus the radioactive decay of ^{238}U can be written as $^{238}\text{U} \rightarrow {}^{234}\text{Th} + \alpha$.

FIGURE 9.23 Rocket propulsion is an example of conservation of momentum.

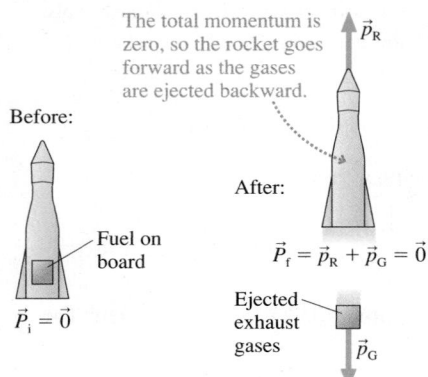

The total momentum is zero, so the rocket goes forward as the gases are ejected backward.

$\vec{p}_R$

Before:

After:

$\vec{P}_f = \vec{p}_R + \vec{p}_G = \vec{0}$

Fuel on board

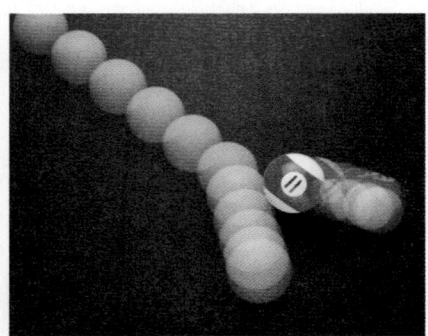

$\vec{P}_i = \vec{0}$

Ejected exhaust gases

$\vec{p}_G$

Much the same reasoning explains how a rocket or jet aircraft accelerates. FIGURE 9.23 shows a rocket with a parcel of fuel on board. Burning converts the fuel to hot gases that are expelled from the rocket motor. If we choose rocket + gases to be the system, the burning and expulsion are both internal forces. There are no other forces, so the total momentum of the rocket + gases system must be conserved. The rocket gains forward velocity and momentum as the exhaust gases are shot out the back, but the *total* momentum of the system remains zero.

The details of rocket propulsion are more complex than we want to handle, because the mass of the rocket is changing, but you should be able to use the law of conservation of momentum to understand the basic principle by which rocket propulsion occurs.

STOP TO THINK 9.5   An explosion in a rigid pipe shoots out three pieces. A 6 g piece comes out the right end. A 4 g piece comes out the left end with twice the speed of the 6 g piece. From which end, left or right, does the third piece emerge?

9.6 Momentum in Two Dimensions

Our examples thus far have been confined to motion along a one-dimensional axis. Many practical examples of momentum conservation involve motion in a plane. The total momentum $\vec{P}$ is a *vector* sum of the momenta $\vec{p} = m\vec{v}$ of the individual particles. Consequently, as we found in Section 9.3, momentum is conserved only if each component of $\vec{P}$ is conserved:

$$(p_{fx})_1 + (p_{fx})_2 + (p_{fx})_3 + \cdots = (p_{ix})_1 + (p_{ix})_2 + (p_{ix})_3 + \cdots$$

$$(p_{fy})_1 + (p_{fy})_2 + (p_{fy})_3 + \cdots = (p_{iy})_1 + (p_{iy})_2 + (p_{iy})_3 + \cdots$$

(9.22)

Collisions and explosions often involve motion in two dimensions.

In this section we'll apply momentum conservation to motion in two dimensions.

EXAMPLE 9.9   **A peregrine falcon strike**

Peregrine falcons often grab their prey from above while both falcon and prey are in flight. A 0.80 kg falcon, flying at 18 m/s, swoops down at a 45° angle from behind a 0.36 kg pigeon flying horizontally at 9.0 m/s. What are the speed and direction of the falcon (now holding the pigeon) immediately after impact?

MODEL The two birds, modeled as particles, are the system. This is a perfectly inelastic collision because after the collision the falcon and pigeon move at a common final velocity. The birds are not a perfectly isolated system because of external forces of the air, but during the brief collision the external impulse delivered by the air resistance will be negligible. Within this approximation, the total momentum of the falcon + pigeon system is conserved during the collision.

VISUALIZE FIGURE 9.24 is a before-and-after pictorial representation. We've used angle ϕ to label the post-collision direction.

SOLVE The initial velocity components of the falcon are $(v_{ix})_F = v_F \cos\theta$ and $(v_{iy})_F = -v_F \sin\theta$. The pigeon's initial velocity is entirely along the x-axis. After the collision, when the falcon and pigeon have the common velocity $\vec{v}_f$, the components are $v_{fx} = v_f \cos\phi$ and $v_{fy} = -v_f \sin\phi$. Conservation of momen-

FIGURE 9.24 Pictorial representation of a falcon catching a pigeon.

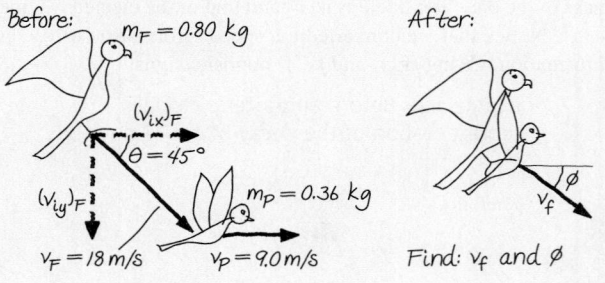

Before: $m_F = 0.80$ kg   After:

$(v_{ix})_F$

$\theta = 45°$

$(v_{iy})_F$ $m_P = 0.36$ kg ϕ

$v_F = 18$ m/s $v_P = 9.0$ m/s v_f

Find: v_f and ϕ

tum in two dimensions requires conservation of both the x- and y-components of momentum. This gives two conservation equations:

$$(m_F + m_P)v_{fx} = (m_F + m_P)v_f \cos\phi$$
$$= m_F(v_{ix})_F + m_P(v_{ix})_P = m_F v_F \cos\theta + m_P v_P$$

$$(m_F + m_P)v_{fy} = -(m_F + m_P)v_f \sin\phi$$
$$= m_F(v_{iy})_F + m_P(v_{iy})_P = -m_F v_F \sin\theta$$

The unknowns are v_f and ϕ. Dividing both equations by the total mass gives

$$v_f \cos\phi = \frac{m_F v_F \cos\theta + m_P v_P}{m_F + m_P} = 11.6 \text{ m/s}$$

$$v_f \sin\phi = \frac{m_F v_F \sin\theta}{m_F + m_P} = 8.78 \text{ m/s}$$

We can eliminate v_f by dividing the second equation by the first to give

$$\frac{v_f \sin\phi}{v_f \cos\phi} = \tan\phi = \frac{8.78 \text{ m/s}}{11.6 \text{ m/s}} = 0.757$$

$$\phi = \tan^{-1}(0.757) = 37°$$

Then $v_f = (11.6 \text{ m/s})/\cos(37°) = 15 \text{ m/s}$. Immediately after impact, the falcon, with its meal, is traveling at 15 m/s at an angle 37° below the horizontal.

ASSESS It makes sense that the falcon would slow down after grabbing the slower-moving pigeon. And Figure 9.24 tells us that the total momentum is at an angle between 0° (the pigeon's momentum) and 45° (the falcon's momentum). Thus our answer seems reasonable.

CHALLENGE EXAMPLE 9.10 **A three-piece explosion**

A 10.0 g projectile is traveling east at 2.0 m/s when it suddenly explodes into three pieces. A 3.0 g fragment is shot due west at 10 m/s while another 3.0 g fragment travels 40° north of east at 12 m/s. What are the speed and direction of the third fragment?

MODEL Although many complex forces are involved in the explosion, they are all internal to the system. There are no external forces, so this is an isolated system and its total momentum is conserved.

VISUALIZE FIGURE 9.25 shows a before-and-after pictorial representation. We'll use uppercase M and V to distinguish the initial object from the three pieces into which it explodes.

FIGURE 9.25 Before-and-after pictorial representation of the three-piece explosion.

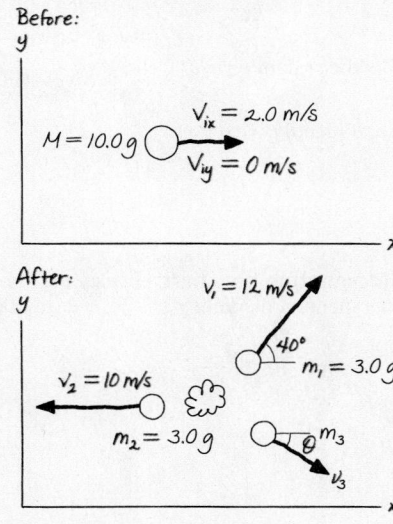

SOLVE The system is the initial object and the subsequent three pieces. Conservation of momentum requires

$$m_1 (v_{fx})_1 + m_2 (v_{fx})_2 + m_3 (v_{fx})_3 = MV_{ix}$$

$$m_1 (v_{fy})_1 + m_2 (v_{fy})_2 + m_3 (v_{fy})_3 = MV_{iy}$$

Conservation of mass implies that

$$m_3 = M - m_1 - m_2 = 4.0 \text{ g}$$

Neither the original object nor m_2 has any momentum along the y-axis. We can use Figure 9.25 to write out the x- and y-components of $\vec{v}_1$ and $\vec{v}_3$, leading to

$$m_1 v_1 \cos 40° - m_2 v_2 + m_3 v_3 \cos\theta = MV$$

$$m_1 v_1 \sin 40° - m_3 v_3 \sin\theta = 0$$

where we used $(v_{fx})_2 = -v_2$ because m_2 is moving in the negative x-direction. Inserting known values in these equations gives us

$$-2.42 + 4v_3 \cos\theta = 20$$

$$23.14 - 4v_3 \sin\theta = 0$$

We can leave the masses in grams in this situation because the conversion factor to kilograms appears on both sides of the equation and thus cancels out. To solve, first use the second equation to write $v_3 = 5.79/\sin\theta$. Substitute this result into the first equation, noting that $\cos\theta/\sin\theta = 1/\tan\theta$, to get

$$-2.42 + 4\left(\frac{5.79}{\sin\theta}\right)\cos\theta = -2.42 + \frac{23.14}{\tan\theta} = 20$$

Now solve for θ:

$$\tan\theta = \frac{23.14}{20 + 2.42} = 1.03$$

$$\theta = \tan^{-1}(1.03) = 45.8°$$

Finally, use this result in the earlier expression for v_3 to find

$$v_3 = \frac{5.79}{\sin 45.8°} = 8.1 \text{ m/s}$$

The third fragment, with a mass of 4.0 g, is shot 46° south of east at a speed of 8.1 m/s.

SUMMARY

The goals of Chapter 9 have been to understand and apply the new concepts of impulse and momentum.

General Principles

Law of Conservation of Momentum

The total momentum $\vec{P} = \vec{p}_1 + \vec{p}_2 + \cdots$ of an isolated system is a constant. Thus

$$\vec{P}_f = \vec{P}_i$$

Newton's Second Law

In terms of momentum, Newton's second law is

$$\vec{F} = \frac{d\vec{p}}{dt}$$

Solving Momentum Conservation Problems

MODEL Choose an isolated system or a system that is isolated during at least part of the problem.

VISUALIZE Draw a pictorial representation of the system before and after the interaction.

SOLVE Write the law of conservation of momentum in terms of vector components:

$$(p_{fx})_1 + (p_{fx})_2 + \cdots = (p_{ix})_1 + (p_{ix})_2 + \cdots$$
$$(p_{fy})_1 + (p_{fy})_2 + \cdots = (p_{iy})_1 + (p_{iy})_2 + \cdots$$

ASSESS Is the result reasonable?

Important Concepts

Momentum $\vec{p} = m\vec{v}$

Impulse $J_x = \displaystyle\int_{t_i}^{t_f} F_x(t)\, dt =$ area under force curve

Impulse and momentum are related by the impulse-momentum theorem

$$\Delta p_x = J_x$$

The impulse delivered to a particle causes the particle's momentum to change. This is an alternative statement of Newton's second law.

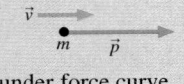

System A group of interacting particles.

Isolated system A system on which there are no external forces or the net external force is zero.

Before-and-after pictorial representation

- Define the system.
- Use two drawings to show the system *before* and *after* the interaction.
- List known information and identify what you are trying to find.

Applications

Collisions Two or more particles come together. In a perfectly inelastic collision, they stick together and move with a common final velocity.

Explosions Two or more particles move away from each other.

Two dimensions No new ideas, but both the x- and y-components of $\vec{P}$ must be conserved, giving two simultaneous equations.

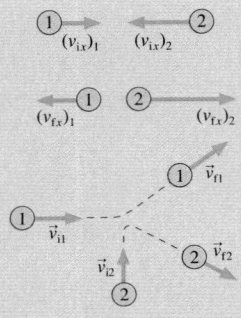

Momentum bar charts display the impulse-momentum theorem $p_{fx} = p_{ix} + J_x$ in graphical form.

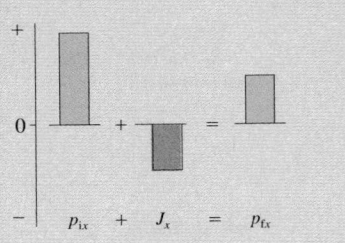

Terms and Notation

collision
impulsive force
momentum, $\vec{p}$
impulse, J_x

impulse-momentum theorem
momentum bar chart
before-and-after pictorial
   representation

impulse approximation
total momentum, $\vec{P}$
isolated system

law of conservation of momentum
perfectly inelastic collision
explosion

CONCEPTUAL QUESTIONS

1. Rank in order, from largest to smallest, the momenta $(p_x)_a$ to $(p_x)_e$ of the objects in **FIGURE Q9.1**.

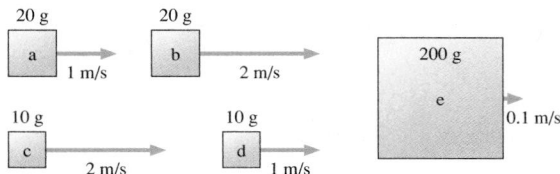

FIGURE Q9.1

2. Explain the concept of *impulse* in nonmathematical language. That is, don't simply put the equation in words to say that "impulse is the time integral of force." Explain it in terms that would make sense to an educated person who had never heard of it.

3. Explain the concept of *isolated system* in nonmathematical language that would make sense to an educated person who had never heard of it.

4. A 0.2 kg plastic cart and a 20 kg lead cart can both roll without friction on a horizontal surface. Equal forces are used to push both carts forward for a time of 1 s, starting from rest. After the force is removed at $t = 1$ s, is the momentum of the plastic cart greater than, less than, or equal to the momentum of the lead cart? Explain.

5. A 0.2 kg plastic cart and a 20 kg lead cart can both roll without friction on a horizontal surface. Equal forces are used to push both carts forward for a distance of 1 m, starting from rest. After traveling 1 m, is the momentum of the plastic cart greater than, less than, or equal to the momentum of the lead cart? Explain.

6. Angie, Brad, and Carlos are discussing a physics problem in which two identical bullets are fired with equal speeds at equal-mass wood and steel blocks resting on a frictionless table. One bullet bounces off the steel block while the second becomes embedded in the wood block. "All the masses and speeds are the same," says Angie, "so I think the blocks will have equal speeds after the collisions." "But what about momentum?" asks Brad. "The bullet hitting the wood block transfers all its momentum and energy to the block, so the wood block should end up going faster than the steel block." "I think the bounce is an important factor," replies Carlos. "The steel block will be faster because the bullet bounces off it and goes back the other direction." Which of these three do you agree with, and why?

7. It feels better to catch a hard ball while wearing a padded glove than to catch it bare handed. Use the ideas of this chapter to explain why.

8. Automobiles are designed with "crumple zones" intended to collapse in a collision. Use the ideas of this chapter to explain why.

9. A 2 kg object is moving to the right with a speed of 1 m/s when it experiences an impulse of 4 N s. What are the object's speed and direction after the impulse?

10. A 2 kg object is moving to the right with a speed of 1 m/s when it experiences an impulse of -4 N s. What are the object's speed and direction after the impulse?

11. A golf club continues forward after hitting the golf ball. Is momentum conserved in the collision? Explain, making sure you are careful to identify "the system."

12. Suppose a rubber ball collides head-on with a steel ball of equal mass traveling in the opposite direction with equal speed. Which ball, if either, receives the larger impulse? Explain.

13. Two particles collide, one of which was initially moving and the other initially at rest.
    a. Is it possible for *both* particles to be at rest after the collision? Give an example in which this happens, or explain why it can't happen.
    b. Is it possible for *one* particle to be at rest after the collision? Give an example in which this happens, or explain why it can't happen.

14. Two ice skaters, Paula and Ricardo, push off from each other. Ricardo weighs more than Paula.
    a. Which skater, if either, has the greater momentum after the push-off? Explain.
    b. Which skater, if either, has the greater speed after the push-off? Explain.

EXERCISES AND PROBLEMS

Problems labeled [ ] integrate material from earlier chapters.

Exercises

Section 9.1 Momentum and Impulse

1. | What is the magnitude of the momentum of
   a. A 3000 kg truck traveling at 15 m/s?
   b. A 200 g baseball thrown at 40 m/s?

2. | At what speed do a bicycle and its rider, with a combined mass of 100 kg, have the same momentum as a 1500 kg car traveling at 5.0 m/s?

3. || What impulse does the force shown in **FIGURE EX9.3** exert on a 250 g particle?

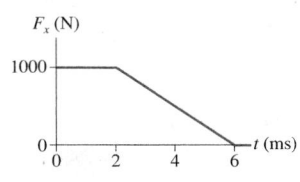

FIGURE EX9.3

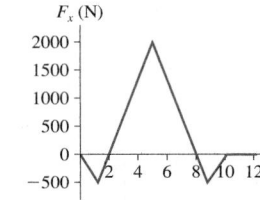

FIGURE EX9.4

4. || What is the impulse on a 3.0 kg particle that experiences the force shown in **FIGURE EX9.4**?

5. ‖ In FIGURE EX9.5, what value of F_{max} gives an impulse of 6.0 N s?

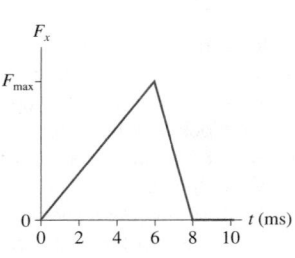

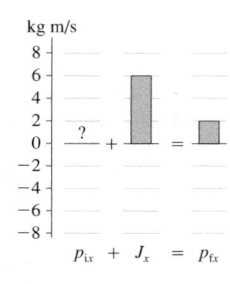

FIGURE EX9.5

FIGURE EX9.6

6. ‖ FIGURE EX9.6 is an incomplete momentum bar chart for a 50 g particle that experiences an impulse lasting 10 ms. What were the speed and direction of the particle before the impulse?

7. ‖ FIGURE EX9.7 is an incomplete momentum bar chart for a collision that lasts 10 ms. What are the magnitude and direction of the average collision force exerted on the object?

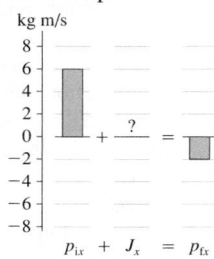

FIGURE EX9.7

Section 9.2 Solving Impulse and Momentum Problems

8. | A 2.0 kg object is moving to the right with a speed of 1.0 m/s when it experiences the force shown in FIGURE EX9.8. What are the object's speed and direction after the force ends?

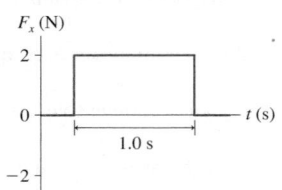

FIGURE EX9.8

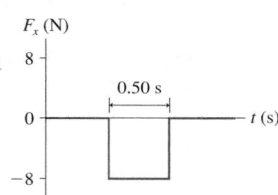

FIGURE EX9.9

9. | A 2.0 kg object is moving to the right with a speed of 1.0 m/s when it experiences the force shown in FIGURE EX9.9. What are the object's speed and direction after the force ends?

10. | A sled slides along a horizontal surface on which the coefficient of kinetic friction is 0.25. Its velocity at point A is 8.0 m/s and at point B is 5.0 m/s. Use the impulse-momentum theorem to find how long the sled takes to travel from A to B.

11. | Far in space, where gravity is negligible, a 425 kg rocket traveling at 75 m/s fires its engines. FIGURE EX9.11 shows the thrust force as a function of time. The mass lost by the rocket during these 30 s is negligible.

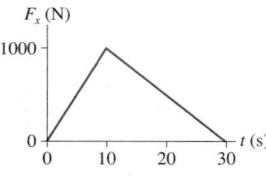

FIGURE EX9.11

a. What impulse does the engine impart to the rocket?

b. At what time does the rocket reach its maximum speed? What is the maximum speed?

12. ‖ A 250 g ball collides with a wall. FIGURE EX9.12 shows the ball's velocity and the force exerted on the ball by the wall. What is v_{fx}, the ball's rebound velocity?

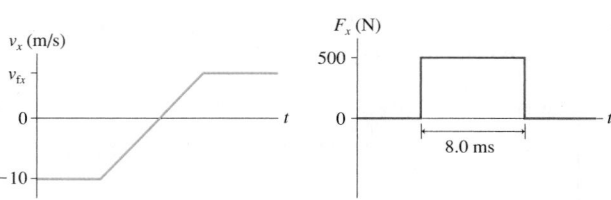

FIGURE EX9.12

13. ‖ A 600 g air-track glider collides with a spring at one end of the track. FIGURE EX9.13 shows the glider's velocity and the force exerted on the glider by the spring. How long is the glider in contact with the spring?

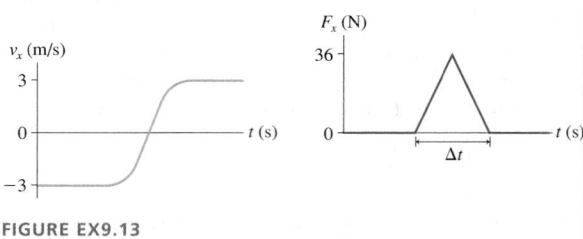

FIGURE EX9.13

Section 9.3 Conservation of Momentum

14. | A 10,000 kg railroad car is rolling at 2.0 m/s when a 4000 kg load of gravel is suddenly dropped in. What is the car's speed just after the gravel is loaded?

15. | A 5000 kg open train car is rolling on frictionless rails at 22 m/s when it starts pouring rain. A few minutes later, the car's speed is 20 m/s. What mass of water has collected in the car?

16. ‖ A 10-m-long glider with a mass of 680 kg (including the passengers) is gliding horizontally through the air at 30 m/s when a 60 kg skydiver drops out by releasing his grip on the glider. What is the glider's velocity just after the skydiver lets go?

Section 9.4 Inelastic Collisions

17. | A 300 g bird flying along at 6.0 m/s sees a 10 g insect heading straight toward it with a speed of 30 m/s. The bird opens its mouth wide and enjoys a nice lunch. What is the bird's speed immediately after swallowing?

18. | The parking brake on a 2000 kg Cadillac has failed, and it is rolling slowly, at 1.0 mph, toward a group of small children. Seeing the situation, you realize you have just enough time to drive your 1000 kg Volkswagen head-on into the Cadillac and save the children. With what speed should you impact the Cadillac to bring it to a halt?

19. | A 1500 kg car is rolling at 2.0 m/s. You would like to stop the car by firing a 10 kg blob of sticky clay at it. How fast should you fire the clay?

Section 9.5 Explosions

20. | A 50 kg archer, standing on frictionless ice, shoots a 100 g arrow at a speed of 100 m/s. What is the recoil speed of the archer?

21. ‖ Dan is gliding on his skateboard at 4.0 m/s. He suddenly jumps backward off the skateboard, kicking the skateboard forward at 8.0 m/s. How fast is Dan going as his feet hit the ground? Dan's mass is 50 kg and the skateboard's mass is 5.0 kg?

22. ‖ A 70.0 kg football player is gliding across very smooth ice at 2.00 m/s. He throws a 0.450 kg football straight forward. What is the player's speed afterward if the ball is thrown at
   a. 15.0 m/s relative to the ground?
   b. 15.0 m/s relative to the player?

Section 9.6 Momentum in Two Dimensions

23. ‖ Two particles collide and bounce apart. **FIGURE EX9.23** shows the initial momenta of both and the final momentum of particle 2. What is the final momentum of particle 1? Write your answer in component form.

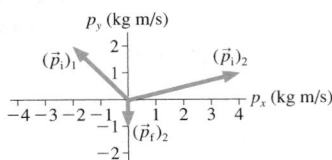

FIGURE EX9.23

24. ‖ An object at rest explodes into three fragments. **FIGURE EX9.24** shows the momentum vectors of two of the fragments. What are p_x and p_y of the third fragment?

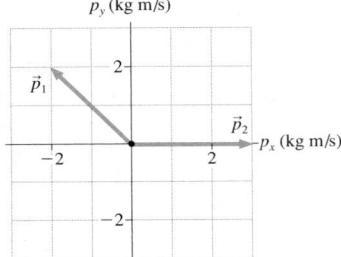

FIGURE EX9.24

25. ‖ A 20 g ball of clay traveling east at 3.0 m/s collides with a 30 g ball of clay traveling north at 2.0 m/s. What are the speed and the direction of the resulting 50 g ball of clay?

Problems

26. ‖ A 60 g tennis ball with an initial speed of 32 m/s hits a wall and rebounds with the same speed. **FIGURE P9.26** shows the force of the wall on the ball during the collision. What is the value of F_{max}, the maximum value of the contact force during the collision?

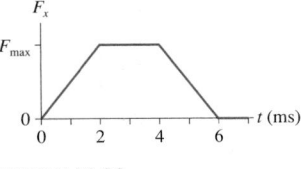

FIGURE P9.26

27. ‖ A tennis player swings her 1000 g racket with a speed of 10 m/s. She hits a 60 g tennis ball that was approaching her at a speed of 20 m/s. The ball rebounds at 40 m/s.
   a. How fast is her racket moving immediately after the impact? You can ignore the interaction of the racket with her hand for the brief duration of the collision.
   b. If the tennis ball and racket are in contact for 10 ms, what is the average force that the racket exerts on the ball? How does this compare to the gravitational force on the ball?

28. ‖‖ A 200 g ball is dropped from a height of 2.0 m, bounces on a hard floor, and rebounds to a height of 1.5 m. **FIGURE P9.28** shows the impulse received from the floor. What maximum force does the floor exert on the ball?

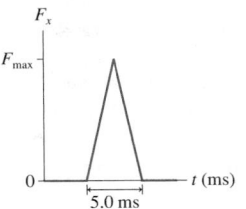

FIGURE P9.28

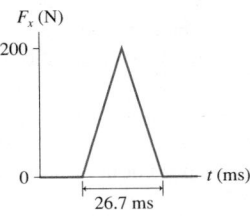

FIGURE P9.29

29. ‖ A 500 g cart is released from rest 1.00 m from the bottom of a frictionless, 30.0° ramp. The cart rolls down the ramp and bounces off a rubber block at the bottom. **FIGURE P9.29** shows the force during the collision. After the cart bounces, how far does it roll back up the ramp?

30. ‖ One week in lab, you're given a spring-loaded bar that can be used to strike a metal ball. Your assignment is to measure what size impulse the bar delivers to the ball. You and your lab partner decide to place several balls of different mass on the edge of the lab table, use the striker to launch them horizontally, and measure the horizontal distance to where each ball hits the floor.
   a. Let the table height be h and the horizontal distance traveled by the ball be its range R. Find an expression for the range. The range depends on h, the ball's mass m, and the impulse J.
   b. What should you graph the measured range against to get a linear graph whose slope is related to J?
   c. After measuring the table height to be 1.5 m, you and your partner acquire the following data:

Mass (g)	Range (cm)
100	247
150	175
200	129
250	98

Draw an appropriate graph of the data and, from the slope of the best-fit line, determine the impulse.

31. | The flowers of the bunchberry plant open with astonishing
BIO force and speed, causing the pollen grains to be ejected out of the flower in a mere 0.30 ms at an acceleration of 2.5×10^4 m/s². If the acceleration is constant, what impulse is delivered to a pollen grain with a mass of 1.0×10^{-7} g?

32. ‖ A particle of mass m is at rest at $t = 0$. Its momentum for $t > 0$ is given by $p_x = 6t^2$ kg m/s, where t is in s. Find an expression for $F_x(t)$, the force exerted on the particle as a function of time.

33. ‖ A small rocket to gather weather data is launched straight up. Several seconds into the flight, its velocity is 120 m/s and it is accelerating at 18 m/s². At this instant, the rocket's mass is 48 kg and it is losing mass at the rate of 0.50 kg/s as it burns fuel. What is the net force on the rocket? **Hint:** Newton's second law was presented in a new form in this chapter.

34. | Three identical train cars, coupled together, are rolling east at speed v_0. A fourth car traveling east at $2v_0$ catches up with the three and couples to make a four-car train. A moment later, the train cars hit a fifth car that was at rest on the tracks, and it couples to make a five-car train. What is the speed of the five-car train?

35. ‖ A clay blob of mass m_1, initially at rest, is pushed across a frictionless surface with constant force F for a distance d. It then hits and sticks to a second clay blob of mass m_2 that is at rest. Find an expression for their speed after the collision.

36. ‖ Air-track gliders with masses 300 g, 400 g, and 200 g are lined up and held in place with lightweight springs compressed between them. All three are released at once. The 200 g glider flies off to the right while the 300 g glider goes left. Their position-versus-time graphs, as measured by motion detectors, are shown in FIGURE P9.36. What are the direction (right or left) and speed of the 400 g glider that was in the middle?

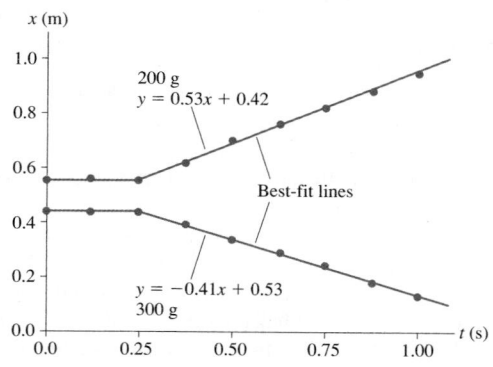

FIGURE P9.36

37. ‖ Most geologists believe that the dinosaurs became extinct 65 million years ago when a large comet or asteroid struck the earth, throwing up so much dust that the sun was blocked out for a period of many months. Suppose an asteroid with a diameter of 2.0 km and a mass of 1.0×10^{13} kg hits the earth with an impact speed of 4.0×10^4 m/s.
   a. What is the earth's recoil speed after such a collision? (Use a reference frame in which the earth was initially at rest.)
   b. What percentage is this of the earth's speed around the sun? (Use the astronomical data inside the back cover.)

38. ‖ At the center of a 50-m-diameter circular ice rink, a 75 kg skater traveling north at 2.5 m/s collides with and holds onto a 60 kg skater who had been heading west at 3.5 m/s.
   a. How long will it take them to glide to the edge of the rink?
   b. Where will they reach it? Give your answer as an angle north of west.

39. ‖ Squids rely on jet propulsion to move around. A 1.5 kg squid
BIO   drifting at 0.40 m/s suddenly expels 0.10 kg of water backward to quickly get itself moving forward at 2.5 m/s. If drag is ignored over the small interval of time needed to expel the water (the impulse approximation), with what speed relative to itself does the squid eject the water?

40. ‖ Two ice skaters, with masses of 50 kg and 75 kg, are at the center of a 60-m-diameter circular rink. The skaters push off against each other and glide to opposite edges of the rink. If the heavier skater reaches the edge in 20 s, how long does the lighter skater take to reach the edge?

41. ‖ A firecracker in a coconut blows the coconut into three pieces. Two pieces of equal mass fly off south and west, perpendicular to each other, at speed v_0. The third piece has twice the mass as the other two. What are the speed and direction of the third piece? Give the direction as an angle east of north.

42. ‖ One billiard ball is shot east at 2.0 m/s. A second, identical billiard ball is shot west at 1.0 m/s. The balls have a glancing collision, not a head-on collision, deflecting the second ball by

90° and sending it north at 1.41 m/s. What are the speed and direction of the first ball after the collision? Give the direction as an angle south of east.

43. ‖ a. A bullet of mass m is fired into a block of mass M that is at rest. The block, with the bullet embedded, slides distance d across a horizontal surface. The coefficient of kinetic friction is μ_k. Find an expression for the bullet's speed v_{bullet}.
   b. What is the speed of a 10 g bullet that, when fired into a 10 kg stationary wood block, causes the block to slide 5.0 cm across a wood table?

44. ‖ Fred (mass 60 kg) is running with the football at a speed of 6.0 m/s when he is met head-on by Brutus (mass 120 kg), who is moving at 4.0 m/s. Brutus grabs Fred in a tight grip, and they fall to the ground. Which way do they slide, and how far? The coefficient of kinetic friction between football uniforms and Astroturf is 0.30.

45. | You are part of a search-and-rescue mission that has been called out to look for a lost explorer. You've found the missing explorer, but, as FIGURE P9.45 shows, you're separated from him by a 200-m-high cliff and a 30-m-wide raging river. To save his life, you need to get a 5.0 kg package of emergency supplies across the river. Unfortunately, you can't throw the package hard enough to make it across. Fortunately, you happen to have a 1.0 kg rocket intended for launching flares. Improvising quickly, you attach a sharpened stick to the front of the rocket, so that it will impale itself into the package of supplies, then fire the rocket at ground level toward the supplies. What minimum speed must the rocket have just before impact in order to save the explorer's life?

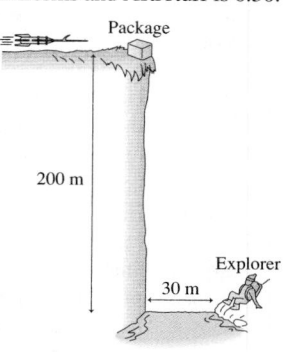

FIGURE P9.45

46. ‖ An object at rest on a flat, horizontal surface explodes into two fragments, one seven times as massive as the other. The heavier fragment slides 8.2 m before stopping. How far does the lighter fragment slide? Assume that both fragments have the same coefficient of kinetic friction.

47. ‖ A 1500 kg weather rocket accelerates upward at 10 m/s². It explodes 2.0 s after liftoff and breaks into two fragments, one twice as massive as the other. Photos reveal that the lighter fragment traveled straight up and reached a maximum height of 530 m. What were the speed and direction of the heavier fragment just after the explosion?

48. ‖ In a ballistics test, a 25 g bullet traveling horizontally at 1200 m/s goes through a 30-cm-thick 350 kg stationary target and emerges with a speed of 900 m/s. The target is free to slide on a smooth horizontal surface. What is the target's speed just after the bullet emerges?

49. | Two 500 g blocks of wood are 2.0 m apart on a frictionless table. A 10 g bullet is fired at 400 m/s toward the blocks. It passes all the way through the first block, then embeds itself in the second block. The speed of the first block immediately afterward is 6.0 m/s. What is the speed of the second block after the bullet stops in it?

50. ‖ The skiing duo of Brian (80 kg) and Ashley (50 kg) is always a crowd pleaser. In one routine, Brian, wearing wood skis, starts at the top of a 200-m-long, 20° slope. Ashley waits for him halfway down. As he skis past, she leaps into his arms and he carries her

the rest of the way down. What is their speed at the bottom of the slope?

51. ‖ The stoplight had just changed and a 2000 kg Cadillac had entered the intersection, heading north at 3.0 m/s, when it was struck by a 1000 kg eastbound Volkswagen. The cars stuck together and slid to a halt, leaving skid marks angled 35° north of east. How fast was the Volkswagen going just before the impact?

52. ‖ Ann (mass 50 kg) is standing at the left end of a 15-m-long, 500 kg cart that has frictionless wheels and rolls on a frictionless track. Initially both Ann and the cart are at rest. Suddenly, Ann starts running along the cart at a speed of 5.0 m/s relative to the cart. How far will Ann have run *relative to the ground* when she reaches the right end of the cart?

53. ‖ A ball of mass m and another ball of mass $3m$ are placed inside a smooth metal tube with a massless spring compressed between them. When the spring is released, the heavier ball flies out of one end of the tube with speed v_0. With what speed does the lighter ball emerge from the other end?

54. ‖‖ Force $F_x = (10 \text{ N})\sin(2\pi t/4.0 \text{ s})$ is exerted on a 250 g particle during the interval $0 \text{ s} \leq t \leq 2.0 \text{ s}$. If the particle starts from rest, what is its speed at $t = 2.0$ s?

55. ‖‖ A 500 g particle has velocity $v_x = -5.0$ m/s at $t = -2$ s. Force $F_x = (4 - t^2)$ N is exerted on the particle between $t = -2$ s and $t = 2$ s. This force increases from 0 N at $t = -2$ s to 4 N at $t = 0$ s and then back to 0 N at $t = 2$ s. What is the particle's velocity at $t = 2$ s?

56. ‖ A 30 ton rail car and a 90 ton rail car, initially at rest, are connected together with a giant but massless compressed spring between them. When released, the 30 ton car is pushed away at a speed of 4.0 m/s relative to the 90 ton car. What is the speed of the 30 ton car relative to the ground?

57. ‖ A 75 kg shell is fired with an initial speed of 125 m/s at an angle 55° above horizontal. Air resistance is negligible. At its highest point, the shell explodes into two fragments, one four times more massive than the other. The heavier fragment lands directly below the point of the explosion. If the explosion exerts forces only in the horizontal direction, how far from the launch point does the lighter fragment land?

58. ‖ A proton (mass 1 u) is shot at a speed of 5.0×10^7 m/s toward a gold target. The nucleus of a gold atom (mass 197 u) repels the proton and deflects it straight back toward the source with 90% of its initial speed. What is the recoil speed of the gold nucleus?

59. ‖ A proton (mass 1 u) is shot toward an unknown target nucleus at a speed of 2.50×10^6 m/s. The proton rebounds with its speed reduced by 25% while the target nucleus acquires a speed of 3.12×10^5 m/s. What is the mass, in atomic mass units, of the target nucleus?

60. ‖ The nucleus of the polonium isotope ^{214}Po (mass 214 u) is radioactive and decays by emitting an alpha particle (a helium nucleus with mass 4 u). Laboratory experiments measure the speed of the alpha particle to be 1.92×10^7 m/s. Assuming the polonium nucleus was initially at rest, what is the recoil speed of the nucleus that remains after the decay?

61. ‖ A neutron is an electrically neutral subatomic particle with a mass just slightly greater than that of a proton. A free neutron is radioactive and decays after a few minutes into other subatomic particles. In one experiment, a neutron at rest was observed to decay into a proton (mass 1.67×10^{-27} kg) and an electron (mass 9.11×10^{-31} kg). The proton and electron were shot out back-to-back. The proton speed was measured to be

1.0×10^5 m/s and the electron speed was 3.0×10^7 m/s. No other decay products were detected.

a. Was momentum conserved in the decay of this neutron?

NOTE ▶ Experiments such as this were first performed in the 1930s and seemed to indicate a failure of the law of conservation of momentum. In 1933, Wolfgang Pauli postulated that the neutron might have a *third* decay product that is virtually impossible to detect. Even so, it can carry away just enough momentum to keep the total momentum conserved. This proposed particle was named the *neutrino,* meaning "little neutral one." Neutrinos were, indeed, discovered nearly 20 years later. ◀

b. If a neutrino was emitted in the above neutron decay, in which direction did it travel? Explain your reasoning.

c. How much momentum did this neutrino "carry away" with it?

62. ‖ A 20 g ball of clay traveling east at 2.0 m/s collides with a 30 g ball of clay traveling 30° south of west at 1.0 m/s. What are the speed and direction of the resulting 50 g blob of clay?

63. ‖ **FIGURE P9.63** shows a collision between three balls of clay. The three hit simultaneously and stick together. What are the speed and direction of the resulting blob of clay?

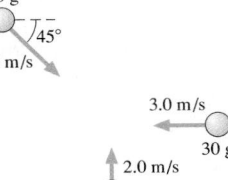

FIGURE P9.63

64. ‖ A 2100 kg truck is traveling east through an intersection at 2.0 m/s when it is hit simultaneously from the side and the rear. (Some people have all the luck!) One car is a 1200 kg compact traveling north at 5.0 m/s. The other is a 1500 kg midsize traveling east at 10 m/s. The three vehicles become entangled and slide as one body. What are their speed and direction just after the collision?

65. ‖ The carbon isotope ^{14}C is used for carbon dating of archeological artifacts. ^{14}C (mass 2.34×10^{-26} kg) decays by the process known as *beta decay* in which the nucleus emits an electron (the beta particle) and a subatomic particle called a neutrino. In one such decay, the electron and the neutrino are emitted at right angles to each other. The electron (mass 9.11×10^{-31} kg) has a speed of 5.0×10^7 m/s and the neutrino has a momentum of 8.0×10^{-24} kg m/s. What is the recoil speed of the nucleus?

In Problems 66 through 69 you are given the equation used to solve a problem. For each of these, you are to

a. Write a realistic problem for which this is the correct equation.

b. Finish the solution of the problem, including a pictorial representation.

66. $(0.10 \text{ kg})(40 \text{ m/s}) - (0.10 \text{ kg})(-30 \text{ m/s}) = \frac{1}{2}(1400 \text{ N}) \Delta t$

67. $(600 \text{ g})(4.0 \text{ m/s}) = (400 \text{ g})(3.0 \text{ m/s}) + (200 \text{ g})(v_{ix})_2$

68. $(3000 \text{ kg})v_{fx} = (2000 \text{ kg})(5.0 \text{ m/s}) + (1000 \text{ kg})(-4.0 \text{ m/s})$

69. $(50 \text{ g})(v_{fx})_1 + (100 \text{ g})(7.5 \text{ m/s}) = (150 \text{ g})(1.0 \text{ m/s})$

Challenge Problems

70. A 1000 kg cart is rolling to the right at 5.0 m/s. A 70 kg man is standing on the right end of the cart. What is the speed of the cart if the man suddenly starts running to the left with a speed of 10 m/s relative to the cart?

71. A spaceship of mass 2.0×10^6 kg is cruising at a speed of 5.0×10^6 m/s when the antimatter reactor fails, blowing the ship into three pieces. One section, having a mass of 5.0×10^5 kg, is blown straight backward with a speed of 2.0×10^6 m/s. A second piece, with mass 8.0×10^5 kg, continues forward at 1.0×10^6 m/s. What are the direction and speed of the third piece?

72. A 20 kg wood ball hangs from a 2.0-m-long wire. The maximum tension the wire can withstand without breaking is 400 N. A 1.0 kg projectile traveling horizontally hits and embeds itself in the wood ball. What is the greatest speed this projectile can have without causing the cable to break?

73. A two-stage rocket is traveling at 1200 m/s with respect to the earth when the first stage runs out of fuel. Explosive bolts release the first stage and push it backward with a speed of 35 m/s relative to the second stage. The first stage is three times as massive as the second stage. What is the speed of the second stage after the separation?

74. You are the ground-control commander of a 2000 kg scientific rocket that is approaching Mars at a speed of 25,000 km/h. It needs to quickly slow to 15,000 km/h to begin a controlled descent to the surface. If the rocket enters the Martian atmosphere too fast it will burn up, and if it enters too slowly, it will use up its maneuvering fuel before reaching the surface and will crash. The rocket has a new braking system: Several 5.0 kg "bullets" on the front of the rocket can be fired straight ahead. Each has a high-explosive charge that fires it at a speed of 139,000 m/s relative to the rocket. You need to send the rocket an instruction to tell it how many bullets to fire. Success will bring you fame and glory, but failure of this $500,000,000 mission will ruin your career. How many bullets will you tell the rocket to fire?

75. You are a world-famous physicist-lawyer defending a client who has been charged with murder. It is alleged that your client,

Mr. Smith, shot the victim, Mr. Wesson. The detective who investigated the scene of the crime found a second bullet, from a shot that missed Mr. Wesson, that had embedded itself into a chair. You arise to cross-examine the detective.

You: In what type of chair did you find the bullet?
Det: A wooden chair.
You: How massive was this chair?
Det: It had a mass of 20 kg.
You: How did the chair respond to being struck with a bullet?
Det: It slid across the floor.
You: How far?
Det: A good three centimeters. The slide marks on the dusty floor are quite distinct.
You: What kind of floor was it?
Det: A wood floor, very nice oak planks.
You: What was the mass of the bullet you retrieved from the chair?
Det: Its mass was 10 g.
You: And how far had it penetrated into the chair?
Det: A distance of 1.5 cm.
You: Have you tested the gun you found in Mr. Smith's possession?
Det: I have.
You: What is the muzzle velocity of bullets fired from that gun?
Det: The muzzle velocity is 450 m/s.
You: And the barrel length?
Det: The gun has a barrel length of 16 cm.

With only a slight hesitation, you turn confidently to the jury and proclaim, "My client's gun did not fire these shots!" How are you going to convince the jury and the judge?

STOP TO THINK ANSWERS

Stop to Think 9.1: f. The cart is initially moving in the negative x-direction, so $p_{ix} = -20$ kg m/s. After it bounces, $p_{fx} = 10$ kg m/s. Thus $\Delta p = (10 \text{ kg m/s}) - (-20 \text{ kg m/s}) = 30$ kg m/s.

Stop to Think 9.2: b. The clay ball goes from $v_{ix} = v$ to $v_{fx} = 0$, so $J_{clay} = \Delta p_x = -mv$. The rubber ball rebounds, going from $v_{ix} = v$ to $v_{fx} = -v$ (same speed, opposite direction). Thus $J_{rubber} = \Delta p_x = -2mv$. The rubber ball has a larger momentum change, and this requires a larger impulse.

Stop to Think 9.3: Less than. The ball's momentum $m_B v_B$ is the same in both cases. Momentum is conserved, so the *total* momentum is the same after both collisions. The ball that rebounds from C has *negative* momentum, so C must have a larger momentum than A.

Stop to Think 9.4: c. Momentum conservation requires $(m_1 + m_2) \times v_f = m_1 v_1 + m_2 v_2$. Because $v_1 > v_2$, it must be that $(m_1 + m_2) \times v_f = m_1 v_1 + m_2 v_2 > m_1 v_2 + m_2 v_2 = (m_1 + m_2) v_2$. Thus $v_f > v_2$. Similarly, $v_2 < v_1$ so $(m_1 + m_2) v_f = m_1 v_1 + m_2 v_2 < m_1 v_1 + m_2 v_1 = (m_1 + m_2) v_1$. Thus $v_f < v_1$. The collision causes m_1 to slow down and m_2 to speed up.

Stop to Think 9.5: Right end. The pieces started at rest, so the total momentum of the system is zero. It's an isolated system, so the total momentum after the explosion is still zero. The 6 g piece has momentum $6v$. The 4 g piece, with velocity $-2v$, has momentum $-8v$. The combined momentum of these two pieces is $-2v$. In order for P to be zero, the third piece must have a *positive* momentum ($+2v$) and thus a positive velocity.

10 Energy

These photovoltaic panels are transforming solar energy into electrical energy.

▶ **Looking Ahead** The goal of Chapter 10 is to introduce the concept of energy and the basic energy model.

Basic Energy Model

Energy is one of the most important concepts in physics. Chapters 10 and 11 will develop the **basic energy model**, a powerful set of ideas for solving problems in mechanics.

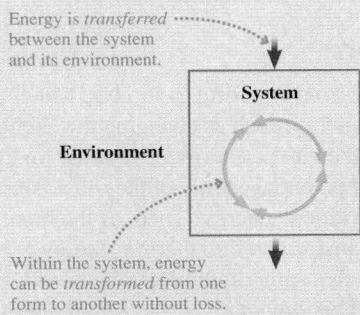

Energy is *transferred* between the system and its environment.

Within the system, energy can be *transformed* from one form to another without loss.

This chapter focuses on energy transformations within the system as one kind of energy is converted to another. Chapter 11 will explore energy transfers to and from the system. For mechanical systems, that transfer is called *work*. Part IV will expand our understanding of energy even further by incorporating the concepts of *heat* and thermodynamics.

Forms of Energy

■ **Kinetic energy** is energy associated with an object's motion.
■ **Potential energy** is stored energy. Potential energy is associated with an object's position.
■ **Thermal energy** is the energy of the random motions of atoms within an object. Thermal energy is associated with temperature.

You will learn about *gravitational potential energy*, the *elastic potential energy* of a stretched or compressed spring, and how these potential energies can be transformed into kinetic energy.

Conservation of Mechanical Energy

Mechanical energy, the sum of kinetic and potential energies, is *conserved* in a system that is both isolated and frictionless. As you learned with momentum, conservation means that

final value = initial value

This will be the basis for a new problem-solving strategy.

◀ **Looking Back**
Sections 9.2–9.3 Before-and-after pictorial representations and conservation of momentum

Energy Diagrams

You'll learn how to interpret an **energy diagram**, a graphical representation for understanding how the speed of a particle changes as it moves through space.

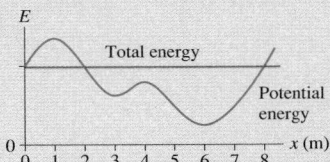

As you'll see, maxima and minima are points of unstable and stable equilibrium, respectively.

Elastic Collisions

A collision that conserves both momentum and mechanical energy is a **perfectly elastic collision.**

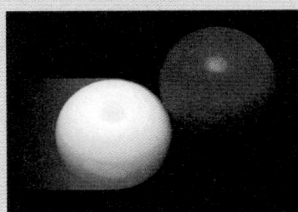

Collisions between two billiard balls or two steel balls come very close to being perfectly elastic.

10.1 The Basic Energy Model

Energy. It's a word you hear all the time. We use chemical energy to heat our homes, electrical energy to power our lights and computers, and solar energy to grow our crops and forests. We're told to use energy wisely and not to waste it.

But just what is energy? The concept of energy has grown and changed with time, and it is not easy to define in a general way just what energy is. Rather than starting with a formal definition, we're going to let the concept of energy expand slowly over the course of several chapters. This chapter introduces three of the most fundamental forms of energy: kinetic energy, potential energy, and thermal energy. Our goal is to understand the characteristics of energy, how energy is used, and, especially important, how energy is transformed from one form to another.

Ultimately we will discover a very powerful conservation law for energy. Some scientists consider the law of conservation of energy to be the most important of all the laws of nature. But all that in due time; first we have to start with the basic ideas.

Some important forms of energy

| Kinetic energy K | Potential energy U | Thermal energy E_{th} |

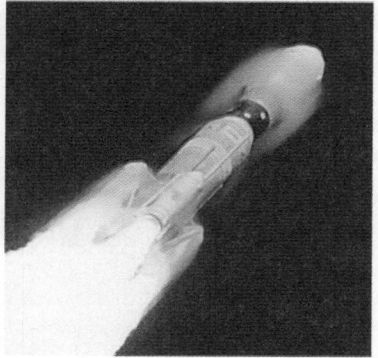  

Kinetic energy is the energy of motion. All moving objects have kinetic energy. The more massive an object or the faster it moves, the larger its kinetic energy.

Potential energy is stored energy associated with an object's position. The roller coaster's gravitational potential energy depends on its height above the ground.

Thermal energy is the sum of the microscopic kinetic and potential energies of all the atoms and bonds that make up the object. An object has more thermal energy when hot than when cold.

Energy Transfer and Transformation

This chapter focuses on the *transformation* of energy from one form to another. Much of modern technology is concerned with transforming energy, such as changing the chemical energy of oil molecules to electrical energy or to the kinetic energy of your car. In the pictures above, you can imagine that the gravitational potential energy of the roller coaster at the top of the hill will soon be transformed into kinetic energy. Then, as the brakes are applied and get hot, that kinetic energy will be transformed into thermal energy.

Remarkably, the total energy of the system—the sum of the various forms of energy—is not changed by these transformations. This *law of conservation of energy* was not recognized until the mid-19th century, long after Newton. The belated discovery of such an important idea was because it took scientists a long time to realize how many types of energy there are and the various ways that energy can be converted from one form into another. As you'll learn, energy ideas go well beyond Newtonian mechanics to include concepts about heat, about chemical energy, and about the energy of the individual atoms that make up a system. All of these forms of energy ultimately have to be included in the law of energy conservation.

Energy not only can be transformed from one kind to another but also can be transferred from one system to another. For example, the roller coaster at the top of the hill acquired its potential energy not through an energy transformation but because an outside force—a chain powered by a motor—dragged it up the hill and in the process *transferred* energy to the roller coaster. This mechanical transfer of energy to a system via forces is called *work*, a topic we'll explore in detail in the next chapter.

As we use energy concepts, we will be "accounting" for energy that is transferred into or out of a system or that is transformed from one form to another within a system. FIGURE 10.1 shows a **basic energy model** that illustrates these ideas. There are many details that must be added to this model, but it's a good starting point. The fact that nature "balances the books" for energy, so that energy is never created or destroyed, is one of the most profound discoveries of science.

There's a lot to say about energy, and energy is an abstract idea, so we'll take it one step at a time. This chapter focuses on the transformations that take place inside the system, especially idealized transformations that don't change the thermal energy. Then, after you've had some practice using the basic concepts of energy, Chapter 11 will introduce energy transfers between the system and the environment and will establish a more rigorous way of defining potential energy. We will extend these ideas even further in Part IV when we reach the study of thermodynamics.

10.2 Kinetic Energy and Gravitational Potential Energy

FIGURE 10.2 is a before-and-after pictorial representation of an object in free fall, as you learned to draw in Chapter 9. We didn't call attention to it in Chapter 2, but one of the free-fall equations also relates "before" and "after." In particular, the free-fall kinematic equation with $a_y = -g$

$$v_{fy}^2 = v_{iy}^2 + 2a_y \Delta y = v_{iy}^2 - 2g(y_f - y_i) \tag{10.1}$$

can easily be rewritten as

$$v_{fy}^2 + 2gy_f = v_{iy}^2 + 2gy_i \tag{10.2}$$

Equation 10.2 is a conservation law for free-fall motion. It tells us that the quantity $v_y^2 + 2gy$ has the same value *after* free fall (regardless of whether the motion is upward or downward) that it had *before* free fall.

Let's introduce a more general technique to arrive at the same result, but a technique that can be extended to other types of motion. Newton's second law for one-dimensional motion along the y-axis is

$$(F_{net})_y = ma_y = m\frac{dv_y}{dt} \tag{10.3}$$

The net force on an object in free fall is $(F_{net})_y = -mg$, so Equation 10.3 becomes

$$m\frac{dv_y}{dt} = -mg \tag{10.4}$$

Recall, from calculus, that we can use the chain rule to write

$$\frac{dv_y}{dt} = \frac{dv_y}{dy}\frac{dy}{dt} = v_y\frac{dv_y}{dy} \tag{10.5}$$

where we used $v_y = dy/dt$. Substituting this into Equation 10.4 gives

$$mv_y\frac{dv_y}{dy} = -mg \tag{10.6}$$

The chain rule has allowed us to change from a derivative of v_y with respect to time to a derivative of v_y with respect to position.

We can rewrite Equation 10.6 as

$$mv_y\,dv_y = -mg\,dy \tag{10.7}$$

Now we can integrate both sides of the equation. However, we have to be careful to make sure the limits of integration match. We want to integrate from "before," when the object

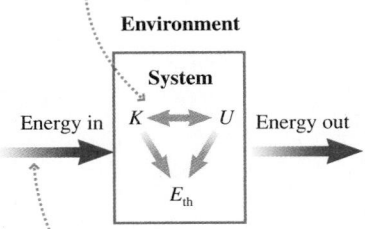

FIGURE 10.1 The basic energy model.

Energy is *transformed* within the system without loss. The energy of an *isolated system* is conserved.

Energy is *transferred* to (and from) the system by forces acting on the system. The forces *do work* on the system.

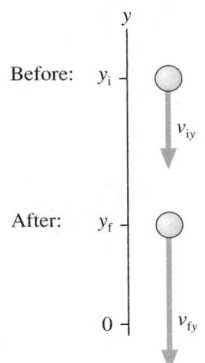

FIGURE 10.2 The before-and-after representation of an object in free fall.

is at position y_i and has velocity v_{iy}, to "after," when the object is at position y_f and has velocity v_{fy}. Figure 10.2 shows these points in the motion. With these limits, the integrals are

$$\int_{v_{iy}}^{v_{fy}} mv_y \, dv_y = -\int_{y_i}^{y_f} mg \, dy \qquad (10.8)$$

Carrying out the integrations, with m and g as constants, we find

$$\frac{1}{2}mv_y^2 \bigg|_{v_{iy}}^{v_{fy}} = \frac{1}{2}mv_{fy}^2 - \frac{1}{2}mv_{iy}^2 = -mgy \bigg|_{y_i}^{y_f} = -mgy_f + mgy_i \qquad (10.9)$$

Because v_y is squared wherever it appears in Equation 10.9, the sign of v_y is not relevant. All we need to know are the initial and final *speeds* v_i and v_f. With this, Equation 10.9 can be written

$$\frac{1}{2}mv_f^2 + mgy_f = \frac{1}{2}mv_i^2 + mgy_i \qquad (10.10)$$

You should recognize that Equation 10.10, other than a constant factor of $\frac{1}{2}m$, is the same as Equation 10.2. This seems like a lot of effort to get to a result we already knew. However, our purpose was not to get the answer but to introduce a *procedure* that will turn out to have other valuable applications.

Kinetic and Potential Energy

The quantity

$$K = \frac{1}{2}mv^2 \qquad \text{(kinetic energy)} \qquad (10.11)$$

is called the **kinetic energy** of the object. **Kinetic energy is an energy of motion.** It depends on the object's speed but not its location. The quantity

$$U_g = mgy \qquad \text{(gravitational potential energy)} \qquad (10.12)$$

is the object's **gravitational potential energy. Potential energy is an energy of position.** It depends on the object's position but not its speed.

The unit of kinetic energy is mass multiplied by velocity squared. In the SI system of units, this is $kg \, m^2/s^2$. The unit of energy is so important that it has been given its own name, the **joule.** We define:

$$1 \text{ joule} = 1 \text{ J} \equiv 1 \text{ kg m}^2/\text{s}^2$$

The unit of potential energy, $kg \times m/s^2 \times m = kg \, m^2/s^2$, is also the joule.

To give you an idea about the size of a joule, consider a 0.5 kg mass (weight on earth $\approx$ 1 lb) moving at 4 m/s ($\approx$ 10 mph). Its kinetic energy is

$$K = \frac{1}{2}mv^2 = \frac{1}{2}(0.5 \text{ kg})(4 \text{ m/s})^2 = 4 \text{ J}$$

Its gravitational potential energy at a height of 1 m ($\approx$ 3 ft) is

$$U_g = mgy = (0.5 \text{ kg})(9.8 \text{ m/s}^2)(1 \text{ m}) \approx 5 \text{ J}$$

This suggests that ordinary-sized objects moving at ordinary speeds will have energies of a fraction of a joule up to, perhaps, a few thousand joules (a running person has $K \approx 1000$ J). A high-speed truck might have $K \approx 10^6$ J.

NOTE ▶ You *must* have masses in kg and velocities in m/s before doing energy calculations. ◀

In terms of energy, Equation 10.10 says that for an object in free fall,

$$K_f + U_{gf} = K_i + U_{gi} \qquad (10.13)$$

In other words, the sum $K + U_g$ of kinetic energy and gravitational potential energy is not changed by free fall. Its value *after* free fall (regardless of whether the motion is upward or downward) is the same as *before* free fall. FIGURE 10.3 illustrates this important idea.

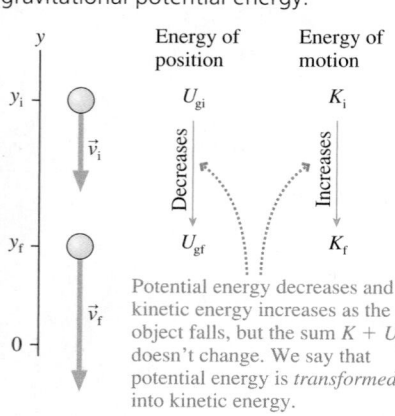

FIGURE 10.3 Kinetic energy and gravitational potential energy.

Potential energy decreases and kinetic energy increases as the object falls, but the sum $K + U_g$ doesn't change. We say that potential energy is *transformed* into kinetic energy.

EXAMPLE 10.1 **Launching a pebble**

Bob uses a slingshot to shoot a 20 g pebble straight up with a speed of 25 m/s. How high does the pebble go?

MODEL This is free-fall motion, so the sum of the kinetic and gravitational potential energy does not change as the pebble rises.

VISUALIZE **FIGURE 10.4** shows a before-and-after pictorial representation. The pictorial representation for energy problems is

FIGURE 10.4 Pictorial representation of a pebble shot upward from a slingshot.

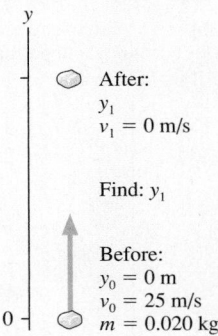

After:
y_1
$v_1 = 0$ m/s

Find: y_1

Before:
$y_0 = 0$ m
$v_0 = 25$ m/s
$m = 0.020$ kg

essentially the same as the pictorial representation you learned in Chapter 9 for momentum problems. We'll use numerical subscripts 0 and 1 for the initial and final points.

SOLVE Equation 10.13,

$$K_1 + U_{g1} = K_0 + U_{g0}$$

tells us that the sum $K + U_g$ is not changed by the motion. Using the definitions of K and U_g, we have

$$\frac{1}{2}mv_1^2 + mgy_1 = \frac{1}{2}mv_0^2 + mgy_0$$

Here $y_0 = 0$ m and $v_1 = 0$ m/s, so the equation simplifies to

$$mgy_1 = \frac{1}{2}mv_0^2$$

This is easily solved for the height y_1:

$$y_1 = \frac{v_0^2}{2g} = \frac{(25 \text{ m/s})^2}{2(9.80 \text{ m/s}^2)} = 32 \text{ m}$$

ASSESS Notice that the mass canceled and wasn't needed, a fact about free fall that you should remember from Chapter 2.

One of the most important characteristics of energy is that it is a scalar, not a vector. Kinetic energy depends on an object's speed v but *not* on the direction of motion. The kinetic energy of a particle is the same whether it moves up or down or left or right. Consequently, the mathematics of using energy is often much easier than the vector mathematics required by force and acceleration.

NOTE ▶ By its definition, kinetic energy can never be a negative number. If you find, in the course of solving a problem, that K is negative—stop! You have made an error somewhere. Don't just "lose" the minus sign and hope that everything turns out OK. ◄

Energy Bar Charts

The pebble of Example 10.1 started with all kinetic energy, an energy of motion. As the pebble ascends, kinetic energy is converted into gravitational potential energy, *but the sum of the two doesn't change.* At the top, the pebble's energy is entirely potential energy. The simple bar chart in **FIGURE 10.5** shows graphically how kinetic energy is transformed into gravitational potential energy as a pebble rises. The potential energy

FIGURE 10.5 Simple energy bar chart for a pebble tossed into the air.

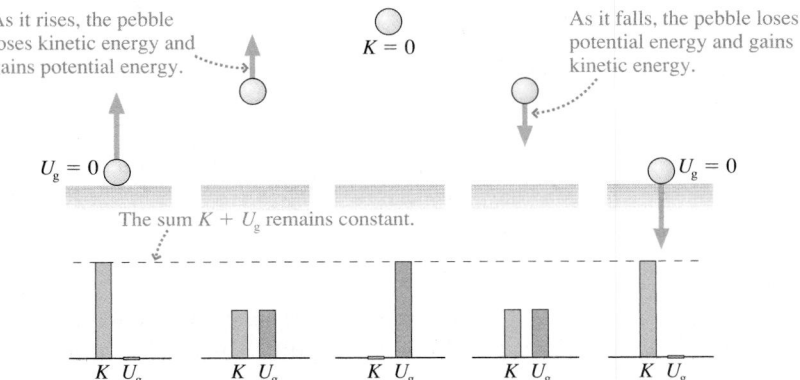

is then transformed back into kinetic energy as the pebble falls. The sum $K + U_g$ remains constant throughout the motion.

FIGURE 10.6a is an energy bar chart more suitable to problem solving. The chart is a graphical representation of the energy equation $K_f + U_{gf} = K_i + U_{gi}$. FIGURE 10.6b applies this to the pebble of Example 10.1. The initial kinetic energy is transformed entirely into potential energy as the pebble reaches its highest point. There are no numerical scales on a bar chart, but you should draw the bar heights proportional to the amount of each type of energy.

FIGURE 10.6 An energy bar chart suitable for problem solving.

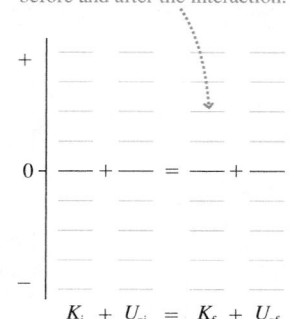 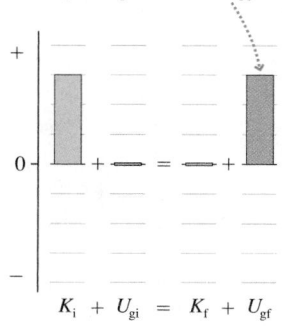

(a) Draw bars to show each energy before and after the interaction.

(b) The initial kinetic energy is transformed entirely into potential energy.

$$K_i + U_{gi} = K_f + U_{gf}$$

$$K_i + U_{gi} = K_f + U_{gf}$$

STOP TO THINK 10.1 Rank in order, from largest to smallest, the gravitational potential energies of balls a to d.

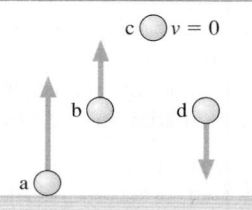

The Zero of Potential Energy

FIGURE 10.7 Amber and Bill use coordinate systems with different origins to determine the potential energy of a rock.

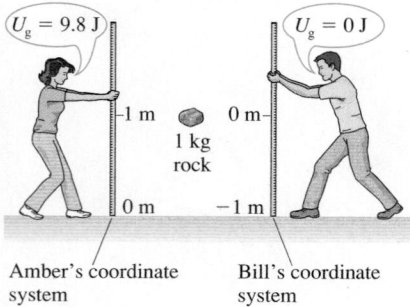

FIGURE 10.7 Amber and Bill use coordinate systems with different origins to determine the potential energy of a rock.

$U_g = 9.8$ J

$U_g = 0$ J

−1 m   0 m−

1 kg
rock

0 m   −1 m

Amber's coordinate system

Bill's coordinate system

Our expression for the gravitational potential energy $U_g = mgy$ seems straightforward. But you might notice, on further reflection, that the value of U_g depends on where you choose to put the origin of your coordinate system. Consider FIGURE 10.7, where Amber and Bill are attempting to determine the potential energy of a 1 kg rock that is 1 m above the ground. Amber chooses to put the origin of her coordinate system on the ground, measures $y_{rock} = 1$ m, and quickly computes $U_g = mgy = 9.8$ J. Bill, on the other hand, read Chapter 1 very carefully and recalls that it is entirely up to him where to locate the origin of his coordinate system. So he places his origin next to the rock, measures $y_{rock} = 0$ m, and declares that $U_g = mgy = 0$ J!

How can the potential energy of one rock at one position in space have two different values? The source of this apparent difficulty comes from our interpretation of Equation 10.9. The integral of $-mg \, dy$ resulted in the expression $-mg(y_f - y_i)$, and this led us to propose that $U_g = mgy$. But all we are *really* justified in concluding is that the potential energy *changes* by $\Delta U = -mg(y_f - y_i)$. To go beyond this and claim $U_g = mgy$ is consistent with $\Delta U = -mg(y_f - y_i)$, but so also would be a claim that $U_g = mgy + C$, where C is any constant.

No matter where the rock is located, Amber's value of y will always equal Bill's value plus 1 m. Consequently, her value of the potential energy will always equal Bill's value plus 9.8 J. That is, their values of U_g differ by a constant. Nonetheless, both will calculate exactly the *same* value for ΔU if the rock changes position.

EXAMPLE 10.2   **The speed of a falling rock**

The 1.0 kg rock shown in Figure 10.7 is released from rest. Use both Amber's and Bill's perspectives to calculate its speed just before it hits the ground.

MODEL This is free-fall motion, so the sum of the kinetic and gravitational potential energy does not change as the rock falls.

VISUALIZE **FIGURE 10.8** shows a before-and-after pictorial representation using both Amber's and Bill's coordinate systems.

FIGURE 10.8 The before-and-after pictorial representation of a falling rock.

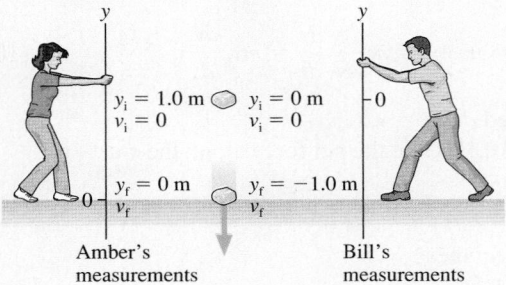

Amber's measurements

Bill's measurements

SOLVE The energy equation is $K_f + U_{gf} = K_i + U_{gi}$. Bill and Amber both agree that $K_i = 0$ because the rock was released from rest, so we have

$$K_f = \frac{1}{2}mv_f^2 = -(U_{gf} - U_{gi}) = -\Delta U$$

According to Amber, $U_{gi} = mgy_i = 9.8$ J and $U_{gf} = mgy_f = 0$ J. Thus

$$\Delta U_{Amber} = U_{gf} - U_{gi} = -9.8 \text{ J}$$

The rock *loses* potential energy as it falls. According to Bill, $U_{gi} = mgy_i = 0$ J and $U_{gf} = mgy_f = -9.8$ J. Thus

$$\Delta U_{Bill} = U_{gf} - U_{gi} = -9.8 \text{ J}$$

Bill has different values for U_{gi} and U_{gf} but the *same* value for ΔU. Thus they both agree that the rock hits the ground with speed

$$v_f = \sqrt{\frac{-2\,\Delta U}{m}} = \sqrt{\frac{-2(-9.8 \text{ J})}{1.0 \text{ kg}}} = 4.4 \text{ m/s}$$

FIGURE 10.9 shows energy bar charts for Amber and Bill. Despite their disagreement over the value of U_g, Amber and Bill arrive at the same value for v_f and their K_f bars are the same height. The reason is that only ΔU has physical significance, not U_g itself, and Amber and Bill found the same value for ΔU. **You can place the origin of your coordinate system, and thus the "zero of potential energy," wherever you choose and be assured of getting the correct answer to a problem.**

FIGURE 10.9 Amber's and Bill's energy bar charts for the falling rock.

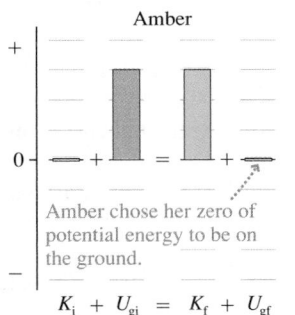

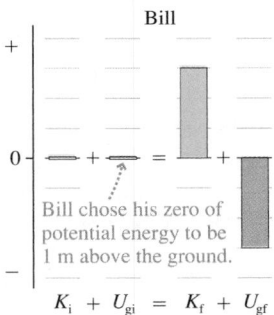

NOTE ▶ Gravitational potential energy can be negative, as U_{gf} is for Bill. **A negative value for U_g means that the particle has less potential for motion at that point than it does at $y = 0$.** But there's nothing wrong with that. Contrast this with kinetic energy, which *cannot* be negative. ◀

10.3 A Closer Look at Gravitational Potential Energy

The concept of energy would be of little interest or use if it applied only to free fall. Let's begin to expand the idea. **FIGURE 10.10a** on the next page shows an object of mass m sliding along a frictionless surface. The only forces acting on the object are gravity

(a)

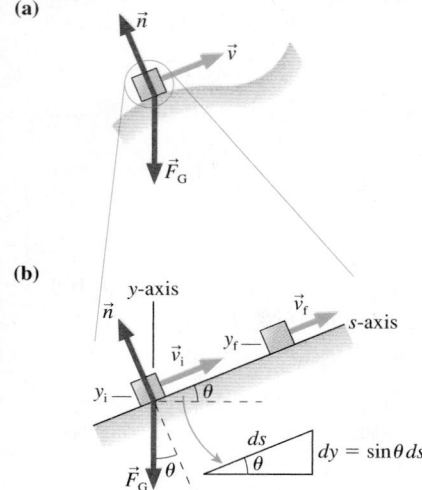

(b)

and the normal force from the surface. If the surface is curved, you know from calculus that we can subdivide the surface into many small (perhaps infinitesimal) straight-line segments. FIGURE 10.10b shows a magnified segment of the surface that, over some small distance, is a straight line at angle θ.

We can analyze the motion along this small segment using the procedure of Equations 10.3 through 10.10. We define an s-axis parallel to the direction of motion. Newton's second law along this axis is

$$(F_{net})_s = ma_s = m\frac{dv_s}{dt} \quad (10.14)$$

Using the chain rule, we can write Equation 10.14 as

$$(F_{net})_s = m\frac{dv_s}{dt} = m\frac{dv_s}{ds}\frac{ds}{dt} = mv_s\frac{dv_s}{ds} \quad (10.15)$$

where, in the last step, we used $ds/dt = v_s$.

You can see from Figure 10.10b that the net force along the s-axis is

$$(F_{net})_s = -F_G\sin\theta = -mg\sin\theta \quad (10.16)$$

Thus Newton's second law becomes

$$-mg\sin\theta = mv_s\frac{dv_s}{ds} \quad (10.17)$$

Multiplying both sides by ds gives

$$mv_s\,dv_s = -mg\sin\theta\,ds \quad (10.18)$$

You can see from the figure that $\sin\theta\,ds$ is dy, so Equation 10.18 becomes

$$mv_s\,dv_s = -mg\,dy \quad (10.19)$$

This is *identical* to Equation 10.7, which we found for free fall. Consequently, integrating this equation from "before" to "after" leads again to Equation 10.10:

$$\frac{1}{2}mv_f^2 + mgy_f = \frac{1}{2}mv_i^2 + mgy_i \quad (10.20)$$

where v_i^2 and v_f^2 are the squares of the *speeds* at the beginning and end of this segment of the motion.

We previously defined the kinetic energy $K = \frac{1}{2}mv^2$ and the gravitational potential energy $U_g = mgy$. Equation 10.20 shows that

$$K_f + U_{gf} = K_i + U_{gi} \quad (10.21)$$

for a particle moving along *any* frictionless surface, regardless of the shape.

NOTE ▶ For energy calculations, the y-axis is specifically a *vertical* axis. Gravitational potential energy depends on the *height* above the earth's surface. A tilted coordinate system, such as we often used in dynamics problems, doesn't work for problems with gravitational potential energy. ◀

STOP TO THINK 10.2 A small child slides down the four frictionless slides a–d. Each has the same height. Rank in order, from largest to smallest, her speeds v_a to v_d at the bottom.

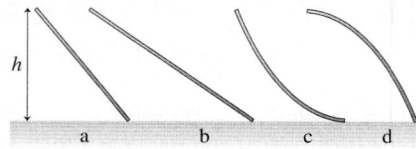

EXAMPLE 10.3 **The speed of a sled**

Christine runs forward with her sled at 2.0 m/s. She hops onto the sled at the top of a 5.0-m-high, very slippery slope. What is her speed at the bottom?

MODEL Model Christine and the sled as a particle. Assume the slope is frictionless. In that case, the sum of her kinetic and gravitational potential energy does not change as she slides down.

VISUALIZE **FIGURE 10.11a** shows a before-and-after pictorial representation. We are not told the angle of the slope, or even if it is a straight slope, but the *change* in potential energy depends only on the height Christine descends and *not* on the shape of the hill. **FIGURE 10.11b** is an energy bar chart in which we see an initial kinetic *and* potential energy being transformed into entirely kinetic energy as she goes down the slope.

SOLVE The quantity $K + U_g$ is the same at the bottom of the hill as it was at the top. Thus

$$\frac{1}{2}mv_1^2 + mgy_1 = \frac{1}{2}mv_0^2 + mgy_0$$

This is easily solved for Christine's speed at the bottom:

$$v_1 = \sqrt{v_0^2 + 2g(y_0 - y_1)} = \sqrt{v_0^2 + 2gh} = 10 \text{ m/s}$$

ASSESS We did not need the mass of either Christine or the sled.

FIGURE 10.11 Pictorial representation and energy bar chart of Christine sliding down the hill.

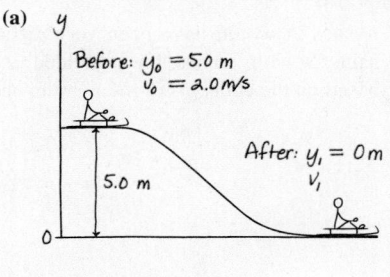

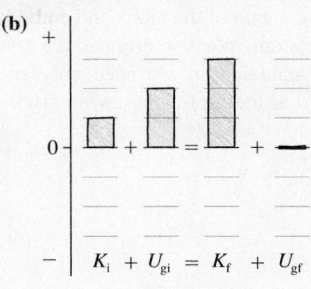

Notice that the normal force $\vec{n}$ doesn't enter an energy analysis. The equation $K_f + U_{gf} = K_i + U_{gi}$ is a statement about how the particle's speed changes as it changes position. $\vec{n}$ does not have a component in the direction of motion, so it cannot change the particle's speed.

The same is true for an object tied to a string and moving in a circle. The tension in the string causes the direction to change, but $\vec{T}$ does not have a component in the direction of motion and does not change the speed of the object. Hence Equation 10.21 also applies to a *pendulum*.

EXAMPLE 10.4 **A ballistic pendulum**

A 10 g bullet is fired into a 1200 g wood block hanging from a 150-cm-long string. The bullet embeds itself into the block, and the block then swings out to an angle of 40°. What was the speed of the bullet? (This is called a *ballistic pendulum*.)

MODEL This is a two-part problem. The impact of the bullet with the block is an inelastic collision. We haven't done any analysis to let us know what happens to energy during a collision, but you learned in Chapter 9 that *momentum* is conserved in an inelastic collision. After the collision is over, the block swings out as a pendulum. The sum of the kinetic and gravitational potential energy does not change as the block swings to its largest angle.

VISUALIZE **FIGURE 10.12** is a pictorial representation in which we've identified before-and-after quantities for both the collision and the swing.

SOLVE The momentum conservation equation $P_f = P_i$ applied to the inelastic collision gives

$$(m_W + m_B)v_{1x} = m_W(v_{0x})_W + m_B(v_{0x})_B$$

FIGURE 10.12 A ballistic pendulum is used to measure the speed of a bullet.

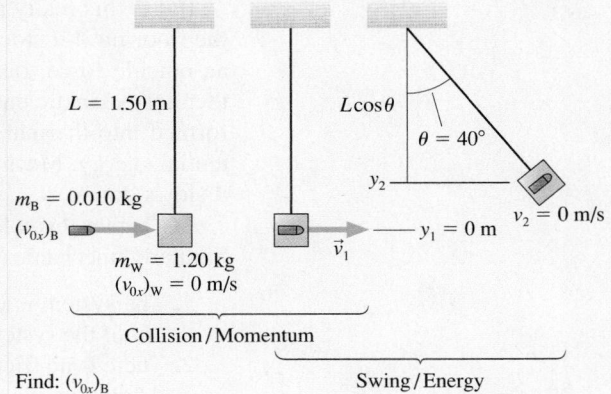

Continued

The wood block is initially at rest, with $(v_{0x})_W = 0$, so the bullet's velocity is

$$(v_{0x})_B = \frac{m_W + m_B}{m_B}v_{1x}$$

where v_{1x} is the velocity of the block + bullet *immediately* after the collision, as the pendulum begins to swing. If we can determine v_{1x} from an analysis of the swing, then we will be able to calculate the speed of the bullet. Turning our attention to the swing, the energy equation $K_f + U_{gf} = K_i + U_{gi}$ is

$$\frac{1}{2}(m_W + m_B)v_2{}^2 + (m_W + m_B)gy_2$$
$$= \frac{1}{2}(m_W + m_B)v_1{}^2 + (m_W + m_B)gy_1$$

We used the *total* mass $(m_W + m_B)$ of the block and embedded bullet, but notice that it cancels out. We also dropped the x-subscript on v_1 because for energy calculations we need only speed, not velocity. The speed is zero at the top of the swing ($v_2 = 0$), and we've defined the y-axis such that $y_1 = 0$ m. Thus

$$v_1 = \sqrt{2gy_2}$$

The initial speed is found simply from the maximum height of the swing. You can see from the geometry of Figure 10.12 that

$$y_2 = L - L\cos\theta = L(1 - \cos\theta) = 0.351 \text{ m}$$

With this, the initial velocity of the pendulum, immediately after the collision, is

$$v_{1x} = v_1 = \sqrt{2gy_2} = \sqrt{2(9.80 \text{ m/s}^2)(0.351 \text{ m})} = 2.62 \text{ m/s}$$

Having found v_{1x} from an energy analysis of the swing, we can now calculate that the speed of the bullet was

$$(v_{0x})_B = \frac{m_W + m_B}{m_B}v_{1x} = \frac{1.210 \text{ kg}}{0.010 \text{ kg}} \times 2.62 \text{ m/s} = 320 \text{ m/s}$$

ASSESS It would have been very difficult to solve this problem using Newton's laws, but it yielded to a straightforward analysis based on the concepts of momentum and energy.

Conservation of Mechanical Energy

The sum of the kinetic energy and the potential energy of a system is called the **mechanical energy:**

$$E_{mech} = K + U \tag{10.22}$$

Here K is the *total* kinetic energy of all the particles in the system and U is the potential energy stored in the system. Our examples thus far suggest that a particle's mechanical energy does not change as it moves under the influence of only gravity. The kinetic energy and the potential energy change, as they are transformed back and forth into each other, but their sum remains constant. We can express the unchanging value of E_{mech} as

$$K_f + U_f = K_i + U_i \tag{10.23}$$

This statement is called the **law of conservation of mechanical energy.**

But is this really a law of nature? Consider shoving a box that then slides along the floor until it stops. The box gains kinetic energy, but it comes from the shove, an outside force, rather than from a transformation of potential energy. The box then loses kinetic energy as it slows down, but in this case kinetic energy is transformed into thermal energy (the box and the floor get hotter) rather than into potential energy. Mechanical energy is conserved neither as the box speeds up nor as it slows down.

In Chapter 9 you learned that momentum is conserved only for an isolated system. Similarly, mechanical energy is conserved only if two requirements are satisfied:

1. The system is isolated, meaning that no external forces transfer energy into or out of the system.
2. There is no friction or drag that would transform kinetic or potential energy into thermal energy.

Fortunately, enough realistic situations satisfy the restrictions, or come very close, that the law of conservation of mechanical energy is an important problem-solving strategy.

STOP TO THINK 10.3   A box slides along the frictionless surface shown in the figure. It is released from rest at the position shown. Is the highest point the box reaches on the other side at level a, level b, or level c?

10.4 Restoring Forces and Hooke's Law

If you stretch a rubber band, a force tries to pull the rubber band back to its equilibrium, or unstretched, length. A force that restores a system to an equilibrium position is called a **restoring force.** Systems that exhibit restoring forces are called **elastic.** The most basic examples of elasticity are things like springs and rubber bands. If you stretch a spring, a tension-like force pulls back. Similarly, a compressed spring tries to re-expand to its equilibrium length. Other examples of elasticity and restoring forces abound. The steel beams bend slightly as you drive your car over a bridge, but they are restored to equilibrium after your car passes by. Nearly everything that stretches, compresses, flexes, bends, or twists exhibits a restoring force and can be called elastic.

We're going to use a simple spring as a prototype of elasticity. Suppose you have a spring whose **equilibrium length** is L_0. This is the length of the spring when it is neither pushing nor pulling. If you now stretch the spring to length L, how hard does it pull back? One way to find out is to attach the spring to a bar, as shown in **FIGURE 10.13**, then to hang a mass m from the spring. The mass stretches the spring to length L. Lengths L_0 and L are easily measured with a meter stick.

The mass hangs in static equilibrium, so the upward spring force $\vec{F}_{sp}$ exactly balances the downward gravitational force $\vec{F}_G$ to give $\vec{F}_{net} = \vec{0}$. That is,

$$F_{sp} = F_G = mg \tag{10.24}$$

By using different masses to stretch the spring to different lengths, we can determine how F_{sp}, the magnitude of the spring's restoring force, depends on the length L.

FIGURE 10.14 shows measured data for the restoring force of a real spring. Notice that the quantity graphed along the horizontal axis is $\Delta s = L - L_0$. This is the distance that the end of the spring has moved, which we call the **displacement from equilibrium.** The graph shows that the restoring force is proportional to the displacement. That is, the data fall along the straight line

$$F_{sp} = k\,\Delta s \tag{10.25}$$

The proportionality constant k, the slope of the force-versus-displacement graph, is called the **spring constant.** The units of the spring constant are N/m.

FIGURE 10.13 A hanging mass stretches a spring of equilibrium length L_0 to length L.

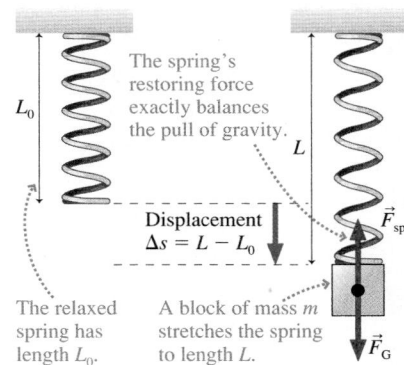

The spring's restoring force exactly balances the pull of gravity.

Displacement $\Delta s = L - L_0$

The relaxed spring has length L_0.

A block of mass m stretches the spring to length L.

FIGURE 10.14 Measured data for the restoring force of a real spring.

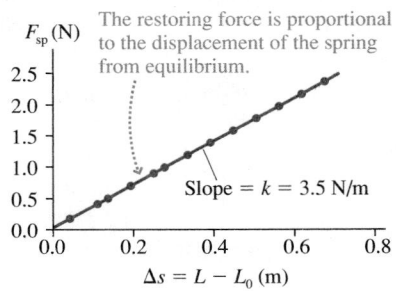

The restoring force is proportional to the displacement of the spring from equilibrium.

Slope = k = 3.5 N/m

NOTE ▶ The force does not depend on the spring's physical length L but, instead, on the *displacement* Δs of the end of the spring. ◀

The spring constant k is a property that characterizes a spring, just as mass m characterizes a particle. If k is large, it takes a large pull to cause a significant stretch, and we call the spring a "stiff" spring. A spring with small k can be stretched with very little force, and we call it a "soft" spring. The spring constant for the spring in Figure 10.14 can be determined from the slope of the straight line to be $k = 3.5$ N/m.

NOTE ▶ Just as we used massless strings, we will adopt the idealization of a *massless spring*. While not a perfect description, it is a good approximation if the mass attached to a spring is much larger than the mass of the spring itself. ◀

Hooke's Law

FIGURE 10.15 The direction of $\vec{F}_{sp}$ is always opposite the displacement $\Delta \vec{s}$.

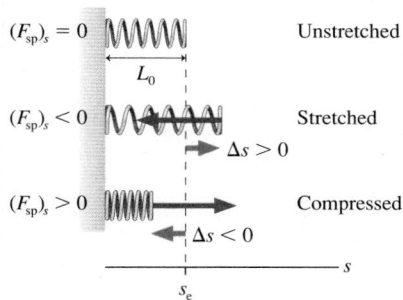

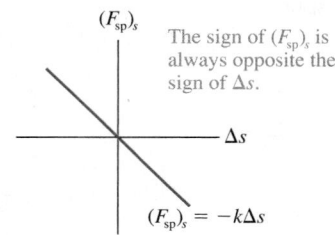

FIGURE 10.15 shows a spring along a generic s-axis. The equilibrium position of the end of the spring is denoted s_e. This is the *position*, or coordinate, of the free end of the spring, *not* the spring's equilibrium length L_0.

When the spring is stretched, the displacement from equilibrium $\Delta s = s - s_e$ is *positive* while $(F_{sp})_s$, the s-component of the restoring force pointing to the left, is *negative*. If the spring is compressed, the displacement from equilibrium Δs is negative while the s-component of $\vec{F}_{sp}$, which now points to the right, is positive. Either way, the sign of the force component $(F_{sp})_s$ is always opposite to the sign of the displacement Δs. We can write this mathematically as

$$(F_{sp})_s = -k\,\Delta s \quad \text{(Hooke's law)} \quad (10.26)$$

where $\Delta s = s - s_e$ is the displacement of the end of the spring from equilibrium. The minus sign is the mathematical indication of a *restoring* force.

Equation 10.26 for the restoring force of a spring is called **Hooke's law.** This "law" was first suggested by Robert Hooke, a contemporary (and sometimes bitter rival) of Newton. Hooke's law is not a true "law of nature," in the sense that Newton's laws are, but is actually just a *model* of a restoring force. It works extremely well for some springs, as in Figure 10.14, but less well for others. Hooke's law will fail for any spring that is compressed or stretched too far.

NOTE ▶ Some of you, in an earlier physics course, may have learned Hooke's law as $F_{sp} = -kx$ (for a spring along the x-axis), rather than as $-k\,\Delta x$. This can be misleading, and it is a common source of errors. The restoring force will be $-kx$ *only* if the coordinate system in the problem is chosen such that the origin is at the equilibrium position of the free end of the spring. That is, $\Delta x = x$ only if $x_e = 0$. This is often done, but in some problems it will be more convenient to locate the origin of the coordinate system elsewhere. So make sure you learn Hooke's law as $(F_{sp})_s = -k\,\Delta s$. ◀

EXAMPLE 10.5 **Pull until it slips**

FIGURE 10.16 shows a spring attached to a 2.0 kg block. The other end of the spring is pulled by a motorized toy train that moves forward at 5.0 cm/s. The spring constant is 50 N/m, and the coefficient of static friction between the block and the surface is 0.60. The spring is at its equilibrium length at $t = 0$ s when the train starts to move. When does the block slip?

MODEL Model the block as a particle and the spring as an ideal spring obeying Hooke's law.

VISUALIZE FIGURE 10.17 is a free-body diagram for the block.

SOLVE Recall that the tension in a massless string pulls equally at *both* ends of the string. The same is true for the spring force: It pulls (or pushes) equally at *both* ends. This is the key to solving the problem. As the right end of the spring moves, stretching the spring, the spring pulls backward on the train *and* forward on the block with equal strength. As the spring stretches, the static friction force on the block increases in

FIGURE 10.16 A toy train stretches the spring until the block slips.

5.0 cm/s

FIGURE 10.17 The free-body diagram.

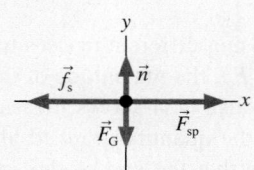

magnitude to keep the block at rest. The block is in static equilibrium, so

$$\sum (F_{net})_x = (F_{sp})_x + (f_s)_x = F_{sp} - f_s = 0$$

where F_{sp} is the *magnitude* of the spring force. The magnitude is $F_{sp} = k \, \Delta x$, where $\Delta x = v_x t$ is the distance the train has moved. Thus

$$f_s = F_{sp} = k \, \Delta x$$

The block slips when the static friction force reaches its maximum

value $f_{s\,max} = \mu_s n = \mu_s mg$. This occurs when the train has moved

$$\Delta x = \frac{f_{s\,max}}{k} = \frac{\mu_s mg}{k} = \frac{(0.60)(2.0 \text{ kg})(9.80 \text{ m/s}^2)}{50 \text{ N/m}}$$

$$= 0.235 \text{ m} = 23.5 \text{ cm}$$

The time at which the block slips is

$$t = \frac{\Delta x}{v_x} = \frac{23.5 \text{ cm}}{5.0 \text{ cm/s}} = 4.7 \text{ s}$$

This example illustrates a class of motion called *stick-slip motion*. Once the block slips, it will shoot forward some distance, then stop and stick again. As the train continues, there will be a recurring sequence of stick, slip, stick, slip, stick. . . .

Earthquakes are an important example of stick-slip motion. The large tectonic plates making up the earth's crust are attempting to slide past each other, but friction causes the edges of the plates to stick together. You may think of rocks as rigid and brittle, but large masses of rock are somewhat elastic and can be "stretched." Eventually the elastic force of the deformed rocks exceeds the friction force between the plates. An earthquake occurs as the plates slip and lurch forward. Once the tension is released, the plates stick together again and the process starts all over.

The slip can range from a few centimeters in a relatively small earthquake to several meters in a very large earthquake.

STOP TO THINK 10.4 The graph shows force versus displacement for three springs. Rank in order, from largest to smallest, the spring constants k_a, k_b, and k_c.

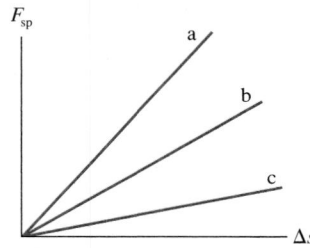

10.5 Elastic Potential Energy

The forces we have worked with thus far—gravity, friction, tension—have been constant forces. That is, their magnitudes do not change as an object moves. That feature has been important because the kinematic equations we developed in Chapter 2 are for motion with constant acceleration. But a spring exerts a *variable* force. The force is zero if $\Delta s = 0$ (no displacement), and it steadily increases as the stretching increases. The "natural motion" of a mass on a spring—think of pulling down on a spring and then releasing it—is an *oscillation*. This is *not* constant-acceleration motion, and we haven't yet developed the kinematics to handle oscillatory motion.

But suppose we're interested not in the time dependence of motion, only in before-and-after situations. For example, FIGURE 10.18 shows a before-and-after situation in which a spring launches a ball. Asking how the compression of the spring (the "before") affects the speed of the ball (the "after") is very different from wanting to know the ball's position as a function of time as the spring expands.

You certainly have a sense that a compressed spring has "stored energy," and Figure 10.18 shows clearly that the stored energy is transformed into the kinetic energy of the ball. Let's analyze this process with the same method we developed for motion under the influence of gravity. Newton's second law for the ball is

$$(F_{net})_s = ma_s = m\frac{dv_s}{dt} \tag{10.27}$$

FIGURE 10.18 Before and after a spring launches a ball.

The compressed spring stores energy.

Before:

After:

The spring's potential energy is transformed into the ball's kinetic energy.

Springs and rubber bands store energy—potential energy—that can be transformed into kinetic energy.

The net force on the ball is given by Hooke's law, $(F_{net})_s = -k(s - s_e)$. Thus

$$m\frac{dv_s}{dt} = -k(s - s_e) \tag{10.28}$$

We'll use a generic s-axis, although it is better in actual problem solving to use x or y, depending on whether the motion is horizontal or vertical.

As before, we use the chain rule to write

$$\frac{dv_s}{dt} = \frac{dv_s}{ds}\frac{ds}{dt} = v_s\frac{dv_s}{ds} \tag{10.29}$$

We substitute this into Equation 10.28 and then multiply both sides by ds to get

$$mv_s\,dv_s = -k(s - s_e)\,ds \tag{10.30}$$

We can integrate both sides of the equation from the initial conditions i to the final conditions f—that is, integrate "from before to after"—to give

$$\int_{v_i}^{v_f} mv_s\,dv_s = \frac{1}{2}mv_f^2 - \frac{1}{2}mv_i^2 = -k\int_{s_i}^{s_f}(s - s_e)\,ds \tag{10.31}$$

The integral on the right is not difficult, but many of you are new to calculus so we'll proceed step by step. The easiest way to get the answer in the most useful form is to make a change of variables. Define $u = (s - s_e)$, in which case $ds = du$. This changes the integrand from $(s - s_e)\,ds$ to $u\,du$.

When we change variables, we also must change the limits of integration. In particular, $s = s_i$ at the lower integration limit makes $u = s_i - s_e = \Delta s_i$, where Δs_i is the initial displacement of the spring from equilibrium. Likewise, $s = s_f$ makes $u = s_f - s_e = \Delta s_f$ at the upper limit. FIGURE 10.19 clarifies the meanings of Δs_i and Δs_f.

With this change of variables, the integral is

$$-k\int_{s_i}^{s_f}(s - s_e)\,ds = -k\int_{\Delta s_i}^{\Delta s_f} u\,du = -\frac{1}{2}ku^2\Big|_{\Delta s_i}^{\Delta s_f}$$
$$= -\frac{1}{2}k(\Delta s_f)^2 + \frac{1}{2}k(\Delta s_i)^2 \tag{10.32}$$

FIGURE 10.19 The initial and final displacements of the spring.

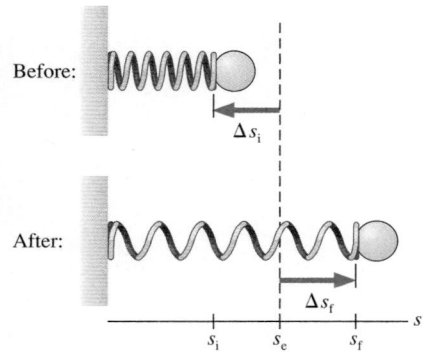

Before:

Δs_i

After:

Δs_f

$s_i \quad s_e \quad s_f$ — s

Using this result makes Equation 10.31 become

$$\frac{1}{2}mv_f^2 - \frac{1}{2}mv_i^2 = -\frac{1}{2}k(\Delta s_f)^2 + \frac{1}{2}k(\Delta s_i)^2 \tag{10.33}$$

which can be rewritten as

$$\frac{1}{2}mv_f^2 + \frac{1}{2}k(\Delta s_f)^2 = \frac{1}{2}mv_i^2 + \frac{1}{2}k(\Delta s_i)^2 \tag{10.34}$$

We've succeeded in our goal of relating before and after. In particular, the quantity

$$\frac{1}{2}mv^2 + \frac{1}{2}k(\Delta s)^2 \tag{10.35}$$

does not change as the spring compresses or expands. You recognize $\frac{1}{2}mv^2$ as the kinetic energy K. Let's define the **elastic potential energy** U_s of a spring to be

$$U_s = \frac{1}{2}k(\Delta s)^2 \quad \text{(elastic potential energy)} \tag{10.36}$$

Then Equation 10.34 tells us that an object moving on a spring obeys

$$K_f + U_{sf} = K_i + U_{si} \tag{10.37}$$

In other words, the mechanical energy $E_{mech} = K + U_s$ is conserved for an object moving *without friction* on an ideal spring.

NOTE ▶ Because Δs is squared, the elastic potential energy is positive for a spring that is either stretched or compressed. U_s is zero when the spring is at its equilibrium length L_0 and $\Delta s = 0$. ◀

EXAMPLE 10.6 A spring-launched plastic ball

A spring-loaded toy gun launches a 10 g plastic ball. The spring, with spring constant 10 N/m, is compressed by 10 cm as the ball is pushed into the barrel. When the trigger is pulled, the spring is released and shoots the ball back out. What is the ball's speed as it leaves the barrel? Assume friction is negligible.

MODEL Assume an ideal spring that obeys Hooke's law. Also assume that the gun is held firmly enough to prevent recoil. There's no friction; hence the mechanical energy $K + U_s$ is conserved.

VISUALIZE FIGURE 10.20a shows a before-and-after pictorial representation. We have chosen to put the origin of the coordinate system at the equilibrium position of the free end of the spring. The bar chart of FIGURE 10.20b shows the potential energy stored in the compressed spring being entirely transformed into the kinetic energy of the ball.

SOLVE The energy conservation equation is $K_2 + U_{s2} = K_1 + U_{s1}$. We can use the elastic potential energy of the spring, Equation 10.36, to write this as

$$\frac{1}{2}mv_2^2 + \frac{1}{2}k(x_2 - x_e)^2 = \frac{1}{2}mv_1^2 + \frac{1}{2}k(x_1 - x_e)^2$$

Notice that we used x, rather than the generic s, and that we explicitly wrote out the meaning of Δx_1 and Δx_2. Using $x_2 = x_e = 0$ m and $v_1 = 0$ m/s simplifies this to

$$\frac{1}{2}mv_2^2 = \frac{1}{2}kx_1^2$$

FIGURE 10.20 Pictorial representation and energy bar chart of a ball being shot from a spring-loaded toy gun.

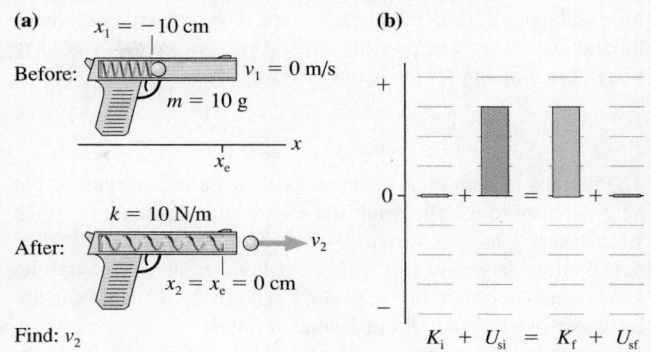

It is now straightforward to solve for the ball's speed:

$$v_2 = \sqrt{\frac{kx_1^2}{m}} = \sqrt{\frac{(10 \text{ N/m})(-0.10 \text{ m})^2}{0.010 \text{ kg}}} = 3.2 \text{ m/s}$$

ASSESS This is a problem that we could *not* have solved with Newton's laws. The acceleration is not constant, and we have not learned how to handle the kinematics of nonconstant acceleration. But with conservation of energy—it's easy! The result, 3.2 m/s, seems reasonable for a toy gun.

EXAMPLE 10.7 A spring-launched projectile

Your lab assignment for the week is to devise a method to determine the spring constant of a spring. You notice several small blocks of different mass lying around, so you decide to measure how high the compressed spring will launch each of the blocks. You and your lab partners quickly realize that you need to compress the spring the same amount each time, so that only the mass is varying, and you choose to use a compression of 4.0 cm. Measuring the height from where you place the mass on the compressed spring generates the following data:

Mass (g)	Height (m)
50	2.07
100	1.11
150	0.65
200	0.51

What value will you report for the spring constant?

MODEL Assume an ideal spring that obeys Hooke's law. There's no friction, and we'll assume no drag; hence the mechanical energy $K + U$ is conserved. However, this system has both elastic *and* gravitational potential energy—two distinct ways of storing energy—and we need to include them both. Thus $U = U_g + U_s$.

VISUALIZE FIGURE 10.21 is a before-and-after pictorial representation. We've chosen to place the origin of the coordinate system at the point of launch, so in this problem the equilibrium position of

FIGURE 10.21 Pictorial representation of a spring-launched projectile.

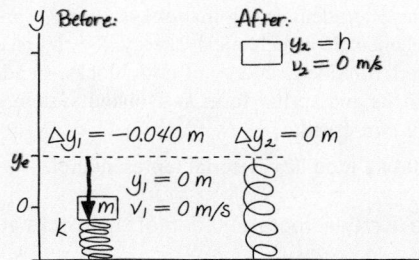

the spring is *not* $y_e = 0$. The projectile reaches height $y_2 = h$, at which point $v_2 = 0$ m/s.

SOLVE Mechanical energy is now $K + U_g + U_s$, so the conservation equation is

$$\frac{1}{2}mv_2^2 + mgy_2 + \frac{1}{2}k(\Delta y_2)^2 = \frac{1}{2}mv_1^2 + mgy_1 + \frac{1}{2}k(\Delta y_1)^2$$

It is important to distinguish between the *position* of the projectile and the *compression* of the spring. While the projectile moves to position y_2, the end of the spring stops at y_e. Thus $\Delta y_2 = 0$, not $\Delta y_2 = y_2$. The initial and final speeds are zero, as is the initial position, so the equation simplifies to

$$mgh = \frac{1}{2}k(\Delta y_1)^2$$

Continued

This equation tells us that the net effect of the launch is to transform the potential energy initially stored in the spring entirely into gravitational potential energy. Kinetic energy is zero at the beginning and again zero at the highest point. The projectile does have kinetic energy as it comes off the spring, but we don't need to know that. Solving for the height, we find

$$h = \frac{k(\Delta y_1)^2}{2mg} = \frac{k(\Delta y_1)^2}{2g} \cdot \frac{1}{m}$$

The first expression for h is correct as an algebraic expression, but here we want to use the result to analyze an experiment in which we measure h as m is varied. By isolating the mass term, we see that plotting h versus $1/m$ (that is, using $1/m$ as the x-variable) should yield a straight line with slope $k(\Delta y_1)^2/2g$. Thus we can use the experimentally determined slope to find k.

FIGURE 10.22 is a graph of h versus $1/m$, with masses first converted to kg. The graph is linear and the best-fit line has a y-intercept very near zero, confirming our analysis of the situation. The experimentally determined slope is 0.105 m kg, with the units determined by rise over run. Thus the experimental value of the spring constant is

$$k = \frac{2g}{(\Delta y_1)^2} \times \text{slope} = 1290 \text{ N/m}$$

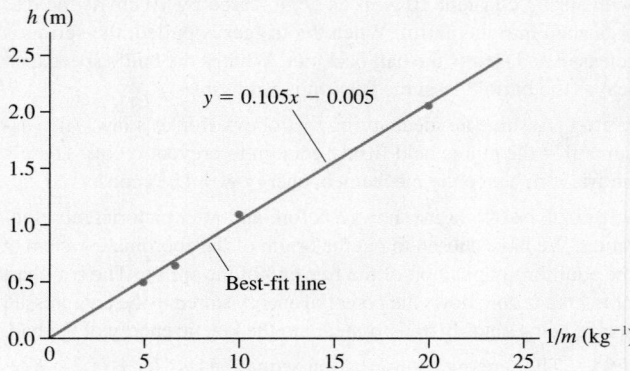

FIGURE 10.22 Graph of the projectile height versus the inverse of its mass.

$y = 0.105x - 0.005$

Best-fit line

ASSESS A spring with spring constant 1290 N/m has potential energy $U_s = \frac{1}{2}k(\Delta y)^2 = 1.0$ J when compressed 4 cm. A 100 g mass has potential energy $U_g = mgy = 1.1$ J at a height of 1.1 m. That these energies are very nearly equal (perfect equality isn't expected with experimental data) gives us confidence in our value for the spring constant.

EXAMPLE 10.8 **Pushing apart**

A spring with spring constant 2000 N/m is sandwiched between a 1.0 kg block and a 2.0 kg block on a frictionless table. The blocks are pushed together to compress the spring by 10 cm, then released. What are the velocities of the blocks as they fly apart?

MODEL Assume an ideal spring that obeys Hooke's law. There's no friction; hence the mechanical energy $K + U_s$ is conserved. Here K is the *total* kinetic energy of both blocks. In addition, because the blocks and spring form an isolated system, their total momentum is conserved.

VISUALIZE FIGURE 10.23 is a pictorial representation.

FIGURE 10.23 Pictorial representation of the blocks and spring.

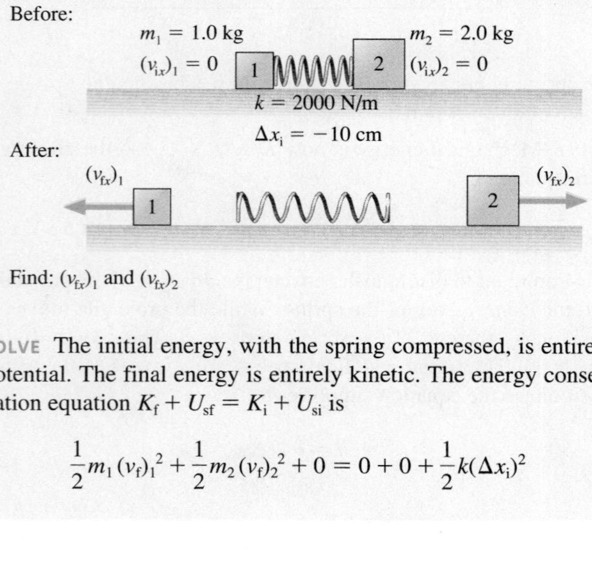

Before:

$m_1 = 1.0$ kg $m_2 = 2.0$ kg
$(v_{ix})_1 = 0$ 1 WWWW 2 $(v_{ix})_2 = 0$
          $k = 2000$ N/m
          $\Delta x_i = -10$ cm

After:

$(v_{fx})_1$ $(v_{fx})_2$

  1   WWWW   2

Find: $(v_{fx})_1$ and $(v_{fx})_2$

SOLVE The initial energy, with the spring compressed, is entirely potential. The final energy is entirely kinetic. The energy conservation equation $K_f + U_{sf} = K_i + U_{si}$ is

$$\frac{1}{2}m_1(v_f)_1^2 + \frac{1}{2}m_2(v_f)_2^2 + 0 = 0 + 0 + \frac{1}{2}k(\Delta x_i)^2$$

Notice that *both* blocks contribute to the kinetic energy. The energy equation has two unknowns, $(v_f)_1$ and $(v_f)_2$, and one equation is not enough to solve the problem. Fortunately, momentum is also conserved. The initial momentum is zero because both blocks are at rest, so the momentum equation is

$$m_1(v_{fx})_1 + m_2(v_{fx})_2 = 0$$

which can be solved to give

$$(v_{fx})_1 = -\frac{m_2}{m_1}(v_{fx})_2$$

The minus sign indicates that the blocks move in opposite directions. The speed $(v_f)_1 = (m_2/m_1)(v_f)_2$ is all we need to calculate the kinetic energy. Substituting $(v_f)_1$ into the energy equation gives

$$\frac{1}{2}m_1\left(\frac{m_2}{m_1}(v_f)_2\right)^2 + \frac{1}{2}m_2(v_f)_2^2 = \frac{1}{2}k(\Delta x_i)^2$$

which simplifies to

$$m_2\left(1 + \frac{m_2}{m_1}\right)(v_f)_2^2 = k(\Delta x_i)^2$$

Solving for $(v_f)_2$, we find

$$(v_f)_2 = \sqrt{\frac{k(\Delta x_i)^2}{m_2(1 + m_2/m_1)}} = 1.8 \text{ m/s}$$

Finally, we can go back to find

$$(v_{fx})_1 = -\frac{m_2}{m_1}(v_{fx})_2 = -3.6 \text{ m/s}$$

The 2.0 kg block moves to the right at 1.8 m/s while the 1.0 kg block goes left at 3.6 m/s.

ASSESS Speeds of a few m/s seem reasonable.

A spring-loaded gun shoots a plastic ball with a speed of 4 m/s. If the spring is compressed twice as far, the ball's speed will be

a. 2 m/s.　　　　　　b. 4 m/s.

c. 8 m/s.　　　　　　d. 16 m/s.

10.6 Energy Diagrams

Potential energy is an energy of position. The gravitational potential energy depends on the height of an object, and the elastic potential energy depends on a spring's displacement. Other potential energies you will meet in the future will depend in some way on position. Functions of position are easy to represent as graphs. A graph showing a system's potential energy and total energy as a function of position is called an **energy diagram.** Energy diagrams allow you to visualize motion based on energy considerations.

FIGURE 10.24 is the energy diagram of a particle in free fall. The gravitational potential energy $U_g = mgy$ is graphed as a line through the origin with slope mg. The *potential-energy curve* is labeled PE. The line labeled TE is the *total energy line,* $E = K + U_g$. It is horizontal because mechanical energy is conserved, meaning that the object's mechanical energy E has the same value at every position.

Suppose the particle is at position y_1. By definition, the distance from the axis to the potential-energy curve is the particle's potential energy U_{g1} at that position. Because $K_1 = E - U_{g1}$, the distance between the potential-energy curve and the total energy line is the particle's kinetic energy.

The four-frame "movie" of FIGURE 10.25 illustrates how an energy diagram is used to visualize motion. The first frame shows a particle projected upward from $y_a = 0$ with kinetic energy K_a. Initially the energy is entirely kinetic, with $U_{ga} = 0$. A pictorial representation and an energy bar chart help to illustrate what the energy diagram is showing.

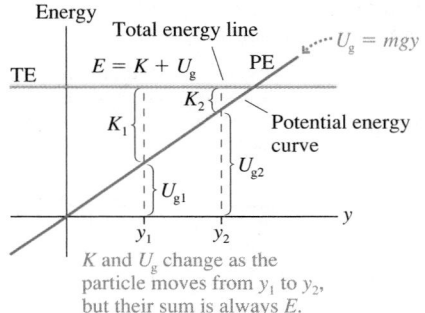

FIGURE 10.24 The energy diagram of a particle in free fall.

K and U_g change as the particle moves from y_1 to y_2, but their sum is always *E*.

FIGURE 10.25 A four-frame movie of a particle in free fall.

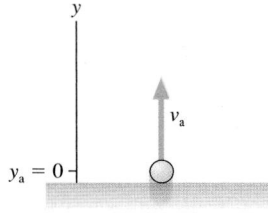 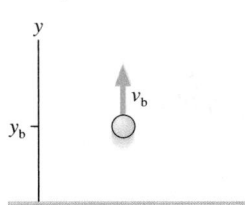 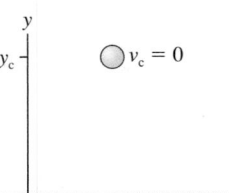 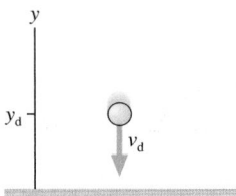

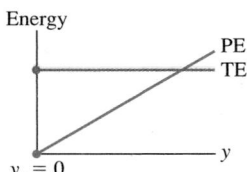 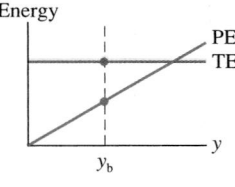 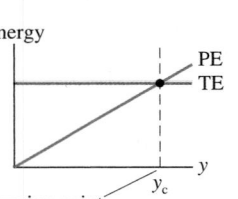 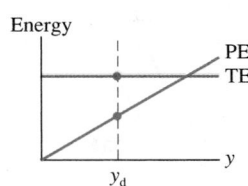

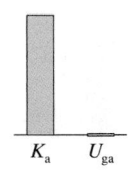 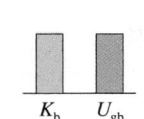 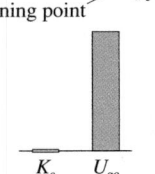 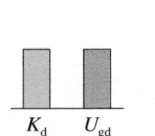

The particle is projected upward. Energy is entirely kinetic.

The particle has gained potential energy and lost kinetic energy.

The energy is entirely potential at the turning point.

The particle gains kinetic energy and loses potential energy as it falls.

In the second frame, the particle has gained height but lost speed. The potential energy U_{gb} is larger, and the distance K_b between the potential-energy curve and the total energy line is less. The particle continues rising and slowing until, in the third frame, it reaches the y-value where the total energy line crosses the potential-energy curve. This point, where $K = 0$ and the energy is entirely potential, is a *turning point* where the particle reverses direction. Finally, we see the particle speeding up as it falls.

A particle with this amount of total energy would need negative kinetic energy to be to the right of the point, at y_c, where the total energy line crosses the potential-energy curve. Negative K is not physically possible, so **the particle cannot be at positions with $U > E$.** Now, it's certainly true that you could make the particle reach a larger value of y simply by throwing it harder. But that would increase E and move the total energy line higher.

> NOTE ▶ The TE line is under your control. You can move the TE line as far up or down as you wish by changing the initial conditions, such as projecting the particle upward with a different speed or dropping it from a different height. Once you've determined the initial conditions, you can use the energy diagram to analyze the motion for that amount of total energy. ◀

FIGURE 10.26 shows the energy diagram of a mass on a horizontal spring. The potential-energy curve $U_s = \frac{1}{2}k(x - x_e)^2$ is a parabola centered at the equilibrium position x_e. The PE curve is determined by the spring constant; you can't change it. But you can set the TE to any height you wish simply by stretching the spring to the proper length. The figure shows one possible TE line.

Suppose you pull the mass out to position x_R and release it. FIGURE 10.27 is a four-frame movie of the subsequent motion. Initially, the energy is entirely potential. The restoring force of the spring pulls the mass toward x_e, increasing the kinetic energy as the potential energy decreases. The mass has maximum speed at position x_e, where $U_s = 0$, and then it slows down as the spring starts to compress.

If the movie were to continue, you should be able to visualize that position x_L is a turning point. The mass will instantaneously have $v_L = 0$ and $K_L = 0$, then reverse direction as the spring starts to expand. The mass will speed up until x_e, then slow

FIGURE 10.26 The energy diagram of a mass on a horizontal spring.

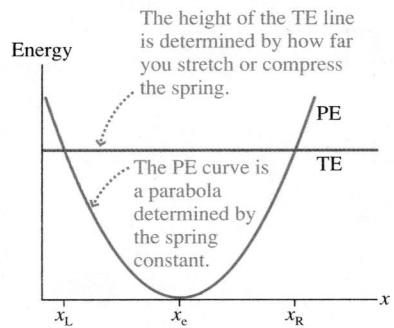

FIGURE 10.27 A four-frame movie of a mass oscillating on a spring.

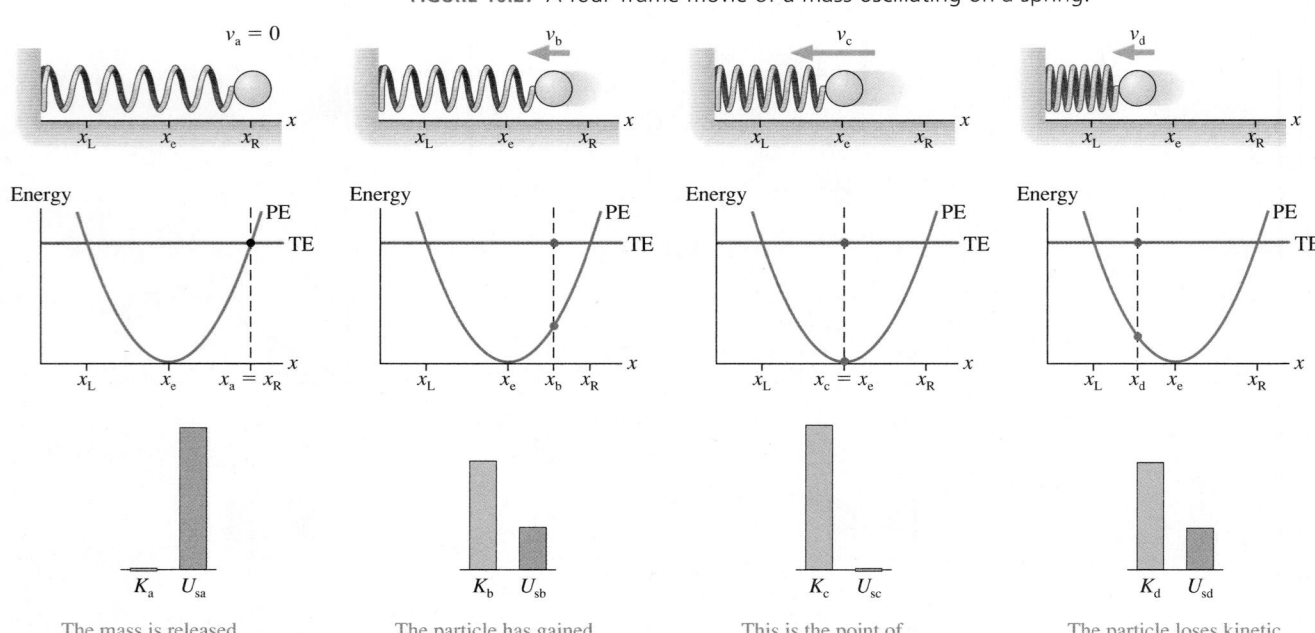

The mass is released from rest. The energy is entirely potential.

The particle has gained kinetic energy as the spring loses potential energy.

This is the point of maximum speed. The energy is entirely kinetic.

The particle loses kinetic energy as it compresses the spring.

down until reaching x_R, where it started. This is another turning point. It will reverse direction again and start the process over. In other words, the mass will *oscillate* back and forth between the left and right turning points at x_L and x_R where the TE line crosses the PE curve.

FIGURE 10.28 applies these ideas to a more general energy diagram. We don't know how this potential energy was created, but we can visualize the motion of a particle that has this potential energy. Suppose the particle is released from rest at position x_1. How will it then move?

The particle's kinetic energy at x_1 is zero; hence the TE line must cross the PE curve at this point. The particle cannot move to the left because $U > E$, so it begins to move toward the right. The particle speeds up from x_1 to x_2 as U decreases and K increases, then slows down from x_2 to x_3 as it goes up the "potential-energy hill." The particle doesn't stop at x_3 because it still has kinetic energy. It speeds up from x_3 to x_4, reaching its maximum speed at x_4, then slows down between x_4 and x_5. Position x_5 is a turning point, a point where the TE line crosses the PE curve. The particle is instantaneously at rest, then reverses direction. The particle will oscillate back and forth between x_1 and x_5, following the pattern of slowing down and speeding up that we've outlined.

Equilibrium Positions

Positions x_2, x_3, and x_4 in Figure 10.28, where the potential energy has a local minimum or maximum, are special positions. Consider a particle with the total energy E_2 shown in FIGURE 10.29. The particle can be at rest at x_2, with $K = 0$, but it cannot move away from x_2. In other words, a particle with energy E_2 is in *static equilibrium* at x_2. If you disturb the particle, giving it a small kinetic energy and a total energy just *slightly* larger than E_2, the particle will undergo a very small oscillation centered on x_2, like a marble in the bottom of a bowl. An equilibrium for which small disturbances cause small oscillations is called a point of **stable equilibrium.** You should recognize that *any* minimum in the PE curve is a point of stable equilibrium. Position x_4 is also a point of stable equilibrium, in this case for a particle with $E = 0$.

Figure 10.29 also shows a particle with energy E_3 that is tangent to the curve at x_3. If a particle is placed *exactly* at x_3, it will stay there at rest ($K = 0$). But if you disturb the particle at x_3, giving it an energy only slightly more than E_3, it will speed up as it moves away from x_3. This is like trying to balance a marble on top of a hill. The slightest displacement will cause the marble to roll down the hill. A point of equilibrium for which a small disturbance causes the particle to move away is called a point of **unstable equilibrium.** Any maximum in the PE curve, such as x_3, is a point of unstable equilibrium.

We can summarize these lessons as follows:

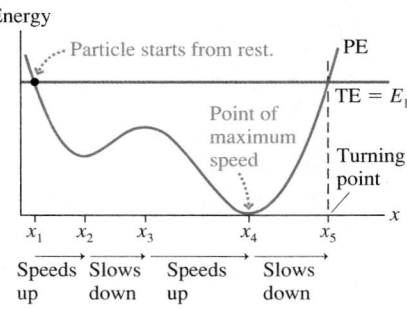

FIGURE 10.28 A more general energy diagram.

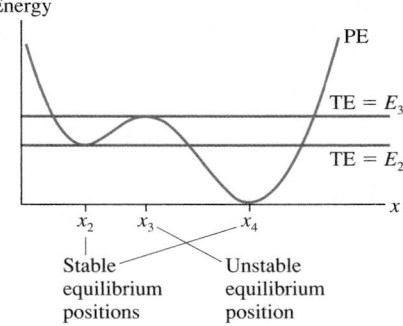

FIGURE 10.29 Points of stable and unstable equilibrium.

TACTICS
BOX 10.1 **Interpreting an energy diagram**

❶ The distance from the axis to the PE curve is the particle's potential energy. The distance from the PE curve to the TE line is its kinetic energy. These are transformed as the position changes, causing the particle to speed up or slow down, but the sum $K + U$ doesn't change.

❷ A point where the TE line crosses the PE curve is a turning point. The particle reverses direction.

❸ The particle cannot be at a point where the PE curve is above the TE line.

❹ The PE curve is determined by the properties of the system—mass, spring constant, and the like. You cannot change the PE curve. However, you can raise or lower the TE line simply by changing the initial conditions to give the particle more or less total energy.

❺ A minimum in the PE curve is a point of stable equilibrium. A maximum in the PE curve is a point of unstable equilibrium.

Exercises 18–20

EXAMPLE 10.9 **Balancing a mass on a spring**

A spring of length L_0 and spring constant k is standing on one end. A block of mass m is placed on the spring, compressing it. What is the length of the compressed spring?

MODEL Assume an ideal spring obeying Hooke's law. The block + spring system has both gravitational potential energy U_g *and* elastic potential energy U_s. The block sitting on top of the spring is at a point of stable equilibrium (small disturbances cause the block to oscillate slightly around the equilibrium position), so we can solve this problem by looking at the energy diagram.

VISUALIZE FIGURE 10.30a is a pictorial representation. We've used a coordinate system with the origin at ground level, so the equilibrium position of the uncompressed spring is $y_e = L_0$.

SOLVE FIGURE 10.30b shows the two potential energies separately and also shows the total potential energy:

$$U_{tot} = U_g + U_s = mgy + \frac{1}{2}k(y - L_0)^2$$

The equilibrium position (the minimum of U_{tot}) has shifted from L_0 to a smaller value of y, closer to the ground. We can find the equilibrium by locating the position of the minimum in the PE curve. You know from calculus that the minimum of a function is at the point where the derivative (or slope) is zero. The derivative of U_{tot} is

$$\frac{dU_{tot}}{dy} = mg + k(y - L_0)$$

The derivative is zero at the point y_{eq}, so we can easily find

$$mg + k(y_{eq} - L_0) = 0$$

$$y_{eq} = L_0 - \frac{mg}{k}$$

The block compresses the spring by the length mg/k from its original length L_0, giving it a new equilibrium length $L_0 - mg/k$.

FIGURE 10.30 The block + spring system has both gravitational and elastic potential energy.

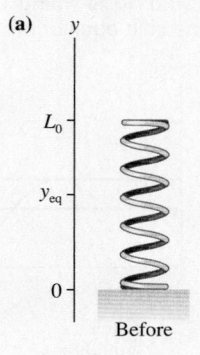

(a)

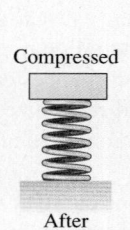

Before

After

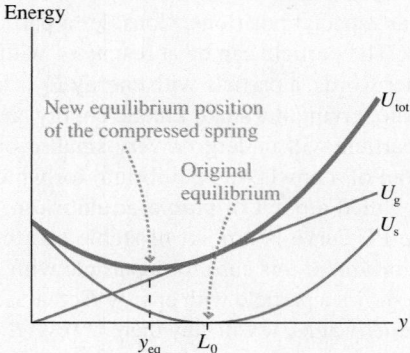

(b) Energy

STOP TO THINK 10.6 A particle with the potential energy shown in the graph is moving to the right. It has 1 J of kinetic energy at $x = 1$ m. Where is the particle's turning point?

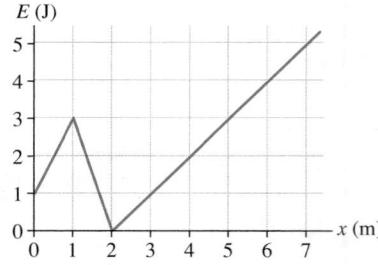

FIGURE 10.31 The energy diagram of the diatomic molecule HCl.

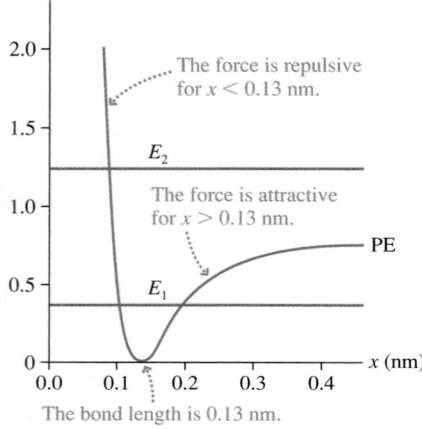

Molecular Bonds

Molecular bonds cause the "springiness" in our atomic models of tension, the normal force, and collisions. A *molecular bond* that holds two atoms together is an electric interaction between the charged electrons and nuclei. FIGURE 10.31 shows the potential-energy diagram for the diatomic molecule HCl (hydrogen chloride) as it has been experimentally determined. Distance x is the *atomic separation*, the distance between the hydrogen and the chlorine atoms. Note the very tiny distances: 1 nm = 10^{-9} m.

Although the potential energy is an electric energy, we can *interpret* the diagram using the steps in Tactics Box 10.1. The molecule has a stable equilibrium at an atomic separation of $x_{eq} = 0.13$ nm. This is the *bond length* of HCl, and you can find this value listed in chemistry books. If we try to push the atoms closer together (smaller x), the potential energy rises very rapidly. Physically, this is the repulsive electric force between the electrons orbiting each atom, preventing the atoms from getting too close.

There is also an attractive force between the atoms, called the *polarization force*. It is similar to the static electricity force by which a comb that has been brushed through your hair attracts small pieces of paper. If you try to pull the atoms apart (larger x), the attractive polarization force resists and is responsible for the increasing potential energy for $x > x_{eq}$. The equilibrium position is where the repulsive force between the electrons and the attractive polarization force are exactly balanced.

The repulsive force keeps getting stronger as you push the atoms together, and thus the potential-energy curve keeps getting steeper on the left. But the attractive polarization force gets *weaker* as the atoms get farther apart. This is why the potential-energy curve becomes *less* steep as the atomic separation increases. This difference between the repulsive and attractive forces leads to an asymmetric curve.

It turns out that, for quantum physics reasons, a molecule cannot have $E = 0$ and thus cannot simply rest at the equilibrium position. By requiring the molecule to have some energy, such as E_1, we see that the atoms oscillate back and forth between two turning points. This is a *molecular vibration*, and atoms held together by molecular bonds are constantly vibrating. For a molecule having an energy $E_1 = 0.35 \times 10^{-18}$ J, as illustrated in Figure 10.31, the bond oscillates in length between roughly 0.10 nm and 0.18 nm.

Suppose we increase the molecule's energy to $E_2 = 1.25 \times 10^{-18}$ J. This could happen if the molecule absorbs some light. You can see from the energy diagram that atoms with this energy are not bound together at large values of x. There is no turning point on the right, so the atoms will keep moving apart. By raising the molecule's energy to E_2 we have *broken the molecular bond*. The breaking of molecular bonds through the absorption of light is called *photodissociation*. It is an important process in making integrated circuits.

10.7 Elastic Collisions

Figure 9.1 showed a molecular-level view of a collision. Billions of spring-like molecular bonds are compressed as two objects collide, then the bonds expand and push the objects apart. In the language of energy, the kinetic energy of the objects is transformed into the elastic potential energy of molecular bonds, then back into kinetic energy as the two objects spring apart.

In some cases, such as the inelastic collisions of Chapter 9, some of the mechanical energy is dissipated inside the objects as thermal energy and not all of the kinetic energy is recovered. We're now interested in collisions in which *all* of the kinetic energy is stored as elastic potential energy in the bonds, and then *all* of the stored energy is transformed back into the post-collision kinetic energy of the objects. A collision in which mechanical energy is conserved is called a **perfectly elastic collision.** Collisions between two very hard objects, such as two billiard balls or two steel balls, come close to being perfectly elastic.

FIGURE 10.32 shows a head-on, perfectly elastic collision of a ball of mass m_1, having initial velocity $(v_{ix})_1$, with a ball of mass m_2 that is initially at rest. The balls' velocities after the collision are $(v_{fx})_1$ and $(v_{fx})_2$. These are velocities, not speeds, and have signs. Ball 1, in particular, might bounce backward and have a negative value for $(v_{fx})_1$.

The collision must obey two conservation laws: conservation of momentum (obeyed in any collision) and conservation of mechanical energy (because the collision is perfectly elastic). Although the energy is transformed into potential energy during the collision, the mechanical energy before and after the collision is purely kinetic energy. Thus

$$\text{momentum conservation:} \quad m_1 (v_{fx})_1 + m_2 (v_{fx})_2 = m_1 (v_{ix})_1 \quad (10.38)$$

$$\text{energy conservation:} \quad \frac{1}{2} m_1 (v_{fx})_1^2 + \frac{1}{2} m_2 (v_{fx})_2^2 = \frac{1}{2} m_1 (v_{ix})_1^2 \quad (10.39)$$

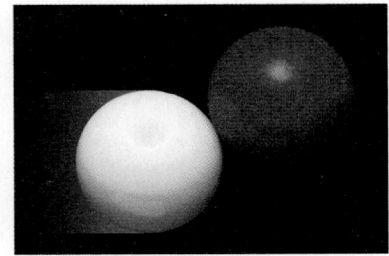

A perfectly elastic collision conserves both momentum and mechanical energy.

FIGURE 10.32 A perfectly elastic collision.

Before: ① $\xrightarrow{(v_{ix})_1}$ ② K_i

During: ①② — Energy is stored in compressed bonds, then released as the bonds re-expand.

After: ① $\xrightarrow{}$ ② $\xrightarrow{}$ $K_f = K_i$
$\quad (v_{fx})_1 \quad (v_{fx})_2$

Momentum conservation alone is not sufficient to analyze the collision because there are two unknowns: the two final velocities. That is why we did not consider perfectly elastic collisions in Chapter 9. Energy conservation gives us another condition. Isolating $(v_{fx})_1$ in Equation 10.38 gives

$$(v_{fx})_1 = (v_{ix})_1 - \frac{m_2}{m_1}(v_{fx})_2 \tag{10.40}$$

We substitute this into Equation 10.39:

$$\frac{1}{2}m_1\left((v_{ix})_1 - \frac{m_2}{m_1}(v_{fx})_2\right)^2 + \frac{1}{2}m_2(v_{fx})_2^2 = \frac{1}{2}m_1(v_{ix})_1^2$$

With a bit of algebra, this can be rearranged to give

$$(v_{fx})_2\left[\left(1 + \frac{m_2}{m_1}\right)(v_{fx})_2 - 2(v_{ix})_1\right] = 0 \tag{10.41}$$

One possible solution to this equation is seen to be $(v_{fx})_2 = 0$. However, this solution is of no interest; it is the case where ball 1 misses ball 2. The other solution is

$$(v_{fx})_2 = \frac{2m_1}{m_1 + m_2}(v_{ix})_1$$

which, finally, can be substituted back into Equation 10.40 to yield $(v_{fx})_1$. The complete solution is

$$(v_{fx})_1 = \frac{m_1 - m_2}{m_1 + m_2}(v_{ix})_1 \qquad \text{(perfectly elastic collision}$$
$$(v_{fx})_2 = \frac{2m_1}{m_1 + m_2}(v_{ix})_1 \qquad \text{with ball 2 initially at rest)} \tag{10.42}$$

Equations 10.42 allow us to compute the final velocity of each ball. These equations are a little difficult to interpret, so let us look at the three special cases shown in **FIGURE 10.33**.

Case 1: $m_1 = m_2$. This is the case of one billiard ball striking another of equal mass. For this case, Equations 10.42 give

$$v_{f1} = 0$$
$$v_{f2} = v_{i1}$$

Case 2: $m_1 \gg m_2$. This is the case of a bowling ball running into a Ping-Pong ball. We do not want an exact solution here, but an approximate solution for the limiting case that $m_1 \rightarrow \infty$. Equations 10.42 in this limit give

$$v_{f1} \approx v_{i1}$$
$$v_{f2} \approx 2v_{i1}$$

Case 3: $m_1 \ll m_2$. Now we have the reverse case of a Ping-Pong ball colliding with a bowling ball. Here we are interested in the limit $m_1 \rightarrow 0$, in which case Equations 10.42 become

$$v_{f1} \approx -v_{i1}$$
$$v_{f2} \approx 0$$

These cases agree well with our expectations and give us confidence that Equations 10.42 accurately describe a perfectly elastic collision.

FIGURE 10.33 Three special elastic collisions.

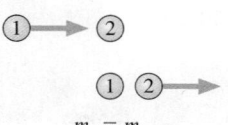

$m_1 = m_2$

Ball 1 stops. Ball 2 goes forward with $v_{f2} = v_{i1}$.

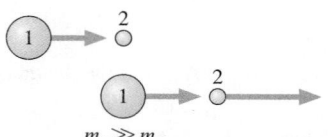

$m_1 \gg m_2$

Ball 1 hardly slows down. Ball 2 is knocked forward at $v_{f2} \approx 2v_{i1}$.

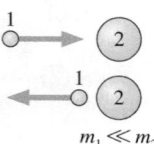

$m_1 \ll m_2$

Ball 1 bounces off ball 2 with almost no loss of speed. Ball 2 hardly moves.

Using Reference Frames

Equations 10.42 assumed that ball 2 was at rest prior to the collision. Suppose, however, you need to analyze the perfectly elastic collision that is just about to take place in FIGURE 10.34. What are the direction and speed of each ball after the collision? You could solve the simultaneous momentum and energy equations, but the mathematics becomes quite messy when both balls have an initial velocity. Fortunately, there's an easier way.

You already know the answer—Equations 10.42—when ball 2 is initially at rest. And in Chapter 4 you learned the Galilean transformation of velocity. This transformation relates an object's velocity as measured in one reference frame to its velocity in a different reference frame that moves with respect to the first. The Galilean transformation provides an elegant and straightforward way to analyze the collision of Figure 10.34.

FIGURE 10.34 A perfectly elastic collision in which both balls have an initial velocity.

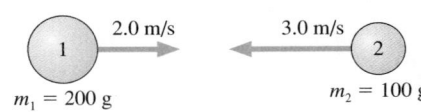

> **TACTICS**
> **BOX 10.2** **Analyzing elastic collisions**
>
> ❶ Use the Galilean transformation to transform the initial velocities of balls 1 and 2 from the "lab frame" to a reference frame in which ball 2 is at rest.
> ❷ Use Equations 10.42 to determine the outcome of the collision in the frame where ball 2 is initially at rest.
> ❸ Transform the final velocities back to the "lab frame."

FIGURE 10.35a shows the situation, just before the collision, in the lab frame L. Ball 1 has initial velocity $(v_{ix})_{1L} = 2.0$ m/s. Recall from Chapter 4 that the subscript notation means "velocity of ball 1 relative to the lab frame L." Because ball 2 is moving to the left, it has $(v_{ix})_{2L} = -3.0$ m/s. We would like to observe the collision from a reference frame in which ball 2 is at rest. That will be true if we choose a moving reference frame M that travels alongside ball 2 with the same velocity: $(v_x)_{ML} = -3.0$ m/s.

FIGURE 10.35 The collision seen in two reference frames: the lab frame L and a moving frame M in which ball 2 is initially at rest.

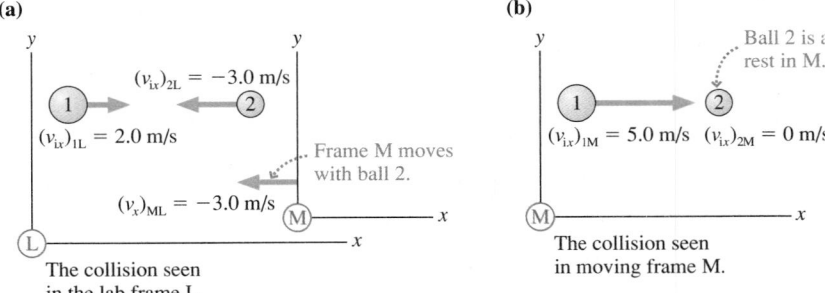

We first need to transform the balls' velocities from the lab frame to the moving reference frame. From Chapter 4, the Galilean transformation of velocity for an object O is

$$(v_x)_{OM} = (v_x)_{OL} + (v_x)_{LM} \qquad (10.43)$$

That is, O's velocity in reference frame M is its velocity in reference frame L plus the velocity of frame L relative to frame M. Because reference frame M is moving to the left relative to L with $(v_x)_{ML} = -3.0$ m/s, reference frame L is moving to the right relative to M with $(v_x)_{LM} = +3.0$ m/s. Applying the transformation to the two initial velocities gives

$$
\begin{aligned}
(v_{ix})_{1M} &= (v_{ix})_{1L} + (v_x)_{LM} = 2.0 \text{ m/s } + 3.0 \text{ m/s } = 5.0 \text{ m/s} \\
(v_{ix})_{2M} &= (v_{ix})_{2L} + (v_x)_{LM} = -3.0 \text{ m/s } + 3.0 \text{ m/s } = 0 \text{ m/s}
\end{aligned}
\qquad (10.44)
$$

$(v_{ix})_{2M} = 0$ m/s, as expected, because we chose a moving reference frame in which ball 2 would be at rest.

FIGURE 10.35b now shows a situation—with ball 2 initially at rest—in which we can use Equations 10.42 to find the post-collision velocities in frame M:

$$(v_{fx})_{1M} = \frac{m_1 - m_2}{m_1 + m_2}(v_{ix})_{1M} = 1.7 \text{ m/s}$$

$$(v_{fx})_{2M} = \frac{2m_1}{m_1 + m_2}(v_{ix})_{1M} = 6.7 \text{ m/s}$$

(10.45)

Reference frame M hasn't changed—it's still moving to the left in the lab frame at 3.0 m/s—but the collision has changed both balls' velocities in frame M.

To finish, we need to transform the post-collision velocities in frame M back to the lab frame L. We can do so with another application of the Galilean transformation:

FIGURE 10.36 The post-collision velocities in the lab frame.

$$(v_{fx})_{1L} = (v_{fx})_{1M} + (v_x)_{ML} = 1.7 \text{ m/s} + (-3.0 \text{ m/s}) = -1.3 \text{ m/s}$$
$$(v_{fx})_{2L} = (v_{fx})_{2M} + (v_x)_{ML} = 6.7 \text{ m/s} + (-3.0 \text{ m/s}) = 3.7 \text{ m/s}$$

(10.46)

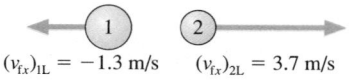

$(v_{fx})_{1L} = -1.3$ m/s $(v_{fx})_{2L} = 3.7$ m/s

FIGURE 10.36 shows the outcome of the collision in the lab frame. It's not hard to confirm that these final velocities do, indeed, conserve both momentum and energy.

CHALLENGE EXAMPLE 10.10 **A rebounding pendulum**

A 200 g steel ball hangs on a 1.0-m-long string. The ball is pulled sideways so that the string is at a 45° angle, then released. At the very bottom of its swing the ball strikes a 500 g steel paperweight that is resting on a frictionless table. To what angle does the ball rebound?

MODEL We can divide this problem into three parts. First the ball swings down as a pendulum. Second, the ball and paperweight have a collision. Steel balls bounce off each other very well, so

we will assume that the collision is perfectly elastic. Third, the ball, after it bounces off the paperweight, swings back up as a pendulum.

VISUALIZE FIGURE 10.37 shows four distinct moments of time: as the ball is released, an instant before the collision, an instant after the collision but before the ball and paperweight have had time to move, and as the ball reaches its highest point on the rebound. Call the ball A and the paperweight B, so $m_A = 0.20$ kg and $m_B = 0.50$ kg.

FIGURE 10.37 Four moments in the collision of a pendulum with a paperweight.

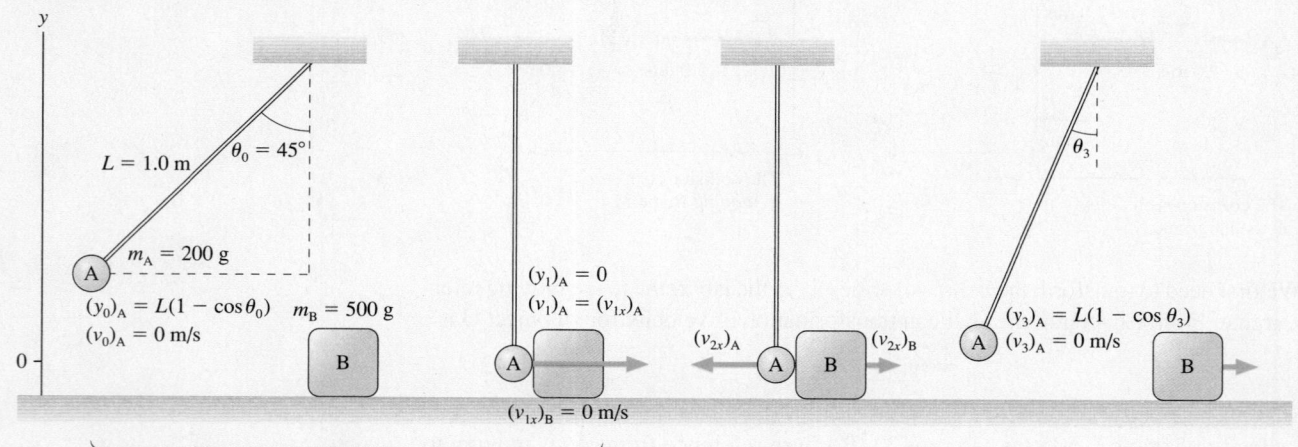

SOLVE Part 1: The first part involves the ball only. Its initial height is

$$(y_0)_A = L - L\cos\theta_0 = L(1 - \cos\theta_0) = 0.293 \text{ m}$$

We can use conservation of mechanical energy to find the ball's velocity at the bottom, just before impact on the paperweight:

$$\frac{1}{2}m_A(v_1)_A^2 + m_A g(y_1)_A = \frac{1}{2}m_A(v_0)_A^2 + m_A g(y_0)_A$$

We know $(v_0)_A = 0$. Solving for the velocity at the bottom, where $(y_1)_A = 0$, gives

$$(v_1)_A = \sqrt{2g(y_0)_A} = 2.40 \text{ m/s}$$

Part 2: The ball and paperweight undergo a perfectly elastic collision in which the paperweight is initially at rest. These are the conditions for which Equations 10.42 were derived. The velocities *immediately* after the collision, prior to any further motion, are

$$(v_{2x})_A = \frac{m_A - m_B}{m_A + m_B}(v_{1x})_A = -1.03 \text{ m/s}$$

$$(v_{2x})_B = \frac{2m_A}{m_A + m_B}(v_{1x})_A = +1.37 \text{ m/s}$$

The ball rebounds toward the left with a speed of 1.03 m/s while the paperweight moves to the right at 1.37 m/s. Kinetic energy has been conserved (you might want to check this), but it is now shared between the ball and the paperweight.

Part 3: Now the ball is a pendulum with an initial speed of 1.03 m/s. Mechanical energy is again conserved, so we can find its maximum height at the point where $(v_3)_A = 0$:

$$\frac{1}{2}m_A(v_3)_A^2 + m_A g(y_3)_A = \frac{1}{2}m_A(v_2)_A^2 + m_A g(y_2)_A$$

Solving for the maximum height gives

$$(y_3)_A = \frac{(v_2)_A^2}{2g} = 0.0541 \text{ m}$$

The height $(y_3)_A$ is related to angle θ_3 by $(y_3)_A = L(1 - \cos\theta_3)$. This can be solved to find the angle of rebound:

$$\theta_3 = \cos^{-1}\left(1 - \frac{(y_3)_A}{L}\right) = 19°$$

The paperweight speeds away at 1.37 m/s and the ball rebounds to an angle of 19°.

ASSESS The ball and the paperweight aren't hugely different in mass, so we expect the ball to transfer a significant fraction of its energy to the paperweight when they collide. Thus a rebound to roughly half the initial angle seems reasonable.

SUMMARY

The goals of Chapter 10 have been to introduce the concept of energy and the basic energy model.

General Principles

Law of Conservation of Mechanical Energy

If a system is isolated and frictionless, then the mechanical energy $E_{mech} = K + U$ of the system is conserved. Thus

$$K_f + U_f = K_i + U_i$$

- K is the sum of the kinetic energies of all particles.
- U is the sum of all potential energies.

Solving Energy Conservation Problems

MODEL Choose an isolated system without friction or other losses of mechanical energy.

VISUALIZE Draw a before-and-after pictorial representation.

SOLVE Use the law of conservation of energy:

$$K_f + U_f = K_i + U_i$$

ASSESS Is the result reasonable?

Important Concepts

Kinetic energy is an energy of motion: $K = \frac{1}{2}mv^2$.

Potential energy is an energy of position.

- **Gravitational:** $U_g = mgy$
- **Elastic:** $U_s = \frac{1}{2}k(\Delta s)^2$

Thermal energy is due to atomic motions. Hotter objects have more thermal energy.

Basic Energy Model

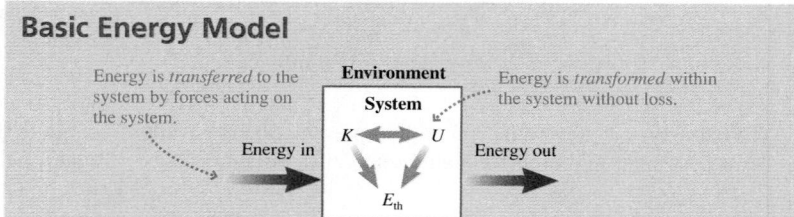

Energy is *transferred* to the system by forces acting on the system.

Energy is *transformed* within the system without loss.

Energy diagrams

These diagrams show the potential-energy curve PE and the total mechanical energy line TE.

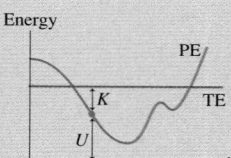

- The distance from the axis to the curve is PE.
- The distance from the curve to the TE line is KE.
- A point where the TE line crosses the PE curve is a **turning point.**
- Minima in the PE curve are points of **stable equilibrium.** Maxima are points of **unstable equilibrium.**
- Regions where PE is greater than TE are forbidden.

Applications

Hooke's law

The restoring force of an ideal spring is

$$(F_{sp})_s = -k\,\Delta s$$

where k is the spring constant and $\Delta s = s - s_e$ is the displacement from equilibrium.

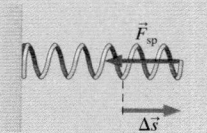

Perfectly elastic collisions

Both mechanical energy and momentum are conserved.

$$(v_{fx})_1 = \frac{m_1 - m_2}{m_1 + m_2}(v_{ix})_1 \qquad (v_{fx})_2 = \frac{2m_1}{m_1 + m_2}(v_{ix})_1$$

If ball 2 is moving, transform to a reference frame in which ball 2 is at rest.

Terms and Notation

energy
basic energy model
kinetic energy, K
gravitational potential energy, U_g
joule, J
mechanical energy
law of conservation of mechanical energy

restoring force
elastic
equilibrium length, L_0
displacement from equilibrium, Δs
spring constant, k
Hooke's law

elastic potential energy, U_s
energy diagram
stable equilibrium
unstable equilibrium
perfectly elastic collision

CONCEPTUAL QUESTIONS

1. Upon what basic quantity does kinetic energy depend? Upon what basic quantity does potential energy depend?
2. Can kinetic energy ever be negative? Can gravitational potential energy ever be negative? For each, give a plausible *reason* for your answer without making use of any equations.
3. If a particle's speed increases by a factor of 3, by what factor does its kinetic energy change?
4. Particle A has half the mass and eight times the kinetic energy of particle B. What is the speed ratio v_A/v_B?
5. A roller-coaster car rolls down a frictionless track, reaching speed v_0 at the bottom. If you want the car to go twice as fast at the bottom, by what factor must you increase the height of the track? Explain.
6. The three balls in FIGURE Q10.6, which have equal masses, are fired with equal speeds from the same height above the ground. Rank in order, from largest to smallest, their speeds v_a, v_b, and v_c as they hit the ground. Explain.

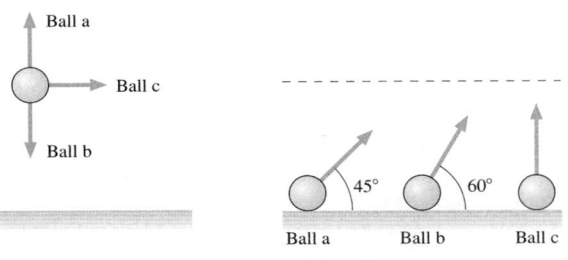

FIGURE Q10.6   **FIGURE Q10.7**

7. The three balls in FIGURE Q10.7, which have equal masses, are fired with equal speeds at the angles shown. Rank in order, from largest to smallest, their speeds v_a, v_b, and v_c as they cross the dashed horizontal line. Explain. (All three are fired with sufficient speed to reach the line.)
8. A spring has an unstretched length of 10 cm. It exerts a restoring force F when stretched to a length of 11 cm.
   a. For what length of the spring is its restoring force $3F$?
   b. At what compressed length is the restoring force $2F$?
9. The left end of a spring is attached to a wall. When Bob pulls on the right end with a 200 N force, he stretches the spring by 20 cm. The same spring is then used for a tug-of-war between Bob and Carlos. Each pulls on his end of the spring with a 200 N force. How far does the spring stretch? Explain.

10. Rank in order, from most to least, the elastic potential energy $(U_s)_a$ to $(U_s)_d$ stored in the springs of FIGURE Q10.10. Explain.

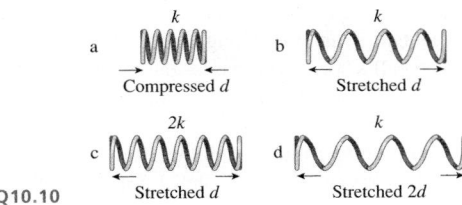

FIGURE Q10.10

11. A spring is compressed 1.0 cm. How far must you compress a spring with twice the spring constant to store the same amount of energy?
12. A spring gun shoots out a plastic ball at speed v_0. The spring is then compressed twice the distance it was on the first shot. By what factor is the ball's speed increased? Explain.
13. A particle with the potential energy shown in FIGURE Q10.13 is moving to the right at $x = 5$ m with total energy E.
   a. At what value or values of x is this particle's speed a maximum?
   b. Does this particle have a turning point or points in the range of x covered by the graph? If so, where?
   c. If E is changed appropriately, could the particle remain at rest at any point or points in the range of x covered by the graph? If so, where?

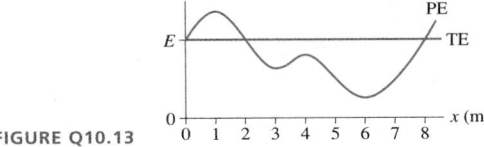

FIGURE Q10.13

14. Two balls of clay of known masses hang from the ceiling on massless strings of equal length. They barely touch when both hang at rest. One ball is pulled back until its string is at 45°, then released. It swings down, collides with the second ball, and they stick together. To determine the angle to which the balls swing on the opposite side, would you invoke (a) conservation of momentum, (b) conservation of mechanical energy, (c) both, (d) either but not both, or (e) these laws alone are not sufficient to find the angle? Explain.

EXERCISES AND PROBLEMS

Problems labeled ▓ integrate material from earlier chapters.

Exercises

Section 10.2 Kinetic Energy and Gravitational Potential Energy

1. | Which has the larger kinetic energy, a 10 g bullet fired at 500 m/s or a 75 kg student running at 5.5 m/s?

2. | The lowest point in Death Valley is 85 m below sea level. The summit of nearby Mt. Whitney has an elevation of 4420 m. What is the change in potential energy of an energetic 65 kg hiker who makes it from the floor of Death Valley to the top of Mt. Whitney?

3. | At what speed does a 1000 kg compact car have the same kinetic energy as a 20,000 kg truck going 25 km/h?

4. | a. What is the kinetic energy of a 1500 kg car traveling at a speed of 30 m/s ($\approx$ 65 mph)?
   b. From what height would the car have to be dropped to have this same amount of kinetic energy just before impact?
   c. Does your answer to part b depend on the car's mass?

5. | A boy reaches out of a window and tosses a ball straight up with a speed of 10 m/s. The ball is 20 m above the ground as he releases it. Use energy to find
   a. The ball's maximum height above the ground.
   b. The ball's speed as it passes the window on its way down.
   c. The speed of impact on the ground.

6. | a. With what minimum speed must you toss a 100 g ball straight up to just touch the 10-m-high roof of the gymnasium if you release the ball 1.5 m above the ground? Solve this problem using energy.
   b. With what speed does the ball hit the ground?

7. || A mother has four times the mass of her young son. Both are running with the same kinetic energy. What is the ratio v_{son}/v_{mother} of their speeds?

Section 10.3 A Closer Look at Gravitational Potential Energy

8. | A 55 kg skateboarder wants to just make it to the upper edge of a "quarter pipe," a track that is one-quarter of a circle with a radius of 3.0 m. What speed does he need at the bottom?

9. || What minimum speed does a 100 g puck need to make it to the top of a 3.0-m-long, 20° frictionless ramp?

10. || A pendulum is made by tying a 500 g ball to a 75-cm-long string. The pendulum is pulled 30° to one side, then released.
    a. What is the ball's speed at the lowest point of its trajectory?
    b. To what angle does the pendulum swing on the other side?

11. || A 20 kg child is on a swing that hangs from 3.0-m-long chains. What is her maximum speed if she swings out to a 45° angle?

12. || A 1500 kg car traveling at 10 m/s suddenly runs out of gas while approaching the valley shown in FIGURE EX10.12. The alert driver immediately puts the car in neutral so that it will roll. What will be the car's speed as it coasts into the gas station on the other side of the valley?

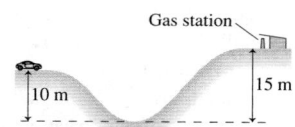

FIGURE EX10.12

Section 10.4 Restoring Forces and Hooke's Law

13. | You need to make a spring scale for measuring mass. You want each 1.0 cm length along the scale to correspond to a mass difference of 100 g. What should be the value of the spring constant?

14. || A 10-cm-long spring is attached to the ceiling. When a 2.0 kg mass is hung from it, the spring stretches to a length of 15 cm.
    a. What is the spring constant k?
    b. How long is the spring when a 3.0 kg mass is suspended from it?

15. || A 60 kg student is standing atop a spring in an elevator as it accelerates upward at 3.0 m/s². The spring constant is 2500 N/m. By how much is the spring compressed?

16. || A spring hanging from the ceiling has equilibrium length L_0. Hanging mass m from the spring stretches its length to L_1. Find an expression for the spring's length L_3 when mass $3m$ hangs from it.

17. || A 5.0 kg mass hanging from a spring scale is slowly lowered onto a vertical spring, as shown in FIGURE EX10.17. The scale reads in newtons.
    a. What does the spring scale read just before the mass touches the lower spring?
    b. The scale reads 20 N when the lower spring has been compressed by 2.0 cm. What is the value of the spring constant for the lower spring?
    c. At what compression length will the scale read zero?

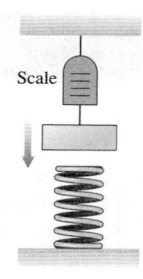

Scale

FIGURE EX10.17

Section 10.5 Elastic Potential Energy

18. | How far must you stretch a spring with $k = 1000$ N/m to store 200 J of energy?

19. | A stretched spring stores 2.0 J of energy. How much energy will be stored if the spring is stretched three times as far?

20. || A student places her 500 g physics book on a frictionless table. She pushes the book against a spring, compressing the spring by 4.0 cm, then releases the book. What is the book's speed as it slides away? The spring constant is 1250 N/m.

21. | A block sliding along a horizontal frictionless surface with speed v collides with a spring and compresses it by 2.0 cm. What will be the compression if the same block collides with the spring at a speed of $2v$?

22. || A 10 kg runaway grocery cart runs into a spring with spring constant 250 N/m and compresses it by 60 cm. What was the speed of the cart just before it hit the spring?

23. | The desperate contestants on a TV survival show are very hungry. The only food they can see is some fruit hanging on a branch high in a tree. Fortunately, they have a spring they can use to launch a rock. The spring constant is 1000 N/m, and they can compress the spring a maximum of 30 cm. All the rocks on the island seem to have a mass of 400 g.
    a. With what speed does the rock leave the spring?
    b. If the fruit hangs 15 m above the ground, will they feast or go hungry?

24. || As a 15,000 kg jet plane lands on an aircraft carrier, its tail hook snags a cable to slow it down. The cable is attached to a spring with spring constant 60,000 N/m. If the spring stretches 30 m to stop the plane, what was the plane's landing speed?

Section 10.6 Energy Diagrams

25. | FIGURE EX10.25 is the potential-energy diagram for a 20 g particle that is released from rest at $x = 1.0$ m.
    a. Will the particle move to the right or to the left? How can you tell?
    b. What is the particle's maximum speed? At what position does it have this speed?
    c. Where are the turning points of the motion?

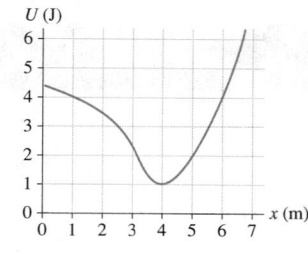

FIGURE EX10.25

26. ‖ **FIGURE EX10.26** is the potential-energy diagram for a 500 g particle that is released from rest at A. What are the particle's speeds at B, C, and D?

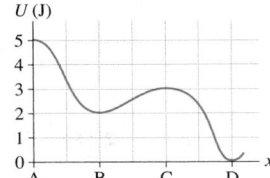

FIGURE EX10.26

27. | a. In **FIGURE EX10.27**, what minimum speed does a 100 g particle need at point A to reach point B?
    b. What minimum speed does a 100 g particle need at point B to reach point A?

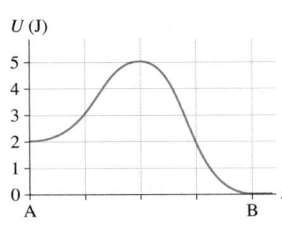

FIGURE EX10.27

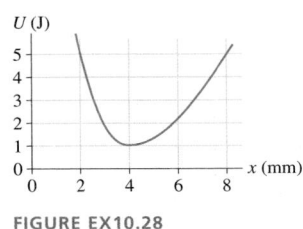

FIGURE EX10.28

28. ‖ In **FIGURE EX10.28**, what is the maximum speed of a 2.0 g particle that oscillates between $x = 2.0$ mm and $x = 8.0$ mm?

Section 10.7 Elastic Collisions

29. | A 50 g marble moving at 2.0 m/s strikes a 20 g marble at rest. What is the speed of each marble immediately after the collision?

30. | A proton is traveling to the right at 2.0×10^7 m/s. It has a head-on perfectly elastic collision with a carbon atom. The mass of the carbon atom is 12 times the mass of the proton. What are the speed and direction of each after the collision?

31. | Ball 1, with a mass of 100 g and traveling at 10 m/s, collides head-on with ball 2, which has a mass of 300 g and is initially at rest. What is the final velocity of each ball if the collision is (a) perfectly elastic? (b) perfectly inelastic?

32. ‖ A 50 g ball of clay traveling at speed v_0 hits and sticks to a 1.0 kg brick sitting at rest on a frictionless surface.
    a. What is the speed of the brick after the collision?
    b. What percentage of the mechanical energy is lost in this collision?

Problems

33. | The maximum energy a bone can absorb without breaking is
BIO surprisingly small. Experimental data show that the leg bones of a healthy, 60 kg human can absorb about 200 J.
    a. From what maximum height could a 60 kg person jump and land rigidly upright on both feet without breaking his legs? Assume that all energy is absorbed by the leg bones in a rigid landing.
    b. People jump safely from much greater heights than this. Explain how this is possible.

34. | You're driving at 35 km/h when the road suddenly descends 15 m into a valley. You take your foot off the accelerator and coast down the hill. Just as you reach the bottom you see the policeman hiding behind the speed limit sign that reads "70 km/h." Are you going to get a speeding ticket?

35. ‖ A cannon tilted up at a 30° angle fires a cannon ball at 80 m/s from atop a 10-m-high fortress wall. What is the ball's impact speed on the ground below?

36. ‖ You have a ball of unknown mass, a spring with spring constant 950 N/m, and a meter stick. You use various compressions of the spring to launch the ball vertically, then use the meter stick to measure the ball's maximum height above the launch point. Your data are as follows:

Compression (cm)	Height (cm)
2.0	32
3.0	65
4.0	115
5.0	189

Use an appropriate graph of the data to determine the ball's mass.

37. ‖ A very slippery ice cube slides in a *vertical* plane around the inside of a smooth, 20-cm-diameter horizontal pipe. The ice cube's speed at the bottom of the circle is 3.0 m/s.
    a. What is the ice cube's speed at the top?
    b. Find an algebraic expression for the ice cube's speed when it is at angle θ, where the angle is measured counterclockwise from the bottom of the circle. Your expression should give 3.0 m/s for $\theta = 0°$ and your answer to part a for $\theta = 180°$.

38. | A 50 g rock is placed in a slingshot and the rubber band is stretched. The force of the rubber band on the rock is shown by the graph in **FIGURE P10.38**.
    a. Is the rubber band stretched to the right or to the left? How can you tell?
    b. Does this rubber band obey Hooke's law? Explain.
    c. What is the rubber band's spring constant k?
    d. The rubber band is stretched 30 cm and then released. What is the speed of the rock?

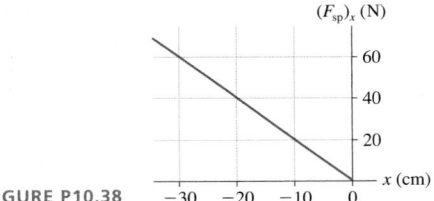

FIGURE P10.38

39. | The elastic energy stored in your tendons can contribute up to
BIO 35% of your energy needs when running. Sports scientists find that (on average) the knee extensor tendons in sprinters stretch 41 mm while those of nonathletes stretch only 33 mm. The spring constant of the tendon is the same for both groups, 33 N/mm. What is the difference in maximum stored energy between the sprinters and the nonathletes?

40. ‖ The spring in **FIGURE P10.40a** is compressed by Δx. It launches the block across a frictionless surface with speed v_0. The two springs in **FIGURE P10.40b** are identical to the spring of Figure P10.40a. They are compressed by the same Δx and used to launch the same block. What is the block's speed now?

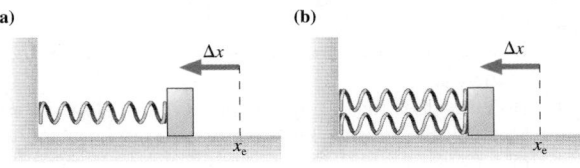

FIGURE P10.40

41. ||| The spring in **FIGURE P10.41a** is compressed by Δx. It launches the block across a frictionless surface with speed v_0. The two springs in **FIGURE P10.41b** are identical to the spring of Figure P10.41a. They are compressed the same *total* Δx and used to launch the same block. What is the block's speed now?

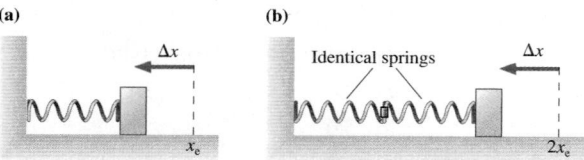

FIGURE P10.41

42. || a. A block of mass m can slide up and down a frictionless slope tilted at angle θ. The block is pressed against a spring at the bottom of the slope, compressing the spring (with spring constant k) by Δx, then released. Find an expression for the block's maximum height h above its starting point.

  b. A 50 g ice cube can slide up and down a frictionless 30° slope. At the bottom, a spring with spring constant 25 N/m is compressed 10 cm and used to launch the ice cube up the slope. How high does it go?

43. || A package of mass m is released from rest at a warehouse loading dock and slides down the 3.0-m-high, frictionless chute of **FIGURE P10.43** to a waiting truck. Unfortunately, the truck driver went on a break without having removed the previous package, of mass $2m$, from the bottom of the chute.

  a. Suppose the packages stick together. What is their common speed after the collision?

  b. Suppose the collision between the packages is perfectly elastic. To what height does the package of mass m rebound?

FIGURE P10.43

44. ||| A 100 g granite cube slides down a 40° frictionless ramp. At the bottom, just as it exits onto a horizontal table, it collides with a 200 g steel cube at rest. How high above the table should the granite cube be released to give the steel cube a speed of 150 cm/s?

45. || A 1000 kg safe is 2.0 m above a heavy-duty spring when the rope holding the safe breaks. The safe hits the spring and compresses it 50 cm. What is the spring constant of the spring?

46. || A vertical spring with $k = 490$ N/m is standing on the ground. You are holding a 5.0 kg block just above the spring, not quite touching it.

  a. How far does the spring compress if you let go of the block suddenly?

  b. How far does the spring compress if you slowly lower the block to the point where you can remove your hand without disturbing it?

  c. Why are your two answers different?

47. || You have been hired to design a spring-launched roller coaster that will carry two passengers per car. The car goes up a 10-m-high hill, then descends 15 m to the track's lowest point. You've determined that the spring can be compressed a maximum of 2.0 m and that a loaded car will have a maximum mass of 400 kg. For safety reasons, the spring constant should be 10% larger than the minimum needed for the car to just make it over the top.

  a. What spring constant should you specify?

  b. What is the maximum speed of a 350 kg car if the spring is compressed the full amount?

48. || It's been a great day of new, frictionless snow. Julie starts at the top of the 60° slope shown in **FIGURE P10.48**. At the bottom, a circular arc carries her through a 90° turn, and she then launches off a 3.0-m-high ramp. How far horizontally is her touchdown point from the end of the ramp?

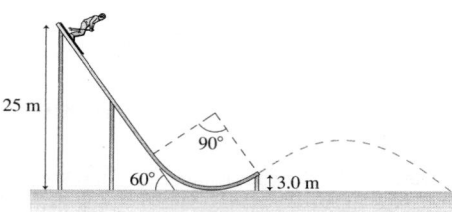

FIGURE P10.48

49. || A 100 g block on a frictionless table is firmly attached to one end of a spring with $k = 20$ N/m. The other end of the spring is anchored to the wall. A 20 g ball is thrown horizontally toward the block with a speed of 5.0 m/s.

  a. If the collision is perfectly elastic, what is the ball's speed immediately after the collision?

  b. What is the maximum compression of the spring?

  c. Repeat parts a and b for the case of a perfectly inelastic collision.

50. || You have been asked to design a "ballistic spring system" to measure the speed of bullets. A bullet of mass m is fired into a block of mass M. The block, with the embedded bullet, then slides across a frictionless table and collides with a horizontal spring whose spring constant is k. The opposite end of the spring is anchored to a wall. The spring's maximum compression d is measured.

  a. Find an expression for the bullet's speed v_B in terms of m, M, k, and d.

  b. What was the speed of a 5.0 g bullet if the block's mass is 2.0 kg and if the spring, with $k = 50$ N/m, was compressed by 10 cm?

  c. What fraction of the bullet's energy is "lost"? Where did it go?

51. ||| You have been asked to design a "ballistic spring system" to measure the speed of bullets. A spring whose spring constant is k is suspended from the ceiling. A block of mass M hangs from the spring. A bullet of mass m is fired vertically upward into the bottom of the block and stops in the block. The spring's maximum compression d is measured.

  a. Find an expression for the bullet's speed v_B in terms of m, M, k, and d.

  b. What was the speed of a 10 g bullet if the block's mass is 2.0 kg and if the spring, with $k = 50$ N/m, was compressed by 45 cm?

52. ‖ In FIGURE P10.52, a block of mass m slides along a frictionless track with speed v_m. It collides with a stationary block of mass M. Find an expression for the minimum value of v_m that will allow the second block to circle the loop-the-loop without falling off if the collision is (a) perfectly inelastic or (b) perfectly elastic.

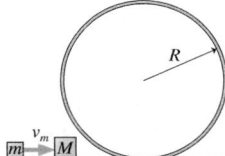

FIGURE P10.52

53. ‖ A block of mass m slides down a frictionless track, then around the inside of a circular loop-the-loop of radius R. From what minimum height h must the block start to make it around without falling off? Give your answer as a multiple of R.

54. ‖ A new event has been proposed for the Winter Olympics. As seen in FIGURE P10.54, an athlete will sprint 100 m, starting from rest, then leap onto a 20 kg bobsled. The person and bobsled will then slide down a 50-m-long ice-covered ramp, sloped at 20°, and into a spring with a carefully calibrated spring constant of 2000 N/m. The athlete who compresses the spring the farthest wins the gold medal. Lisa, whose mass is 40 kg, has been training for this event. She can reach a maximum speed of 12 m/s in the 100 m dash.

   a. How far will Lisa compress the spring?
   b. The Olympic committee has very exact specifications about the shape and angle of the ramp. Is this necessary? What factors about the ramp are important?

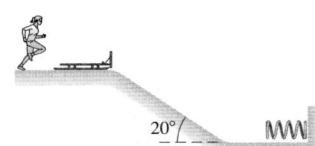

FIGURE P10.54

55. ‖‖ A 20 g ball is fired horizontally with speed v_0 toward a 100 g ball hanging motionless from a 1.0-m-long string. The balls undergo a head-on, perfectly elastic collision, after which the 100 g ball swings out to a maximum angle $\theta_{max} = 50°$. What was v_0?

56. ‖ A 100 g ball moving to the right at 4.0 m/s collides head-on with a 200 g ball that is moving to the left at 3.0 m/s.

   a. If the collision is perfectly elastic, what are the speed and direction of each ball after the collision?
   b. If the collision is perfectly inelastic, what are the speed and direction of the combined balls after the collision?

57. ‖ A 100 g ball moving to the right at 4.0 m/s catches up and collides with a 400 g ball that is moving to the right at 1.0 m/s. If the collision is perfectly elastic, what are the speed and direction of each ball after the collision?

58. ‖ FIGURE P10.58 shows the potential energy of a 500 g particle as it moves along the x-axis. Suppose the particle's mechanical energy is 12 J.

   a. Where are the particle's turning points?
   b. What is the particle's speed when it is at $x = 6.0$ m?
   c. What is the particle's maximum speed? At what position or positions does this occur?
   d. Write a description of the motion of the particle as it moves from the left turning point to the right turning point.
   e. Suppose the particle's energy is lowered to 4.0 J. Describe the possible motions.

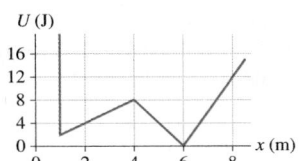

FIGURE P10.58

59. ‖ A particle has potential energy

$$U(x) = x + \sin((2 \text{ rad/m})x)$$

   over the range 0 m $\leq x \leq \pi$ m.

   a. Where are the equilibrium positions in this range?
   b. For each, is it a point of stable or unstable equilibrium?

60. ‖‖ Protons and neutrons (together called *nucleons*) are held together in the nucleus of an atom by a force called the *strong force*. At very small separations, the strong force between two nucleons is larger than the repulsive electrical force between two protons—hence its name. But the strong force quickly weakens as the distance between the protons increases. A well-established model for the potential energy of two nucleons interacting via the strong force is

$$U = U_0[1 - e^{-x/x_0}]$$

   where x is the distance between the centers of the two nucleons, x_0 is a constant having the value $x_0 = 2.0 \times 10^{-15}$ m, and $U_0 = 6.0 \times 10^{-11}$ J.

   a. Calculate and draw an accurate potential-energy curve from $x = 0$ m to $x = 10 \times 10^{-15}$ m. Either use your calculator to compute the value at several points or use computer software.
   b. Quantum effects are essential for a proper understanding of how nucleons behave. Nonetheless, let us innocently consider two neutrons *as if* they were small, hard, electrically neutral spheres of mass 1.67×10^{-27} kg and diameter 1.0×10^{-15} m. (We will consider neutrons rather than protons so as to avoid complications from the electric forces between protons.) You are going to hold two neutrons 5.0×10^{-15} m apart, measured between their centers, then release them. Draw the total energy line for this situation on your diagram of part a.
   c. What is the speed of each neutron as they crash together? Keep in mind that *both* neutrons are moving.

61. ‖‖ A 50 g air-track glider is repelled by a post fixed at one end of the track. It is hypothesized that the glider's potential energy is $U = c/x$, where x is the distance from the post and c is an unknown constant. To test this hypothesis, you launch the glider with the same speed at various distances from the post and then use a motion detector to measure its speed when it is 1.0 m from the post. Your data are as follows:

Initial distance (cm)	Speed at 1.0 m (m/s)
2.0	1.40
4.0	0.98
6.0	0.79
8.0	0.68

   a. Do the data support the hypothesis? To find out, you'll need to compare the shape of an appropriate graph to a theoretical prediction.
   b. Find an experimental value for c. Don't forget to determine the appropriate units.

   Hint: Both the slope *and* the y-intercept of the graph are important.

62. Write a realistic problem for which the energy bar chart shown in FIGURE P10.62 correctly shows the energy at the beginning and end of the problem.

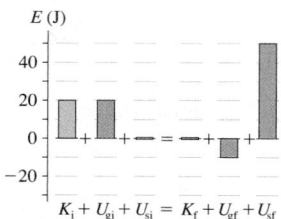

FIGURE P10.62

$$K_i + U_{gi} + U_{si} = K_f + U_{gf} + U_{sf}$$

In Problems 63 through 66 you are given the equation used to solve a problem. For each of these, you are to

a. Write a realistic problem for which this is the correct equation.
b. Draw the before-and-after pictorial representation.
c. Finish the solution of the problem.

63. $\frac{1}{2}(1500 \text{ kg})(5.0 \text{ m/s})^2 + (1500 \text{ kg})(9.80 \text{ m/s}^2)(10 \text{ m})$

$= \frac{1}{2}(1500 \text{ kg})(v_i)^2 + (1500 \text{ kg})(9.80 \text{ m/s}^2)(0 \text{ m})$

64. $\frac{1}{2}(0.20 \text{ kg})(2.0 \text{ m/s})^2 + \frac{1}{2}k(0 \text{ m})^2$

$= \frac{1}{2}(0.20 \text{ kg})(0 \text{ m/s})^2 + \frac{1}{2}k(-0.15 \text{ m})^2$

65. $(0.10 \text{ kg} + 0.20 \text{ kg})v_{1x} = (0.10 \text{ kg})(3.0 \text{ m/s})$

$\frac{1}{2}(0.30 \text{ kg})(0 \text{ m/s})^2 + \frac{1}{2}(3.0 \text{ N/m})(\Delta x_2)^2$

$= \frac{1}{2}(0.30 \text{ kg})(v_{1x})^2 + \frac{1}{2}(3.0 \text{ N/m})(0 \text{ m})^2$

66. $\frac{1}{2}(0.50 \text{ kg})(v_f)^2 + (0.50 \text{ kg})(9.80 \text{ m/s}^2)(0 \text{ m})$

$+ \frac{1}{2}(400 \text{ N/m})(0 \text{ m})^2 = \frac{1}{2}(0.50 \text{ kg})(0 \text{ m/s})^2$

$+ (0.50 \text{ kg})(9.80 \text{ m/s}^2)((-0.10 \text{ m}) \sin 30°)$

$+ \frac{1}{2}(400 \text{ N/m})(-0.10 \text{ m})^2$

Challenge Problems

67. A massless pan hangs from a spring that is suspended from the ceiling. When empty, the pan is 50 cm below the ceiling. If a 100 g clay ball is placed gently on the pan, the pan hangs 60 cm below the ceiling. Suppose the clay ball is dropped from the ceiling onto an empty pan. What is the pan's distance from the ceiling when the spring reaches its maximum length?

68. A pendulum is formed from a small ball of mass m on a string of length L. As FIGURE CP10.68 shows, a peg is height $h = L/3$ above the pendulum's lowest point. From what minimum angle θ must the pendulum be released in order for the ball to go over the top of the peg without the string going slack?

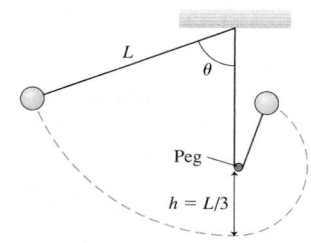

FIGURE CP10.68

69. In a physics lab experiment, a compressed spring launches a 20 g metal ball at a 30° angle. Compressing the spring 20 cm causes the ball to hit the floor 1.5 m below the point at which it leaves the spring after traveling 5.0 m horizontally. What is the spring constant?

70. It's your birthday, and to celebrate you're going to make your first bungee jump. You stand on a bridge 100 m above a raging river and attach a 30-m-long bungee cord to your harness. A bungee cord, for practical purposes, is just a long spring, and this cord has a spring constant of 40 N/m. Assume that your mass is 80 kg. After a long hesitation, you dive off the bridge. How far are you above the water when the cord reaches its maximum elongation?

71. A 10 kg box slides 4.0 m down the frictionless ramp shown in FIGURE CP10.71, then collides with a spring whose spring constant is 250 N/m.

a. What is the maximum compression of the spring?
b. At what compression of the spring does the box have its maximum speed?

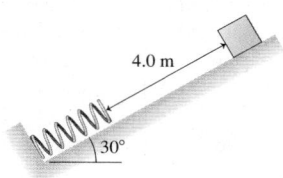

FIGURE CP10.71

72. Old naval ships fired 10 kg cannon balls from a 200 kg cannon. It was very important to stop the recoil of the cannon, since otherwise the heavy cannon would go careening across the deck of the ship. In one design, a large spring with spring constant 20,000 N/m was placed behind the cannon. The other end of the spring braced against a post that was firmly anchored to the ship's frame. What was the speed of the cannon ball if the spring compressed 50 cm when the cannon was fired?

73. A 2.0 kg cart has a spring with $k = 5000$ N/m attached to its front, parallel to the ground. This cart rolls at 4.0 m/s toward a stationary 1.0 kg cart.

a. What is the maximum compression of the spring during the collision?
b. What is the speed of each cart after the collision?

74. The air-track carts in FIGURE CP10.74 are sliding to the right at 1.0 m/s. The spring between them has a spring constant of 120 N/m and is compressed 4.0 cm. The carts slide past a flame that burns through the string holding them together. Afterward, what are the speed and direction of each cart?

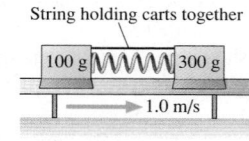

FIGURE CP10.74

75. A 100 g steel ball and a 200 g steel ball each hang from 1.0-m-long strings. At rest, the balls hang side by side, barely touching. The 100 g ball is pulled to the left until the angle between its string and vertical is 45°. The 200 g ball is pulled to a 45° angle on the right. The balls are released so as to collide at the very bottom of their swings. To what angle does each ball rebound?

76. A sled starts from rest at the top of the frictionless, hemispherical, snow-covered hill shown in FIGURE CP10.76.
   a. Find an expression for the sled's speed when it is at angle ϕ.
   b. Use Newton's laws to find the maximum speed the sled can have at angle ϕ without leaving the surface.
   c. At what angle ϕ_{max} does the sled "fly off" the hill?

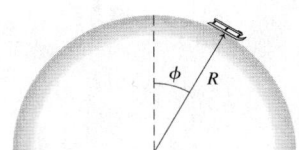

FIGURE CP10.76

STOP TO THINK ANSWERS

Stop to Think 10.1: $(U_g)_c > (U_g)_b = (U_g)_d > (U_g)_a$. Gravitational potential energy depends only on height, not on speed.

Stop to Think 10.2: $v_a = v_b = v_c = v_d$. Her increase in kinetic energy depends only on the vertical height through which she falls, not the shape of the slide.

Stop to Think 10.3: b. Mechanical energy is conserved on a frictionless surface. Because $K_i = 0$ and $K_f = 0$, it must be true that $U_f = U_i$ and thus $y_f = y_i$. The final height matches the initial height.

Stop to Think 10.4: $k_a > k_b > k_c$. The spring constant is the slope of the force-versus-displacement graph.

Stop to Think 10.5: c. U_s depends on $(\Delta s)^2$, so doubling the compression increases U_s by a factor of 4. All the potential energy is converted to kinetic energy, so K increases by a factor of 4. But K depends on v^2, so v increases by only a factor of $(4)^{1/2} = 2$.

Stop to Think 10.6: $x = 6$ m. From the graph, the particle's potential energy at $x = 1$ m is $U = 3$ J. Its total energy is thus $E = K + U = 4$ J. A TE line at 4 J crosses the PE curve at $x = 6$ m.

11 Work

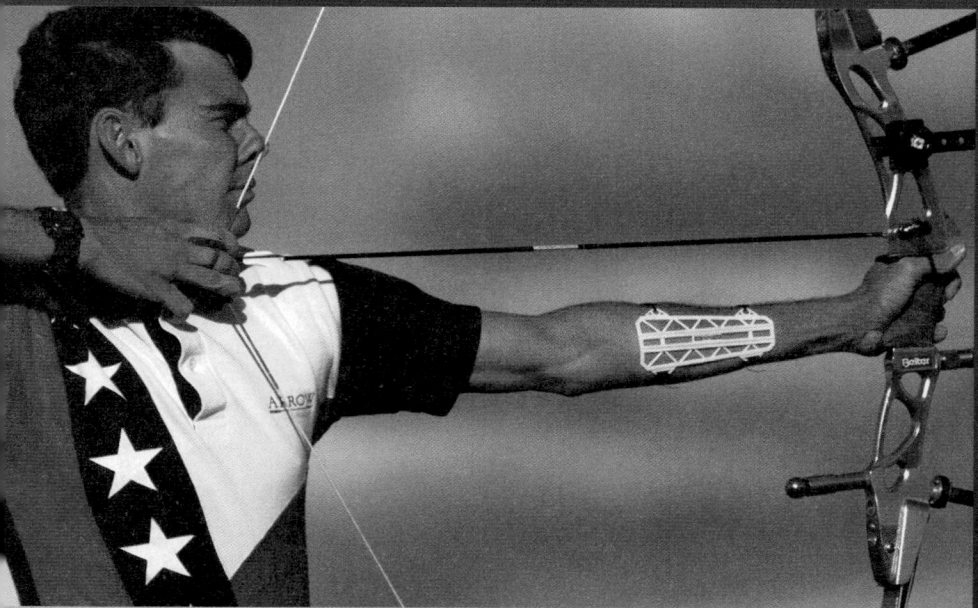

The bow may be very contemporary, but it's still the bow string doing work on the arrow that makes the arrow fly.

▶ **Looking Ahead** The goal of Chapter 11 is to develop a more complete understanding of energy and its conservation.

Unanswered Questions

Chapter 10 introduced energy but left many questions unanswered:

- How does a system gain or lose energy?
- When is a system's energy conserved?
- What role does friction play?

In Chapter 11 we'll answer these questions and introduce powerful new problem-solving tools.

◀ **Looking Back**
Chapter 10 Kinetic energy, potential energy, and energy diagrams

Work

When a force pushes or pulls a particle through a distance, we say that the force does **work** on the particle. The work W changes the particle's kinetic energy by $\Delta K = W$. One of the most important ways to change a system's energy is to do work on the system.

Before: $\vec{F}$ K_i

$\Delta \vec{r}$

After: $\vec{F}$ K_f

You'll learn a simple relationship among work, the force, and the displacement.

Work and Potential Energy

You'll learn that only certain kinds of forces, called **conservative forces**, can be associated with a potential energy. For these forces, the work W done by the force changes the potential energy by $\Delta U = W$.

Gravity is a conservative force. Lifting an object increases its gravitational potential energy.

The Basic Energy Model

We'll expand our basic energy model to include

- Work as an energy *transfer* between the system and the environment, and
- Energy *transformations* within the system.

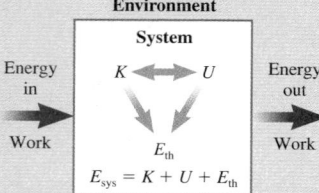

The Energy Equation

The ideas of the basic energy model— transfer and transformation—are captured in the **energy equation**:

$$\Delta E_{sys} = \Delta K + \Delta U + \Delta E_{th} = W_{ext}$$

- An isolated system has $W_{ext} = 0$. The total energy E_{sys} is conserved.
- An isolated system with no friction also has $\Delta E_{th} = 0$, so the mechanical energy $E_{mech} = K + U$ is conserved.

Power

Power is the *rate* at which energy is transferred or transformed. Power is measured in **watts**, where 1 watt is a rate of 1 joule per second.

A 100 W lightbulb transforms electrical energy into light and thermal energy at a rate of 100 J/s.

11.1 The Basic Energy Model Revisited

Chapter 10 introduced the *basic energy model* of FIGURE 11.1 but then focused on isolated systems. We found that kinetic and potential energy could be transformed back and forth without loss in an isolated, frictionless system, which led to the *law of conservation of mechanical energy:* The mechanical energy $E_{mech} = K + U$ is conserved in an isolated, frictionless system.

That was a good start, with many applications, but we need to expand our understanding of energy beyond ideal, isolated systems. Consider the following:

- A speeding car skids to a halt.
- A hand places a book on a high shelf.

Neither of these situations conserves mechanical energy. In the first, kinetic energy is transformed not into potential energy but into **thermal energy** E_{th}, the energy of the random, microscopic motions of the atoms inside the tires, the brakes, and the road. We'll use an arrow $\rightarrow$ as a shorthand way to indicate an energy transformation, writing the energy transformation of the skidding car as $K \rightarrow E_{th}$.

Thermal energy is associated with the system's temperature. Friction raises the temperature—think of rubbing your hands together briskly—so a system with friction transforms kinetic or potential energy into thermal energy. This is usually a one-way process, as we'll discuss when we get to thermodynamics. That is, you can't get your car moving again by transforming the thermal energy of the hot brakes back into the car's kinetic energy. Thus the transformation arrows in Figure 11.1 point only to, not from, E_{th}.

Nonetheless, the energy remains inside the system if we define the system to be all objects that interact—the car and the road in this case. **System energy** E_{sys} is defined to be the sum of the mechanical energy of the objects plus the thermal energy of the atoms *inside* the objects. That is,

$$E_{sys} = E_{mech} + E_{th} = K + U + E_{th} \tag{11.1}$$

Energy transformations within the system do not change the value of E_{sys}, so we can say that **the total energy of an isolated system is conserved.** This generalizes the law of conservation of mechanical energy to include thermal energy.

In the second situation above, an external force—the hand—acts on the book to increase the book's gravitational potential energy. The book is not an isolated system but, instead, is surrounded by a larger environment with which it can exchange energy. Such an exchange is called an **energy transfer.** There are two energy-transfer processes. The first is due to forces—pushes and pulls—exerted on the system by the environment. In this case, the hand gives the book potential energy by lifting it upward. This *mechanical* transfer of energy to or from the system is called **work.** The symbol for work is W.

The second means of transferring energy between a system and its environment is a *nonmechanical* process called *heat.* Heat is a crucial idea that we will add to the energy model when we study thermodynamics, but for now we want to concentrate on the mechanical transfer of energy via work.

As the arrows in Figure 11.1 show, energy can both enter and leave the system. We'll distinguish between the two directions of energy flow by allowing the work W to be either positive or negative. The sign of W is interpreted as follows:

$W > 0$ **The environment does work on the system and the system's energy increases.**

$W < 0$ **The system does work on the environment and the system's energy decreases.**

What is the relationship among the quantities of the basic energy model? Our hypothesis, which is confirmed by experiment, is that

$$\Delta E_{sys} = \Delta K + \Delta U + \Delta E_{th} = W \tag{11.2}$$

FIGURE 11.1 The basic energy model of a system interacting with its environment.

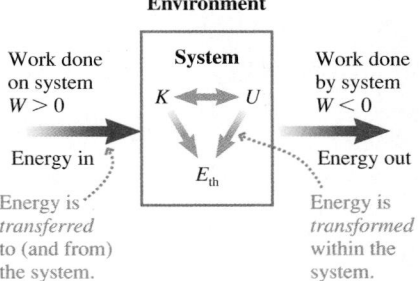

The two essential ideas of the basic energy model and Equation 11.2 are:

1. Energy can be *transferred* to or from a system by doing work on the system. This process changes the energy of the system: $\Delta E_{\text{sys}} = W$.
2. Energy can be *transformed* within the system among K, U, and E_{th}. These processes don't change the energy of the system: $\Delta E_{\text{sys}} = 0$.

This is the essence of the basic energy model. The rest of Chapter 11 will substantiate Equation 11.2 and look at its many implications.

STOP TO THINK 11.1 A child slides down a playground slide at constant speed. The energy transformation is

  a. $U \rightarrow K$ b. $K \rightarrow U$
  c. There is no transformation because energy is conserved.
  d. $U \rightarrow E_{\text{th}}$ e. $K \rightarrow E_{\text{th}}$

11.2 Work and Kinetic Energy

"Work" is a common word in the English language, with many meanings. When you first think of work, you probably think of the first two definitions in this list. After all, we talk about "working out," or we say, "I just got home from work." But that is *not* what work means in physics.

The basic energy model uses "work" in the sense of definition 7: energy transferred to or from a body or system by the application of force. The critical question we must answer is: *How much energy* does a force transfer?

We can answer this question by following the procedure we used in Chapter 10 to find the potential energy of gravity and of a spring. We'll begin, in FIGURE 11.2, with a force $\vec{F}$ acting on a particle of mass m as the particle moves along an s-axis from an initial position s_i, with kinetic energy K_i, to a final position s_f where the kinetic energy is K_f.

The force component F_s parallel to the s-axis causes the particle to speed up or slow down, thus transferring energy to or from the particle. We say that force $\vec{F}$ *does work* on the particle. Our goal is to find a relationship between F_s and ΔK. The s-component of Newton's second law is

$$F_s = ma_s = m\frac{dv_s}{dt} \tag{11.3}$$

where the v_s is the s-component of $\vec{v}$. As we did in Chapter 10, we can use the chain rule to write

$$m\frac{dv_s}{dt} = m\frac{dv_s}{ds}\frac{ds}{dt} = mv_s\frac{dv_s}{ds} \tag{11.4}$$

where $ds/dt = v_s$. Substituting Equation 11.4 into Equation 11.3 gives

$$F_s = mv_s\frac{dv_s}{ds} \tag{11.5}$$

The crucial step here, as it was in Chapter 10, was changing from a derivative with respect to time to a derivative with respect to position. We're going to want to integrate, so we first multiply through by ds to get

$$mv_s\,dv_s = F_s\,ds \tag{11.6}$$

Now we can integrate both sides from "before," where the position is s_i and the speed is v_i, to "after," giving

$$\int_{v_i}^{v_f} mv_s\,dv_s = \frac{1}{2}mv_s^2\Big|_{v_i}^{v_f} = \frac{1}{2}mv_f^2 - \frac{1}{2}mv_i^2 = \int_{s_i}^{s_f} F_s\,ds \tag{11.7}$$

One dictionary defines "work" as:

1. Physical or mental effort; labor.
2. The activity by which one makes a living.
3. A task or duty.
4. Something produced as a result of effort, such as a *work of art.*
5. Plural *works:* A factory or plant where industry is carried on, such as *steel works.*
6. Plural *works:* The essential or operating parts of a mechanism.
7. The transfer of energy to a body by application of a force.

FIGURE 11.2 Force $\vec{F}$ does work as the particle moves from s_i to s_f.

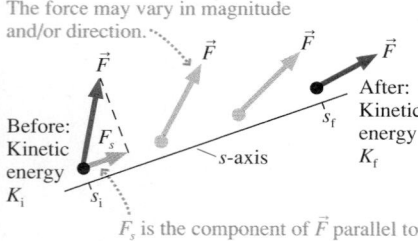

The force may vary in magnitude and/or direction.

Before: Kinetic energy K_i

After: Kinetic energy K_f

s-axis

F_s is the component of $\vec{F}$ parallel to the motion. This component changes the particle's speed and thus its kinetic energy.

The left side of Equation 11.7 is ΔK, the change in the particle's kinetic energy as it moves from s_i to s_f. The integral on the right apparently specifies the extent to which the applied force changes the particle's kinetic energy. We define the *work* done by force $\vec{F}$ as the particle moves from s_i to s_f as

$$W = \int_{s_i}^{s_f} F_s \, ds \tag{11.8}$$

The unit of work, that of force multiplied by distance, is the N m. Using the definition of the newton gives

$$1 \, \text{N m} = 1 \, (\text{kg m/s}^2) \, \text{m} = 1 \, \text{kg m}^2/\text{s}^2 = 1 \, \text{J}$$

Thus the unit of work is really the unit of energy. This is consistent with the idea that work is a transfer of energy. Rather than use N m, we will measure work in joules.

Using Equation 11.8 as the definition of work, we can write Equation 11.7 as

$$\Delta K = W \tag{11.9}$$

This pitcher is increasing the ball's kinetic energy by doing work on it.

Equation 11.9 tells us that a force transfers kinetic energy to a particle by pushing or pulling on it. Furthermore, **Equation 11.8 gives us a specific method to calculate *how much* energy is transferred by the push or pull.** This energy transfer, by mechanical means, is what we mean by the term "work."

Notice that *no* work is done if there is no displacement ($s_f = s_i$) because an integral that spans no interval is zero. **A force does work on a particle only if the particle is displaced.** If you were to hold a 200 lb weight over your head, you might break out in a sweat and your arms would tire. You might "feel" that you had done a lot of work, but you would have done *zero* work in the physics sense because the weight was not displaced while you were holding it and thus you transferred no energy to it.

The Work-Kinetic Energy Theorem

Equation 11.8 is the work done by one force. Because $\vec{F}_{net} = \sum \vec{F}_i$, it's easy to see that the net work done on a particle by several forces is $W_{net} = \sum W_i$, where W_i is the work done by force $\vec{F}_i$. In that case, Equation 11.9 becomes

$$\Delta K = W_{net} \tag{11.10}$$

This basic idea—that the net work done on a particle causes the particle's kinetic energy to change—is a general principle, one worth giving a name:

The work-kinetic energy theorem When one or more forces act on a particle as it is displaced from an initial position to a final position, the net work done on the particle by these forces causes the particle's kinetic energy to *change* by $\Delta K = W_{net}$.

An Analogy with the Impulse-Momentum Theorem

You might have noticed that there is a similarity between the work-kinetic energy theorem and the impulse-momentum theorem of Chapter 9:

$$\text{Work-kinetic energy theorem:} \quad \Delta K = W = \int_{s_i}^{s_f} F_s \, ds$$

$$\tag{11.11}$$

$$\text{Impulse-momentum theorem:} \quad \Delta p_s = J_s = \int_{t_i}^{t_f} F_s \, dt$$

In both cases, a force acting on a particle changes the state of the system. If the force acts over a time interval from t_i to t_f, it creates an *impulse* that changes the particle's momentum. If the force acts over the spatial interval from s_i to s_f, it does *work* that

FIGURE 11.3 Impulse and work are both the area under a force graph, but it's very important to know what the horizontal axis is.

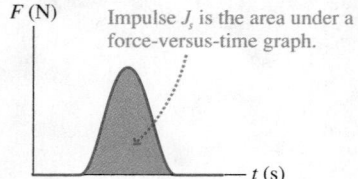

Impulse J_s is the area under a force-versus-time graph.

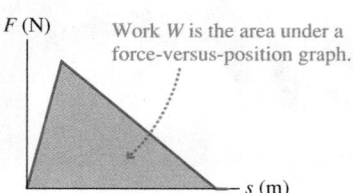

Work W is the area under a force-versus-position graph.

changes the particle's kinetic energy. FIGURE 11.3 shows that the geometric interpretation of impulse as the area under the F-versus-t graph applies equally well to an interpretation of work as the area under the F-versus-s graph.

This does not mean that a force *either* creates an impulse *or* does work but does not do both. Quite the contrary. A force acting on a particle *both* creates an impulse *and* does work, changing both the momentum and the kinetic energy of the particle. Whether you use the work-kinetic energy theorem or the impulse-momentum theorem depends on the question you are trying to answer.

We can, in fact, express the kinetic energy in terms of the momentum as

$$K = \frac{1}{2}mv^2 = \frac{(mv)^2}{2m} = \frac{p^2}{2m} \tag{11.12}$$

You cannot change a particle's kinetic energy without also changing its momentum.

STOP TO THINK 11.2 A particle moving along the x-axis experiences the force shown in the graph. If the particle has 2.0 J of kinetic energy as it passes $x = 0$ m, what is its kinetic energy when it reaches $x = 4$ m?

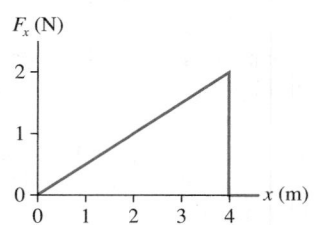

11.3 Calculating and Using Work

In this section we'll practice calculating work and using the work-kinetic energy theorem. We'll also introduce a new mathematical idea, the *dot product* of two vectors, that will allow us to write the work in a compact notation.

Constant Force

FIGURE 11.4 Work being done by a constant force as a particle moves through displacement $\Delta \vec{r}$.

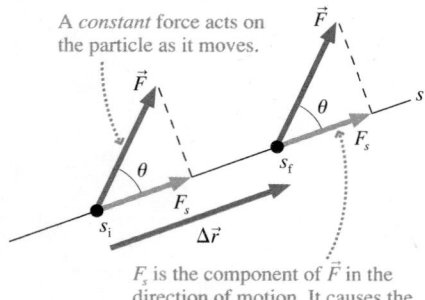

A *constant* force acts on the particle as it moves.

F_s is the component of $\vec{F}$ in the direction of motion. It causes the particle to speed up or slow down.

We'll begin by calculating the work done by a force $\vec{F}$ that acts with a *constant* strength and in a *constant* direction as a particle moves along a straight line through a displacement $\Delta \vec{r}$. FIGURE 11.4 shows the force acting on the particle as it moves along the s-axis. The force vector $\vec{F}$ makes an angle θ with respect to the displacement $\Delta \vec{r}$, so the component of the force vector along the direction of motion is $F_s = F\cos\theta$. According to Equation 11.8, the work done on the particle by this force is

$$W = \int_{s_i}^{s_f} F_s \, ds = \int_{s_i}^{s_f} F\cos\theta \, ds$$

Both F and θ are constant, so they can be taken outside the integral. Thus

$$W = F\cos\theta \int_{s_i}^{s_f} ds = F\cos\theta(s_f - s_i) = F(\Delta r)\cos\theta \tag{11.13}$$

where we used $s_f - s_i = \Delta r$, the magnitude of the particle's displacement. We can use Equation 11.13 to calculate the work done by a constant force if we know the magnitude F of the force, the angle θ of the force from the line of motion, and the distance Δr through which the particle is displaced.

> **NOTE** ▶ You may have learned in an earlier physics course that work is "force times distance." This is *not* the definition of work, merely a special case. Work is "force times distance" only if the force is constant *and* parallel to the displacement (i.e., $\theta = 0°$). ◀

EXAMPLE 11.1 Pulling a suitcase

A rope inclined upward at a 45° angle pulls a suitcase through the airport. The tension in the rope is 20 N. How much work does the tension do if the suitcase is pulled 100 m?

MODEL Model the suitcase as a particle.

VISUALIZE **FIGURE 11.5** shows a pictorial representation.

SOLVE The motion is along the x-axis, so in this case $\Delta r = \Delta x$. We can use Equation 11.13 to find that the tension does work:

$$W = T(\Delta x)\cos\theta = (20 \text{ N})(100 \text{ m})\cos 45° = 1400 \text{ J}$$

FIGURE 11.5 Pictorial representation of a suitcase pulled by a rope.

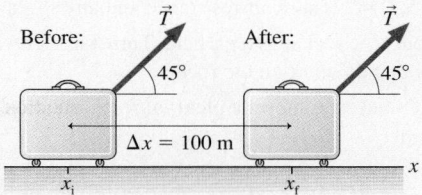

ASSESS Because a person pulls the rope, we would say informally that the person does 1400 J of work on the suitcase.

According to the basic energy model, work can be either positive or negative to indicate energy transfer into or out of the system. The quantities F and Δr are always positive, so **the sign of W is determined entirely by the angle θ between the force $\vec{F}$ and the displacement $\Delta\vec{r}$.**

TACTICS BOX 11.1 Calculating the work done by a constant force

Force and displacement	θ	Work W	Sign	Energy transfer
	0°	$F(\Delta r)$	+	
	<90°	$F(\Delta r)\cos\theta$	+	Energy is transferred into the system. The particle speeds up. K increases.
	90°	0	0	No energy is transferred. Speed and K are constant.
	>90°	$F(\Delta r)\cos\theta$	−	Energy is transferred out of the system. The particle slows down. K decreases.
	180°	$-F(\Delta r)$	−	

Exercises 3–10

NOTE ▶ The sign of W depends on the angle between the force vector and the displacement vector, *not* on the coordinate axes. A force to the left does *positive* work if it pushes a particle to the left (the force and the displacement are in the same direction, so this is a $\theta = 0°$ situation) even though the force component F_x is negative. Think about whether the force is trying to increase the particle's speed ($W > 0$) or decrease the particle's speed ($W < 0$). ◀

EXAMPLE 11.2 **Work during a rocket launch**

A 150,000 kg rocket is launched straight up. The rocket motor generates a thrust of 4.0×10^6 N. What is the rocket's speed at a height of 500 m? Ignore air resistance and any slight mass loss.

MODEL Model the rocket as a particle. Thrust and gravity are constant forces that do work on the rocket.

VISUALIZE **FIGURE 11.6** shows a pictorial representation and a free-body diagram.

FIGURE 11.6 Pictorial representation and free-body diagram of a rocket launch.

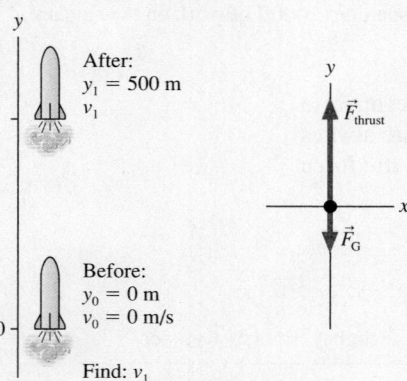

SOLVE We can solve this problem with the work-kinetic energy theorem $\Delta K = W_{\text{net}}$. Both forces do work on the rocket. The thrust is in the direction of motion, with $\theta = 0°$, and thus

$$W_{\text{thrust}} = F_{\text{thrust}}(\Delta r) = (4.0 \times 10^6 \text{ N})(500 \text{ m}) = 2.00 \times 10^9 \text{ J}$$

The gravitational force points downward, opposite the displacement $\Delta \vec{r}$, so $\theta = 180°$. Thus the work done by gravity is

$$W_{\text{grav}} = -F_{\text{G}}(\Delta r) = -mg(\Delta r)$$

$$= -(1.5 \times 10^5 \text{ kg})(9.8 \text{ m/s}^2)(500 \text{ m}) = -0.74 \times 10^9 \text{ J}$$

The work done by the thrust is positive. By itself, the thrust would cause the rocket to speed up. The work done by gravity is negative, not because $\vec{F}_{\text{G}}$ points down but because $\vec{F}_{\text{G}}$ is opposite the displacement. By itself, gravity would cause the rocket to slow down. The work-kinetic energy theorem, using $v_0 = 0$ m/s, is

$$\Delta K = \frac{1}{2}mv_1^2 - 0 = W_{\text{net}} = W_{\text{thrust}} + W_{\text{grav}} = 1.26 \times 10^9 \text{ J}$$

This is easily solved for the speed:

$$v_1 = \sqrt{\frac{2W_{\text{net}}}{m}} = 130 \text{ m/s}$$

ASSESS The net work is positive, meaning that energy is transferred *to* the rocket. In response, the rocket speeds up.

Force Perpendicular to the Direction of Motion

FIGURE 11.7 A perpendicular force does no work.

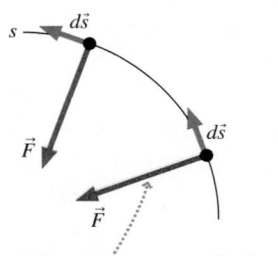

The force is everywhere perpendicular to the displacement, so it does no work.

FIGURE 11.7 shows a particle moving in uniform circular motion. As you learned in Chapter 8, uniform circular motion requires a force pointing toward the center of the circle. How much work does this force do?

Zero! You can see that the force is everywhere perpendicular to the small displacement $d\vec{s}$. Thus F_s, the component of the force parallel to the displacement, is everywhere zero. The force does *no* work on the particle. This shouldn't be surprising. The particle's speed, and hence its kinetic energy, doesn't change in uniform circular motion, so the work-kinetic energy theorem says $W = \Delta K = 0$.

A force everywhere perpendicular to the motion does no work. The friction force on a car turning a corner does no work. Neither does the tension force when a mass on a string is in circular motion.

A crane lowers a steel girder into place. The girder moves with constant speed. Consider the work W_{G} done by gravity and the work W_{T} done by the tension in the cable. Which of the following is correct?

a. W_{G} is positive and W_{T} is positive.
b. W_{G} is positive and W_{T} is negative.
c. W_{G} is negative and W_{T} is positive.
d. W_{G} is negative and W_{T} is negative.
e. W_{G} and W_{T} are both zero.

The Dot Product of Two Vectors

There's something different about the quantity $F(\Delta r)\cos\theta$ in Equation 11.13. We've spent many chapters adding vectors, but this is the first time we've *multiplied* two vectors. Multiplying vectors is not like multiplying scalars. In fact, there is more than one way to multiply vectors. We will introduce one way now, the *dot product*.

FIGURE 11.8 shows two vectors, $\vec{A}$ and $\vec{B}$, with angle α between them. We define the **dot product** of $\vec{A}$ and $\vec{B}$ as

$$\vec{A} \cdot \vec{B} = AB \cos \alpha \qquad (11.14)$$

A dot product *must have* the dot symbol · between the vectors. The notation $\vec{A}\vec{B}$, without the dot, is *not* the same thing as $\vec{A} \cdot \vec{B}$. The dot product is also called the **scalar product** because the value is a scalar. Later, when we need it, we'll introduce a different way to multiply vectors called the *cross product*.

The dot product of two vectors depends on the orientation of the vectors. **FIGURE 11.9** shows five different situations, including the three "special cases" where $\alpha = 0°$, $90°$, and $180°$.

> **NOTE** ▶ The dot product of a vector with itself is well defined. If $\vec{B} = \vec{A}$ (i.e., $\vec{B}$ is a copy of $\vec{A}$), then $\alpha = 0°$. Thus $\vec{A} \cdot \vec{A} = A^2$. ◀

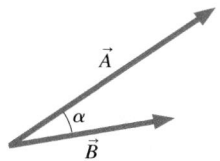

FIGURE 11.8 Vectors $\vec{A}$ and $\vec{B}$, with angle α between them.

FIGURE 11.9 The dot product $\vec{A} \cdot \vec{B}$ as α ranges from 0° to 180°.

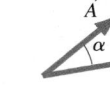

$\alpha = 0$

$\vec{A} \cdot \vec{B} = AB$

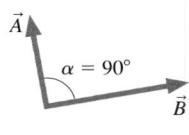

$\alpha < 90°$

$\vec{A} \cdot \vec{B} > 0$

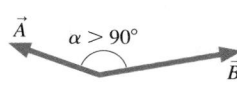

$\alpha = 90°$

$\vec{A} \cdot \vec{B} = 0$

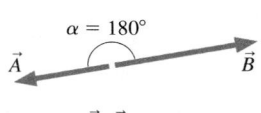

$\alpha > 90°$

$\vec{A} \cdot \vec{B} < 0$

$\alpha = 180°$

$\vec{A} \cdot \vec{B} = -AB$

| EXAMPLE 11.3 | **Calculating a dot product** |

Compute the dot product of the two vectors in **FIGURE 11.10**

SOLVE The angle between the vectors is $\alpha = 30°$, so

$$\vec{A} \cdot \vec{B} = AB \cos \alpha = (3)(4) \cos 30° = 10.4$$

FIGURE 11.10 Vectors $\vec{A}$ and $\vec{B}$ of Example 11.3.

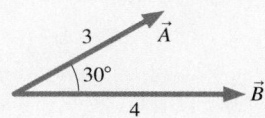

Like vector addition and subtraction, calculating the dot product of two vectors is often performed most easily using vector components. **FIGURE 11.11** reminds you of the unit vectors $\hat{\imath}$ and $\hat{\jmath}$ that point in the positive x-direction and positive y-direction. The two unit vectors are perpendicular to each other, so their dot product is $\hat{\imath} \cdot \hat{\jmath} = 0$. Furthermore, because the magnitudes of $\hat{\imath}$ and $\hat{\jmath}$ are 1, $\hat{\imath} \cdot \hat{\imath} = 1$ and $\hat{\jmath} \cdot \hat{\jmath} = 1$.

In terms of components, we can write the dot product of vectors $\vec{A}$ and $\vec{B}$ as

$$\vec{A} \cdot \vec{B} = (A_x \hat{\imath} + A_y \hat{\jmath}) \cdot (B_x \hat{\imath} + B_y \hat{\jmath})$$

Multiplying this out, and using the results for the dot products of the unit vectors:

$$\vec{A} \cdot \vec{B} = A_x B_x \hat{\imath} \cdot \hat{\imath} + (A_x B_y + A_y B_x)\hat{\imath} \cdot \hat{\jmath} + A_y B_y \hat{\jmath} \cdot \hat{\jmath}$$
$$= A_x B_x + A_y B_y \qquad (11.15)$$

That is, the **dot product is the sum of the products of the components.**

FIGURE 11.11 The unit vectors $\hat{\imath}$ and $\hat{\jmath}$.

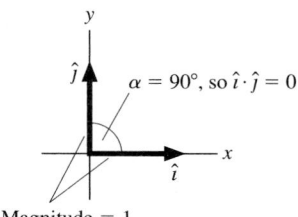

$\alpha = 90°$, so $\hat{\imath} \cdot \hat{\jmath} = 0$

Magnitude $= 1$

| EXAMPLE 11.4 | **Calculating a dot product using components** |

Compute the dot product of $\vec{A} = 3\hat{\imath} + 3\hat{\jmath}$ and $\vec{B} = 4\hat{\imath} - \hat{\jmath}$.

SOLVE FIGURE 11.12 shows vectors $\vec{A}$ and $\vec{B}$. We could calculate the dot product by first doing the geometry needed to find the angle between the vectors and then using Equation 11.14. But calculating the dot product from the vector components is much easier. It is

$$\vec{A} \cdot \vec{B} = A_x B_x + A_y B_y = (3)(4) + (3)(-1) = 9$$

FIGURE 11.12 Vectors $\vec{A}$ and $\vec{B}$.

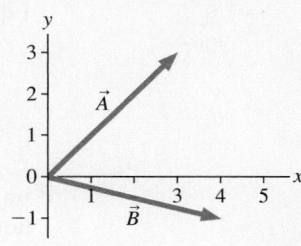

Looking at Equation 11.13, the work done by a constant force, you should recognize that it is the dot product of the force vector and the displacement vector:

$$W = \vec{F} \cdot \Delta \vec{r} \quad \text{(work done by a constant force)} \quad (11.16)$$

This definition of work is valid for a constant force.

EXAMPLE 11.5 **Calculating work using the dot product**

A 70 kg skier is gliding at 2.0 m/s when he starts down a very slippery 50-m-long, 10° slope. What is his speed at the bottom?

MODEL Model the skier as a particle and interpret "very slippery" to mean frictionless. Use the work-kinetic energy theorem to find his final speed.

VISUALIZE **FIGURE 11.13** shows a pictorial representation.

FIGURE 11.13 Pictorial representation of the skier.

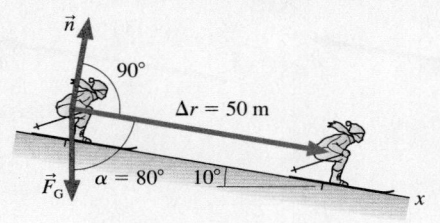

Before:
$x_0 = 0$ m
$v_0 = 2.0$ m/s
$m = 70$ kg

After:
$x_1 = 50$ m
v_1

Find: v_1

SOLVE The only forces on the skier are $\vec{F}_G$ and $\vec{n}$. The normal force is perpendicular to the motion and thus does no work. The work done by gravity is easily calculated as a dot product:

$$W = \vec{F}_G \cdot \Delta \vec{r} = mg(\Delta r)\cos\alpha$$

$$= (70 \text{ kg})(9.8 \text{ m/s}^2)(50 \text{ m})\cos 80° = 5960 \text{ J}$$

Notice that the angle *between* the vectors is 80°, not 10°. Then, from the work-kinetic energy theorem, we find

$$\Delta K = \frac{1}{2}mv_1^2 - \frac{1}{2}mv_0^2 = W$$

$$v_1 = \sqrt{v_0^2 + \frac{2W}{m}} = \sqrt{(2.0 \text{ m/s})^2 + \frac{2(5960 \text{ J})}{70 \text{ kg}}} = 13 \text{ m/s}$$

NOTE ▶ While in the midst of the mathematics of calculating work, do not lose sight of what the work-kinetic energy theorem is all about. It is a statement about *energy transfer:* Work causes a particle's kinetic energy to either increase or decrease. ◀

STOP TO THINK 11.4 Which force does the most work as a particle undergoes displacement $\Delta \vec{r}$?

a. The 10 N force.
b. The 8 N force.
c. The 6 N force.
d. They all do the same amount of work.

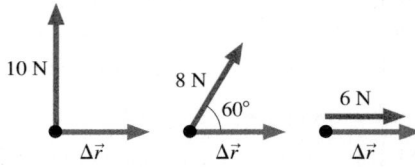

11.4 The Work Done by a Variable Force

We've learned how to calculate the work done on an object by a constant force, but what about a force that changes in either magnitude or direction as the object moves? Equation 11.8, the definition of work, is all we need:

$$W = \int_{s_i}^{s_f} F_s \, ds = \text{area under the force-versus-position graph} \quad (11.17)$$

The integral sums up the small amounts of work $F_s \, ds$ done in each step along the trajectory. The only new feature, because F_s now varies with position, is that we cannot take F_s outside the integral. We must evaluate the integral either geometrically, by finding the area under the curve, or by actually doing the integration.

Using work to find the speed of a car

A 1500 kg car accelerates from rest. **FIGURE 11.14** shows the net force on the car (propulsion force minus any drag forces) as it travels from $x = 0$ m to $x = 200$ m. What is the car's speed after traveling 200 m?

FIGURE 11.14 Force-versus-position graph for a car.

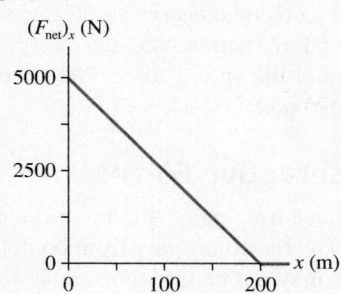

SOLVE The acceleration $a_x = (F_{net})_x/m$ is high as the car starts but decreases as the car picks up speed because of increasing drag. Figure 11.14 is a realistic portrayal of the net force on a car. But a variable force means that we cannot use the familiar constant-acceleration kinematics. Instead, we can use the work-kinetic energy theorem. Because $v_i = 0$ m/s, we have

$$\Delta K = \frac{1}{2}mv_f^2 - 0 = W_{net}$$

Starting from $x_i = 0$ m, the work is

$$W_{net} = \int_{0\,m}^{x_f} (F_{net})_x \, dx$$

$$= \text{area under the } (F_{net})_x\text{-versus-}x \text{ graph from 0 m to } x_f$$

The area under the curve of Figure 11.14 is that of a triangle of width 200 m. Thus

$$W_{net} = \text{area} = \frac{1}{2}(5000 \text{ N})(200 \text{ m}) = 500,000 \text{ J}$$

The work-kinetic energy theorem then gives

$$v_f = \sqrt{\frac{2W_{net}}{m}} = \sqrt{\frac{2(500,000 \text{ J})}{1500 \text{ kg}}} = 26 \text{ m/s}$$

ASSESS 26 m/s $\approx$ 55 mph is a reasonable speed after accelerating for 200 m, so we can have confidence in our answer.

Using the work-kinetic energy theorem for a spring

The "pincube machine" was an ill-fated predecessor of the pinball machine. A 100 g cube is launched by pulling a spring back 20 cm and releasing it. What is the cube's launch speed, as it leaves the spring, if the spring constant is 20 N/m and the surface is frictionless?

MODEL Model the spring as an ideal spring obeying Hooke's law. Use the work-kinetic energy theorem to find the launch speed.

VISUALIZE **FIGURE 11.15** shows a before-and-after pictorial representation and a free-body diagram. We've placed the origin of the x-axis at the equilibrium position of the spring.

FIGURE 11.15 Pictorial representation and free-body diagram for Example 11.7.

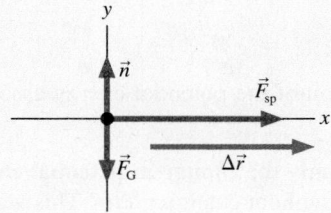

SOLVE The normal force and gravity are perpendicular to the motion and do no work. We can use the work-kinetic energy theorem, with $v_0 = 0$ m/s, to find the launch speed:

$$\Delta K = \frac{1}{2}mv_1^2 - 0 = W_{sp}$$

The spring force is a variable force: $(F_{sp})_x = -k\Delta x = -kx$, where $\Delta x = x - x_e = x$ because we chose a coordinate system with $x_e = 0$ m. Despite the minus sign, $(F_{sp})_x$ is a positive quantity (force pointing to the right) because x is negative throughout the motion. The spring force points in the direction of motion, so W_{sp} is positive. We can use Equation 11.17 to evaluate W_{sp}:

$$W_{sp} = \int_{x_0}^{x_1} (F_{sp})_x \, dx = -k\int_{x_0}^{x_1} x \, dx = -\frac{1}{2}kx^2 \Big|_{x_0}^{x_1}$$

$$= -\left(\frac{1}{2}kx_1^2 - \frac{1}{2}kx_0^2\right)$$

Evaluating W_{sp} for $x_0 = -0.20$ m and $x_1 = 0$ m gives

$$W_{sp} = \frac{1}{2}(20 \text{ N/m})(-0.20 \text{ m})^2 = 0.400 \text{ J}$$

We can now solve for the launch speed, finding

$$v_1 = \sqrt{\frac{2W_{sp}}{m}} = \sqrt{\frac{2(0.400 \text{ J})}{0.100 \text{ kg}}} = 2.8 \text{ m/s}$$

ASSESS You might have noticed that the work done by the spring looks a lot like the spring's potential energy $U_{sp} = \frac{1}{2}k(\Delta x)^2$. The next section will find a connection between work and potential energy.

11.5 Work and Potential Energy

It's time to look more closely at the concept of potential energy. In Chapter 10, the new concept of gravitational potential energy was associated with the gravitational force. Then, after introducing Hooke's law, we used the spring force to "discover" elastic potential energy. In both cases, a force was associated with a potential energy, and we found that it is often easier to solve problems with energy laws rather than force laws.

But that raises the question: Is there a potential energy associated with every force? Is there a "tension potential energy" and a "friction potential energy"? If not, what's special about the gravitational force and the spring force? What conditions must a force meet in order to have an associated potential energy?

Conservative and Nonconservative Forces

FIGURE 11.16 shows a particle that can move from point A to point B along two possible paths while a force $\vec{F}$ is exerted on it. The force may vary from point to point in space, so the force experienced along path 1 may not be the same as the force experienced along path 2. The force changes the particle's speed, as well as its direction, so the particle's speed and kinetic energy when it arrives at B will differ from the speed and kinetic energy it had when it left A.

Let's assume that there is a potential energy associated with force $\vec{F}$, just as the gravitational potential energy $U_g = mgy$ is associated with the gravitational force $\vec{F}_G = -mg\hat{\jmath}$. What restrictions does this assumption place on $\vec{F}$? There are three steps in the logic:

1. Potential energy is an energy of position. The system has one value of potential energy when the particle is at A, a different value when the particle is at B. Thus the overall change in potential energy $\Delta U = U_B - U_A$ is the same whether the particle moves along path 1 or path 2.
2. Potential energy is transformed into kinetic energy, with $\Delta K = -\Delta U$. If ΔU is independent of the path, then ΔK is also independent of the path. The transformation of energy causes the particle to have the same kinetic energy at B no matter which path it follows.
3. The change in a particle's kinetic energy is related to the work done on the particle by force $\vec{F}$. According to the work-kinetic energy theorem, $\Delta K = W$. Because ΔK is independent of the path, it *must* be the case that **the work done by force $\vec{F}$ as the particle moves from A to B is independent of the path followed.**

A force for which the work done on a particle as it moves from an initial to a final position is independent of the path followed is called a **conservative force**. (The name, as you'll soon see, is related to the conditions under which mechanical energy is conserved.) The importance of conservative forces is that **a potential energy can be associated with any conservative force.**

To establish a general connection between work and potential energy, suppose an object moves from initial position i to final position f under the influence of a conservative force $\vec{F}$. We'll denote the work done by the force as $W_c(i \rightarrow f)$, where the notation $i \rightarrow f$ means "as the object moves from position i to position f." Because $\Delta K = W$ and $\Delta K = -\Delta U$, the potential energy difference between these two points must be

$$\Delta U = U_f - U_i = -W_c(i \rightarrow f) \qquad (11.18)$$

Equation 11.18 is a general definition of the potential energy associated with a conservative force.

> **NOTE** ▶ Equation 11.18 defines only the *change* in potential energy ΔU. We can add a constant to both U_f and U_i without changing ΔU. This was the basis for our discussion in Chapter 10 about the zero of potential energy. ◀

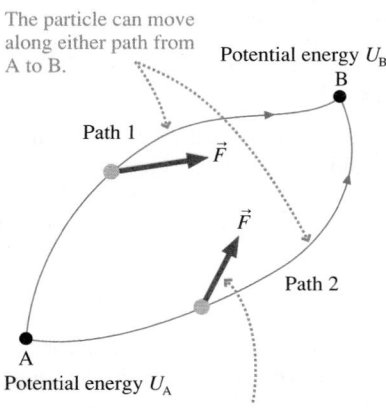

FIGURE 11.16 A particle can move from A to B along either path 1 or path 2.

The particle can move along either path from A to B.

Potential energy U_B

B

Path 1

$\vec{F}$

$\vec{F}$

Path 2

A

Potential energy U_A

The force does work on the particle as it moves from A to B, changing the particle's kinetic energy.

For example, Chapter 10 showed that the kinetic energy gained by an object sliding down a frictionless slope depends only on the vertical distance change Δy and is independent of the shape of the slope. It follows, because $W = \Delta K$, that the work done by gravity does not depend on the path followed from initial height y_i to final height y_f. That is, the gravitational force is a conservative force, and *that* is why we were able to establish a gravitational potential energy. (If you look back, you'll see that the analysis of Chapter 10 that led to $U_g = mgy$ was really a calculation of the work done by the gravitational force, although we didn't call it that at the time.)

What about springs? A homework problem will let you show that Hooke's law is also a conservative force. In Example 11.7 we showed that the work done by a spring is

$$W_{sp}(i \rightarrow f) = \int_{x_i}^{x_f} F_{sp}\, dx = -\left(\frac{1}{2}kx_f^2 - \frac{1}{2}kx_i^2\right)$$

from which it follows that $U_s = \frac{1}{2}kx^2$. Example 11.7 was a "special case" in that we defined the coordinate system to make $x_e = 0$. A more general analysis would give $U_s = \frac{1}{2}k(\Delta s)^2$, as you learned in Chapter 10.

Not all forces are conservative forces. For example, FIGURE 11.17 is a bird's-eye view of two particles sliding across a surface. The friction force always points opposite the direction of motion, 180° from $d\vec{s}$, hence the small amount of work done during displacement $d\vec{s}$ is $dW_{fric} = \vec{f}_k \cdot d\vec{s} = -\mu_k mg\, ds$. Summed over the entire path, the work done by friction as a particle travels total distance Δs is $W_{fric} = -\mu_k mg \Delta s$. We see that the work done by friction depends on Δs, the distance traveled. More work is done on the particle traveling the longer path, so the work done by friction is *not* independent of the path followed.

NOTE ▶ This analysis applies only to the motion of a particle, which has no internal structure and thus no thermal energy. These ideas will be applied to extended objects—such as a car skidding to a halt—in Section 11.7 on thermal energy. The particle equation $W_{fric} = -\mu_k mg \Delta s$ should *not* be used in problem solving. ◀

A force for which the work is *not* independent of the path is called a **nonconservative force.** It is not possible to define a potential energy for a nonconservative force. Friction is a nonconservative force, so we cannot define a potential energy of friction.

This makes sense. If you toss a ball straight up, kinetic energy is transformed into gravitational potential energy. The ball has the potential to transform this energy back into kinetic energy, and it does so as the ball falls. But you cannot recover the kinetic energy lost to friction as a box slides to a halt. There's no "potential" that can be transformed back into kinetic energy.

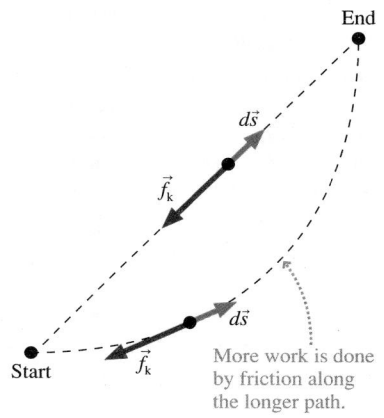

FIGURE 11.17 Top view of two particles sliding across a surface.

End

$d\vec{s}$

$\vec{f}_k$

$d\vec{s}$

$\vec{f}_k$

Start

More work is done by friction along the longer path.

Mechanical Energy

Consider a system of objects interacting via both conservative forces and nonconservative forces. The conservative forces do work W_c as the particles move from initial positions i to final positions f. The nonconservative forces do work W_{nc}. The total work done by *all* forces is $W_{net} = W_c + W_{nc}$. The change in the system's kinetic energy ΔK, as determined by the work-kinetic energy theorem, is

$$\Delta K = W_{net} = W_c(i \rightarrow f) + W_{nc}(i \rightarrow f) \qquad (11.19)$$

The work done by the conservative forces can now be associated with a potential energy U. According to Equation 11.18, $W_c(i \rightarrow f) = -\Delta U$. With this definition, Equation 11.19 becomes

$$\Delta K + \Delta U = \Delta E_{mech} = W_{nc} \qquad (11.20)$$

where, as in Chapter 10, the *mechanical energy* is $E_{mech} = K + U$.

Now we can see that **mechanical energy is conserved if there are no nonconservative forces.** That is,

$$\Delta E_{mech} = 0 \text{ if } W_{nc} = 0 \qquad (11.21)$$

Mechanical energy isn't always conserved. Here, mechanical energy is being transformed into thermal energy.

This important conclusion is what we called the law of conservation of mechanical energy in Chapter 10. There we saw that friction prevents E_{mech} from being conserved, but we really didn't know why. Equation 11.20 tells us that the work done by any nonconservative force causes the mechanical energy to change. Friction and other "dissipative forces" lead to a loss of mechanical energy. Other outside forces, such as the pull of a rope, might increase the mechanical energy.

Equally important, Equation 11.20 tells us what to do if the mechanical energy isn't conserved. You can still use energy concepts to analyze the motion if you compute the work done by the nonconservative forces.

EXAMPLE 11.8 **Using work and potential energy**

The skier from Example 11.5 repeats his run after the wind comes up. Recall that the 70 kg skier was gliding at 2.0 m/s when he started down a 50-m-long, 10°, frictionless slope. What is his speed at the bottom if the wind exerts a steady 50 N retarding force opposite his motion?

MODEL This time let the system be the skier and the earth.

VISUALIZE Figure 11.13 showed the pictorial representation and free-body diagram.

SOLVE In solving this problem with the work-kinetic energy theorem, we had to explicitly calculate the work done by gravity. Now let's use Equation 11.20. Gravity is a conservative force that we can associate with the gravitational potential energy U_g. The retarding force of the wind is nonconservative. Thus

$$\Delta K + \Delta U_g = W_{nc} = W_{wind}$$

The gravitational potential energy is $U_g = mgy$. Because the wind force is opposite the skier's motion, with $\theta = 180°$, it does work $W_{wind} = \vec{F}_{wind} \cdot \Delta \vec{r} = -F_{wind} \Delta r$. Thus the energy equation becomes

$$\frac{1}{2}mv_1^2 - \frac{1}{2}mv_0^2 + mgy_1 - mgy_0 = -F_{wind} \Delta r$$

Using the values given, we find

$$v_1 = \sqrt{v_0^2 + 2gy_0 - 2F_{wind} \Delta r/m} = 10 \text{ m/s}$$

ASSESS What appeared to be a difficult problem, with both gravity and a retarding force, turned out to be straightforward when analyzed with energy and work. The skier's final speed is about 25% slower when the wind is blowing.

Example 11.8 illustrates an important idea. When we associate a potential energy with a conservative force, we

- Enlarge the system to include all objects that interact via conservative forces.
- "Precompute" the work. We can do this because we don't need to know what paths the objects are going to follow. This precomputed work becomes a potential energy and moves from the right side of $\Delta K = W$ to the left side of Equation 11.20.

NOTE ▶ When you use a potential energy, you've already taken the work of that force into account. Don't compute the work explicitly, or you'll double count it! ◀

In Example 11.5, the system consisted of just the skier. We treated the gravitational force as a force from the environment doing work on the system. In Example 11.8, where we revisited the same problem, we brought the earth into the system and represented the conservative earth-skier interaction with a potential energy.

In summary, to analyze a problem using work and energy, you can either

1. Use the work-kinetic energy theorem $\Delta K = W$ and explicitly compute the work done by *every* force. This was the method of Example 11.5. Or
2. Represent the work done by conservative forces as potential energies, then use $\Delta K + \Delta U = W_{nc}$. The only work that must be computed is the work of any nonconservative forces. This was the method of Example 11.8.

In practice, **method 2 is always easier and is the preferred method.**

11.6 Finding Force from Potential Energy

We now know how to find the potential energy due to a conservative force. Can we reverse the procedure and find the force from the potential energy?

FIGURE 11.18a shows an object moving through a *small* displacement Δs while being acted on by a conservative force $\vec{F}$. If Δs is sufficiently small, the force component F_s in the direction of motion is essentially constant during the displacement. The work done on the object as it moves from s to $s + \Delta s$ is

$$W(s \to s + \Delta s) = F_s\, \Delta s \qquad (11.22)$$

Because $\vec{F}$ is a conservative force, the object's potential energy as it moves through Δs changes by

$$\Delta U = -W(s \to s + \Delta s) = -F_s\, \Delta s$$

which we can rewrite as

$$F_s = -\frac{\Delta U}{\Delta s} \qquad (11.23)$$

In the limit $\Delta s \to 0$, we find that the force at position s is

$$F_s = \lim_{\Delta s \to 0}\left(-\frac{\Delta U}{\Delta s}\right) = -\frac{dU}{ds} \qquad (11.24)$$

We see that the force on the object is the *negative* of the derivative of the potential energy with respect to position. **FIGURE 11.18b** shows that we can interpret this result graphically by saying

$$F_s = \text{the negative of the slope of the } U\text{-versus-}s \text{ graph at } s \qquad (11.25)$$

In practice, of course, we will usually use either $F_x = -dU/dx$ or $F_y = -dU/dy$.

As an example, consider the gravitational potential energy $U_g = mgy$. **FIGURE 11.19a** shows the potential-energy diagram U_g versus y. It is simply a straight-line graph passing through the origin. The force on the object at position y, according to Equations 11.24 and 11.25, is simply

$$(F_G)_y = -\frac{dU_g}{dy} = -(\text{slope of } U_g) = -mg$$

The negative sign, as always, indicates that the force points in the negative y-direction. **FIGURE 11.19b** shows the corresponding F-versus-y graph. At each point, the *value* of F is equal to the negative of the *slope* of the U-versus-y graph. This is similar to position and velocity graphs, where the value of v_x at any time t is equal to the slope of the x-versus-t graph.

We already knew that $(F_G)_y = -mg$, of course, so the point of this particular example was to illustrate the meaning of Equation 11.25 rather than to find out anything new. Had we *not* known the gravitational force, we see that it is possible to find it from the potential energy.

FIGURE 11.20a is a more interesting example. The slope of the potential-energy graph is negative between x_1 and x_2. This means that the force on the object, which is the negative of the slope of U, is *positive*. An object between x_1 and x_2 experiences a force toward the right. The force decreases as the magnitude of the slope decreases until, at x_2, $F_x = 0$. This is consistent with our prior identification of x_2 as a point of stable *equilibrium*. The slope is positive (force negative and thus to the left) between x_2 and x_3, zero (zero force) at the unstable equilibrium point x_3, and so on. Point x_4, where the slope is most negative, is the point of maximum force.

FIGURE 11.20b is a plausible graph of F versus x. We don't know the exact shape because we don't have an exact expression for U, but the force graph must look very much like this.

FIGURE 11.18 Relating force and potential energy.

(a)
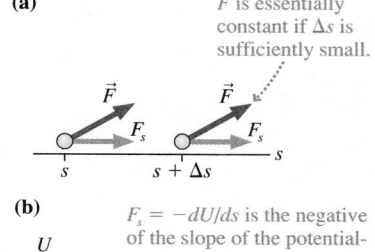

$\vec{F}$ is essentially constant if Δs is sufficiently small.

(b)
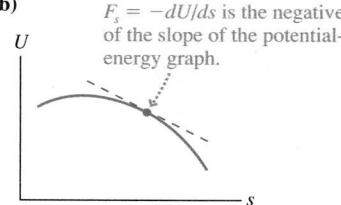

$F_s = -dU/ds$ is the negative of the slope of the potential-energy graph.

FIGURE 11.19 Gravitational potential energy and force diagrams.

(a)
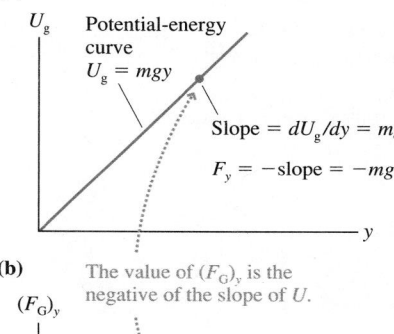

U_g Potential-energy curve
$U_g = mgy$

Slope $= dU_g/dy = mg$
$F_y = -\text{slope} = -mg$

(b)
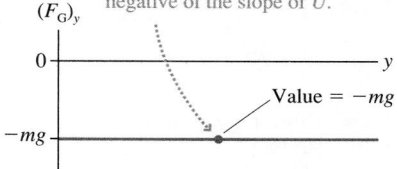

The value of $(F_G)_y$ is the negative of the slope of U.

$(F_G)_y$

Value $= -mg$

FIGURE 11.20 A potential-energy diagram and the corresponding force diagram.

(a)
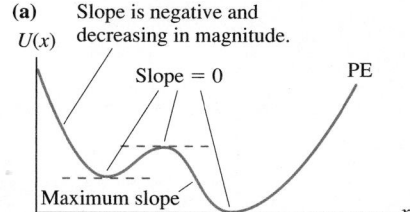

Slope is negative and decreasing in magnitude.
$U(x)$
Slope $= 0$
PE
Maximum slope
x_1 x_2 x_3 x_4 x_5 x_6

(b)
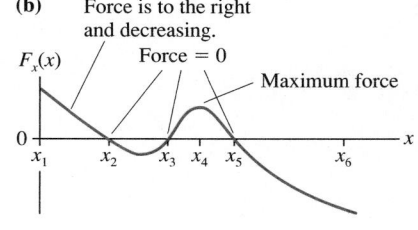

Force is to the right and decreasing.
$F_x(x)$
Force $= 0$
Maximum force
x_1 x_2 x_3 x_4 x_5 x_6

STOP TO THINK 11.5   A particle moves along the x-axis with the potential energy shown. The x-component of the force on the particle when it is at $x = 4$ m is

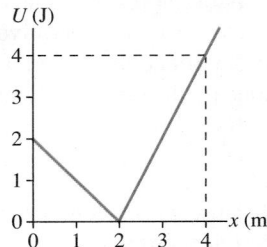

a. 4 N.   b. 2 N.
c. 1 N.   d. −4 N.
e. −2 N.   f. −1 N.

11.7 Thermal Energy

All of the objects we handle and use every day consist of vast numbers of particle-like atoms. We will use the terms **macrophysics** to refer to the motion and dynamics of the object as a whole and **microphysics** to refer to the motion of atoms. You recognize the prefix *micro,* meaning "small." You may not be familiar with *macro,* which means "large" or "large-scale."

Kinetic and Potential Energy at the Microscopic Level

Figure 11.21 shows two different perspectives of an object. In the macrophysics perspective of FIGURE 11.21a you see an object of mass M moving as a whole with velocity v_{obj}. As a consequence of its motion, the object has macroscopic kinetic energy $K_{macro} = \frac{1}{2}Mv_{obj}^2$.

FIGURE 11.21b is a microphysics view of the same object, where now we see a *system of particles.* Each of these atoms is moving about, and in doing so they stretch and compress the spring-like bonds between them. Consequently, there is a *microscopic* kinetic and potential energy associated with the motion of atoms and bonds.

The kinetic energy of one atom is exceedingly small, but there are enormous numbers of atoms in a macroscopic object. The total kinetic energy of all the atoms is what we call the *microscopic kinetic energy K_{micro}.* The total potential energy of all the bonds is the *microscopic potential energy U_{micro}.* These microscopic energies are quite distinct from the energies K_{macro} and U_{macro} of the object as a whole.

Is the microscopic energy worth worrying about? To see, consider a 500 g (≈ 1 lb) iron ball moving at the respectable speed of 20 m/s (≈ 45 mph). Its macroscopic kinetic energy is

$$K_{macro} = \frac{1}{2}Mv_{obj}^2 = 100 \text{ J}$$

A periodic table of the elements shows that iron has atomic mass 56. Recall from chemistry that 56 g of iron is 1 gram-molecular weight and has Avogadro's number ($N_A = 6.02 \times 10^{23}$) of atoms. Thus 500 g of iron is ≈ 9 gram-molecular weights and contains $N \approx 9N_A \approx 5.4 \times 10^{24}$ iron atoms. The mass of each atom is

$$m = \frac{M}{N} \approx \frac{0.50 \text{ kg}}{5.4 \times 10^{24}} \approx 9 \times 10^{-26} \text{ kg}$$

How fast do atoms move? In Part IV you'll learn that $v \approx 500$ m/s is a reasonable estimate. The kinetic energy of one iron atom at this speed is

$$K_{atom} = \frac{1}{2}mv^2 \approx 1.1 \times 10^{-20} \text{ J}$$

This is very tiny, but there are a great many atoms. If all atoms move at roughly this speed, the microscopic kinetic energy is

$$K_{micro} \approx NK_{atom} \approx 60{,}000 \text{ J}$$

FIGURE 11.21 Two perspectives of motion and energy.

(a) The macroscopic motion of the system as a whole

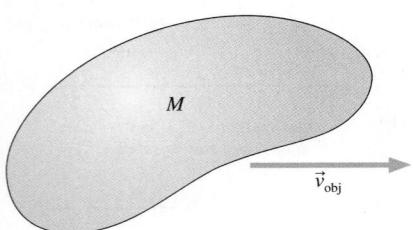

M

$\vec{v}_{obj}$

(b) The microscopic motion of the atoms inside

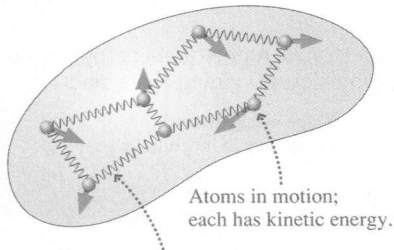

Atoms in motion; each has kinetic energy.

Molecular bonds stretch and compress; each has potential energy.

We'll later see that, on average, U_{micro} for a solid is equal to K_{micro}, so the total microscopic energy is $\approx 120,000$ J. The microscopic energy is much larger than the macroscopic kinetic energy of the object as a whole!

The combined microscopic kinetic and potential energy of the atoms is called the *thermal energy* of the system:

$$E_{th} = K_{micro} + U_{micro} \qquad (11.26)$$

This energy is usually hidden from view in our macrophysics perspective, but it is quite real. We will discover later, when we reach thermodynamics, that the thermal energy is related to the *temperature* of the system. Raising the temperature causes the atoms to move faster and the bonds to stretch more, giving the system more thermal energy.

NOTE ▶ The microscopic energy of atoms is *not* called "heat." The word "heat," like the word "work," has a narrow and precise meaning in physics that is much more restricted than its use in everyday language. We will introduce the concept of heat later, when we need it. For the time being we want to use the correct term "thermal energy" to describe the random, thermal motion of the particles in a system. If the temperature of a system goes up (i.e., it gets hotter), it is because the system's thermal energy has increased. ◀

Dissipative Forces

If you shove a book across the table, it gradually slows down and stops. Where did the energy go? The common answer "It went into heat" isn't quite right.

FIGURE 11.22, the atomic model of friction from Chapter 6, shows why. As two objects slide against each other, atomic interactions at the boundary transform the kinetic energy K_{macro} of the moving object—it's slowing down—into microscopic kinetic and potential energy of vibrating atoms and stretched bonds. The energy transformation is $K \rightarrow E_{th}$, and we perceive this as an increased temperature of *both* objects. Thus the correct answer to What happens to the energy? is "It is transformed into thermal energy."

Forces such as friction and drag cause the macroscopic kinetic energy of a system to be "dissipated" as thermal energy. Hence these are called **dissipative forces.** Dissipative forces are always nonconservative forces. The energy analysis of dissipative forces is a bit subtle. Because friction causes *both* objects to get warmer, we must define the system to include both objects—both the book *and* the table, or both the car *and* the road.

FIGURE 11.23 shows a box being pulled at constant speed across a horizontal surface with friction. As you can imagine, both the surface and the box are getting warmer—increasing thermal energy. But neither the kinetic nor the potential energy of the box is changing, so where is the thermal energy coming from? Recall, from the basic energy model, that work is energy transferred to a system by forces from the environment. If we define the system to be box + surface, then the increasing thermal energy of the system is entirely due to the work being done on the system by tension in the rope: $\Delta E_{th} = W_{tension}$.

The work done by tension in pulling the box a distance Δs is simply $W_{tension} = T\Delta s$; thus $\Delta E_{th} = T\Delta s$. Because the box is moving with constant velocity, Newton's first law $\vec{F}_{net} = \vec{0}$ requires the tension to exactly balance the friction force: $T = f_k$. Consequently, the increase in thermal energy due to the dissipative force of friction is

$$\Delta E_{th} = f_k \Delta s \text{ (increased thermal energy due to friction)} \qquad (11.27)$$

Notice that the increase in thermal energy is directly proportional to the total distance of sliding. **Dissipative forces always increase the thermal energy; they never decrease it.**

You might wonder why we didn't simply calculate the work done by friction. The rather subtle reason is that work is defined only for forces acting on a *particle.*

FIGURE 11.22 The atomic-level view of friction.

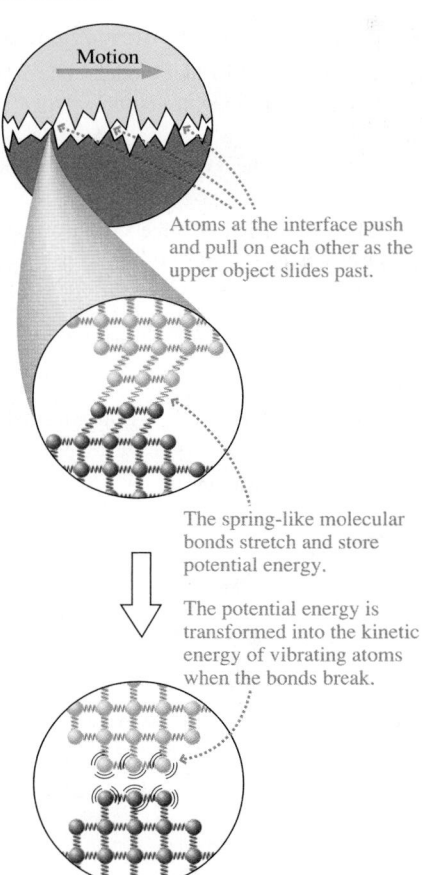

Atoms at the interface push and pull on each other as the upper object slides past.

The spring-like molecular bonds stretch and store potential energy.

The potential energy is transformed into the kinetic energy of vibrating atoms when the bonds break.

FIGURE 11.23 Work done by tension is dissipated as increased thermal energy.

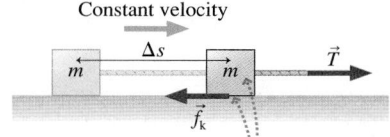

Work done by tension increases the thermal energy of the box *and* the surface.

A particle has no internal structure and thus cannot have thermal energy. Thermal energy appears when we have to deal with extended objects, nonrigid systems of many particle-like atoms.

There is work being done on individual atoms at the boundary as they are pulled this way and that, but we would need a detailed knowledge of atomic-level friction forces to calculate this work. The friction force $\vec{f}_k$ is an average force on the object as a whole; it is not a force on any particular particle, so we cannot use it to calculate work. Further, increasing thermal energy is not an energy transfer from the book to the surface or from the surface to the book. Both book *and* surface are gaining thermal energy at the expense of the macroscopic kinetic energy.

NOTE ▶ The analysis of thermal energy is rather subtle, as we noted above. The considerations that led to Equation 11.27 do allow us to calculate the total increase in thermal energy of the entire system, but we cannot determine what fraction of ΔE_{th} goes to the book and what fraction goes to the surface. ◀

EXAMPLE 11.9 **Calculating the increase in thermal energy**

A rope pulls a 10 kg wooden crate 3.0 m across a wood floor. What is the change in thermal energy? The coefficient of kinetic friction is 0.20.

MODEL Let the system be crate + floor. Assume the floor is horizontal.

SOLVE The friction force on an object moving on a horizontal surface is $F_k = \mu_k n = \mu_k mg$. Thus the change in thermal energy,

given by Equation 11.27, is

$$\Delta E_{th} = f_k \Delta s = \mu_k mg \Delta s$$
$$= (0.20)(10\ \text{kg})(9.80\ \text{m/s}^2)(3.0\ \text{m}) = 59\ \text{J}$$

ASSESS The thermal energy of the crate *and* floor increases by 59 J. We cannot determine ΔE_{th} for the crate (or floor) alone.

11.8 Conservation of Energy

FIGURE 11.24 A system with both internal interaction forces and external forces.

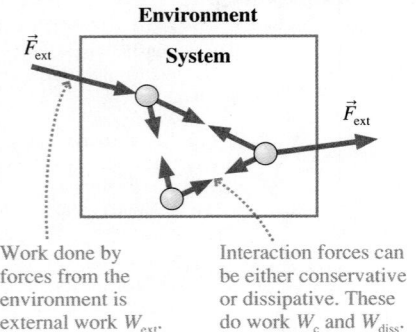

Work done by forces from the environment is external work W_{ext}.

Interaction forces can be either conservative or dissipative. These do work W_c and W_{diss}.

Let's return to the basic energy model and start pulling together the many ideas introduced in this chapter. **FIGURE 11.24** shows a general system consisting of several macroscopic objects. These objects interact with each other, and they may be acted on by external forces from the environment. Both the interaction forces and the external forces do work on the objects. The change in the system's kinetic energy is given by the work-kinetic energy theorem, $\Delta K = W_{net}$.

We previously divided W_{net} into the work W_c done by conservative forces and the work W_{nc} done by nonconservative forces. The work done by the conservative forces can be represented by a potential energy U. Let's now make a further distinction by dividing the nonconservative forces into *dissipative forces* and *external forces*. That is,

$$W_{nc} = W_{diss} + W_{ext} \tag{11.28}$$

To illustrate what we mean by an external force, suppose you pick up a box at rest on the floor and place it at rest on a table. The box gains gravitational potential energy, but $\Delta K = 0$. Or consider pulling the box across the table with a string. The box gains kinetic energy, but not by transforming potential energy. The force of your hand and the tension of the string are forces that "reach in" from the environment to change the system. Thus they are *external forces*.

We have to be careful choosing the system if we want this distinction to be valid. As you can imagine, we're going to associate W_{diss} with ΔE_{th}. We want the thermal energy E_{th} to be an energy *of the system*. Otherwise, it wouldn't make sense to talk about transforming kinetic energy into thermal energy. But for E_{th} to be an energy of the system, *both* objects involved in a dissipative interaction must be part of the system. The book sliding across the table raises the temperature of both the book *and the table*. Consequently, we must include both the book *and the table* in the system. The dissipative forces, like the conservative forces, are atomic-level interaction forces *inside* the system.

With this distinction, the work-kinetic energy theorem is

$$\Delta K = W_c + W_{diss} + W_{ext} \qquad (11.29)$$

As before, we define the potential energy U such that $\Delta U = -W_c$. Remember that potential energy is really just the precomputed work of a conservative force. We've also seen that the work done by dissipative forces—the forces stretching the bonds at the boundary—increases the system's thermal energy: $\Delta E_{th} = -W_{diss}$. With these substitutions, the work-kinetic energy theorem becomes

$$\Delta K = -\Delta U - \Delta E_{th} + W_{ext}$$

We can write this more profitably as

$$\Delta K + \Delta U + \Delta E_{th} = \Delta E_{mech} + \Delta E_{th} = \Delta E_{sys} = W_{ext} \qquad (11.30)$$

where $E_{sys} = E_{mech} + E_{th}$ is the total energy of the system. Equation 11.30 is the **energy equation** of the system.

Equation 11.30 is our most general statement about how the energy of a system changes, but we still need to give a clear interpretation as to what it says. In Chapter 9 we defined an *isolated system* as a system for which the *net* external force is zero. It follows that no external work is done on an isolated system: $W_{ext} = 0$. Thus one conclusion from Equation 11.30 is that **the total energy E_{sys} of an isolated system is conserved**. That is, $\Delta E_{sys} = 0$ for an isolated system. If, in addition, the system is nondissipative (i.e., no friction forces), then $\Delta E_{th} = 0$. In that case, the mechanical energy E_{mech} is conserved.

These conclusions about energy can be summarized as the *law of conservation of energy:*

> **Law of conservation of energy** The total energy $E_{sys} = E_{mech} + E_{th}$ of an isolated system is a constant. The kinetic, potential, and thermal energy within the system can be transformed into each other, but their sum cannot change. Further, the mechanical energy $E_{mech} = K + U$ is conserved if the system is both isolated and nondissipative.

The law of conservation of energy is one of the most powerful statements in physics. **FIGURE 11.25a** redraws the basic energy model of Figure 11.1 Now you can see that this is a pictorial representation of Equation 11.30. E_{sys}, the total energy of the system, changes only if external forces transfer energy into or out of the system by doing work on the system. The kinetic, potential, and thermal energy within the system can be transformed into each other by interaction forces within the system. As **FIGURE 11.25b** shows, $E_{sys} = K + U + E_{th}$ remains constant if the system is isolated. The *transfer* and *transformation* of energy are what the basic energy model is all about.

Energy Bar Charts

The energy bar charts of Chapter 10 can now be expanded to include the thermal energy and the work done by external forces. The energy equation, Equation 11.30, can be written

$$K_i + U_i + W_{ext} = K_f + U_f + \Delta E_{th} \qquad (11.31)$$

The left side is the "before" condition ($K_i + U_i$) plus any energy that is added to or removed from the system. The right side is the "after" situation. The "energy accounting" of Equation 11.31 can be represented by the bar chart of **FIGURE 11.26** on the next page.

NOTE ▶ We don't have any way to determine $(E_{th})_i$ or $(E_{th})_f$, but ΔE_{th} is always positive whenever the system contains dissipative forces. ◀

FIGURE 11.25 The basic energy model is a pictorial representation of the energy equation.

(a) A system interacting with its environment

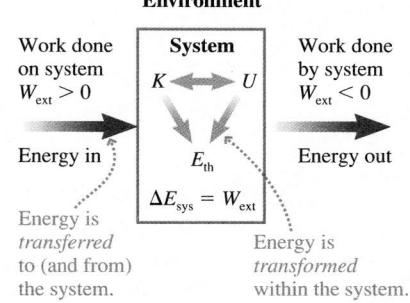

(b) An isolated system

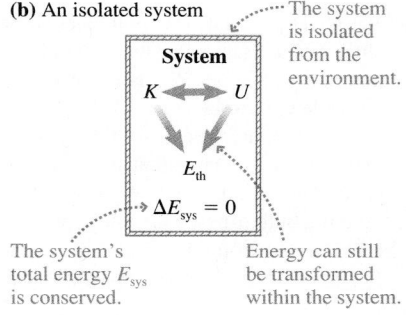

FIGURE 11.26 An energy bar chart shows how all the energy is accounted for.

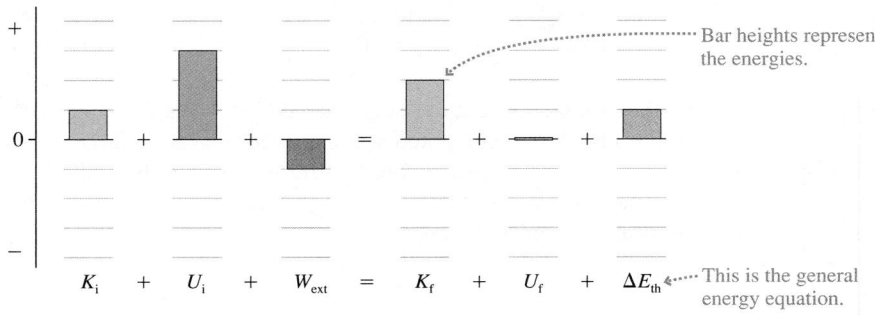

Bar heights represent the energies.

$$K_i \;+\; U_i \;+\; W_{ext} \;=\; K_f \;+\; U_f \;+\; \Delta E_{th}$$

This is the general energy equation.

Let's look at a few examples.

EXAMPLE 11.10 | Energy bar chart I

A speeding car skids to a halt. Show the energy transfers and transformations on an energy bar chart.

SOLVE The car has an initial kinetic energy K_i. That energy is transformed into the thermal energy of the car and the road. The potential energy doesn't change and no work is done by external forces, so the process is an energy transformation $K_i \rightarrow E_{th}$. This is shown in **FIGURE 11.27**. E_{sys} is conserved but E_{mech} is not.

FIGURE 11.27 Energy bar chart for Example 11.10.

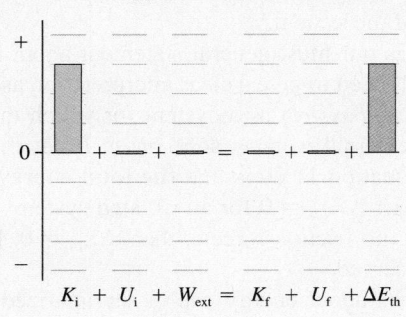

$$K_i \;+\; U_i \;+\; W_{ext} \;=\; K_f \;+\; U_f \;+\; \Delta E_{th}$$

EXAMPLE 11.11 | Energy bar chart II

A rope lifts a box at constant speed. Show the energy transfers and transformations on an energy bar chart.

SOLVE The tension in the rope is an external force that does work on the box, increasing the potential energy of the box. The kinetic energy is unchanged because the speed is constant. The process is an energy transfer $W_{ext} \rightarrow U_f$, as **FIGURE 11.28** shows. This is not an isolated system, so E_{sys} is not conserved.

FIGURE 11.28 Energy bar chart for Example 11.11.

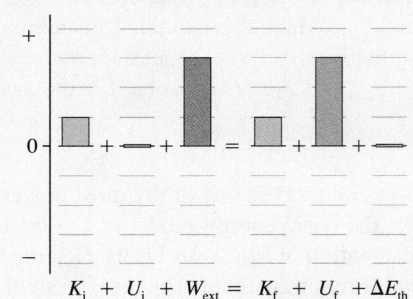

$$K_i \;+\; U_i \;+\; W_{ext} \;=\; K_f \;+\; U_f \;+\; \Delta E_{th}$$

EXAMPLE 11.12 | Energy bar chart III

The box that was lifted in Example 11.11 falls at a steady speed as the rope spins a generator and causes a lightbulb to glow. Air resistance is negligible. Show the energy transfers and transformations on an energy bar chart.

SOLVE The initial potential energy decreases, but K does not change and $\Delta E_{th} = 0$. The tension in the rope is an external force that does work, but W_{ext} is negative in this case because $\vec{T}$ points up while the displacement $\Delta \vec{r}$ is down. Negative work means that energy is transferred from the system to the environment or, in more informal terms, that the *system does work on the environment*. The falling box does work on the generator to spin it and light the bulb. Energy is transferred out of the system and eventually ends up in the lightbulb as electrical energy. The process is $U_i \rightarrow W_{ext}$. This is shown in **FIGURE 11.29**.

FIGURE 11.29 Energy bar chart for Example 11.12.

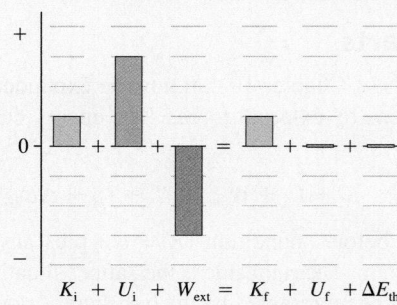

$$K_i \;+\; U_i \;+\; W_{ext} \;=\; K_f \;+\; U_f \;+\; \Delta E_{th}$$

Strategy for Energy Problems

This is a good place to summarize the strategy we have been developing for using the concept of energy.

PROBLEM-SOLVING STRATEGY 11.1 **Solving energy problems**

MODEL Identify which objects are part of the system and which are in the environment. When possible, choose a system without friction or other dissipative forces. Some problems may need to be subdivided into two or more parts.

VISUALIZE Draw a before-and-after pictorial representation and an energy bar chart. A free-body diagram is helpful if you're going to calculate work.

SOLVE If the system is both isolated and nondissipative, then the mechanical energy is conserved:

$$K_f + U_f = K_i + U_i$$

If there are external or dissipative forces, calculate W_{ext} and ΔE_{th}. Then use the more general energy equation

$$K_f + U_f + \Delta E_{th} = K_i + U_i + W_{ext}$$

Kinematics and/or other conservation laws may be needed for some problems.

ASSESS Check that your result has the correct units, is reasonable, and answers the question.

STOP TO THINK 11.6 A child at the playground slides down a pole at constant speed. This is a situation in which

a. $U \rightarrow K$. E_{mech} is not conserved but E_{sys} is.
b. $U \rightarrow E_{th}$. E_{mech} is conserved.
c. $U \rightarrow E_{th}$. E_{mech} is not conserved but E_{sys} is.
d. $K \rightarrow E_{th}$. E_{mech} is not conserved but E_{sys} is.
e. $U \rightarrow W_{ext}$. Neither E_{mech} nor E_{sys} is conserved.

11.9 Power

Work is a transfer of energy between the environment and a system. In many situations we would like to know *how quickly* the energy is transferred. Does the force act quickly and transfer the energy very rapidly, or is it a slow and lazy transfer of energy? If you need to buy a motor to lift 2000 lb of bricks up 50 ft, it makes a *big* difference whether the motor has to do this in 30 s or 30 min!

The question How quickly? implies that we are talking about a *rate*. For example, the velocity of an object—how quickly it is moving—is the *rate of change* of position. So when we raise the issue of how quickly the energy is transferred, we are talking about the *rate of transfer* of energy. The rate at which energy is transferred or transformed is called the **power** P, and it is defined as

$$P \equiv \frac{dE_{sys}}{dt} \tag{11.32}$$

The unit of power is the **watt**, which is defined as 1 watt = 1 W $\equiv$ 1 J/s.

The English unit of power is the *horsepower*. The conversion factor to watts is

1 horsepower = 1 hp = 746 W

Many common appliances, such as motors, are rated in hp.

A force that is doing work (i.e., transferring energy) at a rate of 3 J/s has an "output power" of 3 W. The system gaining energy at the rate of 3 J/s is said to "consume" 3 W of power. Common prefixes used with power are mW (milliwatts), kW (kilowatts), and MW (megawatts).

EXAMPLE 11.13 **Choosing a motor**

What power motor is needed to lift a 2000 kg elevator at a steady 3.0 m/s?

SOLVE The tension in the cable does work on the elevator to lift it. Because the cable is pulled by the motor, we say that the motor does the work of lifting the elevator. The net force is zero because the elevator moves at constant velocity, so the tension is simply $T = mg = 19{,}600$ N. The energy gained by the elevator is

$$\Delta E_{\text{sys}} = W_{\text{ext}} = T\,\Delta y$$

The power required to give the system this much energy in a time interval Δt is

$$P = \frac{\Delta E_{\text{sys}}}{\Delta t} = \frac{T\,\Delta y}{\Delta t}$$

But $\Delta y = v\,\Delta t$, so

$$P = Tv = (19{,}600\ \text{N})(3.0\ \text{m/s}) = 58{,}800\ \text{W}$$

$$= 79\ \text{hp}$$

Highly trained athletes have a tremendous power output.

The idea of power as a *rate* of energy transfer applies no matter what the form of energy. **FIGURE 11.30** shows three examples of the idea of power. For now, we want to focus primarily on *work* as the source of energy transfer. Within this more limited scope, power is simply the **rate of doing work**: $P = dW/dt$. If a particle moves through a small displacement $d\vec{r}$ while acted on by force $\vec{F}$, the force does a small amount of work dW given by

$$dW = \vec{F} \cdot d\vec{r}$$

Dividing both sides by dt, to give a rate of change, yields

$$\frac{dW}{dt} = \vec{F} \cdot \frac{d\vec{r}}{dt}$$

But $d\vec{r}/dt$ is the velocity $\vec{v}$, so we can write the power as

$$P = \vec{F} \cdot \vec{v} = Fv \cos\theta \qquad (11.33)$$

In other words, the power delivered to a particle by a force acting on it is the dot product of the force and the particle's velocity. These ideas will become clearer with some examples.

FIGURE 11.30 Examples of power.

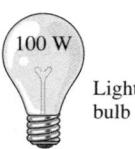

Athlete
$\frac{1}{2}$ hp

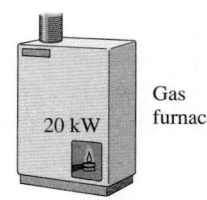

100 W

Light bulb

Gas furnace

20 kW

Electrical energy ⟶ light and thermal energy at 100 J/s

Chemical energy of glucose and fat ⟶ mechanical energy at ≈350 J/s ≈ $\frac{1}{2}$ hp

Chemical energy of gas ⟶ thermal energy at 20,000 J/s

EXAMPLE 11.14 **Power output of a motor**

A factory uses a motor and a cable to drag a 300 kg machine to the proper place on the factory floor. What power must the motor supply to drag the machine at a speed of 0.50 m/s? The coefficient of friction between the machine and the floor is 0.60.

SOLVE The force applied by the motor, through the cable, is the tension force $\vec{T}$. This force does work on the machine with power $P = Tv$. The machine is in dynamic equilibrium because the

motion is at constant velocity, hence the tension in the rope balances the friction and is

$$T = f_k = \mu_k mg$$

The motor's power output is

$$P = Tv = \mu_k mgv = 882 \text{ W}$$

EXAMPLE 11.15 **Power output of a car engine**

A 1500 kg car has a front profile that is 1.6 m wide and 1.4 m high and a drag coefficient of 0.50. The coefficient of rolling friction is 0.02. What power must the engine provide to drive at a steady 30 m/s ($\approx$65 mph) if 25% of the power is "lost" before reaching the drive wheels?

SOLVE The net force on a car moving at a steady speed is zero. The motion is opposed both by rolling friction and by air resistance. The forward force on the car $\vec{F}_{car}$ (recall that this is really $\vec{F}_{ground\ on\ car}$, a reaction to the drive wheels pushing backward on the ground with $\vec{F}_{car\ on\ ground}$) exactly balances the two opposing forces:

$$F_{car} = f_r + D$$

where $\vec{D}$ is the drag due to the air. Using the results of Chapter 6, where both rolling friction and drag were introduced, this becomes

$$F_{car} = \mu_r mg + \frac{1}{2}C\rho Av^2 = 294 \text{ N} + 605 \text{ N} = 899 \text{ N}$$

Here $A = (1.6 \text{ m}) \times (1.4 \text{ m})$ is the front cross-section area of the car, and we used 1.2 kg/m^3 as the density of air. The power required to push the car forward at 30 m/s is

$$P_{car} = F_{car}v = (899 \text{ N})(30 \text{ m/s}) = 27{,}000 \text{ W} = 36 \text{ hp}$$

This is the power *needed* at the drive wheels to push the car against the dissipative forces of friction and air resistance. The power output of the engine is larger because some energy is used to run the water pump, the power steering, and other accessories. In addition, energy is lost to friction in the drive train. If 25% of the power is lost (a typical value), leading to $P_{car} = 0.75P_{engine}$, the engine's power output is

$$P_{engine} = \frac{P_{car}}{0.75} = 31{,}900 \text{ W} = 48 \text{ hp}$$

ASSESS Automobile engines are typically rated at $\approx$200 hp. Most of that power is reserved for fast acceleration and climbing hills.

STOP TO THINK 11.7 Four students run up the stairs in the time shown. Rank in order, from largest to smallest, their power outputs P_a to P_d.

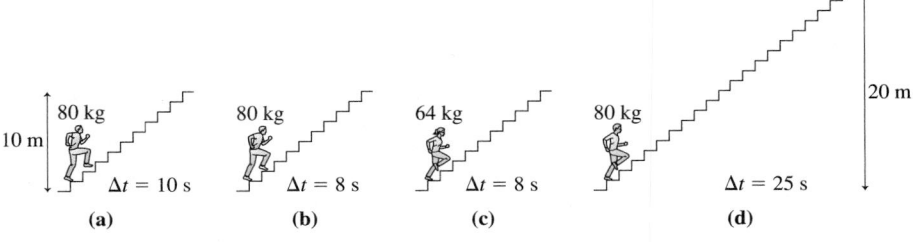

CHALLENGE EXAMPLE 11.16 **Stretching a spring**

A 5.0 kg box is attached to one end of a spring with spring constant 80 N/m. The other end of the spring is anchored to a wall. Initially the box is at rest at the spring's equilibrium position. A rope with a constant tension of 100 N then pulls the box away from the wall. The coefficient of friction between the box and the floor is 0.30. How much power is being supplied by the hand or motor pulling the rope when the box has moved 50 cm?

MODEL This is a complex situation, but one that we can analyze. First, identify the box, the spring, and the floor as the system. We need the floor inside the system because friction increases the temperature of the box *and* the floor. The tension in the rope is an external force. The work W_{ext} done by the rope's tension transfers energy into the system, causing K, U_s, and E_{th} all to increase.

VISUALIZE FIGURE 11.31a is a before-and-after pictorial representation. The energy transfers and transformations are shown in the energy bar chart of FIGURE 11.31b.

SOLVE The power supplied by the rope's tension—the rate at which energy is being delivered to the system—is $P = Tv$. We know the rope's tension, so we need to use energy considerations to find the speed v_1 after the box has moved to $x_1 = 50$ cm. The energy equation $K_f + U_f + \Delta E_{th} = K_i + U_i + W_{ext}$ is

$$\frac{1}{2}mv_1^2 + \frac{1}{2}kx_1^2 + \Delta E_{th} = \frac{1}{2}mv_0^2 + \frac{1}{2}kx_0^2 + W_{ext}$$

We know that $x_0 = 0$ m and $v_0 = 0$ m/s, so the energy equation simplifies to

$$\frac{1}{2}mv_1^2 = W_{ext} - \Delta E_{th} - \frac{1}{2}kx_1^2$$

The external work done by the rope's tension is

$$W_{ext} = \vec{T} \cdot \Delta \vec{r} = T(\Delta x)\cos 0° = (100 \text{ N})(0.50 \text{ m}) = 50.0 \text{ J}$$

The increase in thermal energy is given by Equation 11.27:

$$\Delta E_{th} = f_k \Delta x = \mu_k mg \Delta x$$
$$= (0.30)(5.0 \text{ kg})(9.80 \text{ m/s}^2)(0.50 \text{ m}) = 7.4 \text{ J}$$

Solving for the speed v_1 at $x_1 = 50$ cm = 0.50 m gives

$$v_1 = \sqrt{\frac{2(W_{ext} - \Delta E_{th} - \frac{1}{2}kx_1^2)}{m}} = 3.6 \text{ m/s}$$

The power being supplied at this instant is

$$P_1 = Tv_1 = (100 \text{ N})(3.6 \text{ m/s}) = 360 \text{ W}$$

ASSESS The work done by the rope's tension is energy transferred to the system. Part of the energy increases the speed of the box, part increases the potential energy stored in the spring, and part is transformed into increased thermal energy, increasing the temperature. We had to bring all the energy ideas together to solve this problem.

FIGURE 11.31 Pictorial representation and energy bar chart for Challenge Example 11.16.

(a)
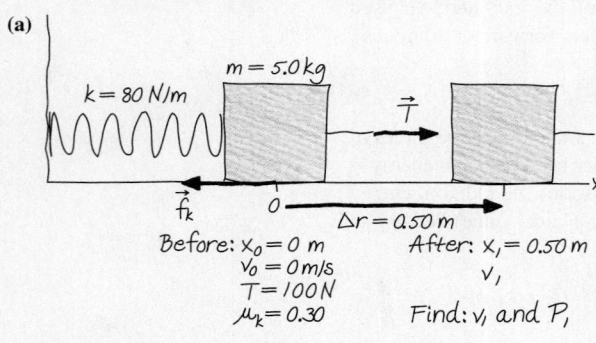

(b)
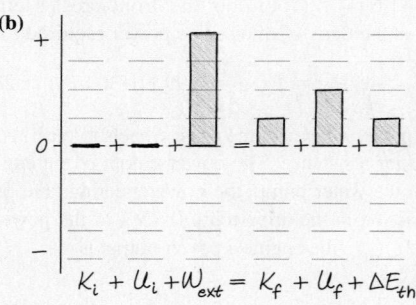

SUMMARY

The goal of Chapter 11 has been to develop a more complete understanding of energy and its conservation.

General Principles

Basic Energy Model

- Energy is *transferred* to or from the system by work.
- Energy is *transformed* within the system.

Two versions of the energy equation are

$$\Delta E_{sys} = \Delta K + \Delta U + \Delta E_{th} = W_{ext}$$

$$K_f + U_f + \Delta E_{th} = K_i + U_i + W_{ext}$$

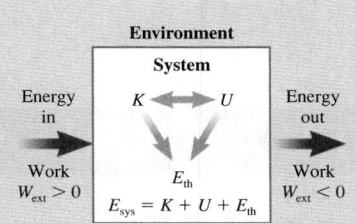

Solving Energy Problems

MODEL Identify objects in the system.

VISUALIZE Draw a before-and-after pictorial representation and an energy bar chart.

SOLVE Use the energy equation

$$K_f + U_f + \Delta E_{th} = K_i + U_i + W_{ext}$$

ASSESS Is the result reasonable?

Law of Conservation of Energy

- **Isolated system:** $W_{ext} = 0$. The total energy $E_{sys} = E_{mech} + E_{th}$ is conserved. $\Delta E_{sys} = 0$.
- **Isolated, nondissipative system:** $W_{ext} = 0$ and $W_{diss} = 0$. The mechanical energy E_{mech} is conserved.

$$\Delta E_{mech} = 0 \text{ or } K_f + U_f = K_i + U_i$$

Important Concepts

The work-kinetic energy theorem is

$$\Delta K = W_{net} = W_c + W_{diss} + W_{ext}$$

With $W_c = -\Delta U$ for conservative forces and $W_{diss} = -\Delta E_{th}$ for dissipative forces, this becomes the energy equation.

The work done by a force on a particle as it moves from s_i to s_f is

$$W = \int_{s_i}^{s_f} F_s \, ds = \text{area under the force curve}$$

$$= \vec{F} \cdot \Delta \vec{r} \text{ if } \vec{F} \text{ is a constant force}$$

Conservative forces are forces for which the work is independent of the path followed. The work done by a conservative force can be represented as a **potential energy:**

$$\Delta U = U_f - U_i = -W_c(i \rightarrow f)$$

A conservative force is found from the potential energy by

$$F_s = -dU/ds = \text{negative of the slope of the PE curve}$$

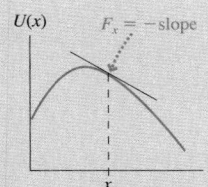

Dissipative forces transform **macroscopic energy** into thermal energy, which is the **microscopic energy** of the atoms and molecules. For friction:

$$\Delta E_{th} = f_k \Delta s$$

Applications

Power is the rate at which energy is transferred or transformed:

$$P = \frac{dE_{sys}}{dt}$$

For a particle moving with velocity $\vec{v}$, the power delivered to the particle by force $\vec{F}$ is $P = \vec{F} \cdot \vec{v} = Fv\cos\theta$.

Dot product

$$\vec{A} \cdot \vec{B} = AB \cos\alpha = A_x B_x + A_y B_y$$

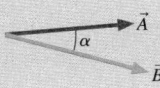

Energy bar charts display the energy equation $K_f + U_f + \Delta E_{th} = K_i + U_i + W_{ext}$ in graphical form.

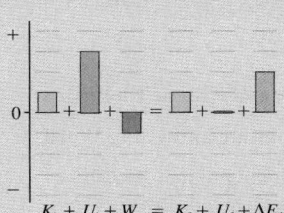

Terms and Notation

thermal energy, E_{th}	dot product	dissipative force
system energy, E_{sys}	scalar product	energy equation
energy transformation	conservative force	law of conservation of energy
energy transfer	nonconservative force	power, P
work, W	macrophysics	watt, W
work-kinetic energy theorem	microphysics	

CONCEPTUAL QUESTIONS

1. A process occurs in which a system's potential energy decreases while the system does work on the environment. Does the system's kinetic energy increase, decrease, or stay the same? Or is there not enough information to tell? Explain.

2. A process occurs in which a system's potential energy increases while the environment does work on the system. Does the system's kinetic energy increase, decrease, or stay the same? Or is there not enough information to tell? Explain.

3. The kinetic energy of a system decreases while its potential energy and thermal energy are unchanged. Does the environment do work on the system, or does the system do work on the environment? Explain.

4. You drop a ball from a high balcony and it falls freely. Does the ball's kinetic energy increase by equal amounts in equal time intervals, or by equal amounts in equal distances? Explain.

5. A particle moves in a vertical plane along the *closed* path seen in FIGURE Q11.5, starting at A and eventually returning to its starting point. How much work is done on the particle by gravity? Explain.

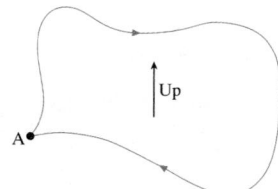

FIGURE Q11.5

6. A 0.2 kg plastic cart and a 20 kg lead cart both roll without friction on a horizontal surface. Equal forces are used to push both carts forward a distance of 1 m, starting from rest. After traveling 1 m, is the kinetic energy of the plastic cart greater than, less than, or equal to the kinetic energy of the lead cart? Explain.

7. You need to raise a heavy block by pulling it with a massless rope. You can either (a) pull the block straight up height h, or (b) pull it up a long, frictionless plane inclined at a 15° angle until its height has increased by h. Assume you will move the block at constant speed either way. Will you do more work in case a or case b? Or is the work the same in both cases? Explain.

8. a. If the force on a particle at some point in space is zero, must its potential energy also be zero at that point? Explain.
   b. If the potential energy of a particle at some point in space is zero, must the force on it also be zero at that point? Explain.

9. A car traveling at 60 mph slams on its brakes and skids to a halt. What happened to the kinetic energy the car had just before stopping?

10. What energy transformations occur as a skier glides down a gentle slope at constant speed?

11. Give a *specific* example of a situation in which
    a. $W_{ext} \rightarrow K$ with $\Delta U = 0$ and $\Delta E_{th} = 0$.
    b. $W_{ext} \rightarrow E_{th}$ with $\Delta K = 0$ and $\Delta U = 0$.

12. The motor of a crane uses power P to lift a steel beam. By what factor must the motor's power increase to lift the beam twice as high in half the time?

EXERCISES AND PROBLEMS

Problems labeled [ ] integrate material from earlier chapters.

Exercises

Section 11.2 Work and Kinetic Energy

Section 11.3 Calculating and Using Work

1. | Evaluate the dot product $\vec{A} \cdot \vec{B}$ if
   a. $\vec{A} = 3\hat{\imath} + 4\hat{\jmath}$ and $\vec{B} = 2\hat{\imath} - 6\hat{\jmath}$.
   b. $\vec{A} = 3\hat{\imath} - 2\hat{\jmath}$ and $\vec{B} = 6\hat{\imath} + 4\hat{\jmath}$.

2. | Evaluate the dot product $\vec{A} \cdot \vec{B}$ if
   a. $\vec{A} = 4\hat{\imath} - 2\hat{\jmath}$ and $\vec{B} = -2\hat{\imath} - 3\hat{\jmath}$.
   b. $\vec{A} = -4\hat{\imath} + 2\hat{\jmath}$ and $\vec{B} = 2\hat{\imath} + 4\hat{\jmath}$.

3. ‖ What is the angle θ between vectors $\vec{A}$ and $\vec{B}$ in each part of Exercise 1?

4. ‖ What is the angle θ between vectors $\vec{A}$ and $\vec{B}$ in each part of Exercise 2?

5. | Evaluate the dot product of the three pairs of vectors in FIGURE EX11.5.

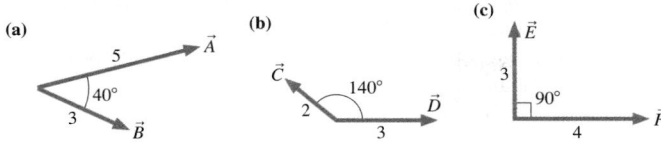

FIGURE EX11.5

6. | Evaluate the dot product of the three pairs of vectors in FIGURE EX11.6.

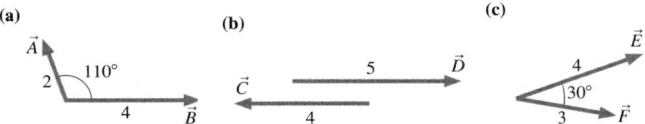

FIGURE EX11.6

7. | How much work is done by the force $\vec{F} = (-3.0\hat{\imath} + 6.0\hat{\jmath})$ N on a particle that moves through displacement (a) $\Delta\vec{r} = 2.0\hat{\imath}$ m and (b) $\Delta\vec{r} = 2.0\hat{\jmath}$ m?

8. | How much work is done by the force $\vec{F} = (-4.0\hat{\imath} - 6.0\hat{\jmath})$ N on a particle that moves through displacement (a) $\Delta\vec{r} = -3.0\hat{\imath}$ m and (b) $\Delta\vec{r} = (3.0\hat{\imath} - 2.0\hat{\jmath})$ m?

9. ‖ A 20 g particle is moving to the left at 30 m/s. How much net work must be done on the particle to cause it to move to the right at 30 m/s?

10. | A 2.0 kg book is lying on a 0.75-m-high table. You pick it up and place it on a bookshelf 2.25 m above the floor.
   a. How much work does gravity do on the book?
   b. How much work does your hand do on the book?

11. ‖ The two ropes seen in FIGURE EX11.11 are used to lower a 255 kg piano 5.00 m from a second-story window to the ground. How much work is done by each of the three forces?

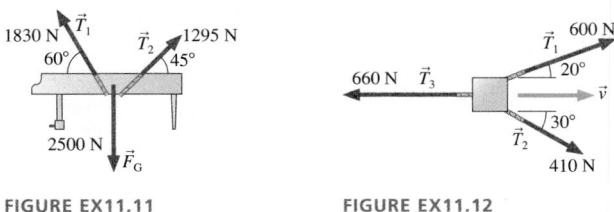

FIGURE EX11.11   **FIGURE EX11.12**

12. ‖ The three ropes shown in the bird's-eye view of FIGURE EX11.12 are used to drag a crate 3.0 m across the floor. How much work is done by each of the three forces?

13. ‖ FIGURE EX11.13 is the velocity-versus-time graph for a 2.0 kg object moving along the x-axis. Determine the work done on the object during each of the four intervals AB, BC, CD, and DE.

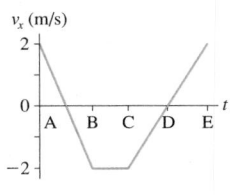

FIGURE EX11.13

Section 11.4 The Work Done by a Variable Force

14. | FIGURE EX11.14 is the force-versus-position graph for a particle moving along the x-axis. Determine the work done on the particle during each of the three intervals 0–1 m, 1–2 m, and 2–3 m.

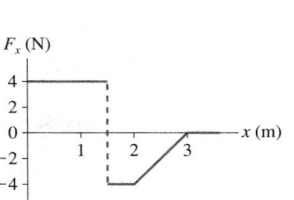

FIGURE EX11.14

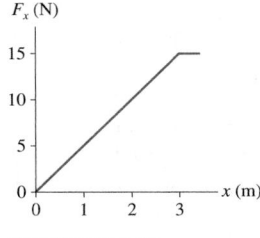

FIGURE EX11.15

15. ‖ A 500 g particle moving along the x-axis experiences the force shown in FIGURE EX11.15. The particle's velocity is 2.0 m/s at $x = 0$ m. What is its velocity at $x = 1$ m, 2 m, and 3 m?

16. ‖ A 2.0 kg particle moving along the x-axis experiences the force shown in FIGURE EX11.16. The particle's velocity is 4.0 m/s at $x = 0$ m. What is its velocity at $x = 2$ m and 4 m?

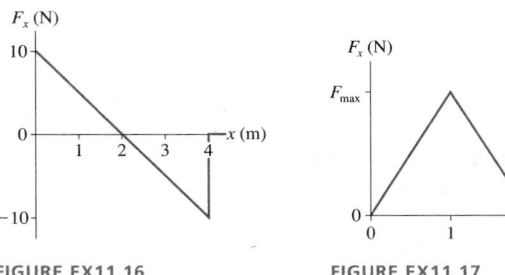

FIGURE EX11.16   **FIGURE EX11.17**

17. ‖ A 500 g particle moving along the x-axis experiences the force shown in FIGURE EX11.17. The particle goes from $v_x = 2.0$ m/s at $x = 0$ m to $v_x = 6.0$ m/s at $x = 2$ m. What is F_{max}?

Section 11.5 Work and Potential Energy

Section 11.6 Finding Force from Potential Energy

18. ‖ A particle has the potential energy shown in FIGURE EX11.18. What is the x-component of the force on the particle at $x = 5$, 15, 25, and 35 cm?

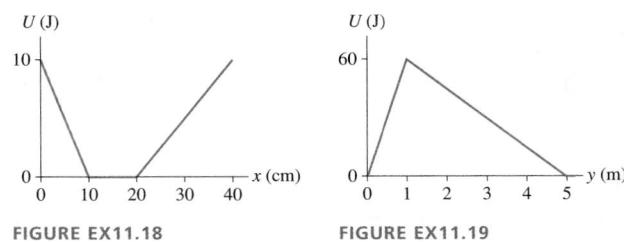

FIGURE EX11.18   **FIGURE EX11.19**

19. ‖ A particle has the potential energy shown in FIGURE EX11.19. What is the y-component of the force on the particle at $y = 0.5$ m and 4 m?

20. ‖ A particle moving along the y-axis has the potential energy $U = 4y^3$ J, where y is in m. What is the y-component of the force on the particle at $y = 0$ m, 1 m, and 2 m?

21. ‖ A particle moving along the x-axis has the potential energy $U = 10/x$ J, where x is in m. What is the x-component of the force on the particle at $x = 2$ m, 5 m, and 8 m?

Section 11.7 Thermal Energy

22. ‖ The mass of a carbon atom is 2.0×10^{-26} kg.
   a. What is the kinetic energy of a carbon atom moving with a speed of 500 m/s?

b. Two carbon atoms are joined by a spring-like carbon-carbon bond. The potential energy stored in the bond has the value you calculated in part a if the bond is stretched 0.050 nm. What is the bond's spring constant?

23. ‖ In Part IV you'll learn to calculate that 1 mole (6.02×10^{23} atoms) of helium atoms in the gas phase has 3700 J of microscopic kinetic energy at room temperature. If we assume that all atoms move with the same speed, what is that speed? The mass of a helium atom is 6.68×10^{-27} kg.

24. ‖ A 20 kg child slides down a 3.0-m-high playground slide. She starts from rest, and her speed at the bottom is 2.0 m/s.
   a. Describe the energy transfers and transformations occurring during the slide.
   b. What is the change in the combined thermal energy of the slide and the seat of her pants?

Section 11.8 Conservation of Energy

25. ‖ A system loses 400 J of potential energy. In the process, it does 400 J of work on the environment and the thermal energy increases by 100 J. Show this process on an energy bar chart.

26. ‖ A system loses 500 J of kinetic energy while gaining 200 J of potential energy. The thermal energy increases 100 J. Show this process on an energy bar chart.

27. | How much work is done by the environment in the process shown in FIGURE EX11.27? Is energy transferred from the environment to the system or from the system to the environment?

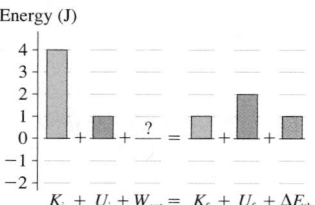

FIGURE EX11.27

28. ‖ A cable with 20.0 N of tension pulls straight up on a 1.02 kg block that is initially at rest. What is the block's speed after being lifted 2.00 m? Solve this problem using work and energy.

Section 11.9 Power

29. | a. How much work does an elevator motor do to lift a 1000 kg elevator a height of 100 m?
   b. How much power must the motor supply to do this in 50 s at constant speed?

30. | a. How much work must you do to push a 10 kg block of steel across a steel table at a steady speed of 1.0 m/s for 3.0 s?
   b. What is your power output while doing so?

31. ‖ At midday, solar energy strikes the earth with an intensity of about 1 kW/m². What is the area of a solar collector that could collect 150 MJ of energy in 1 h? This is roughly the energy content of 1 gallon of gasoline.

32. ‖ Which consumes more energy, a 1.2 kW hair dryer used for 10 min or a 10 W night light left on for 24 h?

33. | A 2.0 hp electric motor on a water well pumps water from 10 m below the surface. The density of water is 1.0 kg per liter. How many liters of water does the motor pump in 1 h?

34. ‖ A 50 kg sprinter, starting from rest, runs 50 m in 7.0 s at constant acceleration.
   a. What is the magnitude of the horizontal force acting on the sprinter?
   b. What is the sprinter's power output at 2.0 s, 4.0 s, and 6.0 s?

35. | a. Estimate the height in meters of the two flights of stairs that go from the first to the third floor of a building.
   b. Estimate how long it takes you to *run* up these two flights of stairs.
   c. Estimate your power output in both watts and horsepower while running up the stairs.

36. | A 70 kg human sprinter can accelerate from rest to 10 m/s in 3.0 s. During the same time interval, a 30 kg greyhound can go from rest to 20 m/s. What is the average power output of each? *Average* power over a time interval Δt is $\Delta E / \Delta t$.

Problems

37. | A particle moves from A to D in FIGURE P11.37 while experiencing force $\vec{F} = (6\hat{\imath} + 8\hat{\jmath})$ N. How much work does the force do if the particle follows path (a) ABD, (b) ACD, and (c) AD? Is this a conservative force? Explain.

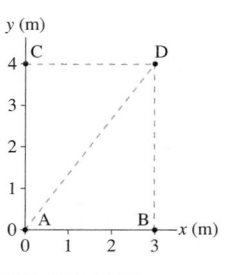

FIGURE P11.37

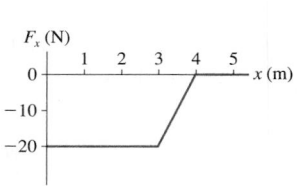

FIGURE P11.38

38. ‖ A 100 g particle experiences the one-dimensional, conservative force F_x shown in FIGURE P11.38.
   a. Draw a graph of the potential energy U from $x = 0$ m to $x = 5$ m. Let the zero of the potential energy be at $x = 0$ m. **Hint:** Think about the definition of potential energy *and* the geometric interpretation of the work done by a varying force.
   b. The particle is shot toward the right from $x = 1.0$ m with a speed of 25 m/s. What is the particle's mechanical energy?
   c. Draw the particle's total energy line on your graph of part a.
   d. Where is the particle's turning point?

39. ‖ A 10 g particle has the potential energy shown in FIGURE P11.39.
   a. Draw a force-versus-position graph from $x = 0$ cm to $x = 8$ cm.
   b. How much work does the force do as the particle moves from $x = 2$ cm to $x = 6$ cm?
   c. What speed does the particle need at $x = 2$ cm to arrive at $x = 6$ cm with a speed of 10 m/s?

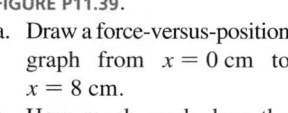

FIGURE P11.39

40. ‖‖ a. FIGURE P11.40a shows the force F_x exerted on a particle that moves along the x-axis. Draw a graph of the particle's potential energy as a function of position x. Let U be zero at $x = 0$ m.
   b. FIGURE P11.40b shows the potential energy U of a particle that moves along the x-axis. Draw a graph of the force F_x as a function of position x.

(a)

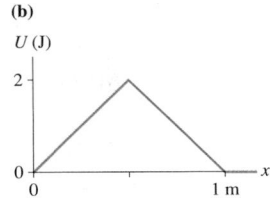

F_x (N)

(b)

U (J)

FIGURE P11.40

41. ‖ FIGURE P11.41 is the velocity-versus-time graph of a 500 g particle that starts at $x = 0$ m and moves along the x-axis. Draw graphs of the following by calculating and plotting numerical values at $t = 0$, 1, 2, 3, and 4 s. Then sketch lines or curves of the appropriate shape between the points. Make sure you include appropriate scales on both axes of each graph.

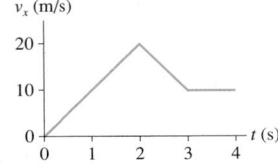

v_x (m/s)

FIGURE P11.41

   a. Acceleration versus time.
   b. Position versus time.
   c. Kinetic energy versus time.
   d. Force versus time.
   e. Use your F_x-versus-t graph to determine the *impulse* delivered to the particle during the time interval 0–2 s and also the interval 2–4 s.
   f. Use the impulse-momentum theorem to determine the particle's velocity at $t = 2$ s and at $t = 4$ s. Do your results agree with the velocity graph?
   g. Now draw a graph of force versus *position*. This requires no calculations; just think carefully about what you learned in parts a to d.
   h. Use your F_x-versus-x graph to determine the *work* done on the particle during the time interval 0–2 s and also the interval 2–4 s.
   i. Use the work-kinetic energy theorem to determine the particle's velocity at $t = 2$ s and at $t = 4$ s. Do your results agree with the velocity graph?

42. ‖ A 1000 kg elevator accelerates upward at 1.0 m/s² for 10 m, starting from rest.
   a. How much work does gravity do on the elevator?
   b. How much work does the tension in the elevator cable do on the elevator?
   c. Use the work-kinetic energy theorem to find the kinetic energy of the elevator as it reaches 10 m.
   d. What is the speed of the elevator as it reaches 10 m?

43. | Bob can throw a 500 g rock with a speed of 30 m/s. He moves his hand forward 1.0 m while doing so.
   a. How much work does Bob do on the rock?
   b. How much force, assumed to be constant, does Bob apply to the rock?
   c. What is Bob's maximum power output as he throws the rock?

44. ‖ a. Starting from rest, a crate of mass m is pushed up a frictionless slope of angle θ by a *horizontal* force of magnitude F. Use work and energy to find an expression for the crate's speed v when it is at height h above the bottom of the slope.
   b. Doug uses a 25 N horizontal force to push a 5.0 kg crate up a 2.0-m-high, 20° frictionless slope. What is the speed of the crate at the top of the slope?

45. ‖ Sam, whose mass is 75 kg, straps on his skis and starts down a 50-m-high, 20° frictionless slope. A strong headwind exerts a *horizontal* force of 200 N on him as he skies. Use work and energy to find Sam's speed at the bottom.

46. ‖ Susan's 10 kg baby brother Paul sits on a mat. Susan pulls the mat across the floor using a rope that is angled 30° above the floor. The tension is a constant 30 N and the coefficient of friction is 0.20. Use work and energy to find Paul's speed after being pulled 3.0 m.

47. ‖ A horizontal spring with spring constant 100 N/m is compressed 20 cm and used to launch a 2.5 kg box across a frictionless, horizontal surface. After the box travels some distance, the surface becomes rough. The coefficient of kinetic friction of the box on the surface is 0.15. Use work and energy to find how far the box slides across the rough surface before stopping.

48. ‖ a. A box of mass m and initial speed v_0 slides distance d across a horizontal floor before coming to rest. Use work and energy to find an expression for the coefficient of kinetic friction.
   b. A baggage handler throws a 15 kg suitcase along the floor of an airplane luggage compartment with a speed of 1.2 m/s. The suitcase slides 2.0 m before stopping. What is the suitcase's coefficient of kinetic friction on the floor?

49. ‖ Truck brakes can fail if they get too hot. In some mountainous areas, ramps of loose gravel are constructed to stop runaway trucks that have lost their brakes. The combination of a slight upward slope and a large coefficient of rolling resistance as the truck tires sink into the gravel brings the truck safely to a halt. Suppose a gravel ramp slopes upward at 6.0° and the coefficient of rolling friction is 0.40. Use work and energy to find the length of a ramp that will stop a 15,000 kg truck that enters the ramp at 35 m/s ($\approx$75 mph).

50. ‖ A freight company uses a compressed spring to shoot 2.0 kg packages up a 1.0-m-high frictionless ramp into a truck, as FIGURE P11.50 shows. The spring constant is 500 N/m and the spring is compressed 30 cm.
   a. What is the speed of the package when it reaches the truck?
   b. A careless worker spills his soda on the ramp. This creates a 50-cm-long sticky spot with a coefficient of kinetic friction 0.30. Will the next package make it into the truck?

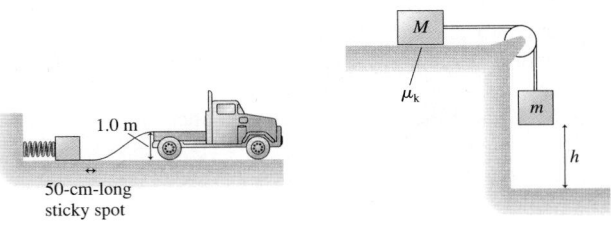

1.0 m

50-cm-long
sticky spot

M

μ_k

m

h

FIGURE P11.50

FIGURE P11.51

51. ‖‖ Use work and energy to find an expression for the speed of the block in FIGURE P11.51 just before it hits the floor if (a) the coefficient of kinetic friction for the block on the table is μ_k and (b) the table is frictionless.

52. ‖‖ An 8.0 kg crate is pulled 5.0 m up a 30° incline by a rope angled 18° above the incline. The tension in the rope is 120 N, and the crate's coefficient of kinetic friction on the incline is 0.25.
   a. How much work is done by tension, by gravity, and by the normal force?
   b. What is the increase in thermal energy of the crate and incline?

53. ‖ You've taken a summer job at a water park. In one stunt, a water skier is going to glide up the 2.0-m-high frictionless ramp shown in **FIGURE P11.53**, then sail over a 5.0-m-wide tank filled with hungry sharks. You

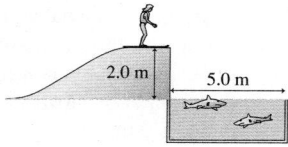

FIGURE P11.53

will be driving the boat that pulls her to the ramp. She'll drop the tow rope at the base of the ramp just as you veer away. What minimum speed must you have as you reach the ramp in order for her to live to do this again tomorrow?

54. ‖ A 50 kg ice skater is gliding along the ice, heading due north at 4.0 m/s. The ice has a small coefficient of static friction, to prevent the skater from slipping sideways, but $\mu_k = 0$. Suddenly, a wind from the northeast exerts a force of 4.0 N on the skater.
   a. Use work and energy to find the skater's speed after gliding 100 m in this wind.
   b. What is the minimum value of μ_s that allows her to continue moving straight north?

55. ‖ a. A 50 g ice cube can slide without friction up and down a 30° slope. The ice cube is pressed against a spring at the bottom of the slope, compressing the spring 10 cm. The spring constant is 25 N/m. When the ice cube is released, what total distance will it travel up the slope before reversing direction?
   b. The ice cube is replaced by a 50 g plastic cube whose coefficient of kinetic friction is 0.20. How far will the plastic cube travel up the slope? Use work and energy.

56. ‖ A 5.0 kg box slides down a 5.0-m-high frictionless hill, starting from rest, across a 2.0-m-wide horizontal surface, then hits a horizontal spring with spring constant 500 N/m. The other end of the spring is anchored against a wall. The ground under the spring is frictionless, but the 2.0-m-wide horizontal surface is rough. The coefficient of kinetic friction of the box on this surface is 0.25.
   a. What is the speed of the box just before reaching the rough surface?
   b. What is the speed of the box just before hitting the spring?
   c. How far is the spring compressed?
   d. Including the first crossing, how many *complete* trips will the box make across the rough surface before coming to rest?

57. ‖ The spring shown in **FIGURE P11.57** is compressed 50 cm and used to launch a 100 kg physics student. The track is frictionless until it starts up the incline. The student's coefficient of kinetic friction on the 30° incline is 0.15.
   a. What is the student's speed just after losing contact with the spring?
   b. How far up the incline does the student go?

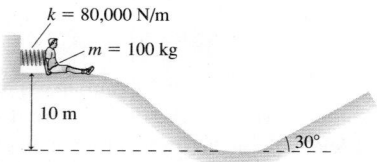

$k = 80,000$ N/m

$m = 100$ kg

10 m

30°

FIGURE P11.57

58. ‖ A block of mass m starts from rest at height h. It slides down a frictionless incline, across a rough horizontal surface of length L, then up a frictionless incline. The coefficient of kinetic friction on the rough surface is μ_k.
   a. What is the block's speed at the bottom of the first incline?
   b. How high does the block go on the second incline?
   Give your answers in terms of m, h, L, μ_k, and g.

59. ‖ Show that Hooke's law for an ideal spring is a conservative force. To do so, first calculate the work done by the spring as it expands from A to B. Then calculate the work done by the spring as it expands from A to point C, which is beyond B, then returns from C to B.

60. ‖ A clever engineer designs a "sprong" that obeys the force law $F_x = -q(x - x_e)^3$, where x_e is the equilibrium position of the end of the sprong and q is the sprong constant. For simplicity, we'll let $x_e = 0$ m. Then $F_x = -qx^3$.
   a. What are the units of q?
   b. Find an expression for the potential energy of a stretched or compressed sprong.
   c. A sprong-loaded toy gun shoots a 20 g plastic ball. What is the launch speed if the sprong constant is 40,000, with the units you found in part a, and the sprong is compressed 10 cm? Assume the barrel is frictionless.

61. ‖ A particle of mass m starts from $x_0 = 0$ m with $v_0 > 0$ m/s. The particle experiences the variable force $F_x = F_0 \sin(cx)$ as it moves to the right along the x-axis, where F_0 and c are constants.
   a. What are the units of F_0?
   b. What are the units of c?
   c. At what position x_{max} does the force first reach a maximum value? Your answer will be in terms of the constants F_0 and c and perhaps other numerical constants.
   d. Sketch a graph of F versus x from x_0 to x_{max}.
   e. What is the particle's velocity as it reaches x_{max}? Give your answer in terms of m, v_0, F_0, and c.

62. ‖ A 5.0 kg cat leaps from the floor to the top of a 95-cm-high table. If the cat pushes against the floor for 0.20 s to accomplish this feat, what was her average power output during the pushoff period?

63. ‖ The human heart pumps the average adult's 6.0 L (6000 cm³) of blood through the body every minute. The heart must do work to overcome frictional forces that resist blood flow. The average adult blood pressure is 1.3×10^4 N/m².
   BIO
   a. How much work does the heart do to move the 6.0 L of blood completely through the body?
   b. What power output must the heart have to do this task once a minute?
   Hint: When the heart contracts, it applies force to the blood. Pressure is force/area. Model the circulatory system as a single closed tube, with cross-section area A and volume $V = 6.0$ L, filled with blood to which the heart applies a force.

64. ‖ When you ride a bicycle at constant speed, nearly all the energy you expend goes into the work you do against the drag force of the air. Model a cyclist as having cross-section area 0.45 m² and, because the human body is not aerodynamically shaped, a drag coefficient of 0.90.
   BIO
   a. What is the cyclist's power output while riding at a steady 7.3 m/s (16 mph)?
   b. *Metabolic power* is the rate at which your body "burns" fuel to power your activities. For many activities, your body is roughly 25% efficient at converting the chemical energy of food into mechanical energy. What is the cyclist's metabolic power while cycling at 7.3 m/s?
   c. The food calorie is equivalent to 4190 J. How many calories does the cyclist burn if he rides over level ground at 7.3 m/s for 1 h?

65. ‖ In a hydroelectric dam, water falls 25 m and then spins a turbine to generate electricity.

a. What is ΔU of 1.0 kg of water?

b. Suppose the dam is 80% efficient at converting the water's potential energy to electrical energy. How many kilograms of water must pass through the turbines each second to generate 50 MW of electricity? This is a typical value for a small hydroelectric dam.

66. ‖ The force required to tow a water skier at speed v is proportional to the speed. That is, $F_{tow} = Av$, where A is a proportionality constant. If a speed of 2.5 mph requires 2 hp, how much power is required to tow a water skier at 7.5 mph?

67. ‖ A Porsche 944 Turbo has a rated engine power of 217 hp. 30% of the power is lost in the drive train, and 70% reaches the wheels. The total mass of the car and driver is 1480 kg, and two-thirds of the weight is over the drive wheels.

a. What is the maximum acceleration of the Porsche on a concrete surface where $\mu_s = 1.00$?
Hint: What force pushes the car forward?

b. If the Porsche accelerates at a_{max}, what is its speed when it reaches maximum power output?

c. How long does it take the Porsche to reach the maximum power output?

In Problems 68 through 71 you are given the equation(s) used to solve a problem. For each of these, you are to
a. Write a realistic problem for which this is the correct equation(s).
b. Draw a pictorial representation.
c. Finish the solution of the problem.

68. $\frac{1}{2}(2.0 \text{ kg})(4.0 \text{ m/s})^2 + 0$
$+ (0.15)(2.0 \text{ kg})(9.8 \text{ m/s}^2)(2.0 \text{ m}) = 0 + 0 + T(2.0 \text{ m})$

69. $\frac{1}{2}(20 \text{ kg})v_1^2 + 0$
$+ (0.15)(20 \text{ kg})(9.8 \text{ m/s}^2)\cos 40°((2.5 \text{ m})/\sin 40°)$
$= 0 + (20 \text{ kg})(9.8 \text{ m/s}^2)(2.5 \text{ m}) + 0$

70. $F_{push} - (0.20)(30 \text{ kg})(9.8 \text{ m/s}^2) = 0$
$75 \text{ W} = F_{push} v$

71. $T - (1500 \text{ kg})(9.8 \text{ m/s}^2) = (1500 \text{ kg})(1.0 \text{ m/s}^2)$
$P = T(2.0 \text{ m/s})$

Challenge Problems

72. A 10.2 kg weather rocket generates a thrust of 200 N. The rocket, pointing upward, is clamped to the top of a vertical spring. The bottom of the spring, whose spring constant is 500 N/m, is anchored to the ground.

a. Initially, before the engine is ignited, the rocket sits at rest on top of the spring. How much is the spring compressed?

b. After the engine is ignited, what is the rocket's speed when the spring has stretched 40 cm? For comparison, what would be the rocket's speed after traveling this distance if it weren't attached to the spring?

73. The spring in FIGURE CP11.73 has a spring constant of 1000 N/m. It is compressed 15 cm, then launches a 200 g block. The horizontal surface is frictionless, but the block's coefficient of kinetic friction on the incline is 0.20. What distance d does the block sail through the air?

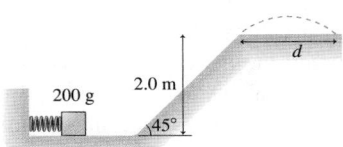

FIGURE CP11.73

74. The equation mgy for gravitational potential energy is valid only for objects near the surface of a planet. Consider two very large objects of mass m_1 and m_2, such as stars or planets, whose centers are separated by the large distance r. These two large objects exert gravitational forces on each other. You'll learn in Chapter 13 that the gravitational potential energy is

$$U = -\frac{Gm_1 m_2}{r}$$

where $G = 6.67 \times 10^{-11} \text{ N m}^2/\text{kg}^2$ is the *gravitational constant*.

a. Sketch a graph of U versus r. The mathematical difficulty at $r = 0$ is not a physically significant difficulty because the masses will collide before they get that close together.

b. What separation r has been chosen as the point of zero potential energy? Does this make sense? Explain.

c. Two stars are at rest 1.0×10^{14} m apart. This is about 10 times the diameter of the solar system. The first star is the size of our sun, with a mass of 2.0×10^{30} kg and a radius of 7.0×10^8 m. The second star has mass 8.0×10^{30} kg and radius of 11.0×10^8 m. Gravitational forces pull the two stars together. What is the speed of each star at the moment of impact?

75. A gardener pushes a 12 kg lawnmower whose handle is tilted up 37° above horizontal. The lawnmower's coefficient of rolling friction is 0.15. How much power does the gardener have to supply to push the lawnmower at a constant speed of 1.2 m/s? Assume his push is parallel to the handle.

STOP TO THINK ANSWERS

Stop to Think 11.1: d. Constant speed means $\Delta K = 0$. Gravitational potential energy is lost, and friction heats up the slide and the child's pants.

Stop to Think 11.2: 6.0 J. $K_f = K_i + W$. W is the area under the curve, which is 4.0 J.

Stop to Think 11.3: b. The gravitational force $\vec{F}_G$ is in the same direction as the displacement. It does positive work. The tension force $\vec{T}$ is opposite the displacement. It does negative work.

Stop to Think 11.4: c. $W = F(\Delta r)\cos\theta$. The 10 N force at 90° does no work at all. $\cos 60° = \frac{1}{2}$, so the 8 N force does less work than the 6 N force.

Stop to Think 11.5: e. Force is the negative of the slope of the potential-energy diagram. At $x = 4$ m the potential energy has risen by 4 J over a distance of 2 m, so the slope is 2 J/m = 2 N.

Stop to Think 11.6: c. Constant speed means $\Delta K = 0$. Gravitational potential energy is lost, and friction heats up the pole and the child's hands.

Stop to Think 11.7: $P_b > P_a = P_c > P_d$. The work done is $mg\Delta y$, so the power is $mg\Delta y/\Delta t$. Runner b does the same work as a but in less time. The ratio $m/\Delta t$ is the same for a and c. Runner d does twice the work of a but takes more than twice as long.

Conservation Laws

In Part II we have discovered that we don't need to know all the details of an interaction to relate the properties of a system "before" an interaction to the system's properties "after" the interaction. Along the way, we found two important quantities, momentum and energy, that characterize a system of particles.

Momentum and energy have specific conditions under which they are conserved. In particular, the total momentum $\vec{P}$ and the total energy E_{sys} are conserved for an *isolated system,* one on which the net external force is zero. Further, the system's mechanical energy is conserved if the system is both isolated and nondissipative (i.e., no friction forces). These ideas are captured in the two most important conservation laws, the law of conservation of momentum and the law of conservation of energy.

Of course, not all systems are isolated. For both momentum and energy, it was useful to develop a *model* of a system interacting with its environment. Interactions between the sys-

tem and the environment change the system's momentum and energy. In particular,

- Impulse is the transfer of momentum to or from the system: $\Delta \vec{P} = \vec{J}$.

- Work is the transfer of energy to or from the system: $\Delta E_{sys} = W_{ext}$.

Interactions within the system do not change $\vec{P}$ or E_{sys}. The kinetic, potential, and thermal energy within the system can be transformed without changing E_{sys}. The basic energy model is built around the twin ideas of the transfer and the transformation of energy.

The table below is a knowledge structure of conservation laws. You should compare this with the knowledge structure of Newtonian mechanics in the Part I Summary. Add the problem-solving strategies, and you now have a very powerful set of tools for understanding motion.

KNOWLEDGE STRUCTURE II Conservation Laws

ESSENTIAL CONCEPTS **BASIC GOALS**	Impulse, momentum, work, energy How is the system "after" an interaction related to the system "before"? What quantities are conserved, and under what conditions?
GENERAL PRINCIPLES	**Impulse-momentum theorem** $\quad \Delta p_s = J_s$ **Work-kinetic energy theorem** $\quad \Delta K = W_{net} = W_c + W_{diss} + W_{ext}$ **Energy equation** $\qquad\qquad\qquad \Delta E_{sys} = \Delta K + \Delta U + \Delta E_{th} = W_{ext}$
CONSERVATION LAWS	For an isolated system, with $\vec{F}_{net} = \vec{0}$ and $W_{net} = 0$ • The total momentum $\vec{P}$ is conserved. • The total energy $E_{sys} = E_{mech} + E_{th}$ is conserved. For an isolated and nondissipative system, with $W_{diss} = 0$ • The mechanical energy $E_{mech} = K + U$ is conserved.

BASIC PROBLEM-SOLVING STRATEGY Draw a before-and-after pictorial representation, then use the momentum or energy equations to relate "before" to "after." Where possible, choose a system for which momentum and/or energy are conserved. If necessary, calculate impulse and/or work.

Basic model of momentum and energy

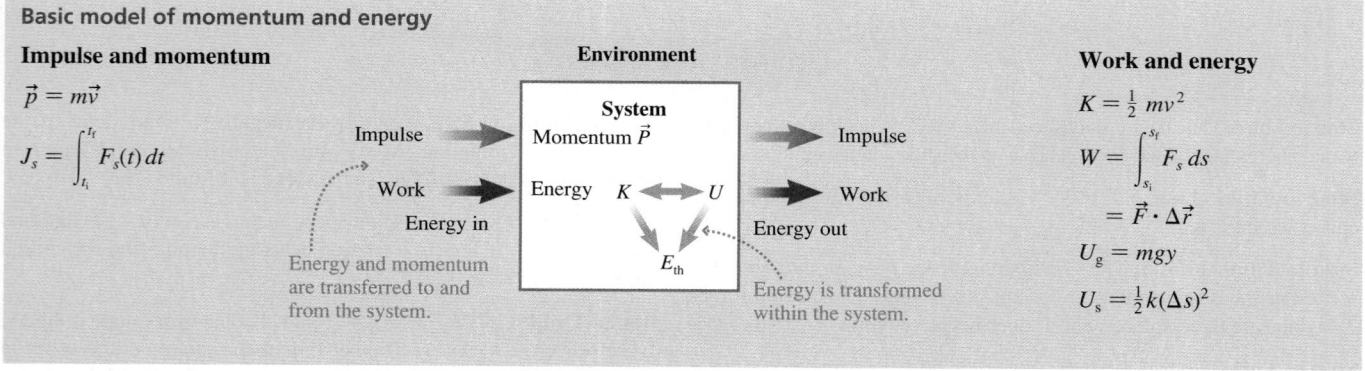

Impulse and momentum

$$\vec{p} = m\vec{v}$$

$$J_s = \int_{t_i}^{t_f} F_s(t)\, dt$$

Impulse → | Work → | Energy in

Energy and momentum are transferred to and from the system.

Environment

System
Momentum $\vec{P}$
Energy $K \leftrightarrow U$
E_{th}

→ Impulse | → Work | Energy out

Energy is transformed within the system.

Work and energy

$$K = \tfrac{1}{2} mv^2$$

$$W = \int_{s_i}^{s_f} F_s\, ds$$

$$= \vec{F} \cdot \Delta \vec{r}$$

$$U_g = mgy$$

$$U_s = \tfrac{1}{2} k(\Delta s)^2$$

ONE STEP BEYOND

Energy Conservation

You hear it all the time. Turn off lights. Buy a more fuel-efficient car. Conserve energy. But why conserve energy if energy is already conserved? Consider the earth as a whole. No work is done on the earth. And while heat energy flows from the sun to the earth, the earth radiates an equal amount of heat back into space. With no work and no net heat flow, the earth's total energy E_{earth} is conserved.

Pumping oil, driving your car, running a nuclear reactor, and turning on the lights are all interactions *within* the earth system. They transform energy from one type to another, but they don't affect the value of E_{earth}. Consider two examples.

- Crude oil, stored in the earth, has chemical energy E_{chem}. Chemical energy, a form of microscopic potential energy, is released when chemical reactions rearrange the bonds. As you burn gasoline in your car engine, the chemical energy is transformed into the kinetic energy of the moving pistons. This kinetic energy, in turn, is transformed into the car's kinetic energy. The car's kinetic energy is ultimately dissipated as thermal energy in the brakes, air, tires, and road because of friction and drag. Overall, the energy process of driving looks like

$$E_{chem} \rightarrow K_{piston} \rightarrow K_{car} \rightarrow E_{th}$$

- Water stored behind a dam has gravitational potential energy U_g. Potential energy is transformed into kinetic energy as the water falls, then into the spinning turbine's kinetic energy. The turbine converts mechanical energy into electric energy E_{elec}. The electric energy reaches a lightbulb where it is transformed partly into thermal energy (lightbulbs are hot!) and partly into light energy. The light is absorbed by surfaces, heating them slightly and thus transforming the light energy into thermal energy. The overall energy process is

$$U_g \rightarrow K_{water} \rightarrow K_{turbine} \rightarrow E_{elec} \rightarrow E_{light} \rightarrow E_{th}$$

Do you notice a trend? Stored energy (fossil fuel, water behind a dam) is transformed through a series of steps, some of which are considered "useful," until the energy is ultimately dissipated as thermal energy. **The total energy has not changed, but its "usefulness" has.**

Thermal energy is rarely "useful" energy. A room full of moving air molecules has a huge thermal energy, but you can't run your lights or your air conditioner with it. You can't turn the thermal energy of your hot brakes back into the kinetic energy of the car. Energy may be conserved, but there's a one-way characteristic of the transformations.

The energy stored in fuels and the energy of the sun are "high-quality energy" because of their potential to be transformed into such useful forms of energy as moving your car and heating your house. But as **FIGURE PSII.1** shows, high-quality energy becomes "degraded" into thermal energy, where it is no longer useful. Thus the phrase "conserve energy" isn't used literally. Instead, it means to conserve or preserve the earth's sources of high-quality energy.

Conserving high-quality energy is important because fossil fuels are a finite resource. Experts may disagree as to how long fossil fuels will last, but all agree that it won't be forever. Oil and natural gas will likely become scarce during your lifetime. In addition, burning fossil fuel generates carbon dioxide, a major contributor to global warming. Energy conservation helps fuels last longer and minimizes their side effects.

There are two paths to conserving energy. One is to use less high-quality energy. Turning off lights and bicycling rather than driving are actions that preserve high-quality energy. A second path is to use energy more efficiently. That is, get more of the useful activity (miles driven, rooms lit) for the same amount of high-quality energy.

Lightbulbs offer a good example. A 100 W incandescent lightbulb actually produces only about 10 W of light energy. Ninety watts of the high-quality electric energy is immediately degraded as thermal energy without doing anything useful. By contrast, a 25 W compact fluorescent bulb generates the same 10 W of light but only 15 W of thermal energy. The same amount of high-quality energy can light four times as many rooms if 100 W incandescent bulbs are replaced by 25 W compact fluorescent bulbs.

So why conserve energy if energy is already conserved? Because technological society needs a dependable and sustainable supply of high-quality energy. Both technology improvement and lifestyle choices will help us achieve a sustainable energy future.

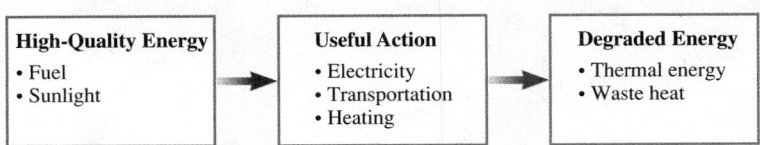

FIGURE PSII.1 "Using" energy transforms high-quality energy into thermal energy.

PART

III

Applications of Newtonian Mechanics

Hurricane Ivan approaches the United States in 2004. A hurricane is a fluid—the air—moving on a rotating sphere—the earth—under the influence of gravity. Understanding hurricanes is very much an application of Newtonian mechanics.

OVERVIEW

Power Over Our Environment

Early humans had to endure whatever nature provided. Only within the last few thousand years have agriculture and technology provided some level of control over the environment. And it has been a mere couple of centuries since machines, and later electronics, began to do much of our work and provide us with "creature comforts."

It's no coincidence that machines began to appear about a century after Galileo, Newton, and others ignited what we now call the *scientific revolution*. The machines and other devices we take for granted today are direct consequences of scientific knowledge and the scientific method.

Parts I and II have established Newton's theory of motion, the foundation of modern science. Most of the applications will be developed in other science and engineering courses, but we're now in a good position to examine a few of the more practical aspects of Newtonian mechanics.

Our goal for Part III is to apply our newfound theory to four important topics:

- **Rotation.** Rotation is a very important form of motion, but to understand rotational motion we'll need to introduce a new model—the *rigid-body model*. We'll then be able to study rolling wheels and spinning space stations. Rotation will also lead to the law of conservation of angular momentum.
- **Gravity.** By adding one more law, Newton's law of gravity, we'll be able to understand much about the physics of the space shuttle, communication satellites, the solar system, and interplanetary travel.
- **Oscillations.** Oscillations are seen in systems ranging from the pendulum in a grandfather clock to the quartz crystal oscillator providing the timing signals in sophisticated electronic circuits. The physics and mathematics of oscillations will later be the starting point for our study of waves.
- **Fluids.** Liquids and gases *flow*. Surprisingly, it takes no new physics to understand the basic mechanical properties of fluids. By applying our understanding of force, we'll be able to understand what pressure is, how a steel ship can float, and how fluids flow through pipes.

Newton's laws of motion and the conservation laws, especially conservation of energy, will be the tools that allow us to analyze and understand a variety of interesting and practical applications.

Science has given us the power to control our environment, but science and engineering are a two-edged sword. Much of the progress of the last two hundred years has come at the expense of the environment. We humans have deforested much of the world, polluted our air and water, and driven many of our fellow travelers on Spaceship Earth to extinction. Now, at the beginning of the 21st century, the evidence is increasingly clear that humans are altering the earth's climate and causing other global changes.

Fortunately, science also gives us the ability to understand the consequences of our actions and to develop better techniques and procedures. It is more important than ever that scientists and engineers in the 21st century distinguish control that is beneficial from control that is harmful. We'll return to some of these ideas in the Summary to Part III.

12 Rotation of a Rigid Body

Not all motion can be described as that of a particle. Rotation requires the idea of an extended object.

▶ **Looking Ahead** The goal of Chapter 12 is to understand the physics of rotating objects.

Rigid Bodies

A **rigid body** is an object whose size and shape don't change as it moves.

This chapter focuses primarily on the rotation of rigid bodies. We'll emphasize two types of rotation:

- Rotation about a fixed axle.
- Rolling without slipping.

The mathematics of circular motion—angular velocity and angular acceleration— will be very important. A review is highly recommended.

◀ **Looking Back**
Sections 4.5–4.7 Kinematics of circular motion

We'll also consider the conditions under which an extended object is in static equilibrium, neither translating nor rotating. Static equilibrium has many important applications.

◀ **Looking Back**
Section 6.1 Equilibrium

Properties of Rigid Bodies

An extended object's motion and stability depend on how its mass is distributed. You'll learn how to calculate an object's **center of mass** and its **moment of inertia.** Moment of inertia is the rotational equivalent of mass.

The center of mass of this lopsided barbell is closer to the heavier end.

Rotating it about the small end is harder than rotating it about the large end because the moment of inertia about the small end is larger.

Conservation Laws

Kinetic energy and linear momentum have their rotational equivalents. A rotating object's **rotational kinetic energy** and **angular momentum** depend on its moment of inertia and its angular velocity.

You'll learn to solve problems using

- Conservation of energy for frictionless, rotating systems, and
- Conservation of angular momentum for isolated systems.

◀ **Looking Back**
Section 9.3 Momentum conservation

Torque

Torque is the tendency or ability of a force to rotate an object around a pivot point.

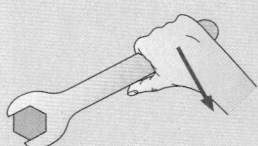

You will learn to calculate torque and will find that the torque depends not only on the magnitude of the force but also on where the force is applied relative to the pivot point. A longer wrench gives a larger torque.

Newton's Second Law

Torque is to rotation what force is to linear motion. Torque τ causes an object with moment of inertia I to undergo angular acceleration $\alpha = \tau/I$.

You'll learn to use Newton's second law to solve problems of rotational dynamics.

◀ **Looking Back**
Section 6.2 Newton's second law

12.1 Rotational Motion

Thus far, our study of physics has focused almost exclusively on the *particle model* in which an object is represented as a mass at a single point in space. The particle model is a perfectly good description of the physics in a vast number of situations, but there are other situations for which we need to consider an *extended object*—a system of particles for which the size and shape *do* make a difference and cannot be neglected.

A **rigid body** is an extended object whose size and shape do not change as it moves. For example, a bicycle wheel can be thought of as a rigid body. FIGURE 12.1 shows a rigid body as a collection of atoms held together by the rigid "massless rods" of molecular bonds.

Real molecular bonds are, of course, not perfectly rigid. That's why an object seemingly as rigid as a bicycle wheel can flex and bend. Thus Figure 12.1 is really a simplified *model* of an extended object, the **rigid-body model**. The rigid-body model is a very good approximation of many real objects of practical interest, such as wheels and axles. Even nonrigid objects can often be modeled as rigid bodies during parts of their motion. For example, a diver is well described as a rotating rigid body while she's in the tuck position.

FIGURE 12.2 illustrates the three basic types of motion of a rigid body: **translational motion, rotational motion,** and **combination motion.**

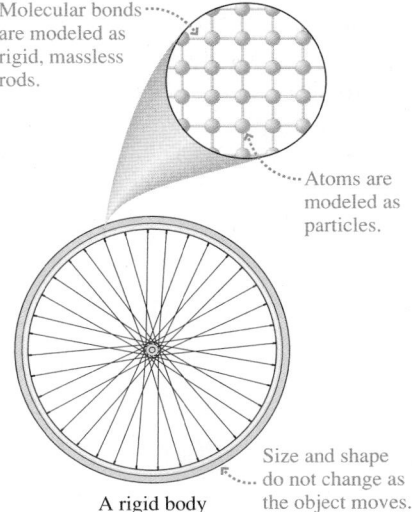

FIGURE 12.1 The rigid-body model.

Molecular bonds are modeled as rigid, massless rods.

Atoms are modeled as particles.

A rigid body

Size and shape do not change as the object moves.

FIGURE 12.2 Three basic types of motion of a rigid body.

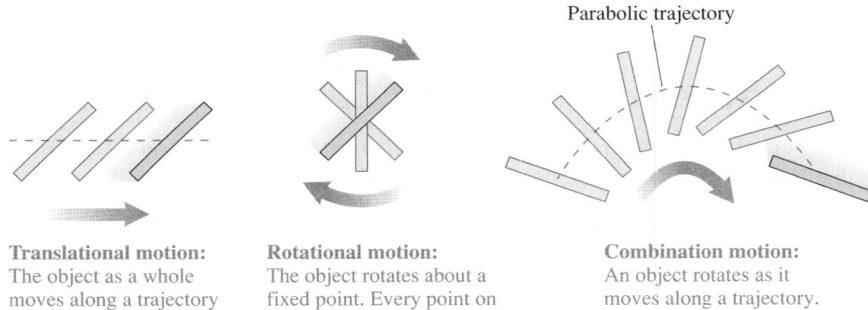

Parabolic trajectory

Translational motion:
The object as a whole moves along a trajectory but does not rotate.

Rotational motion:
The object rotates about a fixed point. Every point on the object moves in a circle.

Combination motion:
An object rotates as it moves along a trajectory.

Brief Review of Rotational Kinematics

Rotation is an extension of circular motion, so we begin with a brief summary of Chapter 4. **A review of Sections 4.5–4.7 is highly recommended.** FIGURE 12.3 shows a wheel rotating on an axle. Its angular velocity

$$\omega = \frac{d\theta}{dt} \tag{12.1}$$

is the rate at which the wheel rotates. The SI units of ω are radians per second (rad/s), but revolutions per second (rev/s) and revolutions per minute (rpm) are frequently used. Notice that all points have equal angular velocities, so we can refer to the angular velocity ω *of the wheel.*

If the wheel is speeding up or slowing down, its angular acceleration is

$$\alpha = \frac{d\omega}{dt} \tag{12.2}$$

The units of angular acceleration are rad/s². Angular acceleration is the *rate* at which the angular velocity ω changes, just as the linear acceleration is the rate at which the linear velocity v changes. Table 12.1 on the next page summarizes the kinematic equations for rotation with constant angular acceleration.

FIGURE 12.3 Two points on a wheel rotate with the same angular velocity.

Every point on the wheel turns through the same angle and thus undergoes circular motion with the same angular velocity ω.

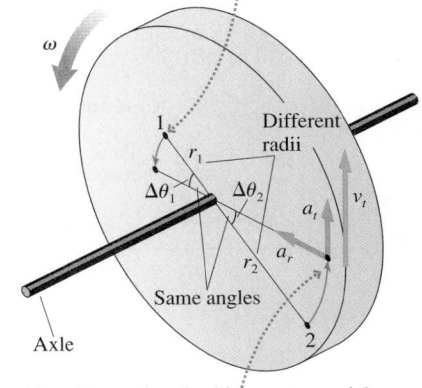

Different radii

r_1

$\Delta\theta_1$ $\Delta\theta_2$ a_t v_t

r_2 a_r

Same angles

Axle

All points on the wheel have a tangential velocity and a radial (centripetal) acceleration. They also have a tangential acceleration if the wheel has angular acceleration.

TABLE 12.1 Rotational kinematics for constant angular acceleration

$\omega_f = \omega_i + \alpha\,\Delta t$

$\theta_f = \theta_i + \omega_i\,\Delta t + \frac{1}{2}\alpha(\Delta t)^2$

$\omega_f^2 = \omega_i^2 + 2\alpha\,\Delta\theta$

FIGURE 12.4 reminds you of the sign conventions for angular velocity and acceleration. They will be especially important in the present chapter. Be careful with the sign of α. Just as with linear acceleration, positive and negative values of α can't be interpreted as simply "speeding up" and "slowing down."

FIGURE 12.4 The signs of angular velocity and angular acceleration.

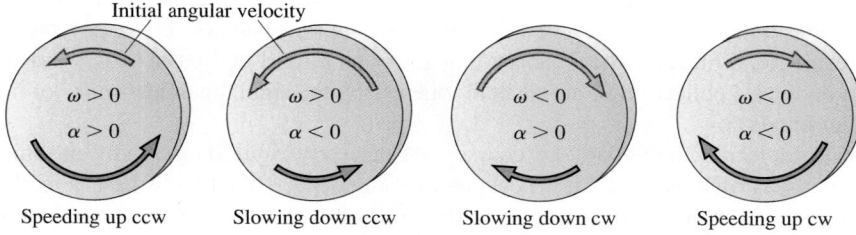

A point at distance r from the rotation axis has instantaneous velocity and acceleration, shown in Figure 12.3, given by

$$v_r = 0 \qquad a_r = \frac{v_t^2}{r} = \omega^2 r$$
$$v_t = r\omega \qquad a_t = r\alpha \tag{12.3}$$

The sign convention for ω implies that v_t and a_t are positive if they point in the counterclockwise (ccw) direction, negative if they point in the clockwise (cw) direction.

12.2 Rotation About the Center of Mass

FIGURE 12.5 Rotation about the center of mass.

(a)

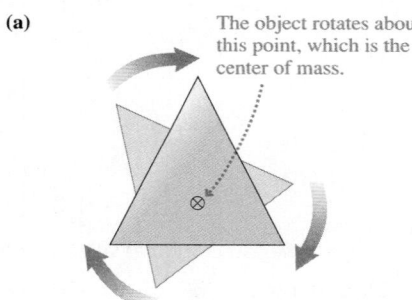

The object rotates about this point, which is the center of mass.

(b)

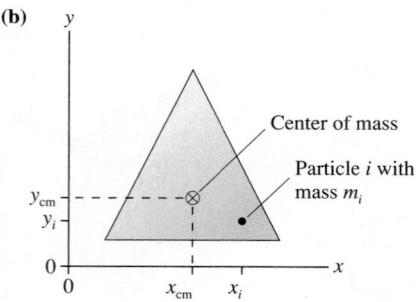

Center of mass

Particle i with mass m_i

Imagine yourself floating in a space capsule deep in space. Suppose you take an object like that shown in **FIGURE 12.5a**, spin it, then let go. The object will rotate, but it will have no translational motion as it floats beside you. *About what point does it rotate?* That is the question we need to answer.

An unconstrained object (i.e., one not on an axle or a pivot) on which there is no net force rotates about a point called the **center of mass.** The center of mass remains motionless while every other point in the object undergoes circular motion around it. You need not go deep into space to demonstrate rotation about the center of mass. If you have an air table, a flat object rotating on the air table rotates about its center of mass.

To locate the center of mass, **FIGURE 12.5b** models the object as if it were constructed from particles numbered $i = 1, 2, 3, \ldots$. Particle i has mass m_i and is located at position (x_i, y_i). We'll prove later in this section that the center of mass is located at position

$$x_{cm} = \frac{1}{M}\sum_i m_i x_i = \frac{m_1 x_1 + m_2 x_2 + m_3 x_3 + \cdots}{m_1 + m_2 + m_3 + \cdots}$$

$$y_{cm} = \frac{1}{M}\sum_i m_i y_i = \frac{m_1 y_1 + m_2 y_2 + m_3 y_3 + \cdots}{m_1 + m_2 + m_3 + \cdots} \tag{12.4}$$

where $M = m_1 + m_2 + m_3 + \cdots$ is the object's total mass.

Let's see if Equations 12.4 make sense. Suppose you have an object consisting of N particles, all with the same mass m. That is, $m_1 = m_2 = \cdots = m_N = m$. We can factor the m out of the numerator, and the denominator becomes simply Nm. The m cancels, and the x-coordinate of the center of mass is

$$x_{cm} = \frac{x_1 + x_2 + \cdots + x_N}{N} = x_{average}$$

In this case, x_{cm} is simply the *average* x-coordinate of all the particles. Likewise, y_{cm} will be the average of all the y-coordinates.

This *does* make sense! If the particle masses are all the same, the center of mass should be at the center of the object. And the "center of the object" is the average position of all the particles. To allow for *unequal* masses, Equations 12.4 are called *weighted averages.* Particles of higher mass count more than particles of lower mass, but the basic idea remains the same. **The center of mass is the mass-weighted center of the object.**

EXAMPLE 12.1   **The center of mass**

A 500 g ball and a 2.0 kg ball are connected by a massless 50-cm-long rod.

a. Where is the center of mass?
b. What is the speed of each ball if they rotate about the center of mass at 40 rpm?

MODEL Model each ball as a particle.

VISUALIZE FIGURE 12.6 shows the two masses. We've chosen a co-ordinate system in which the masses are on the x-axis with the 2.0 kg mass at the origin.

SOLVE a We can use Equations 12.4 to calculate that the center of mass is

$$x_{cm} = \frac{m_1 x_1 + m_2 x_2}{m_1 + m_2}$$

$$= \frac{(2.0 \text{ kg})(0.0 \text{ m}) + (0.50 \text{ kg})(0.50 \text{ m})}{2.0 \text{ kg} + 0.50 \text{ kg}} = 0.10 \text{ m}$$

$y_{cm} = 0$ because all the masses are on the x-axis. The center of mass is 20% of the way from the 2.0 kg ball to the 0.50 kg ball.

FIGURE 12.6 Finding the center of mass.

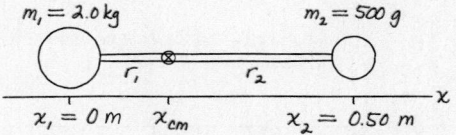

b. Each ball rotates about the center of mass. The radii of the circles are $r_1 = 0.10$ m and $r_2 = 0.40$ m. The tangential velocities are $(v_i)_t = r_i \omega$, but this equation requires ω to be in rad/s. The conversion is

$$\omega = 40 \, \frac{\text{rev}}{\text{min}} \times \frac{1 \text{ min}}{60 \text{ s}} \times \frac{2\pi \text{ rad}}{1 \text{ rev}} = 4.19 \text{ rad/s}$$

Consequently,

$$(v_1)_t = r_1 \omega = (0.10 \text{ m})(4.19 \text{ rad/s}) = 0.42 \text{ m/s}$$

$$(v_2)_t = r_2 \omega = (0.40 \text{ m})(4.19 \text{ rad/s}) = 1.68 \text{ m/s}$$

ASSESS The center of mass is closer to the heavier ball than to the lighter ball. We expected this because x_{cm} is a mass-weighted average of the positions. But the lighter mass moves faster because it is farther from the rotation axis.

For any realistic object, carrying out the summations of Equations 12.4 over all the atoms in the object is not practical. Instead, as **FIGURE 12.7** shows, we can divide an extended object into many small cells or boxes, each with the very small mass Δm. We will number the cells 1, 2, 3, ..., just as we did the particles. Cell i has coordinates (x_i, y_i) and mass $m_i = \Delta m$. The center-of-mass coordinates are then

$$x_{cm} = \frac{1}{M} \sum_i x_i \, \Delta m \qquad \text{and} \qquad y_{cm} = \frac{1}{M} \sum_i y_i \, \Delta m$$

Now, as you might expect, we'll let the cells become smaller and smaller, with the total number increasing. As each cell becomes infinitesimally small, we can replace Δm with dm and the sum by an integral. Then

$$x_{cm} = \frac{1}{M} \int x \, dm \qquad \text{and} \qquad y_{cm} = \frac{1}{M} \int y \, dm \qquad (12.5)$$

Equations 12.5 are a formal definition of the center of mass, but they are *not* ready to integrate in this form. First, integrals are carried out over *coordinates,* not over masses. Before we can integrate, we must replace dm by an equivalent expression involving a coordinate differential such as dx or dy. Second, no limits of integration have been specified. The procedure for using Equations 12.5 is best shown with an example.

FIGURE 12.7 Calculating the center of mass of an extended object.

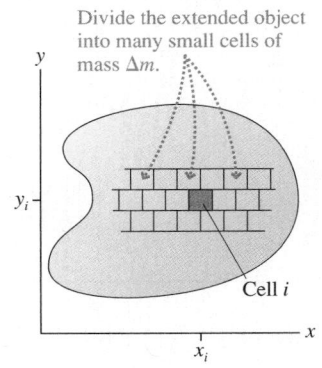

EXAMPLE 12.2 **The center of mass of a rod**

Find the center of mass of a thin, uniform rod of length L and mass M. Use this result to find the tangential acceleration of one tip of a 1.60-m-long rod that rotates about its center of mass with an angular acceleration of 6.0 rad/s^2.

VISUALIZE **FIGURE 12.8** shows the rod. We've chosen a coordinate system such that the rod lies along the x-axis from 0 to L. Because the rod is "thin," we'll assume that $y_{cm} = 0$.

FIGURE 12.8 Finding the center of mass of a long, thin rod.

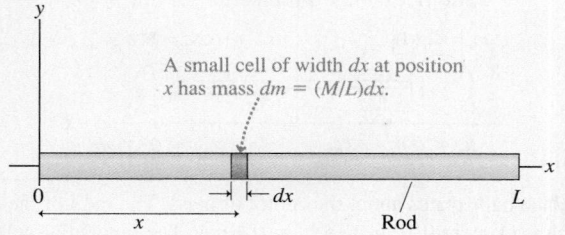

A small cell of width dx at position x has mass $dm = (M/L)dx$.

SOLVE Our first task is to find x_{cm}, which lies somewhere on the x-axis. To do this, we divide the rod into many small cells of mass dm. One such cell, at position x, is shown. The cell's width is dx. Because the rod is *uniform,* the mass of this little cell is the *same fraction* of the total mass M that dx is of the total length L. That is,

$$\frac{dm}{M} = \frac{dx}{L}$$

Consequently, we can express dm in terms of the coordinate differential dx as

$$dm = \frac{M}{L}\,dx$$

NOTE ▶ The change of variables from dm to the differential of a coordinate is *the* key step in calculating the center of mass. ◀

With this expression for dm, Equation 12.5 for x_{cm} becomes

$$x_{cm} = \frac{1}{M}\left(\frac{M}{L}\int x\,dx\right) = \frac{1}{L}\int_0^L x\,dx$$

where in the last step we've noted that summing "all the mass in the rod" means integrating from $x = 0$ to $x = L$. This is a straightforward integral to carry out, giving

$$x_{cm} = \frac{1}{L}\left[\frac{x^2}{2}\right]_0^L = \frac{1}{L}\left[\frac{L^2}{2} - 0\right] = \frac{1}{2}L$$

The center of mass is at the center of the rod. For a 1.60-m-long rod, each tip of the rod rotates in a circle with $r = \frac{1}{2}L = 0.80$ m. The tangential acceleration, the rate at which the tip is speeding up, is

$$a_t = r\alpha = (0.80 \text{ m})(6.0 \text{ rad/s}^2) = 4.8 \text{ m/s}^2$$

ASSESS You could have guessed that the center of mass is at the center of the rod, but now we've shown it rigorously.

NOTE ▶ For any symmetrical object of uniform density, the center of mass is at the physical center of the object. ◀

FIGURE 12.9 Finding the center of mass.

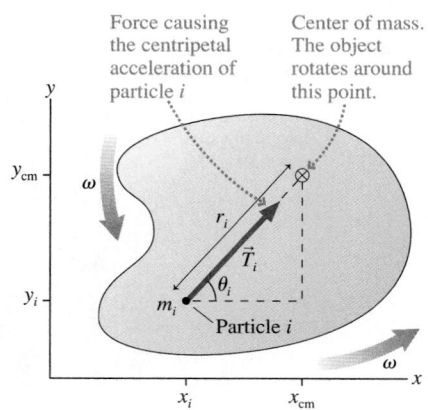

Force causing the centripetal acceleration of particle i

Center of mass. The object rotates around this point.

$\vec{T}_i$

Particle i

To see where the center-of-mass equations come from, **FIGURE 12.9** shows an object rotating about its center of mass. Particle i is moving in a circle, so it *must* have a centripetal acceleration. Acceleration requires a force, and this force is due to tension in the molecular bonds that hold the object together. Force $\vec{T}_i$ on particle i has magnitude

$$T_i = m_i(a_i)_r = m_i r_i \omega^2 \tag{12.6}$$

where r_i is the distance of particle i from the center of mass and we used Equation 12.3 for a_r. All points in a rigid rotating object have the *same* angular velocity, so ω doesn't need a subscript.

At every instant of time, the internal tension forces are all paired as action/reaction forces, equal in magnitude but opposite in direction, so the sum of all the tension forces must be zero. That is, $\sum \vec{T}_i = \vec{0}$. The x-component of this sum is

$$\sum_i (T_i)_x = \sum_i T_i \cos\theta_i = \sum_i (m_i r_i \omega^2)\cos\theta_i = 0 \tag{12.7}$$

You can see from Figure 12.9 that $\cos\theta_i = (x_{cm} - x_i)/r_i$. Thus

$$\sum_i (T_i)_x = \sum_i (m_i r_i \omega^2)\frac{x_{cm} - x_i}{r_i} = \left(\sum_i m_i x_{cm} - \sum_i m_i x_i\right)\omega^2 = 0 \tag{12.8}$$

This equation will be true if the term in parentheses is zero. x_{cm} is a constant, so we can bring it outside the summation to write

$$\sum_i m_i x_{cm} - \sum_i m_i x_i = \left(\sum_i m_i\right)x_{cm} - \sum_i m_i x_i = M x_{cm} - \sum_i m_i x_i = 0 \tag{12.9}$$

where we used the fact that $\sum m_i$ is simply the object's total mass M. Solving for x_{cm}, we find the x-coordinate of the object's center of mass to be

$$x_{cm} = \frac{1}{M}\sum_i m_i x_i = \frac{m_1 x_1 + m_2 x_2 + m_3 x_3 + \cdots}{m_1 + m_2 + m_3 + \cdots} \qquad (12.10)$$

This was Equation 12.4. The y-equation is found similarly.

12.3 Rotational Energy

A rotating rigid body has kinetic energy because all atoms in the object are in motion. The kinetic energy due to rotation is called **rotational kinetic energy.**

FIGURE 12.10 shows a few of the particles making up a solid object that rotates with angular velocity ω. Particle i, which rotates in a circle of radius r_i, moves with speed $v_i = r_i\omega$. The object's rotational kinetic energy is the sum of the kinetic energies of each of the particles:

$$K_{rot} = \frac{1}{2}m_1 v_1^2 + \frac{1}{2}m_2 v_2^2 + \cdots$$

$$= \frac{1}{2}m_1 r_1^2 \omega^2 + \frac{1}{2}m_2 r_2^2 \omega^2 + \cdots = \frac{1}{2}\left(\sum_i m_i r_i^2\right)\omega^2 \qquad (12.11)$$

The quantity $\sum m_i r_i^2$ is called the object's **moment of inertia** I:

$$I = m_1 r_1^2 + m_2 r_2^2 + m_3 r_3^2 + \cdots = \sum_i m_i r_i^2 \qquad (12.12)$$

The units of moment of inertia are $kg\,m^2$. **An object's moment of inertia depends on the axis of rotation.** Once the axis is specified, allowing the values of r_i to be determined, the moment of inertia *about that axis* can be calculated from Equation 12.12.

Written using the moment of inertia I, the rotational kinetic energy is

$$K_{rot} = \frac{1}{2}I\omega^2 \qquad (12.13)$$

Rotational kinetic energy is *not* a new form of energy. This is the familiar kinetic energy of motion, only now expressed in a form that is especially convenient for rotational motion. Notice the analogy with the familiar $\frac{1}{2}mv^2$.

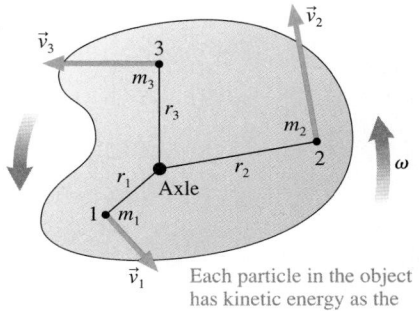

FIGURE 12.10 Rotational kinetic energy is due to the motion of the particles.

Each particle in the object has kinetic energy as the object rotates.

EXAMPLE 12.3 | A rotating widget

Students participating in an engineering project design the triangular widget seen in FIGURE 12.11. The three masses, held together by lightweight plastic rods, rotate in the plane of the page about an axle passing through the right-angle corner. At what angular velocity does the widget have 100 mJ of rotational energy?

FIGURE 12.11 The rotating widget.

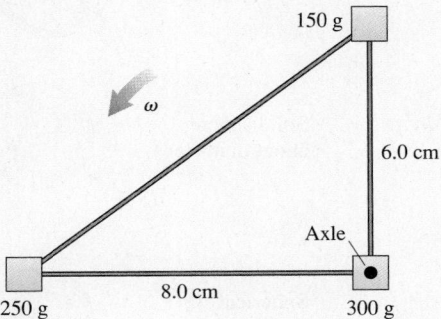

MODEL The widget can be modeled as three particles connected by massless rods.

SOLVE Rotational energy is $K = \frac{1}{2}I\omega^2$. The moment of inertia is measured about the rotation axis, thus

$$I = \sum_i m_i r_i^2 = (0.25\text{ kg})(0.080\text{ m})^2 + (0.15\text{ kg})(0.060\text{ m})^2$$

$$+ (0.30\text{ kg})(0\text{ m})^2$$

$$= 2.14 \times 10^{-3}\text{ kg m}^2$$

The largest mass makes no contribution to I because it is right on the rotation axis with $r = 0$. With I known, the desired angular velocity is

$$\omega = \sqrt{\frac{2K}{I}} = \sqrt{\frac{2(0.10\text{ J})}{2.14 \times 10^{-3}\text{ kg m}^2}}$$

$$= 9.67\text{ rad/s} \times \frac{1\text{ rev}}{2\pi\text{ rad}} = 1.54\text{ rev/s} = 92\text{ rpm}$$

ASSESS The moment of inertia depends on the distance of each mass from the rotation axis. The moment of inertia would be different for an axle passing through either of the other two masses, and thus the required angular velocity would be different.

FIGURE 12.12 Moment of inertia depends on both the mass and how the mass is distributed.

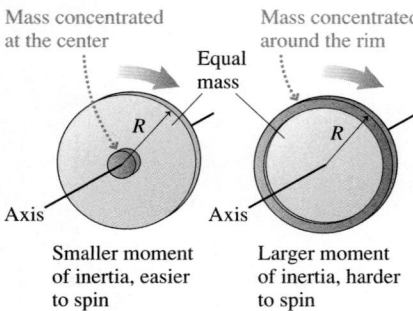

Mass concentrated at the center

Mass concentrated around the rim

Equal mass

Axis

Axis

Smaller moment of inertia, easier to spin

Larger moment of inertia, harder to spin

Before rushing to calculate moments of inertia, let's get a better understanding of the meaning. First, notice that **moment of inertia is the rotational equivalent of mass.** It plays the same role in Equation 12.13 as mass m in the now-familiar $K = \frac{1}{2}mv^2$. Recall that the quantity we call *mass* was actually defined as the *inertial mass.* Objects with larger mass have a larger *inertia,* meaning that they're harder to accelerate. Similarly, an object with a larger moment of inertia is harder to rotate. The fact that *moment of inertia* retains the word "inertia" reminds us of this.

But why does the moment of inertia depend on the distances r_i from the rotation axis? Think about the two wheels shown in FIGURE 12.12. They have the same total mass M and the same radius R. As you probably know from experience, it's much easier to spin the wheel whose mass is concentrated at the center than to spin the one whose mass is concentrated around the rim. This is because having the mass near the center (smaller values of r_i) lowers the moment of inertia.

Moments of inertia for many solid objects are tabulated and found in various science and engineering handbooks. You would need to compute I yourself only for an object of unusual shape. Table 12.2 is a short list of common moments of inertia. We'll see in the next section where these come from, but do notice how I depends on the rotation axis.

If the rotation axis is not through the center of mass, then rotation may cause the center of mass to move up or down. In that case, the object's gravitational potential energy $U_g = Mgy_{cm}$ will change. If there are no dissipative forces (i.e., if the axle is frictionless) and if no work is done by external forces, then the mechanical energy

$$E_{mech} = K_{rot} + U_g = \frac{1}{2}I\omega^2 + Mgy_{cm} \qquad (12.14)$$

is a conserved quantity.

TABLE 12.2 Moments of inertia of objects with uniform density

Object and axis	Picture	I	Object and axis	Picture	I
Thin rod, about center		$\frac{1}{12}ML^2$	Cylinder or disk, about center		$\frac{1}{2}MR^2$
Thin rod, about end		$\frac{1}{3}ML^2$	Cylindrical hoop, about center		MR^2
Plane or slab, about center		$\frac{1}{12}Ma^2$	Solid sphere, about diameter		$\frac{2}{5}MR^2$
Plane or slab, about edge		$\frac{1}{3}Ma^2$	Spherical shell, about diameter		$\frac{2}{3}MR^2$

EXAMPLE 12.4   **The speed of a rotating rod**

A 1.0-m-long, 200 g rod is hinged at one end and connected to a wall. It is held out horizontally, then released. What is the speed of the tip of the rod as it hits the wall?

MODEL The mechanical energy is conserved if we assume the hinge is frictionless. The rod's gravitational potential energy is transformed into rotational kinetic energy as it "falls."

FIGURE 12.13 A before-and-after pictorial representation of the rod.

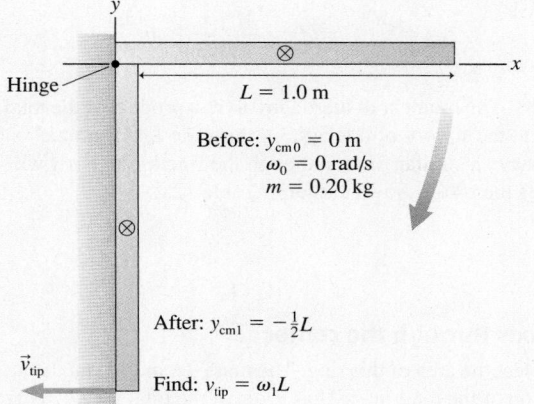

VISUALIZE **FIGURE 12.13** is a familiar before-and-after pictorial representation of the rod. We've placed the origin of the coordinate system at the pivot point.

SOLVE Mechanical energy is conserved, so we can equate the rod's final mechanical energy to its initial mechanical energy:

$$\frac{1}{2}I\omega_1^2 + Mgy_{cm1} = \frac{1}{2}I\omega_0^2 + Mgy_{cm0}$$

The initial conditions are $\omega_0 = 0$ and $y_{cm0} = 0$. The center of mass moves to $y_{cm1} = -\frac{1}{2}L$ as the rod hits the wall. From Table 12.2 we find $I = \frac{1}{3}ML^2$ for a rod rotating about one end. Thus

$$\frac{1}{2}I\omega_1^2 + Mgy_{cm1} = \frac{1}{6}ML^2\omega_1^2 - \frac{1}{2}MgL = 0$$

We can solve this for the rod's angular velocity as it hits the wall:

$$\omega_1 = \sqrt{\frac{3g}{L}}$$

The tip of the rod is moving in a circle with radius $r = L$. Its final speed is

$$v_{tip} = \omega_1 L = \sqrt{3gL} = 5.4 \text{ m/s}$$

ASSESS Energy conservation is a powerful tool for rotational motion, just as it was for translational motion.

12.4 Calculating Moment of Inertia

The equation for rotational energy is easy to write, but we can't make use of it without knowing an object's moment of inertia. Unlike mass, we can't measure moment of inertia by putting an object on a scale. And while we can guess that the center of mass of a symmetrical object is at the physical center of the object, we can *not* guess the moment of inertia of even a simple object. To find I, we really must carry through the calculation.

Equation 12.12 defines the moment of inertia as a sum over all the particles in the system. As we did for the center of mass, we can replace the individual particles with cells 1, 2, 3, . . . of mass Δm. Then the moment of inertia summation can be converted to an integration:

$$I = \sum_i r_i^2 \Delta m \xrightarrow[\Delta m \to 0]{} I = \int r^2 \, dm \qquad (12.15)$$

where r is the distance from the rotation axis. If we let the rotation axis be the z-axis, then we can write the moment of inertia as

$$I = \int (x^2 + y^2) \, dm \qquad (12.16)$$

NOTE ▶ You *must* replace dm by an equivalent expression involving a coordinate differential such as dx or dy before you can carry out the integration of Equation 12.16. ◀

You can use any coordinate system to calculate the coordinates x_{cm} and y_{cm} of the center of mass. But the moment of inertia is defined for rotation about a particular axis, and r is measured from that axis. Thus the coordinate system used for moment-of-inertia calculations *must* have its origin at the pivot point. Two examples will illustrate these ideas.

EXAMPLE 12.5 **Moment of inertia of a rod about a pivot at one end**

Find the moment of inertia of a thin, uniform rod of length L and mass M that rotates about a pivot at one end.

MODEL An object's moment of inertia depends on the axis of rotation. In this case, the rotation axis is at the end of the rod.

VISUALIZE **FIGURE 12.14** defines an x-axis with the origin at the pivot point.

FIGURE 12.14 Finding the moment of inertia about one end of a long, thin rod.

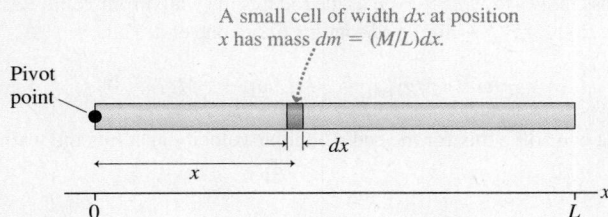

A small cell of width dx at position x has mass $dm = (M/L)dx$.

Pivot point

$$0 \qquad\qquad\qquad\qquad L$$

SOLVE Because the rod is thin, we can assume that $y \approx 0$ for all points on the rod. Thus

$$I = \int x^2 \, dm$$

The small amount of mass dm in the small length dx is $dm = (M/L) \, dx$, as we found in Example 12.2. The rod extends from $x = 0$ to $x = L$, so the moment of inertia for a rod about one end is

$$I_{\text{end}} = \frac{M}{L} \int_0^L x^2 \, dx = \frac{M}{L} \left[\frac{x^3}{3} \right]_0^L = \frac{1}{3} ML^2$$

ASSESS The moment of inertia involves a product of the total mass M with the *square* of a length, in this case L. All moments of inertia have a similar form, although the fraction in front will vary. This is the result shown earlier in Table 12.2.

EXAMPLE 12.6 **Moment of inertia of a circular disk about an axis through the center**

Find the moment of inertia of a circular disk of radius R and mass M that rotates on an axis passing through its center.

VISUALIZE **FIGURE 12.15** shows the disk and defines distance r from the axis.

FIGURE 12.15 Finding the moment of inertia of a disk about an axis through the center.

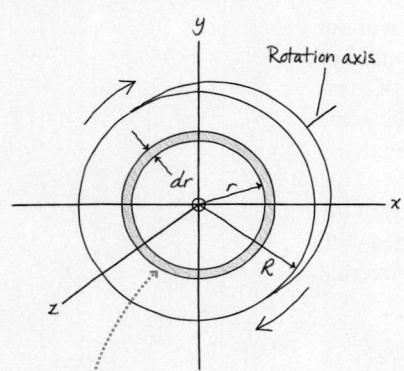

Rotation axis

A narrow ring of width dr has mass $dm = (M/A)dA$. Its area is $dA = $ width $\times$ circumference $= 2\pi r \, dr$.

SOLVE This is a situation of great practical importance. To solve this problem, we need to use a two-dimensional integration scheme that you learned in calculus. Rather than dividing the disk into little boxes, let's divide it into narrow *rings* of mass dm. Figure 12.15 shows one such ring, of radius r and width dr. Let dA

represent the area of this ring. The mass dm in this ring is the same fraction of the total mass M as dA is of the total area A. That is,

$$\frac{dm}{M} = \frac{dA}{A}$$

Thus the mass in the small area dA is

$$dm = \frac{M}{A} dA$$

This is the reasoning we used to find the center of mass of the rod in Example 12.2, only now we're using it in two dimensions.

The total area of the disk is $A = \pi R^2$, but what is dA? If we imagine unrolling the little ring, it would form a long, thin rectangle of length $2\pi r$ and height dr. Thus the *area* of this little ring is $dA = 2\pi r \, dr$. With this information we can write

$$dm = \frac{M}{\pi R^2} (2\pi r \, dr) = \frac{2M}{R^2} r \, dr$$

Now we have an expression for dm in terms of a coordinate differential dr, so we can proceed to carry out the integration for I. Using Equation 12.15, we find

$$I_{\text{disk}} = \int r^2 \, dm = \int r^2 \left(\frac{2M}{R^2} r \, dr \right) = \frac{2M}{R^2} \int_0^R r^3 \, dr$$

where in the last step we have used the fact that the disk extends from $r = 0$ to $r = R$. Performing the integration gives

$$I_{\text{disk}} = \frac{2M}{R^2} \left[\frac{r^4}{4} \right]_0^R = \frac{1}{2} MR^2$$

ASSESS Once again, the moment of inertia involves a product of the total mass M with the *square* of a length, in this case R.

If a complex object can be divided into simpler pieces 1, 2, 3, ... whose moments of inertia I_1, I_2, I_3 ... are already known, the moment of inertia of the entire object is

$$I_{\text{object}} = I_1 + I_2 + I_3 + \cdots \qquad (12.17)$$

This follows from the fact that the sum $I = \sum m_i r_i^2$ can be broken into smaller sums over the simpler objects. Equation 12.17 is useful for solving many problems.

The Parallel-Axis Theorem

The moment of inertia depends on the rotation axis. Suppose you need to know the moment of inertia for rotation about the off-center axis in FIGURE 12.16. You can find this quite easily if you know the moment of inertia for rotation around a *parallel axis* through the center of mass.

If the axis of interest is distance d from a parallel axis through the center of mass, the moment of inertia is

$$I = I_{cm} + Md^2 \qquad (12.18)$$

Equation 12.18 is called the **parallel-axis theorem**. We'll give a proof for the one-dimensional object shown in FIGURE 12.17.

The x-axis has its origin at the rotation axis, and the x'-axis has its origin at the center of mass. You can see that the coordinates of dm along these two axes are related by $x = x' + d$. By definition, the moment of inertia about the rotation axis is

$$I = \int x^2 \, dm = \int (x' + d)^2 \, dm = \int (x')^2 \, dm + 2d \int x' \, dm + d^2 \int dm \qquad (12.19)$$

The first of the three integrals on the right, by definition, is the moment of inertia I_{cm} about the center of mass. The third is simply Md^2 because adding up (integrating) all the dm gives the total mass M.

If you refer back to Equations 12.5, the definition of the center of mass, you'll see that the middle integral on the right is equal to Mx'_{cm}. But $x'_{cm} = 0$ because we specifically chose the x'-axis to have its origin at the center of mass. Thus the second integral is zero and we end up with Equation 12.18. The proof in two dimensions is similar.

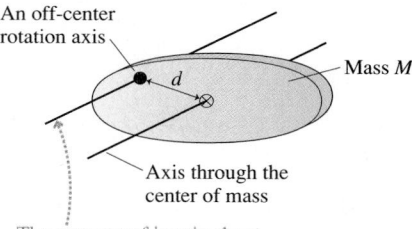

FIGURE 12.16 Rotation about an off-center axis.

An off-center rotation axis

Mass M

Axis through the center of mass

The moment of inertia about this axis is $I = I_{cm} + Md^2$.

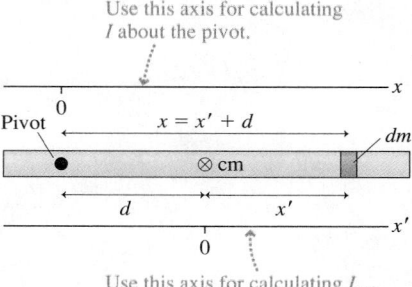

FIGURE 12.17 Proving the parallel-axis theorem.

Use this axis for calculating I about the pivot.

Pivot $x = x' + d$

$\otimes$ cm

Use this axis for calculating I_{cm}.

EXAMPLE 12.7   **The moment of inertia of a thin rod**

Find the moment of inertia of a thin rod with mass M and length L about an axis one-third of the length from one end.

SOLVE From Table 12.2 we know the moment of inertia about the center of mass is $\frac{1}{12}ML^2$. The center of mass is at the center of the rod. An axis $\frac{1}{3}L$ from one end is $d = \frac{1}{6}L$ from the center of mass. Using the parallel-axis theorem, we have

$$I = I_{cm} + Md^2 = \frac{1}{12}ML^2 + M\left(\frac{1}{6}L\right)^2 = \frac{1}{9}ML^2$$

STOP TO THINK 12.1   Four Ts are made from two identical rods of equal mass and length. Rank in order, from largest to smallest, the moments of inertia I_a to I_d for rotation about the dashed line.

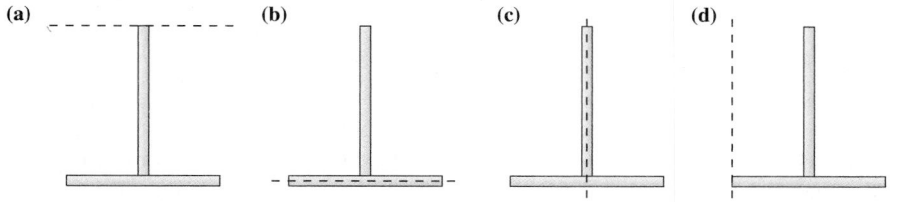

(a)   (b)   (c)   (d)

12.5 Torque

Consider the common experience of pushing open a door. FIGURE 12.18 is a top view of a door hinged on the left. Four pushing forces are shown, all of equal strength. Which of these will be most effective at opening the door?

Force $\vec{F}_1$ will open the door, but force $\vec{F}_2$, which pushes straight at the hinge, will not. Force $\vec{F}_3$ will open the door, but not as easily as $\vec{F}_1$. What about $\vec{F}_4$? It is perpendicular to the door, it has the same magnitude as $\vec{F}_1$, but you know from experience that pushing close to the hinge is not as effective as pushing at the outer edge of the door.

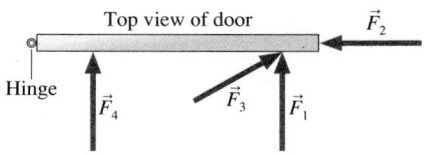

FIGURE 12.18 The four forces have different effects on the swinging door.

Top view of door $\vec{F}_2$

Hinge $\vec{F}_4$ $\vec{F}_3$ $\vec{F}_1$

FIGURE 12.19 Force $\vec{F}$ exerts a torque about the pivot point.

$\vec{F}$ exerts a torque about the pivot point.

Angle ϕ is measured ccw from the radial line.

Point where force is applied

r is measured from the pivot to the point where the force is applied.

Pivot point

Rigid body

Torque is to rotational motion as force is to linear motion.

The ability of a force to cause a rotation depends on three factors:

1. The magnitude F of the force.
2. The distance r from the point of application to the pivot.
3. The angle at which the force is applied.

To make these ideas specific, **FIGURE 12.19** shows a force $\vec{F}$ applied at one point on a rigid body. For example, a string might be pulling on the object at that point, in which case the force would be a tension force.

NOTE ▶ Angle ϕ is measured *counterclockwise* from the dashed line that extends outward along the radial line. This is consistent with our sign convention for the angular position θ. ◀

Let's define a new quantity called the **torque** τ (Greek tau) as

$$\tau \equiv rF \sin \phi \qquad (12.20)$$

Torque depends on the three properties we just listed: the magnitude of the force, its distance from the pivot, and its angle. Loosely speaking, τ measures the "effectiveness" of the force at causing an object to rotate about a pivot. **Torque is the rotational equivalent of force.**

The SI units of torque are newton-meters, abbreviated N m. Although we defined 1 N m = 1 J during our study of energy, torque is not an energy-related quantity and so we do *not* use joules as a measure of torque.

Torque, like force, has a sign. A torque that tries to rotate the object in a ccw direction is positive while a negative torque gives a cw rotation. **FIGURE 12.20** summarizes the signs. Notice that a force pushing straight toward the pivot or pulling straight out from the pivot exerts *no* torque.

FIGURE 12.20 Signs and strengths of the torque.

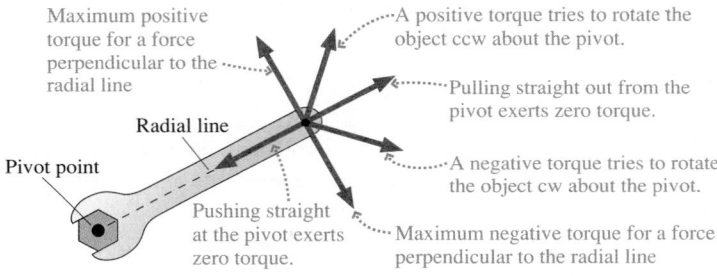

Maximum positive torque for a force perpendicular to the radial line

A positive torque tries to rotate the object ccw about the pivot.

Radial line

Pulling straight out from the pivot exerts zero torque.

Pivot point

A negative torque tries to rotate the object cw about the pivot.

Pushing straight at the pivot exerts zero torque.

Maximum negative torque for a force perpendicular to the radial line

NOTE ▶ Torque differs from force in a very important way. Torque is calculated or measured *about a pivot point*. To say that a torque is 20 N m is meaningless. You need to say that the torque is 20 N m about a particular point. Torque can be calculated about any pivot point, but its value depends on the point chosen. In practice, we measure or calculate torques about the same point from which we measure an object's angular position θ (and thus its angular velocity ω and angular acceleration α). This assumption is built into the equations of rotational dynamics. ◀

Returning to the door of Figure 12.18, you can see that $\vec{F}_1$ is most effective at opening the door because $\vec{F}_1$ exerts the largest torque *about the pivot point*. $\vec{F}_3$ has equal magnitude, but it is applied at an angle less than 90° and thus exerts less torque. $\vec{F}_2$, pushing straight at the hinge with $\phi = 0°$, exerts no torque at all. And $\vec{F}_4$, with a smaller value for r, exerts less torque than $\vec{F}_1$.

Interpreting Torque

Torque can be interpreted from two perspectives. First, FIGURE 12.21a shows that the quantity $F\sin\phi$ is the tangential force component F_t. Consequently, the torque is

$$\tau = rF_t \tag{12.21}$$

In other words, torque is the product of r with the force component F_t that is *perpendicular* to the radial line. This interpretation makes sense because the radial component of $\vec{F}$ points straight at the pivot point and cannot exert a torque.

FIGURE 12.21 Two useful interpretations of the torque.

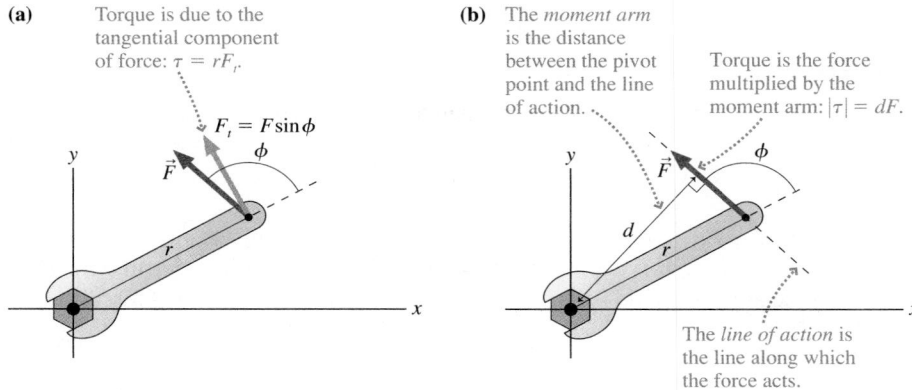

(a) Torque is due to the tangential component of force: $\tau = rF_t$.

$F_t = F\sin\phi$

(b) The *moment arm* is the distance between the pivot point and the line of action.

Torque is the force multiplied by the moment arm: $|\tau| = dF$.

The *line of action* is the line along which the force acts.

A second perspective, widely used in applications, is based on the idea of a *moment arm*. FIGURE 12.21b shows the **line of action**, the line along which the force acts. The minimum distance between the pivot point and the line of action—the length of a line drawn *perpendicular to the line of action*—is called the **moment arm** (or the *lever arm*) d. Because $\sin(180° - \theta) = \sin\phi$, it is easy to see that $d = r\sin\phi$. Thus the torque $rF\sin\theta$ can also be written

$$|\tau| = dF \tag{12.22}$$

NOTE ▶ Equation 12.22 gives only $|\tau|$, the magnitude of the torque; the sign has to be supplied by observing the direction in which the torque acts. ◀

EXAMPLE 12.8   **Applying a torque**

Luis uses a 20-cm-long wrench to turn a nut. The wrench handle is tilted 30° above the horizontal, and Luis pulls straight down on the end with a force of 100 N. How much torque does Luis exert on the nut?

VISUALIZE FIGURE 12.22 shows the situation. The angle is a negative $\phi = -120°$ because it is *clockwise* from the radial line.

FIGURE 12.22 A wrench being used to turn a nut.

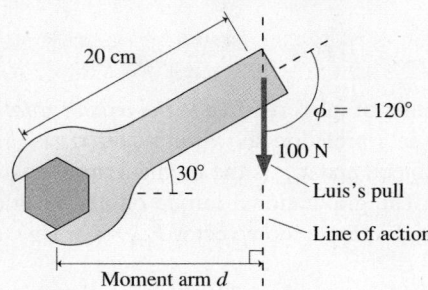

20 cm

$\phi = -120°$

100 N

30°

Luis's pull

Line of action

Moment arm d

SOLVE The tangential component of the force is

$$F_t = F\sin\phi = -86.6\ \text{N}$$

According to our sign convention, F_t is negative because it points in a cw direction. The torque, from Equation 12.21, is

$$\tau = rF_t = (0.20\ \text{m})(-86.6\ \text{N}) = -17\ \text{N m}$$

Alternatively, Figure 12.22 has drawn the *line of action* by extending the force vector forward and backward. The *moment arm*, the distance between the pivot point and the line of action, is

$$d = r\sin(60°) = 0.17\ \text{m}$$

Inserting the moment arm in Equation 12.22 gives

$$|\tau| = dF = (0.17\ \text{m})(100\ \text{N}) = 17\ \text{N m}$$

The torque acts to give a cw rotation, so we insert a minus sign to end up with

$$\tau = -17\ \text{N m}$$

ASSESS Luis could increase the torque by changing the angle so that his pull is perpendicular to the wrench ($\phi = -90°$).

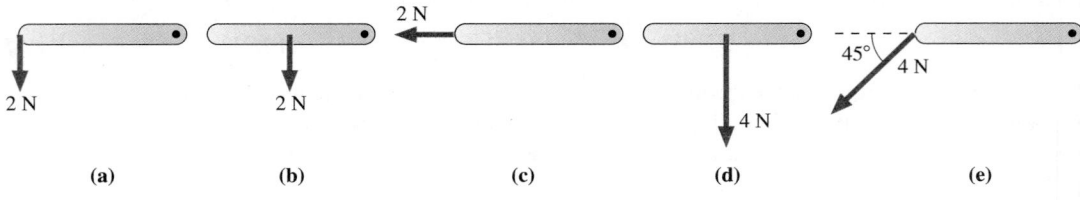

Rank in order, from largest to smallest, the five torques τ_a to τ_e. The rods all have the same length and are pivoted at the dot.

(a)   (b)   (c)   (d)   (e)

Net Torque

FIGURE 12.23 The forces exert a net torque about the pivot point.

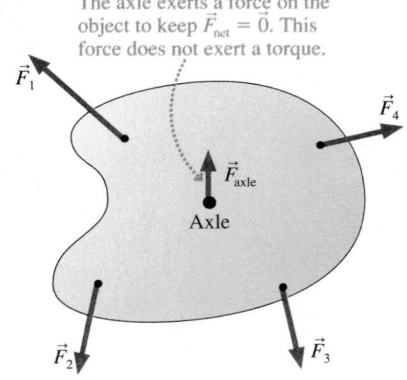

The axle exerts a force on the object to keep $\vec{F}_{net} = \vec{0}$. This force does not exert a torque.

$\vec{F}_1$

$\vec{F}_4$

$\vec{F}_{axle}$

Axle

$\vec{F}_2$ $\vec{F}_3$

FIGURE 12.23 shows forces $\vec{F}_1$, $\vec{F}_2$, $\vec{F}_3$, ... applied to an extended object. The object is free to rotate about the axle, but the axle prevents the object from having any translational motion. It does so by exerting force $\vec{F}_{axle}$ on the object to balance the other forces and keep $\vec{F}_{net} = \vec{0}$.

Forces $\vec{F}_1$, $\vec{F}_2$, $\vec{F}_3$, ... exert torques τ_1, τ_2, τ_3, ... on the object, but $\vec{F}_{axle}$ does *not* exert a torque because it is applied at the pivot point and has zero moment arm. Thus the *net* torque about the axle is the sum of the torques due to the applied forces:

$$\tau_{net} = \tau_1 + \tau_2 + \tau_3 + \cdots = \sum_i \tau_i \qquad (12.23)$$

Gravitational Torque

Gravity exerts a torque on many objects. If the object in **FIGURE 12.24** is released, a torque due to gravity will cause it to rotate around the axle. To calculate the torque about the axle, we start with the fact that gravity acts on *every* particle in the object, exerting a downward force of magnitude $F_i = m_i g$ on particle i. The *magnitude* of the gravitational torque on particle i is $|\tau_i| = d_i m_i g$, where d_i is the moment arm. But we need to be careful with signs.

FIGURE 12.24 Gravitational torque.

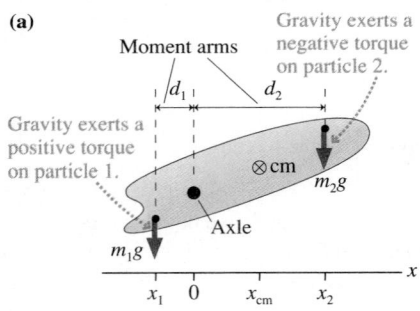

(a)

Moment arms

Gravity exerts a negative torque on particle 2.

Gravity exerts a positive torque on particle 1.

$|d_1|$ d_2

$\otimes$ cm

$m_2 g$

Axle

$m_1 g$

x_1 0 x_{cm} x_2 x

A moment arm must be a positive number because it's a distance. If we establish a coordinate system with the origin at the axle, then you can see from **FIGURE 12.24a** that the moment arm d_i of particle i is $|x_i|$. A particle to the right of the axle (positive x_i) experiences a *negative* torque because gravity tries to rotate this particle in a clockwise direction. Similarly, a particle to the left of the axle (negative x_i) has a positive torque. The torque is opposite in sign to x_i, so we can get the sign right by writing

$$\tau_i = -x_i m_i g = -(m_i x_i) g \qquad (12.24)$$

The net torque due to gravity is found by summing Equation 12.24 over all particles:

$$\tau_{grav} = \sum_i \tau_i = \sum_i (-m_i x_i g) = -\left(\sum_i m_i x_i\right) g \qquad (12.25)$$

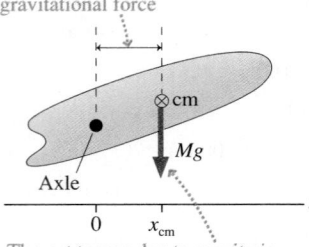

(b)

Moment arm of the net gravitational force

$\otimes$ cm

Mg

Axle

0   x_{cm} x

The net torque due to gravity is found by pretending the object's entire mass is at the center of mass.

But according to the definition of center of mass, Equations 12.4, $\sum m_i x_i = M x_{cm}$. Thus the torque due to gravity is

$$\tau_{grav} = -Mg x_{cm} \qquad (12.26)$$

where x_{cm} is the position of the center of mass *relative to the axis of rotation*.

Equation 12.26 has the simple interpretation shown in **FIGURE 12.24b**. Mg is the net gravitational force on the entire object, and x_{cm} is the moment arm between the rotation axis and the center of mass. The gravitational torque on an extended object of mass M is equivalent to the torque of a *single* force vector $\vec{F}_{grav} = -Mg\,\hat{j}$ acting at the object's center of mass.

In other words, **the gravitational torque is found by treating the object as if all its mass were concentrated at the center of mass.** This is the basis for the well-known technique of finding an object's center of mass by balancing it. An object will balance on a pivot, as shown in FIGURE 12.25, only if the center of mass is directly above the pivot point. If the pivot is *not* under the center of mass, the gravitational torque will cause the object to rotate.

NOTE ▶ The point at which gravity acts is also called the *center of gravity*. As long as gravity is uniform over the object—always true for earthbound objects—there's no difference between center of mass and center of gravity. ◀

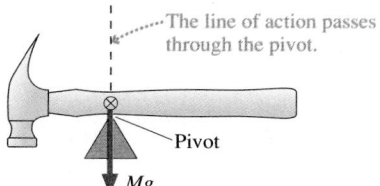

FIGURE 12.25 An object balances on a pivot that is directly under the center of mass.

The line of action passes through the pivot.

Pivot

Mg

EXAMPLE 12.9 **The gravitational torque on a beam**

The 4.00-m-long, 500 kg steel beam shown in FIGURE 12.26 is supported 1.20 m from the right end. What is the gravitational torque about the support?

FIGURE 12.26 A steel beam supported at one point.

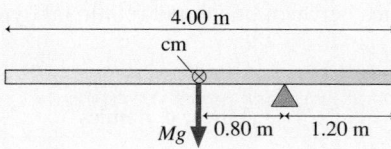

4.00 m

cm

Mg 0.80 m   1.20 m

MODEL The center of mass of the beam is at the midpoint. $x_{cm} = -0.80$ m is measured from the pivot point.

SOLVE This is a straightforward application of Equation 12.26. The gravitational torque is

$$\tau_{grav} = -Mgx_{cm} = -(500 \text{ kg})(9.80 \text{ m/s}^2)(-0.80 \text{ m})$$
$$= 3920 \text{ N m}$$

ASSESS The torque is positive because gravity tries to rotate the beam ccw around the point of support. Notice that the beam in Figure 12.26 is *not* in equilibrium. It will fall over unless other forces, not shown, are supporting it.

12.6 Rotational Dynamics

What does a torque do? **A torque causes an angular acceleration.** To see why, FIGURE 12.27 shows a rigid body undergoing *pure rotational motion* about a fixed and unmoving axis. This might be an unconstrained rotation about the object's center of mass, such as we considered in Section 12.2. Or it might be an object, such as a pulley or a turbine, rotating on an axle.

The forces $\vec{F}_1$, $\vec{F}_2$, $\vec{F}_3$, ... in Figure 12.27 are external forces acting on particles of masses m_1, m_2, m_3, ... that are part of the rigid body. These forces exert torques τ_1, τ_2, τ_3, ... about the rotation axis. The *net* torque on the object is the sum of the torques on all the individual particles in the object:

$$\tau_{net} = \sum_i \tau_i \tag{12.27}$$

Focus on particle i, which is acted on by force $\vec{F}_i$ and undergoes circular motion with radius r_i. In Chapter 8, we found that the radial component of $\vec{F}_i$ is responsible for the centripetal acceleration of circular motion, while the tangential component $(F_i)_t$ causes the particle to speed up or slow down with a tangential acceleration $(a_i)_t$. Newton's second law is

$$(F_i)_t = m_i(a_i)_t = m_i r_i \alpha \tag{12.28}$$

where in the last step we used the relationship between tangential and angular acceleration: $a_t = r\alpha$. The angular acceleration α does not have a subscript because *all particles in the object have the same angular acceleration*. That is, α is the angular acceleration of the entire object.

Multiplying both sides by r_i gives

$$r_i(F_i)_t = m_i r_i^2 \alpha \tag{12.29}$$

But $r_i(F_i)_t$ is the torque τ_i on particle i; hence Newton's second law for a single particle in the object is

$$\tau_i = m_i r_i^2 \alpha \tag{12.30}$$

Returning now to Equation 12.27, we see that the net torque on the object in Figure 12.27 is

$$\tau_{net} = \sum_i \tau_i = \sum_i m_i r_i^2 \alpha = \left(\sum_i m_i r_i^2 \right) \alpha \tag{12.31}$$

FIGURE 12.27 The external forces on a rigid body exert a torque about the rotation axis and thus cause an angular acceleration.

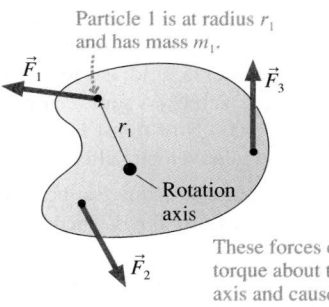

Particle 1 is at radius r_1 and has mass m_1.

$\vec{F}_1$

$\vec{F}_3$

r_1

Rotation axis

$\vec{F}_2$

These forces exert a net torque about the rotation axis and cause the object to have an angular acceleration.

In the last step, we factored out α by using the key idea that every particle in a rotating rigid body has the *same* angular acceleration.

You'll recognize the quantity in parentheses as the moment of inertia I. Substituting the moment of inertia into Equation 12.31 puts the final piece of the puzzle into place. An object that experiences a net torque τ_{net} about the axis of rotation undergoes an angular acceleration

$$\alpha = \frac{\tau_{net}}{I} \qquad \text{(Newton's second law for rotational motion)} \qquad (12.32)$$

where I is the object's moment of inertia *about the rotation axis*. This result, Newton's second law for rotation, is the fundamental equation of rigid-body dynamics.

In practice we often write $\tau_{net} = I\alpha$, but Equation 12.32 better conveys the idea that **torque is the cause of angular acceleration.** In the absence of a net torque ($\tau_{net} = 0$), the object either does not rotate ($\omega = 0$) or rotates with *constant* angular velocity ($\omega = $ constant).

Table 12.3 summarizes the analogies between linear and rotational dynamics.

TABLE 12.3 Rotational and linear dynamics

Rotational dynamics		Linear dynamics	
torque	τ_{net}	force	$\vec{F}_{net}$
moment of inertia	I	mass	m
angular acceleration	α	acceleration	$\vec{a}$
second law	$\alpha = \tau_{net}/I$	second law	$\vec{a} = \vec{F}_{net}/m$

EXAMPLE 12.10   **Rotating rockets**

Far out in space, a 100,000 kg rocket and a 200,000 kg rocket are docked at opposite ends of a motionless 90-m-long connecting tunnel. The tunnel is rigid and its mass is much less than that of either rocket. The rockets start their engines simultaneously, each generating 50,000 N of thrust in opposite directions. What is the structure's angular velocity after 30 s?

MODEL The entire structure can be modeled as two masses at the ends of a massless, rigid rod. There's no net force, so the structure does not undergo translational motion, but the thrusts do create torques that will give the structure angular acceleration and cause it to rotate. We'll assume the thrust forces are perpendicular to the connecting tunnel. This is an unconstrained rotation, so the structure will rotate about its center of mass.

VISUALIZE FIGURE 12.28 shows the rockets and defines distances r_1 and r_2 from the center of mass.

FIGURE 12.28 The thrusts exert a torque on the structure.

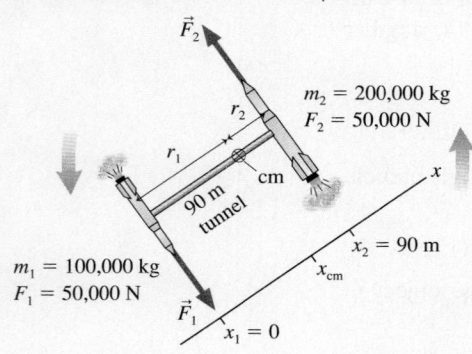

$m_2 = 200,000$ kg
$F_2 = 50,000$ N

$m_1 = 100,000$ kg
$F_1 = 50,000$ N

$x_2 = 90$ m
$x_1 = 0$
90 m tunnel

SOLVE Our strategy will be to use Newton's second law to find the angular acceleration, followed by rotational kinematics to find ω. We'll need to determine the moment of inertia, and that requires knowing the distances of the two rockets from the rotation axis. As we did in Example 12.1, we choose a coordinate system in which the masses are on the x-axis and in which m_1 is at the origin. Then

$$x_{cm} = \frac{m_1 x_1 + m_2 x_2}{m_1 + m_2}$$

$$= \frac{(100,000 \text{ kg})(0 \text{ m}) + (200,000 \text{ kg})(90 \text{ m})}{100,000 \text{ kg} + 200,000 \text{ kg}} = 60 \text{ m}$$

The structure's center of mass is $r_1 = 60$ m from the 100,000 kg rocket and $r_2 = 30$ m from the 200,000 kg rocket. The moment of inertia about the center of mass is

$$I = m_1 r_1^2 + m_2 r_2^2 = 540,000,000 \text{ kg m}^2$$

The two rocket thrusts exert net torque

$$\tau_{net} = r_1 F_1 + r_2 F_2 = (60 \text{ m})(50,000 \text{ N}) + (30 \text{ m})(50,000 \text{ N})$$

$$= 4,500,000 \text{ N m}$$

With I and τ_{net} now known, we can use Newton's second law to find the angular acceleration:

$$\alpha = \frac{\tau}{I} = \frac{4,500,000 \text{ N m}}{540,000,000 \text{ kg m}^2} = 0.00833 \text{ rad/s}^2$$

After 30 seconds, the structure's angular velocity is

$$\omega = \alpha \Delta t = 0.25 \text{ rad/s}$$

ASSESS Few of us have the experience to judge whether or not 0.25 rad/s is a reasonable answer to this problem. The significance of the example is to demonstrate the approach to a rotational dynamics problem.

STOP TO THINK 12.3   Rank in order, from largest to smallest, the angular accelerations α_a to α_e.

(a)

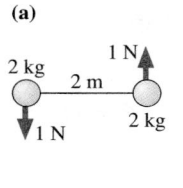

(b)

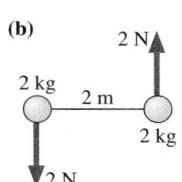

(c)

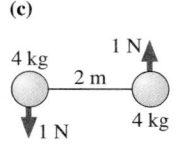

(d)

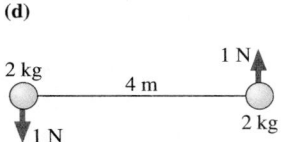

(e)

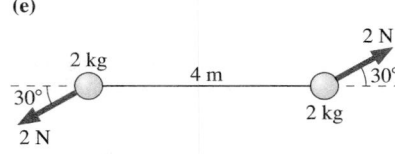

12.7 Rotation About a Fixed Axis

In this section we'll look at rigid bodies that rotate about a fixed axis. The problem-solving strategy for rotational dynamics is very similar to that for linear dynamics.

PROBLEM-SOLVING STRATEGY 12.1   Rotational dynamics problems

MODEL Model the object as a simple shape.

VISUALIZE Draw a pictorial representation to clarify the situation, define coordinates and symbols, and list known information.

■ Identify the axis about which the object rotates.
■ Identify forces and determine their distances from the axis. For most problems it will be useful to draw a free-body diagram.
■ Identify any torques caused by the forces and the signs of the torques.

SOLVE The mathematical representation is based on Newton's second law for rotational motion:

$$\tau_{net} = I\alpha \qquad \text{or} \qquad \alpha = \frac{\tau_{net}}{I}$$

■ Find the moment of inertia in Table 12.2 or, if needed, calculate it as an integral or by using the parallel-axis theorem.
■ Use rotational kinematics to find angles and angular velocities.

ASSESS Check that your result has the correct units, is reasonable, and answers the question.

Exercise 26

EXAMPLE 12.11 **Starting an airplane engine**

The engine in a small airplane is specified to have a torque of 60 N m. This engine drives a 2.0-m-long, 40 kg propeller. On start-up, how long does it take the propeller to reach 200 rpm?

MODEL The propeller can be modeled as a rod that rotates about its center. The engine exerts a torque on the propeller.

VISUALIZE **FIGURE 12.29** shows the propeller and the rotation axis.

FIGURE 12.29 A rotating airplane propeller.

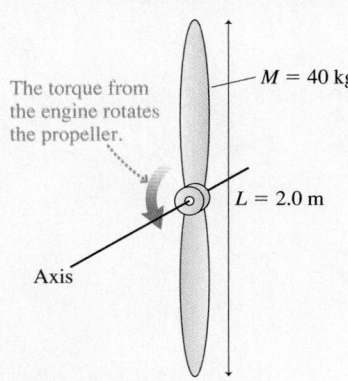

The torque from the engine rotates the propeller.

$M = 40$ kg

$L = 2.0$ m

Axis

SOLVE The moment of inertia of a rod rotating about its center is found from Table 12.2:

$$I = \frac{1}{12}ML^2 = \frac{1}{12}(40 \text{ kg})(2.0 \text{ m})^2 = 13.33 \text{ kg m}^2$$

The 60 N m torque of the engine causes an angular acceleration

$$\alpha = \frac{\tau}{I} = \frac{60 \text{ N m}}{13.33 \text{ kg m}^2} = 4.50 \text{ rad/s}^2$$

The time needed to reach $\omega_f = 200$ rpm $= 3.33$ rev/s $= 20.9$ rad/s is

$$\Delta t = \frac{\Delta \omega}{\alpha} = \frac{\omega_f - \omega_i}{\alpha} = \frac{20.9 \text{ rad/s} - 0 \text{ rad/s}}{4.5 \text{ rad/s}^2} = 4.6 \text{ s}$$

ASSESS We've assumed a constant angular acceleration, which is reasonable for the first few seconds while the propeller is still turning slowly. Eventually, air resistance and friction will cause opposing torques and the angular acceleration will decrease. At full speed, the negative torque due to air resistance and friction cancels the torque of the engine. Then $\tau_{net} = 0$ and the propeller turns at *constant* angular velocity with no angular acceleration.

EXAMPLE 12.12 **An off-center disk**

FIGURE 12.30 shows a piece of a large machine. A 10.0-cm-diameter, 5.0 kg disk turns on an axle. A vertical cable attached to the edge of the disk exerts a 100 N force but, initially, a pin keeps the disk from rotating. What is the initial angular acceleration of the disk when the pin is removed?

FIGURE 12.30 A disk rotates on an off-center axle after the pin is removed.

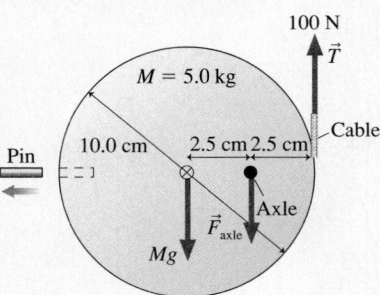

100 N
$\vec{T}$

$M = 5.0$ kg

10.0 cm   2.5 cm 2.5 cm

Cable

Pin

$\vec{F}_{axle}$

Axle

Mg

MODEL The disk has an off-center axle. Gravity and tension exert torques about the axle.

VISUALIZE Both the cable tension and gravity rotate the disk ccw, so their torques are positive.

SOLVE After the pin is removed, the forces on the disk are a downward gravitational force, an upward force from the cable, and a force exerted by the axle. The axle force, which is exerted at the pivot, does not contribute to the torque and doesn't affect the rota-

tion. The center of mass is to the *left* of the axle, at $x_{cm} = -\frac{1}{2}R$; thus the gravitational torque is

$$\tau_{grav} = -Mgx_{cm} = \frac{1}{2}MgR$$

This is a positive torque, as expected. The net torque, including the cable tension, is

$$\tau_{net} = \tau_{grav} + \tau_{cable} = \frac{1}{2}MgR + \frac{1}{2}RT = 3.73 \text{ N m}$$

To find the angular acceleration, we need to know the moment of inertia about the axle. This is where the parallel-axis theorem is useful. We know the moment of inertia about an axis through the center from Table 12.2. The axle is offset by $d = \frac{1}{2}R$. Thus

$$I = I_{cm} + Md^2 = \frac{1}{2}MR^2 + M\left(\frac{1}{2}R\right)^2 = \frac{3}{4}MR^2$$
$$= 9.38 \times 10^{-3} \text{ kg m}^2$$

The torque causes an angular acceleration

$$\alpha = \frac{\tau_{net}}{I} = \frac{3.73 \text{ N m}}{9.38 \times 10^{-3} \text{ kg m}^2} = 400 \text{ rad/s}^2$$

The angular acceleration is positive, indicating that the disk begins rotating in a ccw direction.

ASSESS As the disk rotates, τ_{net} will change as the moment arms change. Consequently, the disk will *not* have constant angular acceleration. This is simply the *initial* value of α.

Constraints Due to Ropes and Pulleys

Many important applications of rotational dynamics involve objects, such as pulleys, that are connected via ropes or belts to other objects. **FIGURE 12.31** shows a rope passing over a pulley and connected to an object in linear motion. If the rope does not slip as the pulley rotates, then the rope's speed v_{rope} must exactly match the speed of the rim of the pulley, which is $v_{\text{rim}} = |\omega|R$. If the pulley has an angular acceleration, the rope's acceleration a_{rope} must match the *tangential* acceleration of the rim of the pulley, $a_t = |\alpha|R$.

The object attached to the other end of the rope has the same speed and acceleration as the rope. Consequently, an object connected to a pulley of radius R by a rope that does not slip must obey the constraints

$$v_{\text{obj}} = |\omega|R$$

$$a_{\text{obj}} = |\alpha|R$$

(motion constraints for a nonslipping rope)   (12.33)

These constraints are very similar to the acceleration constraints introduced in Chapter 7 for two objects connected by a string or rope.

> **NOTE** ▶ The constraints are given as magnitudes. Specific problems will need to introduce signs that depend on the direction of motion and on the choice of coordinate system. ◀

FIGURE 12.31 The rope's motion must match the motion of the rim of the pulley.

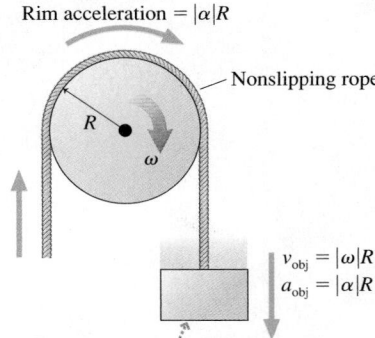

Rim speed $= |\omega|R$
Rim acceleration $= |\alpha|R$

Nonslipping rope

$v_{\text{obj}} = |\omega|R$
$a_{\text{obj}} = |\alpha|R$

The motion of the object must match the motion of the rim.

EXAMPLE 12.13   **Lowering a bucket**

A 2.0 kg bucket is attached to a massless string that is wrapped around a 1.0 kg, 4.0-cm-diameter cylinder, as shown in **FIGURE 12.32a**. The cylinder rotates on an axle through the center. The bucket is released from rest 1.0 m above the floor. How long does it take to reach the floor?

FIGURE 12.32 The falling bucket turns the cylinder.

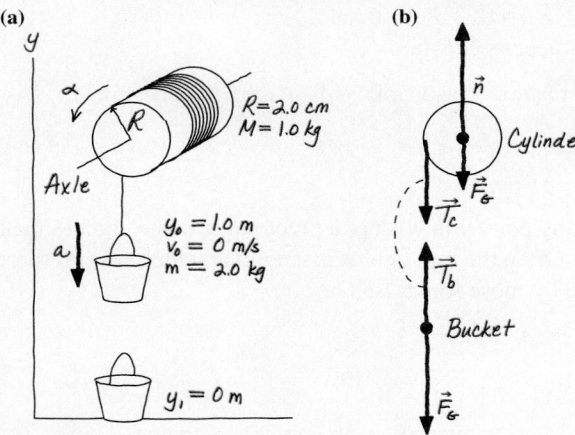

(a)

$R = 2.0$ cm
$M = 1.0$ kg

Axle

$y_0 = 1.0$ m
$v_0 = 0$ m/s
$m = 2.0$ kg

$y_1 = 0$ m

(b)

$\vec{n}$

Cylinder

$\vec{F}_C$

$\vec{T}_c$

$\vec{T}_b$

Bucket

$\vec{F}_G$

MODEL Assume the string does not slip.

VISUALIZE **FIGURE 12.32b** shows the free-body diagram for the cylinder and the bucket. The string tension exerts an upward force on the bucket and a downward force on the outer edge of the cylinder. The string is massless, so these two tension forces act as if they are an action/reaction pair: $T_b = T_c = T$.

SOLVE Newton's second law applied to the linear motion of the bucket is

$$ma_y = T - mg$$

where, as usual, the y-axis points upward. What about the cylinder? The only torque comes from the string tension. The moment

arm for the tension is $d = R$, and the torque is positive because the string turns the cylinder ccw. Thus $\tau_{\text{string}} = TR$ and Newton's second law for the rotational motion is

$$\alpha = \frac{\tau_{\text{net}}}{I} = \frac{TR}{\frac{1}{2}MR^2} = \frac{2T}{MR}$$

The moment of inertia of a cylinder rotating about a center axis was taken from Table 12.2.

The last piece of information we need is the constraint due to the fact that the string doesn't slip. Equation 12.33 relates only the absolute values, but in this problem α is positive (ccw acceleration) while a_y is negative (downward acceleration). Hence

$$a_y = -\alpha R$$

Using α from the cylinder's equation in the constraint, we find

$$a_y = -\alpha R = -\frac{2T}{MR}R = -\frac{2T}{M}$$

Thus the tension is $T = -\frac{1}{2}Ma_y$. If we use this value of the tension in the bucket's equation, we can solve for the acceleration:

$$ma_y = -\frac{1}{2}Ma_y - mg$$

$$a_y = -\frac{g}{(1 + M/2m)} = -7.84 \text{ m/s}^2$$

The time to fall through $\Delta y = -1.0$ m is found from kinematics:

$$\Delta y = \frac{1}{2}a_y(\Delta t)^2$$

$$\Delta t = \sqrt{\frac{2\Delta y}{a_y}} = \sqrt{\frac{2(-1.0 \text{ m})}{-7.84 \text{ m/s}^2}} = 0.50 \text{ s}$$

ASSESS The expression for the acceleration gives $a_y = -g$ if $M = 0$. This makes sense because the bucket would be in free fall if there were no cylinder. When the cylinder has mass, the downward force of gravity on the bucket has to accelerate the bucket *and* spin the cylinder. Consequently, the acceleration is reduced and the bucket takes longer to fall.

12.8 Static Equilibrium

Structures such as bridges are analyzed in engineering statics.

We now have two versions of Newton's second law: $\vec{F}_{net} = M\vec{a}$ for translational motion and $\tau_{net} = I\alpha$ for rotational motion. The condition for a rigid body to be in *static equilibrium* is both $\vec{F}_{net} = \vec{0}$ *and* $\tau_{net} = 0$. That is, no net force *and* no net torque. An important branch of engineering called *statics* analyzes buildings, dams, bridges, and other structures in total static equilibrium.

No matter which pivot point you choose, an object that is not rotating is not rotating about that point. This would seem to be a trivial statement, but it has an important implication: **For a rigid body in total equilibrium, there is no net torque about any point.** This is the basis of a problem-solving strategy.

PROBLEM-SOLVING
STRATEGY 12.2 **Static equilibrium problems**

MODEL Model the object as a simple shape.

VISUALIZE Draw a pictorial representation showing all forces and distances. List known information.

- Pick any point you wish as a pivot point. The net torque about this point is zero.
- Determine the moment arms of all forces about this pivot point.
- Determine the sign of each torque about this pivot point.

SOLVE The mathematical representation is based on the fact that an object in total equilibrium has no net force and no net torque:

$$\vec{F}_{net} = \vec{0} \qquad \text{and} \qquad \tau_{net} = 0$$

- Write equations for $\sum F_x = 0$, $\sum F_y = 0$, and $\sum \tau = 0$.
- Solve the three simultaneous equations.

ASSESS Check that your result is reasonable and answers the question.

Although you can pick any point you wish as a pivot point, some choices make the problem easier than others. Often the best choice is a point at which several forces act because the torques exerted by those forces will be zero.

EXAMPLE 12.14 **Lifting weights**

Weightlifting can exert extremely large forces on the body's joints and tendons. In the *strict curl* event, a standing athlete uses both arms to lift a barbell by moving only his forearms, which pivot at the elbows. The record weight lifted in the strict curl is over 200 pounds (about 900 N). FIGURE 12.33 shows the arm bones and the biceps, the main lifting muscle when the forearm is horizontal. What is the tension in the tendon connecting the biceps muscle to the bone while a 900 N barbell is held stationary in this position?

MODEL Model the arm as two rigid rods connected by a hinge. We'll ignore the arm's weight because it is so much less than that of the barbell. Although the tendon pulls at a slight angle, it is close enough to vertical that we'll treat it as such.

FIGURE 12.33 An arm holding a barbell.

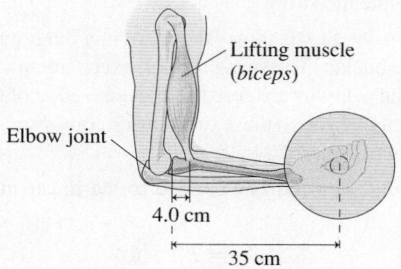

VISUALIZE **FIGURE 12.34** shows the forces acting on our simplified model of the forearm. The biceps pulls the forearm up against the upper arm at the elbow, so the force $\vec{F}_{elbow}$ *on* the forearm at the elbow—a force due to the upper arm—is a downward force.

FIGURE 12.34 A pictorial representation of the forces involved.

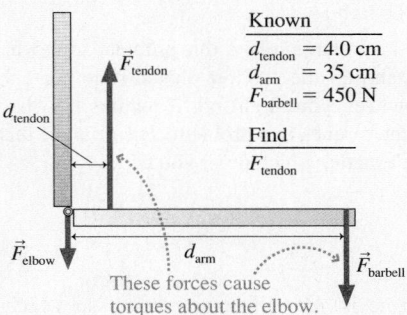

Known
$d_{tendon} = 4.0$ cm
$d_{arm} = 35$ cm
$F_{barbell} = 450$ N

Find
F_{tendon}

These forces cause torques about the elbow.

SOLVE Static equilibrium requires both the net force *and* the net torque on the forearm to be zero. Only the y-component of force is relevant, and setting it to zero gives a first equation:

$$\sum F_y = F_{tendon} - F_{elbow} - F_{barbell} = 0$$

Because each arm supports half the weight of the barbell, $F_{barbell} = 450$ N. We don't know either F_{tendon} or F_{elbow}, nor does the force equation give us enough information to find them. But the fact that the net torque also must be zero gives us that extra information. The torque is zero about *every* point, so we can choose any point we wish to calculate the torque. The elbow joint is a convenient point because force $\vec{F}_{elbow}$ exerts no torque about this point; its moment arm is zero. Thus the torque equation is

$$\tau_{net} = d_{tendon} F_{tendon} - d_{arm} F_{barbell} = 0$$

The tension in the tendon tries to rotate the arm ccw, so it produces a positive torque. Similarly, the torque due to the barbell is negative. We can solve the torque equation for F_{tendon} to find

$$F_{tendon} = F_{barbell} \frac{d_{arm}}{d_{tendon}} = (450 \text{ N}) \frac{35 \text{ cm}}{4.0 \text{ cm}} = 3900 \text{ N}$$

ASSESS The short distance d_{tendon} from the tendon to the elbow joint means that the force supplied by the biceps has to be very large to counter the torque generated by a force applied at the opposite end of the forearm. Although we ended up not needing the force equation in this problem, we could now use it to calculate that the force exerted at the elbow is $F_{elbow} = 3450$ N. These large forces can easily damage the tendon or the elbow.

EXAMPLE 12.15 **Walking the plank**

Adrienne (50 kg) and Bo (90 kg) are playing on a 100 kg rigid plank resting on the supports seen in **FIGURE 12.35**. If Adrienne stands on the left end, can Bo walk all the way to the right end without the plank tipping over? If not, how far can he get past the support on the right?

FIGURE 12.35 Adrienne and Bo on the plank.

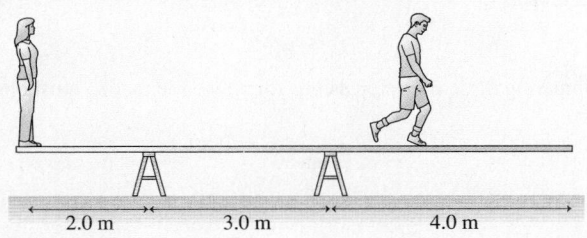

2.0 m   3.0 m   4.0 m

MODEL Model Adrienne and Bo as particles. Assume the plank is uniform, with its center of mass at the center.

VISUALIZE **FIGURE 12.36** shows the forces acting on the plank. Both supports exert upward forces. $\vec{n}_A$ and $\vec{n}_B$ are the normal forces of Adrienne's and Bo's feet pushing down on the board.

SOLVE Because the plank is resting on the supports, not held down, forces $\vec{n}_1$ and $\vec{n}_2$ must point upward. (The supports could pull down if the plank were nailed to them, but that's not the case

FIGURE 12.36 A pictorial representation of the forces on the plank.

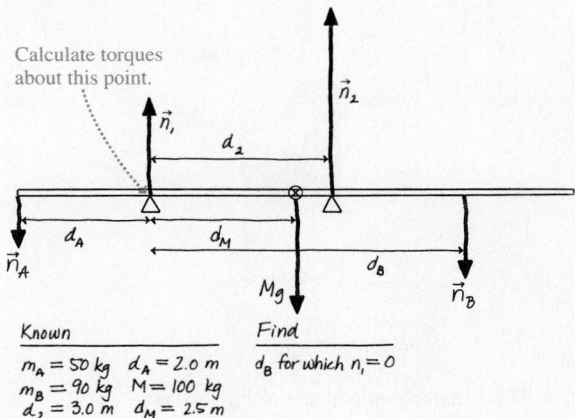

Calculate torques about this point.

Known		Find
$m_A = 50$ kg	$d_A = 2.0$ m	d_B for which $n_1 = 0$
$m_B = 90$ kg	$M = 100$ kg	
$d_2 = 3.0$ m	$d_M = 2.5$ m	

here.) Force $\vec{n}_1$ will decrease as Bo moves to the right, and the tipping point occurs when $n_1 = 0$. The plank remains in static equilibrium right up to the tipping point, so both the net force and the net torque on it are zero. The force equation is

$$\sum F_y = n_1 + n_2 - n_A - n_B - Mg$$
$$= n_1 + n_2 - m_A g - m_B g - Mg = 0$$

Continued

Adrienne is at rest, with zero net force, so her downward force on the board, an action/reaction pair with the upward normal force of the board on her, equals her weight: $n_A = m_A g$. Bo's center of mass oscillates up and down as he walks, so he's *not* in equilibrium and, strictly speaking, $n_B \neq m_B g$. But we'll assume that he edges out onto the board slowly, with minimal bouncing, in which case $n_B = m_B g$ is a reasonable approximation.

We can again choose any point we wish for calculating torque. Let's use the support on the left. Adrienne and the support on the right exert positive torques about this point; the other forces exert negative torques. Force $\vec{n}_1$ exerts no torque, since it acts at the pivot point. Thus the torque equation is

$$\tau_{net} = d_A m_A g - d_B m_B g - d_M M g + d_2 n_2 = 0$$

At the tipping point, where $n_1 = 0$, the force equation gives $n_2 = (m_A + m_B + M)g$. Substituting this into the torque equation and then solving for Bo's position give

$$d_B = \frac{d_A m_A - d_M M + d_2 (m_A + m_B + M)}{m_B} = 6.3 \text{ m}$$

Bo doesn't quite make it to the end. The plank tips when he's 6.3 m past the left support, our pivot point, and thus 3.3 m past the support on the right.

ASSESS We could have solved this problem somewhat more simply had we chosen the support on the right for calculating the torques. However, you might not recognize the "best" point for calculating the torques in a problem. The point of this example is that it doesn't matter which point you choose.

EXAMPLE 12.16 **Will the ladder slip?**

A 3.0-m-long ladder leans against a frictionless wall at an angle of 60°. What is the minimum value of μ_s, the coefficient of static friction with the ground, that prevents the ladder from slipping?

MODEL The ladder is a rigid rod of length L. To not slip, it must be in both translational equilibrium ($\vec{F}_{net} = \vec{0}$) and rotational equilibrium ($\tau_{net} = 0$).

VISUALIZE FIGURE 12.37 shows the ladder and the forces acting on it.

FIGURE 12.37 A ladder in total equilibrium.

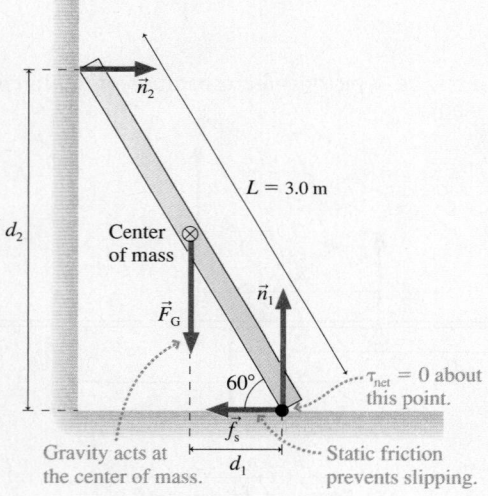

$L = 3.0$ m

Center of mass

d_2

$\vec{n}_2$

$\vec{n}_1$

$\vec{F}_G$

60°

$\tau_{net} = 0$ about this point.

$\vec{f}_s$

d_1

Gravity acts at the center of mass.

Static friction prevents slipping.

SOLVE The x- and y-components of $\vec{F}_{net} = \vec{0}$ are

$$\sum F_x = n_2 - f_s = 0$$
$$\sum F_y = n_1 - Mg = 0$$

The net torque is zero about *any* point, so which should we choose? The bottom corner of the ladder is a good choice because two forces pass through this point and have no torque about it. The torque about the bottom corner is

$$\tau_{net} = d_1 F_G - d_2 n_2 = \frac{1}{2}(L\cos 60°)Mg - (L\sin 60°)n_2 = 0$$

The signs are based on the observation that $\vec{F}_G$ would cause the ladder to rotate ccw while $\vec{n}_2$ would cause it to rotate cw. All together, we have three equations in the three unknowns n_1, n_2, and f_s. If we solve the third for n_2,

$$n_2 = \frac{\frac{1}{2}(L\cos 60°)Mg}{L\sin 60°} = \frac{Mg}{2\tan 60°}$$

we can then substitute this into the first to find

$$f_s = \frac{Mg}{2\tan 60°}$$

Our model of friction is $f_s \leq f_{s\,max} = \mu_s n_1$. We can find n_1 from the second equation: $n_1 = Mg$. Using this, the model of static friction tells us that

$$f_s \leq \mu_s Mg$$

Comparing these two expressions for f_s, we see that μ_s must obey

$$\mu_s \geq \frac{1}{2\tan 60°} = 0.29$$

Thus the minimum value of the coefficient of static friction is 0.29.

ASSESS You know from experience that you can lean a ladder or other object against a wall if the ground is "rough," but it slips if the surface is too smooth. 0.29 is a "medium" value for the coefficient of static friction, which is reasonable.

Balance and Stability

If you tilt a box up on one edge by a small amount and let go, it falls back down. If you tilt it too much, it falls over. And if you tilt "just right," you can get the box to balance on its edge. What determines these three possible outcomes?

FIGURE 12.38 illustrates the idea with a car, but the results are general and apply in many situations. As long as the object's center of mass remains over the base of

FIGURE 12.38 Stability depends on the position of the center of mass.

(a) The torque due to gravity will bring the car back down as long as the center of mass is above the base of support.

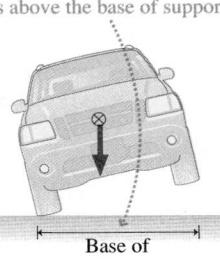

Base of support

(b) The vehicle is at the critical angle θ_c when its center of mass is exactly over the pivot.

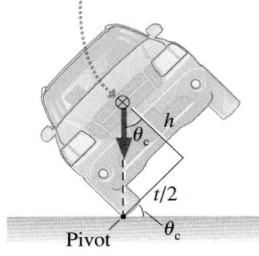

h

$t/2$

Pivot θ_c

(c) Now the center of mass is outside the base of support. Torque due to gravity will cause the car to roll over.

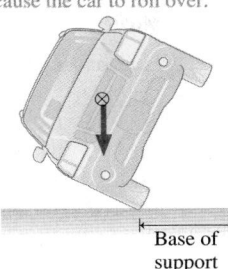

Base of support

support, torque due to gravity will rotate the object back toward its stable equilibrium position. But if the center of mass gets outside the base of support, the torque due to gravity causes a rotation in the opposite direction. Now the box falls over or the car rolls over.

A *critical angle* θ_c is reached when the center of mass is directly over the pivot point. This is the point of balance, with no net torque. For vehicles, the distance between the tires is called the track width t. If the height of the center of mass is h, you can see from **FIGURE 12.38b** that the critical angle is

$$\theta_c = \tan^{-1}\left(\frac{t}{2h}\right)$$

For passenger cars with $h \approx 0.33t$, the critical angle is $\theta_c \approx 57°$. But for a sport utility vehicle (SUV) with $h \approx 0.47t$, a higher center of mass, the critical angle is only $\theta_c \approx 47°$. Loading an SUV with cargo further raises the center of gravity, especially if the roof rack is used, thus reducing θ_c even more. Various automobile safety groups have determined that a vehicle with $\theta_c > 50°$ is unlikely to roll over in an accident. A rollover becomes increasingly likely when θ_c is reduced below 50°. The general rule is that **a wider base of support and/or a lower center of mass improve stability.**

This dancer balances *en pointe* by having her center of mass directly over her toes, her base of support.

EXAMPLE 12.17 **Tilting cans**

A typical can of food is 7.5 cm in diameter. What is the tallest can of food that can rest on a 30° incline without falling over?

MODEL Assume the food inside is uniformly distributed so that the center of mass is at the center of the can.

FIGURE 12.39 A can balanced at the critical angle.

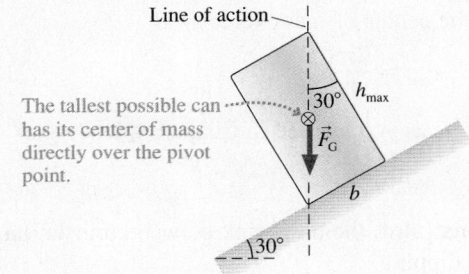

Line of action

The tallest possible can has its center of mass directly over the pivot point.

30° h_{max}

$\vec{F}_G$

b

30°

VISUALIZE **FIGURE 12.39** shows a can at the critical angle. This is the tallest possible can. A shorter can would have its center of mass inside the base of support and would be stable; a taller can would have its center of mass outside the base of support and would fall over.

SOLVE For a can whose height puts it at the critical angle, the line of action is a diagonal through the can. If the height is h_{max} and the diameter of the base b, we see from the figure that $\tan 30° = b/h_{max}$ and thus

$$h_{max} = \frac{b}{\tan 30°} = \frac{7.5 \text{ cm}}{\tan 30°} = 13 \text{ cm}$$

ASSESS A typical can of soup is just under 13 cm tall. It will stand on a 30° incline—try it!—but anything taller will fall over.

A student holds a meter stick straight out with one or more masses dangling from it. Rank in order, from most difficult to least difficult, how hard it will be for the student to keep the meter stick from rotating.

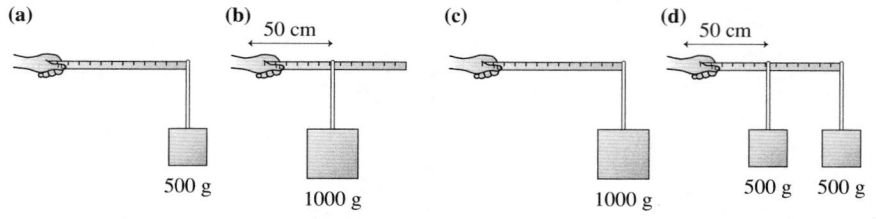

12.9 Rolling Motion

FIGURE 12.40 The trajectories of the center of a wheel and of a point on the rim are seen in a time-exposure photograph.

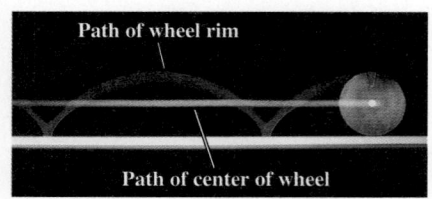

Rolling is a *combination motion* in which an object rotates about an axis that is moving along a straight-line trajectory. For example, FIGURE 12.40 is a time-exposure photo of a rolling wheel with one lightbulb on the axis and a second lightbulb at the edge. The axis light moves straight ahead, but the edge light follows a curve called a *cycloid*. Let's see if we can understand this interesting motion. We'll consider only objects that roll without slipping.

FIGURE 12.41 shows a round object—a wheel or a sphere—that rolls forward exactly one revolution. The point that had been on the bottom follows the cycloid, the curve you saw in Figure 12.40, to the top and back to the bottom. *Because the object doesn't slip,* the center of mass moves forward exactly one circumference: $\Delta x_{cm} = 2\pi R$.

FIGURE 12.41 An object rolling through one revolution.

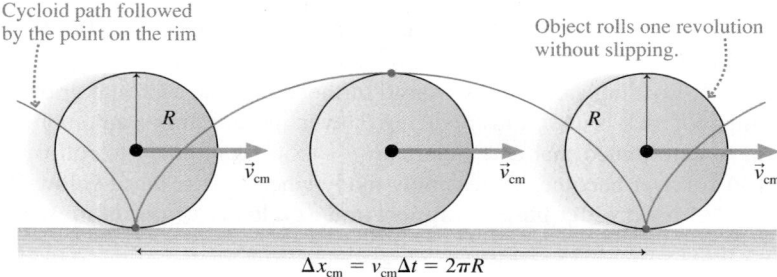

We can also write the distance traveled in terms of the velocity of the center of mass: $\Delta x_{cm} = v_{cm} \Delta t$. But Δt, the time it takes the object to make one complete revolution, is nothing other than the rotation period T. In other words, $\Delta x_{cm} = v_{cm} T$.

These two expressions for Δx_{cm} come from two perspectives on the motion: one looking at the rotation and the other looking at the translation of the center of mass. But it's the same distance no matter how you look at it, so these two expressions must be equal. Consequently,

$$\Delta x_{cm} = 2\pi R = v_{cm} T \tag{12.34}$$

If we divide by T, we can write the center-of-mass velocity as

$$v_{cm} = \frac{2\pi}{T} R \tag{12.35}$$

But $2\pi/T$ is the angular velocity ω, as you learned in Chapter 4, leading to

$$v_{cm} = R\omega \tag{12.36}$$

Equation 12.36 is the **rolling constraint,** the basic link between translation and rotation for objects that roll without slipping.

NOTE ▶ The rolling constraint is equivalent to Equation 12.33 for the speed of a rope that doesn't slip as it passes over a pulley. ◀

Let's look carefully at a particle in the rolling object. As FIGURE 12.42a shows, the position vector $\vec{r}_i$ for particle i is the vector sum $\vec{r}_i = \vec{r}_{cm} + \vec{r}_{i,\,rel}$. Taking the time derivative of this equation, we can write the velocity of particle i as

$$\vec{v}_i = \vec{v}_{cm} + \vec{v}_{i,\,rel} \qquad (12.37)$$

In other words, the velocity of particle i can be divided into two parts: the velocity $\vec{v}_{cm}$ of the object as a whole plus the velocity $\vec{v}_{i,\,rel}$ of particle i relative to the center of mass (i.e., the velocity that particle i would have if the object were only rotating and had no translational motion).

FIGURE 12.42b applies this idea to point P at the very bottom of the rolling object, the point of contact between the object and the surface. This point is moving around the center of the object at angular velocity ω, so $v_{i,\,rel} = -R\omega$. The negative sign indicates that the motion is cw. At the same time, the center-of-mass velocity, Equation 12.36, is $v_{cm} = R\omega$. Adding these, we find that the velocity of point P, the lowest point, is $v_i = 0$. In other words, **the point on the bottom of a rolling object is instantaneously at rest.**

Although this seems surprising, it is really what we mean by "rolling without slipping." If the bottom point had a velocity, it would be moving horizontally relative to the surface. In other words, it would be slipping or sliding across the surface. To roll without slipping, the bottom point, the point touching the surface, must be at rest.

FIGURE 12.43 shows how the velocity vectors at the top, center, and bottom of a rotating wheel are found by adding the rotational velocity vectors to the center-of-mass velocity. You can see that $v_{bottom} = 0$ and that $v_{top} = 2R\omega = 2v_{cm}$.

FIGURE 12.43 Rolling without slipping is a combination of translation and rotation.

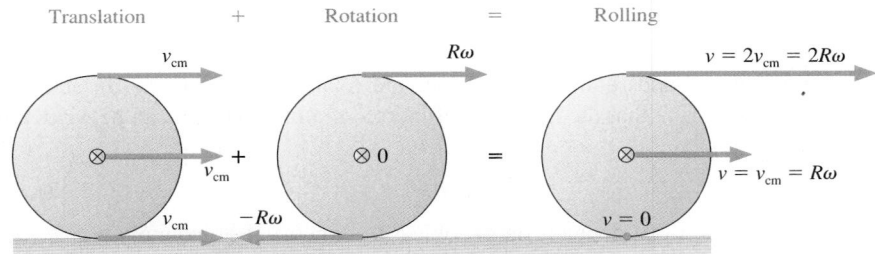

Kinetic Energy of a Rolling Object

We found earlier that the rotational kinetic energy of a rigid body in pure rotational motion is $K_{rot} = \frac{1}{2}I\omega^2$. Now we would like to find the kinetic energy of an object that rolls without slipping, a combination of rotational and translation motion.

We begin with the observation that the bottom point in FIGURE 12.44 is instantaneously at rest. Consequently, we can think of an axis through P as an *instantaneous axis of rotation*. The idea of an instantaneous axis of rotation seems a little far-fetched, but it is confirmed by looking at the instantaneous velocities of the center point and the top point. We found these in Figure 12.43 and they are shown again in Figure 12.44. They are exactly what you would expect as the tangential velocity $v_t = r\omega$ for rotation about P at distances R and $2R$.

From this perspective, the object's motion is pure rotation about point P. Thus the kinetic energy is that of pure rotation:

$$K = K_{\text{rotation about P}} = \frac{1}{2}I_P\omega^2 \qquad (12.38)$$

I_P is the moment of inertia for rotation about point P. We can use the parallel-axis theorem to write I_P in terms of the moment of inertia I_{cm} about the center of mass. Point P is displaced by distance $d = R$; thus

$$I_P = I_{cm} + MR^2$$

FIGURE 12.42 The motion of a particle in the rolling object.

(a)

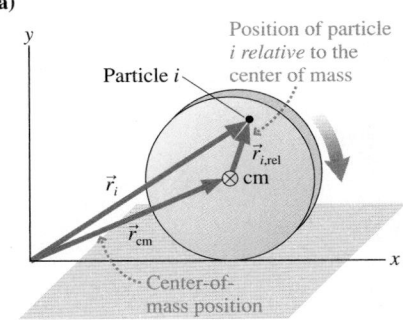

(b)

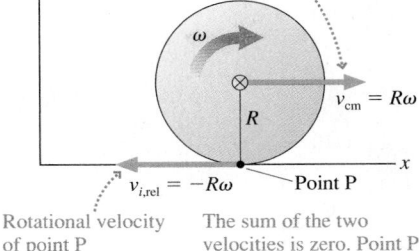

Rotational velocity of point P.   The sum of the two velocities is zero. Point P is instantaneously at rest.

FIGURE 12.44 Rolling motion is an instantaneous rotation about point P.

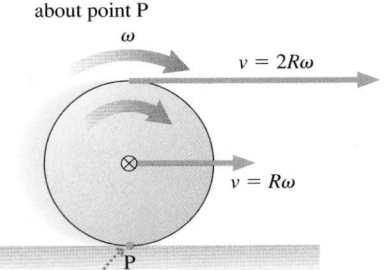

Point P, which is instantaneously at rest, is the pivot point for the entire object.

Using this expression in Equation 12.38 gives us the kinetic energy:

$$K = \frac{1}{2}I_{cm}\omega^2 + \frac{1}{2}M(R\omega)^2 \qquad (12.39)$$

We know from the rolling constraint that $R\omega$ is the center-of-mass velocity v_{cm}. Thus the kinetic energy of a rolling object is

$$K_{rolling} = \frac{1}{2}I_{cm}\omega^2 + \frac{1}{2}Mv_{cm}^2 = K_{rot} + K_{cm} \qquad (12.40)$$

In other words, **the rolling motion of a rigid body can be described as a translation of the center of mass (with kinetic energy K_{cm}) plus a rotation about the center of mass (with kinetic energy K_{rot}).**

The Great Downhill Race

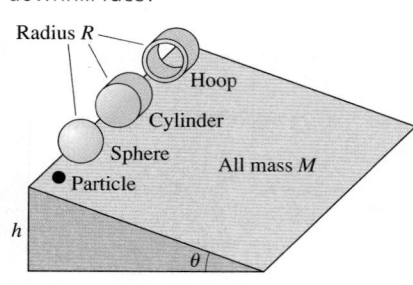

FIGURE 12.45 Which will win the downhill race?

FIGURE 12.45 shows a contest in which a sphere, a cylinder, and a circular hoop, all of mass M and radius R, are placed at height h on a slope of angle θ. All three are released from rest at the same instant of time and roll down the ramp without slipping. To make things more interesting, they are joined by a particle of mass M that slides down the ramp without friction. Which one will win the race to the bottom of the hill? Does rotation affect the outcome?

An object's initial gravitational potential energy is transformed into kinetic energy as it rolls (or slides, in the case of the particle). The kinetic energy, as we just discovered, is a combination of translational and rotational kinetic energy. If we choose the bottom of the ramp as the zero point of potential energy, the statement of energy conservation $K_f = U_i$ can be written

$$\frac{1}{2}I_{cm}\omega^2 + \frac{1}{2}Mv_{cm}^2 = Mgh \qquad (12.41)$$

The translational and rotational velocities are related by $\omega = v_{cm}/R$. In addition, notice from Table 12.2 that the moments of inertia of all the objects can be written in the form

$$I_{cm} = cMR^2 \qquad (12.42)$$

where c is a constant that depends on the object's geometry. For example, $c = \frac{2}{5}$ for a sphere but $c = 1$ for a circular hoop. Even the particle can be represented by $c = 0$, which eliminates the rotational kinetic energy.

With this information, Equation 12.41 becomes

$$\frac{1}{2}(cMR^2)\left(\frac{v_{cm}}{R}\right)^2 + \frac{1}{2}Mv_{cm}^2 = \frac{1}{2}M(1 + c)v_{cm}^2 = Mgh$$

Thus the finishing speed of an object with $I = cMR^2$ is

$$v_{cm} = \sqrt{\frac{2gh}{1 + c}} \qquad (12.43)$$

The final speed is independent of both M and R, but it does depend on the *shape* of the rolling object. The particle, with the smallest value of c, will finish with the highest speed, while the circular hoop, with the largest c, will be the slowest. In other words, the rolling aspect of the motion *does* matter!

We can use Equation 12.43 to find the acceleration a_{cm} of the center of mass. The objects move through distance $\Delta x = h/\sin\theta$, so we can use constant-acceleration kinematics to find

$$v_{cm}^2 = 2a_{cm}\Delta x$$

$$a_{cm} = \frac{v_{cm}^2}{2\Delta x} = \frac{2gh/(1 + c)}{2h/\sin\theta} = \frac{g\sin\theta}{1 + c} \qquad (12.44)$$

Recall, from Chapter 2, that $a_{particle} = g \sin\theta$ is the acceleration of a particle sliding down a frictionless incline. We can use this fact to write Equation 12.44 in an interesting form:

$$a_{cm} = \frac{a_{particle}}{1 + c} \qquad (12.45)$$

This analysis leads us to the conclusion that **the acceleration of a rolling object is less—in some cases significantly less—than the acceleration of a particle.** The reason is that the energy has to be shared between translational kinetic energy and rotational kinetic energy. A particle, by contrast, can put all its energy into translational kinetic energy.

FIGURE 12.46 shows the results of the race. The simple particle wins by a fairly wide margin. Of the solid objects, the sphere has the largest acceleration. Even so, its acceleration is only 71% the acceleration of a particle. The acceleration of the circular hoop, which comes in last, is a mere 50% that of a particle.

> NOTE ▶ The objects having the largest acceleration are those whose mass is most concentrated near the center. Placing the mass far from the center, as in the hoop, increases the moment of inertia. Thus it requires a larger effort to get a hoop rolling than to get a sphere of equal mass rolling. ◀

FIGURE 12.46 And the winner is...

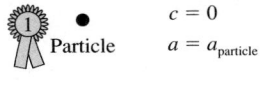
Particle
$c = 0$
$a = a_{particle}$

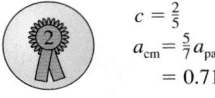
Solid sphere
$c = \frac{2}{5}$
$a_{cm} = \frac{5}{7} a_{particle}$
$= 0.71 a_{particle}$

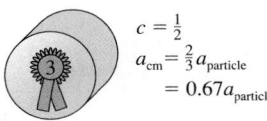
Solid cylinder
$c = \frac{1}{2}$
$a_{cm} = \frac{2}{3} a_{particle}$
$= 0.67 a_{particle}$

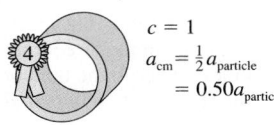
Circular hoop
$c = 1$
$a_{cm} = \frac{1}{2} a_{particle}$
$= 0.50 a_{particle}$

12.10 The Vector Description of Rotational Motion

Rotation about a fixed axis, such as an axle, can be described in terms of a scalar angular velocity ω and a scalar torque τ, using a plus or minus sign to indicate the direction of rotation. This is very much analogous to the one-dimensional kinematics of Chapter 2. For more general rotational motion, angular velocity, torque, and other quantities must be treated as *vectors*. We won't go into much detail because the subject rapidly gets very complicated, but we will sketch some important basic ideas.

The Angular Velocity Vector

FIGURE 12.47 shows a rotating rigid body. We can define an angular velocity vector $\vec{\omega}$ as follows:

- The magnitude of $\vec{\omega}$ is the object's angular velocity ω.
- $\vec{\omega}$ points along the axis of rotation in the direction given by the *right-hand rule* illustrated in Figure 12.47.

If the object rotates in the xy-plane, the vector $\vec{\omega}$ points along the z-axis. The scalar angular velocity $\omega = v_t/r$ that we've been using is now seen to be ω_z, the z-component of the vector $\vec{\omega}$. You should convince yourself that the sign convention for ω (positive for ccw rotation, negative for cw rotation) is equivalent to having the vector $\vec{\omega}$ pointing in the positive z-direction or the negative z-direction.

The Cross Product of Two Vectors

We defined the torque exerted by force $\vec{F}$ to be $\tau = rF \sin\phi$. The quantity F is the magnitude of the force vector $\vec{F}$, and the distance r is really the magnitude of the position vector $\vec{r}$. Hence torque looks very much like a product of the two vectors $\vec{r}$ and $\vec{F}$. Previously, in conjunction with the definition of work, we introduced the dot product of two vectors: $\vec{A} \cdot \vec{B} = AB \cos\alpha$, where α is the angle between the vectors. $\tau = rF \sin\phi$ is a different way of multiplying vectors that depends on the *sine* of the angle between them.

FIGURE 12.47 The angular velocity vector $\vec{\omega}$ is found using the right-hand rule.

1. Using your right hand, curl your fingers in the direction of rotation with your thumb along the rotation axis.

2. Your thumb is then pointing in the direction of $\vec{\omega}$.

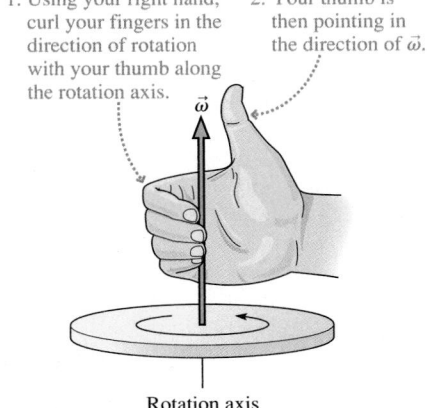

Rotation axis

FIGURE 12.48 The cross product $\vec{A} \times \vec{B}$, is a vector perpendicular to the plane of vectors $\vec{A}$ and $\vec{B}$.

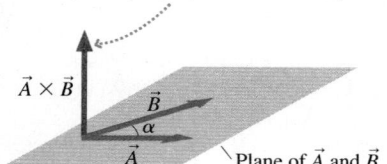

The cross product is perpendicular to the plane.

$\vec{A} \times \vec{B}$
$\vec{B}$
α
$\vec{A}$
Plane of $\vec{A}$ and $\vec{B}$

FIGURE 12.48 shows two vectors, $\vec{A}$ and $\vec{B}$, with angle α between them. We define the **cross product** of $\vec{A}$ and $\vec{B}$ as the vector

$$\vec{A} \times \vec{B} \equiv (AB \sin \alpha, \text{ in the direction given by the right-hand rule}) \qquad (12.46)$$

The symbol $\times$ between the vectors is *required* to indicate a cross product. The cross product is also called the **vector product** because the result is a vector.

The **right-hand rule,** which specifies the direction of $\vec{A} \times \vec{B}$, can be stated in three different but equivalent ways:

Using the right-hand rule

Spread your *right* thumb and index finger apart by angle α. Bend your middle finger so that it is *perpendicular* to your thumb and index finger. Orient your hand so that your thumb points in the direction of $\vec{A}$ and your index finger in the direction of $\vec{B}$. Your middle finger now points in the direction of $\vec{A} \times \vec{B}$.

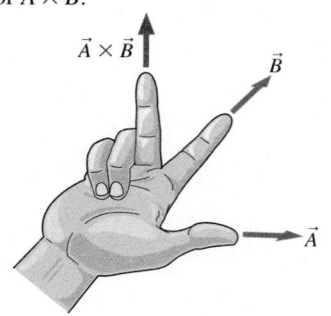

Make a loose fist with your *right* hand with your thumb extended outward. Orient your hand so that your thumb is perpendicular to the plane of $\vec{A}$ and $\vec{B}$ and your fingers are curling *from* the line of vector $\vec{A}$ *toward* the line of vector $\vec{B}$. Your thumb now points in the direction of $\vec{A} \times \vec{B}$.

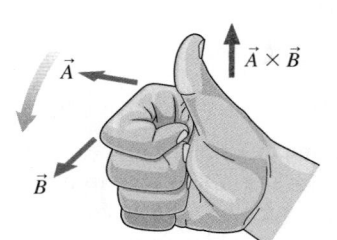

Imagine using a screwdriver to turn the slot in the head of a screw from the direction of $\vec{A}$ to the direction of $\vec{B}$. The screw will move either "in" or "out." The direction in which the screw moves is the direction of $\vec{A} \times \vec{B}$.

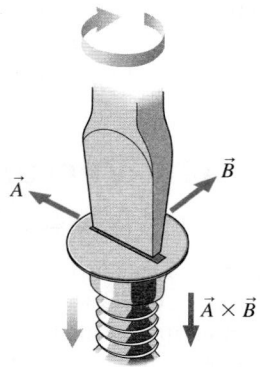

These methods are easier to demonstrate than to describe in words! Your instructor will show you how they work. Some individuals find one method of thinking about the direction of the cross product easier than the others, but they all work, and you'll soon find the method that works best for you.

Referring back to Figure 12.48, you should use the right-hand rule to convince yourself that the cross product $\vec{A} \times \vec{B}$ is a vector that points *upward,* perpendicular to the plane of $\vec{A}$ and $\vec{B}$. **FIGURE 12.49** shows that the cross product, like the dot product, depends on the angle between the two vectors. Notice the two special cases: $\vec{A} \times \vec{B} = \vec{0}$ when $\alpha = 0°$ (parallel vectors) and $\vec{A} \times \vec{B}$ has its maximum magnitude AB when $\alpha = 90°$ (perpendicular vectors).

FIGURE 12.49 The magnitude of the cross-product vector increases from 0 to AB as α increases from 0° to 90°.

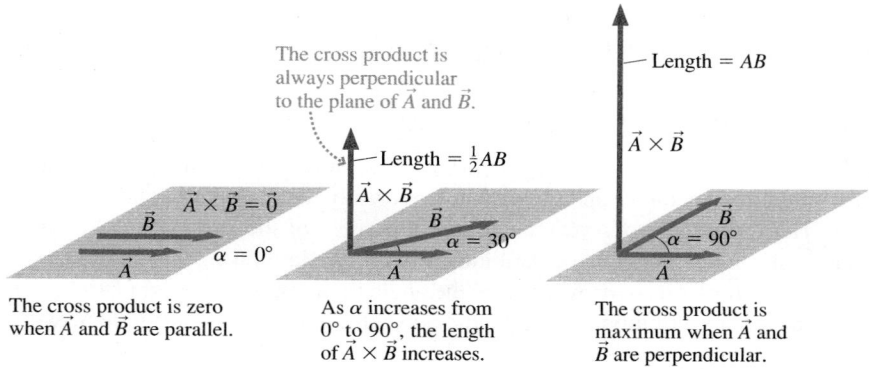

The cross product is always perpendicular to the plane of $\vec{A}$ and $\vec{B}$.

Length $= \frac{1}{2}AB$

Length $= AB$

$\vec{A} \times \vec{B} = \vec{0}$
$\vec{B}$
$\alpha = 0°$
$\vec{A}$

$\vec{A} \times \vec{B}$
$\vec{B}$
$\alpha = 30°$
$\vec{A}$

$\vec{A} \times \vec{B}$
$\vec{B}$
$\alpha = 90°$
$\vec{A}$

The cross product is zero when $\vec{A}$ and $\vec{B}$ are parallel.

As α increases from 0° to 90°, the length of $\vec{A} \times \vec{B}$ increases.

The cross product is maximum when $\vec{A}$ and $\vec{B}$ are perpendicular.

EXAMPLE 12.18 **Calculating a cross product**

FIGURE 12.50 shows vectors $\vec{C}$ and $\vec{D}$ in the plane of the page. What is the cross product $\vec{E} = \vec{C} \times \vec{D}$?

FIGURE 12.50 Vectors $\vec{C}$ and $\vec{D}$.

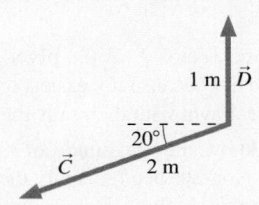

SOLVE The angle between the two vectors is $\alpha = 110°$. Consequently, the magnitude of the cross product is

$$E = CD \sin \alpha = (2 \text{ m})(1 \text{ m}) \sin(110°) = 1.88 \text{ m}^2$$

The direction of $\vec{E}$ is given by the right-hand rule. To curl your right fingers from $\vec{C}$ to $\vec{D}$, you have to point your thumb *into* the page. Alternatively, if you turned a screwdriver from $\vec{C}$ to $\vec{D}$ you would be driving a screw *into* the page. Thus

$$\vec{E} = (1.88 \text{ m}^2, \text{ into page})$$

ASSESS Notice that $\vec{E}$ has units of m^2.

The cross product has three important properties:

1. The product $\vec{A} \times \vec{B}$ is *not* equal to the product $\vec{B} \times \vec{A}$. That is, the cross product does not obey the commutative rule $ab = ba$ that you know from arithmetic. In fact, you can see from the right-hand rule that the product $\vec{B} \times \vec{A}$ points in exactly the opposite direction from $\vec{A} \times \vec{B}$. Thus, as FIGURE 12.51a shows,

$$\vec{B} \times \vec{A} = -\vec{A} \times \vec{B}$$

2. In a *right-handed coordinate system,* which is the standard coordinate system of science and engineering, the z-axis is oriented relative to the xy-plane such that the unit vectors obey $\hat{\imath} \times \hat{\jmath} = \hat{k}$. This is shown in FIGURE 12.51b. You can also see from this figure that $\hat{\jmath} \times \hat{k} = \hat{\imath}$ and $\hat{k} \times \hat{\imath} = \hat{\jmath}$.

3. The derivative of a cross product is

$$\frac{d}{dt}(\vec{A} \times \vec{B}) = \frac{d\vec{A}}{dt} \times \vec{B} + \vec{A} \times \frac{d\vec{B}}{dt} \tag{12.47}$$

FIGURE 12.51 Properties of the cross product.

(a)

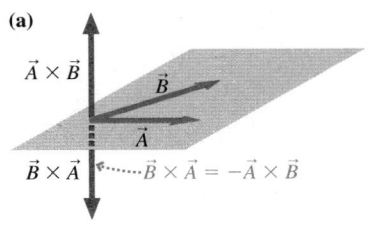

(b)

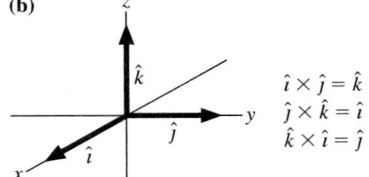

$\hat{\imath} \times \hat{\jmath} = \hat{k}$
$\hat{\jmath} \times \hat{k} = \hat{\imath}$
$\hat{k} \times \hat{\imath} = \hat{\jmath}$

Torque

Now let's return to torque. As a concrete example, FIGURE 12.52 shows a long wrench being used to loosen the nuts holding a car wheel on. Force $\vec{F}$ exerts a torque about the origin. Let's define a *torque vector*

$$\vec{\tau} \equiv \vec{r} \times \vec{F} \tag{12.48}$$

If we place the vector tails together in order to use the right-hand rule, we see that the torque vector is perpendicular to the plane of $\vec{r}$ and $\vec{F}$. The angle between the vectors is ϕ, so the magnitude of the torque is $\tau = rF|\sin \phi|$.

FIGURE 12.52 The torque vector.

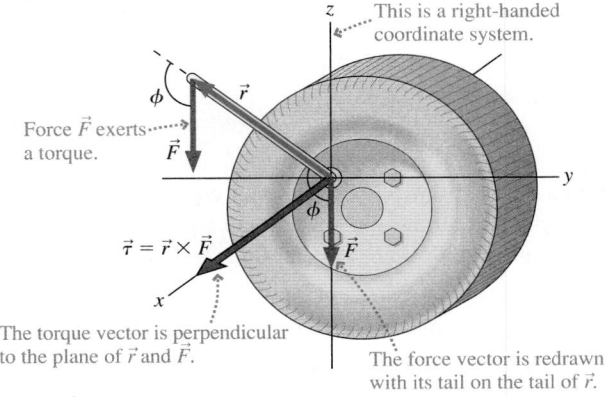

You can see that the scalar torque $\tau = rF\sin\phi$ we've been using is really the component along the rotation axis—in this case τ_x—of the vector $\vec{\tau}$. This is the basis for our earlier sign convention for τ. In Figure 12.52, where the force causes a ccw rotation, the torque vector points in the positive x-direction, and thus τ_x is positive.

EXAMPLE 12.19 | **Wrench torque revisited**

Example 12.8 found the torque that Luis exerts on a nut by pulling on the end of a wrench. What is the torque vector?

VISUALIZE **FIGURE 12.53** shows the position vector $\vec{r}$, drawn from the pivot point to the point where the force is applied. The figure

FIGURE 12.53 Calculating the torque vector.

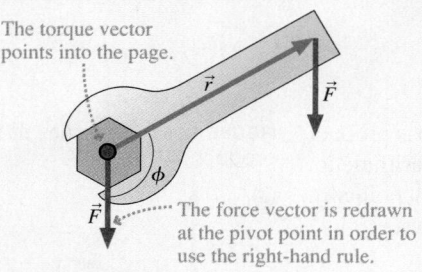

The torque vector points into the page.

ϕ

The force vector is redrawn at the pivot point in order to use the right-hand rule.

also redraws the force vector $\vec{F}$ at the pivot point, not because force is applied there but because it's easiest to use the right-hand rule if the vectors are drawn with their tails together.

SOLVE We already know the magnitude of the torque, 17 N m, from Example 12.8. Now we need to apply the right-hand rule. If you place your right thumb along $\vec{r}$ and your index finger along $\vec{F}$, which is somewhat awkward, you'll see that your middle finger points into the page. Alternatively, make a loose fist of your right hand, then orient your fist so that your fingers curl *from $\vec{r}$ toward $\vec{F}$*. Doing so requires your thumb to point into the page. Using either method, we conclude that

$$\vec{\tau} = (17\text{ N m, into page})$$

12.11 Angular Momentum

FIGURE 12.54 shows a particle that, at this instant, is located at position $\vec{r}$ and is moving with momentum $\vec{p} = m\vec{v}$. Together, $\vec{r}$ and $\vec{p}$ define the *plane of motion*. We define the particle's **angular momentum** $\vec{L}$ relative to the origin to be the vector

$$\vec{L} \equiv \vec{r} \times \vec{p} = (mrv\sin\beta, \text{ direction of right-hand rule}) \qquad (12.49)$$

Because of the cross product, **the angular momentum vector is perpendicular to the plane of motion.** The units of angular momentum are kg m²/s.

FIGURE 12.54 The angular momentum vector $\vec{L}$.

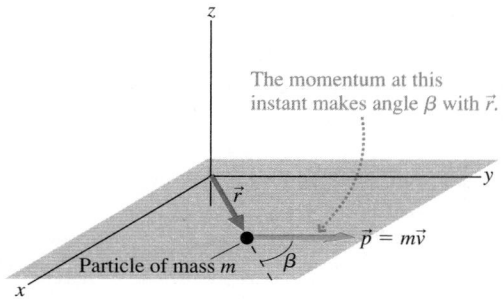

The momentum at this instant makes angle β with $\vec{r}$.

$\vec{p} = m\vec{v}$

Particle of mass m

Vectors $\vec{r}$ and $\vec{p}$ define the plane of motion.

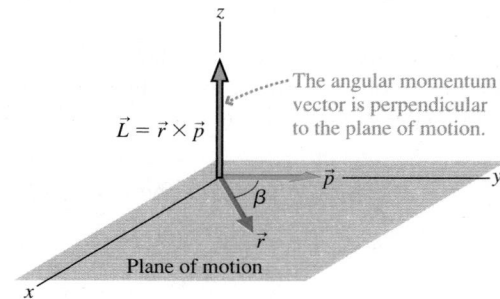

The angular momentum vector is perpendicular to the plane of motion.

$\vec{L} = \vec{r} \times \vec{p}$

Plane of motion

The vector tails are placed together to determine the cross product.

NOTE ▶ Angular momentum is the rotational equivalent of linear momentum in much the same way that torque is the rotational equivalent of force. Notice that the vector definitions are parallel: $\vec{\tau} \equiv \vec{r} \times \vec{F}$ and $\vec{L} \equiv \vec{r} \times \vec{p}$. ◀

Angular momentum, like torque, is *about* the point from which $\vec{r}$ is measured. A different origin would yield a different angular momentum. Angular momentum is especially simple for a particle in circular motion. As **FIGURE 12.55** shows, the angle β between $\vec{p}$ (or $\vec{v}$) and $\vec{r}$ is always 90° if we make the obvious choice of measuring $\vec{r}$ from the center of the circle. For motion in the xy-plane, the angular momentum vector $\vec{L}$ —which must be perpendicular to the plane of motion—is entirely along the z-axis:

$$L_z = mrv_t \qquad \text{(particle in circular motion)} \qquad (12.50)$$

where v_t is the tangential component of velocity. Our sign convention for v_t makes L_z, like ω, positive for a ccw rotation, negative for a cw rotation.

In Chapter 9, we found that Newton's second law for a particle can be written $\vec{F}_{net} = d\vec{p}/dt$. There's a similar connection between torque and angular momentum. To show this, we take the time derivative of $\vec{L}$:

$$\frac{d\vec{L}}{dt} = \frac{d}{dt}(\vec{r} \times \vec{p}) = \frac{d\vec{r}}{dt} \times \vec{p} + \vec{r} \times \frac{d\vec{p}}{dt}$$

$$= \vec{v} \times \vec{p} + \vec{r} \times \vec{F}_{net} \qquad (12.51)$$

where we used Equation 12.47 for the derivative of a cross product. We also used the definitions $\vec{v} = d\vec{r}/dt$ and $\vec{F}_{net} = d\vec{p}/dt$.

Vectors $\vec{v}$ and $\vec{p}$ are parallel, and the cross product of two parallel vectors is $\vec{0}$. Thus the first term in Equation 12.51 vanishes. The second term $\vec{r} \times \vec{F}_{net}$ is the net torque, $\vec{\tau}_{net} = \vec{\tau}_1 + \vec{\tau}_2 + \cdots$, so we arrive at

$$\frac{d\vec{L}}{dt} = \vec{\tau}_{net} \qquad (12.52)$$

Equation 12.52, which says **a net torque causes the particle's angular momentum to change,** is the rotational equivalent of $d\vec{p}/dt = \vec{F}_{net}$.

Angular Momentum of a Rigid Body

Equation 12.52 is the angular momentum of a single particle. The angular momentum of a rigid body composed of particles with individual angular momenta $\vec{L}_1, \vec{L}_2, \vec{L}_3, \ldots$ is the vector sum

$$\vec{L} = \vec{L}_1 + \vec{L}_2 + \vec{L}_3 + \cdots = \sum_i \vec{L}_i \qquad (12.53)$$

We can combine Equations 12.52 and 12.53 to find the rate of change of the system's angular momentum:

$$\frac{d\vec{L}}{dt} = \sum_i \frac{d\vec{L}_i}{dt} = \sum_i \vec{\tau}_i = \vec{\tau}_{net} \qquad (12.54)$$

Because any internal forces are action/reaction pairs of forces, acting with the same strength in opposite directions, the net torque due to internal forces is zero. Thus the only forces that contribute to the net torque are external forces exerted on the system by the environment.

For a system of particles, **the rate of change of the system's angular momentum is the net torque on the system.** Equation 12.54 is analogous to the Chapter 9 result $d\vec{P}/dt = \vec{F}_{net}$, which says that the rate of change of a system's total linear momentum is the net force on the system. Table 12.4 summarizes the analogies between linear and angular momentum and energy.

FIGURE 12.55 Angular momentum of circular motion.

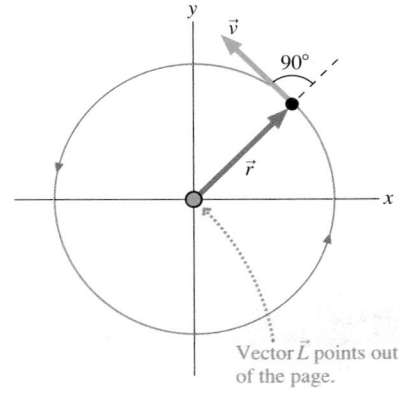

Vector $\vec{L}$ points out of the page.

The spin of an ice skater is determined by her angular momentum.

TABLE 12.4 Angular and linear momentum and energy

Angular motion	Linear motion
$K_{rot} = \frac{1}{2}I\omega^2$	$K_{cm} = \frac{1}{2}Mv_{cm}^2$
$\vec{L} = I\vec{\omega}$ *	$\vec{P} = M\vec{v}_{cm}$
$d\vec{L}/dt = \vec{\tau}_{net}$	$d\vec{P}/dt = \vec{F}_{net}$
The angular momentum of a system is conserved if there is no net torque.	The linear momentum of a system is conserved if there is no net force.

*Rotation about an axis of symmetry.

Conservation of Angular Momentum

A net torque on a rigid body causes its angular momentum to change. Conversely, the angular momentum does *not* change—it is *conserved*—for a system with no net torque. This is the basis of the law of conservation of angular momentum.

> **Law of conservation of angular momentum** The angular momentum $\vec{L}$ of an isolated system ($\vec{\tau}_{net} = \vec{0}$) is conserved. The final angular momentum $\vec{L}_f$ is equal to the initial angular momentum $\vec{L}_i$. Both the magnitude *and* the direction of $\vec{L}$ are unchanged.

EXAMPLE 12.20 | **An expanding rod**

Two equal masses are at the ends of a massless 50-cm-long rod. The rod spins at 2.0 rev/s about an axis through its midpoint. Suddenly, a compressed gas expands the rod out to a length of 160 cm. What is the rotation frequency after the expansion?

MODEL The forces push outward from the pivot and exert no torques. Thus the system's angular momentum is conserved.

VISUALIZE **FIGURE 12.56** is a before-and-after pictorial representation. The angular momentum vectors $\vec{L}_i$ and $\vec{L}_f$ are perpendicular to the plane of motion.

SOLVE The particles are moving in circles, so each has angular momentum $L = mrv_t = mr^2\omega = \frac{1}{4}ml^2\omega$, where we used $r = \frac{1}{2}l$. Thus the initial angular momentum of the system is

$$L_i = \frac{1}{4}ml_i^2\omega_i + \frac{1}{4}ml_i^2\omega_i = \frac{1}{2}ml_i^2\omega_i$$

Similarly, the angular momentum after the expansion is $L_f = \frac{1}{2}ml_f^2\omega_f$. Angular momentum is conserved as the rod expands, thus

$$\frac{1}{2}ml_f^2\omega_f = \frac{1}{2}ml_i^2\omega_i$$

Solving for ω_f, we find

$$\omega_f = \left(\frac{l_i}{l_f}\right)^2 \omega_i = \left(\frac{50 \text{ cm}}{160 \text{ cm}}\right)^2 (2.0 \text{ rev/s}) = 0.20 \text{ rev/s}$$

ASSESS The values of the masses weren't needed. All that matters is the ratio of the lengths.

FIGURE 12.56 The system before and after the rod expands.

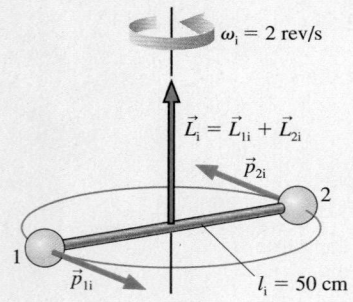

Before:

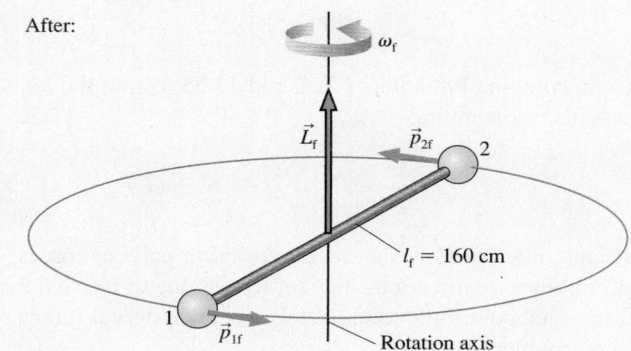

After:

The expansion of the rod in Example 12.20 causes a dramatic slowing of the rotation. Similarly, the rotation would speed up if the weights were pulled in. This is how an ice skater controls her speed as she does a spin. Pulling in her arms decreases her moment of inertia and causes her angular velocity to increase. Similarly, extending her arms increases her moment of inertia, and her angular velocity drops until she can skate out of the spin. It's all a matter of conserving angular momentum.

Angular Momentum and Angular Velocity

The analogy between linear and rotational motion has been so consistent that you might expect one more. The Chapter 9 result $\vec{P} = M\vec{v}_{cm}$ might give us reason to anticipate that angular momentum and angular velocity are related by $\vec{L} = I\vec{\omega}$. Unfortunately, the analogy breaks down here. For an arbitrarily shaped object, the angular momentum vector and the angular velocity vector don't necessarily point in the same direction. The general relationship between $\vec{L}$ and $\vec{\omega}$ is beyond the scope of this text.

The good news is that the analogy *does* continue to hold in two important situations: the rotation of a *symmetrical* object about the symmetry axis and the rotation of any object about a fixed axle. For example, the axis of a cylinder or disk is a symmetry axis, as is any diameter through a sphere. In these two situations—which are all this textbook will consider—the angular momentum and angular velocity are related by

$$\vec{L} = I\vec{\omega} \qquad \text{(rotation about a fixed axle or axis of symmetry)} \qquad (12.55)$$

This relationship is shown for a spinning disk in **FIGURE 12.57**. Equation 12.55 is particularly important for applying the law of conservation of angular momentum.

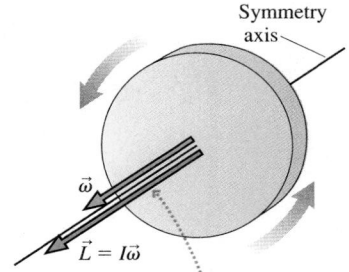

FIGURE 12.57 The angular momentum vector of a rigid body rotating about an axis of symmetry.

Angular velocity and angular momentum vectors point along the rotation axis in the direction determined by the right-hand rule.

EXAMPLE 12.21 **Two interacting disks**

A 20-cm-diameter, 2.0 kg solid disk is rotating at 200 rpm. A 20-cm-diameter, 1.0 kg circular loop is dropped straight down onto the rotating disk. Friction causes the loop to accelerate until it is "riding" on the disk. What is the final angular velocity of the combined system?

MODEL The friction between the two objects creates torques that speed up the loop and slow down the disk. But these torques are internal to the combined disk + loop system, so $\tau_{net} = 0$ and the *total* angular momentum of the disk + loop system is conserved.

VISUALIZE **FIGURE 12.58** is a before-and-after pictorial representation. Initially only the disk is rotating, at angular velocity $\vec{\omega}_i$. The rotation is about an axis of symmetry, so the angular momentum $\vec{L} = I\vec{\omega}$ is parallel to $\vec{\omega}$. At the end of the problem, $\vec{\omega}_{disk} = \vec{\omega}_{loop} = \vec{\omega}_f$.

SOLVE Both angular momentum vectors point along the rotation axis. Conservation of angular momentum tells us that the magnitude of $\vec{L}$ is unchanged. Thus

$$L_f = I_{disk}\omega_f + I_{loop}\omega_f = L_i = I_{disk}\omega_i$$

Solving for ω_f gives

$$\omega_f = \frac{I_{disk}}{I_{disk} + I_{loop}}\omega_i$$

The moments of inertia for a disk and a loop can be found in Table 12.2, leading to

$$\omega_f = \frac{\frac{1}{2}M_{disk}R^2}{\frac{1}{2}M_{disk}R^2 + M_{loop}R^2}\omega_i = 100 \text{ rpm}$$

ASSESS What appeared to be a difficult problem turns out to be fairly easy once you recognize that the total angular momentum is conserved.

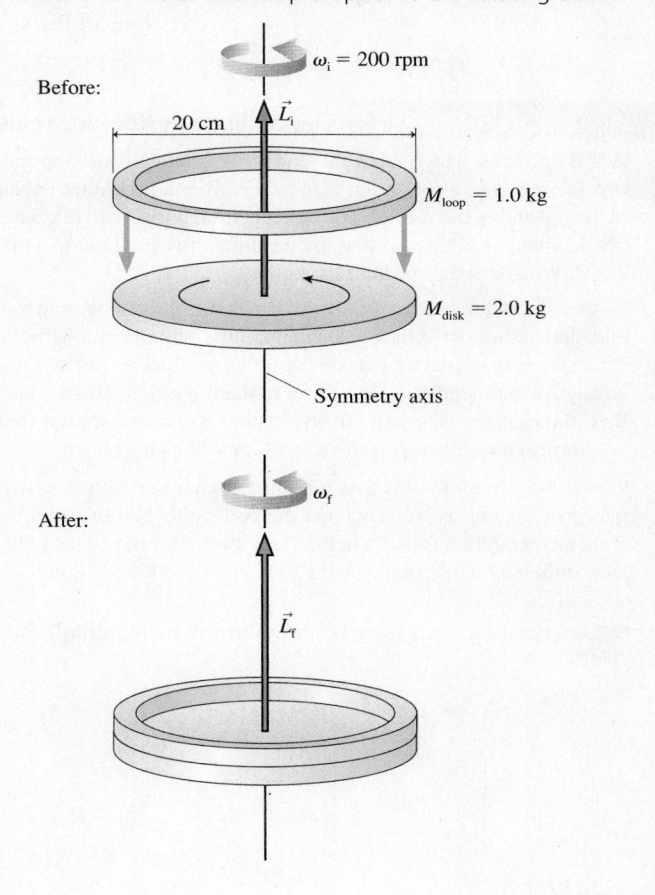

FIGURE 12.58 The circular loop drops onto the rotating disk.

Before:

$\omega_i = 200$ rpm

20 cm

$\vec{L}_i$

$M_{loop} = 1.0$ kg

$M_{disk} = 2.0$ kg

Symmetry axis

After:

ω_f

$\vec{L}_f$

When angular momentum—a vector—is conserved, its direction—the direction of the rotation axis—must remain unchanged. This is often shown with the lecture demonstration illustrated in **FIGURE 12.59** on the next page. A bicycle wheel with two handles is given a spin, then handed to an unsuspecting student. The student is asked to turn the wheel 90°. Surprisingly, this is *very hard to do.*

FIGURE 12.59 The vector nature of
angular momentum makes it difficult to
turn a rapidly spinning wheel.

FIGURE 12.59 The vector nature of
angular momentum makes it difficult to
turn a rapidly spinning wheel.

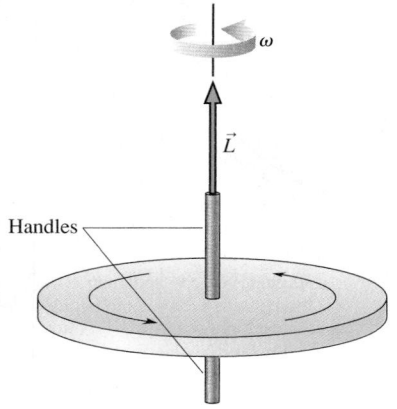

The reason is that the wheel's angular momentum vector, which points straight up, is highly resistant to change. If the wheel is spinning fast, a *large* torque must be supplied to change $\vec{L}$. This directional stability of a rapidly spinning object is why gyroscopes are used as navigational devices on ships and planes. Once the axis of a spinning gyroscope is pointed north, it will maintain that direction as the ship or plane moves.

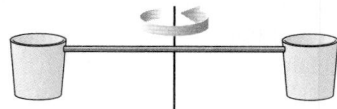

STOP TO THINK 12.5   Two buckets spin around in a horizontal circle on frictionless bearings. Suddenly, it starts to rain. As a result,

a. The buckets continue to rotate at constant angular velocity because the rain is falling vertically while the buckets move in a horizontal plane.
b. The buckets continue to rotate at constant angular velocity because the total mechanical energy of the bucket + rain system is conserved.
c. The buckets speed up because the potential energy of the rain is transformed into kinetic energy.
d. The buckets slow down because the angular momentum of the bucket + rain system is conserved.
e. Both a and b.
f. None of the above.

CHALLENGE EXAMPLE 12.22   **The ballistic pendulum revisited**

A 2.0 kg block hangs from the end of a 1.5 kg, 1.0-m-long rod, together forming a pendulum that swings from a frictionless pivot at the top end of the rod. A 10 g bullet is fired horizontally into the block, where it sticks, causing the pendulum to swing out to a 30° angle. What was the speed of the bullet?

MODEL   Model the rod as a uniform rod that can rotate around one end, and assume the block is small enough to model as a particle. There are no external torques on the bullet + block + rod system, so angular momentum is conserved in the inelastic collision. Further, the mechanical energy of the system is conserved after (but not during) the collision as the pendulum swings outward.

VISUALIZE   FIGURE 12.60 is a pictorial representation. This is a two-part problem, so we've separated the collision's before-and-after from the pendulum swing's before-and-after. The end of the collision is the beginning of the swing.

FIGURE 12.60 Pictorial representation of the bullet hitting the pendulum.

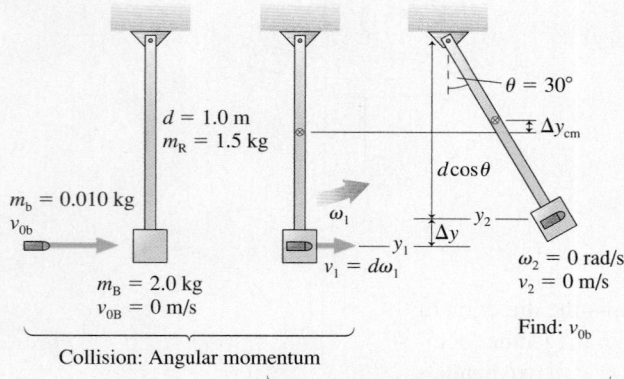

Collision: Angular momentum

Swing: Mechanical energy

SOLVE   This is a *ballistic pendulum*. Example 10.4 considered a simpler ballistic pendulum with a mass on a string, rather than on a rod, and a review of that example is highly recommended. The key to both is that a different conservation law applies to each part of the problem.

Angular momentum is conserved in the collision, thus $L_1 = L_0$. Before the collision, the angular momentum—which we'll measure about the pendulum's pivot point—is entirely that of the bullet. The angular momentum of a particle is $L = mrv \sin \beta$. An instant before the collision, just as the bullet reaches the block, $r = d$ and, because $\vec{v}$ is perpendicular to $\vec{r}$ at that instant, $\beta = 90°$. Thus $L_0 = m_b d v_{0b}$. (This is the magnitude of the angular momentum; from the right-hand rule, the angular momentum vector points out of the page.)

An instant after the collision, but before the pendulum has had time to move, the rod has angular velocity ω_1 and the block, with the embedded bullet, is moving in a circle with speed $v_1 = \omega_1 r = \omega_1 d$. The angular momentum of the block + bullet system is that of a particle, still with $\beta = 90°$, while that of the rod—an object rotating on a fixed axle—is $I_{rod}\omega_1$. Thus the post-collision angular momentum is

$$L_1 = (m_B + m_b)v_1 r + I_{rod}\omega_1 = (m_B + m_b)d^2\omega_1 + \frac{1}{3}m_R d^2\omega_1$$

The moment of inertia of the rod was taken from Table 12.2.

Equating the before-and-after angular momenta, then solving for v_{0b}, gives

$$m_b d v_{0b} = (m_B + m_b)d^2\omega_1 + \frac{1}{3}m_R d^2\omega_1$$

$$v_{0b} = \frac{m_B + m_b + \frac{1}{3}m_R}{m_b}d\omega_1 = 251 d\omega_1$$

Once we know ω_1, which we'll find from energy conservation in the swing, we'll be able to compute the bullet's speed.

Mechanical energy is conserved during the swing, but you must be careful to include all the energies. The kinetic energy has two components: the translational kinetic energy of the block + bullet system and the rotational kinetic energy of the rod. The gravitational potential energy also has two components: the potential energy of the block + bullet system and the potential energy of the rod. The latter changes because the center of mass moves upward as the rod swings. Thus the energy conservation statement is

$$\frac{1}{2}(m_B + m_b)v_2^2 + \frac{1}{2}I_{rod}\omega_2^2 + (m_B + m_b)gy_2 + m_Rgy_{cm2} =$$
$$\frac{1}{2}(m_B + m_b)v_1^2 + \frac{1}{2}I_{rod}\omega_1^2 + (m_B + m_b)gy_1 + m_Rgy_{cm1}$$

Although this looks very complicated, you should convince yourself that we've done nothing more than add up two kinetic energies and two potential energies before and after the swing.

We know that $v_2 = 0$ and $\omega_2 = 0$ at the end of the swing, and that $v_1 = d\omega_1$ at the beginning. We also know the moment of inertia of a rod pivoted at one end. Combining the potential energy terms and using $\Delta y = y_f - y_i$, we thus have

$$\frac{1}{2}\left(m_B + m_b + \frac{1}{3}m_R\right)d^2\omega_1^2 = (m_B + m_b)g\Delta y + m_Rg\Delta y_{cm}$$

We see from Figure 12.60 that the block, at its highest point, is distance $d\cos\theta$ *below* the pivot. It started distance d below the pivot, so the bullet + block system *gained* height $\Delta y = d - d\cos\theta = d(1 - \cos\theta)$. The rod's center of mass started distance $d/2$ below the pivot and rises only half as much as the block, so $\Delta y_{cm} = \frac{1}{2}d(1 - \cos\theta)$. With these, the energy equation becomes

$$\frac{1}{2}\left(m_B + m_b + \frac{1}{3}m_R\right)d^2\omega_1^2 = (m_B + m_b + \frac{1}{2}m_R)gd(1 - \cos\theta)$$

We can now solve for ω_1:

$$\omega_1 = \sqrt{\frac{m_B + m_b + \frac{1}{2}m_R}{m_B + m_b + \frac{1}{3}m_R}\frac{2g(1 - \cos\theta)}{d}} = 1.70 \text{ rad/s}$$

and with that

$$v_{0b} = 251d\omega_1 = 430 \text{ m/s}$$

ASSESS 430 m/s seems a reasonable speed for a bullet. This was a challenging problem, but one that you can solve if you focus on the problem-solving strategies—drawing a careful pictorial representation, defining the system, and thinking about which conservation laws apply—rather than hunting for the "right" equation.

SUMMARY

The goal of Chapter 12 has been to understand the physics of rotating objects.

General Principles

Rotational Dynamics

Every point on a **rigid body** rotating about a fixed axis has the same angular velocity ω and angular acceleration α.

Newton's second law for rotational motion is

$$\alpha = \frac{\tau_{net}}{I}$$

Use rotational kinematics to find angles and angular velocities.

Conservation Laws

Energy is conserved for an isolated system.

- Pure rotation $E = K_{rot} + U_g = \frac{1}{2}I\omega^2 + Mgy_{cm}$
- Rolling $E = K_{rot} + K_{cm} + U_g = \frac{1}{2}I\omega^2 + \frac{1}{2}Mv_{cm}^2 + Mgy_{cm}$

Angular momentum is conserved if $\vec{\tau}_{net} = \vec{0}$.

- Particle $\vec{L} = \vec{r} \times \vec{p}$
- Rotation about a symmetry axis or fixed axle $\vec{L} = I\vec{\omega}$

Important Concepts

Torque is the rotational equivalent of force:

$$\tau = rF\sin\phi = rF_t = dF$$

The vector description of torque is

$$\vec{\tau} = \vec{r} \times \vec{F}$$

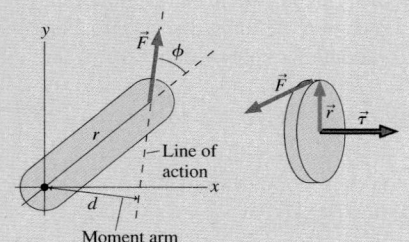

Vector description of rotation

Angular velocity $\vec{\omega}$ points along the rotation axis in the direction of the right-hand rule.

For a rigid body rotating about a fixed axle or an axis of symmetry, the angular momentum is $\vec{L} = I\vec{\omega}$.

Newton's second law is $\dfrac{d\vec{L}}{dt} = \vec{\tau}_{net}$.

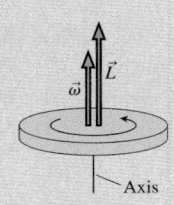

A system of particles on which there is no net force undergoes unconstrained rotation about the center of mass:

$$x_{cm} = \frac{1}{M}\int x\,dm \qquad y_{cm} = \frac{1}{M}\int y\,dm$$

The gravitational torque on a body can be found by treating the body as a particle with all the mass M concentrated at the center of mass.

The moment of inertia

$$I = \sum_i m_i r_i^2 = \int r^2\,dm$$

is the rotational equivalent of mass. The moment of inertia depends on how the mass is distributed around the axis. If I_{cm} is known, I about a parallel axis distance d away is given by the **parallel-axis theorem:** $I = I_{cm} + Md^2$.

Applications

Rotational kinematics

$$\omega_f = \omega_i + \alpha\,\Delta t$$
$$\theta_f = \theta_i + \omega_i\,\Delta t + \tfrac{1}{2}\alpha(\Delta t)^2$$
$$v_t = r\omega \qquad a_t = r\alpha$$

Rigid-body equilibrium

An object is in total equilibrium only if both $\vec{F}_{net} = \vec{0}$ and $\vec{\tau}_{net} = \vec{0}$.

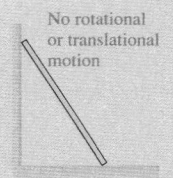

No rotational or translational motion

Rolling motion

For an object that rolls without slipping

$$v_{cm} = R\omega$$
$$K = K_{rot} + K_{cm}$$

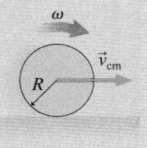

Terms and Notation

rigid body	center of mass	line of action	right-hand rule
rigid-body model	rotational kinetic energy, K_{rot}	moment arm, d	angular momentum, $\vec{L}$
translational motion	moment of inertia, I	rolling constraint	law of conservation of
rotational motion	parallel-axis theorem	cross product	angular momentum
combination motion	torque, τ	vector product	

CONCEPTUAL QUESTIONS

1. Is the center of mass of the dumbbell in FIGURE Q12.1 at point a, b, or c? Explain.

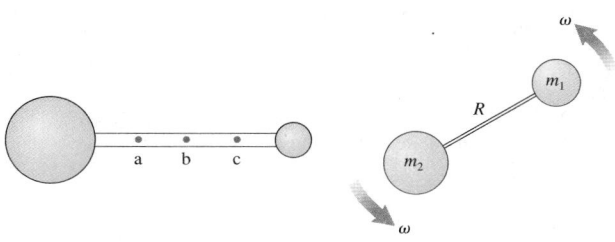

FIGURE Q12.1   FIGURE Q12.2

2. If the angular velocity ω is held constant, by what *factor* must R change to double the rotational kinetic energy of the dumbbell in FIGURE Q12.2?

3. FIGURE Q12.3 shows three rotating disks, all of equal mass. Rank in order, from largest to smallest, their rotational kinetic energies K_a to K_c.

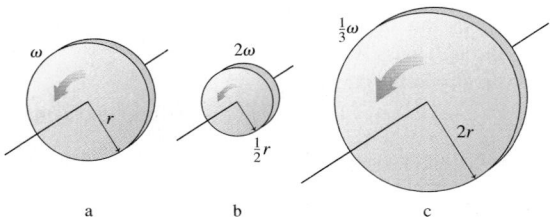

FIGURE Q12.3

4. Must an object be rotating to have a moment of inertia? Explain.

5. The moment of inertia of a uniform rod about an axis through its center is $\frac{1}{12}mL^2$. The moment of inertia about an axis at one end is $\frac{1}{3}mL^2$. Explain *why* the moment of inertia is larger about the end than about the center.

6. You have two steel spheres. Sphere 2 has twice the radius of sphere 1. By what *factor* does the moment of inertia I_2 of sphere 2 exceed the moment of inertia I_1 of sphere 1?

7. The professor hands you two spheres. They have the same mass, the same radius, and the same exterior surface. The professor claims that one is a solid sphere and the other is hollow. Can you determine which is which without cutting them open? If so, how? If not, why not?

8. Six forces are applied to the door in FIGURE Q12.8. Rank in order, from largest to smallest, the six torques τ_a to τ_f about the hinge. Explain.

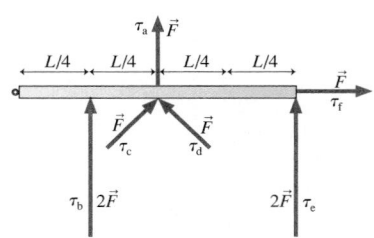

FIGURE Q12.8

9. A student gives a quick push to a ball at the end of a massless, rigid rod, as shown in FIGURE Q12.9, causing the ball to rotate clockwise in a *horizontal* circle. The rod's pivot is frictionless.

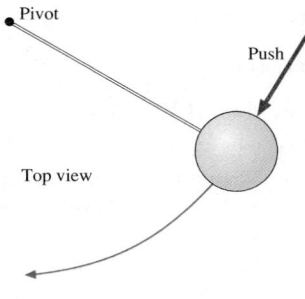

FIGURE Q12.9

   a. As the student is pushing, is the torque about the pivot positive, negative, or zero?

   b. After the push has ended, does the ball's angular velocity (i) steadily increase; (ii) increase for awhile, then hold steady; (iii) hold steady; (iv) decrease for awhile, then hold steady; or (v) steadily decrease? Explain.

   c. Right after the push has ended, is the torque positive, negative, or zero?

10. Rank in order, from largest to smallest, the angular accelerations α_a to α_d in FIGURE Q12.10. Explain.

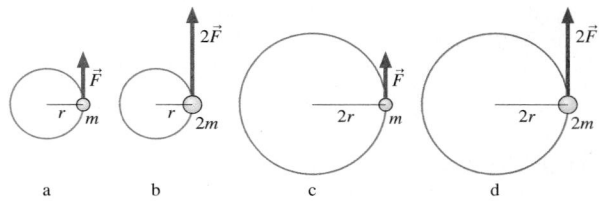

FIGURE Q12.10

11. The solid cylinder and cylindrical shell in FIGURE Q12.11 have the same mass, same radius, and turn on frictionless, horizontal axles. (The cylindrical shell has lightweight spokes connecting the shell to the axle.) A rope is wrapped around each cylinder and tied to a block. The blocks have the same mass and are held the same height above the ground. Both blocks are released simultaneously. Which hits the ground first? Or is it a tie? Explain.

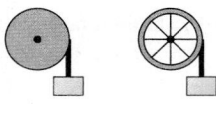

FIGURE Q12.11

12. A diver in the pike position (legs straight, hands on ankles) usually makes only one or one-and-a-half rotations. To make two or three rotations, the diver goes into a tuck position (knees bent, body curled up tight). Why?

13. Is the angular momentum of disk a in FIGURE Q12.13 larger than, smaller than, or equal to the angular momentum of disk b? Explain.

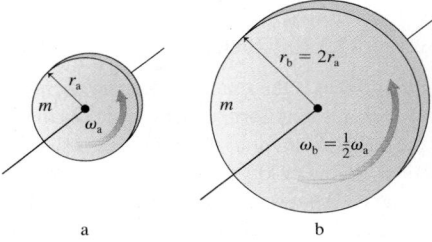

FIGURE Q12.13   a   b

EXERCISES AND PROBLEMS

Problems labeled [ ] integrate material from earlier chapters.

Exercises

Section 12.1 Rotational Motion

1. ‖ A skater holds her arms outstretched as she spins at 180 rpm. What is the speed of her hands if they are 140 cm apart?
2. ‖ A high-speed drill reaches 2000 rpm in 0.50 s.
    a. What is the drill's angular acceleration?
    b. Through how many revolutions does it turn during this first 0.50 s?
3. ‖ A ceiling fan with 80-cm-diameter blades is turning at 60 rpm. Suppose the fan coasts to a stop 25 s after being turned off.
    a. What is the speed of the tip of a blade 10 s after the fan is turned off?
    b. Through how many revolutions does the fan turn while stopping?
4. ‖‖ An 18-cm-long bicycle crank arm, with a pedal at one end, is attached to a 20-cm-diameter sprocket, the toothed disk around which the chain moves. A cyclist riding this bike increases her pedaling rate from 60 rpm to 90 rpm in 10 s.
    a. What is the tangential acceleration of the pedal?
    b. What length of chain passes over the top of the sprocket during this interval?

Section 12.2 Rotation About the Center of Mass

5. ‖ How far from the center of the earth is the center of mass of the earth + moon system? Data for the earth and moon can be found inside the back cover of the book.
6. | The three masses shown in FIGURE EX12.6 are connected by massless, rigid rods. What are the coordinates of the center of mass?

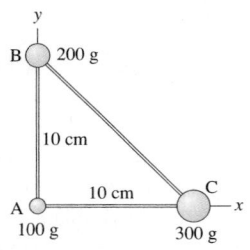

FIGURE EX12.6

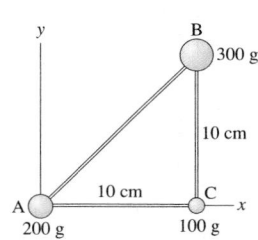

FIGURE EX12.7

7. | The three masses shown in FIGURE EX12.7 are connected by massless, rigid rods. What are the coordinates of the center of mass?
8. ‖ A 100 g ball and a 200 g ball are connected by a 30-cm-long, massless, rigid rod. The balls rotate about their center of mass at 120 rpm. What is the speed of the 100 g ball?

Section 12.3 Rotational Energy

9. ‖ What is the rotational kinetic energy of the earth? Assume the earth is a uniform sphere. Data for the earth can be found inside the back cover of the book.
10. ‖ A thin, 100 g disk with a diameter of 8.0 cm rotates about an axis through its center with 0.15 J of kinetic energy. What is the speed of a point on the rim?

11. ‖‖ The three 200 g masses in FIGURE EX12.11 are connected by massless, rigid rods.
    a. What is the triangle's moment of inertia about the axis through the center?
    b. What is the triangle's kinetic energy if it rotates about the axis at 5.0 rev/s?

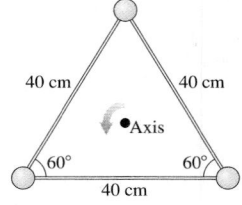

FIGURE EX12.11

12. ‖‖ A drum major twirls a 96-cm-long, 400 g baton about its center of mass at 100 rpm. What is the baton's rotational kinetic energy?

Section 12.4 Calculating Moment of Inertia

13. ‖ The four masses shown in FIGURE EX12.13 are connected by massless, rigid rods.
    a. Find the coordinates of the center of mass.
    b. Find the moment of inertia about an axis that passes through mass A and is perpendicular to the page.

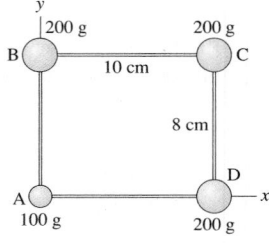

FIGURE EX12.13

14. ‖ The four masses shown in FIGURE EX12.13 are connected by massless, rigid rods.
    a. Find the coordinates of the center of mass.
    b. Find the moment of inertia about a diagonal axis that passes through masses B and D.
15. | The three masses shown in FIGURE EX12.15 are connected by massless, rigid rods.
    a. Find the coordinates of the center of mass.
    b. Find the moment of inertia about an axis that passes through mass A and is perpendicular to the page.
    b. Find the moment of inertia about an axis that passes through masses B and C.

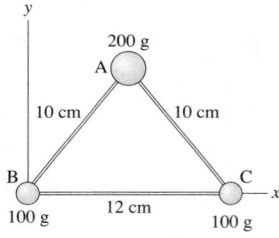

FIGURE EX12.15

16. ‖ A 25 kg solid door is 220 cm tall, 91 cm wide. What is the door's moment of inertia for (a) rotation on its hinges and (b) rotation about a vertical axis inside the door, 15 cm from one edge?
17. ‖ A 12-cm-diameter CD has a mass of 21 g. What is the CD's moment of inertia for rotation about a perpendicular axis (a) through its center and (b) through the edge of the disk?

Section 12.5 Torque

18. | In FIGURE EX12.18, what is the net torque about the axle?

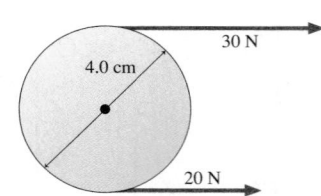

FIGURE EX12.18

19. ‖ In FIGURE EX12.19, what is the net torque about the axle?

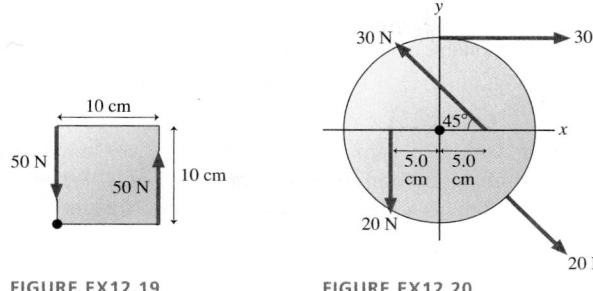

FIGURE EX12.19   FIGURE EX12.20

20. ‖ The 20-cm-diameter disk in FIGURE EX12.20 can rotate on an axle through its center. What is the net torque about the axle?

21. ‖ A 4.0-m-long, 500 kg steel beam extends horizontally from the point where it has been bolted to the framework of a new building under construction. A 70 kg construction worker stands at the far end of the beam. What is the magnitude of the torque about the point where the beam is bolted into place?

22. ‖ An athlete at the gym holds a 3.0 kg steel ball in his hand. His
BIO arm is 70 cm long and has a mass of 4.0 kg. What is the magnitude of the torque about his shoulder if he holds his arm
   a. Straight out to his side, parallel to the floor?
   b. Straight, but 45° below horizontal?

Section 12.6 Rotational Dynamics

Section 12.7 Rotation About a Fixed Axis

23. | An object's moment of inertia is 2.0 $kg\,m^2$. Its angular velocity is increasing at the rate of 4.0 rad/s per second. What is the torque on the object?

24. ‖ An object whose moment of inertia is 4.0 $kg\,m^2$ experiences the torque shown in FIGURE EX12.24. What is the object's angular velocity at $t = 3.0$ s? Assume it starts from rest.

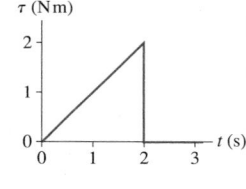

FIGURE EX12.24

25. ‖ A 1.0 kg ball and a 2.0 kg ball are connected by a 1.0-m-long rigid, massless rod. The rod is rotating cw about its center of mass at 20 rpm. What torque will bring the balls to a halt in 5.0 s?

26. ‖ Starting from rest, a 12-cm-diameter compact disk takes 3.0 s to reach its operating angular velocity of 2000 rpm. Assume that the angular acceleration is constant. The disk's moment of inertia is 2.5×10^{-5} $kg\,m^2$.
   a. How much torque is applied to the disk?
   b. How many revolutions does it make before reaching full speed?

27. ‖ A 750 g, 50-cm-long metal rod is free to rotate about a frictionless axle at one end. While at rest, the rod is given a short but sharp 1000 N hammer blow at the center of the rod, aimed in a direction that causes the rod to rotate on the axle. The blow lasts a mere 2.0 ms. What is the rod's angular velocity immediately after the blow?

Section 12.8 Static Equilibrium

28. ‖ How much torque must the pin exert to keep the rod in FIGURE EX12.28 from rotating?

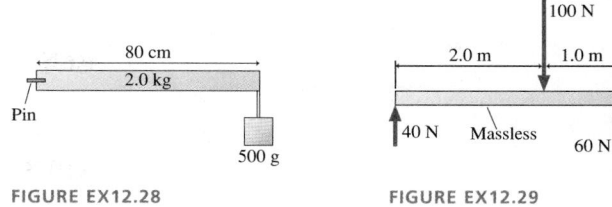

FIGURE EX12.28   FIGURE EX12.29

29. ‖ Is the object in FIGURE EX12.29 in equilibrium? Explain.

30. ‖ The two objects in FIGURE EX12.30 are balanced on the pivot. What is distance d?

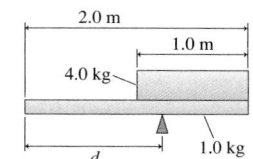

FIGURE EX12.30

31. ‖ A 5.0 kg cat and a 2.0 kg bowl of tuna fish are at opposite ends of the 4.0-m-long seesaw of FIGURE EX12.31. How far to the left of the pivot must a 4.0 kg cat stand to keep the seesaw balanced?

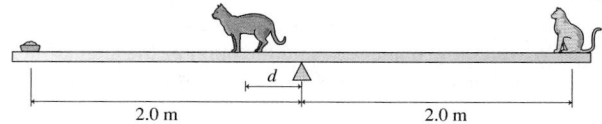

FIGURE EX12.31

Section 12.9 Rolling Motion

32. ‖ A car tire is 60 cm in diameter. The car is traveling at a speed of 20 m/s.
   a. What is the tire's angular velocity, in rpm?
   b. What is the speed of a point at the top edge of the tire?
   c. What is the speed of a point at the bottom edge of the tire?

33. ‖ A 500 g, 8.0-cm-diameter can is filled with uniform, dense food. It rolls across the floor at 1.0 m/s. What is the can's kinetic energy?

34. ‖ An 8.0-cm-diameter, 400 g solid sphere is released from rest at the top of a 2.1-m-long, 25° incline. It rolls, without slipping, to the bottom.
   a. What is the sphere's angular velocity at the bottom of the incline?
   b. What fraction of its kinetic energy is rotational?

35. | A solid sphere of radius R is placed at a height of 30 cm on a 15° slope. It is released and rolls, without slipping, to the bottom. From what height should a circular hoop of radius R be released on the same slope in order to equal the sphere's speed at the bottom?

Section 12.10 The Vector Description of Rotational Motion

36. | Evaluate the cross products $\vec{A} \times \vec{B}$ and $\vec{C} \times \vec{D}$.

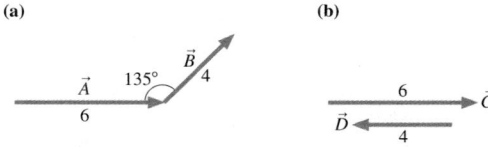

FIGURE EX12.36

37. | Evaluate the cross products $\vec{A} \times \vec{B}$ and $\vec{C} \times \vec{D}$.

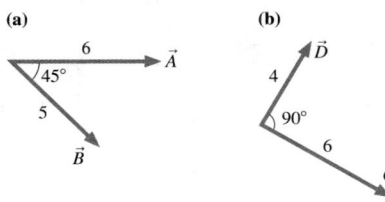

FIGURE EX12.37

38. ‖ a. What is $(\hat{\imath} \times \hat{\jmath}) \times \hat{\imath}$?
    b. What is $\hat{\imath} \times (\hat{\jmath} \times \hat{\imath})$?
39. ‖ a. What is $\hat{\imath} \times (\hat{\imath} \times \hat{\jmath})$?
    b. What is $(\hat{\imath} \times \hat{\jmath}) \times \hat{k}$?
40. ‖ Vector $\vec{A} = 3\hat{\imath} + \hat{\jmath}$ and vector $\vec{B} = 3\hat{\imath} - 2\hat{\jmath} + 2\hat{k}$. What is the cross product $\vec{A} \times \vec{B}$?
41. | Consider the vector $\vec{C} = 3\hat{\imath}$.
    a. What is a vector $\vec{D}$ such that $\vec{C} \times \vec{D} = \vec{0}$?
    b. What is a vector $\vec{E}$ such that $\vec{C} \times \vec{E} = 6\hat{k}$?
    c. What is a vector $\vec{F}$ such that $\vec{C} \times \vec{F} = -3\hat{\jmath}$?
42. ‖ Force $\vec{F} = -10\hat{\jmath}$ N is exerted on a particle at $\vec{r} = (5\hat{\imath} + 5\hat{\jmath})$ m. What is the torque on the particle about the origin?
43. ‖ What are the magnitude and direction of the angular momentum relative to the origin of the 100 g particle in FIGURE EX12.43?

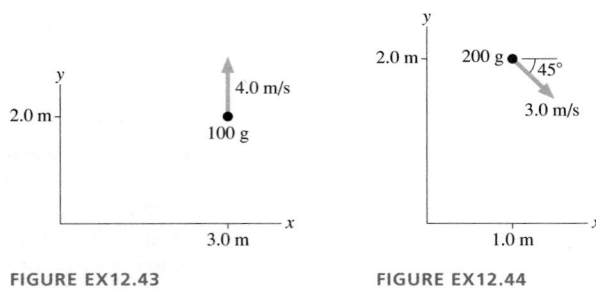

FIGURE EX12.43   **FIGURE EX12.44**

44. ‖ What are the magnitude and direction of the angular momentum relative to the origin of the 200 g particle in FIGURE EX12.44?

Section 12.11 Angular Momentum

45. ‖ What is the angular momentum of the 500 g rotating bar in FIGURE EX12.45?

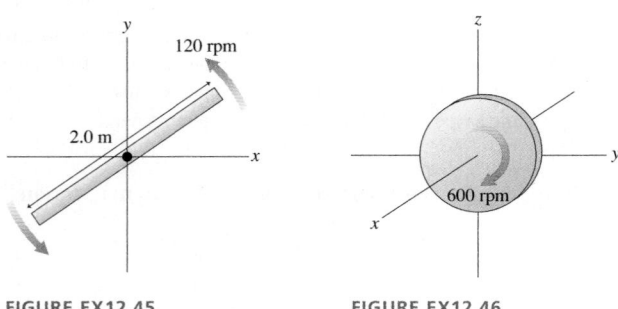

FIGURE EX12.45   **FIGURE EX12.46**

46. ‖ What is the angular momentum of the 2.0 kg, 4.0-cm-diameter rotating disk in FIGURE EX12.46?
47. ‖ How fast, in rpm, would a 5.0 kg, 22-cm-diameter bowling ball have to spin to have an angular momentum of 0.23 kg m²/s?

48. ‖ A 2.0 kg, 20-cm-diameter turntable rotates at 100 rpm on frictionless bearings. Two 500 g blocks fall from above, hit the turntable simultaneously at opposite ends of a diameter, and stick. What is the turntable's angular velocity, in rpm, just after this event?

Problems

49. ‖ A 70 kg man's arm, including the hand, can be modeled as a
BIO 75-cm-long uniform rod with a mass of 3.5 kg. When the man raises both his arms, from hanging down to straight up, by how much does he raise his center of mass?
50. ‖‖ A 300 g ball and a 600 g ball are connected by a 40-cm-long massless, rigid rod. The structure rotates about its center of mass at 100 rpm. What is its rotational kinetic energy?
51. ‖‖ A 60-cm-diameter wheel is rolling along at 20 m/s. What is the speed of a point at the forward edge of the wheel?
52. ‖ An 800 g steel plate has the shape of the isosceles triangle shown in FIGURE P12.52. What are the *x*- and *y*-coordinates of the center of mass?
    Hint: Divide the triangle into vertical strips of width *dx*, then relate the mass *dm* of a strip at position *x* to the values of *x* and *dx*.

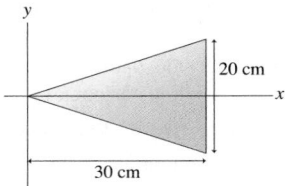

FIGURE P12.52

53. ‖ What is the moment of inertia of a 2.0 kg, 20-cm-diameter disk for rotation about an axis (a) through the center, and (b) through the edge of the disk?
54. ‖ Determine the moment of inertia about the axis of the object shown in FIGURE P12.54.

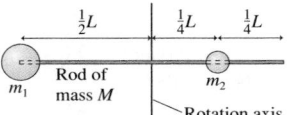

FIGURE P12.54

55. ‖ Calculate by direct integration the moment of inertia for a thin rod of mass M and length L about an axis located distance d from one end. Confirm that your answer agrees with Table 12.2 when $d = 0$ and when $d = L/2$.
56. ‖ a. A disk of mass M and radius R has a hole of radius r centered on the axis. Calculate the moment of inertia of the disk.
    b. Confirm that your answer agrees with Table 12.2 when $r = 0$ and when $r = R$.
    c. A 4.0-cm-diameter disk with a 3.0-cm-diameter hole rolls down a 50-cm-long, 20° ramp. What is its speed at the bottom? What percent is this of the speed of a particle sliding down a frictionless ramp?
57. ‖ Calculate the moment of inertia of the rectangular plate in FIGURE P12.57 for rotation about a perpendicular axis through the center.

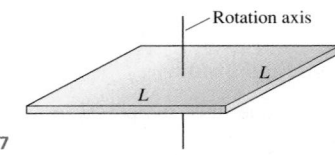

FIGURE P12.57

58. ‖ Calculate the moment of inertia of the steel plate in FIG-URE P12.52 for rotation about a perpendicular axis passing through the origin.

59. ‖ A person's center of mass is easily found by having the person
BIO lie on a *reaction board*. A horizontal, 2.5-m-long, 6.1 kg reaction board is supported only at the ends, with one end resting on a scale and the other on a pivot. A 60 kg woman lies on the reaction board with her feet over the pivot. The scale reads 25 kg. What is the distance from the woman's feet to her center of mass?

60. ‖ A 3.0-m-long ladder, as shown in Figure 12.37, leans against a frictionless wall. The coefficient of static friction between the ladder and the floor is 0.40. What is the minimum angle the ladder can make with the floor without slipping?

61. ‖ The 3.0-m-long, 100 kg rigid beam of FIGURE P12.61 is supported at each end. An 80 kg student stands 2.0 m from support 1. How much upward force does each support exert on the beam?

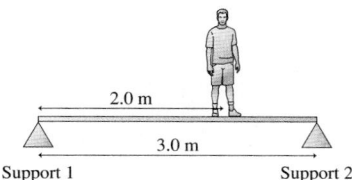

FIGURE P12.61   Support 1   Support 2

2.0 m

3.0 m

62. ‖ In FIGURE P12.62, an 80 kg construction worker sits down 2.0 m from the end of a 1450 kg steel beam to eat his lunch. The cable supporting the beam is rated at 15,000 N. Should the worker be worried?

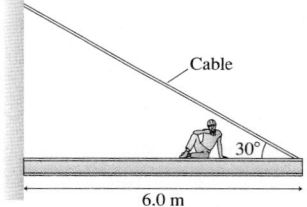

Cable

30°

FIGURE P12.62   6.0 m

63. ‖‖ A 40 kg, 5.0-m-long beam is supported by, but not attached to, the two posts in FIGURE P12.63. A 20 kg boy starts walking along the beam. How close can he get to the right end of the beam without it falling over?

FIGURE P12.63   3.0 m

64. ‖ Your task in a science contest is to stack four identical uniform bricks, each of length L, so that the top brick is as far to the right as possible without the stack falling over. Is it possible, as FIGURE P12.64 shows, to stack the bricks such that no part of the top brick is over the table? Answer this question by determining the maximum possible value of d.

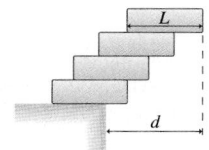

L

d

FIGURE P12.64

65. ‖ A 120-cm-wide sign hangs from a 5.0 kg, 200-cm-long pole. A cable of negligible mass supports the end of the rod as shown in FIGURE P12.65. What is the maximum mass of the sign if the maximum tension in the cable without breaking is 300 N?

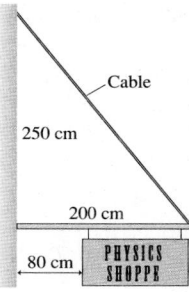

Cable

250 cm

200 cm

80 cm   PHYSICS SHOPPE

FIGURE P12.65

66. ‖ The bunchberry flower has the fastest-moving parts ever ob-
BIO served in a plant. Initially, the stamens are held by the petals in a bent position, storing elastic energy like a coiled spring. When the petals release, the tips of the stamen act like medieval catapults, flipping through a 60° angle in just 0.30 ms to launch pollen from anther sacs at their ends. The human eye just sees a burst of pollen; only high-speed photography reveals the details. As FIGURE P12.66 shows, we can model the stamen tip as a 1.0-mm-long, 10 μg rigid rod with a 10 μg anther sac at the end. Although oversimplifying, we'll assume a constant angular acceleration.
   a. How large is the "straightening torque"?
   b. What is the speed of the anther sac as it releases its pollen?

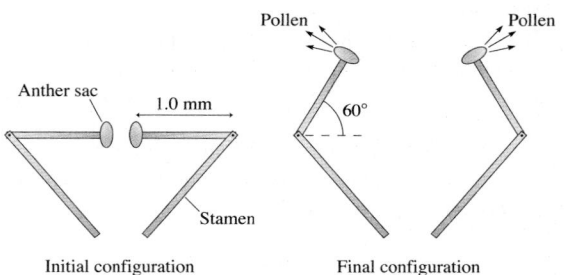

Pollen   Pollen

Anther sac   1.0 mm   60°

Stamen

Initial configuration   Final configuration

FIGURE P12.66

67. ‖ A 60-cm-long, 500 g bar rotates in a horizontal plane on an axle that passes through the center of the bar. Compressed air is fed in through the axle, passes through a small hole down the length of the bar, and escapes as air jets from holes at the ends of the bar. The jets are perpendicular to the bar's axis. Starting from rest, the bar spins up to an angular velocity of 150 rpm at the end of 10 s.
   a. How much force does each jet of escaping air exert on the bar?
   b. If the axle is moved to one end of the bar while the air jets are unchanged, what will be the bar's angular velocity at the end of 10 seconds?

68. ‖‖ Flywheels are large, massive wheels used to store energy. They can be spun up slowly, then the wheel's energy can be released quickly to accomplish a task that demands high power. An industrial flywheel has a 1.5 m diameter and a mass of 250 kg. Its maximum angular velocity is 1200 rpm.
   a. A motor spins up the flywheel with a constant torque of 50 N m. How long does it take the flywheel to reach top speed?
   b. How much energy is stored in the flywheel?
   c. The flywheel is disconnected from the motor and connected to a machine to which it will deliver energy. Half the energy stored in the flywheel is delivered in 2.0 s. What is the average power delivered to the machine?
   d. How much torque does the flywheel exert on the machine?

69. ||| The two blocks in FIGURE P12.69 are connected by a massless rope that passes over a pulley. The pulley is 12 cm in diameter and has a mass of 2.0 kg. As the pulley turns, friction at the axle exerts a torque of magnitude 0.50 N m. If the blocks are released from rest, how long does it take the 4.0 kg block to reach the floor?

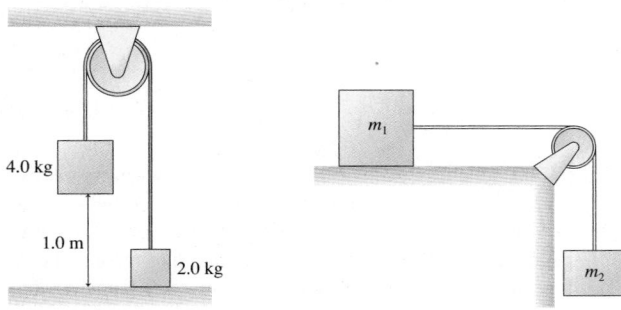

FIGURE P12.69   FIGURE P12.70

70. || Blocks of mass m_1 and m_2 are connected by a massless string that passes over the pulley in FIGURE P12.70. The pulley turns on frictionless bearings. Mass m_1 slides on a horizontal, frictionless surface. Mass m_2 is released while the blocks are at rest.
   a. Assume the pulley is massless. Find the acceleration of m_1 and the tension in the string. This is a Chapter 7 review problem.
   b. Suppose the pulley has mass m_p and radius R. Find the acceleration of m_1 and the tensions in the upper and lower portions of the string. Verify that your answers agree with part a if you set $m_p = 0$.
71. || The 2.0 kg, 30-cm-diameter disk in FIGURE P12.71 is spinning at 300 rpm. How much friction force must the brake apply to the rim to bring the disk to a halt in 3.0 s?

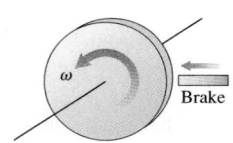

FIGURE P12.71

72. || Your engineering team has been assigned the task of measuring the properties of a new jet-engine turbine. You've previously determined that the turbine's moment of inertia is 2.6 kg m². The next job is to measure the frictional torque of the bearings. Your plan is to run the turbine up to a predetermined rotation speed, cut the power, and time how long it takes the turbine to reduce its rotation speed by 50%. Your data are as follows:

Rotation (rpm)	Time (s)
1500	19
1800	22
2100	25
2400	30
2700	34

Draw an appropriate graph of the data and, from the slope of the best-fit line, determine the frictional torque.

73. || A hollow sphere is rolling along a horizontal floor at 5.0 m/s when it comes to a 30° incline. How far up the incline does it roll before reversing direction?

74. || The 5.0 kg, 60-cm-diameter disk in FIGURE P12.74 rotates on an axle passing through one edge. The axle is parallel to the floor. The cylinder is held with the center of mass at the same height as the axle, then released.
   a. What is the cylinder's initial angular acceleration?
   b. What is the cylinder's angular velocity when it is directly below the axle?

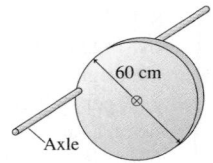

FIGURE P12.74

75. || FIGURE P12.75 shows a hoop of mass M and radius R rotating about an axle at the edge of the hoop. The hoop starts at its highest position and is given a very small push to start it rotating. At its lowest position, what are (a) the angular velocity and (b) the speed of the lowest point on the hoop?

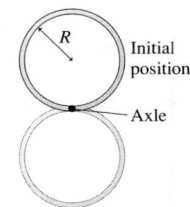

FIGURE P12.75

76. || A long, thin rod of mass M and length L is standing straight up on a table. Its lower end rotates on a frictionless pivot. A very slight push causes the rod to fall over. As it hits the table, what are (a) the angular velocity and (b) the speed of the tip of the rod?

77. || The sphere of mass M and radius R in FIGURE P12.77 is rigidly attached to a thin rod of radius r that passes through the sphere at distance $\frac{1}{2}R$ from the center. A string wrapped around the rod pulls with tension T. Find an expression for the sphere's angular acceleration. The rod's moment of inertia is negligible.

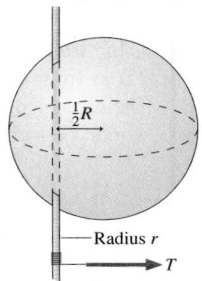

FIGURE P12.77

78. || A satellite follows the elliptical orbit shown in FIGURE P12.78. The only force on the satellite is the gravitational attraction of the planet. The satellite's speed at point a is 8000 m/s.
   a. Does the satellite experience any torque about the center of the planet? Explain.
   b. What is the satellite's speed at point b?
   c. What is the satellite's speed at point c?

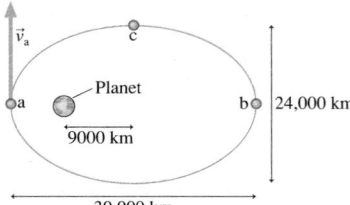

FIGURE P12.78   30,000 km

79. || A 10 g bullet traveling at 400 m/s strikes a 10 kg, 1.0-m-wide door at the edge opposite the hinge. The bullet embeds itself in the door, causing the door to swing open. What is the angular velocity of the door just after impact?
80. || A 200 g, 40-cm-diameter turntable rotates on frictionless bearings at 60 rpm. A 20 g block sits at the center of the turntable. A compressed spring shoots the block radially outward along a frictionless groove in the surface of the turntable. What is the turntable's rotation angular velocity when the block reaches the outer edge?

81. ‖ A merry-go-round is a common piece of playground equipment. A 3.0-m-diameter merry-go-round with a mass of 250 kg is spinning at 20 rpm. John runs tangent to the merry-go-round at 5.0 m/s, in the same direction that it is turning, and jumps onto the outer edge. John's mass is 30 kg. What is the merry-go-round's angular velocity, in rpm, after John jumps on?

82. ‖ A 45 kg figure skater is spinning on the toes of her skates at 1.0 rev/s. Her arms are outstretched as far as they will go. In this orientation, the skater can be modeled as a cylindrical torso (40 kg, 20 cm average diameter, 160 cm tall) plus two rod-like arms (2.5 kg each, 66 cm long) attached to the outside of the torso. The skater then raises her arms straight above her head, where she appears to be a 45 kg, 20-cm-diameter, 200-cm-tall cylinder. What is her new angular velocity, in rev/s?

Challenge Problems

83. In FIGURE CP12.83, a 200 g toy car is placed on a narrow 60-cm-diameter track with wheel grooves that keep the car going in a circle. The 1.0 kg track is free to turn on a frictionless, vertical axis. The spokes have negligible mass. After the car's switch is turned on, it soon reaches a steady speed of 0.75 m/s relative to the track. What then is the track's angular velocity, in rpm?

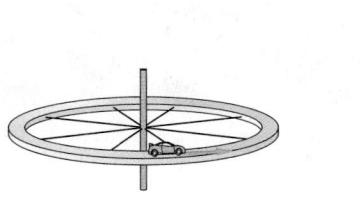

FIGURE CP12.83

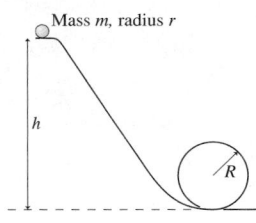

Mass m, radius r

h

R

FIGURE CP12.84

84. The marble rolls down the track shown in FIGURE CP12.84 and around a loop-the-loop of radius R. The marble has mass m and radius r. What minimum height h must the track have for the marble to make it around the loop-the-loop without falling off?

85. FIGURE CP12.85 shows a triangular block of Swiss cheese sitting on a cheese board. You and your friends start to wonder what will happen if you slowly tilt the board, increasing angle θ. Emily thinks the cheese will start to slide before it topples over. Fred thinks it will topple before starting to slide. Some quick Internet research on your part reveals that the coefficient of static friction of Swiss cheese on wood is 0.90. Who is right?

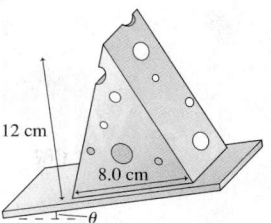

12 cm

8.0 cm

θ

FIGURE CP12.85

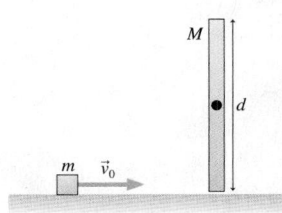

M

d

m $\vec{v}_0$

FIGURE CP12.86

86. FIGURE CP12.86 shows a cube of mass m sliding without friction at speed v_0. It undergoes a perfectly elastic collision with the bottom tip of a rod of length d and mass $M = 2m$. The rod is pivoted about a frictionless axle through its center, and initially it hangs straight down and is at rest. What is the cube's velocity—both speed and direction—after the collision?

87. A 75 g, 30-cm-long rod hangs vertically on a frictionless, horizontal axle passing through its center. A 10 g ball of clay traveling horizontally at 2.5 m/s hits and sticks to the very bottom tip of the rod. To what maximum angle, measured from vertical, does the rod (with the attached ball of clay) rotate?

88. During most of its lifetime, a star maintains an equilibrium size in which the inward force of gravity on each atom is balanced by an outward pressure force due to the heat of the nuclear reactions in the core. But after all the hydrogen "fuel" is consumed by nuclear fusion, the pressure force drops and the star undergoes a *gravitational collapse* until it becomes a *neutron star*. In a neutron star, the electrons and protons of the atoms are squeezed together by gravity until they fuse into neutrons. Neutron stars spin very rapidly and emit intense pulses of radio and light waves, one pulse per rotation. These "pulsing stars" were discovered in the 1960s and are called *pulsars*.

a. A star with the mass ($M = 2.0 \times 10^{30}$ kg) and size ($R = 7.0 \times 10^8$ m) of our sun rotates once every 30 days. After undergoing gravitational collapse, the star forms a pulsar that is observed by astronomers to emit radio pulses every 0.10 s. By treating the neutron star as a solid sphere, deduce its radius.

b. What is the speed of a point on the equator of the neutron star?

Your answers will be somewhat too large because a star cannot be accurately modeled as a solid sphere. Even so, you will be able to show that a star, whose mass is 10^6 larger than the earth's, can be compressed by gravitational forces to a size smaller than a typical state in the United States!

STOP TO THINK ANSWERS

Stop to Think 12.1: $I_a > I_d > I_b > I_c$. The moment of inertia is smaller when the mass is more concentrated near the rotation axis.

Stop to Think 12.2: $\tau_e > \tau_a = \tau_d > \tau_b > \tau_c$. The tangential component in e is larger than 2 N.

Stop to Think 12.3: $\alpha_b > \alpha_a > \alpha_c = \alpha_d = \alpha_e$. Angular acceleration is proportional to torque and inversely proportional to the moment of inertia. The moment of inertia depends on the *square* of the radius. The tangential force component in e is the same as in d.

Stop to Think 12.4: c > d > a = b. To keep the meter stick in equilibrium, the student must supply a torque equal and opposite to the torque due to the hanging masses. Torque depends on the mass *and* on how far the mass is from the pivot point.

Stop to Think 12.5: d. There is no net torque on the bucket + rain system, so the angular momentum is conserved. The addition of mass on the outer edge of the circle increases I, so ω must decrease. Mechanical energy is not conserved because the raindrop collisions are inelastic.

13 Newton's Theory of Gravity

This beautiful galaxy consists of billions of stars orbiting the galactic center exactly as predicted by Newton's theory of gravity.

▶ **Looking Ahead** The goal of Chapter 13 is to use Newton's theory of gravity to understand the motion of satellites and planets.

Copernicus and Galileo

In many ways, modern science began with Copernicus's assertion in 1543 that the planets orbit the sun rather than the sun and planets revolving around the earth.

Copernicus's ideas were confirmed a century later when Galileo, using one of the earliest telescopes, made the first detailed observations of the solar system.

In this chapter you'll learn how Newton's theory of gravity explains the motions of satellites, planets, and even the entire solar system as it revolves around the galactic center.

Newton's Theory

Newton proposed that *any* two masses are attracted toward each other by a gravitational force

$$F_{M \text{ on } m} = F_{m \text{ on } M} = \frac{GMm}{r^2}$$

This is an *inverse-square* force law because the force depends inversely on the square of the distance between the masses.

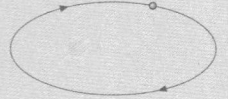

In addition, Newton proposed that his three laws of motion applied to the planets, not just to earthly objects.

We'll use Newton's theory to

- Understand the value of g, and
- Weigh the earth.

Newton's theory explains the orbits of the dust and ice particles that form Saturn's rings.

Gravitational Energy

The gravitational potential energy of Chapter 10, $U_g = mgy$, is valid only very near the surface of a planet. We'll find a more general gravitational potential energy that applies to satellites and planets.

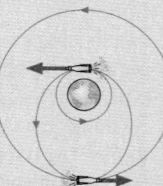

You'll learn to use energy conservation to solve orbit problems, such as how a satellite is transferred from one orbit to another.

◀ **Looking Back**
Chapter 10 Potential energy and energy conservation

Kepler's Laws

Before Galileo and the telescope, Kepler used naked-eye measurements of the planets to make three major discoveries:

- The planets move in elliptical orbits.
- The planets "sweep out" equal areas in equal times.
- The square of the period is proportional to the cube of the orbit's radius.

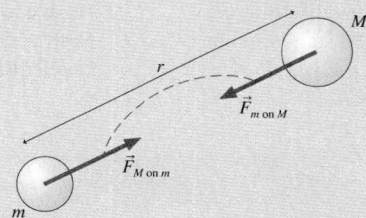

A planet's orbit is an ellipse with the sun at one focus.

Orbits

We'll use Newton's theory to *derive* Kepler's three laws, providing strong evidence in favor of Newton's theory.

Communications satellites are in *geosynchronous orbits* around the earth. You'll learn how to calculate the height of these satellites.

◀ **Looking Back**
Sections 8.2–8.3 Uniform circular motion and orbital motion

13.1 A Little History

The study of the structure of the universe is called **cosmology.** The ancient Greeks developed a cosmological model, with the earth at the center of the universe while the moon, the sun, the planets, and the stars were points of light turning about the earth on large "celestial spheres." This viewpoint was further expanded by the second-century Egyptian astronomer Ptolemy (the P is silent). He developed an elaborate mathematical model of the solar system that quite accurately predicted the complex planetary motions.

Then, in 1543, the medieval world was turned on its head with the publication of Nicholas Copernicus's *De Revolutionibus.* Copernicus argued that it is not the earth at rest in the center of the universe—it is the sun! Furthermore, Copernicus asserted that all of the planets revolve about the sun (hence his title) in circular orbits.

Tycho and Kepler

The greatest medieval astronomer was Tycho Brahe. For 30 years, from 1570 to 1600, Tycho compiled the most accurate astronomical observations the world had known. The invention of the telescope was still to come, but Tycho developed ingenious mechanical sighting devices that allowed him to determine the positions of stars and planets in the sky with unprecedented accuracy.

Tycho had a young mathematical assistant named Johannes Kepler. Kepler had become one of the first outspoken defenders of Copernicus, and his goal was to find evidence for circular planetary orbits in Tycho's records. To appreciate the difficulty of this task, keep in mind that Kepler was working before the development of graphs or of calculus—and certainly before calculators! His mathematical tools were algebra, geometry, and trigonometry, and he was faced with thousands upon thousands of individual observations of planetary positions measured as angles above the horizon.

Many years of work led Kepler to discover that the orbits are not circles, as Copernicus claimed, but *ellipses.* Furthermore, the speed of a planet is not constant but varies as it moves around the ellipse.

Kepler's laws, as we call them today, state that

1. Planets move in elliptical orbits, with the sun at one focus of the ellipse.
2. A line drawn between the sun and a planet sweeps out equal areas during equal intervals of time.
3. The square of a planet's orbital period is proportional to the cube of the semimajor-axis length.

FIGURE 13.1a shows that an ellipse has two *foci* (plural of *focus*), and the sun occupies one of these. The long axis of the ellipse is the *major axis,* and half the length of this axis is called the *semimajor-axis length.* As the planet moves, a line drawn from the sun to the planet "sweeps out" an area. FIGURE 13.1b shows two such areas. Kepler's discovery that the areas are equal for equal Δt implies that the planet moves faster when near the sun, slower when farther away.

All the planets except Mercury have elliptical orbits that are only very slightly distorted circles. As FIGURE 13.2 shows, a circle is an ellipse in which the two foci move to the center, effectively making one focus, and the semimajor-axis length becomes the radius. Because the mathematics of ellipses is difficult, this chapter will focus on circular orbits.

Kepler made an additional contribution that was essential to prepare the way for Newton. For Ptolemy and, later, Copernicus, the role of the sun was merely to light and warm the earth and planets. Kepler was the first to suggest that the sun was a center of force that somehow *caused* the planetary motions. Now, Kepler was working before Galileo and Newton, so he did not speak in terms of forces and centripetal accelerations. The value of his contribution was not the specific mechanism he proposed but his introduction of the idea that the sun somehow exerts forces on the planets to determine their motion.

FIGURE 13.1 The elliptical orbit of a planet about the sun.

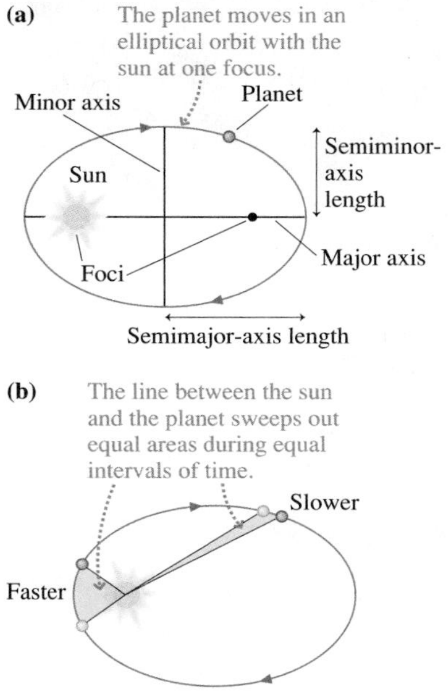

(a) The planet moves in an elliptical orbit with the sun at one focus.

(b) The line between the sun and the planet sweeps out equal areas during equal intervals of time.

FIGURE 13.2 A circular orbit is a special case of an elliptical orbit.

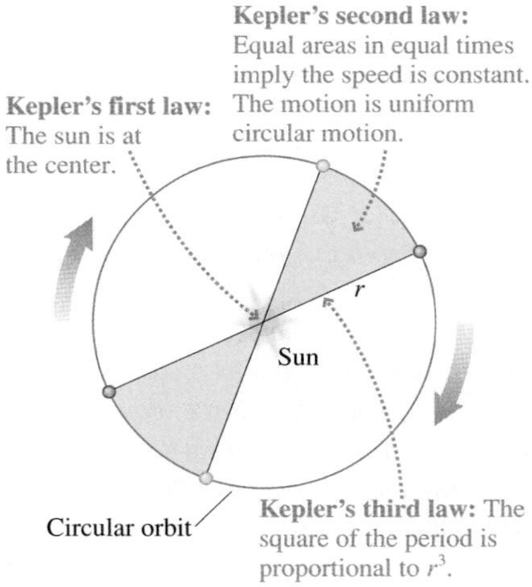

Kepler's second law: Equal areas in equal times imply the speed is constant.

Kepler's first law: The sun is at the center.

The motion is uniform circular motion.

Kepler's third law: The square of the period is proportional to r^3.

Kepler published the first two of his laws in 1609, the same year in which Galileo first turned a telescope to the heavens. Through his telescope Galileo could *see* moons orbiting Jupiter, just as Copernicus had suggested the planets orbit the sun. He could *see* that Venus has phases, like the moon, which implied its orbital motion about the sun. By the time of Galileo's death in 1642, the Copernican revolution was complete.

13.2 Isaac Newton

A popular image has Newton thinking of the idea of gravity after an apple fell on his head. This amusing story is at least close to the truth. Newton himself said that the "notion of gravitation" came to him as he "sat in a contemplative mood" and "was occasioned by the fall of an apple." It occurred to him that, perhaps, the apple was attracted to the *center* of the earth but was prevented from getting there by the earth's surface. And if the apple was so attracted, why not the moon? In other words, perhaps gravitation is a *universal* force between all objects in the universe! This is not shocking today, but no one before Newton had ever thought that the mundane motion of objects on earth had any connection at all with the stately motion of the planets through the heavens.

Newton reasoned along the following lines. Suppose the moon's circular motion around the earth is due to the pull of the earth's gravity. Then, as you learned in Chapter 8 and is shown in **FIGURE 13.3**, the moon must be in *free fall* with the free-fall acceleration $g_{\text{at moon}}$.

Isaac Newton, 1642–1727.

FIGURE 13.3 The moon is in free fall around the earth.

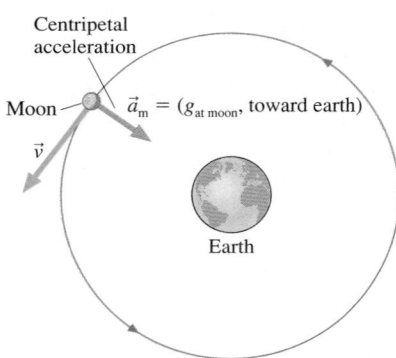

NOTE ▶ We need to be careful with notation. The symbol g_{moon} is the free-fall acceleration caused by the *moon's* gravity—that is, the acceleration of a falling object on the moon. Here we're interested in the acceleration *of* the moon by the earth's gravity, which we'll call $g_{\text{at moon}}$. ◀

The centripetal acceleration of an object in uniform circular motion is

$$a_r = g_{\text{at moon}} = \frac{v_{\text{m}}^2}{r_{\text{m}}} \tag{13.1}$$

The moon's speed is related to the radius r_{m} and period T_{m} of its orbit by $v_{\text{m}} = $ circumference/period $= 2\pi r_{\text{m}}/T_{\text{m}}$. Combining these, Newton found

$$g_{\text{at moon}} = \frac{4\pi^2 r_{\text{m}}}{T_{\text{m}}^2} = \frac{4\pi^2 (3.84 \times 10^8 \text{ m})}{(2.36 \times 10^6 \text{ s})^2} = 0.00272 \text{ m/s}^2$$

Astronomical measurements had established a reasonably good value for r_{m} by the time of Newton, and the period $T_{\text{m}} = 27.3$ days was quite well known.

The moon's centripetal acceleration is significantly less than the free-fall acceleration on the earth's surface. In fact,

$$\frac{g_{\text{at moon}}}{g_{\text{on earth}}} = \frac{0.00272 \text{ m/s}^2}{9.80 \text{ m/s}^2} = \frac{1}{3600}$$

This is an interesting result, but it was Newton's next step that was critical. He compared the radius of the moon's orbit to the radius of the earth:

$$\frac{r_{\text{m}}}{R_{\text{e}}} = \frac{3.84 \times 10^8 \text{ m}}{6.37 \times 10^6 \text{ m}} = 60.2$$

I deduced that the forces which keep the planets in their orbs must be reciprocally as the squares of their distances from the centers about which they revolve; and thereby compared the force requisite to keep the Moon in her orb with the force of gravity at the surface of the Earth; and found them answer pretty nearly.

Isaac Newton

NOTE ▶ We'll use a lowercase r, as in r_{m}, to indicate the radius of an orbit. We'll use an uppercase R, as in R_{e}, to indicate the radius of a star or planet. ◀

Newton recognized that $(60.2)^2$ is almost exactly 3600. Thus, he reasoned:

- If g has the value 9.80 at the earth's surface, and
- If the force of gravity and g decrease in size depending inversely on the square of the distance from the center of the earth,
- Then g will have exactly the value it needs at the distance of the moon to cause the moon to orbit the earth with a period of 27.3 days.

His two ratios were not identical (because the earth isn't a perfect sphere and the moon's orbit isn't a perfect circle), but he found them to "answer pretty nearly" and knew that he had to be on the right track.

A satellite orbits the earth with constant speed at a height above the surface equal to the earth's radius. The magnitude of the satellite's acceleration is

a. $4g_{\text{on earth}}$ b. $2g_{\text{on earth}}$ c. $g_{\text{on earth}}$

d. $\frac{1}{2}g_{\text{on earth}}$ e. $\frac{1}{4}g_{\text{on earth}}$ f. 0

13.3 Newton's Law of Gravity

Newton proposed that *every* object in the universe attracts *every other* object with a force that is

    1. Inversely proportional to the square of the distance between the objects.
    2. Directly proportional to the product of the masses of the two objects.

To make these ideas more specific, **FIGURE 13.4** shows masses m_1 and m_2 separated by distance r. Each mass exerts an attractive force on the other, a force that we call the **gravitational force.** These two forces form an action/reaction pair, so $\vec{F}_{1 \text{ on } 2}$ is equal and opposite to $\vec{F}_{2 \text{ on } 1}$. The magnitude of the forces is given by Newton's law of gravity.

> **Newton's law of gravity** If two objects with masses m_1 and m_2 are a distance r apart, the objects exert attractive forces on each other of magnitude
>
> $$F_{1 \text{ on } 2} = F_{2 \text{ on } 1} = \frac{Gm_1 m_2}{r^2} \qquad (13.2)$$
>
> The forces are directed along the straight line joining the two objects.

The constant G, called the **gravitational constant,** is a proportionality constant necessary to relate the masses, measured in kilograms, to the force, measured in newtons. In the SI system of units, G has the value

$$G = 6.67 \times 10^{-11} \text{ N m}^2/\text{kg}^2$$

FIGURE 13.5 is a graph of the gravitational force as a function of the distance between the two masses. As you can see, an inverse-square force decreases rapidly.

Strictly speaking, Equation 13.2 is valid only for particles. However, Newton was able to show that this equation also applies to spherical objects, such as planets, if r is the distance between their centers. Our intuition and common sense suggest this to us, as they did to Newton. The rather difficult proof is not essential, so we will omit it.

Gravitational Force and Weight

Knowing G, we can calculate the size of the gravitational force. Consider two 1.0 kg masses that are 1.0 m apart. According to Newton's law of gravity, these two masses exert an attractive gravitational force on each other of magnitude

$$F_{1 \text{ on } 2} = F_{2 \text{ on } 1} = \frac{Gm_1 m_2}{r^2}$$

$$= \frac{(6.67 \times 10^{-11} \text{ N m}^2/\text{kg}^2)(1.0 \text{ kg})(1.0 \text{ kg})}{(1.0 \text{ m})^2} = 6.67 \times 10^{-11} \text{ N}$$

This is an exceptionally tiny force, especially when compared to the gravitational force of the entire earth on each mass: $F_G = mg = 9.8$ N.

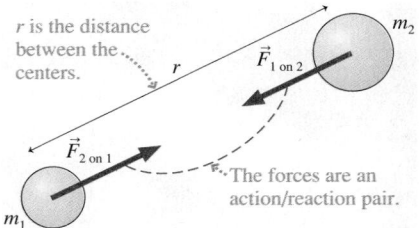

FIGURE 13.4 The gravitational forces on masses m_1 and m_2.

r is the distance between the centers.

$\vec{F}_{1 \text{ on } 2}$

$\vec{F}_{2 \text{ on } 1}$

The forces are an action/reaction pair.

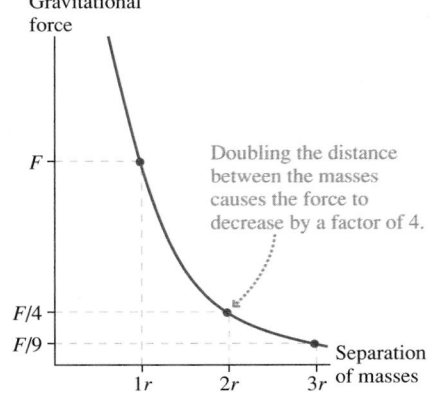

FIGURE 13.5 The gravitational force is an inverse-square force.

Gravitational force

Doubling the distance between the masses causes the force to decrease by a factor of 4.

Separation of masses

The fact that the gravitational force between two ordinary-size objects is so small is the reason we are not aware of it. As you sit there reading, you are being attracted to this book, to the person sitting next to you, and to every object around you, but the forces are so tiny in comparison to the normal forces and friction forces acting on you that they are completely undetectable. Only when one (or both) of the masses is exceptionally large—planet-size—does the force of gravity become important.

We find a more respectable result if we calculate the force *of the earth* on a 1.0 kg mass at the earth's surface:

$$F_{\text{earth on 1 kg}} = \frac{GM_e m_{1\,\text{kg}}}{R_e^2}$$

$$= \frac{(6.67 \times 10^{-11}\,\text{N}\,\text{m}^2/\text{kg}^2)(5.98 \times 10^{24}\,\text{kg})(1.0\,\text{kg})}{(6.37 \times 10^6\,\text{m})^2} = 9.8\,\text{N}$$

The dynamics of stellar motions, spanning many thousands of light years, are governed by Newton's law of gravity.

A galaxy of $\approx 10^{11}$ stars spanning a distance greater than 100,000 light years.

where the distance between the mass and the center of the earth is the earth's radius. The earth's mass M_e and radius R_e were taken from Table 13.2 in Section 13.6. This table, which is also printed inside the back cover of the book, contains astronomical data that will be used for examples and homework.

The force $F_{\text{earth on 1 kg}} = 9.8\,\text{N}$ is exactly the weight of a stationary 1.0 kg mass: $F_G = mg = 9.8\,\text{N}$. Is this a coincidence? Of course not. Weight—the upward force of a spring scale—exactly balances the downward gravitational force, so numerically they must be equal.

Although weak, gravity is a *long-range* force. No matter how far apart two objects may be, there is a gravitational attraction between them given by Equation 13.2. Consequently, gravity is the most ubiquitous force in the universe. It not only keeps your feet on the ground, it also keeps the earth orbiting the sun, the solar system orbiting the center of the Milky Way galaxy, and the entire Milky Way galaxy performing an intricate orbital dance with other galaxies making up what is called the "local cluster" of galaxies.

The Principle of Equivalence

Newton's law of gravity depends on a rather curious assumption. The concept of *mass* was introduced in Chapter 5 by considering the relationship between force and acceleration. The *inertial mass* of an object, which is the mass that appears in Newton's second law, is found by measuring the object's acceleration a in response to force F:

$$m_{\text{inert}} = \text{inertial mass} = \frac{F}{a} \tag{13.3}$$

Gravity plays no role in this definition of mass.

The quantities m_1 and m_2 in Newton's law of gravity are being used in a very different way. Masses m_1 and m_2 govern the strength of the gravitational attraction between two objects. The mass used in Newton's law of gravity is called the **gravitational mass.** The gravitational mass of an object can be determined by measuring the attractive force exerted on it by another mass M a distance r away:

$$m_{\text{grav}} = \text{gravitational mass} = \frac{r^2 F_{M \text{ on } m}}{GM} \tag{13.4}$$

Acceleration does not enter into the definition of the gravitational mass.

These are two very different concepts of mass. Yet Newton, in his theory of gravity, asserts that the inertial mass in his second law is the very same mass that governs the strength of the gravitational attraction between two objects. The assertion that $m_{\text{grav}} = m_{\text{inert}}$ is called the **principle of equivalence.** It says that inertial mass is *equivalent to* gravitational mass.

As a hypothesis about nature, the principle of equivalence is subject to experimental verification or disproof. Many exceptionally clever experiments have looked for any difference between the gravitational mass and the inertial mass, and they have shown that any difference, if it exists at all, is less than 10 parts in a trillion! As far as we know today, the gravitational mass and the inertial mass are exactly the same thing.

But why should a quantity associated with the dynamics of motion, relating force to acceleration, have anything at all to do with the gravitational attraction? This is a question that intrigued Einstein and eventually led to his general theory of relativity, the theory about curved spacetime and black holes. General relativity is beyond the scope of this textbook, but it explains the principle of equivalence as a property of space itself.

Newton's Theory of Gravity

Newton's theory of gravity is more than just Equation 13.2. The *theory* of gravity consists of:

1. A specific force law for gravity, given by Equation 13.2, *and*
2. The principle of equivalence, *and*
3. An assertion that Newton's three laws of motion are universally applicable. These laws are as valid for heavenly bodies, the planets and stars, as for earthly objects.

Consequently, everything we have learned about forces, motion, and energy is relevant to the dynamics of satellites, planets, and galaxies.

STOP TO THINK 13.2 The figure shows a binary star system. The mass of star 2 is twice the mass of star 1. Compared to $\vec{F}_{1\,\text{on}\,2}$, the magnitude of the force $\vec{F}_{2\,\text{on}\,1}$ is

a. Four times as big.
b. Twice as big.
c. The same size.
d. Half as big.
e. One-quarter as big.

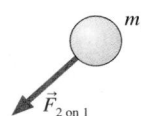

13.4 Little *g* and Big *G*

The familiar equation $F_G = mg$ works well when an object is on the surface of a planet, but *mg* will not help us find the force exerted on the same object if it is in orbit around the planet. Neither can we use *mg* to find the force of attraction between the earth and the moon. Newton's law of gravity provides a more fundamental starting point because it describes a *universal* force that exists between all objects.

To illustrate the connection between Newton's law of gravity and the familiar $F_G = mg$, **FIGURE 13.6** shows an object of mass *m* on the surface of Planet X. Planet X inhabitant Mr. Xhzt, standing on the surface, finds that the downward gravitational force is $F_G = mg_X$, where g_X is the free-fall acceleration on Planet X.

We, taking a more cosmic perspective, reply, "Yes, that is the force *because* of a universal force of attraction between your planet and the object. The size of the force is determined by Newton's law of gravity."

We and Mr. Xhzt are both correct. Whether you think locally or globally, we and Mr. Xhzt must arrive at the *same numerical value* for the magnitude of the force. Suppose an object of mass *m* is on the surface of a planet of mass *M* and radius *R*. The local gravitational force is

$$F_G = mg_{\text{surface}} \tag{13.5}$$

where g_{surface} is the acceleration due to gravity at the planet's surface. The force of gravitational attraction for an object on the surface ($r = R$), as given by Newton's law of gravity, is

$$F_{M\,\text{on}\,m} = \frac{GMm}{R^2} \tag{13.6}$$

FIGURE 13.6 Weighing an object of mass *m* on Planet X.

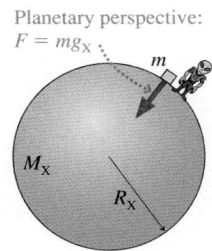

Planetary perspective:
$F = mg_X$

Planet X

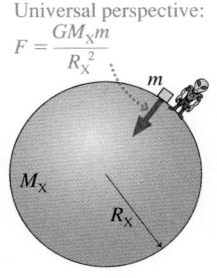

Universal perspective:
$F = \dfrac{GM_X m}{R_X^2}$

Planet X

Because these are two names and two expressions for the same force, we can equate the right-hand sides to find that

$$g_{\text{surface}} = \frac{GM}{R^2} \tag{13.7}$$

We have used Newton's law of gravity to *predict* the value of g at the surface of a planet. The value depends on the mass and radius of the planet as well as on the value of G, which establishes the overall strength of the gravitational force.

The expression for g_{surface} in Equation 13.7 is valid for any planet or star. Using the mass and radius of Mars, we can predict the Martian value of g:

$$g_{\text{Mars}} = \frac{GM_{\text{Mars}}}{R_{\text{Mars}}{}^2} = \frac{(6.67 \times 10^{-11}\,\text{N m}^2/\text{kg}^2)(6.42 \times 10^{23}\,\text{kg})}{(3.37 \times 10^6\,\text{m})^2} = 3.8\,\text{m/s}^2$$

NOTE ▶ We noted in Chapter 6 that measured values of g are very slightly smaller on a rotating planet. We'll ignore rotation in this chapter. ◀

Decrease of *g* with Distance

Equation 13.7 gives g_{surface} at the surface of a planet. More generally, imagine an object of mass m at distance $r > R$ from the center of a planet. Further, suppose that gravity from the planet is the only force acting on the object. Then its acceleration, the free-fall acceleration, is given by Newton's second law:

$$g = \frac{F_{M\,\text{on}\,m}}{m} = \frac{GM}{r^2} \tag{13.8}$$

This more general result agrees with Equation 13.7 if $r = R$, but it allows us to determine the "local" free-fall acceleration at distances $r > R$. Equation 13.8 expresses Newton's discovery, with regard to the moon, that g decreases inversely with the square of the distance.

FIGURE 13.7 shows a satellite orbiting at height h above the earth's surface. Its distance from the center of the earth is $r = R_e + h$. Most people have a mental image that satellites orbit "far" from the earth, but in reality h is typically 200 miles $\approx 3 \times 10^5$ m, while $R_e = 6.37 \times 10^6$ m. Thus the satellite is barely "skimming" the earth at a height only about 5% of the earth's radius!

The value of g at height h above the earth (i.e., above sea level) is

$$g = \frac{GM_e}{(R_e + h)^2} = \frac{GM_e}{R_e^2(1 + h/R_e)^2} = \frac{g_{\text{earth}}}{(1 + h/R_e)^2} \tag{13.9}$$

where $g_{\text{earth}} = 9.83\,\text{m/s}^2$ is the value calculated from Equation 13.7 for $h = 0$ on a nonrotating earth. Table 13.1 shows the value of g evaluated at several values of h.

FIGURE 13.7 A satellite orbits the earth at height h.

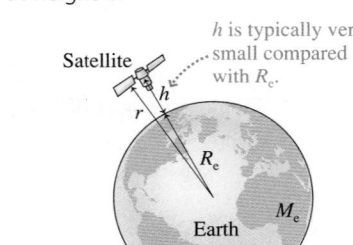

Satellite

h is typically very small compared with R_e.

r

R_e

M_e

Earth

TABLE 13.1 Variation of g with height above the ground

Height h	Example	g (m/s^2)
0 m	ground	9.83
4500 m	Mt. Whitney	9.82
10,000 m	jet airplane	9.80
300,000 m	space shuttle	8.90
35,900,000 m	communications satellite	0.22

NOTE ▶ The free-fall acceleration of a satellite such as the space shuttle is only slightly less than the ground-level value. An object in orbit is not "weightless" because there is no gravity in space but because it is in free fall, as you learned in Chapter 8. ◀

Weighing the Earth

We can predict g if we know the earth's mass. But how do we know the value of M_e? We cannot place the earth on a giant pan balance, so how is its mass known? Furthermore, how do we know the value of G?

Newton did not know the value of G. He could say that the gravitational force is proportional to the product $m_1 m_2$ and inversely proportional to r^2, but he had no means of knowing the value of the proportionality constant.

Determining G requires a *direct* measurement of the gravitational force between two known masses at a known separation. The small size of the gravitational force between ordinary-size objects makes this quite a feat. Yet the English scientist Henry Cavendish came up with an ingenious way of doing so with a device called a *torsion balance*. Two fairly small masses m, typically about 10 g, are placed on the ends of a lightweight rod. The rod is hung from a thin fiber, as shown in FIGURE 13.8a, and allowed to reach equilibrium.

If the rod is then rotated slightly and released, a *restoring force* will return it to equilibrium. This is analogous to displacing a spring from equilibrium, and in fact the restoring force and the angle of displacement obey a version of Hooke's law: $F_{restore} = k\Delta\theta$. The "torsion constant" k can be determined by timing the period of oscillations. Once k is known, a force that twists the rod slightly away from equilibrium can be measured by the product $k\Delta\theta$. It is possible to measure very small angular deflections, so this device can be used to determine very small forces.

Two larger masses M (typically lead spheres with $M \approx 10$ kg) are then brought close to the torsion balance, as shown in FIGURE 13.8b. The gravitational attraction that they exert on the smaller hanging masses causes a very small but measurable twisting of the balance, enough to measure $F_{M \text{ on } m}$. Because m, M, and r are all known, Cavendish was able to determine G from

$$G = \frac{F_{M \text{ on } m}\, r^2}{Mm} \tag{13.10}$$

His first results were not highly accurate, but improvements over the years in this and similar experiments have produced the value of G accepted today.

With an independently determined value of G, we can return to Equation 13.7 to find

$$M_e = \frac{g_{earth} R_e^2}{G} \tag{13.11}$$

We have weighed the earth! The value of g_{earth} at the earth's surface is known with great accuracy from kinematics experiments. The earth's radius R_e is determined by surveying techniques. Combining our knowledge from these very different measurements has given us a way to determine the mass of the earth.

The gravitational constant G is what we call a *universal constant*. Its value establishes the strength of one of the fundamental forces of nature. As far as we know, the gravitational force between two masses would be the same anywhere in the universe. Universal constants tell us something about the most basic and fundamental properties of nature. You will soon meet other universal constants.

FIGURE 13.8 Cavendish's experiment to measure G.

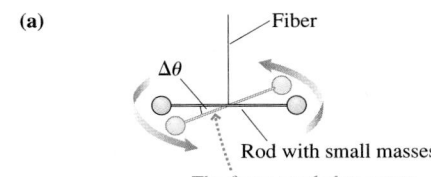

(a)

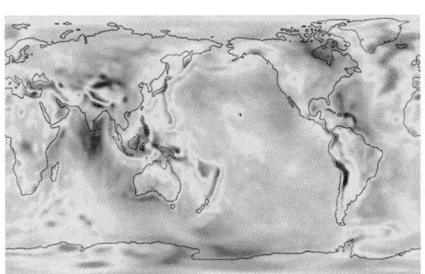

(b)

The free-fall acceleration varies slightly due to mountains and to variation in the density of the earth's crust. This map shows the *gravitational anomaly*, with red regions of slightly stronger gravity and blue regions of slightly weaker gravity. The variation is tiny, less than 0.001 m/s².

STOP TO THINK 13.3 A planet has four times the mass of the earth, but the acceleration due to gravity on the planet's surface is the same as on the earth's surface. The planet's radius is

a. $4R_e$ b. $2R_e$ c. R_e d. $\frac{1}{2}R_e$ e. $\frac{1}{4}R_e$

13.5 Gravitational Potential Energy

Gravitational problems are ideal for the conservation-law tools we developed in Chapters 9 through 11. Because gravity is the only force, and it is a conservative force, both the momentum and the mechanical energy of the system $m_1 + m_2$ are conserved. To employ conservation of energy, however, we need to determine an appropriate form for the gravitational potential energy for two interacting masses.

The definition of potential energy that we developed in Chapter 11 is

$$\Delta U = U_f - U_i = -W_c(\text{i} \rightarrow \text{f}) \tag{13.12}$$

where $W_c(\text{i} \rightarrow \text{f})$ is the work done by a conservative force as a particle moves from position i to position f. For a flat earth, we used $F = -mg$ and the choice that $U = 0$ at the surface ($y = 0$) to arrive at the now-familiar $U_g = mgy$. This result for U_g is valid only for $y \ll R_e$, when the earth's curvature and size are not apparent. We now need to find an expression for the gravitational potential energy of masses that interact over *large* distances.

FIGURE 13.9 shows two particles of mass m_1 and m_2. Let's calculate the work done on mass m_2 by the conservative force $\vec{F}_{1 \text{ on } 2}$ as m_2 moves from an initial position at distance r to a final position very far away. The force, which points to the left, is opposite the displacement; hence this force does *negative* work. Consequently, due to the minus sign in Equation 13.12, ΔU is *positive*. A pair of masses *gains* potential energy as the masses move farther apart, just as a particle near the earth's surface gains potential energy as it moves to a higher altitude.

We can establish a coordinate system with m_1 at the origin and m_2 moving along the x-axis. The gravitational force is a variable force, so we need the full definition of work:

$$W(\text{i} \rightarrow \text{f}) = \int_{x_i}^{x_f} F_x\, dx \tag{13.13}$$

$\vec{F}_{1 \text{ on } 2}$ points toward the left, so its x-component is $(F_{1 \text{ on } 2})_x = -Gm_1 m_2/x^2$. As mass m_2 moves from $x_i = r$ to $x_f = \infty$, the potential energy changes by

$$\Delta U = U_{\text{at } \infty} - U_{\text{at } r} = -\int_r^\infty (F_{1 \text{ on } 2})_x\, dx = -\int_r^\infty \left(\frac{-Gm_1 m_2}{x^2}\right) dx$$

$$= +Gm_1 m_2 \int_r^\infty \frac{dx}{x^2} = -\frac{Gm_1 m_2}{x}\Big|_r^\infty = \frac{Gm_1 m_2}{r} \tag{13.14}$$

NOTE ▶ We chose to integrate along the x-axis, but the fact that gravity is a conservative force means that ΔU will have this value if m_2 moves from r to ∞ along *any* path. ◀

To proceed further, we need to choose the point where $U = 0$. We would like our choice to be valid for any star or planet, regardless of its mass and radius. This will be the case if we set $U = 0$ at the point where the interaction between the masses vanishes. According to Newton's law of gravity, the strength of the interaction is zero only when $r = \infty$. Two masses infinitely far apart will have no tendency, or potential, to move together, so we will *choose* to place the zero point of potential energy at $r = \infty$. That is, $U_{\text{at } \infty} = 0$.

This choice gives us the gravitational potential energy of masses m_1 and m_2:

$$U_g = -\frac{Gm_1 m_2}{r} \tag{13.15}$$

This is the potential energy of masses m_1 and m_2 when their *centers* are separated by distance r. **FIGURE 13.10** is a graph of U_g as a function of the distance r between the masses. Notice that it asymptotically approaches 0 as $r \rightarrow \infty$.

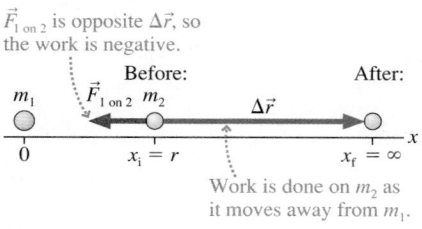

FIGURE 13.9 Calculating the work done by the gravitational force as mass m_2 moves from r to ∞.

$\vec{F}_{1 \text{ on } 2}$ is opposite $\Delta \vec{r}$, so the work is negative.

Before: After:

Work is done on m_2 as it moves away from m_1.

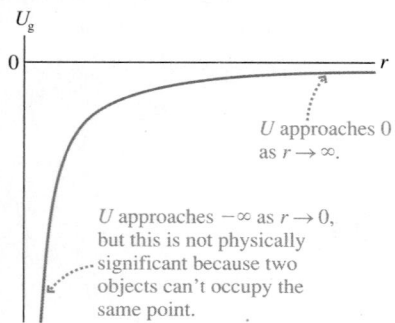

FIGURE 13.10 The gravitational potential-energy curve.

U approaches 0 as $r \rightarrow \infty$.

U approaches $-\infty$ as $r \rightarrow 0$, but this is not physically significant because two objects can't occupy the same point.

NOTE ▶ Although Equation 13.15 looks rather similar to Newton's law of gravity, it depends only on $1/r$, *not* on $1/r^2$. ◀

It may seem disturbing that the potential energy is negative, but we encountered similar situations in Chapter 10. All a negative potential energy means is that the potential energy of the two masses at separation r is *less* than their potential energy at infinite separation. Only the *change* in U has physical significance, and the change will be the same no matter where we place the zero of potential energy.

To illustrate, suppose two masses a distance r_1 apart are released from rest. How will they move? From a force perspective, you would note that each mass experiences an attractive force and accelerates toward the other. The energy perspective of FIGURE 13.11 tells the same thing. By moving toward smaller r (that is, $r_1 \rightarrow r_2$), the system *loses* potential energy and *gains* kinetic energy while conserving E_{mech}. The system is "falling downhill," although in a more general sense than we think about on a flat earth.

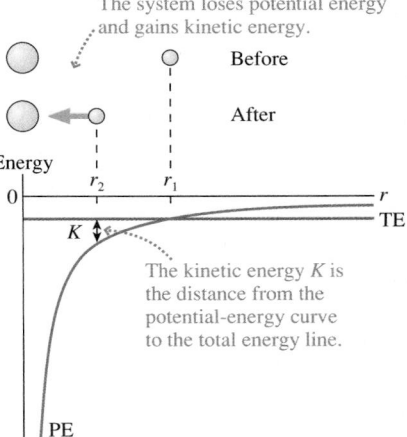

FIGURE 13.11 Two masses gain kinetic energy as their separation decreases.

The system loses potential energy and gains kinetic energy.

The kinetic energy K is the distance from the potential-energy curve to the total energy line.

EXAMPLE 13.1 **Crashing into the sun**

Suppose the earth suddenly came to a halt and ceased revolving around the sun. The gravitational force would then pull it directly into the sun. What would be the earth's speed as it crashed?

MODEL Model the earth and the sun as spherical masses. This is an isolated system, so its mechanical energy is conserved.

VISUALIZE FIGURE 13.12 is a before-and-after pictorial representation for this gruesome cosmic event. The "crash" occurs as the earth touches the sun, at which point the distance between their centers is $r_2 = R_s + R_e$. The initial separation r_1 is the radius of the earth's *orbit* about the sun, not the radius of the earth.

FIGURE 13.12 Before-and-after pictorial representation of the earth crashing into the sun (not to scale).

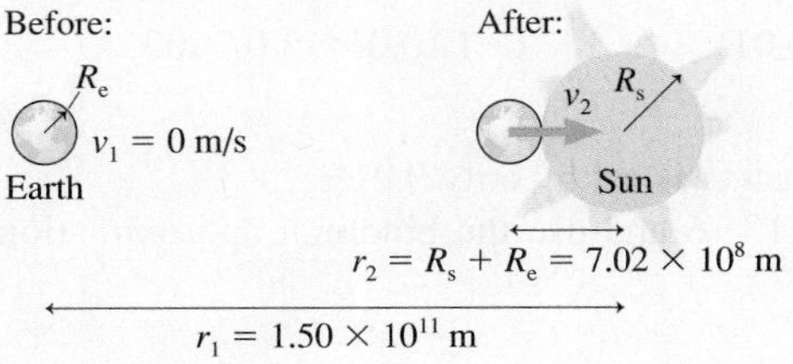

Before: After:

R_e v_2 R_s

Earth $v_1 = 0$ m/s   Sun

$r_2 = R_s + R_e = 7.02 \times 10^8$ m

$r_1 = 1.50 \times 10^{11}$ m

SOLVE Strictly speaking, the kinetic energy is the sum $K = K_{\text{earth}} + K_{\text{sun}}$. However, the sun is so much more massive than the earth that the lightweight earth does almost all of the moving. It is a reasonable approximation to consider the sun as remaining at rest. In that case, the energy conservation equation $K_2 + U_2 = K_1 + U_1$ is

$$\frac{1}{2}M_e v_2^2 - \frac{GM_s M_e}{R_s + R_e} = 0 - \frac{GM_s M_e}{r_1}$$

This is easily solved for the earth's speed at impact. Using data from Table 13.2, we find

$$v_2 = \sqrt{2GM_s \left(\frac{1}{R_s + R_e} - \frac{1}{r_1} \right)} = 6.13 \times 10^5 \text{ m/s}$$

ASSESS The earth would be really flying along at over 1 million miles per hour as it crashed into the sun! It is worth noting that we do not have the mathematical tools to solve this problem using Newton's second law because the acceleration is not constant. But the solution is straightforward when we use energy conservation.

EXAMPLE 13.2 **Escape speed**

A 1000 kg rocket is fired straight away from the surface of the earth. What speed does the rocket need to "escape" from the gravitational pull of the earth and never return? Assume a nonrotating earth.

MODEL In a simple universe, consisting of only the earth and the rocket, an insufficient launch speed will cause the rocket eventually to fall back to earth. Once the rocket finally slows to a halt, gravity will ever so slowly pull it back. The only way the rocket can escape is to never stop ($v = 0$) and thus never have a turning point! That is, the rocket must continue moving away from the earth forever. The *minimum* launch speed for escape, which is called the **escape speed,** will cause the rocket to stop ($v = 0$) only as it reaches $r = \infty$. Now ∞, of course, is not a "place," so a

statement like this means that we want the rocket's speed to approach $v = 0$ asymptotically as $r \rightarrow \infty$.

VISUALIZE FIGURE 13.13 is a before-and-after pictorial representation.

FIGURE 13.13 Pictorial representation of a rocket launched with sufficient speed to escape the earth's gravity.

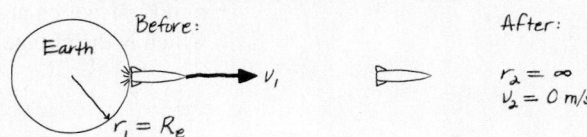

Before: After:

Earth $r_2 = \infty$
                                 $v_2 = 0$ m/s
v_1

$r_1 = R_e$

Continued

SOLVE Energy conservation $K_2 + U_2 = K_1 + U_1$ is

$$0 + 0 = \frac{1}{2}mv_1^2 - \frac{GM_e m}{R_e}$$

where we used the fact that both the kinetic and potential energy are zero at $r = \infty$. Thus the escape speed is

$$v_{escape} = v_1 = \sqrt{\frac{2GM_e}{R_e}} = 11,200 \text{ m/s} \approx 25,000 \text{ mph}$$

ASSESS The problem was mathematically easy; the difficulty was deciding how to interpret it. That is why—as you have now seen many times—the "physics" of a problem consists of thinking, interpreting, and modeling. We will see variations on this problem in the future, with both gravity and electricity, so you might want to review the *reasoning* involved. Notice that the answer does *not* depend on the rocket's mass, so this is the escape speed for any object.

The Flat-Earth Approximation

FIGURE 13.14 We can treat the earth as flat if $y \ll R_e$.

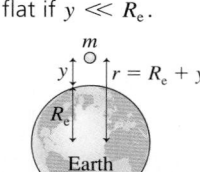

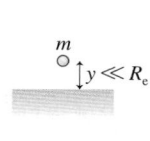

For a spherical earth:
$$U_g = -\frac{GM_e m}{R_e + y}$$

We can treat the earth as flat if $y \ll R_e$:
$$U_g = mgy$$

How does Equation 13.15 for the gravitational potential energy relate to our previous use of $U_g = mgy$ on a flat earth? FIGURE 13.14 shows an object of mass m located at height y above the surface of the earth. The object's distance from the earth's center is $r = R_e + y$ and its gravitational potential energy is

$$U_g = -\frac{GM_e m}{r} = -\frac{GM_e m}{R_e + y} = -\frac{GM_e m}{R_e(1 + y/R_e)} \tag{13.16}$$

where, in the last step, we factored R_e out of the denominator.

Suppose the object is very close to the earth's surface ($y \ll R_e$). In that case, the ratio $y/R_e \ll 1$. There is an approximation you will learn about in calculus, called the *binomial approximation,* that says

$$(1 + x)^n \approx 1 + nx \qquad \text{if } x \ll 1 \tag{13.17}$$

As an illustration, you can easily use your calculator to find that $1/1.01 = 0.9901$, to four significant figures. But suppose you wrote $1.01 = 1 + 0.01$. You could then use the binomial approximation to calculate

$$\frac{1}{1.01} = \frac{1}{1 + 0.01} = (1 + 0.01)^{-1} \approx 1 + (-1)(0.01) = 0.9900$$

You can see that the approximate answer is off by only 0.01%.

If we call $y/R_e = x$ in Equation 13.16 and use the binomial approximation, with $n = -1$, we find

$$U_g(\text{if } y \ll R_e) \approx -\frac{GM_e m}{R_e}\left(1 - \frac{y}{R_e}\right) = -\frac{GM_e m}{R_e} + m\left(\frac{GM_e}{R_e^2}\right)y \tag{13.18}$$

Now the first term is just the gravitational potential energy U_0 when the object is at ground level ($y = 0$). In the second term, you can recognize $GM_e/R_e^2 = g_{earth}$ from the definition of g in Equation 13.7. Thus we can write Equation 13.18 as

$$U_g(\text{if } y \ll R_e) = U_0 + mg_{earth}y \tag{13.19}$$

Although we chose U_g to be zero when $r = \infty$, we are always free to change our minds. If we change the zero point of potential energy to be $U_0 = 0$ at the surface, which is the choice we made in Chapter 10, then Equation 13.19 becomes

$$U_g(\text{if } y \ll R_e) = mg_{earth}y \tag{13.20}$$

We can sleep easier knowing that Equation 13.15 for the gravitational potential energy is consistent with our earlier "flat-earth" expression for the potential energy.

EXAMPLE 13.3   **The speed of a satellite**

A less-than-successful inventor wants to launch small satellites into orbit by launching them straight up from the surface of the earth at very high speed.

a. With what speed should he launch the satellite if it is to have a speed of 500 m/s at a height of 400 km? Ignore air resistance.
b. By what percentage would your answer be in error if you used a flat-earth approximation?

MODEL Mechanical energy is conserved if we ignore drag.

VISUALIZE **FIGURE 13.15** shows a pictorial representation.

FIGURE 13.15 Pictorial representation of a satellite launched straight up.

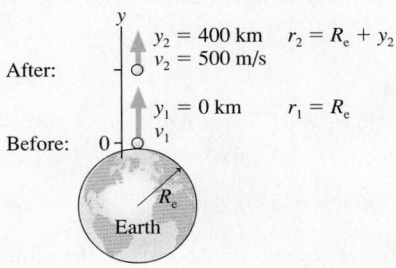

SOLVE a. Although the height is exaggerated in the figure, 400 km = 400,000 m is high enough that we cannot ignore the earth's spherical shape. The energy conservation equation $K_2 + U_2 = K_1 + U_1$ is

$$\frac{1}{2}mv_2^2 - \frac{GM_e m}{R_e + y_2} = \frac{1}{2}mv_1^2 - \frac{GM_e m}{R_e + y_1}$$

where we've written the distance between the satellite and the earth's center as $r = R_e + y$. The initial height is $y_1 = 0$. Notice that the satellite mass m cancels and is not needed. Solving for the launch speed, we have

$$v_1 = \sqrt{v_2^2 + 2GM_e\left(\frac{1}{R_e} - \frac{1}{R_e + y_2}\right)} = 2770 \text{ m/s}$$

This is about 6000 mph, much less than the escape speed.
b. The calculation is the same in the flat-earth approximation except that we use $U_g = mgy$. Thus

$$\frac{1}{2}mv_2^2 + mgy_2 = \frac{1}{2}mv_1^2 + mgy_1$$

$$v_1 = \sqrt{v_2^2 + 2gy_2} = 2840 \text{ m/s}$$

The flat-earth value of 2840 m/s is 70 m/s too big. The error, as a percentage of the correct 2770 m/s, is

$$\text{error} = \frac{70}{2770} \times 100 = 2.5\%$$

ASSESS The true speed is less than the flat-earth approximation because the force of gravity decreases with height. Launching a rocket against a decreasing force takes less effort than it would with the flat-earth force of mg at all heights.

STOP TO THINK 13.4   Rank in order, from largest to smallest, the absolute values of the gravitational potential energies of these pairs of masses. The numbers give the relative masses and distances.

(a) $m_1 = 2$ ◯ - - - $r = 4$ - - - ◯ $m_2 = 2$

(b) $m_1 = 1$ ◯- $r = 1$ -◯ $m_2 = 1$

(c) $m_1 = 1$ ◯- $r = 2$ -◯ $m_2 = 1$

(d) $m_1 = 1$ ◯ - - - $r = 4$ - - - ◯ $m_2 = 4$

(e) $m_1 = 4$ ◯ - - - - - - - $r = 8$ - - - - - - - ◯ $m_2 = 4$

13.6 Satellite Orbits and Energies

Solving Newton's second law to find the trajectory of a mass moving under the influence of gravity is mathematically beyond this textbook. It turns out that the solution is a set of elliptical orbits, which is Kepler's first law. Kepler had no *reason* why orbits should be ellipses rather than some other shape. Newton was able to show that ellipses are a *consequence* of his theory of gravity.

The mathematics of ellipses is rather difficult, so we will restrict most of our analysis to the limiting case in which an ellipse becomes a circle. Most planetary orbits differ only very slightly from being circular. The earth's orbit, for example has a (semiminor axis/semimajor axis) ratio of 0.99986—very close to a true circle!

FIGURE 13.16 shows a massive body M, such as the earth or the sun, with a lighter body m orbiting it. The lighter body is called a **satellite,** even though it may be a planet

FIGURE 13.16 The orbital motion of a satellite due to the force of gravity.

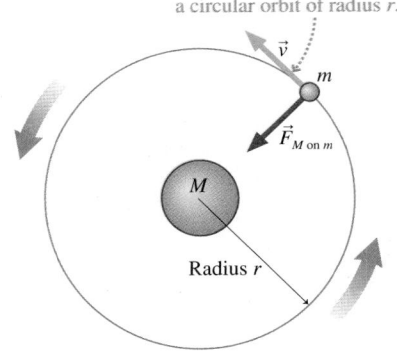

The International Space Station appears to be floating, but it's actually traveling at nearly 8000 m/s as it orbits the earth.

orbiting the sun. For circular motion, the gravitational force must provide the centripetal acceleration v^2/r. Thus Newton's second law for the satellite is

$$F_{M \text{ on } m} = \frac{GMm}{r^2} = ma_r = \frac{mv^2}{r} \tag{13.21}$$

Thus the speed of a satellite in a circular orbit is

$$v = \sqrt{\frac{GM}{r}} \tag{13.22}$$

A satellite must have this specific speed in order to have a circular orbit of radius r about the larger mass M. If the velocity differs from this value, the orbit will become elliptical rather than circular. Notice that the orbital speed does *not* depend on the satellite's mass m. This is consistent with our previous discovery, for motion on a flat earth, that motion due to gravity is independent of the mass.

EXAMPLE 13.4 **The speed of the space shuttle**

The space shuttle in a 300-km-high orbit (≈ 180 mi) wants to capture a smaller satellite for repairs. What are the speeds of the shuttle and the satellite in this orbit?

SOLVE Despite their different masses, the shuttle, the satellite, and the astronaut working in space to make the repairs all travel side by side with the same speed. They are simply in free fall together. Using $r = R_e + h$ with $h = 300$ km $= 3.00 \times 10^5$ m, we find the speed

$$v = \sqrt{\frac{(6.67 \times 10^{-11} \text{ N m}^2/\text{kg}^2)(5.98 \times 10^{24} \text{ kg})}{6.67 \times 10^6 \text{ m}}}$$

$$= 7730 \text{ m/s} \approx 17,000 \text{ mph}$$

ASSESS The answer depends on the mass of the earth but *not* on the mass of the satellite.

Kepler's Third Law

An important parameter of circular motion is the *period*. Recall that the period T is the time to complete one full orbit. The relationship among speed, radius, and period is

$$v = \frac{\text{circumference}}{\text{period}} = \frac{2\pi r}{T} \tag{13.23}$$

We can find a relationship between a satellite's period and the radius of its orbit by using Equation 13.22 for v:

$$v = \frac{2\pi r}{T} = \sqrt{\frac{GM}{r}} \tag{13.24}$$

Squaring both sides and solving for T give

$$T^2 = \left(\frac{4\pi^2}{GM}\right)r^3 \tag{13.25}$$

In other words, the *square* of the period is proportional to the *cube* of the radius. This is Kepler's third law. You can see that Kepler's third law is a direct consequence of Newton's law of gravity.

Table 13.2 contains astronomical information about the solar system. We can use these data to check the validity of Equation 13.25. **FIGURE 13.17** is a graph of log T versus log r for all the planets in Table 13.2 except Mercury. Notice that the scales on each axis are increasing logarithmically—by *factors* of 10—rather than linearly. (Also, the vertical axis has converted T to the SI units of s.) As you can see, the graph is a straight line with a best-fit equation

$$\log T = 1.500 \log r - 9.264$$

TABLE 13.2 Useful astronomical data

Planetary body	Mean distance from sun (m)	Period (years)	Mass (kg)	Mean radius (m)
Sun	—	—	1.99×10^{30}	6.96×10^{8}
Moon	3.84×10^{8}*	27.3 days	7.36×10^{22}	1.74×10^{6}
Mercury	5.79×10^{10}	0.241	3.18×10^{23}	2.43×10^{6}
Venus	1.08×10^{11}	0.615	4.88×10^{24}	6.06×10^{6}
Earth	1.50×10^{11}	1.00	5.98×10^{24}	6.37×10^{6}
Mars	2.28×10^{11}	1.88	6.42×10^{23}	3.37×10^{6}
Jupiter	7.78×10^{11}	11.9	1.90×10^{27}	6.99×10^{7}
Saturn	1.43×10^{12}	29.5	5.68×10^{26}	5.85×10^{7}
Uranus	2.87×10^{12}	84.0	8.68×10^{25}	2.33×10^{7}
Neptune	4.50×10^{12}	165	1.03×10^{26}	2.21×10^{7}

*Distance from earth.

Taking the logarithm of both sides of Equation 13.25, and using the logarithm properties $\log a^n = n \log a$ and $\log(ab) = \log a + \log b$, we have

$$\log T = \frac{3}{2}\log r + \frac{1}{2}\log\left(\frac{4\pi^2}{GM}\right)$$

In other words, theory predicts that the slope of a $\log T$-versus-$\log r$ graph should be exactly $\frac{3}{2}$. As Figure 13.17 shows, the solar-system data agree to an impressive four significant figures. A homework problem will let you use the y-intercept of the graph to determine the mass of the sun.

A particularly interesting application of Equation 13.25 is to communications satellites that are in **geosynchronous orbits** above the earth. These satellites have a period of 24 h = 86,400 s, making their orbital motion synchronous with the earth's rotation. As a result, a satellite in such an orbit appears to remain stationary over one point on the earth's equator. Equation 13.25 allows us to compute the radius of an orbit with this period:

$$r_{\text{geo}} = R_{\text{e}} + h_{\text{geo}} = \left[\left(\frac{GM}{4\pi^2}\right)T^2\right]^{1/3}$$

$$= \left[\left(\frac{(6.67 \times 10^{-11}\,\text{N}\,\text{m}^2/\text{kg}^2)(5.98 \times 10^{24}\,\text{kg})}{4\pi^2}\right)(86,400\,\text{s})^2\right]^{1/3}$$

$$= 4.225 \times 10^7\,\text{m}$$

The height of the orbit is

$$h_{\text{geo}} = r_{\text{geo}} - R_{\text{e}} = 3.59 \times 10^7\,\text{m} = 35,900\,\text{km} \approx 22,300\,\text{mi}$$

NOTE ▶ When you use Equation 13.25, the period *must* be in SI units of s. ◀

Geosynchronous orbits are much higher than the low-earth orbits used by the space shuttle and remote-sensing satellites, where $h \approx 300$ km. Communications satellites in geosynchronous orbits were first proposed in 1948 by science fiction writer Arthur C. Clarke, 10 years before the first artificial satellite of any type!

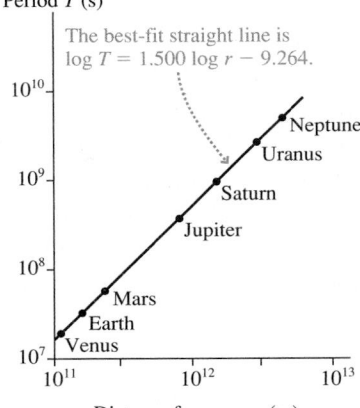

FIGURE 13.17 The graph of $\log T$ versus $\log r$ for the planetary data of Table 13.2.

Extrasolar planets

Astronomers have only recently seen evidence of planets orbiting nearby stars. These are called *extrasolar planets*. Suppose a planet is observed to have a 1200 day period as it orbits a star at the same distance that Jupiter is from the sun. What is the mass of the star in solar masses? (1 *solar mass* is defined to be the mass of the sun.)

SOLVE Here "day" means earth days, as used by astronomers to measure the period. Thus the planet's period in SI units is

$T = 1200$ days $= 1.037 \times 10^8$ s. The orbital radius is that of Jupiter, which we can find in Table 13.2 to be $r = 7.78 \times 10^{11}$ m. Solving Equation 13.25 for the mass of the star gives

$$M = \frac{4\pi^2 r^3}{GT^2} = 2.59 \times 10^{31} \text{ kg} \times \frac{1 \text{ solar mass}}{1.99 \times 10^{30} \text{ kg}}$$

$$= 13 \text{ solar masses}$$

ASSESS This is a large, but not extraordinary, star.

STOP TO THINK 13.5 Two planets orbit a star. Planet 1 has orbital radius r_1 and planet 2 has $r_2 = 4r_1$. Planet 1 orbits with period T_1. Planet 2 orbits with period

a. $T_2 = 8T_1$ b. $T_2 = 4T_1$ c. $T_2 = 2T_1$

d. $T_2 = \frac{1}{2}T_1$ e. $T_2 = \frac{1}{4}T_1$ f. $T_2 = \frac{1}{8}T_1$

Kepler's Second Law

FIGURE 13.18 Angular momentum is conserved for a planet in an elliptical orbit.

(a)

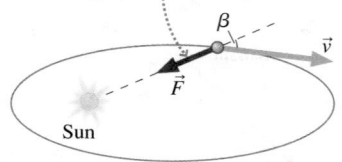

The gravitational force points straight at the sun and exerts no torque.

(b)

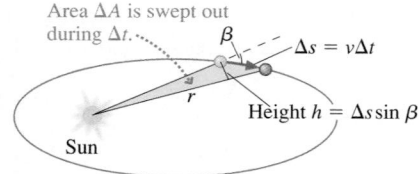

Area ΔA is swept out during Δt.

Height $h = \Delta s \sin \beta$

FIGURE 13.18a shows a planet moving in an elliptical orbit. In Chapter 12 we defined a particle's *angular momentum* to be

$$L = mrv \sin \beta \qquad (13.26)$$

where β is the angle between $\vec{r}$ and $\vec{v}$. For a circular orbit, where β is always 90°, this reduces to simply $L = mrv$.

The only force on the satellite, the gravitational force, points directly toward the star or planet that the satellite is orbiting and exerts no torque; thus **the satellite's angular momentum is conserved as it orbits.**

The satellite moves forward a small distance $\Delta s = v\Delta t$ during the small interval of time Δt. This motion defines the triangle of area ΔA shown in **FIGURE 13.18b**. ΔA is the area "swept out" by the satellite during Δt. You can see that the height of the triangle is $h = \Delta s \sin \beta$, so the triangle's area is

$$\Delta A = \frac{1}{2} \times \text{base} \times \text{height} = \frac{1}{2} \times r \times \Delta s \sin \beta = \frac{1}{2} rv \sin \beta \, \Delta t \quad (13.27)$$

The *rate* at which the area is swept out by the satellite as it moves is

$$\frac{\Delta A}{\Delta t} = \frac{1}{2} rv \sin \beta = \frac{mrv \sin \beta}{2m} = \frac{L}{2m} \qquad (13.28)$$

The angular momentum L is conserved, so it has the same value at every point in the orbit. Consequently, the rate at which the area is swept out by the satellite is constant. This is Kepler's second law, which says that a line drawn between the sun and a planet sweeps out equal areas during equal intervals of time. We see that Kepler's second law is a consequence of the conservation of angular momentum.

Another consequence of angular momentum is that the orbital speed is constant only for a circular orbit. Consider the "ends" of an elliptical orbit, where r is a minimum or maximum. At these points, $\beta = 90°$ and thus $L = mrv$. Because L is constant, the satellite's speed at the farthest point must be less than its speed at the nearest point. In general, a satellite slows as r increases, then speeds up as r decreases, to keep its angular momentum constant.

Kepler's laws summarize observational data about the motions of the planets. They were an outstanding achievement, but they did not form a theory. Newton put forward a *theory,* a specific set of relationships between force and motion that allows *any* motion to be understood and calculated. Newton's theory of gravity has allowed us to *deduce* Kepler's laws and, thus, to understand them at a more fundamental level.

Orbital Energetics

Let us conclude this chapter by thinking about the energetics of orbital motion. We found, with Equation 13.24, that a satellite in a circular orbit must have $v^2 = GM/r$. A satellite's speed is determined entirely by the size of its orbit. The satellite's kinetic energy is thus

$$K = \frac{1}{2}mv^2 = \frac{GMm}{2r} \qquad (13.29)$$

But $-GMm/r$ is the potential energy, U_g, so

$$K = -\frac{1}{2}U_g \qquad (13.30)$$

This is an interesting result. In all our earlier examples, the kinetic and potential energy were two independent parameters. In contrast, a satellite can move in a circular orbit *only* if there is a very specific relationship between K and U. It is not that K and U *have* to have this relationship, but if they do not, the trajectory will be elliptical rather than circular.

Equation 13.30 gives us the mechanical energy of a satellite in a circular orbit:

$$E_{mech} = K + U_g = \frac{1}{2}U_g \qquad (13.31)$$

The gravitational potential energy is negative, hence the *total* mechanical energy is also negative. Negative total energy is characteristic of a **bound system,** a system in which the satellite is bound to the central mass by the gravitational force and cannot get away. The total energy of an unbound system must be ≥ 0 because the satellite can reach infinity, where $U = 0$, while still having kinetic energy. A negative value of E_{mech} tells us that the satellite is unable to escape the central mass.

FIGURE 13.19 shows the energies of a satellite in a circular orbit as a function of the orbit's radius. Notice how $E_{mech} = \frac{1}{2}U_g$. This figure can help us understand the energetics of transferring a satellite from one orbit to another. Suppose a satellite is in an orbit of radius r_1 and we'd like it to be in a larger orbit of radius r_2. The kinetic energy at r_2 is less than at r_1 (the satellite moves more slowly in the larger orbit), but you can see that the total energy *increases* as r increases. Consequently, transferring a satellite to a larger orbit requires a net energy increase $\Delta E > 0$. Where does this increase of energy come from?

Artificial satellites are raised to higher orbits by firing their rocket motors to create a forward thrust. This force does work on the satellite, and the energy equation of Chapter 11 tells us that this work increases the satellite's energy by $\Delta E_{mech} = W_{ext}$. Thus the energy to "lift" a satellite into a higher orbit comes from the chemical energy stored in the rocket fuel.

FIGURE 13.19 The kinetic, potential, and total energy of a satellite in a circular orbit.

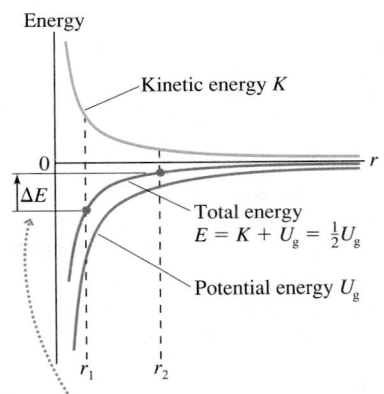

Energy ΔE must be added to move a satellite from an orbit with radius r_1 to radius r_2.

EXAMPLE 13.6 **Raising a satellite**

How much work must be done to boost a 1000 kg communications satellite from a low earth orbit with $h = 300$ km, where it is released by the space shuttle, to a geosynchronous orbit?

SOLVE The required work is $W_{ext} = \Delta E_{mech}$, and from Equation 13.31 we see that $\Delta E_{mech} = \frac{1}{2}\Delta U_g$. The initial orbit has radius $r_{shuttle} = R_e + h = 6.67 \times 10^6$ m. We earlier found the radius of a geosynchronous orbit to be 4.22×10^7 m. Thus

$$W_{ext} = \Delta E_{mech} = \frac{1}{2}\Delta U_g = \frac{1}{2}(-GM_em)\left(\frac{1}{r_{geo}} - \frac{1}{r_{shuttle}}\right) = 2.52 \times 10^{10}\ \text{J}$$

ASSESS It takes a lot of energy to boost satellites to high orbits!

You might think that the way to get a satellite into a larger orbit would be to point the thrusters toward the earth and blast outward. That would work fine *if* the satellite were initially at rest and moved straight out along a linear trajectory. But an orbiting satellite is already moving and has significant inertia. A force directed straight outward would *change* the satellite's velocity vector in that direction but would not cause it to *move* along that line. (Remember all those earlier motion diagrams for motion along curved trajectories.) In addition, a force directed outward would be almost at right angles to the motion and would do essentially zero work. Navigating in space is not as easy as it appears in *Star Wars!*

FIGURE 13.20 Transferring a satellite to a larger circular orbit.

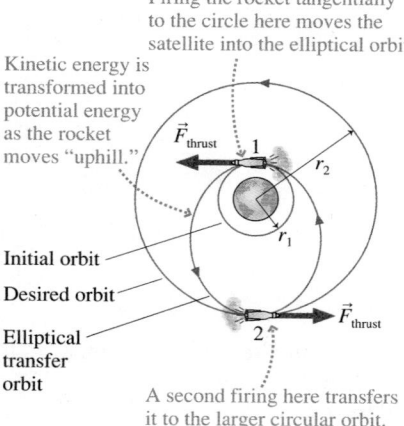

Firing the rocket tangentially to the circle here moves the satellite into the elliptical orbit.

Kinetic energy is transformed into potential energy as the rocket moves "uphill."

$\vec{F}_{thrust}$

r_2

r_1

Initial orbit

Desired orbit

Elliptical transfer orbit

$\vec{F}_{thrust}$

A second firing here transfers it to the larger circular orbit.

To move the satellite in FIGURE 13.20 from the orbit with radius r_1 to the larger circular orbit of radius r_2, the thrusters are turned on at point 1 to apply a brief *forward* thrust force in the direction of motion, *tangent* to the circle. This force does a significant amount of work because the force is parallel to the displacement, so the satellite quickly gains kinetic energy ($\Delta K > 0$). But $\Delta U_g = 0$ because the satellite does not have time to change its distance from the earth during a thrust of short duration. With the kinetic energy increased, but not the potential energy, the satellite no longer meets the requirement $K = -\frac{1}{2}U_g$ for a circular orbit. Instead, it goes into an elliptical orbit.

In the elliptical orbit, the satellite moves "uphill" toward point 2 by transforming kinetic energy into potential energy. At point 2, the satellite has arrived at the desired distance from earth and has the "right" value of the potential energy, but its kinetic energy is now *less* than needed for a circular orbit. (The analysis is more complex than we want to pursue here. It will be left for a homework Challenge Problem.) If no action is taken, the satellite will continue on its elliptical orbit and "fall" back to point 1. But another *forward* thrust at point 2 increases its kinetic energy, without changing U_g, until the kinetic energy reaches the value $K = -\frac{1}{2}U_g$ required for a circular orbit. Presto! The second burn kicks the satellite into the desired circular orbit of radius r_2. The work $W_{ext} = \Delta E_{mech}$ is the *total* work done in both burns. It takes a more extended analysis to see how the work has to be divided between the two burns, but even without those details you now have enough knowledge about orbits and energy to understand the ideas that are involved.

CHALLENGE EXAMPLE 13.7   **A binary star system**

Astronomers discover a binary star system with a period of 90 days. Both stars have a mass twice that of the sun. How far apart are the two stars?

MODEL Model the stars as spherical masses exerting gravitational forces on each other.

VISUALIZE *An isolated system rotates around its center of mass.* FIGURE 13.21 shows the orbits and the forces. If r is the distance of each star to the center of mass—the radius of that star's orbit—then the distance between the stars is $d = 2r$.

FIGURE 13.21 The binary star system.

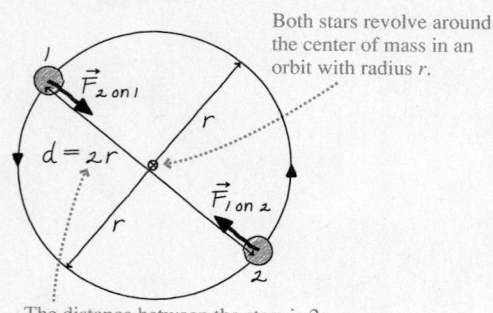

Both stars revolve around the center of mass in an orbit with radius r.

$\vec{F}_{2\,on\,1}$

$d = 2r$

r

r

$\vec{F}_{1\,on\,2}$

The distance between the stars is $2r$.

SOLVE Star 2 has only one force acting on it $\vec{F}_{1\,on\,2}$, and that force has to provide the centripetal acceleration v^2/r of circular motion. Newton's second law for star 2 is

$$F_{1\,on\,2} = \frac{GM_1 M_2}{d^2} = \frac{GM^2}{4r^2} = Ma_r = \frac{Mv^2}{r}$$

where we used $M_1 = M_2 = M$. The equation for star 1 is identical. The star's speed is related to the period and the circumference of its orbit by $v = 2\pi r/T$. With this, the force equation becomes

$$\frac{GM^2}{4r^2} = \frac{4\pi^2 Mr}{T^2}$$

Solving for r gives

$$r = \left[\frac{GMT^2}{16\pi^2}\right]^{1/3}$$

$$= \left[\frac{(6.67 \times 10^{-11}\,\mathrm{N\,m^2/kg^2})(2 \times 1.99 \times 10^{30}\,\mathrm{kg})(7.78 \times 10^6\,\mathrm{s})^2}{16\pi^2}\right]^{1/3}$$

$$= 4.67 \times 10^{10}\,\mathrm{m}$$

The distance between the stars is $d = 2r = 9.3 \times 10^{10}$ m.

ASSESS The result is in the range of solar-system distances and thus is reasonable.

SUMMARY

The goal of Chapter 13 has been to use Newton's theory of gravity to understand the motion of satellites and planets.

General Principles

Newton's Theory of Gravity

1. Two objects with masses M and m a distance r apart exert attractive **gravitational forces** on each other of magnitude

$$F_{M \text{ on } m} = F_{m \text{ on } M} = \frac{GMm}{r^2}$$

where the **gravitational constant** is $G = 6.67 \times 10^{-11} \text{ N m}^2/\text{kg}^2$.

2. Gravitational mass and inertial mass are equivalent.

3. Newton's three laws of motion apply to all objects in the universe.

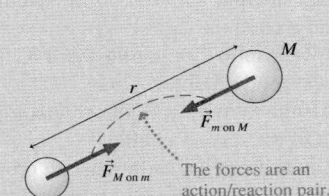

The forces are an action/reaction pair.

Important Concepts

Orbital motion of a planet (or satellite) is described by **Kepler's laws:**

1. Orbits are ellipses with the sun (or planet) at one focus.

2. A line between the sun and the planet sweeps out equal areas during equal intervals of time.

3. The square of the planet's period T is proportional to the cube of the orbit's semimajor axis.

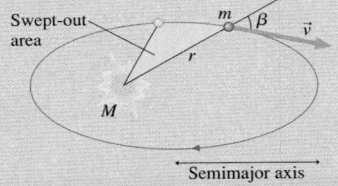

Swept-out area

Semimajor axis

Circular orbits are a special case of an ellipse. For a circular orbit around a mass M,

$$v = \sqrt{\frac{GM}{r}} \qquad \text{and} \qquad T^2 = \left(\frac{4\pi^2}{GM}\right) r^3$$

Conservation of angular momentum

The angular momentum $L = mrv \sin \beta$ remains constant throughout the orbit. Kepler's second law is a consequence of this law.

Orbital energetics

A satellite's mechanical energy $E_{\text{mech}} = K + U_g$ is conserved, where the gravitational potential energy is

$$U_g = -\frac{GMm}{r}$$

For circular orbits, $K = -\frac{1}{2} U_g$ and $E_{\text{mech}} = \frac{1}{2} U_g$. Negative total energy is characteristic of a **bound system.**

Applications

For a planet of mass M and radius R,

- The free-fall acceleration on the surface is $g_{\text{surface}} = \dfrac{GM}{R^2}$

- The escape speed is $v_{\text{escape}} = \sqrt{\dfrac{2GM}{R}}$

- The radius of a geosynchronous orbit is $r_{\text{geo}} = \left(\dfrac{GM}{4\pi^2} T^2\right)^{1/3}$

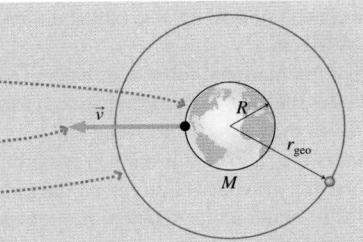

Terms and Notation

cosmology	Newton's law of gravity	principle of equivalence	satellite
Kepler's laws	gravitational constant, G	Newton's theory of gravity	geosynchronous orbit
gravitational force	gravitational mass	escape speed	bound system

CONCEPTUAL QUESTIONS

1. Is the earth's gravitational force on the sun larger than, smaller than, or equal to the sun's gravitational force on the earth? Explain.

2. The gravitational force of a star on orbiting planet 1 is F_1. Planet 2, which is twice as massive as planet 1 and orbits at twice the distance from the star, experiences gravitational force F_2. What is the ratio F_1/F_2?

3. A 1000 kg satellite and a 2000 kg satellite follow exactly the same orbit around the earth.
   a. What is the ratio F_1/F_2 of the force on the first satellite to that on the second satellite?
   b. What is the ratio a_1/a_2 of the acceleration of the first satellite to that of the second satellite?

4. How far away from the earth must an orbiting spacecraft be for the astronauts inside to be weightless? Explain.

5. A space shuttle astronaut is working outside the shuttle as it orbits the earth. If he drops a hammer, will it fall to earth? Explain why or why not.

6. The free-fall acceleration at the surface of planet 1 is 20 m/s². The radius and the mass of planet 2 are twice those of planet 1. What is g on planet 2?

7. *Why* is the gravitational potential energy of two masses negative? Note that saying "because that's what the equation gives" is *not* an explanation.

8. The escape speed from Planet X is 10,000 m/s. Planet Y has the same radius as Planet X but is twice as dense. What is the escape speed from Planet Y?

9. The mass of Jupiter is 300 times the mass of the earth. Jupiter orbits the sun with $T_{\text{Jupiter}} = 11.9$ yr in an orbit with $r_{\text{Jupiter}} = 5.2 r_{\text{earth}}$. Suppose the earth could be moved to the distance of Jupiter and placed in a circular orbit around the sun. Which of the following describes the earth's new period? Explain.
   a. 1 yr
   b. Between 1 yr and 11.9 yr
   c. 11.9 yr
   d. More than 11.9 yr
   e. It would depend on the earth's speed.
   f. It's impossible for a planet of earth's mass to orbit at the distance of Jupiter.

10. Satellites in near-earth orbit experience a very slight drag due to the extremely thin upper atmosphere. These satellites slowly but surely spiral inward, where they finally burn up as they reach the thicker lower levels of the atmosphere. The radius decreases so slowly that you can consider the satellite to have a circular orbit at all times. As a satellite spirals inward, does it speed up, slow down, or maintain the same speed? Explain.

EXERCISES AND PROBLEMS

Problems labeled [ ] integrate material from earlier chapters.

Exercises

Section 13.3 Newton's Law of Gravity

1. ‖ What is the ratio of the sun's gravitational force on you to the earth's gravitational force on you?

2. ‖ The centers of a 10 kg lead ball and a 100 g lead ball are separated by 10 cm.
   a. What gravitational force does each exert on the other?
   b. What is the ratio of this gravitational force to the gravitational force of the earth on the 100 g ball?

3. ‖ What is the ratio of the sun's gravitational force on the moon to the earth's gravitational force on the moon?

4. ‖ A 1.0-m-diameter lead sphere has a mass of 5900 kg. A dust particle rests on the surface. What is the ratio of the gravitational force of the sphere on the dust particle to the gravitational force of the earth on the dust particle?

5. │ Estimate the force of attraction between a 50 kg woman and a 70 kg man sitting 1.0 m apart.

6. ‖ The space shuttle orbits 300 km above the surface of the earth. What is the gravitational force on a 1.0 kg sphere inside the space shuttle?

Section 13.4 Little g and Big G

7. │ a. What is the free-fall acceleration at the surface of the sun?
   b. What is the sun's free-fall acceleration at the distance of the earth?

8. ‖ What is the free-fall acceleration at the surface of (a) the moon and (b) Jupiter?

9. ‖ A sensitive gravimeter at a mountain observatory finds that the free-fall acceleration is 0.0075 m/s² less than that at sea level. What is the observatory's altitude?

10. ‖ Suppose we could shrink the earth without changing its mass. At what fraction of its current radius would the free-fall acceleration at the surface be three times its present value?

11. ‖ Planet Z is 10,000 km in diameter. The free-fall acceleration on Planet Z is 8.0 m/s².
   a. What is the mass of Planet Z?
   b. What is the free-fall acceleration 10,000 km above Planet Z's north pole?

Section 13.5 Gravitational Potential Energy

12. │ An astronaut on earth can throw a ball straight up to a height of 15 m. How high can he throw the ball on Mars?

13. ‖ What is the escape speed from Jupiter?

14. ‖ A rocket is launched straight up from the earth's surface at a speed of 15,000 m/s. What is its speed when it is very far away from the earth?

15. ❙ A space station orbits the sun at the same distance as the earth but on the opposite side of the sun. A small probe is fired away from the station. What minimum speed does the probe need to escape the solar system?

16. ‖ You have been visiting a distant planet. Your measurements have determined that the planet's mass is twice that of earth but the free-fall acceleration at the surface is only one-fourth as large.
    a. What is the planet's radius?
    b. To get back to earth, you need to escape the planet. What minimum speed does your rocket need?

Section 13.6 Satellite Orbits and Energies

17. ❙ The *asteroid belt* circles the sun between the orbits of Mars and Jupiter. One asteroid has a period of 5.0 earth years. What are the asteroid's orbital radius and speed?

18. ❙ Use information about the earth and its orbit to determine the mass of the sun.

19. ‖ Planet X orbits the star Omega with a "year" that is 200 earth days long. Planet Y circles Omega at four times the distance of Planet X. How long is a year on Planet Y?

20. ❙ You are the science officer on a visit to a distant solar system. Prior to landing on a planet you measure its diameter to be 1.8×10^7 m and its rotation period to be 22.3 hours. You have previously determined that the planet orbits 2.2×10^{11} m from its star with a period of 402 earth days. Once on the surface you find that the free-fall acceleration is 12.2 m/s². What is the mass of (a) the planet and (b) the star?

21. ‖ Three satellites orbit a planet of radius R, as shown in FIGURE EX13.21. Satellites S_1 and S_3 have mass m. Satellite S_2 has mass $2m$. Satellite S_1 orbits in 250 minutes and the force on S_1 is 10,000 N.
    a. What are the periods of S_2 and S_3?
    b. What are the forces on S_2 and S_3?
    c. What is the kinetic-energy ratio K_1/K_3 for S_1 and S_3?

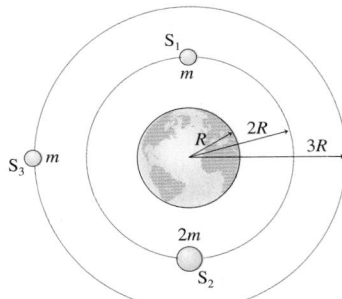

FIGURE EX13.21

22. ‖ A satellite orbits the sun with a period of 1.0 day. What is the radius of its orbit?

23. ‖ An earth satellite moves in a circular orbit at a speed of 5500 m/s. What is its orbital period?

24. ‖ What are the speed and altitude of a geosynchronous satellite orbiting Mars? Mars rotates on its axis once every 24.8 hours.

Problems

25. ‖ Two spherical objects have a combined mass of 150 kg. The gravitational attraction between them is 8.00×10^{-6} N when their centers are 20 cm apart. What is the mass of each?

26. ‖ FIGURE P13.26 shows three masses. What are the magnitude and the direction of the net gravitational force on (a) the 20.0 kg mass and (b) the 5.0 kg mass? Give the direction as an angle cw or ccw from the *y*-axis.

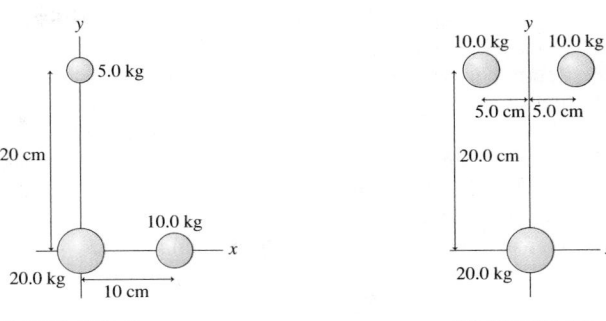

FIGURE P13.26   **FIGURE P13.27**

27. ‖ What are the magnitude and direction of the net gravitational force on the 20.0 kg mass in FIGURE P13.27?

28. ‖ What is the total gravitational potential energy of the three masses in FIGURE P13.26?

29. ‖ What is the total gravitational potential energy of the three masses in FIGURE P13.27?

30. ‖‖ Two 100 kg lead spheres are suspended from 100-m-long massless cables. The tops of the cables have been carefully anchored *exactly* 1 m apart. What is the distance between the centers of the spheres?

31. ‖‖ A 20 kg sphere is at the origin and a 10 kg sphere is at $x = 20$ cm. At what position on the *x*-axis could you place a small mass such that the net gravitational force on it due to the spheres is zero?

32. ‖ a. At what height above the earth is the acceleration due to gravity 10% of its value at the surface?
    b. What is the speed of a satellite orbiting at that height?

33. ‖ A 1.0 kg object is released from rest 500 km ($\approx$ 300 miles) above the earth.
    a. What is its impact speed as it hits the ground? Ignore air resistance.
    b. What would the impact speed be if the earth were flat?
    c. By what percentage is the flat-earth calculation in error?

34. ‖‖ An object of mass m is dropped from height h above a planet of mass M and radius R. Find an expression for the object's speed as it hits the ground.

35. ‖‖ A projectile is shot straight up from the earth's surface at a speed of 10,000 km/h. How high does it go?

36. ‖ Two meteoroids are heading for earth. Their speeds as they cross the moon's orbit are 2.0 km/s.
    a. The first meteoroid is heading straight for earth. What is its speed of impact?
    b. The second misses the earth by 5000 km. What is its speed at its closest point?

37. ‖ A binary star system has two stars, each with the same mass as our sun, separated by 1.0×10^{12} m. A comet is very far away and essentially at rest. Slowly but surely, gravity pulls the comet toward the stars. Suppose the comet travels along a straight line that passes through the midpoint between the two stars. What is the comet's speed at the midpoint?

38. ‖ Suppose that on earth you can jump straight up a distance of 50 cm. Can you escape from a 4.0-km-diameter asteroid with a mass of 1.0×10^{14} kg?

39. ‖ A projectile is fired straight away from the moon from a base on the far side of the moon, away from the earth. What is the projectile's escape speed from the earth-moon system?

40. ‖ Two spherical asteroids have the same radius R. Asteroid 1 has mass M and asteroid 2 has mass $2M$. The two asteroids are released from rest with distance $10R$ between their centers. What is the speed of each asteroid just before they collide?
   Hint: You will need to use two conservation laws.

41. ‖ Two Jupiter-size planets are released from rest 1.0×10^{11} m apart. What are their speeds as they crash together?

42. ‖ A starship is circling a distant planet of radius R. The astronauts find that the free-fall acceleration at their altitude is half the value at the planet's surface. How far above the surface are they orbiting? Your answer will be a multiple of R.

43. ‖ Three stars, each with the mass and radius of our sun, form an equilateral triangle 5.0×10^9 m on a side. If all three are simultaneously released from rest, what are their speeds as they crash together in the center?

44. ‖ The two stars in a binary star system have masses 2.0×10^{30} kg and 6.0×10^{30} kg. They are separated by 2.0×10^{12} m. What are
   a. The system's rotation period, in years?
   b. The speed of each star?

45. ‖ A 4000 kg lunar lander is in orbit 50 km above the surface of the moon. It needs to move out to a 300-km-high orbit in order to link up with the mother ship that will take the astronauts home. How much work must the thrusters do?

46. ‖ The space shuttle is in a 250-km-high circular orbit. It needs to reach a 610-km-high circular orbit to catch the Hubble Space Telescope for repairs. The shuttle's mass is 75,000 kg. How much energy is required to boost it to the new orbit?

47. ‖ In 2000, NASA placed a satellite in orbit around an asteroid. Consider a spherical asteroid with a mass of 1.0×10^{16} kg and a radius of 8.8 km.
   a. What is the speed of a satellite orbiting 5.0 km above the surface?
   b. What is the escape speed from the asteroid?

48. ‖ NASA would like to place a satellite in orbit around the moon such that the satellite always remains in the same position over the lunar surface. What is the satellite's altitude?

49. ‖ A satellite orbiting the earth is directly over a point on the equator at 12:00 midnight every two days. It is not over that point at any time in between. What is the radius of the satellite's orbit?

50. ‖ FIGURE P13.50 shows two planets of mass m orbiting a star of mass M. The planets are in the same orbit, with radius r, but are always at opposite ends of a diameter. Find an exact expression for the orbital period T.
   Hint: Each planet feels two forces.

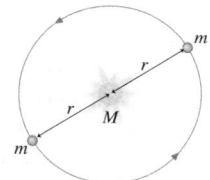

FIGURE P13.50

51. ‖ Figure 13.17 showed a graph of log T versus log r for the planetary data given in Table 13.2. Such a graph is called a *log-log graph*. The scales in Figure 13.17 are logarithmic, not linear, meaning that each division along the axis corresponds to a *factor* of 10 increase in the value. Strictly speaking, the "correct" labels on the y-axis should be 7, 8, 9, and 10 because these are the logarithms of $10^7, \ldots, 10^{10}$.
   a. Consider two quantities u and v that are related by the expression $v^p = Cu^q$, where C is a constant. The exponents p and q are not necessarily integers. Define $x = \log u$ and $y = \log v$. Find an expression for y in terms of x.
   b. What *shape* will a graph of y versus x have? Explain.
   c. What *slope* will a graph of y versus x have? Explain.
   d. Use the experimentally determined "best-fit" line in Figure 13.17 to find the mass of the sun.

52. ‖ Large stars can explode as they finish burning their nuclear fuel, causing a *supernova*. The explosion blows away the outer layers of the star. According to Newton's third law, the forces that push the outer layers away have *reaction forces* that are inwardly directed on the core of the star. These forces compress the core and can cause the core to undergo a *gravitational collapse*. The gravitational forces keep pulling all the matter together tighter and tighter, crushing atoms out of existence. Under these extreme conditions, a proton and an electron can be squeezed together to form a neutron. If the collapse is halted when the neutrons all come into contact with each other, the result is an object called a *neutron star*, an entire star consisting of solid nuclear matter. Many neutron stars rotate about their axis with a period of ≈ 1 s and, as they do so, send out a pulse of electromagnetic waves once a second. These stars were discovered in the 1960s and are called *pulsars*.
   a. Consider a neutron star with a mass equal to the sun, a radius of 10 km, and a rotation period of 1.0 s. What is the speed of a point on the equator of the star?
   b. What is g at the surface of this neutron star?
   c. A stationary 1.0 kg mass has a weight on earth of 9.8 N. What would be its weight on the star?
   d. How many revolutions per minute are made by a satellite orbiting 1.0 km above the surface?
   e. What is the radius of a geosynchronous orbit about the neutron star?

53. ‖ The solar system is 25,000 light years from the center of our Milky Way galaxy. One *light year* is the distance light travels in one year at a speed of 3.0×10^8 m/s. Astronomers have determined that the solar system is orbiting the center of the galaxy at a speed of 230 km/s.
   a. Assuming the orbit is circular, what is the period of the solar system's orbit? Give your answer in years.
   b. Our solar system was formed roughly 5 billion years ago. How many orbits has it completed?
   c. The gravitational force on the solar system is the net force due to all the matter inside our orbit. Most of that matter is concentrated near the center of the galaxy. Assume that the matter has a spherical distribution, like a giant star. What is the approximate mass of the galactic center?
   d. Assume that the sun is a typical star with a typical mass. If galactic matter is made up of stars, approximately how many stars are in the center of the galaxy?
   Astronomers have spent many years trying to determine how many stars there are in the Milky Way. The number of stars seems to be only about 10% of what you found in part d. In other words, about 90% of the mass of the galaxy appears to be in some form other than stars. This is called the *dark matter* of the universe. No one knows what the dark matter is. This is one of the outstanding scientific questions of our day.

54. ‖ Three stars, each with the mass of our sun, form an equilateral triangle with sides 1.0×10^{12} m long. (This triangle would just about fit within the orbit of Jupiter.) The triangle has to rotate, because otherwise the stars would crash together in the center. What is the period of rotation? Give your answer in years.

55. ‖ Pluto moves in a fairly elliptical orbit around the sun. Pluto's speed at its closest approach of 4.43×10^9 km is 6.12 km/s. What is Pluto's speed at the most distant point in its orbit, where it is 7.30×10^9 km from the sun?

56. ‖ Mercury moves in a fairly elliptical orbit around the sun. Mercury's speed is 38.8 km/s when it is at its most distant point, 6.99×10^{10} m from the sun. How far is Mercury from the sun at its closest point, where its speed is 59.0 km/s?

57. ‖ Comets move around the sun in very elliptical orbits. At its closet approach, in 1986, Comet Halley was 8.79×10^7 km from the sun and moving with a speed of 54.6 km/s. What was the comet's speed when it crossed Neptune's orbit in 2006?

58. ‖ A spaceship is in a circular orbit of radius r_0 about a planet of mass M. A brief but intense firing of its engine in the forward direction decreases the spaceship's speed by 50%. This causes the spaceship to move into an elliptical orbit.
   a. What is the spaceship's new speed, just after the rocket burn is complete, in terms of M, G, and r_0?
   b. In terms of r_0, what are the spaceship's maximum and minimum distances from the planet in its new orbit?

In Problems 59 through 61 you are given the equation(s) used to solve a problem. For each of these, you are to
   a. Write a realistic problem for which this is the correct equation(s).
   b. Draw a pictorial representation.
   c. Finish the solution of the problem.

59. $\dfrac{(6.67 \times 10^{-11}\,\text{N}\,\text{m}^2/\text{kg}^2)(5.68 \times 10^{26}\,\text{kg})}{r^2}$

   $= \dfrac{(6.67 \times 10^{-11}\,\text{N}\,\text{m}^2/\text{kg}^2)(5.98 \times 10^{24}\,\text{kg})}{(6.37 \times 10^6\,\text{m})^2}$

60. $\dfrac{(6.67 \times 10^{-11}\,\text{N}\,\text{m}^2/\text{kg}^2)(5.98 \times 10^{24}\,\text{kg})(1000\,\text{kg})}{r^2}$

   $= \dfrac{(1000\,\text{kg})(1997\,\text{m/s})^2}{r}$

61. $\dfrac{1}{2}(100\,\text{kg})v_2^2$

   $- \dfrac{(6.67 \times 10^{-11}\,\text{N}\,\text{m}^2/\text{kg}^2)(7.36 \times 10^{22}\,\text{kg})(100\,\text{kg})}{1.74 \times 10^6\,\text{m}}$

   $= 0 - \dfrac{(6.67 \times 10^{-11}\,\text{N}\,\text{m}^2/\text{kg}^2)(7.36 \times 10^{22}\,\text{kg})(100\,\text{kg})}{3.48 \times 10^6\,\text{m}}$

Challenge Problems

62. A satellite in a circular orbit of radius r has period T. A satellite in a nearby orbit with radius $r + \Delta r$, where $\Delta r \ll r$, has the very slightly different period $T + \Delta T$.
   a. Show that
   $$\frac{\Delta T}{T} = \frac{3}{2}\frac{\Delta r}{r}$$
   b. Two earth satellites are in parallel orbits with radii 6700 km and 6701 km. One day they pass each other, 1 km apart, along a line radially outward from the earth. How long will it be until they are again 1 km apart?

63. In 1996, the Solar and Heliospheric Observatory (SOHO) was "parked" in an orbit slightly inside the earth's orbit, as shown in FIGURE CP13.63. The satellite's period in this orbit is exactly one year, so it remains fixed relative to the earth. At this point, called a *Lagrange point,* the light from the sun is never blocked by the earth, yet the satellite remains "nearby" so that data are easily transmitted to earth. What is SOHO's distance from the earth?

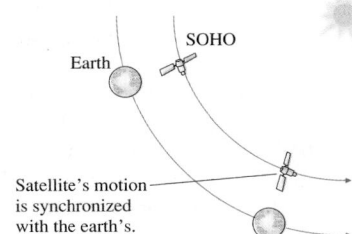

FIGURE CP13.63

Hint: Use the binomial approximation. SOHO's distance from the earth is much less than the earth's distance from the sun.

64. A projectile is fired from the earth in the direction of the earth's motion around the sun. What minimum speed must the projectile have relative to the earth to escape the solar system? Ignore the earth's rotation.
   Hint: This is a three-part problem. First find the speed a projectile at the earth's distance needs to escape the sun. Transform that speed into the earth's reference frame, then determine how fast the projectile must be launched to have this speed when far from the earth.

65. Your job with NASA is to monitor satellite orbits. One day, during a routine survey, you find that a 400 kg satellite in a 1000-km-high circular orbit is going to collide with a smaller 100 kg satellite traveling in the same orbit but in the opposite direction. Knowing the construction of the two satellites, you expect they will become enmeshed into a single piece of space debris. When you notify your boss of this impending collision, he asks you to quickly determine whether the space debris will continue to orbit or crash into the earth. What will the outcome be?

66. While visiting Planet Physics, you toss a rock straight up at 11 m/s and catch it 2.5 s later. While you visit the surface, your cruise ship orbits at an altitude equal to the planet's radius every 230 min. What are the (a) mass and (b) radius of Planet Physics?

67. A moon lander is orbiting the moon at an altitude of 1000 km. By what percentage must it decrease its speed so as to just graze the moon's surface one-half period later?

68. Let's look in more detail at how a satellite is moved from one circular orbit to another. FIGURE CP13.68 shows two circular orbits, of radii r_1 and r_2, and an elliptical orbit that connects them. Points 1 and 2 are at the ends of the semimajor axis of the ellipse.

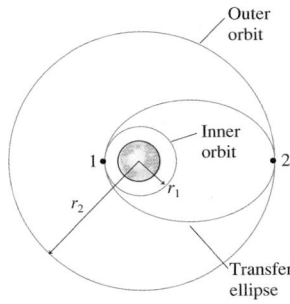

FIGURE CP13.68

   a. A satellite moving along the elliptical orbit has to satisfy two conservation laws. Use these two laws to prove that the velocities at points 1 and 2 are

   $$v_1' = \sqrt{\frac{2GM(r_2/r_1)}{r_1 + r_2}} \quad \text{and} \quad v_2' = \sqrt{\frac{2GM(r_1/r_2)}{r_1 + r_2}}$$

The prime indicates that these are the velocities on the elliptical orbit. Both reduce to Equation 13.22 if $r_1 = r_2 = r$.

b. Consider a 1000 kg communications satellite that needs to be boosted from an orbit 300 km above the earth to a geosynchronous orbit 35,900 km above the earth. Find the velocity v_1 on the inner circular orbit and the velocity v_1' at the low point on the elliptical orbit that spans the two circular orbits.

c. How much work must the rocket motor do to transfer the satellite from the circular orbit to the elliptical orbit?

d. Now find the velocity v_2' at the high point of the elliptical orbit and the velocity v_2 of the outer circular orbit.

e. How much work must the rocket motor do to transfer the satellite from the elliptical orbit to the outer circular orbit?

f. Compute the total work done and compare your answer to the result of Example 13.6.

69. **FIGURE CP13.69** shows a particle of mass m at distance x from the center of a very thin cylinder of mass M and length L. The particle is outside the cylinder, so $x > L/2$.

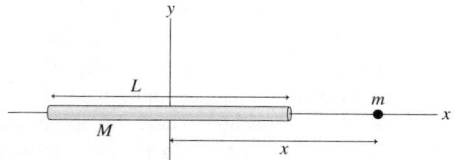

FIGURE CP13.69

a. Calculate the gravitational potential energy of these two masses.

b. Use what you know about the relationship between force and potential energy to find the magnitude of the gravitational force on m when it is at position x.

70. **FIGURE CP13.70** shows a particle of mass m at distance x along the axis of a very thin ring of mass M and radius R.

a. Calculate the gravitational potential energy of these two masses.

b. Use what you know about the relationship between force and potential energy to find the magnitude of the gravitational force on m when it is at position x.

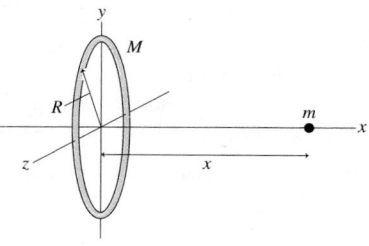

FIGURE CP13.70

STOP TO THINK ANSWERS

Stop to Think 13.1: e. The acceleration decreases inversely with the square of the distance. At height R_e, the distance from the center of the earth is $2R_e$.

Stop to Think 13.2: c. Newton's third law requires $F_{1 \text{ on } 2} = F_{2 \text{ on } 1}$.

Stop to Think 13.3: b. $g_{\text{surface}} = GM/R^2$. Because of the square, a radius twice as large balances a mass four times as large.

Stop to Think 13.4: In absolute value, $U_e > U_a = U_b = U_d > U_c$. $|U_g|$ is proportional to $m_1 m_2/r$.

Stop to Think 13.5: a. T^2 is proportional to r^3, or T is proportional to $r^{3/2}$. $4^{3/2} = 8$.

14 Oscillations

This loudspeaker cone generates sound waves by oscillating back and forth at audio frequencies.

▶ **Looking Ahead** The goal of Chapter 14 is to understand systems that oscillate with simple harmonic motion.

Simple Harmonic Motion

The most basic oscillation, with sinusoidal motion, is called **simple harmonic motion.**

The oscillating cart is an example of simple harmonic motion. You'll learn how to use the mass and the spring constant to determine the frequency of oscillation.

In this chapter you will learn to:

- Represent simple harmonic motion both graphically and mathematically.
- Understand the dynamics of oscillating systems.
- Recognize the similarities among many types of oscillating systems.

Simple harmonic motion has a very close connection to uniform circular motion. You'll learn that an edge-on view of uniform circular motion is none other than simple harmonic motion.

◀ **Looking Back**
Section 4.5 Uniform circular motion

Springs

Simple harmonic motion occurs when there is a **linear restoring force.** The simplest example is a mass on a spring. You will learn how to determine the period of oscillation.

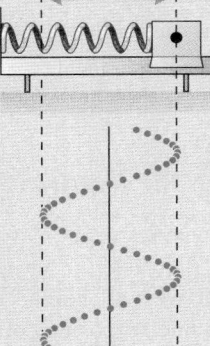

The "bounce" at the bottom of a bungee jump is an exhilarating example of a mass oscillating on a spring.

◀ **Looking Back**
Section 10.4 Restoring forces

Energy of Oscillations

If there is no friction or other dissipation, then the mechanical energy of an oscillator is conserved. Conservation of energy will be an important tool.

The system oscillates between all kinetic energy and all potential energy

◀ **Looking Back**
Section 10.5 Elastic potential energy
Section 10.6 Energy diagrams

Pendulums

A mass swinging at the end of a string or rod is a **pendulum.** Its motion is another example of simple harmonic motion.

The period of a pendulum is determined by the length of the string; neither the mass nor the amplitude matters. Consequently, the pendulum was the basis of time keeping for many centuries.

Damping and Resonance

If there's drag or other dissipation, then the oscillation "runs down." This is called a **damped oscillation.**

The amplitude of a damped oscillation undergoes *exponential decay.*

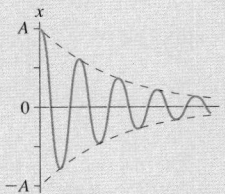

Oscillations can increase in amplitude, sometimes dramatically, when driven at their natural oscillation frequency. This is called **resonance.**

FIGURE 14.1 Examples of position-versus-time graphs for oscillating systems.

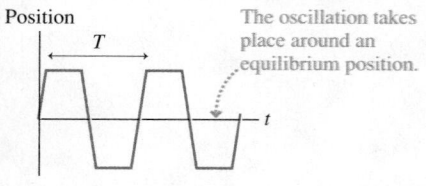

The oscillation takes place around an equilibrium position.

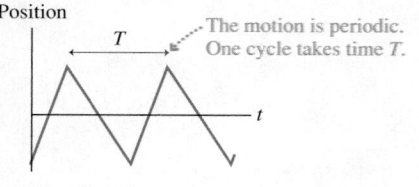

The motion is periodic. One cycle takes time T.

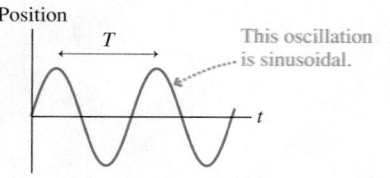

This oscillation is sinusoidal.

14.1 Simple Harmonic Motion

Objects or systems of objects that undergo **oscillatory motion**—a repetitive motion back and forth around an equilibrium position—are called **oscillators.** FIGURE 14.1 shows position-versus-time graphs for three different oscillating systems. Although the shapes of the graphs are different, all these oscillators have two things in common:

1. The oscillation takes place about an equilibrium position, and
2. The motion is *periodic*, repeating at regular intervals of time.

The time to complete one full cycle, or one oscillation, is called the **period** of the motion. Period is represented by the symbol T.

A closely related piece of information is the number of cycles, or oscillations, completed per second. If the period is $\frac{1}{10}$ s, then the oscillator can complete 10 cycles in one second. Conversely, an oscillation period of 10 s allows only $\frac{1}{10}$ of a cycle to be completed per second. In general, T seconds per cycle implies that 1/T cycles will be completed each second. The number of cycles per second is called the **frequency** f of the oscillation. The relationship between frequency and period is

$$f = \frac{1}{T} \quad \text{or} \quad T = \frac{1}{f} \qquad (14.1)$$

The units of frequency are **hertz,** abbreviated Hz, named in honor of the German physicist Heinrich Hertz, who produced the first artificially generated radio waves in 1887. By definition,

$$1 \text{ Hz} \equiv 1 \text{ cycle per second} = 1 \text{ s}^{-1}$$

We will frequently deal with very rapid oscillations and make use of the units shown in Table 14.1.

NOTE ▶ Uppercase and lowercase letters *are* important. 1 MHz is 1 megahertz = 10^6 Hz, but 1 mHz is 1 millihertz = 10^{-3} Hz! ◀

TABLE 14.1 Units of frequency

Frequency	Period
10^3 Hz = 1 kilohertz = 1 kHz	1 ms
10^6 Hz = 1 megahertz = 1 MHz	1 μs
10^9 Hz = 1 gigahertz = 1 GHz	1 ns

EXAMPLE 14.1 **Frequency and period of a loudspeaker cone**

What is the oscillation period of a loudspeaker cone that vibrates back and forth 5000 times per second?

SOLVE The oscillation frequency is $f = 5000$ cycles/s = 5000 Hz = 5.0 kHz. The period is the inverse of the frequency; hence

$$T = \frac{1}{f} = \frac{1}{5000 \text{ Hz}} = 2.0 \times 10^{-4} \text{ s} = 200 \text{ μs}$$

A system can oscillate in many ways, but we will be especially interested in the smooth *sinusoidal* oscillation (i.e., like a sine or cosine) of the third graph in Figure 14.1. This sinusoidal oscillation, the most basic of all oscillatory motions, is called **simple harmonic motion,** often abbreviated SHM. Let's look at a graphical description before we dive into the mathematics of simple harmonic motion.

FIGURE 14.2a shows an air-track glider attached to a spring. If the glider is pulled out a few centimeters and released, it will oscillate back and forth on the nearly frictionless air track. FIGURE 14.2b shows actual results from an experiment in which a computer was used to measure the glider's position 20 times every second. This is a position-versus-time graph that has been rotated 90° from its usual orientation in order for the x-axis to match the motion of the glider.

The object's maximum displacement from equilibrium is called the **amplitude** A of the motion. The object's position oscillates between $x = -A$ and $x = +A$. When using a graph, notice that the amplitude is the distance from the *axis* to the maximum, *not* the distance from the minimum to the maximum.

FIGURE 14.2 A prototype simple-harmonic-motion experiment.

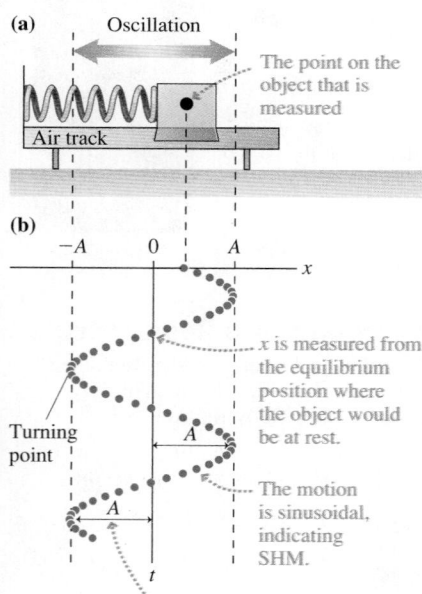

(a) Oscillation

The point on the object that is measured

Air track

(b)

x is measured from the equilibrium position where the object would be at rest.

Turning point

The motion is sinusoidal, indicating SHM.

The motion is symmetrical about the equilibrium position. Maximum distance to the left and to the right is A.

FIGURE 14.3a shows the data with the graph axes in their "normal" positions. You can see that the amplitude in this experiment was $A = 0.17$ m, or 17 cm. You can also measure the period to be $T = 1.60$ s. Thus the oscillation frequency was $f = 1/T = 0.625$ Hz.

FIGURE 14.3b is a velocity-versus-time graph that the computer produced by using $\Delta x/\Delta t$ to find the slope of the position graph at each point. The velocity graph is also sinusoidal, oscillating between $-v_{max}$ (maximum speed to the left) and $+v_{max}$ (maximum speed to the right). As the figure shows,

- The instantaneous velocity is zero at the points where $x = \pm A$. These are the *turning points* in the motion.
- The maximum speed v_{max} is reached as the object passes through the equilibrium position at $x = 0$ m. The *velocity* is positive as the object moves to the right but *negative* as it moves to the left.

We can ask three important questions about this oscillating system:

1. How is the maximum speed v_{max} related to the amplitude A?
2. How are the period and frequency related to the object's mass m, the spring constant k, and the amplitude A?
3. Is the sinusoidal oscillation a consequence of Newton's laws?

A mass oscillating on a spring is the prototype of simple harmonic motion. Our analysis, in which we answer these questions, will be of a spring-mass system. Even so, most of what we learn will be applicable to other types of SHM.

Kinematics of Simple Harmonic Motion

FIGURE 14.4 redraws the position-versus-time graph of Figure 14.3a as a smooth curve. Although these are empirical data (we don't yet have any "theory" of oscillation) the graph for this particular motion is clearly a cosine function. The object's position is

$$x(t) = A \cos\left(\frac{2\pi t}{T}\right) \tag{14.2}$$

where the notation $x(t)$ indicates that the position x is a *function* of time t. Because $\cos(2\pi) = \cos(0)$, it's easy to see that the position at time $t = T$ is the same as the position at $t = 0$. In other words, this is a cosine function with period T. Be sure to convince yourself that this function agrees with the five special points shown in Figure 14.4.

NOTE ▶ The argument of the cosine function is in *radians*. That will be true throughout this chapter. It's especially important to remember to set your calculator to radian mode before working oscillation problems. Leaving it in degree mode will lead to errors. ◀

We can write Equation 14.2 in two alternative forms. Because the oscillation frequency is $f = 1/T$, we can write

$$x(t) = A \cos(2\pi f t) \tag{14.3}$$

Recall from Chapter 4 that a particle in circular motion has an *angular velocity* ω that is related to the period by $\omega = 2\pi/T$, where ω is in rad/s. Now that we've defined the frequency f, you can see that ω and f are related by

$$\omega \text{ (in rad/s)} = \frac{2\pi}{T} = 2\pi f \text{ (in Hz)} \tag{14.4}$$

In this context, ω is called the **angular frequency**. The position can be written in terms of ω as

$$x(t) = A \cos \omega t \tag{14.5}$$

Equations 14.2, 14.3, and 14.5 are equivalent ways to write the position of an object moving in simple harmonic motion.

FIGURE 14.3 Position and velocity graphs of the experimental data.

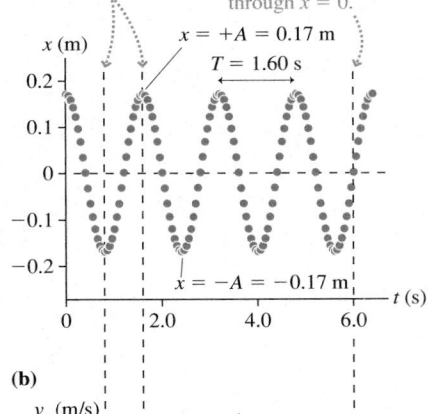

(a) The speed is zero when $x = \pm A$. The speed is maximum as the object passes through $x = 0$.

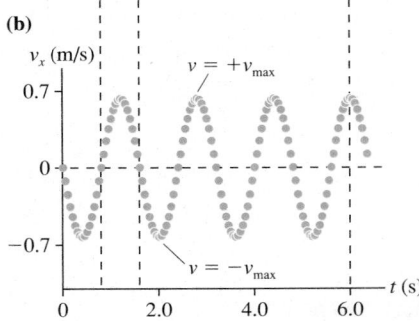

(b)

FIGURE 14.4 The position-versus-time graph for simple harmonic motion.

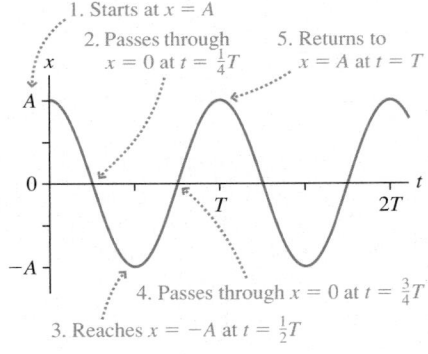

1. Starts at $x = A$
2. Passes through $x = 0$ at $t = \frac{1}{4}T$
5. Returns to $x = A$ at $t = T$
4. Passes through $x = 0$ at $t = \frac{3}{4}T$
3. Reaches $x = -A$ at $t = \frac{1}{2}T$

TABLE 14.2 Derivatives of sine and cosine functions

$$\frac{d}{dt}\big(a\sin(bt + c)\big) = +ab\cos(bt + c)$$

$$\frac{d}{dt}\big(a\cos(bt + c)\big) = -ab\sin(bt + c)$$

FIGURE 14.5 Position and velocity graphs for simple harmonic motion.

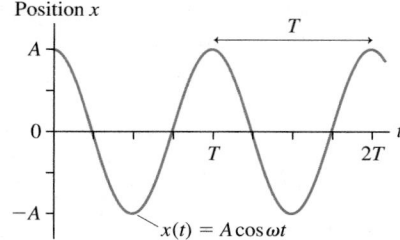

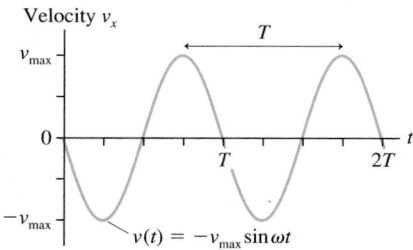

Just as the position graph was clearly a cosine function, the velocity graph shown in FIGURE 14.5 is clearly an "upside-down" sine function with the same period T. The velocity v_x, which is a function of time, can be written

$$v_x(t) = -v_{max} \sin\left(\frac{2\pi t}{T}\right) = -v_{max} \sin(2\pi ft) = -v_{max} \sin \omega t \qquad (14.6)$$

NOTE ▶ v_{max} is the maximum *speed* and thus is a *positive* number. ◀

We deduced Equation 14.6 from the experimental results, but we could equally well find it from the position function of Equation 14.2. After all, velocity is the time derivative of position. Table 14.2 on the previous page reminds you of the derivatives of the sine and cosine functions. Using the derivative of the position function, we find

$$v_x(t) = \frac{dx}{dt} = -\frac{2\pi A}{T} \sin\left(\frac{2\pi t}{T}\right) = -2\pi fA \sin(2\pi ft) = -\omega A \sin \omega t \quad (14.7)$$

Comparing Equation 14.7, the mathematical definition of velocity, to Equation 14.6, the empirical description, we see that the maximum speed of an oscillation is

$$v_{max} = \frac{2\pi A}{T} = 2\pi fA = \omega A \qquad (14.8)$$

Equation 14.8 answers the first question we posed above, which was how the maximum speed v_{max} is related to the amplitude A. Not surprisingly, the object has a greater maximum speed if you stretch the spring farther and give the oscillation a larger amplitude.

EXAMPLE 14.2 **A system in simple harmonic motion**

An air-track glider is attached to a spring, pulled 20.0 cm to the right, and released at $t = 0$ s. It makes 15 oscillations in 10.0 s.

 a. What is the period of oscillation?
 b. What is the object's maximum speed?
 c. What are the position and velocity at $t = 0.800$ s?

MODEL An object oscillating on a spring is in SHM.

SOLVE a. The oscillation frequency is

$$f = \frac{15 \text{ oscillations}}{10.0 \text{ s}} = 1.50 \text{ oscillations/s} = 1.50 \text{ Hz}$$

Thus the period is $T = 1/f = 0.667$ s.
 b. The oscillation amplitude is $A = 0.200$ m. Thus

$$v_{max} = \frac{2\pi A}{T} = \frac{2\pi(0.200 \text{ m})}{0.667 \text{ s}} = 1.88 \text{ m/s}$$

c. The object starts at $x = +A$ at $t = 0$ s. This is exactly the oscillation described by Equations 14.2 and 14.6. The position at $t = 0.800$ s is

$$x = A \cos\left(\frac{2\pi t}{T}\right) = (0.200 \text{ m}) \cos\left(\frac{2\pi(0.800 \text{ s})}{0.667 \text{ s}}\right)$$
$$= (0.200 \text{ m}) \cos(7.54 \text{ rad}) = 0.0625 \text{ m} = 6.25 \text{ cm}$$

The velocity at this instant of time is

$$v_x = -v_{max} \sin\left(\frac{2\pi t}{T}\right) = -(1.88 \text{ m/s}) \sin\left(\frac{2\pi(0.800 \text{ s})}{0.667 \text{ s}}\right)$$
$$= -(1.88 \text{ m/s}) \sin(7.54 \text{ rad}) = -1.79 \text{ m/s} = -179 \text{ cm/s}$$

At $t = 0.800$ s, which is slightly more than one period, the object is 6.25 cm to the right of equilibrium and moving to the *left* at 179 cm/s. Notice the use of radians in the calculations.

EXAMPLE 14.3 **Finding the time**

A mass oscillating in simple harmonic motion starts at $x = A$ and has period T. At what time, as a fraction of T, does the object first pass through $x = \frac{1}{2}A$?

SOLVE Figure 14.4 showed that the object passes through the equilibrium position $x = 0$ at $t = \frac{1}{4}T$. This is one-quarter of the total distance in one-quarter of a period. You might expect it to take $\frac{1}{8}T$ to reach $\frac{1}{2}A$, but this is not the case because the SHM graph is not linear between $x = A$ and $x = 0$. We need to use $x(t) = A \cos(2\pi t/T)$. First, we write the equation with $x = \frac{1}{2}A$:

$$x = \frac{A}{2} = A \cos\left(\frac{2\pi t}{T}\right)$$

Then we solve for the time at which this position is reached:

$$t = \frac{T}{2\pi} \cos^{-1}\left(\frac{1}{2}\right) = \frac{T}{2\pi}\frac{\pi}{3} = \frac{1}{6}T$$

ASSESS The motion is slow at the beginning and then speeds up, so it takes longer to move from $x = A$ to $x = \frac{1}{2}A$ than it does to move from $x = \frac{1}{2}A$ to $x = 0$. Notice that the answer is independent of the amplitude A.

An object moves with simple harmonic motion. If the amplitude and the period are both doubled, the object's maximum speed is

a. Quadrupled.   b. Doubled.   c. Unchanged.
d. Halved.   e. Quartered.

14.2 Simple Harmonic Motion and Circular Motion

The graphs of Figure 14.5 and the position function $x(t) = A \cos \omega t$ are for an oscillation in which the object just happened to be at $x_0 = A$ at $t = 0$. But you will recall that $t = 0$ is an arbitrary choice, the instant of time when you or someone else starts a stopwatch. What if you had started the stopwatch when the object was at $x_0 = -A$, or when the object was somewhere in the middle of an oscillation? In other words, what if the oscillator had different *initial conditions*. The position graph would still show an oscillation, but neither Figure 14.5 nor $x(t) = A \cos \omega t$ would describe the motion correctly.

To learn how to describe the oscillation for other initial conditions it will help to turn to a topic you studied in Chapter 4—circular motion. There's a very close connection between simple harmonic motion and circular motion.

Imagine you have a turntable with a small ball glued to the edge. FIGURE 14.6a shows how to make a "shadow movie" of the ball by projecting a light past the ball and onto a screen. The ball's shadow oscillates back and forth as the turntable rotates. This is certainly periodic motion, with the same period as the turntable, but is it simple harmonic motion?

To find out, you could place a real object on a real spring directly below the shadow, as shown in FIGURE 14.6b. If you did so, and if you adjusted the turntable to have the same period as the spring, you would find that the shadow's motion exactly matches the simple harmonic motion of the object on the spring. **Uniform circular motion projected onto one dimension is simple harmonic motion.**

To understand this, consider the particle in FIGURE 14.7. It is in uniform circular motion, moving *counterclockwise* in a circle with radius A. As in Chapter 4, we can locate the particle by the angle ϕ measured ccw from the x-axis. Projecting the ball's shadow onto a screen in Figure 14.6 is equivalent to observing just the x-component of the particle's motion. Figure 14.7 shows that the x-component, when the particle is at angle ϕ, is

$$x = A \cos \phi \tag{14.9}$$

Recall that the particle's *angular velocity,* in rad/s, is

$$\omega = \frac{d\phi}{dt} \tag{14.10}$$

This is the rate at which the angle ϕ is increasing. If the particle starts from $\phi_0 = 0$ at $t = 0$, its angle at a later time t is simply

$$\phi = \omega t \tag{14.11}$$

As ϕ increases, the particle's x-component is

$$x(t) = A \cos \omega t \tag{14.12}$$

This is identical to Equation 14.5 for the position of a mass on a spring! Thus the x-component of a particle in uniform circular motion is simple harmonic motion.

NOTE ▶ When used to describe oscillatory motion, ω is called the *angular frequency* rather than the angular velocity. The angular frequency of an oscillator has the same numerical value, in rad/s, as the angular velocity of the corresponding particle in circular motion. ◀

FIGURE 14.6 A projection of the circular motion of a rotating ball matches the simple harmonic motion of an object on a spring.

(a)

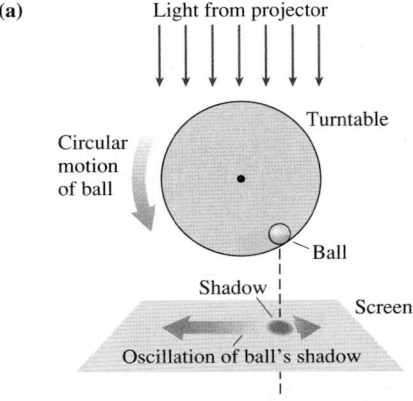

(b)   Simple harmonic motion of block

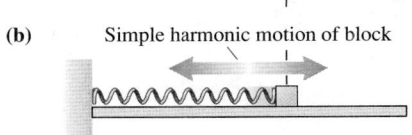

FIGURE 14.7 A particle in uniform circular motion with radius A and angular velocity ω.

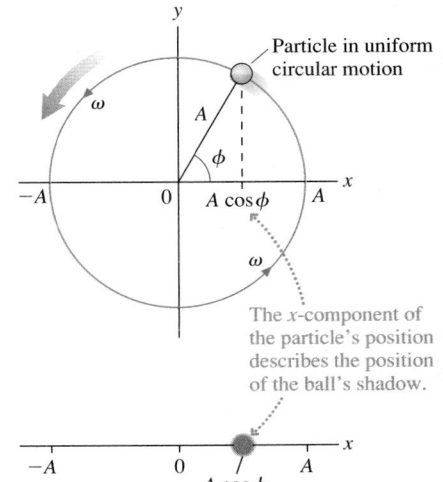

A cup on the turntable in a microwave oven moves in a circle. But from the outside, you see the cup sliding back and forth—in simple harmonic motion!

The names and units can be a bit confusing until you get used to them. It may help to notice that *cycle* and *oscillation* are not true units. Unlike the "standard meter" or the "standard kilogram," to which you could compare a length or a mass, there is no "standard cycle" to which you can compare an oscillation. Cycles and oscillations are simply counted events. Thus the frequency f has units of hertz, where $1 \text{ Hz} = 1 \text{ s}^{-1}$. We may *say* "cycles per second" just to be clear, but the actual units are only "per second."

The radian is the SI unit of angle. However, the radian is a *defined* unit. Further, its definition as a ratio of two lengths ($\theta = s/r$) makes it a pure number without dimensions. As we noted in Chapter 4, the unit of angle, be it radians or degrees, is really just a *name* to remind us that we're dealing with an angle. The 2π in the equation $\omega = 2\pi f$ (and in similar situations), which is stated without units, *means* 2π rad/cycle. When multiplied by the frequency f in cycles/s, it gives the frequency in rad/s. That is why, in this context, ω is called the angular *frequency*.

NOTE ▶ *Hertz* is specifically "cycles per second" or "oscillations per second." It is used for f but *not* for ω. We'll always be careful to use rad/s for ω, but you should be aware that many books give the units of ω as simply s^{-1}. ◀

The Phase Constant

FIGURE 14.8 A particle in uniform circular motion with initial angle ϕ_0.

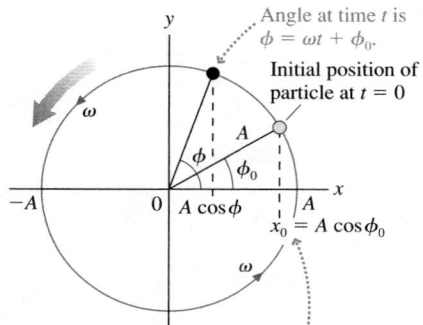

Angle at time t is $\phi = \omega t + \phi_0$.

Initial position of particle at $t = 0$

The initial x-component of the particle's position can be anywhere between $-A$ and A, depending on ϕ_0.

Now we're ready to consider the issue of other initial conditions. The particle in Figure 14.7 started at $\phi_0 = 0$. This was equivalent to an oscillator starting at the far right edge, $x_0 = A$. FIGURE 14.8 shows a more general situation in which the initial angle ϕ_0 can have any value. The angle at a later time t is then

$$\phi = \omega t + \phi_0 \tag{14.13}$$

In this case, the particle's projection onto the x-axis at time t is

$$x(t) = A\cos(\omega t + \phi_0) \tag{14.14}$$

If Equation 14.14 describes the particle's projection, then it must also be the position of an oscillator in simple harmonic motion. The oscillator's velocity v_x is found by taking the derivative dx/dt. The resulting equations,

$$x(t) = A\cos(\omega t + \phi_0)$$
$$v_x(t) = -\omega A\sin(\omega t + \phi_0) = -v_{\max}\sin(\omega t + \phi_0) \tag{14.15}$$

are the two primary kinematic equations of simple harmonic motion.

The quantity $\phi = \omega t + \phi_0$, which steadily increases with time, is called the **phase** of the oscillation. The phase is simply the *angle* of the circular-motion particle whose shadow matches the oscillator. The constant ϕ_0 is called the **phase constant.** It specifies the *initial conditions* of the oscillator.

To see what the phase constant means, set $t = 0$ in Equations 14.15:

$$x_0 = A\cos\phi_0$$
$$v_{0x} = -\omega A\sin\phi_0 \tag{14.16}$$

The position x_0 and velocity v_{0x} at $t = 0$ are the initial conditions. **Different values of the phase constant correspond to different starting points on the circle and thus to different initial conditions.**

The perfect cosine function of Figure 14.5 and the equation $x(t) = A\cos\omega t$ are for an oscillation with $\phi_0 = 0$ rad. You can see from Equations 14.16 that $\phi_0 = 0$ rad implies $x_0 = A$ and $v_0 = 0$. That is, the particle starts from rest at the point of maximum displacement.

FIGURE 14.9 illustrates these ideas by looking at three values of the phase constant: $\phi_0 = \pi/3$ rad (60°), $-\pi/3$ rad ($-60°$), and π rad (180°). Notice that $\phi_0 = \pi/3$ rad and $\phi_0 = -\pi/3$ rad have the same starting position, $x_0 = \frac{1}{2}A$. This is a property of the cosine function in Equation 14.16. But these are *not* the same initial conditions. In one case the oscillator starts at $\frac{1}{2}A$ while moving to the right, in the other case it starts at $\frac{1}{2}A$ while moving to the left. You can distinguish between the two by visualizing the motion.

FIGURE 14.9 Oscillations described by the phase constants $\phi_0 = \pi/3$ rad, $-\pi/3$ rad, and π rad.

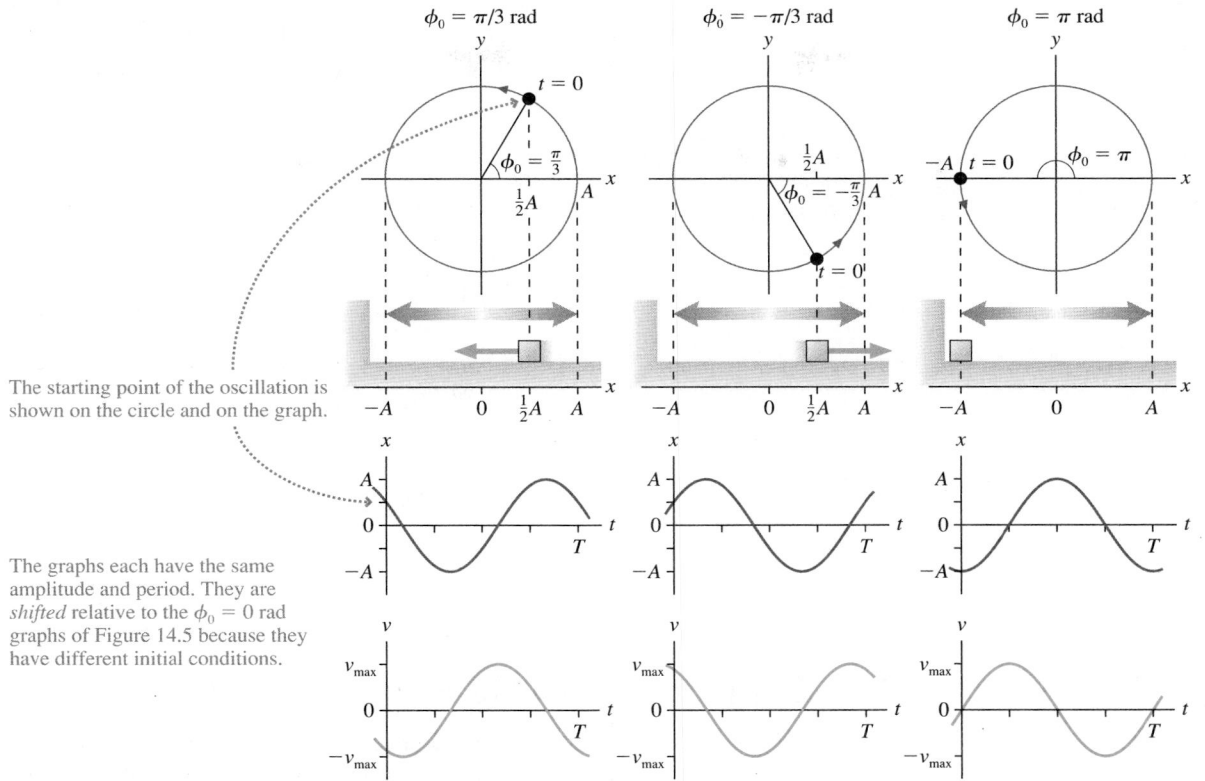

The starting point of the oscillation is shown on the circle and on the graph.

The graphs each have the same amplitude and period. They are *shifted* relative to the $\phi_0 = 0$ rad graphs of Figure 14.5 because they have different initial conditions.

All values of the phase constant ϕ_0 between 0 and π rad correspond to a particle in the upper half of the circle and *moving to the left*. Thus v_{0x} is negative. All values of the phase constant ϕ_0 between π and 2π rad (or, as they are usually stated, between $-\pi$ and 0 rad) have the particle in the lower half of the circle and *moving to the right*. Thus v_{0x} is positive. If you're told that the oscillator is at $x = \frac{1}{2}A$ and moving to the right at $t = 0$, then the phase constant must be $\phi_0 = -\pi/3$ rad, not $+\pi/3$ rad.

EXAMPLE 14.4 **Using the initial conditions**

An object on a spring oscillates with a period of 0.80 s and an amplitude of 10 cm. At $t = 0$ s, it is 5.0 cm to the left of equilibrium and moving to the left. What are its position and direction of motion at $t = 2.0$ s?

MODEL An object oscillating on a spring is in simple harmonic motion.

SOLVE We can find the phase constant ϕ_0 from the initial condition $x_0 = -5.0$ cm $= A\cos\phi_0$. This condition gives

$$\phi_0 = \cos^{-1}\left(\frac{x_0}{A}\right) = \cos^{-1}\left(-\frac{1}{2}\right) = \pm\frac{2}{3}\pi \text{ rad} = \pm 120°$$

Because the oscillator is moving to the *left* at $t = 0$, it is in the upper half of the circular-motion diagram and must have a phase constant between 0 and π rad. Thus ϕ_0 is $\frac{2}{3}\pi$ rad. The angular frequency is

$$\omega = \frac{2\pi}{T} = \frac{2\pi}{0.80 \text{ s}} = 7.85 \text{ rad/s}$$

Thus the object's position at time $t = 2.0$ s is

$$x(t) = A\cos(\omega t + \phi_0)$$
$$= (10 \text{ cm})\cos\left((7.85 \text{ rad/s})(2.0 \text{ s}) + \frac{2}{3}\pi\right)$$
$$= (10 \text{ cm})\cos(17.8 \text{ rad}) = 5.0 \text{ cm}$$

The object is now 5.0 cm to the right of equilibrium. But which way is it moving? There are two ways to find out. The direct way is to calculate the velocity at $t = 2.0$ s:

$$v_x = -\omega A\sin(\omega t + \phi_0) = +68 \text{ cm/s}$$

The velocity is positive, so the motion is to the right. Alternatively, we could note that the phase at $t = 2.0$ s is $\phi = 17.8$ rad. Dividing by π, you can see that

$$\phi = 17.8 \text{ rad} = 5.67\pi \text{ rad} = (4\pi + 1.67\pi) \text{ rad}$$

The 4π rad represents two complete revolutions. The "extra" phase of 1.67π rad falls between π and 2π rad, so the particle in the circular-motion diagram is in the lower half of the circle and moving to the right.

NOTE ▶ The inverse-cosine function $\cos^{-1}$ is a *two-valued* function. Your calculator returns a single value, an angle between 0 rad and π rad. But the negative of this angle is also a solution. As Example 14.4 demonstrates, you must use additional information to choose between them. ◀

STOP TO THINK 14.2   The figure shows four oscillators at $t = 0$. Which one has the phase constant $\phi_0 = \pi/4$ rad?

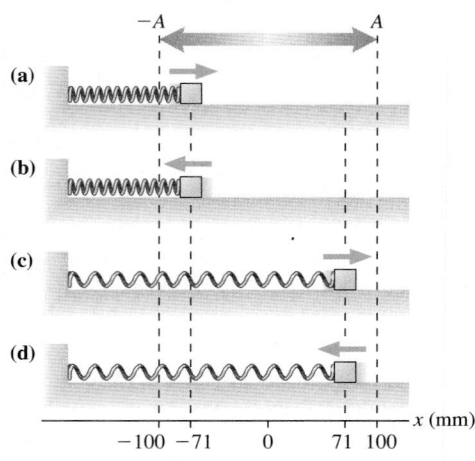

14.3 Energy in Simple Harmonic Motion

FIGURE 14.10 The energy is transformed between kinetic energy and potential energy as the object oscillates, but the mechanical energy $E = K + U$ doesn't change.

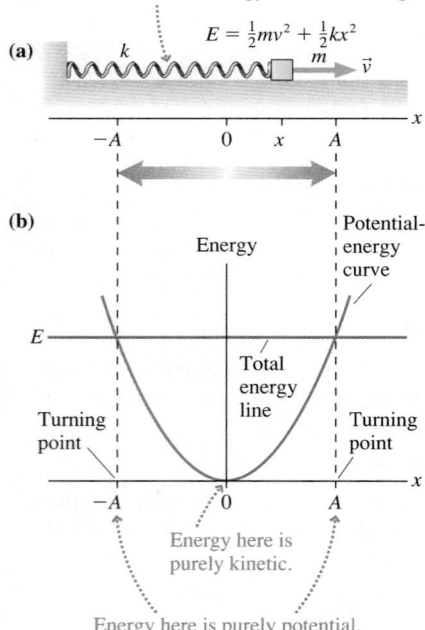

We've begun to develop the mathematical language of simple harmonic motion, but thus far we haven't included any physics. We've made no mention of the mass of the object or the spring constant of the spring. An energy analysis, using the tools of Chapters 10 and 11, is a good starting place.

FIGURE 14.10a shows an object oscillating on a spring, our prototype of simple harmonic motion. Now we'll specify that the object has mass m, the spring has spring constant k, and the motion takes place on a frictionless surface. You learned in Chapter 10 that the elastic potential energy when the object is at position x is $U_s = \frac{1}{2}k(\Delta x)^2$, where $\Delta x = x - x_e$ is the displacement from the equilibrium position x_e. In this chapter we'll always use a coordinate system in which $x_e = 0$, making $\Delta x = x$. There's no chance for confusion with gravitational potential energy, so we can omit the subscript s and write the elastic potential energy as

$$U = \frac{1}{2}kx^2 \qquad (14.17)$$

Thus the mechanical energy of an object oscillating on a spring is

$$E = K + U = \frac{1}{2}mv^2 + \frac{1}{2}kx^2 \qquad (14.18)$$

FIGURE 14.10b is an energy diagram, showing the potential-energy curve $U = \frac{1}{2}kx^2$ as a parabola. Recall that a particle oscillates between the *turning points* where the total energy line E crosses the potential-energy curve. The left turning point is at $x = -A$, and the right turning point is at $x = +A$. To go beyond these points would require a negative kinetic energy, which is physically impossible.

You can see that **the particle has purely potential energy at $x = \pm A$ and purely kinetic energy as it passes through the equilibrium point at $x = 0$.** At maximum displacement, with $x = \pm A$ and $v = 0$, the energy is

$$E(\text{at } x = \pm A) = U = \frac{1}{2}kA^2 \qquad (14.19)$$

At $x = 0$, where $v = \pm v_{\text{max}}$, the energy is

$$E(\text{at } x = 0) = K = \frac{1}{2}m(v_{\text{max}})^2 \qquad (14.20)$$

The system's mechanical energy is conserved because the surface is frictionless and there are no external forces, so the energy at maximum displacement and the energy at maximum speed, Equations 14.19 and 14.20, must be equal. That is

$$\frac{1}{2}m(v_{max})^2 = \frac{1}{2}kA^2 \qquad (14.21)$$

Thus the maximum speed is related to the amplitude by

$$v_{max} = \sqrt{\frac{k}{m}}A \qquad (14.22)$$

This is a relationship based on the physics of the situation.

Earlier, using kinematics, we found that

$$v_{max} = \frac{2\pi A}{T} = 2\pi f A = \omega A \qquad (14.23)$$

Comparing Equations 14.22 and 14.23, we see that frequency and period of an oscillating spring are determined by the spring constant k and the object's mass m:

$$\omega = \sqrt{\frac{k}{m}} \qquad f = \frac{1}{2\pi}\sqrt{\frac{k}{m}} \qquad T = 2\pi\sqrt{\frac{m}{k}} \qquad (14.24)$$

These three expressions are really only one equation. They say the same thing, but each expresses it in slightly different terms.

Equations 14.24 are the answer to the second question we posed at the beginning of the chapter, where we asked how the period and frequency are related to the object's mass m, the spring constant k, and the amplitude A. It is perhaps surprising, but **the period and frequency do not depend on the amplitude A.** A small oscillation and a large oscillation have the same period.

Because energy is conserved, we can combine Equations 14.18, 14.19, and 14.20 to write

$$E = \frac{1}{2}mv^2 + \frac{1}{2}kx^2 = \frac{1}{2}kA^2 = \frac{1}{2}m(v_{max})^2 \quad \text{(conservation of energy)} \quad (14.25)$$

Any pair of these expressions may be useful, depending on the known information. For example, you can use the amplitude A to find the speed at any point x by combining the first and second expressions for E. The speed v at position x is

$$v = \sqrt{\frac{k}{m}(A^2 - x^2)} = \omega\sqrt{A^2 - x^2} \qquad (14.26)$$

FIGURE 14.11 shows graphically how the kinetic and potential energy change with time. They both oscillate but remain *positive* because x and v are squared. Energy is continuously being transformed back and forth between the kinetic energy of the moving block and the stored potential energy of the spring, but their sum remains constant. Notice that K and U both oscillate *twice* each period; make sure you understand why.

FIGURE 14.11 Kinetic energy, potential energy, and the total mechanical energy for simple harmonic motion.

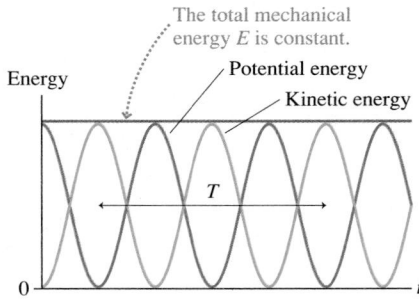

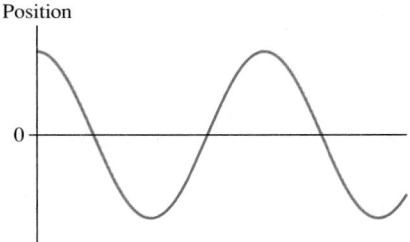

EXAMPLE 14.5	**Using conservation of energy**

A 500 g block on a spring is pulled a distance of 20 cm and released. The subsequent oscillations are measured to have a period of 0.80 s.

a. At what position or positions is the block's speed 1.0 m/s?
b. What is the spring constant?

MODEL The motion is SHM. Energy is conserved.

SOLVE a. The block starts from the point of maximum displacement, where $E = U = \frac{1}{2}kA^2$. At a later time, when the position is x and the speed is v, energy conservation requires

$$\frac{1}{2}mv^2 + \frac{1}{2}kx^2 = \frac{1}{2}kA^2$$

Solving for x, we find

$$x = \sqrt{A^2 - \frac{mv^2}{k}} = \sqrt{A^2 - \left(\frac{v}{\omega}\right)^2}$$

where we used $k/m = \omega^2$ from Equation 14.24. The angular frequency is easily found from the period: $\omega = 2\pi/T = 7.85$ rad/s. Thus

$$x = \sqrt{(0.20\text{ m})^2 - \left(\frac{1.0\text{ m/s}}{7.85\text{ rad/s}}\right)^2} = \pm 0.15\text{ m} = \pm 15\text{ cm}$$

There are two positions because the block has this speed on either side of equilibrium.

b. Although part a did not require that we know the spring constant, it is straightforward to find from Equation 14.24:

$$T = 2\pi\sqrt{\frac{m}{k}}$$

$$k = \frac{4\pi^2 m}{T^2} = \frac{4\pi^2(0.50\text{ kg})}{(0.80\text{ s})^2} = 31\text{ N/m}$$

The four springs shown here have been compressed from their equilibrium position at $x = 0$ cm. When released, the attached mass will start to oscillate. Rank in order, from highest to lowest, the maximum speeds of the masses.

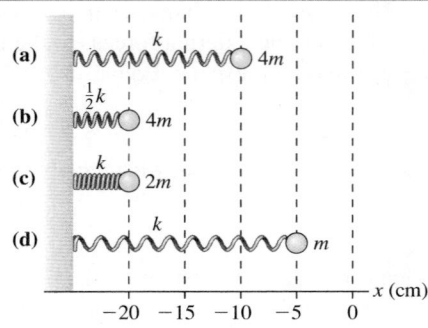

14.4 The Dynamics of Simple Harmonic Motion

Our analysis thus far has been based on the experimental observation that the oscillation of a spring "looks" sinusoidal. It's time to show that Newton's second law *predicts* sinusoidal motion.

A motion diagram will help us visualize the object's acceleration. FIGURE 14.12 shows one cycle of the motion, separating motion to the left and motion to the right to make the diagram clear. As you can see, the object's velocity is large as it passes through the equilibrium point at $x = 0$, but $\vec{v}$ is *not changing* at that point. Acceleration measures the *change* of the velocity; hence $\vec{a} = \vec{0}$ at $x = 0$.

FIGURE 14.12 Motion diagram of simple harmonic motion. The left and right motions are separated vertically for clarity but really occur along the same line.

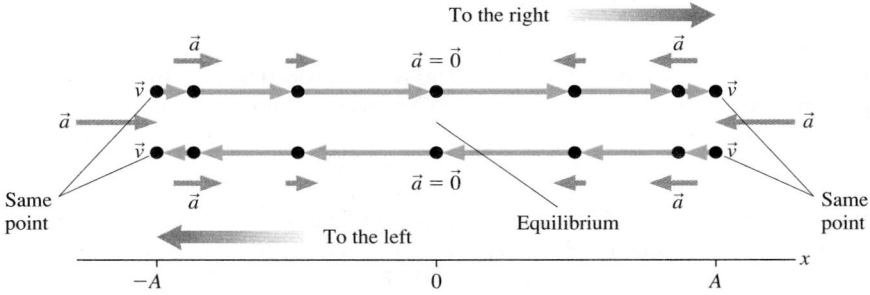

FIGURE 14.13 Position and acceleration graphs for an oscillating spring. We've chosen $\phi_0 = 0$.

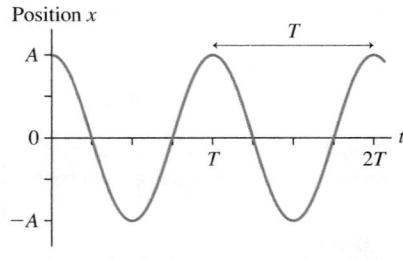

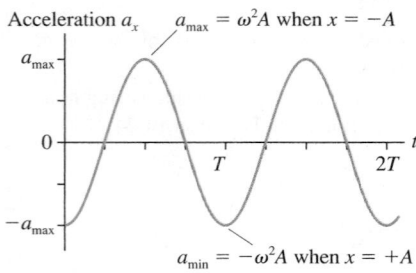

In contrast, the velocity is changing rapidly at the turning points. At the right turning point, $\vec{v}$ changes from a right-pointing vector to a left-pointing vector. Thus the acceleration $\vec{a}$ at the right turning point is large and *to the left*. In one-dimensional motion, the acceleration component a_x has a large *negative* value at the right turning point. Similarly, the acceleration $\vec{a}$ at the left turning point is large and *to the right*. Consequently, a_x has a large positive value at the left turning point.

Our motion-diagram analysis suggests that the acceleration a_x is most positive when the displacement is most negative, most negative when the displacement is a maximum, and zero when $x = 0$. This is confirmed by taking the derivative of the velocity:

$$a_x = \frac{dv_x}{dt} = \frac{d}{dt}(-\omega A \sin \omega t) = -\omega^2 A \cos \omega t \qquad (14.27)$$

then graphing it.

FIGURE 14.13 shows the position graph that we started with in Figure 14.4 and the corresponding acceleration graph. Comparing the two, you can see that the acceleration graph looks like an upside-down position graph. In fact, because $x = A\cos\omega t$, Equation 14.27 for the acceleration can be written

$$a_x = -\omega^2 x \qquad (14.28)$$

That is, **the acceleration is proportional to the negative of the displacement.** The acceleration is, indeed, most positive when the displacement is most negative and is most negative when the displacement is most positive.

Recall that the acceleration is related to the net force by Newton's second law. Consider again our prototype mass on a spring, shown in FIGURE 14.14. This is the simplest possible oscillation, with no distractions due to friction or gravitational forces. We will assume the spring itself to be massless.

As you learned in Chapter 10, the spring force is given by Hooke's law:

$$(F_{sp})_x = -k\,\Delta x \qquad (14.29)$$

The minus sign indicates that the spring force is a **restoring force,** a force that always points back toward the equilibrium position. If we place the origin of the coordinate system at the equilibrium position, as we've done throughout this chapter, then $\Delta x = x$ and Hooke's law is simply $(F_{sp})_x = -kx$.

The x-component of Newton's second law for the object attached to the spring is

$$(F_{net})_x = (F_{sp})_x = -kx = ma_x \qquad (14.30)$$

Equation 14.30 is easily rearranged to read

$$a_x = -\frac{k}{m}x \qquad (14.31)$$

You can see that Equation 14.31 is identical to Equation 14.28 if the system oscillates with angular frequency $\omega = \sqrt{k/m}$. We previously found this expression for ω from an energy analysis. Our experimental observation that the acceleration is proportional to the *negative* of the displacement is exactly what Hooke's law would lead us to expect. That's the good news.

The bad news is that a_x is not a constant. As the object's position changes, so does the acceleration. Nearly all of our kinematic tools have been based on constant acceleration. We can't use those tools to analyze oscillations, so we must go back to the very definition of acceleration:

$$a_x = \frac{dv_x}{dt} = \frac{d^2x}{dt^2}$$

Acceleration is the second derivative of position with respect to time. If we use this definition in Equation 14.31, it becomes

$$\frac{d^2x}{dt^2} = -\frac{k}{m}x \quad \text{(equation of motion for a mass on a spring)} \qquad (14.32)$$

Equation 14.32, which is called the **equation of motion,** is a second-order differential equation. Unlike other equations we've dealt with, Equation 14.32 cannot be solved by direct integration. We'll need to take a different approach.

Solving the Equation of Motion

The solution to an algebraic equation such as $x^2 = 4$ is a number. The solution to a differential equation is a *function*. The x in Equation 14.32 is really $x(t)$, the position as a function of time. The solution to this equation is a function $x(t)$ whose second derivative is the function itself multiplied by $(-k/m)$.

One important property of differential equations that you will learn about in math is that the solutions are *unique*. That is, there is only *one* solution to Equation 14.32 that satisfies the initial conditions. If we were able to *guess* a solution, the uniqueness property would tell us that we had found the *only* solution. That might seem a rather

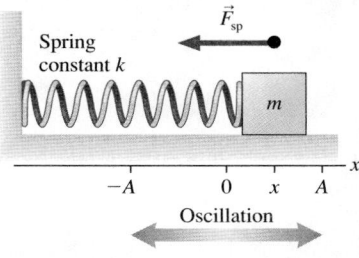

FIGURE 14.14 The prototype of simple harmonic motion: a mass oscillating on a horizontal spring without friction.

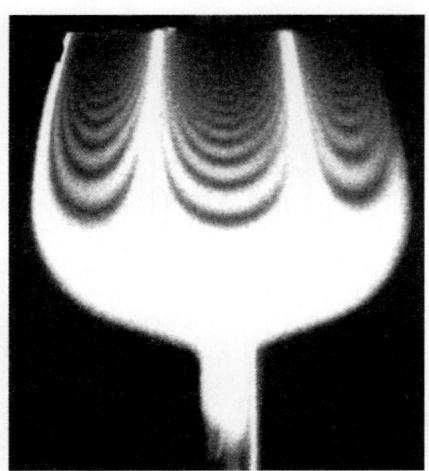

An optical technique called *interferometry* reveals the bell-like vibrations of a wine glass.

strange way to solve equations, but in fact differential equations are frequently solved by using your knowledge of what the solution needs to look like to guess an appropriate function. Let us give it a try!

We know from experimental evidence that the oscillatory motion of a spring appears to be sinusoidal. Let us *guess* that the solution to Equation 14.32 should have the functional form

$$x(t) = A\cos(\omega t + \phi_0) \tag{14.33}$$

where A, ω, and ϕ_0 are unspecified constants that we can adjust to any values that might be necessary to satisfy the differential equation.

If you were to guess that a solution to the algebraic equation $x^2 = 4$ is $x = 2$, you would verify your guess by substituting it into the original equation to see if it works. We need to do the same thing here: Substitute our guess for $x(t)$ into Equation 14.32 to see if, for an appropriate choice of the three constants, it works. To do so, we need the second derivative of $x(t)$. That is straightforward:

$$x(t) = A\cos(\omega t + \phi_0)$$

$$\frac{dx}{dt} = -\omega A\sin(\omega t + \phi_0) \tag{14.34}$$

$$\frac{d^2x}{dt^2} = -\omega^2 A\cos(\omega t + \phi_0)$$

If we now substitute the first and third of Equations 14.34 into Equation 14.32, we find

$$-\omega^2 A\cos(\omega t + \phi_0) = -\frac{k}{m}A\cos(\omega t + \phi_0) \tag{14.35}$$

Equation 14.35 will be true at all instants of time if and only if $\omega^2 = k/m$. There do not seem to be any restrictions on the two constants A and ϕ_0—they are determined by the initial conditions.

So we have found—by guessing!—that *the* solution to the equation of motion for a mass oscillating on a spring is

$$x(t) = A\cos(\omega t + \phi_0) \tag{14.36}$$

where the angular frequency

$$\omega = 2\pi f = \sqrt{\frac{k}{m}} \tag{14.37}$$

is determined by the mass and the spring constant.

NOTE ▶ Once again we see that the oscillation frequency is independent of the amplitude A. ◀

Equations 14.36 and 14.37 seem somewhat anticlimactic because we've been using these results for the last several pages. But keep in mind that we had been *assuming* $x = A\cos\omega t$ simply because the experimental observations "looked" like a cosine function. We've now justified that assumption by showing that Equation 14.36 really is the solution to Newton's second law for a mass on a spring. **The *theory* of oscillation, based on Hooke's law for a spring and Newton's second law, is in good agreement with the experimental observations.** This conclusion gives an affirmative answer to the last of the three questions that we asked early in the chapter, which was whether the sinusoidal oscillation of SHM is a consequence of Newton's laws.

EXAMPLE 14.6 **Analyzing an oscillator**

At $t = 0$ s, a 500 g block oscillating on a spring is observed moving to the right at $x = 15$ cm. It reaches a maximum displacement of 25 cm at $t = 0.30$ s.

a. Draw a position-versus-time graph for one cycle of the motion.
b. At what times during the first cycle does the mass pass through $x = 20$ cm?

MODEL The motion is simple harmonic motion.

SOLVE a. The position equation of the block is $x(t) = A\cos(\omega t + \phi_0)$. We know that the amplitude is $A = 0.25$ m and that $x_0 = 0.15$ m. From these two pieces of information we obtain the phase constant:

$$\phi_0 = \cos^{-1}\left(\frac{x_0}{A}\right) = \cos^{-1}(0.60) = \pm 0.927 \text{ rad}$$

The object is initially moving to the right, which tells us that the phase constant must be between $-\pi$ and 0 rad. Thus $\phi_0 = -0.927$ rad. The block reaches its maximum displacement $x_{max} = A$ at time $t = 0.30$ s. At that instant of time

$$x_{max} = A = A\cos(\omega t + \phi_0)$$

This can be true only if $\cos(\omega t + \phi_0) = 1$, which requires $\omega t + \phi_0 = 0$. Thus

$$\omega = \frac{-\phi_0}{t} = \frac{-(-0.927 \text{ rad})}{0.30 \text{ s}} = 3.09 \text{ rad/s}$$

Now that we know ω, it is straightforward to compute the period:

$$T = \frac{2\pi}{\omega} = 2.0 \text{ s}$$

FIGURE 14.15 graphs $x(t) = (25 \text{ cm})\cos(3.09t - 0.927)$, where t is in s, from $t = 0$ to $t = 2.0$ s.

b. From $x = A\cos(\omega t + \phi_0)$, the time at which the mass reaches position $x = 20$ cm is

FIGURE 14.15 Position-versus-time graph for the oscillator of Example 14.6.

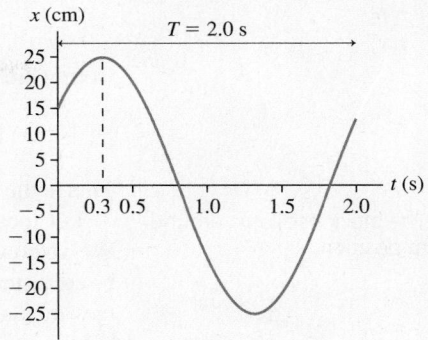

$$t = \frac{1}{\omega}\left(\cos^{-1}\left(\frac{x}{A}\right) - \phi_0\right)$$

$$= \frac{1}{3.09 \text{ rad/s}}\left(\cos^{-1}\left(\frac{20 \text{ cm}}{25 \text{ cm}}\right) + 0.927 \text{ rad}\right) = 0.51 \text{ s}$$

A calculator returns only one value of $\cos^{-1}$, in the range 0 to π rad, but we noted earlier that $\cos^{-1}$ actually has two values. Indeed, you can see in Figure 14.15 that there are two times at which the mass passes $x = 20$ cm. Because they are symmetrical on either side of $t = 0.30$ s, when $x = A$, the first point is $(0.51 \text{ s} - 0.30 \text{ s}) = 0.21$ s *before* the maximum. Thus the mass passes through $x = 20$ cm at $t = 0.09$ s and again at $t = 0.51$ s.

STOP TO THINK 14.4 This is the position graph of a mass on a spring. What can you say about the velocity and the force at the instant indicated by the dashed line?

a. Velocity is positive; force is to the right.
b. Velocity is negative; force is to the right.
c. Velocity is zero; force is to the right.
d. Velocity is positive; force is to the left.
e. Velocity is negative; force is to the left.
f. Velocity is zero; force is to the left.
g. Velocity and force are both zero.

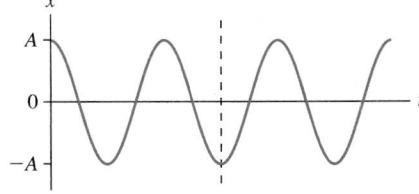

14.5 Vertical Oscillations

We have focused our analysis on a horizontally oscillating spring. But the typical demonstration you'll see in class is a mass bobbing up and down on a spring hung vertically from a support. Is it safe to assume that a vertical oscillation has the same mathematical description as a horizontal oscillation? Or does the additional force of gravity change the motion? Let us look at this more carefully.

FIGURE 14.16 shows a block of mass m hanging from a spring of spring constant k. An important fact to notice is that the equilibrium position of the block is *not* where the spring is at its unstretched length. At the equilibrium position of the block, where it hangs motionless, the spring has stretched by ΔL.

Finding ΔL is a static-equilibrium problem in which the upward spring force balances the downward gravitational force on the block. The y-component of the spring force is given by Hooke's law:

$$(F_{sp})_y = -k\,\Delta y = +k\,\Delta L \tag{14.38}$$

FIGURE 14.16 Gravity stretches the spring.

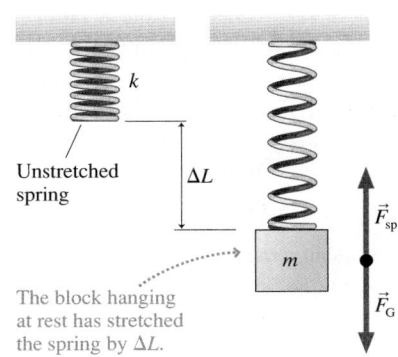

Unstretched spring

ΔL

The block hanging at rest has stretched the spring by ΔL.

Equation 14.38 makes a distinction between ΔL, which is simply a *distance* and is a positive number, and the displacement Δy. The block is displaced downward, so $\Delta y = -\Delta L$. Newton's first law for the block in equilibrium is

$$(F_{net})_y = (F_{sp})_y + (F_G)_y = k\,\Delta L - mg = 0 \qquad (14.39)$$

from which we can find

$$\Delta L = \frac{mg}{k} \qquad (14.40)$$

This is the distance the spring stretches when the block is attached to it.

Let the block oscillate around this equilibrium position, as shown in FIGURE 14.17. We've now placed the origin of the y-axis at the block's equilibrium position in order to be consistent with our analyses of oscillations throughout this chapter. If the block moves upward, as the figure shows, the spring gets shorter compared to its equilibrium length, but the spring is still *stretched* compared to its unstretched length in Figure 14.16. When the block is at position y, the spring is stretched by an amount $\Delta L - y$ and hence exerts an *upward* spring force $F_{sp} = k(\Delta L - y)$. The net force on the block at this point is

$$(F_{net})_y = (F_{sp})_y + (F_G)_y = k(\Delta L - y) - mg = (k\,\Delta L - mg) - ky \qquad (14.41)$$

But $k\,\Delta L - mg$ is zero, from Equation 14.40, so the net force on the block is simply

$$(F_{net})_y = -ky \qquad (14.42)$$

Equation 14.42 for vertical oscillations is *exactly* the same as Equation 14.30 for horizontal oscillations, where we found $(F_{net})_x = -kx$. That is, the restoring force for vertical oscillations is identical to the restoring force for horizontal oscillations. The role of gravity is to determine where the equilibrium position is, but it doesn't affect the oscillatory motion around the equilibrium position.

Because the net force is the same, Newton's second law has exactly the same oscillatory solution:

$$y(t) = A\cos(\omega t + \phi_0) \qquad (14.43)$$

with, again, $\omega = \sqrt{k/m}$. **The vertical oscillations of a mass on a spring are the same simple harmonic motion as those of a block on a horizontal spring.** This is an important finding because it was not obvious that the motion would still be simple harmonic motion when gravity was included.

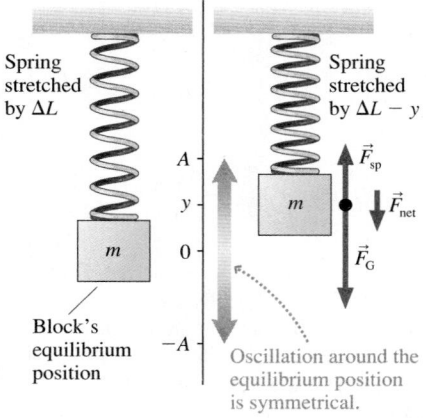

FIGURE 14.17 The block oscillates around the equilibrium position.

Spring stretched by ΔL

Spring stretched by $\Delta L - y$

Block's equilibrium position

Oscillation around the equilibrium position is symmetrical.

EXAMPLE 14.7 **Bungee oscillations**

An 83 kg student hangs from a bungee cord with spring constant 270 N/m. The student is pulled down to a point where the cord is 5.0 m longer than its unstretched length, then released. Where is the student, and what is his velocity 2.0 s later?

MODEL A bungee cord can be modeled as a spring. Vertical oscillations on the bungee cord are SHM.

VISUALIZE FIGURE 14.18 shows the situation.

SOLVE Although the cord is stretched by 5.0 m when the student is released, this is *not* the amplitude of the oscillation. Oscillations occur around the equilibrium position, so we have to begin by finding the equilibrium point where the student hangs motionless. The cord stretch at equilibrium is given by Equation 14.40:

$$\Delta L = \frac{mg}{k} = 3.0 \text{ m}$$

Stretching the cord 5.0 m pulls the student 2.0 m below the equilibrium point, so $A = 2.0$ m. That is, the student oscillates with amplitude $A = 2.0$ m about a point 3.0 m beneath the bungee

FIGURE 14.18 A student on a bungee cord oscillates about the equilibrium position.

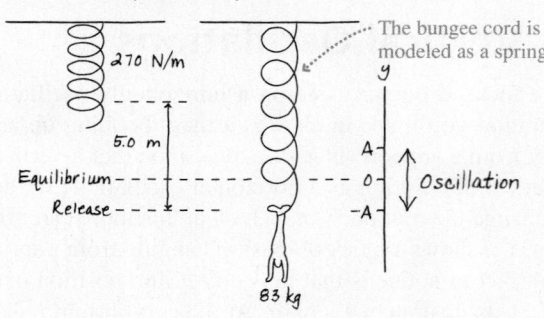

cord's original end point. The student's position as a function of time, as measured from the equilibrium position, is

$$y(t) = (2.0 \text{ m})\cos(\omega t + \phi_0)$$

where $\omega = \sqrt{k/m} = 1.80$ rad/s. The initial condition

$$y_0 = A \cos \phi_0 = -A$$

requires the phase constant to be $\phi_0 = \pi$ rad. At $t = 2.0$ s the student's position and velocity are

$$y = (2.0 \text{ m}) \cos \big((1.80 \text{ rad/s})(2.0 \text{ s}) + \pi \text{ rad} \big) = 1.8 \text{ m}$$

$$v_y = -\omega A \sin(\omega t + \phi_0) = -1.6 \text{ m/s}$$

The student is 1.8 m *above* the equilibrium position, or 1.2m *below* the original end of the cord. Because his velocity is negative, he's passed through the highest point and is heading down.

14.6 The Pendulum

Now let's look at another very common oscillator: a pendulum. FIGURE 14.19a shows a mass m attached to a string of length L and free to swing back and forth. The pendulum's position can be described by the arc of length s, which is zero when the pendulum hangs straight down. Because angles are measured ccw, s and θ are positive when the pendulum is to the right of center, negative when it is to the left.

Two forces are acting on the mass: the string tension $\vec{T}$ and gravity $\vec{F}_G$. It will be convenient to repeat what we did in our study of circular motion: Divide the forces into tangential components, parallel to the motion, and radial components parallel to the string. These are shown on the free-body diagram of FIGURE 14.19b.

Newton's second law for the tangential component, parallel to the motion, is

$$(F_{\text{net}})_t = \sum F_t = (F_G)_t = -mg \sin \theta = ma_t \qquad (14.44)$$

Using $a_t = d^2s/dt^2$ for acceleration "around" the circle, and noting that the mass cancels, we can write Equation 14.44 as

$$\frac{d^2s}{dt^2} = -g \sin \theta \qquad (14.45)$$

This is the equation of motion for an oscillating pendulum. The sine function makes this equation more complicated than the equation of motion for an oscillating spring.

The Small-Angle Approximation

Suppose we restrict the pendulum's oscillations to *small angles* of less than about 10°. This restriction allows us to make use of an interesting and important piece of geometry.

FIGURE 14.20 shows an angle θ and a circular arc of length $s = r\theta$. A right triangle has been constructed by dropping a perpendicular from the top of the arc to the axis. The height of the triangle is $h = r\sin\theta$. Suppose that the angle θ is "small." In that case there is very little difference between h and s. If $h \approx s$, then $r\sin\theta \approx r\theta$. It follows that

$$\sin\theta \approx \theta \quad (\theta \text{ in radians})$$

The result that $\sin\theta \approx \theta$ for small angles is called the **small-angle approximation.** We can similarly note that $l \approx r$ for small angles. Because $l = r\cos\theta$, it follows that $\cos\theta \approx 1$. Finally, we can take the ratio of sine and cosine to find $\tan\theta \approx \sin\theta \approx \theta$. Table 14.3 summarizes the small-angle approximation. We will have other occasions to use the small-angle approximation throughout the remainder of this text.

NOTE ▶ The small-angle approximation is valid *only* if angle θ is in radians! ◀

How small does θ have to be to justify using the small-angle approximation? It's easy to use your calculator to find that the small-angle approximation is good to three

FIGURE 14.19 The motion of a pendulum.

(a)

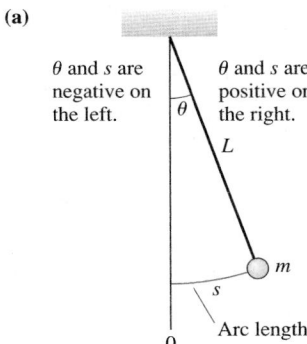

θ and s are negative on the left.

θ and s are positive on the right.

L

m

s

Arc length

0

(b)

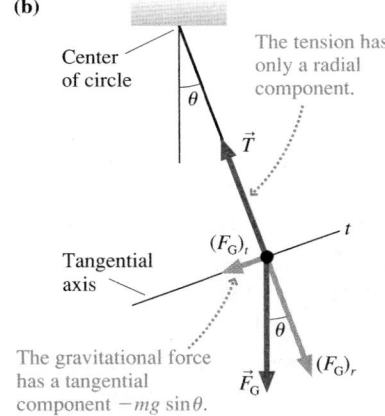

Center of circle

The tension has only a radial component.

$\vec{T}$

$(F_G)_t$

t

Tangential axis

θ

The gravitational force has a tangential component $-mg\sin\theta$.

$\vec{F}_G$ $(F_G)_r$

FIGURE 14.20 The geometrical basis of the small-angle approximation.

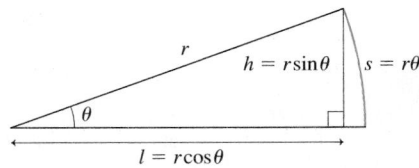

r $h = r\sin\theta$ $s = r\theta$

θ

$l = r\cos\theta$

TABLE 14.3 Small-angle approximations. θ must be in radians.

$\sin\theta \approx \theta$	$\tan\theta \approx \sin\theta \approx \theta$
$\cos\theta \approx 1$	

The pendulum clock has been used for hundreds of years.

significant figures, an error of $\leq 0.1\%$, up to angles of ≈ 0.10 rad ($\approx 5°$). In practice, we will use the approximation up to about 10°, but for angles any larger it rapidly loses validity and produces unacceptable results.

If we restrict the pendulum to $\theta < 10°$, we can use $\sin\theta \approx \theta$. In that case, Equation 14.44 for the net force on the mass is

$$(F_{net})_t = -mg \sin\theta \approx -mg\theta = -\frac{mg}{L}s$$

where, in the last step, we used the fact that angle θ is related to the arc length by $\theta = s/L$. Then the equation of motion becomes

$$\frac{d^2s}{dt^2} = -\frac{g}{L}s \qquad (14.46)$$

This is *exactly* the same as Equation 14.32 for a mass oscillating on a spring. The names are different, with x replaced by s and k/m by g/L, but that does not make it a different equation.

Because we know the solution to the spring problem, we can immediately write the solution to the pendulum problem just by changing variables and constants:

$$s(t) = A\cos(\omega t + \phi_0) \qquad \text{or} \qquad \theta(t) = \theta_{max}\cos(\omega t + \phi_0) \quad (14.47)$$

The angular frequency

$$\omega = 2\pi f = \sqrt{\frac{g}{L}} \qquad (14.48)$$

is determined by the length of the string. The pendulum is interesting in that **the frequency, and hence the period, is independent of the mass.** It depends only on the length of the pendulum. The amplitude A and the phase constant ϕ_0 are determined by the initial conditions, just as they were for an oscillating spring.

EXAMPLE 14.8 **The maximum angle of a pendulum**

A 300 g mass on a 30-cm-long string oscillates as a pendulum. It has a speed of 0.25 m/s as it passes through the lowest point. What maximum angle does the pendulum reach?

MODEL Assume that the angle remains small, in which case the motion is simple harmonic motion.

SOLVE The angular frequency of the pendulum is

$$\omega = \sqrt{\frac{g}{L}} = \sqrt{\frac{9.8 \text{ m/s}^2}{0.30 \text{ m}}} = 5.72 \text{ rad/s}$$

The speed at the lowest point is $v_{max} = \omega A$, so the amplitude is

$$A = s_{max} = \frac{v_{max}}{\omega} = \frac{0.25 \text{ m/s}}{5.72 \text{ rad/s}} = 0.0437 \text{ m}$$

The maximum angle, at the maximum arc length s_{max}, is

$$\theta_{max} = \frac{s_{max}}{L} = \frac{0.0437 \text{ m}}{0.30 \text{ m}} = 0.146 \text{ rad} = 8.3°$$

ASSESS Because the maximum angle is less than 10°, our analysis based on the small-angle approximation is reasonable.

EXAMPLE 14.9 **The gravimeter**

Deposits of minerals and ore can alter the local value of the free-fall acceleration because they tend to be denser than surrounding rocks. Geologists use a *gravimeter*—an instrument that accurately measures the local free-fall acceleration—to search for ore deposits. One of the simplest gravimeters is a pendulum. To achieve the highest accuracy, a stopwatch is used to time 100 oscillations of a pendulum of different lengths. At one location in the field, a geologist makes the following measurements:

Length (m)	Time (s)
0.500	141.7
1.000	200.6
1.500	245.8
2.000	283.5

What is the local value of g?

MODEL Assume the oscillation angle is small, in which case the motion is simple harmonic motion with a period independent of the mass of the pendulum. Because the data are known to four significant figures (± 1 mm on the length and ± 0.1 s on the timing, both of which are easily achievable), we expect to determine g to four significant figures.

SOLVE From Equation 14.48, using $f = 1/T$, we find

$$T^2 = \left(2\pi\sqrt{\frac{L}{g}}\right)^2 = \frac{4\pi^2}{g}L$$

That is, the square of a pendulum's period is proportional to its length. Consequently, a graph of T^2 versus L should be a straight line passing through the origin with slope $4\pi^2/g$. We can use the experimentally measured slope to determine g. FIGURE 14.21 is a graph of the data, with the period found by dividing the measured time by 100.

As expected, the graph is a straight line passing through the origin. The slope of the best-fit line is 4.021 s^2/m. Consequently,

$$g = \frac{4\pi^2}{\text{slope}} = \frac{4\pi^2}{4.021 \text{ s}^2/\text{m}} = 9.818 \text{ m/s}^2$$

ASSESS The fact that the graph is linear and passes through the origin confirms our model of the situation. Had this *not* been the

FIGURE 14.21 Graph of the square of the pendulum's period versus its length.

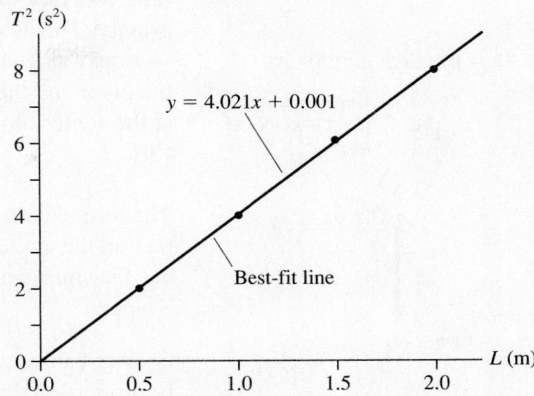

case, we would have had to conclude either that our model of the pendulum as a simple, small-angle pendulum was not valid or that our measurements were bad. This is an important reason for having multiple data points rather than using only one length.

The Conditions for Simple Harmonic Motion

You can begin to see how, in a sense, we have solved *all* simple-harmonic-motion problems once we have solved the problem of the horizontal spring. The restoring force of a spring, $F_{sp} = -kx$, is directly proportional to the displacement x from equilibrium. The pendulum's restoring force, in the small-angle approximation, is directly proportional to the displacement s. A restoring force that is directly proportional to the displacement from equilibrium is called a **linear restoring force.** For *any* linear restoring force, the equation of motion is identical to the spring equation (other than perhaps using different symbols). Consequently, **any system with a linear restoring force will undergo simple harmonic motion around the equilibrium position.**

This is why an oscillating spring is the prototype of SHM. Everything that we learn about an oscillating spring can be applied to the oscillations of any other linear restoring force, ranging from the vibration of airplane wings to the motion of electrons in electric circuits. Let's summarize this information with a Tactics Box.

TACTICS BOX 14.1   **Identifying and analyzing simple harmonic motion**

❶ If the net force acting on a particle is a linear restoring force, the motion will be simple harmonic motion around the equilibrium position.

❷ The position as a function of time is $x(t) = A\cos(\omega t + \phi_0)$. The velocity as a function of time is $v_x(t) = -\omega A \sin(\omega t + \phi_0)$. The maximum speed is $v_{max} = \omega A$. The equations are given here in terms of x, but they can be written in terms of y, θ, or some other parameter if the situation calls for it.

❸ The amplitude A and the phase constant ϕ_0 are determined by the initial conditions through $x_0 = A\cos\phi_0$ and $v_{0x} = -\omega A\sin\phi_0$.

❹ The angular frequency ω (and hence the period $T = 2\pi/\omega$) depends on the physics of the particular situation. But ω does *not* depend on A or ϕ_0.

❺ Mechanical energy is conserved. Thus $\frac{1}{2}mv_x^2 + \frac{1}{2}kx^2 = \frac{1}{2}kA^2 = \frac{1}{2}m(v_{max})^2$. Energy conservation provides a relationship between position and velocity that is independent of time.

Exercises 7–12, 15–19

The Physical Pendulum

A mass on a string is often called a *simple pendulum*. But you can also make a pendulum from any solid object that swings back and forth on a pivot under the influence of gravity. This is called a *physical pendulum*.

FIGURE 14.22 shows a physical pendulum of mass M for which the distance between the pivot and the center of mass is l. The moment arm of the gravitational force acting at the center of mass is $d = l \sin\theta$, so the gravitational torque is

$$\tau = -Mgd = -Mgl\sin\theta$$

The torque is negative because, for positive θ, it's causing a clockwise rotation. If we restrict the angle to being small ($\theta < 10°$), as we did for the simple pendulum, we can use the small-angle approximation to write

$$\tau = -Mgl\theta \qquad (14.49)$$

Gravity causes a linear restoring torque on the pendulum—that is, the torque is directly proportional to the angular displacement θ—so we expect the physical pendulum to undergo SHM.

From Chapter 12, Newton's second law for rotational motion is

$$\alpha = \frac{d^2\theta}{dt^2} = \frac{\tau}{I}$$

where I is the object's moment of inertia about the pivot point. Using Equation 14.49 for the torque, we find

$$\frac{d^2\theta}{dt^2} = \frac{-Mgl}{I}\theta \qquad (14.50)$$

Comparison with Equation 14.32 shows that this is again the SHM equation of motion, this time with angular frequency

$$\omega = 2\pi f = \sqrt{\frac{Mgl}{I}} \qquad (14.51)$$

It appears that the frequency depends on the mass of the pendulum, but recall that the moment of inertia is directly proportional to M. Thus M cancels and the frequency of a physical pendulum, like that of a simple pendulum, is independent of mass.

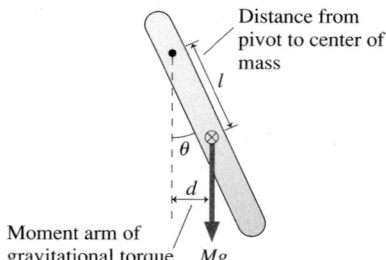

FIGURE 14.22 A physical pendulum.

Distance from pivot to center of mass

l

θ

d

Moment arm of gravitational torque Mg

EXAMPLE 14.10 | **A swinging leg as a pendulum**

A student in a biomechanics lab measures the length of his leg, from hip to heel, to be 0.90 m. What is the frequency of the pendulum motion of the student's leg? What is the period?

MODEL We can model a human leg reasonably well as a rod of uniform cross section, pivoted at one end (the hip) to form a physical pendulum. The center of mass of a uniform leg is at the midpoint, so $l = L/2$.

SOLVE The moment of inertia of a rod pivoted about one end is $I = \frac{1}{3}ML^2$, so the pendulum frequency is

$$f = \frac{1}{2\pi}\sqrt{\frac{Mgl}{I}} = \frac{1}{2\pi}\sqrt{\frac{Mg(L/2)}{ML^2/3}} = \frac{1}{2\pi}\sqrt{\frac{3g}{2L}} = 0.64 \text{ Hz}$$

The corresponding period is $T = 1/f = 1.6$ s. Notice that we didn't need to know the mass.

ASSESS As you walk, your legs do swing as physical pendulums as you bring them forward. The frequency is fixed by the length of your legs and their distribution of mass; it doesn't depend on amplitude. Consequently, you don't increase your walking speed by taking more rapid steps—changing the frequency is difficult. You simply take longer strides, changing the amplitude but not the frequency.

STOP TO THINK 14.5 One person swings on a swing and finds that the period is 3.0 s. A second person of equal mass joins him. With two people swinging, the period is

a. 6.0 s
b. >3.0 s but not necessarily 6.0 s
c. 3.0 s
d. <3.0 s but not necessarily 1.5 s
e. 1.5 s
f. Can't tell without knowing the length

14.7 Damped Oscillations

A pendulum left to itself gradually slows down and stops. The sound of a ringing bell gradually dies away. All real oscillators do run down—some very slowly but others quite quickly—as friction or other dissipative forces transform their mechanical energy into the thermal energy of the oscillator and its environment. An oscillation that runs down and stops is called a **damped oscillation.**

There are many possible reasons for the dissipation of energy, such as air resistance, friction, and internal forces within a metal spring as it flexes. The forces involved in dissipation are complex, but a simple *linear drag* model gives a quite accurate description of most damped oscillations. That is, we'll assume a drag force that depends linearly on the velocity as

$$\vec{D} = -b\vec{v} \quad \text{(model of the drag force)} \qquad (14.52)$$

where the minus sign is the mathematical statement that the force is always opposite in direction to the velocity in order to slow the object.

The **damping constant** b depends in a complicated way on the shape of the object *and* on the viscosity of the air or other medium in which the particle moves. The damping constant plays the same role in our model of air resistance that the coefficient of friction does in our model of friction.

The units of b need to be such that they will give units of force when multiplied by units of velocity. As you can confirm, these units are kg/s. A value $b = 0$ kg/s corresponds to the limiting case of no resistance, in which case the mechanical energy is conserved. A typical value of b for a spring or a pendulum in air is ≤ 0.10 kg/s. Objects moving in a liquid can have significantly larger values of b.

FIGURE 14.23 shows a mass oscillating on a spring in the presence of a drag force. With the drag included, Newton's second law is

$$(F_{\text{net}})_x = (F_{\text{sp}})_x + D_x = -kx - bv_x = ma_x \qquad (14.53)$$

Using $v_x = dx/dt$ and $a_x = d^2x/dt^2$, we can write Equation 14.53 as

$$\frac{d^2x}{dt^2} + \frac{b}{m}\frac{dx}{dt} + \frac{k}{m}x = 0 \qquad (14.54)$$

Equation 14.54 is the equation of motion of a damped oscillator. If you compare it to Equation 14.32, the equation of motion for a block on a frictionless surface, you'll see that it differs by the inclusion of the term involving dx/dt.

Equation 14.54 is another second-order differential equation. We will simply assert (and, as a homework problem, you can confirm) that the solution is

$$x(t) = Ae^{-bt/2m}\cos(\omega t + \phi_0) \quad \text{(damped oscillator)} \qquad (14.55)$$

where the angular frequency is given by

$$\omega = \sqrt{\frac{k}{m} - \frac{b^2}{4m^2}} = \sqrt{\omega_0^2 - \frac{b^2}{4m^2}} \qquad (14.56)$$

Here $\omega_0 = \sqrt{k/m}$ is the angular frequency of an undamped oscillator ($b = 0$). The constant e is the base of natural logarithms, so $e^{-bt/2m}$ is an *exponential function.*

Because $e^0 = 1$, Equation 14.55 reduces to our previous solution, $x(t) = A\cos(\omega t + \phi_0)$, when $b = 0$. This makes sense and gives us confidence in Equation 14.55. A *lightly damped* system, which oscillates many times before stopping, is one for which $b/2m \ll \omega_0$. In that case, $\omega \approx \omega_0$ is a good approximation. That is, light damping does not affect the oscillation frequency.

FIGURE 14.24 is a graph of the position $x(t)$ for a lightly damped oscillator, as given by Equation 14.55. Notice that the term $Ae^{-bt/2m}$, which is shown by the dashed line,

The shock absorbers in cars and trucks are heavily damped springs. The vehicle's vertical motion, after hitting a rock or a pothole, is a damped oscillation.

FIGURE 14.23 An oscillating mass in the presence of a drag force.

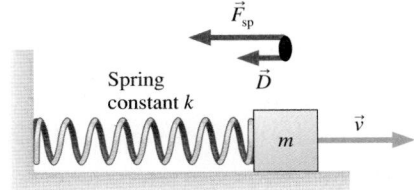

FIGURE 14.24 Position-versus-time graph for a damped oscillator.

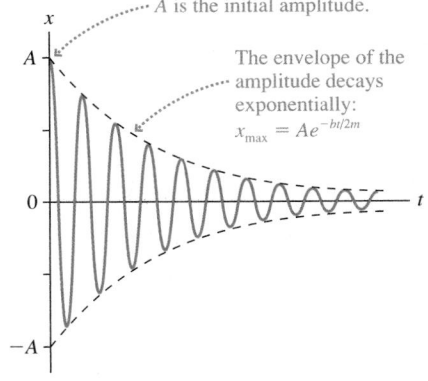

A is the initial amplitude.

The envelope of the amplitude decays exponentially:
$x_{\max} = Ae^{-bt/2m}$

FIGURE 14.25 Several oscillation envelopes, corresponding to different values of the damping constant b.

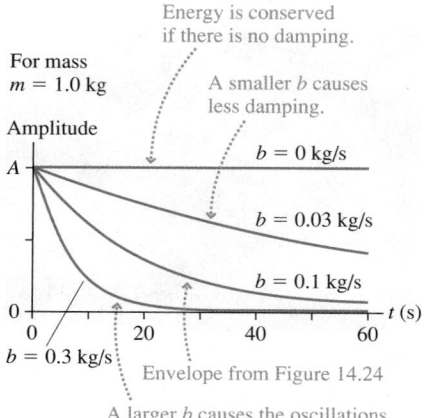

Energy is conserved if there is no damping.

For mass $m = 1.0$ kg

A smaller b causes less damping.

Amplitude

A

$b = 0$ kg/s

$b = 0.03$ kg/s

$b = 0.1$ kg/s

$b = 0.3$ kg/s

Envelope from Figure 14.24

A larger b causes the oscillations to damp more quickly.

acts as a slowly varying amplitude:

$$x_{max}(t) = Ae^{-bt/2m} \tag{14.57}$$

where A is the *initial* amplitude, at $t = 0$. The oscillation keeps bumping up against this line, slowly dying out with time.

A slowly changing line that provides a border to a rapid oscillation is called the **envelope** of the oscillations. In this case, the oscillations have an *exponentially decaying envelope*. Make sure you study Figure 14.24 long enough to see how both the oscillations and the decaying amplitude are related to Equation 14.55.

Changing the amount of damping, by changing the value of b, affects how quickly the oscillations decay. FIGURE 14.25 shows just the envelope $x_{max}(t)$ for several oscillators that are identical except for the value of the damping constant b. (You need to imagine a rapid oscillation within each envelope, as in Figure 14.24.) Increasing b causes the oscillations to damp more quickly, while decreasing b makes them last longer.

MATHEMATICAL ASIDE **Exponential decay**

Exponential decay occurs in a vast number of physical systems of importance in science and engineering. Mechanical vibrations, electric circuits, and nuclear radioactivity all exhibit exponential decay.

The number $e = 2.71828\ldots$ is the base of natural logarithms in the same way that 10 is the base of ordinary logarithms. It arises naturally in calculus from the integral

$$\int \frac{du}{u} = \ln u$$

This integral—which shows up in the analysis of many physical systems—frequently leads to solutions of the form

$$u = Ae^{-v/v_0} = A\exp(-v/v_0)$$

where exp is the *exponential function*.

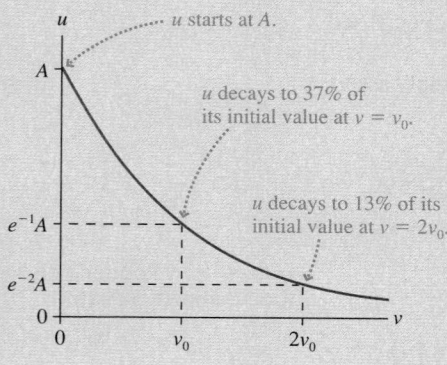

u starts at A.

A

u decays to 37% of its initial value at $v = v_0$.

u decays to 13% of its initial value at $v = 2v_0$.

$e^{-1}A$

$e^{-2}A$

A graph of u illustrates what we mean by *exponential decay*. It starts with $u = A$ at $v = 0$ (because $e^0 = 1$) and then steadily decays, asymptotically approaching zero. The quantity v_0 is called the *decay constant*. When $v = v_0$, $u = e^{-1}A = 0.37A$. When $v = 2v_0$, $u = e^{-2}A = 0.13A$.

Arguments of functions must be pure numbers, without units. That is, we can evaluate e^{-2}, but $e^{-2\,kg}$ makes no sense. If v/v_0 is a pure number, which it must be, then the decay constant v_0 must have the same units as v. If v represents position, then v_0 is a length; if v represents time, then v_0 is a time interval. In a specific situation, v_0 is often called the *decay length* or the *decay time*. It is the length or time in which the quantity decays to 37% of its initial value.

No matter what the process is or what u represents, **a quantity that decays exponentially decays to 37% of its initial value when one decay constant has passed.** Thus exponential decay is a universal behavior. Every time you meet a new system that exhibits exponential decay, its behavior will be exactly the same as every other exponential decay. The decay curve always looks exactly like the figure shown here. Once you've learned the properties of exponential decay, you'll immediately know how to apply this knowledge to a new situation.

Energy in Damped Systems

When considering the oscillator's mechanical energy, it is useful to define the **time constant** τ (also called the *decay time*) to be

$$\tau = \frac{m}{b} \tag{14.58}$$

Because b has units of kg/s, τ has units of seconds. With this definition, we can write the oscillation amplitude as $x_{max}(t) = Ae^{-t/2\tau}$.

Because of the drag force, the mechanical energy is no longer conserved. At any particular time we can compute the mechanical energy from

$$E(t) = \frac{1}{2}k(x_{max})^2 = \frac{1}{2}k(Ae^{-t/2\tau})^2 = \left(\frac{1}{2}kA^2\right)e^{-t/\tau} = E_0e^{-t/\tau} \qquad (14.59)$$

where $E_0 = \frac{1}{2}kA^2$ is the initial energy at $t = 0$ and where we used $(z^m)^2 = z^{2m}$. In other words, **the oscillator's mechanical energy decays exponentially with time constant τ.**

As FIGURE 14.26 shows, the time constant is the amount of time needed for the energy to decay to e^{-1}, or 37%, of its initial value. We say that the time constant τ measures the "characteristic time" during which the energy of the oscillation is dissipated. Roughly two-thirds of the initial energy is gone after one time constant has elapsed, and nearly 90% has dissipated after two time constants have gone by.

For practical purposes, we can speak of the time constant as the *lifetime* of the oscillation—about how long it lasts. Mathematically, there is never a time when the oscillation is "over." The decay approaches zero asymptotically, but it never gets there in any finite time. The best we can do is define a characteristic time when the motion is "almost over," and that is what the time constant τ does.

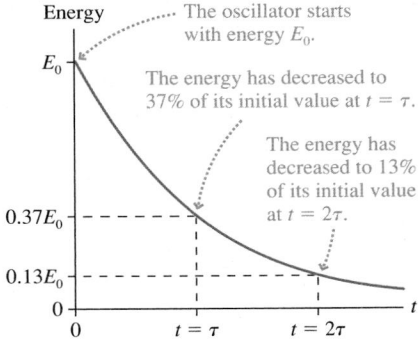

FIGURE 14.26 Exponential decay of the mechanical energy of an oscillator.

EXAMPLE 14.11 **A damped pendulum**

A 500 g mass swings on a 60-cm-string as a pendulum. The amplitude is observed to decay to half its initial value after 35.0 s.

a. What is the time constant for this oscillator?
b. At what time will the *energy* have decayed to half its initial value?

MODEL The motion is a damped oscillation.

SOLVE a. The initial amplitude at $t = 0$ is $x_{max} = A$. At $t = 35.0$ s the amplitude is $x_{max} = \frac{1}{2}A$. The amplitude of oscillation at time t is given by Equation 14.57:

$$x_{max}(t) = Ae^{-bt/2m} = Ae^{-t/2\tau}$$

In this case,

$$\frac{1}{2}A = Ae^{-(35.0\text{ s})/2\tau}$$

Notice that we do not need to know A itself because it cancels out. To solve for τ, we take the natural logarithm of both sides of the equation:

$$\ln\left(\frac{1}{2}\right) = -\ln 2 = \ln e^{-(35.0\text{ s})/2\tau} = -\frac{35.0\text{ s}}{2\tau}$$

This is easily rearranged to give

$$\tau = \frac{35.0\text{ s}}{2\ln 2} = 25.2\text{ s}$$

If desired, we could now determine the damping constant to be $b = m/\tau = 0.020$ kg/s.

b. The energy at time t is given by

$$E(t) = E_0e^{-t/\tau}$$

The time at which an exponential decay is reduced to $\frac{1}{2}E_0$, half its initial value, has a special name. It is called the **half-life** and given the symbol $t_{1/2}$. The concept of the half-life is widely used in applications such as radioactive decay. To relate $t_{1/2}$ to τ, we first write

$$E(\text{at } t = t_{1/2}) = \frac{1}{2}E_0 = E_0e^{-t_{1/2}/\tau}$$

The E_0 cancels, giving

$$\frac{1}{2} = e^{-t_{1/2}/\tau}$$

Again, we take the natural logarithm of both sides:

$$\ln\left(\frac{1}{2}\right) = -\ln 2 = \ln e^{-t_{1/2}/\tau} = -t_{1/2}/\tau$$

Finally, we solve for $t_{1/2}$:

$$t_{1/2} = \tau \ln 2 = 0.693\tau$$

This result that $t_{1/2}$ is 69% of τ is valid for any exponential decay. In this particular problem, half the energy is gone at

$$t_{1/2} = (0.693)(25.2\text{ s}) = 17.5\text{ s}$$

ASSESS The oscillator loses energy faster than it loses amplitude. This is what we should expect because the energy depends on the *square* of the amplitude.

STOP TO THINK 14.6 Rank in order, from largest to smallest, the time constants τ_a to τ_d of the decays shown in the figure. All the graphs have the same scale.

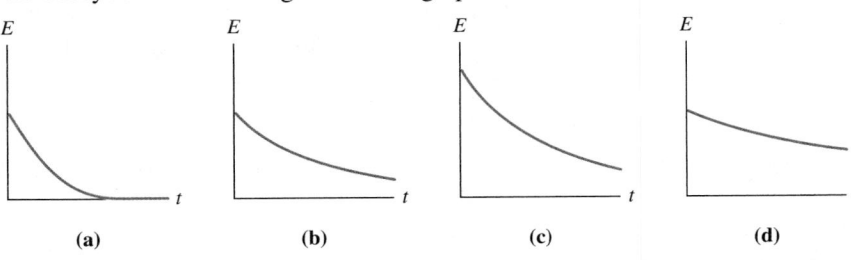

(a)   (b)   (c)   (d)

14.8 Driven Oscillations and Resonance

Thus far we have focused on the free oscillations of an isolated system. Some initial disturbance displaces the system from equilibrium, and it then oscillates freely until its energy is dissipated. These are very important situations, but they do not exhaust the possibilities. Another important situation is an oscillator that is subjected to a periodic external force. Its motion is called a **driven oscillation.**

A simple example of a driven oscillation is pushing a child on a swing, where your push is a periodic external force applied to the swing. A more complex example is a car driving over a series of equally spaced bumps. Each bump causes a periodic upward force on the car's shock absorbers, which are big, heavily damped springs. The electromagnetic coil on the back of a loudspeaker cone provides a periodic magnetic force to drive the cone back and forth, causing it to send out sound waves. Air turbulence moving across the wings of an aircraft can exert periodic forces on the wings and other aerodynamic surfaces, causing them to vibrate if they are not properly designed.

As these examples suggest, driven oscillations have many important applications. However, driven oscillations are a mathematically complex subject. We will simply hint at some of the results, saving the details for more advanced classes.

Consider an oscillating system that, when left to itself, oscillates at a frequency f_0. We will call this the **natural frequency** of the oscillator. The natural frequency for a mass on a spring is $\sqrt{k/m}/2\pi$, but it might be given by some other expression for another type of oscillator. Regardless of the expression, f_0 is simply the frequency of the system if it is displaced from equilibrium and released.

Suppose that this system is subjected to a *periodic* external force of frequency f_{ext}. This frequency, which is called the **driving frequency,** is completely independent of the oscillator's natural frequency f_0. Somebody or something in the environment selects the frequency f_{ext} of the external force, causing the force to push on the system f_{ext} times every second.

Although it is possible to solve Newton's second law with an external driving force, we will be content to look at a graphical representation of the solution. The most important result is that the oscillation amplitude depends very sensitively on the frequency f_{ext} of the driving force. The response to the driving frequency is shown in FIGURE 14.27 for a system with $m = 1.0$ kg, a natural frequency $f_0 = 2.0$ Hz, and a damping constant $b = 0.20$ kg/s. This graph of amplitude versus driving frequency, called the **response curve,** occurs in many different applications.

When the driving frequency is substantially different from the oscillator's natural frequency, at the right and left edges of Figure 14.27, the system oscillates but the amplitude is very small. The system simply does not respond well to a driving frequency that differs much from f_0. As the driving frequency gets closer and closer to the natural frequency, the amplitude of the oscillation rises dramatically. After all, f_0 is the frequency at which the system "wants" to oscillate, so it is quite happy to respond to a driving frequency near f_0. Hence the amplitude reaches a maximum when the driving frequency exactly matches the system's natural frequency: $f_{ext} = f_0$.

The amplitude can become exceedingly large when the frequencies match, especially if the damping constant is very small. FIGURE 14.28 shows the same oscillator with three different values of the damping constant. There's very little response if the damping constant is increased to 0.80 kg/s, but the amplitude for $f_{ext} = f_0$ becomes very large when the damping constant is reduced to 0.08 kg/s. This large-amplitude response to a driving force whose frequency matches the natural frequency of the system is a phenomenon called **resonance.** The condition for resonance is

$$f_{ext} = f_0 \quad \text{(resonance condition)} \tag{14.60}$$

Within the context of driven oscillations, the natural frequency f_0 is often called the **resonance frequency.**

An important feature of Figure 14.28 is how the amplitude and width of the resonance depend on the damping constant. A heavily damped system responds fairly

FIGURE 14.27 The response curve shows the amplitude of a driven oscillator at frequencies near its natural frequency of 2.0 Hz.

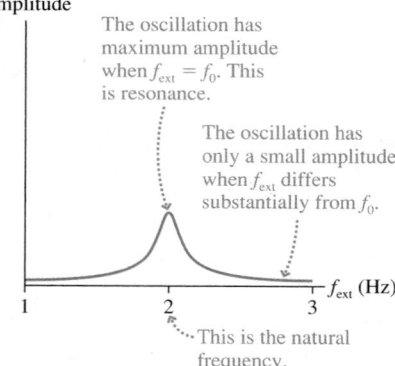

FIGURE 14.28 The resonance amplitude becomes higher and narrower as the damping constant decreases.

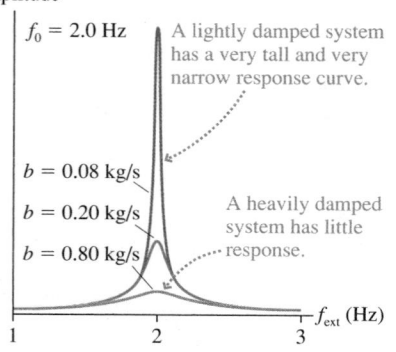

little, even at resonance, but it responds to a wide range of driving frequencies. Very lightly damped systems can reach exceptionally high amplitudes, but notice that the range of frequencies to which the system responds becomes narrower and narrower as b decreases.

This allows us to understand why a few singers can break crystal goblets but not inexpensive, everyday glasses. An inexpensive glass gives a "thud" when tapped, but a fine crystal goblet "rings" for several seconds. In physics terms, the goblet has a much longer time constant than the glass. That, in turn, implies that the goblet is very lightly damped while the ordinary glass is heavily damped (because the internal forces within the glass are not those of a high-quality crystal structure).

The singer causes a sound wave to impinge on the goblet, exerting a small driving force at the frequency of the note she is singing. If the singer's frequency matches the natural frequency of the goblet—resonance! Only the lightly damped goblet, like the top curve in Figure 14.28, can reach amplitudes large enough to shatter. The restriction, though, is that its natural frequency has to be matched very precisely. The sound also has to be very loud.

A singer or musical instrument can shatter a crystal goblet by matching the goblet's natural oscillation frequency.

CHALLENGE EXAMPLE 14.12 | A swinging pendulum

A pendulum consists of a massless, rigid rod with a mass at one end. The other end is pivoted on a frictionless pivot so that the rod can rotate in a complete circle. The pendulum is inverted, with the mass directly above the pivot point, then released. The speed of the mass as it passes through the lowest point is 5.0 m/s. If the pendulum later undergoes small-amplitude oscillations at the bottom of the arc, what will its frequency be?

MODEL This is a simple pendulum because the rod is massless. However, our analysis of a pendulum used the small-angle approximation. It applies only to the small-amplitude oscillations at the end, *not* to the pendulum swinging down from the inverted position. Fortunately, energy is conserved throughout, so we can analyze the big swing using conservation of mechanical energy.

VISUALIZE **FIGURE 14.29** is a pictorial representation of the pendulum swinging down from the inverted position. The pendulum length is L, so the initial height is $2L$.

FIGURE 14.29 Before-and-after pictorial representation of the pendulum swinging down from an inverted position.

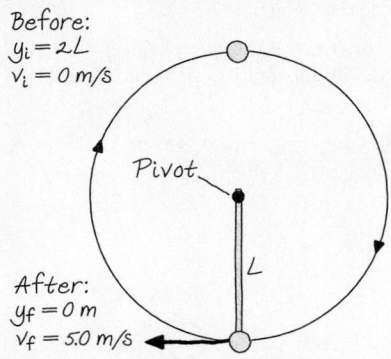

SOLVE The frequency of a simple pendulum is $f = \sqrt{g/L}/2\pi$. We're not given L, but we can find it by analyzing the pendulum's swing down from an inverted position. Mechanical energy is conserved, and the only potential energy is gravitational potential energy. Conservation of mechanical energy $K_f + U_{gf} = K_i + U_{gi}$, with $U_g = mgy$, is

$$\frac{1}{2}mv_f^2 + mgy_f = \frac{1}{2}mv_i^2 + mgy_i$$

The mass cancels, which is good since we don't know it, and two terms are zero. Thus

$$\frac{1}{2}v_f^2 = g(2L) = 2gL$$

Solving for L, we find

$$L = \frac{v_f^2}{4g} = \frac{(5.0 \text{ m/s})^2}{4(9.80 \text{ m/s}^2)} = 0.638 \text{ m}$$

Now we can calculate the frequency:

$$f = \frac{1}{2\pi}\sqrt{\frac{g}{L}} = \frac{1}{2\pi}\sqrt{\frac{9.80 \text{ m/s}^2}{0.638 \text{ m}}} = 0.62 \text{ Hz}$$

ASSESS The frequency corresponds to a period of about 1.5 s, which seems reasonable.

SUMMARY

The goal of Chapter 14 has been to understand systems that oscillate with simple harmonic motion.

General Principles

Dynamics

SHM occurs when a linear restoring force acts to return a system to an equilibrium position.

Horizontal spring

$(F_{\text{net}})_x = -kx$

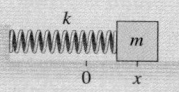

Vertical spring
The origin is at the equilibrium position $\Delta L = mg/k$.

$(F_{\text{net}})_y = -ky$

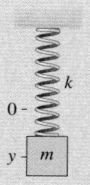

Both: $\omega = \sqrt{\dfrac{k}{m}} \qquad T = 2\pi\sqrt{\dfrac{m}{k}}$

Pendulum

$(F_{\text{net}})_t = -\left(\dfrac{mg}{L}\right)s$

$\omega = \sqrt{\dfrac{g}{L}} \qquad T = 2\pi\sqrt{\dfrac{L}{g}}$

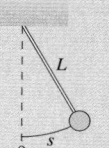

Energy

If there is **no friction** or dissipation, kinetic and potential energy are alternately transformed into each other, but the total mechanical energy $E = K + U$ is conserved.

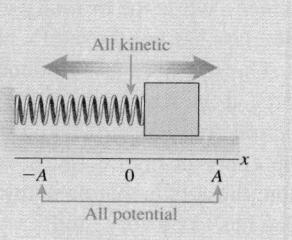

$$E = \frac{1}{2}mv^2 + \frac{1}{2}kx^2$$

$$= \frac{1}{2}m(v_{\text{max}})^2$$

$$= \frac{1}{2}kA^2$$

In a **damped system,** the energy decays exponentially

$$E = E_0 e^{-t/\tau}$$

where τ is the time constant.

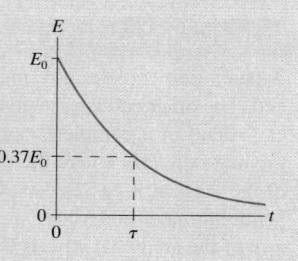

Important Concepts

Simple harmonic motion (SHM) is a sinusoidal oscillation with period T and amplitude A.

Frequency $f = \dfrac{1}{T}$

Angular frequency

$$\omega = 2\pi f = \frac{2\pi}{T}$$

Position $x(t) = A\cos(\omega t + \phi_0)$

$$= A\cos\left(\frac{2\pi t}{T} + \phi_0\right)$$

Velocity $v_x(t) = -v_{\text{max}}\sin(\omega t + \phi_0)$ with maximum speed $v_{\text{max}} = \omega A$

Acceleration $a_x(t) = -\omega^2 x(t) = -\omega^2 A\cos(\omega t + \phi_0)$

SHM is the projection onto the x-axis of **uniform circular motion.**

$\phi = \omega t + \phi_0$ is the phase

The position at time t is

$$x(t) = A\cos\phi$$
$$= A\cos(\omega t + \phi_0)$$

The phase constant ϕ_0 determines the initial conditions:

$$x_0 = A\cos\phi_0 \qquad v_{0x} = -\omega A\sin\phi_0$$

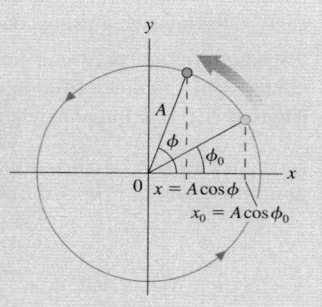

Applications

Resonance

When a system is driven by a periodic external force, it responds with a large-amplitude oscillation if $f_{\text{ext}} \approx f_0$, where f_0 is the system's natural oscillation frequency, or **resonant frequency.**

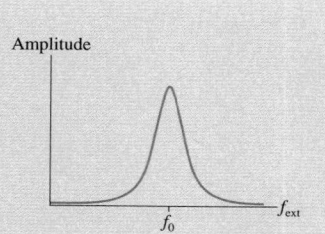

Damping

If there is a drag force $\vec{D} = -b\vec{v}$, where b is the damping constant, then (for lightly damped systems)

$$x(t) = Ae^{-bt/2m}\cos(\omega t + \phi_0)$$

The time constant for energy loss is $\tau = m/b$.

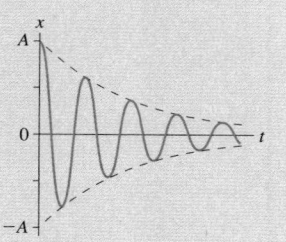

Terms and Notation

oscillatory motion	amplitude, A	linear restoring force	natural frequency, f_0
oscillator	angular frequency, ω	damped oscillation	driving frequency, f_{ext}
period, T	phase, ϕ	damping constant, b	response curve
frequency, f	phase constant, ϕ_0	envelope	resonance
hertz, Hz	restoring force	time constant, τ	resonance frequency, f_0
simple harmonic motion,	equation of motion	half-life, $t_{1/2}$	
SHM	small-angle approximation	driven oscillation	

CONCEPTUAL QUESTIONS

1. A block oscillating on a spring has period $T = 2$ s. What is the period if:
   a. The block's mass is doubled? Explain. Note that you do not know the value of either m or k, so do *not* assume any particular values for them. The required analysis involves thinking about ratios.
   b. The value of the spring constant is quadrupled?
   c. The oscillation amplitude is doubled while m and k are unchanged?

2. A pendulum on Planet X, where the value of g is unknown, oscillates with a period $T = 2$ s. What is the period of this pendulum if:
   a. Its mass is doubled? Explain. Note that you do not know the value of m, L, or g, so do not assume any specific values. The required analysis involves thinking about ratios.
   b. Its length is doubled?
   c. Its oscillation amplitude is doubled?

3. FIGURE Q14.3 shows a position-versus-time graph for a particle in SHM. What are (a) the amplitude A, (b) the angular frequency ω, and (c) the phase constant ϕ_0? Explain.

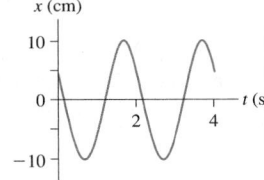

FIGURE Q14.3

4. Equation 14.25 states that $\frac{1}{2}kA^2 = \frac{1}{2}m(v_{max})^2$. What does this mean? Write a couple of sentences explaining how to interpret this equation.

5. A block oscillating on a spring has an amplitude of 20 cm. What will the amplitude be if the total energy is doubled? Explain.

6. A block oscillating on a spring has a maximum speed of 20 cm/s. What will the block's maximum speed be if the total energy is doubled? Explain.

7. FIGURE Q14.7 shows a position-versus-time graph for a particle in SHM.
   a. What is the phase constant ϕ_0? Explain.
   b. What is the phase of the particle at each of the three numbered points on the graph?

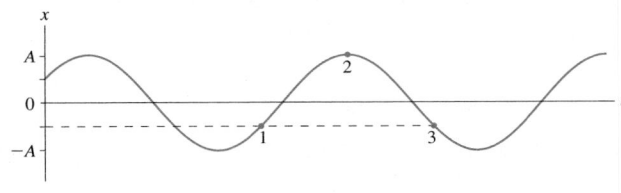

FIGURE Q14.7

8. FIGURE Q14.8 shows a velocity-versus-time graph for a particle in SHM.
   a. What is the phase constant ϕ_0? Explain.
   b. What is the phase of the particle at each of the three numbered points on the graph?

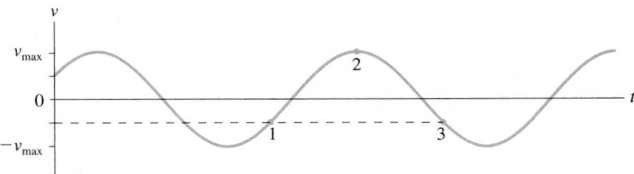

FIGURE Q14.8

9. FIGURE Q14.9 shows the potential-energy diagram and the total energy line of a particle oscillating on a spring.
   a. What is the spring's equilibrium length?
   b. Where are the turning points of the motion? Explain.
   c. What is the particle's maximum kinetic energy?
   d. What will be the turning points if the particle's total energy is doubled?

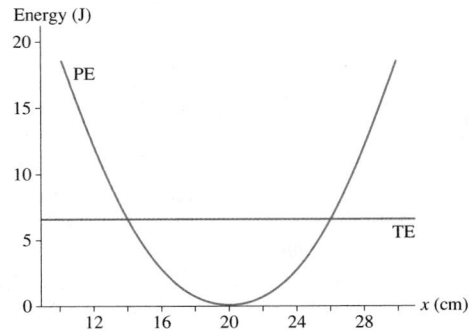

FIGURE Q14.9

10. Suppose the damping constant b of an oscillator increases.
    a. Is the medium more resistive or less resistive?
    b. Do the oscillations damp out more quickly or less quickly?
    c. Is the time constant τ increased or decreased?

11. a. Describe the difference between τ and T. Don't just *name* them; say what is different about the physical concepts they represent.
    b. Describe the difference between τ and $t_{1/2}$.

12. What is the difference between the driving frequency and the natural frequency of an oscillator?

EXERCISES AND PROBLEMS

Problems labeled ▬ integrate material from earlier chapters.

Exercises

Section 14.1 Simple Harmonic Motion

1. | When a guitar string plays the note "A," the string vibrates at 440 Hz. What is the period of the vibration?

2. | An air-track glider attached to a spring oscillates between the 10 cm mark and the 60 cm mark on the track. The glider completes 10 oscillations in 33 s. What are the (a) period, (b) frequency, (c) angular frequency, (d) amplitude, and (e) maximum speed of the glider?

3. ‖ An air-track glider is attached to a spring. The glider is pulled to the right and released from rest at $t = 0$ s. It then oscillates with a period of 2.0 s and a maximum speed of 40 cm/s.
   a. What is the amplitude of the oscillation?
   b. What is the glider's position at $t = 0.25$ s?

Section 14.2 Simple Harmonic Motion and Circular Motion

4. | What are the (a) amplitude, (b) frequency, and (c) phase constant of the oscillation shown in FIGURE EX14.4?

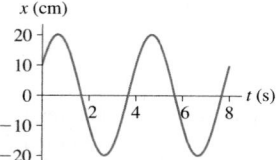

FIGURE EX14.4

5. ‖ What are the (a) amplitude, (b) frequency, and (c) phase constant of the oscillation shown in FIGURE EX14.5?

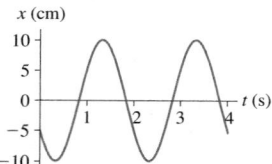

FIGURE EX14.5

6. ‖ An object in simple harmonic motion has an amplitude of 4.0 cm, a frequency of 2.0 Hz, and a phase constant of $2\pi/3$ rad. Draw a position graph showing two cycles of the motion.

7. ‖ An object in simple harmonic motion has an amplitude of 8.0 cm, a frequency of 0.25 Hz, and a phase constant of $-\pi/2$ rad. Draw a position graph showing two cycles of the motion.

8. | An object in simple harmonic motion has amplitude 4.0 cm and frequency 4.0 Hz, and at $t = 0$ s it passes through the equilibrium point moving to the right. Write the function $x(t)$ that describes the object's position.

9. | An object in simple harmonic motion has amplitude 8.0 cm and frequency 0.50 Hz. At $t = 0$ s it has its most negative position. Write the function $x(t)$ that describes the object's position.

10. ‖ An air-track glider attached to a spring oscillates with a period of 1.5 s. At $t = 0$ s the glider is 5.00 cm left of the equilibrium position and moving to the right at 36.3 cm/s.
    a. What is the phase constant?
    b. What is the phase at $t = 0$ s, 0.5 s, 1.0 s, and 1.5 s?

Section 14.3 Energy in Simple Harmonic Motion

Section 14.4 The Dynamics of Simple Harmonic Motion

11. | A block attached to a spring with unknown spring constant oscillates with a period of 2.0 s. What is the period if
    a. The mass is doubled?
    b. The mass is halved?
    c. The amplitude is doubled?
    d. The spring constant is doubled?
    Parts a to d are independent questions, each referring to the initial situation.

12. ‖ A 200 g air-track glider is attached to a spring. The glider is pushed in 10 cm and released. A student with a stopwatch finds that 10 oscillations take 12.0 s. What is the spring constant?

13. ‖ A 200 g mass attached to a horizontal spring oscillates at a frequency of 2.0 Hz. At $t = 0$ s, the mass is at $x = 5.0$ cm and has $v_x = -30$ cm/s. Determine:
    a. The period.   b. The angular frequency.
    c. The amplitude.   d. The phase constant.
    e. The maximum speed.   f. The maximum acceleration.
    g. The total energy.   h. The position at $t = 0.40$ s.

14. | The position of a 50 g oscillating mass is given by $x(t) = (2.0\text{ cm})\cos(10t - \pi/4)$, where t is in s. Determine:
    a. The amplitude.   b. The period.
    c. The spring constant.   d. The phase constant.
    e. The initial conditions. f. The maximum speed.
    g. The total energy.   h. The velocity at $t = 0.40$ s.

15. ‖ A 1.0 kg block is attached to a spring with spring constant 16 N/m. While the block is sitting at rest, a student hits it with a hammer and almost instantaneously gives it a speed of 40 cm/s. What are
    a. The amplitude of the subsequent oscillations?
    b. The block's speed at the point where $x = \frac{1}{2}A$?

Section 14.5 Vertical Oscillations

16. | A spring is hanging from the ceiling. Attaching a 500 g physics book to the spring causes it to stretch 20 cm in order to come to equilibrium.
    a. What is the spring constant?
    b. From equilibrium, the book is pulled down 10 cm and released. What is the period of oscillation?
    c. What is the book's maximum speed?

17. ‖ A spring with spring constant 15 N/m hangs from the ceiling. A ball is attached to the spring and allowed to come to rest. It is then pulled down 6.0 cm and released. If the ball makes 30 oscillations in 20 s, what are its (a) mass and (b) maximum speed?

18. ‖ A spring is hung from the ceiling. When a block is attached to its end, it stretches 2.0 cm before reaching its new equilibrium length. The block is then pulled down slightly and released. What is the frequency of oscillation?

Section 14.6 The Pendulum

19. | A mass on a string of unknown length oscillates as a pendulum with a period of 4.0 s. What is the period if
    a. The mass is doubled?

b. The string length is doubled?

c. The string length is halved?

d. The amplitude is doubled?

Parts a to d are independent questions, each referring to the initial situation.

20. ‖ A 200 g ball is tied to a string. It is pulled to an angle of 8.0° and released to swing as a pendulum. A student with a stopwatch finds that 10 oscillations take 12 s. How long is the string?

21. | What is the period of a 1.0-m-long pendulum on (a) the earth and (b) Venus?

22. | What is the length of a pendulum whose period on the moon matches the period of a 2.0-m-long pendulum on the earth?

23. | Astronauts on the first trip to Mars take along a pendulum that has a period on earth of 1.50 s. The period on Mars turns out to be 2.45 s. What is the free-fall acceleration on Mars?

24. ‖ A uniform steel bar swings from a pivot at one end with a period of 1.2 s. How long is the bar?

Section 14.7 Damped Oscillations

Section 14.8 Driven Oscillations and Resonance

25. | A 2.0 g spider is dangling at the end of a silk thread. You can make the spider bounce up and down on the thread by tapping lightly on his feet with a pencil. You soon discover that you can give the spider the largest amplitude on his little bungee cord if you tap exactly once every second. What is the spring constant of the silk thread?

26. ‖ The amplitude of an oscillator decreases to 36.8% of its initial value in 10.0 s. What is the value of the time constant?

27. ‖ Sketch a position graph from $t = 0$ s to $t = 10$ s of a damped oscillator having a frequency of 1.0 Hz and a time constant of 4.0 s.

28. | In a science museum, a 110 kg brass pendulum bob swings at the end of a 15.0-m-long wire. The pendulum is started at exactly 8:00 A.M. every morning by pulling it 1.5 m to the side and releasing it. Because of its compact shape and smooth surface, the pendulum's damping constant is only 0.010 kg/s. At exactly 12:00 noon, how many oscillations will the pendulum have completed and what is its amplitude?

29. ‖ Vision is blurred if the head is vibrated at 29 Hz because the
BIO vibrations are resonant with the natural frequency of the eyeball in its socket. If the mass of the eyeball is 7.5 g, a typical value, what is the effective spring constant of the musculature that holds the eyeball in the socket?

Problems

30. ‖ FIGURE P14.30 is the velocity-versus-time graph of a particle in simple harmonic motion.

a. What is the amplitude of the oscillation?

b. What is the phase constant?

c. What is the position at $t = 0$ s?

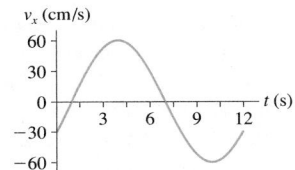

FIGURE P14.30

31. | FIGURE P14.31 is the position-versus-time graph of a particle in simple harmonic motion.

a. What is the phase constant?

b. What is the velocity at $t = 0$ s?

c. What is v_{max}?

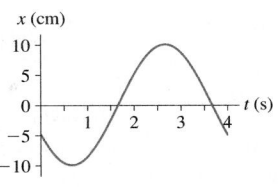   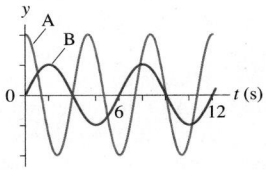

FIGURE P14.31   FIGURE P14.32

32. ‖ The two graphs in FIGURE P14.32 are for two different vertical mass-spring systems. If both systems have the same mass, what is the ratio k_A/k_B of their spring constants?

33. ‖ An object in SHM oscillates with a period of 4.0 s and an amplitude of 10 cm. How long does the object take to move from $x = 0.0$ cm to $x = 6.0$ cm?

34. ‖ A 1.0 kg block oscillates on a spring with spring constant 20 N/m. At $t = 0$ s the block is 20 cm to the right of the equilibrium position and moving to the left at a speed of 100 cm/s. Determine (a) the period and (b) the amplitude.

35. ‖ Astronauts in space cannot weigh themselves by standing on a
BIO bathroom scale. Instead, they determine their mass by oscillating on a large spring. Suppose an astronaut attaches one end of a large spring to her belt and the other end to a hook on the wall of the space capsule. A fellow astronaut then pulls her away from the wall and releases her. The spring's length as a function of time is shown in FIGURE P14.35.

a. What is her mass if the spring constant is 240 N/m?

b. What is her speed when the spring's length is 1.2 m?

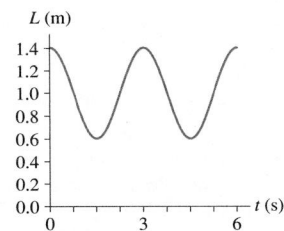

FIGURE P14.35

36. ‖ The motion of a particle is given by $x(t) = (25$ cm$)\cos(10t)$, where t is in s. At what time is the kinetic energy twice the potential energy?

37. ‖ a. When the displacement of a mass on a spring is $\frac{1}{2}A$, what fraction of the energy is kinetic energy and what fraction is potential energy?

b. At what displacement, as a fraction of A, is the energy half kinetic and half potential?

38. ‖ For a particle in simple harmonic motion, show that $v_{max} = (\pi/2)v_{avg}$ where v_{avg} is the average speed during one cycle of the motion.

39. ‖ A 100 g block attached to a spring with spring constant 2.5 N/m oscillates horizontally on a frictionless table. Its velocity is 20 cm/s when $x = -5.0$ cm.

a. What is the amplitude of oscillation?

b. What is the block's maximum acceleration?

c. What is the block's position when the acceleration is maximum?

d. What is the speed of the block when $x = 3.0$ cm?

40. ‖ A block on a spring is pulled to the right and released at $t = 0$ s. It passes $x = 3.00$ cm at $t = 0.685$ s, and it passes $x = -3.00$ cm at $t = 0.886$ s.
    a. What is the angular frequency?
    b. What is the amplitude?
    Hint: $\cos(\pi - \theta) = -\cos\theta$.

41. ‖‖ A 300 g oscillator has a speed of 95.4 cm/s when its displacement is 3.0 cm and 71.4 cm/s when its displacement is 6.0 cm. What is the oscillator's maximum speed?

42. ‖ An ultrasonic transducer, of the type used in medical ultra-
    BIO sound imaging, is a very thin disk ($m = 0.10$ g) driven back and forth in SHM at 1.0 MHz by an electromagnetic coil.
    a. The maximum restoring force that can be applied to the disk without breaking it is 40,000 N. What is the maximum oscillation amplitude that won't rupture the disk?
    b. What is the disk's maximum speed at this amplitude?

43. ‖ A 5.0 kg block hangs from a spring with spring constant 2000 N/m. The block is pulled down 5.0 cm from the equilibrium position and given an initial velocity of 1.0 m/s back toward equilibrium. What are the (a) frequency, (b) amplitude, and (c) total mechanical energy of the motion?

44. ‖ Your lab instructor has asked you to measure a spring constant using a dynamic method—letting it oscillate—rather than a static method of stretching it. You and your lab partner suspend the spring from a hook, hang different masses on the lower end, and start them oscillating. One of you uses a meter stick to measure the amplitude, the other uses a stopwatch to time 10 oscillations. Your data are as follows:

Mass (g)	Amplitude (cm)	Time (s)
100	6.5	7.8
150	5.5	9.8
200	6.0	10.9
250	3.5	12.4

Use the best-fit line of an appropriate graph to determine the spring constant.

45. ‖‖ A 200 g block hangs from a spring with spring constant 10 N/m. At $t = 0$ s the block is 20 cm below the equilibrium point and moving upward with a speed of 100 cm/s. What are the block's
    a. Oscillation frequency?
    b. Distance from equilibrium when the speed is 50 cm/s?
    c. Distance from equilibrium at $t = 1.0$ s?

46. ‖ A spring with spring constant k is suspended vertically from a support and a mass m is attached. The mass is held at the point where the spring is not stretched. Then the mass is released and begins to oscillate. The lowest point in the oscillation is 20 cm below the point where the mass was released. What is the oscillation frequency?

47. ‖ While grocery shopping, you put several apples in the spring scale in the produce department. The scale reads 20 N, and you use your ruler (which you always carry with you) to discover that the pan goes down 9.0 cm when the apples are added. If you tap the bottom of the apple-filled pan to make it bounce up and down a little, what is its oscillation frequency? Ignore the mass of the pan.

48. ‖ A compact car has a mass of 1200 kg. Assume that the car has one spring on each wheel, that the springs are identical, and that the mass is equally distributed over the four springs.

a. What is the spring constant of each spring if the empty car bounces up and down 2.0 times each second?
b. What will be the car's oscillation frequency while carrying four 70 kg passengers?

49. ‖ The two blocks in **FIGURE P14.49** oscillate on a frictionless surface with a period of 1.5 s. The upper block just begins to slip when the amplitude is increased to 40 cm. What is the coefficient of static friction between the two blocks?

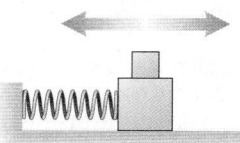

FIGURE P14.49

50. ‖‖ It has recently become possible to "weigh" DNA molecules
    BIO by measuring the influence of their mass on a nano-oscillator.
    FIGURE P14.50 shows a thin rectangular cantilever etched out of silicon (density 2300 kg/m³) with a small gold dot at the end. If pulled down and released, the end of the cantilever vibrates with simple harmonic motion, moving up and down like a diving board after a jump. When bathed with DNA molecules whose ends have been modified to bind with gold, one or more molecules may attach to the gold dot. The addition of their mass causes a very slight—but measurable—decrease in the oscillation frequency.

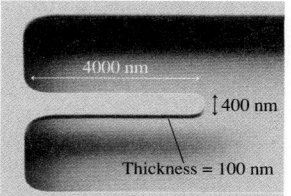

FIGURE P14.50

A vibrating cantilever of mass M can be modeled as a block of mass $\frac{1}{3}M$ attached to a spring. (The factor of $\frac{1}{3}$ arises from the moment of inertia of a bar pivoted at one end.) Neither the mass nor the spring constant can be determined very accurately— perhaps to only two significant figures—but the oscillation frequency can be measured with very high precision simply by counting the oscillations. In one experiment, the cantilever was initially vibrating at exactly 12 MHz. Attachment of a DNA molecule caused the frequency to decrease by 50 Hz. What was the mass of the DNA?

51. ‖ It is said that Galileo discovered a basic principle of the pendulum—that the period is independent of the amplitude—by using his pulse to time the period of swinging lamps in the cathedral as they swayed in the breeze. Suppose that one oscillation of a swinging lamp takes 5.5 s.
    a. How long is the lamp chain?
    b. What maximum speed does the lamp have if its maximum angle from vertical is 3.0°?

52. ‖ A 100 g mass on a 1.0-m-long string is pulled 8.0° to one side and released. How long does it take for the pendulum to reach 4.0° on the opposite side?

53. ‖ Orangutans can move by brachiation, swinging like a pendu-
    BIO lum beneath successive handholds. If an orangutan has arms that are 0.90 m long and repeatedly swings to a 20° angle, taking one swing after another, estimate its speed of forward motion in m/s. While this is somewhat beyond the range of validity of the small-angle approximation, the standard results for a pendulum are adequate for making an estimate.

54. | Show that Equation 14.51 for the angular frequency of a physical pendulum gives Equation 14.48 when applied to a simple pendulum of a mass on a string.

55. ||| A 15-cm-long, 200 g rod is pivoted at one end. A 20 g ball of clay is stuck on the other end. What is the period if the rod and clay swing as a pendulum?

56. ||| A uniform rod of mass M and length L swings as a pendulum on a pivot at distance $L/4$ from one end of the rod. Find an expression for the frequency f of small-angle oscillations.

57. ||| A solid sphere of mass M and radius R is suspended from a thin rod, as shown in FIGURE P14.57. The sphere can swing back and forth at the bottom of the rod. Find an expression for the frequency f of small-angle oscillations.

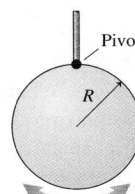

FIGURE P14.57

58. || A geologist needs to determine the local value of g. Unfortunately, his only tools are a meter stick, a saw, and a stopwatch. He starts by hanging the meter stick from one end and measuring its frequency as it swings. He then saws off 20 cm—using the centimeter markings—and measures the frequency again. After two more cuts, these are his data:

Length (cm)	Frequency (Hz)
100	0.61
80	0.67
60	0.79
40	0.96

Use the best-fit line of an appropriate graph to determine the local value of g.

59. || Interestingly, there have been several studies using cadavers to determine the moments of inertia of human body parts, information that is important in biomechanics. In one study, the center of mass of a 5.0 kg lower leg was found to be 18 cm from the knee. When the leg was allowed to pivot at the knee and swing freely as a pendulum, the oscillation frequency was 1.6 Hz. What was the moment of inertia of the lower leg about the knee joint?

60. || A 500 g air-track glider attached to a spring with spring constant 10 N/m is sitting at rest on a frictionless air track. A 250 g glider is pushed toward it from the far end of the track at a speed of 120 cm/s. It collides with and sticks to the 500 g glider. What are the amplitude and period of the subsequent oscillations?

61. || A 200 g block attached to a horizontal spring is oscillating with an amplitude of 2.0 cm and a frequency of 2.0 Hz. Just as it passes through the equilibrium point, moving to the right, a sharp blow directed to the left exerts a 20 N force for 1.0 ms. What are the new (a) frequency and (b) amplitude?

62. || FIGURE P14.62 is a top view of an object of mass m connected between two stretched rubber bands of length L. The object rests on a frictionless surface. At equilibrium, the tension in each rubber band is T. Find an expression for the frequency of oscillations *perpendicular* to the rubber bands. Assume the amplitude is sufficiently small that the magnitude of the tension in the rubber bands is essentially unchanged as the mass oscillates.

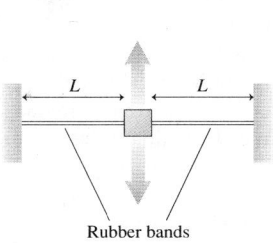

FIGURE P14.62

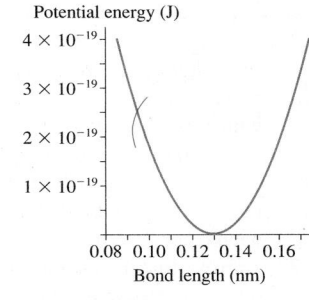

FIGURE P14.63

63. || A molecular bond can be modeled as a spring between two atoms that vibrate with simple harmonic motion. FIGURE P14.63 shows an SHM approximation for the potential energy of an HCl molecule. For $E < 4 \times 10^{-19}$ J it is a good approximation to the more accurate HCl potential-energy curve that was shown in Figure 10.31. Because the chlorine atom is so much more massive than the hydrogen atom, it is reasonable to assume that the hydrogen atom ($m = 1.67 \times 10^{-27}$ kg) vibrates back and forth while the chlorine atom remains at rest. Use the graph to estimate the vibrational frequency of the HCl molecule.

64. || An ice cube can slide around the inside of a vertical circular hoop of radius R. It undergoes small-amplitude oscillations if displaced slightly from the equilibrium position at the lowest point. Find an expression for the period of these small-amplitude oscillations.

65. || A penny rides on top of a piston as it undergoes vertical simple harmonic motion with an amplitude of 4.0 cm. If the frequency is low, the penny rides up and down without difficulty. If the frequency is steadily increased, there comes a point at which the penny leaves the surface.
   a. At what point in the cycle does the penny first lose contact with the piston?
   b. What is the maximum frequency for which the penny just barely remains in place for the full cycle?

66. || On your first trip to Planet X you happen to take along a 200 g mass, a 40-cm-long spring, a meter stick, and a stopwatch. You're curious about the free-fall acceleration on Planet X, where ordinary tasks seem easier than on earth, but you can't find this information in your Visitor's Guide. One night you suspend the spring from the ceiling in your room and hang the mass from it. You find that the mass stretches the spring by 31.2 cm. You then pull the mass down 10.0 cm and release it. With the stopwatch you find that 10 oscillations take 14.5 s. Based on this information, what is g?

67. || The 15 g head of a bobble-head doll oscillates in SHM at a frequency of 4.0 Hz.
   a. What is the spring constant of the spring on which the head is mounted?
   b. The amplitude of the head's oscillations decreases to 0.5 cm in 4.0 s. What is the head's damping constant?

68. || An oscillator with a mass of 500 g and a period of 0.50 s has an amplitude that decreases by 2.0% during each complete oscillation. If the initial amplitude is 10 cm, what will be the amplitude after 25 oscillations?

69. || A spring with spring constant 15.0 N/m hangs from the ceiling. A 500 g ball is attached to the spring and allowed to come to rest. It is then pulled down 6.0 cm and released. What is the time constant if the ball's amplitude has decreased to 3.0 cm after 30 oscillations?

70. ⫾ A 250 g air-track glider is attached to a spring with spring constant 4.0 N/m. The damping constant due to air resistance is 0.015 kg/s. The glider is pulled out 20 cm from equilibrium and released. How many oscillations will it make during the time in which the amplitude decays to e^{-1} of its initial value?

71. ‖ A 200 g oscillator in a vacuum chamber has a frequency of 2.0 Hz. When air is admitted, the oscillation decreases to 60% of its initial amplitude in 50 s. How many oscillations will have been completed when the amplitude is 30% of its initial value?

72. ‖ Prove that the expression for $x(t)$ in Equation 14.55 is a solution to the equation of motion for a damped oscillator, Equation 14.54, if and only if the angular frequency ω is given by the expression in Equation 14.56.

73. ‖ A block on a frictionless table is connected as shown in FIGURE P14.73 to two springs having spring constants k_1 and k_2. Show that the block's oscillation frequency is given by

$$f = \sqrt{f_1^2 + f_2^2}$$

where f_1 and f_2 are the frequencies at which it would oscillate if attached to spring 1 or spring 2 alone.

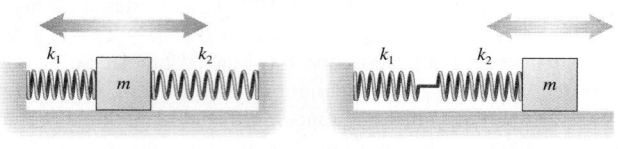

FIGURE P14.73   FIGURE P14.74

74. ‖ A block on a frictionless table is connected as shown in FIGURE P14.74 to two springs having spring constants k_1 and k_2. Find an expression for the block's oscillation frequency f in terms of the frequencies f_1 and f_2 at which it would oscillate if attached to spring 1 or spring 2 alone.

Challenge Problems

75. A block hangs in equilibrium from a vertical spring. When a second identical block is added, the original block sags by 5.0 cm. What is the oscillation frequency of the two-block system?

76. A 1.00 kg block is attached to a horizontal spring with spring constant 2500 N/m. The block is at rest on a frictionless surface. A 10 g bullet is fired into the block, in the face opposite the spring, and sticks. What was the bullet's speed if the subsequent oscillations have an amplitude of 10.0 cm?

77. A spring is standing upright on a table with its bottom end fastened to the table. A block is dropped from a height 3.0 cm above the top of the spring. The block sticks to the top end of the spring and then oscillates with an amplitude of 10 cm. What is the oscillation frequency?

78. The analysis of a simple pendulum assumed that the mass was a particle, with no size. A realistic pendulum is a small, uniform sphere of mass M and radius R at the end of a massless string, with L being the distance from the pivot to the center of the sphere.
    a. Find an expression for the period of this pendulum.
    b. Suppose $M = 25$ g, $R = 1.0$ cm, and $L = 1.0$ m, typical values for a real pendulum. What is the ratio T_{real}/T_{simple}, where T_{real} is your expression from part a and T_{simple} is the expression derived in this chapter?

79. a. A mass m oscillating on a spring has period T. Suppose the mass changes very slightly from m to $m + \Delta m$, where $\Delta m \ll m$. Find an expression for ΔT, the small change in the period. Your expression should involve T, m, and Δm but not the spring constant.
    b. Suppose the period is 2.000 s and the mass increases by 0.1%. What is the new period?

80. FIGURE CP14.80 shows a 200 g uniform rod pivoted at one end. The other end is attached to a horizontal spring. The spring is neither stretched nor compressed when the rod hangs straight down. What is the rod's oscillation period? You can assume that the rod's angle from vertical is always small.

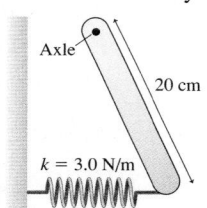

Axle

20 cm

$k = 3.0$ N/m

FIGURE CP14.80

Stop to Think 14.1: c. $v_{max} = 2\pi A/T$. Doubling A and T leaves v_{max} unchanged.

Stop to Think 14.2: d. Think of circular motion. At 45°, the particle is in the first quadrant (positive x) and moving to the left (negative v_x).

Stop to Think 14.3: c > b > a = d. Energy conservation $\frac{1}{2}kA^2 = \frac{1}{2}m(v_{max})^2$ gives $v_{max} = \sqrt{k/m}\,A$. k or m has to be increased or decreased by a factor of 4 to have the same effect as increasing or decreasing A by a factor of 2.

Stop to Think 14.4: c. $v_x = 0$ because the slope of the position graph is zero. The negative value of x shows that the particle is left of the equilibrium position, so the restoring force is to the right.

Stop to Think 14.5: c. The period of a pendulum does not depend on its mass.

Stop to Think 14.6: $\tau_d > \tau_b = \tau_c > \tau_a$. The time constant is the time to decay to 37% of the initial height. The time constant is independent of the initial height.

15 Fluids and Elasticity

This 20,000 pound boat floats while the 130 pound diver sinks. Why?

▶ **Looking Ahead** The goal of Chapter 15 is to understand macroscopic systems that flow or deform.

Pressure

Fluids exert forces on the walls of their container and on other parts of the fluid. **Pressure** is the force-to-area ratio *F/A*.

- Pressure in a liquid is due to gravity and increases with depth.
- Pressure in a gas is primarily thermal. Pressure is constant in a laboratory-size container.

A vacuum—the reduction of pressure below atmospheric pressure—can have serious consequences.

Fluids

A **fluid** is a substance that flows. Both gases and liquids are fluids.

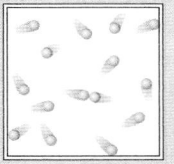 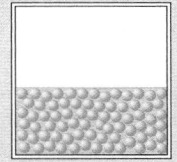

Gas: Freely moving particles, compressible

Liquid: Loosely bound particles, incompressible

You will learn to how to calculate pressure and flow in both gases and liquids.

Fluid Flow

You'll use *Bernoulli's equation*, a statement of energy conservation, to analyze the flow of an ideal fluid.

These *stream-lines* show the fluid flow around the car.

Bernoulli's equation has a wide range of applications, from the flow of liquids through pipes to the generation of lift as air flows over an airplane wing.

Measuring Pressure

Pressure is measured in many ways and in many different units. The SI unit of pressure is the *pascal*, but atmospheres (atm), mm of Hg, and pounds per square inch (psi) are all widely used.

A pressure gauge usually measures *gauge pressure*, which is the pressure in excess of atmospheric pressure.

Buoyancy

A ship floats because of *buoyancy*. You'll learn to use **Archimedes' principle** to calculate the buoyant force as the weight of the displaced fluid.

An object floats if the upward buoyant force is large enough to balance the downward gravitational force.

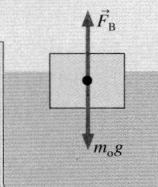

◀ **Looking Back**
Section 6.1 Equilibrium

Elasticity

Elasticity describes the deformation of solids and liquids under stress.

- Pulling a metal wire stretches it slightly.
- Structural concrete in buildings and bridges compresses slightly under the load.

The elastic properties of a material are characterized by a parameter called *Young's modulus*.

◀ **Looking Back**
Section 10.4 Restoring forces

FIGURE 15.1 Simple atomic models of gases and liquids.

(a) A gas

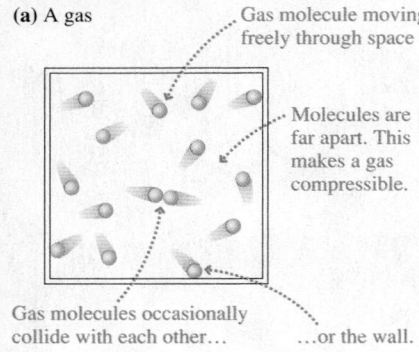

Gas molecule moving freely through space

Molecules are far apart. This makes a gas compressible.

Gas molecules occasionally collide with each other... ...or the wall.

(b) A liquid

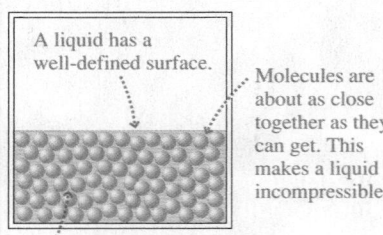

A liquid has a well-defined surface.

Molecules are about as close together as they can get. This makes a liquid incompressible.

Molecules have weak bonds between them, keeping them close together. But the molecules can slide around each other, allowing the liquid to flow and conform to the shape of its container.

FIGURE 15.2 There are 10^6 cm^3 in 1 m^3.

Subdivide the 1 m × 1 m × 1 m cube into little cubes 1 cm on a side. You will get 100 subdivisions along each edge.

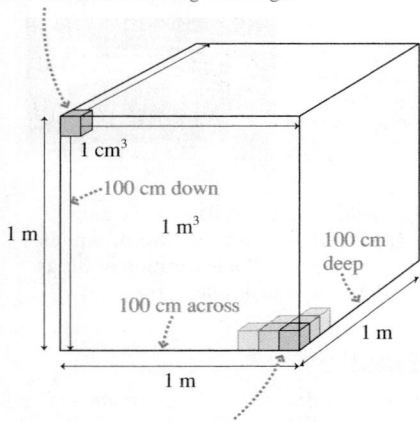

1 cm^3

100 cm down

1 m^3

1 m

100 cm deep

100 cm across

1 m

1 m

There are 100 × 100 × 100 = 10^6 little 1 cm^3 cubes in the big 1 m^3 cube.

15.1 Fluids

Quite simply, a **fluid** is a substance that flows. Because they flow, fluids take the shape of their container rather than retaining a shape of their own. You may think that gases and liquids are quite different, but both are fluids, and their similarities are often more important than their differences.

A **gas**, shown in **FIGURE 15.1a**, is a system in which each molecule moves through space as a free, noninteracting particle until, on occasion, it collides with another molecule or with the wall of the container. The gas you are most familiar with is air, a mixture of mostly nitrogen and oxygen molecules. Gases are fairly simple macroscopic systems, and Part IV of this textbook will delve into the thermal properties of gases. For now, two properties of gases interest us:

1. Gases are *fluids*. They flow, and they exert pressure on the walls of their container.
2. Gases are *compressible*. That is, the volume of a gas is easily increased or decreased, a consequence of the "empty space" between the molecules.

Liquids are more complicated than either gases or solids. Liquids, like solids, are nearly *incompressible*. This property tells us that the molecules in a liquid, as in a solid, are about as close together as they can get without coming into contact with each other. At the same time, a liquid flows and deforms to fit the shape of its container. The fluid nature of a liquid tells us that the molecules are free to move around. These observations suggest the model of a **liquid** shown in **FIGURE 15.1b**.

Volume and Density

One important parameter that characterizes a macroscopic system is its volume V, the amount of space the system occupies. The SI unit of volume is m^3. Nonetheless, both cm^3 and, to some extent, liters (L) are widely used metric units of volume. In most cases, you *must* convert these to m^3 before doing calculations.

While it is true that 1 m = 100 cm, it is *not* true that 1 m^3 = 100 cm^3. **FIGURE 15.2** shows that the volume conversion factor is 1 m^3 = 10^6 cm^3. A liter is 1000 cm^3, so 1 m^3 = 10^3 L. A milliliter (1 mL) is the same as 1 cm^3.

A system is also characterized by its *density*. Suppose you have several blocks of copper, each of different size. Each block has a different mass m and a different volume V. Nonetheless, all the blocks are copper, so there should be some quantity that has the *same* value for all the blocks, telling us, "This is copper, not some other material." The most important such parameter is the *ratio* of mass to volume, which we call the **mass density** ρ (lowercase Greek rho):

$$\rho = \frac{m}{V} \quad \text{(mass density)} \qquad (15.1)$$

Conversely, an object of density ρ has mass $m = \rho V$.

The SI units of mass density are kg/m^3. Nonetheless, units of g/cm^3 are widely used. You need to convert these to SI units before doing most calculations. You must convert both the grams to kilograms and the cubic centimeters to cubic meters. The net result is the conversion factor

$$1 \text{ g/cm}^3 = 1000 \text{ kg/m}^3$$

The mass density is usually called simply "the density" if there is no danger of confusion. However, we will meet other types of density as we go along, and sometimes it is important to be explicit about which density you are using. Table 15.1 on the next page provides a short list of mass densities of various fluids. Notice the enormous difference between the densities of gases and liquids. Gases have lower densities because the molecules in gases are farther apart than in liquids.

What does it *mean* to say that the density of gasoline is 680 kg/m³ or, equivalently, 0.68 g/cm³ ? Density is a mass-to-volume ratio. It is often described as the "mass per unit volume," but for this to make sense you have to know what is meant by "unit volume." Regardless of which system of length units you use, a **unit volume** is one of those units cubed. For example, if you measure lengths in meters, a unit volume is 1 m³. But 1 cm³ is a unit volume if you measure lengths in centimeters, and 1 mi³ is a unit volume if you measure lengths in miles.

Density is the mass of one unit of volume, whatever the units happen to be. To say that the density of gasoline is 680 kg/m³ is to say that the mass of 1 m³ of gasoline is 680 kg. The mass of 1 cm³ of gasoline is 0.68 g, so the density of gasoline in those units is 0.68 g/cm³.

The mass density is independent of the object's size. Mass and volume are parameters that characterize a *specific piece* of some substance—say copper—whereas the mass density characterizes the substance itself. All pieces of copper have the same mass density, which differs from the mass density of any other substance.

TABLE 15.1 Densities of fluids at standard temperature (0°C) and pressure (1 atm)

Substance	ρ (kg/m³)
Air	1.28
Ethyl alcohol	790
Gasoline	680
Glycerin	1260
Helium gas	0.18
Mercury	13,600
Oil (typical)	900
Seawater	1030
Water	1000

EXAMPLE 15.1 **Weighing the air**

What is the mass of air in a living room with dimensions 4.0 m × 6.0 m × 2.5 m?

MODEL Table 15.1 gives air density at a temperature of 0°C. The air density doesn't vary significantly over a small range of temperatures (we'll study this issue in the next chapter), so we'll use this value even though most people keep their living room warmer than 0°C.

SOLVE The room's volume is

$$V = (4.0 \text{ m}) \times (6.0 \text{ m}) \times (2.5 \text{ m}) = 60 \text{ m}^3$$

The mass of the air is

$$m = \rho V = (1.28 \text{ kg/m}^3)(60 \text{ m}^3) = 77 \text{ kg}$$

ASSESS This is perhaps more mass than you might have expected from a substance that hardly seems to be there. For comparison, a swimming pool this size would contain 60,000 kg of water.

STOP TO THINK 15.1 A piece of glass is broken into two pieces of different size. Rank in order, from largest to smallest, the mass densities of pieces a, b, and c.

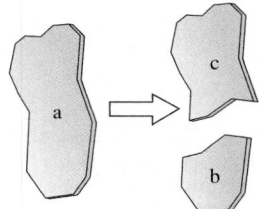

15.2 Pressure

"Pressure" is a word we all know and use. You probably have a commonsense idea of what pressure is. For example, you feel the effects of varying pressure against your eardrums when you swim underwater or take off in an airplane. Cans of whipped cream are "pressurized" to make the contents squirt out when you press the nozzle. It's hard to open a "vacuum sealed" jar of jelly the first time, but easy after the seal is broken.

You've probably seen water squirting out of a hole in the side of a container, as in FIGURE 15.3. Notice that the water emerges at greater speed from a hole at greater depth. And you've probably felt the air squirting out of a hole in a bicycle tire or inflatable air mattress. These observations suggest that

- "Something" pushes the water or air *sideways,* out of the hole.
- In a liquid, the "something" is larger at greater depths. In a gas, the "something" appears to be the same everywhere.

Our goal is to turn these everyday observations into a precise definition of pressure.

FIGURE 15.3 Water pressure pushes the water *sideways*, out of the holes.

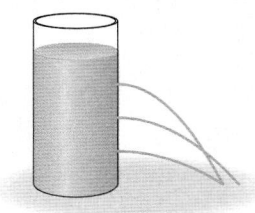

FIGURE 15.4 The fluid presses against area A with force $\vec{F}$.

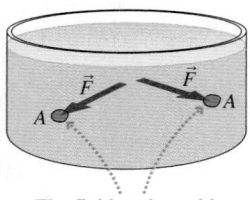

The fluid pushes with force $\vec{F}$ against area A.

FIGURE 15.4 shows a fluid—either a liquid or a gas—pressing against a small area A with force $\vec{F}$. This is the force that pushes the fluid out of a hole. In the absence of a hole, $\vec{F}$ pushes against the wall of the container. Let's define the **pressure** at this point in the fluid to be the ratio of the force to the area on which the force is exerted:

$$p = \frac{F}{A} \tag{15.2}$$

Notice that pressure is a scalar, not a vector. You can see, from Equation 15.2, that a fluid exerts a force of magnitude

$$F = pA \tag{15.3}$$

on a surface of area A. The force is *perpendicular* to the surface.

NOTE ▶ Pressure itself is *not* a force, even though we sometimes talk informally about "the force exerted by the pressure." The correct statement is that the *fluid* exerts a force on a surface. ◀

From its definition, pressure has units of N/m². The SI unit of pressure is the **pascal,** defined as

$$1 \text{ pascal} = 1 \text{ Pa} \equiv 1 \text{ N/m}^2$$

This unit is named for the 17th-century French scientist Blaise Pascal, who was one of the first to study fluids. Large pressures are often given in kilopascals, where 1 kPa = 1000 Pa.

Equation 15.2 is the basis for the simple pressure-measuring device shown in FIGURE 15.5a. Because the spring constant k and the area A are known, we can determine the pressure by measuring the compression of the spring. Once we've built such a device, we can place it in various liquids and gases to learn about pressure. FIGURE 15.5b shows what we can learn from a series of simple experiments.

FIGURE 15.5 Learning about pressure.

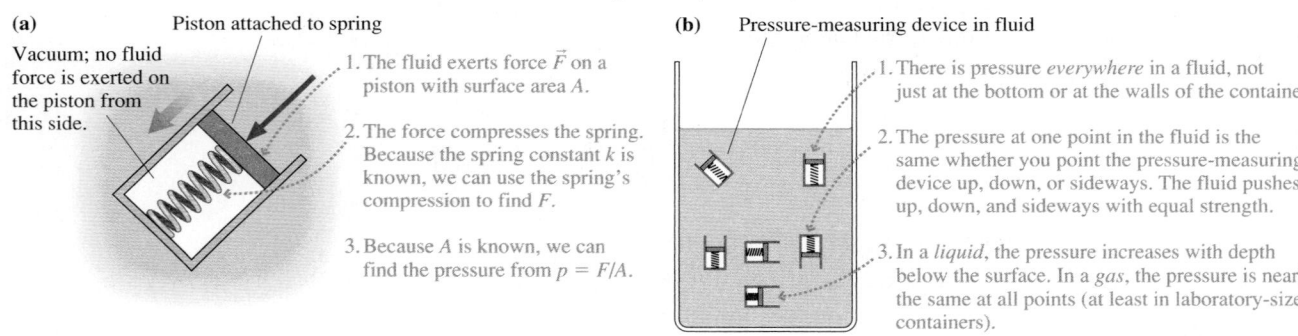

(a) Piston attached to spring

Vacuum; no fluid force is exerted on the piston from this side.

1. The fluid exerts force $\vec{F}$ on a piston with surface area A.

2. The force compresses the spring. Because the spring constant k is known, we can use the spring's compression to find F.

3. Because A is known, we can find the pressure from $p = F/A$.

(b) Pressure-measuring device in fluid

1. There is pressure *everywhere* in a fluid, not just at the bottom or at the walls of the container.

2. The pressure at one point in the fluid is the same whether you point the pressure-measuring device up, down, or sideways. The fluid pushes up, down, and sideways with equal strength.

3. In a *liquid*, the pressure increases with depth below the surface. In a *gas*, the pressure is nearly the same at all points (at least in laboratory-size containers).

The first statement in Figure 15.5b is especially important. Pressure exists at *all* points within a fluid, not just at the walls of the container. You may recall that tension exists at *all* points in a string, not only at its ends where it is tied to an object. We understood tension as the different parts of the string *pulling* against each other. Pressure is an analogous idea, except that the different parts of a fluid are *pushing* against each other.

Causes of Pressure

Gases and liquids are both fluids, but they have some important differences. Liquids are nearly incompressible; gases are highly compressible. The molecules in a liquid attract each other via molecular bonds; the molecules in a gas do not interact other than through occasional collisions. These differences affect how we think about pressure in gases and liquids.

Imagine that you have two sealed jars, each containing a small amount of mercury and nothing else. All the air has been removed from the jars. Suppose you take the two

FIGURE 15.6 A liquid and a gas in a weightless environment.

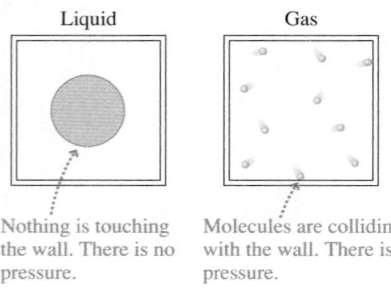

Liquid

Gas

Nothing is touching the wall. There is no pressure.

Molecules are colliding with the wall. There is pressure.

jars into orbit on the space shuttle, where they are weightless. One jar you keep cool, so that the mercury is a liquid. The other you heat until the mercury boils and becomes a gas. What can we say about the pressure in these two jars?

As FIGURE 15.6 shows, molecular bonds hold the liquid mercury together. It might quiver like Jello, but it remains a cohesive drop floating in the center of the jar. The liquid drop exerts no forces on the walls, so there's *no* pressure in the jar containing the liquid. (If we actually did this experiment, a very small fraction of the mercury would be in the vapor phase and create what is called *vapor pressure*.

The gas is different. Figure 15.1 introduced an atomic model of a gas in which a molecule moves freely until it collides with another molecule or with a wall of the container. FIGURE 15.7 shows some of the gas molecules colliding with a wall. Recall, from our study of collisions in Chapter 9, that each molecule as it bounces exerts a tiny impulse on the wall. The impulse from any one collision is extremely small, but there are an extraordinarily large number of collisions every second. These collisions cause the gas to have a pressure. We will do the calculation in Chapter 18.

FIGURE 15.8 shows the jars back on earth. Because of gravity, the liquid now fills the bottom of the jar and exerts a force on the bottom and the sides. Liquid mercury is incompressible, so the volume of liquid in Figure 15.8 is the same as in Figure 15.6. There is still no pressure on the top of the jar (other than the very small vapor pressure).

At first glance, the situation in the gas-filled jar seems unchanged from Figure 15.6. However, the earth's gravitational pull causes the gas density to be *slightly* more at the bottom of the jar than at the top. Because the pressure due to collisions is proportional to the density, the pressure is *slightly* larger at the bottom of the jar than at the top.

Thus there appear to be two contributions to the pressure in a container of fluid:

1. A *gravitational contribution* that arises from gravity pulling down on the fluid. Because a fluid can flow, forces are exerted on both the bottom and sides of the container. The gravitational contribution depends on the strength of the gravitational force.
2. A *thermal contribution* due to the collisions of freely moving gas molecules with the walls. The thermal contribution depends on the absolute temperature of the gas.

A detailed analysis finds that these two contributions are not entirely independent of each other, but the distinction is useful for a basic understanding of pressure. Let's see how these two contributions apply to different situations.

Pressure in Gases

The pressure in a laboratory-size container of gas is due almost entirely to the thermal contribution. A container would have to be ≈ 100 m tall for gravity to cause the pressure at the top to be even 1% less than the pressure at the bottom. Laboratory-size containers are much less than 100 m tall, so we can quite reasonably assume that p has the *same* value at all points in a laboratory-size container of gas.

Decreasing the number of molecules in a container decreases the gas pressure simply because there are fewer collisions with the walls. If a container is completely empty, with no atoms or molecules, then the pressure is $p = 0$ Pa. This is a *perfect vacuum*. No perfect vacuum exists in nature, not even in the most remote depths of outer space, because it is impossible to completely remove every atom from a region of space. In practice, a **vacuum** is an enclosed space in which $p \ll 1$ atm. Using $p = 0$ Pa is then a very good approximation.

Atmospheric Pressure

The earth's atmosphere is *not* a laboratory-size container. The height of the atmosphere is such that the gravitational contribution to pressure *is* important. As FIGURE 15.9 on the next page shows, the density of air slowly decreases with increasing height until approaching zero in the vacuum of space. Consequently, the pressure of the air, what

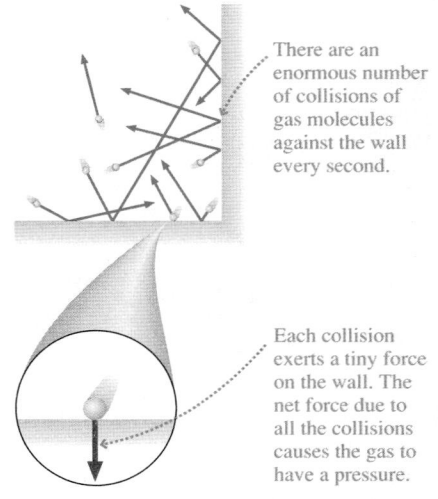

FIGURE 15.7 The pressure in a gas is due to the net force of the molecules colliding with the walls.

There are an enormous number of collisions of gas molecules against the wall every second.

Each collision exerts a tiny force on the wall. The net force due to all the collisions causes the gas to have a pressure.

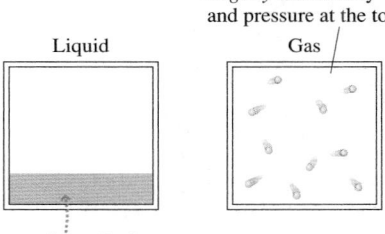

FIGURE 15.8 Gravity affects the pressure of the fluids.

Slightly less density and pressure at the top

Liquid   Gas

As gravity pulls down, the liquid exerts a force on the bottom and sides of its container.

Gravity has little effect on the pressure of the gas.

FIGURE 15.9 The pressure and density decrease with increasing height in the atmosphere.

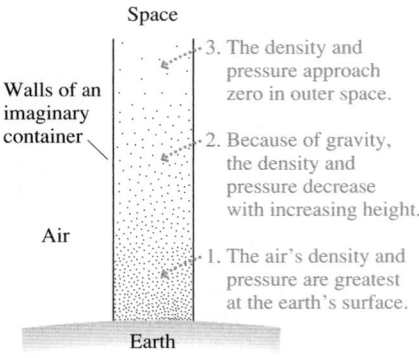

Space

3. The density and pressure approach zero in outer space.

Walls of an imaginary container

2. Because of gravity, the density and pressure decrease with increasing height.

Air

1. The air's density and pressure are greatest at the earth's surface.

Earth

we call the *atmospheric pressure* p_{atmos}, decreases with height. The air pressure is less in Denver than in Miami.

The atmospheric pressure *at sea level* varies slightly with the weather, but the global average sea-level pressure is 101,300 Pa. Consequently, we define the **standard atmosphere** as

$$1 \text{ standard atmosphere} = 1 \text{ atm} \equiv 101,300 \text{ Pa} = 101.3 \text{ kPa}$$

The standard atmosphere, usually referred to simply as "atmospheres," is a commonly used unit of pressure. But it is not an SI unit, so you must convert atmospheres to pascals before doing most calculations with pressure.

NOTE ▶ Unless you happen to live right at sea level, the atmospheric pressure around you is somewhat less than 1 atm. Pressure experiments use a barometer to determine the actual atmospheric pressure. For simplicity, this textbook will always assume that the pressure of the air is $p_{atmos} = 1$ atm unless stated otherwise. ◀

Given that the pressure of the air at sea level is 101.3 kPa, you might wonder why the weight of the air doesn't crush your forearm when you rest it on a table. Your forearm has a surface area of $\approx 200 \text{ cm}^2 = 0.02 \text{ m}^2$, so the force of the air pressing against it is ≈ 2000 N (≈ 450 pounds). How can you even lift your arm?

The reason, as FIGURE 15.10 shows, is that a fluid exerts pressure forces in *all* directions. There *is* a downward force of ≈ 2000 N on your forearm, but the air underneath your arm exerts an upward force of the same magnitude. The *net* force is very close to zero. (To be accurate, there is a net *upward* force called the buoyant force. We'll study buoyancy in Section 15.4. The buoyant force of the air is usually too small to notice.)

But, you say, there isn't any air under my arm if I rest it on a table. Actually, there is. There would be a *vacuum* under your arm if there were no air. Imagine placing your arm on the top of a large vacuum cleaner suction tube. What happens? You feel a downward force as the vacuum cleaner "tries to suck your arm in." However, the downward force you feel is not a *pulling* force from the vacuum cleaner. It is the *pushing* force of the air above your arm *when the air beneath your arm is removed and cannot push back.* Air molecules do not have hooks! They have no ability to "pull" on your arm. The air can only push.

Vacuum cleaners, suction cups, and other similar devices are powerful examples of how strong atmospheric pressure forces can be *if* the air is removed from one side of an object so as to produce an unbalanced force. The fact that we are *surrounded* by the fluid allows us to move around in the air, just as we swim underwater, oblivious of these strong forces.

FIGURE 15.10 Pressure forces in a fluid push with equal strength in all directions.

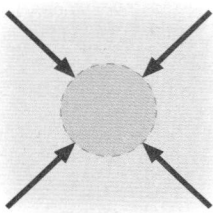

The forces of a fluid push in *all* directions.

EXAMPLE 15.2 **A suction cup**

A 10.0-cm-diameter suction cup is pushed against a smooth ceiling. What is the maximum mass of an object that can be suspended from the suction cup without pulling it off the ceiling? The mass of the suction cup is negligible.

MODEL Pushing the suction cup against the ceiling pushes the air out. We'll assume that the volume enclosed between the suction cup and the ceiling is a perfect vacuum with $p = 0$ Pa. We'll also assume that the pressure in the room is 1 atm.

VISUALIZE FIGURE 15.11 shows a free-body diagram of the suction cup stuck to the ceiling. The downward normal force of the ceiling is distributed around the rim of the suction cup, but in the particle model we can show this as a single force vector.

FIGURE 15.11 A suction cup is held to the ceiling by air pressure pushing upward on the bottom.

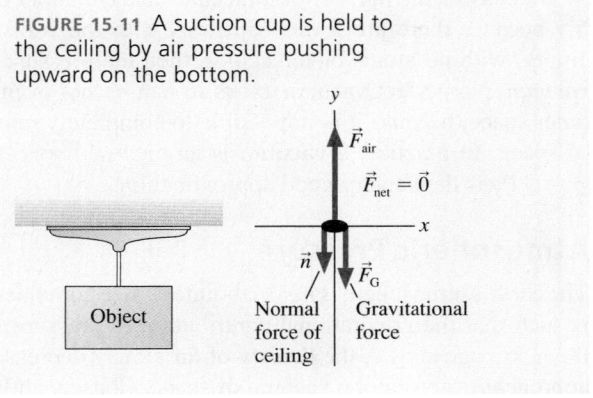

SOLVE The suction cup remains stuck to the ceiling, in static equi-librium, as long as $F_{air} = n + F_G$. The magnitude of the upward force exerted by the air is

$$F_{air} = pA = p\pi r^2 = (101{,}300 \text{ Pa})\pi(0.050 \text{ m})^2 = 796 \text{ N}$$

There is no downward force from the air in this case because there is no air inside the cup. Increasing the hanging mass decreases the normal force n by an equal amount. The maximum weight has been reached when n is reduced to zero. Thus

$$(F_G)_{max} = mg = F_{air} = 796 \text{ N}$$

$$m = \frac{796 \text{ N}}{g} = 81 \text{ kg}$$

ASSESS The suction cup can support a mass of up to 81 kg if all the air is pushed out, leaving a perfect vacuum inside. A real suc-tion cup won't achieve a perfect vacuum, but suction cups can hold substantial weight.

Pressure in Liquids

Gravity causes a liquid to fill the bottom of a container. Thus it's not surprising that the pressure in a liquid is due almost entirely to the gravitational contribution. We'd like to determine the pressure at depth d below the surface of the liquid. We will assume that the liquid is at rest; flowing liquids will be considered later in this chapter.

The shaded cylinder of liquid in FIGURE 15.12 extends from the surface to depth d. This cylinder, like the rest of the liquid, is in static equilibrium with $\vec{F}_{net} = \vec{0}$. Three forces act on this cylinder: the gravitational force mg on the liquid in the cylinder, a downward force p_0A due to the pressure p_0 at the surface of the liquid, and an upward force pA due to the liquid beneath the cylinder pushing up on the bottom of the cylinder. This third force is a consequence of our earlier observation that different parts of a fluid push against each other. Pressure p, which is what we're trying to find, is the pressure at the bottom of the cylinder.

The upward force balances the two downward forces, so

$$pA = p_0A + mg \qquad (15.4)$$

The liquid is a cylinder of cross-section area A and height d. Its volume is $V = Ad$ and its mass is $m = \rho V = \rho Ad$. Substituting this expression for the mass of the liquid into Equation 15.4, we find that the area A cancels from all terms. The pressure at depth d in a liquid is

$$p = p_0 + \rho gd \qquad \text{(hydrostatic pressure at depth } d) \qquad (15.5)$$

where ρ is the liquid's density. Because the fluid is at rest, the pressure given by Equation 15.5 is called the **hydrostatic pressure**. The fact that g appears in Equation 15.5 reminds us that this is a gravitational contribution to the pressure.

As expected, $p = p_0$ at the surface, where $d = 0$. Pressure p_0 is often due to the air or other gas above the liquid. $p_0 = 1$ atm $= 101.3$ kPa for a liquid that is open to the air. However, p_0 can also be the pressure due to a piston or a closed surface pushing down on the top of the liquid.

NOTE ▶ Equation 15.5 assumes that the liquid is *incompressible;* that is, its density ρ doesn't increase with depth. This is an excellent assumption for liquids, but not a good one for a gas, which *is* compressible. ◀

FIGURE 15.12 Measuring the pressure at depth d in a liquid.

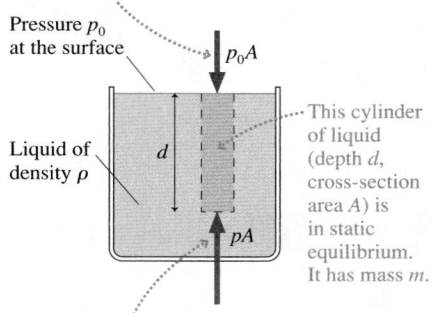

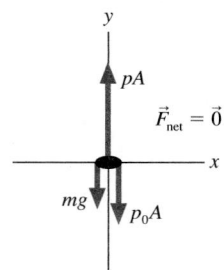

Free-body diagram of the column of liquid

EXAMPLE 15.3 | The pressure on a submarine

A submarine cruises at a depth of 300 m. What is the pressure at this depth? Give the answer in both pascals and atmospheres.

SOLVE The density of seawater, from Table 15.1, is $\rho = 1030$ kg/m³. The pressure at depth $d = 300$ m is found from Equation 15.5 to be

$$p = p_0 + \rho gd = 1.013 \times 10^5 \text{ Pa}$$
$$+ (1030 \text{ kg/m}^3)(9.80 \text{ m/s}^2)(300 \text{ m}) = 3.13 \times 10^6 \text{ Pa}$$

Converting the answer to atmospheres gives

$$p = 3.13 \times 10^6 \text{ Pa} \times \frac{1 \text{ atm}}{1.013 \times 10^5 \text{ Pa}} = 30.9 \text{ atm}$$

ASSESS The pressure deep in the ocean is very large. Windows on submersibles must be very thick to withstand the large forces.

FIGURE 15.13 Some properties of a liquid in hydrostatic equilibrium are not what you might expect.

(a)

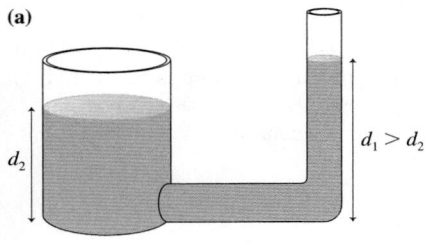

Is this possible?

(b)

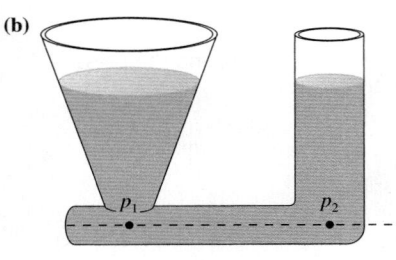

Is $p_1 > p_2$?

The hydrostatic pressure in a liquid depends only on the depth and the pressure at the surface. This observation has some important implications. FIGURE 15.13a shows two connected tubes. It's certainly true that the larger volume of liquid in the wide tube weighs more than the liquid in the narrow tube. You might think that this extra weight would push the liquid in the narrow tube higher than in the wide tube. But it doesn't. If d_1 were larger than d_2, then, according to the hydrostatic pressure equation, the pressure at the bottom of the narrow tube would be higher than the pressure at the bottom of the wide tube. This *pressure difference* would cause the liquid to *flow* from right to left until the heights were equal.

Thus a first conclusion: **A connected liquid in hydrostatic equilibrium rises to the same height in all open regions of the container.**

FIGURE 15.13b shows two connected tubes of different shape. The conical tube holds more liquid above the dotted line, so you might think that $p_1 > p_2$. But it isn't. Both points are at the same depth, thus $p_1 = p_2$. If p_1 were larger than p_2, the pressure at the bottom of the left tube would be larger than the pressure at the bottom of the right tube. This would cause the liquid to flow until the pressures were equal.

Thus a second conclusion: **The pressure is the same at all points on a horizontal line through a connected liquid in hydrostatic equilibrium.**

NOTE ▶ Both of these conclusions are restricted to liquids in hydrostatic equilibrium. The situation is different for flowing fluids, as we'll see later in the chapter. ◀

EXAMPLE 15.4 **Pressure in a closed tube**

Water fills the tube shown in FIGURE 15.14. What is the pressure at the top of the closed tube?

FIGURE 15.14 A water-filled tube.

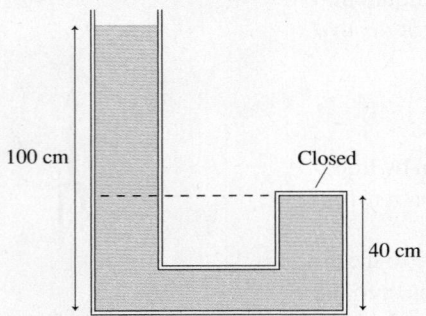

MODEL This is a liquid in hydrostatic equilibrium. The closed tube is not an open region of the container, so the water cannot rise to an equal height. Nevertheless, the pressure is still the same at all points on a horizontal line. In particular, the pressure at the top of the closed tube equals the pressure in the open tube at the height of the dashed line. Assume $p_0 = 1.00$ atm.

SOLVE A point 40 cm above the bottom of the open tube is at a depth of 60 cm. The pressure at this depth is

$$p = p_0 + \rho g d$$

$$= 1.013 \times 10^5 \text{ Pa} + (1000 \text{ kg/m}^3)(9.80 \text{ m/s}^2)(0.60 \text{ m})$$

$$= 1.072 \times 10^5 \text{ Pa} = 1.06 \text{ atm}$$

This is the pressure at the top of the closed tube.

ASSESS The water in the open tube *pushes* the water in the closed tube up against the top of the tube, which is why the pressure is greater than 1 atm.

We can draw one more conclusion from the hydrostatic pressure equation $p = p_0 + \rho g d$. If we change the pressure p_0 at the surface to p_1, the pressure at depth d becomes $p' = p_1 + \rho g d$. The *change* in pressure $\Delta p = p_1 - p_0$ is the same at all points in the fluid, independent of the size or shape of the container. This idea, that **a change in the pressure at one point in an incompressible fluid appears undiminished at all points in the fluid,** was first recognized by Blaise Pascal and is called **Pascal's principle.**

For example, if we compressed the air above the open tube in Example 15.4 to a pressure of 1.5 atm, an increase of 0.5 atm, the pressure at the top of the closed tube would increase to 1.56 atm. Pascal's principle is the basis for hydraulic systems, as we'll see in the next section.

STOP TO THINK 15.2   Water is slowly poured into the container until the water level has risen into tubes A, B, and C. The water doesn't overflow from any of the tubes. How do the water depths in the three columns compare to each other?

a. $d_A > d_B > d_C$
b. $d_A < d_B < d_C$
c. $d_A = d_B = d_C$
d. $d_A = d_C > d_B$
e. $d_A = d_C < d_B$

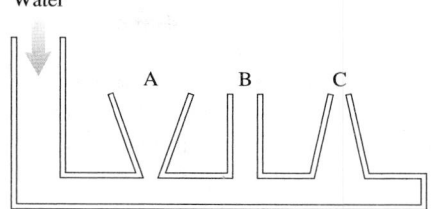
Water

A   B   C

15.3 Measuring and Using Pressure

The pressure in a fluid is measured with a *pressure gauge*. The fluid pushes against some sort of spring, and the spring's displacement is registered by a pointer on a dial.

Many pressure gauges, such as tire gauges and the gauges on air tanks, measure not the actual or absolute pressure p but what is called **gauge pressure**. The gauge pressure, denoted p_g, is the pressure *in excess* of 1 atm. That is,

$$p_g = p - 1 \text{ atm} \tag{15.6}$$

You must add 1 atm = 101.3 kPa to the reading of a pressure gauge to find the absolute pressure p that you need for doing most science or engineering calculations: $p = p_g + 1$ atm.

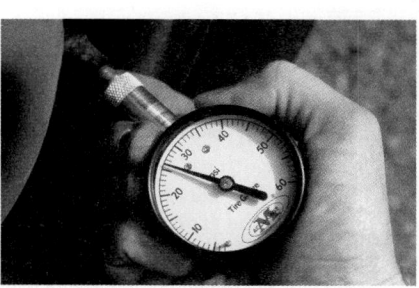
A tire-pressure gauge reads the gauge pressure p_g, not the absolute pressure p.

EXAMPLE 15.5   **An underwater pressure gauge**

An underwater pressure gauge reads 60 kPa. What is its depth?

MODEL   The gauge reads gauge pressure, not absolute pressure.

SOLVE   The hydrostatic pressure at depth d, with $p_0 = 1$ atm, is $p = 1$ atm $+ \rho gd$. Thus the gauge pressure is

$$p_g = p - 1 \text{ atm} = (1 \text{ atm} + \rho gd) - 1 \text{ atm} = \rho gd$$

The term ρgd is the pressure *in excess* of atmospheric pressure and thus *is* the gauge pressure. Solving for d, we find

$$d = \frac{60,000 \text{ Pa}}{(1000 \text{ kg/m}^3)(9.80 \text{ m/s}^2)} = 6.1 \text{ m}$$

Solving Hydrostatic Problems

We now have enough information to formulate a set of rules for thinking about hydrostatic problems.

TACTICS
BOX 15.1   **Hydrostatics**

❶ **Draw a picture.** Show open surfaces, pistons, boundaries, and other features that affect pressure. Include height and area measurements and fluid densities. Identify the points at which you need to find the pressure.
❷ **Determine the pressure at surfaces.**

■ **Surface open to the air:** $p_0 = p_{atmos}$, usually 1 atm.
■ **Surface covered by a gas:** $p_0 = p_{gas}$.
■ **Closed surface:** $p = F/A$, where F is the force the surface, such as a piston, exerts on the fluid.

❸ **Use horizontal lines.** Pressure in a connected fluid is the same at any point along a horizontal line.
❹ **Allow for gauge pressure.** Pressure gauges read $p_g = p - 1$ atm.
❺ **Use the hydrostatic pressure equation.** $p = p_0 + \rho gd$.

Exercises 4–13

FIGURE 15.15 A manometer is used to measure gas pressure.

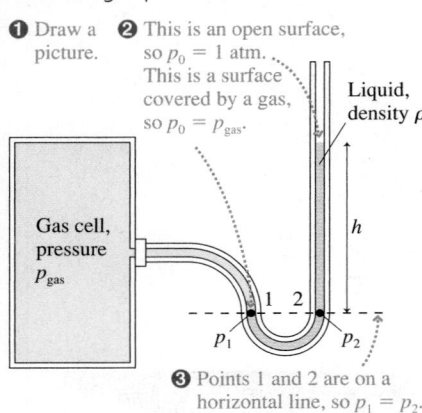

❶ Draw a picture.

❷ This is an open surface, so $p_0 = 1$ atm.
This is a surface covered by a gas, so $p_0 = p_{gas}$.

Liquid, density ρ

Gas cell, pressure p_{gas}

h

1 2
p_1 p_2

❸ Points 1 and 2 are on a horizontal line, so $p_1 = p_2$.

Manometers and Barometers

Gas pressure is sometimes measured with a device called a *manometer*. A manometer, shown in FIGURE 15.15, is a U-shaped tube connected to the gas at one end and open to the air at the other end. The tube is filled with a liquid—usually mercury—of density ρ. The liquid is in static equilibrium. A scale allows the user to measure the height h of the right side above the left side.

Steps 1–3 from Tactics Box 15.1 lead to the conclusion that the pressures p_1 and p_2 must be equal. Pressure p_1, at the surface on the left, is simply the gas pressure: $p_1 = p_{gas}$. Pressure p_2 is the hydrostatic pressure at depth $d = h$ in the liquid on the right: $p_2 = 1$ atm $+ \rho g h$. Equating these two pressures gives

$$p_{gas} = 1 \text{ atm} + \rho g h \qquad (15.7)$$

Figure 15.15 assumed $p_{gas} > 1$ atm, so the right side of the liquid is higher than the left. Equation 15.7 is also valid for $p_{gas} < 1$ atm if the distance of the right side *below* the left side is considered to be a negative value of h.

EXAMPLE 15.6 **Using a manometer**

The pressure of a gas cell is measured with a mercury manometer. The mercury is 36.2 cm higher in the outside arm than in the arm connected to the gas cell.

a. What is the gas pressure?
b. What is the reading of a pressure gauge attached to the gas cell?

SOLVE a. From Table 15.1, the density of mercury is $\rho = 13{,}600$ kg/m^3. Equation 15.7 with $h = 0.362$ m gives

$$p_{gas} = 1 \text{ atm} + \rho g h = 149.5 \text{ kPa}$$

We had to change 1 atm to 101,300 Pa before adding. Converting the result to atmospheres, we have $p_{gas} = 1.476$ atm.

b. The pressure gauge reads gauge pressure: $p_g = p - 1$ atm $= 0.476$ atm or 48.2 kPa.

ASSESS Manometers are useful over a pressure range from near vacuum up to ≈ 2 atm. For higher pressures, the mercury column would be too tall to be practical.

FIGURE 15.16 A barometer.

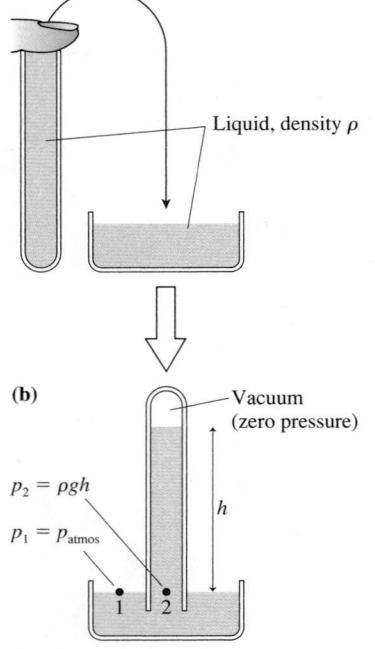

(a) Seal and invert tube.

Liquid, density ρ

(b)

Vacuum (zero pressure)

$p_2 = \rho g h$

$p_1 = p_{atmos}$

h

1 2

Another important pressure-measuring instrument is the *barometer,* which is used to measure the atmospheric pressure p_{atmos}. FIGURE 15.16a shows a glass tube, sealed at the bottom, that has been completely filled with a liquid. If we temporarily seal the top end, we can invert the tube and place it in a beaker of the same liquid. When the temporary seal is removed, some, but not all, of the liquid runs out, leaving a liquid column in the tube that is a height h above the surface of the liquid in the beaker. This device, shown in FIGURE 15.16b, is a barometer. What does it measure? And why doesn't *all* the liquid in the tube run out?

We can analyze the barometer much as we did the manometer. Points 1 and 2 in Figure 15.16b are on a horizontal line drawn even with the surface of the liquid. The liquid is in hydrostatic equilibrium, so the pressure at these two points must be equal. Liquid runs out of the tube only until a balance is reached between the pressure at the base of the tube and the pressure of the air.

You can think of a barometer as rather like a seesaw. If the pressure of the atmosphere increases, it presses down on the liquid in the beaker. This forces liquid up the tube until the pressures at points 1 and 2 are equal. If the atmospheric pressure falls, liquid has to flow out of the tube to keep the pressures equal at these two points.

The pressure at point 2 is the pressure due to the weight of the liquid in the tube plus the pressure of the gas above the liquid. But in this case there is no gas above the liquid! Because the tube had been completely full of liquid when it was inverted, the space left behind when the liquid ran out is a vacuum (ignoring a very slight *vapor pressure* of the liquid, negligible except in extremely precise measurements). Thus pressure p_2 is simply $p_2 = \rho g h$.

Equating p_1 and p_2 gives

$$p_{atmos} = \rho g h \qquad (15.8)$$

Thus we can measure the atmosphere's pressure by measuring the height of the liquid column in a barometer.

The average air pressure at sea level causes a column of mercury in a mercury barometer to stand 760 mm above the surface. Knowing that the density of mercury is 13,600 kg/m³ (at 0°C), we can use Equation 15.8 to find that the average atmospheric pressure is

$$p_{atmos} = \rho_{Hg}gh = (13,600 \text{ kg/m}^3)(9.80 \text{ m/s}^2)(0.760 \text{ m})$$

$$= 1.013 \times 10^5 \text{ Pa} = 101.3 \text{ kPa}$$

This is the value given earlier as "one standard atmosphere."

The barometric pressure varies slightly from day to day as the weather changes. Weather systems are called *high-pressure systems* or *low-pressure systems,* depending on whether the local sea-level pressure is higher or lower than one standard atmosphere. Higher pressure is usually associated with fair weather, while lower pressure portends rain.

Pressure Units

In practice, pressure is measured in several different units. This plethora of units and abbreviations has arisen historically as scientists and engineers working on different subjects (liquids, high-pressure gases, low-pressure gases, weather, etc.) developed what seemed to them the most convenient units. These units continue in use through tradition, so it is necessary to become familiar with converting back and forth between them. Table 15.2 gives the basic conversions.

TABLE 15.2 Pressure units

Unit	Abbreviation	Conversion to 1 atm	Uses
pascal	Pa	101.3 kPa	SI unit: 1 Pa = 1 N/m²
atmosphere	atm	1 atm	general
millimeters of mercury	mm of Hg	760 mm of Hg	gases and barometric pressure
inches of mercury	in	29.92 in	barometric pressure in U.S. weather forecasting
pounds per square inch	psi	14.7 psi	engineering and industry

Blood Pressure

The last time you had a medical checkup, the doctor may have told you something like "Your blood pressure is 120 over 80." What does that mean?

About every 0.8 s, assuming a pulse rate of 75 beats per minute, your heart "beats." The heart muscles contract and push blood out into your aorta. This contraction, like squeezing a balloon, raises the pressure in your heart. The pressure increase, in accordance with Pascal's principle, is transmitted through all your arteries.

FIGURE 15.17 is a pressure graph showing how blood pressure changes during one cycle of the heartbeat. The medical condition of *high blood pressure* usually means that your systolic pressure is higher than necessary for blood circulation. The high pressure causes undue stress and strain on your entire circulatory system, often leading to serious medical problems. Low blood pressure can cause you to get dizzy if you stand up quickly because the pressure isn't adequate to pump the blood up to your brain.

Blood pressure is measured with a cuff that goes around your arm. The doctor or nurse pressurizes the cuff, places a stethoscope over the artery in your arm, then slowly releases the pressure while watching a pressure gauge. Initially, the cuff squeezes the artery shut and cuts off the blood flow. When the cuff pressure drops below the systolic pressure, the pressure pulse during each beat of your heart forces the artery open briefly

FIGURE 15.17 Blood pressure during one cycle of a heartbeat.

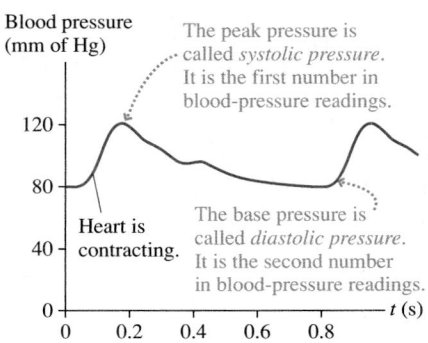

and a squirt of blood goes through. You can feel this, and the doctor or nurse records the pressure when she hears the blood start to flow. This is your systolic pressure.

This pulsing of the blood through your artery lasts until the cuff pressure reaches the diastolic pressure. Then the artery remains open continuously and the blood flows smoothly. This transition is easily heard in the stethoscope, and the doctor or nurse records your diastolic pressure.

Blood pressure is measured in millimeters of mercury. And it is a gauge pressure, the pressure in excess of 1 atm. A fairly typical blood pressure of a healthy young adult is 120/80, meaning that the systolic pressure is $p_g = 120$ mm of Hg (absolute pressure $p = 880$ mm of Hg) and the diastolic pressure is 80 mm of Hg.

The Hydraulic Lift

The use of pressurized liquids to do useful work is a technology known as **hydraulics.** Pascal's principle is the fundamental idea underlying hydraulic devices. If you increase the pressure at one point in a liquid by pushing a piston in, that pressure increase is transmitted to all points in the liquid. A second piston at some other point in the fluid can then push outward and do useful work.

The brake system in your car is a hydraulic system. Stepping on the brake pushes a piston into the *master brake cylinder* and increases the pressure in the *brake fluid.* The fluid itself hardly moves, but the pressure increase is transmitted to the four wheels where it pushes the brake pads against the spinning brake disk. You've used a pressurized liquid to achieve the useful goal of stopping your car.

One advantage of hydraulic systems over simple mechanical linkages is the possibility of *force multiplication.* To see how this works, we'll analyze a *hydraulic lift,* such as the one that lifts your car at the repair shop. FIGURE 15.18a shows force $\vec{F}_2$, perhaps due to the weight of mass m, pressing down on a liquid via a piston of area A_2. A much smaller force $\vec{F}_1$ presses down on a piston of area A_1. Can this system possibly be in equilibrium?

As you now know, the hydrostatic pressure is the same at all points along a horizontal line through a fluid. Consider the line passing through the liquid/piston interface on the left in Figure 15.18a. Pressures p_1 and p_2 must be equal, thus

$$p_0 + \frac{F_1}{A_1} = p_0 + \frac{F_2}{A_2} + \rho g h \qquad (15.9)$$

The atmosphere presses equally on both sides, so p_0 cancels. The system is in static equilibrium if

$$F_2 = \frac{A_2}{A_1} F_1 - \rho g h A_2 \qquad (15.10)$$

If the height h is very small, so that the term $\rho g h A_2$ is negligible, then F_2 (the weight of the heavy object) is larger than F_1 by the factor A_2/A_1. In other words, a small force applied to a small piston really can support a large car because both apply the *same pressure* to the fluid. The ratio A_2/A_1 is a force-multiplying factor.

NOTE ▶ Force $\vec{F}_2$ is the force of the heavy object pushing *down* on the liquid. According to Newton's third law, the liquid pushes *up* on the object with a force of equal magnitude. Thus F_2 in Equation 15.10 is the "lifting force." ◀

Suppose we need to lift the car higher. If piston 1 is pushed down distance d_1, as in FIGURE 15.18b, it displaces volume $V_1 = A_1 d_1$ of liquid. Because the liquid is incompressible, V_1 must equal the volume $V_2 = A_2 d_2$ added beneath piston 2 as it rises distance d_2. That is,

$$d_2 = \frac{d_1}{A_2/A_1} \qquad (15.11)$$

The distance is *divided* by the same factor as that by which force is multiplied. A small force may be able to support a heavy weight, but you have to push the small piston a large distance to raise the heavy weight by a small amount.

FIGURE 15.18 A hydraulic lift.

(a)

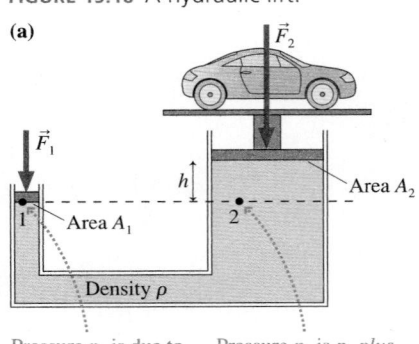

Pressure p_1 is due to atmospheric pressure p_0 *plus* pressure F_1/A_1, due to $\vec{F}_1$.

Pressure p_2 is p_0 *plus* F_2/A_2 *plus* $\rho g h$ from the liquid column of height h.

(b)

Because the fluid is incompressible, $A_1 d_1 = A_2 d_2$.

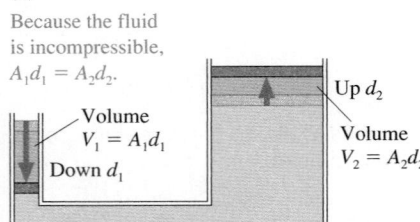

This conclusion is really just a statement of energy conservation. Work is done *on* the liquid by a small force pushing the liquid through a large displacement. Work is done *by* the liquid when it lifts the heavy weight through a small distance. A full analysis must consider the fact that the gravitational potential energy of the liquid is also changing, so we can't simply equate the output work to the input work, but you can see that energy considerations require piston 1 to move farther than piston 2.

EXAMPLE 15.7 **Lifting a car**

The hydraulic lift at a car repair shop is filled with oil. The car rests on a 25-cm-diameter piston. To lift the car, compressed air is used to push down on a 6.0-cm-diameter piston. What does the pressure gauge read when a 1300 kg car is 2.0 m above the compressed-air piston?

MODEL Assume that the oil is incompressible. Its density, from Table 15.1, is 900 kg/m^3.

SOLVE F_2 is the weight of the car pressing down on the piston: $F_2 = mg = 12{,}700$ N. The piston areas are $A_1 = \pi(0.030 \text{ m})^2 = 0.00283 \text{ m}^2$ and $A_2 = \pi(0.125 \text{ m})^2 = 0.0491 \text{m}^2$. The force required to hold the car at height $h = 2.0$ m is found by solving Equation 15.10 for F_1:

$$F_1 = \frac{A_1}{A_2} F_2 + \rho g h A_1$$
$$= \frac{0.00283 \text{ m}^2}{0.0491 \text{ m}^2} \cdot 12{,}700 \text{ N} + (900 \text{ kg/m}^3)(9.8 \text{ m/s}^2)(2.0 \text{ m})(0.00283 \text{ m}^2)$$
$$= 782 \text{ N}$$

The pressure applied to the fluid by the compressed-air piston is

$$p_1 = \frac{F_1}{A_1} = \frac{782 \text{ N}}{0.00283 \text{ m}^2} = 2.76 \times 10^5 \text{ Pa} = 2.7 \text{ atm}$$

This is the pressure *in excess* of atmospheric pressure, which is what a pressure gauge measures, so the gauge reads, depending on its units, 276 kPa or 2.7 atm.

ASSESS 782 N is roughly the weight of an average adult man. The multiplication factor $A_2/A_1 = 17$ makes it quite easy for this much force to lift the car.

STOP TO THINK 15.3 Rank in order, from largest to smallest, the magnitudes of the forces $\vec{F}_a$, $\vec{F}_b$, and $\vec{F}_c$ required to balance the masses. The masses are in kilograms.

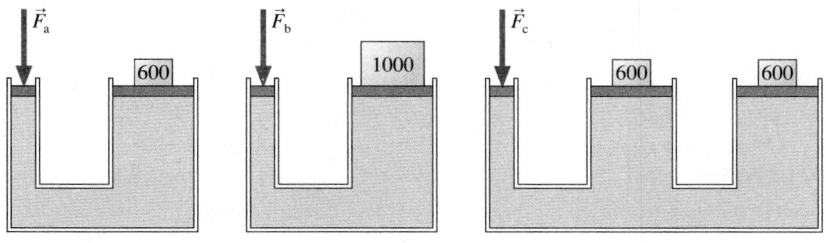

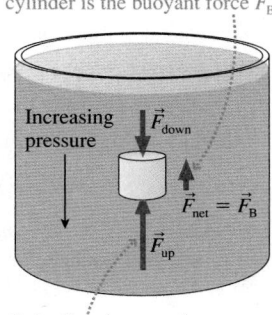

FIGURE 15.19 The buoyant force arises because the fluid pressure at the bottom of the cylinder is larger than at the top.

The net force of the fluid on the cylinder is the buoyant force $\vec{F}_B$.

$F_{up} > F_{down}$ because the pressure is greater at the bottom. Hence the fluid exerts a net upward force.

15.4 Buoyancy

A rock, as you know, sinks like a rock. Wood floats on the surface of a lake. A penny with a mass of a few grams sinks, but a massive steel aircraft carrier floats. How can we understand these diverse phenomena?

An air mattress floats effortlessly on the surface of a swimming pool. But if you've ever tried to push an air mattress underwater, you know it is nearly impossible. As you push down, the water pushes up. This net upward force of a fluid is called the **buoyant force.**

The basic reason for the buoyant force is easy to understand. **FIGURE 15.19** shows a cylinder submerged in a liquid. The pressure in the liquid increases with depth, so the

FIGURE 15.20 The buoyant force on an object is the same as the buoyant force on the fluid it displaces.

(a)

Imaginary boundary around a parcel of fluid

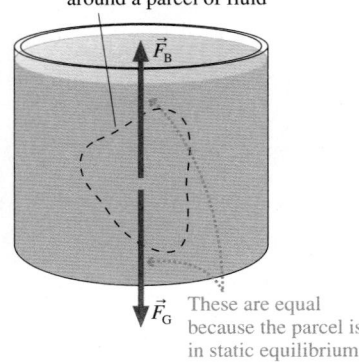

These are equal because the parcel is in static equilibrium.

(b)

Real object with same size and shape as the parcel of fluid

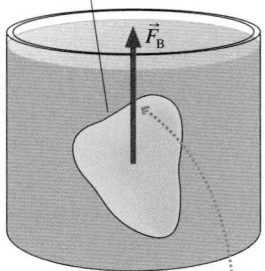

The buoyant force on the object is the same as on the parcel of fluid because the *surrounding* fluid has not changed.

pressure at the bottom of the cylinder is larger than at the top. Both cylinder ends have equal area, so force $\vec{F}_{up}$ is larger than force $\vec{F}_{down}$. (Remember that pressure forces push in *all* directions.) Consequently, the pressure in the liquid exerts a *net upward force* on the cylinder of magnitude $F_{net} = F_{up} - F_{down}$. This is the buoyant force.

The submerged cylinder illustrates the idea in a simple way, but the result is not limited to cylinders or to liquids. Suppose we isolate a parcel of fluid of arbitrary shape and volume by drawing an imaginary boundary around it, as shown in FIGURE 15.20a. This parcel is in static equilibrium. Consequently, the gravitational force pulling down on the parcel must be balanced by an upward force. The upward force, which is exerted on this parcel of fluid by the surrounding fluid, is the buoyant force $\vec{F}_B$. The buoyant force matches the weight of the fluid: $F_B = mg$.

Imagine that we could somehow remove this parcel of fluid and instantaneously replace it with an object of exactly the same shape and size, as shown in FIGURE 15.20b. Because the buoyant force is exerted by the *surrounding* fluid, and the surrounding fluid hasn't changed, the buoyant force on this new object is *exactly the same* as the buoyant force on the parcel of fluid that we removed.

When an object (or a portion of an object) is immersed in a fluid, it *displaces* fluid that would otherwise fill that region of space. This fluid is called the **displaced fluid**. The displaced fluid's volume is exactly the volume of the portion of the object that is immersed in the fluid. Figure 15.20 leads us to conclude that the magnitude of the upward buoyant force matches the weight of this displaced fluid.

This idea was first recognized by the ancient Greek mathematician and scientist Archimedes, perhaps the greatest scientist of antiquity, and today we know it as *Archimedes' principle.*

> **Archimedes' principle** A fluid exerts an upward buoyant force $\vec{F}_B$ on an object immersed in or floating on the fluid. The magnitude of the buoyant force equals the weight of the fluid displaced by the object.

Suppose the fluid has density ρ_f and the object displaces volume V_f of fluid. The mass of the displaced fluid is $m_f = \rho_f V_f$ and so its weight is $m_f g = \rho_f V_f g$. Thus Archimedes' principle in equation form is

$$F_B = \rho_f V_f g \tag{15.12}$$

NOTE ▶ It is important to distinguish the density and volume of the displaced fluid from the density and volume of the object. To do so, we'll use subscript f for the fluid and o for the object. ◀

EXAMPLE 15.8 | **Holding a block of wood underwater**

A 10 cm × 10 cm × 10 cm block of wood with density 700 kg/m³ is held underwater by a string tied to the bottom of the container. What is the tension in the string?

MODEL The buoyant force is given by Archimedes' principle.

VISUALIZE FIGURE 15.21 shows the forces acting on the wood.

SOLVE The block is in static equilibrium, so

$$\sum F_y = F_B - T - m_o g = 0$$

Thus the tension is $T = F_B - m_o g$. The mass of the block is $m_o = \rho_o V_o$, and the buoyant force, given by Equation 15.12, is $F_B = \rho_f V_f g$. Thus

$$T = \rho_f V_f g - \rho_o V_o g = (\rho_f - \rho_o)V_o g$$

FIGURE 15.21 The forces acting on the submerged wood.

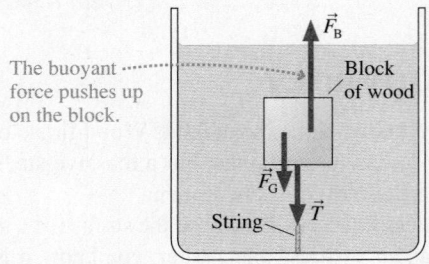

The buoyant force pushes up on the block.

Block of wood

String

where we've used the fact that $V_f = V_o$ for a completely submerged object. The volume is $V_o = 1000 \text{ cm}^3 = 1.0 \times 10^{-3} \text{ m}^3$, and hence the tension in the string is

$$T = \left((1000 \text{ kg/m}^3) - (700 \text{ kg/m}^3) \right)$$
$$\times (1.0 \times 10^{-3} \text{ m}^3)(9.8 \text{ m/s}^2) = 2.9 \text{ N}$$

ASSESS The tension depends on the *difference* in densities. The tension would vanish if the wood density matched the water density.

Float or Sink?

If you *hold* an object underwater and then release it, it floats to the surface, sinks, or remains "hanging" in the water. How can we predict which it will do? The net force on the object an instant after you release it is $\vec{F}_{net} = (F_B - m_o g)\hat{k}$. Whether it heads for the surface or the bottom depends on whether the buoyant force F_B is larger or smaller than the object's weight $m_o g$.

The magnitude of the buoyant force is $\rho_f V_f g$. The weight of a uniform object, such as a block of steel, is simply $\rho_o V_o g$. But a compound object, such as a scuba diver, may have pieces of varying density. If we define the **average density** to be $\rho_{avg} = m_o/V_o$, the weight of a compound object is $\rho_{avg} V_o g$.

Comparing $\rho_f V_f g$ to $\rho_{avg} V_o g$, and noting that $V_f = V_o$ for an object that is fully submerged, we see that an object floats or sinks depending on whether the fluid density ρ_f is larger or smaller than the object's average density ρ_{avg}. If the densities are equal, the object is in static equilibrium and hangs motionless. This is called **neutral buoyancy**. These conditions are summarized in Tactics Box 15.2.

TACTICS BOX 15.2 **Finding whether an object floats or sinks**

❶ Object sinks

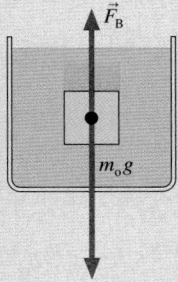

❷ Object floats

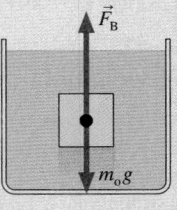

❸ Neutral buoyancy

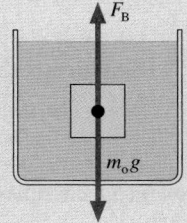

An object sinks if it weighs more than the fluid it displaces—that is, if its average density is greater than the density of the fluid:

$$\rho_{avg} > \rho_f$$

An object floats on the surface if it weighs less than the fluid it displaces—that is, if its average density is less than the density of the fluid:

$$\rho_{avg} < \rho_f$$

An object hangs motionless if it weighs exactly the same as the fluid it displaces—that is, if its average density equals the density of the fluid:

$$\rho_{avg} = \rho_f$$

Exercises 14–18

As an example, steel is denser than water, so a chunk of steel sinks. Oil is less dense than water, so oil floats on water. Fish use *swim bladders* filled with air and scuba divers use weighted belts to adjust their average density to match the density of water. Both are examples of neutral buoyancy.

FIGURE 15.22 A floating object is in static equilibrium.

An object of density ρ_o and volume V_o is floating on a fluid of density ρ_f.

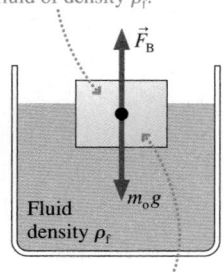

The submerged volume of the object is equal to the volume V_f of displaced fluid.

About 90% of an iceberg is underwater.

If you release a block of wood underwater, the net upward force causes the block to shoot to the surface. Then what? Let's begin with a *uniform* object such as the block shown in **FIGURE 15.22**. This object contains nothing tricky, like indentations or voids. Because it's floating, it must be the case that $\rho_o < \rho_f$.

Now that the object is floating, it's in static equilibrium. The upward buoyant force, given by Archimedes' principle, exactly balances the downward weight of the object. That is,

$$F_B = \rho_f V_f g = m_o g = \rho_o V_o g \qquad (15.13)$$

In this case, the volume of the displaced fluid is *not* the same as the volume of the object. In fact, we can see from Equation 15.13 that the volume of fluid displaced by a floating object of uniform density is

$$V_f = \frac{\rho_o}{\rho_f} V_o < V_o \qquad (15.14)$$

You've often heard it said that "90% of an iceberg is underwater." Equation 15.14 is the basis for that statement. Most icebergs break off glaciers and are fresh-water ice with a density of 917 kg/m^3. The density of seawater is 1030 kg/m^3. Thus

$$V_f = \frac{917 \text{ kg/m}^3}{1030 \text{ kg/m}^3} V_o = 0.89 V_o$$

V_f, the displaced water, is the volume of the iceberg that is underwater. You can see that, indeed, 89% of the volume of an iceberg is underwater.

NOTE ▶ Equation 15.14 applies only to *uniform* objects. It does not apply to boats, hollow spheres, or other objects of nonuniform composition. ◀

EXAMPLE 15.9 **Measuring the density of an unknown liquid**

You need to determine the density of an unknown liquid. You notice that a block floats in this liquid with 4.6 cm of the side of the block submerged. When the block is placed in water, it also floats but with 5.8 cm submerged. What is the density of the unknown liquid?

MODEL The block is an object of uniform composition.

VISUALIZE **FIGURE 15.23** shows the block and defines the cross-section area A and submerged lengths h_u in the unknown liquid and h_w in water.

FIGURE 15.23 More of the block is submerged in water than in an unknown liquid.

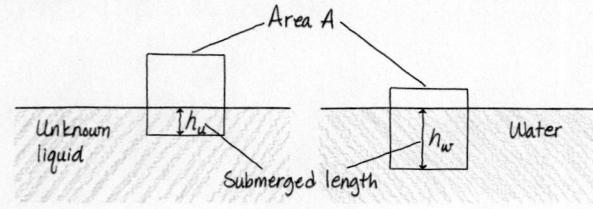

SOLVE The block is floating, so Equation 15.14 applies. The block displaces volume $V_u = Ah_u$ of the unknown liquid. Thus

$$V_u = Ah_u = \frac{\rho_o}{\rho_u} V_o$$

Similarly, the block displaces volume $V_w = Ah_w$ of the water, leading to

$$V_w = Ah_w = \frac{\rho_o}{\rho_w} V_o$$

Because there are two fluids, we've used subscripts w for water and u for the unknown in place of the fluid subscript f. The product $\rho_o V_o$ appears in both equations; hence

$$\rho_u Ah_u = \rho_w Ah_w$$

The unknown area A cancels, and the density of the unknown liquid is

$$\rho_u = \frac{h_w}{h_u} \rho_w = \frac{5.8 \text{ cm}}{4.6 \text{ cm}} \cdot 1000 \text{ kg/m}^3 = 1260 \text{ kg/m}^3$$

ASSESS Comparison with Table 15.1 shows that the unknown liquid is likely to be glycerin.

Boats

We'll conclude by designing a boat. FIGURE 15.24 is a physicist's idea of a boat. Four massless but rigid walls are attached to a solid steel plate of mass m_o and area A. As the steel plate settles down into the water, the sides allow the boat to displace a volume of water much larger than that displaced by the steel alone. The boat will float if the weight of the displaced water equals the weight of the boat.

In terms of density, the boat will float if $\rho_{avg} < \rho_f$. If the sides of the boat are height h, the boat's volume is $V_o = Ah$ and its average density is $\rho_{avg} = m_o/V_o = m_o/Ah$. The boat will float if

$$\rho_{avg} = \frac{m_o}{Ah} < \rho_f \tag{15.15}$$

Thus the minimum height of the sides, a height that would allow the boat to float (in perfectly still water!) with water right up to the rails, is

$$h_{min} = \frac{m_o}{\rho_f A} \tag{15.16}$$

As a quick example, a 5 m × 10 m steel "barge" with a 2-cm-thick floor has an area of 50 m^2 and a mass of 7900 kg. The minimum height of the massless walls, as given by Equation 15.16, is 16 cm.

Real ships and boats are more complicated, but the same idea holds true. Whether it's made of concrete, steel, or lead, **a boat will float if its geometry allows it to displace enough water to equal the weight of the boat.**

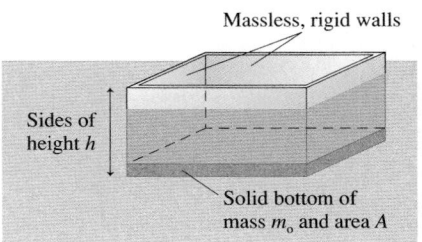

FIGURE 15.24 A physicist's boat.

Massless, rigid walls

Sides of height h

Solid bottom of mass m_o and area A

STOP TO THINK 15.4 An ice cube is floating in a glass of water that is filled entirely to the brim. When the ice cube melts, the water level will

a. Fall.   b. Stay the same, right at the brim.   c. Rise, causing the water to spill.

15.5 Fluid Dynamics

The wind blowing through your hair, a white-water river, and oil gushing from an oil well are examples of fluids in motion. We've focused thus far on fluid statics, but it's time to turn our attention to fluid dynamics.

Fluid flow is a complex subject. Many aspects, especially turbulence and the formation of eddies, are still not well understood and are areas of current science and engineering research. We will avoid these difficulties by using a simplified *model*. The **ideal-fluid model** provides a good, though not perfect, description of fluid flow in many situations. It captures the essence of fluid flow while eliminating unnecessary details.

The ideal-fluid model can be expressed in three assumptions about a fluid:

1. The fluid is *incompressible*. This is a good assumption for liquids, less so for gases.
2. The fluid is *nonviscous*. Water flows much more easily than pancake syrup because the syrup is a very *viscous* fluid. **Viscosity,** a resistance to flow, is analogous to kinetic friction. Assuming that a fluid is nonviscous is equivalent to assuming there's no friction. This is the weakest assumption for many liquids, but assuming a nonviscous liquid avoids major mathematical difficulties.
3. The flow is *steady*. That is, the fluid velocity at each point in the fluid is constant; it does not fluctuate or change with time. Flow under these conditions is called **laminar flow,** and it is distinguished from *turbulent flow*.

The rising smoke in the photograph of FIGURE 15.25 begins as laminar flow, recognizable by the smooth contours, but at some point undergoes a transition to turbulent

FIGURE 15.25 Rising smoke changes from laminar flow to turbulent flow.

Turbulent flow

Laminar flow

flow. A laminar-to-turbulent transition is not uncommon in fluid flow. The ideal-fluid model can be applied to the laminar flow, but not to the turbulent flow.

The Equation of Continuity

FIGURE 15.26 shows smoke being used to help engineers visualize the airflow around a car in a wind tunnel. The smoothness of the flow tells us this is laminar flow. But notice also how the individual smoke trails retain their identity. They don't cross or get mixed together. Each smoke trail represents a *streamline* in the fluid.

FIGURE 15.26 The laminar airflow around a car in a wind tunnel is made visible with smoke. Each smoke trail represents a streamline.

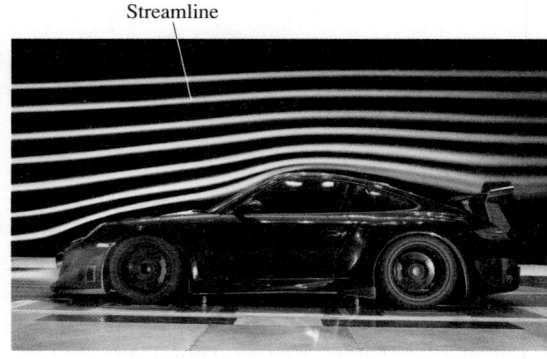

Streamline

FIGURE 15.27 Particles in an ideal fluid move along streamlines.

1. Streamlines never cross.

$\vec{v}$

2. Fluid particle velocity is tangent to the streamline.

3. The speed is higher where the streamlines are closer together.

Imagine that we could inject a colored drop of water into a stream of water flowing as an ideal fluid. Because the flow is steady and frictionless, and the water is incompressible, this colored drop would maintain its identity as it flowed along. Its shape might change, becoming compressed or elongated, but it would not mix with the surrounding water.

The path or trajectory followed by this "particle of fluid" is called a **streamline**. Smoke particles mixed with the air allow you to see the streamlines in the photograph of Figure 15.26. FIGURE 15.27 illustrates three important properties of streamlines.

A bundle of neighboring streamlines, such as those shown in FIGURE 15.28a, form a **flow tube.** Because streamlines never cross, all the streamlines that cross plane 1 within area A_1 later cross plane 2 within area A_2. A flow tube is like an invisible pipe that keeps this portion of the flowing fluid distinct from other portions. Real pipes are also flow tubes.

FIGURE 15.28 A flow tube.

(a)

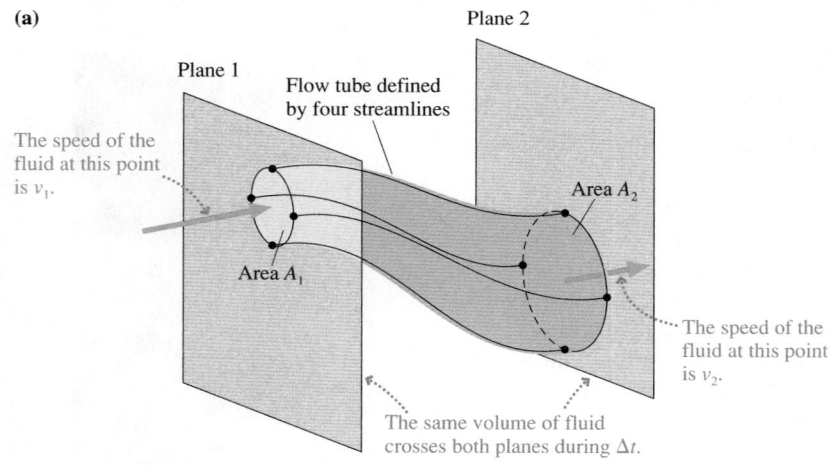

Plane 2

Plane 1

Flow tube defined by four streamlines

The speed of the fluid at this point is v_1.

Area A_2

Area A_1

The speed of the fluid at this point is v_2.

The same volume of fluid crosses both planes during Δt.

(b)

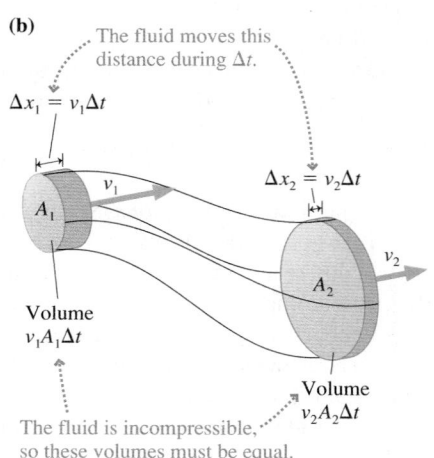

The fluid moves this distance during Δt.

$\Delta x_1 = v_1 \Delta t$

$\Delta x_2 = v_2 \Delta t$

A_1

v_1

A_2

v_2

Volume $v_1 A_1 \Delta t$

Volume $v_2 A_2 \Delta t$

The fluid is incompressible, so these volumes must be equal.

When you squeeze a toothpaste tube, the volume of toothpaste that emerges matches the amount by which you reduce the volume of the tube. An incompressible fluid in a flow tube acts the same way. Fluid is not created or destroyed within the flow tube, and it cannot be stored. If volume V enters the flow tube through area A_1 during some interval of time Δt, then an equal volume V must leave the flow tube through area A_2.

FIGURE 15.28b shows the flow crossing A_1 during a small interval of time Δt. If the fluid speed at this point is v_1, the fluid moves forward a small distance $\Delta x_1 = v_1 \Delta t$ and fills the volume $V_1 = A_1 \Delta x_1 = v_1 A_1 \Delta t$. The same analysis for the fluid crossing A_2 with fluid speed v_2 would find $V_2 = v_2 A_2 \Delta t$. These two volumes must be equal, leading to the conclusion that

FIGURE 15.29 The flow tube diameter changes as the speed increases. This is a consequence of the equation of continuity.

$$v_1 A_1 = v_2 A_2 \qquad (15.17)$$

Equation 15.17 is called the **equation of continuity,** and it is one of two important equations for the flow of an ideal fluid. The equation of continuity says that **the volume of an incompressible fluid entering one part of a flow tube must be matched by an equal volume leaving downstream.**

An important consequence of the equation of continuity is that **flow is faster in narrower parts of a flow tube, slower in wider parts.** You're familiar with this conclusion from many everyday observations. For example, water flowing from the faucet shown in FIGURE 15.29 picks up speed as it falls. As a result, the flow tube "necks down" to a smaller diameter.

The quantity

$$Q = vA \qquad (15.18)$$

is called the **volume flow rate.** The SI units of Q are m^3/s, although in practice Q may be measured in cm^3/s, liters per minute, or, in the United States, gallons per minute. Another way to express the meaning of the equation of continuity is to say that **the volume flow rate is constant at all points in a flow tube.**

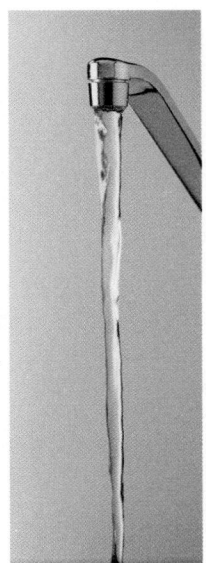

EXAMPLE 15.10 **Blood flow in capillaries**

The heart of a resting adult pumps about 5 L of blood every minute. All this blood must eventually pass through the smallest blood vessels, the capillaries, before returning to the heart. Microscope measurements show that a typical capillary is 6 μm in diameter and 1 mm long and has a blood flow speed of 1 mm/s. Estimate the total surface area of all the capillaries in the body.

MODEL Treat the blood as an ideal fluid.

SOLVE The surface area of one capillary is that of a cylinder:

$$A_1 = 2\pi r L = 2\pi(3 \times 10^{-6} \text{ m})(0.001 \text{ m}) = 1.9 \times 10^{-8} \text{ m}^2$$

The total surface area is $A_{\text{surface}} = NA_1$, where N is the number of capillaries. We can find N by using the equation of continuity: The volume flow rate of blood leaving the heart—$Q = 5$ L/min—must equal the volume flow rate through all N capillaries. In SI units, the volume flow rate is

$$Q = 5 \frac{\text{L}}{\text{min}} \times \frac{1 \text{ m}^3}{1000 \text{ L}} \times \frac{1 \text{ min}}{60 \text{ s}} = 8.3 \times 10^{-5} \text{ m}^3 /s$$

The total cross-section area of all the capillaries together must be

$$A_{\text{total}} = \frac{Q}{v} = \frac{8.3 \times 10^{-5} \text{ m}^3/\text{s}}{10^{-3} \text{ m/s}} = 0.083 \text{ m}^2$$

The cross-section area of one capillary is $A_{\text{cap}} = \pi r^2 = \pi(3 \times 10^{-6} \text{ m})^2 = 2.8 \times 10^{-11} \text{ m}^2$, so the number of capillaries is

$$N = \frac{A_{\text{total}}}{A_{\text{cap}}} = \frac{0.083 \text{ m}^2}{2.8 \times 10^{-11} \text{ m}^2} = 3 \times 10^9$$

Thus the total surface area is

$$A_{\text{surface}} = NA_1 = (3 \times 10^9)(2 \times 10^{-8}) = 60 \text{ m}^2$$

The question asked for an estimate and provided only approximate values, so only a one-significant figure answer is justified.

ASSESS The total surface area is about the area of a two-car garage! Only by having such a large surface area can oxygen and nutrients slowly diffuse into cells. Notice that we had to deal with two types of areas—the cross-section area and the surface area. It is important not to get these confused.

STOP TO THINK 15.5 The figure shows volume flow rates (in cm³/s) for all but one tube. What is the volume flow rate through the unmarked tube? Is the flow direction in or out?

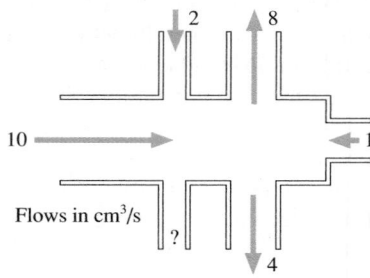

Flows in cm³/s

Bernoulli's Equation

The equation of continuity is one of two important relationships for ideal fluids. The other is a statement of energy conservation. The general statement of energy conservation that you learned in Chapter 11 is

$$\Delta K + \Delta U = W_{\text{ext}} \qquad (15.19)$$

where W_{ext} is the work done by any external forces.

Let's see how this applies to the flow tube of FIGURE 15.30. Our system for analysis is the volume of fluid within the flow tube. Work is done on this volume of fluid by the pressure forces of the *surrounding* fluid. At point 1, the fluid to the left of the flow tube exerts force $\vec{F}_1$ on the system. This force points to the right. At the other end of the flow tube, at point 2, the fluid to the right of the flow tube exerts force $\vec{F}_2$ to the left. The pressure inside the flow tube is not relevant because those forces are internal to the system. Only external forces change the total energy.

FIGURE 15.30 Energy analysis of a flow tube.

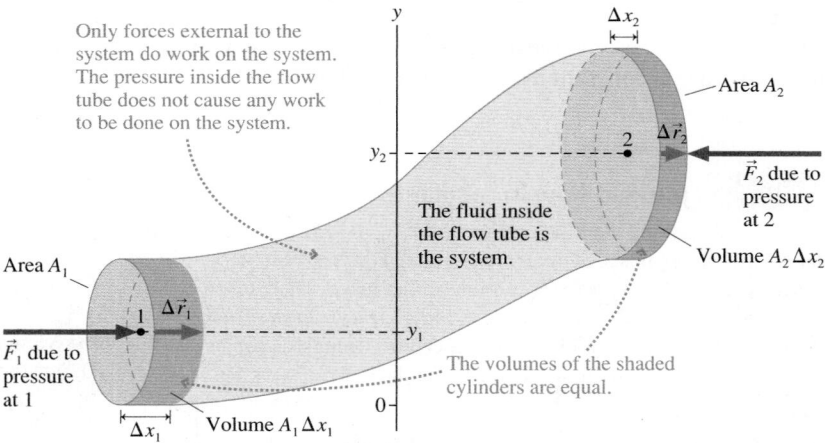

At point 1, force $\vec{F}_1$ pushes the fluid through displacement $\Delta \vec{r}_1$. $\vec{F}_1$ and $\Delta \vec{r}_1$ are parallel, so the work done on the fluid at this point is

$$W_1 = \vec{F}_1 \cdot \Delta \vec{r}_1 = F_1 \Delta r_1 = (p_1 A_1)\Delta x_1 = p_1 V \qquad (15.20)$$

The A_1 and Δx_1 enter the equation from different terms, but they conveniently combine to give the fluid volume $V = A_1 \Delta x_1$.

The situation is much the same at point 2 except that $\vec{F}_2$ points opposite the displacement $\Delta \vec{r}_2$. This introduces a $\cos(180°) = -1$ into the dot product for the work, giving

$$W_2 = \vec{F}_2 \cdot \Delta \vec{r}_2 = -F_2 \Delta r_2 = -(p_2 A_2)\Delta x_2 = -p_2 V \qquad (15.21)$$

The pressure from the left at point 1 pushes the fluid ahead, a positive work. The pressure from the right at point 2 tries to slow the fluid down, a negative work. Together, the work by external forces is

$$W_{\text{ext}} = W_1 + W_2 = p_1 V - p_2 V \qquad (15.22)$$

Now let's see how this work changes the kinetic and potential energy of the system. A small volume of fluid $V = A_1 \Delta x_1$ passes point 1 and, at some later time, arrives at point 2, where the unchanged volume is $V = A_2 \Delta x_2$. The change in gravitational potential energy for this volume of fluid is

$$\Delta U = mgy_2 - mgy_1 = \rho V g y_2 - \rho V g y_1 \qquad (15.23)$$

where ρ is the fluid density. Similarly, the change in kinetic energy is

$$\Delta K = \frac{1}{2}mv_2^2 - \frac{1}{2}mv_1^2 = \frac{1}{2}\rho V v_2^2 - \frac{1}{2}\rho V v_1^2 \qquad (15.24)$$

Combining Equations 15.22, 15.23, and 15.24 gives us the energy equation for the fluid in the flow tube:

$$\frac{1}{2}\rho V v_2^2 - \frac{1}{2}\rho V v_1^2 + \rho V g y_2 - \rho V g y_1 = p_1 V - p_2 V \qquad (15.25)$$

The volume V cancels out of all the terms. If we regroup the terms, the energy equation becomes

$$p_1 + \frac{1}{2}\rho v_1^2 + \rho g y_1 = p_2 + \frac{1}{2}\rho v_2^2 + \rho g y_2 \qquad (15.26)$$

Equation 15.26 is called **Bernoulli's equation.** It is named for the 18th-century Swiss scientist Daniel Bernoulli, who made some of the earliest studies of fluid dynamics.

Bernoulli's equation is really nothing more than a statement about work and energy. It is sometimes useful to express Bernoulli's equation in the alternative form

$$p + \frac{1}{2}\rho v^2 + \rho g y = \text{constant} \qquad (15.27)$$

This version of Bernoulli's equation tells us that the quantity $p + \frac{1}{2}\rho v^2 + \rho g y$ remains constant along a streamline.

One important implication of Bernoulli's equation is easily demonstrated. Before reading the next paragraph, try the simple experiment illustrated in FIGURE 15.31. Really, do try this!

What happened? You probably expected your breath to press the strip of paper down. Instead, the strip *rose*. In fact, the harder you blow, the more nearly the strip becomes parallel to the floor. This counterintuitive result is a consequence of Bernoulli's equation. As the air speed above the strip of paper increases, the pressure has to *decrease* to keep the quantity $p + \frac{1}{2}\rho v^2 + \rho g y$ constant. Consequently, the air pressure above the strip is less than the air pressure beneath the strip, resulting in a net upward force on the paper.

NOTE ▶ Using Bernoulli's equation is very much like using the law of conservation of energy. Rather than identifying a "before" and "after," you want to identify two points on a streamline. As the following examples show, Bernoulli's equation is often used in conjunction with the equation of continuity. ◀

FIGURE 15.31 A simple demonstration of Bernoulli's equation.

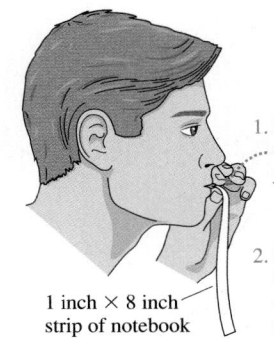

1. Hold strip at lower edge of bottom lip, just touching lip.

2. Pucker lips and blow hard straight out over the *top* of the strip.

1 inch × 8 inch strip of notebook paper

EXAMPLE 15.11   **An irrigation system**

Water flows through the pipes shown in **FIGURE 15.32**. The water's speed through the lower pipe is 5.0 m/s and a pressure gauge reads 75 kPa. What is the reading of the pressure gauge on the upper pipe?

FIGURE 15.32 The water pipes of an irrigation system.

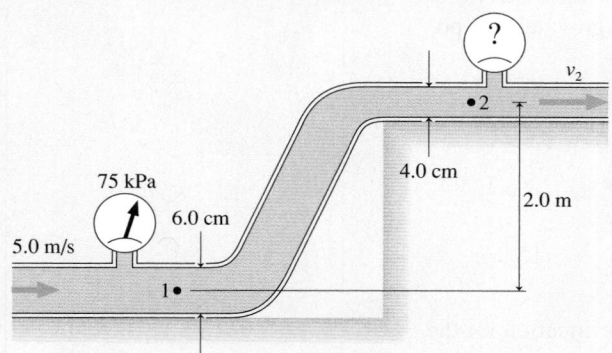

MODEL Treat the water as an ideal fluid obeying Bernoulli's equation. Consider a streamline connecting point 1 in the lower pipe with point 2 in the upper pipe.

SOLVE Bernoulli's equation, Equation 15.26, relates the pressure, fluid speed, and heights at points 1 and 2. It is easily solved for the pressure p_2 at point 2:

$$p_2 = p_1 + \frac{1}{2}\rho v_1^2 - \frac{1}{2}\rho v_2^2 + \rho g y_1 - \rho g y_2$$
$$= p_1 + \frac{1}{2}\rho(v_1^2 - v_2^2) + \rho g(y_1 - y_2)$$

All quantities on the right are known except v_2, and that is where the equation of continuity will be useful. The cross-section areas and water speeds at points 1 and 2 are related by

$$v_1 A_1 = v_2 A_2$$

from which we find

$$v_2 = \frac{A_1}{A_2}v_1 = \frac{r_1^2}{r_2^2}v_1 = \frac{(0.030 \text{ m})^2}{(0.020 \text{ m})^2}(5.0 \text{ m/s}) = 11.25 \text{ m/s}$$

The pressure at point 1 is $p_1 = 75 \text{ kPa} + 1 \text{ atm} = 176{,}300 \text{ Pa}$. We can now use the above expression for p_2 to calculate $p_2 = 105{,}900 \text{ Pa}$. This is the absolute pressure; the pressure gauge on the upper pipe will read

$$p_2 = 105{,}900 \text{ Pa} - 1 \text{ atm} = 4.6 \text{ kPa}$$

ASSESS Reducing the pipe size decreases the pressure because it makes $v_2 > v_1$. Gaining elevation also reduces the pressure.

EXAMPLE 15.12   **Hydroelectric power**

Small hydroelectric plants in the mountains sometimes bring the water from a reservoir down to the power plant through enclosed tubes. In one such plant, the 100-cm-diameter intake tube in the base of the dam is 50 m below the reservoir surface. The water drops 200 m through the tube before flowing into the turbine through a 50-cm-diameter nozzle.

a. What is the water speed into the turbine?
b. By how much does the inlet pressure differ from the hydrostatic pressure at that depth?

MODEL Treat the water as an ideal fluid obeying Bernoulli's equation. Consider a streamline that begins at the surface of the reservoir and ends at the exit of the nozzle. The pressure at the surface is $p_1 = p_{atmos}$ and $v_1 \approx 0$ m/s. The water discharges into air, so $p_3 = p_{atmos}$ at the exit.

VISUALIZE **FIGURE 15.33** is a pictorial representation of the situation.

FIGURE 15.33 Pictorial representation of the water flow to a hydroelectric plant.

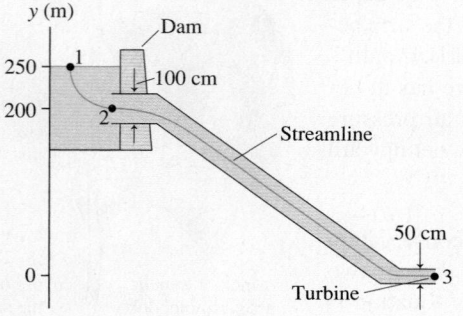

SOLVE a. Bernoulli's equation, with $v_1 = 0$ m/s and $y_3 = 0$ m, is

$$p_{atmos} + \rho g y_1 = p_{atmos} + \frac{1}{2}\rho v_3^2$$

The power plant is in the mountains, where $p_{atmos} < 1$ atm, but p_{atmos} occurs on both sides of Bernoulli's equation and cancels. Solving for v_3 gives

$$v_3 = \sqrt{2g y_1} = \sqrt{2(9.80 \text{ m/s}^2)(250 \text{ m})} = 70 \text{ m/s}$$

b. You might expect the pressure p_2 at the intake to be the hydrostatic pressure $p_{atmos} + \rho g d$ at depth d. But the water is *flowing* into the intake tube, so it's not in static equilibrium. We can find the intake speed v_2 from the equation of continuity:

$$v_2 = \frac{A_3}{A_2}v_3 = \frac{r_3^2}{r_2^2}\sqrt{2g y_1}$$

The intake is along the streamline between points 1 and 3, so we can apply Bernoulli's equation to points 1 and 2:

$$p_{atmos} + \rho g y_1 = p_2 + \frac{1}{2}\rho v_2^2 + \rho g y_2$$

Solving this equation for p_2, and noting that $y_1 - y_2 = d$, we find

$$p_2 = p_{atmos} + \rho g(y_1 - y_2) - \frac{1}{2}\rho v_2^2$$
$$= p_{atmos} + \rho g d - \frac{1}{2}\rho\left(\frac{r_3}{r_2}\right)^4 (2g y_1)$$
$$= p_{static} - \rho g y_1\left(\frac{r_3}{r_2}\right)^4$$

The intake pressure is *less* than hydrostatic pressure by the amount

$$\rho g y_1 \left(\frac{r_3}{r_2}\right)^4 = 153{,}000 \text{ Pa} = 1.5 \text{ atm}$$

ASSESS The water's exit speed from the nozzle is the same as if it fell 250 m from the surface of the reservoir. This isn't surprising because we've assumed a nonviscous (i.e., frictionless) liquid. "Real" water would have less speed but still flow very fast.

Two Applications

The speed of a flowing gas is often measured with a device called a **Venturi tube.** Venturi tubes measure gas speeds in environments as different as chemistry laboratories, wind tunnels, and jet engines.

FIGURE 15.34 shows gas flowing through a tube that changes from cross-section area A_1 to area A_2. A U-shaped glass tube containing liquid of density ρ_{liq} connects the two segments of the flow tube. When gas flows through the horizontal tube, the liquid stands height h higher in the side of the U tube connected to the narrow segment of the flow tube.

Figure 15.34 shows how a Venturi tube works. We can make this analysis quantitative and determine the gas-flow speed from the liquid height h. Two pieces of information we have to work with are Bernoulli's equation

$$p_1 + \frac{1}{2}\rho v_1^2 + \rho g y_1 = p_2 + \frac{1}{2}\rho v_2^2 + \rho g y_2 \tag{15.28}$$

and the equation of continuity

$$v_2 A_2 = v_1 A_1 \tag{15.29}$$

In addition, the hydrostatic equation for the liquid tells us that the pressure p_2 above the right tube differs from the pressure p_1 above the left tube by $\rho_{\text{liq}} gh$. That is,

$$p_2 = p_1 - \rho_{\text{liq}} gh \tag{15.30}$$

First we use Equations 15.29 and 15.30 to eliminate v_2 and p_2 in Bernoulli's equation:

$$p_1 + \frac{1}{2}\rho v_1^2 = (p_1 - \rho_{\text{liq}} gh) + \frac{1}{2}\rho\left(\frac{A_1}{A_2}\right)^2 v_1^2 \tag{15.31}$$

The potential energy terms have disappeared because $y_1 = y_2$ for a horizontal tube. Equation 15.31 can now be solved for v_1, then v_2 is obtained from Equation 15.29. We'll skip a few algebraic steps and go right to the result:

$$
\begin{aligned}
v_1 &= A_2 \sqrt{\frac{2\rho_{\text{liq}} gh}{\rho(A_1^2 - A_2^2)}} \\
v_2 &= A_1 \sqrt{\frac{2\rho_{\text{liq}} gh}{\rho(A_1^2 - A_2^2)}}
\end{aligned}
\tag{15.32}
$$

Equations 15.32 are reasonably accurate as long as the flow speeds are much less than the speed of sound, about 340 m/s. The Venturi tube is an example of the power of Bernoulli's equation.

As a final example, we can use Bernoulli's equation to understand, at least qualitatively, how airplane wings generate *lift.* **FIGURE 15.35** shows the cross section of an airplane wing. This shape is called an *airfoil.*

Although you usually think of an airplane moving through the air, in the airplane's reference frame it is the air that flows across a stationary wing. As it does, the streamlines must separate. The bottom of the wing does not significantly alter the streamlines going under the wing. But the streamlines going over the top of the wing get bunched together. This bunching reduces the cross-section area of a flow tube of streamlines. Consequently, in accordance with the equation of continuity, the air speed must increase as it flows across the top of the wing.

FIGURE 15.34 A Venturi tube measures gas-flow speeds.

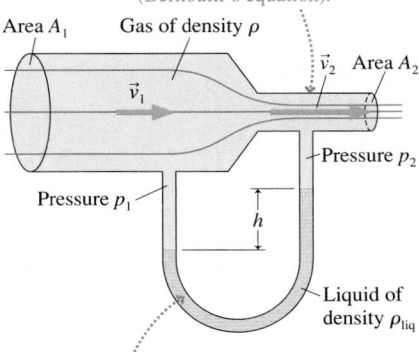

1. As the gas flows into a smaller cross section, it speeds up (equation of continuity). As it speeds up, the pressure decreases (Bernoulli's equation).

Area A_1 Gas of density ρ

$\vec{v}_1$ $\vec{v}_2$ Area A_2

Pressure p_1 Pressure p_2

h

Liquid of density ρ_{liq}

2. The U tube acts like a manometer. The liquid level is higher on the side where the pressure is lower.

FIGURE 15.35 Airflow over a wing generates lift by creating unequal pressures above and below.

1. The streamlines in the flow tube are compressed, indicating that the air speeds up as it flows over the top of the wing. This lowers the pressure to $p < p_{\text{atmos}}$.

2. The pressure difference exerts an upward force on the wing.

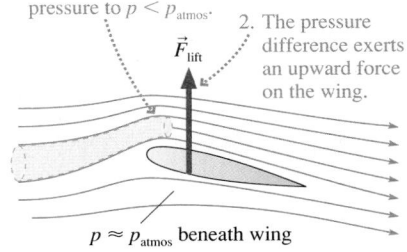

$\vec{F}_{\text{lift}}$

$p \approx p_{\text{atmos}}$ beneath wing

As you've seen several times, an increased air speed implies a decreased air pressure. This is the lesson of Bernoulli's equation. Because the air pressure above the wing is less than the air pressure below, the air exerts a net upward force on the wing, just as it did on the paper strip you blew across. The upward force of the air due to a pressure difference across the wing is called **lift.**

STOP TO THINK 15.6   Rank in order, from highest to lowest, the liquid heights h_a to h_d. The airflow is from left to right.

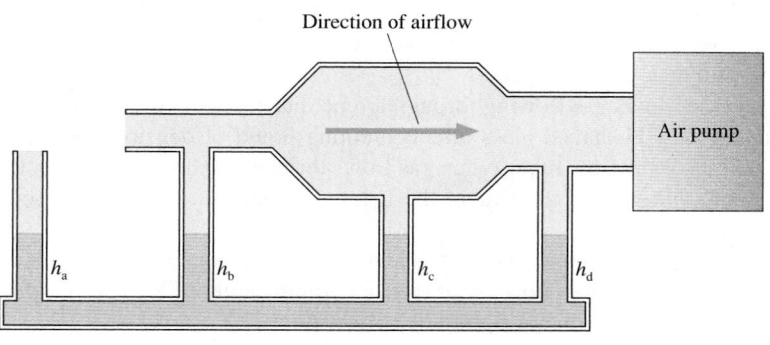

15.6 Elasticity

The final subject to explore in this chapter is elasticity. Although elasticity applies primarily to solids rather than fluids, you will see that similar ideas come into play.

Tensile Stress and Young's Modulus

FIGURE 15.36 Stretching a solid rod.

(a)

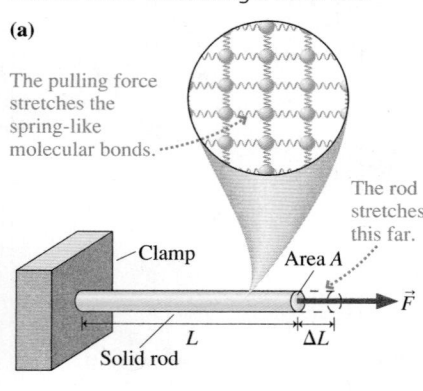

The pulling force stretches the spring-like molecular bonds.

Clamp

Area A

The rod stretches this far.

$\vec{F}$

L ΔL

Solid rod

(b)

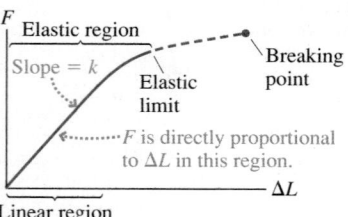

F

Elastic region

Slope = k

Elastic limit

Breaking point

F is directly proportional to ΔL in this region.

ΔL

Linear region

Suppose you clamp one end of a solid rod while using a strong machine to pull on the other with force $\vec{F}$. FIGURE 15.36a shows the experimental arrangement. We usually think of solids as being, well, solid. But any material, be it plastic, concrete, or steel, will stretch as the spring-like molecular bonds expand.

FIGURE 15.36b shows graphically the amount of force needed to stretch the rod by the amount ΔL. This graph contains several regions of interest. First is the *elastic region,* ending at the *elastic limit.* As long as ΔL is less than the elastic limit, the rod will return to its initial length L when the force is removed. Just such a reversible stretch is what we mean when we say a material is *elastic.* A stretch beyond the elastic limit will permanently deform the object; it will not return to its initial length when the force is removed. And, not surprisingly, there comes a point when the rod breaks.

For most materials, the graph begins with a *linear region,* which is where we will focus our attention. If ΔL is within the linear region, the force needed to stretch the rod is

$$F = k \Delta L \qquad (15.33)$$

where k is the slope of the graph. You'll recognize Equation 15.33 as none other than Hooke's law.

The difficulty with Equation 15.33 is that the proportionality constant k depends both on the composition of the rod—whether it is, say, steel or aluminum—and on the rod's length and cross-section area. It would be useful to characterize the elastic properties of steel in general, or aluminum in general, without needing to know the dimensions of a specific rod.

We can meet this goal by thinking about Hooke's law at the atomic scale. The elasticity of a material is directly related to the spring constant of the molecular bonds between neighboring atoms. As FIGURE 15.37 shows, the force pulling each bond is proportional to the quantity F/A. This force causes each bond to stretch by an amount proportional to $\Delta L/L$. We don't know what the proportionality constants are, but we don't need to. Hooke's law applied to a molecular bond tells us that the force pulling on a bond is proportional to the amount that the bond stretches. Thus F/A must be proportional to $\Delta L/L$. We can write their proportionality as

$$\frac{F}{A} = Y \frac{\Delta L}{L} \qquad (15.34)$$

The proportionality constant Y is called **Young's modulus.** It is directly related to the spring constant of the molecular bonds, so it depends on the material from which the object is made but *not* on the object's geometry.

A comparison of Equations 15.33 and 15.34 shows that Young's modulus can be written as $Y = kL/A$. This is not a definition of Young's modulus but simply an expression for making an experimental determination of the value of Young's modulus. This k is the spring constant of the rod seen in Figure 15.36. It is a quantity easily measured in the laboratory.

The quantity F/A, where A is the cross-section area, is called **tensile stress.** Notice that it is essentially the same definition as pressure. Even so, tensile stress differs in that the stress is applied in a particular direction whereas pressure forces are exerted in all directions. Another difference is that stress is measured in N/m² rather than pascals. The quantity $\Delta L/L$, the fractional increase in the length, is called **strain.** Strain is dimensionless. The numerical values of strain are always very small because solids cannot be stretched very much before reaching the breaking point.

With these definitions, Equation 15.34 can be written

$$\text{stress} = Y \times \text{strain} \qquad (15.35)$$

Because strain is dimensionless, Young's modulus Y has the same dimensions as stress, namely N/m². Table 15.3 gives values of Young's modulus for several common materials. Large values of Y characterize materials that are stiff and rigid. "Softer" materials, at least relatively speaking, have smaller values of Y. You can see that steel has a larger Young's modulus than aluminum.

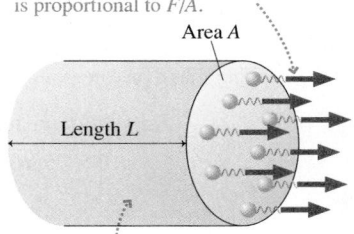

FIGURE 15.37 A material's elasticity is directly related to the spring constant of the molecular bonds.

The number of bonds is proportional to area A. If the rod is pulled with force F, the force pulling on each bond is proportional to F/A.

Area A

Length L

The number of bonds along the rod is proportional to length L. If the rod stretches by ΔL, the stretch of each bond is proportional to $\Delta L/L$.

TABLE 15.3 Elastic properties of various materials

Substance	Young's modulus (N/m²)	Bulk modulus (N/m²)
Aluminum	7×10^{10}	7×10^{10}
Concrete	3×10^{10}	–
Copper	11×10^{10}	14×10^{10}
Mercury	–	3×10^{10}
Plastic (polystyrene)	0.3×10^{10}	–
Steel	20×10^{10}	16×10^{10}
Water	–	0.2×10^{10}
Wood (Douglas fir)	1×10^{10}	–

We introduced Young's modulus by considering how materials stretch. But Equation 15.35 and Young's modulus also apply to the compression of materials. Compression is particularly important in engineering applications, where beams, columns, and support foundations are compressed by the load they bear. Concrete is often compressed, as in columns that support highway overpasses, but rarely stretched.

NOTE ▶ Whether the rod is stretched or compressed, Equation 15.35 is valid only in the linear region of the graph in Figure 15.36b. The breaking point is usually well outside the linear region, so you can't use Young's modulus to compute the maximum possible stretch or compression. ◀

Concrete is a widely used building material because it is relatively inexpensive and, with its large Young's modulus, it has tremendous compressional strength.

EXAMPLE 15.13 **Stretching a wire**

A 2.0-m-long, 1.0-mm-diameter wire is suspended from the ceiling. Hanging a 4.5 kg mass from the wire stretches the wire's length by 1.0 mm. What is Young's modulus for this wire? Can you identify the material?

MODEL The hanging mass creates tensile stress in the wire.

SOLVE The force pulling on the wire, which is simply the weight of the hanging mass, produces tensile stress

$$\frac{F}{A} = \frac{mg}{\pi r^2} = \frac{(4.5 \text{ kg})(9.80 \text{ m/s}^2)}{\pi (0.0005 \text{ m})^2} = 5.6 \times 10^7 \text{ N/m}^2$$

The resulting stretch of 1.0 mm is a strain of $\Delta L/L = (1.0 \text{ mm})/(2000 \text{ mm}) = 5.0 \times 10^{-4}$. Thus Young's modulus for the wire is

$$Y = \frac{F/A}{\Delta L/L} = 11 \times 10^{10} \text{ N/m}^2$$

Referring to Table 15.3, we see that the wire is made of copper.

Volume Stress and the Bulk Modulus

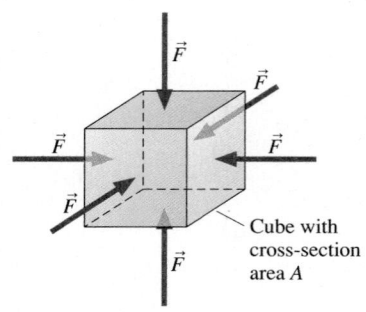

FIGURE 15.38 An object is compressed by pressure forces pushing equally on all sides.

Young's modulus characterizes the response of an object to being pulled in one direction. **FIGURE 15.38** shows an object being squeezed in all directions. For example, objects under water are squeezed from all sides by the water pressure. The force per unit area F/A applied to *all* surfaces of an object is called the **volume stress.** Because the force pushes equally on all sides, the volume stress (unlike the tensile stress) really is the same as pressure p.

No material is perfectly rigid. A volume stress applied to an object compresses its volume slightly. The **volume strain** is defined as $\Delta V/V$. The volume strain is a *negative* number because the volume stress *decreases* the volume.

Volume stress, or pressure, is linearly proportional to the volume strain, much as the tensile stress is linearly proportional to the strain in a rod. That is,

$$\frac{F}{A} = p = -B\frac{\Delta V}{V} \tag{15.36}$$

where B is called the **bulk modulus.** The negative sign in Equation 15.36 ensures that the pressure is a positive number. Table 15.3 gives values of the bulk modulus for several materials. Smaller values of B correspond to materials that are more easily compressed. Both solids and liquids can be compressed and thus have a bulk modulus, whereas Young's modulus applies only to solids.

EXAMPLE 15.14 **Compressing a sphere**

A 1.00-m-diameter solid steel sphere is lowered to a depth of 10,000 m in a deep ocean trench. By how much does its diameter shrink?

MODEL The water pressure applies a volume stress to the sphere.

SOLVE The water pressure at $d = 10,000$ m is

$$p = p_0 + \rho g d = 1.01 \times 10^8 \text{ Pa}$$

where we used the density of seawater. The bulk modulus of steel, taken from Table 15.3, is $16 \times 10^{10} \text{ N/m}^2$. Thus the volume strain is

$$\frac{\Delta V}{V} = -\frac{p}{B} = -\frac{1.01 \times 10^8 \text{ Pa}}{16 \times 10^{10} \text{ Pa}} = -6.3 \times 10^{-4}$$

The volume of a sphere is $V = \frac{4}{3}\pi r^3$. For a very small change, we can use calculus to relate the volume change to the change in radius:

$$\Delta V = \frac{4\pi}{3}\Delta(r^3) = \frac{4\pi}{3} \cdot 3r^2 \Delta r = 4\pi r^2 \Delta r$$

Using this expression for ΔV gives the volume strain:

$$\frac{\Delta V}{V} = \frac{4\pi r^2 \Delta r}{\frac{4}{3}\pi r^3} = \frac{3\Delta r}{r} = -6.3 \times 10^{-4}$$

Solving for Δr gives $\Delta r = -1.05 \times 10^{-4}$ m $= -0.105$ mm. The diameter changes by twice this, decreasing 0.21 mm.

ASSESS The immense pressure of the deep ocean causes only a tiny change in the sphere's diameter. You can see that treating solids and liquids as incompressible is an excellent approximation under nearly all circumstances.

CHALLENGE EXAMPLE 15.15   **Draining a cone**

A conical tank of radius R and height H, pointed end down, is full of water. A small hole of radius r is opened at the bottom of the tank, with $r \ll R$ so that the tank drains slowly. Find an expression for the time T it takes to drain the tank completely.

MODEL   Treat the water as an ideal fluid. We can use Bernoulli's equation to relate the flow speed from the hole to the height of the water in the cone.

VISUALIZE   **FIGURE 15.39** is a pictorial representation. Because the tank drains slowly, we've assumed that the water velocity at the top surface is always very close to zero: $v_1 = 0$. The pressure at the surface is $p_1 = p_{atmos}$. The water discharges into air, so we also have $p_2 = p_{atmos}$ at the exit.

FIGURE 15.39   Pictorial representation of water draining from a tank.

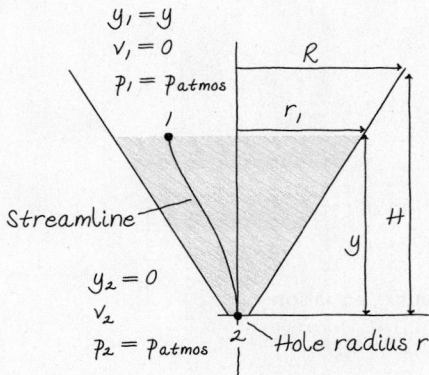

SOLVE   As the tank drains, the water height y decreases from H to 0. If we can find an expression for dy/dt, the *rate* at which the water height changes, we'll be able to find T by integrating from "full tank" at $t = 0$ to "empty tank" at $t = T$. Our starting point is the rate at which water flows out of the hole at the bottom—the volume flow rate $Q = v_2 A_2 = \pi r^2 v_2$, where v_2 is the exit speed. The volume of water inside the tank is changing at the rate

$$\frac{dV_{water}}{dt} = -Q = -\pi r^2 v_2$$

where the minus sign shows that the volume is *decreasing* with time.

We need to relate both V_{water} and v_2 to the height y of the water surface. The volume of a cone is $V = \frac{1}{3} \times$ base $\times$ height, so the cone of water has volume $V_{water} = \frac{1}{3}\pi r_1^2 y$. Based on the similar triangles in Figure 15.39, $r_1/R = y/H$. Thus $r_1 = (R/H)y$ and

$$V_{water} = \frac{\pi R^2}{3H^2}y^3$$

Taking the time derivative, we find

$$\frac{dV_{water}}{dt} = \frac{d}{dt}\left[\frac{\pi R^2}{3H^2}y^3\right] = \frac{\pi R^2}{H^2}y^2\frac{dy}{dt}$$

This relates the rate at which the volume changes to the rate at which the height changes.

We can next relate v_2 to the water height y by using Bernoulli's equation to connect the conditions at the surface (point 1) to conditions at the exit (point 2):

$$p_1 + \frac{1}{2}\rho v_1^2 + \rho g y_1 = p_2 + \frac{1}{2}\rho v_2^2 + \rho g y_2$$

With $p_1 = p_2$, $v_1 = 0$, $y_1 = y$, and $y_2 = 0$ at the bottom, Bernoulli's equation simplifies to $\rho g y = \frac{1}{2}\rho v_2^2$. Thus the exit speed of the water is

$$v_2 = \sqrt{2gy}$$

The exit speed decreases as the water height drops because the pressure at the bottom is less.

With this information, our equation for the rate at which the volume is changing becomes

$$\frac{dV_{water}}{dt} = \frac{\pi R^2}{H^2}y^2\frac{dy}{dt} = -\pi r^2 v_2 = -\pi r^2\sqrt{2gy}$$

In preparation for integration, we need to get all the y's on one side of the equation and dt on the other. Rearranging gives

$$dt = -\frac{R^2}{r^2 H^2\sqrt{2g}}y^{3/2}\,dy$$

We need to integrate this from the beginning, with $y = H$ at $t = 0$, to the moment the tank is empty, with $y = 0$ at $t = T$:

$$\int_0^T dt = T = -\frac{R^2}{r^2 H^2\sqrt{2g}}\int_H^0 y^{3/2}\,dy = \frac{R^2}{r^2 H^2\sqrt{2g}}\int_0^H y^{3/2}\,dy$$

The minus sign was eliminated by reversing the integration limits. Performing the integration gives us the desired result for the time to drain the tank:

$$T = \frac{R^2}{r^2 H^2\sqrt{2g}}\int_0^H y^{3/2}\,dy = \frac{R^2}{r^2 H^2\sqrt{2g}}\left[\frac{2}{5}y^{5/2}\right]_0^H$$

$$= \frac{2}{5}\frac{R^2}{r^2}\sqrt{\frac{H}{2g}}$$

ASSESS   Making the tank larger by increasing R or H increases the time needed to drain. Making the hole at the bottom larger—a larger value of r—decreases the time. These are as we would have expected, giving us confidence in our result.

SUMMARY

The goal of Chapter 15 has been to understand macroscopic systems that flow or deform.

General Principles

Fluid Statics

Gases

- Freely moving particles
- Compressible
- Pressure primarily thermal
- Pressure is constant in a laboratory-size container

Liquids

- Loosely bound particles
- Incompressible
- Pressure primarily gravitational
- Hydrostatic pressure at depth d is $p = p_0 + \rho g d$

Fluid Dynamics

Ideal-fluid model

- Incompressible
- Smooth, laminar flow
- Nonviscous

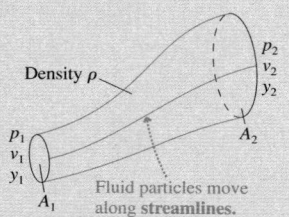

Fluid particles move along **streamlines**.

Equation of continuity

$$v_1 A_1 = v_2 A_2$$

Bernoulli's equation

$$p_1 + \tfrac{1}{2}\rho v_1^2 + \rho g y_1 = p_2 + \tfrac{1}{2}\rho v_2^2 + \rho g y_2$$

Bernoulli's equation is a statement of energy conservation.

Important Concepts

Density $\rho = m/V$, where m is mass and V is volume.

Pressure $p = F/A$, where F is the magnitude of the fluid force and A is the area on which the force acts.

- Pressure exists at all points in a fluid.
- Pressure pushes equally in all directions.
- Pressure is constant along a horizontal line.
- Gauge pressure is $p_g = p - 1$ atm.

Applications

Buoyancy is the upward force of a fluid on an object.

Archimedes' principle

The magnitude of the buoyant force equals the weight of the fluid displaced by the object.

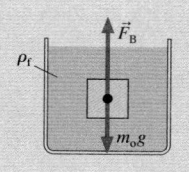

Sink	$\rho_{avg} > \rho_f$	$F_B < m_o g$
Rise to surface	$\rho_{avg} < \rho_f$	$F_B > m_o g$
Neutrally buoyant	$\rho_{avg} = \rho_f$	$F_B = m_o g$

Elasticity describes the deformation of solids and liquids under stress.

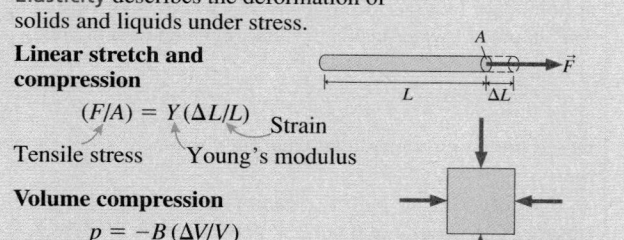

Linear stretch and compression

$$(F/A) = Y(\Delta L/L)$$

Tensile stress — Young's modulus — Strain

Volume compression

$$p = -B(\Delta V/V)$$

Bulk modulus — Volume strain

Terms and Notation

fluid	hydrostatic pressure	ideal-fluid model	Venturi tube
gas	Pascal's principle	viscosity	lift
liquid	gauge pressure, p_g	laminar flow	Young's modulus, Y
mass density, ρ	hydraulics	streamline	tensile stress
unit volume	buoyant force	flow tube	strain
pressure, p	displaced fluid	equation of continuity	volume stress
pascal, Pa	Archimedes' principle	volume flow rate, Q	volume strain
vacuum	average density, ρ_{avg}	Bernoulli's equation	bulk modulus, B
standard atmosphere, atm	neutral buoyancy		

CONCEPTUAL QUESTIONS

1. An object has density ρ.
   a. Suppose each of the object's three dimensions is increased by a factor of 2 without changing the material of which the object is made. Will the density change? If so, by what factor? Explain.
   b. Suppose each of the object's three dimensions is increased by a factor of 2 without changing the object's mass. Will the density change? If so, by what factor? Explain.

2. Rank in order, from largest to smallest, the pressures at a, b, and c in FIGURE Q15.2. Explain.

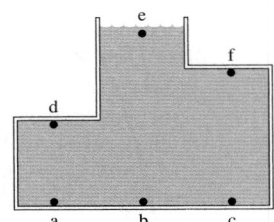

FIGURE Q15.2

3. Rank in order, from largest to smallest, the pressures at d, e, and f in FIGURE Q15.2. Explain.

4. FIGURE Q15.4 shows two rectangular tanks, A and B, full of water. They have equal depths and equal thicknesses (the dimension into the page) but different widths.

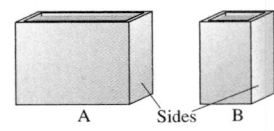

FIGURE Q15.4

   a. Compare the forces the water exerts on the bottoms of the tanks. Is F_A larger than, smaller than, or equal to F_B? Explain.
   b. Compare the forces the water exerts on the sides of the tanks. Is F_A larger than, smaller than, or equal to F_B? Explain.

5. In FIGURE Q15.5, is p_A larger than, smaller than, or equal to p_B? Explain.

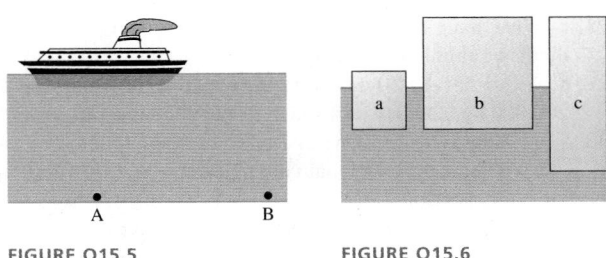

FIGURE Q15.5   FIGURE Q15.6

6. Rank in order, from largest to smallest, the densities of blocks a, b, and c in FIGURE Q15.6. Explain.

7. Blocks a, b, and c in FIGURE Q15.7 have the same volume. Rank in order, from largest to smallest, the sizes of the buoyant forces F_a, F_b, and F_c on a, b, and c. Explain.

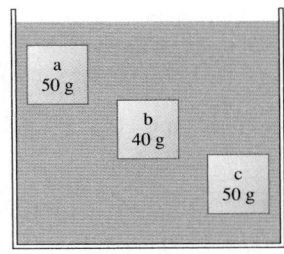

FIGURE Q15.7

8. Blocks a, b, and c in FIGURE Q15.7 have the same density. Rank in order, from largest to smallest, the sizes of the buoyant forces F_a, F_b, and F_c on a, b, and c. Explain.

9. The two identical beakers in FIGURE Q15.9 are filled to the same height with water. Beaker B has a plastic sphere floating in it. Which beaker, with all its contents, weighs more? Or are they equal? Explain.

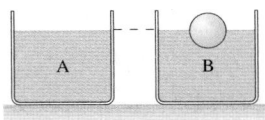   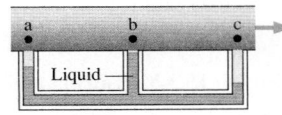

FIGURE Q15.9   FIGURE Q15.10

10. Gas flows through the pipe of FIGURE Q15.10. You can't see into the pipe to know how the inner diameter changes. Rank in order, from largest to smallest, the gas speeds v_a, v_b, and v_c at points a, b, and c. Explain.

11. Wind blows over the house in FIGURE Q15.11. A window on the ground floor is open. Is there an airflow through the house? If so, does the air flow in the window and out the chimney, or in the chimney and out the window? Explain.

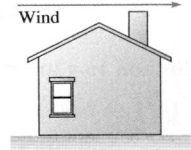

FIGURE Q15.11

12. A 2000 N force stretches a wire by 1 mm. A second wire of the same material is twice as long and has twice the diameter. How much force is needed to stretch it by 1 mm? Explain.

13. A wire is stretched right to the breaking point by a 5000 N force. A longer wire made of the same material has the same diameter. Is the force that will stretch it right to the breaking point larger than, smaller than, or equal to 5000 N? Explain.

EXERCISES AND PROBLEMS

Problems labeled ▓ integrate material from earlier chapters.

Exercises

Section 15.1 Fluids

1. | What is the volume in mL of 55 g of a liquid with density 1100 kg/m³?

2. | Containers A and B have equal volumes. Container A holds helium gas at 1.0 atm pressure and 0°C. Container B is completely filled with a liquid whose mass is 7000 times the mass of helium gas in container A. Identify the liquid in container B.

3. ‖ A 6.0 m × 12.0 m swimming pool slopes linearly from a 1.0 m depth at one end to a 3.0 m depth at the other. What is the mass of water in the pool?

4. ‖ a. 50 g of gasoline are mixed with 50 g of water. What is the average density of the mixture?
   b. 50 cm³ of gasoline are mixed with 50 cm³ of water. What is the average density of the mixture?

Section 15.2 Pressure

5. | The deepest point in the ocean is 11 km below sea level, deeper than Mt. Everest is tall. What is the pressure in atmospheres at this depth?

6. ‖ A 1.0-m-diameter vat of liquid is 2.0 m deep. The pressure at the bottom of the vat is 1.3 atm. What is the mass of the liquid in the vat?

7. ‖ a. What volume of water has the same mass as 8.0 m^3 of ethyl alcohol?
   b. If this volume of water is in a cubic tank, what is the pressure at the bottom?

8. ‖ A 50-cm-thick layer of oil floats on a 120-cm-thick layer of water. What is the pressure at the bottom of the water layer?

9. ‖ A research submarine has a 20-cm-diameter window 8.0 cm thick. The manufacturer says the window can withstand forces up to 1.0×10^6 N. What is the submarine's maximum safe depth? The pressure inside the submarine is maintained at 1.0 atm.

10. ‖ A 20-cm-diameter circular cover is placed over a 10-cm-diameter hole that leads into an evacuated chamber. The pressure in the chamber is 20 kPa. How much force is required to pull the cover off?

Section 15.3 Measuring and Using Pressure

11. | What is the height of a water barometer at atmospheric pressure?

12. | How far must a 2.0-cm-diameter piston be pushed down into one cylinder of a hydraulic lift to raise an 8.0-cm-diameter piston by 20 cm?

13. ‖ What is the minimum hose diameter of an ideal vacuum cleaner that could lift a 10 kg (22 lb) dog off the floor?

Section 15.4 Buoyancy

14. | A 6.00-cm-diameter sphere with a mass of 89.3 g is neutrally buoyant in a liquid. Identify the liquid.

15. | A 2.0 cm × 2.0 cm × 6.0 cm block floats in water with its long axis vertical. The length of the block above water is 2.0 cm. What is the block's mass density?

16. ‖ A sphere completely submerged in water is tethered to the bottom with a string. The tension in the string is one-third the weight of the sphere. What is the density of the sphere?

17. ‖ A 5.0 kg rock whose density is 4800 kg/m^3 is suspended by a string such that half of the rock's volume is under water. What is the tension in the string?

18. | What is the tension of the string in FIGURE EX15.18?

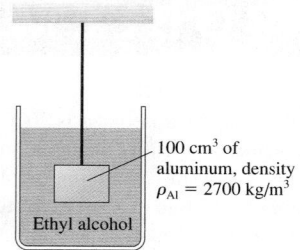

FIGURE EX15.18

100 cm^3 of aluminum, density $\rho_{Al} = 2700$ kg/m^3

Ethyl alcohol

19. ‖ A 10-cm-diameter, 20-cm-tall steel cylinder ($\rho_{steel} = 7900$ kg/m^3) floats in mercury. The axis of the cylinder is perpendicular to the surface. What length of steel is above the surface?

20. ‖ You and your friends are playing in the swimming pool with a 60-cm-diameter beach ball. How much force would be needed to push the ball completely under water?

21. ‖ Styrofoam has a density of 150 kg/m^3. What is the maximum mass that can hang without sinking from a 50-cm-diameter Styrofoam sphere in water? Assume the volume of the mass is negligible compared to that of the sphere.

Section 15.5 Fluid Dynamics

22. ‖ Water flowing through a hose at 4.0 m/s fills a 600 L child's wading pool in 8.0 min. What is the diameter in cm of the hose?

23. ‖ A 1.0-cm-diameter pipe widens to 2.0 cm, then narrows to 5.0 mm. Liquid flows through the first segment at a speed of 4.0 m/s.
   a. What is the speed in the second and third segments?
   b. What is the volume flow rate through the pipe?

24. ‖ A long horizontal tube has a square cross section with sides of width L. A fluid moves through the tube with speed v_0. The tube then changes to a circular cross section with diameter L. What is the fluid's speed in the circular part of the tube?

25. ‖ What does the top pressure gauge read in FIGURE EX15.25?

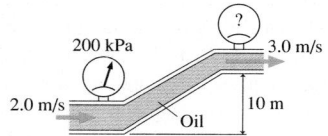

200 kPa   3.0 m/s
2.0 m/s   10 m
Oil

FIGURE EX15.25

Section 15.6 Elasticity

26. | An 80-cm-long, 1.0-mm-diameter steel guitar string must be tightened to a tension of 2000 N by turning the tuning screws. By how much is the string stretched?

27. ‖ A 70 kg mountain climber dangling in a crevasse stretches a 50-m-long, 1.0-cm-diameter rope by 8.0 cm. What is Young's modulus for the rope?

28. ‖ What hanging mass will increase the length of a 1.0-mm-diameter aluminum wire by 1.0%?

29. ‖ A 3.0-m-tall, 50-cm-diameter concrete column supports a 200,000 kg load. By how much is the column compressed?

30. | a. What is the pressure at a depth of 5000 m in the ocean?
   b. What is the fractional volume change $\Delta V/V$ of seawater at this pressure?
   c. What is the density of seawater at this pressure?

Problems

31. ‖ A gymnasium is 16 m high. By what percent is the air pressure at the floor greater than the air pressure at the ceiling?

32. | The two 60-cm-diameter cylinders in FIGURE P15.32, closed at one end, open at the other, are joined to form a single cylinder, then the air inside is removed.
   a. How much force does the atmosphere exert on the flat end of each cylinder?
   b. Suppose one cylinder is bolted to a sturdy ceiling. How many 100 kg football players would need to hang from the lower cylinder to pull the two cylinders apart?

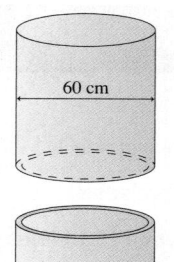

60 cm

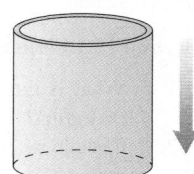

FIGURE P15.32

33. ‖ a. In **FIGURE P15.33**, how much force does the fluid exert on the end of the cylinder at A?

b. How much force does the fluid exert on the end of the cylinder at B?

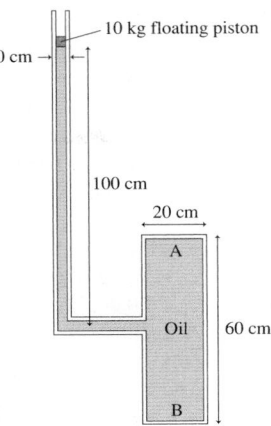

FIGURE P15.33

34. ‖ *Postural hypotension* is the occurrence of low systolic blood pressure when a person stands up too quickly from a reclining position. A brain blood pressure lower than 90 mm of Hg can cause fainting or lightheadedness. In a healthy adult, the automatic constriction and expansion of blood vessels keep the brain blood pressure constant while posture is changing, but disease or aging can weaken this response. If the blood pressure in your brain is 118 mm of Hg while lying down, what would it be when you stand up if this automatic response failed? Assume your brain is 40 cm from your heart and the density of blood is 1060 kg/m³.

35. ‖ A friend asks you how much pressure is in your car tires. You know that the tire manufacturer recommends 30 psi, but it's been a while since you've checked. You can't find a tire gauge in the car, but you do find the owner's manual and a ruler. Fortunately, you've just finished taking physics, so you tell your friend, "I don't know, but I can figure it out." From the owner's manual you find that the car's mass is 1500 kg. It seems reasonable to assume that each tire supports one-fourth of the weight. With the ruler you find that the tires are 15 cm wide and the flattened segment of the tire in contact with the road is 13 cm long. What answer will you give your friend?

36. ‖ A 2.0 mL syringe has an inner diameter of 6.0 mm, a needle inner diameter of 0.25 mm, and a plunger pad diameter (where you place your finger) of 1.2 cm. A nurse uses the syringe to inject medicine into a patient whose blood pressure is 140/100.

a. What is the minimum force the nurse needs to apply to the syringe?

b. The nurse empties the syringe in 2.0 s. What is the flow speed of the medicine through the needle?

37. ‖ What is the total mass of the earth's atmosphere?

38. ‖ Suppose the density of the earth's atmosphere were a constant 1.3 kg/m³, independent of height, until reaching the top. How thick would the atmosphere be?

39. ‖ The container shown in **FIGURE P15.39** is filled with oil. It is open to the atmosphere on the left.

a. What is the pressure at point A?

b. What is the pressure difference between points A and B? Between points A and C?

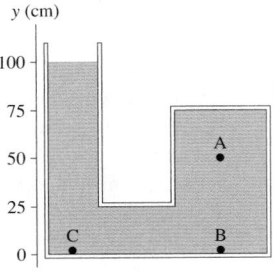

FIGURE P15.39

40. ‖ a. The 70 kg student in **FIGURE P15.40** balances a 1200 kg elephant on a hydraulic lift. What is the diameter of the piston the student is standing on?

b. When a second student joins the first, the piston sinks 35 cm. What is the second student's mass?

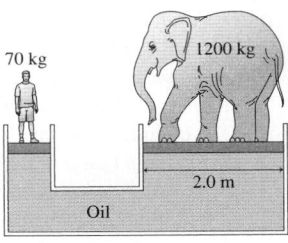

FIGURE P15.40

41. ‖‖ A 55 kg cheerleader uses an oil-filled hydraulic lift to hold four 110 kg football players at a height of 1.0 m. If her piston is 16 cm in diameter, what is the diameter of the football players' piston?

42. ‖ A U-shaped tube, open to the air on both ends, contains mercury. Water is poured into the left arm until the water column is 10.0 cm deep. How far upward from its initial position does the mercury in the right arm rise?

43. ‖ Glycerin is poured into an open U-shaped tube until the height in both sides is 20 cm. Ethyl alcohol is then poured into one arm until the height of the alcohol column is 20 cm. The two liquids do not mix. What is the difference in height between the top surface of the glycerin and the top surface of the alcohol?

44. ‖ Geologists place *tiltmeters* on the sides of volcanoes to measure the displacement of the surface as magma moves inside the volcano. Although most tiltmeters today are electronic, the traditional tilt-meter, used for decades, consisted

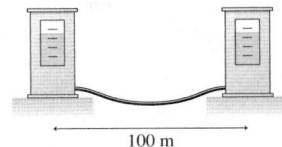

FIGURE P15.44

of two or more water-filled metal cans placed some distance apart and connected by a hose. **FIGURE P15.44** shows two such cans, each having a window to measure the water height. Suppose the cans are placed so that the water level in both is initially at the 5.0 cm mark. A week later, the water level in can 2 is at the 6.5 cm mark.

a. Did can 2 move up or down relative to can 1? By what distance?

b. Where is the water level now in can 1?

45. ‖ An aquarium of length *L*, width (front to back) *W*, and depth *D* is filled to the top with liquid of density ρ.

a. Find an expression for the force of the liquid on the bottom of the aquarium.

b. Find an expression for the force of the liquid on the front window of the aquarium.

c. Evaluate the forces for a 100-cm-long, 35-cm-wide, 40-cm-deep aquarium filled with water.

Hint: This problem requires an integration.

46. ‖ It's possible to use the ideal-gas law to show that the density of the earth's atmosphere decreases exponentially with height. That is, $\rho = \rho_0 \exp(-z/z_0)$, where *z* is the height above sea level, ρ_0 is the density at sea level (you can use the Table 15.1 value), and z_0 is called the *scale height* of the atmosphere.

a. Determine the value of z_0.

b. What is the density of the air in Denver, at an elevation of 1600 m? What percent of sea-level density is this?

Hint: This problem requires an integration. What is the weight of a column of air?

47. ‖ The average density of the body of a fish is 1080 kg/m³. To keep from sinking, a fish increases its volume by inflating an internal air bladder, known as a swim bladder, with air. By what percent must the fish increase its volume to be neutrally buoyant in fresh water? The density of air at 20°C is 1.19 kg/m³.

48. | You need to determine the density of a ceramic statue. If you suspend it from a spring scale, the scale reads 28.4 N. If you then lower the statue into a tub of water, so that it is completely submerged, the scale reads 17.0 N. What is the statue's density?

49. ‖ A cylinder with cross-section area A floats with its long axis vertical in a liquid of density ρ.
    a. Pressing down on the cylinder pushes it deeper into the liquid. Find an expression for the force needed to push the cylinder distance x deeper into the liquid and hold it there.
    b. A 4.0-cm-diameter cylinder floats in water. How much work must be done to push the cylinder 10 cm deeper into the water? **Hint:** An integration is required.

50. ‖ A less-dense liquid of density ρ_1 floats on top of a more-dense liquid of density ρ_2. A uniform cylinder of length l and density ρ, with $\rho_1 < \rho < \rho_2$, floats at the interface with its long axis vertical. What fraction of the length is in the more-dense liquid?

51. ‖ A 30-cm-tall, 4.0-cm-diameter plastic tube has a sealed bottom. 250 g of lead pellets are poured into the bottom of the tube, whose mass is 30 g, then the tube is lowered into a liquid. The tube floats with 5.0 cm extending above the surface. What is the density of the liquid?

52. ‖ One day when you come into physics lab you find several plastic hemispheres floating like boats in a tank of fresh water. Each lab group is challenged to determine the heaviest rock that can be placed in the bottom of a plastic boat without sinking it. You get one try. Sinking the boat gets you no points, and the maximum number of points goes to the group that can place the heaviest rock without sinking. You begin by measuring one of the hemispheres, finding that it has a mass of 21 g and a diameter of 8.0 cm. What is the mass of the heaviest rock that, in perfectly still water, won't sink the plastic boat?

53. ‖‖ A spring with spring constant 35 N/m is attached to the ceiling, and a 5.0-cm-diameter, 1.0 kg metal cylinder is attached to its lower end. The cylinder is held so that the spring is neither stretched nor compressed, then a tank of water is placed underneath with the surface of the water just touching the bottom of the cylinder. When released, the cylinder will oscillate a few times but, damped by the water, quickly reach an equilibrium position. When in equilibrium, what length of the cylinder is submerged?

54. ‖ A plastic "boat" with a 25 cm² square cross section floats in a liquid. One by one, you place 50 g masses inside the boat and measure how far the boat extends below the surface. Your data are as follows:

Mass added (g)	Depth (cm)
50	2.9
100	5.0
150	6.6
200	8.6

Draw an appropriate graph of the data and, from the slope and intercept of the best-fit line, determine the mass of the boat and the density of the liquid.

55. ‖‖ A 355 mL soda can is 6.2 cm in diameter and has a mass of 20 g. Such a soda can half full of water is floating upright in water. What length of the can is above the water level?

56. ‖‖ The bottom of a steel "boat" is a 5.0 m × 10 m × 2.0 cm piece of steel ($\rho_{steel} = 7900$ kg/m³). The sides are made of 0.50-cm-thick steel. What minimum height must the sides have for this boat to float in perfectly calm water?

57. ‖ A nuclear power plant draws 3.0×10^6 L/min of cooling water from the ocean. If the water is drawn in through two parallel, 3.0-m-diameter pipes, what is the water speed in each pipe?

58. ‖ a. A liquid of density ρ flows at speed v_0 through a horizontal pipe that expands smoothly from diameter d_0 to a larger diameter d_1. The pressure in the narrower section is p_0. Find an expression for the pressure p_1 in the wider section.
    b. A pressure gauge reads 50 kPa as water flows at 10.0 m/s through a 16.8-cm-diameter horizontal pipe. What is the reading of a pressure gauge after the pipe has expanded to 20.0 cm in diameter?

59. ‖ A tree loses water to the air by the process of *transpiration* at
    BIO the rate of 110 g/h. This water is replaced by the upward flow of sap through vessels in the trunk. If the trunk contains 2000 vessels, each 100 μm in diameter, what is the upward speed of the sap in each vessel? The density of tree sap is 1040 kg/m³.

60. ‖ Water flows from the pipe shown in FIGURE P15.60 with a speed of 4.0 m/s.
    a. What is the water pressure as it exits into the air?
    b. What is the height h of the standing column of water?

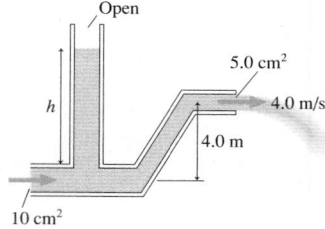

FIGURE P15.60

61. ‖ Water flowing out of a 16-mm-diameter faucet fills a 2.0 L bottle in 10 s. At what distance below the faucet has the water stream narrowed to 10 mm diameter?

62. ‖ A hurricane wind blows across a 6.0 m × 15.0 m flat roof at a speed of 130 km/h.
    a. Is the air pressure above the roof higher or lower than the pressure inside the house? Explain.
    b. What is the pressure difference?
    c. How much force is exerted on the roof? If the roof cannot withstand this much force, will it "blow in" or "blow out"?

63. ‖ Air flows through the tube shown in FIGURE P15.63 at a rate of 1200 cm³/s. Assume that air is an ideal fluid. What is the height h of mercury in the right side of the U-tube?

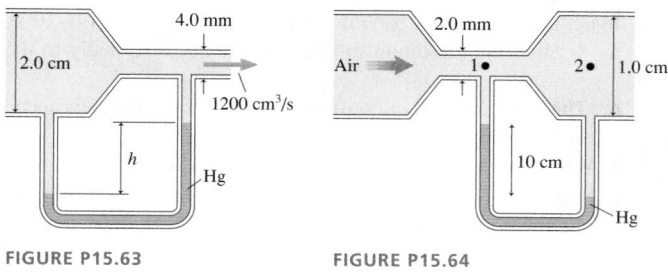

FIGURE P15.63　　　　FIGURE P15.64

64. ‖ Air flows through the tube shown in FIGURE P15.64. Assume that air is an ideal fluid.
    a. What are the air speeds v_1 and v_2 at points 1 and 2?
    b. What is the volume flow rate?

65. ‖ A water tank of height h has a small hole at height y. The water is replenished to keep h from changing. The water squirting from the hole has range x. The range approaches zero as $y \to 0$ because the water squirts right onto the ground. The range also approaches zero as $y \to h$ because the horizontal velocity becomes zero. Thus there must be some height y between 0 and h for which the range is a maximum.

a. Find an algebraic expression for the flow speed v with which the water exits the hole at height y.

b. Find an algebraic expression for the range of a particle shot horizontally from height y with speed v.

c. Combine your expressions from parts a and b. Then find the maximum range x_{max} and the height y of the hole. "Real" water won't achieve quite this range because of viscosity, but it will be close.

66. ‖ a. A cylindrical tank of radius R, filled to the top with a liquid, has a small hole in the side, of radius r, at distance d below the surface. Find an expression for the volume flow rate through the hole.

    b. A 4.0-mm-diameter hole is 1.0 m below the surface of a 2.0-m-diameter tank of water. What is the rate, in mm/min, at which the water level will initially drop if the water is not replenished?

67. ‖ A large 10,000 L aquarium is supported by four wood posts (Douglas fir) at the corners. Each post has a square 4.0 cm × 4.0 cm cross section and is 80 cm tall. By how much is each post compressed by the weight of the aquarium?

68. ‖ BIO There is a disk of cartilage between each pair of vertebrae in your spine. Young's modulus for cartilage is 1.0×10^6 N/m². Suppose a relaxed disk is 4.0 cm in diameter and 5.0 mm thick. If a disk in the lower spine supports half the weight of a 66 kg person, by how many mm does the disk compress?

69. ‖ A cylindrical steel pressure vessel with volume 1.30 m³ is to be tested. The vessel is entirely filled with water, then a piston at one end of the cylinder is pushed in until the pressure inside the vessel has increased by 2000 kPa. Suddenly, a safety plug on the top bursts. How many liters of water come out?

Challenge Problems

70. The 1.0-m-tall cylinder in **FIGURE CP15.70** contains air at a pressure of 1 atm. A very thin, frictionless piston of negligible mass is placed at the top of the cylinder, to prevent any air from escaping, then mercury is slowly poured into the cylinder until no more can be added without the cylinder overflowing. What is the height h of the column of compressed air?

Hint: Boyle's law, which you learned in chemistry, says $p_1V_1 = p_2V_2$ for a gas compressed at constant temperature, which we will assume to be the case.

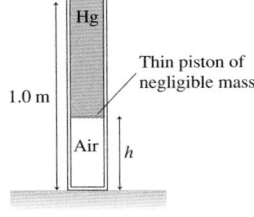

FIGURE CP15.70

71. In **FIGURE CP15.71**, a cone of density ρ_o and total height l floats in a liquid of density ρ_f. The height of the cone above the liquid is h. What is the ratio h/l of the exposed height to the total height?

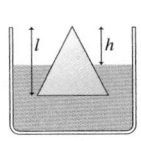
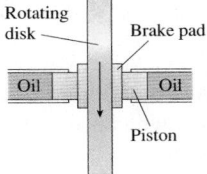

FIGURE CP15.71   **FIGURE CP15.72**

72. Disk brakes, such as those in your car, operate by using pressurized oil to push outward on a piston. The piston, in turn, presses brake pads against a spinning rotor or wheel, as seen in **FIGURE CP15.72**. Consider a 15 kg industrial grinding wheel, 26 cm in diameter, spinning at 900 rpm. The brake pads are actuated by 2.0-cm-diameter pistons, and they contact the wheel an average distance 12 cm from the axis. If the coefficient of kinetic friction between the brake pad and the wheel is 0.60, what oil pressure is needed to stop the wheel in 5.0 s?

73. A cylinder of density ρ_o, length l, and cross-section area A floats in a liquid of density ρ_f with its axis perpendicular to the surface. Length h of the cylinder is submerged when the cylinder floats at rest.

a. Show that $h = (\rho_o / \rho_f)l$.

b. Suppose the cylinder is distance y *above* its equilibrium position. Find an expression for $(F_{net})_y$, the y-component of the net force on the cylinder. Use what you know to cancel terms and write this expression as simply as possible.

c. You should recognize your result of part b as a version of Hooke's law. What is the "spring constant" k?

d. If you push a floating object down and release it, it bobs up and down. So it is like a spring in the sense that it oscillates if displaced from equilibrium. Use your "spring constant" and what you know about simple harmonic motion to show that the cylinder's oscillation period is

$$T = 2\pi \sqrt{\frac{h}{g}}$$

e. What is the oscillation period for a 100-m-tall iceberg ($\rho_{ice} = 917$ kg/m³) in seawater?

74. A cylindrical tank of diameter $2R$ contains water to a depth d. A small hole of diameter $2r$ is opened in the bottom of the tank. $r \ll R$, so the tank drains slowly. Find an expression for the time it takes to drain the tank completely.

STOP TO THINK ANSWERS

Stop to Think 15.1: $\rho_a = \rho_b = \rho_c$. Density depends only on what the object is made of, not how big the pieces are.

Stop to Think 15.2: c. These are all open tubes, so the liquid rises to the same height in all three despite their different shapes.

Stop to Think 15.3: $F_b > F_a = F_c$. The masses in c do not add. The pressure underneath each of the two large pistons is mg/A_2, and the pressure under the small piston must be the same.

Stop to Think 15.4: b. The weight of the displaced water equals the weight of the ice cube. When the ice cube melts and turns into water,

that amount of water will exactly fill the volume that the ice cube is now displacing.

Stop to Think 15.5: 1 cm³/s out. The fluid is incompressible, so the sum of what flows in must match the sum of what flows out. 13 cm³/s is known to be flowing in, while 12 cm³/s flows out. An additional 1 cm³/s must flow out to achieve balance.

Stop to Think 15.6: $h_b > h_d > h_c > h_a$. The liquid level is higher where the pressure is lower. The pressure is lower where the flow speed is higher. The flow speed is highest in the narrowest tube, zero in the open air.

SUMMARY
Applications of Newtonian Mechanics

We have developed two parallel perspectives of motion, each with its own concepts and techniques. We focused on the first of these in Part I, where we dealt with the relationship between force and motion. Newton's second law is the principle most central to the force/motion perspective. Then, in Part II, we developed a before-and-after perspective based on the idea of conservation laws. Newton's laws were essential in the development of conservation laws, but they remain hidden in the background when the conservation laws are applied. Together, these two perspectives form the heart of Newtonian mechanics.

Our goal in Part III has been to see how Newtonian mechanics is applied to several diverse but important topics. We added only one new law of physics in Part III, Newton's law of gravity, and we introduced few completely new concepts. Instead, we've broadened our understanding of the force/motion

perspective and the conservation-law perspective through our investigations of rotational motion, gravity, oscillations, and fluids. In reviewing Part III, pay close attention to the interplay between these two perspectives. Recognizing which is the best tool in a particular situation will help you improve your problem-solving ability.

Our knowledge of mechanics is now essentially complete. We will add a few additional ideas as we need them, but our journey into physics will be taking us in entirely new directions as we continue on. Hence this is an opportune moment to step back a bit to take a look at the "big picture." Newtonian mechanics may seem all very factual and straightforward to us today, but keep in mind that these ideas are all human inventions. There was a time when they did not exist and when our concepts of nature were quite different from what they are today.

KNOWLEDGE STRUCTURE III **Applications of Newtonian Mechanics**

Rotation of a Rigid Body

A rigid body is a system of particles.
Rotational motion is analogous to linear motion.

Rotational motion	**Linear motion**
Angular acceleration α	Acceleration a
Torque τ	Force F
Moment of inertia I	Mass m
Angular momentum L	Momentum p

- **Newton's second law** $\tau_{net} = I\alpha$

- **Rotational kinetic energy** $K = \frac{1}{2}I\omega^2$

NEWTON'S LAWS
+
CONSERVATION LAWS

Newton's Theory of Gravity

Any two masses exert attractive gravitational forces on each other.

Newton's law of gravity is

$$F_{m \text{ on } M} = F_{M \text{ on } m} = \frac{GMm}{r^2}$$

- Kepler's laws describe the elliptical orbits of satellites and planets.

- The gravitational potential energy is

$$U_g = -\frac{GMm}{r}$$

Oscillations

Systems with a linear restoring force exhibit simple harmonic oscillation.

- The **kinematic equations of SHM** are

$$x(t) = A\cos(\omega t + \phi_0)$$
$$v(t) = -v_{max}\sin(\omega t + \phi_0)$$

where $v_{max} = \omega A$ and the phase constant ϕ_0 describes the initial conditions.

- **Energy is transformed between kinetic and potential** as the system oscillates. In an undamped system, the total mechanical energy

$$E = \frac{1}{2}mv^2 + \frac{1}{2}kx^2 = \frac{1}{2}m(v_{max})^2 = \frac{1}{2}kA^2$$

is conserved.

Fluids and Elasticity

Fluids are systems that flow. Gases and liquids are fluids. Fluids are better characterized by density and pressure than by mass and force.

- **Liquids**   Pressure is primarily gravitational. The hydrostatic pressure is

$$p = p_0 + \rho gd$$

- **Gases**   Pressure is primarily thermal. Pressure in a container is constant.

- **Archimedes' principle**   The buoyant force is equal to the weight of the displaced fluid.

For fluid flow, **Bernoulli's equation**

$$p_1 + \frac{1}{2}\rho v_1^2 + \rho gy_1 = p_2 + \frac{1}{2}\rho v_2^2 + \rho gy_2$$

is really a statement of energy conservation.

The Newtonian Synthesis

Newton's achievements, praised by no less than Einstein as "perhaps the greatest advance in thought that a single individual was ever privileged to make," are often called the *Newtonian synthesis*. "Synthesis" means "the uniting or combining of separate elements to form a coherent whole." It is often said of Newton that he "united the heavens and the earth." In doing so, he changed forever the way we view ourselves and our relationship to the universe.

Medieval cosmology considered the heavenly bodies to be perfect, unchanging objects quite unrelated to imperfect and changeable earthly matter. Their perfection and immortality symbolized the perfection of God above, while the material bodies of humans were imperfect and mortal. This cosmology was mirrored in medieval feudal society. The king—ordained by God and whose symbol was the sun—was surrounded by a small circle of nobles and a larger circle of serfs and peasants. Taken together, the ideas and institutions of science, religion, and society of this time form what we call the medieval *worldview*. Their worldview, in its many facets, was hierarchical and authoritarian, reflecting their understanding of "natural order" in the universe.

Copernicus weakened medieval cosmology by questioning the position of the earth in the universe. Galileo, with his telescope, found that the heavens are not perfect and unchanging. Now, at the end of the 17th century, the success of Newton's theories implied that the sun and the planets were merely ordinary matter, obeying the same natural laws as earthly matter. This uniting of earthly motions and heavenly motions—the *synthesis* in the Newtonian synthesis—dealt the final blow to the medieval worldview.

Newton's success changed the way we see and think about the universe. Rather than seeing whirling celestial spheres, people began to think of the universe in terms of the motion of material particles following rigid laws. This Newtonian conception of the cosmos is often called a "clockwork universe." The technology of clocks was progressing rapidly in the 18th century, and people everywhere admired the consistency and predictability of these little machines. The Newtonian universe is a very large machine, but one that is consistent, predictable, and law-abiding. In other words, a perfect clock.

Major thinkers of the 17th and 18th centuries soon concluded that God had created the world by placing all the particles in their original positions, then giving them a push to get them going. God, in this role, was called the "prime mover." But once the universe was started, it went along perfectly well just by obeying Newton's laws. No divine intervention or guidance was needed. This is certainly a very different view of our relationship to God and the universe than was contained in the medieval worldview.

Newton also influenced the way people think about themselves and their society. His theories clearly demonstrated that the universe is not random or capricious but, instead, follows natural laws. Others soon began to apply the concept of natural law to human nature, human behavior, and human institutions. The main protagonist in this school of thought was the English philosopher and political scientist John Locke, a contemporary of Newton. Locke developed a theory of human behavior from the ideas of natural laws and empirical evidence. We cannot go into Locke's theories here, but Newton's success helped to propel Locke's ideas into the mainstream of 18th-century political thought.

Locke's writings had a great influence on a young American named Thomas Jefferson. The concept of natural laws, as they apply to individuals, is very much behind Jefferson's enunciation of "unalienable rights" in the Declaration of Independence. In fact, the first sentence of the Declaration refers explicitly to "the Laws of Nature and of Nature's God." The idea of *checks and balances,* built into the Constitution of the United States, is very much a mechanical and clock-like model of how political institutions function.

Just as medieval feudalism mirrored the medieval understanding of the universe, contemporary constitutional democracy mirrors, in many ways, the Newtonian cosmology. Hierarchy and authority have been replaced by equality and law because they now seem to us the "natural order" of things. Having grown up with this modern worldview, we find it difficult to imagine any other. Nonetheless, it is important to realize that vastly different worldviews have existed at other times and in other cultures.

Science has changed dramatically in the last hundred-odd years. Newton's clockwork universe has been superseded by relativity and quantum physics. Entirely new theories and sciences, such as evolution, ecology, and psychology, have appeared. These new ideas are slowly working their way into other areas of thought and human activity, and bit by bit they are changing the ways in which we see ourselves, our society, and our relationship to nature. A future worldview is in the making.

Thermodynamics

A modern jet engine is a marvel of technical ingenuity. Understanding how a jet engine works requires understanding the thermodynamics of gases and heat engines.

OVERVIEW

It's All About Energy

Thermodynamics—the science of energy in its broadest context—arose hand in hand with the industrial revolution as the systematic study of converting heat energy into mechanical motion and work. Hence the name *thermo + dynamics*. Indeed, the analysis of engines and generators of various kinds remains the focus of engineering thermodynamics. But thermodynamics, as a science, now extends to all forms of energy conversions, including those involving living organisms. For example:

- **Engines** convert the energy of a fuel into the mechanical energy of moving pistons, gears, and wheels.
- **Fuel cells** convert chemical energy into electrical energy.
- **Photovoltaic cells** convert the electromagnetic energy of light into electrical energy.
- **Lasers** convert electrical energy into the electromagnetic energy of light.
- **Organisms** convert the chemical energy of food into a variety of other forms of energy, including kinetic energy, sound energy, and thermal energy.

The major goals of Part IV are to understand both *how* energy transformations such as these take place and *how efficient* they are. We'll discover that the laws of thermodynamics place limits on the efficiency of energy transformations, and understanding these limits is essential for analyzing the very real energy needs of society in the 21st century.

Our ultimate destination in Part IV is an understanding of the thermodynamics of *heat engines*. A heat engine is a device, such as a power plant or an internal combustion engine, that transforms heat energy into useful work. These are the devices that power our modern society.

Understanding how to transform heat into work will be a significant achievement, but we first have many steps to take along the way. We need to understand the concepts of temperature and pressure. We need to learn about the properties of solids, liquids, and gases. Most important, we need to expand our view of energy to include *heat*, the energy that is transferred between two systems at different temperatures.

At a deeper level, we need to see how these concepts are connected to the underlying microphysics of randomly moving molecules. We will find that the familiar concepts of thermodynamics, such as temperature and pressure, have their roots in atomic-level motion and collisions. We will also find it possible to learn a great deal about the properties of molecules, such as their speeds, on the basis of purely macroscopic measurements. This *micro/macro connection* will lead to the second law of thermodynamics, one of the most subtle but also one of the most profound and far-reaching statements in physics.

Only after all these steps have been taken will we be able to analyze a real heat engine. It is an ambitious goal, but one we can achieve.

16 A Macroscopic Description of Matter

The phrase "solid as a rock" is cast in doubt when rocks melt, as they do in this flowing lava.

▶ **Looking Ahead** The goal of Chapter 16 is to learn the characteristics of macroscopic systems.

Temperature

You're familiar with temperature, but what does it actually measure?

We'll start with the simple idea that temperature measures "hotness" and "coldness," but we'll eventually recognize that temperature measures a system's thermal energy.

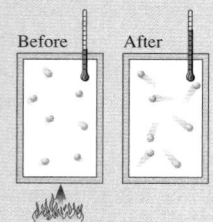

You'll learn to use temperatures in *kelvins*, an absolute temperature scale with absolute zero at 0 K.

Macroscopic Systems

The properties of a macroscopic system are called its **bulk properties**. Examples include a system's volume, density, pressure, and temperature.

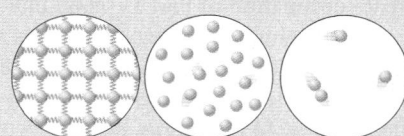

Macroscopic systems can be characterized as solid, liquid, or gas—the three most common *phases* of matter.

Starting with this chapter and continuing throughout Part IV, you'll learn to understand the bulk properties of macroscopic systems in terms of the microscopic motion of their atoms and molecules. This **micro/macro connection** is an important part of our modern understanding of matter.

In this chapter, you will learn to describe the amount of substance in a macroscopic system using moles, mass, and the number of atoms.

◀ **Looking Back**
Sections 15.1–15.3 Fluids and pressure

Ideal Gases

We'll model a gas as consisting of tiny, hard spheres that occasionally collide with each other or the walls of their container but otherwise do not interact. You'll learn to use the *ideal-gas law* to understand the bulk properties of a gas.

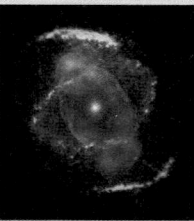

The Cat's Eye Nebula is a huge ball of hot gas ejected by the star in the center.

Ideal-Gas Processes

Heating or compressing a gas changes the *state* of the gas. We'll study three basic ideal-gas processes:
- Constant-pressure process
- Constant-volume process
- Constant-temperature process

You'll learn to represent ideal-gas processes on a *pV* diagram.

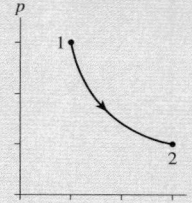

Phase Changes

Melting, freezing, boiling, and condensing are *phase changes*, where a macroscopic system changes from one phase to another. You'll learn to represent phase changes in terms of a *phase diagram*.

These melting ice cubes are undergoing a phase change from solid to liquid.

16.1 Solids, Liquids, and Gases

Each of the elements and most compounds can exist as a solid, liquid, or gas—the three most common **phases** of matter. The change between liquid and solid (freezing or melting) or between liquid and gas (boiling or condensing) is called a **phase change.** We're familiar with only one, or perhaps two, of the phases of most substances because their melting point and/or boiling point are far outside the range of normal human experience. Water is the only substance for which all three phases—ice, liquid, and steam—are everyday occurrences.

NOTE ▶ This use of the word "phase" has no relationship at all to the *phase* or *phase constant* of simple harmonic motion and waves. ◀

Solids, liquids, and gases

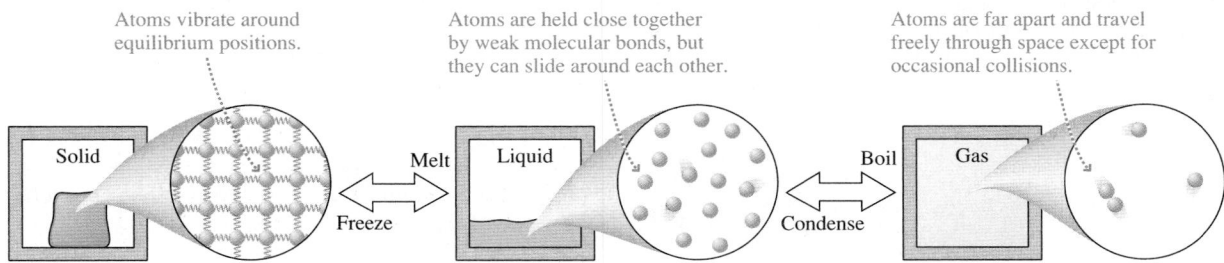

Atoms vibrate around equilibrium positions.

Atoms are held close together by weak molecular bonds, but they can slide around each other.

Atoms are far apart and travel freely through space except for occasional collisions.

A **solid** is a rigid macroscopic system consisting of particle-like atoms connected by spring-like molecular bonds. Each atom vibrates around an equilibrium position but otherwise has a fixed position. Solids are nearly *incompressible,* which tells us that the atoms in a solid are just about as close together as they can get.

The solid shown here is a **crystal,** meaning that the atoms are arranged in a periodic array. The elements and many compounds have a crystal structure when in their solid phase. In other solids, such as glass, the atoms are frozen into random positions. These are called **amorphous solids.**

A **liquid** is more complicated than either a solid or a gas. Like a solid, a liquid is nearly *incompressible.* This tells us that the molecules in a liquid are about as close together as they can get. Like a gas, a liquid flows and deforms to fit the shape of its container. The fluid nature of a liquid tells us that the molecules are free to move around.

Together, these observations suggest a model in which the molecules of the liquid are loosely held together by weak molecular bonds. The bonds are strong enough that the molecules never get far apart but not strong enough to prevent the molecules from sliding around each other.

A **gas** is a system in which each molecule moves through space as a free, noninteracting particle until, on occasion, it collides with another molecule or with the wall of the container. A gas is a *fluid.* A gas is also highly *compressible,* which tells us that there is lots of space between the molecules.

Gases are fairly simple macroscopic systems; hence many of our examples in Part IV will be based on gases.

State Variables

The parameters used to characterize or describe a macroscopic system are known as **state variables** because, taken all together, they describe the *state* of the macroscopic system. You met some state variables in earlier chapters: volume, pressure, mass, mass density, and thermal energy. We'll soon introduce several new state variables: moles, number density, and, most important, the temperature T.

One important state variable, the mass density, is defined as the ratio of two other state variables:

$$\rho = \frac{M}{V} \quad \text{(mass density)} \quad (16.1)$$

In this chapter we'll use an uppercase M for the system mass and a lowercase m for the mass of an atom. Table 16.1 is a short list of mass densities.

TABLE 16.1 Densities of materials

Substance	ρ (kg/m^3)
Air at STP*	1.28
Ethyl alcohol	790
Water (solid)	920
Water (liquid)	1000
Aluminum	2700
Copper	8920
Gold	19,300
Iron	7870
Lead	11,300
Mercury	13,600
Silicon	2330

*$T = 0°C$, $p = 1$ atm

If we change the value of any of the state variables, then we change the state of the system. For example, to *compress* a gas means to decrease its volume. The symbol Δ represents a *change* in the value of a state variable. That is, ΔT is a *change* of temperature and Δp is a *change* of pressure. **For any quantity X, ΔX is always $X_f - X_i$, the final value minus the initial value.**

A system is said to be in **thermal equilibrium** if its state variables are constant and not changing. As an example, a gas is in thermal equilibrium if it has been left undisturbed long enough for p, V, and T to reach steady values.

EXAMPLE 16.1   The mass of a lead pipe

A project on which you are working uses a cylindrical lead pipe with outer and inner diameters of 4.0 cm and 3.5 cm, respectively, and a length of 50 cm. What is its mass?

SOLVE The mass density of lead is $\rho_{\text{lead}} = 11{,}300 \text{ kg/m}^3$. The volume of a circular cylinder of length l is $V = \pi r^2 l$. In this case we need to find the volume of the outer cylinder, of radius r_2, *minus*

the volume of air in the inner cylinder, of radius r_1. The volume of the pipe is

$$V = \pi r_2^2 l - \pi r_1^2 l = \pi(r_2^2 - r_1^2)l = 1.47 \times 10^{-4} \text{ m}^3$$

Hence the pipe's mass is

$$M = \rho_{\text{lead}} V = 1.7 \text{ kg}$$

STOP TO THINK 16.1   The pressure in a system is measured to be 60 kPa. At a later time the pressure is 40 kPa. The value of Δp is

a. 60 kPa   b. 40 kPa   c. 20 kPa   d. -20 kPa

16.2 Atoms and Moles

The mass of a macroscopic system is directly related to the total number of atoms or molecules in the system, denoted N. Because N is determined simply by counting, it is a number with no units. A typical macroscopic system has $N \sim 10^{25}$ atoms, an incredibly large number.

The symbol $\sim$, if you are not familiar with it, stands for "has the order of magnitude." It means that the number is known only to within a factor of 10 or so. The statement $N \sim 10^{25}$, which is read "N is of order 10^{25}," implies that N is somewhere in the range 10^{24} to 10^{26}. It is far less precise than the "approximately equal" symbol $\approx$. Saying $N \sim 10^{25}$ gives us a rough idea of how large N is and allows us to know that it differs significantly from 10^5 or even 10^{15}.

It is often useful to know the number of atoms or molecules per cubic meter in a system. We call this quantity the **number density**. It characterizes how densely the atoms are packed together within the system. In an N-atom system that fills volume V, the number density is

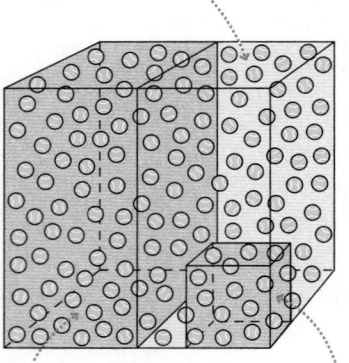

FIGURE 16.1 The number density of a uniform system is independent of the volume.

A 100 m³ room has 10,000 tennis balls bouncing around. The number density of tennis balls in the room is $N/V = 10{,}000/100 \text{ m}^3 = 100 \text{ m}^{-3}$.

If we look at only half the room, we would find 5000 balls in 50 m³, again giving $N/V = 5000/50 \text{ m}^3 = 100 \text{ m}^{-3}$.

In one-tenth of the room, we would find 1000 balls in 10 m³, again giving $N/V = 1000/10 \text{ m}^3 = 100 \text{ m}^{-3}$.

$$\frac{N}{V} \quad \text{(number density)} \tag{16.2}$$

The SI units of number density are m^{-3}. The number density of atoms in a solid is $(N/V)_{\text{solid}} \sim 10^{29} \text{ m}^{-3}$. The number density of a gas depends on the pressure, but is usually less than 10^{27} m^{-3}. As **FIGURE 16.1** shows, **the value of N/V in a *uniform* system is independent of the volume V.** That is, the number density is the same whether you look at the whole system or just a portion of it.

NOTE ▶ While we might say "There are 100 tennis balls per cubic meter," or "There are 10^{29} atoms per cubic meter," tennis balls and atoms are not units. The units of N/V are simply m^{-3}. ◀

Atomic Mass and Atomic Mass Number

You will recall from chemistry that atoms of different elements have different masses. The mass of an atom is determined primarily by its most massive constituents, the protons and neutrons in its nucleus. The *sum* of the number of protons and neutrons is called the **atomic mass number** A:

$$A = \text{number of protons} + \text{number of neutrons}$$

A, which by definition is an integer, is written as a leading superscript on the atomic symbol. For example, the common isotope of hydrogen, with one proton and no neutrons, is ^1H. The "heavy hydrogen" isotope called *deuterium,* which includes one neutron, is ^2H. The primary isotope of carbon, with six protons (which makes it carbon) and six neutrons, is ^{12}C. The radioactive isotope ^{14}C, used for carbon dating of archeological finds, contains six protons and eight neutrons.

The **atomic mass** scale is established by defining the mass of ^{12}C to be exactly 12 u, where u is the symbol for the **atomic mass unit.** That is, $m(^{12}C) = 12$ u. The atomic mass of any other atom is its mass relative to ^{12}C. For example, careful experiments with hydrogen find that the mass *ratio* $m(^1H)/m(^{12}C)$ is 1.0078/12. Thus the atomic mass of hydrogen is $m(^1H) = 1.0078$ u.

The numerical value of the atomic mass of ^1H is close to, but not exactly, its atomic mass number $A = 1$. For our purposes, it will be sufficient to overlook the slight difference and use the integer atomic mass numbers as the values of the atomic mass. That is, we'll use $m(^1H) = 1$ u, $m(^4He) = 4$ u, and $m(^{16}O) = 16$ u. For molecules, the **molecular mass** is the sum of the atomic masses of the atoms forming the molecule. Thus the molecular mass of O_2, the constituent of oxygen gas, is $m(O_2) = 32$ u.

> NOTE ▶ An element's atomic mass number is *not* the same as its atomic number. The *atomic number,* the element's position in the periodic table, is the number of protons in the nucleus. ◀

Table 16.2 shows the atomic mass numbers of some of the elements that we'll use for examples and homework problems. A complete periodic table of the elements, including atomic masses, is found in Appendix B.

Moles and Molar Mass

One way to specify the amount of substance in a macroscopic system is to give its mass. Another is to measure the amount of substance in *moles.* By definition, one **mole** of matter, be it solid, liquid, or gas, is the amount of substance containing as many basic particles as there are atoms in 0.012 kg (12 g) of ^{12}C. Many ingenious experiments have determined that there are 6.02×10^{23} atoms in 0.012 kg of ^{12}C, so we can say that 1 mole of substance, abbreviated 1 mol, is 6.02×10^{23} basic particles.

The basic particle depends on the substance. Helium is a **monatomic gas,** meaning that the basic particle is the helium atom. Thus 6.02×10^{23} helium atoms are 1 mol of helium. But oxygen gas is a **diatomic gas** because the basic particle is the two-atom diatomic molecule O_2. 1 mol of oxygen gas contains 6.02×10^{23} *molecules* of O_2 and thus $2 \times 6.02 \times 10^{23}$ oxygen atoms. Table 16.3 lists the monatomic and diatomic gases that we will use for examples and homework problems.

The number of basic particles per mole of substance is called **Avogadro's number,** N_A. The value of Avogadro's number is

$$N_A = 6.02 \times 10^{23} \text{ mol}^{-1}$$

Despite its name, Avogadro's number is not simply "a number"; it has units. Because there are N_A particles per mole, the number of moles in a substance containing N basic particles is

$$n = \frac{N}{N_A} \quad \text{(moles of substance)} \quad (16.3)$$

One mole of helium, sulfur, copper, and mercury.

TABLE 16.2 Some atomic mass numbers

Element		A
^1H	Hydrogen	1
^4He	Helium	4
^{12}C	Carbon	12
^{14}N	Nitrogen	14
^{16}O	Oxygen	16
^{20}Ne	Neon	20
^{27}Al	Aluminum	27
^{40}Ar	Argon	40
^{207}Pb	Lead	207

TABLE 16.3 Monatomic and diatomic gases

Monatomic		Diatomic	
He	Helium	H_2	Hydrogen
Ne	Neon	N_2	Nitrogen
Ar	Argon	O_2	Oxygen

Avogadro's number allows us to determine atomic masses in kilograms. Knowing that N_A ^{12}C atoms have a mass of 0.012 kg, the mass of one ^{12}C atom must be

$$m(^{12}\text{C}) = \frac{0.012 \text{ kg}}{6.02 \times 10^{23}} = 1.993 \times 10^{-26} \text{ kg}$$

We defined the atomic mass scale such that $m(^{12}\text{C}) = 12$ u. Thus the conversion factor between atomic mass units and kilograms is

$$1 \text{ u} = \frac{m(^{12}\text{C})}{12} = 1.66 \times 10^{-27} \text{ kg}$$

This conversion factor allows us to calculate the mass in kg of any atom. For example, a ^{20}Ne atom has atomic mass $m(^{20}\text{Ne}) = 20$ u. Multiplying by 1.66×10^{-27} kg/u gives $m(^{20}\text{Ne}) = 3.32 \times 10^{-26}$ kg. If the atomic mass is specified in kilograms, the number of atoms in a system of mass M can be found from

$$N = \frac{M}{m} \qquad (16.4)$$

The **molar mass** of a substance is the mass of 1 mol of substance. The molar mass, which we'll designate M_{mol}, has units kg/mol. By definition, the molar mass of ^{12}C is 0.012 kg/mol. For other substances, whose atomic or molecular masses are given relative to ^{12}C, the numerical value of the molar mass is the numerical value of the atomic or molecular mass divided by 1000. For example, the molar mass of He, with $m = 4$ u, is $M_{\text{mol}}(\text{He}) = 0.004$ kg/mol and the molar mass of diatomic O_2 is $M_{\text{mol}}(O_2) = 0.032$ kg/mol.

Equation 16.4 uses the atomic mass to find the number of atoms in a system. Similarly, you can use the molar mass to determine the number of moles. For a system of mass M consisting of atoms or molecules with molar mass M_{mol},

$$n = \frac{M}{M_{\text{mol}}} \qquad (16.5)$$

EXAMPLE 16.2 **Moles of oxygen**

100 g of oxygen gas is how many moles of oxygen?

SOLVE We can do the calculation two ways. First, let's determine the number of molecules in 100 g of oxygen. The diatomic oxygen molecule O_2 has molecular mass $m = 32$ u. Converting this to kg, we get the mass of one molecule:

$$m = 32 \text{ u} \times \frac{1.66 \times 10^{-27} \text{ kg}}{1 \text{ u}} = 5.31 \times 10^{-26} \text{ kg}$$

Thus the number of molecules in 100 g = 0.100 kg is

$$N = \frac{M}{m} = \frac{0.100 \text{ kg}}{5.31 \times 10^{-26} \text{ kg}} = 1.88 \times 10^{24}$$

Knowing the number of molecules gives us the number of moles:

$$n = \frac{N}{N_A} = 3.13 \text{ mol}$$

Alternatively, we can use Equation 16.5 to find

$$n = \frac{M}{M_{\text{mol}}} = \frac{0.100 \text{ kg}}{0.032 \text{ kg/mol}} = 3.13 \text{ mol}$$

STOP TO THINK 16.2 Which system contains more atoms: 5 mol of helium ($A = 4$) or 1 mol of neon ($A = 20$)?

a. Helium.   b. Neon.   c. They have the same number of atoms.

16.3 Temperature

We are all familiar with the idea of temperature. Mass is a measure of the amount of substance in a system. Velocity is a measure of how fast a system moves. What physical property of the system have you determined if you measure its temperature?

We will begin with the commonsense idea that temperature is a measure of how "hot" or "cold" a system is. As we develop these ideas, we'll find that **temperature** T is related to a system's *thermal energy*. We defined thermal energy in Chapter 10 as the kinetic and potential energy of the atoms and molecules in a system as they vibrate (a solid) or move around (a gas). A system has more thermal energy when it is "hot" than when it is "cold." In Chapter 18, we'll replace these vague notions of hot and cold with a precise relationship between temperature and thermal energy.

To start, we need a means to measure the temperature of a system. This is what a *thermometer* does. A thermometer can be any small macroscopic system that undergoes a measurable change as it exchanges thermal energy with its surroundings. It is placed in contact with a larger system whose temperature it will measure. In a common glass-tube thermometer, for example, a small volume of mercury or alcohol expands or contracts when placed in contact with a "hot" or "cold" object. The object's temperature is determined by the length of the column of liquid.

A thermometer needs a *temperature scale* to be a useful measuring device. In 1742, the Swedish astronomer Anders Celsius sealed mercury into a small capillary tube and observed how it moved up and down the tube as the temperature changed. He selected two temperatures that anyone could reproduce, the freezing and boiling points of pure water, and labeled them 0 and 100. He then marked off the glass tube into one hundred equal intervals between these two reference points. By doing so, he invented the temperature scale that we today call the *Celsius scale*. The units of the Celsius temperature scale are "degrees Celsius," which we abbreviate °C. Note that the degree symbol ° is part of the unit, not part of the number.

The *Fahrenheit scale,* still widely used in the United States, is related to the Celsius scale by

$$T_F = \frac{9}{5}T_C + 32° \qquad (16.6)$$

Table 16.4 lists several temperatures measured on the Celsius and Fahrenheit scales and also on the Kelvin scale.

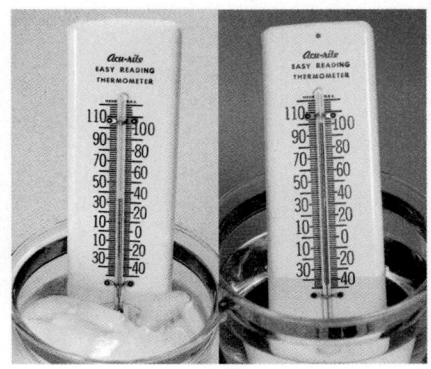

Thermal expansion of the liquid in the thermometer tube pushes it higher in the hot water than in the ice water.

TABLE 16.4 Temperatures measured with different scales

Temperature	T (°C)	T (K)	T (°F)
Melting point of iron	1538	1811	2800
Boiling point of water	100	373	212
Normal body temperature	37	310	99
Room temperature	20	293	68
Freezing point of water	0	273	32
Boiling point of nitrogen	−196	77	−321
Absolute zero	−273	0	−460

Absolute Zero and Absolute Temperature

Any physical property that changes with temperature can be used as a thermometer. In practice, the most useful thermometers have a physical property that changes *linearly* with temperature. One of the most important scientific thermometers is the **constant-volume gas thermometer** shown in FIGURE 16.2a on the next page. This thermometer depends on the fact that the *absolute* pressure (not the gauge pressure) of a gas in a sealed container increases linearly as the temperature increases.

FIGURE 16.2 The pressure in a constant-volume gas thermometer extrapolates to zero at $T_0 = -273°C$. This is the basis for the concept of absolute zero.

(a)

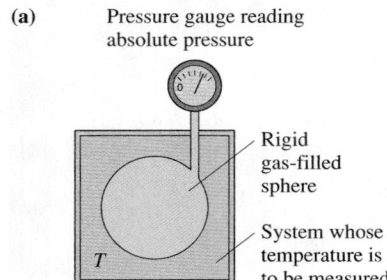

Pressure gauge reading absolute pressure

Rigid gas-filled sphere

System whose temperature is to be measured

T

(b)

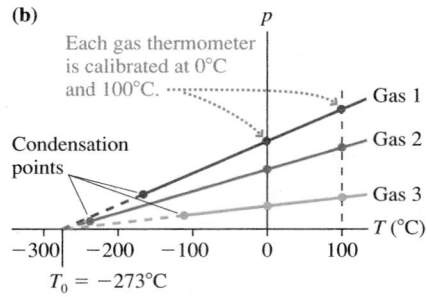

Each gas thermometer is calibrated at 0°C and 100°C.

Condensation points

Gas 1

Gas 2

Gas 3

$T_0 = -273°C$

A gas thermometer is first calibrated by recording the pressure at two reference temperatures, such as the boiling and freezing points of water. These two points are plotted on a pressure-versus-temperature graph and a straight line is drawn through them. The gas bulb is then brought into contact with the system whose temperature is to be measured. The pressure is measured, then the corresponding temperature is read off the graph.

FIGURE 16.2b shows the pressure-temperature relationship for three different gases. Notice two important things about this graph.

1. There is a *linear* relationship between temperature and pressure.
2. All gases extrapolate to *zero pressure* at the same temperature: $T_0 = -273°C$. No gas actually gets that cold without condensing, although helium comes very close, but it is surprising that you get the same zero-pressure temperature for any gas and any starting pressure.

The pressure in a gas is due to collisions of the molecules with each other and the walls of the container. A pressure of zero would mean that all motion, and thus all collisions, had ceased. If there were no atomic motion, the system's thermal energy would be zero. The temperature at which all motion would cease, and at which $E_{th} = 0$, is called **absolute zero**. Because temperature is related to thermal energy, absolute zero is the lowest temperature that has physical meaning. We see from the gas-thermometer data that $T_0 = -273°C$.

It is useful to have a temperature scale with the zero point at absolute zero. Such a temperature scale is called an **absolute temperature scale**. Any system whose temperature is measured on an absolute scale will have $T > 0$. The absolute temperature scale having the same unit size as the Celsius scale is called the *Kelvin scale*. It is the SI scale of temperature. The units of the Kelvin scale are *kelvins,* abbreviated as K. The conversion between the Celsius scale and the Kelvin scale is

$$T_K = T_C + 273 \qquad (16.7)$$

On the Kelvin scale, absolute zero is 0 K, the freezing point of water is 273 K, and the boiling point of water is 373 K.

NOTE ▶ The units are simply "kelvins," *not* "degrees Kelvin." ◀

STOP TO THINK 16.3 The temperature of a glass of water increases from 20°C to 30°C. What is ΔT?

a. 10 K   b. 283 K   c. 293 K   d. 303 K

FIGURE 16.3 The temperature as a function of time as water is transformed from solid to liquid to gas.

(a)

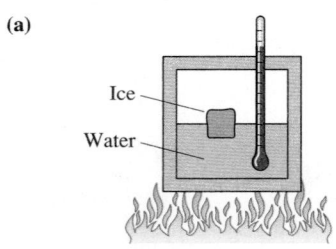

Ice

Water

(b)

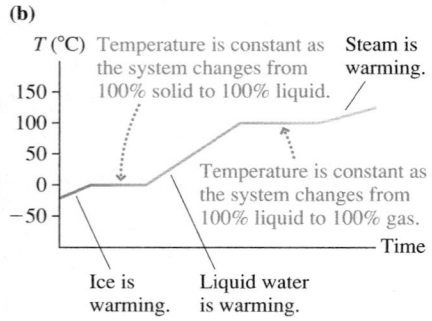

T (°C)   Temperature is constant as the system changes from 100% solid to 100% liquid.   Steam is warming.

150
100
50
0
−50

Temperature is constant as the system changes from 100% liquid to 100% gas.

Time

Ice is warming.   Liquid water is warming.

16.4 Phase Changes

The temperature inside the freezer compartment of a refrigerator is typically about −20°C. Suppose you were to remove a few ice cubes from the freezer, place them in a sealed container with a thermometer, then heat them, as **FIGURE 16.3a** shows. We'll assume that the heating is done so slowly that the inside of the container always has a single, well-defined temperature.

FIGURE 16.3b shows the temperature as a function of time. After steadily rising from the initial −20°C, the temperature remains fixed at 0°C for an extended period of time. This is the interval of time during which the ice melts. As it's melting, the ice temperature is 0°C and the liquid water temperature is 0°C. Even though the system is being heated, the liquid water temperature doesn't begin to rise until all the ice has melted. If you were to turn off the flame at any point, the system would remain a mixture of ice and liquid water at 0°C.

NOTE ▶ In everyday language, the three phases of water are called *ice, water,* and *steam.* That is, the term *water* implies the liquid phase. Scientifically, these are the solid, liquid, and gas phases of the compound called *water.* To be clear, we'll use the term *water* in the scientific sense of a collection of H₂O molecules. We'll say either *liquid* or *liquid water* to denote the liquid phase. ◀

The thermal energy of a solid is the kinetic energy of the vibrating atoms plus the potential energy of the stretched and compressed molecular bonds. Melting occurs when the thermal energy gets so large that molecular bonds begin to break, allowing the atoms to move around. The temperature at which a solid becomes a liquid or, if the thermal energy is reduced, a liquid becomes a solid is called the **melting point** or the **freezing point.** Melting and freezing are *phase changes.*

A system at the melting point is in **phase equilibrium,** meaning that any amount of solid can coexist with any amount of liquid. Raise the temperature ever so slightly and the entire system becomes liquid. Lower it slightly and it all becomes solid. But exactly at the melting point the system has no tendency to move one way or the other. That is why the temperature remains constant at the melting point until the phase change is complete.

You can see the same thing happening in Figure 16.3b at 100°C, the boiling point. This is a phase equilibrium between the liquid phase and the gas phase, and any amount of liquid can coexist with any amount of gas at this temperature. Above this temperature, the thermal energy is too large for bonds to be established between molecules, so the system is a gas. If the thermal energy is reduced, the molecules begin to bond with each other and stick together. In other words, the gas condenses into a liquid. The temperature at which a gas becomes a liquid or, if the thermal energy is increased, a liquid becomes a gas is called the **condensation point** or the **boiling point.**

NOTE ▶ Liquid water becomes solid ice at 0°C, but that doesn't mean the temperature of ice is always 0°C. Ice reaches the temperature of its surroundings. If the air temperature in a freezer is −20°C, then the ice temperature is −20°C. Likewise, steam can be heated to temperatures above 100°C. That doesn't happen when you boil water on the stove because the steam escapes, but steam can be heated far above 100°C in a sealed container. ◀

A **phase diagram** is used to show how the phases and phase changes of a substance vary with both temperature and pressure. **FIGURE 16.4** shows the phase diagrams for water and carbon dioxide. You can see that each diagram is divided into three regions corresponding to the solid, liquid, and gas phases. The boundary lines separating the regions indicate the phase transitions. The system is in phase equilibrium at a pressure-temperature point that falls on one of these lines.

Phase diagrams contain a great deal of information. Notice on the water phase diagram that the dashed line at $p = 1$ atm crosses the solid-liquid boundary at 0°C and the liquid-gas boundary at 100°C. These well-known melting and boiling point temperatures of water apply only at standard atmospheric pressure. You can see that in Denver, where $p_{atmos} < 1$ atm, water melts at slightly above 0°C and boils at a temperature below 100°C. A *pressure cooker* works by allowing the pressure inside to exceed 1 atm. This raises the boiling point, so foods that are in boiling water are at a temperature above 100°C and cook faster.

Crossing the solid-liquid boundary corresponds to melting or freezing while crossing the liquid-gas boundary corresponds to boiling or condensing. But there's another possibility—crossing the solid-gas boundary. The phase change in which a solid becomes a gas is called **sublimation.** It is not an everyday experience with water, but you probably are familiar with the sublimation of dry ice. Dry ice is solid carbon dioxide. You can see on the carbon dioxide phase diagram that the dashed line at $p = 1$ atm crosses the solid-*gas* boundary, rather than the solid-liquid boundary, at $T = -78$°C. This is the *sublimation temperature* of dry ice.

FIGURE 16.4 Phase diagrams (not to scale) for water and carbon dioxide.

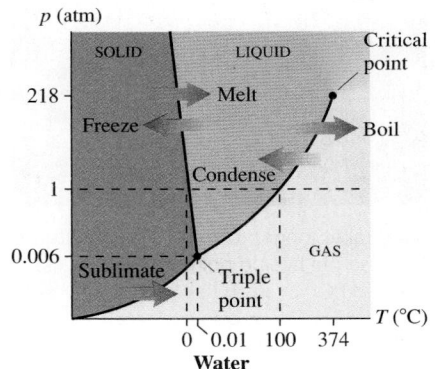

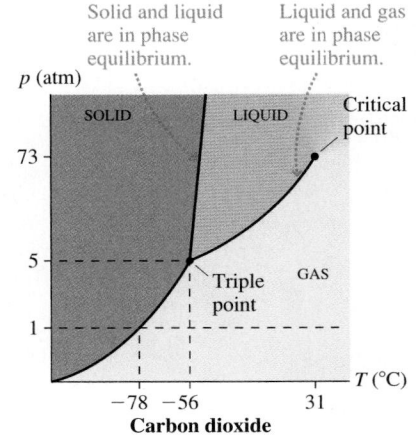

Liquid carbon dioxide does exist, but only at pressures greater than 5 atm and temperatures greater than $-56°C$. A CO_2 fire extinguisher contains *liquid* carbon dioxide under high pressure. (You can hear the liquid slosh if you shake a CO_2 fire extinguisher.)

One important difference between the water and carbon dioxide phase diagrams is the slope of the solid-liquid boundary. For most substances, the solid phase is denser than the liquid phase and the liquid is denser than the gas. Pressurizing the substance compresses it and increases the density. If you start compressing CO_2 gas at room temperature, thus moving upward through the phase diagram along a vertical line, you'll first condense it to a liquid and eventually, if you keep compressing, change it into a solid.

Water is a very unusual substance in that the density of ice is *less* than the density of liquid water. That is why ice floats. If you compress ice, making it denser, you eventually cause a phase transition in which the ice turns to liquid water! Consequently, the solid-liquid boundary for water slopes to the left.

The liquid-gas boundary ends at a point called the **critical point.** Below the critical point, liquid and gas are clearly distinct and there is a phase change if you go from one to the other. But there is no clear distinction between liquid and gas at pressures or temperatures above the critical point. The system is a *fluid,* but it can be varied continuously between high density and low density without a phase change.

The final point of interest on the phase diagram is the **triple point** where the phase boundaries meet. Two phases are in phase equilibrium along the boundaries. The triple point is the *one* value of temperature and pressure for which all three phases can coexist in phase equilibrium. That is, any amounts of solid, liquid, and gas can happily coexist at the triple point. For water, the triple point occurs at $T_3 = 0.01°C$ and $p_3 = 0.006$ atm.

The significance of the triple point of water is its connection to the Kelvin temperature scale. The Celsius scale required two *reference points,* the boiling and melting points of water. We can now see that these are not very satisfactory reference points because their values vary as the pressure changes. In contrast, there's only one temperature at which ice, liquid water, and water vapor will coexist in equilibrium. If you produce this equilibrium in the laboratory, then you *know* the system is at the triple-point temperature.

The triple-point temperature of water is an ideal reference point, hence the Kelvin temperature scale is *defined* to be a linear temperature scale starting from 0 K at absolute zero and passing through 273.16 K at the triple point of water. Because $T_3 = 0.01°C$, absolute zero on the Celsius scale is $T_0 = -273.15°C$.

NOTE ▶ To be consistent with our use of significant figures, $T_0 = -273$ K is the appropriate value to use in calculations *unless* you know other temperatures with an accuracy of better than $1°C$. ◀

Food takes longer to cook at high altitudes because the boiling point of water is less than $100°C$.

STOP TO THINK 16.4 For which is there a sublimation temperature that is higher than a melting temperature?

a. Water   b. Carbon dioxide   c. Both   d. Neither

16.5 Ideal Gases

We noted earlier in the chapter that solids and liquids are nearly incompressible, an observation suggesting that atoms are fairly hard and cannot be pressed together once they come into contact with each other. Based on this observation, suppose we were to model atoms as "hard spheres" that do not interact except for occasional elastic collisions when two atoms come into contact and bounce apart.

This is a *model* of an atom—what we might call the *ideal atom*—because it ignores the weak attractive interactions that hold liquids and solids together. A gas of these noninteracting atoms is called an **ideal gas.** It is a gas of small, hard, randomly moving atoms that bounce off each other and the walls of their container but otherwise do not interact. The ideal gas is a somewhat simplified description of a real gas, but experiments show that the ideal-gas model is quite good for real gases if two conditions are met:

1. The density is low (i.e., the atoms occupy a volume much smaller than that of the container), and
2. The temperature is well above the condensation point.

If the density gets too high, or the temperature too low, then the attractive forces between the atoms begin to play an important role and our model, which ignores those attractive forces, fails. These are the forces that are responsible, under the right conditions, for the gas condensing into a liquid.

We've been using the term "atoms," but many gases, as you know, consist of molecules rather than atoms. Only helium, neon, argon, and the other inert elements in the far-right column of the periodic table of the elements form monatomic gases. Hydrogen (H_2), nitrogen (N_2), and oxygen (O_2) are diatomic gases. As far as translational motion is concerned, the ideal-gas model does not distinguish between a monatomic gas and a diatomic gas; both are considered as simply small, hard spheres. Hence the terms "atoms" and "molecules" can be used interchangeably to mean the basic constituents of the gas.

The Ideal-Gas Law

Section 16.1 introduced the idea of *state variables,* those parameters that describe the state of a macroscopic system. The state variables for an ideal gas are the volume V of its container, the number of moles n of the gas present in the container, the temperature T of the gas and its container, and the pressure p that the gas exerts on the walls of the container. These four state parameters are not independent of each other. If you change the value of one—by, say, raising the temperature—then one or more of the others will change as well. Each change of the parameters is a *change of state* of the system.

Experiments during the 17th and 18th centuries found a very specific relationship between the four state variables. Suppose you change the state of a gas, by heating it or compressing it or doing something else to it, and measure p, V, n, and T. Repeat this many times, changing the state of the gas each time, until you have a large table of p, V, n, and T values.

Then make a graph on which you plot pV, the product of the pressure and volume, on the vertical axis and nT, the product of the number of moles and temperature (in kelvins), on the horizontal axis. The very surprising result is that for *any* gas, whether it is hydrogen or helium or oxygen or methane, **you get exactly the same graph,** the linear graph shown in FIGURE 16.5. In other words, nothing about the graph indicates what gas was used because all gases give the same result.

NOTE ▶ No real gas could extend to $nT = 0$ because it would condense. But an ideal gas never condenses because the only interactions among the molecules are hard-sphere collisions. ◀

As you can see, there is a very clear proportionality between the quantity pV and the quantity nT. If we designate the slope of the line in this graph as R, then we can write the relationship as

$$pV = R \times (nT)$$

It is customary to write this relationship in a slightly different form, namely

$$pV = nRT \qquad \text{(ideal-gas law)} \qquad (16.8)$$

Equation 16.8 is the **ideal-gas law. The ideal-gas law is a relationship among the four state variables—p, V, n, and T—that characterize a gas in thermal equilibrium.**

FIGURE 16.5 A graph of pV versus nT for an ideal gas.

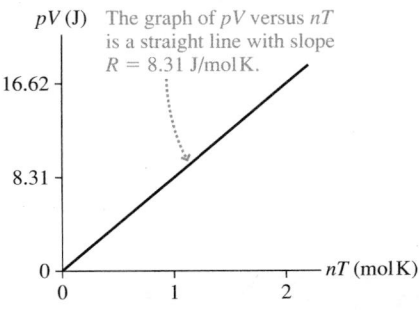

The constant R, which is determined experimentally as the slope of the graph in Figure 16.5, is called the **universal gas constant.** Its value, in SI units, is

$$R = 8.31 \text{ J/mol K}$$

The units of R seem puzzling. The denominator mol K is clear because R multiplies nT. But what about the joules? The left side of the ideal-gas law, pV, has units

$$\text{Pa m}^3 = \frac{\text{N}}{\text{m}^2}\text{m}^3 = \text{N m} = \text{joules}$$

The product pV has units of joules, as shown on the vertical axis in Figure 16.5.

NOTE ▶ You perhaps learned in chemistry to work gas problems using units of atmospheres and liters. To do so, you had a different numerical value of R expressed in those units. In physics, however, we always work gas problems in SI units. Pressures *must* be in Pa, volumes in m^3, and temperatures in K. ◀

The surprising fact, and one worth commenting upon, is that *all* gases have the *same* graph and the *same* value of R. There is no obvious reason a very simple atomic gas such as helium should have the same slope as a more complex gas such as methane (CH$_4$). Nonetheless, both turn out to have the same value for R. The ideal-gas law, within its limits of validity, describes *all* gases with a single value of the constant R.

EXAMPLE 16.3 **Calculating a gas pressure**

100 g of oxygen gas is distilled into an evacuated 600 cm^3 container. What is the gas pressure at a temperature of 150°C?

MODEL The gas can be treated as an ideal gas. Oxygen is a diatomic gas of O$_2$ molecules.

SOLVE From the ideal-gas law, the pressure is $p = nRT/V$. In Example 16.2 we calculated the number of moles in 100 g of O$_2$ and found $n = 3.13$ mol. Gas problems typically involve several conversions to get quantities into the proper units, and this example is no exception. The SI units of V and T are m^3 and K, respectively, thus

$$V = (600 \text{ cm}^3)\left(\frac{1 \text{ m}}{100 \text{ cm}}\right)^3 = 6.00 \times 10^{-4} \text{ m}^3$$

$$T = (150 + 273) \text{ K} = 423 \text{ K}$$

With this information, the pressure is

$$p = \frac{nRT}{V} = \frac{(3.13 \text{ mol})(8.31 \text{ J/mol K})(423 \text{ K})}{6.00 \times 10^{-4} \text{ m}^3}$$

$$= 1.83 \times 10^7 \text{ Pa} = 181 \text{ atm}$$

In this text we will consider only gases in sealed containers. The number of moles (and number of molecules) will not change during a problem. In that case,

$$\frac{pV}{T} = nR = \text{constant} \qquad (16.9)$$

If the gas is initially in state i, characterized by the state variables p_i, V_i, and T_i, and at some later time in a final state f, the state variables for these two states are related by

$$\frac{p_f V_f}{T_f} = \frac{p_i V_i}{T_i} \qquad \text{(ideal gas in a sealed container)} \qquad (16.10)$$

This before-and-after relationship between the two states, reminiscent of a conservation law, will be valuable for many problems.

EXAMPLE 16.4 **Calculating a gas temperature**

A cylinder of gas is at 0°C. A piston compresses the gas to half its original volume and three times its original pressure. What is the final gas temperature?

MODEL Treat the gas as an ideal gas in a sealed container.

SOLVE The before-and-after relationship of Equation 16.10 can be written

$$T_2 = T_1 \frac{p_2}{p_1} \frac{V_2}{V_1}$$

In this problem, the compression of the gas results in $V_2/V_1 = \frac{1}{2}$ and $p_2/p_1 = 3$. The initial temperature is $T_1 = 0°C = 273$ K. With this information,

$$T_2 = 273 \text{ K} \times 3 \times \frac{1}{2} = 409 \text{ K} = 136°C$$

ASSESS We did not need to know actual values of the pressure and volume, just the *ratios* by which they change.

We will often want to refer to the number of molecules N in a gas rather than the number of moles n. This is an easy change to make. Because $n = N/N_A$, the ideal-gas law in terms of N is

$$pV = nRT = \frac{N}{N_A}RT = N\frac{R}{N_A}T \qquad (16.11)$$

R/N_A, the ratio of two known constants, is known as **Boltzmann's constant** k_B:

$$k_B = \frac{R}{N_A} = 1.38 \times 10^{-23} \text{ J/K}$$

The subscript B distinguishes Boltzmann's constant from a spring constant or other uses of the symbol k.

Ludwig Boltzmann was an Austrian physicist who did some of the pioneering work in statistical physics during the mid-19th century. Boltzmann's constant k_B can be thought of as the "gas constant per molecule," whereas R is the "gas constant per mole." With this definition, the ideal-gas law in terms of N is

$$pV = Nk_BT \qquad \text{(ideal-gas law)} \qquad (16.12)$$

Equations 16.8 and 16.12 are both the ideal-gas law, just expressed in terms of different state variables.

Recall that the number density (molecules per m³) was defined as N/V. A re-arrangement of Equation 16.12 gives the number density as

$$\frac{N}{V} = \frac{p}{k_BT} \qquad (16.13)$$

This is a useful consequence of the ideal-gas law, but keep in mind that the pressure *must* be in SI units of pascals and the temperature *must* be in SI units of kelvins.

EXAMPLE 16.5 **The distance between molecules**

"Standard temperature and pressure," abbreviated **STP,** are $T = 0°C$ and $p = 1$ atm. Estimate the average distance between gas molecules at STP.

MODEL Consider the gas to be an ideal gas.

SOLVE Suppose a container of volume V holds N molecules at STP. How do we estimate the distance between them? Imagine placing an imaginary sphere around each molecule, separating it from its neighbors. This divides the total volume V into N little spheres of volume v_i, where $i = 1$ to N. The spheres of two neighboring molecules touch each other, like a crate full of Ping-Pong balls of somewhat different sizes all touching their neighbors, so the distance between two molecules is the sum of the radii of their two spheres. Each of these spheres is somewhat different, but a reasonable *estimate* of the distance between molecules would be twice the *average* radius of a sphere.

The average volume of one of these little spheres is

$$v_{avg} = \frac{V}{N} = \frac{1}{N/V}$$

That is, the average volume per molecule (m³ per molecule) is the inverse of the number density, the number of molecules per m³. This is not the volume of the molecule itself, which is much smaller, but the average volume of space that each molecule can claim as its own. We can use Equation 16.13 to calculate the number density:

$$\frac{N}{V} = \frac{p}{k_BT} = \frac{1.01 \times 10^5 \text{ Pa}}{(1.38 \times 10^{-23} \text{ J/K})(273 \text{ K})}$$
$$= 2.69 \times 10^{25} \text{ molecules/m}^3$$

where we used the definition of STP in SI units. Thus the average volume per molecule is

$$v_{avg} = \frac{1}{N/V} = 3.72 \times 10^{-26} \text{ m}^3$$

The volume of a sphere is $\frac{4}{3}\pi r^3$, so the average radius of a sphere is

$$r_{avg} = \left(\frac{3}{4\pi}v_{avg}\right)^{1/3} = 2.1 \times 10^{-9} \text{ m} = 2.1 \text{ nm}$$

The average distance between two molecules, with their spheres touching, is twice r_{avg}. Thus

$$\text{average distance} = 2r_{avg} \approx 4 \text{ nm}$$

This is a simple estimate, so we've given the answer with only one significant figure.

ASSESS One of the assumptions of the ideal-gas model is that atoms or molecules are "far apart" in comparison to the sizes of atoms and molecules. Chemistry experiments find that small molecules, such as N_2 and O_2, are roughly 0.3 nm in diameter. For a gas at STP, we see that the average distance between molecules is more than 10 times the size of a molecule. Thus the ideal-gas model works very well for a gas at STP.

You have two containers of equal volume. One is full of helium gas. The other holds an equal mass of nitrogen gas. Both gases have the same pressure. How does the temperature of the helium compare to the temperature of the nitrogen?

a. $T_{helium} > T_{nitrogen}$　　　　b. $T_{helium} = T_{nitrogen}$　　　　c. $T_{helium} < T_{nitrogen}$

16.6 Ideal-Gas Processes

The ideal-gas law is the connection between the state variables pressure, temperature, and volume. If the state variables change, as they would from heating or compressing the gas, the state of the gas changes. An **ideal-gas process** is the means by which the gas changes from one state to another.

> NOTE ▶ Even in a sealed container, the ideal-gas law is a relationship among *three* variables. In general, *all three change* during an ideal-gas process. As a result, thinking about cause and effect can be rather tricky. Don't make the mistake of thinking that one variable is constant unless you're sure, beyond a doubt, that it is. ◀

The *pV* Diagram

It will be very useful to represent ideal-gas processes on a graph called a ***pV* diagram.** This is nothing more than a graph of pressure versus volume. The important idea behind the *pV* diagram is that *each point* on the graph represents a single, unique state of the gas. That seems surprising at first, because a point on the graph only directly specifies the values of *p* and *V*. But knowing *p* and *V*, and assuming that *n* is known for a sealed container, we can find the temperature by using the ideal-gas law. Thus each point actually represents a triplet of values (*p*, *V*, *T*) specifying the state of the gas.

For example, FIGURE 16.6a is a *pV* diagram showing three states of a system consisting of 1 mol of gas. The values of *p* and *V* can be read from the axes for each point, then the temperature at that point determined from the ideal-gas law. An ideal-gas process can be represented as a "trajectory" in the *pV* diagram. The trajectory shows all the intermediate states through which the gas passes. FIGURE 16.6b shows two different processes by which the gas of Figure 16.6a can be changed from state 1 to state 3.

There are infinitely many ways to change the gas from state 1 to state 3. Although the initial and final states are the same for each of them, the particular process by which the gas changes—that is, the particular trajectory—will turn out to have very real consequences. For example, you will soon learn that the work done by an expanding gas, a quantity of very practical importance in various devices, depends on the trajectory followed. The *pV* diagram is an important graphical representation of the process.

FIGURE 16.6 The state of the gas and ideal-gas processes can be shown on a *pV* diagram.

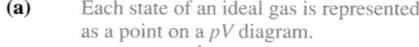
(a) Each state of an ideal gas is represented as a point on a *pV* diagram.

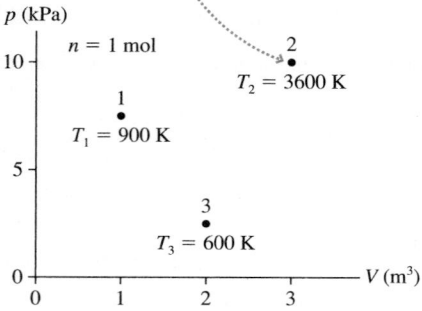

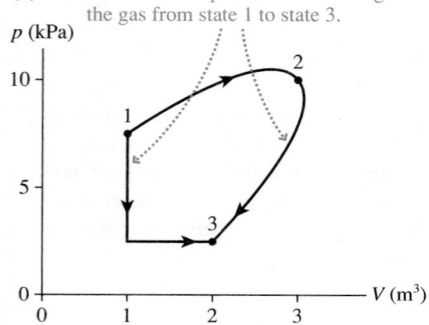
(b) Two different processes that change the gas from state 1 to state 3.

Quasi-Static Processes

Strictly speaking, the ideal-gas law applies only to gases in *thermal equilibrium*, meaning that the state variables are constant and not changing. But, by definition, an ideal-gas process causes some of the state variables to change. The gas is *not* in thermal equilibrium while the process of changing from state 1 to state 2 is under way.

To use the ideal-gas law throughout, we will assume that the process occurs *so slowly* that the system is never far from equilibrium. In other words, the values of *p*, *V*, and *T* at any point in the process are essentially the same as the equilibrium values they would assume if we stopped the process at that point. A process that is essentially in thermal equilibrium at all times is called a **quasi-static process.** It is an idealization, like a frictionless surface, but one that is a very good approximation in many real situations.

An important characteristic of a quasi-static process is that the trajectory through the pV diagram can be *reversed*. If you quasi-statically expand a gas by slowly pulling a piston out, as shown in FIGURE 16.7a, you can reverse the process by slowly pushing the piston in. The gas retraces its pV trajectory until it has returned to its initial state. Contrast this with what happens when the membrane bursts in FIGURE 16.7b. That is a sudden process, not at all quasi-static. The *irreversible* process of Figure 16.7b cannot be represented on a pV diagram.

The critical question is: How slow must a process be to qualify as quasi-static? That is a difficult question to answer. This textbook will always assume that processes are quasi-static. It turns out to be a reasonable assumption for the types of examples and homework problems we will look at. Irreversible processes will be left to more advanced courses.

Constant-Volume Process

Many important gas processes take place in a container of constant, unchanging volume. A constant-volume process is called an **isochoric process,** where *iso* is a prefix meaning "constant" or "equal" while *choric* is from a Greek root meaning "volume." An isochoric process is one for which

$$V_f = V_i \qquad (16.14)$$

For example, suppose that you have a gas in the closed, rigid container shown in FIGURE 16.8a. Warming the gas with a Bunsen burner will raise its pressure without changing its volume. This process is shown as the vertical line $1 \rightarrow 2$ on the pV diagram of FIGURE 16.8b. A constant-volume cooling, by placing the container on a block of ice, would lower the pressure and be represented as the vertical line from 2 to 1. **Any isochoric process appears on a pV diagram as a vertical line.**

FIGURE 16.7 The slow motion of the piston is a quasi-static process. The bursting of the membrane is not.

(a)
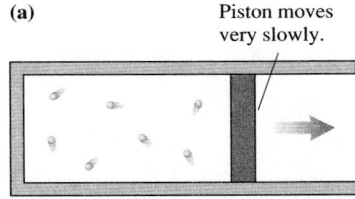
Quasi-static process

(b)
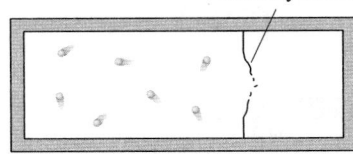
Irreversible process

FIGURE 16.8 A constant-volume (isochoric) process.

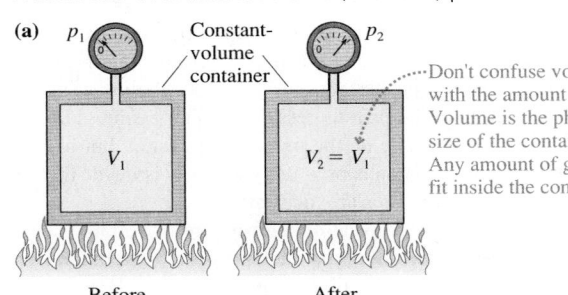

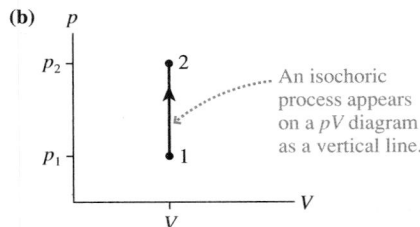

EXAMPLE 16.6 **A constant-volume gas thermometer**

A constant-volume gas thermometer is placed in contact with a reference cell containing water at the triple point. After reaching equilibrium, the gas pressure is recorded as 55.78 kPa. The thermometer is then placed in contact with a sample of unknown temperature. After the thermometer reaches a new equilibrium, the gas pressure is 65.12 kPa. What is the temperature of this sample?

MODEL The thermometer's volume doesn't change, so this is an isochoric process.

SOLVE The temperature at the triple point of water is $T_1 = 0.01°C = 273.16$ K. The ideal-gas law for a closed system

is $p_2V_2/T_2 = p_1V_1/T_1$. The volume doesn't change, so $V_2/V_1 = 1$. Thus

$$T_2 = T_1 \frac{V_2}{V_1}\frac{p_2}{p_1} = T_1 \frac{p_2}{p_1} = (273.16 \text{ K})\frac{65.12 \text{ kPa}}{55.78 \text{ kPa}}$$

$$= 318.90 \text{ K} = 45.75°C$$

The temperature *must* be in kelvins to do this calculation, although it is common to convert the final answer to °C. The fact that the pressures were given to four significant figures justified using $T_K = T_C + 273.15$ rather than the usual $T_C + 273$.

ASSESS $T_2 > T_1$, which we expected from the increase in pressure.

FIGURE 16.9 A constant-pressure (isobaric) process.

(a)

The piston's mass maintains a constant pressure in the cylinder.

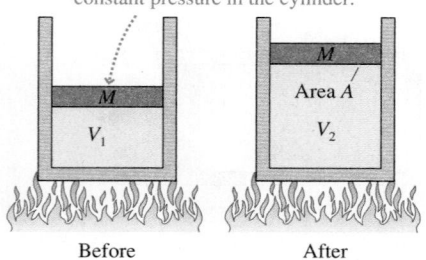

Before   After

(b)

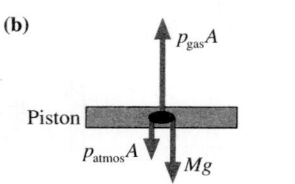

(c)

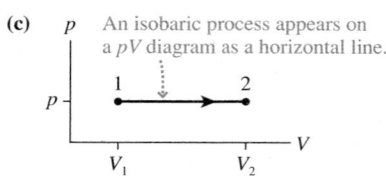

Constant-Pressure Process

Other gas processes take place at a constant, unchanging pressure. A constant-pressure process is called an **isobaric process,** where *baric* is from the same root as "barometer" and means "pressure." An isobaric process is one for which

$$p_f = p_i \tag{16.15}$$

FIGURE 16.9a shows one method of changing the state of a gas while keeping the pressure constant. A cylinder of gas has a tight-fitting piston of mass M that can slide up and down but seals the container so that no atoms enter or escape. As the free-body diagram of FIGURE 16.9b shows, the piston and the air press down with force $p_{atmos}A + Mg$ while the gas inside pushes up with force $p_{gas}A$. In equilibrium, the gas pressure inside the cylinder is

$$p_{gas} = p_{atmos} + \frac{Mg}{A} \tag{16.16}$$

In other words, the gas pressure is determined by the requirement that the gas must support both the mass of the piston and the air pressing inward. **This pressure is independent of the temperature of the gas or the height of the piston, so it stays constant as long as M is unchanged.**

If the cylinder is warmed, the gas will expand and push the piston up. But the pressure, determined by mass M, will not change. This process is shown on the pV diagram of FIGURE 16.9c as the horizontal line $1 \rightarrow 2$. We call this an *isobaric expansion.* An *isobaric compression* occurs if the gas is cooled, lowering the piston. **Any isobaric process appears on a pV diagram as a horizontal line.**

EXAMPLE 16.7 **Comparing pressure**

The two cylinders in FIGURE 16.10 contain ideal gases at 20°C. Each cylinder is sealed by a frictionless piston of mass M.

a. How does the pressure of gas 2 compare to that of gas 1? Is it larger, smaller, or the same?
b. Suppose gas 2 is warmed to 80°C. Describe what happens to the pressure and volume.

FIGURE 16.10 Compare the pressures of the two gases.

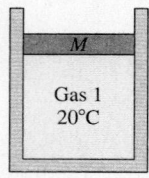 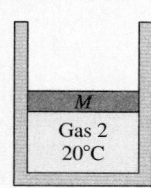

MODEL Treat the gases as ideal gases.

SOLVE a. The pressure in the gas is determined by the requirement that the piston be in mechanical equilibrium. The pressure of the gas inside pushes up on the piston; the air pressure and the weight of the piston press down. The gas pressure $p = p_{atmos} + Mg/A$ depends on the mass of the piston, but not at all on how high the piston is or what type of gas is inside the cylinder. Thus both pressures are the same.

b. Neither does the pressure depend on temperature. Warming the gas increases the temperature, but the pressure—determined by the mass and area of the piston—is unchanged. Because $pV/T = $ constant, and p is constant, it must be true that $V/T = $ constant. As T increases, the volume V also must increase to keep V/T unchanged. In other words, increasing the gas temperature causes the volume to expand—the piston goes up—but with no change in pressure. This is an isobaric process.

EXAMPLE 16.8 **Identifying a gas**

Your lab assistant distilled 50 g of a gas into a cylinder, but he left without writing down what kind of gas it is. The cylinder has a pressure regulator that adjusts a piston to keep the pressure at a constant 2.00 atm. To identify the gas, you measure the cylinder volume at several different temperatures, acquiring the data shown at the right. What is the gas?

Temperature (°C)	Volume (L)
−50	11.6
0	14.0
50	16.2
100	19.4
150	21.8

MODEL The pressure doesn't change, so heating the gas is an isobaric process.

SOLVE The ideal-gas law is $pV = nRT$. Writing this as

$$V = \frac{nR}{p}T$$

we see that a graph of V versus T should be a straight line passing through the origin. Further, we can use the slope of the graph, nR/p, to measure the number of moles of gas, and from that we can identify the gas by determining its molar mass.

FIGURE 16.11 is a graph of the data, with the volumes and temperatures converted to SI units of m^3 ($1\ m^3 = 1000\ L$) and kelvins. The y-intercept of the graph is essentially zero, confirming the behavior of the gas as ideal, and the slope of the best-fit line is $5.16 \times 10^{-5}\ m^3/K$. The number of moles of gas is

$$n = \frac{p}{R} \times \text{slope} = \frac{2 \times 101{,}300\ \text{Pa}}{8.31\ \text{J/mol K}} \times 5.16 \times 10^{-5}\ m^3/K = 1.26\ \text{mol}$$

From this, the molar mass is

FIGURE 16.11 A graph of the gas volume versus its temperature.

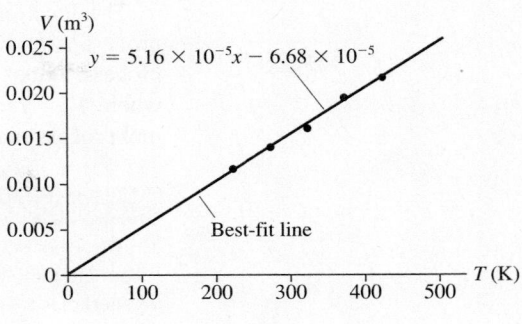

$$M = \frac{0.050\ \text{kg}}{1.26\ \text{mol}} = 0.040\ \text{kg/mol}$$

Thus the atomic mass is 40 u, identifying the gas as argon.

ASSESS The atomic mass is that of a well-known gas, which gives us confidence in the result.

STOP TO THINK 16.6 Two cylinders of equal diameter contain the same number of moles of the same ideal gas. Each cylinder is sealed by a frictionless piston. To have the same pressure in both cylinders, which piston would you use in cylinder 2?

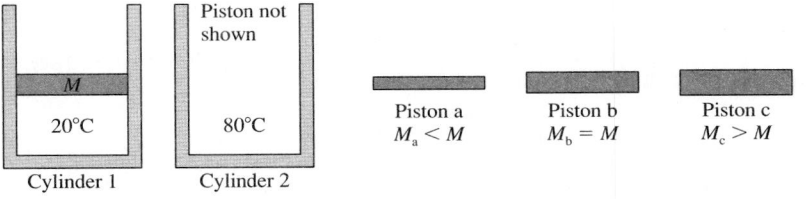

Constant-Temperature Process

The last process we wish to look at for now is one that takes place at a constant temperature. A constant-temperature process is called an **isothermal process.** An isothermal process is one for which $T_f = T_i$. Because $pV = nRT$, a constant-temperature process in a closed system (constant n) is one for which the product pV doesn't change. Thus

$$p_f V_f = p_i V_i \tag{16.17}$$

in an isothermal process.

One possible isothermal process is illustrated in FIGURE 16.12a, where a piston is being pushed down to compress a gas. If the piston is pushed *slowly,* then heat energy transfer through the walls of the cylinder keeps the gas at the same temperature as the surrounding liquid. This is an *isothermal compression.* The reverse process, with the piston slowly pulled out, would be an *isothermal expansion.*

Representing an isothermal process on the pV diagram is a little more complicated than the two previous processes because both p and V change. As long as T remains fixed, we have the relationship

$$p = \frac{nRT}{V} = \frac{\text{constant}}{V} \tag{16.18}$$

The inverse relationship between p and V causes the graph of an isothermal process to be a *hyperbola.* As one state variable goes up, the other goes down.

FIGURE 16.12 A constant-temperature (isothermal) process.

(a)

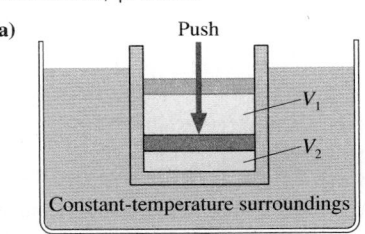

(b)

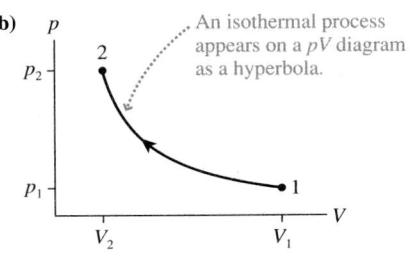

(c)

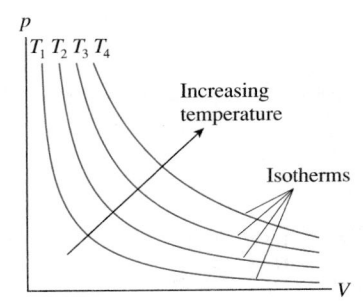

The process shown as $1 \rightarrow 2$ in FIGURE 16.12b represents the *isothermal compression* shown in Figure 16.12a. An *isothermal expansion* would move in the opposite direction along the hyperbola.

The location of the hyperbola depends on the value of T. A lower-temperature process is represented by a hyperbola closer to the origin than a higher-temperature process. FIGURE 16.12c shows four hyperbolas representing the temperatures T_1 to T_4, where $T_4 > T_3 > T_2 > T_1$. These are called **isotherms.** A gas undergoing an isothermal process moves along the isotherm of the appropriate temperature.

EXAMPLE 16.9 | **Compressing air in the lungs**

An ocean snorkeler takes a deep breath at the surface, filling his lungs with 4.0 L of air. He then descends to a depth of 5.0 m. At this depth, what is the volume of air in the snorkeler's lungs?

MODEL At the surface, the pressure in the lungs is 1.00 atm. Because the body cannot sustain large pressure differences between inside and outside, the air pressure in the lungs rises—and the volume decreases—to match the surrounding water pressure as he descends.

SOLVE The ideal-gas law for a sealed container is

$$V_2 = \frac{p_1}{p_2}\frac{T_2}{T_1}V_1$$

Air is quickly warmed to body temperature as it enters through the nose and mouth, and it remains at body temperature as the snorkeler dives, so $T_2/T_1 = 1$. We know $p_1 = 1.00$ atm $= 101,300$ Pa at the surface. We can find p_2 from the hydrostatic pressure equation, using the density of seawater:

$$p_2 = p_1 + \rho g d = 101,300 \text{ Pa} + (1030 \text{ kg/m}^3)(9.80 \text{ m/s}^2)(5.0 \text{ m})$$

$$= 151,800 \text{ Pa}$$

With this, the volume of the lungs at a depth of 5.0 m is

$$V_2 = \frac{101,300 \text{ Pa}}{151,800 \text{ Pa}} \times 1 \times 4.0 \text{ L} = 2.7 \text{ L}$$

ASSESS The air inside your lungs does compress—significantly—as you dive below the surface.

EXAMPLE 16.10 | **A multistep process**

A gas at 2.0 atm pressure and a temperature of 200°C is first expanded isothermally until its volume has doubled. It then undergoes an isobaric compression until it returns to its original volume. First show this process on a pV diagram. Then find the final temperature (in °C) and pressure.

MODEL The final state of the isothermal expansion is the initial state for an isobaric compression.

VISUALIZE FIGURE 16.13 shows the process. As the gas expands isothermally, it moves downward along an isotherm until it reaches volume $V_2 = 2V_1$. The gas is then compressed at constant pressure p_2 until its final volume V_3 equals its original volume V_1. State 3 is on an isotherm closer to the origin, so we expect to find $T_3 < T_1$.

SOLVE $T_2/T_1 = 1$ during the isothermal expansion and $V_2 = 2V_1$, so the pressure at point 2 is

$$p_2 = p_1 \frac{T_2}{T_1}\frac{V_1}{V_2} = p_1 \frac{V_1}{2V_1} = \frac{1}{2}p_1 = 1.0 \text{ atm}$$

We have $p_3/p_2 = 1$ during the isobaric compression and $V_3 = V_1 = \frac{1}{2}V_2$, so

FIGURE 16.13 A pV diagram for the process of Example 16.10.

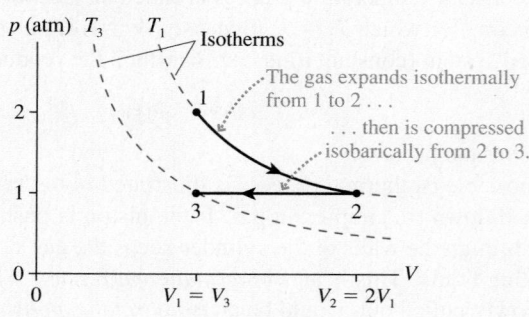

$$T_3 = T_2 \frac{p_3}{p_2}\frac{V_3}{V_2} = T_2 \frac{\frac{1}{2}V_2}{V_2} = \frac{1}{2}T_2 = 236.5 \text{ K} = -36.5°C$$

where we converted T_2 to 473 K before doing calculations and then converted T_3 back to °C. The final state, with $T_3 = -36.5°C$ and $p_3 = 1.0$ atm, is one in which both the pressure and the absolute temperature are half their original values.

What is the ratio T_f/T_i for this process?

a. $\frac{1}{4}$
b. $\frac{1}{2}$
c. 1 (no change)
d. 2
e. 4
f. There's not enough information to tell.

CHALLENGE EXAMPLE 16.11 **Depressing a piston**

A large, 50.0-cm-diameter metal cylinder filled with air supports a 20.0 kg piston that can slide up and down without friction. The piston is 100.0 cm above the bottom when the temperature is 20°C. An 80.0 kg student then stands on the piston. After several minutes have elapsed, by how much has the piston been depressed?

MODEL The metal walls of the cylinder are a good thermal conductor, so after several minutes the gas temperature—even if it initially changed—will return to room temperature. The final temperature matches the initial temperature. Assume that the atmospheric pressure is 1 atm.

VISUALIZE **FIGURE 16.14** shows the cylinder before and after the student stands on it. The volume of the cylinder is $V = Ah$, and only h changes.

SOLVE The ideal-gas law for a sealed container is

$$\frac{p_2 A h_2}{T_2} = \frac{p_1 A h_1}{T_1}$$

Because $T_2 = T_1$, the final height of the piston is

$$h_2 = \frac{p_1}{p_2} h_1$$

The pressure of the gas is determined by the mass of the piston (and anything on the piston) and the pressure of the air above. In equilibrium,

$$p = p_{atmos} + \frac{Mg}{A} = \begin{cases} 1.023 \times 10^5 \text{ Pa} & \text{piston only} \\ 1.063 \times 10^5 \text{ Pa} & \text{piston and student} \end{cases}$$

where we used $p_{atmos} = 1$ atm $= 1.013 \times 10^5$ Pa and $A = \pi r^2 = 0.196$ m^2. The final height of the piston is

FIGURE 16.14 The student compresses the gas.

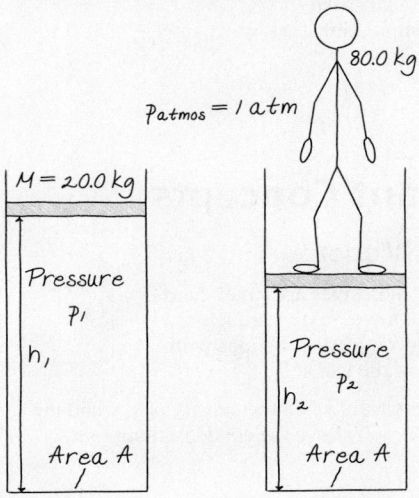

$$h_2 = \frac{1.023 \times 10^5 \text{ Pa}}{1.063 \times 10^5 \text{ Pa}} \times 100.0 \text{ cm} = 96.2 \text{ cm}$$

The question, however, was by how much the piston is depressed. This is $h_1 - h_2 = 3.8$ cm.

ASSESS Neither the piston nor the student increases the gas pressure to much above 1 atm, so it's not surprising that the added weight of the student doesn't push the piston down very far.

SUMMARY

The goal of Chapter 16 has been to learn the characteristics of macroscopic systems.

General Principles

Three Common Phases of Matter

Solid — Rigid, definite shape. Nearly incompressible.

Liquid — Molecules loosely held together by molecular bonds, but able to move around. Nearly incompressible.

Gas — Molecules moving freely through space. Compressible.

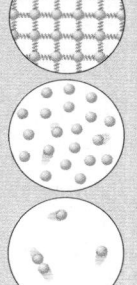

The different phases exist for different conditions of temperature T and pressure p. The boundaries separating the regions of a phase diagram are lines of phase equilibrium. Any amounts of the two phases can coexist in equilibrium. The **triple point** is the one value of temperature and pressure at which all three phases can coexist in equilibrium.

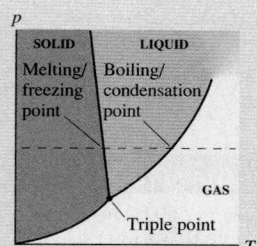

Important Concepts

Ideal-Gas Model

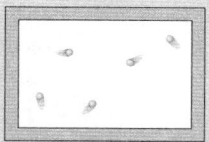

- Atoms and molecules are small, hard spheres that travel freely through space except for occasional collisions with each other or the walls.

- The model is valid when the density is low and the temperature well above the condensation point.

Ideal-Gas Law

The **state variables** of an ideal gas are related by the ideal-gas law

$$pV = nRT \quad \text{or} \quad pV = Nk_BT$$

where $R = 8.31$ J/mol K is the universal gas constant and $k_B = 1.38 \times 10^{-23}$ J/K is Boltzmann's constant. p, V, and T *must* be in SI units of Pa, m³, and K.

For a gas in a sealed container, with constant n:

$$\frac{p_2V_2}{T_2} = \frac{p_1V_1}{T_1}.$$

Counting atoms and moles

A macroscopic sample of matter consists of N atoms (or molecules), each of mass m (the **atomic** or **molecular mass**):

$$N = \frac{M}{m}$$

Alternatively, we can state that the sample consists of n **moles:**

$$n = \frac{N}{N_A} \quad \text{or} \quad \frac{M}{M_{mol}}$$

where $N_A = 6.02 \times 10^{23}$ mol^{-1} is **Avogadro's number.**

The molar mass M_{mol}, in kg/mol, is the numerical value of the atomic or molecular mass in u divided by 1000. The atomic or molecular mass, in atomic mass units u, is well approximated by the **atomic mass number** A. The atomic mass unit is

$$1 \text{ u} = 1.66 \times 10^{-27} \text{ kg}$$

The **number density** of the sample is $\dfrac{N}{V}$.

Volume V

Mass M

Applications

Temperature scales

$$T_F = \frac{9}{5}T_C + 32° \qquad T_K = T_C + 273$$

The Kelvin temperature scale is based on:

- Absolute zero at $T_0 = 0$ K
- The triple point of water at $T_3 = 273.16$ K

Three basic gas processes

1. **Isochoric,** or constant volume
2. **Isobaric,** or constant pressure
3. **Isothermal,** or constant temperature

pV diagram

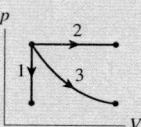

Terms and Notation

bulk properties	atomic mass number, A	absolute zero, T_0	ideal-gas law
micro/macro connection	atomic mass	absolute temperature scale	universal gas constant, R
phase	atomic mass unit, u	melting point	Boltzmann's constant, k_B
phase change	molecular mass	freezing point	STP
solid	mole, n	phase equilibrium	ideal-gas process
crystal	monatomic gas	condensation point	pV diagram
amorphous solid	diatomic gas	boiling point	quasi-static process
liquid	Avogadro's number, N_A	phase diagram	isochoric process
gas	molar mass, M_{mol}	sublimation	isobaric process
state variable	temperature, T	critical point	isothermal process
thermal equilibrium	constant-volume gas	triple point	isotherm
number density, N/V	thermometer	ideal gas	

CONCEPTUAL QUESTIONS

1. Rank in order, from highest to lowest, the temperatures $T_1 = 0$ K, $T_2 = 0°C$, and $T_3 = 0°F$.

2. The sample in an experiment is initially at $10°C$. If the sample's temperature is doubled, what is the new temperature in $°C$?

3. a. Is there a highest temperature at which ice can exist? If so, what is it? If not, why not?
   b. Is there a lowest temperature at which water vapor can exist? If so, what is it? If not, why not?

4. The cylinder in FIGURE Q16.4 is divided into two compartments by a frictionless piston that can slide back and forth. Is the pressure on the left side greater than, less than, or equal to the pressure on the right? Explain.

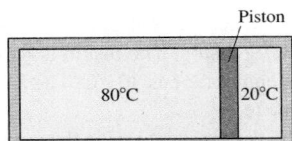

FIGURE Q16.4

5. A gas is in a sealed container. By what factor does the gas temperature change if:
   a. The volume is doubled and the pressure is tripled?
   b. The volume is halved and the pressure is tripled?

6. A gas is in a sealed container. The gas pressure is tripled and the temperature is doubled.
   a. What happens to the number of moles of gas in the container?
   b. What happens to the number density of the gas in the container?

7. An aquanaut lives in an underwater apartment 100 m beneath the surface of the ocean. Compare the freezing and boiling points of water in the aquanaut's apartment to their values at the surface. Are they higher, lower, or the same? Explain.

8. a. A sample of water vapor in an enclosed cylinder has an initial pressure of 500 Pa at an initial temperature of $-0.01°C$. A piston squeezes the sample smaller and smaller, without limit. Describe what happens to the water as the squeezing progresses.
   b. Repeat part a if the initial temperature is $0.03°C$ warmer.

9. A gas is in a sealed container. By what factor does the gas pressure change if:
   a. The volume is doubled and the temperature is tripled?
   b. The volume is halved and the temperature is tripled?

10. A gas undergoes the process shown in FIGURE Q16.10. By what factor does the temperature change?

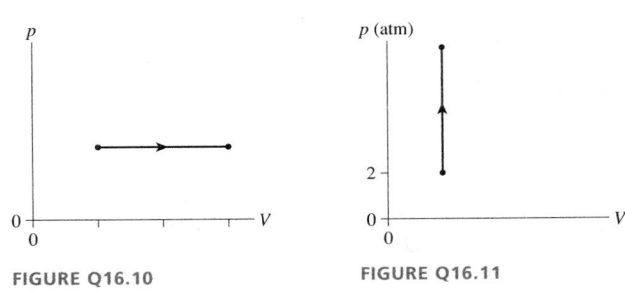

FIGURE Q16.10   FIGURE Q16.11

11. The temperature increases from 300 K to 1200 K as a gas undergoes the process shown in FIGURE Q16.11. What is the final pressure?

12. A student is asked to sketch a pV diagram for a gas that goes through a cycle consisting of (a) an isobaric expansion, (b) a constant-volume reduction in temperature, and (c) an isothermal process that returns the gas to its initial state. The student draws the diagram shown in FIGURE Q16.12. What, if anything, is wrong with the student's diagram?

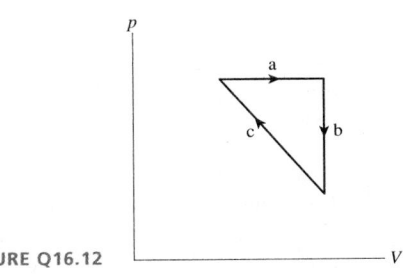

FIGURE Q16.12

EXERCISES AND PROBLEMS

Problems labeled [ ] integrate material from earlier chapters.

Exercises

Section 16.1 Solids, Liquids, and Gases

1. | What volume of water has the same mass as 100 cm^3 of gold?

2. || The nucleus of a uranium atom has a diameter of 1.5×10^{-14} m and a mass of 4.0×10^{-25} kg. What is the density of the nucleus?

3. || A hollow aluminum sphere with outer diameter 10.0 cm has a mass of 690 g. What is the sphere's inner diameter?

4. || What is the diameter of a copper sphere that has the same mass as a 10 cm $\times$ 10 cm $\times$ 10 cm cube of aluminum?

Section 16.2 Atoms and Moles

5. || How many atoms are in a 2.0 cm $\times$ 2.0 cm $\times$ 2.0 cm cube of aluminum?

6. || How many moles are in a 2.0 cm $\times$ 2.0 cm $\times$ 2.0 cm cube of copper?

7. || What is the number density of (a) aluminum and (b) lead?

8. || An element in its solid phase has mass density 1750 kg/m^3 and number density 4.39×10^{28} atoms/m^3. What is the element's atomic mass number?

9. || What volume of aluminum has the same number of atoms as 10 cm^3 of mercury?

10. || 1.0 mol of gold is shaped into a sphere. What is the sphere's diameter?

Section 16.3 Temperature

Section 16.4 Phase Changes

11. | The lowest and highest natural temperatures ever recorded on earth are $-127°F$ in Antarctica and $136°F$ in Libya. What are these temperatures in °C and in K?

12. | At what temperature does the numerical value in °F match the numerical value in °C?

13. || A demented scientist creates a new temperature scale, the "Z scale." He decides to call the boiling point of nitrogen 0°Z and the melting point of iron 1000°Z.
    a. What is the boiling point of water on the Z scale?
    b. Convert 500°Z to degrees Celsius and to kelvins.

14. || What is the temperature in °F and the pressure in Pa at the triple point of (a) water and (b) carbon dioxide?

Section 16.5 Ideal Gases

15. | A cylinder contains nitrogen gas. A piston compresses the gas to half its initial volume. Afterward,
    a. Has the mass density of the gas changed? If so, by what factor? If not, why not?
    b. Has the number of moles of gas changed? If so, by what factor? If not, why not?

16. || 3.0 mol of gas at a temperature of $-120°C$ fills a 2.0 L container. What is the gas pressure?

17. || A gas at 100°C fills volume V_0. If the pressure is held constant, what is the volume if (a) the Celsius temperature is doubled and (b) the Kelvin temperature is doubled?

18. || A rigid container holds 2.0 mol of gas at a pressure of 1.0 atm and a temperature of 30°C.
    a. What is the container's volume?
    b. What is the pressure if the temperature is raised to 130°C?

19. || The total lung capacity of a typical adult is 5.0 L. Approximately 20% of the air is oxygen. At sea level and at a body temperature of 37°C, how many oxygen molecules do the lungs contain at the end of a strong inhalation?

20. || A 20-cm-diameter cylinder that is 40 cm long contains 50 g of oxygen gas at 20°C.
    a. How many moles of oxygen are in the cylinder?
    b. How many oxygen molecules are in the cylinder?
    c. What is the number density of the oxygen?
    d. What is the reading of a pressure gauge attached to the tank?

21. || A 10-cm-diameter cylinder of neon gas is 30 cm long and at 30°C. The pressure gauge reads 120 psi. What is the mass density of the gas?

Section 16.6 Ideal-Gas Processes

22. | A gas with initial state variables p_1, V_1, and T_1 expands isothermally until $V_2 = 2V_1$. What are (a) T_2 and (b) p_2?

23. | A gas with initial state variables p_1, V_1, and T_1 is cooled in an isochoric process until $p_2 = \frac{1}{3}p_1$. What are (a) V_2 and (b) T_2?

24. | A rigid sphere is submerged in boiling water in a room where the air pressure is 1.0 atm. The sphere has an open valve with its inlet just above the water level. After a long period of time has elapsed, the valve is closed. What will be the pressure inside the sphere if it is then placed in (a) a mixture of ice and water and (b) an insulated box filled with dry ice?

25. || A rigid container holds hydrogen gas at a pressure of 3.0 atm and a temperature of 20°C. What will the pressure be if the temperature is lowered to $-20°C$?

26. || A 24-cm-diameter vertical cylinder is sealed at the top by a frictionless 20 kg piston. The piston is 84 cm above the bottom when the gas temperature is 303°C. The air above the piston is at 1.00 atm pressure.
    a. What is the gas pressure inside the cylinder?
    b. What will the height of the piston be if the temperature is lowered to 15°C?

27. || 0.10 mol of argon gas is admitted to an evacuated 50 cm^3 container at 20°C. The gas then undergoes an isochoric heating to a temperature of 300°C.
    a. What is the final pressure of the gas?
    b. Show the process on a pV diagram. Include a proper scale on both axes.

28. | 0.10 mol of argon gas is admitted to an evacuated 50 cm^3 container at 20°C. The gas then undergoes an isobaric heating to a temperature of 300°C.
    a. What is the final volume of the gas?
    b. Show the process on a pV diagram. Include a proper scale on both axes.

29. || 0.10 mol of argon gas is admitted to an evacuated 50 cm^3 container at 20°C. The gas then undergoes an isothermal expansion to a volume of 200 cm^3.
    a. What is the final pressure of the gas?
    b. Show the process on a pV diagram. Include a proper scale on both axes.

30. | 0.0040 mol of gas undergoes the process shown in FIG-URE EX16.30.
   a. What type of process is this?
   b. What are the initial and final temperatures in °C?

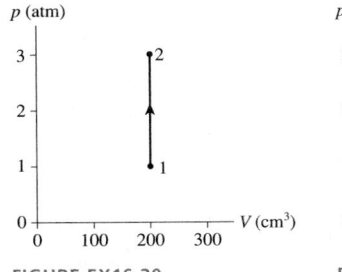

FIGURE EX16.30

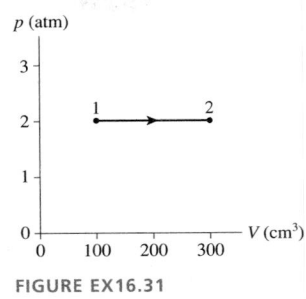

FIGURE EX16.31

31. ‖ A gas with an initial temperature of 900°C undergoes the process shown in FIGURE EX16.31.
   a. What type of process is this?
   b. What is the final temperature in °C?
   c. How many moles of gas are there?

32. | 0.020 mol of gas undergoes the process shown in FIG-URE EX16.32.
   a. What type of process is this?
   b. What is the final temperature in °C?
   c. What is the final volume V_2?

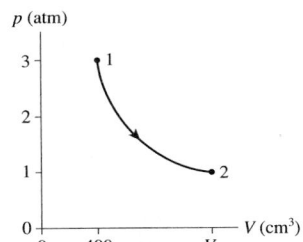

FIGURE EX16.32

Problems

33. ‖ The atomic mass number of copper is $A = 64$. Assume that atoms in solid copper form a cubic crystal lattice. To envision this, imagine that you place atoms at the centers of tiny sugar cubes, then stack the little sugar cubes to form a big cube. If you dissolve the sugar, the atoms left behind are in a cubic crystal lattice. What is the smallest distance between two copper atoms?

34. ‖ An element in its solid phase forms a cubic crystal lattice (see Problem 33) with mass density 7950 kg/m³. The smallest spacing between two adjacent atoms is 0.227 nm. What is the element's atomic mass number?

35. ‖ The molecular mass of water (H_2O) is $A = 18$. How many protons are there in 1.0 L of liquid water?

36. ‖ Estimate the number density of gas molecules in the earth's atmosphere at sea level.

37. | The solar corona is a very hot atmosphere surrounding the visible surface of the sun. X-ray emissions from the corona show that its temperature is about 2×10^6 K. The gas pressure in the corona is about 0.03 Pa. Estimate the number density of particles in the solar corona.

38. ‖ The semiconductor industry manufactures integrated circuits in large vacuum chambers where the pressure is 1.0×10^{-10} mm of Hg.
   a. What fraction is this of atmospheric pressure?
   b. At $T = 20°C$, how many molecules are in a cylindrical chamber 40 cm in diameter and 30 cm tall?

39. ‖ A 6.0-cm-diameter, 10-cm-long cylinder contains 100 mg of oxygen (O_2) at a pressure less than 1 atm. The cap on one end of the cylinder is held in place only by the pressure of the air. One day when the atmospheric pressure is 100 kPa, it takes a 184 N force to pull the cap off. What is the temperature of the gas?

40. ‖ A nebula—a region of the galaxy where new stars are forming—contains a very tenuous gas with 100 atoms/cm³. This gas is heated to 7500 K by ultraviolet radiation from nearby stars. What is the gas pressure in atm?

41. ‖‖ An inflated bicycle inner tube is 2.2 cm in diameter and 200 cm in circumference. A small leak causes the gauge pressure to decrease from 110 psi to 80 psi on a day when the temperature is 20°C. What mass of air is lost? Assume the air is pure nitrogen.

42. ‖ On average, each person in the industrialized world is responsible for the emission of 10,000 kg of carbon dioxide (CO_2) every year. This includes CO_2 that you generate directly, by burning fossil fuels to operate your car or your furnace, as well as CO_2 generated on your behalf by electric generating stations and manufacturing plants. CO_2 is a greenhouse gas that contributes to global warming. If you were to store your yearly CO_2 emissions in a cube at STP, how long would each edge of the cube be?

43. ‖ A gas at temperature T_0 and atmospheric pressure fills a cylinder. The gas is transferred to a new cylinder with three times the volume, after which the pressure is half the original pressure. What is the new temperature of the gas?

44. ‖ To determine the mass of neon contained in a rigid, 2.0 L cylinder, you vary the cylinder's temperature while recording the reading of a pressure gauge. Your data are as follows:

Temperature (°C)	Pressure gauge (atm)
100	6.52
150	7.80
200	8.83
250	9.59

Use the best-fit line of an appropriate graph to determine the mass of the neon.

45. ‖ The 3.0-m-long pipe in FIGURE P16.45 is closed at the top end. It is slowly pushed straight down into the water until the top end of the pipe is level with the water's surface. What is the length L of the trapped volume of air?

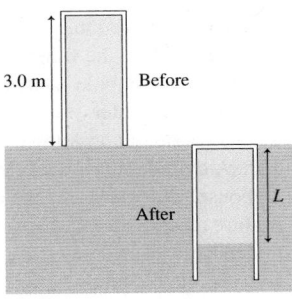

FIGURE P16.45

46. ‖ An electric generating plant boils water to produce high-pressure steam. The steam spins a turbine that is connected to the generator.
   a. How many liters of water must be boiled to fill a 5.0 m³ boiler with 50 atm of steam at 400°C?
   b. The steam has dropped to 2.0 atm pressure at 150°C as it exits the turbine. How much volume does it now occupy?

47. ‖‖ On a cool morning, when the temperature is 15°C, you measure the pressure in your car tires to be 30 psi. After driving 20 mi on the freeway, the temperature of your tires is 45°C. What pressure will your tire gauge now show?

48. ‖ The air temperature and pressure in a laboratory are 20°C and 1.0 atm. A 1.0 L container is open to the air. The container is then sealed and placed in a bath of boiling water. After reaching thermal equilibrium, the container is opened. How many moles of air escape?

49. ‖ A gas cylinder with a tight-fitting, movable piston contains 200 cm³ of air at 1.0 atm. It floats on the surface of a swimming pool filled with 15°C water. The cylinder is then pulled slowly underwater to a depth of 3.0 m. What is the volume of gas at this depth?

50. ‖ The mercury manometer shown in FIGURE P16.50 is attached to a gas cell. The mercury height h is 120 mm when the cell is placed in an ice-water mixture. The mercury height drops to 30 mm when the device is carried into an industrial freezer. What is the freezer temperature?
   Hint: The right tube of the manometer is much narrower than the left tube. What reasonable assumption can you make about the gas volume?

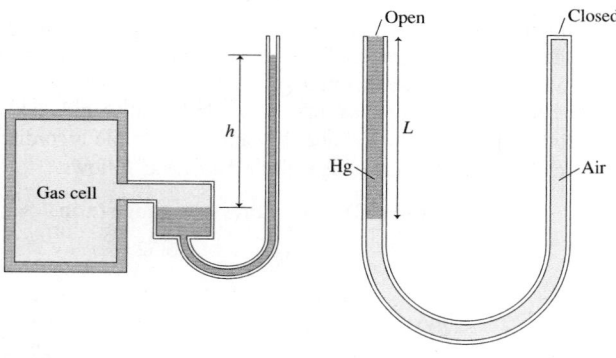

FIGURE P16.50   FIGURE P16.51

51. ‖ The U-shaped tube in FIGURE P16.51 has a total length of 1.0 m. It is open at one end, closed at the other, and is initially filled with air at 20°C and 1.0 atm pressure. Mercury is poured slowly into the open end without letting any air escape, thus compressing the air. This is continued until the open side of the tube is completely filled with mercury. What is the length L of the column of mercury?

52. ‖ A diver 50 m deep in 10°C fresh water exhales a 1.0-cm-diameter bubble. What is the bubble's diameter just as it reaches the surface of the lake, where the water temperature is 20°C?
   Hint: Assume that the air bubble is always in thermal equilibrium with the surrounding water.

53. ‖ A compressed-air cylinder is known to fail if the pressure exceeds 110 atm. A cylinder that was filled to 25 atm at 20°C is stored in a warehouse. Unfortunately, the warehouse catches fire and the temperature reaches 950°C. Does the cylinder blow?

54. ‖ Reproduce FIGURE P16.54 on a piece of paper. A gas starts with pressure p_1 and volume V_1. Show on the figure the process in which the gas undergoes an isochoric process that doubles the pressure, then an isobaric process that doubles the volume, followed by an isothermal process that doubles the volume again. Label each of the three processes.

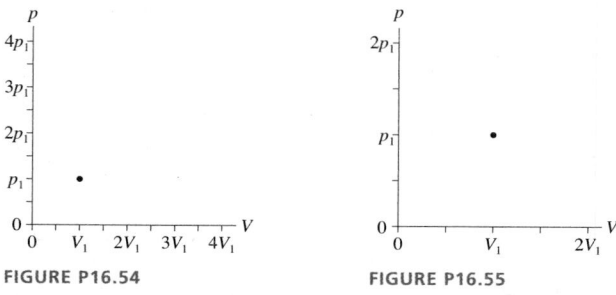

FIGURE P16.54   FIGURE P16.55

55. ‖ Reproduce FIGURE P16.55 on a piece of paper. A gas starts with pressure p_1 and volume V_1. Show on the figure the process in which the gas undergoes an isothermal process during which the volume is halved, then an isochoric process during which the pressure is halved, followed by an isobaric process during which the volume is doubled. Label each of the three processes.

56. ‖ 8.0 g of helium gas follows the process $1 \rightarrow 2 \rightarrow 3$ shown in FIGURE P16.56. Find the values of V_1, V_3, p_2, and T_3.

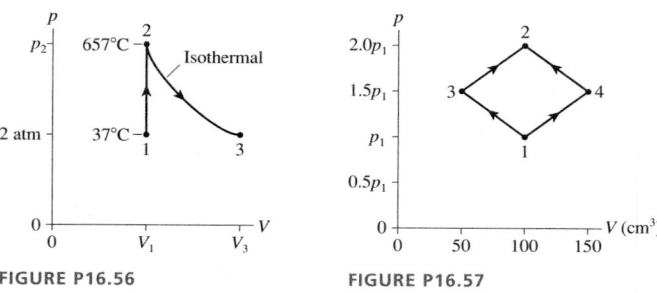

FIGURE P16.56   FIGURE P16.57

57. ‖ FIGURE P16.57 shows two different processes by which 1.0 g of nitrogen gas moves from state 1 to state 2. The temperature of state 1 is 25°C. What are (a) pressure p_1 and (b) temperatures (in °C) T_2, T_3, and T_4?

58. ‖ FIGURE P16.58 shows two different processes by which 80 mol of gas move from state 1 to state 2. The dashed line is an isotherm.
   a. What is the temperature of the isothermal process?
   b. What maximum temperature is reached along the straight-line process?

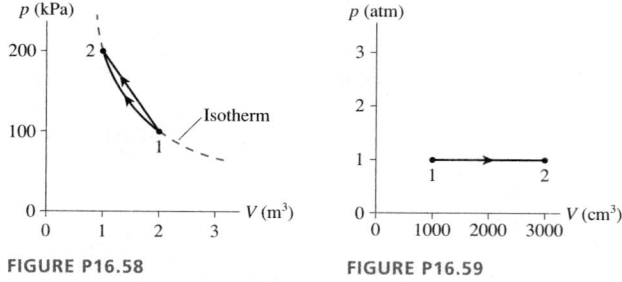

FIGURE P16.58   FIGURE P16.59

59. ‖ 0.10 mol of gas undergoes the process $1 \rightarrow 2$ shown in FIGURE P16.59.
   a. What are temperatures T_1 and T_2 (in °C)?
   b. What type of process is this?
   c. The gas undergoes an isothermal compression from point 2 until the volume is restored to the value it had at point 1. What is the final pressure of the gas?

60. | 0.0050 mol of gas undergoes the process $1 \rightarrow 2 \rightarrow 3$ shown in FIGURE P16.60. What are (a) temperature T_1, (b) pressure p_2, and (c) volume V_3?

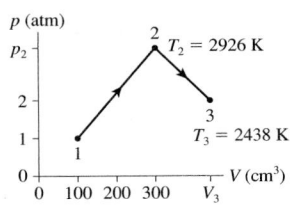
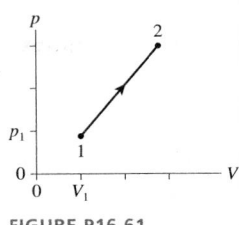

FIGURE P16.60   FIGURE P16.61

61. || 4.0 g of oxygen gas, starting at 20°C, follow the process $1 \rightarrow 2$ shown in FIGURE P16.61. What is temperature T_2 (in °C)?

62. || 10 g of dry ice (solid CO_2) is placed in a 10,000 cm³ container, then all the air is quickly pumped out and the container sealed. The container is warmed to 0°C, a temperature at which CO_2 is a gas.
    a. What is the gas pressure? Give your answer in atm.
    The gas then undergoes an isothermal compression until the pressure is 3.0 atm, immediately followed by an isobaric compression until the volume is 1000 cm³.
    b. What is the final temperature of the gas (in °C)?
    c. Show the process on a pV diagram.

63. || A container of gas at 2.0 atm pressure and 127°C is compressed at constant temperature until the volume is halved. It is then further compressed at constant pressure until the volume is halved again.
    a. What are the final pressure and temperature of the gas?
    b. Show this process on a pV diagram.

64. || Five grams of nitrogen gas at an initial pressure of 3.0 atm and at 20°C undergo an isobaric expansion until the volume has tripled.
    a. What is the gas volume after the expansion?
    b. What is the gas temperature after the expansion (in °C)?
    The gas pressure is then decreased at constant volume until the original temperature is reached.
    c. What is the gas pressure after the decrease?
    Finally, the gas is isothermally compressed until it returns to its initial volume.
    d. What is the final gas pressure?
    e. Show the full three-step process on a pV diagram. Use appropriate scales on both axes.

In Problems 65 through 68 you are given the equation(s) used to solve a problem. For each of these, you are to
    a. Write a realistic problem for which this is the correct equation(s).
    b. Draw a pV diagram.
    c. Finish the solution of the problem.

65. $p_2 = \dfrac{300 \text{ cm}^3}{100 \text{ cm}^3} \times 1 \times 2 \text{ atm}$

66. $(T_2 + 273) \text{ K} = \dfrac{200 \text{ kPa}}{500 \text{ kPa}} \times 1 \times (400 + 273) \text{ K}$

67. $V_2 = \dfrac{(400 + 273) \text{ K}}{(50 + 273) \text{ K}} \times 1 \times 200 \text{ cm}^3$

68. $(2.0 \times 101,300 \text{ Pa})(100 \times 10^{-6} \text{ m}^3) = n(8.31 \text{ J/mol K})T_1$

$n = \dfrac{0.12 \text{ g}}{20 \text{ g/mol}}$

$T_2 = \dfrac{200 \text{ cm}^3}{100 \text{ cm}^3} \times 1 \times T_1$

Challenge Problems

69. The 50 kg lead piston shown in FIGURE CP16.69 floats on 0.12 mol of compressed air.
    a. What is the piston height h if the temperature is 30°C?
    b. How far does the piston move if the temperature is increased by 100°C?

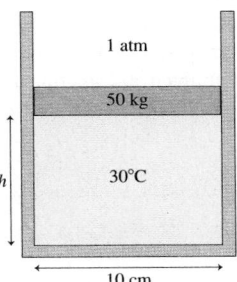

FIGURE CP16.69

70. A diving bell is a 3.0-m-tall cylinder closed at the upper end but open at the lower end. The temperature of the air in the bell is 20°C. The bell is lowered into the ocean until its lower end is 100 m deep. The temperature at that depth is 10°C.
    a. How high does the water rise in the bell after enough time has passed for the air inside to reach thermal equilibrium?
    b. A compressed-air hose from the surface is used to expel all the water from the bell. What minimum air pressure is needed to do this?

71. 10,000 cm³ of 200°C steam at a pressure of 20 atm is cooled until it condenses. What is the volume of the liquid water? Give your answer in cm³.

72. The cylinder in FIGURE CP16.72 has a moveable piston attached to a spring. The cylinder's cross-section area is 10 cm², it contains 0.0040 mol of gas, and the spring constant is 1500 N/m. At 20°C the spring is neither compressed nor stretched. How far is the spring compressed if the gas temperature is raised to 100°C?

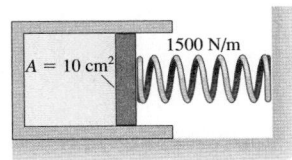

FIGURE CP16.72

73. Containers A and B in FIGURE CP16.73 hold the same gas. The volume of B is four times the volume of A. The two containers are connected by a thin tube (negligible volume) and a valve that is closed. The gas in A is at 300 K and pressure of 1.0×10^5 Pa. The gas in B is at 400 K and pressure of 5.0×10^5 Pa. Heaters will maintain the temperatures of A and B even after the valve is opened.
    a. After the valve is opened, gas will flow one way or the other until A and B have equal pressure. What is this final pressure?
    b. Is this a reversible or an irreversible process? Explain.

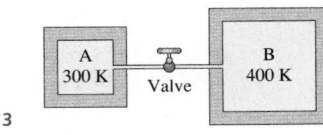

FIGURE CP16.73

74. The closed cylinder of FIGURE CP16.74 has a tight-fitting but frictionless piston of mass M. The piston is in equilibrium when the left chamber has pressure p_0 and length L_0 while the spring on the right is compressed by ΔL.
    a. What is ΔL in terms of p_0, L_0, A, M, and k?
    b. Suppose the piston is moved a small distance x to the right. Find an expression for the net force $(F_x)_{net}$ on the piston. Assume all motions are slow enough for the gas to remain at the same temperature as its surroundings.
    c. If released, the piston will oscillate around the equilibrium position. Assuming $x \ll L_0$ find an expression for the oscillation period T.

Hint: Use the binomial approximation.

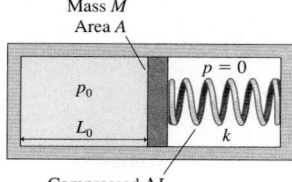

Mass M
Area A

$p = 0$

p_0

L_0

k

FIGURE CP16.74   Compressed ΔL

STOP TO THINK ANSWERS

Stop to Think 16.1: d. The pressure *decreases* by 20 kPa.

Stop to Think 16.2: a. The number of atoms depends only on the number of moles, not the substance.

Stop to Think 16.3: a. The step size on the Kelvin scale is the same as the step size on the Celsius scale. A *change* of 10°C is a *change* of 10 K.

Stop to Think 16.4: a. On the water phase diagram, you can see that for a pressure just slightly below the triple-point pressure, the solid/gas transition occurs at a higher temperature than does the solid/liquid transition at high pressures. This is not true for carbon dioxide.

Stop to Think 16.5: c. $T = pV/nR$. Pressure and volume are the same, but n differs. The number of moles in mass M is $n = M/M_{mol}$. Helium, with the smaller molar mass, has a larger number of moles and thus a lower temperature.

Stop to Think 16.6: b. The pressure is determined entirely by the weight of the piston pressing down. Changing the temperature changes the volume of the gas, but not its pressure.

Stop to Think 16.7: b. The temperature decreases by a factor of 4 during the isochoric process, where $p_f/p_i = \frac{1}{4}$. The temperature then increases by a factor of 2 during the isobaric expansion, where $V_f/V_i = 2$.

17 Work, Heat, and the First Law of Thermodynamics

This false-color thermal image—an infrared photo—shows where heat is escaping from the house.

▶ **Looking Ahead** The goal of Chapter 17 is to develop and apply the first law of thermodynamics.

Energy Transfers

There are two ways to transfer energy between a system and its environment: work and heat.

Work is the transfer of energy in a *mechanical interaction*—when external forces push or pull on the system.

You will learn to calculate the work done in an ideal-gas process as the negative of the area under a *pV* curve.

Heat is the transfer of energy in a *thermal interaction*—when the system and its environment have different temperatures.

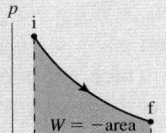

You will learn to calculate the heat energy required for various processes. For example, it takes just over 4000 J of heat to raise the temperature of 1 kg of water by 1°C.

◀ **Looking Back**
Section 11.4 Work

The First Law of Thermodynamics

The first law of thermodynamics is a very general statement of the idea that energy can be transferred and transformed but not created or destroyed.

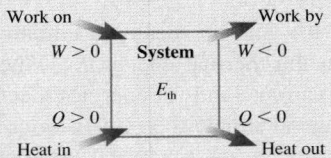

- The system's thermal energy increases if energy is transferred into the system as heat or work.
- The system's thermal energy decreases if energy is transferred out of the system as heat or work.

In a jet engine, part of the heat from burning fuel is used to do work—pushing the aircraft forward. The remainder becomes increased thermal energy of the hot exhaust gases. No energy is destroyed.

◀ **Looking Back**
Sections 11.7–11.8 Conservation of energy

Thermal Properties of Matter

Changing the thermal energy can cause

- A temperature change, or
- A phase change.

A material's response to heat is governed by its *specific heat*, its *heat of fusion*, and its *heat of vaporization*.

You'll learn how to do practical calorimetry calculations to determine the final temperature of two or more interacting systems.

◀ **Looking Back**
Sections 16.4–16.6 Phase changes and ideal gases

Heat Transfer

Heat energy can be transferred between a system and its environment by

- Conduction
- Radiation
- Convection
- Evaporation

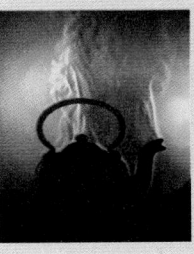

Air heated by the hot teakettle streams upward, an example of heat transfer by convection.

17.1 It's All About Energy

A key idea of Chapter 11 was the work-kinetic energy theorem in the form

$$\Delta K = W_c + W_{diss} + W_{ext} \qquad (17.1)$$

Equation 17.1 tells us that the kinetic energy of a system of particles is changed when forces do work on the particles by pushing or pulling them through a distance. Here

1. W_c is the work done by conservative forces. This work can be represented as a change in the system's potential energy: $\Delta U = -W_c$.
2. W_{diss} is the work done by friction-like dissipative forces within the system. This work increases the system's thermal energy: $\Delta E_{th} = -W_{diss}$.
3. W_{ext} is the work done by external forces that originate in the environment. The push of a piston rod would be an external force.

With these definitions, Equation 17.1 becomes

$$\Delta K + \Delta U + \Delta E_{th} = W_{ext} \qquad (17.2)$$

The system's *mechanical energy* was defined as $E_{mech} = K + U$. **FIGURE 17.1** reminds you that the mechanical energy is associated with the motion of the system as a whole, while E_{th} is associated with the motion of the atoms and molecules within the system. E_{mech} is the *macroscopic* energy of the system as a whole while E_{th} is the *microscopic* energy of the particle-like atoms and spring-like molecular bonds. This led to our final energy statement of Chapter 11:

$$\Delta E_{sys} = \Delta E_{mech} + \Delta E_{th} = W_{ext} \qquad (17.3)$$

Thus the total energy of an *isolated system,* for which $W_{ext} = 0$, is constant. This was the essence of the law of conservation of energy as stated in Chapter 11.

The emphasis in Chapters 10 and 11 was on isolated systems. There we were interested in learning how kinetic and potential energy were *transformed* into each other and, where there is friction, into thermal energy. Now we want to focus on how energy is *transferred* between the system and its environment, when W_{ext} is *not* zero.

> **NOTE** ▶ Strictly speaking, Equation 17.3 should use the *internal energy* E_{int} rather than the thermal energy E_{th}, where $E_{int} = E_{th} + E_{chem} + E_{nuc} + \cdots$ includes all the various kinds of energies that can be stored inside a system. This textbook will focus on simple thermodynamics systems in which the internal energy is entirely thermal: $E_{int} = E_{th}$. We'll leave other forms of internal energy to more advanced courses. ◀

Energy Transfer

Doing work on a system can have very different consequences. **FIGURE 17.2a** shows an object being lifted at steady speed by a rope. The rope's tension is an external force doing work W_{ext} on the system. In this case, the energy transferred into the system goes entirely to increasing the system's macroscopic potential energy U_{grav}, part of the mechanical energy. The energy-transfer process $W_{ext} \rightarrow E_{mech}$ is shown graphically in the energy bar chart of Figure 17.2a.

Contrast this with **FIGURE 17.2b**, where the same rope with the same tension now drags the object at steady speed across a rough surface. The tension does the same amount of work, but the mechanical energy does not change. Instead, friction increases the thermal energy of the object + surface system. The energy-transfer process $W_{ext} \rightarrow E_{th}$ is shown in the energy bar chart of Figure 17.2b.

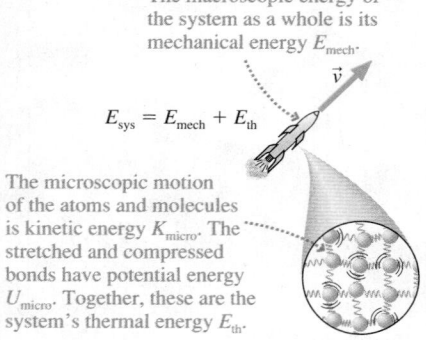

FIGURE 17.1 The total energy of a system consists of the macroscopic energy plus the microscopic thermal energy.

The macroscopic energy of the system as a whole is its mechanical energy E_{mech}.

$$E_{sys} = E_{mech} + E_{th}$$

The microscopic motion of the atoms and molecules is kinetic energy K_{micro}. The stretched and compressed bonds have potential energy U_{micro}. Together, these are the system's thermal energy E_{th}.

FIGURE 17.2 The work done by tension can have very different consequences.

(a) Lift at steady speed

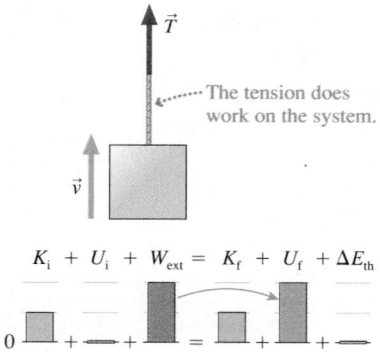

The tension does work on the system.

$$K_i + U_i + W_{ext} = K_f + U_f + \Delta E_{th}$$

The energy transferred to the system goes entirely to the system's mechanical energy.

(b) Drag at steady speed

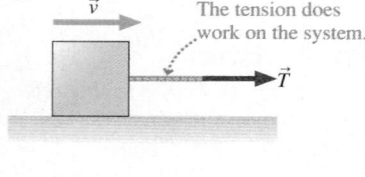

The tension does work on the system.

$$K_i + U_i + W_{ext} = K_f + U_f + \Delta E_{th}$$

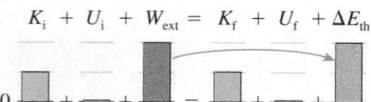

The energy transferred to the system goes entirely to the system's thermal energy.

The point of this example is that the energy transferred to a system can go entirely to the system's mechanical energy, entirely to its thermal energy, or (imagine dragging the object up an incline) some combination of the two. The energy isn't lost, but where it ends up depends on the circumstances.

That Can't Be All

You can transfer energy into a system by the mechanical process of doing work on the system. But that can't be all there is to energy transfer. What happens when you place a pan of water on the stove and light the burner? The water temperature increases, so $\Delta E_{th} > 0$. But no work is done ($W_{ext} = 0$) and there is no change in the water's mechanical energy ($\Delta E_{mech} = 0$). This process clearly violates the energy equation $\Delta E_{mech} + \Delta E_{th} = W_{ext}$. What's wrong?

Nothing is wrong. The energy equation is correct as far as it goes, but it is incomplete. Work is energy transferred in a mechanical interaction, but that is not the only way a system can interact with its environment. Energy can also be transferred between the system and the environment if they have a *thermal interaction*. The energy transferred in a thermal interaction is called *heat*.

The symbol for heat is Q. When heat is included, the energy equation becomes

$$\Delta E_{sys} = \Delta E_{mech} + \Delta E_{th} = W + Q \qquad (17.4)$$

Heat and work are both energy transferred between the system and the environment.

NOTE ▶ We've dropped the subscript "ext" from W. The work that we consider in thermodynamics is *always* the work done by the environment on the system. We won't need to distinguish this work from W_c or W_{diss}, so the subscript is superfluous. ◀

We'll return to Equation 17.4 in Section 17.4 after we look at how work is calculated for ideal-gas processes and at what heat is.

STOP TO THINK 17.1 A gas cylinder and piston are covered with heavy insulation. The piston is pushed into the cylinder, compressing the gas. In this process the gas temperature

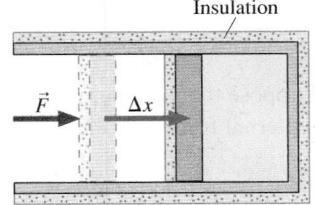

a. Increases.
b. Decreases.
c. Doesn't change.
d. There's not sufficient information to tell.

17.2 Work in Ideal-Gas Processes

We introduced the idea of *work* in Chapter 11. **Work** is the energy transferred between a system and the environment when a net force acts on the system over a distance. The process itself is a **mechanical interaction,** meaning that the system and the environment interact via macroscopic pushes and pulls. Loosely speaking, we say that the environment (or a particular force from the environment) "does work" on the system. A system is in **mechanical equilibrium** if there is no net force on the system.

FIGURE 17.3 on the next page reminds you that work can be either positive or negative. **The sign of the work is *not* just an arbitrary convention, nor does it have anything to do with the choice of coordinate system.** The sign of the work tells us which way energy is being transferred.

The pistons in a car engine do work on the air-fuel mixture by compressing it.

FIGURE 17.3 The sign of work.

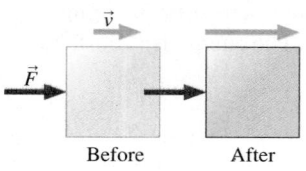

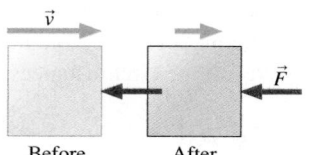

Work is *positive* when the force is in the direction of motion.
- The force causes the object to speed up.
- Energy is transferred from the environment to the system.
- The system's energy increases.

Work is *negative* when the force is opposite to the motion.
- The force causes the object to slow down.
- Energy is transferred from the system to the environment.
- The system's energy decreases.

In contrast to the mechanical energy or the thermal energy, **work is not a state variable.** That is, work is not a number characterizing the system. Instead, work is the amount of energy that moves between the system and the environment during a mechanical interaction. We can measure the *change* in a state variable, such as a temperature change $\Delta T = T_f - T_i$, but it would make no sense to talk about a "change of work." Consequently, work always appears as W, never as ΔW.

You learned in Chapter 11 how to calculate work. The small amount of work dW done by force $\vec{F}$ as a system moves through the small displacement $d\vec{s}$ is $dW = \vec{F} \cdot d\vec{s}$. If we restrict ourselves to situations where $\vec{F}$ is either parallel or opposite to $d\vec{s}$, then the total work done on the system as it moves from s_i to s_f is

$$W = \int_{s_i}^{s_f} F_s \, ds \qquad (17.5)$$

FIGURE 17.4 The external force does work on the gas as the piston moves.

(a) The gas pushes on the piston with force $\vec{F}_{gas}$.

Pressure p

To keep the piston in place, an external force must be equal and opposite to $\vec{F}_{gas}$.

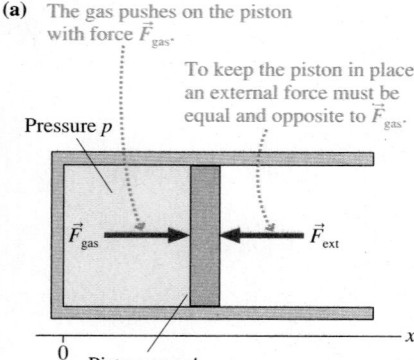

$\vec{F}_{gas}$ $\vec{F}_{ext}$

0   Piston area A

(b) As the piston moves dx, the external force does work $(F_{ext})_x \, dx$ on the gas.

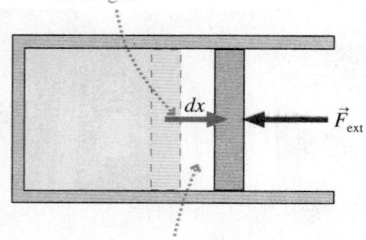

dx $\vec{F}_{ext}$

The volume changes by $dV = A \, dx$ as the piston moves dx.

Let's apply this definition to a gas as it expands or is compressed. FIGURE 17.4a shows a gas cylinder sealed at one end by a movable piston. Force $\vec{F}_{ext}$, perhaps a force supplied by a piston rod, is equal in magnitude and opposite in direction to $\vec{F}_{gas}$. The gas pressure would blow the piston out of the cylinder if the external force weren't there! Using the coordinate system of Figure 17.4a,

$$(F_{ext})_x = -(F_{gas})_x = -pA \qquad (17.6)$$

Suppose the piston moves the small distance dx shown in FIGURE 17.4b. As it does so, the external force (i.e., the environment) does work

$$dW = (F_{ext})_x \, dx = -pA \, dx \qquad (17.7)$$

If dx is positive (the gas expands), then dW is negative. This is because the external force is opposite the displacement. dW is positive if the gas is slightly compressed (negative dx) because the force and the displacement are in the same direction. This is an important idea.

NOTE ▶ The force $\vec{F}_{gas}$ due to the gas pressure inside the cylinder also does work. Because $\vec{F}_{gas} = -\vec{F}_{ext}$, by Newton's third law, the work done by the gas is simply $W_{gas} = -W_{ext}$. To compress the gas, the environment does positive work and the gas does negative work. As the gas expands, W_{gas} is positive and W_{ext} is negative. But the work that appeared in the work-kinetic energy theorem, and now appears in the laws of thermodynamics, is the work done *on* the system by external forces, not the work done *by* the system. It is W_{ext} that tells us whether energy enters the system or leaves the system—by whether it is positive or negative—and that is why we focus our attention on W_{ext} rather than on W_{gas}. ◀

As the piston moves dx, the volume of the gas changes by $dV = A \, dx$. Consequently, Equation 17.7 can be written in terms of the cylinder's volume as

$$dW = -p \, dV \qquad (17.8)$$

If we let the piston move in a slow quasi-static process from initial volume V_i to final volume V_f, the total work done by the environment on the gas is found by integrating Equation 17.8:

$$W = -\int_{V_i}^{V_f} p\,dV \qquad \text{(work done on a gas)} \qquad (17.9)$$

Equation 17.9 is a key result of thermodynamics. Although we used a cylinder to derive Equation 17.9, it turns out to be true for a container of any shape.

NOTE ▶ The pressure of a gas usually changes as the gas expands or contracts. Consequently, p is *not* a constant that can be brought outside the integral. You need to know how the pressure changes with volume before you can carry out the integration. ◀

We can give the work done on a gas a nice geometric interpretation. You learned in Chapter 16 how to represent an ideal-gas process as a curve in the pV diagram. FIGURE 17.5 shows that the work done on a gas is the negative of the area under the pV curve as the volume changes from V_i to V_f. That is

W = the negative of the area under the pV curve between V_i and V_f

Figure 17.5a shows a process in which a gas *expands* from V_i to a larger volume V_f. The area under the curve is positive, so the environment does a negative amount of work on an expanding gas. Figure 17.5b shows a process in which a gas is compressed to a smaller volume. This one is a little trickier because we have to integrate "backward" along the V-axis. You learned in calculus that integrating from a larger limit to a smaller limit gives a negative result, so the area in Figure 17.5b is a negative area. Consequently, as the minus sign in Equation 17.9 indicates, the environment does positive work on a gas to compress it.

FIGURE 17.5 The work done on a gas is the negative of the area under the curve.

(a)

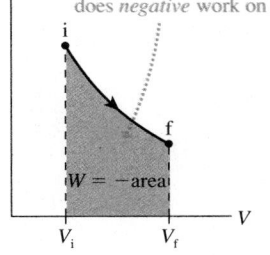

For an *expanding* gas ($V_f > V_i$), the area under the pV curve is positive (integration direction is to the right). Thus the environment does *negative* work on the gas.

Integration direction

(b)

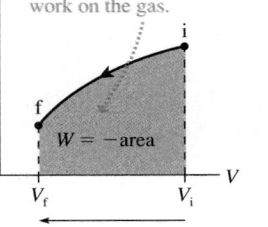

For a *compressed* gas ($V_f < V_i$), the area is negative because the integration direction is to the left. Thus the environment does *positive* work on the gas.

Integration direction

EXAMPLE 17.1 **The work done on an expanding gas**

How much work is done on the gas in the ideal-gas process of FIGURE 17.6?

FIGURE 17.6 The ideal-gas process of Example 17.1.

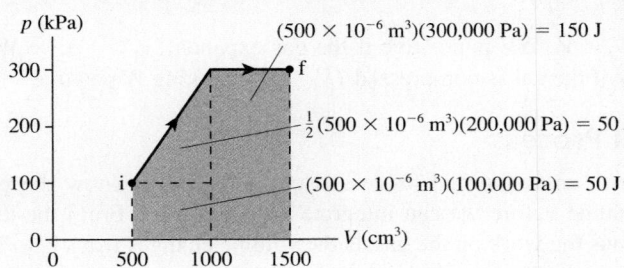

$(500 \times 10^{-6}\,\text{m}^3)(300{,}000\,\text{Pa}) = 150\,\text{J}$

$\frac{1}{2}(500 \times 10^{-6}\,\text{m}^3)(200{,}000\,\text{Pa}) = 50\,\text{J}$

$(500 \times 10^{-6}\,\text{m}^3)(100{,}000\,\text{Pa}) = 50\,\text{J}$

MODEL The work done on a gas is the negative of the area under the pV curve. The gas is *expanding*, so we expect the work to be negative.

SOLVE As Figure 17.6 shows, the area under the curve can be divided into two rectangles and a triangle. Volumes *must* be converted to SI units of m^3. The total area under the curve is 250 J, so the work done on the gas as it expands is

$$W = -(\text{area under the } pV \text{ curve}) = -250\,\text{J}$$

ASSESS We noted previously that the product Pa m^3 is equivalent to joules. The work is negative, as expected, because the external force pushing on the piston is opposite the direction of the piston's displacement.

Equation 17.9 is the basis for a problem-solving strategy.

PROBLEM-SOLVING
STRATEGY 17.1 **Work in ideal-gas processes**

MODEL Assume the gas is ideal and the process is quasi-static.

VISUALIZE Show the process on a pV diagram. Note whether it happens to be one of the basic gas processes: isochoric, isobaric, or isothermal.

SOLVE Calculate the work as the area under the pV curve either geometrically or by carrying out the integration:

$$\text{Work done on the gas } W = -\int_{V_i}^{V_f} p\, dV = -(\text{area under } pV \text{ curve})$$

ASSESS Check your signs.

- $W > 0$ when the gas is compressed. Energy is transferred from the environment to the gas.
- $W < 0$ when the gas expands. Energy is transferred from the gas to the environment.
- No work is done if the volume doesn't change. $W = 0$.

Exercise 4

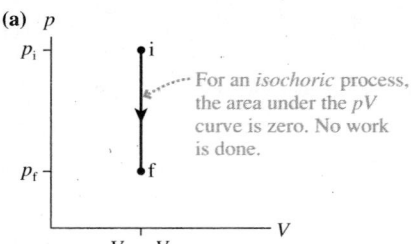

(a) p

For an *isochoric* process, the area under the pV curve is zero. No work is done.

$V_i = V_f$

Isochoric Process

The isochoric process in FIGURE 17.7a is one in which the volume does not change. Consequently,

$$W = 0 \quad \text{(isochoric process)} \qquad (17.10)$$

An isochoric process is the *only* ideal-gas process in which no work is done.

Isobaric Process

FIGURE 17.7b shows an isobaric process in which the volume changes from V_i to V_f. The rectangular area under the curve is $p\Delta V$, so the work done during this process is

$$W = -p\,\Delta V \quad \text{(isobaric process)} \qquad (17.11)$$

(b) p

For an *isobaric* process, the area is $p\Delta V$. The work done on the gas is $-p\Delta V$.

ΔV

$V_i \qquad V_f$

where $\Delta V = V_f - V_i$. ΔV is positive if the gas expands ($V_f > V_i$), so W is negative. ΔV is *negative* if the gas is compressed ($V_f < V_i$), making W positive.

Isothermal Process

FIGURE 17.8 shows an isothermal process. Here we need to know the pressure as a function of volume before we can integrate Equation 17.9. From the ideal-gas law, $p = nRT/V$. Thus the work on the gas as the volume changes from V_i to V_f is

FIGURE 17.8 An isothermal process.

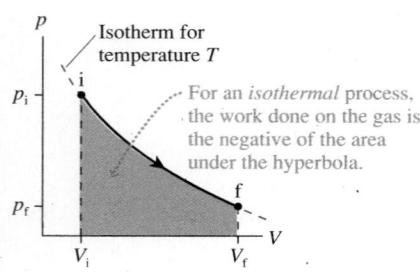

$$W = -\int_{V_i}^{V_f} p\, dV = -\int_{V_i}^{V_f} \frac{nRT}{V} dV = -nRT \int_{V_i}^{V_f} \frac{dV}{V} \qquad (17.12)$$

Isotherm for temperature T

For an *isothermal* process, the work done on the gas is the negative of the area under the hyperbola.

where we could take the T outside the integral because temperature is constant during an isothermal process. This is a straightforward integration, giving

$$W = -nRT \int_{V_i}^{V_f} \frac{dV}{V} = -nRT \ln V \Big|_{V_i}^{V_f}$$

$$= -nRT(\ln V_f - \ln V_i) = -nRT \ln\left(\frac{V_f}{V_i}\right) \qquad (17.13)$$

Because $nRT = p_iV_i = p_fV_f$ during an isothermal process, the work is:

$$W = -nRT \ln\left(\frac{V_f}{V_i}\right) = -p_iV_i \ln\left(\frac{V_f}{V_i}\right) = -p_fV_f \ln\left(\frac{V_f}{V_i}\right) \qquad (17.14)$$
(isothermal process)

Which version of Equation 17.14 is easiest to use will depend on the information you're given. The pressure, volume, and temperature *must* be in SI units.

EXAMPLE 17.2 **The work of an isothermal compression**

A cylinder contains 7.0 g of nitrogen gas. How much work must be done to compress the gas at a constant temperature of 80°C until the volume is halved?

MODEL This is an isothermal ideal-gas process.

SOLVE Nitrogen gas is N_2, with molar mass $M_{mol} = 28$ g/mol, so 7.0 g is 0.25 mol of gas. The temperature is $T = 353$ K. Although we don't know the actual volume, we do know that $V_f = \frac{1}{2}V_i$. The volume ratio is all we need to calculate the work:

$$W = -nRT \ln\left(\frac{V_f}{V_i}\right)$$

$$= -(0.25 \text{ mol})(8.31 \text{ J/mol K})(353 \text{ K})\ln(1/2) = 508 \text{ J}$$

ASSESS The work is positive because a force from the environment pushes the piston inward to compress the gas.

Work Depends on the Path

FIGURE 17.9a shows two different processes that take a gas from an initial state i to a final state f. Although the initial and final states are the same, the work done during these two processes is *not* the same. **The work done during an ideal-gas process depends on the path followed through the pV diagram.**

You may recall that "work is independent of the path," but that referred to a different situation. In Chapter 11, we found that the work done by a conservative force is independent of the physical path of the object through space. For an ideal-gas process, the "path" is a sequence of thermodynamic states on a pV diagram. It is a figurative path because we can draw a picture of it on a pV diagram, but it is not a literal path.

The path dependence of work has an important implication for multistep processes such as the one shown in FIGURE 17.9b. The total work done on the gas during the process $1 \rightarrow 2 \rightarrow 3$ must be calculated as $W_{1 \text{ to } 3} = W_{1 \text{ to } 2} + W_{2 \text{ to } 3}$. In this case, $W_{1 \text{ to } 2}$ is negative and $W_{2 \text{ to } 3}$ is positive. Trying to compute the work in a single step, using $\Delta V = V_3 - V_1$, would give you the work of a process that goes directly from 1 to 3. The initial and final states are the same, but the work is *not* the same because work depends on the path followed through the pV diagram.

FIGURE 17.9 The work done during an ideal-gas process depends on the path.

(a)

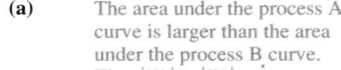
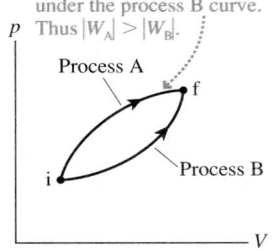

The area under the process A curve is larger than the area under the process B curve. Thus $|W_A| > |W_B|$.

(b)

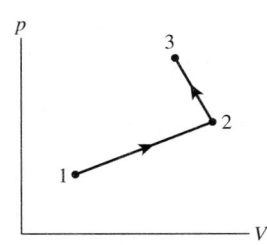

STOP TO THINK 17.2 Two processes take an ideal gas from state 1 to state 3. Compare the work done by process A to the work done by process B.

a. $W_A = W_B = 0$
b. $W_A = W_B$ but neither is zero
c. $W_A > W_B$
d. $W_A < W_B$

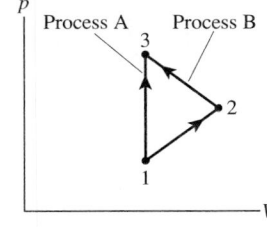

17.3 Heat

Heat is a more elusive concept than work. We use the word "heat" very loosely in the English language, often as synonymous with *hot*. We might say on a very hot day, "This heat is oppressive." If your apartment is cold, you may say, "Turn up the heat." These expressions date to a time long ago when it was thought that heat was a *substance* with fluid-like properties.

FIGURE 17.10 Joule's experiments to show the equivalence of heat and work.

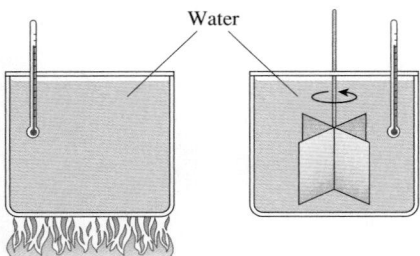

Water

The flame heats the water. The temperature increases.

The spinning paddle does work on the water. The temperature increases.

Our concept of heat changed with the work of British physicist James Joule in the 1840s. Joule was the first to carry out careful experiments to learn how it is that systems change their temperature. Using experiments like those shown in FIGURE 17.10, Joule found that you can raise the temperature of a beaker of water by two entirely different means:

1. Heating it with a flame, or
2. Doing work on it with a rapidly spinning paddle wheel.

The final state of the water is *exactly* the same in both cases. This implies that heat and work are essentially equivalent. In other words, heat is not a substance. Instead, heat is *energy.* Heat and work, which previously had been regarded as two completely different phenomena, were now seen to be simply two different ways of transferring energy to or from a system.

Thermal Interactions

To be specific, **heat** is the energy transferred between a system and the environment as a consequence of a *temperature difference* between them. Unlike a mechanical interaction in which work is done, heat requires no macroscopic motion of the system. Instead (we'll look at the details in Chapter 18), energy is transferred when the *faster* molecules in the hotter object collide with the *slower* molecules in the cooler object. On average, these collisions cause the faster molecules to lose energy and the slower molecules to gain energy. The net result is that energy is transferred from the hotter object to the colder object. The process itself, whereby energy is transferred between the system and the environment via atomic-level collisions, is called a **thermal interaction.**

When you place a pan of water on the stove, heat is the energy transferred *from* the hotter flame *to* the cooler water. If you place the water in a freezer, heat is the energy transferred from the warmer water to the colder air in the freezer. A system is in **thermal equilibrium** with the environment, or two systems are in thermal equilibrium with each other, if there is no temperature difference.

Like work, **heat is not a state variable. That is, heat is not a property of the system.** Instead, heat is the amount of energy that moves between the system and the environment during a thermal interaction. It would not be meaningful to talk about a "change of heat." Thus heat appears in the energy equation simply as a value Q, never as ΔQ. FIGURE 17.11 shows how to interpret the sign of Q.

> **NOTE** ▶ For both heat and work, a positive value indicates energy being transferred from the environment to the system. Table 17.1 summarizes the similarities and differences between work and heat. ◀

FIGURE 17.11 The sign of heat.

(a) Positive heat

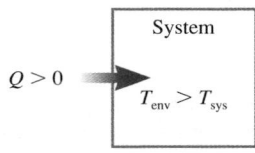

$Q > 0$

System

$T_{env} > T_{sys}$

(b) Negative heat

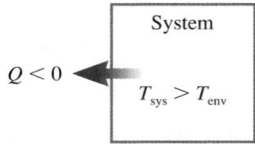

$Q < 0$

System

$T_{sys} > T_{env}$

(c) Thermal equilibrium

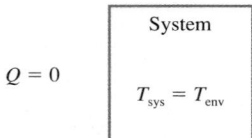

$Q = 0$

System

$T_{sys} = T_{env}$

TABLE 17.1 Understanding work and heat

	Work	Heat
Interaction:	Mechanical	Thermal
Requires:	Force and displacement	Temperature difference
Process:	Macroscopic pushes and pulls	Microscopic collisions
Positive value:	$W > 0$ when a gas is compressed. Energy is transferred in.	$Q > 0$ when the environment is at a higher temperature than the system. Energy is transferred in.
Negative value:	$W < 0$ when a gas expands. Energy is transferred out.	$Q < 0$ when the system is at a higher temperature than the environment. Energy is transferred out.
Equilibrium:	A system is in mechanical equilibrium when there is no net force or torque on it.	A system is in thermal equilibrium when it is at the same temperature as the environment.

Units of Heat

Heat is energy transferred between the system and the environment. Consequently, the SI unit of heat is the joule. Historically, before the connection between heat and work had been recognized, a unit for measuring heat, the calorie, had been defined as

> 1 calorie = 1 cal = the quantity of heat needed to change
> the temperature of 1 g of water by 1°C

Once Joule established that heat is energy, it was apparent that the calorie is really a unit of energy. In today's SI units, the conversion is

$$1 \text{ cal} = 4.186 \text{ J}$$

The calorie you know in relation to food is not the same as the heat calorie. The *food calorie,* abbreviated Cal with a capital C, is

> 1 food calorie = 1 Cal = 1000 cal = 1 kcal = 4186 J

We will not use calories in this textbook, but there are some fields of science and engineering where calories are still widely used. All the calculations you learn to do with joules can equally well be done with calories.

Heat, Temperature, and Thermal Energy

It is important to distinguish among *heat, temperature,* and *thermal energy.* These three ideas are related, but the distinctions among them are crucial. In brief,

- Thermal energy is an energy *of the system* due to the motion of its atoms and molecules. It is a *form* of energy. Thermal energy is a state variable, and it makes sense to talk about how E_{th} changes during a process. The system's thermal energy continues to exist even if the system is isolated and not interacting thermally with its environment.
- Heat is energy transferred *between the system* and the environment as they interact. Heat is *not* a particular form of energy, nor is it a state variable. It makes no sense to talk about how heat changes. $Q = 0$ if a system does not interact thermally with its environment. Heat may cause the system's thermal energy to change, but that doesn't make heat and thermal energy the same.
- Temperature is a state variable that quantifies the "hotness" or "coldness" of a system. We haven't given a precise definition of temperature, but it is related to the thermal energy *per molecule.* A temperature difference is a requirement for a thermal interaction in which heat energy is transferred between the system and the environment.

Heat is the energy transferred in a thermal interaction.

It is especially important not to associate an observed temperature increase with heat. Heating a system is one way to change its temperature, but, as Joule showed, not the only way. You can also change the system's temperature by doing work on the system or, as is the case with friction, transforming mechanical energy into thermal energy. **Observing the system tells us nothing about the process by which energy enters or leaves the system.**

STOP TO THINK 17.3   Which one or more of the following processes involves heat?

- a. The brakes in your car get hot when you stop.
- b. A steel block is held over a candle.
- c. You push a rigid cylinder of gas across a frictionless surface.
- d. You push a piston into a cylinder of gas, increasing the temperature of the gas.
- e. You place a cylinder of gas in hot water. The gas expands, causing a piston to rise and lift a weight. The temperature of the gas does not change.

17.4 The First Law of Thermodynamics

Heat was the missing piece that we needed to arrive at a completely general statement of the law of conservation of energy. Restating Equation 17.4, we have

$$\Delta E_{sys} = \Delta E_{mech} + \Delta E_{th} = W + Q$$

Work and heat, two ways of transferring energy between a system and the environment, cause the system's energy to change.

At this point in the text we are not interested in systems that have a macroscopic motion of the system as a whole. Moving macroscopic systems were important to us for many chapters, but now, as we investigate the thermal properties of a system, we would like the system as a whole to rest peacefully on the laboratory bench while we study it. So we will assume, throughout the remainder of Part IV, that $\Delta E_{mech} = 0$.

With this assumption clearly stated, the law of conservation of energy becomes

$$\Delta E_{th} = W + Q \qquad \text{(first law of thermodynamics)} \qquad (17.15)$$

The energy equation, in this form, is called the **first law of thermodynamics** or simply "the first law." The first law is a very general statement about the conservation of energy.

Chapters 10 and 11 introduced the basic energy model. It was called *basic* because it included work but not heat. The first law of thermodynamics has included heat, but it excludes situations where the mechanical energy changes. FIGURE 17.12 is a pictorial representation of the **thermodynamic energy model** described by the first law. Work and heat are energies transferred between the system and the environment. Energy added to the system (W or Q positive) increases the system's thermal energy ($\Delta E_{th} > 0$). Likewise, the thermal energy decreases when energy is removed from the system.

Two comments are worthwhile:

1. The first law of thermodynamics doesn't tell us anything about the value of E_{th}, only how E_{th} changes. Doing 1 J of work changes the thermal energy by $\Delta E_{th} = 1$ J regardless of whether $E_{th} = 10$ J or 10,000 J.
2. The system's thermal energy isn't the only thing that changes. Work or heat that changes the thermal energy also changes the pressure, volume, temperature, and other state variables. The first law tells us only about ΔE_{th}. Other laws and relationships must be used to learn how the other state variables change.

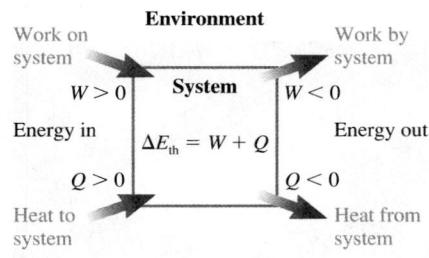

FIGURE 17.12 The thermodynamic energy model.

Three Special Ideal-Gas Processes

There are three ideal-gas processes in which one of the terms in the first law—ΔE_{th}, W, or Q—is zero. To investigate these processes, FIGURE 17.13 shows a gas cylinder with three special properties:

- You can keep the gas volume from changing by inserting the locking pin into the piston. Without the pin, the piston can slide up or down. The piston is massless, frictionless, and insulated.
- You can change the gas pressure by adding or removing masses on top of the piston. Work is done as the piston moves the masses up and down.
- You can warm or cool the gas by placing the cylinder above a flame or on a block of ice. The thin bottom of the cylinder is the only surface through which heat energy can be transferred.

You learned in Chapter 16 (see Figure 16.9) that the gas pressure when the piston "floats" is determined by the atmospheric pressure and by the total mass M on the piston:

$$p_{gas} = p_{atmos} + \frac{Mg}{A} \qquad (17.16)$$

The pressure doesn't change as the piston moves unless you change the mass. This is a particularly important point to understand. Equation 17.16 is *not* valid when the piston is locked. The pressure with the piston locked could be either higher or lower than the value found with Equation 17.16.

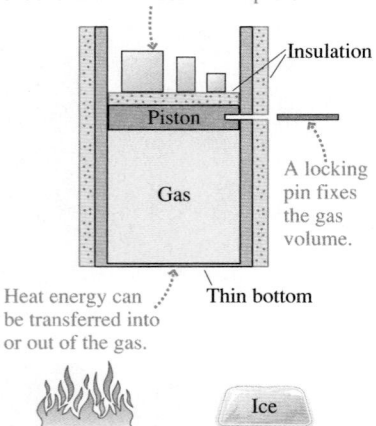

FIGURE 17.13 The gas can be heated and have work done on it.

An isochoric cooling process ($W = 0$): No work is done in an isochoric (constant volume) process because the piston doesn't move. To cool the gas without doing work:

■ Insert the locking pin so that the volume cannot change.
■ Place the cylinder on the block of ice. Heat energy will be transferred from the gas to the ice, causing the gas temperature and pressure to fall.
■ Remove the cylinder from the ice when the desired pressure is reached.
■ Remove masses from the piston until the total mass M balances the new gas pressure. This step must be done before removing the locking pin; otherwise, the piston will move when the pin is removed.
■ Remove the locking pin.

Figure 17.7a showed the pV diagram. The final point is on a lower isotherm than the initial point, so $T_f < T_i$. No work was done, but heat energy was transferred out of the gas ($Q < 0$) and the thermal energy of the gas decreased ($\Delta E_{th} < 0$) as the temperature fell. FIGURE 17.14 shows this result on a first-law bar chart.

An isothermal expansion ($\Delta E_{th} = 0$): The thermal energy does not change in an isothermal process because the temperature of the gas doesn't change. To expand the gas without changing its thermal energy:

■ Place the cylinder over the flame. Heat energy will be transferred to the gas, and the gas will begin to expand.
■ The product pV must remain constant during an isothermal process. Slowly remove masses from the piston to reduce the pressure as the volume increases. The temperature remains constant as heat energy from the flame balances the negative work done on the gas as it expands.
■ Remove the cylinder from the flame when the gas reaches the desired volume.

Figure 17.8 showed the pV diagram, and FIGURE 17.15 is the first-law bar chart. The temperature doesn't change in an isothermal process ($\Delta T = 0$), hence the thermal energy cannot change ($\Delta E_{th} = 0$). Heat energy is transferred to the gas, but that energy is used to do work (the piston lifts the masses) rather than to increase the temperature. Here "do work" means that the gas is doing work ($W_{gas} > 0$), so the external work *on* the gas—the W in the first law—is negative.

> **NOTE** ▶ It is surprising, but true, that we can heat the system without changing its temperature. But to do so, we must have a process in which the energy coming into the system as heat is exactly balanced by the energy leaving the system as work. **The important point is that $\Delta T = 0$ does *not* mean $Q = 0$.** ◄

An adiabatic compression ($Q = 0$): A process in which no heat energy is transferred between the system and the environment is called an **adiabatic process**. Although the system cannot have thermal interactions with its environment, it can still have mechanical interactions as the insulated piston pushes or pulls on the gas. To compress the gas without heat:

■ Add insulation beneath the cylinder.
■ Slowly add masses to the piston, increasing the pressure. The piston will slowly descend, compressing the gas and decreasing its volume.
■ Stop adding masses when the gas reaches the desired volume.

$Q = 0$ in an adiabatic process, so the first law $\Delta E_{th} = W + Q$ can be satisfied only if $\Delta E_{th} = W$. This information is shown on the first-law bar chart of FIGURE 17.16.

> **NOTE** ▶ Just because the system is well insulated—thermally isolated from the environment—does not mean its temperature remains constant. An adiabatic compression uses work to increase the temperature of the gas. Similarly, an adiabatic expansion lowers the temperature of the gas. **The important point is that $Q = 0$ does *not* mean $\Delta T = 0$.** ◄

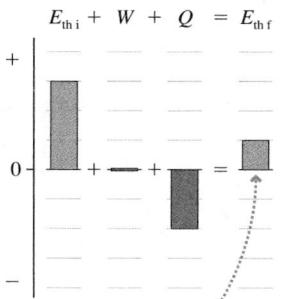

FIGURE 17.14 A first-law bar chart for a process that does no work.

$$E_{th\,i} + W + Q = E_{th\,f}$$

Thermal energy has decreased by the amount of energy that left the system as heat.

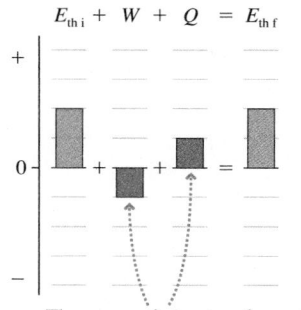

FIGURE 17.15 A first-law bar chart for a process that doesn't change the thermal energy.

$$E_{th\,i} + W + Q = E_{th\,f}$$

The energy that enters the system as heat leaves as work. The thermal energy is unchanged.

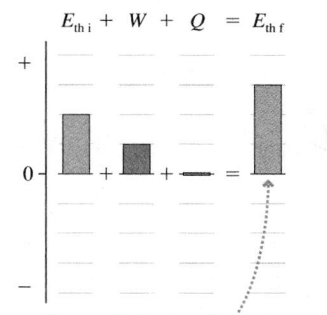

FIGURE 17.16 A first-law bar chart for a process that transfers no heat energy.

$$E_{th\,i} + W + Q = E_{th\,f}$$

Energy that enters the system as work increases the thermal energy —and thus the temperature.

We'll examine adiabatic gas processes and their pV curve later in the chapter. For now, make sure you understand which quantities are zero and which aren't in these three special processes.

STOP TO THINK 17.4 Which first-law bar chart describes the process shown in the pV diagram?

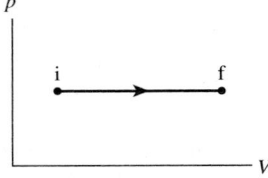

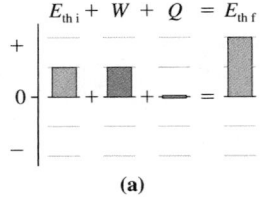

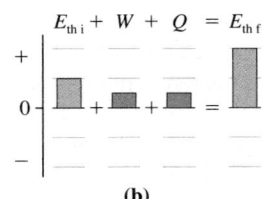

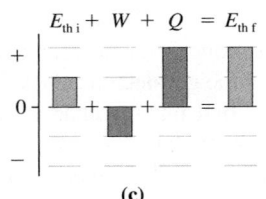

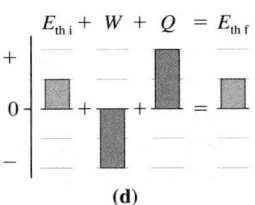

     (a)   **(b)**   **(c)**   **(d)**

17.5 Thermal Properties of Matter

Heat and work are equivalent in the sense that the change of the system is *exactly the same* whether you transfer heat energy to it or do an equal amount of work on it. Adding energy to the system, or removing it, changes the system's thermal energy.

What happens to a system when you change its thermal energy? In this section we'll consider two distinct possibilities:

- The temperature of the system changes.
- The system undergoes a phase change, such as melting or freezing.

Temperature Change and Specific Heat

TABLE 17.2 Specific heats and molar specific heats of solids and liquids

Substance	c (J/kg K)	C (J/mol K)
Solids		
Aluminum	900	24.3
Copper	385	24.4
Iron	449	25.1
Gold	129	25.4
Lead	128	26.5
Ice	2090	37.6
Liquids		
Ethyl alcohol	2400	110.4
Mercury	140	28.1
Water	4190	75.4

Suppose you do an experiment in which you add energy to water, either by doing work on it or by transferring heat to it. Either way, you will find that adding 4190 J of energy raises the temperature of 1 kg of water by 1 K. If you were fortunate enough to have 1 kg of gold, you would need to add only 129 J of energy to raise its temperature by 1 K.

The amount of energy that raises the temperature of 1 kg of a substance by 1 K is called the **specific heat** of that substance. The symbol for specific heat is c. Water has specific heat $c_{\text{water}} = 4190$ J/kg K. The specific heat of gold is $c_{\text{gold}} = 129$ J/kg K. Specific heat depends only on the material from which an object is made. Table 17.2 provides some specific heats for common liquids and solids.

NOTE ▶ The term *specific heat* does not use the word "heat" in the way that we have defined it. Specific heat is an old idea, dating back to the days of the caloric theory when heat was thought to be a substance contained in the object. The term has continued in use even though our understanding of heat has changed. ◀

If energy c is required to raise the temperature of 1 kg of a substance by 1 K, then energy Mc is needed to raise the temperature of mass M by 1 K and $(Mc)\Delta T$ is needed to raise the temperature of mass M by ΔT. In other words, the thermal energy of the system changes by

$$\Delta E_{\text{th}} = Mc\Delta T \qquad \text{(temperature change)} \qquad (17.17)$$

when its temperature changes by ΔT. ΔE_{th} can be either positive (thermal energy increases as the temperature goes up) or negative (thermal energy decreases as the temperature goes down). Recall that uppercase M is used for the mass of an entire system while lowercase m is reserved for the mass of an atom or molecule.

NOTE ▶ In practice, ΔT is usually measured in °C. But the Kelvin and the Celsius temperature scales have the same step size, so ΔT in K has exactly the same numerical value as ΔT in °C. Thus

- You do not need to convert temperatures from °C to K if you need only a temperature *change* ΔT.
- You do need to convert anytime you need the actual temperature T. ◀

The first law of thermodynamics, $\Delta E_{th} = W + Q$, allows us to write Equation 17.17 as $Mc\Delta T = W + Q$. In other words, **we can change the system's temperature either by heating it or by doing an equivalent amount of work on it.** In working with solids and liquids, we almost always change the temperature by heating. If $W = 0$, which we will assume for the rest of this section, then the heat energy needed to bring about a temperature change ΔT is

$$Q = Mc\Delta T \qquad \text{(temperature change)} \qquad (17.18)$$

Because $\Delta T = \Delta E_{th}/Mc$, it takes more energy to change the temperature of a substance with a large specific heat than to change the temperature of a substance with a small specific heat. You can think of specific heat as measuring the *thermal inertia* of a substance. Metals, with small specific heats, warm up and cool down quickly. A piece of aluminum foil can be safely held within seconds of removing it from a hot oven. Water, with a very large specific heat, is slow to warm up and slow to cool down. This is fortunate for us. The large thermal inertia of water is essential for the biological processes of life. We wouldn't be here studying physics if water had a small specific heat!

EXAMPLE 17.3 **Running a fever**

A 70 kg student catches the flu, and his body temperature increases from 37.0°C (98.6°F) to 39.0°C (102.2°F). How much energy is required to raise his body's temperature? The specific heat of a mammalian body is 3400 J/kg K, nearly that of water because mammals are mostly water.

MODEL Energy is supplied by the chemical reactions of the body's metabolism. These exothermic reactions transfer heat to the body. Normal metabolism provides enough heat energy to offset energy losses (radiation, evaporation, etc.) while maintaining

a normal body temperature of 37°C. We need to calculate the additional energy needed to raise the body's temperature by 2.0°C, or 2.0 K.

SOLVE The necessary heat energy is

$$Q = Mc\Delta T = (70 \text{ kg})(3400 \text{ J/kg K})(2.0 \text{ K}) = 4.8 \times 10^5 \text{ J}$$

ASSESS This appears to be a lot of energy, but a joule is actually a very small amount of energy. It is only 110 Cal, approximately the energy gained by eating an apple.

The **molar specific heat** is the amount of energy that raises the temperature of 1 mol of a substance by 1 K. We'll use an uppercase C for the molar specific heat. The heat energy needed to bring about a temperature change ΔT of n moles of substance is

$$Q = nC\Delta T \qquad (17.19)$$

Molar specific heats are listed in Table 17.2. Look at the five elemental solids (excluding ice). All have C very near 25 J/mol K. If we were to expand the table, we would find that most elemental solids have $C \approx 25$ J/mol K. This can't be a coincidence, but what is it telling us? This is a puzzle we will address in Chapter 18, where we will explore thermal energy at the atomic level.

Phase Change and Heat of Transformation

FIGURE 17.17 The temperature of a system that is heated at a steady rate.

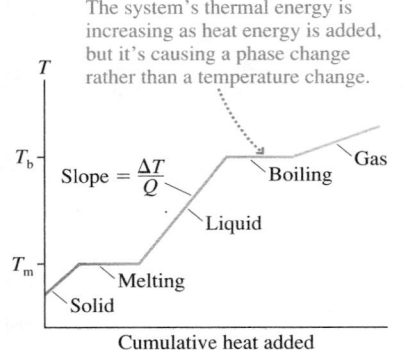

The system's thermal energy is increasing as heat energy is added, but it's causing a phase change rather than a temperature change.

T

T_b Slope = $\dfrac{\Delta T}{Q}$ Boiling Gas

Liquid

T_m Melting

Solid

Cumulative heat added

Suppose you start with a system in its solid phase and heat it at a steady rate. **FIGURE 17.17**, which you saw in Chapter 16, shows how the system's temperature changes. At first, the temperature increases linearly. This is not hard to understand because Equation 17.18 can be written

$$\text{slope of the } T\text{-versus-}Q \text{ graph} = \frac{\Delta T}{Q} = \frac{1}{Mc} \qquad (17.20)$$

The slope of the graph depends inversely on the system's specific heat. A constant specific heat implies a constant slope and thus a linear graph. In fact, you can measure c from such a graph.

NOTE ▶ The different slopes indicate that the solid, liquid, and gas phases of a substance have different specific heats. ◀

But there are times, shown as horizontal line segments, during which heat is being transferred to the system but the temperature isn't changing. These are *phase changes*. The thermal energy continues to increase during a phase change, but the additional energy goes into breaking molecular bonds rather than speeding up the molecules. **A phase change is characterized by a change in thermal energy without a change in temperature.**

The amount of heat energy that causes 1 kg of a substance to undergo a phase change is called the **heat of transformation** of that substance. For example, laboratory experiments show that 333,000 J of heat are needed to melt 1 kg of ice at 0°C. The symbol for heat of transformation is L. The heat required for the entire system of mass M to undergo a phase change is

$$Q = ML \qquad \text{(phase change)} \qquad (17.21)$$

Heat of transformation is a generic term that refers to any phase change. Two specific heats of transformation are the **heat of fusion** L_f, the heat of transformation between a solid and a liquid, and the **heat of vaporization** L_v, the heat of transformation between a liquid and a gas. The heat needed for these phase changes is

$$Q = \begin{cases} \pm ML_f & \text{melt/freeze} \\ \pm ML_v & \text{boil/condense} \end{cases} \qquad (17.22)$$

Lava—molten rock—undergoes a phase change when it contacts the much colder water. This is one way in which new islands are formed.

where the $\pm$ indicates that heat must be *added* to the system during melting or boiling but *removed* from the system during freezing or condensing. **You must explicitly include the minus sign when it is needed.**

Table 17.3 gives the heats of transformation of a few substances. Notice that the heat of vaporization is always much larger than the heat of fusion. We can understand this. Melting breaks just enough molecular bonds to allow the system to lose rigidity and flow. Even so, the molecules in a liquid remain close together and loosely bonded. Vaporization breaks all bonds completely and sends the molecules flying apart. This process requires a larger increase in the thermal energy and thus a larger quantity of heat.

TABLE 17.3 Melting/boiling temperatures and heats of transformation

Substance	T_m (°C)	L_f (J/kg)	T_b (°C)	L_v (J/kg)
Nitrogen (N_2)	−210	0.26×10^5	−196	1.99×10^5
Ethyl alcohol	−114	1.09×10^5	78	8.79×10^5
Mercury	−39	0.11×10^5	357	2.96×10^5
Water	0	3.33×10^5	100	22.6×10^5
Lead	328	0.25×10^5	1750	8.58×10^5

| EXAMPLE 17.4 | **Melting wax** |

An insulated jar containing 200 g of solid candle wax is placed on a hot plate that supplies heat energy to the wax at the rate of 220 J/s. The wax temperature is measured every 30 s, yielding the following data:

Time (s)	Temperature (°C)	Time (s)	Temperature (°C)
0	20.0	180	70.5
30	31.7	210	70.5
60	42.2	240	70.6
90	55.0	270	70.5
120	64.7	300	70.4
150	70.4	330	74.5

What are the specific heat of the solid wax, the melting point, and the wax's heat of fusion?

MODEL The wax is in an insulated jar, so assume that heat loss to the environment is negligible.

VISUALIZE Heat energy is being supplied at the rate of 220 J/s, so the total heat energy that has been transferred into the wax at time t is $Q = 220t$ J. **FIGURE 17.18** shows the temperature graphed against

the cumulative heat Q, although notice that the horizontal axis is in kJ, not J. The initial linear slope corresponds to raising the wax's temperature to the melting point. Temperature remains constant during a phase change, even though the sample is still being heated, so the horizontal section of the graph is when the wax is melting. The temperature increase at the end shows that the temperature of the liquid wax is beginning to rise after melting is complete.

SOLVE From $Q = Mc\Delta T$, the slope of the T-versus-Q graph is $\Delta T/Q = 1/Mc$. The experimental slope of the best-fit line is $1.708°C/kJ = 0.001708$ K/J. Thus the specific heat of the solid wax is

$$c = \frac{1}{M \times \text{slope}} = \frac{1}{(0.200 \text{ kg})(0.001708 \text{ K/J})} = 2930 \text{ J/kg K}$$

From the table, we see that the melting temperature—which remains constant during the phase change—is 70.5°C. The heat required for the phase change is $Q = ML_f$, so the heat of fusion is $L_f = Q/M$. With data recorded only every 30 s, it's not exactly clear when the melting began and when it ended. The extension of the initial slope shows that the temperature reached the melting point about halfway between 120 s and 150 s, so the melting started at about 135 s. We'll assume it was complete about halfway between 300 s and 330 s, or at about 315 s. Thus the melting took 180 s, during which, at 220 J/s, 39,600 J of heat energy was transferred from the hot plate to the wax. With this value of Q, the heat of fusion is

$$L_f = \frac{Q}{M} = \frac{39,600 \text{ J}}{0.200 \text{ kg}} = 2.0 \times 10^5 \text{ J/kg}$$

ASSESS Both the specific heat and the heat of fusion are similar to values in Tables 17.2 and 17.3, which gives us confidence in our results.

FIGURE 17.18 The heating curve of the wax.

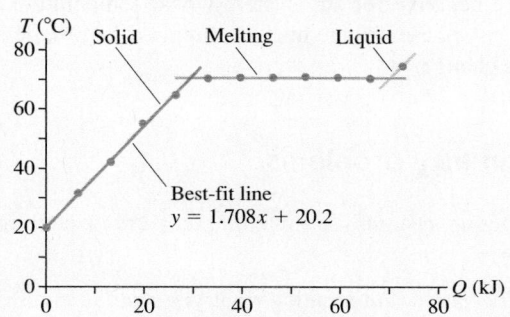

STOP TO THINK 17.5   Objects A and B are brought into close thermal contact with each other, but they are well isolated from their surroundings. Initially $T_A = 0°C$ and $T_B = 100°C$. The specific heat of A is less than the specific heat of B. The two objects will soon reach a common final temperature T_f. The final temperature is

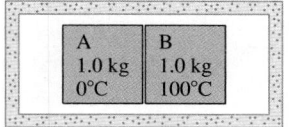

a. $T_f > 50°C$ b. $T_f = 50°C$ c. $T_f < 50°C$

17.6 Calorimetry

At one time or another you've probably put an ice cube into a hot drink to cool it quickly. You were engaged, in a somewhat trial-and-error way, in a practical aspect of heat transfer known as **calorimetry.**

FIGURE 17.19 on the next page shows two systems thermally interacting with each other but isolated from everything else. Suppose they start at different temperatures T_1 and T_2. As you know from experience, heat energy will be transferred from the hotter

FIGURE 17.19 Two systems interact thermally.

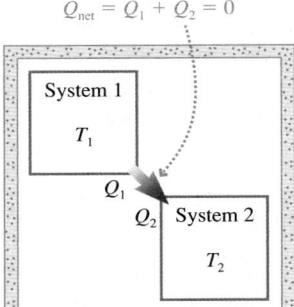

Heat energy is transferred from system 1 to system 2. Energy conservation requires

$$|Q_1| = |Q_2|$$

Opposite signs mean that

$$Q_{net} = Q_1 + Q_2 = 0$$

to the colder system until they reach a common final temperature T_f. The systems will then be in thermal equilibrium and the temperature will not change further.

The insulation prevents any heat energy from being transferred to or from the environment, so energy conservation tells us that any energy leaving the hotter system must enter the colder system. That is, the systems *exchange* energy with no net loss or gain. The concept is straightforward, but to state the idea mathematically we need to be careful with signs.

Let Q_1 be the energy transferred to system 1 as heat. Similarly, Q_2 is the energy transferred to system 2. The fact that the systems are merely exchanging energy can be written $|Q_1| = |Q_2|$. That is, the energy *lost* by the hotter system is the energy *gained* by the colder system. Thus Q_1 and Q_2 have opposite signs: $Q_1 = -Q_2$. No energy is exchanged with the environment, hence it makes more sense to write this relationship as

$$Q_{net} = Q_1 + Q_2 = 0 \tag{17.23}$$

This idea is not limited to the interaction of only two systems. If three or more systems are combined in isolation from the rest of their environment, each at a different initial temperature, they will all come to a common final temperature that can be found from the relationship

$$Q_{net} = Q_1 + Q_2 + Q_3 + \cdots = 0 \tag{17.24}$$

NOTE ▶ The signs are very important in calorimetry problems. ΔT is always $T_f - T_i$, so ΔT and Q are negative for any system whose temperature decreases. The proper sign of Q for any phase change must be supplied *by you*, depending on the direction of the phase change. ◀

PROBLEM-SOLVING STRATEGY 17.2 **Calorimetry problems**

MODEL Identify the interacting systems. Assume that they are isolated from the larger environment.

VISUALIZE List known information and identify what you need to find. Convert all quantities to SI units.

SOLVE The mathematical representation, which is a statement of energy conservation, is

$$Q_{net} = Q_1 + Q_2 + \cdots = 0$$

- For systems that undergo a temperature change, $Q = Mc(T_f - T_i)$. Be sure to have the temperatures T_i and T_f in the correct order.
- For systems that undergo a phase change, $Q = \pm ML$. Supply the correct sign by observing whether energy enters or leaves the system.
- Some systems may undergo a temperature change *and* a phase change. Treat the changes separately. The heat energy is $Q = Q_{\Delta T} + Q_{phase}$.

ASSESS Is the final temperature in the middle? T_f that is higher or lower than all initial temperatures is an indication that something is wrong, usually a sign error.

Exercise 15

NOTE ▶ You may have learned to solve calorimetry problems in other courses by writing $Q_{gained} = Q_{lost}$. That is, by balancing heat gained with heat lost. That approach works in simple problems, but it has two drawbacks. First, you often have to "fudge" the signs to make them work. Second, and more serious, you can't extend this approach to a problem with three or more interacting systems. Using $Q_{net} = 0$ is much preferred. ◀

EXAMPLE 17.5 **Calorimetry with a phase change**

Your 500 mL soda is at 20°C, room temperature, so you add 100 g of ice from the −20°C freezer. Does all the ice melt? If so, what is the final temperature? If not, what fraction of the ice melts? Assume that you have a well-insulated cup.

MODEL We have a thermal interaction between the soda, which is essentially water, and the ice. We need to distinguish between three possible outcomes. If all the ice melts, then $T_f > 0°C$. It's also possible that the soda will cool to 0°C before all the ice has melted, leaving the ice and liquid in equilibrium at 0°C. A third possibility is that the soda will freeze solid before the ice warms up to 0°C. That seems unlikely here, but there are situations, such as the pouring of molten metal out of furnaces, when all the liquid does solidify. We need to distinguish between these before knowing how to proceed.

VISUALIZE All the initial temperatures, masses, and specific heats are known. The final temperature of the combined soda + ice system is unknown.

SOLVE Let's first calculate the heat needed to melt all the ice and leave it as liquid water at 0°C. To do so, we must warm the ice to 0°C, then change it to water. The heat input for this two-stage process is

$$Q_{melt} = M_i c_i (20\ \text{K}) + M_i L_f = 37,500\ \text{J}$$

where L_f is the heat of fusion of water. It is used as a *positive* quantity because we must *add* heat to melt the ice. Next, let's calculate how much heat energy will leave the soda if it cools all the

way to 0°C. The volume is $V = 500\ \text{mL} = 5.00 \times 10^{-4}\ \text{m}^3$ and thus the mass is $M_s = \rho V = 0.500\ \text{kg}$. The heat is

$$Q_{cool} = M_s c_w (-20\ \text{K}) = -41,900\ \text{J}$$

where $\Delta T = -20$ K because the temperature decreases. Because $|Q_{cool}| > Q_{melt}$, the soda has sufficient energy to melt all the ice. Hence the final state will be all liquid at $T_f > 0$. (Had we found $|Q_{cool}| < Q_{melt}$, then the final state would have been an ice-liquid mixture at 0°C.)

Energy conservation requires $Q_{ice} + Q_{soda} = 0$. The heat Q_{ice} consists of three terms: warming the ice to 0°C, melting the ice to water at 0°C, then warming the 0°C water to T_f. The mass will still be M_i in the last of these steps because it is the "ice system," but we need to use the specific heat of *liquid water*. Thus

$$Q_{ice} + Q_{soda} = [M_i c_i (20\ \text{K}) + M_i L_f + M_i c_w (T_f - 0°C)]$$
$$+ M_s c_w (T_f - 20°C) = 0$$

We've already done part of the calculation, allowing us to write

$$37,500\ \text{J} + M_i c_w (T_f - 0°C) + M_s c_w (T_f - 20°C) = 0$$

Solving for T_f gives

$$T_f = \frac{20 M_s c_w - 37,500}{M_i c_w + M_s c_w} = 1.7°C$$

ASSESS As expected, the soda has been cooled to nearly the freezing point.

EXAMPLE 17.6 **Three interacting systems**

A 200 g piece of iron at 120°C and a 150 g piece of copper at −50°C are dropped into an insulated beaker containing 300 g of ethyl alcohol at 20°C. What is the final temperature?

MODEL Here you can't use a simple $Q_{gained} = Q_{lost}$ approach because you don't know whether the alcohol is going to warm up or cool down.

VISUALIZE All the initial temperatures, masses, and specific heats are known. We need to find the final temperature.

SOLVE Energy conservation requires

$$Q_i + Q_c + Q_e = M_i c_i (T_f - 120°C) + M_c c_c (T_f - (-50°C))$$
$$+ M_e c_e (T_f - 20°C) = 0$$

Solving for T_f gives

$$T_f = \frac{120 M_i c_i - 50 M_c c_c + 20 M_e c_e}{M_i c_i + M_c c_c + M_e c_e} = 25.7°C$$

ASSESS The temperature is between the initial iron and copper temperatures, as expected. It turns out that the alcohol warms up ($Q_e > 0$), but we had no way to know this without doing the calculation.

17.7 The Specific Heats of Gases

Specific heats are given in Table 17.2 for solids and liquids. Gases are harder to characterize because the heat required to cause a specified temperature change depends on the *process* by which the gas changes state.

FIGURE 17.20 shows two isotherms on the pV diagram for a gas. Processes A and B, which start on the T_i isotherm and end on the T_f isotherm, have the *same* temperature change $\Delta T = T_f - T_i$. But process A, which takes place at constant volume, requires a *different* amount of heat than does process B, which occurs at constant pressure. The reason is that work is done in process B but not in process A. This is a situation that we are now equipped to analyze.

It is useful to define two different versions of the specific heat of gases, one for constant-volume (isochoric) processes and one for constant-pressure (isobaric) processes. We will define these as molar specific heats because we usually do gas

FIGURE 17.20 Processes A and B have the same ΔT and the same ΔE_{th}, but they require different amounts of heat.

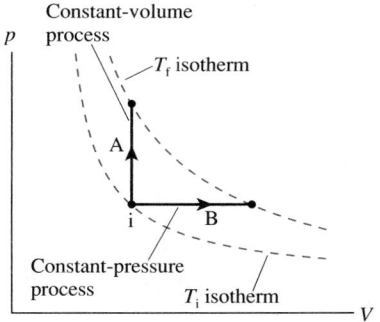

TABLE 17.4 Molar specific heats of gases (J/mol K)

Gas	C_P	C_V	$C_P - C_V$
Monatomic Gases			
He	20.8	12.5	8.3
Ne	20.8	12.5	8.3
Ar	20.8	12.5	8.3
Diatomic Gases			
H_2	28.7	20.4	8.3
N_2	29.1	20.8	8.3
O_2	29.2	20.9	8.3

calculations using moles instead of mass. The quantity of heat needed to change the temperature of n moles of gas by ΔT is

$$Q = nC_V\Delta T \quad \text{(temperature change at constant volume)}$$
$$Q = nC_P\Delta T \quad \text{(temperature change at constant pressure)}$$
(17.25)

where C_V is the **molar specific heat at constant volume** and C_P is the **molar specific heat at constant pressure**. Table 17.4 gives the values of C_V and C_P for a few common monatomic and diatomic gases. The units are J/mol K.

NOTE ▶ Equations 17.25 apply to two specific ideal-gas processes. In a general gas process, for which neither p nor V is constant, we have no direct way to relate Q to ΔT. In that case, the heat must be found indirectly from the first law as $Q = \Delta E_{th} - W$. ◀

EXAMPLE 17.7 **Heating and cooling a gas**

Three moles of O_2 gas are at 20.0°C. 600 J of heat energy are transferred to the gas at constant pressure, then 600 J are removed at constant volume. What is the final temperature? Show the process on a pV diagram.

MODEL O_2 is a diatomic ideal gas. The gas is heated as an isobaric process, then cooled as an isochoric process.

SOLVE The heat transferred during the constant-pressure process causes a temperature rise

$$\Delta T = T_2 - T_1 = \frac{Q}{nC_P} = \frac{600 \text{ J}}{(3.0 \text{ mol})(29.2 \text{ J/mol K})} = 6.8°C$$

where C_P for oxygen was taken from Table 17.4. Heating leaves the gas at temperature $T_2 = T_1 + \Delta T = 26.8°C$. The temperature then falls as heat is removed during the constant-volume process:

$$\Delta T = T_3 - T_2 = \frac{Q}{nC_V} = \frac{-600 \text{ J}}{(3.0 \text{ mol})(20.9 \text{ J/mol K})} = -9.5°C$$

We used a *negative* value for Q because heat energy is transferred from the gas to the environment. The final temperature of the gas is $T_3 = T_2 + \Delta T = 17.3°C$. FIGURE 17.21 shows the process on a

pV diagram. The gas expands (moves horizontally on the diagram) as heat is added, then cools at constant volume (moves vertically on the diagram) as heat is removed.

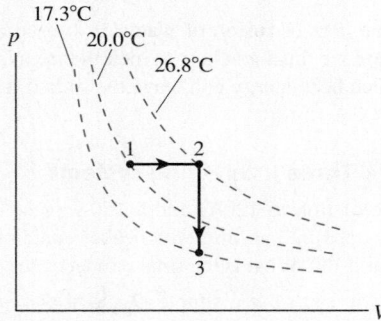

FIGURE 17.21 The pV diagram for Example 17.7.

ASSESS The final temperature is lower than the initial temperature because $C_P > C_V$.

EXAMPLE 17.8 **Calorimetry with a gas and a solid**

The interior volume of a 200 g hollow aluminum box is 800 cm³. The box contains nitrogen gas at STP. A 20 cm³ block of copper at a temperature of 300°C is placed inside the box, then the box is sealed. What is the final temperature?

MODEL This example has three interacting systems: the aluminum box, the nitrogen gas, and the copper block. They must all come to a common final temperature T_f.

VISUALIZE The box and gas have the same initial temperature: $T_{Al} = T_{N2} = 0°C$. The box doesn't change size, so this is a constant-volume process. The final temperature is unknown.

SOLVE Although one of the systems is now a gas, the calorimetry equation $Q_{net} = Q_{Al} + Q_{N2} + Q_{Cu} = 0$ is still appropriate. In this case,

$$Q_{net} = m_{Al}c_{Al}(T_f - T_{Al}) + n_{N2}C_V(T_f - T_{N2})$$
$$+ m_{Cu}c_{Cu}(T_f - T_{Cu}) = 0$$

Notice that we used masses and specific heats for the solids but moles and the molar specific heat for the gas. We used C_V because this is a constant-volume process. Solving for T_f gives

$$T_f = \frac{m_{Al}c_{Al}T_{Al} + n_{N2}C_VT_{N2} + m_{Cu}c_{Cu}T_{Cu}}{m_{Al}c_{Al} + n_{N2}C_V + m_{Cu}c_{Cu}}$$

The specific heat values are found in Tables 17.2 and 17.4. The mass of the copper is

$$m_{Cu} = \rho_{Cu}V_{Cu} = (8920 \text{ kg/cm}^3)(20 \times 10^{-6} \text{ m}^3) = 0.178 \text{ kg}$$

The number of moles of the gas is found from the ideal-gas law, using the initial conditions. Notice that inserting the copper block *displaces* 20 cm³ of gas; hence the gas volume is only $V = 780 \text{ cm}^3 = 7.80 \times 10^{-4} \text{ m}^3$. Thus

$$n_{N2} = \frac{pV}{RT} = 0.0348 \text{ mol}$$

Computing the final temperature gives $T_f = 83°C$.

C_P and C_V

You may have noticed two curious features in Table 17.4. First, the molar specific heats of monatomic gases are *all alike*. And the molar specific heats of diatomic gases, while different from monatomic gases, are again *very nearly alike*. We saw a similar feature in Table 17.2 for the molar specific heats of solids. Second, the *difference* $C_P - C_V = 8.3$ J/mol K is the same in every case. And, most puzzling of all, the value of $C_P - C_V$ appears to be equal to the universal gas constant R! Why should this be?

The relationship between C_V and C_P hinges on one crucial idea: ΔE_{th}, **the change in the thermal energy of a gas, is the same for *any* two processes that have the same ΔT.** The thermal energy of a gas is associated with temperature, so any process that changes the gas temperature from T_i to T_f has the same ΔE_{th} as any other process that goes from T_i to T_f. Furthermore, the first law $\Delta E_{th} = Q + W$ tells us that a gas cannot distinguish between heat and work. The system's thermal energy changes in response to energy added to or removed from the system, but the response of the gas is the same whether you heat the system, do work on the system, or do some combination of both. Thus **any two processes that change the thermal energy of the gas by ΔE_{th} will cause the same temperature change ΔT.**

With that in mind, look back at Figure 17.20. Both gas processes have the same ΔT, so both have the same value of ΔE_{th}. Process A is an isochoric process in which no work is done (the piston doesn't move), so the first law for this process is

$$(\Delta E_{th})_A = W + Q = 0 + Q_{\text{const vol}} = nC_V \Delta T \qquad (17.26)$$

Process B is an isobaric process. You learned earlier that the work done on the gas during an isobaric process is $W = -p\Delta V$. Thus

$$(\Delta E_{th})_B = W + Q = -p\Delta V + Q_{\text{const press}} = -p\Delta V + nC_P \Delta T \quad (17.27)$$

$(\Delta E_{th})_B = (\Delta E_{th})_A$ because both have the same ΔT, so we can equate the right sides of Equations 17.26 and 17.27:

$$-p\Delta V + nC_P \Delta T = nC_V \Delta T \qquad (17.28)$$

For the final step, we can use the ideal-gas law $pV = nRT$ to relate ΔV and ΔT during process B. For any gas process,

$$\Delta(pV) = \Delta(nRT) \qquad (17.29)$$

For a constant-pressure process, where p is constant, Equation 17.29 becomes

$$p\Delta V = nR\Delta T \qquad (17.30)$$

Substituting this expression for $p\Delta V$ into Equation 17.28 gives

$$-nR\Delta T + nC_P \Delta T = nC_V \Delta T \qquad (17.31)$$

The $n\Delta T$ cancels, and we are left with

$$C_P = C_V + R \qquad (17.32)$$

This result, which applies to all ideal gases, is exactly what we see in the data of Table 17.4.

But that's not the only conclusion we can draw. Equation 17.26 found that $\Delta E_{th} = nC_V \Delta T$ for a constant-volume process. But we had just noted that ΔE_{th} is the same for *all* gas processes that have the same ΔT. Consequently, this expression for ΔE_{th} is equally true for any other process. That is

$$\Delta E_{th} = nC_V \Delta T \qquad \text{(any ideal-gas process)} \qquad (17.33)$$

Compare this result to Equations 17.25. We first made a distinction between constant-volume and constant-pressure processes, but now we're saying that Equation 17.33 is true for any process. Are we contradicting ourselves? No, the difference lies in what you need to calculate.

- The change in thermal energy when the temperature changes by ΔT is the same for any process. That's Equation 17.33.
- The *heat* required to bring about the temperature change depends on what the process is. That's Equations 17.25. An isobaric process requires more heat than an isochoric process that produces the same ΔT.

The reason for the difference is seen by writing the first law as $Q = \Delta E_{th} - W$. In an isochoric process, where $W = 0$, *all* the heat input is used to increase the gas temperature. But in an isobaric process, some of the energy that enters the system as heat then leaves the system as work ($W < 0$) done by the expanding gas. Thus more heat is needed to produce the same ΔT.

Heat Depends on the Path

FIGURE 17.22 Is the heat input along these two paths the same or different?

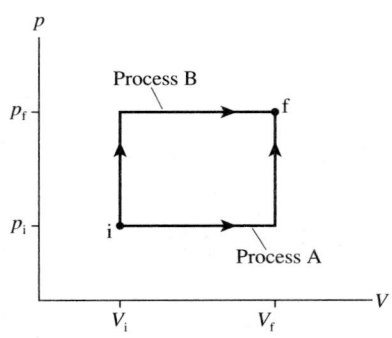

Consider the two ideal-gas processes shown in FIGURE 17.22. Even though the initial and final states are the same, the heat added during these two processes is *not* the same. We can use the first law $\Delta E_{th} = W + Q$ to see why.

The thermal energy is a state variable. That is, its value depends on the state of the gas, not the process by which the gas arrived at that state. Thus $\Delta E_{th} = E_{th\,f} - E_{th\,i}$ is the same for both processes. If ΔE_{th} is the same for processes A and B, then $W_A + Q_A = W_B + Q_B$.

You learned in Section 17.2 that the work done during an ideal-gas process depends on the path in the pV diagram. There's more area under the process B curve, so $|W_B| > |W_A|$. Both values of W are negative because the gas expands, so W_B is more negative than W_A. Consequently, $W_A + Q_A$ can equal $W_B + Q_B$ only if $Q_B > Q_A$. **The heat added or removed during an ideal-gas process depends on the path followed through the pV diagram.**

Adiabatic Processes

FIGURE 17.23 The relationship of three important processes to the first law of thermodynamics.

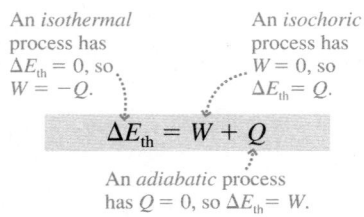

Section 17.4 introduced the idea of an *adiabatic process,* a process in which no heat energy is transferred ($Q = 0$). FIGURE 17.23 compares an adiabatic process with isothermal and isochoric processes. We're now prepared to look at adiabatic processes in more detail.

In practice, there are two ways that an adiabatic process can come about. First, a gas cylinder can be completely surrounded by thermal insulation, such as thick pieces of Styrofoam. The environment can interact mechanically with the gas by pushing or pulling on the insulated piston, but there is no thermal interaction.

Second, the gas can be expanded or compressed very rapidly in what we call an *adiabatic expansion* or an *adiabatic compression.* In a rapid process there is essentially no time for heat to be transferred between the gas and the environment. We've already alluded to the idea that heat is transferred via atomic-level collisions. These collisions take time. If you stick one end of a copper rod into a flame, the other end will eventually get too hot to hold—but not instantly. Some amount of time is required for heat to be transferred from one end to the other. A process that takes place faster than the heat can be transferred is adiabatic.

NOTE ▶ You may recall reading in Chapter 16 that we are going to study only quasi-static processes, processes that proceed slowly enough to remain essentially in equilibrium at all times. Now we're proposing to study processes that take place very rapidly. Isn't this a contradiction? Yes, to some extent it is. What we need to establish are the appropriate time scales. How slow must a process go to be quasi-static? How fast must it go to be adiabatic? These types of calculations must be deferred to a more advanced course. It turns out—fortunately!—that many practical applications, such as the compression strokes in gasoline and diesel engines, are fast enough to be adiabatic yet slow enough to still be considered quasi-static. ◀

For an adiabatic process, with $Q = 0$, the first law of thermodynamics is $\Delta E_{th} = W$. Compressing a gas adiabatically ($W > 0$) increases the thermal energy. Thus **an adiabatic compression raises the temperature of a gas.** A gas that expands adiabatically ($W < 0$) gets colder as its thermal energy decreases. Thus **an adiabatic expansion lowers the temperature of a gas.** You can use an adiabatic process to change the gas temperature without using heat!

The work done in an adiabatic process goes entirely to changing the thermal energy of the gas. But we just found that $\Delta E_{th} = nC_V \Delta T$ for *any* process. Thus

$$W = nC_V \Delta T \qquad \text{(adiabatic process)} \qquad (17.34)$$

Equation 17.34 joins with the equations we derived earlier for the work done in isochoric, isobaric, and isothermal processes.

Gas processes can be represented as trajectories in the pV diagram. For example, a gas moves along a hyperbola during an isothermal process. How does an adiabatic process appear in a pV diagram? The result is more important than the derivation, which is a bit tedious, so we'll begin with the answer and then, at the end of this section, show where it comes from.

First, we define the **specific heat ratio** γ (lowercase Greek gamma) to be

$$\gamma = \frac{C_P}{C_V} = \begin{cases} 1.67 & \text{monatomic gas} \\ 1.40 & \text{diatomic gas} \end{cases} \qquad (17.35)$$

The specific heat ratio has many uses in thermodynamics. Notice that γ is dimensionless.

An adiabatic process is one in which

$$pV^\gamma = \text{constant} \qquad \text{or} \qquad p_f V_f^\gamma = p_i V_i^\gamma \qquad (17.36)$$

This is similar to the isothermal $pV = \text{constant}$, but somewhat more complex due to the exponent γ.

The curves found by graphing $p = \text{constant}/V^\gamma$ are called **adiabats.** In FIGURE 17.24 you see that the two adiabats are steeper than the hyperbolic isotherms. An adiabatic process moves along an adiabat in the same way that an isothermal process moves along an isotherm. You can see that the temperature falls during an adiabatic expansion and rises during an adiabatic compression.

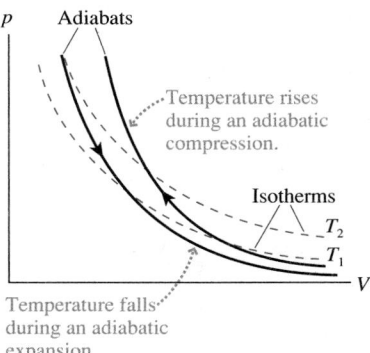

FIGURE 17.24 An adiabatic process moves along pV curves called *adiabats.*

EXAMPLE 17.9 | An adiabatic compression

Air containing gasoline vapor is admitted into the cylinder of an internal combustion engine at 1.00 atm pressure and 30°C. The piston rapidly compresses the gas from 500 cm³ to 50 cm³, a *compression ratio* of 10.

  a. What are the final temperature and pressure of the gas?
  b. Show the compression process on a pV diagram.
  c. How much work is done to compress the gas?

MODEL The compression is rapid, with insufficient time for heat to be transferred from the gas to the environment, so we will model it as an adiabatic compression. We'll treat the gas as if it were 100% air.

SOLVE a. We know the initial pressure and volume, and we know the volume after the compression. For an adiabatic process, where pV^γ remains constant, the final pressure is

$$p_f = p_i \left(\frac{V_i}{V_f}\right)^\gamma = (1.00 \text{ atm})(10)^{1.40} = 25.1 \text{ atm}$$

Air is a mixture of N_2 and O_2, diatomic gases, so we used $\gamma = 1.40$. We can now find the temperature by using the ideal-gas law:

$$T_f = T_i \frac{p_f}{p_i} \frac{V_f}{V_i} = (303 \text{ K})(25.1)\left(\frac{1}{10}\right) = 761 \text{ K} = 488°C$$

Temperature *must* be in kelvins for doing gas calculations such as these.

b. FIGURE 17.25 shows the pV diagram. The 30°C and 488°C isotherms are included to show how the temperature changes during the process.

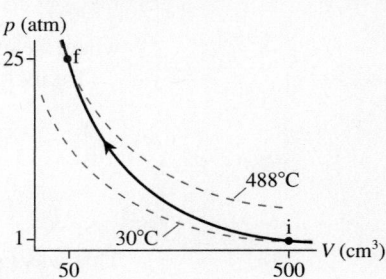

FIGURE 17.25 The adiabatic compression of the gas in an internal combustion engine.

Continued

c. The work done is $W = nC_V \Delta T$, with $\Delta T = 458$ K. The number of moles is found from the ideal-gas law and the initial conditions:

$$n = \frac{p_i V_i}{RT_i} = 0.0201 \text{ mol}$$

Thus the work done to compress the gas is

$$W = nC_V \Delta T = (0.0201 \text{ mol})(20.8 \text{ J/mol K})(458 \text{ K}) = 192 \text{ J}$$

ASSESS The temperature rises dramatically during the compression stroke of an engine. But the higher temperature has nothing to do with heat! **The temperature and thermal energy of the gas are increased not by heating the gas but by doing work on it.** This is an important idea to understand.

If we use the ideal-gas-law expression $p = nRT/V$ in the adiabatic equation $pV^\gamma = $ constant, we see that $TV^{\gamma-1}$ is also constant during an adiabatic process. Thus another useful equation for adiabatic processes is

$$T_f V_f^{\gamma-1} = T_i V_i^{\gamma-1} \tag{17.37}$$

Proof of Equation 17.36

Now let's see where Equation 17.36 comes from. Consider an adiabatic process in which an infinitesimal amount of work dW done on a gas causes an infinitesimal change in the thermal energy. For an adiabatic process, with $dQ = 0$, the first law of thermodynamics is

$$dE_{th} = dW \tag{17.38}$$

We can use Equation 17.33, which is valid for *any* gas process, to write $dE_{th} = nC_V dT$. Earlier in the chapter we found that the work done during a small volume change is $dW = -p\,dV$. With these substitutions, Equation 17.38 becomes

$$nC_V dT = -p\,dV \tag{17.39}$$

The ideal-gas law can now be used to write $p = nRT/V$. The n cancels, and the C_V can be moved to the other side of the equation to give

$$\frac{dT}{T} = -\frac{R}{C_V} \frac{dV}{V} \tag{17.40}$$

We're going to integrate Equation 17.40, but anticipating the need for $\gamma = C_P/C_V$ we can first use the fact that $C_P = C_V + R$ to write

$$\frac{R}{C_V} = \frac{C_P - C_V}{C_V} = \frac{C_P}{C_V} - 1 = \gamma - 1 \tag{17.41}$$

Now we integrate Equation 17.40 from the initial state i to the final state f:

$$\int_{T_i}^{T_f} \frac{dT}{T} = -(\gamma - 1) \int_{V_i}^{V_f} \frac{dV}{V} \tag{17.42}$$

Carrying out the integration gives

$$\ln\left(\frac{T_f}{T_i}\right) = \ln\left(\frac{V_i}{V_f}\right)^{\gamma-1} \tag{17.43}$$

where we used the logarithm properties $\log a - \log b = \log(a/b)$ and $c \log a = \log(a^c)$. Taking the exponential of both sides now gives

$$\left(\frac{T_f}{T_i}\right) = \left(\frac{V_i}{V_f}\right)^{\gamma-1} \tag{17.44}$$

$$T_f V_f^{\gamma-1} = T_i V_i^{\gamma-1}$$

This was Equation 17.37. Writing $T = pV/nR$ and canceling $1/nR$ from both sides of the equation give Equation 17.36:

$$p_f V_f^\gamma = p_i V_i^\gamma \tag{17.45}$$

This was a lengthy derivation, but it is good practice at seeing how the ideal-gas law and the first law of thermodynamics can work together to yield results of great importance.

> **STOP TO THINK 17.6** For the two processes shown, which of the following is true:
>
> a. $Q_A > Q_B$
> b. $Q_A = Q_B$
> c. $Q_A < Q_B$

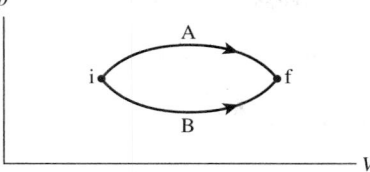

17.8 Heat-Transfer Mechanisms

You feel warmer when the sun is shining on you, colder when sitting on a metal bench or when the wind is blowing, especially if your skin is wet. This is due to the transfer of heat. Although we've talked about heat a lot in this chapter, we haven't said much about *how* heat is transferred from a hotter object to a colder object. There are four basic mechanisms by which objects exchange heat with their surroundings. Evaporation was treated in Section 17.5; in this section, we will consider the other mechanisms.

Heat-transfer mechanisms

When two objects are in direct contact, such as the soldering iron and the circuit board, heat is transferred by *conduction.*

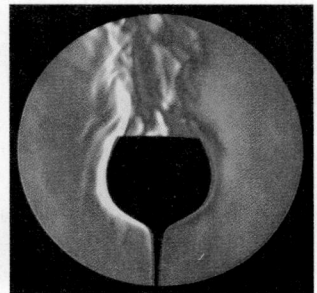

Air currents near a warm glass of water rise, taking thermal energy with them in a process known as *convection.*

The lamp at the top shines on the lambs huddled below, warming them. The energy is transferred by *radiation.*

Blowing on a hot cup of tea or coffee cools it by *evaporation.*

Conduction

FIGURE 17.26 shows an object sandwiched between a higher temperature T_H and a lower temperature T_C. The temperature *difference* causes heat energy to be transferred from the hot side to the cold side in a process known as **conduction.**

It is not surprising that more heat is transferred if the temperature difference ΔT is larger. A material with a larger cross section A (a fatter pipe) transfers more heat, while a thicker material, increasing the distance L between the hot and cold sources, decreases the rate of heat transfer.

These observations about heat conduction can be summarized in a single formula. If heat Q is transferred in a time interval Δt, the *rate* of heat transfer is $Q/\Delta t$. For a material of cross-section area A and length L, spanning a temperature difference $\Delta T = T_H - T_C$, the rate of heat transfer is

$$\frac{Q}{\Delta t} = k\frac{A}{L}\Delta T \qquad (17.46)$$

FIGURE 17.26 Conduction of heat through a solid.

This material is conducting heat across the temperature difference.

TABLE 17.5 Thermal conductivities

Material	k (W/m K)
Diamond	2000
Silver	430
Copper	400
Aluminum	240
Iron	80
Stainless steel	14
Ice	1.7
Concrete	0.8
Glass	0.8
Styrofoam	0.035
Air (20°C, 1 atm)	0.023

The quantity k, which characterizes whether the material is a good conductor of heat or a poor conductor, is called the **thermal conductivity** of the material. Because the heat-transfer rate J/s is a *power,* measured in watts, the units of k are W/m K. Values of k for common materials are given in Table 17.5; a material with a larger value of k is a better conductor of heat.

Most good heat conductors are metals, which are also good conductors of electricity. One exception is diamond, in which the strong bonds among atoms that make diamond such a hard material lead to a rapid transfer of thermal energy. Air and other gases are poor conductors of heat because there are no bonds between adjacent molecules.

Some of our perceptions of hot and cold have more to do with thermal conductivity than with temperature. For example, a metal chair feels colder to your bare skin than a wooden chair not because it has a lower temperature—both are at room temperature— but because it has a much larger thermal conductivity that conducts heat away from your body at a much higher rate.

EXAMPLE 17.10 **Keeping a freezer cold**

A 1.8-m-wide by 1.0-m-tall by 0.65-m-deep home freezer is insulated with 5.0-cm-thick Styrofoam insulation. At what rate must the compressor remove heat from the freezer to keep the inside at −20°C in a room where the air temperature is 25°C?

MODEL Heat is transferred through each of the six sides by conduction. The compressor must remove heat at the same rate it enters to maintain a steady temperature inside. The heat conduction is determined primarily by the thick insulation, so we'll neglect the thin inner and outer panels.

SOLVE Each of the six sides is a slab of Styrofoam with cross-section area A_i and thickness $L = 5.0$ cm. The total rate of heat transfer is

$$\frac{Q}{\Delta t} = \sum_{i=1}^{6} k \frac{A_i}{L} \Delta T = \frac{k \Delta T}{L} \sum_{i=1}^{6} A_i = \frac{k \Delta T}{L} A_{\text{total}}$$

The total surface area is

$$A_{\text{total}} = 2 \times (1.8 \text{ m} \times 1.0 \text{ m} + 1.8 \text{ m} \times 0.65 \text{ m}$$
$$+ 1.0 \text{ m} \times 0.65 \text{ m}) = 7.24 \text{ m}^2$$

Using $k = 0.035$ W/m K from Table 17.5, we find

$$\frac{Q}{\Delta t} = \frac{k \Delta t}{L} A_{\text{total}} = \frac{(0.035 \text{ W/m K})(45 \text{ K})(7.24 \text{ m}^2)}{0.050 \text{ m}} = 230 \text{ W}$$

Heat enters the freezer through the walls at the rate 230 J/s; thus the compressor must remove 230 J of heat energy every second to keep the temperature at −20°C.

ASSESS We'll learn in Chapter 19 how the compressor does this and how much work it must do. A typical freezer uses electric energy at a rate of about 150 W, so our result seems reasonable.

Convection

Warm water (colored) moves by convection.

Air is a poor conductor of heat, but thermal energy is easily transferred through air, water, and other fluids because the air and water can flow. A pan of water on the stove is heated at the bottom. This heated water expands, becomes less dense than the water above it, and thus rises to the surface, while cooler, denser water sinks to take its place. The same thing happens to air. This transfer of thermal energy by the motion of a fluid—the well-known idea that "heat rises"—is called **convection.**

Convection is usually the main mechanism for heat transfer in fluid systems. On a small scale, convection mixes the pan of water that you heat on the stove; on a large scale, convection is responsible for making the wind blow and ocean currents circulate. Air is a very poor thermal conductor, but it is very effective at transferring energy by convection. To use air for thermal insulation, it is necessary to trap the air in small pockets to limit convection. And that's exactly what feathers, fur, double-paned windows, and fiberglass insulation do. Convection is much more rapid in water than in air, which is why people can die of hypothermia in 68°F (20°C) water but can live quite happily in 68°F air.

Because convection involves the often-turbulent motion of fluids, there is no simple equation for energy transfer by convection. Our description must remain qualitative.

Radiation

The sun *radiates* energy to earth through the vacuum of space. Similarly, you feel the warmth from the glowing red coals in a fireplace.

All objects emit energy in the form of **radiation,** electromagnetic waves generated by oscillating electric charges in the atoms that form the object. These waves transfer energy from the object that emits the radiation to the object that absorbs it. Electromagnetic waves carry energy from the sun; this energy is absorbed when sunlight falls on your skin, warming you by increasing your thermal energy. Your skin also emits electromagnetic radiation, helping to keep your body cool by decreasing your thermal energy. Radiation is a significant part of the *energy balance* that keeps your body at the proper temperature.

> NOTE ▶ The word "radiation" comes from "radiate," meaning "to beam." Radiation can refer to x rays or to the radioactive decay of nuclei, but it also can refer simply to light and other forms of electromagnetic waves that "beam" from an object. Here we are using this second meaning of the term. ◀

You are familiar with radiation from objects hot enough to glow "red hot" or, at a high enough temperature, "white hot." The sun is simply a very hot ball of glowing gas, and the white light from an incandescent lightbulb is radiation emitted by a thin wire filament heated to a very high temperature by an electric current. Objects at lower temperatures also radiate, but at infrared wavelengths. You can't see this radiation (although you can sometimes feel it), but infrared-sensitive detectors can measure it and are used to make thermal images.

The energy radiated by an object depends strongly on temperature. If heat energy Q is radiated in a time interval Δt by an object with surface area A and absolute temperature T, the *rate* of heat transfer is found to be

$$\frac{Q}{\Delta t} = e\sigma A T^4 \qquad (17.47)$$

Because the rate of energy transfer is power (1 J/s = 1 W), $Q/\Delta t$ is often called the *radiated power.* Notice the very strong fourth-power dependence on temperature. Doubling the absolute temperature of an object increases the radiated power by a factor of 16!

The parameter e in Equation 17.47 is the **emissivity** of the surface, a measure of how effectively it radiates. The value of e ranges from 0 to 1. σ is a constant, known as the Stefan-Boltzmann constant, with the value

$$\sigma = 5.67 \times 10^{-8} \text{ W/m}^2 \text{ K}^4$$

> NOTE ▶ Just as in the ideal-gas law, the temperature in Equation 17.47 *must* be in kelvins. ◀

Objects not only emit radiation, they also *absorb* radiation emitted by their surroundings. Suppose an object at temperature T is surrounded by an environment at temperature T_0. The *net* rate at which the object radiates heat energy—that is, radiation emitted minus radiation absorbed—is

$$\frac{Q_{net}}{\Delta t} = e\sigma A (T^4 - T_0^4) \qquad (17.48)$$

This makes sense. An object should have no *net* radiation if it's in thermal equilibrium ($T = T_0$) with its surroundings.

Notice that the emissivity e appears for absorption as well as emission; good emitters are also good absorbers. A perfect absorber ($e = 1$), one absorbing all light and radiation impinging on it but reflecting none, would appear completely black. Thus a perfect absorber is sometimes called a **black body.** But a perfect absorber would also be a perfect emitter, so thermal radiation from an ideal emitter is called **black-body radiation.** It seems strange that black objects are perfect emitters, but think of black charcoal glowing bright red in a fire. At room temperature, it "glows" equally bright with infrared.

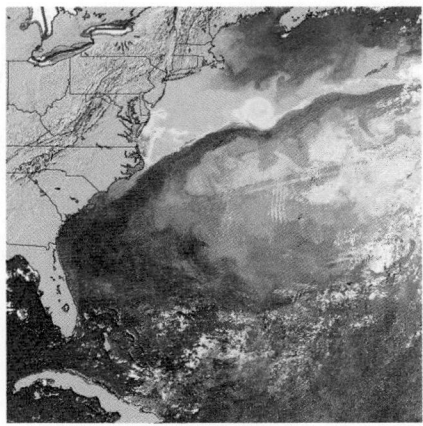

This satellite image shows radiation emitted by the ocean waters off the east coast of the United States. You can clearly see the warm waters of the Gulf Stream, a large-scale convection that transfers heat to northern latitudes.

EXAMPLE 17.11 **Taking the sun's temperature**

The radius of the sun is 6.96×10^8 m. At the distance of the earth, 1.50×10^{11} m, the intensity of solar radiation (measured by satellites above the atmosphere) is 1370 W/m². What is the temperature of the sun's surface?

MODEL Assume the sun to be an ideal radiator with $e = 1$.

SOLVE The total power radiated by the sun is the power per m² multiplied by the surface area of a sphere extending to the earth:

$$P = \frac{1370 \text{ W}}{1 \text{ m}^2} \times 4\pi (1.50 \times 10^{11} \text{ m})^2 = 3.87 \times 10^{26} \text{ W}$$

That is, the sun radiates energy at the rate $Q/\Delta t = 3.87 \times 10^{26}$ J/s. That's a lot of power! This energy is radiated from the surface of a

sphere of radius R_S. Using this information in Equation 17.47, we find that the sun's surface temperature is

$$T = \left[\frac{Q/\Delta t}{e\sigma(4\pi R_S^2)} \right]^{1/4}$$

$$= \left[\frac{3.87 \times 10^{26} \text{ W}}{(1)(5.67 \times 10^{-8} \text{ W/m}^2 \text{ K}^4)4\pi(6.96 \times 10^8 \text{ m})^2} \right]^{1/4}$$

$$= 5790 \text{ K}$$

ASSESS This temperature is confirmed by measurements of the solar spectrum, a topic we'll explore in Part VII.

Thermal radiation plays a prominent role in climate and global warming. The earth as a whole is in thermal equilibrium. Consequently, it must radiate back into space exactly as much energy as it receives from the sun. The incoming radiation from the hot sun is mostly visible light. The earth's atmosphere is transparent to visible light, so this radiation reaches the surface and is absorbed. The cooler earth radiates infrared radiation, but the atmosphere is *not* completely transparent to infrared. Some components of the atmosphere, notably water vapor and carbon dioxide, are strong absorbers of infrared radiation. They hinder the emission of radiation and, rather like a blanket, keep the earth's surface warmer than it would be without these gases in the atmosphere.

The **greenhouse effect,** as it's called, is a natural part of the earth's climate. The earth would be much colder and mostly frozen were it not for naturally occurring carbon dioxide in the atmosphere. But carbon dioxide also results from the burning of fossil fuels, and human activities since the beginning of the industrial revolution have increased the atmospheric concentration of carbon dioxide by nearly 50%. This human contribution has amplified the greenhouse effect and is the primary cause of global warming.

STOP TO THINK 17.7 Suppose you are an astronaut in space, hard at work in your sealed spacesuit. The only way that you can transfer excess heat to the environment is by

a. Conduction.   b. Convection.   c. Radiation.   d. Evaporation.

CHALLENGE EXAMPLE 17.12 **Boiling water**

400 mL of water is poured into an 8.0-cm-diameter, 150 g glass beaker with a 2.0-mm-thick bottom; then the beaker is placed on a 400°C hot plate. Once the water reaches the boiling point, how long will it take to boil away all the water?

MODEL The bottom of the beaker is a heat-conducting material transferring heat energy from the 400°C hot plate to the 100°C boiling water. The temperature of both the water and the beaker remains constant until the water has boiled away. We'll assume that heat losses due to convection and radiation are negligible, in which case the heat energy entering the system is used entirely for the phase change of the water. The beaker's mass isn't relevant because its temperature isn't changing.

SOLVE The heat energy required to boil mass M of water is

$$Q = ML_v$$

where $L_v = 2.26 \times 10^6$ J/kg is the heat of vaporization. The heat energy transferred through the bottom of the beaker during a time interval Δt is

$$Q = k\frac{A}{L}\Delta T \Delta t$$

where $k = 0.80$ W/m K is the thermal conductivity of glass. Because the heat transferred by conduction is used entirely for boiling the water, we can combine these two expressions:

$$k\frac{A}{L}\Delta T \Delta t = ML_v$$

and then solve for Δt:

$$\Delta t = \frac{MLL_v}{kA\Delta T} = \frac{(0.40 \text{ kg})(0.0020 \text{ m})(2.26 \times 10^6 \text{ J/kg})}{(0.80 \text{ W/m K})(0.0050 \text{ m}^2)(300 \text{ K})}$$

$$= 1500 \text{ s} = 25 \text{ min}$$

We used the density of water to find that $M = 400$ g $= 0.40$ kg and calculated $A = \pi r^2 = 0.0050$ m² as the area through which heat conduction occurs.

ASSESS 400 mL is roughly 2 cups, a small hot plate can bring 2 cups of water to a boil in 5 min or so, and boiling the water away takes quite a bit longer than bringing it to a boil. 25 min is probably a slight underestimate since we neglected energy losses due to convection and radiation, but it seems reasonable. A stove could boil the water away much faster because the burner temperature (gas flame or red-hot heating coil) is much higher.

SUMMARY

The goal of Chapter 17 has been to develop and apply the first law of thermodynamics.

General Principles

First Law of Thermodynamics

$$\Delta E_{th} = W + Q$$

The first law is a general statement of energy conservation.

Work W and heat Q depend on the process by which the system is changed.

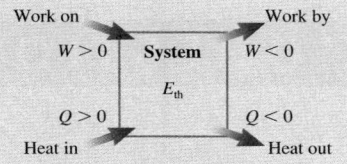

Work on ← $W > 0$ **System** E_{th} $W < 0$ → Work by

$Q > 0$ Heat in $Q < 0$ Heat out

The change in the system depends only on the total energy exchanged $W + Q$, not on the process.

Energy

Thermal energy E_{th} Microscopic energy of moving molecules and stretched molecular bonds. ΔE_{th} depends on the initial/final states but is independent of the process.

Work W Energy transferred to the system by forces in a mechanical interaction.

Heat Q Energy transferred to the system via atomic-level collisions when there is a temperature difference. A thermal interaction.

Important Concepts

The work done on a gas is

$$W = -\int_{V_i}^{V_f} p\, dV$$

$$= -(\text{area under the } pV \text{ curve})$$

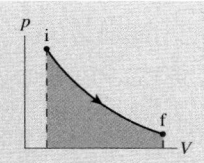

An adiabatic process is one for which $Q = 0$. Gases move along an **adiabat** for which $pV^{\gamma} = $ constant, where $\gamma = C_P/C_V$ is the **specific heat ratio.** An adiabatic process changes the temperature of the gas without heating it.

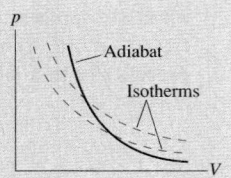

Calorimetry When two or more systems interact thermally, they come to a common final temperature determined by

$$Q_{net} = Q_1 + Q_2 + \cdots = 0$$

The heat of transformation L is the energy needed to cause 1 kg of substance to undergo a phase change

$$Q = \pm ML$$

The specific heat c of a substance is the energy needed to raise the temperature of 1 kg by 1 K:

$$Q = Mc\Delta T$$

The molar specific heat C is the energy needed to raise the temperature of 1 mol by 1 K:

$$Q = nC\Delta T$$

The molar specific heat of gases depends on the *process* by which the temperature is changed:

$C_V = $ molar specific heat at **constant volume**

$C_P = C_V + R = $ molar specific heat at **constant pressure**

Heat is transferred by conduction, convection, radiation, and evaporation.

Conduction: $Q/\Delta t = (kA/L)\Delta T$

Radiation: $Q/\Delta t = e\sigma A T^4$

Summary of Basic Gas Processes

Process	Definition	Stays constant	Work	Heat
Isochoric	$\Delta V = 0$	V and p/T	$W = 0$	$Q = nC_V\Delta T$
Isobaric	$\Delta p = 0$	p and V/T	$W = -p\Delta V$	$Q = nC_P\Delta T$
Isothermal	$\Delta T = 0$	T and pV	$W = -nRT\ln(V_f/V_i)$	$\Delta E_{th} = 0$
Adiabatic	$Q = 0$	pV^{γ}	$W = \Delta E_{th}$	$Q = 0$
All gas processes	First law $\Delta E_{th} = W + Q = nC_V\Delta T$		Ideal-gas law $pV = nRT$	

Terms and Notation

work, W	adiabatic process	molar specific heat at	thermal conductivity, k
mechanical interaction	specific heat, c	constant volume, C_V	convection
mechanical equilibrium	molar specific heat, C	molar specific heat at	radiation
heat, Q	heat of transformation, L	constant pressure, C_P	emissivity, e
thermal interaction	heat of fusion, L_f	specific heat ratio, γ	black body
thermal equilibrium	heat of vaporization, L_v	adiabat	black-body radiation
first law of thermodynamics	calorimetry	conduction	greenhouse effect
thermodynamic energy model			

CONCEPTUAL QUESTIONS

1. When the space shuttle returns to earth, its surfaces get very hot as it passes through the atmosphere at high speed. Has the space shuttle been heated? If so, what was the source of the heat? If not, why is it hot?

2. Do (a) temperature, (b) heat, and (c) thermal energy describe a property of a system, an interaction of the system with its environment, or both? Explain.

3. Two containers hold equal masses of nitrogen gas at equal temperatures. You supply 10 J of heat to container A while not allowing its volume to change, and you supply 10 J of heat to container B while not allowing its pressure to change. Afterward, is temperature T_A greater than, less than, or equal to T_B? Explain.

4. You need to raise the temperature of a gas by 10°C. To use the least amount of heat energy, should you heat the gas at constant pressure or at constant volume? Explain.

5. *Why* is the molar specific heat of a gas at constant pressure larger than the molar specific heat at constant volume?

6. FIGURE Q17.6 shows an adiabatic process.
   a. Is the final temperature higher than, lower than, or equal to the initial temperature?
   b. Is any heat energy added to or removed from the system in this process? Explain.

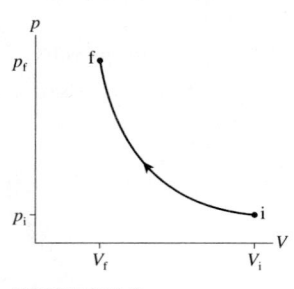

FIGURE Q17.6   **FIGURE Q17.7**

7. FIGURE Q17.7 shows two different processes taking an ideal gas from state i to state f. Is the work done on the gas in process A greater than, less than, or equal to the work done in process B? Explain.

8. FIGURE Q17.8 shows two different processes taking an ideal gas from state i to state f.
   a. Is the temperature *change* ΔT during process A larger than, smaller than, or equal to the change during process B? Explain.
   b. Is the heat energy added during process A greater than, less than, or equal to the heat added during process B? Explain.

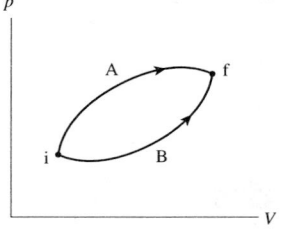

FIGURE Q17.8   **FIGURE Q17.9**

9. Describe a series of steps in which you use the cylinder of Figure 17.13 to implement the ideal-gas process shown in FIGURE Q17.9. Then show the process as a first-law bar chart.

10. Describe a series of steps in which you use the cylinder of Figure 17.13 to implement the ideal-gas process shown in FIGURE Q17.10. Then show the process as a first-law bar chart.

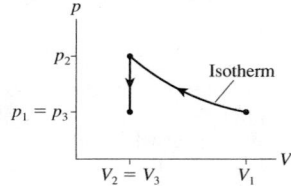

FIGURE Q17.10

11. The gas cylinder in FIGURE Q17.11, similar to the cylinder shown in Figure 17.13, is placed on a block of ice. The initial gas temperature is $> 0°C$.
    a. During the process that occurs until the gas reaches a new equilibrium, are (i) ΔT, (ii) W, and (iii) Q greater than, less than, or equal to zero? Explain.
    b. Draw a pV diagram showing the process.

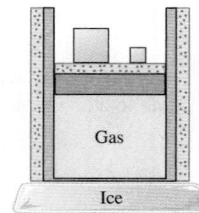

FIGURE Q17.11

12. The gas cylinder in FIGURE Q17.12 is similar to the cylinder described earlier in Figure 17.13, except that the bottom is insulated. Masses are slowly removed from the top of the piston until the total mass is reduced by 50%.
    a. During this process, are (i) ΔT, (ii) W, and (iii) Q greater than, less than, or equal to zero? Explain.
    b. Draw a pV diagram showing the process.

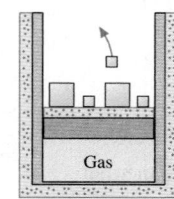

FIGURE Q17.12

EXERCISES AND PROBLEMS

Problems labeled [ ] integrate material from earlier chapters.

Exercises

Section 17.1 It's All About Energy

Section 17.2 Work in Ideal-Gas Processes

1. ‖ How much work is done on the gas in the process shown in FIGURE EX17.1?

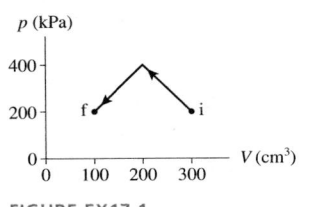

FIGURE EX17.1 **FIGURE EX17.2**

2. ‖ How much work is done on the gas in the process shown in FIGURE EX17.2?

3. ‖ 80 J of work are done on the gas in the process shown in FIGURE EX17.3. What is V_1 in cm^3?

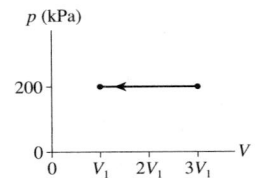

FIGURE EX17.3

4. ‖ A 2000 cm^3 container holds 0.10 mol of helium gas at 300°C. How much work must be done to compress the gas to 1000 cm^3 at (a) constant pressure and (b) constant temperature?

Section 17.3 Heat

Section 17.4 The First Law of Thermodynamics

5. | Draw a first-law bar chart (see Figure 17.14) for the gas process in FIGURE EX17.5.

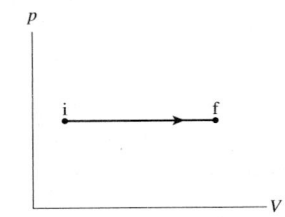

FIGURE EX17.5

6. | Draw a first-law bar chart (see Figure 17.14) for the gas process in FIGURE EX17.6.

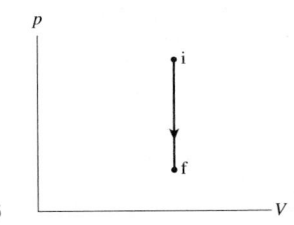

FIGURE EX17.6

7. | Draw a first-law bar chart (see Figure 17.14) for the gas process in FIGURE EX17.7.

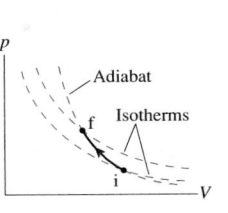

FIGURE EX17.7 **FIGURE EX17.8**

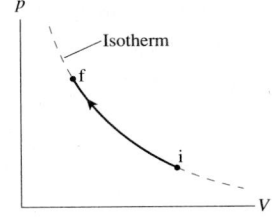

8. | Draw a first-law bar chart (see Figure 17.14) for the gas process in FIGURE EX17.8.

9. ‖ A gas is compressed from 600 cm^3 to 200 cm^3 at a constant pressure of 400 kPa. At the same time, 100 J of heat energy is transferred out of the gas. What is the change in thermal energy of the gas during this process?

10. | 500 J of work are done on a system in a process that decreases the system's thermal energy by 200 J. How much heat energy is transferred to or from the system?

Section 17.5 Thermal Properties of Matter

11. ‖ How much heat energy must be added to a 6.0 cm × 6.0 cm × 6.0 cm block of aluminum to raise its temperature from −50°C to 50°C?

12. ‖ A rapidly spinning paddle wheel raises the temperature of 200 mL of water from 21°C to 25°C. How much (a) heat is transferred and (b) work is done in this process?

13. | a. 100 J of heat energy are transferred to 20 g of mercury. By how much does the temperature increase?
    b. How much heat is needed to raise the temperature of 20 g of water by the same amount?

14. ‖ How much heat is needed to change 20 g of mercury at 20°C into mercury vapor at the boiling point?

15. ‖ What is the maximum mass of ethyl alcohol you could boil with 1000 J of heat, starting from 20°C?

Section 17.6 Calorimetry

16. ‖ 30 g of copper pellets are removed from a 300°C oven and immediately dropped into 100 mL of water at 20°C in an insulated cup. What will the new water temperature be?

17. ‖ A 750 g aluminum pan is removed from the stove and plunged into a sink filled with 10.0 L of water at 20.0°C. The water temperature quickly rises to 24.0°C. What was the initial temperature of the pan in °C and in °F?

18. ‖ A 50.0 g thermometer is used to measure the temperature of 200 mL of water. The specific heat of the thermometer, which is mostly glass, is 750 J/kg K, and it reads 20.0°C while lying on the table. After being completely immersed in the water, the thermometer's reading stabilizes at 71.2°C. What was the actual water temperature before it was measured?

19. ‖ A 500 g metal sphere is heated to 300°C, then dropped into a beaker containing 300 cm^3 of mercury at 20.0°C. A short time later the mercury temperature stabilizes at 99.0°C. Identify the metal.

20. ‖ A 65 cm^3 block of iron is removed from an 800°C furnace and immediately dropped into 200 mL of 20°C water. What fraction of the water boils away?

Section 17.7 The Specific Heats of Gases

21. ‖ A container holds 1.0 g of argon at a pressure of 8.0 atm.
   a. How much heat is required to increase the temperature by 100°C at constant volume?
   b. How much will the temperature increase if this amount of heat energy is transferred to the gas at constant pressure?

22. ‖ A container holds 1.0 g of oxygen at a pressure of 8.0 atm.
   a. How much heat is required to increase the temperature by 100°C at constant pressure?
   b. How much will the temperature increase if this amount of heat energy is transferred to the gas at constant volume?

23. ‖ A rigid cylinder contains 7.0 g of nitrogen at 20°C. What is the minimum amount of heat energy that must be removed to liquify the nitrogen?

24. ‖ The volume of a gas is halved during an adiabatic compression that increases the pressure by a factor of 2.5.
   a. What is the specific heat ratio γ?
   b. By what factor does the temperature increase?

25. ‖ A gas cylinder holds 0.10 mol of O$_2$ at 150°C and a pressure of 3.0 atm. The gas expands adiabatically until the pressure is halved. What are the final (a) volume and (b) temperature?

26. ‖ A gas cylinder holds 0.10 mol of O$_2$ at 150°C and a pressure of 3.0 atm. The gas expands adiabatically until the volume is doubled. What are the final (a) pressure and (b) temperature?

Section 17.8 Heat-Transfer Mechanisms

27. ‖ A 10 m × 14 m house is built on a 12-cm-thick concrete slab. What is the heat-loss rate through the slab if the ground temperature is 5°C while the interior of the house is 22°C?

28. ‖ The ends of a 20-cm-long, 2.0-cm-diameter rod are maintained at 0°C and 100°C by immersion in an ice-water bath and boiling water. Heat is conducted through the rod at 4.5×10^4 J per hour. Of what material is the rod made?

29. ‖ What maximum power can be radiated by a 10-cm-diameter solid lead sphere? Assume an emissivity of 1.

30. ‖ BIO Radiation from the head is a major source of heat loss from the human body. Model a head as a 20-cm-diameter, 20-cm-tall cylinder with a flat top. If the body's surface temperature is 35°C, what is the net rate of heat loss on a chilly 5°C day? All skin, regardless of color, is effectively black in the infrared where the radiation occurs, so use an emissivity of 0.95.

Problems

31. ‖ A 5.0 g ice cube at −20°C is in a rigid, sealed container from which all the air has been evacuated. How much heat is required to change this ice cube into steam at 200°C?

32. ‖ A 5.0-m-diameter garden pond is 30 cm deep. Solar energy is incident on the pond at an average rate of 400 W/m^2. If the water absorbs all the solar energy and does not exchange energy with its surroundings, how many hours will it take to warm from 15°C to 25°C?

33. ‖ An 11 kg bowling ball at 0°C is dropped into a tub containing a mixture of ice and water. A short time later, when a new equilibrium has been established, there are 5.0 g less ice. From what height was the ball dropped? Assume no water or ice splashes out.

34. ‖ The burner on an electric stove has a power output of 2.0 kW. A 750 g stainless steel teakettle is filled with 20°C water and placed on the already hot burner. If it takes 3.0 min for the water to reach a boil, what volume of water, in cm^3, was in the kettle? Stainless steel is mostly iron, so you can assume its specific heat is that of iron.

35. ‖ BIO Reptiles don't use enough metabolic energy to keep their body temperature constant. They cool off at night and must warm up in the morning sun. Suppose a 2.9-m-long, 60-cm-wide, 350 kg alligator is basking in the sun. If the sun's intensity on the back of the alligator is 500 W/m^2, and if energy losses can be ignored, how long will it take the alligator to warm up from 23°C to a more favorable 30°C? The average specific heat of body tissue is 3400 J/kg K.

36. ‖ BIO One way you keep from overheating is by perspiring. Evaporation—a phase change—requires heat, and the heat energy is removed from your body. Evaporation is much like boiling, only water's heat of vaporization at 35°C is a somewhat larger 24×10^5 J/kg because at lower temperatures more energy is required to break the molecular bonds. Very strenuous activity can cause an adult human to produce 30 g of perspiration per minute. If all the perspiration evaporates, rather than dripping off, at what rate (in J/s) is it possible to exhaust heat by perspiring?

37. ‖ BIO When air is inhaled, it quickly becomes saturated with water vapor as it passes through the moist airways. Consequently, an adult human exhales about 25 mg of evaporated water with each breath. Evaporation—a phase change—requires heat, and the heat energy is removed from your body. Evaporation is much like boiling, only water's heat of vaporization at 35°C is a somewhat larger 24×10^5 J/kg because at lower temperatures more energy is required to break the molecular bonds. At 12 breaths/min, on a dry day when the inhaled air has almost no water content, what is the body's rate of energy loss (in J/s) due to exhaled water? (For comparison, the energy loss from radiation, usually the largest loss on a cool day, is about 100 J/s.)

38. ‖ Two cars collide head-on while each is traveling at 80 km/h. Suppose all their kinetic energy is transformed into the thermal energy of the wrecks. What is the temperature increase of each car? You can assume that each car's specific heat is that of iron.

39. ‖ 10 g of aluminum at 200°C and 20 g of copper are dropped into 50 cm^3 of ethyl alcohol at 15°C. The temperature quickly comes to 25°C. What was the initial temperature of the copper?

40. ‖ A 100 g ice cube at −10°C is placed in an aluminum bucket whose initial temperature is 70°C. The system comes to an equilibrium temperature of 20°C. What is the mass of the bucket?

41. ‖ 512 g of an unknown metal at a temperature of 15°C is dropped into a 100 g aluminum container holding 325 g of water at 98°C. A short time later, the container of water and metal stabilizes at a new temperature of 78°C. Identify the metal.

42. ‖ A 150 L ($\approx$ 40 gal) electric hot-water tank has a 5.0 kW heater. How many minutes will it take to raise the water temperature from 65°F to 140°F?

43. ‖ What is oxygen's specific heat at constant volume in J/kg K?

44. ‖ BIO Suppose you take and hold a deep breath on a chilly day, inhaling 3.0 L of air at 0°C and 1 atm.
   a. How much heat must your body supply to warm the air to your internal body temperature of 37°C?
   b. By how much does the air's volume increase as it warms?

45. | An experiment measures the temperature of a 500 g substance while steadily supplying heat to it. FIGURE P17.45 shows the results of the experiment. What are the (a) specific heat of the solid phase, (b) specific heat of the liquid phase, (c) melting and boiling temperatures, and (d) heats of fusion and vaporization?

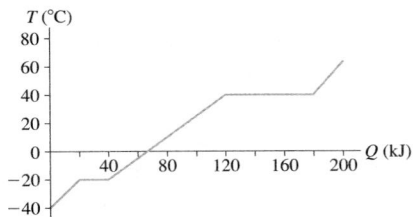

FIGURE P17.45

46. ‖ Your 300 mL cup of coffee is too hot to drink when served at 90°C. What is the mass of an ice cube, taken from a −20°C freezer, that will cool your coffee to a pleasant 60°C?

47. ‖ A typical nuclear reactor generates 1000 MW (1000 MJ/s) of electrical energy. In doing so, it produces 2000 MW of "waste heat" that must be removed from the reactor to keep it from melting down. Many reactors are sited next to large bodies of water so that they can use the water for cooling. Consider a reactor where the intake water is at 18°C. State regulations limit the temperature of the output water to 30°C so as not to harm aquatic organisms. How many liters of cooling water have to be pumped through the reactor each minute?

48. ‖ A beaker with a metal bottom is filled with 20 g of water at 20°C. It is brought into good thermal contact with a 4000 cm³ container holding 0.40 mol of a monatomic gas at 10 atm pressure. Both containers are well insulated from their surroundings. What is the gas pressure after a long time has elapsed? You can assume that the containers themselves are nearly massless and do not affect the outcome.

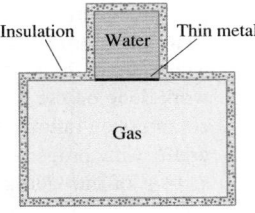

FIGURE P17.48

49. ‖ 2.0 mol of gas are at 30°C and a pressure of 1.5 atm. How much work must be done on the gas to compress it to one third of its initial volume at (a) constant temperature and (b) constant pressure? (c) Show both processes on a single pV diagram.

50. ‖ 500 J of work must be done to compress a gas to half its initial volume at constant temperature. How much work must be done to compress the gas by a factor of 10, starting from its initial volume?

51. ‖ A 6.0-cm-diameter cylinder of nitrogen gas has a 4.0-cm-thick movable copper piston. The cylinder is oriented vertically, as shown in FIGURE P17.51, and the air above the piston is evacuated. When the gas temperature is 20°C, the piston floats 20 cm above the bottom of the cylinder.
a. What is the gas pressure?
b. How many gas molecules are in the cylinder?
Then 2.0 J of heat energy are transferred to the gas.
c. What is the new equilibrium temperature of the gas?
d. What is the final height of the piston?
e. How much work is done on the gas as the piston rises?

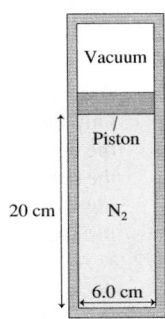

FIGURE P17.51

52. ‖ A 10-cm-diameter cylinder contains argon gas at 10 atm pressure and a temperature of 50°C. A piston can slide in and out of the cylinder. The cylinder's initial length is 20 cm. 2500 J of heat are transferred to the gas, causing the gas to expand at constant pressure. What are (a) the final temperature and (b) the final length of the cylinder?

53. ‖ A cube 20 cm on each side contains 3.0 g of helium at 20°C. 1000 J of heat energy are transferred to this gas. What are (a) the final pressure if the process is at constant volume and (b) the final volume if the process is at constant pressure? (c) Show and label both processes on a single pV diagram.

54. ‖ An 8.0-cm-diameter, well-insulated vertical cylinder containing nitrogen gas is sealed at the top by a 5.1 kg frictionless piston. The air pressure above the piston is 100 kPa.
a. What is the gas pressure inside the cylinder?
b. Initially, the piston height above the bottom of the cylinder is 26 cm. What will be the piston height if an additional 3.5 kg are placed on top of the piston?

55. ‖ n moles of an ideal gas at temperature T_1 and volume V_1 expand isothermally until the volume has doubled. In terms of n, T_1, and V_1, what are (a) the final temperature, (b) the work done on the gas, and (c) the heat energy transferred to the gas?

56. ‖ 5.0 g of nitrogen gas at 20°C and an initial pressure of 3.0 atm undergo an isobaric expansion until the volume has tripled.
a. What are the gas volume and temperature after the expansion?
b. How much heat energy is transferred to the gas to cause this expansion?
The gas pressure is then decreased at constant volume until the original temperature is reached.
c. What is the gas pressure after the decrease?
d. What amount of heat energy is transferred from the gas as its pressure decreases?
e. Show the total process on a pV diagram. Provide an appropriate scale on both axes.

57. ‖ FIGURE P17.57 shows two processes that take a gas from state i to state f. Show that $Q_A - Q_B = p_iV_i$.

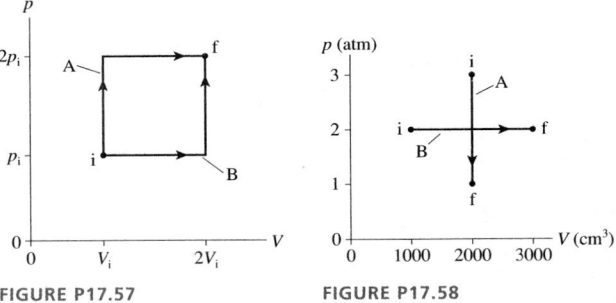

FIGURE P17.57 · · · FIGURE P17.58

58. | 0.10 mol of nitrogen gas follow the two processes shown in FIGURE P17.58. How much heat is required for each?

59. ‖ 0.10 mol of nitrogen gas follow the two processes shown in FIGURE P17.59. How much heat is required for each?

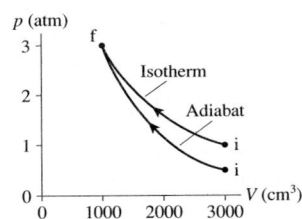

FIGURE P17.59

60. ‖ 0.10 mol of a monatomic gas follow the process shown in FIGURE P17.60.
    a. How much heat energy is transferred to or from the gas during process 1 → 2?
    b. How much heat energy is transferred to or from the gas during process 2 → 3?
    c. What is the total change in thermal energy of the gas?

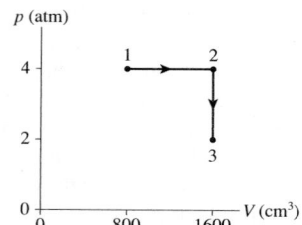

FIGURE P17.60

61. ‖‖ Your laboratory assignment for the week is to measure the specific heat ratio γ of carbon dioxide. The gas is contained in a cylinder with a movable piston and a thermometer. When the piston is withdrawn as far as possible, the cylinder's length is 20 cm. You decide to push the piston in very rapidly by various amounts and, for each push, to measure the temperature of the carbon dioxide. Before each push, you withdraw the piston all the way and wait several minutes for the gas to come to the room temperature of 21°C. Your data are as follows:

Push (cm)	Temperature (°C)
5	35
10	68
13	110
15	150

Use the best-fit line of an appropriate graph to determine γ for carbon dioxide.

62. ‖ Two cylinders each contain 0.10 mol of a diatomic gas at 300 K and a pressure of 3.0 atm. Cylinder A expands isothermally and cylinder B expands adiabatically until the pressure of each is 1.0 atm.
    a. What are the final temperature and volume of each?
    b. Show both processes on a single pV diagram. Use an appropriate scale on both axes.

63. ‖‖ A monatomic gas follows the process 1 → 2 → 3 shown in FIGURE P17.63. How much heat is needed for (a) process 1 → 2 and (b) process 2 → 3?

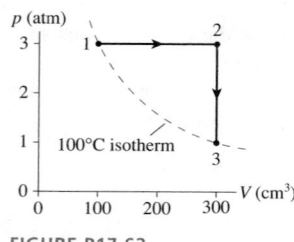

FIGURE P17.63

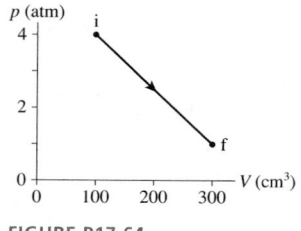

FIGURE P17.64

64. ‖ FIGURE P17.64 shows a thermodynamic process followed by 0.015 mol of hydrogen. How much heat energy is transferred to the gas?

65. ‖ FIGURE P17.65 shows a thermodynamic process followed by 120 mg of helium.
    a. Determine the pressure (in atm), temperature (in °C), and volume (in cm³) of the gas at points 1, 2, and 3. Put your results in a table for easy reading.
    b. How much work is done on the gas during each of the three segments?
    c. How much heat energy is transferred to or from the gas during each of the three segments?

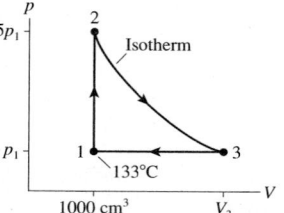

FIGURE P17.65

66. ‖ a. What compression ratio V_{max}/V_{min} will raise the air temperature from 20°C to 1000°C in an adiabatic process?
    b. What pressure ratio p_{max}/p_{min} does this process have?

67. ‖ Two containers of a diatomic gas have the same initial conditions. One container, heated at constant pressure, has a temperature increase of 20°C. The other container receives the same quantity of heat energy, but at constant volume. What is its temperature increase?

68. ‖ 14 g of nitrogen gas at STP are adiabatically compressed to a pressure of 20 atm. What are (a) the final temperature, (b) the work done on the gas, (c) the heat input to the gas, and (d) the compression ratio V_{max}/V_{min} ? (e) Show the process on a pV diagram, using proper scales on both axes.

69. ‖ 14 g of nitrogen gas at STP are pressurized in an isochoric process to a pressure of 20 atm. What are (a) the final temperature, (b) the work done on the gas, (c) the heat input to the gas, and (d) the pressure ratio p_{max}/p_{min} ? (e) Show the process on a pV diagram, using proper scales on both axes.

70. ‖ When strong winds rapidly carry air down from mountains to a lower elevation, the air has no time to exchange heat with its surroundings. The air is compressed as the pressure rises, and its temperature can increase dramatically. These warm winds are called Chinook winds in the Rocky Mountains and Santa Ana winds in California. Suppose the air temperature high in the mountains behind Los Angeles is 0°C at an elevation where the air pressure is 60 kPa. What will the air temperature be, in °C and °F, when the Santa Ana winds have carried this air down to an elevation near sea level where the air pressure is 100 kPa?

71. ‖ You would like to put a solar hot water system on your roof, but you're not sure it's feasible. A reference book on solar energy shows that the ground-level solar intensity in your city is 800 W/m² for at least 5 hours a day throughout most of the year. Assuming that a completely black collector plate loses energy only by radiation, and that the air temperature is 20°C, what is the equilibrium temperature of a collector plate directly facing the sun? Note that while a plate has two sides, only the side facing the sun will radiate because the opposite side will be well insulated.

72. ‖ A cubical box 20 cm on a side is constructed from 1.2-cm-thick concrete panels. A 100 W lightbulb is sealed inside the box. What is the air temperature inside the box when the light is on if the surrounding air temperature is 20°C?

73. ‖ The sun's intensity at the distance of the earth is 1370 W/m². 30% of this energy is reflected by water and clouds; 70% is absorbed. What would be the earth's average temperature (in °C) if the earth had no atmosphere? The emissivity of the surface is very close to 1. (The actual average temperature of the earth, about 15°C, is higher than your calculation because of the greenhouse effect.)

In Problems 74 through 76 you are given the equation used to solve a problem. For each of these, you are to
  a. Write a realistic problem for which this is the correct equation.
  b. Finish the solution of the problem.

74. $50 \text{ J} = -n(8.31 \text{ J/mol K})(350 \text{ K})\ln\left(\frac{1}{3}\right)$

75. $(200 \times 10^{-6} \text{ m}^3)(13{,}600 \text{ kg/m}^3)$
$\times (140 \text{ J/kg K})(90°C - 15°C)$
$+ (0.50 \text{ kg})(449 \text{ J/kg K})(90°C - T_i) = 0$

76. $(10 \text{ atm})V_2^{1.40} = (1.0 \text{ atm})V_1^{1.40}$

Challenge Problems

77. **FIGURE CP17.77** shows a thermodynamic process followed by 120 mg of helium.
  a. Determine the pressure (in atm), temperature (in °C), and volume (in cm³) of the gas at points 1, 2, and 3. Put your results in a table for easy reading.

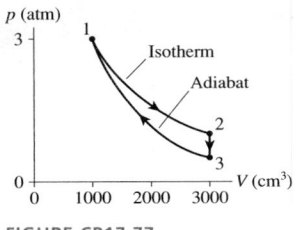

FIGURE CP17.77

  b. How much work is done on the gas during each of the three segments?
  c. How much heat is transferred to or from the gas during each of the three segments?

78. One cylinder in the diesel engine of a truck has an initial volume of 600 cm³. Air is admitted to the cylinder at 30°C and a pressure of 1.0 atm. The piston rod then does 400 J of work to rapidly compress the air. What are its final temperature and volume?

79. You come into lab one day and find a well-insulated 2000 mL thermos bottle containing 500 mL of boiling liquid nitrogen. The remainder of the thermos has nitrogen gas at a pressure of 1.0 atm. The gas and liquid are in thermal equilibrium. While waiting for lab to start, you notice a piece of iron on the table with "197 g" written on it. Just for fun, you drop the iron into the thermos and seal the cap tightly so that no gas can escape. After a few seconds have passed, what is the pressure inside the thermos? The density of liquid nitrogen is 810 kg/m³.

80. A cylindrical copper rod and an iron rod with exactly the same dimensions are welded together end to end. The outside end of the copper rod is held at 100°C, and the outside end of the iron rod is held at 0°C. What is the temperature at the midpoint where the rods are joined together?

81. 0.020 mol of a diatomic gas, with initial temperature 20°C, are compressed from 1500 cm³ to 500 cm³ in a process in which $pV^2 = $ constant. How much heat is added during this process?

82. A monatomic gas fills the left end of the cylinder in **FIGURE CP17.82**. At 300 K, the gas cylinder length is 10.0 cm and the spring is compressed by 2.0 cm. How much heat energy must be added to the gas to expand the cylinder length to 16.0 cm?

FIGURE CP17.82

Stop to Think 17.1: a. The piston does work W on the gas. There's no heat because of the insulation, and $\Delta E_{mech} = 0$ because the gas as a whole doesn't move. Thus $\Delta E_{th} = W > 0$. The work increases the system's thermal energy and thus raises its temperature.

Stop to Think 17.2: d. $W_A = 0$ because A is an isochoric process. $W_B = W_{1 \text{ to } 2} + W_{2 \text{ to } 3}$. $|W_{2 \text{ to } 3}| > |W_{1 \text{ to } 2}|$ because there's more area under the curve, and $W_{2 \text{ to } 3}$ is positive whereas $W_{1 \text{ to } 2}$ is negative. Thus W_B is positive.

Stop to Think 17.3: b and e. The temperature rises in d from doing work on the gas ($\Delta E_{th} = W$), not from heat. e involves heat because there is a temperature difference. The temperature of the gas doesn't change because the heat is used to do the work of lifting a weight.

Stop to Think 17.4: c. The temperature increases so E_{th} must increase. W is negative in an expansion, so Q must be positive and larger than $|W|$.

Stop to Think 17.5: a. A has a smaller specific heat and thus less thermal inertia. The temperature of A will change more than the temperature of B.

Stop to Think 17.6: a. $W_A + Q_A = W_B + Q_B$. The area under process A is larger than the area under B, so W_A is *more negative* than W_B. Q_A has to be more positive than Q_B to maintain the equality.

Stop to Think 17.7: c. Conduction, convection, and evaporation require matter. Only radiation transfers energy through the vacuum of space.

18 The Micro/Macro Connection

Heating the air in a hot-air balloon increases the thermal energy of the molecules. This causes the gas to expand, lowering its density and allowing it to float in the cooler surrounding air.

▶ **Looking Ahead** The goal of Chapter 18 is to understand a macroscopic system in terms of the microscopic behavior of its molecules.

Macro Puzzles Micro Explanations

- Why does the ideal-gas law work for every gas?
- Why is the molar specific heat for every monatomic gas the same? And for every diatomic gas and for every elemental solid?
- What does temperature measure?

These are puzzles we uncovered in the last two chapters. In this chapter, you will learn that we can resolve these puzzles and understand many of the properties of macroscopic systems by investigating the microscopic behavior of its atoms and molecules.

Macro	Micro
A container of an ideal gas	N molecules of gas with number density N/V

Macroscopic properties include pressure, temperature and thermal energy.

The molecules have kinetic energy. They collide with each other and the walls of the container.

Collisions

You will learn how to understand the pressure of a gas in terms of atomic collisions with the walls of the container.

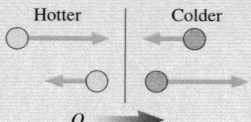

You'll also learn that heat transfer occurs via collisions at the boundary between two systems. More-energetic molecules on one side transfer their energy to less-energetic molecules on the other.

Energy and Temperature

We'll find that the average energy of a molecule depends only on temperature.

This will allow us to interpret temperature in terms of thermal energy—the microscopic energy of the moving molecules—and then to understand why every monatomic gas has the same molar specific heat.

◀ **Looking Back**
Sections 17.3–17.7 Heat, the first law of thermodynamics, and specific heat

The Second Law of Thermodynamics

You will learn a new law of nature, the second law of thermodynamics, that governs how systems evolve in time.

One statement of the second law is that heat energy is transferred spontaneously from a hotter system to a colder system, never from colder to hotter. Heat transfer is an *irreversible process.*

Heat goes from the burner to the teakettle, making the water hotter and the burner a little cooler. It would not violate energy conservation for heat to go from the kettle to the burner, making the water colder and the burner hotter. But it doesn't happen.

You'll learn to use the concept of **entropy** to understand why all macroscopic interactions are irreversible. Entropy explains why the future is different from the past and why there are theoretical limits to the efficiency of using energy in practical ways.

18.1 Molecular Speeds and Collisions

Let us begin by thinking about gases at the atomic level. If gases really are composed of atoms and molecules in motion, how fast are the molecules moving? Do all molecules move with the same speed, or is there a range of speeds?

To answer these questions, FIGURE 18.1 shows an experiment to measure the speeds of molecules in a gas. The two rotating disks form a *velocity selector*. Once every revolution, the slot in the first disk allows a small pulse of molecules to pass through. By the time these molecules reach the second disk, the slots have rotated. The molecules can pass through the second slot and be detected *only if* they have exactly the right speed $v = L/\Delta t$ to travel between the two disks during time interval Δt it takes the axle to complete one revolution. Molecules having any other speed are blocked by the second disk. By changing the rotation period of the axle, this apparatus can measure how many molecules have each of many possible speeds.

FIGURE 18.2 shows the results for nitrogen gas (N_2) at $T = 20°C$. The data are presented in the form of a **histogram,** a bar chart in which the height of each bar tells how many (or, in this case, what percentage) of the molecules have a speed in the *range* of speeds shown below the bar. For example, 16% of the molecules have speeds in the range from 600 m/s to 700 m/s. All the bars sum to 100%, showing that this histogram describes *all* of the molecules leaving the source.

It turns out that the molecules have what is called a *distribution* of speeds, ranging from as low as ≈ 100 m/s to as high as ≈ 1200 m/s. But not all speeds are equally likely; there is a *most likely speed* of ≈ 550 m/s. This is really fast, ≈ 1200 mph! Changing the temperature or changing to a different gas changes the most likely speed, as we'll learn later in the chapter, but it does not change the *shape* of the distribution.

If you were to repeat the experiment, you would again find the most likely speed to be ≈ 550 m/s and that 16% of the molecules have speeds between 600 m/s and 700 m/s. Think about what this means. The "molecular deck of cards" is constantly being reshuffled by molecular collisions, yet 16% of the molecules always have speeds between 600 m/s and 700 m/s.

This is an important lesson. Although a gas consists of a vast number of molecules, each moving randomly, *averages,* such as the average number of molecules in the speed range 600 to 700 m/s, have precise, predictable values. **The micro/macro connection is built on the idea that the macroscopic properties of a system, such as temperature or pressure, are related to the *average* behavior of the atoms and molecules.**

Mean Free Path

Imagine someone opening a bottle of strong perfume a few feet away from you. If molecular speeds are hundreds of meters per second, you might expect to smell the perfume almost instantly. But that isn't what happens. As you know, it takes many seconds for the molecules to *diffuse* across the room. Let's see why this is.

FIGURE 18.3 shows a "movie" of one molecule. Instead of zipping along in a straight line, as it would in a vacuum, the molecule follows a convoluted zig-zag path in which it frequently collides with other molecules. A molecule may have traveled hundreds of meters by the time it manages to get 1 or 2 m away from its starting point.

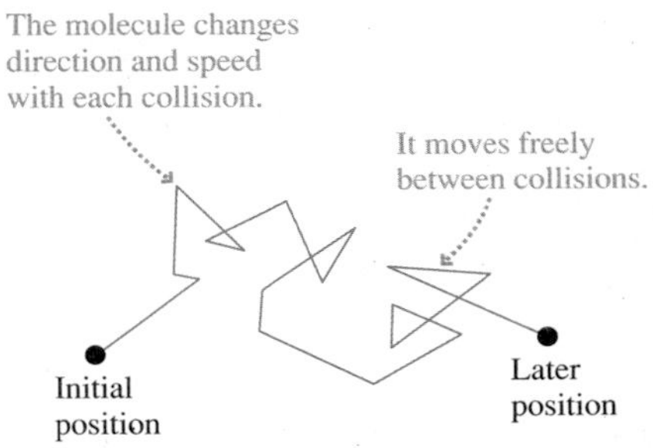

The random distribution of the molecules in the gas causes the straight-line segments between collisions to be of unequal lengths. A question we could ask is: What is the *average* distance between collisions? If a molecule has N_{coll} collisions as it travels distance L, the average distance between collisions, which is called the **mean free path** λ (lowercase Greek lambda), is

$$\lambda = \frac{L}{N_{coll}} \qquad (18.1)$$

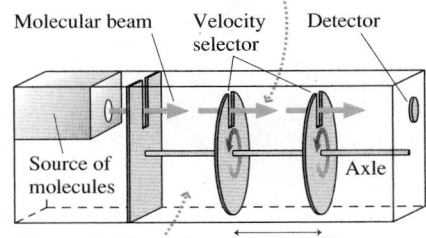

FIGURE 18.1 An experiment to measure the speeds of molecules in a gas.

The only molecules that reach the detector are those whose speed allows them to travel distance L during the time it takes the disks to make one full revolution.

Vacuum inside, so that molecules travel without collisions.

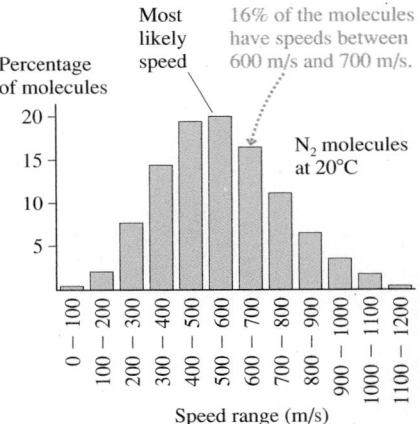

FIGURE 18.2 The distribution of molecular speeds in a sample of nitrogen gas.

16% of the molecules have speeds between 600 m/s and 700 m/s.

N_2 molecules at 20°C

Percentage of molecules

Speed range (m/s)

FIGURE 18.3 A single molecule follows a zig-zag path through a gas as it collides with other molecules.

The molecule changes direction and speed with each collision.

It moves freely between collisions.

Initial position

Later position

FIGURE 18.4 A sample molecule will collide with all target molecules whose centers are within a bent cylinder of radius 2r centered on its path.

(a)

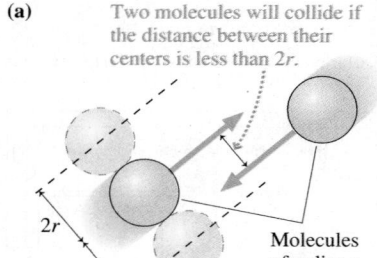

Two molecules will collide if the distance between their centers is less than 2r.

2r

2r

Molecules of radius r

(b) Target molecules

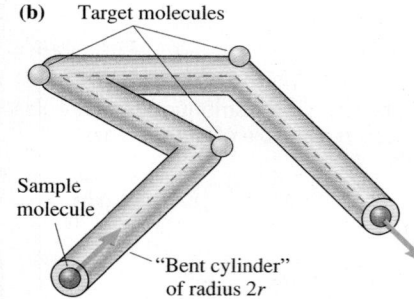

Sample molecule

"Bent cylinder" of radius 2r

The concept of mean free path is used not only in gases but also to describe electrons moving through conductors and light passing through a medium that scatters the photons.

Our task is to determine the number of collisions. FIGURE 18.4a shows two molecules approaching each other. We will assume that the molecules are spherical and of radius r. We will also continue the ideal-gas assumption that the molecules undergo hard-sphere collisions, like billiard balls. In that case, the molecules will collide if the distance between their *centers* is less than $2r$. They will miss if the distance is greater than $2r$.

FIGURE 18.4b shows a cylinder of radius $2r$ centered on the trajectory of a "sample" molecule. The sample molecule collides with any "target" molecule whose center is located within the cylinder, causing the cylinder to bend at that point. Hence the number of collisions N_{coll} is equal to the number of molecules in a cylindrical volume of length L.

The volume of a cylinder is $V_{cyl} = AL = \pi(2r)^2 L$. If the number density of the gas is N/V particles per m³, then the number of collisions along a trajectory of length L is

$$N_{coll} = \frac{N}{V} V_{cyl} = \frac{N}{V}\pi(2r)^2 L = 4\pi \frac{N}{V}r^2 L \qquad (18.2)$$

Thus the mean free path between collisions is

$$\lambda = \frac{L}{N_{coll}} = \frac{1}{4\pi(N/V)r^2}$$

We made a tacit assumption in this derivation that the target molecules are at rest. While the general idea behind our analysis is correct, a more detailed calculation with all the molecules moving introduces an extra factor of $\sqrt{2}$, giving

$$\lambda = \frac{1}{4\sqrt{2}\,\pi(N/V)r^2} \qquad \text{(mean free path)} \qquad (18.3)$$

Laboratory measurements are necessary to determine atomic and molecular radii, but a reasonable rule of thumb is to assume that atoms in a monatomic gas have $r \approx 0.5 \times 10^{-10}$ m and diatomic molecules have $r \approx 1.0 \times 10^{-10}$ m.

EXAMPLE 18.1 **The mean free path at room temperature**

What is the mean free path of a nitrogen molecule at 1.0 atm pressure and room temperature (20°C)?

SOLVE Nitrogen is a diatomic molecule, so $r \approx 1.0 \times 10^{-10}$ m. We can use the ideal-gas law in the form $pV = Nk_B T$ to determine the number density:

$$\frac{N}{V} = \frac{p}{k_B T} = \frac{101,300 \text{ Pa}}{(1.38 \times 10^{-23} \text{ J/K})(293 \text{ K})} = 2.5 \times 10^{25} \text{ m}^{-3}$$

Thus the mean free path is

$$\lambda = \frac{1}{4\sqrt{2}\,\pi(N/V)r^2}$$

$$= \frac{1}{4\sqrt{2}\,\pi(2.5 \times 10^{25} \text{ m}^{-3})(1.0 \times 10^{-10} \text{ m})^2}$$

$$= 2.3 \times 10^{-7} \text{ m} = 230 \text{ nm}$$

ASSESS You learned in Example 16.5 that the average separation between gas molecules at STP is ≈ 4 nm. It seems that any given molecule can slip between its neighbors, which are spread out in three dimensions, and travel—on average—about 60 times the average spacing before it collides with another molecule.

STOP TO THINK 18.1 The table shows the properties of four gases, each having the same number of molecules. Rank in order, from largest to smallest, the mean free paths λ_A to λ_D of molecules in these gases.

Gas	A	B	C	D
Volume	V	2V	V	V
Atomic mass	m	m	2m	m
Atomic radius	r	r	r	2r

18.2 Pressure in a Gas

Why does a gas have pressure? In Chapter 15, where pressure was introduced, we suggested that the pressure in a gas is due to collisions of the molecules with the walls of its container. The force due to one such collision may be unmeasurably tiny, but the steady rain of a vast number of molecules striking a wall each second exerts a measurable macroscopic force. The gas pressure is the force per unit area ($p = F/A$) resulting from these molecular collisions.

Our task in this section is to calculate the pressure by doing the appropriate averaging over molecular motions and collisions. This task can be divided into three main pieces:

1. Calculate the impulse a single molecule exerts on the wall during a collision.
2. Find the force due to all collisions.
3. Introduce an appropriate average speed.

Force Due to a Single Collision

FIGURE 18.5 shows a molecule with an x-component of velocity v_x colliding with a wall and rebounding with its x-component of velocity changed from $+v_x$ to $-v_x$. This molecule experiences an impulse. We can use the impulse-momentum theorem from Chapter 9 to write

$$(J_x)_{\text{wall on molecule}} = \Delta p = m(-v_x) - mv_x = -2mv_x \qquad (18.4)$$

According to Newton's third law, the wall experiences the equal but opposite impulse

$$(J_x)_{\text{molecule on wall}} = +2mv_x \qquad (18.5)$$

as a result of this single collision.

Suppose there are N_{coll} such collisions during a very small time interval Δt. If we assume for the moment that all molecules have the *same* x-component velocity v_x, the net impulse of these collisions on the wall is

$$J_{\text{wall}} = N_{\text{coll}} \times (J_x)_{\text{molecule on wall}} = 2N_{\text{coll}}mv_x \qquad (18.6)$$

FIGURE 18.6 reminds you that impulse is the area under the force-versus-time curve and thus $J_{\text{wall}} = F_{\text{avg}}\Delta t$, where F_{avg} is the *average* force exerted on the wall. Using this in Equation 18.6, we see that the average force on the wall due to many molecular collisions is

$$F_{\text{avg}} = 2\frac{N_{\text{coll}}}{\Delta t}mv_x \qquad (18.7)$$

The quantity $N_{\text{coll}}/\Delta t$ is the *rate* of collisions with the wall—that is, the number of collisions per second. FIGURE 18.7 shows how to determine the rate of collisions. Let the time interval Δt be much less than the average time between molecular collisions, so no collisions alter the molecular speeds during this interval. (This assumption about Δt isn't really necessary, but it makes it easier to think about what's going on.) During Δt, all molecules travel distance $\Delta x = v_x \Delta t$ along the x-axis. This distance is shaded in the figure. *Every one* of the molecules in this shaded region that is moving to the right will reach and collide with the wall during time Δt. Molecules outside this region will not reach the wall during Δt and will not collide.

The shaded region has volume $A\Delta x$, where A is the surface area of the wall. Because of their random motions, only half the molecules are moving to the right, hence the number of collisions during Δt is

$$N_{\text{coll}} = \frac{1}{2}\frac{N}{V}A\Delta x = \frac{1}{2}\frac{N}{V}Av_x\Delta t \qquad (18.8)$$

and thus the rate of collisions is

$$\frac{N_{\text{coll}}}{\Delta t} = \frac{1}{2}\frac{N}{V}Av_x \qquad (18.9)$$

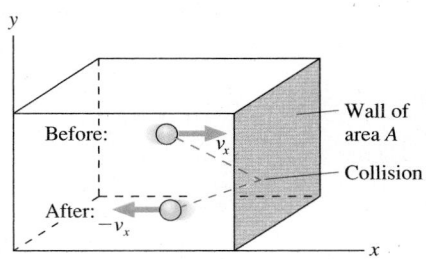

FIGURE 18.5 A molecule colliding with the wall exerts an impulse on it.

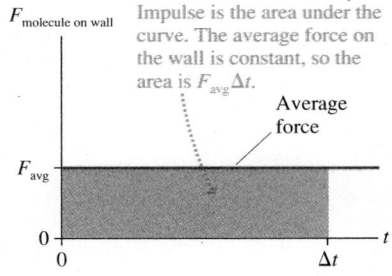

FIGURE 18.6 Impulse is the area under the force-versus-time curve.

Impulse is the area under the curve. The average force on the wall is constant, so the area is $F_{\text{avg}}\Delta t$.

Average force

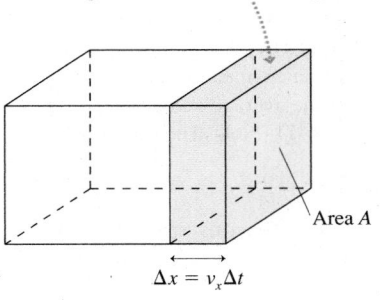

FIGURE 18.7 Determining the rate of collisions.

Only molecules moving to the right in the shaded region will hit the wall during Δt.

The average force on the wall is found by substituting $N_{coll}/\Delta t$ from Equation 18.9 into Equation 18.7:

$$F_{avg} = 2\left(\frac{1}{2}\frac{N}{V}Av_x\right)mv_x = \frac{N}{V}mv_x^2 A \qquad (18.10)$$

Notice that this expression for F_{avg} does not depend on any details of the molecular collisions.

We can relax the assumption that all molecules have the same speed by replacing the squared velocity v_x^2 in Equation 18.10 with its average value. That is,

$$F_{avg} = \frac{N}{V}m(v_x^2)_{avg}A \qquad (18.11)$$

where $(v_x^2)_{avg}$ is the quantity v_x^2 averaged over all the molecules in the container.

The Root-Mean-Square Speed

We need to be somewhat careful when averaging velocities. The velocity component v_x has a sign. At any instant of time, half the molecules in a container move to the right and have positive v_x while the other half move to the left and have negative v_x. Thus the *average velocity* is $(v_x)_{avg} = 0$. If this weren't true, the entire container of gas would move away!

The speed of a molecule is $v = (v_x^2 + v_y^2 + v_z^2)^{1/2}$. Thus the average of the speed squared is

$$(v^2)_{avg} = (v_x^2 + v_y^2 + v_z^2)_{avg} = (v_x^2)_{avg} + (v_y^2)_{avg} + (v_z^2)_{avg} \qquad (18.12)$$

The square root of $(v^2)_{avg}$ is called the **root-mean-square speed** v_{rms}:

$$v_{rms} = \sqrt{(v^2)_{avg}} \qquad \text{(root-mean-square speed)} \qquad (18.13)$$

This is usually called the *rms speed*. You can remember its definition by noting that its name is the *opposite* of the sequence of operations: First you square all the speeds, then you average the squares (find the mean), then you take the square root. Because the square root "undoes" the square, v_{rms} must, in some sense, give an average speed.

> **NOTE** ▶ We could compute a true average speed v_{avg}, but that calculation is difficult. More important, the root-mean-square speed tends to arise naturally in many scientific and engineering calculations. It turns out that v_{rms} differs from v_{avg} by less than 10%, so for practical purposes we can interpret v_{rms} as being essentially the average speed of a molecule in a gas. ◀

EXAMPLE 18.2 **Calculating the root-mean-square speed**

FIGURE 18.8 shows the velocities of all the molecules in a six-molecule, two-dimensional gas. Calculate and compare the average velocity $\vec{v}_{avg}$, the average speed v_{avg}, and the rms speed v_{rms}.

SOLVE Table 18.1 on the next page shows the velocity components v_x and v_y for each molecule, the squares v_x^2 and v_y^2, their sum $v^2 = v_x^2 + v_y^2$, and the speed $v = (v_x^2 + v_y^2)^{1/2}$. Averages of all the values in each column are shown at the bottom. You can see that the average velocity is $\vec{v}_{avg} = \vec{0}$ m/s and the average speed is $v_{avg} = 11.9$ m/s. The rms speed is

$$v_{rms} = \sqrt{(v^2)_{avg}} = \sqrt{148.3 \text{ m}^2/\text{s}^2} = 12.2 \text{ m/s}$$

ASSESS The rms speed is only 2.5% greater than the average speed.

FIGURE 18.8 The molecular velocities of Example 18.2. Units are m/s.

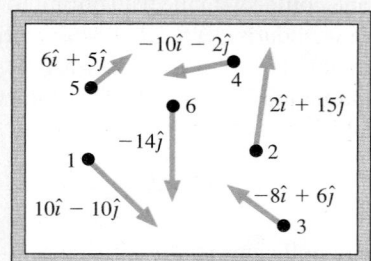

TABLE 18.1 Calculation of rms speed and average speed for the molecules of Example 18.2

Molecule	v_x	v_y	v_x^2	v_y^2	v^2	v
1	10	-10	100	100	200	14.1
2	2	15	4	225	229	15.1
3	-8	6	64	36	100	10.0
4	-10	-2	100	4	104	10.2
5	6	5	36	25	61	7.8
6	0	-14	0	196	196	14.0
Average	0	0			148.3	11.9

There's nothing special about the x-axis. The coordinate system is something that *we* impose on the problem, so *on average* it must be the case that

$$(v_x^2)_{\text{avg}} = (v_y^2)_{\text{avg}} = (v_z^2)_{\text{avg}} \qquad (18.14)$$

Hence we can use Equation 18.12 and the definition of v_{rms} to write

$$v_{\text{rms}}^2 = (v_x^2)_{\text{avg}} + (v_y^2)_{\text{avg}} + (v_z^2)_{\text{avg}} = 3(v_x^2)_{\text{avg}} \qquad (18.15)$$

Consequently, $(v_x^2)_{\text{avg}}$ is

$$(v_x^2)_{\text{avg}} = \frac{1}{3} v_{\text{rms}}^2 \qquad (18.16)$$

Using this result in Equation 18.11 gives us the net force on the wall of the container:

$$F_{\text{net}} = \frac{1}{3} \frac{N}{V} m v_{\text{rms}}^2 A \qquad (18.17)$$

Thus the pressure on the wall of the container due to all the molecular collisions is

$$p = \frac{F}{A} = \frac{1}{3} \frac{N}{V} m v_{\text{rms}}^2 \qquad (18.18)$$

We have met our goal. Equation 18.18 expresses the macroscopic pressure in terms of the microscopic physics. The pressure depends on the number density of molecules in the container and on how fast, on average, the molecules are moving.

EXAMPLE 18.3 **The rms speed of helium atoms**

A container holds helium at a pressure of 200 kPa and a temperature of 60.0°C. What is the rms speed of the helium atoms?

SOLVE The rms speed can be found from the pressure and the number density. Using the ideal-gas law gives us the number density:

$$\frac{N}{V} = \frac{p}{k_B T} = \frac{200,000 \text{ Pa}}{(1.38 \times 10^{-23} \text{ J/K})(333 \text{ K})} = 4.35 \times 10^{25} \text{ m}^{-3}$$

The mass of a helium atom is $m = 4 \text{ u} = 6.64 \times 10^{-27} \text{ kg}$. Thus

$$v_{\text{rms}} = \sqrt{\frac{3p}{(N/V)m}} = 1440 \text{ m/s}$$

STOP TO THINK 18.2 The speed of every molecule in a gas is suddenly increased by a factor of 4. As a result, v_{rms} increases by a factor of

a. 2.
c. 4.
e. 16.

b. <4 but not necessarily 2.
d. >4 but not necessarily 16.
f. v_{rms} doesn't change.

18.3 Temperature

A molecule of mass m and velocity v has translational kinetic energy

$$\epsilon = \frac{1}{2}mv^2 \qquad (18.19)$$

We'll use ϵ (lowercase Greek epsilon) to distinguish the energy of a molecule from the system energy E. Thus the average translational kinetic energy is

$$\epsilon_{avg} = \text{average translational kinetic energy of a molecule}$$

$$= \frac{1}{2}m(v^2)_{avg} = \frac{1}{2}mv_{rms}^2 \qquad (18.20)$$

We've included the word "translational" to distinguish ϵ from rotational kinetic energy, which we will consider later in this chapter.

We can write the gas pressure, Equation 18.18, in terms of the average translational kinetic energy as

$$p = \frac{2}{3}\frac{N}{V}\left(\frac{1}{2}mv_{rms}^2\right) = \frac{2}{3}\frac{N}{V}\epsilon_{avg} \qquad (18.21)$$

The pressure is directly proportional to the average molecular translational kinetic energy. This makes sense. More-energetic molecules will hit the walls harder as they bounce and thus exert more force on the walls.

It's instructive to write Equation 18.21 as

$$pV = \frac{2}{3}N\epsilon_{avg} \qquad (18.22)$$

We know, from the ideal-gas law, that

$$pV = Nk_BT \qquad (18.23)$$

Comparing these two equations, we reach the significant conclusion that the average translational kinetic energy per molecule is

$$\epsilon_{avg} = \frac{3}{2}k_BT \qquad \text{(average translational kinetic energy)} \qquad (18.24)$$

where the temperature T is in kelvins. For example, the average translational kinetic energy of a molecule at room temperature (20°C) is

$$\epsilon_{avg} = \frac{3}{2}(1.38 \times 10^{-23} \text{ J/K})(293 \text{ K}) = 6.1 \times 10^{-21} \text{ J}$$

NOTE ▶ A molecule's average translational kinetic energy depends *only* on the temperature, not on the molecule's mass. If two gases have the same temperature, their molecules have the same average translational kinetic energy. ◀

Equation 18.24 is especially satisfying because it finally gives real meaning to the concept of temperature. Writing it as

$$T = \frac{2}{3k_B}\epsilon_{avg} \qquad (18.25)$$

we can see that, for a gas, **this thing we call *temperature*** measures the average **translational kinetic energy.** A higher temperature corresponds to a larger value of ϵ_{avg} and thus to higher molecular speeds. This concept of temperature also gives meaning to *absolute zero* as the temperature at which $\epsilon_{avg} = 0$ and all molecular motion ceases. (Quantum effects at very low temperatures prevent the motions from actually stopping, but our classical theory predicts that they would.) FIGURE 18.9 summarizes what we've learned thus far about the micro/macro connection.

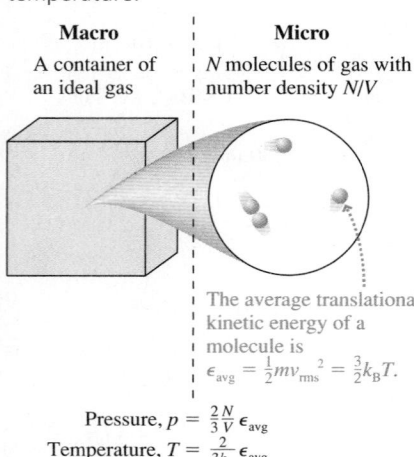

FIGURE 18.9 The micro/macro connection for pressure and temperature.

We can now justify our assumption that molecular collisions are perfectly elastic. Suppose they were not. If kinetic energy was lost in collisions, the average translational kinetic energy ϵ_{avg} of the gas would decrease and we would see a steadily decreasing temperature. But that doesn't happen. The temperature of an isolated system remains constant, indicating that ϵ_{avg} is not changing with time. Consequently, the collisions must be perfectly elastic.

EXAMPLE 18.4   **Total microscopic kinetic energy**

What is the total translational kinetic energy of the molecules in 1.0 mol of gas at STP?

SOLVE The average translational kinetic energy of each molecule is

$$\epsilon_{avg} = \frac{3}{2}k_BT = \frac{3}{2}(1.38 \times 10^{-23} \text{ J/K})(273 \text{ K})$$

$$= 5.65 \times 10^{-21} \text{ J}$$

1.0 mol of gas contains N_A molecules; hence the total kinetic energy is

$$K_{micro} = N_A\epsilon_{avg} = 3400 \text{ J}$$

ASSESS The energy of any one molecule is incredibly small. Nonetheless, a macroscopic system has substantial thermal energy because it consists of an incredibly large number of molecules.

By definition, $\epsilon_{avg} = \frac{1}{2}mv_{rms}^2$. Using the ideal-gas law, we found $\epsilon_{avg} = \frac{3}{2}k_BT$. By equating these expressions we find that the rms speed of molecules in a gas is

$$v_{rms} = \sqrt{\frac{3k_BT}{m}} \qquad (18.26)$$

The rms speed depends on the square root of the temperature and inversely on the square root of the molecular mass.

EXAMPLE 18.5   **Calculating an rms speed**

What is the rms speed of nitrogen molecules at room temperature (20°C)?

SOLVE The molecular mass is $m = 28$ u $= 4.68 \times 10^{-26}$ kg and $T = 20°C = 293$ K. It is then a simple calculation to find

$$v_{rms} = \sqrt{\frac{3(1.38 \times 10^{-23} \text{ J/K})(293 \text{ K})}{4.68 \times 10^{-26} \text{ kg}}} = 509 \text{ m/s}$$

Some speeds will be greater than this and others smaller, but 509 m/s will be a typical or fairly average speed. This is in excellent agreement with the experimental results of Figure 18.2.

EXAMPLE 18.6   **Mean time between collisions**

Estimate the mean time between collisions for a nitrogen molecule at 1.0 atm pressure and room temperature (20°C).

MODEL Because v_{rms} is essentially the average molecular speed, the *mean time between collisions* is simply the time needed to travel distance λ, the mean free path, at speed v_{rms}.

SOLVE We found $\lambda = 2.3 \times 10^{-7}$ m in Example 18.1 and $v_{rms} = 509$ m/s in Example 18.5. Thus the mean time between collisions is

$$\tau_{coll} = \frac{\lambda}{v_{rms}} = \frac{2.3 \times 10^{-7} \text{ m}}{509 \text{ m/s}} = 4.5 \times 10^{-10} \text{ s}$$

ASSESS The air molecules around us move very fast, they collide with their neighbors about two billion times every second, and they manage to move, on average, only about 230 nm between collisions.

STOP TO THINK 18.3   The speed of every molecule in a gas is suddenly increased by a factor of 4. As a result, T increases by a factor of

a. 2.

c. 4.

e. 16.

b. <4 but not necessarily 2.

d. >4 but not necessarily 16.

f. T doesn't change.

18.4 Thermal Energy and Specific Heat

We defined the thermal energy of a system to be $E_{th} = K_{micro} + U_{micro}$, where K_{micro} is the microscopic kinetic energy of the moving molecules and U_{micro} is the potential energy of the stretched and compressed molecular bonds. We're now ready to take a microscopic look at thermal energy.

Monatomic Gases

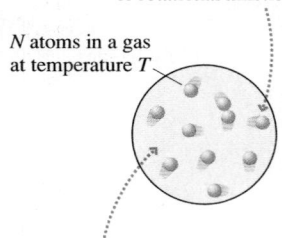

FIGURE 18.10 The atoms in a monatomic gas have only translational kinetic energy.

Atom i has translational kinetic energy ϵ_i but no potential energy or rotational kinetic energy.

N atoms in a gas at temperature T

The thermal energy of the gas is $E_{th} = \epsilon_1 + \epsilon_2 + \epsilon_3 + \cdots = N\epsilon_{avg}$.

FIGURE 18.10 shows a monatomic gas such as helium or neon. The atoms in an ideal gas have no molecular bonds with their neighbors; hence $U_{micro} = 0$. Furthermore, the kinetic energy of a monatomic gas particle is entirely translational kinetic energy ϵ. Thus the thermal energy of a monatomic gas of N atoms is

$$E_{th} = K_{micro} = \epsilon_1 + \epsilon_2 + \epsilon_3 + \cdots + \epsilon_N = N\epsilon_{avg} \tag{18.27}$$

where ϵ_i is the translational kinetic energy of atom i. We found that $\epsilon_{avg} = \frac{3}{2}k_B T$; hence the thermal energy is

$$E_{th} = \frac{3}{2}Nk_B T = \frac{3}{2}nRT \quad \text{(thermal energy of a monatomic gas)} \tag{18.28}$$

where we used $N = nN_A$ and the definition of Boltzmann's constant, $k_B = R/N_A$.

We've noted for the last two chapters that thermal energy is associated with temperature. Now we have an explicit result for a monatomic gas: E_{th} is directly proportional to the temperature. Notice that E_{th} is independent of the atomic mass. Any two monatomic gases will have the same thermal energy if they have the same temperature and the same number of atoms (or moles).

If the temperature of a monatomic gas changes by ΔT, its thermal energy changes by

$$\Delta E_{th} = \frac{3}{2}nR\Delta T \tag{18.29}$$

In Chapter 17 we found that the change in thermal energy for *any* ideal-gas process is related to the molar specific heat at constant volume by

$$\Delta E_{th} = nC_V \Delta T \tag{18.30}$$

Equation 18.29 is a microscopic result that we obtained by relating the temperature to the average translational kinetic energy of the atoms. Equation 18.30 is a macroscopic result that we arrived at from the first law of thermodynamics. We can make a micro/macro connection by combining these two equations. Doing so gives us a *prediction* for the molar specific heat:

$$C_V = \frac{3}{2}R = 12.5 \text{ J/mol K} \quad \text{(monatomic gas)} \tag{18.31}$$

This was exactly the value of C_V for all three monatomic gases in Table 17.4. The perfect agreement of theory and experiment is strong evidence that gases really do consist of moving, colliding molecules.

The Equipartition Theorem

The particles of a monatomic gas are atoms. Their energy consists exclusively of their translational kinetic energy. A particle's translational kinetic energy can be written

$$\epsilon = \frac{1}{2}mv^2 = \frac{1}{2}mv_x^2 + \frac{1}{2}mv_y^2 + \frac{1}{2}mv_z^2 = \epsilon_x + \epsilon_y + \epsilon_z \tag{18.32}$$

where we have written separately the energy associated with translational motion along the three axes. Because each axis in space is independent, we can think of ϵ_x, ϵ_y, and ϵ_z as independent *modes* of storing energy within the system.

Other systems have additional modes of energy storage. For example,

- Two atoms joined by a spring-like molecular bond can vibrate back and forth. Both kinetic and potential energy are associated with this vibration.
- A diatomic molecule, in addition to translational kinetic energy, has rotational kinetic energy if it rotates end-over-end like a dumbbell.

We define the number of **degrees of freedom** as the number of distinct and independent modes of energy storage. A monatomic gas has three degrees of freedom, the three modes of translational kinetic energy. Systems that can vibrate or rotate have more degrees of freedom.

An important result of statistical physics says that the energy in a system is distributed so that all modes of energy storage have equal amounts of energy. This conclusion is known as the *equipartition theorem,* meaning that the energy is equally divided. The proof is beyond what we can do in this textbook, so we will state the theorem without proof:

> **Equipartition theorem** The thermal energy of a system of particles is equally divided among all the possible degrees of freedom. For a system of N particles at temperature T, the energy stored in each mode (each degree of freedom) is $\frac{1}{2}Nk_BT$ or, in terms of moles, $\frac{1}{2}nRT$.

A monatomic gas has three degrees of freedom and thus, as we found above, $E_{th} = \frac{3}{2}Nk_BT$.

Solids

FIGURE 18.11 reminds you of our "bedspring model" of a solid with particle-like atoms connected by a lattice of spring-like molecular bonds. How many degrees of freedom does a solid have? Three degrees of freedom are associated with the kinetic energy, just as in a monatomic gas. In addition, the molecular bonds can be compressed or stretched independently along the x-, y-, and z-axes. Three additional degrees of freedom are associated with these three modes of potential energy. Altogether, a solid has six degrees of freedom.

The energy stored in each of these six degrees of freedom is $\frac{1}{2}Nk_BT$. The thermal energy of a solid is the total energy stored in all six modes, or

$$E_{th} = 3Nk_BT = 3nRT \qquad \text{(thermal energy of a solid)} \qquad (18.33)$$

We can use this result to predict the molar specific heat of a solid. If the temperature changes by ΔT, then the thermal energy changes by

$$\Delta E_{th} = 3nR\Delta T \qquad (18.34)$$

In Chapter 17 we defined the molar specific heat of a solid such that

$$\Delta E_{th} = nC\Delta T \qquad (18.35)$$

By comparing Equations 18.34 and 18.35 we can predict that the molar specific heat of a solid is

$$C = 3R = 25.0 \text{ J/mol K} \quad \text{(solid)} \qquad (18.36)$$

Not bad. The five elemental solids in Table 17.2 had molar specific heats clustered right around 25 J/mol K. They ranged from 24.3 J/mol K for aluminum to 26.5 J/mol K for lead. There are two reasons the agreement between theory and experiment isn't quite as perfect as it was for monatomic gases. First, our simple bedspring model of a solid isn't quite as accurate as our model of a monatomic gas. Second, quantum effects

FIGURE 18.11 A simple model of a solid.

Each atom has microscopic translational kinetic energy *and* microscopic potential energy along all three axes.

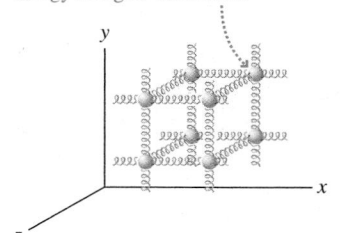

are beginning to make their appearance. More on this shortly. Nonetheless, our ability to predict C to within a few percent from a simple model of a solid is further evidence for the atomic structure of matter.

Diatomic Molecules

Diatomic molecules are a bigger challenge. How many degrees of freedom does a diatomic molecule have? FIGURE 18.12 shows a diatomic molecule, such as molecular nitrogen N_2, oriented along the x-axis. Three degrees of freedom are associated with the molecule's translational kinetic energy. The molecule can have a dumbbell-like end-over-end rotation about either the y-axis or the z-axis. It can also rotate about its own axis. These are three rotational degrees of freedom. The two atoms can also vibrate back and forth, stretching and compressing the molecular bond. This vibrational motion has both kinetic and potential energy—thus two more degrees of freedom.

Altogether, then, a diatomic molecule has eight degrees of freedom, and we would expect the thermal energy of a gas of diatomic molecules to be $E_{th} = 4k_BT$. The analysis we followed for a monatomic gas would then lead to the prediction $C_V = 4R = 33.2$ J/mol K. As compelling as this reasoning seems to be, this is *not* the experimental value of C_V that was reported for diatomic gases in Table 17.4. Instead, we found $C_V = 20.8$ J/mol K.

Why should a theory that works so well for monatomic gases and solids fail so miserably for diatomic molecules? To see what's going on, notice that 20.8 J/mol K = $\frac{5}{2}R$. A monatomic gas, with three degrees of freedom, has $C_V = \frac{3}{2}R$. A solid, with six degrees of freedom, has $C = 3R$. A diatomic gas would have $C_V = \frac{5}{2}R$ if it had five degrees of freedom, not eight.

This discrepancy was a major conundrum as statistical physics developed in the late 19th century. Although it was not recognized as such at the time, we are here seeing our first evidence for the breakdown of classical Newtonian physics. Classically, a diatomic molecule has eight degrees of freedom. The equipartition theorem doesn't distinguish between them; all eight should have the same energy. But atoms and molecules are not classical particles. It took the development of quantum theory in the 1920s to accurately characterize the behavior of atoms and molecules. We don't yet have the tools needed to see why, but quantum effects prevent three of the modes—the two vibrational modes and the rotation of the molecule about its own axis—from being active at room temperature.

FIGURE 18.13 shows C_V as a function of temperature for hydrogen gas. C_V is right at $\frac{5}{2}R$ for temperatures from ≈ 200 K up to ≈ 800 K. But at very low temperatures C_V drops to the monatomic-gas value $\frac{3}{2}R$. The two rotational modes become "frozen out" and the nonrotating molecule has only translational kinetic energy. Quantum physics can explain this, but not Newtonian physics. You can also see that the two vibrational modes *do* become active at very high temperatures, where C_V rises to

FIGURE 18.12 A diatomic molecule can rotate or vibrate.

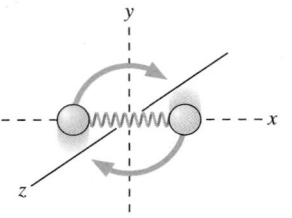

Rotation end-over-end about the z-axis

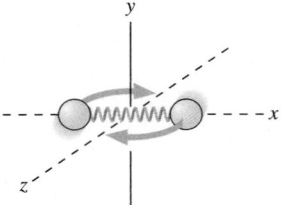

Rotation end-over-end about the y-axis

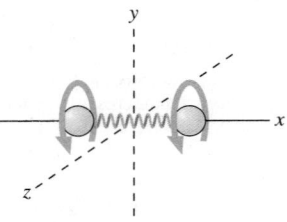

Rotation about the x-axis

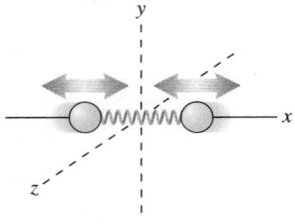

Vibration back and forth along the x-axis

FIGURE 18.13 Hydrogen molar specific heat at constant volume as a function of temperature. The temperature scale is logarithmic.

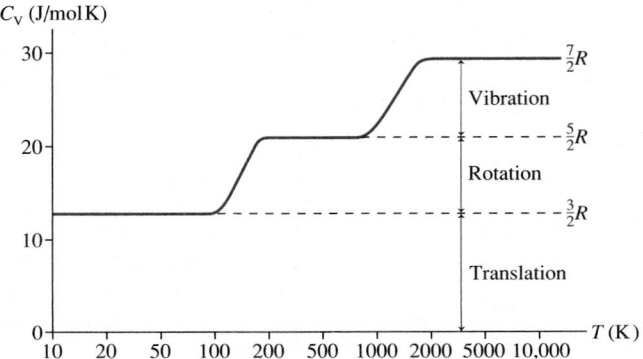

$\frac{7}{2}R$. Thus the real answer to What's wrong? is that Newtonian physics is not the right physics for describing atoms and molecules. We are somewhat fortunate that Newtonian physics is adequate to understand monatomic gases and solids, at least at room temperature.

Accepting the quantum result that a diatomic gas has only five degrees of freedom at commonly used temperatures (the translational degrees of freedom and the two end-over-end rotations), we find

$$E_{th} = \frac{5}{2}Nk_BT = \frac{5}{2}nRT$$

$$\text{(diatomic gases)} \qquad (18.37)$$

$$C_V = \frac{5}{2}R = 20.8 \text{ J/mol K}$$

A diatomic gas has more thermal energy than a monatomic gas at the same temperature because the molecules have rotational as well as translational kinetic energy.

While the micro/macro connection firmly establishes the atomic structure of matter, it also heralds the need for a new theory of matter at the atomic level. That is a task we will take up in Part VII. For now, Table 18.2 summarizes what we have learned from kinetic theory about thermal energy and molar specific heats.

TABLE 18.2 Kinetic theory predictions for the thermal energy and the molar specific heat

System	Degrees of freedom	E_{th}	C_V
Monatomic gas	3	$\frac{3}{2}Nk_BT = \frac{3}{2}nRT$	$\frac{3}{2}R = 12.5$ J/mol K
Diatomic gas	5	$\frac{5}{2}Nk_BT = \frac{5}{2}nRT$	$\frac{5}{2}R = 20.8$ J/mol K
Elemental solid	6	$3Nk_BT = 3nRT$	$3R = 25.0$ J/mol K

EXAMPLE 18.7 **The rotational frequency of a molecule**

The nitrogen molecule N_2 has a bond length of 0.12 nm. Estimate the rotational frequency of N_2 at 20°C.

MODEL The molecule can be modeled as a rigid dumbbell of length $L = 0.12$ nm rotating about its center.

SOLVE The rotational kinetic energy of the molecule is $\epsilon_{rot} = \frac{1}{2}I\omega^2$, where I is the moment of inertia about the center. Because we have two point masses each moving in a circle of radius $r = L/2$, the moment of inertia is

$$I = mr^2 + mr^2 = 2m\left(\frac{L}{2}\right)^2 = \frac{mL^2}{2}$$

Thus the rotational kinetic energy is

$$\epsilon_{rot} = \frac{1}{2}\frac{mL^2}{2}\omega^2 = \frac{mL^2\omega^2}{4} = \pi^2 mL^2 f^2$$

where we used $\omega = 2\pi f$ to relate the rotational frequency f to the angular frequency ω. From the equipartition theorem, the energy

associated with this mode is $\frac{1}{2}Nk_BT$, so the *average* rotational kinetic energy per molecule is

$$(\epsilon_{rot})_{avg} = \frac{1}{2}k_BT$$

Equating these two expressions for ϵ_{rot} gives us

$$\pi^2 mL^2 f^2 = \frac{1}{2}k_BT$$

Thus the rotational frequency is

$$f = \sqrt{\frac{k_BT}{2\pi^2 mL^2}} = 7.8 \times 10^{11} \text{ rev/s}$$

We evaluated f at $T = 293$ K, using $m = 14$ u $= 2.34 \times 10^{-26}$ kg for each *atom*.

ASSESS This is a *very* high frequency, but these values are typical of molecular rotations.

STOP TO THINK 18.4 How many degrees of freedom does a bead on a rigid rod have?

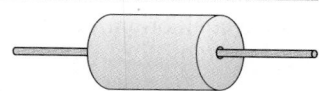

a. 1   b. 2   c. 3   d. 4   e. 5   f. 6

18.5 Thermal Interactions and Heat

FIGURE 18.14 Two gases can interact thermally through a very thin barrier.

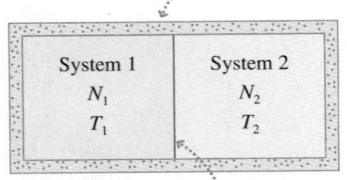

Insulation prevents heat from entering or leaving the container.

| System 1 N_1 T_1 | System 2 N_2 T_2 |

A thin barrier prevents atoms from moving from system 1 to 2 but still allows them to collide. The barrier is clamped in place and cannot move.

We can now look in more detail at what happens when two systems at different temperatures interact with each other. **FIGURE 18.14** shows a rigid, insulated container divided into two sections by a very thin, stiff membrane. The left side, which we'll call system 1, has N_1 atoms at an initial temperature T_{1i}. System 2 on the right has N_2 atoms at an initial temperature T_{2i}. The membrane is so thin that atoms can collide at the boundary as if the membrane were not there, yet it is a barrier that prevents atoms from moving from one side to the other. The situation is analogous, on an atomic scale, to basketballs colliding through a shower curtain.

Suppose that system 1 is initially at a higher temperature: $T_{1i} > T_{2i}$. This is not an equilibrium situation. The temperatures will change with time until the systems eventually reach a common final temperature T_f. If you *watch* the gases as one warms and the other cools, you see nothing happening. This interaction is quite different from a mechanical interaction in which, for example, you might see a piston move from one side toward the other. The only way in which the gases can interact is via molecular collisions at the boundary. This is a *thermal interaction,* and our goal is to understand how thermal interactions bring the systems to thermal equilibrium.

System 1 and system 2 begin with thermal energies

$$E_{1i} = \frac{3}{2}N_1 k_B T_{1i} = \frac{3}{2}n_1 R T_{1i}$$

$$E_{2i} = \frac{3}{2}N_2 k_B T_{2i} = \frac{3}{2}n_2 R T_{2i}$$

(18.38)

We've written the energies for monatomic gases; you could do the same calculation if one or both of the gases is diatomic by replacing the $\frac{3}{2}$ with $\frac{5}{2}$. Notice that we've omitted the subscript "th" to keep the notation manageable.

The total energy of the combined systems is $E_{tot} = E_{1i} + E_{2i}$. As systems 1 and 2 interact, their individual thermal energies E_1 and E_2 can change but their sum E_{tot} remains constant. The system will have reached thermal equilibrium when the individual thermal energies reach final values E_{1f} and E_{2f} that no longer change.

The Systems Exchange Energy

FIGURE 18.15 Collisions at the barrier transfer energy from faster molecules to slower molecules.

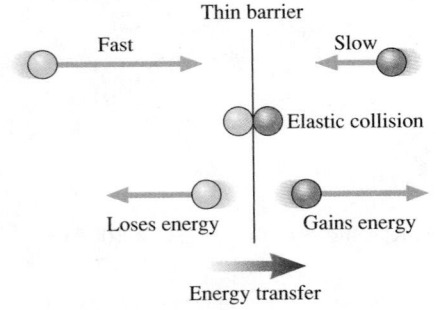

Thin barrier

Fast   Slow

Elastic collision

Loses energy   Gains energy

Energy transfer

FIGURE 18.15 shows a fast atom and a slow atom approaching the barrier from opposite sides. They undergo a perfectly elastic collision at the barrier. Although no net energy is lost in a perfectly elastic collision, the faster atom loses energy while the slower one gains energy. In other words, there is an energy *transfer* from the faster atom's side to the slower atom's side.

The average translational kinetic energy per molecule is directly proportional to the temperature: $\epsilon_{avg} = \frac{3}{2}k_B T$. Because $T_{1i} > T_{2i}$, the atoms in system 1 are, on average, more energetic than the atoms in system 2. Thus *on average* the collisions transfer energy from system 1 to system 2. Not in every collision: sometimes a fast atom in system 2 collides with a slow atom in system 1, transferring energy from 2 to 1. But the net energy transfer, from all collisions, is from the warmer system 1 to the cooler system 2. In other words, **heat is the energy transferred *via collisions* between the more-energetic (warmer) atoms on one side and the less-energetic (cooler) atoms on the other.**

How do the systems "know" when they've reached thermal equilibrium? Energy transfer continues until the atoms on both sides of the barrier have the *same average translational kinetic energy.* Once the average translational kinetic energies are the same, there is no tendency for energy to flow in either direction. This is the state of thermal equilibrium, so the condition for thermal equilibrium is

$$(\epsilon_1)_{avg} = (\epsilon_2)_{avg} \quad \text{(thermal equilibrium)}$$

(18.39)

where, as before, ϵ is the translational kinetic energy of an atom.

Because the average energies are directly proportional to the final temperatures, $\epsilon_{avg} = \frac{3}{2}k_B T_f$, thermal equilibrium is characterized by the macroscopic condition

$$T_{1f} = T_{2f} = T_f \qquad \text{(thermal equilibrium)} \qquad (18.40)$$

In other words, **two thermally interacting systems reach a common final temperature *because* they exchange energy via collisions until the atoms on each side have, on average, equal translational kinetic energies.** This is a very important idea.

Equation 18.40 can be used to determine the equilibrium thermal energies. Because these are monatomic gases, $E_{th} = N\epsilon_{avg}$. Thus the equilibrium condition $(\epsilon_1)_{avg} = (\epsilon_2)_{avg} = (\epsilon_{tot})_{avg}$ implies

$$\frac{E_{1f}}{N_1} = \frac{E_{2f}}{N_2} = \frac{E_{tot}}{N_1 + N_2} \qquad (18.41)$$

from which we can conclude

$$E_{1f} = \frac{N_1}{N_1 + N_2}E_{tot} = \frac{n_1}{n_1 + n_2}E_{tot}$$

$$E_{2f} = \frac{N_2}{N_1 + N_2}E_{tot} = \frac{n_2}{n_1 + n_2}E_{tot} \qquad (18.42)$$

where in the last step we used moles rather than molecules.

Notice that $E_{1f} + E_{2f} = E_{tot}$, verifying that energy has been conserved even while being redistributed between the systems.

No work is done on either system because the barrier has no macroscopic displacement, so the first law of thermodynamics is

$$Q_1 = \Delta E_1 = E_{1f} - E_{1i}$$

$$Q_2 = \Delta E_2 = E_{2f} - E_{2i} \qquad (18.43)$$

As a homework problem you can show that $Q_1 = -Q_2$, as required by energy conservation. That is, the heat lost by one system is gained by the other. $|Q_1|$ is the quantity of heat that is transferred from the warmer gas to the cooler gas during the thermal interaction.

NOTE ▶ In general, the equilibrium thermal energies of the system are *not* equal. That is, $E_{1f} \neq E_{2f}$. They will be equal only if $N_1 = N_2$. Equilibrium is reached when the average translational kinetic energies in the two systems are equal—that is, when $(\epsilon_1)_{avg} = (\epsilon_2)_{avg}$, not when $E_{1f} = E_{2f}$. The distinction is important. FIGURE 18.16 summarizes these ideas. ◀

FIGURE 18.16 Equilibrium is reached when the atoms on each side have, on average, equal energies.

Collisions transfer energy from the warmer system to the cooler system as more-energetic atoms lose energy to less-energetic atoms.

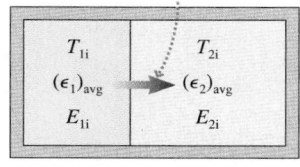

Thermal equilibrium occurs when the systems have the same average translational kinetic energy and thus the same temperature.

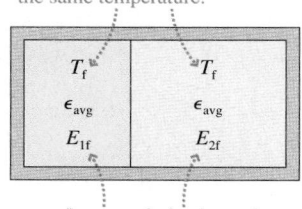

In general, the thermal energies E_{1f} and E_{2f} are *not* equal.

EXAMPLE 18.8 **A thermal interaction**

A sealed, insulated container has 2.0 g of helium at an initial temperature of 300 K on one side of a barrier and 10.0 g of argon at an initial temperature of 600 K on the other side.

a. How much heat energy is transferred, and in which direction?
b. What is the final temperature?

MODEL The systems start with different temperatures, so they are not in thermal equilibrium. Energy will be transferred via collisions from the argon to the helium until both systems have the same average molecular energy.

SOLVE a. Let the helium be system 1. Helium has molar mass $M_{mol} = 0.004$ kg/mol, so $n_1 = M/M_{mol} = 0.50$ mol. Similarly, argon has $M_{mol} = 0.040$ kg/mol, so $n_2 = 0.25$ mol. The initial thermal energies of the two monatomic gases are

$$E_{1i} = \frac{3}{2}n_1 RT_{1i} = 225R = 1870 \text{ J}$$

$$E_{2i} = \frac{3}{2}n_2 RT_{2i} = 225R = 1870 \text{ J}$$

The systems start with *equal* thermal energies, but they are not in thermal equilibrium. The total energy is $E_{tot} = 3740$ J.

Continued

In equilibrium, this energy is distributed between the two systems as

$$E_{1f} = \frac{n_1}{n_1 + n_2} E_{tot} = \frac{0.50}{0.75} 3740 \text{ J} = 2493 \text{ J}$$

$$E_{2f} = \frac{n_2}{n_1 + n_2} E_{tot} = \frac{0.25}{0.75} 3740 \text{ J} = 1247 \text{ J}$$

The heat entering or leaving each system is

$$Q_1 = Q_{He} = E_{1f} - E_{1i} = 623 \text{ J}$$

$$Q_2 = Q_{Ar} = E_{2f} - E_{2i} = -623 \text{ J}$$

The helium and the argon interact thermally via collisions at the boundary, causing 623 J of heat to be transferred from the warmer argon to the cooler helium.

b. These are constant-volume processes, thus $Q = nC_V\Delta T$. $C_V = \frac{3}{2}R$ for monatomic gases, so the temperature changes are

$$\Delta T_{He} = \frac{Q_{He}}{\frac{3}{2}nR} = \frac{623 \text{ J}}{1.5(0.50 \text{ mol})(8.31 \text{ J/mol K})} = 100 \text{ K}$$

$$\Delta T_{Ar} = \frac{Q_{Ar}}{\frac{3}{2}nR} = \frac{-623 \text{ J}}{1.5(0.25 \text{ mol})(8.31 \text{ J/mol K})} = -200 \text{ K}$$

Both gases reach the common final temperature $T_f = 400$ K.

ASSESS $E_{1f} = 2E_{2f}$ because there are twice as many atoms in system 1.

The main idea of this section is that two systems reach a common final temperature not by magic or by a prearranged agreement but simply from the energy exchange of vast numbers of molecular collisions. Real interacting systems, of course, are separated by walls rather than our unrealistic thin membrane. As the systems interact, the energy is first transferred via collisions from system 1 into the wall and subsequently, as the cooler molecules collide with a warm wall, into system 2. That is, the energy transfer is $E_1 \rightarrow E_{wall} \rightarrow E_2$. This is still heat because the energy transfer is occurring via molecular collisions rather than mechanical motion.

STOP TO THINK 18.5 Systems A and B are interacting thermally. At this instant of time,

a. $T_A > T_B$
b. $T_A = T_B$
c. $T_A < T_B$

A	B
$N = 1000$	$N = 2000$
$\epsilon_{avg} = 1.0 \times 10^{-20}$ J	$\epsilon_{avg} = 0.5 \times 10^{-20}$ J
$E_{th} = 1.0 \times 10^{-17}$ J	$E_{th} = 1.0 \times 10^{-17}$ J

18.6 Irreversible Processes and the Second Law of Thermodynamics

The preceding section looked at the thermal interaction between a warm gas and a cold gas. Heat energy is transferred from the warm gas to the cold gas until they reach a common final temperature. But why isn't heat transferred from the cold gas to the warm gas, making the cold side colder and the warm side warmer? Such a process could still conserve energy, but it never happens. The transfer of heat energy from hot to cold is an example of an **irreversible process,** a process that can happen only in one direction.

Examples of irreversible processes abound. Stirring the cream in your coffee mixes the cream and coffee together. No amount of stirring ever unmixes them. If you shake a jar that has red marbles on the top and blue marbles on the bottom, the two colors are quickly mixed together. No amount of shaking ever separates them again. If you watched a movie of someone shaking a jar and saw the red and blue marbles separating, you would be certain that the movie was running backward. In fact, a reasonable definition of an irreversible process is one for which a backward-running movie shows a physically impossible process.

FIGURE 18.17a is a two-frame movie of a collision between two particles, perhaps two gas molecules. Suppose that sometime after the collision is over we could reach in and reverse the velocities of both particles. That is, replace vector $\vec{v}$ with vector $-\vec{v}$. Then, as in a movie playing backward, the collision would happen in reverse. This is the movie of **FIGURE 18.17b**.

FIGURE 18.17 Molecular collisions are reversible.

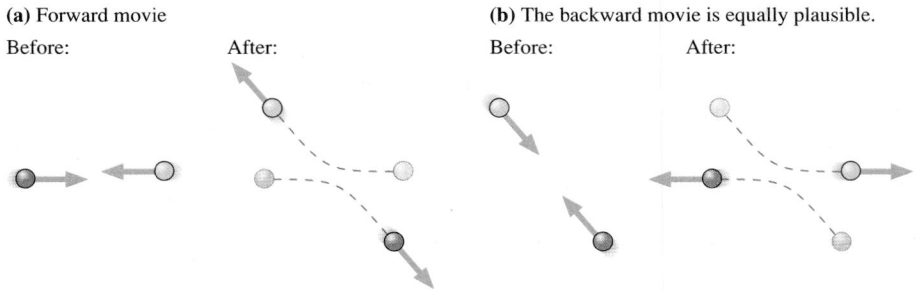

(a) Forward movie

Before: After:

(b) The backward movie is equally plausible.

Before: After:

You cannot tell, just by looking at the two movies, which is really going forward and which is being played backward. Maybe Figure 18.17b was the original collision and Figure 18.17a is the backward version. Nothing in either collision looks wrong, and no measurements you might make on either would reveal any violations of Newton's laws. Interactions at the molecular level are reversible processes.

Contrast this with the two-frame car crash movies in **FIGURE 18.18**. Past and future are clearly distinct in an irreversible process, and the backward movie of Figure 18.18b is obviously wrong. But what has been violated in the backward movie? To have the crumpled car spring away from the wall would not violate any laws of physics we have so far discovered. It would simply require transforming the thermal energy of the car and wall back into the macroscopic center-of-mass energy of the car as a whole.

The paradox stems from our assertion that macroscopic phenomena can be understood on the basis of microscopic molecular motions. If the microscopic motions are all reversible, how can the macroscopic phenomena end up being irreversible? If reversible collisions can cause heat to be transferred from hot to cold, why do they never cause heat to be transferred from cold to hot? There must be another law of physics preventing it. The law we seek must, in some sense, be able to distinguish the past from the future.

FIGURE 18.18 A car crash is irreversible.

(a) Forward movie

Before: After:

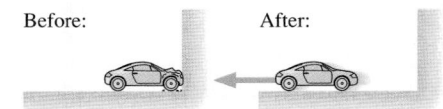

(b) The backward movie is physically impossible.

Before: After:

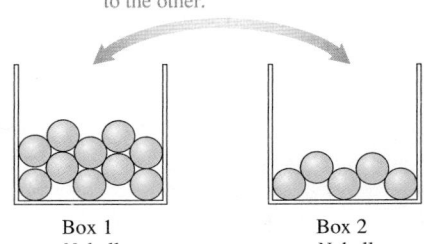

Which Way to Equilibrium?

Stated another way, how do two systems initially at different temperatures "know" which way to go to reach equilibrium? Perhaps an analogy will help.

FIGURE 18.19 shows two boxes, numbered 1 and 2, containing identical balls. Box 1 starts with more balls than box 2, so $N_{1i} > N_{2i}$. Once every second, one ball is chosen at random and moved to the other box. This is a reversible process because a ball can move from box 2 to box 1 just as easily as from box 1 to box 2. What do you expect to see if you return several hours later?

Because balls are chosen at random, and because $N_{1i} > N_{2i}$, it's initially more likely that a ball will move from box 1 to box 2 than from box 2 to box 1. Sometimes a ball will move "backward" from box 2 to box 1, but overall there's a net movement of balls from box 1 to box 2. The system will evolve until $N_1 \approx N_2$. This is a stable situation—equilibrium!—with an equal number of balls moving in both directions.

But couldn't it go the other way, with N_1 getting even larger while N_2 decreases? In principle, any possible arrangement of the balls is possible in the same way that any number of heads are possible if you throw N coins in the air and let them fall. If you

FIGURE 18.19 Two interacting systems. Balls are chosen at random and moved to the other box.

Balls are chosen at random and moved from one box to the other.

Box 1
N_1 balls

Box 2
N_2 balls

throw four coins, the odds are 1 in 2^4, or 1 in 16, of getting four heads. With four balls, the odds are 1 in 16 that, at a randomly chosen instant of time, you would find $N_1 = 4$. You wouldn't find that to be terribly surprising.

With 10 balls, the probability that $N_1 = 10$ is $0.5^{10} \approx 1/1000$. With 100 balls, the probability that $N_1 = 100$ has dropped to $\approx 10^{-30}$. With 10^{20} balls, the odds of finding all of them, or even most of them, in one box are so staggeringly small that it's safe to say it will "never" happen. Although each transfer is reversible, **the statistics of large numbers make it overwhelmingly more likely that the system will evolve toward a state in which $N_1 \approx N_2$ than toward a state in which $N_1 > N_2$.**

The balls in our analogy represent energy. The total energy, like the total number of balls, is conserved, but molecular collisions can move energy between system 1 and system 2. Each collision is reversible, just as likely to transfer energy from 1 to 2 as from 2 to 1. But if $(\epsilon_{1i})_{avg} > (\epsilon_{2i})_{avg}$, and if we're dealing with two macroscopic systems where $N > 10^{20}$, then it's overwhelmingly likely that the net result of many, many collisions will be to transfer energy from system 1 to system 2 until $(\epsilon_{1f})_{avg} = (\epsilon_{2f})_{avg}$—in other words, for heat energy to be transferred from hot to cold.

The system reaches thermal equilibrium not by any plan or by outside intervention, but simply because **equilibrium is the *most probable* state in which to be.** It is *possible* that the system will move away from equilibrium, with heat moving from cold to hot, but remotely improbable in any realistic system. The consequence of a vast number of random events is that the system evolves in one direction, toward equilibrium, and not the other. **Reversible microscopic events lead to irreversible macroscopic behavior because some macroscopic states are vastly more probable than others.**

Order, Disorder, and Entropy

FIGURE 18.20 Ordered and disordered systems.

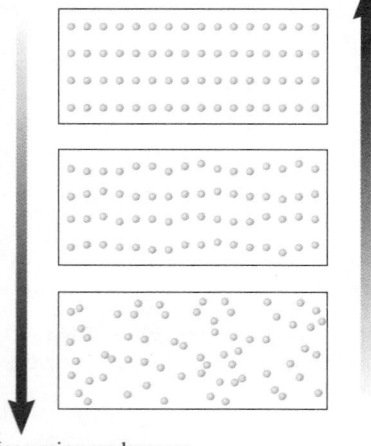

Increasing order
Decreasing entropy
Decreasing probability

Increasing randomness
Increasing entropy
Increasing probability

FIGURE 18.20 shows three different systems. At the top is a group of atoms arranged in a crystal-like lattice. This is a highly ordered and nonrandom system, with each atom's position precisely specified. Contrast this with the system on the bottom, where there is no order at all. The position of every atom was assigned entirely at random.

It is extremely improbable that the atoms in a container would *spontaneously* arrange themselves into the ordered pattern of the top picture. In a system of, say, 10^{20} atoms, the probability of this happening is similar to the probability that 10^{20} tossed coins will all be heads. We can safely say that it will never happen. By contrast, there are a vast number of arrangements like the one on the bottom that randomly fill the container.

The middle picture of Figure 18.20 is an in-between situation. This situation might arise as a solid melts. The positions of the atoms are clearly not completely random, so the system preserves some degree of order. This in-between situation is more likely to occur spontaneously than the highly ordered lattice on the top, but is less likely to occur than the completely random system on the bottom.

Scientists and engineers use a state variable called **entropy** to measure the probability that a macroscopic state will occur spontaneously. The ordered lattice, which has a very small probability of spontaneous occurrence, has a very low entropy. The entropy of the randomly filled container is high. The entropy of the middle picture is somewhere in between. It is often said that entropy measures the amount of *disorder* in a system. The entropy in Figure 18.20 increases as you move from the ordered system on the top to the disordered system on the bottom.

Similarly, two thermally interacting systems with different temperatures have a low entropy. These systems are ordered in the sense that the faster atoms are on one side of the barrier, the slower atoms on the other. The most random possible distribution of energy, and hence the least ordered system, corresponds to the situation where the two systems are in thermal equilibrium with equal temperatures. Entropy increases as two systems with initially different temperatures move toward equilibrium. Entropy would decrease if heat energy moved from cold to hot, making the hot system hotter and the cold system colder.

Entropy can be calculated, but we'll leave that to more advanced courses. For our purposes, the *concept* of entropy as a measure of the disorder in a system, or of the probability that a macroscopic state will occur, is more important than a numerical value.

The Second Law of Thermodynamics

The fact that macroscopic systems evolve irreversibly toward equilibrium is a statement about nature that is not contained in any of the laws of physics we have encountered. It is, in fact, a new law of physics, one known as the **second law of thermodynamics.**

The formal statement of the second law of thermodynamics is given in terms of entropy:

> **Second law, formal statement** The entropy of an isolated system (or group of systems) never decreases. The entropy either increases, until the system reaches equilibrium, or, if the system began in equilibrium, stays the same.

Tossing all heads, while not impossible, is extremely unlikely, and the probability of doing so rapidly decreases as the number of coins increases.

The qualifier "isolated" is most important. We can order the system by reaching in from the outside, perhaps using tiny tweezers to place the atoms in a lattice. Similarly, we can transfer heat from cold to hot by using a refrigerator. The second law is about what a system can or cannot do *spontaneously,* on its own, without outside intervention.

The second law of thermodynamics tells us that an isolated system evolves such that

- Order turns into disorder and randomness.
- Information is lost rather than gained.
- The system "runs down."

An isolated system never spontaneously generates order out of randomness. It is not that the system "knows" about order or randomness, but rather that there are vastly more states corresponding to randomness than there are corresponding to order. As collisions occur at the microscopic level, the laws of probability dictate that the system will, on average, move inexorably toward the most probable and thus most random macroscopic state.

The second law of thermodynamics is often stated in several equivalent but more informal versions. One of these, and the one most relevant to our discussion, is

> **Second law, informal statement #1** When two systems at different temperatures interact, heat energy is transferred spontaneously from the hotter to the colder system, never from the colder to the hotter.

The second law of thermodynamics is an independent statement about nature, separate from the first law. The first law is a precise statement about energy conservation. The second law, by contrast, is a *probabilistic* statement, based on the statistics of very large numbers. While it is conceivable that heat could spontaneously move from cold to hot, it will never occur in any realistic macroscopic system.

The irreversible evolution from less-likely macroscopic states to more-likely macroscopic states is what gives us a macroscopic direction of time. Stirring blends your coffee and cream, it never unmixes them. Friction causes an object to stop while increasing its thermal energy; the random atomic motions of thermal energy never spontaneously organize themselves into a macroscopic motion of the entire object.

A plant in a sealed jar dies and decomposes to carbon and various gases; the gases and carbon never spontaneously assemble themselves into a flower. These are all examples of irreversible processes. They each show a clear direction of time, a distinct difference between past and future.

Thus another statement of the second law is

> **Second law, informal statement #2** The time direction in which the entropy of an isolated macroscopic system increases is "the future."

Establishing the "arrow of time" is one of the most profound implications of the second law of thermodynamics.

The second law of thermodynamics has important implications for issues ranging from how we as a society use energy and resources to biological evolution and the future of the universe. We'll return to some of these issues in the Summary to Part IV. In the meantime, the second law will be used in Chapter 19 to understand some of the practical aspects of the thermodynamics of engines.

STOP TO THINK 18.6 Two identical boxes each contain 1,000,000 molecules. In box A, 750,000 molecules happen to be in the left half of the box while 250,000 are in the right half. In box B, 499,900 molecules happen to be in the left half of the box while 500,100 are in the right half. At this instant of time,

  a. The entropy of box A is larger than the entropy of box B.
  b. The entropy of box A is equal to the entropy of box B.
  c. The entropy of box A is smaller than the entropy of box B.

SUMMARY

The goal of Chapter 18 has been to understand a macroscopic system in terms of the microscopic behavior of its molecules.

General Principles

The **micro/macro connection** relates the macroscopic properties of a system to the motion and collisions of its atoms and molecules.

The Equipartition Theorem

Tells us how collisions distribute the energy in the system. The energy stored in each mode of the system (each **degree of freedom**) is $\frac{1}{2}Nk_BT$ or, in terms of moles, $\frac{1}{2}nRT$.

The Second Law of Thermodynamics

Tells us how collisions move a system toward equilibrium. The entropy of an isolated system can only increase or, in equilibrium, stay the same.

- Order turns into disorder and randomness.

- Systems run down.

- Heat energy is transferred spontaneously from a hotter to a colder system, never from colder to hotter.

Important Concepts

Pressure is due to the force of the molecules colliding with the walls:

$$p = \frac{1}{3}\frac{N}{V}mv_{rms}{}^2 = \frac{2}{3}\frac{N}{V}\epsilon_{avg}$$

The average translational kinetic energy of a molecule is $\epsilon_{avg} = \frac{3}{2}k_BT$. The temperature of the gas $T = \frac{2}{3k_B}\epsilon_{avg}$ measures the average translational kinetic energy.

Entropy measures the probability that a macroscopic state will occur or, equivalently, the amount of disorder in a system.

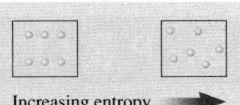

Increasing entropy ➡

The thermal energy of a system is

E_{th} = translational kinetic energy + rotational kinetic energy + vibrational energy

- **Monatomic gas** $E_{th} = \frac{3}{2}Nk_BT = \frac{3}{2}nRT$
- **Diatomic gas** $E_{th} = \frac{5}{2}Nk_BT = \frac{5}{2}nRT$
- **Elemental solid** $E_{th} = 3Nk_BT = 3nRT$

Heat is energy transferred via collisions from more-energetic molecules on one side to less-energetic molecules on the other. Equilibrium is reached when $(\epsilon_1)_{avg} = (\epsilon_2)_{avg}$, which implies $T_{1f} = T_{2f}$.

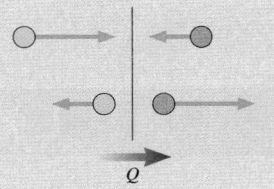

Applications

The root-mean-square speed v_{rms} is the square root of the average of the squares of the molecular speeds:

$$v_{rms} = \sqrt{(v^2)_{avg}}$$

For molecules of mass m at temperature T, $v_{rms} = \sqrt{\dfrac{3k_BT}{m}}$

Molar specific heats can be predicted from the thermal energy because $\Delta E_{th} = nC\Delta T$.

- **Monatomic gas** $C_V = \frac{3}{2}R$
- **Diatomic gas** $C_V = \frac{5}{2}R$
- **Elemental solid** $C = 3R$

Terms and Notation

histogram	degrees of freedom	entropy
mean free path, λ	equipartition theorem	second law of thermodynamics
root-mean-square speed, v_{rms}	irreversible process	

CONCEPTUAL QUESTIONS

1. Solids and liquids resist being compressed. They are not totally incompressible, but it takes large forces to compress them even slightly. If it is true that matter consists of atoms, what can you infer about the microscopic nature of solids and liquids from their incompressibility?

2. Gases, in contrast with solids and liquids, are very compressible. What can you infer from this observation about the microscopic nature of gases?

3. The density of air at STP is about $\frac{1}{1000}$ the density of water. How does the average distance between air molecules compare to the average distance between water molecules? Explain.

4. The mean free path of molecules in a gas is 200 nm.
   a. What will be the mean free path if the pressure is doubled while all other state variables are held constant?
   b. What will be the mean free path if the absolute temperature is doubled while all other state variables are held constant?

5. If the pressure of a gas is really due to the *random* collisions of molecules with the walls of the container, why do pressure gauges—even very sensitive ones—give perfectly steady readings? Shouldn't the gauge be continually jiggling and fluctuating? Explain.

6. Suppose you could suddenly increase the speed of every molecule in a gas by a factor of 2.
   a. Would the rms speed of the molecules increase by a factor of $2^{1/2}$, 2, or 2^2? Explain.
   b. Would the gas pressure increase by a factor of $2^{1/2}$, 2, or 2^2? Explain.

7. Suppose you could suddenly increase the speed of every molecule in a gas by a factor of 2.
   a. Would the temperature of the gas increase by a factor of $2^{1/2}$, 2, or 2^2? Explain.

   b. Would the molar specific heat at constant volume change? If so, by what factor? If not, why not?

8. The two containers of gas in **FIGURE Q18.8** are in good thermal contact with each other but well insulated from the environment. They have been in contact for a long time and are in thermal equilibrium.
   a. Is v_{rms} of helium greater than, less than, or equal to v_{rms} of argon? Explain.
   b. Does the helium have more thermal energy, less thermal energy, or the same amount of thermal energy as the argon? Explain.

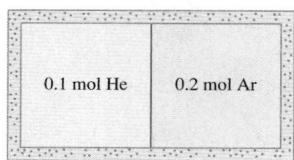

FIGURE Q18.8

9. Suppose you place an ice cube in a beaker of room-temperature water, then seal them in a rigid, well-insulated container. No energy can enter or leave the container.
   a. If you open the container an hour later, will you find a beaker of water slightly cooler than room temperature, or a large ice cube and some 100°C steam?
   b. Finding a large ice cube and some 100°C steam would not violate the first law of thermodynamics. $W = 0$ J and $Q = 0$ J because the container is sealed, and $\Delta E_{th} = 0$ J because the increase in thermal energy of the water molecules that became steam is offset by the decrease in thermal energy of the water molecules that turned to ice. Energy would be conserved, yet we never see an outcome like this. Why not?

EXERCISES AND PROBLEMS

Problems labeled [ ] integrate material from earlier chapters.

Exercises

Section 18.1 Molecular Speeds and Collisions

1. | A 1.0 m × 1.0 m × 1.0 m cube of nitrogen gas is at 20°C and 1.0 atm. Estimate the number of molecules in the cube with a speed between 700 m/s and 1000 m/s.

2. | The number density of an ideal gas at STP is called the *Loschmidt number*. Calculate the Loschmidt number.

3. ‖ At what pressure will the mean free path in room-temperature (20°C) nitrogen be 1.0 m?

4. ‖ Integrated circuits are manufactured in vacuum chambers in which the air pressure is 1.0×10^{-10} mm of Hg. What are (a) the number density and (b) the mean free path of a molecule? Assume $T = 20°C$.

5. | The mean free path of a molecule in a gas is 300 nm. What will the mean free path be if the gas temperature is doubled at (a) constant volume and (b) constant pressure?

6. ‖ For a monatomic gas, what is the ratio of the volume per atom (V/N) to the volume *of* an atom when the mean free path is ten times the atomic diameter?

7. ‖ A lottery machine uses blowing air to keep 2000 Ping-Pong balls bouncing around inside a 1.0 m × 1.0 m × 1.0 m box. The diameter of a Ping-Pong ball is 3.0 cm. What is the mean free path between collisions? Give your answer in cm.

Section 18.2 Pressure in a Gas

8. | Eleven molecules have speeds 15, 16, 17, . . . , 25 m/s. Calculate (a) v_{avg} and (b) v_{rms}.

9. ‖ The molecules in a six-particle gas have velocities

$$\vec{v}_1 = (20\hat{\imath} - 30\hat{\jmath}) \text{ m/s} \qquad \vec{v}_4 = 30\hat{\imath} \text{ m/s}$$

$$\vec{v}_2 = (40\hat{\imath} + 70\hat{\jmath}) \text{ m/s} \qquad \vec{v}_5 = (40\hat{\imath} - 40\hat{\jmath}) \text{ m/s}$$

$$\vec{v}_3 = (-80\hat{\imath} + 20\hat{\jmath}) \text{ m/s} \qquad \vec{v}_6 = (-50\hat{\imath} - 20\hat{\jmath}) \text{ m/s}$$

Calculate (a) $\vec{v}_{avg}$, (b) v_{avg}, and (c) v_{rms}.

10. | **FIGURE EX18.10** is a histogram showing the speeds of the molecules in a very small gas. What are (a) the most probable speed, (b) the average speed, and (c) the rms speed?

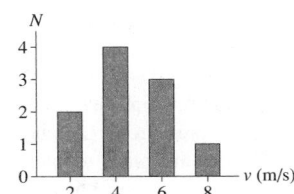

FIGURE EX18.10

11. ‖ The number density in a container of neon gas is 5.00×10^{25} m^{-3}. The atoms are moving with an rms speed of 660 m/s. What are (a) the temperature and (b) the pressure inside the container?

12. ‖ A cylinder contains gas at a pressure of 2.0 atm and a number density of 4.2×10^{25} m^{-3}. The rms speed of the atoms is 660 m/s. Identify the gas.

13. ‖ At 100°C the rms speed of nitrogen molecules is 576 m/s. Nitrogen at 100°C and a pressure of 2.0 atm is held in a container with a 10 cm × 10 cm square wall. Estimate the rate of molecular collisions (collisions/s) on this wall.

Section 18.3 Temperature

14. ‖ What are the rms speeds of (a) argon atoms and (b) hydrogen molecules at 800°C?

15. ‖ A gas consists of a mixture of neon and argon. The rms speed of the neon atoms is 400 m/s. What is the rms speed of the argon atoms?

16. | 1.5 m/s is a typical walking speed. At what temperature (in °C) would nitrogen molecules have an rms speed of 1.5 m/s?

17. ‖ At what temperature (in °C) do hydrogen molecules have the same rms speed as nitrogen molecules at 100°C?

18. | At what temperature (in °C) is the rms speed of helium atoms (a) half and (b) twice its value at STP?

19. | The rms speed of molecules in a gas is 600 m/s. What will be the rms speed if the gas pressure and volume are both halved?

20. ‖ By what factor does the rms speed of a molecule change if the temperature is increased from 10°C to 1000°C?

21. ‖ Atoms can be "cooled" to incredibly low temperatures by letting them interact with a laser beam. Various novel quantum phenomena appear at these temperatures. What is the rms speed of cesium atoms that have been cooled to a temperature of 100 nK?

22. | At STP, what is the total translational kinetic energy of the molecules in 1.0 mol of (a) hydrogen, (b) helium, and (c) oxygen?

23. | Suppose you double the temperature of a gas at constant volume. Do the following change? If so, by what factor?
   a. The average translational kinetic energy of a molecule.
   b. The rms speed of a molecule.
   c. The mean free path.

24. ‖ During a physics experiment, helium gas is cooled to a temperature of 10 K at a pressure of 0.10 atm. What are (a) the mean free path in the gas, (b) the rms speed of the atoms, and (c) the average energy per atom?

25. | What are (a) the average kinetic energy and (b) the rms speed of a proton in the center of the sun, where the temperature is 2.0×10^7 K?

26. | The atmosphere of the sun consists mostly of hydrogen *atoms* (not molecules) at a temperature of 6000 K. What are (a) the average translational kinetic energy per atom and (b) the rms speed of the atoms?

Section 18.4 Thermal Energy and Specific Heat

27. ‖ A 10 g sample of neon gas has 1700 J of thermal energy. Estimate the average speed of a neon atom.

28. ‖ The rms speed of the atoms in a 2.0 g sample of helium gas is 700 m/s. What is the thermal energy of the gas?

29. ‖ A 6.0 m × 8.0 m × 3.0 m room contains air at 20°C. What is the room's thermal energy?

30. | The thermal energy of 1.0 mol of a substance is increased by 1.0 J. What is the temperature change if the system is (a) a monatomic gas, (b) a diatomic gas, and (c) a solid?

31. ‖ What is the thermal energy of 100 cm^3 of aluminum at 100°C?

32. | 1.0 mol of a monatomic gas interacts thermally with 1.0 mol of an elemental solid. The gas temperature decreases by 50°C at constant volume. What is the temperature change of the solid?

33. ‖ A cylinder of nitrogen gas has a volume of 15,000 cm^3 and a pressure of 100 atm.
   a. What is the thermal energy of this gas at room temperature (20°C)?
   b. What is the mean free path in the gas?
   c. The valve is opened and the gas is allowed to expand slowly and isothermally until it reaches a pressure of 1.0 atm. What is the change in the thermal energy of the gas?

34. ‖ A rigid container holds 0.20 g of hydrogen gas. How much heat is needed to change the temperature of the gas
   a. From 50 K to 100 K?
   b. From 250 K to 300 K?
   c. From 2250 K to 2300 K?

Section 18.5 Thermal Interactions and Heat

35. ‖ 4.0 mol of monatomic gas A interacts with 3.0 mol of monatomic gas B. Gas A initially has 9000 J of thermal energy, but in the process of coming to thermal equilibrium it transfers 1000 J of heat energy to gas B. How much thermal energy did gas B have initially?

36. | 2.0 mol of monatomic gas A initially has 5000 J of thermal energy. It interacts with 3.0 mol of monatomic gas B, which initially has 8000 J of thermal energy.
   a. Which gas has the higher initial temperature?
   b. What is the final thermal energy of each gas?

Problems

37. ‖‖ The pressure inside a tank of neon is 150 atm. The temperature is 25°C. On average, how many atomic diameters does a neon atom move between collisions?

38. ‖ From what height must an oxygen molecule fall in a vacuum so that its kinetic energy at the bottom equals the average energy of an oxygen molecule at 300 K?

39. ‖ A gas at $p = 50$ kPa and $T = 300$ K has a mass density of 0.0802 kg/m^3.
   a. Identify the gas.
   b. What is the rms speed of the atoms in this gas?
   c. What is the mean free path of the atoms in the gas?

40. ‖ Dust particles are $\approx 10 \ \mu m$ in diameter. They are pulverized rock, with $\rho \approx 2500 \ kg/m^3$. If you treat dust as an ideal gas, what is the rms speed of a dust particle at 20°C?

41. ‖ Interstellar space, far from any stars, is filled with a very low density of hydrogen atoms (H, not H_2). The number density is about 1 atom/cm³ and the temperature is about 3 K.
    a. Estimate the pressure in interstellar space. Give your answer in Pa and in atm.
    b. What is the rms speed of the atoms?
    c. What is the edge length L of an $L \times L \times L$ cube of gas with 1.0 J of thermal energy?

42. ‖ Equation 18.3 is the mean free path of a particle through a gas of identical particles of equal radius. An electron can be thought of as a point particle with zero radius.
    a. Find an expression for the mean free path of an electron through a gas.
    b. Electrons travel 3 km through the Stanford Linear Accelerator. In order for scattering losses to be negligible, the pressure inside the accelerator tube must be reduced to the point where the mean free path is at least 50 km. What is the maximum possible pressure inside the accelerator tube, assuming $T = 20°C$? Give your answer in both Pa and atm.

43. ‖ Uranium has two naturally occurring isotopes. ^{238}U has a natural abundance of 99.3% and ^{235}U has an abundance of 0.7%. It is the rarer ^{235}U that is needed for nuclear reactors. The isotopes are separated by forming uranium hexafluoride, UF_6, which is a gas, then allowing it to diffuse through a series of porous membranes. $^{235}UF_6$ has a slightly larger rms speed than $^{238}UF_6$ and diffuses slightly faster. Many repetitions of this procedure gradually separate the two isotopes. What is the ratio of the rms speed of $^{235}UF_6$ to that of $^{238}UF_6$?

44. ‖ On earth, STP is based on the average atmospheric pressure at the surface and on a phase change of water that occurs at an easily produced temperature, being only slightly cooler than the average air temperature. The atmosphere of Venus is almost entirely carbon dioxide (CO_2), the pressure at the surface is a staggering 93 atm, and the average temperature is 470°C. Venusian scientists, if they existed, would certainly use the surface pressure as part of their definition of STP. To complete the definition, they would seek a phase change that occurs near the average temperature. Conveniently, the melting point of the element tellurium is 450°C. What are (a) the rms speed and (b) the mean free path of carbon dioxide molecules at Venusian STP based on this phase change in tellurium. The radius of a CO_2 molecule is 1.5×10^{-10} m.

45. ‖‖ 5.0×10^{23} nitrogen molecules collide with a 10 cm² wall each second. Assume that the molecules all travel with a speed of 400 m/s and strike the wall head-on. What is the pressure on the wall?

46. ‖‖ A 10 cm × 10 cm × 10 cm box contains 0.010 mol of nitrogen at 20°C. What is the rate of collisions (collisions/s) on one wall of the box?

47. ‖ FIGURE P18.47 shows the thermal energy of 0.14 mol of gas as a function of temperature. What is C_V for this gas?

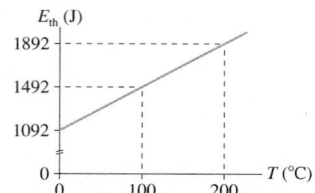

E_{th} (J)

FIGURE P18.47

48. ‖ A 100 cm³ box contains helium at a pressure of 2.0 atm and a temperature of 100°C. It is placed in thermal contact with a 200 cm³ box containing argon at a pressure of 4.0 atm and a temperature of 400°C.
    a. What is the initial thermal energy of each gas?
    b. What is the final thermal energy of each gas?
    c. How much heat energy is transferred, and in which direction?
    d. What is the final temperature?
    e. What is the final pressure in each box?

49. ‖ 2.0 g of helium at an initial temperature of 300 K interacts thermally with 8.0 g of oxygen at an initial temperature of 600 K.
    a. What is the initial thermal energy of each gas?
    b. What is the final thermal energy of each gas?
    c. How much heat energy is transferred, and in which direction?
    d. What is the final temperature?

50. ‖ A gas of 1.0×10^{20} atoms or molecules has 1.0 J of thermal energy. Its molar specific heat at constant pressure is 20.8 J/mol K. What is the temperature of the gas?

51. ‖ How many degrees of freedom does a system have if $\gamma = 1.29$?

52. ‖ A monatomic gas and a diatomic gas have equal numbers of moles and equal temperatures. Both are heated at constant pressure until their volume doubles. What is the ratio $Q_{diatomic}/Q_{monatomic}$?

53. ‖ In the discussion following Equation 18.43 it was said that $Q_1 = -Q_2$. Prove that this is so.

54. ‖ A monatomic gas is adiabatically compressed to $\frac{1}{8}$ of its initial volume. Does each of the following quantities change? If so, does it increase or decrease, and by what factor? If not, why not?
    a. The rms speed.
    b. The mean free path.
    c. The thermal energy of the gas.
    d. The molar specific heat at constant volume.

55. ‖ A diatomic gas is isobarically expanded to four times its initial volume. Does each of the following quantities change? If so, does it increase or decrease, and by what factor? If not, why not?
    a. The rms speed.
    b. The mean free path.
    c. The thermal energy of the gas.
    d. The molar heat capacity at constant volume.

56. ‖ The 2010 Nobel Prize in Physics was awarded for the discovery of graphene, a two-dimensional form of carbon in which the atoms form a two-dimensional crystal-lattice sheet only one atom thick. Predict the molar specific heat of graphene. Give your answer as a multiple of R.

57. ‖ Equal masses of hydrogen gas and oxygen gas are mixed together in a container and held at constant temperature. What is the hydrogen/oxygen ratio of (a) v_{rms}, (b) ϵ_{avg}, and (c) E_{th}?

58. ‖ The rms speed of the molecules in 1.0 g of hydrogen gas is 1800 m/s.
    a. What is the total translational kinetic energy of the gas molecules?
    b. What is the thermal energy of the gas?
    c. 500 J of work are done to compress the gas while, in the same process, 1200 J of heat energy are transferred from the gas to the environment. Afterward, what is the rms speed of the molecules?

59. ‖ At what temperature does the rms speed of (a) a nitrogen molecule and (b) a hydrogen molecule equal the escape speed from the earth's surface? (c) You'll find that these temperatures are very high, so you might think that the earth's gravity could easily contain both gases. But not all molecules move with v_{rms}. There

is a distribution of speeds, and a small percentage of molecules have speeds several times v_{rms}. Bit by bit, a gas can slowly leak out of the atmosphere as its fastest molecules escape. A reasonable rule of thumb is that the earth's gravity can contain a gas only if the average translational kinetic energy per molecule is less than 1% of the kinetic energy needed to escape. Use this rule to show why the earth's atmosphere contains nitrogen but not hydrogen, even though hydrogen is the most abundant element in the universe.

60. ‖ n_1 moles of a monatomic gas and n_2 moles of a diatomic gas are mixed together in a container.
   a. Derive an expression for the molar specific heat at constant volume of the mixture.
   b. Show that your expression has the expected behavior if $n_1 \rightarrow 0$ or $n_2 \rightarrow 0$.

61. ‖ A 1.0 kg ball is at rest on the floor in a 2.0 m × 2.0 m × 2.0 m room of air at STP. Air is 80% nitrogen (N_2) and 20% oxygen (O_2) by volume.
   a. What is the thermal energy of the air in the room?
   b. What fraction of the thermal energy would have to be conveyed to the ball for it to be spontaneously launched to a height of 1.0 m?
   c. By how much would the air temperature have to decrease to launch the ball?
   d. Your answer to part c is so small as to be unnoticeable, yet this event never happens. Why not?

62. ‖ An inventor wants you to invest money with his company, offering you 10% of all future profits. He reminds you that the brakes on cars get extremely hot when they stop and that there is a large quantity of thermal energy in the brakes. He has invented a device, he tells you, that converts that thermal energy into the forward motion of the car. This device will take over from the engine after a stop and accelerate the car back up to its original speed, thereby saving a tremendous amount of gasoline. Now, you're a smart person, so he admits up front that this device is not 100% efficient, that there is some unavoidable heat loss to the air and to friction within the device, but the upcoming research for which he needs your investment will make those losses extremely small. You do also have to start the car with cold brakes after it has been parked awhile, so you'll still need a

gasoline engine for that. Nonetheless, he tells you, his prototype car gets 500 miles to the gallon and he expects to be at well over 1000 miles to the gallon after the next phase of research. Should you invest? Base your answer on an analysis of the *physics* of the situation.

Challenge Problems

63. n moles of a diatomic gas with $C_V = \frac{5}{2}R$ has initial pressure p_i and volume V_i. The gas undergoes a process in which the pressure is directly proportional to the volume until the rms speed of the molecules has doubled.
   a. Show this process on a pV diagram.
   b. How much heat does this process require? Give your answer in terms of n, p_i, and V_i.

64. An experiment you're designing needs a gas with $\gamma = 1.50$. You recall from your physics class that no individual gas has this value, but it occurs to you that you could produce a gas with $\gamma = 1.50$ by mixing together a monatomic gas and a diatomic gas. What fraction of the molecules need to be monatomic?

65. Consider a container like that shown in Figure 18.14, with n_1 moles of a monatomic gas on one side and n_2 moles of a diatomic gas on the other. The monatomic gas has initial temperature T_{1i}. The diatomic gas has initial temperature T_{2i}.
   a. Show that the equilibrium thermal energies are

   $$E_{1f} = \frac{3n_1}{3n_1 + 5n_2}(E_{1i} + E_{2i})$$

   $$E_{2f} = \frac{5n_2}{3n_1 + 5n_2}(E_{1i} + E_{2i})$$

   b. Show that the equilibrium temperature is

   $$T_f = \frac{3n_1 T_{1i} + 5n_2 T_{2i}}{3n_1 + 5n_2}$$

   c. 2.0 g of helium at an initial temperature of 300 K interacts thermally with 8.0 g of oxygen at an initial temperature of 600 K. What is the final temperature? How much heat energy is transferred, and in which direction?

STOP TO THINK ANSWERS

Stop to Think 18.1: $\lambda_B > \lambda_A = \lambda_C > \lambda_D$. Increasing the volume makes the gas less dense, so λ increases. Increasing the radius makes the targets larger, so λ decreases. The mean free path doesn't depend on the atomic mass.

Stop to Think 18.2: c. Each v^2 increases by a factor of 16 but, after averaging, v_{rms} takes the square root.

Stop to Think 18.3: e. Temperature is proportional to the average energy. The energy of a gas molecule is kinetic, proportional to v^2. The average energy, and thus T, increases by 4^2.

Stop to Think 18.4: b. The bead can slide along the wire (one degree of translational motion) and rotate around the wire (one degree of rotational motion).

Stop to Think 18.5: a. Temperature measures the average translational kinetic energy *per molecule,* not the thermal energy of the entire system.

Stop to Think 18.6: c. With 1,000,000 molecules, it's highly unlikely that 750,000 of them would spontaneously move into one side of the box. A state with a very small probability of occurrence has a very low entropy. Having an imbalance of only 100 out of 1,000,000 is well within what you might expect for random fluctuations. This is a highly probable situation and thus one of large entropy.

19 Heat Engines and Refrigerators

This power plant is generating electricity by turning heat into work—but not very efficiently. The cooling towers dissipate roughly two-thirds of the fuel's energy into the air as "waste heat."

▶ **Looking Ahead** The goal of Chapter 19 is to study the physical principles that govern heat engines and refrigerators.

Turning Heat into Work

Modern society is powered by devices that transform the heat energy of burning fuel into useful work, such as

- Pumping water.
- Propelling cars and airplanes.
- Generating electricity.

Our goal in this chapter is not to study specific devices but to look at the underlying physics that governs all such devices

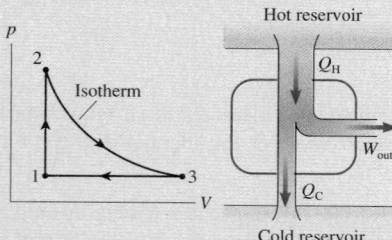

You will learn that heat engines

- Follow a cyclical process that can be shown on pV diagrams and on *energy-transfer diagrams*.
- Require not only a source of heat but also a source of cooling. These are called the *hot reservoir* and the *cold reservoir*.
- Are governed by the first and second laws of thermodynamics.

◀ **Looking Back**
Sections 17.2–17.4 Work, heat, and the first law of thermodynamics

Heat Engines

Heat engine is the generic name for any device that uses a cyclical process to transform heat energy into work.

The heat from burning fuel boils water to make high-pressure steam that then does work by spinning this turbine at an electric generating station.

Efficiency

How good is a heat engine at transforming heat into work? We'll define an engine's **thermal efficiency** as

$$\text{efficiency} = \frac{\text{work done}}{\text{heat required}}$$

You'll learn that the laws of thermodynamics set limits on the maximum possible efficiency. The fact that no heat engine can have an efficiency of 100% prevents us from extracting and using the vast thermal energy in the air and water around us.

Refrigerators

A **refrigerator** is a heat engine in reverse, using work to "pump energy uphill" from cold to hot.

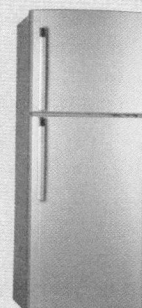

In a refrigerator, a compressor does work to pump heat energy from the colder inside to the warmer room. Air conditioners are "refrigerators" pumping heat energy from the cool inside of a house to the hot outside.

The Carnot Engine

We'll use the second law of thermodynamics to show that a *perfectly reversible heat engine*—called a **Carnot engine**—has the maximum possible thermal efficiency.

You'll learn that the efficiency of a Carnot engine depends only on the temperatures of the hot and cold reservoirs. Any real engine's efficiency will be less—often much less.

◀ **Looking Back**
Section 18.6 The second law of thermodynamics

19.1 Turning Heat into Work

Thermodynamics is the branch of physics that studies the transformation of energy. Many practical devices are designed to transform energy from one form, such as the heat from burning fuel, into another, such as work. Chapters 17 and 18 established two laws of thermodynamics that any such device must obey:

First law Energy is conserved; that is, $\Delta E_{th} = W + Q$.

Second law Most macroscopic processes are irreversible. In particular, heat energy is transferred spontaneously from a hotter to a colder system but never from a colder system to a hotter system.

Our goal in this chapter is to discover what these two laws, especially the second law, imply about devices that turn heat into work. In particular:

- How does a practical device transform heat into work?
- What are the limitations and restrictions on these energy transformations?

Work Done by the System

The work W in the first law is the work done *on* the system by external forces from the environment. However, it makes more sense in "practical thermodynamics" to use the work done *by* the system. For example, you want to know how much useful work you can obtain from an expanding gas. The work done by the system is called W_s.

Work done by the environment and work done by the system are not mutually exclusive. In fact, they are very simply related by $W_s = -W$. In FIGURE 19.1, force $\vec{F}_{gas}$ due to the gas pressure does work when the piston moves. This is W_s, the work done *by* the system. At the same time, some object in the environment, such as a piston rod, must be pushing inward with force $\vec{F}_{ext} = -\vec{F}_{gas}$ to keep the gas pressure from blowing the piston out. This force does the work W *on* the system, work that you've learned is the negative of the area under the pV curve of the process.

Because the forces are equal but opposite, we see that

$$W_s = -W = \text{the area under the } pV \text{ curve} \qquad (19.1)$$

When a gas expands and pushes the piston out, transferring energy out of the system, we say "the system does work on the environment." While this may seem to imply that the environment is doing no work on the system, all the phrase means is that W_s is positive and W is negative.

Similarly, "the environment does work on the system" means that $W > 0$ (energy is transferred into the system) and thus $W_s < 0$. Whether we use W or W_s is a matter of convenience. They are always opposite to each other rather than one being zero.

The first law of thermodynamics $\Delta E_{th} = W + Q$ can be written in terms of W_s as

$$Q = W_s + \Delta E_{th} \qquad \text{(first law of thermodynamics)} \qquad (19.2)$$

Any energy transferred into a system as heat is either used to do work or stored within the system as an increased thermal energy.

Energy-Transfer Diagrams

Suppose you drop a hot rock into the ocean. Heat is transferred from the rock to the ocean until the rock and ocean are at the same temperature. Although the ocean warms up ever so slightly, ΔT_{ocean} is so small as to be completely insignificant. For all practical purposes, the ocean is infinite and unchangeable.

An **energy reservoir** is an object or a part of the environment so large that its temperature does not change when heat is transferred between the system and the reservoir. A reservoir at a higher temperature than the system is called a *hot reservoir*.

FIGURE 19.1 Forces $\vec{F}_{gas}$ and $\vec{F}_{ext}$ both do work as the piston moves.

(a)

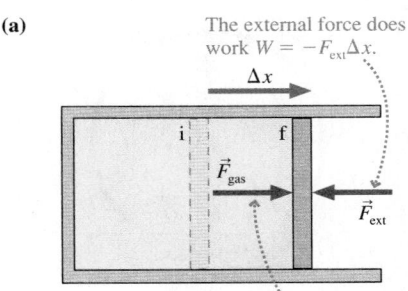

The external force does work $W = -F_{ext}\Delta x$.

The system does work $W_s = F_{gas}\Delta x$.
$W_s = -W$ because $\vec{F}_{gas} = -\vec{F}_{ext}$.

(b)

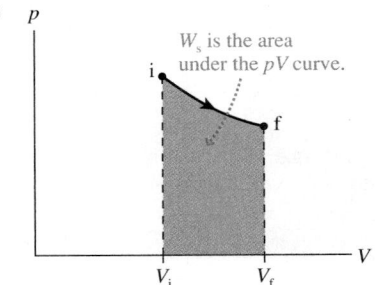

W_s is the area under the pV curve.

FIGURE 19.2 Energy-transfer diagrams.

(a)

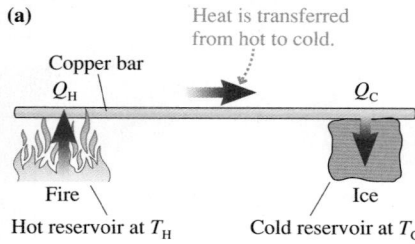

Heat is transferred from hot to cold.

Copper bar

Q_H

Q_C

Fire

Ice

Hot reservoir at T_H

Cold reservoir at T_C

(b) Heat energy is transferred from a hot reservoir to a cold reservoir. Energy conservation requires $Q_C = Q_H$.

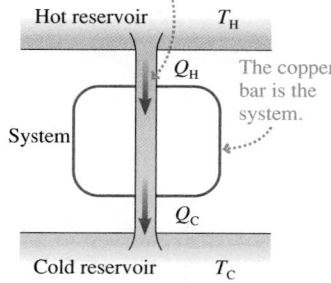

Hot reservoir $\quad T_H$

Q_H

The copper bar is the system.

System

Q_C

Cold reservoir $\quad T_C$

(c) The second law forbids a process in which heat is spontaneously transferred from a colder object to a hotter object.

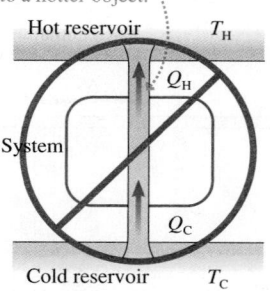

Hot reservoir $\quad T_H$

Q_H

System

Q_C

Cold reservoir $\quad T_C$

FIGURE 19.3 Work can be transformed into heat with 100% efficiency.

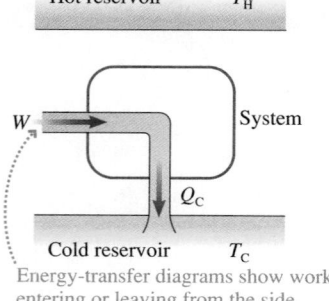

Hot reservoir $\quad T_H$

W_s

System

Q_C

Cold reservoir $\quad T_C$

Energy-transfer diagrams show work entering or leaving from the side.

A vigorously burning flame is a hot reservoir for small objects placed in the flame. A reservoir at a lower temperature than the system is called a *cold reservoir*. The ocean is a cold reservoir for the hot rock. We will use T_H and T_C to designate the temperatures of the hot and cold reservoirs.

Hot and cold reservoirs are idealizations, in the same category as frictionless surfaces and massless strings. No real object can maintain a perfectly constant temperature as heat is transferred in or out. Even so, an object can be modeled as a reservoir if it is much larger than the system that thermally interacts with it.

Heat energy is transferred between a system and a reservoir if they have different temperatures. We will define

Q_H = amount of heat transferred to or from a hot reservoir

Q_C = amount of heat transferred to or from a cold reservoir

By definition, Q_H and Q_C are *positive* quantities. The direction of heat transfer, which determines the sign of Q in the first law, will always be clear as we deal with thermodynamic devices.

FIGURE 19.2a shows a heavy copper bar between a hot reservoir (at temperature T_H) and a cold reservoir (at temperature T_C). Heat Q_H is transferred from the hot reservoir into the copper and heat Q_C is transferred from the copper to the cold reservoir. FIGURE 19.2b is an **energy-transfer diagram** for this process. The hot reservoir is always drawn at the top, the cold reservoir at the bottom, and the system—the copper bar in this case—between them. Figure 19.2b shows heat Q_H being transferred into the system and Q_C being transferred out.

The first law of thermodynamics $Q = W_s + \Delta E_{th}$ refers to the *system*. Q is the *net* heat to the system, which, in this case, is $Q = Q_H - Q_C$. The copper bar does no work, so $W_s = 0$. The bar warms up when first placed between the two reservoirs, but it soon comes to a steady state where its temperature no longer changes. Then $\Delta E_{th} = 0$. Thus the first law tells us that $Q = Q_H - Q_C = 0$, from which we conclude that $Q_C = Q_H$.

In other words, all of the heat transferred into the hot end of the rod is subsequently transferred out of the cold end. This isn't surprising. After all, we know that heat is transferred spontaneously from a hotter object to a colder object. Even so, there has to be some *means* by which the heat energy gets from the hotter object to the colder. The copper bar provides a route for the energy transfer, and $Q_C = Q_H$ is the statement that energy is conserved as it moves through the bar.

Contrast Figure 19.2b with FIGURE 19.2c. Figure 19.2c shows a system in which heat is being transferred from the cold reservoir to the hot reservoir. The first law of thermodynamics is not violated, because $Q_H = Q_C$, but the second law is. If there were such a system, it would allow the spontaneous (i.e., with no outside input or assistance) transfer of heat from a colder object to a hotter object. The process of Figure 19.2c is forbidden by the second law of thermodynamics.

Work into Heat and Heat into Work

Turning work into heat is easy—just rub two objects together. Work from the friction force increases the objects' thermal energy and their temperature. Heat energy is then transferred from the warmer objects to the cooler environment. FIGURE 19.3 is the energy-transfer diagram for this process. The conversion of work into heat is 100% efficient in that *all* the energy supplied to the system as work is ultimately transferred to the environment as heat. Notice that the objects have returned to their initial state at the end of this process, ready to repeat the process for as long as there's a source of motion.

The reverse—transforming heat into work—isn't so easy. Heat can be transformed into work in a one-time process, such as an isothermal expansion of a gas, but at the end the system is not restored to its initial state. To be practical, **a device that transforms heat into work must return to its initial state at the end of the process and be ready for continued use.** You want your car engine to turn over and over for as long as there's fuel.

Interestingly, no one has ever invented a "perfect engine" that transforms heat into work with 100% efficiency *and returns to its initial state* so that it can continue to do work as long as there is fuel. Of course, that such a device has not been invented is not a proof that it can't be done. We'll provide a proof shortly, but for now we'll make the hypothesis that the process of FIGURE 19.4 is somehow forbidden.

Notice the asymmetry between Figures 19.3 and 19.4. The perfect transformation of work into heat is permitted, but the perfect transformation of heat into work is forbidden. This asymmetry parallels the asymmetry of the two processes in Figure 19.2. In fact, we'll soon see that the "perfect engine" of Figure 19.4 is forbidden for exactly the same reason: the second law of thermodynamics.

FIGURE 19.4 There are no perfect engines that turn heat into work with 100% efficiency.

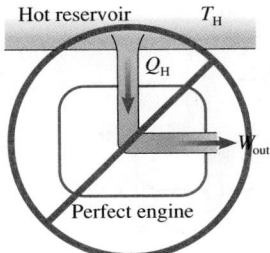

19.2 Heat Engines and Refrigerators

The steam generator at your local electric power plant works by boiling water to produce high-pressure steam that spins a turbine (which then spins a generator to produce electricity). That is, the steam pressure is doing work. The steam is then condensed to liquid water and pumped back to the boiler to start the process again. There are two crucial ideas here. First, the device works in a cycle, with the water returning to its initial conditions once a cycle. Second, heat is transferred to the water in the boiler, but heat is transferred *out* of the water in the condenser.

Car engines and steam generators are examples of what we call *heat engines*. A **heat engine** is any closed-cycle device that extracts heat Q_H from a hot reservoir, does useful work, and exhausts heat Q_C to a cold reservoir. A **closed-cycle device** is one that periodically *returns to its initial conditions*, repeating the same process over and over. That is, all state variables (pressure, temperature, thermal energy, and so on) return to their initial values once every cycle. Consequently, a heat engine can continue to do useful work for as long as it is attached to the reservoirs.

FIGURE 19.5 is the energy-transfer diagram of a heat engine. Unlike the forbidden "perfect engine" of Figure 19.4, a heat engine is connected to both a hot reservoir *and* a cold reservoir. You can think of a heat engine as "siphoning off" some of the heat that moves from the hot reservoir to the cold reservoir and transforming that heat into work—*some* of the heat, but not all.

Because the temperature and thermal energy of a heat engine return to their initial values at the end of each cycle, there is no *net* change in E_{th}:

$$(\Delta E_{th})_{net} = 0 \quad \text{(any heat engine, over one full cycle)} \quad (19.3)$$

Consequently, the first law of thermodynamics *for a full cycle* of a heat engine is $(\Delta E_{th})_{net} = Q - W_s = 0$.

Let's define W_{out} to be the useful work done by the heat engine *per cycle*. The first law applied to a heat engine is

$$W_{out} = Q_{net} = Q_H - Q_C \quad \text{(work per cycle done by a heat engine)} \quad (19.4)$$

This is just energy conservation. The energy-transfer diagram of Figure 19.5 is a pictorial representation of Equation 19.4.

NOTE ▶ Equations 19.3 and 19.4 apply only to a *full cycle* of the heat engine. They are *not* valid for any of the individual processes that make up a cycle. ◀

For practical reasons, we would like an engine to do the maximum amount of work with the minimum amount of fuel. We can measure the performance of a heat engine in terms of its **thermal efficiency** η (lowercase Greek eta), defined as

$$\eta = \frac{W_{out}}{Q_H} = \frac{\text{what you get}}{\text{what you had to pay}} \quad (19.5)$$

The steam turbine in a modern power plant is an enormous device. Expanding steam does work by spinning the turbine.

FIGURE 19.5 The energy-transfer diagram of a heat engine.

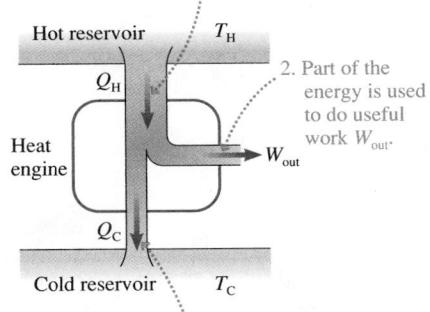

1. Heat energy Q_H is transferred from the hot reservoir (typically burning fuel) to the system.

2. Part of the energy is used to do useful work W_{out}.

3. The remaining energy $Q_C = Q_H - W_{out}$ is exhausted to the cold reservoir (cooling water or the air) as waste heat.

FIGURE 19.6 η is the fraction of heat energy that is transformed into useful work.

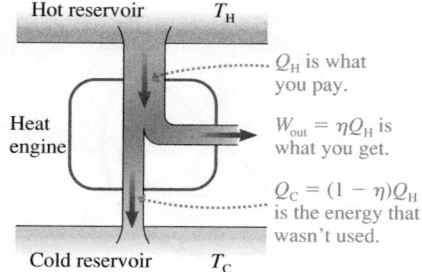

Using Equation 19.4 for W_{out}, we can also write the thermal efficiency as

$$\eta = 1 - \frac{Q_C}{Q_H} \qquad (19.6)$$

FIGURE 19.6 illustrates the idea of thermal efficiency.

A *perfect* heat engine would have $\eta_{perfect} = 1$. That is, it would be 100% efficient at converting heat from the hot reservoir (the burning fuel) into work. You can see from Equation 19.6 that a perfect engine would have no exhaust ($Q_C = 0$) and would not need a cold reservoir. Figure 19.4 has already suggested that there are no perfect heat engines, that an engine with $\eta = 1$ is impossible. A heat engine *must* exhaust **waste heat** to a cold reservoir. It is energy that was extracted from the hot reservoir but *not* transformed to useful work.

Practical heat engines, such as car engines and steam generators, have thermal efficiencies in the range $\eta \approx 0.1 - 0.5$. This is not large. Can a clever designer do better, or is this some kind of physical limitation?

STOP TO THINK 19.1 Rank in order, from largest to smallest, the work W_{out} performed by these four heat engines.

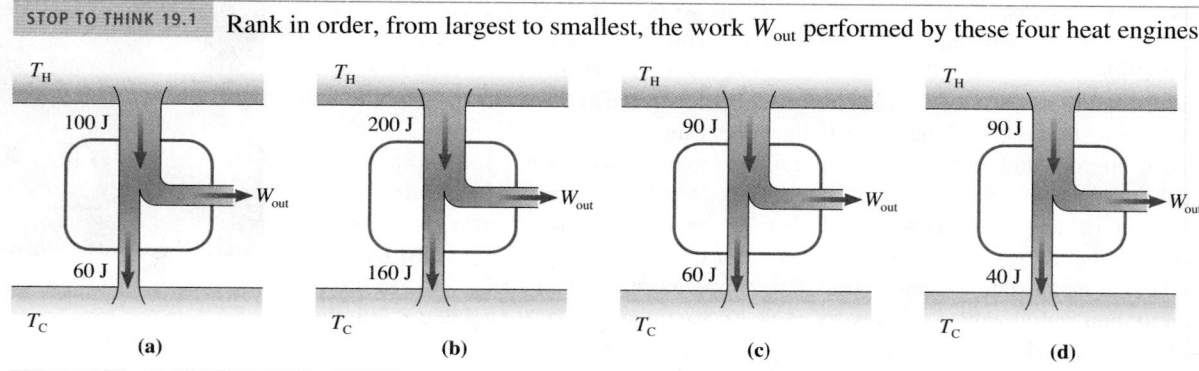

A Heat-Engine Example

To illustrate how these ideas actually work, **FIGURE 19.7** shows a simple engine that converts heat into the work of lifting mass M.

FIGURE 19.7 A simple heat engine transforms heat into work.

(a) Heat is transferred into the gas from the burning fuel.

(b) The gas does work by lifting the mass in an isobaric expansion.

(c) The piston is locked and the mass is removed. The heat is turned off.

(d) The gas cools back to room temperature at constant volume. Then the piston is unlocked.

(e) A steadily increasing external force steadily raises the pressure in an isothermal compression until the pressure has been restored to its initial value.

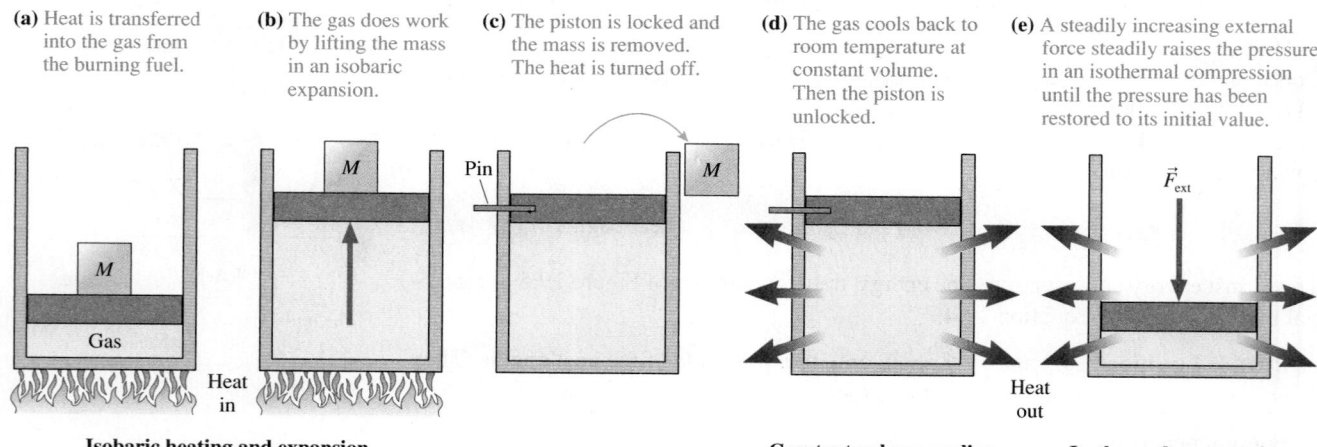

Isobaric heating and expansion   **Constant-volume cooling**   **Isothermal compression**

The net effect of this multistep process is to convert some of the fuel's energy into the useful work of lifting the mass. There has been no net change in the gas, which has returned to its initial pressure, volume, and temperature at the end of step (e). We can start the whole process over again and continue lifting masses (doing work) as long as we have fuel.

FIGURE 19.8 shows the heat-engine process on a *pV* diagram. It is a *closed cycle* because the gas returns to its initial conditions. No work is done during the isochoric process, and, as you can see from the areas under the curve, the work done *by* the gas to lift the mass is greater than the work the environment must do *on* the gas to recompress it. Thus this heat engine, by burning fuel, does *net* work per cycle:

$$W_{net} = W_{lift} - W_{ext} = (W_s)_{1 \to 2} + (W_s)_{3 \to 1}.$$

Notice that the cyclical process of Figure 19.8 involves two cooling processes in which heat is transferred *from* the gas to the environment. Heat energy is transferred from hotter objects to colder objects, so the system *must* be connected to a cold reservoir with $T_C < T_{gas}$ during these two processes. A key to understanding heat engines is that they require both a heat source (burning fuel) *and* a heat sink (cooling water, the air, or something at a lower temperature than the system).

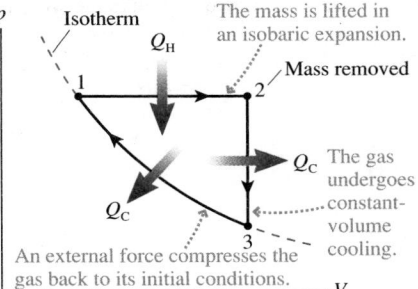

FIGURE 19.8 The closed-cycle *pV* diagram for the heat engine of Figure 19.7.

The mass is lifted in an isobaric expansion.
Mass removed
The gas undergoes constant-volume cooling.
An external force compresses the gas back to its initial conditions.

EXAMPLE 19.1 **Analyzing a heat engine I**

Analyze the heat engine of FIGURE 19.9 to determine (a) the net work done per cycle, (b) the engine's thermal efficiency, and (c) the engine's power output if it runs at 600 rpm. Assume the gas is monatomic.

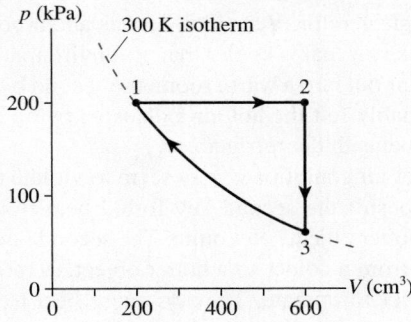

FIGURE 19.9 The heat engine of Example 19.1.

MODEL The gas follows a closed cycle consisting of three distinct processes, each of which was studied in Chapters 16 and 17. For each of the three we need to determine the work done and the heat transferred.

SOLVE To begin, we can use the initial conditions at state 1 and the ideal-gas law to determine the number of moles of gas:

$$n = \frac{p_1 V_1}{R T_1} = \frac{(200 \times 10^3 \text{ Pa})(2.0 \times 10^{-4} \text{ m}^3)}{(8.31 \text{ J/mol K})(300 \text{ K})} = 0.0160 \text{ mol}$$

Process $1 \to 2$: The work done *by* the gas in the isobaric expansion is

$$(W_s)_{12} = p\Delta V = (200 \times 10^3 \text{ Pa})\big((6.0 - 2.0) \times 10^{-4} \text{ m}^3\big) = 80 \text{ J}$$

We can use the ideal-gas law at constant pressure to find $T_2 = (V_2/V_1)T_1 = 3T_1 = 900$ K. The heat transfer during a constant-pressure process is

$$Q_{12} = nC_P \Delta T$$
$$= (0.0160 \text{ mol})(20.8 \text{ J/mol K})(900 \text{ K} - 300 \text{ K}) = 200 \text{ J}$$

where we used $C_P = \frac{5}{2}R$ for a monatomic ideal gas.

Process $2 \to 3$: No work is done in an isochoric process, so $(W_s)_{23} = 0$. The temperature drops back to 300 K, so the heat transfer, with $C_V = \frac{3}{2}R$, is

$$Q_{23} = nC_V \Delta T$$
$$= (0.0160 \text{ mol})(12.5 \text{ J/mol K})(300 \text{ K} - 900 \text{ K}) = -120 \text{ J}$$

Process $3 \to 1$: The gas returns to its initial state with volume V_1. The work done *by* the gas during an isothermal process is

$$(W_s)_{31} = nRT \ln\left(\frac{V_1}{V_3}\right)$$
$$= (0.0160 \text{ mol})(8.31 \text{ J/mol K})(300 \text{ K}) \ln\left(\frac{1}{3}\right) = -44 \text{ J}$$

W_s is negative because the environment does work on the gas to compress it. An isothermal process has $\Delta E_{th} = 0$ and hence, from the first law,

$$Q_{31} = (W_s)_{31} = -44 \text{ J}$$

Q is negative because the gas must be cooled as it is compressed to keep the temperature constant.

a. The *net* work done by the engine during one cycle is

$$W_{out} = (W_s)_{12} + (W_s)_{23} + (W_s)_{31} = 36 \text{ J}$$

As a consistency check, notice that the net heat transfer is

$$Q_{net} = Q_{12} + Q_{23} + Q_{31} = 36 \text{ J}$$

Equation 19.4 told us that a heat engine *must* have $W_{out} = Q_{net}$, and we see that it does.

b. The efficiency depends not on the net heat transfer but on the heat Q_H transferred into the engine from the flame. Heat enters during process $1 \to 2$, where Q is positive, and exits during processes $2 \to 3$ and $3 \to 1$, where Q is negative. Thus

$$Q_H = Q_{12} = 200 \text{ J}$$
$$Q_C = |Q_{23}| + |Q_{31}| = 164 \text{ J}$$

Notice that $Q_H - Q_C = 36$ J $= W_{out}$. In this heat engine, 200 J of heat from the hot reservoir does 36 J of useful work. Thus the thermal efficiency is

$$\eta = \frac{W_{out}}{Q_H} = \frac{36 \text{ J}}{200 \text{ J}} = 0.18 \text{ or } 18\%$$

This heat engine is far from being a perfect engine!

Continued

c. An engine running at 600 rpm goes through 10 cycles per second. The power output is the work done *per second*:

$$P_{out} = (\text{work per cycle}) \times (\text{cycles per second})$$
$$= 360 \text{ J/s} = 360 \text{ W}$$

ASSESS Although we didn't need Q_{net}, verifying that $Q_{net} = W_{out}$ was a check of self-consistency. Heat-engine analysis requires many calculations and offers many opportunities to get signs wrong. However, there are a sufficient number of self-consistency checks so that you can almost always spot calculational errors *if you check for them*.

Let's think about this example a bit more before going on. We've said that a heat engine operates between a hot reservoir and a cold reservoir. Figure 19.9 doesn't explicitly show the reservoirs. Nonetheless, we know that heat is transferred from a hotter object to a colder object. Heat Q_H is transferred into the system during process $1 \rightarrow 2$ as the gas warms from 300 K to 900 K. For this to be true, the hot-reservoir temperature T_H must be ≥ 900 K. Likewise, heat Q_C is transferred from the system to the cold reservoir as the temperature drops from 900 K to 300 K in process $2 \rightarrow 3$. For this to be true, the cold-reservoir temperature T_C must be ≤ 300 K.

So, while we really don't know what the reservoirs are or their exact temperatures, we can say with certainty that the hot-reservoir temperature T_H must exceed the highest temperature reached by the system and the cold-reservoir temperature T_C must be less than the coldest system temperature.

Refrigerators

Your house or apartment has a refrigerator. Very likely it has an air conditioner. The purpose of these devices is to make air that is cooler than its environment even colder. The first does so by blowing hot air out into a warm room, the second by blowing it out to the hot outdoors. You've probably felt the hot air exhausted by an air conditioner compressor or coming out from beneath the refrigerator.

At first glance, a refrigerator or air conditioner may seem to violate the second law of thermodynamics. After all, doesn't the second law forbid heat from being transferred from a colder object to a hotter object? Not quite: The second law says that heat is not *spontaneously* transferred from a colder to a hotter object. A refrigerator or air conditioner requires electric power to operate. They do cause heat to be transferred from cold to hot, but the transfer is "assisted" rather than spontaneous.

A **refrigerator** is any closed-cycle device that uses external work W_{in} to remove heat Q_C from a cold reservoir and exhaust heat Q_H to a hot reservoir. FIGURE 19.10 is the energy-transfer diagram of a refrigerator. The cold reservoir is the air inside the refrigerator or the air inside your house on a summer day. To keep the air cold, in the face of inevitable "heat leaks," the refrigerator or air conditioner compressor continuously removes heat from the cold reservoir and exhausts heat into the room or outdoors. You can think of a refrigerator as "pumping heat uphill," much as a water pump lifts water uphill.

Because a refrigerator, like a heat engine, is a cyclical device, $\Delta E_{th} = 0$. Conservation of energy requires

$$Q_H = Q_C + W_{in} \tag{19.7}$$

To move energy from a colder to a hotter reservoir, a refrigerator must exhaust *more* heat to the outside than it removes from the inside. This has significant implications for whether or not you can cool a room by leaving the refrigerator door open.

The thermal efficiency of a heat engine was defined as "what you get (useful work W_{out})" versus "what you had to pay (fuel to supply Q_H)." By analogy, we define the **coefficient of performance** K of a refrigerator to be

$$K = \frac{Q_C}{W_{in}} = \frac{\text{what you get}}{\text{what you had to pay}} \tag{19.8}$$

What you get, in this case, is the removal of heat from the cold reservoir. But you have to pay the electric company for the work needed to run the refrigerator. A better

This air conditioner transfers heat energy *from* the cool indoors *to* the hot exterior.

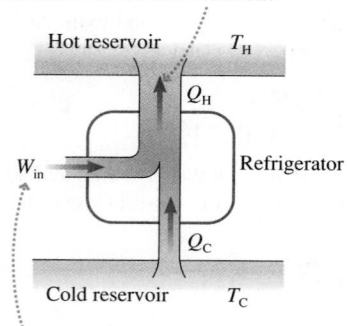
FIGURE 19.10 The energy-transfer diagram of a refrigerator.

The amount of heat exhausted to the hot reservoir is larger than the amount of heat extracted from the cold reservoir.

External work is used to remove heat from a cold reservoir and exhaust heat to a hot reservoir.

refrigerator will require less work to remove a given amount of heat, thus having a larger coefficient of performance.

A perfect refrigerator would require no work ($W_{in} = 0$) and would have $K_{perfect} = \infty$. But if Figure 19.10 had no work input, it would look like Figure 19.2c. That device was forbidden by the second law of thermodynamics because, with no work input, heat would move *spontaneously* from cold to hot.

We noted in Chapter 18 that the second law of thermodynamics can be stated several different but equivalent ways. We can now give a third statement:

> **Second law, informal statement #3** There are no perfect refrigerators with coefficient of performance $K = \infty$.

Any real refrigerator or air conditioner *must* use work to move energy from the cold reservoir to the hot reservoir, hence $K < \infty$.

No Perfect Heat Engines

We hypothesized above that there are no perfect heat engines—that is, no heat engines like the one shown in Figure 19.4 with $Q_C = 0$ and $\eta = 1$. Now we're ready to prove this hypothesis. **FIGURE 19.11** shows a hot reservoir at temperature T_H and a cold reservoir at temperature T_C. An ordinary refrigerator, one that obeys all the laws of physics, is operating between these two reservoirs.

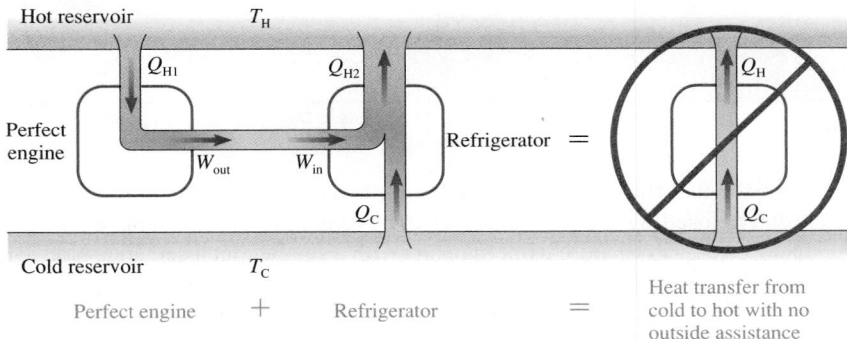

FIGURE 19.11 A perfect engine driving an ordinary refrigerator would be able to violate the second law of thermodynamics.

Suppose we had a perfect heat engine, one that takes in heat Q_H from the high-temperature reservoir and transforms that energy entirely into work W_{out}. If we had such a heat engine, we could use its output to provide the work input to the refrigerator. The two devices combined have no connection to the external world. That is, there's no net input or net output of work.

If we built a box around the heat engine and refrigerator, so that you couldn't see what was inside, the only thing you would observe is heat being transferred *with no outside assistance* from the cold reservoir to the hot reservoir. But a spontaneous or unassisted transfer of heat from a colder to a hotter object is exactly what the second law of thermodynamics forbids. Consequently, our assumption of a perfect heat engine must be wrong. Hence another statement of the second law of thermodynamics is:

> **Second law, informal statement #4** There are no perfect heat engines with efficiency $\eta = 1$.

Any real heat engine *must* exhaust waste heat Q_C to a cold reservoir.

STOP TO THINK 19.2 It's a hot day and your air conditioner is broken. Your roommate says, "Let's open the refrigerator door and cool this place off." Will this work?

a. Yes.   b. No.   c. It might, but it will depend on how hot the room is.

19.3 Ideal-Gas Heat Engines

We will focus on heat engines that use a gas as the *working substance*. The gasoline or diesel engine in your car is an engine that alternately compresses and expands a gaseous fuel-air mixture. A discussion of engines such as steam generators that rely on phase changes will be deferred to more advanced courses.

A gas heat engine can be represented by a closed-cycle trajectory in the pV diagram, such as the one shown in FIGURE 19.12a. This observation leads to an important geometric interpretation of the work done by the system during one full cycle. You learned in Section 19.1 that the work done *by* the system is the area under the curve of a pV trajectory. As FIGURE 19.12b shows, the net work done during a full cycle is

$$W_{out} = W_{expand} - |W_{compress}| = \text{area } inside \text{ the closed curve} \qquad (19.9)$$

FIGURE 19.12 The work W_{out} done by the system during one full cycle is the area enclosed within the curve.

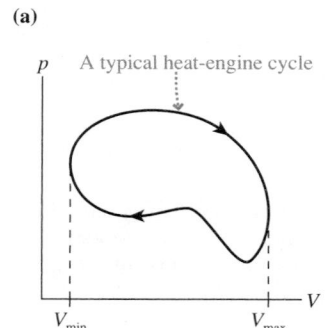

(a)

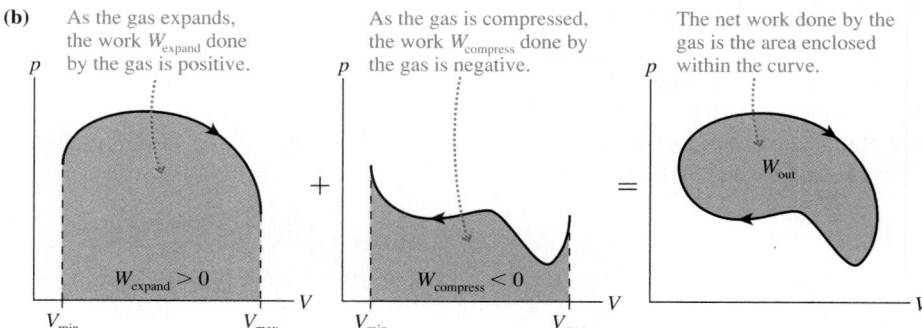

(b)

You can see that **the net work done by a gas heat engine during one full cycle is the area enclosed by the pV curve for the cycle.** A thermodynamic cycle with a larger enclosed area does more work than one with a smaller enclosed area. Notice that the gas must go around the pV trajectory in a *clockwise* direction for W_{out} to be positive. We'll see later that a refrigerator uses a counterclockwise (ccw) cycle.

Ideal-Gas Summary

We've learned a lot about ideal gases in the last three chapters. All gas processes obey the ideal-gas law $pV = nRT$ and the first law of thermodynamics $\Delta E_{th} = Q - W_s$. Table 19.1 summarizes the results for specific gas processes. This table shows W_s, the work done *by* the system, so the signs are opposite those in Chapter 17.

TABLE 19.1 Summary of ideal-gas processes

Process	Gas law	Work W_s	Heat Q	Thermal energy
Isochoric	$p_i/T_i = p_f/T_f$	0	$nC_V \Delta T$	$\Delta E_{th} = Q$
Isobaric	$V_i/T_i = V_f/T_f$	$p \Delta V$	$nC_P \Delta T$	$\Delta E_{th} = Q - W_s$
Isothermal	$p_i V_i = p_f V_f$	$nRT \ln(V_f/V_i)$ $pV \ln(V_f/V_i)$	$Q = W_s$	$\Delta E_{th} = 0$
Adiabatic	$p_i V_i^\gamma = p_f V_f^\gamma$ $T_i V_i^{\gamma-1} = T_f V_f^{\gamma-1}$	$(p_f V_f - p_i V_i)/(1 - \gamma)$ $-nC_V \Delta T$	0	$\Delta E_{th} = -W_s$
Any	$p_i V_i/T_i = p_f V_f/T_f$	area under curve		$\Delta E_{th} = nC_V \Delta T$

There is one entry in this table that you haven't seen before. The expression

$$W_s = \frac{p_f V_f - p_i V_i}{1 - \gamma} \qquad \text{(work in an adiabatic process)} \qquad (19.10)$$

for the work done in an adiabatic process follows from writing $W_s = -\Delta E_{th} = -nC_V \Delta T$, which you learned in Chapter 17, then using $\Delta T = \Delta(pV)/nR$ and the definition of γ. The proof will be left for a homework problem.

You learned in Chapter 18 that the thermal energy of an ideal gas depends only on its temperature. Table 19.2 lists the thermal energy, molar specific heats, and specific heat ratio $\gamma = C_P/C_V$ for monatomic and diatomic gases.

A Strategy for Heat-Engine Problems

The engine of Example 19.1 was not a realistic heat engine, but it did illustrate the kinds of reasoning and computations involved in the analysis of a heat engine.

TABLE 19.2 Properties of monatomic and diatomic gases

	Monatomic	Diatomic
E_{th}	$\frac{3}{2}nRT$	$\frac{5}{2}nRT$
C_V	$\frac{3}{2}R$	$\frac{5}{2}R$
C_P	$\frac{5}{2}R$	$\frac{7}{2}R$
γ	$\frac{5}{3} = 1.67$	$\frac{7}{5} = 1.40$

PROBLEM-SOLVING
STRATEGY 19.1   **Heat-engine problems**   (MP)

MODEL Identify each process in the cycle.

VISUALIZE Draw the pV diagram of the cycle.

SOLVE There are several steps in the mathematical analysis.

- Use the ideal-gas law to complete your knowledge of n, p, V, and T at one point in the cycle.
- Use the ideal-gas law and equations for specific gas processes to determine p, V, and T at the beginning and end of each process.
- Calculate Q, W_s, and ΔE_{th} for each process.
- Find W_{out} by adding W_s for each process in the cycle. If the geometry is simple, you can confirm this value by finding the area enclosed within the pV curve.
- Add just the *positive* values of Q to find Q_H.
- Verify that $(\Delta E_{th})_{net} = 0$. This is a self-consistency check to verify that you haven't made any mistakes.
- Calculate the thermal efficiency η and any other quantities you need to complete the solution.

ASSESS Is $(\Delta E_{th})_{net} = 0$? Do all the signs of W_s and Q make sense? Does η have a reasonable value? Have you answered the question?

EXAMPLE 19.2   **Analyzing a heat engine II**

A heat engine with a diatomic gas as the working substance uses the closed cycle shown in FIGURE 19.13. How much work does this engine do per cycle, and what is its thermal efficiency?

FIGURE 19.13 The pV diagram for the heat engine of Example 19.2.

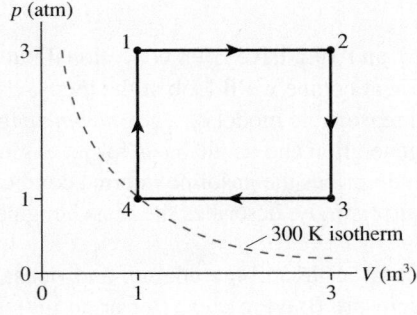

MODEL Processes $1 \to 2$ and $3 \to 4$ are isobaric. Processes $2 \to 3$ and $4 \to 1$ are isochoric.

VISUALIZE The pV diagram has already been drawn.

SOLVE We know the pressure, volume, and temperature at state 4. The number of moles of gas in the heat engine is

$$n = \frac{p_4 V_4}{RT_4} = \frac{(101,300 \text{ Pa})(1.0 \text{ m}^3)}{(8.31 \text{ J/mol K})(300 \text{ K})} = 40.6 \text{ mol}$$

$p/T =$ constant during an isochoric process and $V/T =$ constant during an isobaric process. These allow us to find that $T_1 = T_3 = 900$ K and $T_2 = 2700$ K. This completes our knowledge of the state variables at all four corners of the diagram.

Process $1 \to 2$ is an isobaric expansion, so

$$(W_s)_{12} = p\Delta V = (3.0 \times 101,300 \text{ Pa})(2.0 \text{ m}^3) = 6.08 \times 10^5 \text{ J}$$

Continued

where we converted the pressure to pascals. The heat transfer during an isobaric expansion is

$$Q_{12} = nC_P\Delta T = (40.6 \text{ mol})(29.1 \text{ J/mol K})(1800 \text{ K})$$

$$= 21.27 \times 10^5 \text{ J}$$

where $C_P = \frac{7}{2}R$ for a diatomic gas. Then, using the first law,

$$\Delta E_{12} = Q_{12} - (W_s)_{12} = 15.19 \times 10^5 \text{ J}$$

Process $2 \rightarrow 3$ is an isochoric process, so $(W_s)_{23} = 0$ and

$$\Delta E_{23} = Q_{23} = nC_V\Delta T = -15.19 \times 10^5 \text{ J}$$

Notice that ΔT is *negative*.

Process $3 \rightarrow 4$ is an isobaric compression. Now ΔV is negative, so

$$(W_s)_{34} = p\Delta V = -2.03 \times 10^5 \text{ J}$$

and

$$Q_{34} = nC_P\Delta T = -7.09 \times 10^5 \text{ J}$$

Then $\Delta E_{th} = Q_{34} - (W_s)_{34} = -5.06 \times 10^5$ J.

Process $4 \rightarrow 1$ is another constant-volume process, so again $(W_s)_{41} = 0$ and

$$\Delta E_{41} = Q_{41} = nC_V\Delta T = 5.06 \times 10^5 \text{ J}$$

The results of all four processes are shown in Table 19.3. The net results for W_{out}, Q_{net}, and $(\Delta E_{th})_{net}$ are found by summing the columns. As expected, $W_{out} = Q_{net}$ and $(\Delta E_{th})_{net} = 0$.

TABLE 19.3 Energy transfers in Example 19.2. All energies $\times 10^5$ J

Process	W_s	Q	ΔE_{th}
$1 \rightarrow 2$	6.08	21.27	15.19
$2 \rightarrow 3$	0	−15.19	−15.19
$3 \rightarrow 4$	−2.03	−7.09	−5.06
$4 \rightarrow 1$	0	5.06	5.06
Net	4.05	4.05	0

The work done during one cycle is $W_{out} = 4.05 \times 10^5$ J. Heat enters the system from the hot reservoir during processes $1 \rightarrow 2$ and $4 \rightarrow 1$, where Q is positive. Summing these gives $Q_H = 26.33 \times 10^5$ J. Thus the thermal efficiency of this engine is

$$\eta = \frac{W_{out}}{Q_H} = \frac{4.05 \times 10^5 \text{ J}}{26.33 \times 10^5 \text{ J}} = 0.15 = 15\%$$

ASSESS The verification that $W_{out} = Q_{net}$ and $(\Delta E_{th})_{net} = 0$ gives us great confidence that we didn't make any calculational errors. This engine may not seem very efficient, but η is quite typical of many real engines.

We noted in Example 19.1 that a heat engine's hot-reservoir temperature T_H must exceed the highest temperature reached by the system and the cold-reservoir temperature T_C must be less than the coldest system temperature. Although we don't know what the reservoirs are in Example 19.2, we can be sure that $T_H > 2700$ K and $T_C < 300$ K.

STOP TO THINK 19.3 What is the thermal efficiency of this heat engine?

a. 0.10
b. 0.50
c. 0.25
d. 4
e. Can't tell without knowing Q_C

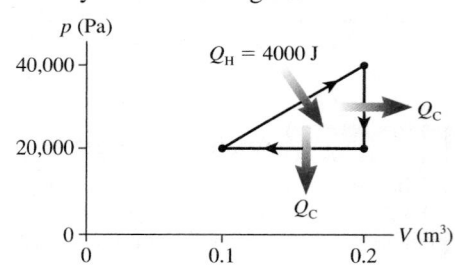

The Brayton Cycle

The heat engines of Examples 19.1 and 19.2 have been educational but not realistic. As an example of a more realistic heat engine we'll look at the thermodynamic cycle known as the *Brayton cycle*. It is a reasonable model of a *gas turbine engine*. Gas turbines are used for electric power generation and as the basis for jet engines in aircraft and rockets. The *Otto cycle*, which describes the gasoline internal combustion engine, and the *Diesel cycle*, which, not surprisingly, describes the diesel engine, will be the subject of homework problems.

FIGURE 19.14a is a schematic look at a gas turbine engine, and FIGURE 19.14b is the corresponding pV diagram. To begin the Brayton cycle, air at an initial pressure p_1 is rapidly compressed in a *compressor*. This is an *adiabatic process*, with $Q = 0$,

A jet engine uses a modified Brayton cycle.

because there is no time for heat to be exchanged with the surroundings. Recall that an adiabatic compression raises the temperature of a gas by doing work on it, not by heating it, so the air leaving the compressor is very hot.

The hot gas flows into a combustion chamber. Fuel is continuously admitted to the combustion chamber where it mixes with the hot gas and is ignited, transferring heat to the gas at constant pressure and raising the gas temperature yet further. The high-pressure gas then expands, spinning a turbine that does some form of useful work. This adiabatic expansion, with $Q = 0$, drops the temperature and pressure of the gas. The pressure at the end of the expansion through the turbine is back to p_1, but the gas is still quite hot. The gas completes the cycle by flowing through a device called a **heat exchanger** that transfers heat energy to a cooling fluid. Large power plants are often sited on rivers or oceans in order to use the water for the cooling fluid in the heat exchanger.

This thermodynamic cycle, called a Brayton cycle, has two adiabatic processes—the compression and the expansion through the turbine—plus a constant-pressure heating and a constant-pressure cooling. There's no heat transfer during the adiabatic processes. The hot-reservoir temperature must be $T_H \geq T_3$ for heat to be transferred into the gas during process $2 \rightarrow 3$. Similarly, the heat exchanger will remove heat from the gas only if $T_C \leq T_1$.

The thermal efficiency of any heat engine is

$$\eta = \frac{W_{out}}{Q_H} = 1 - \frac{Q_C}{Q_H}$$

Heat is transferred into the gas only during process $2 \rightarrow 3$. This is an isobaric process, so $Q_H = nC_P \Delta T = nC_P(T_3 - T_2)$. Similarly, heat is transferred out only during the isobaric process $4 \rightarrow 1$.

We have to be careful with signs. Q_{41} is negative because the temperature decreases, but Q_C was defined as the *amount* of heat exchanged with the cold reservoir, a positive quantity. Thus

$$Q_C = |Q_{41}| = |nC_P(T_1 - T_4)| = nC_P(T_4 - T_1) \tag{19.11}$$

With these expressions for Q_H and Q_C, the thermal efficiency is

$$\eta_{Brayton} = 1 - \frac{T_4 - T_1}{T_3 - T_2} \tag{19.12}$$

This expression isn't useful unless we compute all four temperatures. Fortunately, we can cast Equation 19.12 into a more useful form.

You learned in Chapter 17 that $pV^\gamma = $ constant during an adiabatic process, where $\gamma = C_P/C_V$ is the specific heat ratio. If we use $V = nRT/p$ from the ideal-gas law, $V^\gamma = (nR)^\gamma T^\gamma p^{-\gamma}$. $(nR)^\gamma$ is a constant, so we can write $pV^\gamma = $ constant as

$$p^{(1-\gamma)}T^\gamma = \text{constant} \tag{19.13}$$

Equation 19.13 is a pressure-temperature relationship for an adiabatic process. Because $(T^\gamma)^{1/\gamma} = T$, we can simplify Equation 19.13 by raising both sides to the power $1/\gamma$. Doing so gives

$$p^{(1-\gamma)/\gamma}T = \text{constant} \tag{19.14}$$

during an adiabatic process.

Process $1 \rightarrow 2$ is an adiabatic process; hence

$$p_1^{(1-\gamma)/\gamma}T_1 = p_2^{(1-\gamma)/\gamma}T_2 \tag{19.15}$$

Isolating T_1 gives

$$T_1 = \frac{p_2^{(1-\gamma)/\gamma}}{p_1^{(1-\gamma)/\gamma}}T_2 = \left(\frac{p_2}{p_1}\right)^{(1-\gamma)/\gamma}T_2 = \left(\frac{p_{max}}{p_{min}}\right)^{(1-\gamma)/\gamma}T_2 \tag{19.16}$$

If we define the **pressure ratio** r_p as $r_p = p_{max}/p_{min}$, then T_1 and T_2 are related by

$$T_1 = r_p^{(1-\gamma)/\gamma}T_2 \tag{19.17}$$

FIGURE 19.14 A gas turbine engine follows a Brayton cycle.

(a)

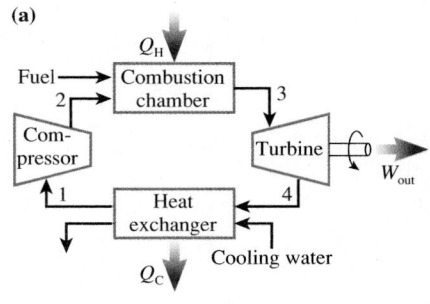

(b)

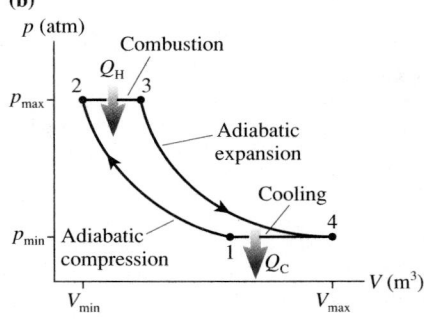

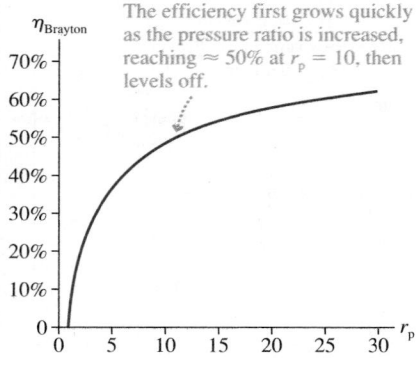

The efficiency first grows quickly as the pressure ratio is increased, reaching $\approx 50\%$ at $r_p = 10$, then levels off.

Any increase in efficiency beyond $\approx 50\%$ has to be weighed against the higher costs of a better compressor that can achieve a much higher pressure ratio.

The algebra of getting to Equation 19.17 was a bit tricky, but the final result is fairly simple.

Process $3 \rightarrow 4$ is also an adiabatic process. The same reasoning leads to

$$T_4 = r_p^{(1-\gamma)/\gamma} T_3 \tag{19.18}$$

If we substitute these expressions for T_1 and T_4 into Equation 19.12, the efficiency is

$$\eta_B = 1 - \frac{T_4 - T_1}{T_3 - T_2} = 1 - \frac{r_p^{(1-\gamma)/\gamma} T_3 - r_p^{(1-\gamma)/\gamma} T_2}{T_3 - T_2} = 1 - \frac{r_p^{(1-\gamma)/\gamma}(T_3 - T_2)}{T_3 - T_2}$$

$$= 1 - r_p^{(1-\gamma)/\gamma}$$

Remarkably, all the temperatures cancel and we're left with an expression that depends only on the pressure ratio. Noting that $(1 - \gamma)$ is negative, we can make one final change and write

$$\eta_B = 1 - \frac{1}{r_p^{(\gamma-1)/\gamma}} \tag{19.19}$$

FIGURE 19.15 is a graph of the efficiency of the Brayton cycle as a function of the pressure ratio, assuming $\gamma = 1.40$ for a diatomic gas such as air.

19.4 Ideal-Gas Refrigerators

Suppose we were to operate a Brayton heat engine backward, going ccw rather than cw in the pV diagram. **FIGURE 19.16a**, (which you should compare to Figure 19.14a) shows a device for doing this. **FIGURE 19.16b** is its pV diagram, and **FIGURE 19.16c** is the energy-transfer diagram. Starting from point 4, the gas is adiabatically compressed to increase its temperature and pressure. It then flows through a high-temperature heat exchanger where the gas *cools* at constant pressure from temperature T_3 to T_2. The gas then expands adiabatically, leaving it significantly colder at T_1 than it started at T_4. It completes the cycle by flowing through a low-temperature heat exchanger, where it *warms* back to its starting temperature.

(a)

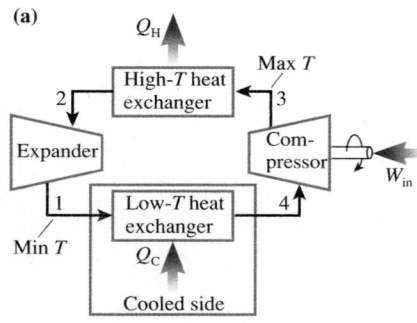

(b)

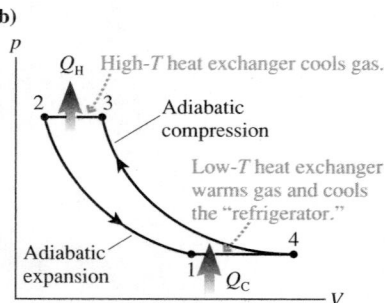

(c)

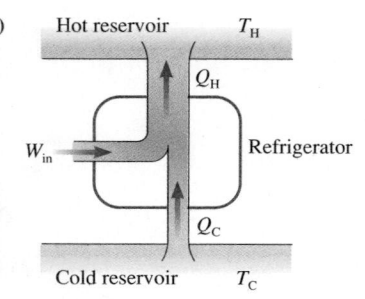

Suppose that the low-temperature heat exchanger is a closed container of air surrounding a pipe through which the engine's cold gas is flowing. The heat-exchange process $1 \rightarrow 4$ *cools* the air in the container as it warms the gas flowing through the pipe. If you were to place eggs and milk inside this closed container, you would call it a refrigerator!

Going around a closed pV cycle in a ccw direction reverses the sign of W for each process in the cycle. Consequently, the area inside the curve of Figure 19.16b is W_{in}, the work done *on* the system. Here work is used to extract heat Q_C from the cold reservoir and exhaust a larger amount of heat $Q_H = Q_C + W_{in}$ to the hot reservoir. But where, in this situation, are the energy reservoirs?

Understanding a refrigerator is a little harder than understanding a heat engine. The key is to remember that **heat is always transferred from a hotter object to a colder object.** In particular,

- The gas in a refrigerator can extract heat Q_C *from* the cold reservoir only if the gas temperature is *lower* than the cold-reservoir temperature T_C. Heat energy is then transferred *from* the cold reservoir *into* the colder gas.
- The gas in a refrigerator can exhaust heat Q_H *to* the hot reservoir only if the gas temperature is *higher* than the hot-reservoir temperature T_H. Heat energy is then transferred *from* the warmer gas *into* the hot reservoir.

These two requirements place severe constraints on the thermodynamics of a refrigerator. Because there is no reservoir colder than T_C, the gas cannot reach a temperature lower than T_C by heat exchange. The gas in a refrigerator *must* use an adiabatic expansion ($Q = 0$) to lower the temperature below T_C. Likewise, a gas refrigerator requires an adiabatic compression to raise the gas temperature above T_H.

EXAMPLE 19.3 **Analyzing a refrigerator**

A refrigerator using helium gas operates on a reversed Brayton cycle with a pressure ratio of 5.0. Prior to compression, the gas occupies 100 cm³ at a pressure of 150 kPa and a temperature of −23°C. Its volume at the end of the expansion is 80 cm³. What are the refrigerator's coefficient of performance and its power input if it operates at 60 cycles per second?

MODEL The Brayton cycle has two adiabatic processes and two isobaric processes. The work per cycle needed to run the refrigerator is $W_{in} = Q_H - Q_C$; hence we can determine both the coefficient of performance and the power requirements from Q_H and Q_C. Heat energy is transferred only during the two isobaric processes.

VISUALIZE **FIGURE 19.17** shows the pV cycle. We know from the pressure ratio of 5.0 that the maximum pressure is 750 kPa. Neither V_2 nor V_3 is known.

FIGURE 19.17 A Brayton-cycle refrigerator.

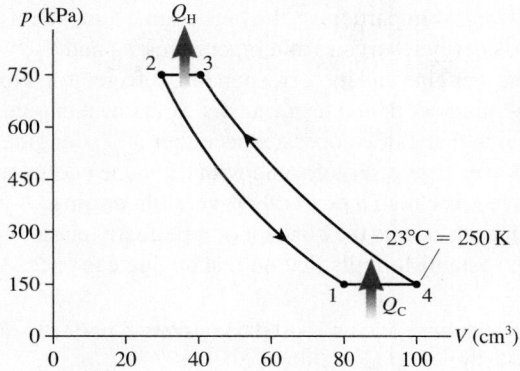

SOLVE To calculate heat we're going to need the temperatures at the four corners of the cycle. First, we can use the conditions of state 4 to find the number of moles of helium:

$$n = \frac{p_4 V_4}{RT_4} = 0.00722 \text{ mol}$$

Process $1 \rightarrow 4$ is isobaric; hence temperature T_1 is

$$T_1 = \frac{V_1}{V_4} T_4 = (0.80)(250 \text{ K}) = 200 \text{ K} = -73°C$$

With Equation 19.14 we found that the quantity $p^{(1-\gamma)/\gamma}T$ remains constant during an adiabatic process. Helium is a monatomic gas with $\gamma = \frac{5}{3}$, so $(1 - \gamma)/\gamma = -\frac{2}{5} = -0.40$. For the adiabatic compression $4 \rightarrow 3$,

$$p_3^{-0.40}T_3 = p_4^{-0.40}T_4$$

Solving for T_3 gives

$$T_3 = \left(\frac{p_4}{p_3}\right)^{-0.40} T_4 = \left(\frac{1}{5}\right)^{-0.40}(250 \text{ K}) = 476 \text{ K} = 203°C$$

The same analysis applied to the $2 \rightarrow 1$ adiabatic expansion gives

$$T_2 = \left(\frac{p_1}{p_2}\right)^{-0.40} T_1 = \left(\frac{1}{5}\right)^{-0.40}(200 \text{ K}) = 381 \text{ K} = 108°C$$

Now we can use $C_P = \frac{5}{2}R = 20.8$ J/mol K for a monatomic gas to compute the heat transfers:

$$Q_H = |Q_{32}| = nC_P(T_3 - T_2)$$
$$= (0.00722 \text{ mol})(20.8 \text{ J/mol K})(95 \text{ K}) = 14.3 \text{ J}$$
$$Q_C = |Q_{14}| = nC_P(T_4 - T_1)$$
$$= (0.00722 \text{ mol})(20.8 \text{ J/mol K})(50 \text{ K}) = 7.5 \text{ J}$$

Thus the work *input* to the refrigerator is $W_{in} = Q_H - Q_C = 6.8$ J. During each cycle, 6.8 J of work are done *on* the gas to extract 7.5 J of heat from the cold reservoir. Then 14.3 J of heat are exhausted into the hot reservoir.

The refrigerator's coefficient of performance is

$$K = \frac{Q_C}{W_{in}} = \frac{7.5 \text{ J}}{6.8 \text{ J}} = 1.1$$

The power input needed to run the refrigerator is

$$P_{in} = 6.8 \frac{\text{J}}{\text{cycle}} \times 60 \frac{\text{cycles}}{\text{s}} = 410 \frac{\text{J}}{\text{s}} = 410 \text{ W}$$

ASSESS These are fairly realistic values for a kitchen refrigerator. You pay your electric company for providing the work W_{in} that operates the refrigerator. The cold reservoir is the freezer compartment. The cold temperature T_C must be higher than T_4 ($T_C > -23°C$) in order for heat to be transferred *from* the cold reservoir *to* the gas. A typical freezer temperature is $-15°C$, so this condition is satisfied. The hot reservoir is the air in the room. The back and underside of a refrigerator have heat-exchanger coils where the hot gas, after compression, transfers heat to the air. The hot temperature T_H must be less than T_2 ($T_H < 108°C$) in order for heat to be transferred *from* the gas *to* the air. An air temperature $\approx 25°C$ under a refrigerator satisfies this condition.

STOP TO THINK 19.4 What, if anything, is wrong with this refrigerator?

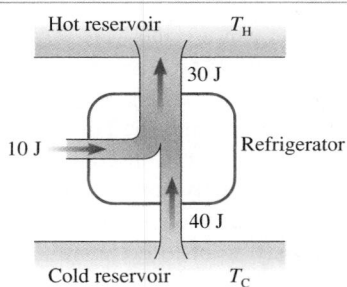

19.5 The Limits of Efficiency

Thermodynamics has its historical roots in the development of the steam engine and other machines of the early industrial revolution. Early steam engines, built on the basis of experience rather than scientific understanding, were not very efficient at converting fuel energy into work. The first major theoretical analysis of heat engines was published by the French engineer Sadi Carnot in 1824. The question that Carnot raised was: Can we make a heat engine whose thermal efficiency η approaches 1, or is there an upper limit η_{max} that cannot be exceeded? To frame the question more clearly, imagine we have a hot reservoir at temperature T_H and a cold reservoir at T_C. What is the most efficient heat engine (maximum η) that can operate between these two energy reservoirs? Similarly, what is the most efficient refrigerator (maximum K) that can operate between the two reservoirs?

We just saw that a refrigerator is, in some sense, a heat engine running backward. We might thus suspect that the most efficient heat engine is related to the most efficient refrigerator. Suppose we have a heat engine that we can turn into a refrigerator by reversing the direction of operation, thus changing the direction of the energy transfers, and with *no other changes*. In particular, the heat engine and the refrigerator operate between the same two energy reservoirs at temperatures T_H and T_C.

FIGURE 19.18a shows such a heat engine and its corresponding refrigerator. Notice that the refrigerator has *exactly the same* work and heat transfer as the heat engine, only in the opposite directions. A device that can be operated as either a heat engine or a refrigerator between the same two energy reservoirs and with the same energy transfers, with only their direction changed, is called a **perfectly reversible engine**. A perfectly reversible engine is an idealization, as was the concept of a perfectly elastic collision. Nonetheless, it will allow us to establish limits that no real engine can exceed.

FIGURE 19.18 If a perfectly reversible heat engine is used to operate a perfectly reversible refrigerator, the two devices exactly cancel each other.

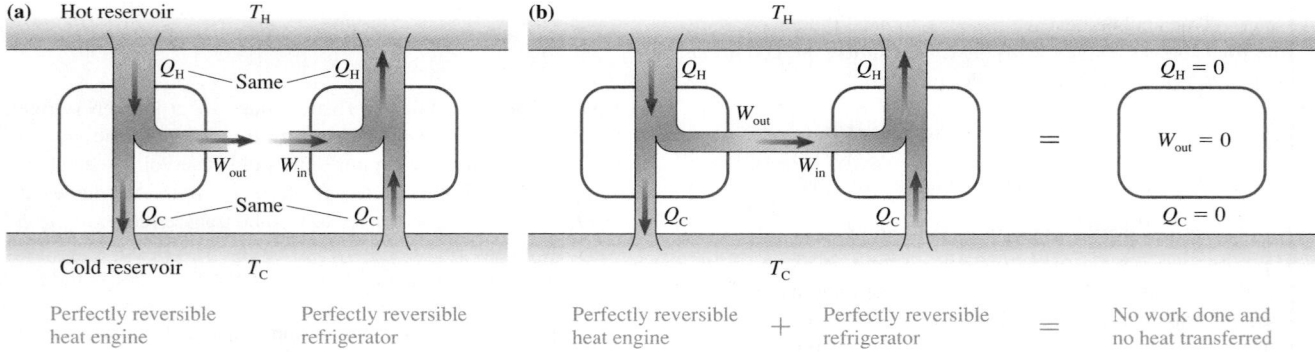

Suppose we have a perfectly reversible heat engine and a perfectly reversible refrigerator (the same device running backward) operating between a hot reservoir at temperature T_H and a cold reservoir at temperature T_C. Because the work W_{in} needed to operate the refrigerator is exactly the same as the useful work W_{out} done by the heat engine, we can use the heat engine, as shown in **FIGURE 19.18b**, to drive the refrigerator. The heat Q_C the engine exhausts to the cold reservoir is exactly the same as the heat Q_C the refrigerator extracts from the cold reservoir. Similarly, the heat Q_H the engine extracts from the hot reservoir matches the heat Q_H the refrigerator exhausts to the hot reservoir. Consequently, there is no net heat transfer in either direction. The refrigerator exactly replaces all the heat energy that had been transferred out of the hot reservoir by the heat engine.

You may want to compare the reasoning used here with the reasoning we used with Figure 19.11. There we tried to use the output of a "perfect" heat engine to run a refrigerator but did *not* succeed.

A Perfectly Reversible Engine Has Maximum Efficiency

Now we've arrived at the critical step in the reasoning. Suppose I claim to have a heat engine that can operate between temperatures T_H and T_C with *more* efficiency than a perfectly reversible engine. FIGURE 19.19 shows the output of this heat engine operating the same perfectly reversible refrigerator that we used in Figure 19.18b.

FIGURE 19.19 A heat engine more efficient than a perfectly reversible engine could be used to violate the second law of thermodynamics.

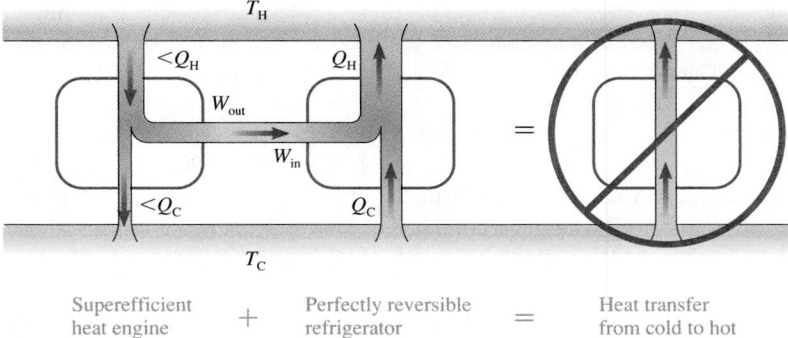

Superefficient heat engine + Perfectly reversible refrigerator = Heat transfer from cold to hot

Recall that the thermal efficiency and the work of a heat engine are

$$\eta = \frac{W_{out}}{Q_H} \quad \text{and} \quad W_{out} = Q_H - Q_C$$

If the new heat engine is more efficient than the perfectly reversible engine it replaces, it needs *less* heat Q_H from the hot reservoir to perform the *same* work W_{out}. If Q_H is less while W_{out} is the same, then Q_C must also be less. That is, the new heat engine exhausts less heat to the cold reservoir than does the perfectly reversible heat engine.

When this new heat engine drives the perfectly reversible refrigerator, the heat it exhausts to the cold reservoir is *less* than the heat extracted from the cold reservoir by the refrigerator. Similarly, this engine extracts *less* heat from the hot reservoir than the refrigerator exhausts. Thus the net result of using this superefficient heat engine to operate a perfectly reversible refrigerator is that heat is transferred from the cold reservoir to the hot reservoir *without outside assistance*.

But this can't happen. It would violate the second law of thermodynamics. Hence we have to conclude that no heat engine operating between reservoirs at temperatures T_H and T_C can be more efficient than a perfectly reversible engine. This very important conclusion is another version of the second law:

> **Second law, informal statement #5** No heat engine operating between reservoirs at temperatures T_H and T_C can be more efficient than a perfectly reversible engine operating between these temperatures.

The answer to our question "Is there a maximum η that cannot be exceeded?" is a clear Yes! The maximum possible efficiency η_{max} is that of a perfectly reversible engine. Because the perfectly reversible engine is an idealization, any real engine will have an efficiency less than η_{max}.

A similar argument shows that no refrigerator can be more efficient than a perfectly reversible refrigerator. If we had such a refrigerator, and if we ran it with the output of a perfectly reversible heat engine, we could transfer heat from cold to hot with no outside assistance. Thus:

> **Second law, informal statement #6** No refrigerator operating between reservoirs at temperatures T_H and T_C can have a coefficient of performance larger than that of a perfectly reversible refrigerator operating between these temperatures.

Conditions for a Perfectly Reversible Engine

This argument tells us that η_{max} and K_{max} exist, but it doesn't tell us what they are. Our final task will be to "design" and analyze a perfectly reversible engine. Under what conditions is an engine reversible?

An engine transfers energy by both mechanical and thermal interactions. Mechanical interactions are pushes and pulls. The environment does work on the system, transferring energy into the system by pushing in on a piston. The system transfers energy back to the environment by pushing out on the piston.

The energy transferred by a moving piston is perfectly reversible, returning the system to its initial state, with no change of temperature or pressure, only if the motion is *frictionless*. The slightest bit of friction will prevent the mechanical transfer of energy from being perfectly reversible.

The circumstances under which heat transfer can be *completely* reversed aren't quite so obvious. After all, Chapter 18 emphasized the *irreversible* nature of heat transfer. If objects A and B are in thermal contact, with $T_A > T_B$, then heat energy is transferred from A to B. But the second law of thermodynamics prohibits a heat transfer from B back to A. Heat transfer through a temperature *difference* is an irreversible process.

But suppose $T_A = T_B$. With no temperature difference, any heat that is transferred from A to B can, at a later time, be transferred from B back to A. This transfer wouldn't violate the second law, which prohibits only heat transfer from a colder object to a hotter object. Now you might object, and rightly so, that heat *can't* move from A to B if they are at the same temperature because heat, by definition, is the energy transferred between two objects at different temperatures.

This is true, so let's consider a limiting case in which $T_A = T_B + dT$. The temperature difference is infinitesimal. Heat is transferred from A to B, but *very slowly!* If you later try to make the heat move from B back to A, the second law will prevent you from doing so with perfect precision. But because the temperature difference is infinitesimal, you'll be missing only an infinitesimal amount dQ of heat. You can transfer heat reversibly in the limit $dT \rightarrow 0$, but you must be prepared to spend an infinite amount of time doing so.

Thus the thermal transfer of energy is reversible if the heat is transferred infinitely slowly in an isothermal process. This is an idealization, but so are completely frictionless processes. Nonetheless, we can now say that a perfectly reversible engine must use only two types of processes:

1. Frictionless mechanical interactions with no heat transfer ($Q = 0$), and
2. Thermal interactions in which heat is transferred in an isothermal process ($\Delta E_{th} = 0$).

Any engine that uses only these two types of processes is called a **Carnot engine.** A Carnot engine is a perfectly reversible engine; thus it has the maximum possible thermal efficiency η_{max} and, if operated as a refrigerator, the maximum possible coefficient of performance K_{max}.

19.6 The Carnot Cycle

The definition of a Carnot engine does not specify whether the engine's working substance is a gas or a liquid. It makes no difference. Our argument that a perfectly reversible engine is the most efficient possible heat engine depended only on the engine's reversibility. Consequently, **any Carnot engine operating between T_H and T_C must have exactly the same efficiency as any other Carnot engine operating between the same two energy reservoirs.** If we can determine the thermal efficiency of one Carnot engine, we'll know the efficiency of all Carnot engines. Because liquids and phase changes are complicated, we'll analyze a Carnot engine that uses an ideal gas.

Designing a Carnot Engine

The **Carnot cycle** is an ideal-gas cycle that consists of the two adiabatic processes ($Q = 0$) and two isothermal processes ($\Delta E_{th} = 0$) shown in FIGURE 19.20. These are the two types of processes allowed in a perfectly reversible gas engine. As a Carnot cycle operates,

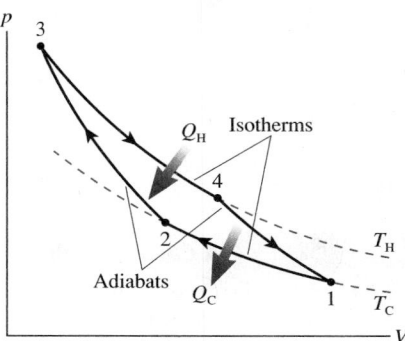

FIGURE 19.20 The Carnot cycle is perfectly reversible.

1. The gas is isothermally compressed while in thermal contact with the cold reservoir at temperature T_C. Heat energy $Q_C = |Q_{12}|$ is removed from the gas as it is compressed in order to keep the temperature constant. The compression must take place extremely slowly because there can be only an infinitesimal temperature difference between the gas and the reservoir.
2. The gas is adiabatically compressed while thermally isolated from the environment. This compression increases the gas temperature until it matches temperature T_H of the hot reservoir. No heat is transferred during this process.
3. After reaching maximum compression, the gas expands isothermally at temperature T_H. Heat $Q_H = Q_{34}$ is transferred from the hot reservoir into the gas as it expands in order to keep the temperature constant.
4. Finally, the gas expands adiabatically, with $Q = 0$, until the temperature decreases back to T_C.

Work is done in all four processes of the Carnot cycle, but heat is transferred only during the two isothermal processes.

The thermal efficiency of any heat engine is

$$\eta = \frac{W_{out}}{Q_H} = 1 - \frac{Q_C}{Q_H}$$

We can determine η_{Carnot} by finding the heat transfer in the two isothermal processes.

Process $1 \rightarrow 2$: Table 19.1 gives us the heat transfer in an isothermal process at temperature T_C:

$$Q_{12} = (W_s)_{12} = nRT_C \ln\left(\frac{V_2}{V_1}\right) = -nRT_C \ln\left(\frac{V_1}{V_2}\right) \tag{19.20}$$

$V_1 > V_2$, so the logarithm on the right is positive. Q_{12} is negative because heat is transferred out of the system, but Q_C is simply the *amount* of heat transferred to the cold reservoir:

$$Q_C = |Q_{12}| = nRT_C \ln\left(\frac{V_1}{V_2}\right) \tag{19.21}$$

Process $3 \rightarrow 4$: Similarly, the heat transferred in the isothermal expansion at temperature T_H is

$$Q_H = Q_{34} = (W_s)_{34} = nRT_H \ln\left(\frac{V_4}{V_3}\right) \tag{19.22}$$

Thus the thermal efficiency of the Carnot cycle is

$$\eta_{Carnot} = 1 - \frac{Q_C}{Q_H} = 1 - \frac{T_C \ln(V_1/V_2)}{T_H \ln(V_4/V_3)} \tag{19.23}$$

We can simplify this expression. During the two adiabatic processes,

$$T_C V_2^{\gamma-1} = T_H V_3^{\gamma-1} \qquad \text{and} \qquad T_C V_1^{\gamma-1} = T_H V_4^{\gamma-1} \tag{19.24}$$

An algebraic rearrangement gives

$$V_2 = V_3 \left(\frac{T_H}{T_C}\right)^{1/(\gamma-1)} \qquad \text{and} \qquad V_1 = V_4 \left(\frac{T_H}{T_C}\right)^{1/(\gamma-1)} \tag{19.25}$$

from which it follows that

$$\frac{V_1}{V_2} = \frac{V_4}{V_3} \tag{19.26}$$

Consequently, the two logarithms in Equation 19.23 cancel and we're left with the result that the thermal efficiency of a Carnot engine operating between a hot reservoir at temperature T_H and a cold reservoir at temperature T_C is

$$\eta_{Carnot} = 1 - \frac{T_C}{T_H} \quad \text{(Carnot thermal efficiency)} \quad (19.27)$$

This remarkably simple result, an efficiency that depends only on the ratio of the temperatures of the hot and cold reservoirs, is Carnot's legacy to thermodynamics.

NOTE ▶ Temperatures T_H and T_C are *absolute* temperatures. ◀

EXAMPLE 19.4 A Carnot engine

A Carnot engine is cooled by water at $T_C = 10°C$. What temperature must be maintained in the hot reservoir of the engine to have a thermal efficiency of 70%?

MODEL The efficiency of a Carnot engine depends only on the temperatures of the hot and cold reservoirs.

SOLVE The thermal efficiency $\eta_{Carnot} = 1 - T_C/T_H$ can be rearranged to give

$$T_H = \frac{T_C}{1 - \eta_{Carnot}} = 943 \text{ K} = 670°C$$

where we used $T_C = 283$ K.

ASSESS A "real" engine would need a higher temperature than this to provide 70% efficiency because no real engine will match the Carnot efficiency.

EXAMPLE 19.5 A real engine

The heat engine of Example 19.2 had a highest temperature of 2700 K, a lowest temperature of 300 K, and a thermal efficiency of 15%. What is the efficiency of a Carnot engine operating between these two temperatures?

SOLVE The Carnot efficiency is

$$\eta_{Carnot} = 1 - \frac{T_C}{T_H} = 1 - \frac{300 \text{ K}}{2700 \text{ K}} = 0.89 = 89\%$$

ASSESS The thermodynamic cycle used in Example 19.2 doesn't come anywhere close to the Carnot efficiency.

The Maximum Efficiency

In Section 19.2 we tried to invent a perfect engine with $\eta = 1$ and $Q_C = 0$. We found that we could not do so without violating the second law, so no engine can have $\eta = 1$. However, that example didn't rule out an engine with $\eta = 0.9999$. Further analysis has now shown that no heat engine operating between energy reservoirs at temperatures T_H and T_C can be more efficient than a perfectly reversible engine operating between these temperatures.

We've now reached the endpoint of this line of reasoning by establishing an exact result for the thermal efficiency of a perfectly reversible engine, the Carnot engine. We can summarize our conclusions:

Second law, informal statement #7 No heat engine operating between energy reservoirs at temperatures T_H and T_C can exceed the Carnot efficiency

$$\eta_{Carnot} = 1 - \frac{T_C}{T_H}$$

As Example 19.5 showed, real engines usually fall well short of the Carnot limit.

We also found that no refrigerator can exceed the coefficient of performance of a perfectly reversible refrigerator. We'll leave the proof as a homework problem, but an analysis very similar to that above shows that the coefficient of performance of a Carnot refrigerator is

$$K_{Carnot} = \frac{T_C}{T_H - T_C} \quad \text{(Carnot coefficient of performance)} \quad (19.28)$$

Thus we can state:

> **Second law, informal statement #8** No refrigerator operating between energy reservoirs at temperatures T_H and T_C can exceed the Carnot coefficient of performance
>
> $$K_{Carnot} = \frac{T_C}{T_H - T_C}$$

EXAMPLE 19.6 | Brayton versus Carnot

The Brayton-cycle refrigerator of Example 19.3 had coefficient of performance $K = 1.1$. Compare this to the limit set by the second law of thermodynamics.

SOLVE Example 19.3 found that the reservoir temperatures had to be $T_C \geq 250$ K and $T_H \leq 381$ K. A Carnot refrigerator operating between 250 K and 381 K has

$$K_{Carnot} = \frac{T_C}{T_H - T_C} = \frac{250\ \text{K}}{381\ \text{K} - 250\ \text{K}} = 1.9$$

ASSESS This is the minimum value of K_{Carnot}. It will be even higher if $T_C > 250$ K or $T_H < 381$ K. The coefficient of performance of the reasonably realistic refrigerator of Example 19.3 is less than 60% of the limiting value.

Statements #7 and #8 of the second law are a major result of this chapter, one with profound implications. The efficiency limit of a heat engine is set by the temperatures of the hot and cold reservoirs. High efficiency requires $T_C/T_H \ll 1$ and thus $T_H \gg T_C$. However, practical realities often prevent T_H from being significantly larger than T_C, in which case the engine cannot possibly have a large efficiency. This limit on the efficiency of heat engines is a consequence of the second law of thermodynamics.

EXAMPLE 19.7 | Generating electricity

An electric power plant boils water to produce high-pressure steam at 400°C. The high-pressure steam spins a turbine as it expands, then the turbine spins the generator. The steam is then condensed back to water in an ocean-cooled heat exchanger at 25°C. What is the *maximum* possible efficiency with which heat energy can be converted to electric energy?

MODEL The maximum possible efficiency is that of a Carnot engine operating between these temperatures.

SOLVE The Carnot efficiency depends on absolute temperatures, so we must use $T_H = 400°C = 673$ K and $T_C = 25°C = 298$ K. Then

$$\eta_{max} = 1 - \frac{298}{673} = 0.56 = 56\%$$

ASSESS This is an upper limit. Real coal-, oil-, gas-, and nuclear-heated steam generators actually operate at $\approx 35\%$ thermal efficiency, converting only about one-third of the fuel energy to electric energy while exhausting about two-thirds of the energy to the environment as waste heat. (The heat *source* has nothing to do with the efficiency. All it does is boil water.) Not much can be done to alter the low-temperature limit. The high-temperature limit is determined by the maximum temperature and pressure the boiler and turbine can withstand. The efficiency of electricity generation is far less than most people imagine, but it is an unavoidable consequence of the second law of thermodynamics.

A limit on the efficiency of heat engines was not expected. We are used to thinking in terms of energy conservation, so it comes as no surprise that we cannot make an engine with $\eta > 1$. But the limits arising from the second law were not anticipated, nor are they obvious. Nonetheless, they are a very real fact of life and a very real constraint on any practical device. No one has ever invented a machine that exceeds the second-law limits, and we have seen that the maximum efficiency for realistic engines is surprisingly low.

STOP TO THINK 19.5 Could this heat engine be built?

a. Yes.
b. No.
c. It's impossible to tell without knowing what kind of cycle it uses.

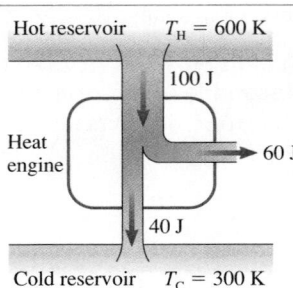

Hot reservoir $T_H = 600$ K
100 J
Heat engine
60 J
40 J
Cold reservoir $T_C = 300$ K

CHALLENGE EXAMPLE 19.8 Calculating efficiency

A heat engine using a monatomic ideal gas goes through the following closed cycle:

- Isochoric heating until the pressure is doubled.

- Isothermal expansion until the pressure is restored to its initial value.

- Isobaric compression until the volume is restored to its initial value.

What is the thermal efficiency of this heat engine? What would be the thermal efficiency of a Carnot engine operating between the highest and lowest temperatures reached by this engine?

MODEL The cycle consists of three familiar processes; we'll need to analyze each. The amount of work and heat will depend on the quantity of gas, which we don't know, but efficiency is a work-to-heat ratio that is independent of the amount of gas.

VISUALIZE FIGURE 19.21 shows the cycle. The initial pressure, volume, and temperature are p, V, and T. The isochoric process increases the pressure to $2p$ and, because the ratio p/T is constant in an isochoric process, increases the temperature to $2T$. The isothermal expansion is along the $2T$ isotherm. The product pV is constant in an isothermal process, so the volume doubles to $2V$ as the pressure returns to p.

FIGURE 19.21 The pV cycle of the heat engine.

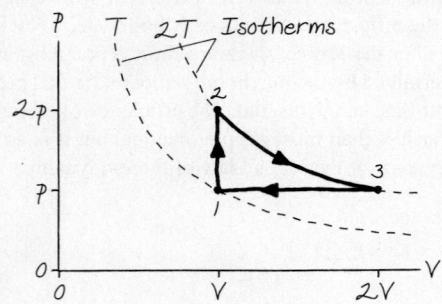

SOLVE We know, symbolically, the state variables at each corner of the pV diagram. That is sufficient for calculating W_s, Q, and ΔE_{th}.

Process $1 \rightarrow 2$: An isochoric process has $W_s = 0$ and

$$Q = \Delta E_{th} = nC_V \Delta T = \tfrac{3}{2} nRT$$

where we used $C_V = \tfrac{3}{2}R$ for a monatomic gas and $\Delta T = 2T - T = T$.

Process $2 \rightarrow 3$: An isothermal process has $\Delta E_{th} = 0$ and

$$Q = W_s = nR(2T)\ln(2V/V) = (2\ln 2)nRT$$

Here we used the Table 19.1 result for the work done in an isothermal process.

Process $3 \rightarrow 1$: The work done by the gas is the area under the curve, which is negative because $\Delta V = V - 2V = -V$ in the compression:

$$W_s = \text{area} = p\Delta V = -pV = -nRT$$

We used the ideal-gas law in the last step to express the result in terms of n and T. The heat transfer is also negative because $\Delta T = T - 2T = -T$:

$$Q = nC_P \Delta T = -\tfrac{5}{2}nRT$$

where we used $C_P = \tfrac{5}{2}R$ for a monatomic gas. Based on the first law, $\Delta E_{th} = Q - W_s = -\tfrac{3}{2}nRT$.

Summing over the three processes, we see that $(\Delta E_{th})_{net} = 0$, as expected, and

$$W_{out} = (2\ln 2 - 1)nRT$$

Heat energy is supplied to the gas ($Q > 0$) in processes $1 \rightarrow 2$ and $2 \rightarrow 3$, so

$$Q_H = (2\ln 2 + \tfrac{3}{2})nRT$$

Thus the thermal efficiency of this heat engine is

$$\eta = \frac{W_{out}}{Q_H} = \frac{(2\ln 2 - 1)nRT}{(2\ln 2 + \tfrac{3}{2})nRT} = 0.134 = 13.4\%$$

A Carnot engine would be able to operate between a high temperature $T_H = 2T$ and a low temperature $T_C = T$. Its efficiency would be

$$\eta_{Carnot} = 1 - \frac{T_C}{T_H} = 1 - \frac{T}{2T} = 0.500 = 50.0\%$$

ASSESS As we anticipated, the thermal efficiency depends on the *shape* of the pV cycle but not on the quantity of gas or even on the values of p, V, or T. The heat engine's 13.4% efficiency is considerably less than the 50% maximum possible efficiency set by the second law of thermodynamics.

SUMMARY

The goal of Chapter 19 has been to study the physical principles that govern heat engines and refrigerators.

General Principles

Heat Engines

Devices that transform heat into work. They require two energy reservoirs at different temperatures.

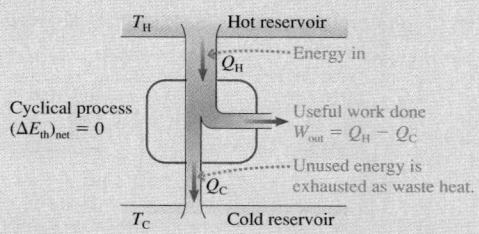

Thermal efficiency

$$\eta = \frac{W_{out}}{Q_H} = \frac{\text{what you get}}{\text{what you pay}}$$

Second-law limit:

$$\eta \leq 1 - \frac{T_C}{T_H}$$

Refrigerators

Devices that use work to transfer heat from a colder object to a hotter object.

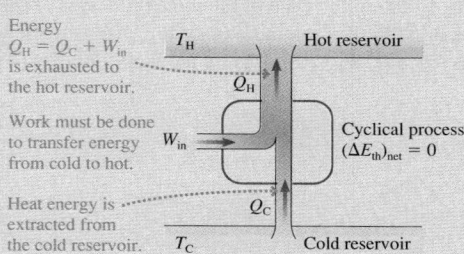

Coefficient of performance

$$K = \frac{Q_C}{W_{in}} = \frac{\text{what you get}}{\text{what you pay}}$$

Second-law limit:

$$K \leq \frac{T_C}{T_H - T_C}$$

Important Concepts

A perfectly reversible engine (a **Carnot engine**) can be operated as either a heat engine or a refrigerator between the same two energy reservoirs by reversing the cycle and with no other changes.

- A **Carnot heat engine** has the maximum possible thermal efficiency of any heat engine operating between T_H and T_C:

$$\eta_{\text{Carnot}} = 1 - \frac{T_C}{T_H}$$

- A **Carnot refrigerator** has the maximum possible coefficient of performance of any refrigerator operating between T_H and T_C:

$$K_{\text{Carnot}} = \frac{T_C}{T_H - T_C}$$

The Carnot cycle for a gas engine consists of two isothermal processes and two adiabatic processes.

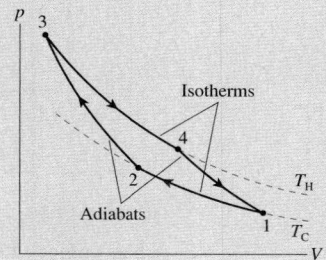

An energy reservoir is a part of the environment so large in comparison to the system that its temperature doesn't change as the system extracts heat energy from or exhausts heat energy to the reservoir. All heat engines and refrigerators operate between two energy reservoirs at different temperatures T_H and T_C.

The **work** W_s done *by* the system has the opposite sign to the work done *on* the system.

$W_s =$ area under pV curve

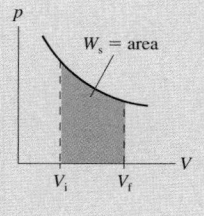

Applications

To analyze a heat engine or refrigerator:

MODEL Identify each process in the cycle.

VISUALIZE Draw the pV diagram of the cycle.

SOLVE There are several steps:

- Determine p, V, and T at the beginning and end of each process.
- Calculate ΔE_{th}, W_s, and Q for each process.
- Determine W_{in} or W_{out}, Q_H, and Q_C.
- Calculate $\eta = W_{out}/Q_H$ or $K = Q_C/W_{in}$.

ASSESS Verify $(\Delta E_{th})_{net} = 0$. Check signs.

Terms and Notation

thermodynamics	closed-cycle device	coefficient of performance, K	Carnot engine
energy reservoir	thermal efficiency, η	heat exchanger	Carnot cycle
energy-transfer diagram	waste heat	pressure ratio, r_p	
heat engine	refrigerator	perfectly reversible engine	

CONCEPTUAL QUESTIONS

1. In going from i to f in each of the three processes of FIGURE Q19.1, is work done *by* the system ($W < 0$, $W_s > 0$), is work done *on* the system ($W > 0$, $W_s < 0$), or is *no* net work done?

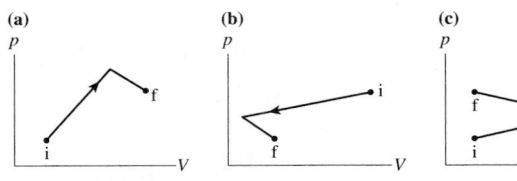

FIGURE Q19.1

2. Rank in order, from largest to smallest, the amount of work $(W_s)_1$ to $(W_s)_4$ done by the gas in each of the cycles shown in FIGURE Q19.2. Explain.

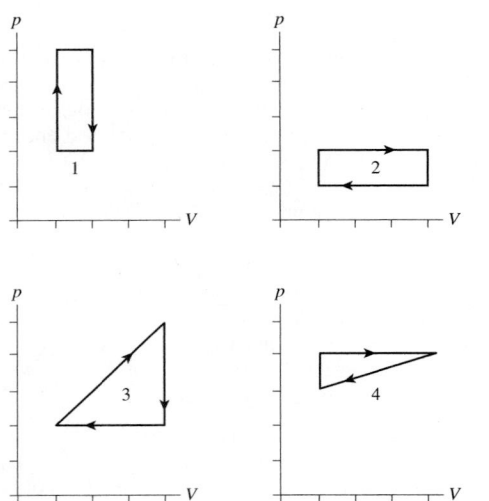

FIGURE Q19.2

3. Could you have a heat engine with $\eta > 1$? Explain.

4. FIGURE Q19.4 shows the pV diagram of a heat engine. During which stage or stages is (a) heat added to the gas, (b) heat removed from the gas, (c) work done on the gas, and (d) work done by the gas?

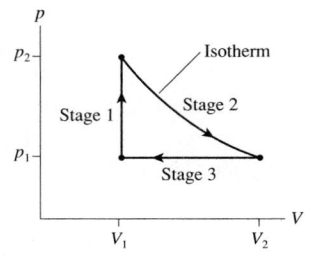

FIGURE Q19.4

5. Rank in order, from largest to smallest, the thermal efficiencies η_1 to η_4 of the four heat engines in FIGURE Q19.5. Explain.

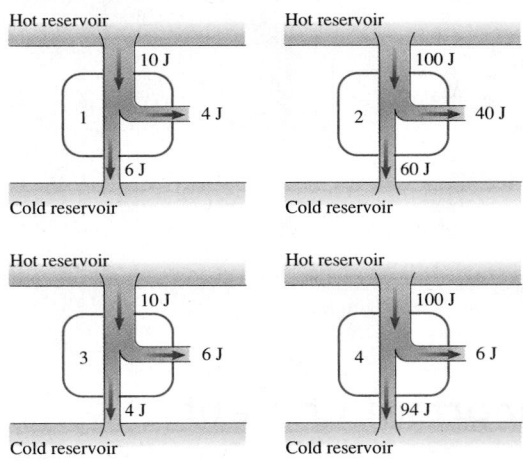

FIGURE Q19.5

6. FIGURE Q19.6 shows the thermodynamic cycles of two heat engines. Which heat engine has the larger thermal efficiency? Or are they the same? Explain.

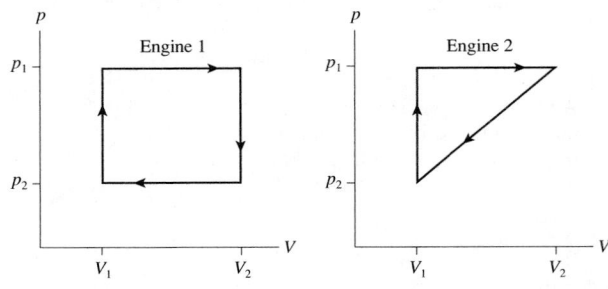

FIGURE Q19.6

7. A heat engine satisfies $W_{out} = Q_{net}$. Why is there no ΔE_{th} term in this relationship?

8. Do the energy-transfer diagrams in FIGURE Q19.8 represent possible heat engines? If not, what is wrong?

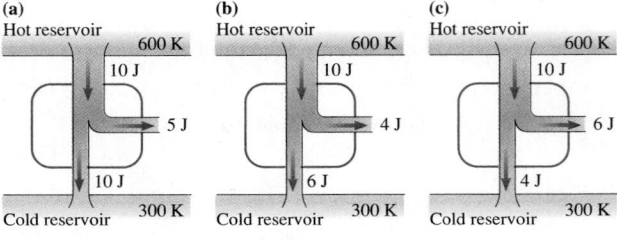

FIGURE Q19.8

9. Do the energy-transfer diagrams in **FIGURE Q19.9** represent possible refrigerators? If not, what is wrong?

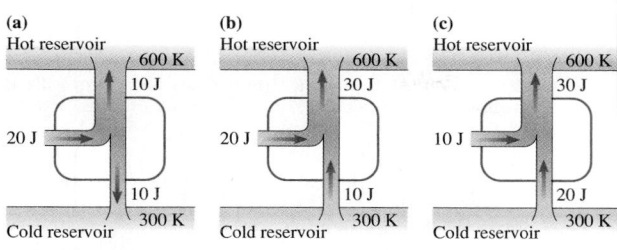

FIGURE Q19.9

10. It gets pretty hot in your apartment. In browsing the Internet, you find a company selling small "room air conditioners." You place the air conditioner on the floor, plug it in, and—the advertisement says—it will lower the room temperature up to 10°F. Should you order one? Explain.

11. The first and second laws of thermodynamics are sometimes stated as "You can't win" and "You can't even break even." Do these sayings accurately characterize the laws of thermodynamics as applied to heat engines? Why or why not?

EXERCISES AND PROBLEMS

Problems labeled integrate material from earlier chapters.

Exercises

Section 19.1 Turning Heat into Work

Section 19.2 Heat Engines and Refrigerators

1. | A heat engine with a thermal efficiency of 40% does 100 J of work per cycle. How much heat is (a) extracted from the hot reservoir and (b) exhausted to the cold reservoir per cycle?

2. ‖ A heat engine does 200 J of work per cycle while exhausting 400 J of waste heat. What is the engine's thermal efficiency?

3. | A heat engine extracts 55 kJ of heat from the hot reservoir each cycle and exhausts 40 kJ of heat. What are (a) the thermal efficiency and (b) the work done per cycle?

4. ‖ A refrigerator requires 200 J of work and exhausts 600 J of heat per cycle. What is the refrigerator's coefficient of performance?

5. | 50 J of work are done per cycle on a refrigerator with a coefficient of performance of 4.0. How much heat is (a) extracted from the cold reservoir and (b) exhausted to the hot reservoir per cycle?

6. ‖ The power output of a car engine running at 2400 rpm is 500 kW. How much (a) work is done and (b) heat is exhausted per cycle if the engine's thermal efficiency is 20%? Give your answers in kJ.

7. ‖ A 32%-efficient electric power plant produces 900 MW of electric power and discharges waste heat into 20°C ocean water. Suppose the waste heat could be used to heat homes during the winter instead of being discharged into the ocean. A typical American house requires an average 20 kW for heating. How many homes could be heated with the waste heat of this one power plant?

8. ‖ 1.0 L of 20°C water is placed in a refrigerator. The refrigerator's motor must supply an extra 8.0 W of power to chill the water to 5°C in 1.0 h. What is the refrigerator's coefficient of performance?

Section 19.3 Ideal-Gas Heat Engines

Section 19.4 Ideal-Gas Refrigerators

9. ‖ The cycle of **FIGURE EX19.9** consists of four processes. Make a table with rows labeled A to D and columns labeled ΔE_{th}, W_s, and Q. Fill each box in the table with +, −, or 0 to indicate whether the quantity increases, decreases, or stays the same during that process.

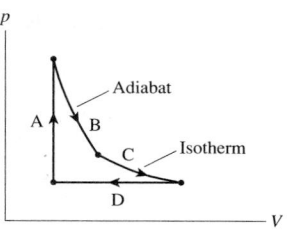

FIGURE EX19.9   **FIGURE EX19.10**

10. ‖ The cycle of **FIGURE EX19.10** consists of three processes. Make a table with rows labeled A–C and columns labeled ΔE_{th}, W_s, and Q. Fill each box in the table with +, −, or 0 to indicate whether the quantity increases, decreases, or stays the same during that process.

11. ‖ How much work is done per cycle by a gas following the pV trajectory of **FIGURE EX19.11**?

FIGURE EX19.11   **FIGURE EX19.12**

12. ‖ A gas following the pV trajectory of **FIGURE EX19.12** does 60 J of work per cycle. What is p_{max}?

13. ‖ What are (a) W_{out} and Q_H and (b) the thermal efficiency for the heat engine shown in **FIGURE EX19.13**?

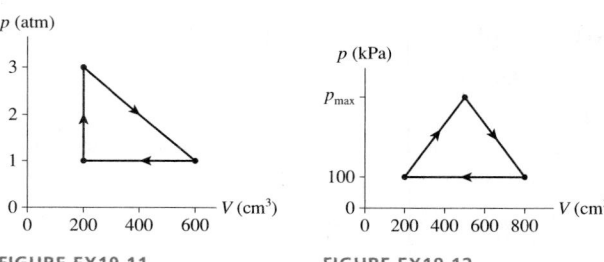

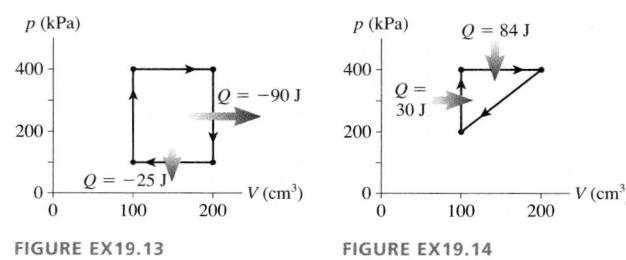

FIGURE EX19.13   **FIGURE EX19.14**

14. ‖ What are (a) W_{out} and Q_C and (b) the thermal efficiency for the heat engine shown in **FIGURE EX19.14**?

15. ‖ How much heat is exhausted to the cold reservoir by the heat engine shown in FIGURE EX19.15?

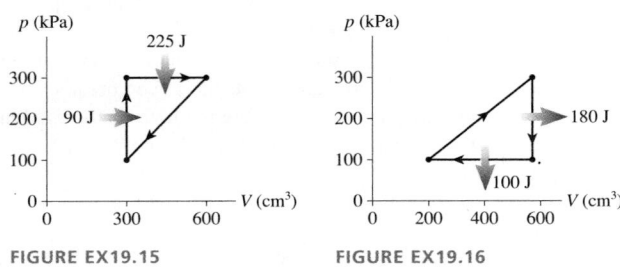

FIGURE EX19.15   FIGURE EX19.16

16. ‖ What are (a) the thermal efficiency and (b) the heat extracted from the hot reservoir for the heat engine shown in FIGURE EX19.16?
17. ‖ A heat engine uses a diatomic gas in a Brayton cycle. What is the engine's thermal efficiency if the gas volume is halved during the adiabatic compression?
18. ‖ What are (a) the heat extracted from the cold reservoir and (b) the coefficient of performance for the refrigerator shown in FIGURE EX19.18?

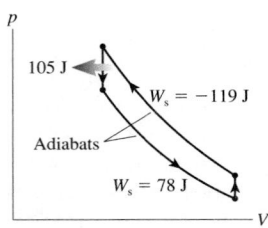

FIGURE EX19.18

Section 19.5 The Limits of Efficiency

Section 19.6 The Carnot Cycle

19. | Which, if any, of the heat engines in FIGURE EX19.19 violate (a) the first law of thermodynamics or (b) the second law of thermodynamics? Explain.

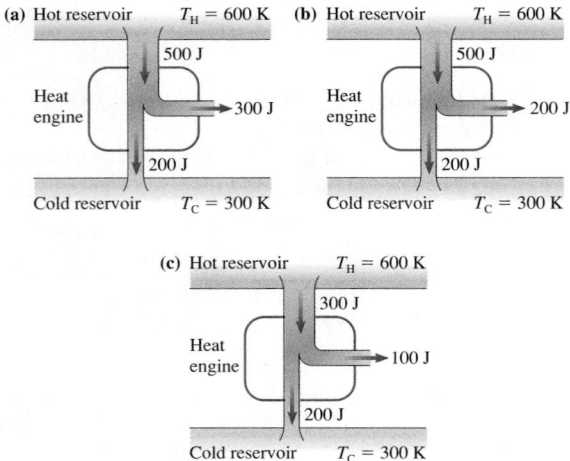

FIGURE EX19.19

20. | Which, if any, of the refrigerators in FIGURE EX19.20 violate (a) the first law of thermodynamics or (b) the second law of thermodynamics? Explain.

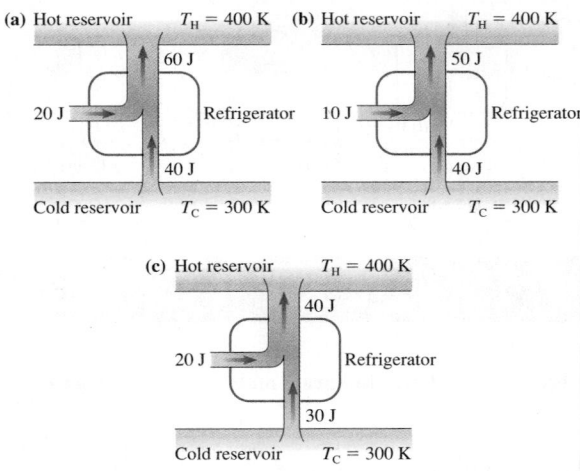

FIGURE EX19.20

21. ‖ At what cold-reservoir temperature (in °C) would a Carnot engine with a hot-reservoir temperature of 427°C have an efficiency of 60%?
22. ‖ A heat engine does 10 J of work and exhausts 15 J of waste heat during each cycle.
    a. What is the engine's thermal efficiency?
    b. If the cold-reservoir temperature is 20°C, what is the minimum possible temperature in °C of the hot reservoir?
23. | a. A heat engine does 200 J of work per cycle while exhausting 600 J of heat to the cold reservoir. What is the engine's thermal efficiency?
    b. A Carnot engine with a hot-reservoir temperature of 400°C has the same thermal efficiency. What is the cold-reservoir temperature in °C?
24. | A Carnot engine operating between energy reservoirs at temperatures 300 K and 500 K produces a power output of 1000 W. What are (a) the thermal efficiency of this engine, (b) the rate of heat input, in W, and (c) the rate of heat output, in W?
25. ‖ A Carnot engine whose hot-reservoir temperature is 400°C has a thermal efficiency of 40%. By how many degrees should the temperature of the cold reservoir be decreased to raise the engine's efficiency to 60%?
26. ‖ A heat engine operating between energy reservoirs at 20°C and 600°C has 30% of the maximum possible efficiency. How much energy must this engine extract from the hot reservoir to do 1000 J of work?
27. ‖ A heat engine operating between a hot reservoir at 500°C and a cold reservoir at 0°C is 60% as efficient as a Carnot engine. If this heat engine and the Carnot engine do the same amount of work, what is the ratio $Q_H/(Q_H)_{\text{Carnot}}$?
28. ‖ A Carnot refrigerator operating between −20°C and +20°C extracts heat from the cold reservoir at the rate 200 J/s. What are (a) the coefficient of performance of this refrigerator, (b) the rate at which work is done on the refrigerator, and (c) the rate at which heat is exhausted to the hot side?

29. ‖ The coefficient of performance of a refrigerator is 5.0. The compressor uses 10 J of energy per cycle.
    a. How much heat energy is exhausted per cycle?
    b. If the hot-reservoir temperature is 27°C, what is the lowest possible temperature in °C of the cold reservoir?

30. ‖ A Carnot heat engine with thermal efficiency $\frac{1}{3}$ is run backward as a Carnot refrigerator. What is the refrigerator's coefficient of performance?

Problems

31. ‖ The engine that powers a crane burns fuel at a flame temperature of 2000°C. It is cooled by 20°C air. The crane lifts a 2000 kg steel girder 30 m upward. How much heat energy is transferred to the engine by burning fuel if the engine is 40% as efficient as a Carnot engine?

32. ‖‖ 100 mL of water at 15°C is placed in the freezer compartment of a refrigerator with a coefficient of performance of 4.0. How much heat energy is exhausted into the room as the water is changed to ice at 15°C?

33. ‖ Prove that the work done in an adiabatic process $i \rightarrow f$ is $W_s = (p_f V_f - p_i V_i)/(1 - \gamma)$.

34. ‖ A Carnot refrigerator operates between reservoirs at $-20°C$ and 50°C in a 25°C room. The refrigerator is a 40 cm × 40 cm × 40 cm box. Five of the walls are perfect insulators, but the sixth is a 1.0-cm-thick piece of stainless steel. What electric power does the refrigerator require to maintain the inside temperature at –20°C?

35. ‖ Prove that the coefficient of performance of a Carnot refrigerator is $K_{\text{Carnot}} = T_C/(T_H - T_C)$.

36. ‖ An ideal refrigerator utilizes a Carnot cycle operating between 0°C and 25°C. To turn 10 kg of liquid water at 0°C into 10 kg of ice at 0°C, (a) how much heat is exhausted into the room and (b) how much energy must be supplied to the refrigerator?

37. ‖ There has long been an interest in using the vast quantities of thermal energy in the oceans to run heat engines. A heat engine needs a temperature *difference*, a hot side and a cold side. Conveniently, the ocean surface waters are warmer than the deep ocean waters. Suppose you build a floating power plant in the tropics where the surface water temperature is ≈ 30°C. This would be the hot reservoir of the engine. For the cold reservoir, water would be pumped up from the ocean bottom where it is always ≈ 5°C. What is the maximum possible efficiency of such a power plant?

38. ‖ The ideal gas in a Carnot engine extracts 1000 J of heat energy during the isothermal expansion at 300°C. How much heat energy is exhausted during the isothermal compression at 50°C?

39. ‖ The hot-reservoir temperature of a Carnot engine with 25% efficiency is 80°C higher than the cold-reservoir temperature. What are the reservoir temperatures, in °C?

40. ‖ A Carnot heat engine operates between reservoirs at 182°C and 0°C. If the engine extracts 25 J of energy from the hot reservoir per cycle, how many cycles will it take to lift a 10 kg mass a height of 10 m?

41. ‖ A Carnot refrigerator operates between reservoirs at 55°C and –20°C. If the engine exhausts 250 J of energy to the hot reservoir per cycle, how many cycles will it take to cool a 500 mL soda from 25°C to 5°C?

42. ‖ FIGURE P19.42 shows a Carnot heat engine driving a Carnot refrigerator.

a. Determine Q_1, Q_2, Q_3, and Q_4.
b. Is Q_3 greater than, less than, or equal to Q_1?
c. Do these two devices, when operated together in this way, violate the second law?

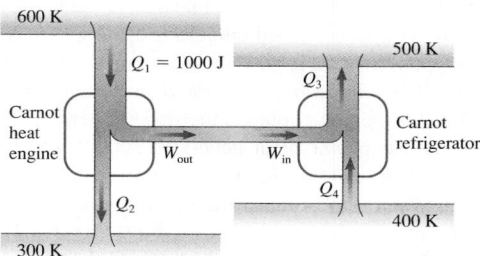

FIGURE P19.42

43. ‖ A Carnot heat engine and an ordinary refrigerator with coefficient of performance 2.00 operate between reservoirs at 350 K and 250 K. The work done by the Carnot heat engine drives the refrigerator. If the heat engine extracts 10.0 J of energy from the hot reservoir, how much energy does the refrigerator exhaust to the hot reservoir?

44. ‖ A heat engine running backward is called a refrigerator if its purpose is to extract heat from a cold reservoir. The same engine running backward is called a *heat pump* if its purpose is to exhaust warm air into the hot reservoir. Heat pumps are widely used for home heating. You can think of a heat pump as a refrigerator that is cooling the already cold outdoors and, with its exhaust heat Q_H, warming the indoors. Perhaps this seems a little silly, but consider the following. Electricity can be directly used to heat a home by passing an electric current through a heating coil. This is a direct, 100% conversion of work to heat. That is, 15 kW of electric power (generated by doing work at the rate of 15 kJ/s at the power plant) produces heat energy inside the home at a rate of 15 kJ/s. Suppose that the neighbor's home has a heat pump with a coefficient of performance of 5.0, a realistic value.
    a. How much electric power (in kW) does the heat pump use to deliver 15 kJ/s of heat energy to the house?
    b. An average price for electricity is about 40 MJ per dollar. A furnace or heat pump will run typically 200 hours per month during the winter. What does one month's heating cost in the home with a 15 kW electric heater and in the home of the neighbor who uses an equivalent heat pump?

45. ‖ You and your roommates need a new refrigerator. At the appliance store, the salesman shows you the DreamFridge. According to its sticker, the DreamFridge uses a mere 100 W of power to remove 100 kJ of heat per minute from the 2°C interior. According to the fine print on the sticker, this claim is true in a 22°C kitchen. Should you buy? Explain.

46. ‖ Three engineering students submit their solutions to a design problem in which they were asked to design an engine that operates between temperatures 300 K and 500 K. The heat input/output and work done by their designs are shown in the following table:

Student	Q_H	Q_C	W_{out}
1	250 J	140 J	110 J
2	250 J	170 J	90 J
3	250 J	160 J	90 J

Critique their designs. Are they acceptable or not? Is one better than the others? Explain.

47. ‖ A typical coal-fired power plant burns 300 metric tons of coal *every hour* to generate 750 MW of electricity. 1 metric ton = 1000 kg. The density of coal is 1500 kg/m³ and its heat of combustion is 28 MJ/kg. Assume that *all* heat is transferred from the fuel to the boiler and that *all* the work done in spinning the turbine is transformed into electric energy.
    a. Suppose the coal is piled up in a 10 m × 10 m room. How tall must the pile be to operate the plant for one day?
    b. What is the power plant's thermal efficiency?

48. ‖ A nuclear power plant generates 3000 MW of heat energy from nuclear reactions in the reactor's core. This energy is used to boil water and produce high-pressure steam at 300°C. The steam spins a turbine, which produces 1000 MW of electric power, then the steam is condensed and the water is cooled to 25°C before starting the cycle again.
    a. What is the maximum possible thermal efficiency of the power plant?
    b. What is the plant's actual efficiency?
    c. Cooling water from a river flows through the condenser (the low-temperature heat exchanger) at the rate of 1.2×10^8 L/h (≈ 30 million gallons per hour). If the river water enters the condenser at 18°C, what is its exit temperature?

49. ‖ The electric output of a power plant is 750 MW. Cooling water flows through the power plant at the rate 1.0×10^8 L/h. The cooling water enters the plant at 16°C and exits at 27°C. What is the power plant's thermal efficiency?

50. ‖ a. A large nuclear power plant has a power output of 1000 MW. In other words, it generates electric energy at the rate 1000 MJ/s. How much energy does this power plant supply in one day?
    b. The oceans are vast. How much energy could be extracted from 1 km³ of water if its temperature were decreased by 1°C? For simplicity, assume fresh water.
    c. A friend of yours who is an inventor comes to you with an idea. He has done the calculations that you just did in parts a and b, and he's concluded that a few cubic kilometers of ocean water could meet most of the energy needs of the United States. This is an insignificant fraction of the U.S. coastal waters. In addition, the oceans are constantly being reheated by the sun, so energy obtained from the ocean is essentially solar energy. He has sketched out some design plans—highly secret, of course, because they're not patented—and now he needs some investors to provide money for a prototype. A working prototype will lead to a patent. As an initial investor, you'll receive a fraction of all future royalties. Time is of the essence because a rival inventor is working on the same idea. He needs $10,000 from you right away. You could make millions if it works out. Will you invest? If so, explain why. If not, why not? Either way, your explanation should be based on scientific principles. Sketches and diagrams are a reasonable part of an explanation.

51. ‖ An air conditioner removes 5.0×10^5 J/min of heat from a house and exhausts 8.0×10^5 J/min to the hot outdoors.
    a. How much power does the air conditioner's compressor require?
    b. What is the air conditioner's coefficient of performance?

52. ‖ A heat engine using 1.0 mol of a monatomic gas follows the cycle shown in FIGURE P19.52. 3750 J of heat energy is transferred to the gas during process $1 \rightarrow 2$.
    a. Determine W_s, Q, and ΔE_{th} for each of the four processes in this cycle. Display your results in a table.
    b. What is the thermal efficiency of this heat engine?

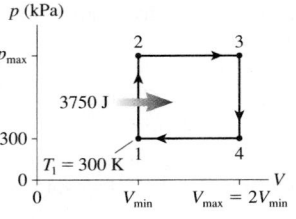

FIGURE P19.52

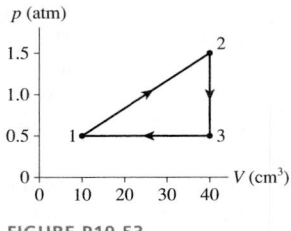

FIGURE P19.53

53. ‖ A heat engine using a diatomic gas follows the cycle shown in FIGURE P19.53. Its temperature at point 1 is 20°C.
    a. Determine W_s, Q, and ΔE_{th} for each of the three processes in this cycle. Display your results in a table.
    b. What is the thermal efficiency of this heat engine?
    c. What is the power output of the engine if it runs at 500 rpm?

54. ‖ FIGURE P19.54 shows the cycle for a heat engine that uses a gas having $\gamma = 1.25$. The initial temperature is $T_1 = 300$ K, and this engine operates at 20 cycles per second.
    a. What is the power output of the engine?
    b. What is the engine's thermal efficiency?

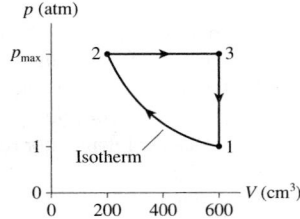

FIGURE P19.54

55. ‖ A heat engine using a monatomic gas follows the cycle shown in FIGURE P19.55.
    a. Find W_s, Q, and ΔE_{th} for each process in the cycle. Display your results in a table.
    b. What is the thermal efficiency of this heat engine?

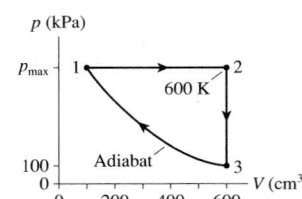

FIGURE P19.55

56. ‖ A heat engine uses a diatomic gas that follows the pV cycle in FIGURE P19.56.
    a. Determine the pressure, volume, and temperature at point 2.
    b. Determine ΔE_{th}, W_s, and Q for each of the three processes. Put your results in a table for easy reading.
    c. How much work does this engine do per cycle and what is its thermal efficiency?

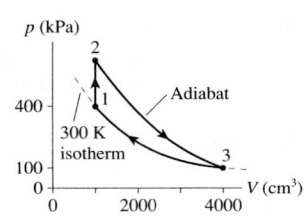

FIGURE P19.56

57. ‖ A heat engine uses a diatomic gas that follows the *pV* cycle in **FIGURE P19.57**.
    a. Determine the pressure, volume, and temperature at point 1.
    b. Determine ΔE_{th}, W_s, and Q for each of the three processes. Put your results in a table for easy reading.
    c. How much work does this engine do per cycle and what is its thermal efficiency?

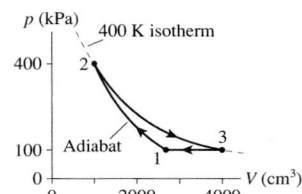

FIGURE P19.57

58. ‖ A refrigerator using helium gas operates on the reversed cycle shown in **FIGURE P19.58**. What are the refrigerator's (a) coefficient of performance and (b) power input if it operates at 60 cycles per second?

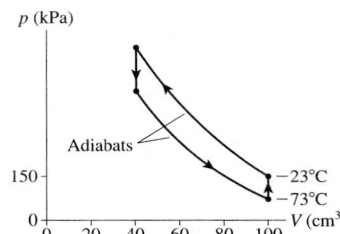

FIGURE P19.58

59. ‖ A heat engine using 120 mg of helium as the working substance follows the cycle shown in **FIGURE P19.59**.
    a. Determine the pressure, temperature, and volume of the gas at points 1, 2, and 3.
    b. What is the engine's thermal efficiency?
    c. What is the maximum possible efficiency of a heat engine that operates between T_{max} and T_{min}?

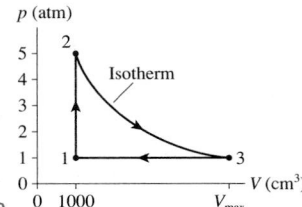

FIGURE P19.59

60. ‖ The heat engine shown in **FIGURE P19.60** uses 2.0 mol of a monatomic gas as the working substance.
    a. Determine T_1, T_2, and T_3.
    b. Make a table that shows ΔE_{th}, W_s, and Q for each of the three processes.
    c. What is the engine's thermal efficiency?

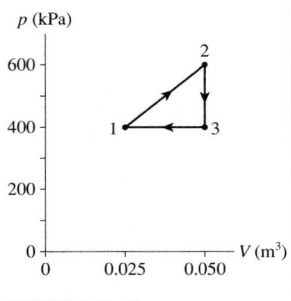

FIGURE P19.60

61. ‖ The heat engine shown in **FIGURE P19.61** uses 0.020 mol of a diatomic gas as the working substance.
    a. Determine T_1, T_2, and T_3.
    b. Make a table that shows ΔE_{th}, W_s, and Q for each of the three processes.
    c. What is the engine's thermal efficiency?

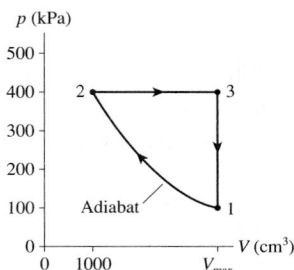

FIGURE P19.61

62. ‖‖ A heat engine using a diatomic ideal gas goes through the following closed cycle:
    ■ Isothermal compression until the volume is halved.
    ■ Isobaric expansion until the volume is restored to its initial value.
    ■ Isochoric cooling until the pressure is restored to its initial value.

    What are the thermal efficiencies of (a) this heat engine and (b) a Carnot engine operating between the highest and lowest temperatures reached by this engine?

63. ‖‖ A heat engine with 0.20 mol of a monatomic ideal gas initially fills a 2000 cm³ cylinder at 600 K. The gas goes through the following closed cycle:
    ■ Isothermal expansion to 4000 cm³.
    ■ Isochoric cooling to 300 K.
    ■ Isothermal compression to 2000 cm³.
    ■ Isochoric heating to 600 K.

    How much work does this engine do per cycle and what is its thermal efficiency?

64. ‖ **FIGURE P19.64** is the *pV* diagram of Example 19.2, but now the device is operated in reverse.
    a. During which processes is heat transferred into the gas?
    b. Is this Q_H, heat extracted from a hot reservoir, or Q_C, heat extracted from a cold reservoir? Explain.
    c. Determine the values of Q_H and Q_C.
    Hint: The calculations have been done in Example 19.2 and do not need to be repeated. Instead, you need to determine which processes now contribute to Q_H and which to Q_C.
    d. Is the area inside the curve W_{in} or W_{out}? What is its value?
    e. The device is now being operated in a ccw cycle. Is it a refrigerator? Explain.

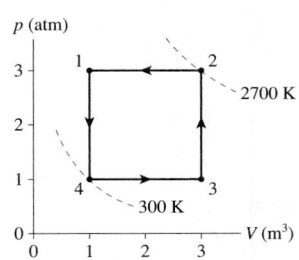

FIGURE P19.64

In Problems 65 through 68 you are given the equation(s) used to solve a problem. For each of these, you are to
  a. Write a realistic problem for which this is the correct equation(s).
  b. Finish the solution of the problem.

65. $0.80 = 1 - (0°C + 273)/(T_H + 273)$

66. $4.0 = Q_C/W_{in}$
    $Q_H = 100$ J

67. $0.20 = 1 - Q_C/Q_H$
    $W_{out} = Q_H - Q_C = 20$ J

68. 400 kJ $= \frac{1}{2}(p_{max} - 100$ kPa$)(3.0$ m$^3 - 1.0$ m$^3)$

Challenge Problems

69. **FIGURE CP19.69** shows a heat engine going through one cycle. The gas is diatomic. The masses are such that when the pin is removed, in steps 3 and 6, the piston does not move.
    a. Draw the pV diagram for this heat engine.
    b. How much work is done per cycle?
    c. What is this engine's thermal efficiency?

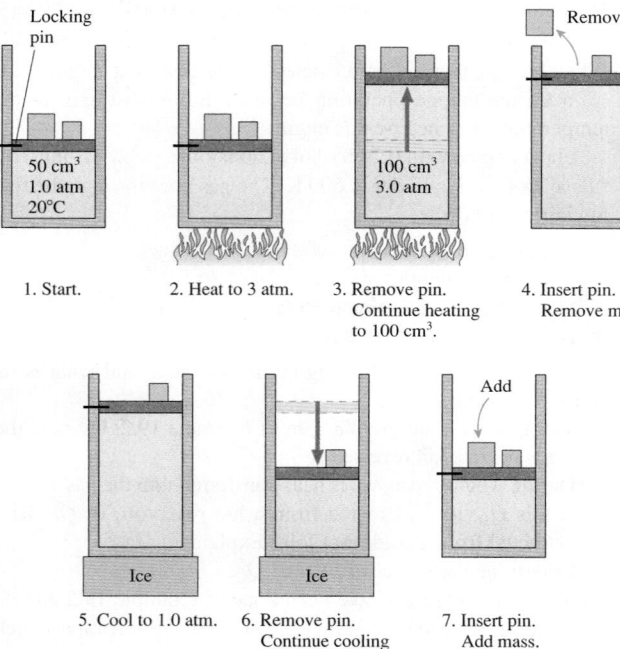

1. Start.   2. Heat to 3 atm.   3. Remove pin.
                                     Continue heating
                                     to 100 cm³.   4. Insert pin.
                                                       Remove mass.

5. Cool to 1.0 atm.   6. Remove pin.
                          Continue cooling
                          to 50 cm³.   7. Insert pin.
                                           Add mass.
                                           Start again.

FIGURE CP19.69

70. **FIGURE CP19.70** shows two insulated compartments separated by a thin wall. The left side contains 0.060 mol of helium at an initial temperature of 600 K and the right side contains 0.030 mol of helium at an initial temperature of 300 K. The compartment on the right is attached to a vertical cylinder, above which the air pressure is 1.0 atm. A 10-cm-diameter, 2.0 kg piston can slide without friction up and down the cylinder. Neither the cylinder diameter nor the volumes of the compartments are known.

  a. What is the final temperature?
  b. How much heat is transferred from the left side to the right side?
  c. How high is the piston lifted due to this heat transfer?
  d. What fraction of the heat is converted into work?

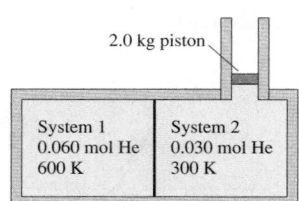

FIGURE CP19.70

71. The gasoline engine in your car can be modeled as the Otto cycle shown in **FIGURE CP19.71**. A fuel-air mixture is sprayed into the cylinder at point 1, where the piston is at its farthest distance from the spark plug. This mixture is compressed as the piston moves toward the spark plug during the adiabatic *compression stroke*. The spark plug fires at point 2, releasing heat energy that had been stored in the gasoline. The fuel burns so quickly that the piston doesn't have time to move, so the heating is an isochoric process. The hot, high-pressure gas then pushes the piston outward during the *power stroke*. Finally, an exhaust value opens to allow the gas temperature and pressure to drop back to their initial values before starting the cycle over again.
    a. Analyze the Otto cycle and show that the work done per cycle is

$$W_{out} = \frac{nR}{1 - \gamma}(T_2 - T_1 + T_4 - T_3)$$

   b. Use the adiabatic connection between T_1 and T_2 and also between T_3 and T_4 to show that the thermal efficiency of the Otto cycle is

$$\eta = 1 - \frac{1}{r^{(\gamma-1)}}$$

   where $r = V_{max}/V_{min}$ is the engine's *compression ratio*.
   c. Graph η versus r out to $r = 30$ for a diatomic gas.

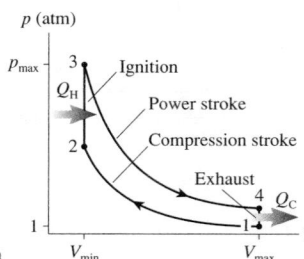

FIGURE CP19.71

72. **FIGURE CP19.72** shows the Diesel cycle. It is similar to the Otto cycle (see Problem CP19.71), but there are two important differences. First, the fuel is not admitted until the air is fully compressed at point 2. Because of the high temperature at the end of an adiabatic compression, the fuel begins to burn spontaneously. (There are no spark plugs in a diesel engine!) Second,

combustion takes place more slowly, with fuel continuing to be injected. This makes the ignition stage a constant-pressure process. The cycle shown, for one cylinder of a diesel engine, has a *displacement* $V_{max} - V_{min}$ of 1000 cm^3 and a compression ratio $r = V_{max}/V_{min} = 21$. These are typical values for a diesel truck. The engine operates with intake air ($\gamma = 1.40$) at 25°C and 1.0 atm pressure. The quantity of fuel injected into the cylinder has a heat of combustion of 1000 J.

a. Find p, V, and T at each of the four corners of the cycle. Display your results in a table.

b. What is the net work done by the cylinder during one full cycle?

c. What is the thermal efficiency of this engine?

d. What is the power output in kW and horsepower (1 hp = 746 W) of an eight-cylinder diesel engine running at 2400 rpm?

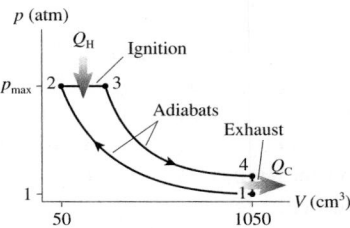

FIGURE CP19.72

STOP TO THINK ANSWERS

Stop to Think 19.1: $W_d > W_a = W_b > W_c$. $W_{out} = Q_H - Q_C$.

Stop to Think 19.2: b. Energy conservation requires $Q_H = Q_C + W_{in}$. The refrigerator will exhaust more heat out the back than it removes from the front. A refrigerator with an open door will heat the room rather than cool it.

Stop to Think 19.3: c. W_{out} = area inside triangle = 1000 J. $\eta = W_{out}/Q_H = (1000\ \text{J})/(4000\ \text{J}) = 0.25$.

Stop to Think 19.4: To conserve energy, the heat Q_H exhausted to the hot reservoir needs to be $Q_H = Q_C + W_{in} = 40\ \text{J} + 10\ \text{J} = 50\ \text{J}$, not 30 J. The numbers shown here would be appropriate to a heat engine if the energy-transfer arrows were all reversed.

Stop to Think 19.5: b. The efficiency of this engine would be $\eta = W_{out}/Q_H = 0.6$. That exceeds the Carnot efficiency $\eta_{Carnot} = 1 - T_C/T_H = 0.5$, so it is not possible.

Thermodynamics

Part IV had two important goals: first, to learn how energy is transformed; second, to establish a micro/macro connection in which we can understand the macroscopic properties of solids, liquids, and gases in terms of the microscopic motions of atoms and molecules. We have been quite successful. You have learned that:

- Temperature is a measure of the thermal energy of the molecules in a system, and the average energy per molecule is simply $\frac{1}{2}k_B T$ per degree of freedom.

- The pressure of a gas is due to collisions of the molecules with the walls of the container.

- Heat is the energy transferred between two systems that have different temperatures. An important mechanism of heat transfer is molecular collisions at the boundary between the two systems.

- Work, heat, and thermal energy can be transformed into each other in accord with the first law of thermodynamics, $\Delta E_{th} = W + Q$. This is a statement that energy is conserved.

- Practical devices for turning heat into work, called heat engines, are limited in their efficiency by the second law of thermodynamics.

The knowledge structure of thermodynamics below summarizes the basic laws, diagramming our energy model and presenting our model of a heat engine in pictorial form. Thermodynamics, more than most topics in physics, can seem very "equation oriented." It's undeniable that there are more equations than we used in earlier parts of this text and more things to remember. But focusing on the equations is seeing only the trees, not the forest. A better strategy is to focus on the ideas embedded in the knowledge structure. You can find the necessary equations if you know how the ideas are connected, but memorizing all the equations won't help if you don't know which are relevant to different situations.

KNOWLEDGE STRUCTURE IV **Thermodynamics**

ESSENTIAL CONCEPTS	Work, heat, and thermal energy
BASIC GOALS	How is energy converted from one form to another?
	How are macroscopic properties related to microscopic behavior?

GENERAL PRINCIPLES	**First law of thermodynamics**	Energy is conserved, $\Delta E_{th} = W + Q$.
	Second law of thermodynamics	Heat is not spontaneously transferred from a colder object to a hotter object.

GAS LAWS AND PROCESSES Ideal-gas law $pV = nRT = Nk_B T$

- Isochoric process $V = $ constant and $W = 0$
- Isothermal process $T = $ constant and $\Delta E_{th} = 0$

- Isobaric process $p = $ constant
- Adiabatic process $Q = 0$

Energy Transformation

Work on system

Environment

Work by system

$W > 0$

System
Thermal energy E_{th}
+
Other state variables
$p, V, T, n, M, \ldots$

$W < 0$

Energy in

Energy out

$Q > 0$ First law: $\Delta E_{th} = W + Q$ $Q < 0$

Heat to system

Heat out of system

Work

Requires volume change

Gas: $W = -\int p\,dV$

$= -$(area under pV curve)

Thermal Energy

$E_{th} = \frac{1}{2}Nk_B T$ per degree of freedom

Heat

Requires temperature difference

$Q = Mc\Delta T$ or $nC\Delta T$

$Q = \pm ML$ for phase changes

Heat Engines

$W_{out} = $ area inside pV curve

$= Q_H - Q_C$

$\eta = \dfrac{W_{out}}{Q_H}$

$\eta_{max} = \eta_{Carnot} = 1 - \dfrac{T_C}{T_H}$

Hot reservoir T_H

Q_H

Heat engine

W_{out}

Q_C

Cold reservoir T_C

Order Out of Chaos

The second law predicts that systems will run down, that order will evolve toward disorder and randomness, and that complexity will give way to simplicity. But just look around you!

■ Plants grow from simple seeds to complex entities.

■ Single-cell fertilized eggs grow into complex adult organisms.

■ Electric current passing through a "soup" of simple random molecules produces such complex chemicals as amino acids.

■ Over the last billion or so years life has evolved from simple unicellular organisms to very complex forms.

■ Knowledge and information seem to grow every year, not to fade away.

Everywhere we look, it seems, the second law is being violated. How can this be?

There is an important qualification in the second law of thermodynamics: It applies only to *isolated* systems, systems that do not exchange energy with their environment. The situation is entirely different if energy is transferred into or out of the system, and we cannot predict what will happen to the entropy of a nonisolated system. The popular-science literature is full of arguments and predictions that make incorrect use of the second law by trying to apply it to systems that are not isolated.

Systems that become *more* ordered as time passes, and in which the entropy decreases, are called *self-organizing systems.* All the examples listed above are self-organizing systems. One of the major characteristics of self-organizing systems is a substantial flow of energy *through* the system. For example, plants and animals take in energy from the sun or chemical energy from food, make use of that energy, and then give waste heat back to the environment via evaporation, decay, and other means. It is this energy flow that allows the systems to maintain, or even increase, a high degree of order and a very low entropy.

But—and this is the important point—the entropy of the *entire* system, including the earth and the sun, undergoes a significant *increase* so as to let selected subsystems decrease their entropy and become more ordered. The second law is not violated at all, but you must apply the second law to the combined systems that are interacting and not just to a single subsystem.

The snowflake in the photo is a beautiful example. As water freezes, the random motion of water molecules is transformed into a highly ordered crystal. The entropy of

A snowflake is a highly ordered arrangement of water molecules. The creation of a snowflake decreases the entropy of the water, but the second law of thermodynamics is not violated because the water molecules are not an isolated system.

the water molecules certainly decreases, but water doesn't freeze as an isolated system. For it to freeze, heat energy must be transferred from the water to the surrounding air. The entropy of the air increases by *more* than the entropy of the water decreases. Thus the *total* entropy of the water + air system increases when a snowflake is formed, just as the second law predicts.

Self-organization is closely related to nonlinear mechanics, chaos, and the geometry of fractals. It has important applications in fields ranging from ecology to computer science to aeronautical engineering. For example, the airflow across a wing gives rise to large-scale turbulence—eddies and whirlpools—in the wake behind an airplane. Their formation affects the aerodynamics of the plane and can also create hazards for following aircraft. Whirlpools are ordered, large-scale macroscopic structures with low entropy, but they are produced from disordered, random collisions of the air molecules.

Self-organizing systems are a very active field of research in both science and engineering. The 1977 Nobel Prize in chemistry was awarded to the Belgian scientist Ilya Prigogine for his studies of *nonequilibrium thermodynamics,* the basic science underlying self-organizing systems. Prigogine and others have shown how energy flow through a system can, when the conditions are right, "bring order out of chaos."

Waves and Optics

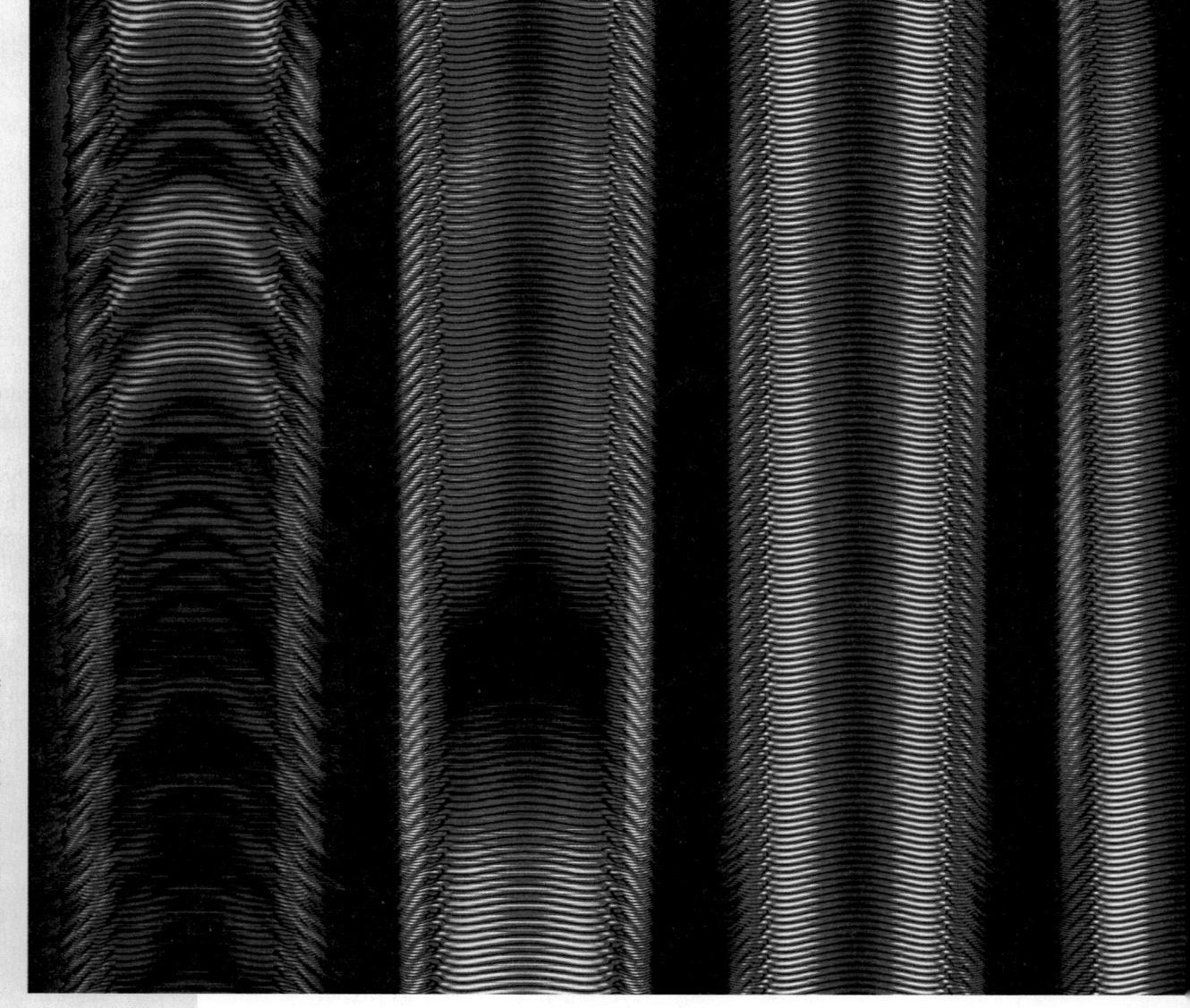

The song of a humpback whale can travel hundreds of kilometers underwater. This graph uses a procedure called wavelet analysis to study the frequency structure of a humpback whale song.

OVERVIEW

The Wave Model

Parts I–IV of this text have been primarily about the physics of particles. You've seen that macroscopic systems ranging from balls and rockets to a gas of molecules can be thought of as particles or as systems of particles. A *particle* is one of the two fundamental models of classical physics. The other, to which we now turn our attention, is a *wave*.

Waves are ubiquitous in nature. Familiar examples of waves include

- Undulating ripples on a pond.
- The swaying ground of an earthquake.
- A vibrating guitar string.
- The sweet sound of a flute.
- The colors of the rainbow.

The physics of waves is the subject of Part V, the next stage of our journey. Despite the great diversity of types and sources of waves, a single, elegant physical theory is capable of describing them all. Our exploration of wave phenomena will call upon sound waves, light waves, and vibrating strings for examples, but our goal is to emphasize the unity and coherence of the ideas that are common to *all* types of waves.

A wave, in contrast with a particle, is diffuse, spread out, not to be found at a single point in space. We will start with waves traveling outward through some medium, like the spreading ripples after a pebble hits a pool of water. These are called *traveling waves.* An investigation of what happens when waves travel through each other will lead us to *standing waves,* which are essential for understanding phenomena ranging from those as common as musical instruments and water sloshing in a tub to as complex as lasers and the electrons in atoms. We'll also study one of the most important defining characteristics of waves—their ability to exhibit *interference.*

Three chapters will be devoted to light and optics, perhaps the most important application of waves. Although light is an electromagnetic wave, your understanding of these chapters will depend on nothing more than the "waviness" of light. You can study these chapters either before or after your study of electricity and magnetism in Part VI. The electromagnetic aspects of light waves will be taken up in Chapter 34.

Our investigation of light will be aided by a second model, the *ray model,* in which light travels in straight lines, reflects from mirrors, and is focused by lenses. Many practical applications of optics, from the camera to the telescope, are best understood with the ray model of light.

In fact, that you're able to read this book at all is due to the first optical instrument you ever used—your eyes. We will investigate the optics of the eye, learn how the cornea and lens form an image on the retina, and see how glasses or contact lenses can be used to correct the image if it is out of focus.

20 Traveling Waves

This surfer is "catching a wave." At the same time, he's seeing light waves and hearing sound waves.

▶ **Looking Ahead** The goal of Chapter 20 is to learn the basic properties of traveling waves.

The Wave Model

A **wave** is a disturbance traveling through a medium. Our goal is to develop a model —the wave model—that describes the basic properties of all waves.

The wave propagates, but the particles of the medium don't. The water molecules simply oscillate up and down as the ripples spread outward.

Two Types of Waves

You'll find that waves come in two basic types:

Transverse waves: The displacement is perpendicular to the direction of travel.

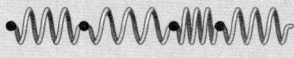

Longitudinal waves: The displacement is parallel to the direction of travel.

Sound and Light

Two types of waves are especially important: sound and light.
- Sound waves are longitudinal waves.
- Light waves are transverse waves.

You'll learn that the colors of visible light correspond to different wavelengths.

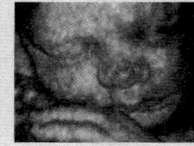

Ultrasound images are made with very-high-frequency sound waves.

Wave Properties

You'll learn that a wave is characterized by three basic properties:
- **Wave speed**: How fast it travels through the medium.
- **Wavelength**: The distance between two neighboring crests.
- **Frequency**: The number of oscillations per second.

You'll also see that wave motion is closely related to simple harmonic motion.

◀ **Looking Back**
Sections 14.1 and 14.2 Properties of simple harmonic motion

Intensity and Loudness

Waves carry energy. The rate at which a wave delivers energy to a surface is the **intensity** of the wave.

Your ears are sensitive to a remarkable range of intensities. You'll learn to use the logarithmic **decibel** scale to characterize the loudness of a sound.

Focusing the sun's light into a smaller area increases its intensity.

The Doppler Effect

The frequency and wavelength of a wave are shifted when there is relative motion between the source and the observer of the waves. This is called the **Doppler effect.**

The pitch of the ambulance siren drops as it races past you. The frequency is shifted up as it approaches, then shifted down as it recedes.

20.1 The Wave Model

Balls, cars, and rockets obviously differ from one another, but the general features of their motions are well described by the *particle model* of Parts I–IV. In Part V we will explore the basic properties of waves with a **wave model,** emphasizing those aspects of wave behavior common to all waves. Although water waves, sound waves, and light waves are clearly different, the wave model will allow us to understand many of the important features they have in common.

The wave model is built around the idea of a **traveling wave,** which is an organized disturbance traveling with a well-defined wave speed. We'll begin our study of traveling waves by looking at two distinct wave motions.

Two types of traveling waves

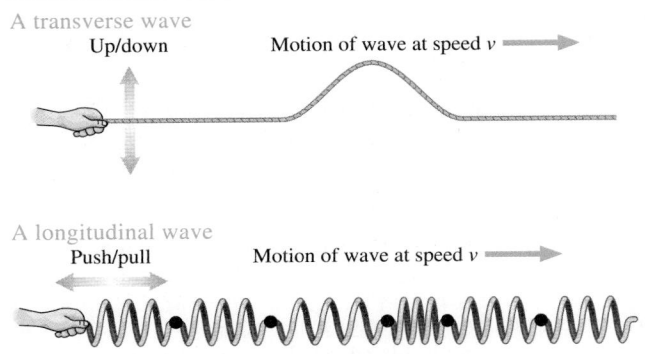

A transverse wave
Up/down   Motion of wave at speed v ⟶

A **transverse wave** is a wave in which the displacement is *perpendicular* to the direction in which the wave travels. For example, a wave travels along a string in a horizontal direction while the particles that make up the string oscillate vertically. Electromagnetic waves are also transverse waves because the electromagnetic fields oscillate perpendicular to the direction in which the wave travels.

A longitudinal wave
Push/pull   Motion of wave at speed v ⟶

In a **longitudinal wave,** the particles in the medium move *parallel* to the direction in which the wave travels. Here we see a chain of masses connected by springs. If you give the first mass in the chain a sharp push, a disturbance travels down the chain by compressing and expanding the springs. Sound waves in gases and liquids are the most well known examples of longitudinal waves.

We can also classify waves on the basis of what is "waving":

1. **Mechanical waves** travel only within a material *medium*, such as air or water. Two familiar mechanical waves are sound waves and water waves.
2. **Electromagnetic waves,** from radio waves to visible light to x rays, are a self-sustaining oscillation of the *electromagnetic field*. Electromagnetic waves require no material medium and can travel through a vacuum.

The **medium** of a mechanical wave is the substance through or along which the wave moves. For example, the medium of a water wave is the water, the medium of a sound wave is the air, and the medium of a wave on a stretched string is the string. A medium must be *elastic*. That is, a restoring force of some sort brings the medium back to equilibrium after it has been displaced or disturbed. The tension in a stretched string pulls the string back straight after you pluck it. Gravity restores the level surface of a lake after the wave generated by a boat has passed by.

As a wave passes through a medium, the atoms of the medium—we'll simply call them the particles of the medium—are displaced from equilibrium. This is a **disturbance** of the medium. The water ripples of **FIGURE 20.1** are a disturbance of the water's surface. A pulse traveling down a string is a disturbance, as are the wake of a boat and the sonic boom created by a jet traveling faster than the speed of sound. **The disturbance of a wave is an *organized* motion of the particles in the medium,** in contrast to the *random* molecular motions of thermal energy.

FIGURE 20.1 Ripples on a pond are a traveling wave.

⋯⋯The disturbance is the rippling of the water's surface.

The water is the medium.⋯⋯

Wave Speed

A wave disturbance is created by a *source*. The source of a wave might be a rock thrown into water, your hand plucking a stretched string, or an oscillating loudspeaker cone pushing on the air. Once created, the disturbance travels outward through the medium at the **wave speed** v. This is the speed with which a ripple moves across the water or a pulse travels down a string.

NOTE ▶ The disturbance propagates through the medium, but **the medium as a whole does not move!** The ripples on the pond (the disturbance) move outward from the splash of the rock, but there is no outward flow of water from the splash. Likewise, the particles of a string oscillate up and down but do not move in the direction of a pulse traveling along the string. **A wave transfers energy, but it does not transfer any material or substance outward from the source.** ◀

As an example, we'll prove in Section 20.3 that the wave speed on a string stretched with tension T_s is

$$v_{string} = \sqrt{\frac{T_s}{\mu}} \qquad \text{(wave speed on a stretched string)} \qquad (20.1)$$

where μ is the string's **linear density,** its mass-to-length ratio:

$$\mu = \frac{m}{L} \qquad (20.2)$$

The SI unit of linear density is kg/m. A fat string has a larger value of μ than a skinny string made of the same material. Similarly, a steel wire has a larger value of μ than a plastic string of the same diameter. We'll assume that strings are *uniform,* meaning the linear density is the same everywhere along the length of the string.

NOTE ▶ The subscript s on the symbol T_s for the string's tension distinguishes it from the symbol T for the *period* of oscillation. ◀

Equation 20.1 is the wave *speed,* not the wave velocity, so v_{string} always has a positive value. Every point on a wave travels with this speed. You can increase the wave speed either by *increasing* the string's tension (make it tighter) or by *decreasing* the string's linear density (make it skinnier). We'll examine the implications for stringed musical instruments in Chapter 21.

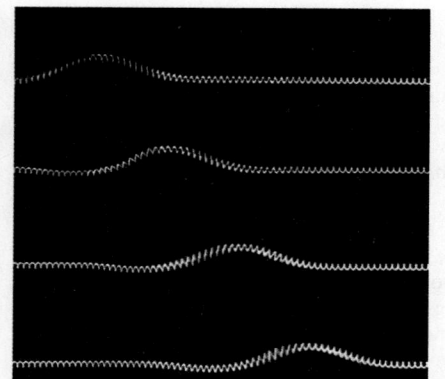

This sequence of photographs shows a wave pulse traveling along a spring.

EXAMPLE 20.1   Measuring the linear density

In a laboratory experiment, one end of a metal wire is connected to a motion sensor. The wire is stretched horizontally to a pulley 1.50 m away, then attached to a hanging mass that provides tension. A mechanical pick plucks the horizontal segment of the wire right at the pulley, creating a small wave pulse that travels along the wire. The plucking motion starts a timer that is stopped by the motion sensor when the pulse reaches the end of the wire. Changing the hanging mass changes the time required for the pulse to travel the length of the wire. The data are as follows:

Mass (kg)	Time (ms)
0.50	31
1.00	23
1.50	18
2.00	15
2.50	14

Use the data to determine the wire's linear density.

MODEL The wave pulse is a traveling wave on a stretched string. The hanging mass is in static equilibrium.

VISUALIZE **FIGURE 20.2** is a pictorial representation.

SOLVE The wave speed on the wire is determined by the wire's linear density μ and tension T_s. The hanging mass is in static

FIGURE 20.2 A wave pulse on the wire.

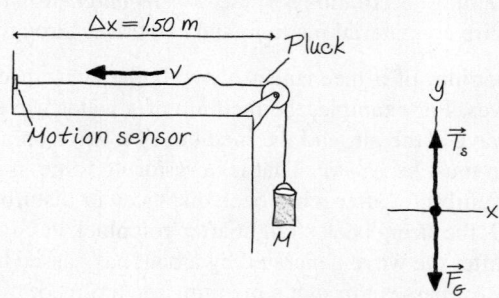

equilibrium, with no net force, so we see from the free-body diagram that the tension in the wire is $T_s = F_G = Mg$. Squaring both sides of Equation 20.1 gives

$$v^2 = \left(\frac{\Delta x}{\Delta t}\right)^2 = \frac{T_s}{\mu} = \frac{Mg}{\mu}$$

Mass M is the independent variable that we've changed, each time measuring the pulse travel time Δt, so we can rearrange the wave-speed equation as

$$(\Delta t)^2 = \frac{\mu (\Delta x)^2}{g} \frac{1}{M}$$

Theory predicts that a graph of the *square* of the travel time versus the *inverse* of the hanging mass should be a straight line passing through the origin with slope $\mu(\Delta x)^2/g$. The graph of FIGURE 20.3, with the times converted from ms to s, is indeed linear with a *y*-intercept of zero. The slope of the best-fit line is seen to be 4.85×10^{-4} kg s^2 (recall that spreadsheets and graphing calculators display this as 4.85E–04), from which we find the wire's linear density:

$$\mu = \frac{g \times \text{slope}}{(\Delta x)^2} = 0.0021 \text{ kg/m} = 2.1 \text{ g/m}$$

ASSESS A meter of thin wire is likely to have a mass of a few grams, so a linear density of a few g/m seems reasonable.

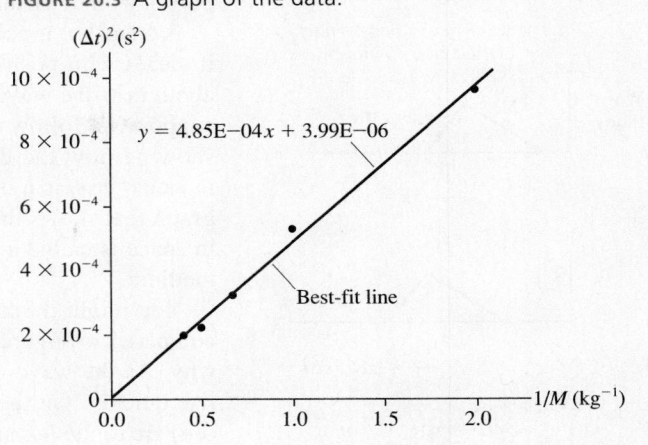

FIGURE 20.3 A graph of the data.

$y = 4.85\text{E}{-}04x + 3.99\text{E}{-}06$

Best-fit line

The wave speed on a string is a property of the string—its tension and linear density. In general, **the wave speed is a property *of the medium*.** The wave speed depends on the restoring forces within the medium but not at all on the shape or size of the pulse, how the pulse was generated, or how far it has traveled.

STOP TO THINK 20.1 Which of the following actions would make a pulse travel faster along a stretched string? More than one answer may be correct. If so, give all that are correct.

a. Move your hand up and down more quickly as you generate the pulse.
b. Move your hand up and down a larger distance as you generate the pulse.
c. Use a heavier string of the same length, under the same tension.
d. Use a lighter string of the same length, under the same tension.
e. Stretch the string tighter to increase the tension.
f. Loosen the string to decrease the tension.
g. Put more force into the wave.

20.2 One-Dimensional Waves

To understand waves we must deal with functions of *two* variables. Until now, we have been concerned with quantities that depend only on time, such as $x(t)$ or $v(t)$. Functions of the one variable *t* are all right for a particle because a particle is only in one place at a time, but a wave is not localized. It is spread out through space at each instant of time. To describe a wave mathematically requires a function that specifies not only an instant of time (when) but also a point in space (where).

Rather than leaping into mathematics, we will start by thinking about waves graphically. Consider the wave pulse shown moving along a stretched string in FIGURE 20.4. (We will consider somewhat artificial triangular and square-shaped pulses in this section to make clear where the edges of the pulse are.) The graph shows the string's displacement Δy at a particular instant of time t_1 as a function of position *x* along the string. This is a "snapshot" of the wave, much like what you might make with a camera whose shutter is opened briefly at t_1. A graph that shows the wave's displacement as a function of position at a single instant of time is called a **snapshot graph.** For a wave on a string, a snapshot graph is literally a picture of the wave at this instant.

FIGURE 20.5 shows a sequence of snapshot graphs as the wave of Figure 20.4 continues to move. These are like successive frames from a movie. Notice that the wave

FIGURE 20.4 A snapshot graph of a wave pulse on a string.

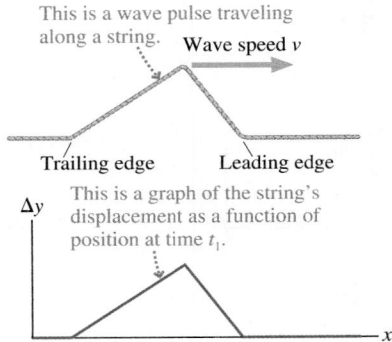

This is a wave pulse traveling along a string. Wave speed *v*

Trailing edge   Leading edge

This is a graph of the string's displacement as a function of position at time t_1.

Δy

FIGURE 20.5 A sequence of snapshot graphs shows the wave in motion.

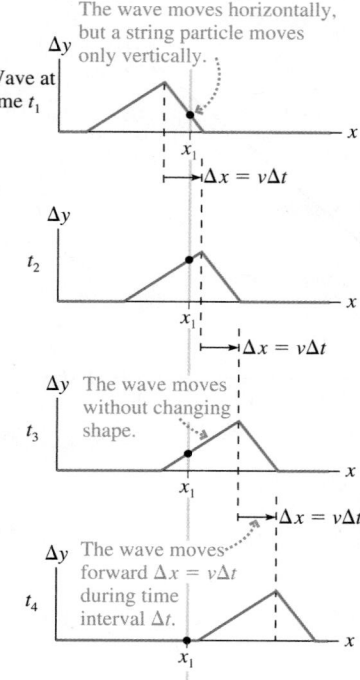

FIGURE 20.5 A sequence of snapshot graphs shows the wave in motion.

pulse moves forward distance $\Delta x = v \Delta t$ during the time interval Δt. That is, the wave moves with constant speed.

A snapshot graph tells only half the story. It tells us *where* the wave is and how it varies with position, but only at one instant of time. It gives us no information about how the wave *changes* with time. As a different way of portraying the wave, suppose we follow the dot marked on the string in Figure 20.5 and produce a graph showing how the displacement of this dot changes with time. The result, shown in FIGURE 20.6, is a displacement-versus-time graph at a single position in space. A graph that shows the wave's displacement as a function of time at a single position in space is called a **history graph**. It tells the history of that particular point in the medium.

You might think we have made a mistake; the graph of Figure 20.6 is reversed compared to Figure 20.5. It is not a mistake, but it requires careful thought to see why. As the wave moves toward the dot, the steep *leading edge* causes the dot to rise quickly. On the displacement-versus-time graph, *earlier* times (smaller values of *t*) are to the *left* and later times (larger *t*) to the right. Thus the leading edge of the wave is on the *left* side of the Figure 20.6 history graph. As you move to the right on Figure 20.6 you see the slowly falling *trailing edge* of the wave as it moves past the dot at later times.

The snapshot graph of Figure 20.4 and the history graph of Figure 20.6 portray complementary information. The snapshot graph tells us how things look throughout all of space, but at only one instant of time. The history graph tells us how things look at all times, but at only one position in space. We need them both to have the full story of the wave. An alternative representation of the wave is the series of graphs in FIGURE 20.7, where we can get a clearer sense of the wave moving forward. But graphs like these are essentially impossible to draw by hand, so it is necessary to move back and forth between snapshot graphs and history graphs.

FIGURE 20.6 A history graph for the dot on the string in Figure 20.5.

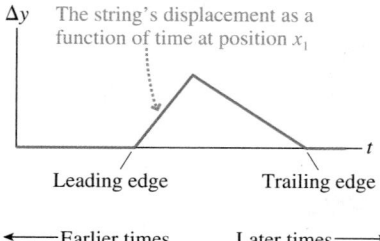

FIGURE 20.7 An alternative look at a traveling wave.

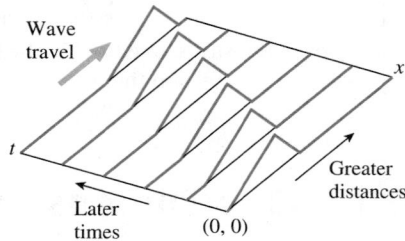

EXAMPLE 20.2 **Finding a history graph from a snapshot graph**

FIGURE 20.8 is a snapshot graph at $t = 0$ s of a wave moving to the right at a speed of 2.0 m/s. Draw a history graph for the position $x = 8.0$ m.

FIGURE 20.8 A snapshot graph at $t = 0$ s.

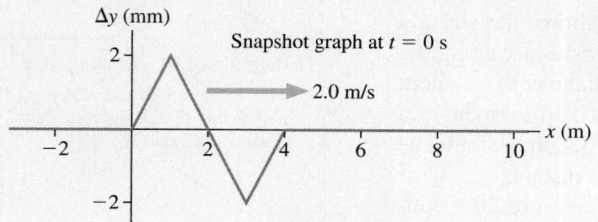

MODEL This is a wave traveling at constant speed. The pulse moves 2.0 m to the right every second.

VISUALIZE The snapshot graph of Figure 20.8 shows the wave at all points on the *x*-axis at $t = 0$ s. You can see that nothing is happening at $x = 8.0$ m at this instant of time because the wave has not yet reached $x = 8.0$ m. In fact, at $t = 0$ s the leading edge of the wave is still 4.0 m away from $x = 8.0$ m. Because the wave is traveling at 2.0 m/s, it will take 2.0 s for the leading edge to reach $x = 8.0$ m. Thus the history graph for $x = 8.0$ m will be zero until $t = 2.0$ s. The first part of the wave causes a *downward* displacement of the medium, so immediately after $t = 2.0$ s the displacement at $x = 8.0$ m will be negative. The negative portion of the wave pulse is 2.0 m wide and takes 1.0 s to pass $x = 8.0$ m, so the midpoint of the pulse reaches $x = 8.0$ m at $t = 3.0$ s. The positive portion takes another 1.0 s to go past, so the trailing edge of the pulse arrives at $t = 4.0$ s. You could also note that the trailing edge was initially 8.0 m away from $x = 8.0$ m and needed 4.0 s to travel that distance at 2.0 m/s. The displacement at $x = 8.0$ m returns to zero at $t = 4.0$ s and remains zero for all later times. This information is all portrayed on the history graph of FIGURE 20.9.

FIGURE 20.9 The corresponding history graph at $x = 8.0$ m.

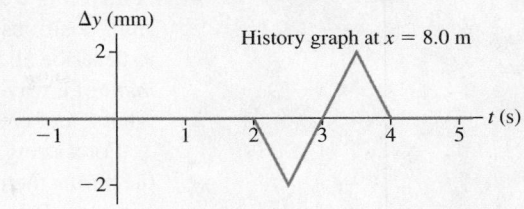

STOP TO THINK 20.2 The graph at the right is the history graph at $x = 4.0$ m of a wave traveling to the right at a speed of 2.0 m/s. Which is the history graph of this wave at $x = 0$ m?

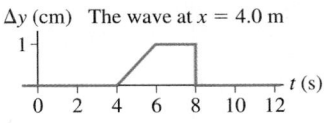

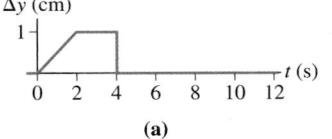

(a)

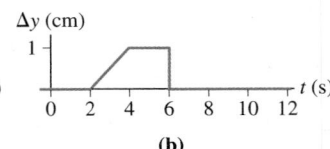

(b)

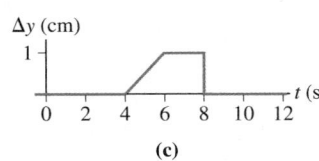

(c)

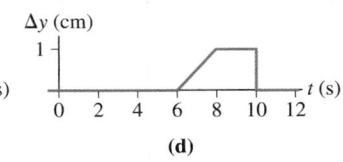

(d)

Longitudinal Waves

For a wave on a string, a transverse wave, the snapshot graph is literally a picture of the wave. Not so for a longitudinal wave, where the particles in the medium are displaced parallel to the direction in which the wave is traveling. Thus the displacement is Δx rather than Δy, and a snapshot graph is a graph of Δx versus x.

FIGURE 20.10a is a snapshot graph of a longitudinal wave, such as a sound wave. It's purposefully drawn to have the same shape as the string wave in Example 20.2. Without practice, it's not clear what this graph tells us about the particles in the medium.

FIGURE 20.10 Visualizing a longitudinal wave.

(a)

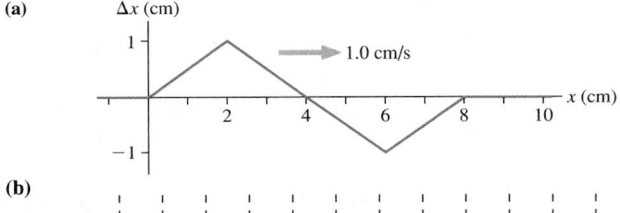

Snapshot graph of a longitudinal wave at $t_1 = 0$ s

(b)

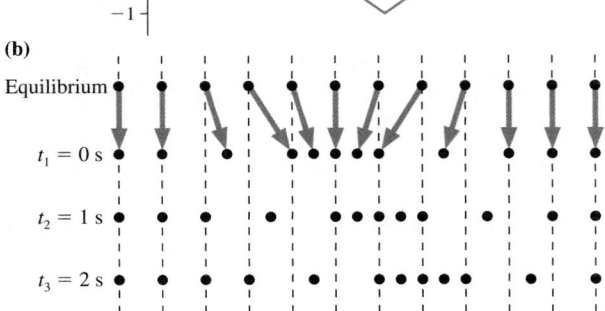

1. Draw a series of equally spaced vertical lines to represent the equilibrium positions of particles before the wave arrives.

2. Use information from the graph to displace the particles in the medium to the right or left.

3. The wave propagates to the right at 1.0 cm/s.

To help you find out, FIGURE 20.10b provides a tool for visualizing longitudinal waves. In the second row, we've used information from the graph to displace the particles in the medium to the right or to the left of their equilibrium positions. For example, the particle at $x = 1.0$ cm has been displaced 0.5 cm to the right because the snapshot graph shows $\Delta x = 0.5$ cm at $x = 1.0$ cm. We now have a picture of the longitudinal wave pulse at $t_1 = 0$ s. You can see that the medium is compressed to higher density at the center of the pulse and, to compensate, expanded to lower density at the leading and trailing edges. Two more lines show the medium at $t_2 = 1$ s and $t_3 = 2$ s so that you can see the wave propagating through the medium at 1.0 cm/s.

You've probably seen or participated in "the wave" at a sporting event. The wave moves around the stadium, but the people (the medium) simply undergo small displacements from their equilibrium positions.

The Displacement

A traveling wave causes the particles of the medium to be displaced from their equilibrium positions. Because one of our goals is to develop a mathematical representation to describe all types of waves, we'll use the generic symbol D to stand for the *displacement* of a wave of any type. But what do we mean by a "particle" in the medium? And what about electromagnetic waves, for which there is no medium?

For a string, where the atoms stay fixed relative to each other, you can think of either the atoms themselves or very small segments of the string as being the particles of the medium. D is then the perpendicular displacement Δy of a point on the string. For a sound wave, D is the longitudinal displacement Δx of a small volume of fluid. For any other mechanical wave, D is the appropriate displacement. Even electromagnetic waves can be described within the same mathematical representation if D is interpreted as a yet-undefined *electromagnetic field strength,* a "displacement" in a more abstract sense as an electromagnetic wave passes through a region of space.

Because the displacement of a particle in the medium depends both on *where* the particle is (position x) and on *when* you observe it (time t), D must be a function of the two variables x and t. That is,

$D(x, t) =$ the displacement at time t of a particle at position x

The values of *both* variables—where and when—must be specified before you can evaluate the displacement D.

20.3 Sinusoidal Waves

A wave source that oscillates with simple harmonic motion (SHM) generates a **sinusoidal wave.** For example, a loudspeaker cone that oscillates in SHM radiates a sinusoidal sound wave. The sinusoidal electromagnetic waves broadcast by television and FM radio stations are generated by electrons oscillating back and forth in the antenna wire with SHM. **The frequency f of the wave is the frequency of the oscillating source.**

FIGURE 20.11 shows a sinusoidal wave moving through a medium. The source of the wave, which is undergoing vertical SHM, is located at $x = 0$. Notice how the wave crests move with steady speed toward larger values of x at later times t.

FIGURE 20.12a is a history graph for a sinusoidal wave, showing the displacement of the medium at one point in space. Each particle in the medium undergoes simple harmonic motion with frequency f, so this graph of SHM is identical to the graphs you learned to work with in Chapter 14. The *period* of the wave, shown on the graph, is the time interval for one cycle of the motion. The period is related to the wave frequency f by

$$T = \frac{1}{f} \tag{20.3}$$

exactly as in simple harmonic motion. The **amplitude** A of the wave is the maximum value of the displacement. The crests of the wave have displacement $D_{\text{crest}} = A$ and the troughs have displacement $D_{\text{trough}} = -A$.

FIGURE 20.11 A sinusoidal wave moving along the x-axis.

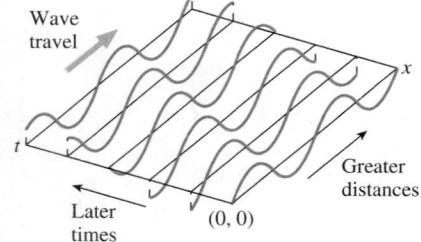

FIGURE 20.12 History and snapshot graphs for a sinusoidal wave.

(a) A history graph at one point in space

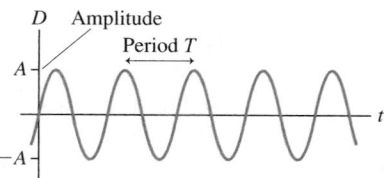

(b) A snapshot graph at one instant of time

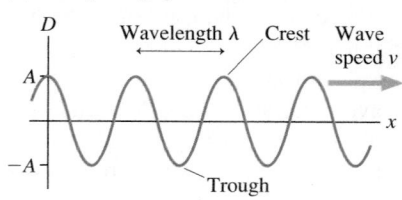

Displacement versus time is only half the story. FIGURE 20.12b shows a snapshot graph for the same wave at one instant in time. Here we see the wave stretched out in space, moving to the right with speed v. An important characteristic of a sinusoidal wave is that it is periodic *in space* as well as in time. As you move from left to right along the "frozen" wave in the snapshot graph, the disturbance repeats itself over and over. The distance spanned by one cycle of the motion is called the **wavelength** of the wave. Wavelength is symbolized by λ (lowercase Greek lambda) and, because it is a length, it is measured in units of meters. The wavelength is shown in Figure 20.12b as the distance between two crests, but it could equally well be the distance between two troughs.

NOTE ▶ Wavelength is the spatial analog of period. The period T is the *time* in which the disturbance at a single point in space repeats itself. The wavelength λ is the *distance* in which the disturbance at one instant of time repeats itself. ◀

The Fundamental Relationship for Sinusoidal Waves

There is an important relationship between the wavelength and the period of a wave. FIGURE 20.13 shows this relationship through five snapshot graphs of a sinusoidal wave at time increments of one-quarter of the period T. One full period has elapsed between the first graph and the last, which you can see by observing the motion at a fixed point on the x-axis. Each point in the medium has undergone exactly one complete oscillation.

The critical observation is that the wave crest marked by an arrow has moved one full wavelength between the first graph and the last. That is, **during a time interval of exactly one period T, each crest of a sinusoidal wave travels forward a distance of exactly one wavelength λ.** Because speed is distance divided by time, the wave speed must be

$$v = \frac{\text{distance}}{\text{time}} = \frac{\lambda}{T} \tag{20.4}$$

Because $f = 1/T$, it is customary to write Equation 20.4 in the form

$$v = \lambda f \tag{20.5}$$

Although Equation 20.5 has no special name, it is *the* fundamental relationship for periodic waves. When using it, keep in mind the *physical* meaning that **a wave moves forward a distance of one wavelength during a time interval of one period.**

NOTE ▶ Wavelength and period are defined only for *periodic* waves, so Equations 20.4 and 20.5 apply only to periodic waves. A wave pulse has a wave speed, but it doesn't have a wavelength or a period. Hence Equations 20.4 and 20.5 cannot be applied to wave pulses. ◀

Because the wave speed is a property of the medium while the wave frequency is a property of the source, it is often useful to write Equation 20.5 as

$$\lambda = \frac{v}{f} = \frac{\text{property of the medium}}{\text{property of the source}} \tag{20.6}$$

The wavelength is a *consequence* of a wave of frequency f traveling through a medium in which the wave speed is v.

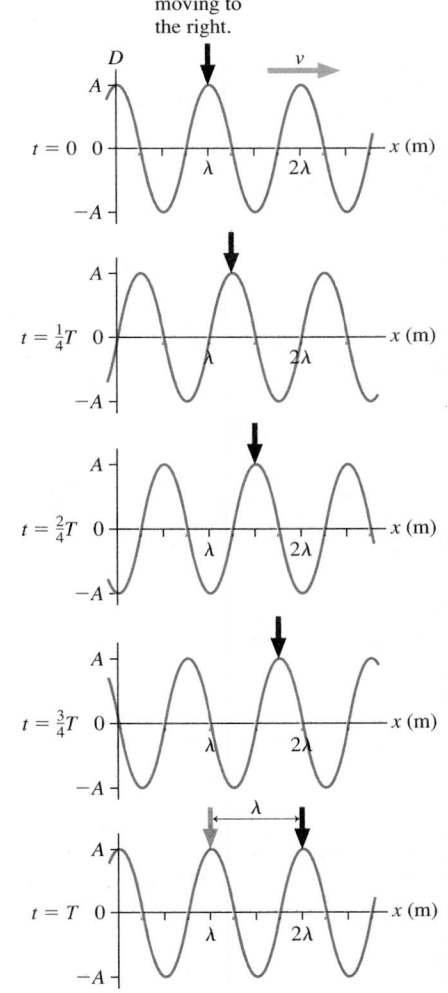

FIGURE 20.13 A series of snapshot graphs at time increments of one-quarter of the period T.

During a time interval of exactly one period, the crest has moved forward exactly one wavelength.

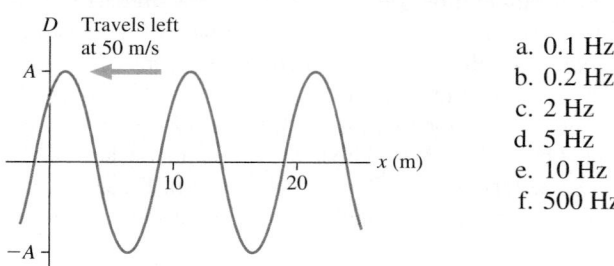

STOP TO THINK 20.3   What is the frequency of this traveling wave?

D Travels left at 50 m/s

a. 0.1 Hz
b. 0.2 Hz
c. 2 Hz
d. 5 Hz
e. 10 Hz
f. 500 Hz

The Mathematics of Sinusoidal Waves

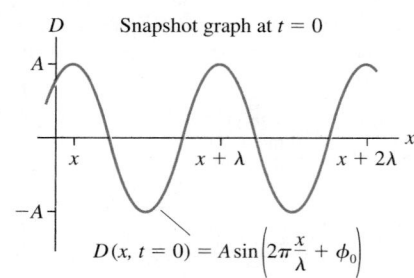

FIGURE 20.14 A sinusoidal wave is "frozen" at $t = 0$.

FIGURE 20.14 shows a snapshot graph at $t = 0$ of a sinusoidal wave. The sinusoidal function that describes the displacement of this wave is

$$D(x, t = 0) = A \sin\left(2\pi \frac{x}{\lambda} + \phi_0\right) \tag{20.7}$$

where the notation $D(x, t = 0)$ means that we've frozen the time at $t = 0$ to make the displacement a function of only x. The term ϕ_0 is a *phase constant* that characterizes the initial conditions. (We'll return to the phase constant momentarily.)

The function of Equation 20.7 is periodic with period λ. We can see this by writing

$$D(x + \lambda) = A \sin\left(2\pi \frac{(x + \lambda)}{\lambda} + \phi_0\right) = A \sin\left(2\pi \frac{x}{\lambda} + \phi_0 + 2\pi \text{ rad}\right)$$

$$= A \sin\left(2\pi \frac{x}{\lambda} + \phi_0\right) = D(x)$$

where we used the fact that $\sin(a + 2\pi \text{ rad}) = \sin a$. In other words, the disturbance created by the wave at $x + \lambda$ is exactly the same as the disturbance at x.

The next step—and it's an important step to graph—is to set the wave in motion. We can do this by replacing x in Equation 20.7 with $x - vt$. To see why this works, recall that the wave moves distance vt during time t. In other words, whatever displacement the wave has at position x at time t, the wave must have had that same displacement at position $x - vt$ at the earlier time $t = 0$. Mathematically, this idea can be captured by writing

$$D(x, t) = D(x - vt, t = 0) \tag{20.8}$$

Make sure you understand how this statement describes a wave moving in the positive *x*-direction at speed v.

This is what we were looking for. $D(x, t)$ is the general function describing the traveling wave. It's found by taking the function that describes the wave at $t = 0$—the function of Equation 20.7—and replacing x with $x - vt$. Thus the displacement equation of a sinusoidal wave traveling in the positive *x*-direction at speed v is

$$D(x, t) = A \sin\left(2\pi \frac{x - vt}{\lambda} + \phi_0\right) = A \sin\left(2\pi \left(\frac{x}{\lambda} - \frac{t}{T}\right) + \phi_0\right) \tag{20.9}$$

In the last step we used $v = \lambda f = \lambda/T$ to write $v/\lambda = 1/T$. The function of Equation 20.9 is not only periodic in space with period λ, it is also periodic in time with period T. That is, $D(x, t + T) = D(x, t)$.

It will be useful to introduce two new quantities. First, recall from simple harmonic motion the *angular frequency*

$$\omega = 2\pi f = \frac{2\pi}{T} \tag{20.10}$$

The units of ω are rad/s, although many textbooks use simply s^{-1}.

You can see that ω is 2π times the reciprocal of the period in time. This suggests that we define an analogous quantity, called the **wave number** k, that is 2π times the reciprocal of the period in space:

$$k = \frac{2\pi}{\lambda} \tag{20.11}$$

The units of k are rad/m, although many textbooks use simply m^{-1}.

NOTE ▶ The wave number k is *not* a spring constant, even though it uses the same symbol. This is a most unfortunate use of symbols, but every major textbook and professional tradition uses the same symbol k for these two very different meanings, so we have little choice but to follow along. ◀

We can use the fundamental relationship $v = \lambda f$ to find an analogous relationship between ω and k:

$$v = \lambda f = \frac{2\pi}{k}\frac{\omega}{2\pi} = \frac{\omega}{k} \tag{20.12}$$

which is usually written

$$\omega = vk \tag{20.13}$$

Equation 20.13 contains no new information. It is a variation of Equation 20.5, but one that is convenient when working with k and ω.

If we use the definitions of Equations 20.10 and 20.11, Equation 20.9 for the displacement can be written

$$D(x, t) = A\sin(kx - \omega t + \phi_0) \tag{20.14}$$
(sinusoidal wave traveling in the positive x-direction)

A sinusoidal wave traveling in the negative x-direction is $A\sin(kx + \omega t + \phi_0)$. Equation 20.14 is graphed versus x and t in FIGURE 20.15.

Just as it did for simple harmonic motion, the phase constant ϕ_0 characterizes the initial conditions. At $(x, t) = (0\text{ m}, 0\text{ s})$ Equation 20.14 becomes

$$D(0\text{ m}, 0\text{ s}) = A\sin\phi \tag{20.15}$$

Different values of ϕ_0 describe different initial conditions for the wave.

FIGURE 20.15 Interpreting the equation of a sinusoidal traveling wave.

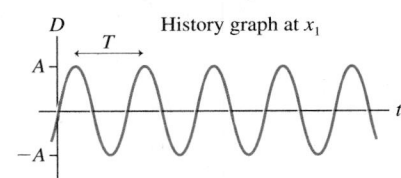

If x is fixed, $D(x_1, t) = A\sin(kx_1 - \omega t + \phi_0)$ gives a sinusoidal history graph at one point in space, x_1. It repeats every T s.

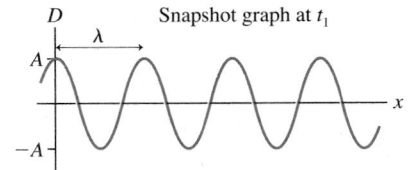

If t is fixed, $D(x, t_1) = A\sin(kx - \omega t_1 + \phi_0)$ gives a sinusoidal snapshot graph at one instant of time, t_1. It repeats every λ m.

EXAMPLE 20.3 **Analyzing a sinusoidal wave**

A sinusoidal wave with an amplitude of 1.00 cm and a frequency of 100 Hz travels at 200 m/s in the positive x-direction. At $t = 0$ s, the point $x = 1.00$ m is on a crest of the wave.

a. Determine the values of A, v, λ, k, f, ω, T, and ϕ_0 for this wave.
b. Write the equation for the wave's displacement as it travels.
c. Draw a snapshot graph of the wave at $t = 0$ s.

VISUALIZE The snapshot graph will be sinusoidal, but we must do some numerical analysis before we know how to draw it.

SOLVE a. There are several numerical values associated with a sinusoidal traveling wave, but they are not all independent. From the problem statement itself we learn that

$$A = 1.00\text{ cm} \qquad v = 200\text{ m/s} \qquad f = 100\text{ Hz}$$

We can then find:

$$\lambda = v/f = 2.00\text{ m}$$
$$k = 2\pi/\lambda = \pi\text{ rad/m or 3.14 rad/m}$$

Continued

$$\omega = 2\pi f = 628 \text{ rad/s}$$

$$T = 1/f = 0.0100 \text{ s} = 10.0 \text{ ms}$$

The phase constant ϕ_0 is determined by the initial conditions. We know that a wave crest, with displacement $D = A$, is passing $x_0 = 1.00$ m at $t_0 = 0$ s. Equation 20.14 at x_0 and t_0 is

$$D(x_0, t_0) = A = A \sin\left(k(1.00 \text{ m}) + \phi_0\right)$$

This equation is true only if $\sin\left(k(1.00 \text{ m}) + \phi_0\right) = 1$, which requires

$$k(1.00 \text{ m}) + \phi_0 = \frac{\pi}{2} \text{ rad}$$

Solving for the phase constant gives

$$\phi_0 = \frac{\pi}{2} \text{ rad} - (\pi \text{ rad/m})(1.00 \text{ m}) = -\frac{\pi}{2} \text{ rad}$$

b. With the information gleaned from part a, the wave's displacement is

$$D(x, t) = 1.00 \text{ cm} \times$$

$$\sin\left[(3.14 \text{ rad/m})x - (628 \text{ rad/s})t - \pi/2 \text{ rad}\right]$$

Notice that we included units with A, k, ω, and ϕ_0.

c. We know that $x = 1.00$ m is a wave crest at $t = 0$ s and that the wavelength is $\lambda = 2.00$ m. Because the origin is $\lambda/2$ away from the crest at $x = 1.00$ m, we expect to find a wave trough at $x = 0$. This is confirmed by calculating $D(0 \text{ m}, 0 \text{ s}) = (1.00 \text{ cm}) \sin(-\pi/2 \text{ rad}) = -1.00$ cm. FIGURE 20.16 is a snapshot graph that portrays this information.

FIGURE 20.16 A snapshot graph at $t = 0$ s of the sinusoidal wave of Example 20.3.

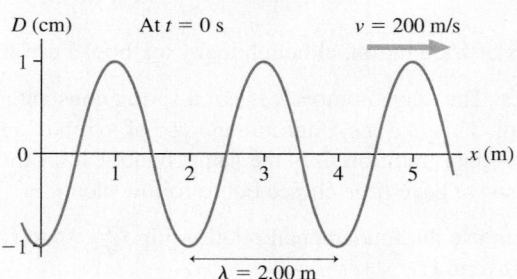

Wave Motion on a String

The displacement equation, Equation 20.14, allows us to learn more about wave motion on a string. As a wave travels along the x-axis, the points on the string oscillate back and forth in the y-direction. The displacement D of a point on the string is simply that point's y-coordinate, so Equation 20.14 for a string wave is

$$y(x, t) = A \sin(kx - \omega t + \phi_0) \qquad (20.16)$$

The velocity of a particle on the string—**which is not the same as the velocity of the wave along the string**—is the time derivative of Equation 20.16:

$$v_y = \frac{dy}{dt} = -\omega A \cos(kx - \omega t + \phi_0) \qquad (20.17)$$

The maximum velocity of a small segment of the string is $v_{\text{max}} = \omega A$. This is the same result we found for simple harmonic motion because the motion of the string particles is simple harmonic motion. FIGURE 20.17 shows velocity vectors *of the particles* at different points on a sinusoidal wave.

NOTE ▶ Creating a wave of larger amplitude increases the speed of particles in the medium, but it does *not* change the speed of the wave *through* the medium. ◀

Pursuing this line of thought, we can derive an expression for the wave speed along the string. FIGURE 20.18 shows a small segment of the string with length $\Delta x \ll \lambda$ right at a crest of the wave. You can see that the string's tension exerts a downward force on this piece of the string, pulling it back to equilibrium. Newton's second law for this small segment of string is

$$(F_{\text{net}})_y = ma_y = (\mu \Delta x)a_y \qquad (20.18)$$

where we used the string's linear density μ to write the mass as $m = \mu \Delta x$.

FIGURE 20.17 A snapshot graph of a wave on a string with vectors showing the velocity *of the string* at various points.

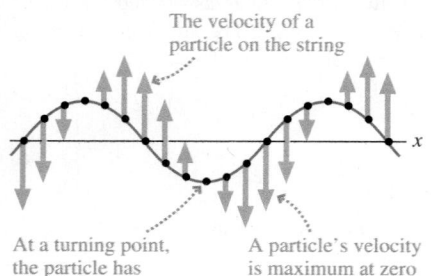

The velocity of the wave ⟶

The velocity of a particle on the string

At a turning point, the particle has zero velocity.

A particle's velocity is maximum at zero displacement.

From simple harmonic motion, we know that this point of maximum displacement is also the point of maximum acceleration. The acceleration of a point on the string is the time derivative of Equation 20.17:

$$a_y = \frac{dv_y}{dt} = -\omega^2 A \sin(kx - \omega t + \phi_0) \tag{20.19}$$

Thus the acceleration at the crest of the wave is $a_y = -\omega^2 A$. But the angular frequency ω with which the particles of the string oscillate is related to the wave's speed v along the string by Equation 20.13, $\omega = vk$. Thus

$$a_y = -\omega^2 A = -v^2 k^2 A \tag{20.20}$$

A large wave speed causes the particles of the string to oscillate more quickly and thus to have a larger acceleration.

You can see from Figure 20.18 that the y-component of the tension is $T_s \sin\theta$, where θ is the angle of the string at $x = \frac{1}{2}\Delta x$. θ is a *negative* angle because it is below the x-axis. This segment of string is pulled from both ends, so

$$(F_{net})_y = 2T_s \sin\theta \tag{20.21}$$

The angle θ is very small because $\Delta x \ll \lambda$, so we can use the small-angle approximation ($\sin u \approx \tan u$ if $u \ll 1$) to write

$$(F_{net})_y \approx 2T_s \tan\theta \tag{20.22}$$

where $\tan\theta$ is the slope of the string at $x = \frac{1}{2}\Delta x$.

At this specific instant, with the crest of the wave at $x = 0$, the equation of the string is

$$y = A\cos(kx)$$

The slope of the string at $x = \frac{1}{2}\Delta x$ is the derivative evaluated at that point:

$$\tan\theta = \frac{dy}{dx}\bigg|_{\text{at } \Delta x/2} = -kA\sin(kx)\big|_{\text{at } \Delta x/2} = -kA\sin\left(\frac{k\Delta x}{2}\right)$$

Now $\Delta x \ll \lambda$, so $k\Delta x/2 = \pi\Delta x/\lambda \ll 1$. Thus the small-angle approximation ($\sin u \approx u$ if $u \ll 1$) of the slope is

$$\tan\theta \approx -kA\left(\frac{k\Delta x}{2}\right) = -\frac{k^2 A\Delta x}{2} \tag{20.23}$$

If we substitute this expression for $\tan\theta$ into Equation 20.22, we find that the net force on this little piece of string is

$$(F_{net})_y = -k^2 A T_s \Delta x \tag{20.24}$$

Now we can use Equation 20.20 for a_y and Equation 20.24 for $(F_{net})_y$ in Newton's second law. With these substitutions, Equation 20.18 becomes

$$(F_{net})_y = -k^2 A T_s \Delta x = (\mu\Delta x)a_y = -v^2 k^2 A\mu\Delta x \tag{20.25}$$

The term $-k^2 A\Delta x$ cancels, and we're left with

$$v = \sqrt{\frac{T_s}{\mu}} \tag{20.26}$$

This was the result that we stated, without proof, in Equation 20.1. Although we've derived Equation 20.26 with the assumption of a sinusoidal wave, the wave speed does not depend on the shape of the wave. Thus any wave on a stretched string will have this wave speed.

FIGURE 20.18 A small segment of string at the crest of a wave.

A small segment of the string at the crest of the wave. Because of the curvature of the string, the tension forces exert a net downward force on this segment.

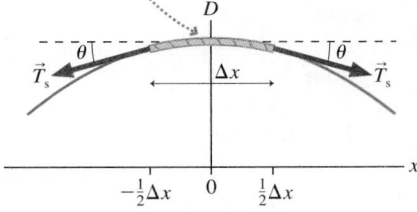

EXAMPLE 20.4   **Generating a sinusoidal wave**

A very long string with $\mu = 2.0$ g/m is stretched along the x-axis with a tension of 5.0 N. At $x = 0$ m it is tied to a 100 Hz simple harmonic oscillator that vibrates perpendicular to the string with an amplitude of 2.0 mm. The oscillator is at its maximum positive displacement at $t = 0$ s.

a. Write the displacement equation for the traveling wave on the string.

b. At $t = 5.0$ ms, what is the string's displacement at a point 2.7 m from the oscillator?

MODEL   The oscillator generates a sinusoidal traveling wave on a string. The displacement of the wave has to match the displacement of the oscillator at $x = 0$ m.

SOLVE   a. The equation for the displacement is

$$D(x, t) = A \sin(kx - \omega t + \phi_0)$$

with A, k, ω, and ϕ_0 to be determined. The wave amplitude is the same as the amplitude of the oscillator that generates the wave, so $A = 2.0$ mm. The oscillator has its maximum displacement $y_{osc} = A = 2.0$ mm at $t = 0$ s, thus

$$D(0 \text{ m}, 0 \text{ s}) = A \sin(\phi_0) = A$$

This requires the phase constant to be $\phi_0 = \pi/2$ rad. The wave's frequency is $f = 100$ Hz, the frequency of the source;

therefore the angular frequency is $\omega = 2\pi f = 200\pi$ rad/s. We still need $k = 2\pi/\lambda$, but we do not know the wavelength. However, we have enough information to determine the wave speed, and we can then use either $\lambda = v/f$ or $k = \omega/v$. The speed is

$$v = \sqrt{\frac{T_s}{\mu}} = \sqrt{\frac{5.0 \text{ N}}{0.0020 \text{ kg/m}}} = 50 \text{ m/s}$$

Using v, we find $\lambda = 0.50$ m and $k = 2\pi/\lambda = 4\pi$ rad/m. Thus the wave's displacement equation is

$$D(x, t) = (2.0 \text{ mm}) \times$$
$$\sin\left[2\pi\left((2.0 \text{ m}^{-1})x - (100 \text{ s}^{-1})t\right) + \pi/2 \text{ rad}\right]$$

Notice that we have separated out the 2π. This step is not essential, but for some problems it makes subsequent steps easier.

b. The wave's displacement at $t = 5.0$ ms $= 0.0050$ s is

$$D(x, t = 5.0 \text{ ms}) = (2.0 \text{ mm}) \sin(4\pi x - \pi \text{ rad} + \pi/2 \text{ rad})$$
$$= (2.0 \text{ mm}) \sin(4\pi x - \pi/2 \text{ rad})$$

At $x = 2.7$ m (calculator set to radians!), the displacement is

$$D(2.7 \text{ m}, 5.0 \text{ ms}) = 1.6 \text{ mm}$$

20.4 Waves in Two and Three Dimensions

Suppose you were to take a photograph of ripples spreading on a pond. If you mark the location of the *crests* on the photo, your picture would look like FIGURE 20.19a. The lines that locate the crests are called **wave fronts,** and they are spaced precisely one wavelength apart. The diagram shows only a single instant of time, but you can imagine a movie in which you would see the wave fronts moving outward from the source at speed v. A wave like this is called a **circular wave.** It is a two-dimensional wave that spreads across a surface.

Although the wave fronts are circles, you would hardly notice the curvature if you observed a small section of the wave front very, very far away from the source. The wave fronts would appear to be parallel lines, still spaced one wavelength apart and traveling at speed v. A good example is an ocean wave reaching a beach. Ocean waves are generated by storms and wind far out at sea, hundreds or thousands of miles away. By the time they reach the beach where you are working on your tan, the crests appear to be straight lines. An aerial view of the ocean would show a wave diagram like FIGURE 20.19b.

Many waves of interest, such as sound waves or light waves, move in three dimensions. For example, loudspeakers and lightbulbs emit **spherical waves.** That is, the crests of the wave form a series of concentric spherical shells separated by the wavelength λ. In essence, the waves are three-dimensional ripples. It will still be useful to draw wave-front diagrams such as Figure 20.19, but now the circles are slices through the spherical shells locating the wave crests.

If you observe a spherical wave very, very far from its source, the small piece of the wave front that you can see is a little patch on the surface of a very large sphere. If the radius of the sphere is sufficiently large, you will not notice the curvature and this little patch of the wave front appears to be a plane. FIGURE 20.20 illustrates the idea of a **plane wave.**

FIGURE 20.19 The wave fronts of a circular or spherical wave.

(a)

Wave fronts are the crests of the wave. They are spaced one wavelength apart.

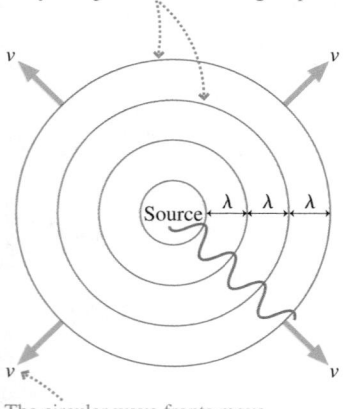

The circular wave fronts move outward from the source at speed v.

(b)

Very far away from the source, small sections of the wave fronts appear to be straight lines.

To visualize a plane wave, imagine standing on the x-axis facing a sound wave as it comes toward you from a very distant loudspeaker. Sound is a longitudinal wave, so the particles of medium oscillate toward you and away from you. If you were to locate all of the particles that, at one instant of time, were at their maximum displacement toward you, they would all be located in a plane perpendicular to the travel direction. This is one of the wave fronts in Figure 20.20, and all the particles in this plane are doing exactly the same thing at that instant of time. This plane is moving toward you at speed v. There is another plane one wavelength behind it where the molecules are also at maximum displacement, yet another two wavelengths behind the first, and so on.

Because a plane wave's displacement depends on x but not on y or z, the displacement function $D(x, t)$ describes a plane wave just as readily as it does a one-dimensional wave. Once you specify a value for x, the displacement is the same at every point in the yz-plane that slices the x-axis at that value (i.e., one of the planes shown in Figure 20.20).

NOTE ▶ There are no perfect plane waves in nature, but many waves of practical interest can be modeled as plane waves. ◀

We can describe a circular wave or a spherical wave by changing the mathematical description from $D(x, t)$ to $D(r, t)$, where r is the radial distance measured outward from the source. Then the displacement of the medium will be the same at every point on a spherical surface. In particular, a sinusoidal spherical wave with wave number k and angular frequency ω is written

$$D(r, t) = A(r)\sin(kr - \omega t + \phi_0) \qquad (20.27)$$

Other than the change of x to r, the only difference is that the amplitude is now a function of r. A one-dimensional wave propagates with no change in the wave amplitude. But circular and spherical waves spread out to fill larger and larger volumes of space. To conserve energy, an issue we'll look at later in the chapter, the wave's amplitude has to decrease with increasing distance r. This is why sound and light decrease in intensity as you get farther from the source. We don't need to specify exactly how the amplitude decreases with distance, but you should be aware that it does.

Phase and Phase Difference

The quantity $(kx - \omega t + \phi_0)$ is called the **phase** of the wave, denoted ϕ. The phase of a wave will be an important concept in Chapters 21 and 22, where we will explore the consequences of adding various waves together. For now, we can note that the wave fronts seen in Figures 20.19 and 20.20 are "surfaces of constant phase." To see this, use the phase to write the displacement as simply $D(x, t) = A\sin\phi$. Because each point on a wave front has the same displacement, the phase must be the same at every point.

It will be useful to know the *phase difference* $\Delta\phi$ between two different points on a sinusoidal wave. FIGURE 20.21 shows two points on a sinusoidal wave at time t. The phase difference between these points is

$$\Delta\phi = \phi_2 - \phi_1 = (kx_2 - \omega t + \phi_0) - (kx_1 - \omega t + \phi_0)$$
$$= k(x_2 - x_1) = k\Delta x = 2\pi\frac{\Delta x}{\lambda} \qquad (20.28)$$

That is, **the phase difference between two points on a wave depends on only the ratio of their separation Δx to the wavelength λ.** For example, two points on a wave separated by $\Delta x = \frac{1}{2}\lambda$ have a phase difference $\Delta\phi = \pi$ rad.

An important consequence of Equation 20.28 is that **the phase difference between two adjacent wave fronts is $\Delta\phi = 2\pi$ rad.** This follows from the fact that two adjacent wave fronts are separated by $\Delta x = \lambda$. This is an important idea. Moving from one crest of the wave to the next corresponds to changing the *distance* by λ and changing the *phase* by 2π rad.

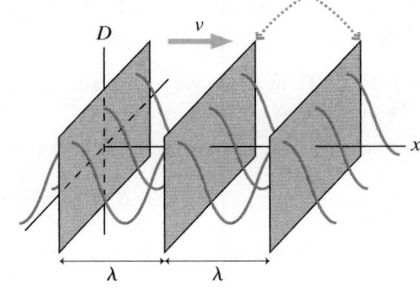

FIGURE 20.20 A plane wave.

Very far from the source, small segments of spherical wave fronts appear to be planes. The wave is cresting at every point in these planes.

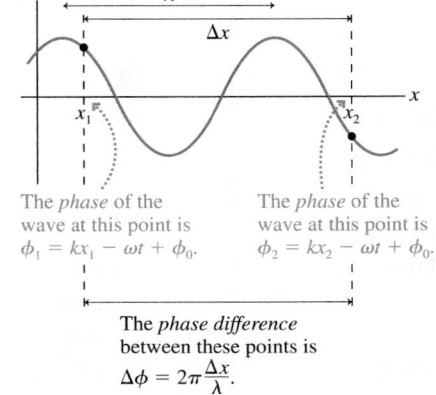

FIGURE 20.21 The phase difference between two points on a wave.

The *phase* of the wave at this point is $\phi_1 = kx_1 - \omega t + \phi_0$.

The *phase* of the wave at this point is $\phi_2 = kx_2 - \omega t + \phi_0$.

The *phase difference* between these points is $\Delta\phi = 2\pi\frac{\Delta x}{\lambda}$.

EXAMPLE 20.5 **The phase difference between two points on a sound wave**

A 100 Hz sound wave travels with a wave speed of 343 m/s.

a. What is the phase difference between two points 60.0 cm apart along the direction the wave is traveling?

b. How far apart are two points whose phase differs by 90°?

MODEL Treat the wave as a plane wave traveling in the positive x-direction.

SOLVE a. The phase difference between two points is

$$\Delta\phi = 2\pi \frac{\Delta x}{\lambda}$$

In this case, $\Delta x = 60.0$ cm $= 0.600$ m. The wavelength is

$$\lambda = \frac{v}{f} = \frac{343 \text{ m/s}}{100 \text{ Hz}} = 3.43 \text{ m}$$

and thus

$$\Delta\phi = 2\pi \frac{0.600 \text{ m}}{3.43 \text{ m}} = 0.350\pi \text{ rad} = 63.0°$$

b. A phase difference $\Delta\phi = 90°$ is $\pi/2$ rad. This will be the phase difference between two points when $\Delta x/\lambda = \frac{1}{4}$, or when $\Delta x = \lambda/4$. Here, with $\lambda = 3.43$ m, $\Delta x = 85.8$ cm.

ASSESS The phase difference increases as Δx increases, so we expect the answer to part b to be larger than 60 cm.

STOP TO THINK 20.4 What is the phase difference between the crest of a wave and the adjacent trough?

a. -2π rad b. 0 rad c. $\pi/4$ rad
d. $\pi/2$ rad e. π rad f. 3π rad

FIGURE 20.22 A sound wave in a fluid is a sequence of compressions and rarefactions. The variation in density and the amount of motion have been greatly exaggerated.

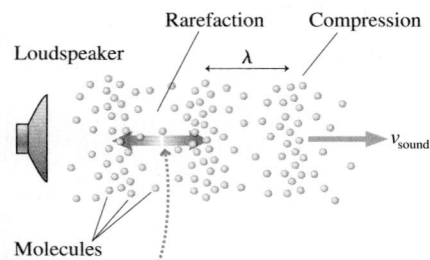

Individual molecules oscillate back and forth with displacement D. As they do so, the compressions propagate forward at speed v_{sound}. Because compressions are regions of higher pressure, a sound wave can be thought of as a pressure wave.

TABLE 20.1 The speed of sound

Medium	Speed (m/s)
Air (0°C)	331
Air (20°C)	343
Helium (0°C)	970
Ethyl alcohol	1170
Water	1480
Granite	6000
Aluminum	6420

20.5 Sound and Light

Although there are many kinds of waves in nature, two are especially significant for us as humans. These are sound waves and light waves, the basis of hearing and seeing.

Sound Waves

We usually think of sound waves traveling in air, but sound can travel through any gas, through liquids, and even through solids. FIGURE 20.22 shows a loudspeaker cone vibrating back and forth in a fluid such as air or water. Each time the cone moves forward, it collides with the molecules and pushes them closer together. A half cycle later, as the cone moves backward, the fluid has room to expand and the density decreases a little. These regions of higher and lower density (and thus higher and lower pressure) are called **compressions** and **rarefactions**.

This periodic sequence of compressions and rarefactions travels outward from the loudspeaker as a longitudinal sound wave. When the wave reaches your ear, the oscillating pressure causes your eardrum to vibrate. These vibrations are transferred into your inner ear and perceived as sound.

Your ears are able to detect sinusoidal sound waves with frequencies between about 20 Hz and about 20,000 Hz, or 20 kHz. Low frequencies are perceived as "low pitch" bass notes, while high frequencies are heard as "high pitch" treble notes. Your high-frequency range of hearing can deteriorate either with age or as a result of exposure to loud sounds that damage the ear.

The speed of sound waves depends on the properties of the medium. A thermodynamic analysis of the compressions and expansions shows that the wave speed in a gas depends on the temperature and on the molecular mass of the gas. For air at room temperature (20°C),

$$v_{\text{sound}} = 343 \text{ m/s} \qquad \text{(sound speed in air at 20°C)}$$

The speed of sound is a little lower at lower temperatures and a little higher at higher temperatures. Liquids and solids are less compressible than air, and that makes the speed of sound in those media higher than in air. Table 20.1 gives the speed of sound in several substances.

A speed of 343 m/s is high, but not extraordinarily so. A distance as small as 100 m is enough to notice a slight delay between when you see something, such as a person hammering a nail, and when you hear it. The time required for sound to travel 1 km is $t = (1000 \text{ m})/(343 \text{ m/s}) \approx 3$ s. You may have learned to estimate the distance to a bolt of lightning by timing the number of seconds between when you see the flash and when you hear the thunder. Because sound takes 3 s to travel 1 km, the time divided by 3 gives the distance in kilometers. Or, in English units, the time divided by 5 gives the distance in miles.

Sound waves exist at frequencies well above 20 kHz, even though humans can't hear them. These are called *ultrasonic* frequencies. Oscillators vibrating at frequencies of many MHz generate the ultrasonic waves used in ultrasound medical imaging. A 3 MHz wave traveling through water (which is basically what your body is) at a sound speed of 1480 m/s has a wavelength of about 0.5 mm. It is this very small wavelength that allows ultrasound to image very small objects. We'll see why when we study *diffraction* in Chapter 22.

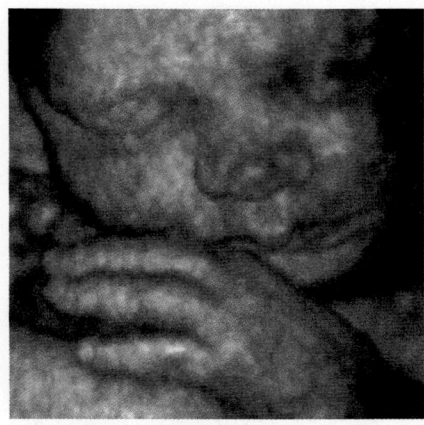

This ultrasound image is an example of using high-frequency sound waves to "see" within the human body.

EXAMPLE 20.6   **Sound wavelengths**

What are the wavelengths of sound waves at the limits of human hearing and at the midrange frequency of 500 Hz? Notes sung by human voices are near 500 Hz, as are notes played by striking keys near the center of a piano keyboard.

MODEL Assume a room temperature of 20°C.

SOLVE We can use the fundamental relationship $\lambda = v/f$ to find the wavelengths for sounds of various frequencies:

$$f = 20 \text{ Hz} \qquad \lambda = \frac{343 \text{ m/s}}{20 \text{ Hz}} = 17 \text{ m}$$

$$f = 500 \text{ Hz} \qquad \lambda = \frac{343 \text{ m/s}}{500 \text{ Hz}} = 0.69 \text{ m}$$

$$f = 20,000 \text{ Hz} \qquad \lambda = \frac{343 \text{ m/s}}{20,000 \text{ Hz}} = 0.017 \text{ m} = 1.7 \text{ cm}$$

ASSESS The wavelength of a 20 kHz note is a small 1.7 cm while, at the other extreme, a 20 Hz note has a huge wavelength of 17 m! This is because a wave moves forward one wavelength during a time interval of one period, and a wave traveling at 343 m/s can move 17 m during the $\frac{1}{20}$ s period of a 20 Hz note. The 69 cm wavelength of a 500 Hz note is more of a "human scale." You might note that most musical instruments are a meter or a little less in size. This is not a coincidence. You will see in the next chapter how the wavelength produced by a musical instrument is related to its size.

Electromagnetic Waves

A light wave is an *electromagnetic wave,* an oscillation of the electromagnetic field. Other electromagnetic waves, such as radio waves, microwaves, and ultraviolet light, have the same physical characteristics as light waves even though we cannot sense them with our eyes. It is easy to demonstrate that light will pass unaffected through a container from which all the air has been removed, and light reaches us from distant stars through the vacuum of interstellar space. Such observations raise interesting but difficult questions. If light can travel through a region in which there is no matter, then what is the *medium* of a light wave? What is it that is waving?

It took scientists over 50 years, most of the 19th century, to answer this question. We will examine the answers in more detail in Part IV after we introduce the ideas of electric and magnetic fields. For now we can say that light waves are a "self-sustaining oscillation of the electromagnetic field." That is, the displacement D is an electric or magnetic field. Being self-sustaining means that electromagnetic waves require *no material medium* in order to travel; hence electromagnetic waves are not mechanical waves. Fortunately, we can learn about the wave properties of light without having to understand electromagnetic fields.

It was predicted theoretically in the late 19th century, and has been subsequently confirmed, that all electromagnetic waves travel through vacuum with the same speed, called the *speed of light.* The value of the speed of light is

$$v_{\text{light}} = c = 299,792,458 \text{ m/s} \qquad \text{(electromagnetic wave speed in vacuum)}$$

where the special symbol c is used to designate the speed of light. (This is the c in Einstein's famous formula $E = mc^2$.) Now *this* is really moving—about one million times faster than the speed of sound in air!

NOTE ▶ $c = 3.00 \times 10^8$ m/s is the appropriate value to use in calculations. ◀

The wavelengths of light are extremely small. You will learn in Chapter 22 how these wavelengths are determined, but for now we will note that visible light is an electromagnetic wave with a wavelength (in air) in the range of roughly 400 nm (400×10^{-9} m) to 700 nm (700×10^{-9} m). Each wavelength is perceived as a different color, with the longer wavelengths seen as orange or red light and the shorter wavelengths seen as blue or violet light. A prism is able to spread the different wavelengths apart, from which we learn that "white light" is all the colors, or wavelengths, combined. The spread of colors seen with a prism, or seen in a rainbow, is called the *visible spectrum.*

If the wavelengths of light are unbelievably small, the oscillation frequencies are unbelievably large. The frequency for a 600 nm wavelength of light (orange) is

$$f = \frac{v}{\lambda} = \frac{3.00 \times 10^8 \text{ m/s}}{600 \times 10^{-9} \text{ m}} = 5.00 \times 10^{14} \text{ Hz}$$

The frequencies of light waves are roughly a factor of a trillion (10^{12}) higher than sound frequencies.

Electromagnetic waves exist at many frequencies other than the rather limited range that our eyes detect. One of the major technological advances of the 20th century was learning to generate and detect electromagnetic waves at many frequencies, ranging from low-frequency radio waves to the extraordinarily high frequencies of x rays. FIGURE 20.23 shows that the visible spectrum is a small slice of the much broader **electromagnetic spectrum.**

White light passing through a prism is spread out into a band of colors called the *visible spectrum.*

FIGURE 20.23 The electromagnetic spectrum from 10^6 Hz to 10^{18} Hz.

Increasing frequency (Hz) ⟶

10^6	10^8	10^{10}	10^{12}	10^{14}	10^{16}	10^{18}
AM radio	FM radio/TV	Microwaves	Infrared		Ultraviolet	X rays
300	3	0.03	3×10^{-4}	3×10^{-6}	3×10^{-8}	3×10^{-10}

⟵ Increasing wavelength (m)

Visible light

700 nm 600 nm 500 nm 400 nm

EXAMPLE 20.7 **Traveling at the speed of light**

A satellite exploring Jupiter transmits data to the earth as a radio wave with a frequency of 200 MHz. What is the wavelength of the electromagnetic wave, and how long does it take the signal to travel 800 million kilometers from Jupiter to the earth?

SOLVE Radio waves are sinusoidal electromagnetic waves traveling with speed c. Thus

$$\lambda = \frac{c}{f} = \frac{3.00 \times 10^8 \text{ m/s}}{2.00 \times 10^8 \text{ Hz}} = 1.5 \text{ m}$$

The time needed to travel 800×10^6 km $= 8.0 \times 10^{11}$ m is

$$\Delta t = \frac{\Delta x}{c} = \frac{8.0 \times 10^{11} \text{ m}}{3.00 \times 10^8 \text{ m/s}} = 2700 \text{ s} = 45 \text{ min}$$

The Index of Refraction

Light waves travel with speed c in a vacuum, but they slow down as they pass through transparent materials such as water or glass or even, to a very slight extent, air. The slowdown is a consequence of interactions between the electromagnetic field of the wave and the electrons in the material. The speed of light in a material is characterized by the material's **index of refraction** n, defined as

$$n = \frac{\text{speed of light in a vacuum}}{\text{speed of light in the material}} = \frac{c}{v} \tag{20.29}$$

The index of refraction of a material is always greater than 1 because $v < c$. A vacuum has $n = 1$ exactly. Table 20.2 shows the index of refraction for several materials. You can see that liquids and solids have larger indices of refraction than gases.

NOTE ▶ An accurate value for the index of refraction of air is relevant only in very precise measurements. We will assume $n_{air} = 1.00$ in this text. ◀

If the speed of a light wave changes as it enters into a transparent material, such as glass, what happens to the light's frequency and wavelength? Because $v = \lambda f$, either λ or f or both have to change when v changes.

As an analogy, think of a sound wave in the air as it impinges on the surface of a pool of water. As the air oscillates back and forth, it periodically pushes on the surface of the water. These pushes generate the compressions of the sound wave that continues on into the water. Because each push of the air causes one compression of the water, the frequency of the sound wave in the water must be *exactly the same* as the frequency of the sound wave in the air. In other words, **the frequency of a wave is the frequency of the source. It does not change as the wave moves from one medium to another.**

The same is true for electromagnetic waves; the frequency does not change as the wave moves from one material to another.

FIGURE 20.24 shows a light wave passing through a transparent material with index of refraction n. As the wave travels through vacuum it has wavelength λ_{vac} and frequency f_{vac} such that $\lambda_{vac} f_{vac} = c$. In the material, $\lambda_{mat} f_{mat} = v = c/n$. The frequency does not change as the wave enters ($f_{mat} = f_{vac}$), so the wavelength must. The wavelength in the material is

$$\lambda_{mat} = \frac{v}{f_{mat}} = \frac{c}{n f_{mat}} = \frac{c}{n f_{vac}} = \frac{\lambda_{vac}}{n} \qquad (20.30)$$

The wavelength in the transparent material is less than the wavelength in vacuum. This makes sense. Suppose a marching band is marching at one step per second at a speed of 1 m/s. Suddenly they slow their speed to $\frac{1}{2}$ m/s but maintain their march at one step per second. The only way to go slower while marching at the same pace is to take *smaller steps*. When a light wave enters a material, the only way it can go slower while oscillating at the same frequency is to have a *smaller wavelength*.

TABLE 20.2 Typical indices of refraction

Material	Index of refraction
Vacuum	1 exactly
Air	1.0003
Water	1.33
Glass	1.50
Diamond	2.42

FIGURE 20.24 Light passing through a transparent material with index of refraction n.

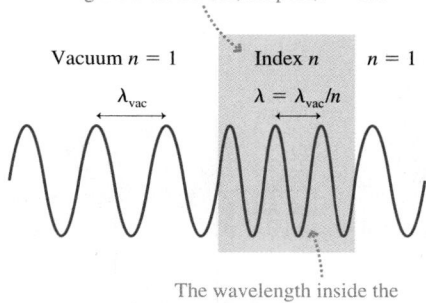

A transparent material in which light travels slower, at speed $v = c/n$

The wavelength inside the material decreases, but the frequency doesn't change.

EXAMPLE 20.8 **Light traveling through glass**

Orange light with a wavelength of 600 nm is incident upon a 1.00-mm-thick glass microscope slide.

a. What is the light speed in the glass?
b. How many wavelengths of the light are inside the slide?

SOLVE a. From Table 20.2 we see that the index of refraction of glass is $n_{glass} = 1.50$. Thus the speed of light in glass is

$$v_{glass} = \frac{c}{n_{glass}} = \frac{3.00 \times 10^8 \text{ m/s}}{1.50} = 2.00 \times 10^8 \text{ m/s}$$

b. The wavelength inside the glass is

$$\lambda_{glass} = \frac{\lambda_{vac}}{n_{glass}} = \frac{600 \text{ nm}}{1.50} = 400 \text{ nm} = 4.00 \times 10^{-7} \text{ m}$$

N wavelengths span a distance $d = N\lambda$, so the number of wavelengths in $d = 1.00$ mm is

$$N = \frac{d}{\lambda} = \frac{1.00 \times 10^{-3} \text{ m}}{4.00 \times 10^{-7} \text{ m}} = 2500$$

ASSESS The fact that 2500 wavelengths fit within 1 mm shows how small the wavelengths of light are.

STOP TO THINK 20.5 A light wave travels through three transparent materials of equal thickness. Rank in order, from largest to smallest, the indices of refraction n_a, n_b, and n_c.

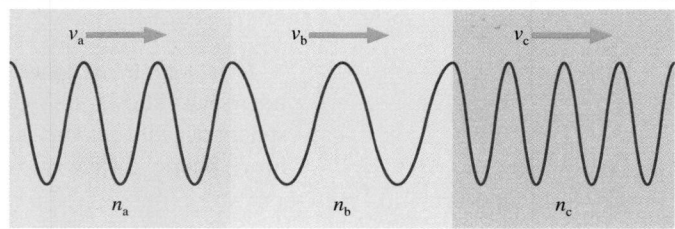

20.6 Power, Intensity, and Decibels

A traveling wave transfers energy from one point to another. The sound wave from a loudspeaker sets your eardrum into motion. Light waves from the sun warm the earth. The *power* of a wave is the rate, in joules per second, at which the wave transfers energy. As you learned in Chapter 11, power is measured in watts. A loudspeaker might emit 2 W of power, meaning that energy in the form of sound waves is radiated at the rate of 2 joules per second. A lightbulb might emit 5 W, or 5 J/s, of visible light. (In fact, this is about right for a so-called 100 watt bulb, with the other 95 W of power being emitted as heat, or infrared radiation, rather than as visible light.)

Imagine doing two experiments with a lightbulb that emits 5 W of visible light. In the first, you hang the bulb in the center of a room and allow the light to illuminate the walls. In the second experiment, you use mirrors and lenses to "capture" the bulb's light and focus it onto a small spot on one wall. This is what a computer projector does. The energy emitted by the bulb is the same in both cases, but, as you know, the light is much brighter when focused onto a small area. We would say that the focused light is more *intense* than the diffuse light that goes in all directions. Similarly, a loudspeaker that beams its sound forward into a small area produces a louder sound in that area than a speaker of equal power that radiates the sound in all directions. Quantities such as brightness and loudness depend not only on the rate of energy transfer, or power, but also on the *area* that receives that power.

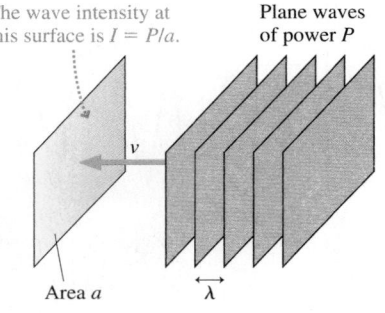

FIGURE 20.25 Plane waves of power P impinge on area a with intensity $I = P/a$.

The wave intensity at this surface is $I = P/a$.

Plane waves of power P

v

Area a

λ

FIGURE 20.25 shows a wave impinging on a surface of area a. The surface is perpendicular to the direction in which the wave is traveling. This might be a real, physical surface, such as your eardrum or a photovoltaic cell, but it could equally well be a mathematical surface in space that the wave passes right through. If the wave has power P, we define the **intensity** I of the wave to be

$$I = \frac{P}{a} = \text{power-to-area ratio} \qquad (20.31)$$

The SI units of intensity are W/m². Because intensity is a power-to-area ratio, a wave focused into a small area will have a larger intensity than a wave of equal power that is spread out over a large area.

EXAMPLE 20.9 **The intensity of a laser beam**

A helium-neon laser, the kind that provides the familiar red light of classroom demonstrations and supermarket checkout scanners, emits 1.0 mW of light power into a 1.0-mm-diameter laser beam. What is the intensity of the laser beam?

MODEL The laser beam is a light wave.

SOLVE The light waves of the laser beam pass through a mathematical surface that is a circle of diameter 1.0 mm. The intensity of the laser beam is

$$I = \frac{P}{a} = \frac{P}{\pi r^2} = \frac{0.0010 \text{ W}}{\pi (0.00050 \text{ m})^2} = 1300 \text{ W/m}^2$$

ASSESS This is roughly the intensity of sunlight at noon on a summer day. The difference between the sun and a small laser is not their intensities, which are about the same, but their powers. The laser has a small power of 1 mW. It can produce a very intense wave only because the area through which the wave passes is very small. The sun, by contrast, radiates a total power $P_{\text{sun}} \approx 4 \times 10^{26}$ W. This immense power is spread through *all* of space, producing an intensity of 1400 W/m² at a distance of 1.5×10^{11} m, the radius of the earth's orbit.

If a source of spherical waves radiates uniformly in all directions, then, as **FIGURE 20.26** shows, the power at distance r is spread uniformly over the surface of a sphere of radius r. The surface area of a sphere is $a = 4\pi r^2$, so the intensity of a uniform spherical wave is

$$I = \frac{P_{\text{source}}}{4\pi r^2} \qquad \text{(intensity of a uniform spherical wave)} \qquad (20.32)$$

The inverse-square dependence of r is really just a statement of energy conservation. The source emits energy at the rate P joules per second. The energy is spread over a larger and larger area as the wave moves outward. Consequently, the energy *per unit area* must decrease in proportion to the surface area of a sphere.

If the intensity at distance r_1 is $I_1 = P_{\text{source}}/4\pi r_1^2$ and the intensity at r_2 is $I_2 = P_{\text{source}}/4\pi r_2^2$, then you can see that the intensity *ratio* is

$$\frac{I_1}{I_2} = \frac{r_2^2}{r_1^2} \tag{20.33}$$

You can use Equation 20.33 to compare the intensities at two distances from a source without needing to know the power of the source.

> NOTE ▶ Wave intensities are strongly affected by reflections and absorption. Equations 20.32 and 20.33 apply to situations such as the light from a star or the sound from a firework exploding high in the air. Indoor sound does *not* obey a simple inverse-square law because of the many reflecting surfaces. ◀

For a sinusoidal wave, each particle in the medium oscillates back and forth in simple harmonic motion. You learned in Chapter 14 that a particle in SHM with amplitude A has energy $E = \frac{1}{2}kA^2$, where k is the spring constant of the medium, not the wave number. It is this oscillatory energy of the medium that is transferred, particle to particle, as the wave moves through the medium.

Because a wave's intensity is proportional to the rate at which energy is transferred through the medium, and because the oscillatory energy in the medium is proportional to the *square* of the amplitude, we can infer that

$$I \propto A^2 \tag{20.34}$$

That is, **the intensity of a wave is proportional to the square of its amplitude.** If you double the amplitude of a wave, you increase its intensity by a factor of 4.

Human hearing spans an extremely wide range of intensities, from the *threshold of hearing* at $\approx 1 \times 10^{-12}$ W/m^2 (at midrange frequencies) to the *threshold of pain* at ≈ 10 W/m^2. If we want to make a scale of loudness, it's convenient and logical to place the zero of our scale at the threshold of hearing. To do so, we define the **sound intensity level,** expressed in **decibels** (dB), as

$$\beta = (10 \text{ dB}) \log_{10}\left(\frac{I}{I_0}\right) \tag{20.35}$$

where $I_0 = 1.0 \times 10^{-12}$ W/m^2. The symbol β is the Greek letter beta. Notice that β is computed as a base-10 logarithm, not a natural logarithm.

The decibel is named after Alexander Graham Bell, inventor of the telephone. Sound intensity level is actually dimensionless because it's formed from the ratio of two intensities, so decibels are just a *name* to remind us that we're dealing with an intensity *level* rather than a true intensity.

Right at the threshold of hearing, where $I = I_0$, the sound intensity level is

$$\beta = (10 \text{ dB}) \log_{10}\left(\frac{I_0}{I_0}\right) = (10 \text{ dB}) \log_{10}(1) = 0 \text{ dB}$$

Note that 0 dB doesn't mean no sound; it means that, for most people, no sound is heard. Dogs have more sensitive hearing than humans, and most dogs can easily perceive a 0 dB sound. The sound intensity level at the pain threshold is

$$\beta = (10 \text{ dB}) \log_{10}\left(\frac{10 \text{ W/m}^2}{10^{-12} \text{ W/m}^2}\right) = (10 \text{ dB}) \log_{10}(10^{13}) = 130 \text{ dB}$$

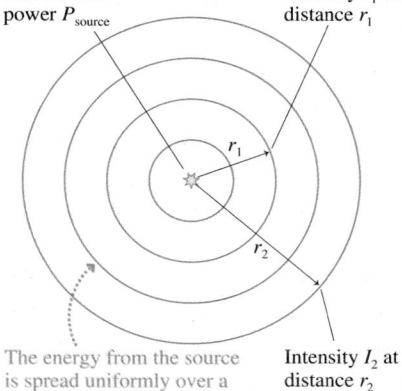

FIGURE 20.26 A source emitting uniform spherical waves.

Source with power P_{source}

Intensity I_1 at distance r_1

r_1

r_2

The energy from the source is spread uniformly over a spherical surface of area $4\pi r^2$.

Intensity I_2 at distance r_2

TABLE 20.3 Sound intensity levels of common sounds

Sound	β (dB)
Threshold of hearing	0
Person breathing, at 3 m	10
A whisper, at 1 m	20
Quiet room	30
Outdoors, no traffic	40
Quiet restaurant	50
Normal conversation, at 1 m	60
Busy traffic	70
Vacuum cleaner, for user	80
Niagara Falls, at viewpoint	90
Snowblower, at 2 m	100
Stereo, at maximum volume	110
Rock concert	120
Threshold of pain	130

The major point to notice is that the sound intensity level increases by 10 dB each time the actual intensity increases by a *factor* of 10. For example, the sound intensity level increases from 70 dB to 80 dB when the sound intensity increases from 10^{-5} W/m^2 to 10^{-4} W/m^2. Perception experiments find that sound is perceived as "twice as loud" when the intensity increases by a factor of 10. In terms of decibels, we can say that the perceived loudness of a sound doubles with each increase in the sound intensity level by 10 dB.

Table 20.3 gives the sound intensity levels for a number of sounds. Although 130 dB is the threshold of pain, quieter sounds can damage your hearing. A fairly short exposure to 120 dB can cause damage to the hair cells in the ear, but lengthy exposure to sound intensity levels of over 85 dB can produce damage as well.

EXAMPLE 20.10 **Blender noise**

The blender making a smoothie produces a sound intensity level of 83 dB. What is the intensity of the sound? What will the sound intensity level be if a second blender is turned on?

SOLVE We can solve Equation 20.35 for the sound intensity, finding $I = I_0 \times 10^{\beta/10\,dB}$. Here we used the fact that 10 raised to a power is an "antilogarithm." In this case,

$$I = (1.0 \times 10^{-12}\ \text{W/m}^2) \times 10^{8.3} = 2.0 \times 10^{-4}\ \text{W/m}^2$$

A second blender doubles the sound power and thus raises the intensity to $I = 4.0 \times 10^{-4}$ W/m^2. The new sound intensity level is

$$\beta = (10\ \text{dB}) \log_{10}\left(\frac{4.0 \times 10^{-4}\ \text{W/m}^2}{1.0 \times 10^{-12}\ \text{W/m}^2}\right) = 86\ \text{dB}$$

ASSESS In general, doubling the actual sound intensity increases the decibel level by 3 dB.

STOP TO THINK 20.6 Four trumpet players are playing the same note. If three of them suddenly stop, the sound intensity level decreases by

a. 40 dB   b. 12 dB   c. 6 dB   d. 4 dB

20.7 The Doppler Effect

Our final topic for this chapter is an interesting effect that occurs when you are in motion relative to a wave source. It is called the *Doppler effect.* You've likely noticed that the pitch of an ambulance's siren drops as it goes past you. Why?

FIGURE 20.27a shows a source of sound waves moving away from Pablo and toward Nancy at a steady speed v_s. The subscript s indicates that this is the speed of the source, not the speed of the waves. The source is emitting sound waves of frequency f_0 as it travels. The figure is a motion diagram showing the position of the source at times $t = 0, T, 2T,$ and $3T$, where $T = 1/f_0$ is the period of the waves.

Nancy measures the frequency of the wave emitted by the *approaching source* to be f_+. At the same time, Pablo measures the frequency of the wave emitted by the *receding source* to be f_-. Our task is to relate f_+ and f_- to the source frequency f_0 and speed v_s.

After a wave crest leaves the source, its motion is governed by the properties of the medium. That is, the motion of the source cannot affect a wave that has already been emitted. Thus each circular wave front in FIGURE 20.27b is centered on the point from which it was emitted. The wave crest from point 3 was emitted just as this figure was made, but it hasn't yet had time to travel any distance.

The wave crests are bunched up in the direction the source is moving, stretched out behind it. The distance between one crest and the next is one wavelength, so the wavelength λ_+ Nancy measures is *less* than the wavelength $\lambda_0 = v/f_0$ that would be emitted if the source were at rest. Similarly, λ_- behind the source is larger than λ_0.

FIGURE 20.27 A motion diagram showing the wave fronts emitted by a source as it moves to the right at speed v_s.

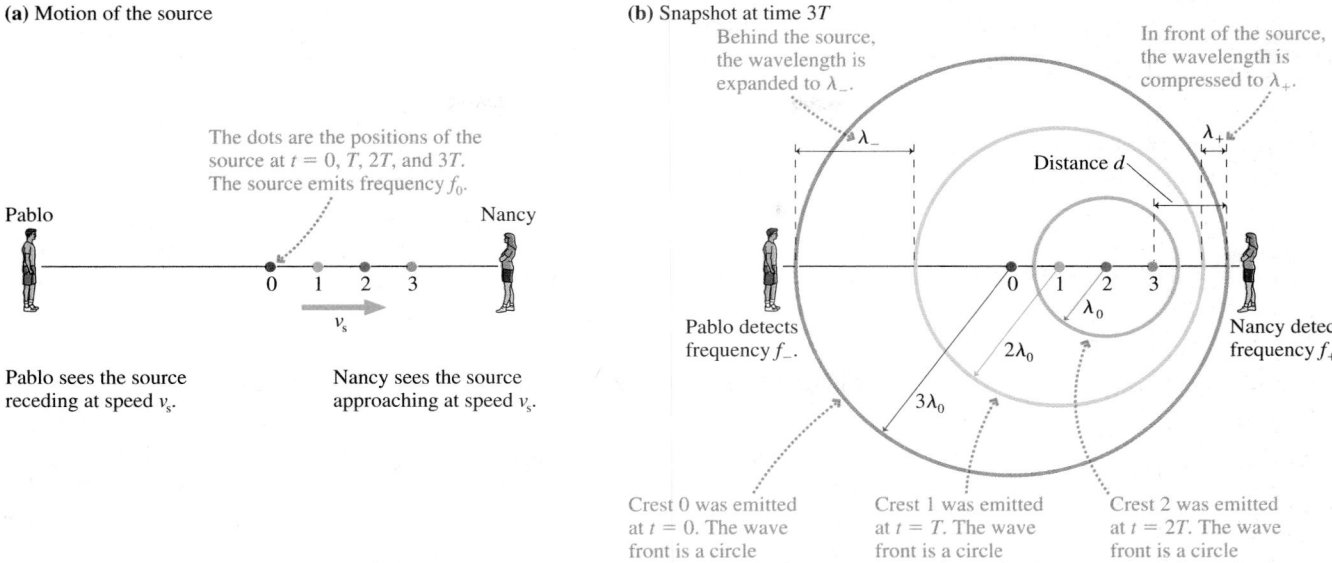

(a) Motion of the source

The dots are the positions of the source at $t = 0$, T, $2T$, and $3T$. The source emits frequency f_0.

Pablo

Nancy

0 1 2 3

v_s

Pablo sees the source receding at speed v_s.

Nancy sees the source approaching at speed v_s.

(b) Snapshot at time $3T$

Behind the source, the wavelength is expanded to λ_-.

In front of the source, the wavelength is compressed to λ_+.

λ_-

λ_+

Distance d

Pablo detects frequency f_-.

λ_0

$2\lambda_0$

$3\lambda_0$

Nancy detects frequency f_+.

Crest 0 was emitted at $t = 0$. The wave front is a circle centered on point 0.

Crest 1 was emitted at $t = T$. The wave front is a circle centered on point 1.

Crest 2 was emitted at $t = 2T$. The wave front is a circle centered on point 2.

These crests move through the medium at the wave speed v. Consequently, the frequency $f_+ = v/\lambda_+$ detected by the observer whom the source is approaching is *higher* than the frequency f_0 emitted by the source. Similarly, $f_- = v/\lambda_-$ detected behind the source is *lower* than frequency f_0. This change of frequency when a source moves relative to an observer is called the **Doppler effect**.

The distance labeled d in Figure 20.27b is the difference between how far the wave has moved and how far the source has moved at time $t = 3T$. These distances are

$$\Delta x_{\text{wave}} = vt = 3vT$$
$$\Delta x_{\text{source}} = v_s t = 3v_s T$$

(20.36)

The distance d spans three wavelengths; thus the wavelength of the wave emitted by an approaching source is

$$\lambda_+ = \frac{d}{3} = \frac{\Delta x_{\text{wave}} - \Delta x_{\text{source}}}{3} = \frac{3vT - 3v_s T}{3} = (v - v_s)T \qquad (20.37)$$

You can see that our arbitrary choice of three periods was not relevant because the 3 cancels. The frequency detected in Nancy's direction is

$$f_+ = \frac{v}{\lambda_+} = \frac{v}{(v - v_s)T} = \frac{v}{(v - v_s)}f_0 \qquad (20.38)$$

where $f_0 = 1/T$ is the frequency of the source and is the frequency you would detect if the source were at rest. We'll find it convenient to write the detected frequency as

$$f_+ = \frac{f_0}{1 - v_s/v} \qquad \text{(Doppler effect for an approaching source)}$$

(20.39)

$$f_- = \frac{f_0}{1 + v_s/v} \qquad \text{(Doppler effect for a receding source)}$$

Proof of the second version, for the frequency f_- of a receding source, is similar. You can see that $f_+ > f_0$ in front of the source, because the denominator is less than 1, and $f_- < f_0$ behind the source.

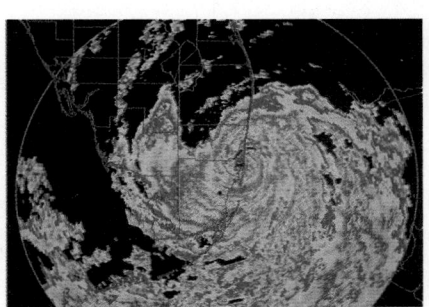

Doppler weather radar uses the Doppler shift of reflected radar signals to measure wind speeds and thus better gauge the severity of a storm.

EXAMPLE 20.11 **How fast are the police traveling?**

A police siren has a frequency of 550 Hz as the police car approaches you, 450 Hz after it has passed you and is receding. How fast are the police traveling? The temperature is 20°C.

MODEL The siren's frequency is altered by the Doppler effect. The frequency is f_+ as the car approaches and f_- as it moves away.

SOLVE To find v_s, we rewrite Equations 20.39 as

$$f_0 = (1 + v_s/v)f_-$$

$$f_0 = (1 - v_s/v)f_+$$

We subtract the second equation from the first, giving

$$0 = f_- - f_+ + \frac{v_s}{v}(f_- + f_+)$$

This is easily solved to give

$$v_s = \frac{f_+ - f_-}{f_+ + f_-}v = \frac{100 \text{ Hz}}{1000 \text{ Hz}} 343 \text{ m/s} = 34.3 \text{ m/s}$$

ASSESS If you now solve for the siren frequency when at rest, you will find $f_0 = 495$ Hz. Surprisingly, the at-rest frequency is not halfway between f_- and f_+.

A Stationary Source and a Moving Observer

Suppose the police car in Example 20.11 is at rest while you drive toward it at 34.3 m/s. You might think that this is equivalent to having the police car move toward you at 34.3 m/s, but it isn't. Mechanical waves move through a medium, and the Doppler effect depends not just on how the source and the observer move with respect to each other but also on how they move with respect to the medium. We'll omit the proof, but it's not hard to show that the frequencies heard by an observer moving at speed v_o relative to a stationary source emitting frequency f_0 are

$$f_+ = (1 + v_o/v)f_0 \quad \text{(observer approaching a source)}$$

$$f_- = (1 - v_o/v)f_0 \quad \text{(observer receding from a source)}$$

(20.40)

A quick calculation shows that the frequency of the police siren as you approach it at 34.3 m/s is 545 Hz, not the 550 Hz you heard as it approached you at 34.3 m/s.

The Doppler Effect for Light Waves

The Doppler effect is observed for all types of waves, not just sound waves. If a source of light waves is receding from you, the wavelength λ_- that you detect is longer than the wavelength λ_0 emitted by the source.

Although the reason for the Doppler shift for light is the same as for sound waves, there is one fundamental difference. We derived Equations 20.39 for the Doppler-shifted frequencies by measuring the wave speed v relative to the medium. For electromagnetic waves in empty space, there is no medium. Consequently, we need to turn to Einstein's theory of relativity to determine the frequency of light waves from a moving source. The result, which we state without proof, is

$$\lambda_- = \sqrt{\frac{1 + v_s/c}{1 - v_s/c}} \lambda_0 \quad \text{(receding source)}$$

(20.41)

$$\lambda_+ = \sqrt{\frac{1 - v_s/c}{1 + v_s/c}} \lambda_0 \quad \text{(approaching source)}$$

Here v_s is the speed of the source *relative to* the observer.

The light waves from a receding source are shifted to longer wavelengths ($\lambda_- > \lambda_0$). Because the longest visible wavelengths are perceived as the color red, the light from a receding source is **red shifted.** That is *not* to say that the light is red, simply that its wavelength is shifted toward the red end of the spectrum. If $\lambda_0 = 470$ nm (blue) light emitted by a rapidly receding source is detected at $\lambda_- = 520$ nm (green), we would say that the light has been red shifted. Similarly, light from an approaching source is **blue shifted,** meaning that the detected wavelengths are shorter than the emitted wavelengths ($\lambda_+ < \lambda_0$) and thus are shifted toward the blue end of the spectrum.

EXAMPLE 20.12 Measuring the velocity of a galaxy

Hydrogen atoms in the laboratory emit red light with wavelength 656 nm. In the light from a distant galaxy, this "spectral line" is observed at 691 nm. What is the speed of this galaxy relative to the earth?

MODEL The observed wavelength is longer than the wavelength emitted by atoms at rest with respect to the observer (i.e., red shifted), so we are looking at light emitted from a galaxy that is receding from us.

SOLVE Squaring the expression for λ_- in Equations 20.41 and solving for v_s give

$$v_s = \frac{(\lambda_-/\lambda_0)^2 - 1}{(\lambda_-/\lambda_0)^2 + 1}c$$

$$= \frac{(691 \text{ nm}/656 \text{ nm})^2 - 1}{(691 \text{ nm}/656 \text{ nm})^2 + 1}c$$

$$= 0.052c = 1.56 \times 10^7 \text{ m/s}$$

ASSESS The galaxy is moving away from the earth at about 5% of the speed of light!

In the 1920s, an analysis of the red shifts of many galaxies led the astronomer Edwin Hubble to the conclusion that the galaxies of the universe are *all* moving apart from each other. Extrapolating backward in time must bring us to a point when all the matter of the universe—and even space itself, according to the theory of relativity—began rushing out of a primordial fireball. Many observations and measurements since have given support to the idea that the universe began in a *Big Bang* about 14 billion years ago.

As an example, **FIGURE 20.28** is a Hubble Space Telescope picture of a *quasar*, short for *quasistellar object.* Quasars are extraordinarily powerful sources of light and radio waves. The light reaching us from quasars is highly red shifted, corresponding in some cases to objects that are moving away from us at greater than 90% of the speed of light. Astronomers have determined that some quasars are 10 to 12 *billion* light years away from the earth, hence the light we see was emitted when the universe was only about 25% of its present age. Today, the red shifts of distant quasars and supernovae (exploding stars) are being used to refine our understanding of the structure and evolution of the universe.

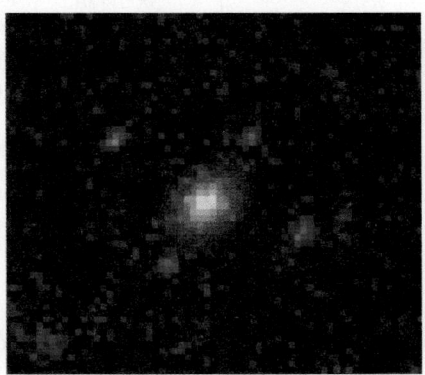

FIGURE 20.28 A Hubble Space Telescope picture of a quasar.

STOP TO THINK 20.7 Amy and Zack are both listening to the source of sound waves that is moving to the right. Compare the frequencies each hears.

a. $f_{Amy} > f_{Zack}$
b. $f_{Amy} = f_{Zack}$
c. $f_{Amy} < f_{Zack}$

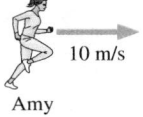

Amy   10 m/s f_0 10 m/s   10 m/s   Zack

CHALLENGE EXAMPLE 20.13 Decreasing the sound

The loudspeaker on a homecoming float—mounted on a pole—is stuck playing an annoying 210 Hz tone. When the speaker is 10 m away, you measure the sound to be a loud 95 dB at 208 Hz. How long will it take for the sound intensity level to drop to a tolerable 55 dB?

MODEL The source is on a pole, so model the sound waves as uniform spherical waves. Assume a temperature of 20°C.

SOLVE The 208 Hz frequency you measure is less than the 210 Hz frequency that was emitted, so the float must be moving away from you. The Doppler effect for a receding source is

$$f_- = \frac{f_0}{1 + v_s/v}$$

We can solve this to find the speed of the float:

$$v_s = \left(\frac{f_0}{f_-} - 1\right)v = \left(\frac{210 \text{ Hz}}{208 \text{ Hz}} - 1\right) \times 343 \text{ m/s} = 3.3 \text{ m/s}$$

The sound intensity of a spherical wave decreases with the inverse square of the distance from the source. A sound intensity level β corresponds to an intensity $I = I_0 \times 10^{\beta/10 \text{ dB}}$, where $I_0 = 1.0 \times 10^{-12} \text{ W/m}^2$. At the initial 95 dB, the intensity is

$$I_1 = I_0 \times 10^{9.5} = 3.2 \times 10^{-3} \text{ W/m}^2$$

At the desired 55 dB, the intensity will have dropped to

$$I_2 = I_0 \times 10^{5.5} = 3.2 \times 10^{-7} \text{ W/m}^2$$

The intensity ratio is related to the distances by

$$\frac{I_1}{I_2} = \frac{r_2^2}{r_1^2}$$

Thus the sound will have dropped to 55 dB when the distance to the speaker is

$$r_2 = \sqrt{\frac{I_1}{I_2}}\, r_1 = \sqrt{10^4} \times 10 \text{ m} = 1000 \text{ m}$$

The float has to travel $\Delta x = 990$ m, which will take

$$\Delta t = \frac{\Delta x}{v_s} = \frac{990 \text{ m}}{3.3 \text{ m/s}} = 300 \text{ s} = 5.0 \text{ min}$$

ASSESS To drop the sound intensity level by 40 dB requires decreasing the intensity by a factor of 10^4. And with the intensity depending on the inverse square of the distance, that requires increasing the distance by a factor of 100. Floats don't move very fast—3.3 m/s is about 7 mph—so needing several minutes to travel the ≈ 1000 m seems reasonable.

SUMMARY

The goal of Chapter 20 has been to learn the basic properties of traveling waves.

General Principles

The Wave Model

This model is based on the idea of a traveling wave, which is an organized disturbance traveling at a well-defined **wave speed** v.

- In transverse waves the displacement is perpendicular to the direction in which the wave travels.

- In longitudinal waves the particles of the medium move parallel to the direction in which the wave travels.

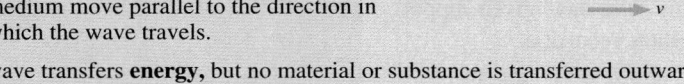

A wave transfers **energy**, but no material or substance is transferred outward from the source.

Two basic types of waves:

- Mechanical waves travel through a material medium such as water or air.

- Electromagnetic waves require no material medium and can travel through a vacuum.

For mechanical waves, such as sound waves and waves on strings, the speed of the wave is a property of the medium. Speed does not depend on the size or shape of the wave.

Important Concepts

The **displacement** D of a wave is a function of both position (where) and time (when).

- A snapshot graph shows the wave's displacement as a function of position at a single instant of time.

- A history graph shows the wave's displacement as a function of time at a single point in space.

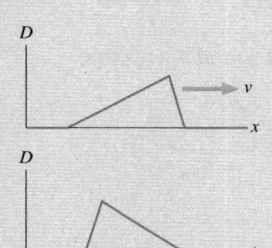

For a transverse wave on a string, the snapshot graph is a picture of the wave. The displacement of a longitudinal wave is parallel to the motion; thus the snapshot graph of a longitudinal sound wave is *not* a picture of the wave.

Sinusoidal waves are periodic in both time (period T) and space (wavelength λ):

$$D(x, t) = A \sin\left[2\pi(x/\lambda - t/T) + \phi_0\right]$$
$$= A \sin(kx - \omega t + \phi_0)$$

where A is the **amplitude**, $k = 2\pi/\lambda$ is the **wave number,** $\omega = 2\pi f = 2\pi/T$ is the **angular frequency,** and ϕ_0 is the **phase constant** that describes initial conditions.

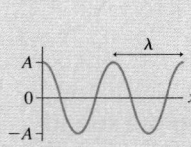

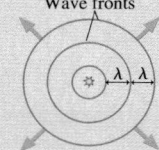

One-dimensional waves   Two- and three-dimensional waves

The fundamental relationship for any sinusoidal wave is $v = \lambda f$.

Applications

- **String** (transverse): $v = \sqrt{T_s/\mu}$
- **Sound** (longitudinal): $v = 343$ m/s in 20°C air
- **Light** (transverse): $v = c/n$, where $c = 3.00 \times 10^8$ m/s is the speed of light in a vacuum and n is the material's **index of refraction**

The wave intensity is the power-to-area ratio: $I = P/a$

For a circular or spherical wave: $I = P_{\text{source}}/4\pi r^2$

The sound intensity level is

$$\beta = (10 \text{ dB}) \log_{10}(I/1.0 \times 10^{-12} \text{ W/m}^2)$$

The Doppler effect occurs when a wave source and detector are moving with respect to each other: the frequency detected differs from the frequency f_0 emitted.

Approaching source

$$f_+ = \frac{f_0}{1 - v_s/v}$$

Observer approaching a source

$$f_+ = (1 + v_o/v)f_0$$

Receding source

$$f_- = \frac{f_0}{1 + v_s/v}$$

Observer receding from a source

$$f_- = (1 - v_o/v)f_0$$

The Doppler effect for light uses a result derived from the theory of relativity.

Terms and Notation

wave model	wave speed, v	wave front	index of refraction, n
traveling wave	linear density, μ	circular wave	intensity, I
transverse wave	snapshot graph	spherical wave	sound intensity level, β
longitudinal wave	history graph	plane wave	decibels
mechanical waves	sinusoidal wave	phase, ϕ	Doppler effect
electromagnetic waves	amplitude, A	compression	red shifted
medium	wavelength, λ	rarefaction	blue shifted
disturbance	wave number, k	electromagnetic spectrum	

CONCEPTUAL QUESTIONS

1. The three wave pulses in FIGURE Q20.1 travel along the same stretched string. Rank in order, from largest to smallest, their wave speeds v_a, v_b, and v_c. Explain.

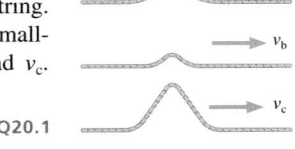

FIGURE Q20.1

2. A wave pulse travels along a stretched string at a speed of 200 cm/s. What will be the speed if:
   a. The string's tension is doubled?
   b. The string's mass is quadrupled (but its length is unchanged)?
   c. The string's length is quadrupled (but its mass is unchanged)?
   Note: Each part is independent and refers to changes made to the original string.

3. FIGURE Q20.3 is a history graph showing the displacement as a function of time at one point on a string. Did the displacement at this point reach its maximum of 2 mm *before* or *after* the interval of time when the displacement was a constant 1 mm?

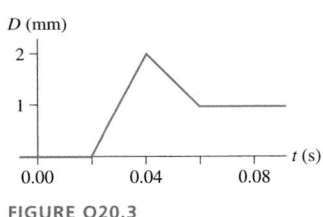

FIGURE Q20.3

4. FIGURE Q20.4 shows a snapshot graph *and* a history graph for a wave pulse on a stretched string. They describe the same wave from two perspectives.
   a. In which direction is the wave traveling? Explain.
   b. What is the speed of this wave?

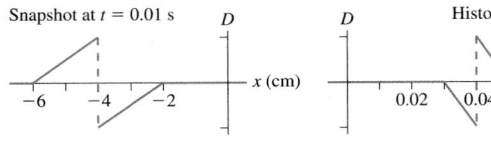

FIGURE Q20.4

5. Rank in order, from largest to smallest, the wavelengths λ_a, λ_b, and λ_c for sound waves having frequencies $f_a = 100$ Hz, $f_b = 1000$ Hz, and $f_c = 10,000$ Hz. Explain.

6. A sound wave with wavelength λ_0 and frequency f_0 moves into a new medium in which the speed of sound is $v_1 = 2v_0$. What are the new wavelength λ_1 and frequency f_1? Explain.

7. What are the amplitude, wavelength, frequency, and phase constant of the traveling wave in FIGURE Q20.7?

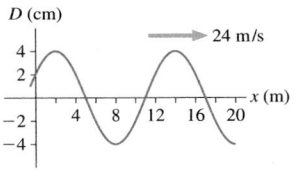

Snapshot graph at $t = 0$ s

FIGURE Q20.7

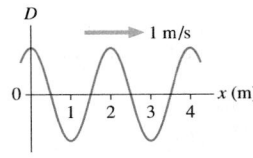

Snapshot graph at $t = 1.0$ s

FIGURE Q20.8

8. FIGURE Q20.8 is a snapshot graph of a sinusoidal wave at $t = 1.0$ s. What is the phase constant of this wave?

9. FIGURE Q20.9 shows the wave fronts of a circular wave. What is the phase difference between (a) points A and B, (b) points C and D, and (c) points E and F?

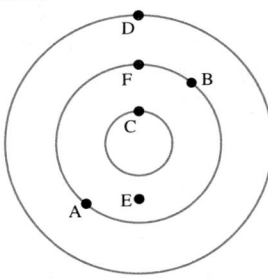

FIGURE Q20.9

10. Sound wave A delivers 2 J of energy in 2 s. Sound wave B delivers 10 J of energy in 5 s. Sound wave C delivers 2 mJ of energy in 1 ms. Rank in order, from largest to smallest, the sound powers P_A, P_B, and P_C of these three sound waves. Explain.

11. One physics professor talking produces a sound intensity level of 52 dB. It's a frightening idea, but what would be the sound intensity level of 100 physics professors talking simultaneously?

12. You are standing at $x = 0$ m, listening to a sound that is emitted at frequency f_0. The graph of FIGURE Q20.12 shows the frequency you hear during a 4-second interval. Which of the following describes the sound source? Explain your choice.

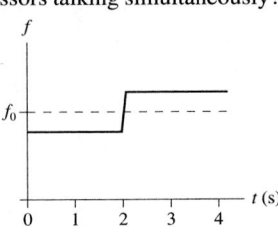

FIGURE Q20.12

   A. It moves from left to right and passes you at $t = 2$ s.
   B. It moves from right to left and passes you at $t = 2$ s.
   C. It moves toward you but doesn't reach you. It then reverses direction at $t = 2$ s.
   D. It moves away from you until $t = 2$ s. It then reverses direction and moves toward you but doesn't reach you.

EXERCISES AND PROBLEMS

Problems labeled [ ] integrate material from earlier chapters.

Exercises

Section 20.1 The Wave Model

1. | The wave speed on a string is 150 m/s when the tension is 75 N. What tension will give a speed of 180 m/s?
2. | The wave speed on a string under tension is 200 m/s. What is the speed if the tension is halved?
3. || A 25 g string is under 20 N of tension. A pulse travels the length of the string in 50 ms. How long is the string?

Section 20.2 One-Dimensional Waves

4. || Draw the history graph $D(x = 0 \text{ m}, t)$ at $x = 0$ m for the wave shown in FIGURE EX20.4.

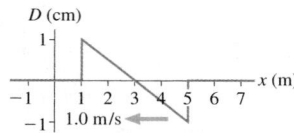

FIGURE EX20.4 — Snapshot graph of a wave at $t = 2$ s

Snapshot graph of a wave at $t = 0$ s

FIGURE EX20.5

5. || Draw the history graph $D(x = 5.0 \text{ m}, t)$ at $x = 5.0$ m for the wave shown in FIGURE EX20.5.
6. || Draw the snapshot graph $D(x, t = 0 \text{ s})$ at $t = 0$ s for the wave shown in FIGURE EX20.6.

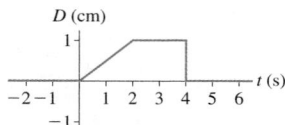 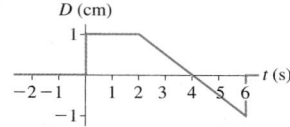

FIGURE EX20.6 — History graph of a wave at $x = 2$ m. Wave moving to the left at 1.0 m/s

FIGURE EX20.7 — History graph of a wave at $x = 0$ m. Wave moving to the right at 1.0 m/s

7. || Draw the snapshot graph $D(x, t = 1.0 \text{ s})$ at $t = 1.0$ s for the wave shown in FIGURE EX20.7.
8. || FIGURE EX20.8 is the snapshot graph at $t = 0$ s of a *longitudinal* wave. Draw the corresponding picture of the particle positions, as was done in Figure 20.10b. Let the equilibrium spacing between the particles be 1.0 cm.

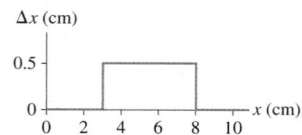

FIGURE EX20.8

9. || FIGURE EX20.9 is a picture at $t = 0$ s of the particles in a medium as a longitudinal wave is passing through. The equilibrium spacing between the particles is 1.0 cm. Draw the snapshot graph $D(x, t = 0 \text{ s})$ of this wave at $t = 0$ s.

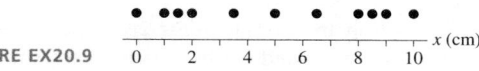

FIGURE EX20.9

Section 20.3 Sinusoidal Waves

10. | A wave travels with speed 200 m/s. Its wave number is 1.5 rad/m. What are its (a) wavelength and (b) frequency?
11. | A wave has angular frequency 30 rad/s and wavelength 2.0 m. What are its (a) wave number and (b) wave speed?
12. | The displacement of a wave traveling in the positive x-direction is $D(x, t) = (3.5 \text{ cm}) \sin(2.7x - 124t)$, where x is in m and t is in s. What are the (a) frequency, (b) wavelength, and (c) speed of this wave?
13. | The displacement of a wave traveling in the negative y-direction is $D(y, t) = (5.2 \text{ cm}) \sin(5.5y + 72t)$, where y is in m and t is in s. What are the (a) frequency, (b) wavelength, and (c) speed of this wave?
14. | What are the amplitude, frequency, and wavelength of the wave in FIGURE EX20.14?

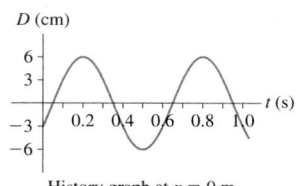

FIGURE EX20.14 — History graph at $x = 0$ m. Wave traveling left at 2.0 m/s

Section 20.4 Waves in Two and Three Dimensions

15. | A spherical wave with a wavelength of 2.0 m is emitted from the origin. At one instant of time, the phase at $r = 4.0$ m is π rad. At that instant, what is the phase at $r = 3.5$ m and at $r = 4.5$ m?
16. | A circular wave travels outward from the origin. At one instant of time, the phase at $r_1 = 20$ cm is 0 rad and the phase at $r_2 = 80$ cm is 3π rad. What is the wavelength of the wave?
17. || A loudspeaker at the origin emits a 120 Hz tone on a day when the speed of sound is 340 m/s. The phase difference between two points on the x-axis is 5.5 rad. What is the distance between these two points?
18. || A sound source is located somewhere along the x-axis. Experiments show that the same wave front simultaneously reaches listeners at $x = -7.0$ m and $x = +3.0$ m.
    a. What is the x-coordinate of the source?
    b. A third listener is positioned along the positive y-axis. What is her y-coordinate if the same wave front reaches her at the same instant it does the first two listeners?

Section 20.5 Sound and Light

19. || A hammer taps on the end of a 4.00-m-long metal bar at room temperature. A microphone at the other end of the bar picks up two pulses of sound, one that travels through the metal and one that travels through the air. The pulses are separated in time by 9.00 ms. What is the speed of sound in this metal?
20. || a. What is the wavelength of a 2.0 MHz ultrasound wave traveling through aluminum?
    b. What frequency of electromagnetic wave would have the same wavelength as the ultrasound wave of part a?

21. | a. What is the frequency of an electromagnetic wave with a wavelength of 20 cm?
    b. What would be the wavelength of a sound wave in water with the same frequency as the electromagnetic wave of part a?

22. | a. What is the frequency of blue light that has a wavelength of 450 nm?
    b. What is the frequency of red light that has a wavelength of 650 nm?
    c. What is the index of refraction of a material in which the red-light wavelength is 450 nm?

23. | a. An FM radio station broadcasts at a frequency of 101.3 MHz. What is the wavelength?
    b. What is the frequency of a sound source that produces the same wavelength in 20°C air?

24. | a. Telephone signals are often transmitted over long distances by microwaves. What is the frequency of microwave radiation with a wavelength of 3.0 cm?
    b. Microwave signals are beamed between two mountaintops 50 km apart. How long does it take a signal to travel from one mountaintop to the other?

25. ‖ a. How long does it take light to travel through a 3.0-mm-thick piece of window glass?
    b. Through what thickness of water could light travel in the same amount of time?

26. ‖ Cell phone conversations are transmitted by high-frequency radio waves. Suppose the signal has wavelength 35 cm while traveling through air. What are the (a) frequency and (b) wavelength as the signal travels through 3-mm-thick window glass into your room?

27. | A light wave has a 670 nm wavelength in air. Its wavelength in a transparent solid is 420 nm.
    a. What is the speed of light in this solid?
    b. What is the light's frequency in the solid?

Section 20.6 Power, Intensity, and Decibels

28. ‖ A sound wave with intensity 2.0×10^{-3} W/m^2 is perceived to
    BIO   be modestly loud. Your eardrum is 6.0 mm in diameter. How much energy will be transferred to your eardrum while listening to this sound for 1.0 min?

29. ‖ The intensity of electromagnetic waves from the sun is 1.4 kW/m^2 just above the earth's atmosphere. Eighty percent of this reaches the surface at noon on a clear summer day. Suppose you think of your back as a 30 cm × 50 cm rectangle. How many joules of solar energy fall on your back as you work on your tan for 1.0 h?

30. ‖ A concert loudspeaker suspended high above the ground emits 35 W of sound power. A small microphone with a 1.0 cm^2 area is 50 m from the speaker.
    a. What is the sound intensity at the position of the microphone?
    b. How much sound energy impinges on the microphone each second?

31. ‖ During takeoff, the sound intensity level of a jet engine is 140 dB at a distance of 30 m. What is the sound intensity level at a distance of 1.0 km?

32. | The sun emits electromagnetic waves with a power of 4.0×10^{26} W. Determine the intensity of electromagnetic waves from the sun just outside the atmospheres of Venus, the earth, and Mars.

33. | What are the sound intensity levels for sound waves of intensity (a) 3.0×10^{-6} W/m^2 and (b) 3.0×10^{-2} W/m^2?

34. | What are the intensities of sound waves with sound intensity levels (a) 46 dB and (b) 103 dB?

35. ‖ A loudspeaker on a tall pole broadcasts sound waves equally in all directions. What is the speaker's power output if the sound intensity level is 90 dB at a distance of 20 m?

Section 20.7 The Doppler Effect

36. | A friend of yours is loudly singing a single note at 400 Hz while racing toward you at 25.0 m/s on a day when the speed of sound is 340 m/s.
    a. What frequency do you hear?
    b. What frequency does your friend hear if you suddenly start singing at 400 Hz?

37. | An opera singer in a convertible sings a note at 600 Hz while cruising down the highway at 90 km/h. What is the frequency heard by
    a. A person standing beside the road in front of the car?
    b. A person on the ground behind the car?

38. ‖ A bat locates insects by emitting ultrasonic "chirps" and then
    BIO   listening for echoes from the bugs. Suppose a bat chirp has a frequency of 25 kHz. How fast would the bat have to fly, and in what direction, for you to just barely be able to hear the chirp at 20 kHz?

39. | A mother hawk screeches as she dives at you. You recall from biology that female hawks screech at 800 Hz, but you hear the screech at 900 Hz. How fast is the hawk approaching?

Problems

40. ‖ The displacement of a traveling wave is

$$D(x, t) = \begin{cases} 1 \text{ cm} & \text{if } |x - 3t| \le 1 \\ 0 \text{ cm} & \text{if } |x - 3t| > 1 \end{cases}$$

    where x is in m and t in s.
    a. Draw displacement-versus-position graphs from $x = -2$ m to $x = 12$ m at 1 s intervals from $t = 0$ s to $t = 3$ s.
    b. Determine the wave speed from the graphs. Explain.
    c. Determine the wave speed from the equation for $D(x, t)$. Does it agree with your answer to part b?

41. ‖ FIGURE P20.41 is a history graph at $x = 0$ m of a wave traveling in the positive x-direction at 4.0 m/s.
    a. What is the wavelength?
    b. What is the phase constant of the wave?
    c. Write the displacement equation for this wave.

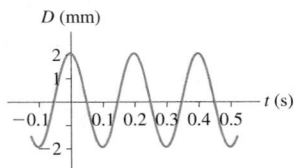

History graph at $x = 0$ m
Wave traveling right at 4.0 m/s

FIGURE P20.41

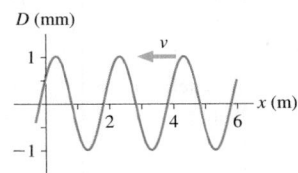

Snapshot graph at $t = 0$ s

FIGURE P20.42

42. ‖ FIGURE P20.42 is a snapshot graph at $t = 0$ s of a 5.0 Hz wave traveling to the left.
    a. What is the wave speed?
    b. What is the phase constant of the wave?
    c. Write the displacement equation for this wave.

43. ‖ A wave travels along a string at speed v_0. What will be the speed if the string is replaced by one made of the same material and under the same tension but having twice the radius?

44. | String 1 in FIGURE P20.44 has linear density 2.0 g/m and string 2 has linear density 4.0 g/m. A student sends pulses in both directions by quickly pulling up on the knot, then releasing it. What should the string lengths L_1 and L_2 be if the pulses are to reach the ends of the strings simultaneously?

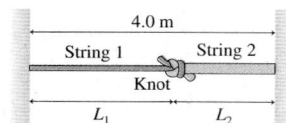

FIGURE P20.44

45. ‖ Ships measure the distance to the ocean bottom with sonar. A pulse of sound waves is aimed at the ocean bottom, then sensitive microphones listen for the echo. FIGURE P20.45 shows the delay time as a function of the ship's position as it crosses 60 km of ocean. Draw a graph of the ocean bottom. Let the ocean surface define $y = 0$ and ocean bottom have negative values of y. This way your graph will be a picture of the ocean bottom. The speed of sound in ocean water varies slightly with temperature, but you can use 1500 m/s as an average value.

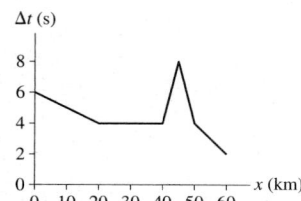

FIGURE P20.45

46. ‖ Oil explorers set off explosives to make loud sounds, then listen for the echoes from underground oil deposits. Geologists suspect that there is oil under 500-m-deep Lake Physics. It's known that Lake Physics is carved out of a granite basin. Explorers detect a weak echo 0.94 s after exploding dynamite at the lake surface. If it's really oil, how deep will they have to drill into the granite to reach it?

47. ‖ One cue your hearing system uses to localize a sound (i.e., to tell where a sound is coming from) is the slight difference in the arrival times of the sound at your ears. Your ears are spaced approximately 20 cm apart. Consider a sound source 5.0 m from the center of your head along a line 45° to your right. What is the difference in arrival times? Give your answer in microseconds.
Hint: You are looking for the difference between two numbers that are nearly the same. What does this near equality imply about the necessary precision during intermediate stages of the calculation?

48. ‖ A helium-neon laser beam has a wavelength in air of 633 nm. It takes 1.38 ns for the light to travel through 30 cm of an unknown liquid. What is the wavelength of the laser beam in the liquid?

49. | A 440 Hz sound wave in 20°C air propagates into the water of a swimming pool. What are the wave's (a) frequency and (b) wavelength in the water?

50. ‖ Earthquakes are essentially sound waves—called seismic waves—traveling through the earth. Because the earth is solid, it can support both longitudinal and transverse seismic waves. The speed of longitudinal waves, called P waves, is 8000 m/s. Transverse waves, called S waves, travel at a slower 4500 m/s.

A seismograph records the two waves from a distant earthquake. If the S wave arrives 2.0 min after the P wave, how far away was the earthquake? You can assume that the waves travel in straight lines, although actual seismic waves follow more complex routes.

51. ‖ A sound wave is described by $D(y, t) = (0.0200 \text{ mm}) \times \sin[(8.96 \text{ rad/m})y + (3140 \text{ rad/s})t + \pi/4 \text{ rad}]$, where y is in m and t is in s.
    a. In what direction is this wave traveling?
    b. Along which axis is the air oscillating?
    c. What are the wavelength, the wave speed, and the period of oscillation?

52. ‖ A wave on a string is described by $D(x, t) = (3.0 \text{ cm}) \times \sin[2\pi(x/(2.4 \text{ m}) + t/(0.20 \text{ s}) + 1)]$, where x is in m and t is in s.
    a. In what direction is this wave traveling?
    b. What are the wave speed, the frequency, and the wave number?
    c. At $t = 0.50$ s, what is the displacement of the string at $x = 0.20$ m?

53. ‖ A wave on a string is described by $D(x, t) = (2.00 \text{ cm}) \times \sin[(12.57 \text{ rad/m})x - (638 \text{ rad/s})t]$, where x is in m and t in s. The linear density of the string is 5.00 g/m. What are
    a. The string tension?
    b. The maximum displacement of a point on the string?
    c. The maximum speed of a point on the string?

54. | Write the displacement equation for a sinusoidal wave that is traveling in the negative y-direction with wavelength 50 cm, speed 4.0 m/s, and amplitude 5.0 cm. Assume $\phi_0 = 0$.

55. | Write the displacement equation for a sinusoidal wave that is traveling in the positive x-direction with frequency 200 Hz, speed 400 m/s, amplitude 0.010 mm, and phase constant $\pi/2$ rad.

56. | A string with linear density 2.0 g/m is stretched along the positive x-axis with tension 20 N. One end of the string, at $x = 0$ m, is tied to a hook that oscillates up and down at a frequency of 100 Hz with a maximum displacement of 1.0 mm. At $t = 0$ s, the hook is at its lowest point.
    a. What are the wave speed on the string and the wavelength?
    b. What are the amplitude and phase constant of the wave?
    c. Write the equation for the displacement $D(x, t)$ of the traveling wave.
    d. What is the string's displacement at $x = 0.50$ m and $t = 15$ ms?

57. ‖ FIGURE P20.57 shows a snapshot graph of a wave traveling to the right along a string at 45 m/s. At this instant, what is the velocity of points 1, 2, and 3 on the string?

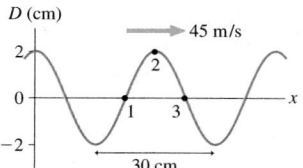

FIGURE P20.57

58. ‖ FIGURE P20.58 shows two masses hanging from a steel wire. The mass of the wire is 60.0 g. A wave pulse travels along the wire from point 1 to point 2 in 24.0 ms. What is mass m?

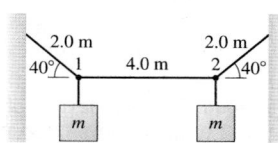

FIGURE P20.58

59. ‖ A wire is made by welding together two metals having different densities. FIGURE P20.59 shows a 2.00-m-long section of wire centered on the junction, but the wire extends much farther in both directions. The wire is placed under 2250 N tension, then a 1500 Hz wave with an amplitude of 3.00 mm is sent down the wire. How many wavelengths (complete cycles) of the wave are in this 2.00-m-long section of the wire?

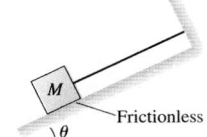

FIGURE P20.59

2250 N ← 1.00 m → ← 1.00 m → 2250 N
$\mu_1 = 9.00$ g/m $\mu_2 = 25.0$ g/m

60. ‖ The string in FIGURE P20.60 has linear density μ. Find an expression in terms of M, μ and θ for the speed of waves on the string.

FIGURE P20.60

M — Frictionless — θ

61. ‖ A string that is under 50.0 N of tension has linear density 5.0 g/m. A sinusoidal wave with amplitude 3.0 cm and wavelength 2.0 m travels along the string. What is the maximum speed of a particle on the string?

62. ‖ A sinusoidal wave travels along a stretched string. A particle on the string has a maximum speed of 2.0 m/s and a maximum acceleration of 200 m/s². What are the frequency and amplitude of the wave?

63. ‖ a. A 100 W lightbulb produces 5.0 W of visible light. (The other 95 W are dissipated as heat and infrared radiation.) What is the light intensity on a wall 2.0 m away from the lightbulb?
    b. A krypton laser produces a cylindrical red laser beam 2.0 mm in diameter with 5.0 W of power. What is the light intensity on a wall 2.0 m away from the laser?

64. ‖ An AM radio station broadcasts with a power of 25 kW at a frequency of 920 kHz. Estimate the intensity of the radio wave at a point 10 km from the broadcast antenna.

65. ‖ LASIK eye surgery uses pulses of laser light to shave off
BIO tissue from the cornea, reshaping it. A typical LASIK laser emits a 1.0-mm-diameter laser beam with a wavelength of 193 nm. Each laser pulse lasts 15 ns and contains 1.0 mJ of light energy
    a. What is the power of one laser pulse?
    b. During the very brief time of the pulse, what is the intensity of the light wave?

66. ‖ The sound intensity 50 m from a wailing tornado siren is 0.10 W/m².
    a. What is the intensity at 1000 m?
    b. The weakest intensity likely to be heard over background noise is $\approx 1\ \mu$W/m². Estimate the maximum distance at which the siren can be heard.

67. ‖ The sound intensity level 5.0 m from a large power saw is 100 dB. At what distance will the sound be a more tolerable 80 dB?

68. ‖ Two loudspeakers on elevated platforms are at opposite ends of a field. Each broadcasts equally in all directions. The sound intensity level at a point halfway between the loudspeakers is 75.0 dB. What is the sound intensity level at a point one-quarter of the way from one speaker to the other along the line joining them?

69. ‖ Your ears are sensitive to differences in pitch, but they are not
BIO very sensitive to differences in intensity. You are not capable of detecting a difference in sound intensity level of less than 1 dB. By what factor does the sound intensity increase if the sound intensity level increases from 60 dB to 61 dB?

70. ‖‖ The intensity of a sound source is described by an inverse-square law only if the source is very small (a point source) and only if the waves can travel unimpeded in all directions. For an extended source or in a situation where obstacles absorb or reflect the waves, the intensity at distance r can often be expressed as $I = cP_{source}/r^x$, where c is a constant and the exponent x—which would be 2 for an ideal spherical wave—depends on the situation. In one such situation, you use a sound meter to measure the sound intensity level at different distances from a source, acquiring the following data:

Distance (m)	Intensity level (dB)
1	100
3	93
10	85
30	78
100	70

Use the best-fit line of an appropriate graph to determine the exponent x that characterizes this sound source.

71. ‖‖ A mad doctor believes that baldness can be cured by warming the scalp with sound waves. His patients sit underneath the Bald-o-Matic loudspeakers, where their heads are bathed with 93 dB of soothing 800 Hz sound waves. Suppose we model a bald head as a 16-cm-diameter hemisphere. If 0.10 J of sound energy is considered an appropriate "dose," how many minutes should each therapy session last?

72. ‖ A physics professor demonstrates the Doppler effect by tying a 600 Hz sound generator to a 1.0-m-long rope and whirling it around her head in a horizontal circle at 100 rpm. What are the highest and lowest frequencies heard by a student in the classroom?

73. ‖ Show that the Doppler frequency f_- of a receding source is $f_- = f_0/(1 + v_s/v)$.

74. | A starship approaches its home planet at a speed of $0.1c$. When it is 54×10^6 km away, it uses its green laser beam ($\lambda = 540$ nm) to signal its approach.
    a. How long does the signal take to travel to the home planet?
    b. At what wavelength is the signal detected on the home planet?

75. | Wavelengths of light from a distant galaxy are found to be 0.5% longer than the corresponding wavelengths measured in a terrestrial laboratory. Is the galaxy approaching or receding from the earth? At what speed?

76. | You have just been pulled over for running a red light, and the police officer has informed you that the fine will be $250. In desperation, you suddenly recall an idea that your physics professor recently discussed in class. In your calmest voice, you tell the officer that the laws of physics prevented you from knowing that the light was red. In fact, as you drove toward it, the light was Doppler shifted to where it appeared green to you. "OK," says the officer, "Then I'll ticket you for speeding. The fine is $1 for every 1 km/h over the posted speed limit of 50 km/h." How big is your fine? Use 650 nm as the wavelength of red light and 540 nm as the wavelength of green light.

Challenge Problems

77. One way to monitor global warming is to measure the average temperature of the ocean. Researchers are doing this by measuring the time it takes sound pulses to travel underwater over large distances. At a depth of 1000 m, where ocean temperatures hold steady near 4°C, the average sound speed is 1480 m/s. It's known from laboratory measurements that the sound speed increases 4.0 m/s for every 1.0°C increase in temperature. In one experiment, where sounds generated near California are detected in the South Pacific, the sound waves travel 8000 km. If the smallest time change that can be reliably detected is 1.0 s, what is the smallest change in average temperature that can be measured?

78. The G string on a guitar is a 0.46-mm-diameter steel string with a linear density of 1.3 g/m. When the string is properly tuned to 196 Hz, the wave speed on the string is 250 m/s. Tuning is done by turning the tuning screw, which slowly tightens—and stretches—the string. By how many mm does a 75-cm-long G string stretch when it's first tuned?

79. A rope of mass m and length L hangs from a ceiling.
    a. Show that the wave speed on the rope a distance y above the lower end is $v = \sqrt{gy}$.
    b. Show that the time for a pulse to travel the length of the string is $\Delta t = 2\sqrt{L/g}$.

80. Some modern optical devices are made with glass whose index of refraction changes with distance from the front surface. FIGURE CP20.80 shows the index of refraction as a function of the distance into a slab of glass of thickness L. The index of refraction increases linearly from n_1 at the front surface to n_2 at the rear surface.

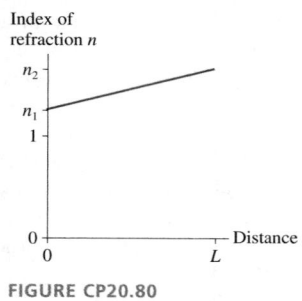

Index of refraction n

FIGURE CP20.80

a. Find an expression for the time light takes to travel through this piece of glass.
b. Evaluate your expression for a 1.0-cm-thick piece of glass for which $n_1 = 1.50$ and $n_2 = 1.60$.

81. A water wave is a *shallow-water wave* if the water depth d is less than $\approx \lambda/10$. It is shown in hydrodynamics that the speed of a shallow-water wave is $v = \sqrt{gd}$, so waves slow down as they move into shallower water. Ocean waves, with wavelengths of typically 100 m, are shallow-water waves when the water depth is less than ≈ 10 m. Consider a beach where the depth increases linearly with distance from the shore until reaching a depth of 5.0 m at a distance of 100 m. How long does it take a wave to move the last 100 m to the shore? Assume that the waves are so small that they don't break before reaching the shore.

82. An important characteristic of the heart, one used to diagnose heart disease, is the *pressure difference* between the blood pressure inside the heart and the blood pressure in the aorta, the large artery leading away from the heart. The blood inside the heart is essentially at rest, but it speeds up significantly as it enters the aorta—and its speed can be measured by using the Doppler shift of reflected ultrasound.

    a. The Doppler effect enters twice in calculating the frequency of the reflection from a moving object. Suppose the object's speed v_o is very small compared to the wave speed v. Show that a good approximation for the *Doppler shift*—the difference between the reflected frequency and the incident frequency—is

    $$\Delta f = 2f_0 \frac{v_o}{v}$$

    b. A doctor using 2.5 MHz ultrasound measures a 6000 Hz Doppler shift as the ultrasound reflects from blood ejected from the heart into the aorta. What is the blood pressure difference, in mm of Hg, between the inside of the heart and the aorta? Assume the patient is lying down so that there is no height difference between the heart and the aorta. The density of blood is 1060 kg/m³.

STOP TO THINK ANSWERS

Stop to Think 20.1: d and e. The wave speed depends on properties of the medium, not on how you generate the wave. For a string, $v = \sqrt{T_s/\mu}$. Increasing the tension or decreasing the linear density (lighter string) will increase the wave speed.

Stop to Think 20.2: b. The wave is traveling to the right at 2.0 m/s, so each point on the wave passes $x = 0$ m, the point of interest, 2.0 s before reaching $x = 4.0$ m. The graph has the same shape, but everything happens 2.0 s earlier.

Stop to Think 20.3: d. The wavelength—the distance between two crests—is seen to be 10 m. The frequency is $f = v/\lambda = (50 \text{ m/s})/(10 \text{ m}) = 5$ Hz.

Stop to Think 20.4: e. A crest and an adjacent trough are separated by $\lambda/2$. This is a phase difference of π rad.

Stop to Think 20.5: $n_c > n_a > n_b$. $\lambda = \lambda_{vac}/n$, so a shorter wavelength corresponds to a larger index of refraction.

Stop to Think 20.6: c. Any factor-of-2 change in intensity changes the sound intensity level by 3 dB. One trumpet is $\frac{1}{4}$ the original number, so the intensity has decreased by two factors of 2.

Stop to Think 20.7: c. Zack hears a higher frequency as he and the source approach. Amy is moving with the source, so $f_{Amy} = f_0$.

21 Superposition

This swirl of colors is due to a very thin layer of oil. Oil is clear. The colors arise from the interference of reflected light waves.

▶ **Looking Ahead** The goal of Chapter 21 is to understand and use the idea of superposition.

Standing Waves

Standing waves are created from the superposition of traveling waves bouncing back and forth between the edges of the medium.

Standing waves occur in well-defined patterns called **modes,** each with its own distinct frequency. Some points on the wave, called **nodes,** do not oscillate at all.

You'll learn how to calculate the frequencies and wavelengths of standing waves on strings and in air.

Applications

You'll learn how standing waves determine the notes of a guitar and other musical instruments...

...and how interference is used to design antireflection coatings for lenses.

◀ **Looking Back**
Section 20.5 Sound waves

Superposition

Waves can pass through each other—a characteristic that distinguishes waves from particles. As they do, their displacements add together. This is called the **principle of superposition**.

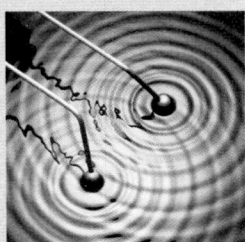

These water waves are exhibiting superposition as the ripples pass through each other.

You'll learn to analyze the response of the medium when two waves pass through each other.

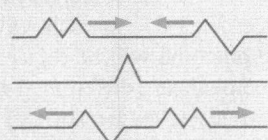

◀ **Looking Back**
Sections 20.2–20.4 Properties of traveling waves

Interference

When two sources emit waves of the same wavelength and frequency, the overlapped waves create an **interference pattern.**

You'll learn to interpret interference patterns such as this one. The two black dots are the sources of the waves.

Constructive interference occurs where the waves add to make a larger wave. *Destructive interference* is where the waves cancel to make a smaller wave.

Beats

The superposition of two waves of slightly different frequency produces a soft-loud-soft-loud-... modulation of the intensity called **beats.**

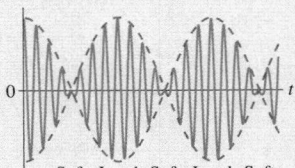

Beats are easily demonstrated with sound waves, but the concept is used in applications from ultrasonics to telecommunications.

21.1 The Principle of Superposition

FIGURE 21.1a shows two baseball players, Alan and Bill, at batting practice. Unfortunately, someone has turned the pitching machines so that pitching machine A throws baseballs toward Bill while machine B throws toward Alan. If two baseballs are launched at the same time, and with the same speed, they collide at the crossing point. Two particles cannot occupy the same point of space at the same time.

FIGURE 21.1 Unlike particles, two waves can pass directly through each other.

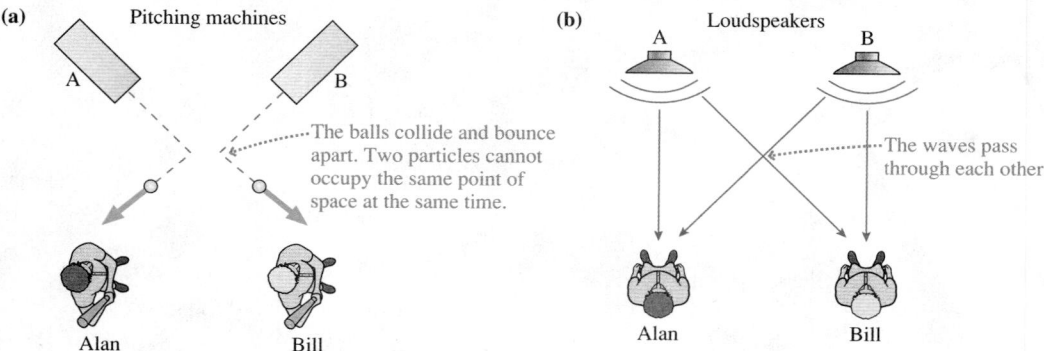

But waves, unlike particles, can pass directly through each other. In FIGURE 21.1b Alan and Bill are listening to the stereo system in the locker room after practice. Because both hear the music quite well, the sound wave that travels from loudspeaker A toward Bill must pass through the wave traveling from loudspeaker B toward Alan. What happens to the medium at a point where two waves are present simultaneously?

If wave 1 displaces a particle in the medium by D_1 and wave 2 *simultaneously* displaces it by D_2, the net displacement of the particle is simply $D_1 + D_2$. This is a very important idea because it tells us how to combine waves. It is known as the *principle of superposition.*

> **Principle of superposition** When two or more waves are *simultaneously* present at a single point in space, the displacement of the medium at that point is the sum of the displacements due to each individual wave.

When one object is placed on top of another, the two are said to be *superimposed.* But through some quirk in the English language, the result of superimposing objects is called a *superposition,* without the syllable "im." When one wave is "placed" on top of another wave, we have a superposition of waves.

Mathematically, the net displacement of a particle in the medium is

$$D_{\text{net}} = D_1 + D_2 + \cdots = \sum_i D_i \tag{21.1}$$

where D_i is the displacement that would be caused by wave *i* alone. We will make the simplifying assumption that the displacements of the individual waves are along the same line so that we can add displacements as scalars rather than vectors.

To use the principle of superposition you must know the displacement caused by each wave if traveling alone. Then you go through the medium *point by point* and add the displacements due to each wave *at that point* to find the net displacement at that point. The outcome will be different at each and every point in the medium because the displacements are different at each point.

To illustrate, FIGURE 21.2 shows snapshot graphs taken 1 s apart of two waves traveling at the same speed (1 m/s) in opposite directions along a string. The principle of superposition comes into play wherever the waves overlap. The solid line is the sum *at each point* of the two displacements at that point. This is the displacement that you would actually observe as the two waves pass through each other.

FIGURE 21.2 The superposition of two waves on a string as they pass through each other.

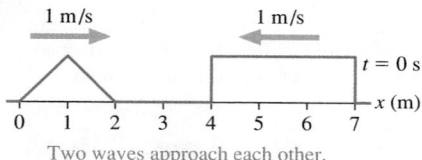

Two waves approach each other.

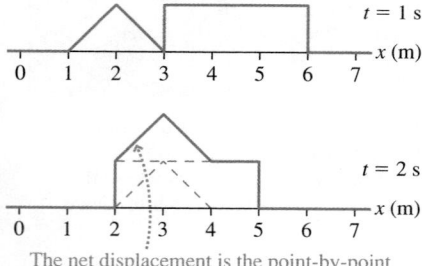

The net displacement is the point-by-point summation of the individual waves.

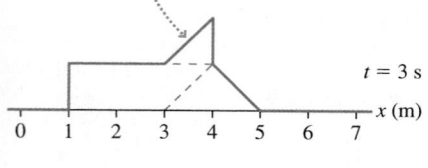

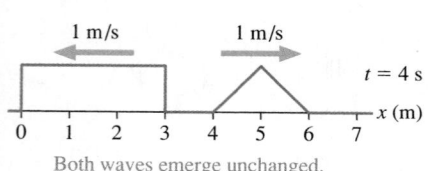

Both waves emerge unchanged.

STOP TO THINK 21.1 Two pulses on a string approach each other at speeds of 1 m/s. What is the shape of the string at $t = 6$ s?

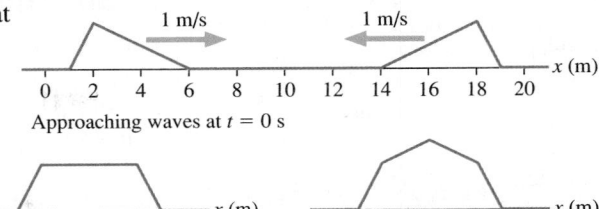

Approaching waves at $t = 0$ s

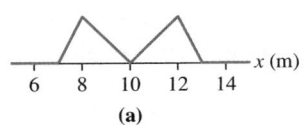

(a)

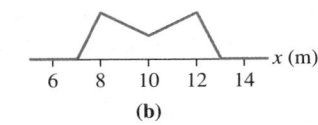

(b)

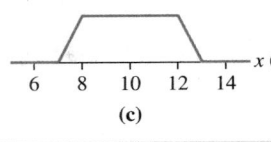

(c)

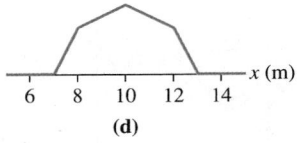

(d)

21.2 Standing Waves

FIGURE 21.3 is a time-lapse photograph of a *standing wave* on a vibrating string. It's not obvious from the photograph, but this is actually a superposition of two waves. To understand this, consider two sinusoidal waves **with the same frequency, wavelength, and amplitude** traveling in opposite directions. For example, **FIGURE 21.4a** shows two waves on a string, and **FIGURE 21.4b** shows nine snapshot graphs, at intervals of $\frac{1}{8}T$. The dots identify two of the crests to help you visualize the wave movement.

At *each point,* the net displacement—the superposition—is found by adding the red displacement and the green displacement. **FIGURE 21.4c** shows the result. It is the wave you would actually observe. The blue dot shows that the blue wave is moving neither right nor left. The wave of Figure 21.4c is called a **standing wave** because the crests and troughs "stand in place" as the wave oscillates.

FIGURE 21.3 A vibrating string is an example of a standing wave.

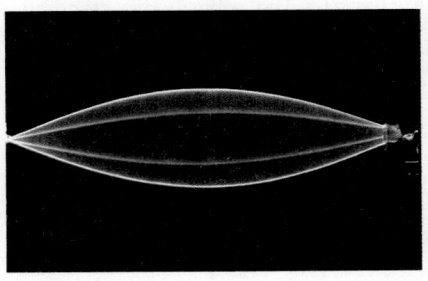

FIGURE 21.4 The superposition of two sinusoidal waves traveling in opposite directions.

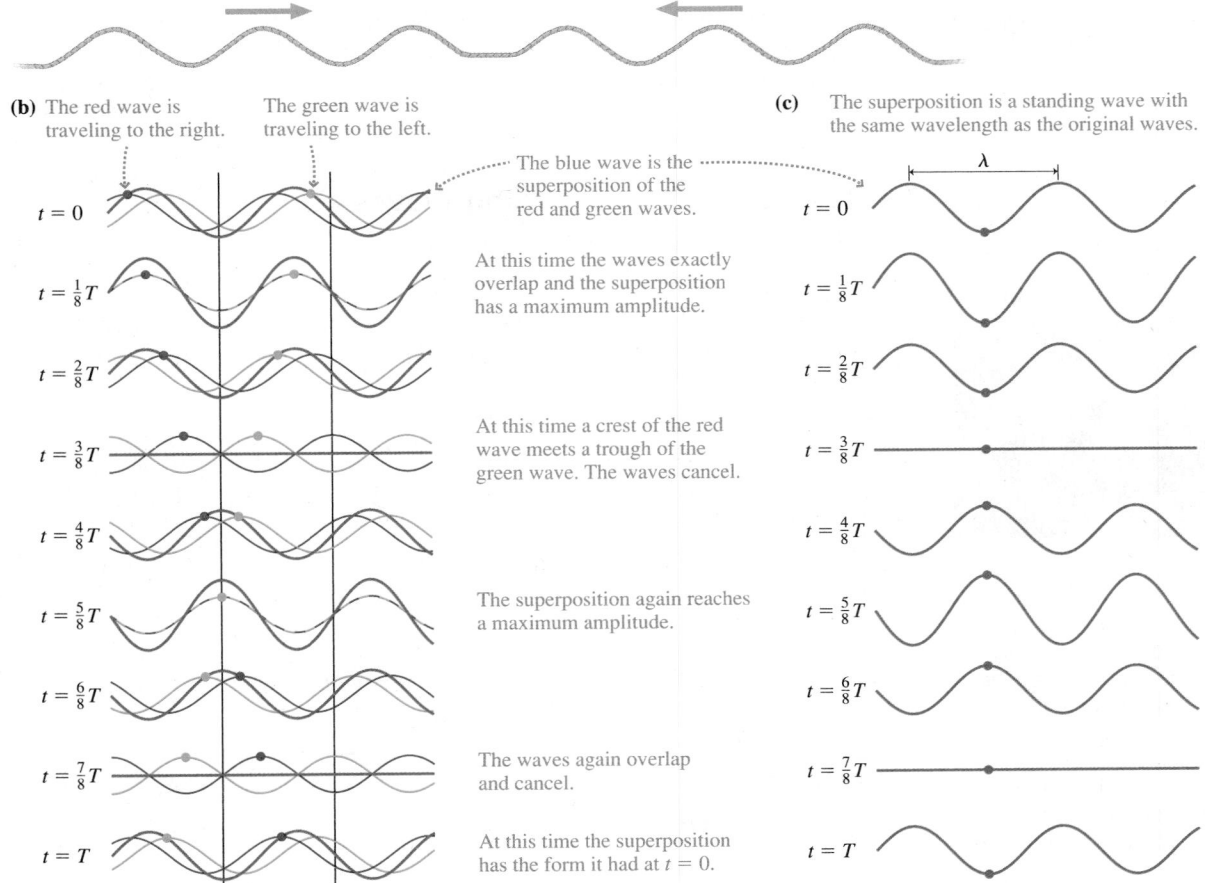

(a) A string is carrying two waves moving in opposite directions.

(b) The red wave is traveling to the right. The green wave is traveling to the left.

The blue wave is the superposition of the red and green waves.

$t = 0$

$t = \frac{1}{8}T$ At this time the waves exactly overlap and the superposition has a maximum amplitude.

$t = \frac{2}{8}T$

$t = \frac{3}{8}T$ At this time a crest of the red wave meets a trough of the green wave. The waves cancel.

$t = \frac{4}{8}T$

$t = \frac{5}{8}T$ The superposition again reaches a maximum amplitude.

$t = \frac{6}{8}T$

$t = \frac{7}{8}T$ The waves again overlap and cancel.

$t = T$ At this time the superposition has the form it had at $t = 0$.

Antinode   Node

(c) The superposition is a standing wave with the same wavelength as the original waves.

λ

$t = 0$

$t = \frac{1}{8}T$

$t = \frac{2}{8}T$

$t = \frac{3}{8}T$

$t = \frac{4}{8}T$

$t = \frac{5}{8}T$

$t = \frac{6}{8}T$

$t = \frac{7}{8}T$

$t = T$

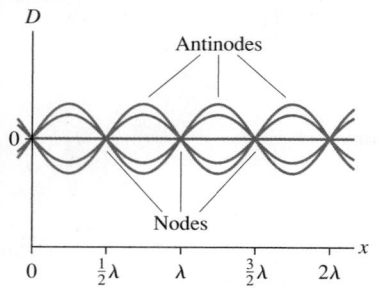

The nodes and antinodes are spaced λ/2 apart.

FIGURE 21.6 The intensity of a standing wave is maximum at the antinodes, zero at the nodes.

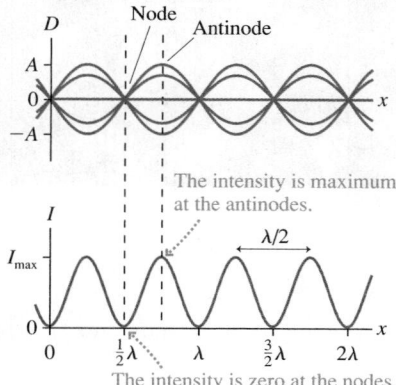

The intensity is maximum at the antinodes.

The intensity is zero at the nodes.

This photograph shows the Tacoma Narrows suspension bridge on the day in 1940 when it experienced a catastrophic standing-wave oscillation that led to its collapse. Aerodynamic forces caused the amplitude of a particular resonant mode of the bridge to increase dramatically until the bridge failed. In this photo, the red line shows the original line of the deck of the bridge. You can clearly see the large amplitude of the oscillation and the node at the center of the span.

Nodes and Antinodes

FIGURE 21.5 has collapsed the nine graphs of Figure 21.4b into a single graphical representation of a standing wave. Compare this to the Figure 21.3 photograph of a vibrating string. A striking feature of a standing-wave pattern is the existence of **nodes,** points that *never move!* **The nodes are spaced λ/2 apart.** Halfway between the nodes are the points where the particles in the medium oscillate with maximum displacement. These points of maximum amplitude are called **antinodes,** and you can see that they are also spaced λ/2 apart.

It seems surprising and counterintuitive that some particles in the medium have no motion at all. To understand how this happens, look carefully at the two traveling waves in Figure 21.4a. You will see that the nodes occur at points where at *every instant* of time the displacements of the two traveling waves have equal magnitudes but *opposite signs.* Thus the superposition of the displacements at these points is always zero. The antinodes correspond to points where the two displacements have equal magnitudes and the *same sign* at all times.

Two waves 1 and 2 are said to be *in phase* at a point where D_1 is *always* equal to D_2. The superposition at that point yields a wave whose amplitude is twice that of the individual waves. This is called a point of *constructive interference.* The antinodes of a standing wave are points of constructive interference between the two traveling waves.

In contrast, two waves are said to be *out of phase* at points where D_1 is *always* equal to $-D_2$. Their superposition gives a wave with zero amplitude—no wave at all! This is a point of *destructive interference.* The nodes of a standing wave are points of destructive interference. We will defer the main discussion of constructive and destructive interference until later in this chapter, but you'll then recognize that you're seeing constructive and destructive interference at the antinodes and nodes of a standing wave.

In Chapter 20 you learned that the *intensity* of a wave is proportional to the square of the amplitude: $I \propto A^2$. You can see in **FIGURE 21.6** that maximum intensity occurs at the antinodes and that the intensity is zero at the nodes. If this is a sound wave, the loudness is maximum at the antinodes and zero at the nodes. A standing light wave is bright at the antinodes, dark at the nodes. The key idea is that **the intensity is maximum at points of constructive interference and zero at points of destructive interference.**

The Mathematics of Standing Waves

A sinusoidal wave traveling to the right along the x-axis with angular frequency $\omega = 2\pi f$, wave number $k = 2\pi/\lambda$, and amplitude a is

$$D_R = a\sin(kx - \omega t) \tag{21.2}$$

An equivalent wave traveling to the left is

$$D_L = a\sin(kx + \omega t) \tag{21.3}$$

We previously used the symbol A for the wave amplitude, but here we will use a lowercase a to represent the amplitude of each individual wave and reserve A for the amplitude of the net wave.

According to the principle of superposition, the net displacement of the medium when both waves are present is the sum of D_R and D_L:

$$D(x, t) = D_R + D_L = a\sin(kx - \omega t) + a\sin(kx + \omega t) \tag{21.4}$$

We can simplify Equation 21.4 by using the trigonometric identity

$$\sin(\alpha \pm \beta) = \sin\alpha\cos\beta \pm \cos\alpha\sin\beta$$

Doing so gives

$$D(x, t) = a(\sin kx \cos\omega t - \cos kx \sin\omega t) + a(\sin kx \cos\omega t + \cos kx \sin\omega t)$$
$$= (2a\sin kx)\cos\omega t \tag{21.5}$$

It is useful to write Equation 21.5 as

$$D(x, t) = A(x) \cos \omega t \qquad (21.6)$$

where the **amplitude function** $A(x)$ is defined as

$$A(x) = 2a \sin kx \qquad (21.7)$$

The amplitude reaches a maximum value $A_{max} = 2a$ at points where $\sin kx = 1$.

The displacement $D(x, t)$ given by Equation 21.6 is neither a function of $x - vt$ nor a function of $x + vt$; hence it is *not* a traveling wave. Instead, the $\cos \omega t$ term in Equation 21.6 describes a medium in which each point oscillates in simple harmonic motion with frequency $f = \omega/2\pi$. The function $A(x) = 2a \sin kx$ gives the amplitude of the oscillation for a particle at position x.

FIGURE 21.7 graphs Equation 21.6 at several different instants of time. Notice that the graphs are identical to those of Figure 21.5, showing us that Equation 21.6 is the mathematical description of a standing wave.

The nodes of the standing wave are the points at which the amplitude is zero. They are located at positions x for which

$$A(x) = 2a \sin kx = 0 \qquad (21.8)$$

The sine function is zero if the angle is an integer multiple of π rad, so Equation 21.8 is satisfied if

$$kx_m = \frac{2\pi x_m}{\lambda} = m\pi \qquad m = 0, 1, 2, 3, \ldots \qquad (21.9)$$

Thus the position x_m of the mth node is

$$x_m = m\frac{\lambda}{2} \qquad m = 0, 1, 2, 3, \ldots \qquad (21.10)$$

You can see that the spacing between two adjacent nodes is $\lambda/2$, in agreement with Figure 21.6. The nodes are *not* spaced by λ, as you might have expected.

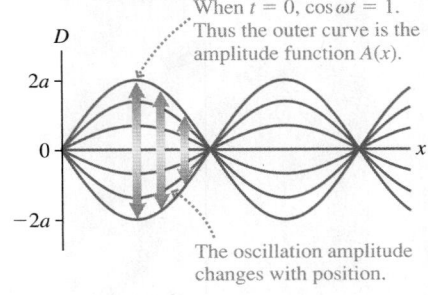

FIGURE 21.7 The net displacement resulting from two counter-propagating sinusoidal waves.

When $t = 0$, $\cos \omega t = 1$. Thus the outer curve is the amplitude function $A(x)$.

The oscillation amplitude changes with position.

EXAMPLE 21.1 **Node spacing on a string**

A very long string has a linear density of 5.0 g/m and is stretched with a tension of 8.0 N. 100 Hz waves with amplitudes of 2.0 mm are generated at the ends of the string.

a. What is the node spacing along the resulting standing wave?
b. What is the maximum displacement of the string?

MODEL Two counter-propagating waves of equal frequency create a standing wave.

VISUALIZE The standing wave will look like Figure 21.5.

SOLVE a. The speed of the waves on the string is

$$v = \sqrt{\frac{T_s}{\mu}} = \sqrt{\frac{8.0 \text{ N}}{0.0050 \text{ kg/m}}} = 40 \text{ m/s}$$

and the wavelength is

$$\lambda = \frac{v}{f} = \frac{40 \text{ m/s}}{100 \text{ Hz}} = 0.40 \text{ m} = 40 \text{ cm}$$

Thus the spacing between adjacent nodes is $\lambda/2 = 20$ cm.

b. The maximum displacement is $A_{max} = 2a = 4.0$ mm.

21.3 Standing Waves on a String

Wiggling both ends of a very long string is not a practical way to generate standing waves. Instead, as in the photograph in Figure 21.3, standing waves are usually seen on a string that is fixed at both ends. To understand why this condition causes standing waves, we need to examine what happens when a traveling wave encounters a discontinuity.

FIGURE 21.8a on the next page shows a *discontinuity* between a string with a larger linear density and one with a smaller linear density. The tension is the same in both strings, so the wave speed is slower on the left, faster on the right. Whenever a wave encounters a discontinuity, some of the wave's energy is *transmitted* forward and some is *reflected*.

FIGURE 21.8 A wave reflects when it encounters a discontinuity or a boundary.

(a)

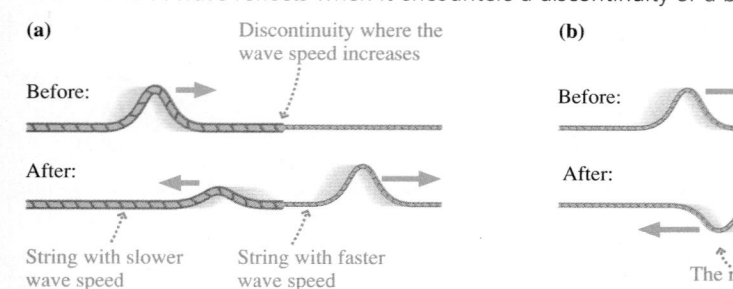

Discontinuity where the wave speed increases

Before:

After:

String with slower wave speed   String with faster wave speed

(b)

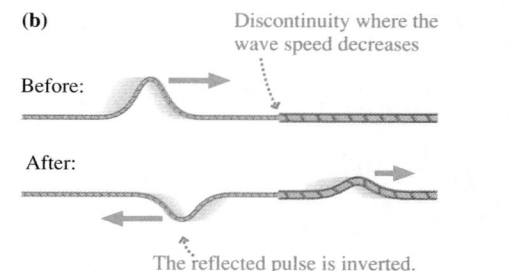

Discontinuity where the wave speed decreases

Before:

After:

The reflected pulse is inverted.

(c)

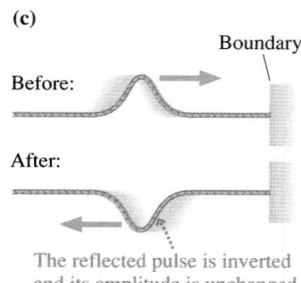

Boundary

Before:

After:

The reflected pulse is inverted and its amplitude is unchanged.

FIGURE 21.9 A strobe photo of a pulse traveling along a rope-like spring.

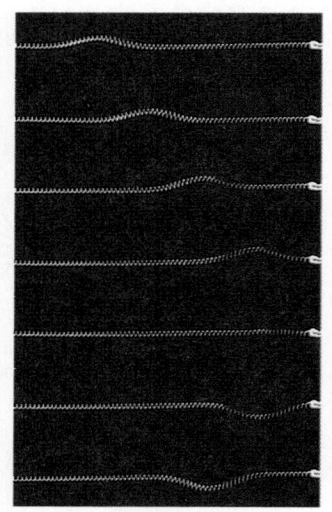

FIGURE 21.10 Reflections at the two boundaries cause a standing wave on the string.

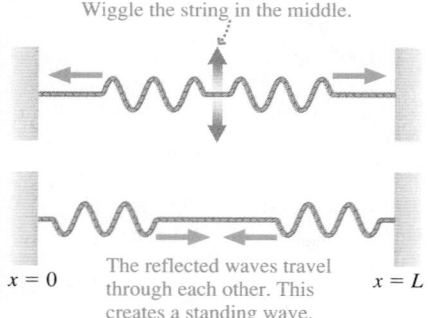

Wiggle the string in the middle.

$x = 0$ The reflected waves travel through each other. This creates a standing wave. $x = L$

Light waves exhibit an analogous behavior when they encounter a piece of glass. Most of the light wave's energy is transmitted through the glass, which is why glass is transparent, but a small amount of energy is reflected. That is how you see your reflection dimly in a storefront window.

In **FIGURE 21.8b**, an incident wave encounters a discontinuity at which the wave speed decreases. In this case, the reflected pulse is *inverted*. A positive displacement of the incident wave becomes a negative displacement of the reflected wave. Because $\sin(\phi + \pi) = -\sin \phi$, we say that the reflected wave has a *phase change of π upon reflection*. This aspect of reflection will be important later in the chapter when we look at the interference of light waves.

The wave in **FIGURE 21.8c** reflects from a *boundary*. You can think of this as Figure 21.8b in the limit that the string on the right becomes infinitely massive. Thus the reflection in Figure 21.8c looks like that of Figure 21.8b with one exception: Because there is no transmitted wave, *all* the wave's energy is reflected. Hence **the amplitude of a wave reflected from a boundary is unchanged.** **FIGURE 21.9** is a sequence of strobe photos in which you see a pulse on a rope-like spring reflecting from a boundary at the right of the photo. The reflected pulse is inverted but otherwise unchanged.

Creating Standing Waves

FIGURE 21.10 shows a string of length L tied at $x = 0$ and $x = L$. If you wiggle the string in the middle, sinusoidal waves travel outward in both directions and soon reach the boundaries. Because the speed of a reflected wave does not change, **the wavelength and frequency of a reflected sinusoidal wave are unchanged.** Consequently, reflections at the ends of the string cause two waves of *equal amplitude and wavelength* to travel in opposite directions along the string. As we've just seen, these are the conditions that cause a standing wave!

To connect the mathematical analysis of standing waves in Section 21.2 with the physical reality of a string tied down at the ends, we need to impose *boundary conditions*. A **boundary condition** is a mathematical statement of any constraint that *must* be obeyed at the boundary or edge of a medium. Because the string is tied down at the ends, the displacements at $x = 0$ and $x = L$ must be zero at all times. Thus the standing-wave boundary conditions are $D(x = 0, t) = 0$ and $D(x = L, t) = 0$. Stated another way, we require nodes at both ends of the string.

We found that the displacement of a standing wave is $D(x, t) = (2a \sin kx) \cos \omega t$. This equation already satisfies the boundary condition $D(x = 0, t) = 0$. That is, the origin has already been located at a node. The second boundary condition, at $x = L$, requires $D(x = L, t) = 0$. This condition will be met at all times if

$$2a \sin kL = 0 \qquad \text{(boundary condition at } x = L) \qquad (21.11)$$

Equation 21.11 will be true if $\sin kL = 0$, which in turn requires

$$kL = \frac{2\pi L}{\lambda} = m\pi \qquad m = 1, 2, 3, 4, \ldots \qquad (21.12)$$

kL must be a multiple of $m\pi$, but $m = 0$ is excluded because L can't be zero.

For a string of fixed length L, the only quantity in Equation 21.12 that can vary is λ. That is, the boundary condition is satisfied only if the wavelength has one of the values

$$\lambda_m = \frac{2L}{m} \qquad m = 1, 2, 3, 4, \ldots \qquad (21.13)$$

A standing wave can exist on the string *only* if its wavelength is one of the values given by Equation 21.13. The mth possible wavelength $\lambda_m = 2L/m$ is just the right size so that its mth node is located at the end of the string (at $x = L$).

NOTE ▶ Other wavelengths, which would be perfectly acceptable wavelengths for a traveling wave, cannot exist as a *standing* wave of length L because they cannot meet the boundary conditions requiring a node at each end of the string. ◀

If standing waves are possible only for certain wavelengths, then only a few specific oscillation frequencies are allowed. Because $\lambda f = v$ for a sinusoidal wave, the oscillation frequency corresponding to wavelength λ_m is

$$f_m = \frac{v}{\lambda_m} = \frac{v}{2L/m} = m\frac{v}{2L} \qquad m = 1, 2, 3, 4, \ldots \qquad (21.14)$$

The lowest allowed frequency

$$f_1 = \frac{v}{2L} \qquad \text{(fundamental frequency)} \qquad (21.15)$$

which corresponds to wavelength $\lambda_1 = 2L$, is called the **fundamental frequency** of the string. The allowed frequencies can be written in terms of the fundamental frequency as

$$f_m = mf_1 \qquad m = 1, 2, 3, 4, \ldots \qquad (21.16)$$

The allowed standing-wave frequencies are all integer multiples of the fundamental frequency. The higher-frequency standing waves are called **harmonics,** with the $m = 2$ wave at frequency f_2 called the *second harmonic,* the $m = 3$ wave called the *third harmonic,* and so on.

FIGURE 21.11 graphs the first four possible standing waves on a string of fixed length L. These possible standing waves are called the **modes** of the string, or sometimes the *normal modes.* Each mode, numbered by the integer m, has a unique wavelength and frequency. Keep in mind that these drawings simply show the *envelope,* or outer edge, of the oscillations. The string is continuously oscillating at all positions between these edges, as we showed in more detail in Figure 21.5.

There are three things to note about the modes of a string.

1. m is the number of *antinodes* on the standing wave, not the number of nodes. You can tell a string's mode of oscillation by counting the number of antinodes.
2. The *fundamental mode,* with $m = 1$, has $\lambda_1 = 2L$, not $\lambda_1 = L$. Only half of a wavelength is contained between the boundaries, a direct consequence of the fact that the spacing between nodes is $\lambda/2$.
3. The frequencies of the normal modes form a series: $f_1, 2f_1, 3f_1, 4f_1, \ldots$. The fundamental frequency f_1 can be found as the *difference* between the frequencies of any two adjacent modes. That is, $f_1 = \Delta f = f_{m+1} - f_m$.

FIGURE 21.12 is a time-exposure photograph of the $m = 3$ standing wave on a string. The nodes and antinodes are quite distinct. The string vibrates three times faster for the $m = 3$ mode than for the fundamental $m = 1$ mode.

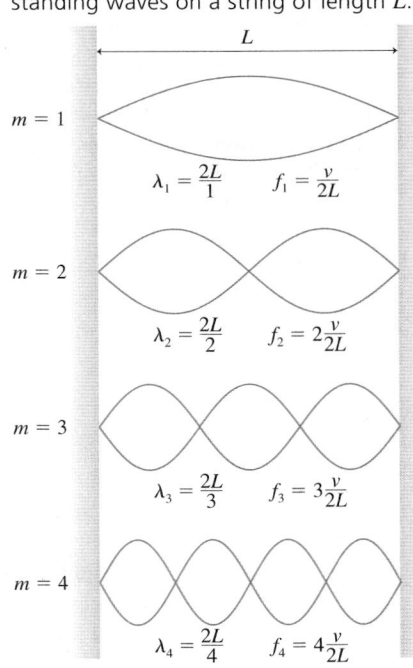

FIGURE 21.11 The first four modes for standing waves on a string of length L.

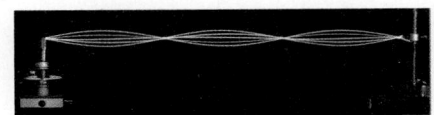

FIGURE 21.12 Time-exposure photograph of the $m = 3$ standing-wave mode on a stretched string.

Measuring g

Standing-wave frequencies can be measured very accurately. Consequently, standing waves are often used in experiments to make accurate measurements of other quantities. One such experiment, shown in **FIGURE 21.13**, uses standing waves to measure the free-fall acceleration g. A heavy mass is suspended from a 1.65-m-long, 5.85 g steel wire; then an oscillating magnetic field (because steel is magnetic) is used to excite the $m = 3$ standing wave on the wire. Measuring the frequency for different masses yields the following data:

FIGURE 21.13 An experiment to measure g with standing waves.

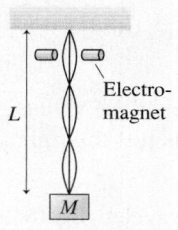

Mass (kg)	f_3 (Hz)
2.00	68
4.00	97
6.00	117
8.00	135
10.00	152

Analyze these data to determine the local value of g.

MODEL The hanging mass creates tension in the wire. This establishes the wave speed along the wire and thus the frequencies of standing waves. Masses of a few kg might stretch the wire a mm or so, but that doesn't change the length L until the third decimal place. The mass of the wire itself is insignificant in comparison to that of the hanging mass. We'll be justified in determining g to three significant figures.

SOLVE The frequency of the third harmonic is

$$f_3 = \frac{3}{2}\frac{v}{L}$$

The wave speed on the wire is

$$v = \sqrt{\frac{T_s}{\mu}} = \sqrt{\frac{Mg}{m/L}} = \sqrt{\frac{MgL}{m}}$$

where Mg is the weight of the hanging mass, and thus the tension in the wire, while m is the mass of the wire. Combining these two equations, we have

$$f_3 = \frac{3}{2}\sqrt{\frac{Mg}{mL}} = \frac{3}{2}\sqrt{\frac{g}{mL}}\sqrt{M}$$

Squaring both sides gives

$$f_3^2 = \frac{9g}{4mL}M$$

A graph of the square of the standing-wave frequency versus mass M should be a straight line passing through the origin with slope $9g/4mL$. We can use the experimental slope to determine g.

FIGURE 21.14 is a graph of f_3^2 versus M. The slope of the best-fit line is 2289 $kg^{-1}s^{-2}$, from which we find

$$g = \text{slope} \times \frac{4mL}{9}$$

$$= 2289 \text{ kg}^{-1}\text{s}^{-2} \times \frac{4(0.00585 \text{ kg})(1.65 \text{ m})}{9} = 9.82 \text{ m/s}^2$$

FIGURE 21.14 Graph of the data.

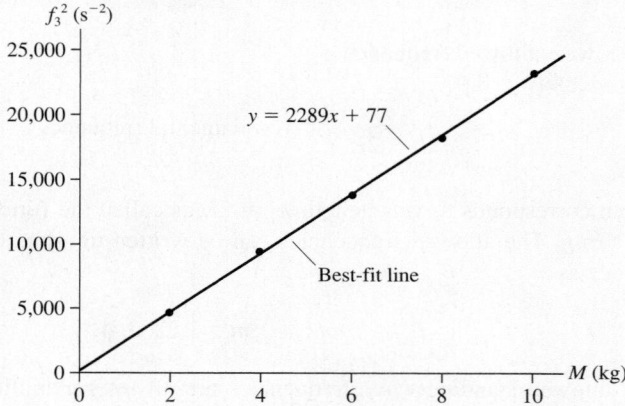

ASSESS The fact that the graph is linear and passes through the origin confirms our model. This is an important reason for having multiple data points rather than using only one mass.

A standing wave on a string vibrates as shown at the right. Suppose the string tension is quadrupled while the frequency and the length of the string are held constant. Which standing-wave pattern is produced?

Original standing wave

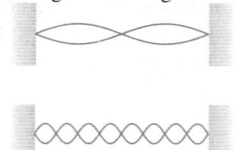

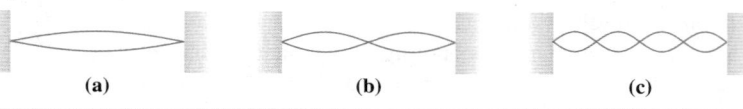

(a) (b) (c) (d)

Standing Electromagnetic Waves

Because electromagnetic waves are transverse waves, a standing electromagnetic wave is very much like a standing wave on a string. Standing electromagnetic waves can be established between two parallel mirrors that reflect light back and forth. The mirrors are boundaries, analogous to the boundaries at the ends of a string. In fact, this is exactly how a laser operates. The two facing mirrors in **FIGURE 21.15** form what is called a *laser cavity*.

Because the mirrors act like the points to which a string is tied, the light wave must have a node at the surface of each mirror. One of the mirrors is only partially reflective, to allow some light to escape and form the laser beam, but this doesn't affect the boundary condition.

Because the boundary conditions are the same, Equations 21.13 and 21.14 for λ_m and f_m apply to a laser just as they do to a vibrating string. The primary difference is the size of the wavelength. A typical laser cavity has a length $L \approx 30$ cm, and visible light has a wavelength $\lambda \approx 600$ nm. The standing light wave in a laser cavity has a mode number m that is approximately

$$m = \frac{2L}{\lambda} \approx \frac{2 \times 0.30 \text{ m}}{6.00 \times 10^{-7} \text{ m}} = 1{,}000{,}000$$

In other words, the standing light wave inside a laser cavity has approximately one million antinodes! This is a consequence of the very short wavelength of light.

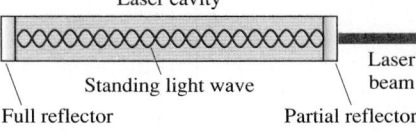

FIGURE 21.15 A laser contains a standing light wave between two parallel mirrors.

Laser cavity

Standing light wave

Full reflector   Partial reflector

Laser beam

EXAMPLE 21.3 **The standing light wave inside a laser**

Helium-neon lasers emit the red laser light commonly used in classroom demonstrations and supermarket checkout scanners. A helium-neon laser operates at a wavelength of precisely 632.9924 nm when the spacing between the mirrors is 310.372 mm.

a. In which mode does this laser operate?
b. What is the next longest wavelength that could form a standing wave in this laser cavity?

MODEL The light wave forms a standing wave between the two mirrors.

VISUALIZE The standing wave looks like Figure 21.15.

SOLVE a. We can use $\lambda_m = 2L/m$ to find that m (the mode) is

$$m = \frac{2L}{\lambda_m} = \frac{2(0.310372 \text{ m})}{6.329924 \times 10^{-7} \text{ m}} = 980{,}650$$

There are 980,650 antinodes in the standing light wave.

b. The next longest wavelength that can fit in this laser cavity will have one fewer node. It will be the $m = 980{,}649$ mode and its wavelength will be

$$\lambda = \frac{2L}{m} = \frac{2(0.310372 \text{ m})}{980{,}649} = 632.9930 \text{ nm}$$

ASSESS The wavelength increases by a mere 0.0006 nm when the mode number is decreased by 1.

Microwaves, with a wavelength of a few centimeters, can also set up standing waves. This is not always good. If the microwaves in a microwave oven form a standing wave, there are nodes where the electromagnetic field intensity is always zero. These nodes cause cold spots where the food does not heat. Although designers of microwave ovens try to prevent standing waves, ovens usually do have cold spots spaced $\lambda/2$ apart at nodes in the microwave field. A turntable in a microwave oven keeps the food moving so that no part of your dinner remains at a node.

21.4 Standing Sound Waves and Musical Acoustics

A long, narrow column of air, such as the air in a tube or pipe, can support a *longitudinal* standing sound wave. Longitudinal waves are somewhat trickier than string waves because a graph—showing displacement *parallel* to the tube—is not a picture of the wave.

To illustrate the ideas, **FIGURE 21.16** on the next page is a series of three graphs and pictures that show the $m = 2$ standing wave inside a column of air closed at both ends. We call this a *closed-closed tube*. The air at the closed ends cannot oscillate because the air molecules are pressed up against the wall, unable to move; hence **a closed end of a column of air must be a displacement node.** Thus the boundary conditions— nodes at the ends—are the same as for a standing wave on a string.

Although the graph looks familiar, it is now a graph of *longitudinal* displacement. At $t = 0$, positive displacements in the left half and negative displacements in the right half cause all the air molecules to converge at the center of the tube. The density and pressure rise at the center and fall at the ends—a *compression* and *rarefaction* in the terminology of Chapter 20. A half cycle later, the molecules have rushed to the ends

FIGURE 21.16 This time sequence of graphs and pictures illustrates the $m = 2$ standing sound wave in a closed-closed tube of air.

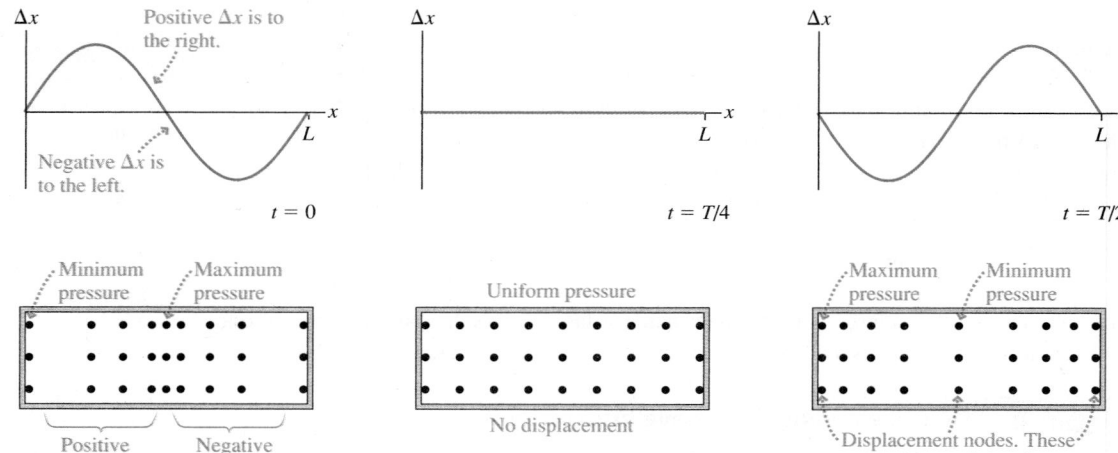

FIGURE 21.17 The $m = 2$ longitudinal standing wave can be represented as a displacement wave or as a pressure wave.

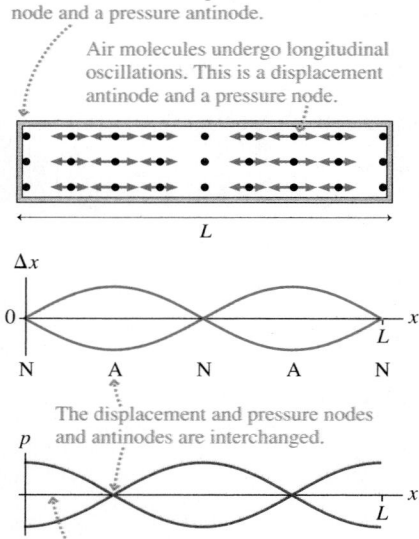

The closed end is a displacement node and a pressure antinode.

Air molecules undergo longitudinal oscillations. This is a displacement antinode and a pressure node.

The displacement and pressure nodes and antinodes are interchanged.

The pressure is oscillating around atmospheric pressure p_{atmos}.

of the tube. Now the pressure is maximum at the ends, minimum in the center. Try to visualize the air molecules sloshing back and forth this way.

FIGURE 21.17 combines these illustrations into a single picture showing where the molecules are oscillating (antinodes) and where they're not (nodes). A graph of the displacement Δx looks just like the $m = 2$ graph of a standing wave on a string. Because the boundary conditions are the same, the possible wavelengths and frequencies of standing waves in a closed-closed tube are the same as for a string of the same length.

It is often useful to think of sound as a *pressure wave* rather than a displacement wave, and the bottom graph in Figure 21.17 shows the $m = 2$ pressure standing wave in a closed-closed tube. Notice that the pressure is oscillating around p_{atmos}, its equilibrium value. **The nodes and antinodes of the pressure wave are interchanged with those of the displacement wave,** and a careful study of Figure 21.16 reveals why. The gas molecules are alternately pushed up against the ends of the tube, then pulled away, causing the pressure at the closed ends to oscillate with maximum amplitude—an antinode.

EXAMPLE 21.4 **Singing in the shower**

A shower stall is 2.45 m (8 ft) tall. For what frequencies less than 500 Hz are there standing sound waves in the shower stall?

MODEL The shower stall, to a first approximation, is a column of air 2.45 m long. It is closed at the ends by the ceiling and floor. Assume a 20°C speed of sound.

VISUALIZE A standing sound wave will have nodes at the ceiling and the floor. The $m = 2$ mode will look like Figure 21.17 rotated 90°.

SOLVE The fundamental frequency for a standing sound wave in this air column is

$$f_1 = \frac{v}{2L} = \frac{343 \text{ m/s}}{2(2.45 \text{ m})} = 70 \text{ Hz}$$

The possible standing-wave frequencies are integer multiples of the fundamental frequency. These are 70 Hz, 140 Hz, 210 Hz, 280 Hz, 350 Hz, 420 Hz, and 490 Hz.

ASSESS The many possible standing waves in a shower cause the sound to *resonate*, which helps explain why some people like to sing in the shower. Our approximation of the shower stall as a one-dimensional tube is actually a bit too simplistic. A three-dimensional analysis would find additional modes, making the "sound spectrum" even richer.

Air columns closed at both ends are of limited interest unless, as in Example 21.4, you are inside the column. Columns of air that *emit* sound are open at one or both ends. Many musical instruments fit this description. For example, a flute is a tube of air open at both ends. The flutist blows across one end to create a standing wave inside the tube, and a note of that frequency is emitted from both ends of the flute. (The blown end of a flute is open on the side, rather than across the tube. That is necessary for practical reasons of how flutes are played, but from a physics perspective this is the "end" of the tube because it opens the tube to the atmosphere.) A trumpet, however, is open at the bell end but is *closed* by the player's lips at the other end.

You saw earlier that a wave is partially transmitted and partially reflected at a discontinuity. When a sound wave traveling through a tube of air reaches an open end, some of the wave's energy is transmitted out of the tube to become the sound that you hear and some portion of the wave is reflected back into the tube. These reflections, analogous to the reflection of a string wave from a boundary, allow standing sound waves to exist in a tube of air that is open at one or both ends.

Not surprisingly, the *boundary condition* at the open end of a column of air is not the same as the boundary condition at a closed end. The air pressure at the open end of a tube is constrained to match the atmospheric pressure of the surrounding air. Consequently, the open end of a tube must be a pressure node. Because pressure nodes and antinodes are interchanged with those of the displacement wave, **an open end of an air column is required to be a displacement antinode.** (A careful analysis shows that the antinode is actually just outside the open end, but for our purposes we'll assume the antinode is exactly at the open end.)

FIGURE 21.18 shows displacement and pressure graphs of the first three standing-wave modes of a tube closed at both ends (a *closed-closed tube*), a tube open at both ends (an *open-open tube*), and a tube open at one end but closed at the other (an *open-closed tube*), all with the same length L. Notice the pressure and displacement boundary conditions. The standing wave in the open-open tube looks like the closed-closed tube except that the positions of the nodes and antinodes are interchanged. In both cases there are m half-wavelength segments between the ends; thus the wavelengths and frequencies of an open-open tube and a closed-closed tube are the same as those of a string tied at both ends:

FIGURE 21.18 The first three standing sound wave modes in columns of air with different boundary conditions.

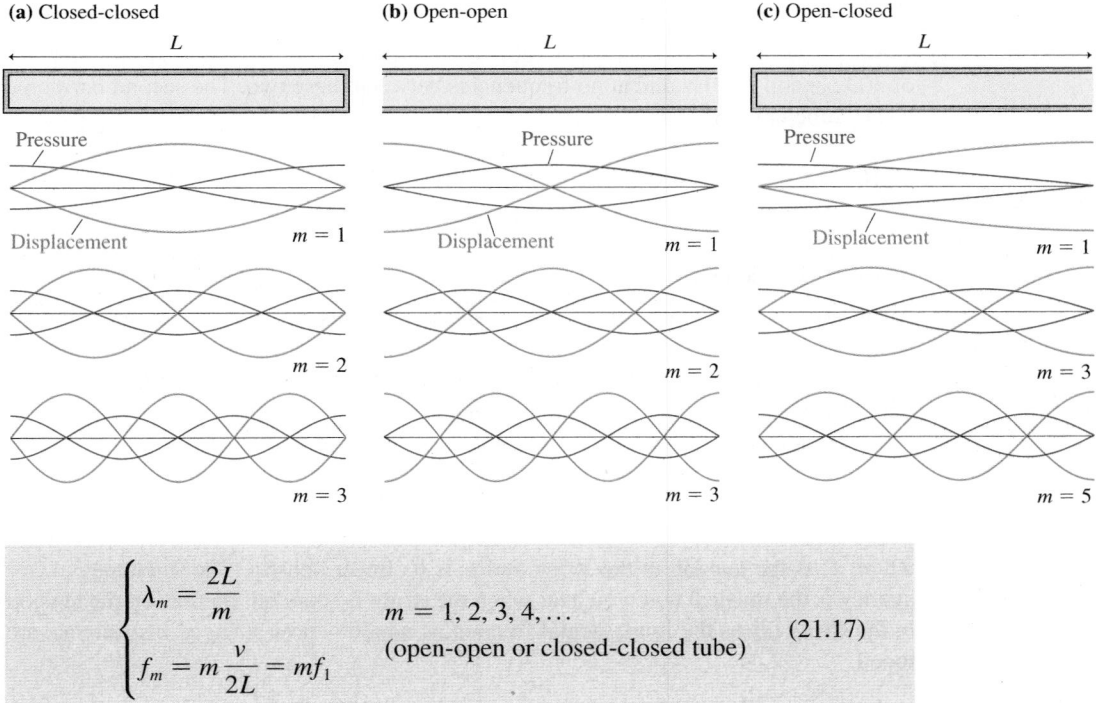

$$\begin{cases} \lambda_m = \dfrac{2L}{m} \\[2mm] f_m = m\dfrac{v}{2L} = mf_1 \end{cases} \quad \begin{aligned} &m = 1, 2, 3, 4, \ldots \\ &\text{(open-open or closed-closed tube)} \end{aligned} \quad (21.17)$$

The open-closed tube is different. The fundamental mode has only one-quarter of a wavelength in a tube of length L; hence the $m = 1$ wavelength is $\lambda_1 = 4L$. This is twice the λ_1 wavelength of an open-open or a closed-closed tube. Consequently, **the fundamental frequency of an open-closed tube is half that of an open-open or a closed-closed tube of the same length.** It will be left as a homework problem for you to show that the possible wavelengths and frequencies of an open-closed tube of length L are

$$\begin{cases} \lambda_m = \dfrac{4L}{m} \\ f_m = m\dfrac{v}{4L} = mf_1 \end{cases} \qquad \begin{array}{l} m = 1, 3, 5, 7, \dots \\ \text{(open-closed tube)} \end{array} \qquad (21.18)$$

Notice that m in Equation 21.18 takes on only *odd* values.

EXAMPLE 21.5 **Resonances of the ear canal**

The eardrum, which transmits sounds vibrations to the sensory organs of the inner ear, lies at the end of the ear canal. For adults, the ear canal is about 2.5 cm in length. What frequency standing waves can occur in the ear canal that are within the range of human hearing? The speed of sound in the warm air of the ear canal is 350 m/s.

MODEL The ear canal is open to the air at one end, closed by the eardrum at the other. We can model it as an open-closed tube. The standing waves will be those of Figure 21.18c.

SOLVE The lowest standing-wave frequency is the fundamental frequency for a 2.5-cm-long open-closed tube:

$$f_1 = \frac{v}{4L} = \frac{350 \text{ m/s}}{4(0.025 \text{ m})} = 3500 \text{ Hz}$$

Standing waves also occur at the harmonics, but an open-closed tube has only odd harmonics. These are

$$f_3 = 3f_1 = 10{,}500 \text{ Hz}$$
$$f_5 = 5f_1 = 17{,}500 \text{ Hz}$$

Higher harmonics are beyond the range of human hearing, as discussed in Section 20.5.

ASSESS The ear canal is short, so we expected the standing-wave frequencies to be relatively high. The air in your ear canal responds readily to sounds at these frequencies—what we call a *resonance* of the ear canal—and transmits theses sounds to the eardrum. Consequently, your ear actually is slightly more sensitive to sounds with frequencies around 3500 Hz and 10,500 Hz than to sounds at nearby frequencies.

STOP TO THINK 21.3 An open-open tube of air supports standing waves at frequencies of 300 Hz and 400 Hz and at no frequencies between these two. The second harmonic of this tube has frequency

a. 100 Hz   b. 200 Hz   c. 400 Hz   d. 600 Hz   e. 800 Hz

Musical Instruments

An important application of standing waves is to musical instruments. Instruments such as the guitar, the piano, and the violin have strings fixed at the ends and tightened to create tension. A disturbance generated on the string by plucking, striking, or bowing it creates a standing wave on the string.

The fundamental frequency of a vibrating string is

$$f_1 = \frac{v}{2L} = \frac{1}{2L}\sqrt{\frac{T_s}{\mu}}$$

where T_s is the tension in the string and μ is its linear density. The fundamental frequency is the musical note you hear when the string is sounded. Increasing the tension in the string raises the fundamental frequency, which is how stringed instruments are tuned.

NOTE ▶ v is the wave speed *on the string,* not the speed of sound in air. ◀

For the guitar or the violin, the strings are all the same length and under approximately the same tension. Were that not the case, the neck of the instrument would tend to twist toward the side of higher tension. The strings have different frequencies because they differ in linear density: The lower-pitched strings are "fat" while the higher-pitched strings are "skinny." This difference changes the frequency by changing the wave speed. *Small* adjustments are then made in the tension to bring each string to the exact desired frequency. Once the instrument is tuned, you play it by using your fingertips to alter the effective length of the string. As you shorten the string's length, the frequency and pitch go up.

A piano covers a much wider range of frequencies than a guitar or violin. This range cannot be produced by changing only the linear densities of the strings. The high end would have strings too thin to use without breaking, and the low end would have solid rods rather than flexible wires! So a piano is tuned through a combination of changing the linear density *and* the length of the strings. The bass note strings are not only fatter, they are also longer.

With a wind instrument, blowing into the mouthpiece creates a standing sound wave inside a tube of air. The player changes the notes by using her fingers to cover holes or open valves, changing the length of the tube and thus its frequency. The fact that the holes are on the side makes very little difference; the first open hole becomes an antinode because the air is free to oscillate in and out of the opening.

A wind instrument's frequency depends on the speed of sound *inside* the instrument. But the speed of sound depends on the temperature of the air. When a wind player first blows into the instrument, the air inside starts to rise in temperature. This increases the sound speed, which in turn raises the instrument's frequency for each note until the air temperature reaches a steady state. Consequently, wind players must "warm up" before tuning their instrument.

Many wind instruments have a "buzzer" at one end of the tube, such as a vibrating reed on a saxophone or vibrating lips on a trombone. Buzzers generate a continuous range of frequencies rather than single notes, which is why they sound like a "squawk" if you play on just the mouthpiece without the rest of the instrument. When a buzzer is connected to the body of the instrument, most of those frequencies cause no response of the air molecules. But the frequency from the buzzer that matches the fundamental frequency of the instrument causes the buildup of a large-amplitude response at just that frequency—a standing-wave resonance. This is the energy input that generates and sustains the musical note.

The strings on a harp vibrate as standing waves. Their frequencies determine the notes that you hear.

EXAMPLE 21.6 **Flutes and clarinets**

A clarinet is 66.0 cm long. A flute is nearly the same length, with 63.5 cm between the hole the player blows across and the end of the flute. What are the frequencies of the lowest note and the next higher harmonic on a flute and on a clarinet? The speed of sound in warm air is 350 m/s.

MODEL The flute is an open-open tube, open at the end as well as at the hole the player blows across. A clarinet is an open-closed tube because the player's lips and the reed seal the tube at the upper end.

SOLVE The lowest frequency is the fundamental frequency. For the flute, an open-open tube, this is

$$f_1 = \frac{v}{2L} = \frac{350 \text{ m/s}}{2(0.635 \text{ m})} = 275 \text{ Hz}$$

The clarinet, an open-closed tube, has

$$f_1 = \frac{v}{4L} = \frac{350 \text{ m/s}}{4(0.660 \text{ m})} = 133 \text{ Hz}$$

The next higher harmonic on the flute's open-open tube is $m = 2$ with frequency $f_2 = 2f_1 = 550$ Hz. An open-closed tube has only odd harmonics, so the next higher harmonic of the clarinet is $f_3 = 3f_1 = 399$ Hz.

ASSESS The clarinet plays a much lower note than the flute—musically, about an octave lower—because it is an open-closed tube. It's worth noting that neither of our fundamental frequencies is exactly correct because our open-open and open-closed tube models are a bit too simplified to adequately describe a real instrument. However, both calculated frequencies are close because our models do capture the essence of the physics.

A vibrating string plays the musical note corresponding to the fundamental frequency f_1, so stringed instruments must use several strings to obtain a reasonable range of notes. In contrast, wind instruments can sound at the second or third harmonic of the tube of air (f_2 or f_3). These higher frequencies are sounded by *overblowing* (flutes, brass instruments) or with keys that open small holes to encourage the formation of an antinode at that point (clarinets, saxophones). The controlled use of these higher harmonics gives wind instruments a wide range of notes.

21.5 Interference in One Dimension

One of the most basic characteristics of waves is the ability of two waves to combine into a single wave whose displacement is given by the principle of superposition. The pattern resulting from the superposition of two waves is often called **interference.** A standing wave is the interference pattern produced when two waves of equal frequency travel in opposite directions. In this section we will look at the interference of two waves traveling in the *same* direction.

FIGURE 21.19 Two overlapped waves travel along the *x*-axis.

(a) Two overlapped light waves

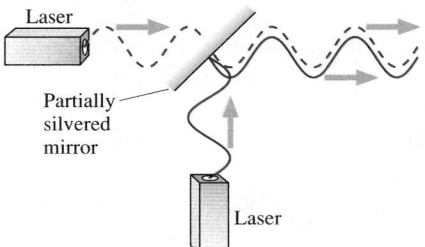

Laser

Partially
silvered
mirror

Laser

(b) Two overlapped sound waves

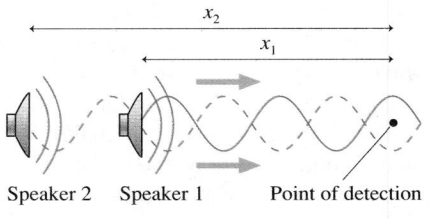

x_2

x_1

Speaker 2   Speaker 1   Point of detection

FIGURE 21.19a shows two light waves impinging on a partially silvered mirror. Such a mirror partially transmits and partially reflects each wave, causing two *overlapped* light waves to travel along the *x*-axis to the right of the mirror. Or consider the two loudspeakers in **FIGURE 21.19b**. The sound wave from loudspeaker 2 passes just to the side of loudspeaker 1; hence two overlapped sound waves travel to the right along the *x*-axis. We want to find out what happens when two overlapped waves travel in the same direction along the same axis.

Figure 21.19b shows a point on the *x*-axis where the overlapped waves are detected, either by your ear or by a microphone. This point is distance x_1 from loudspeaker 1 and distance x_2 from loudspeaker 2. (We will use loudspeakers and sound waves for most of our examples, but our analysis is valid for any wave.) What is the amplitude of the combined waves at this point?

Throughout this section, **we will assume that the waves are sinusoidal, have the same frequency and amplitude, and travel to the right along the *x*-axis.** Thus we can write the displacements of the two waves as

$$D_1(x_1, t) = a\sin(kx_1 - \omega t + \phi_{10}) = a\sin\phi_1$$
$$D_2(x_2, t) = a\sin(kx_2 - \omega t + \phi_{20}) = a\sin\phi_2$$

(21.19)

where ϕ_1 and ϕ_2 are the *phases* of the waves. Both waves have the same wave number $k = 2\pi/\lambda$ and the same angular frequency $\omega = 2\pi f$.

The phase constants ϕ_{10} and ϕ_{20} are characteristics of *the sources,* not the medium. **FIGURE 21.20** shows snapshot graphs at $t = 0$ of waves emitted by three sources with phase constants $\phi_0 = 0$ rad, $\phi_0 = \pi/2$ rad, and $\phi_0 = \pi$ rad. You can see that **the phase constant tells us what the source is doing at $t = 0$.** For example, a loudspeaker at its center position and moving backward at $t = 0$ has $\phi_0 = 0$ rad. Looking back at Figure 21.19b, you can see that loudspeaker 1 has phase constant $\phi_{10} = 0$ rad and loudspeaker 2 has $\phi_{20} = \pi$ rad.

NOTE ▶ We will often consider *identical sources,* by which we mean that $\phi_{20} = \phi_{10}$. That is, the sources oscillate in phase. ◀

Let's examine overlapped waves graphically before diving into the mathematics. **FIGURE 21.21** shows two important situations. In part a, the crests of the two waves are aligned as they travel along the *x*-axis. In part b, the crests of one wave align with the troughs of the other wave. The graphs and the wave fronts are slightly displaced from

FIGURE 21.20 Waves from three sources having phase constants $\phi_0 = 0$ rad, $\phi_0 = \pi/2$ rad, and $\phi_0 = \pi$ rad.

(a) Snapshot graph at $t = 0$ for $\phi_0 = 0$ rad

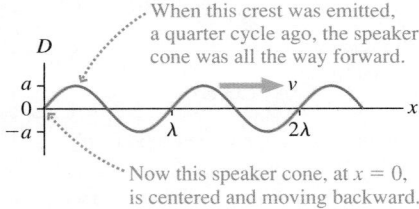

When this crest was emitted, a quarter cycle ago, the speaker cone was all the way forward.

Now this speaker cone, at $x = 0$, is centered and moving backward.

(b) Snapshot graph at $t = 0$ for $\phi_0 = \pi/2$ rad

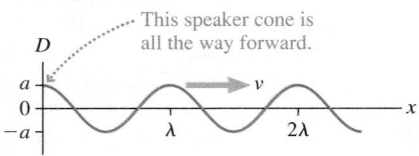

This speaker cone is all the way forward.

(c) Snapshot graph at $t = 0$ for $\phi_0 = \pi$ rad

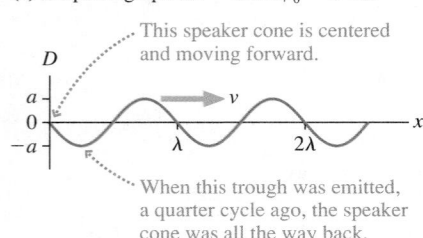

This speaker cone is centered and moving forward.

When this trough was emitted, a quarter cycle ago, the speaker cone was all the way back.

each other so that you can see what each wave is doing, but the *physical situation* is one in which the waves are traveling *on top of* each other. Recall, from Chapter 20, that the wave fronts shown in the middle panel locate the crests of the waves.

The two waves of FIGURE 21.21a have the same displacement at every point: $D_1(x) = D_2(x)$. Two waves that are aligned crest to crest and trough to trough are said to be **in phase**. Waves that are in phase march along "in step" with each other.

When we combine two in-phase waves, using the principle of superposition, the net displacement at each point is twice the displacement of each individual wave. The superposition of two waves to create a traveling wave with an amplitude *larger* than either individual wave is called **constructive interference**. When the waves are exactly in phase, giving $A = 2a$, we have *maximum constructive interference*.

In FIGURE 21.21b, where the crests of one wave align with the troughs of the other, the waves march along "out of step" with $D_1(x) = -D_2(x)$ at every point. Two waves that are aligned crest to trough are said to be *180° out of phase* or, more generally, just **out of phase**. A superposition of two waves to create a wave with an amplitude smaller than either individual wave is called **destructive interference**. In this case, because $D_1 = -D_2$, the net displacement is *zero* at *every point* along the axis. The combination of two waves that cancel each other to give no wave is called *perfect destructive interference*.

> NOTE ▶ Perfect destructive interference occurs only if the two waves have equal wavelengths and amplitudes, as we're assuming. Two waves of unequal amplitudes can interfere destructively, but the cancellation won't be perfect. ◀

The Phase Difference

To understand interference, we need to focus on the *phases* of the two waves, which are

$$\phi_1 = kx_1 - \omega t + \phi_{10}$$
$$\phi_2 = kx_2 - \omega t + \phi_{20} \qquad (21.20)$$

The difference between the two phases ϕ_1 and ϕ_2, called the **phase difference** $\Delta\phi$, is

$$\Delta\phi = \phi_2 - \phi_1 = (kx_2 - \omega t + \phi_{20}) - (kx_1 - \omega t + \phi_{10})$$
$$= k(x_2 - x_1) + (\phi_{20} - \phi_{10}) \qquad (21.21)$$
$$= 2\pi\frac{\Delta x}{\lambda} + \Delta\phi_0$$

You can see that there are two contributions to the phase difference. $\Delta x = x_2 - x_1$, the distance between the two sources, is called **path-length difference**. It is the extra distance traveled by wave 2 on the way to the point where the two waves are combined. $\Delta\phi_0 = \phi_{20} - \phi_{10}$ is the *inherent phase difference* between the sources.

The condition of being in phase, where crests are aligned with crests and troughs with troughs, is $\Delta\phi = 0$, 2π, 4π, or any integer multiple of 2π rad. Thus the condition for maximum constructive interference is

Maximum constructive interference:

$$\Delta\phi = 2\pi\frac{\Delta x}{\lambda} + \Delta\phi_0 = m \cdot 2\pi \text{ rad} \qquad m = 0, 1, 2, 3, \ldots \qquad (21.22)$$

For identical sources, which have $\Delta\phi_0 = 0$ rad, maximum constructive interference occurs when $\Delta x = m\lambda$. That is, **two identical sources produce maximum constructive interference when the path-length difference is an integer number of wavelengths**.

FIGURE 21.21 Constructive and destructive interference of two waves traveling along the *x*-axis.

(a) Constructive interference

These two waves are in phase. Their crests are aligned.

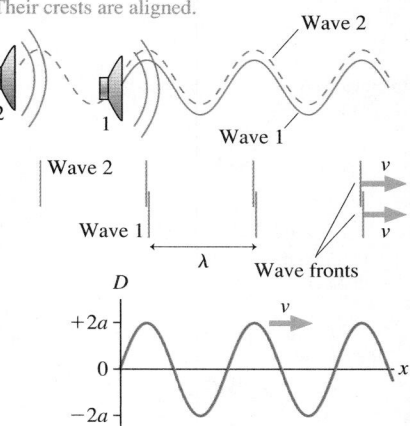

Their superposition produces a traveling wave moving to the right with amplitude 2*a*. This is maximum constructive interference.

(b) Destructive interference

These two waves are out of phase. The crests of one wave are aligned with the troughs of the other.

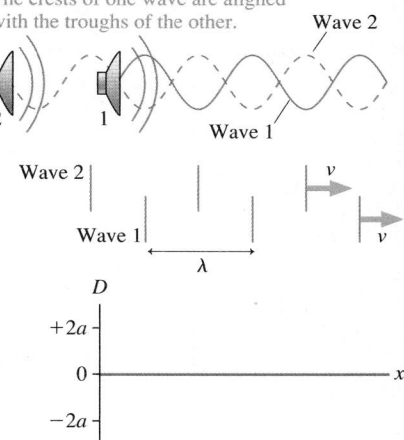

Their superposition produces a wave with zero amplitude. This is perfect destructive interference.

FIGURE 21.22 Two identical sources one wavelength apart.

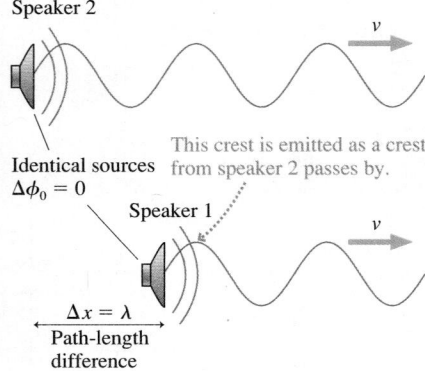

Speaker 2

This crest is emitted as a crest from speaker 2 passes by.

Identical sources $\Delta\phi_0 = 0$

Speaker 1

$\Delta x = \lambda$
Path-length difference

v

v

The two waves are in phase ($\Delta\phi = 2\pi$ rad) and interfere constructively.

FIGURE 21.22 shows two identical sources (i.e., the two loudspeakers are doing the same thing at the same time), so $\Delta\phi_0 = 0$ rad. The path-length difference Δx is the extra distance traveled by the wave from loudspeaker 2 before it combines with loudspeaker 1. In this case, $\Delta x = \lambda$. Because a wave moves forward exactly one wavelength during one period, loudspeaker 1 emits a crest exactly as a crest of wave 2 passes by. The two waves are "in step," with $\Delta\phi = 2\pi$ rad, so the two waves interfere constructively to produce a wave of amplitude $2a$.

Perfect destructive interference, where the crests of one wave are aligned with the troughs of the other, occurs when two waves are *out of phase*, meaning that $\Delta\phi = \pi$, 3π, 5π, or any odd multiple of π rad. Thus the condition for perfect destructive interference is

Perfect destructive interference:

$$\Delta\phi = 2\pi\frac{\Delta x}{\lambda} + \Delta\phi_0 = \left(m + \frac{1}{2}\right)\cdot 2\pi \text{ rad} \qquad m = 0, 1, 2, 3, \ldots \qquad (21.23)$$

For identical sources, which have $\Delta\phi_0 = 0$ rad, perfect destructive interference occurs when $\Delta x = (m + \frac{1}{2})\lambda$. **That is, two identical sources produce perfect destructive interference when the path-length difference is a half-integer number of wavelengths.**

Two waves can be out of phase because the sources are located at different positions, because the sources themselves are out of phase, or because of a combination of these two. FIGURE 21.23 illustrates these ideas by showing three different ways in which two waves interfere destructively. Each of these three arrangements creates waves with $\Delta\phi = \pi$ rad.

FIGURE 21.23 Destructive interference three ways.

(a) The sources are out of phase.

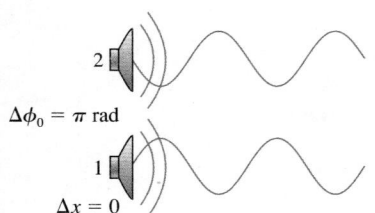

2

$\Delta\phi_0 = \pi$ rad

1

$\Delta x = 0$

(b) Identical sources are separated by half a wavelength.

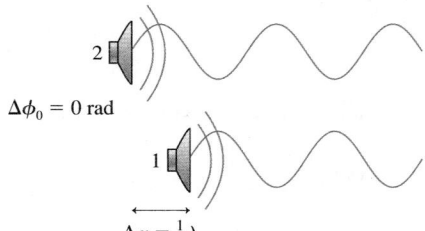

2

$\Delta\phi_0 = 0$ rad

1

$\Delta x = \frac{1}{2}\lambda$

(c) The sources are both separated and partially out of phase.

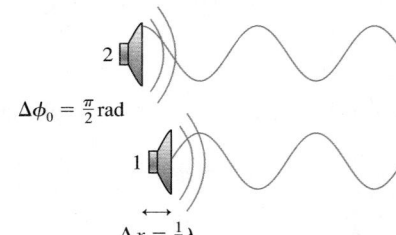

2

$\Delta\phi_0 = \frac{\pi}{2}$ rad

1

$\Delta x = \frac{1}{4}\lambda$

NOTE ▶ Don't confuse the phase difference of the waves ($\Delta\phi$) with the phase difference of the sources ($\Delta\phi_0$). It is $\Delta\phi$, the phase difference of the waves, that governs interference. ◀

EXAMPLE 21.7 **Interference between two sound waves**

You are standing in front of two side-by-side loudspeakers playing sounds of the same frequency. Initially there is almost no sound at all. Then one of the speakers is moved slowly away from you. The sound intensity increases as the separation between the speakers increases, reaching a maximum when the speakers are 0.75 m apart. Then, as the speaker continues to move, the intensity starts to decrease. What is the distance between the speakers when the sound intensity is again a minimum?

MODEL The changing sound intensity is due to the interference of two overlapped sound waves.

VISUALIZE Moving one speaker relative to the other changes the phase difference between the waves.

SOLVE A minimum sound intensity implies that the two sound waves are interfering destructively. Initially the loudspeakers are side by side, so the situation is as shown in Figure 21.23a with $\Delta x = 0$ and $\Delta\phi_0 = \pi$ rad. That is, the speakers themselves are out of phase. Moving one of the speakers does not change $\Delta\phi_0$, but it does change the path-length difference Δx and thus increases the overall phase difference $\Delta\phi$. Constructive interference, causing maximum intensity, is reached when

$$\Delta\phi = 2\pi\frac{\Delta x}{\lambda} + \Delta\phi_0 = 2\pi\frac{\Delta x}{\lambda} + \pi = 2\pi \text{ rad}$$

where we used $m = 1$ because this is the first separation giving constructive interference. The speaker separation at which this occurs is $\Delta x = \lambda/2$. This is the situation shown in **FIGURE 21.24**.

Because $\Delta x = 0.75$ m is $\lambda/2$, the sound's wavelength is $\lambda = 1.50$ m. The next point of destructive interference, with $m = 1$, occurs when

$$\Delta\phi = 2\pi\frac{\Delta x}{\lambda} + \Delta\phi_0 = 2\pi\frac{\Delta x}{\lambda} + \pi = 3\pi \text{ rad}$$

Thus the distance between the speakers when the sound intensity is again a minimum is

$$\Delta x = \lambda = 1.50 \text{ m}$$

ASSESS A separation of λ gives constructive interference for two *identical* speakers ($\Delta\phi_0 = 0$). Here the phase difference of π rad between the speakers (one is pushing forward as the other pulls back) gives destructive interference at this separation.

FIGURE 21.24 Two out-of-phase sources generate waves that are in phase if the sources are one half-wavelength apart.

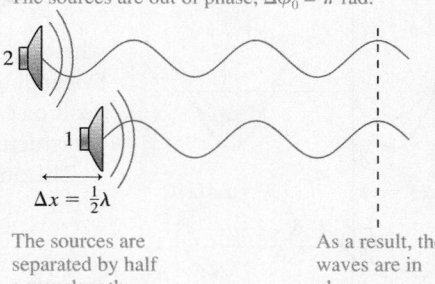

The sources are out of phase, $\Delta\phi_0 = \pi$ rad.

$\Delta x = \frac{1}{2}\lambda$

The sources are separated by half a wavelength.

As a result, the waves are in phase.

STOP TO THINK 21.4 Two loudspeakers emit waves with $\lambda = 2.0$ m. Speaker 2 is 1.0 m in front of speaker 1. What, if anything, can be done to cause constructive interference between the two waves?

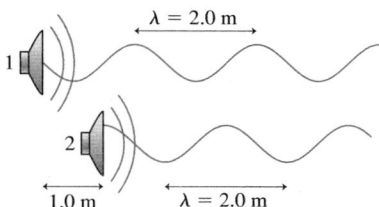

$\lambda = 2.0$ m

1.0 m $\lambda = 2.0$ m

a. Move speaker 1 forward (to the right) 1.0 m.
b. Move speaker 1 forward (to the right) 0.5 m.
c. Move speaker 1 backward (to the left) 0.5 m.
d. Move speaker 1 backward (to the left) 1.0 m.
e. Nothing. The situation shown already causes constructive interference.
f. Constructive interference is not possible for any placement of the speakers.

21.6 The Mathematics of Interference

Let's look more closely at the superposition of two waves. As two waves of equal amplitude and frequency travel together along the x-axis, the net displacement of the medium is

$$D = D_1 + D_2 = a\sin(kx_1 - \omega t + \phi_{10}) + a\sin(kx_2 - \omega t + \phi_{20})$$
$$= a\sin\phi_1 + a\sin\phi_2 \tag{21.24}$$

where the phases ϕ_1 and ϕ_2 were defined in Equation 21.20.

A useful trigonometric identity is

$$\sin\alpha + \sin\beta = 2\cos\left[\tfrac{1}{2}(\alpha - \beta)\right]\sin\left[\tfrac{1}{2}(\alpha + \beta)\right] \tag{21.25}$$

This identity is certainly not obvious, although it is easily proven by working backward from the right side. We can use this identity to write the net displacement of Equation 21.24 as

$$D = \left[2a\cos\left(\frac{\Delta\phi}{2}\right)\right]\sin(kx_{\text{avg}} - \omega t + (\phi_0)_{\text{avg}}) \tag{21.26}$$

where $\Delta\phi = \phi_2 - \phi_1$ is the phase difference between the two waves, exactly as in Equation 21.21. $x_{\text{avg}} = (x_1 + x_2)/2$ is the average distance to the two sources and $(\phi_0)_{\text{avg}} = (\phi_{10} + \phi_{20})/2$ is the average phase constant of the sources.

The sine term shows that the superposition of the two waves is still a traveling wave. An observer would see a sinusoidal wave moving along the x-axis with the *same* wavelength and frequency as the original waves.

FIGURE 21.25 The interference of two waves for three different values of the phase difference.

For $\Delta\phi = 40°$, the interference is constructive but not maximum constructive.

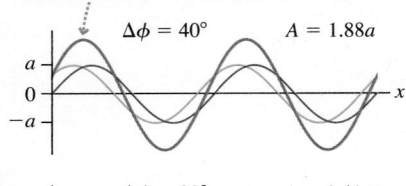

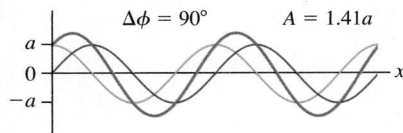

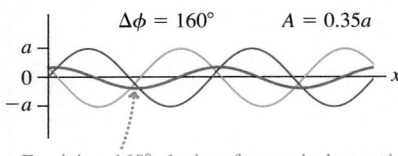

For $\Delta\phi = 160°$, the interference is destructive but not perfect destructive.

But how *big* is this wave compared to the two original waves? They each had amplitude a, but the amplitude of their superposition is

$$A = \left| 2a\cos\left(\frac{\Delta\phi}{2}\right) \right| \tag{21.27}$$

where we have used an absolute value sign because amplitudes must be positive. Depending upon the phase difference of the two waves, the amplitude of their superposition can be anywhere from zero (perfect destructive interference) to $2a$ (maximum constructive interference).

The amplitude has its maximum value $A = 2a$ if $\cos(\Delta\phi/2) = \pm 1$. This occurs when

$$\Delta\phi = m \cdot 2\pi \qquad \text{(maximum amplitude } A = 2a\text{)} \tag{21.28}$$

where m is an integer. Similarly, the amplitude is zero if $\cos(\Delta\phi/2) = 0$, which occurs when

$$\Delta\phi = \left(m + \tfrac{1}{2}\right) \cdot 2\pi \qquad \text{(minimum amplitude } A = 0\text{)} \tag{21.29}$$

Equations 21.28 and 21.29 are identical to the conditions of Equations 21.22 and 21.23 for constructive and destructive interference. We initially found these conditions by considering the alignment of the crests and troughs. Now we have confirmed them with an algebraic addition of the waves.

It is entirely possible, of course, that the two waves are neither exactly in phase nor exactly out of phase. Equation 21.27 allows us to calculate the amplitude of the superposition for any value of the phase difference. As an example, FIGURE 21.25 shows the calculated interference of two waves that differ in phase by 40°, by 90°, and by 160°.

EXAMPLE 21.8 **More interference of sound waves**

Two loudspeakers emit 500 Hz sound waves with an amplitude of 0.10 mm. Speaker 2 is 1.00 m behind speaker 1, and the phase difference between the speakers is 90°. What is the amplitude of the sound wave at a point 2.00 m in front of speaker 1?

MODEL The amplitude is determined by the interference of the two waves. Assume that the speed of sound has a room-temperature (20°C) value of 343 m/s.

SOLVE The amplitude of the sound wave is

$$A = |2a\cos(\Delta\phi/2)|$$

where $a = 0.10$ mm and the phase difference between the waves is

$$\Delta\phi = \phi_2 - \phi_1 = 2\pi\frac{\Delta x}{\lambda} + \Delta\phi_0$$

The sound's wavelength is

$$\lambda = \frac{v}{f} = \frac{343 \text{ m/s}}{500 \text{ Hz}} = 0.686 \text{ m}$$

Distances $x_1 = 2.00$ m and $x_2 = 3.00$ m are measured from the speakers, so the path-length difference is $\Delta x = 1.00$ m. We're given that the inherent phase difference between the speakers is $\Delta\phi_0 = \pi/2$ rad. Thus the phase difference at the observation point is

$$\Delta\phi = 2\pi\frac{\Delta x}{\lambda} + \Delta\phi_0 = 2\pi\frac{1.00 \text{ m}}{0.686 \text{ m}} + \frac{\pi}{2} \text{ rad} = 10.73 \text{ rad}$$

and the amplitude of the wave at this point is

$$A = \left| 2a\cos\left(\frac{\Delta\phi}{2}\right) \right| = \left| (0.200 \text{ mm})\cos\left(\frac{10.73}{2}\right) \right| = 0.121 \text{ mm}$$

ASSESS The interference is constructive because $A > a$, but less than maximum constructive interference.

Application: Thin-Film Optical Coatings

The shimmering colors of soap bubbles and oil slicks, as seen in the photo at the beginning of the chapter, are due to the interference of light waves. In fact, the idea of light-wave interference in one dimension has an important application in the optics industry, namely the use of **thin-film optical coatings**. These films, less than 1 μm (10^{-6} m) thick, are placed on glass surfaces, such as lenses, to control reflections from the glass. Antireflection coatings on the lenses in cameras, microscopes, and other optical equipment are examples of thin-film coatings.

FIGURE 21.26 shows a light wave of wavelength λ approaching a piece of glass that has been coated with a transparent film of thickness d whose index of refraction is n. The air-film boundary is a discontinuity at which the wave speed suddenly decreases, and you saw earlier, in Figure 21.8, that a discontinuity causes a reflection. Most of the light is transmitted into the film, but a little bit is reflected.

Furthermore, you saw in Figure 21.8 that the wave reflected from a discontinuity at which the speed decreases is *inverted* with respect to the incident wave. For a sinusoidal wave, which we're now assuming, the inversion is represented mathematically as a phase shift of π rad. The speed of a light wave decreases when it enters a material with a *larger* index of refraction. Thus **a light wave that reflects from a boundary at which the index of refraction increases has a phase shift of π rad.** There is no phase shift for the reflection from a boundary at which the index of refraction decreases. The reflection in Figure 21.26 is from a boundary between air ($n_{air} = 1.00$) and a transparent film with $n_{film} > n_{air}$, so the reflected wave is inverted due to the phase shift of π rad.

When the transmitted wave reaches the glass, most of it continues on into the glass but a portion is reflected back to the left. We'll assume that the index of refraction of the glass is larger than that of the film, $n_{glass} > n_{film}$, so this reflection also has a phase shift of π rad. This second reflection, after traveling back through the film, passes back into the air. There are now *two* equal-frequency waves traveling to the left, and these waves will interfere. If the two reflected waves are *in phase,* they will interfere constructively to cause a *strong reflection.* If the two reflected waves are *out of phase,* they will interfere destructively to cause a *weak reflection* or, if their amplitudes are equal, *no reflection* at all.

This suggests practical uses for thin-film optical coatings. The reflections from glass surfaces, even if weak, are often undesirable. For example, reflections degrade the performance of optical equipment. These reflections can be eliminated by coating the glass with a film whose thickness is chosen to cause *destructive* interference of the two reflected waves. This is an *antireflection coating.*

The amplitude of the reflected light depends on the phase difference between the two reflected waves. This phase difference is

$$\Delta\phi = \phi_2 - \phi_1 = (kx_2 + \phi_{20} + \pi \text{ rad}) - (kx_1 + \phi_{10} + \pi \text{ rad})$$
$$= 2\pi\frac{\Delta x}{\lambda_f} + \Delta\phi_0 \tag{21.30}$$

where we explicitly included the reflection phase shift of each wave. In this case, because *both* waves had a phase shift of π rad, the reflection phase shifts cancel.

The wavelength λ_f is the wavelength *in the film* because that's where the path-length difference Δx occurs. You learned in Chapter 20 that the wavelength in a transparent material with index of refraction n is $\lambda_f = \lambda/n$, where the unsubscripted λ is the wavelength in vacuum or air. That is, λ is the wavelength that we measure on "our" side of the air-film boundary.

The path-length difference between the two waves is $\Delta x = 2d$ because wave 2 travels through the film *twice* before rejoining wave 1. The two waves have a common origin—the initial division of the incident wave at the front surface of the film—so the inherent phase difference is $\Delta\phi_0 = 0$. Thus the phase difference of the two reflected waves is

$$\Delta\phi = 2\pi\frac{2d}{\lambda/n} = 2\pi\frac{2nd}{\lambda} \tag{21.31}$$

The interference is constructive, causing a strong reflection, when $\Delta\phi = m \cdot 2\pi$ rad. So when both reflected waves have a phase of π rad, constructive interference occurs for wavelengths

$$\lambda_C = \frac{2nd}{m} \qquad m = 1, 2, 3, \dots \qquad \text{(constructive interference)} \tag{21.32}$$

FIGURE 21.26 The two reflections, one from the coating and one from the glass, interfere.

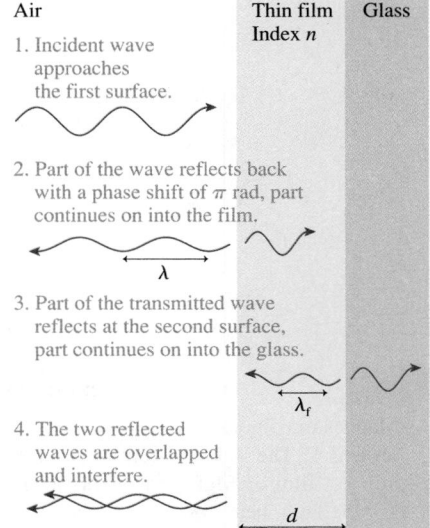

1. Incident wave approaches the first surface.

2. Part of the wave reflects back with a phase shift of π rad, part continues on into the film.

3. Part of the transmitted wave reflects at the second surface, part continues on into the glass.

4. The two reflected waves are overlapped and interfere.

Antireflection coatings use the interference of light waves to nearly eliminate reflections from glass surfaces.

You will notice that m starts with 1, rather than 0, in order to give meaningful results. Destructive interference, with minimum reflection, requires $\Delta\phi = \left(m - \frac{1}{2}\right) \cdot 2\pi$ rad. This—again, when both waves have a phase shift of π rad—occurs for wavelengths

$$\lambda_{\mathrm{D}} = \frac{2nd}{m - \frac{1}{2}} \qquad m = 1, 2, 3, \ldots \qquad \text{(destructive interference)} \quad (21.33)$$

We've used $m - \frac{1}{2}$, rather than $m + \frac{1}{2}$, so that m can start with 1 to match the condition for constructive interference.

NOTE ▶ The exact condition for constructive or destructive interference is satisfied for only a few discrete wavelengths λ. Nonetheless, reflections are strongly enhanced (nearly constructive interference) for a range of wavelengths near λ_{C}. Likewise, there is a range of wavelengths near λ_{D} for which the reflection is nearly canceled. ◀

EXAMPLE 21.9 | **Designing an antireflection coating**

Magnesium fluoride (MgF_2) is used as an antireflection coating on lenses. The index of refraction of MgF_2 is 1.39. What is the thinnest film of MgF_2 that works as an antireflection coating at $\lambda = 510$ nm, near the center of the visible spectrum?

MODEL Reflection is minimized if the two reflected waves interfere destructively.

SOLVE The film thicknesses that cause destructive interference at wavelength λ are

$$d = \left(m - \frac{1}{2}\right)\frac{\lambda}{2n}$$

The thinnest film has $m = 1$. Its thickness is

$$d = \frac{\lambda}{4n} = \frac{510 \text{ nm}}{4(1.39)} = 92 \text{ nm}$$

The film thickness is significantly less than the wavelength of visible light!

ASSESS The reflected light is completely eliminated (perfect destructive interference) only if the two reflected waves have equal amplitudes. In practice, they don't. Nonetheless, the reflection is reduced from $\approx 4\%$ of the incident intensity for "bare glass" to well under 1%. Furthermore, the intensity of reflected light is much reduced across most of the visible spectrum (400–700 nm), even though the phase difference deviates more and more from π rad as the wavelength moves away from 510 nm. It is the increasing reflection at the ends of the visible spectrum ($\lambda \approx 400$ nm and $\lambda \approx 700$ nm), where $\Delta\phi$ deviates significantly from π rad, that gives a reddish-purple tinge to the lenses on cameras and binoculars. Homework problems will let you explore situations where only one of the two reflections has a reflection phase shift of π rad.

21.7 Interference in Two and Three Dimensions

Ripples on a lake move in two dimensions. The glow from a lightbulb spreads outward as a spherical wave. A circular or spherical wave can be written

$$D(r, t) = a\sin(kr - \omega t + \phi_0) \qquad (21.34)$$

where r is the distance measured outward from the source. Equation 21.34 is our familiar wave equation with the one-dimensional coordinate x replaced by a more general radial coordinate r. Strictly speaking, the amplitude a of a circular or spherical wave diminishes as r increases. However, we will assume that a remains essentially constant over the region in which we study the wave. FIGURE 21.27 shows the wave-front diagram for a circular or spherical wave. Recall that the wave fronts represent the *crests* of the wave and are spaced by the wavelength λ.

What happens when two circular or spherical waves overlap? For example, imagine two paddles oscillating up and down on the surface of a pond. We will assume that the two paddles oscillate with the same frequency and amplitude and that they are in phase. FIGURE 21.28 shows the wave fronts of the two waves. The ripples overlap as they travel, and, as was the case in one dimension, this causes interference.

Constructive interference with $A = 2a$ occurs where two crests align or two troughs align. Several locations of constructive interference are marked in Figure 21.28. Intersecting wave fronts are points where two crests are aligned. It's a bit harder to

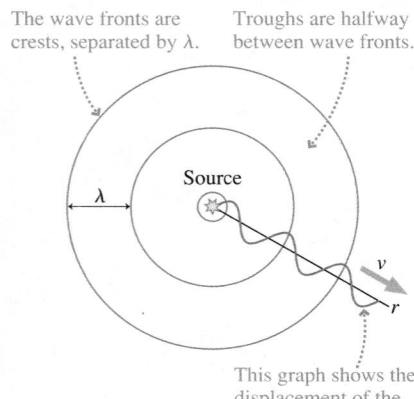

FIGURE 21.27 A circular or spherical wave.

The wave fronts are crests, separated by λ.

Troughs are halfway between wave fronts.

λ

Source

v

r

This graph shows the displacement of the medium.

visualize, but two troughs are aligned when a midpoint between two wave fronts is overlapped with another midpoint between two wave fronts. Destructive interference with $A = 0$ occurs where the crest of one wave aligns with a trough of the other wave. Several points of destructive interference are also indicated in Figure 21.28.

A picture on a page is static, but **the wave fronts are in motion.** Try to imagine the wave fronts of Figure 21.28 expanding outward as new circular rings are born at the sources. The waves will move forward half a wavelength during half a period, causing the crests in Figure 21.28 to be replaced by troughs while the troughs become crests.

The important point to recognize is that **the motion of the waves does not affect the points of constructive and destructive interference.** Points in the figure where two crests overlap will become points where two troughs overlap, but this overlap is still constructive interference. Similarly, points in the figure where a crest and a trough overlap will become a point where a trough and a crest overlap—still destructive interference.

The mathematical description of interference in two or three dimensions is very similar to that of one-dimensional interference. The net displacement of a particle in the medium is

$$D = D_1 + D_2 = a \sin(kr_1 - \omega t + \phi_{10}) + a \sin(kr_2 - \omega t + \phi_{20}) \quad (21.35)$$

The only difference between Equation 21.35 and the earlier one-dimensional Equation 21.24 is that the linear coordinates x_1 and x_2 have been changed to radial coordinates r_1 and r_2. Thus our conclusions are unchanged. The superposition of the two waves yields a wave traveling outward with amplitude

$$A = \left| 2a \cos\left(\frac{\Delta\phi}{2}\right) \right| \quad (21.36)$$

where the phase difference, with x replaced by r, is now

$$\Delta\phi = 2\pi \frac{\Delta r}{\lambda} + \Delta\phi_0 \quad (21.37)$$

The term $2\pi(\Delta r/\lambda)$ is the phase difference that arises when the waves travel different distances from the sources to the point at which they combine. Δr itself is the *path-length difference.* As before, $\Delta\phi_0$ is any inherent phase difference of the sources themselves.

Maximum constructive interference with $A = 2a$ occurs, just as in one dimension, at those points where $\cos(\Delta\phi/2) = \pm 1$. Similarly, perfect destructive interference occurs at points where $\cos(\Delta\phi/2) = 0$. The conditions for constructive and destructive interference are

> Maximum constructive interference:
>
> $$\Delta\phi = 2\pi \frac{\Delta r}{\lambda} + \Delta\phi_0 = m \cdot 2\pi$$
>
> $$m = 0, 1, 2, \dots \quad (21.38)$$
>
> Perfect destructive interference:
>
> $$\Delta\phi = 2\pi \frac{\Delta r}{\lambda} + \Delta\phi_0 = \left(m + \frac{1}{2}\right) \cdot 2\pi$$

For two identical sources (i.e., sources that oscillate in phase with $\Delta\phi_0 = 0$), the conditions for constructive and destructive interference are simple:

Constructive: $\Delta r = m\lambda$

Destructive: $\Delta r = \left(m + \frac{1}{2}\right)\lambda$ (identical sources)   (21.39)

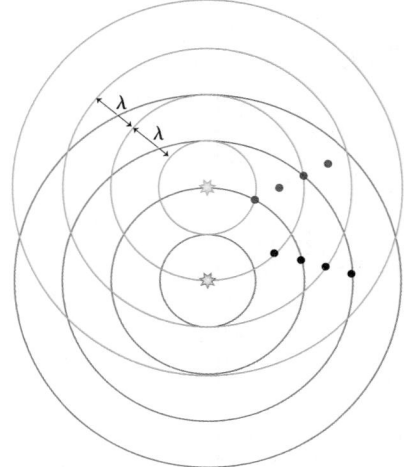

FIGURE 21.28 The overlapping ripple patterns of two sources. Several points of constructive and destructive interference are noted.

Two in-phase sources emit circular or spherical waves.

• Points of constructive interference. A crest is aligned with a crest, or a trough with a trough.

• Points of destructive interference. A crest is aligned with a trough of another wave.

Two overlapping water waves create an interference pattern.

FIGURE 21.29 The path-length difference Δr determines whether the interference at a particular point is constructive or destructive.

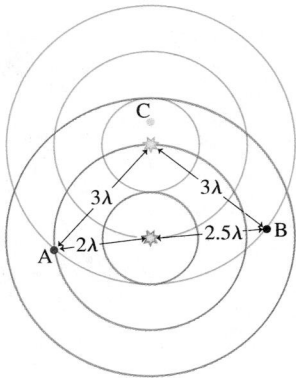

- At A, $\Delta r_A = \lambda$, so this is a point of constructive interference.

- At B, $\Delta r_B = \frac{1}{2}\lambda$, so this is a point of destructive interference.

The waves from two identical sources interfere constructively at points where the path-length difference is an integer number of wavelengths because, for these values of Δr, crests are aligned with crests and troughs with troughs. **The waves interfere destructively at points where the path-length difference is a half-integer number of wavelengths** because, for these values of Δr, crests are aligned with troughs. These two statements are the essence of interference.

> **NOTE** ▶ Equation 21.39 applies only if the sources are in phase. If the sources are not in phase, you must use the more general Equation 21.38 to locate the points of constructive and destructive interference. ◀

Wave fronts are spaced exactly one wavelength apart; hence we can measure the distances r_1 and r_2 simply by counting the rings in the wave-front pattern. In FIGURE 21.29, which is based on Figure 21.28, point A is distance $r_1 = 3\lambda$ from the first source and $r_2 = 2\lambda$ from the second. The path-length difference is $\Delta r_A = 1\lambda$, the condition for the maximum constructive interference of identical sources. Point B has $\Delta r_B = \frac{1}{2}\lambda$, so it is a point of perfect destructive interference.

> **NOTE** ▶ Interference is determined by Δr, the path-length *difference,* rather than by r_1 or r_2. ◀

STOP TO THINK 21.5   The interference at point C in Figure 21.29 is

a. Maximum constructive.
b. Constructive, but less than maximum.
c. Perfect destructive.
d. Destructive, but not perfect.
e. There is no interference at point C.

We can now locate the points of maximum constructive interference, for which $\Delta r = m\lambda$, by drawing a line through *all* the points at which $\Delta r = 0$, another line through all the points at which $\Delta r = \lambda$, and so on. These lines, shown in red in FIGURE 21.30, are called **antinodal lines.** They are analogous to the antinodes of a standing wave, hence the name. An antinode is a *point* of maximum constructive interference; for circular waves, oscillation at maximum amplitude occurs along a continuous *line.* Similarly, destructive interference occurs along lines called **nodal lines.** The displacement is *always zero* along these lines, just as it is at a node in a standing-wave pattern.

FIGURE 21.30 The points of constructive and destructive interference fall along antinodal and nodal lines.

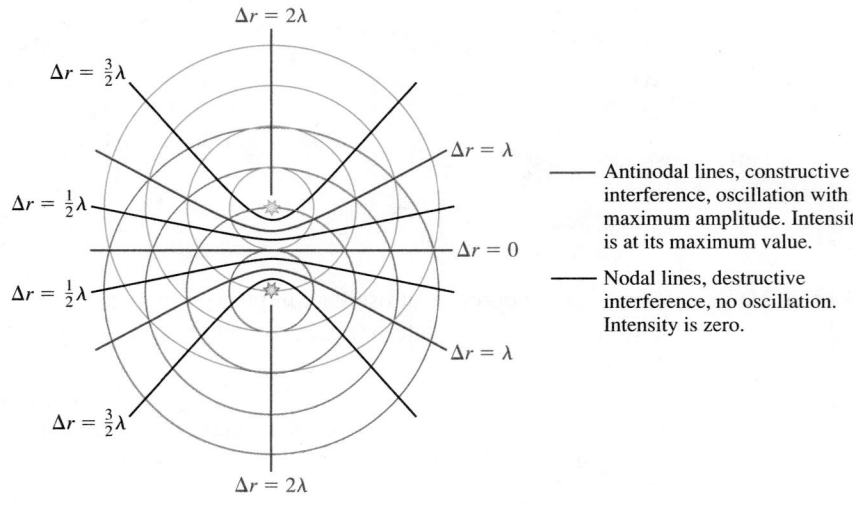

—— Antinodal lines, constructive interference, oscillation with maximum amplitude. Intensity is at its maximum value.

—— Nodal lines, destructive interference, no oscillation. Intensity is zero.

A Problem-Solving Strategy for Interference Problems

The information in this section is the basis of a strategy for solving interference problems. This strategy applies equally well to interference in one dimension if you use Δx instead of Δr.

PROBLEM-SOLVING STRATEGY 21.1 **Interference of two waves**

MODEL Make simplifying assumptions, such as assuming waves are circular and of equal amplitude.

VISUALIZE Draw a picture showing the sources of the waves and the point where the waves interfere. Give relevant dimensions. Identify the distances r_1 and r_2 from the sources to the point. Note any phase difference $\Delta\phi_0$ between the two sources.

SOLVE The interference depends on the path-length difference $\Delta r = r_2 - r_1$ and the source phase difference $\Delta\phi_0$.

Constructive: $\Delta\phi = 2\pi\dfrac{\Delta r}{\lambda} + \Delta\phi_0 = m \cdot 2\pi$

$m = 0, 1, 2, \ldots$

Destructive: $\Delta\phi = 2\pi\dfrac{\Delta r}{\lambda} + \Delta\phi_0 = \left(m + \dfrac{1}{2}\right) \cdot 2\pi$

For identical sources ($\Delta\phi_0 = 0$), the interference is maximum constructive if $\Delta r = m\lambda$, perfect destructive if $\Delta r = \left(m + \frac{1}{2}\right)\lambda$.

ASSESS Check that your result has the correct units, is reasonable, and answers the question.

Exercise 18

EXAMPLE 21.10 **Two-dimensional interference between two loudspeakers**

Two loudspeakers in a plane are 2.0 m apart and in phase with each other. Both emit 700 Hz sound waves into a room where the speed of sound is 341 m/s. A listener stands 5.0 m in front of the loudspeakers and 2.0 m to one side of the center. Is the interference at this point maximum constructive, perfect destructive, or in between? How will the situation differ if the loudspeakers are out of phase?

MODEL The two speakers are sources of in-phase, spherical waves. The overlap of these waves causes interference.

VISUALIZE **FIGURE 21.31** shows the loudspeakers and defines the distances r_1 and r_2 to the point of observation. The figure includes dimensions and notes that $\Delta\phi_0 = 0$ rad.

FIGURE 21.31 Pictorial representation of the interference between two loudspeakers.

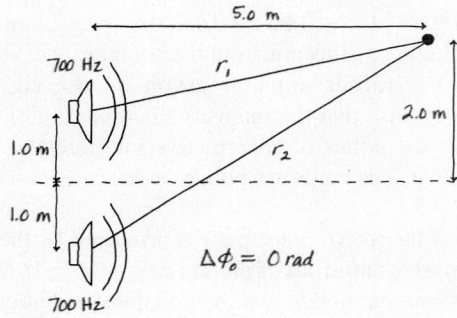

SOLVE It's not r_1 and r_2 that matter, but the *difference* Δr between them. From the geometry of the figure we can calculate that

$$r_1 = \sqrt{(5.0\text{ m})^2 + (1.0\text{ m})^2} = 5.10\text{ m}$$
$$r_2 = \sqrt{(5.0\text{ m})^2 + (3.0\text{ m})^2} = 5.83\text{ m}$$

Thus the path-length difference is $\Delta r = r_2 - r_1 = 0.73$ m. The wavelength of the sound waves is

$$\lambda = \frac{v}{f} = \frac{341\text{ m/s}}{700\text{ Hz}} = 0.487\text{ m}$$

In terms of wavelengths, the path-length difference is $\Delta r/\lambda = 1.50$, or

$$\Delta r = \frac{3}{2}\lambda$$

Because the sources are in phase ($\Delta\phi_0 = 0$), this is the condition for *destructive* interference. If the sources were out of phase ($\Delta\phi_0 = \pi$ rad), then the phase difference of the waves at the listener would be

$$\Delta\phi = 2\pi\frac{\Delta r}{\lambda} + \Delta\phi_0 = 2\pi\left(\frac{3}{2}\right) + \pi\text{ rad} = 4\pi\text{ rad}$$

This is an integer multiple of 2π rad, so in this case the interference would be *constructive*.

ASSESS Both the path-length difference *and* any inherent phase difference of the sources must be considered when evaluating interference.

Picturing Interference

A *contour map* is a useful way to visualize an interference pattern. FIGURE 21.32a shows the superposition of the waves from two identical sources ($\Delta\phi_0 = 0$) emitting waves with $\lambda = 1$ m. The sources, indicated with black dots, are located two wavelengths apart at $y = \pm 1$ m. Positive displacements are shown in red, with the deepest red representing the maximum displacement of the wave at this instant in time. These are the points where the crests of the individual waves interfere constructively to give $D = 2a$. Negative displacements are blue, with the darkest blue being the most negative displacement of the wave. These are also points of constructive interference, with two troughs overlapping to give $D = -2a$.

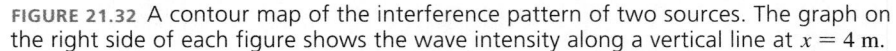

FIGURE 21.32 A contour map of the interference pattern of two sources. The graph on the right side of each figure shows the wave intensity along a vertical line at $x = 4$ m.

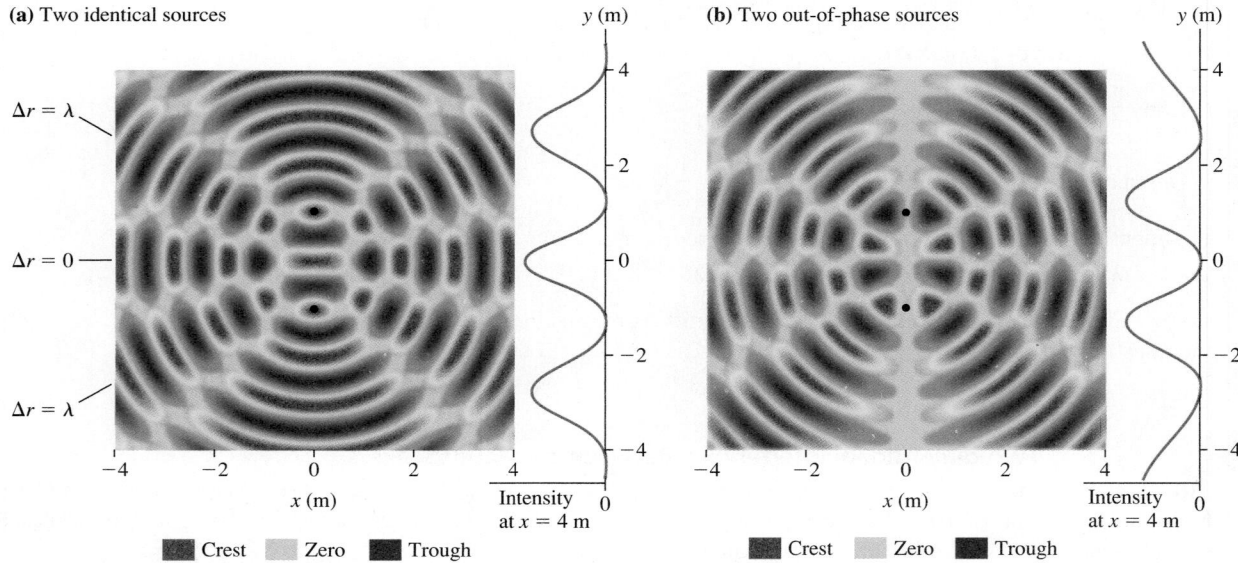

To understand this figure, try to visualize the waves expanding outward from the center. The red-blue-red-blue-red-··· pattern of crests and troughs moves outward along the antinodal lines as a *traveling wave* of amplitude $A = 2a$. Nothing ever happens along the nodal lines, where the amplitude is always zero.

Suppose you were to observe the *intensity* of the wave as it crosses the vertical line at $x = 4$ m on the right edge of the figure. If, for example, these are sound waves, you could listen to (or measure with a microphone) the sound intensity as you walk from $(x, y) = (4$ m, -4 m$)$ at the bottom of the figure to $(x, y) = (4$ m, 4 m$)$ at the top. The intensity is zero as you cross the nodal lines at $y \approx \pm 1$ m $\left(\Delta r = \frac{1}{2}\lambda\right)$. The intensity is maximum at the antinodal lines at $y = 0$ ($\Delta r = 0$) and $y \approx \pm 2.5$ m ($\Delta r = \lambda$), where a wave of maximum amplitude streams out from the sources.

The intensity is shown in the rather unusual graph on the right side of Figure 21.32a. It is unusual in the sense that the intensity, the quantity of interest, is graphed to the left. The peaks are the points of constructive interference, where you would measure maximum amplitude. The zeros are points of destructive interference, where the intensity is zero.

FIGURE 21.32b is a contour map of the interference pattern produced by the same two sources but with the sources themselves now out of phase ($\Delta\phi_0 = \pi$ rad). We'll leave the investigation of this figure to you, but notice that the nodal and antinodal lines are reversed from those of Figure 21.32a.

EXAMPLE 21.11 **The intensity of two interfering loudspeakers**

Two loudspeakers in a plane are 6.0 m apart and in phase. They emit equal-amplitude sound waves with a wavelength of 1.0 m. Each speaker alone creates sound with intensity I_0. An observer at point A is 10 m in front of the plane containing the two loudspeakers and centered between them. A second observer at point B is 10 m directly in front of one of the speakers. In terms of I_0, what are the intensity I_A at point A and the intensity I_B at point B?

MODEL The two speakers are sources of in-phase waves. The overlap of these waves causes interference.

VISUALIZE FIGURE 21.33 shows the two loudspeakers and the two points of observation. Distances r_1 and r_2 are defined for point B.

FIGURE 21.33 Pictorial representation of the interference between two loudspeakers.

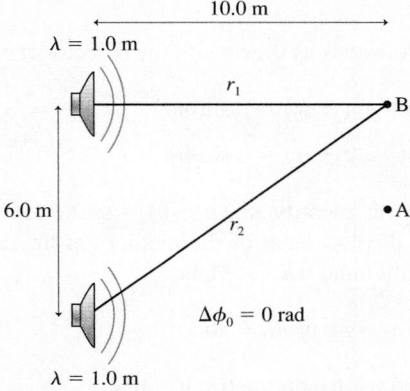

SOLVE Let the amplitude of the wave from each speaker be a. The intensity of a wave is proportional to the square of the amplitude,

so the intensity of each speaker alone is $I_0 = ca^2$, where c is an unknown proportionality constant. Point A is a point of constructive interference because the speakers are in phase ($\Delta\phi_0 = 0$) and the path-length difference is $\Delta r = 0$. The amplitude at this point is given by Equation 21.36:

$$A_A = \left| 2a\cos\left(\frac{\Delta\phi}{2}\right) \right| = 2a\cos(0) = 2a$$

Consequently, the intensity at this point is

$$I_A = cA_A^2 = c(2a)^2 = 4ca^2 = 4I_0$$

The intensity at A is four times that of either speaker played alone. At point B, the path-length difference is

$$\Delta r = \sqrt{(10.0\ \text{m})^2 + (6.0\ \text{m})^2} - 10.0\ \text{m} = 1.662\ \text{m}$$

The phase difference of the waves at this point is

$$\Delta\phi = 2\pi\frac{\Delta r}{\lambda} = 2\pi\frac{1.662\ \text{m}}{1.0\ \text{m}} = 10.44\ \text{rad}$$

Consequently, the amplitude at B is

$$A_B = \left| 2a\cos\left(\frac{\Delta\phi}{2}\right) \right| = |2a\cos(5.22\ \text{rad})| = 0.972a$$

Thus the intensity at this point is

$$I_B = cA_B^2 = c(0.972a)^2 = 0.95ca^2 = 0.95I_0$$

ASSESS Although B is directly in front of one of the speakers, superposition of the two waves results in an intensity that is less than it would be if this speaker played alone.

STOP TO THINK 21.6 These two loudspeakers are in phase. They emit equal-amplitude sound waves with a wavelength of 1.0 m. At the point indicated, is the interference maximum constructive, perfect destructive, or something in between?

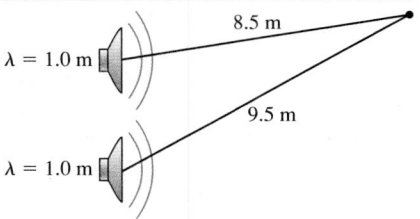

21.8 Beats

Thus far we have looked at the superposition of sources having the same wavelength and frequency. We can also use the principle of superposition to investigate a phenomenon that is easily demonstrated with two sources of slightly different frequency.

If you listen to two sounds with very different frequencies, such as a high note and a low note, you hear two distinct tones. But if the frequency difference is very small, just one or two hertz, then you hear a single tone whose intensity is *modulated* once or twice every second. That is, the sound goes up and down in volume, loud, soft, loud, soft,..., making a distinctive sound pattern called **beats**.

Consider two sinusoidal waves traveling along the x-axis with angular frequencies $\omega_1 = 2\pi f_1$ and $\omega_2 = 2\pi f_2$. The two waves are

$$D_1 = a\sin(k_1 x - \omega_1 t + \phi_{10})$$
$$D_2 = a\sin(k_2 x - \omega_2 t + \phi_{20})$$

(21.40)

where the subscripts 1 and 2 indicate that the frequencies, wave numbers, and phase constants of the two waves may be different.

To simplify the analysis, let's make several assumptions:

1. The two waves have the same amplitude a,
2. A detector, such as your ear, is located at the origin ($x = 0$),
3. The two sources are in phase ($\phi_{10} = \phi_{20}$), and
4. The source phases happen to be $\phi_{10} = \phi_{20} = \pi$ rad.

None of these assumptions is essential to the outcome. All could be otherwise and we would still come to basically the same conclusion, but the mathematics would be far messier. Making these assumptions allows us to emphasize the physics with the least amount of mathematics.

With these assumptions, the two waves as they reach the detector at $x = 0$ are

$$D_1 = a\sin(-\omega_1 t + \pi) = a\sin\omega_1 t$$
$$D_2 = a\sin(-\omega_2 t + \pi) = a\sin\omega_2 t$$

(21.41)

where we've used the trigonometric identity $\sin(\pi - \theta) = \sin\theta$. The principle of superposition tells us that the *net* displacement of the medium at the detector is the sum of the displacements of the individual waves. Thus

$$D = D_1 + D_2 = a(\sin\omega_1 t + \sin\omega_2 t)$$

(21.42)

Earlier, for interference, we used the trigonometric identity

$$\sin\alpha + \sin\beta = 2\cos\left[\tfrac{1}{2}(\alpha - \beta)\right]\sin\left[\tfrac{1}{2}(\alpha + \beta)\right]$$

We can use this identity again to write Equation 21.42 as

$$D = 2a\cos\left[\tfrac{1}{2}(\omega_1 - \omega_2)t\right]\sin\left[\tfrac{1}{2}(\omega_1 + \omega_2)t\right]$$
$$= \left[2a\cos(\omega_{mod}t)\right]\sin(\omega_{avg}t)$$

(21.43)

where $\omega_{avg} = \tfrac{1}{2}(\omega_1 + \omega_2)$ is the *average* angular frequency and $\omega_{mod} = \tfrac{1}{2}(\omega_1 - \omega_2)$ is called the *modulation frequency*.

We are interested in the situation when the two frequencies are very nearly equal: $\omega_1 \approx \omega_2$. In that case, ω_{avg} hardly differs from either ω_1 or ω_2 while ω_{mod} is very near to—but not exactly—zero. When ω_{mod} is very small, the term $\cos(\omega_{mod}t)$ oscillates *very* slowly. We have grouped it with the $2a$ term because, together, they provide a slowly changing "amplitude" for the rapid oscillation at frequency ω_{avg}.

FIGURE 21.34 is a history graph of the wave at the detector ($x = 0$). It shows the oscillation of the air against your eardrum at frequency $f_{avg} = \omega_{avg}/2\pi = \tfrac{1}{2}(f_1 + f_2)$. This oscillation determines the note you hear; it differs little from the two notes at frequencies f_1 and f_2. We are especially interested in the time-dependent amplitude, shown as a dashed line, that is given by the term $2a\cos(\omega_{mod}t)$. This periodically varying amplitude is called a **modulation** of the wave, which is where ω_{mod} gets its name.

As the amplitude rises and falls, the sound alternates as loud, soft, loud, soft, and so on. But that is exactly what you hear when you listen to beats! The alternating loud and soft sounds arise from the two waves being alternately in phase and out of phase, causing constructive and then destructive interference.

FIGURE 21.34 Beats are caused by the superposition of two waves of nearly identical frequency.

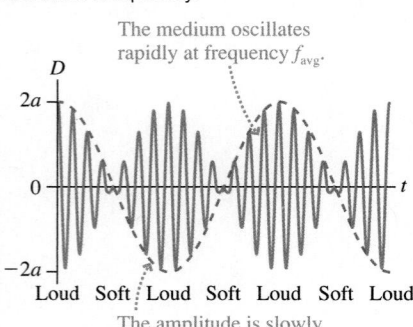

The medium oscillates rapidly at frequency f_{avg}.

The amplitude is slowly modulated as $2a\cos(\omega_{mod}t)$.

Imagine two people walking side by side at just slightly different paces. Initially both of their right feet hit the ground together, but after a while they get out of step. A little bit later they are back in step and the process alternates. The sound waves are doing the same. Initially the crests of each wave, of amplitude a, arrive together at your ear and the net displacement is doubled to $2a$. But after a while the two waves, being of slightly different frequency, get out of step and a crest of one arrives with a trough of the other. When this happens, the two waves cancel each other to give a net displacement of zero. This process alternates over and over, loud and soft.

Notice, from the figure, that the sound intensity rises and falls *twice* during one cycle of the modulation envelope. Each "loud-soft-loud" is one beat, so the **beat frequency** f_{beat}, which is the number of beats per second, is *twice* the modulation frequency $f_{mod} = \omega_{mod}/2\pi$. From the above definition of ω_{mod}, the beat frequency is

$$f_{beat} = 2f_{mod} = 2\frac{\omega_{mod}}{2\pi} = 2 \cdot \frac{1}{2}\left(\frac{\omega_1}{2\pi} - \frac{\omega_2}{2\pi}\right) = f_1 - f_2 \qquad (21.44)$$

where, to keep f_{beat} from being negative, we will always let f_1 be the larger of the two frequencies. The beat frequency is simply the *difference* between the two individual frequencies.

EXAMPLE 21.12   **Detecting bats with beats**

The little brown bat is a common species in North America. It emits echolocation pulses at a frequency of 40 kHz, well above the range of human hearing. To allow researchers to "hear" these bats, the bat detector shown in FIGURE 21.35 combines the bat's sound wave at frequency f_1 with a wave of frequency f_2 from a tunable oscillator. The resulting beat frequency is then amplified and sent to a loudspeaker. To what frequency should the tunable oscillator be set to produce an audible beat frequency of 3 kHz?

SOLVE Combining two waves with different frequencies gives a beat frequency

$$f_{beat} = f_1 - f_2$$

A beat frequency will be generated at 3 kHz if the oscillator frequency and the bat frequency *differ* by 3 kHz. An oscillator frequency of either 37 kHz or 43 kHz will work nicely.

ASSESS The electronic circuitry of radios, televisions, and cell phones makes extensive use of *mixers* to generate difference frequencies.

FIGURE 21.35 The operation of a bat detector.

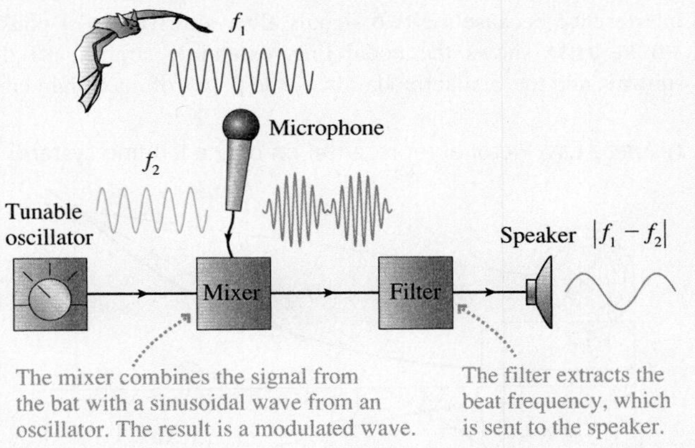

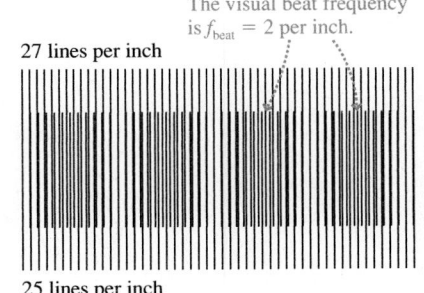

The mixer combines the signal from the bat with a sinusoidal wave from an oscillator. The result is a modulated wave.

The filter extracts the beat frequency, which is sent to the speaker.

Beats aren't limited to sound waves. FIGURE 21.36 shows a graphical example of beats. Two "fences" of slightly different frequencies are superimposed on each other. The difference in the two frequencies is two lines per inch. You can confirm, with a ruler, that the figure has two "beats" per inch, in agreement with Equation 21.44.

Beats are important in many other situations. For example, you have probably seen movies where rotating wheels seem to turn slowly backward. Why is this? Suppose the movie camera is shooting at 30 frames per second but the wheel is rotating 32 times per second. The combination of the two produces a "beat" of 2 Hz, meaning that the wheel appears to rotate only twice per second. The same is true if the wheel is rotating 28 times per second, but in this case, where the wheel frequency slightly lags the camera frequency, it appears to rotate *backward* twice per second!

FIGURE 21.36 A graphical example of beats.

The visual beat frequency is $f_{beat} = 2$ per inch.

27 lines per inch

25 lines per inch

You hear three beats per second when two sound tones are generated. The frequency of one tone is 610 Hz. The frequency of the other is

a. 604 Hz   b. 607 Hz   c. 613 Hz
d. 616 Hz   e. Either a or d.   f. Either b or c.

CHALLENGE EXAMPLE 21.13 **An airplane landing system**

Your firm has been hired to design a system that allows airplane pilots to make instrument landings in rain or fog. You've decided to place two radio transmitters 50 m apart on either side of the runway. These two transmitters will broadcast the same frequency, but out of phase with each other. This will cause a nodal line to extend straight off the end of the runway. As long as the airplane's receiver is silent, the pilot knows she's directly in line with the runway. If she drifts to one side or the other, the radio will pick up a signal and sound a warning beep. To have sufficient accuracy, the first intensity maxima need to be 60 m on either side of the nodal line at a distance of 3.0 km. What frequency should you specify for the transmitters?

MODEL The two transmitters are sources of out-of-phase, circular waves. The overlap of these waves produces an interference pattern.

VISUALIZE For out-of-phase sources, the center line—with zero path-length difference—is a nodal line of perfect destructive interference because the two signals always arrive out of phase. FIGURE 21.37 shows the nodal line, extending straight off the runway, and the first antinodal line—the points of maximum con-

FIGURE 21.37 Pictorial representation of the landing system.

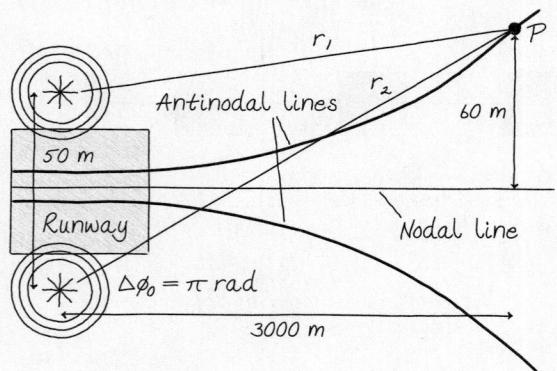

structive interference—on either side. Comparing this to Figure 21.30, where the two sources were in phase, you can see that the nodal and antinodal lines have been reversed.

SOLVE Point P, 60 m to the side at a distance of 3000 m, needs to be a point of maximum constructive interference. The distances are

$$r_1 = \sqrt{(3000 \text{ m})^2 + (60 \text{ m} - 25 \text{ m})^2} = 3000.204 \text{ m}$$
$$r_2 = \sqrt{(3000 \text{ m})^2 + (60 \text{ m} + 25 \text{ m})^2} = 3001.204 \text{ m}$$

We needed to keep several extra significant figures because we're looking for the difference between two numbers that are almost the same. The path-length difference at P is

$$\Delta r = r_2 - r_1 = 1.000 \text{ m}$$

We know, for out-of-phase transmitters, that the phase difference of the sources is $\Delta\phi_0 = \pi$ rad. The first maximum will occur where the phase difference between the waves is $\Delta\phi = 1 \cdot 2\pi$ rad. Thus the condition that we must satisfy at P is

$$\Delta\phi = 2\pi \text{ rad} = 2\pi \frac{\Delta r}{\lambda} + \pi \text{ rad}$$

Solving for λ, we find

$$\lambda = 2\,\Delta r = 2.00 \text{ m}$$

Consequently, the required frequency is

$$f = \frac{c}{\lambda} = \frac{3.00 \times 10^8 \text{ m/s}}{2.00 \text{ m}} = 1.50 \times 10^8 \text{ Hz} = 150 \text{ MHz}$$

ASSESS 150 MHz is slightly higher than the frequencies of FM radio (≈ 100 MHz) but is well within the radio frequency range. Notice that the condition to be satisfied at P is that the path-length difference must be $\frac{1}{2}\lambda$. This makes sense. A path-length difference of $\frac{1}{2}\lambda$ contributes π rad to the phase difference. When combined with the π rad from the out-of-phase sources, the total phase difference of 2π rad creates constructive interference.

SUMMARY

The goal of Chapter 21 has been to understand and use the idea of superposition.

General Principles

Principle of Superposition

The displacement of a medium when more than one wave is present is the sum of the displacements due to each individual wave.

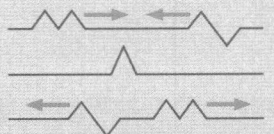

Important Concepts

Standing waves are due to the superposition of two traveling waves moving in opposite directions.

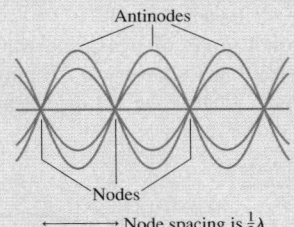

Antinodes

Nodes

Node spacing is $\frac{1}{2}\lambda$.

The amplitude at position x is

$$A(x) = 2a\sin kx$$

where a is the amplitude of each wave.

The boundary conditions determine which standing-wave frequencies and wavelengths are allowed. The allowed standing waves are **modes** of the system.

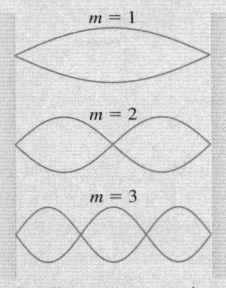

$m = 1$

$m = 2$

$m = 3$

Standing waves on a string

Interference

In general, the superposition of two or more waves into a single wave is called interference.

Maximum constructive interference occurs where crests are aligned with crests and troughs with troughs. These waves are in phase. The maximum displacement is $A = 2a$.

Perfect destructive interference occurs where crests are aligned with troughs. These waves are out of phase. The amplitude is $A = 0$.

Interference depends on the phase difference $\Delta\phi$ between the two waves.

Constructive: $\Delta\phi = 2\pi\dfrac{\Delta r}{\lambda} + \Delta\phi_0 = m \cdot 2\pi$

Destructive: $\Delta\phi = 2\pi\dfrac{\Delta r}{\lambda} + \Delta\phi_0 = \left(m + \dfrac{1}{2}\right) \cdot 2\pi$

Δr is the path-length difference of the two waves, and $\Delta\phi_0$ is any phase difference between the sources. For identical sources (in phase, $\Delta\phi_0 = 0$):

Interference is constructive if the path-length difference $\Delta r = m\lambda$.

Interference is destructive if the path-length difference $\Delta r = \left(m + \frac{1}{2}\right)\lambda$.

The amplitude at a point where the phase difference is $\Delta\phi$ is $A = \left| 2a\cos\left(\dfrac{\Delta\phi}{2}\right) \right|$.

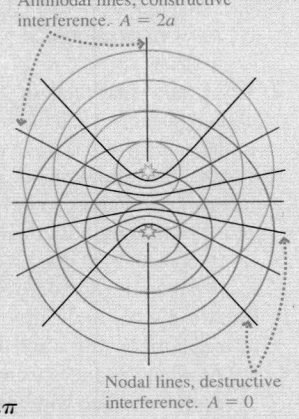

Antinodal lines, constructive interference. $A = 2a$

Nodal lines, destructive interference. $A = 0$

Applications

Boundary conditions

Strings, electromagnetic waves, and sound waves in closed-closed tubes must have nodes at both ends:

$$\lambda_m = \frac{2L}{m} \qquad f_m = m\frac{v}{2L} = mf_1$$

where $m = 1, 2, 3, \ldots$.

The frequencies and wavelengths are the same for a sound wave in an open-open tube, which has antinodes at both ends.

A sound wave in an open-closed tube must have a node at the closed end but an antinode at the open end. This leads to

$$\lambda_m = \frac{4L}{m} \qquad f_m = m\frac{v}{4L} = mf_1$$

where $m = 1, 3, 5, 7, \ldots$.

Beats (loud-soft-loud-soft modulations of intensity) occur when two waves of slightly different frequency are superimposed.

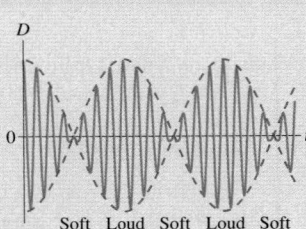

Soft Loud Soft Loud Soft

The beat frequency between waves of frequencies f_1 and f_2 is

$$f_{\text{beat}} = f_1 - f_2$$

Terms and Notation

principle of superposition	mode	path-length difference, Δx or Δr
standing wave	interference	thin-film optical coating
node	in phase	antinodal line
antinode	constructive interference	nodal line
amplitude function, $A(x)$	out of phase	beats
boundary condition	destructive interference	modulation
fundamental frequency, f_1	phase difference, $\Delta\phi$	beat frequency, f_{beat}
harmonic		

CONCEPTUAL QUESTIONS

1. **FIGURE Q21.1** shows a standing wave oscillating on a string at frequency f_0.

   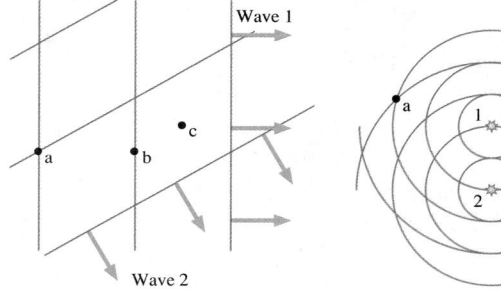
   FIGURE Q21.1

   a. What mode (m-value) is this?
   b. How many antinodes will there be if the frequency is doubled to $2f_0$?

2. If you take snapshots of a standing wave on a string, there are certain instants when the string is totally flat. What has happened to the energy of the wave at those instants?

3. **FIGURE Q21.3** shows the displacement of a standing sound wave in a 32-cm-long horizontal tube of air open at both ends.

   a. What mode (m-value) is this?
   b. Are the air molecules moving horizontally or vertically? Explain.
   c. At what distances from the left end of the tube do the molecules oscillate with maximum amplitude?
   d. At what distances from the left end of the tube does the air pressure oscillate with maximum amplitude?

4. An organ pipe is tuned to exactly 384 Hz when the room temperature is 20°C. If the room temperature later increases to 22°C, does the pipe's frequency increase, decrease, or stay the same? Explain.

5. If you pour liquid into a tall, narrow glass, you may hear sound with a steadily rising pitch. What is the source of the sound? And why does the pitch rise as the glass fills?

6. A flute filled with helium will, until the helium escapes, play notes at a much higher pitch than normal. Why?

7. In music, two notes are said to be an *octave* apart when one note is exactly twice the frequency of the other. Suppose you have a guitar string playing frequency f_0. To increase the frequency by an octave, to $2f_0$, by what factor would you have to (a) increase the tension or (b) decrease the length?

8. **FIGURE Q21.8** is a snapshot graph of two plane waves passing through a region of space. Each wave has a 2.0 mm amplitude and the same wavelength. What is the net displacement of the medium at points a, b, and c?

   FIGURE Q21.8   **FIGURE Q21.9**

9. **FIGURE Q21.9** shows the circular waves emitted by two in-phase sources. Are points a, b, and c points of maximum constructive interference or perfect destructive interference? Explain.

10. A trumpet player hears 5 beats per second when she plays a note and simultaneously sounds a 440 Hz tuning fork. After pulling her tuning valve out to slightly increase the length of her trumpet, she hears 3 beats per second against the tuning fork. Was her initial frequency 435 Hz or 445 Hz? Explain.

EXERCISES AND PROBLEMS

Problems labeled ▇ integrate material from earlier chapters.

Exercises

Section 21.1 The Principle of Superposition

1. | **FIGURE EX21.1** is a snapshot graph at $t = 0$ s of two waves approaching each other at 1.0 m/s. Draw six snapshot graphs, stacked vertically, showing the string at 1 s intervals from $t = 1$ s to $t = 6$ s.

   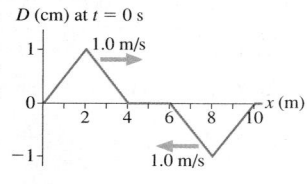
   FIGURE EX21.1

2. | **FIGURE EX21.2** is a snapshot graph at $t = 0$ s of two waves approaching each other at 1.0 m/s. Draw six snapshot graphs, stacked vertically, showing the string at 1 s intervals from $t = 1$ s to $t = 6$ s.

   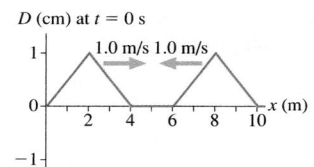
   FIGURE EX21.2

3. ‖ **FIGURE EX21.3** is a snapshot graph at $t = 0$ s of two waves approaching each other at 1.0 m/s. Draw four snapshot graphs, stacked vertically, showing the string at $t = 2, 4, 6,$ and 8 s.

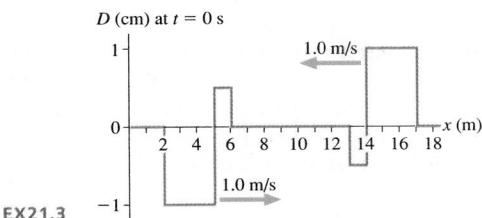

FIGURE EX21.3

4. ‖ **FIGURE EX21.4a** is a snapshot graph at $t = 0$ s of two waves approaching each other at 1.0 m/s.
   a. At what time was the snapshot graph in **FIGURE EX21.4b** taken?
   b. Draw a history graph of the string at $x = 5.0$ m from $t = 0$ s to $t = 6$ s.

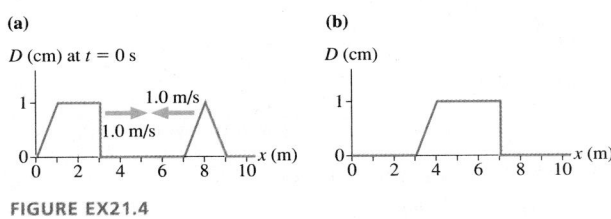

FIGURE EX21.4

Section 21.2 Standing Waves

Section 21.3 Standing Waves on a String

5. ‖ **FIGURE EX21.5** is a snapshot graph at $t = 0$ s of two waves moving to the right at 1.0 m/s. The string is fixed at $x = 8.0$ m. Draw four snapshot graphs, stacked vertically, showing the string at $t = 2, 4, 6,$ and 8 s.

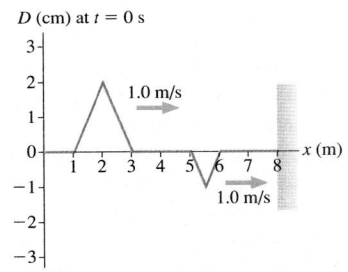

FIGURE EX21.5

6. | **FIGURE EX21.6** shows a standing wave oscillating at 100 Hz on a string. What is the wave speed?

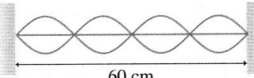

FIGURE EX21.6

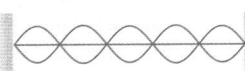

FIGURE EX21.7

7. ‖ **FIGURE EX21.7** shows a standing wave on a 2.0-m-long string that has been fixed at both ends and tightened until the wave speed is 40 m/s. What is the frequency?

8. ‖ **FIGURE EX21.8** shows a standing wave that is oscillating at frequency f_0.
   a. How many antinodes will there be if the frequency is doubled to $2f_0$? Explain.

FIGURE EX21.8

b. If the tension in the string is increased by a factor of four, for what frequency, in terms of f_0, will the string continue to oscillate as a standing wave with four antinodes?

9. | a. What are the three longest wavelengths for standing waves on a 240-cm-long string that is fixed at both ends?
   b. If the frequency of the second-longest wavelength is 50 Hz, what is the frequency of the third-longest wavelength?

10. | Standing waves on a 1.0-m-long string that is fixed at both ends are seen at successive frequencies of 36 Hz and 48 Hz.
    a. What are the fundamental frequency and the wave speed?
    b. Draw the standing-wave pattern when the string oscillates at 48 Hz.

11. ‖ A heavy piece of hanging sculpture is suspended by a 90-cm-long, 5.0 g steel wire. When the wind blows hard, the wire hums at its fundamental frequency of 80 Hz. What is the mass of the sculpture?

12. | A carbon dioxide laser is an infrared laser. A CO_2 laser with a cavity length of 53.00 cm oscillates in the $m = 100,000$ mode. What are the wavelength and frequency of the laser beam?

Section 21.4 Standing Sound Waves and Musical Acoustics

13. | What are the three longest wavelengths for standing sound waves in a 121-cm-long tube that is (a) open at both ends and (b) open at one end, closed at the other?

14. | **FIGURE EX21.14** shows a standing sound wave in an 80-cm-long tube. The tube is filled with an unknown gas. What is the speed of sound in this gas?

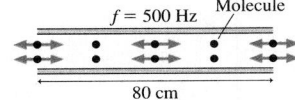

FIGURE EX21.14

15. ‖ The fundamental frequency of an open-open tube is 1500 Hz when the tube is filled with 0°C helium. What is its frequency when filled with 0°C air?

16. | We can make a simple model of the human vocal tract as an
BIO open-closed tube extending from the opening of the mouth to the diaphragm. What is the length of this tube if its fundamental frequency equals a typical speech frequency of 250 Hz? The speed of sound in the warm air is 350 m/s.

17. ‖ The lowest note on a grand piano has a frequency of 27.5 Hz. The entire string is 2.00 m long and has a mass of 400 g. The vibrating section of the string is 1.90 m long. What tension is needed to tune this string properly?

18. ‖ A violin string is 30 cm long. It sounds the musical note A (440 Hz) when played without fingering. How far from the end of the string should you place your finger to play the note C (523 Hz)?

Section 21.5 Interference in One Dimension

Section 21.6 The Mathematics of Interference

19. ‖ Two loudspeakers emit sound waves along the x-axis. The sound has maximum intensity when the speakers are 20 cm apart. The sound intensity decreases as the distance between the speakers is increased, reaching zero at a separation of 60 cm.
    a. What is the wavelength of the sound?
    b. If the distance between the speakers continues to increase, at what separation will the sound intensity again be a maximum?

20. ‖ Two loudspeakers in a 20°C room emit 686 Hz sound waves along the *x*-axis.
   a. If the speakers are in phase, what is the smallest distance between the speakers for which the interference of the sound waves is perfectly destructive?
   b. If the speakers are out of phase, what is the smallest distance between the speakers for which the interference of the sound waves is maximum constructive?

21. | What is the thinnest film of MgF_2 ($n = 1.39$) on glass that produces a strong reflection for orange light with a wavelength of 600 nm?

22. ‖ A very thin oil film ($n = 1.25$) floats on water ($n = 1.33$). What is the thinnest film that produces a strong reflection for green light with a wavelength of 500 nm?

Section 21.7 Interference in Two and Three Dimensions

23. ‖ FIGURE EX21.23 shows the circular wave fronts emitted by two wave sources.
   a. Are these sources in phase or out of phase? Explain.
   b. Make a table with rows labeled P, Q, and R and columns labeled r_1, r_2, Δr, and C/D. Fill in the table for points P, Q, and R, giving the distances as multiples of λ and indicating, with a C or a D, whether the interference at that point is constructive or destructive.

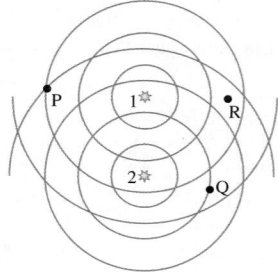
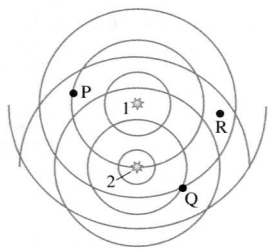

FIGURE EX21.23   FIGURE EX21.24

24. ‖ FIGURE EX21.24 shows the circular wave fronts emitted by two wave sources.
   a. Are these sources in phase or out of phase? Explain.
   b. Make a table with rows labeled P, Q, and R and columns labeled r_1, r_2, Δr, and C/D. Fill in the table for points P, Q, and R, giving the distances as multiples of λ and indicating, with a C or a D, whether the interference at that point is constructive or destructive.

25. ‖ Two in-phase speakers 2.0 m apart in a plane are emitting 1800 Hz sound waves into a room where the speed of sound is 340 m/s. Is the point 4.0 m in front of one of the speakers, perpendicular to the plane of the speakers, a point of maximum constructive interference, perfect destructive interference, or something in between?

26. ‖ Two out-of-phase radio antennas at $x = \pm 300$ m on the *x*-axis are emitting 3.0 MHz radio waves. Is the point $(x, y) = (300\ \text{m}, 800\ \text{m})$ a point of maximum constructive interference, perfect destructive interference, or something in between?

Section 21.8 Beats

27. | Two strings are adjusted to vibrate at exactly 200 Hz. Then the tension in one string is increased slightly. Afterward, three beats per second are heard when the strings vibrate at the same time. What is the new frequency of the string that was tightened?

28. | A flute player hears four beats per second when she compares her note to a 523 Hz tuning fork (the note C). She can match the frequency of the tuning fork by pulling out the "tuning joint" to lengthen her flute slightly. What was her initial frequency?

29. | Two microwave signals of nearly equal wavelengths can generate a beat frequency if both are directed onto the same microwave detector. In an experiment, the beat frequency is 100 MHz. One microwave generator is set to emit microwaves with a wavelength of 1.250 cm. If the second generator emits the longer wavelength, what is that wavelength?

Problems

30. ‖ Two waves on a string travel in opposite directions at 100 m/s. FIGURE P21.30 shows a snapshot graph of the string at $t = 0$ s, when the two waves are overlapped, and a snapshot graph of the left-traveling wave at $t = 0.050$ s. Draw a snapshot graph of the right-traveling wave at $t = 0.050$ s.

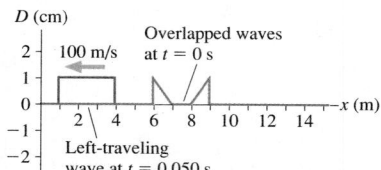

FIGURE P21.30

31. | A 2.0-m-long string vibrates at its second-harmonic frequency with a maximum amplitude of 2.0 cm. One end of the string is at $x = 0$ cm. Find the oscillation amplitude at $x = 10$, 20, 30, 40, and 50 cm.

32. ‖ A string vibrates at its third-harmonic frequency. The amplitude at a point 30 cm from one end is half the maximum amplitude. How long is the string?

33. ‖ A string of length L vibrates at its fundamental frequency. The amplitude at a point $\frac{1}{4}L$ from one end is 2.0 cm. What is the amplitude of each of the traveling waves that form this standing wave?

34. ‖ Two sinusoidal waves with equal wavelengths travel along a string in opposite directions at 3.0 m/s. The time between two successive instants when the antinodes are at maximum height is 0.25 s. What is the wavelength?

35. ‖ BIO Tendons are, essentially, elastic cords stretched between two fixed ends. As such, they can support standing waves. A woman has a 20-cm-long Achilles tendon—connecting the heel to a muscle in the calf—with a cross-section area of 90 mm^2. The density of tendon tissue is 1100 kg/m^3. For a reasonable tension of 500 N, what will be the fundamental frequency of her Achilles tendon?

36. ‖ BIO Biologists think that some spiders "tune" strands of their web to give enhanced response at frequencies corresponding to those at which desirable prey might struggle. Orb spider web silk has a typical diameter of 20 μm, and spider silk has a density of 1300 kg/m^3. To have a fundamental frequency at 100 Hz, to what tension must a spider adjust a 12-cm-long strand of silk?

37. ‖ A particularly beautiful note reaching your ear from a rare Stradivarius violin has a wavelength of 39.1 cm. The room is slightly warm, so the speed of sound is 344 m/s. If the string's linear density is 0.600 g/m and the tension is 150 N, how long is the vibrating section of the violin string?

38. ‖ A violinist places her finger so that the vibrating section of a 1.0 g/m string has a length of 30 cm, then she draws her bow across it. A listener nearby in a 20°C room hears a note with a wavelength of 40 cm. What is the tension in the string?

39. ‖ A steel wire is used to stretch the spring of FIGURE P21.39. An oscillating magnetic field drives the steel wire back and forth. A standing wave with three antinodes is created when the spring is stretched 8.0 cm. What stretch of the spring produces a standing wave with two antinodes?

FIGURE P21.39

40. ‖ Astronauts visiting Planet X have a 250-cm-long string whose mass is 5.00 g. They tie the string to a support, stretch it horizontally over a pulley 2.00 m away, and hang a 4.00 kg mass on the free end. Then the astronauts begin to excite standing waves on the horizontal portion of the string. Their data are as follows:

m	Frequency (Hz)
1	31
2	66
3	95
4	130
5	162

Use the best-fit line of an appropriate graph to determine the value of g, the free-fall acceleration on Planet X.

41. ‖ A 75 g bungee cord has an equilibrium length of 1.20 m. The cord is stretched to a length of 1.80 m, then vibrated at 20 Hz. This produces a standing wave with two antinodes. What is the spring constant of the bungee cord?

42. ‖ A metal wire under tension T_0 vibrates at its fundamental frequency. For what tension will the second-harmonic frequency be the same as the fundamental frequency at tension T_0?

43. ‖‖ In a laboratory experiment, one end of a horizontal string is tied to a support while the other end passes over a frictionless pulley and is tied to a 1.5 kg sphere. Students determine the frequencies of standing waves on the horizontal segment of the string, then they raise a beaker of water until the hanging 1.5 kg sphere is completely submerged. The frequency of the fifth harmonic with the sphere submerged exactly matches the frequency of the third harmonic before the sphere was submerged. What is the diameter of the sphere?

44. ‖‖ What is the fundamental frequency of the steel wire in FIGURE P21.44?

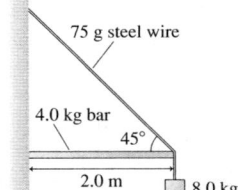

FIGURE P21.44

45. ‖ The two strings in FIGURE P21.45 are of equal length and are being driven at equal frequencies. The linear density of the left string is μ_0. What is the linear density of the right string?

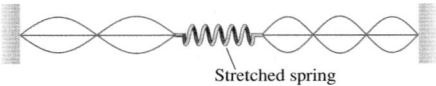

Stretched spring

FIGURE P21.45

46. ‖ Microwaves pass through a small hole into the "microwave cavity" of FIGURE P21.46. What frequencies between 10 GHz and 20 GHz will create standing waves in the cavity?

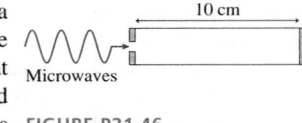

FIGURE P21.46

47. ‖ An open-open organ pipe is 78.0 cm long. An open-closed pipe has a fundamental frequency equal to the third harmonic of the open-open pipe. How long is the open-closed pipe?

48. ‖ A narrow column of 20°C air is found to have standing waves at frequencies of 390 Hz, 520 Hz, and 650 Hz and at no frequencies in between these. The behavior of the tube at frequencies less than 390 Hz or greater than 650 Hz is not known.
    a. Is this an open-open tube or an open-closed tube? Explain.
    b. How long is the tube?

49. ‖ BIO Deep-sea divers often breathe a mixture of helium and oxygen to avoid getting the "bends" from breathing high-pressure nitrogen. The helium has the side effect of making the divers' voices sound odd. Although your vocal tract can be roughly described as an open-closed tube, the way you hold your mouth and position your lips greatly affects the standing-wave frequencies of the vocal tract. This is what allows different vowels to sound different. The "ee" sound is made by shaping your vocal tract to have standing-wave frequencies at, normally, 270 Hz and 2300 Hz. What will these frequencies be for a helium-oxygen mixture in which the speed of sound at body temperature is 750 m/s? The speed of sound in air at body temperature is 350 m/s.

50. ‖ In 1866, the German scientist Adolph Kundt developed a technique for accurately measuring the speed of sound in various gases. A long glass tube, known today as a Kundt's tube, has a vibrating piston at one end and is closed at the other. Very finely ground particles of cork are sprinkled in the bottom of the tube before the piston is inserted. As the vibrating piston is slowly moved forward, there are a few positions that cause the cork particles to collect in small, regularly spaced piles along the bottom. FIGURE P21.50 shows an experiment in which the tube is filled with pure oxygen and the piston is driven at 400 Hz. What is the speed of sound in oxygen?

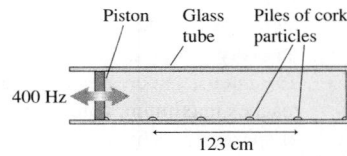

FIGURE P21.50

51. ‖ The 40-cm-long tube of FIGURE P21.51 has a 40-cm-long insert that can be pulled in and out. A vibrating tuning fork is held next to the tube. As the insert is slowly pulled out, the sound from the tuning fork creates standing waves in the tube when the total length L is 42.5 cm, 56.7 cm, and 70.9 cm. What is the frequency of the tuning fork? Assume $v_{sound} = 343$ m/s.

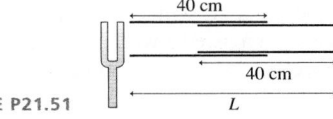

FIGURE P21.51

52. ‖ A 1.0-m-tall vertical tube is filled with 20°C water. A tuning fork vibrating at 580 Hz is held just over the top of the tube as the water is slowly drained from the bottom. At what water heights, measured from the bottom of the tube, will there be a standing wave in the tube above the water?

53. ‖ A 25-cm-long wire with a linear density of 20 g/m passes across the open end of an 85-cm-long open-closed tube of air. If the wire, which is fixed at both ends, vibrates at its fundamental frequency, the sound wave it generates excites the second vibrational mode of the tube of air. What is the tension in the wire? Assume $v_{sound} = 340$ m/s.

54. ‖ A longitudinal standing wave can be created in a long, thin aluminum rod by stroking the rod with very dry fingers. This is often done as a physics demonstration, creating a high-pitched, very annoying whine. From a wave perspective, the standing wave is equivalent to a sound standing wave in an open-open tube. As FIGURE P21.54 shows, both ends of the rod are anti-nodes. What is the fundamental frequency of a 2.0-m-long aluminum rod?

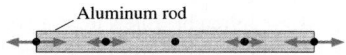

Aluminum rod

FIGURE P21.54

55. ‖ An old mining tunnel disappears into a hillside. You would like to know how long the tunnel is, but it's too dangerous to go inside. Recalling your recent physics class, you decide to try setting up standing-wave resonances inside the tunnel. Using your subsonic amplifier and loudspeaker, you find resonances at 4.5 Hz and 6.3 Hz, and at no frequencies between these. It's rather chilly inside the tunnel, so you estimate the sound speed to be 335 m/s. Based on your measurements, how far is it to the end of the tunnel?

56. ‖ Analyze the standing sound waves in an open-closed tube to show that the possible wavelengths and frequencies are given by Equation 21.18.

57. ‖‖ Two in-phase loudspeakers emit identical 1000 Hz sound waves along the x-axis. What distance should one speaker be placed behind the other for the sound to have an amplitude 1.5 times that of each speaker alone?

58. ‖ Two loudspeakers emit sound waves of the same frequency along the x-axis. The amplitude of each wave is a. The sound intensity is minimum when speaker 2 is 10 cm behind speaker 1. The intensity increases as speaker 2 is moved forward and first reaches maximum, with amplitude $2a$, when it is 30 cm in front of speaker 1. What is
   a. The wavelength of the sound?
   b. The phase difference between the two loudspeakers?
   c. The amplitude of the sound (as a multiple of a) if the speakers are placed side by side?

59. ‖‖ Two loudspeakers emit sound waves along the x-axis. A listener in front of both speakers hears a maximum sound intensity when speaker 2 is at the origin and speaker 1 is at $x = 0.50$ m. If speaker 1 is slowly moved forward, the sound intensity decreases and then increases, reaching another maximum when speaker 1 is at $x = 0.90$ m.
   a. What is the frequency of the sound? Assume $v_{sound} = 340$ m/s.
   b. What is the phase difference between the speakers?

60. ‖ A sheet of glass is coated with a 500-nm-thick layer of oil ($n = 1.42$).
   a. For what *visible* wavelengths of light do the reflected waves interfere constructively?
   b. For what *visible* wavelengths of light do the reflected waves interfere destructively?
   c. What is the color of reflected light? What is the color of transmitted light?

61. ‖ A manufacturing firm has hired your company, Acoustical Consulting, to help with a problem. Their employees are complaining about the annoying hum from a piece of machinery. Using a frequency meter, you quickly determine that the machine emits a rather loud sound at 1200 Hz. After investigating, you tell the owner that you cannot solve the problem entirely, but you can at least improve the situation by eliminating reflections of this sound from the walls. You propose to do this by installing mesh screens in front of the walls. A portion of the sound will reflect from the mesh; the rest will pass through the mesh and reflect from the wall. How far should the mesh be placed in front of the wall for this scheme to work?

62. ‖ A soap bubble is essentially a very thin film of water ($n = 1.33$) surrounded by air. The colors that you see in soap bubbles are produced by interference.
   a. Derive an expression for the wavelengths λ_C for which constructive interference causes a strong reflection from a soap bubble of thickness d.
   Hint: Think about the reflection phase shifts at both boundaries.
   b. What visible wavelengths of light are strongly reflected from a 390-nm-thick soap bubble? What color would such a soap bubble appear to be?

63. ‖ Two radio antennas are separated by 2.0 m. Both broadcast identical 750 MHz waves. If you walk around the antennas in a circle of radius 10 m, how many maxima will you detect?

64. ‖ You are standing 2.5 m directly in front of one of the two loudspeakers shown in FIGURE P21.64. They are 3.0 m apart and both are playing a 686 Hz tone in phase. As you begin to walk directly away from the speaker, at what distances from the speaker do you hear a *minimum* sound intensity? The room temperature is 20°C.

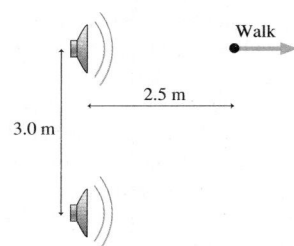

Walk

2.5 m

3.0 m

FIGURE P21.64

65. ‖ Two loudspeakers in a plane, 5.0 apart, are playing the same frequency. If you stand 12.0 m in front of the plane of the speakers, centered between them, you hear a sound of maximum intensity. As you walk parallel to the plane of the speakers, staying 12.0 m in front of them, you first hear a minimum of sound intensity when you are directly in front of one of the speakers. What is the frequency of the sound? Assume a sound speed of 340 m/s.

66. ‖ Two in-phase loudspeakers are located at (x, y) coordinates $(-3.0$ m, $+2.0$ m$)$ and $(-3.0$ m, -2.0 m$)$. They emit identical sound waves with a 2.0 m wavelength and amplitude a. Determine the amplitude of the sound at the five positions on the y-axis ($x = 0$) with $y = 0.0$ m, 0.5 m, 1.0 m, 1.5 m, and 2.0 m.

67. ‖ Two identical loudspeakers separated by distance Δx each emit sound waves of wavelength λ and amplitude a along the x-axis. What is the minimum value of the ratio $\Delta x/\lambda$ for which the amplitude of their superposition is also a?

68. ‖ Two radio antennas are 100 m apart along a north-south line. They broadcast identical radio waves at a frequency of 3.0 MHz. Your job is to monitor the signal strength with a hand-held receiver. To get to your first measuring point, you walk 800 m east from the midpoint between the antennas, then 600 m north.
    a. What is the phase difference between the waves at this point?
    b. Is the interference at this point maximum constructive, perfect destructive, or somewhere in between? Explain.
    c. If you now begin to walk farther north, does the signal strength increase, decrease, or stay the same? Explain.

69. ‖ The three identical loudspeakers in FIGURE P21.69 play a 170 Hz tone in a room where the speed of sound is 340 m/s. You are standing 4.0 m in front of the middle speaker. At this point, the amplitude of the wave from each speaker is a.
    a. What is the amplitude at this point?
    b. How far must speaker 2 be moved to the left to produce a maximum amplitude at the point where you are standing?
    c. When the amplitude is maximum, by what factor is the sound intensity greater than the sound intensity from a single speaker?

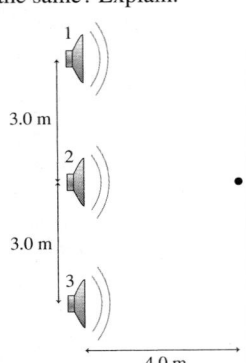

FIGURE P21.69

70. | Piano tuners tune pianos by listening to the beats between the *harmonics* of two different strings. When properly tuned, the note A should have a frequency of 440 Hz and the note E should be at 659 Hz.
    a. What is the frequency difference between the third harmonic of the A and the second harmonic of the E?
    b. A tuner first tunes the A string very precisely by matching it to a 440 Hz tuning fork. She then strikes the A and E strings simultaneously and listens for beats between the harmonics. What beat frequency indicates that the E string is properly tuned?
    c. The tuner starts with the tension in the E string a little low, then tightens it. What is the frequency of the E string when she hears four beats per second?

71. ‖ A flutist assembles her flute in a room where the speed of sound is 342 m/s. When she plays the note A, it is in perfect tune with a 440 Hz tuning fork. After a few minutes, the air inside her flute has warmed to where the speed of sound is 346 m/s.
    a. How many beats per second will she hear if she now plays the note A as the tuning fork is sounded?
    b. How far does she need to extend the "tuning joint" of her flute to be in tune with the tuning fork?

72. ‖ Two loudspeakers face each other from opposite walls of a room. Both are playing exactly the same frequency, thus setting up a standing wave with distance $\lambda/2$ between antinodes. Assume that λ is much less than the room width, so there are many antinodes.
    a. Yvette starts at one speaker and runs toward the other at speed v_Y. As the does so, she hears a loud-soft-loud modulation of the sound intensity. From your perspective, as you sit at rest in the room, Yvette is running through the nodes and antinodes of the standing wave. Find an expression for the number of sound maxima she hears per second.

b. From Yvette's perspective, the two sound waves are Doppler shifted. They're not the same frequency, so they don't create a standing wave. Instead, she hears a loud-soft-loud modulation of the sound intensity because of beats. Find an expression for the beat frequency that Yvette hears.
    c. Are your answers to parts a and b the same or different? *Should* they be the same or different?

73. ‖ Two loudspeakers emit 400 Hz notes. One speaker sits on the ground. The other speaker is in the back of a pickup truck. You hear eight beats per second as the truck drives away from you. What is the truck's speed?

Challenge Problems

74. a. The frequency of a standing wave on a string is f when the string's tension is T_s. If the tension is changed by the *small* amount ΔT_s, without changing the length, show that the frequency changes by an amount Δf such that

$$\frac{\Delta f}{f} = \frac{1}{2}\frac{\Delta T_s}{T_s}$$

    b. Two identical strings vibrate at 500 Hz when stretched with the same tension. What percentage increase in the tension of one of the strings will cause five beats per second when both strings vibrate simultaneously?

75. A 280 Hz sound wave is directed into one end of the trombone slide seen in FIGURE CP21.75. A microphone is placed at the other end to record the intensity of sound waves that are transmitted through the tube. The straight sides of the slide are 80 cm in length and 10 cm apart with a semicircular bend at the end. For what slide extensions s will the microphone detect a maximum of sound intensity?

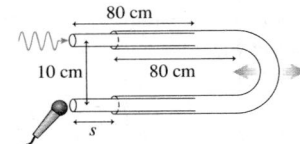

FIGURE CP21.75

76. As the captain of the scientific team sent to Planet Physics, one of your tasks is to measure g. You have a long, thin wire labeled 1.00 g/m and a 1.25 kg weight. You have your accurate space cadet chronometer but, unfortunately, you seem to have forgotten a meter stick. Undeterred, you first find the midpoint of the wire by folding it in half. You then attach one end of the wire to the wall of your laboratory, stretch it horizontally to pass over a pulley at the midpoint of the wire, then tie the 1.25 kg weight to the end hanging over the pulley. By vibrating the wire, and measuring time with your chronometer, you find that the wire's second-harmonic frequency is 100 Hz. Next, with the 1.25 kg weight still tied to one end of the wire, you attach the other end to the ceiling to make a pendulum. You find that the pendulum requires 314 s to complete 100 oscillations. Pulling out your trusty calculator, you get to work. What value of g will you report back to headquarters?

77. When mass M is tied to the bottom of a long, thin wire suspended from the ceiling, the wire's second-harmonic frequency is 200 Hz. Adding an additional 1.0 kg to the hanging mass increases the second-harmonic frequency to 245 Hz. What is M?

78. Ultrasound has many medical applications, one of which is to
    BIO monitor fetal heartbeats by reflecting ultrasound off a fetus in the
    womb.

    a. Consider an object moving at speed v_o toward an at-rest
       source that is emitting sound waves of frequency f_0. Show
       that the reflected wave (i.e., the echo) that returns to the
       source has a Doppler-shifted frequency

    $$f_{echo} = \left(\frac{v + v_o}{v - v_o}\right) f_0$$

       where v is the speed of sound in the medium.

    b. Suppose the object's speed is much less than the wave speed:
       $v_o \ll v$. Then $f_{echo} \approx f_0$, and a microphone that is sensitive to
       these frequencies will detect a beat frequency if it listens to
       f_0 and f_{echo} simultaneously. Use the binomial approximation
       and other appropriate approximations to show that the beat
       frequency is $f_{beat} \approx (2v_o/v)f_0$.

    c. The reflection of 2.40 MHz ultrasound waves from the surface
       of a fetus's beating heart is combined with the 2.40 MHz wave
       to produce a beat frequency that reaches a maximum of 65 Hz.
       What is the maximum speed of the surface of the heart? The
       speed of ultrasound waves within the body is 1540 m/s.

    d. Suppose the surface of the heart moves in simple harmonic
       motion at 90 beats/min. What is the amplitude in mm of the
       heartbeat?

79. A water wave is called a *deep-water wave* if the water's depth
    is more than one-quarter of the wavelength. Unlike the waves
    we've considered in this chapter, the speed of a deep-water wave
    depends on its wavelength:

    $$v = \sqrt{\frac{g\lambda}{2\pi}}$$

    Longer wavelengths travel faster. Let's apply this to standing
    waves. Consider a diving pool that is 5.0 m deep and 10.0 m
    wide. Standing water waves can set up across the width of the
    pool. Because water sloshes up and down at the sides of the pool,
    the boundary conditions require antinodes at $x = 0$ and $x = L$.
    Thus a standing water wave resembles a standing sound wave in
    an open-open tube.

    a. What are the wavelengths of the first three standing-wave
       modes for water in the pool? Do they satisfy the condition for
       being deep-water waves? Draw a graph of each.

    b. What are the wave speeds for each of these waves?

    c. Derive a general expression for the frequencies f_m of the pos-
       sible standing waves. Your expression should be in terms of
       m, g, and L.

    d. What are the oscillation *periods* of the first three standing-
       wave modes?

80. The broadcast antenna of an
    AM radio station is located at
    the edge of town. The station
    owners would like to beam all
    of the energy into town and
    none into the countryside, but
    a single antenna radiates en-
    ergy equally in all directions.
    FIGURE CP21.80 shows two par-
    allel antennas separated by dis-

    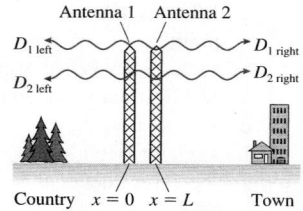
    FIGURE CP21.80

    tance L. Both antennas broadcast a signal at wavelength λ, but
    antenna 2 can delay its broadcast relative to antenna 1 by a time
    interval Δt in order to create a phase difference $\Delta\phi_0$ between
    the sources. Your task is to find values of L and Δt such that the
    waves interfere constructively on the town side and destructively
    on the country side.

    Let antenna 1 be at $x = 0$. The wave that travels to the right is
    $a\sin[2\pi(x/\lambda - t/T)]$. The left wave is $a\sin[2\pi(-x/\lambda - t/T)]$.
    (It must be this, rather than $a\sin[2\pi(x/\lambda + t/T)]$, so that the
    two waves match at $x = 0$.) Antenna 2 is at $x = L$. It broadcasts
    wave $a\sin[2\pi((x-L)/\lambda - t/T) + \phi_{20}]$ to the right and wave
    $a\sin[2\pi(-(x-L)/\lambda - t/T) + \phi_{20}]$ to the left.

    a. What is the smallest value of L for which you can create per-
       fect constructive interference on the town side and perfect
       destructive interference on the country side? Your answer
       will be a multiple or fraction of the wavelength λ.

    b. What phase constant ϕ_{20} of antenna 2 is needed?

    c. What fraction of the oscillation period T must Δt be to pro-
       duce the proper value of ϕ_{20}?

    d. Evaluate both L and Δt for the realistic AM radio frequency
       of 1000 KHz.

    Comment: This is a simple example of what is called a *phased
    array,* where phase differences between identical emitters are
    used to "steer" the radiation in a particular direction. Phased ar-
    rays are widely used in radar technology.

STOP TO THINK ANSWERS

Stop to Think 21.1: c. The figure shows the two waves at $t = 6$ s and
their superposition. The superposition is the *point-by-point* addition
of the displacements of the two individual waves.

$$\text{(graph with }x\text{ (m) axis marked } 0\ 2\ 4\ 6\ 8\ 10\ 12\ 14\ 16\ 18\ 20)$$

Stop to Think 21.2: a. The allowed standing-wave frequencies are
$f_m = m(v/2L)$, so the mode number of a standing wave of frequency
f is $m = 2Lf/v$. Quadrupling T_s increases the wave speed v by a factor
of 2. The initial mode number was 2, so the new mode number is 1.

Stop to Think 21.3: b. 300 Hz and 400 Hz are allowed standing
waves, but they are not f_1 and f_2 because 400 Hz $\neq$ 2 × 300 Hz.
Because there's a 100 Hz difference between them, these must be

$f_3 = 3 \times 100$ Hz and $f_4 = 4 \times 100$ Hz, with a fundamental frequency
$f_1 = 100$ Hz. Thus the second harmonic is $f_2 = 2 \times 100$ Hz =
200 Hz.

Stop to Think 21.4: c. Shifting the top wave 0.5 m to the left aligns
crest with crest and trough with trough.

Stop to Think 21.5: a. $r_1 = 0.5\lambda$ and $r_2 = 2.5\lambda$, so $\Delta r = 2.0\lambda$. This
is the condition for maximum constructive interference.

Stop to Think 21.6: Maximum constructive. The path-length difference is $\Delta r = 1.0$ m $= \lambda$. For identical sources, interference is constructive when Δr is an integer multiple of λ.

Stop to Think 21.7: f. The beat frequency is the difference between
the two frequencies.

22 Wave Optics

The vivid colors of this peacock—which change as you see the feathers from different angles—are not due to pigments. Instead, the colors are due to the interference of light waves.

▶ **Looking Ahead** The goal of Chapter 22 is to understand and apply the wave model of light.

Models of Light

You'll learn that light has aspects of both waves and particles. We'll introduce three models of light:

The **wave model** of light—the subject of this chapter—allows us to understand the colors of a soap bubble.

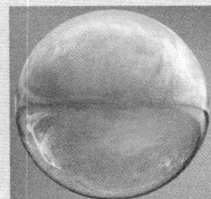

To understand the focusing of light by a contact lens, Chapter 23 will introduce a **ray model** in which light travels in particle-like straight lines.

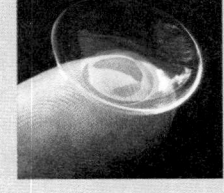

Solar cells generate electricity from sunlight. The **photon model** of Part VII will be most appropriate for understanding this aspect of light.

◀ **Looking Back**
Sections 20.4–20.6 Wave fronts, phase, and intensity

Diffraction

Diffraction is the ability of waves to spread out after going through small holes or around corners. The diffraction of light indicates that light is a wave.

The "ripples" around the edges of this razor blade—back lit with a blue laser beam—are due to the diffraction of light.

The Diffraction Grating

A **diffraction grating** is a periodic array of closely spaced holes or slits or grooves. You'll learn how a diffraction grating sends different wavelengths off at different angles.

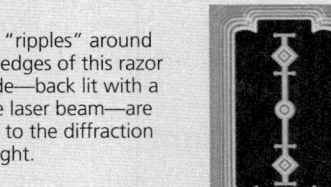

The microscopic pits in this DVD act as a diffraction grating, breaking white light into its component colors.

Diffraction gratings are the basis for *spectroscopy*, an important tool for determining the composition of materials by the wavelengths they emit.

Double-Slit Interference

You'll learn that an interference pattern is formed when light shines on an opaque screen with two narrow, closely spaced slits. This also shows that light is a wave.

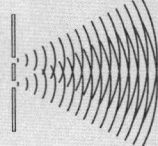

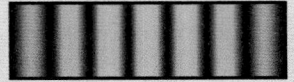

Interference fringes from green light passing through two closely spaced slits

◀ **Looking Back**
Section 21.7 Interference

Interferometry

Today, the controlled interference of light has applications that include optical computing, precision measurements in engineering, holography, and observing movements of the earth's crust.

Interference fringes such as these can be used to monitor vibrations and displacements of only a few nanometers.

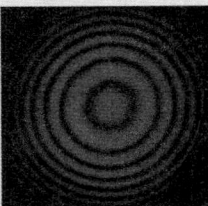

22.1 Light and Optics

The study of light is called **optics.** But what is light? The first Greek scientists did not make a distinction between light and vision. Light, to them, was inseparable from seeing. But gradually there arose a view that light actually "exists," that light is some sort of physical entity that is present regardless of whether or not someone is looking. But if light is a physical entity, what is it? What are its characteristics? Is it a wave, similar to sound? Or is light a collection of small particles that blows by like the wind?

Newton, in addition to his pioneering work in mathematics and mechanics in the 1660s, investigated the nature of light. Newton knew that a water wave, after passing through an opening, *spreads out* to fill the space behind the opening. You can see this in FIGURE 22.1a, where plane waves, approaching from the left, spread out in circular arcs after passing through a hole in a barrier. This inexorable spreading of waves is the phenomenon called **diffraction.** Diffraction is a sure sign that whatever is passing through the hole is a wave.

In contrast, FIGURE 22.1b shows that sunlight makes a sharp-edged shadow after passing through a door. We don't see sunlight light spreading out in circular arcs. This behavior is exactly what you would expect if light consists of particles traveling in straight lines. Some particles would pass through the door to make a bright area on the floor, others would be blocked and cause the well-defined shadow. This reasoning led Newton to the conclusion that light consists of very small, light, fast particles that he called *corpuscles.*

The situation changed dramatically in 1801, when the English scientist Thomas Young announced that he had produced *interference* between two waves of light. Young's experiment, which we will analyze in the next section, was painstakingly difficult with the technology of his era. Nonetheless, Young's experiment quickly settled the debate in favor of a wave theory of light because interference is a distinctly wave-like phenomenon.

But if light is a wave, what is waving? This was the question that Young posed to the 19th century. It was ultimately established that light is an *electromagnetic wave,* an oscillation of the electromagnetic field requiring no material medium in which to travel. Further, as we have already seen, visible light is just one small slice out of a vastly broader *electromagnetic spectrum.*

But this satisfying conclusion was soon undermined by new discoveries at the start of the 20th century. Albert Einstein's introduction of the concept of the *photon*—a wave having certain particle-like characteristics—marked the end of *classical physics* and the beginning of a new era called *quantum physics.* Equally important, Einstein's theory marked yet another shift in our age-old effort to understand light.

FIGURE 22.1 Water waves spread out behind a small hole in a barrier, but light passing through a doorway makes a sharp-edged shadow.

(a) Plane waves approach from the left.

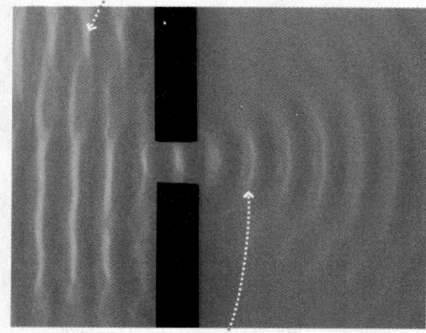

Circular waves spread out on the right.

(b)

A beam of sunlight has a sharp edge.

Models of Light

Light is a real physical entity, but the nature of light is elusive. Light is the chameleon of the physical world. Under some circumstances, light acts like particles traveling in straight lines. But change the circumstances, and light shows the same kinds of wave-like behavior as sound waves or water waves. Change the circumstances yet again, and light exhibits behavior that is neither wave-like nor particle-like but has characteristics of both.

Rather than an all-encompassing "theory of light," it will be better to develop three **models of light.** Each model successfully explains the behavior of light within a certain domain—that is, within a certain range of physical situations. Our task will be twofold:

1. To develop clear and distinct models of light.
2. To learn the conditions and circumstances for which each model is valid.

We'll begin with a brief summary of all three models.

Three models of light

The Wave Model

The wave model of light is responsible for the widely known "fact" that light is a wave. Indeed, under many circumstances light exhibits the same behavior as sound or water waves. Lasers and electro-optical devices are best described by the wave model of light. Some aspects of the wave model were introduced in Chapters 20 and 21, and it is the primary focus of this chapter.

The Ray Model

An equally well-known "fact" is that light travels in straight lines. These straight-line paths are called *light rays*. The properties of prisms, mirrors, and lenses are best understood in terms of light rays. Unfortunately, it's difficult to reconcile "light travels in straight lines" with "light is a wave." For the most part, waves and rays are mutually exclusive models of light. One of our important tasks will be to learn when each model is appropriate. Ray optics is the subject of Chapters 23 and 24.

The Photon Model

Modern technology is increasingly reliant on quantum physics. In the quantum world, light behaves like neither a wave nor a particle. Instead, light consists of *photons* that have both wave-like and particle-like properties. Much of the quantum theory of light is beyond the scope of this textbook, but we will take a peek at some of the important ideas in Part VII.

22.2 The Interference of Light

Newton might have reached a different conclusion had he seen the experiment depicted in **FIGURE 22.2**. Here light of a single wavelength (or color) passes through a "window"—a narrow slit—that is only 0.1 mm wide, about twice the width of a human hair. The image shows how the light appears on a viewing screen 2 m behind the slit. If light consists of corpuscles traveling in straight lines, as Newton thought, we should see a narrow strip of light, about 0.1 mm wide, with dark shadows on either side. Instead, we see a band of light extending over about 2.5 cm, a distance much wider than the aperture, with dimmer patches of light extending even farther on either side.

If you compare Figure 22.2 to the water wave of Figure 22.1, you see that *the light is spreading out* behind the 0.1-mm-wide hole. The light is exhibiting diffraction, the sure signature of waviness. We will look at diffraction in more detail later in the chapter. For now, we merely need the *observation* that light does, indeed, spread out behind a hole that is sufficiently small.

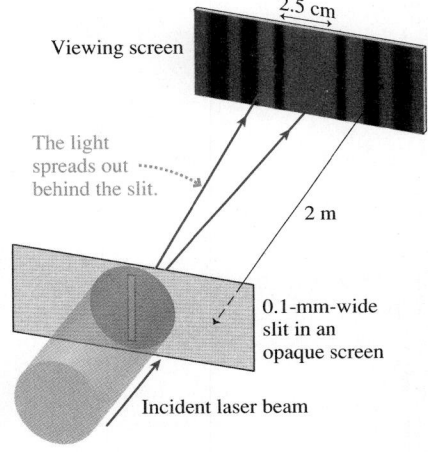

FIGURE 22.2 Light, just like a water wave, does spread out behind a hole *if* the hole is sufficiently small.

Young's Double-Slit Experiment

Rather than one small hole, suppose we use two. **FIGURE 22.3a** shows an experiment in which a laser beam is aimed at an opaque screen containing two long, narrow slits that are very close together. This pair of slits is called a **double slit**, and in a typical

FIGURE 22.3 A double-slit interference experiment.

(a)

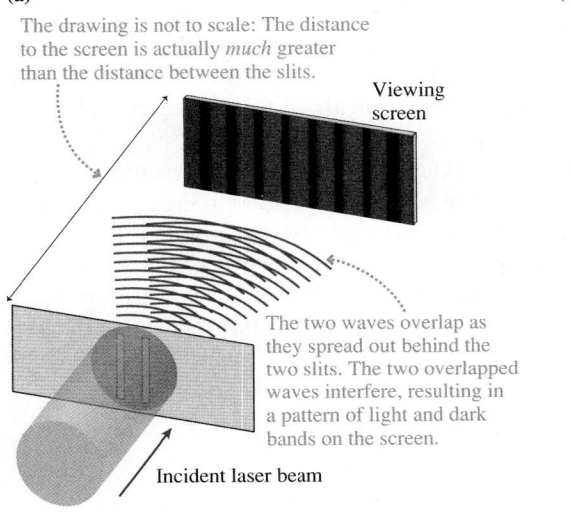

(b)

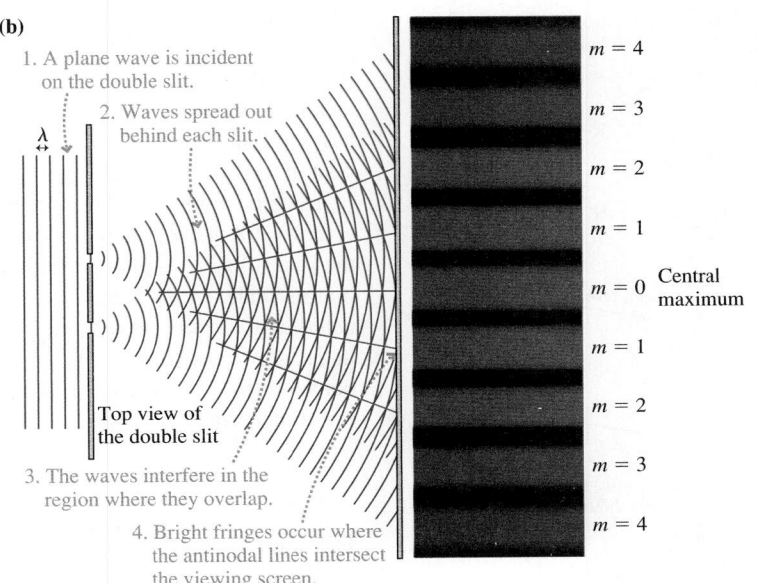

experiment they are ≈ 0.1 mm wide and spaced ≈ 0.5 mm apart. We will assume that the laser beam illuminates both slits equally, and any light passing through the slits impinges on a viewing screen. This is the essence of Young's experiment of 1801, although he used sunlight rather than a laser.

What should we expect to see on the screen? FIGURE 22.3b is a view from above the experiment, looking down on the top ends of the slits and the top edge of the viewing screen. Because the slits are very narrow, **light spreads out behind each slit** as it did in Figure 22.2, and these two spreading waves overlap in the region between the slits and the screen.

The primary conclusion of Chapter 21 was that two overlapped waves of equal wavelength produce interference. In fact, Figure 22.3b is equivalent to the waves emitted by two loudspeakers, a situation we analyzed in Section 21.7. (It is very useful to compare Figure 22.3b with Figures 21.30 and 21.32a.) Nothing in that analysis depended on what type of wave it was, so the conclusions apply equally well to two overlapped light waves. If light really is a wave, we should see interference between the two light waves over the small region, typically a few centimeters wide, where they overlap on the viewing screen.

The image in Figure 22.3b shows how the screen looks. As expected, the light is intense at points where an antinodal line intersects the screen. There is no light at all at points where a nodal line intersects the screen. These alternating bright and dark bands of light, due to constructive and destructive interference, are called **interference fringes**. The fringes are numbered $m = 0, 1, 2, 3, \ldots$, going outward from the center. The brightest fringe, at the midpoint of the viewing screen, with $m = 0$, is called the **central maximum.**

STOP TO THINK 22.1   Suppose the viewing screen in Figure 22.3 is moved closer to the double slit. What happens to the interference fringes?

a. They get brighter but otherwise do not change.
b. They get brighter and closer together.
c. They get brighter and farther apart.
d. They get out of focus.
e. They fade out and disappear.

Analyzing Double-Slit Interference

Figure 22.3 showed qualitatively how interference is produced behind a double slit by the overlap of the light waves spreading out behind each slit. Now let's analyze the experiment more carefully. FIGURE 22.4 shows a double-slit experiment in which the spacing between the two slits is d and the distance to the viewing screen is L. We will assume that L is *very* much larger than d. Consequently, we don't see the individual slits in the upper part of Figure 22.4.

Let P be a point on the screen at angle θ. Our goal is to determine whether the interference at P is constructive, destructive, or in between. The insert to Figure 22.4 shows the individual slits and the paths from these slits to point P. Because P is so far away on this scale, the two paths are virtually parallel, both at angle θ. Both slits are illuminated by the *same* wave front from the laser; hence the slits act as sources of identical, in-phase waves ($\Delta\phi_0 = 0$). You learned in Chapter 21 that constructive interference between the waves from in-phase sources occurs at points for which the path-length difference $\Delta r = r_2 - r_1$ is an integer number of wavelengths:

$$\Delta r = m\lambda \qquad m = 0, 1, 2, 3, \ldots \qquad \text{(constructive interference)} \qquad (22.1)$$

Thus the interference at point P is constructive, producing a bright fringe, if $\Delta r = m\lambda$ at that point.

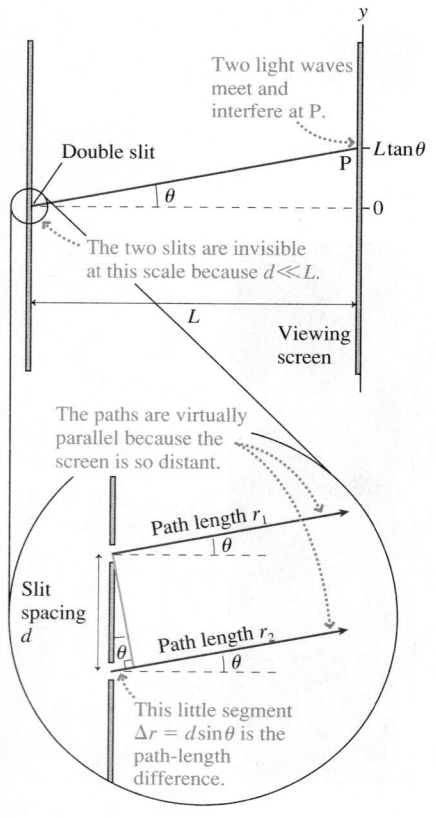

FIGURE 22.4 Geometry of the double-slit experiment.

The midpoint on the viewing screen at $y = 0$ is equally distant from both slits ($\Delta r = 0$) and thus is a point of constructive interference. This is the bright fringe identified as the central maximum in Figure 22.3b. The path-length difference increases as you move away from the center of the screen, and the $m = 1$ fringes occur at the points where $\Delta r = 1\lambda$—that is, where one wave has traveled exactly one wavelength farther than the other. In general, **the mth bright fringe occurs where the wave from one slit travels m wavelengths farther than the wave from the other slit and thus $\Delta r = m\lambda$.**

You can see from the magnified portion of Figure 22.4 that the wave from the lower slit travels an extra distance

$$\Delta r = d \sin\theta \tag{22.2}$$

If we use this in Equation 22.1, we find that bright fringes (constructive interference) occur at angles θ_m such that

$$\Delta r = d\sin\theta_m = m\lambda \qquad m = 0, 1, 2, 3, \ldots \tag{22.3}$$

We added the subscript m to denote that θ_m is the angle of the mth bright fringe, starting with $m = 0$ at the center.

In practice, the angle θ in a double-slit experiment is very small ($<1°$). We can use the small-angle approximation $\sin\theta \approx \theta$, where θ must be in radians, to write Equation 22.3 as

$$\theta_m = m\frac{\lambda}{d} \qquad m = 0, 1, 2, 3, \ldots \qquad \text{(angles of bright fringes)} \tag{22.4}$$

This gives the angular positions *in radians* of the bright fringes in the interference pattern.

It's usually easier to measure distances rather than angles, so we can also specify point P by its position on a y-axis with the origin directly across from the midpoint between the slits. You can see from Figure 22.4 that

$$y = L\tan\theta \tag{22.5}$$

Using the small-angle approximation once again, this time in the form $\tan\theta \approx \theta$, we can substitute θ_m from Equation 22.4 for $\tan\theta_m$ in Equation 22.1 to find that the mth bright fringe occurs at position

$$y_m = \frac{m\lambda L}{d} \qquad m = 0, 1, 2, 3, \ldots \qquad \text{(positions of bright fringes)} \tag{22.6}$$

The interference pattern is symmetrical, so there is an mth bright fringe at the same distance on both sides of the center. You can see this in Figure 22.3b. As we've noted, **the $m = 1$ fringes occur at points on the screen where the light from one slit travels exactly one wavelength farther than the light from the other slit.**

NOTE ▶ Equations 22.4 and 22.6 do *not* apply to the interference of sound waves from two loudspeakers. The approximations we've used (small angles, $L \gg d$) are usually not valid for the much longer wavelengths of sound waves. ◀

Equation 22.6 predicts that **the interference pattern is a series of equally spaced bright lines** on the screen, exactly as shown in Figure 22.3b. How do we know the fringes are equally spaced? The **fringe spacing** between the m fringe and the $m + 1$ fringe is

$$\Delta y = y_{m+1} - y_m = \frac{(m+1)\lambda L}{d} - \frac{m\lambda L}{d} = \frac{\lambda L}{d} \tag{22.7}$$

Because Δy is independent of m, *any* two adjacent bright fringes have the same spacing.

The dark fringes in the image are bands of destructive interference. You learned in Chapter 21 that destructive interference occurs at positions where the path-length difference of the waves is a half-integer number of wavelengths:

$$\Delta r = \left(m + \frac{1}{2}\right)\lambda \qquad m = 0, 1, 2, \ldots \qquad \text{(destructive interference)} \qquad (22.8)$$

We can use Equation 22.2 for Δr and the small-angle approximation to find that the dark fringes are located at positions

$$y'_m = \left(m + \frac{1}{2}\right)\frac{\lambda L}{d} \qquad m = 0, 1, 2, \ldots \qquad \text{(positions of dark fringes)} \qquad (22.9)$$

We have used y'_m, with a prime, to distinguish the location of the mth minimum from the mth maximum at y_m. You can see from Equation 22.9 that **the dark fringes are located exactly halfway between the bright fringes.**

EXAMPLE 22.1 | **Double-slit interference of a laser beam**

Light from a helium-neon laser ($\lambda = 633$ nm) illuminates two slits spaced 0.40 mm apart. A viewing screen is 2.0 m behind the slits. What are the distances between the two $m = 2$ bright fringes and between the two $m = 2$ dark fringes?

MODEL Two closely spaced slits produce a double-slit interference pattern.

VISUALIZE The interference pattern looks like the image of Figure 22.3b. It is symmetrical, with $m = 2$ bright fringes at equal distances on both sides of the central maximum.

SOLVE The positions of the bright fringes are given by Equation 22.6. The $m = 2$ bright fringe is located at position

$$y_m = \frac{m\lambda L}{d} = \frac{2(633 \times 10^{-9}\,\text{m})(2.0\,\text{m})}{4.0 \times 10^{-4}\,\text{m}} = 6.3\,\text{mm}$$

Each of the $m = 2$ fringes is 6.3 mm from the central maximum; so the distance between the two $m = 2$ bright fringes is 12.6 mm. The $m = 2$ dark fringe is located at

$$y'_m = \left(m + \frac{1}{2}\right)\frac{\lambda L}{d} = 7.9\,\text{mm}$$

Thus the distance between the two $m = 2$ dark fringes is 15.8 mm.

ASSESS Because the fringes are counted outward from the center, the $m = 2$ bright fringe occurs *before* the $m = 2$ dark fringe.

EXAMPLE 22.2 | **Measuring the wavelength of light**

A double-slit interference pattern is observed on a screen 1.0 m behind two slits spaced 0.30 mm apart. Ten bright fringes span a distance of 1.7 cm. What is the wavelength of the light?

MODEL It is not always obvious which fringe is the central maximum. Slight imperfections in the slits can make the interference fringe pattern less than ideal. However, you do not need to identify the $m = 0$ fringe because you can make use of the fact that the fringe spacing Δy is uniform. Ten bright fringes have *nine* spaces between them (not ten—be careful!).

VISUALIZE The interference pattern looks like the image of Figure 22.3b.

SOLVE The fringe spacing is

$$\Delta y = \frac{1.7\,\text{cm}}{9} = 1.89 \times 10^{-3}\,\text{m}$$

Using this fringe spacing in Equation 22.7, we find that the wavelength is

$$\lambda = \frac{d}{L}\Delta y = 5.7 \times 10^{-7}\,\text{m} = 570\,\text{nm}$$

It is customary to express the wavelengths of visible light in nanometers. Be sure to do this as you solve problems.

ASSESS Young's double-slit experiment not only demonstrated that light is a wave, it provided a means for measuring the wavelength. You learned in Chapter 20 that the wavelengths of visible light span the range 400–700 nm. These lengths are smaller than we can easily comprehend. A wavelength of 570 nm, which is in the middle of the visible spectrum, is only about 1% of the diameter of a human hair.

STOP TO THINK 22.2 Light of wavelength λ_1 illuminates a double slit, and interference fringes are observed on a screen behind the slits. When the wavelength is changed to λ_2, the fringes get closer together. Is λ_2 larger or smaller than λ_1?

Intensity of the Double-Slit Interference Pattern

Equations 22.6 and 22.9 locate the positions of maximum and zero intensity. To complete our analysis we need to calculate the light *intensity* at every point on the screen. All the tools we need to do this calculation were developed in Chapters 20 and 21.

You learned in Chapter 20 that the wave intensity I is proportional to the square of the wave's amplitude. The light spreading out behind a *single* slit produces the wide band of light that you saw in Figure 22.2. The intensity in this band of light is $I_1 = ca^2$, where a is the light-wave amplitude at the screen due to *one* wave and c is a proportionality constant.

If there were no interference, the light intensity due to two slits would be twice the intensity of one slit: $I_2 = 2I_1 = 2ca^2$. In other words, two slits would cause the broad band of light on the screen to be twice as bright. But that's not what happens. Instead, the superposition of the two light waves creates bright and dark interference fringes.

We found in Chapter 21 (Equation 21.36) that the net amplitude of two superimposed waves is

$$A = \left| 2a \cos\left(\frac{\Delta\phi}{2}\right) \right| \tag{22.10}$$

where a is the amplitude of each individual wave. Because the sources (i.e., the two slits) are in phase, the phase difference $\Delta\phi$ at the point where the two waves are combined is due only to the path-length difference: $\Delta\phi = 2\pi(\Delta r/\lambda)$. Using Equation 22.2 for Δr, along with the small-angle approximation and Equation 22.5 for y, we find the phase difference at position y on the screen to be

$$\Delta\phi = 2\pi\frac{\Delta r}{\lambda} = 2\pi\frac{d\sin\theta}{\lambda} \approx 2\pi\frac{d\tan\theta}{\lambda} = \frac{2\pi d}{\lambda L}y \tag{22.11}$$

Substituting Equation 22.11 into Equation 22.10, we find the wave amplitude at position y to be

$$A = \left| 2a \cos\left(\frac{\pi d}{\lambda L}y\right) \right| \tag{22.12}$$

Consequently, the light intensity at position y on the screen is

$$I = cA^2 = 4ca^2 \cos^2\left(\frac{\pi d}{\lambda L}y\right) \tag{22.13}$$

But ca^2 is I_1, the light intensity of a single slit. Thus the intensity of the double-slit interference pattern at position y is

$$I_{\text{double}} = 4I_1 \cos^2\left(\frac{\pi d}{\lambda L}y\right) \tag{22.14}$$

FIGURE 22.5a is a graph of the double-slit intensity versus position y. Notice the unusual orientation of the graph, with the intensity increasing toward the *left* so that the y-axis can match the experimental layout. You can see that the intensity oscillates between dark fringes ($I_{\text{double}} = 0$) and bright fringes ($I_{\text{double}} = 4I_1$). The maxima occur at points where $y_m = m\lambda L/d$. This is what we found earlier for the positions of the bright fringes, so Equation 22.14 is consistent with our initial analysis.

One curious feature is that the light intensity at the maxima is $I = 4I_1$, four times the intensity of the light from each slit alone. You might think that two slits would make the light twice as intense as one slit, but interference leads to a different result. Mathematically, two slits make the *amplitude* twice as big at points of constructive interference ($A = 2a$), so the intensity increases by a factor of $2^2 = 4$. Physically, this is conservation of energy. The line labeled $2I_1$ in Figure 22.5a is the uniform intensity that two slits would produce *if* the waves did not interfere. Interference does not change the amount of light energy coming through the two slits, but it does redistribute the light energy on the viewing screen. You can see that the *average* intensity of the

FIGURE 22.5 Intensity of the interference fringes in a double-slit experiment.

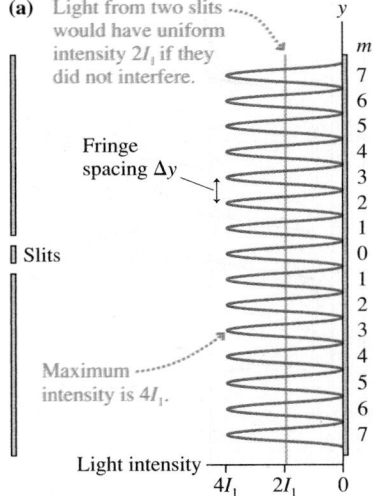

(a) Light from two slits would have uniform intensity $2I_1$ if they did not interfere.

Fringe spacing Δy

Slits

Maximum intensity is $4I_1$.

Light intensity $4I_1$ $2I_1$ 0

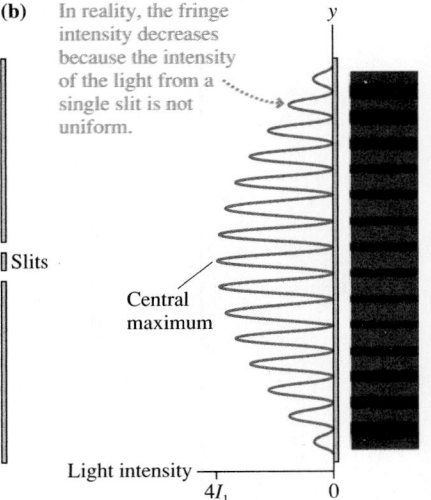

(b) In reality, the fringe intensity decreases because the intensity of the light from a single slit is not uniform.

Slits

Central maximum

Light intensity $4I_1$ 0

oscillating curve is $2I_1$, but the intensity of the bright fringes gets pushed up from $2I_1$ to $4I_1$ in order for the intensity of the dark fringes to drop from $2I_1$ to 0.

There is still one problem. Equation 22.14 predicts that all interference fringes are equally bright, but you saw in Figure 22.3b that the fringes decrease in brightness as you move away from the center. The erroneous prediction stems from our assumption that the amplitude a of the wave from each slit is constant across the screen. This isn't really true. A more detailed calculation, in which the amplitude gradually decreases as you move away from the center, finds that Equation 22.14 is correct if I_1 slowly decreases as y increases.

FIGURE 22.5b summarizes this analysis by graphing the light intensity (Equation 22.14) with I_1 slowly decreasing as y increases. Comparing this graph to the image, you can see that the wave model of light has provided an excellent description of Young's double-slit interference experiment.

22.3 The Diffraction Grating

Suppose we were to replace the double slit with an opaque screen that has N closely spaced slits. When illuminated from one side, each of these slits becomes the source of a light wave that diffracts, or spreads out, behind the slit. Such a multi-slit device is called a **diffraction grating.** The light intensity pattern on a screen behind a diffraction grating is due to the interference of N overlapped waves.

FIGURE 22.6 shows a diffraction grating in which N slits are equally spaced a distance d apart. This is a top view of the grating, as we look down on the experiment, and the slits extend above and below the page. Only 10 slits are shown here, but a practical grating will have hundreds or even thousands of slits. Suppose a plane wave of wavelength λ approaches from the left. The crest of a plane wave arrives *simultaneously* at each of the slits, causing the wave emerging from each slit to be in phase with the wave emerging from every other slit. Each of these emerging waves spreads out, just like the light wave in Figure 22.2, and after a short distance they all overlap with each other and interfere.

We want to know how the interference pattern will appear on a screen behind the grating. The light wave at the screen is the superposition of N waves, from N slits, as they spread and overlap. As we did with the double slit, we'll assume that the distance L to the screen is very large in comparison with the slit spacing d; hence the path followed by the light from one slit to a point on the screen is *very nearly* parallel to the path followed by the light from neighboring slits. The paths cannot be perfectly parallel, of course, or they would never meet to interfere, but the slight deviation from perfect parallelism is too small to notice. You can see in Figure 22.6 that the wave from one slit travels distance $\Delta r = d \sin \theta$ more than the wave from the slit above it and $\Delta r = d \sin \theta$ less than the wave below it. This is the same reasoning we used in Figure 22.4 to analyze the double-slit experiment.

Figure 22.6 is a magnified view of the slits. FIGURE 22.7 steps back to where we can see the viewing screen. If the angle θ is such that $\Delta r = d \sin \theta = m\lambda$, where m is an integer, then the light wave arriving at the screen from one slit will be *exactly in phase* with the light waves arriving from the two slits next to it. But each of those waves is in phase with waves from the slits next to them, and so on until we reach the end of the grating. In other words, *N* light waves, from *N* different slits, will *all* be in phase with each other when they arrive at a point on the screen at angle θ_m such that

$$d \sin \theta_m = m\lambda \qquad m = 0, 1, 2, 3, \dots \qquad (22.15)$$

The screen will have bright constructive-interference fringes at the values of θ_m given by Equation 22.15. We say that the light is "diffracted at angle θ_m."

Because it's usually easier to measure distances rather than angles, the position y_m of the mth maximum is

$$y_m = L \tan \theta_m \qquad \text{(positions of bright fringes)} \qquad (22.16)$$

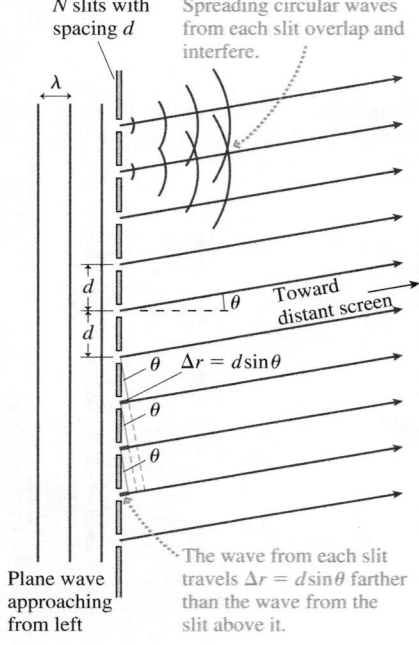

FIGURE 22.6 Top view of a diffraction grating with $N = 10$ slits.

N slits with spacing d

Spreading circular waves from each slit overlap and interfere.

λ

d

d

θ Toward distant screen

θ $\Delta r = d \sin \theta$

θ

θ

The wave from each slit travels $\Delta r = d \sin \theta$ farther than the wave from the slit above it.

Plane wave approaching from left

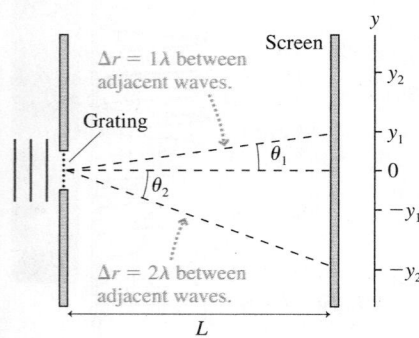

FIGURE 22.7 Angles of constructive interference.

$\Delta r = 1\lambda$ between adjacent waves.

Screen

Grating

θ_1

θ_2

$\Delta r = 2\lambda$ between adjacent waves.

L

y

y_2

y_1

0

$-y_1$

$-y_2$

The integer m is called the **order** of the diffraction. For example, light diffracted at θ_2 would be the second-order diffraction. Practical gratings, with very small values for d, display only a few orders. Because d is usually very small, it is customary to characterize a grating by the number of *lines per millimeter*. Here "line" is synonymous with "slit," so the number of lines per millimeter is simply the inverse of the slit spacing d in millimeters.

NOTE ▶ The condition for constructive interference in a grating of N slits is identical to Equation 22.4 for just two slits. Equation 22.15 is simply the requirement that the path-length difference between adjacent slits, be they two or N, is $m\lambda$. But unlike the angles in double-slit interference, the angles of constructive interference from a diffraction grating are generally *not* small angles. The reason is that the slit spacing d in a diffraction grating is so small that λ/d is not a small number. Thus you *cannot* use the small-angle approximation to simplify Equations 22.15 and 22.16. ◀

The wave amplitude at the points of constructive interference is Na because N waves of amplitude a combine in phase. Because the intensity depends on the square of the amplitude, the intensities of the bright fringes of a diffraction grating are

$$I_{max} = N^2 I_1 \qquad (22.17)$$

where, as before, I_1 is the intensity of the wave from a single slit. Equation 22.17 is consistent with our prior conclusion that the intensity of a bright fringe in a double-slit interference experiment is four times the intensity of the light from each slit alone. You can see that the fringe intensities increase rapidly as the number of slits increases.

Not only do the fringes get brighter as N increases, they also get narrower. This is again a matter of conservation of energy. If the light waves did not interfere, the intensity from N slits would be NI_1. Interference increases the intensity of the bright fringes by an extra factor of N, so to conserve energy the width of the bright fringes must be proportional to $1/N$. For a realistic diffraction grating, with $N > 100$, the interference pattern consists of a small number of *very* bright and *very* narrow fringes while most of the screen remains dark. FIGURE 22.8a shows the interference pattern behind a diffraction grating both graphically and with a simulation of the viewing screen. A comparison with Figure 22.5b shows that the bright fringes of a diffraction grating are much sharper and more distinct than the fringes of a double slit.

Because the bright fringes are so distinct, diffraction gratings are used for measuring the wavelengths of light. Suppose the incident light consists of two slightly different wavelengths. Each wavelength will be diffracted at a slightly different angle and, if N is sufficiently large, we'll see two distinct fringes on the screen. FIGURE 22.8b illustrates this idea. By contrast, the bright fringes in a double-slit experiment are too broad to distinguish the fringes of one wavelength from those of the other.

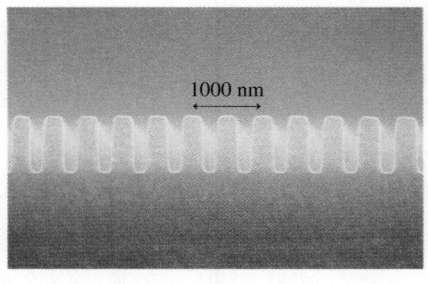

A microscopic side-on look at a diffraction grating.

FIGURE 22.8 The interference pattern behind a diffraction grating.

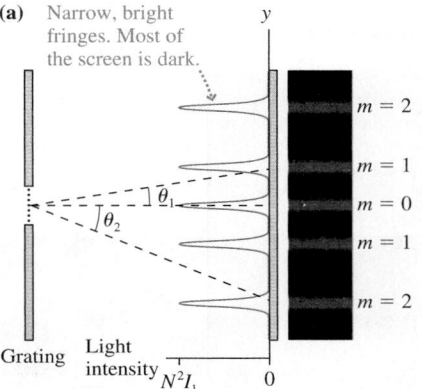

(a) Narrow, bright fringes. Most of the screen is dark.

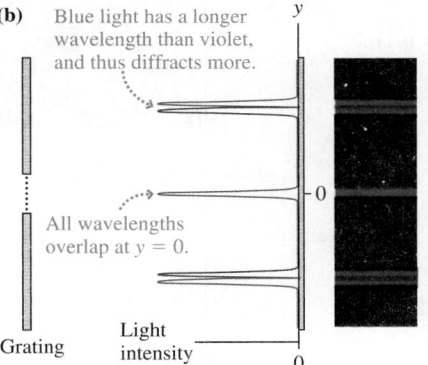

(b) Blue light has a longer wavelength than violet, and thus diffracts more.

All wavelengths overlap at $y = 0$.

EXAMPLE 22.3 **Measuring wavelengths emitted by sodium atoms**

Light from a sodium lamp passes through a diffraction grating having 1000 slits per millimeter. The interference pattern is viewed on a screen 1.000 m behind the grating. Two bright yellow fringes are visible 72.88 cm and 73.00 cm from the central maximum. What are the wavelengths of these two fringes?

VISUALIZE This is the situation shown in Figure 22.8b. The two fringes are very close together, so we expect the wavelengths to be only slightly different. No other yellow fringes are mentioned, so we will assume these two fringes are the first-order diffraction ($m = 1$).

SOLVE The distance y_m of a bright fringe from the central maximum is related to the diffraction angle by $y_m = L \tan\theta_m$. Thus the diffraction angles of these two fringes are

$$\theta_1 = \tan^{-1}\left(\frac{y_1}{L}\right) = \begin{cases} 36.08° & \text{fringe at 72.88 cm} \\ 36.13° & \text{fringe at 73.00 cm} \end{cases}$$

These angles must satisfy the interference condition $d \sin\theta_1 = \lambda$, so the wavelengths are $\lambda = d \sin\theta_1$. What is d? If a 1 mm length of the grating has 1000 slits, then the spacing from one slit to the next must be 1/1000 mm, or $d = 1.000 \times 10^{-6}$ m. Thus the wavelengths creating the two bright fringes are

$$\lambda = d \sin\theta_1 = \begin{cases} 589.0 \text{ nm} & \text{fringe at 72.88 cm} \\ 589.6 \text{ nm} & \text{fringe at 73.00 cm} \end{cases}$$

ASSESS We had data accurate to four significant figures, and all four were necessary to distinguish the two wavelengths.

The science of measuring the wavelengths of atomic and molecular emissions is called **spectroscopy.** The two sodium wavelengths in this example are called the *sodium doublet,* a name given to two closely spaced wavelengths emitted by the atoms of one element. This doublet is an identifying characteristic of sodium. Because no other element emits these two wavelengths, the doublet can be used to identify the presence of sodium in a sample of unknown composition, even if sodium is only a very minor constituent. This procedure is called *spectral analysis.*

Reflection Gratings

We have analyzed what is called a *transmission grating,* with many parallel slits. In practice, most diffraction gratings are manufactured as *reflection gratings.* The simplest reflection grating, shown in **FIGURE 22.9a**, is a mirror with hundreds or thousands of narrow, parallel grooves cut into the surface. The grooves divide the surface into many parallel reflective stripes, each of which, when illuminated, becomes the source of a spreading wave. Thus an incident light wave is divided into *N* overlapped waves. The interference pattern is exactly the same as the interference pattern of light transmitted through *N* parallel slits.

Naturally occurring reflection gratings are responsible for some forms of color in nature. As the micrograph of **FIGURE 22.9b** shows, a peacock feather consists of nearly parallel rods of melanin. These act as a reflection grating and create the ever-changing, multicolored hues of iridescence as the angle between the grating and your eye changes. The iridescence of some insects is due to diffraction from parallel microscopic ridges on the shell.

The rainbow of colors reflected from the surface of a DVD is a similar display of interference. The surface of a DVD is smooth plastic with a mirror-like reflective coating in which millions of microscopic holes, each about 1 μm in diameter, encode digital information. From an optical perspective, the array of holes in a shiny surface is a two-dimensional version of the reflection grating shown in Figure 22.9a. Reflection gratings can be manufactured at very low cost simply by stamping holes or grooves into a reflective surface, and these are widely sold as toys and novelty items. Rainbows of color are seen as each wavelength of white light is diffracted at a unique angle.

STOP TO THINK 22.3 White light passes through a diffraction grating and forms rainbow patterns on a screen behind the grating. For each rainbow,

a. The red side is on the right, the violet side on the left.
b. The red side is on the left, the violet side on the right.
c. The red side is closest to the center of the screen, the violet side is farthest from the center.
d. The red side is farthest from the center of the screen, the violet side is closest to the center.

22.4 Single-Slit Diffraction

We opened this chapter with a photograph (Figure 22.1a) of a water wave passing through a hole in a barrier, then spreading out on the other side. You then saw an image (Figure 22.2) showing that light, after passing through a very narrow slit, also spreads out on the other side. This phenomenon is called *diffraction.* We're now ready to look at the details of diffraction.

FIGURE 22.10 shows the experimental arrangement for observing the diffraction of light through a narrow slit of width *a.* Diffraction through a tall, narrow slit is known as **single-slit diffraction.** A viewing screen is placed distance *L* behind the slit, and we will assume that $L \gg a$. The light pattern on the viewing screen consists of a *central maximum*

FIGURE 22.9 Reflection gratings.

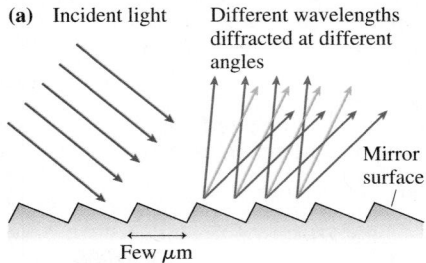

(a) Incident light   Different wavelengths diffracted at different angles

Mirror surface

Few μm

A reflection grating can be made by cutting parallel grooves in a mirror surface. These can be very precise, for scientific use, or mass produced in plastic.

(b)

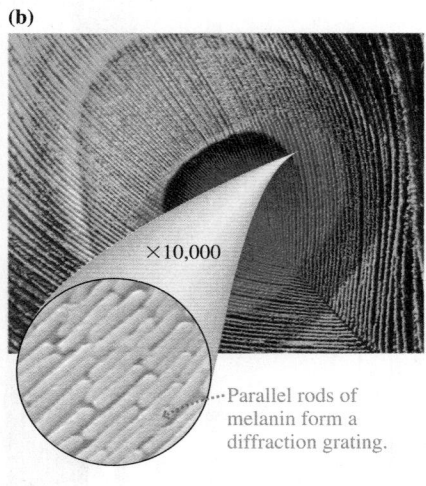

×10,000

Parallel rods of melanin form a diffraction grating.

FIGURE 22.10 A single-slit diffraction experiment.

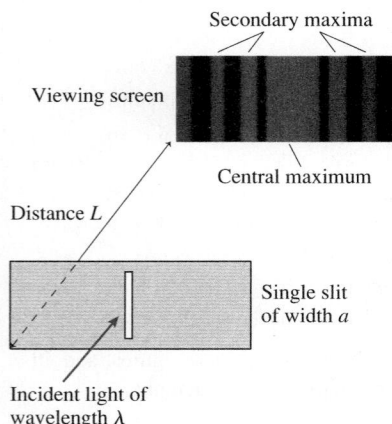

Secondary maxima

Viewing screen

Central maximum

Distance *L*

Single slit of width *a*

Incident light of wavelength λ

flanked by a series of weaker **secondary maxima** and dark fringes. Notice that the central maximum is significantly broader than the secondary maxima. It is also significantly brighter than the secondary maxima, although that is hard to tell here because this image has been overexposed to make the secondary maxima show up better.

Huygens' Principle

Our analysis of the superposition of waves from distinct sources, such as two loud-speakers or the two slits in a double-slit experiment, has tacitly assumed that the sources are *point sources,* with no measurable extent. To understand diffraction, we need to think about the propagation of an *extended* wave front. This is a problem first considered by the Dutch scientist Christiaan Huygens, a contemporary of Newton who argued that light is a wave.

Huygens lived before a mathematical theory of waves had been developed, so he developed a geometrical model of wave propagation. His idea, which we now call **Huygens' principle,** has two steps:

1. Each point on a wave front is the source of a spherical *wavelet* that spreads out at the wave speed.
2. At a later time, the shape of the wave front is the line tangent to all the wavelets.

FIGURE 22.11 illustrates Huygens' principle for a plane wave and a spherical wave. As you can see, the line tangent to the wavelets of a plane wave is a plane that has propagated to the right. The line tangent to the wavelets of a spherical wave is a larger sphere.

Huygens' principle is a visual device, not a theory of waves. Nonetheless, the full mathematical theory of waves, as it developed in the 19th century, justifies Huygens' basic idea, although it is beyond the scope of this textbook to prove it.

Analyzing Single-Slit Diffraction

FIGURE 22.12a shows a wave front passing through a narrow slit of width a. According to Huygens' principle, each point on the wave front can be thought of as the source of a spherical wavelet. These wavelets overlap and interfere, producing the diffraction pattern seen on the viewing screen. The full mathematical analysis, using *every* point on the wave front, is a fairly difficult problem in calculus. We'll be satisfied with a geometrical analysis based on just a few wavelets.

FIGURE 22.12b shows the paths of several wavelets that travel straight ahead to the central point on the screen. (The screen is *very* far to the right in this magnified view of the slit.) The paths are very nearly parallel to each other, thus all the wavelets travel the same distance and arrive at the screen *in phase* with each other. The *constructive interference* between these wavelets produces the central maximum of the diffraction pattern at $\theta = 0$.

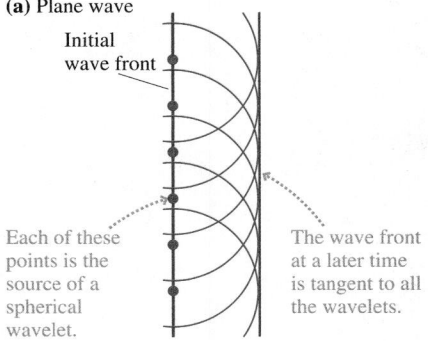

FIGURE 22.11 Huygens' principle applied to the propagation of plane waves and spherical waves.

(a) Plane wave

Initial wave front

Each of these points is the source of a spherical wavelet.

The wave front at a later time is tangent to all the wavelets.

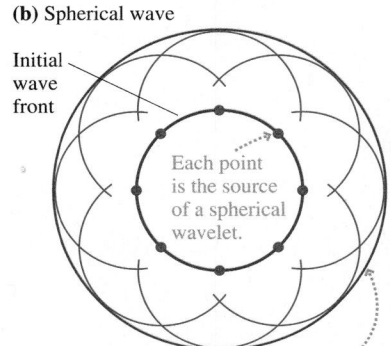

(b) Spherical wave

Initial wave front

Each point is the source of a spherical wavelet.

The wave front at a later time is tangent to all the wavelets.

FIGURE 22.12 Each point on the wave front is a source of spherical wavelets. The superposition of these wavelets produces the diffraction pattern on the screen.

(a) Greatly magnified view of slit

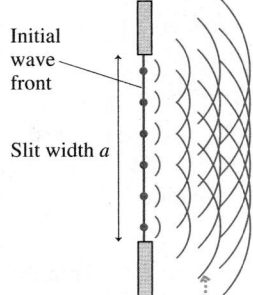

Initial wave front

Slit width a

The wavelets from each point on the initial wave front overlap and interfere, creating a diffraction pattern on the screen.

(b)

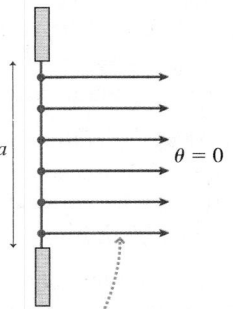

a

$\theta = 0$

The wavelets going straight forward all travel the same distance to the screen. Thus they arrive in phase and interfere constructively to produce the central maximum.

(c)

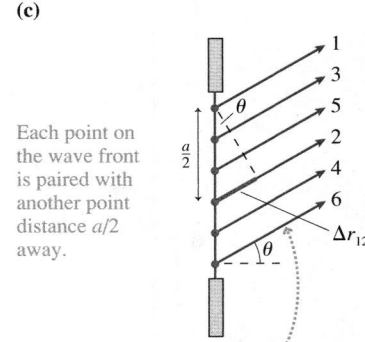

Each point on the wave front is paired with another point distance $a/2$ away.

$\frac{a}{2}$

θ

Δr_{12}

These wavelets all meet on the screen at angle θ. Wavelet 2 travels distance $\Delta r_{12} = (a/2)\sin\theta$ farther than wavelet 1.

The situation is different at points away from the center. Wavelets 1 and 2 in **FIGURE 22.12c** start from points that are distance $a/2$ apart. If the angle is such that Δr_{12}, the extra distance traveled by wavelet 2, happens to be $\lambda/2$, then wavelets 1 and 2 arrive out of phase and interfere destructively. But if Δr_{12} is $\lambda/2$, then the difference Δr_{34} between paths 3 and 4 and the difference Δr_{56} between paths 5 and 6 are also $\lambda/2$. Those pairs of wavelets also interfere destructively. The superposition of all the wavelets produces perfect destructive interference.

Figure 22.12c shows six wavelets, but our conclusion is valid for any number of wavelets. The key idea is that **every point on the wave front can be paired with another point distance $a/2$ away.** If the path-length difference is $\lambda/2$, the wavelets originating at these two points arrive at the screen out of phase and interfere destructively. When we sum the displacements of all N wavelets, they will—pair by pair—add to zero. The viewing screen at this position will be dark. This is the main idea of the analysis, one worth thinking about carefully.

You can see from Figure 22.12c that $\Delta r_{12} = (a/2) \sin \theta$. This path-length difference will be $\lambda/2$, the condition for destructive interference, if

$$\Delta r_{12} = \frac{a}{2} \sin \theta_1 = \frac{\lambda}{2} \qquad (22.18)$$

or, equivalently, if $a \sin \theta_1 = \lambda$.

> **NOTE** ▶ Equation 22.18 cannot be satisfied if the slit width a is less than the wavelength λ. If a wave passes through an opening smaller than the wavelength, the central maximum of the diffraction pattern expands to where it *completely* fills the space behind the opening. There are no minima or dark spots at any angle. This situation is uncommon for light waves, because λ is so small, but quite common in the diffraction of sound and water waves. ◀

We can extend this idea to find other angles of perfect destructive interference. Suppose each wavelet is paired with another wavelet from a point $a/4$ away. If Δr between these wavelets is $\lambda/2$, then all N wavelets will again cancel in pairs to give complete destructive interference. The angle θ_2 at which this occurs is found by replacing $a/2$ in Equation 22.18 with $a/4$, leading to the condition $a \sin \theta_2 = 2\lambda$. This process can be continued, and we find that the general condition for complete destructive interference is

$$a \sin \theta_p = p\lambda \qquad p = 1, 2, 3, \ldots \qquad (22.19)$$

When $\theta_p \ll 1$ rad, which is almost always true for light waves, we can use the small-angle approximation to write

$$\theta_p = p \frac{\lambda}{a} \qquad p = 1, 2, 3, \ldots \qquad \text{(angles of dark fringes)} \qquad (22.20)$$

Equation 22.20 gives the angles *in radians* to the dark minima in the diffraction pattern of Figure 22.10. Notice that $p = 0$ is explicitly *excluded*. $p = 0$ corresponds to the straight-ahead position at $\theta = 0$, but you saw in Figures 22.10 and 22.12b that $\theta = 0$ is the central *maximum*, not a minimum.

> **NOTE** ▶ It is perhaps surprising that Equations 22.19 and 22.20 are *mathematically* the same as the condition for the mth *maximum* of the double-slit interference pattern. But the physical meaning here is quite different. Equation 22.20 locates the *minima* (dark fringes) of the single-slit diffraction pattern. ◀

You might think that we could use this method of pairing wavelets from different points on the wave front to find the maxima in the diffraction pattern. Why not take two points on the wave front that are distance $a/2$ apart, find the angle at which their wavelets are in phase and interfere constructively, then sum over all points on the wave front? There is a subtle but important distinction. **FIGURE 22.13** shows six vector

FIGURE 22.13 Destructive interference by pairs leads to net destructive interference, but constructive interference by pairs does *not* necessarily lead to net constructive interference.

(a)

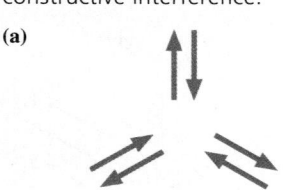

Each pair of vectors interferes destructively. The vector sum of all six vectors is zero.

(b)

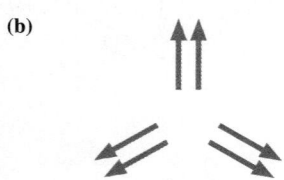

Each pair of vectors interferes constructively. Even so, the vector sum of all six vectors is zero.

arrows. The arrows in FIGURE 22.13a are arranged in pairs such that the two members of each pair cancel. The sum of all six vectors is clearly the zero vector $\vec{0}$, representing destructive interference. This is the procedure we used in Figure 22.12c to arrive at Equation 22.18.

The arrows in FIGURE 22.13b are arranged in pairs such that the two members of each pair point in the same direction—constructive interference! Nonetheless, the sum of all six vectors is still $\vec{0}$. To have N waves interfere constructively requires more than simply having constructive interference between pairs. Each pair must also be in phase with every other pair, a condition not satisfied in Figure 22.13b. Constructive interference by pairs does *not* necessarily lead to net constructive interference. It turns out that there is no simple formula to locate the maxima of a single-slit diffraction pattern.

It is possible, although beyond the scope of this textbook, to calculate the entire light intensity pattern. The results of such a calculation are shown graphically in FIGURE 22.14. You can see the bright central maximum at $\theta = 0$, the weaker secondary maxima, and the dark points of destructive interference at the angles given by Equation 22.20. Compare this graph to the image of Figure 22.10 and make sure you see the agreement between the two.

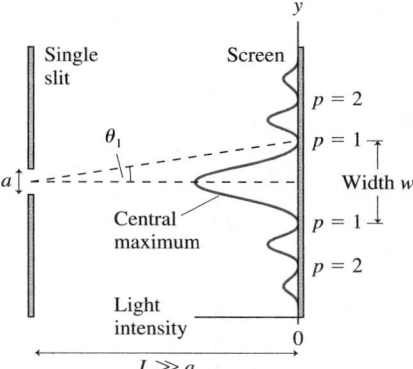

FIGURE 22.14 A graph of the intensity of a single-slit diffraction pattern.

| EXAMPLE 22.4 | **Diffraction of a laser through a slit** |

Light from a helium-neon laser ($\lambda = 633$ nm) passes through a narrow slit and is seen on a screen 2.0 m behind the slit. The first minimum in the diffraction pattern is 1.2 cm from the central maximum. How wide is the slit?

MODEL A narrow slit produces a single-slit diffraction pattern. A displacement of only 1.2 cm in a distance of 200 cm means that angle θ_1 is certainly a small angle.

VISUALIZE The intensity pattern will look like Figure 22.14.

SOLVE We can use the small-angle approximation to find that the angle to the first minimum is

$$\theta_1 = \frac{1.2 \text{ cm}}{200 \text{ cm}} = 0.00600 \text{ rad} = 0.344°$$

The first minimum is at angle $\theta_1 = \lambda/a$, from which we find that the slit width is

$$a = \frac{\lambda}{\theta_1} = \frac{633 \times 10^{-9} \text{ m}}{6.00 \times 10^{-3} \text{ rad}} = 1.1 \times 10^{-4} \text{ m} = 0.11 \text{ mm}$$

ASSESS This is typical of the slit widths used to observe single-slit diffraction. You can see that the small-angle approximation is well satisfied.

The Width of a Single-Slit Diffraction Pattern

We'll find it useful, as we did for the double slit, to measure positions on the screen rather than angles. The position of the pth dark fringe, at angle θ_p, is $y_p = L \tan \theta_p$, where L is the distance from the slit to the viewing screen. Using Equation 22.20 for θ_p and the small-angle approximation $\tan \theta_p \approx \theta_p$, we find that the dark fringes in the single-slit diffraction pattern are located at

$$y_p = \frac{p\lambda L}{a} \qquad p = 1, 2, 3, \dots \qquad \text{(positions of dark fringes)} \qquad (22.21)$$

A diffraction pattern is dominated by the central maximum, which is much brighter than the secondary maxima. The width w of the central maximum, shown in Figure 22.14, is defined as the distance between the two $p = 1$ minima on either side of the central maximum. Because the pattern is symmetrical, the width is simply $w = 2y_1$. This is

$$w = \frac{2\lambda L}{a} \qquad \text{(single slit)} \qquad (22.22)$$

The width of the central maximum is *twice* the spacing $\lambda L/a$ between the dark fringes on either side. The farther away the screen (larger L), the wider the pattern of light on it becomes. In other words, the light waves are *spreading out* behind the slit, and they fill a wider and wider region as they travel farther.

An important implication of Equation 22.22, one contrary to common sense, is that a narrower slit (smaller a) causes a *wider* diffraction pattern. **The smaller the opening you squeeze a wave through, the *more* it spreads out on the other side.**

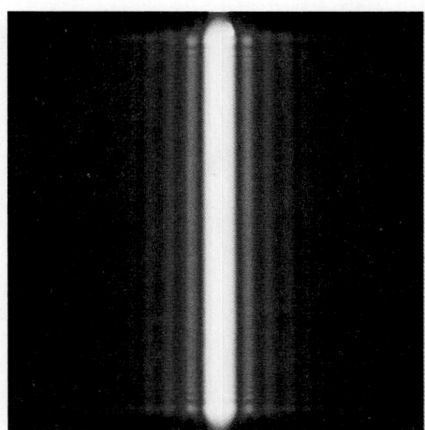

The central maximum of this single-slit diffraction pattern appears white because it is overexposed. The width of the central maximum is clear.

Determining the wavelength

Light passes through a 0.12-mm-wide slit and forms a diffraction pattern on a screen 1.00 m behind the slit. The width of the central maximum is 0.85 cm. What is the wavelength of the light?

SOLVE From Equation 22.22, the wavelength is

$$\lambda = \frac{aw}{2L} = \frac{(1.2 \times 10^{-4} \text{ m})(0.0085 \text{ m})}{2(1.00 \text{ m})}$$

$$= 5.1 \times 10^{-7} \text{ m} = 510 \text{ nm}$$

STOP TO THINK 22.4 The figure shows two single-slit diffraction patterns. The distance between the slit and the viewing screen is the same in both cases. Which of the following (perhaps more than one) could be true?

a. The slits are the same for both; $\lambda_1 > \lambda_2$.
b. The slits are the same for both; $\lambda_2 > \lambda_1$.
c. The wavelengths are the same for both; $a_1 > a_2$.
d. The wavelengths are the same for both; $a_2 > a_1$.
e. The slits and the wavelengths are the same for both; $p_1 > p_2$.
f. The slits and the wavelengths are the same for both; $p_2 > p_1$.

λ_1

λ_2

22.5 Circular-Aperture Diffraction

Diffraction occurs if a wave passes through an opening of any shape. Diffraction by a single slit establishes the basic ideas of diffraction, but a common situation of practical importance is diffraction of a wave by a **circular aperture**. Circular diffraction is mathematically more complex than diffraction from a slit, and we will present results without derivation.

Consider some examples. A loudspeaker cone generates sound by the rapid oscillation of a diaphragm, but the sound wave must pass through the circular aperture defined by the outer edge of the speaker cone before it travels into the room beyond. This is diffraction by a circular aperture. Telescopes and microscopes are the reverse. Light waves from outside need to enter the instrument. To do so, they must pass through a circular lens. In fact, the performance limit of optical instruments is determined by the diffraction of the circular openings through which the waves must pass. This is an issue we'll look at in Chapter 24.

FIGURE 22.15 shows a circular aperture of diameter D. Light waves passing through this aperture spread out to generate a *circular* diffraction pattern. You should compare this to Figure 22.10 for a single slit to note the similarities and differences. The diffraction pattern still has a *central maximum*, now circular, and it is surrounded by a series of secondary bright fringes.

FIGURE 22.15 The diffraction of light by a circular opening.

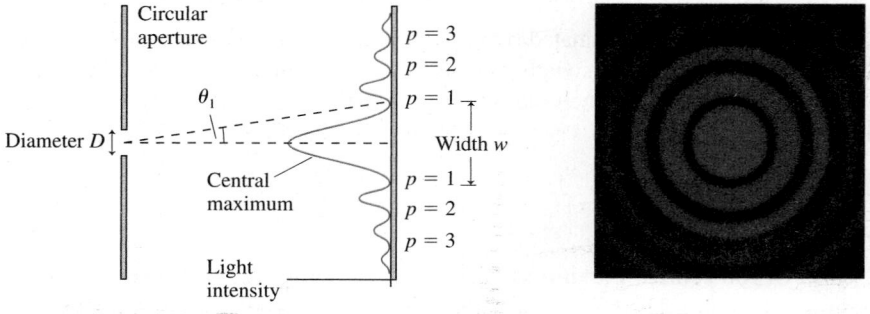

Angle θ_1 locates the first minimum in the intensity, where there is perfect destructive interference. A mathematical analysis of circular diffraction finds

$$\theta_1 = \frac{1.22\lambda}{D} \tag{22.23}$$

where D is the *diameter* of the circular opening. Equation 22.23 has assumed the small-angle approximation, which is almost always valid for the diffraction of light but usually is *not* valid for the diffraction of longer-wavelength sound waves.

Within the small-angle approximation, the width of the central maximum is

$$w = 2y_1 = 2L\tan\theta_1 \approx \frac{2.44\lambda L}{D} \quad \text{(circular aperture)} \quad (22.24)$$

The diameter of the diffraction pattern increases with distance L, showing that light spreads out behind a circular aperture, but it decreases if the size D of the aperture is increased.

EXAMPLE 22.6 **Shining a laser through a circular hole**

Light from a helium-neon laser ($\lambda = 633$ nm) passes through a 0.50-mm-diameter hole. How far away should a viewing screen be placed to observe a diffraction pattern whose central maximum is 3.0 mm in diameter?

SOLVE Equation 22.24 gives us the appropriate screen distance:

$$L = \frac{wD}{2.44\lambda} = \frac{(3.0 \times 10^{-3}\,\text{m})(5.0 \times 10^{-4}\,\text{m})}{2.44(633 \times 10^{-9}\,\text{m})} = 0.97\,\text{m}$$

The Wave and Ray Models of Light

We opened this chapter by noting that there are three models of light, each useful within a certain range of circumstances. We are now at a point where we can establish an important condition that separates the wave model of light from the ray model of light.

When light passes through an opening of size a, the angle of the first diffraction minimum is

$$\theta_1 = \sin^{-1}\left(\frac{\lambda}{a}\right) \quad (22.25)$$

Equation 22.25 is for a slit, but the result is very nearly the same if a is the diameter of a circular aperture. Regardless of the shape of the opening, **the factor that determines how much a wave spreads out behind an opening is the ratio λ/a, the size of the wavelength compared to the size of the opening.**

FIGURE 22.16 illustrates the difference between a wave whose wavelength is much smaller than the size of the opening and a second wave whose wavelength is comparable to the opening. A wave with $\lambda/a \approx 1$ quickly spreads to fill the region behind the opening. Light waves, because of their very short wavelength, almost always have $\lambda/a \ll 1$ and diffract to produce a slowly spreading "beam" of light.

Now we can better appreciate Newton's dilemma. With everyday-sized openings, sound and water waves have $\lambda/a \approx 1$ and diffract to fill the space behind the opening. Consequently, this is what we come to expect for the behavior of waves. Newton saw no evidence of this for light passing through openings. We see now that light really does spread out behind an opening, but the very small λ/a ratio usually makes the diffraction pattern too small to see. Diffraction begins to be discernible only when the size of the opening is a fraction of a millimeter or less. If we wanted the diffracted light wave to *fill* the space behind the opening ($\theta_1 \approx 90°$), as a sound wave does, we would need to reduce the size of the opening to $a \approx 0.001$ mm! Although holes this small can be made today, with the processes used to make integrated circuits, the light passing through such a small opening is too weak to be seen by the eye.

FIGURE 22.17 shows light passing through a hole of diameter D. According to the ray model, light rays passing through the hole travel straight ahead to create a bright circular spot of diameter D on a viewing screen. This is the *geometric image* of the slit. In reality, diffraction causes the light to spread out behind the slit, but—and this is the important point—**we will not notice the spreading if it is less than the diameter D of the geometric image.** That is, we will not be aware of diffraction unless the bright spot on the screen increases in diameter.

FIGURE 22.16 The diffraction of a long-wavelength wave and a short-wavelength wave through the same opening.

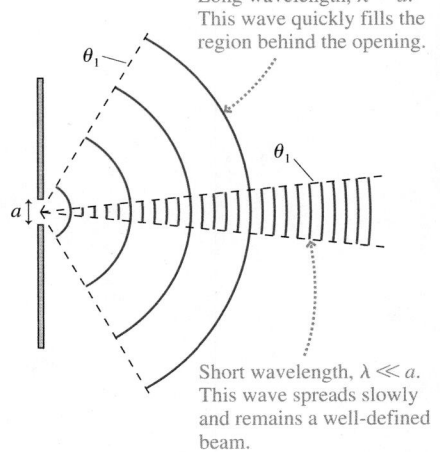

Long wavelength, $\lambda \approx a$. This wave quickly fills the region behind the opening.

Short wavelength, $\lambda \ll a$. This wave spreads slowly and remains a well-defined beam.

FIGURE 22.17 Diffraction will be noticed only if the bright spot on the screen is wider than D.

If light travels in straight lines, the image on the screen is the same size as the hole. Diffraction will not be noticed unless the light spreads over a diameter larger than D.

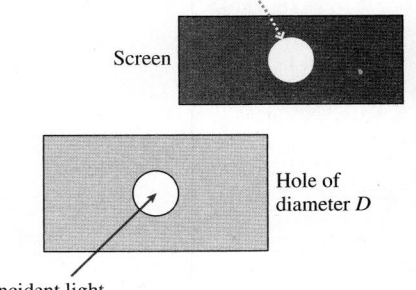

Screen

Hole of diameter D

Incident light

This idea provides a reasonable criterion for when to use ray optics and when to use wave optics:

- If the spreading due to diffraction is less than the size of the opening, use the ray model and think of light as traveling in straight lines.
- If the spreading due to diffraction is greater than the size of the opening, use the wave model of light.

The crossover point between these two regimes occurs when the spreading due to diffraction is equal to the size of the opening. The central-maximum width of a circular-aperture diffraction pattern is $w = 2.44\lambda L/D$. If we equate this diffraction width to the diameter of the aperture itself, we have

$$\frac{2.44\lambda L}{D_c} = D_c \qquad (22.26)$$

where the subscript c on D_c indicates that this is the crossover between the ray model and the wave model. Because we're making an estimate—the change from the ray model to the wave model is gradual, not sudden—to one significant figure, we find

$$D_c \approx \sqrt{2\lambda L} \qquad (22.27)$$

This is the diameter of a circular aperture whose diffraction pattern, at distance L, has width $w = D$. We know that visible light has $\lambda \approx 500$ nm, and a typical distance in laboratory work is $L \approx 1$ m. For these values,

$$D_c \approx 1 \text{ mm}$$

This brings us to an important and very practical conclusion, presented in Tactics Box 22.1.

TACTICS BOX 22.1 **Choosing a model of light**

❶ When visible light passes through openings smaller than about 1 mm in size, diffraction effects are usually important. Use the wave model of light.
❷ When visible light passes through openings larger than about 1 mm in size, diffraction effects are usually not important. Use the ray model of light.

Openings ≈ 1 mm in size are a gray area. Whether one should use a ray model or a wave model will depend on the precise values of λ and L. We'll avoid such ambiguous cases in this book, sticking with examples and homework that fall clearly within the wave model or the ray model. Lenses and mirrors, in particular, are almost always >1 mm in size. We will study the optics of lenses and mirrors in the chapter on ray optics. This chapter on wave optics deals with objects and openings <1 mm in size.

22.6 Interferometers

Scientists and engineers have devised many ingenious methods for using interference to control the flow of light and to make very precise measurements with light waves. A device that makes practical use of interference is called an **interferometer.**

Interference requires two waves of *exactly* the same wavelength. One way of guaranteeing that two waves have exactly equal wavelengths is to divide one wave into two parts of smaller amplitude. Later, at a different point in space, the two parts are recombined. Interferometers are based on the division and recombination of a single wave.

To illustrate the idea, FIGURE 22.18 shows an *acoustical interferometer*. A sound wave is sent into the left end of the tube. The wave splits into two parts at the junction, and waves of smaller amplitude travel around each side. Distance L can be changed by sliding the upper tube in and out like a trombone. After traveling distances r_1 and r_2, the waves recombine and their superposition travels out to the microphone. The sound emerging from the right end has maximum intensity, zero intensity, or somewhere in between depending on the phase difference between the two waves as they recombine.

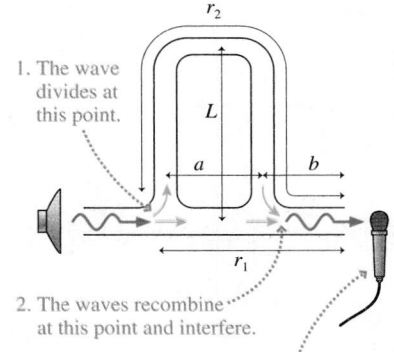

FIGURE 22.18 An acoustical interferometer.

1. The wave divides at this point.

2. The waves recombine at this point and interfere.

3. The microphone detects the superposition of the two waves that traveled different distances.

The two waves traveling through the interferometer started from the *same* source, the loudspeaker; hence the phase difference $\Delta\phi_0$ between the wave sources is automatically zero. The phase difference $\Delta\phi$ between the recombined waves is due entirely to the different distances they travel. Consequently, the conditions for constructive and destructive interference are those we found in Chapter 21 for identical sources:

Constructive: $\Delta r = m\lambda$

Destructive: $\Delta r = \left(m + \dfrac{1}{2}\right)\lambda$ $m = 0, 1, 2, \ldots$ (22.28)

The distance each wave travels is easily found from Figure 22.18:

$$r_1 = a + b$$
$$r_2 = L + a + L + b = 2L + a + b$$

Thus the path-length difference between the waves is $\Delta r = r_2 - r_1 = 2L$, and the conditions for constructive and destructive interference are

Constructive: $L = m\dfrac{\lambda}{2}$

Destructive: $L = \left(m + \dfrac{1}{2}\right)\dfrac{\lambda}{2}$ $m = 0, 1, 2, \ldots$ (22.29)

The interference conditions involve $\lambda/2$ rather than just λ because the wave following the upper path travels distance L *twice*, once up and once down. The upper wave travels a full wavelength λ farther than the lower wave when $L = \lambda/2$.

The interferometer is used by recording the alternating maxima and minima in the sound as the top tube is pulled out and L changes. The interference changes from a maximum to a minimum and back to a maximum every time L increases by half a wavelength. FIGURE 22.19 is a graph of the sound intensity at the microphone as L is increased. You can see, from Equation 22.29, that the number Δm of maxima appearing as the length changes by ΔL is

$$\Delta m = \frac{\Delta L}{\lambda/2}$$ (22.30)

Equation 22.30 is the basis for measuring wavelengths very accurately.

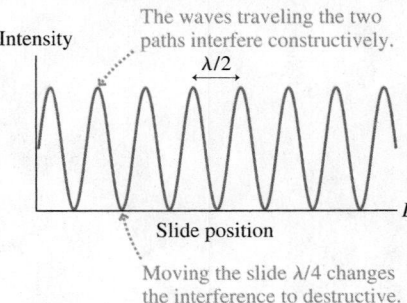

FIGURE 22.19 Interference maxima and minima alternate as the slide on an acoustical interferometer is withdrawn.

The waves traveling the two paths interfere constructively.

Moving the slide $\lambda/4$ changes the interference to destructive.

EXAMPLE 22.7 Measuring the wavelength of sound

A loudspeaker broadcasts a sound wave into an acoustical interferometer. The interferometer is adjusted so that the output sound intensity is a maximum, then the slide is slowly withdrawn. Exactly 10 new maxima appear as the slide moves 31.52 cm. What is the wavelength of the sound wave?

MODEL An interferometer produces a new maximum each time L increases by $\lambda/2$, causing the path-length difference Δr to increase by λ.

SOLVE Using Equation 22.30, we have

$$\lambda = \frac{2\Delta L}{\Delta m} = \frac{2(31.52 \text{ cm})}{10} = 6.304 \text{ cm}$$

ASSESS The wavelength can be determined to four significant figures because the distance was measured to four significant figures.

The Michelson Interferometer

Albert Michelson, the first American scientist to receive a Nobel Prize, invented an optical interferometer analogous to the acoustical interferometer. In the Michelson interferometer of FIGURE 22.20, the light wave is divided by a **beam splitter,** a partially silvered mirror that reflects half the light but transmits the other half. The two waves then travel toward mirrors M_1 and M_2. Half of the wave reflected from M_1 is transmitted through the beam splitter, where it recombines with the reflected half of the wave returning from M_2. The superimposed waves travel on to a light detector, originally a human observer but now more likely an electronic photodetector.

FIGURE 22.20 A Michelson interferometer.

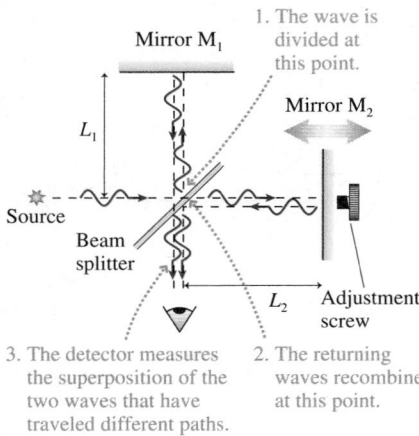

1. The wave is divided at this point.

Mirror M_1

L_1

Mirror M_2

Source

Beam splitter

L_2 Adjustment screw

3. The detector measures the superposition of the two waves that have traveled different paths.

2. The returning waves recombine at this point.

Mirror M_2 can be moved forward or backward by turning a precision screw. This is equivalent to pulling out the slide on the acoustical interferometer. The waves travel distances $r_1 = 2L_1$ and $r_2 = 2L_2$, with the factors of 2 appearing because the waves travel to the mirrors and back again. Thus the path-length difference between the two waves is

$$\Delta r = 2L_2 - 2L_1 \qquad (22.31)$$

The condition for constructive interference is $\Delta r = m\lambda$; hence constructive interference occurs when

$$\text{Constructive:} \qquad L_2 - L_1 = m\frac{\lambda}{2} \qquad m = 0, 1, 2, \ldots \qquad (22.32)$$

This result is essentially identical to Equation 22.29 for an acoustical interferometer. Both divide a wave, send the two smaller waves along two paths that differ in length by Δr, then recombine the two waves at a detector.

You might expect the interferometer output to be either "bright" or "dark." Instead, a viewing screen shows the pattern of circular interference fringes seen in FIGURE 22.21. Our analysis was for light waves that impinge on the mirrors exactly perpendicular to the surface. In an actual experiment, some of the light waves enter the interferometer at slightly different angles and, as a result, the recombined waves have slightly altered path-length differences Δr. These waves cause the alternating bright and dark fringes as you move outward from the center of the pattern. Their analysis will be left to more advanced courses in optics. Equation 22.32 is valid at the *center* of the circular pattern; thus there is a bright central spot when Equation 22.32 is true.

FIGURE 22.21 Photograph of the interference fringes produced by a Michelson interferometer.

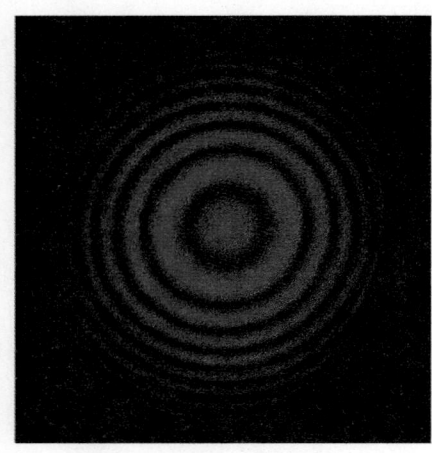

If mirror M_2 is moved by turning the screw, the central spot in the fringe pattern alternates between bright and dark. The output recorded by a detector looks exactly like the alternating loud and soft sounds shown in Figure 22.19. Suppose the interferometer is adjusted to produce a bright central spot. The next bright spot will appear when M_2 has moved half a wavelength, increasing the path-length difference by one full wavelength. The number Δm of maxima appearing as M_2 moves through distance ΔL_2 is

$$\Delta m = \frac{\Delta L_2}{\lambda/2} \qquad (22.33)$$

Very precise wavelength measurements can be made by moving the mirror while counting the number of new bright spots appearing at the center of the pattern. The number Δm is counted and known exactly. The only limitation on how precisely λ can be measured this way is the precision with which distance ΔL_2 can be measured. Unlike λ, which is microscopic, ΔL_2 is typically a few millimeters, a macroscopic distance that can be measured very accurately using precision screws, micrometers, and other techniques. Michelson's invention provided a way to transfer the precision of macroscopic distance measurements to an equal precision for the wavelength of light.

EXAMPLE 22.8 **Measuring the wavelength of light**

An experimenter uses a Michelson interferometer to measure one of the wavelengths of light emitted by neon atoms. She slowly moves mirror M_2 until 10,000 new bright central spots have appeared. (In a modern experiment, a photodetector and computer would eliminate the possibility of experimenter error while counting.) She then measures that the mirror has moved a distance of 3.164 mm. What is the wavelength of the light?

MODEL An interferometer produces a new maximum each time L_2 increases by $\lambda/2$.

SOLVE The mirror moves $\Delta L_2 = 3.164$ mm $= 3.164 \times 10^{-3}$ m. We can use Equation 22.33 to find

$$\lambda = \frac{2\Delta L_2}{\Delta m} = 6.328 \times 10^{-7} \text{ m} = 632.8 \text{ nm}$$

ASSESS A measurement of ΔL_2 accurate to four significant figures allowed us to determine λ to four significant figures. This happens to be the neon wavelength that is emitted as the laser beam in a helium-neon laser.

STOP TO THINK 22.5 A Michelson interferometer using light of wavelength λ has been adjusted to produce a bright spot at the center of the interference pattern. Mirror M_1 is then moved distance λ toward the beam splitter while M_2 is moved distance λ away from the beam splitter. How many bright-dark-bright fringe shifts are seen?

a. 0   b. 1   c. 2   d. 4
e. 8   f. It's not possible to say without knowing λ.

Holography

No discussion of wave optics would be complete without mentioning holography, which has both scientific and artistic applications. The basic idea is a simple extension of interferometry.

FIGURE 22.22a shows how a **hologram** is made. A beam splitter divides a laser beam into two waves. One wave illuminates the object of interest. The light scattered by this object is a very complex wave, but it is the wave you would see if you looked at the object from the position of the film. The other wave, called the *reference beam,* is reflected directly toward the film. The scattered light and the reference beam meet at the film and interfere. The film records their interference pattern.

The interference patterns we've looked at in this chapter have been simple patterns of stripes and circles because the light waves have been well-behaved plane waves and spherical waves. The light wave scattered by the object in Figure 22.22a is exceedingly complex. As a result, the interference pattern recorded on the film—the hologram—is a seemingly random pattern of whorls and blotches. FIGURE 22.22b is an enlarged photograph of a portion of a hologram. It's certainly not obvious that information is stored in this pattern, but it is.

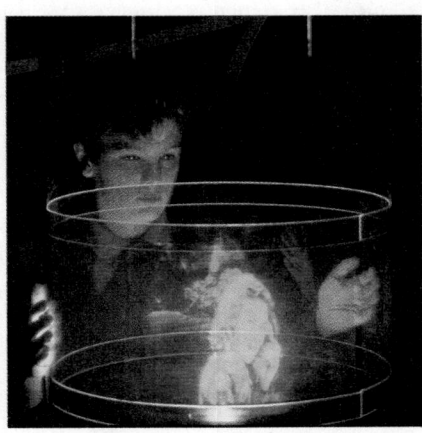

A hologram.

FIGURE 22.22 Holography is an important application of wave optics.

(a) Recording a hologram

The interference between the scattered light and the reference beam is recorded on the film.

Film
Plane waves
Reference beam
Laser
Beam splitter
Object beam
The scattered light has a complex wave front.
Object

(b) A hologram

An enlarged photo of the developed film. This is the hologram.

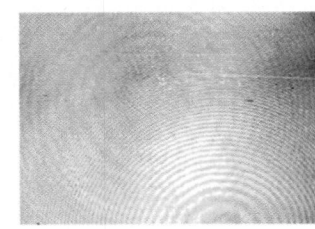

(c) Playing a hologram

The diffraction of the laser beam through the light and dark patches of the film reconstructs the original scattered wave.

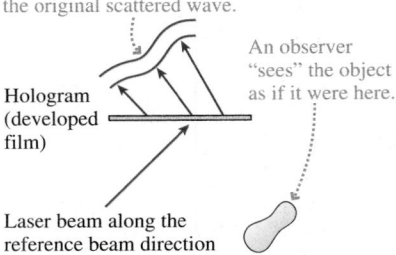

Hologram (developed film)
An observer "sees" the object as if it were here.
Laser beam along the reference beam direction

The hologram is "played" by sending just the reference beam through it, as seen in FIGURE 22.2c. The reference beam diffracts through the transparent parts of the hologram, just as it would through the slits of a diffraction grating. Amazingly, the diffracted wave is *exactly the same* as the light wave that had been scattered by the object! In other words, the diffracted reference beam *reconstructs* the original scattered wave. As you look at this diffracted wave, from the far side of the hologram, you "see" the object exactly as if it were there. The view is three dimensional because, by moving your head with respect to the hologram, you can see different portions of the wave front.

CHALLENGE EXAMPLE 22.9   **Measuring the index of refraction of a gas**

A Michelson interferometer uses a helium-neon laser with wavelength $\lambda_{vac} = 633$ nm. In one arm, the light passes through a 4.00-cm-thick glass cell. Initially the cell is evacuated, and the interferometer is adjusted so that the central spot is a bright fringe. The cell is then slowly filled to atmospheric pressure with a gas. As the cell fills, 43 bright-dark-bright fringe shifts are seen and counted. What is the index of refraction of the gas at this wavelength?

MODEL   Adding one additional wavelength to the round trip causes one bright-dark-bright fringe shift. Changing the length of the arm is one way to add wavelengths, but not the only way. Increasing the index of refraction also adds wavelengths because light has a shorter wavelength when traveling through a material with a larger index of refraction.

VISUALIZE   FIGURE 22.23 shows a Michelson interferometer with a cell of thickness d in one arm.

FIGURE 22.23 Measuring the index of refraction.

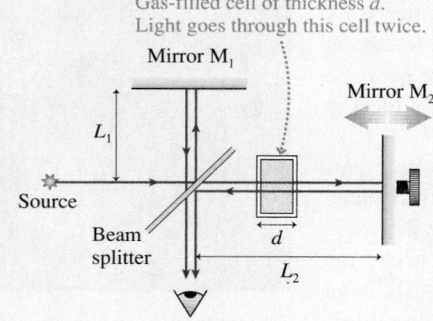

Gas-filled cell of thickness d.
Light goes through this cell twice.

Mirror M_1

Mirror M_2

L_1

Source

Beam splitter

d

L_2

SOLVE   To begin, all the air is pumped out of the cell. As light travels from the beam splitter to the mirror and back, the number of wavelengths inside the cell is

$$m_1 = \frac{2d}{\lambda_{vac}}$$

where the 2 appears because the light passes through the cell twice.

The cell is then filled with gas at 1 atm pressure. Light travels slower in the gas, $v = c/n$, and you learned in Chapter 20 that the reduction in speed decreases the wavelength to λ_{vac}/n. With the cell filled, the number of wavelengths spanning distance d is

$$m_2 = \frac{2d}{\lambda} = \frac{2d}{\lambda_{vac}/n}$$

The physical distance has not changed, but the number of wavelengths along the path has. Filling the cell has increased the path by

$$\Delta m = m_2 - m_1 = (n-1)\frac{2d}{\lambda_{vac}}$$

wavelengths. Each increase of one wavelength causes one bright-dark-bright fringe shift at the output. Solving for n, we find

$$n = 1 + \frac{\lambda_{vac}\Delta m}{2d} = 1 + \frac{(6.33 \times 10^{-7} \text{ m})(43)}{2(0.0400 \text{ m})} = 1.00034$$

ASSESS   This may seem like a six-significant-figure result, but there are really only two. What we're measuring is not n but $n - 1$. We know the fringe count to two significant figures, and that has allowed us to compute $n - 1 = \lambda_{vac}\Delta m/2d = 3.4 \times 10^{-4}$.

SUMMARY

The goal of Chapter 22 has been to understand and apply the wave model of light.

General Principles

Huygens' principle says that each point on a wave front is the source of a spherical wavelet. The wave front at a later time is tangent to all the wavelets.

Diffraction is the spreading of a wave after it passes through an opening.

Constructive and destructive interference are due to the overlap of two or more waves as they spread behind openings.

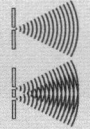

Important Concepts

The wave model of light considers light to be a wave propagating through space. Diffraction and interference are important.

The ray model of light considers light to travel in straight lines like little particles. Diffraction and interference are not important.

Diffraction is important when the width of the diffraction pattern of an aperture equals or exceeds the size of the aperture. For a circular aperture, the crossover between the ray and wave models occurs for an opening of diameter $D_c \approx \sqrt{2\lambda L}$.

In practice, $D_c \approx 1$ mm for visible light. Thus

- Use the wave model when light passes through openings < 1 mm in size. Diffraction effects are usually important.

- Use the ray model when light passes through openings > 1 mm in size. Diffraction is usually not important.

Applications

Single slit of width a. A bright **central maximum** of width

$$w = \frac{2\lambda L}{a}$$

is flanked by weaker **secondary maxima.** Dark fringes are located at angles such that

$$a\sin\theta_p = p\lambda \qquad p = 1, 2, 3, \ldots$$

If $\lambda/a \ll 1$, then from the small-angle approximation

$$\theta_p = \frac{p\lambda}{a} \qquad y_p = \frac{p\lambda L}{a}$$

Interference due to wave-front division

Waves overlap as they spread out behind slits. Constructive interference occurs along antinodal lines. Bright fringes are seen where the antinodal lines intersect the viewing screen.

Double slit with separation d. Equally spaced bright fringes are located at

$$\theta_m = \frac{m\lambda}{d} \qquad y_m = \frac{m\lambda L}{d} \qquad m = 0, 1, 2, \ldots$$

The **fringe spacing** is $\Delta y = \dfrac{\lambda L}{d}$

Diffraction grating with slit spacing d. Very bright and narrow fringes are located at angles and positions

$$d\sin\theta_m = m\lambda \qquad y_m = L\tan\theta_m$$

Circular aperture of diameter D. A bright central maximum of diameter

$$w = \frac{2.44\lambda L}{D}$$

is surrounded by circular secondary maxima. The first dark fringe is located at

$$\theta_1 = \frac{1.22\lambda}{D} \qquad y_1 = \frac{1.22\lambda L}{D}$$

For an aperture of any shape, a smaller opening causes a more rapid spreading of the wave behind the opening.

Interference due to amplitude division

An interferometer divides a wave, lets the two waves travel different paths, then recombines them. Interference is constructive if one wave travels an integer number of wavelengths more or less than the other wave. The difference can be due to an actual path-length difference or to a different index of refraction.

Michelson interferometer

The number of bright-dark-bright fringe shifts as mirror M_2 moves distance ΔL_2 is

$$\Delta m = \frac{\Delta L_2}{\lambda/2}$$

Terms and Notation

optics	photon model	diffraction grating	Huygens' principle
diffraction	double slit	order, m	circular aperture
models of light	interference fringes	spectroscopy	interferometer
wave model	central maximum	single-slit diffraction	beam splitter
ray model	fringe spacing, Δy	secondary maxima	hologram

CONCEPTUAL QUESTIONS

1. FIGURE Q22.1 shows light waves passing through two closely spaced, narrow slits. The graph shows the intensity of light on a screen behind the slits. Reproduce these graph axes, including the zero and the tick marks locating the double-slit fringes, then draw a graph to show how the light-intensity pattern will appear if the right slit is blocked, allowing light to go through only the left slit. Explain your reasoning.

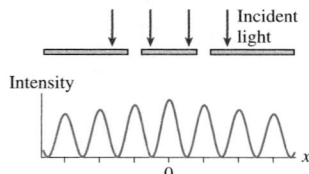

FIGURE Q22.1

2. In a double-slit interference experiment, which of the following actions (perhaps more than one) would cause the fringe spacing to increase? (a) Increasing the wavelength of the light. (b) Increasing the slit spacing. (c) Increasing the distance to the viewing screen. (d) Submerging the entire experiment in water.

3. FIGURE Q22.3 shows the viewing screen in a double-slit experiment. Fringe C is the central maximum. What will happen to the fringe spacing if
   a. The wavelength of the light is decreased?
   b. The spacing between the slits is decreased?
   c. The distance to the screen is decreased?
   d. Suppose the wavelength of the light is 500 nm. How much farther is it from the dot on the screen in the center of fringe E to the left slit than it is from the dot to the right slit?

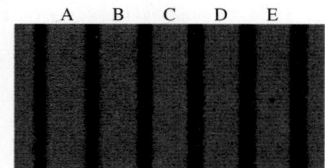

FIGURE Q22.3

4. FIGURE Q22.3 is the interference pattern seen on a viewing screen behind 2 slits. Suppose the 2 slits were replaced by 20 slits having the same spacing d between adjacent slits.
   a. Would the number of fringes on the screen increase, decrease, or stay the same?
   b. Would the fringe spacing increase, decrease, or stay the same?
   c. Would the width of each fringe increase, decrease, or stay the same?
   d. Would the brightness of each fringe increase, decrease, or stay the same?

5. FIGURE Q22.5 shows the light intensity on a viewing screen behind a single slit of width a. The light's wavelength is λ. Is $\lambda < a$, $\lambda = a$, $\lambda > a$, or is it not possible to tell? Explain.

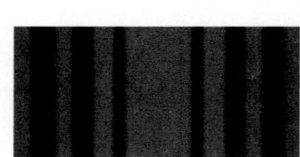

FIGURE Q22.5

FIGURE Q22.6

6. FIGURE Q22.6 shows the light intensity on a viewing screen behind a circular aperture. What happens to the width of the central maximum if
   a. The wavelength of the light is increased?
   b. The diameter of the aperture is increased?
   c. How will the screen appear if the aperture diameter is less than the light wavelength?

7. Narrow, bright fringes are observed on a screen behind a diffraction grating. The entire experiment is then immersed in water. Do the fringes on the screen get closer together, get farther apart, remain the same, or disappear? Explain.

8. a. Green light shines through a 100-mm-diameter hole and is observed on a screen. If the hole diameter is increased by 20%, does the circular spot of light on the screen decrease in diameter, increase in diameter, or stay the same? Explain.
   b. Green light shines through a 100-μm-diameter hole and is observed on a screen. If the hole diameter is increased by 20%, does the circular spot of light on the screen decrease in diameter, increase in diameter, or stay the same? Explain.

9. A Michelson interferometer using 800 nm light is adjusted to have a bright central spot. One mirror is then moved 200 nm forward, the other 200 nm back. Afterward, is the central spot bright, dark, or in between? Explain.

10. A Michelson interferometer is set up to display constructive interference (a bright central spot in the fringe pattern of Figure 22.21) using light of wavelength λ. If the wavelength is changed to $\lambda/2$, does the central spot remain bright, does the central spot become dark, or do the fringes disappear? Explain. Assume the fringes are viewed by a detector sensitive to both wavelengths.

EXERCISES AND PROBLEMS

Problems labeled ▨ integrate material from earlier chapters.

Exercises

Section 22.2 The Interference of Light

1. | Two narrow slits 80 μm apart are illuminated with light of wavelength 600 nm. What is the angle of the $m = 3$ bright fringe in radians? In degrees?

2. | A double slit is illuminated simultaneously with orange light of wavelength 600 nm and light of an unknown wavelength. The $m = 4$ bright fringe of the unknown wavelength overlaps the $m = 3$ bright orange fringe. What is the unknown wavelength?

3. | Light of wavelength 500 nm illuminates a double slit, and the interference pattern is observed on a screen. At the position of the $m = 2$ bright fringe, how much farther is it to the more distant slit than to the nearer slit?

4. | A double-slit experiment is performed with light of wavelength 600 nm. The bright interference fringes are spaced 1.8 mm apart on the viewing screen. What will the fringe spacing be if the light is changed to a wavelength of 400 nm?

5. ‖ Light of 600 nm wavelength illuminates a double slit. The intensity pattern shown in FIGURE EX22.5 is seen on a screen 2.0 m behind the slits. What is the spacing (in mm) between the slits?

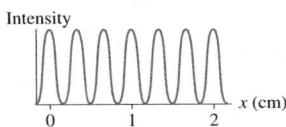

FIGURE EX22.5

6. ‖ Light from a sodium lamp ($\lambda = 589$ nm) illuminates two narrow slits. The fringe spacing on a screen 150 cm behind the slits is 4.0 mm. What is the spacing (in mm) between the two slits?

7. ‖ In a double-slit experiment, the slit separation is 200 times the wavelength of the light. What is the angular separation (in degrees) between two adjacent bright fringes?

8. ‖ A double-slit interference pattern is created by two narrow slits spaced 0.20 mm apart. The distance between the first and the fifth minimum on a screen 60 cm behind the slits is 6.0 mm. What is the wavelength (in nm) of the light used in this experiment?

Section 22.3 The Diffraction Grating

9. | A 4.0-cm-wide diffraction grating has 2000 slits. It is illuminated by light of wavelength 550 nm. What are the angles (in degrees) of the first two diffraction orders?

10. ‖ A diffraction grating produces a first-order maximum at an angle of 20.0°. What is the angle of the second-order maximum?

11. ‖ Light of wavelength 600 nm illuminates a diffraction grating. The second-order maximum is at angle 39.5°. How many lines per millimeter does this grating have?

12. ‖ The two most prominent wavelengths in the light emitted by a hydrogen discharge lamp are 656 nm (red) and 486 nm (blue). Light from a hydrogen lamp illuminates a diffraction grating with 500 lines/mm, and the light is observed on a screen 1.5 m behind the grating. What is the distance between the first-order red and blue fringes?

13. ‖ A helium-neon laser ($\lambda = 633$ nm) illuminates a diffraction grating. The distance between the two $m = 1$ bright fringes is 32 cm on a screen 2.0 m behind the grating. What is the spacing between slits of the grating?

14. | A diffraction grating is illuminated simultaneously with red light of wavelength 660 nm and light of an unknown wavelength. The fifth-order maximum of the unknown wavelength exactly overlaps the third-order maximum of the red light. What is the unknown wavelength?

Section 22.4 Single-Slit Diffraction

15. | A helium-neon laser ($\lambda = 633$ nm) illuminates a single slit and is observed on a screen 1.5 m behind the slit. The distance between the first and second minima in the diffraction pattern is 4.75 mm. What is the width (in mm) of the slit?

16. | In a single-slit experiment, the slit width is 200 times the wavelength of the light. What is the width (in mm) of the central maximum on a screen 2.0 m behind the slit?

17. | The central maximum of a single slit has width 4000λ when viewed on a screen 1.0 m behind the slit. How wide (in mm) is the slit?

18. ‖ Light of 600 nm wavelength illuminates a single slit. The intensity pattern shown in FIGURE EX22.18 is seen on a screen 2.0 m behind the slits. What is the width (in mm) of the slit?

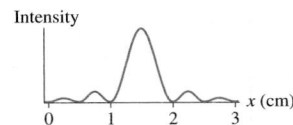

FIGURE EX22.18

19. ‖ A 0.50-mm-wide slit is illuminated by light of wavelength 500 nm. What is the width (in mm) of the central maximum on a screen 2.0 m behind the slit?

20. ‖ You need to use your cell phone, which broadcasts an 800 MHz signal, but you're behind two massive, radio-wave-absorbing buildings that have only a 15 m space between them. What is the angular width, in degrees, of the electromagnetic wave after it emerges from between the buildings?

21. | The opening to a cave is a tall, 30-cm-wide crack. A bat that is preparing to leave the cave emits a 30 kHz ultrasonic chirp. How wide is the "sound beam" 100 m outside the cave opening? Use $v_{\text{sound}} = 340$ m/s.

Section 22.5 Circular-Aperture Diffraction

22. ‖ A 0.50-mm-diameter hole is illuminated by light of wavelength 500 nm. What is the width (in mm) of the central maximum on a screen 2.0 m behind the slit?

23. | Infrared light of wavelength 2.5 μm illuminates a 0.20-mm-diameter hole. What is the angle of the first dark fringe in radians? In degrees?

24. | You want to photograph a circular diffraction pattern whose central maximum has a diameter of 1.0 cm. You have a helium-neon laser ($\lambda = 633$ nm) and a 0.12-mm-diameter pinhole. How far behind the pinhole should you place the screen that's to be photographed?

25. ‖ Light from a helium-neon laser ($\lambda = 633$ nm) passes through a circular aperture and is observed on a screen 4.0 m behind the aperture. The width of the central maximum is 2.5 cm. What is the diameter (in mm) of the hole?

Section 22.6 Interferometers

26. | A Michelson interferometer uses red light with a wavelength of 656.45 nm from a hydrogen discharge lamp. How many bright-dark-bright fringe shifts are observed if mirror M_2 is moved exactly 1 cm?

27. | Moving mirror M_2 of a Michelson interferometer a distance of 100 μm causes 500 bright-dark-bright fringe shifts. What is the wavelength of the light?

28. ‖ A Michelson interferometer uses light whose wavelength is known to be 602.446 nm. Mirror M_2 is slowly moved while exactly 33,198 bright-dark-bright fringe shifts are observed. What distance has M_2 moved? Be sure to give your answer to an appropriate number of significant figures.

29. | A Michelson interferometer uses light from a sodium lamp. Sodium atoms emit light having wavelengths 589.0 nm and 589.6 nm. The interferometer is initially set up with both arms of equal length ($L_1 = L_2$), producing a bright spot at the center of the interference pattern. How far must mirror M_2 be moved so that one wavelength has produced one more new maximum than the other wavelength?

Problems

30. | FIGURE P22.30 shows the light intensity on a screen 2.5 m behind an aperture. The aperture is illuminated with light of wavelength 600 nm.
    a. Is the aperture a single slit or a double slit? Explain.
    b. If the aperture is a single slit, what is its width? If it is a double slit, what is the spacing between the slits?

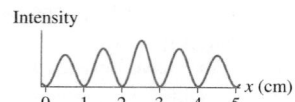

Intensity

FIGURE P22.30   0 1 2 3 4 5 x (cm)

31. | FIGURE P22.31 shows the light intensity on a screen 2.5 m behind an aperture. The aperture is illuminated with light of wavelength 600 nm.
    a. Is the aperture a single slit or a double slit? Explain.
    b. If the aperture is a single slit, what is its width? If it is a double slit, what is the spacing between the slits?

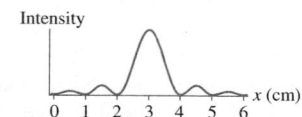

Intensity

FIGURE P22.31   0 1 2 3 4 5 6 x (cm)

32. ‖ Light from a helium-neon laser ($\lambda = 633$ nm) is used to illuminate two narrow slits. The interference pattern is observed on a screen 3.0 m behind the slits. Twelve bright fringes are seen, spanning a distance of 52 mm. What is the spacing (in mm) between the slits?

33. ‖ FIGURE P22.33 shows the light intensity on a screen behind a double slit. The slit spacing is 0.20 mm and the wavelength of the light is 600 nm. What is the distance from the slits to the screen?

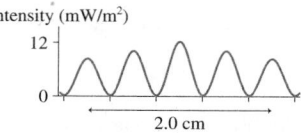

Intensity (mW/m²)

12

0

FIGURE P22.33   2.0 cm

34. ‖ FIGURE P22.33 shows the light intensity on a screen behind a double slit. The slit spacing is 0.20 mm and the screen is 2.0 m behind the slits. What is the wavelength (in nm) of the light?

35. | FIGURE P22.33 shows the light intensity on a screen behind a double slit. Suppose one slit is covered. What will be the light intensity at the center of the screen due to the remaining slit?

36. ‖‖ A laser beam with a wavelength of 524 nm is exactly perpendicular to a screen having two narrow slits spaced 0.150 mm apart. Interference fringes, including a central maximum, are observed on a viewing screen 1.00 m away. The direction of the laser beam is then slowly rotated by 1.0° around an axis parallel to the slits until it makes an 89.0° angle with the screen. How far does the central maximum move on the viewing screen?

37. ‖‖ A double-slit experiment is set up using a helium-neon laser ($\lambda = 633$ nm). Then a very thin piece of glass ($n = 1.50$) is placed over one of the slits. Afterward, the central point on the screen is occupied by what had been the $m = 10$ dark fringe. How thick is the glass?

38. ‖ A diffraction grating having 500 lines/mm diffracts visible light at 30°. What is the light's wavelength?

39. ‖ Helium atoms emit light at several wavelengths. Light from a helium lamp illuminates a diffraction grating and is observed on a screen 50.0 cm behind the grating. The emission at wavelength 501.5 nm creates a first-order bright fringe 21.90 cm from the central maximum. What is the wavelength of the bright fringe that is 31.60 cm from the central maximum?

40. ‖ A triple-slit experiment consists of three narrow slits, equally spaced by distance d and illuminated by light of wavelength λ. Each slit alone produces intensity I_1 on the viewing screen at distance L.
    a. Consider a point on the distant viewing screen such that the path-length difference between any two adjacent slits is λ. What is the intensity at this point?
    b. What is the intensity at a point where the path-length difference between any two adjacent slits is $\lambda/2$?

41. ‖ Because sound is a wave, it's possible to make a diffraction grating for sound from a large board of sound-absorbing material with several parallel slits cut for sound to go through. When 10 kHz sound waves pass through such a grating, listeners 10 m from the grating report "loud spots" 1.4 m on both sides of center. What is the spacing between the slits? Use 340 m/s for the speed of sound.

42. ‖ A diffraction grating with 600 lines/mm is illuminated with light of wavelength 500 nm. A very wide viewing screen is 2.0 m behind the grating.
    a. What is the distance between the two $m = 1$ bright fringes?
    b. How many bright fringes can be seen on the screen?

43. ‖ A 500 line/mm diffraction grating is illuminated by light of wavelength 510 nm. How many bright fringes are seen on a 2.0-m-wide screen located 2.0 m behind the grating?

44. ‖ White light (400–700 nm) incident on a 600 line/mm diffraction grating produces rainbows of diffracted light. What is the width of the first-order rainbow on a screen 2.0 m behind the grating?

45. ‖ For your science fair project you need to design a diffraction grating that will disperse the visible spectrum (400–700 nm) over 30.0° in first order.
   a. How many lines per millimeter does your grating need?
   b. What is the first-order diffraction angle of light from a sodium lamp ($\lambda = 589$ nm)?

46. ‖ FIGURE P22.46 shows the interference pattern on a screen 1.0 m behind an 800 line/mm diffraction grating. What is the wavelength (in nm) of the light?

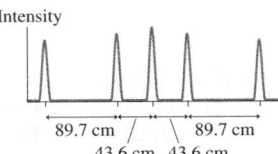

Intensity

| 89.7 cm | 89.7 cm |
| 43.6 cm | 43.6 cm |

FIGURE P22.46

47. ‖ FIGURE P22.46 shows the interference pattern on a screen 1.0 m behind a diffraction grating. The wavelength of the light is 600 nm. How many lines per millimeter does the grating have?

48. ‖ Light from a sodium lamp ($\lambda = 589$ nm) illuminates a narrow slit and is observed on a screen 75 cm behind the slit. The distance between the first and third dark fringes is 7.5 mm. What is the width (in mm) of the slit?

49. | The wings of some beetles
BIO have closely spaced parallel lines of melanin, causing the wing to act as a reflection grating. Suppose sunlight shines straight onto a beetle wing. If the melanin lines on the wing are spaced 2.0 μm apart, what is the first-order diffraction angle for green light ($\lambda = 550$ nm)?

50. | If sunlight shines straight onto a peacock feather, the feather
BIO appears bright blue when viewed from 15° on either side of the incident beam of light. The blue color is due to diffraction from parallel rods of melanin in the feather barbules, as was shown in the photograph on page 636. Other wavelengths in the incident light are diffracted at different angles, leaving only the blue light to be seen. The average wavelength of blue light is 470 nm. Assuming this to be the first-order diffraction, what is the spacing of the melanin rods in the feather?

51. ‖ You've found an unlabeled diffraction grating. Before you can use it, you need to know how many lines per mm it has. To find out, you illuminate the grating with light of several different wavelengths and then measure the distance between the two first-order bright fringes on a viewing screen 150 cm behind the grating. Your data are as follows:

Wavelength (nm)	Distance (cm)
430	109.6
480	125.4
530	139.8
580	157.2
630	174.4
680	194.8

Use the best-fit line of an appropriate graph to determine the number of lines per mm.

52. ‖ A diffraction grating has slit spacing d. Fringes are viewed on a screen at distance L. What wavelength of light produces a first-order fringe on the viewing screen at distance L from the center of the screen?

53. | For what slit-width-to-wavelength ratio does the first minimum of a single-slit diffraction pattern appear at (a) 30°, (b) 60°, and (c) 90°?

54. ‖ Light from a helium-neon laser ($\lambda = 633$ nm) is incident on a single slit. What is the largest slit width for which there are no minima in the diffraction pattern?

55. ‖ FIGURE P22.55 shows the light intensity on a screen behind a single slit. The slit width is 0.20 mm and the screen is 1.5 m behind the slit. What is the wavelength (in nm) of the light?

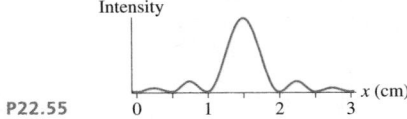

Intensity

FIGURE P22.55

56. ‖ FIGURE P22.55 shows the light intensity on a screen behind a single slit. The wavelength of the light is 600 nm and the slit width is 0.15 mm. What is the distance from the slit to the screen?

57. ‖ FIGURE P22.55 shows the light intensity on a screen behind a circular aperture. The wavelength of the light is 500 nm and the screen is 1.0 m behind the slit. What is the diameter (in mm) of the aperture?

58. ‖ Light from a helium-neon laser ($\lambda = 633$ nm) illuminates a circular aperture. It is noted that the diameter of the central maximum on a screen 50 cm behind the aperture matches the diameter of the geometric image. What is the aperture's diameter (in mm)?

59. ‖ One day, after pulling down your window shade, you notice that sunlight is passing through a pinhole in the shade and making a small patch of light on the far wall. Having recently studied optics in your physics class, you're not too surprised to see that the patch of light seems to be a circular diffraction pattern. It appears that the central maximum is about 1 cm across, and you estimate that the distance from the window shade to the wall is about 3 m. Estimate (a) the average wavelength of the sunlight (in nm) and (b) the diameter of the pinhole (in mm).

60. | A radar for tracking aircraft broadcasts a 12 GHz microwave beam from a 2.0-m-diameter circular radar antenna. From a wave perspective, the antenna is a circular aperture through which the microwaves diffract.
   a. What is the diameter of the radar beam at a distance of 30 km?
   b. If the antenna emits 100 kW of power, what is the average microwave intensity at 30 km?

61. ‖ Scientists use *laser range-finding* to measure the distance to the moon with great accuracy. A brief laser pulse is fired at the moon, then the time interval is measured until the "echo" is seen by a telescope. A laser beam spreads out as it travels because it diffracts through a circular exit as it leaves the laser. In order for the reflected light to be bright enough to detect, the laser spot on the moon must be no more than 1.0 km in diameter. Staying within this diameter is accomplished by using a special large-diameter laser. If $\lambda = 532$ nm, what is the minimum diameter of the circular opening from which the laser beam emerges? The earth-moon distance is 384,000 km.

62. ‖ Light of wavelength 600 nm passes though two slits separated by 0.20 mm and is observed on a screen 1.0 m behind the slits. The location of the central maximum is marked on the screen and labeled $y = 0$.
    a. At what distance, on either side of $y = 0$, are the $m = 1$ bright fringes?
    b. A very thin piece of glass is then placed in one slit. Because light travels slower in glass than in air, the wave passing through the glass is delayed by 5.0×10^{-16} s in comparison to the wave going through the other slit. What fraction of the period of the light wave is this delay?
    c. With the glass in place, what is the phase difference $\Delta\phi_0$ between the two waves as they leave the slits?
    d. The glass causes the interference fringe pattern on the screen to shift sideways. Which way does the central maximum move (toward or away from the slit with the glass) and by how far?

63. ‖ A 600 line/mm diffraction grating is in an empty aquarium tank. The index of refraction of the glass walls is $n_{glass} = 1.50$. A helium-neon laser ($\lambda = 633$ nm) is outside the aquarium. The laser beam passes through the glass wall and illuminates the diffraction grating.
    a. What is the first-order diffraction angle of the laser beam?
    b. What is the first-order diffraction angle of the laser beam after the aquarium is filled with water ($n_{water} = 1.33$)?

64. | You've set up a Michelson interferometer with a helium-neon laser ($\lambda = 632.8$ nm). After adjusting mirror M_2 to produce a bright spot at the center of the pattern, you carefully move M_2 away from the beam splitter while counting 1200 new bright spots at the center. Then you put the laser away. Later another student wants to restore the interferometer to its starting condition, but he mistakenly sets up a hydrogen discharge lamp and uses the 656.5 nm emission from hydrogen atoms. He then counts 1200 new bright spots while slowly moving M_2 back toward the beam splitter. What is the net displacement of M_2 when he is done? Is M_2 now closer to or farther from the beam splitter?

65. ‖ A Michelson interferometer operating at a 600 nm wavelength has a 2.00-cm-long glass cell in one arm. To begin, the air is pumped out of the cell and mirror M_2 is adjusted to produce a bright spot at the center of the interference pattern. Then a valve is opened and air is slowly admitted into the cell. The index of refraction of air at 1.00 atm pressure is 1.00028. How many bright-dark-bright fringe shifts are observed as the cell fills with air?

66. | A 0.10-mm-thick piece of glass is inserted into one arm of a Michelson interferometer that is using light of wavelength 500 nm. This causes the fringe pattern to shift by 200 fringes. What is the index of refraction of this piece of glass?

67. ‖ Optical computers require microscopic optical switches to turn signals on and off. One device for doing so, which can be implemented in an integrated circuit, is the *Mach-Zender interferometer* seen in FIGURE P22.67. Light from an on-chip infrared laser ($\lambda = 1.000\ \mu$m) is split into two waves that travel equal distances around the arms of the interferometer. One arm passes through an *electro-optic crystal,* a transparent material that can change its index of refraction in response to an applied voltage. Suppose both arms are exactly the same length and the crystal's index of refraction with no applied voltage is 1.522.
    a. With no voltage applied, is the output bright (switch closed, optical signal passing through) or dark (switch open, no signal)? Explain.

b. What is the first index of refraction of the electro-optic crystal larger than 1.522 that changes the optical switch to the state opposite the state you found in part a?

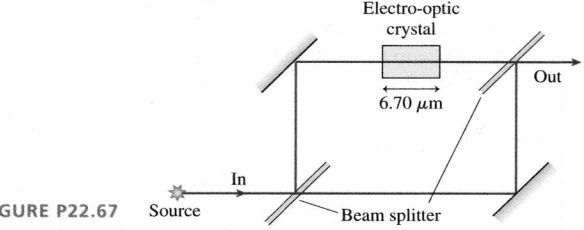

FIGURE P22.67

68. ‖ To illustrate one of the ideas of holography in a simple way, consider a diffraction grating with slit spacing d. The small-angle approximation is usually not valid for diffraction gratings, because d is only slightly larger than λ, but assume that the λ/d ratio of this grating is small enough to make the small-angle approximation valid.
    a. Use the small-angle approximation to find an expression for the fringe spacing on a screen at distance L behind the grating.
    b. Rather than a screen, suppose you place a piece of film at distance L behind the grating. The bright fringes will expose the film, but the dark spaces in between will leave the film unexposed. After being developed, the film will be a series of alternating light and dark stripes. What if you were to now "play" the film by using it as a diffraction grating? In other words, what happens if you shine the same laser through the film and look at the film's diffraction pattern on a screen at the same distance L? Demonstrate that the film's diffraction pattern is a reproduction of the original diffraction grating.

Challenge Problems

69. A helium-neon laser ($\lambda = 633$ nm) is built with a glass tube of inside diameter 1.0 mm, as shown in FIGURE CP22.69. One mirror is partially transmitting to allow the laser beam out. An electrical discharge in the tube causes it to glow like a neon light. From an optical perspective, the laser beam is a light wave that diffracts out through a 1.0-mm-diameter circular opening.
    a. Can a laser beam be *perfectly* parallel, with no spreading? Why or why not?
    b. The angle θ_1 to the first minimum is called the *divergence angle* of a laser beam. What is the divergence angle of this laser beam?
    c. What is the diameter (in mm) of the laser beam after it travels 3.0 m?
    d. What is the diameter of the laser beam after it travels 1.0 km?

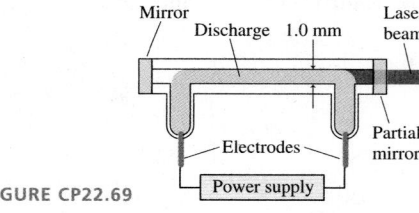

FIGURE CP22.69

70. The intensity at the central maximum of a double-slit interference pattern is $4I_1$. The intensity at the first minimum is zero. At what fraction of the distance from the central maximum to the first minimum is the intensity I_1?

71. Light consisting of two nearly equal wavelengths $\lambda + \Delta\lambda$ and λ, where $\Delta\lambda \ll \lambda$, is incident on a diffraction grating. The slit separation of the grating is d.
    a. Show that the angular separation of these two wavelengths in the mth order is

    $$\Delta\theta = \frac{\Delta\lambda}{\sqrt{(d/m)^2 - \lambda^2}}$$

    b. Sodium atoms emit light at 589.0 nm and 589.6 nm. What are the first-order and second-order angular separations (in degrees) of these two wavelengths for a 600 line/mm grating?
72. **FIGURE CP22.72** shows two nearly overlapped intensity peaks of the sort you might produce with a diffraction grating (see Figure 22.8b). As a practical matter, two peaks can just barely be resolved if their spacing Δy equals the width w of each peak, where w is measured at half of the peak's height. Two peaks closer together than w will merge into a single peak. We can use this idea to understand the *resolution* of a diffraction grating.
    a. In the small-angle approximation, the position of the $m = 1$ peak of a diffraction grating falls at the same location as the $m = 1$ fringe of a double slit: $y_1 = \lambda L/d$. Suppose two wavelengths differing by $\Delta\lambda$ pass through a grating at the same time. Find an expression for Δy, the separation of their first-order peaks.
    b. We noted that the widths of the bright fringes are proportional to $1/N$, where N is the number of slits in the grating. Let's hypothesize that the fringe width is $w = y_1/N$. Show that this is true for the double-slit pattern. We'll then assume it to be true as N increases.
    c. Use your results from parts a and b together with the idea that $\Delta y_{min} = w$ to find an expression for $\Delta\lambda_{min}$, the minimum wavelength separation (in first order) for which the diffraction fringes can barely be resolved.
    d. Ordinary hydrogen atoms emit red light with a wavelength of 656.45 nm. In deuterium, which is a "heavy" isotope of hydrogen, the wavelength is 656.27 nm. What is the minimum number of slits in a diffraction grating that can barely resolve these two wavelengths in the first-order diffraction pattern?

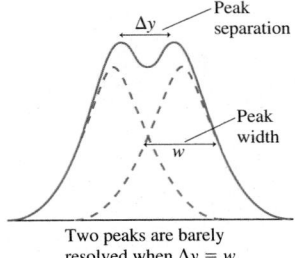

FIGURE CP22.72   Two peaks are barely resolved when $\Delta y = w$.

73. The diffraction grating analysis in this chapter assumed that the incident light is normal to the grating. **FIGURE CP22.73** shows a plane wave approaching a diffraction grating at angle ϕ.
    a. Show that the angles θ_m for constructive interference are given by the grating equation

    $$d(\sin\theta_m + \sin\phi) = m\lambda$$

    where $m = 0, \pm 1, \pm 2, \ldots$. Angles are considered positive if they are above the horizontal line, negative if below it.
    b. The two first-order maxima, $m = +1$ and $m = -1$, are no longer symmetrical about the center. Find θ_1 and θ_{-1} for 500 nm light incident on a 600 line/mm grating at $\phi = 30°$.

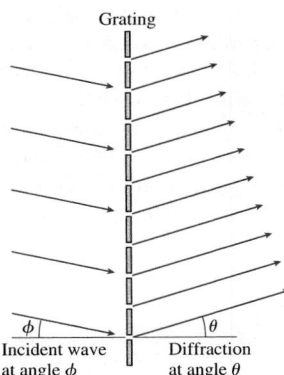

FIGURE CP22.73   Incident wave at angle ϕ Diffraction at angle θ

74. **FIGURE CP22.74** shows light of wavelength λ incident at angle ϕ on a *reflection* grating of spacing d. We want to find the angles θ_m at which constructive interference occurs.
    a. The figure shows paths 1 and 2 along which two waves travel and interfere. Find an expression for the path-length difference $\Delta r = r_2 - r_1$.
    b. Using your result from part a, find an equation (analogous to Equation 22.15) for the angles θ_m at which diffraction occurs when the light is incident at angle ϕ. Notice that m can be a negative integer in your expression, indicating that path 2 is shorter than path 1.
    c. Show that the zeroth-order diffraction is simply a "reflection." That is, $\theta_0 = \phi$.
    d. Light of wavelength 500 nm is incident at $\phi = 40°$ on a reflection grating having 700 reflection lines/mm. Find all angles θ_m at which light is diffracted. Negative values of θ_m are interpreted as an angle left of the vertical.
    e. Draw a picture showing a *single* 500 nm light ray incident at $\phi = 40°$ and showing all the diffracted waves at the correct angles.

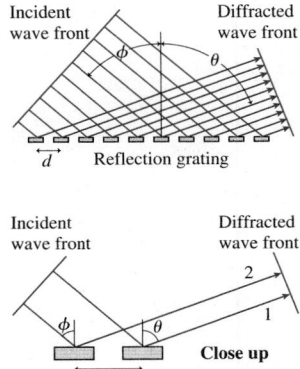

FIGURE CP22.74

75. The pinhole camera of **FIGURE CP22.75** images distant objects by allowing only a narrow bundle of light rays to pass through the hole and strike the film. If light consist-

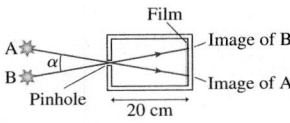

FIGURE CP22.75

ed of particles, you could make the image sharper and sharper (at the expense of getting dimmer and dimmer) by making the aperture smaller and smaller. In practice, diffraction of light by the circular aperture limits the maximum sharpness that can be obtained. Consider two distant points of light, such as two distant streetlights. Each will produce a circular diffraction pattern on the film. The two images can just barely be resolved if the central maximum of one image falls on the first dark fringe of the other image. (This is called Rayleigh's criterion, and we will explore its implication for optical instruments in Chapter 24.)

a. Optimum sharpness of one image occurs when the diameter of the central maximum equals the diameter of the pinhole. What is the optimum hole size for a pinhole camera in which the film is 20 cm behind the hole? Assume $\lambda = 550$ nm, an average value for visible light.
b. For this hole size, what is the angle α (in degrees) between two distant sources that can barely be resolved?
c. What is the distance between two street lights 1 km away that can barely be resolved?

STOP TO THINK ANSWERS

Stop to Think 22.1: b. The antinodal lines seen in Figure 22.3b are diverging.

Stop to Think 22.2: Smaller. Shorter-wavelength light doesn't spread as rapidly as longer-wavelength light. The fringe spacing Δy is directly proportional to the wavelength λ.

Stop to Think 22.3: d. Larger wavelengths have larger diffraction angles. Red light has a larger wavelength than violet light, so red light is diffracted farther from the center.

Stop to Think 22.4: b or c. The width of the central maximum, which is proportional to λ/a, has increased. This could occur either because the wavelength has increased or because the slit width has decreased.

Stop to Think 22.5: d. Moving M_1 in by λ decreases r_1 by 2λ. Moving M_2 out by λ increases r_2 by 2λ. These two actions together change the path length by $\Delta r = 4\lambda$.

23 Ray Optics

The observation that light travels in straight lines—*light rays*—will help us understand the physics of lenses and prisms.

▶ **Looking Ahead** The goals of Chapter 23 are to understand and apply the ray model of light.

The Ray Model of Light

The ray model applies when light interacts with objects that are very large compared to the wavelength. You'll learn that...

...light rays travel in straight lines unless they are...

...reflected by a surface or...

...refracted at a boundary.

Light rays can also be *scattered* or *absorbed* by the medium they travel through.

Images Formed by Lenses and Mirrors

You'll discover how lenses and mirrors form **images.** We'll start with a graphical method called **ray tracing.**

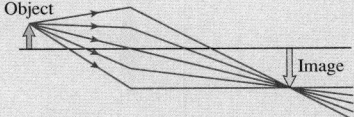

Ray tracing shows how this lens forms a *real image* on the opposite side of the lens from the object.

We'll then develop the **thin-lens equation** for more quantitative results.

A magnifying glass creates a *virtual image* that you see by looking through the lens.

We'll use the same graphical and mathematical techniques to understand how curved mirrors create images.

The passenger-side rearview mirror is curved, allowing you to see a wider field of view.

Reflection

Light rays can bounce, or **reflect,** off a surface. There are two important cases:

Specular reflection, like from a mirror.

Diffuse reflection, like from the page of this book.

You'll learn to use the *law of reflection.*

Refraction

When light rays travel from one medium to another, they change directions, or **refract,** at the boundary.

Refraction causes the laser beam to change direction as it goes through the prism.

You'll learn to use *Snell's law* to find the angles on both sides.

◀ **Looking Back**
Section 20.5 Index of refraction

23.1 The Ray Model of Light

A flashlight makes a beam of light through the night's darkness. Sunbeams stream into a darkened room through a small hole in the shade. Laser beams are even more well defined. Our everyday experience that light travels in straight lines is the basis of the *ray model* of light.

The ray model is an oversimplification of reality but nonetheless is very useful within its range of validity. In particular, the ray model of light is valid as long as any apertures through which the light passes (lenses, mirrors, and holes) are very large compared to the wavelength of light. In that case, diffraction and other wave aspects of light are negligible and can be ignored. The analysis of Section 22.5 found that the crossover between wave optics and ray optics occurs for apertures ≈ 1 mm in diameter. Lenses and mirrors are almost always larger than 1 mm, so the ray model of light is an excellent basis for the practical optics of image formation.

To begin, let us define a **light ray** as a line in the direction along which light energy is flowing. A light ray is an abstract idea, not a physical entity or a "thing." Any narrow beam of light, such as the laser beam in **FIGURE 23.1**, is actually a bundle of many parallel light rays. You can think of a single light ray as the limiting case of a laser beam whose diameter approaches zero. Laser beams are good approximations of light rays, certainly adequate for demonstrating ray behavior, but any real laser beam is a bundle of many parallel rays.

The following table outlines five basic ideas and assumptions of the ray model of light.

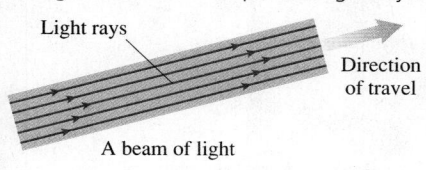

FIGURE 23.1 A laser beam or beam of sunlight is a bundle of parallel light rays.

Light rays

Direction of travel

A beam of light

The ray model of light

Light rays travel in straight lines.

Light travels through a transparent material in straight lines called light rays. The speed of light is $v = c/n$, where n is the index of refraction of the material.

Light rays can cross.

Light rays do not interact with each other. Two rays can cross without either being affected in any way.

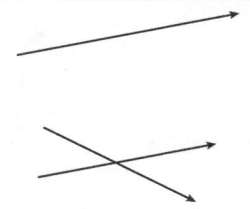

A light ray travels forever unless it interacts with matter.

A light ray continues forever unless it has an interaction with matter that causes the ray to change direction or to be absorbed. Light interacts with matter in four different ways:

- At an interface between two materials, light can be either *reflected* or *refracted*.
- Within a material, light can be either *scattered* or *absorbed*.

Material 1   Material 2

Reflection

Refraction

Scattering

Absorption

These interactions are discussed later in the chapter.

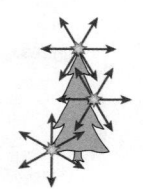

An object is a source of light rays.

An **object** is a source of light rays. Rays originate from *every* point on the object, and each point sends rays in *all* directions. We make no distinction between self-luminous objects and reflective objects.

Diverging bundle of rays

The eye sees by focusing a diverging bundle of rays.

The eye "sees" an object when *diverging* bundles of rays from each point on the object enter the pupil and are focused to an image on the retina. (Imaging is discussed later in the chapter.) From the movements the eye's lens has to make to focus the image, your brain determines the point from which the rays originated, and you perceive the object as being at that point.

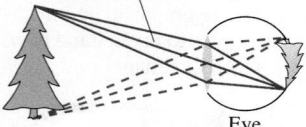

Eye

Objects

FIGURE 23.2 illustrates the idea that objects can be either *self-luminous,* such as the sun, flames, and lightbulbs, or *reflective.* Most objects are reflective. A tree, unless it is on fire, is seen or photographed by virtue of reflected sunlight or reflected skylight. People, houses, and this page in the book reflect light from self-luminous sources. In this chapter we are concerned not with how the light originates but with how it behaves after leaving the object.

Light rays from an object are emitted in all directions, but you are not *aware* of light rays unless they enter the pupil of your eye. Consequently, most light rays go completely unnoticed. For example, light rays travel from the sun to the tree in Figure 23.2, but you're not aware of these unless the tree reflects some of them into your eye. Or consider a laser beam. You've probably noticed that it's almost impossible to see a laser beam from the side unless there's dust in the air. The dust scatters a few of the light rays toward your eye, but in the absence of dust you would be completely unaware of a very powerful light beam traveling past you. **Light rays exist independently of whether you are seeing them.**

FIGURE 23.3 shows two idealized sets of light rays. The diverging rays from a **point source** are emitted in all directions. It is useful to think of each point on an object as a point source of light rays. A **parallel bundle** of rays could be a laser beam. Alternatively it could represent a *distant object,* an object such as a star so far away that the rays arriving at the observer are essentially parallel to each other.

Ray Diagrams

Rays originate from *every* point on an object and travel outward in *all* directions, but a diagram trying to show all these rays would be hopelessly messy and confusing. To simplify the picture, we usually use a **ray diagram** showing only a few rays. For example, **FIGURE 23.4** is a ray diagram showing only a few rays leaving the top and bottom points of the object and traveling to the right. These rays will be sufficient to show us how the object is imaged by lenses or mirrors.

> **NOTE** ▸ Ray diagrams are the basis for a *pictorial representation* that we'll use throughout this chapter. Be careful not to think that a ray diagram shows all of the rays. The rays shown on the diagram are just a subset of the infinitely many rays leaving the object. ◂

Apertures

A popular form of entertainment during ancient Roman times was a visit to a **camera obscura,** Latin for "dark room." As **FIGURE 23.5a** shows, a camera obscura was a darkened room with a single, small hole to the outside world. After their eyes became dark adapted, visitors could see a dim but full-color image of the outside world displayed on the back wall of the room. However, the image was upside down! The *pinhole camera* is a miniature version of the camera obscura.

A hole through which light passes is called an **aperture.** **FIGURE 23.5b** uses the ray model of light passing through a small aperture to explain how the camera obscura works. Each point on an object emits light rays in all directions, but only a very few of these rays pass through the aperture and reach the back wall. As the figure illustrates, the geometry of the rays causes the image to be upside down.

Actually, as you may have realized, each *point* on the object illuminates a small but extended *patch* on the wall. This is because the non-zero size of the aperture—needed for the image to be bright enough to see—allows several rays from each point on the object to pass through at slightly different angles. As a result, the image is slightly blurred and out of focus. (Diffraction also becomes an issue if the hole gets too small.) We'll later discover how a modern camera, with a lens, improves on the camera obscura.

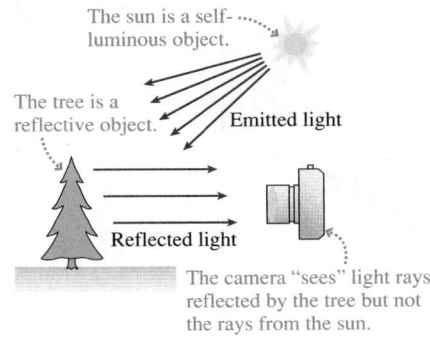

FIGURE 23.2 Self-luminous and reflective objects.

The sun is a self-luminous object.

The tree is a reflective object.

Emitted light

Reflected light

The camera "sees" light rays reflected by the tree but not the rays from the sun.

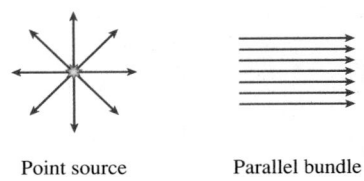

FIGURE 23.3 Point sources and parallel bundles represent idealized objects.

Point source   Parallel bundle

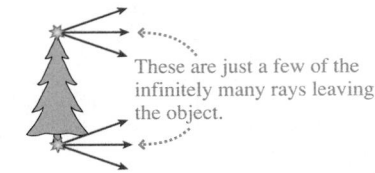

FIGURE 23.4 A ray diagram simplifies the situation by showing only a few rays.

These are just a few of the infinitely many rays leaving the object.

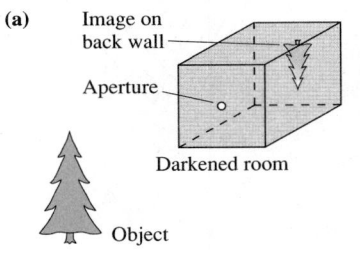

FIGURE 23.5 A camera obscura.

(a)   Image on back wall

Aperture

Darkened room

Object

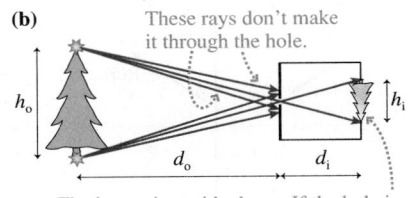

(b)   These rays don't make it through the hole.

h_o h_i

d_o d_i

The image is upside down. If the hole is sufficiently small, each point on the image corresponds to one point on the object.

FIGURE 23.6 Light through an aperture.

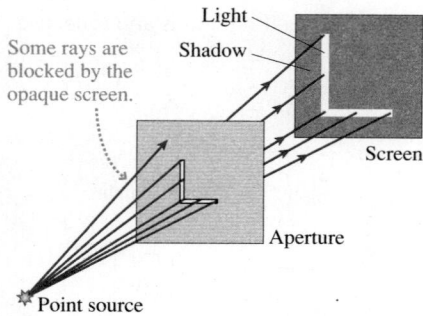

You can see from the similar triangles in Figure 23.5b that the object and image heights are related by

$$\frac{h_i}{h_o} = \frac{d_i}{d_o} \qquad (23.1)$$

where d_o is the distance to the object and d_i is the depth of the camera obscura. Any realistic camera obscura has $d_i < d_o$; thus the image is smaller than the object.

We can apply the ray model to more complex apertures, such as the L-shaped aperture in FIGURE 23.6. The pattern of light on the screen is found by tracing all the straight-line paths—the ray trajectories—that start from the point source and pass through the aperture. We will see an enlarged L on the screen, with a sharp boundary between the image and the dark shadow.

STOP TO THINK 23.1   A long, thin lightbulb illuminates a vertical aperture. Which pattern of light do you see on a viewing screen behind the aperture?

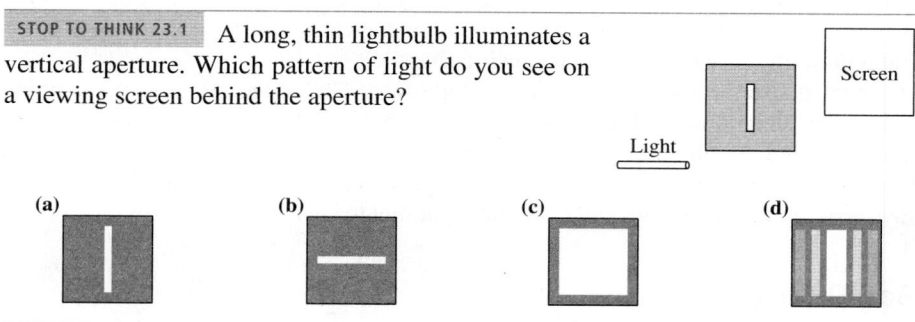

23.2 Reflection

Reflection of light is a familiar, everyday experience. You see your reflection in the bathroom mirror first thing every morning, reflections in your car's rearview mirror as you drive to school, and the sky reflected in puddles of standing water. Reflection from a flat, smooth surface, such as a mirror or a piece of polished metal, is called **specular reflection,** from *speculum,* the Latin word for "mirror."

FIGURE 23.7a shows a bundle of parallel light rays reflecting from a mirror-like surface. You can see that the incident and reflected rays are both in a plane that is normal, or perpendicular, to the reflective surface. A three-dimensional perspective accurately shows the relationship between the light rays and the surface, but figures such as this are hard to draw by hand. Instead, it is customary to represent reflection with the simpler pictorial representation of FIGURE 23.7b. In this figure,

FIGURE 23.7 Specular reflection of light.

(a) The incident and reflected rays lie in the plane of incidence, a plane perpendicular to the surface.

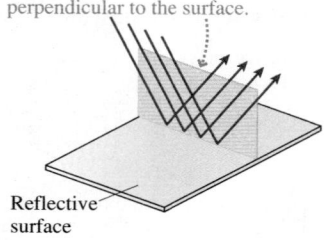

Reflective surface

(b)

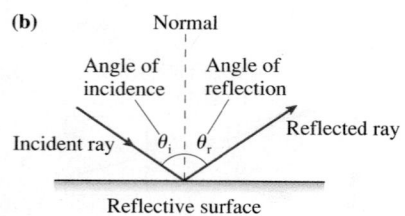

- The plane of the page is the *plane of incidence*, the plane containing both incident and reflected rays. The reflective surface extends into the page.
- A *single* light ray represents the entire bundle of parallel rays. This is oversimplified, but it keeps the figure and the analysis clear.

The angle θ_i between the ray and a line perpendicular to the surface—the *normal* to the surface—is called the **angle of incidence.** Similarly, the **angle of reflection** θ_r is the angle between the reflected ray and the normal to the surface. The **law of reflection,** easily demonstrated with simple experiments, states that

1. The incident ray and the reflected ray are in the same plane normal to the surface, and
2. The angle of reflection equals the angle of incidence: $\theta_r = \theta_i$.

NOTE ▶ Optics calculations *always* use the angle measured from the normal, not the angle between the ray and the surface. ◀

| EXAMPLE 23.1 | **Light reflecting from a mirror** |

A dressing mirror on a closet door is 1.50 m tall. The bottom is 0.50 m above the floor. A bare lightbulb hangs 1.00 m from the closet door, 2.50 m above the floor. How long is the streak of reflected light across the floor?

MODEL Treat the lightbulb as a point source and use the ray model of light.

FIGURE 23.8 Pictorial representation of the light rays reflecting from a mirror.

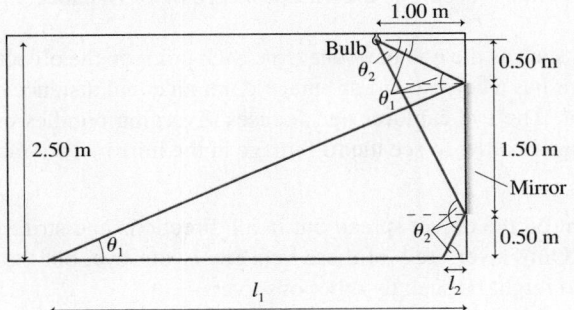

VISUALIZE FIGURE 23.8 is a pictorial representation of the light rays. We need to consider only the two rays that strike the edges of the mirror. All other reflected rays will fall between these two.

SOLVE Figure 23.8 has used the law of reflection to set the angles of reflection equal to the angles of incidence. Other angles have been identified with simple geometry. The two angles of incidence are

$$\theta_1 = \tan^{-1}\left(\frac{0.50 \text{ m}}{1.00 \text{ m}}\right) = 26.6°$$

$$\theta_2 = \tan^{-1}\left(\frac{2.00 \text{ m}}{1.00 \text{ m}}\right) = 63.4°$$

The distances to the points where the rays strike the floor are then

$$l_1 = \frac{2.00 \text{ m}}{\tan\theta_1} = 4.00 \text{ m}$$

$$l_2 = \frac{0.50 \text{ m}}{\tan\theta_2} = 0.25 \text{ m}$$

Thus the length of the light streak is $l_1 - l_2 = 3.75$ m.

Diffuse Reflection

Most objects are seen by virtue of their reflected light. For a "rough" surface, the law of reflection $\theta_r = \theta_i$ is obeyed at each point but the irregularities of the surface cause the reflected rays to leave in many random directions. This situation, shown in FIGURE 23.9, is called **diffuse reflection.** It is how you see this page, the wall, your hand, your friend, and so on.

By a "rough" surface, we mean a surface that is rough or irregular in comparison to the wavelength of light. Because visible-light wavelengths are $\approx 0.5 \ \mu$m, any surface with texture, scratches, or other irregularities larger than 1 μm will cause diffuse reflection rather than specular reflection. A piece of paper may feel quite smooth to your hand, but a microscope would show that the surface consists of distinct fibers much larger than 1 μm. By contrast, the irregularities on a mirror or a piece of polished metal are much smaller than 1 μm.

FIGURE 23.9 Diffuse reflection from an irregular surface.

Each ray obeys the law of reflection at that point, but the irregular surface causes the reflected rays to leave in many random directions.

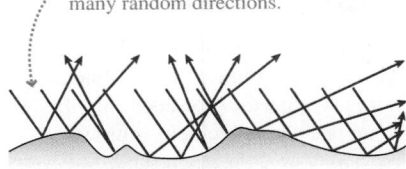

Magnified view of surface

The Plane Mirror

One of the most commonplace observations is that you can see yourself in a mirror. How? FIGURE 23.10a shows rays from point source P reflecting from a mirror. Consider the particular ray shown in FIGURE 23.10b. The reflected ray travels along a line that passes through point P′ on the "back side" of the mirror. Because $\theta_r = \theta_i$, simple geometry dictates that P′ is the same distance behind the mirror as P is in front of the mirror. That is, $s' = s$.

FIGURE 23.10 The light rays reflecting from a plane mirror.

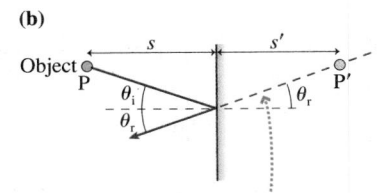

(a)

Rays from P reflect from the mirror. Each ray obeys the law of reflection.

(b)

This reflected ray appears to have been traveling along a line that passed through point P′.

(c)   Object distance Image distance

The reflected rays *all* diverge from P′, which appears to be the source of the reflected rays. Your eye collects the bundle of diverging rays and "sees" the light coming from P′.

The location of point P′ in Figure 23.10b is independent of the value of θ_i. Consequently, as FIGURE 23.10c shows, **the reflected rays all *appear* to be coming from point P′.** For a plane mirror, the distance s' to point P′ is equal to the object distance s:

$$s' = s \quad \text{(plane mirror)} \tag{23.2}$$

If rays diverge from an object point P and interact with a mirror so that the reflected rays diverge from point P′ and *appear* to come from P′, then we call P′ a **virtual image** of point P. The image is "virtual" in the sense that no rays actually leave P′, which is in darkness behind the mirror. But as far as your eye is concerned, the light rays act exactly *as if* the light really originated at P′. So while you may say "I see P in the mirror," what you are actually seeing is the virtual image of P. Distance s' is the *image distance*.

For an extended object, such as the one in FIGURE 23.11, each point on the object from which rays strike the mirror has a corresponding image point an equal distance on the opposite side of the mirror. The eye captures and focuses diverging bundles of rays from each point of the image in order to see the full image in the mirror. Two facts are worth noting:

1. Rays from each point on the object spread out in all directions and strike *every point* on the mirror. Only a very few of these rays enter your eye, but the other rays are very real and might be seen by other observers.
2. Rays from points P and Q enter your eye after reflecting from *different* areas of the mirror. This is why you can't always see the full image of an object in a very small mirror.

FIGURE 23.11 Each point on the extended object has a corresponding image point an equal distance on the opposite side of the mirror.

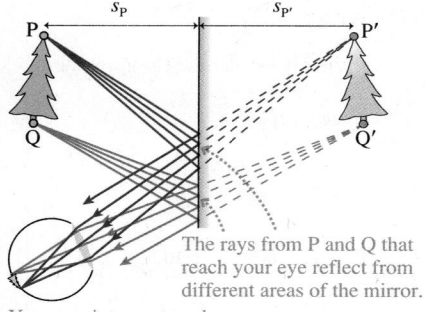

The rays from P and Q that reach your eye reflect from different areas of the mirror.

Your eye intercepts only a very small fraction of all the reflected rays.

EXAMPLE 23.2 **How high is the mirror?**

If your height is h, what is the shortest mirror on the wall in which you can see your full image? Where must the top of the mirror be hung?

MODEL Use the ray model of light.

VISUALIZE FIGURE 23.12 is a pictorial representation of the light rays. We need to consider only the two rays that leave your head and feet and reflect into your eye.

SOLVE Let the distance from your eyes to the top of your head be l_1 and the distance to your feet be l_2. Your height is $h = l_1 + l_2$. A light ray from the top of your head that reflects from the mirror at $\theta_r = \theta_i$ and enters your eye must, by congruent triangles, strike the mirror a distance $\frac{1}{2} l_1$ above your eyes. Similarly, a ray from your foot to your eye strikes the mirror a distance $\frac{1}{2} l_2$ below your eyes. The distance between these two points on the mirror is $\frac{1}{2} l_1 + \frac{1}{2} l_2 = \frac{1}{2} h$. A ray from anywhere else on your body will reach your eye if it strikes the mirror between these two points. Pieces of the mirror outside these two points are irrelevant, not because rays don't strike them but because the reflected rays don't reach your

FIGURE 23.12 Pictorial representation of light rays from your head and feet reflecting into your eye.

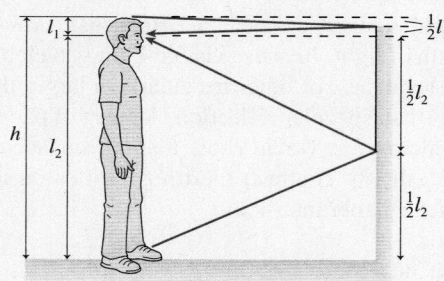

eye. Thus the shortest mirror in which you can see your full reflection is $\frac{1}{2} h$. But this will work only if the top of the mirror is hung midway between your eyes and the top of your head.

ASSESS It is interesting that the answer does not depend on how far you are from the mirror.

STOP TO THINK 23.2 Two plane mirrors form a right angle. How many images of the ball can you see in the mirrors?

a. 1
b. 2
c. 3
d. 4

Observer

23.3 Refraction

Two things happen when a light ray is incident on a smooth boundary between two transparent materials, such as the boundary between air and glass:

1. Part of the light *reflects* from the boundary, obeying the law of reflection. This is how you see reflections from pools of water or storefront windows, even though water and glass are transparent.

2. Part of the light continues into the second medium. It is *transmitted* rather than reflected, but the transmitted ray changes direction as it crosses the boundary. The transmission of light from one medium to another, but with a change in direction, is called **refraction.**

The photograph of FIGURE 23.13 shows the refraction of a light beam as it passes through a glass prism. Notice that the ray direction changes as the light enters and leaves the glass. Our goal in this section is to understand refraction, so we will usually ignore the weak reflection and focus on the transmitted light.

> NOTE ▶ A transparent material through which light travels is called the *medium.* This term has to be used with caution. The material does affect the light speed, but a transparent material differs from the medium of a sound or water wave in that particles of the medium do *not* oscillate as a light wave passes through. For a light wave it is the electromagnetic field that oscillates. ◀

FIGURE 23.14a shows the refraction of light rays in a parallel beam of light, such as a laser beam, and rays from a point source. It's good to remember that an infinite number of rays are incident on the boundary, but our analysis will be simplified if we focus on a single light ray. FIGURE 23.14b is a ray diagram showing the refraction of a single ray at a boundary between medium 1 and medium 2. Let the angle between the ray and the normal be θ_1 in medium 1 and θ_2 in medium 2. For the medium in which the ray is approaching the boundary, this is the *angle of incidence* as we've previously defined it. The angle on the transmitted side, *measured from the normal,* is called the **angle of refraction.** Notice that θ_1 is the angle of incidence in Figure 23.14b and the angle of refraction in FIGURE 23.14c, where the ray is traveling in the opposite direction, even though the value of θ_1 has not changed.

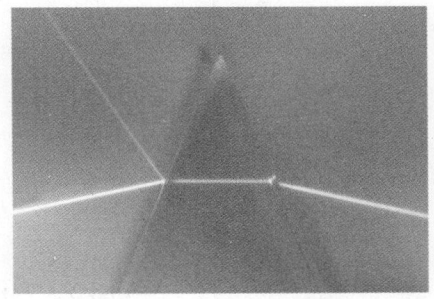

FIGURE 23.13 A light beam refracts twice in passing through a glass prism. You can see a weak reflection from the left surface of the prism.

FIGURE 23.14 Refraction of light rays.

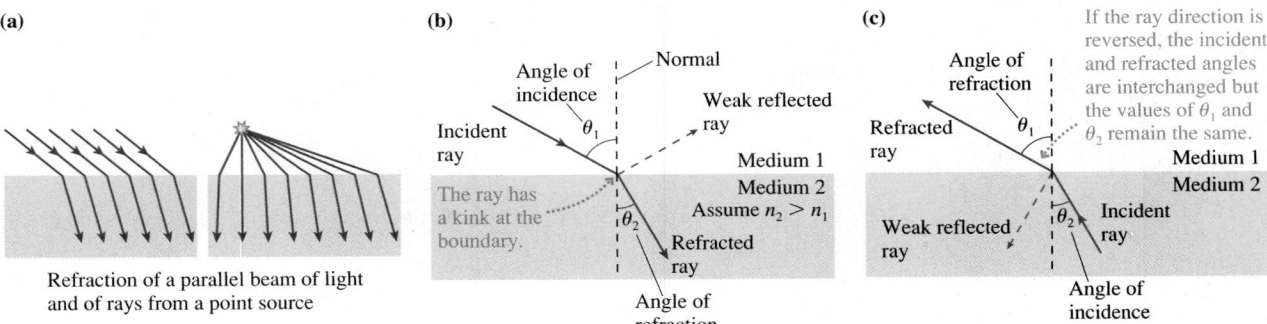

(a)

Refraction of a parallel beam of light and of rays from a point source

(b)

Angle of incidence · Normal
Incident ray θ_1 · Weak reflected ray
The ray has a kink at the boundary. · Medium 1
θ_2 · Medium 2 Assume $n_2 > n_1$
Refracted ray
Angle of refraction

(c)

Angle of refraction · If the ray direction is reversed, the incident and refracted angles are interchanged but the values of θ_1 and θ_2 remain the same.
Refracted ray θ_1
· Medium 1
Weak reflected ray · Medium 2
θ_2 Incident ray
Angle of incidence

Refraction was first studied experimentally by the Arab scientist Ibn Al-Haitham, in about the year 1000, and later by the Dutch scientist Willebrord Snell. **Snell's law** says that when a ray refracts between medium 1 and medium 2, having indices of refraction n_1 and n_2, the ray angles θ_1 and θ_2 in the two media are related by

$$n_1 \sin\theta_1 = n_2 \sin\theta_2 \qquad \text{(Snell's law of refraction)} \qquad (23.3)$$

Notice that Snell's law does not mention which is the incident angle and which the refracted angle.

The Index of Refraction

To Snell and his contemporaries, n was simply an "index of the refractive power" of a transparent substance. The relationship between the index of refraction and the speed of light was not recognized until the development of a wave theory of light in the 19th century. Theory predicts, and experiment confirms, that light travels through a transparent medium, such as glass or water, at a speed *less* than its speed c in vacuum. In Section 20.5, we defined the *index of refraction n* of a transparent medium as

$$n = \frac{c}{v_{medium}} \tag{23.4}$$

where v_{medium} is the light speed in the medium. This implies, of course, that $v_{medium} = c/n$. The index of refraction of a medium is always $n > 1$ except for vacuum, which has $n = 1$ exactly.

Table 23.1 shows measured values of n for several materials. There are many types of glass, each with a slightly different index of refraction, so we will keep things simple by accepting $n = 1.50$ as a typical value. Notice that cubic zirconia, used to make costume jewelry, has an index of refraction much higher than glass, although not equal to diamond.

We can accept Snell's law as simply an empirical discovery about light. Alternatively, and perhaps surprisingly, we can use the wave model of light to justify Snell's law. The key ideas we need are:

■ Wave fronts represent the crests of waves. They are spaced one wavelength apart.
■ The wavelength in a medium with index of refraction n is $\lambda = \lambda_{vac}/n$, where λ_{vac} is the vacuum wavelength.
■ Wave fronts are perpendicular to the wave's direction of travel. Consequently, wave fronts are perpendicular to rays.
■ The wave fronts have to stay lined up as a wave crosses from one medium into another.

FIGURE 23.15 shows what happens as a wave crosses the boundary between two media, where we're assuming $n_2 > n_1$. **Because the wavelengths differ on opposite sides of the boundary, the wave fronts can stay lined up only if the waves in the two media are traveling in different directions.** In other words, the wave must refract at the boundary to keep the crests of the wave aligned.

To analyze Figure 23.15, consider the segment of boundary of length l between the two dots. This segment is the common hypotenuse of two right triangles. From the upper triangle, which has one side of length λ_1, we see

$$l = \frac{\lambda_1}{\sin\theta_1} \tag{23.5}$$

where θ_1 is the angle of incidence. Similarly, the lower triangle, where θ_2 is the angle of refraction, gives

$$l = \frac{\lambda_2}{\sin\theta_2} \tag{23.6}$$

Equating these two expressions for l, and using $\lambda_1 = \lambda_{vac}/n_1$ and $\lambda_2 = \lambda_{vac}/n_2$, we find

$$\frac{\lambda_{vac}}{n_1\sin\theta_1} = \frac{\lambda_{vac}}{n_2\sin\theta_2} \tag{23.7}$$

Equation 23.7 can be true only if

$$n_1\sin\theta_1 = n_2\sin\theta_2 \tag{23.8}$$

which is Snell's law.

TABLE 23.1 Indices of refraction

Medium	n
Vacuum	1.00 exactly
Air (actual)	1.0003
Air (accepted)	1.00
Water	1.33
Ethyl alcohol	1.36
Oil	1.46
Glass (typical)	1.50
Polystyrene plastic	1.59
Cubic zirconia	2.18
Diamond	2.41
Silicon (infrared)	3.50

FIGURE 23.15 Snell's law is a consequence of the wave model of light.

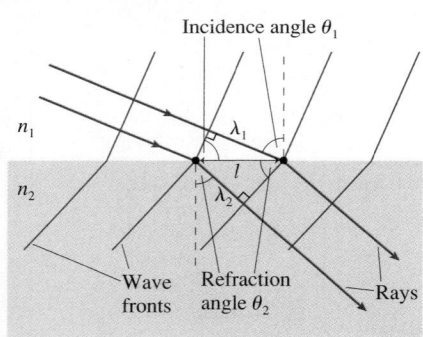

Examples of Refraction

Look back at Figure 23.14. As the ray in Figure 23.14b moves from medium 1 to medium 2, where $n_2 > n_1$, it bends closer to the normal. In Figure 23.14c, where the ray moves from medium 2 to medium 1, it bends away from the normal. This is a general conclusion that follows from Snell's law:

- When a ray is transmitted into a material with a higher index of refraction, it bends toward the normal.
- When a ray is transmitted into a material with a lower index of refraction, it bends away from the normal.

This rule becomes a central idea in a procedure for analyzing refraction problems.

TACTICS
BOX 23.1 **Analyzing refraction**

❶ **Draw a ray diagram.** Represent the light beam with one ray.
❷ **Draw a line normal to the boundary.** Do this at each point where the ray intersects a boundary.
❸ **Show the ray bending in the correct direction.** The angle is larger on the side with the smaller index of refraction. This is the qualitative application of Snell's law.
❹ **Label angles of incidence and refraction.** Measure all angles from the normal.
❺ **Use Snell's law.** Calculate the unknown angle or unknown index of refraction.

Exercises 11–15

EXAMPLE 23.3 **Deflecting a laser beam**

A laser beam is aimed at a 1.0-cm-thick sheet of glass at an angle 30° above the glass.

a. What is the laser beam's direction of travel in the glass?
b. What is its direction in the air on the other side?
c. By what distance is the laser beam displaced?

MODEL Represent the laser beam with a single ray and use the ray model of light.

VISUALIZE **FIGURE 23.16** is a pictorial representation in which the first four steps of Tactics Box 23.1 are identified. Notice that the angle of incidence is $\theta_1 = 60°$, not the 30° value given in the problem.

FIGURE 23.16 The ray diagram of a laser beam passing through a sheet of glass.

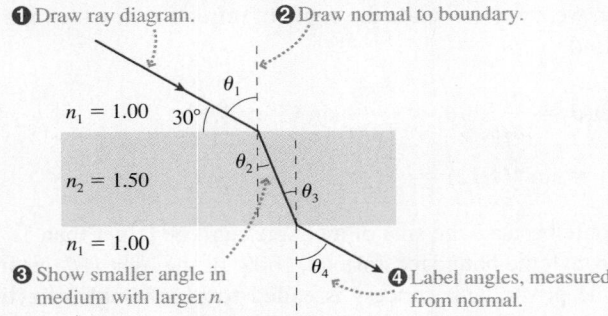

❶ Draw ray diagram. ❷ Draw normal to boundary.
$n_1 = 1.00$ 30° θ_1
$n_2 = 1.50$ θ_2 θ_3
$n_1 = 1.00$
❸ Show smaller angle in medium with larger n. θ_4 ❹ Label angles, measured from normal.

SOLVE a. Snell's law, the final step in the Tactics Box, is $n_1 \sin\theta_1 = n_2 \sin\theta_2$. Using $\theta_1 = 60°$, we find that the direction of travel in the glass is

$$\theta_2 = \sin^{-1}\left(\frac{n_1 \sin\theta_1}{n_2}\right) = \sin^{-1}\left(\frac{\sin 60°}{1.5}\right)$$

$$= \sin^{-1}(0.577) = 35.3°$$

b. Snell's law at the second boundary is $n_2 \sin\theta_3 = n_1 \sin\theta_4$. You can see from Figure 23.16 that the interior angles are equal: $\theta_3 = \theta_2 = 35.3°$. Thus the ray emerges back into the air traveling at angle

$$\theta_4 = \sin^{-1}\left(\frac{n_2 \sin\theta_3}{n_1}\right) = \sin^{-1}(1.5 \sin 35.3°)$$

$$= \sin^{-1}(0.867) = 60°$$

This is the same as θ_1, the original angle of incidence. The glass doesn't change the direction of the laser beam.

c. Although the exiting laser beam is parallel to the initial laser beam, it has been displaced sideways by distance d. **FIGURE 23.17** on the next page shows the geometry for finding d. From trigonometry, $d = l \sin\phi$. Further, $\phi = \theta_1 - \theta_2$ and $l = t/\cos\theta_2$, where t is the thickness of the glass. Combining these gives

Continued

$$d = l \sin \phi = \frac{t}{\cos \theta_2} \sin(\theta_1 - \theta_2)$$

$$= \frac{(1.0 \text{ cm}) \sin 24.7°}{\cos 35.3°} = 0.51 \text{ cm}$$

The glass causes the laser beam to be displaced sideways by 0.51 cm.

ASSESS The laser beam exits the glass still traveling in the same direction as it entered. This is a general result for light traveling through a medium with parallel sides. Notice that the displacement d becomes zero in the limit $t \to 0$. This will be an important observation when we get to lenses.

FIGURE 23.17 The laser beam is deflected sideways by distance d.

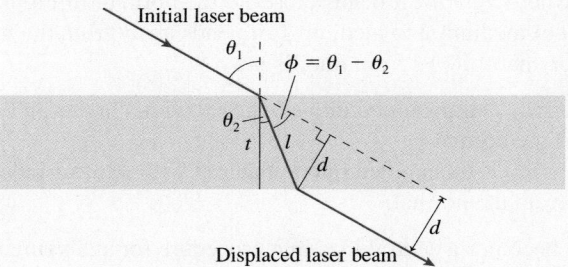

Initial laser beam

Displaced laser beam

EXAMPLE 23.4 **Measuring the index of refraction**

FIGURE 23.18 shows a laser beam deflected by a 30°-60°-90° prism. What is the prism's index of refraction?

FIGURE 23.18 A prism deflects a laser beam.

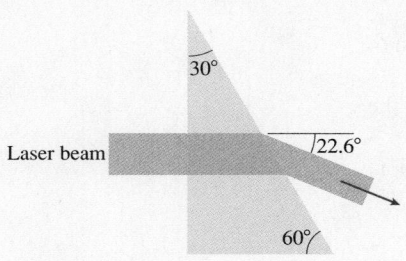

MODEL Represent the laser beam with a single ray and use the ray model of light.

VISUALIZE **FIGURE 23.19** uses the steps of Tactics Box 23.1 to draw a ray diagram. The ray is incident perpendicular to the front face of the prism ($\theta_{\text{incident}} = 0°$), thus it is transmitted through the first boundary without deflection. At the second boundary it is especially important to *draw the normal to the surface* at the point of incidence and to *measure angles from the normal*.

SOLVE From the geometry of the triangle you can find that the laser's angle of incidence on the hypotenuse of the prism is

FIGURE 23.19 Pictorial representation of a laser beam passing through the prism.

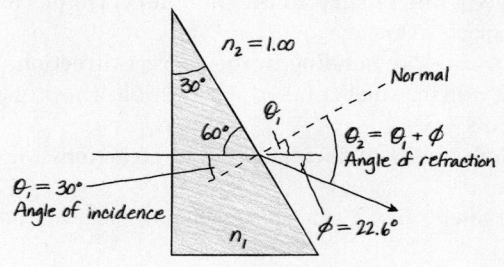

θ_1 and θ_2 are measured from the normal.

$\theta_1 = 30°$, the same as the apex angle of the prism. The ray exits the prism at angle θ_2 such that the deflection is $\phi = \theta_2 - \theta_1 = 22.6°$. Thus $\theta_2 = 52.6°$. Knowing both angles and $n_2 = 1.00$ for air, we can use Snell's law to find n_1:

$$n_1 = \frac{n_2 \sin \theta_2}{\sin \theta_1} = \frac{1.00 \sin 52.6°}{\sin 30°} = 1.59$$

ASSESS Referring to the indices of refraction in Table 23.1, we see that the prism is made of plastic.

Total Internal Reflection

FIGURE 23.20 The blue laser beam undergoes total internal reflection inside the prism.

What would have happened in Example 23.4 if the prism angle had been 45° rather than 30°? The light rays would approach the rear surface of the prism at an angle of incidence $\theta_1 = 45°$. When we try to calculate the angle of refraction at which the ray emerges into the air, we find

$$\sin \theta_2 = \frac{n_1}{n_2} \sin \theta_1 = \frac{1.59}{1.00} \sin 45° = 1.12$$

$$\theta_2 = \sin^{-1}(1.12) = \text{???}$$

Angle θ_2 doesn't compute because the sine of an angle can't be larger than 1. The ray is unable to refract through the boundary. Instead, 100% of the light *reflects* from the boundary back into the prism. This process is called **total internal reflection,** often abbreviated TIR. That it really happens is illustrated in **FIGURE 23.20.** Here three laser beams enter a prism from the left. The bottom two refract out through the right side of the prism. The blue beam, which is incident on the prism's top face, undergoes total internal reflection and then emerges through the right surface.

FIGURE 23.21 shows several rays leaving a point source in a medium with index of refraction n_1. The medium on the other side of the boundary has $n_2 < n_1$. As we've seen, crossing a boundary into a material with a lower index of refraction causes the ray to bend away from the normal. Two things happen as angle θ_1 increases. First, the refraction angle θ_2 approaches 90°. Second, the fraction of the light energy transmitted decreases while the fraction reflected increases.

A **critical angle** is reached when $\theta_2 = 90°$. Because $\sin 90° = 1$, Snell's law $n_1 \sin \theta_c = n_2 \sin 90°$ gives the critical angle of incidence as

$$\theta_c = \sin^{-1}\left(\frac{n_2}{n_1}\right) \qquad (23.9)$$

The refracted light vanishes at the critical angle and the reflection becomes 100% for any angle $\theta_1 \geq \theta_c$. The critical angle is well defined because of our assumption that $n_2 < n_1$. **There is no critical angle and no total internal reflection if $n_2 > n_1$.**

As a quick example, the critical angle in a typical piece of glass at the glass-air boundary is

$$\theta_{c\,glass} = \sin^{-1}\left(\frac{1.00}{1.50}\right) = 42°$$

The fact that the critical angle is less than 45° has important applications. For example, FIGURE 23.22 shows a pair of binoculars. The lenses are much farther apart than your eyes, so the light rays need to be brought together before exiting the eyepieces. Rather than using mirrors, which get dirty and require alignment, binoculars use a pair of prisms on each side. Thus the light undergoes two total internal reflections and emerges from the eyepiece. (The actual arrangement is a little more complex than in Figure 23.22, to avoid left-right reversals, but this illustrates the basic idea.)

FIGURE 23.21 Refraction and reflection of rays as the angle of incidence increases.

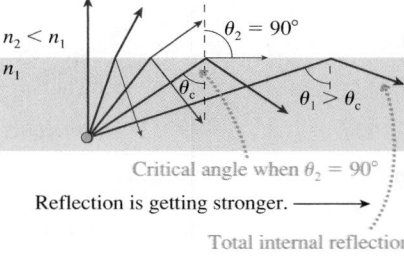

Critical angle when $\theta_2 = 90°$

Total internal reflection occurs when $\theta_1 \geq \theta_c$.

FIGURE 23.22 Binoculars and other optical instruments make use of total internal reflection.

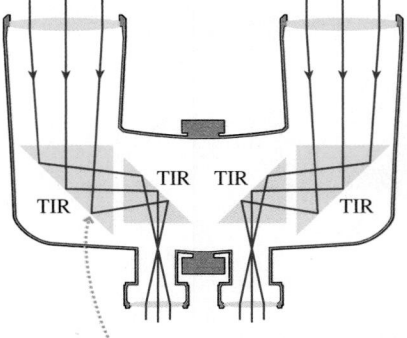

Angles of incidence exceed the critical angle.

EXAMPLE 23.5 | **Total internal reflection**

A lightbulb is set in the bottom of a 3.0-m-deep swimming pool. What is the diameter of the circle of light seen on the water's surface from above?

MODEL Represent the lightbulb as a point source and use the ray model of light.

VISUALIZE FIGURE 23.23 is a pictorial representation of the light rays. The lightbulb emits rays at all angles, but only some of the rays refract into the air where they can be seen from above. Rays striking the surface at greater than the critical angle undergo TIR and remain within the water. The diameter of the circle of light is the distance between the two points at which rays strike the surface at the critical angle.

SOLVE From trigonometry, the circle diameter is $D = 2h \tan\theta_c$, where h is the depth of the water. The critical angle for a water-air boundary is $\theta_c = \sin^{-1}(1.00/1.33) = 48.7°$. Thus

$$D = 2(3.0 \text{ m}) \tan 48.7° = 6.8 \text{ m}$$

FIGURE 23.23 Pictorial representation of the rays leaving a lightbulb at the bottom of a swimming pool.

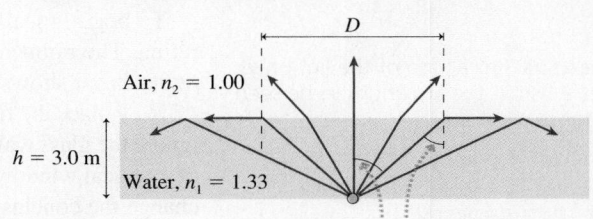

Rays at the critical angle θ_c form the edge of the circle of light seen from above.

Fiber Optics

The most important modern application of total internal reflection is the transmission of light through optical fibers. FIGURE 23.24a on the next page shows a laser beam shining into the end of a long, narrow-diameter glass tube. The light rays pass easily from the air into the glass, but they then impinge on the inside wall of the glass tube at an angle

FIGURE 23.24 Light rays are confined within an optical fiber by total internal reflection.

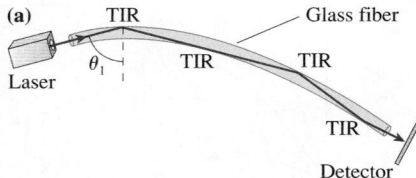

(a)

Laser
TIR
θ_1
TIR
TIR
Glass fiber
TIR
Detector

(b)

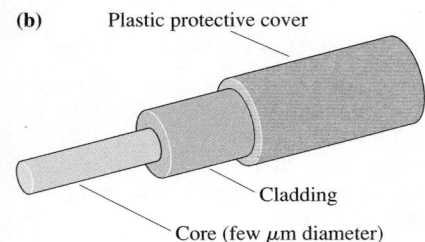

Plastic protective cover
Cladding
Core (few μm diameter)

of incidence θ_1 approaching 90°. This is well above the critical angle, so the laser beam undergoes TIR and remains inside the glass. The laser beam continues to "bounce" its way down the tube as if the light were inside a pipe. Indeed, optical fibers are sometimes called "light pipes." The rays are *below* the critical angle ($\theta_1 \approx 0$) when they finally reach the end of the fiber, thus they refract out without difficulty and can be detected.

While a simple glass tube can transmit light, reliance on a glass-air boundary is not sufficiently reliable for commercial use. Any small scratch on the side of the tube alters the rays' angle of incidence and allows leakage of light. **FIGURE 23.24b** shows the construction of a practical optical fiber. A small-diameter glass *core* is surrounded by a layer of glass *cladding*. The glasses used for the core and the cladding have $n_{core} > n_{cladding}$; thus light undergoes TIR at the core-cladding boundary and remains confined within the core. This boundary is not exposed to the environment and hence retains its integrity even under adverse conditions.

Even glass of the highest purity is not perfectly transparent. Absorption in the glass, even if very small, causes a gradual decrease in light intensity. The glass used for the core of optical fibers has a minimum absorption at a wavelength of 1.3 μm, in the infrared, so this is the laser wavelength used for long-distance signal transmission. Light at this wavelength can travel hundreds of kilometers through a fiber without significant loss.

STOP TO THINK 23.3 A light ray travels from medium 1 to medium 3 as shown. For these media,

a. $n_3 > n_1$ b. $n_3 = n_1$ c. $n_3 < n_1$
d. We can't compare n_1 to n_3 without knowing n_2.

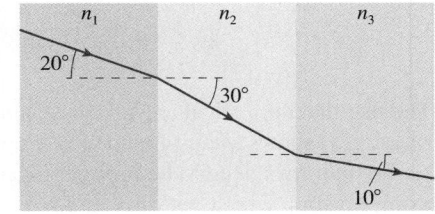

n_1 n_2 n_3
20° 30° 10°

23.4 Image Formation by Refraction

If you see a fish that appears to be swimming close to the front window of the aquarium, but then look through the side of the aquarium, you'll find that the fish is actually farther from the window than you thought. Why is this?

To begin, recall that vision works by focusing a diverging bundle of rays onto the retina. The point from which the rays diverge is where you perceive the object to be. **FIGURE 23.25a** shows how you would see a fish out of water at distance d.

Now place the fish back into the aquarium at the same distance d. For simplicity, we'll ignore the glass wall of the aquarium and consider the water-air boundary. (The thin glass of a typical window has only a very small effect on the refraction of the rays and doesn't change the conclusions.) Light rays again leave the fish, but this time they refract at the water-air boundary. Because they're going from a higher to a lower index of refraction, the rays refract *away from* the normal. **FIGURE 23.25b** shows the consequences.

A bundle of diverging rays still enters your eye, but now these rays are diverging from a closer point, at distance d'. As far as your eye and brain are concerned, it's exactly *as if* the rays really originate at distance d', and this is the location at which you "see" the fish. **The object appears closer than it really is because of the refraction of light at the boundary.**

We found that the rays reflected from a mirror diverge from a point that is not the object point. We called that point a *virtual image*. Similarly, if rays from an object point P refract at a boundary between two media such that the rays then diverge from a point P′ and *appear* to come from P′, we call P′ a virtual image of point P. The virtual image of the fish is what you see.

Let's examine this image formation a bit more carefully. **FIGURE 23.26** shows a boundary between two transparent media having indices of refraction n_1 and n_2. Point P, a source of light rays, is the object. Point P′, from which the rays *appear* to diverge, is

FIGURE 23.25 Refraction of the light rays causes a fish in the aquarium to be seen at distance d'.

(a) A fish out of water

The rays that reach the eye are diverging from this point, the object.

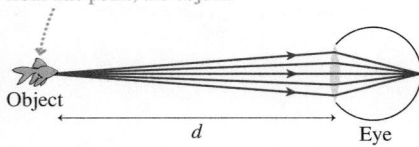

Object
d
Eye

(b) A fish in the aquarium

Refraction causes the rays to bend at the boundary.

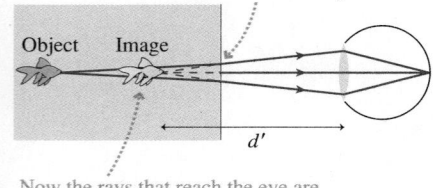

Object Image
d'

Now the rays that reach the eye are diverging from this point, the image.

the virtual image of P. Distance s is called the **object distance.** Our goal is to determine distance s', the **image distance. Both are measured from the boundary.**

A line perpendicular to the boundary is called the **optical axis.** Consider a ray leaving the object at angle θ_1 with respect to the optical axis. θ_1 is also the angle of incidence at the boundary, where the ray refracts into the second medium at angle θ_2. By tracing the refracted ray backward, you can see that θ_2 is also the angle between the refracted ray and the optical axis at point P′.

The distance l is common to both the incident and the refracted rays, and you can see that $l = s \tan \theta_1 = s' \tan \theta_2$. Thus

$$s' = \frac{\tan \theta_1}{\tan \theta_2} s \tag{23.10}$$

Snell's law relates the sines of angles θ_1 and θ_2; that is,

$$\frac{\sin \theta_1}{\sin \theta_2} = \frac{n_2}{n_1} \tag{23.11}$$

In practice, the angle between any of these rays and the optical axis is very small because the size of the pupil of your eye is very much less than the distance between the object and your eye. (The angles in the figure have been greatly exaggerated.) Rays that are nearly *parallel* to the *axis* are called **paraxial rays.** The small-angle approximation $\sin \theta \approx \tan \theta \approx \theta$, where θ is in radians, can be applied to paraxial rays. Consequently,

$$\frac{\tan \theta_1}{\tan \theta_2} \approx \frac{\sin \theta_1}{\sin \theta_2} = \frac{n_2}{n_1} \tag{23.12}$$

Using this result in Equation 23.10, we find that the image distance is

$$s' = \frac{n_2}{n_1} s \tag{23.13}$$

NOTE ▶ The fact that the result for s' is independent of θ_1 implies that *all* paraxial rays appear to diverge from the same point P′. This property of the diverging rays is essential in order to have a well-defined image. ◀

This section has given us a first look at image formation via refraction. We will extend this idea to image formation with lenses in Section 23.6.

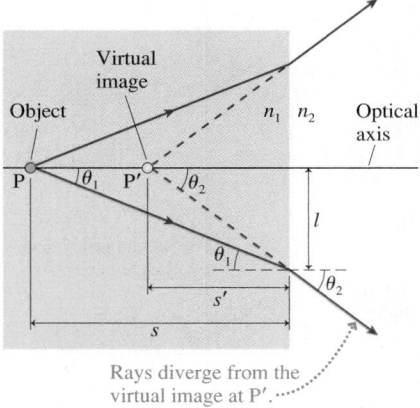

FIGURE 23.26 Finding the virtual image P′ of an object at P. We've assumed $n_1 > n_2$.

Rays diverge from the virtual image at P′.

EXAMPLE 23.6 **An air bubble in a window**

A fish and a sailor look at each other through a 5.0-cm-thick glass porthole in a submarine. There happens to be an air bubble right in the center of the glass. How far behind the surface of the glass does the air bubble appear to the fish? To the sailor?

MODEL Represent the air bubble as a point source and use the ray model of light.

VISUALIZE Paraxial light rays from the bubble refract into the air on one side and into the water on the other. The ray diagram looks like Figure 23.26.

SOLVE The index of refraction of the glass is $n_1 = 1.50$. The bubble is in the center of the window, so the object distance from

either side of the window is $s = 2.5$ cm. From the water side, the fish sees the bubble at an image distance

$$s' = \frac{n_2}{n_1} s = \frac{1.33}{1.50}(2.5 \text{ cm}) = 2.2 \text{ cm}$$

This is the apparent depth of the bubble. The sailor, in air, sees the bubble at an image distance

$$s' = \frac{n_2}{n_1} s = \frac{1.00}{1.50}(2.5 \text{ cm}) = 1.7 \text{ cm}$$

ASSESS The image distance is *less* for the sailor because of the *larger* difference between the two indices of refraction.

23.5 Color and Dispersion

One of the most obvious visual aspects of light is the phenomenon of color. Yet color, for all its vivid sensation, is not inherent in the light itself. Color is a *perception,* not a physical quantity. Color is associated with the wavelength of light, but the fact that we see light with a wavelength of 650 nm as "red" tells us how our visual system responds

FIGURE 23.27 Newton used prisms to study color.

(a)

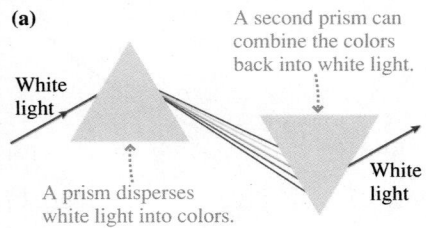

White light

A prism disperses white light into colors.

A second prism can combine the colors back into white light.

White light

(b)

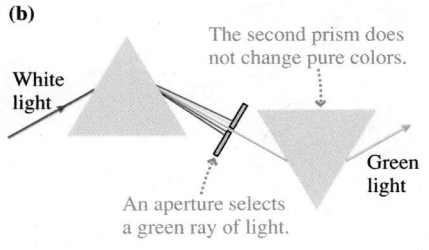

White light

The second prism does not change pure colors.

An aperture selects a green ray of light.

Green light

TABLE 23.2 A brief summary of the visible spectrum of light

Color	Approximate wavelength
Deepest red	700 nm
Red	650 nm
Green	550 nm
Blue	450 nm
Deepest violet	400 nm

FIGURE 23.28 Dispersion curves show how the index of refraction varies with wavelength.

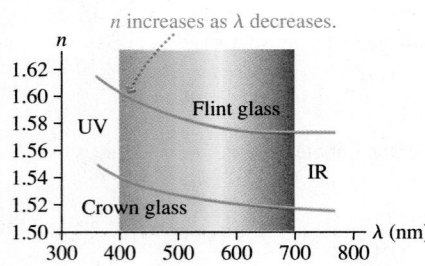

n increases as λ decreases.

to electromagnetic waves of this wavelength. There is no "redness" associated with the light wave itself.

Most of the results of optics do not depend on color. We generally don't need to know the color of light—or, to be more precise, its wavelength—to use the laws of reflection and refraction. Nonetheless, color is an interesting subject, one worthy of a short digression.

Color

It has been known since antiquity that irregularly shaped glass and crystals cause sunlight to be broken into various colors. A common idea was that the glass or crystal somehow altered the properties of the light by *adding* color to the light. Newton suggested a different explanation. He first passed a sunbeam through a prism, producing the familiar rainbow of light. We say that the prism *disperses* the light. Newton's novel idea, shown in FIGURE 23.27a, was to use a second prism, inverted with respect to the first, to "reassemble" the colors. He found that the light emerging from the second prism was a beam of pure, white light.

But the emerging light beam is white only if *all* the rays are allowed to move between the two prisms. Blocking some of the rays with small obstacles, as in FIGURE 23.27b, causes the emerging light beam to have color. This suggests that color is associated with the light itself, not with anything that the prism is "doing" to the light. Newton tested this idea by inserting a small aperture between the prisms to pass only the rays of a particular color, such as green. If the prism alters the properties of light, then the second prism should change the green light to other colors. Instead, the light emerging from the second prism is unchanged from the green light entering the prism.

These and similar experiments show that

1. What we perceive as white light is a mixture of all colors. White light can be dispersed into its various colors and, equally important, mixing all the colors produces white light.
2. The index of refraction of a transparent material differs slightly for different colors of light. Glass has a slightly larger index of refraction for violet light than for green light or red light. Consequently, different colors of light refract at slightly different angles. A prism does not alter the light or add anything to the light; it simply causes the different colors that are inherent in white light to follow slightly different trajectories.

Dispersion

It was Thomas Young, with his two-slit interference experiment, who showed that different colors are associated with light of different wavelengths. The longest wavelengths are perceived as red light and the shortest as violet light. Table 23.2 is a brief summary of the *visible spectrum* of light. Visible-light wavelengths are used so frequently that it is well worth committing this short table to memory.

The slight variation of index of refraction with wavelength is known as **dispersion**. FIGURE 23.28 shows the *dispersion curves* of two common glasses. Notice that **n is larger when the wavelength is shorter,** thus violet light refracts more than red light.

EXAMPLE 23.7 **Dispersing light with a prism**

Example 23.4 found that a ray incident on a 30° prism is deflected by 22.6° if the prism's index of refraction is 1.59. Suppose this is the index of refraction of deep violet light and deep red light has an index of refraction of 1.54.

a. What is the deflection angle for deep red light?
b. If a beam of white light is dispersed by this prism, how wide is the rainbow spectrum on a screen 2.0 m away?

VISUALIZE Figure 23.19 showed the geometry. A ray of any wavelength is incident on the hypotenuse of the prism at $\theta_1 = 30°$.

SOLVE a. If $n_1 = 1.54$ for deep red light, the refraction angle is

$$\theta_2 = \sin^{-1}\left(\frac{n_1 \sin\theta_1}{n_2}\right) = \sin^{-1}\left(\frac{1.54 \sin 30°}{1.00}\right) = 50.4°$$

Example 23.4 showed that the deflection angle is $\phi = \theta_2 - \theta_1$, so deep red light is deflected by $\phi_{red} = 20.4°$. This angle is slightly smaller than the previously observed $\phi_{violet} = 22.6°$.

b. The entire spectrum is spread between $\phi_{red} = 20.4°$ and $\phi_{violet} = 22.6°$. The angular spread is

$$\delta = \phi_{violet} - \phi_{red} = 2.2° = 0.038 \text{ rad}$$

At distance r, the spectrum spans an arc length

$$s = r\delta = (2.0 \text{ m})(0.038 \text{ rad}) = 0.076 \text{ m} = 7.6 \text{ cm}$$

ASSESS The angle is so small that there's no appreciable difference between arc length and a straight line. The spectrum will be 7.6 cm wide at a distance of 2.0 m.

Rainbows

One of the most interesting sources of color in nature is the rainbow. The details get somewhat complicated, but FIGURE 23.29a shows that the basic cause of the rainbow is a combination of refraction, reflection, and dispersion.

Figure 23.29a might lead you to think that the top edge of a rainbow is violet. In fact, the top edge is red, and violet is on the bottom. The rays leaving the drop in Figure 23.29a are spreading apart, so they can't all reach your eye. As FIGURE 23.29b shows, a ray of red light reaching your eye comes from a drop *higher* in the sky than a ray of violet light. In other words, the colors you see in a rainbow refract toward your eye from different raindrops, not from the same drop. You have to look higher in the sky to see the red light than to see the violet light.

FIGURE 23.29 Light seen in a rainbow has undergone refraction + reflection + refraction in a raindrop.

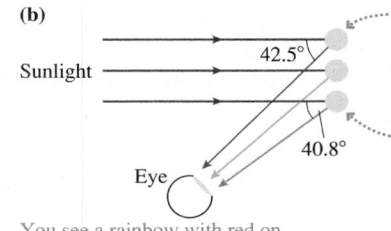

(a)
2. Dispersion causes different colors to refract at different angles.
Sunlight
1. The sun is behind your back when you see a rainbow.
3. Most of the light refracts into the air at this point, but a little reflects back into the drop.
4. Dispersion separates the colors even more as the rays refract back into the air.

(b)
Sunlight
42.5°
40.8°
Eye
You see a rainbow with red on the top, violet on the bottom.

Red light is refracted predominantly at 42.5°. The red light reaching your eye comes from drops higher in the sky.

Violet light is refracted predominantly at 40.8°. The violet light reaching your eye comes from drops lower in the sky.

Colored Filters and Colored Objects

White light passing through a piece of green glass emerges as green light. A possible explanation would be that the green glass *adds* "greenness" to the white light, but Newton found otherwise. Green glass is green because it *absorbs* any light that is "not green." We can think of a piece of colored glass or plastic as a *filter* that removes all wavelengths except a chosen few.

If a green filter and a red filter are overlapped, as in FIGURE 23.30, *no* light gets through. The green filter transmits only green light, which is then absorbed by the red filter because it is "not red."

This behavior is true not just for glass filters, which transmit light, but for *pigments* that absorb light of some wavelengths but *reflect* light at other wavelengths. For example, red paint contains pigments reflecting light at wavelengths near 650 nm while absorbing all other wavelengths. Pigments in paints, inks, and natural objects are responsible for most of the color we observe in the world, from the red of lipstick to the blue of a bluebird's feathers.

As an example, FIGURE 23.31 on the next page shows the absorption curve of *chlorophyll*. Chlorophyll is essential for photosynthesis in green plants. The chemical reactions of photosynthesis are able to use red light and blue/violet light, thus chlorophyll absorbs red light and blue/violet light from sunlight and puts it to use. But

FIGURE 23.30 No light at all passes through both a green and a red filter.

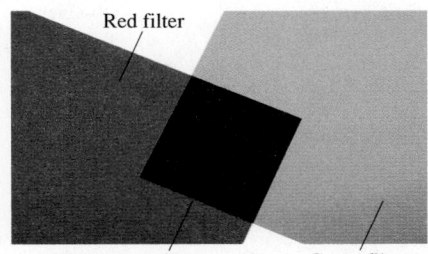

Red filter
Black where filters overlap
Green filter

FIGURE 23.31 The absorption curve of chlorophyll.

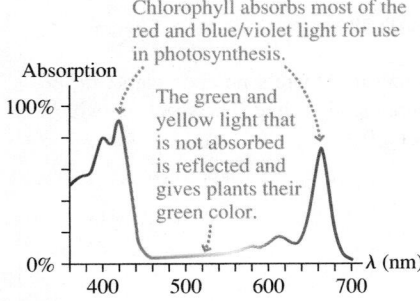

Chlorophyll absorbs most of the red and blue/violet light for use in photosynthesis.

The green and yellow light that is not absorbed is reflected and gives plants their green color.

Sunsets are red because all the blue light has scattered as the sunlight passes through the atmosphere.

FIGURE 23.32 Rayleigh scattering by molecules in the air gives the sky and sunsets their color.

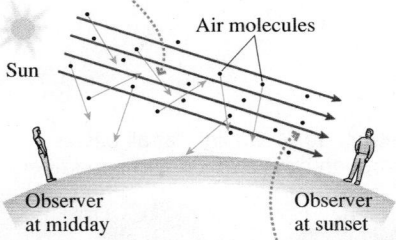

At midday the scattered light is mostly blue because molecules preferentially scatter shorter wavelengths.

Air molecules

Sun

Observer at midday

Observer at sunset

At sunset, when the light has traveled much farther through the atmosphere, the light is mostly red because the shorter wavelengths have been lost to scattering.

green and yellow light are not absorbed. Instead, to conserve energy, these wavelengths are mostly *reflected* to give the object a greenish-yellow color. When you look at the green leaves on a tree, you're seeing the light that was reflected because it *wasn't* needed for photosynthesis.

Light Scattering: Blue Skies and Red Sunsets

In the ray model of Section 23.1 we noted that light within a medium can be scattered or absorbed. As we've now seen, the absorption of light can be wavelength dependent and can create color in objects. What are the effects of scattering?

Light can scatter from small particles that are suspended in a medium. If the particles are large compared to the wavelengths of light—even though they may be microscopic and not readily visible to the naked eye—the light essentially reflects off the particles. The law of reflection doesn't depend on wavelength, so all colors are scattered equally. White light scattered from many small particles makes the medium appear cloudy and white. Two well-known examples are clouds, where micrometer-size water droplets scatter the light, and milk, which is a colloidal suspension of microscopic droplets of fats and proteins.

A more interesting aspect of scattering occurs at the atomic level. The atoms and molecules of a transparent medium are much smaller than the wavelengths of light, so they can't scatter light simply by reflection. Instead, the oscillating electric field of the light wave interacts with the electrons in each atom in such a way that the light is scattered. This atomic-level scattering is called **Rayleigh scattering.**

Unlike the scattering by small particles, Rayleigh scattering from atoms and molecules *does* depend on the wavelength. A detailed analysis shows that the intensity of scattered light depends inversely on the fourth power of the wavelength: $I_{scattered} \propto \lambda^{-4}$. This wavelength dependence explains why the sky is blue and sunsets are red.

As sunlight travels through the atmosphere, the λ^{-4} dependence of Rayleigh scattering causes the shorter wavelengths to be preferentially scattered. If we take 650 nm as a typical wavelength for red light and 450 nm for blue light, the intensity of scattered blue light relative to scattered red light is

$$\frac{I_{blue}}{I_{red}} = \left(\frac{650}{450}\right)^4 \approx 4$$

Four times more blue light is scattered toward us than red light and thus, as FIGURE 23.32 shows, the sky appears blue.

Because of the earth's curvature, sunlight has to travel much farther through the atmosphere when we see it at sunrise or sunset than it does during the midday hours. In fact, the path length through the atmosphere at sunset is so long that essentially all the short wavelengths have been lost due to Rayleigh scattering. Only the longer wavelengths remain—orange and red—and they make the colors of the sunset.

23.6 Thin Lenses: Ray Tracing

A camera obscura or a pinhole camera forms images on a screen, but the images are faint and not perfectly focused. The ability to create a bright, well-focused image is vastly improved by using a lens. A **lens** is a transparent material that uses refraction at *curved* surfaces to form an image from diverging light rays. We will defer a mathematical analysis of the refraction of lenses until the next section. First, we want to establish a pictorial method of understanding image formation. This method is called **ray tracing.**

FIGURE 23.33 Parallel light rays pass through a converging lens and a diverging lens.

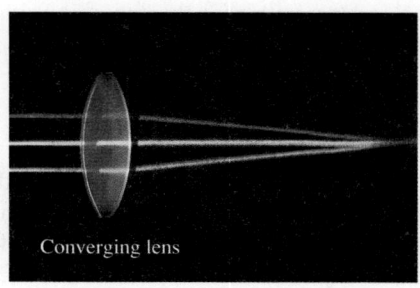 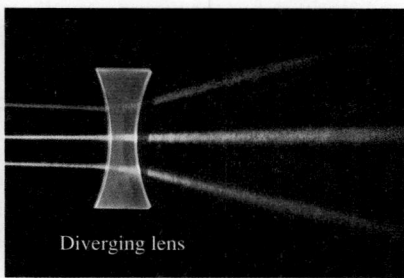

Converging lens   Diverging lens

FIGURE 23.33 shows parallel light rays entering two different lenses. The left lens, called a **converging lens,** causes the rays to refract *toward* the optical axis. The common point through which initially parallel rays pass is called the **focal point** of the lens. The distance of the focal point from the lens is called the **focal length** f of the lens. The right lens, called a **diverging lens,** refracts parallel rays *away from* the optical axis. This lens also has a focal point, but it is not as obvious.

NOTE ▶ A converging lens is thicker in the center than at the edges. A diverging lens is thicker at the edges than at the center. ◀

FIGURE 23.34 clarifies the situation. In the case of a diverging lens, a backward projection of the diverging rays shows that they *appear* to have started from the same point. This is the focal point of a diverging lens, and its distance from the lens is the focal length of the lens. In the next section we'll relate the focal length to the curvature and index of refraction of the lens, but now we'll use the practical definition that **the focal length is the distance from the lens at which rays parallel to the optical axis converge or from which they diverge.**

NOTE ▶ The focal length f is a property *of the lens,* independent of how the lens is used. The focal length characterizes a lens in much the same way that a mass m characterizes an object or a spring constant k characterizes a spring. ◀

Converging Lenses

These basic observations about lenses are enough to understand image formation by a thin lens. A **thin lens** is a lens whose thickness is very small in comparison to its focal length and in comparison to the object and image distances. We'll make the approximation that the thickness of a thin lens is zero and that the lens lies in a plane called the **lens plane.** Within this approximation, **all refraction occurs as the rays cross the lens plane, and all distances are measured from the lens plane.** Fortunately, the thin-lens approximation is quite good for most practical applications of lenses.

NOTE ▶ We'll *draw* lenses as if they have a thickness, because that is how we expect lenses to look, but our analysis will not depend on the shape or thickness of a lens. ◀

FIGURE 23.35 shows three important situations of light rays passing through a thin converging lens. Part a is familiar from Figure 23.34. If the direction of each of the rays in **FIGURE 23.35a** is reversed, Snell's law tells us that each ray will exactly retrace its path and emerge from the lens parallel to the optical axis. This leads to **FIGURE 23.35b,** which is the "mirror image" of part (a). Notice that the lens actually has *two* focal points, located at distances f on either side of the lens.

FIGURE 23.35c shows three rays passing through the *center* of the lens. At the center, the two sides of a lens are very nearly parallel to each other. Earlier, in Example 23.3, we found that a ray passing through a piece of glass with parallel sides is *displaced*

FIGURE 23.34 The focal lengths of converging and diverging lenses.

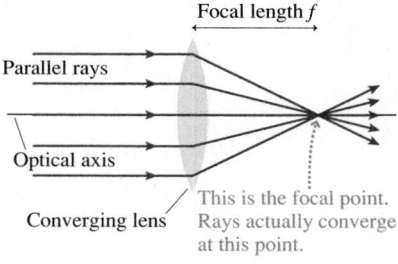

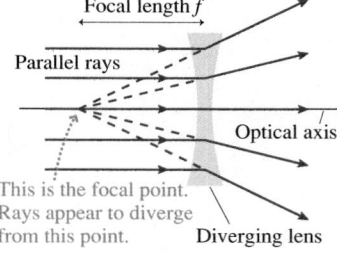

FIGURE 23.35 Three important sets of rays passing through a thin converging lens.

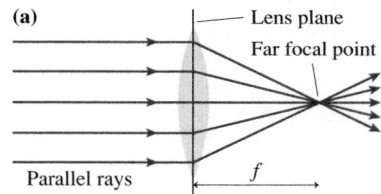

Any ray initially parallel to the optical axis will refract through the focal point on the far side of the lens.

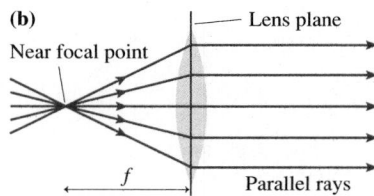

Any ray passing through the near focal point emerges from the lens parallel to the optical axis.

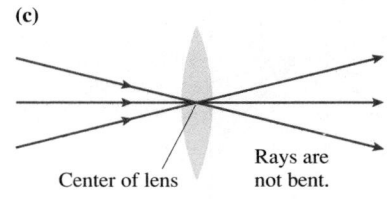

Any ray directed at the center of the lens passes through in a straight line.

but *not bent* and that the displacement becomes zero as the thickness approaches zero. Consequently, a ray through the center of a thin lens, with zero thickness, is neither bent nor displaced but travels in a straight line.

These three situations form the basis for ray tracing.

Real Images

FIGURE 23.36 shows a lens and an object whose distance from the lens is larger than the focal length. Rays from point P on the object are refracted by the lens so as to converge at point P′ on the opposite side of the lens. If rays diverge from an object point P and interact with a lens such that the refracted rays *converge* at point P′, actually meeting at P′, then we call P′ a **real image** of point P. Contrast this with our prior definition of a *virtual image* as a point from which rays—which never meet—appear to *diverge*.

FIGURE 23.36 Rays from an object point P are refracted by the lens and converge to a real image at point P′.

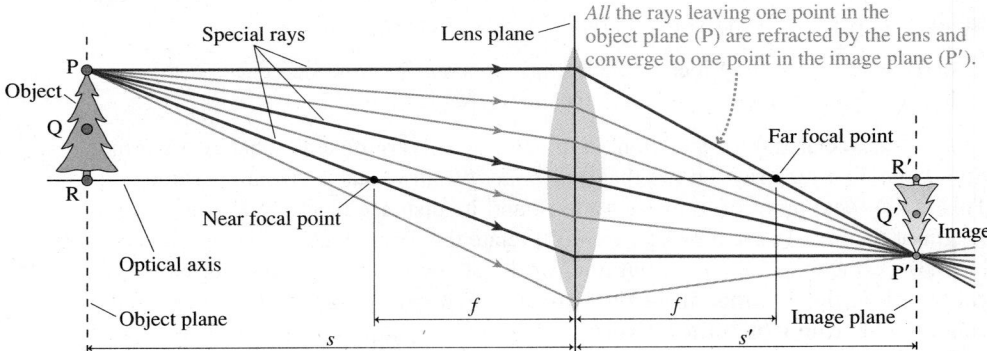

All points on the object that are in the same plane, the **object plane,** converge to image points in the **image plane.** Points Q and R in the object plane of Figure 23.36 have image points Q′ and R′ in the same plane as point P′. Once we locate *one* point in the image plane, such as point P′, we know that the full image lies in the same plane.

There are two important observations to make about Figure 23.36. First, the image is upside down with respect to the object. This is called an **inverted image,** and it is a standard characteristic of real-image formation with a converging lens. Second, rays from point P *fill* the entire lens surface, and all portions of the lens contribute to the image. A larger lens will "collect" more rays and thus make a brighter image.

FIGURE 23.37 is a close-up view of the rays very near the image plane. The rays don't stop at P′ unless we place a screen in the image plane. When we do so, we see a sharp, well-focused image on the screen. To focus an image, you must either move the screen to coincide with the image plane or move the lens or object to make the image plane coincide with the screen. For example, the focus knob on a projector moves the lens forward or backward until the image plane matches the screen position.

NOTE ▶ The ability to see a real image on a screen sets real images apart from *virtual* images. But keep in mind that we need not *see* a real image in order to *have* an image. A real image exists at a point in space where the rays converge even if there's no viewing screen in the image plane. ◀

Figure 23.36 highlights three "special rays" based on the three situations of Figure 23.35. These three rays alone are sufficient to locate the image point P′. That is, we don't need to draw all the rays shown in Figure 23.36. The procedure known as *ray tracing* consists of locating the image by the use of just these three rays.

FIGURE 23.37 A close-up look at the rays near the image plane.

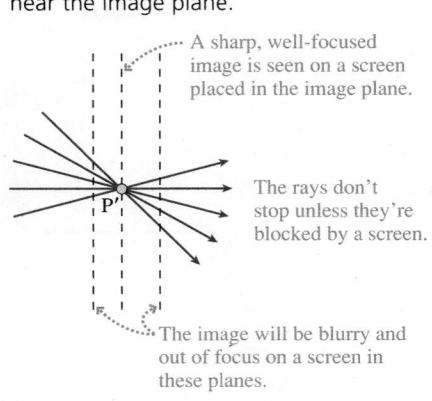

A sharp, well-focused image is seen on a screen placed in the image plane.

The rays don't stop unless they're blocked by a screen.

The image will be blurry and out of focus on a screen in these planes.

TACTICS
BOX 23.2   **Ray tracing for a converging lens**

❶ **Draw an optical axis.** Use graph paper or a ruler! Establish an appropriate scale.

❷ **Center the lens on the axis.** Mark and label the focal points at distance f on either side.

❸ **Represent the object with an upright arrow at distance s.** It's usually best to place the base of the arrow on the axis and to draw the arrow about half the radius of the lens.

❹ **Draw the three "special rays" from the tip of the arrow.** Use a straight edge.

    a. A ray parallel to the axis refracts through the far focal point.
    b. A ray that enters the lens along a line through the near focal point emerges parallel to the axis.
    c. A ray through the center of the lens does not bend.

❺ **Extend the rays until they converge.** This is the image point. Draw the rest of the image in the image plane. If the base of the object is on the axis, then the base of the image will also be on the axis.

❻ **Measure the image distance s'.** Also, if needed, measure the image height relative to the object height.

Exercises 22–27

EXAMPLE 23.8   **Finding the image of a flower**

A 4.0-cm-diameter flower is 200 cm from the 50-cm-focal-length lens of a camera. How far should the light detector be placed behind the lens to record a well-focused image? What is the diameter of the image on the detector?

MODEL   The flower is in the object plane. Use ray tracing to locate the image.

VISUALIZE   **FIGURE 23.38** shows the ray-tracing diagram and the steps of Tactics Box 23.2. The image has been drawn in the plane where the three special rays converge. You can see *from the drawing* that the image distance is $s' \approx 67$ cm. This is where the detector needs to be placed to record a focused image.

The heights of the object and image are labeled h and h'. The ray through the center of the lens is a straight line, thus the object and image both subtend the same angle θ. Using similar triangles,

$$\frac{h'}{s'} = \frac{h}{s}$$

Solving for h' gives

$$h' = h\frac{s'}{s} = (4.0 \text{ cm})\frac{67 \text{ cm}}{200 \text{ cm}} = 1.3 \text{ cm}$$

The flower's image has a diameter of 1.3 cm.

ASSESS   We've been able to learn a great deal about the image from a simple geometric procedure.

FIGURE 23.38 Ray-tracing diagram for Example 23.8.

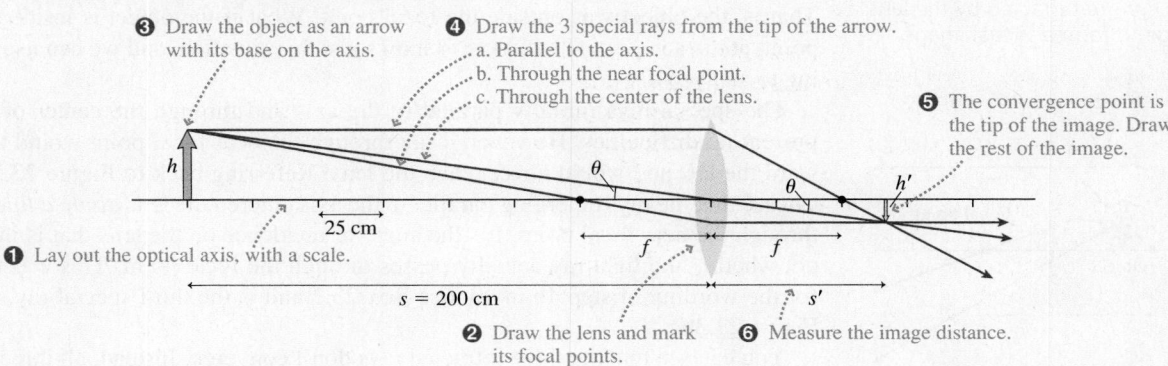

Lateral Magnification

The image can be either larger or smaller than the object, depending on the location and focal length of the lens. But there's more to a description of the image than just its size. We also want to know its *orientation* relative to the object. That is, is the image upright or inverted? It is customary to combine size and orientation information into a single number. The **lateral magnification** m is defined as

$$m = -\frac{s'}{s} \qquad (23.14)$$

You just saw in Example 23.8 that the image-to-object height ratio is $h'/h = s'/s$. Consequently, we interpret the lateral magnification m as follows:

1. A positive value of m indicates that the image is upright relative to the object. A negative value of m indicates that the image is inverted relative to the object.
2. The absolute value of m gives the size ratio of the image and object: $h'/h = |m|$.

The lateral magnification in Example 23.8 would be $m = -0.33$, indicating that the image is inverted and 33% the size of the object.

NOTE ▶ The image-to-object height ratio is called *lateral* magnification to distinguish it from angular magnification, which we'll introduce in the next chapter. In practice, m is simply called "magnification" when there's no chance of confusion. Magnification can be less than 1, meaning that the image is smaller than the object. ◀

STOP TO THINK 23.4 A lens produces a sharply focused, inverted image on a screen. What will you see on the screen if the lens is removed?

a. The image will be inverted and blurry.
b. The image will be upright and sharp.
c. The image will be upright and blurry.
d. The image will be much dimmer but otherwise unchanged.
e. There will be no image at all.

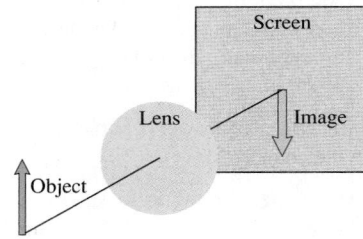

Virtual Images

The previous section considered a converging lens with the object at distance $s > f$. That is, the object was outside the focal point. What if the object is inside the focal point, at distance $s < f$? FIGURE 23.39 shows just this situation, and we can use ray tracing to analyze it.

The special rays initially parallel to the axis and through the center of the lens present no difficulties. However, a ray through the near focal point would travel toward the left and would never reach the lens! Referring back to Figure 23.35b, you can see that the rays emerging parallel to the axis entered the lens *along a line* passing through the near focal point. It's the angle of incidence on the lens that is important, not whether the light ray actually passes through the focal point. This was the basis for the wording of step 4b in Tactics Box 23.2 and is the third special ray shown in Figure 23.39.

You can see that the three refracted rays don't converge. Instead, all three rays appear to *diverge* from point P'. This is the situation we found for rays reflecting from

FIGURE 23.39 Rays from an object at distance $s < f$ are refracted by the lens and diverge to form a virtual image.

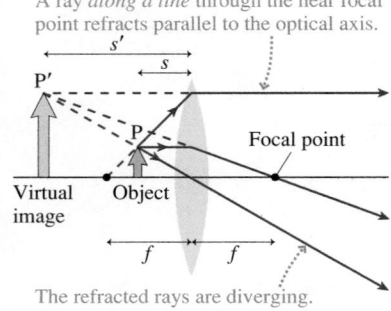

A ray *along a line* through the near focal point refracts parallel to the optical axis.

The refracted rays are diverging. They appear to come from point P'.

a mirror and for the rays refracting out of an aquarium. Point P' is a *virtual image* of the object point P. Furthermore, it is an **upright image,** having the same orientation as the object.

The refracted rays, which are all to the right of the lens, *appear* to come from P', but none of the rays were ever at that point. No image would appear on a screen placed in the image plane at P'. So what good is a virtual image?

Your eye collects and focuses bundles of diverging rays; thus, as FIGURE 23.40a shows, you can "see" a virtual image by looking *through* the lens. This is exactly what you do with a magnifying glass, producing a scene like the one in FIGURE 23.40b. In fact, you view a virtual image anytime you look *through* the eyepiece of an optical instrument such as a microscope or binoculars.

The image distance s' for a virtual image is defined to be a *negative number* (s' < 0), indicating that the image is on the opposite side of the lens from a real image. With this choice of sign, the definition of magnification, $m = -s'/s$, is still valid. A virtual image with negative s' has $m > 0$, thus the image is upright. This agrees with the rays in Figure 23.39 and the photograph of Figure 23.40b.

NOTE ▶ A lens thicker in the middle than at the edges is classified as a converging lens. The light rays from an object *can* converge to form a real image after passing through such a lens, but only if the object distance is larger than the focal length of the lens: $s > f$. If $s < f$, the rays leaving a converging lens are diverging to produce a virtual image. ◀

FIGURE 23.40 A converging lens is a magnifying glass when the object distance is less than f.

(a)

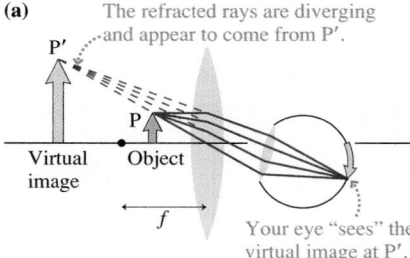

(b)

EXAMPLE 23.9 **Magnifying a flower**

To see a flower better, a naturalist holds a 6.0-cm-focal-length magnifying glass 4.0 cm from the flower. What is the magnification?

MODEL The flower is in the object plane. Use ray tracing to locate the image.

VISUALIZE FIGURE 23.41 shows the ray-tracing diagram. The three special rays diverge from the lens, but we can use a straightedge to extend the rays backward to the point from which they diverge. This point, the image point, is seen to be 12 cm to the left of the lens. Because this is a virtual image, the image distance is $s' = -12$ cm. Thus the magnification is

$$m = -\frac{s'}{s} = -\frac{-12 \text{ cm}}{4.0 \text{ cm}} = 3.0$$

The image is three times as large as the object and, because m is positive, upright.

FIGURE 23.41 Ray-tracing diagram for Example 23.9.

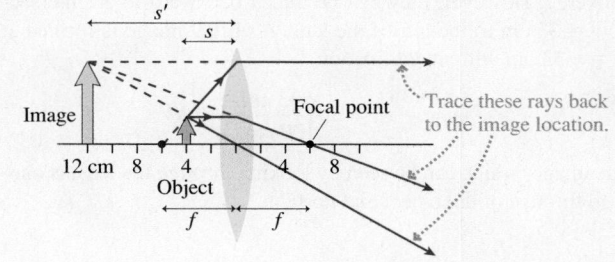

Diverging Lenses

A lens thicker at the edges than in the middle is called a *diverging lens.* FIGURE 23.42 shows three important sets of rays passing through a diverging lens. These are based on Figures 23.33 and 23.34, where you saw that rays initially parallel to the axis diverge after passing through a diverging lens.

FIGURE 23.42 Three important sets of rays passing through a thin diverging lens.

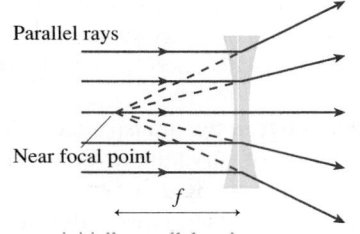

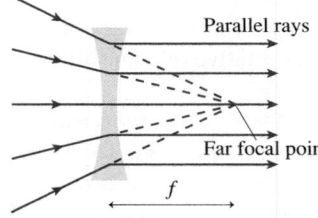

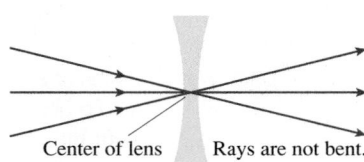

Any ray initially parallel to the optical axis diverges along a line through the near focal point.

Any ray directed along a line toward the far focal point emerges from the lens parallel to the optical axis.

Any ray directed at the center of the lens passes through in a straight line.

Ray tracing follows the steps of Tactics Box 23.2 for a converging lens *except* that two of the three special rays in step 4 are different.

TACTICS
BOX 23.3 **Ray tracing for a diverging lens**

①–③ Follow steps 1 through 3 of Tactics Box 23.2.
④ Draw the three "special rays" from the tip of the arrow. Use a straight-edge.

   a. A ray parallel to the axis diverges along a line through the near focal point.
   b. A ray along a line toward the far focal point emerges parallel to the axis.
   c. A ray through the center of the lens does not bend.

⑤ Trace the diverging rays backward. The point from which they are diverging is the image point, which is always a virtual image.
⑥ Measure the image distance s′. This will be a negative number.

Exercise 28

EXAMPLE 23.10 **Demagnifying a flower**

A diverging lens with a focal length of 50 cm is placed 100 cm from a flower. Where is the image? What is its magnification?

MODEL The flower is in the object plane. Use ray tracing to locate the image.

VISUALIZE **FIGURE 23.43** shows the ray-tracing diagram. The three special rays (labeled a, b, and c to match the Tactics Box) do not converge. However, they can be traced backward to an intersection ≈ 33 cm to the left of the lens. A virtual image is formed at $s' = -33$ cm with magnification

$$m = -\frac{s'}{s} = -\frac{-33 \text{ cm}}{100 \text{ cm}} = 0.33$$

The image, which can be seen by looking *through* the lens, is one-third the size of the object and upright.

FIGURE 23.43 Ray-tracing diagram for Example 23.10.

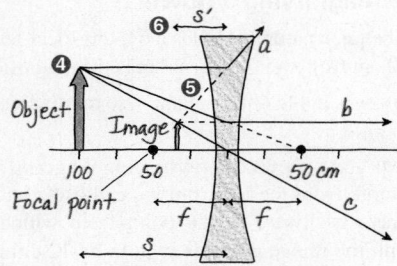

ASSESS Ray tracing with a diverging lens is somewhat trickier than with a converging lens, so this example is worth careful study.

Diverging lenses *always* make virtual images and, for this reason, are rarely used alone. However, they have important applications when used in combination with other lenses. Cameras, eyepieces, and eyeglasses often incorporate diverging lenses.

23.7 Thin Lenses: Refraction Theory

Ray tracing is a powerful visual approach for understanding image formation, but it doesn't provide precise information about the image location or image properties. We need to develop a quantitative relationship between the object distance s and the image distance s'.

To begin, **FIGURE 23.44** shows a *spherical* boundary between two transparent media with indices of refraction n_1 and n_2. The sphere has radius of curvature R. Consider a ray that leaves object point P at angle α and later, after refracting, reaches point P′. Figure 23.44 has exaggerated the angles to make the picture clear, but we will restrict our analysis to *paraxial rays* traveling nearly parallel to the axis. For paraxial rays, all the angles are small and we can use the small-angle approximation.

FIGURE 23.44 Image formation due to refraction at a spherical surface. The angles are exaggerated.

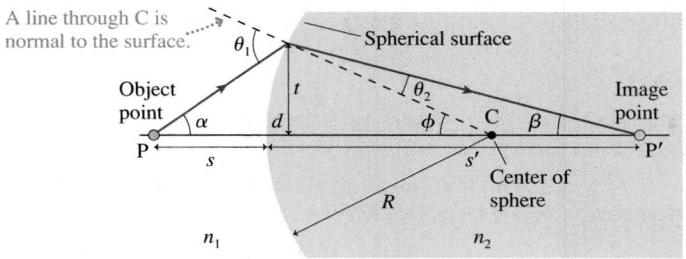

The ray from P is incident on the boundary at angle θ_1 and refracts into medium n_2 at angle θ_2, both measured from the normal to the surface at the point of incidence. Snell's law is $n_1 \sin \theta_1 = n_2 \sin \theta_2$, which in the small-angle approximation is

$$n_1 \theta_1 = n_2 \theta_2 \tag{23.15}$$

You can see from the geometry of Figure 23.44 that angles α, β, and ϕ are related by

$$\begin{aligned} \theta_1 &= \alpha + \phi \\ \theta_2 &= \phi - \beta \end{aligned} \tag{23.16}$$

Using these expressions in Equation 23.15, we can write Snell's law as

$$n_1(\alpha + \phi) = n_2(\phi - \beta) \tag{23.17}$$

This is one important relationship between the angles.

The line of height t, from the axis to the point of incidence, is the vertical leg of three different right triangles having vertices at points P, C, and P′. Consequently,

$$\tan \alpha \approx \alpha = \frac{t}{s+d} \qquad \tan \beta \approx \beta = \frac{t}{s'-d} \qquad \tan \phi \approx \phi = \frac{t}{R-d} \tag{23.18}$$

But $d \to 0$ for paraxial rays, thus

$$\alpha = \frac{t}{s} \qquad \beta = \frac{t}{s'} \qquad \phi = \frac{t}{R} \tag{23.19}$$

This is the second important relationship that comes from Figure 23.44.

If we use the angles of Equation 23.19 in Equation 23.17, we find

$$n_1 \left(\frac{t}{s} + \frac{t}{R} \right) = n_2 \left(\frac{t}{R} - \frac{t}{s'} \right) \tag{23.20}$$

The t cancels, and we can rearrange Equation 23.20 to read

$$\frac{n_1}{s} + \frac{n_2}{s'} = \frac{n_2 - n_1}{R} \tag{23.21}$$

Equation 23.21 is independent of angle α. Consequently, **all paraxial rays leaving point P later converge at point P′**. If an object is located at distance s from a spherical refracting surface, an image will be formed at distance s' given by Equation 23.21.

Equation 23.21 was derived for a surface that is convex toward the object point, and the image is real. However, the result is also valid for virtual images or for surfaces that are concave toward the object point as long as we adopt the *sign convention* shown in Table 23.3.

Section 23.4 considered image formation due to refraction by a plane surface. There we found (in Equation 23.13) an image distance $s' = (n_2/n_1)s$. A plane can be thought of as a sphere in the limit $R \to \infty$, so we should be able to reach the same conclusion from Equation 23.21. As $R \to \infty$, the term $(n_2 - n_1)/R \to 0$ and Equation 23.21 becomes $s' = -(n_2/n_1)s$. This seems to differ from Equation 23.13, but it

TABLE 23.3 Sign convention for refracting surfaces

	Positive	Negative
R	Convex toward the object	Concave toward the object
s'	Real image, opposite side from object	Virtual image, same side as object

doesn't really. Equation 23.13 gives the actual distance to the image. Equation 23.21 is based on a sign convention in which virtual images have negative image distances, hence the minus sign.

EXAMPLE 23.11 | Image formation inside a glass rod

One end of a 4.0-cm-diameter glass rod is shaped like a hemisphere. A small lightbulb is 6.0 cm from the end of the rod. Where is the bulb's image located?

MODEL Model the lightbulb as a point source of light and consider the paraxial rays that refract into the glass rod.

FIGURE 23.45 The curved surface refracts the light to form a real image.

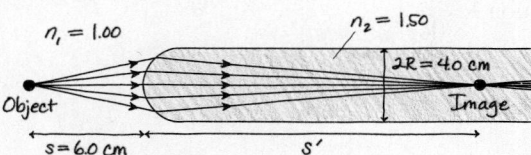

VISUALIZE FIGURE 23.45 shows the situation. $n_1 = 1.00$ for air and $n_2 = 1.50$ for glass.

SOLVE The radius of the surface is half the rod diameter, so $R = 2.0$ cm. Equation 23.21 is

$$\frac{1.00}{6.0 \text{ cm}} + \frac{1.50}{s'} = \frac{1.50 - 1.00}{2.0 \text{ cm}} = \frac{0.50}{2.0 \text{ cm}}$$

Solving for the image distance s' gives

$$\frac{1.50}{s'} = \frac{0.50}{2.0 \text{ cm}} - \frac{1.00}{6.0 \text{ cm}} = 0.0833 \text{ cm}^{-1}$$

$$s' = \frac{1.50}{0.0833} = 18 \text{ cm}$$

ASSESS This is a real image located 18 cm inside the glass rod.

EXAMPLE 23.12 | A goldfish in a bowl

A goldfish lives in a spherical fish bowl 50 cm in diameter. If the fish is 10 cm from the near edge of the bowl, where does the fish appear when viewed from the outside?

MODEL Model the fish as a point source and consider the paraxial rays that refract from the water into the air. The thin glass wall has little effect and will be ignored.

FIGURE 23.46 The curved surface of a fish bowl produces a virtual image of the fish.

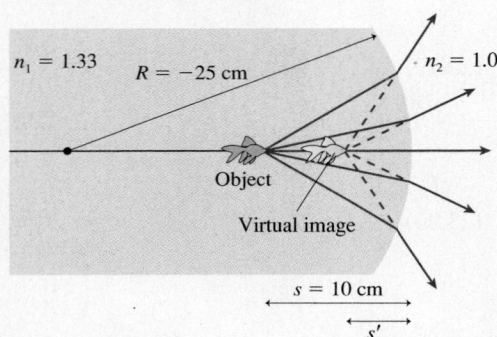

VISUALIZE FIGURE 23.46 shows the rays refracting *away* from the normal as they move from the water into the air. We expect to find a virtual image at a distance less than 10 cm.

SOLVE The object is in the water, so $n_1 = 1.33$ and $n_2 = 1.00$. The inner surface is concave (you can remember "concave" because it's like looking into a cave), so $R = -25$ cm. The object distance is $s = 10$ cm. Thus Equation 23.21 is

$$\frac{1.33}{10 \text{ cm}} + \frac{1.00}{s'} = \frac{1.00 - 1.33}{-25 \text{ cm}} = \frac{0.33}{25 \text{ cm}}$$

Solving for the image distance s' gives

$$\frac{1.00}{s'} = \frac{0.33}{25 \text{ cm}} - \frac{1.33}{10 \text{ cm}} = -0.12 \text{ cm}^{-1}$$

$$s' = \frac{1.00}{-0.12 \text{ cm}^{-1}} = -8.3 \text{ cm}$$

ASSESS The image is virtual, located to the left of the boundary. A person looking into the bowl will see a fish that appears to be 8.3 cm from the edge of the bowl.

STOP TO THINK 23.5 Which of these actions will move the real image point P′ farther from the boundary? More than one may work.

a. Increase the radius of curvature R.
b. Increase the index of refraction n.
c. Increase the object distance s.
d. Decrease the radius of curvature R.
e. Decrease the index of refraction n.
f. Decrease the object distance s.

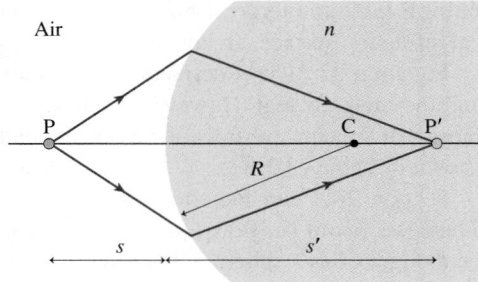

Lenses

The thin-lens approximation assumes rays refract one time, at the lens plane. In fact, as **FIGURE 23.47** shows, rays refract *twice,* at spherical surfaces having radii of curvature R_1 and R_2. Let the lens have thickness t and be made of a material with index of refraction n. For simplicity, we'll assume that the lens is surrounded by air.

FIGURE 23.47 Image formation by a lens.

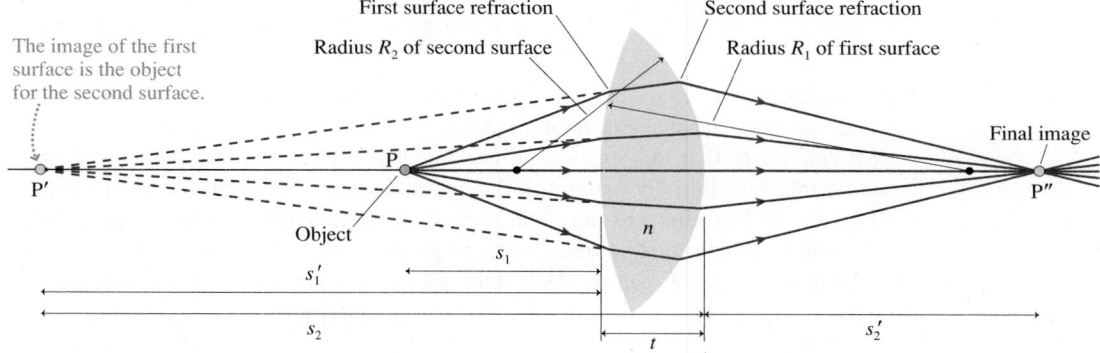

The image of the first surface is the object for the second surface.

First surface refraction   Second surface refraction

Radius R_2 of second surface   Radius R_1 of first surface

Final image

P′

Object s_1 n

s_1'

s_2 t s_2'

P″

The object at point P is distance s_1 to the left of the lens. The first surface of the lens, of radius R_1, refracts the rays from P to create an image at point P′. We can use Equation 23.21 for a spherical surface to find the image distance s_1':

$$\frac{1}{s_1} + \frac{n}{s_1'} = \frac{n-1}{R_1} \tag{23.22}$$

where we used $n_1 = 1$ for the air and $n_2 = n$ for the lens. We'll assume that the image P′ is a virtual image, but this assumption isn't essential to the outcome.

With two refracting surfaces, the image P′ of the first surface becomes the object for the second surface. That is, the rays refracting at the second surface appear to have come from P′. Object distance s_2 from P′ to the second surface looks like it should be $s_2 = s_1' + t$, but P′ is a virtual image, so s_1' is a *negative* number. Thus the distance to the second surface is $s_2 = |s_1'| + t = t - s_1'$. We can find the image of P′ by a second application of Equation 23.21, but with a switch. The rays are incident on the surface from within the lens, so this time $n_1 = n$ and $n_2 = 1$. Consequently,

$$\frac{n}{t - s_1'} + \frac{1}{s_2'} = \frac{1-n}{R_2} \tag{23.23}$$

For a *thin lens,* which has $t \to 0$, Equation 23.23 becomes

$$-\frac{n}{s_1'} + \frac{1}{s_2'} = \frac{1-n}{R_2} = -\frac{n-1}{R_2} \tag{23.24}$$

Our goal is to find the distance s_2' to point P″, the image produced by the lens as a whole. This goal is easily reached if we simply add Equations 23.22 and 23.24, eliminating s_1' and giving

$$\frac{1}{s_1} + \frac{1}{s_2'} = \frac{n-1}{R_1} - \frac{n-1}{R_2} = (n-1)\left(\frac{1}{R_1} - \frac{1}{R_2}\right) \tag{23.25}$$

The numerical subscripts on s_1 and s_2' no longer serve a purpose. If we replace s_1 by s, the object distance from the lens, and s_2' by s', the image distance, Equation 23.25 becomes the *thin-lens equation:*

$$\frac{1}{s} + \frac{1}{s'} = \frac{1}{f} \qquad \text{(thin-lens equation)} \qquad (23.26)$$

where the *focal length* of the lens is

$$\frac{1}{f} = (n-1)\left(\frac{1}{R_1} - \frac{1}{R_2}\right) \qquad \text{(lens maker's equation)} \qquad (23.27)$$

Equation 23.27 is known as the *lens maker's equation.* It allows you to determine the focal length from the shape of a thin lens and the material used to make it.

We can verify that this expression for f really is the focal length of the lens by recalling that rays initially parallel to the optical axis pass through the focal point on the far side. In fact, this was our *definition* of the focal length of a lens. Parallel rays must come from an object extremely far away, with object distance $s \rightarrow \infty$ and thus $1/s = 0$. In that case, Equation 23.26 tells us that the parallel rays will converge at distance $s' = f$ on the far side of the lens, exactly as expected.

We derived the thin-lens equation and the lens maker's equation from the specific lens geometry shown in Figure 23.47, but the results are valid for any lens as long as all quantities are given appropriate signs. The sign convention used with Equations 23.26 and 23.27 is given in Table 23.4.

TABLE 23.4 Sign convention for thin lenses

	Positive	Negative
R_1, R_2	Convex toward the object	Concave toward the object
f	Converging lens, thicker in center	Diverging lens, thinner in center
s'	Real image, opposite side from object	Virtual image, same side as object

NOTE ▶ For a *thick lens,* where the thickness t is not negligible, we can solve Equations 23.22 and 23.23 in sequence to find the position of the image point P''. ◀

EXAMPLE 23.13 **Focal length of a meniscus lens**

What is the focal length of the glass *meniscus lens* shown in FIGURE 23.48? Is this a converging or diverging lens?

FIGURE 23.48 A meniscus lens.

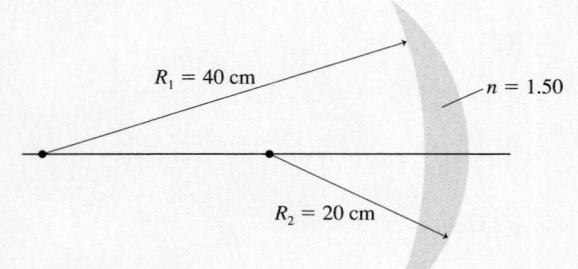

$R_1 = 40$ cm
$n = 1.50$
$R_2 = 20$ cm

SOLVE If the object is on the left, then the first surface has $R_1 = -40$ cm (concave toward the object) and the second surface has $R_2 = -20$ cm (also concave toward the object). The index of refraction of glass is $n = 1.50$, so the lens maker's equation is

$$\frac{1}{f} = (n-1)\left(\frac{1}{R_1} - \frac{1}{R_2}\right) = (1.50 - 1)\left(\frac{1}{-40 \text{ cm}} - \frac{1}{-20 \text{ cm}}\right)$$
$$= 0.0125 \text{ cm}^{-1}$$

Inverting this expression gives $f = 80$ cm. This is a converging lens, as seen both from the positive value of f and from the fact that the lens is thicker in the center.

Thin-Lens Image Formation

Although the thin-lens equation allows precise calculations, the lessons of ray tracing should not be forgotten. The most powerful tool of optical analysis is a combination of ray tracing, to gain an intuitive understanding of the ray trajectories, and the thin-lens equation.

EXAMPLE 23.14 | Designing a lens

The objective lens of a microscope uses a planoconvex glass lens with the flat side facing the specimen. A real image is formed 160 mm behind the lens when the lens is 8.0 mm from the specimen. What is the radius of the lens's curved surface?

MODEL Treat the lens as a thin lens with the specimen as the object. The lens's focal length is given by the lens maker's equation.

VISUALIZE FIGURE 23.49 clarifies the shape of the lens and defines R_2. The index of refraction was taken from Table 23.1.

FIGURE 23.49 A planoconvex microscope lens.

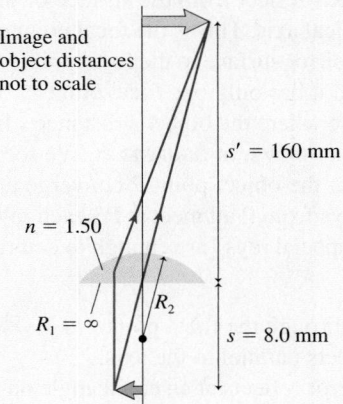

SOLVE We can use the lens maker's equation to solve for R_2 if we know the lens's focal length. Because we know both the object and image distances, we can use the thin-lens equation to find

$$\frac{1}{f} = \frac{1}{s} + \frac{1}{s'} = \frac{1}{8.0 \text{ mm}} + \frac{1}{160 \text{ mm}} = 0.131 \text{ mm}^{-1}$$

The focal length is $f = 1/(0.131 \text{ mm}^{-1}) = 7.6 \text{ mm}$, but $1/f$ is all we need for the lens maker's equation. The front surface of the lens is planar, which we can consider a portion of a sphere with $R_1 \rightarrow \infty$. Consequently $1/R_1 = 0$. With this, we can solve the lens maker's equation for R_2:

$$\frac{1}{R_2} = \frac{1}{R_1} - \frac{1}{n-1} \frac{1}{f} = 0 - \left(\frac{1}{1.50 - 1}\right)(0.131 \text{ mm}^{-1})$$

$$= -0.262 \text{ mm}^{-1}$$

$$R_2 = -3.8 \text{ mm}$$

The minus sign appears because the curved surface is concave toward the object. Physically, the radius of the curved surface is 3.8 mm.

ASSESS The actual thickness of the lens has to be less than R_2, probably no more than about 1.0 mm. This thickness is significantly less than the object and image distances, so the thin-lens approximation is justified.

EXAMPLE 23.15 | A magnifying lens

A stamp collector uses a magnifying lens that sits 2.0 cm above the stamp. The magnification is 4.0. What is the focal length of the lens?

FIGURE 23.50 Pictorial representation of a magnifying lens.

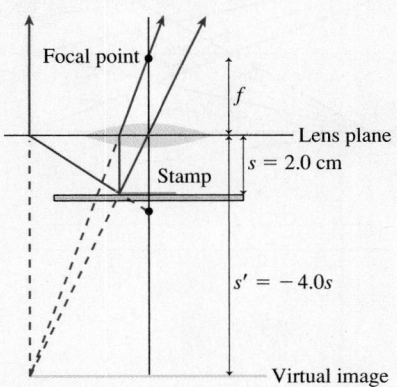

MODEL A magnifying lens is a converging lens with the object distance less than the focal length ($s < f$). Assume it is a thin lens.

VISUALIZE FIGURE 23.50 shows the lens and a ray-tracing diagram. We do not need to know the actual shape of the lens, so the figure shows a generic converging lens.

SOLVE A virtual image is upright, so $m = +4.0$. The magnification is $m = -s'/s$, thus

$$s' = -4.0s = -(4.0)(2.0 \text{ cm}) = -8.0 \text{ cm}$$

We can use s and s' in the thin-lens equation to find the focal length:

$$\frac{1}{f} = \frac{1}{s} + \frac{1}{s'} = \frac{1}{2.0 \text{ cm}} + \frac{1}{-8.0 \text{ cm}} = 0.375 \text{ cm}^{-1}$$

$$f = 2.7 \text{ cm}$$

ASSESS $f > 2$ cm, as expected.

STOP TO THINK 23.6   A lens forms a real image of a lightbulb, but the image of the bulb on a viewing screen is blurry because the screen is slightly in front of the image plane. To focus the image, should you move the lens toward the bulb or away from the bulb?

23.8 Image Formation with Spherical Mirrors

Curved mirrors—such as those used in telescopes, security and rearview mirrors, and searchlights—can be used to form images, and their images can be analyzed with ray diagrams similar to those used with lenses. We'll consider only the important case of **spherical mirrors,** whose surface is a section of a sphere.

Concave Mirrors

FIGURE 23.51 The focal point and focal length of a concave mirror.

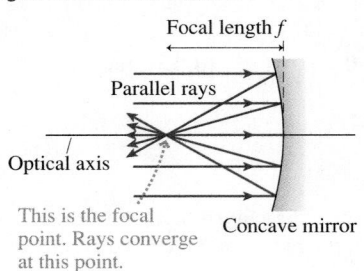

FIGURE 23.51 shows a **concave mirror,** a mirror in which the edges curve *toward* the light source. Rays parallel to the optical axis reflect from the surface of the mirror so as to pass through a single point on the optical axis. This is the focal point of the mirror. The focal length is the distance from the mirror surface to the focal point. A concave mirror is analogous to a converging lens, but it has only one focal point.

Let's begin by considering the case where the object's distance s from the mirror is greater than the focal length ($s > f$), as shown in **FIGURE 23.52**. We see that the image is *real* (and inverted) because rays from the object point P converge at the image point P'. Although an infinite number of rays from P all meet at P', each ray obeying the law of reflection, you can see that three "special rays" are enough to determine the position and size of the image:

■ A ray parallel to the axis reflects through the focal point.
■ A ray through the focal point reflects parallel to the axis.
■ A ray striking the center of the mirror reflects at an equal angle on the opposite side of the axis.

These three rays also locate the image if $s < f$, but in that case the image is *virtual* and behind the mirror.

FIGURE 23.52 A real image formed by a concave mirror.

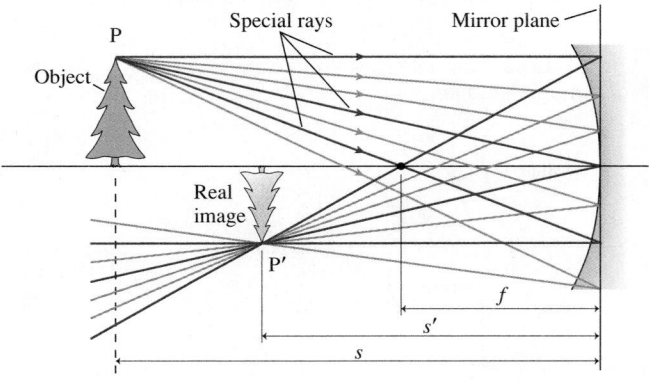

Convex Mirrors

FIGURE 23.53 The focal point and focal length of a convex mirror.

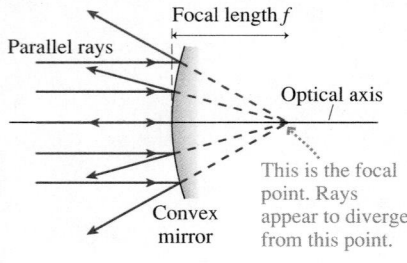

FIGURE 23.53 shows parallel light rays approaching a mirror in which the edges curve *away from* the light source. This is called a **convex mirror.** In this case, the reflected rays appear to come from a point behind the mirror. This is the focal point for a convex mirror.

A common example of a convex mirror is a silvered ball, such as a tree ornament. You may have noticed that if you look at your reflection in such a ball, your image appears right-side-up but is quite small. As another example, **FIGURE 23.54** shows a city skyline reflected in a polished metal sphere. Let's use ray tracing to understand why the skyscrapers all appear to be so small.

FIGURE 23.55 shows an object in front of a convex mirror. In this case, the reflected rays—each obeying the law of reflection—create an upright image of reduced height behind the mirror. We see that the image is virtual because no rays actually converge at the image point P′. Instead, diverging rays *appear* to come from this point. Once again, three special rays are enough to find the image.

FIGURE 23.54 A city skyline is reflected in this polished sphere.

FIGURE 23.55 A virtual image formed by a convex mirror.

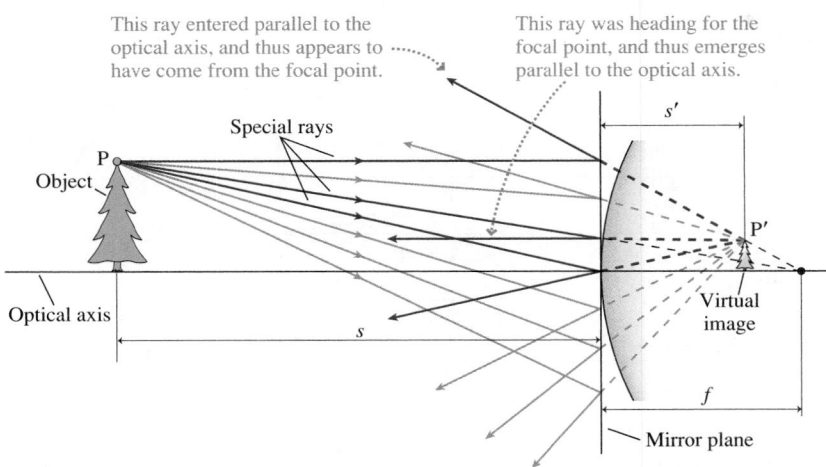

This ray entered parallel to the optical axis, and thus appears to have come from the focal point.

This ray was heading for the focal point, and thus emerges parallel to the optical axis.

Special rays

Object
P
Optical axis
s
s'
P′
Virtual image
f
Mirror plane

Convex mirrors are used for a variety of safety and monitoring applications, such as passenger-side rearview mirrors and the round mirrors used in stores to keep an eye on the customers. When an object is reflected in a convex mirror, the image appears smaller than the object itself. Because the image is, in a sense, a miniature version of the object, you can *see much more of it* within the edges of the mirror than you could with an equal-sized flat mirror.

TACTICS BOX 23.4 Ray tracing for a spherical mirror

❶ **Draw an optical axis.** Use graph paper or a ruler! Establish an appropriate scale.

❷ **Center the mirror on the axis.** Mark and label the focal point at distance f from the mirror's surface.

❸ **Represent the object with an upright arrow at distance s.** It's usually best to place the base of the arrow on the axis and to draw the arrow about half the radius of the mirror.

❹ **Draw the three "special rays" from the tip of the arrow.** Use a straight-edge.

    a. A ray parallel to the axis reflects through (concave) or away from (convex) the focal point.

    b. An incoming ray passing through (concave) or heading toward (convex) the focal point reflects parallel to the axis.

    c. A ray that strikes the center of the mirror reflects at an equal angle on the opposite side of the optical axis.

❺ **Extend the rays forward or backward until they converge.** This is the image point. Draw the rest of the image in the image plane. If the base of the object is on the axis, then the base of the image will also be on the axis.

❻ **Measure the image distance s'.** Also, if needed, measure the image height relative to the object height.

Exercises 32–33

EXAMPLE 23.16 **Analyzing a concave mirror**

A 3.0-cm-high object is located 60 cm from a concave mirror. The mirror's focal length is 40 cm. Use ray tracing to find the position and height of the image.

MODEL Use the ray-tracing steps of Tactics Box 23.4.

VISUALIZE **FIGURE 23.56** shows the steps of Tactics Box 23.4.

SOLVE We can use a ruler to find that the image position is $s' \approx 120$ cm in front of the mirror and its height is $h' \approx 6$ cm.

ASSESS The image is a *real* image because light rays converge at the image point.

FIGURE 23.56 Ray-tracing diagram for a concave mirror.

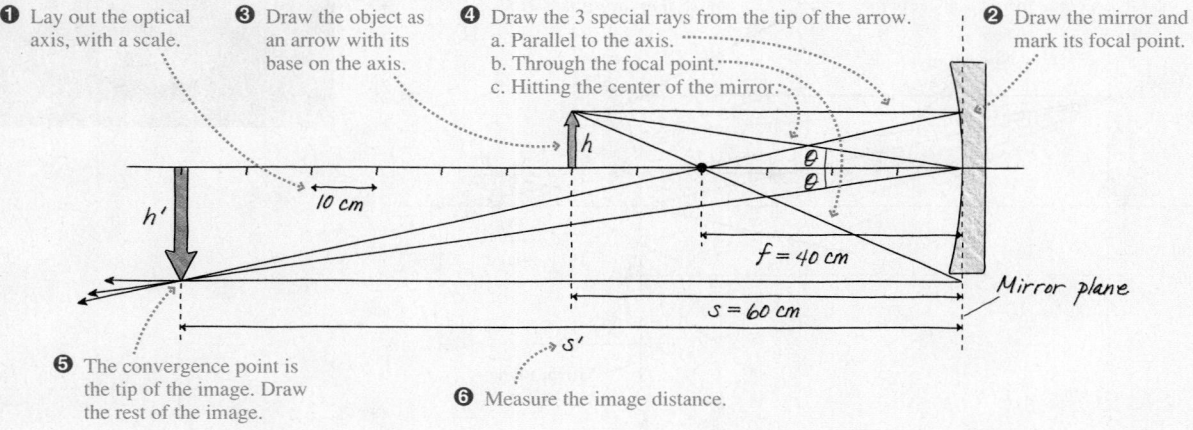

❶ Lay out the optical axis, with a scale.

❸ Draw the object as an arrow with its base on the axis.

❹ Draw the 3 special rays from the tip of the arrow.
  a. Parallel to the axis.
  b. Through the focal point.
  c. Hitting the center of the mirror.

❷ Draw the mirror and mark its focal point.

❺ The convergence point is the tip of the image. Draw the rest of the image.

❻ Measure the image distance.

The Mirror Equation

The thin-lens equation assumes lenses have negligible thickness (so a single refraction occurs in the lens plane) and the rays are nearly parallel to the optical axis (paraxial rays). If we make the same assumptions about spherical mirrors—the mirror has negligible thickness and so paraxial rays reflect at the mirror plane—then the object and image distances are related exactly as they were for thin lenses:

$$\frac{1}{s} + \frac{1}{s'} = \frac{1}{f} \qquad \text{(mirror equation)} \qquad (23.28)$$

The focal length of the mirror, as you can show as a homework problem, is related to the mirror's radius of curvature by

$$f = \frac{R}{2} \qquad (23.29)$$

TABLE 23.5 Sign convention for spherical mirrors

	Positive	Negative
R, f	Concave toward the object	Convex toward the object
s'	Real image, same side as object	Virtual image, opposite side from object

Table 23.5 shows the sign convention used with spherical mirrors. It differs from the convention for lenses, so you'll want to carefully compare this table to Table 23.4. A concave mirror (analogous to a converging lens) has a positive focal length while a convex mirror (analogous to a diverging lens) has a negative focal length. The lateral magnification of a spherical mirror is computed exactly as for a lens:

$$m = -\frac{s'}{s} \qquad (23.30)$$

EXAMPLE 23.17 **Analyzing a concave mirror**

A 3.0-cm-high object is located 20 cm from a concave mirror. The mirror's radius of curvature is 80 cm. Determine the position, orientation, and height of the image.

MODEL Treat the mirror as a thin mirror.

VISUALIZE The mirror's focal length is $f = R/2 = +40$ cm, where we used the sign convention from Table 23.5. With the focal length known, the three special rays in **FIGURE 23.57** show that the image is a magnified, virtual image behind the mirror.

FIGURE 23.57 Pictorial representation of Example 23.17.

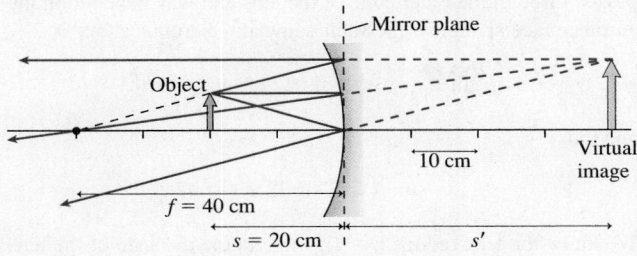

SOLVE The thin-mirror equation is

$$\frac{1}{20 \text{ cm}} + \frac{1}{s'} = \frac{1}{40 \text{ cm}}$$

This is easily solved to give $s' = -40$ cm, in agreement with the ray tracing. The negative sign tells us this is a virtual image behind the mirror. The magnification is

$$m = -\frac{-40 \text{ cm}}{20 \text{ cm}} = +2.0$$

Consequently, the image is 6.0 cm tall and upright.

ASSESS This is a virtual image because light rays diverge from the image point. You could see this enlarged image by standing behind the object and looking into the mirror. In fact, this is how magnifying cosmetic mirrors work.

STOP TO THINK 23.7 A concave mirror of focal length f forms an image of the moon. Where is the image located?

a. At the mirror's surface
b. Almost exactly a distance f behind the mirror
c. Almost exactly a distance f in front of the mirror
d. At a distance behind the mirror equal to the distance of the moon in front of the mirror

---

CHALLENGE EXAMPLE 23.18 **Optical fiber imaging**

An *endoscope* is a thin bundle of optical fibers that can be inserted through a bodily opening or small incision to view the interior of the body. As FIGURE 23.58 shows, an *objective* lens forms a real image on the entrance face of the fiber bundle. Individual fibers, using total internal reflection, transport the light to the exit face, where it emerges. The doctor (or a TV camera) observes the object by viewing the exit face through an *eyepiece* lens.

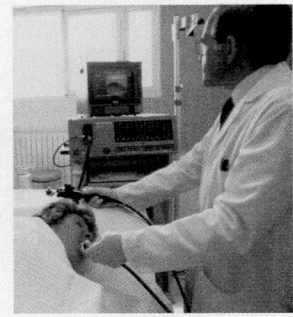

FIGURE 23.58 An endoscope.

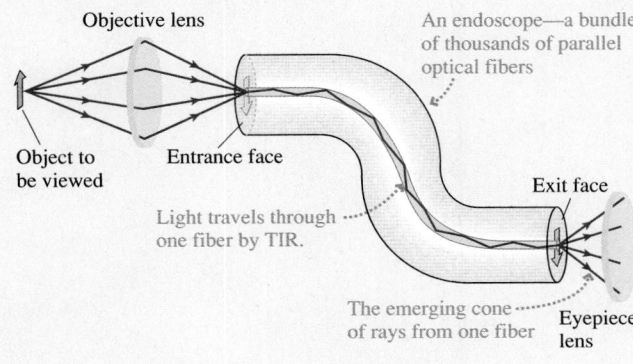

Consider an endoscope having a 3.0-mm-diameter objective lens with a focal length of 1.1 mm. These are typical values. The indices of refraction of the core and the cladding of the optical fibers are 1.62 and 1.50, respectively. To give maximum brightness, the objective lens is positioned so that, for an on-axis object, rays passing through the outer edge of the lens have the maximum angle of incidence for undergoing TIR in the fiber. How far should the objective lens be placed from the object the doctor wishes to view?

MODEL Represent the object as an on-axis point source and use the ray model of light.

VISUALIZE FIGURE 23.59 on the next page shows the real image being focused on the entrance face of the endoscope. Inside the fiber, rays that strike the cladding at an angle of incidence greater than the critical angle θ_c undergo TIR and stay in the fiber; rays are lost if their angle of incidence is less than θ_c. For maximum brightness, the lens is positioned so that a ray passing through the outer edge refracts into the fiber at the maximum angle of incidence θ_{max} for which TIR is possible. A smaller-diameter lens would sacrifice light-gathering power, whereas the outer rays from a larger-diameter lens would impinge on the core-cladding boundary at less than θ_c and would not undergo TIR. Note that the lens-to-fiber distance, although unknown, is fixed by the manufacturer and cannot be changed. Only object distance is under the doctor's control.

Continued

FIGURE 23.59 Magnified view of the entrance of an optical fiber.

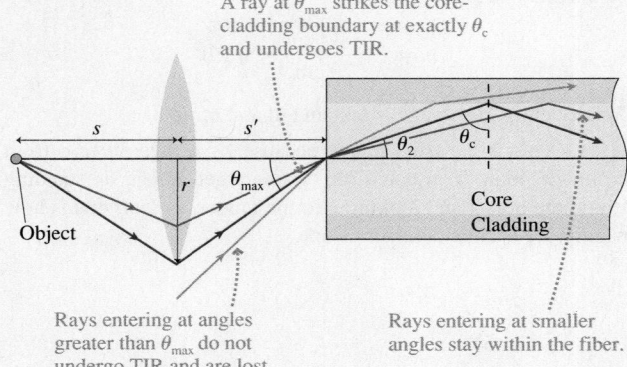

A ray at θ_{max} strikes the core-cladding boundary at exactly θ_c and undergoes TIR.

s   *s'*

r θ_{max}

Object

θ_2 θ_c

Core
Cladding

Rays entering at angles greater than θ_{max} do not undergo TIR and are lost.

Rays entering at smaller angles stay within the fiber.

SOLVE We know the focal length of the lens. We can use the geometry of the ray at the critical angle to find the image distance s', then use the thin-lens equation to find the object distance s. The critical angle for TIR inside the fiber is

$$\theta_c = \sin^{-1}\left(\frac{n_{cladding}}{n_{core}}\right) = \sin^{-1}\left(\frac{1.50}{1.62}\right) = 67.8°$$

A ray incident on the core-cladding boundary at exactly the critical angle must have entered the fiber, at the entrance face, at angle

$\theta_2 = 90° - \theta_c = 22.2°$. For optimum lens placement, this ray passed through the outer edge of the lens and was incident on the entrance face at angle θ_{max}. Snell's law at the entrance face is

$$n_{air}\sin\theta_{max} = 1.0 \cdot \sin\theta_{max} = n_{core}\sin\theta_2$$

and thus

$$\theta_{max} = \sin^{-1}(1.62\sin 22.2°) = 37.7°$$

We know the lens radius, $r = 1.5\,\text{mm}$, so the distance of the lens from the fiber—the image distance s'—is

$$s' = \frac{r}{\tan\theta_{max}} = \frac{1.5\ \text{mm}}{\tan(37.7°)} = 1.9\ \text{mm}$$

Now we can use the thin-lens equation to locate the object:

$$\frac{1}{s} = \frac{1}{f} - \frac{1}{s'} = \frac{1}{1.1\ \text{mm}} - \frac{1}{1.9\ \text{mm}}$$

$$s = 2.6\ \text{mm}$$

The doctor, viewing the exit face of the fiber bundle, will see a focused image when the objective lens is 2.6 mm from the object she wishes to view.

ASSESS The object and image distances are both greater than the focal length, which is correct for forming a real image.

SUMMARY

The goals of Chapter 23 have been to understand and apply the ray model of light.

General Principles

Reflection

Law of reflection: $\theta_r = \theta_i$

Reflection can be **specular** (mirror-like) or **diffuse** (from rough surfaces).

Plane mirrors: A virtual image is formed at P' with $s' = s$.

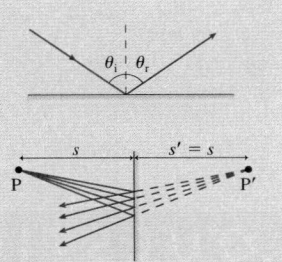

Refraction

Snell's law of refraction:

$$n_1 \sin\theta_1 = n_2 \sin\theta_2$$

Index of refraction is $n = c/v$. The ray is closer to the normal on the side with the larger index of refraction.

If $n_2 < n_1$, **total internal reflection** (TIR) occurs when the angle of incidence $\theta_1 \geq \theta_c = \sin^{-1}(n_2/n_1)$.

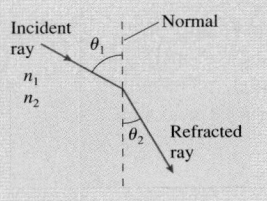

Important Concepts

The ray model of light

Light travels along straight lines, called **light rays,** at speed $v = c/n$.

A light ray continues forever unless an interaction with matter causes it to reflect, refract, scatter, or be absorbed.

Light rays come from **objects.** Each point on the object sends rays in all directions.

The eye sees an object (or an image) when diverging rays are collected by the pupil and focused on the retina.

▶ Ray optics is valid when lenses, mirrors, and apertures are larger than ≈ 1 mm.

Image formation

If rays diverge from P and interact with a lens or mirror so that the refracted rays *converge* at P', then P' is a real image of P.

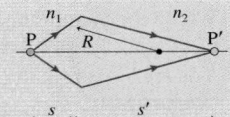

If rays diverge from P and interact with a lens or mirror so that the refracted/reflected rays *diverge* from P' and appear to come from P', then P' is a virtual image of P.

Spherical surface: Object and image distances are related by

$$\frac{n_1}{s} + \frac{n_2}{s'} = \frac{n_2 - n_1}{R}$$

Plane surface: $R \rightarrow \infty$, so $|s'/s| = n_2/n_1$.

Applications

Ray tracing

3 special rays in 3 basic situations:

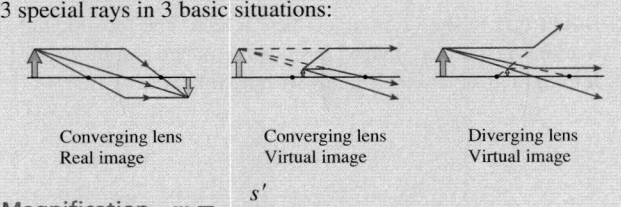

| Converging lens Real image | Converging lens Virtual image | Diverging lens Virtual image |

Magnification $m = -\dfrac{s'}{s}$

m is $+$ for an upright image, $-$ for inverted.
The height ratio is $h'/h = |m|$.

Thin lenses

The image and object distances are related by

$$\frac{1}{s} + \frac{1}{s'} = \frac{1}{f}$$

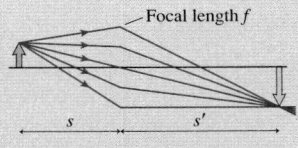

where the focal length is given by the lens maker's equation:

$$\frac{1}{f} = (n - 1)\left(\frac{1}{R_1} - \frac{1}{R_2}\right)$$

R $+$ for surface convex toward object	$-$ for concave
f $+$ for a converging lens	$-$ for diverging
s' $+$ for a real image	$-$ for virtual

Spherical mirrors

The image and object distances are related by

$$\frac{1}{s} + \frac{1}{s'} = \frac{1}{f}$$

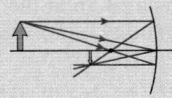

R, f $+$ for concave mirror	$-$ for convex
s' $+$ for a real image	$-$ for virtual

Focal length $f = R/2$

Terms and Notation

light ray	diffuse reflection	dispersion	object plane
object	virtual image	Rayleigh scattering	image plane
point source	refraction	lens	inverted image
parallel bundle	angle of refraction	ray tracing	lateral magnification, m
ray diagram	Snell's law	converging lens	upright image
camera obscura	total internal reflection (TIR)	focal point	spherical mirror
aperture	critical angle, θ_c	focal length, f	concave mirror
specular reflection	object distance, s	diverging lens	convex mirror
angle of incidence	image distance, s'	thin lens	
angle of reflection	optical axis	lens plane	
law of reflection	paraxial rays	real image	

CONCEPTUAL QUESTIONS

1. If you turn on your car headlights during the day, the road ahead of you doesn't appear to get brighter. Why not?

2. Suppose you have two pinhole cameras. The first has a small round hole in the front. The second is identical except it has a square hole of the same area as the round hole in the first camera. Would the pictures taken by these two cameras, under the same conditions, be different in any obvious way? Explain.

3. You are looking at the image of a pencil in a mirror, as shown in FIGURE Q23.3.
   a. What happens to the image if the top half of the mirror, down to the midpoint, is covered with a piece of cardboard? Explain.
   b. What happens to the image if the bottom half of the mirror is covered with a piece of cardboard? Explain.

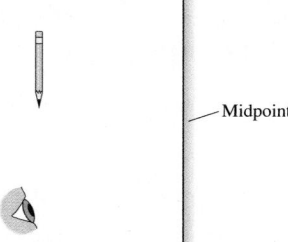

FIGURE Q23.3

4. One problem with using optical fibers for communication is that a light ray passing directly down the center of the fiber takes less time to travel from one end to the other than a ray taking a longer, zig-zag path. Thus light rays starting at the same time but traveling in slightly different directions reach the end of the fiber at different times. This problem can be solved by making the refractive index of the glass change gradually from a higher value in the center to a lower value near the edges of the fiber. Explain how this reduces the difference in travel times.

5. Suppose you looked at the sky on a clear day through pieces of red and blue plastic oriented as shown in FIGURE Q23.5. Describe the color and brightness of the light coming through sections 1, 2, and 3.

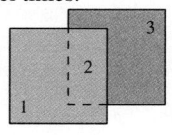

FIGURE Q23.5

6. A red card is illuminated by red light. What color will the card appear? What if it's illuminated by blue light?

7. The center of the galaxy is filled with low-density hydrogen gas. An astronomer wants to take a picture of the center of the galaxy. Will the view be better using ultraviolet light, visible light, or infrared light? (High-quality telescopes are available in all three spectral regions.) Explain.

8. Consider *one* point on an object near a lens.
   a. What is the minimum number of rays needed to locate its image point? Explain.
   b. How many rays from this point actually strike the lens and refract to the image point?

9. The object and lens in FIGURE Q23.9 are positioned to form a well-focused, inverted image on a viewing screen. Then a piece of cardboard is lowered just in front of the lens to cover the top half of the lens. Describe what you see on the screen when the cardboard is in place.

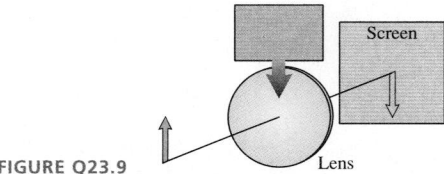

FIGURE Q23.9

10. FIGURE Q23.10 shows an object near a lens. The focal points are marked. Is there an image? If so, is the image real or virtual? Is it upright or inverted? If not, why not? Explain.

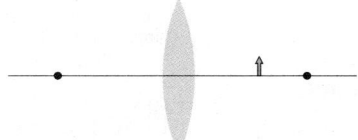

FIGURE Q23.10

11. A concave mirror brings the sun's rays to a focus in front of the mirror. Suppose the mirror is submerged in a swimming pool but still pointed up at the sun. Will the sun's rays be focused nearer to, farther from, or at the same distance from the mirror? Explain.

12. When you look at your reflection in the bowl of a spoon, it is upside down. Why?

EXERCISES AND PROBLEMS

Exercises

Section 23.1 The Ray Model of Light

1. ‖ a. How long (in ns) does it take light to travel 1.0 m in vacuum?
   b. What distance does light travel in water, glass, and cubic zirconia during the time that it travels 1.0 m in vacuum?

2. ‖ A point source of light illuminates an aperture 2.0 m away. A 12.0-cm-wide bright patch of light appears on a screen 1.0 m behind the aperture. How wide is the aperture?

3. ‖ A 5.0-cm-thick layer of oil is sandwiched between a 1.0-cm-thick sheet of glass and a 2.0-cm-thick sheet of polystyrene plastic. How long (in ns) does it take light incident perpendicular to the glass to pass through this 8.0-cm-thick sandwich?

4. ‖ A student has built a 15-cm-long pinhole camera for a science fair project. She wants to photograph her 180-cm-tall friend and have the image on the film be 5.0 cm high. How far should the front of the camera be from her friend?

Section 23.2 Reflection

5. | The mirror in FIGURE EX23.5 deflects a horizontal laser beam by 60°. What is the angle ϕ?

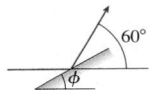

FIGURE EX23.5

6. | A light ray leaves point A in FIGURE EX23.6, reflects from the mirror, and reaches point B. How far below the top edge does the ray strike the mirror?

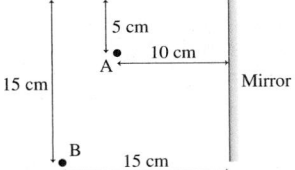

FIGURE EX23.6

7. ‖ The laser beam in FIGURE EX23.7 is aimed at the center of a rotating hexagonal mirror. How long is the streak of laser light as the reflected laser beam sweeps across the wall behind the laser?

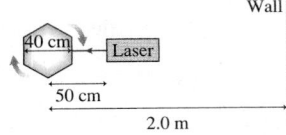

FIGURE EX23.7

8. ‖ At what angle ϕ should the laser beam in FIGURE EX23.8 be aimed at the mirrored ceiling in order to hit the midpoint of the far wall?

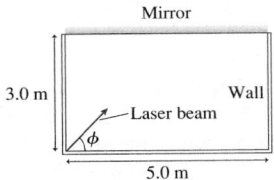

FIGURE EX23.8

9. ‖ It is 165 cm from your eyes to your toes. You're standing 200 cm in front of a tall mirror. How far is it from your eyes to the image of your toes?

Section 23.3 Refraction

10. ‖ A 1.0-cm-thick layer of water stands on a horizontal slab of glass. A light ray in the air is incident on the water 60° from the normal. What is the ray's direction of travel in the glass?

11. ‖ A costume jewelry pendant made of cubic zirconia is submerged in oil. A light ray strikes one face of the zirconia crystal at an angle of incidence of 25°. Once inside, what is the ray's angle with respect to the face of the crystal?

12. ‖ An underwater diver sees the sun 50° above horizontal. How high is the sun above the horizon to a fisherman in a boat above the diver?

13. | A laser beam in air is incident on a liquid at an angle of 53° with respect to the normal. The laser beam's angle in the liquid is 35°. What is the liquid's index of refraction?

14. ‖ The glass core of an optical fiber has an index of refraction 1.60. The index of refraction of the cladding is 1.48. What is the maximum angle a light ray can make with the wall of the core if it is to remain inside the fiber?

15. ‖ A thin glass rod is submerged in oil. What is the critical angle for light traveling inside the rod?

Section 23.4 Image Formation by Refraction

16. ‖ A fish in a flat-sided aquarium sees a can of fish food on the counter. To the fish's eye, the can looks to be 30 cm outside the aquarium. What is the actual distance between the can and the aquarium? (You can ignore the thin glass wall of the aquarium.)

17. | A biologist keeps a specimen of his favorite beetle embedded in a cube of polystyrene plastic. The hapless bug appears to be 2.0 cm within the plastic. What is the beetle's actual distance beneath the surface?

18. | A 150-cm-tall diver is standing completely submerged on the bottom of a swimming pool full of water. You are sitting on the end of the diving board, almost directly over her. How tall does the diver appear to be?

19. ‖ To a fish in an aquarium, the 4.00-mm-thick walls appear to be only 3.50 mm thick. What is the index of refraction of the walls?

Section 23.5 Color and Dispersion

20. ‖ A sheet of glass has $n_{red} = 1.52$ and $n_{violet} = 1.55$. A narrow beam of white light is incident on the glass at 30°. What is the angular spread of the light inside the glass?

21. | A narrow beam of white light is incident on a sheet of quartz. The beam disperses in the quartz, with red light ($\lambda \approx 700$ nm) traveling at an angle of 26.3° with respect to the normal and violet light ($\lambda \approx 400$ nm) traveling at 25.7°. The index of refraction of quartz for red light is 1.45. What is the index of refraction of quartz for violet light?

22. ‖ A hydrogen discharge lamp emits light with two prominent wavelengths: 656 nm (red) and 486 nm (blue). The light enters a flint-glass prism perpendicular to one face and then refracts through the hypotenuse back into the air. The angle between these two faces is 35°.
   a. Use Figure 23.28 to estimate to ±0.002 the index of refraction of flint glass at these two wavelengths.
   b. What is the angle (in degrees) between the red and blue light as it leaves the prism?

23. ‖ Infrared telescopes, which use special infrared detectors, are able to peer farther into star-forming regions of the galaxy because infrared light is not scattered as strongly as is visible light by the tenuous clouds of hydrogen gas from which new stars are created. For what wavelength of light is the scattering only 1% that of light with a visible wavelength of 500 nm?

Section 23.6 Thin Lenses: Ray Tracing

24. ‖ An object is 20 cm in front of a converging lens with a focal length of 10 cm. Use ray tracing to determine the location of the image. Is the image upright or inverted?

25. ‖ An object is 30 cm in front of a converging lens with a focal length of 5 cm. Use ray tracing to determine the location of the image. Is the image upright or inverted?

26. ‖ An object is 6 cm in front of a converging lens with a focal length of 10 cm. Use ray tracing to determine the location of the image. Is the image upright or inverted?

27. ‖ An object is 15 cm in front of a diverging lens with a focal length of −15 cm. Use ray tracing to determine the location of the image. Is the image upright or inverted?

Section 23.7 Thin Lenses: Refraction Theory

28. ‖ Find the focal length of the glass lens in FIGURE EX23.28.

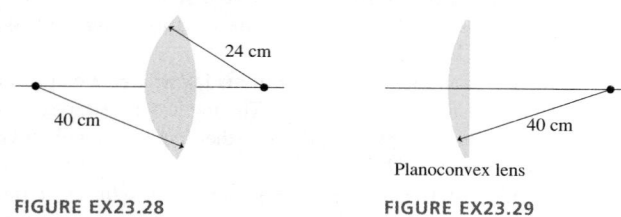

FIGURE EX23.28   FIGURE EX23.29

29. ‖ Find the focal length of the planoconvex polystyrene plastic lens in FIGURE EX23.29.

30. ‖ Find the focal length of the glass lens in FIGURE EX23.30.

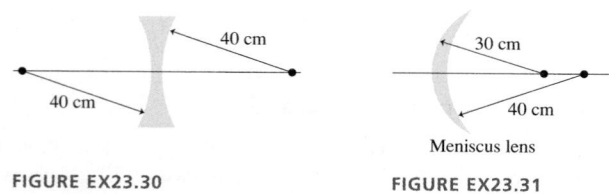

FIGURE EX23.30   FIGURE EX23.31

31. ‖ Find the focal length of the meniscus polystyrene plastic lens in FIGURE EX23.31.

32. ‖ An air bubble inside an 8.0-cm-diameter plastic ball is 2.0 cm from the surface. As you look at the ball with the bubble turned toward you, how far beneath the surface does the bubble appear to be?

33. ‖ A goldfish lives in a 50-cm-diameter spherical fish bowl. The fish sees a cat watching it. If the cat's face is 20 cm from the edge of the bowl, how far from the edge does the fish see it as being? (You can ignore the thin glass wall of the bowl.)

34. ‖ A 1.0-cm-tall candle flame is 60 cm from a lens with a focal length of 20 cm. What are the image distance and the height of the flame's image?

Section 23.8 Image Formation with Spherical Mirrors

35. ‖ An object is 40 cm in front of a concave mirror with a focal length of 20 cm. Use ray tracing to locate the image. Is the image upright or inverted?

36. ‖ An object is 12 cm in front of a concave mirror with a focal length of 20 cm. Use ray tracing to locate the image. Is the image upright or inverted?

37. ‖ An object is 30 cm in front of a convex mirror with a focal length of −20 cm. Use ray tracing to locate the image. Is the image upright or inverted?

Problems

38. ‖ An advanced computer sends information to its various parts via infrared light pulses traveling through silicon fibers. To acquire data from memory, the central processing unit sends a light-pulse request to the memory unit. The memory unit processes the request, then sends a data pulse back to the central processing unit. The memory unit takes 0.5 ns to process a request. If the information has to be obtained from memory in 2.0 ns, what is the maximum distance the memory unit can be from the central processing unit?

39. ‖ A red ball is placed at point A in FIGURE P23.39.
   a. How many images are seen by an observer at point O?
   b. What are the (x, y) coordinates of each image?

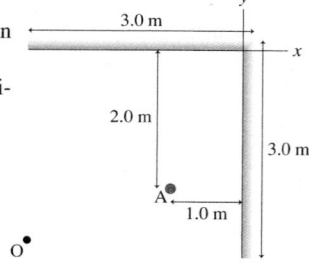

FIGURE P23.39

40. ‖ A laser beam is incident on the left mirror in FIGURE P23.40. Its initial direction is parallel to a line that bisects the mirrors. What is the angle φ of the reflected laser beam?

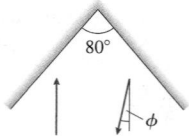

FIGURE P23.40

41. ‖ The place you get your hair cut has two nearly parallel mirrors 5.0 m apart. As you sit in the chair, your head is 2.0 m from the nearer mirror. Looking toward this mirror, you first see your face and then, farther away, the back of your head. (The mirrors need to be slightly nonparallel for you to be able to see the back of your head, but you can treat them as parallel in this problem.) How far away does the back of your head appear to be? Neglect the thickness of your head.

42. ‖ You're helping with an experiment in which a vertical cylinder will rotate about its axis by a very small angle. You need to devise a way to measure this angle. You decide to use what is called an *optical lever.* You begin by mounting a small mirror

on top of the cylinder. A laser 5.0 m away shoots a laser beam at the mirror. Before the experiment starts, the mirror is adjusted to reflect the laser beam directly back to the laser. Later, you measure that the reflected laser beam, when it returns to the laser, has been deflected sideways by 2.0 mm. Through how many degrees has the cylinder rotated?

43. ‖ A microscope is focused on a black dot. When a 1.00-cm-thick piece of plastic is placed over the dot, the microscope objective has to be raised 0.40 cm to bring the dot back into focus. What is the index of refraction of the plastic?

44. ‖ A light ray in air is incident on a transparent material whose index of refraction is n.
    a. Find an expression for the (non-zero) angle of incidence whose angle of refraction is half the angle of incidence.
    b. Evaluate your expression for light incident on glass.

45. ‖ The meter stick in **FIGURE P23.45** lies on the bottom of a 100-cm-long tank with its zero mark against the left edge. You look into the tank at a 30° angle, with your line of sight just grazing the upper left edge of the tank. What mark do you see on the meter stick if the tank is (a) empty, (b) half full of water, and (c) completely full of water?

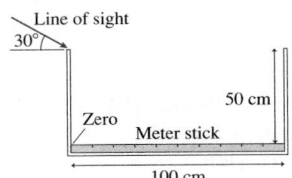

FIGURE P23.45

46. ‖ The 80-cm-tall, 65-cm-wide tank shown in **FIGURE P23.46** is completely filled with water. The tank has marks every 10 cm along one wall, and the 0 cm mark is barely submerged. As you stand beside the opposite wall, your eye is level with the top of the water.
    a. Can you see the marks from the top of the tank (the 0 cm mark) going down, or from the bottom of the tank (the 80 cm mark) coming up? Explain.
    b. Which is the lowest or highest mark, depending on your answer to part a, that you can see?

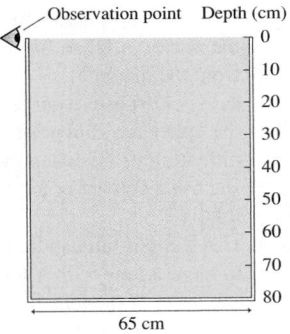

FIGURE P23.46

47. ‖ A 4.0-m-wide swimming pool is filled to the top. The bottom of the pool becomes completely shaded in the afternoon when the sun is 20° above the horizon. How deep is the pool?

48. ‖ It's nighttime, and you've dropped your goggles into a 3.0-m-deep swimming pool. If you hold a laser pointer 1.0 m above the edge of the pool, you can illuminate the goggles if the laser beam enters the water 2.0 m from the edge. How far are the goggles from the edge of the pool?

49. ‖ Shown from above in **FIGURE P23.49** is one corner of a rectangular box filled with water. A laser beam starts 10 cm from side A of the container and enters the water at position x. You can ignore the thin walls of the container.
    a. If $x = 15$ cm, does the laser beam refract back into the air through side B or reflect from side B back into the water? Determine the angle of refraction or reflection.

b. Repeat part a for $x = 25$ cm.
c. Find the minimum value of x for which the laser beam passes through side B and emerges into the air.

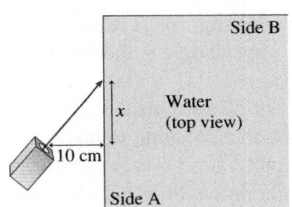

FIGURE P23.49

50. ‖ A fish is 20 m from the shore of a lake. A bonfire is burning on the edge of the lake nearest the fish.
    a. Does the fish need to be shallow (just below the surface) or very deep to see the light from the bonfire? Explain.
    b. What is the deepest or shallowest, depending on your answer to part a, that the fish can be and still see light from the fire?

51. ‖ Your supervisor asks you to measure the index of refraction of a piece of plastic. You notice that, because of scattering of the light, you can see the path of a laser beam through the plastic. You decide to shoot a laser beam toward the plastic at several different incident angles and measure the refraction angle in the plastic. Your data are as follows:

Incident angle	Refraction angle
15°	9°
30°	19°
45°	26°
60°	34°
75°	37°

Use the best-fit line of an appropriate graph to determine the plastic's index of refraction.

52. ‖‖ One of the contests at the school carnival is to throw a spear at an underwater target lying flat on the bottom of a pool. The water is 1.0 m deep. You're standing on a small stool that places your eyes 3.0 m above the bottom of the pool. As you look at the target, your gaze is 30° below horizontal. At what angle below horizontal should you throw the spear in order to hit the target? Your raised arm brings the spear point to the level of your eyes as you throw it, and over this short distance you can assume that the spear travels in a straight line rather than a parabolic trajectory.

53. ‖ White light is incident onto a 30° prism at the 40° angle shown in **FIGURE P23.53**. Violet light emerges perpendicular to the rear face of the prism. The index of refraction of violet light in this glass is 2.0% larger than the index of refraction of red light. At what angle ϕ does red light emerge from the rear face?

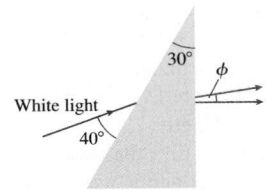

FIGURE P23.53

54. ‖ There's one angle of incidence β onto a prism for which the light inside an isosceles prism travels parallel to the base and emerges at angle β.

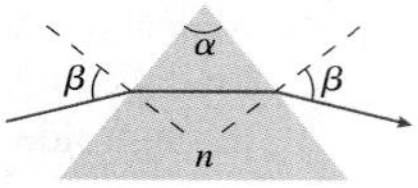

    FIGURE P23.54

    a. Find an expression for β in terms of the prism's apex angle α and index of refraction n.
    b. A laboratory measurement finds that $\beta = 52.2°$ for a prism shaped like an equilateral triangle. What is the prism's index of refraction?

55. ‖ Paraxial light rays approach a transparent sphere parallel to an optical axis passing through the center of the sphere. The rays come to a focus on the far surface of the sphere. What is the sphere's index of refraction?

56. ‖ A 6.0-cm-diameter cubic zirconia sphere has an air bubble exactly in the center. As you look into the sphere, how far beneath the surface does the bubble appear to be?

57. ‖ A 1.0-cm-tall object is 10 cm in front of a converging lens that has a 30 cm focal length.
    a. Use ray tracing to find the position and height of the image. To do this accurately, use a ruler or paper with a grid. Determine the image distance and image height by making measurements on your diagram.
    b. Calculate the image position and height. Compare with your ray-tracing answers in part a.

58. ‖ A 2.0-cm-tall object is 40 cm in front of a converging lens that has a 20 cm focal length.
    a. Use ray tracing to find the position and height of the image. To do this accurately, use a ruler or paper with a grid. Determine the image distance and image height by making measurements on your diagram.
    b. Calculate the image position and height. Compare with your ray-tracing answers in part a.

59. ‖ A 1.0-cm-tall object is 75 cm in front of a converging lens that has a 30 cm focal length.
    a. Use ray tracing to find the position and height of the image. To do this accurately, use a ruler or paper with a grid. Determine the image distance and image height by making measurements on your diagram.
    b. Calculate the image position and height. Compare with your ray-tracing answers in part a.

60. ‖ A 2.0-cm-tall object is 15 cm in front of a converging lens that has a 20 cm focal length.
    a. Use ray tracing to find the position and height of the image. To do this accurately, use a ruler or paper with a grid. Determine the image distance and image height by making measurements on your diagram.
    b. Calculate the image position and height. Compare with your ray-tracing answers in part a.

61. ‖ A 1.0-cm-tall object is 60 cm in front of a diverging lens that has a −30 cm focal length.
    a. Use ray tracing to find the position and height of the image. To do this accurately, use a ruler or paper with a grid. Determine the image distance and image height by making measurements on your diagram.
    b. Calculate the image position and height. Compare with your ray-tracing answers in part a.

62. ‖ A 2.0-cm-tall object is 15 cm in front of a diverging lens that has a −20 cm focal length.
    a. Use ray tracing to find the position and height of the image. To do this accurately, use a ruler or paper with a grid.

Determine the image distance and image height by making measurements on your diagram.
    b. Calculate the image position and height. Compare with your ray-tracing answers in part a.

63. ‖ To determine the focal length of a lens, you place the lens in front of a small lightbulb and then adjust a viewing screen to get a sharply focused image. Varying the lens position produces the following data:

Bulb to lens (cm)	Lens to screen (cm)
20	61
22	47
24	39
26	37
28	32

Use the best-fit line of an appropriate graph to determine the focal length of the lens.

64. | A 1.0-cm-tall object is 20 cm in front of a concave mirror that has a 60 cm focal length. Calculate the position and height of the image. State whether the image is in front of or behind the mirror, and whether the image is upright or inverted.

65. | A 1.0-cm-tall object is 20 cm in front of a convex mirror that has a −60 cm focal length. Calculate the position and height of the image. State whether the image is in front of or behind the mirror, and whether the image is upright or inverted.

66. ‖ The illumination lights in an operating room use a concave
BIO mirror to focus an image of a bright lamp onto the surgical site. One such light uses a mirror with a 30 cm radius of curvature. If the mirror is 1.2 m from the patient, how far should the lamp be from the mirror?

67. ‖ A dentist uses a curved mirror to view the back side of teeth in
BIO the upper jaw. Suppose she wants an upright image with a magnification of 1.5 when the mirror is 1.2 cm from a tooth. Should she use a convex or a concave mirror? What focal length should it have?

68. ‖ A 2.0-cm-tall candle flame is 2.0 m from a wall. You happen to have a lens with a focal length of 32 cm. How many places can you put the lens to form a well-focused image of the candle flame on the wall? For each location, what are the height and orientation of the image?

69. ‖ A lightbulb is 3.0 m from a wall. What are the focal length and the position (measured from the bulb) of a lens that will form an image on the wall that is twice the size of the lightbulb?

70. ‖ a. Estimate the diameter of your eyeball.
BIO b. Bring this page up to the closest distance at which the text is sharp—not the closest at which you can still read it, but the closest at which the letters remain sharp. If you wear glasses or contact lenses, leave them on. This distance is called the *near point* of your (possibly corrected) eye. Measure it.
    c. Estimate the effective focal length of your eye. The effective focal length includes the focusing due to the lens, the curvature of the cornea, and any corrections you wear. Ignore the effects of the fluid in your eye.

71. ‖ A slide projector needs to create a 98-cm-high image of a 2.0-cm-tall slide. The screen is 300 cm from the slide.
    a. What focal length does the lens need? Assume that it is a thin lens.
    b. How far should you place the lens from the slide?

72. ‖ A lens placed 10 cm in front of an object creates an upright image twice the height of the object. The lens is then moved along the optical axis until it creates an inverted image twice the height of the object. How far did the lens move?

73. ‖ An object is 60 cm from a screen. What are the radii of a symmetric converging plastic lens (i.e., two equally curved surfaces) that will form an image on the screen twice the height of the object?

74. ‖ A sports photographer has a 150-mm-focal-length lens on his camera. The photographer wants to photograph a sprinter running straight away from him at 5.0 m/s. What is the speed (in mm/s) of the sprinter's image at the instant the sprinter is 10 m in front of the lens?

75. ‖ A concave mirror has a 40 cm radius of curvature. How far from the mirror must an object be placed to create an upright image three times the height of the object?

76. ‖‖ A 2.0-cm-tall object is placed in front of a mirror. A 1.0-cm-tall upright image is formed behind the mirror, 150 cm from the object. What is the focal length of the mirror?

77. ‖ A spherical mirror of radius R has its center at C, as shown in FIGURE P23.77. A ray parallel to the axis reflects through F, the focal point. Prove that $f = R/2$ if $\phi \ll 1$ rad.

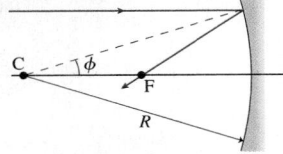

FIGURE P23.77

Challenge Problems

78. Consider a lens having index of refraction n_2 and surfaces with radii R_1 and R_2. The lens is immersed in a fluid that has index of refraction n_1.
    a. Derive a generalized lens maker's equation to replace Equation 23.27 when the lens is surrounded by a medium other than air. That is, when $n_1 \neq 1$.
    b. A symmetric converging glass lens (i.e., two equally curved surfaces) has two surfaces with radii of 40 cm. Find the focal length of this lens in air and the focal length of this lens in water.

79. FIGURE CP23.79 shows a light ray that travels from point A to point B. The ray crosses the boundary at position x, making angles θ_1 and θ_2 in the two media. Suppose that you did *not* know Snell's law.
    a. Write an expression for the *time t* it takes the light ray to travel from A to B. Your expression should be in terms of the distances a, b, and w; the variable x; and the indices of refraction n_1 and n_2.

b. The time depends on x. There's one value of x for which the light travels from A to B in the shortest possible time. We'll call it x_{min}. Write an expression (but don't try to solve it!) from which x_{min} could be found.

c. Now, by using the geometry of the figure, derive Snell's law from your answer to part b.

You've proven that Snell's law is equivalent to the statement that "light traveling between two points follows the path that requires the shortest time." This interesting way of thinking about refraction is called *Fermat's principle.*

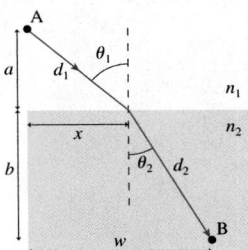

FIGURE P23.79

80. A fortune teller's "crystal ball" (actually just glass) is 10 cm in diameter. Her secret ring is placed 6.0 cm from the edge of the ball.
    a. An image of the ring appears on the opposite side of the crystal ball. How far is the image from the center of the ball?
    b. Draw a ray diagram showing the formation of the image.
    c. The crystal ball is removed and a thin lens is placed where the center of the ball had been. If the image is still in the same position, what is the focal length of the lens?

81. A beam of white light enters a transparent material. Wavelengths for which the index of refraction is n are refracted at angle θ_2. Wavelengths for which the index of refraction is $n + \delta n$, where $\delta n \ll n$, are refracted at angle $\theta_2 + \delta \theta$.
    a. Show that the angular separation in radians is $\delta\theta = -(\delta n/n)\tan\theta_2$.
    b. A beam of white light is incident on a piece of glass at 30.0°. Deep violet light is refracted 0.28° more than deep red light. The index of refraction for deep red light is known to be 1.552. What is the index of refraction for deep violet light?

82. Consider an object of thickness ds (parallel to the axis) in front of a lens or mirror. The image of the object has thickness ds'. Define the *longitudinal magnification* as $M = ds'/ds$. Prove that $M = -m^2$, where m is the lateral magnification.

STOP TO THINK ANSWERS

Stop to Think 23.1: c. The light spreads vertically as it goes through the vertical aperture. The light spreads horizontally due to different points on the horizontal lightbulb.

Stop to Think 23.2: c. There's one image behind the vertical mirror and a second behind the horizontal mirror. A third image in the corner arises from rays that reflect twice, once off each mirror.

Stop to Think 23.3: a. The ray travels closer to the normal in both media 1 and 3 than in medium 2, so n_1 and n_3 are both larger than n_2. The angle is smaller in medium 3 than in medium 1, so $n_3 > n_1$.

Stop to Think 23.4: e. The rays from the object are diverging. Without a lens, the rays cannot converge to form any kind of image on the screen.

Stop to Think 23.5: a, e, or f. Any of these will increase the angle of refraction θ_2.

Stop to Think 23.6: Away from. You need to decrease s' to bring the image plane onto the screen. s' is decreased by increasing s.

Stop to Think 23.7: c. A concave mirror forms a real image in front of the mirror. Because the object distance is $s \approx \infty$, the image distance is $s' \approx f$.

24 Optical Instruments

The world's greatest collection of telescopes is on the summit of Mauna Kea on the Big Island of Hawaii, towering 4200 m (13,800 ft) over the Pacific Ocean.

▶ **Looking Ahead** The goal of Chapter 24 is to understand some common optical instruments and their limitations.

Lenses in Combination

The "lenses" of optical instruments are always built with several individual lenses to give better optical performance.

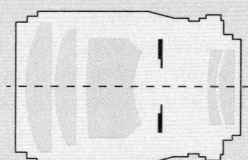

A cross section of a typical camera lens shows that it is built of 5 individual lenses and an adjustable iris.

You'll learn how to analyze a system with multiple lenses.

Optical Systems That Magnify

Lenses and mirrors can be used to magnify objects both near and far. Optical instruments open a realm far beyond what the unaided eye can see.

A simple magnifying glass has a low magnification of only 2× or 3×.

A microscope uses two sets of lenses in combination to produce magnifications of up to 1000×.

Small telescopes use lenses; larger telescopes use a curved mirror as the primary optical element.

The Camera

A camera uses a lens to project a real image onto a light-sensitive detector.

Although a modern digital camera is very complex, at its heart it's just a light-tight box with a lens to focus the image.

You'll learn about focusing, zoom, and exposure.

◀ **Looking Back**
Sections 23.6–23.7 Ray tracing and image formation by lenses

The Human Eye

The human eye is much like a camera: The cornea and lens together focus a real image onto the retina.

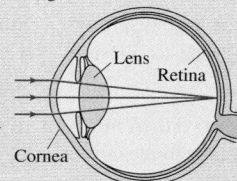

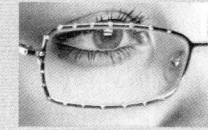

You'll discover how eyeglasses and contact lenses are used to correct defects of vision.

Resolution of Lenses

Light passing through a lens undergoes diffraction, just like light passing through a circular hole. Diffraction limits a lens's ability to form a perfectly focused image.

An ideal lens would have focused the light to two points. Instead, we get two overlapped diffraction patterns.

You'll learn about *Rayleigh's criterion* for when two images can be resolved.

◀ **Looking Back**
Section 22.5 Circular diffraction

24.1 Lenses in Combination

Only the simplest magnifiers are built with a single lens of the sort we analyzed in Chapter 23. Optical instruments, such as microscopes and cameras, are invariably built with multiple lenses. The reason, as we'll see, is to improve the image quality.

The analysis of multi-lens systems requires only one new rule: **The image of the first lens acts as the object for the second lens.** To see why this is so, FIGURE 24.1 shows a simple telescope consisting of a large-diameter converging lens, called the *objective,* and a smaller converging lens used as the *eyepiece.* (We'll analyze telescopes more thoroughly later in the chapter.) Highlighted are the three special rays you learned to use in Chapter 23:

■ A ray parallel to the optical axis refracts through the focal point.
■ A ray through the focal point refracts parallel to the optical axis.
■ A ray through the center of the lens is undeviated.

FIGURE 24.1 Ray-tracing diagram of a simple astronomical telescope.

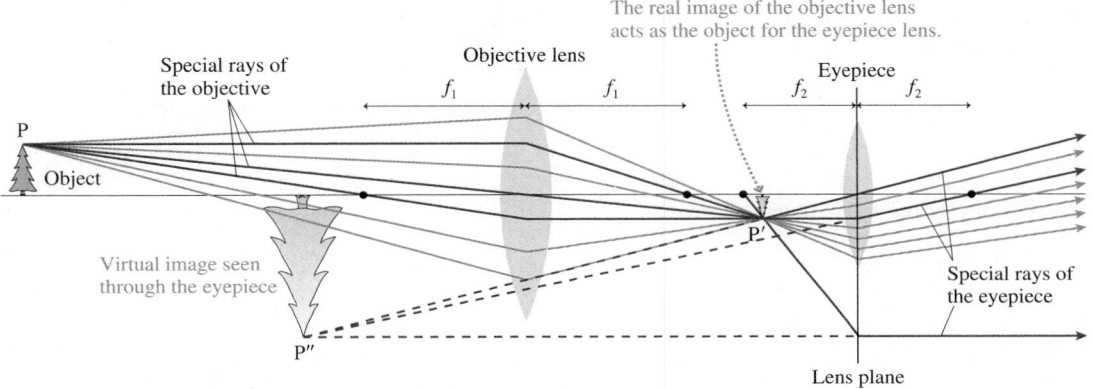

The rays passing through the objective converge to a real image at P′, but they don't stop there. Instead, light rays *diverge* from P′ as they approach the second lens. **As far as the eyepiece is concerned, the rays are coming from P′, and thus P′ acts as the object for the second lens.** The three special rays passing through the objective lens are sufficient to locate the image P′, but these rays are generally *not* the special rays for the second lens. However, other rays converging at P′ leave at the correct angles to be the special rays for the eyepiece. That is, a new set of special rays is drawn from P′ to the second lens and used to find the final image point P″.

NOTE ▶ One ray seems to "miss" the eyepiece lens, but this isn't a problem. All rays passing through the lens converge to (or diverge from) a single point, and the purpose of the special rays is to locate that point. To do so, we can let the special rays refract as they cross the *lens plane,* regardless of whether the physical lens really extends that far. ◀

EXAMPLE 24.1 **A camera lens**

The "lens" on a camera is usually a combination of two or more single lenses. Consider a camera in which light passes first through a diverging lens, with $f_1 = -120$ mm, then a converging lens, with $f_2 = 42$ mm, spaced 60 mm apart. A reasonable definition of the *effective focal length* of this lens combination is the focal length of a *single* lens that could produce an image in the same location if placed at the midpoint of the lens combination. A 10-cm-tall object is 500 mm from the first lens.

a. What are the location, size, and orientation of the image?
b. What is the effective focal length of the double-lens system used in this camera?

MODEL Each lens is a thin lens. The image of the first lens is the object for the second.

VISUALIZE The ray-tracing diagram of FIGURE 24.2 shows the production of a real, inverted image ≈ 55 mm behind the second lens.

Continued

SOLVE

a. $s_1 = 500$ mm is the object distance of the first lens. Its image, a virtual image, is found from the thin-lens equation:

$$\frac{1}{s_1'} = \frac{1}{f_1} - \frac{1}{s_1} = \frac{1}{-120 \text{ mm}} - \frac{1}{500 \text{ mm}} = -0.0103 \text{ mm}^{-1}$$

$$s_1' = -97 \text{ mm}$$

This is consistent with the ray-tracing diagram. The image of the first lens now acts as the object for the second lens. Because the lenses are 60 mm apart, the object distance is $s_2 = 97$ mm + 60 mm = 157 mm. A second application of the thin-lens equation yields

$$\frac{1}{s_2'} = \frac{1}{f_2} - \frac{1}{s_2} = \frac{1}{42 \text{ mm}} - \frac{1}{157 \text{ mm}} = 0.0174 \text{ mm}^{-1}$$

$$s_2' = 57 \text{ mm}$$

The image of the lens combination is 57 mm behind the second lens. The lateral magnifications of the two lenses are

$$m_1 = -\frac{s_1'}{s_1} = -\frac{-97 \text{ cm}}{500 \text{ cm}} = 0.194$$

$$m_2 = -\frac{s_2'}{s_2} = -\frac{57 \text{ cm}}{157 \text{ cm}} = -0.363$$

The second lens magnifies the image of the first lens, which magnifies the object, so **the total magnification is the product of the individual magnifications:**

$$m = m_1 m_2 = -0.070$$

Thus the image is 57 mm behind the second lens, inverted (m is negative), and 0.70 cm tall.

b. If a single lens midway between these two lenses produced an image in the same plane, its object and image distances would be $s = 500$ mm + 30 mm = 530 mm and $s' = 57$ mm + 30 mm = 87 mm. A final application of the thin-lens equation gives the effective focal length:

$$\frac{1}{f_{\text{eff}}} = \frac{1}{s} + \frac{1}{s'} = \frac{1}{530 \text{ mm}} + \frac{1}{87 \text{ mm}} = 0.0134 \text{ mm}^{-1}$$

$$f_{\text{eff}} = 75 \text{ mm}$$

ASSESS This combination lens would be sold as a "75 mm lens."

FIGURE 24.2 Pictorial representation of a combination lens.

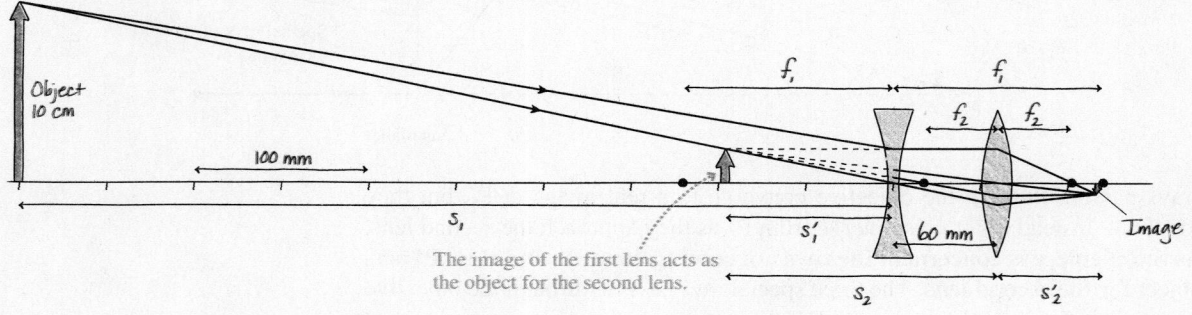

STOP TO THINK 24.1 The second lens in this optical instrument

a. Causes the light rays to focus closer than they would with the first lens acting alone.
b. Causes the light rays to focus farther away than they would with the first lens acting alone.
c. Inverts the image but does not change where the light rays focus.
d. Prevents the light rays from reaching a focus.

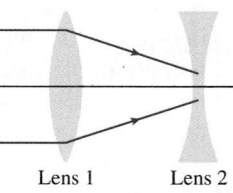

24.2 The Camera

A **camera,** shown in FIGURE 24.3, "takes a picture" by using a lens to form a real, inverted image on a light-sensitive detector in a light-tight box. Film was the detector of choice for well over a hundred years, but today's digital cameras use an electronic detector called a *charge-coupled device,* or CCD.

FIGURE 24.3 A camera.

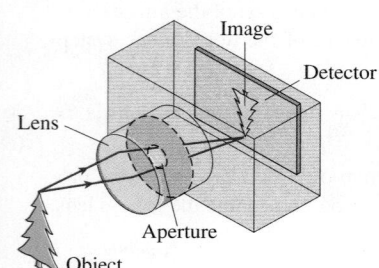

The camera "lens" is always a combination of two or more individual lenses. The simplest such lens, shown in FIGURE 24.4, consists of a converging lens and a somewhat weaker diverging lens. This combination of positive and negative lenses corrects some of the defects inherent in single lenses, as we'll discuss later in the chapter. As Example 24.1 suggested, we can model a combination lens as a single lens with an **effective focal length** (usually called simply "the focal length") f. A *zoom lens* changes the effective focal length by changing the spacing between the converging lens and the diverging lens; this is what happens when the lens barrel on your digital camera moves in and out as you use the zoom. A typical digital camera has a lens whose effective focal length can be varied from 6 mm to 18 mm, giving, as we'll see, a 3× zoom.

FIGURE 24.4 A simple camera lens is a combination lens.

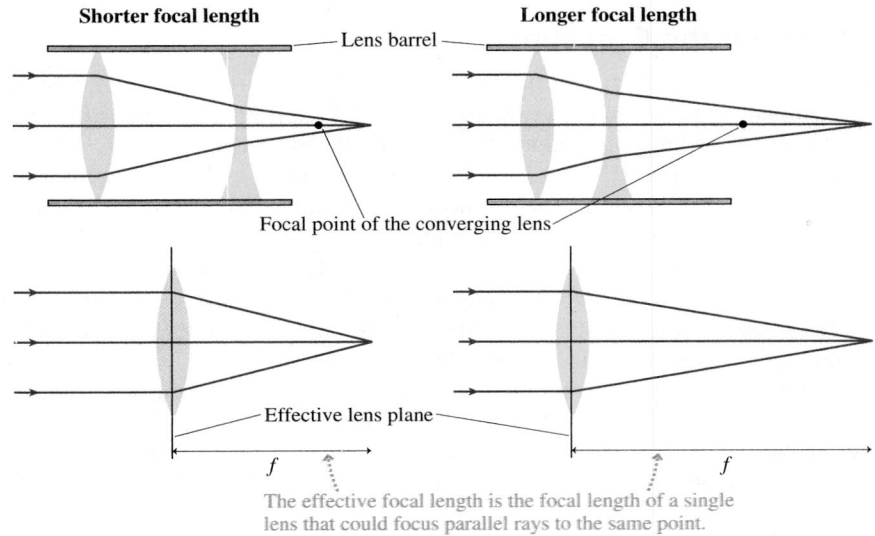

The effective focal length is the focal length of a single lens that could focus parallel rays to the same point.

A camera must carry out two important functions: focus the image on the detector and control the exposure. Cameras are focused by moving the lens forward or backward until the image is well focused on the detector. Most modern cameras do this automatically, but older cameras required manual focusing.

EXAMPLE 24.2 **Focusing a camera**

Your digital camera lens, with an effective focal length of 10.0 mm, is focused on a flower 20.0 cm away. You then turn to take a picture of a distant landscape. How far, and in which direction, must the lens move to bring the landscape into focus?

MODEL Model the camera's combination lens as a single thin lens with $f = 10.0$ mm. Image and object distances are measured from the effective lens plane. Assume all the lenses in the combination move together as the camera refocuses.

SOLVE The flower is at object distance $s = 20.0$ cm $= 200$ mm. When the camera is focused, the image distance between the

effective lens plane and the detector is found by solving the thin-lens equation $1/s + 1/s' = 1/f$ to give

$$s' = \left(\frac{1}{f} - \frac{1}{s}\right)^{-1} = \left(\frac{1}{10.0 \text{ mm}} - \frac{1}{200 \text{ mm}}\right)^{-1} = 10.5 \text{ mm}$$

The distant landscape is effectively at object distance $s = \infty$, so its image distance is $s' = f = 10.0$ mm. To refocus as you shift scenes, the lens must move 0.5 mm closer to the detector.

ASSESS The required motion of the lens is very small, about the diameter of the lead used in a mechanical pencil.

Zoom Lenses

For objects more than 10 focal lengths from the lens (roughly $s > 20$ cm for a typical digital camera), the approximation $s \gg f$ (and thus $1/s \ll 1/f$) leads to $s' \approx f$. In other words, objects more than about 10 focal lengths away are essentially "at infinity," and we know that the parallel rays from an infinitely distant object are focused one focal length behind the lens. For such an object, the lateral magnification of the image is

$$m = -\frac{s'}{s} \approx -\frac{f}{s} \tag{24.1}$$

The magnification is much less than 1, because $s \gg f$, so the image on the detector is much smaller than the object itself. This comes as no surprise. More important, **the size of the image is directly proportional to the focal length of the lens.** We saw in Figure 24.4 that the effective focal length of a combination lens is easily changed by varying the distance between the individual lenses, and this is exactly how a zoom lens works. A lens that can be varied from $f_{min} = 6$ mm to $f_{max} = 18$ mm gives magnifications spanning a factor of 3, and that is why you see it specified as a $3\times$ zoom lens.

Controlling the Exposure

The camera also must control the amount of light reaching the detector. Too little light results in photos that are *underexposed;* too much light gives *overexposed* pictures. Both the shutter and the lens diameter help control the exposure.

The *shutter* is "opened" for a selected amount of time as the image is recorded. Older cameras used a spring-loaded mechanical shutter that literally opened and closed; digital cameras electronically control the amount of time the detector is active. Either way, the exposure—the amount of light captured by the detector—is directly proportional to the time the shutter is open. Typical exposure times range from 1/1000 s or less for a sunny scene to 1/30 s or more for dimly lit or indoor scenes. The exposure time is generally referred to as the *shutter speed.*

The amount of light passing through the lens is controlled by an adjustable **aperture,** also called an *iris* because it functions much like the iris of your eye. The aperture sets the effective diameter D of the lens. The full area of the lens is used when the aperture is fully open, but a *stopped-down* aperture allows light to pass through only the central portion of the lens.

The light intensity on the detector is directly proportional to the area of the lens; a lens with twice as much area will collect and focus twice as many light rays from the object to make an image twice as bright. The lens area is proportional to the square of its diameter, so the intensity I is proportional to D^2. The light intensity—power per square meter—is also *inversely* proportional to the area of the image. That is, the light reaching the detector is more intense if the rays collected from the object are focused into a small area than if they are spread out over a large area. The lateral size of the image is proportional to the focal length of the lens, as we saw in Equation 24.1, so the *area* of the image is proportional to f^2 and thus I is proportional to $1/f^2$. Altogether, $I \propto D^2/f^2$.

By long tradition, the light-gathering ability of a lens is specified by its **f-number,** defined as

$$f\text{-number} = \frac{f}{D} \tag{24.2}$$

The *f*-number of a lens may be written either as *f*/4.0, to mean that the *f*-number is 4.0, or as F4.0. The instruction manuals with some digital cameras call this the *aperture value* rather than the *f*-number. A digital camera in fully automatic mode does not display shutter speed or *f*-number, but that information is displayed if you set your camera to any of the other modes. For example, the display 1/125 F5.6 means that your camera is going to achieve the correct exposure by adjusting the diameter of the lens aperture to give $f/D = 5.6$ and by opening the shutter for 1/125 s. If your lens's effective focal length is 10 mm, the diameter of the lens aperture will be

$$D = \frac{f}{f\text{-number}} = \frac{10 \text{ mm}}{5.6} = 1.8 \text{ mm}$$

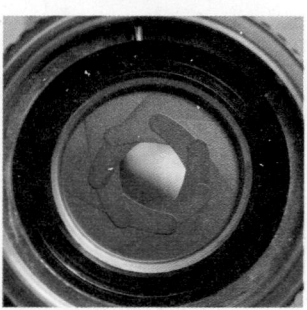

An iris can change the effective diameter of a lens and thus the amount of light reaching the detector.

NOTE ▶ The f in f-number is not the focal length f; it's just a name. And the / in $f/4$ does not mean division; it's just a notation. These both derive from the long history of photography. ◀

Because the aperture diameter is in the denominator of the f-number, a *larger-diameter* aperture, which gathers more light and makes a brighter image, has a *smaller* f-number. The light intensity on the detector is related to the lens's f-number by

$$I \propto \frac{D^2}{f^2} = \frac{1}{(f\text{-number})^2} \tag{24.3}$$

Historically, a lens's f-numbers could be adjusted in the sequence 2.0, 2.8, 4.0, 5.8, 8.0, 11, 16. Each differs from its neighbor by a factor of $\sqrt{2}$, so changing the lens by one "f stop" changed the light intensity by a factor of 2. A modern digital camera is able to adjust the f-number continuously.

The exposure, the total light reaching the detector while the shutter is open, depends on the product $I\Delta t_{\text{shutter}}$. A small f-number (large aperture diameter D) and short $\Delta t_{\text{shutter}}$ can produce the same exposure as a larger f-number (smaller aperture) and a longer $\Delta t_{\text{shutter}}$. It might not make any difference for taking a picture of a distant mountain, but action photography needs very short shutter times to "freeze" the action. Thus action photography requires a large-diameter lens with a small f-number.

Focal length and f-number information is stamped on a camera lens. This lens is labeled 5.8–23.2 mm 1:2.6–5.5. The first numbers are the range of focal lengths. They span a factor of 4, so this is a 4× zoom lens. The second numbers show that the minimum f-number ranges from $f/2.6$ (for the $f = 5.8$ mm focal length) to $f/5.5$ (for the $f = 23.2$ mm focal length).

EXAMPLE 24.3 | **Capturing the action**

Before a race, a photographer finds that she can make a perfectly exposed photo of the track while using a shutter speed of 1/250 s and a lens setting of $f/8.0$. To freeze the sprinters as they go past, she plans to use a shutter speed of 1/1000 s. To what f-number must she set her lens?

MODEL The exposure depends on $I\Delta t_{\text{shutter}}$, and the light intensity depends inversely on the square of the f-number.

SOLVE Changing the shutter speed from 1/250 s to 1/1000 s will reduce the light reaching the detector by a factor of 4. To compensate, she needs to let 4 times as much light through the lens. Because $I \propto 1/(f\text{-number})^2$, the intensity will increase by a factor of 4 if she *decreases* the f-number by a factor of 2. Thus the correct lens setting is $f/4.0$.

ASSESS To keep the photo properly exposed, a decreased shutter time must be balanced by an increased lens aperture diameter.

The Detector

For traditional cameras, the light-sensitive detector is film. Today's digital cameras use an electronic light-sensitive surface called a *charge-coupled device* or **CCD**. A CCD consists of a rectangular array of many millions of small detectors called **pixels**. When light hits one of these pixels, it generates an electric charge proportional to the light intensity. Thus an image is recorded on the CCD in terms of little packets of charge. After the CCD has been exposed, the charges are read out, the signal levels are digitized, and the picture is stored in the digital memory of the camera.

FIGURE 24.5a shows a CCD "chip" and, schematically, the magnified appearance of the pixels on its surface. To record color information, different pixels are covered by red, green, or blue filters. A pixel covered by a green filter, for instance, records only the intensity of the green light hitting it. Later, the camera's microprocessor interpolates nearby colors to give each pixel an overall true color. The pixels are so small that the picture looks "smooth" even after some enlargement, but, as you can see in FIGURE 24.5b, sufficient magnification reveals the individual pixels.

FIGURE 24.5 The CCD detector used in a digital camera.

(a) 2500 × 2000 pixels

1 pixel

(b)

STOP TO THINK 24.2 A photographer has adjusted his camera for a correct exposure with a short-focal-length lens. He then decides to zoom in by increasing the focal length. To maintain a correct exposure without changing the shutter speed, the diameter of the lens aperture should

a. Be increased.   b. Be decreased.   c. Stay the same.

24.3 Vision

The human eye is a marvelous and intricate organ. If we leave the biological details to biologists and focus on the eye's optical properties, we find that it functions very much like a camera. Like a camera, the eye has refracting surfaces that focus incoming light rays, an adjustable iris to control the light intensity, and a light-sensitive detector.

FIGURE 24.6 shows the basic structure of the eye. It is roughly spherical, about 2.4 cm in diameter. The transparent **cornea,** which is somewhat more sharply curved, and the *lens* are the eye's refractive elements. The eye is filled with a clear, jellylike fluid called the *aqueous humor* (in front of the lens) and the *vitreous humor* (behind the lens). The indices of refraction of the aqueous and vitreous humors are 1.34, only slightly different from water. The lens, although not uniform, has an average index of 1.44. The **pupil,** a variable-diameter aperture in the **iris,** automatically opens and closes to control the light intensity. A fully dark-adapted eye can open to ≈ 8 mm, and the pupil closes down to ≈ 1.5 mm in bright sun. This corresponds to *f*-numbers from roughly *f*/3 to *f*/16, very similar to a camera.

FIGURE 24.6 The human eye.

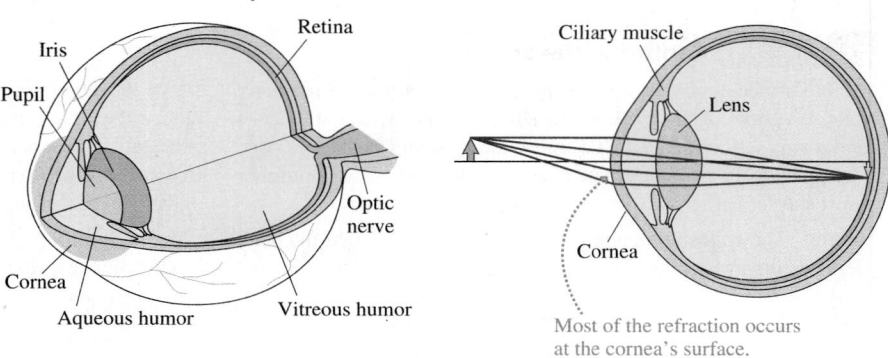

The eye's detector, the **retina,** consists of specialized light-sensitive cells called *rods* and *cones.* The rods, sensitive mostly to light and dark, are most important in very dim lighting. Color vision, which requires somewhat more light, is due to the cones, of which there are three types. FIGURE 24.7 shows the wavelength responses of the cones. They have overlapping ranges, especially the red- and green-sensitive cones, so two or even all three cones respond to light of any particular wavelength. The relative response of the different cones is interpreted by your brain as light of a particular color. Color is a *perception,* a response of our sensory and nervous systems, not something inherent in the light itself. Other animals, with slightly different retinal cells, can see ultraviolet or infrared wavelengths that we cannot see.

FIGURE 24.7 Wavelength sensitivity of the three types of cones in the human retina.

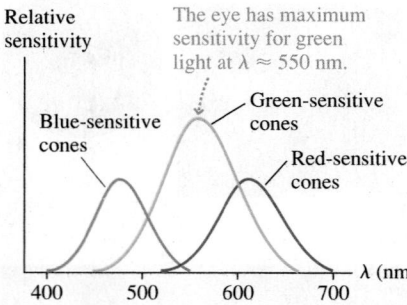

Focusing and Accommodation

The eye, like a camera, focuses light rays to an inverted image on the retina. Perhaps surprisingly, most of the refractive power of the eye is due to the cornea, not the lens. The cornea is a sharply curved, spherical surface, and you learned in Chapter 23 that images are formed by refraction at a spherical surface. The rather large difference between the index of refraction of air and that of the aqueous humor causes a significant refraction of light rays at the cornea. In contrast, there is much less difference between the indices of the lens and its surrounding fluid, so refraction at the lens surfaces is weak. The lens is important for fine-tuning, but the air-cornea boundary is responsible for the majority of the refraction.

You can recognize the power of the cornea if you open your eyes underwater. Everything is very blurry! When light enters the cornea through water, rather than through air, there's almost no difference in the indices of refraction at the surface. Light rays pass through the cornea with almost no refraction, so what little focusing ability you have while underwater is due to the lens alone.

A camera focuses by moving the lens. The eye focuses by changing the focal length of the lens, a feat it accomplishes by using the *ciliary muscles* to change the curvature of the lens surface. The ciliary muscles are relaxed when you look at a distant scene. Thus the lens surface is relatively flat and the lens has its longest focal length. As you shift your gaze to a nearby object, the ciliary muscles contract and cause the lens to bulge. This process, called **accommodation,** decreases the lens's radius of curvature and thus decreases its focal length.

The farthest distance at which a relaxed eye can focus is called the eye's **far point** (FP). The far point of a normal eye is infinity; that is, the eye can focus on objects extremely far away. The closest distance at which an eye can focus, using maximum accommodation, is the eye's **near point** (NP). (Objects can be *seen* closer than the near point, but they're not sharply focused on the retina.) Both situations are shown in FIGURE 24.8.

FIGURE 24.8 Normal vision of far and near objects.

The ciliary muscles are relaxed for distant vision.

FP = ∞

NP = 25 cm

The ciliary muscles are contracted for near vision, causing the lens to curve more.

Vision Defects and Their Correction

The near point of normal vision is considered to be 25 cm, but the near point of any individual changes with age. The near point of young children can be as little as 10 cm. The "normal" 25 cm near point is characteristic of young adults, but the near point of most individuals begins to move outward by age 40 or 45 and can reach 200 cm by age 60. This loss of accommodation, which arises because the lens loses flexibility, is called **presbyopia.** Even if their vision is otherwise normal, individuals with presbyopia need reading glasses to bring their near point back to 25 or 30 cm, a comfortable distance for reading.

Presbyopia is known as a *refractive error* of the eye. Two other common refractive errors are *hyperopia* and *myopia.* All three can be corrected with lenses—either eyeglasses or contact lenses—that assist the eye's focusing. Corrective lenses are prescribed not by their focal length but by their **power.** The power of a lens is the inverse of its focal length:

$$\text{Power of a lens} = P = \frac{1}{f} \tag{24.4}$$

A lens with more power (shorter focal length) causes light rays to refract through a larger angle. The SI unit of lens power is the **diopter,** abbreviated D, defined as $1\ \text{D} = 1\ \text{m}^{-1}$. Thus a lens with $f = 50\ \text{cm} = 0.50\ \text{m}$ has power $P = 2.0\ \text{D}$.

A person who is *farsighted* can see faraway objects (but even then must use some accommodation rather than a relaxed eye), but his near point is larger than 25 cm, often much larger, so he cannot focus on nearby objects. The cause of farsightedness—called **hyperopia**—is an eyeball that is too short for the refractive power of the cornea and lens. As FIGURES 24.9a and b on the next page show, no amount of accommodation allows the eye to focus on an object 25 cm away, the normal near point.

With hyperopia, the eye needs assistance to focus the rays from a near object onto the closer-than-normal retina. This assistance is obtained by adding refractive power with the positive (i.e., converging) lens shown in FIGURE 24.9c. To understand why this works, recall that the image of a first lens acts as the object for a second lens. The goal is to allow the person to focus on an object 25 cm away. If a corrective lens forms an upright, virtual image at the person's actual near point, that virtual image acts as an object for the eye itself and, with maximum accommodation, the eye can focus these rays onto the retina. Presbyopia, the loss of accommodation with age, is corrected in the same way.

The optometrist's prescription is −2.25 D for the right eye (top) and −2.50 D for the left (bottom), the minus sign indicating that these are diverging lenses. The optometrist doesn't write the D because the lens maker already knows that prescriptions are in diopters. Most people's eyes are not exactly the same, so each eye usually gets a different lens.

FIGURE 24.9 Hyperopia.

(a)

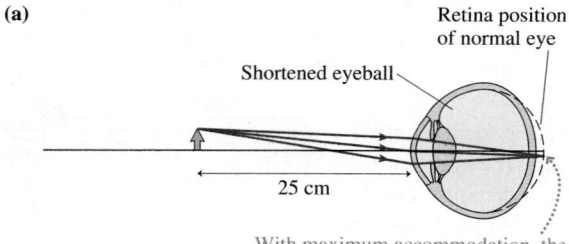

Shortened eyeball

Retina position of normal eye

25 cm

With maximum accommodation, the eye tries to focus the image behind the actual retina. Thus the image is blurry.

(b)

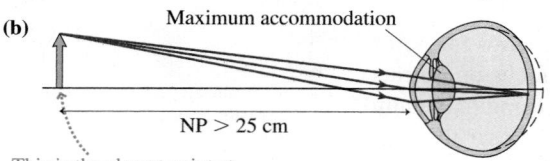

Maximum accommodation

NP > 25 cm

This is the closest point at which the eye can focus.

(c)

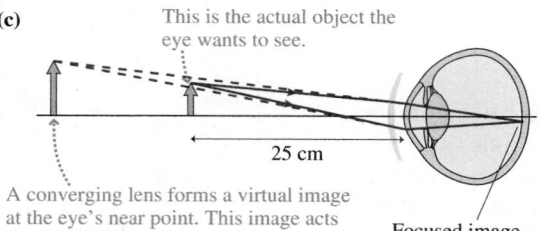

This is the actual object the eye wants to see.

25 cm

A converging lens forms a virtual image at the eye's near point. This image acts as the object for the eye and is what the eye actually focuses on.

Focused image

FIGURE 24.10 Myopia.

(a)

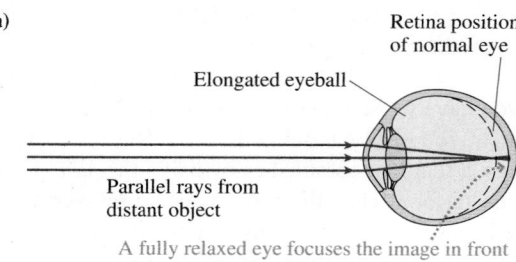

Elongated eyeball

Retina position of normal eye

Parallel rays from distant object

A fully relaxed eye focuses the image in front of the actual retina. The image is blurry.

(b)

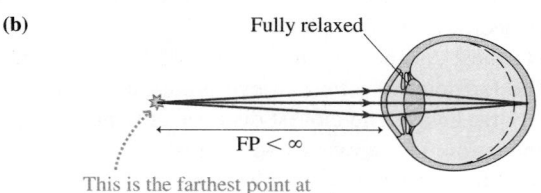

Fully relaxed

FP < ∞

This is the farthest point at which the eye can focus.

(c)

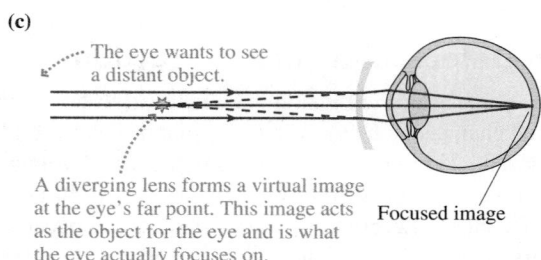

The eye wants to see a distant object.

A diverging lens forms a virtual image at the eye's far point. This image acts as the object for the eye and is what the eye actually focuses on.

Focused image

NOTE ▶ Figures 24.9 and 24.10 show the corrective lenses as they are actually shaped—called *meniscus lenses*—rather than with our usual lens shape. Nonetheless, the lens in Figure 24.9c is a converging lens because it's thicker in the center than at the edges. The lens in Figure 24.10c is a diverging lens because it's thicker at the edges than in the center. ◀

A person who is *nearsighted* can clearly see nearby objects when the eye is relaxed (and extremely close objects by using accommodation), but no amount of relaxation allows her to see distant objects. Nearsightedness—called **myopia**—is caused by an eyeball that is too long. As FIGURE 24.10a shows, rays from a distant object come to a focus in front of the retina and have begun to diverge by the time they reach the retina. The eye's far point, shown in FIGURE 24.10b, is less than infinity.

To correct myopia, we needed a diverging lens, as shown in FIGURE 24.10c, to slightly defocus the rays and move the image point back to the retina. To focus on a very distant object, the person needs a corrective lens that forms an upright, virtual image at her actual far point. That virtual image acts as an object for the eye itself and, when fully relaxed, the eye can focus these rays onto the retina.

EXAMPLE 24.4 | **Correcting hyperopia**

Sanjay has hyperopia. The near point of his left eye is 150 cm. What prescription lens will restore normal vision?

MODEL Normal vision will allow Sanjay to focus on an object 25 cm away. In measuring distances, we'll ignore the small space between the lens and his eye.

SOLVE Because Sanjay can see objects at 150 cm, using maximum accommodation, we want a lens that creates a virtual image

at position $s' = -150$ cm (negative because it's a virtual image) of an object held at $s = 25$ cm. From the thin-lens equation,

$$\frac{1}{f} = \frac{1}{s} + \frac{1}{s'} = \frac{1}{0.25 \text{ m}} + \frac{1}{-1.50 \text{ m}} = 3.3 \text{ m}^{-1}$$

$1/f$ is the lens power, and m^{-1} are diopters. Thus the prescription is for a lens with power $P = 3.3$ D.

ASSESS Hyperopia is always corrected with a converging lens.

EXAMPLE 24.5 **Correcting myopia**

Martina has myopia. The far point of her left eye is 200 cm. What prescription lens will restore normal vision?

MODEL Normal vision will allow Martina to focus on a very distant object. In measuring distances, we'll ignore the small space between the lens and her eye.

SOLVE Because Martina can see objects at 200 cm with a fully relaxed eye, we want a lens that will create a virtual image at

position $s' = -200$ cm (negative because it's a virtual image) of a distant object at $s = \infty$ cm. From the thin-lens equation,

$$\frac{1}{f} = \frac{1}{s} + \frac{1}{s'} = \frac{1}{\infty \text{ m}} + \frac{1}{-2.0 \text{ m}} = -0.5 \text{ m}^{-1}$$

Thus the prescription is for a lens with power $P = -0.5$ D.

ASSESS Myopia is always corrected with a diverging lens.

STOP TO THINK 24.3 You need to improvise a magnifying glass to read some very tiny print. Should you borrow the eyeglasses from your hyperopic friend or from your myopic friend?

a. The hyperopic friend
c. Either will do.

b. The myopic friend
d. Neither will work.

24.4 Optical Systems That Magnify

The camera, with its fast shutter speed, allows us to capture images of events that take place too quickly for our unaided eye to resolve. Another use of optical systems is to magnify—to see objects smaller or closer together than our eye can see.

The easiest way to magnify an object requires no extra optics at all; simply get closer! The closer you get, the bigger the object appears. Obviously the actual size of the object is unchanged as you approach it, so what exactly is getting "bigger"? Consider the green arrow in FIGURE 24.11a. We can determine the size of its image on the retina by tracing the ray that is undeviated as it passes through the center of a lens. (Here we're modeling the eye's optical system as one thin lens.) If we get closer to the arrow, now shown as red, we find the arrow makes a larger image on the retina. Our brain interprets the larger image as a larger-appearing object. The object's actual size doesn't change, but its *apparent size* gets larger as it gets closer.

Technically, we say that closer objects look larger because they subtend a larger angle θ, called the **angular size** of the object. The red arrow has a larger angular size than the green arrow, $\theta_2 > \theta_1$, so the red arrow looks larger and we can see more detail. But you can't keep increasing an object's angular size because you can't focus on the object if it's closer than your near point, which we'll take to be a normal 25 cm. FIGURE 24.11b defines the angular size θ_{NP} of an object at your near point. If the object's height is h and if we assume the small-angle approximation $\tan \theta \approx \theta$, the maximum angular size viewable by your unaided eye is

$$\theta_{NP} = \frac{h}{25 \text{ cm}} \tag{24.5}$$

Suppose we view the same object, of height h, through the single converging lens in FIGURE 24.12 on the next page. If the object's distance from the lens is less than the lens's focal length, we'll see an enlarged, upright image. Used in this way, the lens is called a **magnifier** or *magnifying glass.* The eye sees the virtual image subtending angle θ, and it can focus on this virtual image as long as the image distance is more than 25 cm. Within the small-angle approximation, the image subtends angle $\theta = h/s$. In practice, we usually want the image to be at distance $s' \approx \infty$ so that we can view it with a relaxed eye as a "distant object." This will be true if the object is very near the focal point: $s \approx f$. In this case, the image subtends angle

$$\theta = \frac{h}{s} \approx \frac{h}{f} \tag{24.6}$$

FIGURE 24.11 Angular size.

(a) Same object at two different distances

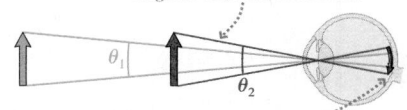

As the object gets closer, the angle it subtends becomes larger. Its *angular size* has increased.

Further, the size of the image on the retina gets larger. The object's *apparent size* has increased.

(b)

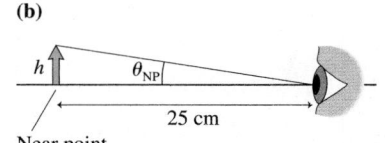

Near point

FIGURE 24.12 The magnifier.

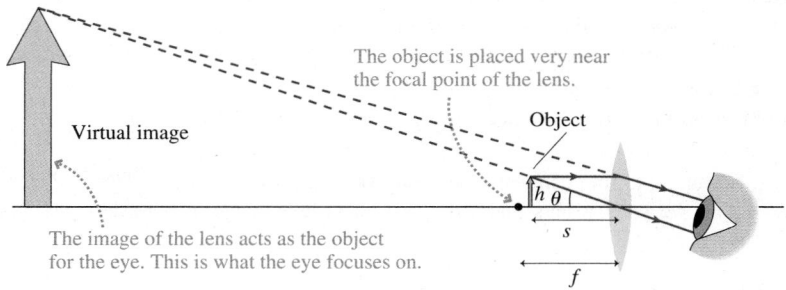

Let's define the **angular magnification** M as

$$M = \frac{\theta}{\theta_{NP}} \qquad (24.7)$$

Angular magnification is the increase in the *apparent size* of the object that you achieve by using a magnifying lens rather than simply holding the object at your near point. Substituting from Equations 24.5 and 24.6, we find the angular magnification of a magnifying glass is

$$M = \frac{25 \text{ cm}}{f} \qquad (24.8)$$

The angular magnification depends on the focal length of the lens but not on the size of the object. Although it would appear we could increase angular magnification without limit by using lenses with shorter and shorter focal lengths, the inherent limitations of lenses we discuss in the next section limit the magnification of a simple lens to about 4×. Slightly more complex magnifiers with two lenses reach 20×, but beyond that one would use a microscope.

NOTE ▶ Don't confuse angular magnification with lateral magnification. Lateral magnification m compares the height of an object to the height of its image. The lateral magnification of a magnifying glass is $\approx \infty$ because the virtual image is at $s' \approx \infty$, but that doesn't make the object seem infinitely big. Its apparent size is determined by the angle subtended on your retina, and that angle remains finite. Thus angular magnification tells us how much bigger things appear. ◀

The Microscope

A microscope, whose major parts are shown in FIGURE 24.13a, can attain a magnification of up to 1000× by a *two-step* magnification process. A specimen to be observed is placed on the *stage* of the microscope, directly beneath the **objective,** a converging lens with a relatively short focal length. The objective creates a magnified real image that is further enlarged by the **eyepiece.** Both the objective and the eyepiece are complex combination lenses, but we'll model them as single thin lenses. It's common for a prism to bend the rays so that the eyepiece is at a comfortable viewing angle. However, we'll consider a simplified version of a microscope in which the light travels along a straight tube.

FIGURE 24.13b shows the optics in more detail. The object is placed just outside the focal point of the objective, which then creates a highly magnified real image with lateral magnification $m = -s'/s$. The object is so close to the focal point that $s \approx f_{obj}$ is an excellent approximation. In addition, the focal lengths of the objective and the eyepiece are much less than the tube length L, so $s' \approx L$ is another good approximation. With these approximations, the lateral magnification of the objective is

$$m_{obj} = -\frac{s'}{s} \approx -\frac{L}{f_{obj}} \qquad (24.9)$$

FIGURE 24.13 The microscope.

(a)

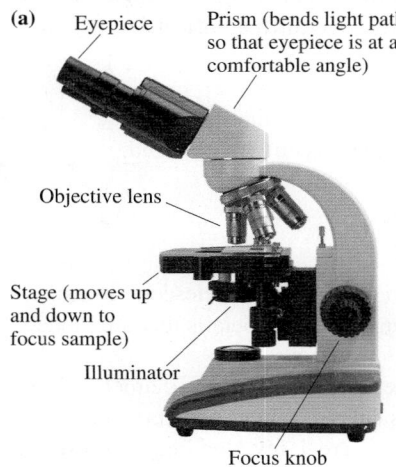

Eyepiece

Prism (bends light path so that eyepiece is at a comfortable angle)

Objective lens

Stage (moves up and down to focus sample)

Illuminator

Focus knob

(b)

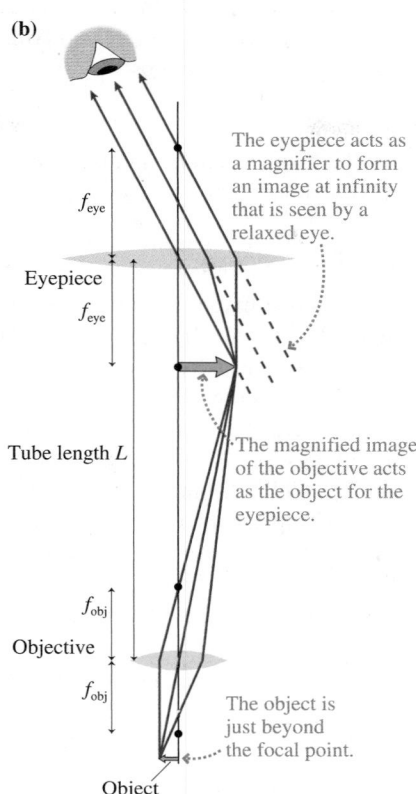

The eyepiece acts as a magnifier to form an image at infinity that is seen by a relaxed eye.

f_{eye}

Eyepiece

f_{eye}

Tube length L

The magnified image of the objective acts as the object for the eyepiece.

f_{obj}

Objective

f_{obj}

The object is just beyond the focal point.

Object

The image of the objective acts as the object for the eyepiece, which functions as a simple magnifier. The angular magnification of the eyepiece is given by Equation 24.8, $M_{eye} = (25 \text{ cm})/f_{eye}$. Together, the objective and eyepiece produce a total angular magnification

$$M = m_{obj}M_{eye} = -\frac{L}{f_{obj}}\frac{25 \text{ cm}}{f_{eye}} \tag{24.10}$$

The minus sign shows that the image seen in a microscope is inverted.

In practice, the magnifications of the objective (without the minus sign) and the eyepiece are stamped on the barrels. A set of objectives on a rotating turret might include 10×, 20×, 40×, and 100×. When combined with a 10× eyepiece, the microscope's total angular magnification ranges from 100× to 1000×. In addition, most biological microscopes are standardized with a tube length $L = 160$ mm. Thus a 40× objective has focal length $f_{obj} = 160$ mm/40 = 4.0 mm.

EXAMPLE 24.6 **Viewing blood cells**

A pathologist inspects a sample of 7-μm-diameter human blood cells under a microscope. She selects a 40× objective and a 10× eyepiece. What size object, viewed from 25 cm, has the same apparent size as a blood cell seen through the microscope?

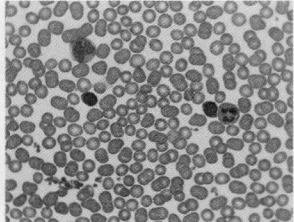

MODEL Angular magnification compares the magnified angular size to the angular size seen at the near-point distance of 25 cm.

SOLVE The microscope's angular magnification is $M = -(40) \times (10) = -400$. The magnified cells will have the same apparent size as an object $400 \times 7 \ \mu$m ≈ 3 mm in diameter seen from a distance of 25 cm.

ASSESS 3 mm is about the size of a capital O in this textbook, so a blood cell seen through the microscope will have about the same apparent size as an O seen from a comfortable reading distance.

STOP TO THINK 24.4 A biologist rotates the turret of a microscope to replace a 20× objective with a 10× objective. To keep the same overall magnification, the focal length of the eyepiece must be

a. Doubled.　　　　b. Halved.　　　　c. Kept the same.

d. The magnification cannot be kept the same if the objective is changed.

The Telescope

A microscope magnifies small, nearby objects to look large. A telescope magnifies distant objects, which might be quite large, so that we can see details that are blended together when seen by eye.

FIGURE 24.14 shows the optical layout of a simple telescope. A large-diameter objective lens (larger lenses collect more light and thus can see fainter objects) collects the parallel rays from a distant object ($s = \infty$) and forms a real, inverted image at distance $s' = f_{obj}$. Unlike a microscope, which uses a short-focal-length objective, the focal length of a telescope objective is very nearly the length of the telescope tube. Then, just as in the microscope, the eyepiece functions as a simple magnifier. The viewer observes an inverted image, but that's not a serious problem in astronomy. Terrestrial telescopes use a different design to obtain an upright image.

FIGURE 24.14 A refracting telescope.

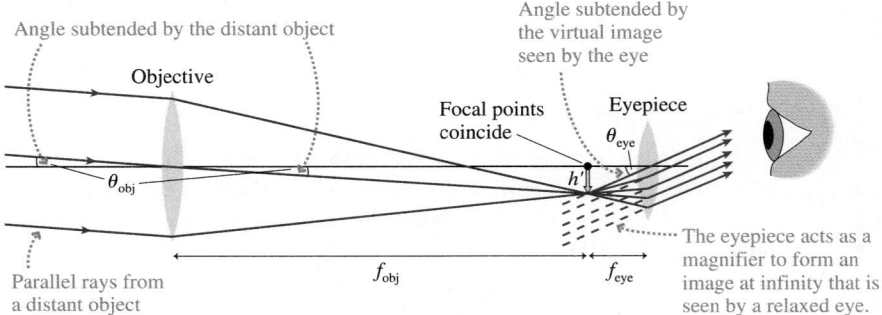

Suppose the distant object, as seen by the objective lens, subtends angle θ_{obj}. If the image seen through the eyepiece subtends a larger angle θ_{eye}, then the angular magnification is $M = \theta_{eye}/\theta_{obj}$. We can see from the undeviated ray passing through the center of the objective lens that (using the small-angle approximation)

$$\theta_{obj} \approx -\frac{h'}{f_{obj}}$$

where the minus sign indicates the inverted image. The image of height h' acts as the object for the eyepiece, and we can see that the final image observed by the viewer subtends angle

$$\theta_{eye} = \frac{h'}{f_{eye}}$$

Consequently, the angular magnification of a telescope is

$$M = \frac{\theta_{eye}}{\theta_{obj}} = -\frac{f_{obj}}{f_{eye}} \tag{24.11}$$

The angular magnification is simply the ratio of the objective focal length to the eyepiece focal length.

Because the stars and galaxies are so distant, light-gathering power is more important to astronomers than magnification. Large light-gathering power requires a large-diameter objective lens, but large lenses are not practical; they begin to sag under their own weight. Thus **refracting telescopes,** with two lenses, are relatively small. Serious astronomy is done with a **reflecting telescope,** such as the one shown in FIGURE 24.15.

A large-diameter mirror (the *primary mirror*) focuses the rays to form a real image, but, for practical reasons, a small flat mirror (the *secondary mirror*) reflects the rays sideways before they reach a focus. This moves the primary mirror's image out to the edge of the telescope where it can be viewed by an eyepiece on the side. None of these changes affects the overall analysis of the telescope, and its angular magnification is given by Equation 24.11 if f_{obj} is replaced by f_{pri}, the focal length of the primary mirror.

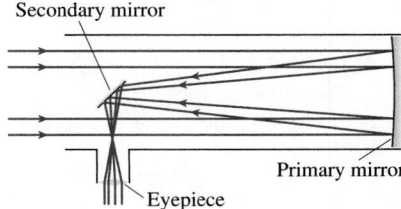

FIGURE 24.15 A reflecting telescope.

Secondary mirror

Primary mirror

Eyepiece

24.5 **The Resolution of Optical Instruments**

A camera *could* focus light with a single lens. A microscope objective *could* be built with a single lens. So why would anyone ever use a lens combination in place of a single lens? There are two primary reasons.

First, any lens has dispersion. That is, its index of refraction varies slightly with wavelength. Because the index of refraction for violet light is larger than for red light, a lens's focal length is shorter for violet light than for red light. Consequently, different colors of light come to a focus at slightly different distances from the lens. If red light is sharply focused on a viewing screen, then blue and violet wavelengths are not well focused. This imaging error, illustrated in FIGURE 24.16a, is called **chromatic aberration.**

Second, our analysis of thin lenses was based on paraxial rays traveling nearly parallel to the optical axis. A more exact analysis, taking all the rays into account, finds that rays incident on the outer edges of a spherical surface are not focused at exactly the same point as rays incident near the center. This imaging error, shown in FIGURE 24.16b, is called **spherical aberration.** Spherical aberration, which causes the image to be slightly blurred, gets worse as the lens diameter increases.

FIGURE 24.16 Chromatic aberration and spherical aberration prevent simple lenses from forming perfect images.

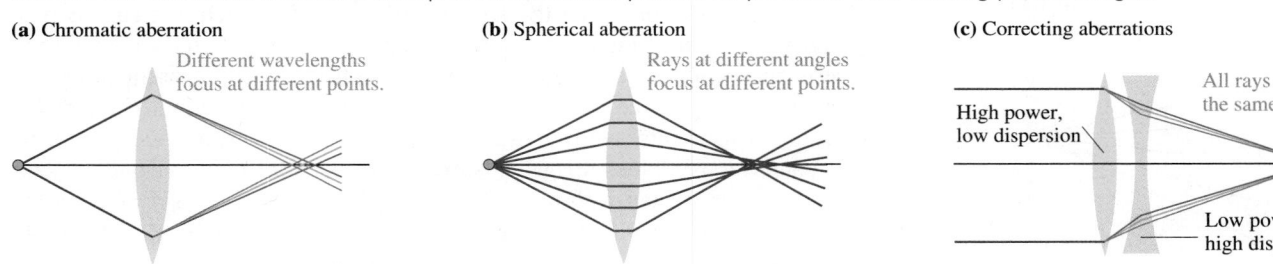

(a) Chromatic aberration

Different wavelengths focus at different points.

(b) Spherical aberration

Rays at different angles focus at different points.

(c) Correcting aberrations

High power, low dispersion

All rays meet at the same focus.

Low power, high dispersion

Fortunately, the chromatic and spherical aberrations of a converging lens and a diverging lens are in opposite directions. When a converging lens and a diverging lens are used in combination, their aberrations tend to cancel. A combination lens, such as the one in FIGURE 24.16c, can produce a much sharper focus than a single lens with the equivalent focal length. Consequently, most optical instruments use combination lenses rather than single lenses.

Diffraction Again

According to the ray model of light, a perfect lens (one with no aberrations) should be able to form a perfect image. But the ray model of light, though a very good model for lenses, is not an absolutely correct description of light. If we look closely, the wave aspects of light haven't entirely disappeared. In fact, the performance of optical equipment is limited by the diffraction of light.

FIGURE 24.17 A lens both focuses and diffracts the light passing through.

(a) A lens acts as a circular aperture.

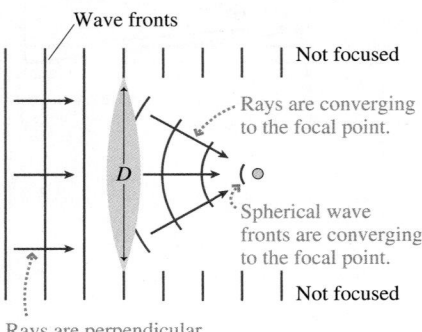

Wave fronts

Not focused

Rays are converging to the focal point.

Spherical wave fronts are converging to the focal point.

Not focused

Rays are perpendicular to the wave fronts.

(b) The aperture and focusing effects can be separated.

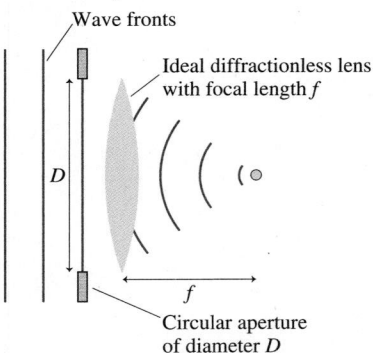

Wave fronts

Ideal diffractionless lens with focal length f

Circular aperture of diameter D

(c) The lens focuses the diffraction pattern in the focal plane.

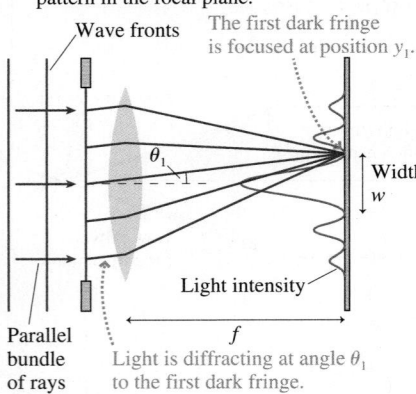

Wave fronts

The first dark fringe is focused at position y_1.

Width w

Light intensity

Parallel bundle of rays

Light is diffracting at angle θ_1 to the first dark fringe.

FIGURE 24.17a shows a plane wave, with parallel light rays, being focused by a lens of diameter D. According to the ray model of light, a perfect lens would focus parallel rays to a perfect point. Notice, though, that only a piece of each wave front passes *through* the lens and gets focused. In effect, **the lens itself acts as a circular aperture** in an opaque barrier, allowing through only a portion of each wave front. Consequently, **the lens diffracts the light wave.** The diffraction is usually very small because D is usually much greater than the wavelength of the light; nonetheless, this small amount of diffraction is the limiting factor in how well the lens can focus the light.

FIGURE 24.17b separates the diffraction from the focusing by modeling the lens as an actual aperture of diameter D followed by an "ideal" diffractionless lens. You learned in Chapter 22 that a circular aperture produces a diffraction pattern with a bright central maximum surrounded by dimmer fringes. A converging lens brings this diffraction pattern to a focus in the image plane, as shown in FIGURE 24.17c. As a result, a perfect lens focuses parallel light rays not to a perfect point of light, as we expected, but to a small, circular diffraction pattern.

The angle to the first minimum of a circular diffraction pattern is $\theta_1 = 1.22\lambda/D$. The ray that passes through the center of a lens is not bent, so Figure 24.17c uses this ray to show that the position of the dark fringe is $y_1 = f\tan\theta_1 \approx f\theta_1$. Thus the width of the central maximum in the focal plane is

$$w_{min} \approx 2f\theta_1 = \frac{2.44\lambda f}{D} \qquad \text{(minimum spot size)} \qquad (24.12)$$

This is the **minimum spot size** to which a lens can focus light.

Lenses are often limited by aberrations, so not all lenses can focus parallel light rays to a spot this small. A well-crafted lens, for which Equation 24.12 is the minimum spot size, is called a *diffraction-limited lens.* No optical design can overcome the spreading of light due to diffraction, and it is because of this spreading that the image point has a minimum spot size. The image of an actual object, rather than of parallel rays, becomes a mosaic of overlapping diffraction patterns, so even the most perfect lens inevitably forms an image that is slightly fuzzy.

For various reasons, it is difficult to produce a diffraction-limited lens having a focal length that is much less than its diameter. The very best microscope objectives have $f \approx 0.5D$. This implies that **the smallest diameter to which you can focus a spot of light, no matter how hard you try, is $w_{min} \approx \lambda$.** This is a fundamental limit on the performance of optical equipment. Diffraction has very real consequences!

One example of these consequences is found in the manufacturing of integrated circuits. Integrated circuits are made by creating a "mask" showing all the components and their connections. A lens images this mask onto the surface of a semiconductor wafer that has been coated with a substance called *photoresist.* Bright areas in the mask expose the photoresist, and subsequent processing steps chemically etch away the exposed areas while leaving behind areas that had been in the shadows of the mask. This process is called *photolithography.*

The power of a microprocessor and the amount of memory in a memory chip depend on how small the circuit elements can be made. Diffraction dictates that a circuit element can be no smaller than the smallest spot to which light can be focused, which is roughly the wavelength of the light. If the mask is projected with ultraviolet light having $\lambda \approx 200$ nm, then the smallest elements on a chip are about 200 nm wide. This is, in fact, just about the current limit of technology.

EXAMPLE 24.7 **Seeing stars**

A 12-cm-diameter telescope lens has a focal length of 1.0 m. What is the diameter of the image of a star in the focal plane if the lens is diffraction limited *and* if the earth's atmosphere is not a limitation?

MODEL Stars are so far away that they appear as points in space. An ideal diffractionless lens would focus their light to arbitrarily small points. Diffraction prevents this. Model the telescope lens as a 12-cm-diameter aperture in front of an ideal lens with a 1.0 m focal length.

SOLVE The minimum spot size in the focal plane of this lens is

$$w = \frac{2.44\lambda f}{D}$$

where D is the lens diameter. What is λ? Because stars emit white light, the *longest* wavelengths spread the most and determine the size of the image that is seen. If we use $\lambda = 700$ nm as the approximate upper limit of visible wavelengths, we find $w = 1.4 \times 10^{-5}$ m $= 14$ μm.

ASSESS This is certainly small, and it would appear as a point to your unaided eye. Nonetheless, the spot size would be easily noticed if it were recorded on film and enlarged. Turbulence and temperature effects in the atmosphere, the causes of the "twinkling" of stars, prevent ground-based telescopes from being this good, but space-based telescopes really are diffraction limited.

Resolution

Suppose you point a telescope at two nearby stars in a galaxy far, far away. If you use the best possible detector, will you be able to distinguish separate images for the two stars, or will they blur into a single blob of light? A similar question could be asked of a microscope. Can two microscopic objects, very close together, be distinguished if sufficient magnification is used? Or is there some size limit at which their images will blur together and never be separated? These are important questions about the *resolution* of optical instruments.

Because of diffraction, the image of a distant star is not a point but a circular diffraction pattern. Our question, then, really is: How close together can two diffraction patterns be before you can no longer distinguish them? One of the major scientists of the 19th century, Lord Rayleigh, studied this problem and suggested a reasonable rule that today is called **Rayleigh's criterion.**

FIGURE 24.18 shows two distant point sources being imaged by a lens of diameter D. The angular separation between the objects, as seen from the lens, is α. Rayleigh's criterion states that

- The two objects are resolvable if $\alpha > \theta_{min}$, where $\theta_{min} = \theta_1 = 1.22\lambda/D$ is the angle of the first dark fringe in the circular diffraction pattern.
- The two objects are not resolvable if $\alpha < \theta_{min}$ because their diffraction patterns are too overlapped.
- The two objects are marginally resolvable if $\alpha = \theta_{min}$. The central maximum of one image falls exactly on top of the first dark fringe of the other image. This is the situation shown in the figure.

FIGURE 24.19 shows enlarged photographs of the images of two point sources. The images are circular diffraction patterns, not points. The two images are close but distinct where the objects are separated by $\alpha > \theta_{min}$. Two objects really were recorded in the photo at the bottom, but their separation is $\alpha < \theta_{min}$ and their images have blended together. In the middle photo, with $\alpha = \theta_{min}$, you can see that the two images are just barely resolved.

The angle

$$\theta_{min} = \frac{1.22\lambda}{D} \quad \text{(angular resolution of a lens)} \quad (24.13)$$

is called the **angular resolution** of a lens. The angular resolution of a telescope depends on the diameter of the objective lens (or the primary mirror) and the wavelength of the light; magnification is not a factor. Two images will remain overlapped and unresolved no matter what the magnification if their angular separation is less than θ_{min}. For visible light, where λ is pretty much fixed, the only parameter over which the astronomer has any control is the diameter of the lens or mirror of the

FIGURE 24.18 Two images that are marginally resolved.

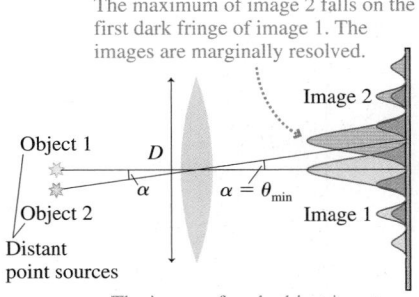

The maximum of image 2 falls on the first dark fringe of image 1. The images are marginally resolved.

The image of each object is not a perfect point, but a small circular diffraction pattern.

FIGURE 24.19 Enlarged photographs of the images of two point sources.

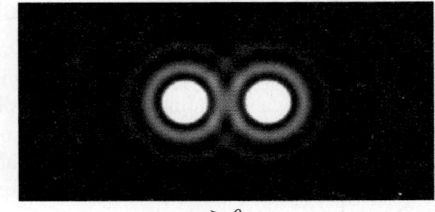

$\alpha > \theta_{min}$
Resolved

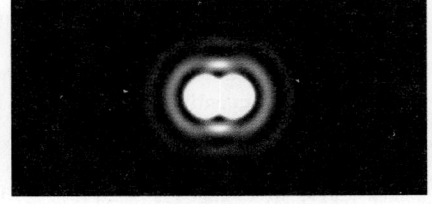

$\alpha = \theta_{min}$
Marginally resolved

$\alpha < \theta_{min}$
Not resolved

The size of the features in an integrated circuit is limited by the diffraction of light.

telescope. The urge to build ever-larger telescopes is motivated, in part, by a desire to improve the angular resolution. (Another motivation is to increase the light-gathering power so as to see objects farther away.)

The performance of a microscope is also limited by the diffraction of light passing through the objective lens. Just as light cannot be focused to a spot smaller than about a wavelength, the most perfect microscope cannot resolve the features of objects that are smaller than a wavelength. Similarly, two objects separated by less than one wavelength—roughly 500 nm—will blur into a single object and cannot be resolved. Because atoms are approximately 0.1 nm in diameter, vastly smaller than the wavelength of visible or even ultraviolet light, there is no hope of ever seeing atoms with an optical microscope. This limitation is not simply a matter of needing a better design or more precise components; it is a fundamental limit set by the wave nature of the light with which we see.

STOP TO THINK 24.5 Four diffraction-limited lenses focus plane waves of light with the same wavelength λ. Rank in order, from largest to smallest, the spot sizes w_a to w_d.

$f = 10$ mm

a | 2 mm

$f = 5$ mm

b | 2 mm

$f = 10$ mm

c | 4 mm

$f = 24$ mm

d | 8 mm

CHALLENGE EXAMPLE 24.8 **Visual acuity**

The normal human eye has maximum visual acuity with a pupil diameter of about 3 mm. For larger pupils, acuity decreases due to increasing aberrations; for smaller pupils, acuity decreases due to increasing diffraction. If your pupil diameter is 2.0 mm, as it would be in bright light, what is the smallest-diameter circle that you should be able to see as a circle, rather than just an unresolved blob, on an eye chart at the standard distance of 20 ft? The index of refraction inside the eye is 1.33.

MODEL Assume that a 2.0-mm-diameter pupil is diffraction limited. Then the angular resolution is given by Rayleigh's criterion. Diffraction increases with wavelength, so the eye's acuity will be affected more by longer wavelengths than by shorter wavelengths. Consequently, assume that the light's wavelength in air is 600 nm.

VISUALIZE Let the diameter of the circle be d. **FIGURE 24.20** shows the circle at distance $s = 20$ ft = 6.1 m. "Seeing the circle," shown edge-on, requires resolving the top and bottom lines as distinct.

FIGURE 24.20 Viewing a circle of diameter d.

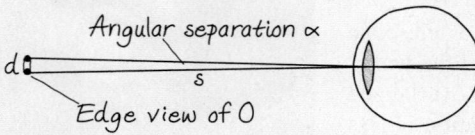

SOLVE The angular separation of the top and bottom lines of the circle is $\alpha = d/s$. Rayleigh's criterion says that a perfect lens with aperture D can just barely resolve these two lines if

$$\alpha = \frac{d}{s} = \theta_{min} = \frac{1.22\lambda_{eye}}{D} = \frac{1.22\lambda_{air}}{n_{eye}D}$$

The diffraction takes place inside the eye, where the wavelength is shortened to $\lambda_{eye} = \lambda_{air}/n_{eye}$. Thus the circle diameter that can barely be resolved with perfect vision is

$$d = \frac{1.22\lambda_{air}s}{n_{eye}D} = \frac{1.22(600 \times 10^{-9} \text{ m})(6.1 \text{ m})}{(1.33)(0.0020 \text{ m})} \approx 2 \text{ mm}$$

That's about the height of a capital O in this book, so in principle you should—in very bright light—just barely be able to recognize it as an O at 20 feet.

ASSESS On an eye chart, the O on the line for 20/20 vision—the standard of excellent vision—is about 7 mm tall, so the calculated 2 mm, although in the right range, is a bit too small. There are three reasons. First, eye tests are done with medium-bright indoor lighting. Your acuity really does improve in light bright enough to reduce your pupil diameter to 2.0 mm. Second, although aberrations of the eye are reduced with a smaller pupil, they haven't vanished. And third, for a 2-mm-tall object at 20 ft, the size of the image on the retina is barely larger than the spacing between the cone cells, so the resolution of the "detector" is also a factor. Your eye is a very good optical instrument, but not perfect.

SUMMARY

The goal of Chapter 24 has been to understand some common optical instruments and their limitations.

Important Concepts

Lens Combinations

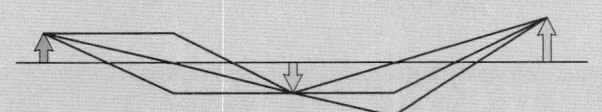

The image of the first lens acts as the object for the second lens.

Lens power: $P = \dfrac{1}{f}$ diopters, $1\text{ D} = 1\text{ m}^{-1}$

Resolution

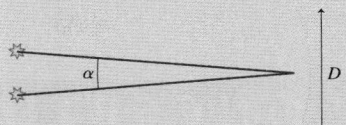 

The angular resolution of a lens of diameter D is

$$\theta_{min} = 1.22\lambda/D$$

Rayleigh's criterion states that two objects separated by an angle α are marginally resolvable if $\alpha = \theta_{min}$.

Applications

Cameras

Forms a real, inverted image on a detector. The lens's **f-number** is

$$f\text{-number} = \frac{f}{D}$$

The light intensity on the detector is

$$I \propto \frac{1}{(f\text{-number})^2}$$

Magnifiers

For relaxed-eye viewing, the angular magnification is

$$M = \frac{25\text{ cm}}{f}$$

For microscopes and telescopes, angular magnification, not lateral magnification, is the important characteristic. The eyepiece acts as a magnifier to view the image formed by the objective lens.

Vision

Refraction at the cornea is responsible for most of the focusing. The lens provides fine-tuning by changing its shape **(accommodation).**

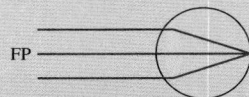 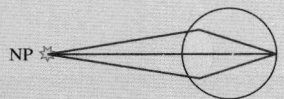

In normal vision, the eye can focus from a far point (FP) at ∞ (relaxed eye) to a near point (NP) at ≈ 25 cm (maximum accommodation).

- **Hyperopia** (farsightedness) is corrected with a converging lens.
- **Myopia** (nearsightedness) is corrected with a diverging lens.

Microscopes

The object is very close to the focal point of the objective. The total angular magnification is

$$M = -\frac{L}{f_{obj}} \frac{25\text{ cm}}{f_{eye}}$$

The best possible spatial resolution of a microscope, limited by diffraction, is about one wavelength of light.

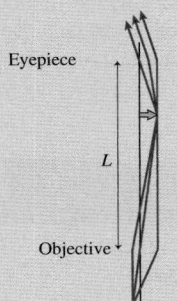

Focusing and spatial resolution

The minimum spot size to which a lens of diameter D can focus light is limited by diffraction to

$$w_{min} = \frac{2.44\lambda f}{D}$$

With the best lenses that can be manufactured, $w_{min} \approx \lambda$.

Telescopes

The object is very far from the objective.

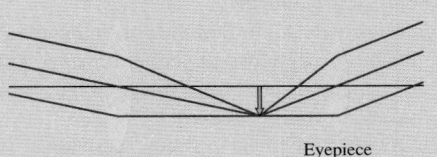

The total angular magnification is $M = -\dfrac{f_{obj}}{f_{eye}}$.

Terms and Notation

camera	iris	hyperopia	reflecting telescope
effective focal length, f	retina	myopia	chromatic aberration
aperture	accommodation	angular size	spherical aberration
f-number	far point	magnifier	minimum spot size, w_{min}
CCD	near point	angular magnification, M	Rayleigh's criterion
pixel	presbyopia	objective	angular resolution
cornea	power, P	eyepiece	
pupil	diopter, D	refracting telescope	

CONCEPTUAL QUESTIONS

1. Suppose a camera's exposure is correct when the lens has a focal length of 8.0 mm. Will the picture be overexposed, underexposed, or still correct if the focal length is "zoomed" to 16.0 mm without changing the diameter of the lens aperture? Explain.

2. A camera has a circular aperture immediately behind the lens. Reducing the aperture diameter to half its initial value will
   A. Make the image blurry.
   B. Cut off the outer half of the image and leave the inner half unchanged.
   C. Make the image less bright.
   D. All the above.
   Explain your choice.

3. Suppose you wanted special glasses designed to let you see underwater without a face mask. Should the glasses use a converging or diverging lens? Explain.

4. A friend lends you the eyepiece of his microscope to use on your own microscope. He claims the spatial resolution of your microscope will be halved, since his eyepiece has the same diameter as yours but twice the magnification. Is his claim valid? Explain.

5. A diffraction-limited lens can focus light to a 10-μm-diameter spot on a screen. Do the following actions make the spot diameter larger, make it smaller, or leave it unchanged?
   A. Decreasing the wavelength of the light.
   B. Decreasing the lens diameter.
   C. Decreasing the lens focal length.
   D. Decreasing the lens-to-screen distance.

6. To focus parallel light rays to the smallest possible spot, should you use a lens with a small f-number or a large f-number? Explain.

7. An astronomer is trying to observe two distant stars. The stars are marginally resolved when she looks at them through a filter that passes green light with a wavelength near 550 nm. Which of the following actions would improve the resolution? Assume that the resolution is not limited by the atmosphere.
   A. Changing the filter to a different wavelength. If so, should she use a shorter or a longer wavelength?
   B. Using a telescope with an objective lens of the same diameter but a different focal length. If so, should she select a shorter or a longer focal length?
   C. Using a telescope with an objective lens of the same focal length but a different diameter. If so, should she select a larger or a smaller diameter?
   D. Using an eyepiece with a different magnification. If so, should she select an eyepiece with more or less magnification?

EXERCISES AND PROBLEMS

Exercises

Section 24.1 Lenses in Combination

1. ‖ Two converging lenses with focal lengths of 40 cm and 20 cm are 10 cm apart. A 2.0-cm-tall object is 15 cm in front of the 40-cm-focal-length lens.
   a. Use ray tracing to find the position and height of the image. Do this accurately with a ruler or paper with a grid. Estimate the image distance and image height by making measurements on your diagram.
   b. Calculate the image position and height. Compare with your ray-tracing answers in part a.

2. ‖ A converging lens with a focal length of 40 cm and a diverging lens with a focal length of −40 cm are 160 cm apart. A 2.0-cm-tall object is 60 cm in front of the converging lens.
   a. Use ray tracing to find the position and height of the image. Do this accurately with a ruler or paper with a grid. Estimate the image distance and image height by making measurements on your diagram.
   b. Calculate the image position and height. Compare with your ray-tracing answers in part a.

3. ‖ A 2.0-cm-tall object is 20 cm to the left of a lens with a focal length of 10 cm. A second lens with a focal length of 15 cm is 30 cm to the right of the first lens.
   a. Use ray tracing to find the position and height of the image. Do this accurately with a ruler or paper with a grid. Estimate the image distance and image height by making measurements on your diagram.
   b. Calculate the image position and height. Compare with your ray-tracing answers in part a.

4. ‖ A 2.0-cm-tall object is 20 cm to the left of a lens with a focal length of 10 cm. A second lens with a focal length of 5 cm is 30 cm to the right of the first lens.
   a. Use ray tracing to find the position and height of the image. Do this accurately with a ruler or paper with a grid. Estimate the image distance and image height by making measurements on your diagram.
   b. Calculate the image position and height. Compare with your ray-tracing answers in part a.

5. ‖‖ A 2.0-cm-tall object is 20 cm to the left of a lens with a focal length of 10 cm. A second lens with a focal length of −5 cm is 30 cm to the right of the first lens.
   a. Use ray tracing to find the position and height of the image. Do this accurately with a ruler or paper with a grid. Estimate the image distance and image height by making measurements on your diagram.
   b. Calculate the image position and height. Compare with your ray-tracing answers in part a.

Section 24.2 The Camera

6. | A 2.0-m-tall man is 10 m in front of a camera with a 15-mm-focal-length lens. How tall is his image on the detector?

7. | What is the f-number of a lens with a 35 mm focal length and a 7.0-mm-diameter aperture?

8. | A 12-mm-focal-length lens has a 4.0-mm-diameter aperture. What is the aperture diameter of an 18-mm-focal-length lens with the same f-number?

9. | What is the aperture diameter of a 12-mm-focal-length lens set to $f/4.0$?

10. | A camera takes a properly exposed photo at $f/5.6$ and 1/125 s. What shutter speed should be used if the lens is changed to $f/4.0$?

11. ‖‖ A camera takes a properly exposed photo with a 3.0-mm-diameter aperture and a shutter speed of 1/125 s. What is the appropriate aperture diameter for a 1/500 s shutter speed?

Section 24.3 Vision

12. ‖ Ramon has contact lenses with the prescription +2.0 D.
   BIO
   a. What eye condition does Ramon have?
   b. What is his near point without the lenses?

13. | Ellen wears eyeglasses with the prescription −1.0 D.
   BIO
   a. What eye condition does Ellen have?
   b. What is her far point without the glasses?

14. | What is the f-number of a relaxed eye with the pupil fully
   BIO dilated to 8.0 mm? Model the eye as a single lens 2.4 cm in front of the retina.

Section 24.4 Optical Systems That Magnify

15. | A magnifier has a magnification of 5×. How far from the lens should an object be held so that its image is at the near-point distance of 25 cm?

16. ‖ A microscope has a 20 cm tube length. What focal-length objective will give total magnification 500× when used with a eyepiece having a focal length of 5.0 cm?

17. ‖ A standardized biological microscope has an 8.0-mm-focal-length objective. What focal-length eyepiece should be used to achieve a total magnification of 100×?

18. ‖ A 6.0-mm-diameter microscope objective has a focal length of 9.0 mm. What object distance gives a lateral magnification of −40?

19. | A 20× telescope has a 12-cm-diameter objective lens. What minimum diameter must the eyepiece lens have to collect all the light rays from an on-axis distant source?

20. ‖ A reflecting telescope is built with a 20-cm-diameter mirror having a 1.00 m focal length. It is used with a 10× eyepiece. What are (a) the magnification and (b) the f-number of the telescope?

Section 24.5 The Resolution of Optical Instruments

21. ‖ A scientist needs to focus a helium-neon laser beam ($\lambda = 633$ nm) to a 10-μm-diameter spot 8.0 cm behind a lens.
   a. What focal-length lens should she use?
   b. What minimum diameter must the lens have?

22. ‖ Two lightbulbs are 1.0 m apart. From what distance can these lightbulbs be marginally resolved by a small telescope with a 4.0-cm-diameter objective lens? Assume that the lens is diffraction limited and $\lambda = 600$ nm.

Problems

23. ‖ A 1.0-cm-tall object is located 4.0 cm to the left of a converging lens with a focal length of 5.0 cm. A diverging lens, of focal length −8.0 cm, is 12 cm to the right of the first lens. Find the position, size, and orientation of the final image.

24. | In FIGURE P24.24, are parallel rays from the left focused to a point? If so, on which side of the lens and at what distance?

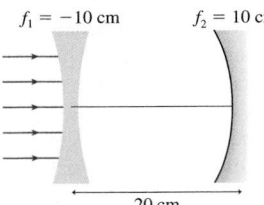

$f_1 = -10$ cm $f_2 = 10$ cm $f = 10$ cm $f = 10$ cm

|← 20 cm →| 5.0 cm   5.0 cm
                                      Mirror   Lens

FIGURE P24.24   **FIGURE P24.25**

25. ‖ The rays leaving the two-component optical system of FIGURE P24.25 produce two distinct images of the 1.0-cm-tall object.
   a. What are the position (relative to the lens), orientation, and height of each image?
   b. Draw two ray diagrams, one for each image, showing how the images are formed.

26. | A common optical instrument in a laser laboratory is a *beam expander*. One type of beam expander is shown in FIGURE P24.26. The parallel rays of a laser beam of width w_1 enter from the left.
   a. For what lens spacing d does a parallel laser beam exit from the right?
   b. What is the width w_2 of the exiting laser beam?

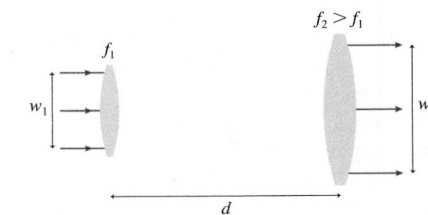

FIGURE P24.26

27. | A common optical instrument in a laser laboratory is a *beam expander*. One type of beam expander is shown in FIGURE P24.27. The parallel rays of a laser beam of width w_1 enter from the left.
    a. For what lens spacing d does a parallel laser beam exit from the right?
    b. What is the width w_2 of the exiting laser beam?

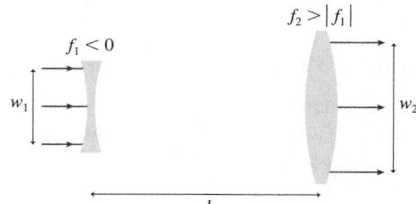

FIGURE P24.27

28. ||| A 15-cm-focal-length converging lens is 20 cm to the right of a 7.0-cm-focal-length converging lens. A 1.0-cm-tall object is distance L to the left of the first lens.
    a. For what value of L is the final image of this two-lens system halfway between the two lenses?
    b. What are the height and orientation of the final image?
29. | A 1.0-cm-tall object is 110 cm from a screen. A diverging lens with focal length -20 cm is 20 cm in front of the object. What are the focal length and distance from the screen of a second lens that will produce a well-focused, 2.0-cm-tall image on the screen?
30. ||| You use your 8× binoculars to focus on a 14-cm-long bird in a tree 18 m away from you. What angle (in degrees) does the image of the warbler subtend on your retina?
31. || Yang can focus on objects 150 cm away with a relaxed eye.
    BIO With full accommodation, she can focus on objects 20 cm away. After her eyesight is corrected for distance vision, what will her near point be while wearing her glasses?
32. ||| The cornea, a boundary between the air and the aqueous hu-
    BIO mor, has a 3.0 cm focal length when acting alone. What is its radius of curvature?
33. | The objective lens of a telescope is a symmetric glass lens with 100 cm radii of curvature. The eyepiece lens is also a symmetric glass lens. What are the radii of curvature of the eyepiece lens if the telescope's magnification is 20×?
34. || You've been asked to build a telescope from a 2.0× magnifying lens and a 5.0× magnifying lens.
    a. What is the maximum magnification you can achieve?
    b. Which lens should be used as the objective? Explain.
    c. What will be the length of your telescope?
35. | Marooned on a desert island and with a lot of time on your hands, you decide to disassemble your glasses to make a crude telescope with which you can scan the horizon for rescuers. Luckily you're farsighted and, like most people, your two eyes have different lens prescriptions. Your left eye uses a lens of power +4.5 D, and your right eye's lens is +3.0 D.
    a. Which lens should you use for the objective and which for the eyepiece? Explain.
    b. What will be the magnification of your telescope?
    c. How far apart should the two lenses be when you focus on distant objects?
36. || You've been asked to build a 12× microscope from a 2.0× magnifying lens and a 4.0× magnifying lens.
    a. Which lens should be used as the objective?
    b. What will be the tube length of your microscope?

37. || A microscope with a tube length of 180 mm achieves a total magnification of 800× with a 40× objective and a 20× eyepiece. The microscope is focused for viewing with a relaxed eye. How far is the sample from the objective lens?
38. | High-power lasers are used to cut and weld materials by focusing the laser beam to a very small spot. This is like using a magnifying lens to focus the sun's light to a small spot that can burn things. As an engineer, you have designed a laser cutting device in which the material to be cut is placed 5.0 cm behind the lens. You have selected a high-power laser with a wavelength of 1.06 μm. Your calculations indicate that the laser must be focused to a 5.0-μm-diameter spot in order to have sufficient power to make the cut. What is the minimum diameter of the lens you must install?
39. ||| Once dark adapted, the pupil of your eye is approximately
    BIO 7 mm in diameter. The headlights of an oncoming car are 120 cm apart. If the lens of your eye is diffraction limited, at what distance are the two headlights marginally resolved? Assume a wavelength of 600 nm and that the index of refraction inside the eye is 1.33. (Your eye is not really good enough to resolve headlights at this distance, due both to aberrations in the lens and to the size of the receptors in your retina, but it comes reasonably close.)
40. || The Hubble Space Telescope has a mirror diameter of 2.4 m. Suppose the telescope is used to photograph stars near the center of our galaxy, 30,000 light years away, using red light with a wavelength of 650 nm.
    a. What's the distance (in km) between two stars that are marginally resolved? The resolution of a reflecting telescope is calculated exactly the same as for a refracting telescope.
    b. For comparison, what is this distance as a multiple of the distance of Jupiter from the sun?
41. || Alpha Centauri, the nearest star to our solar system, is 4.3 light years away. Assume that Alpha Centauri has a planet with an advanced civilization. Professor Dhg, at the planet's Astronomical Institute, wants to build a telescope with which he can find out whether any planets are orbiting our sun.
    a. What is the minimum diameter for an objective lens that will just barely resolve Jupiter and the sun? The radius of Jupiter's orbit is 780 million km. Assume $\lambda = 600$ nm.
    b. Building a telescope of the necessary size does not appear to be a major problem. What practical difficulties might prevent Professor Dhg's experiment from succeeding?

Challenge Problems

42. In FIGURE CP24.42, what are the position, height, and orientation of the final image? Give the position as a distance to the right or left of the lens.

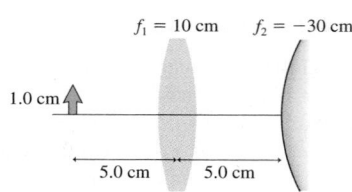

FIGURE CP24.42

43. Mars (6800 km diameter) is viewed through a telescope on a night when it is 1.1×10^8 km from the earth. Its angular size as seen through the eyepiece is 0.50°, the same size as the full moon seen by the naked eye. If the eyepiece focal length is 25 mm, how long is the telescope?

44. Your task in physics laboratory is to make a microscope from two lenses. One lens has a focal length of 2.0 cm, the other 1.0 cm. You plan to use the more powerful lens as the objective, and you want the eyepiece to be 16 cm from the objective.
    a. For viewing with a relaxed eye, how far should the sample be from the objective lens?
    b. What is the magnification of your microscope?

45. The lens shown in FIGURE CP24.45 is called an *achromatic doublet,* meaning that it has no chromatic aberration. The left side is flat, and all other surfaces have radii of curvature R.
    a. For parallel light rays coming from the left, show that the effective focal length of this two-lens system is $f = R/(2n_2 - n_1 - 1)$, where n_1 and n_2 are, respectively, the indices of refraction of the diverging and the converging lenses. Don't forget to make the thin-lens approximation.
    b. Because of dispersion, either lens alone would focus red rays and blue rays at different points. Define Δn_1 and Δn_2 as $n_{blue} - n_{red}$ for the two lenses. Find an expression for Δn_2 in terms of Δn_1 that makes $f_{blue} = f_{red}$ for the two-lens system. That is, the two-lens system does *not* exhibit chromatic aberration.
    c. Indices of refraction for two types of glass are given in the table. To make an achromatic doublet, which glass should you use for the converging lens and which for the diverging lens? Explain.

FIGURE CP24.45

	n_{blue}	n_{red}
Crown glass	1.525	1.517
Flint glass	1.632	1.616

    d. What value of R gives a focal length of 10.0 cm?

46. FIGURE CP24.46 shows a simple zoom lens in which the magnitudes of both focal lengths are f. If the spacing $d < f$, the image of the converging lens falls on the right side of the diverging lens. Our procedure of letting the image of the first lens act as the object of the second lens will continue to work in this case if we use a *negative* object distance for the second lens. This is called a *virtual object.* Consider a very distant object ($s \approx \infty$ for the first lens) and define the effective focal length as the distance from the midpoint between the lenses to the final image.
    a. Show that the effective focal length is

$$f_{eff} = \frac{f^2 - fd + \frac{1}{2}d^2}{d}$$

    b. What is the zoom for a lens that can be adjusted from $d = \frac{1}{2}f$ to $d = \frac{1}{4}f$?

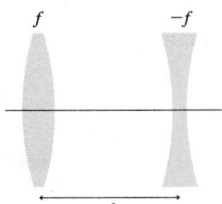

FIGURE CP24.46

STOP TO THINK ANSWERS

Stop to Think 24.1: b. A diverging lens refracts rays away from the optical axis, so the rays will travel farther down the axis before converging.

Stop to Think 24.2: a. Because the shutter speed doesn't change, the f-number must remain unchanged. The f-number is f/D, so increasing f requires increasing D.

Stop to Think 24.3: a. A magnifier is a converging lens. Converging lenses are used to correct hyperopia.

Stop to Think 24.4: b. If the objective magnification is halved, the eyepiece magnification must be doubled. $M_{eye} = 25 \text{ cm}/f_{eye}$, so doubling M_{eye} requires halving f_{eye}.

Stop to Think 24.5: $w_a > w_d > w_b = w_c$. The spot size is proportional to f/D.

V Waves and Optics

We end our study of waves a long distance from where we started. Who would have guessed, as we examined our first pulse on a string, that we would end up discussing the resolution of microscopes? But despite the wide disparity between string waves, sound waves, and light waves, a few key ideas have stayed with us throughout Part V: the principle of superposition, interference and diffraction, and standing waves. As part of your final study of waves, you should trace the influence of these ideas through the chapters of Part V.

One point we have tried to emphasize is the *unity* of wave physics. We did not need separate theories of string waves and sound waves and light waves. Instead, a few basic ideas enabled us to understand waves of all types. By focusing on similarities, we have been able to analyze vibrating guitar strings and anti-reflection coatings on lenses in a single part of this book.

Unfortunately, the physics of waves is not as easily summarized as the physics of particles. Newton's laws and the conservation laws are two very general sets of principles about particles, principles that allowed us to develop the powerful problem-solving strategies of Parts I and II. You probably noticed that we have not found any general problem-solving strategies for wave problems.

This is not to say that wave physics has no structure. Rather, the knowledge structure of waves and optics rests more heavily on *phenomena* than on general principles. Unlike the knowledge structure of Newtonian mechanics, which was a "pyramid of ideas," the knowledge structure of waves is a logical grouping of the major topics you studied. This is a different way of structuring knowledge, but it still provides you with a mental framework for analyzing and thinking about wave problems.

KNOWLEDGE STRUCTURE V Waves and Optics

ESSENTIAL CONCEPTS	Wave speed, wavelength, frequency, phase, wave front, and ray
BASIC GOALS	What are the distinguishing features of waves?
	How does a wave travel through a medium?
	How does a medium respond to the presence of more than one wave?
	What is light and what are its properties?
GENERAL PRINCIPLES	Principle of superposition
	$v = \lambda f$ for periodic waves

Traveling Waves

- The wave speed v is a property of the medium.
- The motion of particles in the medium is distinct from the motion of the wave.
- Snapshot graphs and history graphs show the same wave from different perspectives.
- The Doppler effect of shifted frequencies is observed whenever the wave source or the detector is moving.

Standing Waves

- Standing waves are the superposition of waves moving in opposite directions.
- Nodes and antinodes are spaced by $\lambda/2$.
- Only certain discrete frequencies are allowed, depending on the boundary conditions.

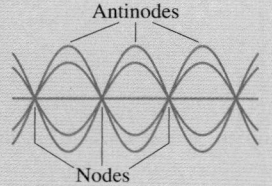
Antinodes

Nodes

Interference

- Interference is constructive—crests align with crests—if two waves are in phase: $\Delta\phi = 0, 2\pi, 4\pi, \ldots$. The wave is enhanced.
- Interference is destructive—crests align with troughs—if two waves are out of phase: $\Delta\phi = \pi, 3\pi, 5\pi, \ldots$. The wave is reduced.
- The phase difference depends on the path-length difference Δr and on any phase difference between the sources.
- Beats occur when $f_1 \neq f_2$.

Light and Optics

- The wave model, used for interference and diffraction, is appropriate when apertures are comparable in size to the wavelength.

Single-slit diffraction: Double-slit interference:

- The ray model, used for mirrors and lenses, is appropriate when apertures are much larger than the wavelength.
- Diffraction, a wave effect, limits the best possible resolution of a lens.

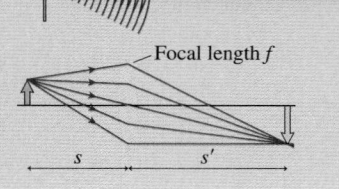
Focal length f

s s'

ONE STEP BEYOND

Tsunami!

In December 2004, an earthquake off the Indonesian coast produced a devastating water wave, a *tsunami,* that caused tremendous destruction and loss of life around the edges of the Indian Ocean, often thousands of miles from the earthquake's epicenter. The tsunami was a dramatic reminder of the power of the earth's forces and an impressive illustration of the energy carried by waves.

The Indian Ocean tsunami of 2004 was caused when a very large earthquake disrupted the seafloor along a fault line, pushing one side of the fault up several meters. This dramatic shift in the seafloor produced an almost instantaneous rise in the surface of the ocean above, much like giving a quick shake to one end of a rope. This was the disturbance that produced the tsunami. And just as shaking one end of rope causes a pulse to travel along it, the resulting water wave propagated throughout the Indian Ocean, as we see in the figure, carrying energy from the earthquake.

This computer simulation of the tsunami looks much like the ripples that spread out when you drop a pebble into a pond, but on an immensely larger scale. The individual wave pulses are up to 100 km wide, and the leading wave front spans more than 5000 km.

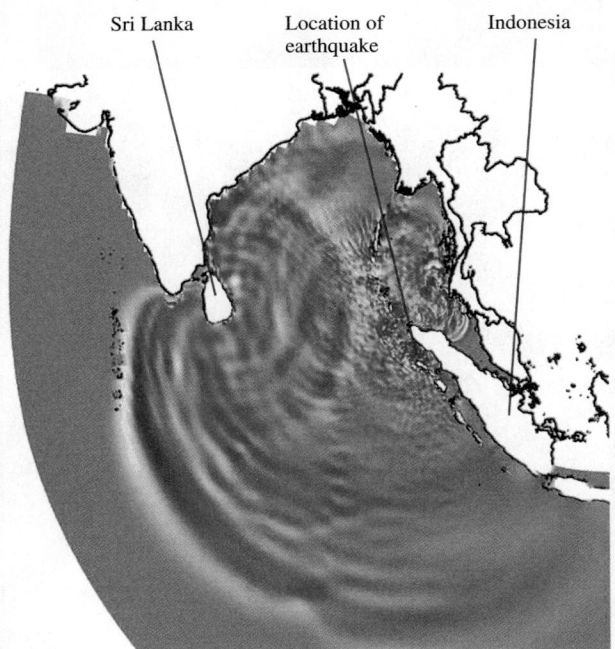

A frame from a computer simulation of the tsunami, showing the Indian Ocean about three hours after the earthquake. Notice the interference pattern to the east of Sri Lanka, where incoming waves and reflected waves are superimposed.

Technically, a tsunami is a "shallow-water wave," even in the deep ocean, because the scale of the wave (roughly 100 km) is much larger than the depth of the ocean (typically 4 km). Consequently, a tsunami travels differently than normal ocean waves. Unlike normal waves on the surface, whose speed is independent of depth, the speed of a shallow-water wave is determined by the depth of the ocean: The greater the depth, the greater the speed. In the deep ocean, a tsunami travels at hundreds of kilometers per hour, about the speed of a jet plane. This great speed allows a tsunami to cross oceans in only a few hours.

The height of the tsunami as it raced across the open ocean was about half a meter. Why should such a small wave—one that ships didn't even notice as it passed—be so fearsome? It's the width of the wave that matters. The wave pulse may have been only half a meter high, but it was about 100 km wide. In other words, the tsunami far from land was a half-meter-high, 100-km-wide wall of water. This is a tremendous amount of water displaced upward, and thus the tsunami was carrying a tremendous amount of energy.

As a tsunami nears shore, the ocean depth decreases and—because its speed is determined by depth—the tsunami begins to slow. This is when the awesome power of a tsunami begins to become apparent. As the leading edge of the wave slows, the trailing edge, still 100 km away and traveling much faster in deeper water, quickly begins to catch up. Water is nearly incompressible. As the width of the wave pulse decreases, the water begins to pile up higher and higher and the wave increases dramatically in height. The Indian Ocean tsunami had a height of up to 15 m (50 ft) as it came ashore.

Despite its height, a tsunami doesn't break and crash on the beach like a normal wave. The wave pulse may have narrowed dramatically from its 100 km width in the open ocean, but it is still several kilometers wide. Thus a tsunami reaching shore is more like a huge water surge than a typical wave—a wall of water that moves onto the shore and just keeps on coming. In many places, the Indian Ocean tsunami reached 2 km inland.

The impact of the Indian Ocean tsunami was devastating, but it was the first tsunami for which scientists were able to use satellites and ocean sensors to make planet-wide measurements. An analysis of the data, including computer simulations like the one seen here, has helped us better understand the physics of these ocean waves. We won't be able to stop future tsunamis, but with a better knowledge of how they are formed and how they travel, we will be better able to warn people to get out of their way.

Electricity and Magnetism

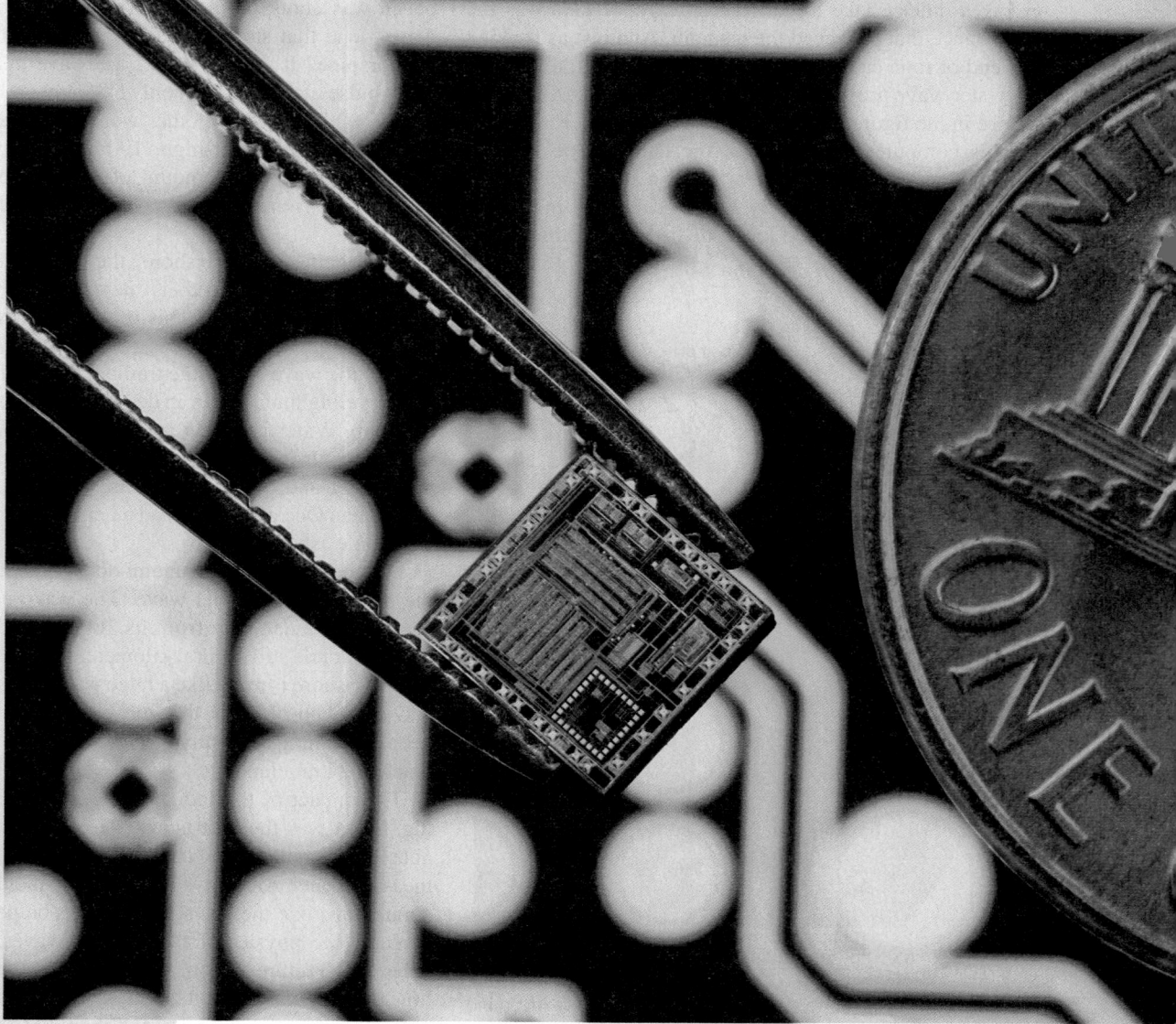

This integrated circuit contains millions of circuit elements. The density of circuit elements in integrated circuits has doubled about every 18 months for the past 30 years. Whether this trend continues depends on whether scientists and engineers can understand the physics of nanoscale electric circuits.

OVERVIEW

Phenomena and Theories

Amber, or fossilized tree resin, has long been prized for its beauty. Amber is of scientific interest today because biologists have learned how to recover DNA strands from million-year-old insects trapped in the resin. But amber has an ancient scientific connection as well. The Greek word for amber is *elektron*.

It has been known since antiquity that a piece of amber rubbed with fur can attract feathers or straw—seemingly magical powers to a pre-scientific society. It was also known to the ancient Greeks that certain stones from the region they called *Magnesia* could pick up pieces of iron. It is from these humble beginnings that we today have high-speed computers, lasers, and magnetic resonance imaging as well as such mundane modern-day miracles as the lightbulb.

The basic phenomena of electricity and magnetism are not as familiar as those of mechanics. You have spent your entire life exerting forces on objects and watching them move, but your experience with electricity and magnetism is probably much more limited. We will deal with this lack of experience by placing a large emphasis on the *phenomena* of electricity and magnetism.

We will begin by looking in detail at *electric charge* and the process of *charging* an object. It is easy to make systematic observations of how charges behave, and we will consider the forces between charges and how charges behave in different materials. Similarly, we will begin our study of magnetism by observing how magnets stick to some metals but not others and how magnets affect compass needles. But our most important observation will be that an electric current affects a compass needle in exactly the same way as a magnet. This observation, suggesting a close connection between electricity and magnetism, will eventually lead us to the discovery of electromagnetic waves.

Our goal in Part VI is to develop a theory to explain the phenomena of electricity and magnetism. The linchpin of our theory will be the entirely new concept of a *field*. Electricity and magnetism are about the long-range interactions of charges, both static charges and moving charges, and the field concept will help us understand how these interactions take place. We will want to know how fields are created by charges and how charges, in return, respond to the fields. Bit by bit, we will assemble a theory—based on the new concepts of electric and magnetic fields—that will allow us to understand, explain, and predict a wide range of electromagnetic behavior.

The story of electricity and magnetism is vast. The 19th-century formulation of the theory of electromagnetism, which led to sweeping revolutions in science and technology, has been called by no less than Einstein "the most important event in physics since Newton's time." Not surprisingly, all we can do in this text is develop some of the basic ideas and concepts, leaving many details and applications to later courses. Even so, our study of electricity and magnetism will explore some of the most exciting and important topics in physics.

25 Electric Charges and Forces

Electricity is one of the fundamental forces of nature. Lightning is a vivid manifestation of electric charges and forces.

▶ **Looking Ahead** The goal of Chapter 25 is to describe electric phenomena in terms of charges, forces, and fields.

Charge Model

Electric phenomena seem mysterious at first, but we'll find that we can understand them in terms of a **charge model:**

- There are two kinds of charge, called *positive* and *negative.*
- Two charges of the same kind repel; two opposite charges attract.
- Small neutral objects are attracted to a charge of either sign.

You'll learn how a comb rubbed through your hair picks up small pieces of paper.

Coulomb's Law

The law governing the electric force is called **Coulomb's law.** It tells us how the force between charged particles depends on their charge and on the distance between them.

You'll find that Coulomb's law, like Newton's law of gravity, is an *inverse-square law*.

◀ **Looking Back**

Sections 3.2–3.4 Vector addition
Sections 13.3–13.4 Newton's theory of gravity

Field Model

How is a long-range force transmitted from one charge to another? We'll develop the idea that every charge alters the space around it by creating an **electric field.** It is the electric field that then exerts forces on other charges.

The liquid crystal displays (LCD) of your calculator, your digital watch, and your computer screen use electric fields to turn the pixels on and off.

Charges and Atoms

Electrons and protons—the constituents of atoms—are the basic charges of ordinary matter.

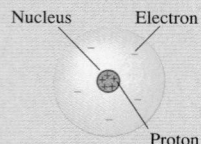

You'll learn that *charging* an object can be understood as the transfer of electrons from one material to another.

An object that is negative has an excess of electrons; a positively charged object is missing electrons.

Conductors and Insulators

There are two types of materials with very different electrical properties:

- **Conductors** are materials through or along which charge easily moves.
- **Insulators** are materials on or in which charge is immobile.

The metal wire—a conductor—carries a current of moving charges. It is separated from the support by a ceramic insulator.

Point Charges

A charged particle, with no physical size, is called a **point charge.** You'll learn that real objects can be modeled as point charges if they are very small compared to the distances between them.

The electric field of a point charge will be important throughout our study of electricity.

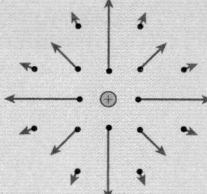

25.1 Developing a Charge Model

You can receive a mildly unpleasant shock and produce a little spark if you touch a metal doorknob after walking across a carpet. Vigorously brushing your freshly washed hair makes all the hairs fly apart. A plastic comb that you've run through your hair will pick up bits of paper and other small objects, but a metal comb won't.

The common factor in these observations is that two objects are *rubbed* together. Why should rubbing an object cause forces and sparks? What kind of forces are these? Why do metallic objects behave differently from nonmetallic? These are the questions with which we begin our study of electricity.

Our first goal is to develop a model for understanding electric phenomena in terms of *charges* and *forces*. We will later use our contemporary knowledge of atoms to understand electricity on a microscopic level, but the basic concepts of electricity make *no* reference to atoms or electrons. The theory of electricity was well established long before the electron was discovered.

Experimenting with Charges

Let us enter a laboratory where we can make observations of electric phenomena. The major tools in the lab are:

- A variety of plastic and glass rods, each several centimeters long.
- A few metal rods with wood handles.
- Pieces of wool and silk.
- Small metal spheres, an inch or two in diameter, on wood stands.

Let's see what we can learn with these tools.

Discovering electricity I

Experiment 1	Experiment 2	Experiment 3	Experiment 4
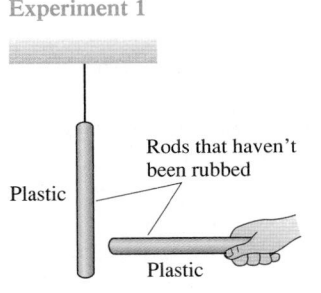		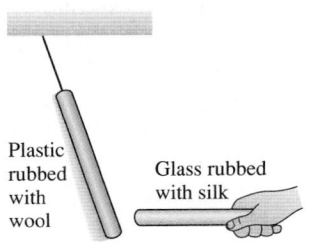	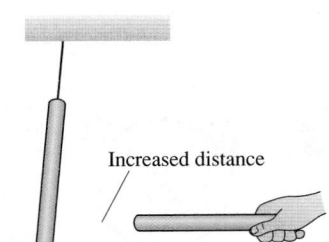

Experiment 1: Plastic. Rods that haven't been rubbed. Plastic.

Experiment 2: Plastic rubbed with wool.

Experiment 3: Plastic rubbed with wool. Glass rubbed with silk.

Experiment 4: Increased distance.

Take a plastic rod that has been undisturbed for a long period of time and hang it by a thread. Pick up another undisturbed plastic rod and bring it close to the hanging rod. Nothing happens to either rod.

Rub both plastic rods with wool. Now the hanging rod tries to move away from the handheld rod when you bring the two close together. Two glass rods rubbed with silk also repel each other.

Bring a glass rod that has been rubbed with silk close to a hanging plastic rod that has been rubbed with wool. These two rods *attract* each other.

Further observations show that:

- These forces are greater for rods that have been rubbed more vigorously.
- The strength of the forces decreases as the separation between the rods increases.

No forces were observed in Experiment 1. We will say that the original objects are **neutral**. Rubbing the rods (Experiments 2 and 3) somehow causes forces to be exerted between them. We will call the rubbing process **charging** and say that a rubbed rod is *charged*. For now, these are simply descriptive terms. The terms don't tell us anything about the process itself.

Experiment 2 shows that there is a *long-range repulsive force*, requiring no contact, between two identical objects that have been charged in the *same* way. Furthermore, Experiment 4 shows that the force between two charged objects depends on the distance between them. This is the first long-range force we've encountered since gravity was introduced in Chapter 5. It is also the first time we've observed a repulsive force, so right away we see that new ideas will be needed to understand electricity.

Experiment 3 is a puzzle. Two rods *seem* to have been charged in the same way, by rubbing, but these two rods *attract* each other rather than repel. Why does the outcome of Experiment 3 differ from that of Experiment 2? Back to the lab.

Discovering electricity II

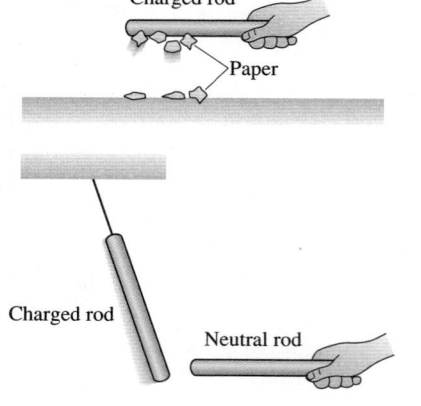

Experiment 5

Hold a charged (i.e., rubbed) plastic rod over small pieces of paper on the table. The pieces of paper leap up and stick to the rod. A charged glass rod does the same. However, a neutral rod has no effect on the pieces of paper.

Experiment 6

Rub a plastic rod with wool and a glass rod with silk. Hang both by threads, some distance apart. Both rods are attracted to a *neutral* (i.e., unrubbed) plastic rod that is held close. Interestingly, both are also attracted to a *neutral* glass rod. In fact, the charged rods are attracted to *any* neutral object, such as a finger, a piece of paper, or a metal rod.

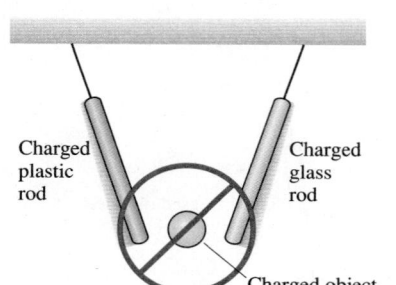

Experiment 7

Rub a hanging plastic rod with wool and then hold the *wool* close to the rod. The rod is weakly *attracted* to the wool. The plastic rod is *repelled* by a piece of silk that has been used to rub glass.

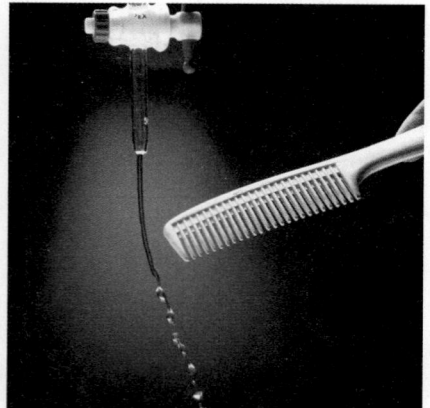

Experiment 8

Further experiments show that:

- Other objects, after being rubbed, attract one of the hanging charged rods (plastic or glass) and repel the other. These objects always pick up small pieces of paper.
- There appear to be *no* objects that, after being rubbed, pick up pieces of paper and attract *both* the charged plastic and glass rods.

Our first set of experiments found that charged objects exert forces on each other. The forces are sometimes attractive, sometimes repulsive. Experiments 5 and 6 show that there is an attractive force between a charged object and a *neutral* (uncharged) object. This discovery presents us with a problem: How can we tell if an object is charged or neutral? Because of the attractive force between a charged and a neutral object, simply observing an electric force does *not* imply that an object is charged.

However, an important characteristic of any *charged* object appears to be that **a charged object picks up small pieces of paper.** This behavior provides a straightforward test to answer the question, Is this object charged? An object that passes the test by picking up paper is charged; an object that fails the test is neutral.

These observations let us tentatively advance the first stages of a **charge model.**

Charge model, part I The basic postulates of our model are:

1. Frictional forces, such as rubbing, add something called **charge** to an object or remove it from the object. The process itself is called *charging*. More vigorous rubbing produces a larger quantity of charge.

A plastic comb that has been charged by running it through your hair attracts neutral objects—here drops of water.

2. There are two and only two kinds of charge. For now we will call these "plastic charge" and "glass charge." Other objects can sometimes be charged by rubbing, but the charge they receive is either "plastic charge" or "glass charge."

3. Two **like charges** (plastic/plastic or glass/glass) exert repulsive forces on each other. Two **opposite charges** (plastic/glass) attract each other.

4. The force between two charges is a long-range force. The size of the force increases as the quantity of charge increases and decreases as the distance between the charges increases.

5. *Neutral* objects have an *equal mixture* of both "plastic charge" and "glass charge." The rubbing process somehow manages to separate the two.

Postulate 2 is based on Experiment 8. If an object is charged (i.e., picks up paper), it always attracts one charged rod and repels the other. That is, it acts either "like plastic" or "like glass." If there were a third kind of charge, different from the first two, an object with that charge should pick up paper and attract *both* the charged plastic and glass rods. No such objects have ever been found.

The basis for postulate 5 is the observation in Experiment 7 that a charged plastic rod is attracted to the wool used to rub it but repelled by silk that has rubbed glass. It appears that rubbing glass causes the silk to acquire "plastic charge." The easiest way to explain this is to hypothesize that the silk starts out with equal amounts of "glass charge" and "plastic charge" and that the rubbing somehow transfers "glass charge" from the silk to the rod. This leaves an excess of "glass charge" on the rod and an excess of "plastic charge" on the silk.

While the charge model is *consistent* with the observations, it is by no means proved. One could easily imagine other hypotheses that are just as consistent with the limited observations we have made so far. We still have some large unexplained puzzles, such as why charged objects exert attractive forces on neutral objects.

Electric Properties of Materials

We still need to clarify how different types of materials respond to charges.

Discovering electricity III

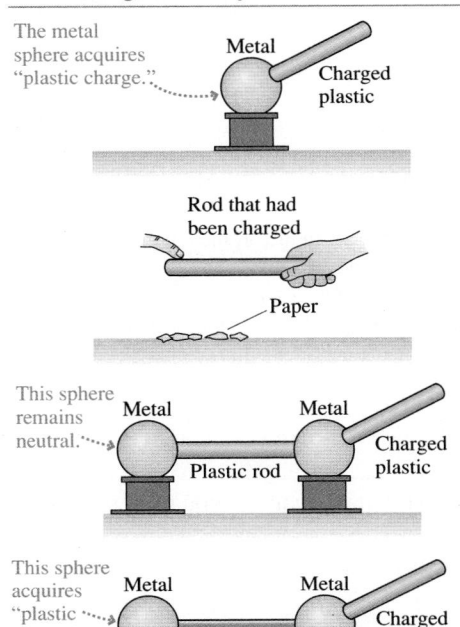

The metal sphere acquires "plastic charge."

Metal

Charged plastic

Rod that had been charged

Paper

This sphere remains neutral.

Metal

Metal

Plastic rod

Charged plastic

This sphere acquires "plastic charge."

Metal

Metal

Metal rod

Charged plastic

Experiment 9

Charge a plastic rod by rubbing it with wool. Touch a neutral metal sphere with the rubbed area of the rod. The metal sphere then picks up small pieces of paper and repels a charged, hanging plastic rod. The metal sphere appears to have acquired "plastic charge."

Experiment 10

Charge a plastic rod, then run your finger along it. After you've done so, the rod no longer picks up small pieces of paper or repels a charged, hanging plastic rod. Similarly, the metal sphere of Experiment 9 no longer repels the plastic rod after you touch it with your finger.

Experiment 11

Place two metal spheres close together with a plastic rod connecting them. Charge a second plastic rod, by rubbing, and touch it to one of the metal spheres. Afterward, the metal sphere that was touched picks up small pieces of paper and repels a charged, hanging plastic rod. The other metal sphere does neither.

Experiment 12

Repeat Experiment 11 with a metal rod connecting the two metal spheres. Touch one metal sphere with a charged plastic rod. Afterward, *both* metal spheres pick up small pieces of paper and repel a charged, hanging plastic rod.

Our final set of experiments has shown that

- Charge can be *transferred* from one object to another, but only when the objects *touch*. Contact is required. Removing charge from an object, which you can do by touching it, is called **discharging.**
- There are two types or classes of materials with very different electric properties. We call these *conductors* and *insulators*.

Experiment 12, in which a metal rod is used, is in sharp contrast to Experiment 11. Charge somehow *moves through* or along a metal rod, from one sphere to the other, but remains *fixed in place* on a plastic or glass rod. Let us define **conductors** as those materials through or along which charge easily moves and **insulators** as those materials on or in which charges remain immobile. Glass and plastic are insulators; metal is a conductor.

This information lets us add two more postulates to our charge model:

Charge model, part II

6. There are two types of materials. Conductors are materials through or along which charge easily moves. Insulators are materials on or in which charges remain fixed in place.

7. Charge can be transferred from one object to another by contact.

NOTE ▶ Both insulators and conductors can be charged. They differ in the *mobility* of the charge. ◀

We have by no means exhausted the number of experiments and observations we might try. Early scientific investigators were faced with all of these results, plus many others. Moreover, many of these experiments are hard to reproduce with much accuracy. How should we make sense of it all? The charge model seems promising, but certainly not proven. We have not yet explained how charged objects exert attractive forces on *neutral* objects, nor have we explained what charge is, how it is transferred, or *why* it moves through some objects but not others. Nonetheless, we will take advantage of our historical hindsight and continue to pursue this model. Homework problems will let you practice using the model to explain other observations.

EXAMPLE 25.1 **Transferring charge**

In Experiment 12, touching one metal sphere with a charged plastic rod caused a second metal sphere to become charged with the same type of charge as the rod. Use the postulates of the charge model to explain this.

SOLVE We need the following ideas from the charge model:

1. Charge is transferred upon contact.
2. Metal is a conductor.
3. Like charges repel.

The plastic rod was charged by rubbing with wool. The charge doesn't move around on the rod, because it is an insulator, but some of the "plastic charge" is transferred to the metal upon contact. Once in the metal, which is a conductor, the charges are free to move around. Furthermore, because like charges repel, these plastic charges quickly move as far apart as they possibly can. Some move through the connecting metal rod to the second sphere. Consequently, the second sphere acquires "plastic charge."

STOP TO THINK 25.1 To determine if an object has "glass charge," you need to

a. See if the object attracts a charged plastic rod.
b. See if the object repels a charged glass rod.
c. Do both a and b.
d. Do either a or b.

25.2 Charge

As you probably know, the modern names for the two types of charge are *positive charge* and *negative charge*. You may be surprised to learn that the names were coined by Benjamin Franklin. Franklin found that charge behaves like positive and negative numbers. If a plastic rod is charged twice, by rubbing, and twice transfers charge to a metal sphere, the electric forces exerted by the sphere are doubled. That is, $2 + 2 = 4$. But the sphere is found to be neutral after receiving equal amounts of "plastic charge" and "glass charge." This is like $2 + (-2) = 0$.

So what is positive and what is negative? It's entirely up to us! Franklin established the convention that **a glass rod that has been rubbed with silk is *positively* charged.** That's it. Any other object that repels a charged glass rod is also positively charged. Any charged object that attracts a charged glass rod is negatively charged. Thus **a plastic rod rubbed with wool is negative.** It was only long afterward, with the discovery of electrons and protons, that electrons were found to be attracted to a charged glass rod while protons were repelled. Thus *by convention* electrons have a negative charge and protons a positive charge.

Atoms and Electricity

Now let's fast forward to the 21st century. The theory of electricity was developed without knowledge of atoms, but there is no reason for us to continue to overlook this important part of our contemporary perspective. **FIGURE 25.1** shows that an atom consists of a very small and dense *nucleus* (diameter $\sim 10^{-14}$ m) surrounded by much less massive orbiting *electrons*. The electron orbital frequencies are so enormous ($\sim 10^{15}$ revolutions per second) that the electrons seem to form an **electron cloud** of diameter $\sim 10^{-10}$ m, a factor 10^4 larger than the nucleus. In fact, the wave–particle duality of quantum physics destroys any notion of a well-defined electron trajectory, and *all* we know about the electrons is the size and shape of the electron cloud.

Experiments at the end of the 19th century revealed that electrons are particles with both mass and a negative charge. The nucleus is a composite structure consisting of *protons,* positively charged particles, and neutral *neutrons.* The atom is held together by the attractive electric force between the positive nucleus and the negative electrons.

One of the most important discoveries is that **charge, like mass, is an inherent property of electrons and protons.** It's no more possible to have an electron without charge than it is to have an electron without mass. As far as we know today, electrons and protons have charges of opposite sign but *exactly* equal magnitude. (Very careful experiments have never found any difference.) This atomic-level unit of charge, called the **fundamental unit of charge,** is represented by the symbol e. Table 25.1 shows the masses and charges of protons and electrons. We need to define a unit of charge, which we will do in Section 25.4, before we can specify how much charge e is.

The Micro/Macro Connection

Electrons and protons are the basic charges of ordinary matter. Consequently, the various observations we made in Section 25.1 need to be explained in terms of electrons and protons.

> NOTE ▶ Electrons and protons are particles of matter. Their motion is governed by Newton's laws. Electrons can move from one object to another when the objects are in contact, but neither electrons nor protons can leap through the air from one object to another. An object does not become charged simply from being close to a charged object. ◀

FIGURE 25.1 An atom.

The nucleus, exaggerated for clarity, contains positive protons.

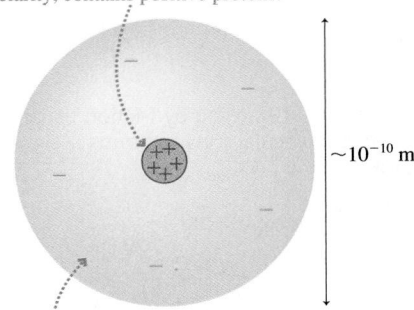

$\sim 10^{-10}$ m

The electron cloud is negatively charged.

TABLE 25.1 Protons and electrons

Particle	Mass (kg)	Charge
Proton	1.67×10^{-27}	$+e$
Electron	9.11×10^{-31}	$-e$

Charge is represented by the symbol q (or sometimes Q). A macroscopic object, such as a plastic rod, has charge

$$q = N_p e - N_e e = (N_p - N_e)e \qquad (25.1)$$

where N_p and N_e are the number of protons and electrons contained in the object. Most macroscopic objects have an *equal number* of protons and electrons and therefore have $q = 0$. An object with no *net* charge (i.e., $q = 0$) is said to be *electrically neutral*.

NOTE ▶ *Neutral* does *not* mean "no charges" but, instead, means that there is no *net* charge. ◀

A charged object has an unequal number of protons and electrons. An object is positively charged if $N_p > N_e$. It is negatively charged if $N_p < N_e$. Notice that an object's charge is always an integer multiple of e. That is, the amount of charge on an object varies by small but discrete steps, not continuously. This is called **charge quantization.**

In practice, objects acquire a positive charge not by gaining protons, as you might expect, but by losing electrons. Protons are *extremely* tightly bound within the nucleus and cannot be added to or removed from atoms. Electrons, on the other hand, are bound rather loosely and can be removed without great difficulty. The process of removing an electron from the electron cloud of an atom is called **ionization.** An atom that is missing an electron is called a *positive ion*. Its *net* charge is $q = +e$.

Some atoms can accommodate an *extra* electron and thus become a *negative ion* with net charge $q = -e$. A saltwater solution is a good example. When table salt (the chemical sodium chloride, NaCl) dissolves, it separates into positive sodium ions Na^+ and negative chlorine ions Cl^-. FIGURE 25.2 shows positive and negative ions.

All the charging processes we observed in Section 25.1 involved rubbing and friction. The forces of friction cause molecular bonds at the surface to break as the two materials slide past each other. Molecules are electrically neutral, but FIGURE 25.3 shows that *molecular ions* can be created when one of the bonds in a large molecule is broken. The positive molecular ions remain on one material and the negative ions on the other, so one of the objects being rubbed ends up with a net positive charge and the other with a net negative charge. This is the way in which a plastic rod is charged by rubbing with wool or a comb is charged by passing through your hair.

Charge Conservation and Charge Diagrams

One of the important discoveries about charge is the **law of conservation of charge:** Charge is neither created nor destroyed. Charge can be transferred from one object to another as electrons and ions move about, but the *total* amount of charge remains constant. For example, charging a plastic rod by rubbing it with wool transfers electrons from the wool to the plastic as the molecular bonds break. The wool is left with a positive charge equal in magnitude but opposite in sign to the negative charge of the rod: $q_{wool} = -q_{plastic}$. The *net* charge remains zero.

Diagrams are going to be an important tool for understanding and explaining charges and the forces on charged objects. As you begin to use diagrams, it will be important to make explicit use of charge conservation. The net number of plusses and minuses drawn on your diagrams should *not* change as you show them moving around.

FIGURE 25.2 Positive and negative ions.

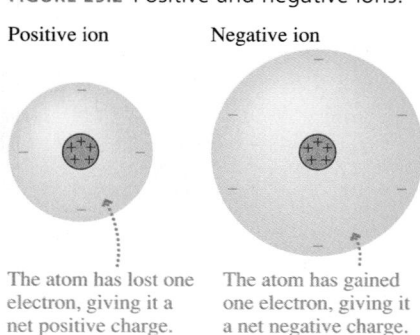

Positive ion   Negative ion

The atom has lost one electron, giving it a net positive charge.

The atom has gained one electron, giving it a net negative charge.

FIGURE 25.3 Charging by friction usually creates molecular ions as bonds are broken.

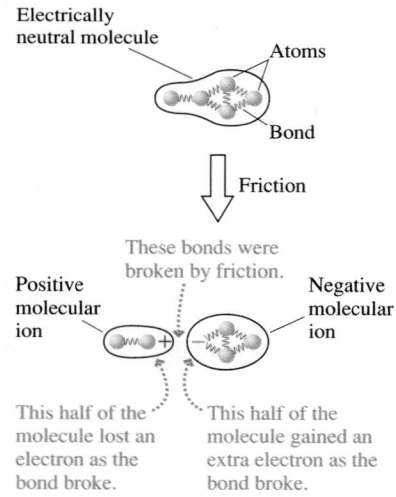

Electrically neutral molecule

Atoms

Bond

Friction

These bonds were broken by friction.

Positive molecular ion

Negative molecular ion

This half of the molecule lost an electron as the bond broke.

This half of the molecule gained an extra electron as the bond broke.

STOP TO THINK 25.2 Rank in order, from most positive to most negative, the charges q_a to q_e of these five systems.

Proton	Electron	17 protons 19 electrons	1,000,000 protons 1,000,000 electrons	Glass ball missing 3 electrons
•	•			○
(a)	**(b)**	**(c)**	**(d)**	**(e)**

25.3 Insulators and Conductors

You have seen that there are two classes of materials as defined by their electrical properties: insulators and conductors. It's time for a closer look at these materials.

FIGURE 25.4 looks inside an insulator and a metallic conductor. The electrons in the insulator are all tightly bound to the positive nuclei and not free to move around. Charging an insulator by friction leaves patches of molecular ions on the surface, but these patches are immobile.

In metals, the outer atomic electrons (called the *valence electrons* in chemistry) are only weakly bound to the nuclei. As the atoms come together to form a solid, these outer electrons become detached from their parent nuclei and are free to wander about through the entire solid. The solid *as a whole* remains electrically neutral, because we have not added or removed any electrons, but the electrons are now rather like a negatively charged gas or liquid—what physicists like to call a **sea of electrons**—permeating an array of positively charged **ion cores.**

The primary consequence of this structure is that electrons in a metal are highly mobile. They can quickly and easily move through the metal in response to electric forces. The motion of charges through a material is what we will later call a **current,** and the charges that physically move are called the **charge carriers.** The charge carriers in metals are electrons.

Metals aren't the only conductors. Ionic solutions, such as salt water, are also good conductors. But the charge carriers in an ionic solution are the ions, not electrons. We'll focus on metallic conductors because of their importance in applications of electricity.

FIGURE 25.4 A microscopic look at insulators and conductors.

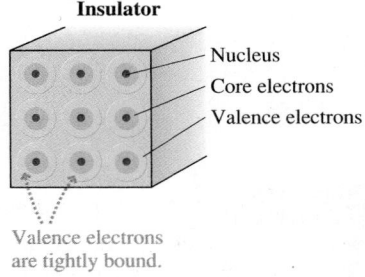

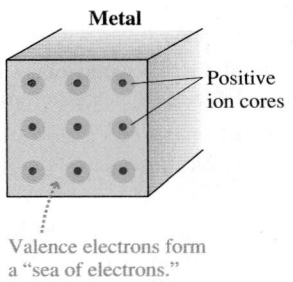

Charging

Insulators are often charged by rubbing. The charge diagrams of FIGURE 25.5 show that the charges on the rod are on the surface and that charge is conserved. The charge can be transferred to another object upon contact, but it doesn't move around on the rod.

FIGURE 25.5 An insulating rod is charged by rubbing.

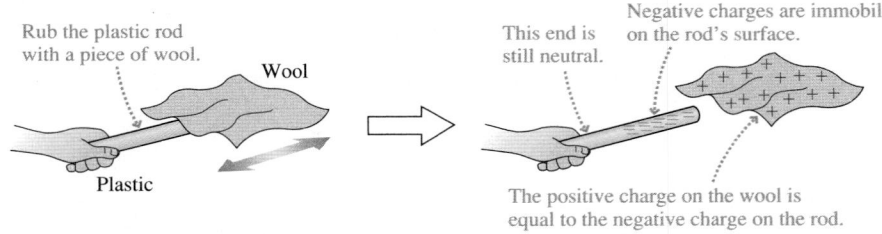

Metals usually cannot be charged by rubbing, but Experiment 9 showed that a metal sphere can be charged by contact with a charged plastic rod. FIGURE 25.6 gives a pictorial explanation. An essential idea is that **the electrons in a conductor are free to move.** Once charge is transferred to the metal, repulsive forces between the negative charges cause the electrons to move apart from each other.

Note that the newly added electrons do not themselves need to move to the far corners of the metal. Because of the repulsive forces, the newcomers simply "shove" the entire electron sea a little to the side. The electron sea takes an extremely short time to adjust itself to the presence of the added charge, typically less than 10^{-9} s. For all practical purposes, a conductor responds *instantaneously* to the addition or removal of charge.

Other than this very brief interval during which the electron sea is adjusting, the charges in an *isolated* conductor are in static equilibrium. That is, the charges are at rest and there is no net force on any charge. This condition is called **electrostatic equilibrium.** If there *were* a net force on one of the charges, it would quickly move to an equilibrium point at which the force was zero.

FIGURE 25.6 A conductor is charged by contact with a charged plastic rod.

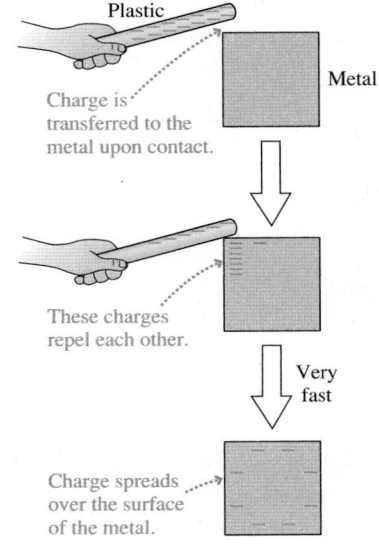

Electrostatic equilibrium has an important consequence:

In an isolated conductor, any excess charge is located on the surface of the conductor.

To see this, suppose there *were* an excess electron in the interior of an isolated conductor. The extra electron would upset the electrical neutrality of the interior and exert forces on nearby electrons, causing them to move. But their motion would violate the assumption of static equilibrium, so we're forced to conclude that there cannot be any excess electrons in the interior. Any excess electrons push each other apart until they're all on the surface.

EXAMPLE 25.2 Charging an electroscope

Many electricity demonstrations are carried out with the help of an *electroscope* like the one shown in FIGURE 25.7. Touching the sphere at the top of an electroscope with a charged plastic rod causes the leaves to fly apart and remain hanging at an angle. Use charge diagrams to explain why.

MODEL We'll use the charge model and the model of a conductor as a material through which electrons move.

VISUALIZE FIGURE 25.8 uses a series of charge diagrams to show the charging of an electroscope.

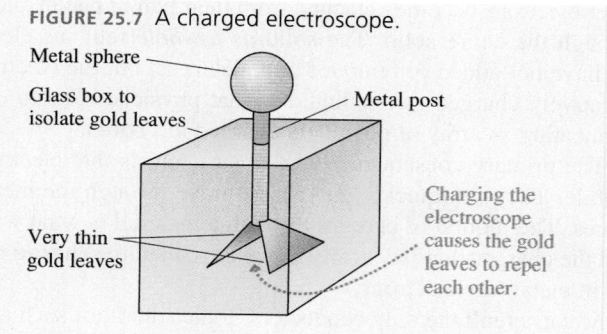

FIGURE 25.7 A charged electroscope.

Metal sphere

Glass box to isolate gold leaves

Metal post

Very thin gold leaves

Charging the electroscope causes the gold leaves to repel each other.

FIGURE 25.8 The process by which an electroscope is charged.

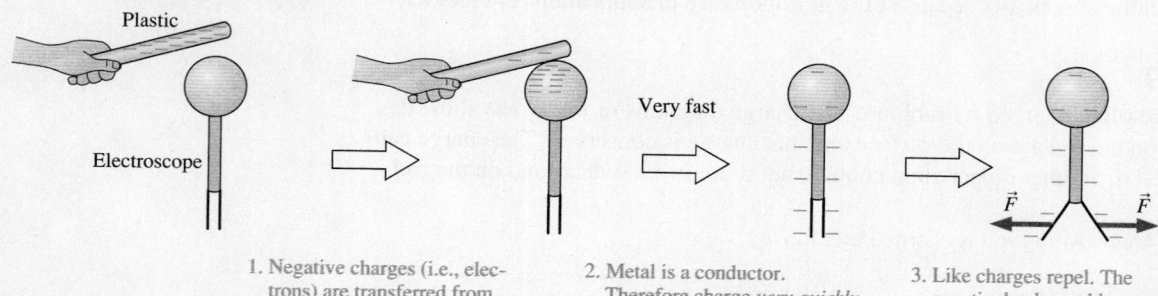

Plastic

Electroscope

Very fast

$\vec{F}$ $\vec{F}$

1. Negative charges (i.e., electrons) are transferred from the rod to the metal sphere upon contact.

2. Metal is a conductor. Therefore charge *very quickly* spreads throughout the entire electroscope.

3. Like charges repel. The negatively charged leaves exert repulsive forces on each other, causing them to spread apart.

FIGURE 25.9 Touching a charged metal discharges it.

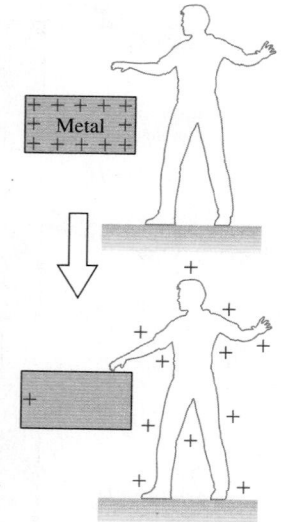

Metal

Charges spread through the metal + human system. Very little charge is left on the metal.

Discharging

Pure water is not a terribly good conductor, but nearly all water contains a variety of dissolved minerals that float around as ions. Dissolved table salt, as we noted previously, separates into Na^+ and Cl^- ions. These ions are the charge carriers, allowing salt water to be a fairly good conductor.

The human body consists largely of salt water. Consequently, and occasionally tragically, humans are reasonably good conductors. This fact allows us to understand how it is that *touching* a charged object discharges it, as we observed in Experiment 10. As FIGURE 25.9 shows, the net effect of touching a charged metal is that it and the conducting human together become a much larger conductor than the metal alone. Any excess charge that was initially confined to the metal can now spread over the larger metal + human conductor. This may not entirely discharge the metal, but in typical circumstances, where the human is much larger than the metal, the residual charge remaining on the metal is much reduced from the original charge. The metal, for most practical purposes, is discharged. In essence, two conductors in contact "share" the charge that was originally on just one of them.

Moist air is a conductor, although a rather poor one. Charged objects in air slowly lose their charge as the object shares its charge with the air. The earth itself is a giant conductor because of its water, moist soil, and a variety of ions. Any object that is physically connected to the earth through a conductor is said to be **grounded.** The effect of being grounded is that the object shares any excess charge it has with the entire earth! But the earth is so enormous that any conductor attached to the earth will be completely discharged.

The purpose of *grounding* objects, such as circuits and appliances, is to prevent the buildup of any charge on the objects. The third prong on appliances and electronics that have a three-prong plug is the ground connection. The building wiring physically connects that third wire deep into the ground somewhere just outside the building, often by attaching it to a metal water pipe that goes underground.

Charge Polarization

One observation from Section 25.1 still needs an explanation. How do charged objects of either sign exert an attractive force on a *neutral* object? To begin answering this question, FIGURE 25.10 shows a positively charged rod held close to—but not touching— a *neutral* electroscope. The leaves move apart and stay apart as long as you hold the rod near, but they quickly collapse when it is removed. Can we understand this behavior?

We can, and FIGURE 25.11a shows how. Although the metal as a whole is still electrically neutral, we say that the object has been *polarized.* **Charge polarization** is a slight separation of the positive and negative charges in a neutral object. Charge polarization produces an excess positive charge on the leaves of the electroscope shown in FIGURE 25.11b, so they repel each other. But because the electroscope has no *net* charge, the electron sea quickly readjusts once the rod is removed.

FIGURE 25.10 A charged rod held close to an electroscope causes the leaves to repel each other.

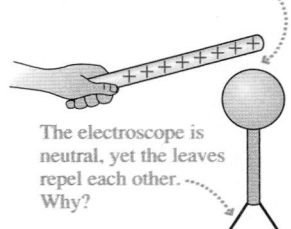

Bring a positively charged glass rod close to an electroscope without touching the sphere.

The electroscope is neutral, yet the leaves repel each other. Why?

FIGURE 25.11 A charged rod polarizes a metal.

(a) The sea of electrons is attracted to the rod and shifts so that there is excess negative charge on the near surface.

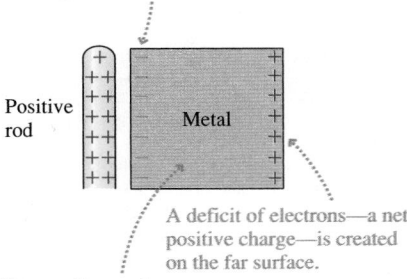

Positive rod

Metal

A deficit of electrons—a net positive charge—is created on the far surface.

The metal's net charge is still zero, but it has been *polarized* by the charged rod.

(b) The electroscope is polarized by the charged rod. The sea of electrons shifts toward the positive rod.

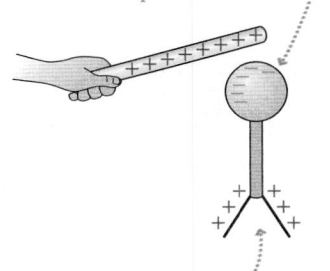

Although the net charge on the electroscope is still zero, the leaves have excess positive charge and repel each other.

Why don't *all* the electrons in Figure 25.11a rush to the side near the positive charge? Once the electron sea shifts slightly, the stationary positive ions begin to exert a force, a restoring force, pulling the electrons back to the right. The equilibrium position for the sea of electrons is just far enough to the left that the forces due to the external charge and the positive ions are in balance. In practice, the displacement of the electron sea is usually *less than* 10^{-15} *m*!

Charge polarization explains not only why the electroscope leaves deflect but also how a charged object exerts an attractive force on a neutral object. FIGURE 25.12 on the next page shows a positively charged rod near a neutral piece of metal. Because the electric force decreases with distance, the attractive force on the electrons at the top surface is *slightly greater* than the repulsive force on the ions at the bottom. The net force toward the charged rod is called a **polarization force.** The polarization force arises because the charges in the metal are separated, *not* because the rod and metal are oppositely charged.

FIGURE 25.12 The polarization force on a neutral piece of metal is due to the slight charge separation.

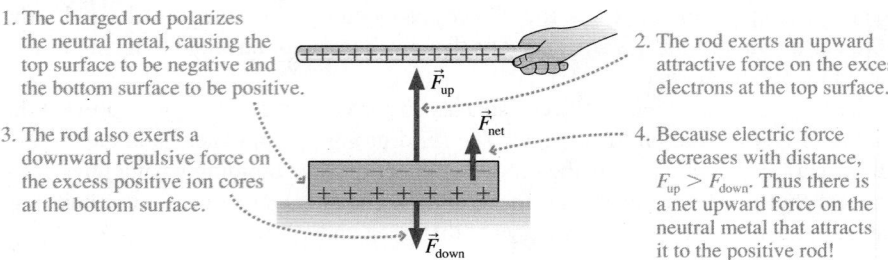

1. The charged rod polarizes the neutral metal, causing the top surface to be negative and the bottom surface to be positive.

2. The rod exerts an upward attractive force on the excess electrons at the top surface.

3. The rod also exerts a downward repulsive force on the excess positive ion cores at the bottom surface.

4. Because electric force decreases with distance, $F_{up} > F_{down}$. Thus there is a net upward force on the neutral metal that attracts it to the positive rod!

A negatively charged rod would push the electron sea slightly away, polarizing the metal to have a positive upper surface charge and a negative lower surface charge. Once again, these are the conditions for the charge to exert a *net attractive force* on the metal. Thus our charge model explains how a charged object of *either* sign attracts neutral pieces of metal.

The Electric Dipole

Now let's consider a slightly trickier situation. Why does a charged rod pick up paper, which is an insulator rather than a metal? First consider what happens when we bring a positive charge near an atom. As FIGURE 25.13a shows, the charge polarizes the atom. The electron cloud doesn't move far, because the force from the positive nucleus pulls it back, but the center of positive charge and the center of negative charge are now slightly separated.

FIGURE 25.13 A neutral atom is polarized by an external charge, forming an *electric dipole*.

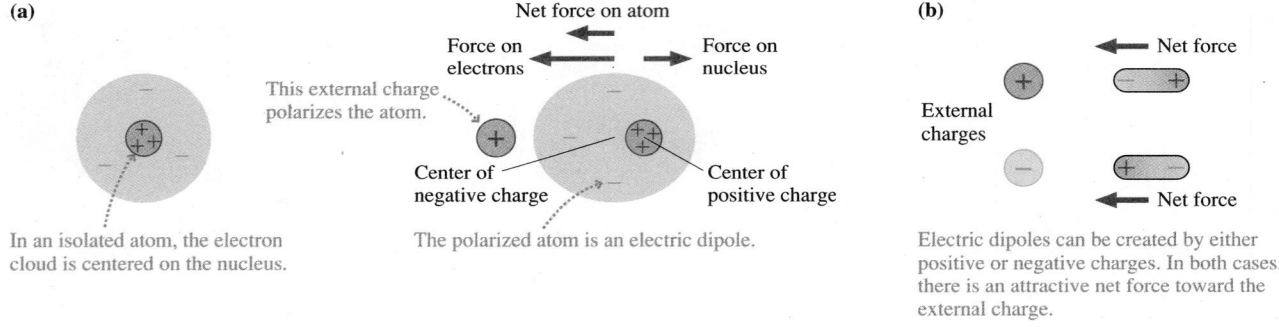

(a)

In an isolated atom, the electron cloud is centered on the nucleus.

Net force on atom

Force on electrons

Force on nucleus

This external charge polarizes the atom.

Center of negative charge

Center of positive charge

The polarized atom is an electric dipole.

(b)

External charges

Net force

Net force

Electric dipoles can be created by either positive or negative charges. In both cases, there is an attractive net force toward the external charge.

FIGURE 25.14 The atoms in an insulator are polarized by an external charge.

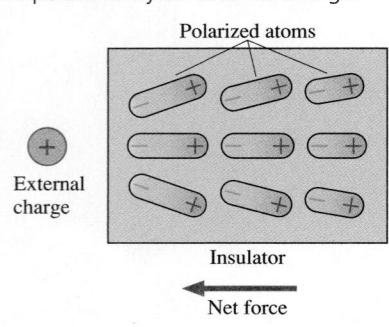

Polarized atoms

External charge

Insulator

Net force

Two opposite charges with a slight separation between them form what is called an **electric dipole.** FIGURE 25.13b shows that an external charge of either sign polarizes the atom to produce an electric dipole with the near end opposite in sign to the charge. (The actual distortion from a perfect sphere is minuscule, nothing like the distortion shown in the figure.) The attractive force on the dipole's near end *slightly* exceeds the repulsive force on its far end because the near end is closer to the charge. The net force, an *attractive* force between the charge and the atom, is another example of a polarization force.

An insulator has no sea of electrons to shift if an external charge is brought close. Instead, as FIGURE 25.14 shows, all the individual atoms inside the insulator become polarized. The polarization force acting *on each atom* produces a net polarization force toward the external charge. This solves the puzzle. A charged rod picks up pieces of paper by

- Polarizing the atoms in the paper,
- Then exerting an attractive polarization force on each atom.

This is important. Make sure you understand all the steps in the reasoning.

STOP TO THINK 25.3   An electroscope is positively charged by *touching* it with a positive glass rod. The electroscope leaves spread apart and the glass rod is removed. Then a negatively charged plastic rod is brought close to the top of the electroscope, but it doesn't touch. What happens to the leaves?

a. The leaves get closer together.
b. The leaves spread farther apart.
c. One leaf moves higher, the other lower.
d. The leaves don't move.

Charging by Induction

Charge polarization is responsible for an interesting and counterintuitive way of charging an electroscope. FIGURE 25.15 shows a positively charged glass rod held near an electroscope but not touching it, while a person touches the electroscope with a finger. Unlike what happens in Figure 25.10, the electroscope leaves hardly move.

FIGURE 25.15 Charging by induction.

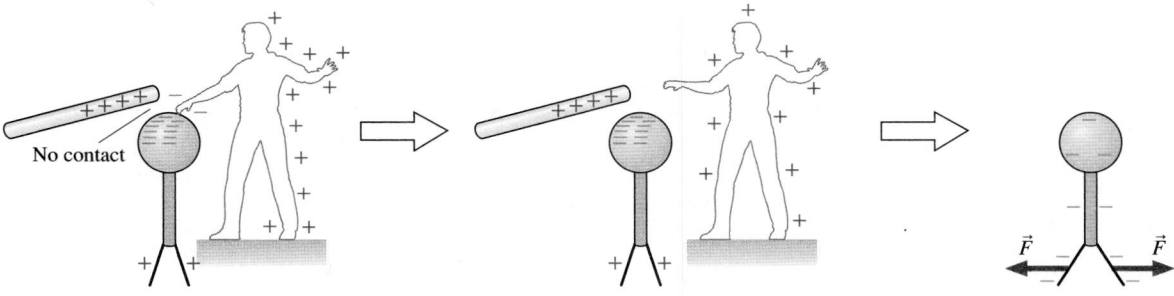

1. The charged rod polarizes the electroscope + person conductor. The leaves repel slightly due to polarization, but overall the electroscope has an excess of electrons and the person has a deficit of electrons.

2. The negative charge on the electroscope is isolated when contact is broken.

3. When the rod is removed, the leaves first collapse as the polarization vanishes, then repel as the excess negative charge spreads out. The electroscope has been *negatively* charged.

Charge polarization occurs, as it did in Figure 25.10, but this time in the much larger electroscope + person conductor. If the person removes his or her finger while the system is polarized, the electroscope is left with a *net* negative charge and the person has a net positive charge. The electroscope has been charged *opposite to the rod* in a process called **charging by induction.**

25.4 Coulomb's Law

The first three sections have established a *model* of charges and electric forces. This model has successfully provided a qualitative explanation of electric phenomena; now it's time to become quantitative. Experiment 4 in Section 25.1 found that the electric force decreases with distance. The force law that describes this behavior is known as *Coulomb's law*.

Charles Coulomb was one of many scientists investigating electricity in the late 18th century. Coulomb had the idea of studying electric forces using the torsion balance scheme by which Cavendish had measured the value of the gravitational constant *G* (see Section 13.4). This was a difficult experiment. Despite many obstacles, Coulomb announced in 1785 that the electric force obeys an *inverse-square law* analogous to Newton's law of gravity. Today we know it as **Coulomb's law.**

Coulomb's law:

1. If two charged particles having charges q_1 and q_2 are a distance r apart, the particles exert forces on each other of magnitude

$$F_{1 \text{ on } 2} = F_{2 \text{ on } 1} = \frac{K|q_1||q_2|}{r^2} \qquad (25.2)$$

where K is called the **electrostatic constant.** These forces are an action/reaction pair, equal in magnitude and opposite in direction.

2. The forces are directed along the line joining the two particles. The forces are *repulsive* for two like charges and *attractive* for two opposite charges.

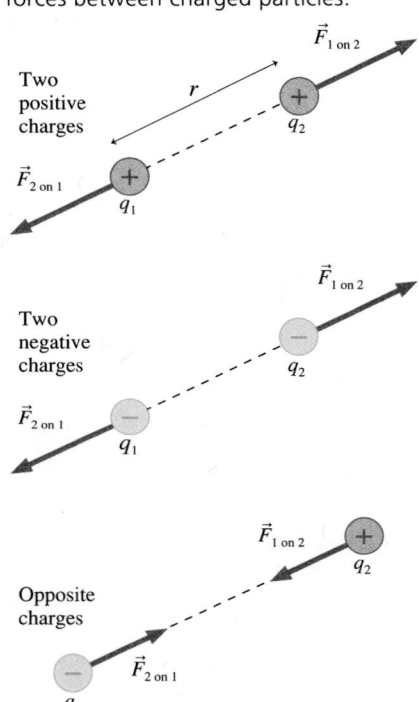

FIGURE 25.16 Attractive and repulsive forces between charged particles.

We sometimes speak of the "force between charge q_1 and charge q_2," but keep in mind that we are really dealing with charged *objects* that also have a mass, a size, and other properties. Charge is not some disembodied entity that exists apart from matter. Coulomb's law describes the force between charged *particles,* which are also called **point charges.** A charged particle, which is an extension of the particle model we used in Part I, has a mass and a charge but has no size.

Coulomb's law looks much like Newton's law of gravity, but there is one important difference: The charge q can be either positive or negative. Consequently, the absolute value signs in Equation 25.2 are especially important. The first part of Coulomb's law gives only the *magnitude* of the force, which is always positive. The direction must be determined from the second part of the law. **FIGURE 25.16** shows the forces between different combinations of positive and negative charges.

Units of Charge

Coulomb had no *unit* of charge, so he was unable to determine a value for K, whose numerical value depends on the units of both charge and distance. The SI unit of charge, the **coulomb** (C), is derived from the SI unit of *current,* so we'll have to await the study of current in Chapter 30 before giving a precise definition. For now we'll note that the fundamental unit of charge e has been measured to have the value

$$e = 1.60 \times 10^{-19} \text{ C}$$

This is a very small amount of charge. Stated another way, 1 C is the net charge of roughly 6.25×10^{18} protons.

> **NOTE** ▶ The amount of charge produced by rubbing plastic or glass rods is typically in the range 1 nC (10^{-9} C) to 100 nC (10^{-7} C). This corresponds to an excess or deficit of 10^{10} to 10^{12} electrons. ◀

Once the unit of charge is established, torsion balance experiments such as Coulomb's can be used to measure the electrostatic constant K. In SI units

$$K = 8.99 \times 10^9 \text{ N m}^2/\text{C}^2$$

It is customary to round this to $K = 9.0 \times 10^9 \text{ N m}^2/\text{C}^2$ for all but extremely precise calculations, and we will do so.

Surprisingly, we will find that Coulomb's law is not explicitly used in much of the theory of electricity. While it *is* the basic force law, most of our future discussion and calculations will be of things called *fields* and *potentials.* It turns out that we can make many future equations easier to use if we rewrite Coulomb's law in a somewhat more complicated way. Let's define a new constant, called the **permittivity constant** ϵ_0 (pronounced "epsilon zero" or "epsilon naught"), as

$$\epsilon_0 = \frac{1}{4\pi K} = 8.85 \times 10^{-12}\ \text{C}^2/\text{N m}^2$$

Rewriting Coulomb's law in terms of ϵ_0 gives us

$$F = \frac{1}{4\pi\epsilon_0}\frac{|q_1||q_2|}{r^2} \tag{25.3}$$

It will be easiest when using Coulomb's law directly to use the electrostatic constant K. However, in later chapters we will switch to the second version with ϵ_0.

Using Coulomb's Law

Coulomb's law is a force law, and forces are vectors. It has been many chapters since we made much use of vectors and vector addition, but these mathematical techniques will be essential in our study of electricity and magnetism.

There are two important observations regarding Coulomb's law:

1. **Coulomb's law applies only to point charges.** A point charge is an idealized material object with charge and mass but with no size or extension. For practical purposes, two charged objects can be modeled as point charges if they are much smaller than the separation between them.
2. **Electric forces, like other forces, can be superimposed.** If multiple charges 1, 2, 3, . . . are present, the *net* electric force on charge j due to all other charges is

$$\vec{F}_{\text{net}} = \vec{F}_{1\ \text{on}\ j} + \vec{F}_{2\ \text{on}\ j} + \vec{F}_{3\ \text{on}\ j} + \cdots \tag{25.4}$$

where each of the $\vec{F}_{i\ \text{on}\ j}$ is given by Equation 25.2 or 25.3.

These conditions are the basis of a strategy for using Coulomb's law to solve electrostatic force problems.

PROBLEM-SOLVING STRATEGY 25.1   **Electrostatic forces and Coulomb's law**

MODEL Identify point charges or objects that can be modeled as point charges.

VISUALIZE Use a *pictorial representation* to establish a coordinate system, show the positions of the charges, show the force vectors on the charges, define distances and angles, and identify what the problem is trying to find. This is the process of translating words to symbols.

SOLVE The mathematical representation is based on Coulomb's law:

$$F_{1\ \text{on}\ 2} = F_{2\ \text{on}\ 1} = \frac{K|q_1||q_2|}{r^2}$$

- Show the directions of the forces—repulsive for like charges, attractive for opposite charges—on the pictorial representation.
- When possible, do graphical vector addition on the pictorial representation. While not exact, it tells you the type of answer you should expect.
- Write each force vector in terms of its *x*- and *y*-components, then add the components to find the net force. Use the pictorial representation to determine which components are positive and which are negative.

ASSESS Check that your result has the correct units, is reasonable, and answers the question.

Exercise 26

EXAMPLE 25.3 **The point of zero force**

Two positively charged particles q_1 and $q_2 = 3q_1$ are 10.0 cm apart on the x-axis. Where (other than at infinity) could a third charge q_3 be placed so as to experience no net force?

MODEL Model the charged particles as point charges.

VISUALIZE **FIGURE 25.17** establishes a coordinate system with q_1 at the origin. We first need to identify the region of space in which q_3 must be located. We have no information about the sign of q_3, so apparently the position for which we are looking will work for either sign. You can see from the figure that the forces at point A, above the axis, and at point B, outside the charges, cannot possibly add to zero. However, at point C on the x-axis *between* the charges, the two forces are oppositely directed.

FIGURE 25.17 A pictorial representation of the charges and forces.

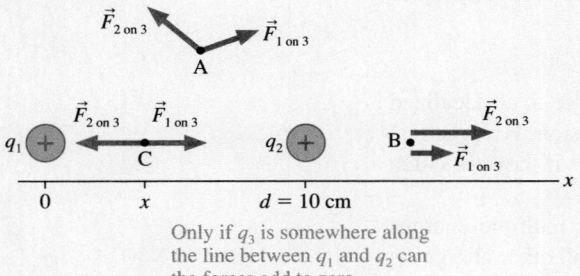

Only if q_3 is somewhere along the line between q_1 and q_2 can the forces add to zero.

SOLVE The mathematical problem is to find the position for which the forces $\vec{F}_{1\,\text{on}\,3}$ and $\vec{F}_{2\,\text{on}\,3}$ are equal in magnitude. If q_3 is distance x from q_1, it is distance $d - x$ from q_2. The *magnitudes* of the forces are

$$F_{1\,\text{on}\,3} = \frac{Kq_1|q_3|}{r_{13}^{\,2}} = \frac{Kq_1|q_3|}{x^2}$$

$$F_{2\,\text{on}\,3} = \frac{Kq_2|q_3|}{r_{23}^{\,2}} = \frac{K(3q_1)|q_3|}{(d-x)^2}$$

Charges q_1 and q_2 are positive and do not need absolute value signs. Equating the two forces gives

$$\frac{Kq_1|q_3|}{x^2} = \frac{3Kq_1|q_3|}{(d-x)^2}$$

The term $Kq_1|q_3|$ cancels. Multiplying by $x^2(d-x)^2$ gives

$$(d-x)^2 = 3x^2$$

which can be rearranged into the quadratic equation

$$2x^2 + 2dx - d^2 = 2x^2 + 20x - 100 = 0$$

where we used $d = 10$ cm and x is in cm. The solutions to this equation are

$$x = +3.66 \text{ cm and } -13.66 \text{ cm}$$

Both are points where the *magnitudes* of the two forces are equal, but $x = -13.66$ cm is a point where the magnitudes are equal while the directions are the same. The solution we want, which is between the charges, is $x = 3.66$ cm. Thus the point to place q_3 is 3.66 cm from q_1 along the line joining q_1 and q_2.

ASSESS q_1 is smaller than q_2, so we expect the point at which the forces balance to be closer to q_1 than to q_2. The solution seems reasonable. Note that the problem statement has no coordinates, so "$x = 3.66$ cm" is *not* an acceptable answer. You need to describe the position relative to q_1 and q_2.

EXAMPLE 25.4 **Three charges**

Three charged particles with $q_1 = -50$ nC, $q_2 = +50$ nC, and $q_3 = +30$ nC are placed on the corners of the 5.0 cm × 10.0 cm rectangle shown in **FIGURE 25.18**. What is the net force on charge q_3 due to the other two charges? Give your answer both in component form and as a magnitude and direction.

MODEL Model the charged particles as point charges.

VISUALIZE The pictorial representation of **FIGURE 25.19** establishes a coordinate system. q_1 and q_3 are opposite charges, so force vector $\vec{F}_{1\,\text{on}\,3}$ is an attractive force toward q_1. q_2 and q_3 are like charges, so force vector $\vec{F}_{2\,\text{on}\,3}$ is a repulsive force away from q_2. q_1 and q_2 have equal magnitudes, but $\vec{F}_{2\,\text{on}\,3}$ has been drawn shorter than $\vec{F}_{1\,\text{on}\,3}$ because q_2 is farther from q_3. Vector addition has been used to draw the net force vector and to define the angle ϕ.

SOLVE The question asks for a *force*, so our answer will be the vector sum $\vec{F}_3 = \vec{F}_{1\,\text{on}\,3} + \vec{F}_{2\,\text{on}\,3}$. We need to write $\vec{F}_{1\,\text{on}\,3}$ and $\vec{F}_{2\,\text{on}\,3}$ in component form. The magnitude of force $\vec{F}_{1\,\text{on}\,3}$ can be found using Coulomb's law:

FIGURE 25.18 The three charges.

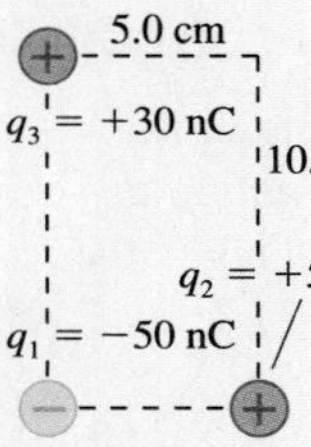

$q_3 = +30$ nC
5.0 cm
10.0 cm
$q_2 = +50$ nC
$q_1 = -50$ nC

FIGURE 25.19 A pictorial representation of the charges and forces.

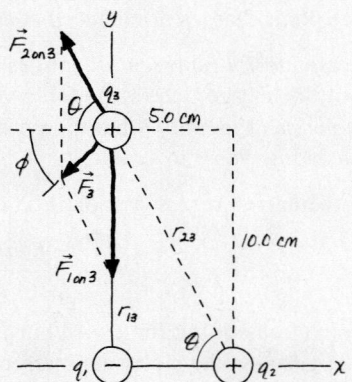

$$F_{1\,\text{on}\,3} = \frac{K|q_1||q_3|}{r_{13}^{\,2}}$$

$$= \frac{(9.0 \times 10^9 \text{ N m}^2/\text{C}^2)(50 \times 10^{-9} \text{ C})(30 \times 10^{-9} \text{ C})}{(0.100 \text{ m})^2}$$

$$= 1.35 \times 10^{-3} \text{ N}$$

where we used $r_{13} = 10.0$ cm.

The pictorial representation shows that $\vec{F}_{1 \text{ on } 3}$ points downward, in the negative y-direction, so

$$\vec{F}_{1 \text{ on } 3} = -1.35 \times 10^{-3}\hat{\jmath} \text{ N}$$

To calculate $\vec{F}_{2 \text{ on } 3}$ we first need the distance r_{23} between the charges:

$$r_{23} = \sqrt{(5.0 \text{ cm})^2 + (10.0 \text{ cm})^2} = 11.2 \text{ cm}$$

The magnitude of $\vec{F}_{2 \text{ on } 3}$ is thus

$$F_{2 \text{ on } 3} = \frac{K|q_2||q_3|}{r_{23}^2}$$

$$= \frac{(9.0 \times 10^9 \text{ N m}^2/\text{C}^2)(50 \times 10^{-9} \text{ C})(30 \times 10^{-9} \text{ C})}{(0.112 \text{ m})^2}$$

$$= 1.08 \times 10^{-3} \text{ N}$$

This is only a magnitude. The *vector* $\vec{F}_{2 \text{ on } 3}$ is

$$\vec{F}_{2 \text{ on } 3} = -F_{2 \text{ on } 3}\cos\theta\,\hat{\imath} + F_{2 \text{ on } 3}\sin\theta\,\hat{\jmath}$$

where angle θ is defined in the figure and the signs (negative x-component, positive y-component) were determined from the pictorial representation. From the geometry of the rectangle,

$$\theta = \tan^{-1}\left(\frac{10.0 \text{ cm}}{5.0 \text{ cm}}\right) = \tan^{-1}(2.0) = 63.4°$$

Thus $\vec{F}_{2 \text{ on } 3} = (-4.83\hat{\imath} + 9.66\hat{\jmath}) \times 10^{-4} \text{ N}$. Now we can add $\vec{F}_{1 \text{ on } 3}$ and $\vec{F}_{2 \text{ on } 3}$ to find

$$\vec{F}_3 = \vec{F}_{1 \text{ on } 3} + \vec{F}_{2 \text{ on } 3} = (-4.83\hat{\imath} - 3.84\hat{\jmath}) \times 10^{-4} \text{ N}$$

This would be an acceptable answer for many problems, but sometimes we need the net force as a magnitude and direction. With angle ϕ as defined in the figure, these are

$$F_3 = \sqrt{F_{3x}^2 + F_{3y}^2} = 6.2 \times 10^{-4} \text{ N}$$

$$\phi = \tan^{-1}\left|\frac{F_{3y}}{F_{3x}}\right| = 38°$$

Thus $\vec{F}_3 = (6.2 \times 10^{-4} \text{ N}, 38°$ below the negative x-axis).

ASSESS The forces are not large, but they are typical of electrostatic forces. Even so, you'll soon see that these forces can produce very large accelerations because the masses of the charged objects are usually very small.

EXAMPLE 25.5 **Lifting a glass bead**

A small plastic sphere charged to -10 nC is held 1.0 cm above a small glass bead at rest on a table. The bead has a mass of 15 mg and a charge of $+10$ nC. Will the glass bead "leap up" to the plastic sphere?

MODEL Model the plastic sphere and glass bead as point charges.

VISUALIZE FIGURE 25.20 establishes a y-axis, identifies the plastic sphere as q_1 and the glass bead as q_2, and shows a free-body diagram.

SOLVE If $F_{1 \text{ on } 2}$ is less than the gravitational force $F_G = m_{\text{bead}}g$, then the bead will remain at rest on the table with $\vec{F}_{1 \text{ on } 2} + \vec{F}_G + \vec{n} = \vec{0}$. But if $F_{1 \text{ on } 2}$ is greater than $m_{\text{bead}}g$, the glass bead will accelerate upward from the table. Using the values provided, we have

$$F_{1 \text{ on } 2} = \frac{K|q_1||q_2|}{r^2} = 9.0 \times 10^{-3} \text{ N}$$

$$F_G = m_{\text{bead}}g = 1.5 \times 10^{-4} \text{ N}$$

$F_{1 \text{ on } 2}$ exceeds $m_{\text{bead}}g$ by a factor of 60, so the glass bead will leap upward.

FIGURE 25.20 A pictorial representation of the charges and forces.

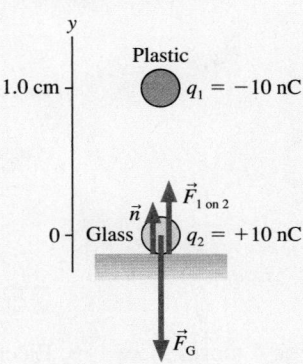

ASSESS The values used in this example are realistic for spheres ≈ 2 mm in diameter. In general, as in this example, electric forces are *significantly* larger than gravitational forces. Consequently, we can neglect gravity when working electric-force problems unless the particles are fairly massive.

EXAMPLE 25.6 **A point charge and a charged wire**

Coulomb's inverse-square law applies only to the electric force between two point charges. Your lab assignment for the week is to discover the law describing the force between a point charge and a long, straight, charged metal wire. It is postulated that the force on a point charge q is characterized by a *power law* $F \propto qr^n$, where r is the distance from the wire. To test this hypothesis and, if it is correct, to determine the exponent n, you first set up a long, straight metal wire and charge it by connecting it to a high-voltage

Continued

power supply. You then charge a small plastic ball and, using a sensitive force probe, measure the force on the ball at different distances from the wire. Your data are as follows:

Distance (cm)	Force (μN)
2.0	895
4.0	455
6.0	310
8.0	215
10.0	185

Is the force described by a power law? And if so, what is the exponent?

MODEL Model the small plastic ball as a point charge.

SOLVE A power law is represented by a linear log-log graph. To see why, we can write the postulated force law as

$$F = cqr^n$$

where c is an unknown proportionality constant. If we take the logarithm of both sides, applying the rules $\log(ab) = \log a + \log b$ and $\log a^n = n \log a$, we get

$$\log F = \log(cqr^n) = \log(cq) + \log r^n = \log(cq) + n \log r$$

If we plot $\log F$ on the y-axis against $\log r$ on the x-axis—a log-log graph—it should be a straight line with slope n. A nonlinear log-log graph would disprove the hypothesis that the force is characterized by a power law.

FIGURE 25.21 is a graph of $\log F$ versus $\log r$. It is clearly linear, which validates the postulated power-law force. And because distances were measured to only two significant figures, the experimental slope of -0.997 is consistent with the simpler $n = -1$. Thus our data show that the force between a point charge and a long, charged wire can be characterized as $F \propto q/r$.

FIGURE 25.21 A log-log graph of force versus distance.

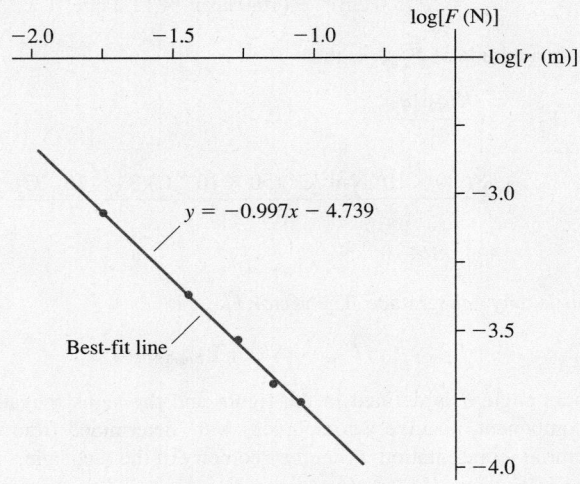

$$y = -0.997x - 4.739$$

Best-fit line

ASSESS The force depends inversely on the distance. The inverse-square dependence of Coulomb's law describes only the force between two point charges.

STOP TO THINK 25.4 Charged spheres A and B exert repulsive forces on each other. $q_A = 4q_B$. Which statement is true?

A   B

a. $F_{A \text{ on } B} > F_{B \text{ on } A}$ b. $F_{A \text{ on } B} = F_{B \text{ on } A}$ c. $F_{A \text{ on } B} < F_{B \text{ on } A}$

25.5 The Field Model

FIGURE 25.22 If charge A moves, how long does it take the force vector on B to respond?

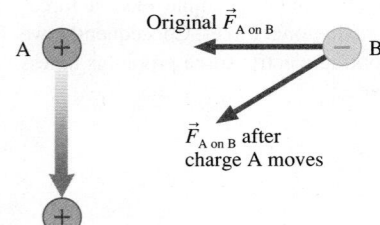

Electric and magnetic forces, like gravity, are *long-range* forces; no contact is required for one charged particle to exert a force on another. But this raises some troubling issues. For example, consider the charged particles A and B in FIGURE 25.22. If A suddenly starts moving, as shown by the arrow, the force vector on B must pivot to follow A. Does this happen *instantly*? Or is there some *delay* between when A moves and when the force $\vec{F}_{A \text{ on } B}$ responds?

Neither Coulomb's law nor Newton's law of gravity is dependent on time, so the answer from the perspective of Newtonian physics has to be "instantly." Yet most scientists found this troubling. What if A is 100,000 light years from B? Will B respond *instantly* to an event 100,000 light years away? The idea of instantaneous transmission of forces had become unbelievable to most scientists by the beginning of the 19th century. But if there is a delay, how long is it? How does the information to "change force" get sent from A to B? These were the issues when a young Michael Faraday appeared on the scene.

Michael Faraday is one of the most interesting figures in the history of science. Because of the late age at which he started his education—he was a teenager—he never became fluent in mathematics. In place of equations, Faraday's brilliant and insightful mind developed many ingenious *pictorial* methods for thinking about and describing physical phenomena. By far the most important of these was the field.

The Concept of a Field

Faraday was particularly impressed with the pattern that iron filings make when sprinkled around a magnet, as seen in **FIGURE 25.23**. The pattern's regularity and the curved lines suggested to Faraday that the *space itself* around the magnet is filled with some kind of magnetic influence. Perhaps the magnet in some way alters the space around it. In this view, a piece of iron near the magnet responds not directly to the magnet but, instead, to the alteration of space caused by the magnet. This space alteration, whatever it is, is the *mechanism* by which the long-range force is exerted.

FIGURE 25.24 illustrates Faraday's idea. The Newtonian view was that A and B interact directly. In Faraday's view, A first alters or modifies the space around it, and particle B then comes along and interacts with this altered space. The alteration of space becomes the *agent* by which A and B interact. Furthermore, this alteration could easily be imagined to take a finite time to propagate outward from A, perhaps in a wave-like fashion. If A changes, B responds only when the new alteration of space reaches it. The interaction between B and this alteration of space is a *local* interaction, rather like a contact force.

Faraday's idea came to be called a **field.** The term "field," which comes from mathematics, describes a function that assigns a vector to every point in space. When used in physics, a field conveys the idea that the physical entity exists at every point in space. That is, indeed, what Faraday was suggesting about how long-range forces operate. The charge makes an alteration *everywhere* in space. Other charges then respond to the alteration at their position. The alteration of the space around a mass is called the *gravitational field.* Similarly, the space around a charge is altered to create the **electric field.**

> **NOTE** ▶ The concept of a field is in sharp contrast to the concept of a particle. A particle exists at *one* point in space. The purpose of Newton's laws of motion is to determine how the particle moves from point to point along a trajectory. A field exists simultaneously at *all* points in space. ◀

Faraday's idea was not taken seriously at first; it seemed too vague and nonmathematical to scientists steeped in the Newtonian tradition of particles and forces. But the significance of the concept of field grew as electromagnetic theory developed during the first half of the 19th century. What seemed at first a pictorial "gimmick" came to be seen as more and more essential for understanding electric and magnetic forces.

Faraday's field ideas were finally placed on a mathematical foundation in 1865 by James Clerk Maxwell. Maxwell was able to describe completely all the known behaviors of electric and magnetic fields in four equations, known today as Maxwell's equations. We will explore aspects of Maxwell's theory as we go along, then look at the full implications of Maxwell's equations in Chapter 34.

The Electric Field

We begin our investigation of electric fields by postulating a **field model** that describes how charges interact:

1. Some charges, which we will call the **source charges,** alter the space around them by creating an *electric field $\vec{E}$.*
2. A separate charge *in* the electric field experiences a force $\vec{F}$ exerted *by the field.*

FIGURE 25.23 Iron filings sprinkled around the ends of a magnet suggest that the influence of the magnet extends into the space around it.

FIGURE 25.24 Newton's and Faraday's ideas about long-range forces.

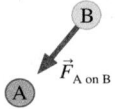

In the Newtonian view, A exerts a force directly on B.

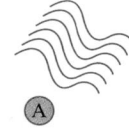

In Faraday's view, A alters the space around it. (The wavy lines are poetic license. We don't know what the alteration looks like.)

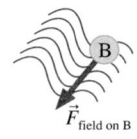

Particle B then responds to the altered space. The altered space is the agent that exerts the force on B.

Suppose some set of charges—the source charges—have created an electric field. We can learn about the field by using a different charge q as a *probe charge*. As we move the probe charge around from point to point in space, it experiences a changing electric force $\vec{F}_{on\,q}$ due to the other charges. This suggests that "something" is present at each point in space to cause the force that charge q experiences. We can use the force on the probe charge to define the electric field $\vec{E}$ at the point (x, y, z) as

$$\vec{E}(x, y, z) \equiv \frac{\vec{F}_{on\,q} \text{ at } (x, y, z)}{q} \qquad (25.5)$$

We're *defining* the electric field as a force-to-charge ratio; hence the units of the electric field are newtons per coulomb, or N/C. The magnitude E of the electric field is called the **electric field strength.**

If a probe charge q experiences an electric force at a point in space, as FIGURE 25.25a shows, we say that there is an electric field at that point causing the force. Further, we *define* the electric field at that point to be the vector given by Equation 25.5. FIGURE 25.25b shows the electric field only at two points, but you can imagine "mapping out" the electric field by moving charge q all through space.

NOTE ▶ Probe charge q also creates an electric field. But charges don't exert forces on themselves, so q is measuring only the electric field of *other* charges. ◀

The basic idea of the field model is that **the field is the agent that exerts an electric force on a charged particle.** Notice three important ideas about the field:

1. Equation 25.5 assigns a *vector* to *every point* in space. That is, the electric field is a *vector field*. Electric field diagrams will show a sample of the vectors, but there is an electric field vector at every point whether one is shown or not.
2. If q is positive, the electric field vector points in the same direction as the force on the charge.
3. Because q appears in Equation 25.5, it may seem that the electric field depends on the size of the charge used to probe the field. It doesn't. We know from Coulomb's law that the force $\vec{F}_{on\,q}$ is proportional to q. Thus the electric field defined in Equation 25.5 is *independent* of the charge q that probes the field. The electric field depends only on the source charges that create the field.

In practice we often want to turn Equation 25.5 around and find the force exerted by a known field. That is, a charged particle with charge q at a point in space where the electric field is $\vec{E}$ experiences an electric force

$$\vec{F}_{on\,q} = q\vec{E} \qquad (25.6)$$

If q is positive, the force on the particle is in the direction of $\vec{E}$. The force on a negative charge is *opposite* the direction of $\vec{E}$.

FIGURE 25.25 Charge q is a probe of the electric field.

(a)

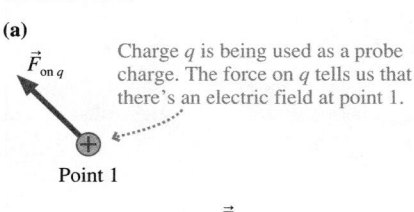

Charge q is being used as a probe charge. The force on q tells us that there's an electric field at point 1.

Point 1

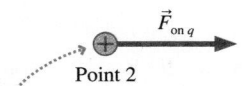

Point 2

Now charge q is placed at point 2. There's also an electric field here that differs from the field at point 1.

(b)

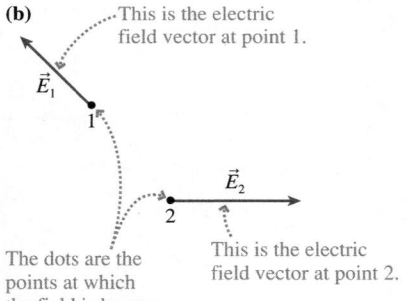

This is the electric field vector at point 1.

This is the electric field vector at point 2.

The dots are the points at which the field is known.

EXAMPLE 25.7 **Electric forces in a cell**

Every cell in your body is electrically active in various ways. For example, nerve propagation occurs when large electric fields in the cell membranes of neurons cause ions to move through the cell walls. The field strength in a typical cell membrane is 1.0×10^7 N/C. What is the magnitude of the electric force on a singly charged calcium ion?

MODEL The ion is a point charge in an electric field. A singly charged ion is missing one electron and has net charge $q = +e$.

SOLVE A charged particle in an electric field experiences an electric force $\vec{F}_{on\,q} = q\vec{E}$. In this case, the magnitude of the force is

$$F = eE = (1.6 \times 10^{-19}\,C)(1.0 \times 10^7\,N/C) = 1.6 \times 10^{-12}\,N$$

ASSESS This may seem like an incredibly tiny force, but it is applied to a particle with mass $m \sim 10^{-26}$ kg. The ion would have an unimaginable acceleration ($F/m \sim 10^{14}$ m/s²) were it not for resistive forces as it moves through the membrane. Even so, an ion can cross the cell wall in less than 1 μs.

STOP TO THINK 25.5 An electron is placed at the position marked by the dot. The force on the electron is

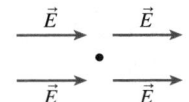

a. Zero.   b. To the right.   c. To the left.
d. There's not enough information to tell.

The Electric Field of a Point Charge

We will begin to put the definition of the electric field to full use in the next chapter. For now, to develop the ideas, we will determine the electric field of a single point charge q. FIGURE 25.26a shows charge q and a point in space at which we would like to know the electric field. We need a second charge, shown as q' in FIGURE 25.26b, to serve as a probe of the electric field.

For the moment, assume both charges are positive. The force on q', which is repulsive and directed straight away from q, is given by Coulomb's law:

$$\vec{F}_{on\,q'} = \left(\frac{1}{4\pi\epsilon_0} \frac{qq'}{r^2},\ \text{away from } q \right) \tag{25.7}$$

It's customary to use $1/4\pi\epsilon_0$ rather than K for field calculations. Equation 25.5 defined the electric field in terms of the force on a probe charge, thus the electric field at this point is

$$\vec{E} = \frac{\vec{F}_{on\,q'}}{q'} = \left(\frac{1}{4\pi\epsilon_0} \frac{q}{r^2},\ \text{away from } q \right) \tag{25.8}$$

The electric field is shown in FIGURE 25.26c.

NOTE ▶ The expression for the electric field is similar to Coulomb's law. To distinguish the two, remember that Coulomb's law has a product of two charges in the numerator. It describes the force between *two* charges. The electric field has a single charge in the numerator. It is the field of *a* charge. ◀

If we calculate the field at a sufficient number of points in space, we can draw a **field diagram** such as the one shown in FIGURE 25.27. Notice that the field vectors all point straight away from charge q. Also notice how quickly the arrows decrease in length due to the inverse-square dependence on r.

Keep these three important points in mind when using field diagrams:

1. The diagram is just a representative sample of electric field vectors. The field exists at all the other points. A well-drawn diagram can tell you fairly well what the field would be like at a neighboring point.
2. The arrow indicates the direction and the strength of the electric field *at the point to which it is attached*—that is, at the point where the *tail* of the vector is placed. In this chapter, we indicate the point at which the electric field is measured with

FIGURE 25.26 Charge q' is used to probe the electric field of point charge q.

(a) What is the electric field of q at this point?

Point charge q

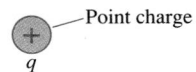

(b) 1. Place q' at the point to probe the field.
2. Measure the force on q'.

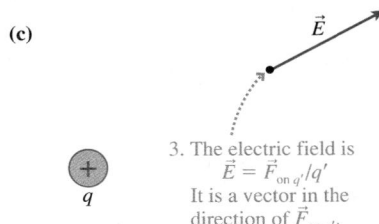

(c) 3. The electric field is $\vec{E} = \vec{F}_{on\,q'}/q'$. It is a vector in the direction of $\vec{F}_{on\,q'}$.

FIGURE 25.27 The electric field of a positive point charge.

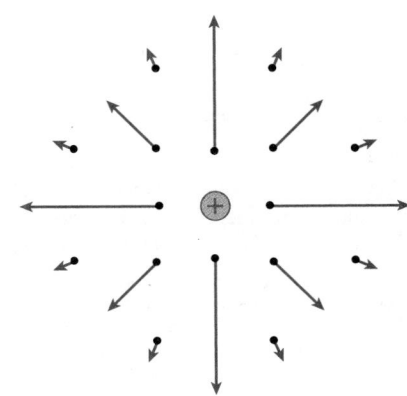

a dot. The length of any vector is significant only relative to the lengths of other vectors.

3. Although we have to draw a vector across the page, from one point to another, an electric field vector is *not* a spatial quantity. It does not "stretch" from one point to another. Each vector represents the electric field at *one point* in space.

Unit Vector Notation

FIGURE 25.28 Using the unit vector $\hat{r}$.

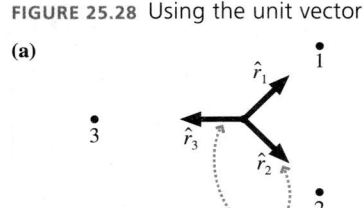

The unit vectors specify the directions to the points.

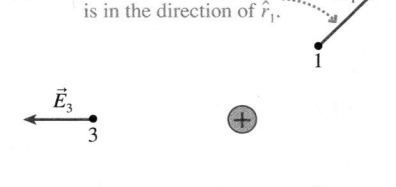

Electric field at point 1 is in the direction of $\hat{r}_1$.

$\vec{E}_2$ is in the direction of $\hat{r}_2$.

Equation 25.8 is precise, but it's not terribly convenient. Furthermore, what happens if the source charge q is negative? We need a more concise notation to write the electric field, a notation that will allow q to be either positive or negative.

The basic need is to express "away from q" in mathematical notation. "Away from q" is a *direction* in space. To guide us, recall that we already have a notation for expressing certain directions—namely, the unit vectors $\hat{\imath}$, $\hat{\jmath}$, and $\hat{k}$. For example, unit vector $\hat{\imath}$ means "in the direction of the positive x-axis." With a minus sign, $-\hat{\imath}$ means "in the direction of the negative x-axis." Unit vectors, with a magnitude of 1 and no units, provide purely directional information.

With this in mind, let's define the unit vector $\hat{r}$ to be a vector of length 1 that points from the origin to a point of interest. Unit vector $\hat{r}$ provides no information at all about the distance to the point; it merely specifies the direction.

FIGURE 25.28a shows unit vectors $\hat{r}_1$, $\hat{r}_2$, and $\hat{r}_3$ pointing toward points 1, 2, and 3. Unlike $\hat{\imath}$ and $\hat{\jmath}$, unit vector $\hat{r}$ does not have a fixed direction. Instead, unit vector $\hat{r}$ specifies the direction "straight outward from this point." But that's just what we need to describe the electric field vector. FIGURE 25.28b shows the electric fields at points 1, 2, and 3 due to a positive charge at the origin. No matter which point you choose, the electric field at that point is "straight outward" from the charge. In other words, the electric field $\vec{E}$ points in the direction of the unit vector $\hat{r}$.

With this notation, the electric field at distance r from a point charge q is

$$\vec{E} = \frac{1}{4\pi\epsilon_0} \frac{q}{r^2} \hat{r} \quad \text{(electric field of a point charge)} \qquad (25.9)$$

FIGURE 25.29 The electric field of a negative point charge.

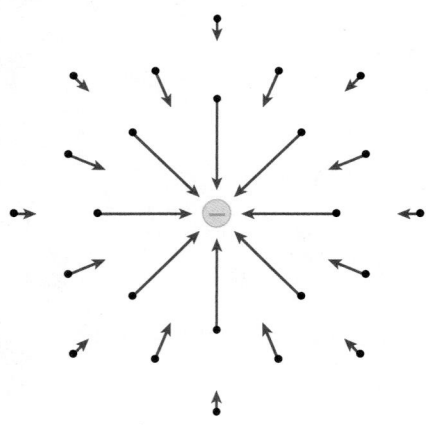

where $\hat{r}$ is the unit vector from the charge toward the point at which we want to know the field. Equation 25.9 is identical to Equation 25.8, but written in a notation in which the unit vector $\hat{r}$ expresses the idea "away from q."

Equation 25.9 works equally well if q is negative. A negative sign in front of a vector simply reverses its direction, so the unit vector $-\hat{r}$ points *toward* charge q. FIGURE 25.29 shows the electric field of a negative point charge. It looks like the electric field of a positive point charge except that the vectors point inward, toward the charge, instead of outward.

We'll end this chapter with three examples of the electric field of a point charge. Chapter 26 will expand these ideas to the electric fields of multiple charges and of extended objects.

EXAMPLE 25.8 **Calculating the electric field**

A -1.0 nC charged particle is located at the origin. Points 1, 2, and 3 have (x, y) coordinates $(1 \text{ cm}, 0 \text{ cm})$, $(0 \text{ cm}, 1 \text{ cm})$, and $(1 \text{ cm}, 1 \text{ cm})$, respectively. Determine the electric field $\vec{E}$ at these points, then show the vectors on an electric field diagram.

MODEL The electric field is that of a negative point charge.

VISUALIZE The electric field points straight *toward* the origin. It will be weaker at $(1 \text{ cm}, 1 \text{ cm})$, which is farther from the charge.

SOLVE The electric field is

$$\vec{E} = \frac{1}{4\pi\epsilon_0} \frac{q}{r^2} \hat{r}$$

where $q = -1.0 \text{ nC} = -1.0 \times 10^{-9}$ C. The distance r is 1.0 cm $=$ 0.010 m for points 1 and 2 and $\left(\sqrt{2} \times 1.0 \text{ cm}\right) = 0.0141$ m for point 3. The *magnitude* of $\vec{E}$ at the three points is

$$E_1 = E_2 = \frac{1}{4\pi\epsilon_0} \frac{|q|}{r_1^2}$$

$$= \frac{(9.0 \times 10^9 \text{ N m}^2/\text{C}^2)(1.0 \times 10^{-9} \text{ C})}{(0.010 \text{ m})^2} = 90,000 \text{ N/C}$$

$$E_3 = \frac{1}{4\pi\epsilon_0} \frac{|q|}{r_3^2}$$

$$= \frac{(9.0 \times 10^9 \text{ N m}^2/\text{C}^2)(1.0 \times 10^{-9} \text{ C})}{(0.0141 \text{ m})^2} = 45,000 \text{ N/C}$$

Because q is negative, the field at each of these positions points directly at charge q. The electric field vectors, in component form, are

$$\vec{E}_1 = -90,000\hat{\imath} \text{ N/C}$$

$$\vec{E}_2 = -90,000\hat{\jmath} \text{ N/C}$$

$$\vec{E}_3 = -E_3 \cos 45°\hat{\imath} - E_3 \sin 45°\hat{\jmath}$$

$$= (-31,800\hat{\imath} - 31,800\hat{\jmath}) \text{ N/C}$$

These vectors are shown on the electric field diagram of FIGURE 25.30.

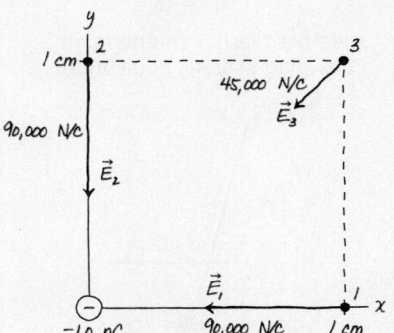

FIGURE 25.30 The electric field diagram of a -1.0 nC charged particle.

EXAMPLE 25.9 **The electric field of a proton**

The electron in a hydrogen atom orbits the proton at a radius of 0.053 nm.

a. What is the proton's electric field strength at the position of the electron?
b. What is the magnitude of the electric force on the electron?

SOLVE a. The proton's charge is $q = e$. Its electric field strength at the distance of the electron is

$$E = \frac{1}{4\pi\epsilon_0} \frac{e}{r^2} = \frac{1}{4\pi\epsilon_0} \frac{1.6 \times 10^{-19} \text{ C}}{(5.3 \times 10^{-11} \text{ m})^2} = 5.1 \times 10^{11} \text{ N/C}$$

Notice how large this field is in comparison to the field of Example 25.8.

b. We could use Coulomb's law to find the force on the electron, but the whole point of knowing the electric field is that we can use it directly to find the force on a charge in the field. The magnitude of the force on the electron is

$$F_{\text{on elec}} = |q_e| E_{\text{of proton}}$$

$$= (1.60 \times 10^{-19} \text{ C})(5.1 \times 10^{11} \text{ N/C})$$

$$= 8.2 \times 10^{-8} \text{ N}$$

STOP TO THINK 25.6   Rank in order, from largest to smallest, the electric field strengths E_a to E_d at points a to d.

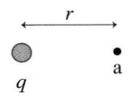 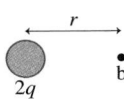 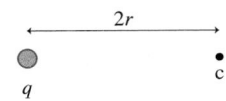 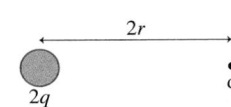

CHALLENGE EXAMPLE 25.10   **A charge in static equilibrium**

A horizontal electric field causes the charged ball in FIGURE 25.31 to hang at a 15° angle, as shown. The spring is plastic, so it doesn't discharge the ball, and in its equilibrium position the spring extends only to the vertical dashed line. What is the electric field strength?

FIGURE 25.31 A charged ball hanging in static equilibrium.

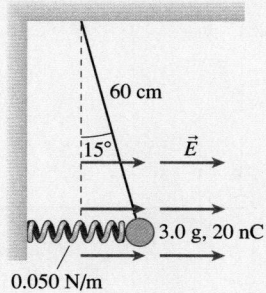

MODEL Model the ball as a point charge in static equilibrium. The electric force on the ball is $\vec{F}_E = q\vec{E}$. The charge is positive, so the force is in the same direction as the field.

VISUALIZE FIGURE 25.32 is a free-body diagram for the ball.

SOLVE The ball is in static equilibrium, so the net force on the ball must be zero. With the field applied, the spring is stretched by $\Delta x = L \sin\theta = (0.60 \text{ m})(\sin 15°) = 0.16 \text{ m}$, where L is the string length, and exerts a pulling force $F_{sp} = k\Delta x$ to the left.

Newton's first law, which we've not used in quite some time, is

$$\sum F_x = F_E - F_{sp} - T\sin\theta = 0$$

$$\sum F_y = T\cos\theta - F_G = T\cos\theta - mg = 0$$

FIGURE 25.32 The free-body diagram.

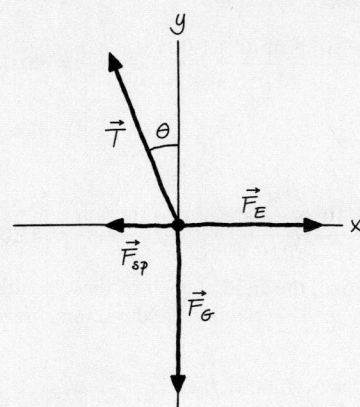

From the y-equation,

$$T = \frac{mg}{\cos\theta}$$

The x-equation is then

$$qE - k\Delta x - mg\tan\theta = 0$$

We can solve this for the electric field strength:

$$E = \frac{mg\tan\theta + k\Delta x}{q}$$

$$= \frac{(0.0030 \text{ kg})(9.8 \text{ m/s}^2)\tan 15° + (0.050 \text{ N/m})(0.16 \text{ m})}{20 \times 10^{-9} \text{ C}}$$

$$= 7.9 \times 10^5 \text{ N/C}$$

ASSESS We don't yet have a way of judging whether this is a reasonable field strength, but we'll see in the next chapter that this is typical of the electric field strength near an object that has been charged by rubbing.

SUMMARY

The goal of Chapter 25 has been to describe electric phenomena in terms of charges, forces, and fields.

General Principles

Coulomb's Law

The forces between two charged particles q_1 and q_2 separated by distance r are

$$F_{1\text{ on }2} = F_{2\text{ on }1} = \frac{K|q_1||q_2|}{r^2}$$

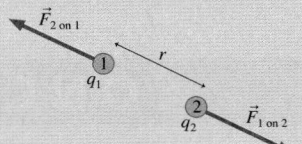

These forces are an action/reaction pair directed along the line joining the particles.

- The forces are repulsive for two like charges, attractive for two opposite charges.
- The net force on a charge is the sum of the forces from all other charges.
- The unit of charge is the coulomb (C).
- The electrostatic constant is $K = 9.0 \times 10^9 \text{ N m}^2/\text{C}^2$.

Important Concepts

The Charge Model

There are two kinds of charge, positive and negative.

- Fundamental charges are protons and electrons, with charge $\pm e$ where $e = 1.60 \times 10^{-19}$ C.
- Objects are charged by adding or removing electrons.
- The amount of charge is $q = (N_p - N_e)e$.
- An object with an equal number of protons and electrons is neutral, meaning no *net* charge.

Charged objects exert electric forces on each other.

- Like charges repel, opposite charges attract.
- The force increases as the charge increases.
- The force decreases as the distance increases.

There are two types of material, insulators and conductors.

- Charge remains fixed in or on an insulator.
- Charge moves easily through or along conductors.
- Charge is transferred by contact between objects.

Charged objects attract neutral objects.

- Charge polarizes metal by shifting the electron sea.
- Charge polarizes atoms, creating electric dipoles.
- The polarization force is always an attractive force.

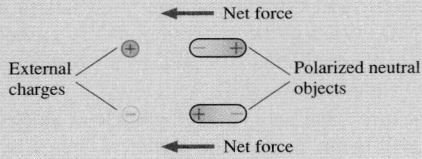

The Field Model

Charges interact with each other via the electric field $\vec{E}$.

- Charge A alters the space around it by creating an electric field.

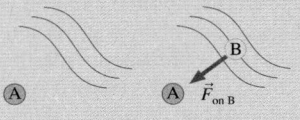

- The field is the agent that exerts a force. The force on charge q_B is $\vec{F}_{\text{on B}} = q_B\vec{E}$.

An electric field is identified and measured in terms of the force on a **probe charge** q:

$$\vec{E} = \vec{F}_{\text{on }q}/q$$

- The electric field exists at all points in space.
- An electric field vector shows the field only at one point, the point at the tail of the vector.

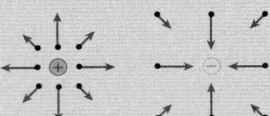

The electric field of a **point charge** is

$$\vec{E} = \frac{1}{4\pi\epsilon_0}\frac{q}{r^2}\hat{r}$$

Terms and Notation

neutral	electron cloud	electrostatic equilibrium	coulomb, C
charging	fundamental unit of charge, e	grounded	permittivity constant, ϵ_0
charge model	charge quantization	charge polarization	field
charge, q or Q	ionization	polarization force	electric field, $\vec{E}$
like charges	law of conservation of charge	electric dipole	field model
opposite charges	sea of electrons	charging by induction	source charge
discharging	ion core	Coulomb's law	electric field strength, E
conductor	current	electrostatic constant, K	field diagram
insulator	charge carriers	point charge	

CONCEPTUAL QUESTIONS

1. Can an insulator be charged? If so, how would you charge an insulator? If not, why not?

2. Can a conductor be charged? If so, how would you charge a conductor? If not, why not?

3. Four lightweight balls A, B, C, and D are suspended by threads. Ball A has been touched by a plastic rod that was rubbed with wool. When the balls are brought close together, without touching, the following observations are made:
   • Balls B, C, and D are attracted to ball A.
   • Balls B and D have no effect on each other.
   • Ball B is attracted to ball C.
   What are the charge states (glass, plastic, or neutral) of balls A, B, C, and D? Explain.

4. Charged plastic and glass rods hang by threads.
   a. An object repels the plastic rod. Can you predict what it will do to the glass rod? If so, what? If not, why not?
   b. A different object attracts the plastic rod. Can you predict what it will do to the glass rod? If so, what? If not, why not?

5. A lightweight metal ball hangs by a thread. When a charged rod is held near, the ball moves toward the rod, touches the rod, then quickly "flies away" from the rod. Explain this behavior.

6. Suppose there exists a third type of charge in addition to the two types we've called glass and plastic. Call this third type X charge. What experiment or series of experiments would you use to test whether an object has X charge? State clearly how each possible outcome of the experiments is to be interpreted.

7. A negatively charged electroscope has separated leaves.
   a. Suppose you bring a negatively charged rod close to the top of the electroscope, but not touching. How will the leaves respond? Use both charge diagrams and words to explain.
   b. How will the leaves respond if you bring a positively charged rod close to the top of the electroscope, but not touching? Use both charge diagrams and words to explain.

8. The two oppositely charged metal spheres in FIGURE Q25.8 have equal quantities of charge. They are brought into contact with a neutral metal rod. What is the final charge state of each sphere and of the rod? Use both charge diagrams and words to explain.

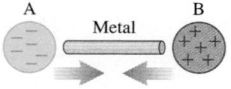

FIGURE Q25.8

FIGURE Q25.9

9. Metal sphere A in FIGURE Q25.9 has 4 units of negative charge and metal sphere B has 2 units of positive charge. The two spheres are brought into contact. What is the final charge state of each sphere? Explain.

10. Metal spheres A and B in FIGURE Q25.10 are initially neutral and are touching. A positively charged rod is brought near A, but not touching. Is A now positive, negative, or neutral? Use both charge diagrams and words to explain.

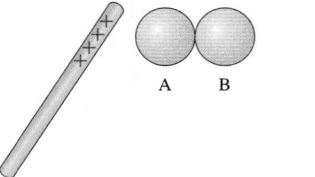

FIGURE Q25.10 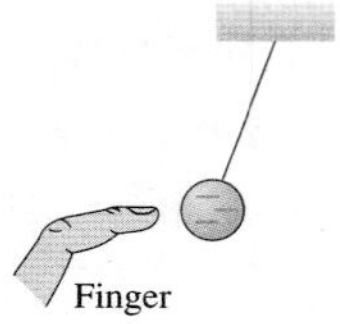   **FIGURE Q25.11**

11. If you bring your finger near a lightweight, negatively charged hanging ball, the ball swings over toward your finger as shown in FIGURE Q25.11. Use charge diagrams and words to explain this observation.

12. Reproduce FIGURE Q25.12 on your paper. Then draw a dot (or dots) on the figure to show the position (or positions) where an electron would experience no net force.

| | | | $\oplus$ | | | $\ominus$ | | | |

FIGURE Q25.12

13. Charges A and B in FIGURE Q25.13 are equal. Each charge exerts a force on the other of magnitude F. Suppose the charge of B is increased by a factor of 4, but everything else is unchanged. In terms of F, (a) what is the magnitude of the force on A, and (b) what is the magnitude of the force on B?

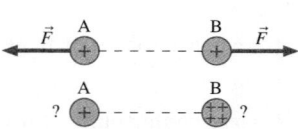

FIGURE Q25.13

14. The electric field strength at one point near a point charge is 1000 N/C. What is the field strength if (a) the distance from the point charge is doubled, and (b) the distance from the point charge is halved?

15. The electric force on a charged particle in an electric field is F. What will be the force if the particle's charge is tripled and the electric field strength is halved?

EXERCISES AND PROBLEMS

Problems labeled ▓ integrate material from earlier chapters.

Exercises

Section 25.1 Developing a Charge Model

Section 25.2 Charge

1. | A plastic rod is charged to −12 nC by rubbing.
   a. Have electrons been added to the rod or protons removed? Explain.
   b. How many electrons have been added or protons removed?
2. | A glass rod is charged to +8.0 nC by rubbing.
   a. Have electrons been removed from the rod or protons added? Explain.
   b. How many electrons have been removed or protons added?
3. | A glass rod that has been charged to +12 nC touches a metal sphere. Afterward, the rod's charge is +8.0 nC.
   a. What kind of charged particle was transferred between the rod and the sphere, and in which direction? That is, did it move from the rod to the sphere or from the sphere to the rod?
   b. How many charged particles were transferred?
4. | A plastic rod that has been charged to −15 nC touches a metal sphere. Afterward, the rod's charge is −10 nC.
   a. What kind of charged particle was transferred between the rod and the sphere, and in which direction? That is, did it move from the rod to the sphere or from the sphere to the rod?
   b. How many charged particles were transferred?
5. ‖ What is the total charge of all the protons in 1.0 mol of He gas?
6. ‖‖ What is the total charge of all the electrons in 1.0 L of liquid water?

Section 25.3 Insulators and Conductors

7. | Figure 25.8 showed how an electroscope becomes negatively charged. The leaves will also repel each other if you touch the electroscope with a positively charged glass rod. Use a series of charge diagrams to explain what happens and why the leaves repel each other.
8. | A plastic balloon that has been rubbed with wool will stick to a wall.
   a. Can you conclude that the wall is charged? If not, why not? If so, where does the charge come from?
   b. Draw a series of charge diagrams showing how the balloon is held to the wall.
9. | Two neutral metal spheres on wood stands are touching. A negatively charged rod is held directly above the top of the left sphere, not quite touching it. While the rod is there, the right sphere is moved so that the spheres no longer touch. Then the rod is withdrawn. Afterward, what is the charge state of each sphere? Use charge diagrams to explain your answer.
10. ‖ You have two neutral metal spheres on wood stands. Devise a procedure for charging the spheres so that they will have like charges of *exactly* equal magnitude. Use charge diagrams to explain your procedure.
11. ‖ You have two neutral metal spheres on wood stands. Devise a procedure for charging the spheres so that they will have opposite charges of *exactly* equal magnitude. Use charge diagrams to explain your procedure.

Section 25.4 Coulomb's Law

12. ‖ Two 1.0 kg masses are 1.0 m apart (center to center) on a frictionless table. Each has +10 μC of charge.
    a. What is the magnitude of the electric force on one of the masses?
    b. What is the initial acceleration of this mass if it is released and allowed to move?
13. ‖ Two small plastic spheres each have a mass of 2.0 g and a charge of −50.0 nC. They are placed 2.0 cm apart (center to center).
    a. What is the magnitude of the electric force on each sphere?
    b. By what factor is the electric force on a sphere larger than its weight?
14. ‖ A small glass bead has been charged to +20 nC. A metal ball bearing 1.0 cm above the bead feels a 0.018 N downward electric force. What is the charge on the ball bearing?
15. | Two protons are 2.0 fm apart.
    a. What is the magnitude of the electric force on one proton due to the other proton?
    b. What is the magnitude of the gravitational force on one proton due to the other proton?
    c. What is the ratio of the electric force to the gravitational force?
16. | What is the net electric force on charge A in FIGURE EX25.16?

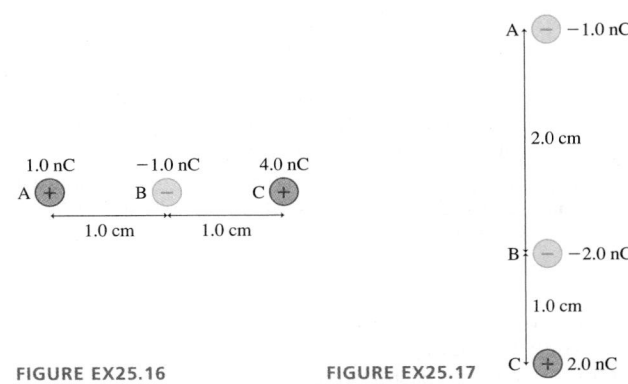

FIGURE EX25.16   FIGURE EX25.17

17. | What is the net electric force on charge B in FIGURE EX25.17?
18. | Object A, which has been charged to +4.0 nC, is at the origin. Object B, which has been charged to −8.0 nC, is at $(x, y) = (0.0$ cm, 2.0 cm$)$. Determine the electric force on each object. Write each force vector in component form.
19. | A small plastic bead has been charged to −15 nC. What are the magnitude and direction of the acceleration of (a) a proton and (b) an electron that is 1.0 cm from the center of the bead?

Section 25.5 The Field Model

20. | What are the strength and direction of the electric field 1.0 mm from (a) a proton and (b) an electron?
21. | The electric field at a point in space is $\vec{E} = (400\,\hat{i} + 100\,\hat{j})$ N/C.
    a. What is the electric force on a proton at this point? Give your answer in component form.
    b. What is the electric force on an electron at this point? Give your answer in component form.
    c. What is the magnitude of the proton's acceleration?
    d. What is the magnitude of the electron's acceleration?

22. ‖ What magnitude charge creates a 1.0 N/C electric field at a point 1.0 m away?

23. | What are the strength and direction of the electric field 4.0 cm from a small plastic bead that has been charged to −8.0 nC?

24. ‖ The electric field 2.0 cm from a small object points away from the object with a strength of 270,000 N/C. What is the object's charge?

25. ‖ What are the strength and direction of an electric field that will balance the weight of a 1.0 g plastic sphere that has been charged to −3.0 nC?

26. ‖ A +12 nC charge is located at the origin.
    a. What are the electric fields at the positions $(x, y) =$ (5.0 cm, 0 cm), (−5.0 cm, 5.0 cm), and (−5.0 cm, −5.0 cm)? Write each electric field vector in component form.
    b. Draw a field diagram showing the electric field vectors at these points.

27. ‖ A −12 nC charge is located at $(x, y) = (1.0$ cm, 0 cm). What are the electric fields at the positions $(x, y) =$ (5.0 cm, 0 cm), (−5.0 cm, 0 cm), and (0 cm, 5.0 cm)? Write each electric field vector in component form.

Problems

28. ‖‖ Pennies today are copper-covered zinc, but older pennies are 3.1 g of solid copper. What are the total positive charge and total negative charge in a solid copper penny that is electrically neutral?

29. | A 2.0 g plastic bead charged to −4.0 nC and a 4.0 g glass bead charged to +8.0 nC are 2.0 cm apart (center to center). What are the accelerations of (a) the plastic bead and (b) the glass bead?

30. ‖ The nucleus of a ^{125}Xe atom (an isotope of the element xenon with mass 125 u) is 6.0 fm in diameter. It has 54 protons and charge $q = +54e$.
    a. What is the electric force on a proton 2.0 fm from the surface of the nucleus?
    b. What is the proton's acceleration?
    Hint: Treat the spherical nucleus as a point charge.

31. ‖ Two 1.0 g spheres are charged equally and placed 2.0 cm apart. When released, they begin to accelerate at 150 m/s². What is the magnitude of the charge on each sphere?

32. ‖ Objects A and B are both positively charged. Both have a mass of 100 g, but A has twice the charge of B. When A and B are placed 10 cm apart, B experiences an electric force of 0.45 N.
    a. What is the charge on A?
    b. If the objects are released, what is the initial acceleration of A?

33. ‖ What is the force $\vec{F}$ on the 1.0 nC charge in FIGURE P25.33? Give your answer as a magnitude and a direction.

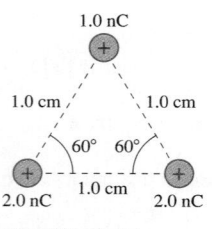

FIGURE P25.33

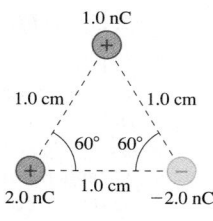

FIGURE P25.34

34. ‖ What is the force $\vec{F}$ on the 1.0 nC charge in FIGURE P25.34? Give your answer as a magnitude and a direction.

35. ‖ What is the force $\vec{F}$ on the −10 nC charge in FIGURE P25.35? Give your answer as a magnitude and an angle measured cw or ccw (specify which) from the +x-axis.

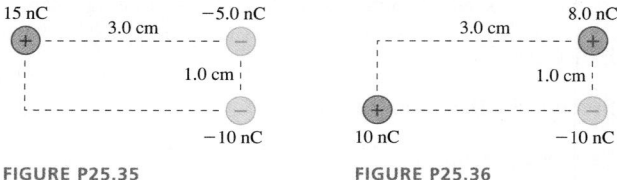

FIGURE P25.35   **FIGURE P25.36**

36. ‖ What is the force $\vec{F}$ on the −10 nC charge in FIGURE P25.36? Give your answer as a magnitude and an angle measured cw or ccw (specify which) from the +x-axis.

37. ‖ What is the force $\vec{F}$ on the 5.0 nC charge in FIGURE P25.37? Give your answer as a magnitude and an angle measured cw or ccw (specify which) from the +x-axis.

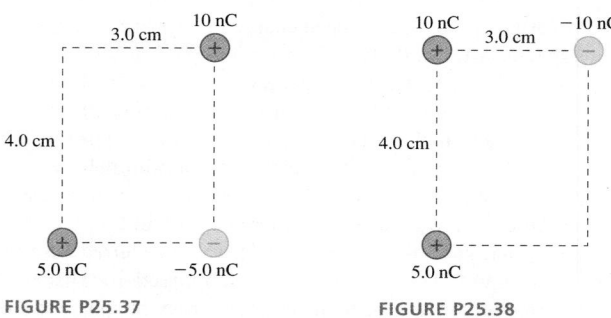

FIGURE P25.37   **FIGURE P25.38**

38. ‖ What is the force $\vec{F}$ on the 5.0 nC charge in FIGURE P25.38? Give your answer as a magnitude and an angle measured cw or ccw (specify which) from the +x-axis.

39. ‖‖ What is the force $\vec{F}$ on the 1.0 nC charge in the middle of FIGURE P25.39 due to the four other charges? Give your answer in component form.

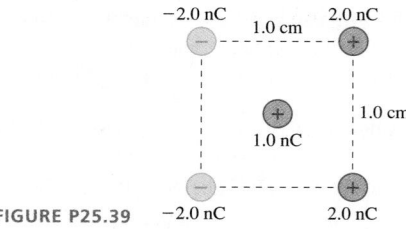

FIGURE P25.39

40. ‖ What is the force $\vec{F}$ on the 1.0 nC charge at the bottom in FIGURE P25.40? Give your answer in component form.

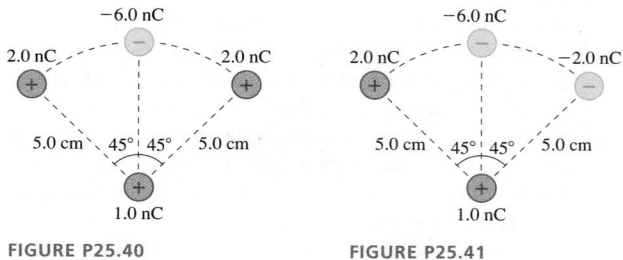

FIGURE P25.40   **FIGURE P25.41**

41. ‖ What is the force $\vec{F}$ on the 1.0 nC charge at the bottom in FIGURE P25.41? Give your answer in component form.

42. ⫴ A +2.0 nC charge is at the origin and a −4.0 nC charge is at $x = 1.0$ cm.
    a. At what x-coordinate could you place a proton so that it would experience no net force?
    b. Would the net force be zero for an electron placed at the same position? Explain.

43. ‖ The net force on the 1.0 nC charge in FIGURE P25.43 is zero. What is q?

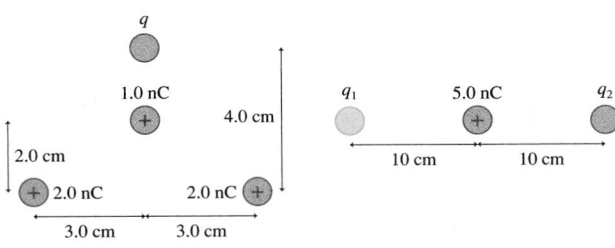

FIGURE P25.43   FIGURE P25.44

44. ‖ Charge q_2 in FIGURE P25.44 is in static equilibrium. What is q_1?

45. ‖ A positive point charge Q is located at $x = a$ and a negative point charge $-Q$ is at $x = -a$. A positive charge q can be placed anywhere on the y-axis. Find an expression for $(F_{net})_x$, the x-component of the net force on q.

46. ‖ A positive point charge Q is located at $x = a$ and a negative point charge $-Q$ is at $x = -a$. A positive charge q can be placed anywhere on the x-axis. Find an expression for $(F_{net})_x$, the x-component of the net force on q, when (a) $|x| < a$ and (b) $|x| > a$.

47. ⫴ FIGURE P25.47 shows four charges at the corners of a square of side L. What is the magnitude of the net force on q?

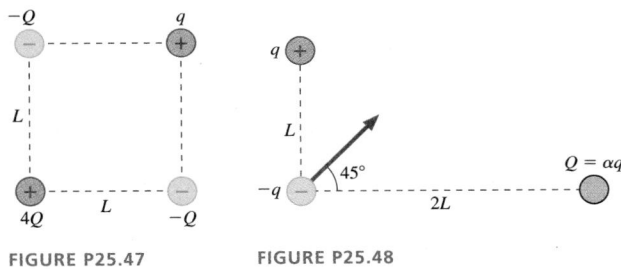

FIGURE P25.47   FIGURE P25.48

48. ‖ FIGURE P25.48 shows three charges and the net force on charge $-q$. Charge Q is some multiple α of q. What is α?

49. ‖ Two positive point charges q and $4q$ are at $x = 0$ and $x = L$, respectively, and free to move. A third charge is placed so that the entire three-charge system is in static equilibrium. What are the magnitude, sign, and x-coordinate of the third charge?

50. ⫴ Suppose the magnitude of the proton charge differs from the magnitude of the electron charge by a mere 1 part in 10^9.
    a. What would be the force between two 2.0-mm-diameter copper spheres 1.0 cm apart? Assume that each copper atom has an equal number of electrons and protons.
    b. Would this amount of force be detectable? What can you conclude from the fact that no such forces are observed?

51. ‖ In a simple model of the hydrogen atom, the electron moves in a circular orbit of radius 0.053 nm around a stationary proton. How many revolutions per second does the electron make?

52. ‖ You have two small, 2.0 g balls that have been given equal but opposite charges, but you don't know the magnitude of the charge. To find out, you place the balls distance d apart on a slippery horizontal surface, release them, and use a motion detector to measure the initial acceleration of one of the balls toward the other. After repeating this for several different separation distances, your data are as follows:

Distance (cm)	Acceleration (m/s²)
2.0	0.74
3.0	0.30
4.0	0.19
5.0	0.10

Use an appropriate graph of the data to determine the magnitude of the charge.

53. ‖ A 0.10 g honeybee acquires a charge of +23 pC while flying.
BIO a. The earth's electric field near the surface is typically (100 N/C, downward). What is the ratio of the electric force on the bee to the bee's weight?
    b. What electric field (strength and direction) would allow the bee to hang suspended in the air?

54. ‖ As a science project, you've invented an "electron pump" that moves electrons from one object to another. To demonstrate your invention, you bolt a small metal plate to the ceiling, connect the pump between the metal plate and yourself, and start pumping electrons from the metal plate to you. How many electrons must be moved from the metal plate to you in order for you to hang suspended in the air 2.0 m below the ceiling? Your mass is 60 kg. **Hint:** Assume that both you and the plate can be modeled as point charges.

55. ‖ You have a lightweight spring whose unstretched length is 4.0 cm. First, you attach one end of the spring to the ceiling and hang a 1.0 g mass from it. This stretches the spring to a length of 5.0 cm. You then attach two small plastic beads to the opposite ends of the spring, lay the spring on a frictionless table, and give each plastic bead the same charge. This stretches the spring to a length of 4.5 cm. What is the magnitude of the charge (in nC) on each bead?

56. ‖ An electric dipole consists of two opposite charges $\pm q$ separated by a small distance s. The product $p = qs$ is called the *dipole moment*. FIGURE P25.56 shows an electric dipole perpendicular to an electric field $\vec{E}$. Find an expression in terms of p and E for the magnitude of the torque that the electric field exerts on the dipole.

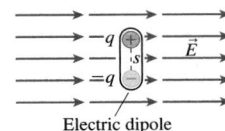

FIGURE P25.56   Electric dipole

57. ‖ You sometimes create a spark when you touch a doorknob after shuffling your feet on a carpet. Why? The air always has a few free electrons that have been kicked out of atoms by cosmic rays. If an electric field is present, a free electron is accelerated until it collides with an air molecule. It will transfer its kinetic energy to the molecule, then accelerate, then collide, then accelerate, collide, and so on. If the electron's kinetic energy just before a collision is 2.0×10^{-18} J or more, it has sufficient energy to kick an electron out of the molecule it hits. Where there was one free electron, now there are two! Each of these can then

accelerate, hit a molecule, and kick out another electron. Then there will be four free electrons. In other words, as FIGURE P25.57 shows, a sufficiently strong electric field causes a "chain reaction" of electron production. This is called a *breakdown* of the air. The current of moving electrons is what gives you the shock, and a spark is generated when the electrons recombine with the positive ions and give off excess energy as a burst of light.

a. The average distance an electron travels between collisions is 2.0 μm. What acceleration must an electron have to gain 2.0×10^{-18} J of kinetic energy in this distance?

b. What force must act on an electron to give it the acceleration found in part a?

c. What strength electric field will exert this much force on an electron? This is the *breakdown field strength*. **Note:** The measured breakdown field strength is a little less than your calculated value because our model of the process is a bit too simple. Even so, your calculated value is close.

d. Suppose a free electron in air is 1.0 cm away from a point charge. What minimum charge must this point charge have to cause a breakdown of the air and create a spark?

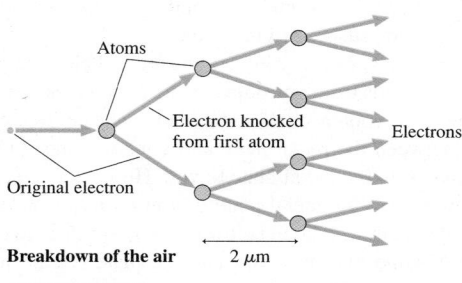

Breakdown of the air

FIGURE P25.57

58. ‖ Two 5.0 g point charges on 1.0-m-long threads repel each other after being charged to $+100$ nC, as shown in FIGURE P25.58. What is the angle θ? You can assume that θ is a small angle.

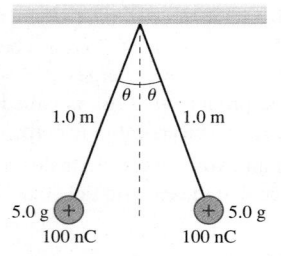

FIGURE P25.58

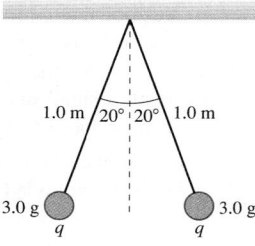

FIGURE P25.59

59. ‖ Two 3.0 g point charges on 1.0-m-long threads repel each other after being equally charged, as shown in FIGURE P25.59. What is the charge q?

60. ‖ What are the electric fields at points 1, 2, and 3 in FIGURE P25.60? Give your answer in component form.

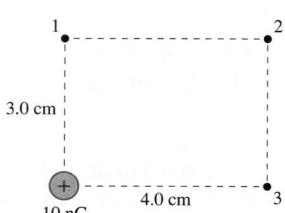

FIGURE P25.60

61. ‖ What are the electric fields at points 1 and 2 in FIGURE P25.61? Give your answer as a magnitude and direction.

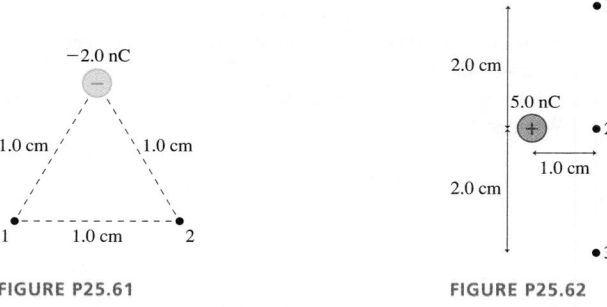

FIGURE P25.61   **FIGURE P25.62**

62. ‖ What are the electric fields at points 1, 2, and 3 in FIGURE P25.62? Give your answer in component form.

63. ‖ A -10.0 nC charge is located at position $(x, y) = (2.0$ cm, 1.0 cm$)$. At what (x, y) position(s) is the electric field
   a. $-225,000\,\hat{\imath}$ N/C?
   b. $(161,000\,\hat{\imath} - 80,500\,\hat{\jmath})$ N/C?
   c. $(28,800\,\hat{\imath} + 21,600\,\hat{\jmath})$ N/C?

64. ‖ A 10.0 nC charge is located at position $(x, y) = (1.0$ cm, 2.0 cm$)$. At what (x, y) position(s) is the electric field
   a. $-225,000\,\hat{\imath}$ N/C?
   b. $(161,000\,\hat{\imath} + 80,500\,\hat{\jmath})$ N/C?
   c. $(21,600\,\hat{\imath} - 28,800\,\hat{\jmath})$ N/C?

65. ‖ Three 1.0 nC charges are placed as shown in FIGURE P25.65. Each of these charges creates an electric field $\vec{E}$ at a point 3.0 cm in front of the middle charge.

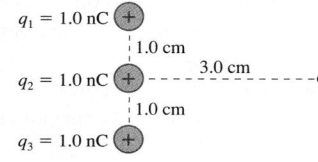

FIGURE P25.65

   a. What are the three fields $\vec{E}_1$, $\vec{E}_2$, and $\vec{E}_3$ created by the three charges? Write your answer for each as a vector in component form.

   b. Do you think that electric fields obey a principle of superposition? That is, is there a "net field" at this point given by $\vec{E}_{net} = \vec{E}_1 + \vec{E}_2 + \vec{E}_3$? Use what you learned in this chapter and previously in our study of forces to argue why this is or is not true.

   c. If it is true, what is $\vec{E}_{net}$?

66. ‖ An electric field $\vec{E} = 100,000\,\hat{\imath}$ N/C causes the 5.0 g point charge in FIGURE P25.66 to hang at a 20° angle. What is the charge on the ball?

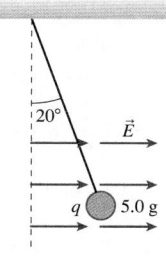

FIGURE P25.66

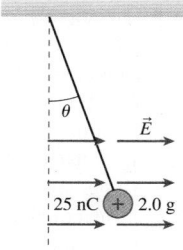

FIGURE P25.67

67. ‖ An electric field $\vec{E} = 200,000\,\hat{\imath}$ N/C causes the point charge in FIGURE P25.67 to hang at an angle. What is θ?

In Problems 68 through 71 you are given the equation(s) used to solve a problem. For each of these,

a. Write a realistic problem for which this is the correct equation(s).
b. Finish the solution of the problem.

68. $\dfrac{(9.0 \times 10^9 \, \text{N m}^2/\text{C}^2) \times N \times (1.60 \times 10^{-19} \, \text{C})}{(1.0 \times 10^{-6} \, \text{m})^2}$

$= 1.5 \times 10^6 \, \text{N/C}$

69. $\dfrac{(9.0 \times 10^9 \, \text{N m}^2/\text{C}^2) q^2}{(0.0150 \, \text{m})^2} = 0.020 \, \text{N}$

70. $\dfrac{(9.0 \times 10^9 \, \text{N m}^2/\text{C}^2)(15 \times 10^{-9} \, \text{C})}{r^2} = 54{,}000 \, \text{N/C}$

71. $\sum F_x = 2 \times \dfrac{(9.0 \times 10^9 \, \text{N m}^2/\text{C}^2)(1.0 \times 10^{-9} \, \text{C})q}{\left((0.020 \, \text{m})/\sin 30°\right)^2} \times \cos 30°$

$= 5.0 \times 10^{-5} \, \text{N}$

$\sum F_y = 0 \, \text{N}$

Challenge Problems

72. A 2.0-mm-diameter copper ball is charged to $+50$ nC. What fraction of its electrons have been removed?
73. Three 3.0 g balls are tied to 80-cm-long threads and hung from a *single* fixed point. Each of the balls is given the same charge q. At equilibrium, the three balls form an equilateral triangle in a horizontal plane with 20 cm sides. What is q?
74. The identical small spheres shown in FIGURE CP25.74 are charged to $+100$ nC and -100 nC. They hang as shown in a 100,000 N/C electric field. What is the mass of each sphere?

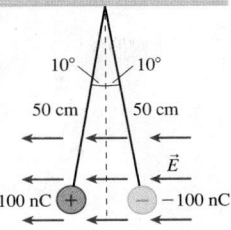

FIGURE CP25.74

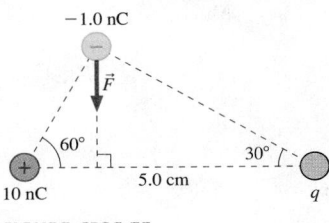

FIGURE CP25.75

75. The force on the -1.0 nC charge is as shown in FIGURE CP25.75. What is the magnitude of this force?
76. In Section 25.3 we claimed that a charged object exerts a net attractive force on an electric dipole. Let's investigate this. FIGURE CP25.76 shows a *permanent* electric dipole consisting of charges $+q$ and $-q$ separated by the fixed distance s. Charge $+Q$ is distance r from the center of the dipole. We'll assume, as is usually the case in practice, that $s \ll r$.
   a. Write an expression for the net force exerted on the dipole by charge $+Q$.
   b. Is this force toward $+Q$ or away from $+Q$? Explain.
   c. Use the *binomial approximation* $(1 + x)^{-n} \approx 1 - nx$ if $x \ll 1$ to show that your expression from part a can be written $F_{\text{net}} = 2KqQs/r^3$.
   d. How can an electric force have an inverse-cube dependence? Doesn't Coulomb's law say that the electric force depends on the inverse square of the distance? Explain.

$$\underset{q \quad -q}{\boxed{+}\ \boxed{-}} \xleftarrow{\ s\ } \qquad \qquad \overset{Q}{\boxed{+}}$$
$$\xleftarrow{\qquad\qquad r \qquad\qquad}$$

FIGURE CP25.76

STOP TO THINK ANSWERS

Stop to Think 25.1: b. Charged objects are attracted to neutral objects, so an attractive force is inconclusive. Repulsion is the only sure test.

Stop to Think 25.2: $q_e(+3e) > q_a(+1e) > q_d(0) > q_b(-1e) > q_c(-2e)$.

Stop to Think 25.3: a. The negative plastic rod will polarize the electroscope by pushing electrons down toward the leaves. This will partially neutralize the positive charge the leaves had acquired from the glass rod.

Stop to Think 25.4: b. The two forces are an action/reaction pair, opposite in direction but *equal* in magnitude.

Stop to Think 25.5: c. There's an electric field at *all* points, whether an $\vec{E}$ vector is shown or not. The electric field at the dot is to the right. But an electron is a negative charge, so the force of the electric field on the electron is to the left.

Stop to Think 25.6: $E_b > E_a > E_d > E_c$.

26 The Electric Field

In a plasma ball, electrons follow the electric field lines outward from the center electrode. The streamers appear where gas atoms emit light after the high-speed electrons collide with them.

▶ **Looking Ahead** The goal of Chapter 26 is to learn how to calculate and use the electric field.

Fields of Multiple Charges

You'll learn that the electric field due to several point charges is the vector sum of the individual fields.

You'll also learn to use **electric field lines.** This figure shows the electric field lines of a *dipole*, two equal but opposite point charges.

◀ **Looking Back**
Section 25.5 The electric field of a point charge

The Field of a Continuous Distribution of Charge

You'll learn a strategy for computing the electric field of a macroscopic charged object, such as a charged rod or a disk of charge.

- A charged object can be described by its **charge density,** the charge per unit length, area, or volume.
- The vector sum of electric fields will become an integral. We'll develop a step-by-step approach to setting up and evaluating these integrals.

We'll calculate the electric field of charged wires, charged disks, planes of charge, and spheres of charge.

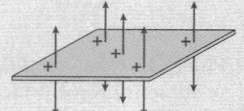

The electric field of a plane of charge is perpendicular to the plane. Many practical devices can be modeled as planes or lines of charge.

Uniform Electric Fields

Two parallel conducting plates with equal but opposite charges are called a **parallel-plate capacitor.**

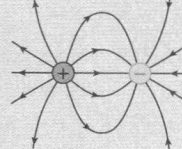

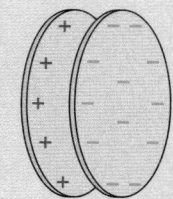

You'll learn that parallel-plate capacitors are important for creating a **uniform electric field.**

Charges in Electric Fields

Electric fields exert forces on charged particles. You'll learn to calculate the trajectories of charged particles moving in electric fields.

Older televisions and computer monitors use a *cathode-ray tube*. The picture is formed as a changing electric field sweeps an electron beam back and forth across the screen.

◀ **Looking Back**
Section 4.3 Projectile motion

Dipoles in Electric Fields

You learned in Chapter 25 that charged objects of either sign attract a neutral object. We'll understand better why this happens.

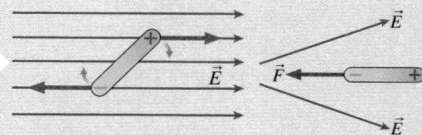

An electric field exerts a *torque* on a dipole, causing it to align with the field.

A nonuniform field exerts a force on a dipole, drawing it toward the stronger field.

26.1 Electric Field Models

Chapter 25 made a distinction between those charged particles that are the *sources* of an electric field and other charged particles that *experience* and move in the electric field. This is a very important distinction. Most of this chapter will be concerned with the *sources* of the electric field. Only at the end, once we know how to calculate the electric field, will we look at what happens to charges that find themselves *in* an electric field.

The electric fields used in science and engineering are often caused by fairly complicated distributions of charge. Sometimes these fields require exact calculations, but much of the time we can understand the essential physics on the basis of simplified *models* of the electric field.

FIGURE 26.1 Four basic electric field models.

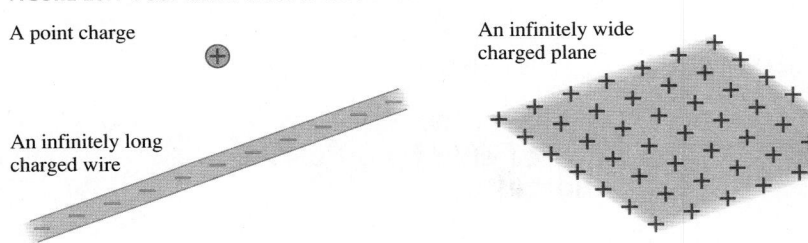

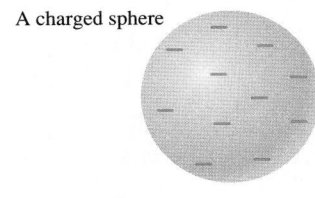

Four widely used electric field models, illustrated in **FIGURE 26.1**, are:

- The electric field of a point charge.
- The electric field of an infinitely long charged wire.
- The electric field of an infinitely wide charged plane.
- The electric field of a charged sphere.

Small charged objects can often be modeled as point charges or charged spheres. Real wires aren't infinitely long, but in many practical situations this approximation is perfectly reasonable. As we derive and use these electric fields, we'll consider the conditions under which they are appropriate models.

Our starting point is the electric field of a point charge q:

$$\vec{E} = \frac{1}{4\pi\epsilon_0} \frac{q}{r^2} \hat{r} \qquad \text{(electric field of a point charge)} \qquad (26.1)$$

where $\hat{r}$ is a unit vector pointing away from q and $\epsilon_0 = 8.85 \times 10^{-12}$ C^2/N m^2 is the permittivity constant. **FIGURE 26.2** reminds you of the electric fields of point charges. Although we have to give each vector we draw a length, keep in mind that each arrow represents the electric field *at a point*. The electric field is not a spatial quantity that "stretches" from one end of the arrow to the other.

The electric field was defined as $\vec{E} = \vec{F}_{\text{on }q}/q$, where $\vec{F}_{\text{on }q}$ is the electric force on charge q. Forces add as vectors, so the net force on q due to a group of point charges is the vector sum

$$\vec{F}_{\text{on }q} = \vec{F}_{1\text{ on }q} + \vec{F}_{2\text{ on }q} + \cdots$$

Consequently, the net electric field due to a group of point charges is

$$\vec{E}_{\text{net}} = \frac{\vec{F}_{\text{on }q}}{q} = \frac{\vec{F}_{1\text{ on }q}}{q} + \frac{\vec{F}_{2\text{ on }q}}{q} + \cdots = \vec{E}_1 + \vec{E}_2 + \cdots = \sum_i \vec{E}_i \qquad (26.2)$$

where $\vec{E}_i$ is the field from point charge i.

Equation 26.2, which is the primary tool for calculating electric fields, tells us that **the net electric field is the *vector sum* of the electric fields due to each charge.** In other words, electric fields obey the *principle of superposition*.

Knowing typical electric field strengths will also be helpful. The values in Table 26.1 on the next page will help you judge the reasonableness of your solutions to problems.

FIGURE 26.2 The electric field of a positive and a negative point charge.

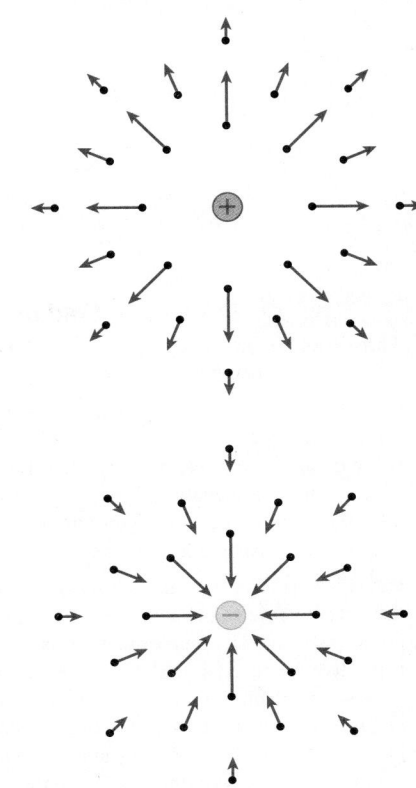

TABLE 26.1 Typical electric field strengths

Field location	Field strength (N/C)
Inside a current-carrying wire	$10^{-3} - 10^{-1}$
Near the earth's surface	$10^2 - 10^4$
Near objects charged by rubbing	$10^3 - 10^6$
Electric breakdown in air, causing a spark	3×10^6
Inside an atom	10^{11}

26.2 The Electric Field of Multiple Point Charges

Suppose the source of an electric field is a group of point charges $q_1, q_2, \ldots$. According to Equation 26.2, the net electric field $\vec{E}_{net}$ at each point in space is a superposition of the electric fields due to each individual charge:

$$(E_{net})_x = (E_1)_x + (E_2)_x + \cdots = \sum (E_i)_x$$
$$(E_{net})_y = (E_1)_y + (E_2)_y + \cdots = \sum (E_i)_y \qquad (26.3)$$
$$(E_{net})_z = (E_1)_z + (E_2)_z + \cdots = \sum (E_i)_z$$

Sometimes you'll want to write $\vec{E}_{net}$ in component form:

$$\vec{E}_{net} = (E_{net})_x \hat{\imath} + (E_{net})_y \hat{\jmath} + (E_{net})_z \hat{k}$$

At other times you will give $\vec{E}_{net}$ as a magnitude and a direction.

PROBLEM-SOLVING STRATEGY 26.1 **The electric field of multiple point charges**

MODEL Model charged objects as point charges.

VISUALIZE For the pictorial representation:

- Establish a coordinate system and show the locations of the charges.
- Identify the point P at which you want to calculate the electric field.
- Draw the electric field of each charge at P.
- Use symmetry to determine if any components of $\vec{E}_{net}$ are zero.

SOLVE The mathematical representation is $\vec{E}_{net} = \sum \vec{E}_i$.

- For each charge, determine its distance from P and the angle of $\vec{E}_i$ from the axes.
- Calculate the field strength of each charge's electric field.
- Write each vector $\vec{E}_i$ in component form.
- Sum the vector components to determine $\vec{E}_{net}$.
- If needed, determine the magnitude and direction of $\vec{E}_{net}$.

ASSESS Check that your result has the correct units, is reasonable, and agrees with any known limiting cases.

Exercise 16

EXAMPLE 26.1 **The electric field of three equal point charges**

Three equal point charges q are located on the y-axis at $y = 0$ and at $y = \pm d$. What is the electric field at a point on the x-axis?

MODEL This problem is a step along the way to understanding the electric field of a charged wire. We'll assume that q is positive when drawing pictures, but the solution should allow for the possibility that q is negative. The question does not ask about any specific point, so we will be looking for a symbolic expression in terms of the unspecified position x.

VISUALIZE FIGURE 26.3 shows the charges, the coordinate system, and the three electric field vectors $\vec{E}_1$, $\vec{E}_2$, and $\vec{E}_3$. Each of these fields points *away from* its source charge because of the assumption that q is positive. We need to find the vector sum $\vec{E}_{net} = \vec{E}_1 + \vec{E}_2 + \vec{E}_3$.

Before rushing into a calculation, we can make our task *much* easier by first thinking qualitatively about the situation. For example, the fields $\vec{E}_1$, $\vec{E}_2$, and $\vec{E}_3$ all lie in the xy-plane, hence we can conclude without any calculations that $(E_{net})_z = 0$. Next,

FIGURE 26.3 Calculating the electric field of three equal point charges.

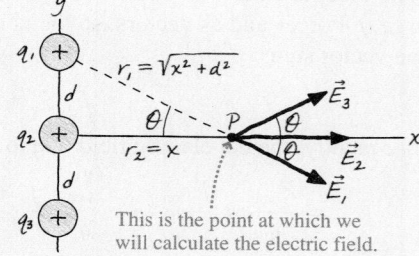

look at the y-components of the fields. The fields $\vec{E}_1$ and $\vec{E}_3$ have equal magnitudes and are tilted away from the x-axis by the same angle θ. Consequently, the y-components of $\vec{E}_1$ and $\vec{E}_3$ will *cancel* when added. $\vec{E}_2$ has no y-component, so we can conclude that $(E_{net})_y = 0$. The only component we need to calculate is $(E_{net})_x$.

SOLVE We're ready to calculate. The x-component of the field is

$$(E_{\text{net}})_x = (E_1)_x + (E_2)_x + (E_3)_x = 2(E_1)_x + (E_2)_x$$

where we used the fact that fields $\vec{E}_1$ and $\vec{E}_3$ have *equal* x-components. Vector $\vec{E}_2$ has *only* the x-component

$$(E_2)_x = E_2 = \frac{1}{4\pi\epsilon_0}\frac{q_2}{r_2^2} = \frac{1}{4\pi\epsilon_0}\frac{q}{x^2}$$

where $r_2 = x$ is the distance from q_2 to the point at which we are calculating the field. Vector $\vec{E}_1$ is at angle θ from the x-axis, so its x-component is

$$(E_1)_x = E_1\cos\theta = \frac{1}{4\pi\epsilon_0}\frac{q_1}{r_1^2}\cos\theta$$

where r_1 is the distance from q_1. This expression for $(E_1)_x$ is correct, but it is not yet sufficient. Both the distance r_1 and the angle θ vary with the position x and need to be expressed as functions of x. From the Pythagorean theorem, $r_1 = (x^2 + d^2)^{1/2}$. Then from trigonometry,

$$\cos\theta = \frac{x}{r_1} = \frac{x}{(x^2 + d^2)^{1/2}}$$

By combining these pieces, we see that $(E_1)_x$ is

$$(E_1)_x = \frac{1}{4\pi\epsilon_0}\frac{q}{x^2 + d^2}\frac{x}{(x^2 + d^2)^{1/2}} = \frac{1}{4\pi\epsilon_0}\frac{xq}{(x^2 + d^2)^{3/2}}$$

This expression is a bit complex, but notice that the dimensions of $x/(x^2 + d^2)^{3/2}$ are $1/\text{m}^2$, as they *must* be for the field of a point charge. Checking dimensions is a good way to verify that you haven't made algebra errors.

We can now combine $(E_1)_x$ and $(E_2)_x$ to write the x-component of $\vec{E}_{\text{net}}$ as

$$(E_{\text{net}})_x = 2(E_1)_x + (E_2)_x = \frac{q}{4\pi\epsilon_0}\left[\frac{1}{x^2} + \frac{2x}{(x^2 + d^2)^{3/2}}\right]$$

The other two components of $\vec{E}_{\text{net}}$ are zero, hence the electric field of the three charges at a point on the x-axis is

$$\vec{E}_{\text{net}} = \frac{q}{4\pi\epsilon_0}\left[\frac{1}{x^2} + \frac{2x}{(x^2 + d^2)^{3/2}}\right]\hat{\imath}$$

ASSESS This is the electric field only at points *on the x-axis*. Furthermore, this expression is valid only for $x > 0$. The electric field to the left of the charges points in the opposite direction, but our expression doesn't change sign for negative x. (This is a consequence of how we wrote $(E_2)_x$.) We would need to modify this expression to use it for negative values of x. The good news, though, is that our expression *is* valid for both positive and negative q. A negative value of q makes $(E_{\text{net}})_x$ negative, which would be an electric field pointing to the left, toward the negative charges.

Let's explore this example a bit more. There are two limiting cases for which we know what the result should be. First, let x become really, really small. As the point in Figure 26.3 approaches the origin, the fields $\vec{E}_1$ and $\vec{E}_3$ become opposite to each other and cancel. Thus as $x \to 0$, the field *should* be that of the single point charge q at the origin, a field we already know. Is it? Notice that

$$\lim_{x\to 0}\frac{2x}{(x^2 + d^2)^{3/2}} = 0 \tag{26.4}$$

Thus $E_{\text{net}} \to q/4\pi\epsilon_0 x^2$ as $x \to 0$, the expected field of a single point charge.

Now consider the opposite situation, where x becomes extremely large. From very far away, the three source charges will seem to merge into a single charge of size $3q$, just as three very distant lightbulbs appear to be a single light. Thus the field for $x \gg d$ *should* be that of a point charge $3q$. Is it?

The field is zero in the limit $x \to \infty$. That doesn't tell us much, so we don't want to go *that* far away. We simply want x to be very large in comparison to the spacing d between the source charges. If $x \gg d$, then the denominator of the second term of $\vec{E}_{\text{net}}$ is well approximated by $(x^2 + d^2)^{3/2} \approx (x^2)^{3/2} = x^3$. Thus

$$\lim_{x\gg d}\left[\frac{1}{x^2} + \frac{2x}{(x^2 + d^2)^{3/2}}\right] = \frac{1}{x^2} + \frac{2x}{x^3} = \frac{3}{x^2} \tag{26.5}$$

Consequently, the net electric field far from the source charges is

$$\vec{E}_{\text{net}}(x \gg d) = \frac{1}{4\pi\epsilon_0}\frac{(3q)}{x^2}\hat{\imath} \tag{26.6}$$

As expected, this is the field of a point charge $3q$. These checks of limiting cases provide confidence in the result of the calculation.

FIGURE 26.4 is a graph of the field strength E_{net} for the three charges of Example 26.1. Although we don't have any numerical values, we can specify x as a multiple of the charge separation d. Notice how the graph matches the field of a single point charge when $x \ll d$ and matches the field of a point charge $3q$ when $x \gg d$.

FIGURE 26.4 The electric field strength along a line perpendicular to three equal point charges.

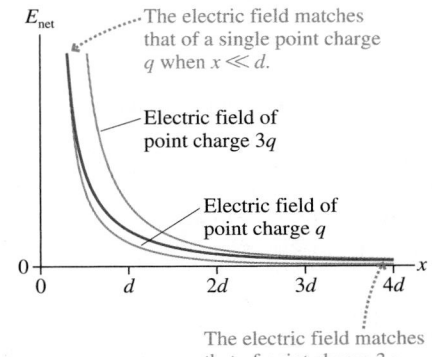

The electric field matches that of a single point charge q when $x \ll d$.

Electric field of point charge $3q$

Electric field of point charge q

The electric field matches that of point charge $3q$ when $x \gg d$.

FIGURE 26.5 Permanent and induced electric dipoles.

A water molecule is a *permanent* dipole because the negative electrons spend more time with the oxygen atom.

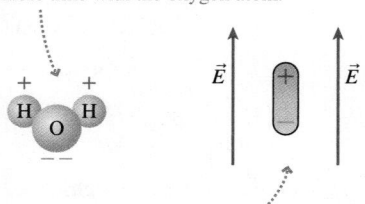

This dipole is *induced*, or stretched, by the electric field acting on the + and − charges.

The Electric Field of a Dipole

Two equal but opposite charges separated by a small distance form an *electric dipole*. FIGURE 26.5 shows two examples. In a *permanent electric dipole,* such as the water molecule, the oppositely charged particles maintain a small permanent separation. We can also create an electric dipole, as you learned in Chapter 25, by polarizing a neutral atom with an external electric field. This is an *induced electric dipole*.

FIGURE 26.6 shows that we can represent an electric dipole, whether permanent or induced, by two opposite charges $\pm q$ separated by the small distance s. The dipole has zero net charge, but it *does* have an electric field. Consider a point on the positive y-axis. This point is slightly closer to $+q$ than to $-q$, so the fields of the two charges do not cancel. You can see in the figure that $\vec{E}_{\text{dipole}}$ points in the positive y-direction. Similarly, vector addition shows that $\vec{E}_{\text{dipole}}$ points in the negative y-direction at points along the x-axis.

Let's calculate the electric field of a dipole at a point on the axis of the dipole. This is the y-axis in Figure 26.6. The point is distance $r_+ = y - s/2$ from the positive charge and $r_- = y + s/2$ from the negative charge. The net electric field at this point has only a y-component, and the sum of the fields of the two point charges gives

$$(E_{\text{dipole}})_y = (E_+)_y + (E_-)_y = \frac{1}{4\pi\epsilon_0}\frac{q}{(y - \frac{1}{2}s)^2} + \frac{1}{4\pi\epsilon_0}\frac{(-q)}{(y + \frac{1}{2}s)^2}$$

$$= \frac{q}{4\pi\epsilon_0}\left[\frac{1}{(y - \frac{1}{2}s)^2} - \frac{1}{(y + \frac{1}{2}s)^2}\right] \quad (26.7)$$

FIGURE 26.6 The dipole electric field at two points.

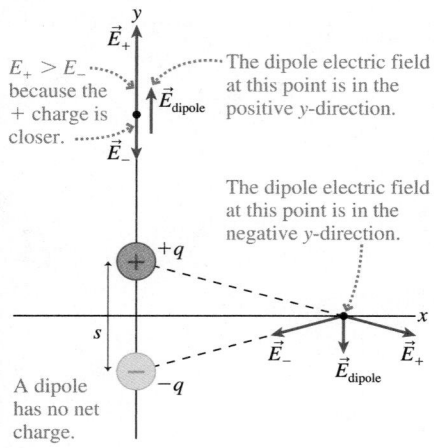

$E_+ > E_-$ because the + charge is closer.

The dipole electric field at this point is in the positive y-direction.

The dipole electric field at this point is in the negative y-direction.

A dipole has no net charge.

Combining the two terms over a common denominator, we find

$$(E_{\text{dipole}})_y = \frac{q}{4\pi\epsilon_0}\left[\frac{2ys}{(y - \frac{1}{2}s)^2(y + \frac{1}{2}s)^2}\right] \quad (26.8)$$

We omitted some of the algebraic steps, but be sure you can do this yourself. Some of the homework problems will require similar algebra.

In practice, we almost always observe the electric field of a dipole only for distances $y \gg s$—that is, for distances much larger than the charge separation. In such cases, the denominator can be approximated $(y - \frac{1}{2}s)^2(y + \frac{1}{2}s)^2 \approx y^4$. With this approximation, Equation 26.8 becomes

$$(E_{\text{dipole}})_y \approx \frac{1}{4\pi\epsilon_0}\frac{2qs}{y^3} \quad (26.9)$$

It is useful to define the **dipole moment** $\vec{p}$, shown in FIGURE 26.7, as the vector

$$\vec{p} = (qs, \text{ from the negative to the positive charge}) \quad (26.10)$$

The direction of $\vec{p}$ identifies the orientation of the dipole, and the dipole-moment magnitude $p = qs$ determines the electric field strength. The SI units of the dipole moment are C m.

We can use the dipole moment to write a succinct expression for the electric field at a point on the axis of a dipole:

$$\vec{E}_{\text{dipole}} \approx \frac{1}{4\pi\epsilon_0}\frac{2\vec{p}}{r^3} \quad \text{(on the axis of an electric dipole)} \quad (26.11)$$

FIGURE 26.7 The dipole moment.

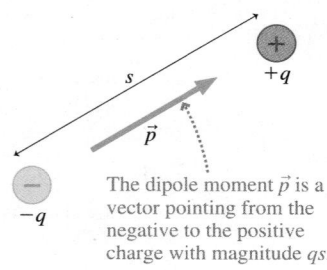

The dipole moment $\vec{p}$ is a vector pointing from the negative to the positive charge with magnitude qs.

where r is the distance measured from the *center* of the dipole. We've switched from y to r because we've now specified that Equation 26.11 is valid only along the axis of the dipole. Notice that the electric field along the axis points in the direction of the dipole moment $\vec{p}$.

A homework problem will let you calculate the electric field in the plane that bisects the dipole. This is the field shown on the x-axis in Figure 26.6, but it could equally well be the field on the z-axis as it comes out of the page. The field, for $r \gg s$, is

$$\vec{E}_{\text{dipole}} \approx -\frac{1}{4\pi\epsilon_0}\frac{\vec{p}}{r^3} \quad \text{(bisecting plane)} \qquad (26.12)$$

This field is *opposite* to $\vec{p}$, and it is only half the strength of the on-axis field at the same distance.

NOTE ▶ Do these inverse-cube equations violate Coulomb's law? Not at all. Coulomb's law describes the force between two *point charges,* and from Coulomb's law we found that the electric field of a *point charge* varies with the inverse square of the distance. But a dipole is not a point charge. The field of a dipole decreases more rapidly than that of a point charge, which is to be expected because the dipole is, after all, electrically neutral. ◀

EXAMPLE 26.2 The electric field of a water molecule

The water molecule H_2O has a permanent dipole moment of magnitude $6.2 \times 10^{-30}\,\mathrm{C\,m}$. What is the electric field strength 1.0 nm from a water molecule at a point on the dipole's axis?

MODEL The size of a molecule is ≈ 0.1 nm. Thus $r \gg s$, and we can use Equation 26.11 for the on-axis electric field of the molecule's dipole moment.

SOLVE The on-axis electric field strength at $r = 1.0$ nm is

$$E \approx \frac{1}{4\pi\epsilon_0}\frac{2p}{r^3} = (9.0 \times 10^9\,\mathrm{N\,m^2/C^2})\frac{2(6.2 \times 10^{-30}\,\mathrm{C\,m})}{(1.0 \times 10^{-9}\,\mathrm{m})^3}$$
$$= 1.1 \times 10^8\,\mathrm{N/C}$$

ASSESS By referring to Table 26.1 you can see that the field strength is "strong" compared to our everyday experience with charged objects but "weak" compared to the electric field inside the atoms themselves. This seems reasonable.

Picturing the Electric Field

We can't see the electric field. Consequently, we need pictorial tools to help us visualize it in a region of space. One method, introduced in Chapter 25, is to picture the electric field by drawing electric field vectors at various points in space. Another way to picture the field is to draw **electric field lines.**

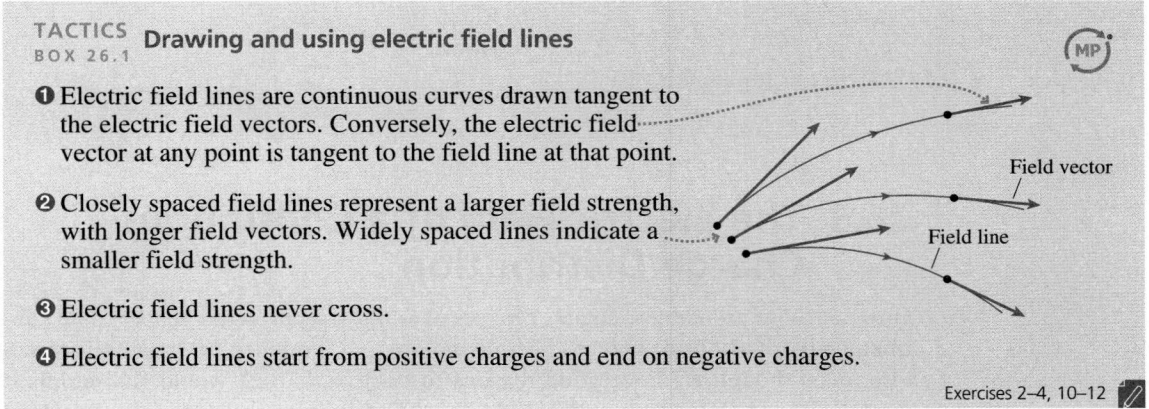

TACTICS BOX 26.1 Drawing and using electric field lines

❶ Electric field lines are continuous curves drawn tangent to the electric field vectors. Conversely, the electric field vector at any point is tangent to the field line at that point.

❷ Closely spaced field lines represent a larger field strength, with longer field vectors. Widely spaced lines indicate a smaller field strength.

❸ Electric field lines never cross.

❹ Electric field lines start from positive charges and end on negative charges.

Field vector

Field line

Exercises 2–4, 10–12

Step 3 is required to make sure that $\vec{E}$ has a unique direction at every point in space. Step 4 follows from the fact that electric fields are created by charges. However, we will have to modify step 4 in Chapter 33 when we find another way to create an electric field.

FIGURE 26.8a on the next page represents the electric field of a dipole as a field-vector diagram. FIGURE 26.8b shows the same field using electric field lines. Notice how the on-axis field points in the direction of $\vec{p}$, both above and below the dipole, while the field in the bisecting plane points opposite to $\vec{p}$. At most points, however, $\vec{E}$ has components both parallel to $\vec{p}$ and perpendicular to $\vec{p}$.

FIGURE 26.8 The electric field of a dipole: (a) field vectors, (b) field lines.

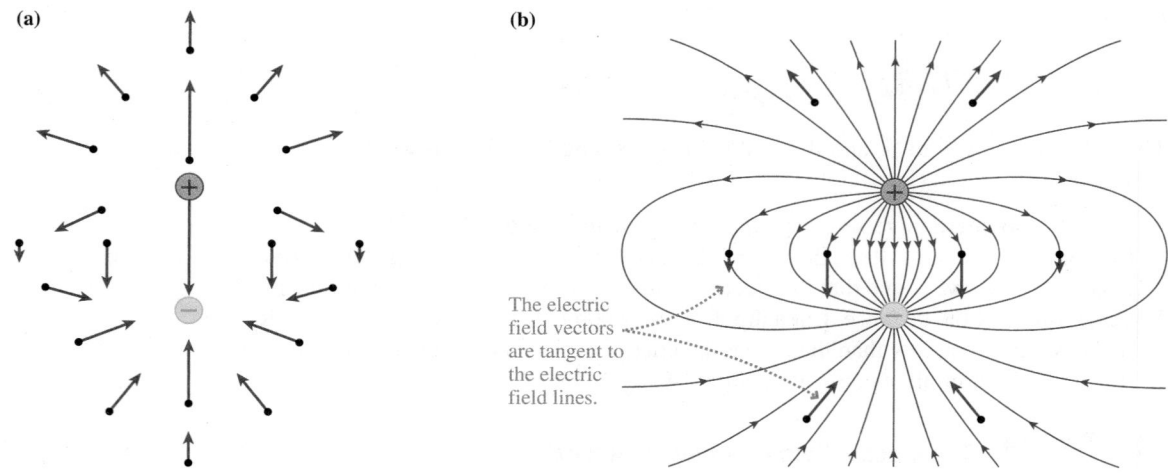

(a)

(b)

The electric field vectors are tangent to the electric field lines.

FIGURE 26.9 The electric field of two equal positive charges.

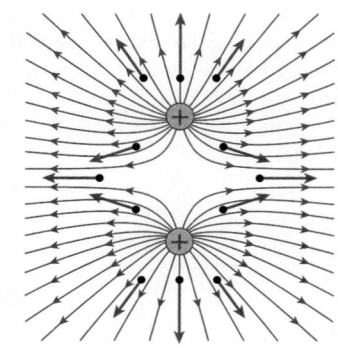

FIGURE 26.9 shows the electric field of two same-sign charges. A careful comparison of Figures 26.8b and 26.9 is worthwhile. Make sure you can explain the similarities and differences.

Neither field-vector diagrams nor field-line diagrams are perfect pictorial representations of an electric field. The field vectors are somewhat harder to draw, and they show the field at only a few points, but they do clearly indicate the direction and strength of the electric field at those points. Field-line diagrams perhaps look more elegant, and they're sometimes easier to sketch, but there's no formula for knowing where to draw the lines. We'll use both field-vector diagrams and field-line diagrams, depending on the circumstances.

STOP TO THINK 26.1 At the dot, the electric field points

a. Left.   b. Right.
c. Up.   d. Down.
e. The electric field is zero.

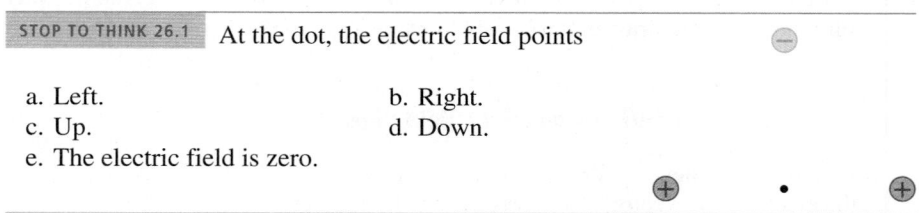

26.3 The Electric Field of a Continuous Charge Distribution

Ordinary objects—tables, chairs, beakers of water—seem to our senses to be continuous distributions of matter. There is no obvious evidence for an atomic structure, even though we have good reasons to believe that we would find atoms if we subdivided the matter sufficiently far. Thus it is easier, for many practical purposes, to consider matter to be continuous and to talk about the *density* of matter. Density—the number of kilograms of matter per cubic meter—allows us to describe the distribution of matter *as if* the matter were continuous rather than atomic.

Much the same situation occurs with charge. If a charged object contains a large number of excess electrons—for example, 10^{12} extra electrons on a metal rod—it is not practical to track every electron. It makes more sense to consider the charge to be *continuous* and to describe how it is distributed over the object.

FIGURE 26.10a shows an object of length L, such as a plastic rod or a metal wire, with charge Q spread uniformly along it. (We will use an uppercase Q for the total charge of an object, reserving lowercase q for individual point charges.) The **linear charge density** λ is defined to be

$$\lambda = \frac{Q}{L} \qquad (26.13)$$

Linear charge density, which has units of C/m, is the amount of charge *per meter* of length. The linear charge density of a 20-cm-long wire with 40 nC of charge is 2.0 nC/cm or 2.0×10^{-7} C/m.

NOTE ▶ The linear charge density λ is analogous to the linear mass density μ that you used in Chapter 20 to find the speed of a wave on a string. ◀

We'll also be interested in charged surfaces. FIGURE 26.10b shows a two-dimensional distribution of charge across a surface of area A. We define the **surface charge density** η (lowercase Greek eta) to be

$$\eta = \frac{Q}{A} \qquad (26.14)$$

Surface charge density, with units of C/m², is the amount of charge *per square meter*. A 1.0 mm × 1.0 mm square on a surface with $\eta = 2.0 \times 10^{-4}$ C/m² contains 2.0×10^{-10} C or 0.20 nC of charge. (The volume charge density $\rho = Q/V$, measured in C/m³, will be used in Chapter 27.)

Figure 26.10 and the definitions of Equations 26.13 and 26.14 assume that the object is **uniformly charged,** meaning that the charges are evenly spread over the object. We will assume objects are uniformly charged unless noted otherwise.

NOTE ▶ Some textbooks represent the surface charge density with the symbol σ. Because σ is also used to represent *conductivity,* an idea we'll introduce in Chapter 30, we've selected a different symbol for surface charge density. ◀

FIGURE 26.10 One-dimensional and two-dimensional continuous charge distributions.

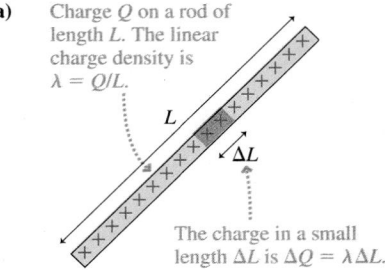

(a) Charge Q on a rod of length L. The linear charge density is $\lambda = Q/L$.

L

ΔL

The charge in a small length ΔL is $\Delta Q = \lambda \Delta L$.

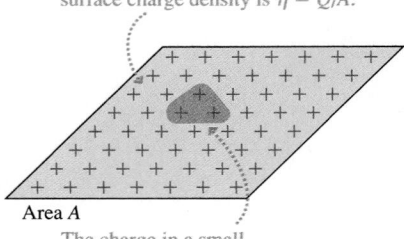

(b) Charge Q on a surface of area A. The surface charge density is $\eta = Q/A$.

Area A

The charge in a small area ΔA is $\Delta Q = \eta \Delta A$.

STOP TO THINK 26.2 A piece of plastic is uniformly charged with surface charge density η_a. The plastic is then broken into a large piece with surface charge density η_b and a small piece with surface charge density η_c. Rank in order, from largest to smallest, the surface charge densities η_a to η_c.

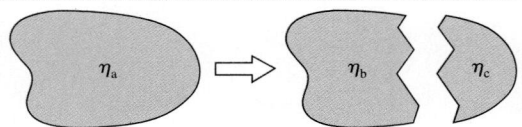

A Problem-Solving Strategy

Our goal is to find the electric field of a continuous distribution of charge, such as a charged rod or a charged disk. We have two basic tools to work with:

- The electric field of a point charge, and
- The principle of superposition.

We can apply these tools to a continuous distribution of charge if we follow a three-step strategy:

1. Divide the total charge Q into many small point-like charges ΔQ.
2. Use our knowledge of the electric field of a point charge to find the electric field of each ΔQ.
3. Calculate the net field $\vec{E}_{net}$ by summing the fields of all the ΔQ.

In practice, as you may have guessed, we'll let the sum become an integral.

The difficulty with electric field calculations is not the summation or integration itself, which is the last step, but setting up the calculation and knowing *what* to integrate. We will go step by step through several examples to illustrate the procedures. However, we first need to flesh out the steps of the problem-solving strategy. The aim of this problem-solving strategy is to break a difficult problem down into small steps that are individually manageable.

PROBLEM-SOLVING
STRATEGY 26.2

The electric field of a continuous distribution of charge

MODEL Model the distribution as a simple shape, such as a line of charge or a disk of charge. Assume the charge is uniformly distributed.

VISUALIZE For the pictorial representation:

❶ Draw a picture and establish a coordinate system.
❷ Identify the point P at which you want to calculate the electric field.
❸ Divide the total charge Q into small pieces of charge ΔQ, using shapes for which you *already know* how to determine $\vec{E}$. This is often, but not always, a division into point charges.
❹ Draw the electric field vector at P for one or two small pieces of charge. This will help you identify distances and angles that need to be calculated.
❺ Look for symmetries of the charge distribution that simplify the field. You may conclude that some components of $\vec{E}$ are zero.

SOLVE The mathematical representation is $\vec{E}_{\text{net}} = \sum \vec{E}_i$.

- Use superposition to form an algebraic expression for *each* of the three components of $\vec{E}$ (unless you are sure one or more is zero) at point P.
- Let the (x, y, z) coordinates of the point remain variables.
- Replace the small charge ΔQ with an equivalent expression involving a charge density and a coordinate, such as dx, that describes the shape of charge ΔQ. **This is the critical step in making the transition from a sum to an integral** because you need a coordinate to serve as the integration variable.
- Express all angles and distances in terms of the coordinates.
- Let the sum become an integral. The integration will be over the *one* coordinate variable that is related to ΔQ. The integration limits for this variable must "cover" the entire charged object.

ASSESS Check that your result is consistent with any limits for which you know what the field should be.

EXAMPLE 26.3 **The electric field of a line of charge**

FIGURE 26.11 shows a thin, uniformly charged rod of length L with total charge Q. Find the electric field strength at radial distance r in the plane that bisects the rod.

FIGURE 26.11 A thin, uniformly charged rod.

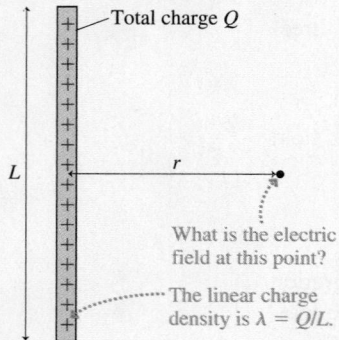

Total charge Q

L

r

What is the electric field at this point?

The linear charge density is $\lambda = Q/L$.

MODEL The rod is thin, so we'll assume the charge lies along a line and forms what we call a *line of charge*. This is an important

charge distribution that models the electric field of a charged rod or a charged metal wire. The rod's linear charge density is $\lambda = Q/L$.

VISUALIZE FIGURE 26.12 illustrates the five steps of the problem-solving strategy. We've chosen a coordinate system in which the rod lies along the y-axis and point P, in the bisecting plane, is on the x-axis. We've then divided the rod into N small segments of charge ΔQ, each of which can be modeled as a point charge. For every ΔQ in the bottom half of the wire with a field that points to the right and up, there's a matching ΔQ in the top half whose field points to the right and down. The y-components of these two fields cancel, hence the net electric field on the x-axis points straight away from the rod. The only component we need to calculate is E_x. (This is the same reasoning on the basis of symmetry that we used in Example 26.1.)

SOLVE Each of the little segments of charge can be modeled as a point charge. We know the electric field of a point charge, so we can write the x-component of $\vec{E}_i$, the electric field of segment i, as

$$(E_i)_x = E_i \cos\theta_i = \frac{1}{4\pi\epsilon_0} \frac{\Delta Q}{r_i^2} \cos\theta_i$$

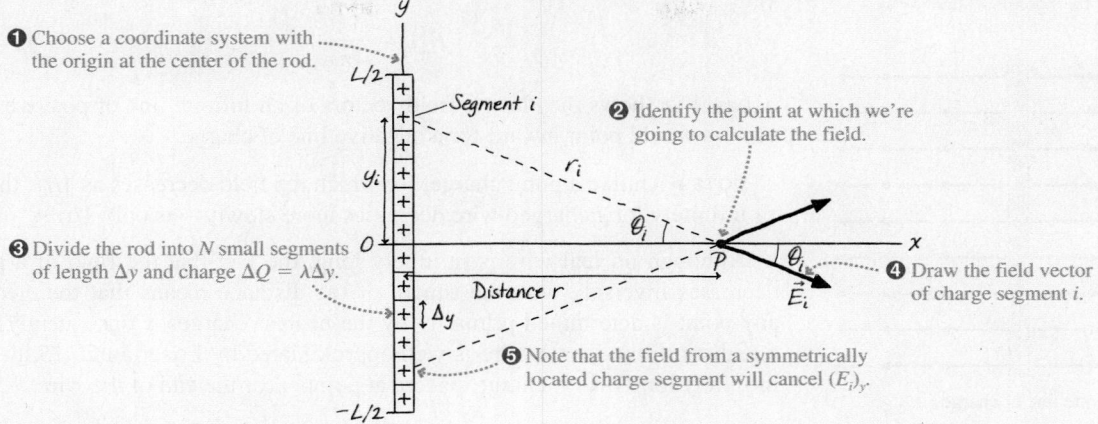

FIGURE 26.12 Calculating the electric field of a line of charge.

❶ Choose a coordinate system with the origin at the center of the rod.

Segment i

❷ Identify the point at which we're going to calculate the field.

❸ Divide the rod into N small segments of length Δy and charge $\Delta Q = \lambda \Delta y$.

❹ Draw the field vector of charge segment i.

Distance r

❺ Note that the field from a symmetrically located charge segment will cancel $(E_i)_y$.

where r_i is the distance from charge i to point P. You can see from the figure that $r_i = (y_i^2 + r^2)^{1/2}$ and $\cos\theta_i = r/r_i = r/(y_i^2 + r^2)^{1/2}$. With these, $(E_i)_x$ is

$$(E_i)_x = \frac{1}{4\pi\epsilon_0} \frac{\Delta Q}{y_i^2 + r^2} \frac{r}{\sqrt{y_i^2 + r^2}}$$

$$= \frac{1}{4\pi\epsilon_0} \frac{r\,\Delta Q}{(y_i^2 + r^2)^{3/2}}$$

Compare this result to the very similar calculation we did in Example 26.1. If we now sum this expression over all the charge segments, the net x-component of the electric field is

$$E_x = \sum_{i=1}^{N} (E_i)_x = \frac{1}{4\pi\epsilon_0} \sum_{i=1}^{N} \frac{r\,\Delta Q}{(y_i^2 + r^2)^{3/2}}$$

This is the same superposition we did for the $N = 3$ case in Example 26.1. The only difference is that we have now written the result as an explicit summation so that N can have any value. We want to let $N \rightarrow \infty$ and to replace the sum with an integral, but we can't integrate over Q; it's not a geometric quantity. This is where the linear charge density enters. The quantity of charge in each segment is related to its length Δy by $\Delta Q = \lambda \Delta y = (Q/L)\Delta y$. In terms of the linear charge density, the electric field is

$$E_x = \frac{Q/L}{4\pi\epsilon_0} \sum_{i=1}^{N} \frac{r\,\Delta y}{(y_i^2 + r^2)^{3/2}}$$

Now we're ready to let the sum become an integral. If we let $N \rightarrow \infty$, then each segment becomes an infinitesimal length $\Delta y \rightarrow dy$ while the discrete position variable y_i becomes the continuous integration variable y. The sum from $i = 1$ at the bottom end of the line of charge to $i = N$ at the top end will be replaced with an integral from $y = -L/2$ to $y = +L/2$. Thus in the limit $N \rightarrow \infty$,

$$E_x = \frac{Q/L}{4\pi\epsilon_0} \int_{-L/2}^{L/2} \frac{r\,dy}{(y^2 + r^2)^{3/2}}$$

This is a standard integral that you have learned to do in calculus and that can be found in Appendix A. Note that r is a *constant* as far as this integral is concerned. Integrating gives

$$E_x = \frac{Q/L}{4\pi\epsilon_0} \frac{y}{r\sqrt{y^2 + r^2}} \Big|_{-L/2}^{L/2}$$

$$= \frac{Q/L}{4\pi\epsilon_0} \left[\frac{L/2}{r\sqrt{(L/2)^2 + r^2}} - \frac{-L/2}{r\sqrt{(-L/2)^2 + r^2}} \right]$$

$$= \frac{1}{4\pi\epsilon_0} \frac{Q}{r\sqrt{r^2 + (L/2)^2}}$$

Because E_x is the *only* component of the field, the electric field strength E_{rod} at distance r from the center of a charged rod is

$$E_{\text{rod}} = \frac{1}{4\pi\epsilon_0} \frac{|Q|}{r\sqrt{r^2 + (L/2)^2}}$$

The field strength must be positive, so we added absolute value signs to Q to allow for the possibility that the charge could be negative. The only restriction is to remember that this is the electric field at a point in the plane that bisects the rod.

ASSESS Suppose we are at a point *very* far from the rod. If $r \gg L$, the length of the rod is not relevant and the rod appears to be a point charge Q in the distance. Thus in the *limiting case $r \gg L$*, we expect the rod's electric field to be that of a point charge. If $r \gg L$, the square root becomes $(r^2 + (L/2)^2)^{1/2} \approx (r^2)^{1/2} = r$ and the electric field strength at distance r becomes $E_{\text{rod}} \approx Q/4\pi\epsilon_0 r^2$, the field of a point charge. The fact that our expression of E_{rod} has the correct limiting behavior gives us confidence that we haven't made any mistakes in its derivation.

An Infinite Line of Charge

What if the rod or wire becomes very long, becoming a **line of charge,** while the linear charge density λ remains constant? To answer this question, we can rewrite the expression for E_{rod} by factoring $(L/2)^2$ out of the denominator:

$$E_{\text{rod}} = \frac{1}{4\pi\epsilon_0} \frac{|Q|}{r \cdot L/2} \frac{1}{\sqrt{1 + 4r^2/L^2}} = \frac{1}{4\pi\epsilon_0} \frac{2|\lambda|}{r} \frac{1}{\sqrt{1 + 4r^2/L^2}}$$

FIGURE 26.13 The electric field of an infinite line of charge.

The field points straight away from the line at all points.

Infinite line of charge

The field strength decreases with distance.

where $|\lambda| = |Q|/L$ is the magnitude of the linear charge density. If we now let $L \to \infty$, the last term becomes simply 1 and we're left with

$$E_{\text{line}} = \frac{1}{4\pi\epsilon_0} \frac{2|\lambda|}{r} \qquad (26.15)$$

FIGURE 26.13 shows the electric field vectors of an infinite line of positive charge. The vectors would point inward for a negative line of charge.

NOTE ▶ Unlike a point charge, for which the field decreases as $1/r^2$, the field of an infinitely long charged wire decreases more slowly—as only $1/r$. ◀

Although no real wire is infinitely long, the fact that the field of a point charge decreases inversely with the square of the distance means that the electric field at any point is determined primarily by the nearest charges. Consequently, the field of a realistic finite-length wire is well approximated by Equation 26.15, the field of an infinitely long line of charge, except at points near the end of the wire.

STOP TO THINK 26.3 Which of the following actions will increase the electric field strength at the position of the dot?

a. Make the rod longer without changing the charge.
b. Make the rod shorter without changing the charge.
c. Make the rod wider without changing the charge.
d. Make the rod narrower without changing the charge.
e. Add charge to the rod.
f. Remove charge from the rod.
g. Move the dot farther from the rod.
h. Move the dot closer to the rod.

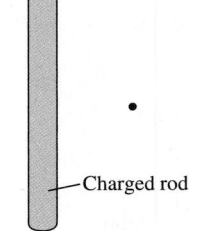

Charged rod

26.4 The Electric Fields of Rings, Disks, Planes, and Spheres

In this section we'll derive the electric fields for several important charge distributions.

EXAMPLE 26.4 **The electric field of a ring of charge**

A thin ring of radius R is uniformly charged with total charge Q. Find the electric field at a point on the axis of the ring (perpendicular to the ring).

MODEL Because the ring is thin, we'll assume the charge lies along a circle of radius R. You can think of this as a line of charge of length $2\pi R$ wrapped into a circle. The linear charge density along the ring is $\lambda = Q/2\pi R$.

VISUALIZE **FIGURE 26.14** shows the ring and illustrates the five steps of the problem-solving strategy. We've chosen a coordinate system in which the ring lies in the xy-plane and point P is on the z-axis. We've then divided the ring into N small segments of charge ΔQ, each of which can be modeled as a point charge. As you can see from the figure, the component of the field perpendicular to the axis cancels for two diametrically opposite segments. Thus we need to calculate only the z-component E_z.

SOLVE The z-component of the electric field due to segment i is

$$(E_i)_z = E_i \cos\theta_i = \frac{1}{4\pi\epsilon_0} \frac{\Delta Q}{r_i^2} \cos\theta_i$$

FIGURE 26.14 Calculating the on-axis electric field of a ring of charge.

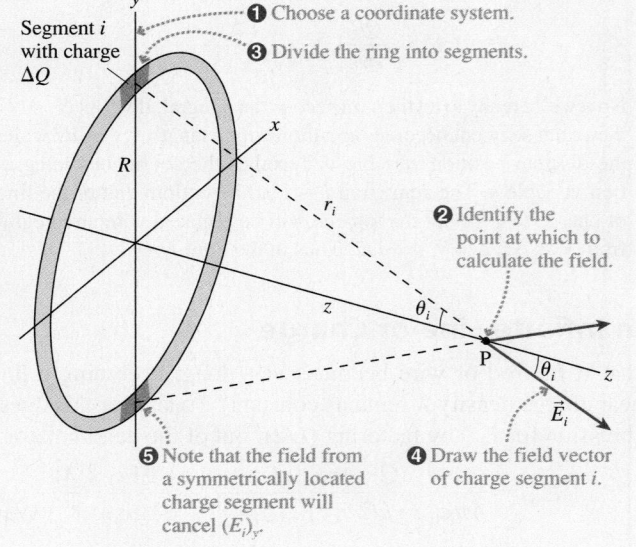

❶ Choose a coordinate system.

Segment i with charge ΔQ

❸ Divide the ring into segments.

❷ Identify the point at which to calculate the field.

❹ Draw the field vector of charge segment i.

❺ Note that the field from a symmetrically located charge segment will cancel $(E_i)_y$.

You can see from the figure that *every* segment of the ring, independent of i, has

$$r_i = \sqrt{z^2 + R^2}$$

$$\cos\theta_i = \frac{z}{r_i} = \frac{z}{\sqrt{z^2 + R^2}}$$

Consequently, the field of segment i is

$$(E_i)_z = \frac{1}{4\pi\epsilon_0} \frac{\Delta Q}{z^2 + R^2} \frac{z}{\sqrt{z^2 + R^2}} = \frac{1}{4\pi\epsilon_0} \frac{z}{(z^2 + R^2)^{3/2}} \Delta Q$$

The net electric field is found by summing $(E_i)_z$ due to all N segments:

$$E_z = \sum_{i=1}^{N} (E_i)_z = \frac{1}{4\pi\epsilon_0} \frac{z}{(z^2 + R^2)^{3/2}} \sum_{i=1}^{N} \Delta Q$$

We were able to bring all terms involving z to the front because z is a constant as far as the summation is concerned. Surprisingly, we don't need to convert the sum to an integral to complete this calculation. The sum of all the ΔQ around the ring is simply the ring's total charge, $\sum \Delta Q = Q$, hence the field on the axis is

$$(E_{\text{ring}})_z = \frac{1}{4\pi\epsilon_0} \frac{zQ}{(z^2 + R^2)^{3/2}}$$

This expression is valid for both positive and negative z (i.e., on either side of the ring) and for both positive and negative charge.

ASSESS It will be left as a homework problem to show that this result gives the expected limit when $z \gg R$.

FIGURE 26.15 shows two representations of the on-axis electric field of a positively charged ring. FIGURE 26.15a shows that the electric field vectors point away from the ring, increasing in length until reaching a maximum when $|z| \approx R$, then decreasing. The graph of $(E_{\text{ring}})_z$ in FIGURE 26.15b confirms that the field strength has a maximum on either side of the ring. Notice that the electric field at the center of the ring is zero, even though this point is surrounded by charge. You might want to spend a minute thinking about why this has to be the case.

A Disk of Charge

FIGURE 26.16 shows a disk of radius R that is uniformly charged with charge Q. This is a mathematical disk, with no thickness, and its surface charge density is

$$\eta = \frac{Q}{A} = \frac{Q}{\pi R^2} \tag{26.16}$$

We would like to calculate the on-axis electric field of this disk. Our problem-solving strategy tells us to divide a continuous charge into segments for which we already know how to find $\vec{E}$. Because we now know the on-axis electric field of a ring of charge, let's divide the disk into N very narrow rings of radius r and width Δr. One such ring, with radius r_i and charge ΔQ_i, is shown.

We need to be careful with notation. The R in Example 26.4 was the radius of the ring. Now we have many rings, and the radius of ring i is r_i. Similarly, Q was the charge on the ring. Now the charge on ring i is ΔQ_i, a small fraction of the total charge on the disk. With these changes, the electric field of ring i, with radius r_i, is

$$(E_i)_z = \frac{1}{4\pi\epsilon_0} \frac{z \Delta Q_i}{(z^2 + r_i^2)^{3/2}} \tag{26.17}$$

The on-axis electric field of the charged disk is the sum of the electric fields of all of the rings:

$$(E_{\text{disk}})_z = \sum_{i=1}^{N} (E_i)_z = \frac{z}{4\pi\epsilon_0} \sum_{i=1}^{N} \frac{\Delta Q_i}{(z^2 + r_i^2)^{3/2}} \tag{26.18}$$

The critical step, as always, is to relate ΔQ to a coordinate. Because we now have a surface, rather than a line, the charge in ring i is $\Delta Q = \eta \Delta A_i$, where ΔA_i is the area of ring i. We can find ΔA_i, as you've learned to do in calculus, by "unrolling" the ring to form a narrow rectangle of length $2\pi r_i$ and height Δr. Thus the area of ring i is $\Delta A_i = 2\pi r_i \Delta r$ and the charge is $\Delta Q_i = 2\pi \eta r_i \Delta r$. With this substitution, Equation 26.18 becomes

$$(E_{\text{disk}})_z = \frac{\eta z}{2\epsilon_0} \sum_{i=1}^{N} \frac{r_i \Delta r}{(z^2 + r_i^2)^{3/2}} \tag{26.19}$$

FIGURE 26.15 The on-axis electric field of a ring of charge.

(a)

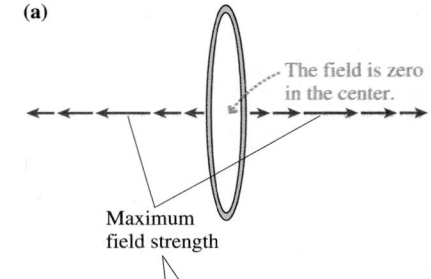

The field is zero in the center.

Maximum field strength

(b)

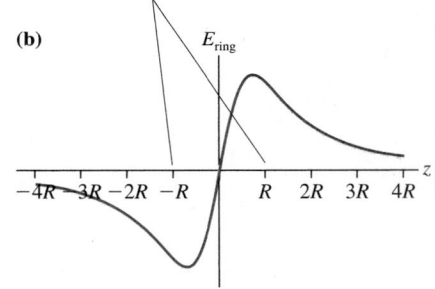

FIGURE 26.16 Calculating the on-axis field of a charged disk.

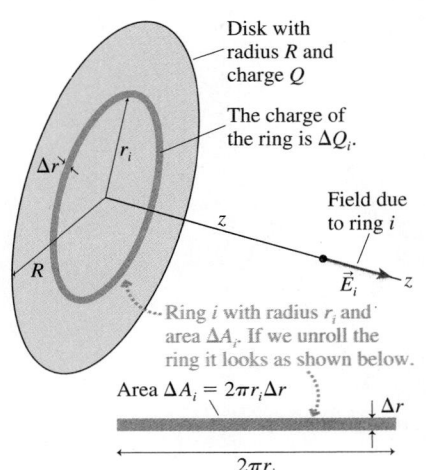

Disk with radius R and charge Q

The charge of the ring is ΔQ_i.

Field due to ring i

Ring i with radius r_i and area ΔA_i. If we unroll the ring it looks as shown below.

Area $\Delta A_i = 2\pi r_i \Delta r$

As $N \rightarrow \infty$, $\Delta r \rightarrow dr$ and the sum becomes an integral. Adding all the rings means integrating from $r = 0$ to $r = R$; thus

$$(E_{\text{disk}})_z = \frac{\eta z}{2\epsilon_0} \int_0^R \frac{r \, dr}{(z^2 + r^2)^{3/2}} \qquad (26.20)$$

All that remains is to carry out the integration. This is straightforward if we make the variable change $u = z^2 + r^2$. Then $du = 2r \, dr$ or, equivalently, $r \, dr = \frac{1}{2} du$. At the lower integration limit $r = 0$, our new variable is $u = z^2$. At the upper limit $r = R$, the new variable is $u = z^2 + R^2$.

NOTE ▶ When changing variables in a definite integral, you *must* also change the limits of integration. ◀

With this variable change the integral becomes

$$(E_{\text{disk}})_z = \frac{\eta z}{2\epsilon_0} \frac{1}{2} \int_{z^2}^{z^2+R^2} \frac{du}{u^{3/2}} = \frac{\eta z}{4\epsilon_0} \frac{-2}{u^{1/2}} \Big|_{z^2}^{z^2+R^2} = \frac{\eta z}{2\epsilon_0} \left[\frac{1}{z} - \frac{1}{\sqrt{z^2 + R^2}} \right] \qquad (26.21)$$

If we multiply through by z, the on-axis electric field of a charged disk with surface charge density $\eta = Q/\pi R^2$ is

$$(E_{\text{disk}})_z = \frac{\eta}{2\epsilon_0} \left[1 - \frac{z}{\sqrt{z^2 + R^2}} \right] \qquad (26.22)$$

NOTE ▶ This expression is valid only for $z > 0$. The field for $z < 0$ has the same magnitude but points in the opposite direction. ◀

It's a bit difficult see what Equation 26.22 is telling us, so let's compare it to what we already know. First, you can see that the quantity in square brackets is dimensionless. The surface charge density $\eta = Q/A$ has the same units as q/r^2, so $\eta/2\epsilon_0$ has the same units as $q/4\pi\epsilon_0 r^2$. This tells us that $\eta/2\epsilon_0$ really is an electric field.

Next, let's move very far away from the disk. At distance $z \gg R$, the disk appears to be a point charge Q in the distance and the field of the disk should approach that of a point charge. If we let $z \rightarrow \infty$ in Equation 26.22, so that $z^2 + R^2 \approx z^2$, we find $(E_{\text{disk}})_z \rightarrow 0$. This is true, but not quite what we wanted. We need to let z be very large in comparison to R, but not so large as to make E_{disk} vanish. That requires a little more care in taking the limit.

We can cast Equation 26.22 into a somewhat more useful form by factoring the z^2 out of the square root to give

$$(E_{\text{disk}})_z = \frac{\eta}{2\epsilon_0} \left[1 - \frac{1}{\sqrt{1 + R^2/z^2}} \right] \qquad (26.23)$$

Now $R^2/z^2 \ll 1$ if $z \gg R$, so the second term in the square brackets is of the form $(1 + x)^{-1/2}$ where $x \ll 1$. We can then use the *binomial approximation*

$$(1 + x)^n \approx 1 + nx \quad \text{if} \quad x \ll 1 \qquad \text{(binomial approximation)}$$

to simplify the expression in square brackets:

$$1 - \frac{1}{\sqrt{1 + R^2/z^2}} = 1 - (1 + R^2/z^2)^{-1/2} \approx 1 - \left(1 + \left(-\frac{1}{2} \right) \frac{R^2}{z^2} \right) = \frac{R^2}{2z^2} \qquad (26.24)$$

This is a good approximation when $z \gg R$. Substituting this approximation into Equation 26.23, we find that the electric field of the disk for $z \gg R$ is

$$(E_{\text{disk}})_z \approx \frac{\eta}{2\epsilon_0} \frac{R^2}{2z^2} = \frac{Q/\pi R^2}{4\epsilon_0} \frac{R^2}{z^2} = \frac{1}{4\pi\epsilon_0} \frac{Q}{z^2} \quad \text{if} \quad z \gg R \qquad (26.25)$$

This is, indeed, the field of a point charge Q, giving us confidence in Equation 26.22 for the on-axis electric field of a disk of charge.

NOTE ▶ The binomial approximation is an important tool for looking at the limiting cases of electric fields. ◀

EXAMPLE 26.5 **The electric field of a charged disk**

A 10-cm-diameter plastic disk is charged uniformly with an extra 10^{11} electrons. What is the electric field 1.0 mm above the surface at a point near the center?

MODEL Model the plastic disk as a uniformly charged disk. We are seeking the on-axis electric field. Because the charge is negative, the field will point *toward* the disk.

SOLVE The total charge on the plastic square is $Q = N(-e) = -1.60 \times 10^{-8}$ C. The surface charge density is

$$\eta = \frac{Q}{A} = \frac{Q}{\pi R^2} = \frac{-1.60 \times 10^{-8} \text{ C}}{\pi (0.050 \text{ m})^2} = -2.04 \times 10^{-6} \text{ C/m}^2$$

The electric field at $z = 0.0010$ m, given by Equation 26.23, is

$$E_z = \frac{\eta}{2\epsilon_0} \left[1 - \frac{1}{\sqrt{1 + R^2/z^2}} \right] = -1.1 \times 10^5 \text{ N/C}$$

The minus sign indicates that the field points *toward*, rather than away from, the disk. As a vector,

$$\vec{E} = (1.1 \times 10^5 \text{ N/C, toward the disk})$$

ASSESS The total charge, -16 nC, is typical of the amount of charge produced on a small plastic object by rubbing or friction. Thus 10^5 N/C is a typical electric field strength near an object that has been charged by rubbing.

A Plane of Charge

Many electronic devices use charged, flat surfaces—disks, squares, rectangles, and so on—to steer electrons along the proper paths. These charged surfaces are called **electrodes.** Although any real electrode is finite in extent, we can often model an electrode as an infinite **plane of charge.** As long as the distance z to the electrode is small in comparison to the distance to the edges, we can reasonably treat the edges *as if* they are infinitely far away.

The electric field of a plane of charge is found from the on-axis field of a charged disk by letting the radius $R \to \infty$. That is, a disk with infinite radius is an infinite plane. From Equation 26.22, we see that the electric field of a plane of charge with surface charge density η is:

$$E_{\text{plane}} = \frac{\eta}{2\epsilon_0} = \text{constant} \tag{26.26}$$

This is a simple result, but what does it tell us? First, the field strength is directly proportional to the charge density η: More charge, bigger field. Second, and more interesting, the field strength is the same at *all* points in space, independent of the distance z. The field strength 1000 m from the plane is the same as the field strength 1 mm from the plane.

How can this be? It seems that the field should get weaker as you move away from the plane of charge. But remember that we are dealing with an *infinite* plane of charge. What does it mean to be "close to" or "far from" an infinite object? For a disk of finite radius R, whether a point at distance z is "close to" or "far from" the disk is a comparison of z to R. If $z \ll R$, the point is close to the disk. If $z \gg R$, the point is far from the disk. But as $R \to \infty$, we have no *scale* for distinguishing near and far. In essence, *every* point in space is "close to" a disk of infinite radius.

No real plane is infinite in extent, but we can interpret Equation 26.26 as saying that the field of a surface of charge, regardless of its shape, is a constant $\eta/2\epsilon_0$ for those points whose distance z to the surface is much smaller than their distance to the edge. Eventually, when $z \gg R$, the charged surface will begin to look like a point charge Q and the field will have to decrease as $1/z^2$.

We do need to note that the derivation leading to Equation 26.26 considered only $z > 0$. For a positively charged plane, with $\eta > 0$, the electric field points *away from* the plane on both sides of the plane. This requires $E_z < 0$ ($\vec{E}$ pointing in the negative

FIGURE 26.17 Two views of the electric field of a plane of charge.

Perspective view

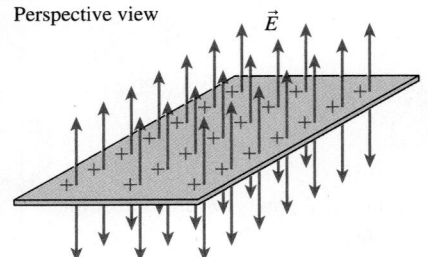

Edge view

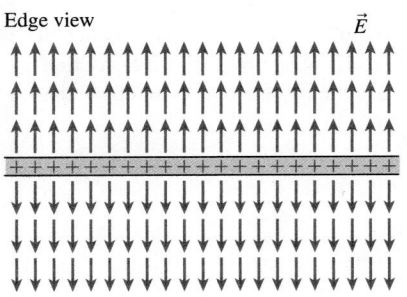

FIGURE 26.18 The electric field of a sphere of positive charge.

The electric field outside a sphere or spherical shell is the same as the field of a point charge Q at the center.

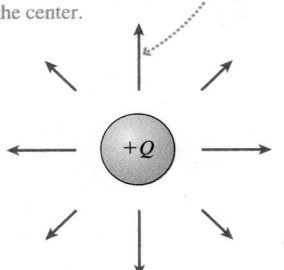

z-direction) on the side with $z < 0$. Thus a complete description of the electric field, valid for both sides of the plane and for either sign of η, is

$$(E_{\text{plane}})_z = \begin{cases} +\dfrac{\eta}{2\epsilon_0} & z > 0 \\[2mm] -\dfrac{\eta}{2\epsilon_0} & z < 0 \end{cases} \qquad (26.27)$$

FIGURE 26.17 shows two views of the electric field of a positively charged plane. All the arrows would be reversed for a negatively charged plane. It would have been very difficult to anticipate this result from Coulomb's law or from the electric field of a single point charge, but step by step we have been able to use the concept of the electric field to look at increasingly complex distributions of charge.

A Sphere of Charge

The one last charge distribution for which we need to know the electric field is a **sphere of charge.** This problem is analogous to wanting to know the gravitational field of a spherical planet or star. The procedure for calculating the field of a sphere of charge is the same as we used for lines and planes, but the integrations are significantly more difficult. We will skip the details of the calculations and, for now, simply assert the result without proof. In Chapter 27 we'll use an alternative procedure to find the field of a sphere of charge.

A sphere of charge Q and radius R, be it a uniformly charged sphere or just a spherical shell, has an electric field *outside* the sphere ($r \geq R$) that is exactly the same as that of a point charge Q located at the center of the sphere:

$$\vec{E}_{\text{sphere}} = \frac{Q}{4\pi\epsilon_0 r^2}\hat{r} \qquad \text{for } r \geq R \qquad (26.28)$$

This assertion is analogous to our earlier assertion that the gravitational force between stars and planets can be computed as if all the mass is at the center.

FIGURE 26.18 shows the electric field of a sphere of positive charge. The field of a negative sphere would point inward.

STOP TO THINK 26.4 Rank in order, from largest to smallest, the electric field strengths E_a to E_e at these five points near a plane of charge.

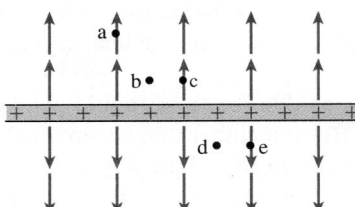

26.5 The Parallel-Plate Capacitor

FIGURE 26.19 shows two electrodes, one with charge $+Q$ and the other with $-Q$, placed face-to-face a distance d apart. This arrangement of two electrodes, charged equally but oppositely, is called a **parallel-plate capacitor.** Capacitors play important roles in many electric circuits. Our goal is to find the electric field both inside the capacitor (i.e., between the plates) and outside the capacitor.

NOTE ▶ The *net* charge of a capacitor is zero. Capacitors are charged by transferring electrons from one plate to the other. The plate that gains N electrons has charge $-Q = N(-e)$. The plate that loses electrons has charge $+Q$. ◀

FIGURE 26.19 A parallel-plate capacitor.

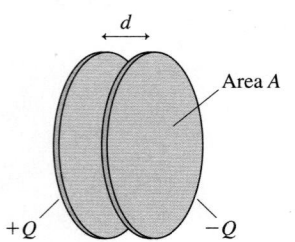

Let's begin with a qualitative investigation. FIGURE 26.20 is an enlarged view of the capacitor plates, seen from the side. Because opposite charges attract, all of the charge is on the *inner* surfaces of the two plates. Thus the inner surfaces of the plates can be modeled as two planes of charge with equal but opposite surface charge densities. As you can see from the figure, at all points in space the electric field $\vec{E}_+$ of the positive plate points *away from* the plane of positive charges. Similarly, the field $\vec{E}_-$ of the negative plate everywhere points *toward* the plane of negative charges.

NOTE ▶ You might think the right capacitor plate would somehow "block" the electric field created by the positive plate and prevent the presence of an $\vec{E}_+$ field to the right of the capacitor. To see that it doesn't, consider an analogous situation with gravity. The strength of gravity above a table is the same as its strength below it. Just as the table doesn't block the earth's gravitational field, intervening matter or charges do not alter or block an object's electric field. ◀

Inside the capacitor, $\vec{E}_+$ and $\vec{E}_-$ are parallel and of equal strength. Their superposition creates a net electric field inside the capacitor that points from the positive plate to the negative plate. Outside the capacitor, $\vec{E}_+$ and $\vec{E}_-$ point in opposite directions and, because the field of a plane of charge is independent of the distance from the plane, have equal magnitudes. Consequently, the fields $\vec{E}_+$ and $\vec{E}_-$ add to zero outside the capacitor plates.

We can calculate the fields between the capacitor plates from the field of an infinite charged plane. Between the electrodes, $\vec{E}_+$ is of magnitude $\eta/2\epsilon_0$ and points from the positive toward the negative side. The field $\vec{E}_-$ is *also* of magnitude $\eta/2\epsilon_0$ and *also* points from positive to negative. Thus the electric field inside the capacitor is

$$\vec{E}_{\text{capacitor}} = \vec{E}_+ + \vec{E}_- = \left(\frac{\eta}{\epsilon_0}, \text{from positive to negative}\right)$$

$$= \left(\frac{Q}{\epsilon_0 A}, \text{from positive to negative}\right)$$

(26.29)

where A is the surface area of each electrode. Outside the capacitor plates, where $\vec{E}_+$ and $\vec{E}_-$ have equal magnitudes but *opposite* directions, $\vec{E} = \vec{0}$.

FIGURE 26.21a shows the electric field of an ideal parallel-plate capacitor constructed from two infinite charged planes. Now, it's true that no real capacitor is infinite in extent, but the ideal parallel-plate capacitor is a very good approximation for all but the most precise calculations as long as the electrode separation d is much smaller than the electrodes' size. FIGURE 26.21b shows that the interior field of a real capacitor is virtually identical to that of an ideal capacitor but that the exterior field isn't quite zero. This weak field outside the capacitor is called the **fringe field**. We will keep things simple by always assuming the plates are very close together and using Equation 26.29 for the field inside a parallel-plate capacitor.

NOTE ▶ The shape of the electrodes—circular or square or any other shape—is not relevant as long as the electrodes are very close together. ◀

FIGURE 26.20 The electric fields inside and outside a parallel-plate capacitor.

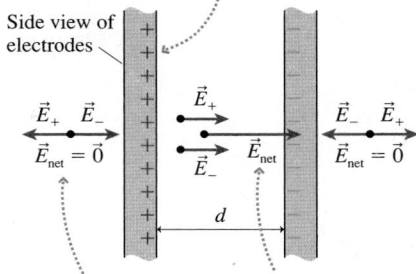

The capacitor's charge resides on the inner surfaces as planes of charge.

Outside the capacitor, $\vec{E}_+$ and $\vec{E}_-$ are opposite, so the net field is zero.   Inside the capacitor, $\vec{E}_+$ and $\vec{E}_-$ are parallel, so the net field is large.

FIGURE 26.21 The electric field of a capacitor.

(a) Ideal capacitor

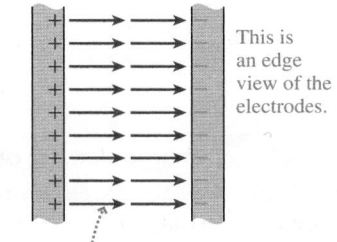

This is an edge view of the electrodes.

The field is uniform, pointing from the positive to the negative electrode.

(b) Real capacitor

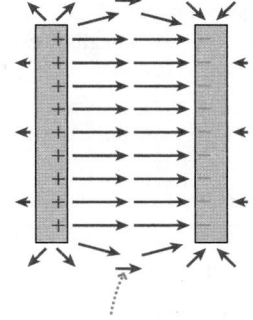

A weak fringe field extends outside the electrodes.

EXAMPLE 26.6 **The electric field inside a capacitor**

Two 1.0 cm × 2.0 cm rectangular electrodes are 1.0 mm apart. What charge must be placed on each electrode to create a uniform electric field of strength 2.0×10^6 N/C? How many electrons must be moved from one electrode to the other to accomplish this?

MODEL The electrodes can be modeled as a parallel-plate capacitor because the spacing between them is much smaller than their lateral dimensions.

SOLVE The electric field strength inside the capacitor is $E = Q/\epsilon_0 A$. Thus the charge to produce a field of strength E is

$$Q = (8.85 \times 10^{-12} \text{ C}^2/\text{N m}^2)(2.0 \times 10^{-4} \text{ m}^2)(2.0 \times 10^6 \text{ N/C})$$

$$= 3.5 \times 10^{-9} \text{ C} = 3.5 \text{ nC}$$

The positive plate must be charged to $+3.5$ nC and the negative plate to -3.5 nC. In practice, the plates are charged by using a

Continued

battery to move electrons from one plate to the other. The number of electrons in 3.5 nC is

$$N = \frac{Q}{e} = \frac{3.5 \times 10^{-9} \text{ C}}{1.60 \times 10^{-19} \text{ C/electron}} = 2.2 \times 10^{10} \text{ electrons}$$

Thus 2.2×10^{10} electrons are moved from one electrode to the other. Note that the capacitor *as a whole* has no net charge.

ASSESS The plate spacing does not enter the result. As long as the spacing is much smaller than the plate dimensions, as is true in this example, the field is independent of the spacing.

Uniform Electric Fields

FIGURE 26.22 A uniform electric field.

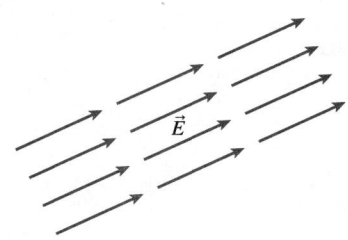

FIGURE 26.22 shows an electric field that is the *same*—in strength and direction—at every point in a region of space. This is called a **uniform electric field.** A uniform electric field is analogous to the uniform gravitational field near the surface of the earth. Uniform fields are of great practical significance because, as you will see in the next section, computing the trajectory of a charged particle moving in a uniform electric field is a straightforward process.

The easiest way to produce a uniform electric field is with a parallel-plate capacitor, as you can see in Figure 26.21a. Indeed, our interest in capacitors is due in large measure to the fact that the electric field is uniform. Many electric field problems refer to a uniform electric field. Such problems carry an implicit assumption that the action is taking place *inside* a parallel-plate capacitor.

EXAMPLE 26.7   **Charge density on a cell wall**

Example 25.7 noted that the electric field strength in the cell wall of a neuron is typically 1.0×10^7 N/C. This electric field is established because the outer surface of the cell wall is positive and the inner surface negative. What is a typical surface charge density on the surface of a cell wall?

MODEL Although cells are roughly spherical, the wall thickness is much less than the radius of the cell. Locally, at a point inside the cell wall, the curvature is negligible, so we can model the cell wall as a parallel-plate capacitor.

VISUALIZE FIGURE 26.23 shows a section of the cell wall. The charges are due to ions, not electrons, but that doesn't affect our analysis.

SOLVE The electric field strength inside a capacitor is $E = \eta/\epsilon_0$. The surface charge density needed to produce a known field is

$$\eta = \epsilon_0 E = (8.85 \times 10^{-12} \text{ C}^2/\text{N m}^2)(1.0 \times 10^7 \text{ N/C})$$

$$= 8.9 \times 10^{-5} \text{ C/m}^2$$

FIGURE 26.23 The electric field inside the cell wall is due to charges on the surfaces.

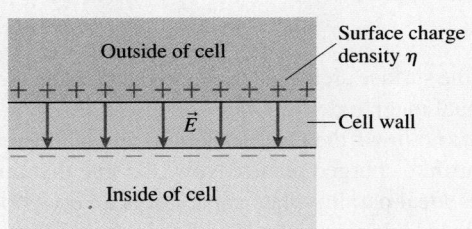

ASSESS The charge density may seem rather large, but cells are very small. A typical cell is $\approx 10 \ \mu\text{m}$ in diameter, with a surface area of $\approx 3 \times 10^{-10} \text{ m}^2$. At a surface charge density of $9 \times 10^{-5} \text{ C/m}^2$, the total charge on the outer surface of the cell is $\approx 3 \times 10^{-14} \text{ C}$, or $\approx 200,000$ ions.

STOP TO THINK 26.5   Rank in order, from largest to smallest, the forces F_a to F_e a proton would experience if placed at points a to e in this parallel-plate capacitor.

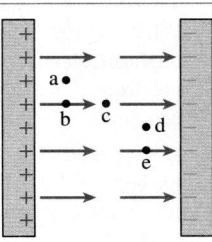

26.6 Motion of a Charged Particle in an Electric Field

Our motivation for introducing the concept of the electric field was to understand the long-range electric interaction of charges. We said that some charges, the *source charges,* create an electric field. Other charges then respond to that electric field. The first five sections of this chapter have focused on the electric field of the source charges. Now we turn our attention to the second half of the interaction.

FIGURE 26.24 shows a particle of charge q and mass m at a point where an electric field $\vec{E}$ has been produced by *other* charges, the source charges. The electric field exerts a force

$$\vec{F}_{\text{on } q} = q\vec{E}$$

on the charged particle. Notice that the force on a negatively charged particle is *opposite* in direction to the electric field vector. Signs are important!

FIGURE 26.24 The electric field exerts a force on a charged particle.

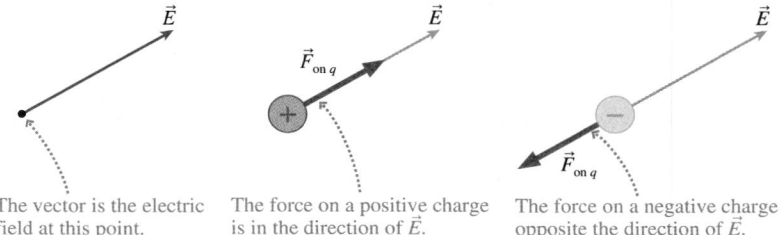

The vector is the electric field at this point.

The force on a positive charge is in the direction of $\vec{E}$.

The force on a negative charge is opposite the direction of $\vec{E}$.

If $\vec{F}_{\text{on } q}$ is the only force acting on q, it causes the charged particle to accelerate with

$$\vec{a} = \frac{\vec{F}_{\text{on } q}}{m} = \frac{q}{m}\vec{E} \qquad (26.30)$$

This acceleration is the *response* of the charged particle to the source charges that created the electric field. The ratio q/m is especially important for the dynamics of charged-particle motion. It is called the **charge-to-mass ratio.** Two *equal* charges, say a proton and a Na^+ ion, will experience *equal* forces $\vec{F} = q\vec{E}$ if placed at the same point in an electric field, but their accelerations will be *different* because they have different masses and thus different charge-to-mass ratios. Two particles with different charges and masses *but* with the same charge-to-mass ratio will undergo the same acceleration and follow the same trajectory.

Motion in a Uniform Field

The motion of a charged particle in a *uniform* electric field is especially important for its basic simplicity and because of its many valuable applications. A uniform field is *constant* at all points—constant in both magnitude and direction—within the region of space where the charged particle is moving. It follows, from Equation 26.30, that **a charged particle in a uniform electric field will move with constant acceleration.** The magnitude of the acceleration is

$$a = \frac{qE}{m} = \text{constant} \qquad (26.31)$$

where E is the electric field strength, and the direction of $\vec{a}$ is parallel or antiparallel to $\vec{E}$, depending on the sign of q.

"DNA fingerprints" are measured with the technique of *gel electrophoresis.* A solution of DNA fragments is placed in a well at one end of a plate covered with gel. The fragments are negatively charged when in solution, and they begin to migrate through the gel when a uniform electric field is established parallel to the surface of the plate. Because the gel exerts a drag force, the fragments move at a terminal speed inversely proportional to their size. Thus gel electrophoresis sorts the DNA fragments by size, and fluorescent markers allow the results to be seen.

Identifying the motion of a charged particle in a uniform field as being one of constant acceleration brings into play all the kinematic machinery that we developed in Chapters 2 and 4 for constant-acceleration motion. The basic trajectory of a charged particle in a uniform field is a *parabola,* analogous to the projectile motion of a mass in the near-earth uniform gravitational field. In the special case of a charged particle moving parallel to the electric field vectors, the motion is one-dimensional, analogous to the one-dimensional vertical motion of a mass tossed straight up or falling straight down.

NOTE ▶ The gravitational acceleration $\vec{a}_{grav}$ always points straight down. The electric field acceleration $\vec{a}_{elec}$ can point in *any* direction. You must determine the electric field $\vec{E}$ in order to learn the direction of $\vec{a}$. ◀

EXAMPLE 26.8 | An electron moving across a capacitor

Two 6.0-cm-diameter electrodes are spaced 5.0 mm apart. They are charged by transferring 1.0×10^{11} electrons from one electrode to the other. An electron is released from rest at the surface of the negative electrode. How long does it take the electron to cross to the positive electrode? What is its speed as it collides with the positive electrode? Assume the space between the electrodes is a vacuum.

MODEL The electrodes form a parallel-plate capacitor. The electric field inside a parallel-plate capacitor is a uniform field, so the electron will have constant acceleration.

VISUALIZE FIGURE 26.25 shows an edge view of the capacitor and the electron. The force on the negative electron is *opposite* the electric field, so the electron is repelled by the negative electrode as it accelerates across the gap of width d.

FIGURE 26.25 An electron accelerates across a capacitor (plate separation exaggerated).

The capacitor was charged by transferring 10^{11} electrons from the right electrode to the left electrode.

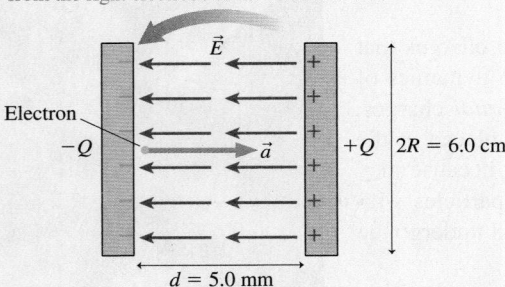

SOLVE The electrodes are not point charges, so we cannot use Coulomb's law to find the force on the electron. Instead, we must analyze the electron's motion in terms of the electric field inside the capacitor. The field is the agent that exerts the force on the electron, causing it to accelerate. The electric field strength inside a parallel-plate capacitor with charge $Q = Ne$ is

$$E = \frac{\eta}{\epsilon_0} = \frac{Q}{\epsilon_0 A} = \frac{Ne}{\epsilon_0 \pi R^2} = 639,000 \text{ N/C}$$

The electron's acceleration in this field is

$$a = \frac{eE}{m} = 1.1 \times 10^{17} \text{ m/s}^2$$

where we used the electron mass $m = 9.11 \times 10^{-31}$ kg. This is an enormous acceleration compared to accelerations we're familiar with for macroscopic objects. We can use one-dimensional kinematics, with $x_i = 0$ and $v_i = 0$, to find the time required for the electron to cross the capacitor:

$$x_f = d = \frac{1}{2}a(\Delta t)^2$$

$$\Delta t = \sqrt{\frac{2d}{a}} = 3.0 \times 10^{-10} \text{ s} = 0.30 \text{ ns}$$

The electron's speed as it reaches the positive electrode is

$$v = a\,\Delta t = 3.3 \times 10^7 \text{ m/s}$$

ASSESS We used e rather than $-e$ to find the acceleration because we already knew the direction; we needed only the magnitude. The electron's speed, after traveling a mere 5 mm, is approximately 10% the speed of light.

Parallel electrodes such as those in Example 26.8 are often used to accelerate charged particles. If the positive plate has a small hole in the center, a *beam* of electrons will pass through the hole, after accelerating across the capacitor gap, and emerge with a speed of 3.3×10^7 m/s. This is the basic idea of the *electron gun* used until quite recently in *cathode-ray tube* (CRT) devices such as televisions and computer display terminals. (A negatively charged electrode is called a *cathode,* so the physicists who first learned to produce electron beams in the late 19th century called them *cathode rays.*) The following example shows that parallel electrodes can also be used to deflect charged particles sideways.

EXAMPLE 26.9 **Deflecting an electron beam**

An electron gun creates a beam of electrons moving horizontally with a speed of 3.3×10^7 m/s. The electrons enter a 2.0-cm-long gap between two parallel electrodes where the electric field is $\vec{E} = (5.0 \times 10^4$ N/C, down). In which direction, and by what angle, is the electron beam deflected by these electrodes?

MODEL The electric field between the electrodes is uniform. Assume that the electric field outside the electrodes is zero.

VISUALIZE **FIGURE 26.26** shows an electron moving through the electric field. The electric field points down, so the force on the (negative) electrons is upward. The electrons will follow a parabolic trajectory, analogous to that of a ball thrown horizontally, except that the electrons "fall up" rather than down.

FIGURE 26.26 The deflection of an electron beam in a uniform electric field.

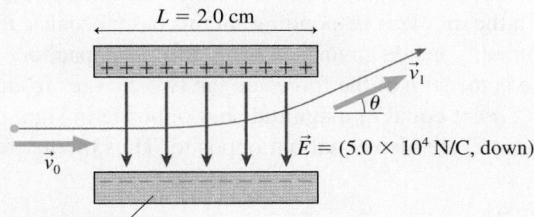

Deflection plates

SOLVE This is a two-dimensional motion problem. The electron enters the capacitor with velocity *vector* $\vec{v}_0 = v_{0x}\hat{\imath} = 3.3 \times 10^7 \hat{\imath}$ m/s and leaves with velocity $\vec{v}_1 = v_{1x}\hat{\imath} + v_{1y}\hat{\jmath}$. The electron's angle of travel upon leaving the electric field is

$$\theta = \tan^{-1}\left(\frac{v_{1y}}{v_{1x}}\right)$$

This is the *deflection angle*. To find θ we must compute the final velocity vector $\vec{v}_1$.

There is no horizontal force on the electron, so $v_{1x} = v_{0x} = 3.3 \times 10^7$ m/s. The electron's upward acceleration has magnitude

$$a = \frac{eE}{m} = \frac{(1.60 \times 10^{-19} \text{ C})(5.0 \times 10^4 \text{ N/C})}{9.11 \times 10^{-31} \text{ kg}}$$

$$= 8.78 \times 10^{15} \text{ m/s}^2$$

We can use the fact that the horizontal velocity is constant to determine the time interval Δt needed to travel length 2.0 cm:

$$\Delta t = \frac{L}{v_{0x}} = \frac{0.020 \text{ m}}{3.3 \times 10^7 \text{ m/s}} = 6.06 \times 10^{-10} \text{ s}$$

Vertical acceleration will occur during this time interval, resulting in a final vertical velocity

$$v_{1y} = v_{0y} + a \Delta t = 5.3 \times 10^6 \text{ m/s}$$

The electron's velocity as it leaves the capacitor is thus

$$\vec{v}_1 = (3.3 \times 10^7 \hat{\imath} + 5.3 \times 10^6 \hat{\jmath}) \text{ m/s}$$

and the deflection angle θ is

$$\theta = \tan^{-1}\left(\frac{v_{1y}}{v_{1x}}\right) = 9.1°$$

ASSESS We know that the electron beam in a cathode-ray tube can be deflected enough to cover the screen, so a deflection angle of 9° seems reasonable. Our neglect of the gravitational force is seen to be justified because the acceleration of the electrons is enormous in comparison to the free-fall acceleration g.

Example 26.9 demonstrates how an electron beam is steered to a point on the screen of a cathode-ray tube. First, a high-speed electron beam is created by an electron gun like that of Example 26.8. The beam then passes first through a set of *vertical deflection plates,* as in Example 26.9, then through a second set of *horizontal deflection plates.* After leaving the deflection plates, it travels in a straight line (through vacuum, to eliminate collisions with air molecules) to the screen of the CRT, where it strikes a phosphor coating on the inside surface and makes a dot of light. Properly choosing the electric fields within the deflection plates steers the electron beam to any point on the screen.

STOP TO THINK 26.6 Which electric field is responsible for the proton's trajectory?

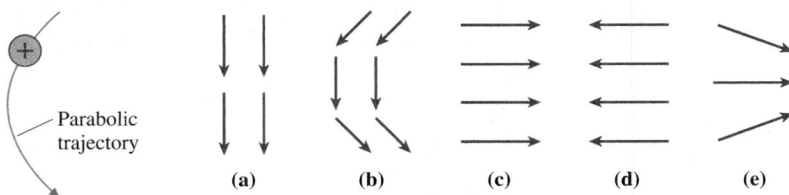

26.7 Motion of a Dipole in an Electric Field

Let us conclude this chapter by returning to one of the more striking puzzles we faced when making the observations at the beginning of Chapter 25. There you found that charged objects of *either* sign exert forces on neutral objects, such as when a comb used to brush your hair picks up pieces of paper. Our qualitative understanding of the *polarization force* was that it required two steps:

■ The charge polarizes the neutral object, creating an induced electric dipole.
■ The charge then exerts an attractive force on the near end of the dipole that is slightly stronger than the repulsive force on the far end.

We are now in a position to make that understanding more quantitative.

Dipoles in a Uniform Field

FIGURE 26.27a shows an electric dipole in a *uniform* external electric field $\vec{E}$ that has been created by source charges we do not see. That is, $\vec{E}$ is *not* the field of the dipole but, instead, is a field to which the dipole is responding. In this case, because the field is uniform, the dipole is presumably inside an unseen parallel-plate capacitor.

The net force on the dipole is the sum of the forces on the two charges forming the dipole. Because the charges $\pm q$ are equal in magnitude but opposite in sign, the two forces $\vec{F}_+ = +q\vec{E}$ and $\vec{F}_- = -q\vec{E}$, are also equal but opposite. Thus the net force on the dipole is

$$\vec{F}_{net} = \vec{F}_+ + \vec{F}_- = \vec{0} \tag{26.32}$$

There is no net force on a dipole in a uniform electric field.

There may be no net force, but the electric field *does* affect the dipole. Because the two forces in Figure 26.27a are in opposite directions but not aligned with each other, the electric field exerts a *torque* on the dipole and causes the dipole to *rotate*.

The torque causes the dipole to rotate until it is aligned with the electric field, as shown in **FIGURE 26.27b**. In this position, the dipole experiences not only no net force but also no torque. Thus Figure 26.27b represents the *equilibrium position* for a dipole in a uniform electric field. Notice that the positive end of the dipole is in the direction in which $\vec{E}$ points.

FIGURE 26.28 shows a sample of permanent dipoles, such as water molecules, in an external electric field. All the dipoles rotate until they are aligned with the electric field. This is the mechanism by which the sample becomes *polarized*. Once the dipoles are aligned, there is an excess of positive charge at one end of the sample and an excess of negative charge at the other end. The excess charges at the ends of the sample are the basis of the polarization forces we discussed in Section 25.3.

It's not hard to calculate the torque. Recall from Chapter 12 that the magnitude of a torque is the product of the force and the moment arm. **FIGURE 26.29** shows that there are two forces of the same magnitude ($F_+ = F_- = qE$), each with the same moment arm ($d = \frac{1}{2}s\sin\theta$). Thus the torque on the dipole is

$$\tau = 2 \times dF_+ = 2(\tfrac{1}{2}s\sin\theta)(qE) = pE\sin\theta \tag{26.33}$$

where $p = qs$ was our definition of the dipole moment. The torque is zero when the dipole is aligned with the field, making $\theta = 0$.

Recall from Chapter 12 that the torque can be written in a compact mathematical form as the cross product between two vectors. The terms p and E in Equation 26.33 are the magnitudes of vectors, and θ is the angle between them. Thus in vector notation, the torque exerted on a dipole moment $\vec{p}$ by an electric field $\vec{E}$ is

$$\vec{\tau} = \vec{p} \times \vec{E} \tag{26.34}$$

The torque is greatest when $\vec{p}$ is perpendicular to $\vec{E}$, zero when $\vec{p}$ is aligned with or opposite to $\vec{E}$.

FIGURE 26.27 A dipole in a uniform electric field.

(a)

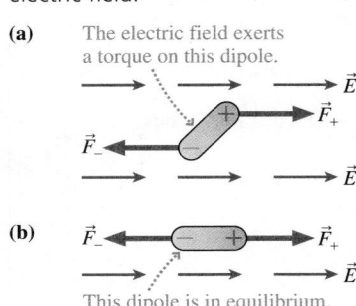

The electric field exerts a torque on this dipole.

(b)

This dipole is in equilibrium.

FIGURE 26.28 A sample of permanent dipoles is *polarized* in an electric field.

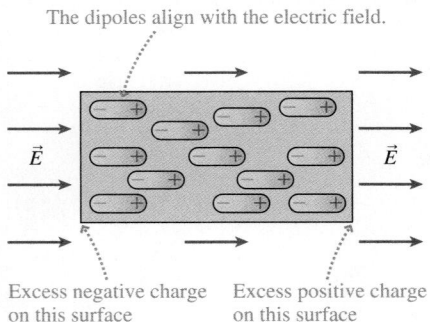

The dipoles align with the electric field.

Excess negative charge on this surface   Excess positive charge on this surface

FIGURE 26.29 The torque on a dipole.

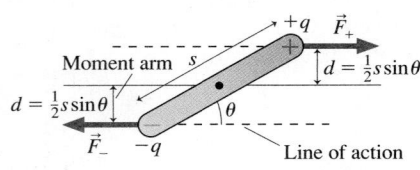

Moment arm s

$d = \frac{1}{2}s\sin\theta$

$d = \frac{1}{2}s\sin\theta$

Line of action

In terms of vectors, $\vec{\tau} = \vec{p} \times \vec{E}$.

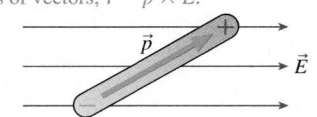

EXAMPLE 26.10   **The angular acceleration of a dipole dumbbell**

Two 1.0 g balls are connected by a 2.0-cm-long insulating rod of negligible mass. One ball has a charge of $+10$ nC, the other a charge of -10 nC. The rod is held in a 1.0×10^4 N/C uniform electric field at an angle of $30°$ with respect to the field, then released. What is its initial angular acceleration?

MODEL The two oppositely charged balls form an electric dipole. The electric field exerts a torque on the dipole, causing an angular acceleration.

VISUALIZE **FIGURE 26.30** shows the dipole in the electric field.

FIGURE 26.30 The dipole of Example 26.10.

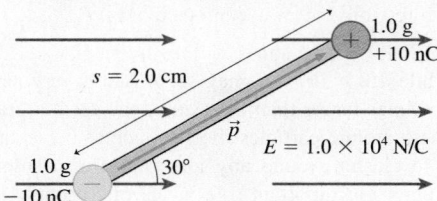

SOLVE The dipole moment is $p = qs = (1.0 \times 10^{-8}$ C$) \times (0.020$ m$) = 2.0 \times 10^{-10}$ C m. The torque exerted on the dipole moment by the electric field is

$$\tau = pE \sin\theta = (2.0 \times 10^{-10} \text{ C m})(1.0 \times 10^4 \text{ N/C}) \sin 30°$$
$$= 1.0 \times 10^{-6} \text{ N m}$$

You learned in Chapter 12 that a torque causes an angular acceleration $\alpha = \tau/I$, where I is the moment of inertia. The dipole rotates about its center of mass, which is at the center of the rod, so the moment of inertia is

$$I = m_1 r_1^2 + m_2 r_2^2 = 2m\left(\frac{1}{2}s\right)^2 = \frac{1}{2}ms^2 = 2.0 \times 10^{-7} \text{ kg m}^2$$

Thus the rod's angular acceleration is

$$\alpha = \frac{\tau}{I} = \frac{1.0 \times 10^{-6} \text{ N m}}{2.0 \times 10^{-7} \text{ kg m}^2} = 5.0 \text{ rad/s}^2$$

ASSESS This value of α is the initial angular acceleration, when the rod is first released. The torque and the angular acceleration will decrease as the rod rotates toward alignment with $\vec{E}$.

Dipoles in a Nonuniform Field

Suppose that a dipole is placed in a nonuniform electric field, one in which the field strength changes with position. For example, **FIGURE 26.31** shows a dipole in the non-uniform field of a point charge. The first response of the dipole is to rotate until it is aligned with the field, with the dipole's positive end pointing in the same direction as the field. Now, however, there is a slight difference between the forces acting on the two ends of the dipole. This difference occurs because the electric field, which depends on the distance from the point charge, is stronger at the end of the dipole nearest the charge. This causes a net force to be exerted on the dipole.

Which way does the force point? **FIGURE 26.31a** shows a positive point charge. Once the dipole is aligned, the leftward attractive force on its negative end is slightly stronger than the rightward repulsive force on its positive end. This causes a net force to the *left,* toward the point charge. The dipole in **FIGURE 26.31b** aligns in the opposite orientation in the field of a negative point charge, but the net force is still to the left.

As you can see, **the net force on a dipole is toward the direction of the strongest field.** Because any finite-size charged object, such as a charged rod or a charged disk, has a field strength that increases as you get closer to the object, we can conclude that **a dipole will experience a net force toward any charged object.**

FIGURE 26.31 An aligned dipole is drawn toward a point charge.

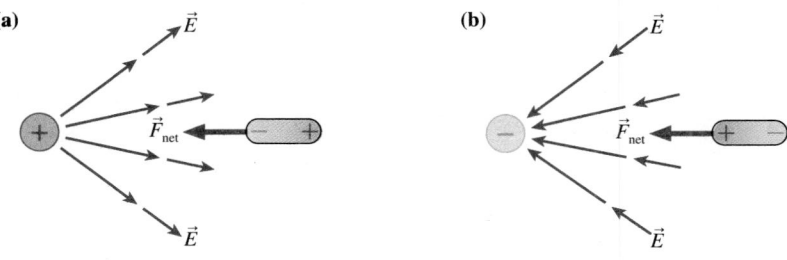

EXAMPLE 26.11   **The force on a water molecule**

The water molecule H_2O has a permanent dipole moment of magnitude 6.2×10^{-30} C m. A water molecule is located 10 nm from a Na^+ ion in a saltwater solution. What force does the ion exert on the water molecule?

VISUALIZE **FIGURE 26.32** shows the ion and the dipole. The forces are an action/reaction pair.

FIGURE 26.32 The interaction between an ion and a permanent dipole.

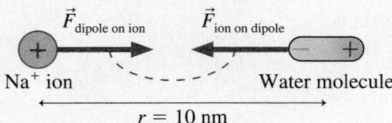

Na$^+$ ion   Water molecule

$r = 10$ nm

SOLVE A Na^+ ion has charge $q = +e$. The electric field of the ion aligns the water's dipole moment and exerts a net force on it. We could calculate the net force on the dipole as the small difference between the attractive force on its negative end and the repulsive force on its positive end. Alternatively, we know from Newton's third law that the force $\vec{F}_{\text{dipole on ion}}$ has the same mag-

nitude as the force $\vec{F}_{\text{ion on dipole}}$ that we are seeking. We calculated the on-axis field of a dipole in Section 26.2. An ion of charge $q = e$ will experience a force of magnitude $F = qE_{\text{dipole}} = eE_{\text{dipole}}$ when placed in that field. The dipole's electric field, which we found in Equation 26.11, is

$$E_{\text{dipole}} = \frac{1}{4\pi\epsilon_0}\frac{2p}{r^3}$$

The force on the ion at distance $r = 1.0 \times 10^{-8}$ m is

$$F_{\text{dipole on ion}} = eE_{\text{dipole}} = \frac{1}{4\pi\epsilon_0}\frac{2ep}{r^3} = 1.8 \times 10^{-14}\,\text{N}$$

Thus the force on the water molecule is $F_{\text{ion on dipole}} = 1.8 \times 10^{-14}$ N.

ASSESS While 1.8×10^{-14} N may seem like a very small force, it is $\approx 10^{11}$ times larger than the size of the earth's gravitational force on these atomic particles. Forces such as these cause water molecules to cluster around any ions that are in solution. This clustering plays an important role in the microscopic physics of solutions studied in chemistry and biochemistry.

An orbiting proton

In a vacuum chamber, a proton orbits a 1.0-cm-diameter metal ball 1.0 mm above the surface with a period of 1.0 μs. What is the charge on the ball?

MODEL Model the ball as a charged sphere. The electric field of a charged sphere is the same as that of a point charge at the center, so the radius of the ball is irrelevant. Assume that the gravitational force on the proton is extremely small compared to the electric force and can be neglected.

VISUALIZE **FIGURE 26.33** shows the orbit and the force on the proton.

FIGURE 26.33 An orbiting proton.

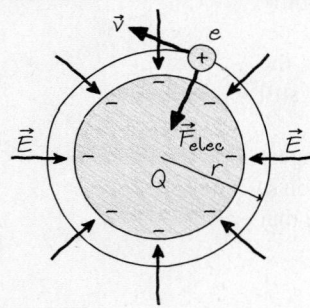

SOLVE The ball must be negative, with an inward electric field exerting an inward electric force on the positive proton. This is

exactly the necessary condition for uniform circular motion. Recall from Chapter 8 that Newton's second law for uniform circular motion is $(F_{\text{net}})_r = mv^2/r$. Here the only radial force has magnitude $F_{\text{elec}} = eE$, so the proton will move in a circular orbit if

$$eE = \frac{mv^2}{r}$$

The electric field strength of a sphere of charge Q at distance r is $E = Q/4\pi\epsilon_0 r^2$. From Chapter 4, orbital speed and period are related by $v = \text{circumference/period} = 2\pi r/T$. With these substitutions, Newton's second law becomes

$$\frac{eQ}{4\pi\epsilon_0 r^2} = \frac{4\pi^2 m}{T^2}r$$

Solving for Q, we find

$$Q = \frac{16\pi^3\epsilon_0 m r^3}{eT^2} = 9.9 \times 10^{-12}\,\text{C}$$

where we used $r = 6.0$ mm as the radius of the proton's orbit. Q is the *magnitude* of the charge on the ball. Including the sign, we have

$$Q_{\text{ball}} = -9.9 \times 10^{-12}\,\text{C}$$

ASSESS This is not a lot of charge, but it shouldn't take much charge to affect the motion of something as light as a proton.

SUMMARY

The goal of Chapter 26 has been to learn how to calculate and use the electric field.

General Principles

Sources of $\vec{E}$

Electric fields are created by charges.

Two major tools for calculating $\vec{E}$ are

- The field of a point charge:

$$\vec{E} = \frac{1}{4\pi\epsilon_0} \frac{q}{r^2} \hat{r}$$

- The principle of superposition

Multiple point charges
Use superposition: $\vec{E} = \vec{E}_1 + \vec{E}_2 + \vec{E}_3 + \cdots$

Continuous distribution of charge

- Divide the charge into segments ΔQ for which you already know the field.
- Find the field of each ΔQ.
- Find $\vec{E}$ by summing the fields of all ΔQ.

The summation usually becomes an integral. A critical step is replacing ΔQ with an expression involving a **charge density** (λ or η) and an integration coordinate.

Consequences of $\vec{E}$

The electric field exerts a force on a charged particle:

$$\vec{F} = q\vec{E}$$

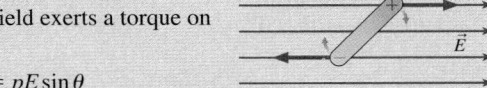

The force causes acceleration:

$$\vec{a} = (q/m)\vec{E}$$

Trajectories of charged particles are calculated with kinematics.

The electric field exerts a torque on a dipole:

$$\tau = pE\sin\theta$$

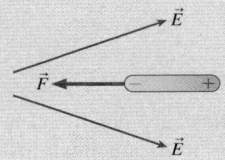

The torque tends to align the dipoles with the field.

In a nonuniform electric field, a dipole has a net force in the direction of increasing field strength.

Applications

Electric dipole

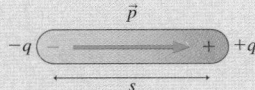

The electric dipole moment is

$\vec{p} = (qs,$ from negative to positive$)$

Field on axis: $\vec{E} = \dfrac{1}{4\pi\epsilon_0} \dfrac{2\vec{p}}{r^3}$

Field in bisecting plane: $\vec{E} = -\dfrac{1}{4\pi\epsilon_0} \dfrac{\vec{p}}{r^3}$

Infinite line of charge with linear charge density λ

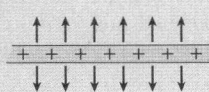

$\vec{E} = \left(\dfrac{1}{4\pi\epsilon_0} \dfrac{2\lambda}{r}, \text{perpendicular to line} \right)$

Infinite plane of charge with surface charge density η

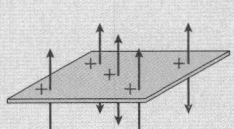

$\vec{E} = \left(\dfrac{\eta}{2\epsilon_0}, \text{perpendicular to plane} \right)$

Sphere of charge

Same as a point charge Q for $r > R$

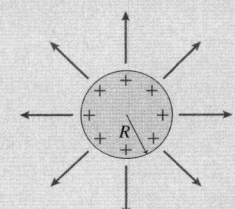

Parallel-plate capacitor

The electric field inside an ideal capacitor is a **uniform electric field:**

$$\vec{E} = \left(\frac{\eta}{\epsilon_0}, \text{from positive to negative} \right)$$

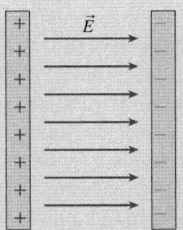

A real capacitor has a weak **fringe field** around it.

Terms and Notation

dipole moment, $\vec{p}$	uniformly charged	plane of charge	fringe field
electric field line	line of charge	sphere of charge	uniform electric field
linear charge density, λ	electrode	parallel-plate capacitor	charge-to-mass ratio, q/m
surface charge density, η			

CONCEPTUAL QUESTIONS

1. You've been assigned the task of determining the magnitude and direction of the electric field at a point in space. Give a step-by-step procedure of how you will do so. List any objects you will use, any measurements you will make, and any calculations you will need to perform. Make sure that your measurements do not disturb the charges that are creating the field.

2. Reproduce FIGURE Q26.2 on your paper. For each part, draw a dot or dots on the figure to show any position or positions (other than infinity) where $\vec{E} = \vec{0}$.

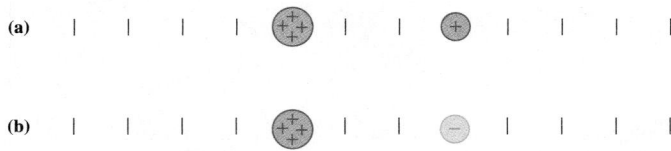

(a)

(b)

FIGURE Q26.2

3. Rank in order, from largest to smallest, the electric field strengths E_1 to E_4 at points 1 to 4 in FIGURE Q26.3. Explain.

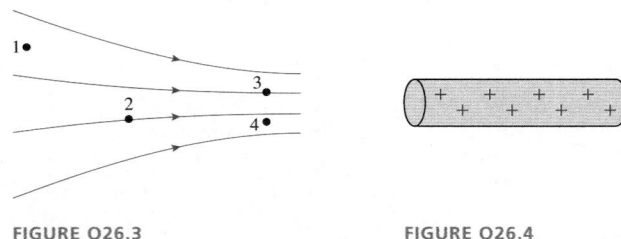

FIGURE Q26.3   **FIGURE Q26.4**

4. A small segment of wire in FIGURE Q26.4 contains 10 nC of charge.
   a. The segment is shrunk to one-third of its original length. What is the ratio λ_f/λ_i, where λ_i and λ_f are the initial and final linear charge densities?
   b. A proton is very far from the wire. What is the ratio F_f/F_i of the electric force on the proton after the segment is shrunk to the force before the segment was shrunk?
   c. Suppose the original segment of wire is stretched to 10 times its original length. How much charge must be *added* to the wire to keep the linear charge density unchanged?

5. An electron experiences a force of magnitude F when it is 1 cm from a very long, charged wire with linear charge density λ. If the charge density is doubled, at what distance from the wire will a proton experience a force of the same magnitude F?

6. FIGURE Q26.6 shows a hollow soda straw that has been uniformly charged with positive charge. What is the electric field at the center (inside) of the straw? Explain.

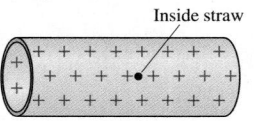

Inside straw

FIGURE Q26.6

7. The irregularly shaped area of charge in FIGURE Q26.7 has surface charge density η_i. Each dimension (x and y) of the area is reduced by a factor of 3.163.
   a. What is the ratio η_f/η_i, where η_f is the final surface charge density?
   b. An electron is very far from the area. What is the ratio F_f/F_i of the electric force on the electron after the area is reduced to the force before the area was reduced?

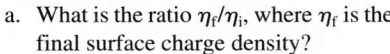

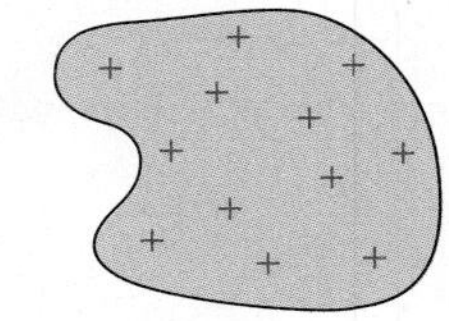

FIGURE Q26.7

8. A circular disk has surface charge density 8 nC/cm². What will the surface charge density be if the radius of the disk is doubled?

9. A sphere of radius R has charge Q. The electric field strength at distance $r > R$ is E_i. What is the ratio E_f/E_i of the final to initial electric field strengths if (a) Q is halved, (b) R is halved, and (c) r is halved (but is still $> R$)? Each part changes only one quantity; the other quantities have their initial values.

10. The ball in FIGURE Q26.10 is suspended from a large, uniformly charged positive plate. It swings with period T. If the ball is discharged, will the period increase, decrease, or stay the same? Explain.

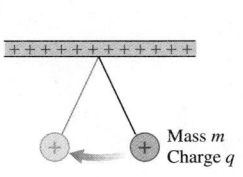

FIGURE Q26.10

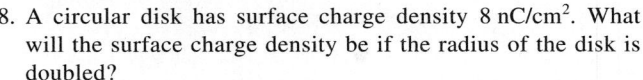

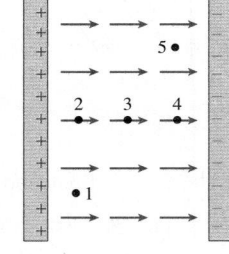

FIGURE Q26.11

11. Rank in order, from largest to smallest, the electric field strengths E_1 to E_5 at the five points in FIGURE Q26.11. Explain.

12. A parallel-plate capacitor consists of two square plates, size $L \times L$, separated by distance d. The plates are given charge $\pm Q$. What is the ratio E_f/E_i of the final to initial electric field strengths if (a) Q is doubled, (b) L is doubled, and (c) d is doubled? Each part changes only one quantity; the other quantities have their initial values.

13. A small object is released at point 3 in the center of the capacitor in FIGURE Q26.11. For each situation, does the object move to the right, to the left, or remain in place? If it moves, does it accelerate or move at constant speed?
   a. A positive object is released from rest.
   b. A neutral but polarizable object is released from rest.
   c. A negative object is released from rest.

14. A proton and an electron are released from rest in the center of a capacitor.
    a. Is the force ratio F_p/F_e greater than 1, less than 1, or equal to 1? Explain.
    b. Is the acceleration ratio a_p/a_e greater than 1, less than 1, or equal to 1? Explain.

15. Three charges are placed at the corners of the triangle in FIGURE Q26.15. The ++ charge has twice the quantity of charge of the two − charges; the net charge is zero. Is the triangle in equilibrium? If so, explain why. If not, draw the equilibrium orientation.

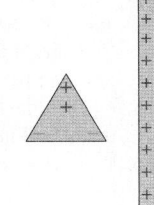

FIGURE Q26.15

EXERCISES AND PROBLEMS

Problems labeled ▨ integrate material from earlier chapters.

Exercises

Section 26.2 The Electric Field of Multiple Point Charges

1. ‖ What are the strength and direction of the electric field at the position indicated by the dot in FIGURE EX26.1? Specify the direction as an angle above or below horizontal.

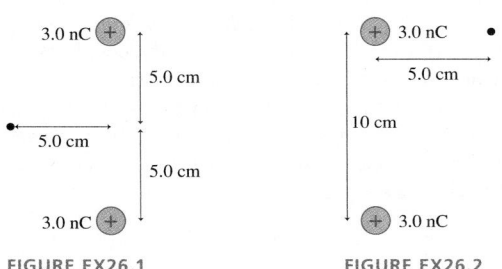

FIGURE EX26.1   FIGURE EX26.2

2. ‖ What are the strength and direction of the electric field at the position indicated by the dot in FIGURE EX26.2? Specify the direction as an angle above or below horizontal.

3. ‖ What are the strength and direction of the electric field at the position indicated by the dot in FIGURE EX26.3? Specify the direction as an angle above or below horizontal.

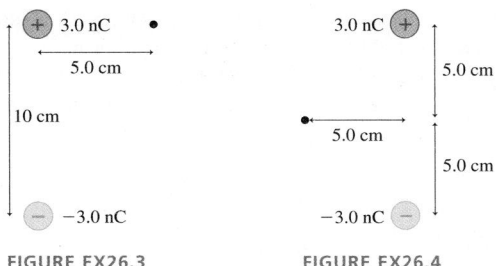

FIGURE EX26.3   FIGURE EX26.4

4. ‖ What are the strength and direction of the electric field at the position indicated by the dot in FIGURE EX26.4? Specify the direction as an angle above or below horizontal.

5. ‖ An electric dipole is formed from ± 1.0 nC charges spaced 2.0 mm apart. The dipole is at the origin, oriented along the x-axis. What is the electric field strength at the points (a) $(x, y) = $ (10 cm, 0 cm) and (b) $(x, y) = (0$ cm, 10 cm)?

6. ‖ An electric dipole is formed from two charges, $\pm q$, spaced 1.0 cm apart. The dipole is at the origin, oriented along the y-axis.

The electric field strength at the point $(x, y) = (0$ cm, 10 cm) is 360 N/C.
a. What is the charge q? Give your answer in nC.
b. What is the electric field strength at the point $(x, y) = $ (10 cm, 0 cm)?

Section 26.3 The Electric Field of a Continuous Charge Distribution

7. | The electric field strength 10.0 cm from a very long charged wire is 2000 N/C. What is the electric field strength 5.0 cm from the wire?

8. ‖ A 10-cm-long thin glass rod uniformly charged to +10 nC and a 10-cm-long thin plastic rod uniformly charged to −10 nC are placed side by side, 4.0 cm apart. What are the electric field strengths E_1 to E_3 at distances 1.0 cm, 2.0 cm, and 3.0 cm from the glass rod along the line connecting the midpoints of the two rods?

9. ‖ Two 10-cm-long thin glass rods uniformly charged to +10 nC are placed side by side, 4.0 cm apart. What are the electric field strengths E_1 to E_3 at distances 1.0 cm, 2.0 cm, and 3.0 cm to the right of the rod on the left along the line connecting the midpoints of the two rods?

10. ‖ A small glass bead charged to +6.0 nC is 4.0 cm from a thin, uniformly charged, 10-cm-long glass rod. The bead is repelled from the rod with a force of 840 μN. What is the total charge on the rod?

Section 26.4 The Electric Fields of Rings, Disks, Planes, and Spheres

11. | Two 10-cm-diameter charged rings face each other, 20 cm apart. The left ring is charged to −20 nC and the right ring is charged to +20 nC.
    a. What is the electric field $\vec{E}$, both magnitude and direction, at the midpoint between the two rings?
    b. What is the force $\vec{F}$ on a −1.0 nC charge placed at the midpoint?

12. ‖ Two 10-cm-diameter charged rings face each other, 20 cm apart. Both rings are charged to +20 nC. What is the electric field strength at (a) the midpoint between the two rings and (b) the center of the left ring?

13. ‖ Two 10-cm-diameter charged disks face each other, 20 cm apart. The left disk is charged to −50 nC and the right disk is charged to +50 nC.
    a. What is the electric field $\vec{E}$, both magnitude and direction, at the midpoint between the two disks?
    b. What is the force $\vec{F}$ on a −1.0 nC charge placed at the midpoint?

14. ‖ Two 10-cm-diameter charged disks face each other, 20 cm apart. Both disks are charged to $+50$ nC. What is the electric field strength at (a) the midpoint between the two disks and (b) a point on the axis 5.0 cm from one disk?

15. ‖ The electric field strength 2.0 cm from a 10-cm-diameter metal ball is 50,000 N/C. What is the charge (in nC) on the ball?

16. ‖ A 20 cm × 20 cm horizontal metal electrode is uniformly charged to $+80$ nC. What is the electric field strength 2.0 mm above the center of the electrode?

Section 26.5 The Parallel-Plate Capacitor

17. ‖ Two circular disks spaced 0.50 mm apart form a parallel-plate capacitor. Transferring 3.0×10^9 electrons from one disk to the other causes the electric field strength to be 2.0×10^5 N/C. What are the diameters of the disks?

18. ‖ A parallel-plate capacitor is formed from two 6.0-cm-diameter electrodes spaced 2.0 mm apart. The electric field strength inside the capacitor is 1.0×10^6 N/C. What is the charge (in nC) on each electrode?

19. ‖ Air "breaks down" when the electric field strength reaches 3.0×10^6 N/C, causing a spark. A parallel-plate capacitor is made from two 4.0 cm × 4.0 cm disks. How many electrons must be transferred from one disk to the other to create a spark between the disks?

Section 26.6 Motion of a Charged Particle in an Electric Field

20. ‖ A 0.10 g glass bead is charged by the removal of 1.0×10^{10} electrons. What electric field $\vec{E}$ (strength and direction) will cause the bead to hang suspended in the air?

21. ‖ Two 2.0-cm-diameter disks face each other, 1.0 mm apart. They are charged to ± 10 nC.
   a. What is the electric field strength between the disks?
   b. A proton is shot from the negative disk toward the positive disk. What launch speed must the proton have to just barely reach the positive disk?

22. ‖ An electron in a uniform electric field increases its speed from 2.0×10^7 m/s to 4.0×10^7 m/s over a distance of 1.2 cm. What is the electric field strength?

23. ‖ The surface charge density on an infinite charged plane is -2.0×10^{-6} C/m². A proton is shot straight away from the plane at 2.0×10^6 m/s. How far does the proton travel before reaching its turning point?

24. ‖ A 1.0-μm-diameter oil droplet (density 900 kg/m³) is negatively charged with the addition of 25 extra electrons. It is released from rest 2.0 mm from a very wide plane of positive charge, after which it accelerates toward the plane and collides with a speed of 3.5 m/s. What is the surface charge density of the plane?

Section 26.7 Motion of a Dipole in an Electric Field

25. ‖ The permanent electric dipole moment of the water molecule (H_2O) is 6.2×10^{-30} C m. What is the maximum possible torque on a water molecule in a 5.0×10^8 N/C electric field?

26. ‖ A point charge Q is distance r from the center of a dipole consisting of charges $\pm q$ separated by distance s. The charge is located in the plane that bisects the dipole. At this instant, what are (a) the force (magnitude and direction) and (b) the magnitude of the torque on the dipole? You can assume $r \gg s$.

27. ‖ An ammonia molecule (NH_3) has a permanent electric dipole moment 5.0×10^{-30} C m. A proton is 2.0 nm from the molecule in the plane that bisects the dipole. What is the electric force of the molecule on the proton?

Problems

28. ‖ What are the strength and direction of the electric field at the position indicated by the dot in FIGURE P26.28? Give your answer (a) in component form and (b) as a magnitude and angle measured cw or ccw (specify which) from the positive x-axis.

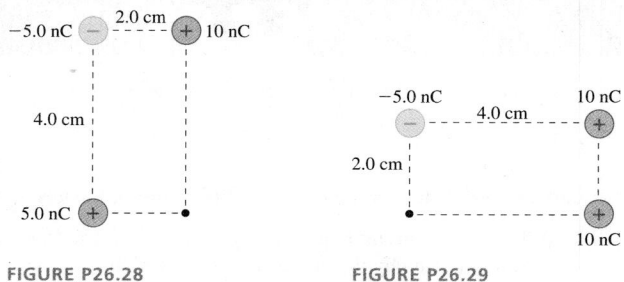

FIGURE P26.28   FIGURE P26.29

29. ‖ What are the strength and direction of the electric field at the position indicated by the dot in FIGURE P26.29? Give your answer (a) in component form and (b) as a magnitude and angle measured cw or ccw (specify which) from the positive x-axis.

30. ‖ What are the strength and direction of the electric field at the position indicated by the dot in FIGURE P26.30? Give your answer (a) in component form and (b) as a magnitude and angle measured cw or ccw (specify which) from the positive x-axis.

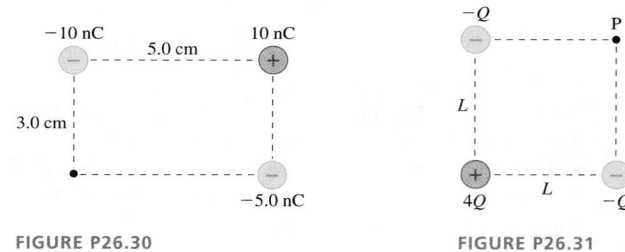

FIGURE P26.30   FIGURE P26.31

31. ‖ FIGURE P26.31 shows three charges at the corners of a square. Write the electric field at point P in component form.

32. ‖ Charges $-q$ and $+2q$ in FIGURE P26.32 are located at $x = \pm a$. Determine the electric field at points 1 to 4. Write each field in component form.

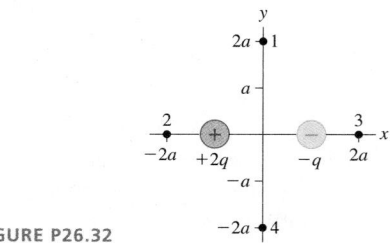

FIGURE P26.32

33. ‖ Two positive charges q are on the y-axis at $y = \pm\frac{1}{2}s$.
   a. Find an expression for the electric field strength at distance x on the axis that bisects the two charges.
   b. For $q = 1.0$ nC and $s = 6.0$ mm, evaluate E at $x = 0, 2, 4, 6,$ and 10 mm.

34. ‖ Derive Equation 26.12 for the field $\vec{E}_{dipole}$ in the plane that bisects an electric dipole.

35. ‖ Three charges are on the y-axis. Charges $-q$ are at $y = \pm d$ and charge $+2q$ is at $y = 0$.
    a. Find an expression for the electric field $\vec{E}$ at a point on the x-axis.
    b. Verify that your answer to part a has the expected behavior as x becomes very small and very large.

36. ‖ FIGURE P26.36 is a cross section of two infinite lines of charge that extend out of the page. Both have linear charge density λ. Find an expression for the electric field strength E at height y above the midpoint between the lines.

37. ‖‖ FIGURE P26.37 is a cross section of two infinite lines of charge that extend out of the page. The linear charge densities are $\pm\lambda$. Find an expression for the electric field strength E at height y above the midpoint between the lines.

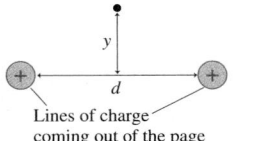

FIGURE P26.36

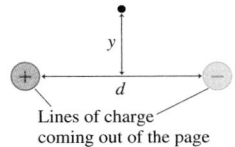

FIGURE P26.37

38. ‖ Two infinite lines of charge, each with linear charge density λ, lie along the x- and y-axes, crossing at the origin. What is the electric field strength at position (x, y)?

39. ‖ The electric field 5.0 cm from a very long charged wire is (2000 N/C, toward the wire). What is the charge (in nC) on a 1.0-cm-long segment of the wire?

40. ‖ FIGURE P26.40 shows a thin rod of length L with total charge Q.
    a. Find an expression for the electric field strength at point P on the axis of the rod at distance r from the center.
    b. Verify that your expression has the expected behavior if $r \gg L$.
    c. Evaluate E at $r = 3.0$ cm if $L = 5.0$ cm and $Q = 3.0$ nC.

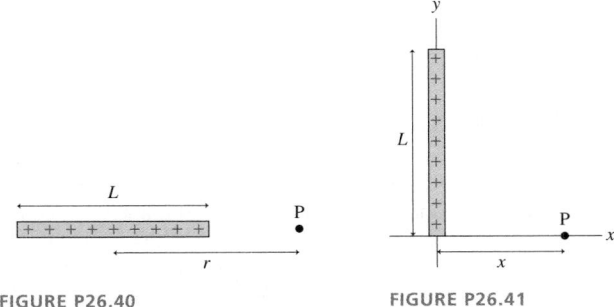

FIGURE P26.40

FIGURE P26.41

41. ‖‖ FIGURE P26.41 shows a thin rod of length L with total charge Q. Find an expression for the electric field $\vec{E}$ at point P. Give your answer in component form.

42. ‖ Show that the on-axis electric field of a ring of charge has the expected behavior when $z \ll R$ and when $z \gg R$.

43. ‖ A ring of radius R has total charge Q.
    a. At what distance along the z-axis is the electric field strength a maximum?
    b. What is the electric field strength at this point?

44. ‖ Charge Q is uniformly distributed along a thin, flexible rod of length L. The rod is then bent into the semicircle shown in FIGURE P26.44.

    a. Find an expression for the electric field $\vec{E}$ at the center of the semicircle.
    Hint: A small piece of arc length Δs spans a small angle $\Delta\theta = \Delta s/R$, where R is the radius.
    b. Evaluate the field strength if $L = 10$ cm and $Q = 30$ nC.

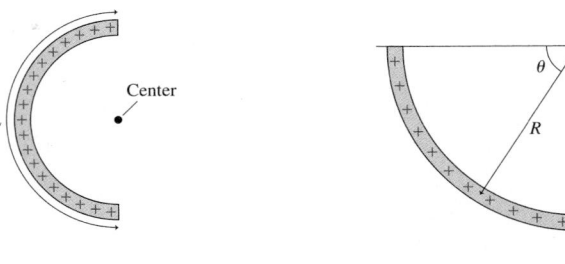

FIGURE P26.44

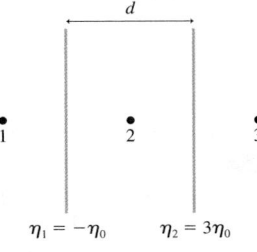

FIGURE P26.45

45. ‖ A plastic rod with linear charge density λ is bent into the quarter circle shown in FIGURE P26.45. We want to find the electric field at the origin.
    a. Write expressions for the x- and y-components of the electric field at the origin due to a small piece of charge at angle θ.
    b. Write, but do not evaluate, definite integrals for the x- and y-components of the net electric field at the origin.
    c. Evaluate the integrals and write $\vec{E}_{net}$ in component form.

46. ‖ You've hung two very large sheets of plastic facing each other with distance d between them, as shown in FIGURE P26.46. By rubbing them with wool and silk, you've managed to give one sheet a uniform surface charge density $\eta_1 = -\eta_0$ and the other a uniform surface charge density $\eta_2 = +3\eta_0$. What are the electric field vectors at points 1, 2, and 3?

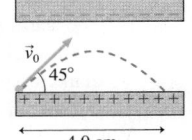

FIGURE P26.46

47. ‖ Two 2.0-cm-diameter insulating spheres have a 6.0 cm space between them. One sphere is charged to $+10$ nC, the other to -15 nC. What is the electric field strength at the midpoint between the two spheres?

48. ‖ Two parallel plates 1.0 cm apart are equally and oppositely charged. An electron is released from rest at the surface of the negative plate and simultaneously a proton is released from rest at the surface of the positive plate. How far from the negative plate is the point at which the electron and proton pass each other?

49. ‖ A parallel-plate capacitor has 2.0 cm $\times$ 2.0 cm electrodes with surface charge densities $\pm 1.0 \times 10^{-6}$ C/m². A proton traveling parallel to the electrodes at 1.0×10^6 m/s enters the center of the gap between them. By what distance has the proton been deflected sideways when it reaches the far edge of the capacitor? Assume the field is uniform inside the capacitor and zero outside the capacitor.

50. ‖ An electron is launched at a 45° angle and a speed of 5.0×10^6 m/s from the positive plate of the parallel-plate capacitor shown in FIGURE P26.50. The electron lands 4.0 cm away.
    a. What is the electric field strength inside the capacitor?
    b. What is the smallest possible spacing between the plates?

FIGURE P26.50

51. ‖‖ The two parallel plates in FIGURE P26.51 are 2.0 cm apart and the electric field strength between them is 1.0×10^4 N/C. An electron is launched at a 45° angle from the positive plate. What is the maximum initial speed v_0 the electron can have without hitting the negative plate?

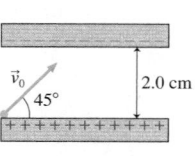

2.0 cm

v_0

45°

+++++++++++++

FIGURE P26.51

FIGURE P26.52

52. ‖ A problem of practical interest is to make a beam of electrons turn a 90° corner. This can be done with the parallel-plate capacitor shown in FIGURE P26.52. An electron with kinetic energy 3.0×10^{-17} J enters through a small hole in the bottom plate of the capacitor.
   a. Should the bottom plate be charged positive or negative relative to the top plate if you want the electron to turn to the right? Explain.
   b. What strength electric field is needed if the electron is to emerge from an exit hole 1.0 cm away from the entrance hole, traveling at right angles to its original direction?
   Hint: The difficulty of this problem depends on how you choose your coordinate system.
   c. What minimum separation d_{min} must the capacitor plates have?

53. ‖ The combustion of fossil fuels produces micron-sized particles of soot, one of the major components of air pollution. The terminal speeds of these particles are extremely small, so they remain suspended in air for very long periods of time. Furthermore, very small particles almost always acquire small amounts of charge from cosmic rays and various atmospheric effects, so their motion is influenced not only by gravity but also by the earth's weak electric field. Consider a small spherical particle of radius r, density ρ, and charge q. A small sphere moving with speed v experiences a drag force $F_{drag} = 6\pi\eta r v$, where η is the viscosity of the air. (This differs from the drag force you learned in Chapter 6 because there we considered macroscopic rather than microscopic objects.)
   a. A particle falling at its terminal speed v_{term} is in dynamic equilibrium with no net force. Write Newton's first law for this particle falling in the presence of a *downward* electric field of strength E, then solve to find an expression for v_{term}.
   b. Soot is primarily carbon, and carbon in the form of graphite has a density of 2200 kg/m³. In the absence of an electric field, what is the terminal speed in mm/s of a 1.0-μm-diameter graphite particle? The viscosity of air at 20°C is 1.8×10^{-5} kg/m s.
   c. The earth's electric field is typically (150 N/C, downward). In this field, what is the terminal speed in mm/s of a 1.0-μm-diameter graphite particle that has acquired 250 extra electrons?

54. ‖ A 2.0-mm-diameter glass sphere has a charge of +1.0 nC. What speed does an electron need to orbit the sphere 1.0 mm above the surface?

55. ‖ In a classical model of the hydrogen atom, the electron orbits the proton in a circular orbit of radius 0.053 nm. What is the orbital frequency? The proton is so much more massive than the electron that you can assume the proton is at rest.

56. ‖‖ In a classical model of the hydrogen atom, the electron orbits a stationary proton in a circular orbit. What is the radius of the orbit for which the orbital frequency is 1.0×10^{12} s⁻¹?

57. ‖ An electric field can *induce* an electric dipole in a neutral atom or molecule by pushing the positive and negative charges in opposite directions. The dipole moment of an induced dipole is directly proportional to the electric field. That is, $\vec{p} = \alpha\vec{E}$, where α is called the *polarizability* of the molecule. A bigger field stretches the molecule farther and causes a larger dipole moment.
   a. What are the units of α?
   b. An ion with charge q is distance r from a molecule with polarizability α. Find an expression for the force $\vec{F}_{ion\ on\ dipole}$.

58. ‖ Show that an infinite line of charge with linear charge density λ exerts an attractive force on an electric dipole with magnitude $F = 2\lambda p/4\pi\epsilon_0 r^2$. Assume that r is much larger than the charge separation in the dipole.

In Problems 59 through 62 you are given the equation(s) used to solve a problem. For each of these
   a. Write a realistic problem for which this is the correct equation(s).
   b. Finish the solution of the problem.

59. $(9.0 \times 10^9 \text{ N m}^2/\text{C}^2) \dfrac{(2.0 \times 10^{-9} \text{ C})s}{(0.025 \text{ m})^3} = 1150$ N/C

60. $(9.0 \times 10^9 \text{ N m}^2/\text{C}^2) \dfrac{2(2.0 \times 10^{-7} \text{ C/m})}{r} = 25{,}000$ N/C

61. $\dfrac{\eta}{2\epsilon_0}\left[1 - \dfrac{z}{\sqrt{z^2 + R^2}}\right] = \dfrac{1}{2}\dfrac{\eta}{2\epsilon_0}$

62. 2.0×10^{12} m/s² $= \dfrac{(1.60 \times 10^{-19} \text{ C})E}{(1.67 \times 10^{-27} \text{ kg})}$

$E = \dfrac{Q}{(8.85 \times 10^{-12} \text{ C}^2/\text{N m}^2)(0.020 \text{ m})^2}$

Challenge Problems

63. Your physics assignment is to figure out a way to use electricity to launch a small 6.0-cm-long plastic drink stirrer. You decide that you'll charge the little plastic rod by rubbing it with fur, then hold it near a long, charged wire, as shown in FIGURE CP26.63. When you let go, the electric force of the wire on the plastic rod will shoot it away. Suppose you can uniformly charge the plastic stirrer to 10 nC and that the linear charge density of the long wire is 1.0×10^{-7} C/m. What is the net electric force on the plastic stirrer if the end closest to the wire is 2.0 cm away?

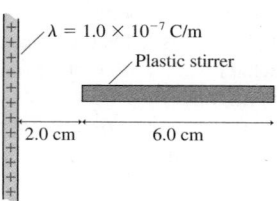

$\lambda = 1.0 \times 10^{-7}$ C/m

Plastic stirrer

2.0 cm   6.0 cm

FIGURE CP26.63

64. Three 10-cm-long rods form an equilateral triangle in a plane. Two of the rods are charged to $+10$ nC, the third to -10 nC. What is the electric field strength at the center of the triangle?

65. A rod of length L lies along the y-axis with its center at the origin. The rod has a nonuniform linear charge density $\lambda = a|y|$, where a is a constant with the units C/m^2.
    a. Draw a graph of λ versus y over the length of the rod.
    b. Determine the constant a in terms of L and the rod's total charge Q.
    Hint: This requires an integration. Think about how to handle the absolute value sign.
    c. Find the electric field strength of the rod at distance x on the x-axis.

66. a. An infinitely long *sheet* of charge of width L lies in the xy-plane between $x = -L/2$ and $x = L/2$. The surface charge density is η. Derive an expression for the electric field $\vec{E}$ at height z above the centerline of the sheet.
    b. Verify that your expression has the expected behavior if $z \ll L$ and if $z \gg L$.
    c. Draw a graph of field strength E versus z.

67. a. An infinitely long *sheet* of charge of width L lies in the xy-plane between $x = -L/2$ and $x = L/2$. The surface charge density is η. Derive an expression for the electric field $\vec{E}$ along the x-axis for points outside the sheet ($x > L/2$).
    b. Verify that your expression has the expected behavior if $x \gg L$.
    Hint: $\ln(1 + u) \approx u$ if $u \ll 1$.
    c. Draw a graph of field strength E versus x for $x > L/2$.

68. One type of ink-jet printer, called an electrostatic ink-jet printer, forms the letters by using deflecting electrodes to steer charged ink drops up and down vertically as the ink jet sweeps horizontally across the page. The ink jet forms 30-μm-diameter drops of ink, charges them by spraying 800,000 electrons on the surface, and shoots them toward the page at a speed of 20 m/s. Along the way, the drops pass through two horizontal, parallel electrodes that are 6.0 mm long, 4.0 mm wide, and spaced 1.0 mm

apart. The distance from the center of the electrodes to the paper is 2.0 cm. To form the tallest letters, which have a height of 6.0 mm, the drops need to be deflected upward (or downward) by 3.0 mm. What electric field strength is needed between the electrodes to achieve this deflection? Ink, which consists of dye particles suspended in alcohol, has a density of 800 kg/m³.

69. A proton orbits a long charged wire, making 1.0×10^6 revolutions per second. The radius of the orbit is 1.0 cm. What is the wire's linear charge density?

70. A *positron* is an elementary particle identical to an electron except that its charge is $+e$. An electron and a positron can rotate about their center of mass as if they were a dumbbell connected by a massless rod. What is the orbital frequency for an electron and a positron 1.0 nm apart?

71. You have a summer intern position with a company that designs and builds nanomachines. An engineer with the company is designing a microscopic oscillator to help keep time, and you've been assigned to help him analyze the design. He wants to place a negative charge at the center of a very small, positively charged metal ring. His claim is that the negative charge will undergo simple harmonic motion at a frequency determined by the amount of charge on the ring.
    a. Consider a negative charge near the center of a positively charged ring centered on the z-axis. Show that there is a restoring force on the charge if it moves along the z-axis but stays close to the center of the ring. That is, show there's a force that tries to keep the charge at $z = 0$.
    b. Show that for *small* oscillations, with amplitude $\ll R$, a particle of mass m with charge $-q$ undergoes simple harmonic motion with frequency

    $$f = \frac{1}{2\pi} \sqrt{\frac{qQ}{4\pi\epsilon_0 mR^3}}$$

    R and Q are the radius and charge of the ring.
    c. Evaluate the oscillation frequency for an electron at the center of a 2.0-μm-diameter ring charged to 1.0×10^{-13} C.

STOP TO THINK ANSWERS

Stop to Think 26.1: c. From symmetry, the fields of the positive charges cancel. The net field is that of the negative charge, which is toward the charge.

Stop to Think 26.2: $\eta_c = \eta_b = \eta_a$. All pieces of a uniformly charged surface have the same surface charge density.

Stop to Think 26.3: b, e, and h. b and e both increase the linear charge density λ.

Stop to Think 26.4: $E_a = E_b = E_c = E_d = E_e$. The field strength of a charged plane is the same at all distances from the plane. An electric field diagram shows the electric field vectors at only a few points; the field exists at all points.

Stop to Think 26.5: $F_a = F_b = F_c = F_d = F_e$. The field strength inside a capacitor is the same at all points, hence the force on a charge is the same at all points. The electric field exists at all points whether or not a vector is shown at that point.

Stop to Think 26.6: c. Parabolic trajectories require *constant* acceleration and thus a *uniform* electric field. The proton has an initial velocity component to the left, but it's being pushed back to the right.

27 Gauss's Law

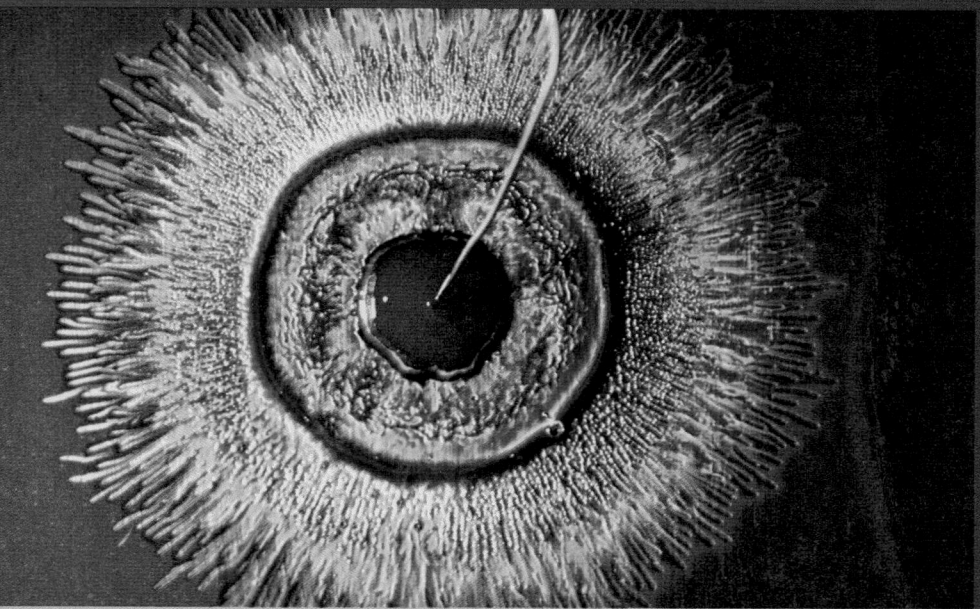

An electric field image of blood plasma from healthy blood. The wire in the center creates the electric field. Variations in the shape and color of the pattern can give early warning of cancer.

▶ **Looking Ahead** The goal of Chapter 27 is to understand and apply Gauss's law.

Symmetry

You'll learn how the shape of some important electric fields, those with a high degree of **symmetry,** can be deduced from the shape of the charge distribution. The idea of symmetry plays an important role in science and mathematics.

An infinitely long charged wire has *cylindrical symmetry.* The electric field of the wire must have the same symmetry.

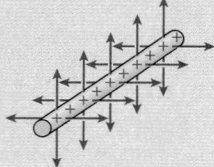

Electric Flux

The amount of electric field passing through a surface is called the **electric flux.** You'll learn how to calculate the flux through open and closed surfaces.

The electric flux is analogous to the amount of air or water flowing through a loop.

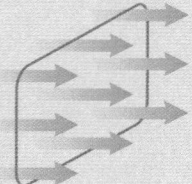

◀ **Looking Back**
Section 11.3 The vector dot product

Gauss's Law

In Chapter 26, you learned to calculate electric fields based on the superposition of the fields of point charges. In this chapter, you'll learn a different way to calculate electric fields based on the idea of electric flux.

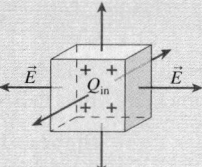

Gauss's law says that the electric flux through a *closed surface* is proportional to the charge Q_{in} enclosed within the surface. This will be the basis of a powerful problem-solving strategy for finding the electric fields of highly symmetric charge distributions.

Gauss's law is a more general statement about the nature of electric fields than is Coulomb's law. It is the first of the four equations that we'll later call *Maxwell's equations*, the governing equations of electricity and magnetism.

◀ **Looking Back**
Section 25.5 The field of a point charge
Section 26.2 Electric field lines

Using Gauss's Law

You'll learn how Gauss's law can be used to find the electric field both inside and outside of charged spheres, cylinders, and planes. In these highly symmetric situations, Gauss's law is much easier to use than superposition.

To find the field of a sphere of charge, you'll draw a *Gaussian surface* around the sphere and then calculate the electric flux through the surface.

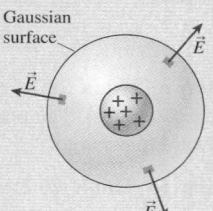

Conductors

Gauss's law can be used to establish several important properties of conductors in **electrostatic equilibrium.**

- Excess charge is on the surface.
- The interior electric field is zero.

The metal grid in the door of a microwave oven shields the room because the electric field inside the metal must be zero. It turns out that the holes don't matter because they are very small compared to the wavelength of the microwaves.

27.1 Symmetry

Suppose we knew only two things about electric fields:

1. The field points away from positive charges, toward negative charges, and
2. An electric field exerts a force on a charged particle.

From this information alone, what can we deduce about the electric field of the infinitely long charged cylinder shown in FIGURE 27.1?

We don't know if the cylinder's diameter is large or small. We don't know if the charge density is the same at the outer edge as along the axis. All we know is that the charge is positive and the charge distribution has *cylindrical symmetry*. We say that a charge distribution is **symmetric** if there is a group of *geometric transformations* that don't cause any *physical* change.

To make this idea concrete, suppose you close your eyes while a friend transforms a charge distribution in one of the following three ways. He or she can

- *Translate* (that is, displace) the charge parallel to an axis,
- *Rotate* the charge about an axis, or
- *Reflect* the charge in a mirror.

When you open your eyes, will you be able to tell if the charge distribution has been changed? You might tell by observing a visual difference in the distribution. Or the results of an experiment with charged particles could reveal that the distribution has changed. If nothing you can see or do reveals any change, then we say that the charge distribution is symmetric under that particular transformation.

FIGURE 27.2 shows that the charge distribution of Figure 27.1 is symmetric with respect to

- Translation parallel to the cylinder axis. Shifting an infinitely long cylinder by 1 mm or 1000 m makes no noticeable or measurable change.
- Rotation by any angle about the cylinder axis. Turning a cylinder about its axis by 1° or 100° makes no detectable change.
- Reflections in any plane containing or perpendicular to the cylinder axis. Exchanging top and bottom, front and back, or left and right makes no detectable change.

A charge distribution that is symmetric under these three groups of geometric transformations is said to be *cylindrically symmetric*. Other charge distributions have other types of symmetries. Some charge distributions have no symmetry at all.

Our interest in symmetry can be summed up in a single statement:

The symmetry of the electric field must match the symmetry of the charge distribution.

If this were not true, you could use the electric field to test whether the charge distribution had undergone a transformation.

Now we're ready to see what we can learn about the electric field in Figure 27.1. Could the field look like FIGURE 27.3a? (Imagine this picture rotated about the axis.) That is, is this a *possible* field? This field looks the same if it's translated parallel to the

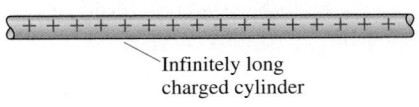

FIGURE 27.1 A charge distribution with cylindrical symmetry.

Infinitely long charged cylinder

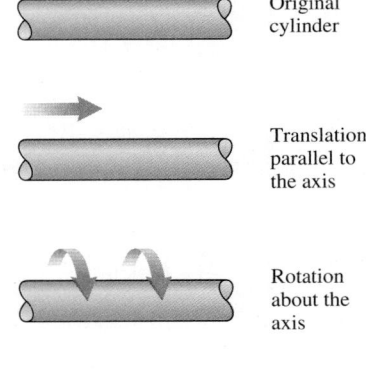

FIGURE 27.2 Transformations that don't change an infinite cylinder of charge.

Original cylinder

Translation parallel to the axis

Rotation about the axis

Reflection in plane containing the axis

Reflection perpendicular to the axis

FIGURE 27.3 Could the field of a cylindrical charge distribution look like this?

(a) Is this a possible electric field of an infinitely long charged cylinder? Suppose the charge and the field are reflected in a plane perpendicular to the axis.

Reflection plane

$\vec{E}$

$\vec{E}$

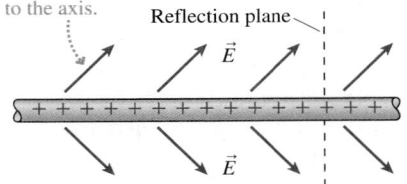

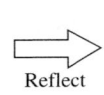

Reflect

(b) The charge distribution is not changed by the reflection, but the field is. This field doesn't match the symmetry of the cylinder, so the cylinder's field can't look like this.

$\vec{E}$

$\vec{E}$

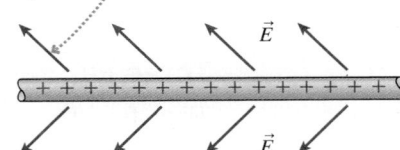

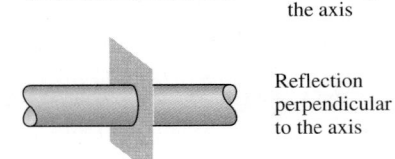

FIGURE 27.4 Or might the field of a cylindrical charge distribution look like this?

(a)

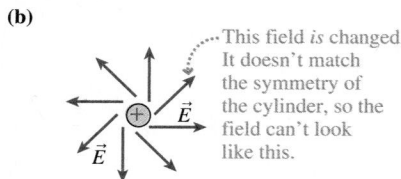

End view of cylinder — Reflection plane

The charge distribution is not changed by reflecting it in a plane containing the axis.

$\vec{E}$

$\vec{E}$

Reflect

(b)

This field *is* changed. It doesn't match the symmetry of the cylinder, so the field can't look like this.

$\vec{E}$

$\vec{E}$

cylinder axis, if up and down are exchanged by reflecting the field in a plane coming out of the page, or if you rotate the cylinder about its axis.

But the proposed field fails one test: reflection in a plane perpendicular to the axis, a reflection that exchanges left and right. This reflection, which would *not* make any change in the charge distribution itself, produces the field shown in FIGURE 27.3b. This change in the field is detectable because a positively charged particle would now have a component of motion to the left instead of to the right.

The field of Figure 27.3a, which makes a distinction between left and right, is not cylindrically symmetric and thus is *not* a possible field. In general, **the electric field of a cylindrically symmetric charge distribution cannot have a component parallel to the cylinder axis.**

Well then, what about the electric field shown in FIGURE 27.4a? Here we're looking down the axis of the cylinder. The electric field vectors are restricted to planes perpendicular to the cylinder and thus do not have any component parallel to the cylinder axis. This field is symmetric for rotations about the axis, but it's *not* symmetric for a reflection in a plane containing the axis.

The field of FIGURE 27.4b, after this reflection, is easily distinguishable from the field of Figure 27.4a. Thus **the electric field of a cylindrically symmetric charge distribution cannot have a component tangent to the circular cross section.**

FIGURE 27.5 shows the only remaining possible field shape. The electric field is radial, pointing straight out from the cylinder like the bristles on a bottle brush. This is the one electric field shape matching the symmetry of the charge distribution.

FIGURE 27.5 This is the only shape for the electric field that matches the symmetry of the charge distribution.

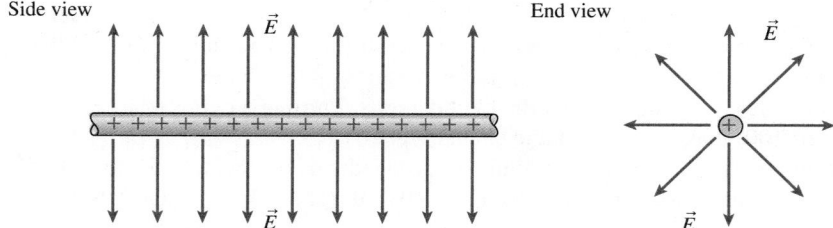

Side view $\vec{E}$

End view $\vec{E}$

$\vec{E}$

$\vec{E}$

$\vec{E}$

What Good Is Symmetry?

Given how little we assumed about Figure 27.1—that the charge distribution is cylindrically symmetric and that electric fields point away from positive charges—we've been able to deduce a great deal about the electric field. In particular, we've deduced the *shape* of the electric field.

Now, shape is not everything. We've learned nothing about the strength of the field or how strength changes with distance. Is E constant? Does it decrease like $1/r$ or $1/r^2$? We don't yet have a complete description of the field, but knowing what shape the field *has* to have will make finding the field strength a much easier task.

That's the good of symmetry. Symmetry arguments allow us to *rule out* many conceivable field shapes as simply being incompatible with the symmetry of the charge distribution. Knowing what doesn't happen, or can't happen, is often as useful as knowing what does happen. By the process of elimination, we're led to the one and only shape the field can possibly have. Reasoning on the basis of symmetry is a sometimes subtle but always powerful means of reasoning.

Three Fundamental Symmetries

Three fundamental symmetries appear frequently in electrostatics. The first row of FIGURE 27.6 shows the simplest form of each symmetry. The second row shows a more complex, but more realistic, situation with the same symmetry.

FIGURE 27.6 Three fundamental symmetries.

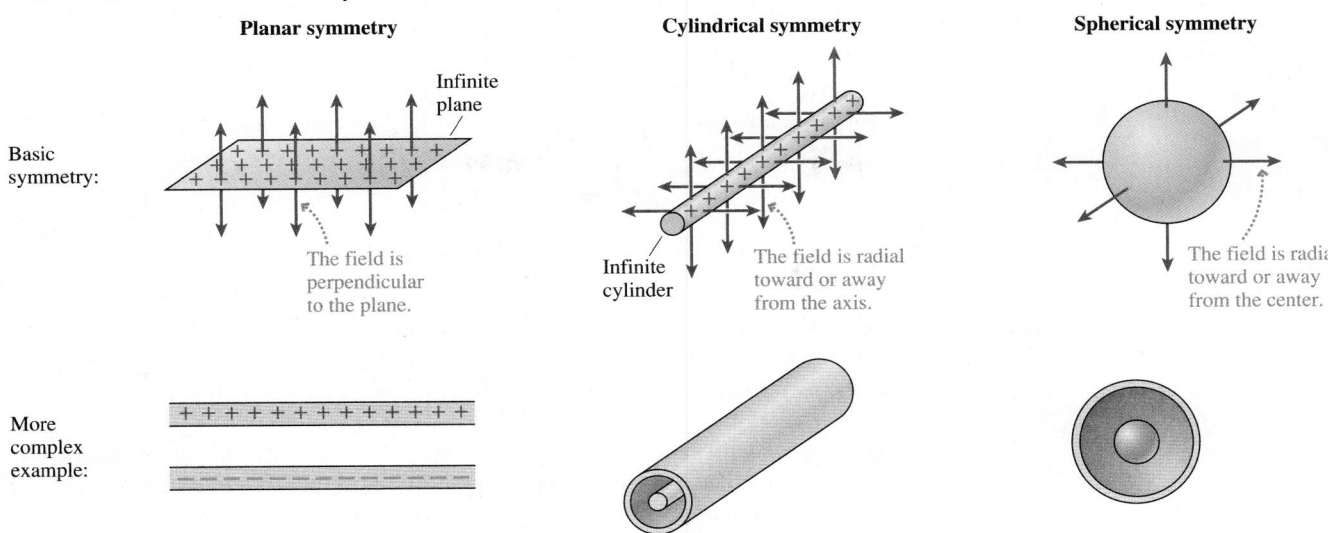

NOTE ▶ Figures must be finite in extent, but the planes and cylinders in Figure 27.6 are assumed to be infinite. ◀

Objects do exist that are extremely close to being perfect spheres, but no real cylinder or plane can be infinite in extent. Even so, the fields of infinite planes and cylinders are good models for the fields of finite planes and cylinders at points not too close to an edge or an end. The fields that we'll study in this chapter, even if idealized, have many important applications.

STOP TO THINK 27.1 A uniformly charged rod has a *finite* length L. The rod is symmetric under rotations about the axis and under reflection in any plane containing the axis. It is *not* symmetric under translations or under reflections in a plane perpendicular to the axis unless that plane bisects the rod. Which field shape or shapes match the symmetry of the rod?

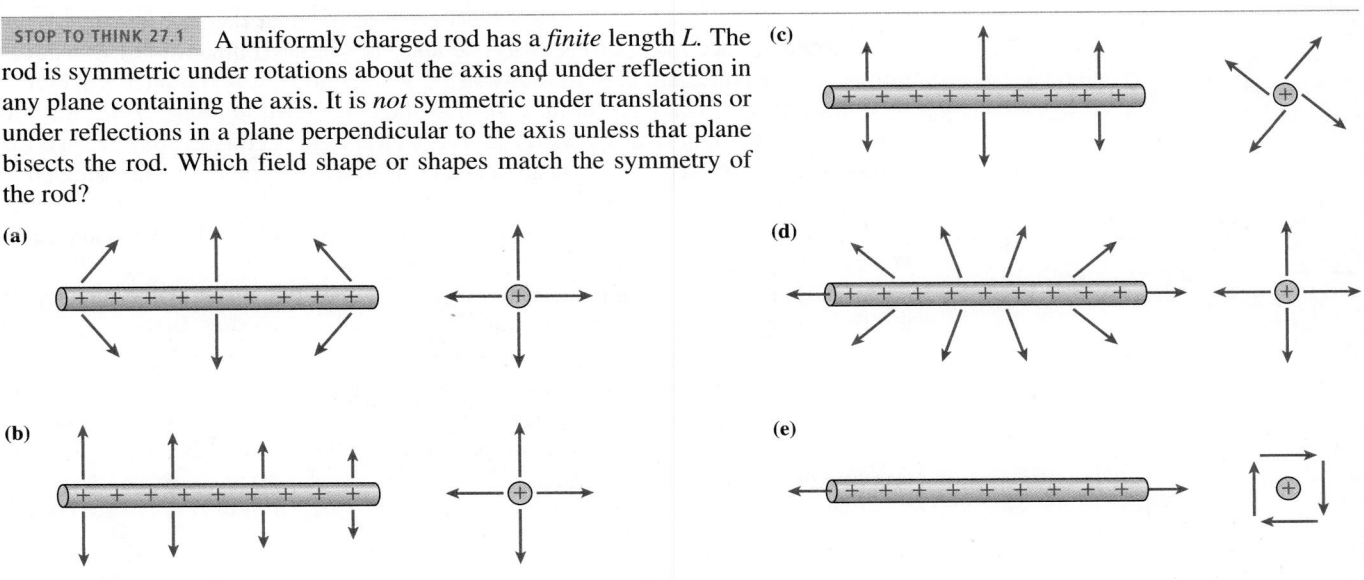

27.2 The Concept of Flux

FIGURE 27.7a on the next page shows an opaque box surrounding a region of space. We can't see what's in the box, but there's an electric field vector coming out of each face of the box. Can you figure out what's in the box?

(a) The field is coming out of each face of the box. There must be a positive charge in the box.

(b) The field is going into each face of the box. There must be a negative charge in the box.

(c) A field passing through the box implies there's no net charge in the box.

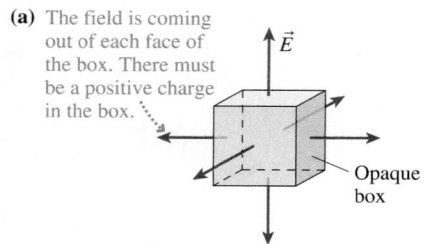 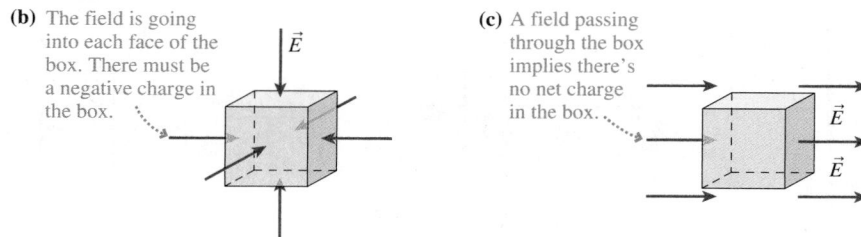

Opaque box

Of course you can. Because electric fields point away from positive charges, and the electric field is coming out of every face of the box, it seems clear that the box contains a positive charge or charges. Similarly, the box in **FIGURE 27.7b** certainly contains a negative charge.

What can we tell about the box in **FIGURE 27.7c**? The electric field points into the box on the left. An equal electric field points out on the right. This might be the electric field between a large positive electrode somewhere out of sight on the left and a large negative electrode off to the right. An electric field passes through the box, but we see no evidence there's any charge (or at least any net charge) inside the box.

These examples suggest that the electric field as it passes into, out of, or through the box is in some way connected to the charge within the box. However, these simple pictures don't tell us how much charge there is or where within the box the charge is located. Perhaps a better box would be more informative.

Suppose we surround a region of space with a *closed surface,* a surface that divides space into distinct inside and outside regions. Within the context of electrostatics, a closed surface through which an electric field passes is called a **Gaussian surface,** named after the 19th-century mathematician Karl Gauss who developed the mathematical foundations of geometry. This is an imaginary, mathematical surface, not a physical surface, although it might coincide with a physical surface. For example, **FIGURE 27.8a** shows a spherical Gaussian surface surrounding a charge.

A closed surface must, of necessity, be a surface in three dimensions. But three-dimensional pictures are hard to draw, so we'll often look at two-dimensional cross sections through a Gaussian surface, such as the one shown in **FIGURE 27.8b**. Now we can tell from the *spherical symmetry* of the electric field vectors poking through the surface that the positive charge inside must be spherically symmetric and centered at the *center* of the sphere. Notice two features that will soon be important: The electric field is everywhere *perpendicular* to the spherical surface and has the *same magnitude* at each point on the surface.

A Gaussian surface is most useful when it matches the shape and symmetry of the field. For example, **FIGURE 27.9a** shows a cylindrical Gaussian surface—a *closed* cylinder—surrounding some kind of cylindrical charge distribution, such as a charged wire. **FIGURE 27.9b** simplifies the drawing by showing two-dimensional end and side views. Because the Gaussian surface matches the symmetry of the charge distribution, the electric field is everywhere *perpendicular* to the side wall and *no* field passes through the top and bottom surfaces.

FIGURE 27.8 Gaussian surface surrounding a charge. A two-dimensional cross section is usually easier to draw.

(a)

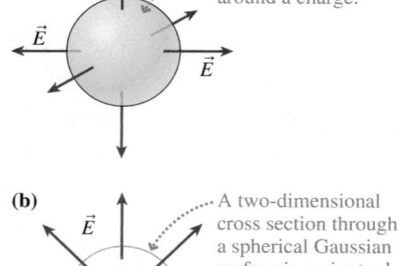

A Gaussian surface is a closed surface around a charge.

(b)

A two-dimensional cross section through a spherical Gaussian surface is easier to draw.

FIGURE 27.9 A Gaussian surface is most useful when it matches the shape of the field.

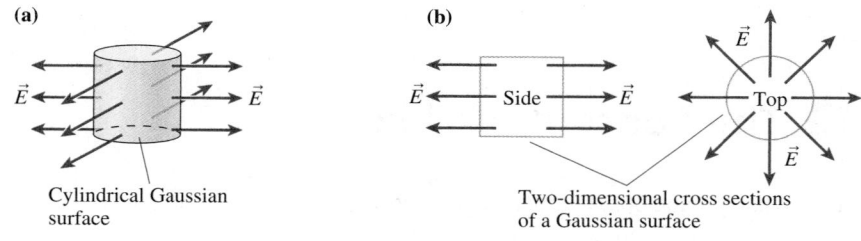

(a)

Cylindrical Gaussian surface

(b)

Side

Top

Two-dimensional cross sections of a Gaussian surface

For contrast, consider the spherical surface in FIGURE 27.10a. This is also a Gaussian surface, and the protruding electric field tells us there's a positive charge inside. It might be a point charge located on the left side, but we can't really say. A Gaussian surface that doesn't match the symmetry of the charge distribution isn't terribly useful.

The nonclosed surface of FIGURE 27.10b doesn't provide much help either. What appears to be a uniform electric field to the right could be due to a large positive plate on the left, a large negative plate on the right, or both. A nonclosed surface doesn't provide enough information.

These examples lead us to two conclusions:

1. The electric field, in some sense, "flows" *out of* a closed surface surrounding a region of space containing a net positive charge and *into* a closed surface surrounding a net negative charge. The electric field may flow *through* a closed surface surrounding a region of space in which there is no net charge, but the *net flow* is zero.
2. The electric field pattern through the surface is particularly simple if the closed surface matches the symmetry of the charge distribution inside.

The electric field doesn't really flow like a fluid, but the metaphor is a useful one. The Latin word for flow is *flux,* and the amount of electric field passing through a surface is called the **electric flux.** Our first conclusions, stated in terms of electric flux, are

- There is an outward flux through a closed surface around a net positive charge.
- There is an inward flux through a closed surface around a net negative charge.
- There is no net flux through a closed surface around a region of space in which there is no net charge.

This chapter has been entirely qualitative thus far as we've established pictorially what we mean by symmetry, the idea of flux, and the fact that the electric flux through a closed surface has something to do with the charge inside. Understanding these qualitative ideas is essential, but to go further we need to make these ideas quantitative and precise. In the next section, you'll learn how to calculate the electric flux through a surface. Then, in the section following that, we'll establish a precise relationship between the net flux through a Gaussian surface and the enclosed charge. That relationship, Gauss's law, will allow us to determine the electric fields of some interesting and useful charge distributions.

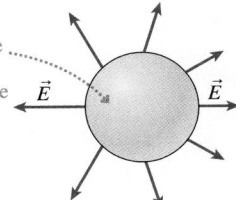

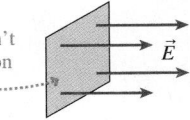

FIGURE 27.10 Not every surface is useful for learning about charge.

(a)

A Gaussian surface that doesn't match the symmetry of the electric field isn't very useful. $\vec{E}$ $\vec{E}$

(b)

A nonclosed surface doesn't provide enough information about the charges. $\vec{E}$

STOP TO THINK 27.2 This box contains

a. A positive charge.
b. A negative charge.
c. No charge.
d. A net positive charge.
e. A net negative charge.
f. No net charge.

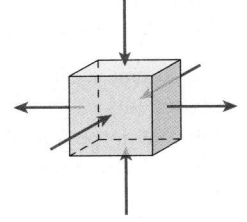

27.3 Calculating Electric Flux

Let's start with a brief overview of where this section will take us. We'll begin with a definition of flux that is easy to understand, then we'll turn that simple definition into a formidable-looking integral. We need the integral because the simple definition applies only to uniform electric fields and flat surfaces. Those are good starting points, but we'll soon need to calculate the flux of nonuniform fields through curved surfaces.

Mathematically, the flux of a nonuniform field through a curved surface is described by a special kind of integral called a *surface integral.* It's quite possible that you have not yet encountered surface integrals in your calculus course, and the "novelty factor" contributes to making this integral look worse than it really is. We will emphasize over and over the idea that an integral is just a fancy way of doing a sum, in this case the sum of the small amounts of flux through many small pieces of a surface.

The good news is that *every* surface integral we need to evaluate in this chapter, or that you will need to evaluate for the homework problems, is either zero or is so easy that you will be able to do it in your head. This seems like an astounding claim, but you will soon see it is true. The key will be to make effective use of the *symmetry* of the electric field.

Now that you've been warned, you needn't panic at the sight of the mathematical notation that will be introduced. We'll go step by step, and you'll see that, at least as far as electrostatics is concerned, calculating the electric flux is not difficult.

The Basic Definition of Flux

Imagine holding a rectangular wire loop of area A in front of a fan. As FIGURE 27.11 shows, the volume of air flowing through the loop each second depends on the angle between the loop and the direction of flow. The flow is maximum through a loop that is perpendicular to the airflow; no air goes through the same loop if it lies parallel to the flow.

FIGURE 27.11 The amount of air flowing through a loop depends on the angle between $\vec{v}$ and $\hat{n}$.

(a)

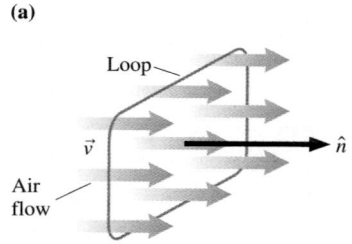

The air flowing through the loop is maximum when $\theta = 0°$.

(b)

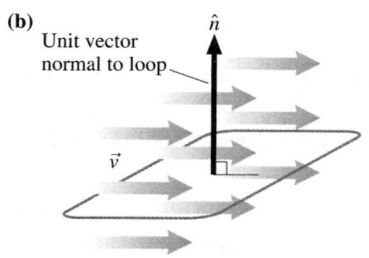

No air flows through the loop when $\theta = 90°$.

(c) The loop is tilted by angle θ.

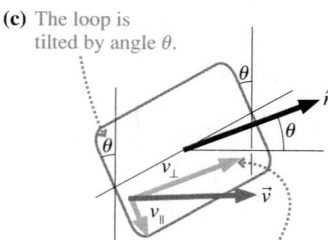

$v_\perp = v\cos\theta$ is the component of the air velocity perpendicular to the loop.

The flow direction is identified by the velocity vector $\vec{v}$. We can identify the loop's orientation by defining a unit vector $\hat{n}$ normal to the plane of the loop. Angle θ is then the angle between $\vec{v}$ and $\hat{n}$. The loop perpendicular to the flow in FIGURE 27.11a has $\theta = 0°$; the loop parallel to the flow in FIGURE 27.11b has $\theta = 90°$. You can think of θ as the angle by which a loop has been tilted away from perpendicular.

NOTE ▶ A surface has two sides, so $\hat{n}$ could point either way. We'll choose the side that makes $\theta \leq 90°$. ◀

You can see from FIGURE 27.11c that the velocity vector $\vec{v}$ can be decomposed into components $v_\perp = v\cos\theta$ perpendicular to the loop and $v_\parallel = v\sin\theta$ parallel to the loop. Only the perpendicular component $v_\perp$ carries air *through* the loop. Consequently, the volume of air flowing through the loop each second is

$$\text{volume of air per second (m}^3\text{/s)} = v_\perp A = vA\cos\theta \qquad (27.1)$$

$\theta = 0°$ is the orientation for maximum flow through the loop, as expected, and no air flows through the loop if it is tilted $90°$.

An electric field doesn't flow in a literal sense, but we can apply the same idea to an electric field passing through a surface. FIGURE 27.12 shows a surface of area A in a uniform electric field $\vec{E}$. Unit vector $\hat{n}$ is normal to the surface, and θ is the angle between $\hat{n}$ and $\vec{E}$. Only the component $E_\perp = E\cos\theta$ passes *through* the surface.

With this in mind, and using Equation 27.1 as an analog, let's define the *electric flux* Φ_e (uppercase Greek phi) as

$$\Phi_e = E_\perp A = EA\cos\theta \qquad (27.2)$$

The electric flux measures the amount of electric field passing through a surface of area A if the normal to the surface is tilted at angle θ from the field.

FIGURE 27.12 An electric field passing through a surface.

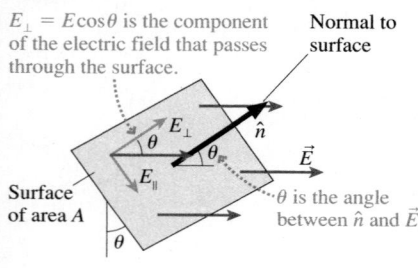

$E_\perp = E\cos\theta$ is the component of the electric field that passes through the surface.

Normal to surface

Surface of area A

θ is the angle between $\hat{n}$ and $\vec{E}$.

Equation 27.2 looks very much like a vector dot product: $\vec{E} \cdot \vec{A} = EA \cos \theta$. For this idea to work, let's define an **area vector** $\vec{A} = A\hat{n}$ to be a vector in the direction of $\hat{n}$—that is, *perpendicular* to the surface—with a magnitude A equal to the area of the surface. Vector $\vec{A}$ has units of m². FIGURE 27.13a shows two area vectors.

FIGURE 27.13 The electric flux can be defined in terms of the area vector $\vec{A}$.

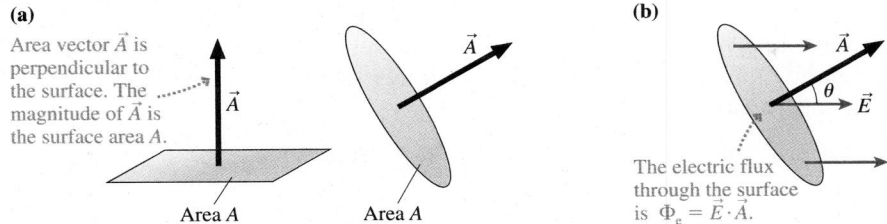

(a)

Area vector $\vec{A}$ is perpendicular to the surface. The magnitude of $\vec{A}$ is the surface area A.

$\vec{A}$

Area A

$\vec{A}$

Area A

(b)

$\vec{A}$

θ

$\vec{E}$

The electric flux through the surface is $\Phi_e = \vec{E} \cdot \vec{A}$.

FIGURE 27.13b shows an electric field passing through a surface of area A. The angle between vectors $\vec{A}$ and $\vec{E}$ is the same angle used in Equation 27.2 to define the electric flux, so Equation 27.2 really is a dot product. We can define the electric flux more concisely as

$$\Phi_e = \vec{E} \cdot \vec{A} \quad \text{(electric flux of a constant electric field)} \qquad (27.3)$$

Writing the flux as a dot product helps make clear how angle θ is defined: θ is the angle between the electric field and a line *perpendicular* to the plane of the surface.

NOTE ▶ Figure 27.13b shows a circular area, but the shape of the surface is not relevant. However, Equation 27.3 is restricted to a *constant* electric field passing through a *planar* surface. ◀

EXAMPLE 27.1 **The electric flux inside a parallel-plate capacitor**

Two 100 cm² parallel electrodes are spaced 2.0 cm apart. One is charged to +5.0 nC, the other to −5.0 nC. A 1.0 cm × 1.0 cm surface between the electrodes is tilted to where its normal makes a 45° angle with the electric field. What is the electric flux through this surface?

MODEL Assume the surface is located near the center of the capacitor where the electric field is uniform. The electric flux doesn't depend on the shape of the surface.

VISUALIZE The surface is square, rather than circular, but otherwise the situation looks like Figure 27.13b.

SOLVE In Chapter 26, we found the electric field inside a parallel-plate capacitor to be

$$E = \frac{Q}{\epsilon_0 A_{\text{plates}}} = \frac{5.0 \times 10^{-9} \text{ C}}{(8.85 \times 10^{-12} \text{ C}^2/\text{N m}^2)(1.0 \times 10^{-2} \text{ m}^2)}$$

$$= 5.65 \times 10^4 \text{ N/C}$$

A 1.0 cm × 1.0 cm surface has $A = 1.0 \times 10^{-4}$ m². The electric flux through this surface is

$$\Phi_e = \vec{E} \cdot \vec{A} = EA \cos \theta$$

$$= (5.65 \times 10^4 \text{ N/C})(1.0 \times 10^{-4} \text{ m}^2) \cos 45°$$

$$= 4.0 \text{ N m}^2/\text{C}$$

ASSESS The units of electric flux are the product of electric field and area units: N m²/C.

The Electric Flux of a Nonuniform Electric Field

Our initial definition of the electric flux assumed that the electric field $\vec{E}$ was constant over the surface. How should we calculate the electric flux if $\vec{E}$ varies from point to point on the surface? We can answer this question by returning to the analogy of air flowing through a loop. Suppose the airflow varies from point to point. We can still find the total volume of air passing through the loop each second by dividing the loop into many small areas, finding the flow through each small area, then adding them. Similarly, **the electric flux through a surface can be calculated as the sum of the fluxes through smaller pieces of the surface.** Because flux is a scalar, adding fluxes is easier than adding electric fields.

FIGURE 27.14 A surface in a nonuniform electric field.

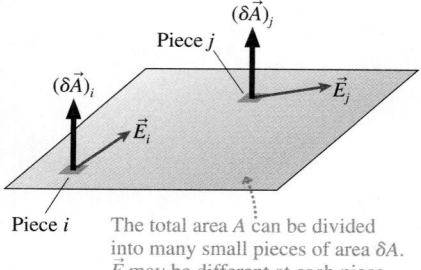

Piece *i*   The total area *A* can be divided into many small pieces of area δA. $\vec{E}$ may be different at each piece.

FIGURE 27.14 shows a surface in a nonuniform electric field. Imagine dividing the surface into many small pieces of area δA. Each little area has an area vector $\delta \vec{A}$ perpendicular to the surface. Two of the little pieces are shown in the figure. The electric fluxes through these two pieces differ because the electric fields are different.

Consider the small piece *i* where the electric field is $\vec{E}_i$. The small electric flux $\delta \Phi_i$ through area $(\delta \vec{A})_i$ is

$$\delta \Phi_i = \vec{E}_i \cdot (\delta \vec{A})_i \tag{27.4}$$

The flux through every other little piece of the surface is found the same way. The total electric flux through the entire surface is then the sum of the fluxes through each of the small areas:

$$\Phi_e = \sum_i \delta \Phi_i = \sum_i \vec{E}_i \cdot (\delta \vec{A})_i \tag{27.5}$$

Now let's go to the limit $\delta \vec{A} \rightarrow d\vec{A}$. That is, the little areas become infinitesimally small, and there are infinitely many of them. Then the sum becomes an integral, and the electric flux through the surface is

$$\Phi_e = \int_{\text{surface}} \vec{E} \cdot d\vec{A} \tag{27.6}$$

The integral in Equation 27.6 is called a **surface integral.**

Equation 27.6 may look rather frightening if you haven't seen surface integrals before. Despite its appearance, a surface integral is no more complicated than integrals you know from calculus. After all, what does $\int f(x)\, dx$ really mean? This expression is a shorthand way to say "Divide the *x*-axis into many little segments of length δx, evaluate the function $f(x)$ in each of them, then add up $f(x)\,\delta x$ for all the segments along the line." The integral in Equation 27.6 differs only in that we're dividing a surface into little pieces instead of a line into little segments. In particular, we're summing the fluxes through a vast number of very tiny pieces.

You may be thinking, "OK, I understand the idea, but I don't know what to *do*. In calculus, I learned formulas for evaluating integrals such as $\int x^2\, dx$. How do I evaluate a surface integral?" This is a good question. We'll deal with evaluation shortly, and it will turn out that the surface integrals in electrostatics are quite easy to evaluate. But don't confuse *evaluating* the integral with understanding what the integral *means*. The surface integral in Equation 27.6 is simply a shorthand notation for the summation of the electric fluxes through a vast number of very tiny pieces of a surface.

The electric field might be different at every point on the surface, but suppose it isn't. That is, suppose the surface is in a uniform electric field $\vec{E}$. A field that is the same at every single point on a surface is a constant as far as the integration of Equation 27.6 is concerned, so we can take it outside the integral. In that case,

$$\Phi_e = \int_{\text{surface}} \vec{E} \cdot d\vec{A} = \int_{\text{surface}} E \cos\theta \, dA = E \cos\theta \int_{\text{surface}} dA \tag{27.7}$$

The integral that remains in Equation 27.7 tells us to add up all the little areas into which the full surface was subdivided. But the sum of all the little areas is simply the area of the surface:

$$\int_{\text{surface}} dA = A \tag{27.8}$$

This idea—that the surface integral of dA is the area of the surface—is one we'll use to evaluate most of the surface integrals of electrostatics. If we substitute Equation 27.8 into Equation 27.7, we find that the electric flux in a uniform electric field is $\Phi_e = EA \cos\theta$. We already knew this, from Equation 27.2, but it was important to see that the surface integral of Equation 27.6 gives the correct result for the case of a uniform electric field.

The Flux Through a Curved Surface

Most of the Gaussian surfaces we considered in the last section were curved surfaces. FIGURE 27.15 shows an electric field passing through a curved surface. How do we find the electric flux through this surface? Just as we did for a flat surface!

Divide the surface into many small pieces of area δA. For each, define the area vector $\delta \vec{A}$ perpendicular to the surface *at that point.* Compared to Figure 27.14, the only difference that the curvature of the surface makes is that the $\delta \vec{A}$ are no longer parallel to each other. Find the small electric flux $\delta \Phi_i = \vec{E}_i \cdot (\delta \vec{A})_i$ through each little area, then add them all up. The result, once again, is

$$\Phi_e = \int_{\text{surface}} \vec{E} \cdot d\vec{A} \qquad (27.9)$$

We *assumed,* in deriving this expression the first time, that the surface was flat and that all the $\delta \vec{A}$ were parallel to each other. But that assumption wasn't necessary. The *meaning* of Equation 27.9—a summation of the fluxes through a vast number of very tiny pieces—is unchanged if the pieces lie on a curved surface.

We seem to be getting more and more complex, using surface integrals first for nonuniform fields and now for curved surfaces. But consider the two situations shown in FIGURE 27.16. The electric field $\vec{E}$ in FIGURE 27.16a is everywhere tangent, or parallel, to the curved surface. We don't need to know the magnitude of $\vec{E}$ to recognize that $\vec{E} \cdot d\vec{A}$ is *zero at every point* on the surface because $\vec{E}$ is perpendicular to $d\vec{A}$ at every point. Thus $\Phi_e = 0$. A tangent electric field never pokes through the surface, so it has no flux through the surface.

The electric field in FIGURE 27.16b is everywhere perpendicular to the surface *and* has the same magnitude E at every point. $\vec{E}$ differs in direction at different points on a curved surface, but at any particular point $\vec{E}$ is parallel to $d\vec{A}$ and $\vec{E} \cdot d\vec{A}$ is simply $E\,dA$. In this case,

$$\Phi_e = \int_{\text{surface}} \vec{E} \cdot d\vec{A} = \int_{\text{surface}} E\,dA = E \int_{\text{surface}} dA = EA \qquad (27.10)$$

As we evaluated the integral, the fact that E has the same magnitude at every point on the surface allowed us to bring the constant value outside the integral. We then used the fact that the integral of dA over the surface is the surface area A.

We can summarize these two situations with a Tactics Box.

> **TACTICS BOX 27.1 Evaluating surface integrals** (MP)
>
> ❶ If the electric field is everywhere tangent to a surface, the electric flux through the surface is $\Phi_e = 0$.
> ❷ If the electric field is everywhere perpendicular to a surface *and* has the same magnitude E at every point, the electric flux through the surface is $\Phi_e = EA$.

These two results will be of immeasurable value for using Gauss's law because *every* flux we'll need to calculate will be one of these situations. This is the basis for our earlier claim that the evaluation of surface integrals is not going to be difficult.

The Electric Flux Through a Closed Surface

Our final step, to calculate the electric flux through a closed surface such as a box, a cylinder, or a sphere, requires nothing new. We've already learned how to calculate the electric flux through flat and curved surfaces, and a closed surface is nothing more than a surface that happens to be closed.

However, the mathematical notation for the surface integral over a closed surface differs slightly from what we've been using. It is customary to use a little circle on

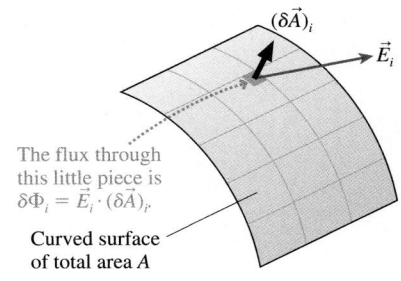

FIGURE 27.15 A curved surface in an electric field.

The flux through this little piece is $\delta \Phi_i = \vec{E}_i \cdot (\delta \vec{A})_i$.

Curved surface of total area A

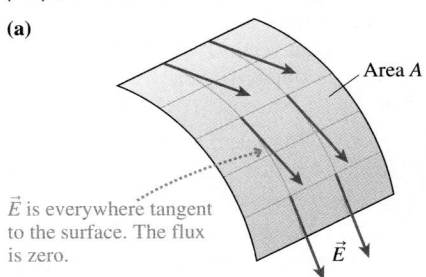

FIGURE 27.16 Electric fields that are everywhere tangent to or everywhere perpendicular to a curved surface.

(a)

Area A

$\vec{E}$ is everywhere tangent to the surface. The flux is zero.

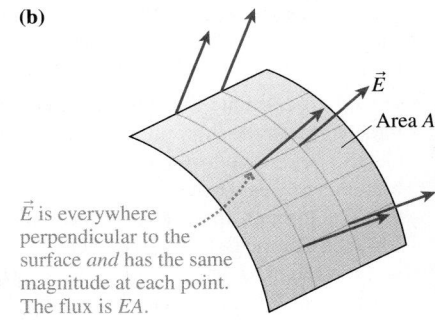

(b)

$\vec{E}$

Area A

$\vec{E}$ is everywhere perpendicular to the surface *and* has the same magnitude at each point. The flux is EA.

the integral sign to indicate that the surface integral is to be performed over a closed surface. With this notation, the electric flux through a closed surface is

$$\Phi_e = \oint \vec{E} \cdot d\vec{A} \tag{27.11}$$

Only the notation has changed. The electric flux is still the summation of the fluxes through a vast number of tiny pieces, pieces that now cover a closed surface.

> NOTE ▶ A closed surface has a distinct inside and outside. The area vector $d\vec{A}$ is defined to always point *toward the outside*. This removes an ambiguity that was present for a single surface, where $d\vec{A}$ could point to either side. ◀

EXAMPLE 27.2 | **Calculating the electric flux through a closed cylinder**

A charge distribution with cylindrical symmetry has created the electric field $\vec{E} = E_0(r^2/r_0^2)\hat{r}$, where E_0 and r_0 are constants and where unit vector $\hat{r}$ lies in the xy-plane. Calculate the electric flux through a closed cylinder of length L and radius R that is centered along the z-axis.

MODEL The electric field extends radially outward from the z-axis with cylindrical symmetry. The z-component is $E_z = 0$. The cylinder is a Gaussian surface.

VISUALIZE FIGURE 27.17a is a view of the electric field looking along the z-axis. The field strength increases with increasing radial distance, and it's symmetric about the z-axis. FIGURE 27.17b is the closed Gaussian surface for which we need to calculate the electric flux. We can place the cylinder anywhere along the z-axis because the electric field extends forever in that direction.

SOLVE To calculate the flux, we divide the closed cylinder into three surfaces: the top, the bottom, and the cylindrical wall. The electric field is tangent to the surface at every point on the top and bottom surfaces. Hence, according to step 1 in Tactics Box 27.1, the flux through those two surfaces is zero. For the cylindrical wall, the electric field is perpendicular to the surface at every point *and* has the constant magnitude $E = E_0(R^2/r_0^2)$ at every point on the surface. Thus, from step 2 in Tactics Box 27.1,

$$\Phi_{wall} = EA_{wall}$$

If we add the three pieces, the net flux through the closed surface is

$$\Phi_e = \oint \vec{E} \cdot d\vec{A} = \Phi_{top} + \Phi_{bottom} + \Phi_{wall} = 0 + 0 + EA_{wall}$$

$$= EA_{wall}$$

We've evaluated the surface integral, using the two steps in Tactics Box 27.1, and there was nothing to it! To finish, all we need to recall is that the surface area of a cylindrical wall is circumference × height, or $A_{wall} = 2\pi RL$. Thus

$$\Phi_e = \left(E_0 \frac{R^2}{r_0^2}\right)(2\pi RL) = \frac{2\pi LR^3}{r_0^2}E_0$$

ASSESS LR^3/r_0^2 has units of m^2, an area, so this expression for Φ_e has units of $N\,m^2/C$. These are the correct units for electric flux, giving us confidence in our answer. Notice the important role played by symmetry. The electric field was perpendicular to the wall and of constant value at every point on the wall *because* the Gaussian surface had the same symmetry as the charge distribution. We would not have been able to evaluate the surface integral in such an easy way for a surface of any other shape. Symmetry is the key.

FIGURE 27.17 The electric field and the closed surface through which we will calculate the electric flux.

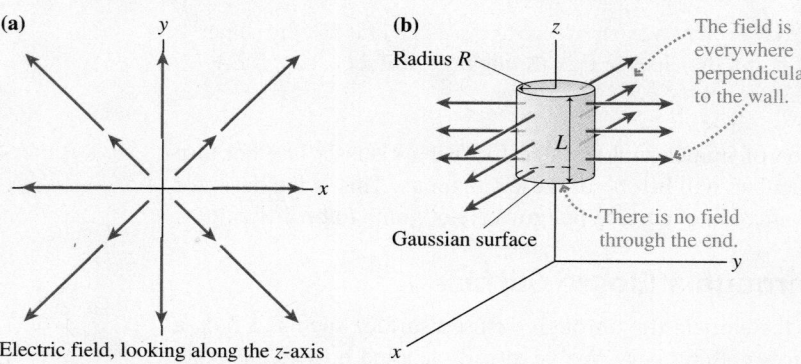

(a)

Electric field, looking along the z-axis

(b)

Radius R

Gaussian surface

L

The field is everywhere perpendicular to the wall.

There is no field through the end.

Example 27.2 illustrated a two-step approach to performing a flux integral over a closed surface. In summary:

Finding the flux through a closed surface

❶ Divide the closed surface into pieces that are everywhere tangent to the electric field and everywhere perpendicular to the electric field.
❷ Use Tactics Box 27.1 to evaluate the surface integrals over these surfaces, then add the results.

Exercise 11

STOP TO THINK 27.3 The total electric flux through this box is

a. $0 \, \text{N}\,\text{m}^2/\text{C}$
b. $1 \, \text{N}\,\text{m}^2/\text{C}$
c. $2 \, \text{N}\,\text{m}^2/\text{C}$
d. $4 \, \text{N}\,\text{m}^2/\text{C}$
e. $6 \, \text{N}\,\text{m}^2/\text{C}$
f. $8 \, \text{N}\,\text{m}^2/\text{C}$

$\vec{E} = (1 \, \text{N/C, up})$

Plane of charge

Cross section of a $1 \, \text{m} \times 1 \, \text{m} \times 1 \, \text{m}$ box

$\vec{E} = (1 \, \text{N/C, down})$

27.4 Gauss's Law

The last section was long, but knowing how to calculate the electric flux through a closed surface is essential for the main topic of this chapter: Gauss's law. Gauss's law is equivalent to Coulomb's law for static charges, although Gauss's law will look very different.

The purpose of learning Gauss's law is twofold:

■ Gauss's law allows the electric fields of some continuous distributions of charge to be found much more easily than does Coulomb's law.
■ Gauss's law is valid for *moving* charges, but Coulomb's law is not (although it's a very good approximation for velocities that are much less than the speed of light). Thus Gauss's law is ultimately a more fundamental statement about electric fields than is Coulomb's law.

Let's start with Coulomb's law for the electric field of a point charge. FIGURE 27.18 shows a spherical Gaussian surface of radius r centered on a positive charge q. Keep in mind that this is an imaginary, mathematical surface, not a physical surface. There is a net flux through this surface because the electric field points outward at every point on the surface. To evaluate the flux, given formally by the surface integral of Equation 27.11, notice that the electric field is perpendicular to the surface at every point on the surface *and,* from Coulomb's law, it has the same magnitude $E = q/4\pi\epsilon_0 r^2$ at every point on the surface. This simple situation arises because **the Gaussian surface has the same symmetry as the electric field.**

Thus we know, without having to do any hard work, that the flux integral is

$$\Phi_e = \oint \vec{E} \cdot d\vec{A} = EA_{\text{sphere}} \tag{27.12}$$

The surface area of a sphere of radius r is $A_{\text{sphere}} = 4\pi r^2$. If we use A_{sphere} and the Coulomb-law expression for E in Equation 27.12, we find that the electric flux through the spherical surface is

$$\Phi_e = \frac{q}{4\pi\epsilon_0 r^2} \, 4\pi r^2 = \frac{q}{\epsilon_0} \tag{27.13}$$

FIGURE 27.18 A spherical Gaussian surface surrounding a point charge.

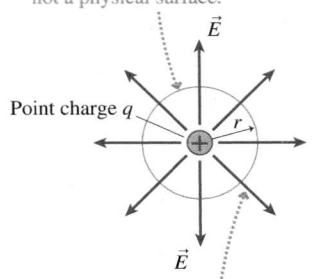

Cross section of a Gaussian sphere of radius r. This is a mathematical surface, not a physical surface.

Point charge q

The electric field is everywhere perpendicular to the surface *and* has the same magnitude at every point.

You should examine the logic of this calculation closely. We really did evaluate the surface integral of Equation 27.11, although it may appear, at first, as if we didn't do much. The integral was easily evaluated, we reiterate for emphasis, because the closed surface on which we performed the integration matched the *symmetry* of the charge distribution.

NOTE ▶ We found Equation 27.13 for a positive charge, but it applies equally to negative charges. According to Equation 27.13, Φ_e is negative if q is negative. And that's what we would expect from the basic definition of flux, $\vec{E} \cdot \vec{A}$. The electric field of a negative charge points inward, while the area vector of a closed surface points outward, making the dot product negative. ◀

Electric Flux Is Independent of Surface Shape and Radius

Notice something interesting about Equation 27.13. The electric flux depends on the amount of charge but *not* on the radius of the sphere. Although this may seem a bit surprising, it's really a direct consequence of what we *mean* by flux. Think of the fluid analogy with which we introduced the term "flux." If fluid flows outward from a central point, all the fluid crossing a small-radius spherical surface will, at some later time, cross a large-radius spherical surface. No fluid is lost along the way, and no new fluid is created. Similarly, the point charge in FIGURE 27.19 is the only source of electric field. Every electric field line passing through a small-radius spherical surface also passes through a large-radius spherical surface. Hence the electric flux is independent of r.

NOTE ▶ This argument hinges on the fact that Coulomb's law is an inverse-square force law. The electric field strength, which is proportional to $1/r^2$, decreases with distance. But the surface area, which increases in proportion to r^2, exactly compensates for this decrease. Consequently, the electric flux of a point charge through a spherical surface is independent of the radius of the sphere. ◀

This conclusion about the flux has an extremely important generalization. FIGURE 27.20a shows a point charge and a closed Gaussian surface of arbitrary shape and dimensions. All we know is that the charge is *inside* the surface. What is the electric flux through this arbitrary surface?

One way to answer the question is to approximate the surface as a patchwork of spherical and radial pieces. The spherical pieces are centered on the charge and the radial pieces lie along lines extending outward from the charge. (Figure 27.20 is a two-dimensional drawing so you need to imagine these arcs as actually being pieces of a spherical shell.) The figure, of necessity, shows fairly large pieces that don't match the actual surface all that well. However, we can make this approximation as good as we want by letting the pieces become sufficiently small.

The electric field is everywhere tangent to the radial pieces. Hence the electric flux through the radial pieces is zero. The spherical pieces, although at varying distances from the charge, form a *complete sphere*. That is, any line drawn radially outward from the charge will pass through exactly one spherical piece, and no radial lines can avoid passing through a spherical piece. You could even imagine, as FIGURE 27.20b shows, sliding the spherical pieces in and out *without changing the angle they subtend* until they come together to form a complete sphere.

Consequently, the electric flux through these spherical pieces that, when assembled, form a complete sphere must be exactly the same as the flux q/ϵ_0 through a spherical Gaussian surface. In other words, **the flux through *any* closed surface surrounding a point charge q is**

$$\Phi_e = \oint \vec{E} \cdot d\vec{A} = \frac{q}{\epsilon_0} \qquad (27.14)$$

This surprisingly simple result is a consequence of the fact that Coulomb's law is an inverse-square force law. Even so, the reasoning that got us to Equation 27.14 is rather subtle and well worth reviewing.

FIGURE 27.19 The electric flux is the same through *every* sphere centered on a point charge.

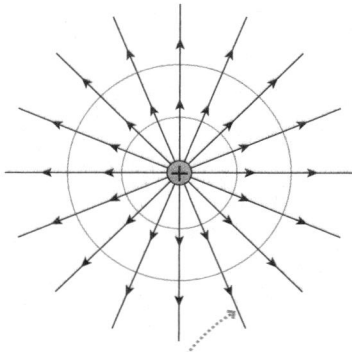

Every field line passing through the smaller sphere also passes through the larger sphere. Hence the flux through the two spheres is the same.

FIGURE 27.20 An arbitrary Gaussian surface can be approximated with spherical and radial pieces.

(a) Point charge   The spherical pieces are centered on the charge.

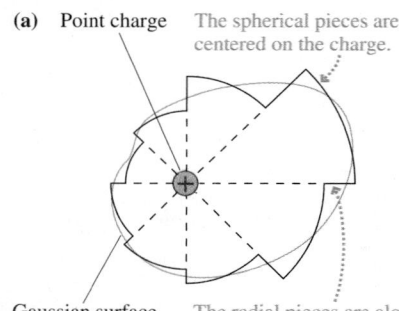

Gaussian surface of arbitrary shape

The radial pieces are along lines extending out from the charge. There's no flux through these.

(b)

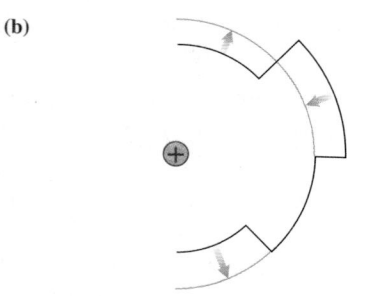

The spherical pieces can slide in or out to form a complete sphere. Hence the flux through the pieces is the same as the flux through a sphere.

Charge Outside the Surface

The closed surface shown in FIGURE 27.21a has a point charge q outside the surface but no charges inside. Now what can we say about the flux? By approximating this surface with spherical and radial pieces *centered on the charge,* as we did in Figure 27.20, we can reassemble the surface into the equivalent surface of FIGURE 27.21b. This closed surface consists of sections of two spherical shells, and it is equivalent in the sense that the electric flux through this surface is the same as the electric flux through the original surface of Figure 27.21a.

FIGURE 27.21 A point charge outside a Gaussian surface.

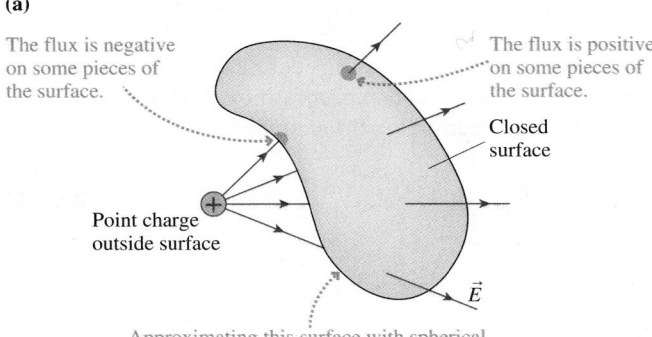

(a)

The flux is negative on some pieces of the surface.

The flux is positive on some pieces of the surface.

Closed surface

Point charge outside surface

$\vec{E}$

Approximating this surface with spherical and radial pieces allows it to be reassembled as the surface in part (b) that has the same flux.

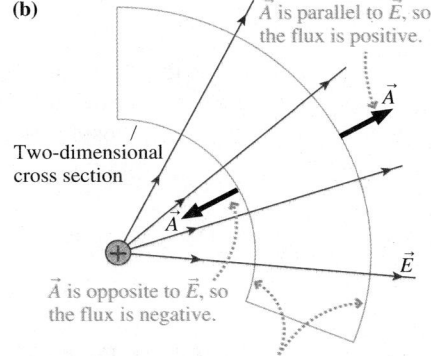

(b)

$\vec{A}$ is parallel to $\vec{E}$, so the flux is positive.

$\vec{A}$

Two-dimensional cross section

$\vec{A}$

$\vec{A}$ is opposite to $\vec{E}$, so the flux is negative.

$\vec{E}$

The fluxes through these surfaces are equal but opposite. The net flux is zero.

If the electric field were a fluid flowing outward from the charge, all the fluid *entering* the closed region through the first spherical surface would later *exit* the region through the second spherical surface. There is no *net* flow into or out of the closed region. Similarly, every electric field line entering this closed volume through one spherical surface exits through the other spherical surface.

Mathematically, the electric fluxes through the two spherical surfaces have the same magnitude because Φ_e is independent of r. But they have *opposite signs* because the outward-pointing area vector $\vec{A}$ is parallel to $\vec{E}$ on one surface but opposite to $\vec{E}$ on the other. The sum of the fluxes through the two surfaces is zero, and we are led to the conclusion that **the net electric flux is zero through a closed surface that does not contain any net charge.** Charges outside the surface do not produce a net flux through the surface.

This isn't to say that the flux through a small piece of the surface is zero. In fact, as Figure 27.21a shows, nearly every piece of the surface has an electric field either entering or leaving and thus has a nonzero flux. But some of these are positive and some are negative. When summed over the *entire* surface, the positive and negative contributions exactly cancel to give no *net* flux.

Multiple Charges

Finally, consider an arbitrary Gaussian surface and a group of charges $q_1, q_2, q_3, \ldots$ such as those shown in FIGURE 27.22. What is the net electric flux through the surface?

By definition, the net flux is

$$\Phi_e = \oint \vec{E} \cdot d\vec{A}$$

From the principle of superposition, the electric field is $\vec{E} = \vec{E}_1 + \vec{E}_2 + \vec{E}_3 + \cdots$, where $\vec{E}_1, \vec{E}_2, \vec{E}_3, \ldots$ are the fields of the individual charges. Thus the flux is

$$\Phi_e = \oint \vec{E}_1 \cdot d\vec{A} + \oint \vec{E}_2 \cdot d\vec{A} + \oint \vec{E}_3 \cdot d\vec{A} + \cdots \qquad (27.15)$$

$$= \Phi_1 + \Phi_2 + \Phi_3 + \cdots$$

FIGURE 27.22 Charges both inside and outside a Gaussian surface.

The fluxes due to charges outside the surface are all zero.

q_3

q_2

q_1

Two-dimensional cross section of a Gaussian surface

Total charge inside is Q_{in}.

The fluxes due to charges inside the surface add.

where Φ_1, Φ_2, Φ_3, ... are the fluxes through the Gaussian surface due to the individual charges. That is, the net flux is the sum of the fluxes due to individual charges. But we know what those are: q/ϵ_0 for the charges inside the surface and zero for the charges outside. Thus

$$\Phi_e = \left(\frac{q_1}{\epsilon_0} + \frac{q_2}{\epsilon_0} + \cdots + \frac{q_i}{\epsilon_0} \text{ for all charges inside the surface} \right) \qquad (27.16)$$
$$+(0 + 0 + \cdots + 0 \text{ for all charges outside the surface})$$

We define

$$Q_{in} = q_1 + q_2 + \cdots + q_i \text{ for all charges inside the surface} \qquad (27.17)$$

as the total charge enclosed *within* the surface. With this definition, we can write our result for the net electric flux in a very neat and compact fashion. For any *closed* surface enclosing total charge Q_{in}, the net electric flux through the surface is

$$\Phi_e = \oint \vec{E} \cdot d\vec{A} = \frac{Q_{in}}{\epsilon_0} \qquad (27.18)$$

This result for the electric flux is known as **Gauss's law.**

What Does Gauss's Law Tell Us?

In one sense, Gauss's law doesn't say anything new or anything that we didn't already know from Coulomb's law. After all, we derived Gauss's law from Coulomb's law. But in another sense, Gauss's law is more important than Coulomb's law. Gauss's law states a very general property of electric fields—namely, that charges create electric fields in just such a way that the net flux of the field is the same through *any* surface surrounding the charges, no matter what its size and shape may be. This fact may have been implied by Coulomb's law, but it was by no means obvious. And Gauss's law will turn out to be particularly useful later when we combine it with other electric and magnetic field equations.

Gauss's law is the mathematical statement of our observations in Section 27.2. There we noticed a net "flow" of electric field out of a closed surface containing charges. Gauss's law quantifies this idea by making a specific connection between the "flow," now measured as electric flux, and the amount of charge.

But is it useful? Although to some extent Gauss's law is a formal statement about electric fields, not a tool for solving practical problems, there are exceptions: Gauss's law will allow us to find the electric fields of some very important and very practical charge distributions much more easily than if we had to rely on Coulomb's law. We'll consider some examples in the next section.

STOP TO THINK 27.4 These are two-dimensional cross sections through three-dimensional closed spheres and a cube. Rank in order, from largest to smallest, the electric fluxes Φ_a to Φ_e through surfaces a to e.

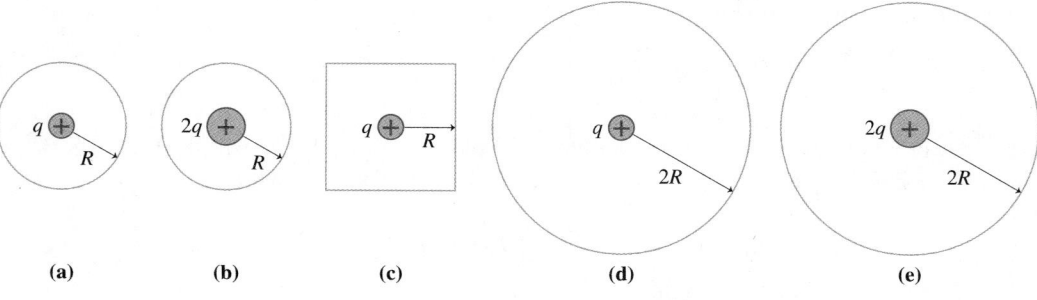

(a) (b) (c) (d) (e)

27.5 Using Gauss's Law

In this section, we'll use Gauss's law to determine the electric fields of several important charge distributions. Some of these you already know, from Chapter 26; others will be new. Three important observations can be made about using Gauss's law:

1. Gauss's law applies only to a *closed* surface, called a Gaussian surface.
2. A Gaussian surface is not a physical surface. It need not coincide with the boundary of any physical object (although it could if we wished). It is an imaginary, mathematical surface in the space surrounding one or more charges.
3. We can't find the electric field from Gauss's law alone. We need to apply Gauss's law in situations where, from symmetry and superposition, we already can guess the *shape* of the field.

These observations and our previous discussion of symmetry and flux lead to the following strategy for solving electric field problems with Gauss's law.

PROBLEM-SOLVING STRATEGY 27.1   Gauss's law

MODEL Model the charge distribution as a distribution with symmetry.

VISUALIZE Draw a picture of the charge distribution.

- Determine the symmetry of its electric field.
- Choose and draw a Gaussian surface with the *same symmetry*.
- You need not enclose all the charge within the Gaussian surface.
- Be sure every part of the Gaussian surface is either tangent to or perpendicular to the electric field.

SOLVE The mathematical representation is based on Gauss's law

$$\Phi_e = \oint \vec{E} \cdot d\vec{A} = \frac{Q_{in}}{\epsilon_0}$$

- Use Tactics Boxes 27.1 and 27.2 to evaluate the surface integral.

ASSESS Check that your result has the correct units, is reasonable, and answers the question.

Exercise 19

EXAMPLE 27.3   **Outside a sphere of charge**

In Chapter 26 we asserted, without proof, that the electric field outside a sphere of total charge Q is the same as the field of a point charge Q at the center. Use Gauss's law to prove this result.

MODEL The charge distribution within the sphere need not be uniform (i.e., the charge density might increase or decrease with r), but it must have spherical symmetry in order for us to use Gauss's law. We will assume that it does.

VISUALIZE FIGURE 27.23 shows a sphere of charge Q and radius R. We want to find $\vec{E}$ outside this sphere, for distances $r > R$. The spherical symmetry of the charge distribution tells us that the electric field must point *radially outward* from the sphere. Although Gauss's law is true for any surface surrounding the charged sphere, it is useful only if we choose a Gaussian surface to match the spherical symmetry of the charge distribution and the field. Thus a spherical surface of radius $r > R$ *concentric with*

FIGURE 27.23 A spherical Gaussian surface surrounding a sphere of charge.

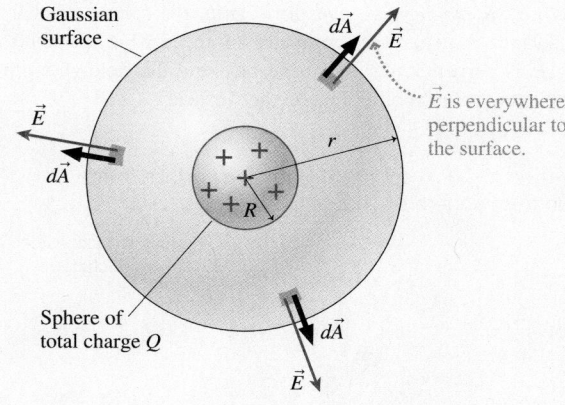

Continued

the charged sphere will be our Gaussian surface. Because this surface surrounds the entire sphere of charge, the enclosed charge is simply $Q_{in} = Q$.

SOLVE Gauss's law is

$$\Phi_e = \oint \vec{E} \cdot d\vec{A} = \frac{Q_{in}}{\epsilon_0} = \frac{Q}{\epsilon_0}$$

To calculate the flux, notice that the electric field is everywhere perpendicular to the spherical surface. And although we don't know the electric field magnitude E, spherical symmetry dictates that E must have the same value at all points equally distant from the center of the sphere. Thus we have the simple result that the net flux through the Gaussian surface is

$$\Phi_e = EA_{sphere} = 4\pi r^2 E$$

where we used the fact that the surface area of a sphere is $A_{sphere} = 4\pi r^2$. With this result for the flux, Gauss's law is

$$4\pi r^2 E = \frac{Q}{\epsilon_0}$$

Thus the electric field at distance r outside a sphere of charge is

$$E_{outside} = \frac{1}{4\pi\epsilon_0} \frac{Q}{r^2}$$

Or in vector form, making use of the fact that $\vec{E}$ is radially outward,

$$\vec{E}_{outside} = \frac{1}{4\pi\epsilon_0} \frac{Q}{r^2} \hat{r}$$

where $\hat{r}$ is a radial unit vector.

ASSESS The field is exactly that of a point charge Q, which is what we wanted to show.

The derivation of the electric field of a sphere of charge depended crucially on a proper choice of the Gaussian surface. We would not have been able to evaluate the flux integral so simply for any other choice of surface. It's worth noting that the result of Example 27.3 can also be proven by the superposition of point-charge fields, but it requires a difficult three-dimensional integral and about a page of algebra. We obtained the answer using Gauss's law in just a few lines. Where Gauss's law works, it works *extremely* well! However, it works only in situations, such as this, with a very high degree of symmetry.

EXAMPLE 27.4 **Inside a sphere of charge**

What is the electric field *inside* a uniformly charged sphere?

MODEL We haven't considered a situation like this before. To begin, we don't know if the field strength is increasing or decreasing as we move outward from the center of the sphere. But the field inside must have spherical symmetry. That is, the field must point radially inward or outward, and the field strength can depend only on r. This is sufficient information to solve the problem because it allows us to choose a Gaussian surface.

VISUALIZE **FIGURE 27.24** shows a spherical Gaussian surface with radius $r \leq R$ inside, and *concentric with,* the sphere of charge. This surface matches the symmetry of the charge distribution, hence $\vec{E}$ is perpendicular to this surface and the field strength E has the same value at all points on the surface.

FIGURE 27.24 A spherical Gaussian surface inside a uniform sphere of charge.

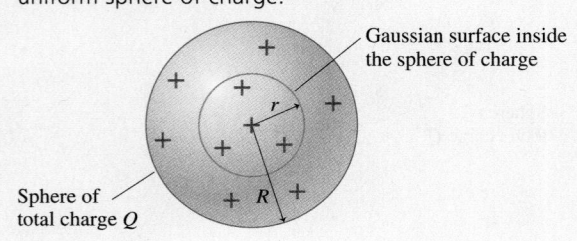

Sphere of total charge Q

Gaussian surface inside the sphere of charge

SOLVE The flux integral is identical to that of Example 27.3:

$$\Phi_e = EA_{sphere} = 4\pi r^2 E$$

Consequently, Gauss's law is

$$\Phi_e = 4\pi r^2 E = \frac{Q_{in}}{\epsilon_0}$$

The difference between this example and Example 27.3 is that Q_{in} is no longer the total charge of the sphere. Instead, Q_{in} is the amount of charge *inside* the Gaussian sphere of radius r. Because the charge distribution is *uniform,* the volume charge density is

$$\rho = \frac{Q}{V_R} = \frac{Q}{\frac{4}{3}\pi R^3}$$

The charge enclosed in a sphere of radius r is thus

$$Q_{in} = \rho V_r = \left(\frac{Q}{\frac{4}{3}\pi R^3}\right)\left(\frac{4}{3}\pi r^3\right) = \frac{r^3}{R^3}Q$$

The amount of enclosed charge increases with the cube of the distance r from the center and, as expected, $Q_{in} = Q$ if $r = R$. With this expression for Q_{in}, Gauss's law is

$$4\pi r^2 E = \frac{(r^3/R^3)Q}{\epsilon_0}$$

Thus the electric field at radius r inside a uniformly charged sphere is

$$E_{\text{inside}} = \frac{1}{4\pi\epsilon_0}\frac{Q}{R^3}r$$

The electric field strength inside the sphere increases *linearly* with the distance r from the center.

ASSESS The field inside and the field outside a sphere of charge match at the boundary of the sphere, $r = R$, where both give $E = Q/4\pi\epsilon_0 R^2$. In other words, the field strength is *continuous* as we cross the boundary of the sphere. These results are shown graphically in FIGURE 27.25.

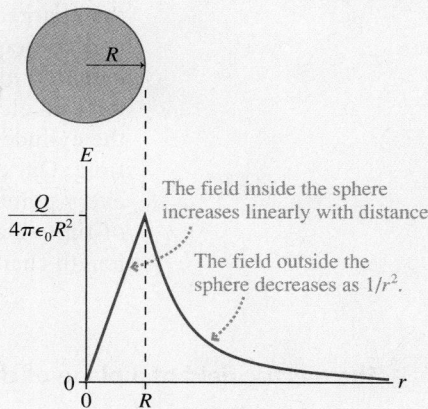

FIGURE 27.25 The electric field strength of a uniform sphere of charge of radius R.

The field inside the sphere increases linearly with distance.

The field outside the sphere decreases as $1/r^2$.

EXAMPLE 27.5 **The electric field of a long, charged wire**

In Chapter 26, we used superposition to find the electric field of an infinitely long line of charge with linear charge density (C/m) λ. It was not an easy derivation. Find the electric field using Gauss's law.

MODEL A long, charged wire can be modeled as an infinitely long line of charge.

VISUALIZE FIGURE 27.26 shows an infinitely long line of charge. We can use the symmetry of the situation to see that the only possible shape of the electric field is to point straight into or out from the wire, rather like the bristles on a bottle brush. The shape of the field suggests that we choose our Gaussian surface to be a cylinder of radius r and length L, centered on the wire. Because Gauss's law refers to *closed* surfaces, we must include the ends of the cylinder as part of the surface.

FIGURE 27.26 A Gaussian surface around a charged wire.

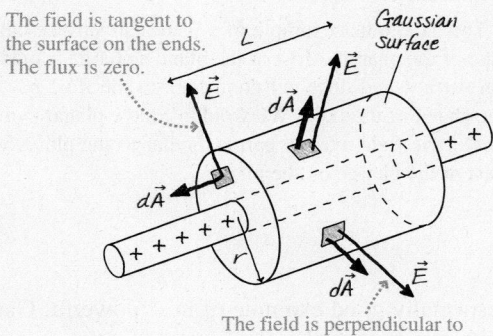

The field is tangent to the surface on the ends. The flux is zero.

Gaussian surface

The field is perpendicular to the surface on the cylinder wall.

SOLVE Gauss's law is

$$\Phi_e = \oint \vec{E} \cdot d\vec{A} = \frac{Q_{\text{in}}}{\epsilon_0}$$

where Q_{in} is the charge *inside* the closed cylinder. We have two tasks: to evaluate the flux integral, and to determine how much

charge is inside the closed surface. The wire has linear charge density λ, so the amount of charge inside a cylinder of length L is simply

$$Q_{\text{in}} = \lambda L$$

Finding the net flux is just as straightforward. We can divide the flux through the entire closed surface into the flux through each end plus the flux through the cylindrical wall. The electric field $\vec{E}$, pointing straight out from the wire, is tangent to the end surfaces at every point. Thus the flux through these two surfaces is zero. On the wall, $\vec{E}$ is perpendicular to the surface and has the same strength E at every point. Thus

$$\Phi_e = \Phi_{\text{top}} + \Phi_{\text{bottom}} + \Phi_{\text{wall}} = 0 + 0 + EA_{\text{cyl}} = 2\pi r L E$$

where we used $A_{\text{cyl}} = 2\pi r L$ as the surface area of a cylindrical wall of radius r and length L. Once again, the proper choice of the Gaussian surface reduces the flux integral merely to finding a surface area. With these expressions for Q_{in} and Φ_e, Gauss's law is

$$\Phi_e = 2\pi r L E = \frac{Q_{\text{in}}}{\epsilon_0} = \frac{\lambda L}{\epsilon_0}$$

Thus the electric field at distance r from a long, charged wire is

$$E_{\text{wire}} = \frac{\lambda}{2\pi\epsilon_0 r}$$

ASSESS This agrees exactly with the result of the more complex derivation in Chapter 26. Notice that the result does not depend on our choice of L. A Gaussian surface is an imaginary device, not a physical object. We needed a finite-length cylinder to do the flux calculation, but the electric field of an *infinitely* long wire can't depend on the length of an imaginary cylinder.

Example 27.5, for the electric field of a long, charged wire, contains a subtle but important idea, one that often occurs when using Gauss's law. The Gaussian cylinder of length L encloses only some of the wire's charge. The pieces of the charged wire outside the cylinder are not enclosed by the Gaussian surface and consequently do not contribute anything to the net flux. Even so, *they are essential* to the use of Gauss's law because it takes the *entire* charged wire to produce an electric field with cylindrical symmetry. In other words, the wire outside the cylinder may not contribute to the flux, but it affects the *shape* of the electric field. Our ability to write $\Phi_e = EA_{cyl}$ depended on knowing that E is the same at every point on the wall of the cylinder. That would not be true for a charged wire of finite length, so we cannot use Gauss's law to find the electric field of a finite-length charged wire.

EXAMPLE 27.6 The electric field of a plane of charge

Use Gauss's law to find the electric field of an infinite plane of charge with surface charge density (C/m^2) η.

MODEL A uniformly charged flat electrode can be modeled as an infinite plane of charge.

VISUALIZE FIGURE 27.27 shows a uniformly charged plane with surface charge density η. We will assume that the plane extends infinitely far in all directions, although we obviously have to show "edges" in our drawing. The planar symmetry allows the electric field to point only straight toward or away from the plane. With this in mind, choose as a Gaussian surface a cylinder with length L and faces of area A centered on the plane of charge. Although we've drawn them as circular, the shape of the faces is not relevant.

FIGURE 27.27 The Gaussian surface extends to both sides of a plane of charge.

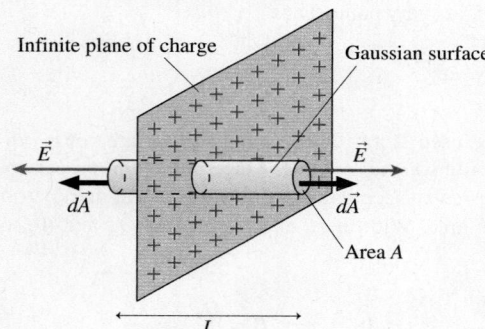

Infinite plane of charge
Gaussian surface
$\vec{E}$
$d\vec{A}$
$\vec{E}$
$d\vec{A}$
Area A
L

SOLVE The electric field is perpendicular to both faces of the cylinder, so the total flux through both faces is $\Phi_{faces} = 2EA$. (The fluxes add rather than cancel because the area vector $\vec{A}$ points *outward* on each face.) There's *no* flux through the wall of the cylinder because the field vectors are tangent to the wall. Thus the net flux is simply

$$\Phi_e = 2EA$$

The charge inside the cylinder is the charge contained in area A of the plane. This is

$$Q_{in} = \eta A$$

With these expressions for Q_{in} and Φ_e, Gauss's law is

$$\Phi_e = 2EA = \frac{Q_{in}}{\epsilon_0} = \frac{\eta A}{\epsilon_0}$$

Thus the electric field of an infinite charged plane is

$$E_{plane} = \frac{\eta}{2\epsilon_0}$$

This agrees with the result in Chapter 26.

ASSESS This is another example of a Gaussian surface enclosing only some of the charge. Most of the plane's charge is outside the Gaussian surface and does not contribute to the flux, but it does affect the shape of the field. We wouldn't have planar symmetry, with the electric field exactly perpendicular to the plane, without all the rest of the charge on the plane.

The plane of charge is an especially good example of how powerful Gauss's law can be. Finding the electric field of a plane of charge via superposition was a difficult and tedious derivation. With Gauss's law, once you see how to apply it, the problem is simple enough to solve in your head!

You might wonder, then, why we bothered with superposition at all. The reason is that Gauss's law, powerful though it may be, is effective only in a limited number of situations where the field is highly symmetric. Superposition always works, even if the derivation is messy, because superposition goes directly back to the fields of individual point charges. It's good to use Gauss's law when you can, but superposition is often the only way to attack real-world charge distributions.

STOP TO THINK 27.5 Which Gaussian surface would allow you to use Gauss's law to determine the electric field outside a uniformly charged cube?

a. A sphere whose center coincides with the center of the charged cube
b. A cube whose center coincides with the center of the charged cube and that has parallel faces
c. Either a or b
d. Neither a nor b

27.6 Conductors in Electrostatic Equilibrium

Consider a charged conductor, such as a charged metal electrode, in electrostatic equilibrium. That is, there is no current through the conductor and the charges are all stationary. One very important conclusion is that **the electric field is zero at all points within a conductor in electrostatic equilibrium.** That is, $\vec{E}_{in} = \vec{0}$. If this weren't true, the electric field would cause the charge carriers to move and thus violate the assumption that all the charges are at rest. Let's use Gauss's law to see what else we can learn.

At the Surface of a Conductor

FIGURE 27.28 shows a Gaussian surface just barely inside the physical surface of a conductor that's in electrostatic equilibrium. The electric field is zero at all points within the conductor, hence the electric flux Φ_e through this Gaussian surface must be zero. But if $\Phi_e = 0$, Gauss's law tells us that $Q_{in} = 0$. That is, there's no net charge within this surface. There are charges—electrons and positive ions—but no *net* charge.

If there's no net charge in the interior of a conductor in electrostatic equilibrium, then **all the excess charge on a charged conductor resides on the exterior surface of the conductor.** Any charges added to a conductor quickly spread across the surface until reaching positions of electrostatic equilibrium, but there is no net charge *within* the conductor.

There may be no electric field within a charged conductor, but the presence of net charge requires an exterior electric field in the space outside the conductor. **FIGURE 27.29** shows that **the electric field right at the surface of the conductor has to be perpendicular to the surface.** To see that this is so, suppose $\vec{E}_{surface}$ had a component tangent to the surface. This component of $\vec{E}_{surface}$ would exert a force on the surface charges and cause a surface current, thus violating the assumption that all charges are at rest. The only exterior electric field consistent with electrostatic equilibrium is one that is perpendicular to the surface.

We can use Gauss's law to relate the field strength at the surface to the charge density on the surface. **FIGURE 27.30** shows a small Gaussian cylinder with faces very slightly above and below the surface of a charged conductor. The charge inside this Gaussian cylinder is ηA, where η is the surface charge density at this point on the conductor. There's a flux $\Phi = AE_{surface}$ through the outside face of this cylinder but, unlike Example 27.6 for the plane of charge, *no* flux through the inside face because $\vec{E}_{in} = \vec{0}$ within the conductor. Furthermore, there's no flux through the wall of the cylinder because $\vec{E}_{surface}$ is perpendicular to the surface. Thus the net flux is $\Phi_e = AE_{surface}$. Gauss's law is

$$\Phi_e = AE_{surface} = \frac{Q_{in}}{\epsilon_0} = \frac{\eta A}{\epsilon_0} \qquad (27.19)$$

from which we can conclude that the electric field at the surface of a charged conductor is

$$\vec{E}_{surface} = \left(\frac{\eta}{\epsilon_0}, \text{ perpendicular to surface} \right) \qquad (27.20)$$

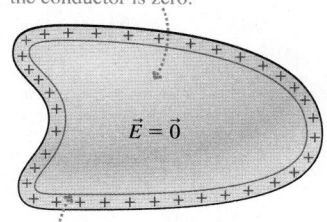

FIGURE 27.28 A Gaussian surface just inside a conductor that's in electrostatic equilibrium.

The electric field inside the conductor is zero.

$\vec{E} = \vec{0}$

The flux through the Gaussian surface is zero. There's no net charge inside the conductor. Hence all the excess charge is on the surface.

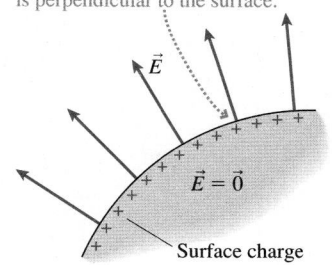

FIGURE 27.29 The electric field at the surface of a charged conductor.

The electric field at the surface is perpendicular to the surface.

$\vec{E}$

$\vec{E} = \vec{0}$

Surface charge

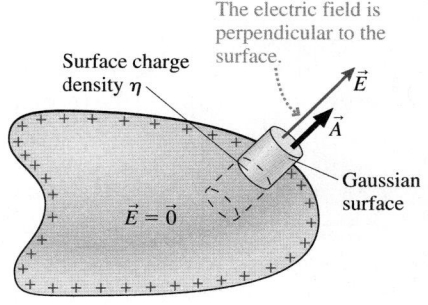

FIGURE 27.30 A Gaussian surface extending through the surface of the conductor has a flux only through the outer face.

The electric field is perpendicular to the surface.

Surface charge density η

$\vec{E}$

$\vec{A}$

Gaussian surface

$\vec{E} = \vec{0}$

In general, the surface charge density η is *not* constant on the surface of a conductor but depends on the shape of the conductor. If we can determine η, by either calculating it or measuring it, then Equation 27.20 tells us the electric field at that point on the surface. Alternatively, we can use Equation 27.20 to deduce the charge density on the conductor's surface if we know the electric field just outside the conductor.

Charges and Fields Within a Conductor

FIGURE 27.31 shows a charged conductor with a hole inside. Can there be charge on this interior surface? To find out, we place a Gaussian surface around the hole, infinitesimally close but entirely within the conductor. The electric flux Φ_e through this Gaussian surface is zero because the electric field is zero everywhere inside the conductor. Thus we must conclude that $Q_{in} = 0$. There's no net charge inside this Gaussian surface and thus no charge on the surface of the hole. Any excess charge resides on the *exterior* surface of the conductor, not on any interior surfaces.

Furthermore, because there's no electric field inside the conductor and no charge inside the hole, the electric field inside the hole must also be zero. This conclusion has an important practical application. For example, suppose we need to exclude the electric field from the region in FIGURE 27.32a enclosed within dashed lines. We can do so by surrounding this region with the neutral conducting box of FIGURE 27.32b.

FIGURE 27.31 A Gaussian surface surrounding a hole inside a conductor in electrostatic equilibrium.

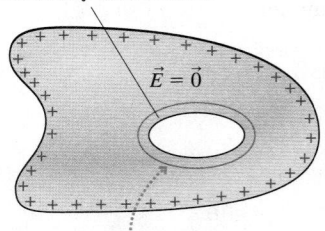

A hollow completely enclosed by the conductor

$\vec{E} = \vec{0}$

The flux through the Gaussian surface is zero. There's no net charge inside, hence no charge on this interior surface.

FIGURE 27.32 The electric field can be excluded from a region of space by surrounding it with a conducting box.

(a) Parallel-plate capacitor

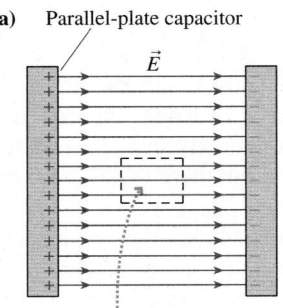

$\vec{E}$

We want to exclude the electric field from this region.

(b) The conducting box has been polarized and has induced surface charges.

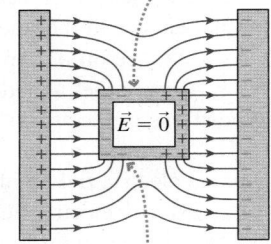

$\vec{E} = \vec{0}$

The electric field is perpendicular to all conducting surfaces.

This region of space is now a hole inside a conductor, thus the interior electric field is zero. The use of a conducting box to exclude electric fields from a region of space is called **screening**. Solid metal walls are ideal, but in practice wire screen or wire mesh—sometimes called a *Faraday cage*—provides sufficient screening for all but the most sensitive applications. The price we pay is that the exterior field is now very complicated.

Finally, FIGURE 27.33 shows a charge q inside a hole within a neutral conductor. The electric field *within* the conductor is still zero, hence the electric flux through the Gaussian surface is zero. But $\Phi_e = 0$ requires $Q_{in} = 0$. Consequently, the charge inside the hole attracts an equal charge of opposite sign, and charge $-q$ now lines the inner surface of the hole.

The conductor as a whole is neutral, so moving $-q$ to the surface of the hole must leave $+q$ behind somewhere else. Where is it? It can't be in the interior of the conductor, as we've seen, and that leaves only the exterior surface. In essence, an internal charge polarizes the conductor just as an external charge would. Net charge $-q$ moves to the inner surface and net charge $+q$ is left behind on the exterior surface.

In summary, conductors in electrostatic equilibrium have the properties described in Tactics Box 27.3.

FIGURE 27.33 A charge in the hole causes a net charge on the interior and exterior surfaces.

The flux through the Gaussian surface is zero, hence there's no *net* charge inside this surface. There must be charge $-q$ on the inside surface to balance point charge q.

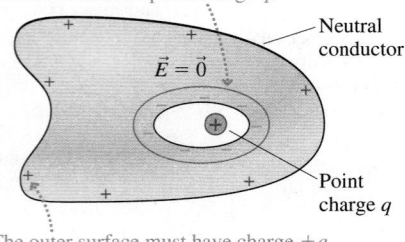

Neutral conductor

$\vec{E} = \vec{0}$

Point charge q

The outer surface must have charge $+q$ so that the conductor remains neutral.

TACTICS
BOX 27.3
Finding the electric field of a conductor in electrostatic equilibrium

❶ The electric field is zero at all points within the volume of the conductor.
❷ Any excess charge resides entirely on the *exterior* surface.
❸ The external electric field at the surface of a charged conductor is perpendicular to the surface and of magnitude η/ϵ_0, where η is the surface charge density at that point.
❹ The electric field is zero inside any hole within a conductor unless there is a charge in the hole.

Exercises 20–24

EXAMPLE 27.7 The electric field at the surface of a charged metal sphere

A 2.0-cm-diameter brass sphere has been given a charge of 2.0 nC. What is the electric field strength at the surface?

MODEL Brass is a conductor. The excess charge resides on the surface.

VISUALIZE The charge distribution has spherical symmetry. The electric field points radially outward from the surface.

SOLVE We can solve this problem in two ways. One uses the fact that a sphere is the one shape for which any excess charge will spread out to a *uniform* surface charge density. Thus

$$\eta = \frac{q}{A_{\text{sphere}}} = \frac{q}{4\pi R^2} = \frac{2.0 \times 10^{-9}\,\text{C}}{4\pi(0.010\,\text{m})^2} = 1.59 \times 10^{-6}\,\text{C/m}^2$$

From Equation 27.20, we know the electric field at the surface has strength

$$E_{\text{surface}} = \frac{\eta}{\epsilon_0} = \frac{1.59 \times 10^{-6}\,\text{C/m}^2}{8.85 \times 10^{-12}\,\text{C}^2/\text{N m}^2} = 1.8 \times 10^5\,\text{N/C}$$

Alternatively, we could have used the result, obtained earlier in the chapter, that the electric field strength outside a sphere of charge Q is $E_{\text{outside}} = Q_{\text{in}}/(4\pi\epsilon_0 r^2)$. But $Q_{\text{in}} = q$ and, at the surface, $r = R$. Thus

$$E_{\text{surface}} = \frac{1}{4\pi\epsilon_0}\frac{q}{R^2} = (9.0 \times 10^9\,\text{N m}^2/\text{C}^2)\frac{2.0 \times 10^{-9}\,\text{C}}{(0.010\,\text{m})^2}$$

$$= 1.8 \times 10^5\,\text{N/C}$$

As we can see, both methods lead to the same result.

CHALLENGE EXAMPLE 27.8 The electric field of a slab of charge

An infinite slab of charge of thickness $2a$ is centered in the xy-plane. The charge density is $\rho = \rho_0(1 - |z|/a)$. Find the electric field strengths inside and outside this slab of charge.

MODEL The charge density is not uniform. Starting at ρ_0 in the xy-plane, it decreases linearly with distance above and below the xy-plane until reaching zero at $z = \pm a$, the edges of the slab.

VISUALIZE **FIGURE 27.34** shows an edge view of the slab of charge and, as Gaussian surfaces, side views of two cylinders with cross-section area A. By symmetry, the electric field must point away from the xy-plane; the field cannot have an x- or y-component.

FIGURE 27.34 Two cylindrical Gaussian surfaces for an infinite slab of charge.

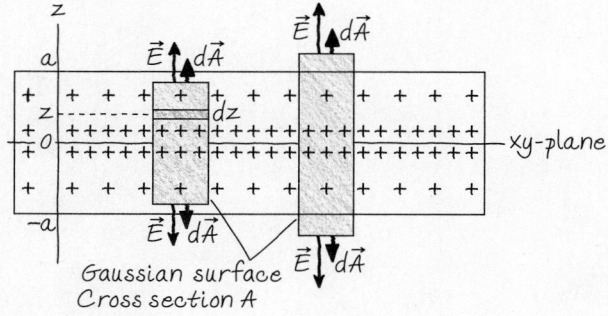

SOLVE Gauss's law is

$$\Phi_e = \oint \vec{E} \cdot d\vec{A} = \frac{Q_{\text{in}}}{\epsilon_0}$$

With symmetry, finding the net flux is straightforward. The electric field is perpendicular to the faces of the cylinders and pointing outward, so the total flux through the faces is $\Phi_{\text{faces}} = 2EA$, where E may depend on distance z. The field is parallel to the walls of the cylinders, so $\Phi_{\text{wall}} = 0$. Thus the net flux is simply

$$\Phi_e = 2EA$$

Because the charge density is not uniform, we need to integrate to find Q_{in}, the charge *inside* the cylinder. We can slice the cylinder into small slabs of infinitesimal thickness dz and volume $dV = A\,dz$. Figure 27.34 shows one such little slab at distance z from the xy-plane. The charge in this little slab is

$$dq = \rho\,dV = \rho_0\left(1 - \frac{z}{a}\right)A\,dz$$

where we assumed that z is positive. Because the charge is symmetric about $z = 0$, we can avoid difficulties with the absolute value sign in the charge density by integrating from 0 and

Continued

multiplying by 2. For the Gaussian cylinder that ends inside the slab of charge, at distance z, the total charge inside is

$$Q_{in} = \int dq = 2 \int_0^z \rho_0 \left(1 - \frac{z}{a}\right) A \, dz$$

$$= 2\rho_0 A \left[z \Big|_0^z - \frac{1}{2a} z^2 \Big|_0^z \right]$$

$$= 2\rho_0 A z \left(1 - \frac{z}{2a}\right)$$

Gauss's law inside the slab is then

$$\Phi_e = 2E_{inside} A = \frac{Q_{in}}{\epsilon_0} = \frac{2\rho_0 A z}{\epsilon_0} \left(1 - \frac{z}{2a}\right)$$

The area A cancels, as it must because it was an arbitrary choice, leaving

$$E_{inside} = \frac{\rho_0 z}{\epsilon_0} \left(1 - \frac{z}{2a}\right)$$

The field strength is zero at $z = 0$, then increases as z increases. This expression is valid only above the xy-plane, for $z > 0$, but the field strength is symmetric on the other side.

For the Gaussian cylinder that extends outside the slab of charge, the integral for Q has to end at $z = a$. Thus

$$Q_{in} = 2\rho_0 A a \left(1 - \frac{a}{2a}\right) = \rho_0 A a$$

independent of distance z. With this, Gauss's law gives

$$E_{outside} = \frac{Q_{in}}{2\epsilon_0 A} = \frac{\rho_0 a}{2\epsilon_0}$$

This matches E_{inside} at the surface, $z = a$, so the field is continuous as it crosses the boundary.

ASSESS Outside a sphere of charge, the field is the same as that of a point charge at the center. Similarly, the field outside an infinite slab of charge should be the same as that of an infinite charged plane. We found, by integration, that the total charge in an area A of the slab is $Q = \rho_0 A a$. If we squished this charge into a plane, the surface charge density would be $\eta = Q/A = \rho_0 a$. Thus our expression for $E_{outside}$ could be written $\eta/2\epsilon_0$, which matches the field we found in Example 27.6 for a plane of charge.

SUMMARY

The goal of Chapter 27 has been to understand and apply Gauss's law.

General Principles

Gauss's Law

For any *closed* surface enclosing net charge Q_{in}, the net electric flux through the surface is

$$\Phi_e = \oint \vec{E} \cdot d\vec{A} = \frac{Q_{in}}{\epsilon_0}$$

The electric flux Φ_e is the same for *any* closed surface enclosing charge Q_{in}.

Symmetry

The symmetry of the electric field must match the symmetry of the charge distribution.

In practice, Φ_e is computable only if the symmetry of the Gaussian surface matches the symmetry of the charge distribution.

Important Concepts

Charge creates the electric field that is responsible for the electric flux.

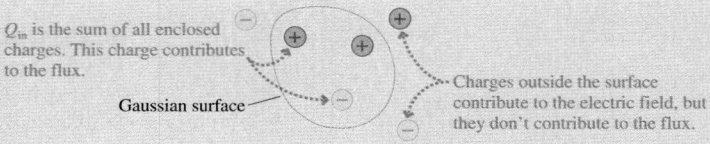

Q_{in} is the sum of all enclosed charges. This charge contributes to the flux.

Gaussian surface

Charges outside the surface contribute to the electric field, but they don't contribute to the flux.

Flux is the amount of electric field passing through a surface of area A:

$$\Phi_e = \vec{E} \cdot \vec{A}$$

where $\vec{A}$ is the **area vector.**

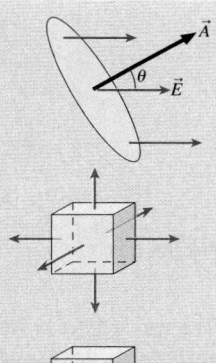

For closed surfaces:
A net flux in or out indicates that the surface encloses a net charge.

Field lines through but with no *net* flux mean that the surface encloses no *net* charge.

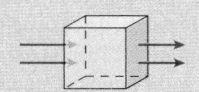

Surface integrals calculate the flux by summing the fluxes through many small pieces of the surface:

$$\Phi_e = \sum \vec{E} \cdot \delta\vec{A}$$

$$\rightarrow \int \vec{E} \cdot d\vec{A}$$

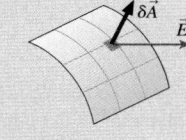

Two important situations:
If the electric field is everywhere tangent to the surface, then

$$\Phi_e = 0$$

If the electric field is everywhere perpendicular to the surface *and* has the same strength E at all points, then

$$\Phi_e = EA$$

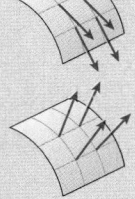

Applications

Conductors in electrostatic equilibrium

- The electric field is zero at all points within the conductor.
- Any excess charge resides entirely on the exterior surface.
- The external electric field is perpendicular to the surface and of magnitude η/ϵ_0, where η is the surface charge density.
- The electric field is zero inside any hole within a conductor unless there is a charge in the hole.

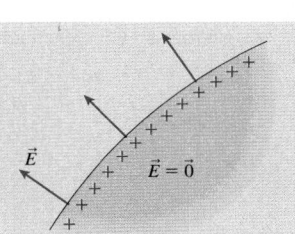

Terms and Notation

CONCEPTUAL QUESTIONS

1. Suppose you have the uniformly charged cube in FIGURE Q27.1. Can you use symmetry alone to deduce the *shape* of the cube's electric field? If so, sketch and describe the field shape. If not, why not?

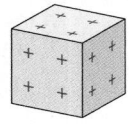

FIGURE Q27.1

2. FIGURE Q27.2 shows cross sections of three-dimensional closed surfaces. They have a flat top and bottom surface above and below the plane of the page. However, the electric field is everywhere parallel to the page, so there is no flux through the top or bottom surface. The electric field is uniform over each face of the surface. For each, does the surface enclose a net positive charge, a net negative charge, or no net charge? Explain.

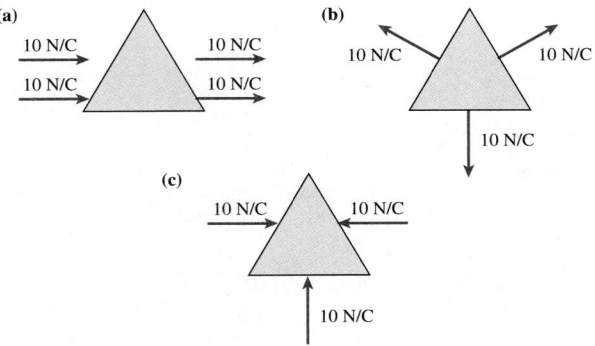

FIGURE Q27.2

3. The square and circle in FIGURE Q27.3 are in the same uniform field. The diameter of the circle equals the edge length of the square. Is Φ_{square} larger than, smaller than, or equal to Φ_{circle}? Explain.

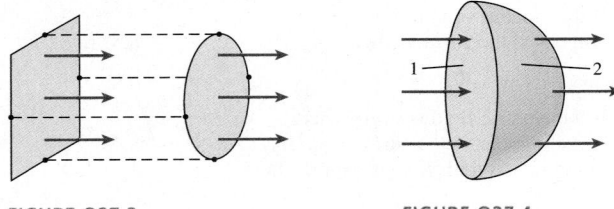

FIGURE Q27.3   **FIGURE Q27.4**

4. In FIGURE Q27.4, where the field is uniform, is Φ_1 larger than, smaller than, or equal to Φ_2? Explain.

5. What is the electric flux through each of the surfaces in FIGURE Q27.5? Give each answer as a multiple of q/ϵ_0.

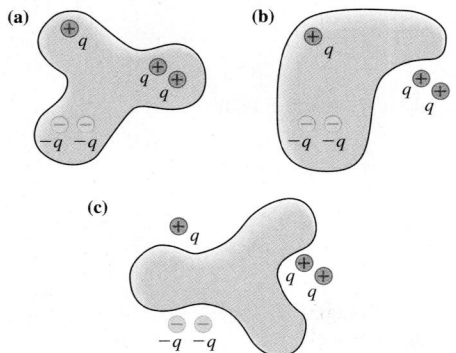

FIGURE Q27.5

6. What is the electric flux through each of the surfaces A to E in FIGURE Q27.6? Give each answer as a multiple of q/ϵ_0.

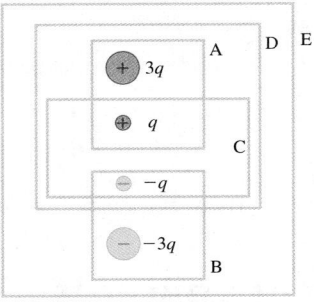

FIGURE Q27.6

7. The charged balloon in FIGURE Q27.7 expands as it is blown up, increasing in size from the initial to final diameters shown. Do the electric field strengths at points 1, 2, and 3 increase, decrease, or stay the same? Explain your reasoning for each.

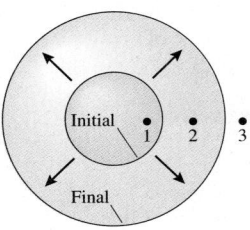

FIGURE Q27.7

8. The two spheres in FIGURE Q27.8 surround equal charges. Three students are discussing the situation.
   Student 1: The fluxes through spheres A and B are equal because they enclose equal charges.
   Student 2: But the electric field on sphere B is weaker than the electric field on sphere A. The flux depends on the electric field strength, so the flux through A is larger than the flux through B.
   Student 3: I thought we learned that flux was about surface area. Sphere B is larger than sphere A, so I think the flux through B is larger than the flux through A.
   Which of these students, if any, do you agree with? Explain.

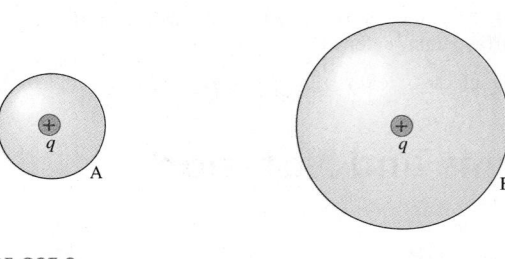

FIGURE Q27.8

9. The sphere and ellipsoid in **FIGURE Q27.9** surround equal charges. Four students are discussing the situation.

Student 1: The fluxes through A and B are equal because the average radius is the same.

Student 2: I agree that the fluxes are equal, but that's because they enclose equal charges.

Student 3: The electric field is not perpendicular to the surface for B, and that makes the flux through B less than the flux through A.

Student 4: I don't think that Gauss's law even applies to a situation like B, so we can't compare the fluxes through A and B.

Which of these students, if any, do you agree with? Explain.

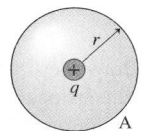
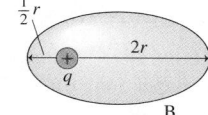

FIGURE Q27.9

10. A small, metal sphere hangs by an insulating thread within the larger, hollow conducting sphere of **FIGURE Q27.10**. A conducting wire extends from the small sphere through, but not touching, a small hole in the hollow sphere. A charged rod is used to transfer positive charge to the protruding wire. After the charged rod has touched the wire and been removed, are the following surfaces positive, negative, or not charged? Explain.

a. The small sphere.
b. The inner surface of the hollow sphere.
c. The outer surface of the hollow sphere.

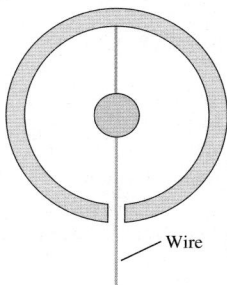

FIGURE Q27.10

EXERCISES AND PROBLEMS

Exercises

Section 27.1 Symmetry

1. | **FIGURE EX27.1** shows two cross sections of two infinitely long coaxial cylinders. The inner cylinder has a positive charge, the outer cylinder has an equal negative charge. Draw this figure on your paper, then draw electric field vectors showing the shape of the electric field.

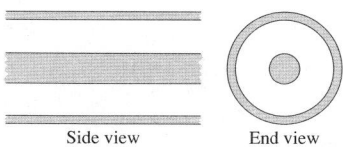

FIGURE EX27.1

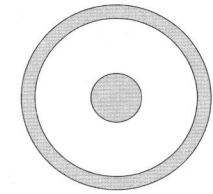

FIGURE EX27.2

2. | **FIGURE EX27.2** shows a cross section of two concentric spheres. The inner sphere has a negative charge. The outer sphere has a positive charge larger in magnitude than the charge on the inner sphere. Draw this figure on your paper, then draw electric field vectors showing the shape of the electric field.

3. | **FIGURE EX27.3** shows a cross section of two infinite parallel planes of charge. Draw this figure on your paper, then draw electric field vectors showing the shape of the electric field.

FIGURE EX27.3

Section 27.2 The Concept of Flux

4. | The electric field is constant over each face of the cube shown in **FIGURE EX27.4**. Does the box contain positive charge, negative charge, or no charge? Explain.

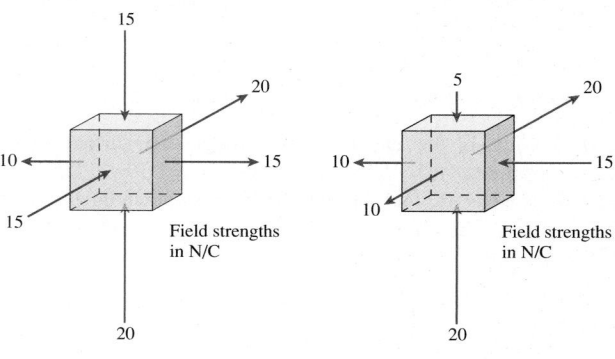

FIGURE EX27.4   **FIGURE EX27.5**

5. | The electric field is constant over each face of the cube shown in **FIGURE EX27.5**. Does the box contain positive charge, negative charge, or no charge? Explain.

6. | The cube in **FIGURE EX27.6** contains negative charge. The electric field is constant over each face of the cube. Does the missing electric field vector on the front face point in or out? What strength must this field exceed?

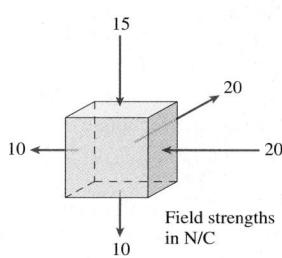

FIGURE EX27.6

7. | The cube in FIGURE EX27.7 contains negative charge. The electric field is constant over each face of the cube. Does the missing electric field vector on the front face point in or out? What strength must this field exceed?

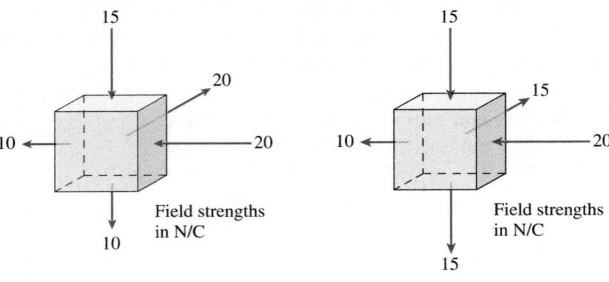

FIGURE EX27.7   FIGURE EX27.8

8. | The cube in FIGURE EX27.8 contains no net charge. The electric field is constant over each face of the cube. Does the missing electric field vector on the front face point in or out? What is the field strength?

Section 27.3 Calculating Electric Flux

9. ‖ What is the electric flux through the surface shown in FIGURE EX27.9?

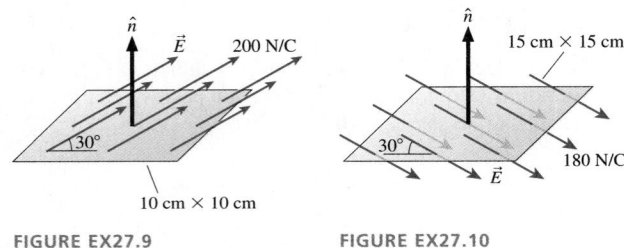

FIGURE EX27.9   FIGURE EX27.10

10. ‖ What is the electric flux through the surface shown in FIGURE EX27.10?

11. ‖ The electric flux through the surface shown in FIGURE EX27.11 is 25 N m^2/C. What is the electric field strength?

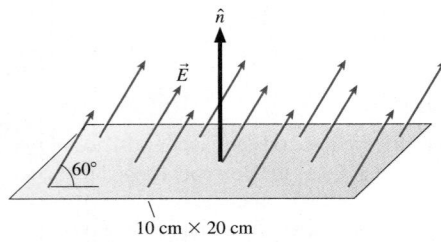

FIGURE EX27.11

12. ‖ A 2.0 cm × 3.0 cm rectangle lies in the xy-plane. What is the electric flux through the rectangle if
a. $\vec{E} = (100\hat{i} + 50\hat{k})$ N/C?
b. $\vec{E} = (100\hat{i} + 50\hat{j})$ N/C?

13. ‖ A 2.0 cm × 3.0 cm rectangle lies in the xz-plane. What is the electric flux through the rectangle if
a. $\vec{E} = (100\hat{i} + 50\hat{k})$ N/C?
b. $\vec{E} = (100\hat{i} + 50\hat{j})$ N/C?

14. ‖ A 3.0-cm-diameter circle lies in the xz-plane in a region where the electric field is $\vec{E} = (1500\hat{i} + 1500\hat{j} - 1500\hat{k})$ N/C. What is the electric flux through the circle?

15. ‖ A 1.0 cm × 1.0 cm × 1.0 cm box with its edges aligned with the xyz-axes is in the electric field $\vec{E} = (350x + 150)\hat{i}$ N/C, where x is in meters. What is the net electric flux through the box?

16. | What is the net electric flux through the two cylinders shown in FIGURE EX27.16? Give your answer in terms of R and E.

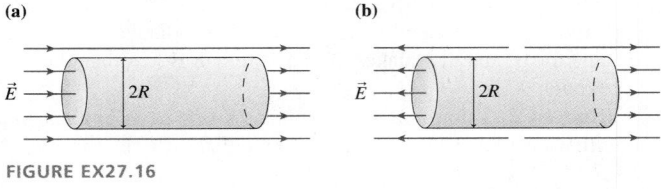

FIGURE EX27.16

Section 27.4 Gauss's Law

Section 27.5 Using Gauss's Law

17. | FIGURE EX27.17 shows three charges. Draw these charges on your paper four times. Then draw two-dimensional cross sections of three-dimensional closed surfaces through which the electric flux is (a) $2q/\epsilon_0$, (b) q/ϵ_0, (c) 0, and (d) $5q/\epsilon_0$.

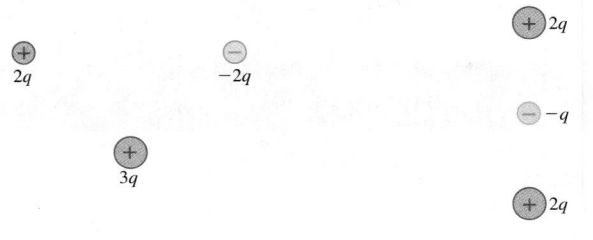

FIGURE EX27.17   FIGURE EX27.18

18. | FIGURE EX27.18 shows three charges. Draw these charges on your paper four times. Then draw two-dimensional cross sections of three-dimensional closed surfaces through which the electric flux is (a) $-q/\epsilon_0$, (b) q/ϵ_0, (c) $3q/\epsilon_0$, and (d) $4q/\epsilon_0$.

19. | FIGURE EX27.19 shows three Gaussian surfaces and the electric flux through each. What are the three charges q_1, q_2, and q_3?

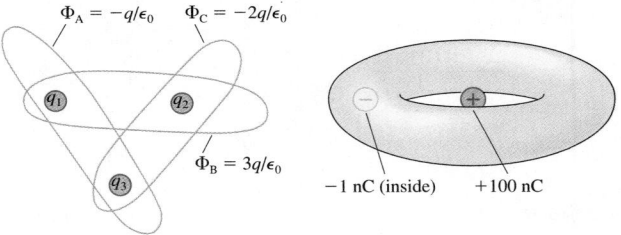

FIGURE EX27.19   FIGURE EX27.20

20. ‖ What is the net electric flux through the torus (i.e., doughnut shape) of FIGURE EX27.20?

21. | What is the net electric flux through the cylinder of FIGURE EX27.21?

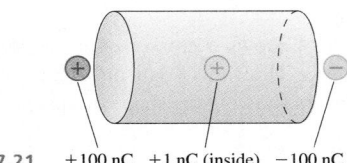

FIGURE EX27.21   +100 nC   +1 nC (inside)   −100 nC

22. ‖ The net electric flux through an octahedron is -1000 N m^2/C. How much charge is enclosed within the octahedron?

23. ‖ 55.3 million excess electrons are inside a closed surface. What is the net electric flux through the surface?

Section 27.6 Conductors in Electrostatic Equilibrium

24. ‖ The electric field strength just above one face of a copper penny is 2000 N/C. What is the surface charge density on this face of the penny?

25. | A spark occurs at the tip of a metal needle if the electric field strength exceeds 3.0×10^6 N/C, the field strength at which air breaks down. What is the minimum surface charge density for producing a spark?

26. | The conducting box in FIGURE EX27.26 has been given an excess negative charge. The surface density of excess electrons at the center of the top surface is 5.0×10^{10} electrons/m². What are the electric field strengths E_1 to E_3 at points 1 to 3?

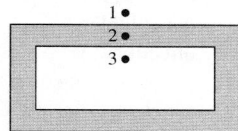

FIGURE EX27.26

27. | A thin, horizontal, 10-cm-diameter copper plate is charged to 3.5 nC. If the electrons are uniformly distributed on the surface, what are the strength and direction of the electric field
   a. 0.1 mm above the center of the top surface of the plate?
   b. at the plate's center of mass?
   c. 0.1 mm below the center of the bottom surface of the plate?

28. ‖ FIGURE EX27.28 shows a hollow cavity within a neutral conductor. A point charge Q is inside the cavity. What is the net electric flux through the closed surface that surrounds the conductor?

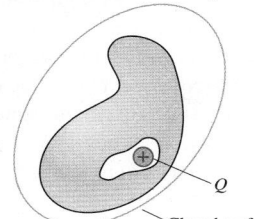

FIGURE EX27.28

Closed surface

Problems

29. | FIGURE P27.29 shows four sides of a 3.0 cm × 3.0 cm ×3.0 cm cube.
   a. What are the electric fluxes Φ_1 to Φ_4 through sides 1 to 4?
   b. What is the net flux through these four sides?

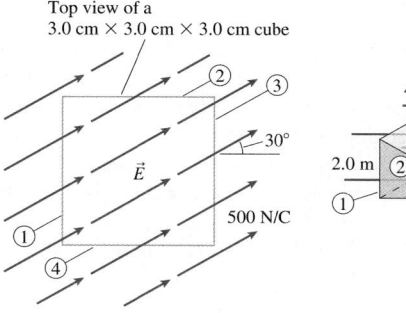

FIGURE P27.29

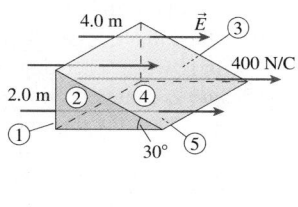

FIGURE P27.30

30. ‖‖ Find the electric fluxes Φ_1 to Φ_5 through surfaces 1 to 5 in FIGURE P27.30.

31. ‖ A tetrahedron has an equilateral triangle base with 20-cm-long edges and three equilateral triangle sides. The base is parallel to the ground, and a vertical uniform electric field of strength 200 N/C passes upward through the tetrahedron.
   a. What is the electric flux through the base?
   b. What is the electric flux through each of the three sides?

32. | Charges $q_1 = -4Q$ and $q_2 = +2Q$ are located at $x = -a$ and $x = +a$, respectively. What is the net electric flux through a sphere of radius $2a$ centered (a) at the origin and (b) at $x = 2a$?

33. ‖ A 10 nC point charge is at the center of a 2.0 m × 2.0 m × 2.0 m cube. What is the electric flux through the top surface of the cube?

34. ‖ The electric flux is 300 N m²/C through two opposing faces of a 2.0 cm × 2.0 cm × 2.0 cm box. The flux through each of the other faces is 100 N m²/C. How much charge is inside the box?

35. ‖ A spherically symmetric charge distribution produces the electric field $\vec{E} = (200/r)\hat{r}$ N/C, where r is in m.
   a. What is the electric field strength at $r = 10$ cm?
   b. What is the electric flux through a 20-cm-diameter spherical surface that is concentric with the charge distribution?
   c. How much charge is inside this 20-cm-diameter spherical surface?

36. ‖ A spherically symmetric charge distribution produces the electric field $\vec{E} = (5000r^2)\hat{r}$ N/C, where r is in m.
   a. What is the electric field strength at $r = 20$ cm?
   b. What is the electric flux through a 40-cm-diameter spherical surface that is concentric with the charge distribution?
   c. How much charge is inside this 40-cm-diameter spherical surface?

37. ‖ A neutral conductor contains a hollow cavity in which there is a +100 nC point charge. A charged rod then transfers −50 nC to the conductor. Afterward, what is the charge (a) on the inner wall of the cavity wall, and (b) on the exterior surface of the conductor?

38. ‖ A hollow metal sphere has inner radius a and outer radius b. The hollow sphere has charge $+2Q$. A point charge $+Q$ sits at the center of the hollow sphere.
   a. Determine the electric fields in the three regions $r \le a$, $a < r < b$, and $r \ge b$.
   b. How much charge is on the inside surface of the hollow sphere? On the exterior surface?

39. ‖ A 20-cm-radius ball is uniformly charged to 80 nC.
   a. What is the ball's volume charge density (C/m³)?
   b. How much charge is enclosed by spheres of radii 5, 10, and 20 cm?
   c. What is the electric field strength at points 5, 10, and 20 cm from the center?

40. ‖ FIGURE P27.40 shows a solid metal sphere at the center of a hollow metal sphere. What is the total charge on (a) the exterior of the inner sphere, (b) the inside surface of the hollow sphere, and (c) the exterior surface of the hollow sphere?

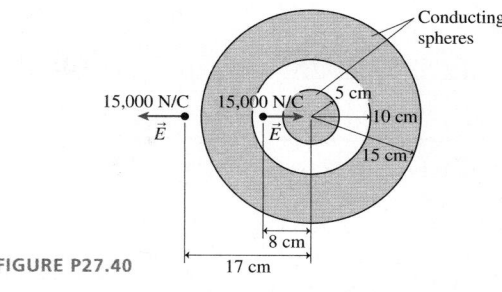

FIGURE P27.40

41. ‖ The earth has a vertical electric field at the surface, pointing down, that averages 100 N/C. This field is maintained by various atmospheric processes, including lightning. What is the excess charge on the surface of the earth?

42. ‖ Figure 27.32b showed a conducting box inside a parallel-plate capacitor. The electric field inside the box is $\vec{E} = \vec{0}$. Suppose the surface charge on the exterior of the box could be frozen. Draw a picture of the electric field inside the box after the box, with its frozen charge, is removed from the capacitor.
    Hint: Superposition.

43. ‖ A hollow metal sphere has 6 cm and 10 cm inner and outer radii, respectively. The surface charge density on the inside surface is -100 nC/m². The surface charge density on the exterior surface is $+100$ nC/m². What are the strength and direction of the electric field at points 4, 8, and 12 cm from the center?

44. ‖ A positive point charge q sits at the center of a hollow spherical shell. The shell, with radius R and negligible thickness, has net charge $-2q$. Find an expression for the electric field strength (a) inside the sphere, $r < R$, and (b) outside the sphere, $r > R$. In what direction does the electric field point in each case?

45. ‖ Find the electric field inside and outside a hollow plastic ball of radius R that has charge Q uniformly distributed on its outer surface.

46. ‖ A uniformly charged ball of radius a and charge $-Q$ is at the center of a hollow metal shell with inner radius b and outer radius c. The hollow sphere has net charge $+2Q$. Determine the electric field strength in the four regions $r \le a$, $a < r < b$, $b \le r \le c$, and $r > c$.

47. ‖ The three parallel planes of charge shown in **FIGURE P27.47** have surface charge densities $-\frac{1}{2}\eta$, η, and $-\frac{1}{2}\eta$. Find the electric fields $\vec{E}_1$ to $\vec{E}_4$ in regions 1 to 4.

$$-\tfrac{1}{2}\eta \; - \; - \; - \; - \; - \; - \; - \; \overset{1}{-} \; - \; -$$
$$\overset{2}{}$$
$$\eta \; +$$
$$\overset{3}{}$$
$$-\tfrac{1}{2}\eta \; - \; - \; - \; - \; - \; - \; - \; - \; - \; -$$
$$\overset{4}{}$$

FIGURE P27.47

48. ‖ An infinite slab of charge of thickness $2z_0$ lies in the xy-plane between $z = -z_0$ and $z = +z_0$. The volume charge density ρ (C/m³) is a constant.
    a. Use Gauss's law to find an expression for the electric field strength inside the slab ($-z_0 \le z \le z_0$).
    b. Find an expression for the electric field strength above the slab ($z \ge z_0$).
    c. Draw a graph of E from $z = 0$ to $z = 3z_0$.

49. ‖ **FIGURE P27.49** shows an infinitely wide conductor parallel to and distance d from an infinitely wide plane of charge with surface charge density η. What are the electric fields $\vec{E}_1$ to $\vec{E}_4$ in regions 1 to 4?

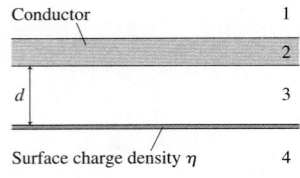

FIGURE P27.49

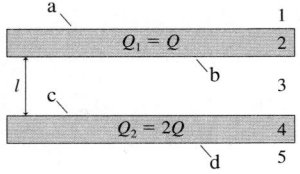

FIGURE P27.50

50. ‖ **FIGURE P27.50** shows two very large slabs of metal that are parallel and distance l apart. Each slab has a total surface area (top + bottom) A. The thickness of each slab is so small in comparison to its lateral dimensions that the surface area around the sides is negligible. Metal 1 has total charge $Q_1 = Q$ and metal 2

has total charge $Q_2 = 2Q$. Assume Q is positive. In terms of Q and A, determine
    a. The electric field strengths E_1 to E_5 in regions 1 to 5.
    b. The surface charge densities η_a to η_d on the four surfaces a to d.

51. ‖ A long, thin straight wire with linear charge density λ runs down the center of a thin, hollow metal cylinder of radius R. The cylinder has a net linear charge density 2λ. Assume λ is positive. Find expressions for the electric field strength (a) inside the cylinder, $r < R$, and (b) outside the cylinder, $r > R$. In what direction does the electric field point in each of the cases?

52. ‖ A very long, uniformly charged cylinder has radius R and linear charge density λ. Find the cylinder's electric field (a) outside the cylinder, $r \ge R$, and (b) inside the cylinder, $r \le R$. (c) Show that your answers to parts a and b match at the boundary, $r = R$.

53. ‖ A spherical shell has inner radius R_{in} and outer radius R_{out}. The shell contains total charge Q, uniformly distributed. The interior of the shell is empty of charge and matter.
    a. Find the electric field outside the shell, $r \ge R_{out}$.
    b. Find the electric field in the interior of the shell, $r \le R_{in}$.
    c. Find the electric field within the shell, $R_{in} \le r \le R_{out}$.
    d. Show that your solutions match at both the inner and outer boundaries.

54. ‖ An early model of the atom, proposed by Rutherford after his discovery of the atomic nucleus, had a positive point charge $+Ze$ (the nucleus) at the center of a sphere of radius R with uniformly distributed negative charge $-Ze$. Z is the atomic number, the number of protons in the nucleus and the number of electrons in the negative sphere.
    a. Show that the electric field inside this atom is

$$E_{in} = \frac{Ze}{4\pi\epsilon_0}\left(\frac{1}{r^2} - \frac{r}{R^3}\right)$$

    b. What is E at the surface of the atom? Is this the expected value? Explain.
    c. A uranium atom has $Z = 92$ and $R = 0.10$ nm. What is the electric field strength at $r = \frac{1}{2}R$?

Challenge Problems

55. All examples of Gauss's law have used highly symmetric surfaces where the flux integral is either zero or EA. Yet we've claimed that the net $\Phi_e = Q_{in}/\epsilon_0$ is independent of the surface. This is worth checking. **FIGURE CP27.55** shows a cube of edge length L centered on a long thin wire with linear charge density λ. The flux through one face of the cube is *not* simply EA because, in this case, the electric field varies in both strength and direction. But you can calculate the flux by actually doing the flux integral.

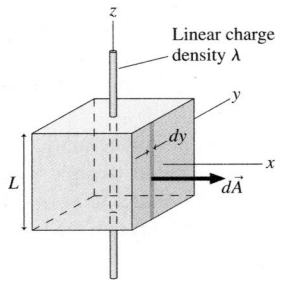

FIGURE CP27.55

a. Consider the face parallel to the yz-plane. Define area $d\vec{A}$ as a strip of width dy and height L with the vector pointing in the x-direction. One such strip is located at position y. Use the known electric field of a wire to calculate the electric flux $d\Phi$ through this little area. Your expression should be written in terms of y, which is a variable, and various constants. It should not explicitly contain any angles.

b. Now integrate $d\Phi$ to find the total flux through this face.

c. Finally, show that the net flux through the cube is $\Phi_e = Q_{in}/\epsilon_0$.

56. An infinite cylinder of radius R has a linear charge density λ. The volume charge density (C/m^3) within the cylinder ($r \le R$) is $\rho(r) = r\rho_0/R$, where ρ_0 is a constant to be determined.

a. Draw a graph of ρ versus x for an x-axis that crosses the cylinder perpendicular to the cylinder axis. Let x range from $-2R$ to $2R$.

b. The charge within a small volume dV is $dq = \rho\,dV$. The integral of $\rho\,dV$ over a cylinder of length L is the total charge $Q = \lambda L$ within the cylinder. Use this fact to show that $\rho_0 = 3\lambda/2\pi R^2$.

Hint: Let dV be a cylindrical shell of length L, radius r, and thickness dr. What is the volume of such a shell?

c. Use Gauss's law to find an expression for the electric field E inside the cylinder, $r \le R$.

d. Does your expression have the expected value at the surface, $r = R$? Explain.

57. A sphere of radius R has total charge Q. The volume charge density (C/m^3) within the sphere is $\rho(r) = C/r^2$, where C is a constant to be determined.

a. The charge within a small volume dV is $dq = \rho\,dV$. The integral of $\rho\,dV$ over the entire volume of the sphere is the total charge Q. Use this fact to determine the constant C in terms of Q and R.

Hint: Let dV be a spherical shell of radius r and thickness dr. What is the volume of such a shell?

b. Use Gauss's law to find an expression for the electric field E inside the sphere, $r \le R$.

c. Does your expression have the expected value at the surface, $r = R$? Explain.

58. A sphere of radius R has total charge Q. The volume charge density (C/m^3) within the sphere is

$$\rho = \rho_0\left(1 - \frac{r}{R}\right)$$

This charge density decreases linearly from ρ_0 at the center to zero at the edge of the sphere.

a. Show that $\rho_0 = 3Q/\pi R^3$.

b. Show that the electric field inside the sphere points radially outward with magnitude

$$E = \frac{Qr}{4\pi\epsilon_0 R^3}\left(4 - 3\frac{r}{R}\right)$$

c. Show that your result of part b has the expected value at $r = R$.

59. A spherical ball of charge has radius R and total charge Q. The electric field strength inside the ball ($r \le R$) is $E(r) = E_{max}(r^4/R^4)$.

a. What is E_{max} in terms of Q and R?

b. Find an expression for the volume charge density $\rho(r)$ inside the ball as a function of r.

c. Verify that your charge density gives the total charge Q when integrated over the volume of the ball.

STOP TO THINK ANSWERS

Stop to Think 27.1: a and d. Symmetry requires the electric field to be unchanged if front and back are reversed, if left and right are reversed, or if the field is rotated about the wire's axis. Fields a and d both have the proper symmetry. Other factors would now need to be considered to determine the correct field.

Stop to Think 27.2: e. The net flux is into the box.

Stop to Think 27.3: c. There's no flux through the four sides. The flux is positive 1 N m^2/C through both the top and bottom because $\vec{E}$ and $\vec{A}$ both point outward.

Stop to Think 27.4: $\Phi_b = \Phi_e > \Phi_a = \Phi_c = \Phi_d$. The flux through a closed surface depends only on the amount of enclosed charge, not the size or shape of the surface.

Stop to Think 27.5: d. A cube doesn't have enough symmetry to use Gauss's law. The electric field of a charged cube is *not* constant over the face of a cubic Gaussian surface, so we can't evaluate the surface integral for the flux.

28 The Electric Potential

City lights seen from space show where millions of lightbulbs are transforming electric energy into light and thermal energy.

▶ **Looking Ahead** The goals of Chapter 28 are to calculate and use the electric potential and electric potential energy.

Electric Energy

Energy allows things to happen. You want your lights to light, your computer to compute, and your stereo to keep your neighbors awake. All these require energy—*electric* energy.

This is the first of two chapters that explore electric energy and its connection to electric forces and fields.

Lightning is a dramatic example of the transformation of electric energy into light, sound, and thermal energy.

You'll learn to calculate the electric potential energy of charged particles and to solve problems using conservation of mechanical energy.

There's a close connection between electric potential energy and gravitational potential energy because both forces obey inverse-square laws.

◀ **Looking Back**
Sections 10.2–10.5 Kinetic energy, potential energy, and conservation

◀ **Looking Back**
Sections 11.2–11.5 Work and potential energy

The Electric Potential

Just as source charges create an electric field, they also create an **electric potential.** A charge moving in an electric potential has an electric potential energy.

The unit of electric potential is the **volt,** perhaps the most well known of all electrical units. A voltmeter reads the *potential difference* between two points.

Using Electric Potential

Charged particles *accelerate* as they move through a potential difference.

You'll learn to use the electric potential and a conservation of energy problem-solving strategy to solve problems about the motion of charged particles.

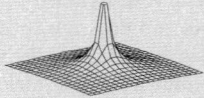
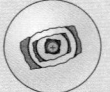

0 V   500 V

◀ **Looking Back**
Section 10.6 Energy diagrams

Calculating Electric Potential

You'll learn how to calculate the electric potential for several important charge distributions.

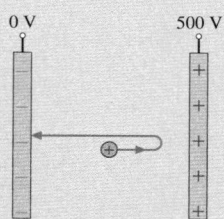

Elevation graph   **Equipotential surfaces**

You'll also learn to use several different representations of the electric potential.

◀ **Looking Back**
Section 26.3 Calculating electric fields

Sources of Electric Potential

In practice, electric potential is created by separating positive and negative charges—an idea we'll explore more thoroughly in Chapter 29.

A battery is the most common source of electric potential. As you'll learn, its *voltage* is the potential difference between separated charges—the plus and minus terminals.

28.1 Electric Potential Energy

In electricity, just as in mechanics, it takes energy to make things happen. It's been many chapters since we dealt much with work and energy, but these ideas will now be *essential* to our story. Consequently, the Looking Back recommendations in the chapter preview are especially important. You will recall that a system's mechanical energy $E_{mech} = K + U$ is conserved for particles that interact with each other via *conservative forces*, where K and U are the kinetic and potential energy. That is,

$$\Delta E_{mech} = \Delta K + \Delta U = 0 \qquad (28.1)$$

We need to be careful with notation because we are now using E to represent the electric field strength. To avoid confusion, we will represent mechanical energy either as the explicit sum $K + U$ or as E_{mech}, with an explicit subscript.

NOTE ▶ Recall that for any X, the *change* in X is $\Delta X = X_{final} - X_{initial}$. ◀

The kinetic energy $K = \sum K_i$, where $K_i = \frac{1}{2} m_i v_i^2$, is the sum of the kinetic energies of all the particles in the system. The potential energy U is the *interaction energy* of the system. In particular, we defined the *change* in potential energy in terms of the work W done by the forces of interaction as the system moves from an initial position or configuration i to a final position or configuration f:

$$\Delta U = U_f - U_i = -W_{interaction\ forces} \qquad \text{(position i} \rightarrow \text{position f)} \qquad (28.2)$$

This formal definition of ΔU is rather abstract and will make more sense when we see specific applications.

A *constant* force does work

$$W = \vec{F} \cdot \Delta \vec{r} = F \Delta r \cos\theta \qquad (28.3)$$

on a particle that undergoes a linear displacement $\Delta \vec{r}$, where θ is the angle between the force $\vec{F}$ and $\Delta \vec{r}$. FIGURE 28.1 reminds you of the three special cases $\theta = 0°$, $90°$, and $180°$. It also shows that, in general, the work is done by the force component F_r in the direction of motion.

NOTE ▶ Work is *not* the oft-remembered "force times distance." Work is force times distance only in the one very special case in which the force is both constant *and* parallel to the displacement. ◀

If the force is *not* constant or the displacement is *not* along a linear path, we can calculate the work by dividing the path into many small segments. FIGURE 28.2 shows how this is done. The work done as the particle moves distance ds is $F_s\,ds$, where F_s is the force component parallel to ds (i.e., the component in the direction of motion). The total work done on the particle is

$$W = \sum_j (F_s)_j \Delta s_j \rightarrow \int_{s_i}^{s_f} F_s\,ds = \int_i^f \vec{F} \cdot d\vec{s} \qquad (28.4)$$

The second integral recognizes that $F_s\,ds = F\cos\theta\,ds$ is equivalent to the dot product $\vec{F} \cdot d\vec{s}$, allowing us to write the work in vector notation. As with Gauss's law, this integral looks more formidable than it really is. We'll look at examples shortly.

Finally, recall that a *conservative force* is one for which the work done as a particle moves from position i to position f is *independent of the path followed*. In other words, the integral in Equation 28.4 gives the same value for *any* path between points i and f. We'll assert for now, and prove later, that **the electric force is a conservative force.**

Uniform Fields

Gravity, like electricity, is a long-range force. Much as we defined the electric field $\vec{E} = \vec{F}_{on\ q}/q$, we can also define a gravitational field—the agent that exerts gravitational forces on masses—as $\vec{F}_{on\ m}/m$. But $\vec{F}_{on\ m} = m\vec{g}$ near the earth's surface; thus

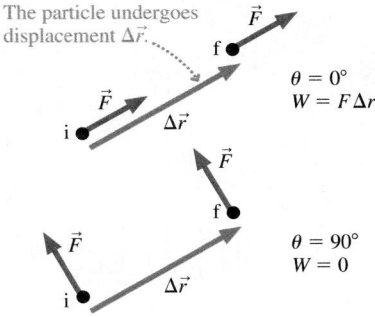

FIGURE 28.1 The work done by a constant force.

The particle undergoes displacement $\Delta \vec{r}$.

$\theta = 0°$
$W = F\Delta r$

$\theta = 90°$
$W = 0$

$\theta = 180°$
$W = -F\Delta r$

General case
$W = F_r\Delta r$
$= F\Delta r\cos\theta$

The work is done by the component of $\vec{F}$ in the direction of motion.

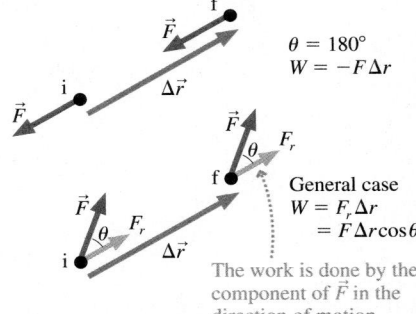

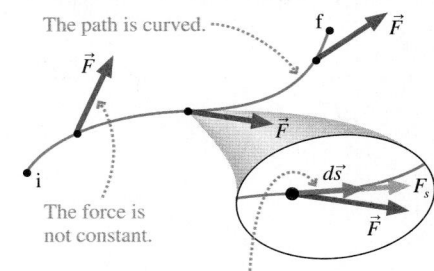

FIGURE 28.2 The work done along a curved path or by a variable force.

The path is curved.

The force is not constant.

$d\vec{s}$

F_s

$\vec{F}$

The work done in this small segment of the motion is $F_s\,ds = \vec{F} \cdot d\vec{s}$.

the familiar $\vec{g} = (9.80 \text{ N/kg, down})$ is really the gravitational field! Notice how we've written the units of $\vec{g}$ as N/kg, as is appropriate for a field, but you can easily show that N/kg = m/s^2. The gravitational field near the earth's surface is a *uniform* field in the downward direction.

FIGURE 28.3 shows a particle of mass m falling in the gravitational field. The gravitational force is in the same direction as the particle's displacement, so the gravitational field does a *positive* amount of work on the particle. The gravitational force is constant, hence the work done by gravity is

$$W_{\text{grav}} = F_G \, \Delta r \cos 0° = mg |y_f - y_i| = mgy_i - mgy_f \qquad (28.5)$$

We have to be careful with signs because Δr, the magnitude of the displacement vector, must be a positive number.

Now we can see how the definition of ΔU in Equation 28.2 makes sense. The *change* in gravitational potential energy is

$$\Delta U_{\text{grav}} = U_f - U_i = -W_{\text{grav}}(\text{i} \rightarrow \text{f}) = mgy_f - mgy_i \qquad (28.6)$$

Comparing the initial and final terms on the two sides of the equation, we see that the gravitational potential energy near the earth is the familiar quantity

$$U_{\text{grav}} = U_0 + mgy \qquad (28.7)$$

where U_0 is the value of U_{grav} at $y = 0$. We usually choose $U_0 = 0$, in which case $U_{\text{grav}} = mgy$, but such a choice is not necessary. The zero point of potential energy is an arbitrary choice because we have defined ΔU rather than U.

The uniform electric field between the plates of the parallel-plate capacitor of FIGURE 28.4 looks very much like the uniform gravitational field near the earth's surface. The one difference is that $\vec{g}$ always points down whereas the positive-to-negative electric field can point in any direction. To deal with this, let's define a coordinate axis s that points *from* the negative plate, which we define to be $s = 0$, *toward* the positive plate. The electric field $\vec{E}$ then points in the negative s-direction, just as the gravitational field $\vec{g}$ points in the negative y-direction. This s-axis, which is valid no matter how the capacitor is oriented, is analogous to the y-axis used for gravitational potential energy.

A positive charge q inside the capacitor speeds up and gains kinetic energy as it "falls" toward the negative plate. Is the charge losing potential energy as it gains kinetic energy? Indeed it is, and the calculation of the potential energy is just like the calculation of gravitational potential energy. The electric field exerts a *constant* force $F = qE$ on the charge in the direction of motion; thus the work done on the charge by the electric field is

$$W_{\text{elec}} = F \, \Delta r \cos 0° = qE |s_f - s_i| = qEs_i - qEs_f \qquad (28.8)$$

where we again have to be careful with the signs because $s_f < s_i$.

The work done by the electric field causes the charge to experience a change in *electric* potential energy given by

$$\Delta U_{\text{elec}} = U_f - U_i = -W_{\text{elec}}(\text{i} \rightarrow \text{f}) = qEs_f - qEs_i \qquad (28.9)$$

Comparing the initial and final terms on the two sides of the equation, we see that the **electric potential energy** of charge q in a uniform electric field is

$$U_{\text{elec}} = U_0 + qEs \qquad (28.10)$$

where s is measured from the negative plate and U_0 is the potential energy at the negative plate ($s = 0$). It will often be convenient to choose $U_0 = 0$, but the choice has no physical consequences because it doesn't affect ΔU_{elec}, the *change* in the electric potential energy. Only the *change* is significant.

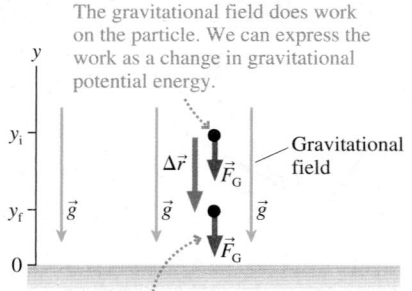

FIGURE 28.3 Potential energy is transformed into kinetic energy as a particle moves in a gravitational field.

The gravitational field does work on the particle. We can express the work as a change in gravitational potential energy.

The net force on the particle is down. It gains kinetic energy (i.e., speeds up) as it loses potential energy.

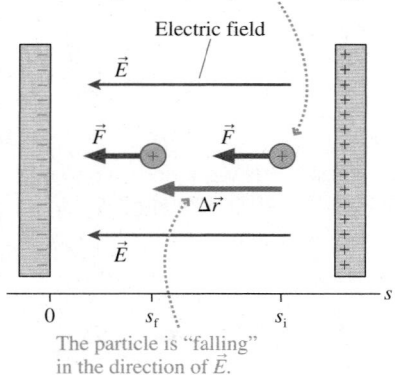

FIGURE 28.4 The electric field does work on the charged particle.

The electric field does work on the particle. We can express the work as a change in electric potential energy.

Electric field

The particle is "falling" in the direction of $\vec{E}$.

Equation 28.10 was derived with the assumption that q is positive, but it is valid for either sign of q. A negative value for q in Equation 28.10 causes the potential energy U_{elec} to become *more negative* as s increases. As FIGURE 28.5 shows, a negative charge gains kinetic energy as it moves *away from* the negative plate of the capacitor.

FIGURE 28.5 A charged particle of either sign gains kinetic energy as it moves in the direction of decreasing potential energy.

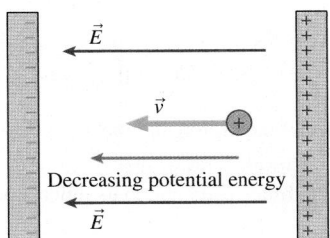

The potential energy of a positive charge decreases in the direction of $\vec{E}$. The charge gains kinetic energy as it moves toward the negative plate.

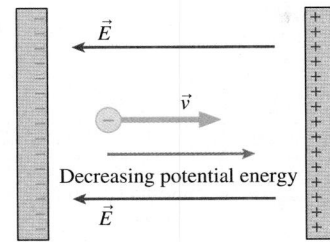

The potential energy of a negative charge decreases in the direction opposite to $\vec{E}$. The charge gains kinetic energy as it moves away from the negative plate.

FIGURE 28.6 The energy diagram for a positively charged particle in a uniform electric field.

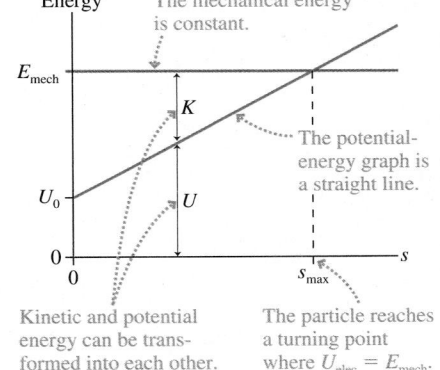

The mechanical energy is constant.

The potential-energy graph is a straight line.

Kinetic and potential energy can be transformed into each other.

The particle reaches a turning point where $U_{elec} = E_{mech}$.

NOTE ▶ Although Equation 28.10 is often called "the potential energy of charge q," it is really the potential energy of the charge + capacitor system. To the extent that the charges on the capacitor plate stay fixed, we're justified in thinking of this as the potential energy of just the charge q. ◀

FIGURE 28.6 is the *energy diagram* for a positively charged particle in a uniform electric field. Recall that an energy diagram is a graphical representation of how the kinetic and potential energy are transformed as a particle moves. The potential energy, given by Equation 28.10, increases linearly with distance, but the particle's total mechanical energy E_{mech} is fixed. If a positively charged particle is projected against a uniform field, it gradually slows (transforming kinetic to potential energy) until reaching the *turning point* where $U_{elec} = E_{mech}$.

EXAMPLE 28.1   **Conservation of energy**

A 2.0 cm × 2.0 cm parallel-plate capacitor with a 2.0 mm spacing is charged to ±1.0 nC. First a proton, then an electron are released from rest at the midpoint of the capacitor.

a. What is each particle's change in electric potential energy from its release until it collides with one of the plates?

b. What is each particle's speed as it reaches the plate?

MODEL The mechanical energy of each particle is conserved. A parallel-plate capacitor has a uniform electric field.

VISUALIZE FIGURE 28.7 is a before-and-after pictorial representation, as you learned to draw in Part II. On the energy diagram of Figure 28.6, each particle is released at the turning point ($K = 0$) and moves toward lower potential energy. Thus the proton moves toward the negative plate, the electron toward the positive plate.

SOLVE a. The s-axis was defined to point from the negative toward the positive plate of the capacitor. Both charged particles have $s_i = \frac{1}{2}d$, where $d = 2.0$ mm is the plate separation. The positive proton loses potential energy and gains kinetic energy

FIGURE 28.7 A proton and an electron in a capacitor.

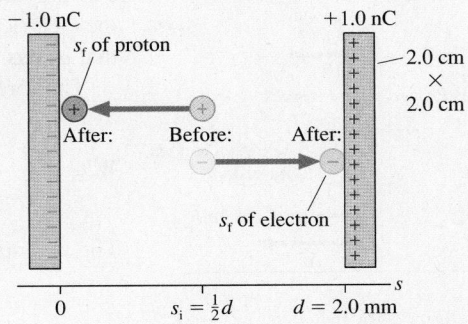

as it moves toward the negative plate. For the proton, with $q = +e$ and $s_f = 0$, the change in potential energy is

$$\Delta U_p = U_f - U_i = (U_0 + 0) - \left(U_0 + eE\frac{d}{2}\right) = -\frac{1}{2}eEd$$

Continued

where we used the electric potential energy for a charge in a uniform electric field. ΔU_p is negative, as expected. Notice that U_0 cancels when ΔU is calculated.

The electron moves toward the positive plate, which is the direction of decreasing potential energy for a negative charge. The electron has $q = -e$ and ends at $s_f = d$. Thus

$$\Delta U_e = U_f - U_i = (U_0 + (-e)Ed) - \left(U_0 + (-e)E\frac{d}{2}\right)$$

$$= -\frac{1}{2}eEd$$

Both particles have the *same* change in potential energy. The capacitor's electric field is

$$E = \frac{\eta}{\epsilon_0} = \frac{Q}{\epsilon_0 A} = 2.82 \times 10^5 \text{ N/C}$$

Using $d = 0.0020$ m, we find

$$\Delta U_p = \Delta U_e = -4.5 \times 10^{-17} \text{ J}$$

b. The law of conservation of energy is $\Delta K + \Delta U = 0$. Both particles are released from rest; hence $\Delta K = K_f - 0 = \frac{1}{2}mv_f^2$. Thus $\frac{1}{2}mv_f^2 = -\Delta U$, or

$$v_f = \sqrt{\frac{-2\,\Delta U}{m}} = \begin{cases} 2.3 \times 10^5 \text{ m/s for the proton} \\ 1.0 \times 10^7 \text{ m/s for the electron} \end{cases}$$

where we used the masses of the proton and the electron.

ASSESS Even though both particles have the same ΔU, the electron reaches a much faster final speed due to its much smaller mass.

STOP TO THINK 28.1 A glass rod is positively charged. The figure shows an end view of the rod. A negatively charged particle moves in a circular arc around the glass rod. Is the work done on the charged particle by the rod's electric field positive, negative, or zero?

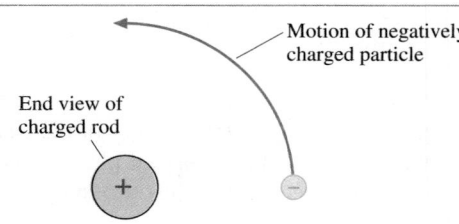

28.2 The Potential Energy of Point Charges

Now that we've introduced the idea of electric potential energy, let's look at *the* fundamental interaction of electricity—the force between two point charges. This force, given by Coulomb's law, varies with the distance between the two charges; hence we need to use the integral expression of Equation 28.4 to calculate the work done.

FIGURE 28.8a shows two charges q_1 and q_2, which we will assume to be like charges. The potential energy of their interaction can be found by calculating the work done by the electric field of q_1 on q_2 as q_2 moves from position x_i to position x_f. We'll assume that q_1 has been glued down and is unable to move, as shown in FIGURE 28.8b.

The force is entirely in the direction of motion, so $F_s\,ds = F_{1 \text{ on } 2}\,dx$. Thus

FIGURE 28.8 The interaction between two point charges.

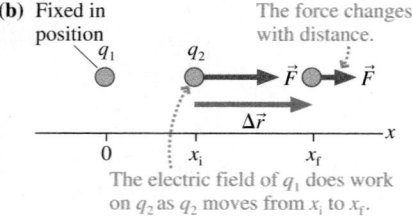

$$W_{\text{elec}} = \int_{x_i}^{x_f} F_{1 \text{ on } 2}\,dx = \int_{x_i}^{x_f} \frac{Kq_1q_2}{x^2}\,dx = Kq_1q_2 \frac{-1}{x}\Big|_{x_i}^{x_f} = -\frac{Kq_1q_2}{x_f} + \frac{Kq_1q_2}{x_i} \quad (28.11)$$

The potential energy of the two charges is related to the work done by

$$\Delta U_{\text{elec}} = U_f - U_i = -W_{\text{elec}}(i \rightarrow f) = \frac{Kq_1q_2}{x_f} - \frac{Kq_1q_2}{x_i} \quad (28.12)$$

By comparing the left and right sides of the equation we see that the potential energy of the two-point-charge system is

$$U_{\text{elec}} = \frac{Kq_1q_2}{x} \quad (28.13)$$

We could include a constant U_0, as we did in Equation 28.10, for the potential energy of a charge in a uniform electric field, but it is customary to set $U_0 = 0$.

We chose to integrate along the *x*-axis for convenience, but what is really important is the *distance* between the charges. Thus a more general expression for the electric potential energy is

$$U_{elec} = \frac{Kq_1q_2}{r} = \frac{1}{4\pi\epsilon_0}\frac{q_1q_2}{r} \qquad \text{(two point charges)} \qquad (28.14)$$

This is explicitly the energy *of the system,* not the energy of just q_1 or q_2.

NOTE ▶ The electric potential energy of two point charges looks *almost* the same as the force between the charges. The difference is the r in the denominator of the potential energy compared to the r^2 in Coulomb's law. ◀

Three important points need to be noted:

■ The choice $U_0 = 0$ is equivalent to saying that the potential energy of two charged particles is zero only when they are infinitely far apart. This makes sense because two charged particles cease interacting only when they are infinitely far apart.

■ We derived Equation 28.14 for two like charges, but it is equally valid for two opposite charges. The potential energy of two like charges is *positive* and of two opposite charges is *negative.*

■ Because the electric field outside a *sphere of charge* is the same as that of a point charge at the center, Equation 28.14 is also the electric potential energy of two charged spheres. Distance r is the distance between their centers.

FIGURE 28.9a shows the potential-energy curve—a hyperbola—for two like charges as a function of the distance r between them. Distances must be positive numbers, so the graph shows only $r > 0$. Also shown is the total energy line for two charged particles shot toward each other with equal but opposite momenta. Recall, from Chapter 10, that the total energy line is horizontal because the mechanical energy is conserved.

You can see that the total energy line crosses the potential-energy curve at r_{min}. This is a turning point. The two charges gradually slow down, because of the repulsive force between them, until the distance separating them is r_{min}. At this point, the kinetic energy is zero and both particles are instantaneously at rest. Both then reverse direction and move apart, speeding up as they go. r_{min} is the *distance of closest approach.*

Two opposite charges are a little trickier because of the negative energies. Negative total energies seem troubling at first, but they characterize *bound systems.* FIGURE 28.9b shows two oppositely charged particles shot apart from each other with equal but opposite momenta. If $E_{mech} < 0$, as shown, then their total energy line crosses the potential-energy curve at r_{max}. That is, the particles slow down, lose kinetic energy, reverse directions at *maximum separation* r_{max}, and then "fall" back together. They cannot escape from each other. Although moving in three dimensions rather than one, the electron and proton of a hydrogen atom are a realistic example of a bound system, and their mechanical energy is negative.

Two oppositely charged particles *can* escape from each other if $E_{mech} > 0$. They'll slow down, but eventually the potential energy vanishes and the particles still have kinetic energy. The threshold condition for escape is $E_{mech} = 0$, which will allow the particles to reach infinite separation ($U \to 0$) at infinitesimally slow speed ($K \to 0$). The initial speed that gives $E_{mech} = 0$ is called the *escape speed.*

NOTE ▶ Real particles can't be infinitely far apart, but because U_{elec} decreases with distance, there comes a point when $U_{elec} = 0$ is an excellent approximation. Two charged particles for which $U_{elec} \approx 0$ are sometimes described as "far apart" or "far away." ◀

FIGURE 28.9 The potential-energy diagrams for two like charges and two opposite charges.

(a) Like charges

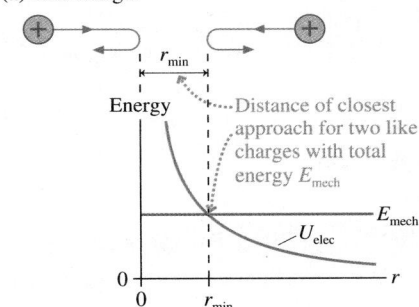

(b) Opposite charges

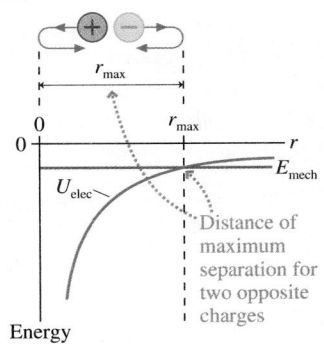

The Electric Force Is a Conservative Force

Potential energy can be defined only if the force is *conservative*, meaning that the work done on the particle as it moves from position i to position f is independent of the path followed between i and f. FIGURE 28.10 demonstrates that electric force is indeed conservative.

FIGURE 28.10 The work done on q_2 is independent of the path from i to f.

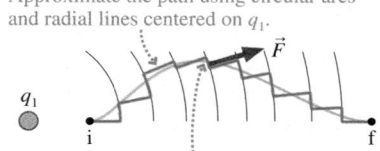

Consider an alternative path for q_2 to move from i to f.

Approximate the path using circular arcs and radial lines centered on q_1.

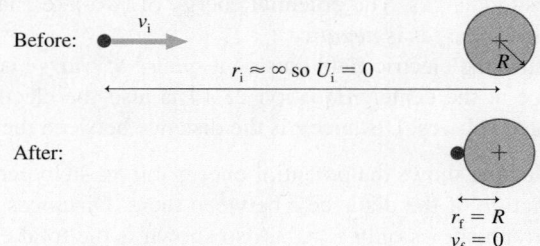

The electric force is a *central force*. As a result, zero work is done as q_2 moves along a circular arc because the force is perpendicular to the displacement.

All the work is done along the radial line segments, which are equivalent to a straight line from i to f. This is the work that was calculated in Equation 28.11.

EXAMPLE 28.2 Approaching a charged sphere

A proton is fired from far away at a 1.0-mm-diameter glass sphere that has been charged to +100 nC. What initial speed must the proton have to just reach the surface of the glass?

MODEL Energy is conserved. The glass sphere can be treated as a charged particle, so the potential energy is that of two point charges. The proton starts "far away," which we interpret as sufficiently far to make $U_i \approx 0$.

VISUALIZE FIGURE 28.11 shows the before-and-after pictorial representation. To "just reach" the glass sphere means that the proton comes to rest, $v_f = 0$, as it reaches $r_f = 0.50$ mm, the *radius* of the sphere.

SOLVE Conservation of energy $K_f + U_f = K_i + U_i$ is

$$0 + \frac{Kq_p q_{sphere}}{r_f} = \frac{1}{2}mv_i^2 + 0$$

FIGURE 28.11 A proton approaching a glass sphere.

Before: ●→ v_i

$r_i \approx \infty$ so $U_i = 0$ ⊕ R

After: ⊕

$r_f = R$
$v_f = 0$

The proton charge is $q_p = e$. With this, we can solve for the proton's initial speed:

$$v_i = \sqrt{\frac{2Keq_{sphere}}{mr_f}} = 1.86 \times 10^7 \text{ m/s}$$

EXAMPLE 28.3 Escape velocity

An interaction between two elementary particles causes an electron and a positron (a positive electron) to be shot out back to back with equal speeds. What minimum speed must each have when they are 100 fm apart in order to escape each other?

MODEL Energy is conserved. The particles end "far apart," which we interpret as sufficiently far to make $U_f \approx 0$.

VISUALIZE FIGURE 28.12 shows the before-and-after pictorial representation. The minimum speed to escape is the speed that allows the particles to reach $r_f = \infty$ with $v_f = 0$.

FIGURE 28.12 An electron and a positron flying apart.

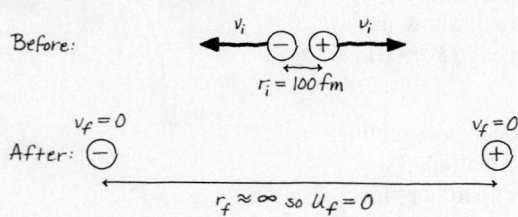

SOLVE Here it is essential to interpret U_{elec} as the potential energy of the electron + positron system. Similarly, K is the *total* kinetic energy of the system. The electron and the positron, with equal masses and equal speeds, have equal kinetic energies. Conservation of energy $K_f + U_f = K_i + U_i$ is

$$0 + 0 + 0 = \frac{1}{2}mv_i^2 + \frac{1}{2}mv_i^2 + \frac{Kq_e q_p}{r_i} = mv_i^2 - \frac{Ke^2}{r_i}$$

Using $r_i = 100$ fm $= 1.0 \times 10^{-13}$ m, we can calculate the minimum initial speed to be

$$v_i = \sqrt{\frac{Ke^2}{mr_i}} = 5.0 \times 10^7 \text{ m/s}$$

ASSESS v_i is a little more than 10% the speed of light, just about the limit of what a "classical" calculation can predict. We would need to use the theory of relativity if v_i were any larger.

Multiple Point Charges

If more than two charges are present, the potential energy is the sum of the potential energies due to all pairs of charges:

$$U_{\text{elec}} = \sum_{i<j} \frac{Kq_iq_j}{r_{ij}} \tag{28.15}$$

where r_{ij} is the distance between q_i and q_j. The summation contains the $i < j$ restriction to ensure that each pair of charges is counted only once.

NOTE ▶ For energy conservation problems, it's necessary to calculate only the potential energy for those pairs of charges for which the distance r_{ij} changes. The potential energy of charges that don't move is an additive constant with no physical consequences. ◀

EXAMPLE 28.4 | **Launching an electron**

Three electrons are spaced 1.0 mm apart along a vertical line. The outer two electrons are fixed in position.

a. Is the center electron at a point of stable or unstable equilibrium?
b. If the center electron is displaced horizontally by a small distance, what will its speed be when it is very far away?

MODEL Energy is conserved. The outer two electrons don't move, so we don't need to include the potential energy of their interaction.

VISUALIZE FIGURE 28.13 shows the situation.

FIGURE 28.13 Three electrons.

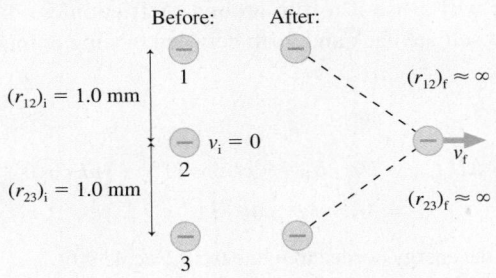

$(r_{12})_i = 1.0$ mm

$v_i = 0$

$(r_{23})_i = 1.0$ mm

SOLVE a. The center electron is in equilibrium *exactly* in the center because the two electric forces on it balance. But if it moves a little to the right or left, no matter how little, then the horizontal components of the forces from both outer electrons will push the center electron farther away. This is an unstable equilibrium for horizontal displacements, like being on the top of a hill.

b. A small displacement will cause the electron to move away. If the displacement is only infinitesimal, the initial conditions are $(r_{12})_i = (r_{23})_i = 1.0$ mm and $v_i = 0$. "Far away" is interpreted as $r_f \to \infty$, where $U_f \approx 0$. There are now *two* terms in the potential energy, so conservation of energy $K_f + U_f = K_i + U_i$ gives

$$\frac{1}{2}mv_f^2 + 0 + 0 = 0 + \left[\frac{Kq_1q_2}{(r_{12})_i} + \frac{Kq_2q_3}{(r_{23})_i}\right]$$

$$= \left[\frac{Ke^2}{(r_{12})_i} + \frac{Ke^2}{(r_{23})_i}\right]$$

This is easily solved to give

$$v_f = \sqrt{\frac{2}{m}\left[\frac{Ke^2}{(r_{12})_i} + \frac{Ke^2}{(r_{23})_i}\right]} = 1000 \text{ m/s}$$

STOP TO THINK 28.2 Rank in order, from largest to smallest, the potential energies U_a to U_d of these four pairs of charges. Each + symbol represents the same amount of charge.

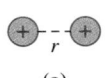   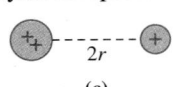   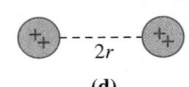

(a)   (b)   (c)   (d)

28.3 The Potential Energy of a Dipole

The electric dipole has been our model for understanding how charged objects interact with neutral objects. In Chapter 26 we found that an electric field exerts a *torque* on a dipole. We can complete the picture by calculating the potential energy of an electric dipole in a uniform electric field.

FIGURE 28.14 shows a dipole in an electric field $\vec{E}$. Recall that the dipole moment $\vec{p}$ is a vector that points from $-q$ to q with magnitude $p = qs$. The forces $\vec{F}_+$ and $\vec{F}_-$ exert a torque on the dipole, but now we're interested in calculating the *work* done by these forces as the dipole rotates from angle ϕ_i to angle ϕ_f.

FIGURE 28.14 The electric field does work as a dipole rotates.

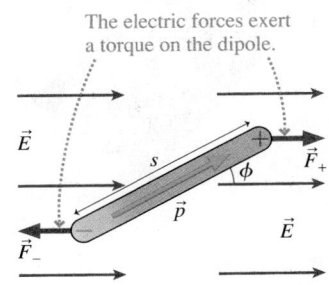

The electric forces exert a torque on the dipole.

When a force component F_s acts through a small displacement ds, the force does work $dW = F_s\, ds$. If we exploit the rotational-linear motion analogy from Chapter 12, where torque τ is the analog of force and angular displacement $\Delta\phi$ is the analog of linear displacement, then a torque acting through a small angular displacement $d\phi$ does work $dW = \tau\, d\phi$. From Chapter 26, the torque on the dipole in Figure 28.14 is $\tau = -pE\sin\phi$, where the minus sign is due to the torque trying to cause a clockwise rotation. Thus the work done by the electric field on the dipole as it rotates through the small angle $d\phi$ is

$$dW_{\text{elec}} = -pE\sin\phi\, d\phi \tag{28.16}$$

The total work done by the electric field as the dipole turns from ϕ_i to ϕ_f is

$$W_{\text{elec}} = -pE\int_{\phi_i}^{\phi_f} \sin\phi\, d\phi = pE\cos\phi_f - pE\cos\phi_i \tag{28.17}$$

The potential energy associated with the work done on the dipole is

$$\Delta U_{\text{dipole}} = U_f - U_i = -W_{\text{elec}}(\text{i}\to\text{f}) = -pE\cos\phi_f + pE\cos\phi_i \tag{28.18}$$

By comparing the left and right sides of Equation 28.18, we see that the potential energy of an electric dipole $\vec{p}$ in a uniform electric field $\vec{E}$ is

$$U_{\text{dipole}} = -pE\cos\phi = -\vec{p}\cdot\vec{E} \tag{28.19}$$

FIGURE 28.15 shows the energy diagram of a dipole. The potential energy is minimum at $\phi = 0°$ where the dipole is aligned with the electric field. This is a point of stable equilibrium. A dipole exactly opposite $\vec{E}$, at $\phi = \pm 180°$, is at a point of unstable equilibrium. Any disturbance will cause it to flip around. A frictionless dipole with mechanical energy E_{mech} will oscillate back and forth between turning points on either side of $\phi = 0°$.

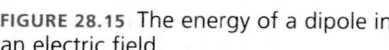

FIGURE 28.15 The energy of a dipole in an electric field.

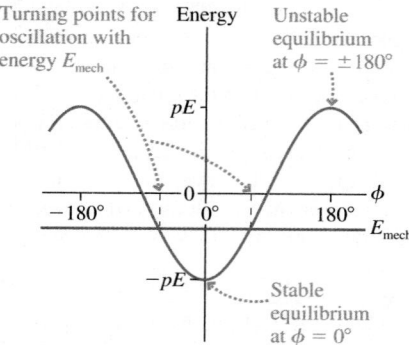

EXAMPLE 28.5 **Rotating a molecule**

The water molecule is a permanent electric dipole with dipole moment 6.2×10^{-30} C m. A water molecule is aligned in an electric field with field strength 1.0×10^7 N/C. How much energy is needed to rotate the molecule 90°?

MODEL The molecule is at the point of minimum energy. It won't spontaneously rotate 90°. However, an external force that supplies energy, such as a collision with another molecule, can cause the water molecule to rotate.

SOLVE The molecule starts at $\phi_i = 0°$ and ends at $\phi_f = 90°$. The increase in potential energy is

$$\Delta U_{\text{dipole}} = U_f - U_i = -pE\cos 90° - (-pE\cos 0°)$$
$$= pE = 6.2 \times 10^{-23}\,\text{J}$$

This is the energy needed to rotate the molecule 90°.

ASSESS ΔU_{dipole} is significantly less than k_BT at room temperature. Thus collisions with other molecules can easily supply the energy to rotate the water molecules and keep them from staying aligned with the electric field.

28.4 The Electric Potential

We introduced the concept of the *electric field* in Chapter 25 because action at a distance raised concerns and difficulties. The field provides an intermediary through which two charges exert forces on each other. Charge q_1 somehow alters the space around it by creating an electric field $\vec{E}_1$. Charge q_2 then responds to the field, experiencing force $\vec{F} = q_2\vec{E}_1$.

We face the same kinds of difficulties when we try to understand electric potential energy. For a mass on a spring, we can *see* how the energy is stored in the stretched or compressed spring. But when we say two charged particles have a potential energy, an energy that can be converted to a tangible kinetic energy of motion, *where is the energy?* It's indisputable that two positive charges fly apart when you release them, gaining kinetic energy, but there's no obvious place that the energy had been stored.

This battery is labeled 1.5 Volts. As we'll soon see, a battery is a source of electric potential.

In defining the electric field, we chose to separate the charges that are the *source* of the field from the charge *in* the field. The force on charge q is related to the electric field of the source charges by

$$\text{force on } q \text{ by sources} = [\text{charge } q] \times [\text{alteration of space by the source charges}]$$

Let's try a similar procedure for the potential energy. The electric potential energy is due to the interaction of charge q with other charges, so let's write

$$\text{potential energy of } q + \text{sources}$$

$$= [\text{charge } q] \times [\textit{potential} \text{ for interaction of the source charges}]$$

FIGURE 28.16 shows this idea schematically.

In analogy with the electric field, we will define the **electric potential** V (or, for brevity, just *the potential*) as

$$V \equiv \frac{U_{q+\text{sources}}}{q} \tag{28.20}$$

Charge q is used as a probe to determine the electric potential, but the value of V is *independent of q*. **The electric potential, like the electric field, is a property of the source charges.**

In practice, we're usually more interested in knowing the potential energy if a charge q happens to be at a point in space where the electric potential of the source charges is V. Turning Equation 28.20 around, we see that the electric potential energy is

$$U_{q+\text{sources}} = qV \tag{28.21}$$

Once the potential has been determined, it's very easy to find the potential energy.

The unit of electric potential is the joule per coulomb, which is called the **volt** V:

$$1 \text{ volt} = 1 \text{ V} \equiv 1 \text{ J/C}$$

This unit is named for Alessandro Volta, who invented the electric battery in the year 1800. Microvolts (μV), millivolts (mV), and kilovolts (kV) are commonly used units.

NOTE ▶ Once again, commonly used symbols are in conflict. The symbol V is widely used to represent *volume,* and now we're introducing the same symbol to represent *potential*. To make matters more confusing, V is the abbreviation for *volts.* In printed text, V for potential is italicized and V for volts is not, but you can't make such a distinction in handwritten work. This is not a pleasant state of affairs, but these are the commonly accepted symbols. It's incumbent upon you to be especially alert to the *context* in which a symbol is used. ◀

Using the Electric Potential

The electric potential is an abstract idea, and it will take some practice to see just what it means and how it is useful. We'll use multiple representations—words, pictures, graphs, and analogies—to explain and describe the electric potential.

NOTE ▶ It is unfortunate that the terms *potential* and *potential energy* are so much alike. Despite the similar names, they are very different concepts and are not interchangeable. Table 28.1 will help you to distinguish between the two. ◀

Basically, knowing the electric potential in a region of space allows us to determine whether a charged particle speeds up or slows down as it moves through that region. FIGURE 28.17 on the next page illustrates this idea. Here a group of source charges, which remains hidden offstage, has created an electric potential V that increases from left to right. A charged particle q, which for now we'll assume to be positive, has electric

FIGURE 28.16 Source charges alter the space around them by creating an electric potential.

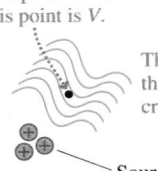

The potential at this point is V.

The source charges alter the space around them by creating an electric potential.

Source charges

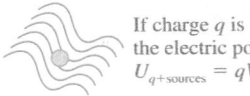

If charge q is in the potential, the electric potential energy is $U_{q+\text{sources}} = qV$.

TABLE 28.1 Distinguishing electric potential and potential energy

The *electric potential* is a property of the source charges and, as you'll soon see, is related to the electric field. The electric potential is present whether or not a charged particle is there to experience it. Potential is measured in J/C, or V.

The *electric potential energy* is the interaction energy of a charged particle with the source charges. Potential energy is measured in J.

FIGURE 28.17 A charged particle speeds up or slows down as it moves through a potential difference.

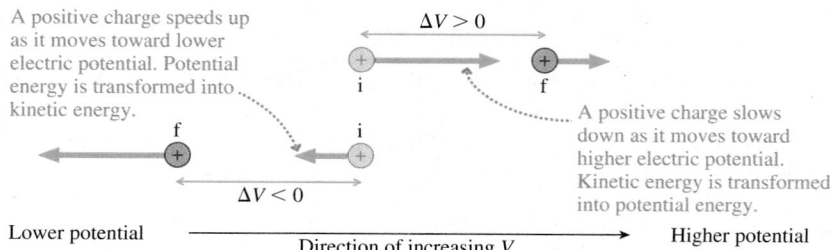

A positive charge speeds up as it moves toward lower electric potential. Potential energy is transformed into kinetic energy.

$\Delta V > 0$

A positive charge slows down as it moves toward higher electric potential. Kinetic energy is transformed into potential energy.

$\Delta V < 0$

Lower potential — Direction of increasing V → Higher potential

potential energy $U = qV$. If the particle moves to the right, its potential energy increases and so, by energy conservation, its kinetic energy must decrease. **A positive charge slows down as it moves into a region of higher electric potential.**

It is customary to say that the particle moves through a **potential difference** $\Delta V = V_f - V_i$. The potential difference between two points is often called the **voltage.** The particle moving to the right moves through a positive potential difference ($\Delta V > 0$ because $V_f > V_i$), so we can say that a positively charged particle slows down as it moves through a positive potential difference.

The particle moving to the left in Figure 28.17 travels in the direction of decreasing electric potential—through a negative potential difference—and is losing potential energy. It speeds up as it transforms potential energy into kinetic energy. A negatively charged particle would slow down because its potential energy qV would increase as V decreases. Table 28.2 summarizes these ideas.

If a particle moves through a potential difference ΔV, its electric potential energy changes by $\Delta U = q \Delta V$. We can write the conservation of energy equation in terms of the electric potential as $\Delta K + \Delta U = \Delta K + q \Delta V = 0$ or, as is often more practical,

$$K_f + qV_f = K_i + qV_i \qquad (28.22)$$

Conservation of energy is the basis of a powerful problem-solving strategy.

TABLE 28.2 Charged particles moving in an electric potential

| | Electric potential | |
	Increasing ($\Delta V > 0$)	Decreasing ($\Delta V < 0$)
+ charge	Slows down	Speeds up
− charge	Speeds up	Slows down

PROBLEM-SOLVING STRATEGY 28.1 **Conservation of energy in charge interactions**

MODEL Check whether there are any dissipative forces that would keep the mechanical energy from being conserved.

VISUALIZE Draw a before-and-after pictorial representation. Define symbols that will be used in the problem, list known values, and identify what you're trying to find.

SOLVE The mathematical representation is based on the law of conservation of mechanical energy:

$$K_f + qV_f = K_i + qV_i$$

- Is the electric potential given in the problem statement? If not, you'll need to use a known potential, such as that of a point charge, or calculate the potential using the procedure given later, in Problem-Solving Strategy 28.2.
- K_i and K_f are the sums of the kinetic energies of all moving particles.
- Some problems may need additional conservation laws, such as conservation of charge or conservation of momentum.

ASSESS Check that your result has the correct units, is reasonable, and answers the question.

Exercise 22

EXAMPLE 28.6 **Moving through a potential difference**

A proton with a speed of 2.0×10^5 m/s enters a region of space in which source charges have created an electric potential. What is the proton's speed after it moves through a potential difference of 100 V? What will be the final speed if the proton is replaced by an electron?

MODEL Energy is conserved. The electric potential determines the potential energy.

VISUALIZE FIGURE 28.18 is a before-and-after pictorial representation of a charged particle moving through a potential difference. A positive charge *slows down* as it moves into a region of higher potential ($K \rightarrow U$). A negative charge *speeds up* ($U \rightarrow K$).

FIGURE 28.18 A charged particle moving through a potential difference.

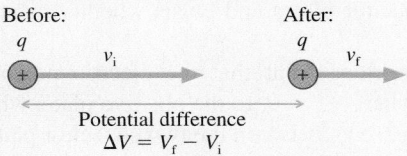

Before:

After:

$$\Delta V = V_f - V_i$$

Potential difference

SOLVE The potential energy of charge q is $U = qV$. Conservation of energy, now expressed in terms of the electric potential V, is $K_f + qV_f = K_i + qV_i$, or

$$K_f = K_i - q\,\Delta V$$

where $\Delta V = V_f - V_i$ is the potential difference through which the particle moves. In terms of the speeds, energy conservation is

$$\frac{1}{2}mv_f^2 = \frac{1}{2}mv_i^2 - q\,\Delta V$$

We can solve this for the final speed:

$$v_f = \sqrt{v_i^2 - \frac{2q}{m}\Delta V}$$

For a proton, with $q = e$, the final speed is

$$(v_f)_p = \sqrt{(2.0 \times 10^5 \text{ m/s})^2 - \frac{2(1.60 \times 10^{-19} \text{ C})(100 \text{ V})}{1.67 \times 10^{-27} \text{ kg}}}$$

$$= 1.4 \times 10^5 \text{ m/s}$$

An electron, though, with $q = -e$ and a different mass, speeds up to $(v_f)_e = 5.9 \times 10^6$ m/s.

ASSESS The electric potential *already existed* in space due to other charges that are not explicitly seen in the problem. The electron and proton have nothing to do with creating the potential. Instead, they *respond* to the potential by having potential energy $U = qV$.

STOP TO THINK 28.3   A proton is released from rest at point B, where the potential is 0 V. Afterward, the proton

a. Remains at rest at B.
b. Moves toward A with a steady speed.
c. Moves toward A with an increasing speed.
d. Moves toward C with a steady speed.
e. Moves toward C with an increasing speed.

$$-100 \text{ V} \qquad 0 \text{ V} \qquad +100 \text{ V}$$

A•   B•   C•

28.5 The Electric Potential Inside a Parallel-Plate Capacitor

We began this chapter with the potential energy of a charge inside a parallel-plate capacitor. Now let's investigate the electric potential. FIGURE 28.19 shows two parallel electrodes, separated by distance d, with surface charge density $\pm \eta$. As a specific example, we'll let $d = 3.00$ mm and $\eta = 4.42 \times 10^{-9}$ C/m^2. The electric field inside the capacitor, as you learned in Chapter 26, is

$$\vec{E} = \left(\frac{\eta}{\epsilon_0}, \text{ from positive toward negative}\right) \tag{28.23}$$

$$= (500 \text{ N/C, from right to left})$$

This electric field is due to the *source charges* on the capacitor plates.

FIGURE 28.19 A parallel-plate capacitor.

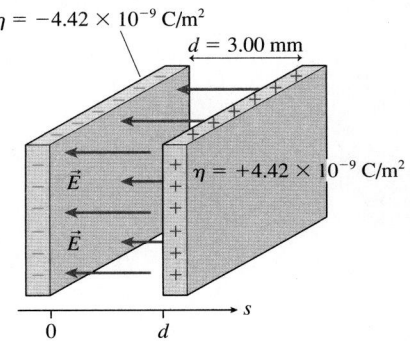

$\eta = -4.42 \times 10^{-9}$ C/m^2

$d = 3.00$ mm

$\eta = +4.42 \times 10^{-9}$ C/m^2

In Section 28.1, we found that the electric potential energy of a charge q in the uniform electric field of a parallel-plate capacitor is

$$U_{elec} = U_{q+sources} = qEs \qquad (28.24)$$

We've set the constant term U_0 to zero. U_{elec} is the energy of q interacting with the source charges on the capacitor plates.

Our new view of the interaction is to separate the role of charge q from the role of the source charges by defining the electric potential $V = U_{q+sources}/q$. Thus the electric potential inside a parallel-plate capacitor is

$$V = Es \qquad \text{(electric potential inside a parallel-plate capacitor)} \qquad (28.25)$$

where s **is the distance from the** ***negative*** **electrode.** The electric potential, like the electric field, exists at *all points* inside the capacitor. The electric potential is created by the source charges on the capacitor plates and exists whether or not charge q is inside the capacitor.

FIGURE 28.20 illustrates the important point that the electric potential increases linearly from the negative plate, where $V_- = 0$, to the positive plate, where $V_+ = Ed$. Let's define the *potential difference* ΔV_C between the two capacitor plates to be

$$\Delta V_C = V_+ - V_- = Ed \qquad (28.26)$$

In our specific example, $\Delta V_C = (500 \text{ N/C})(0.0030 \text{ m}) = 1.5 \text{ V}$. The units work out because $1.5 \text{ (N m)/C} = 1.5 \text{ J/C} = 1.5 \text{ V}$.

NOTE ▶ People who work with circuits would call ΔV_C "the voltage across the capacitor" or simply "the capacitor voltage." ◀

Equation 28.26 has an interesting implication. Thus far, we've determined the electric field inside a capacitor by specifying the surface charge density η on the plates. Alternatively, we could specify the capacitor voltage ΔV_C (i.e., the potential difference between the capacitor plates) and then determine the electric field strength as

$$E = \frac{\Delta V_C}{d} \qquad (28.27)$$

In fact, this is how E is determined in practical applications because it's easy to measure ΔV_C with a voltmeter but difficult, in practice, to know the value of η.

Equation 28.27 implies that the units of electric field are volts per meter, or V/m. We have been using electric field units of newtons per coulomb. In fact, as you can show as a homework problem, these units are equivalent to each other. That is,

$$1 \text{ N/C} = 1 \text{ V/m}$$

NOTE ▶ Volts per meter are the electric field units used by scientists and engineers in practice. We will now adopt them as our standard electric field units. ◀

Returning to the electric potential, we can substitute Equation 28.27 for E into Equation 28.25 for V. Thus the electric potential inside the capacitor is

$$V = Es = \frac{s}{d}\Delta V_C \qquad (28.28)$$

The potential increases linearly from $V_- = 0 \text{ V}$ at the negative plate ($s = 0$) to $V_+ = \Delta V_C$ at the positive plate ($s = d$).

Let's explore the electric potential inside the capacitor by looking at several different, but related, ways that the potential can be represented graphically.

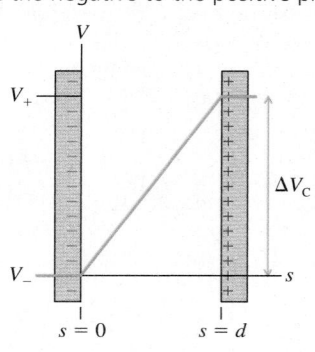

FIGURE 28.20 The electric potential of a parallel-plate capacitor increases linearly from the negative to the positive plate.

Graphical representations of the electric potential inside a capacitor

| A graph of potential versus s. You can see the potential increasing from 0.0 V at the negative plate to 1.5 V at the positive plate. | A three-dimensional view showing **equipotential surfaces.** These are mathematical surfaces, not physical surfaces, with the same value of V at every point. The equipotential surfaces of a capacitor are planes parallel to the capacitor plates. The capacitor plates are also equipotential surfaces. | A two-dimensional **contour map.** The capacitor plates and the equipotential surfaces are seen edge-on, so you need to imagine them extending above and below the plane of the page. | A three-dimensional **elevation graph.** The potential is graphed vertically versus the s-coordinate on one axis and a generalized "yz-coordinate" on the other axis. Viewing the right face of the elevation graph gives you the potential graph. |

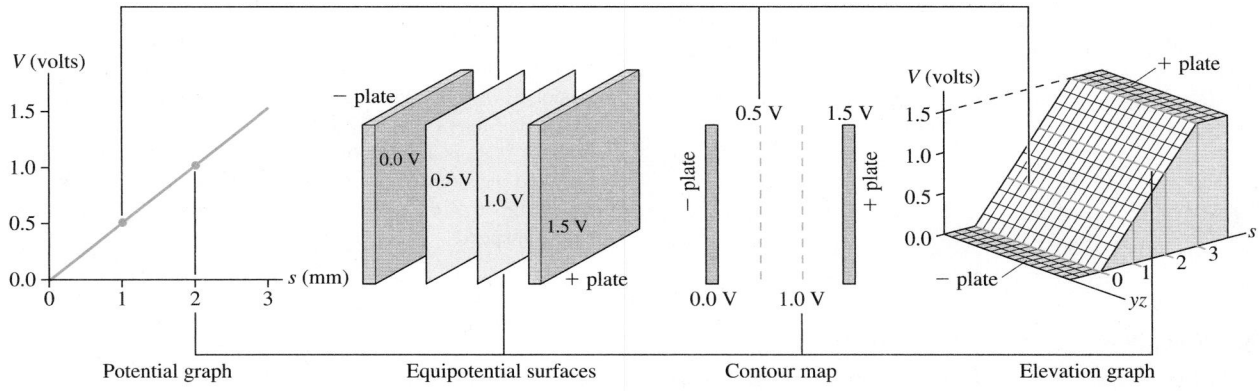

| Potential graph | Equipotential surfaces | Contour map | Elevation graph |

These four graphical representations show the same information from different perspectives, and the connecting lines help you see how they are related. If you think of the elevation graph as a "mountain," then the contour lines on the contour map are like the lines of a topographic map.

The potential graph and the contour map are the two representations most widely used in practice because they are easy to draw. Their limitation is that they are trying to convey three-dimensional information in a two-dimensional presentation. When you see graphs or contour maps, you need to imagine the three-dimensional equipotential surfaces or the three-dimensional elevation graph.

There's nothing special about showing equipotential surfaces or contour lines every 0.5 V. We chose these intervals because they were convenient. As an alternative, **FIGURE 28.21** shows how the contour map looks if the contour lines are spaced every 0.3 V. Contour lines and equipotential surfaces are *imaginary* lines and surfaces drawn to help us visualize how the potential changes in space. Drawing the map more than one way reinforces the idea that there is an electric potential at *every* point inside the capacitor, not just at the points where we happened to draw a contour line or an equipotential surface.

Figure 28.21 also shows the electric field vectors. Notice that

- The electric field vectors are perpendicular to the equipotential surfaces.
- The electric field points in the direction of decreasing potential. In other words, the electric field points "downhill" on a graph or map of the electric potential.

Chapter 29 will present a more in-depth exploration of the connection between the electric field and the electric potential. There you will find that these observations are always true. They are not unique to the parallel-plate capacitor.

Finally, you might wonder how we can arrange a capacitor to have a surface charge density of precisely 4.42×10^{-9} C/m^2. Simple! As **FIGURE 28.22** shows, we use wires to attach the capacitor plates to a 1.5 V battery. This is another topic that we'll explore in Chapter 29, but it's worth noting now that **a battery is a source of potential.** That's why batteries are labeled in volts, and it's a major reason we need to thoroughly understand the concept of potential.

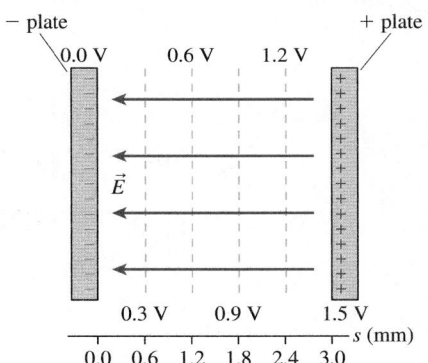

FIGURE 28.21 The contour lines of the electric potential and the electric field vectors inside a parallel-plate capacitor.

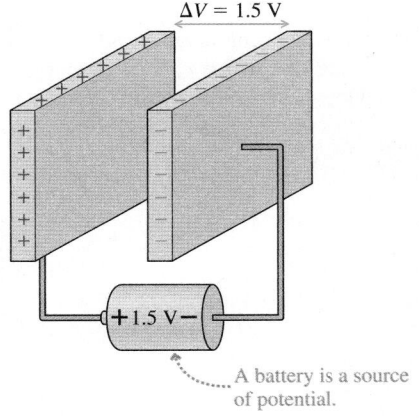

FIGURE 28.22 Using a battery to charge a capacitor to a precise value of ΔV_C.

EXAMPLE 28.7 **Measuring the speed of a proton**

The lab in which you work has a small proton accelerator. You've been assigned the task of measuring the speed of the protons as they emerge from the accelerator. To do so, you decide to measure how much voltage is needed across a parallel-plate capacitor to stop the protons. The capacitor you choose has a 2.0 mm plate separation and a small hole in one plate that you shoot the protons through. By filling the space between the plates with a low-density gas, you can see (with a microscope) a slight glow from the region where the protons collide with and excite the gas molecules. The width of the glow tells you how far the protons travel before being stopped and reversing direction. Varying the voltage across the capacitor gives the following data:

Capacitor voltage (V)	Glow width (mm)
1000	1.7
1250	1.3
1500	1.1
1750	1.0
2000	0.8

What value will you report for the speed of the protons?

MODEL Energy is conserved. The proton's potential energy can be found from the capacitor's electric potential.

VISUALIZE FIGURE 28.23 shows a before-and-after pictorial representation of the proton entering the capacitor with speed v_i, which we want to find, and later reaching a turning point with $v_f = 0$ m/s after traveling distance $s_f =$ glow width. For the protons to slow and stop, the hole through which they pass has to be in the negative plate. We've established an s-axis with $s = 0$ at this point.

FIGURE 28.23 A proton being stopped in a capacitor.

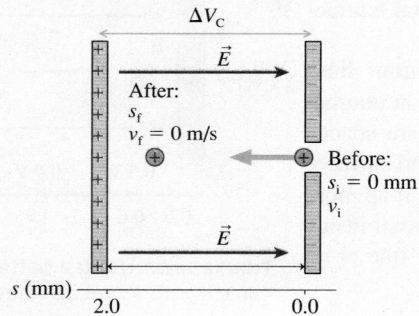

SOLVE The conservation of energy equation, with the proton having charge $q = e$, is $K_f + eV_f = K_i + eV_i$. The initial potential energy is zero, because the capacitor's electric potential is zero at $s_i = 0$, and the final kinetic energy is zero. Using Equation 28.28 for the potential inside the capacitor, we have

$$eV_f = e\left(\frac{s_f}{d}\Delta V_C\right) = K_i = \frac{1}{2}mv_i^2$$

Solving for the distance traveled, we find

$$s_f = \frac{dmv_i^2}{2e}\frac{1}{\Delta V_C}$$

Thus a graph of the distance traveled versus the *inverse* of the capacitor voltage should be a straight line with zero y-intercept and slope $dmv_i^2/2e$. We can use the experimentally determined slope to find the proton speed.

FIGURE 28.24 is a graph of s_f versus $1/\Delta V_C$. It has the expected shape, and the slope of the best-fit line is seen to be 1.72 V m. The units are those of the rise-over-run. Using the slope, we calculate the proton speed:

$$v_i = \sqrt{\frac{2e}{dm}\times\text{slope}} = \sqrt{\frac{2(1.60\times10^{-19}\text{ C})(1.72\text{ V m})}{(0.0020\text{ m})(1.67\times10^{-27}\text{ kg})}}$$

$$= 4.1\times10^5\text{ m/s}$$

FIGURE 28.24 A graph of the data.

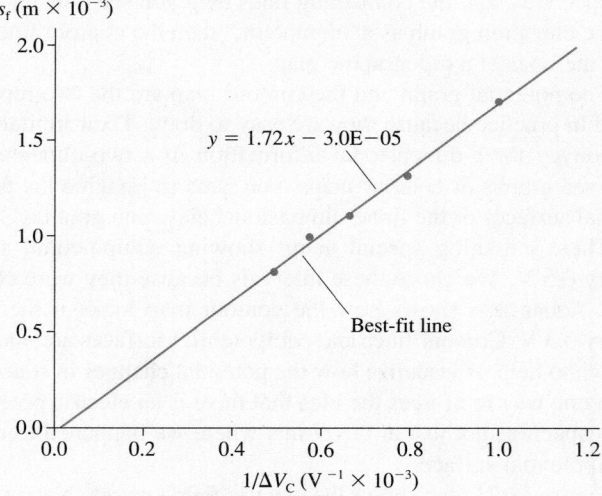

ASSESS This would be a very high speed for a macroscopic object but is quite typical of the speeds of charged particles.

In writing the electric potential inside a parallel-plate capacitor, we made the choice that $V_- = 0$ V at the negative plate. But that is not the only possible choice. FIGURE 28.25 shows three parallel-plate capacitors, each having the same capacitor voltage $\Delta V_C = V_+ - V_- = 100$ V, but each with a different choice for the location of the zero point of the electric potential. Notice the *terminal symbols* (lines with small circles at the end) showing how the potential, from a battery or a power supply, is applied to each plate; these symbols are common in electronics.

FIGURE 28.25 These three choices for $V = 0$ represent the same physical situation. These are contour maps, showing the edges of the equipotential surfaces.

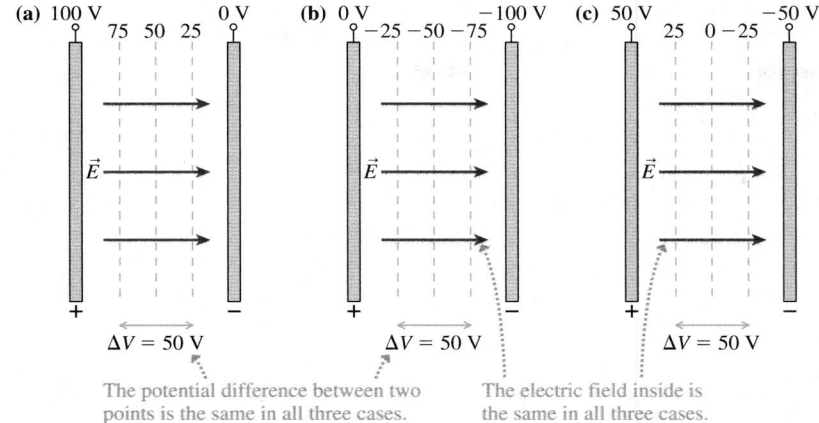

The potential difference between two points is the same in all three cases.

The electric field inside is the same in all three cases.

The important thing to notice is that the three contour maps in Figure 28.25 represent the *same physical situation*. The potential difference between any two points is the same in all three maps. The electric field is the same in all three. We may *prefer* one of these figures over the others, but there is no measurable physical difference between them.

EXAMPLE 28.8 **The force on an ion**

Example 26.7 noted that a cell wall can be modeled as a parallel-plate capacitor, with the outer surface of the cell wall being positive while the inner surface is negative. The potential difference between the inside of the cell and the outside is called the *membrane potential*. Suppose a molecular ion with charge $5e$ is embedded within the 5.0-nm-thick wall of a cell with a membrane potential of -70 mV, typical for a nerve cell in its resting state. What is the force on the molecular ion?

MODEL Model the cell wall as a parallel-plate capacitor with the inner surface being the negative plate. Although the walls are actually curved, and not large flat planes, the parallel-plate approximation is valid if the wall thickness is much less than the radius of the cell. The capacitor voltage is $\Delta V_C = 70$ mV $= 0.070$ V. The membrane potential is negative because the potential inside the cell is less than the potential outside, but ΔV_C, the capacitor voltage, is the *magnitude* of the potential difference and thus always positive.

SOLVE The force on a charged particle is $\vec{F} = q\vec{E}$. The electric field strength inside the parallel-plate capacitor of the cell wall is

$$E = \frac{\Delta V_C}{d} = \frac{0.070 \text{ V}}{5.0 \times 10^{-9} \text{ m}} = 1.4 \times 10^7 \text{ V/m}$$

Notice that we're now using V/m rather than N/C as the units of electric field. Because the field points from positive to negative, the field vector is $\vec{E} = (1.4 \times 10^7$ V/m, toward inside). Thus the force on an ion with $q = 5e = 8.0 \times 10^{-19}$ C is

$$\vec{F} = q\vec{E} = (1.1 \times 10^{-11} \text{ N, toward inside})$$

ASSESS For cells to function, a steady flow of molecules must pass back and forth through the cell wall. Although the details of how this happens are very complex, a key idea is that a potential difference between the inside and outside of the cell creates an electric field that pushes positive ions toward the inside, negative ions toward the outside.

STOP TO THINK 28.4 Rank in order, from largest to smallest, the potentials V_a to V_e at the points a to e.

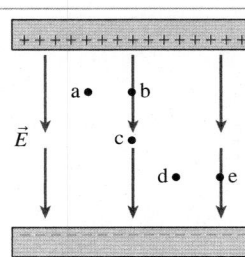

28.6 The Electric Potential of a Point Charge

FIGURE 28.26 Measuring the electric potential of charge q.

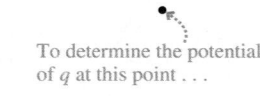

To determine the potential of q at this point . . .

q

q'

r

. . . place charge q' at the point as a probe and measure the potential energy $U_{q'+q}$.

q

Another important electric potential is that of a point charge. Let q in FIGURE 28.26 be the source charge, and let a second charge q' probe the electric potential of q. The potential energy of the two point charges is

$$U_{q'+q} = \frac{1}{4\pi\epsilon_0} \frac{qq'}{r} \qquad (28.29)$$

Thus, by definition, the electric potential of charge q is

$$V = \frac{U_{q'+q}}{q'} = \frac{1}{4\pi\epsilon_0} \frac{q}{r} \qquad \text{(electric potential of a point charge)} \quad (28.30)$$

The potential of Equation 28.30 extends through all of space, showing the influence of charge q, but it weakens with distance as $1/r$. This expression for V assumes that we have chosen $V = 0$ V to be at $r = \infty$. This is the most logical choice for a point charge because the influence of charge q ends at infinity.

The expression for the electric potential of charge q is similar to that for the electric field of charge q. The difference most quickly seen is that V depends on $1/r$ whereas $\vec{E}$ depends on $1/r^2$. But it is also important to notice that **the potential is a scalar** whereas the field is a vector. Thus the mathematics of using the potential are much easier than the vector mathematics using the electric field requires.

EXAMPLE 28.9 **Calculating the potential of a point charge**

What is the electric potential 1.0 cm from a $+1.0$ nC charge? What is the potential difference between a point 1.0 cm away and a second point 3.0 cm away?

SOLVE The potential at $r = 1.0$ cm is

$$V_{1\,cm} = \frac{1}{4\pi\epsilon_0} \frac{q}{r} = (9.0 \times 10^9 \text{ N m}^2/\text{C}^2) \frac{1.0 \times 10^{-9} \text{ C}}{0.010 \text{ m}}$$

$$= 900 \text{ V}$$

We can similarly calculate $V_{3\,cm} = 300$ V. Thus the potential difference between these two points is $\Delta V = V_{1\,cm} - V_{3\,cm} = 600$ V.

ASSESS 1 nC is typical of the electrostatic charge produced by rubbing, and you can see that such a charge creates a fairly large potential nearby. Why are we not shocked and injured when working with the "high voltages" of such charges? The sensation of being shocked is a result of current, not potential. Some high-potential sources simply do not have the ability to generate much current. We will look at this issue in Chapter 31.

Visualizing the Potential of a Point Charge

FIGURE 28.27 shows four graphical representations of the electric potential of a point charge. These match the four representations of the electric potential inside a capacitor, and a comparison of the two is worthwhile. This figure assumes that q is positive; you may want to think about how the representations would change if q were negative.

FIGURE 28.27 Four graphical representations of the electric potential of a point charge.

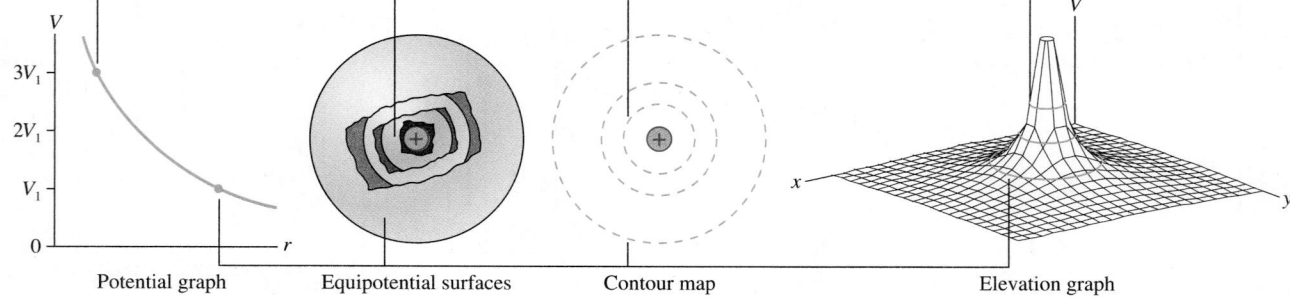

Potential graph   Equipotential surfaces   Contour map   Elevation graph

STOP TO THINK 28.5 Rank in order, from largest to smallest, the potential differences ΔV_{ab}, ΔV_{ac}, and ΔV_{bc} between points a and b, points a and c, and points b and c.

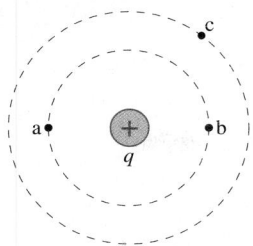

The Electric Potential of a Charged Sphere

In practice, you are more likely to work with a charged sphere, of radius R and total charge Q, than with a point charge. Outside a uniformly charged sphere, the electric potential is identical to that of a point charge Q at the center. That is,

$$V = \frac{1}{4\pi\epsilon_0} \frac{Q}{r} \qquad \text{(sphere of charge, } r \geq R\text{)} \qquad (28.31)$$

We can cast this result in a more useful form. It is customary to speak of charging an electrode, such as a sphere, "to" a certain potential, as in "Bob charged the sphere to a potential of 3000 volts." This potential, which we will call V_0, is the potential right on the surface of the sphere. We can see from Equation 28.31 that

$$V_0 = V(\text{at } r = R) = \frac{Q}{4\pi\epsilon_0 R} \qquad (28.32)$$

Consequently, a sphere of radius R that is charged to potential V_0 has total charge

$$Q = 4\pi\epsilon_0 R V_0 \qquad (28.33)$$

If we substitute this expression for Q into Equation 28.31, we can write the potential outside a sphere that is charged to potential V_0 as

$$V = \frac{R}{r} V_0 \qquad \text{(sphere charged to potential } V_0\text{)} \qquad (28.34)$$

Equation 28.34 tells us that the potential of a sphere is V_0 on the surface and decreases inversely with the distance. The potential at $r = 3R$ is $\frac{1}{3} V_0$.

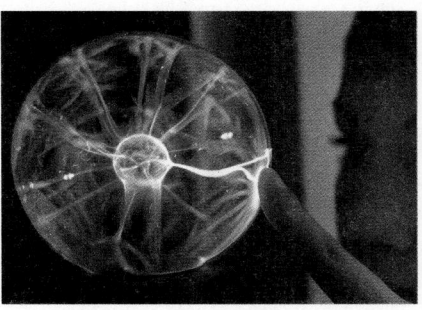

A *plasma ball* consists of a small metal ball charged to a potential of about 2000 V inside a hollow glass sphere. The glass sphere is filled with gas—typically neon or argon because of the colors they produce—at a pressure of about 0.01 atm. The electric field of the high-voltage ball is sufficient to cause a gas breakdown at this pressure, creating "lightning bolts" between the ball and the glass sphere.

EXAMPLE 28.10   **A proton and a charged sphere**

A proton is released from rest at the surface of a 1.0-cm-diameter sphere that has been charged to +1000 V.

a. What is the charge of the sphere?
b. What is the proton's speed at 1.0 cm from the sphere?

MODEL Energy is conserved. The potential outside the charged sphere is the same as the potential of a point charge at the center.

VISUALIZE FIGURE 28.28 shows the situation.

FIGURE 28.28 A sphere and a proton.

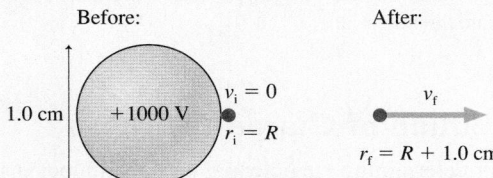

SOLVE a. The charge of the sphere is

$$Q = 4\pi\epsilon_0 R V_0 = 0.56 \times 10^{-9} \text{ C} = 0.56 \text{ nC}$$

b. A sphere charged to $V_0 = +1000$ V is positively charged. The proton will be repelled by this charge and move away from the sphere. The conservation of energy equation $K_f + eV_f = K_i + eV_i$, with Equation 28.34 for the potential of a sphere, is

$$\frac{1}{2}mv_f^2 + \frac{eR}{r_f}V_0 = \frac{1}{2}mv_i^2 + \frac{eR}{r_i}V_0$$

The proton starts from the surface of the sphere, $r_i = R$, with $v_i = 0$. When the proton is 1.0 cm from the *surface* of the sphere, it has $r_f = 1.0$ cm $+ R = 1.5$ cm. Using these, we can solve for v_f:

$$v_f = \sqrt{\frac{2eV_0}{m}\left(1 - \frac{R}{r_f}\right)} = 3.6 \times 10^5 \text{ m/s}$$

ASSESS This example illustrates how the ideas of electric potential and potential energy work together, yet they are *not* the same thing.

28.7 The Electric Potential of Many Charges

Suppose there are many source charges $q_1, q_2, \ldots$. The electric potential V at a point in space is the sum of the potentials due to each charge:

$$V = \sum_i \frac{1}{4\pi\epsilon_0} \frac{q_i}{r_i} \qquad (28.35)$$

where r_i is the distance from charge q_i to the point in space where the potential is being calculated. In other words, **the electric potential, like the electric field, obeys the principle of superposition.**

As an example, the contour map and elevation graph in FIGURE 28.29 show that the potential of an electric dipole is the sum of the potentials of the positive and negative charges. Potentials such as these have many practical applications. For example, electrical activity within the body can be monitored by measuring equipotential lines on the skin. Figure 28.29c shows that the equipotentials near the heart are a slightly distorted but recognizable electric dipole.

FIGURE 28.29 The electric potential of an electric dipole.

(a) Contour map

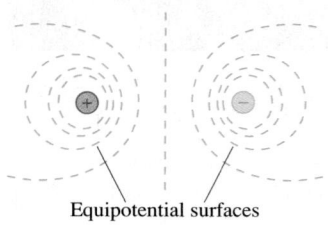

Equipotential surfaces

(b) Elevation graph

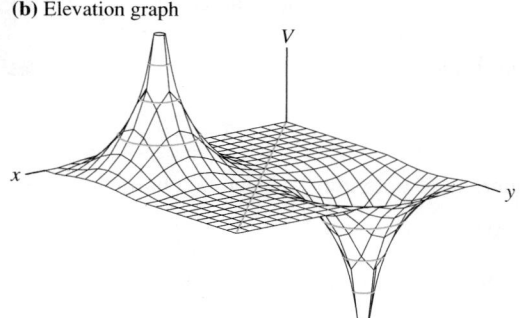

(c)

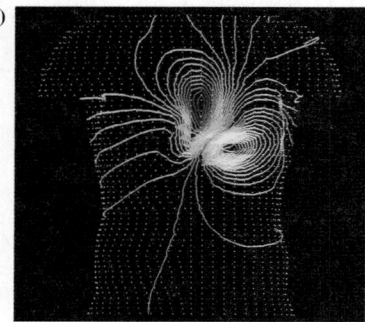

Equipotentials on the chest of a human are a slightly distorted electric dipole.

EXAMPLE 28.11 **The potential of two charges**

What is the electric potential at the point indicated in FIGURE 28.30?

FIGURE 28.30 Finding the potential of two charges.

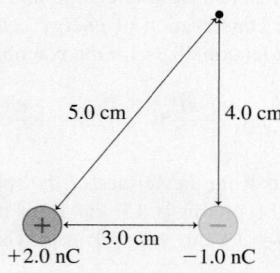

MODEL The potential is the sum of the potentials due to each charge.

SOLVE The potential at the indicated point is

$$V = \frac{1}{4\pi\epsilon_0} \frac{q_1}{r_1} + \frac{1}{4\pi\epsilon_0} \frac{q_2}{r_2}$$

$$= (9.0 \times 10^9 \text{ N m}^2/\text{C}^2)\left(\frac{2.0 \times 10^{-9} \text{ C}}{0.050 \text{ m}} + \frac{-1.0 \times 10^{-9} \text{ C}}{0.040 \text{ m}}\right)$$

$$= 135 \text{ V}$$

ASSESS The potential is a *scalar*, so we found the net potential by adding two numbers. We don't need any angles or components to calculate the potential.

A Continuous Distribution of Charge

Equation 28.35 is the basis for determining the potential of a continuous distribution of charge, such as a charged rod or a charged disk. The procedure is much like the one you learned in Chapter 26 for calculating the electric field of a continuous distribution of charge, but *easier* because the potential is a scalar. We will continue to assume that the object is *uniformly charged,* meaning that the charges are evenly spaced over the object.

PROBLEM-SOLVING STRATEGY 28.2   **The electric potential of a continuous distribution of charge**

MODEL Model the charges as a simple shape, such as a line or a disk. Assume the charge is uniformly distributed.

VISUALIZE For the pictorial representation:

❶ Draw a picture and establish a coordinate system.
❷ Identify the point P at which you want to calculate the electric potential.
❸ Divide the total charge Q into small pieces of charge ΔQ, using shapes for which you *already know* how to determine V. This division is often, but not always, into point charges.
❹ Identify distances that need to be calculated.

SOLVE The mathematical representation is $V = \sum V_i$.

■ Use superposition to form an algebraic expression for the potential at P.
■ Let the (x, y, z) coordinates of the point remain as variables.
■ Replace the small charge ΔQ with an equivalent expression involving a *charge density* and a *coordinate,* such as dx, that describes the shape of charge ΔQ. **This is the critical step in making the transition from a sum to an integral** because you need a coordinate to serve as the integration variable.
■ All distances must be expressed in terms of the coordinates.
■ Let the sum become an integral. The integration will be over the coordinate variable that is related to ΔQ. The integration limits for this variable will depend on the coordinate system you have chosen. Carry out the integration and simplify the result.

ASSESS Check that your result is consistent with any limits for which you know what the potential should be.

Exercise 29

EXAMPLE 28.12   **The potential of a ring of charge**

A thin, uniformly charged ring of radius R has total charge Q. Find the potential at distance z on the axis of the ring.

MODEL Because the ring is thin, we'll assume the charge lies along a circle of radius R.

VISUALIZE **FIGURE 28.31** illustrates the four steps of the problem-solving strategy. We've chosen a coordinate system in which the ring lies in the xy-plane and point P is on the z-axis. We've then divided the ring into N small segments of charge ΔQ, each of which can be modeled as a point charge. The distance r_i between segment i and point P is

$$r_i = \sqrt{R^2 + z^2}$$

Note that r_i is a constant distance, the same for every charge segment.

SOLVE The potential V at P is the sum of the potentials due to each segment of charge:

$$V = \sum_{i=1}^{N} V_i = \sum_{i=1}^{N} \frac{1}{4\pi\epsilon_0} \frac{\Delta Q}{r_i} = \frac{1}{4\pi\epsilon_0} \frac{1}{\sqrt{R^2 + z^2}} \sum_{i=1}^{N} \Delta Q$$

FIGURE 28.31 Finding the potential of a ring of charge.

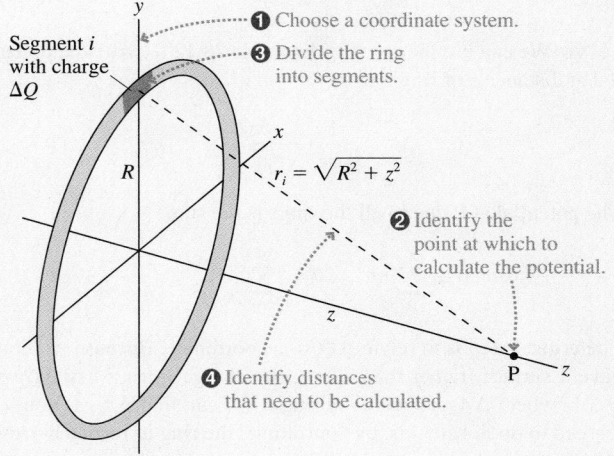

Continued

We were able to bring all terms involving z to the front because z is a constant as far as the summation is concerned. Surprisingly, we don't need to convert the sum to an integral to complete this calculation. The sum of all the ΔQ charge segments around the ring is simply the ring's total charge, $\Sigma(\Delta Q) = Q$; hence the electric potential on the axis of a charged ring is

$$V_{\text{ring on axis}} = \frac{1}{4\pi\epsilon_0}\frac{Q}{\sqrt{R^2 + z^2}}$$

ASSESS From far away, the ring appears as a point charge Q in the distance. Thus we expect the potential of the ring to be that of a point charge when $z \gg R$. You can see that $V_{\text{ring}} \approx Q/4\pi\epsilon_0 z$ when $z \gg R$, which is, indeed, the potential of a point charge Q.

CHALLENGE EXAMPLE 28.13 **The potential of a charged dime**

A 17.5-mm-diameter dime is charged to +5.00 nC.

a. What is the potential of the dime?

b. What is the potential energy of an electron 1.00 cm above the dime?

MODEL Model the dime as a thin, uniformly charged disk of radius R and charge Q. The disk has uniform surface charge density $\eta = Q/A = Q/\pi R^2$. We can take advantage of now knowing the on-axis potential of a ring of charge.

VISUALIZE Orient the disk in the xy-plane, as shown in FIGURE 28.32, with point P at distance z. Then divide the disk into *rings* of equal width Δr. Ring i has radius r_i and charge ΔQ_i.

FIGURE 28.32 Finding the potential of a disk of charge.

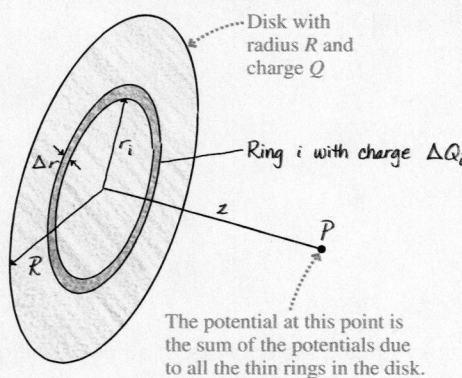

The potential at this point is the sum of the potentials due to all the thin rings in the disk.

SOLVE We can use the result of Example 28.12 to write the potential at distance z of ring i as

$$V_i = \frac{1}{4\pi\epsilon_0}\frac{\Delta Q_i}{\sqrt{r_i^2 + z^2}}$$

The potential at P due to all the rings is the sum

$$V = \sum_i V_i = \frac{1}{4\pi\epsilon_0}\sum_{i=1}^{N}\frac{\Delta Q_i}{\sqrt{r_i^2 + z^2}}$$

The critical step is to relate ΔQ_i to a coordinate. Because we now have a surface, rather than a line, the charge in ring i is $\Delta Q_i = \eta\,\Delta A_i$, where ΔA_i is the area of ring i. We can find ΔA_i, as you've learned to do in calculus, by "unrolling" the ring to form a narrow

rectangle of length $2\pi r_i$ and height Δr. Thus the area of ring i is $\Delta A_i = 2\pi r_i\,\Delta r$ and the charge is

$$\Delta Q_i = \eta\,\Delta A_i = \frac{Q}{\pi R^2}2\pi r_i\,\Delta r = \frac{2Q}{R^2}r_i\,\Delta r$$

With this substitution, the potential at P is

$$V = \frac{1}{4\pi\epsilon_0}\sum_{i=1}^{N}\frac{2Q}{R^2}\frac{r_i\,\Delta r_i}{\sqrt{r_i^2 + z^2}} \rightarrow \frac{Q}{2\pi\epsilon_0 R^2}\int_0^R\frac{r\,dr}{\sqrt{r^2 + z^2}}$$

where, in the last step, we let $N \rightarrow \infty$ and the sum become an integral. This integral can be found in Appendix A, but it's not hard to evaluate with a change of variables. Let $u = r^2 + z^2$, in which case $r\,dr = \frac{1}{2}du$. Changing variables requires that we also change the integration limits. You can see that $u = z^2$ when $r = 0$, and $u = R^2 + z^2$ when $r = R$. With these changes, the on-axis potential of a charged disk is

$$V_{\text{disk on axis}} = \frac{Q}{2\pi\epsilon_0 R^2}\int_{z^2}^{R^2+z^2}\frac{\frac{1}{2}du}{u^{1/2}} = \frac{Q}{2\pi\epsilon_0 R^2}u^{1/2}\Big|_{z^2}^{R^2+z^2}$$

$$= \frac{Q}{2\pi\epsilon_0 R^2}\left(\sqrt{R^2 + z^2} - z\right)$$

We can find the potential V_0 of the disk itself by setting $z = 0$, giving $V_0 = Q/2\pi\epsilon_0 R$. In other words, placing charge Q on a disk of radius R charges it to potential V_0. The on-axis potential of the disk can be written in terms of V_0 as

$$V_{\text{disk on axis}} = V_0\left[\sqrt{1 + (z/R)^2} - (z/R)\right]$$

Now we can evaluate the case of the charged dime.

a. The potential of the dime is the potential of a disk at $z = 0$:

$$V_0 = \frac{Q}{2\pi\epsilon_0 R} = 10,300 \text{ V}$$

b. To calculate the potential energy $U = qV$ of charge q, we first need to determine the potential of the disk at $z = 1.0$ cm. This is

$$V = V_0\left[\sqrt{1 + (z/R)^2} - (z/R)\right] = 3870 \text{ V}$$

The electron's charge is $q = -e = -1.60 \times 10^{-19}$ C, so its potential energy at $z = 1.00$ cm is $U = qV = -6.19 \times 10^{-16}$ J.

ASSESS Although we had to go through a number of steps, this procedure is easier than evaluating the electric field because we do not have to worry about vector components.

SUMMARY

The goals of Chapter 28 have been to calculate and use the electric potential and electric potential energy.

General Principles

Sources of V

The electric potential, like the electric field, is created by charges.

Two major tools for calculating V are

- The potential of a point charge $V = \dfrac{1}{4\pi\epsilon_0}\dfrac{q}{r}$

- The principle of superposition

Multiple point charges

Use superposition: $V = V_1 + V_2 + V_3 + \cdots$

Continuous distribution of charge

- Divide the charge into point-like ΔQ.

- Find the potential of each ΔQ.

- Find V by summing the potentials of all ΔQ.

The summation usually becomes an integral. A critical step is replacing ΔQ with an expression involving a charge density and an integration coordinate. Calculating V is usually easier than calculating $\vec{E}$ because the potential is a scalar.

Consequences of V

A charged particle has potential energy

$$U = qV$$

at a point where source charges have created an electric potential V.

The electric force is a conservative force, so the mechanical energy is conserved for a charged particle in an electric potential:

$$K_f + qV_f = K_i + qV_i$$

The potential energy of **two point charges** separated by distance r is

$$U_{q_1 + q_2} = \frac{Kq_1q_2}{r} = \frac{1}{4\pi\epsilon_0}\frac{q_1q_2}{r}$$

The **zero point** of potential and potential energy is chosen to be convenient. For point charges, we let $U = 0$ when $r \to \infty$.

The potential energy in an electric field of an **electric dipole** with dipole moment $\vec{p}$ is

$$U_{\text{dipole}} = -pE\cos\theta = -\vec{p}\cdot\vec{E}$$

Applications

Graphical representations of the potential:

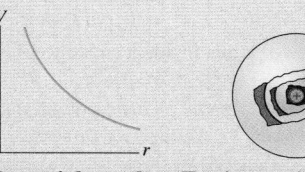

Potential graph   **Equipotential surfaces**

   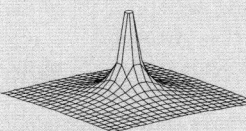

Contour map   **Elevation graph**

Sphere of charge Q

Same as a point charge if $r \geq R$

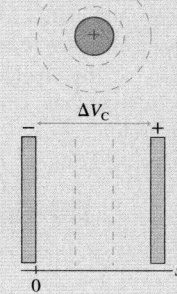

Parallel-plate capacitor

$V = Es$, where s is measured from the negative plate. The electric field inside is

$$E = \frac{\Delta V_C}{d}$$

Units

Electric potential: 1 V = 1 J/C

Electric field: 1 V/m = 1 N/C

Terms and Notation

electric potential energy, U	potential difference, ΔV	contour map
electric potential, V	voltage, ΔV	elevation graph
volt, V	equipotential surface	

CONCEPTUAL QUESTIONS

1. a. Charge q_1 is distance r from a positive point charge Q. Charge $q_2 = q_1/3$ is distance $2r$ from Q. What is the ratio U_1/U_2 of their potential energies due to their interactions with Q?

   b. Charge q_1 is distance s from the negative plate of a parallel-plate capacitor. Charge $q_2 = q_1/3$ is distance $2s$ from the negative plate. What is the ratio U_1/U_2 of their potential energies?

2. FIGURE Q28.2 shows the potential energy of a proton ($q = +e$) and a lead nucleus ($q = +82e$). The horizontal scale is in units of *femtometers*, where 1 fm $= 10^{-15}$ m.

   a. A proton is fired toward a lead nucleus from very far away. How much initial kinetic energy does the proton need to reach a turning point 10 fm from the nucleus? Explain.

   b. How much kinetic energy does the proton of part a have when it is 20 fm from the nucleus and moving toward it, before the collision?

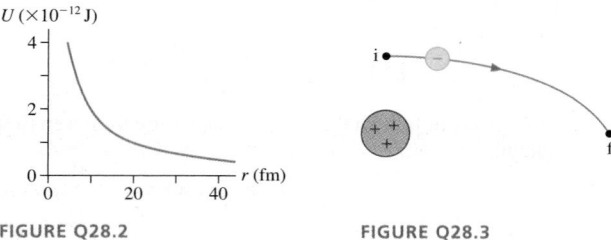

FIGURE Q28.2   FIGURE Q28.3

3. An electron moves along the trajectory of FIGURE Q28.3 from i to f.

   a. Does the electric potential energy increase, decrease, or stay the same? Explain.

   b. Is the electron's speed at f greater than, less than, or equal to its speed at i? Explain.

4. Two protons are launched with the same speed from point 1 inside the parallel-plate capacitor of FIGURE Q28.4. Points 2 and 3 are the same distance from the negative plate.

   a. Is $\Delta U_{1\to2}$, the change in potential energy along the path $1 \to 2$, larger than, smaller than, or equal to $\Delta U_{1\to3}$?

   b. Is the proton's speed v_2 at point 2 larger than, smaller than, or equal to v_3? Explain.

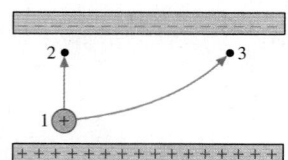

FIGURE Q28.4

5. Rank in order, from most positive to most negative, the potential energies U_a to U_f of the six electric dipoles in the uniform electric field of FIGURE Q28.5. Explain.

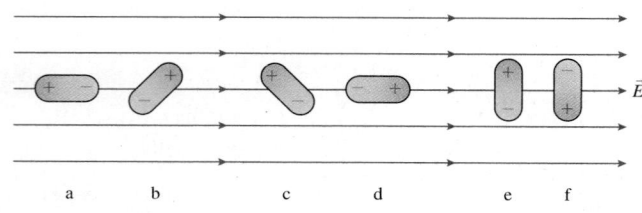

FIGURE Q28.5

6. FIGURE Q28.6 shows the electric potential along the x-axis.

   a. Draw a graph of the potential energy of a 0.1 C charged particle. Provide a numerical scale for both axes.

   b. If the charged particle is shot toward the right from $x = 1$ m with 1.0 J of kinetic energy, where is its turning point? Use your graph to explain.

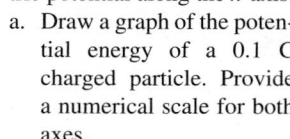
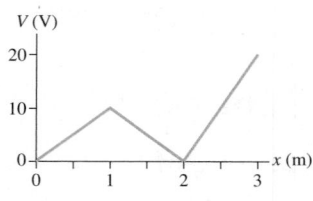

FIGURE Q28.6

7. A capacitor with plates separated by distance d is charged to a potential difference ΔV_C. All wires and batteries are disconnected, then the two plates are pulled apart (with insulated handles) to a new separation of distance $2d$.

   a. Does the capacitor charge Q change as the separation increases? If so, by what factor? If not, why not?

   b. Does the electric field strength E change as the separation increases? If so, by what factor? If not, why not?

   c. Does the potential difference ΔV_C change as the separation increases? If so, by what factor? If not, why not?

8. Rank in order, from largest to smallest, the electric potentials V_a to V_e at points a to e in FIGURE Q28.8. Explain.

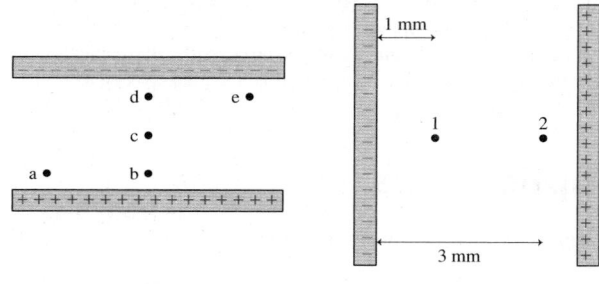

FIGURE Q28.8   FIGURE Q28.9

9. FIGURE Q28.9 shows two points inside a capacitor. Let $V = 0$ V at the negative plate.

   a. What is the ratio V_2/V_1 of the electric potentials? Explain.

   b. What is the ratio E_2/E_1 of the electric field strengths?

10. FIGURE Q28.10 shows two points near a positive point charge.

    a. What is the ratio V_2/V_1 of the electric potentials? Explain.

    b. What is the ratio E_2/E_1 of the electric field strengths?

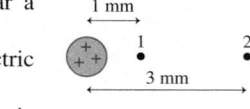

FIGURE Q28.10

11. FIGURE Q28.11 shows three points in the vicinity of two point charges. The charges have equal magnitudes. Rank in order, from most positive to most negative, the potentials V_a to V_c.

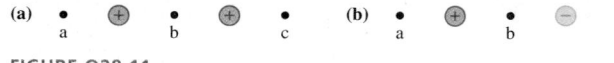

FIGURE Q28.11

12. Reproduce FIGURE Q28.12 on your paper. Then draw a dot (or dots) on the figure to show the position (or positions) at which the electric potential is zero.

FIGURE Q28.12

EXERCISES AND PROBLEMS

Problems labeled ▓ integrate material from earlier chapters.

Exercises

Section 28.1 Electric Potential Energy

1. ‖ The electric field strength is 50,000 N/C inside a parallel-plate capacitor with a 2.0 mm spacing. A proton is released from rest at the positive plate. What is the proton's speed when it reaches the negative plate?
2. ‖ The electric field strength is 20,000 N/C inside a parallel-plate capacitor with a 1.0 mm spacing. An electron is released from rest at the negative plate. What is the electron's speed when it reaches the positive plate?
3. ‖ A proton is released from rest at the positive plate of a parallel-plate capacitor. It crosses the capacitor and reaches the negative plate with a speed of 50,000 m/s. What will be the final speed of an electron released from rest at the negative plate?
4. | A proton is released from rest at the positive plate of a parallel-plate capacitor. It crosses the capacitor and reaches the negative plate with a speed of 50,000 m/s. The experiment is repeated with a He$^+$ ion (charge e, mass 4 u). What is the ion's speed at the negative plate?

Section 28.2 The Potential Energy of Point Charges

5. ‖ What is the electric potential energy of the proton in FIGURE EX28.5? The electrons are fixed and cannot move.

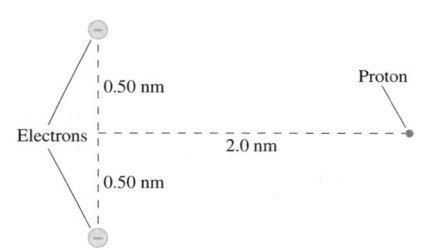

FIGURE EX28.5

6. ‖ What is the electric potential energy of the group of charges in FIGURE EX28.6?

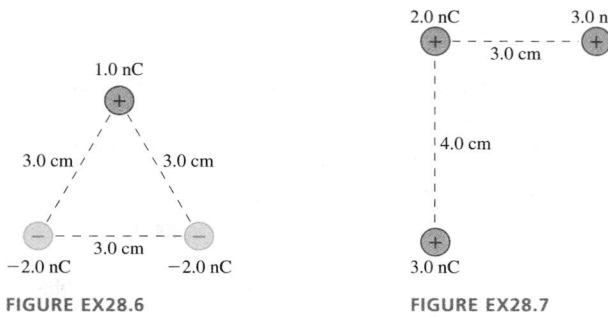

FIGURE EX28.6   **FIGURE EX28.7**

7. ‖ What is the electric potential energy of the group of charges in FIGURE EX28.7?

Section 28.3 The Potential Energy of a Dipole

8. | A water molecule perpendicular to an electric field has 1.0×10^{-21} J more potential energy than a water molecule aligned with the field. The dipole moment of a water molecule is 6.2×10^{-30} C m. What is the strength of the electric field?
9. | FIGURE EX28.9 shows the potential energy of an electric dipole. Consider a dipole that oscillates between $\pm 60°$.
   a. What is the dipole's mechanical energy?
   b. What is the dipole's kinetic energy when it is aligned with the electric field?

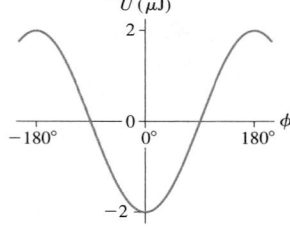

FIGURE EX28.9

Section 28.4 The Electric Potential

10. | What is the speed of a proton that has been accelerated from rest through a potential difference of -1000 V?
11. | What is the speed of an electron that has been accelerated from rest through a potential difference of 1000 V?
12. ‖ What potential difference is needed to accelerate an electron from rest to a speed of 2.0×10^6 m/s?
13. ‖ What potential difference is needed to accelerate a He$^+$ ion (charge $+e$, mass 4 u) from rest to a speed of 2.0×10^6 m/s?
14. | A proton with an initial speed of 800,000 m/s is brought to rest by an electric field.
   a. Did the proton move into a region of higher potential or lower potential?
   b. What was the potential difference that stopped the proton?
15. ‖ An electron with an initial speed of 500,000 m/s is brought to rest by an electric field.
   a. Did the electron move into a region of higher potential or lower potential?
   b. What was the potential difference that stopped the electron?

Section 28.5 The Electric Potential Inside a Parallel-Plate Capacitor

16. | Show that 1 V/m = 1 N/C.
17. | a. What is the potential of an ordinary AA or AAA battery? (If you're not sure, find one and look at the label.)
   b. An AA battery is connected to a parallel-plate capacitor having 4.0 cm × 4.0 cm plates spaced 1.0 mm apart. How much charge does the battery supply to each plate?
18. | Two 2.00 cm × 2.00 cm plates that form a parallel-plate capacitor are charged to ± 0.708 nC. What are the electric field strength inside and the potential difference across the capacitor if the spacing between the plates is (a) 1.00 mm and (b) 2.00 mm?
19. | A 3.0-cm-diameter parallel-plate capacitor has a 2.0 mm spacing. The electric field strength inside the capacitor is 1.0×10^5 V/m.
   a. What is the potential difference across the capacitor?
   b. How much charge is on each plate?

20. ‖ Two 2.0-cm-diameter disks spaced 2.0 mm apart form a parallel-plate capacitor. The electric field between the disks is 5.0×10^5 V/m.
    a. What is the voltage across the capacitor?
    b. An electron is launched from the negative plate. It strikes the positive plate at a speed of 2.0×10^7 m/s. What was the electron's speed as it left the negative plate?

Section 28.6 The Electric Potential of a Point Charge

21. ‖ a. What is the electric potential at points A, B, and C in FIGURE EX28.21?
    b. What are the potential differences ΔV_{AB} and ΔV_{BC}?

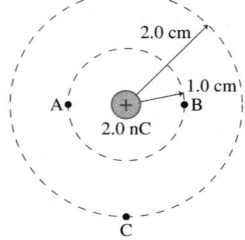

FIGURE EX28.21

22. ‖ A 1.0-mm-diameter ball bearing has 2.0×10^9 excess electrons. What is the ball bearing's potential?
23. ‖ In a semiclassical model of the hydrogen atom, the electron orbits the proton at a distance of 0.053 nm.
    a. What is the electric potential of the proton at the position of the electron?
    b. What is the electron's potential energy?

Section 28.7 The Electric Potential of Many Charges

24. ‖ What is the electric potential at the point indicated with the dot in FIGURE EX28.24?

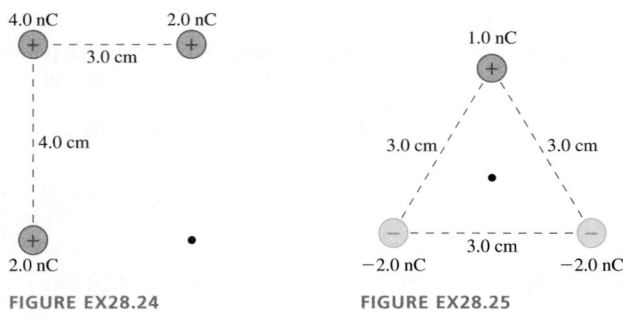

FIGURE EX28.24   FIGURE EX28.25

25. ‖ What is the electric potential at the point indicated with the dot in FIGURE EX28.25?
26. ‖ The electric potential at the dot in FIGURE EX28.26 is 3140 V. What is charge q?

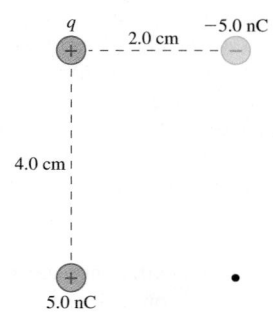

FIGURE EX28.26

27. ‖ A -2.0 nC charge and a $+2.0$ nC charge are located on the x-axis at $x = -1.0$ cm and $x = +1.0$ cm, respectively.
    a. Other than at infinity, is there a position or positions on the x-axis where the electric field is zero? If so, where?
    b. Other than at infinity, at what position or positions on the x-axis is the electric potential zero?
    c. Sketch graphs of the electric field strength and the electric potential along the x-axis.

28. ‖ Two point charges q_a and q_b are located on the x-axis at $x = a$ and $x = b$. FIGURE EX28.28 is a graph of E_x, the x-component of the electric field.
    a. What are the signs of q_a and q_b?
    b. What is the ratio $|q_a/q_b|$?
    c. Draw a graph of V, the electric potential, as a function of x.

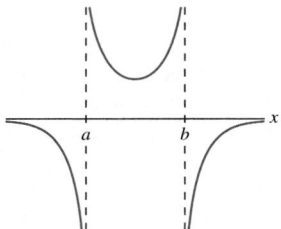

FIGURE EX28.28

29. ‖ Two point charges q_a and q_b are located on the x-axis at $x = a$ and $x = b$. FIGURE EX28.29 is a graph of V, the electric potential.
    a. What are the signs of q_a and q_b?
    b. What is the ratio $|q_a/q_b|$?
    c. Draw a graph of E_x, the x-component of the electric field, as a function of x.

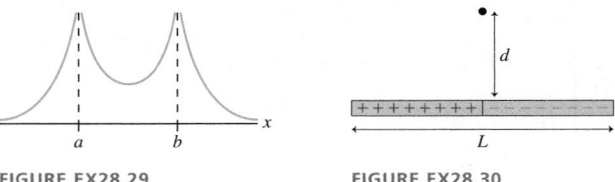

FIGURE EX28.29   FIGURE EX28.30

30. ‖ The two halves of the rod in FIGURE EX28.30 are uniformly charged to $\pm Q$. What is the electric potential at the point indicated by the dot?

Problems

31. ‖ Two positive point charges are 5.0 cm apart. If the electric potential energy is 72 μJ, what is the magnitude of the force between the two charges?
32. ‖ Two point charges 2.0 cm apart have an electric potential energy -180 μJ. The total charge is 30 nC. What are the two charges?
33. ‖ A -10.0 nC point charge and a $+20.0$ nC point charge are 15.0 cm apart on the x-axis.
    a. What is the electric potential at the point on the x-axis where the electric field is zero?
    b. What is the magnitude of the electric field at the point on the x-axis, between the charges, where the electric potential is zero?
34. ‖ A $+3.0$ nC charge is at $x = 0$ cm and a -1.0 nC charge is at $x = 4$ cm. At what point or points on the x-axis is the electric potential zero?
35. ‖ A -3.0 nC charge is on the x-axis at $x = -9$ cm and a $+4.0$ nC charge is on the x-axis at $x = 16$ cm. At what point or points on the y-axis is the electric potential zero?

36. ‖ Two small metal cubes with masses 2.0 g and 4.0 g are tied together by a 5.0-cm-long massless string and are at rest on a frictionless surface. Each is charged to $+2.0 \ \mu C$.
    a. What is the energy of this system?
    b. What is the tension in the string?
    c. The string is cut. What is the speed of each cube when they are far apart?
    Hint: There are *two* conserved quantities. Make use of both.

37. ‖ The four 1.0 g spheres shown in FIGURE P28.37 are released simultaneously and allowed to move away from each other. What is the speed of each sphere when they are very far apart?

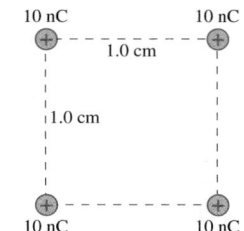

FIGURE P28.37

38. ‖ A proton's speed as it passes point A is 50,000 m/s. It follows the trajectory shown in FIGURE P28.38. What is the proton's speed at point B?

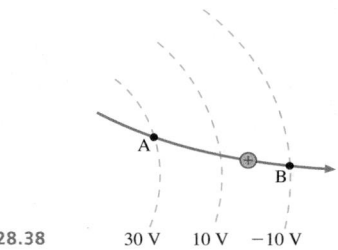

FIGURE P28.38   30 V   10 V   −10 V

39. ‖ Living cells "pump" singly ionized sodium ions, Na^+, from
BIO the inside of the cell to the outside to maintain a membrane potential $\Delta V_{membrane} = V_{in} - V_{out} = -70$ mV. It is called *pumping* because work must be done to move a positive ion from the negative inside of the cell to the positive outside, and it must go on continuously because sodium ions "leak" back through the cell wall by diffusion.
    a. How much work must be done to move one sodium ion from the inside of the cell to the outside?
    b. At rest, the human body uses energy at the rate of approximately 100 W to maintain basic metabolic functions. It has been estimated that 20% of this energy is used to operate the sodium pumps of the body. Estimate—to one significant figure—the number of sodium ions pumped per second.

40. ‖ An arrangement of source charges produces the electric potential $V = 5000x^2$ along the x-axis, where V is in volts and x is in meters. What is the maximum speed of a 1.0 g, 10 nC charged particle that moves in this potential with turning points at ± 8.0 cm?

41. ‖ A proton moves along the x-axis, where an arrangement of source charges has created the electric potential $V = 6000x^2$, where V is in volts and x is in meters. By exploiting the analogy with the potential energy of a mass on a spring, determine the proton's oscillation frequency.

42. ‖ In FIGURE P28.42, a proton is fired with a speed of 200,000 m/s from the midpoint of the capacitor toward the positive plate.
    a. Show that this is insufficient speed to reach the positive plate.
    b. What is the proton's speed as it collides with the negative plate?

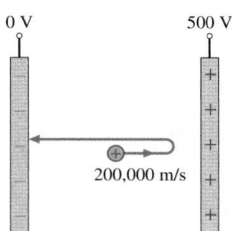

FIGURE P28.42

43. ‖ The electron gun in an old TV picture tube accelerates electrons between two parallel plates 1.2 cm apart with a 25 kV potential difference between them. The electrons enter through a small hole in the negative plate, accelerate, then exit through a small hole in the positive plate. Assume that the holes are small enough not to affect the electric field or potential.
    a. What is the electric field strength between the plates?
    b. With what speed does an electron exit the electron gun if its entry speed is close to zero?
    NOTE ▶ The exit speed is so fast that we really need to use the theory of relativity to compute an accurate value. Your answer to part b is in the right range but a little too big. ◀

44. ‖ An uncharged parallel-plate capacitor with spacing d is horizontal. A small bead with mass m and positive charge q is shot straight up from the bottom plate with speed v_0. It reaches maximum height y_{max} before falling back. Then the capacitor is charged with the bottom plate negative. Find an expression for the capacitor voltage ΔV_C for which the bead's maximum height is reduced to $\frac{1}{2} y_{max}$. Ignore air resistance.

45. ‖ A room with 3.0-m-high ceilings has a metal plate on the floor with $V = 0$ V and a separate metal plate on the ceiling. A 1.0 g glass ball charged to $+4.9$ nC is shot straight up at 5.0 m/s. How high does the ball go if the ceiling voltage is (a) $+3.0 \times 10^6$ V and (b) -3.0×10^6 V?

46. ‖ In *proton-beam therapy*, a high-energy beam of protons is
BIO fired at a tumor. As the protons stop in the tumor, their kinetic energy breaks apart the tumor's DNA, thus killing the tumor cells. For one patient, it is desired to deposit 0.10 J of proton energy in the tumor. To create the proton beam, protons are accelerated from rest through a 10,000 kV potential difference. What is the total charge of the protons that must be fired at the tumor?

47. ‖ What is the escape speed of an electron launched from the surface of a 1.0-cm-diameter glass sphere that has been charged to 10 nC?

48. ‖ An electric dipole consists of 1.0 g spheres charged to ± 2.0 nC at the ends of a 10-cm-long massless rod. The dipole rotates on a frictionless pivot at its center. The dipole is held perpendicular to a uniform electric field with field strength 1000 V/m, then released. What is the dipole's angular velocity at the instant it is aligned with the electric field?

49. ‖‖ Three electrons form an equilateral triangle 1.0 nm on each side. A proton is at the center of the triangle. What is the potential energy of this group of charges?

50. ‖‖ A 2.0-mm-diameter glass bead is positively charged. The potential difference between a point 2.0 mm from the bead and a point 4.0 mm from the bead is 500 V. What is the charge on the bead?

51. ▌▌▌ Your lab assignment for the week is to measure the amount of charge on the 6.0-cm-diameter metal sphere of a Van de Graaff generator. To do so, you're going to use a spring with spring constant 0.65 N/m to launch a small, 1.5 g bead horizontally toward the sphere. You can reliably charge the bead to 2.5 nC, and your plan is to use a video camera to measure the bead's closest approach to the sphere as you change the compression of the spring. Your data are as follows:

Compression (cm)	Closest approach (cm)
1.6	5.5
1.9	2.6
2.2	1.6
2.5	0.4

Use an appropriate graph of the data to determine the sphere's charge in nC. You can assume that the bead's motion is entirely horizontal and that the spring is so far away that the bead has no interaction with the sphere as it's launched.

52. ▌▌ A proton is fired from far away toward the nucleus of an iron atom. Iron is element number 26, and the diameter of the nucleus is 9.0 fm. What initial speed does the proton need to just reach the surface of the nucleus? Assume the nucleus remains at rest.

53. ▌▌ A proton is fired from far away toward the nucleus of a mercury atom. Mercury is element number 80, and the diameter of the nucleus is 14.0 fm. If the proton is fired at a speed of 4.0×10^7 m/s, what is its closest approach to the surface of the nucleus? Assume the nucleus remains at rest.

54. ▌▌ In the form of radioactive decay known as *alpha decay,* an unstable nucleus emits a helium-atom nucleus, which is called an *alpha particle.* An alpha particle contains two protons and two neutrons, thus having mass $m = 4$ u and charge $q = 2e$. Suppose a uranium nucleus with 92 protons decays into thorium, with 90 protons, and an alpha particle. The alpha particle is initially at rest at the surface of the thorium nucleus, which is 15 fm in diameter. What is the speed of the alpha particle when it is detected in the laboratory? Assume the thorium nucleus remains at rest.

55. ▌▌ One form of nuclear radiation, *beta decay,* occurs when a neutron changes into a proton, an electron, and a neutral particle called a *neutrino:* $n \rightarrow p^+ + e^- + \nu$ where ν is the symbol for a neutrino. When this change happens to a neutron within the nucleus of an atom, the proton remains behind in the nucleus while the electron and neutrino are ejected from the nucleus. The ejected electron is called a *beta particle.* One nucleus that exhibits beta decay is the isotope of hydrogen ^{3}H, called *tritium,* whose nucleus consists of one proton (making it hydrogen) and two neutrons (giving tritium an atomic mass $m = 3$ u). Tritium is radioactive, and it decays to helium: ^{3}H $\rightarrow$ ^{3}He $+ e^- + \nu$.
   a. Is charge conserved in the beta decay process? Explain.
   b. Why is the final product a helium atom? Explain.
   c. The nuclei of both ^{3}H and ^{3}He have radii of 1.5×10^{-15} m. With what minimum speed must the electron be ejected if it is to escape from the nucleus and not fall back?

56. ▌▌ The sun is powered by *fusion,* with four protons fusing together to form a helium nucleus (two of the protons turn into neutrons) and, in the process, releasing a large amount of thermal energy. The process happens in several steps, not all at once. In one step, two protons fuse together, with one proton then becoming a neutron, to form the "heavy hydrogen" isotope *deuterium* (^{2}H).

A proton is essentially a 2.4-fm-diameter sphere of charge, and fusion occurs only if two protons come into contact with each other. This requires extraordinarily high temperatures due to the strong repulsion between the protons. Recall that the average kinetic energy of a gas particle is $\frac{3}{2} k_B T$.
   a. Suppose two protons, each with exactly the average kinetic energy, have a head-on collision. What is the minimum temperature for fusion to occur?
   b. Your answer to part a is much hotter than the 15 million K in the core of the sun. If the temperature were as high as you calculated, every proton in the sun would fuse almost instantly and the sun would explode. For the sun to last for billions of years, fusion can occur only in collisions between two protons with kinetic energies much higher than average. Only a very tiny fraction of the protons have enough kinetic energy to fuse when they collide, but that fraction is enough to keep the sun going. Suppose two protons with the same kinetic energy collide head-on and just barely manage to fuse. By what factor does each proton's energy exceed the average kinetic energy at 15 million K?

57. ▌▌ Two 10-cm-diameter electrodes 0.50 cm apart form a parallel-plate capacitor. The electrodes are attached by metal wires to the terminals of a 15 V battery. After a long time, the capacitor is disconnected from the battery but is not discharged. What are the charge on each electrode, the electric field strength inside the capacitor, and the potential difference between the electrodes
   a. Right after the battery is disconnected?
   b. After insulating handles are used to pull the electrodes away from each other until they are 1.0 cm apart?
   c. After the original electrodes (not the modified electrodes of part b) are expanded until they are 20 cm in diameter?

58. ▌▌ Two 10-cm-diameter electrodes 0.50 cm apart form a parallel-plate capacitor. The electrodes are attached by metal wires to the terminals of a 15 V battery. What are the charge on each electrode, the electric field strength inside the capacitor, and the potential difference between the electrodes
   a. While the capacitor is attached to the battery?
   b. After insulating handles are used to pull the electrodes away from each other until they are 1.0 cm apart? The electrodes remain connected to the battery during this process.
   c. After the original electrodes (not the modified electrodes of part b) are expanded until they are 20 cm in diameter while remaining connected to the battery?

59. ▌▌ a. Find an algebraic expression for the electric field strength E_0 at the surface of a charged sphere in terms of the sphere's potential V_0 and radius R.
   b. What is the electric field strength at the surface of a 1.0-cm-diameter marble charged to 500 V?

60. ▌▌ Two spherical drops of mercury each have a charge of 0.10 nC and a potential of 300 V at the surface. The two drops merge to form a single drop. What is the potential at the surface of the new drop?

61. ▌▌ A Van de Graaff generator is a device for generating a large electric potential by building up charge on a hollow metal sphere. A typical classroom-demonstration model has a diameter of 30 cm.
   a. How much charge is needed on the sphere for its potential to be 500,000 V?
   b. What is the electric field strength just outside the surface of the sphere when it is charged to 500,000 V?

62. ▌▌ A thin spherical shell of radius R has total charge Q. What is the electric potential at the center of the shell?

63. | **FIGURE P28.63** shows two uniformly charged spheres. What is the potential difference between points a and b? Which point is at the higher potential?

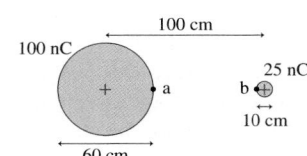

FIGURE P28.63

Hint: The potential at any point is the superposition of the potentials due to *all* charges.

64. ‖ An electric dipole with dipole moment p is oriented along the y-axis.
   a. Find an expression for the electric potential on the y-axis at a point where y is much larger than the charge spacing s. Write your expression in terms of the dipole moment p.
   b. The dipole moment of a water molecule is 6.2×10^{-30} C m. What is the electric potential 1.0 nm from a water molecule along the axis of the dipole?

65. ‖ Two positive point charges q are located on the y-axis at $y = \pm\frac{1}{2}s$.
   a. Find an expression for the potential along the x-axis.
   b. Draw a graph of V versus x for $-\infty < x < \infty$. For comparison, use a dotted line to show the potential of a point charge $2q$ located at the origin.

66. ‖ The arrangement of charges shown in **FIGURE P28.66** is called a *linear electric quadrupole*. The positive charges are located at $y = \pm s$. Notice that the net charge is zero. Find an expression for the electric potential on the y-axis at distances $y \gg s$.

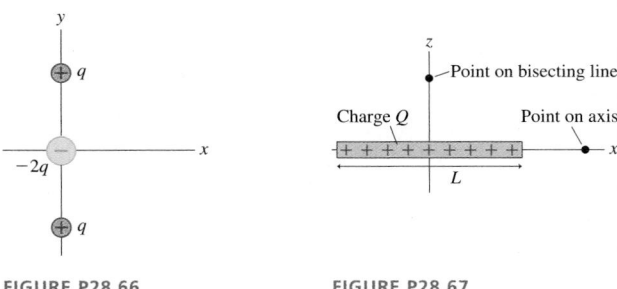

FIGURE P28.66   **FIGURE P28.67**

67. ‖ **FIGURE P28.67** shows a thin rod of length L and charge Q. Find an expression for the electric potential a distance x away from the center of the rod on the axis of the rod.

68. ‖‖ **FIGURE P28.67** showed a thin rod of length L and charge Q. Find an expression for the electric potential a distance z away from the center of rod on the line that bisects the rod.

69. | **FIGURE P28.69** shows a thin rod with charge Q that has been bent into a semicircle of radius R. Find an expression for the electric potential at the center.

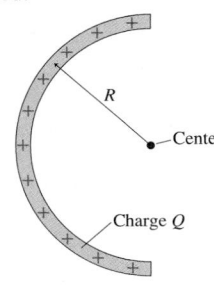

FIGURE P28.69

70. ‖ A disk with a hole has inner radius R_{in} and outer radius R_{out}. The disk is uniformly charged with total charge Q. Find an expression for the on-axis electric potential at distance z from the center of the disk. Verify that your expression has the correct behavior when $R_{in} \to 0$.

In Problems 71 through 73 you are given the equation(s) used to solve a problem. For each of these,
   a. Write a realistic problem for which this is the correct equation(s).
   b. Finish the solution of the problem.

71. $\dfrac{(9.0 \times 10^9 \text{ N m}^2/\text{C}^2)q_1q_2}{0.030 \text{ m}} = 90 \times 10^{-6} \text{ J}$

   $q_1 + q_2 = 40 \text{ nC}$

72. $\frac{1}{2}(1.67 \times 10^{-27} \text{ kg})(2.5 \times 10^6 \text{ m/s})^2 + 0 =$

   $\frac{1}{2}(1.67 \times 10^{-27} \text{ kg})v_i^2 +$

   $\dfrac{(9.0 \times 10^9 \text{ N m}^2/\text{C}^2)(2.0 \times 10^{-9} \text{ C})(1.60 \times 10^{-19} \text{ C})}{0.0010 \text{ m}}$

73. $\dfrac{(9.0 \times 10^9 \text{ N m}^2/\text{C}^2)(3.0 \times 10^{-9} \text{ C})}{0.030 \text{ m}} +$

   $\dfrac{(9.0 \times 10^9 \text{ N m}^2/\text{C}^2)(3.0 \times 10^{-9} \text{ C})}{(0.030 \text{ m}) + d} = 1200 \text{ V}$

Challenge Problems

74. A proton and an alpha particle ($q = +2e$, $m = 4$ u) are fired directly toward each other from far away, each with an initial speed of $0.010c$. What is their distance of closest approach, as measured between their centers?

75. Bead A has a mass of 15 g and a charge of -5.0 nC. Bead B has a mass of 25 g and a charge of -10.0 nC. The beads are held 12 cm apart (measured between their centers) and released. What maximum speed is achieved by each bead?

76. Two 2.0-mm-diameter beads, C and D, are 10 mm apart, measured between their centers. Bead C has mass 1.0 g and charge 2.0 nC. Bead D has mass 2.0 g and charge -1.0 nC. If the beads are released from rest, what are the speeds v_C and v_D at the instant the beads collide?

77. An electric dipole has dipole moment p. If $r \gg s$, where s is the separation between the charges, show that the electric potential of the dipole can be written

$$V = \frac{1}{4\pi\epsilon_0}\frac{p\cos\theta}{r^2}$$

where r is the distance from the center of the dipole and θ is the angle from the dipole axis.

78. Electrodes of area A are spaced distance d apart to form a parallel-plate capacitor. The electrodes are charged to $\pm q$.
   a. What is the infinitesimal increase in electric potential energy dU if an infinitesimal amount of charge dq is moved from the negative electrode to the positive electrode?
   b. An uncharged capacitor can be charged to $\pm Q$ by transferring charge dq over and over and over. Use your answer to part a to show that the potential energy of a capacitor charged to $\pm Q$ is $U_{cap} = \frac{1}{2}Q\,\Delta V_C$.

79. A sphere of radius R has charge q.
    a. What is the infinitesimal increase in electric potential energy dU if an infinitesimal amount of charge dq is brought from infinity to the surface of the sphere?
    b. An uncharged sphere can acquire total charge Q by the transfer of charge dq over and over and over. Use your answer to part a to find an expression for the potential energy of a sphere of radius R with total charge Q.
    c. Your answer to part b is the amount of energy needed to assemble a charged sphere. It is often called the *self-energy* of the sphere. What is the self-energy of a proton, assuming it to be a charged sphere with a diameter of 1.0×10^{-15} m?

80. The wire in FIGURE CP28.80 has linear charge density λ. What is the electric potential at the center of the semicircle?

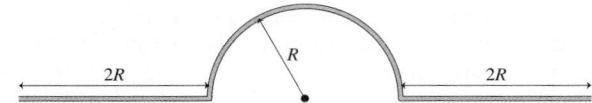

FIGURE CP28.80

81. A circular disk of radius R and total charge Q has the charge distributed with surface charge density $\eta = cr$, where c is a constant. Find an expression for the electric potential at distance z on the axis of the disk. Your expression should include R and Q, but not c.

82. A hollow cylindrical shell of length L and radius R has charge Q uniformly distributed along its length. What is the electric potential at the center of the cylinder?

STOP TO THINK ANSWERS

Stop to Think 28.1: Zero. The motion is always perpendicular to the electric force.

Stop to Think 28.2: $U_b = U_d > U_a = U_c$. The potential energy depends inversely on r. The effects of doubling the charge and doubling the distance cancel each other.

Stop to Think 28.3: c. The proton gains speed by losing potential energy. It loses potential energy by moving in the direction of decreasing electric potential.

Stop to Think 28.4: $V_a = V_b > V_c > V_d = V_e$. The potential decreases steadily from the positive to the negative plate. It depends only on the distance from the positive plate.

Stop to Think 28.5: $\Delta V_{ac} = \Delta V_{bc} > \Delta V_{ab}$. The potential depends only on the *distance* from the charge, not the direction. $\Delta V_{ab} = 0$ because these points are at the same distance.

29 Potential and Field

These solar cells are *photovoltaic* cells, meaning that light creates a voltage—a potential difference.

▶ **Looking Ahead** The goal of Chapter 29 is to understand how the electric potential is related to the electric field.

Field and Potential

The electric potential and the electric field are intimately connected. They are two different perspectives of how source charges alter the space around them.

You'll learn to:

- Use the electric potential to find the electric field.
- Use the electric field to find the electric potential.

The mathematical connection is analogous to that between force and potential energy.

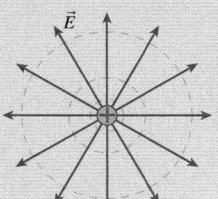

The electric field and the electric potential can be related to each other geometrically.

You'll also learn that:

- The electric field is always perpendicular to equipotential surfaces.
- The electric field points "downhill" in the direction of decreasing potential.
- The electric field is stronger where equipotential lines are closer together.

◀ **Looking Back**

Sections 28.4–28.6 Electric potential and its graphical representations

Sources of Potential

A potential difference—a *voltage*—is created by separating positive and negative charges.

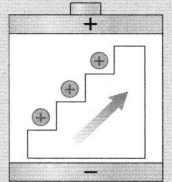

We'll develop a *charge escalator* model of a battery in which chemical reactions separate charge to create a potential difference.

You'll learn that any nonelectrical means of separating charge—in batteries, photocells, and generators—does work and develops what we'll call an **emf.**

Conductors

You'll learn several important characteristics of conductors in **electrostatic equilibrium,** with stationary charges.

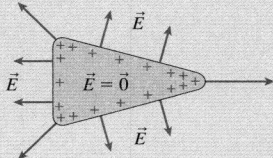

- Any excess charge is on the surface.
- The interior electric field is zero.
- The exterior electric field is perpendicular to the surface.
- The entire conductor is an equipotential.

Capacitors

Capacitors are circuit elements that store charge and energy. They are used in devices ranging from high-speed computers to heart defibrillators.

The flash on your camera uses energy stored in a capacitor. The capacitor can discharge in a few microseconds, much faster than a battery can provide energy.

You'll learn to:

- Work with combinations of capacitors called *in series* and *in parallel*.
- Calculate the energy stored in a capacitor's electric field.
- Understand capacitors with dielectrics.

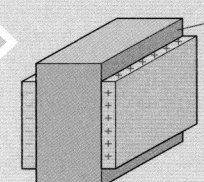

Dielectric

An insulator between the capacitor plates is called a **dielectric.** It changes the capacitor properties in many useful ways.

◀ **Looking Back**

Section 26.5 Parallel-plate capacitors
Section 26.7 Dipoles in electric fields

29.1 Connecting Potential and Field

FIGURE 29.1 The four key ideas.

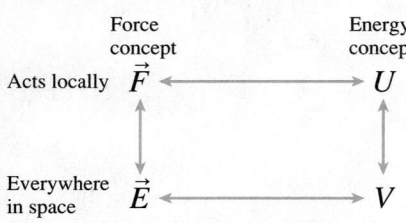

FIGURE 29.1 shows the four key ideas of force, field, potential energy, and potential. The electric field and the electric potential were based on force and potential energy. We know, from Chapters 10 and 11, that force and potential energy are closely related. The focus of this chapter is to establish a similar relationship between the electric field and the electric potential. **The electric potential and electric field are not two distinct entities but, instead, two different perspectives or two different mathematical representations of how source charges alter the space around them.**

If this is true, we should be able to find the electric potential from the electric field. Chapter 28 introduced all the pieces we need to do so. We used the potential energy of charge q and the source charges to define the electric potential as

$$V \equiv \frac{U_{q+\text{sources}}}{q} \tag{29.1}$$

Potential energy is defined in terms of the work done by force $\vec{F}$ on charge q as it moves from position i to position f:

$$\Delta U = -W(\text{i} \rightarrow \text{f}) = -\int_{s_i}^{s_f} F_s \, ds = -\int_i^f \vec{F} \cdot d\vec{s} \tag{29.2}$$

But the force exerted on charge q by the electric field is $\vec{F} = q\vec{E}$. Putting these three pieces together, you can see that the charge q cancels out and the potential difference between two points in space is

$$\Delta V = V_f - V_i = -\int_{s_i}^{s_f} E_s \, ds = -\int_i^f \vec{E} \cdot d\vec{s} \tag{29.3}$$

where s is the position along a line from point i to point f. That is, we can find the potential difference between two points if we know the electric field.

We can think of an integral as an area under a curve. Thus a graphical interpretation of Equation 29.3 is

$$V_f = V_i - (\text{area under the } E_s\text{-versus-}s \text{ curve between } s_i \text{ and } s_f) \tag{29.4}$$

Notice, because of the minus sign in Equation 29.3, that the area is *subtracted* from V_i.

EXAMPLE 29.1 **Finding the potential**

FIGURE 29.2 is a graph of E_x, the x-component of the electric field, versus position along the x-axis. Find and graph $V(x)$. Assume $V = 0$ V at $x = 0$ m.

FIGURE 29.2 Graph of E_x versus x.

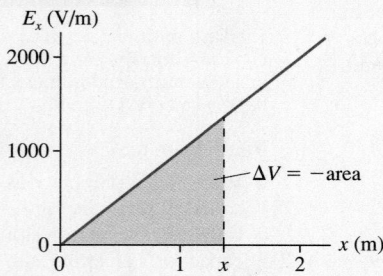

MODEL The potential difference is the *negative* of the area under the curve.

VISUALIZE E_x is positive throughout this region of space, meaning that $\vec{E}$ points in the positive x-direction.

SOLVE If we integrate from $x = 0$, then $V_i = V(x = 0) = 0$. The potential for $x > 0$ is the negative of the triangular area under the

E_x curve. We can see that $E_x = 1000x$ V/m, where x is in m. Thus

$$V_f = V(x) = 0 - (\text{area under the } E_x \text{ curve})$$
$$= -\tfrac{1}{2} \times \text{base} \times \text{height} = -\tfrac{1}{2}(x)(1000x) = -500x^2 \text{ V}$$

FIGURE 29.3 shows that the electric potential in this region of space is parabolic, decreasing from 0 V at $x = 0$ m to -2000 V at $x = 2$ m.

FIGURE 29.3 Graph of V versus x.

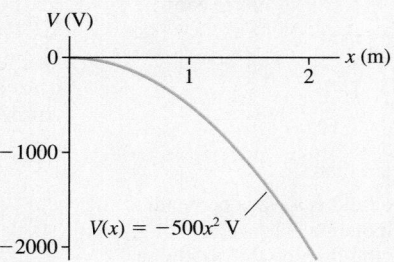

ASSESS The electric field points in the direction in which V is *decreasing*. We'll soon see that this is a general rule.

Finding the potential from the electric field

❶ Draw a picture and identify the point at which you wish to find the potential. Call this position f.
❷ Choose the zero point of the potential, often at infinity. Call this position i.
❸ Establish a coordinate axis from i to f along which you already know or can easily determine the electric field component E_s.
❹ Carry out the integration of Equation 29.3 to find the potential.

To see how this works, let's use the electric field of a point charge to find its electric potential. FIGURE 29.4 identifies a point P at $s_f = r$ at which we want to know the potential and calls this position f. We've chosen position i to be at $s_i = \infty$ and identified that as the zero point of the potential. The integration of Equation 29.3 is straight inward along the radial line from i to f:

$$\Delta V = V(r) - V(\infty) = -\int_\infty^r E_s \, ds = \int_r^\infty E_s \, ds \tag{29.5}$$

The electric field is radially outward. Its s-component is

$$E_s = \frac{1}{4\pi\epsilon_0} \frac{q}{s^2}$$

Thus the potential at distance r from a point charge q is

$$V(r) = V(\infty) + \frac{q}{4\pi\epsilon_0} \int_r^\infty \frac{ds}{s^2} = V(\infty) + \frac{q}{4\pi\epsilon_0} \frac{-1}{s}\Big|_r^\infty = 0 + \frac{1}{4\pi\epsilon_0} \frac{q}{r} \tag{29.6}$$

We've rediscovered the potential of a point charge that you learned in Chapter 28:

$$V_{\text{point charge}} = \frac{1}{4\pi\epsilon_0} \frac{q}{r} \tag{29.7}$$

FIGURE 29.4 Finding the potential of a point charge.

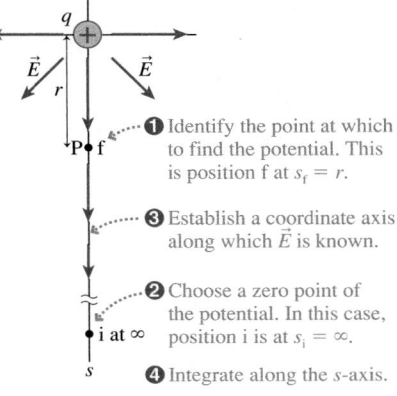

❶ Identify the point at which to find the potential. This is position f at $s_f = r$.
❸ Establish a coordinate axis along which $\vec{E}$ is known.
❷ Choose a zero point of the potential. In this case, position i is at $s_i = \infty$.
❹ Integrate along the s-axis.

EXAMPLE 29.2 **The potential of a parallel-plate capacitor**

In Chapter 26, the electric field inside a capacitor was found to be

$$\vec{E} = \left(\frac{Q}{\epsilon_0 A}, \text{ from positive to negative}\right)$$

Find the electric potential inside the capacitor. Let $V = 0$ V at the negative plate.

MODEL The electric field inside a capacitor is a uniform field.

VISUALIZE FIGURE 29.5 shows the capacitor and establishes a point P where we want to find the potential. We've chosen an s-axis measured from the negative plate, which is the zero point of the potential.

SOLVE We'll integrate along the s-axis from $s_i = 0$ (where $V_f = 0$ V) to $s_f = s$. Notice that $\vec{E}$ points in the negative s-direction, so $E_s = -Q/\epsilon_0 A$. $Q/\epsilon_0 A$ is a constant, so

$$V(s) = V_f = V_i - \int_0^s E_s \, ds = -\left(-\frac{Q}{\epsilon_0 A}\right)\int_0^s ds = \frac{Q}{\epsilon_0 A} s = Es$$

ASSESS $V = Es$ is the capacitor potential we deduced in Chapter 28 by working directly with the potential energy. The

FIGURE 29.5 Finding the potential inside a capacitor.

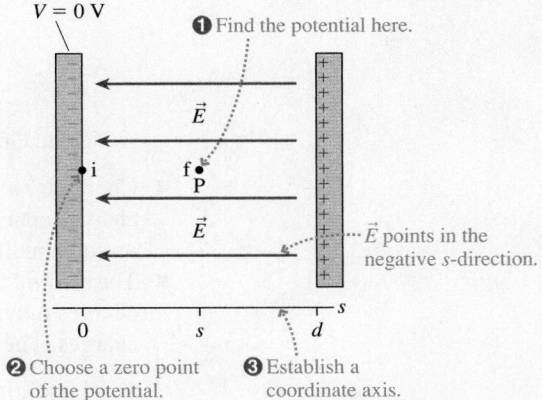

$V = 0$ V
❶ Find the potential here.
$\vec{E}$ points in the negative s-direction.
❷ Choose a zero point of the potential.
❸ Establish a coordinate axis.

potential increases linearly from $V = 0$ at the negative plate to $V = Ed$ at the positive plate. Here we found the potential by explicitly recognizing the connection between the potential and the field.

29.2 Sources of Electric Potential

A *separation of charge* creates an electric potential difference. Shuffling your feet on the carpet transfers electrons from the carpet to you, creating a potential difference between you and a doorknob that causes a spark and a shock as you touch it. Charging a capacitor by moving electrons from one plate to the other creates a potential difference across the capacitor.

In fact, as **FIGURE 29.6** shows, *any* separation of charge causes a potential difference. The charge separation between the two electrodes creates an electric field $\vec{E}$ pointing from the positive toward the negative electrode. As a consequence, there is a potential difference between the electrodes that is given by

$$\Delta V = V_{\text{pos}} - V_{\text{neg}} = -\int_{\text{neg}}^{\text{pos}} E_s \, ds$$

where the integral runs from any point on the negative electrode to any point on the positive. The key idea is that **we can create a potential difference by creating a charge separation.**

The **Van de Graaff generator** shown in **FIGURE 29.7a** is a mechanical charge separator—essentially a fancy foot shuffler. A moving plastic or leather belt is charged, then the charge is mechanically transported via the conveyor belt to the spherical electrode at the top of the insulating column. The charging of the belt could be done by friction, but in practice a *corona discharge* created by the strong electric field at the tip of a needle is more efficient and reliable.

FIGURE 29.6 A charge separation creates a potential difference.

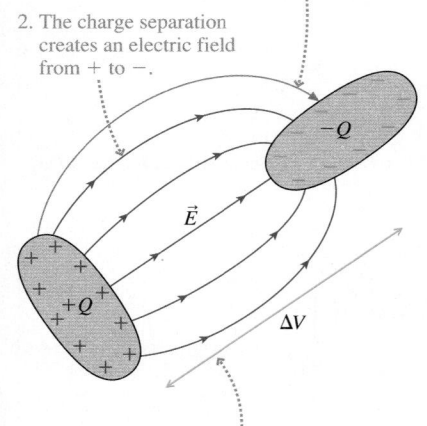

1. Charge is separated by moving electrons from one electrode to the other.

2. The charge separation creates an electric field from + to −.

3. Because of the electric field, there's a potential difference between the electrodes.

FIGURE 29.7 A Van de Graaff generator.

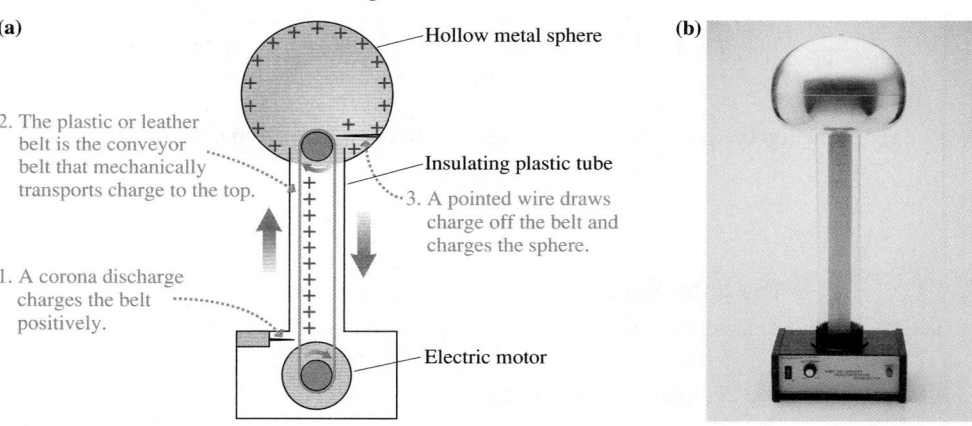

(a)

Hollow metal sphere

2. The plastic or leather belt is the conveyor belt that mechanically transports charge to the top.

Insulating plastic tube

3. A pointed wire draws charge off the belt and charges the sphere.

1. A corona discharge charges the belt positively.

Electric motor

(b)

A Van de Graaff generator has two noteworthy features:

- Charge is *mechanically* transported from the negative side to the positive side. This charge separation creates a potential difference between the spherical electrode and its surroundings.
- The electric field of the spherical electrode exerts a downward force on the positive charges moving up the belt. Consequently, *work must be done* to "lift" the positive charges. The work is done by the electric motor that runs the belt.

A classroom-demonstration Van de Graaff generator like the one shown in **FIGURE 29.7b** creates a potential difference of several hundred thousand volts between the upper sphere and its surroundings. The maximum potential is reached when the electric field near the sphere becomes large enough to cause a breakdown of the air. This produces a spark and temporarily discharges the sphere. A large Van de Graaff generator surrounded by vacuum can reach a potential of 20 MV or more. These generators are used to accelerate protons for nuclear physics experiments.

Batteries and emf

The most common source of electric potential is a **battery.** A battery consists of chemicals, called *electrolytes,* sandwiched between two electrodes made of different metals. Chemical reactions in the electrolytes separate charge by moving positive ions to one electrode and negative ions to the other. In other words, chemical reactions, rather than a mechanical conveyor belt, transport charge from one electrode to the other. The procedure is different, but the outcome is the same: a potential difference.

We can sidestep the chemistry details by introducing the **charge escalator model** of a battery shown in FIGURE 29.8. The escalator separates charge by "lifting" positive charges from the negative terminal to the positive terminal. Lifting positive charges to a positive terminal requires that work be done, and the chemical reactions within the battery provide the energy to do this work. When the chemicals are used up, the reactions cease, and the battery is dead.

By separating the charge, the charge escalator establishes a potential difference ΔV_{bat} between the terminals. The value of ΔV_{bat} is determined by the specific chemical reactions employed by the battery. To see how, suppose the chemical reactions do work W_{chem} to move charge q from the negative to the positive terminal. In an **ideal battery,** in which there are no internal energy losses, the charge gains electric potential energy $\Delta U = W_{chem}$. This is analogous to a book gaining gravitational potential energy as you do work to lift it from the floor to a shelf.

The quantity W_{chem}/q, which is the work done *per charge* by the charge escalator, is called the **emf** of the battery, pronounced as the sequence of three letters "e-m-f." The symbol for emf is $\mathcal{E}$, a script E, and the units are those of the electric potential: joules per coulomb, or volts. The *rating* of a battery, such as 1.5 V or 9 V, is the battery's emf. Originally the term emf was an abbreviation of "electromotive force." That is an outdated term (work per charge is not a force!), so today we just call it emf and it is not an abbreviation of anything.

By definition, the electric potential is related to the electric potential energy of charge q by $\Delta V = \Delta U/q$. But $\Delta U = W_{chem}$ for the charges in a battery, hence the potential difference between the terminals of an ideal battery is

$$\Delta V_{bat} = \frac{W_{chem}}{q} = \mathcal{E} \qquad \text{(ideal battery)} \qquad (29.8)$$

In other words, a battery constructed to have an emf of 1.5 V (i.e., the chemical reactions do 1.5 J of work to separate 1 C of charge) creates a 1.5 V potential difference between its positive and negative terminals. In practice, the measured potential difference ΔV_{bat} between the terminals of a real battery, called the **terminal voltage,** is usually slightly less than $\mathcal{E}$. You will learn the reason for this in Chapter 31.

Many consumer goods, from flashlights to digital cameras, use more than one battery. Why? A particular type of battery, such as an AA or AAA battery, produces a fixed emf determined by the chemical reactions inside. The emf of one battery, often 1.5 V, is not sufficient to light a lightbulb or power a camera. But just as you can reach the third floor of a building by taking three escalators in succession, we can produce a larger potential difference by placing two or more batteries *in series.* FIGURE 29.9 shows two batteries with the positive terminal of one literally touching the negative terminal of the next. Flashlight batteries usually are arranged like this. Other devices, such as cameras, achieve the same effect by using conducting metal wires between one battery and the next. Either way, the total potential difference of batteries in series is simply the sum of their individual terminal voltages:

$$\Delta V_{series} = \Delta V_1 + \Delta V_2 + \cdots \qquad \text{(batteries in series)} \qquad (29.9)$$

Electric generators, photocells, and other sources of potential difference use different means to separate charges, but otherwise they function exactly the same as a battery. The common feature of all such devices is that they use a *nonelectrical* means to separate charge and, thus, to create a potential difference. The emf $\mathcal{E}$ of any device is the work done per charge to separate the charge.

FIGURE 29.8 The charge escalator model of a battery.

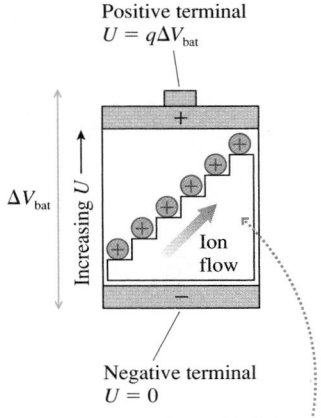

The charge escalator "lifts" charge from the negative side to the positive side. Charge q gains energy $\Delta U = q\Delta V_{bat}$.

Flashlight batteries are placed in series to create twice the potential difference of one battery.

FIGURE 29.9 Batteries in series.

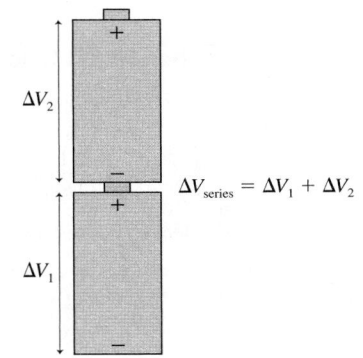

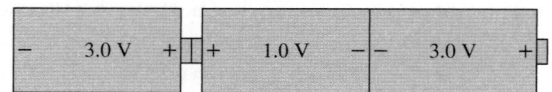

STOP TO THINK 29.1 What total potential difference is created by these three batteries?

| − 3.0 V + | + 1.0 V − | − 3.0 V + |

29.3 Finding the Electric Field from the Potential

FIGURE 29.10 The electric field does work on charge q.

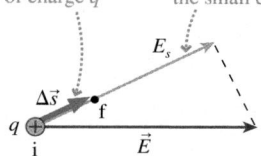

A very small displacement of charge q

E_s, the component of $\vec{E}$ in the direction of motion, is essentially constant over the small distance Δs.

FIGURE 29.10 shows two points i and f separated by a very small distance Δs, so small that the electric field is essentially constant over this very short distance. The work done by the electric field as a charge q moves through this small distance is $W = F_s \Delta s = qE_s \Delta s$. Consequently, the potential difference between these two points is

$$\Delta V = \frac{\Delta U_{q+\text{sources}}}{q} = \frac{-W}{q} = -E_s \Delta s \qquad (29.10)$$

In terms of the potential, the component of the electric field in the s-direction is $E_s = -\Delta V/\Delta s$. In the limit $\Delta s \to 0$,

$$E_s = -\frac{dV}{ds} \qquad (29.11)$$

Now we have reversed Equation 29.3 and can find the electric field from the potential. We'll begin with examples where the field is parallel to a coordinate axis, then we'll look at what Equation 29.11 tells us about the geometry of the field and the potential.

Field Parallel to a Coordinate Axis

The derivative in Equation 29.11 gives E_s, the component of the electric field parallel to the displacement $\Delta \vec{s}$. It doesn't tell us about the electric field component perpendicular to $\Delta \vec{s}$. Thus Equation 29.11 is most useful if we can use symmetry to select a coordinate axis that is parallel to $\vec{E}$ and along which the perpendicular component of $\vec{E}$ is known to be zero.

For example, suppose we knew the potential of a point charge to be $V = q/4\pi\epsilon_0 r$ but didn't remember the electric field. Symmetry requires that the field point straight outward from the charge, with only a radial component E_r. If we choose the s-axis to be in the radial direction, parallel to $\vec{E}$, we can use Equation 29.11 to find

$$E_r = -\frac{dV}{dr} = -\frac{d}{dr}\left(\frac{q}{4\pi\epsilon_0 r}\right) = \frac{1}{4\pi\epsilon_0}\frac{q}{r^2} \qquad (29.12)$$

This is, indeed, the well-known electric field of a point charge.

Equation 29.11 is especially useful for a continuous distribution of charge because calculating V, which is a scalar, is usually much easier than calculating the vector $\vec{E}$ directly from the charge. Once V is known, $\vec{E}$ is found simply by taking a derivative.

EXAMPLE 29.3 **The electric field of a ring of charge**

In Chapter 28, we found the on-axis potential of a ring of radius R and charge Q to be

$$V_{\text{ring}} = \frac{1}{4\pi\epsilon_0}\frac{Q}{\sqrt{z^2 + R^2}}$$

Find the on-axis electric field of a ring of charge.

SOLVE Symmetry requires the electric field along the axis to point straight outward from the ring with only a z-component E_z. The electric field at position z is

$$E_z = -\frac{dV}{dz} = -\frac{d}{dz}\left(\frac{1}{4\pi\epsilon_0}\frac{Q}{\sqrt{z^2 + R^2}}\right)$$

$$= \frac{1}{4\pi\epsilon_0}\frac{zQ}{(z^2 + R^2)^{3/2}}$$

ASSESS This result is in perfect agreement with the electric field we found in Chapter 26, but this calculation was easier because we didn't have to deal with angles.

A geometric interpretation of Equation 29.11 is that the electric field is the negative of the *slope* of the *V*-versus-*s* graph. This interpretation should be familiar. You learned in Chapter 11 that the force on a particle is the negative of the slope of the potential-energy graph: $F = -dU/ds$. In fact, Equation 29.11 is simply $F = -dU/ds$ with both sides divided by q to yield E and V. This geometric interpretation is an important step in developing an understanding of potential.

EXAMPLE 29.4   Finding *E* from the slope of *V*

FIGURE 29.11 is a graph of the electric potential in a region of space where $\vec{E}$ is parallel to the *x*-axis. Draw a graph of E_x versus *x*.

FIGURE 29.11 Graph of *V* versus position *x*.

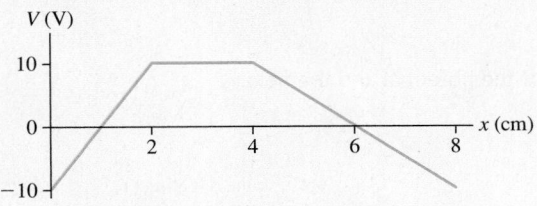

$$V (V)$$

MODEL The electric field is the *negative* of the slope of the potential graph.

SOLVE There are three regions of different slope:

$$0 < x < 2 \text{ cm} \quad \begin{cases} \Delta V/\Delta x = (20 \text{ V})/(0.020 \text{ m}) = 1000 \text{ V/m} \\ E_x = -1000 \text{ V/m} \end{cases}$$

$$2 < x < 4 \text{ cm} \quad \begin{cases} \Delta V/\Delta x = 0 \text{ V/m} \\ E_x = 0 \text{ V/m} \end{cases}$$

$$4 < x < 8 \text{ cm} \quad \begin{cases} \Delta V/\Delta x = (-20 \text{ V})/(0.040 \text{ m}) = -500 \text{ V/m} \\ E_x = 500 \text{ V/m} \end{cases}$$

The results are shown in **FIGURE 29.12**.

FIGURE 29.12 Graph of E_x versus position *x*.

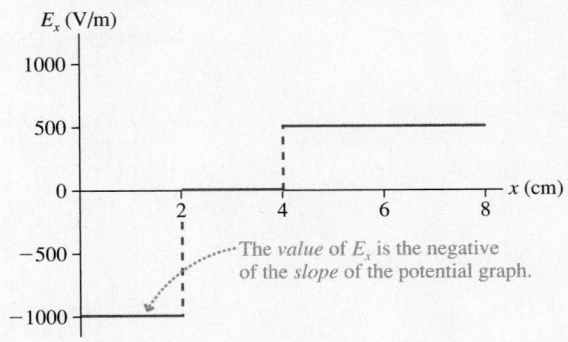

The *value* of E_x is the negative of the *slope* of the potential graph.

ASSESS The electric field $\vec{E}$ points to the left (E_x is negative) for $0 < x < 2$ cm and to the right (E_x is positive) for $4 < x < 8$ cm. Notice that **the electric field is zero in a region of space where the potential is not changing.**

STOP TO THINK 29.2 Which potential graph describes the electric field at the left?

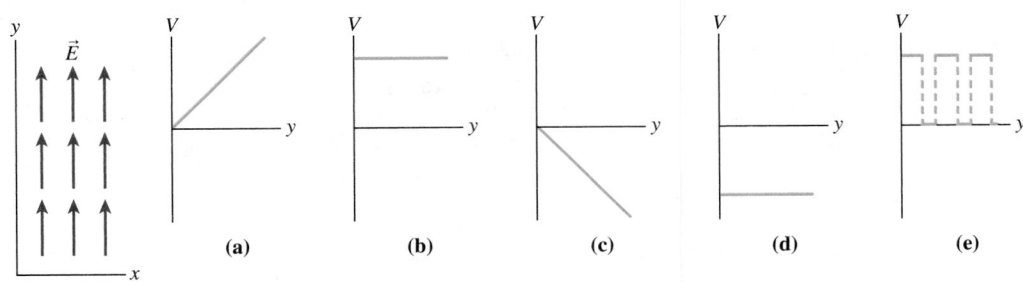

The Geometry of Potential and Field

Equations 29.3 for *V* in terms of E_s and 29.11 for E_s in terms of *V* have profound implications for the geometry of the potential and the field. **FIGURE 29.13** shows two equipotential surfaces, with V_+ positive relative to V_-. To learn about the electric field $\vec{E}$ at point P, allow a charge to move through the two displacements $\Delta \vec{s}_1$ and $\Delta \vec{s}_2$. Displacement $\Delta \vec{s}_1$ is *tangent* to the equipotential surface, hence a charge moving in this direction experiences *no* potential difference. According to Equation 29.11, the electric field component along a direction of *constant* potential is $E_s = -dV/ds = 0$. In other words, the electric field component tangent to the equipotential is $E_{\parallel} = 0$.

FIGURE 29.13 The electric field at P is related to the shape of the equipotential surfaces.

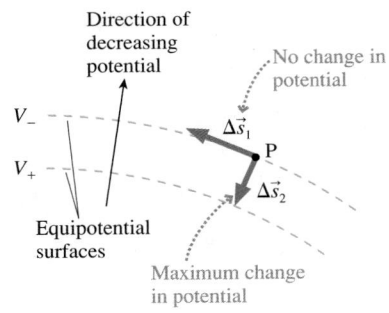

Displacement $\Delta\vec{s}_2$ is *perpendicular* to the equipotential surface. There is a potential difference along $\Delta\vec{s}_2$, hence the electric field component is

$$E_{\perp} = -\frac{dV}{ds} \approx -\frac{\Delta V}{\Delta s} = -\frac{V_+ - V_-}{\Delta s_2}$$

You can, see that the electric field is inversely proportional to Δs_2, the spacing between the equipotential surfaces. Furthermore, because $(V_+ - V_-) > 0$, the minus sign tells us that the electric field is *opposite* in direction to $\Delta\vec{s}_2$. In other words, $\vec{E}$ is **perpendicular to the equipotential surfaces and points "downhill" in the direction of *decreasing* potential.**

These important ideas about the geometry of the potential and the field are summarized in **FIGURE 29.14**.

FIGURE 29.14 The geometry of the potential and the field.

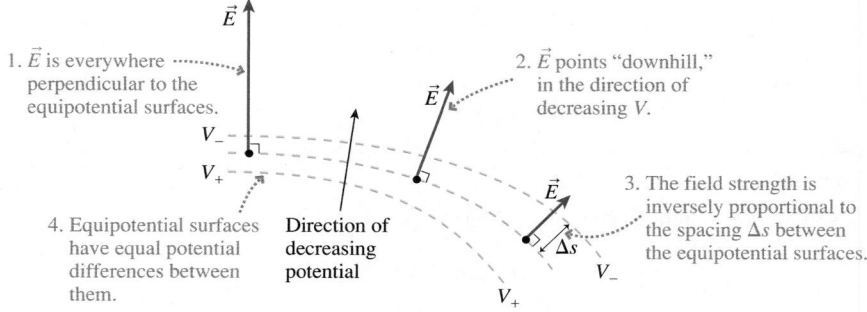

1. $\vec{E}$ is everywhere perpendicular to the equipotential surfaces.

2. $\vec{E}$ points "downhill," in the direction of decreasing V.

3. The field strength is inversely proportional to the spacing Δs between the equipotential surfaces.

4. Equipotential surfaces have equal potential differences between them.

Direction of decreasing potential

Mathematically, we can calculate the individual components of $\vec{E}$ at any point by extending Equation 29.11 to three dimensions:

$$\vec{E} = E_x\hat{\imath} + E_y\hat{\jmath} + E_z\hat{k} = -\left(\frac{\partial V}{\partial x}\hat{\imath} + \frac{\partial V}{\partial y}\hat{\jmath} + \frac{\partial V}{\partial z}\hat{k}\right) \qquad (29.13)$$

where $\partial V/\partial x$ is the partial derivative of V with respect to x while y and z are held constant. You may recognize from calculus that the expression in parentheses is the *gradient* of V, written ∇V. Thus, $\vec{E} = -\nabla V$. More advanced treatments of the electric field make extensive use of this mathematical relationship, but for the most part we'll limit our investigations to those we can analyze graphically.

EXAMPLE 29.5 **Finding the electric field from the equipotential surfaces**

In **FIGURE 29.15** a 1 cm × 1 cm grid is superimposed on a contour map of the potential. Estimate the strength and direction of the electric field at points 1, 2, and 3. Show your results graphically by drawing the electric field vectors on the contour map.

FIGURE 29.15 Equipotential lines.

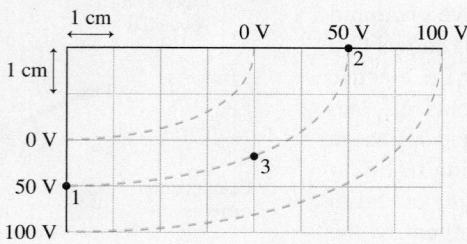

MODEL The electric field is perpendicular to the equipotential lines, points "downhill," and depends on the slope of the potential hill.

VISUALIZE The potential is highest on the bottom and the right. An elevation graph of the potential would look like the lower-right quarter of a bowl or a football stadium.

SOLVE Some distant but unseen source charges have created an electric field and potential. We do not need to see the source charges to relate the field to the potential. Because $E \approx -\Delta V/\Delta s$, the electric field is stronger where the equipotential lines are closer together and weaker where they are farther apart. If Figure 29.15 were a topographic map, you would interpret the closely spaced contour lines at the bottom of the figure as a steep slope.

FIGURE 29.16 shows how measurements of Δs from the grid are combined with values of ΔV to determine $\vec{E}$. Point 3 requires an estimate of the spacing between the 0 V and the 100 V surfaces. Notice that we're using the 0 V and 100 V equipotential surfaces to determine $\vec{E}$ at a point on the 50 V equipotential.

ASSESS The *directions* of $\vec{E}$ are found by drawing downhill vectors perpendicular to the equipotentials. The distances between the equipotential surfaces are needed to determine the field strengths.

FIGURE 29.16 The electric field at points 1 to 3.

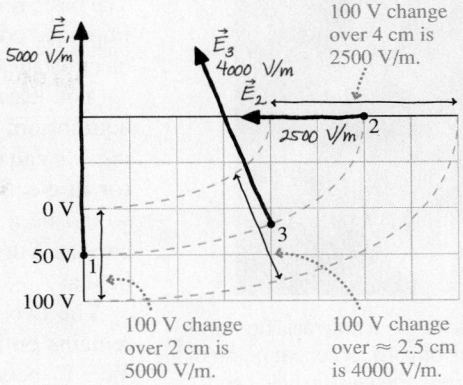

Kirchhoff's Loop Law

FIGURE 29.17 shows two points, 1 and 2, in a region of electric field and potential. You learned in Chapter 28 that the work done in moving a charge between points 1 and 2 is *independent of the path*. Consequently, the potential difference between points 1 and 2 along any two paths that join them is $\Delta V = 20$ V. This must be true in order for the idea of an equipotential surface to make sense.

Now consider the path 1–a–b–c–2–d–1 that ends where it started. What is the potential difference "around" this closed path? The potential increases by 20 V in moving from 1 to 2, but then decreases by 20 V in moving from 2 back to 1. Thus $\Delta V = 0$ V around the closed path.

The numbers are specific to this example, but the idea applies to any loop (i.e., a closed path) through an electric field. The situation is analogous to hiking on the side of a mountain. You may walk uphill during parts of your hike and downhill during other parts, but if you return to your starting point your *net* change of elevation is zero. So for any path that starts and ends at the same point, we can conclude that

FIGURE 29.17 The potential difference between points 1 and 2 is the same along either path.

The potential difference along path 1-a-b-c-2 is $\Delta V = 0$ V + 10 V + 0 V + 10 V = 20 V.

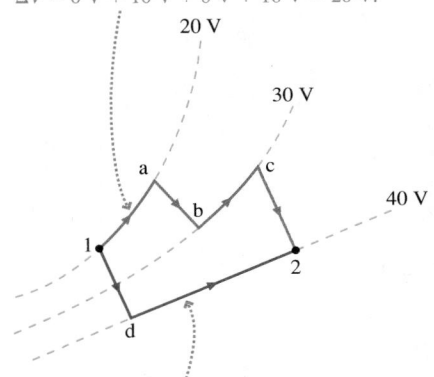

The potential difference along path 1-d-2 is $\Delta V = 20$ V + 0 V = 20 V.

$$\Delta V_{\text{loop}} = \sum_i (\Delta V)_i = 0 \qquad (29.14)$$

Stated in words, **the sum of all the potential differences encountered while moving around a loop or closed path is zero.** This statement is known as **Kirchhoff's loop law.**

Kirchhoff's loop law is a statement of energy conservation because a charge that moves around a loop and returns to its starting point has $\Delta U = q\,\Delta V = 0$. Kirchhoff's loop law and a second Kirchhoff's law you'll meet in the next chapter will turn out to be the two fundamental principles of circuit analysis.

STOP TO THINK 29.3 Which set of equipotential surfaces matches this electric field?

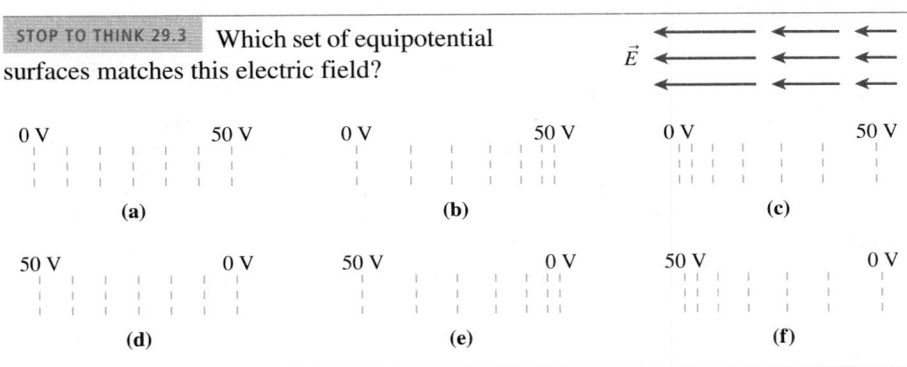

A corona discharge, with crackling noises and glimmers of light, occurs at pointed metal tips where the electric field can be very strong.

FIGURE 29.18 All points inside a conductor in electrostatic equilibrium are at the same potential.

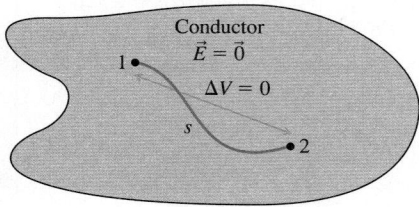

29.4 A Conductor in Electrostatic Equilibrium

The basic relationships between potential and field allow us to draw some interesting and important conclusions about conductors. Consider a conductor, such as a metal, that is in electrostatic equilibrium. The conductor may be charged, but all the charges are at rest.

You learned in Chapter 25 that any excess charges on a conductor in electrostatic equilibrium are always located on the *surface* of the conductor. Using similar reasoning, we can conclude that **the electric field is zero at any interior point of a conductor in electrostatic equilibrium.** Why? If the field were other than zero, then there would be a force $\vec{F} = q\vec{E}$ on the charge carriers and they would move, creating a current. But there are no currents in a conductor in electrostatic equilibrium, so it must be that $\vec{E} = \vec{0}$ at all interior points.

The two points inside the conductor in **FIGURE 29.18** are connected by a line that remains entirely inside the conductor. We can find the potential difference $\Delta V = V_2 - V_1$ between these points by using Equation 29.3 to integrate E_s along the line from 1 to 2. But $E_s = 0$ at all points along the line, because $\vec{E} = \vec{0}$; thus the value of the integral is zero and $\Delta V = 0$. In other words, **any two points inside a conductor in electrostatic equilibrium are at the same potential.**

When a conductor is in electrostatic equilibrium, the *entire conductor* is at the same potential. If we charge a metal sphere, then the entire sphere is at a single potential. Similarly, a charged metal rod or wire is at a single potential *if* it is in electrostatic equilibrium.

If $\vec{E} = \vec{0}$ inside a charged conductor but $\vec{E} \neq \vec{0}$ outside, what happens right at the surface? If the entire conductor is at the same potential, then the surface is an equipotential surface. You have seen that the electric field is always perpendicular to an equipotential surface, hence **the exterior electric field $\vec{E}$ of a charged conductor is perpendicular to the surface.**

We can also conclude that the electric field, and thus the surface charge density, is largest at sharp points. This follows from our earlier discovery that the field at the surface of a sphere of radius R can be written $E = V_0/R$. If we approximate the rounded corners of a conductor with sections of spheres, all of which are at the same potential V_0, the field strength will be largest at the corners with the smallest radii of curvature—the sharpest points.

FIGURE 29.20 Estimating the field and potential between two charged conductors.

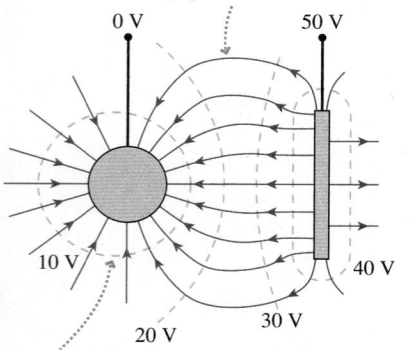

FIGURE 29.19 Electric properties of a conductor in electrostatic equilibrium.

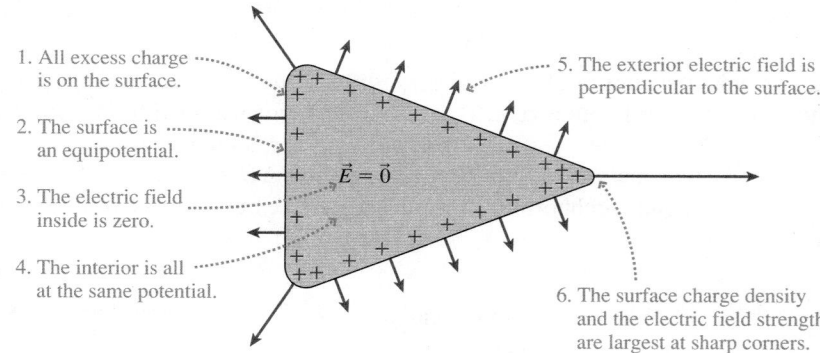

1. All excess charge is on the surface.

2. The surface is an equipotential.

3. The electric field inside is zero.

4. The interior is all at the same potential.

5. The exterior electric field is perpendicular to the surface.

$\vec{E} = \vec{0}$

6. The surface charge density and the electric field strength are largest at sharp corners.

FIGURE 29.19 summarizes what we know about conductors in electrostatic equilibrium. These are important and practical conclusions because conductors are the primary components of electrical devices.

We can use similar reasoning to estimate the electric field and potential between two charged conductors. As an example, **FIGURE 29.20** shows a negatively charged metal sphere near a flat metal plate. The surfaces of the sphere and the flat plate are equipotentials, hence the electric field must be perpendicular to both. Close to a surface, the electric field is still *nearly* perpendicular to the surface. Consequently, **an equipotential surface close to an electrode must roughly match the shape of the electrode.**

In between, the equipotential surfaces *gradually* change as they "morph" from one electrode shape to the other. It's not hard to sketch a contour map showing a plausible set of equipotential surfaces. You can then draw electric field lines (field lines are easier to draw than field vectors) that are perpendicular to the equipotentials, point "downhill," and are closer together where the contour line spacing is smaller.

STOP TO THINK 29.4 Three charged metal spheres of different radii are connected by a thin metal wire. The potential and electric field at the surface of each sphere are V and E. Which of the following is true?

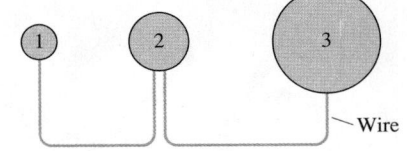

a. $V_1 = V_2 = V_3$ and $E_1 = E_2 = E_3$ b. $V_1 = V_2 = V_3$ and $E_1 > E_2 > E_3$

c. $V_1 > V_2 > V_3$ and $E_1 = E_2 = E_3$ d. $V_1 > V_2 > V_3$ and $E_1 > E_2 > E_3$

e. $V_3 > V_2 > V_1$ and $E_3 = E_2 = E_1$ f. $V_3 > V_2 > V_1$ and $E_3 > E_2 > E_1$

29.5 Capacitance and Capacitors

We introduced the parallel-plate capacitor in Chapter 26 and have made frequent use of it since. We've assumed that the capacitor is charged, but we haven't really addressed the issue of *how* it gets charged. FIGURE 29.21 shows the two plates of a capacitor connected with conducting wires to the two terminals of a battery. What happens? And how is the potential difference ΔV_C across the capacitor related to the battery's potential difference ΔV_{bat}?

FIGURE 29.21a shows the situation shortly after the capacitor is connected to the battery and before it is fully charged. The battery's charge escalator is moving charge from one capacitor plate to the other, and it is this work done by the battery that charges the capacitor. (The connecting wires are conductors, and you learned in Chapter 25 that charges can move through conductors as a *current*.) The capacitor voltage ΔV_C steadily increases as the charge separation continues.

Capacitors are important elements in electric circuits. They come in a variety of sizes and shapes.

FIGURE 29.21 A parallel-plate capacitor is charged by a battery.

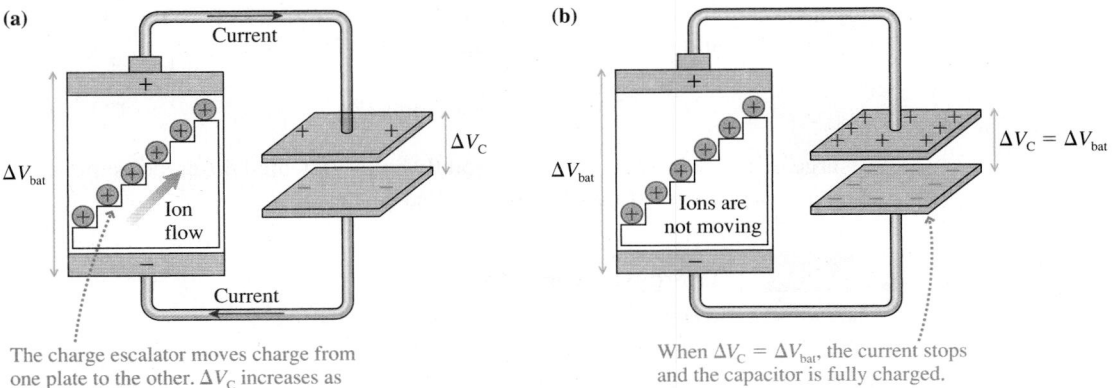

The charge escalator moves charge from one plate to the other. ΔV_C increases as the charge separation increases.

When $\Delta V_C = \Delta V_{bat}$, the current stops and the capacitor is fully charged.

But this process cannot continue forever. The growing positive charge on the upper capacitor plate exerts a repulsive force on new charges coming up the escalator, and eventually the capacitor charge gets so large that no new charges can arrive. The capacitor in FIGURE 29.21b is now *fully charged*. In Chapter 31 we'll analyze how long the charging process takes, but it is typically less than a nanosecond for a capacitor connected directly to a battery with copper wires.

Once the capacitor is fully charged, with charges no longer in motion, the positive capacitor plate, the upper wire, and the positive terminal of the battery form a single

The keys on most computer keyboards are capacitor switches. Pressing the key pushes two capacitor plates closer together, increasing their capacitance. A larger capacitor can hold more charge, so a momentary current carries charge from the battery (or power supply) to the capacitor. This current is sensed, and the keystroke is then recorded. Capacitor switches are much more reliable than make-and-break contact switches.

conductor in electrostatic equilibrium. This is an important idea, and it wasn't true while the capacitor was charging. As you just learned, any two points in a conductor in electrostatic equilibrium are at the same potential. Thus the positive plate of a fully charged capacitor is at the same potential as the positive terminal of the battery.

Similarly, the negative plate of a fully charged capacitor is at the same potential as the negative terminal of the battery. Consequently, the potential difference ΔV_C between the capacitor plates exactly matches the potential difference ΔV_{bat} between the battery terminals. **A capacitor attached to a battery charges until $\Delta V_C = \Delta V_{bat}$.** Once the capacitor is charged, you can disconnect it from the battery; it will maintain this charge and potential difference until and unless something—a current—allows positive charge to move back to the negative plate. An ideal capacitor in vacuum would stay charged forever.

You learned in Chapter 28 that a parallel-plate capacitor's potential difference is related to the electric field inside by $\Delta V_C = Ed$, where d is the separation between the plates. And you know from Chapter 26 that a capacitor's electric field is

$$E = \frac{Q}{\epsilon_0 A} \qquad (29.15)$$

where A is the surface area of the plates. Combining these gives

$$Q = \frac{\epsilon_0 A}{d} \Delta V_C \qquad (29.16)$$

In other words, **the charge on the capacitor plates is directly proportional to the potential difference between the plates.**

The ratio of the charge Q to the potential difference ΔV_C is called the **capacitance C**:

$$C \equiv \frac{Q}{\Delta V_C} = \frac{\epsilon_0 A}{d} \qquad \text{(parallel-plate capacitor)} \qquad (29.17)$$

Capacitance is a purely *geometric* property of two electrodes because it depends only on their surface area and spacing. The SI unit of capacitance is the **farad,** named in honor of Michael Faraday. One farad is defined as

$$1 \text{ farad} = 1 \text{ F} \equiv 1 \text{ C/V}$$

One farad turns out to be an enormous amount of capacitance. Practical capacitors are usually measured in units of microfarads (μF) or picofarads (1 pF $= 10^{-12}$ F).

With this definition of capacitance, Equation 29.17 can be written

$$Q = C\Delta V_C \qquad \text{(charge on a capacitor)} \qquad (29.18)$$

The charge on a capacitor is determined jointly by the potential difference supplied by a battery *and* a property of the electrodes called capacitance.

EXAMPLE 29.6 **Charging a capacitor**

The spacing between the plates of a 1.0 μF capacitor is 0.050 mm.

a. What is the surface area of the plates?
b. How much charge is on the plates if this capacitor is attached to a 1.5 V battery?

MODEL Assume the battery is ideal and the capacitor is a parallel-plate capacitor.

SOLVE a. From the definition of capacitance,

$$A = \frac{dC}{\epsilon_0} = 5.65 \text{ m}^2$$

b. The charge is $Q = C\Delta V_C = 1.5 \times 10^{-6} \text{ C} = 1.5 \ \mu\text{C}$.

ASSESS The surface area needed to construct a 1.0 μF capacitor (a fairly typical value) is enormous. We'll see in Section 29.7 how the area can be reduced by inserting an insulator between the capacitor plates.

Forming a Capacitor

The parallel-plate capacitor is important because it is straightforward to analyze and it produces a uniform electric field. But capacitors and capacitance are not limited to flat, parallel electrodes. *Any* two electrodes, regardless of their shape, form a capacitor.

FIGURE 29.22 shows two arbitrary electrodes charged to $\pm Q$. The net charge, as was the case with a parallel-plate capacitor, is zero. By definition, the capacitance of the two electrodes is

$$C = \frac{Q}{\Delta V_C} \qquad (29.19)$$

where ΔV_C is the potential difference between the positive and negative electrodes. It might appear that the capacitance depends on the amount of charge, but the potential difference is proportional to Q. Consequently, **the capacitance depends only on the geometry of the electrodes.**

To make use of Equation 29.19, you must be able to determine the potential difference between the electrodes when they are charged to $\pm Q$. You can do so if you know the electric field—say from Gauss's law—by carrying out the integration of Equation 29.3. Several homework problems will let you try this to calculate the capacitance of electrodes with other geometries.

FIGURE 29.22 Any two electrodes form a capacitor.

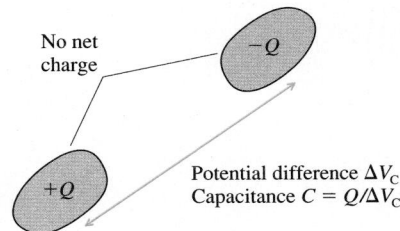

No net charge

$-Q$

$+Q$

Potential difference ΔV_C
Capacitance $C = Q/\Delta V_C$

Combinations of Capacitors

In practice, two or more capacitors are sometimes joined together. FIGURE 29.23 illustrates two basic combinations: **parallel capacitors** and **series capacitors.** Notice that a capacitor, no matter what its actual geometric shape, is represented in *circuit diagrams* by two parallel lines.

FIGURE 29.23 Parallel and series capacitors.

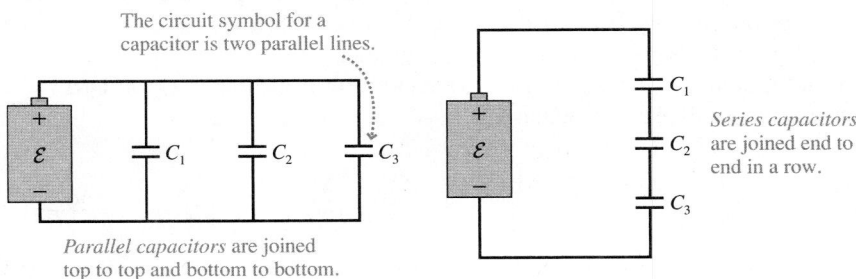

The circuit symbol for a capacitor is two parallel lines.

$\mathcal{E}$ C_1 C_2 C_3

Parallel capacitors are joined top to top and bottom to bottom.

$\mathcal{E}$ C_1 C_2 C_3

Series capacitors are joined end to end in a row.

NOTE ▶ The terms "parallel capacitors" and "parallel-plate capacitor" do not describe the same thing. The former term describes how two or more capacitors are connected to each other, the latter describes how a particular capacitor is constructed. ◀

As we'll show, parallel or series capacitors (or, as is sometimes said, capacitors "in parallel" or "in series") can be represented by a single **equivalent capacitance.** We'll demonstrate this first with the two parallel capacitors C_1 and C_2 of FIGURE 29.24a. Because the two top electrodes are connected by a conducting wire, they form a single conductor in electrostatic equilibrium. Thus the two top electrodes are at the same potential. Similarly, the two connected bottom electrodes are at the same potential. Consequently, two (or more) capacitors in parallel each have the *same* potential difference ΔV_C between the two electrodes.

The charges on the two capacitors are $Q_1 = C_1 \Delta V_C$ and $Q_2 = C_2 \Delta V_C$. Altogether, the battery's charge escalator moved total charge $Q = Q_1 + Q_2$ from the negative

FIGURE 29.24 Replacing two parallel capacitors with an equivalent capacitor.

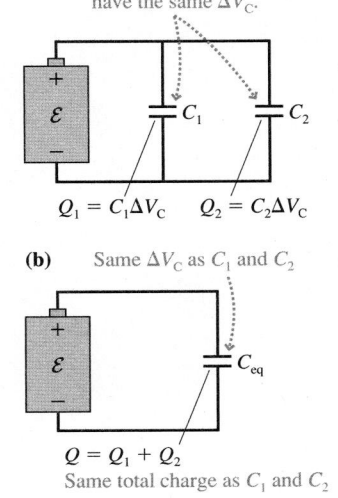

(a)   Parallel capacitors have the same ΔV_C.

$\mathcal{E}$ C_1 C_2

$Q_1 = C_1 \Delta V_C$ $Q_2 = C_2 \Delta V_C$

(b)   Same ΔV_C as C_1 and C_2

$\mathcal{E}$ C_{eq}

$Q = Q_1 + Q_2$
Same total charge as C_1 and C_2

electrodes to the positive electrodes. Suppose, as in FIGURE 29.24b, we replaced the two capacitors with a single capacitor having charge $Q = Q_1 + Q_2$ and potential difference ΔV_C. This capacitor is equivalent to the original two in the sense that the battery can't tell the difference. In either case, the battery has to establish the same potential difference and move the same amount of charge.

By definition, the capacitance of this equivalent capacitor is

$$C_{eq} = \frac{Q}{\Delta V_C} = \frac{Q_1 + Q_2}{\Delta V_C} = \frac{Q_1}{\Delta V_C} + \frac{Q_2}{\Delta V_C} = C_1 + C_2 \qquad (29.20)$$

This analysis hinges on the fact that **parallel capacitors each have the same potential difference ΔV_C.** We could easily extend this analysis to more than two capacitors. If capacitors C_1, C_2, C_3, ... are in parallel, their equivalent capacitance is

$$C_{eq} = C_1 + C_2 + C_3 + \cdots \qquad \text{(parallel capacitors)} \qquad (29.21)$$

Neither the battery nor any other part of a circuit can tell if the parallel capacitors are replaced by a single capacitor having capacitance C_{eq}.

Now consider the two series capacitors in FIGURE 29.25a. The center section, consisting of the bottom plate of C_1, the top plate of C_2, and the connecting wire, is electrically isolated. The battery cannot remove charge from or add charge to this section. If it starts out with no net charge, it must end up with no net charge. As a consequence, the two capacitors in series have equal charges $\pm Q$. The battery transfers Q from the bottom of C_2 to the top of C_1. This transfer polarizes the center section, as shown, but it still has $Q_{net} = 0$.

The potential differences across the two capacitors are $\Delta V_1 = Q/C_1$ and $\Delta V_2 = Q/C_2$. The total potential difference across both capacitors is $\Delta V_C = \Delta V_1 + \Delta V_2$. Suppose, as in FIGURE 29.25b, we replaced the two capacitors with a single capacitor having charge Q and potential difference $\Delta V_C = \Delta V_1 + \Delta V_2$. This capacitor is equivalent to the original two because the battery has to establish the same potential difference and move the same amount of charge in either case.

By definition, the capacitance of this equivalent capacitor is $C_{eq} = Q/\Delta V_C$. The inverse of the equivalent capacitance is thus

$$\frac{1}{C_{eq}} = \frac{\Delta V_C}{Q} = \frac{\Delta V_1 + \Delta V_2}{Q} = \frac{\Delta V_1}{Q} + \frac{\Delta V_2}{Q} = \frac{1}{C_1} + \frac{1}{C_2} \qquad (29.22)$$

This analysis hinges on the fact that **series capacitors each have the same charge Q.** We could easily extend this analysis to more than two capacitors. If capacitors C_1, C_2, C_3, ... are in series, their equivalent capacitance is

$$C_{eq} = \left(\frac{1}{C_1} + \frac{1}{C_2} + \frac{1}{C_3} + \cdots \right)^{-1} \qquad \text{(series capacitors)} \qquad (29.23)$$

NOTE ▶ Be careful to avoid the common error of adding the inverses but forgetting to invert the sum. ◀

Let's summarize the key facts before looking at a numerical example:

- Parallel capacitors all have the same potential difference ΔV_C. Series capacitors all have the same amount of charge $\pm Q$.
- The equivalent capacitance of a parallel combination of capacitors is *larger* than any single capacitor in the group. The equivalent capacitance of a series combination of capacitors is *smaller* than any single capacitor in the group.

FIGURE 29.25 Replacing two series capacitors with an equivalent capacitor.

(a) Series capacitors have the same Q.

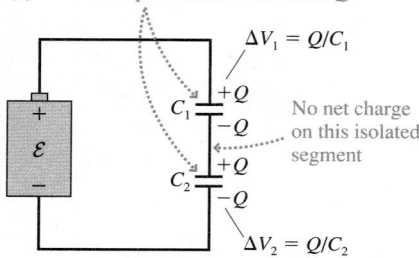

$\Delta V_1 = Q/C_1$

$+Q$
$-Q$
C_1

No net charge on this isolated segment

$+Q$
$-Q$
C_2

$\Delta V_2 = Q/C_2$

(b) Same Q as C_1 and C_2

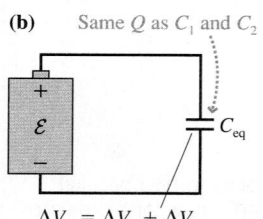

C_{eq}

$\Delta V_C = \Delta V_1 + \Delta V_2$

Same total potential difference as C_1 and C_2

EXAMPLE 29.7   **A capacitor circuit**

Find the charge on and the potential difference across each of the three capacitors in **FIGURE 29.26**.

FIGURE 29.26 A capacitor circuit.

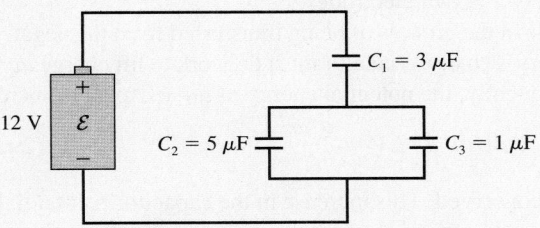

MODEL Assume the battery is ideal, with $\Delta V_{bat} = \mathcal{E} = 12$ V. Use the results for parallel and series capacitors.

SOLVE The three capacitors are neither in parallel nor in series, but we can break them down into smaller groups that are. A useful method of *circuit analysis* is first to combine elements until reaching a single equivalent element, then to reverse the process and calculate values for each element. **FIGURE 29.27** shows the analysis of this circuit. Notice that we redraw the circuit after every step. The equivalent capacitance of the 3 μF and 6 μF capacitors in series is found from

$$C_{eq} = \left(\frac{1}{3\ \mu F} + \frac{1}{6\ \mu F} \right)^{-1} = \left(\frac{2}{6} + \frac{1}{6} \right)^{-1} \mu F = 2\ \mu F$$

FIGURE 29.27 Analyzing the capacitor circuit.

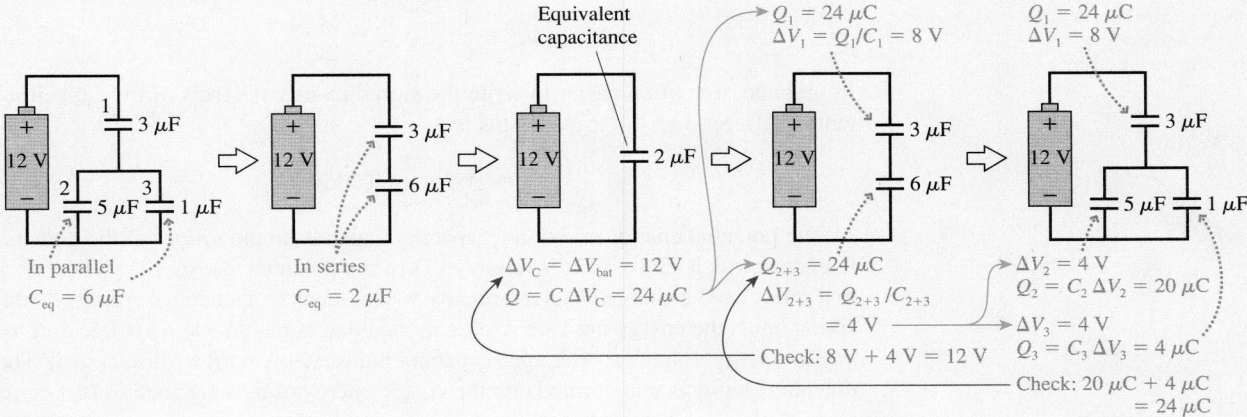

Once we get to the single equivalent capacitance, we find that $\Delta V_C = \Delta V_{bat} = 12$ V and $Q = C\Delta V_C = 24\ \mu$C. Now we can reverse direction. Capacitors in series all have the same charge, so the charge on C_1 and on C_{2+3} is $\pm 24\ \mu$C. This is enough to determine that $\Delta V_1 = 8$ V and $\Delta V_{2+3} = 4$ V. Capacitors in parallel all have the same potential difference, so $\Delta V_2 = \Delta V_3 = 4$ V. This is enough to find that $Q_2 = 20\ \mu$C and $Q_3 = 4\ \mu$C. The charge on and the potential difference across each of the three capacitors are shown in the final step of Figure 29.27.

ASSESS Notice that we had two important checks of internal consistency. $\Delta V_1 + \Delta V_{2+3} = 8$ V + 4 V add up to the 12 V we had found for the 2 μF equivalent capacitor. Then $Q_2 + Q_3 = 20\ \mu$C + 4 μC add up to the 24 μC we had found for the 6 μF equivalent capacitor. We'll do much more circuit analysis of this type in the next chapter, but it's worth noting now that circuit analysis becomes nearly foolproof *if* you make use of these checks of internal consistency.

STOP TO THINK 29.5   Rank in order, from largest to smallest, the equivalent capacitance $(C_{eq})_a$ to $(C_{eq})_d$ of circuits a to d.

5 μF

3 μF   3 μF

3 μF
3 μF

3 μF   4 μF
         4 μF

(a)   (b)   (c)   (d)

29.6 The Energy Stored in a Capacitor

FIGURE 29.28 The charge escalator does work on charge *dq* as the capacitor is being charged.

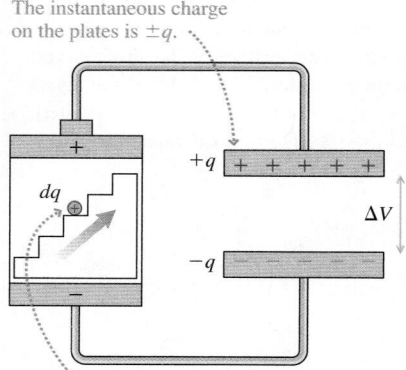

The instantaneous charge on the plates is ±*q*.

dq

+*q*

−*q*

ΔV

The charge escalator does work *dq* ΔV to move charge *dq* from the negative plate to the positive plate.

Capacitors are important elements in electric circuits because of their ability to store energy. **FIGURE 29.28** shows a capacitor being charged. The instantaneous value of the charge on the two plates is ±*q*, and this charge separation has established a potential difference $\Delta V = q/C$ between the two electrodes.

An additional charge *dq* is in the process of being transferred from the negative to the positive electrode. The battery's charge escalator must do work to lift charge *dq* "uphill" to a higher potential. Consequently, the potential energy of *dq* + capacitor increases by

$$dU = dq\,\Delta V = \frac{q\,dq}{C} \tag{29.24}$$

NOTE ▶ Energy must be conserved. This increase in the capacitor's potential energy is provided by the battery. ◀

The total energy transferred from the battery to the capacitor is found by integrating Equation 29.24 from the start of charging, when $q = 0$, until the end, when $q = Q$. Thus we find that the energy stored in a charged capacitor is

$$U_C = \frac{1}{C}\int_0^Q q\,dq = \frac{Q^2}{2C} \tag{29.25}$$

In practice, it is often easier to write the stored energy in terms of the capacitor's potential difference $\Delta V_C = Q/C$. This is

$$U_C = \frac{Q^2}{2C} = \frac{1}{2}C(\Delta V_C)^2 \tag{29.26}$$

The potential energy stored in a capacitor depends on the *square* of the potential difference across it. This result is reminiscent of the potential energy $U = \frac{1}{2}k(\Delta x)^2$ stored in a spring, and a charged capacitor really is analogous to a stretched spring. A stretched spring holds the energy until we release it, then that potential energy is transformed into kinetic energy. Likewise, a charged capacitor holds energy until we discharge it. Then the potential energy is transformed into the kinetic energy of moving charges (the current).

EXAMPLE 29.8 **Storing energy in a capacitor**

How much energy is stored in a 2.0 μF capacitor that has been charged to 5000 V? What is the average power dissipation if this capacitor is discharged in 10 μs?

SOLVE The energy stored in the charged capacitor is

$$U_C = \frac{1}{2}C(\Delta V_C)^2 = \frac{1}{2}(2.0 \times 10^{-6}\ \text{F})(5000\ \text{V})^2 = 25\ \text{J}$$

If this energy is released in 10 μs, the average power dissipation is

$$P = \frac{\Delta E}{\Delta t} = \frac{25\ \text{J}}{1.0 \times 10^{-5}\ \text{s}} = 2.5 \times 10^6\ \text{W} = 2.5\ \text{MW}$$

ASSESS The stored energy is equivalent to raising a 1 kg mass 2.5 m. This is a rather large amount of energy, which you can see by imagining the damage a 1 kg mass could do after falling 2.5 m. When this energy is released very quickly, which is possible in an electric circuit, it provides an *enormous* amount of power.

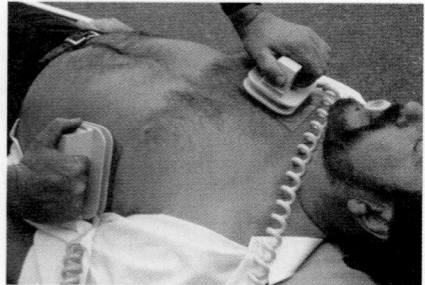

A defibrillator, which can restore a normal heartbeat, discharges a capacitor through the patient's chest.

The usefulness of a capacitor stems from the fact that it can be charged slowly (the charging rate is usually limited by the battery's ability to transfer charge) but then can release the energy very quickly. A mechanical analogy would be using a crank to slowly stretch the spring of a catapult, then quickly releasing the energy to launch a massive rock.

The capacitor described in Example 29.8 is typical of the capacitors used in high-power pulsed lasers. The capacitor is charged relatively slowly, in about 0.1 s, then quickly discharged into the laser tube to generate a high-power laser pulse. Exactly the same thing occurs, only on a smaller scale, in the flash unit of a camera. The camera batteries charge a capacitor, then the energy stored in the capacitor is quickly discharged into a *flashlamp*. The charging process in a camera takes several seconds, which is why you can't fire a camera flash twice in quick succession.

An important medical application of capacitors is the *defibrillator*. A heart attack or a serious injury can cause the heart to enter a state known as *fibrillation* in which the heart muscles twitch randomly and cannot pump blood. A strong electric shock through the chest completely stops the heart, giving the cells that control the heart's rhythm a chance to restore the proper heartbeat. A defibrillator has a large capacitor that can store up to 360 J of energy. This energy is released in about 2 ms through two "paddles" pressed against the patient's chest. It takes several seconds to charge the capacitor, which is why, on television medical shows, you hear an emergency room doctor or nurse shout "Charging!"

The Energy in the Electric Field

We can "see" the potential energy of a stretched spring in the tension of the coils. If a charged capacitor is analogous to a stretched spring, where is the stored energy? It's in the electric field!

FIGURE 29.29 shows a parallel-plate capacitor in which the plates have area A and are separated by distance d. The potential difference across the capacitor is related to the electric field inside the capacitor by $\Delta V_C = Ed$. The capacitance, which we found in Equation 29.17, is $C = \epsilon_0 A/d$. Substituting these into Equation 29.26, we find that the energy stored in the capacitor is

$$U_C = \frac{1}{2} C (\Delta V_C)^2 = \frac{1}{2} \frac{\epsilon_0 A}{d} (Ed)^2 = \frac{\epsilon_0}{2} (Ad) E^2 \qquad (29.27)$$

The quantity Ad is the volume *inside* the capacitor, the region in which the capacitor's electric field exists. (Recall that an ideal capacitor has $\vec{E} = \vec{0}$ everywhere except between the plates.) Although we talk about "the energy stored in the capacitor," Equation 29.27 suggests that, strictly speaking, **the energy is stored in the capacitor's electric field.**

Because Ad is the volume in which the energy is stored, we can define an **energy density** u_E of the electric field:

$$u_E = \frac{\text{energy stored}}{\text{volume in which it is stored}} = \frac{U_C}{Ad} = \frac{\epsilon_0}{2} E^2 \qquad (29.28)$$

The energy density has units J/m³. We've derived Equation 29.28 for a parallel-plate capacitor, but it turns out to be the correct expression for any electric field.

From this perspective, charging a capacitor stores energy in the capacitor's electric field as the field grows in strength. Later, when the capacitor is discharged, the energy is released as the field collapses.

We first introduced the electric field as a way to visualize how a long-range force operates. But if the field can store energy, the field must be real, not merely a pictorial device. We'll explore this idea further in Chapter 34, where we'll find that the energy transported by a light wave—the very real energy of warm sunshine—is the energy of electric and magnetic fields.

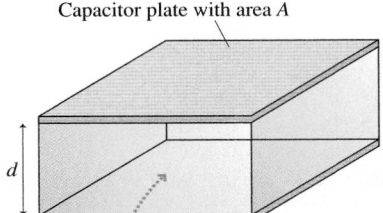

FIGURE 29.29 A capacitor's energy is stored in the electric field.

Capacitor plate with area A

The capacitor's energy is stored in the electric field in volume Ad between the plates.

29.7 Dielectrics

FIGURE 29.30a shows a parallel-plate capacitor with the plates separated by vacuum, the perfect insulator. Suppose the capacitor is charged to voltage $(\Delta V_C)_0$, then disconnected from the battery. The charge on the plates will be $\pm Q_0$, where $Q_0 = C_0 (\Delta V_C)_0$. We'll use a subscript 0 in this section to refer to a vacuum-insulated capacitor.

FIGURE 29.30 Vacuum-insulated and dielectric-filled capacitors.

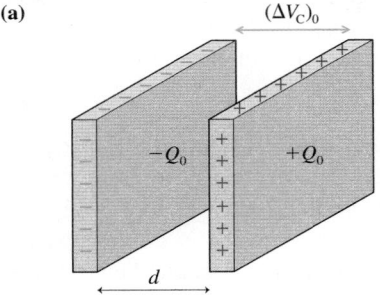

(a)

$(\Delta V_C)_0$

$-Q_0$ $+Q_0$

d

Capacitance C_0 in vacuum

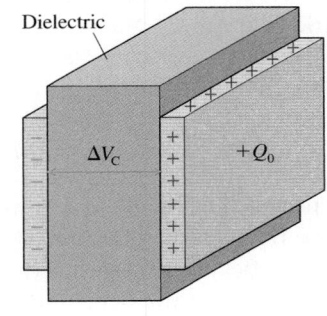

(b)

Dielectric

ΔV_C $+Q_0$

Capacitance $C > C_0$

Now suppose, as in FIGURE 29.30b, an insulating material, such as oil or glass or plastic, is slipped between the capacitor plates. We'll assume for now that the insulator is of thickness d and completely fills the space. An insulator in an electric field is called a **dielectric,** for reasons that will soon become clear, so we call this a *dielectric-filled capacitor.* How does a dielectric-filled capacitor differ from the vacuum-insulated capacitor?

The charge on the capacitor plates does not change. The insulator doesn't allow charge to move through it, and the capacitor has been disconnected from the battery, so no charge can be added to or removed from either plate. That is, $Q = Q_0$. Nonetheless, measurements of the capacitor voltage with a voltmeter would find that the voltage has decreased: $\Delta V_C < (\Delta V_C)_0$. Consequently, based on the definition of capacitance, the capacitance has increased:

$$C = \frac{Q}{\Delta V_C} > \frac{Q_0}{(\Delta V_C)_0} = C_0$$

Example 29.6 found that the plate size needed to make a 1 μF capacitor is unreasonably large. It appears that we can get more capacitance *with the same plates* by filling the capacitor with an insulator.

We can utilize two tools you learned in Chapter 26, superposition and polarization, to understand the properties of dielectric-filled capacitors. Figure 26.28 showed how an insulating material becomes *polarized* in an external electric field. FIGURE 29.31a reproduces the basic ideas from that earlier figure. The electric dipoles in Figure 29.31a could be permanent dipoles, such as water molecules, or simply induced dipoles due to a slight charge separation in the atoms. However the dipoles originate, their alignment in the electric field—the *polarization* of the material—produces an excess positive charge on one surface, an excess negative charge on the other. The insulator as a whole is still neutral, but the external electric field separates positive and negative charge.

FIGURE 29.31b represents the polarized insulator as simply two sheets of charge with surface charge densities $\pm \eta_{induced}$. The size of $\eta_{induced}$ depends both on the strength of the electric field and on the properties of the insulator. These two sheets of charge create an electric field—a situation we analyzed in Chapter 26. In essence, the two sheets of induced charge act just like the two charged plates of a parallel-plate capacitor. The **induced electric field** (keep in mind that this field is due to the insulator responding to the external electric field) is

$$\vec{E}_{induced} = \begin{cases} \left(\dfrac{\eta_{induced}}{\epsilon_0}, \text{ from positive to negative} \right) & \text{inside the insulator} \\ \vec{0} & \text{outside the insulator} \end{cases} \quad (29.29)$$

It is because an insulator in an electric field has *two* sheets of induced *electric* charge that we call it a *dielectric*, with the prefix *di*, meaning *two*, the same as in "diatomic" and "dipole."

FIGURE 29.32 shows what happens when you insert a dielectric into a capacitor. The capacitor plates have their own surface charge density $\eta_0 = Q_0/A$. This creates the electric field $\vec{E}_0 = (\eta_0/\epsilon_0$, from positive to negative) into which the dielectric is placed. The dielectric responds with induced surface charge density $\eta_{induced}$ and the induced electric field $\vec{E}_{induced}$. Notice that $\vec{E}_{induced}$ points *opposite* to $\vec{E}_0$. By the principle of superposition, another important lesson from Chapter 26, the net electric field between the capacitor plates is the *vector* sum of these two fields:

$$\vec{E} = \vec{E}_0 + \vec{E}_{induced} = (E_0 - E_{induced}, \text{ from positive to negative}) \quad (29.30)$$

The presence of the dielectric weakens the electric field, from E_0 to $E_0 - E_{induced}$, but the field still points from the positive capacitor plate to the negative capacitor plate. The field is weakened because the induced surface charge in the dielectric acts to counter the electric field of the capacitor plates.

FIGURE 29.31 An insulator in an external electric field.

(a) The insulator is polarized.

Excess positive charge on this surface   Excess negative charge on this surface

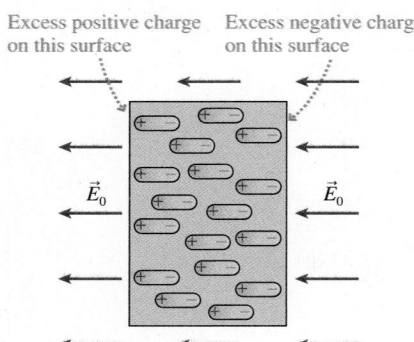

(b) The polarized insulator—a dielectric—can be represented as two sheets of surface charge. This surface charge creates an electric field inside the insulator.

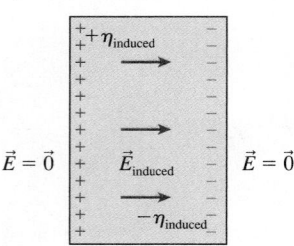

FIGURE 29.32 The consequences of filling a capacitor with a dielectric.

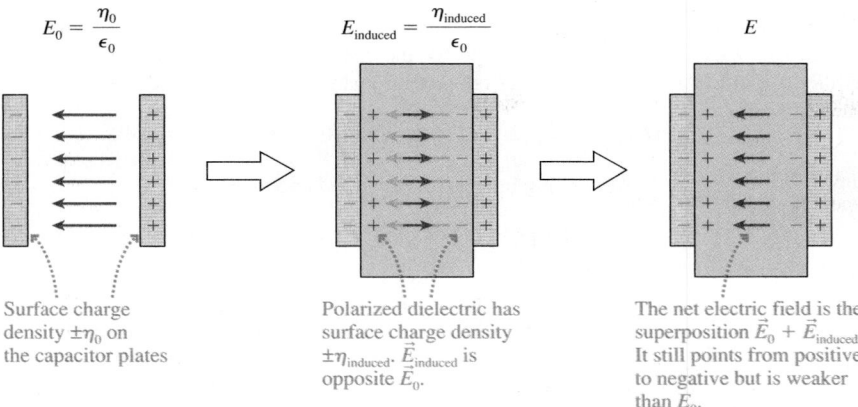

Surface charge density $\pm\eta_0$ on the capacitor plates

Polarized dielectric has surface charge density $\pm\eta_{\text{induced}}$. $\vec{E}_{\text{induced}}$ is opposite $\vec{E}_0$.

The net electric field is the superposition $\vec{E}_0 + \vec{E}_{\text{induced}}$. It still points from positive to negative but is weaker than $\vec{E}_0$.

Let's define the **dielectric constant** κ (Greek *kappa*) as

$$\kappa \equiv \frac{E_0}{E} \tag{29.31}$$

Equivalently, the field strength inside a dielectric in an external field is $E = E_0/\kappa$. The dielectric constant is the factor by which a dielectric *weakens* an electric field, so $\kappa \geq 1$. You can see from the definition that κ is a pure number with no units.

The dielectric constant, like density or specific heat, is a property of a material. Easily polarized materials have larger dielectric constants than materials not easily polarized. Vacuum has $\kappa = 1$ exactly, and low-pressure gases have $\kappa \approx 1$. (Air has $\kappa_{\text{air}} = 1.00$ to three significant figures, so we won't worry about the very slight effect air has on capacitors.) Table 29.1 lists the dielectric constants for different materials.

The electric field inside the capacitor, although weakened, is still uniform. Consequently, the potential difference across the capacitor is

$$\Delta V_{\text{C}} = Ed = \frac{E_0}{\kappa}d = \frac{(\Delta V_{\text{C}})_0}{\kappa} \tag{29.32}$$

where $(\Delta V_{\text{C}})_0 = E_0 d$ was the voltage of the vacuum-insulated capacitor. The presence of a dielectric reduces the capacitor voltage, the observation with which we started this section. Now we see why; it is due to the polarization of the material. Further, the new capacitance is

$$C = \frac{Q}{\Delta V_{\text{C}}} = \frac{Q_0}{(\Delta V_{\text{C}})_0/\kappa} = \kappa \frac{Q_0}{(\Delta V_{\text{C}})_0} = \kappa C_0 \tag{29.33}$$

Filling a capacitor with a dielectric increases the capacitance by a factor equal to the dielectric constant. This ranges from virtually no increase for an air-filled capacitor to a capacitance 300 times larger if the capacitor is filled with strontium titanate.

We'll leave it as a homework problem to show that the induced surface charge density is

$$\eta_{\text{induced}} = \eta_0\left(1 - \frac{1}{\kappa}\right) \tag{29.34}$$

η_{induced} ranges from nearly zero when $\kappa \approx 1$ to $\approx \eta_0$ when $\kappa \gg 1$.

NOTE ▶ We assumed that the capacitor was disconnected from the battery after being charged, so Q couldn't change. If you insert a dielectric while a capacitor is attached to a battery, then it will be ΔV_{C}, fixed at the battery voltage, that can't change. In this case, more charge will flow from the battery until $Q = \kappa Q_0$. In both cases, the capacitance increases to $C = \kappa C_0$. ◀

TABLE 29.1 Properties of dielectrics

Material	Dielectric constant κ	Dielectric strength $E_{\text{max}}(10^6$ V/m)
Vacuum	1	—
Air (1 atm)	1.0006	3
Teflon	2.1	60
Polystyrene plastic	2.6	24
Mylar	3.1	7
Paper	3.7	16
Pyrex glass	4.7	14
Pure water (20°C)	80	—
Titanium dioxide	110	6
Strontium titanate	300	8

EXAMPLE 29.9 A water-filled capacitor

A 5.0 nF parallel-plate capacitor is charged to 160 V. It is then disconnected from the battery and immersed in distilled water. What are (a) the capacitance and voltage of the water-filled capacitor and (b) the energy stored in the capacitor before and after its immersion?

MODEL Pure distilled water is a good insulator. (The conductivity of tap water is due to dissolved ions.) Thus the immersed capacitor has a dielectric between the electrodes.

SOLVE a. From Table 29.1, the dielectric constant of water is $\kappa = 80$. The presence of the dielectric increases the capacitance to

$$C = \kappa C_0 = 80 \times 5.0 \text{ nF} = 400 \text{ nF}$$

At the same time, the voltage decreases to

$$\Delta V_C = \frac{(\Delta V_C)_0}{\kappa} = \frac{160 \text{ V}}{80} = 2.0 \text{ V}$$

b. The presence of a dielectric does not alter the derivation leading to Equation 29.26 for the energy stored in a capacitor. Right after being disconnected from the battery, the stored energy was

$$(U_C)_0 = \frac{1}{2} C_0 (\Delta V_C)_0^2 = \frac{1}{2}(5.0 \times 10^{-9} \text{ F})(160 \text{ V})^2 = 6.4 \times 10^{-5} \text{ J}$$

After being immersed, the stored energy is

$$U_C = \frac{1}{2} C (\Delta V_C)^2 = \frac{1}{2}(400 \times 10^{-9} \text{ F})(2.0 \text{ V})^2 = 8.0 \times 10^{-7} \text{ J}$$

ASSESS Water, with its large dielectric constant, has a *big* effect on the capacitor. But where did the energy go? We learned in Chapter 26 that a dipole is drawn into a region of stronger electric field. The electric field inside the capacitor is much stronger than just outside the capacitor, so the polarized dielectric is actually *pulled* into the capacitor. The "lost" energy is the work the capacitor's electric field did pulling in the dielectric.

EXAMPLE 29.10 Energy density of a defibrillator

A defibrillator unit contains a 150 μF capacitor that is charged to 2100 V. The capacitor plates are separated by a 0.050-mm-thick insulator with dielectric constant 120.

a. What is the area of the capacitor plates?
b. What are the stored energy and the energy density in the electric field when the capacitor is charged?

MODEL Model the defibrillator as a parallel-plate capacitor with a dielectric.

SOLVE a. The capacitance of a parallel-plate capacitor in a vacuum is $C_0 = \epsilon_0 A/d$. A dielectric increases the capacitance by the factor κ, to $C = \kappa C_0$, so the area of the capacitor plates is

$$A = \frac{Cd}{\kappa \epsilon_0} = \frac{(150 \times 10^{-6} \text{ F})(5.0 \times 10^{-5} \text{ m})}{120 \,(8.85 \times 10^{-12} \text{ C}^2/\text{N m}^2)} = 7.1 \text{ m}^2$$

Although the surface area is very large, Figure 29.33 below shows how very large sheets of very thin metal can be rolled up into capacitors that you hold in your hand.

b. The energy stored in the capacitor is

$$U_c = \frac{1}{2} C (\Delta V_C)^2 = \frac{1}{2}(150 \times 10^{-6} \text{ F})(2100 \text{ V})^2 = 330 \text{ J}$$

Because the dielectric has increased C by a factor of κ, the energy density of Equation 29.28 is increased by a factor of κ to $u_E = \frac{1}{2}\kappa\epsilon_0 E^2$. The electric field strength in the capacitor is

$$E = \frac{\Delta V_C}{d} = \frac{2100 \text{ V}}{5.0 \times 10^{-5} \text{ m}} = 4.2 \times 10^7 \text{ V/m}$$

Consequently, the energy density is

$$u_E = \frac{1}{2}(120)(8.85 \times 10^{-12} \text{ C}^2/\text{N m}^2)(4.2 \times 10^7 \text{ V/m})^2$$
$$= 9.4 \times 10^5 \text{ J/m}^3$$

ASSESS 330 J is a substantial amount of energy—equivalent to that of a 1 kg mass traveling at 25 m/s. And it can be delivered very quickly as the capacitor is discharged through the patient's chest.

FIGURE 29.33 A practical capacitor.

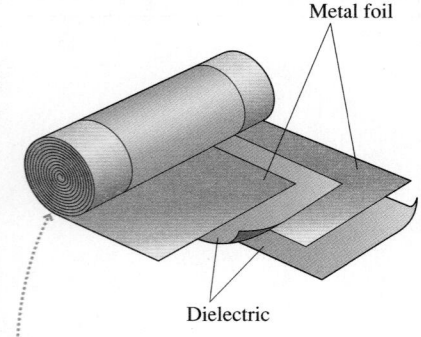

Metal foil

Dielectric

Many real capacitors are a rolled-up sandwich of metal foils and thin, insulating dielectrics.

Solid or liquid dielectrics allow a set of electrodes to have more capacitance than they would if filled with air. Not surprisingly, as **FIGURE 29.33** shows, this is important in the production of practical capacitors. In addition, dielectrics allow capacitors to be charged to higher voltages. All materials have a maximum electric field they can sustain without *breakdown*—the production of a spark. The breakdown electric field of air, as we've noted previously, is about 3×10^6 V/m. In general, a material's maximum sustainable electric field is called its **dielectric strength**. Table 29.1 includes dielectric strengths for air and the solid dielectrics. (The breakdown of water is extremely sensitive to ions and impurities in the water, so water doesn't have a well-defined dielectric strength.)

Many materials have dielectric strengths much larger than air. Teflon, for an example, has a dielectric strength 20 times that of air. Consequently, a Teflon-filled capacitor can be safely charged to a voltage 20 times larger than an air-filled capacitor with the same plate separation. An air-filled capacitor with a plate separation of 0.2 mm can be charged only to 600 V, but a capacitor with a 0.2-mm-thick Teflon sheet could be charged to 12,000 V.

CHALLENGE EXAMPLE 29.11 | A Geiger counter

The radiation detector known as a *Geiger counter* consists of a 25-mm-diameter cylindrical metal tube, sealed at the ends, with a 1.0-mm-diameter wire along its axis. The wire and cylinder are separated by a low-pressure gas whose dielectric strength is 1.0×10^6 V/m. What is the maximum potential difference between the wire and the tube?

MODEL Model the Geiger counter as two long, concentric, conducting cylinders. To avoid breakdown of the gas, the field strength at the surface of the wire—the point of maximum field strength—must not exceed the dielectric strength.

VISUALIZE FIGURE 29.34 shows a cross section of the Geiger counter tube. Applying a potential difference between the inner and outer cylinders charges it like a capacitor; indeed, it *is* a cylindrical capacitor. We've chosen to let the outer cylinder be positive, with an inward-pointing electric field, but a negative outer cylinder would lead to the same answer since it's only the field strength that we're interested in.

FIGURE 29.34 Cross section of a Geiger counter tube.

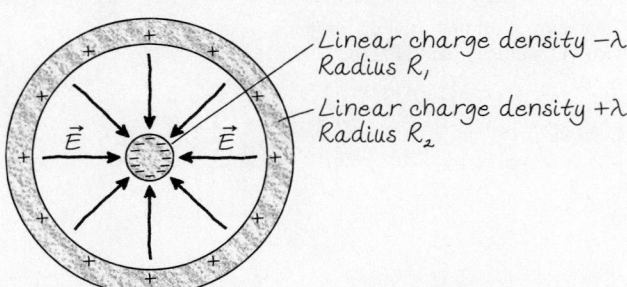

SOLVE Gauss's law tells us that the electric field between the cylinders is due only to the charge on the inner cylinder. Thus $\vec{E}$ is the field of a long, charged wire—a field we found in Chapter 26 using superposition and again in Chapter 27 using Gauss's law. It is

$$\vec{E} = \left(\frac{\lambda}{2\pi\epsilon_0 r}, \text{ inward} \right)$$

where λ is the magnitude of the linear charge density. We need to connect this field to the potential difference between the wire and the outer cylinder. For that, we need to use Equation 29.3:

$$\Delta V = V_f - V_i = -\int_{s_i}^{s_f} E_s \, ds$$

We'll integrate along a radial line from $s_i = R_1$ on the surface of the inner cylinder to $s_f = R_2$ at the outer cylinder. The field component E_s is negative because the field points inward. Thus the potential difference is

$$\Delta V = -\int_{R_1}^{R_2} \left(-\frac{\lambda}{2\pi\epsilon_0 s} \right) ds = \frac{\lambda}{2\pi\epsilon_0} \int_{R_1}^{R_2} \frac{ds}{s}$$

$$= \frac{\lambda}{2\pi\epsilon_0} \ln s \Big|_{R_1}^{R_2} = \frac{\lambda}{2\pi\epsilon_0} \ln\left(\frac{R_2}{R_1}\right)$$

We see that the applied potential difference and the linear charge density are related by

$$\frac{\lambda}{2\pi\epsilon_0} = \frac{\Delta V}{\ln(R_2/R_1)}$$

Using this in the expression for $\vec{E}$, we find the electric field strength at distance r is

$$E = \frac{\Delta V}{r \ln(R_2/R_1)}$$

The field strength is a maximum at the surface of the wire, where it reaches

$$E_{max} = \frac{\Delta V}{R_1 \ln(R_2/R_1)}$$

The maximum applied voltage will bring E_{max} to the dielectric strength, $E_{max} = 1.0 \times 10^6$ V/m. Thus the maximum potential difference between the wire and the tube is

$$\Delta V_{max} = R_1 E_{max} \ln\left(\frac{R_2}{R_1}\right)$$

$$= (5.0 \times 10^{-4} \text{ m})(1.0 \times 10^6 \text{ V/m}) \ln(25)$$

$$= 1600 \text{ V}$$

ASSESS This is the *maximum* possible voltage, but it's not practical to operate right at the maximum. Real Geiger counters operate with typically a 1000 V potential difference to avoid an accidental breakdown of the gas. If a high-speed charged particle from a radioactive decay then happens to pass through the tube, it will collide with and ionize a number of the gas atoms. Because the tube is already very close to breakdown, the addition of these extra ions and electrons is enough to push it over the edge: A breakdown of the gas occurs, with a spark jumping across the tube. The "clicking" sounds of a Geiger counter are made by amplifying the current pulses associated with the sparks.

SUMMARY

The goal of Chapter 29 has been to understand how the electric potential is related to the electric field.

General Principles

Connecting V and $\vec{E}$

The electric potential and the electric field are two different perspectives of how source charges alter the space around them. V and $\vec{E}$ are related by

$$\Delta V = V_f - V_i = -\int_{s_i}^{s_f} E_s \, ds$$

where s is measured from point i to point f and E_s is the component of $\vec{E}$ parallel to the line of integration.

Graphically

$\Delta V =$ the negative of the area under the E_s graph

and

$$E_s = -\frac{dV}{ds}$$

$\quad =$ the negative of the slope of the potential graph

The Geometry of Potential and Field

The electric field

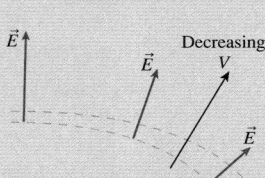

- Is perpendicular to the equipotential surfaces.

- Points "downhill" in the direction of decreasing V.

- Is inversely proportional to the spacing Δs between the equipotential surfaces.

Conservation of Energy

The sum of all potential differences around a closed path is zero.

$$\sum (\Delta V)_i = 0$$

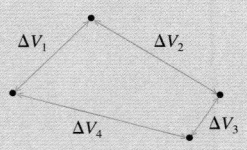

Important Concepts

A **battery** is a source of potential. The charge escalator in a battery uses chemical reactions to move charges from the negative terminal to the positive terminal:

$$\Delta V_{\text{bat}} = \mathcal{E}$$

where the emf $\mathcal{E}$ is the work per charge done by the charge escalator.

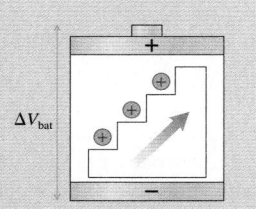

For a conductor in electrostatic equilibrium

- The interior electric field is zero.

- The exterior electric field is perpendicular to the surface.

- The surface is an equipotential.

- The interior is at the same potential as the surface.

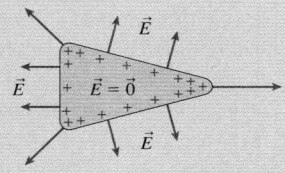

Applications

Capacitors

The **capacitance** of two conductors charged to $\pm Q$ is

$$C = \frac{Q}{\Delta V_C}$$

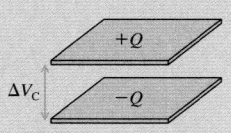

A parallel-plate capacitor has

$$C = \frac{\epsilon_0 A}{d}$$

Filling the space between the plates with a dielectric of dielectric constant κ increases the capacitance to $C = \kappa C_0$.

The energy stored in a capacitor is $u_C = \frac{1}{2} C (\Delta V_C)^2$.

This energy is stored in the electric field at density $u_E = \frac{1}{2} \kappa \epsilon_0 E^2$.

Combinations of capacitors

Series capacitors

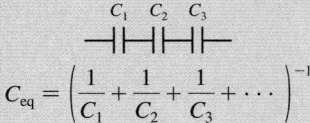

$$C_{\text{eq}} = \left(\frac{1}{C_1} + \frac{1}{C_2} + \frac{1}{C_3} + \cdots \right)^{-1}$$

Parallel capacitors

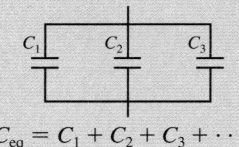

$$C_{\text{eq}} = C_1 + C_2 + C_3 + \cdots$$

Terms and Notation

Van de Graaff generator	terminal voltage, ΔV_{bat}	series capacitors	dielectric constant, κ
battery	Kirchhoff's loop law	equivalent capacitance, C_{eq}	dielectric strength
charge escalator model	capacitance, C	energy density, u_E	
ideal battery	farad, F	dielectric	
emf, $\mathcal{E}$	parallel capacitors	induced electric field	

CONCEPTUAL QUESTIONS

1. **FIGURE Q29.1** shows the x-component of $\vec{E}$ as a function of x. Draw a graph of V versus x in this same region of space. Let $V = 0$ V at $x = 0$ m and include an appropriate vertical scale.

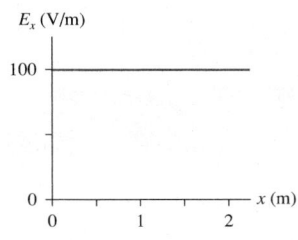

FIGURE Q29.1

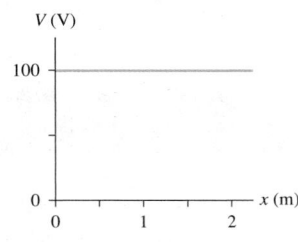

FIGURE Q29.2

2. **FIGURE Q29.2** shows the electric potential as a function of x. Draw a graph of E_x versus x in this same region of space.

3. a. Suppose that $\vec{E} = \vec{0}$ V/m throughout some region of space. Can you conclude that $V = 0$ V in this region? Explain.
   b. Suppose that $V = 0$ V throughout some region of space. Can you conclude that $\vec{E} = \vec{0}$ V/m in this region? Explain.

4. For each contour map in **FIGURE Q29.4**, estimate the electric fields $\vec{E}_1$ and $\vec{E}_2$ at points 1 and 2. Don't forget that $\vec{E}$ is a vector.

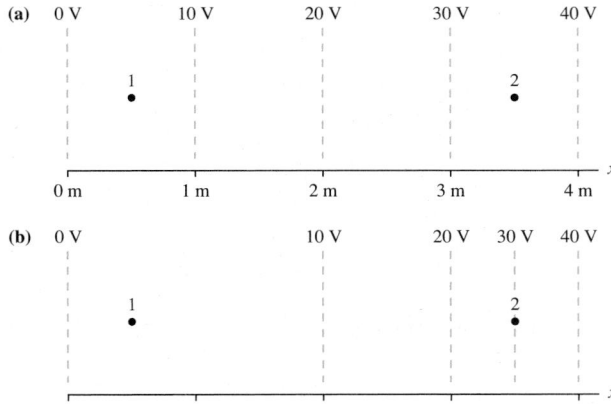

FIGURE Q29.4

5. An electron is released from rest at $x = 2$ m in the potential shown in **FIGURE Q29.5**. Does it move? If so, to the left or to the right? Explain.

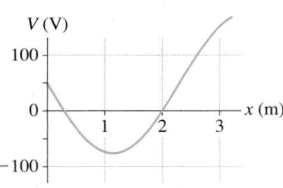

FIGURE Q29.5

6. **FIGURE Q29.6** shows an electric field diagram. Dashed lines 1 and 2 are two surfaces in space, not physical objects.
   a. Is the electric potential at point a higher than, lower than, or equal to the electric potential at point b? Explain.
   b. Rank in order, from largest to smallest, the magnitudes of the potential differences ΔV_{ab}, ΔV_{cd}, and ΔV_{ef}.
   c. Is surface 1 an equipotential surface? What about surface 2? Explain why or why not.

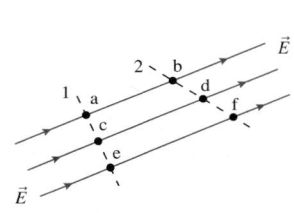

FIGURE Q29.6

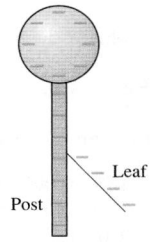

FIGURE Q29.7

7. **FIGURE Q29.7** shows a negatively charged electroscope. The gold leaf stands away from the rigid metal post. Is the electric potential of the leaf higher than, lower than, or equal to the potential of the post? Explain.

8. The two metal spheres in **FIGURE Q29.8** are connected by a metal wire with a switch in the middle. Initially the switch is open. Sphere 1, with the larger radius, is given a positive charge. Sphere 2, with the smaller radius, is neutral. Then the switch is closed. Afterward, sphere 1 has charge Q_1, is at potential V_1, and the electric field strength at its surface is E_1. The values for sphere 2 are Q_2, V_2, and E_2.
   a. Is V_1 larger than, smaller than, or equal to V_2? Explain.
   b. Is Q_1 larger than, smaller than, or equal to Q_2? Explain.
   c. Is E_1 larger than, smaller than, or equal to E_2? Explain.

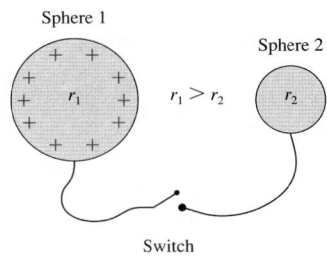

FIGURE Q29.8

9. **FIGURE Q29.9** shows a 3 V battery with metal wires attached to each end. What are the potential differences ΔV_{12}, ΔV_{23}, ΔV_{34}, and ΔV_{14}?

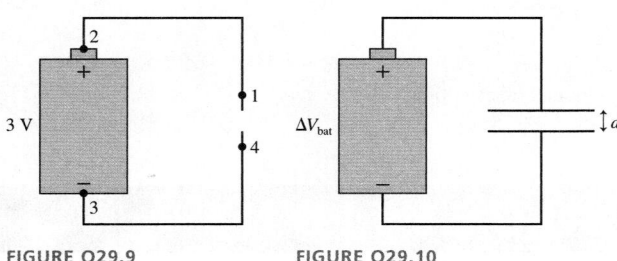

FIGURE Q29.9   **FIGURE Q29.10**

10. The parallel-plate capacitor in **FIGURE Q29.10** is connected to a battery having potential difference ΔV_{bat}. Without breaking any of the connections, insulating handles are used to increase the plate separation to $2d$.

a. Does the potential difference ΔV_C change as the separation increases? If so, by what factor? If not, why not?
b. Does the capacitance change? If so, by what factor? If not, why not?
c. Does the capacitor charge Q change? If so, by what factor? If not, why not?

11. Rank in order, from largest to smallest, the potential differences $(\Delta V_C)_1$ to $(\Delta V_C)_4$ of the four capacitors in **FIGURE Q29.11**. Explain.

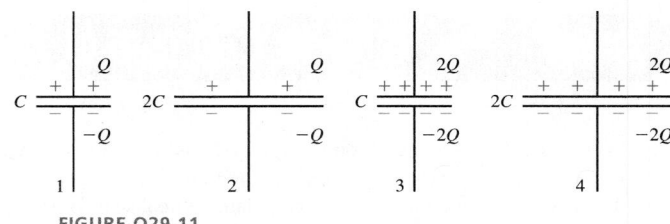

FIGURE Q29.11

EXERCISES AND PROBLEMS

Problems labeled ▓ integrate material from earlier chapters.

Exercises

Section 29.1 Connecting Potential and Field

1. ‖ What is the potential difference between $x_i = 10$ cm and $x_f = 30$ cm in the uniform electric field $E_x = 1000$ V/m?
2. ‖ What is the potential difference between $y_i = -5$ cm and $y_f = 5$ cm in the uniform electric field $\vec{E} = (20,000\hat{i} - 50,000\hat{j})$ V/m?
3. ‖ **FIGURE EX29.3** is a graph of E_x. What is the potential difference between $x_i = 1.0$ m and $x_f = 3.0$ m?

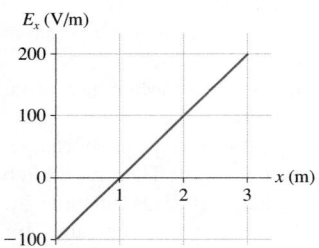

FIGURE EX29.3   **FIGURE EX29.4**

4. ‖ **FIGURE EX29.4** is a graph of E_x. The potential at the origin is -50 V. What is the potential at $x = 3.0$ m?

Section 29.2 Sources of Electric Potential

5. | How much work does the charge escalator do to move 1.0 μC of charge from the negative terminal to the positive terminal of a 1.5 V battery?
6. ‖ How much work does the electric motor of a Van de Graaff generator do to lift a positive ion ($q = e$) if the potential of the spherical electrode is 1.0 MV?
7. ‖ How much charge does a 9.0 V battery transfer from the negative to the positive terminal while doing 27 J of work?

8. | Light from the sun allows a solar cell to move electrons from the positive to the negative terminal, doing 2.4×10^{-19} J of work per electron. What is the emf of this solar cell?

Section 29.3 Finding the Electric Field from the Potential

9. | What are the magnitude and direction of the electric field at the dot in **FIGURE EX29.9**?

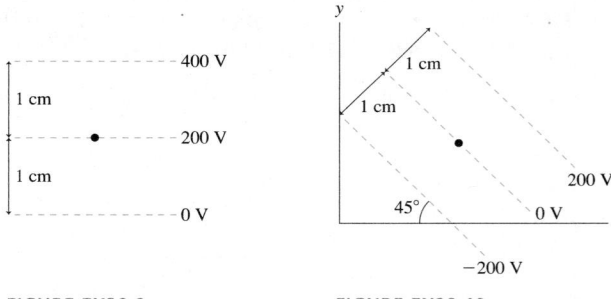

FIGURE EX29.9   **FIGURE EX29.10**

10. | What are the magnitude and direction of the electric field at the dot in **FIGURE EX29.10**?
11. ‖ **FIGURE EX29.11** is a graph of V versus x. Draw the corresponding graph of E_x versus x.

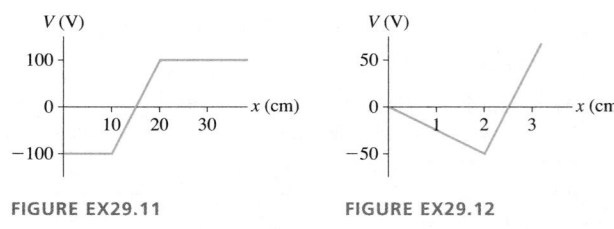

FIGURE EX29.11   **FIGURE EX29.12**

12. ‖ **FIGURE EX29.12** is a graph of V versus x. Draw the corresponding graph of E_x versus x.

13. ‖ The electric potential in a region of uniform electric field is -1000 V at $x = -1.0$ m and $+1000$ V at $x = +1.0$ m. What is E_x?

14. ‖ The electric potential along the x-axis is $V = 100x^2$ V, where x is in meters. What is E_x at (a) $x = 0$ m and (b) $x = 1$ m?

15. ‖ The electric potential along the x-axis is $V = 100e^{-2x}$ V, where x is in meters. What is E_x at (a) $x = 1.0$ m and (b) $x = 2.0$ m?

16. | What is the potential difference ΔV_{34} in FIGURE EX29.16?

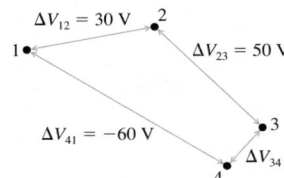

FIGURE EX29.16

Section 29.5 Capacitance and Capacitors

17. | Two 3.0-cm-diameter aluminum electrodes are spaced 0.50 mm apart. The electrodes are connected to a 100 V battery.
    a. What is the capacitance?
    b. What is the magnitude of the charge on each electrode?

18. ‖ You need to construct a 100 pF capacitor for a science project. You plan to cut two $L \times L$ metal squares and insert small spacers between their corners. The thinnest spacers you have are 0.20 mm thick. What is the proper value of L?

19. | A switch that connects a battery to a $10\ \mu$F capacitor is closed. Several seconds later you find that the capacitor plates are charged to $\pm 30\ \mu$C. What is the emf of the battery?

20. | A $6\ \mu$F capacitor, a $10\ \mu$F capacitor, and a $16\ \mu$F capacitor are connected in series. What is their equivalent capacitance?

21. | A $6\ \mu$F capacitor, a $10\ \mu$F capacitor, and a $16\ \mu$F capacitor are connected in parallel. What is their equivalent capacitance?

22. | You need a capacitance of $50\ \mu$F, but you don't happen to have a $50\ \mu$F capacitor. You do have a $30\ \mu$F capacitor. What additional capacitor do you need to produce a total capacitance of $50\ \mu$F? Should you join the two capacitors in parallel or in series?

23. | You need a capacitance of $50\ \mu$F, but you don't happen to have a $50\ \mu$F capacitor. You do have a $75\ \mu$F capacitor. What additional capacitor do you need to produce a total capacitance of $50\ \mu$F? Should you join the two capacitors in parallel or in series?

24. ‖ What is the capacitance of the two metal spheres shown in FIGURE EX29.24?

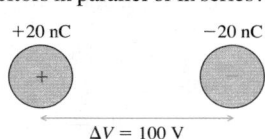

FIGURE EX29.24

Section 29.6 The Energy Stored in a Capacitor

25. ‖ To what potential should you charge a $1.0\ \mu$F capacitor to store 1.0 J of energy?

26. ‖ FIGURE EX29.26 shows Q versus t for a $2.0\ \mu$F capacitor. Draw a graph showing U_C versus t.

FIGURE EX29.26

27. | Capacitor 2 has half the capacitance and twice the potential difference as capacitor 1. What is the ratio U_{C1}/U_{C2}?

28. ‖ 50 pJ of energy is stored in a 2.0 cm $\times$ 2.0 cm $\times$ 2.0 cm region of uniform electric field. What is the electric field strength?

29. ‖ A 2.0-cm-diameter parallel-plate capacitor with a spacing of 0.50 mm is charged to 200 V. What are (a) the total energy stored in the electric field and (b) the energy density?

Section 29.7 Dielectrics

30. ‖ Two 4.0 cm $\times$ 4.0 cm metal plates are separated by a 0.20-mm-thick piece of Teflon.
    a. What is the capacitance?
    b. What is the maximum potential difference between the plates?

31. ‖ Two 5.0 mm $\times$ 5.0 mm electrodes with a 0.10-mm-thick sheet of Mylar between them are attached to a 9.0 V battery. Without disconnecting the battery, the Mylar is withdrawn. (Very small spacers keep the electrode separation unchanged.) What are the charge, potential difference, and electric field (a) before and (b) after the Mylar is withdrawn?

32. ‖ A typical cell has a layer of negative charge on the inner
    BIO surface of the cell wall and a layer of positive charge on the outside surface, thus making the cell wall a capacitor. What is the capacitance of a 50-μm-diameter cell with a 7.0-nm-thick cell wall whose dielectric constant is 9.0? Because the cell's diameter is much larger than the wall thickness, it is reasonable to ignore the curvature of the cell and think of it as a parallel-plate capacitor.

Problems

33. ‖ a. Which point in FIGURE P29.33, A or B, has a larger electric potential?
    b. What is the potential difference between A and B?

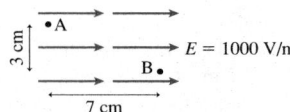

FIGURE P29.33

34. ‖‖ The electric field in a region of space is $E_x = -1000x$ V/m, where x is in meters.
    a. Graph E_x versus x over the region -1 m $\le x \le 1$ m.
    b. What is the potential difference between $x_i = -20$ cm and $x_f = 30$ cm?

35. ‖ The electric field in a region of space is $E_x = 5000x$ V/m, where x is in meters.
    a. Graph E_x versus x over the region -1 m $\le x \le 1$ m.
    b. Find an expression for the potential V at position x. As a reference, let $V = 0$ V at the origin.
    c. Graph V versus x over the region -1 m $\le x \le 1$ m.

36. ‖ An infinitely long cylinder of radius R has linear charge density λ. The potential on the surface of the cylinder is V_0, and the electric field outside the cylinder is $E_r = \lambda/2\pi\epsilon_0 r$. Find the potential relative to the surface at a point that is distance r from the axis, assuming $r > R$.

37. ‖ FIGURE P29.37 is an edge view of three charged metal electrodes. Let the left electrode be the zero point of the electric potential. What are V and $\vec{E}$ at (a) $x = 0.5$ cm, (b) $x = 1.5$ cm, and (c) $x = 2.5$ cm?

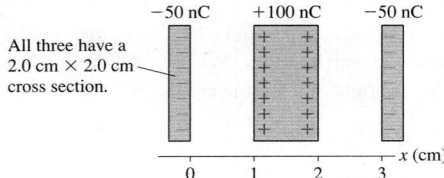

All three have a 2.0 cm × 2.0 cm cross section.

FIGURE P29.37

38. ‖ FIGURE P29.38 shows a graph of V versus x in a region of space. The potential is independent of y and z. What is E_x at (a) $x = -2$ cm, (b) $x = 0$ cm, and (c) $x = 2$ cm?

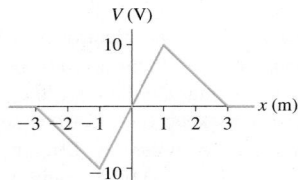

FIGURE P29.38

39. ‖ Use the on-axis potential of a charged disk from Chapter 28 to find the on-axis electric field of a charged disk.

40. ‖ a. Use the methods of Chapter 28 to find the potential at distance x on the axis of the charged rod shown in FIGURE P29.40.
   b. Use the result of part a to find the electric field at distance x on the axis of a rod.

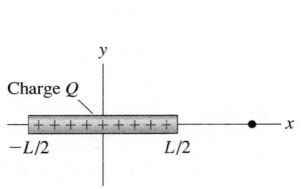

FIGURE P29.40

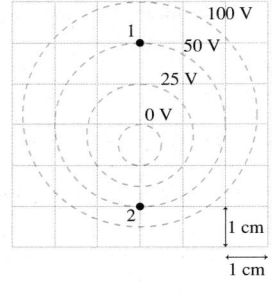

FIGURE P29.41

41. | Determine the magnitude and direction of the electric field at points 1 and 2 in FIGURE P29.41.

42. ‖ It is postulated that the radial electric field of a group of charges falls off as $E_r = C/r^n$, where r is the distance from the center of the group and n is an unknown exponent. To test this hypothesis, you make a *field probe* consisting of two needle tips spaced 1.00 mm apart. You orient the needles so that a line between the tips points to the center of the charges, then use a voltmeter to read the potential difference between the tips. After you take measurements at several distances from the center of the group, your data are as follows:

Distance (cm)	Potential difference (mV)
2.0	34.7
4.0	6.6
6.0	2.1
8.0	1.2
10.0	0.6

Use an appropriate graph of the data to determine the constants C and n.

43. ‖ The electric potential in a region of space is $V = (150x^2 - 200y^2)$ V, where x and y are in meters. What are the strength and direction of the electric field at $(x, y) = (2.0$ m, 2.0 m)? Give the direction as an angle cw or ccw (specify which) from the positive x-axis.

44. ‖ The electric potential in a region of space is $V = 200/\sqrt{x^2 + y^2}$, where x and y are in meters. What are the strength and direction of the electric field at $(x, y) = (2.0$ m, 1.0 m)? Give the direction as an angle cw or ccw (specify which) from the positive x-axis.

45. ‖ Metal sphere 1 has a positive charge of 6.0 nC. Metal sphere 2, which is twice the diameter of sphere 1, is initially uncharged. The spheres are then connected together by a long, thin metal wire. What are the final charges on each sphere?

46. ‖ The metal spheres in FIGURE P29.46 are charged to ±300 V. Draw this figure on your paper, then draw a plausible contour map of the potential, showing and labeling the −300 V, −200 V, −100 V, ... , 300 V equipotential surfaces.

FIGURE P29.46

47. ‖ The potential at the center of a 4.0-cm-diameter copper sphere is 500 V, relative to $V = 0$ V at infinity. How much excess charge is on the sphere?

48. ‖ Two 2.0 cm × 2.0 cm metal electrodes are spaced 1.0 mm apart and connected by wires to the terminals of a 9.0 V battery.
   a. What are the charge on each electrode and the potential difference between them?
   The wires are disconnected, and insulated handles are used to pull the plates apart to a new spacing of 2.0 mm.
   b. What are the charge on each electrode and the potential difference between them?

49. | Two 2.0 cm × 2.0 cm metal electrodes are spaced 1.0 mm apart and connected by wires to the terminals of a 9.0 V battery.
   a. What are the charge on each electrode and the potential difference between them?
   While the plates are still connected to the battery, insulated handles are used to pull them apart to a new spacing of 2.0 mm.
   b. What are the charge on each electrode and the potential difference between them?

50. | Find expressions for the equivalent capacitance of (a) N identical capacitors C in parallel and (b) N identical capacitors C in series.

51. | What is the equivalent capacitance of the three capacitors in FIGURE P29.51?

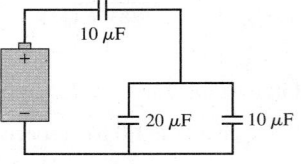

FIGURE P29.51

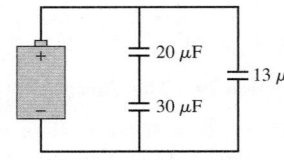

FIGURE P29.52

52. | What is the equivalent capacitance of the three capacitors in FIGURE P29.52?

53. | What are the charge on and the potential difference across each capacitor in FIGURE P29.53?

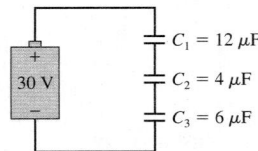

FIGURE P29.53

54. ‖ What are the charge on and the potential difference across each capacitor in FIGURE P29.54?

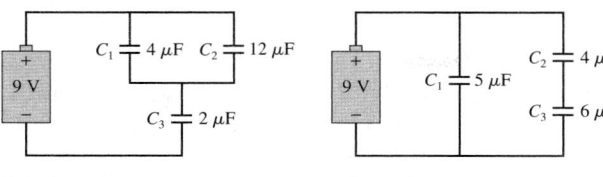

FIGURE P29.54   FIGURE P29.55

55. ‖ What are the charge on and the potential difference across each capacitor in FIGURE P29.55?

56. ‖ You have three 12 μF capacitors. Draw diagrams showing how you could arrange all three so that their equivalent capacitance is (a) 4.0 μF, (b) 8.0 μF, (c) 18 μF, and (d) 36 μF.

57. ‖ Six identical capacitors with capacitance C are connected as shown in FIGURE P29.57.
   a. What is the equivalent capacitance of these six capacitors?
   b. What is the potential difference between points a and b?

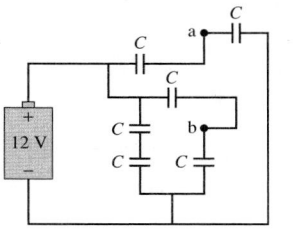

FIGURE P29.57   FIGURE P29.58

58. ‖ What is the capacitance of the two electrodes in FIGURE P29.58? **Hint:** Can you think of this as a combination of capacitors?

59. ‖ Initially, the switch in FIGURE P29.59 is in position A and capacitors C_2 and C_3 are uncharged. Then the switch is flipped to position B. Afterward, what are the charge on and the potential difference across each capacitor?

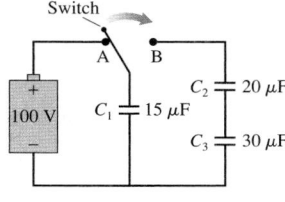
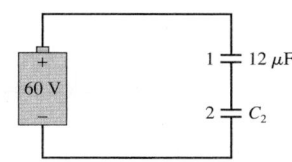

FIGURE P29.59   FIGURE P29.60

60. ‖ A battery with an emf of 60 V is connected to the two capacitors shown in FIGURE P29.60. Afterward, the charge on capacitor 2 is 450 μC. What is the capacitance of capacitor 2?

61. ‖ Capacitors $C_1 = 10 \mu$F and $C_2 = 20 \mu$F are each charged to 10 V, then disconnected from the battery without changing the charge on the capacitor plates. The two capacitors are then connected in parallel, with the positive plate of C_1 connected to the negative plate of C_2 and vice versa. Afterward, what are the charge on and the potential difference across each capacitor?

62. ‖ An isolated 5.0 μF parallel-plate capacitor has 4.0 mC of charge. An external force changes the distance between the electrodes until the capacitance is 2.0 μF. How much work is done by the external force?

63. ‖ A parallel-plate capacitor is constructed from two 10 cm × 10 cm electrodes spaced 1.0 mm apart. The capacitor plates are charged to ± 10 nC, then disconnected from the battery.
   a. How much energy is stored in the capacitor?
   b. Insulating handles are used to pull the capacitor plates apart until the spacing is 2.0 mm. Now how much energy is stored in the capacitor?
   c. Energy must be conserved. How do you account for the difference between a and b?

64. ‖ What is the energy density in the electric field at the surface of a 1.0-cm-diameter sphere charged to a potential of 1000 V?

65. ‖ The 90 μF capacitor in a defibrillator unit supplies an average
   BIO of 6500 W of power to the chest of the patient during a discharge lasting 5.0 ms. To what voltage is the capacitor charged?

66. ‖ The flash unit in a camera uses a 3.0 V battery to charge a capacitor. The capacitor is then discharged through a flashlamp. The discharge takes 10 μs, and the average power dissipated in the flashlamp is 10 W. What is the capacitance of the capacitor?

67. ‖ You need to use a motor and lightweight cable to lift a 2.0 kg copper weight to a height of 3.0 m. To do so, you've decided to use a 1000 V power supply to charge a capacitor, then run the motor by letting the capacitor discharge through it. If the motor is 90% efficient (that is, 10% of the energy supplied to the motor is dissipated as heat), what minimum capacitance do you need?

68. ‖ Two 5.0-cm-diameter metal disks separated by a 0.50-mm-thick piece of Pyrex glass are charged to a potential difference of 1000 V. What are (a) the surface charge density on the disks and (b) the surface charge density on the glass?

69. ‖ A typical cell has a membrane potential of −70 mV, mean-
   BIO ing that the potential inside the cell is 70 mV less than the potential outside due to a layer of negative charge on the inner surface of the cell wall and a layer of positive charge on the outer surface. This effectively makes the cell wall a charged capacitor. Because a cell's diameter is much larger than the wall thickness, it is reasonable to ignore the curvature of the cell and think of it as a parallel-plate capacitor. How much energy is stored in the electric field of a 50-μm-diameter cell with a 7.0-nm-thick cell wall whose dielectric constant is 9.0?

70. ‖‖ A nerve cell in its resting state has a membrane potential of
   BIO −70 mV, meaning that the potential inside the cell is 70 mV less than the potential outside due to a layer of negative charge on the inner surface of the cell wall and a layer of positive charge on the outer surface. This effectively makes the cell wall a charged capacitor. When the nerve cell fires, sodium ions, Na^+, flood through the cell wall to briefly switch the membrane potential to +40 mV. Model the central body of a nerve cell—the *soma*—as a 50-μm-diameter sphere with a 7.0-nm-thick cell wall whose dielectric constant is 9.0. Because a cell's diameter is much larger than the wall thickness, it is reasonable to ignore the curvature of the cell and think of it as a parallel-plate capacitor. How many sodium ions enter the cell as it fires?

71. ‖ Derive Equation 29.34 for the induced surface charge density on the dielectric in a capacitor.

72. ‖ A vacuum-insulated parallel-plate capacitor with plate separation d has capacitance C_0. What is the capacitance if an insulator with dielectric constant κ and thickness is $d/2$ slipped between the electrodes?

In Problems 73 through 75 you are given the equation(s) used to solve a problem. For each of these, you are to
 a. Write a realistic problem for which this is the correct equation(s).
 b. Finish the solution of the problem.

73. $2az$ V/m $= -\dfrac{dV}{dz}$, where a is a constant with units of V/m^2

$V(z = 0) = 10$ V

74. 400 nC $= (100$ V$)C$

$C = \dfrac{(8.85 \times 10^{-12} \text{ C}^2/\text{N m}^2)(0.10 \text{ m} \times 0.10 \text{ m})}{d}$

75. $\left(\dfrac{1}{3 \ \mu\text{F}} + \dfrac{1}{6 \ \mu\text{F}}\right)^{-1} + C = 4 \ \mu\text{F}$

Challenge Problems

76. The electric potential in a region of space is $V = 100(x^2 - y^2)$ V, where x and y are in meters.
 a. Draw a contour map of the potential, showing and labeling the -400 V, -100 V, 0 V, $+100$ V, and $+400$ V equipotential surfaces.
 b. Find an expression for the electric field $\vec{E}$ at position (x, y).
 c. Draw the electric field lines on your diagram of part a.
77. An electric dipole at the origin consists of two charges $\pm q$ spaced distance s apart along the y-axis.
 a. Find an expression for the potential $V(x, y)$ at an arbitrary point in the xy-plane. Your answer will be in terms of q, s, x, and y.
 b. Use the binomial approximation to simplify your result of part a when $s \ll x$ and $s \ll y$.
 c. Assuming $s \ll x$ and y, find expressions for E_x and E_y, the components of $\vec{E}$ for a dipole.
 d. What is the on-axis field $\vec{E}$? Does your result agree with Equation 26.11?
 e. What is the field $\vec{E}$ on the bisecting axis? Does your result agree with Equation 26.12?
78. Charge is uniformly distributed with charge density ρ inside a very long cylinder of radius R. Find the potential difference between the surface and the axis of the cylinder.
79. Consider a uniformly charged sphere of radius R and total charge Q. The electric field E_{out} outside the sphere ($r \geq R$) is simply that of a point charge Q. In Chapter 27, we used Gauss's law to find that the electric field E_{in} inside the sphere ($r \leq R$) is radially outward with field strength

$$E_{\text{in}} = \frac{1}{4\pi\epsilon_0}\frac{Q}{R^3}r$$

 a. The electric potential V_{out} outside the sphere is that of a point charge Q. Find an expression for the electric potential V_{in} at position r inside the sphere. As a reference, let $V_{\text{in}} = V_{\text{out}}$ at the surface of the sphere.
 b. What is the ratio $V_{\text{center}}/V_{\text{surface}}$?
 c. Graph V versus r for $0 \leq r \leq 3R$.
80. a. Find an expression for the capacitance of a *spherical capacitor*, consisting of concentric spherical shells of radii R_1 (inner shell) and R_2 (outer shell).
 b. A spherical capacitor with a 1.0 mm gap between the shells has a capacitance of 100 pF. What are the diameters of the two spheres?
81. High-frequency signals are often transmitted along a *coaxial cable*, such as the one shown in FIGURE CP29.81. For example, the cable TV hookup coming into your home is a coaxial cable. The signal is carried on a wire of radius R_1 while the outer conductor of radius R_2 is grounded (i.e., at $V = 0$ V). An insulating material fills the space between them, and an insulating plastic coating goes around the outside.

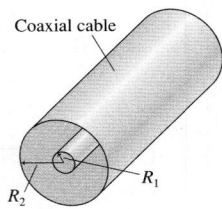
FIGURE CP29.81

 a. Find an expression for the capacitance per meter of a coaxial cable. Assume that the insulating material between the cylinders is air.
 b. Evaluate the capacitance per meter of a cable having $R_1 = 0.50$ mm and $R_2 = 3.0$ mm.
82. Each capacitor in FIGURE CP29.82 has capacitance C. What is the equivalent capacitance between points a and b?

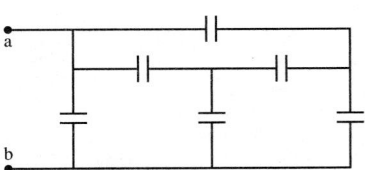
FIGURE CP29.82

Stop to Think 29.1: 5.0 V. The potentials add, but $\Delta V_2 = -1.0$ V because the charge escalator goes *down* by 1.0 V.

Stop to Think 29.2: c. E_y is the negative of the slope of the V-versus-y graph. E_y is positive because $\vec{E}$ points up, so the graph has a negative slope. E_y has constant magnitude, so the slope has a constant value.

Stop to Think 29.3: c. $\vec{E}$ points "downhill," so V must decrease from right to left. E is larger on the left than on the right, so the contour lines must be closer together on the left.

Stop to Think 29.4: b. Because of the connecting wire, the three spheres form a single conductor in electrostatic equilibrium. Thus all points are at the same potential. The electric field of a sphere is related to the sphere's potential by $E = V/R$, so a smaller-radius sphere has a larger E.

Stop to Think 29.5: $(C_{\text{eq}})_b > (C_{\text{eq}})_a = (C_{\text{eq}})_d > (C_{\text{eq}})_c$. $(C_{\text{eq}})_b = 3 \ \mu\text{F} + 3 \ \mu\text{F} = 6 \ \mu\text{F}$. The equivalent capacitance of series capacitors is less than any capacitor in the group, so $(C_{\text{eq}})_c < 3 \ \mu\text{F}$. Only d requires any real calculation. The two 4 μF capacitors are in series and are equivalent to a single 2 μF capacitor. The 2 μF equivalent capacitor is in parallel with 3 μF, so $(C_{\text{eq}})_d = 5 \ \mu\text{F}$.

30 Current and Resistance

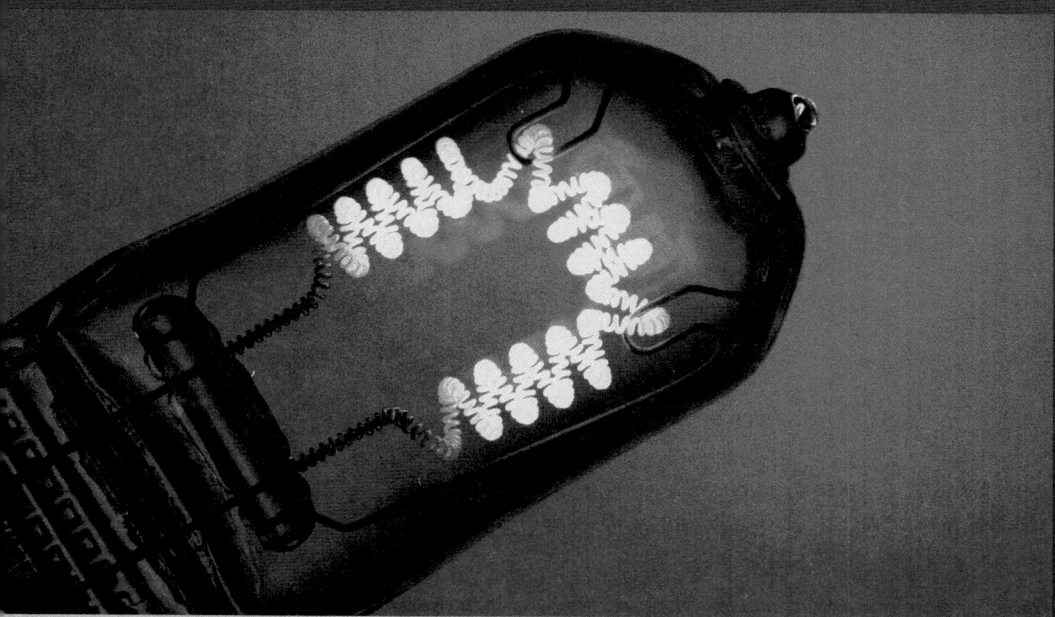

A lightbulb filament is a very thin tungsten wire—coiled repeatedly to increase its length—heated until it glows by passing a current through it.

▶ **Looking Ahead** The goal of Chapter 30 is to learn how and why charge moves through a conductor as what we call a current.

A Model of Conduction

You'll learn to use a model of conduction to understand many of the properties of current.

A nonuniform surface charge distribution, typically established when the ends of a wire are connected to the terminals of a battery, creates an electric field in the wire.

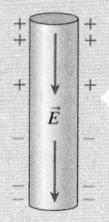

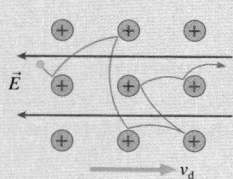

The electric field pushes the **sea of electrons** opposite the field direction, but the electrons undergo frequent collisions with the positive ions of the crystal lattice. The net result is a slow but sustained flow of charges at the **drift speed v_d.** This is the **electron current.**

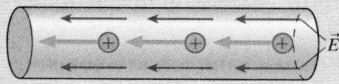

For historical reasons, current is defined to be in the direction that positive charges would move. Current is measured in **amperes,** where one ampere (or one *amp*) is a charge flow rate of 1 coulomb per second.

Current

Current is the flow of charge through a conductor. But we can't see the charges moving, so how do we know they do?

You'll learn that the flow of charge can be recognized by its effects. These include heating wires and deflecting compass needles. These are *indicators* of a current.

◀ **Looking Back**
Section 26.6 The motion of charge in an electric field

Conservation of Current

Any charge entering one end of a wire must be balanced by an equal charge leaving the other end.

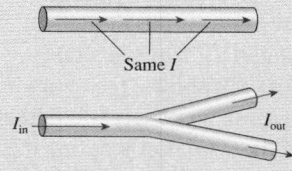

As a consequence, you'll learn that the current is the same from one end of a wire to the other. At a junction, the sum of the currents entering must equal the sum of the currents leaving.

Resistance

Collisions of electrons with the crystal lattice cause conductors to resist the flow of charges. You'll learn to use:

- **Resistivity,** an electric property of a material, such as copper.
- **Resistance,** a property of a specific wire based on its geometry and the material of which it is made.

Heater wires, such as those in toasters, are made of an alloy called *nichrome* because its resistivity is larger than that of ordinary metals.

Ohm's Law

You'll discover that the current I through a conductor is determined by the potential difference ΔV across the conductor and the conductor's resistance R.

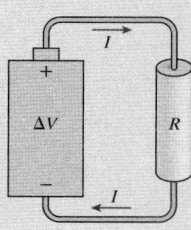

Ohm's law is $I = \dfrac{\Delta V}{R}$

◀ **Looking Back**
Section 29.2 Sources of potential

30.1 The Electron Current

We've focused thus far on situations in which charges are in static equilibrium. Now it's time to explore the *controlled* motion of charges—currents. Let's begin with a simple question: How does a capacitor get discharged? FIGURE 30.1a shows a charged capacitor. If, as in FIGURE 30.1b, we connect the two capacitor plates with a metal wire, a conductor, the plates quickly become neutral; that is, the capacitor has been *discharged*. Charge has somehow moved from one plate to the other.

FIGURE 30.1 A capacitor is discharged by a metal wire.

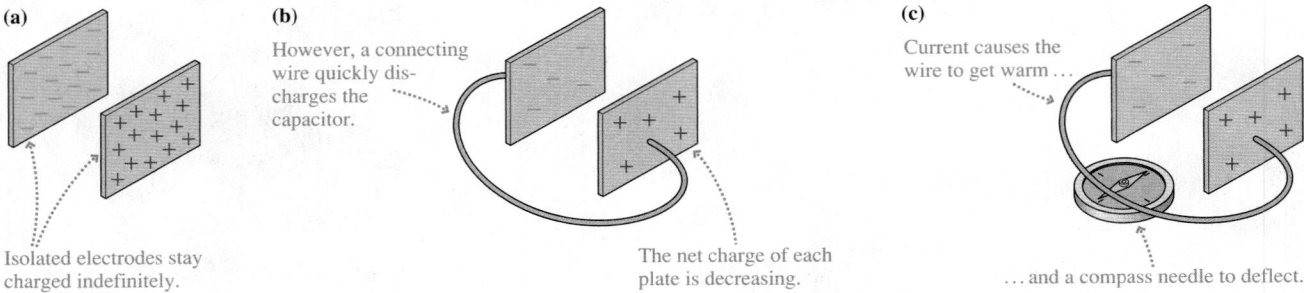

(a)

Isolated electrodes stay charged indefinitely.

(b)

However, a connecting wire quickly discharges the capacitor.

The net charge of each plate is decreasing.

(c)

Current causes the wire to get warm …

… and a compass needle to deflect.

In Chapter 25, we defined **current** as the motion of charges. It would seem that the capacitor is discharged by a current in the connecting wire. Let's see what else we can observe. FIGURE 30.1c shows that the connecting wire gets warm. If the wire is very thin in places, such as the thin filament in a lightbulb, the wire gets hot enough to glow. The current-carrying wire also deflects a compass needle, an observation we'll explore further in Chapter 32. For now, we will use "makes the wire warm" and "deflects a compass needle" as *indicators* that a current is present in a wire.

Charge Carriers

The charges that move in a conductor are called the *charge carriers*. FIGURE 30.2 reminds you of the microscopic model of a metallic conductor that we introduced in Chapter 25. The outer electrons of metal atoms—the valence electrons—are only weakly bound to the nuclei. When the atoms come together to form a solid, the outer electrons become detached from their parent nuclei to form a fluid-like *sea of electrons* that can move through the solid. That is, **electrons are the charge carriers in metals.** Notice that the metal as a whole remains electrically neutral. This is not a perfect model because it overlooks some quantum effects, but it provides a reasonably good description of current in a metal.

FIGURE 30.2 The sea of electrons is a model of how conduction electrons behave in a metal.

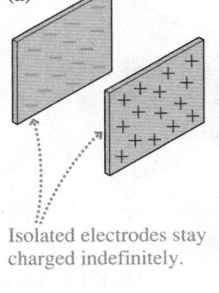

Ions (the metal atoms minus valence electrons) occupy fixed positions in the lattice.

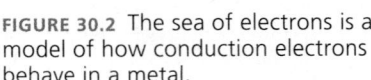

The metal as a whole is electrically neutral.

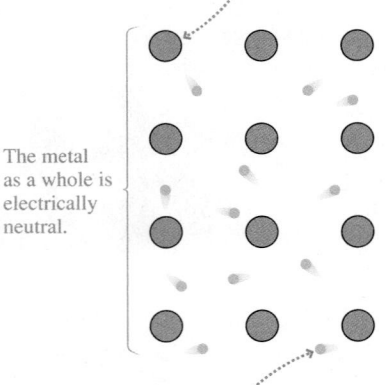

The conduction electrons are free to move around. They are bound to the solid as a whole, not to any particular atom.

NOTE ▶ Electrons are the charge carriers in *metals*. Other conductors, such as ionic solutions or semiconductors, have different charge carriers. We will focus on metals because of their importance to circuits, but don't think that electrons are *always* the charge carrier. ◀

The conduction electrons in a metal, like molecules in a gas, undergo random thermal motions, but there is no *net* motion. We can change that by pushing on the sea of electrons with an electric field, causing the entire sea of electrons to move in one direction like a gas or liquid flowing through a pipe. This net motion, which takes place at what we'll call the **drift speed** v_d, is superimposed on top of the random thermal motions of the individual electrons. The drift speed is quite small. As we'll establish later, 10^{-4} m/s is a fairly typical value for v_d.

As FIGURE 30.3 shows, the entire sea of electrons moves from left to right at the drift speed. Suppose an observer could count the electrons as they pass through this cross section of the wire. Let's define the **electron current** i_e to be the number of electrons *per second* that pass through a cross section of a wire or other conductor. The units of electron current are s^{-1}. Stated another way, the number N_e of electrons that pass through the cross section during the time interval Δt is

$$N_e = i_e \, \Delta t \qquad (30.1)$$

Increasing the drift speed will increase the number of electrons passing through a wire each second—that is, will increase the electron current. To quantify this idea, FIGURE 30.4 shows the sea of electrons moving through a wire at the drift speed v_d. The electrons passing through a particular cross section of the wire during the interval Δt are shaded. How many of them are there?

FIGURE 30.4 The sea of electrons moves to the right with drift speed v_d.

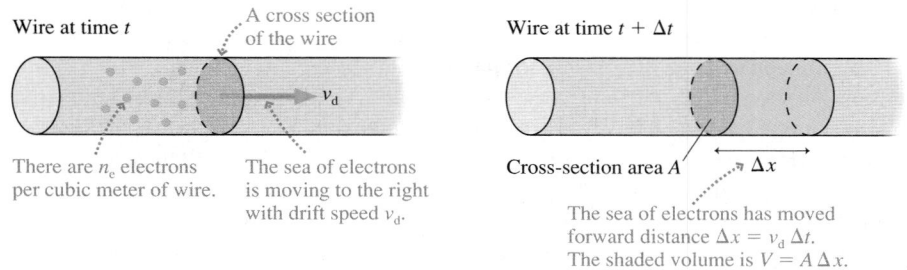

The electrons travel distance $\Delta x = v_d \, \Delta t$ to the right during the interval Δt, forming a cylinder of charge with volume $V = A \, \Delta x$. If the *number density* of conduction electrons is n_e electrons per cubic meter, then the total number of electrons in the cylinder is

$$N_e = n_e V = n_e A \, \Delta x = n_e A v_d \, \Delta t \qquad (30.2)$$

Comparing Equations 30.2 and 30.1, you can see that the electron current in the wire is

$$i_e = n_e A v_d \qquad (30.3)$$

You can increase the electron current—the number of electrons per second moving through the wire—by making them move faster, by having more of them per cubic meter, or by increasing the size of the pipe they're flowing through. That all makes sense.

In most metals, each atom contributes one valence electron to the sea of electrons. Thus the number of conduction electrons per cubic meter is the same as the number of atoms per cubic meter, a quantity that can be determined from the metal's mass density. Table 30.1 gives values of the conduction-electron density n_e for several metals.

FIGURE 30.3 The electron current.

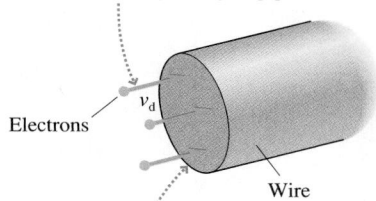

The sea of electrons flows through a wire at the drift speed v_d, much like a fluid flowing through a pipe.

Wire

The electron current i_e is the number of electrons passing through this cross section of the wire per second.

TABLE 30.1 Conduction-electron density in metals

Metal	Electron density (m^{-3})
Aluminum	6.0×10^{28}
Copper	8.5×10^{28}
Iron	8.5×10^{28}
Gold	5.9×10^{28}
Silver	5.8×10^{28}

EXAMPLE 30.1 | **The size of the electron current**

What is the electron current in a 2.0-mm-diameter copper wire if the electron drift speed is 1.0×10^{-4} m/s?

SOLVE This is a straightforward calculation. The wire's cross-section area is $A = \pi r^2 = 3.14 \times 10^{-6} \, m^2$. Table 30.1 gives the electron density for copper as $8.5 \times 10^{28} \, m^{-3}$. Thus we find

$$i_e = n_e A v_d = 2.7 \times 10^{19} \, s^{-1}$$

ASSESS This is an incredible number of electrons to pass through a section of the wire every second. The number is high not because the sea of electrons moves fast—in fact, it moves at literally a snail's pace—but because the density of electrons is so enormous. This is a fairly typical electron current.

STOP TO THINK 30.1 These four wires are made of the same metal. Rank in order, from largest to smallest, the electron currents i_a to i_d.

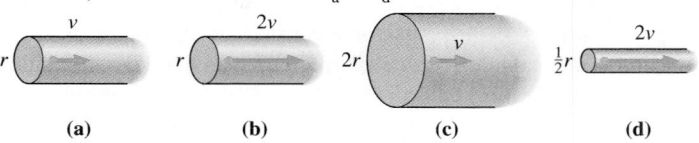

FIGURE 30.5 How long does it take to discharge this capacitor?

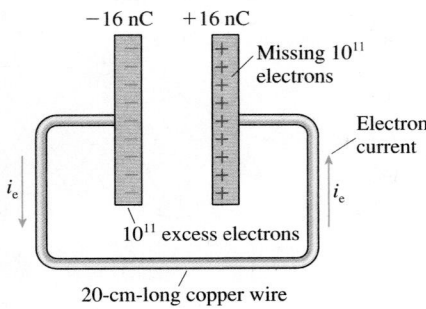

Discharging a Capacitor

FIGURE 30.5 shows a capacitor charged to ± 16 nC as it is being discharged by a 2.0-mm-diameter, 20-cm-long copper wire. *How long does it take* to discharge the capacitor? We've noted that a fairly typical drift speed of the electron current through a wire is 10^{-4} m/s. At this rate, it would take 2000 s, or about a half hour, for an electron to travel 20 cm. We should have time to go for a cup of coffee while we wait for the discharge to occur!

But this isn't what happens. As far as our senses are concerned, the discharge of a capacitor by a copper wire is instantaneous. So what's wrong with our simple calculation?

The important point we overlooked is that the wire is *already full* of electrons. As an analogy, think of water in a hose. If the hose is already full of water, adding a drop to one end immediately (or very nearly so) pushes a drop out the other end. Likewise with the wire. As soon as the excess electrons move from the negative capacitor plate into the wire, they immediately (or very nearly so) push an equal number of electrons out the other end of the wire and onto the positive plate, thus neutralizing it. We don't have to wait for electrons to move all the way through the wire from one plate to the other. Instead, we just need to slightly rearrange the charges on the plates *and* in the wire.

Let's do a rough estimate of how much rearrangement is needed and how long the discharge takes. Using the conduction-electron density of copper in Table 30.1, we can calculate that there are 5×10^{22} conduction electrons in the wire. The negative plate in **FIGURE 30.6**, with $Q = -16$ nC, has 10^{11} excess electrons, far fewer than in the wire. In fact, the length of copper wire needed to hold 10^{11} electrons is a mere 4×10^{-13} m, only about 1% the diameter of an atom.

FIGURE 30.6 The sea of electrons needs only a minuscule rearrangement to discharge the capacitor.

1. The 10^{11} excess electrons on the negative plate move into the wire. The length of wire needed to accommodate these electrons is only 4×10^{-13} m.

3. 10^{11} electrons are pushed out of the wire and onto the positive plate. This plate is now neutral.

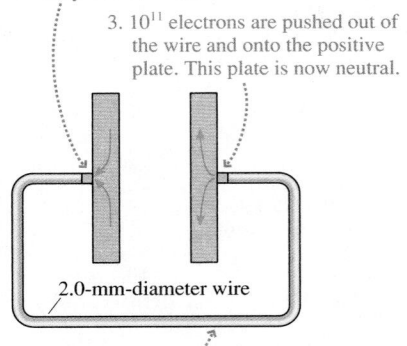

2. The sea of 5×10^{22} electrons in the wire is pushed to the side. It moves only 4×10^{-13} m, taking almost no time.

The instant the wire joins the capacitor plates together, the repulsive forces between the excess 10^{11} electrons on the negative plate cause them to push their way into the wire. As they do, 10^{11} electrons are squeezed out of the final 4×10^{-13} m of the wire and onto the positive plate. If the electrons all move together, and if they move at the typical drift speed of 10^{-4} m/s—both less than perfect assumptions but fine for making an estimate—it takes 4×10^{-9} s, or 4 ns, to move 4×10^{-13} m and discharge the capacitor. And, indeed, this is the right order of magnitude for how long the electrons take to rearrange themselves so that the capacitor plates are neutral.

STOP TO THINK 30.2 Why does the light in a room come on instantly when you flip a switch several meters away?

30.2 Creating a Current

Suppose you want to slide a book across the table to your friend. You give it a quick push to start it moving, but it begins slowing down because of friction as soon as you take your hand off. The book's kinetic energy is transformed into thermal energy, leaving the book and the table slightly warmer. The only way to keep the book moving at a constant speed is to *continue pushing it*.

As **FIGURE 30.7** shows, the sea of electrons is similar to the book. If you push the sea of electrons, you create a current of electrons moving through the conductor. But the electrons aren't moving in a vacuum. Collisions between the electrons and the atoms of the metal transform the electrons' kinetic energy into the thermal energy of the metal, making the metal warmer. (Recall that "makes the wire warm" is one of our indicators of a current.) Consequently, the sea of electrons will quickly slow down and stop *unless you continue pushing*. How do you push on electrons? With an electric field!

One of the important conclusions of Chapter 27 was that $\vec{E} = \vec{0}$ inside a conductor in electrostatic equilibrium. But a conductor with electrons moving through it is *not* in electrostatic equilibrium. **An electron current is a nonequilibrium motion of charges sustained by an internal electric field.**

FIGURE 30.7 An electron current is sustained by pushing on the sea of electrons with an electric field.

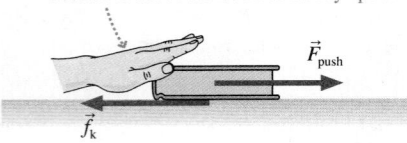

Because of friction, a steady push is needed to move the book at steady speed.

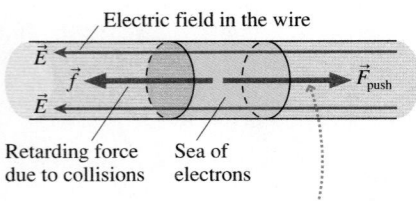

Electric field in the wire

Retarding force due to collisions | Sea of electrons

Because of collisions with atoms, a steady push is needed to move the sea of electrons at steady speed.

Thus the quick answer to "What creates a current?" is "An electric field." But why is there an electric field in a current-carrying wire?

Establishing the Electric Field in a Wire

FIGURE 30.8a shows two metal wires attached to the plates of a charged capacitor. The wires are conductors, so some of the charges on the capacitor plates become spread out along the wires as a surface charge. (Remember that all excess charge on a conductor is located on the surface.)

This is an electrostatic situation, with no current and no charges in motion. Consequently—because this is always true in electrostatic equilibrium—the electric field inside the wire is zero. Symmetry requires there to be equal amounts of charge to either side of each point to make $\vec{E} = \vec{0}$ at that point; hence the surface charge density must be uniform along each wire except near the ends (where the details need not concern us). We implied this uniform density in Figure 30.8a by drawing equally spaced + and − symbols along the wire. Remember that a positively charged surface is a surface that is *missing* electrons.

FIGURE 30.8 The surface charge on the wires before and after they are connected.

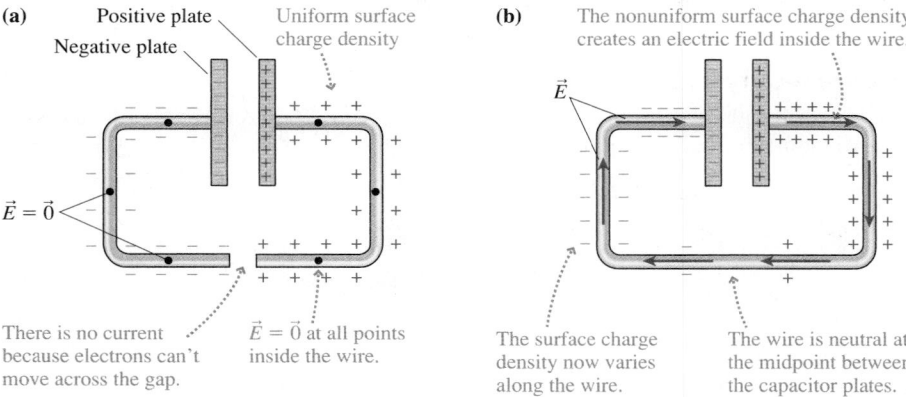

(a) Positive plate
Negative plate
Uniform surface charge density
$\vec{E} = \vec{0}$

There is no current because electrons can't move across the gap.
$\vec{E} = \vec{0}$ at all points inside the wire.

(b) The nonuniform surface charge density creates an electric field inside the wire.
$\vec{E}$

The surface charge density now varies along the wire.
The wire is neutral at the midpoint between the capacitor plates.

Now we connect the ends of the wires together. What happens? The excess electrons on the negative wire suddenly have an opportunity to move onto the positive wire that is missing electrons. Within a *very* brief interval of time ($\approx 10^{-9}$ s), the sea of electrons shifts slightly and the surface charge is rearranged into a *nonuniform* distribution like that shown in FIGURE 30.8b. The surface charge near the positive and negative plates remains strongly positive and negative because of the large amount of charge on the capacitor plates, but the midpoint of the wire, halfway between the positive and negative plates, is now electrically neutral. The new surface charge density on the wire varies from positive at the positive capacitor plate through zero at the midpoint to negative at the negative plate.

This nonuniform distribution of surface charge has an *extremely* important consequence. FIGURE 30.9 shows a section from a wire on which the surface charge density becomes more positive toward the left and more negative toward the right. Calculating

FIGURE 30.9 A varying surface charge distribution creates an internal electric field inside the wire.

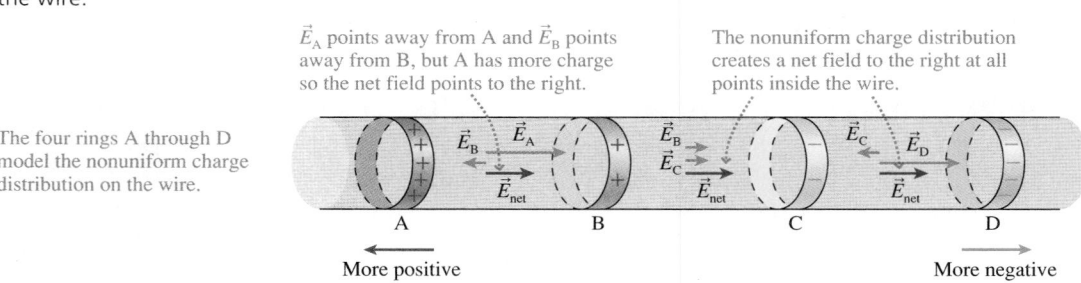

The four rings A through D model the nonuniform charge distribution on the wire.

$\vec{E}_A$ points away from A and $\vec{E}_B$ points away from B, but A has more charge so the net field points to the right.

The nonuniform charge distribution creates a net field to the right at all points inside the wire.

$\vec{E}_B$ $\vec{E}_A$ $\vec{E}_{net}$ A
$\vec{E}_B$ $\vec{E}_C$ $\vec{E}_{net}$ B
$\vec{E}_C$ $\vec{E}_D$ $\vec{E}_{net}$ C
D

More positive More negative

the exact electric field is complicated, but we can understand the basic idea if we *model* this section of wire with four circular rings of charge.

In Chapter 26, we found that the on-axis field of a ring of charge

- Points away from a positive ring, toward a negative ring;
- Is proportional to the amount of charge on the ring; and
- Decreases with distance away from the ring.

The field at the midpoint between rings A and B is well approximated as $\vec{E}_{net} \approx \vec{E}_A + \vec{E}_B$. Ring A has more charge than ring B, so $\vec{E}_{net}$ points away from A.

The analysis of Figure 30.9 leads to a very important conclusion:

> A *nonuniform* distribution of surface charges along a wire creates a net electric field *inside* the wire that points from the more positive end of the wire toward the more negative end of the wire. This is the internal electric field $\vec{E}$ that pushes the electron current through the wire.

Note that the surface charges are *not* the moving charges of the current. Further, the current—the moving charges—is *inside* the wire, not on the surface. In fact, as the next example shows, the electric field inside a current-carrying wire can be established with an extremely small amount of surface charge.

EXAMPLE 30.2 **The surface charge on a current-carrying wire**

Table 26.1 in Chapter 26 gave a typical electric field strength in a current-carrying wire as 0.01 N/C or, as we would now say, 0.01 V/m. (We'll verify this value later in this chapter.) Two 2.0-mm-diameter rings are 2.0 mm apart. They are charged to $\pm Q$. What value of Q causes the electric field at the midpoint to be 0.010 V/m?

MODEL Use the on-axis electric field of a ring of charge from Chapter 26.

VISUALIZE **FIGURE 30.10** shows the two rings. Both contribute equally to the field strength, so the electric field strength of the

FIGURE 30.10 The electric field of two charged rings.

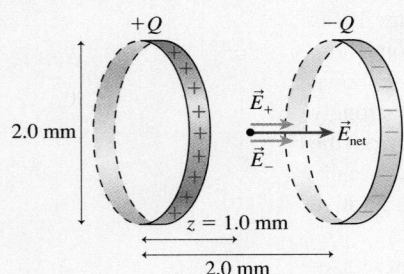

positive ring is $E_+ = 0.0050$ V/m. The distance $z = 1.0$ mm is half the ring spacing.

SOLVE Chapter 26 found the on-axis electric field of a ring of charge Q to be

$$E_+ = \frac{1}{4\pi\epsilon_0} \frac{zQ}{(z^2 + R^2)^{3/2}}$$

Thus the charge needed to produce the desired field is

$$Q = \frac{4\pi\epsilon_0 (z^2 + R^2)^{3/2}}{z} E_+$$

$$= \frac{((0.0010 \text{ m})^2 + (0.0010 \text{ m})^2)^{3/2}}{(9.0 \times 10^9 \text{ N m}^2/\text{C}^2)(0.0010 \text{ m})} (0.0050 \text{ V/m})$$

$$= 1.6 \times 10^{-18} \text{ C}$$

ASSESS The electric field of a ring of charge is largest at $z \approx R$, so these two rings are a simple but reasonable model for estimating the electric field inside a 2.0-mm-diameter wire. We find that the surface charge needed to establish the electric field is *very small*. A mere 10 electrons have to be moved from one ring to the other to charge them to $\pm 1.6 \times 10^{-18}$ C. The resulting electric field is sufficient to drive a sizable electron current through the wire.

STOP TO THINK 30.3 The two charged rings are a model of the surface charge distribution along a wire. Rank in order, from largest to smallest, the electron currents E_a to E_e at the midpoint between the rings.

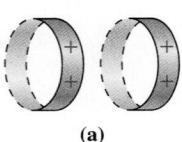

(a)

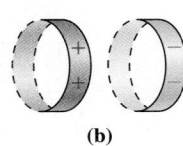

(b)

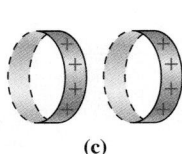

(c)

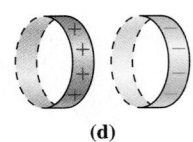

(d)

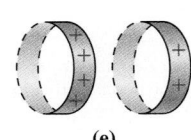

(e)

A Model of Conduction

Electrons don't just magically move through a wire as a current. They move because an electric field inside the wire—a field created by a nonuniform surface charge density on the wire—pushes on the sea of electrons to create the electron current. The field has to *keep* pushing because the electrons continuously lose energy in collisions with the positive ions that form the structure of the solid. These collisions provide a drag force, much like friction.

We will model the conduction electrons—those electrons that make up the sea of electrons—as free particles moving through the lattice of the metal. In the absence of an electric field, the electrons, like the molecules in a gas, move randomly in all directions with a distribution of speeds. If we assume that the average thermal energy of the electrons is given by the same $\frac{3}{2}k_BT$ that applies to an ideal gas, we can calculate that the average electron speed at room temperature is $\approx 10^5$ m/s. This estimate turns out, for quantum physics reasons, to be not quite right, but it correctly indicates that the conduction electrons are moving very fast.

However, an individual electron does not travel far before colliding with an ion and being scattered to a new direction. FIGURE 30.11a shows that an electron bounces back and forth between collisions, but its *average* velocity is zero, and it undergoes no *net* displacement. This is similar to molecules in a container of gas.

Suppose we now turn on an electric field. FIGURE 30.11b shows that the steady electric force causes the electrons to move along *parabolic trajectories* between collisions. Because of the curvature of the trajectories, the negatively charged electrons begin to drift slowly in the direction opposite the electric field. The motion is similar to a ball moving in a pinball machine with a slight downward tilt. An individual electron ricochets back and forth between the ions at a high rate of speed, but now there is a slow *net* motion in the "downhill" direction. Even so, this net displacement is a *very* small effect superimposed on top of the much larger thermal motion. Figure 30.11b has greatly exaggerated the rate at which the drift would occur.

Suppose an electron just had a collision with an ion and has rebounded with velocity $\vec{v}_0$. The acceleration of the electron between collisions is

$$a_x = \frac{F}{m} = \frac{eE}{m} \tag{30.4}$$

where E is the electric field strength inside the wire and m is the mass of the electron. (We'll assume that $\vec{E}$ points in the negative x-direction.) The field causes the x-component of the electron's velocity to increase linearly with time:

$$v_x = v_{0x} + a_x\,\Delta t = v_{0x} + \frac{eE}{m}\Delta t \tag{30.5}$$

The electron speeds up, with increasing kinetic energy, until its next collision with an ion. The collision transfers much of the electron's kinetic energy to the ion and thus to the thermal energy of the metal. **This energy transfer is the "friction" that raises the temperature of the wire.** The electron then rebounds, in a random direction, with a new initial velocity $\vec{v}_0$, and starts the process all over.

FIGURE 30.12a on the next page shows how the velocity abruptly changes due to a collision. Notice that the acceleration (the slope of the line) is the same before and after the collision. FIGURE 30.12b follows an electron through a series of collisions. You can see that each collision "resets" the velocity. The primary observation we can make from Figure 30.12b is that this repeated process of speeding up and colliding gives the electron a nonzero *average* velocity. **The magnitude of the electron's average velocity, due to the electric field, is the *drift speed* v_d of the electron.**

If we observe all the electrons in the metal at one instant of time, their average velocity is

$$v_d = \overline{v_x} = \overline{v_{0x}} + \frac{eE}{m}\overline{\Delta t} \tag{30.6}$$

FIGURE 30.11 A microscopic view of a conduction electron moving through a metal.

(a) No electric field

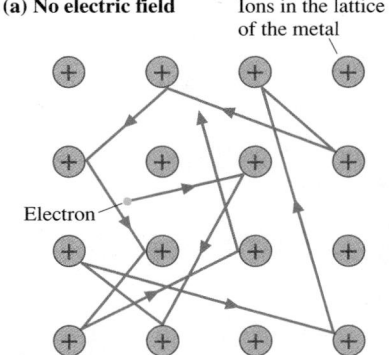

The electron has frequent collisions with ions, but it undergoes no net displacement.

(b) With an electric field

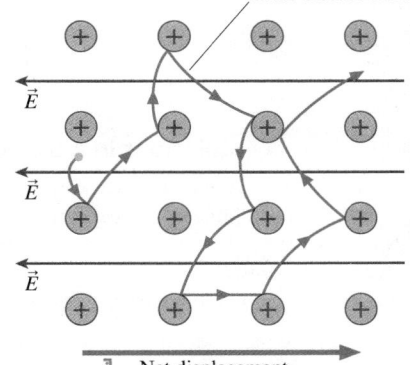

A net displacement in the direction opposite to $\vec{E}$ is superimposed on the random thermal motion.

FIGURE 30.12 The electron velocity as a function of time.

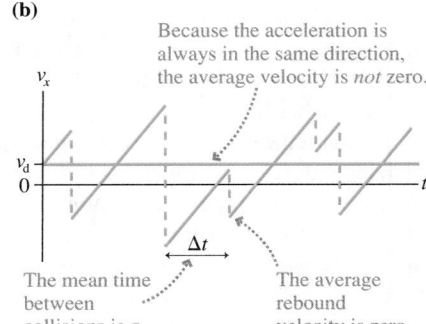

(a) The acceleration between collisions (the slope of the line) is $a = eE/m$.

The electron velocity as it collides with an ion

Collision

After the collision the acceleration is again eE/m.

The electron rebounds with velocity v_{0x}.

(b) Because the acceleration is always in the same direction, the average velocity is *not* zero.

The mean time between collisions is τ.

The average rebound velocity is zero.

where a bar over a quantity indicates an average value. The average value of v_{0x}, the velocity with which an electron rebounds after a collision, is zero. We know this because, in the absence of an electric field, the sea of electrons moves neither right nor left.

The quantity Δt is the time between collisions, so the average value of Δt is the **mean time between collisions,** which we designate τ. The mean time between collisions, analogous to the mean free path between collisions in the kinetic theory of gases, depends on the metal's temperature but can be considered a constant in the equations below.

Thus the average speed at which the electrons are pushed along by the electric field is

$$v_d = \frac{e\tau}{m}E \qquad (30.7)$$

We can complete our model of conduction by using Equation 30.7 for v_d in the electron-current equation $i_e = n_e A v_d$. Upon doing so, we find that an electric field strength E in a wire of cross-section area A causes an electron current

$$i_e = \frac{n_e e \tau A}{m}E \qquad (30.8)$$

The electron density n_e and the mean time between collisions τ are properties of the metal.

Equation 30.8 is the main result of this model of conduction. We've found that **the electron current is directly proportional to the electric field strength.** A stronger electric field pushes the electrons faster and thus increases the electron current.

EXAMPLE 30.3 **Collisions in a copper wire**

Example 30.1 found the electron current to be 2.7×10^{19} s^{-1} for a 2.0-mm-diameter copper wire in which the electron drift speed is 1.0×10^{-4} m/s. If an internal electric field of 0.020 V/m is needed to sustain this current, a typical value, how many collisions per second, on average, do electrons in copper undergo?

MODEL Use the model of conduction.

SOLVE From Equation 30.7, the mean time between collisions is

$$\tau = \frac{mv_d}{eE} = 2.8 \times 10^{-14} \text{ s}$$

The average number of collisions per second is the inverse:

$$\text{Collision rate} = \frac{1}{\tau} = 3.5 \times 10^{13} \text{ s}^{-1}$$

ASSESS This was another straightforward calculation simply to illustrate the incredibly large collision rate of conduction electrons.

30.3 Current and Current Density

We have developed the idea of a current as the motion of electrons through metals. But the properties of currents were known and used for a century before the discovery that electrons are the charge carriers in metals. We need to connect our ideas about the electron current to the conventional definition of current.

Because the coulomb is the unit of charge, and because currents are charges in motion, it seemed quite natural in the 19th century to define current as the *rate,* in coulombs per second, at which charge moves through a wire. If Q is the total amount of charge that has moved past a point in the wire, we define the current I in the wire to be the rate of charge flow:

$$I \equiv \frac{dQ}{dt} \tag{30.9}$$

For a *steady current,* which will be our primary focus, the amount of charge delivered by current I during the time interval Δt is

$$Q = I\,\Delta t \tag{30.10}$$

The SI unit for current is the coulomb per second, which is called the **ampere** A:

$$1 \text{ ampere} = 1 \text{ A} \equiv 1 \text{ coulomb per second} = 1 \text{ C/s}$$

The current unit is named after the French scientist André Marie Ampère, who made major contributions to the study of electricity and magnetism in the early 19th century. The *amp* is an informal abbreviation of ampere. Household currents are typically ≈ 1 A. For example, the current through a 100 watt lightbulb is 0.85 A, meaning that 0.85 C of charge flow through the bulb every second. Currents in consumer electronics, such as stereos and computers, are much less. They are typically measured in milliamps ($1 \text{ mA} = 10^{-3}$ A) or microamps ($1 \text{ }\mu\text{A} = 10^{-6}$ A).

Equation 30.10 is closely related to Equation 30.1, which said that the number of electrons delivered during a time interval Δt is $N_e = i_e\,\Delta t$. Each electron has charge of magnitude e; hence the total charge of N_e electrons is $Q = eN_e$. Consequently, the conventional current I and the electron current i_e are related by

$$I = \frac{Q}{\Delta t} = \frac{eN_e}{\Delta t} = ei_e \tag{30.11}$$

Because electrons are the charge carriers, the rate at which charge moves is e times the rate at which the electrons move.

EXAMPLE 30.4 **The current in a copper wire**

The electron current in the copper wire of Examples 30.1 and 30.3 was 2.7×10^{19} electrons/s. What is the current I? How much charge flows through a cross section of the wire each hour?

SOLVE The current in the wire is

$$I = ei_e = (1.60 \times 10^{-19} \text{ C})(2.7 \times 10^{19} \text{ s}^{-1}) = 4.3 \text{ A}$$

The amount of charge passing through the wire in 1 h $= 3600$ s is

$$Q = I\,\Delta t = (4.3 \text{ A})(3600 \text{ s}) = 16,000 \text{ C}$$

In one sense, the current I and the electron current i_e differ by only a scale factor. The electron current i_e, the rate at which electrons move through a wire, is more *fundamental* because it looks directly at the charge carriers. The current I, the rate at which the charge of the electrons moves through the wire, is more *practical* because we can measure charge more easily than we can count electrons.

Despite the close connection between i_e and I, there's one extremely important distinction. Because currents were known and studied before it was known what the charge carriers are, **the direction of current is *defined* to be the direction in which positive charges *seem* to move.** Thus the direction of the current I is the same as that of the internal electric field $\vec{E}$. But because the charge carriers turned out to be negative, at least for a metal, **the direction of the current I in a metal is opposite the direction of motion of the electrons.**

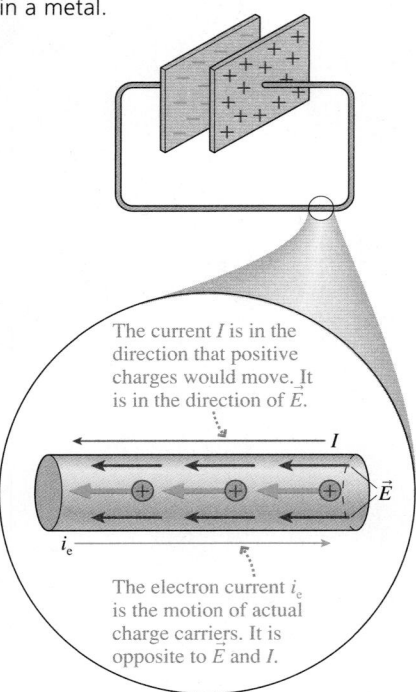

The current I is in the direction that positive charges would move. It is in the direction of $\vec{E}$.

The electron current i_e is the motion of actual charge carriers. It is opposite to $\vec{E}$ and I.

The situation shown in **FIGURE 30.13** may seem disturbing, but it makes no real difference. A capacitor is discharged regardless of whether positive charges move toward the negative plate or negative charges move toward the positive plate. The primary application of current is the analysis of circuits, and in a circuit—a macroscopic device—we simply can't tell what is moving through the wires. All of our calculations will be correct and all of our circuits will work perfectly well if we choose to think of current as the flow of positive charge. The distinction is important only at the microscopic level.

The Current Density in a Wire

We found the electron current in a wire of cross-section area A to be $i_e = n_e A v_d$. Thus the current I is

$$I = e i_e = n_e e v_d A \tag{30.12}$$

The quantity $n_e e v_d$ depends on the charge carriers and on the internal electric field that determines the drift speed, whereas A is simply a physical dimension of the wire. It will be useful to separate these quantities by defining the **current density** J in a wire as the current per square meter of cross section:

$$J = \text{current density} \equiv \frac{I}{A} = n_e e v_d \tag{30.13}$$

The current density has units of A/m². A specific piece of metal, shaped into a wire with cross-section area A, carries current $I = JA$.

EXAMPLE 30.5 | **Finding the electron drift speed**

A 1.0 A current passes through a 1.0-mm-diameter aluminum wire. What are the current density and the drift speed of the electrons in the wire?

SOLVE We can find the drift speed from the current density. The current density is

$$J = \frac{I}{A} = \frac{I}{\pi r^2} = \frac{1.0 \text{ A}}{\pi (0.00050 \text{ m})^2} = 1.3 \times 10^6 \text{ A/m}^2$$

The electron drift speed is thus

$$v_d = \frac{J}{n_e e} = 1.3 \times 10^{-4} \text{ m/s} = 0.13 \text{ mm/s}$$

where the conduction-electron density for aluminum was taken from Table 30.1.

ASSESS We earlier used 1.0×10^{-4} m/s as a typical electron drift speed. This example shows where that value comes from.

Conservation of Current

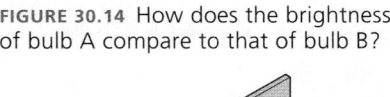
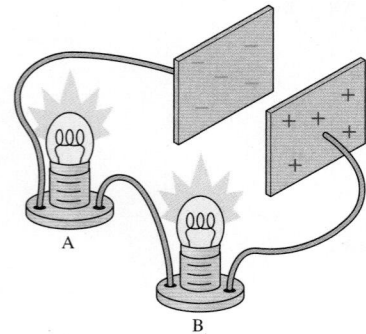

FIGURE 30.14 shows two lightbulbs in the wire connecting two charged capacitor plates. Both bulbs glow as the capacitor is discharged. How do you think the brightness of bulb A compares to that of bulb B? Is one brighter than the other? Or are they equally bright? Think about this before going on.

You might have predicted that B is brighter than A because the current I, which carries positive charges from plus to minus, reaches B first. In order to be glowing, B must use up some of the current, leaving less for A. Or perhaps you realized that the actual charge carriers are electrons, moving from minus to plus. The conventional current I may be mathematically equivalent, but physically it's the negative electrons rather than positive charge that actually move. Because the electron current gets to A first, you might have predicted that A is brighter than B.

In fact, both bulbs are equally bright. This is an important observation, one that demands an explanation. After all, "something" gets used up to make the bulb glow, so why don't we observe a decrease in the current? Current is the amount of charge moving through the wire per second. There are only two ways to decrease I: either decrease the amount of charge, or decrease the charge's drift speed through the wire. Electrons, the charge carriers, are charged particles. The lightbulb can't destroy electrons without violating both the law of conservation of mass and the law of conservation of charge. Thus the amount of charge (i.e., the *number* of electrons) cannot be changed by a lightbulb.

Do charges slow down after passing through the bulb? This is a little trickier, so consider the fluid analogy shown in FIGURE 30.15. Suppose the water flows into one end at a rate of 2.0 kg/s. Is it possible that the water, after turning a paddle wheel, flows out the other end at a rate of only 1.5 kg/s? That is, does turning the paddle wheel cause the water current to decrease?

We can't destroy water molecules any more than we can destroy electrons, we can't increase the density of water by pushing the molecules closer together, and there's nowhere to store extra water inside the pipe. Each drop of water entering the left end pushes a drop out the right end; hence water flows out at the exactly the same rate it flows in.

The same is true for electrons in a wire. **The rate of electrons leaving a lightbulb (or any other device) is exactly the same as the rate of electrons entering the lightbulb. The current does not change.** A lightbulb doesn't "use up" current, but it *does*—like the paddlewheel in the fluid analogy—use energy. The kinetic energy of the electrons is dissipated by their collisions with the ions in the lattice of the metal (the atomic-level friction) as the electrons move through the atoms, making the wire hotter until, in the case of the lightbulb filament, it glows. The lightbulb affects the amount of current *everywhere* in the wire, a process we'll examine later in the chapter, but the current doesn't change as it passes through the bulb.

There are many issues that we'll need to look at before we can say that we understand how currents work, and we'll take them one at a time. For now, we draw a first important conclusion:

Law of conservation of current The current is the same at all points in a current-carrying wire.

The law of conservation of current is really a practical application of the law of conservation of charge.

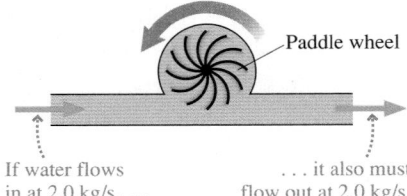

FIGURE 30.15 Water flowing through a pipe.

If water flows in at 2.0 kg/s . . .

Paddle wheel

. . . it also must flow out at 2.0 kg/s.

FIGURE 30.16 The sum of the currents into a junction must equal the sum of the currents leaving the junction.

(a)

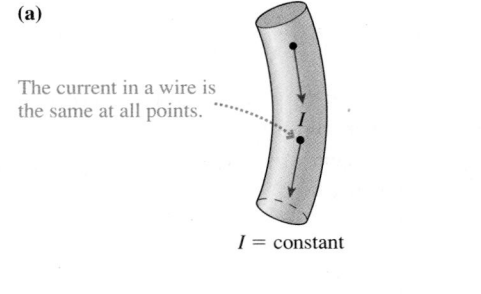

The current in a wire is the same at all points.

I

$I = \text{constant}$

(b)

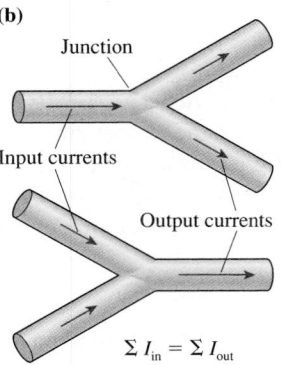

Junction

Input currents

Output currents

$\sum I_{\text{in}} = \sum I_{\text{out}}$

FIGURE 30.16a summarizes the law of conservation in a single wire. But what about FIGURE 30.16b, where two wires merge into one and another wire splits into two? A point where a wire branches is called a **junction.** The presence of a junction doesn't change our basic reasoning. We cannot create or destroy electrons in the wire, and neither can we store them in the junction. The rate at which electrons flow into one *or many* wires must be exactly balanced by the rate at which they flow out of others. For a *junction,* the law of conservation of charge requires that

$$\sum I_{\text{in}} = \sum I_{\text{out}} \qquad (30.14)$$

where, as usual, the Σ symbol means summation.

This basic conservation statement—that the sum of the currents into a junction equals the sum of the currents leaving—is called **Kirchhoff's junction law.** The junction law, together with *Kirchhoff's loop law* that you met in Chapter 29, will play an important role in circuit analysis in the next chapter.

STOP TO THINK 30.4 What are the magnitude and the direction of the current in the fifth wire?

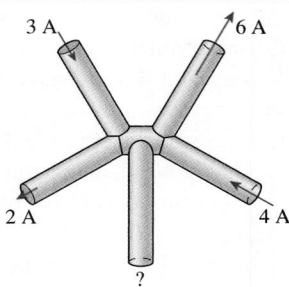

30.4 Conductivity and Resistivity

The current density $J = n_e e v_d$ is directly proportional to the electron drift speed v_d. We earlier used the microscopic model of conduction to find that the drift speed is $v_d = e\tau E/m$, where τ is the mean time between collisions and m is the mass of an electron. Combining these, we find the current density is

$$J = n_e e v_d = n_e e \left(\frac{e\tau E}{m} \right) = \frac{n_e e^2 \tau}{m} E \qquad (30.15)$$

The quantity $n_e e^2 \tau/m$ depends *only* on the conducting material. According to Equation 30.15, a given electric field strength will generate a larger current density in a material with a larger electron density n_e or longer times τ between collisions than in materials with smaller values. In other words, such a material is a *better conductor* of current.

It makes sense, then, to define the **conductivity** σ of a material as

$$\sigma = \text{conductivity} = \frac{n_e e^2 \tau}{m} \qquad (30.16)$$

Conductivity, like density, characterizes a material as a whole. All pieces of copper (at the same temperature) have the same value of σ, but the conductivity of copper is different from that of aluminum. Notice that the mean time between collisions τ can be inferred from measured values of the conductivity.

With this definition of conductivity, Equation 30.15 becomes

$$J = \sigma E \qquad (30.17)$$

This is a result of fundamental importance. Equation 30.17 tells us three things:

1. Current is caused by an electric field exerting forces on the charge carriers.
2. The current density, and hence the current $I = JA$, depends linearly on the strength of the electric field. To double the current, you must double the strength of the electric field that pushes the charges along.
3. The current density also depends on the *conductivity* of the material. Different conducting materials have different conductivities because they have different values of the electron density and, especially, different values of the mean time between electron collisions with the lattice of atoms.

The value of the conductivity is affected by the structure of a metal, by any impurities, and by the temperature. As the temperature increases, so do the thermal vibrations of the lattice atoms. This makes them "bigger targets" and causes collisions to be more frequent, thus lowering τ and decreasing the conductivity. Metals conduct better at low temperatures than at high temperatures.

For many practical applications of current it will be convenient to use the inverse of the conductivity, called the **resistivity:**

$$\rho = \text{resistivity} = \frac{1}{\sigma} = \frac{m}{n_e e^2 \tau} \qquad (30.18)$$

The resistivity of a material tells us how reluctantly the electrons move in response to an electric field. Table 30.2 gives measured values of the resistivity and conductivity for several metals and for carbon. You can see that they vary quite a bit, with copper and silver being the best two conductors.

The units of conductivity, from Equation 30.17, are those of J/E, namely $A\,C/N\,m^2$. These are clearly awkward. In the next section we will introduce a new unit called the *ohm,* symbolized by Ω (uppercase Greek omega). It will then turn out that resistivity has units of $\Omega\,m$ and conductivity has units of $\Omega^{-1}\,m^{-1}$.

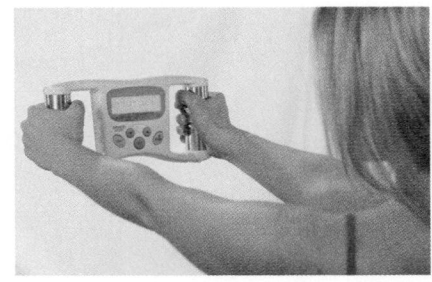

This woman is measuring her percentage body fat by gripping a device that sends a small electric current through her body. Because muscle and fat have different resistivities, the amount of current allows the fat-to-muscle ratio to be determined.

EXAMPLE 30.6 **The electric field in a wire**

A 2.0-mm-diameter aluminum wire carries a current of 800 mA. What is the electric field strength inside the wire?

SOLVE The electric field strength is

$$E = \frac{J}{\sigma} = \frac{I}{\sigma \pi r^2} = \frac{0.80\ A}{(3.5 \times 10^7\ \Omega^{-1}\,m^{-1})\pi(0.0010\ m)^2} = 0.0072\ V/m$$

where the conductivity of aluminum was taken from Table 30.2.

ASSESS This is a *very* small field in comparison with those we calculated in Chapters 25 and 26. This calculation justifies the claim in Table 26.1 that a typical electric field strength inside a current-carrying wire is ≈ 0.01 V/m. It takes *very few* surface charges on a wire to create the weak electric field necessary to push a considerable current through the wire. The reason, once again, is the enormous value of the charge-carrier density n_e. Even though the electric field is very tiny and the drift speed is agonizingly slow, a wire can carry a substantial current due to the vast number of charge carriers able to move.

TABLE 30.2 Resistivity and conductivity of conducting materials

Material	Resistivity $(\Omega\,m)$	Conductivity $(\Omega^{-1}\,m^{-1})$
Aluminum	2.8×10^{-8}	3.5×10^7
Copper	1.7×10^{-8}	6.0×10^7
Gold	2.4×10^{-8}	4.1×10^7
Iron	9.7×10^{-8}	1.0×10^7
Silver	1.6×10^{-8}	6.2×10^7
Tungsten	5.6×10^{-8}	1.8×10^7
Nichrome*	1.5×10^{-6}	6.7×10^5
Carbon	3.5×10^{-5}	2.9×10^4

*Nickel-chromium alloy used for heating wires.

Superconductivity

In 1911, the Dutch physicist Kamerlingh Onnes was studying the conductivity of metals at very low temperatures. Scientists had just recently discovered how to liquefy helium, and this opened a whole new field of *low-temperature physics.* As we noted above, metals become better conductors (i.e., they have higher conductivity and lower resistivity) at lower temperatures. But the effect is gradual. Onnes, however, found that mercury suddenly and dramatically loses *all* resistance to current when cooled below a temperature of 4.2 K. This complete loss of resistance at low temperatures is called **superconductivity.**

Later experiments established that the resistivity of a superconducting metal is not just small, it is truly zero. The electrons are moving in a frictionless environment, and charge will continue to move through a superconductor *without an electric field.* Superconductivity was not understood until the 1950s, when it was explained as being a specific quantum effect.

Superconducting wires can carry enormous currents because the wires are not heated by electrons colliding with the atoms. Very strong magnetic fields can be created with superconducting electromagnets, but applications remained limited for many decades because all known superconductors required temperatures less than 20 K. This situation changed dramatically in 1986 with the discovery of *high-temperature superconductors.* These ceramic-like materials are superconductors at temperatures as "high" as 125 K. Although $-150°C$ may not seem like a high tem-

Superconductors have unusual magnetic properties. Here a small permanent magnet levitates above a disk of the high-temperature superconductor $YBa_2Cu_3O_7$ that has been cooled to liquid-nitrogen temperature.

perature to you, the technology for producing such temperatures is simple and inexpensive. Thus many new superconductor applications are likely to appear in coming years.

STOP TO THINK 30.5 Rank in order, from largest to smallest, the current densities J_a to J_d in these four wires.

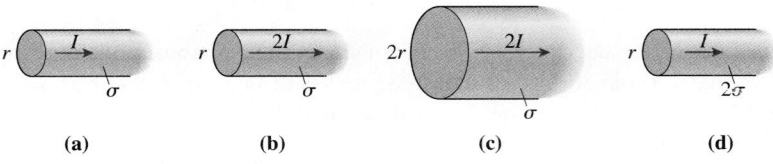

(a)   (b)   (c)   (d)

30.5 Resistance and Ohm's Law

FIGURE 30.17 The current I is related to the potential difference ΔV.

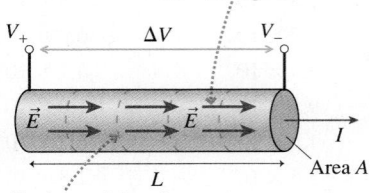

The potential difference creates an electric field inside the conductor and causes charges to flow through it.

Equipotential surfaces are perpendicular to the electric field.

FIGURE 30.17 shows a section of a conductor in which an electric field $\vec{E}$ is creating current I by pushing the charge carriers. We found in Chapter 29 that an electric field requires a potential difference. Further, the electric field points "downhill" and is perpendicular to the equipotential surfaces. Thus it should come as no surprise that current is related to potential difference.

Recall that the electric field component E_s is related to the potential by $E_s = -dV/ds$. We're interested in only the electric field strength $E = |E_s|$, so the minus sign isn't relevant. The field strength is constant inside a constant-diameter conductor (a consequence of conservation of current); thus

$$E = \frac{\Delta V}{\Delta s} = \frac{\Delta V}{L} \qquad (30.19)$$

where $\Delta V = V_+ - V_-$ is the potential difference between the ends of a conductor of length L. Equation 30.19 is an important result: The electric field strength inside a constant-diameter conductor—the field that drives the current forward—is simply the potential difference between the ends of the conductor divided by its length.

Now we can use E to find the current I in the conductor. We found earlier that the current density is $J = \sigma E$, and the current in a wire of cross-section area A is related to the current density by $I = JA$. Thus

$$I = JA = A\sigma E = \frac{A}{\rho}E \qquad (30.20)$$

where $\rho = 1/\sigma$ is the resistivity.

Combining Equations 30.19 and 30.20, we see that the current is

$$I = \frac{A}{\rho L}\Delta V \qquad (30.21)$$

That is, **the current is proportional to the potential difference between the ends of a conductor.** We can cast Equation 30.21 into a more useful form if we define the **resistance** of a conductor to be

$$R = \frac{\rho L}{A} \qquad (30.22)$$

The resistance is a property of a *specific* conductor because it depends on the conductor's length and diameter as well as on the resistivity of the material from which it is made.

The SI unit of resistance is the **ohm,** defined as

$$1 \text{ ohm} = 1 \ \Omega \equiv 1 \text{ V/A}$$

The ohm is the basic unit of resistance, although kilohms (1 k$\Omega = 10^3 \ \Omega$) and megohms (1 M$\Omega = 10^6 \ \Omega$) are widely used. You can now see from Equation 30.22 why the resistivity ρ has units of Ω m while the units of conductivity σ are $\Omega^{-1}\text{m}^{-1}$.

The resistance of a wire or conductor increases as the length increases. This seems reasonable because it should be harder to push electrons through a longer wire than a shorter one. Decreasing the cross-section area also increases the resistance. This again seems reasonable because the same electric field can push more electrons through a fat wire than a skinny one.

NOTE ▶ It is important to distinguish between resistivity and resistance. *Resistivity* describes just the *material,* not any particular piece of it. *Resistance* characterizes a specific piece of the conductor with a specific geometry. The relationship between resistivity and resistance is analogous to that between mass density and mass. ◀

The definition of resistance allows us to write the current through a conductor as

$$I = \frac{\Delta V}{R} \qquad \text{(Ohm's law)} \qquad (30.23)$$

In other words, establishing a potential difference ΔV between the ends of a conductor of resistance R creates an electric field that, in turn, causes a current $I = \Delta V/R$ through the conductor. The smaller the resistance, the larger the current. This simple relationship between potential difference and current is known as **Ohm's law.**

EXAMPLE 30.7 **The resistivity of a leaf**

Resistivity measurements on the leaves of corn plants are a good way to assess stress and the plant's overall health. To determine resistivity, the current is measured when a voltage is applied between two electrodes placed 20 cm apart on a leaf that is 2.5 cm wide and 0.20 mm thick. The following data are obtained by using several different voltages:

Voltage (V)	Current (μA)
5.0	2.3
10.0	5.1
15.0	7.5
20.0	10.3
25.0	12.2

What is the resistivity of the leaf tissue?

MODEL Model the leaf as a bar of length $L = 0.20$ m with a rectangular cross-section area $A = (0.025 \text{ m})(2.0 \times 10^{-4} \text{ m}) = 5.0 \times 10^{-6} \text{ m}^2$. The potential difference creates an electric field inside the leaf and causes a current. The current and the potential difference are related by Ohm's law.

SOLVE We can find the leaf's resistivity ρ from its resistance R. Ohm's law

$$I = \frac{1}{R} \Delta V$$

tells us that a graph of current versus potential difference should be a straight line through the origin with slope $1/R$. The graph of

the data in FIGURE 30.18 is as expected. Using the slope of the best-fit line, $0.50\ \mu$A/V, we find the leaf's resistance to be

$$R = \frac{1}{0.50\ \mu\text{A/V}} = 2.0 \times 10^6\ \frac{\text{V}}{\text{A}} = 2.0 \times 10^6\ \Omega$$

We can now use Equation 30.22 to find the resistivity:

$$\rho = \frac{AR}{L} = \frac{(5.0 \times 10^{-6} \text{ m}^2)(2.0 \times 10^6\ \Omega)}{0.20 \text{ m}} = 50\ \Omega \text{ m}$$

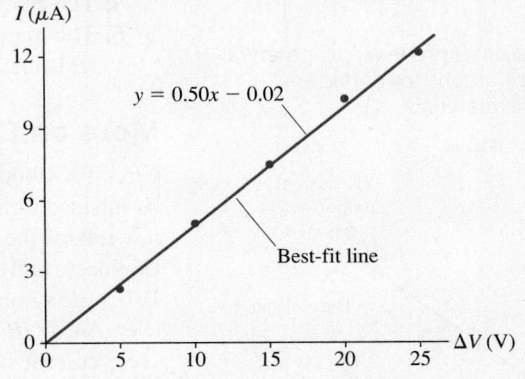

FIGURE 30.18 A graph of current versus potential difference.

ASSESS This is a huge resistivity compared to metals, but that's not surprising; the conductivity of the salty fluids in a leaf is certainly much less than that of a metal. In fact, this value is typical of the resistivities of plant and animal tissues.

Batteries and Current

Our study of current has focused on the discharge of a capacitor because we can understand where all the charges are and how they move. By contrast, we can't easily see what's happening to the charges inside a battery. Nonetheless, current in most "real" circuits is driven by a battery rather than by a capacitor. Just like the wire

FIGURE 30.19 A battery's charge escalator causes a sustained current in a wire.

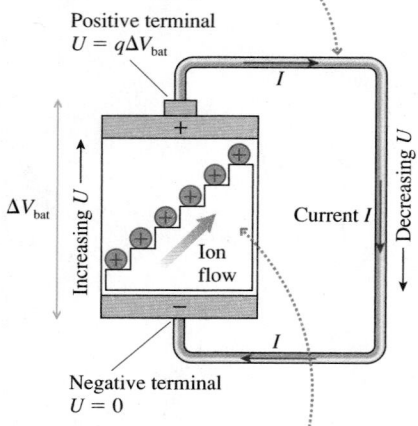

The charge "falls downhill" through the wire, but a current can be sustained because of the charge escalator.

Positive terminal
$U = q\Delta V_{bat}$

ΔV_{bat}

Increasing U

Decreasing U

Current I

I

Ion flow

I

Negative terminal
$U = 0$

The charge escalator "lifts" charge from the negative side to the positive side. Charge q gains energy $\Delta U = q\Delta V_{bat}$.

FIGURE 30.20 Current-versus-potential-difference graphs for ohmic and nonohmic materials.

(a) Ohmic material

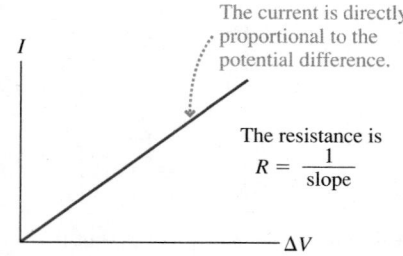

The current is directly proportional to the potential difference.

I

The resistance is
$R = \dfrac{1}{slope}$

ΔV

(b) Nonohmic materials

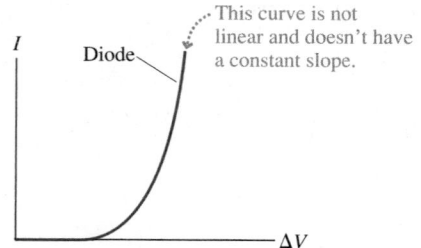

I

Diode

This curve is not linear and doesn't have a constant slope.

ΔV

discharging a capacitor, a wire connecting two battery terminals gets warm, deflects a compass needle, and makes a lightbulb glow brightly. These indicators tell us that charges flow through the wire from one terminal to the other.

The one major difference between a capacitor and a battery is the duration of the current. The current discharging a capacitor is transient, ceasing as soon as the excess charge on the capacitor plates is removed. In contrast, the current supplied by a battery is *sustained.*

We can use the charge escalator model of a battery to understand why. FIGURE 30.19 shows the charge escalator creating a potential difference ΔV_{bat} by lifting positive charge from the negative terminal to the positive terminal. Once at the positive terminal, positive charges can move *through the wire* as current I. In essence, the charges are "falling downhill" through the wire, losing the energy they gained on the escalator. This energy transfer to the wire warms the wire.

Eventually the charges find themselves back at the negative terminal of the battery, where they can ride the escalator back up and repeat the journey. A battery, unlike a charged capacitor, has an internal source of energy (the chemical reactions) that keeps the charge escalator running. It is the charge escalator that *sustains* the current in the wire by providing a continually renewed supply of charge at the battery terminals.

An important consequence of the charge escalator model, one you learned in the previous chapter, is that **a battery is a source of potential difference.** It is true that charges flow through a wire connecting the battery terminals, but current is a *consequence* of the battery's potential difference. The battery's emf is the *cause;* current, heat, light, sound, and so on are all *effects* that happen when the battery is used in certain ways.

Distinguishing cause and effect will be vitally important for understanding how a battery functions in a circuit. The reasoning is as follows:

1. A battery is a source of potential difference ΔV_{bat}. An ideal battery has $\Delta V_{bat} = \mathcal{E}$.
2. The battery creates a potential difference $\Delta V_{wire} = \Delta V_{bat}$ between the ends of a wire.
3. The potential difference ΔV_{wire} causes an electric field $E = \Delta V_{wire}/L$ in the wire.
4. The electric field establishes a current $I = JA = \sigma AE$ in the wire.
5. The magnitude of the current is determined *jointly* by the battery and the wire's resistance R to be $I = \Delta V_{wire}/R$.

More on Ohm's Law

Circuit textbooks often write Ohm's law as $V = IR$ rather than $I = \Delta V/R$. This can be misleading until you have sufficient experience with circuit analysis. First, Ohm's law relates the current to the potential *difference* between the ends of the conductor. Engineers and circuit designers *mean* "potential difference" when they use the symbol V, but the symbol is easily misinterpreted as simply "the potential." Second, $V = IR$ or even $\Delta V = IR$ suggests that a current I causes a potential difference ΔV. As you have seen, current is a *consequence* of a potential difference; hence $I = \Delta V/R$ is a better description of cause and effect.

Despite its name, Ohm's law is *not* a law of nature. It is limited to those materials whose resistance R remains constant—or very nearly so—during use. The materials to which Ohm's law applies are called *ohmic.* FIGURE 30.20a shows that the current through an ohmic material is directly proportional to the potential difference. Doubling the potential difference doubles the current. Metal and other conductors are ohmic devices.

Because the resistance of metals is small, a circuit made exclusively of metal wires would have enormous currents and would quickly deplete the battery. It is useful to limit the current in a circuit with ohmic devices, called **resistors,** whose resistance is significantly larger than the metal wires. Resistors are made with poorly conducting materials, such as carbon, or by depositing very thin metal films on an insulating substrate.

Some materials and devices are *nonohmic*, meaning that the current through the device is *not* directly proportional to the potential difference. For example, FIGURE 30.20b shows the *I*-versus-ΔV graph of a commonly used semiconductor device called a *diode*. Diodes do not have a well-defined resistance. Batteries, where $\Delta V = \mathcal{E}$ is determined by chemical reactions, and capacitors, where the relationship between *I* and ΔV differs from that of a resistor, are important nonohmic devices.

We can identify three important classes of ohmic circuit materials:

1. *Wires* are metals with very small resistivities ρ and thus very small resistances ($R \ll 1 \Omega$). An **ideal wire** has $R = 0 \Omega$; hence the potential difference between the ends of an ideal wire is $\Delta V = 0$ V *even if there is a current in it.* We will usually adopt the *ideal-wire model* of assuming that any connecting wires in a circuit are ideal.

2. *Resistors* are poor conductors with resistances usually in the range 10^1 to $10^6 \Omega$. They are used to control the current in a circuit. Most resistors in a circuit have a specified value of *R*, such as 500 Ω. The filament in a lightbulb (a tungsten wire with a high resistance due to an extremely small cross-section area *A*) functions as a resistor as long as it is glowing, but the filament is slightly nonohmic because the value of its resistance when hot is larger than its room-temperature value.

3. *Insulators* are materials such as glass, plastic, or air. An **ideal insulator** has $R = \infty \Omega$; hence there is no current in an insulator even if there is a potential difference across it ($I = \Delta V/R = 0$ A). This is why insulators can be used to hold apart two conductors at different potentials. All practical insulators have $R \gg 10^9 \Omega$ and can be treated, for our purposes, as ideal.

NOTE ▶ Ohm's law will be an important part of circuit analysis in the next chapter because resistors are essential components of almost any circuit. However, it is important that you apply Ohm's law *only* to the resistors and not to anything else. ◀

FIGURE 30.21a shows a resistor connected to a battery with current-carrying wires. Current must be conserved; hence the current *I* through the resistor is the same as the current in each wire. Because the wire's resistance is *much* less than that of the resistor, $R_{wire} \ll R_{resist}$, the potential difference $\Delta V_{wire} = IR_{wire}$ between the ends of each wire is *much* less than the potential difference $\Delta V_{resist} = IR_{resist}$ across the resistor. FIGURE 30.21b shows the potential along the wire-resistor-wire combination. You can see the large *voltage drop*, or potential difference, across the resistor. The voltage drops across the two wires are much smaller.

If we assume ideal wires with $R_{wire} = 0 \Omega$, then $\Delta V_{wire} = 0$ V and *all* the voltage drop occurs across the resistor. In this *ideal-wire model*, shown in FIGURE 30.21c, the segments of the graph corresponding to the wires are horizontal. As we begin circuit analysis in the next chapter, we will assume that all wires are ideal unless stated otherwise. Thus our analysis will be focused on the resistors.

EXAMPLE 30.8 A battery and a resistor

What resistor would have a 15 mA current if connected across the terminals of a 9.0 V battery?

MODEL Assume the resistor is connected to the battery with ideal wires.

SOLVE Connecting the resistor to the battery with ideal wires makes $\Delta V_{resist} = \Delta V_{bat} = 9.0$ V. From Ohm's law, the resistance giving a 15 mA current is

$$R = \frac{\Delta V_{resist}}{I} = \frac{9.0 \text{ V}}{0.015 \text{ A}} = 600 \Omega$$

The resistors used in circuits range from a few ohms to millions of ohms of resistance.

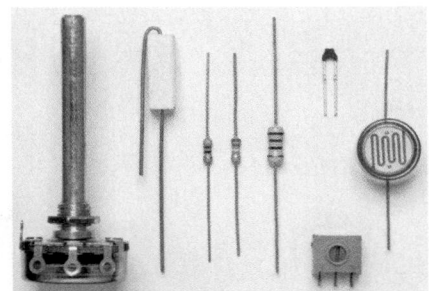

FIGURE 30.21 The potential along a wire-resistor-wire combination.

(a)

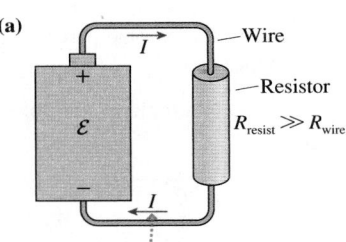

The current is constant along the wire-resistor-wire combination.

(b) The voltage drop along the wires is much less than across the resistor because the wires have much less resistance.

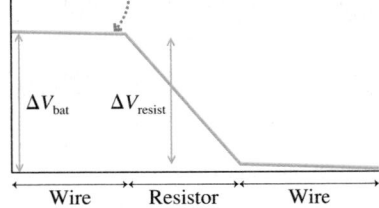

(c) In the ideal-wire model, with $R_{wire} = 0 \Omega$, there is no voltage drop along the wires. All the voltage drop is across the resistor; thus $\Delta V_{resist} = \Delta V_{bat}$.

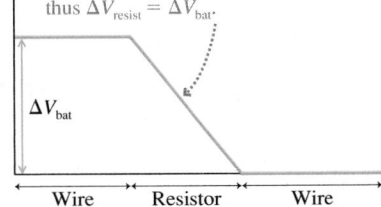

STOP TO THINK 30.6 A wire connects the positive and negative terminals of a battery. Two identical wires connect the positive and negative terminals of an identical battery. Rank in order, from largest to smallest, the currents I_a to I_d at points a to d.

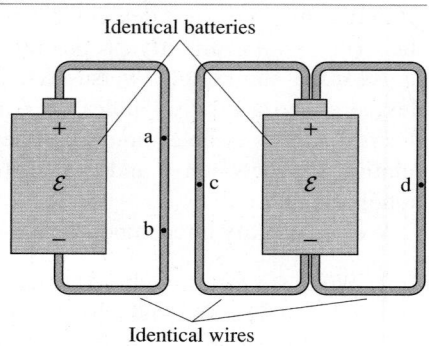

CHALLENGE EXAMPLE 30.9 **Measuring body composition**

The woman in the photo on page 879 is gripping a device that measures body fat. To illustrate how this works, FIGURE 30.22 models an upper arm as part muscle and part fat, showing the resistivities of each. Nonconductive elements, such as skin and bone, have been ignored. This is obviously not a picture of the actual structure, but gathering all the fat tissue together and all the muscle tissue together is a model that predicts the arm's electrical character quite well.

A 0.87 mA current is recorded when a 0.60 V potential difference is applied across an upper arm having the dimensions shown in the figure. What are the percentages of muscle and fat in this person's upper arm?

FIGURE 30.22 A simple model for the resistance of an arm.

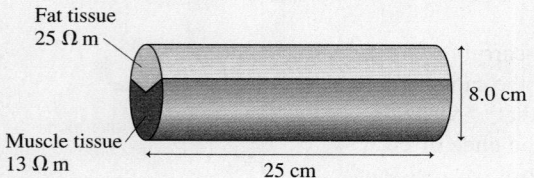

MODEL Model the muscle and the fat as separate resistors connected to a 0.60 V battery. Assume the connecting wires to be ideal, with no "loss" of potential along the wires.

VISUALIZE FIGURE 30.23 shows the circuit, with the side-by-side muscle and fat resistors connected to the two terminals of the battery.

FIGURE 30.23 Circuit for passing current through the upper arm.

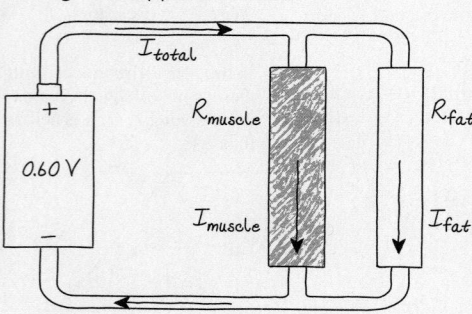

SOLVE The measured current of 0.87 mA is I_{total}, the current traveling from the battery to the arm and later back to the battery. This current splits at the junction between the two resistors. Kirchhoff's junction law, for the conservation of current, requires

$$I_{total} = I_{muscle} + I_{fat}$$

The current through each resistor can be found from Ohm's law: $I = \Delta V/R$. Each resistor has $\Delta V = 0.60$ V because each is connected to the battery terminals by lossless, ideal wires, but they have different resistances.

Let the fraction of muscle tissue be x; the fraction of fat is then $1 - x$. If the cross-section area of the upper arm is $A = \pi r^2$, then the muscle resistor has $A_{muscle} = xA$ while the fat resistor has $A_{fat} = (1 - x)A$. The resistances are related to the resistivities and the geometry by

$$R_{muscle} = \frac{\rho_{muscle} L}{A_{muscle}} = \frac{\rho_{muscle} L}{x\pi r^2}$$

$$R_{fat} = \frac{\rho_{fat} L}{A_{fat}} = \frac{\rho_{fat} L}{(1 - x)\pi r^2}$$

The currents are thus

$$I_{muscle} = \frac{\Delta V}{R_{muscle}} = \frac{x\pi r^2 \Delta V}{\rho_{muscle} L} = 0.93x \text{ mA}$$

$$I_{fat} = \frac{\Delta V}{R_{fat}} = \frac{(1 - x)\pi r^2 \Delta V}{\rho_{fat} L} = 0.48(1 - x) \text{ mA}$$

The sum of these is the total current:

$$I_{total} = 0.87 \text{mA} = 0.93x \text{ mA} + 0.48(1 - x) \text{ mA}$$

$$= (0.48 + 0.45x) \text{ mA}$$

Solving, we find $x = 0.87$. This subject's upper arm is 87% muscle tissue, 13% fat tissue.

ASSESS The percentages seem reasonable for a healthy adult. A real measurement of body fat requires a more detailed model of the human body, because the current passes through both arms and across the chest, but the principles are the same.

SUMMARY

The goal of Chapter 30 has been to learn how and why charge moves through a conductor as what we call a current.

General Principles

Current is a nonequilibrium motion of charges sustained by an electric field. Nonuniform surface charge density creates an electric field in a wire. The electric field pushes the electron current i_e in a direction opposite to $\vec{E}$. The conventional current I is in the direction in which positive charge *seems* to move.

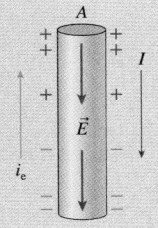

Conservation of Current

The current is the same at any two points in a wire.
At a junction,

$$\sum I_{in} = \sum I_{out}$$

This is **Kirchhoff's junction law.**

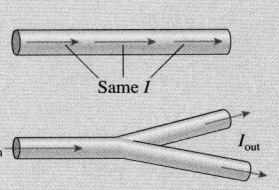

Electron current

i_e = rate of electron flow
$N_e = i_e \Delta t$

Conventional current

I = rate of charge flow = ei_e
$Q = I \Delta t$

Current density

$J = I/A$

Important Concepts

Sea of electrons

Conduction electrons move freely around the positive ions that form the atomic lattice.

Conduction

An electric field causes a slow drift at speed v_d to be superimposed on the rapid but random thermal motions of the electrons.

Collisions of electrons with the ions transfer energy to the atoms. This makes the wire warm and lightbulbs glow. More collisions mean a higher resistivity ρ and a lower conductivity σ.

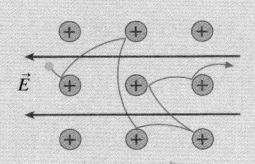

The drift speed is $v_d = \dfrac{e\tau}{m} E$, where τ is the mean time between collisions.

The electron current is related to the drift speed by

$$i_e = n_e A v_d$$

where n_e is the electron density.

An electric field E in a conductor causes a current density $J = n_e e v_d = \sigma E$, where the conductivity is

$$\sigma = \frac{n_e e^2 \tau}{m}$$

The resistivity is $\rho = 1/\sigma$.

Applications

Resistors
A potential difference ΔV_{wire} between the ends of a wire creates an electric field inside the wire:

$$E_{wire} = \frac{\Delta V_{wire}}{L}$$

The electric field causes a current in the direction of decreasing potential.

The size of the current is

$$I = \frac{\Delta V_{wire}}{R}$$

where $R = \dfrac{\rho L}{A}$ is the wire's **resistance.**

This is **Ohm's law.**

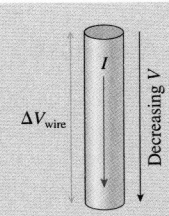

Terms and Notation

current, I	ampere, A	conductivity, σ	Ohm's law
drift speed, v_d	current density, J	resistivity, ρ	resistor
electron current, i_e	law of conservation of current	superconductivity	ideal wire
mean time between collisions, τ	junction	resistance, R	ideal insulator
	Kirchhoff's junction law	ohm, Ω	

CONCEPTUAL QUESTIONS

1. Suppose a time machine has just brought you forward from 1750 (post-Newton but pre-electricity) and you've been shown the lightbulb demonstration of FIGURE Q30.1. Do observations or *simple* measurements you might make—measurements that must make sense to you with your 1700s knowledge—prove that something is *flowing* through the wires? Or might you advance an alternative hypothesis for why the bulb is glowing? If your answer to the first question is yes, state what observations and/or measurements are relevant and the reasoning from which you can infer that something must be flowing. If not, can you offer an alternative hypothesis about why the bulb glows that could be tested?

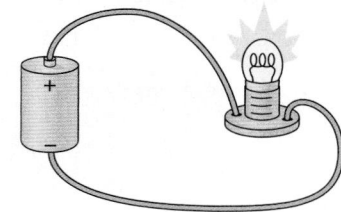

FIGURE Q30.1

2. Consider a lightbulb circuit such as the one in FIGURE Q30.1.
   a. From the simple observations and measurements you can make on this circuit, can you distinguish a current composed of positive charge carriers from a current composed of negative charge carriers? If so, describe how you can tell which it is. If not, why not?
   b. One model of current is the motion of discrete charged particles. Another model is that current is the flow of a continuous charged fluid. Do simple observations and measurements on this circuit provide evidence in favor of either one of these models? If so, describe how.

3. The electron drift speed in a wire is exceedingly slow—typically only a fraction of a millimeter per second. Yet when you turn on a flashlight switch, the light comes on almost instantly. Resolve this apparent paradox.

4. Is FIGURE Q30.4 a possible surface charge distribution for a current-carrying wire? If so, in which direction is the current? If not, why not?

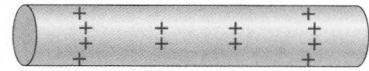

FIGURE Q30.4

5. What is the difference between current and current density?

6. All the wires in FIGURE Q30.6 are made of the same material and have the same diameter. Rank in order, from largest to smallest, the currents I_a to I_d. Explain.

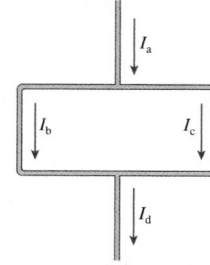

FIGURE Q30.6

7. Both batteries in FIGURE Q30.7 are identical and all lightbulbs are the same. Rank in order, from brightest to least bright, the brightness of bulbs a to c. Explain.

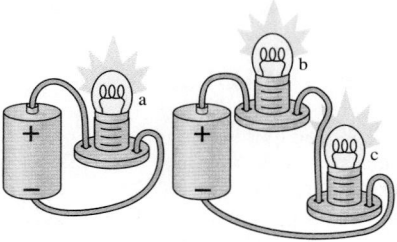

FIGURE Q30.7

8. Both batteries in FIGURE Q30.8 are identical and all lightbulbs are the same. Rank in order, from brightest to least bright, the brightness of bulbs a to c. Explain.

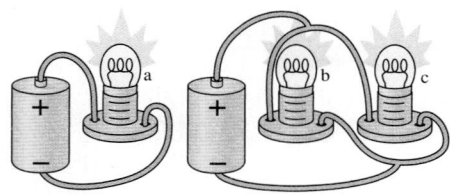

FIGURE Q30.8

9. The wire in FIGURE Q30.9 consists of two segments of different diameters but made from the same metal. The current in segment 1 is I_1.

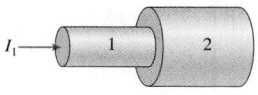

FIGURE Q30.9

   a. Compare the currents in the two segments. That is, is I_2 greater than, less than, or equal to I_1? Explain.
   b. Compare the current densities J_1 and J_2 in the two segments.
   c. Compare the electric field strengths E_1 and E_2 in the two segments.
   d. Compare the drift speeds $(v_d)_1$ and $(v_d)_2$ in the two segments.

10. The current in a wire is doubled. What happens to (a) the current density, (b) the conduction-electron density, (c) the mean time between collisions, and (d) the electron drift speed? Are each of these doubled, halved, or unchanged? Explain.

11. The wires in FIGURE Q30.11 are all made of the same material. Rank in order, from largest to smallest, the resistances R_a to R_e of these wires. Explain.

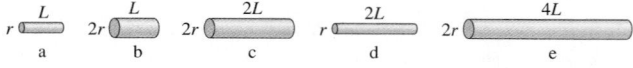

FIGURE Q30.11

12. Which, if any, of these statements are true? (More than one may be true.) Explain.
   a. A battery supplies the energy to a circuit.
   b. A battery is a source of potential difference; the potential difference between the terminals of the battery is always the same.
   c. A battery is a source of current; the current leaving the battery is always the same.

EXERCISES AND PROBLEMS

Problems labeled ▧ integrate material from earlier chapters.

Exercises

Section 30.1 The Electron Current

1. ‖ The electron drift speed in a 1.0-mm-diameter gold wire is 5.0×10^{-5} m/s. How long does it take 1 mole of electrons to flow through a cross section of the wire?

2. ‖ 1.0×10^{20} electrons flow through a cross section of a 2.0-mm-diameter iron wire in 5.0 s. What is the electron drift speed?

3. ‖ Electrons flow through a 1.6-mm-diameter aluminum wire at 2.0×10^{-4} m/s. How many electrons move through a cross section of the wire each day?

4. ‖ 1.0×10^{16} electrons flow through a cross section of silver wire in 320 μs with a drift speed of 8.0×10^{-4} m/s. What is the diameter of the wire?

Section 30.2 Creating a Current

5. | The electron drift speed is 2.0×10^{-4} m/s in a metal with a mean time between collisions of 5.0×10^{-14} s. What is the electric field strength?

6. ‖ a. How many conduction electrons are there in a 1.0-mm-diameter gold wire that is 10 cm long?
   b. How far must the sea of electrons in the wire move to deliver -32 nC of charge to an electrode?

7. ‖ The mean time between collisions in iron is 4.2×10^{-15} s. What electron current is driven through a 1.8-mm-diameter iron wire by a 0.065 V/m electric field?

8. ‖ A 2.0×10^{-3} V/m electric field creates a 3.5×10^{17} electrons/s current in a 1.0-mm-diameter aluminum wire. What are (a) the drift speed and (b) the mean time between collisions for electrons in this wire?

Section 30.3 Current and Current Density

9. | The wires leading to and from a 0.12-mm-diameter lightbulb filament are 1.5 mm in diameter. The wire to the filament carries a current with a current density of 4.5×10^5 A/m^2. What are (a) the current and (b) the current density in the filament?

10. ‖ The current in a 100 watt lightbulb is 0.85 A. The filament inside the bulb is 0.25 mm in diameter.
    a. What is the current density in the filament?
    b. What is the electron current in the filament?

11. ‖ In an integrated circuit, the current density in a 2.5-μm-thick $\times$ 75-μm-wide gold film is 7.5×10^5 A/m^2. How much charge flows through the film in 15 min?

12. | When a nerve cell fires, charge is transferred across the cell BIO membrane to change the cell's potential from negative to positive. For a typical nerve cell, 9.0 pC of charge flows in a time of 0.50 ms. What is the average current through the cell membrane?

13. | The current in an electric hair dryer is 10.0 A. How many electrons flow through the hair dryer in 5.0 min?

14. ‖ 2.0×10^{13} electrons flow through a transistor in 1.0 ms. What is the current through the transistor?

15. | In an ionic solution, 5.0×10^{15} positive ions with charge $+2e$ pass to the right each second while 6.0×10^{15} negative ions with charge $-e$ pass to the left. What is the current in the solution?

16. | A hollow copper wire with an inner diameter of 1.0 mm and an outer diameter of 2.0 mm carries a current of 10 A. What is the current density in the wire?

17. ‖ The current in a 2.0 mm $\times$ 2.0 mm square aluminum wire is 2.5 A. What are (a) the current density and (b) the electron drift speed?

Section 30.4 Conductivity and Resistivity

18. | What is the mean time between collisions for electrons in an aluminum wire and in an iron wire?

19. | The electric field in a 2.0 mm $\times$ 2.0 mm square aluminum wire is 0.012 V/m. What is the current in the wire?

20. | A 15-cm-long nichrome wire is connected across the terminals of a 1.5 V battery.
    a. What is the electric field inside the wire?
    b. What is the current density inside the wire?
    c. If the current in the wire is 2.0 A, what is the wire's diameter?

21. ‖ A 3.0-mm-diameter wire carries a 12 A current when the electric field is 0.085 V/m. What is the wire's resistivity?

22. | A 0.0075 V/m electric field creates a 3.9 mA current in a 1.0-mm-diameter wire. What material is the wire made of?

23. ‖ A 0.50-mm-diameter silver wire carries a 20 mA current. What are (a) the electric field and (b) the electron drift speed in the wire?

24. | The two segments of the wire in **FIGURE EX30.24** have equal diameters but different conductivities σ_1 and σ_2. Current I passes through this wire. If the conductivities have the ratio $\sigma_2/\sigma_1 = 2$, what is the ratio E_2/E_1 of the electric field strengths in the two segments of the wire?

FIGURE EX30.24

25. | A metal cube 1.0 cm on each side is sandwiched between two electrodes. The electrodes create a 0.0050 V/m electric field in the metal. A current of 9.0 A passes through the cube, from the positive electrode to the negative electrode. Identify the metal.

Section 30.5 Resistance and Ohm's Law

26. | A 1.5 V battery provides 0.50 A of current.
    a. At what rate (C/s) is charge lifted by the charge escalator?
    b. How much work does the charge escalator do to lift 1.0 C of charge?
    c. What is the power output of the charge escalator?

27. ‖ Wires 1 and 2 are made of the same metal. Wire 2 has twice the length and twice the diameter of wire 1. What are the ratios (a) ρ_2/ρ_1 of the resistivities and (b) R_2/R_1 of the resistances of the two wires?

28. | What is the resistance of
    a. A 2.0-m-long gold wire that is 0.20 mm in diameter?
    b. A 10-cm-long piece of carbon with a 1.0 mm $\times$ 1.0 mm square cross section?

29. ‖ A 10-m-long wire with a diameter of 0.80 mm has a resistance of 1.1 Ω. Of what material is the wire made?

30. | The electric field inside a 30-cm-long copper wire is 5.0 mV/m. What is the potential difference between the ends of the wire?

31. | a. How long must a 0.60-mm-diameter aluminum wire be to have a 0.50 A current when connected to the terminals of a 1.5 V flashlight battery?
    b. What is the current if the wire is half this length?

32. ‖ The terminals of a 0.70 V watch battery are connected by a 100-m-long gold wire with a diameter of 0.10 mm. What is the current in the wire?

33. | The femoral artery is the large artery that carries blood to the
BIO leg. What is the resistance of a 20-cm-long column of blood in a 1.0-cm-diameter femoral artery? The conductivity of blood is 0.63 $\Omega^{-1}\,m^{-1}$.

34. ‖ Pencil "lead" is actually carbon. What is the current if a 9.0 V potential difference is applied between the ends of a 0.70-mm-diameter, 6.0-cm-long lead from a mechanical pencil?

35. ‖ The resistance of a very fine aluminum wire with a 10 μm × 10 μm square cross section is 1000 Ω. A 1000 Ω resistor is made by wrapping this wire in a spiral around a 3.0-mm-diameter glass core. How many turns of wire are needed?

36. | **FIGURE EX30.36** is a current-versus-potential-difference graph for a material. What is the material's resistance?

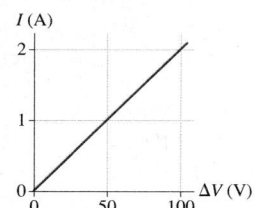

FIGURE EX30.36

37. ‖ A circuit calls for a 0.50-mm-diameter copper wire to be stretched between two points. You don't have any copper wire, but you do have aluminum wire in a wide variety of diameters. What diameter aluminum wire will provide the same resistance?

Problems

38. ‖ For what electric field strength would the current in a 2.0-mm-diameter nichrome wire be the same as the current in a 1.0-mm-diameter aluminum wire in which the electric field strength is 0.0080 V/m?

39. ‖ You've been asked to determine whether a new material your company has made is ohmic and, if so, to measure its electrical conductivity. Taking a 0.50 mm × 1.0 mm × 45 mm sample, you wire the ends of the long axis to a power supply and then measure the current for several different potential differences. Your data are as follows:

Voltage (V)	Current (A)
0.200	0.47
0.400	1.06
0.600	1.53
0.800	1.97

Use an appropriate graph of the data to determine whether the material is ohmic and, if so, its conductivity.

40. ‖ The electron beam inside a television picture tube is 0.40 mm in diameter and carries a current of 50 μA. This electron beam impinges on the inside of the picture tube screen.
    a. How many electrons strike the screen each second?
    b. What is the current density in the electron beam?

c. The electrons move with a velocity of 4.0×10^7 m/s. What electric field strength is needed to accelerate electrons from rest to this velocity in a distance of 5.0 mm?

d. Each electron transfers its kinetic energy to the picture tube screen upon impact. What is the *power* delivered to the screen by the electron beam?

41. ‖ **FIGURE P30.41** shows a 4.0-cm-wide plastic film being wrapped onto a 2.0-cm-diameter roller that turns at 90 rpm. The plastic has a uniform surface charge density of -2.0 nC/cm^2.

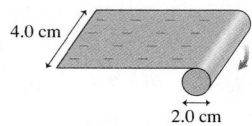

FIGURE P30.41

a. What is the current of the moving film?
b. How long does it take the roller to accumulate a charge of $-10\ \mu$C?

42. ‖ A sculptor has asked you to help electroplate gold onto a brass statue. You know that the charge carriers in the ionic solution are gold ions, and you've calculated that you must deposit 0.50 g of gold to reach the necessary thickness. How much current do you need, in mA, to plate the statue in 3.0 hours?

43. ‖ In a classic model of the hydrogen atom, the electron moves around the proton in a circular orbit of radius 0.053 nm.
    a. What is the electron's orbital frequency?
    b. What is the effective current of the electron?

44. | The biochemistry that takes place inside cells depends on vari-
BIO ous elements, such as sodium, potassium, and calcium, that are dissolved in water as ions. These ions enter cells through narrow pores in the cell membrane known as *ion channels*. Each ion channel, which is formed from a specialized protein molecule, is selective for one type of ion. Measurements with microelectrodes have shown that a 0.30-nm-diameter potassium ion (K$^+$) channel carries a current of 1.8 pA.
    a. How many potassium ions pass through if the ion channel opens for 1.0 ms?
    b. What is the current density in the ion channel?

45. ‖ The starter motor of a car engine draws a current of 150 A from the battery. The copper wire to the motor is 5.0 mm in diameter and 1.2 m long. The starter motor runs for 0.80 s until the car engine starts.
    a. How much charge passes through the starter motor?
    b. How far does an electron travel along the wire while the starter motor is on?

46. | A car battery is rated at 90 A h, meaning that it can supply a 90 A current for 1 h before being completely discharged. If you leave your headlights on until the battery is completely dead, how much charge leaves the battery?

47. ‖ Variations in the resistivity of blood can give valuable clues
BIO about changes in various properties of the blood. Suppose a medical device attaches two electrodes into a 1.5-mm-diameter vein at positions 5.0 cm apart. What is the blood resistivity if a 9.0 V potential difference causes a 230 μA current through the blood in the vein?

48. ‖ The conducting path between the right hand and the left hand
BIO can be modeled as a 10-cm-diameter, 160-cm-long cylinder. The average resistivity of the interior of the human body is 5.0 Ω m. Dry skin has a much higher resistivity, but skin resistance can be made negligible by soaking the hands in salt water. If skin resistance is neglected, what potential difference between the hands is needed for a lethal shock of 100 mA across the chest? Your result shows that even small potential differences can produce dangerous currents when the skin is wet.

49. ‖ You need to design a 1.0 A fuse that "blows" if the current exceeds 1.0 A. The fuse material in your stockroom melts at a current density of 500 A/cm². What diameter wire of this material will do the job?

50. ‖ A hollow metal cylinder has inner radius a, outer radius b, length L, and conductivity σ. The current I is *radially* outward from the inner surface to the outer surface.
   a. Find an expression for the electric field strength inside the metal as a function of the radius r from the cylinder's axis.
   b. Evaluate the electric field strength at the inner and outer surfaces of an iron cylinder if $a = 1.0$ cm, $b = 2.5$ cm, $L = 10$ cm, and $I = 25$ A.

51. ‖ A hollow metal sphere has inner radius a, outer radius b, and conductivity σ. The current I is *radially* outward from the inner surface to the outer surface.
   a. Find an expression for the electric field strength inside the metal as a function of the radius r from the center.
   b. Evaluate the electric field strength at the inner and outer surfaces of a copper sphere if $a = 1.0$ cm, $b = 2.5$ cm, and $I = 25$ A.

52. ‖ The total amount of charge in coulombs that has entered a wire at time t is given by the expression $Q = 4t - t^2$, where t is in seconds and $t \geq 0$.
   a. Find an expression for the current in the wire at time t.
   b. Graph I versus t for the interval $0 \leq t \leq 4$ s.

53. ‖ The total amount of charge that has entered a wire at time t is given by the expression $Q = (20 \text{ C})(1 - e^{-t/(2.0 \text{ s})})$, where t is in seconds and $t \geq 0$.
   a. Find an expression for the current in the wire at time t.
   b. What is the maximum value of the current?
   c. Graph I versus t for the interval $0 \leq t \leq 10$ s.

54. ‖ The current in a wire at time t is given by the expression $I = (2.0 \text{ A})e^{-t/(2.0 \text{ }\mu\text{s})}$, where t is in microseconds and $t \geq 0$.
   a. Find an expression for the total amount of charge (in coulombs) that has entered the wire at time t. The initial conditions are $Q = 0$ C at $t = 0$ μs.
   b. Graph Q versus t for the interval $0 \leq t \leq 10$ μs.

55. ‖ The two wires in FIGURE P30.55 are made of the same material. What are the current and the electron drift speed in the 2.0-mm-diameter segment of the wire?

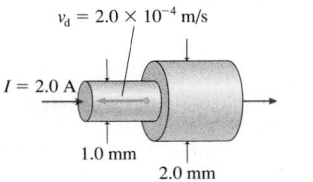

FIGURE P30.55

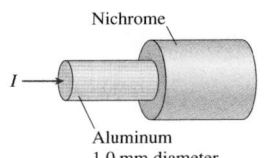

FIGURE P30.57

56. ‖ What is the electron drift speed at the 3.0-mm-diameter end (the left end) of the wire in FIGURE P30.56?

57. | What diameter should the nichrome wire in FIGURE P30.57 be in order for the electric field strength to be the same in both wires?

58. ‖ An aluminum wire consists of the three segments shown in FIGURE P30.58. The current in the top segment is 10 A. For each of these three segments, find the
   a. Current I.
   b. Current density J.
   c. Electric field E.
   d. Drift velocity v_d.
   e. Electron current i.
   Place your results in a table for easy viewing.

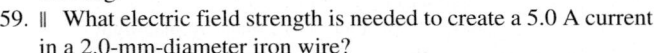

FIGURE P30.58

59. ‖ What electric field strength is needed to create a 5.0 A current in a 2.0-mm-diameter iron wire?

60. ‖ A 20-cm-long hollow nichrome tube of inner diameter 2.8 mm, outer diameter 3.0 mm is connected to a 3.0 V battery. What is the current in the tube?

61. ‖ The batteries in FIGURE P30.61 are identical. Both resistors have equal currents. What is the resistance of the resistor on the right?

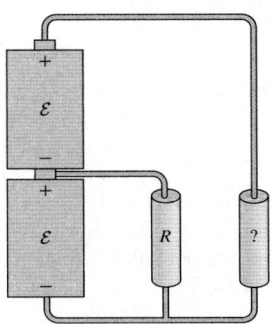

FIGURE P30.61

62. ‖ A 1.5 V flashlight battery is connected to a wire with a resistance of 3.0 Ω. FIGURE P30.62 shows the battery's potential difference as a function of time. What is the total charge lifted by the charge escalator?

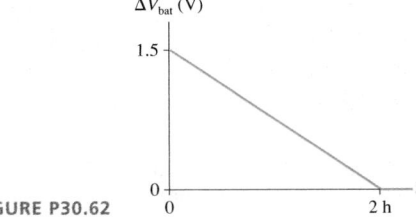

FIGURE P30.62

63. ‖ Two 10-cm-diameter metal plates 1.0 cm apart are charged to ± 12.5 nC. They are suddenly connected together by a 0.224-mm-diameter copper wire stretched taut from the center of one plate to the center of the other.
   a. What is the maximum current in the wire?
   b. Does the current increase with time, decrease with time, or remain steady? Explain.
   c. What is the total amount of energy dissipated in the wire?

64. ‖ A long, round wire has resistance R. What will the wire's resistance be if you stretch it to twice its initial length?

65. ‖ **FIGURE P30.65** shows the potential along a tungsten wire. What is the current density in the wire?

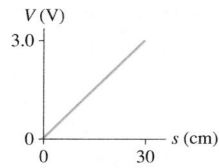

FIGURE P30.65

66. ‖ Household wiring often uses 2.0-mm-diameter copper wires. The wires can get rather long as they snake through the walls from the fuse box to the farthest corners of your house. What is the potential difference across a 20-m-long, 2.0-mm-diameter copper wire carrying an 8.0 A current?

67. ‖ You've decided to protect your house by placing a 5.0-m-tall iron lightning rod next to the house. The top is sharpened to a point and the bottom is in good contact with the ground. From your research, you've learned that lightning bolts can carry up to 50 kA of current and last up to 50 μs.
   a. How much charge is delivered by a lightning bolt with these parameters?
   b. You don't want the potential difference between the top and bottom of the lightning rod to exceed 100 V. What minimum diameter must the rod have?

Challenge Problems

68. The conductive tissues of the upper leg can be modeled as a 40-cm-long, 12-cm-diameter cylinder of muscle and fat. The resistivities of muscle and fat are 13 Ω m and 25 Ω m, respectively. One person's upper leg is 82% muscle, 18% fat. What current is measured if a 1.5 V potential difference is applied between the person's hip and knee?
   BIO

69. The current supplied by a battery slowly decreases as the battery runs down. Suppose that the current as a function of time is $I = (0.75 \text{ A})e^{-t/(6\text{ h})}$. What is the total number of electrons transported from the positive electrode to the negative electrode by the charge escalator from the time the battery is first used until it is completely dead?

70. The electric field in a current-carrying wire can be modeled as the electric field at the midpoint between two charged rings. Model a 3.0-mm-diameter aluminum wire as two 3.0-mm-diameter rings 2.0 mm apart. What is the current in the wire after 20 electrons are transferred from one ring to the other?

71. A 5.0-mm-diameter proton beam carries a total current of 1.5 mA. The current density in the proton beam, which increases with distance from the center, is given by $J = J_{\text{edge}}(r/R)$, where R is the radius of the beam and J_{edge} is the current density at the edge.
   a. How many protons per second are delivered by this proton beam?
   b. Determine the value of J_{edge}.

72. A metal wire connecting the terminals of a battery with potential difference ΔV_{bat} gets warm as it draws a current I.
   a. What is ΔU, the change in potential energy of charge Q as it passes through the wire?
   b. Where does this energy go?
   c. Power is the *rate* of transfer of energy. Based on your answer to part a, find an expression for the power supplied by the battery to warm the wire.
   d. What power does a 1.5 V battery supply to a wire drawing a 1.2 A current?

73. **FIGURE CP30.73** shows a wire that is made of two equal-diameter segments with conductivities σ_1 and σ_2. When current I passes through the wire, a thin layer of charge appears at the boundary between the segments.

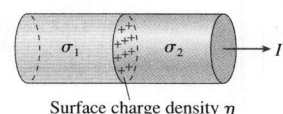

FIGURE CP30.73

   a. Find an expression for the surface charge density η on the boundary. Give your result in terms of I, σ_1, σ_2, and the wire's cross-section area A.
   b. A 1.0-mm-diameter wire made of copper and iron segments carries a 5.0 A current. How much charge accumulates at the boundary between the segments?

STOP TO THINK ANSWERS

Stop to Think 30.1: $i_c > i_b > i_a > i_d$. The electron current is proportional to $r^2 v_d$. Changing r by a factor of 2 has more influence than changing v_d by a factor of 2.

Stop to Think 30.2: The electrons don't have to move from the switch to the bulb, which could take hours. Because the wire between the switch and the bulb is already full of electrons, a flow of electrons from the switch into the wire immediately causes electrons to flow from the other end of the wire into the lightbulb.

Stop to Think 30.3: $E_d > E_b > E_e > E_a = E_c$. The electric field strength depends on the *difference* in the charge on the two wires. The electric fields of the rings in a and c are opposed to each other, so the net field is zero. The rings in d have the largest charge *difference*.

Stop to Think 30.4: 1 A into the junction. The total current entering the junction must equal the total current leaving the junction.

Stop to Think 30.5: $J_b > J_a = J_d > J_c$. The current density $J = I/\pi r^2$ is independent of the conductivity σ, so a and d are the same. Changing r by a factor of 2 has more influence than changing I by a factor of 2.

Stop to Think 30.6: $I_a = I_b = I_c = I_d$. Conservation of current requires $I_a = I_b$. The current in each wire is $I = \Delta V_{\text{wire}}/R$. All the wires have the same resistance because they are identical, and they all have the same potential difference because each is connected directly to the battery, which is a *source of potential*.

31 Fundamentals of Circuits

A microprocessor, the heart of a powerful computer, is a complex device. Even so, a microprocessor operates on the basis of just a few fundamental physical principles.

▶ **Looking Ahead** The goal of Chapter 31 is to understand the fundamental physical principles that govern electric circuits.

DC Circuits

Circuits—from a simple lightbulb to a supercomputer—are based on the controlled motion of charges. You will learn about the fundamental physical principles by which circuits operate.

This chapter will focus on **DC circuits,** meaning *direct current*, in which potentials and currents are steady. Chapter 35 will extend these ideas to AC circuits in which the potential difference oscillates sinusoidally.

◀ **Looking Back**
Section 29.2 Sources of potential

Analyzing Circuits

Circuits consist of many elements—batteries, resistors, capacitors, and more—connected together. Two basic tools will help you find the potential difference across and current through each element:

- Kirchhoff's junction law.
- Kirchhoff's loop law.

Also important will be Ohm's law, for resistors, and the properties of batteries and capacitors.

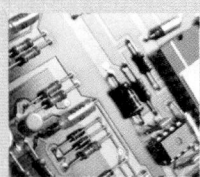

Energy and Power

Circuits do things by using energy. You'll learn to calculate *power*, the rate at which the battery supplies energy to a circuit and the rate at which a resistor dissipates it.

The power delivered by these photovoltaic cells is the product of their emf and the current they deliver. One solar panel provides about 200 W at midday on a sunny day.

Circuit Diagrams

You will learn how to use symbols of circuit elements to draw a **circuit diagram.** This is a logical picture of how the circuit elements are related rather than a literal picture of how they look.

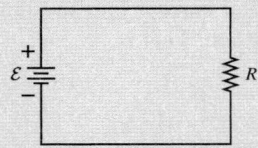

This is the circuit diagram of a simple circuit in which a resistor is connected to a battery.

Combining Resistors

Resistors often occur **in series** or **in parallel.**

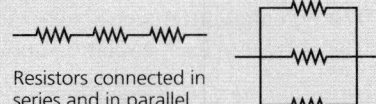

Resistors connected in series and in parallel

You'll learn that these combinations of resistors can be "simplified" by replacing them with one **equivalent resistor.**

◀ **Looking Back**
Sections 30.3–30.5 Current, resistance, and Ohm's law

RC Circuits

A capacitor is charged or discharged by current through a resistor. These important circuits are called *RC* **circuits.** Applications range from defibrillators to timing circuits.

You'll learn that the capacitor charge decays exponentially. The time to decay to e^{-1} of the initial value is called the time constant τ.

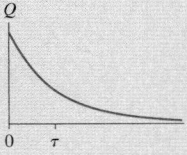

◀ **Looking Back**
Section 29.5 Capacitors

31.1 Circuit Elements and Diagrams

FIGURE 31.1 An electric circuit.

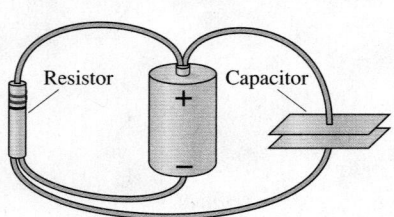

FIGURE 31.1 shows an electric circuit in which a resistor and a capacitor are connected by wires to a battery. To understand the functioning of this circuit, we do not need to know whether the wires are bent or straight, or whether the battery is to the right or to the left of the resistor. The literal picture of Figure 31.1 provides many irrelevant details. It is customary when describing or analyzing circuits to use a more abstract picture called a **circuit diagram**. A circuit diagram is a *logical* picture of what is connected to what.

A circuit diagram also replaces pictures of the circuit elements with symbols. FIGURE 31.2 shows the basic symbols that we will need. The longer line at one end of the battery symbol represents the positive terminal of the battery. Notice that a lightbulb, like a wire or a resistor, has two "ends," and current passes *through* the bulb. It is often useful to think of a lightbulb as a resistor that gives off light when a current is present. A lightbulb filament is not a perfectly ohmic material, but the resistance of a *glowing* lightbulb remains reasonably constant if you don't change ΔV by much.

FIGURE 31.2 A library of basic symbols used for electric circuit drawings.

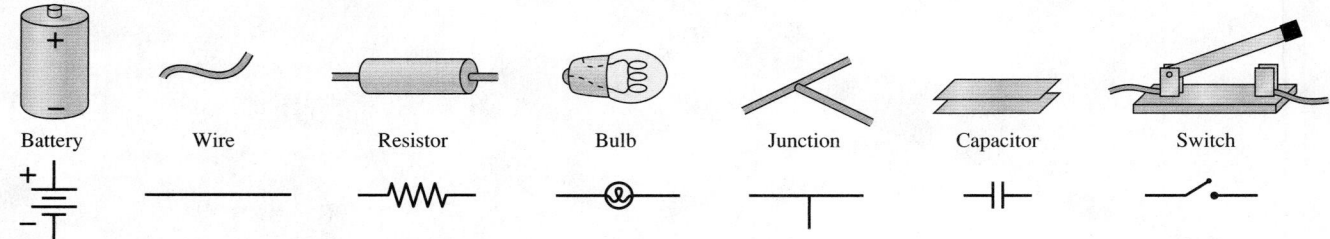

FIGURE 31.3 A circuit diagram for the circuit of Figure 31.1.

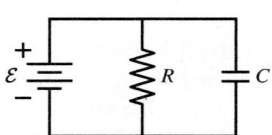

FIGURE 31.3 is a circuit diagram of the circuit shown in Figure 31.1. Notice how the circuit elements are labeled. The battery's emf $\mathcal{E}$ is shown beside the battery, and $+$ and $-$ symbols, even though somewhat redundant, are shown beside the terminals. We would use numerical values for $\mathcal{E}$, R, and C if we knew them. The wires, which in practice may bend and curve, are shown as straight-line connections between the circuit elements.

STOP TO THINK 31.1 Which of these diagrams represent the same circuit?

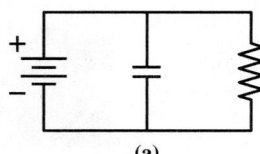
(a)

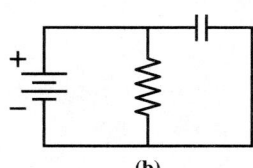
(b)

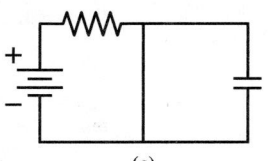
(c)

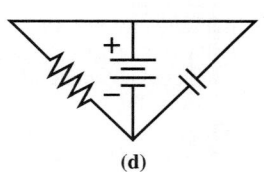
(d)

31.2 Kirchhoff's Laws and the Basic Circuit

We are now ready to begin analyzing circuits. To analyze a circuit means to find:

1. The potential difference across each circuit component.
2. The current in each circuit component.

FIGURE 31.4 Kirchhoff's junction law.

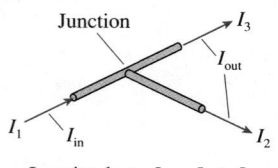

Junction law: $I_1 = I_2 + I_3$

Because charge and current are conserved, the total current into the junction of FIGURE 31.4 must equal the total current leaving the junction. That is,

$$\sum I_{\text{in}} = \sum I_{\text{out}} \qquad (31.1)$$

This statement, which you met in Chapter 30, is **Kirchhoff's junction law.**

An important property of the electric potential is that the sum of the potential differences around any loop or closed path is zero. This is a statement of energy conservation, because a charge that moves around a closed path and returns to its starting point has $\Delta U = 0$. We apply this idea to the circuit of FIGURE 31.5 by adding all of the potential differences *around* the loop formed by the circuit. Doing so gives

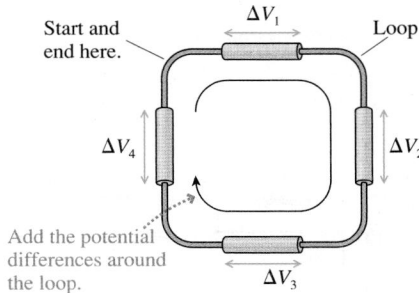

FIGURE 31.5 Kirchhoff's loop law.

$$\Delta V_{\text{loop}} = \sum (\Delta V)_i = 0 \qquad (31.2)$$

where $(\Delta V)_i$ is the potential difference of the *i*th component in the loop. This statement, introduced in Chapter 29, is **Kirchhoff's loop law.**

Kirchhoff's loop law can be true only if at least one of the $(\Delta V)_i$ is negative. To apply the loop law, we need to explicitly identify which potential differences are positive and which are negative.

Loop law: $\Delta V_1 + \Delta V_2 + \Delta V_3 + \Delta V_4 = 0$

TACTICS BOX 31.1 **Using Kirchhoff's loop law** (MP)

❶ **Draw a circuit diagram.** Label all known and unknown quantities.
❷ **Assign a direction to the current.** Draw and label a current arrow *I* to show your choice.

- ■ If you know the actual current direction, choose that direction.
- ■ If you don't know the actual current direction, make an arbitrary choice. All that will happen if you choose wrong is that your value for *I* will end up negative.

❸ **"Travel" around the loop.** Start at any point in the circuit, then go all the way around the loop in the direction you assigned to the current in step 2. As you go through each circuit element, ΔV is interpreted to mean

$$\Delta V = V_{\text{downstream}} - V_{\text{upstream}}$$

- ■ For an ideal battery in the negative-to-positive direction:

$$\Delta V_{\text{bat}} = +\mathcal{E}$$

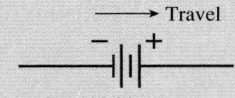

Potential increases

- ■ For an ideal battery in the positive-to-negative direction:

$$\Delta V_{\text{bat}} = -\mathcal{E}$$

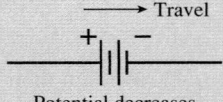

Potential decreases

- ■ For a resistor: $\Delta V_{\text{res}} = -\Delta V_R = -IR$

Potential decreases

❹ **Apply the loop law:** $\sum (\Delta V)_i = 0$

Exercises 4–7

NOTE ▶ Ohm's law gives us only the *magnitude* $\Delta V_R = IR$ of the potential difference across a resistor. Kirchhoff's law requires us to recognize that the electric potential inside a resistor *decreases* in the direction of the current. Thus $\Delta V_{\text{res}} = V_{\text{downstream}} - V_{\text{upstream}} = -\Delta V_R.$ ◀

The Basic Circuit

The most basic electric circuit is a single resistor connected to the two terminals of a battery. FIGURE 31.6a on the next page shows a literal picture of the circuit elements and the connecting wires; FIGURE 31.6b is the circuit diagram. Notice that this is a **complete circuit,** forming a continuous path between the battery terminals.

FIGURE 31.6 The basic circuit of a resistor connected to a battery.

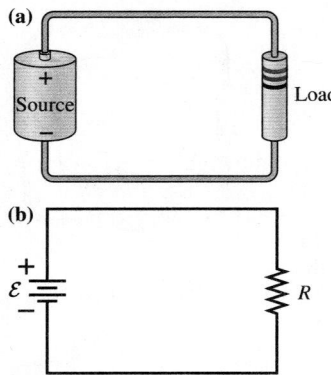

The resistor might be a known resistor, such as "a 10 Ω resistor," or it might be some other resistive device, such as a lightbulb. Regardless of what the resistor is, it is called the **load**. The battery is called the **source.**

FIGURE 31.7 shows the use of Kirchhoff's loop law to analyze this circuit. Two things are worth noting:

1. This circuit has no junctions, so the current I is the same in all four sides of the circuit. Kirchhoff's junction law is not needed.
2. We're assuming the ideal-wire model, in which there are *no* potential differences along the connecting wires.

Kirchhoff's loop law, with two circuit elements, is

$$\Delta V_{loop} = \sum(\Delta V)_i = \Delta V_{bat} + \Delta V_{res} = 0 \tag{31.3}$$

Let's look at each of the two terms in Equation 31.3:

1. The potential *increases* as we travel through the battery on our clockwise journey around the loop. We enter the negative terminal and, farther downstream, exit the positive terminal after having gained potential $\mathcal{E}$. Thus

$$\Delta V_{bat} = +\mathcal{E}$$

2. The potential of a conductor *decreases* in the direction of the current, which we've indicated with the + and − signs in Figure 31.7. Thus

$$\Delta V_{res} = V_{downstream} - V_{upstream} = -IR$$

FIGURE 31.7 Analysis of the basic circuit using Kirchhoff's loop law.

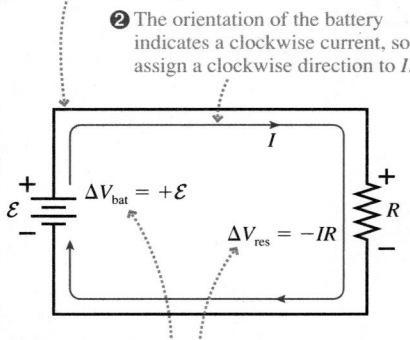

❶ Draw a circuit diagram.

❷ The orientation of the battery indicates a clockwise current, so assign a clockwise direction to I.

❸ Determine ΔV for each circuit element.

NOTE ▶ Determining which potential differences are positive and which are negative is perhaps *the* most important step in circuit analysis. ◀

With this information, the loop equation becomes

$$\mathcal{E} - IR = 0 \tag{31.4}$$

We can solve the loop equation to find that the current in the circuit is

$$I = \frac{\mathcal{E}}{R} \tag{31.5}$$

We can then use the current to find that the magnitude of the resistor's potential difference is

$$\Delta V_R = IR = \mathcal{E} \tag{31.6}$$

This result should come as no surprise. The potential energy that the charges gain in the battery is subsequently lost as they "fall" through the resistor.

NOTE ▶ The current that the battery delivers depends jointly on the emf of the battery and the resistance of the load. ◀

Two resistors and two batteries

Analyze the circuit shown in FIGURE 31.8.

a. Find the current in and the potential difference across each resistor.
b. Draw a graph showing how the potential changes around the circuit, starting from $V = 0$ V at the negative terminal of the 6 V battery.

MODEL Assume ideal connecting wires and ideal batteries, for which $\Delta V_{bat} = \mathcal{E}$.

FIGURE 31.8 Circuit for Example 31.1.

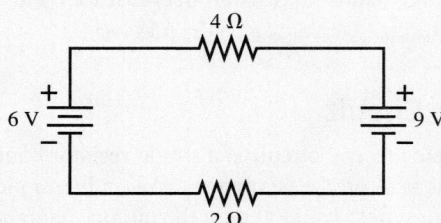

VISUALIZE In **FIGURE 31.9**, we've redrawn the circuit and defined $\mathcal{E}_1$, $\mathcal{E}_2$, R_1, and R_2. Because there are no junctions, the current is the same through *each* component in the circuit. With some thought, we might deduce whether the current is cw or ccw, but we do not need to know in advance of our analysis. We will choose a clockwise direction and solve for the value of I. If our solution is positive, then the current really is cw. If the solution should turn out to be negative, we will know that the current is ccw.

FIGURE 31.9 Analyzing the circuit.

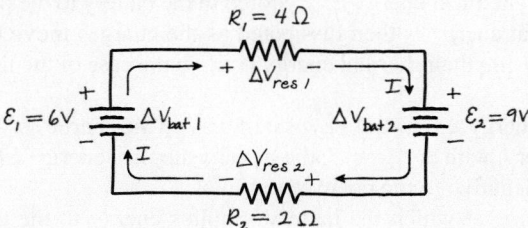

SOLVE a. How do we deal with *two* batteries? Can charge flow "backward" through a battery, from positive to negative? Consider the charge escalator analogy. Left to itself, a charge escalator lifts charge from lower to higher potential. But it *is* possible to run down an up escalator, as many of you have probably done. If two escalators are placed "head to head," whichever is stronger will, indeed, force the charge to run down the up escalator of the other battery. The current in a battery *can* be from positive to negative if driven in that direction by a larger emf from a second battery. Indeed, this is how rechargeable batteries are recharged.

Kirchhoff's loop law, going clockwise from the negative terminal of battery 1, is

$$\Delta V_{\text{closed loop}} = \sum (\Delta V)_i = \Delta V_{\text{bat 1}} + \Delta V_{\text{res 1}}$$

$$+ \Delta V_{\text{bat 2}} + \Delta V_{\text{res 2}} = 0$$

All the signs are + because this is a formal statement of *adding* potential differences around the loop. Next we can evaluate each ΔV. As we go cw, the charges *gain* potential in battery 1 but *lose* potential in battery 2. Thus $\Delta V_{\text{bat 1}} = +\mathcal{E}_1$ and $\Delta V_{\text{bat 2}} = -\mathcal{E}_2$. There is a *loss* of potential in traveling through each resistor, because we're traversing them in the

direction we assigned to the current, so $\Delta V_{\text{res 1}} = -IR_1$ and $\Delta V_{\text{res 2}} = -IR_2$. Thus Kirchhoff's loop law becomes

$$\sum (\Delta V)_i = \mathcal{E}_1 - IR_1 - \mathcal{E}_2 - IR_2$$

$$= \mathcal{E}_1 - \mathcal{E}_2 - I(R_1 + R_2) = 0$$

We can solve this equation to find the current in the loop:

$$I = \frac{\mathcal{E}_1 - \mathcal{E}_2}{R_1 + R_2} = \frac{6\text{ V} - 9\text{ V}}{4\ \Omega + 2\ \Omega} = -0.50\text{ A}$$

The value of I is negative; hence the actual current in this circuit is 0.50 A *counterclockwise*. You perhaps anticipated this from the orientation of the 9 V battery with its larger emf.

b. The potential difference across the 4 Ω resistor is

$$\Delta V_{\text{res 1}} = -IR_1 = -(-0.50\text{ A})(4\ \Omega) = +2.0\text{ V}$$

Because the current is actually ccw, the resistor's potential *increases* in the cw direction of our travel around the loop. Similarly, the potential difference across the 2 Ω resistor is $\Delta V_{\text{res 2}} = 1.0$ V. **FIGURE 31.10** shows the potential experienced by charges flowing around the circuit. The distance s is measured from the 6 V battery's negative terminal, and we have chosen to let $V = 0$ V at that point. The potential ends at the value from which it started.

FIGURE 31.10 A graphical presentation of how the potential changes around the loop.

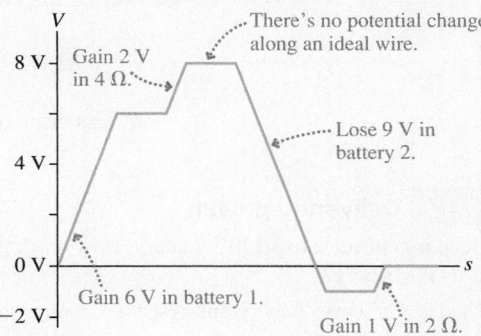

ASSESS Notice how the potential *drops* 9 V upon passing through battery 2 in the cw direction. It then gains 1 V upon passing through R_2 to end at the starting potential.

What is ΔV across the unspecified circuit element? Does the potential increase or decrease when traveling through this element in the direction assigned to I?

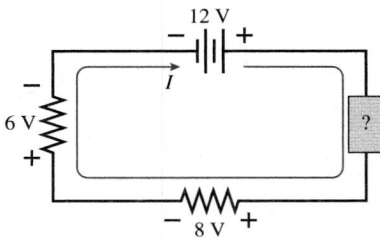

31.3 Energy and Power

FIGURE 31.11 Which lightbulb is brighter?

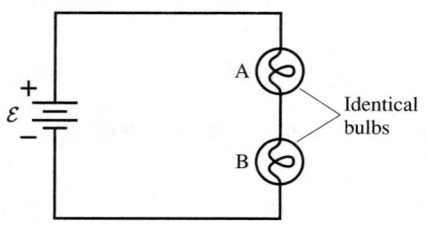

The circuit of FIGURE 31.11 has two identical lightbulbs, A and B. Which is brighter? Or are they equally bright? Think about this before going on.

You might have been tempted to say that A is brighter. After all, the current gets to A first, so A might "use up" some of the current and leave less for B. But this would violate the laws of conservation of charge and conservation of current. There are no junctions between A and B, so the current through the two bulbs must be the same. Hence the bulbs are equally bright.

It's not current that the bulbs use up, it's *energy*. Because a battery supplies a potential difference, it also supplies energy to a circuit. The charge escalator is an energy-transfer process, transferring chemical energy E_{chem} stored in the battery to the potential energy U of the charges. That energy is then dissipated as the charges move through the wires and resistors, increasing their thermal energy until, in the case of the lightbulb filaments, they glow.

A charge gains potential energy $\Delta U = q\Delta V_{bat}$ as it moves up the charge escalator in the battery. For an ideal battery, with $\Delta V_{bat} = \mathcal{E}$, the battery supplies energy $\Delta U = q\mathcal{E}$ as it lifts charge q from the negative to the positive terminal.

It is useful to know the *rate* at which the battery supplies energy to the charges. Recall from Chapter 11 that the rate at which energy is transferred is *power,* measured in joules per second or *watts.* If energy $\Delta U = q\mathcal{E}$ is transferred to charge q, then the *rate* at which energy is transferred from the battery to the moving charges is

$$P_{bat} = \text{rate of energy transfer} = \frac{dU}{dt} = \frac{dq}{dt}\mathcal{E} \qquad (31.7)$$

But dq/dt, the rate at which charge moves through the battery, is the current I. Hence the power supplied by a battery, or the rate at which the battery (or any other source of emf) transfers energy to the charges passing through it, is

$$P_{bat} = I\mathcal{E} \qquad \text{(power delivered by an emf)} \qquad (31.8)$$

$I\mathcal{E}$ has units of J/s, or W.

EXAMPLE 31.2 **Delivering power**

A 90 Ω load is connected to a 120 V battery. How much power is delivered by the battery?

SOLVE This is our basic battery-and-resistor circuit, which we analyzed earlier. In this case

$$I = \frac{\mathcal{E}}{R} = \frac{120\text{ V}}{90\ \Omega} = 1.33\text{ A}$$

Thus the power delivered by the battery is

$$P_{bat} = I\mathcal{E} = (1.33\text{ A})(120\text{ V}) = 160\text{ W}$$

FIGURE 31.12 A current-carrying resistor dissipates power because the electric force does work on the charges.

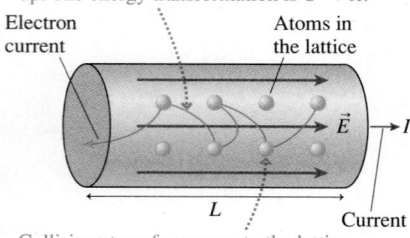

The electric field causes electrons to speed up. The energy transformation is $U \rightarrow K$.

Electron current
Atoms in the lattice

$\vec{E}$ → I

L
Current

Collisions transfer energy to the lattice.
The energy transformation is $K \rightarrow E_{th}$.

P_{bat} is the energy transferred per second from the battery's store of chemicals to the moving charges that make up the current. But what happens to this energy? Where does it end up? FIGURE 31.12, a section of a current-carrying resistor, reminds you of our microscopic model of conduction. The electrons accelerate in the electric field, then collide with atoms in the lattice. The acceleration phase is a transformation of potential to kinetic energy. The collisions then transfer the electron's kinetic energy to the *thermal* energy of the lattice. The potential energy was acquired in the battery, from the conversion of chemical energy, so the entire energy-transfer process looks like

$$E_{chem} \rightarrow U \rightarrow K \rightarrow E_{th}$$

The net result is that **the battery's chemical energy is transferred to the thermal energy of the resistors,** raising their temperature.

Suppose the average distance between collisions is d. The electric force $\vec{F} = q\vec{E}$ exerted on charge q does work as it pushes the charge through distance d. The field is constant inside the resistor, so the work is simply

$$W = F \Delta s = qEd \tag{31.9}$$

According to the work-kinetic energy theorem, this work increases the kinetic energy of charge q by $\Delta K = W = qEd$. This kinetic energy is transferred to the lattice when charge q collides with a lattice atom, causing the energy of the lattice to increase by

$$\Delta E_{\text{per collision}} = \Delta K = qEd$$

Collisions occur over and over as the charge makes its way through the resistor. After many such collisions, the total energy that charge q transfers while traveling distance L, the length of the resistor, is

$$\Delta E_{\text{th}} = qEL \tag{31.10}$$

But EL is the potential difference ΔV_{R} between the two ends of the resistor. Thus *each* charge q, as it travels the length of the resistor, transfers energy to the atomic lattice in the amount

$$\Delta E_{\text{th}} = q \Delta V_{\text{R}} \tag{31.11}$$

The *rate* at which energy is transferred from the current to the resistor is thus

$$P_{\text{R}} = \frac{dE_{\text{th}}}{dt} = \frac{dq}{dt} \Delta V_{\text{R}} = I \Delta V_{\text{R}} \tag{31.12}$$

We say that this power—so many joules per second—is *dissipated* by the resistor as charge flows through it. The resistor, in turn, transfers this energy to the air and to the circuit board on which it is mounted, causing the circuit and all its surroundings to heat up.

From our analysis of the basic circuit, in which a single resistor is connected to a battery, we learned that $\Delta V_{\text{R}} = \mathcal{E}$. That is, the potential difference across the resistor is exactly the emf supplied by the battery. But then Equations 31.8 and 31.12, for P_{bat} and P_{R}, are numerically equal, and we find that

$$P_{\text{R}} = P_{\text{bat}} \tag{31.13}$$

The answer to the question "What happens to the energy supplied by the battery?" is "The battery's chemical energy is transformed into the thermal energy of the resistor." The *rate* at which the battery supplies energy is exactly equal to the *rate* at which the resistor dissipates energy. This is, of course, exactly what we would have expected from energy conservation.

EXAMPLE 31.3 **The power of light**

How much current is "drawn" by a 100 W lightbulb connected to a 120 V outlet?

MODEL Most household appliances, such as a 100 W lightbulb or a 1500 W hair dryer, have a power rating. The rating does *not* mean that these appliances *always* dissipate that much power. These appliances are intended for use at a standard household voltage of 120 V, and their rating is the power they will dissipate *if* operated with a potential difference of 120 V. Their power consumption will differ from the rating if they are operated at any other potential difference.

SOLVE Because the lightbulb is operating as intended, it will dissipate 100 W of power. Thus

$$I = \frac{P_{\text{R}}}{\Delta V_{\text{R}}} = \frac{100 \text{ W}}{120 \text{ V}} = 0.833 \text{ A}$$

ASSESS A current of 0.833 A in this lightbulb transfers 100 J/s to the thermal energy of the filament, which, in turn, dissipates 100 J/s as heat and light to its surroundings.

A resistor obeys Ohm's law, $\Delta V_R = IR$. (Remember that Ohm's law gives only the *magnitude* of ΔV_R.) This gives us two alternative ways of writing the power dissipated by a resistor. We can either substitute IR for ΔV_R or substitute $\Delta V_R/R$ for I. Thus

$$P_R = I\Delta V_R = I^2 R = \frac{(\Delta V_R)^2}{R} \qquad \text{(power dissipated by a resistor)} \qquad (31.14)$$

If the same current I passes through several resistors in series, then $P_R = I^2 R$ tells us that most of the power will be dissipated by the largest resistance. This is why a lightbulb filament glows but the connecting wires do not. Essentially *all* of the power supplied by the battery is dissipated by the high-resistance lightbulb filament and essentially no power is dissipated by the low-resistance wires. The filament gets very hot, but the wires do not.

EXAMPLE 31.4 | **The power of sound**

Most loudspeakers are designed to have a resistance of 8 Ω. If an 8 Ω loudspeaker is connected to a stereo amplifier with a rating of 100 W, what is the maximum possible current to the loudspeaker?

MODEL The rating of an amplifier is the *maximum* power it can deliver. Most of the time it delivers far less, but the maximum might be reached for brief, intense sounds like cymbal crashes.

SOLVE The loudspeaker is a resistive load. The maximum current to the loudspeaker occurs when the amplifier delivers maximum power $P_{max} = (I_{max})^2 R$. Thus

$$I_{max} = \sqrt{\frac{P_{max}}{R}} = \sqrt{\frac{100 \text{ W}}{8 \text{ }\Omega}} = 3.5 \text{ A}$$

Kilowatt Hours

The electric meter on the side of your house or apartment records the kilowatt hours of electric energy that you use.

The energy dissipated (i.e., transformed into thermal energy) by a resistor during time Δt is $E_{th} = P_R \Delta t$. The product of watts and seconds is joules, the SI unit of energy. However, your local electric company prefers to use a different unit, the *kilowatt hour,* to measure the energy you use each month.

A load that consumes P_R kW of electricity for Δt hours has used $P_R \Delta t$ **kilowatt hours** of energy, abbreviated kWh. For example, a 4000 W electric water heater uses 40 kWh of energy in 10 hours. A 1500 W hair dryer uses 0.25 kWh of energy in 10 minutes. Despite the rather unusual name, a kilowatt hour is a unit of energy. A homework problem will let you find the conversion factor from kilowatt hours to joules.

Your monthly electric bill specifies the number of kilowatt hours you used last month. This is the amount of energy that the electric company delivered to you, via an electric current, and that you transformed into light and thermal energy inside your home. The cost of electricity varies throughout the country, but the average cost of electricity in the United States is approximately 10¢ per kWh ($0.10/kWh). Thus it costs about $4.00 to run your water heater for 10 hours, about 2.5¢ to dry your hair.

STOP TO THINK 31.3 Rank in order, from largest to smallest, the powers P_a to P_d dissipated in resistors a to d.

(a) $+ \; \Delta V \; -$
     R

(b) $+ \; 2\Delta V \; -$
     R

(c) $+ \; \Delta V \; -$
     $2R$

(d) $+ \; \Delta V \; -$
     $\frac{1}{2}R$

31.4 Series Resistors

Consider the three lightbulbs in **FIGURE 31.13**. The batteries are identical and the bulbs are identical. You learned in the previous section that B and C are equally bright, because of conservation of current, but how does the brightness of B compare to that of A? Think about this before going on.

FIGURE 31.14a shows two resistors placed end to end between points a and b. Resistors that are aligned end to end, *with no junctions between them,* are called **series resistors** or, sometimes, resistors "in series." Because there are no junctions and because current is conserved, the current I must be the same through each of these resistors. That is, the current out of the last resistor in a series is equal to the current into the first resistor.

The potential differences across the two resistors are $\Delta V_1 = IR_1$ and $\Delta V_2 = IR_2$. The total potential difference ΔV_{ab} between points a and b is the sum of the individual potential differences:

$$\Delta V_{ab} = \Delta V_1 + \Delta V_2 = IR_1 + IR_2 = I(R_1 + R_2) \qquad (31.15)$$

FIGURE 31.14 Replacing two series resistors with an equivalent resistor.

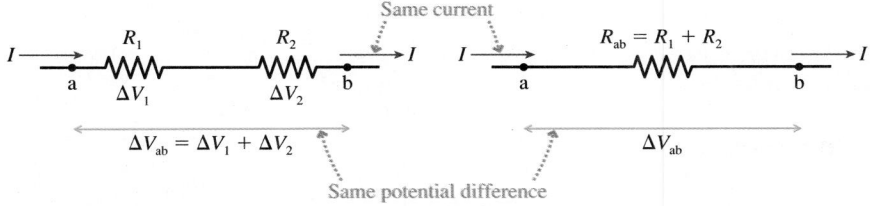

(a) Two resistors in series   **(b)** An equivalent resistor

Suppose, as in FIGURE 31.14b, we replaced the two resistors with a single resistor having current I and potential difference $\Delta V_{ab} = \Delta V_1 + \Delta V_2$. We can then use Ohm's law to find that the resistance R_{ab} between points a and b is

$$R_{ab} = \frac{\Delta V_{ab}}{I} = \frac{I(R_1 + R_2)}{I} = R_1 + R_2 \qquad (31.16)$$

Because the battery has to establish the same potential difference across the load and provide the same current in both cases, the two resistors R_1 and R_2 act exactly the same as a *single* resistor of value $R_1 + R_2$. We can say that the single resistor R_{ab} is *equivalent* to the two resistors in series.

There was nothing special about having only two resistors. If we have N resistors in series, their **equivalent resistance** is

$$R_{eq} = R_1 + R_2 + \cdots + R_N \qquad \text{(series resistors)} \qquad (31.17)$$

The current and the power output of the battery will be unchanged if the N series resistors are replaced by the single resistor R_{eq}. The key idea in this analysis is that **resistors in series all have the same current.**

NOTE ▶ Compare this idea to what you learned in Chapter 29 about capacitors in series. The end-to-end connections are the same, but the equivalent capacitance is *not* the sum of the individual capacitances. ◀

Now we can answer the lightbulb question posed at the beginning of this section. Suppose the resistance of each lightbulb is R. The battery drives current $I_A = \mathcal{E}/R$ through bulb A. Bulbs B and C are in series, with an equivalent resistance $R_{eq} = 2R$, but the battery has the same emf $\mathcal{E}$. Thus the current through bulbs B and C is $I_{B+C} = \mathcal{E}/R_{eq} = \mathcal{E}/2R = \frac{1}{2}I_A$. Bulb B has only half the current of bulb A, so B is dimmer.

Many people predict that A and B should be equally bright. It's the same battery, so shouldn't it provide the same current to both circuits? No! A battery is a source of emf, *not* a source of current. In other words, the battery's emf is the same no matter how the battery is used. When you buy a 1.5 V battery you're buying a device that provides a specified amount of potential difference, not a specified amount of current. The battery does provide the current to the circuit, but the *amount* of current depends on the resistance of the load. Your 1.5 V battery causes 1 A to pass through a 1.5 Ω load but only 0.1 A to pass through a 15 Ω load. As an analogy, think about a water

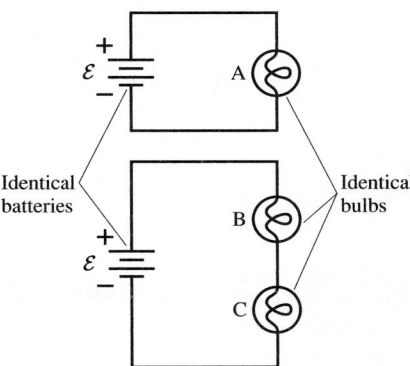

FIGURE 31.13 How does the brightness of bulb B compare to that of A?

faucet. The pressure in the water main underneath the street is a fixed and unvarying quantity set by the water company, but the amount of water coming out of a faucet depends on how far you open it. A faucet opened slightly has a "high resistance," so only a little water flows. A wide-open faucet has a "low resistance," and the water flow is large.

In summary, **a battery provides a fixed and unvarying emf (potential difference). It does *not* provide a fixed and unvarying current. The amount of current depends jointly on the battery's emf *and* the resistance of the circuit attached to the battery.**

EXAMPLE 31.5 **A series resistor circuit**

a. What is the current in the circuit of FIGURE 31.15a?
b. Draw a graph of potential versus position in the circuit, going cw from $V = 0$ V at the battery's negative terminal.

MODEL The three resistors are end to end, with no junctions between them, and thus are in series. Assume ideal connecting wires and an ideal battery.

SOLVE a. The battery "acts" the same—it provides the same current at the same potential difference—if we replace the three series resistors by their equivalent resistance

$$R_{eq} = 15\ \Omega + 4\ \Omega + 8\ \Omega = 27\ \Omega$$

This is shown as an equivalent circuit in FIGURE 31.15b. Now we have a circuit with a single battery and a single resistor, for which we know the current to be

$$I = \frac{\mathcal{E}}{R_{eq}} = \frac{9\ \text{V}}{27\ \Omega} = 0.333\ \text{A}$$

b. $I = 0.333$ A is the current in each of the three resistors in the original circuit. Thus the potential differences across the resistors are $\Delta V_{res\,1} = -IR_1 = -5.0$ V, $\Delta V_{res\,2} = -IR_2 = -1.3$ V, and $\Delta V_{res\,3} = -IR_3 = -2.7$ V for the 15 Ω, the 4 Ω, and the 8 Ω resistors, respectively. FIGURE 31.15c shows that the potential increases by 9 V due to the battery's emf, then decreases by 9 V in three steps.

FIGURE 31.15 Analyzing a circuit with series resistors.

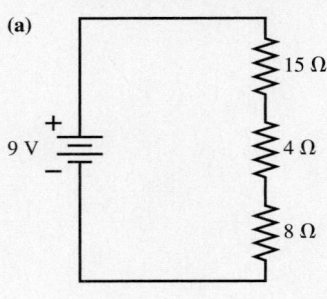

(a)

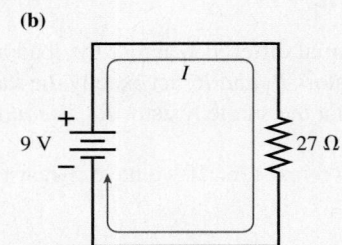

(b)

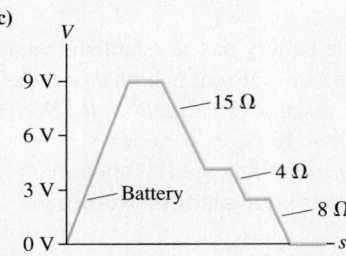

(c)

FIGURE 31.16 An ammeter measures the current in a circuit element.

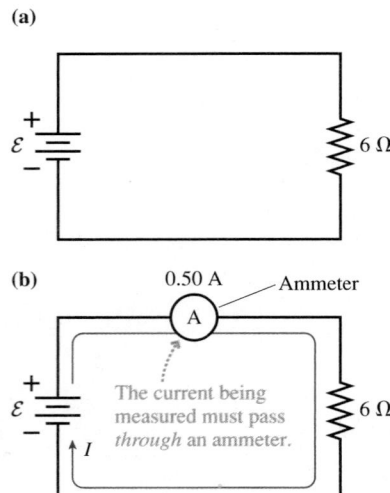

Ammeters

A device that measures the current in a circuit element is called an **ammeter.** Because charge flows *through* circuit elements, an ammeter must be placed *in series* with the circuit element whose current is to be measured.

FIGURE 31.16a shows a simple one-resistor circuit with an unknown emf $\mathcal{E}$. We can measure the current in the circuit by inserting the ammeter as shown in FIGURE 31.16b. Notice that we have to *break the connection* between the battery and the resistor in order to insert the ammeter. Now the current in the resistor has to first pass through the ammeter.

Because the ammeter is now in series with the resistor, the total resistance seen by the battery is $R_{eq} = 6\ \Omega + R_{ammeter}$. In order that the ammeter measure the current without changing the current, the ammeter's resistance must, in this case, be $\ll 6\ \Omega$. Indeed, an ideal ammeter has $R_{ammeter} = 0\ \Omega$ and thus has no effect on the current. Real ammeters come very close to this ideal.

The ammeter in Figure 31.16b reads 0.50 A, meaning that the current through the 6 Ω resistor is $I = 0.50$ A. Thus the resistor's potential difference is $\Delta V_R = IR = 3.0$ V. If the ammeter is ideal, with no resistance and thus no potential difference across it, then, from Kirchhoff's loop law, the battery's emf is $\mathcal{E} = \Delta V_R = 3.0$ V.

STOP TO THINK 31.4   What are the current and the potential at points a to e?

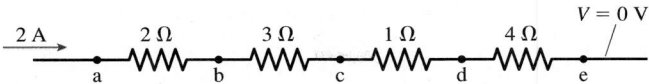

31.5 Real Batteries

Let's look at how real batteries differ from the ideal battery we have been assuming. Real batteries, like ideal batteries, separate charge and create a potential difference. However, real batteries also provide a slight resistance to the charges on the charge escalator. They have what is called an **internal resistance,** which is symbolized by r. FIGURE 31.17 shows both an ideal and a real battery.

From our vantage point outside a battery, we cannot see $\mathcal{E}$ and r separately. To the user, the battery provides a potential difference ΔV_{bat} called the **terminal voltage.** $\Delta V_{bat} = \mathcal{E}$ for an ideal battery, but the presence of the internal resistance affects ΔV_{bat}. Suppose the current in the battery is I. As charges travel from the negative to the positive terminal, they gain potential $\mathcal{E}$ but *lose* potential $\Delta V_{int} = -Ir$ due to the internal resistance. Thus the terminal voltage of the battery is

$$\Delta V_{bat} = \mathcal{E} - Ir \leq \mathcal{E} \qquad (31.18)$$

Only when $I = 0$, meaning that the battery is not being used, is $\Delta V_{bat} = \mathcal{E}$.

FIGURE 31.18 shows a single resistor R connected to the terminals of a battery having emf $\mathcal{E}$ and internal resistance r. Resistances R and r are in series, so we can replace them, for the purpose of circuit analysis, with a single equivalent resistor $R_{eq} = R + r$. Hence the current in the circuit is

$$I = \frac{\mathcal{E}}{R_{eq}} = \frac{\mathcal{E}}{R + r} \qquad (31.19)$$

If $r \ll R$, so that the internal resistance of the battery is negligible, then $I \approx \mathcal{E}/R$, exactly the result we found before. But the current decreases significantly as r increases.

FIGURE 31.17 An ideal battery and a real battery.

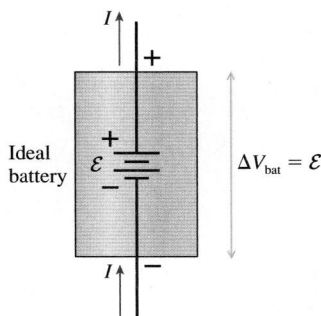

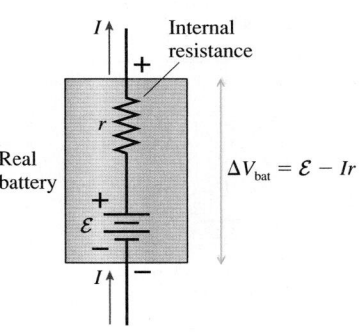

FIGURE 31.18 A single resistor connected to a real battery is in series with the battery's internal resistance, giving $R_{eq} = R + r$.

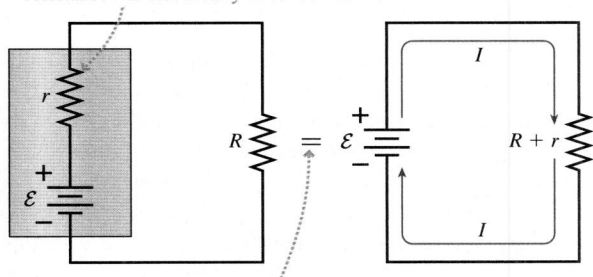

Although physically separated, the internal resistance r is electrically in series with R.

This means the two circuits are equivalent.

We can use Ohm's law to find that the potential difference across the load resistor R is

$$\Delta V_R = IR = \frac{R}{R + r}\mathcal{E} \qquad (31.20)$$

Similarly, the potential difference across the terminals of the battery is

$$\Delta V_{bat} = \mathcal{E} - Ir = \mathcal{E} - \frac{r}{R + r}\mathcal{E} = \frac{R}{R + r}\mathcal{E} \qquad (31.21)$$

The potential difference across the resistor is equal to the potential difference between the *terminals* of the battery, where the resistor is attached, *not* equal to the battery's emf. Notice that $\Delta V_{bat} = \mathcal{E}$ only if $r = 0$ (an ideal battery with no internal resistance).

EXAMPLE 31.6 | **Lighting up a flashlight**

A 6 Ω flashlight bulb is powered by a 3 V battery with an internal resistance of 1 Ω. What are the power dissipation of the bulb and the terminal voltage of the battery?

MODEL Assume ideal connecting wires but not an ideal battery.

VISUALIZE The circuit diagram looks like Figure 31.18. R is the resistance of the bulb's filament.

SOLVE Equation 31.19 gives us the current:

$$I = \frac{\mathcal{E}}{R + r} = \frac{3\text{ V}}{6\text{ Ω} + 1\text{ Ω}} = 0.43\text{ A}$$

This is 15% less than the 0.5 A an ideal battery would supply. The potential difference across the resistor is $\Delta V_R = IR = 2.6$ V, thus the power dissipation is

$$P_R = I\Delta V_R = 1.1\text{ W}$$

The battery's terminal voltage is

$$\Delta V_{bat} = \frac{R}{R + r}\mathcal{E} = \frac{6\text{ Ω}}{6\text{ Ω} + 1\text{ Ω}}3\text{ V} = 2.6\text{ V}$$

ASSESS 1 Ω is a typical internal resistance for a flashlight battery. The internal resistance causes the battery's terminal voltage to be 0.4 V less than its emf in this circuit.

A Short Circuit

FIGURE 31.19 The short-circuit current of a battery.

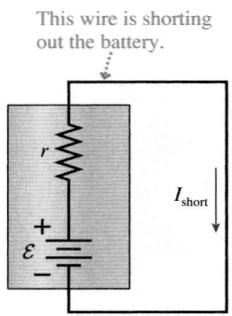

This wire is shorting out the battery.

In FIGURE 31.19 we've replaced the resistor with an ideal wire having $R_{wire} = 0$ Ω. When a connection of very low or zero resistance is made between two points in a circuit that are normally separated by a higher resistance, we have what is called a **short circuit.** The wire in Figure 31.19 is *shorting out* the battery.

If the battery were ideal, shorting it with an ideal wire ($R = 0$ Ω) would cause the current to be $I = \mathcal{E}/0 = \infty$. The current, of course, cannot really become infinite. Instead, the battery's internal resistance r becomes the only resistance in the circuit. If we use $R = 0$ Ω in Equation 31.19, we find that the *short-circuit current* is

$$I_{short} = \frac{\mathcal{E}}{r} \qquad (31.22)$$

A 3 V battery with 1 Ω internal resistance generates a short circuit current of 3 A. This is the *maximum possible current* that this battery can produce. Adding any external resistance R will decrease the current to a value less than 3 A.

EXAMPLE 31.7 | **A short-circuited battery**

What is the short-circuit current of a 12 V car battery with an internal resistance of 0.020 Ω? What happens to the power supplied by the battery?

SOLVE The short-circuit current is

$$I_{short} = \frac{\mathcal{E}}{r} = \frac{12\text{ V}}{0.02\text{ Ω}} = 600\text{ A}$$

Power is generated by chemical reactions in the battery and dissipated by the load resistance. But with a short-circuited battery, the load resistance is *inside* the battery! The "shorted" battery has to dissipate power $P = I^2r = 7200$ W *internally.*

ASSESS This value is realistic. Car batteries are designed to drive the starter motor, which has a very small resistance and can draw a current of a few hundred amps. That is why the battery cables are so thick. A shorted car battery can produce an *enormous* amount of current. The normal response of a shorted car battery is to explode; it simply cannot dissipate this much power. Shorting a flashlight battery can make it rather hot, but your life is not in danger. Although the voltage of a car battery is relatively small, a car battery can be dangerous and should be treated with great respect.

Most of the time a battery is used under conditions in which $r \ll R$ and the internal resistance is negligible. The ideal battery model is fully justified in that case. Thus we will assume that batteries are ideal *unless stated otherwise.* But keep in mind that batteries (and other sources of emf) do have an internal resistance, and this internal resistance limits the current of the battery.

31.6 Parallel Resistors

FIGURE 31.20 is another lightbulb puzzle. Initially the switch is open. The current is the same through bulbs A and B, because of conservation of current, and they are equally bright. Bulb C is not glowing. What happens to the brightness of A and B when the switch is closed? And how does the brightness of C then compare to that of A and B? Think about this before going on.

FIGURE 31.21a shows two resistors aligned side by side with their ends connected at c and d. Resistors connected *at both ends* are called **parallel resistors** or, sometimes, resistors "in parallel." The left ends of both resistors are at the same potential V_c. Likewise, the right ends are at the same potential V_d. Thus the potential *differences* ΔV_1 and ΔV_2 are the *same* and are simply ΔV_{cd}.

Kirchhoff's junction law applies at the junctions. The input current I splits into currents I_1 and I_2 at the left junction. On the right, the two currents are recombined into current I. According to the junction law,

$$I = I_1 + I_2 \tag{31.23}$$

We can apply Ohm's law to each resistor, along with $\Delta V_1 = \Delta V_2 = \Delta V_{cd}$, to find that the current is

$$I = \frac{\Delta V_1}{R_1} + \frac{\Delta V_2}{R_2} = \frac{\Delta V_{cd}}{R_1} + \frac{\Delta V_{cd}}{R_2} = \Delta V_{cd}\left(\frac{1}{R_1} + \frac{1}{R_2}\right) \tag{31.24}$$

Suppose, as in FIGURE 31.21b, we replaced the two resistors with a single resistor having current I and potential difference ΔV_{cd}. This resistor is equivalent to the original two because the battery has to establish the same potential difference and provide the same current in either case. A second application of Ohm's law shows that the resistance between points c and d is

$$R_{cd} = \frac{\Delta V_{cd}}{I} = \left(\frac{1}{R_1} + \frac{1}{R_2}\right)^{-1} \tag{31.25}$$

The single resistor R_{cd} draws the same current as resistors R_1 and R_2, so, as far as the battery is concerned, resistor R_{cd} is *equivalent* to the two resistors in parallel.

There is nothing special about having chosen two resistors to be in parallel. If we have N resistors in parallel, the *equivalent resistance* is

$$R_{eq} = \left(\frac{1}{R_1} + \frac{1}{R_2} + \cdots + \frac{1}{R_N}\right)^{-1} \quad \text{(parallel resistors)} \tag{31.26}$$

The behavior of the circuit will be unchanged if the N parallel resistors are replaced by the single resistor R_{eq}. The key idea of this analysis is that **resistors in parallel all have the same potential difference.**

NOTE ▶ Don't forget to take the inverse—the -1 exponent in Equation 31.26—after adding the inverses of all the resistances. ◀

FIGURE 31.20 What happens to the brightness of the bulbs when the switch is closed?

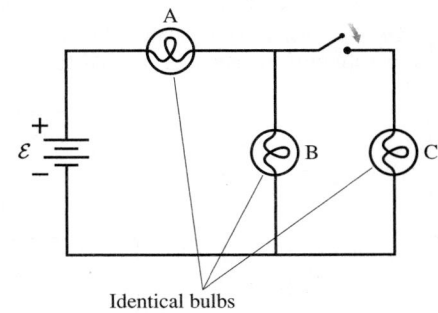

Identical bulbs

FIGURE 31.21 Replacing two parallel resistors with an equivalent resistor.

(a) Two resistors in parallel

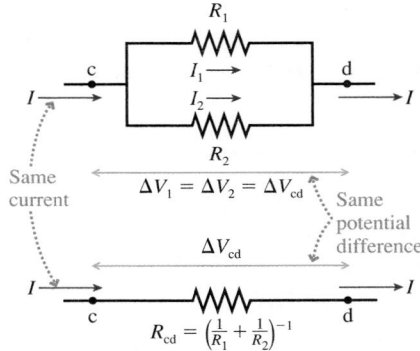

(b) An equivalent resistor

Two identical resistors*

In series	$R_{eq} = 2R$
In parallel	$R_{eq} = \dfrac{R}{2}$

$*R_1 = R_2 = R$

EXAMPLE 31.8 **A parallel resistor circuit**

The three resistors of FIGURE 31.22 are connected to a 9 V battery. Find the potential difference across and the current through each resistor.

FIGURE 31.22 Parallel resistor circuit of Example 31.8.

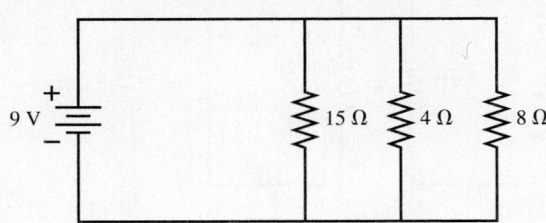

MODEL The resistors are in parallel. Assume an ideal battery and ideal connecting wires.

SOLVE The three parallel resistors can be replaced by a single equivalent resistor

$$R_{eq} = \left(\frac{1}{15\ \Omega} + \frac{1}{4\ \Omega} + \frac{1}{8\ \Omega}\right)^{-1} = (0.4417\ \Omega^{-1})^{-1} = 2.26\ \Omega$$

The equivalent circuit is shown in FIGURE 31.23a on the next page, from which we find the current to be

$$I = \frac{\mathcal{E}}{R_{eq}} = \frac{9\ \text{V}}{2.26\ \Omega} = 3.98\ \text{A}$$

Continued

The potential difference across R_{eq} is $\Delta V_{eq} = \mathcal{E} = 9.0$ V. Now we have to be careful. Current I divides at the junction into the smaller currents I_1, I_2, and I_3 shown in FIGURE 31.23b. However, the division is *not* into three equal currents. According to Ohm's law, resistor i has current $I_i = \Delta V_i / R_i$. Because the three resistors are each connected to the battery by ideal wires, as is the equivalent resistor, their potential differences are equal:

$$\Delta V_1 = \Delta V_2 = \Delta V_3 = \Delta V_{eq} = 9.0 \text{ V}$$

Thus the currents are

$$I_1 = \frac{9 \text{ V}}{15 \text{ }\Omega} = 0.60 \text{ A} \qquad I_2 = \frac{9 \text{ V}}{4 \text{ }\Omega} = 2.25 \text{ A}$$

$$I_3 = \frac{9 \text{ V}}{8 \text{ }\Omega} = 1.13 \text{ A}$$

ASSESS The *sum* of the three currents is 3.98 A, as required by Kirchhoff's junction law.

FIGURE 31.23 The parallel resistors can be replaced by a single equivalent resistor.

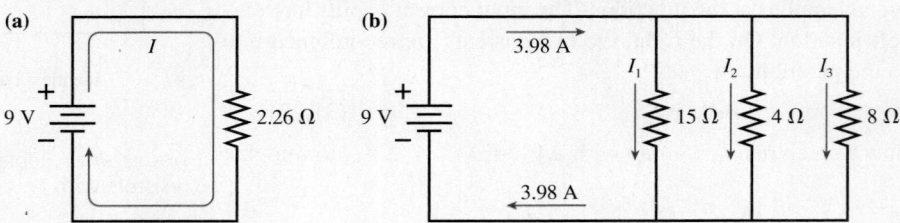

Summary of series and parallel resistors

	I	ΔV
Series	Same	Add
Parallel	Add	Same

The result of Example 31.8 seems surprising. The equivalent of a parallel combination of 15 Ω, 4 Ω, and 8 Ω was found to be 2.26 Ω. How can the equivalent of a group of resistors be *less* than any single resistance in the group? Shouldn't more resistors imply more resistance? The answer is yes for resistors in series but not for resistors in parallel. Even though a resistor is an obstacle to the flow of charge, parallel resistors provide more pathways for charge to get through. Consequently, the equivalent of several resistors in parallel is always *less* than any single resistor in the group.

Complex combinations of resistors can often be reduced to a single equivalent resistance through a step-by-step application of the series and parallel rules. The final example in this section illustrates this idea.

EXAMPLE 31.9 | **A combination of resistors**

What is the equivalent resistance of the group of resistors shown in FIGURE 31.24?

FIGURE 31.24 A combination of resistors.

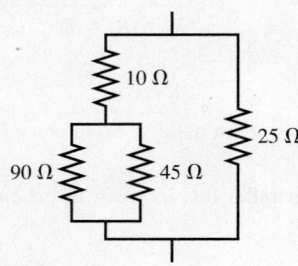

MODEL This circuit contains both series and parallel resistors.

SOLVE Reduction to a single equivalent resistance is best done in a series of steps, with the circuit being redrawn after each step. The procedure is shown in FIGURE 31.25. Note that the 10 Ω and 25 Ω

resistors are *not* in parallel. They are connected at their top ends but not at their bottom ends. Resistors must be connected to each other at *both* ends to be in parallel. Similarly, the 10 Ω and 45 Ω resistors are *not* in series because of the junction between them. If the original group of four resistors occurred within a larger circuit, they could be replaced with a single 15.4 Ω resistor without having any effect on the rest of the circuit.

FIGURE 31.25 The combination is reduced to a single equivalent resistor.

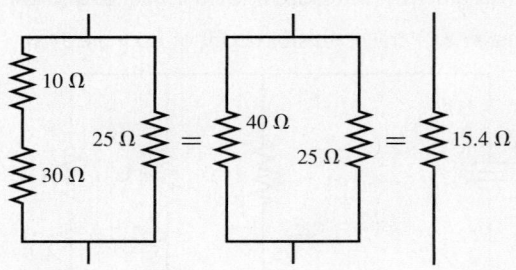

To return to the lightbulb question that began this section, FIGURE 31.26 has redrawn the circuit with each bulb shown as a resistance R. Initially, before the switch is closed, bulbs A and B are in series with equivalent resistance $2R$. The current from the battery is

$$I_{\text{before}} = \frac{\mathcal{E}}{2R} = \frac{1}{2}\frac{\mathcal{E}}{R}$$

This is the current in both bulbs.

Closing the switch places bulbs B and C in parallel. The equivalent resistance of two identical resistors in parallel is $R_{\text{eq}} = \frac{1}{2}R$. This equivalent resistance of B and C is in series with bulb A; hence the total resistance of the circuit is $\frac{3}{2}R$ and the current leaving the battery is

$$I_{\text{after}} = \frac{\mathcal{E}}{3R/2} = \frac{2}{3}\frac{\mathcal{E}}{R} > I_{\text{before}}$$

Closing the switch *decreases* the circuit resistance and thus *increases* the current leaving the battery.

All the charge flows through A, so A *increases* in brightness when the switch is closed. The current I_{after} then splits at the junction. Bulbs B and C have equal resistance, so the current splits equally. The current in B is $\frac{1}{3}(\mathcal{E}/R)$, which is *less* than I_{before}. Thus B *decreases* in brightness when the switch is closed. Bulb C has the same brightness as bulb B.

Voltmeters

A device that measures the potential difference across a circuit element is called a **voltmeter.** Because potential difference is measured *across* a circuit element, from one side to the other, a voltmeter is placed in *parallel* with the circuit element whose potential difference is to be measured.

FIGURE 31.27a shows a simple circuit in which a 17 Ω resistor is connected across a 9 V battery with an unknown internal resistance. By connecting a voltmeter across the resistor, as shown in FIGURE 31.27b, we can measure the potential difference across the resistor. Unlike an ammeter, using a voltmeter does *not* require us to break the connections.

Because the voltmeter is now in parallel with the resistor, the total resistance seen by the battery is $R_{\text{eq}} = (1/17\ \Omega + 1/R_{\text{voltmeter}})^{-1}$. In order that the voltmeter measure the voltage without changing the voltage, the voltmeter's resistance must, in this case, be $\gg 17\ \Omega$. Indeed, an *ideal voltmeter* has $R_{\text{voltmeter}} = \infty\ \Omega$, and thus has no effect on the voltage. Real voltmeters come very close to this ideal, and we will always assume them to be so.

The voltmeter in Figure 31.27b reads 8.5 V. This is less than $\mathcal{E}$ because of the battery's internal resistance. Equation 31.20 found an expression for the resistor's potential difference ΔV_R. That equation is easily solved for the internal resistance r:

$$r = \frac{\mathcal{E} - \Delta V_R}{\Delta V_R}R = \frac{0.5\ \text{V}}{8.5\ \text{V}}17\ \Omega = 1.0\ \Omega$$

Here a voltmeter reading was the one piece of experimental data we needed in order to determine the battery's internal resistance.

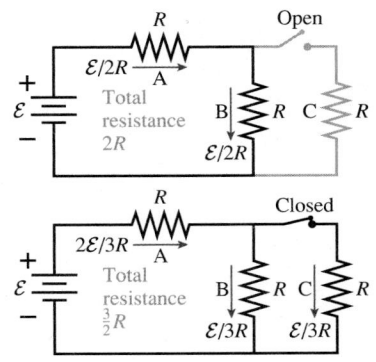

FIGURE 31.26 The lightbulbs of Figure 31.20 with the switch open and closed.

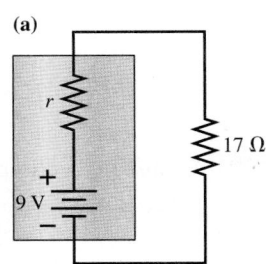

FIGURE 31.27 A voltmeter measures the potential difference across an element.

(a)

(b)

STOP TO THINK 31.5 Rank in order, from brightest to dimmest, the identical bulbs A to D.

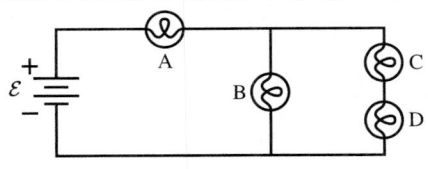

31.7 Resistor Circuits

We can use the information in this chapter to analyze a variety of more complex but more realistic circuits.

PROBLEM-SOLVING
STRATEGY 31.1 **Resistor circuits**

MODEL Assume that wires are ideal and, where appropriate, that batteries are ideal.

VISUALIZE Draw a circuit diagram. Label all known and unknown quantities.

SOLVE Base your mathematical analysis on Kirchhoff's laws and on the rules for series and parallel resistors.

■ Step by step, reduce the circuit to the smallest possible number of equivalent resistors.
■ Write Kirchhoff's loop law for each independent loop in the circuit.
■ Determine the current through and the potential difference across the equivalent resistors.
■ Rebuild the circuit, using the facts that the current is the same through all resistors in series and the potential difference is the same for all parallel resistors.

ASSESS Use two important checks as you rebuild the circuit.

■ Verify that the sum of the potential differences across series resistors matches ΔV for the equivalent resistor.
■ Verify that the sum of the currents through parallel resistors matches I for the equivalent resistor.

Exercise 26

EXAMPLE 31.10 **Analyzing a complex circuit**

Find the current through and the potential difference across each of the four resistors in the circuit shown in FIGURE 31.28.

FIGURE 31.28 A complex resistor circuit.

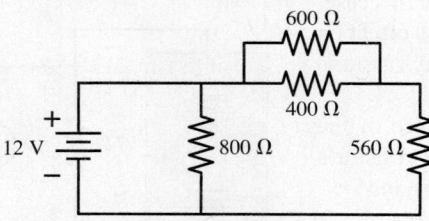

MODEL Assume an ideal battery, with no internal resistance, and ideal connecting wires.

VISUALIZE Figure 31.28 shows the circuit diagram. We'll keep redrawing the diagram as we analyze the circuit.

SOLVE First, we break the circuit down, step by step, into one with a single resistor. FIGURE 31.29a shows this done in three steps. The final battery-and-resistor circuit is our basic circuit, with current

$$I = \frac{\mathcal{E}}{R} = \frac{12 \text{ V}}{400 \ \Omega} = 0.030 \text{ A} = 30 \text{ mA}$$

The potential difference across the 400 Ω resistor is $\Delta V_{400} = \Delta V_{\text{bat}} = \mathcal{E} = 12$ V.

Second, we rebuild the circuit, step by step, finding the currents and potential differences at each step. FIGURE 31.29b repeats the steps of Figure 31.29a exactly, but in reverse order. The 400 Ω resistor came from two 800 Ω resistors in parallel. Because $\Delta V_{400} = 12$ V, it must be true that each $\Delta V_{800} = 12$ V. The current through each 800 Ω is then $I = \Delta V/R = 15$ mA. The checkpoint is to note that 15 mA + 15 mA = 30 mA.

The right 800 Ω resistor was formed by 240 Ω and 560 Ω in series. Because $I_{800} = 15$ mA, it must be true that $I_{240} = I_{560} = 15$ mA. The potential difference across each is $\Delta V = IR$, so $\Delta V_{240} = 3.6$ V and $\Delta V_{560} = 8.4$ V. Here the checkpoint is to note that 3.6 V + 8.4 V = 12 V = ΔV_{800}, so the potential differences add as they should.

Finally, the 240 Ω resistor came from 600 Ω and 400 Ω in parallel, so they each have the same 3.6 V potential difference as their 240 Ω equivalent. The currents are $I_{600} = 6$ mA and $I_{400} = 9$ mA. Note that 6 mA + 9 mA = 15 mA, which is our third checkpoint. We now know all currents and potential differences.

ASSESS We *checked our work* at each step of the rebuilding process by verifying that currents summed properly at junctions and that potential differences summed properly along a series of resistances. This "check as you go" procedure is extremely important. It provides you, the problem solver, with a built-in error finder that will immediately inform you if a mistake has been made.

FIGURE 31.29 The step-by-step circuit analysis.

(a) Break down the circuit.

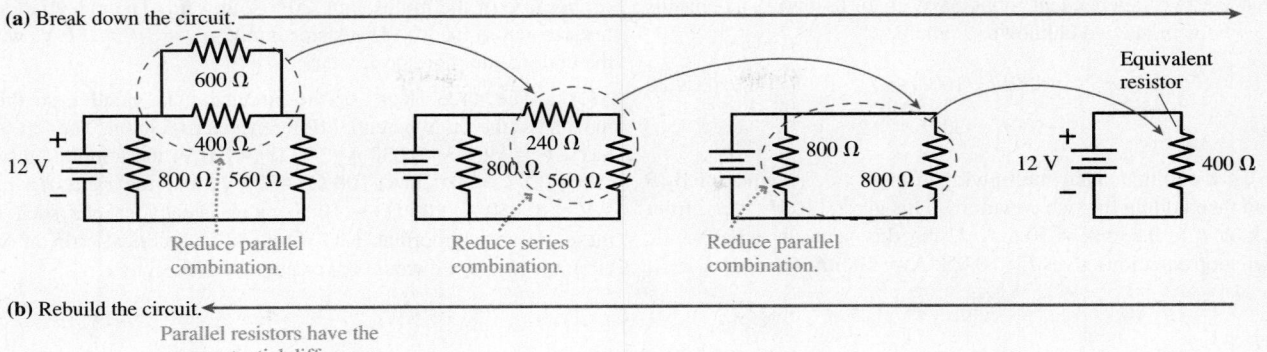

Reduce parallel combination. Reduce series combination. Reduce parallel combination.

Equivalent resistor

(b) Rebuild the circuit.

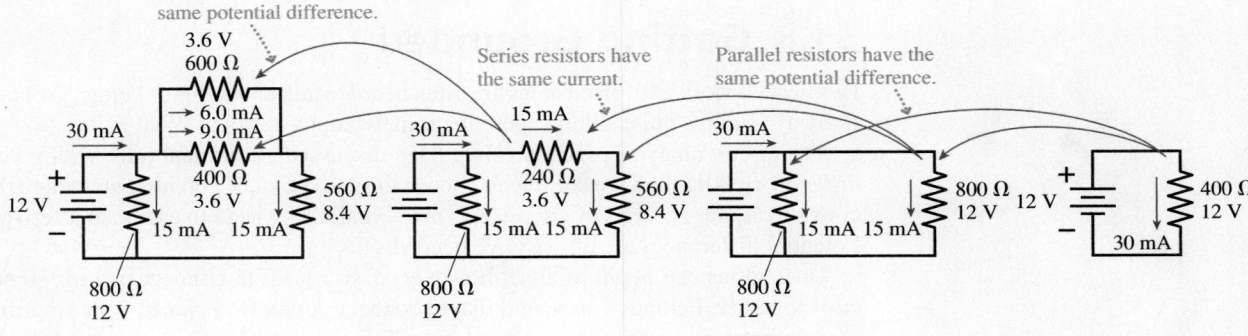

Parallel resistors have the same potential difference.

Series resistors have the same current.

Parallel resistors have the same potential difference.

EXAMPLE 31.11 **Analyzing a two-loop circuit**

Find the current through and the potential difference across the 100 Ω resistor in the circuit of FIGURE 31.30.

FIGURE 31.30 A two-loop circuit.

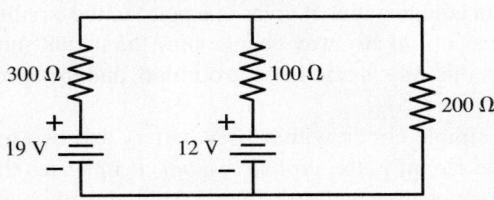

MODEL Assume ideal batteries and ideal connecting wires.

VISUALIZE Figure 31.30 shows the circuit diagram. None of the resistors are connected in series or in parallel, so this circuit cannot be reduced to a simpler circuit.

SOLVE Kirchhoff's loop law applies to *any* loop. To analyze a multiloop problem, we need to write a loop-law equation for each loop. FIGURE 31.31 redraws the circuit and defines clockwise currents I_1 in the left loop and I_2 in the right loop. But what about the middle branch? Let's assign a downward current I_3 to the middle branch. If we apply Kirchhoff's junction law $\sum I_{in} = \sum I_{out}$ to the junction above the 100 Ω resistor, as shown in the blow-up of Figure 31.31, we see that $I_1 = I_2 + I_3$ and thus $I_3 = I_1 - I_2$. If I_3 ends up being a positive number, then the current in the middle branch really is downward. A negative I_3 will signify an upward current.

FIGURE 31.31 Applying Kirchhoff's laws.

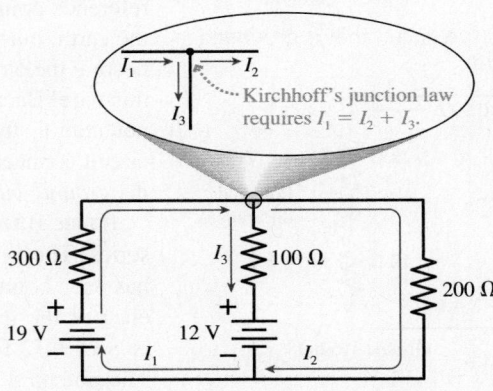

Kirchhoff's junction law requires $I_1 = I_2 + I_3$.

Kirchhoff's loop law for the left loop, going clockwise from the lower-left corner, is

$$\sum (\Delta V)_i = 19\ \text{V} - (300\ \Omega)I_1 - (100\ \Omega)I_3 - 12\ \text{V} = 0$$

We're traveling through the 100 Ω resistor in the direction of I_3, the "downhill" direction, so the potential decreases. The 12 V battery is traversed positive to negative, so there we have $\Delta V = -\mathcal{E} = -12\ \text{V}$. For the right loop, we're going to travel "uphill" through the 100 Ω resistor, opposite to I_3, and gain potential. Thus the loop law for the right loop is

$$\sum (\Delta V)_i = 12\ \text{V} + (100\ \Omega)I_3 - (200\ \Omega)I_2 = 0$$

Continued

If we substitute $I_3 = I_1 - I_2$ and then rearrange the terms, we find that the two independent loops have given us two simultaneous equations in the two unknowns I_1 and I_2:

$$400I_1 - 100I_2 = 7$$

$$-100I_1 + 300I_2 = 12$$

We can eliminate I_2 by multiplying through the first equation by 3 and then adding the two equations. This gives $1100I_1 = 33$, from which $I_1 = 0.030$ A $= 30$ mA. Using this value in either of the two loop equations gives $I_2 = 0.050$ A $= 50$ mA. Because $I_2 > I_1$,

the current through the 100 Ω resistor is $I_3 = I_1 - I_2 = -20$ mA, or, because of the minus sign, 20 mA upward. The potential difference across the 100 Ω resistor is $\Delta V_{100\,\Omega} = I_3R = 2.0$ V, with the bottom end more positive.

ASSESS The three "legs" of the circuit are in parallel, so they must have the same potential difference across them. The left leg has $\Delta V = 19$ V $- (0.030$ A$)(300\,\Omega) = 10$ V, the middle leg has $\Delta V = 12$ V $- (0.020$ A$)(100\,\Omega) = 10$ V, and the right leg has $\Delta V = (0.050$ A$)(200\,\Omega) = 10$ V. Consistency checks such as these are very important. Had we made a numerical error in our circuit analysis, we would have caught it at this point.

31.8 Getting Grounded

People who work with electronics are often heard to talk about things being "grounded." It always sounds quite serious, perhaps somewhat mysterious. What is it?

The circuit analysis procedures we have discussed so far deal only with potential *differences.* Although we are free to choose the zero point of potential anywhere that is convenient, our analysis of circuits has not revealed any need to establish a zero point. Potential differences are all we have needed.

Difficulties can begin to arise, however, if you want to connect two *different* circuits together. Perhaps you would like to connect your CD player to your amplifier or your computer monitor to the computer itself. Incompatibilities can arise unless all the circuits to be connected have a *common* reference point for the potential.

You learned previously that the earth itself is a conductor. Suppose we have two circuits. If we connect *one* point of each circuit to the earth by an ideal wire, and we also agree to call the potential of the earth $V_{\text{earth}} = 0$ V, then both circuits have a common reference point. But notice something very important: *one* wire connects the circuit to the earth, but there is not a second wire returning to the circuit. That is, the wire connecting the circuit to the earth is not part of a complete circuit, so there is *no current* in this wire! Because the wire is an equipotential, it gives one point in the circuit the same potential as the earth, but it does *not* in any way change how the circuit functions. A circuit connected to the earth in this way is said to be **grounded,** and the wire is called the *ground wire.*

FIGURE 31.32a shows a fairly simple circuit with a 10 V battery and two resistors in series. The symbol beneath the circuit is the *ground symbol.* It indicates that a wire has been connected between the negative battery terminal and the earth, but the presence of the ground wire does not affect the circuit's behavior. The total resistance is $8\,\Omega + 12\,\Omega = 20\,\Omega$, so the current in the loop is $I = (10$ V$)/(20\,\Omega) = 0.50$ A. The potential differences across the two resistors are found, using Ohm's law, to be $\Delta V_8 = 4$ V and $\Delta V_{12} = 6$ V. These are the same values that we would find if the ground wire were *not* present. So what has grounding the circuit accomplished?

FIGURE 31.32b shows the actual potential at several points in the circuit. By definition, $V_{\text{earth}} = 0$ V. The negative battery terminal and the bottom of the 12 Ω resistor are connected by ideal wires to the earth, so *the* potential at these two points must also be zero. The positive terminal of the battery is 10 V more positive than the negative terminal, so $V_{\text{neg}} = 0$ V implies $V_{\text{pos}} = +10$ V. Similarly, the fact that the potential *decreases* by 6 V as charge flows through the 12 Ω resistor now implies that *the* potential at the junction of the resistors must be $+6$ V. The potential difference across the 8 Ω resistor is 4 V, so the top has to be at $+10$ V. This agrees with the potential at the positive battery terminal, as it must because these two points are connected by an ideal wire.

All that grounding the circuit does is allow us to have *specific values* for the potential at each point in the circuit. Now we can say "The voltage at the resistor junction is 6 V," whereas before all we could say was "There is a 6 V potential difference across the 12 Ω resistor."

The circular prong of a three-prong plug is a connection to ground.

FIGURE 31.32 A circuit that is grounded at one point.

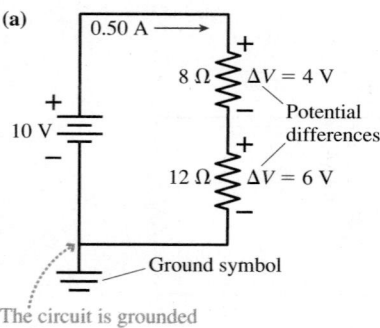

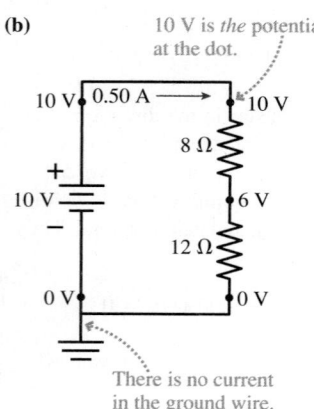

There is one important lesson from this: **Being grounded does not affect the circuit's behavior under normal conditions.** You cannot use "because it is grounded" to *explain* anything about a circuit's behavior.

We added "under normal conditions" because there is one exception. Most circuits are enclosed in a case of some sort that is held away from the circuit with insulators. Sometimes a circuit breaks or malfunctions in such a way that the case comes into electrical contact with the circuit. If the circuit uses high voltage, or even ordinary 120 V household voltage, anyone touching the case could be injured or killed by electrocution. To prevent this, many appliances or electrical instruments have the case itself grounded. Grounding ensures that the potential of the case will always remain at 0 V and be safe. If a malfunction occurs that connects the case to the circuit, a large current will pass through the ground wire to the earth and cause a fuse to blow. This is the *only* time the ground wire would ever have a current, and it is *not* a normal operation of the circuit.

EXAMPLE 31.12   A grounded circuit

Suppose the circuit of Figure 31.32 were grounded at the junction between the two resistors instead of at the bottom. Find the potential at each corner of the circuit.

VISUALIZE FIGURE 31.33 shows the new circuit. (It is customary to draw the ground symbol so that its "point" is always down.)

FIGURE 31.33 Circuit of Figure 31.32 grounded at the point between the resistors.

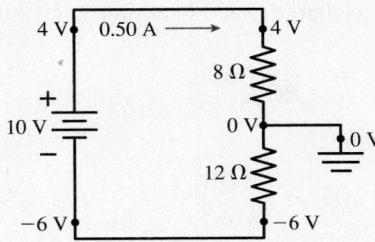

SOLVE Changing the ground point does not affect the circuit's behavior. The current is still 0.50 A, and the potential differences

across the two resistors are still 4 V and 6 V. All that has happened is that we have moved the $V = 0$ V reference point. Because the earth has $V_{earth} = 0$ V, the junction itself now has a potential of 0 V. The potential decreases by 4 V as charge flows through the 8 Ω resistor. Because it *ends* at 0 V, the potential at the top of the 8 Ω resistor must be +4 V. Similarly, the potential decreases by 6 V through the 12 Ω resistor. Because it *starts* at 0 V, the bottom of the 12 Ω resistor must be at −6 V. The negative battery terminal is at the same potential as the bottom of the 12 Ω resistor, because they are connected by a wire, so $V_{neg} = -6$ V. Finally, the potential increases by 10 V as the charge flows through the battery, so $V_{pos} = +4$ V, in agreement, as it should be, with the potential at the top of the 8 Ω resistor.

ASSESS A negative voltage means only that the potential at that point is less than the potential at some other point that we chose to call $V = 0$ V. Only potential *differences* are physically meaningful, and only potential differences enter into Ohm's law: $I = \Delta V/R$. The potential difference across the 12 Ω resistor in this example is 6 V, decreasing from top to bottom, regardless of which point we choose to call $V = 0$ V.

31.9 *RC* Circuits

Thus far we've considered only circuits in which the current is steady and continuous. There are many circuits in which the time dependence of the current is a crucial feature. Charging and discharging a capacitor is an important example.

FIGURE 31.34a shows a charged capacitor, a switch, and a resistor. The capacitor has charge Q_0 and potential difference $\Delta V_0 = Q_0/C$. There is no current, so the potential difference across the resistor is zero. Then, at $t = 0$, the switch closes and the capacitor begins to discharge through the resistor. A circuit such as this, with resistors and capacitors, is called an **RC circuit.**

How long does the capacitor take to discharge? How does the current through the resistor vary as a function of time? To answer these questions, **FIGURE 31.34b** shows the circuit at some point in time after the switch was closed.

Kirchhoff's loop law is valid for any circuit, not just circuits with batteries. The loop law applied to the circuit of Figure 31.34b, going around the loop cw, is

$$\Delta V_{cap} + \Delta V_{res} = \frac{Q}{C} - IR = 0 \qquad (31.27)$$

Q and I in this equation are the *instantaneous* values of the capacitor charge and resistor current.

FIGURE 31.34 An *RC* circuit.

(a) Before the switch closes

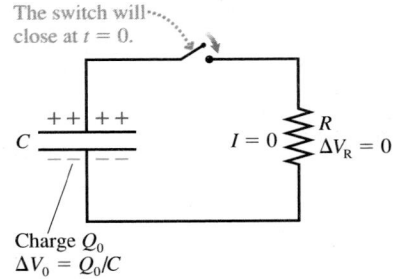

Charge Q_0
$\Delta V_0 = Q_0/C$

(b) After the switch closes

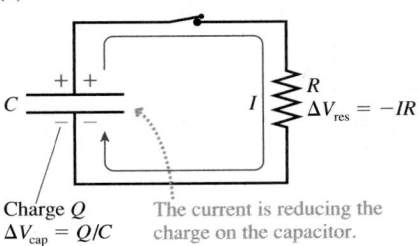

Charge Q The current is reducing the
$\Delta V_{cap} = Q/C$ charge on the capacitor.

The rear flasher on a bike helmet flashes on and off. The timing is controlled by an *RC* circuit.

The current I is the rate at which charge flows through the resistor: $I = dq/dt$. But the charge flowing through the resistor is charge that was *removed* from the capacitor. That is, an infinitesimal charge dq flows through the resistor when the capacitor charge *decreases* by dQ. Thus $dq = -dQ$, and the resistor current is related to the instantaneous capacitor charge by

$$I = -\frac{dQ}{dt} \tag{31.28}$$

Now I is positive when Q is decreasing, as we would expect. The reasoning that has led to Equation 31.28 is rather subtle but very important. You'll see the same reasoning later in other contexts.

If we substitute Equation 31.28 into Equation 31.27 and then divide by R, the loop law for the *RC* circuit becomes

$$\frac{dQ}{dt} + \frac{Q}{RC} = 0 \tag{31.29}$$

Equation 31.29 is a first-order differential equation for the capacitor charge Q, but one that we can solve by direct integration. First, we rearrange Equation 31.29 to get all the charge terms on one side of the equation:

$$\frac{dQ}{Q} = -\frac{1}{RC} dt$$

The product RC is a constant for any particular circuit.

The capacitor charge was Q_0 at $t = 0$ when the switch was closed. We want to integrate from these starting conditions to charge Q at a later time t. That is,

$$\int_{Q_0}^{Q} \frac{dQ}{Q} = -\frac{1}{RC} \int_0^t dt \tag{31.30}$$

Both are well-known integrals, giving

$$\ln Q \Big|_{Q_0}^{Q} = \ln Q - \ln Q_0 = \ln\left(\frac{Q}{Q_0}\right) = -\frac{t}{RC}$$

We can solve for the capacitor charge Q by taking the exponential of both sides, then multiplying by Q_0. Doing so gives

$$Q = Q_0 e^{-t/RC} \tag{31.31}$$

Notice that $Q = Q_0$ at $t = 0$, as expected.

The argument of an exponential function must be dimensionless, so the quantity RC must have dimensions of time. It is useful to define the **time constant** τ of the *RC* circuit as

$$\tau = RC \tag{31.32}$$

We can then write Equation 31.31 as

$$Q = Q_0 e^{-t/\tau} \tag{31.33}$$

And because the capacitor voltage is directly proportional to the charge, it also decays exponentially as

$$\Delta V_C = \Delta V_0 e^{-t/\tau} \tag{31.34}$$

The meaning of Equation 31.33 is easier to understand if we portray it graphically. FIGURE 31.35a shows the capacitor charge as a function of time. The charge decays exponentially, starting from Q_0 at $t = 0$ and asymptotically approaching zero as $t \to \infty$. The time constant τ is the time at which the charge has decreased to e^{-1} (about 37%) of its initial value. At time $t = 2\tau$, the charge has decreased to e^{-2} (about 13%) of its initial value. A voltage graph would have the same shape.

> **NOTE** ▶ The *shape* of the graph of Q is always the same, regardless of the specific value of the time constant τ. ◀

FIGURE 31.35 The decay curves of the capacitor charge and the resistor current.

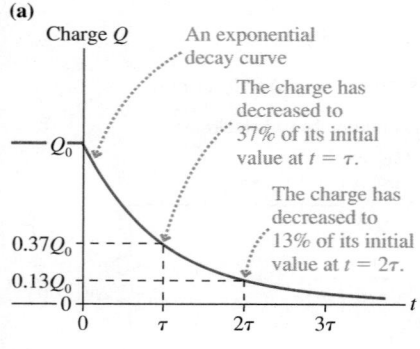

(a)

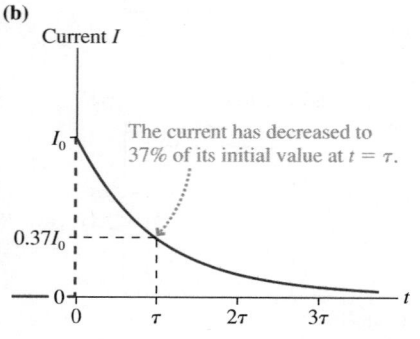

(b)

We find the resistor current by using Equation 31.28:

$$I = -\frac{dQ}{dt} = \frac{Q_0}{\tau}e^{-t/\tau} = \frac{Q_0}{RC}e^{-t/\tau} = \frac{\Delta V_0}{R}e^{-t/\tau} = I_0 e^{-t/\tau} \qquad (31.35)$$

where $I_0 = \Delta V_0/R$ is the initial current, immediately after the switch closes. **FIGURE 31.35b** is a graph of the resistor current versus t. You can see that the current undergoes the same exponential decay, with the same time constant, as the capacitor charge.

NOTE ▶ There's no specific time at which the capacitor has been discharged, because Q approaches zero asymptotically, but the charge and current have dropped to less than 1% of their initial values at $t = 5\tau$. Thus 5τ is a reasonable answer to the question "How long does it take to discharge a capacitor?" ◀

EXAMPLE 31.13 | **Measuring capacitance**

To determine the capacitance of an unmarked capacitor, you set up the circuit shown in **FIGURE 31.36**. After holding the switch in position a for several seconds, you suddenly flip it—at a time you choose to call $t = 0$ s—to position b while monitoring the resistor voltage with a voltmeter. Your measurements are as follows:

Time (s)	Voltage (V)
0.0	9.0
2.0	5.4
4.0	2.7
6.0	1.6
8.0	1.0

FIGURE 31.36 An RC circuit for measuring capacitance.

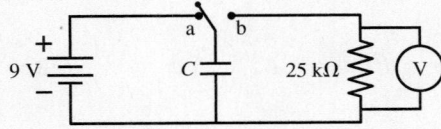

What is the capacitance? And what was the resistor current 5.0 s after the switch changed position?

MODEL The battery charges the capacitor to 9.0 V. Then, when the switch is flipped to position b, the capacitor discharges through the 25,000 Ω resistor with time constant $\tau = RC$.

SOLVE With the switch in position b, the resistor is in parallel with the capacitor and both have the same potential difference $\Delta V_R = \Delta V_C = Q/C$ at all times. The capacitor charge decays exponentially as

$$Q = Q_0 e^{-t/\tau}$$

Consequently, the resistor (and capacitor) voltage also decays exponentially:

$$\Delta V_R = \frac{Q_0}{C}e^{-t/\tau} = \Delta V_0 e^{-t/\tau}$$

where $\Delta V_0 = 9.0$ V is the potential difference at the instant the switch closes. To analyze exponential decays, we take the natural logarithm of both sides. This gives

$$\ln(\Delta V_R) = \ln(\Delta V_0) + \ln(e^{-t/\tau}) = \ln(\Delta V_0) - \frac{1}{\tau}t$$

This result tells us that a graph of $\ln(\Delta V_R)$ versus t—a *semi-log graph*—should be linear with y-intercept $\ln(\Delta V_0)$ and slope $-1/\tau$. If this turns out to be true, we can determine τ and hence C from an experimental measurement of the slope.

FIGURE 31.37 is a graph of $\ln(\Delta V_R)$ versus t. It is, indeed, linear with a negative slope. From the y-intercept of the best-fit line, we find $\Delta V_0 = e^{2.20} = 9.0$ V, as expected. This gives us confidence in our analysis. Using the slope, we find

$$\tau = -\frac{1}{\text{slope}} = -\frac{1}{-0.28 \text{ s}^{-1}} = 3.6 \text{ s}$$

With this, we can calculate

$$C = \frac{\tau}{R} = \frac{3.6 \text{ s}}{25,000 \text{ }\Omega} = 1.4 \times 10^{-4} \text{ F} = 140 \text{ }\mu\text{F}$$

The initial current is $I_0 = (9.0 \text{ V})/(25,000 \text{ }\Omega) = 360 \text{ }\mu\text{A}$. Current also decays exponentially with the same time constant, so the current after 5.0 s is

$$I = I_0 e^{-t/\tau} = (360 \text{ }\mu\text{A})e^{-(5.0 \text{ s})/(3.6 \text{ s})} = 90 \text{ }\mu\text{A}$$

FIGURE 31.37 A semi-log graph of the data.

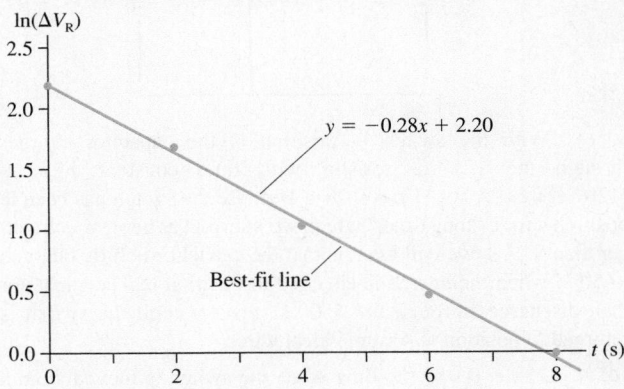

ASSESS The time constant of an exponential decay can be estimated as the time required to decay to one-third of the initial value. Looking at the data, we see that the voltage drops to one-third of the initial 9.0 V in just under 4 s. This is consistent with the more precise $\tau = 3.6$ s, so we have confidence in our results.

FIGURE 31.38 A circuit for charging a capacitor.

(a)

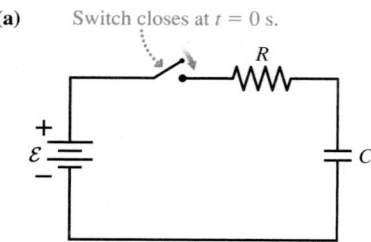

Switch closes at $t = 0$ s.

(b)

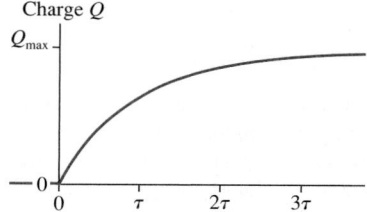

Charge Q

Charging a Capacitor

FIGURE 31.38a shows a circuit that charges a capacitor. After the switch is closed, the battery's charge escalator moves charge from the bottom electrode of the capacitor to the top electrode. The resistor, by limiting the current, slows the process but doesn't stop it. The capacitor charges until $\Delta V_C = \mathcal{E}$; then the charging current ceases. The full charge of the capacitor is $Q_{max} = C(\Delta V_C)_{max} = C\mathcal{E}$.

As a homework problem, you can show that the capacitor charge at time t is

$$Q = Q_{max}(1 - e^{-t/\tau}) \tag{31.36}$$

where again $\tau = RC$. This "upside-down decay" to Q_{max} is shown graphically in **FIGURE 31.38b**. RC circuits that alternately charge and discharge a capacitor are at the heart of time-keeping circuits in computers and other digital electronics.

STOP TO THINK 31.6 The time constant for the discharge of this capacitor is

a. 5 s.
b. 4 s.
c. 2 s.
d. 1 s.
e. The capacitor doesn't discharge because the resistors cancel each other.

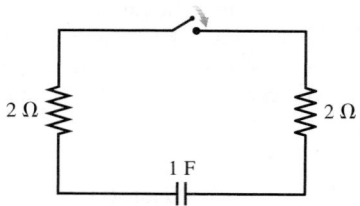

CHALLENGE EXAMPLE 31.14 **Energy dissipated during a capacitor discharge**

The switch in **FIGURE 31.39** has been in position a for a long time. It is suddenly switched to position b for 1.0 s, then back to a. How much energy is dissipated by the 5500 Ω resistor?

FIGURE 31.39 Circuit of a switched capacitor.

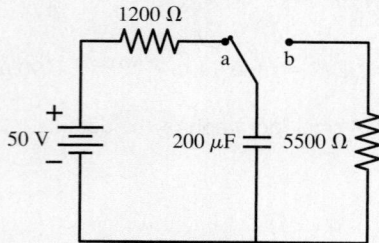

MODEL With the switch in position a, the capacitor charges through the 1200 Ω resistor with time constant $\tau_{charge} = (1200\ \Omega)(2.0 \times 10^{-4}\ F) = 0.24$ s. Because the switch has been in position a for a "long time," which we interpret as being much longer than 0.24 s, we will assume that the capacitor is fully charged to 50 V when the switch is changed to position b. The capacitor then discharges through the 5500 Ω resistor until the switch is returned to position a. Assume ideal wires.

SOLVE Let $t = 0$ s be the time when the switch is moved from a to b, initiating the discharge. The battery and 1200 Ω resistor are irrelevant during the discharge, so the circuit looks like that of Figure 31.34b. The time constant is $\tau = (5500\ \Omega)(2.0 \times 10^{-4}\ F) = 1.1$ s, so the capacitor voltage decreases from 50 V at $t = 0$ s to

$$\Delta V_C = (50\ V)e^{-(1.0\ s)/(1.1\ s)} = 20\ V$$

at $t = 1.0$ s.

There are two ways to determine the energy dissipated in the resistor. We learned in Section 31.3 that a resistor dissipates energy at the rate $dE/dt = P_R = I^2R$. The current decays exponentially as $I = I_0 \exp(-t/\tau)$, with $I_0 = \Delta V_0/R = 9.09$ mA. We can find the energy dissipated during a time T by integrating:

$$\Delta E = \int_0^T I^2R\, dt = I_0^2 R \int_0^T e^{-2t/\tau}\, dt = -\frac{1}{2}\tau I_0^2 Re^{-2t/\tau}\Big|_0^T$$

$$= \frac{1}{2}\tau I_0^2 R\left(1 - e^{-2T/\tau}\right)$$

The 2 in the exponent appears because we squared the expression for I. Evaluating for $T = 1.0$ s, we find

$$\Delta E = \frac{1}{2}(1.1\ s)(0.00909\ A)^2 (5500\ \Omega)\left(1 - e^{-(2.0\ s)/(1.1\ s)}\right) = 0.21\ J$$

Alternatively, we can use the known capacitor voltages at $t = 0$ s and $t = 1.0$ s and $U_C = \frac{1}{2}C(\Delta V_C)^2$ to calculate the energy stored in the capacitor at these times:

$$U_C\ (t = 0.0\ s) = \frac{1}{2}(2.0 \times 10^{-4}\ F)(50\ V)^2 = 0.25\ J$$

$$U_C\ (t = 1.0\ s) = \frac{1}{2}(2.0 \times 10^{-4}\ F)(20\ V)^2 = 0.04\ J$$

The capacitor has lost $\Delta E = 0.21$ J of energy, and this energy was dissipated by the current through the resistor.

ASSESS Not every problem can be solved two ways, but doing so when it's possible gives us great confidence in our result.

SUMMARY

The goal of Chapter 31 has been to understand the fundamental physical principles that govern electric circuits.

General Strategy

MODEL Assume that wires and, where appropriate, batteries are ideal.

VISUALIZE Draw a circuit diagram. Label all known and unknown quantities.

SOLVE Base the solution on Kirchhoff's laws.

- Reduce the circuit to the smallest possible number of equivalent resistors.
- Write one loop equation for each independent loop.
- Find the current and the potential difference.
- Rebuild the circuit to find I and ΔV for each resistor.

ASSESS Verify that

- The sum of potential differences across series resistors matches ΔV for the equivalent resistor.
- The sum of the currents through parallel resistors matches I for the equivalent resistor.

Kirchhoff's loop law

For a closed loop:

- Assign a direction to the current I.
- $\sum (\Delta V)_i = 0$

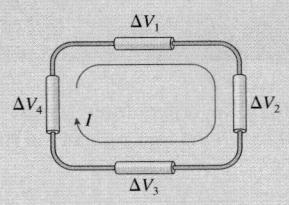

Kirchhoff's junction law

For a junction:

- $\sum I_{in} = \sum I_{out}$

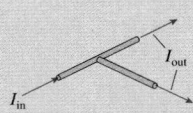

Important Concepts

Ohm's Law

A potential difference ΔV between the ends of a conductor with resistance R creates a current

$$I = \frac{\Delta V}{R}$$

Signs of ΔV for Kirchhoff's loop law

$$\Delta V_{bat} = +\mathcal{E} \qquad \Delta V_{bat} = -\mathcal{E} \qquad \Delta V_{res} = -IR$$

The energy used by a circuit is supplied by the emf $\mathcal{E}$ of the battery through the energy transformations

$$E_{chem} \rightarrow U \rightarrow K \rightarrow E_{th}$$

The battery *supplies* energy at the rate

$$P_{bat} = I\mathcal{E}$$

The resistors *dissipate* energy at the rate

$$P_R = I\Delta V_R = I^2R = \frac{(\Delta V_R)^2}{R}$$

Applications

Series resistors

$$R_{eq} = R_1 + R_2 + R_3 + \cdots$$

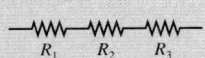

Parallel resistors

$$R_{eq} = \left(\frac{1}{R_1} + \frac{1}{R_2} + \frac{1}{R_3} + \cdots \right)^{-1}$$

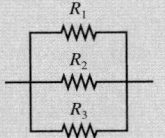

RC circuits

The discharge of a capacitor through a resistor satisfies:

$$Q = Q_0 e^{-t/\tau}$$

$$I = -\frac{dQ}{dt} = \frac{Q_0}{\tau} e^{-t/\tau} = I_0 e^{-t/\tau}$$

where $\tau = RC$ is the **time constant.**

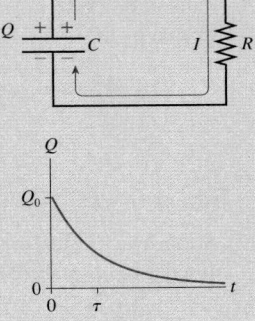

Terms and Notation

circuit diagram	source	internal resistance, r	grounded
Kirchhoff's junction law	kilowatt hour, kWh	terminal voltage, ΔV_{bat}	RC circuit
Kirchhoff's loop law	series resistors	short circuit	time constant, τ
complete circuit	equivalent resistance, R_{eq}	parallel resistors	
load	ammeter	voltmeter	

CONCEPTUAL QUESTIONS

1. Rank in order, from largest to smallest, the currents I_a to I_d through the four resistors in FIGURE Q31.1.

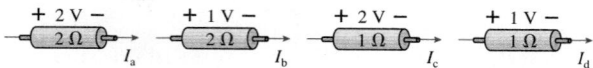

FIGURE Q31.1

2. The tip of a flashlight bulb is touching the top of the 3 V battery in FIGURE Q31.2. Does the bulb light? Why or why not?

FIGURE Q31.2

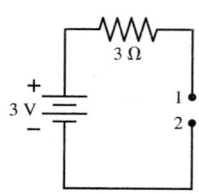

FIGURE Q31.3

3. The wire is broken on the right side of the circuit in FIGURE Q31.3. What is the potential difference ΔV_{12} between points 1 and 2? Explain.

4. The circuit of FIGURE Q31.4 has two resistors, with $R_1 > R_2$. Which of the two resistors dissipates the larger amount of power? Explain.

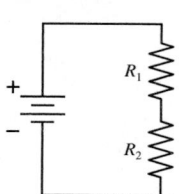

FIGURE Q31.4

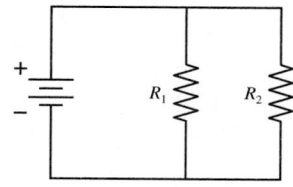

FIGURE Q31.5

5. The circuit of FIGURE Q31.5 has two resistors, with $R_1 > R_2$. Which of the two resistors dissipates the larger amount of power? Explain.

6. Rank in order, from largest to smallest, the powers P_a to P_d dissipated by the four resistors in FIGURE Q31.6.

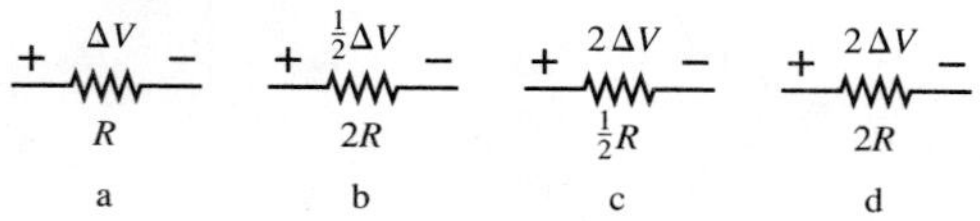

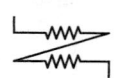

FIGURE Q31.6

7. Are the two resistors in FIGURE Q31.7 in series or in parallel? Explain.

FIGURE Q31.7

8. A battery with internal resistance r is connected to a load resistance R. If R is increased, does the terminal voltage of the battery increase, decrease, or stay the same? Explain.

9. Initially bulbs A and B in FIGURE Q31.9 are glowing. What happens to each bulb if the switch is closed? Does it get brighter, stay the same, get dimmer, or go out? Explain.

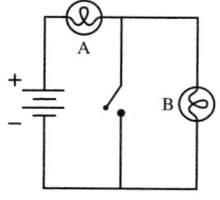

FIGURE Q31.9

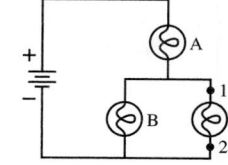

FIGURE Q31.10

10. Bulbs A, B, and C in FIGURE Q31.10 are identical, and all are glowing.
    a. Rank in order, from most to least, the brightnesses of the three bulbs. Explain.
    b. Suppose a wire is connected between points 1 and 2. What happens to each bulb? Does it get brighter, stay the same, get dimmer, or go out? Explain.

11. Bulbs A and B in FIGURE Q31.11 are identical, and both are glowing. Bulb B is removed from its socket. Does the potential difference ΔV_{12} between points 1 and 2 increase, stay the same, decrease, or become zero? Explain.

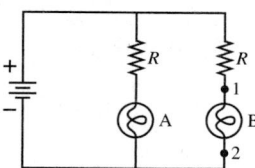

FIGURE Q31.11

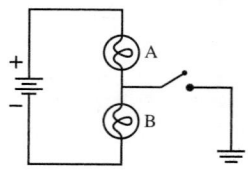

FIGURE Q31.12

12. Bulbs A and B in FIGURE Q31.12 are identical, and both are glowing. What happens to each bulb when the switch is closed? Does its brightness increase, stay the same, decrease, or go out? Explain.

13. FIGURE Q31.13 shows the voltage as a function of time of a capacitor as it is discharged (separately) through three different resistors. Rank in order, from largest to smallest, the values of the resistances R_1 to R_3.

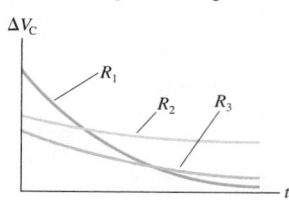

FIGURE Q31.13

EXERCISES AND PROBLEMS

Problems labeled [ ] integrate material from earlier chapters.

Exercises

Section 31.1 Circuit Elements and Diagrams

1. | Draw a circuit diagram for the circuit of FIGURE EX31.1.

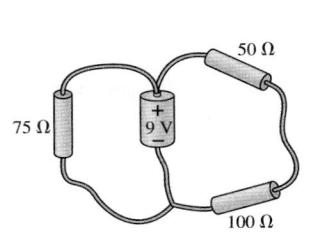

FIGURE EX31.1

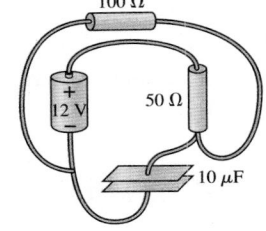

FIGURE EX31.2

2. | Draw a circuit diagram for the circuit of FIGURE EX31.2.

Section 31.2 Kirchhoff's Laws and the Basic Circuit

3. || In FIGURE EX31.3, what is the current in the wire to the right of the junction? Does the charge in this wire flow to the right or to the left?

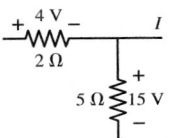

FIGURE EX31.3

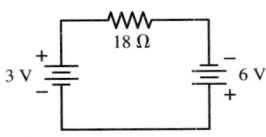

FIGURE EX31.4

4. | a. What are the magnitude and direction of the current in the 18 Ω resistor in FIGURE EX31.4?
   b. Draw a graph of the potential as a function of the distance traveled through the circuit, traveling cw from $V = 0$ V at the lower left corner.

5. | a. What are the magnitude and direction of the current in the 10 Ω resistor in FIGURE EX31.5?
   b. Draw a graph of the potential as a function of the distance traveled through the circuit, traveling cw from $V = 0$ V at the lower left corner.

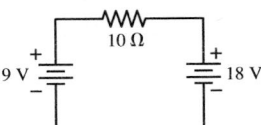

FIGURE EX31.5

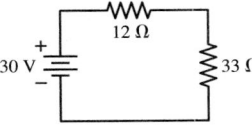

FIGURE EX31.6

6. | What is the potential difference across each resistor in FIGURE EX31.6?

Section 31.3 Energy and Power

7. | What is the resistance of a 1500 W (120 V) hair dryer? What is the current in the hair dryer when it is used?

8. | How much power is dissipated by each resistor in FIGURE EX31.8?

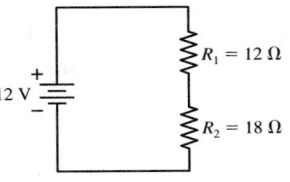

FIGURE EX31.8

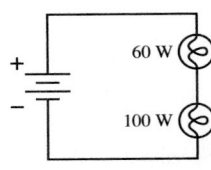

FIGURE EX31.9

9. || A 60 W lightbulb and a 100 W lightbulb are placed one after the other in the circuit of FIGURE EX31.9. The battery's emf is large enough that both bulbs are glowing. Which is the true statement?
   A. The 60 W bulb is brighter.
   B. Both bulbs are equally bright.
   C. The 100 W bulb is brighter.
   D. There's not enough information to tell which bulb is brighter.

10. || A standard 100 W (120 V) lightbulb contains a 7.0-cm-long tungsten filament. The high-temperature resistivity of tungsten is 9.0×10^{-7} Ω m. What is the diameter of the filament?

11. || A typical American family uses 1000 kWh of electricity a month.
    a. What is the average current in the 120 V power line to the house?
    b. On average, what is the resistance of a household?

12. | A waterbed heater uses 450 W of power. It is on 25% of the time, off 75%. What is the annual cost of electricity at a billing rate of $0.12/kWh?

Section 31.4 Series Resistors

Section 31.5 Real Batteries

13. | Two of the three resistors in FIGURE EX31.13 are unknown but equal. The total resistance between points a and b is 200 Ω. What is the value of R?

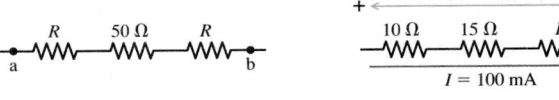

FIGURE EX31.13   **FIGURE EX31.14**

14. | What is the value of resistor R in FIGURE EX31.14?

15. | The battery in FIGURE EX31.15 is short-circuited by an ideal ammeter having zero resistance.
    a. What is the battery's internal resistance?
    b. How much power is dissipated inside the battery?

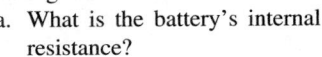
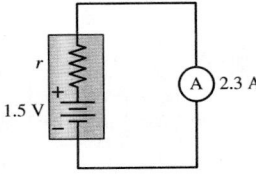

FIGURE EX31.15

16. || The voltage across the terminals of a 9.0 V battery is 8.5 V when the battery is connected to a 20 Ω load. What is the battery's internal resistance?

17. ‖ Compared to an ideal battery, by what percentage does the battery's internal resistance reduce the potential difference across the 20 Ω resistor in FIGURE EX31.17?

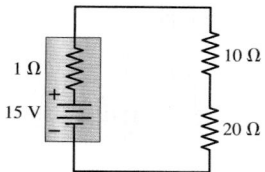

FIGURE EX31.17

Section 31.6 Parallel Resistors

18. ‖ A metal wire of resistance R is cut into two pieces of equal length. The two pieces are connected together side by side. What is the resistance of the two connected wires?

19. | Two of the three resistors in FIGURE EX31.19 are unknown but equal. The total resistance between points a and b is 75 Ω. What is the value of R?

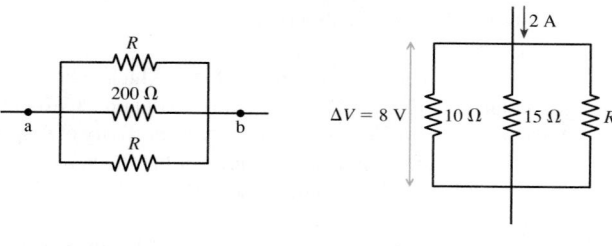

FIGURE EX31.19   FIGURE EX31.20

20. | What is the value of resistor R in FIGURE EX31.20?
21. | What is the equivalent resistance between points a and b in FIGURE EX31.21?

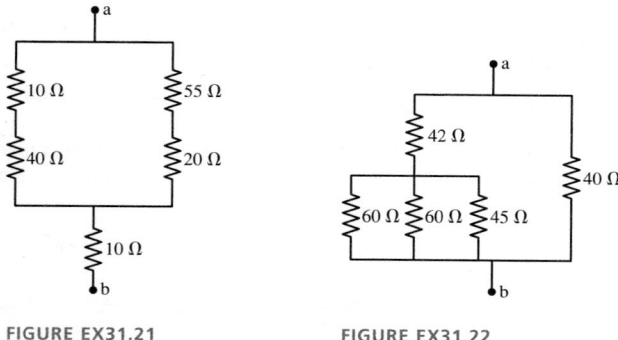

FIGURE EX31.21   FIGURE EX31.22

22. | What is the equivalent resistance between points a and b in FIGURE EX31.22?
23. | What is the equivalent resistance between points a and b in FIGURE EX31.23?

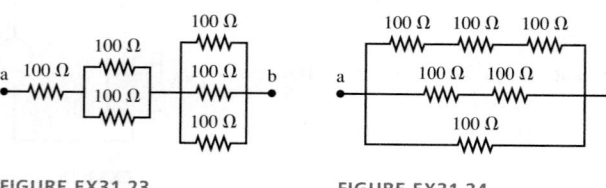

FIGURE EX31.23   FIGURE EX31.24

24. | What is the equivalent resistance between points a and b in FIGURE EX31.24?

Section 31.8 Getting Grounded

25. ‖ In FIGURE EX31.25, what is the value of the potential at points a and b?

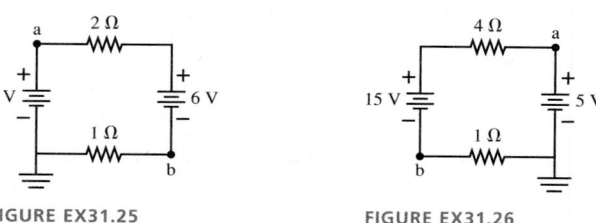

FIGURE EX31.25   FIGURE EX31.26

26. ‖‖ In FIGURE EX31.26, what is the value of the potential at points a and b?

Section 31.9 *RC* Circuits

27. | Show that the product RC has units of s.
28. | What is the time constant for the discharge of the capacitors in FIGURE EX31.28?

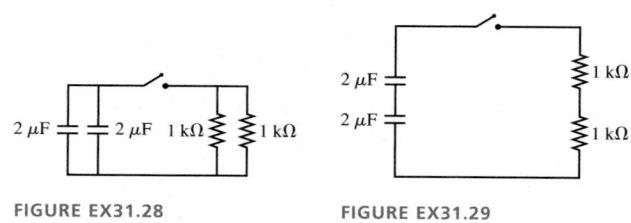

FIGURE EX31.28   FIGURE EX31.29

29. ‖ What is the time constant for the discharge of the capacitors in FIGURE EX31.29?
30. ‖ A 10 μF capacitor initially charged to 20 μC is discharged through a 1.0 kΩ resistor. How long does it take to reduce the capacitor's charge to 10 μC?
31. | The switch in FIGURE EX31.31 has been in position a for a long time. It is changed to position b at $t = 0$ s. What are the charge Q on the capacitor and the current I through the resistor (a) immediately after the switch is closed? (b) at $t = 50$ μs? (c) at $t = 200$ μs?

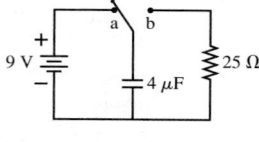

FIGURE EX31.31

32. ‖ What value resistor will discharge a 1.0 μF capacitor to 10% of its initial charge in 2.0 ms?
33. ‖ A capacitor is discharged through a 100 Ω resistor. The discharge current decreases to 25% of its initial value in 2.5 ms. What is the value of the capacitor?

Problems

34. ‖ The five identical bulbs in FIGURE P31.34 are all glowing. The battery is ideal. What is the order of brightness of the bulbs, from brightest to dimmest? Some may be equal.
    A. $P = S > Q = R = T$
    B. $P = S = T > Q = R$
    C. $P > S = T > Q = R$
    D. $P > Q = R > S = T$

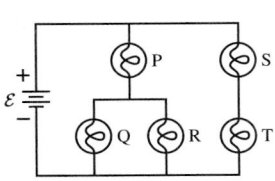

FIGURE P31.34

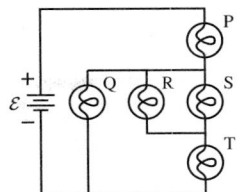

FIGURE P31.35

35. ‖ The five identical bulbs in FIGURE P31.35 are all glowing. The battery is ideal. What is the order of brightness of the bulbs, from brightest to dimmest? Some may be equal.
   A. $P = T > Q = R = S$
   B. $P > Q = R = S > T$
   C. $P = T > Q > R = S$
   D. $P > Q > T > R = S$

36. ‖‖ Two 75 W (120 V) lightbulbs are wired in series, then the combination is connected to a 120 V supply. How much power is dissipated by each bulb?

37. ‖‖ The corroded contacts in a lightbulb socket have 5.0 Ω resistance. How much actual power is dissipated by a 100 W (120 V) lightbulb screwed into this socket?

38. ‖ An electric eel develops a 450 V potential difference between
BIO its head and tail. The eel can stun a fish or other prey by using this potential difference to drive a 0.80 A current pulse for 1.0 ms. What are (a) the energy delivered by this pulse and (b) the total charge that flows?

39. ‖ You have a 2.0 Ω resistor, a 3.0 Ω resistor, a 6.0 Ω resistor, and a 6.0 V battery. Draw a diagram of a circuit in which all three resistors are used and the battery delivers 9.0 W of power.

40. | You have three 12 Ω resistors. Draw diagrams showing how you could arrange all three so that their equivalent resistance is (a) 4.0 Ω, (b) 8.0 Ω, (c) 18 Ω, and (d) 36 Ω.

41. ‖ What is the equivalent resistance between points a and b in FIGURE P31.41?

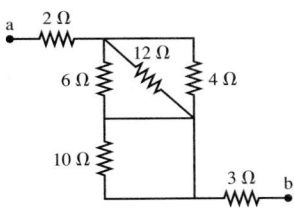

FIGURE P31.41

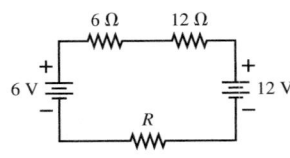

FIGURE P31.42

42. | There is a current of 0.25 A in the circuit of FIGURE P31.42. What is the power dissipated by R?

43. ‖ A variable resistor R is connected across the terminals of a battery. FIGURE P31.43 shows the current in the circuit as R is varied. What are the emf and internal resistance of the battery?

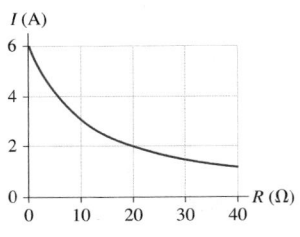

FIGURE P31.43

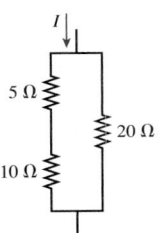

FIGURE P31.44

44. | The 10 Ω resistor in FIGURE P31.44 is dissipating 40 W of power. How much power are the other two resistors dissipating?

45. ‖ What are the emf and internal resistance of the battery in FIGURE P31.45?

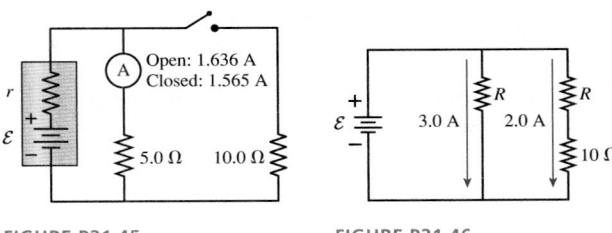

FIGURE P31.45

FIGURE P31.46

46. | What is the emf of the battery in FIGURE P31.46?

47. ‖ A 2.5 V battery with 0.70 Ω internal resistance is connected in parallel with a 1.5 V battery having 0.30 Ω internal resistance. That is, their positive terminals are connected by a wire and their negative terminals are connected by a wire. What is the terminal voltage of the 2.5 V battery?

48. ‖ a. Load resistor R is attached to a battery of emf $\mathcal{E}$ and internal resistance r. For what value of the resistance R, in terms of $\mathcal{E}$ and r, will the power dissipated by the load resistor be a maximum?
   b. What is the maximum power that the load can dissipate if the battery has $\mathcal{E} = 9.0$ V and $r = 1.0$ Ω?
   c. *Why* should the power dissipated by the load have a maximum value? Explain.
   Hint: What happens to the power dissipation when R is either very small or very large?

49. | The ammeter in FIGURE P31.49 reads 3.0 A. Find I_1, I_2, and $\mathcal{E}$.

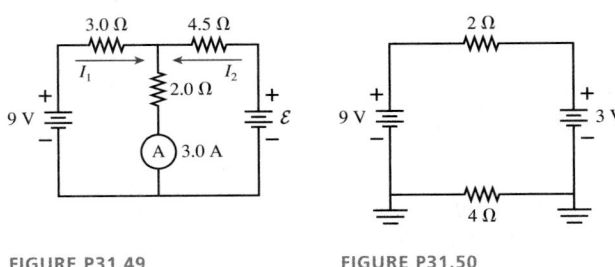

FIGURE P31.49

FIGURE P31.50

50. ‖ What is the current in the 2 Ω resistor in FIGURE P31.50?

51. | It seems hard to justify spending $5 for a compact fluorescent lightbulb when an ordinary incandescent bulb costs 50¢. To see if this makes sense, compare a 60 W incandescent bulb lasting 1000 hours to a 15 W compact fluorescent bulb having a lifetime of 10,000 hours. Both bulbs produce the same amount of visible light and are interchangeable. If electricity costs $0.10/kWh, what is the total cost—purchase plus energy—to obtain 10,000 hours of light from each type of bulb? This is called the *life-cycle cost*.

52. ‖ A refrigerator has a 1000 W compressor, but the compressor runs only 20% of the time.
   a. If electricity costs $0.10/kWh, what is the monthly (30 day) cost of running the refrigerator?
   b. A more energy-efficient refrigerator with an 800 W compressor costs $100 more. If you buy the more expensive refrigerator, how many months will it take to recover your additional cost?

53. | For an ideal battery ($r = 0\ \Omega$), closing the switch in FIGURE P31.53 does not affect the brightness of bulb A. In practice, bulb A dims *just a little* when the switch closes. To see why, assume that the 1.50 V battery has an internal resistance $r = 0.50\ \Omega$ and that the resistance of a glowing bulb is $R = 6.00\ \Omega$.

   a. What is the current through bulb A when the switch is open?
   b. What is the current through bulb A after the switch has closed?
   c. By what percentage does the current through A change when the switch is closed?

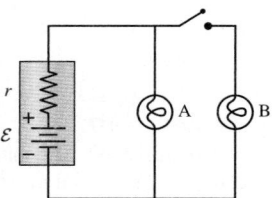

FIGURE P31.53

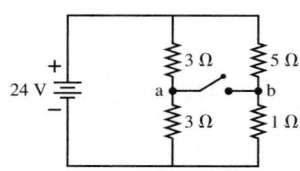

FIGURE P31.54

54. | What are the battery current I_{bat} and the potential difference ΔV_{ab} between points a and b when the switch in FIGURE P31.54 is (a) open and (b) closed?

55. || The circuit in FIGURE P31.55 is called a *voltage divider*. What value of R will make $V_{out} = V_{in}/10$?

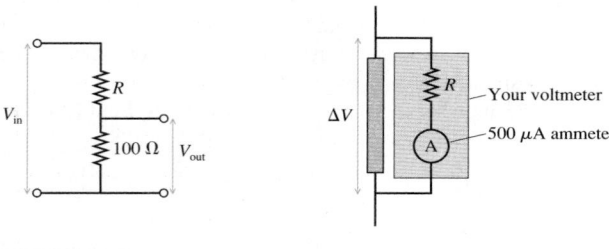

FIGURE P31.55   FIGURE P31.56

56. || A circuit you're building needs a voltmeter that goes from 0 V to a full-scale reading of 5.0 V. Unfortunately, the only meter in the storeroom is an *ammeter* that goes from 0 μA to a full-scale reading of 500 μA. Fortunately, you've just finished a physics class, and you realize that you can convert this meter to a voltmeter by putting a resistor in series with it, as shown in FIGURE P31.56. You've measured that the resistance of the ammeter is 50.0 Ω, not the 0 Ω of an ideal ammeter. What value of R must you use so that the meter will go to full scale when the potential difference across the object being measured is 5.0 V?

57. || A circuit you're building needs an ammeter that goes from 0 mA to a full-scale reading of 50 mA. Unfortunately, the only ammeter in the storeroom goes from 0 μA to a full-scale reading of only 500 μA. Fortunately, you've just finished a physics class, and you realize that you can make this ammeter work by putting a resistor in parallel with it, as shown in FIGURE P31.57. You've measured that the resistance of the ammeter is 50.0 Ω, not the 0 Ω of an ideal ammeter.

   a. What value of R must you use so that the meter will go to full scale when the current I is 50 mA?
   b. What is the effective resistance of your ammeter?

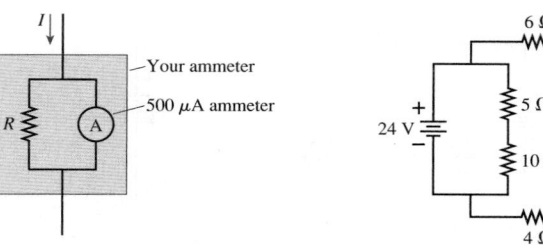

FIGURE P31.57   FIGURE P31.58

58. || For the circuit shown in FIGURE P31.58, find the current through and the potential difference across each resistor. Place your results in a table for ease of reading.

59. || For the circuit shown in FIGURE P31.59, find the current through and the potential difference across each resistor. Place your results in a table for ease of reading.

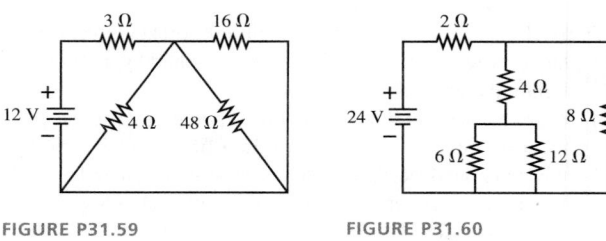

FIGURE P31.59   FIGURE P31.60

60. || For the circuit shown in FIGURE P31.60, find the current through and the potential difference across each resistor. Place your results in a table for ease of reading.

61. || For the circuit shown in FIGURE P31.61, find the current through and the potential difference across each resistor. Place your results in a table for ease of reading.

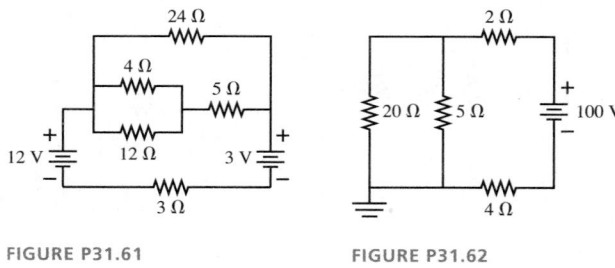

FIGURE P31.61   FIGURE P31.62

62. || What is the current through the 20 Ω resistor in FIGURE P31.62?
63. || What is the current through the 10 Ω resistor in FIGURE P31.63? Is the current from left to right or right to left?

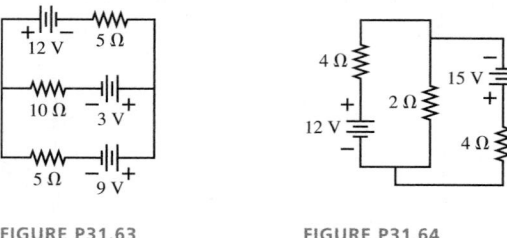

FIGURE P31.63   FIGURE P31.64

64. || What power is dissipated by the 2 Ω resistor in FIGURE P31.64?

65. ‖ For what emf $\mathcal{E}$ does the 200 Ω resistor in FIGURE P31.65 dissipate no power? Should the emf be oriented with its positive terminal at the top or at the bottom?

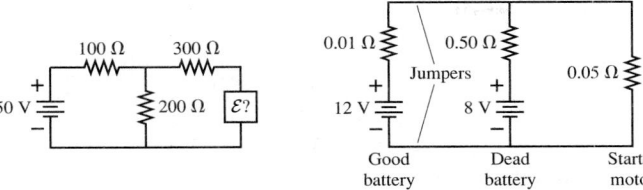

FIGURE P31.65   FIGURE P31.66

66. ‖ A 12 V car battery dies not so much because its voltage drops but because chemical reactions increase its internal resistance. A good battery connected with jumper cables can both start the engine and recharge the dead battery. Consider the automotive circuit of FIGURE P31.66.
   a. How much current could the good battery alone drive through the starter motor?
   b. How much current is the dead battery alone able to drive through the starter motor?
   c. With the jumper cables attached, how much current passes through the starter motor?
   d. With the jumper cables attached, how much current passes through the dead battery, and in which direction?

67. ‖ How much current flows through the bottom wire in FIGURE P31.67, and in which direction?

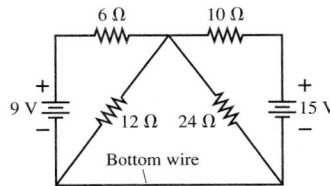

FIGURE P31.67

68. ‖ The capacitor in an RC circuit is discharged with a time constant of 10 ms. At what time after the discharge begins are (a) the charge on the capacitor reduced to half its initial value and (b) the energy stored in the capacitor reduced to half its initial value?

69. ‖ A circuit you're using discharges a 20 μF capacitor through an unknown resistor. After charging the capacitor, you close a switch at $t = 0$ s and then monitor the resistor current with an ammeter. Your data are as follows:

Time (s)	Current (μA)
0.5	890
1.0	640
1.5	440
2.0	270
2.5	200

Use an appropriate graph of the data to determine (a) the resistance and (b) the initial capacitor voltage.

70. ‖ A 150 μF defibrillator capacitor is charged to 1500 V. When
BIO fired through a patient's chest, it loses 95% of its charge in 40 ms. What is the resistance of the patient's chest?

71. ‖ A 50 μF capacitor that had been charged to 30 V is discharged through a resistor. FIGURE P31.71 shows the capacitor voltage as a function of time. What is the value of the resistance?

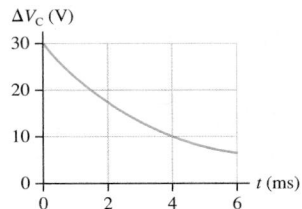

FIGURE P31.71

72. ‖ A 0.25 μF capacitor is charged to 50 V. It is then connected in series with a 25 Ω resistor and a 100 Ω resistor and allowed to discharge completely. How much energy is dissipated by the 25 Ω resistor?

73. ‖ The capacitor in FIGURE P31.73 begins to charge after the switch closes at $t = 0$ s.
   a. What is ΔV_C a very long time after the switch has closed?
   b. What is Q_{max} in terms of $\mathcal{E}$, R, and C?
   c. In this circuit, does $I = +dQ/dt$ or $-dQ/dt$? Explain.
   d. Find an expression for the current I at time t. Graph I from $t = 0$ to $t = 5\tau$.

FIGURE P31.73

74. ‖ The capacitors in FIGURE P31.74 are charged and the switch closes at $t = 0$ s. At what time has the current in the 8 Ω resistor decayed to half the value it had immediately after the switch was closed?

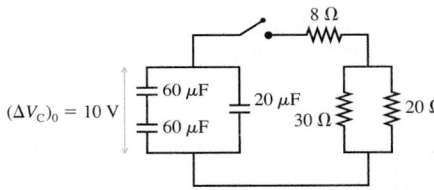

FIGURE P31.74

Challenge Problems

75. You've made the finals of the Science Olympics! As one of your tasks, you're given 1.0 g of aluminum and asked to make a wire, using all the aluminum, that will dissipate 7.5 W when connected to a 1.5 V battery. What length and diameter will you choose for your wire?

76. The switch in FIGURE CP31.76 has been closed for a very long time.
   a. What is the charge on the capacitor?
   b. The switch is opened at $t = 0$ s. At what time has the charge on the capacitor decreased to 10% of its initial value?

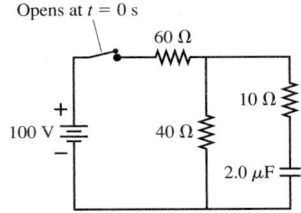

FIGURE CP31.76

77. A capacitor-charging circuit has a time constant of 40 ms. When the switch is closed, the initial current to the 50 μF capacitor is 65 mA. What is the capacitor's voltage after 20 ms? Assume the capacitor was completely uncharged when the switch closed.

78. The capacitor in Figure 31.38a begins to charge after the switch closes at $t = 0$ s. Analyze this circuit and show that $Q = Q_{max}(1 - e^{-t/\tau})$, where $Q_{max} = C\mathcal{E}$.

79. The switch in Figure 31.38a closes at $t = 0$ s and, after a very long time, the capacitor is fully charged. Find expressions for (a) the total energy supplied by the battery as the capacitor is being charged, (b) total energy dissipated by the resistor as the capacitor is being charged, and (c) the energy stored in the capacitor when it is fully charged. Your expressions will be in terms of $\mathcal{E}$, R, and C. (d) Do your results for parts a to c show that energy is conserved? Explain.

80. An *oscillator circuit* is important to many applications. A simple oscillator circuit can be built by adding a neon gas tube to an *RC* circuit, as shown in **FIGURE CP31.80**. Gas is normally a good insulator, and the resistance of the gas tube is essentially infinite when the light is off. This allows the capacitor to charge. When the capacitor voltage reaches a value V_{on}, the electric field inside the tube becomes strong enough to ionize the neon gas. Visually, the tube lights with an orange glow. Electrically, the ionization of the gas provides a very-low-resistance path through the tube. The capacitor very rapidly (we can think of it as instantaneously) discharges through the tube and the capacitor voltage drops. When the capacitor voltage has dropped to a value V_{off}, the electric field inside the tube becomes too weak to sustain the ionization and the neon light turns off. The capacitor then starts to charge again. The capacitor voltage oscillates between V_{off}, when it starts charging, and V_{on}, when the light comes on to discharge it.

a. Show that the oscillation period is

$$T = RC\ln\left(\frac{\mathcal{E} - V_{off}}{\mathcal{E} - V_{on}}\right)$$

b. A neon gas tube has $V_{on} = 80$ V and $V_{off} = 20$ V. What resistor value should you choose to go with a 10 μF capacitor and a 90 V battery to make a 10 Hz oscillator?

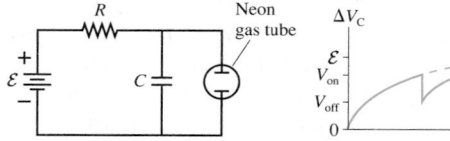

FIGURE CP31.80

81. A 2.0-m-long, 1.0-mm-diameter wire has a variable resistivity given by

$$\rho(x) = (2.5 \times 10^{-6})\left[1 + \left(\frac{x}{1.0 \text{ m}}\right)^2\right] \Omega \text{ m}$$

where x is measured from one end of the wire. What is the current if this wire is connected to the terminals of a 9.0 V battery?

STOP TO THINK ANSWERS

Stop to Think 31.1: a, b, and d. These three are the same circuit because the logic of the connections is the same. In c, the functioning of the circuit is changed by the extra wire connecting the two sides of the capacitor.

Stop to Think 31.2: ΔV increases by 2 V in the direction of I. Kirchhoff's loop law, starting on the left side of the battery, is then $+12$ V $+ 2$ V $- 8$ V $- 6$ V $= 0$ V.

Stop to Think 31.3: $P_b > P_d > P_a > P_c$. The power dissipated by a resistor is $P_R = (\Delta V_R)^2/R$. Increasing R decreases P_R; increasing ΔV_R increases P_R. But the potential has a larger effect because P_R depends on the square of ΔV_R.

Stop to Think 31.4: $I = 2$ A for all. $V_a = 20$ V, $V_b = 16$ V, $V_c = 10$ V, $V_d = 8$ V, $V_e = 0$ V. Current is conserved. The potential is 0 V on the right and increases by IR for each resistor going to the left.

Stop to Think 31.5: A > B > C = D. All the current from the battery goes through A, so it is brightest. The current divides at the junction, but not equally. Because B is in parallel with C + D but has half the resistance, twice as much current travels through B as through C + D. So B is dimmer than A but brighter than C and D. C and D are equal because of conservation of current.

Stop to Think 31.6: b. The two 2 Ω resistors are in series and equivalent to a 4 Ω resistor. Thus $\tau = RC = 4$ s.

32 The Magnetic Field

The aurora occurs when high-energy charged particles from the sun are steered into the upper atmosphere by the earth's magnetic field.

▶ **Looking Ahead** The goal of Chapter 32 is to learn how to calculate and use the magnetic field.

Magnetic Fields

Magnetism has been known since antiquity. Whereas electricity is understood in terms of electric charges, magnetism is based on **magnetic poles.** You'll learn how to use the **magnetic field,** with symbol $\vec{B}$, to work with the long-range interactions of magnetism.

Iron filings, like little compasses, show the shape of the magnetic field.

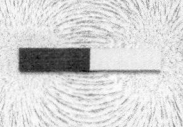

This bar magnet— a *dipole*, with a north and a south pole—is a permanent magnet.

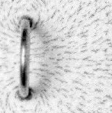

A loop of current also creates a dipole magnetic field.

One of our key tasks will be to understand the connection between electromagnets and permanent magnets.

Compasses work because the earth is a large magnet. It is an electromagnet, with circulating currents in its molten iron core.

Magnetic Forces

Magnetic fields exert forces on *moving* charged particles. The force is perpendicular to the plane of $\vec{v}$ and $\vec{B}$.

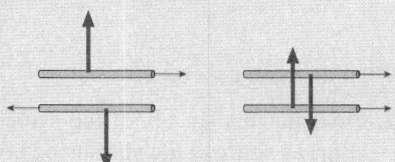

Currents are moving charged particles. You'll learn that currents create magnetic fields, and currents exert magnetic forces on each other. Opposite currents repel, parallel currents attract.

Magnetic Torque

Magnetic forces exert a *torque* on a current traveling around a closed loop.

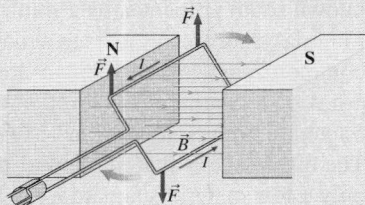

You'll learn that motors work because of magnetic torque.

◀ **Looking Back**
Sections 12.5 and 12.10 Torque and the vector cross product

Motion of Charges

The magnetic force causes charged particles to move in circular orbits in a magnetic field. This **cyclotron motion** has many important applications, from particle accelerators to the aurora.

Magnetism is three dimensional. You'll learn how to represent vectors perpendicular to a plane. Here the ×'s show a magnetic field into the page.

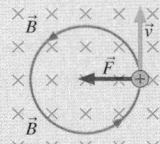

◀ **Looking Back**
Sections 8.2–8.3 Circular motion

Magnetic Materials

Iron and a few other materials exhibit pronounced magnetic properties, including the ability to form permanent magnets. You'll learn that **ferromagnetism** arises because electrons have an inherent magnetic moment called **electron spin.**

This hard disk is made of nickel, a magnetic material. It stores digital data—1's and 0's—in the alignment of microscopic **magnetic domains.**

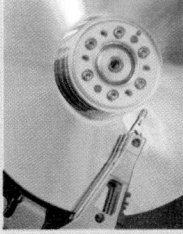

32.1 Magnetism

We began our investigation of electricity in Chapter 25 by looking at the results of simple experiments with charged rods. We'll do the same with magnetism.

Discovering magnetism

Experiment 1

If a bar magnet is taped to a piece of cork and allowed to float in a dish of water, it always turns to align itself in an approximate north-south direction. The end of a magnet that points north is called the *north-seeking pole,* or simply the **north pole.** The other end is the **south pole.**

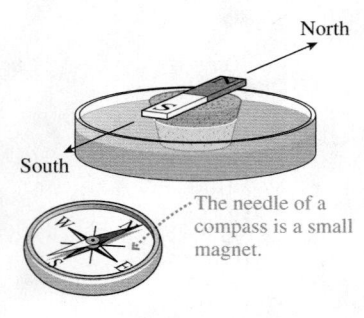

North

South

The needle of a compass is a small magnet.

Experiment 2

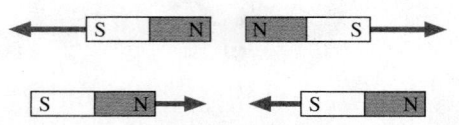

If the north pole of one magnet is brought near the north pole of another magnet, they repel each other. Two south poles also repel each other, but the north pole of one magnet exerts an attractive force on the south pole of another magnet.

Experiment 3

The north pole of a bar magnet attracts one end of a compass needle and repels the other. Apparently the compass needle itself is a little bar magnet with a north pole and a south pole.

Experiment 4

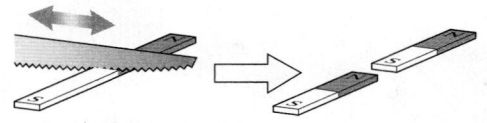

Cutting a bar magnet in half produces two weaker but still complete magnets, each with a north pole and a south pole. No matter how small the magnets are cut, even down to microscopic sizes, each piece remains a complete magnet with two poles.

Experiment 5

Magnets can pick up some objects, such as paper clips, but not all. If an object is attracted to one end of a magnet, it is also attracted to the other end. Most materials, including copper (a penny), aluminum, glass, and plastic, experience no force from a magnet.

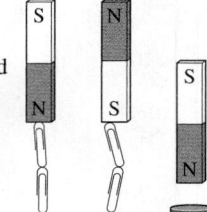

Experiment 6

A magnet does not affect an electroscope. A charged rod exerts a weak *attractive* force on *both* ends of a magnet. However, the force is the same as the force on a metal bar that isn't a magnet, so it is simply a polarization force like the ones we studied in Chapter 25. Other than polarization forces, charges have *no effects* on magnets.

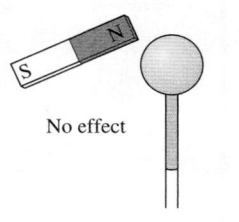

No effect

What do these experiments tell us?

1. First, magnetism is not the same as electricity. **Magnetic poles and electric charges share some similar behavior, but they are not the same.**
2. Magnetism is a long-range force. Paper clips leap upward to a magnet.
3. Magnets have two poles, called north and south poles, and thus are **magnetic dipoles.** Two like poles exert repulsive forces on each other; two opposite poles attract. The behavior is *analogous* to electric charges, but, as noted, magnetic poles and electric charges are *not* the same. Unlike charges, isolated north or south poles do not exist.
4. The poles of a bar magnet can be identified by using it as a compass. The poles of other magnets, such as flat refrigerator magnets, can be identified by testing them against a bar magnet. A pole that attracts a known north pole and repels a known south pole must be a south magnetic pole.
5. Materials that are attracted to a magnet are called **magnetic materials.** The most common magnetic material is iron. Magnetic materials are attracted to *both* poles of a magnet. This attraction is analogous to how neutral objects are attracted to both positively and negatively charged rods by the polarization force. The difference is that *all* neutral objects are attracted to a charged rod whereas only a few materials are attracted to a magnet.

Our goal is to develop a theory of magnetism to explain these observations.

Compasses and Geomagnetism

The north pole of a compass needle is attracted toward the geographic north pole of the earth. Apparently the earth itself is a large magnet, as shown in FIGURE 32.1. The reasons for the earth's magnetism are complex, but geophysicists think that the earth's magnetic poles arise from currents in its molten iron core. Two interesting facts about the earth's magnetic field are (1) that the magnetic poles are offset slightly from the geographic poles of the earth's rotation axis, and (2) that the geographic north pole is actually a *south* magnetic pole! You should be able to use what you have learned thus far to convince yourself that this is the case.

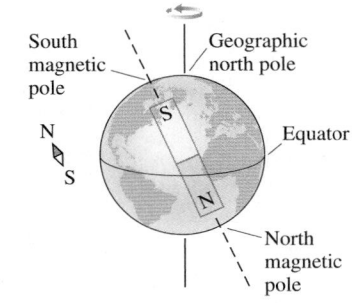

FIGURE 32.1 The earth is a large magnet.

STOP TO THINK 32.1   Does the compass needle rotate clockwise (cw), counterclockwise (ccw), or not at all?

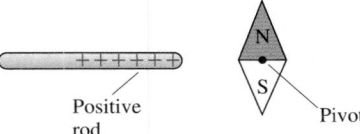

32.2 The Discovery of the Magnetic Field

As electricity began to be seriously studied in the 18th century, some scientists specu-lated that there might be a connection between electricity and magnetism. Interest-ingly, the link between electricity and magnetism was discovered *in the midst of a classroom lecture demonstration* in 1819 by the Danish scientist Hans Christian Oersted. Oersted was using a battery—a fairly recent invention—to produce a large current in a wire. By chance, a compass was sitting next to the wire, and Oersted no-ticed that the current caused the compass needle to turn. In other words, the compass responded as if a magnet had been brought near.

Oersted had long been interested in a possible connection between electricity and magnetism, so the significance of this serendipitous observation was immediately ap-parent to him. Oersted's discovery that **magnetism is caused by an electric current** is illustrated in FIGURE 32.2. Part c of the figure demonstrates an important **right-hand rule** that relates the orientation of the compass needles to the direction of the current.

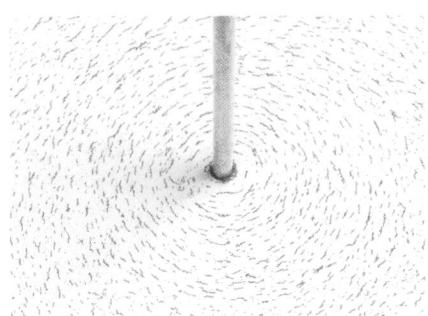

The magnetic field is revealed by the pattern of iron filings around a current-carrying wire.

FIGURE 32.2 Response of compass needles to a current in a straight wire.

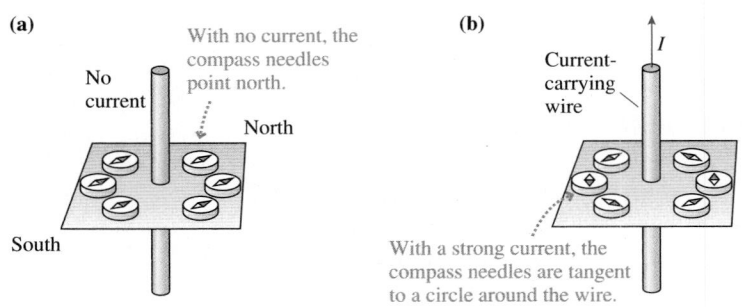

Magnetism is more demanding than electricity in requiring a three-dimensional perspective of the sort shown in Figure 32.2. But since two-dimensional figures are easier to draw, we will make as much use of them as we can. Consequently, we will often need to indicate field vectors or currents that are perpendicular to the page. FIGURE 32.3 shows the notation we will use. FIGURE 32.4 on the next page demonstrates this notation by showing the compasses around a current that is directed into the page. To use the right-hand rule, point your right thumb in the direction of the current (into the page). Your fingers will curl cw, and that is the direction in which the north poles of the compass needles point.

FIGURE 32.3 The notation for vectors and currents perpendicular to the page.

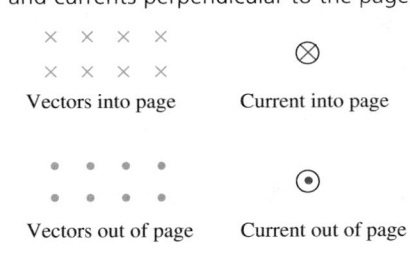

FIGURE 32.4 The orientation of the compasses around a current is given by the right-hand rule.

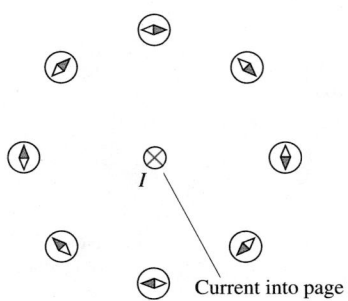

FIGURE 32.4 The orientation of the compasses around a current is given by the right-hand rule.

Current into page

FIGURE 32.5 The magnetic field exerts forces on the poles of a compass, causing the needle to align with the field.

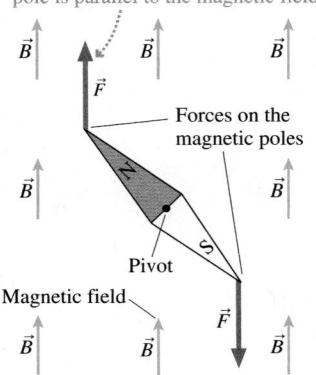

The magnetic force on the north pole is parallel to the magnetic field.

Forces on the magnetic poles

Pivot

Magnetic field

FIGURE 32.6 The magnetic field around a current-carrying wire.

(a) The magnetic field vectors are tangent to circles around the wire, pointing in the direction given by the right-hand rule. The field is weaker farther from the wire.

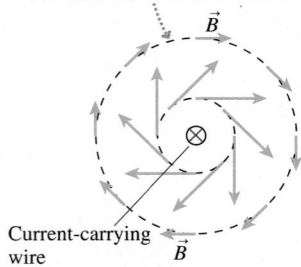

Current-carrying wire

(b) Magnetic field lines are circles.

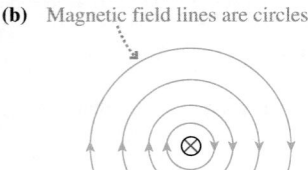

The Magnetic Field

We introduced the idea of a *field* as a way to understand the long-range electric force. Although this idea appeared rather far-fetched, it turned out to be very useful. We need a similar idea to understand the long-range force exerted by a current on a compass needle.

Let us define the **magnetic field** $\vec{B}$ as having the following properties:

1. A magnetic field is created at *all* points in space surrounding a current-carrying wire.
2. The magnetic field at each point is a vector. It has both a magnitude, which we call the *magnetic field strength B,* and a direction.
3. The magnetic field exerts forces on magnetic poles. The force on a north pole is parallel to $\vec{B}$; the force on a south pole is opposite $\vec{B}$.

FIGURE 32.5 shows a compass needle in a magnetic field. The field vectors are shown at several points, but keep in mind that the field is present at *all* points in space. A magnetic force is exerted on each of the two poles of the compass, parallel to $\vec{B}$ for the north pole and opposite $\vec{B}$ for the south pole. This pair of opposite forces exerts a torque on the needle, rotating the needle until it is parallel to the magnetic field at that point.

Notice that the north pole of the compass needle, when it reaches the equilibrium position, is in the direction of the magnetic field. Thus a compass needle can be used as a probe of the magnetic field, just as a charge was a probe of the electric field. **Magnetic forces cause a compass needle to become aligned parallel to a magnetic field, with the north pole of the compass showing the direction of the magnetic field at that point.**

Look back at the compass alignments around the current-carrying wire in Figure 32.4. Because compass needles align with the magnetic field, the magnetic field at each point must be tangent to a circle around the wire. **FIGURE 32.6a** shows the magnetic field by drawing field vectors. Notice that the field is weaker (shorter vectors) at greater distances from the wire.

Another way to picture the field is with the use of **magnetic field lines.** These are imaginary lines drawn through a region of space so that

- A tangent to a field line is in the direction of the magnetic field, and
- The field lines are closer together where the magnetic field strength is larger.

FIGURE 32.6b shows the magnetic field lines around a current-carrying wire. Notice that magnetic field lines form loops, with no beginning or ending point. This is in contrast to electric field lines, which stop and start on charges.

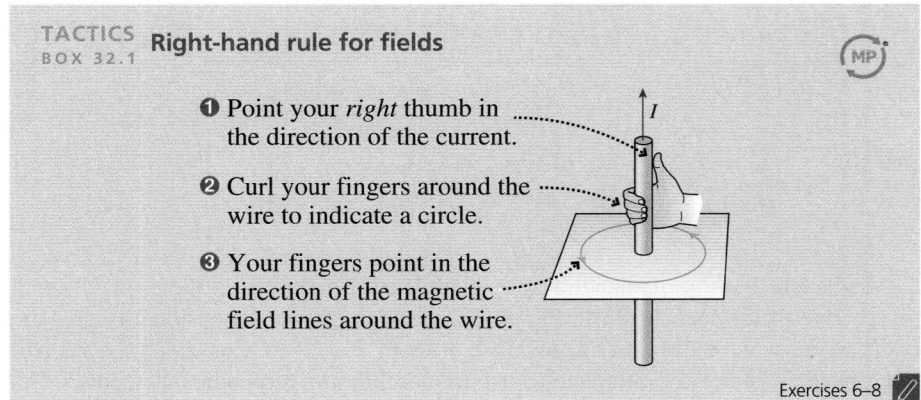

TACTICS
BOX 32.1 **Right-hand rule for fields**

❶ Point your *right* thumb in the direction of the current.

❷ Curl your fingers around the wire to indicate a circle.

❸ Your fingers point in the direction of the magnetic field lines around the wire.

I

Exercises 6–8

NOTE ▶ The magnetic field of a current-carrying wire is very different from the electric field of a charged wire. The electric field of a charged wire points radially outward (positive wire) or inward (negative wire). ◀

Two Kinds of Magnetism?

You might be concerned that we have introduced two kinds of magnetism. We opened this chapter discussing permanent magnets and their forces. Then, without warning, we switched to the magnetic forces caused by a current. It is not at all obvious that these forces are the same kind of magnetism as that exhibited by stationary chunks of metal called "magnets." Perhaps there are two different types of magnetic forces, one having to do with currents and the other being responsible for permanent magnets. One of the major goals for our study of magnetism is to see that these two quite different ways of producing magnetic effects are really just two different aspects of a *single* magnetic force.

STOP TO THINK 32.2   The magnetic field at position P points

•P

a. Up.
c. Into the page.

b. Down.
d. Out of the page.

⊗▭▭▭▭▭▭▭▭▭▭▶ *I*

32.3 The Source of the Magnetic Field: Moving Charges

Figure 32.6 is a qualitative picture of the wire's magnetic field. Our first task is to turn that picture into a quantitative description. Because current in a wire generates a magnetic field, and a current is a collection of moving charges, our starting point is the idea that **moving charges are the source of the magnetic field.** FIGURE 32.7 shows a charged particle q moving with velocity $\vec{v}$. The magnetic field of this moving charge is found to be

$$\vec{B}_{\text{point charge}} = \left(\frac{\mu_0}{4\pi} \frac{qv\sin\theta}{r^2}, \text{ direction given by the right-hand rule} \right) \quad (32.1)$$

where r is the distance from the charge and θ is the angle between $\vec{v}$ and $\vec{r}$.

Equation 32.1 is called the **Biot-Savart law** for a point charge (rhymes with *Leo* and *bazaar*), named for two French scientists whose investigations were motivated by Oersted's observations. It is analogous to Coulomb's law for the electric field of a point charge. Notice that the Biot-Savart law, like Coulomb's law, is an inverse-square law. However, the Biot-Savart law is somewhat more complex than Coulomb's law because the magnetic field depends on the angle θ between the charge's velocity and the line to the point where the field is evaluated.

NOTE ▶ A moving charge has both a magnetic field *and* an electric field. What you know about electric fields has not changed. ◀

The SI unit of magnetic field strength is the **tesla,** abbreviated as T. The tesla is defined as

$$1 \text{ tesla} = 1 \text{ T} \equiv 1 \text{ N/A m}$$

You will see later in the chapter that this definition is based on the magnetic force on a current-carrying wire. One tesla is quite a large field; most magnetic fields are a small fraction of a tesla. Table 32.1 lists a few magnetic field strengths.

The constant μ_0 in Equation 32.1 is called the **permeability constant.** Its value is

$$\mu_0 = 4\pi \times 10^{-7} \text{ T m/A} = 1.257 \times 10^{-6} \text{ T m/A}$$

This constant plays a role in magnetism similar to that of the permittivity constant ϵ_0 in electricity.

FIGURE 32.7 The magnetic field of a moving point charge.

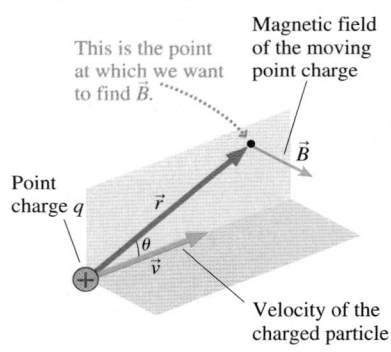

TABLE 32.1 Typical magnetic field strengths

Field source	Field strength (T)
Surface of the earth	5×10^{-5}
Refrigerator magnet	5×10^{-3}
Laboratory magnet	0.1 to 1
Superconducting magnet	10

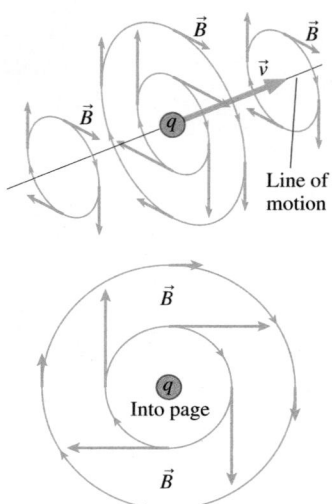

◀ FIGURE 32.8 Two views of the magnetic field of a moving positive charge.

The right-hand rule for finding the direction of $\vec{B}$ is similar to the rule used for a current-carrying wire: Point your right thumb in the direction of $\vec{v}$. The magnetic field vector $\vec{B}$ is perpendicular to the plane of $\vec{r}$ and $\vec{v}$, pointing in the direction in which your fingers curl. In other words, the $\vec{B}$ vectors are tangent to circles drawn about the charge's line of motion. FIGURE 32.8 shows a more complete view of the magnetic field of a moving positive charge. Notice that $\vec{B}$ is zero along the line of motion, where $\theta = 0°$ or $180°$, due to the $\sin\theta$ term in Equation 32.1.

NOTE ▶ The vector arrows in Figure 32.8 would have the same lengths but be reversed in direction for a negative charge. ◀

The requirement that a charge be moving to generate a magnetic field is explicit in Equation 32.1. If the speed v of the particle is zero, the magnetic field (but not the electric field!) is zero. This helps to emphasize a fundamental distinction between electric and magnetic fields: **All charges create electric fields, but only *moving* charges create magnetic fields.**

EXAMPLE 32.1 **The magnetic field of a proton**

A proton moves with velocity $\vec{v} = 1.0 \times 10^7\,\hat{\imath}$ m/s. As it passes the origin, what is the magnetic field at the (x, y, z) positions (1 mm, 0 mm, 0 mm), (0 mm, 1 mm, 0 mm), and (1 mm, 1 mm, 0 mm)?

MODEL The magnetic field is that of a moving charged particle.

VISUALIZE FIGURE 32.9 shows the geometry. The first point is on the x-axis, directly in front of the proton, with $\theta_1 = 0°$. The second point is on the y-axis, with $\theta_2 = 90°$, and the third is in the xy-plane.

FIGURE 32.9 The magnetic field of Example 32.1.

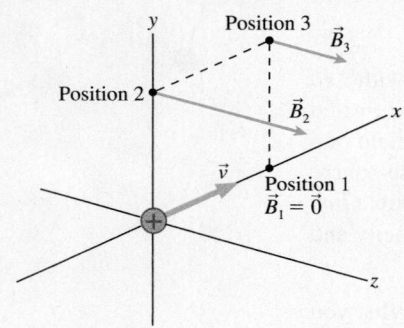

SOLVE Position 1, which is along the line of motion, has $\theta_1 = 0°$. Thus $\vec{B}_1 = \vec{0}$. Position 2 (at 0 mm, 1 mm, 0 mm) is at distance $r_2 = 1$ mm $= 0.001$ m. Equation 32.1, the Biot-Savart law, gives us the magnetic field strength at this point as

$$B = \frac{\mu_0}{4\pi}\frac{qv\sin\theta_2}{r_2^2}$$
$$= \frac{4\pi \times 10^{-7}\,\text{T m/A}(1.60 \times 10^{-19}\,\text{C})(1.0 \times 10^7\,\text{m/s})\sin 90°}{4\pi\,(0.0010\,\text{m})^2}$$
$$= 1.60 \times 10^{-13}\,\text{T}$$

According to the right-hand rule, the field points in the positive z-direction. Thus

$$\vec{B}_2 = 1.60 \times 10^{-13}\,\hat{k}\,\text{T}$$

where $\hat{k}$ is the unit vector in the positive z-direction. The field at position 3, at (1 mm, 1 mm, 0 mm), also points in the z-direction, but it is weaker than at position 2 both because r is larger *and* because θ is smaller. From geometry we know $r_3 = \sqrt{2}$ mm $= 0.00141$ m and $\theta_3 = 45°$. Another calculation using Equation 32.1 gives

$$\vec{B}_3 = 0.57 \times 10^{-13}\,\hat{k}\,\text{T}$$

ASSESS The magnetic field of a single moving charge is *very* small.

Superposition

The Biot-Savart law is the starting point for generating all magnetic fields, just as our earlier expression for the electric field of a point charge was the starting point for generating all electric fields. Magnetic fields, like electric fields, have been found experimentally to obey the principle of superposition. If there are n moving point charges, the net magnetic field is given by the vector sum

$$\vec{B}_{\text{total}} = \vec{B}_1 + \vec{B}_2 + \cdots + \vec{B}_n \tag{32.2}$$

where each individual $\vec{B}$ is calculated with Equation 32.1. The principle of superposition will be the basis for calculating the magnetic fields of several important current distributions.

The Vector Cross Product

In Chapter 25, we found that the electric field of a point charge could be written concisely and accurately as

$$\vec{E} = \frac{1}{4\pi\epsilon_0} \frac{q}{r^2} \hat{r}$$

where $\hat{r}$ is a *unit vector* that points from the charge to the point at which we wish to calculate the field. Unit vector $\hat{r}$ expresses the idea "away from q."

The unit vector $\hat{r}$ also allows us to write the Biot-Savart law more concisely, but we'll need to use the form of vector multiplication called the *cross product*. To remind you, FIGURE 32.10 shows two vectors, $\vec{C}$ and $\vec{D}$, with angle α between them. The **cross product** of $\vec{C}$ and $\vec{D}$ is defined to be the vector

$$\vec{C} \times \vec{D} = (CD\sin\alpha, \text{ direction given by the right-hand rule}) \qquad (32.3)$$

The symbol $\times$ between the vectors is *required* to indicate a cross product.

NOTE ▶ The cross product of two vectors and the right-hand rule used to determine the direction of the cross product were introduced in Section 12.10 to describe torque and angular momentum. If you omitted that section, you will want to turn to it now to read about the cross product. A review is worthwhile even if you did learn about the cross product earlier. ◀

The Biot-Savart law, Equation 32.1, can be written in terms of the cross product as

$$\vec{B}_{\text{point charge}} = \frac{\mu_0}{4\pi} \frac{q\vec{v} \times \hat{r}}{r^2} \qquad \text{(magnetic field of a point charge)} \qquad (32.4)$$

where unit vector $\hat{r}$, shown in FIGURE 32.11, points from charge q to the point at which we want to evaluate the field. This expression for the magnetic field $\vec{B}$ has magnitude $(\mu_0/4\pi)qv\sin\theta/r^2$ (because the magnitude of $\hat{r}$ is 1) and points in the correct direction (given by the right-hand rule), so it agrees completely with Equation 32.1.

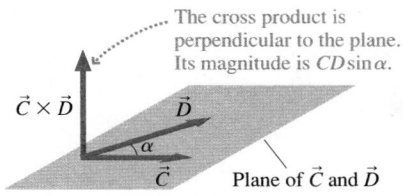

FIGURE 32.10 The cross product $\vec{C} \times \vec{D}$ is a vector perpendicular to the plane of vectors $\vec{C}$ and $\vec{D}$.

The cross product is perpendicular to the plane. Its magnitude is $CD\sin\alpha$.

Plane of $\vec{C}$ and $\vec{D}$

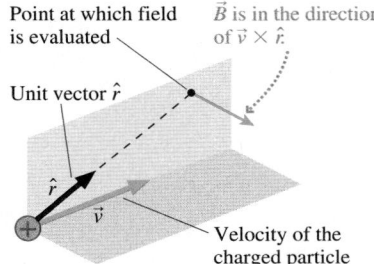

FIGURE 32.11 Unit vector $\hat{r}$ defines the direction from the moving charge to the point at which we want to evaluate the magnetic field.

Point at which field is evaluated

$\vec{B}$ is in the direction of $\vec{v} \times \hat{r}$.

Unit vector $\hat{r}$

Velocity of the charged particle

EXAMPLE 32.2 | **The magnetic field direction of a moving electron**

The electron in FIGURE 32.12 is moving to the right. What is the direction of the electron's magnetic field at the position indicated with a dot?

FIGURE 32.12 A moving electron.

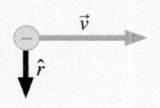

VISUALIZE Because the charge is negative, the magnetic field points in the direction of $-(\vec{v} \times \hat{r})$, or opposite the direction of $\vec{v} \times \hat{r}$. Unit vector $\hat{r}$ points from the charge toward the dot. We can use the right-hand rule to find that $\vec{v} \times \hat{r}$ points *into* the page. Thus the electron's magnetic field at the dot points *out* of the page.

STOP TO THINK 32.3 The positive charge is moving straight out of the page. What is the direction of the magnetic field at the position of the dot?

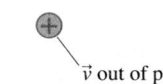

$\vec{v}$ out of page

a. Up   b. Down   c. Left   d. Right

32.4 The Magnetic Field of a Current

In practice we're more interested in the magnetic field of a current—a collection of moving charges—than in the very small magnetic fields of individual charges. The Biot-Savart law and the principle of superposition will be our primary tools for calculating magnetic fields. First, however, it will be useful to rewrite the Biot-Savart law in terms of current.

FIGURE 32.13 Relating the charge velocity $\vec{v}$ to the current I.

(a) Charge ΔQ in a small length Δs of a current-carrying wire

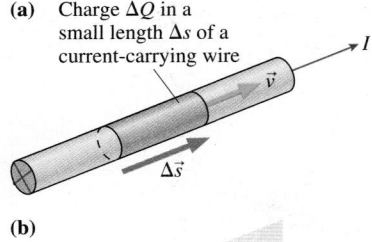

(b)

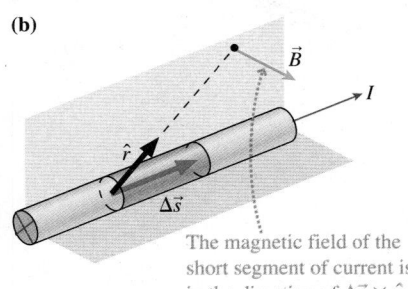

The magnetic field of the short segment of current is in the direction of $\Delta \vec{s} \times \hat{r}$.

FIGURE 32.13a shows a current-carrying wire. The wire as a whole is electrically neutral, but current I represents the motion of positive charge carriers through the wire. Suppose the small amount of moving charge ΔQ spans the small length Δs. The charge has velocity $\vec{v} = \Delta \vec{s}/\Delta t$, where the vector $\Delta \vec{s}$, which is parallel to $\vec{v}$, is the charge's displacement vector. If ΔQ is small enough to treat as a point charge, the magnetic field it creates at a point in space is proportional to $(\Delta Q)\vec{v}$. We can write $(\Delta Q)\vec{v}$ in terms of the wire's current I as

$$(\Delta Q)\vec{v} = \Delta Q \frac{\Delta \vec{s}}{\Delta t} = \frac{\Delta Q}{\Delta t} \Delta \vec{s} = I \Delta \vec{s} \qquad (32.5)$$

where we used the definition of current, $I = \Delta Q/\Delta t$.

If we replace $q\vec{v}$ in the Biot-Savart law with $I \Delta \vec{s}$, we find that the magnetic field of a very short segment of wire carrying current I is

$$\vec{B}_{\text{current segment}} = \frac{\mu_0}{4\pi} \frac{I \Delta \vec{s} \times \hat{r}}{r^2} \qquad (32.6)$$

(magnetic field of a very short segment of current)

Equation 32.6 is still the Biot-Savart law, only now written in terms of current rather than the motion of an individual charge. **FIGURE 32.13b** shows the direction of the current segment's magnetic field as determined by using the right-hand rule.

Equation 32.6 is the basis of a strategy for calculating the magnetic field of a current-carrying wire. You will recognize that it is the same basic strategy you learned for calculating the electric field of a continuous distribution of charge. The goal is to break a problem down into small steps that are individually manageable.

PROBLEM-SOLVING
STRATEGY 32.1 **The magnetic field of a current**

MODEL Model the wire as a simple shape, such as a straight line or a loop.

VISUALIZE For the pictorial representation:

❶ Draw a picture and establish a coordinate system.
❷ Identify the point P at which you want to calculate the magnetic field.
❸ Divide the current-carrying wire into small segments for which you *already know* how to determine $\vec{B}$. This is usually, though not always, a division into very short segments of length Δs.
❹ Draw the magnetic field vector for one or two segments. This will help you identify distances and angles that need to be calculated.
❺ Look for symmetries that simplify the field. You may conclude that some components of $\vec{B}$ are zero.

SOLVE The mathematical representation is $\vec{B}_{\text{net}} = \Sigma \vec{B}_i$.

■ Use superposition to form an algebraic expression for *each* of the three components of $\vec{B}$ (unless you are sure one or more is zero) at point P.
■ Let the (x, y, z)-coordinates of the point remain as variables.
■ Express all angles and distances in terms of the coordinates.
■ Let $\Delta s \rightarrow ds$ and the sum become an integral. Think carefully about the integration limits for this variable; they will depend on the boundaries of the wire and on the coordinate system you have chosen to use. Carry out the integration and simplify the results as much as possible.

ASSESS Check that your result is consistent with any limits for which you know what the field should be.

EXAMPLE 32.3   **The magnetic field of a long, straight wire**

A long, straight wire carries current I in the positive x-direction. Find the magnetic field at a point that is distance d from the wire.

MODEL Because the wire is "long," let's model it as being infinitely long.

VISUALIZE **FIGURE 32.14** illustrates the steps in the problem-solving strategy. We've chosen a coordinate system with point P on the y-axis. We've then divided the wire into small segments, labeled with index i, each containing a small amount ΔQ of *moving charge*. Unit vector $\hat{r}$ and angle θ_i are shown for segment i. You should use the right-hand rule to convince yourself that $\vec{B}_i$ points *out of the page*, in the positive z-direction. This is the direction no matter where segment i happens to be along the x-axis. Consequently, B_x (the component of $\vec{B}$ parallel to the wire) and B_y (the component of $\vec{B}$ straight away from the wire) are zero. The only component of $\vec{B}$ we need to evaluate is B_z, the component tangent to a circle around the wire.

FIGURE 32.14 Calculating the magnetic field of a long, straight wire carrying current I.

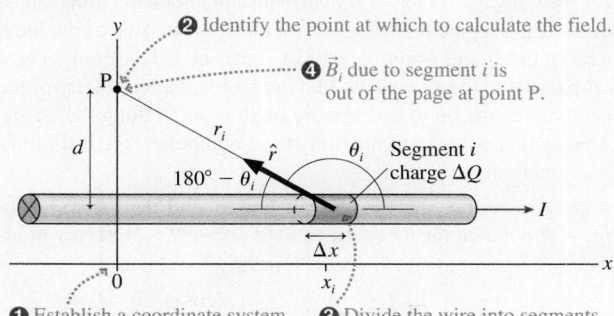

❷ Identify the point at which to calculate the field.

❹ $\vec{B}_i$ due to segment i is out of the page at point P.

Segment i charge ΔQ

❶ Establish a coordinate system. ❸ Divide the wire into segments.

SOLVE We can use the Biot-Savart law to find the field $(B_i)_z$ of segment i. The cross product $\Delta\vec{s}_i \times \hat{r}$ has magnitude $(\Delta x)(1)\sin\theta_i$, hence

$$(B_i)_z = \frac{\mu_0}{4\pi}\frac{I\,\Delta x \sin\theta_i}{r_i^2} = \frac{\mu_0}{4\pi}\frac{I\sin\theta_i}{r_i^2}\Delta x = \frac{\mu_0}{4\pi}\frac{I\sin\theta_i}{x_i^2 + d^2}\Delta x$$

where we wrote the distance r_i in terms of x_i and d. We also need to express θ_i in terms of x_i and d. Because $\sin(180° - \theta) = \sin\theta$, this is

$$\sin\theta_i = \sin(180° - \theta_i) = \frac{d}{r_i} = \frac{d}{\sqrt{x_i^2 + d^2}}$$

With this expression for $\sin\theta_i$, the magnetic field of segment i is

$$(B_i)_z = \frac{\mu_0}{4\pi}\frac{Id}{(x_i^2 + d^2)^{3/2}}\Delta x$$

Now we're ready to sum the magnetic fields of all the segments. The superposition is a vector sum, but in this case only the z-components are nonzero. Summing all the $(B_i)_z$ gives

$$B_{\text{wire}} = \frac{\mu_0 Id}{4\pi}\sum_i \frac{\Delta x}{(x_i^2 + d^2)^{3/2}} \to \frac{\mu_0 Id}{4\pi}\int_{-\infty}^{\infty}\frac{dx}{(x^2 + d^2)^{3/2}}$$

Only at the very last step did we convert the sum to an integral. Then our model of the wire as being infinitely long sets the integration limits at $\pm\infty$. This is a standard integral that can be found in Appendix A or with integration software. Evaluation gives

$$B_{\text{wire}} = \frac{\mu_0 Id}{4\pi}\frac{x}{d^2(x^2 + d^2)^{1/2}}\bigg|_{-\infty}^{\infty} = \frac{\mu_0}{2\pi}\frac{I}{d}$$

This is the magnitude of the field. The field direction is determined by using the right-hand rule. We can combine these two pieces of information to write

$$\vec{B}_{\text{wire}} = \left(\frac{\mu_0}{2\pi}\frac{I}{d}, \text{ tangent to a circle around the wire in the right-hand direction}\right)$$

ASSESS **FIGURE 32.15** shows the magnetic field of a current-carrying wire. Compare this to Figure 32.2 and convince yourself that the direction shown agrees with the right-hand rule.

FIGURE 32.15 The magnetic field of a long, straight wire carrying current I.

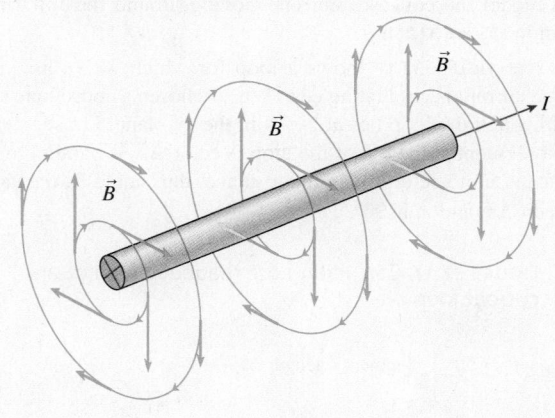

NOTE ▶ The difficulty magnetic field calculations present is not doing the integration itself, which is the last step, but setting up the calculation and knowing *what* to integrate. The purpose of the problem-solving strategy is to guide you through the process of setting up the integral. ◀

EXAMPLE 32.4   **The magnetic field strength near a heater wire**

A 1.0-m-long, 1.0-mm-diameter nichrome heater wire is connected to a 12 V battery. What is the magnetic field strength 1.0 cm away from the wire?

MODEL 1 cm is much less than the 1 m length of the wire, so model the wire as infinitely long.

SOLVE The current through the wire is $I = \Delta V_{\text{bat}}/R$, where the wire's resistance R is

$$R = \frac{\rho L}{A} = \frac{\rho L}{\pi r^2} = 1.91\ \Omega$$

Continued

The nichrome resistivity $\rho = 1.50 \times 10^{-6}$ Ω m was taken from Table 30.2. Thus the current is $I = (12 \text{ V})/(1.91 \text{ Ω}) = 6.28$ A. The magnetic field strength at distance $d = 1.0$ cm $= 0.010$ m from the wire is

$$B_{\text{wire}} = \frac{\mu_0}{2\pi} \frac{I}{d} = (2.0 \times 10^{-7} \text{ T m/A}) \frac{6.28 \text{ A}}{0.010 \text{ m}}$$
$$= 1.3 \times 10^{-4} \text{ T}$$

ASSESS The magnetic field of the wire is slightly more than twice the strength of the earth's magnetic field.

Motors, loudspeakers, metal detectors, and many other devices generate magnetic fields with *coils* of wire. The simplest coil is a single-turn circular loop of wire. A circular loop of wire with a circulating current is called a **current loop**.

EXAMPLE 32.5 The magnetic field of a current loop

FIGURE 32.16a shows a current loop, a circular loop of wire with radius R that carries current I. Find the magnetic field of the current loop at distance z on the axis of the loop.

FIGURE 32.16 A current loop.

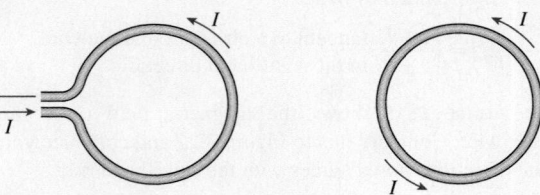

(a) A practical current loop

(b) An ideal current loop

MODEL Real coils need wires to bring the current in and out, but we'll model the coil as a current moving around the full circle shown in FIGURE 32.16b.

VISUALIZE FIGURE 32.17 shows a loop for which we've assumed that the current is circulating ccw. We've chosen a coordinate system in which the loop lies at $z = 0$ in the xy-plane. Let segment i be the segment at the top of the loop. Vector $\Delta \vec{s}_i$ is parallel to the x-axis and unit vector $\hat{r}$ is in the yz-plane, thus angle θ_i, the angle between $\Delta \vec{s}_i$ and $\hat{r}$, is 90°.

FIGURE 32.17 Calculating the magnetic field of a current loop.

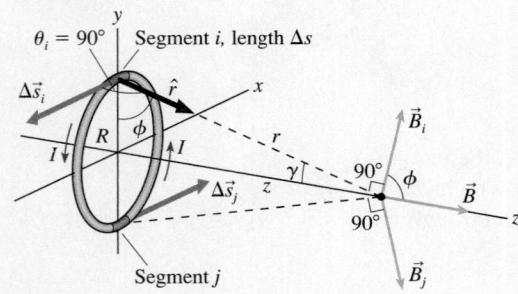

The direction of $\vec{B}_i$, the magnetic field due to the current in segment i, is given by the cross product $\Delta \vec{s}_i \times \hat{r}$. $\vec{B}_i$ must be perpendicular to $\Delta \vec{s}_i$ *and* perpendicular to $\hat{r}$. You should convince yourself that $\vec{B}_i$ in Figure 32.17 points in the correct direction. Notice that the y-component of $\vec{B}_i$ is canceled by the y-component of magnetic field $\vec{B}_j$ due to the current segment at the bottom of the loop, 180° away. In fact, *every* current segment on the loop can be paired with a segment 180° away, on the opposite side of the loop, such that the x- and y-components of $\vec{B}$ cancel and the components of $\vec{B}$ parallel to the z-axis add. In other words, the symmetry of the loop requires the on-axis magnetic field to point along the z-axis. Knowing that we need to sum only the z-components will simplify our calculation.

SOLVE We can use the Biot-Savart law to find the z-component $(B_i)_z = B_i \cos\phi$ of the magnetic field of segment i. The cross product $\Delta \vec{s}_i \times \hat{r}$ has magnitude $(\Delta s)(1) \sin 90° = \Delta s$, thus

$$(B_i)_z = \frac{\mu_0}{4\pi} \frac{I \Delta s}{r^2} \cos\phi = \frac{\mu_0 I \cos\phi}{4\pi (z^2 + R^2)} \Delta s$$

where we used $r = (z^2 + R^2)^{1/2}$. You can see, because $\phi + \gamma = 90°$, that angle ϕ is also the angle between $\hat{r}$ and the radius of the loop. Hence $\cos\phi = R/r$, and

$$(B_i)_z = \frac{\mu_0 IR}{4\pi (z^2 + R^2)^{3/2}} \Delta s$$

The final step is to sum the magnetic fields due to all the segments:

$$B_{\text{loop}} = \sum_i (B_i)_z = \frac{\mu_0 IR}{4\pi (z^2 + R^2)^{3/2}} \sum_i \Delta s$$

In this case, unlike the straight wire, none of the terms multiplying Δs depends on the position of segment i, so all these terms can be factored out of the summation. We're left with a summation that adds up the lengths of all the small segments. But this is just the total length of the wire, which is the circumference $2\pi R$. Thus the on-axis magnetic field of a current loop is

$$B_{\text{loop}} = \frac{\mu_0 IR}{4\pi (z^2 + R^2)^{3/2}} 2\pi R = \frac{\mu_0}{2} \frac{IR^2}{(z^2 + R^2)^{3/2}}$$

In practice, current often passes through a *coil* consisting of N *turns* of wire. If the turns are all very close together, so that the magnetic field of each is essentially the same, then the magnetic field of a coil is N times the magnetic field of a current loop. The magnetic field at the center ($z = 0$) of an N-turn coil is

$$B_{\text{coil center}} = \frac{\mu_0}{2} \frac{NI}{R} \tag{32.7}$$

EXAMPLE 32.6   **Matching the earth's magnetic field**

What current is needed in a 5-turn, 10-cm-diameter coil to cancel the earth's magnetic field at the center of the coil?

MODEL One way to create a zero-field region of space is to generate a magnetic field equal to the earth's field but pointing in the opposite direction. The vector sum of the two fields is zero.

VISUALIZE FIGURE 32.18 shows a five-turn coil of wire. The magnetic field is five times that of a single current loop.

SOLVE The earth's magnetic field, from Table 32.1, is 5×10^{-5} T. We can use Equation 32.7 to find that the current needed to generate a 5×10^{-5} T field is

$$I = \frac{2RB}{\mu_0 N} = \frac{2(0.050 \text{ m})(5.0 \times 10^{-5} \text{ T})}{5(4\pi \times 10^{-7} \text{ T m/A})} = 0.80 \text{ A}$$

FIGURE 32.18 A coil of wire.

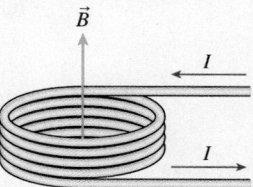

ASSESS A 0.80 A current is easily produced. Although there are better ways to cancel the earth's field than using a simple coil, this illustrates the idea.

32.5 Magnetic Dipoles

We were able to calculate the on-axis magnetic field of a current loop, but determining the field at off-axis points requires either numerical integrations or an experimental mapping of the field. FIGURE 32.19 shows the full magnetic field of a current loop. This is a field with *rotational symmetry,* so to picture the full three-dimensional field, imagine FIGURE 32.19a rotated about the axis of the loop. FIGURE 32.19b shows the magnetic field in the plane of the loop as seen from the right. There is a clear sense, seen in the photo of FIGURE 32.19c, that the magnetic field leaves the loop on one side, "flows" around the outside, then returns to the loop.

FIGURE 32.19 The magnetic field of a current loop.

(a) Cross section through the current loop

(b) The current loop seen from the right

(c) A photo of iron filings

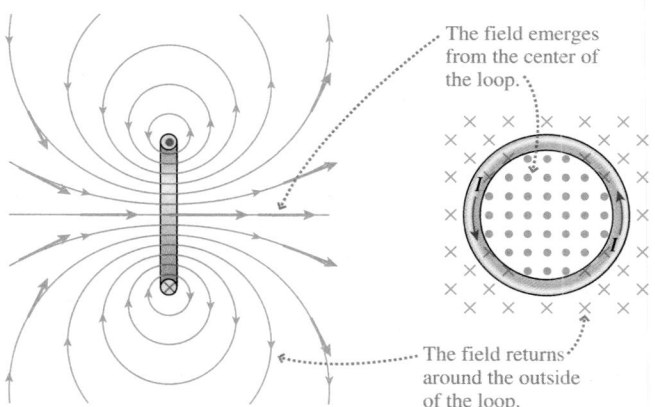

The field emerges from the center of the loop.

The field returns around the outside of the loop.

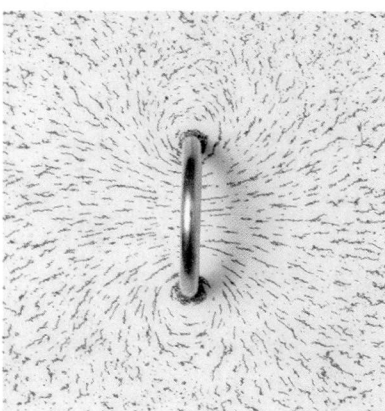

There are two versions of the right-hand rule that you can use to determine which way a loop's field points. Try these in Figure 32.19. Being able to quickly ascertain the field direction of a current loop is an important skill.

TACTICS
BOX 32.2   **Finding the magnetic field direction of a current loop**   (MP)

Use either of the following methods to find the magnetic field direction:

❶ Point your right thumb in the direction of the current at any point on the loop and let your fingers curl through the center of the loop. Your fingers are then pointing in the direction in which $\vec{B}$ leaves the loop.

❷ Curl the fingers of your right hand around the loop in the direction of the current. Your thumb is then pointing in the direction in which $\vec{B}$ leaves the loop.

Exercises 18–20

A Current Loop Is a Magnetic Dipole

A current loop has two distinct sides. Bar magnets and flat refrigerator magnets also have two distinct sides or ends, so you might wonder if current loops are related to these permanent magnets. Consider the following experiments with a current loop. Notice that we're using a simplified picture that shows the magnetic field only in the plane of the loop.

Investigating current loops

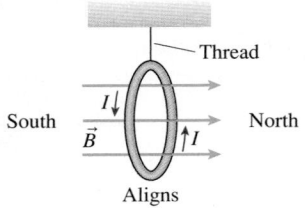

A current loop hung by a thread aligns itself with the magnetic field pointing north.

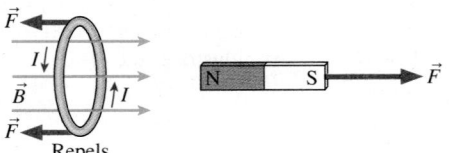

The north pole of a permanent magnet repels the side of a current loop from which the magnetic field is emerging.

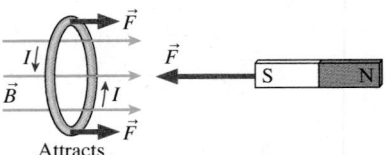

The south pole of a permanent magnet attracts the side of a current loop from which the magnetic field is emerging.

These investigations show that **a current loop is a magnet,** just like a permanent magnet. A magnet created by a current in a coil of wire is called an **electromagnet.** An electromagnet picks up small pieces of iron, influences a compass needle, and acts in every way like a permanent magnet.

In fact, FIGURE 32.20 shows that a flat permanent magnet and a current loop generate the same magnetic field—the field of a magnetic dipole. For both, **you can identify the north pole as the face or end** *from which* **the magnetic field emerges.** The magnetic fields of both point *into* the south pole.

FIGURE 32.20 A current loop has magnetic poles and generates the same magnetic field as a flat permanent magnet.

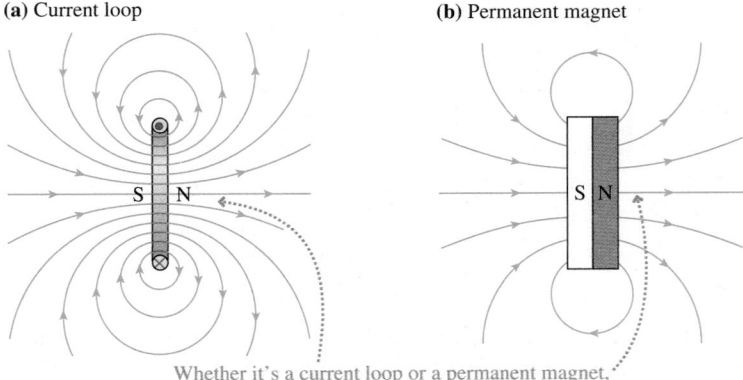

Whether it's a current loop or a permanent magnet, the magnetic field emerges from the north pole.

One of the goals of this chapter is to show that magnetic forces exerted by currents and magnetic forces exerted by permanent magnets are just two different aspects of a single magnetism. We've now found a strong connection between permanent magnets and current loops, and this connection will turn out to be a big piece of the puzzle.

The Magnetic Dipole Moment

The expression for the electric field of an electric dipole was considerably simplified when we considered the field at distances significantly larger than the size of the charge separation s. The on-axis field of an electric dipole when $z \gg s$ is

$$\vec{E}_{\text{dipole}} = \frac{1}{4\pi\epsilon_0} \frac{2\vec{p}}{z^3}$$

where the electric dipole moment $\vec{p} = (qs,$ from negative to positive charge$)$.

The on-axis magnetic field of a current loop is

$$B_{\text{loop}} = \frac{\mu_0}{2} \frac{IR^2}{(z^2 + R^2)^{3/2}}$$

If z is much larger than the diameter of the current loop, $z \gg R$, we can make the approximation $(z^2 + R^2)^{3/2} \to z^3$. Then the loop's field is

$$B_{\text{loop}} \approx \frac{\mu_0}{2} \frac{IR^2}{z^3} = \frac{\mu_0}{4\pi} \frac{2(\pi R^2)I}{z^3} = \frac{\mu_0}{4\pi} \frac{2AI}{z^3} \qquad (32.8)$$

where $A = \pi R^2$ is the area of the loop.

A more advanced treatment of current loops shows that, if z is much larger than the size of the loop, Equation 32.8 is the on-axis magnetic field of a current loop of *any* shape, not just a circular loop. The shape of the loop affects the nearby field, but the distant field depends only on the current I and the area A enclosed within the loop. With this in mind, let's define the **magnetic dipole moment** $\vec{\mu}$ of a current loop enclosing area A to be

$$\vec{\mu} = (AI, \text{ from the south pole to the north pole})$$

The SI units of the magnetic dipole moment are $A\,m^2$.

NOTE ▶ Don't confuse the magnetic dipole moment $\vec{\mu}$ with the constant μ_0 in the Biot-Savart law. ◀

The magnetic dipole moment, like the electric dipole moment, is a vector. It has the same direction as the on-axis magnetic field. Thus the right-hand rule for determining the direction of $\vec{B}$ also shows the direction of $\vec{\mu}$. **FIGURE 32.21** shows the magnetic dipole moment of a circular current loop.

Because the on-axis magnetic field of a current loop points in the same direction as $\vec{\mu}$, we can combine Equation 32.8 and the definition of $\vec{\mu}$ to write the on-axis field of a magnetic dipole as

$$\vec{B}_{\text{dipole}} = \frac{\mu_0}{4\pi} \frac{2\vec{\mu}}{z^3} \quad \text{(on the axis of a magnetic dipole)} \qquad (32.9)$$

If you compare $\vec{B}_{\text{dipole}}$ to $\vec{E}_{\text{dipole}}$, you can see that the magnetic field of a magnetic dipole has the same basic shape as the electric field of an electric dipole.

A permanent magnet also has a magnetic dipole moment and its on-axis magnetic field is given by Equation 32.9 when z is much larger than the size of the magnet. Equation 32.9 and laboratory measurements of the on-axis magnetic field can be used to determine a permanent magnet's dipole moment.

FIGURE 32.21 The magnetic dipole moment of a circular current loop.

The magnetic dipole moment is perpendicular to the loop, in the direction of the right-hand rule. The magnitude of $\vec{\mu}$ is AI.

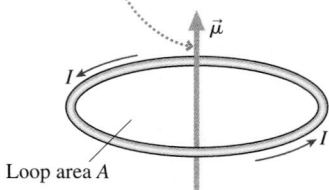

Loop area A

| EXAMPLE 32.7 | **Measuring current in a superconducting loop** |

You'll learn in Chapter 33 that a current can be *induced* in a closed loop of wire. If the loop happens to be made of a superconducting material, with zero resistance, the induced current will—in principle—persist forever. The current cannot be measured with an ammeter because any real ammeter has resistance that will quickly stop the current. Instead, physicists measure the persistent current in a superconducting loop by measuring its magnetic field. In one experiment, the axial magnetic field of a 3.0-mm-diameter superconducting loop is measured at several distances from the center of the loop, yielding the following data:

Distance (cm)	Magnetic field (μT)
1.0	130
1.5	35
2.0	19
2.5	9
3.0	5

Continued

To measure such small magnetic fields requires careful shielding from the earth's magnetic field. Based on these data, what is the current in the loop?

MODEL The measurements are made far enough from the loop in comparison to its radius ($z \gg R$) that we can approximate the loop as a magnetic dipole rather than using the exact expression for the on-axis field of a current loop.

SOLVE The axial magnetic field strength of a dipole is

$$B = \frac{\mu_0}{4\pi}\frac{2\mu}{z^3} = \frac{\mu_0}{4\pi}\frac{2\pi R^2 I}{z^3} = \frac{\mu_0 R^2 I}{2}\frac{1}{z^3}$$

where we used $\mu = AI = \pi R^2 I$ for the magnetic dipole moment of a circular loop of radius R. If we graph B versus $1/z^3$ the result should be a straight line whose slope can be used to determine I.

The graph of **FIGURE 32.22** is a straight line passing through the origin, as expected. The best-fit line has slope $129\,\mu\text{T cm}^3$, where the rather unusual units are determined by rise over run. Converting to SI units, we find

$$\text{slope} = 129\,\mu\text{T cm}^3 \times \frac{10^{-6}\,\text{T}}{1\,\mu\text{T}} \times \left(\frac{1\,\text{m}}{100\,\text{cm}}\right)^3$$

$$= 1.29 \times 10^{-10}\,\text{T m}^3$$

With this, we can determine the current:

$$I = \frac{2}{\mu_0 R^2} \times \text{slope} = 91\,\text{A}$$

FIGURE 32.22 A graph of the data.

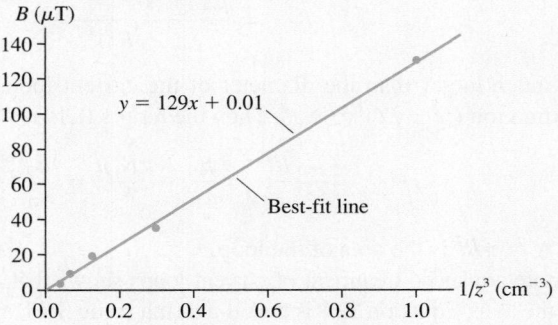

$$y = 129x + 0.01$$

Best-fit line

ASSESS This would be a very large current for ordinary wire. An important property of superconducting wires is their ability to carry current that would melt an ordinary wire.

STOP TO THINK 32.4 What is the current direction in this loop? And which side of the loop is the north pole?

a. Current cw; north pole on top
b. Current cw; north pole on bottom
c. Current ccw; north pole on top
d. Current ccw; north pole on bottom

32.6 Ampère's Law and Solenoids

In principle, the Biot-Savart law can be used to calculate the magnetic field of any current distribution. In practice, the integrals are difficult to evaluate for anything other than very simple situations. We faced a similar situation for calculating electric fields, but we discovered an alternative method—Gauss's law—for calculating the electric field of charge distributions with a high degree of symmetry.

Likewise, there's an alternative method, called *Ampère's law,* for calculating the magnetic fields of current distributions with a high degree of symmetry. Ampère's law, like Gauss's law, doesn't work in all situations, but it is simple and elegant where it does. Whereas Gauss's law is written in terms of a surface integral, Ampère's law is based on the mathematical procedure called a *line integral.*

Line Integrals

We've flirted with the idea of a line integral ever since introducing the concept of work in Chapter 11, but now we need to take a more serious look at what a line integral represents and how it is used. **FIGURE 32.23a** shows a curved line that goes from an initial point i to a final point f.

FIGURE 32.23 Integrating along a line from i to f.

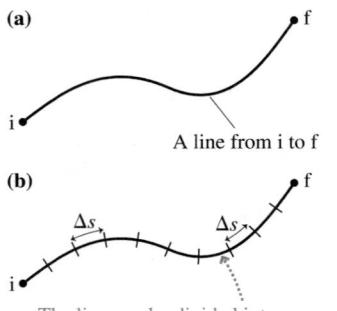

(a)

A line from i to f

(b)

The line can be divided into many small segments. The sum of all the Δs's is the length l of the line.

Suppose, as shown in FIGURE 32.23b, we divide the line into many small segments of length Δs. The first segment is Δs_1, the second is Δs_2, and so on. The sum of all the Δs's is the length l of the line between i and f. We can write this mathematically as

$$l = \sum_k \Delta s_k \rightarrow \int_i^f ds \qquad (32.10)$$

where, in the last step, we let $\Delta s \rightarrow ds$ and the sum become an integral.

This integral is called a **line integral**. All we've done is to subdivide a line into infinitely many infinitesimal pieces, then add them up. This is exactly what you do in calculus when you evaluate an integral such as $\int x \, dx$. In fact, an integration along the x-axis *is* a line integral, one that happens to be along a straight line. Figure 32.23 differs only in that the line is curved. The underlying idea in both cases is that an integral is just a fancy way of doing a sum.

The line integral of Equation 32.10 is not terribly exciting. FIGURE 32.24a makes things more interesting by allowing the line to pass through a magnetic field. FIGURE 32.24b again divides the line into small segments, but this time $\Delta \vec{s}_k$ is the displacement vector of segment k. The magnetic field at this point in space is $\vec{B}_k$.

Suppose we were to evaluate the dot product $\vec{B}_k \cdot \Delta \vec{s}_k$ at each segment, then add the values of $\vec{B}_k \cdot \Delta \vec{s}_k$ due to every segment. Doing so, and again letting the sum become an integral, we have

$$\sum_k \vec{B}_k \cdot \Delta \vec{s}_k \rightarrow \int_i^f \vec{B} \cdot d\vec{s} = \text{the line integral of } \vec{B} \text{ from i to f}$$

Once again, the integral is just a shorthand way to say: Divide the line into lots of little pieces, evaluate $\vec{B}_k \cdot \Delta \vec{s}_k$ for each piece, then add them up.

Although this process of evaluating the integral could be difficult, the only line integrals we'll need to deal with fall into two simple cases. If the magnetic field is *everywhere perpendicular* to the line, then $\vec{B} \cdot d\vec{s} = 0$ at every point along the line and the integral is zero. If the magnetic field is *everywhere tangent* to the line *and* has the same magnitude B at every point, then $\vec{B} \cdot d\vec{s} = B \, ds$ at every point and

$$\int_i^f \vec{B} \cdot d\vec{s} = \int_i^f B \, ds = B \int_i^f ds = Bl \qquad (32.11)$$

We used Equation 32.10 in the last step to integrate ds along the line.

Tactics Box 32.3 summarizes these two situations.

FIGURE 32.24 Integrating $\vec{B}$ along a line from i to f.

(a)

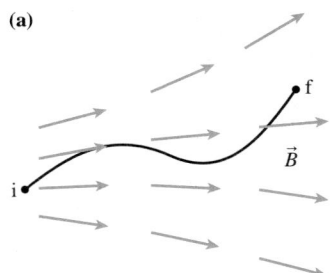

The line passes through a magnetic field.

(b)

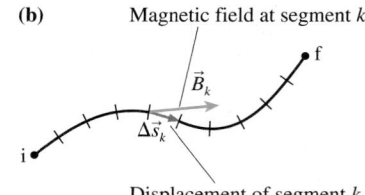

TACTICS
BOX 32.3   **Evaluating line integrals**   (MP)

❶ If $\vec{B}$ is everywhere perpendicular to a line, the line integral of $\vec{B}$ is

$$\int_i^f \vec{B} \cdot d\vec{s} = 0$$

❷ If $\vec{B}$ is everywhere tangent to a line of length l *and* has the same magnitude B at every point, then

$$\int_i^f \vec{B} \cdot d\vec{s} = Bl$$

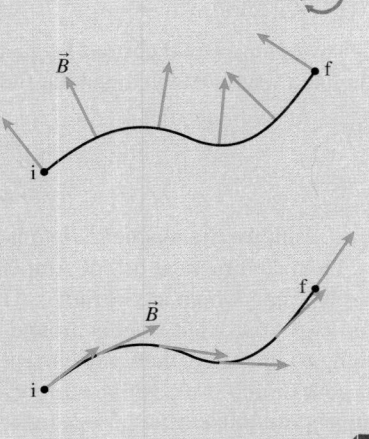

Exercises 23–24

Ampère's Law

FIGURE 32.25 Integrating the magnetic field around a wire.

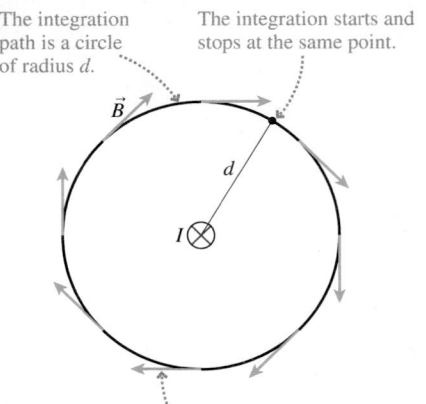

The integration path is a circle of radius *d*.

The integration starts and stops at the same point.

$\vec{B}$ is everywhere tangent to the integration path and has constant magnitude.

FIGURE 32.25 shows a wire carrying current I into the page and the magnetic field at distance d. The magnetic field of a current-carrying wire is everywhere tangent to a circle around the wire *and* has the same magnitude $\mu_0 I/2\pi d$ at all points on the circle. According to Tactics Box 32.3, these conditions allow us to easily evaluate the line integral of $\vec{B}$ along a circular path around the wire. Suppose we were to integrate the magnetic field *all the way around* the circle. That is, the initial point i of the integration path and the final point f will be the same point. This would be a line integral around a *closed curve,* which is denoted

$$\oint \vec{B} \cdot d\vec{s}$$

The little circle on the integral sign indicates that the integration is performed around a closed curve. The notation has changed, but the meaning has not.

Because $\vec{B}$ is tangent to the circle *and* of constant magnitude at every point on the circle, we can use Option 2 from Tactics Box 32.3 to write

$$\oint \vec{B} \cdot d\vec{s} = Bl = B(2\pi d) \qquad (32.12)$$

where, in this case, the path length l is the circumference $2\pi d$ of the circle. The magnetic field strength of a current-carrying wire is $B = \mu_0 I/2\pi d$, thus

$$\oint \vec{B} \cdot d\vec{s} = \mu_0 I \qquad (32.13)$$

The interesting result is that the line integral of $\vec{B}$ around the current-carrying wire is independent of the radius of the circle. Any circle, from one touching the wire to one far away, would give the same result. The integral depends only on the amount of current passing *through* the circle that we integrated around.

This is reminiscent of Gauss's law. In our investigation of Gauss's law, we started with the observation that the electric flux Φ_e through a sphere surrounding a point charge depends only on the amount of charge inside, not on the radius of the sphere. After examining several cases, we concluded that the shape of the surface wasn't relevant. The electric flux through *any* closed surface enclosing total charge Q_{in} turned out to be $\Phi_e = Q_{in}/\epsilon_0$.

Although we'll skip the details, the same type of reasoning that we used to prove Gauss's law shows that the result of Equation 32.13

■ Is independent of the shape of the curve around the current.
■ Is independent of where the current passes through the curve.
■ Depends only on the total amount of current through the area enclosed by the integration path.

FIGURE 32.26 Using Ampère's law.

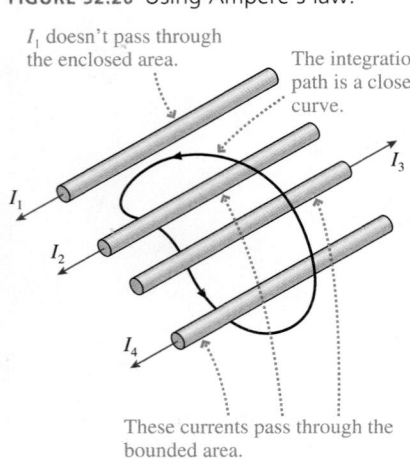

I_1 doesn't pass through the enclosed area.

The integration path is a closed curve.

These currents pass through the bounded area.

Thus whenever total current $I_{through}$ passes through an area bounded by a *closed curve,* the line integral of the magnetic field around the curve is

$$\oint \vec{B} \cdot d\vec{s} = \mu_0 I_{through} \qquad (32.14)$$

This result for the magnetic field is known as **Ampère's law.**

To make practical use of Ampère's law, we need to determine which currents are positive and which are negative. The right-hand rule is once again the proper tool. If you curl your right fingers around the closed path in the direction in which you are going to integrate, then any current passing though the bounded area in the direction of your thumb is a positive current. Any current in the opposite direction is a negative current. In FIGURE 32.26, for example, currents I_2 and I_4 are positive, I_3 is negative. Thus $I_{through} = I_2 - I_3 + I_4$.

NOTE ▶ The integration path of Ampère's law is a mathematical curve through space. It does not have to match a physical surface or boundary, although it could if we want it to. ◀

In one sense, Ampère's law doesn't tell us anything new. After all, we derived Ampère's law from the Biot-Savart law for the magnetic field of a current. But in another sense, Ampère's law is more important than the Biot-Savart law because it states a very general property about magnetic fields. We will use Ampère's law to find the magnetic fields of some important current distributions that have a high degree of symmetry.

EXAMPLE 32.8   **The magnetic field inside a current-carrying wire**

A wire of radius R carries current I. Find the magnetic field inside the wire at distance $r < R$ from the axis.

MODEL Assume the current density is uniform over the cross section of the wire.

VISUALIZE **FIGURE 32.27** shows a cross section through the wire. The wire has cylindrical symmetry, with all the charges moving parallel to the wire, so the magnetic field *must* be tangent to circles that are concentric with the wire. We don't know how the strength of the magnetic field depends on the distance from the center—that's what we're going to find—but the symmetry of the situation dictates the *shape* of the magnetic field.

FIGURE 32.27 Using Ampère's law inside a current-carrying wire.

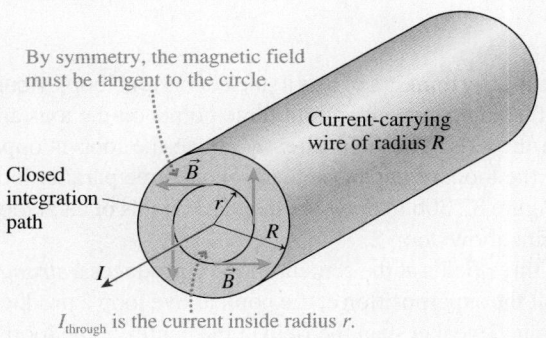

By symmetry, the magnetic field must be tangent to the circle.

Closed integration path

Current-carrying wire of radius R

I_{through} is the current inside radius r.

SOLVE To find the field strength at radius r, we draw a circle of radius r. The amount of current passing through this circle is

$$I_{\text{through}} = JA_{\text{circle}} = \pi r^2 J$$

where J is the current density. Our assumption of a uniform current density allows us to use the full current I passing through a wire of radius R to find that

$$J = \frac{I}{A} = \frac{I}{\pi R^2}$$

Thus the current through the circle of radius r is

$$I_{\text{through}} = \frac{r^2}{R^2}I$$

Let's integrate $\vec{B}$ around the circumference of this circle. According to Ampère's law,

$$\oint \vec{B} \cdot d\vec{s} = \mu_0 I_{\text{through}} = \frac{\mu_0 r^2}{R^2}I$$

We know from the symmetry of the wire that $\vec{B}$ is everywhere tangent to the circle *and* has the same magnitude at all points on the circle. Consequently, the line integral of $\vec{B}$ around the circle can be evaluated using Option 2 of Tactics Box 32.3:

$$\oint \vec{B} \cdot d\vec{s} = Bl = 2\pi rB$$

where $l = 2\pi r$ is the path length. If we substitute this expression into Ampère's law, we find that

$$2\pi rB = \frac{\mu_0 r^2}{R^2}I$$

Solving for B, we find that the magnetic field strength at radius r *inside* a current-carrying wire is

$$B = \frac{\mu_0 I}{2\pi R^2}r$$

ASSESS The magnetic field strength increases linearly with distance from the center of the wire until, at the surface of the wire, $B = \mu_0 I/2\pi R$ matches our earlier solution for the magnetic field outside a current-carrying wire. This agreement at $r = R$ gives us confidence in our result. The magnetic field strength both inside and outside the wire is shown graphically in **FIGURE 32.28**.

FIGURE 32.28 Graphical representation of the magnetic field of a current-carrying wire.

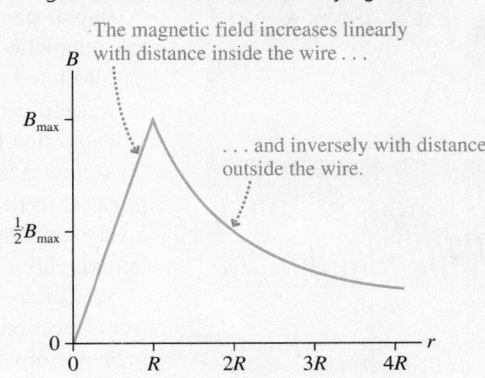

The magnetic field increases linearly with distance inside the wire . . .

. . . and inversely with distance outside the wire.

The Magnetic Field of a Solenoid

FIGURE 32.29 A solenoid.

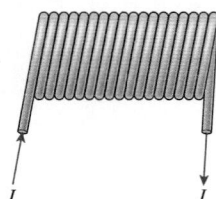

In our study of electricity, we made extensive use of the idea of a uniform electric field: a field that is the same at every point in space. We found that two closely spaced, parallel charged plates generate a uniform electric field between them, and this was one reason we focused so much attention on the parallel-plate capacitor.

Similarly, there are many applications of magnetism for which we would like to generate a **uniform magnetic field,** a field having the same magnitude and the same direction at every point within some region of space. None of the sources we have looked at thus far produces a uniform magnetic field.

In practice, a uniform magnetic field is generated with a **solenoid.** A solenoid, shown in FIGURE 32.29, is a helical coil of wire with the same current I passing through each loop in the coil. Solenoids may have hundreds or thousands of coils, often called *turns,* sometimes wrapped in several layers.

FIGURE 32.30 Using superposition to find the magnetic field of a stack of current loops.

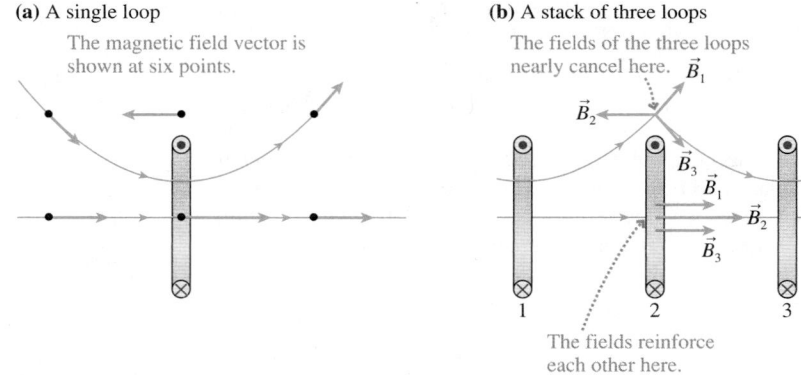

(a) A single loop

The magnetic field vector is shown at six points.

(b) A stack of three loops

The fields of the three loops nearly cancel here.

The fields reinforce each other here.

We can understand a solenoid by thinking of it as a stack of current loops. FIGURE 32.30a shows the magnetic field of a single current loop at three points on the axis and three points equally distant from the axis. The field directly above the loop is opposite in direction to the field inside the loop. FIGURE 32.30b then shows three parallel loops. We can use information from Figure 32.30b to draw the magnetic fields of each loop at the center of loop 2 and at a point above loop 2.

The superposition of the three fields at the center of loop 2 produces a *stronger* field than that of loop 2 alone. But the superposition at the point above loop 2 produces a net magnetic field that is very much weaker than the field at the center of the loop. We've used only three current loops to illustrate the idea, but these tendencies are reinforced by including more loops. With many current loops along the same axis, **the field in the center is strong and roughly parallel to the axis, whereas the field outside the loops is very close to zero.**

FIGURE 32.31a is a photo of the magnetic field of a short solenoid. You can see that the magnetic field inside the coils is nearly uniform (i.e., the field lines are nearly parallel) and the field outside is much weaker. Our goal of producing a uniform magnetic field can be achieved by increasing the number of coils until we have an *ideal solenoid* that is infinitely long and in which the coils are as close together as possible. As FIGURE 32.31b shows, **the magnetic field inside an ideal solenoid is uniform and parallel to the axis; the magnetic field outside is zero.** No real solenoid is ideal, but a very uniform magnetic field can be produced near the center of a tightly wound solenoid whose length is much larger than its diameter.

We can use Ampère's law to calculate the field of an ideal solenoid. FIGURE 32.32 shows a cross section through an infinitely long solenoid. The integration path that we'll use is a rectangle of width l, enclosing N turns of the solenoid coil. Because this is a mathematical curve, not a physical boundary, there's no difficulty with letting it protrude through the

FIGURE 32.31 The magnetic field of a solenoid.

(a) A short solenoid

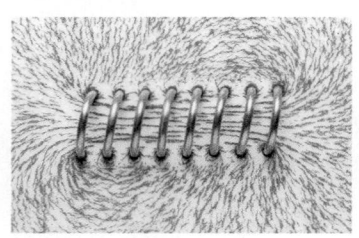

(b)

$\vec{B} = \vec{0}$

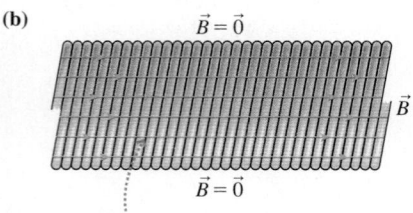

$\vec{B}$

$\vec{B} = \vec{0}$

The magnetic field is uniform inside this section of an ideal, infinitely long solenoid. The magnetic field outside the solenoid is zero.

wall of the solenoid wherever we wish. The solenoid's magnetic field direction, given by the right-hand rule, is left to right, so we'll integrate around this path in the ccw direction.

Each of the N wires enclosed by the integration path carries current I, so the total current passing through the rectangle is $I_{through} = NI$. Ampère's law is thus

$$\oint \vec{B} \cdot d\vec{s} = \mu_0 I_{through} = \mu_0 NI \qquad (32.15)$$

The line integral around this path is the sum of the line integrals along each side. Along the bottom, where $\vec{B}$ is parallel to $d\vec{s}$ and of constant value B, the integral is simply Bl. The integral along the top is zero because the magnetic field outside an ideal solenoid is zero.

The left and right sides sample the magnetic field both inside and outside the solenoid. The magnetic field outside is zero, and the interior magnetic field is everywhere *perpendicular* to the line of integration. Consequently, as we recognized in Option 1 of Tactics Box 32.3, the line integral is zero.

Only the integral along the bottom path is nonzero, leading to

$$\oint \vec{B} \cdot d\vec{s} = Bl = \mu_0 NI$$

Thus the strength of the uniform magnetic field inside a solenoid is

$$B_{solenoid} = \frac{\mu_0 NI}{l} = \mu_0 nI \qquad (32.16)$$

where $n = N/l$ is the number of turns per unit length. Measurements that need a uniform magnetic field are often conducted inside a solenoid, which can be built quite large.

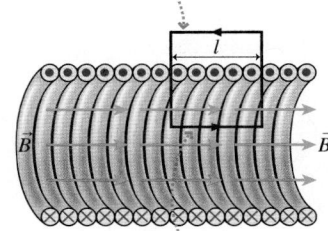

FIGURE 32.32 A closed path inside and outside an ideal solenoid.

This is the integration path for Ampère's law. There are N turns inside.

$\vec{B}$ is tangent to the integration path along the bottom edge.

EXAMPLE 32.9 | **Generating an MRI magnetic field**

A 1.0-m-long MRI solenoid generates a 1.2 T magnetic field. To produce such a large field, the solenoid is wrapped with superconducting wire that can carry a 100 A current. How many turns of wire does the solenoid need?

MODEL Assume that the solenoid is ideal.

SOLVE Generating a magnetic field with a solenoid is a trade-off between current and turns of wire. A larger current requires fewer turns, but the resistance of ordinary wires causes them to overheat if the current is too large. For a superconducting wire that can carry 100 A with no resistance, we can use Equation 32.16 to find the required number of turns:

$$N = \frac{lB}{\mu_0 I} = \frac{(1.0 \text{ m})(1.2 \text{ T})}{(4\pi \times 10^{-7} \text{ T m/A})(100 \text{ A})} = 9500 \text{ turns}$$

ASSESS The solenoid coil requires a large number of turns, but that's not surprising for generating a very strong field. If the wires are 1 mm in diameter, there would be 10 layers with approximately 1000 turns per layer.

The magnetic field of a finite-length solenoid is approximately uniform *inside* the solenoid and weak, but not zero, outside. As **FIGURE 32.33** shows, the magnetic field outside the solenoid looks like that of a bar magnet. Thus a solenoid is an electromagnet, and you can use the right-hand rule to identify the north-pole end. A solenoid with many turns and a large current can be a very powerful magnet.

FIGURE 32.33 The magnetic fields of a finite-length solenoid and of a bar magnet.

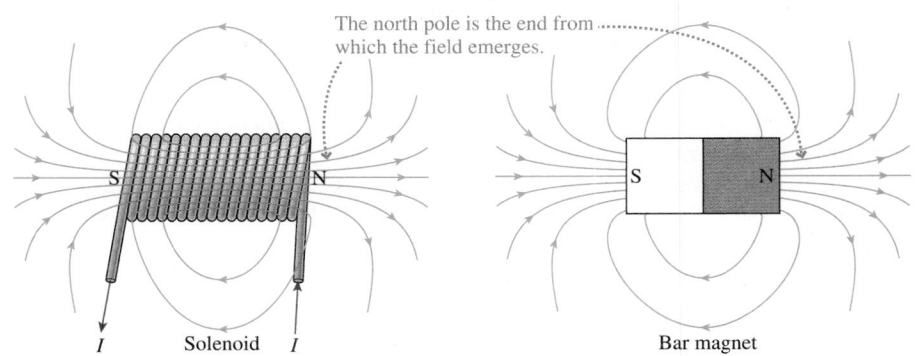

The north pole is the end from which the field emerges.

S   Solenoid   N

S   Bar magnet   N

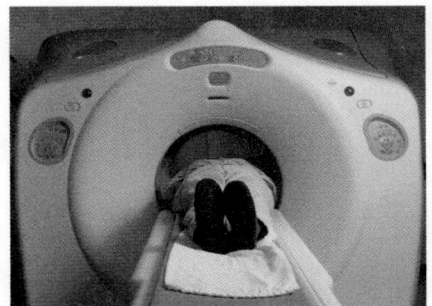

This patient is undergoing magnetic resonance imaging (MRI). The large cylinder surrounding the patient contains a solenoid that is wound with superconducting wire to generate a strong uniform magnetic field.

32.7 The Magnetic Force on a Moving Charge

FIGURE 32.34 Ampère's experiment to observe the forces between parallel current-carrying wires.

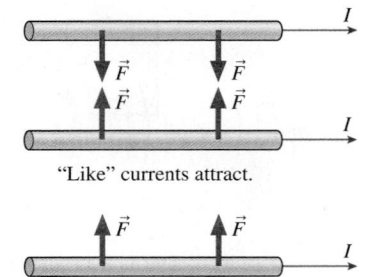

"Like" currents attract.

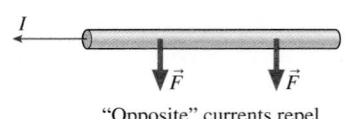

"Opposite" currents repel.

It's time to switch our attention from how magnetic fields are generated to how magnetic fields exert forces and torques. Oersted discovered that a current passing through a wire causes a magnetic torque to be exerted on a nearby compass needle. Upon hearing of Oersted's discovery, André-Marie Ampère, for whom the SI unit of current is named, reasoned that the current was acting like a magnet and, if this were true, that two current-carrying wires should exert magnetic forces on each other.

To find out, Ampère set up two parallel wires that could carry large currents either in the same direction or in opposite (or "antiparallel") directions. FIGURE 32.34 shows the outcome of his experiment. Notice that, for currents, "likes" attract and "opposites" repel. This is the opposite of what would have happened had the wires been charged and thus exerting electric forces on each other. Ampère's experiment showed that **a magnetic field exerts a force on a current.**

Magnetic Force

Because a current consists of moving charges, Ampère's experiment implied that a magnetic field exerts a force on a *moving* charge. This is true, although the exact form of the force law was not discovered until later in the 19th century. The magnetic force turns out to depend not only on the charge and the charge's velocity, but also on how the velocity vector is oriented relative to the magnetic field. FIGURE 32.35 shows the outcome of three experiments to observe the magnetic force.

FIGURE 32.35 The relationship among $\vec{v}$, $\vec{B}$, and $\vec{F}$.

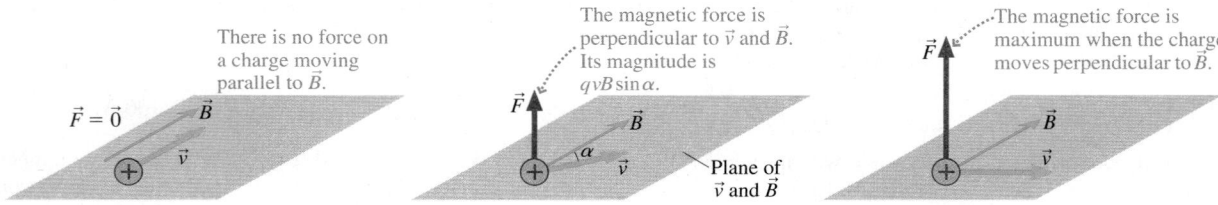

There is no force on a charge moving parallel to $\vec{B}$.

$\vec{F} = \vec{0}$

The magnetic force is perpendicular to $\vec{v}$ and $\vec{B}$. Its magnitude is $qvB \sin \alpha$.

Plane of $\vec{v}$ and $\vec{B}$

The magnetic force is maximum when the charge moves perpendicular to $\vec{B}$.

If you compare the experiment in the middle of Figure 32.35 to Figure 32.10, you'll see that the relationship among $\vec{v}$, $\vec{B}$, and $\vec{F}$ is exactly the same as the geometric relationship among $\vec{C}$, $\vec{D}$, and $\vec{C} \times \vec{D}$. The magnetic force on a charge q as it moves through a magnetic field $\vec{B}$ with velocity $\vec{v}$ can be written

$$\vec{F}_{\text{on } q} = q\vec{v} \times \vec{B} = (qvB \sin \alpha, \text{ direction of right-hand rule}) \quad (32.17)$$

where α is the angle between $\vec{v}$ and $\vec{B}$.

FIGURE 32.36 The right-hand rule for magnetic forces.

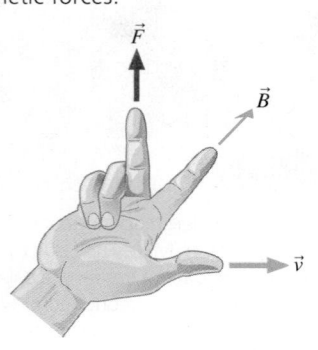

The right-hand rule is that of the cross product, shown in FIGURE 32.36. Notice that **the magnetic force on a moving charged particle is perpendicular to both $\vec{v}$ and $\vec{B}$.**

The magnetic force has several important properties:

1. Only a *moving* charge experiences a magnetic force. There is no magnetic force on a charge at rest ($\vec{v} = \vec{0}$) in a magnetic field.
2. There is no force on a charge moving parallel ($\alpha = 0°$) or antiparallel ($\alpha = 180°$) to a magnetic field.
3. When there is a force, the force is perpendicular to *both* $\vec{v}$ and $\vec{B}$.
4. The force on a negative charge is in the direction *opposite* to $\vec{v} \times \vec{B}$.
5. For a charge moving perpendicular to $\vec{B}$ ($\alpha = 90°$), the magnitude of the magnetic force is $F = |q|vB$.

FIGURE 32.37 shows the relationship among $\vec{v}$, $\vec{B}$, and $\vec{F}$ for four moving charges. (The *source* of the magnetic field isn't shown, only the field itself.) You can see the inherent three-dimensionality of magnetism, with the force perpendicular to both $\vec{v}$ and $\vec{B}$. The magnetic force is very different from the electric force, which is parallel to the electric field.

FIGURE 32.37 Magnetic forces on moving charges.

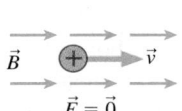 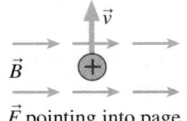 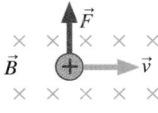

| EXAMPLE 32.10 | **The magnetic force on an electron** |

A long wire carries a 10 A current from left to right. An electron 1.0 cm above the wire is traveling to the right at a speed of 1.0×10^7 m/s. What are the magnitude and the direction of the magnetic force on the electron?

MODEL The magnetic field is that of a long, straight wire.

VISUALIZE FIGURE 32.38 shows the current and an electron moving to the right. The right-hand rule tells us that the wire's magnetic

FIGURE 32.38 An electron moving parallel to a current-carrying wire.

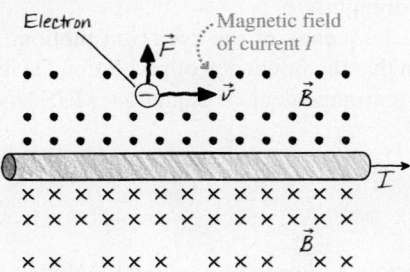

field above the wire is out of the page, so the electron is moving perpendicular to the field.

SOLVE The electron charge is negative, thus the direction of the force is opposite the direction of $\vec{v} \times \vec{B}$. The right-hand rule shows that $\vec{v} \times \vec{B}$ points down, toward the wire, so $\vec{F}$ points up, away from the wire. The magnitude of the force is $|q|vB = evB$. The field is that of a long, straight wire:

$$B = \frac{\mu_0 I}{2\pi d} = 2.0 \times 10^{-4} \text{ T}$$

Thus the magnitude of the force on the electron is

$$F = evB = (1.60 \times 10^{-19} \text{ C})(1.0 \times 10^7 \text{ m/s})(2.0 \times 10^{-4} \text{ T})$$
$$= 3.2 \times 10^{-16} \text{ N}$$

The force on the electron is $\vec{F} = (3.2 \times 10^{-16} \text{ N, up})$.

ASSESS This force will cause the electron to curve away from the wire.

We can draw an interesting and important conclusion at this point. You have seen that the magnetic field is *created by* moving charges. Now you also see that magnetic forces are *exerted on* moving charges. Thus it appears that **magnetism is an interaction between moving charges.** Any two charges, whether moving or stationary, interact with each other through the electric field. In addition, two *moving* charges interact with each other through the magnetic field.

Cyclotron Motion

Many important applications of magnetism involve the motion of charged particles in a magnetic field. Older television picture tubes use magnetic fields to steer electrons through a vacuum from the electron gun to the screen. Microwave generators, which are used in applications ranging from ovens to radar, use a device called a *magnetron* in which electrons oscillate rapidly in a magnetic field.

You've just seen that there is no force on a charge that has velocity $\vec{v}$ parallel or antiparallel to a magnetic field. Consequently, **a magnetic field has no effect on a charge moving parallel or antiparallel to the field.** To understand the motion of charged particles in magnetic fields, we need to consider only motion *perpendicular* to the field.

FIGURE 32.39 shows a positive charge q moving with a velocity $\vec{v}$ in a plane that is perpendicular to a *uniform* magnetic field $\vec{B}$. According to the right-hand rule, the

FIGURE 32.39 Cyclotron motion of a charged particle moving in a uniform magnetic field.

The magnetic force is always perpendicular to $\vec{v}$, causing the particle to move in a circle.

magnetic force on this particle is *perpendicular* to the velocity $\vec{v}$. A force that is always perpendicular to $\vec{v}$ changes the *direction* of motion, by deflecting the particle sideways, but it cannot change the particle's speed. Thus **a particle moving perpendicular to a uniform magnetic field undergoes uniform circular motion at constant speed.** This motion is called the **cyclotron motion** of a charged particle in a magnetic field.

NOTE ▶ A negative charge will orbit in the opposite direction from that shown in Figure 32.39 for a positive charge. ◀

You've seen many analogies to cyclotron motion earlier in this text. For a mass moving in a circle at the end of a string, the tension force is always perpendicular to $\vec{v}$. For a satellite moving in a circular orbit, the gravitational force is always perpendicular to $\vec{v}$. Now, for a charged particle moving in a magnetic field, it is the magnetic force of strength $F = qvB$ that points toward the center of the circle and causes the particle to have a centripetal acceleration.

Newton's second law for circular motion, which you learned in Chapter 8, is

$$F = qvB = ma_r = \frac{mv^2}{r} \tag{32.18}$$

Thus the radius of the cyclotron orbit is

$$r_{cyc} = \frac{mv}{qB} \tag{32.19}$$

The inverse dependence on B indicates that the size of the orbit can be decreased by increasing the magnetic field strength.

We can also determine the frequency of the cyclotron motion. Recall from your earlier study of circular motion that the frequency of revolution f is related to the speed and radius by $f = v/2\pi r$. A rearrangement of Equation 32.19 gives the **cyclotron frequency:**

$$f_{cyc} = \frac{qB}{2\pi m} \tag{32.20}$$

where the ratio q/m is the particle's *charge-to-mass ratio*. Notice that the cyclotron frequency depends on the charge-to-mass ratio and the magnetic field strength but *not* on the charge's velocity.

Electrons undergoing circular motion in a magnetic field. You can see the electrons' path because they collide with a low-density gas that then emits light.

EXAMPLE 32.11 **The radius of cyclotron motion**

In FIGURE 32.40, an electron is accelerated from rest through a potential difference of 500 V, then injected into a uniform magnetic field. Once in the magnetic field, it completes half a revolution in 2.0 ns. What is the radius of its orbit?

FIGURE 32.40 An electron is accelerated, then injected into a magnetic field.

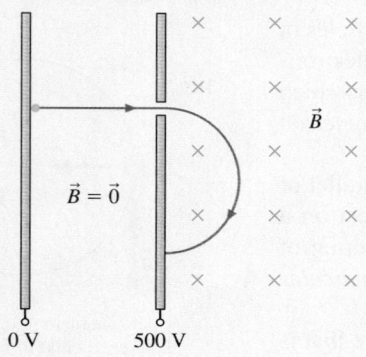

MODEL Energy is conserved as the electron is accelerated by the potential difference. The electron then undergoes cyclotron motion in the magnetic field, although it completes only half a revolution before hitting the back of the acceleration electrode.

SOLVE The electron accelerates from rest ($v_i = 0$ m/s) at $V_i = 0$ V to speed v_f at $V_f = 500$ V. We can use conservation of energy $K_f + qV_f = K_i + qV_i$ to find the speed v_f with which it enters the magnetic field:

$$\frac{1}{2}mv_f^2 + (-e)V_f = 0 + 0$$

$$v_f = \sqrt{\frac{2eV_f}{m}} = \sqrt{\frac{2(1.60 \times 10^{-19}\,\text{C})(500\,\text{V})}{9.11 \times 10^{-31}\,\text{kg}}}$$

$$= 1.33 \times 10^7\,\text{m/s}$$

The cyclotron radius in the magnetic field is $r_{cyc} = mv/eB$, but we first need to determine the field strength. Were it not for the electrode, the electron would undergo circular motion with period $T = 4.0$ ns. Hence the cyclotron frequency is $f = 1/T = 2.5 \times 10^8$ Hz. We can use the cyclotron frequency to determine that the magnetic field strength is

$$B = \frac{2\pi m f_{cyc}}{e} = \frac{2\pi(9.11 \times 10^{-31}\ \text{kg})(2.50 \times 10^8\ \text{Hz})}{1.60 \times 10^{-19}\ \text{C}}$$

$$= 8.94 \times 10^{-3}\ \text{T}$$

Thus the radius of the electron's orbit is

$$r_{cyc} = \frac{mv}{qB} = 8.5 \times 10^{-3}\ \text{m} = 8.5\ \text{mm}$$

FIGURE 32.41a shows a more general situation in which the charged particle's velocity $\vec{v}$ is neither parallel nor perpendicular to $\vec{B}$. The component of $\vec{v}$ parallel to $\vec{B}$ is not affected by the field, so the charged particle spirals around the magnetic field lines in a helical trajectory. The radius of the helix is determined by $\vec{v}_\perp$, the component of $\vec{v}$ perpendicular to $\vec{B}$.

FIGURE 32.41 In general, charged particles spiral along helical trajectories around the magnetic field lines. This motion is responsible for the earth's aurora.

(a) Charged particles spiral around the magnetic field lines.

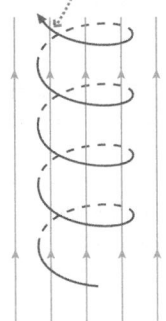

(b) The earth's magnetic field leads particles into the atmosphere near the poles, causing the aurora.

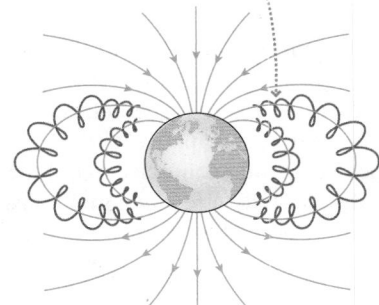

(c) The aurora

The motion of charged particles in a magnetic field is responsible for the earth's aurora. High-energy particles and radiation streaming out from the sun, called the *solar wind,* create ions and electrons as they strike molecules high in the atmosphere. Some of these charged particles become trapped in the earth's magnetic field, creating what is known as the *Van Allen radiation belt.*

As FIGURE 32.41b shows, the electrons spiral along the magnetic field lines until the field leads them into the atmosphere. The shape of the earth's magnetic field is such that most electrons enter the atmosphere near the magnetic poles. There they collide with oxygen and nitrogen atoms, exciting the atoms and causing them to emit auroral light seen in FIGURE 32.41c.

STOP TO THINK 32.5 An electron moves perpendicular to a magnetic field. What is the direction of $\vec{B}$?

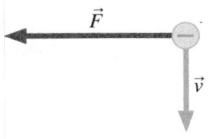

a. Left   b. Up   c. Into the page
d. Right   e. Down   f. Out of the page

The Cyclotron

Physicists studying the structure of the atomic nucleus and of elementary particles usually use a device called a *particle accelerator.* The first practical particle accelerator,

FIGURE 32.42 A cyclotron.

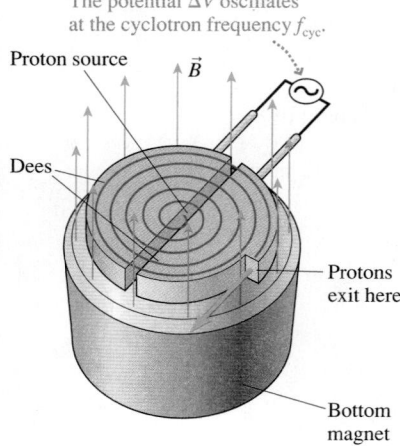

The potential ΔV oscillates at the cyclotron frequency f_{cyc}.

Proton source

$\vec{B}$

Dees

Protons exit here.

Bottom magnet

invented in the 1930s, was the **cyclotron.** Cyclotrons remain important for many applications of nuclear physics, such as the creation of radioisotopes for medicine.

A cyclotron, shown in FIGURE 32.42, consists of an evacuated chamber within a large, uniform magnetic field. Inside the chamber are two hollow conductors shaped like the letter D and hence called "dees." The dees are made of copper, which doesn't affect the magnetic field; are open along the straight sides; and are separated by a small gap. A charged particle, typically a proton, is injected into the magnetic field from a source near the center of the cyclotron, and it begins to move in and out of the dees in a circular cyclotron orbit.

The cyclotron operates by taking advantage of the fact that the cyclotron frequency f_{cyc} of a charged particle is independent of the particle's speed. An *oscillating* potential difference ΔV is connected across the dees and adjusted until its frequency is exactly the cyclotron frequency. There is almost no electric field inside the dees (you learned in Chapter 27 that the electric field inside a hollow conductor is zero), but a strong electric field points from the positive to the negative dee in the gap between them.

Suppose the proton emerges into the gap from the positive dee. The electric field in the gap *accelerates* the proton across the gap into the negative dee, and it gains kinetic energy $e\Delta V$. A half cycle later, when it next emerges into the gap, the potential of the dees (whose potential difference is oscillating at f_{cyc}) will have changed sign. The proton will *again* be emerging from the positive dee and will *again* accelerate across the gap and gain kinetic energy $e\Delta V$.

This pattern will continue orbit after orbit. The proton's kinetic energy increases by $2e\Delta V$ every orbit, so after N orbits its kinetic energy is $K = 2Ne\Delta V$ (assuming that its initial kinetic energy was near zero). The radius of its orbit increases as it speeds up; hence the proton follows the *spiral* path shown in Figure 32.42 until it finally reaches the outer edge of the dee. It is then directed out of the cyclotron and aimed at a target. Although ΔV is modest, usually a few hundred volts, the fact that the proton can undergo many thousands of orbits before reaching the outer edge allows it to acquire a very large kinetic energy.

The Hall Effect

A charged particle moving through a vacuum is deflected sideways, perpendicular to $\vec{v}$, by a magnetic field. In 1879, a graduate student named Edwin Hall showed that the same is true for the charges moving through a conductor as part of a current. This phenomenon—now called the **Hall effect**—is used to gain information about the charge carriers in a conductor. It is also the basis of a widely used technique for measuring magnetic field strengths.

FIGURE 32.43a shows a magnetic field perpendicular to a flat, current-carrying conductor. You learned in Chapter 30 that the charge carriers move through a conductor at the drift speed v_d. Their motion is perpendicular to $\vec{B}$, so each charge carrier experiences a magnetic force $F_B = ev_dB$ perpendicular to both $\vec{B}$ and the current I. However, for the first time we have a situation in which it *does* matter whether the charge carriers are positive or negative.

FIGURE 32.43 The charge carriers in a current are deflected to one surface of a conductor, creating the Hall voltage ΔV_H.

(a)

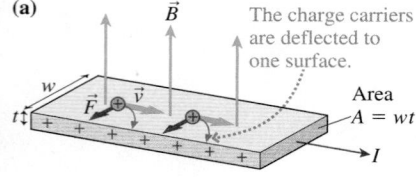

$\vec{B}$

The charge carriers are deflected to one surface.

w

$\vec{F}$ $\vec{v}$

t

Area $A = wt$

I

(b) Top surface is negative.

Conventional current of positive charge carriers

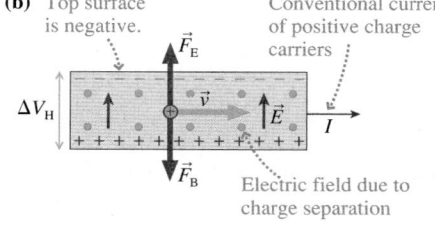

ΔV_H

$\vec{F}_E$

$\vec{v}$

$\vec{E}$

I

$\vec{F}_B$

Electric field due to charge separation

(c) Electron current

Top surface is positive.

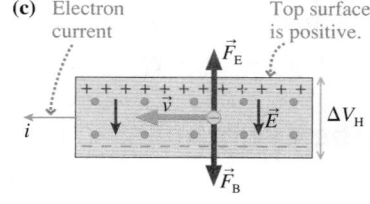

$\vec{F}_E$

$\vec{v}$

$\vec{E}$

ΔV_H

i

$\vec{F}_B$

FIGURE 32.43b, with the field out of the page, shows that positive charge carriers moving in the direction of I are pushed toward the bottom surface of the conductor. This creates an excess positive charge on the bottom surface and leaves an excess negative charge on the top. **FIGURE 32.43c**, where the electrons in an electron current i move opposite the direction of I, shows that electrons would be pushed toward the bottom surface. (Be sure to use the right-hand rule and the sign of the electron charge to confirm the deflections shown in these figures.) Thus the sign of the excess charge on the bottom surface is the same as the sign of the charge carriers. Experimentally, the bottom surface is negative when the conductor is a metal, and this is one more piece of evidence that the charge carriers in metals are electrons.

Electrons are deflected toward the bottom surface once the current starts flowing, but the process can't continue indefinitely. As excess charge accumulates on the top and bottom surfaces, it acts like the charge on the plates of a capacitor, creating a potential difference ΔV between the two surfaces and an electric field $E = \Delta V/w$ inside the conductor of width w. This electric field increases until the upward electric force $\vec{F}_E$ on the charge carriers exactly balances the downward magnetic force $\vec{F}_B$. Once the forces are balanced, a steady state is reached in which the charge carriers move in the direction of the current and no additional charge is deflected to the surface.

The steady-state condition, in which $F_B = F_E$, is

$$F_B = ev_d B = F_E = eE = e\frac{\Delta V}{w} \qquad (32.21)$$

Thus the steady-state potential difference between the two surfaces of the conductor, which is called the **Hall voltage** ΔV_H, is

$$\Delta V_H = wv_d B \qquad (32.22)$$

You learned in Chapter 30 that the drift speed is related to the current density J by $J = nev_d$, where n is the charge-carrier density (charge carriers per m³). Thus

$$v_d = \frac{J}{ne} = \frac{I/A}{ne} = \frac{I}{wtne} \qquad (32.23)$$

where $A = wt$ is the cross-section area of the conductor. If we use this expression for v_d in Equation 32.22, we find that the Hall voltage is

$$\Delta V_H = \frac{IB}{tne} \qquad (32.24)$$

The Hall voltage is very small for metals in laboratory-sized magnetic fields, typically in the microvolt range. Even so, measurements of the Hall voltage in a known magnetic field are used to determine the charge-carrier density n. Interestingly, the Hall voltage is larger for *poor* conductors that have smaller charge-carrier densities. A laboratory probe for measuring magnetic field strengths, called a *Hall probe*, measures ΔV_H for a poor conductor whose charge-carrier density is known. The magnetic field is then determined from Equation 32.24.

EXAMPLE 32.12 **Measuring the magnetic field**

A Hall probe consists of a strip of the metal bismuth that is 0.15 mm thick and 5.0 mm wide. Bismuth is a poor conductor with charge-carrier density 1.35×10^{25} m⁻³. The Hall voltage on the probe is 2.5 mV when the current through it is 1.5 A. What is the strength of the magnetic field, and what is the electric field strength inside the bismuth?

VISUALIZE The bismuth strip looks like Figure 32.43a. The thickness is $t = 1.5 \times 10^{-4}$ m and the width is $w = 5.0 \times 10^{-3}$ m.

SOLVE Equation 32.24 gives the Hall voltage. We can rearrange the equation to find that the magnetic field is

$$B = \frac{tne}{I}\Delta V_H$$
$$= \frac{(1.5 \times 10^{-4}\text{ m})(1.35 \times 10^{25}\text{ m}^{-3})(1.60 \times 10^{-19}\text{ C})}{1.5\text{ A}}\, 0.0025\text{ V}$$
$$= 0.54\text{ T}$$

The electric field created inside the bismuth by the excess charge on the surface is

$$E = \frac{\Delta V_H}{w} = \frac{0.0025\text{ V}}{5.0 \times 10^{-3}\text{ m}} = 0.50\text{ V/m}$$

ASSESS 0.54 T is a fairly typical strength for a laboratory magnet.

32.8 Magnetic Forces on Current-Carrying Wires

FIGURE 32.44 Magnetic force on a current-carrying wire.

(a)   **(b)**

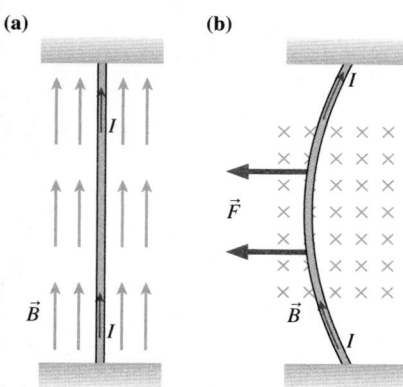

There's no force on a current parallel to a magnetic field.

A current perpendicular to the field experiences a force in the direction of the right-hand rule.

Ampère's observation of magnetic forces between current-carrying wires motivated us to look at the magnetic forces on moving charges. We're now ready to apply that knowledge to Ampère's experiment. As a first step, let us find the force exerted by a uniform magnetic field on a long, straight wire carrying current I through the field. As **FIGURE 32.44a** shows, there's *no* force on a current-carrying wire *parallel* to a magnetic field. This shouldn't be surprising; it follows from the fact that there is no force on a charged particle moving parallel to $\vec{B}$.

FIGURE 32.44b shows a wire *perpendicular* to the magnetic field. By the right-hand rule, each charge in the current has a force of magnitude qvB directed to the left. Consequently, the entire length of wire within the magnetic field experiences a force to the left, perpendicular to both the current direction and the field direction.

To find the magnitude of the force, we must relate the current I in the wire to the charge q moving through the wire. **FIGURE 32.45** shows a segment of wire of length l carrying current I. The current I, by definition, is the amount of moving charge q in this segment of wire divided by the time Δt it takes the charge to flow through the segment: $I = q/\Delta t$. The time required is $\Delta t = l/v$, giving

$$q = I\Delta t = I\frac{l}{v}$$

FIGURE 32.45 Two ways to think of a current.

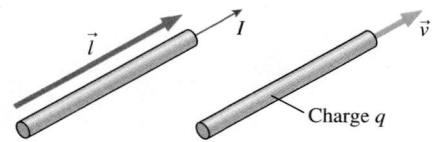

A current consists of charge carriers q moving with velocity $\vec{v}$.

Thus $Il = qv$. If we define vector $\vec{l}$ to have magnitude l and point in the direction of $\vec{v}$, the direction of current, then $I\vec{l} = q\vec{v}$. Substituting this for $q\vec{v}$ in the force equation $\vec{F} = q\vec{v} \times \vec{B}$, we find that the magnetic force on a current-carrying wire is

$$\vec{F}_{\text{wire}} = I\vec{l} \times \vec{B} = (IlB\sin\alpha, \text{ direction of right-hand rule}) \qquad (32.25)$$

where α is the angle between $\vec{l}$ (the direction of the current) and $\vec{B}$. As an aside, you can see from Equation 32.25 that the magnetic field B must have units of N/A m. This is why we defined 1 T = 1 N/A m in Section 32.3.

NOTE ▶ The familiar right-hand rule applies to a current-carrying wire. Point your right thumb in the direction of the current (parallel to $\vec{l}$) and your index finger in the direction of $\vec{B}$. Your middle finger is then pointing in the direction of the force $\vec{F}$ on the wire. ◀

EXAMPLE 32.13 | **Magnetic levitation**

The 0.10 T uniform magnetic field of **FIGURE 32.46** is horizontal, parallel to the floor. A straight segment of 1.0-mm-diameter copper wire, also parallel to the floor, is perpendicular to the magnetic field. What current through the wire, and in which direction, will allow the wire to "float" in the magnetic field?

FIGURE 32.46 Magnetic levitation.

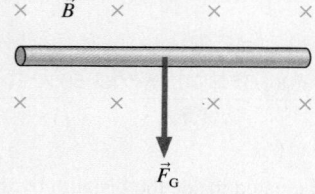

MODEL The wire will float in the magnetic field if the magnetic force on the wire points upward and has magnitude mg, allowing it to balance the downward gravitational force.

SOLVE We can use the right-hand rule to determine which current direction experiences an upward force. With $\vec{B}$ pointing away from us, the direction of the current needs to be from left to right. The forces will balance when

$$F = IlB = mg = \rho(\pi r^2 l)g$$

where $\rho = 8920$ kg/m³ is the density of copper. The length of the wire cancels, leading to

$$I = \frac{\rho\pi r^2 g}{B} = \frac{(8920 \text{ kg/m}^3)\pi(0.00050 \text{ m})^2 (9.80 \text{ m/s}^2)}{0.10 \text{ T}}$$
$$= 0.69 \text{ A}$$

A 0.69 A current from left to right will levitate the wire in the magnetic field.

ASSESS A 0.69 A current is quite reasonable, but this idea is useful only if we can get the current into and out of this segment of wire. In practice, we could do so with wires that come in from below the page. These input and output wires would be parallel to $\vec{B}$ and not experience a magnetic force. Although this example is very simple, it is the basis for applications such as magnetic levitation trains.

Force Between Two Parallel Wires

Now consider Ampère's experimental arrangement of two parallel wires of length l, distance d apart. FIGURE 32.47a shows the currents I_1 and I_2 in the same direction; FIGURE 32.47b shows the currents in opposite directions. We will assume that the wires are sufficiently long to allow us to use the earlier result for the magnetic field of a long, straight wire: $B = \mu_0 I/2\pi d$.

FIGURE 32.47 Magnetic forces between parallel current-carrying wires.

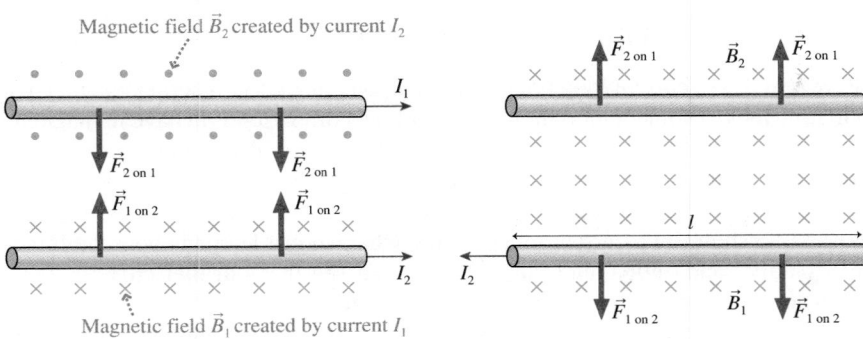

(a) Currents in same direction

(b) Currents in opposite directions

As Figure 32.47a shows, the current I_2 in the lower wire creates a magnetic field $\vec{B}_2$ at the position of the upper wire. $\vec{B}_2$ points out of the page, perpendicular to current I_1. **It is field $\vec{B}_2$, due to the lower wire, that exerts a magnetic force on the upper wire.** Using the right-hand rule, you can see that the force on the upper wire is downward, thus attracting it toward the lower wire. The field of the lower current is not a uniform field, but it is the *same* at all points along the upper wire because the two wires are parallel. Consequently, we can use the field of a long, straight wire to determine the magnetic force exerted by the lower wire on the upper wire when they are separated by distance d:

$$F_{\text{parallel wires}} = I_1 l B_2 = I_1 l \frac{\mu_0 I_2}{2\pi d} = \frac{\mu_0 l I_1 I_2}{2\pi d} \qquad (32.26)$$

As an exercise, you should convince yourself that the current in the upper wire exerts an upward-directed magnetic force on the lower wire with exactly the same magnitude. You should also convince yourself, using the right-hand rule, that the forces are repulsive and tend to push the wires apart if the two currents are in opposite directions.

Thus two parallel wires exert equal but opposite forces on each other, as required by Newton's third law. **Parallel wires carrying currents in the same direction attract each other; parallel wires carrying currents in opposite directions repel each other.**

EXAMPLE 32.14 | **A current balance**

Two stiff, 50-cm-long, parallel wires are connected at the ends by metal springs. Each spring has an unstretched length of 5.0 cm and a spring constant of 0.025 N/m. The wires push each other apart when a current travels around the loop. How much current is required to stretch the springs to lengths of 6.0 cm?

MODEL Two parallel wires carrying currents in opposite directions exert repulsive magnetic forces on each other.

VISUALIZE FIGURE 32.48 shows the "circuit." The springs are conductors, allowing a current to travel around the loop. In equilibrium, the repulsive magnetic forces between the wires are balanced by the restoring forces $F_{sp} = k\Delta y$ of the springs.

FIGURE 32.48 The current-carrying wires of Example 32.14.

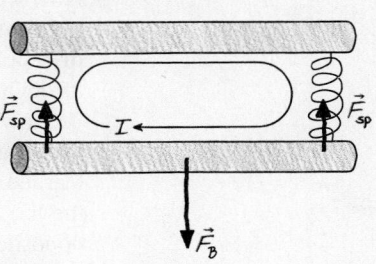

Continued

SOLVE Figure 32.48 shows the forces on the lower wire. The net force is zero, hence the magnetic force is $F_B = 2F_{sp}$. The force between the wires is given by Equation 32.26 with $I_1 = I_2 = I$:

$$F_B = \frac{\mu_0 l I^2}{2\pi d} = 2F_{sp} = 2k\Delta y$$

where k is the spring constant and $\Delta y = 1.0$ cm is the amount by which each spring stretches. Solving for the current, we find

$$I = \sqrt{\frac{4\pi k d \Delta y}{\mu_0 l}} = 17 \text{ A}$$

ASSESS Devices in which a magnetic force balances a mechanical force are called *current balances*. They can be used to make very accurate current measurements.

32.9 Forces and Torques on Current Loops

You have seen that a current loop is a magnetic dipole, much like a permanent magnet. We will now look at some important features of how current loops behave in magnetic fields. This discussion will be largely qualitative, but it will highlight some of the important properties of magnets and magnetic fields. We will use these ideas in the next section to make the connection between electromagnets and permanent magnets.

FIGURE 32.49a shows two current loops. Using what we just learned about the forces between parallel and antiparallel currents, you can see that **parallel current loops exert attractive magnetic forces on each other if the currents circulate in the same direction; they repel each other when the currents circulate in opposite directions.**

FIGURE 32.49 Two alternative but equivalent ways to view magnetic forces.

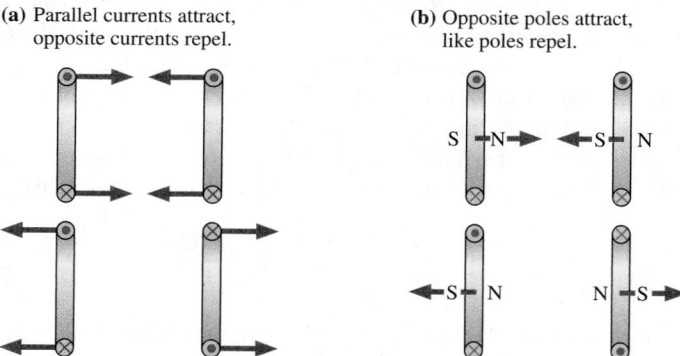

(a) Parallel currents attract, opposite currents repel.

(b) Opposite poles attract, like poles repel.

We can think of these forces in terms of magnetic poles. Recall that the north pole of a current loop is the side from which the magnetic field emerges, which you can determine with the right-hand rule. FIGURE 32.49b shows the north and south magnetic poles of the current loops. When the currents circulate in the same direction, a north and a south pole face each other and exert attractive forces on each other. When the currents circulate in opposite directions, the two like poles repel each other.

Here, at last, we have a real connection to the behavior of magnets that opened our discussion of magnetism—namely, that like poles repel and opposite poles attract. Now we have an *explanation* for this behavior, at least for electromagnets. **Magnetic poles attract or repel because the moving charges in one current exert attractive or repulsive magnetic forces on the moving charges in the other current.** Our tour through interacting moving charges is finally starting to show some practical results!

Now let's consider what happens to a current loop in a magnetic field. FIGURE 32.50 shows a square current loop in a uniform magnetic field along the z-axis. As we've learned, the field exerts magnetic forces on the currents in each of the four sides of the loop. Their directions are given by the right-hand rule. Forces $\vec{F}_{\text{front}}$ and $\vec{F}_{\text{back}}$ are opposite to each other and cancel. Forces $\vec{F}_{\text{top}}$ and $\vec{F}_{\text{bottom}}$ also add to give no net force, but because $\vec{F}_{\text{top}}$ and $\vec{F}_{\text{bottom}}$ don't act along the same line they will *rotate* the loop by exerting a torque on it.

Recall that torque is the magnitude of the force F multiplied by the moment arm d, the distance between the pivot point and the line of action. Both forces have the same moment arm $d = \frac{1}{2}l\sin\theta$, hence the torque on the loop—a torque exerted by the magnetic field—is

$$\tau = 2Fd = 2(IlB)(\tfrac{1}{2}l\sin\theta) = (Il^2)B\sin\theta = \mu B\sin\theta \qquad (32.27)$$

where $\mu = Il^2 = IA$ is the loop's magnetic dipole moment.

Although we derived Equation 32.27 for a square loop, the result is valid for a current loop of any shape. Notice that Equation 32.27 looks like another example of a cross product. We earlier defined the magnetic dipole moment vector $\vec{\mu}$ to be a vector perpendicular to the current loop in a direction given by the right-hand rule. Figure 32.50 shows that θ is the angle between $\vec{B}$ and $\vec{\mu}$, hence the torque on a magnetic dipole is

$$\vec{\tau} = \vec{\mu} \times \vec{B} \qquad (32.28)$$

The torque is zero when the magnetic dipole moment $\vec{\mu}$ is aligned parallel or antiparallel to the magnetic field, and is maximum when $\vec{\mu}$ is perpendicular to the field. It is this magnetic torque that causes a compass needle—a magnetic moment—to rotate until it is aligned with the magnetic field.

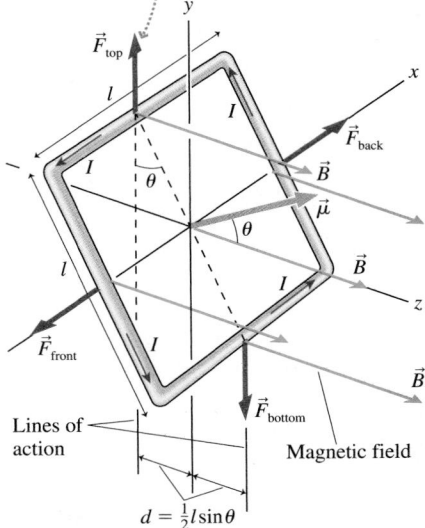

FIGURE 32.50 A uniform magnetic field exerts a torque on a current loop.

$\vec{F}_{top}$ and $\vec{F}_{bottom}$ exert a torque that rotates the loop about the x-axis.

An Electric Motor

The torque on a current loop in a magnetic field is the basis for how an electric motor works. As FIGURE 32.51 shows, the *armature* of a motor is a coil of wire wound on an axle. When a current passes through the coil, the magnetic field exerts a torque on the armature and causes it to rotate. If the current were steady, the armature would oscillate back and forth around the equilibrium position until (assuming there's some friction or damping) it stopped with the plane of the coil perpendicular to the field. To keep the motor turning, a device called a *commutator* reverses the current direction in the coils every 180°. (Notice that the commutator is split, so the positive terminal of the battery sends current into whichever wire touches the right half of the commutator.) The current reversal prevents the armature from ever reaching an equilibrium position, so the magnetic torque keeps the motor spinning as long as there is a current.

FIGURE 32.51 A simple electric motor.

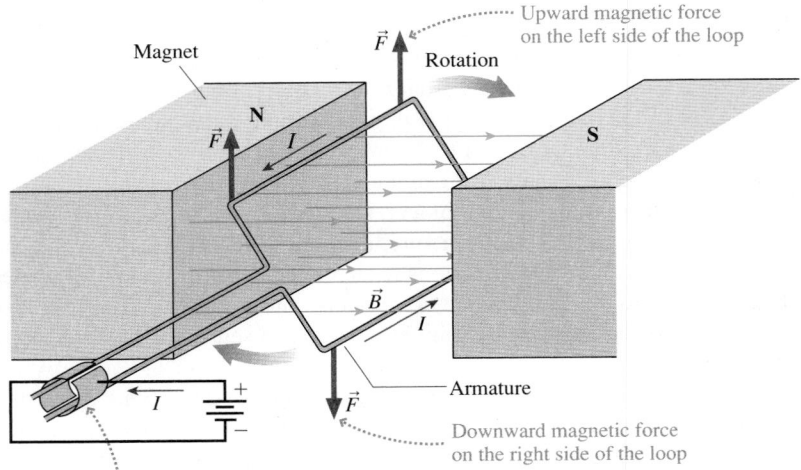

Upward magnetic force on the left side of the loop

Magnet

Rotation

N

S

Armature

Downward magnetic force on the right side of the loop

The commutator reverses the current in the loop every half cycle so that the force is always upward on the left side of the loop.

What is the current direction in the loop?

a. Out of the page at the top of the loop, into the page at the bottom
b. Out of the page at the bottom of the loop, into the page at the top

Repel

32.10 Magnetic Properties of Matter

Our theory has focused mostly on the magnetic properties of currents, yet our everyday experience is mostly with permanent magnets. We have seen that current loops and solenoids have magnetic poles and exhibit behaviors like those of permanent magnets, but we still lack a specific connection between electromagnets and permanent magnets. The goal of this section is to complete our understanding by developing an atomic-level view of the magnetic properties of matter.

Atomic Magnets

FIGURE 32.52 A classical orbiting electron is a tiny magnetic dipole.

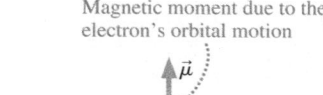

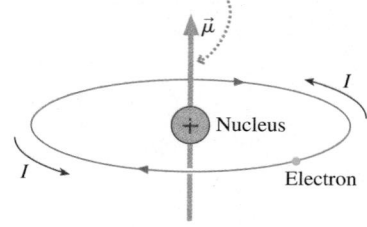

A plausible explanation for the magnetic properties of materials is the orbital motion of the atomic electrons. **FIGURE 32.52** shows a simple, classical model of an atom in which a negative electron orbits a positive nucleus. In this picture of the atom, the electron's motion is that of a current loop! It is a microscopic current loop, to be sure, but a current loop nonetheless. Consequently, an orbiting electron acts as a tiny magnetic dipole, with a north pole and a south pole. You can think of the magnetic dipole as an atomic-size magnet.

However, the atoms of most elements contain many electrons. Unlike the solar system, where all of the planets orbit in the same direction, electron orbits are arranged to oppose each other: one electron moves counterclockwise for every electron that moves clockwise. Thus the magnetic moments of individual orbits tend to cancel each other and the *net* magnetic moment is either zero or very small.

The cancellation continues as the atoms are joined into molecules and the molecules into solids. When all is said and done, the net magnetic moment of any bulk matter due to the orbiting electrons is so small as to be negligible. There are various subtle magnetic effects that can be observed under laboratory conditions, but orbiting electrons cannot explain the very strong magnetic effects of a piece of iron.

The Electron Spin

The key to understanding atomic magnetism was the 1922 discovery that electrons have an *inherent magnetic moment*. Perhaps this shouldn't be surprising. An electron has a *mass,* which allows it to interact with gravitational fields, and a *charge,* which allows it to interact with electric fields. There's no reason an electron shouldn't also interact with magnetic fields, and to do so it comes with a magnetic moment.

FIGURE 32.53 Magnetic moment of the electron.

The arrow represents the inherent magnetic moment of the electron.

An electron's inherent magnetic moment, shown in **FIGURE 32.53**, is often called the electron *spin* because, in a classical picture, a spinning ball of charge would have a magnetic moment. This classical picture is not a realistic portrayal of how the electron really behaves, but its inherent magnetic moment makes it seem *as if* the electron were spinning. While it may not be spinning in a literal sense, an electron really is a microscopic magnet.

We must appeal to the results of quantum physics to find out what happens in an atom with many electrons. The spin magnetic moments, like the orbital magnetic moments, tend to oppose each other as the electrons are placed into their shells, causing the net magnetic moment of a *filled* shell to be zero. However, atoms containing an odd number of electrons must have at least one valence electron with an unpaired spin. These atoms have net magnetic moment due to the electron's spin.

But atoms with magnetic moments don't necessarily form a solid with magnetic properties. For most elements, the magnetic moments of the atoms are randomly arranged when the atoms join together to form a solid. As FIGURE 32.54 shows, this random arrangement produces a solid whose net magnetic moment is very close to zero. This agrees with our common experience that most materials are not magnetic.

Ferromagnetism

It happens that in iron, and a few other substances, the spins interact with each other in such a way that atomic magnetic moments tend to all line up in the *same* direction, as shown in FIGURE 32.55. Materials that behave in this fashion are called **ferromagnetic,** with the prefix *ferro* meaning "iron-like."

In ferromagnetic materials, the individual magnetic moments add together to create a *macroscopic* magnetic dipole. The material has a north and a south magnetic pole, generates a magnetic field, and aligns parallel to an external magnetic field. In other words, it is a magnet!

FIGURE 32.54 The random magnetic moments of the atoms in a typical solid.

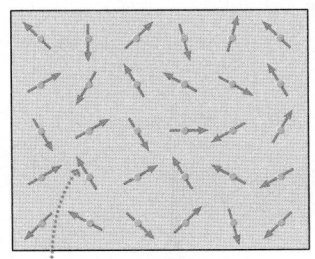

The atomic magnetic moments due to unpaired spins point in random directions. The sample has no net magnetic moment.

FIGURE 32.55 The aligned atomic magnetic moments in a ferromagnetic material.

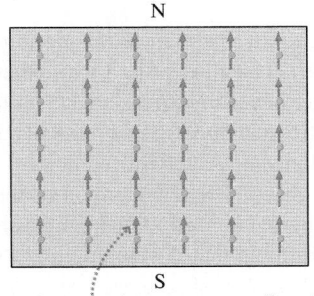

The atomic magnetic moments are aligned. The sample has north and south magnetic poles.

FIGURE 32.56 Magnetic domains in a ferromagnetic material. The net magnetic dipole is nearly zero.

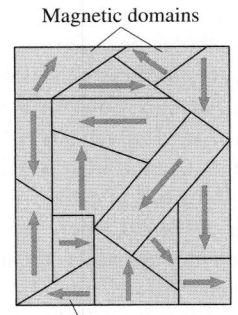

Magnetic domains

Magnetic moment of the domain

Although iron is a magnetic material, a typical piece of iron is not a strong permanent magnet. You need not worry that a steel nail, which is mostly iron and is easily lifted with a magnet, will leap from your hands and pin itself against the hammer because of its own magnetism. It turns out, as shown in FIGURE 32.56, that a piece of iron is divided into small regions, typically less than 100 μm in size, called **magnetic domains.** The magnetic moments of all the iron atoms within each domain are perfectly aligned, so each individual domain, like Figure 32.55, is a strong magnet.

However, the various magnetic domains that form a larger solid, such as you might hold in your hand, are randomly arranged. Their magnetic dipoles largely cancel, much like the cancellation that occurs on the atomic scale for nonferromagnetic substances, so the solid as a whole has only a small magnetic moment. That is why the nail is not a strong permanent magnet.

Induced Magnetic Dipoles

If a ferromagnetic substance is subjected to an *external* magnetic field, the external field exerts a torque on the magnetic dipole of each domain. The torque causes many of the domains to rotate and become aligned with the external field, just as a compass needle aligns with a magnetic field, although internal forces between the domains generally prevent the alignment from being perfect. In addition, atomic-level forces between the spins can cause the *domain boundaries* to move. Domains that are aligned along the external field become larger at the expense of domains that are opposed to the field. These changes in the size and orientation of the domains cause the material to develop a *net magnetic dipole* that is aligned with the external field. This magnetic dipole has been *induced* by the external field, so it is called an **induced magnetic dipole.**

NOTE ▶ The induced magnetic dipole is analogous to the polarization forces and induced electric dipoles that you studied in Chapter 26. ◀

FIGURE 32.57 The magnetic field of the solenoid creates an induced magnetic dipole in the iron.

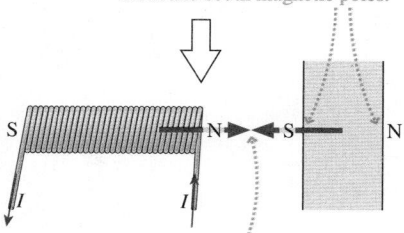

The magnetic domains align with the solenoid's magnetic field.

Ferromagnetic material

The induced magnetic dipole has north and south magnetic poles.

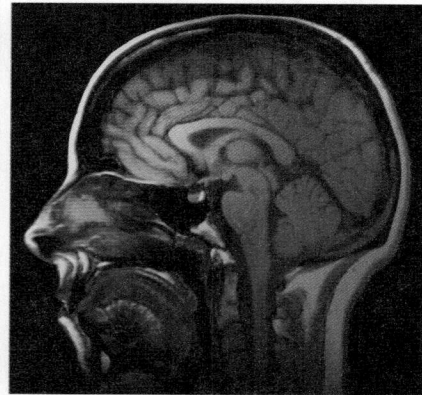

The attractive force between the opposite poles pulls the ferromagnetic material toward the solenoid.

FIGURE 32.57 shows a ferromagnetic material near the end of a solenoid. The magnetic moments of the domains align with the solenoid's field, creating an induced magnetic dipole whose south pole faces the solenoid's north pole. Consequently, the magnetic force between the poles pulls the ferromagnetic object to the electromagnet.

The fact that a magnet attracts and picks up ferromagnetic objects was one of the basic observations about magnetism with which we started the chapter. Now we have an *explanation* of how it works, based on three ideas:

1. Electrons are microscopic magnets due to their spin.
2. A ferromagnetic material in which the spins are aligned is organized into magnetic domains.
3. The individual domains align with an external magnetic field to produce an induced magnetic dipole moment for the entire object.

The object's magnetic dipole may not return to zero when the external field is removed because some domains remain "frozen" in the alignment they had in the external field. Thus a ferromagnetic object that has been in an external field may be left with a net magnetic dipole moment after the field is removed. In other words, the object has become a **permanent magnet.** A permanent magnet is simply a ferromagnetic material in which a majority of the magnetic domains are aligned with each other to produce a net magnetic dipole moment.

Whether or not a ferromagnetic material can be made into a permanent magnet depends on the internal crystalline structure of the material. *Steel* is an alloy of iron with other elements. An alloy of mostly iron with the right percentages of chromium and nickel produces *stainless steel,* which has virtually no magnetic properties at all because its particular crystalline structure is not conducive to the formation of domains. A very different steel alloy called Alnico V is made with 51% iron, 24% cobalt, 14% nickel, 8% aluminum, and 3% copper. It has extremely prominent magnetic properties and is used to make high-quality permanent magnets. You can see from the complex formula that developing good magnetic materials requires a lot of engineering skill as well as a lot of patience!

So we've come full circle. One of our initial observations about magnetism was that a permanent magnet can exert forces on some materials but not others. The *theory* of magnetism that we then proceeded to develop was about the interactions between moving charges. What moving charges had to do with permanent magnets was not obvious. But finally, by considering magnetic effects at the atomic level, we found that properties of permanent magnets and magnetic materials can be traced to the interactions of vast numbers of electron spins.

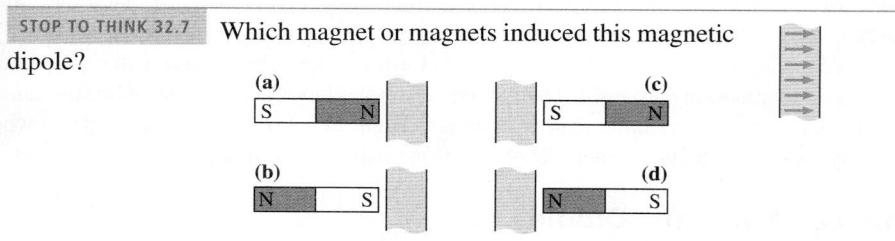

STOP TO THINK 32.7 Which magnet or magnets induced this magnetic dipole?

(a)

(b)

(c)

(d)

Magnetic resonance imaging, or MRI, uses the magnetic properties of atoms as a noninvasive probe of the human body.

CHALLENGE EXAMPLE 32.15 **Designing a loudspeaker**

A loudspeaker consists of a paper cone wrapped at the bottom with several turns of fine wire. As **FIGURE 32.58** shows, this coil sits in a narrow gap between the poles of a circular magnet. To produce sound, the amplifier drives a current through the coil. The magnetic field then exerts a force on this current, pushing the cone and thus pushing the air to create a sound wave. An ideal speaker would experience only forces from the magnetic field, thus responding only to the current from the amplifier. Real speakers are balanced so as to come close to this ideal unless driven very hard.

FIGURE 32.58 The coil and magnet of a loudspeaker.

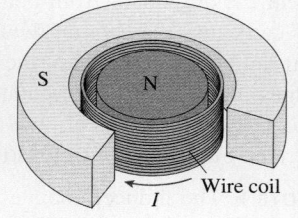

Consider a 5.5 g loudspeaker cone with a 5.0-cm-diameter, 20-turn coil having a resistance of 8.0 Ω. There is a 0.18 T field in the gap between the poles. These values are typical of the loudspeakers found in car stereo systems. What is the oscillation amplitude of this speaker if driven by a 100 Hz oscillatory voltage from the amplifier with a peak value of 12 V?

MODEL Model the loudspeaker as ideal, responding only to magnetic forces. These forces cause the cone to accelerate. We'll use kinematics to relate the acceleration to the displacement.

VISUALIZE FIGURE 32.59 shows the coil in the gap between the magnet poles. Magnetic fields go from north to south poles, so the field is radially outward. Consequently, the field at all points is perpendicular to the circular current. According to the right-hand rule, the magnetic force on the current is into or out of the page, depending on whether the current is counterclockwise or clockwise, respectively.

FIGURE 32.59 The magnetic field in the gap, from north to south, is perpendicular to the current.

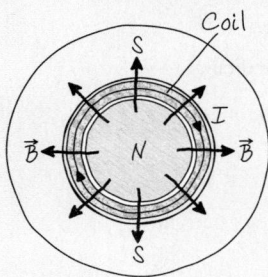

SOLVE We can write the output voltage of the amplifier as $\Delta V = V_0 \cos \omega t$, where $V_0 = 12$ V is the peak voltage and $\omega = 2\pi f = 628$ rad/s is the angular frequency at 100 Hz. The voltage drives current

$$I = \frac{\Delta V}{R} = \frac{V_0 \cos \omega t}{R}$$

through the coil, where R is the coil's resistance. This causes the oscillating in-and-out force that drives the speaker cone back and forth. Even though the coil isn't a straight wire, the fact that the magnetic field is everywhere perpendicular to the current means that we can calculate the magnetic force as $F = IlB$ where l is the total length of the wire in the coil. The circumference of the coil is $\pi(0.050 \text{ m}) = 0.157$ m so 20 turns gives $l = 3.1$ m. The cone responds to the force by accelerating with $a = F/m$. Combining these pieces, we find the cone's acceleration is

$$a = \frac{IlB}{m} = \frac{V_0 lB \cos \omega t}{mR} = a_{max} \cos \omega t$$

It is straightforward to evaluate $a_{max} = 152$ m/s^2.

From kinematics, $a = dv/dt$ and $v = dx/dt$. We need to integrate twice to find the displacement. First,

$$v = \int a \, dt = a_{max} \int \cos \omega t \, dt = \frac{a_{max}}{\omega} \sin \omega t$$

The integration constant is zero because we know, from simple harmonic motion, that the average velocity is zero. Integrating again, we get

$$x = \int v \, dt = \frac{a_{max}}{\omega} \int \sin \omega t \, dt = -\frac{a_{max}}{\omega^2} \cos \omega t$$

where the integration constant is again zero if we assume the oscillation takes place around the origin. The minus sign tells us that the displacement and acceleration are out of phase. The *amplitude* of the oscillation, which we seek, is

$$A = \frac{a_{max}}{\omega^2} = \frac{152 \text{ m/s}^2}{(628 \text{ rad/s})^2} = 3.8 \times 10^{-4} \text{ m} = 0.38 \text{ mm}$$

ASSESS If you've ever placed your hand on a loudspeaker cone, you know that you can feel a slight vibration. An amplitude of 0.38 mm is consistent with this observation. The fact that the amplitude increases with the inverse square of the frequency explains why you can sometimes *see* the cone vibrating with an amplitude of several millimeters for low-frequency bass notes.

SUMMARY

The goal of Chapter 32 has been to learn how to calculate and use the magnetic field.

General Principles

At its most fundamental level, magnetism is an interaction between moving charges. The magnetic field of one moving charge exerts a force on another moving charge.

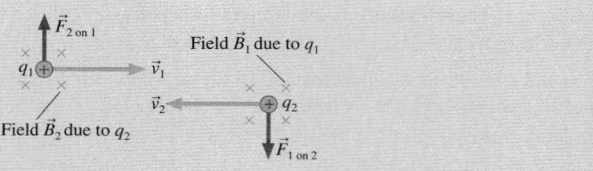

Magnetic Fields

The Biot-Savart law

- A **point charge**, $\vec{B} = \dfrac{\mu_0}{4\pi} \dfrac{q\vec{v} \times \hat{r}}{r^2}$

- A **short current element**, $\vec{B} = \dfrac{\mu_0}{4\pi} \dfrac{I\Delta\vec{s} \times \hat{r}}{r^2}$

To find the magnetic field of a current:

- Divide the wire into many short segments.
- Find the field of each segment Δs.
- Find $\vec{B}$ by summing the fields of all Δs, usually as an integral.

An alternative method for fields with a high degree of symmetry is Ampère's law:

$$\oint \vec{B} \cdot d\vec{s} = \mu_0 I_{\text{through}}$$

where I_{through} is the current through the area bounded by the integration path.

Magnetic Forces

The magnetic force on a moving charge is

$$\vec{F} = q\vec{v} \times \vec{B}$$

The force is perpendicular to $\vec{v}$ and $\vec{B}$.

The magnetic force on a current-carrying wire is

$$\vec{F} = I\vec{l} \times \vec{B}$$

$\vec{F} = \vec{0}$ for a charge or current moving parallel to $\vec{B}$.

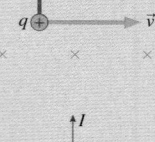

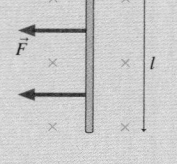

The magnetic torque on a magnetic dipole is

$$\vec{\tau} = \vec{\mu} \times \vec{B}$$

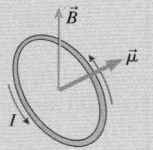

Applications

Wire

$$B = \dfrac{\mu_0}{2\pi} \dfrac{I}{d}$$

Loop

Solenoid

$$B = \dfrac{\mu_0 NI}{l}$$

Flat magnet

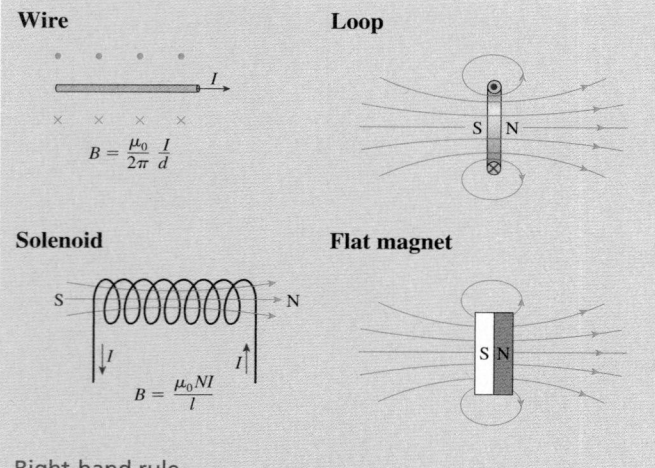

Right-hand rule

Point your right thumb in the direction of I. Your fingers curl in the direction of $\vec{B}$. For a dipole, $\vec{B}$ emerges from the side that is the north pole.

Charged-particle motion

No force if $\vec{v}$ is parallel to $\vec{B}$

Circular motion at the cyclotron frequency $f_{\text{cyc}} = qB/2\pi m$ if $\vec{v}$ is perpendicular to $\vec{B}$

Parallel wires and current loops

Parallel currents attract.
Opposite currents repel.

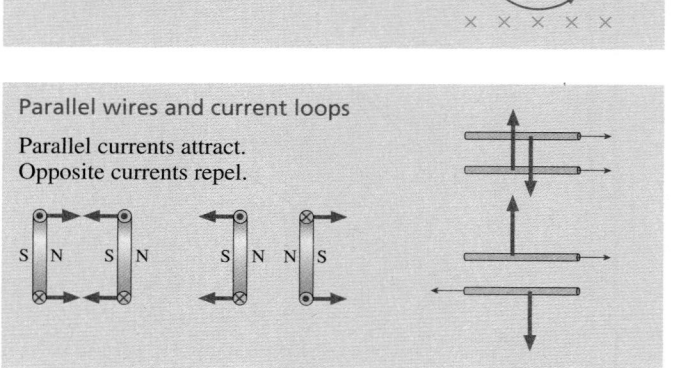

Terms and Notation

north pole	Biot-Savart law	line integral	Hall effect
south pole	tesla, T	Ampère's law	Hall voltage, ΔV_H
magnetic dipole	permeability constant, μ_0	uniform magnetic field	ferromagnetic
magnetic material	cross product	solenoid	magnetic domain
right-hand rule	current loop	cyclotron motion	induced magnetic dipole
magnetic field, $\vec{B}$	electromagnet	cyclotron frequency, f_{cyc}	permanent magnet
magnetic field lines	magnetic dipole moment, $\vec{\mu}$	cyclotron	

CONCEPTUAL QUESTIONS

1. The lightweight glass sphere in FIGURE Q32.1 hangs by a thread. The north pole of a bar magnet is brought near the sphere.
   a. Suppose the sphere is electrically neutral. Is it attracted to, repelled by, or not affected by the magnet? Explain.
   b. Answer the same question if the sphere is positively charged.

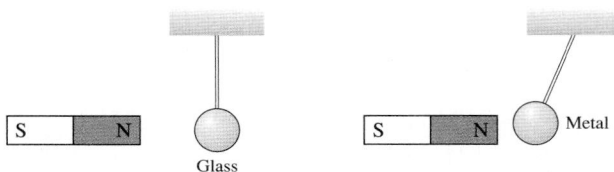

FIGURE Q32.1   FIGURE Q32.2

2. The metal sphere in FIGURE Q32.2 hangs by a thread. When the north pole of a magnet is brought near, the sphere is strongly attracted to the magnet. Then the magnet is reversed and its south pole is brought near the sphere. How does the sphere respond? Explain.
3. You have two electrically neutral metal cylinders that exert strong attractive forces on each other. You have no other metal objects. Can you determine if *both* of the cylinders are magnets, or if one is a magnet and the other is just a piece of iron? If so, how? If not, why not?
4. What is the current direction in the wire of FIGURE Q32.4? Explain.

FIGURE Q32.4   FIGURE Q32.5

5. What is the current direction in the wire of FIGURE Q32.5? Explain.
6. What is the *initial* direction of deflection for the charged particles entering the magnetic fields shown in FIGURE Q32.6?

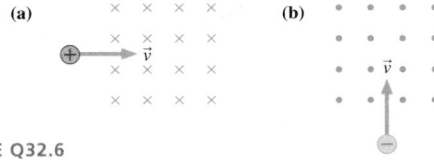

FIGURE Q32.6

7. What is the *initial* direction of deflection for the charged particles entering the magnetic fields shown in FIGURE Q32.7?

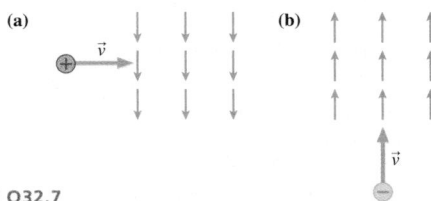

FIGURE Q32.7

8. Determine the magnetic field direction that causes the charged particles shown in FIGURE Q32.8 to experience the indicated magnetic force.

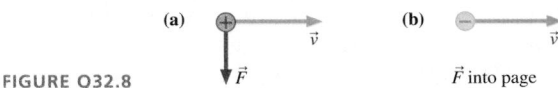

FIGURE Q32.8

9. Determine the magnetic field direction that causes the charged particles shown in FIGURE Q32.9 to experience the indicated magnetic force.

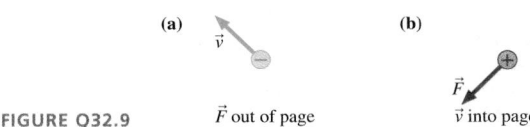

FIGURE Q32.9 $\vec{F}$ out of page $\vec{v}$ into page

10. You have a horizontal cathode-ray tube (CRT) for which the controls have been adjusted such that the electron beam *should* make a single spot of light exactly in the center of the screen. You observe, however, that the spot is deflected to the right. It is possible that the CRT is broken. But as a clever scientist, you realize that your laboratory might be in either an electric or a magnetic field. Assuming that you do not have a compass, any magnets, or any charged rods, how can you use the CRT itself to determine whether the CRT is broken, is in an electric field, or is in a magnetic field? You cannot remove the CRT from the room.
11. The south pole of a bar magnet is brought toward the current loop of FIGURE Q32.11. Does the bar magnet attract, repel, or have no effect on the loop? Explain.

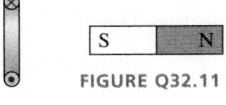

FIGURE Q32.11

12. Give a step-by-step explanation, using both words and pictures, of how a permanent magnet can pick up a piece of nonmagnetized iron.

EXERCISES AND PROBLEMS

Problems labeled ▓▓▓ integrate material from earlier chapters.

Exercises

Section 32.3 The Source of the Magnetic Field: Moving Charges

1. | Points 1 and 2 in FIGURE EX32.1 are the same distance from the wires as the point where $B = 2.0$ mT. What are the strength and direction of $\vec{B}$ at points 1 and 2?

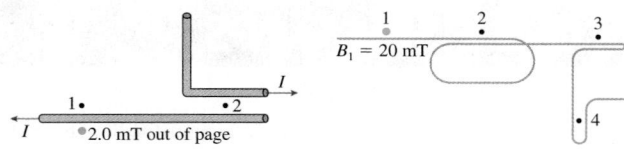

FIGURE EX32.1   FIGURE EX32.2

2. | What is the magnetic field strength at points 2 to 4 in FIGURE EX32.2? Assume that the wires overlap closely and that points 1 to 4 are equally distant from the wires.

3. ‖ A proton moves along the x-axis with $v_x = 1.0 \times 10^7$ m/s. As it passes the origin, what are the strength and direction of the magnetic field at the (x, y, z) positions (a) (1 cm, 0 cm, 0 cm), (b) (0 cm, 1 cm, 0 cm), and (c) (0 cm, −2 cm, 0 cm)?

4. ‖ An electron moves along the z-axis with $v_z = 2.0 \times 10^7$ m/s. As it passes the origin, what are the strength and direction of the magnetic field at the (x, y, z) positions (a) (1 cm, 0 cm, 0 cm), (b) (0 cm, 0 cm, 1 cm), and (c) (0 cm, 1 cm, 1 cm)?

5. ‖ What is the magnetic field at the position of the dot in FIGURE EX32.5? Give your answer as a vector.

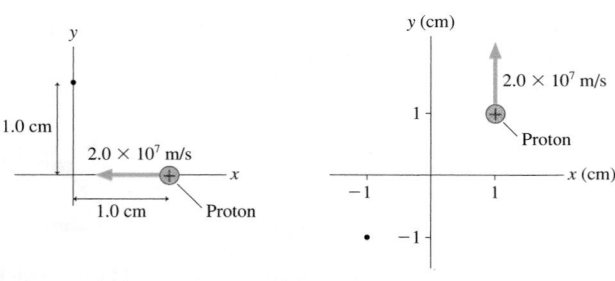

FIGURE EX32.5   FIGURE EX32.6

6. ‖ What is the magnetic field at the position of the dot in FIGURE EX32.6? Give your answer as a vector.

7. ‖ A proton is passing the origin. The magnetic field at the (x, y, z) position (1 mm, 0 mm, 0 mm) is $1.0 \times 10^{-13} \hat{j}$ T. The field at (0 mm, 1 mm, 0 mm) is $-1.0 \times 10^{-13} \hat{i}$ T. What are the speed and direction of the proton?

Section 32.4 The Magnetic Field of a Current

8. | What currents are needed to generate the magnetic field strengths of Table 32.1 at a point 1.0 cm from a long, straight wire?

9. | At what distances from a very thin, straight wire carrying a 10 A current would the magnetic field strengths of Table 32.1 be generated?

10. ‖ The element niobium, which is a metal, is a superconductor (i.e., no electrical resistance) at temperatures below 9 K. However, the superconductivity is destroyed if the magnetic field at the surface of the metal reaches or exceeds 0.10 T. What is the maximum current in a straight, 3.0-mm-diameter superconducting niobium wire?

11. ‖ The magnetic field at the center of a 1.0-cm-diameter loop is 2.5 mT.
    a. What is the current in the loop?
    b. A long straight wire carries the same current you found in part a. At what distance from the wire is the magnetic field 2.5 mT?

12. | A wire carries current I into the junction shown in FIGURE EX32.12. What is the magnetic field at the dot?

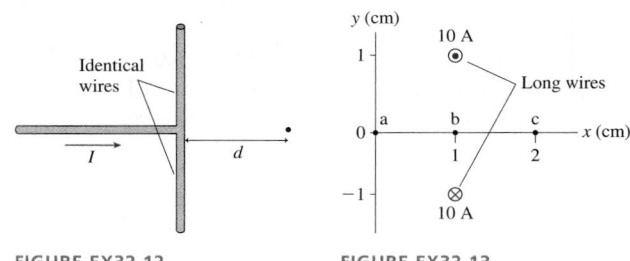

FIGURE EX32.12   FIGURE EX32.13

13. | What are the magnetic fields at points a to c in FIGURE EX32.13? Give your answers as vectors.

14. ‖ What are the magnetic field strength and direction at points a to c in FIGURE EX32.14?

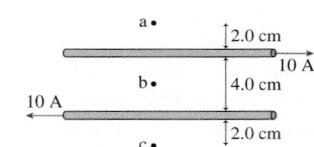

FIGURE EX32.14

Section 32.5 Magnetic Dipoles

15. | The on-axis magnetic field strength 10 cm from a small bar magnet is 5.0 μT.
    a. What is the bar magnet's magnetic dipole moment?
    b. What is the on-axis field strength 15 cm from the magnet?

16. ‖ A 100 A current circulates around a 2.0-mm-diameter superconducting ring.
    a. What is the ring's magnetic dipole moment?
    b. What is the on-axis magnetic field strength 5.0 cm from the ring?

17. ‖ A small, square loop carries a 25 A current. The on-axis magnetic field strength 50 cm from the loop is 7.5 nT. What is the edge length of the square?

18. ‖ The earth's magnetic dipole moment is 8.0×10^{22} A m².
    a. What is the magnetic field strength on the surface of the earth at the earth's north magnetic pole? How does this compare to the value in Table 32.1? You can assume that the current loop is deep inside the earth.
    b. Astronauts discover an earth-size planet without a magnetic field. To create a magnetic field with the same strength as earth's, they propose running a current through a wire around the equator. What size current would be needed?

Section 32.6 Ampère's Law and Solenoids

19. | What is the line integral of $\vec{B}$ between points i and f in **FIGURE EX32.19**?

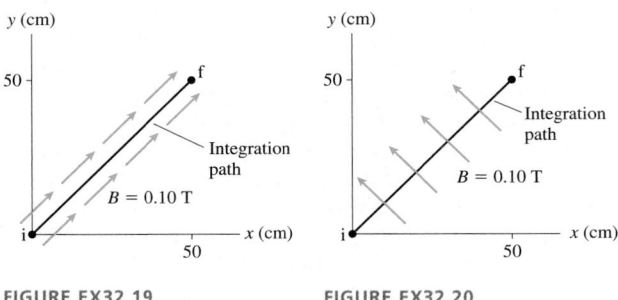

FIGURE EX32.19   **FIGURE EX32.20**

20. | What is the line integral of $\vec{B}$ between points i and f in **FIGURE EX32.20**?

21. ‖ The value of the line integral of $\vec{B}$ around the closed path in **FIGURE EX32.21** is 3.77×10^{-6} T m. What is I_3?

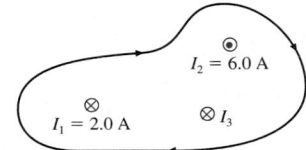

FIGURE EX32.21

22. ‖ The value of the line integral of $\vec{B}$ around the closed path in **FIGURE EX32.22** is 1.38×10^{-5} T m. What are the direction (in or out of the page) and magnitude of I_3?

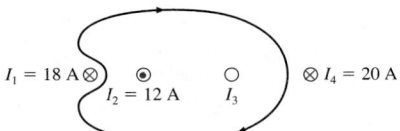

FIGURE EX32.22

23. ‖ What is the line integral of $\vec{B}$ between points i and f in **FIGURE EX32.23**?

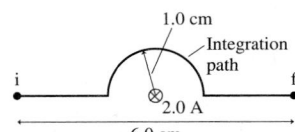

FIGURE EX32.23

24. ‖ Magnetic resonance imaging needs a magnetic field strength
BIO of 1.5 T. The solenoid is 1.8 m long and 75 cm in diameter. It is tightly wound with a single layer of 2.0-mm-diameter superconducting wire. What size current is needed?

25. ‖ A 2.0-cm-diameter, 15-cm-long solenoid is tightly wound with one layer of wire. A 2.5 A current through the wire generates a 3.0 mT magnetic field inside the solenoid. What is the diameter of the wire, in mm?

Section 32.7 The Magnetic Force on a Moving Charge

26. | A proton moves in the magnetic field $\vec{B} = 0.50\,\hat{\imath}$ T with a speed of 1.0×10^7 m/s in the directions shown in **FIGURE EX32.26**. For each, what is magnetic force $\vec{F}$ on the proton? Give your answers in component form.

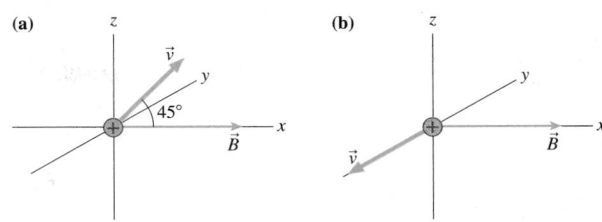

FIGURE EX32.26

27. ‖ An electron moves in the magnetic field $\vec{B} = 0.50\,\hat{\imath}$ T with a speed of 1.0×10^7 m/s in the directions shown in **FIGURE EX32.27**. For each, what is magnetic force $\vec{F}$ on the electron? Give your answers in component form.

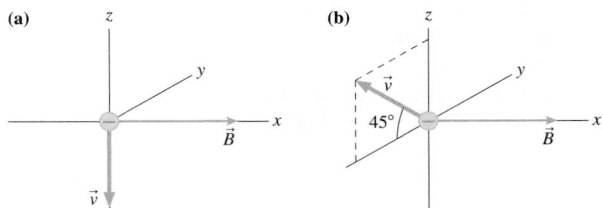

FIGURE EX32.27

28. ‖ To five significant figures, what are the cyclotron frequencies in a 3.0000 T magnetic field of the ions (a) O_2^+, (b) N_2^+, and (c) CO^+? The atomic masses are shown in the table; the mass of the missing electron is less than 0.001 u and is not relevant at this level of precision. Although N_2^+ and CO^+ both have a nominal molecular mass of 28, they are easily distinguished by virtue of their slightly different cyclotron frequencies. Use the following constants: $1\,u = 1.6605 \times 10^{-27}$ kg, $e = 1.6022 \times 10^{-19}$ C.

Atomic masses	
^{12}C	12.000
^{14}N	14.003
^{16}O	15.995

29. | Radio astronomers detect electromagnetic radiation at 45 MHz from an interstellar gas cloud. They suspect this radiation is emitted by electrons spiraling in a magnetic field. What is the magnetic field strength inside the gas cloud?

30. | For your senior project, you would like to build a cyclotron that will accelerate protons to 10% of the speed of light. The largest vacuum chamber you can find is 50 cm in diameter. What magnetic field strength will you need?

31. | The Hall voltage across a conductor in a 55 mT magnetic field is 1.9 μV. When used with the same current in a different magnetic field, the voltage across the conductor is 2.8 μV. What is the strength of the second field?

32. | Test instruments to measure magnetic field strengths are often based on the Hall effect. In one instrument, the "probe" is a 1.0-mm-thick, 6.0-mm-wide semiconductor with a charge-carrier density of 2.1×10^{21} m^{-3}, much less than the charge-carrier density in a conductor. Passing a 60 mA current through the probe generates a Hall voltage of 120 mV. What is the magnetic field strength?

Section 32.8 Magnetic Forces on Current-Carrying Wires

33. | What magnetic field strength and direction will levitate the 2.0 g wire in FIGURE EX32.33?

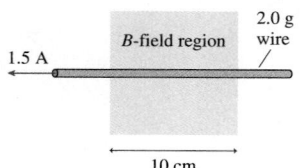

FIGURE EX32.33

34. | The right edge of the circuit in FIGURE EX32.34 extends into a 50 mT uniform magnetic field. What are the magnitude and direction of the net force on the circuit?

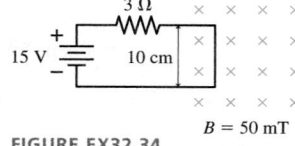

FIGURE EX32.34

35. ‖ The two 10-cm-long parallel wires in FIGURE EX32.35 are separated by 5.0 mm. For what value of the resistor R will the force between the two wires be 5.4×10^{-5} N?

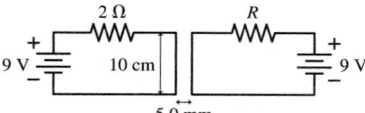

FIGURE EX32.35

36. | What is the net force (magnitude and direction) on each wire in FIGURE EX32.36?

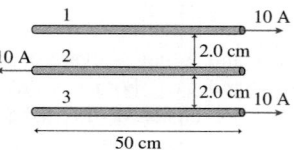

FIGURE EX32.36

Section 32.9 Forces and Torques on Current Loops

37. ‖ A square current loop 5.0 cm on each side carries a 500 mA current. The loop is in a 1.2 T uniform magnetic field. The axis of the loop, perpendicular to the plane of the loop, is 30° away from the field direction. What is the magnitude of the torque on the current loop?

38. | A small bar magnet experiences a 0.020 N m torque when the axis of the magnet is at 45° to a 0.10 T magnetic field. What is the magnitude of its magnetic dipole moment?

39. ‖ a. What is the magnitude of the torque on the current loop in FIGURE EX32.39?
b. What is the loop's equilibrium orientation?

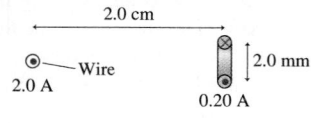

FIGURE EX32.39

Problems

40. | Although the evidence is weak, there has been concern in recent years over possible health effects from the magnetic fields generated by electric transmission lines. A typical high-voltage transmission line is 20 m above the ground and carries a 200 A current at a potential of 110 kV.
a. What is the magnetic field strength on the ground directly under such a transmission line?
b. What percentage is this of the earth's magnetic field of 50 μT?

41. | A biophysics experiment uses a very sensitive magnetic field probe to determine the current associated with a nerve impulse traveling along an axon. If the peak field strength 1.0 mm from an axon is 8.0 pT, what is the peak current carried by the axon?

42. ‖ A long wire carrying a 5.0 A current perpendicular to the xy-plane intersects the x-axis at $x = -2.0$ cm. A second, parallel wire carrying a 3.0 A current intersects the x-axis at $x = +2.0$ cm. At what point or points on the x-axis is the magnetic field zero if (a) the two currents are in the same direction and (b) the two currents are in opposite directions?

43. ‖ The two insulated wires in FIGURE P32.43 cross at a 30° angle but do not make electrical contact. Each wire carries a 5.0 A current. Points 1 and 2 are each 4.0 cm from the intersection and equally distant from both wires. What are the magnitude and direction of the magnetic fields at points 1 and 2?

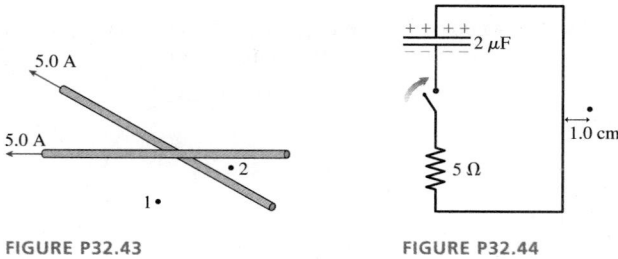

FIGURE P32.43   FIGURE P32.44

44. ‖ The capacitor in FIGURE P32.44 is charged to 50 V. The switch closes at $t = 0$ s. Draw a graph showing the magnetic field strength as a function of time at the position of the dot. On your graph indicate the maximum field strength, and provide an appropriate numerical scale on the horizontal axis.

45. ‖ At what distance on the axis of a current loop is the magnetic field half the strength of the field at the center of the loop? Give your answer as a multiple of R.

46. ‖ Find an expression for the magnetic field strength at the center (point P) of the circular arc in FIGURE P32.46.

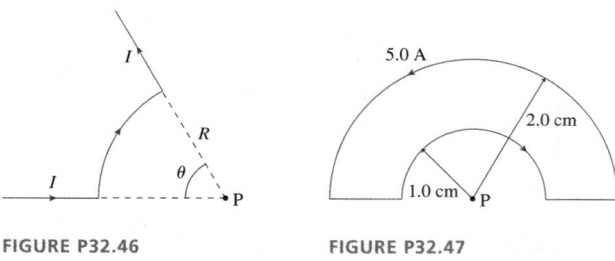

FIGURE P32.46   FIGURE P32.47

47. ‖ What are the strength and direction of the magnetic field at point P in FIGURE P32.47?

48. ‖ What are the strength and direction of the magnetic field at the center of the loop in FIGURE P32.48?

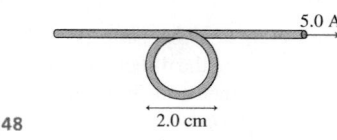

FIGURE P32.48

49. ‖ Your employer asks you to build a 20-cm-long solenoid with an interior field of 5.0 mT. The specifications call for a single layer of wire, wound with the coils as close together as possible. You have two spools of wire available. Wire with a #18 gauge has a diameter of 1.02 mm and has a maximum current rating of 6 A. Wire with a #26 gauge is 0.41 mm in diameter and can carry up to 1 A. Which wire should you use, and what current will you need?

50. ‖ The magnetic field strength at the north pole of a 2.0-cm-diameter, 8-cm-long Alnico magnet is 0.10 T. To produce the same field with a solenoid of the same size, carrying a current of 2.0 A, how many turns of wire would you need? Does this seem feasible?

51. ‖ The earth's magnetic field, with a magnetic dipole moment of 8.0×10^{22} A m^2, is generated by currents within the molten iron of the earth's outer core. Suppose we model the core current as a 3000-km-diameter current loop made from a 1000-km-diameter "wire." The loop diameter is measured from the centers of this very fat wire.
  a. What is the current in the current loop?
  b. What is the current density J in the current loop?
  c. To decide whether this is a large or a small current density, compare it to the current density of a 1.0 A current in a 1.0-mm-diameter wire.

52. ‖ Weak magnetic fields can be measured at the surface of the
BIO brain. Although the currents causing these fields are quite complicated, we can estimate their size by modeling them as a current loop around the equator of a 16-cm-diameter (the width of a typical head) sphere. What current is needed to produce a 3.0 pT field— the strength measured for one subject—at the pole of this sphere?

53. ‖ The heart produces a weak magnetic field that can be used to
BIO diagnose certain heart problems. It is a dipole field produced by a current loop in the outer layers of the heart.
  a. It is estimated that the field at the center of the heart is 90 pT. What current must circulate around an 8.0-cm-diameter loop, about the size of a human heart, to produce this field?
  b. What is the magnitude of the heart's magnetic dipole moment?

54. ‖‖ Two identical coils are parallel to each other on the same axis. They are separated by a distance equal to their radius. They each have N turns and carry equal currents I in the same direction.
  a. Find an expression for the magnetic field strength at the midpoint between the loops.
  b. Calculate the field strength if the loops are 10 cm in diameter, have 10 turns, and carry a 1.0 A current.

55. ‖ Use the Biot-Savart law to find the magnetic field strength at the center of the semicircle in FIGURE P32.55.

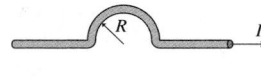

FIGURE P32.55

56. ‖ The *toroid* of FIGURE P32.56 is a coil of wire wrapped around a doughnut-shaped ring (a *torus*) made of nonconducting material. Toroidal magnetic fields are used to confine fusion plasmas.
  a. From symmetry, what must be the *shape* of the magnetic field in this toroid? Explain.
  b. Consider a toroid with N closely spaced turns carrying current I. Use Ampère's law to find an expression for the magnetic field strength at a point inside the torus at distance r from the axis.
  c. Is a toroidal magnetic field a uniform field? Explain.

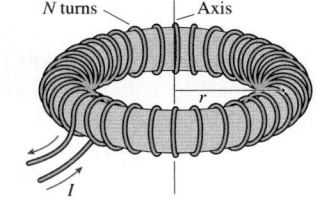

FIGURE P32.56

57. ‖ A long, hollow wire has inner radius R_1 and outer radius R_2. The wire carries current I uniformly distributed across the area of the wire. Use Ampère's law to find an expression for the magnetic field strength in the three regions $0 < r < R_1$, $R_1 < r < R_2$, and $R_2 < r$.

58. ‖ An electron orbits in a 5.0 mT field with angular momentum 8.0×10^{-26} kg m^2/s. What is the diameter of the orbit?

59. ‖ A proton moving in a uniform magnetic field with $\vec{v}_1 = 1.00 \times 10^6 \hat{\imath}$ m/s experiences force $\vec{F}_1 = 1.20 \times 10^{-16} \hat{k}$ N. A second proton with $\vec{v}_2 = 2.00 \times 10^6 \hat{\jmath}$ m/s experiences $\vec{F}_2 = -4.16 \times 10^{-16} \hat{k}$ N in the same field. What is $\vec{B}$? Give your answer as a magnitude and an angle measured ccw from the $+x$-axis.

60. ‖ An electron travels with speed 1.0×10^7 m/s between the two parallel charged plates shown in FIGURE P32.60. The plates are separated by 1.0 cm and are charged by a 200 V battery. What magnetic field strength and direction will allow the electron to pass between the plates without being deflected?

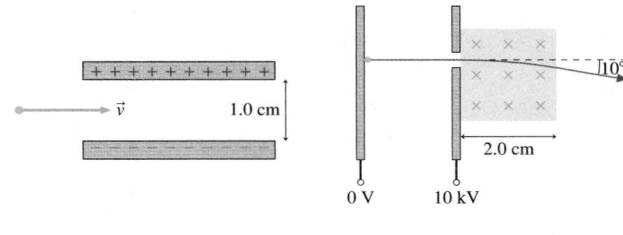

FIGURE P32.60   FIGURE P32.61

61. ‖ An electron in a cathode-ray tube is accelerated through a potential difference of 10 kV, then passes through the 2.0-cm-wide region of uniform magnetic field in FIGURE P32.61. What field strength will deflect the electron by 10°?

62. ‖ The microwaves in a microwave oven are produced in a special tube called a *magnetron*. The electrons orbit the magnetic field at 2.4 GHz, and as they do so they emit 2.4 GHz electromagnetic waves.
  a. What is the magnetic field strength?
  b. If the maximum diameter of the electron orbit before the electron hits the wall of the tube is 2.5 cm, what is the maximum electron kinetic energy?

63. ‖ An antiproton (same properties as a proton except that $q = -e$) is moving in the combined electric and magnetic fields of FIGURE P32.63. What are the magnitude and direction of the antiproton's acceleration at this instant?

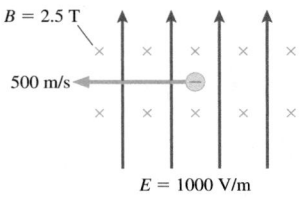

FIGURE P32.63 $E = 1000$ V/m

64. ‖ a. A 65-cm-diameter cyclotron uses a 500 V oscillating potential difference between the dees. What is the maximum kinetic energy of a proton if the magnetic field strength is 0.75 T?
  b. How many revolutions does the proton make before leaving the cyclotron?

65. ‖ **FIGURE P32.65** shows a *mass spectrometer,* an analytical instrument used to identify the various molecules in a sample by measuring their charge-to-mass ratio q/m. The sample is ionized, the positive ions are accelerated (starting from rest) through a potential difference ΔV, and they then enter a region of uniform magnetic field. The field bends the ions into circular trajectories, but after just half a circle they either strike the wall or pass through a small opening to a detector. As the accelerating voltage is slowly increased, different ions reach the detector and are measured. Consider a mass spectrometer with a 200.00 mT magnetic field and an 8.0000 cm spacing between the entrance and exit holes. To five significant figures, what accelerating potential differences ΔV are required to detect the ions (a) O_2^+, (b) N_2^+, and (c) CO^+? See Exercise 28 for atomic masses; the mass of the missing electron is less than 0.001 u and is not relevant at this level of precision. Although N_2^+ and CO^+ both have a nominal molecular mass of 28, they are easily distinguished by virtue of their slightly different accelerating voltages. Use the following constants: $1\ u = 1.6605 \times 10^{-27}$ kg, $e = 1.6022 \times 10^{-19}$ C.

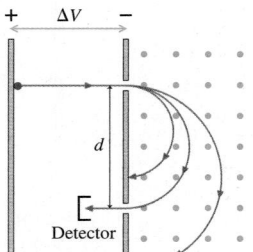

FIGURE P32.65

66. ‖ A Hall-effect probe to measure magnetic field strengths needs to be calibrated in a known magnetic field. Although it is not easy to do, magnetic fields can be precisely measured by measuring the cyclotron frequency of protons. A testing laboratory adjusts a magnetic field until the proton's cyclotron frequency is 10.0 MHz. At this field strength, the Hall voltage on the probe is 0.543 mV when the current through the probe is 0.150 mA. Later, when an unknown magnetic field is measured, the Hall voltage at the same current is 1.735 mV. What is the strength of this magnetic field?

67. ‖ The 10-turn loop of wire shown in **FIGURE P32.67** lies in a horizontal plane, parallel to a uniform horizontal magnetic field, and carries a 2.0 A current. The loop is free to rotate about a nonmagnetic axle through the center. A 50 g mass hangs from one edge of the loop. What magnetic field strength will prevent the loop from rotating about the axle?

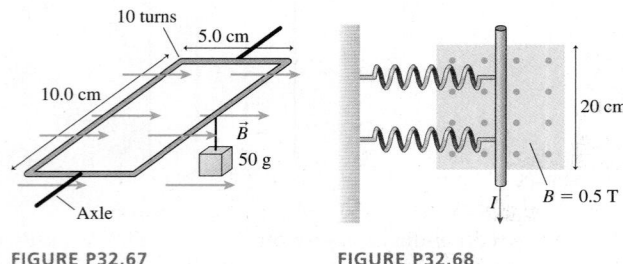

FIGURE P32.67   FIGURE P32.68

68. ‖ The two springs in **FIGURE P32.68** each have a spring constant of 10 N/m. They are compressed by 1.0 cm when a current passes through the wire. How big is the current?

69. ‖ Magnetic fields are sometimes measured by balancing magnetic forces against known mechanical forces. Your task is to measure the strength of a horizontal magnetic field using a 12-cm-long rigid metal rod that hangs from two nonmagnetic springs, one at each end, with spring constants 1.3 N/m. You first position the rod to be level and perpendicular to the field, whose direction you determined with a compass. You then connect the ends of the rod to wires that run parallel to the field and thus experience no forces. Finally, you measure the downward deflection of the rod, stretching the springs, as you pass current through it. Your data are as follows:

Current (A)	Deflection (mm)
1.0	4
2.0	9
3.0	12
4.0	15
5.0	21

Use an appropriate graph of the data to determine the magnetic field strength.

70. ‖ A conducting bar of length l and mass m rests at the left end of the two frictionless rails of length d in **FIGURE P32.70**. A uniform magnetic field of strength B points upward.
   a. In which direction, into or out of the page, will a current through the conducting bar cause the bar to experience a force to the right?
   b. Find an expression for the bar's speed as it leaves the rails at the right end.

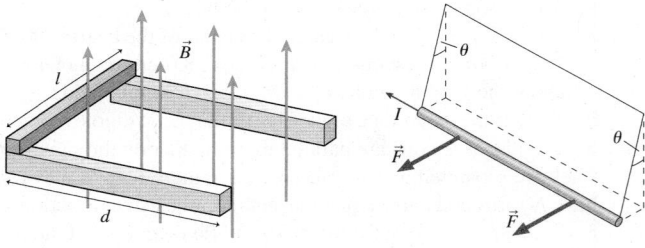

FIGURE P32.70   FIGURE P32.71

71. ‖ a. In **FIGURE P32.71**, a long, straight, current-carrying wire of linear mass density μ is suspended by threads. A magnetic field perpendicular to the wire exerts a horizontal force that deflects the wire to an equilibrium angle θ. Find an expression for the strength and direction of the magnetic field $\vec{B}$.
   b. What $\vec{B}$ deflects a 55 g/m wire to a 12° angle when the current is 10 A?

72. ‖ **FIGURE P32.72** is a cross section through three long wires with linear mass density 50 g/m. They each carry equal currents in the directions shown. The lower two wires are 4.0 cm apart and are attached to a table. What current I will allow the upper wire to "float" so as to form an equilateral triangle with the lower wires?

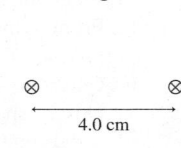

FIGURE P32.72

73. ‖ In the semiclassical Bohr model of the hydrogen atom, the electron moves in a circular orbit of radius 5.3×10^{-11} m with speed 2.2×10^6 m/s. According to this model, what is the magnetic field at the center of a hydrogen atom?
Hint: Determine the *average* current of the orbiting electron.

74. ‖ A wire along the x-axis carries current I in the negative x-direction through the magnetic field

$$\vec{B} = \begin{cases} B_0 \dfrac{x}{l}\, \hat{k} & 0 \le x \le l \\ 0 & \text{elsewhere} \end{cases}$$

  a. Draw a graph of B versus x over the interval $-\tfrac{3}{2}l < x < \tfrac{3}{2}l$.
  b. Find an expression for the net force $\vec{F}_{\text{net}}$ on the wire.
  c. Find an expression for the net torque on the wire about the point $x = 0$.

75. ‖ A *nonuniform* magnetic field exerts a net force on a current loop of radius R. FIGURE P32.75 shows a magnetic field that is diverging from the end of a bar magnet. The magnetic field at the position of the current loop makes an angle θ with respect to the vertical.
  a. Find an expression for the net magnetic force on the current.
  b. Calculate the force if $R = 2.0$ cm, $I = 0.50$ A, $B = 200$ mT, and $\theta = 20°$.

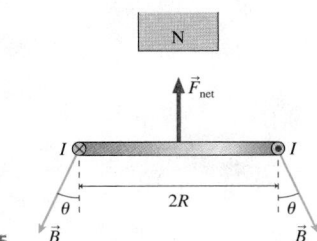

FIGURE P32.75

Challenge Problems

76. You have a 1.0-m-long copper wire. You want to make an N-turn current loop that generates a 1.0 mT magnetic field at the center when the current is 1.0 A. You must use the entire wire. What will be the diameter of your coil?

77. a. Derive an expression for the magnetic field strength at distance d from the center of a straight wire of finite length l that carries current I.
  b. Determine the field strength at the center of a current-carrying *square* loop having sides of length $2R$.
  c. Compare your answer to part b to the field at the center of a *circular* loop of diameter $2R$. Do so by computing the ratio $B_{\text{square}}/B_{\text{circle}}$.

78. A flat, circular disk of radius R is uniformly charged with total charge Q. The disk spins at angular velocity ω about an axis through its center. What is the magnetic field strength at the center of the disk?

79. A long, straight conducting wire of radius R has a nonuniform current density $J = J_0 r/R$, where J_0 is a constant. The wire carries total current I.
  a. Find an expression for J_0 in terms of I and R.
  b. Find an expression for the magnetic field strength inside the wire at radius r.
  c. At the boundary, $r = R$, does your solution match the known field outside a long, straight current-carrying wire?

80. The coaxial cable shown in FIGURE CP32.80 consists of a solid inner conductor of radius R_1 surrounded by a hollow, very thin outer conductor of radius R_2. The two carry equal currents I, but in *opposite* directions. The current density is uniformly distributed over each conductor.
  a. Find expressions for three magnetic fields: within the inner conductor, in the space between the conductors, and outside the outer conductor.
  b. Draw a graph of B versus r from $r = 0$ to $r = 2R_2$ if $R_1 = \tfrac{1}{3}R_2$.

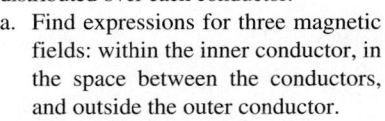

FIGURE CP32.80

81. An infinitely wide flat sheet of charge flows out of the page in FIGURE CP32.81. The current per unit width along the sheet (amps per meter) is given by the linear current density J_s.
  a. What is the *shape* of the magnetic field? To answer this question, you may find it helpful to approximate the current sheet as many parallel, closely spaced current-carrying wires. Give your answer as a picture showing magnetic field vectors.
  b. Find the magnetic field strength at distance d above or below the current sheet.

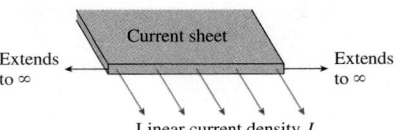

FIGURE CP32.81

82. The uniform 30 mT magnetic field in FIGURE CP32.82 points in the positive z-direction. An electron enters the region of magnetic field with a speed of 5.0×10^6 m/s and at an angle of $30°$ above the xy-plane. Find the radius r and the pitch p of the electron's spiral trajectory.

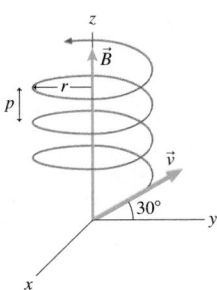

FIGURE CP32.82

STOP TO THINK ANSWERS

Stop to Think 32.1: Not at all. The charge exerts weak, attractive polarization forces on both ends of the compass needle, but in this configuration the forces will balance and have no net effect.

Stop to Think 32.2: d. Point your right thumb in the direction of the current and curl your fingers around the wire.

Stop to Think 32.3: b. Point your right thumb out of the page, in the direction of $\vec{v}$. Your fingers are pointing down as they curl around the left side.

Stop to Think 32.4: b. The right-hand rule gives a downward $\vec{B}$ for a clockwise current. The north pole is on the side from which the field emerges.

Stop to Think 32.5: c. For a field pointing into the page, $\vec{v} \times \vec{B}$ is to the right. But the electron is negative, so the force is in the direction of $-(\vec{v} \times \vec{B})$.

Stop to Think 32.6: b. Repulsion indicates that the south pole of the loop is on the right, facing the bar magnet; the north pole is on the left. Then the right-hand rule gives the current direction.

Stop to Think 32.7: a or c. Any magnetic field to the right, whether leaving a north pole or entering a south pole, will align the magnetic domains as shown.

33 Electromagnetic Induction

Electromagnetic induction is the physics that underlies many modern technologies, from the generation of electricity to data storage.

▶ **Looking Ahead** The goal of Chapter 33 is to understand and apply electromagnetic induction.

Magnetic Flux

A key idea will be the amount of magnetic field passing through a loop. This is called the **magnetic flux.**

You'll find that the magnetic flux depends on the strength of the magnetic field, the area of the loop, and the angle between them.

◀ **Looking Back**
Sections 27.2–27.3 Electric flux

Connecting *E* and *B*

We previously found that a current generates a magnetic field. In fact, the connection between electric and magnetic fields is much more profound.

You'll learn that pushing a magnet into a coil of wire, or pulling it out, causes an **induced current** in the wire. The process is called **electromagnetic induction.**

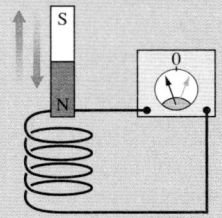

◀ **Looking Back**
Chapter 32 Magnetic fields and forces

Applications

Electromagnetic induction has many applications. You'll learn about using **inductors**—coils of wire that store magnetic energy—in circuits.

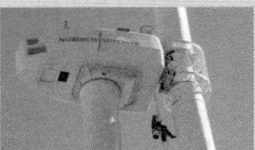

You'll also learn how a generator, such as this **generator** turned by windmill blades, transforms mechanical energy into electric energy.

Lenz's Law

Lenz's law says that a current is induced in a closed loop *if and only if the magnetic flux through the loop is changing*. Simply having a magnetic flux doesn't do anything; the flux has to *change*.

You'll learn how to use Lenz's law to determine the direction of an induced current.

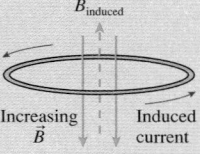

◀ **Looking Back**
Section 29.2 Sources of potential

Faraday's Law

You'll learn to use **Faraday's law,** the most important law connecting electric and magnetic fields.

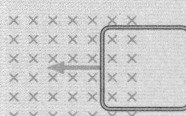

The magnetic flux through this loop is increasing as the loop moves into the field. Faraday's law allows us to compute the induced emf and the induced current. The current direction is given by Lenz's law.

Induced Fields

At its most fundamental level, Faraday's law tells us that a changing magnetic field creates an **induced electric field.** It is the induced electric field that then creates the induced current in a conducting loop.

An increasing magnetic field (blue) creates an electric field (red) that circulates in closed loops. You'll learn to calculate the strength of the induced field.

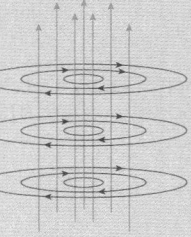

33.1 Induced Currents

Oersted's 1820 discovery that a current creates a magnetic field generated enormous excitement. One question scientists hoped to answer was whether the converse of Oersted's discovery was true: that is, can a magnet be used to create a current?

The breakthrough came in 1831 when the American science teacher Joseph Henry and the English scientist Michael Faraday each discovered the process we now call *electromagnetic induction*. Faraday—whom you met in Chapter 25 as the inventor of the concept of a *field*—was the first to publish his findings, so today we study Faraday's law rather than Henry's law.

Faraday's 1831 discovery, like Oersted's, was a happy combination of an unplanned event and a mind that was ready to recognize its significance. Faraday was experimenting with two coils of wire wrapped around an iron ring, as shown in FIGURE 33.1. He had hoped that the magnetic field generated in the coil on the left would induce a magnetic field in the iron, and that the magnetic field in the iron might then somehow create a current in the circuit on the right.

Like all his previous attempts, this technique failed to generate a current. But Faraday happened to notice that the needle of the current meter jumped ever so slightly at the instant he closed the switch in the circuit on the left. After the switch was closed, the needle immediately returned to zero. The needle again jumped when he later opened the switch, but this time in the opposite direction. Faraday recognized that the motion of the needle indicated a current in the circuit on the right, but a momentary current only during the brief interval when the current on the left was starting or stopping.

Faraday's observations, coupled with his mental picture of field lines, led him to suggest that a current is generated only if the magnetic field through the coil is *changing*. This explains why all the previous attempts to generate a current with static magnetic fields had been unsuccessful. Faraday set out to test this hypothesis.

FIGURE 33.1 Faraday's discovery.

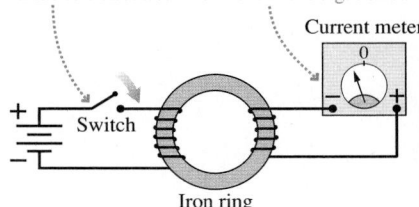

Closing the switch in the left circuit... ...causes a momentary current in the right circuit.

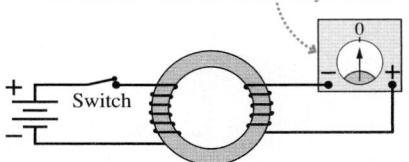

No current while the switch stays closed

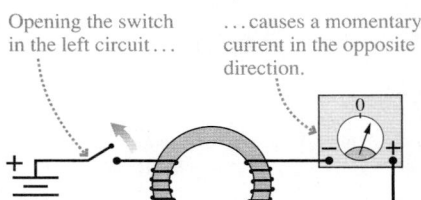

Opening the switch in the left circuit... ...causes a momentary current in the opposite direction.

Faraday investigates electromagnetic induction

Faraday placed one coil directly above the other, without the iron ring. There was no current in the lower circuit while the switch was in the closed position, but a momentary current appeared whenever the switch was opened or closed.

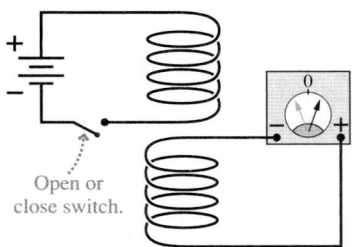

Opening or closing the switch creates a momentary current.

He pushed a bar magnet into a coil of wire. This action caused a momentary deflection of the current-meter needle, although *holding* the magnet inside the coil had no effect. A quick withdrawal of the magnet deflected the needle in the other direction.

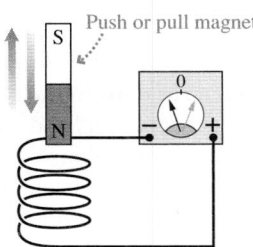

Pushing the magnet into the coil or pulling it out creates a momentary current.

Must the magnet move? Faraday created a momentary current by rapidly pulling a coil of wire out of a magnetic field. Pushing the coil *into* the magnet caused the needle to deflect in the opposite direction.

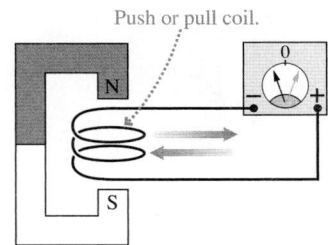

Pushing the coil into the magnet or pulling it out creates a momentary current.

Faraday found that there is a current in a coil of wire if and only if the magnetic field passing through the coil is *changing*. This is an informal statement of what we'll soon call *Faraday's law*. The current in a circuit due to a changing magnetic field is called an **induced current.** An induced current is not caused by a battery; it is a completely new way to generate a current.

33.2 Motional emf

We'll start our investigation of electromagnetic induction by looking at situations in which the magnetic field is fixed while the circuit moves or changes. Consider a conductor of length l that moves with velocity $\vec{v}$ through a perpendicular uniform magnetic field $\vec{B}$, as shown in FIGURE 33.2. The charge carriers inside the wire—assumed to be positive—also move with velocity $\vec{v}$, so they each experience a magnetic force $\vec{F}_B = q\vec{v} \times \vec{B}$ of strength $F_B = qvB$. This force causes the charge carriers to move, separating the positive and negative charges. The separated charges then create an electric field inside the conductor.

FIGURE 33.2 The magnetic force on the charge carriers in a moving conductor creates an electric field inside the conductor.

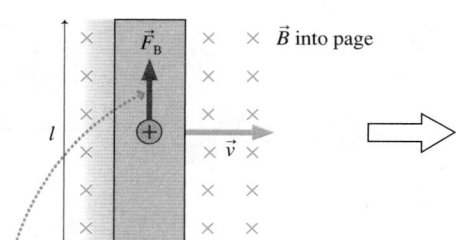

Charge carriers in the wire experience an upward force of magnitude $F_B = qvB$. Being free to move, positive charges flow upward (or, if you prefer, negative charges downward).

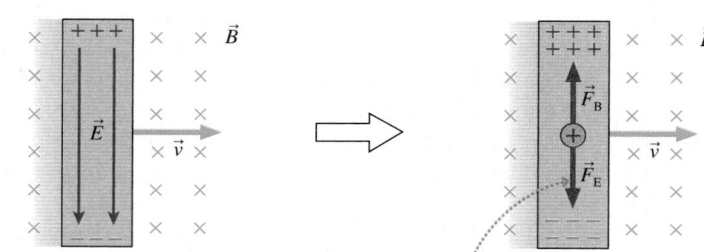

The charge separation creates an electric field in the conductor. $\vec{E}$ increases as more charge flows.

The charge flow continues until the downward electric force $\vec{F}_E$ is large enough to balance the upward magnetic force $\vec{F}_B$. Then the net force on a charge is zero and the current ceases.

The charge carriers continue to separate until the electric force $F_E = qE$ exactly balances the magnetic force $F_B = qvB$, creating an equilibrium situation. This balance happens when the electric field strength is

$$E = vB \tag{33.1}$$

In other words, **the magnetic force on the charge carriers in a moving conductor creates an electric field $E = vB$ inside the conductor.**

The electric field, in turn, creates an electric potential difference between the two ends of the moving conductor. FIGURE 33.3a defines a coordinate system in which $\vec{E} = -vB\hat{\jmath}$. Using the connection between the electric field and the electric potential,

$$\Delta V = V_{\text{top}} - V_{\text{bottom}} = -\int_0^l E_y \, dy = -\int_0^l (-vB) \, dy = vlB \tag{33.2}$$

Thus **the motion of the wire through a magnetic field *induces* a potential difference vlB between the ends of the conductor.** The potential difference depends on the strength of the magnetic field and on the wire's speed through the field.

There's an important analogy between this potential difference and the potential difference of a battery. FIGURE 33.3b reminds you that a battery uses a nonelectric force—the charge escalator—to separate positive and negative charges. The emf $\mathcal{E}$ of the battery was defined as the work performed per charge (W/q) to separate the charges. An isolated battery, with no current, has a potential difference $\Delta V_{\text{bat}} = \mathcal{E}$. We could refer to a battery, where the charges are separated by chemical reactions, as a source of *chemical emf*.

The moving conductor develops a potential difference because of the work done by magnetic forces to separate the charges. You can think of the moving conductor as a "battery" that stays charged only as long as it keeps moving but "runs down" if it stops. The emf of the conductor is due to its motion, rather than to chemical reactions inside, so we can define the **motional emf** of a conductor moving with velocity $\vec{v}$ perpendicular to a magnetic field $\vec{B}$ to be

$$\mathcal{E} = vlB \tag{33.3}$$

FIGURE 33.3 Generating an emf.

(a) Magnetic forces separate the charges and cause a potential difference between the ends. This is a motional emf.

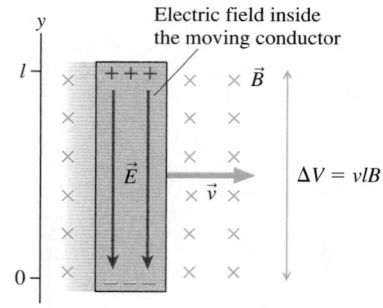

(b) Chemical reactions separate the charges and cause a potential difference between the ends. This is a chemical emf.

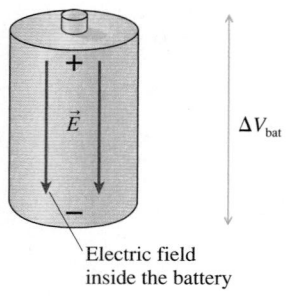

Electric field inside the battery

STOP TO THINK 33.1 A square conductor moves through a uniform magnetic field. Which of the figures shows the correct charge distribution on the conductor?

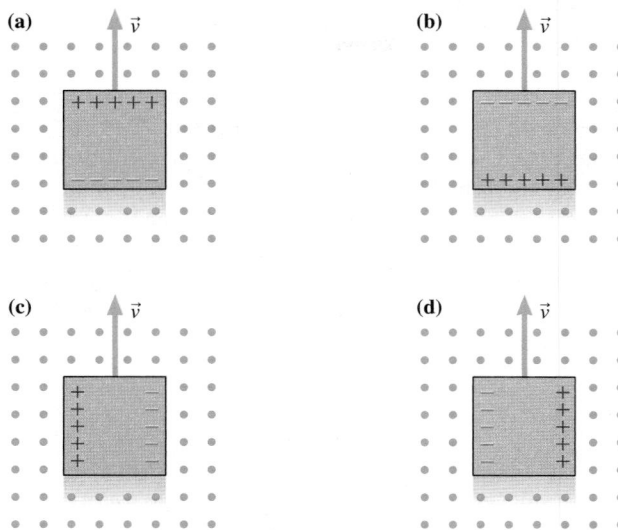

$\vec{B}$ out of page

EXAMPLE 33.1 Measuring the earth's magnetic field

It is known that the earth's magnetic field over northern Canada points straight down. The crew of a Boeing 747 aircraft flying at 260 m/s over northern Canada finds a 0.95 V potential difference between the wing tips. The wing span of a Boeing 747 is 65 m. What is the magnetic field strength there?

MODEL The wing is a conductor moving through a magnetic field, so there is a motional emf.

SOLVE The magnetic field is perpendicular to the velocity, so we can use Equation 33.3 to find

$$B = \frac{\mathcal{E}}{vL} = \frac{0.95 \text{ V}}{(260 \text{ m/s})(65 \text{ m})} = 5.6 \times 10^{-5} \text{ T}$$

ASSESS Chapter 32 noted that the earth's magnetic field is roughly 5×10^{-5} T. The field is somewhat stronger than this near the magnetic poles, somewhat weaker near the equator.

EXAMPLE 33.2 Potential difference along a rotating bar

A metal bar of length l rotates with angular velocity ω about a pivot at one end of the bar. A uniform magnetic field $\vec{B}$ is perpendicular to the plane of rotation. What is the potential difference between the ends of the bar?

VISUALIZE **FIGURE 33.4** is a pictorial representation of the bar. The magnetic forces on the charge carriers will cause the outer end to be positive with respect to the pivot.

FIGURE 33.4 Pictorial representation of a metal bar rotating in a magnetic field.

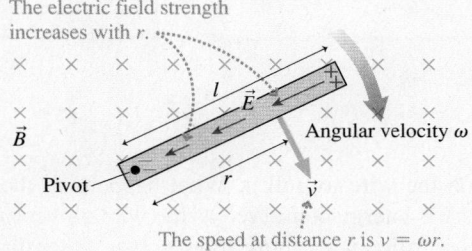

The electric field strength increases with r.

The speed at distance r is $v = \omega r$.

SOLVE Even though the bar is rotating, rather than moving in a straight line, the velocity of each charge carrier is perpendicular to $\vec{B}$. Consequently, the electric field created inside the bar is exactly that given in Equation 33.1, $E = vB$. But v, the speed of the charge carrier, now depends on its distance from the pivot. Recall that in rotational motion the tangential speed at radius r from the center of rotation is $v = \omega r$. Thus the electric field at distance r from the pivot is $E = \omega r B$. The electric field increases in strength as you move outward along the bar.

The electric field $\vec{E}$ points toward the pivot, so its radial component is $E_r = -\omega r B$. If we integrate outward from the center, the potential difference between the ends of the bar is

$$\Delta V = V_{\text{tip}} - V_{\text{pivot}} = -\int_0^l E_r \, dr$$

$$= -\int_0^l (-\omega r B) \, dr = \omega B \int_0^l r \, dr = \frac{1}{2}\omega l^2 B$$

ASSESS $\frac{1}{2}\omega l$ is the speed at the midpoint of the bar. Thus ΔV is $v_{\text{mid}} l B$, which seems reasonable.

FIGURE 33.5 A current is induced in the circuit as the wire moves through a magnetic field.

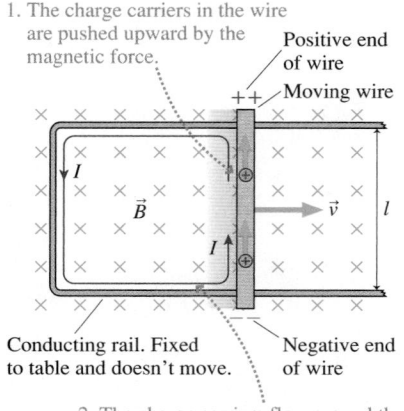

FIGURE 33.5 A current is induced in the circuit as the wire moves through a magnetic field.

1. The charge carriers in the wire are pushed upward by the magnetic force.

Positive end of wire

Moving wire

$\vec{B}$

$\vec{v}$

l

I

Conducting rail. Fixed to table and doesn't move.

Negative end of wire

2. The charge carriers flow around the conducting loop as an induced current.

FIGURE 33.6 A pulling force is needed to move the wire to the right.

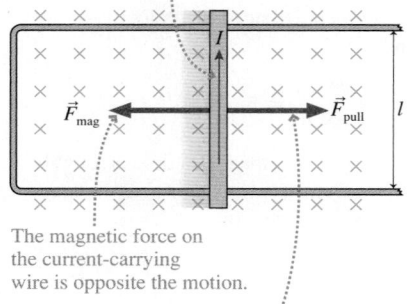

The induced current flows through the moving wire.

$\vec{F}_{mag}$

$\vec{F}_{pull}$

l

The magnetic force on the current-carrying wire is opposite the motion.

A pulling force to the right must balance the magnetic force to keep the wire moving at constant speed. This force does work on the wire.

Induced Current in a Circuit

The moving conductor of Figure 33.2 had an emf, but it couldn't sustain a current because the charges had nowhere to go. It's like a battery that is disconnected from a circuit. We can change this by including the moving conductor in a circuit.

FIGURE 33.5 shows a conducting wire sliding with speed v along a U-shaped conducting rail. We'll assume that the rail is attached to a table and cannot move. The wire and the rail together form a closed conducting loop—a circuit.

Suppose a magnetic field $\vec{B}$ is perpendicular to the plane of the circuit. Charges in the moving wire will be pushed to the ends of the wire by the magnetic force, just as they were in Figure 33.2, but now the charges can continue to flow around the circuit. That is, the moving wire acts like a battery in a circuit.

The current in the circuit is an *induced current*. In this example, the induced current is counterclockwise (ccw). If the total resistance of the circuit is R, the induced current is given by Ohm's law as

$$I = \frac{\mathcal{E}}{R} = \frac{vlB}{R} \qquad (33.4)$$

In this situation, the induced current is due to magnetic forces on moving charges.

We've assumed that the wire is moving along the rail at constant speed. It turns out that we must apply a continuous pulling force $\vec{F}_{pull}$ to make this happen. FIGURE 33.6 shows why. The moving wire, which now carries induced current I, is in a magnetic field. You learned in Chapter 32 that a magnetic field exerts a force on a current-carrying wire. According to the right-hand rule, the magnetic force $\vec{F}_{mag}$ on the moving wire points to the left. This "magnetic drag" will cause the wire to slow down and stop *unless* we exert an equal but opposite pulling force $\vec{F}_{pull}$ to keep the wire moving.

The magnitude of the magnetic force on a current-carrying wire was found in Chapter 32 to be $F_{mag} = IlB$. Using that result, along with Equation 33.4 for the induced current, we find that the force required to pull the wire with a constant speed v is

$$F_{pull} = F_{mag} = IlB = \left(\frac{vlB}{R}\right)lB = \frac{vl^2B^2}{R} \qquad (33.5)$$

STOP TO THINK 33.2 Is there an induced current in this circuit? If so, what is its direction?

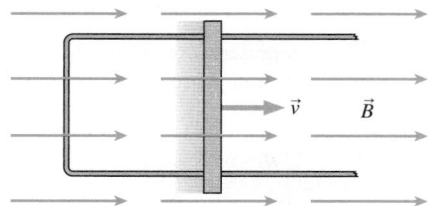

$\vec{v}$ $\vec{B}$

Energy Considerations

The environment must do work on the wire to pull it. What happens to the energy transferred to the wire by this work? Is energy conserved as the wire moves along the rail? It will be easier to answer this question if we think about power rather than work. Power is the *rate* at which work is done on the wire. You learned in Chapter 11 that the power exerted by a force pushing or pulling an object with velocity v is $P = Fv$. The power provided to the circuit by pulling on the wire is

$$P_{\text{input}} = F_{\text{pull}}v = \frac{v^2l^2B^2}{R} \qquad (33.6)$$

This is the rate at which energy is added to the circuit by the pulling force.

But the circuit also dissipates energy by transforming electric energy into the thermal energy of the wires and components, heating them up. The power dissipated by current I as it passes through resistance R is $P = I^2R$. Equation 33.4 for the induced current I gives us the power dissipated by the circuit of Figure 33.5:

$$P_{\text{dissipated}} = I^2R = \frac{v^2l^2B^2}{R} \qquad (33.7)$$

You can see that Equations 33.6 and 33.7 are identical. **The rate at which work is done on the circuit exactly balances the rate at which energy is dissipated.** Thus *energy is conserved.*

If you have to *pull* on the wire to get it to move to the right, you might think that it would spring back to the left on its own. FIGURE 33.7 shows the same circuit with the wire moving to the left. In this case, you must *push* the wire to the left to keep it moving. The magnetic force is always opposite to the wire's direction of motion.

In both Figure 33.6, where the wire is pulled, and Figure 33.7, where it is pushed, a mechanical force is used to create a current. In other words, we have a conversion of *mechanical* energy to *electric* energy. A device that converts mechanical energy to electric energy is called a **generator.** The slide-wire circuits of Figures 33.6 and 33.7 are simple examples of a generator. We will look at more practical examples of generators later in the chapter.

We can summarize our analysis as follows:

1. Pulling or pushing the wire through the magnetic field at speed v creates a motional emf $\mathcal{E}$ in the wire and induces a current $I = \mathcal{E}/R$ in the circuit.
2. To keep the wire moving at constant speed, a pulling or pushing force must balance the magnetic force on the wire. This force does work on the circuit.
3. The work done by the pulling or pushing force exactly balances the energy dissipated by the current as it passes through the resistance of the circuit.

FIGURE 33.7 A pushing force is needed to move the wire to the left.

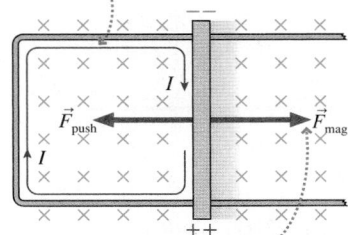

1. The magnetic force on the charge carriers is down, so the induced current flows clockwise.

2. The magnetic force on the current-carrying wire is to the right.

EXAMPLE 33.3 | **Lighting a bulb**

FIGURE 33.8 shows a circuit consisting of a flashlight bulb, rated 3.0 V/1.5 W, and ideal wires with no resistance. The right wire of the circuit, which is 10 cm long, is pulled at constant speed v through a perpendicular magnetic field of strength 0.10 T.

a. What speed must the wire have to light the bulb to full brightness?

b. What force is needed to keep the wire moving?

FIGURE 33.8 Circuit of Example 33.3.

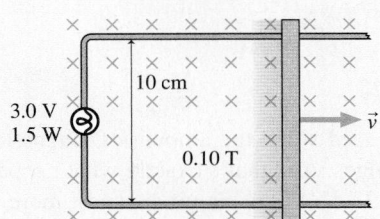

3.0 V
1.5 W
10 cm
0.10 T
$\vec{v}$

MODEL Treat the moving wire as a source of motional emf.

VISUALIZE The magnetic force on the charge carriers, $\vec{F}_B = q\vec{v} \times \vec{B}$, causes a counterclockwise (ccw) induced current.

SOLVE a. The bulb's rating of 3.0 V/1.5 W means that at full brightness it will dissipate 1.5 W at a potential difference of 3.0 V. Because the power is related to the voltage and current by $P = I\Delta V$, the current causing full brightness is

$$I = \frac{P}{\Delta V} = \frac{1.5 \text{ W}}{3.0 \text{ V}} = 0.50 \text{ A}$$

The bulb's resistance—the total resistance of the circuit—is

$$R = \frac{\Delta V}{I} = \frac{3.0 \text{ V}}{0.50 \text{ A}} = 6.0 \text{ } \Omega$$

Equation 33.4 gives the speed needed to induce this current:

$$v = \frac{IR}{lB} = \frac{(0.50 \text{ A})(6.0 \text{ } \Omega)}{(0.10 \text{ m})(0.10 \text{ T})} = 300 \text{ m/s}$$

You can confirm from Equation 33.6 that the input power at this speed is 1.5 W.

Continued

b. From Equation 33.5, the pulling force must be

$$F_{pull} = \frac{vl^2B^2}{R} = 5.0 \times 10^{-3} \text{ N}$$

You can also obtain this result from $F_{pull} = P/v$.

ASSESS Example 33.1 showed that high speeds are needed to produce significant potential difference. Thus 300 m/s is not surprising. The pulling force is not very large, but even a small force can deliver large amounts of power $P = Fv$ when v is large.

FIGURE 33.9 Eddy currents.

(a) Eddy currents are induced when a metal sheet is pulled through a magnetic field.

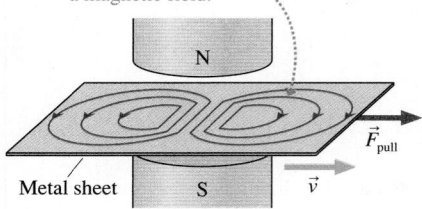

Metal sheet

(b) The magnetic force on the eddy currents is opposite in direction to $\vec{v}$.

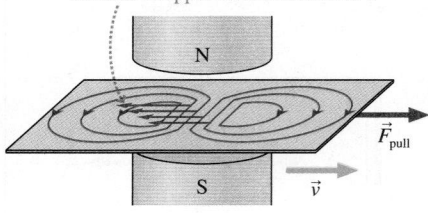

FIGURE 33.10 Magnetic braking system.

The electromagnets are part of the moving train car.

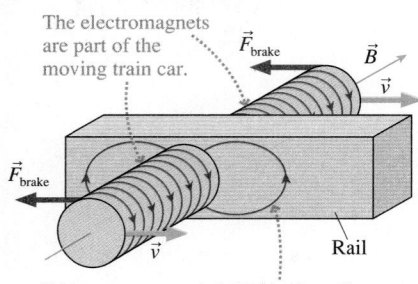

Rail

Eddy currents are induced in the rail. Magnetic forces between the eddy currents and the electromagnets slow the train.

Eddy Currents

These ideas have interesting implications. Consider pulling a *sheet* of metal through a magnetic field, as shown in FIGURE 33.9a. The metal, we will assume, is not a magnetic material, so it experiences no magnetic force if it is at rest. The charge carriers in the metal experience a magnetic force as the sheet is dragged between the pole tips of the magnet. A current is induced, just as in the loop of wire, but here the currents do not have wires to define their path. As a consequence, two "whirlpools" of current begin to circulate in the metal. These spread-out current whirlpools in a solid metal are called **eddy currents.**

FIGURE 33.9b shows the magnetic force on the eddy current as it passes between the pole tips. This force is to the left, acting as a retarding force. Thus **an external force is required to pull a metal through a magnetic field.** If the pulling force ceases, the retarding magnetic force quickly causes the metal to decelerate until it stops. Similarly, a force is required to push a sheet of metal *into* a magnetic field.

Eddy currents are often undesirable. The power dissipation of eddy currents can cause unwanted heating, and the magnetic forces on eddy currents mean that extra energy must be expended to move metals in magnetic fields. But eddy currents also have important useful applications. A good example is magnetic braking.

The moving train car has an electromagnet that straddles the rail, as shown in FIGURE 33.10. During normal travel, there is no current through the electromagnet and no magnetic field. To stop the car, a current is switched into the electromagnet. The current creates a strong magnetic field that passes *through* the rail, and the motion of the rail relative to the magnet induces eddy currents in the rail. The magnetic force between the electromagnet and the eddy currents acts as a braking force on the magnet and, thus, on the car. Magnetic braking systems are very efficient, and they have the added advantage that they heat the rail rather than the brakes.

STOP TO THINK 33.3 A square loop of copper wire is pulled through a region of magnetic field. Rank in order, from strongest to weakest, the pulling forces $\vec{F}_a$, $\vec{F}_b$, $\vec{F}_c$, and $\vec{F}_d$ that must be applied to keep the loop moving at constant speed.

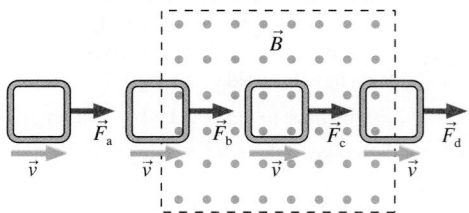

33.3 Magnetic Flux

Faraday found that a current is induced when the amount of magnetic field passing through a coil or a loop of wire changes. And that's exactly what happens as the slide wire moves down the rail in Figure 33.5! As the circuit expands, more magnetic field passes through. It's time to define more clearly what we mean by "the amount of field passing through a loop."

Imagine holding a rectangular loop of wire in front of a fan, as shown in FIGURE 33.11. The amount of air that flows through the loop depends on the effective area of the loop as seen along the direction of flow. You can see from the figure that the effective area (i.e., as seen facing the fan) is

$$A_{\text{eff}} = ab \cos \theta = A \cos \theta \qquad (33.8)$$

where $A = ab$ is the area of the loop and θ is the tilt angle of the loop. A loop perpendicular to the flow, with $\theta = 0°$, has $A_{\text{eff}} = A$, the full area of the loop. No air at all flows through the loop if it is tilted 90°, and you can see that $A_{\text{eff}} = 0$ in this case.

FIGURE 33.11 The amount of air flowing through a loop depends on the effective area of the loop.

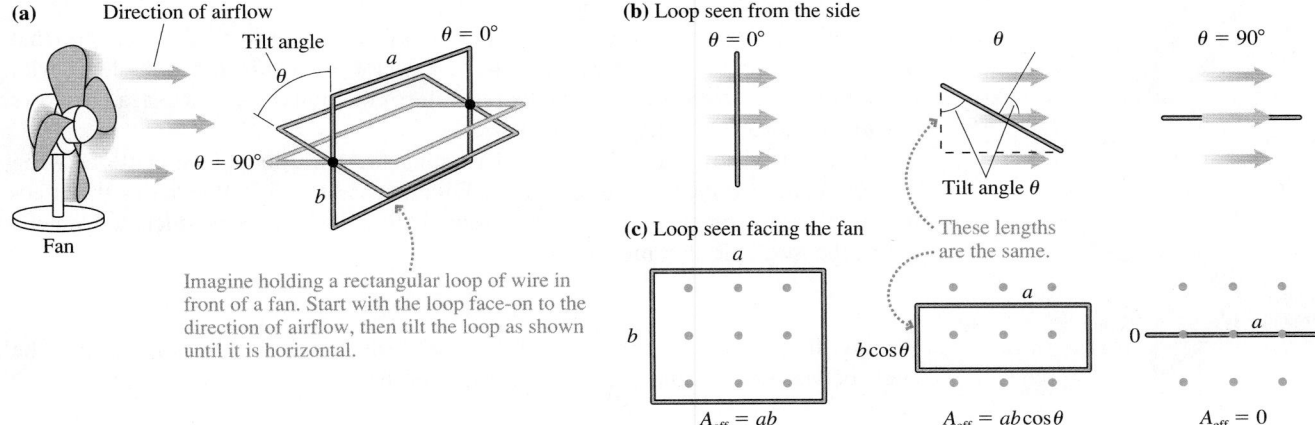

We can apply this idea to a magnetic field passing through a loop. FIGURE 33.12 shows a loop of area $A = ab$ in a uniform magnetic field. Think of the field vectors, seen here from behind, as if they were arrows shot into the page. The density of arrows (arrows per m²) is proportional to the strength B of the magnetic field; a stronger field would be represented by arrows packed closer together. The number of arrows passing through a loop of wire depends on two factors:

1. The density of arrows, which is proportional to B, and
2. The effective area $A_{\text{eff}} = A \cos \theta$ of the loop.

FIGURE 33.12 Magnetic field through a loop that is tilted at various angles.

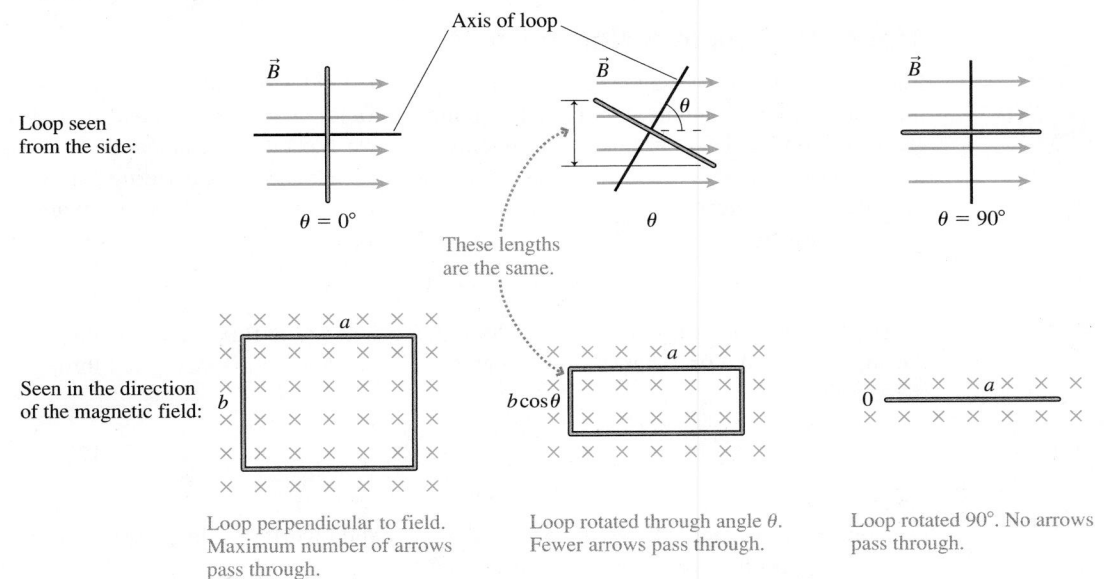

The angle θ is the angle between the magnetic field and the axis of the loop. The maximum number of arrows passes through the loop when it is perpendicular to the magnetic field ($\theta = 0°$). No arrows pass through the loop if it is tilted 90°.

With this in mind, let's define the **magnetic flux** Φ_m as

$$\Phi_m = A_{eff}B = AB\cos\theta \qquad (33.9)$$

The magnetic flux measures the amount of magnetic field passing through a loop of area A if the loop is tilted at angle θ from the field. The SI unit of magnetic flux is the **weber.** From Equation 33.9 you can see that

$$1 \text{ weber} = 1 \text{ Wb} = 1 \text{ T m}^2$$

Equation 33.9 is reminiscent of the vector dot product: $\vec{A} \cdot \vec{B} = AB\cos\theta$. With that in mind, let's define an **area vector** $\vec{A}$ to be a vector *perpendicular* to the loop, with magnitude equal to the area A of the loop. Vector $\vec{A}$ has units of m². FIGURE 33.13a shows the area vector $\vec{A}$ for a circular loop of area A.

FIGURE 33.13b shows a magnetic field passing through a loop. The angle between vectors $\vec{A}$ and $\vec{B}$ is the same angle used in Equations 33.8 and 33.9 to define the effective area and the magnetic flux. So Equation 33.9 really is a dot product, and we can define the magnetic flux more concisely as

$$\Phi_m = \vec{A} \cdot \vec{B} \qquad (33.10)$$

Writing the flux as a dot product helps make clear how angle θ is defined: θ is the angle between the magnetic field and the axis of the loop.

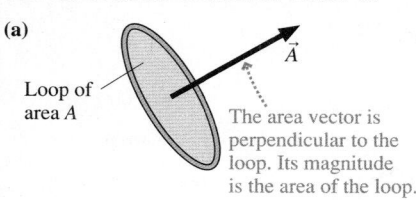

FIGURE 33.13 Magnetic flux can be defined in terms of an area vector $\vec{A}$.

(a)

Loop of area A

The area vector is perpendicular to the loop. Its magnitude is the area of the loop.

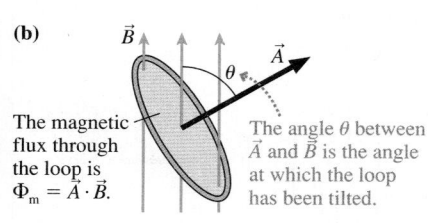

(b)

The magnetic flux through the loop is $\Phi_m = \vec{A} \cdot \vec{B}$.

The angle θ between $\vec{A}$ and $\vec{B}$ is the angle at which the loop has been tilted.

EXAMPLE 33.4 | **A circular loop in a magnetic field**

FIGURE 33.14 is an edge view of a 10-cm-diameter circular loop in a uniform 0.050 T magnetic field. What is the magnetic flux through the loop?

SOLVE Angle θ is the angle between the loop's area vector $\vec{A}$, which is perpendicular to the plane of the loop, and the magnetic field $\vec{B}$. In this case, $\theta = 60°$, not the 30° angle shown in the figure. Vector $\vec{A}$ has magnitude $A = \pi r^2 = 7.85 \times 10^{-3}$ m². Thus the magnetic flux is

$$\Phi_m = \vec{A} \cdot \vec{B} = AB\cos\theta = 2.0 \times 10^{-4} \text{ Wb}$$

FIGURE 33.14 A circular loop in a magnetic field.

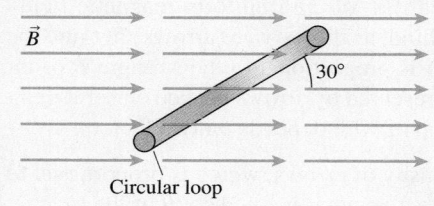

Circular loop

Magnetic Flux in a Nonuniform Field

Equation 33.10 for the magnetic flux assumes that the field is uniform over the area of the loop. We can calculate the flux in a nonuniform field, one where the field strength changes from one edge of the loop to the other, but we'll need to use calculus.

FIGURE 33.15 shows a loop in a nonuniform magnetic field. Imagine dividing the loop into many small pieces of area dA. The infinitesimal flux $d\Phi_m$ through one such area, where the magnetic field is $\vec{B}$, is

$$d\Phi_m = \vec{B} \cdot d\vec{A} \qquad (33.11)$$

The total magnetic flux through the loop is the sum of the fluxes through each of the small areas. We find that sum by integrating. Thus the total magnetic flux through the loop is

$$\Phi_m = \int_{\text{area of loop}} \vec{B} \cdot d\vec{A} \qquad (33.12)$$

Equation 33.12 is a more general definition of magnetic flux. It may look rather formidable, so we'll illustrate its use with an example.

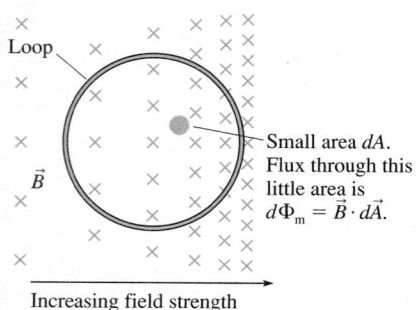

FIGURE 33.15 A loop in a nonuniform magnetic field.

Loop

Small area dA. Flux through this little area is $d\Phi_m = \vec{B} \cdot d\vec{A}$.

$\vec{B}$

Increasing field strength

EXAMPLE 33.5 **Magnetic flux from the current in a long straight wire**

The 1.0 cm × 4.0 cm rectangular loop of FIGURE 33.16 is 1.0 cm away from a long straight wire. The wire carries a current of 1.0 A. What is the magnetic flux through the loop?

FIGURE 33.16 A loop next to a current carrying wire.

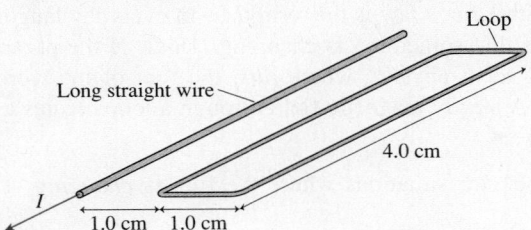

MODEL We'll treat the wire as if it were infinitely long. The magnetic field strength of a wire decreases with distance from the wire, so the field is *not* uniform over the area of the loop.

FIGURE 33.17 Calculating the magnetic flux through the loop.

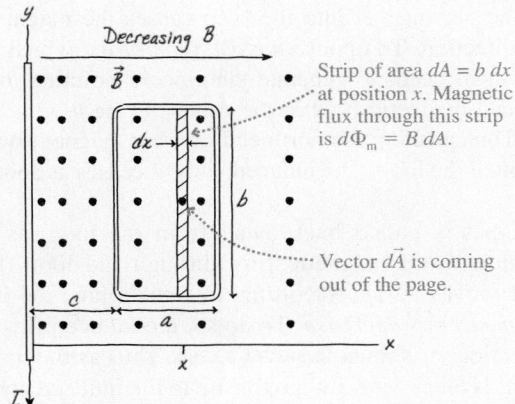

Strip of area $dA = b\,dx$ at position x. Magnetic flux through this strip is $d\Phi_m = B\,dA$.

Vector $d\vec{A}$ is coming out of the page.

VISUALIZE Using the right-hand rule, we see that the field, as it circles the wire, is perpendicular to the plane of the loop. FIGURE 33.17 redraws the loop with the field coming out of the page and establishes a coordinate system.

SOLVE Let the loop have dimensions a and b, as shown, with the near edge at distance c from the wire. The magnetic field varies with distance x from the wire, but the field is constant along a line parallel to the wire. This suggests dividing the loop into many narrow rectangular strips of length b and width dx, each forming a small area $dA = b\,dx$. The magnetic field has the same strength at all points within this small area. One such strip is shown in the figure at position x.

The area vector $d\vec{A}$ is perpendicular to the strip (coming out of the page), which makes it parallel to $\vec{B}$ ($\theta = 0°$). Thus the infinitesimal flux through this little area is

$$d\Phi_m = \vec{B} \cdot d\vec{A} = B\,dA = B(b\,dx) = \frac{\mu_0 Ib}{2\pi x}dx$$

where, from Chapter 32, we've used $B = \mu_0 I/2\pi x$ as the magnetic field at distance x from a long straight wire. Integrating "over the area of the loop" means to integrate from the near edge of the loop at $x = c$ to the far edge at $x = c + a$. Thus

$$\Phi_m = \frac{\mu_0 Ib}{2\pi}\int_c^{c+a}\frac{dx}{x} = \frac{\mu_0 Ib}{2\pi}\ln x\Big|_c^{c+a} = \frac{\mu_0 Ib}{2\pi}\ln\left(\frac{c+a}{c}\right)$$

Evaluating for $a = c = 0.010$ m, $b = 0.040$ m, and $I = 1.0$ A gives

$$\Phi_m = 5.5 \times 10^{-9} \text{ Wb}$$

ASSESS The flux measures how much of the wire's magnetic field passes through the loop, but we had to integrate, rather than simply using Equation 33.10, because the field is stronger at the near edge of the loop than at the far edge.

33.4 Lenz's Law

We started out by looking at a situation in which a moving wire caused a loop to expand in a magnetic field. This is one way to change the magnetic flux through the loop. But Faraday found that a current can be induced by any change in the magnetic flux, no matter how it's accomplished.

For example, a momentary current is induced in the loop of FIGURE 33.18 as the bar magnet is pushed toward the loop, increasing the flux through the loop. Pulling the magnet back out of the loop causes the current meter to deflect in the opposite direction. The conducting wires aren't moving, so this is not a motional emf. Nonetheless, the induced current is very real.

The German physicist Heinrich Lenz began to study electromagnetic induction after learning of Faraday's discovery. Three years later, in 1834, Lenz announced a rule for determining the direction of the induced current. We now call his rule **Lenz's law,** and it can be stated as follows:

FIGURE 33.18 Pushing a bar magnet toward the loop induces a current.

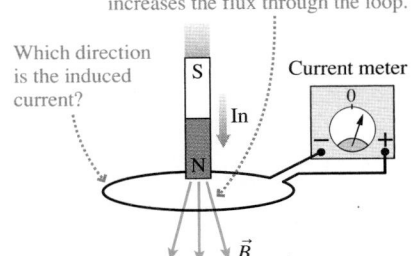

A bar magnet pushed into a loop increases the flux through the loop.

Which direction is the induced current?

Current meter

Lenz's law There is an induced current in a closed, conducting loop if and only if the magnetic flux through the loop is changing. The direction of the induced current is such that the induced magnetic field opposes the *change* in the flux.

Lenz's law is rather subtle, and it takes some practice to see how to apply it.

NOTE ▶ One difficulty with Lenz's law is the term *flux*. In everyday language, the word *flux* already implies that something is changing. Think of the phrase, "The situation is in flux." Not so in physics, where *flux*, the root of the word *flow*, means "passes through." A steady magnetic field through a loop creates a steady, *un*changing magnetic flux. ◀

Lenz's law tells us to look for situations where the flux is *changing*. This can happen in three ways.

1. The magnetic field through the loop changes (increases or decreases),
2. The loop changes in area or angle, or
3. The loop moves into or out of a magnetic field.

FIGURE 33.19 The induced current is ccw.

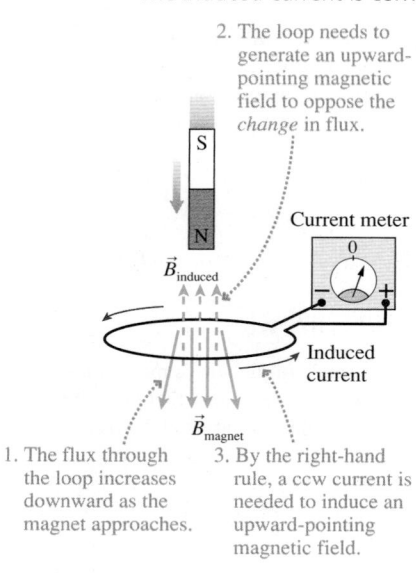

2. The loop needs to generate an upward-pointing magnetic field to oppose the *change* in flux.

Current meter

$\vec{B}_{induced}$

Induced current

$\vec{B}_{magnet}$

1. The flux through the loop increases downward as the magnet approaches.

3. By the right-hand rule, a ccw current is needed to induce an upward-pointing magnetic field.

Lenz's law depends on the idea that an induced current generates its own magnetic field $\vec{B}_{induced}$. This is the *induced magnetic field* of Lenz's law. You learned in Chapter 32 how to use the right-hand rule to determine the direction of this induced magnetic field.

In Figure 33.18, pushing the bar magnet into the loop causes the magnetic flux to *increase* in the downward direction. To oppose the *change* in flux, which is what Lenz's law requires, the loop itself needs to generate the *upward*-pointing magnetic field of FIGURE 33.19. The induced magnetic field at the center of the loop will point upward if the current is ccw. Thus pushing the north end of a bar magnet toward the loop induces a ccw current around the loop. The induced current ceases as soon as the magnet stops moving.

Now suppose the bar magnet is pulled back away from the loop, as shown in FIGURE 33.20a. There is a downward magnetic flux through the loop, but the flux *decreases* as the magnet moves away. According to Lenz's law, the induced magnetic field of the loop *opposes this decrease*. To do so, the induced field needs to point in the *downward* direction, as shown in FIGURE 33.20b. Thus as the magnet is withdrawn, the induced current is clockwise (cw), opposite to the induced current of Figure 33.19.

FIGURE 33.20 Pulling the magnet away induces a cw current.

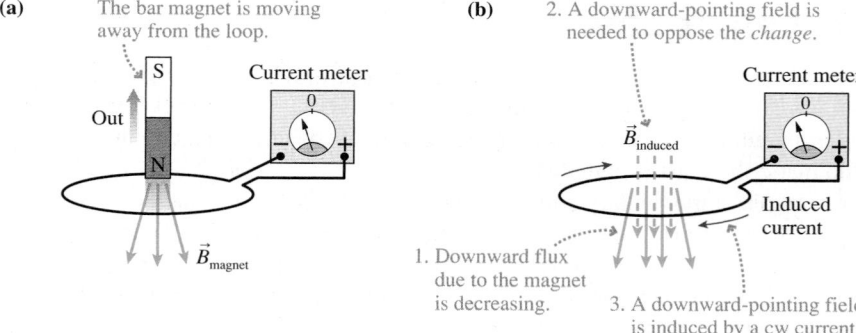

(a) The bar magnet is moving away from the loop.

Out

Current meter

$\vec{B}_{magnet}$

(b) 2. A downward-pointing field is needed to oppose the *change*.

Current meter

$\vec{B}_{induced}$

Induced current

1. Downward flux due to the magnet is decreasing.

3. A downward-pointing field is induced by a cw current.

NOTE ▶ Notice that the magnetic field of the bar magnet is pointing downward in both Figures 33.19 and 33.20. It is not the *flux* due to the magnet that the induced current opposes, but the *change* in the flux. This is a subtle but critical distinction.

If the induced current opposed the flux itself, the current in both Figures 33.19 and 33.20 would be ccw to generate an upward magnetic field. But that's not what happens. When the field of the magnet points down and is increasing, the induced current opposes the increase by generating an upward field. When the field of the magnet points down but is decreasing, the induced current opposes the decrease by generating a downward field. ◄

FIGURE 33.21 shows six basic situations. The magnetic field can point either up or down through the loop. For each, the flux can either increase, hold steady, or decrease in strength. These observations form the basis for a set of rules about using Lenz's law.

FIGURE 33.21 The induced current for six different situations.

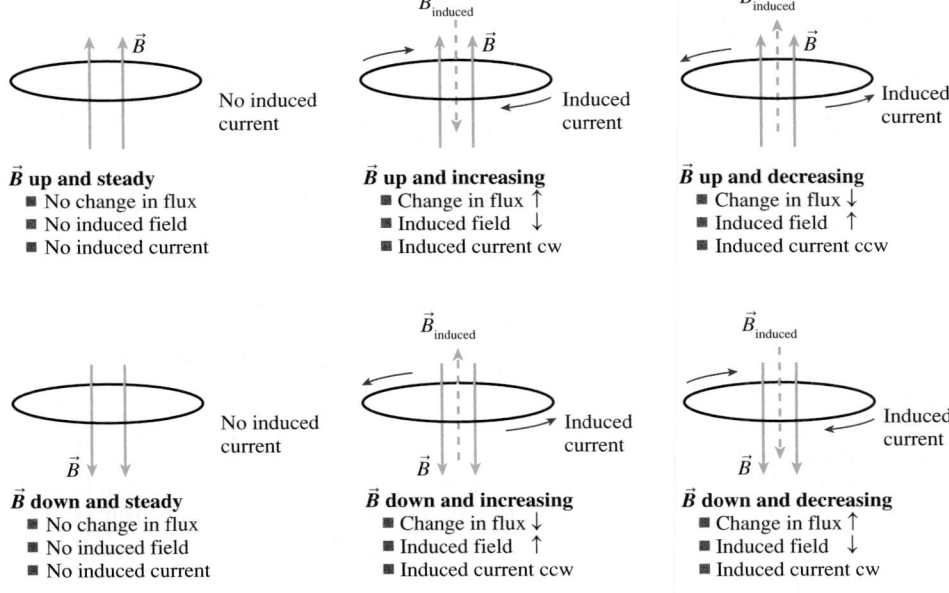

$\vec{B}$ up and steady
- No change in flux
- No induced field
- No induced current

$\vec{B}$ up and increasing
- Change in flux ↑
- Induced field ↓
- Induced current cw

$\vec{B}$ up and decreasing
- Change in flux ↓
- Induced field ↑
- Induced current ccw

$\vec{B}$ down and steady
- No change in flux
- No induced field
- No induced current

$\vec{B}$ down and increasing
- Change in flux ↓
- Induced field ↑
- Induced current ccw

$\vec{B}$ down and decreasing
- Change in flux ↑
- Induced field ↓
- Induced current cw

TACTICS
BOX 33.1 **Using Lenz's law**

❶ **Determine the direction of the applied magnetic field.** The field must pass through the loop.
❷ **Determine how the flux is changing.** Is it increasing, decreasing, or staying the same?
❸ **Determine the direction of an induced magnetic field that will oppose the *change* in the flux.**

- Increasing flux: the induced magnetic field points opposite the applied magnetic field.
- Decreasing flux: the induced magnetic field points in the same direction as the applied magnetic field.
- Steady flux: there is no induced magnetic field.

❹ **Determine the direction of the induced current.** Use the right-hand rule to determine the current direction in the loop that generates the induced magnetic field you found in step 3.

Exercises 10–14

Let's look at some examples.

EXAMPLE 33.6 **Lenz's law 1**

FIGURE 33.22 shows two loops, one above the other. The upper loop has a battery and a switch that has been closed for a long time. How does the lower loop respond when the switch is opened in the upper loop?

MODEL We'll use the right-hand rule to find the magnetic fields of current loops.

SOLVE FIGURE 33.23 shows the four steps of using Lenz's law. Opening the switch induces a ccw current in the lower loop. This is a momentary current, lasting only until the magnetic field of the upper loop drops to zero.

ASSESS The conclusion is consistent with Figure 33.21.

FIGURE 33.22 The two loops of Example 33.6.

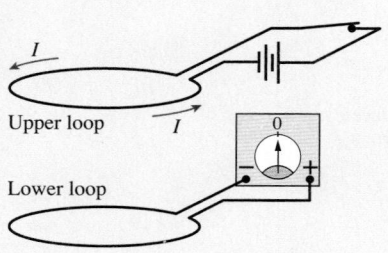

FIGURE 33.23 Applying Lenz's law.

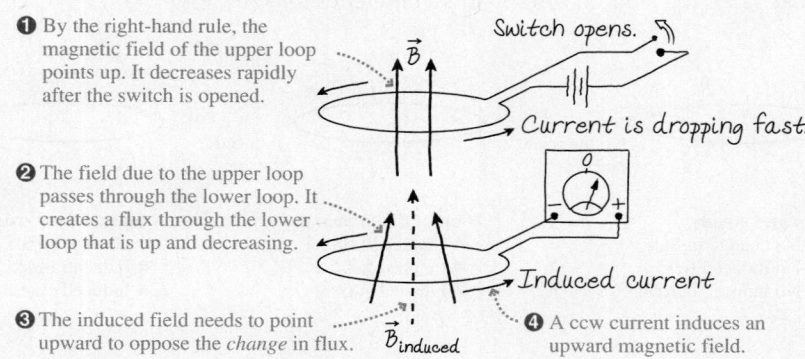

❶ By the right-hand rule, the magnetic field of the upper loop points up. It decreases rapidly after the switch is opened.

❷ The field due to the upper loop passes through the lower loop. It creates a flux through the lower loop that is up and decreasing.

❸ The induced field needs to point upward to oppose the *change* in flux.

❹ A ccw current induces an upward magnetic field.

Switch opens.

Current is dropping fast.

Induced current

EXAMPLE 33.7 **Lenz's law 2**

FIGURE 33.24 shows two coils wrapped side by side on a cylinder. When the switch for coil 1 is closed, does the induced current in coil 2 pass from right to left or from left to right through the current meter?

MODEL We'll use the right-hand rule to find the magnetic field of a coil.

VISUALIZE It is very important to look at the *direction* in which a coil is wound around the cylinder. Notice that the two coils in Figure 33.24 are wound in opposite directions.

SOLVE FIGURE 33.25 shows the four steps of using Lenz's law. Closing the switch induces a current that passes from right to left through the current meter. The induced current is only momentary. It lasts only until the field from coil 1 reaches full strength and is no longer changing.

ASSESS The conclusion is consistent with Figure 33.21.

FIGURE 33.24 The two solenoids of Example 33.7.

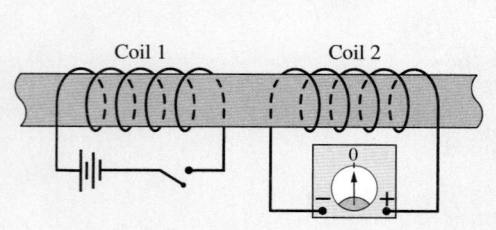

FIGURE 33.25 Applying Lenz's law.

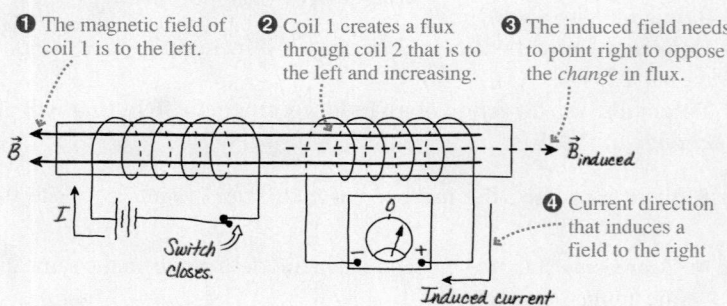

❶ The magnetic field of coil 1 is to the left.

❷ Coil 1 creates a flux through coil 2 that is to the left and increasing.

❸ The induced field needs to point right to oppose the *change* in flux.

❹ Current direction that induces a field to the right

Switch closes.

Induced current

STOP TO THINK 33.4 A current-carrying wire is pulled away from a conducting loop in the direction shown. As the wire is moving, is there a cw current around the loop, a ccw current, or no current?

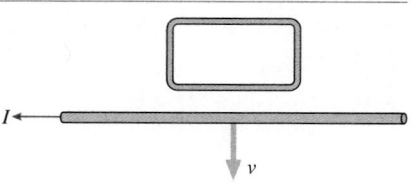

33.5 Faraday's Law

Charges don't start moving spontaneously. A current requires an emf to provide the energy. We started our analysis of induced currents with circuits in which a *motional emf* can be understood in terms of magnetic forces on moving charges. But we've also seen that a current can be induced by changing the magnetic field through a stationary circuit, a circuit in which there is no motion. There *must* be an emf in this circuit, even though the mechanism for this emf is not yet clear.

The emf associated with a changing magnetic flux, regardless of what causes the change, is called an **induced emf** $\mathcal{E}$. Then, if there is a complete circuit having resistance R, a current

$$I_{\text{induced}} = \frac{\mathcal{E}}{R} \tag{33.13}$$

is established in the wire as a *consequence* of the induced emf. The direction of the current is given by Lenz's law. The last piece of information we need is the size of the induced emf $\mathcal{E}$.

The research of Faraday and others eventually led to the discovery of the basic law of electromagnetic induction, which we now call **Faraday's law.** It states:

> **Faraday's law** An emf $\mathcal{E}$ is induced around a closed loop if the magnetic flux through the loop changes. The magnitude of the emf is
>
> $$\mathcal{E} = \left| \frac{d\Phi_m}{dt} \right| \tag{33.14}$$
>
> and the direction of the emf is such as to drive an induced current in the direction given by Lenz's law.

In other words, the induced emf is the *rate of change* of the magnetic flux through the loop.

As a corollary to Faraday's law, an N-turn coil of wire in a changing magnetic field acts like N batteries in series. The induced emf of each of the coils adds, so the induced emf of the entire coil is

$$\mathcal{E}_{\text{coil}} = N \left| \frac{d\Phi_{\text{per coil}}}{dt} \right| \quad \text{(Faraday's law for an } N\text{-turn coil)} \tag{33.15}$$

As a first example of using Faraday's law, return to the situation of Figure 33.5, where a wire moves through a magnetic field by sliding on a U-shaped conducting rail. FIGURE 33.26 shows the circuit again. The magnetic field $\vec{B}$ is perpendicular to the plane of the conducting loop, so $\theta = 0°$ and the magnetic flux is $\Phi = AB$, where A is the area of the loop. If the slide wire is distance x from the end, the area is $A = xl$ and the flux at that instant of time is

$$\Phi_m = AB = xlB \tag{33.16}$$

The flux through the loop increases as the wire moves. According to Faraday's law, the induced emf is

$$\mathcal{E} = \left| \frac{d\Phi_m}{dt} \right| = \frac{d}{dt}(xlB) = \frac{dx}{dt}lB = vlB \tag{33.17}$$

where the wire's velocity is $v = dx/dt$. We can now use Equation 33.13 to find that the induced current is

$$I = \frac{\mathcal{E}}{R} = \frac{vlB}{R} \tag{33.18}$$

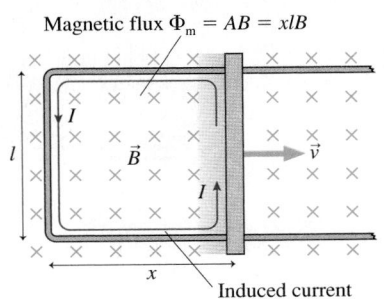

FIGURE 33.26 The magnetic flux through the loop increases as the slide wire moves.

Magnetic flux $\Phi_m = AB = xlB$

The flux is increasing into the loop, so the induced magnetic field opposes this increase by pointing out of the loop. This requires a ccw induced current in the loop. Faraday's law leads us to the conclusion that the loop will have a ccw induced current $I = vlB/R$. This is exactly the conclusion we reached in Section 33.2, where we analyzed the situation from the perspective of magnetic forces on moving charge carriers. Faraday's law confirms what we already knew but, at least in this case, doesn't seem to offer anything new.

Using Faraday's Law

Most electromagnetic induction problems can be solved with a four-step strategy.

PROBLEM-SOLVING
STRATEGY 33.1 **Electromagnetic induction**

MODEL Make simplifying assumptions about wires and magnetic fields.

VISUALIZE Draw a picture or a circuit diagram. Use Lenz's law to determine the direction of the induced current.

SOLVE The mathematical representation is based on Faraday's law

$$\mathcal{E} = \left| \frac{d\Phi_m}{dt} \right|$$

For an N-turn coil, multiply by N. The size of the induced current is $I = \mathcal{E}/R$.

ASSESS Check that your result has the correct units, is reasonable, and answers the question.

Exercise 18

EXAMPLE 33.8 **Electromagnetic induction in a solenoid**

A 2.0-cm-diameter loop of wire with a resistance of 0.010 Ω is placed in the center of the solenoid seen in FIGURE 33.27a. The solenoid is 4.0 cm in diameter, 20 cm long, and wrapped with 1000 turns of wire. FIGURE 33.27b shows the current through the

solenoid as a function of time as the solenoid is "powered up." A positive current is defined to be cw when seen from the left. Find the current in the loop as a function of time and show the result as a graph.

MODEL The solenoid's length is much greater than its diameter, so the field near the center should be nearly uniform.

VISUALIZE The magnetic field of the solenoid creates a magnetic flux through the loop of wire. The solenoid current is always positive, meaning that it is cw as seen from the left. Consequently, from the right-hand rule, the magnetic field inside the solenoid always points to the right. During the first second, while the solenoid current is increasing, the flux through the loop is to the right and increasing. To oppose the *change* in the flux, the loop's induced magnetic field must point to the left. Thus, again using the right-hand rule, the induced current must flow ccw as seen from the left. This is a *negative* current. There's no *change* in the flux for $t > 1$ s, so the induced current is zero.

SOLVE Now we're ready to use Faraday's law to find the magnitude of the current. Because the field is uniform inside the solenoid and perpendicular to the loop ($\theta = 0°$), the flux is $\Phi_m = AB$, where $A = \pi r^2 = 3.14 \times 10^{-4}$ m^2 is the area of the loop (*not* the area of the solenoid). The field of a long solenoid of length l was found in Chapter 32 to be

$$B = \frac{\mu_0 N I_{sol}}{l}$$

FIGURE 33.27 A loop inside a solenoid.

(a)

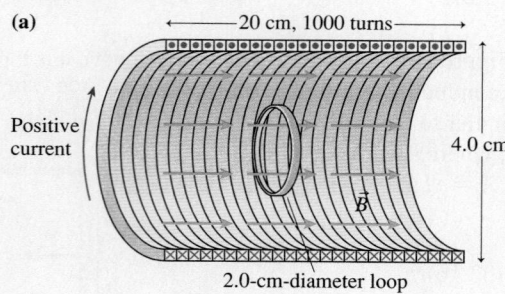

(b)

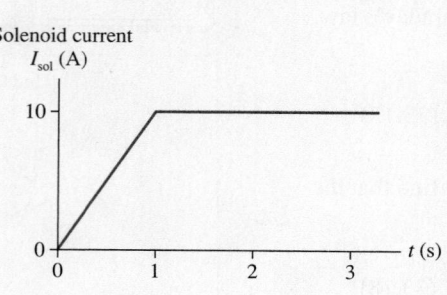

The flux when the solenoid current is I_{sol} is thus

$$\Phi_m = \frac{\mu_0 A N I_{sol}}{l}$$

The changing flux creates an induced emf $\mathcal{E}$ that is given by Faraday's law:

$$\mathcal{E} = \left|\frac{d\Phi_m}{dt}\right| = \frac{\mu_0 A N}{l}\left|\frac{dI_{sol}}{dt}\right| = 2.0 \times 10^{-6}\left|\frac{dI_{sol}}{dt}\right|$$

From the slope of the graph, we find

$$\left|\frac{dI_{sol}}{dt}\right| = \begin{cases} 10 \text{ A/s} & 0.0 \text{ s} < t < 1.0 \text{ s} \\ 0 & 1.0 \text{ s} < t < 3.0 \text{ s} \end{cases}$$

Thus the induced emf is

$$\mathcal{E} = \begin{cases} 2.0 \times 10^{-5} \text{ V} & 0.0 \text{ s} < t < 1.0 \text{ s} \\ 0 \text{ V} & 1.0 \text{ s} < t < 3.0 \text{ s} \end{cases}$$

Finally, the current induced in the loop is

$$I_{loop} = \frac{\mathcal{E}}{R} = \begin{cases} -2.0 \text{ mA} & 0.0 \text{ s} < t < 1.0 \text{ s} \\ 0 \text{ mA} & 1.0 \text{ s} < t < 3.0 \text{ s} \end{cases}$$

where the negative sign comes from Lenz's law. This result is shown in **FIGURE 33.28**.

FIGURE 33.28 The induced current in the loop.

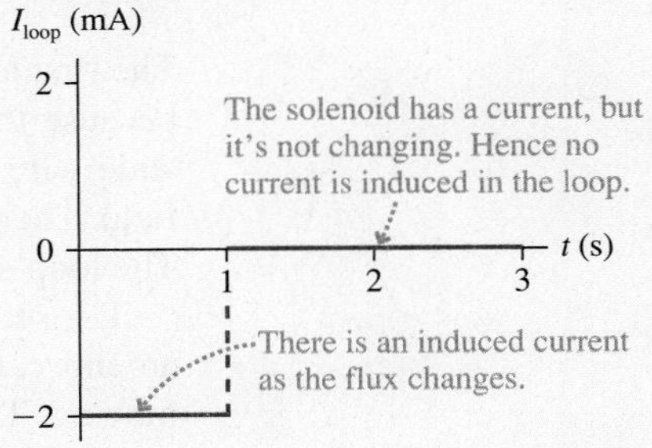

I_{loop} (mA)

> The solenoid has a current, but it's not changing. Hence no current is induced in the loop.

> There is an induced current as the flux changes.

EXAMPLE 33.9 **Current induced by an MRI machine**

The body is a conductor, so rapid magnetic field changes in an MRI machine can induce currents in the body. To estimate the size of these currents, and any biological hazard they might impose, consider the "loop" of muscle tissue shown in **FIGURE 33.29**. This might be muscle circling the bone of your arm or thigh. Although muscle is not a great conductor—its resistivity is $1.5 \ \Omega\,\text{m}$—we can consider it to be a conducting loop with a rather high resistance. Suppose the magnetic field along the axis of the loop drops from 1.6 T to 0 T in 0.30 s, which is about the largest possible rate of change for an MRI solenoid. What current will be induced?

FIGURE 33.29 Edge view of a loop of muscle tissue in a magnetic field.

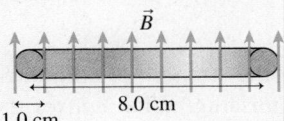

$\vec{B}$

8.0 cm

1.0 cm

MODEL Model the muscle as a current loop. Assume that B decreases linearly with time.

SOLVE The magnetic field is parallel to the axis of the loop, with $\theta = 0°$, so the magnetic flux through the loop is $\Phi_m = AB = \pi r^2 B$. The flux changes with time because B changes. According to Faraday's law, the magnitude of the induced emf is

$$\mathcal{E} = \left|\frac{d\Phi_m}{dt}\right| = \pi r^2 \left|\frac{dB}{dt}\right|$$

The rate at which the magnetic field changes is

$$\frac{dB}{dt} = \frac{\Delta B}{\Delta t} = \frac{-1.60 \text{ T}}{0.30 \text{ s}} = -5.3 \text{ T/s}$$

dB/dt is negative because the field is decreasing, but all we need for Faraday's law is the absolute value. Thus

$$\mathcal{E} = \pi r^2 \left|\frac{dB}{dt}\right| = \pi(0.040 \text{ m})^2(5.3 \text{ T/s}) = 0.027 \text{ V}$$

To find the current, we need to know the resistance of the loop. Recall, from Chapter 30, that a conductor with resistivity ρ, length L, and cross-section area A has resistance $R = \rho L/A$. The length is the circumference of the loop, $L = 0.25$ m, and we can use the 1.0 cm diameter of the "wire" to find $A = 7.9 \times 10^{-5} \text{ m}^2$. With these values, we can compute $R = 4700 \ \Omega$. As a result, the induced current is

$$I = \frac{\mathcal{E}}{R} = \frac{0.027 \text{ V}}{4700 \ \Omega} = 5.7 \times 10^{-6} \text{ A} = 5.7 \ \mu\text{A}$$

ASSESS This is a very small current. Power—the rate of energy dissipation in the muscle—is

$$P = I^2 R = (5.7 \times 10^{-6} \text{ A})^2(4700 \ \Omega) = 1.5 \times 10^{-7} \text{ W}$$

The current is far too small to notice, and the tiny energy dissipation will certainly not heat the tissue.

What Does Faraday's Law Tell Us?

The induced current in the slide-wire circuit of Figure 33.26 can be understood as a motional emf due to magnetic forces on moving charges. We had not anticipated this kind of current in Chapter 32, but it takes no new laws of physics to understand it. The induced currents in Examples 33.8 and 33.9 are different. We cannot explain these induced currents on the basis of previous laws or principles. This is new physics.

Faraday recognized that all induced currents are associated with a changing magnetic flux. There are two fundamentally different ways to change the magnetic flux through a conducting loop:

1. The loop can expand, contract, or rotate, creating a motional emf.
2. The magnetic field can change.

We can see both of these if we write Faraday's law as

$$\mathcal{E} = \left| \frac{d\Phi_m}{dt} \right| = \left| \vec{B} \cdot \frac{d\vec{A}}{dt} + \vec{A} \cdot \frac{d\vec{B}}{dt} \right| \qquad (33.19)$$

The first term on the right side represents a motional emf. The magnetic flux changes because the loop itself is changing. This term includes not only situations like the slide-wire circuit, where the area A changes, but also loops that rotate in a magnetic field. The physical area of a rotating loop does not change, but the area *vector* $\vec{A}$ does. The loop's motion causes magnetic forces on the charge carriers in the loop.

The second term on the right side is the new physics in Faraday's law. It says that an emf can also be created simply by changing a magnetic field, even if nothing is moving. This was the case in Examples 33.8 and 33.9. Faraday's law tells us that the induced emf is simply the rate of change of the magnetic flux through the loop, *regardless* of what causes the flux to change.

STOP TO THINK 33.5 A conducting loop is halfway into a magnetic field. Suppose the magnetic field begins to increase rapidly in strength. What happens to the loop?

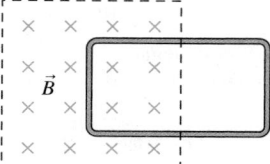

a. The loop is pushed upward, toward the top of the page.
b. The loop is pushed downward, toward the bottom of the page.
c. The loop is pulled to the left, into the magnetic field.
d. The loop is pushed to the right, out of the magnetic field.
e. The tension in the wires increases but the loop does not move.

FIGURE 33.30 An induced electric field creates a current in the loop.

(a)

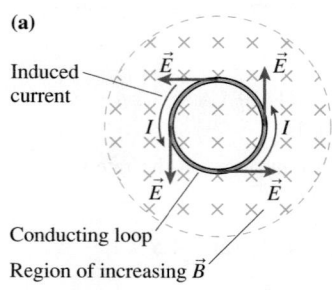

Induced current

Conducting loop
Region of increasing $\vec{B}$

(b)

Induced electric field $\vec{E}$

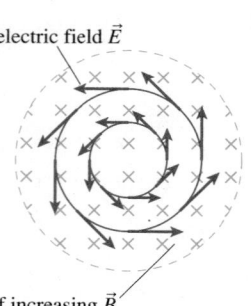

Region of increasing $\vec{B}$

33.6 Induced Fields

Faraday's law is a tool for calculating the strength of an induced current, but one important piece of the puzzle is still missing. What *causes* the current? That is, what *force* pushes the charges around the loop against the resistive forces of the metal? The agents that exert forces on charges are electric fields and magnetic fields. Magnetic forces are responsible for motional emfs, but magnetic forces cannot explain the current induced in a *stationary* loop by a changing magnetic field.

FIGURE 33.30a shows a conducting loop in an increasing magnetic field. According to Lenz's law, there is an induced current in the ccw direction. Something has to act on the charge carriers to make them move, so we infer that there must be an *electric* field tangent to the loop at all points. This electric field is *caused* by the changing magnetic field and is called an **induced electric field**. The induced electric field is the mechanism that creates a current inside a stationary loop when there's a changing magnetic field.

The conducting loop isn't necessary. The space in which the magnetic field is changing is filled with the pinwheel pattern of induced electric fields shown in FIGURE 33.30b. Charges will move if a conducting path is present, but the induced electric field is there as a direct consequence of the changing magnetic field.

But this is a rather peculiar electric field. All the electric fields we have examined until now have been created by charges. Electric field vectors pointed away from positive charges and toward negative charges. An electric field created by charges is called a **Coulomb electric field.** The induced electric field of Figure 33.30b is caused not by charges but by a changing magnetic field. It is called a **non-Coulomb electric field.**

So it appears that there are two different ways to create an electric field:

1. A Coulomb electric field is created by positive and negative charges.
2. A non-Coulomb electric field is created by a changing magnetic field.

Both exert a force $\vec{F} = q\vec{E}$ on a charge, and both create a current in a conductor. However, the origins of the fields are very different. FIGURE 33.31 is a quick summary of the two ways to create an electric field.

We first introduced the idea of a field as a way of thinking about how two charges exert long-range forces on each other through the emptiness of space. The field may have seemed like a useful pictorial representation of charge interactions, but we had little evidence that fields are *real,* that they actually exist. Now we do. The electric field has shown up in a completely different context, independent of charges, as the explanation of the very real existence of induced currents.

The electric field is not just a pictorial representation; it is real.

Calculating the Induced Field

The induced electric field is peculiar in another way: It is nonconservative. Recall that a force is conservative if it does no net work on a particle moving around a closed path. "Uphills" are balanced by "downhills." We can associate a potential energy with a conservative force, hence we have gravitational potential energy for the conservative gravitational force and electric potential energy for the conservative electric force of charges (a Coulomb electric field).

But a charge moving around a closed path in the induced electric field of Figure 33.30 is always being pushed *in the same direction* by the electric force $\vec{F} = q\vec{E}$. There's never any negative work to balance the positive work, so the net work done in going around a closed path is not zero. Because it's nonconservative, we cannot associate an electric potential with an induced electric field. Only the Coulomb field of charges has an electric potential.

However, we can associate the induced field with the emf of Faraday's law. The emf was defined as the work required per unit charge to separate the charge. That is,

$$\mathcal{E} = \frac{W}{q} \tag{33.20}$$

In batteries, a familiar source of emf, this work is done by chemical forces. But the emf that appears in Faraday's law arises when work is done by the force of an induced electric field.

If a charge q moves through a small displacement $d\vec{s}$, the small amount of work done by the electric field is $dW = \vec{F} \cdot d\vec{s} = q\vec{E} \cdot d\vec{s}$. The emf of Faraday's law is an emf around a *closed curve* through which the magnetic flux Φ_m is changing. The work done by the induced electric field as charge q moves around a closed curve is

$$W_{\text{closed curve}} = q \oint \vec{E} \cdot d\vec{s} \tag{33.21}$$

where the integration symbol with the circle is the same as the one we used in Ampère's law to indicate an integral around a closed curve. If we use this work in Equation 33.20, we find that the emf around a closed loop is

$$\mathcal{E} = \frac{W_{\text{closed curve}}}{q} = \oint \vec{E} \cdot d\vec{s} \tag{33.22}$$

FIGURE 33.31 Two ways to create an electric field.

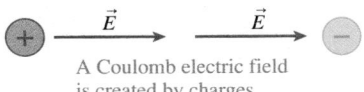

A Coulomb electric field is created by charges.

$\vec{B}$ increasing or decreasing

A non-Coulomb electric field is created by a changing magnetic field.

If we restrict ourselves to situations such as Figure 33.30 where the loop is perpendicular to the magnetic field and only the field is changing, we can write Faraday's law as $\mathcal{E} = |d\Phi_m/dt| = A|dB/dt|$. Consequently

$$\oint \vec{E} \cdot d\vec{s} = A \left| \frac{dB}{dt} \right| \qquad (33.23)$$

Equation 33.23 is an alternative statement of Faraday's law that relates the induced electric field to the changing magnetic field.

The solenoid in FIGURE 33.32a provides a good example of the connection between $\vec{E}$ and $\vec{B}$. If there were a conducting loop inside the solenoid, we could use Lenz's law to determine that the direction of the induced current would be clockwise. But Faraday's law, in the form of Equation 33.23, tells us that **an induced electric field is present whether there's a conducting loop or not.** The electric field is induced simply due to the fact that $\vec{B}$ is changing.

FIGURE 33.32 The induced electric field circulates around the changing magnetic field inside a solenoid.

(a) The current through the solenoid is increasing.

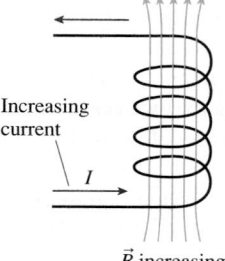

(b) The induced electric field circulates around the magnetic field lines.

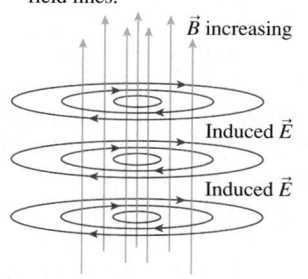

(c) Top view into the solenoid. $\vec{B}$ is coming out of the page.

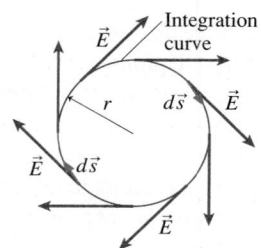

The shape and direction of the induced electric field have to be such that it *could* drive a current around a conducting loop, if one were present, and it has to be consistent with the cylindrical symmetry of the solenoid. The only possible choice, shown in FIGURE 33.32b, is an electric field that circulates clockwise around the magnetic field lines.

NOTE ▶ Circular electric field lines violate the Chapter 26 rule that electric field lines have to start and stop on charges. However, that rule applied only to Coulomb fields created by source charges. An induced electric field is a non-Coulomb field created not by source charges but by a changing magnetic field. Without source charges, induced electric field lines *must* form closed loops. ◀

To use Faraday's law, choose a *clockwise* circle of radius r as the closed curve for evaluating the integral. FIGURE 33.32c shows that the electric field vectors are everywhere tangent to the curve, so the line integral of $\vec{E}$ is

$$\oint \vec{E} \cdot d\vec{s} = El = 2\pi r E \qquad (33.24)$$

where $l = 2\pi r$ is the length of the closed curve. This is exactly like the integrals we did for Ampère's law in Chapter 32.

If we stay inside the solenoid ($r < R$), the flux passes through area $A = \pi r^2$ and Equation 33.24 becomes

$$\oint \vec{E} \cdot d\vec{s} = 2\pi r E = A \left| \frac{dB}{dt} \right| = \pi r^2 \left| \frac{dB}{dt} \right| \qquad (33.25)$$

Thus the strength of the induced electric field inside the solenoid is

$$E_{inside} = \frac{r}{2}\left|\frac{dB}{dt}\right|$$ (33.26)

This result shows very directly that the induced electric field is created by a *changing* magnetic field. A constant $\vec{B}$, with $dB/dt = 0$, would give $E = 0$.

EXAMPLE 33.10   **An induced electric field**

A 4.0-cm-diameter solenoid is wound with 2000 turns per meter. The current through the solenoid oscillates at 60 Hz with an amplitude of 2.0 A. What is the maximum strength of the induced electric field inside the solenoid?

MODEL Assume that the magnetic field inside the solenoid is uniform.

VISUALIZE The electric field lines are concentric circles around the magnetic field lines, as was shown in Figure 33.32b. They reverse direction twice every period as the current oscillates.

SOLVE You learned in Chapter 32 that the magnetic field strength inside a solenoid with n turns per meter is $B = \mu_0 nI$. In this case, the current through the solenoid is $I = I_0 \sin \omega t$, where $I_0 = 2.0$ A is the peak current and $\omega = 2\pi(60 \text{ Hz}) = 377$ rad/s. Thus the induced electric field strength at radius r is

$$E = \frac{r}{2}\left|\frac{dB}{dt}\right| = \frac{r}{2}\frac{d}{dt}(\mu_0 nI_0 \sin \omega t) = \frac{1}{2}\mu_0 nr\omega I_0 \cos \omega t$$

The field strength is maximum at maximum radius ($r = R$) *and* at the instant when $\cos \omega t = 1$. That is,

$$E_{max} = \frac{1}{2}\mu_0 nR\omega I_0 = 0.019 \text{ V/m}$$

ASSESS This field strength, although not large, is similar to the field strength that the emf of a battery creates in a wire. Hence this induced electric field can drive a substantial induced current through a conducting loop *if* a loop is present. But the induced electric field exists inside the solenoid whether or not there is a conducting loop.

Occasionally it is useful to have a version of Faraday's law without the absolute value signs. The essence of Lenz's law is that the emf $\mathcal{E}$ opposes the *change* in Φ_m. Mathematically, this means that $\mathcal{E}$ must be opposite in sign to dB/dt. Consequently, we can write Faraday's law as

$$\mathcal{E} = \oint \vec{E} \cdot d\vec{s} = -\frac{d\Phi_m}{dt}$$ (33.27)

For practical applications, it's always easier to calculate just the magnitude of the emf with Faraday's law and to use Lenz's law to find the direction of the emf or the induced current. However, the mathematically rigorous version of Faraday's law in Equation 33.27 will prove to be useful when we combine it with other equations, in Chapter 34, to predict the existence of electromagnetic waves.

Maxwell's Theory of Electromagnetic Waves

In 1855, less than two years after receiving his undergraduate degree, the Scottish physicist James Clerk Maxwell presented a paper titled "On Faraday's Lines of Force." In this paper, he began to sketch out how Faraday's pictorial ideas about fields could be given a rigorous mathematical basis. Maxwell was troubled by a certain lack of symmetry. Faraday had found that a changing magnetic field creates an induced electric field, a non-Coulomb electric field not tied to charges. But what, Maxwell began to wonder, about a changing *electric* field?

To complete the symmetry, Maxwell proposed that a changing electric field creates an **induced magnetic field,** a new kind of magnetic field not tied to the existence of currents. **FIGURE 33.33** shows a region of space where the *electric* field is increasing. This region of space, according to Maxwell, is filled with a pinwheel pattern of induced magnetic fields. The induced magnetic field looks like the induced electric field, with $\vec{E}$ and $\vec{B}$ interchanged, except that—for technical reasons explored in the

FIGURE 33.33 Maxwell hypothesized the existence of induced magnetic fields.

A changing magnetic field creates an induced electric field.

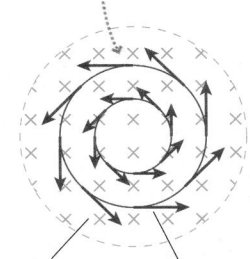

Region of increasing $\vec{B}$ — Induced electric field $\vec{E}$

A changing electric field creates an induced magnetic field.

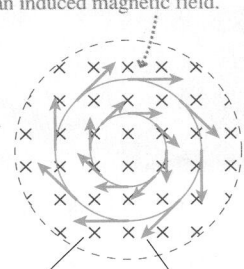

Region of increasing $\vec{E}$ — Induced magnetic field $\vec{B}$

next chapter—the induced $\vec{B}$ points the opposite way from the induced $\vec{E}$. Although there was no experimental evidence that induced magnetic fields existed, Maxwell went ahead and included them in his electromagnetic field theory. This was an inspired hunch, soon to be vindicated.

Maxwell soon realized that it might be possible to establish self-sustaining electric and magnetic fields that would be entirely independent of any charges or currents. That is, a changing electric field $\vec{E}$ creates a magnetic field $\vec{B}$, which then changes in just the right way to recreate the electric field, which then changes in just the right way to again recreate the magnetic field, and so on. The fields are continually recreated through electromagnetic induction without any reliance on charges or currents.

Maxwell was able to predict that electric and magnetic fields would be able to sustain themselves, free from charges and currents, if they took the form of an **electromagnetic wave.** The wave would have to have a very specific geometry, shown in FIGURE 33.34, in which $\vec{E}$ and $\vec{B}$ are perpendicular to each other as well as perpendicular to the direction of travel. That is, an electromagnetic wave would be a *transverse* wave.

Furthermore, Maxwell's theory predicted that the wave would travel with speed

$$v_{\text{em wave}} = \frac{1}{\sqrt{\epsilon_0 \mu_0}}$$

where ϵ_0 is the permittivity constant from Coulomb's law and μ_0 is the permeability constant from the law of Biot and Savart. Maxwell computed that an electromagnetic wave, if it existed, would travel with speed $v_{\text{em wave}} = 3.00 \times 10^8$ m/s.

We don't know Maxwell's immediate reaction, but it must have been both shock and excitement. His predicted speed for electromagnetic waves, a prediction that came directly from his theory, was none other than the speed of light! This agreement could be just a coincidence, but Maxwell didn't think so. Making a bold leap of imagination, Maxwell concluded that **light is an electromagnetic wave.**

It took 25 more years for Maxwell's predictions to be tested. In 1886, the German physicist Heinrich Hertz discovered how to generate and transmit radio waves. Two years later, in 1888, he was able to show that radio waves travel at the speed of light. Maxwell, unfortunately, did not live to see his triumph. He had died in 1879, at the age of 48.

Chapter 34 will develop some of the mathematical details of Maxwell's theory and show how the ideas contained in Faraday's law lead to electromagnetic waves.

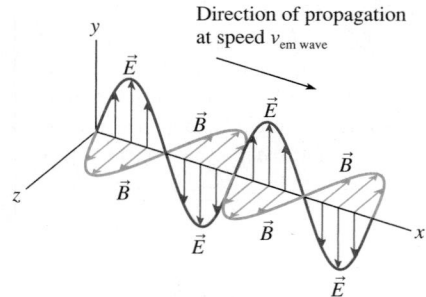

FIGURE 33.34 A self-sustaining electromagnetic wave.

Direction of propagation at speed $v_{\text{em wave}}$

33.7 Induced Currents: Three Applications

There are many applications of Faraday's law and induced currents in modern technology. In this section we will look at three: generators, transformers, and metal detectors.

Generators

A generator is a device that transforms mechanical energy into electric energy. FIGURE 33.35 shows a generator in which a coil of wire, perhaps spun by a windmill, rotates in a magnetic field. Both the field and the area of the loop are constant, but the magnetic flux through the loop changes continuously as the loop rotates. The induced current is removed from the rotating loop by *brushes* that press up against rotating *slip rings.*

The flux through the coil is

$$\Phi_{\text{m}} = \vec{A} \cdot \vec{B} = AB \cos\theta = AB \cos\omega t \qquad (33.28)$$

A generator inside a hydroelectric dam uses electromagnetic induction to convert the mechanical energy of a spinning turbine into electric energy.

where ω is the angular frequency ($\omega = 2\pi f$) with which the coil rotates. The induced emf is given by Faraday's law,

$$\mathcal{E}_{\text{coil}} = -N\frac{d\Phi_m}{dt} = -ABN\frac{d}{dt}(\cos\omega t) = \omega ABN\sin\omega t \qquad (33.29)$$

where N is the number of turns on the coil. Here it's best to use the signed version of Faraday's law to see how $\mathcal{E}_{\text{coil}}$ alternates between positive and negative.

Because the emf alternates in sign, the current through resistor R alternates back and forth in direction. Hence the generator of Figure 33.35 is an alternating-current generator, producing what we call an *AC voltage*.

FIGURE 33.35 An alternating-current generator.

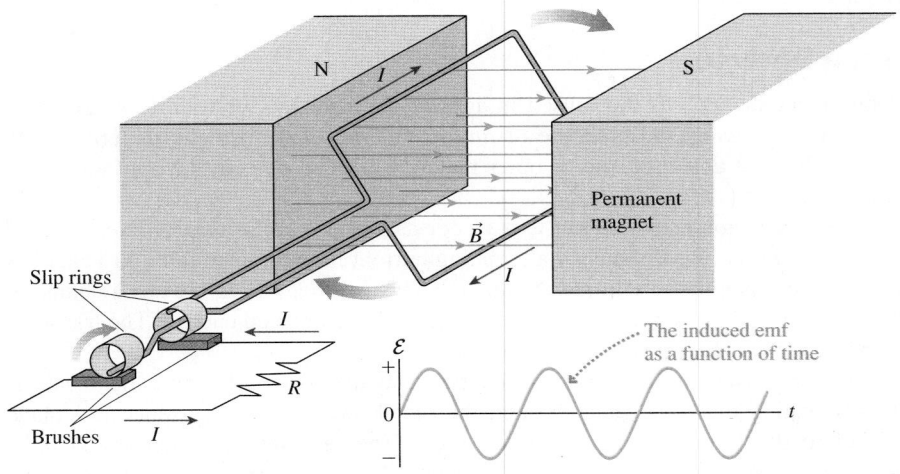

EXAMPLE 33.11 **An AC generator**

A coil with area 2.0 m² rotates in a 0.010 T magnetic field at a frequency of 60 Hz. How many turns are needed to generate a peak voltage of 160 V?

SOLVE The coil's maximum voltage is found from Equation 33.29:

$$\mathcal{E}_{\text{max}} = \omega ABN = 2\pi f ABN$$

The number of turns needed to generate $\mathcal{E}_{\text{max}} = 160$ V is

$$N = \frac{\mathcal{E}_{\text{max}}}{2\pi fAB} = \frac{160\text{ V}}{2\pi(60\text{ Hz})(2.0\text{ m}^2)(0.010\text{ T})} = 21\text{ turns}$$

ASSESS A 0.010 T field is modest, so you can see that generating large voltages is not difficult with large (2 m²) coils. Commercial generators use water flowing through a dam, rotating windmill blades, or turbines spun by expanding steam to rotate the generator coils. Work is required to rotate the coil, just as work was required to pull the slide wire in Section 33.2, because the magnetic field exerts retarding forces on the currents in the coil. Thus a generator is a device that turns motion (mechanical energy) into a current (electric energy). A generator is the opposite of a motor, which turns a current into motion.

Transformers

FIGURE 33.36 shows two coils wrapped on an iron core. The left coil is called the **primary coil**. It has N_1 turns and is driven by an oscillating voltage $V_1\cos\omega t$. The magnetic field of the primary follows the iron core and passes through the right coil, which has N_2 turns and is called the **secondary coil**. The alternating current through the primary coil causes an oscillating magnetic flux through the secondary coil and, hence, an induced emf. The induced emf of the secondary coil is delivered to the load as the oscillating voltage $V_2\cos\omega t$.

The changing magnetic field inside the iron core is inversely proportional to the number of turns on the primary coil: $B \propto 1/N_1$. (This relation is a consequence of the coil's inductance, an idea discussed in the next section.) According to Faraday's law, the emf induced in the secondary coil is directly proportional to its number of turns:

FIGURE 33.36 A transformer.

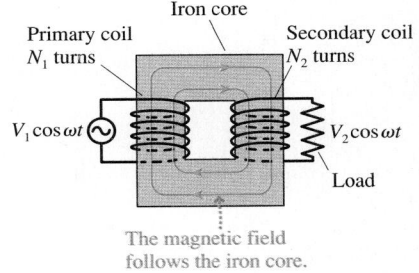

Transformers are essential for transporting electric energy from the power plant to cities and homes.

FIGURE 33.37 A metal detector.

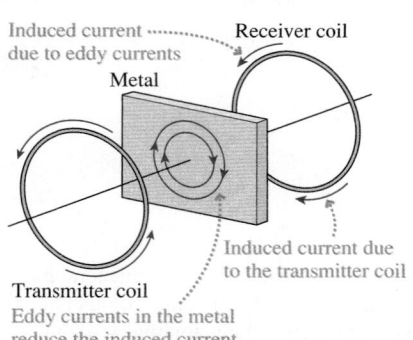

$\mathcal{E}_{\text{sec}} \propto N_2$. Combining these two proportionalities, the secondary voltage of an ideal transformer is related to the primary voltage by

$$V_2 = \frac{N_2}{N_1} V_1 \qquad (33.30)$$

Depending on the ratio N_2/N_1, the voltage V_2 across the load can be *transformed* to a higher or a lower voltage than V_1. Consequently, this device is called a **transformer.** Transformers are widely used in the commercial generation and transmission of electricity. A *step-up transformer,* with $N_2 \gg N_1$, boosts the voltage of a generator up to several hundred thousand volts. Delivering power with smaller currents at higher voltages reduces losses due to the resistance of the wires. High-voltage transmission lines carry electric power to urban areas, where *step-down transformers* ($N_2 \ll N_1$) lower the voltage to 120 V.

Metal Detectors

Metal detectors, such as those used in airports for security, seem fairly mysterious. How can they detect the presence of *any* metal—not just magnetic materials such as iron—but not detect plastic or other materials? Metal detectors work because of induced currents.

A metal detector, shown in FIGURE 33.37, consists of two coils: a *transmitter coil* and a *receiver coil.* A high-frequency alternating current in the transmitter coil generates an alternating magnetic field along the axis. This magnetic field creates a changing flux through the receiver coil and causes an alternating induced current. The transmitter and receiver are similar to a transformer.

Suppose a piece of metal is placed between the transmitter and the receiver. The alternating magnetic field through the metal induces eddy currents in a plane parallel to the transmitter and receiver coils. The receiver coil then responds to the *superposition* of the transmitter's magnetic field and the magnetic field of the eddy currents. Because the eddy currents attempt to prevent the flux from changing, in accordance with Lenz's law, the net field at the receiver *decreases* when a piece of metal is inserted between the coils. Electronic circuits detect the current decrease in the receiver coil and set off an alarm. Eddy currents can't flow in an insulator, so this device detects only metals.

33.8 Inductors

Capacitors are useful circuit elements because they store potential energy U_C in the electric field. Similarly, a coil of wire can be a useful circuit element because it stores energy in the magnetic field. Using as an analogy the definition of capacitance as the charge-to-voltage ratio, $C = Q/\Delta V$, let's define the **inductance** L of a coil as its flux-to-current ratio:

$$L = \frac{\Phi_{\text{m}}}{I} \qquad (33.31)$$

Strictly speaking, this is called *self-inductance* because the flux we're considering is the magnetic flux the solenoid creates in itself when there is a current.

The SI unit of inductance is the **henry,** named in honor of Joseph Henry, defined as

$$1 \text{ henry} = 1 \text{ H} \equiv 1 \text{ Wb/A} = 1 \text{ T m}^2/\text{A}$$

Practical inductances are typically millihenries (mH) or microhenries (μH).

A coil of wire used in a circuit for the purpose of providing inductance is called an **inductor.** An *ideal inductor* is one for which the wire forming the coil has no electric resistance. The circuit symbol for an inductor is —⟲⟲⟲⟲—.

It's not hard to find the inductance of a solenoid. In Chapter 32 we found that the magnetic field inside an ideal solenoid having N turns and length l is

$$B = \frac{\mu_0 NI}{l}$$

The magnetic flux through one turn of the coil is $\Phi_{\text{per turn}} = AB$, where A is the cross-section area of the solenoid. The total magnetic flux through all N turns is

$$\Phi_{\text{m}} = N\Phi_{\text{per turn}} = \frac{\mu_0 N^2 A}{l} I \qquad (33.32)$$

Thus the inductance of the solenoid, using the definition of Equation 33.31, is

$$L_{\text{solenoid}} = \frac{\Phi_{\text{m}}}{I} = \frac{\mu_0 N^2 A}{l} \qquad (33.33)$$

The inductance of a solenoid depends only on its geometry, not at all on the current. You may recall that the capacitance of two parallel plates depends only on their geometry, not at all on their potential difference.

EXAMPLE 33.12 | **The length of an inductor**

An inductor is made by tightly wrapping 0.30-mm-diameter wire around a 4.0-mm-diameter cylinder. What length cylinder has an inductance of 10 μH?

SOLVE The cross-section area of the solenoid is $A = \pi r^2$. If the wire diameter is d, the number of turns of wire on a cylinder of length l is $N = l/d$. Thus the inductance is

$$L = \frac{\mu_0 N^2 A}{l} = \frac{\mu_0 (l/d)^2 \pi r^2}{l} = \frac{\mu_0 \pi r^2 l}{d^2}$$

The length needed to give inductance $L = 1.0 \times 10^{-5}$ H is

$$l = \frac{d^2 L}{\mu_0 \pi r^2} = \frac{(0.00030 \text{ m})^2 (1.0 \times 10^{-5} \text{ H})}{(4\pi \times 10^{-7} \text{ T m/A})\pi (0.0020 \text{ m})^2}$$

$$= 0.057 \text{ m} = 5.7 \text{ cm}$$

The Potential Difference Across an Inductor

An inductor is not very interesting when the current through it is steady. If the inductor is ideal, with $R = 0\ \Omega$, the potential difference due to a steady current is zero. Inductors become important circuit elements when currents are changing. **FIGURE 33.38a** shows a steady current into the left side of an inductor. The solenoid's magnetic field passes through the coils of the solenoid, establishing a flux.

In **FIGURE 33.38b**, the current into the solenoid is increasing. This creates an increasing flux to the left. According to Lenz's law, an induced current in the coils will oppose this increase by creating an induced magnetic field pointing to the right. This requires the induced current to be *opposite* the current into the solenoid. This induced current will carry positive charge carriers to the left until a potential difference is established across the solenoid.

You saw a similar situation in Section 33.2. The induced current in a conductor moving through a magnetic field carried positive charge carriers to the top of the wire and established a potential difference across the conductor. The induced current in the moving wire was due to magnetic forces on the moving charges. Now, in Figure 33.38b, the induced current is due to the non-Coulomb electric field induced by the changing magnetic field. Nonetheless, the outcome is the same: a potential difference across the conductor.

We can use Faraday's law to find the potential difference. The emf induced in a coil is

$$\mathcal{E}_{\text{coil}} = N \left| \frac{d\Phi_{\text{per turn}}}{dt} \right| = \left| \frac{d\Phi_{\text{m}}}{dt} \right| \qquad (33.34)$$

FIGURE 33.38 Increasing the current through an inductor.

(a)

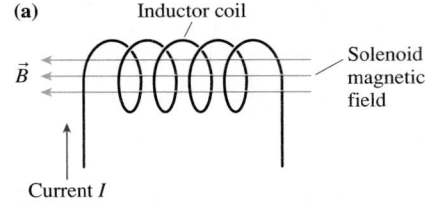

Current I

(b) The induced current is opposite the solenoid current.

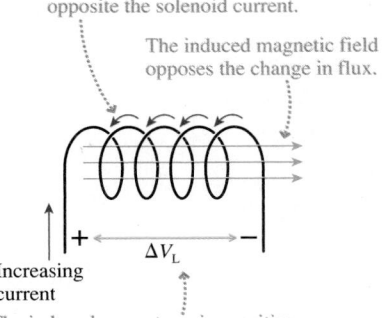

The induced current carries positive charge carriers to the left and establishes a potential difference across the inductor.

FIGURE 33.39 Decreasing the current through an inductor.

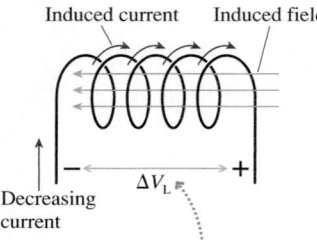

Induced current   Induced field

Decreasing current ΔV_{L} $-$ $+$

The induced current carries positive charge carriers to the right. The potential difference is opposite that of Figure 33.38b.

where $\Phi_{\text{m}} = N\Phi_{\text{per turn}}$ is the total flux through all the coils. The inductance was defined such that $\Phi_{\text{m}} = LI$, so Equation 33.34 becomes

$$\mathcal{E}_{\text{coil}} = L\left|\frac{dI}{dt}\right| \tag{33.35}$$

The induced emf is directly proportional to the *rate of change* of current through the coil. We'll consider the appropriate sign in a moment, but Equation 33.35 gives us the size of the potential difference that is developed across a coil as the current through the coil changes. Note that $\mathcal{E}_{\text{coil}} = 0$ for a steady, unchanging current.

FIGURE 33.39 shows the same inductor, but now the current (still *in* to the left side) is decreasing. To oppose the decrease in flux, the induced current is in the *same* direction as the input current. The induced current carries charge to the right and establishes a potential difference opposite that in Figure 33.38b.

NOTE ▶ Notice that the induced current does not oppose the current through the inductor, which is from left to right in both Figures 33.38 and 33.39. Instead, in accordance with Lenz's law, the induced current opposes the *change* in the current in the solenoid. The practical result is that it is hard to change the current through an inductor. Any effort to increase or decrease the current is met with opposition in the form of an opposing induced current. You can think of the current in an inductor as having inertia, trying to continue what it was doing without change. ◀

Before we can use inductors in a circuit we need to establish a rule about signs that is consistent with our earlier circuit analysis. **FIGURE 33.40** first shows current I passing through a resistor. You learned in Chapter 31 that the potential difference across a resistor is $\Delta V_{\text{res}} = -\Delta V_{\text{R}} = -IR$, where the minus sign indicates that the potential *decreases* in the direction of the current.

We'll use the same convention for an inductor. The potential difference across an inductor, *measured along the direction of the current*, is

$$\Delta V_{\text{L}} = -L\frac{dI}{dt} \tag{33.36}$$

FIGURE 33.40 The potential difference across a resistor and an inductor.

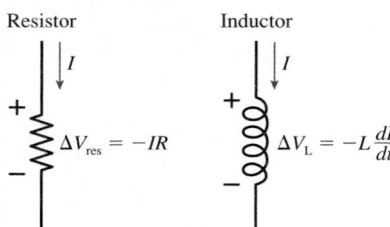

Resistor   Inductor

$+$ I $+$ I

$\Delta V_{\text{res}} = -IR$ $\Delta V_{\text{L}} = -L\frac{dI}{dt}$

$-$ $-$

The potential always decreases.   The potential decreases if the current is increasing.

The potential increases if the current is decreasing.

If the current is increasing ($dI/dt > 0$), the input side of the inductor is more positive than the output side and the potential decreases in the direction of the current ($\Delta V_{\text{L}} < 0$). This was the situation in Figure 33.38b. If the current is decreasing ($dI/dt < 0$), the input side is more negative and the potential increases in the direction of the current ($\Delta V_{\text{L}} > 0$). This was the situation in Figure 33.39.

The potential difference across an inductor can be very large if the current changes very abruptly (large dI/dt). **FIGURE 33.41** shows an inductor connected across a battery. There is a large current through the inductor, limited only by the internal resistance of the battery. Suppose the switch is suddenly opened. A very large induced voltage is created across the inductor as the current rapidly drops to zero. This potential difference (plus ΔV_{bat}) appears across the gap of the switch as it is opened. A large potential difference across a small gap often creates a spark.

FIGURE 33.41 Creating sparks.

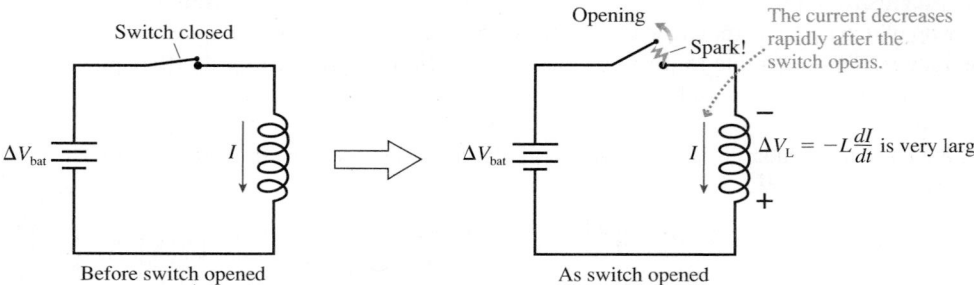

Switch closed   Opening   Spark!   The current decreases rapidly after the switch opens.

ΔV_{bat} I ΔV_{bat} I $\Delta V_{\text{L}} = -L\frac{dI}{dt}$ is very large.

Before switch opened   As switch opened

Indeed, this is exactly how the spark plugs in your car work. The car's generator sends a current through the *coil,* which is a big inductor. When a switch is suddenly opened, breaking the current, the induced voltage, typically a few thousand volts, appears across the terminals of the spark plug, creating the spark that ignites the gasoline. Older cars use a *distributor* to open and close an actual switch; more recent cars have *electronic ignition* in which the mechanical switch has been replaced by a transistor.

EXAMPLE 33.13 **Large voltage across an inductor**

A 1.0 A current passes through a 10 mH inductor coil. What potential difference is induced across the coil if the current drops to zero in 5.0 μs?

MODEL Assume this is an ideal inductor, with $R = 0 \ \Omega$, and that the current decrease is linear with time.

SOLVE The rate of current decrease is

$$\frac{dI}{dt} \approx \frac{\Delta I}{\Delta t} = \frac{-1.0 \ A}{5.0 \times 10^{-6} \ s} = -2.0 \times 10^5 \ A/s$$

The induced voltage is

$$\Delta V_L = -L\frac{dI}{dt} \approx -(0.010 \ H)(-2.0 \times 10^5 \ A/s) = 2000 \ V$$

ASSESS Inductors may be physically small, but they can pack a punch if you try to change the current through them too quickly.

STOP TO THINK 33.6   The potential at a is higher than the potential at b. Which of the following statements about the inductor current I could be true?

a. I is from a to b and steady.
b. I is from a to b and increasing.
c. I is from a to b and decreasing.
d. I is from b to a and steady.
e. I is from b to a and increasing.
f. I is from b to a and decreasing.

$V_a > V_b$

a ⟶〰〰〰 b

Energy in Inductors and Magnetic Fields

Recall that electric power is $P_{elec} = I\Delta V$. As current passes through an inductor, for which $\Delta V_L = -L(dI/dt)$, the electric power is

$$P_{elec} = I\Delta V_L = -LI\frac{dI}{dt} \tag{33.37}$$

P_{elec} is negative because the current is *losing* electric energy. That energy is being transferred to the inductor, which is *storing* energy U_L at the rate

$$\frac{dU_L}{dt} = +LI\frac{dI}{dt} \tag{33.38}$$

where we've noted that power is the rate of change of energy.

We can find the total energy stored in an inductor by integrating Equation 33.38 from $I = 0$, where $U_L = 0$, to a final current I. Doing so gives

$$U_L = L\int_0^I I \ dI = \frac{1}{2}LI^2 \tag{33.39}$$

The potential energy stored in an inductor depends on the square of the current through it. Notice the analogy with the energy $U_C = \frac{1}{2}C(\Delta V)^2$ stored in a capacitor.

In working with circuits we say that the energy is "stored in the inductor." Strictly speaking, the energy is stored in the inductor's magnetic field, analogous to how a capacitor stores energy in the electric field. We can use the inductance of a solenoid, Equation 33.33, to relate the inductor's energy to the magnetic field strength:

$$U_L = \frac{1}{2}LI^2 = \frac{\mu_0 N^2 A}{2l}I^2 = \frac{1}{2\mu_0}Al\left(\frac{\mu_0 NI}{l}\right)^2 \tag{33.40}$$

We made the last rearrangement in Equation 33.40 because $\mu_0 NI/l$ is the magnetic field inside the solenoid. Thus

$$U_L = \frac{1}{2\mu_0}AlB^2 \tag{33.41}$$

But Al is the volume inside the solenoid. Dividing by Al, the magnetic field *energy density* inside the solenoid (energy per m³) is

$$u_B = \frac{1}{2\mu_0}B^2 \tag{33.42}$$

We've derived this expression for energy density based on the properties of a solenoid, but it turns out to be the correct expression for the energy density anywhere there's a magnetic field. Compare this to the energy density of an electric field $u_E = \frac{1}{2}\epsilon_0 E^2$ that we found in Chapter 29.

Energy in electric and magnetic fields

Electric fields	Magnetic fields
A capacitor stores energy	An inductor stores energy
$U_C = \frac{1}{2}C(\Delta V)^2$	$U_L = \frac{1}{2}LI^2$
Energy density in the field is	Energy density in the field is
$u_E = \frac{\epsilon_0}{2}E^2$	$u_B = \frac{1}{2\mu_0}B^2$

EXAMPLE 33.14 **Energy stored in an inductor**

The 10 μH inductor of Example 33.12 was 5.7 cm long and 4.0 mm in diameter. Suppose it carries a 100 mA current. What are the energy stored in the inductor, the magnetic energy density, and the magnetic field strength?

SOLVE The stored energy is

$$U_L = \frac{1}{2}LI^2 = \frac{1}{2}(1.0 \times 10^{-5} \text{ H})(0.10 \text{ A})^2 = 5.0 \times 10^{-8} \text{ J}$$

The solenoid volume is $(\pi r^2)l = 7.16 \times 10^{-7}$ m³. Using this gives the energy density of the magnetic field:

$$u_B = \frac{5.0 \times 10^{-8} \text{ J}}{7.16 \times 10^{-7} \text{ m}^3} = 0.070 \text{ J/m}^3$$

From Equation 33.42, the magnetic field with this energy density is

$$B = \sqrt{2\mu_0 u_B} = 4.2 \times 10^{-4} \text{ T}$$

33.9 *LC* Circuits

Telecommunication—radios, televisions, cell phones—is based on electromagnetic signals that *oscillate* at a well-defined frequency. These oscillations are generated and detected by a simple circuit consisting of an inductor and a capacitor in parallel. This is called an **LC circuit**. In this section we will learn why an *LC* circuit oscillates and determine the oscillation frequency.

FIGURE 33.42 shows a capacitor with initial charge Q_0, an inductor, and a switch. The switch has been open for a long time, so there is no current in the circuit. Then, at $t = 0$, the switch is closed. How does the circuit respond? Let's think it through qualitatively before getting into the mathematics.

As FIGURE 33.43 shows, the inductor provides a conducting path for discharging the capacitor. However, the discharge current has to pass through the inductor, and, as we've seen, an inductor resists changes in current. Consequently, the current doesn't stop when the capacitor charge reaches zero.

A block attached to a stretched spring is a useful mechanical analogy. Closing the switch to discharge the capacitor is like releasing the block. The block doesn't stop when it reaches the origin; its momentum keeps it going until the spring is fully compressed. Likewise, the current continues until it has recharged the capacitor with the opposite polarization. This process repeats over and over, charging the capacitor first one way, then the other. That is, the charge and current *oscillate*.

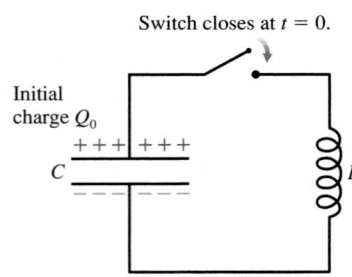

FIGURE 33.42 An *LC* circuit.

FIGURE 33.43 The capacitor charge oscillates much like a block attached to a spring.

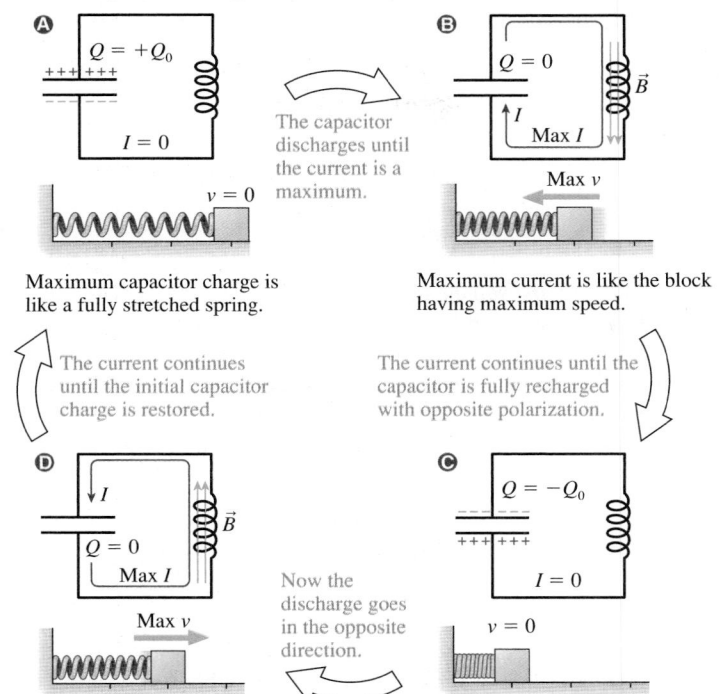

The goal of our circuit analysis will be to find expressions showing how the capacitor charge Q and the inductor current I change with time. As always, our starting point for circuit analysis is Kirchhoff's voltage law, which says that all the potential differences around a closed loop must sum to zero. Choosing a cw direction for I, Kirchhoff's law is

$$\Delta V_C + \Delta V_L = 0 \tag{33.43}$$

The potential difference across a capacitor is $\Delta V_C = Q/C$, and we found the potential difference across an inductor in Equation 33.36. Using these, Kirchhoff's law becomes

$$\frac{Q}{C} - L\frac{dI}{dt} = 0 \tag{33.44}$$

Equation 33.44 has two unknowns, Q and I. We can eliminate one of the unknowns by finding another relation between Q and I. Current is the rate at which charge moves, $I = dq/dt$, but the charge flowing through the inductor is charge that was *removed* from the capacitor. That is, an infinitesimal charge dq flows through the inductor when the capacitor charge changes by $dQ = -dq$. Thus the current through the inductor is related to the charge on the capacitor by

$$I = -\frac{dQ}{dt} \tag{33.45}$$

Now I is positive when Q is decreasing, as we would expect. This is a subtle but important step in the reasoning.

Equations 33.44 and 33.45 are two equations in two unknowns. To solve them, we'll first take the time derivative of Equation 33.45:

$$\frac{dI}{dt} = \frac{d}{dt}\left(-\frac{dQ}{dt}\right) = -\frac{d^2Q}{dt^2} \tag{33.46}$$

We can substitute this result into Equation 33.44:

$$\frac{Q}{C} + L\frac{d^2Q}{dt^2} = 0 \tag{33.47}$$

Now we have an equation for the capacitor charge Q.

A cell phone is actually a very sophisticated two-way radio that communicates with the nearest base station via high-frequency radio waves—roughly 1000 MHz. As in any radio or communications device, the transmission frequency is established by the oscillating current in an *LC* circuit.

Equation 33.47 is a second-order differential equation for Q. Fortunately, it is an equation we've seen before and already know how to solve. To see this, we rewrite Equation 33.47 as

$$\frac{d^2Q}{dt^2} = -\frac{1}{LC}Q \qquad (33.48)$$

Recall, from Chapter 14, that the equation of motion for an undamped mass on a spring is

$$\frac{d^2x}{dt^2} = -\frac{k}{m}x \qquad (33.49)$$

Equation 33.48 is *exactly the same equation,* with x replaced by Q and k/m replaced by $1/LC$. This should be no surprise because we've already seen that a mass on a spring is a mechanical analog of the LC circuit.

We know the solution to Equation 33.49. It is simple harmonic motion $x(t) = x_0 \cos \omega t$ with angular frequency $\omega = \sqrt{k/m}$. Thus the solution to Equation 33.48 must be

$$Q(t) = Q_0 \cos \omega t \qquad (33.50)$$

where Q_0 is the initial charge, at $t = 0$, and the angular frequency is

$$\omega = \sqrt{\frac{1}{LC}} \qquad (33.51)$$

The charge on the upper plate of the capacitor oscillates back and forth between $+Q_0$ and $-Q_0$ (the opposite polarization) with period $T = 2\pi/\omega$.

As the capacitor charge oscillates, so does the current through the inductor. Using Equation 33.45 gives the current through the inductor:

$$I = -\frac{dQ}{dt} = \omega Q_0 \sin \omega t = I_{max} \sin \omega t \qquad (33.52)$$

where $I_{max} = \omega Q_0$ is the maximum current.

An LC circuit is an *electric oscillator,* oscillating at frequency $f = \omega/2\pi$. FIGURE 33.44 shows graphs of the capacitor charge Q and the inductor current I as functions of time. Notice that Q and I are 90° out of phase. The current is zero when the capacitor is fully charged, as expected, and the charge is zero when the current is maximum.

FIGURE 33.44 The oscillations of an LC circuit.

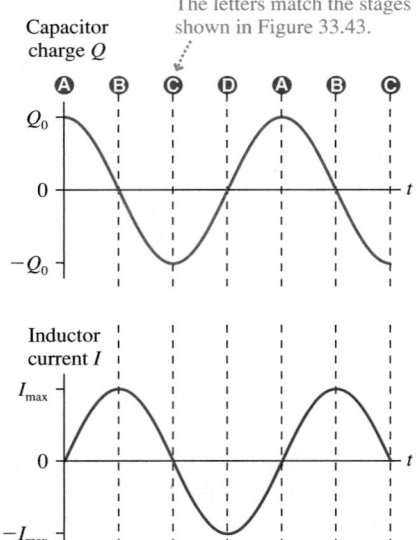

The letters match the stages shown in Figure 33.43.

EXAMPLE 33.15 **An AM radio oscillator**

You have a 1.0 mH inductor. What capacitor should you choose to make an oscillator with a frequency of 920 kHz? (This frequency is near the center of the AM radio band.)

SOLVE The angular frequency is $\omega = 2\pi f = 5.78 \times 10^6$ rad/s. Using Equation 33.51 for ω gives the required capacitor:

$$C = \frac{1}{\omega^2 L} = \frac{1}{(5.78 \times 10^6 \text{ rad/s})^2 (0.0010 \text{ H})}$$

$$= 3.0 \times 10^{-11} \text{ F} = 30 \text{ pF}$$

An LC circuit, like a mass on a spring, wants to respond only at its natural oscillation frequency $\omega = 1/\sqrt{LC}$. In Chapter 14 we defined a strong response at the natural frequency as a *resonance,* and resonance is the basis for all telecommunications. The input circuit in radios, televisions, and cell phones is an LC circuit driven by the signal picked up by the antenna. This signal is the superposition of hundreds of sinusoidal waves at different frequencies, one from each transmitter in the area, but the circuit responds only to the *one* signal that matches the circuit's natural frequency. That particular signal generates a large-amplitude current that can be further amplified and decoded to become the output that you hear.

33.10 *LR* Circuits

A circuit consisting of an inductor, a resistor, and (perhaps) a battery is called an **LR circuit**. FIGURE 33.45a is an example of an *LR* circuit. We'll assume that the switch has been in position a for such a long time that the current is steady and unchanging. There's no potential difference across the inductor, because $dI/dt = 0$, so it simply acts like a piece of wire. The current flowing around the circuit is determined entirely by the battery and the resistor: $I_0 = \Delta V_{bat}/R$.

What happens if, at $t = 0$, the switch is suddenly moved to position b? With the battery no longer in the circuit, you might expect the current to stop immediately. But the inductor won't let that happen. The current will continue for some period of time as the inductor's magnetic field drops to zero. In essence, the energy stored in the inductor allows it to act like a battery for a short period of time. Our goal is to determine how the current decays after the switch is moved.

NOTE ▶ It's important not to open switches in inductor circuits because they'll spark, as Figure 33.41 showed. The unusual switch in Figure 33.45 is designed to make the new contact just before breaking the old one. ◀

FIGURE 33.45b shows the circuit after the switch is changed. Our starting point, once again, is Kirchhoff's voltage law. The potential differences around a closed loop must sum to zero. For this circuit, Kirchhoff's law is

$$\Delta V_{res} + \Delta V_L = 0 \tag{33.53}$$

The potential differences in the direction of the current are $\Delta V_{res} = -IR$ for the resistor and $\Delta V_L = -L(dI/dt)$ for the inductor. Substituting these into Equation 33.53 gives

$$-RI - L\frac{dI}{dt} = 0 \tag{33.54}$$

We're going to need to integrate to find the current I as a function of time. Before doing so, we rearrange Equation 33.54 to get all the current terms on one side of the equation and all the time terms on the other:

$$\frac{dI}{I} = -\frac{R}{L}dt = -\frac{dt}{(L/R)} \tag{33.55}$$

We know that the current at $t = 0$, when the switch was moved, was I_0. We want to integrate from these starting conditions to current I at the unspecified time t. That is,

$$\int_{I_0}^{I}\frac{dI}{I} = -\frac{1}{(L/R)}\int_0^t dt \tag{33.56}$$

Both are common integrals, giving

$$\ln I \Big|_{I_0}^{I} = \ln I - \ln I_0 = \ln\left(\frac{I}{I_0}\right) = -\frac{t}{(L/R)} \tag{33.57}$$

We can solve for the current I by taking the exponential of both sides, then multiplying by I_0. Doing so gives I, the current as a function of time:

$$I = I_0 e^{-t/(L/R)} \tag{33.58}$$

Notice that $I = I_0$ at $t = 0$, as expected.

The argument of the exponential function must be dimensionless, so L/R must have dimensions of time. If we define the **time constant** τ of the *LR* circuit to be

$$\tau = \frac{L}{R} \tag{33.59}$$

FIGURE 33.45 An *LR* circuit.

(a)

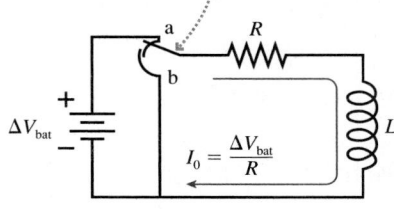

The switch has been in this position for a long time. At $t = 0$ it is moved to position b.

(b)

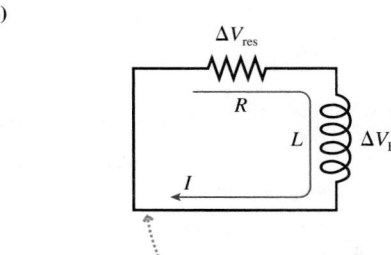

This is the circuit with the switch in position b. The inductor prevents the current from stopping instantly.

FIGURE 33.46 The current decay in an *LR* circuit.

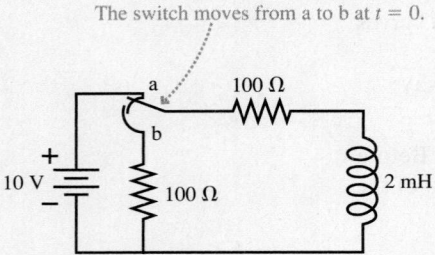

Current *I*

I_0

The current has decreased to 37% of its initial value at $t = \tau$.

The current has decreased to 13% of its initial value at $t = 2\tau$.

$0.50I_0$

$0.37I_0$

$0.13I_0$

0

0   τ 2τ 3τ *t*

then we can write Equation 33.58 as

$$I = I_0e^{-t/\tau} \qquad (33.60)$$

The time constant is the time at which the current has decreased to e^{-1} (about 37%) of its initial value. We can see this by computing the current at the time $t = \tau$:

$$I(\text{at } t = \tau) = I_0e^{-\tau/\tau} = e^{-1}I_0 = 0.37I_0 \qquad (33.61)$$

Thus the time constant for an *LR* circuit functions in exactly the same way as the time constant for the *RC* circuit we analyzed in Chapter 31. At time $t = 2\tau$, the current has decreased to $e^{-2}I_0$, or about 13% of its initial value.

The current is graphed in **FIGURE 33.46**. You can see that the current decays exponentially. The *shape* of the graph is always the same, regardless of the specific value of the time constant τ.

EXAMPLE 33.16 **Exponential decay in an *LR* circuit**

The switch in **FIGURE 33.47** has been in position a for a long time. It is changed to position b at $t = 0$ s.

a. What is the current in the circuit at $t = 5.0 \; \mu s$?
b. At what time has the current decayed to 1% of its initial value?

FIGURE 33.47 The *LR* circuit of Example 33.16.

The switch moves from a to b at $t = 0$.

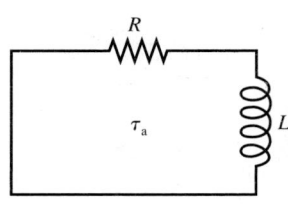

a
100 Ω
b
10 V
100 Ω
2 mH

MODEL This is an *LR* circuit. We'll assume ideal wires and an ideal inductor.

VISUALIZE The two resistors will be in series after the switch is thrown.

SOLVE Before the switch is thrown, while $\Delta V_L = 0$, the current is $I_0 = (10 \text{ V})/(100 \; \Omega) = 0.10 \text{ A} = 100 \text{ mA}$. This will be the initial current after the switch is thrown because the current through an

inductor can't change instantaneously. The circuit resistance after the switch is thrown is $R = 200 \; \Omega$, so the time constant is

$$\tau = \frac{L}{R} = \frac{2.0 \times 10^{-3} \text{ H}}{200 \; \Omega} = 1.0 \times 10^{-5} \text{ s} = 10 \; \mu s$$

a. The current at $t = 5.0 \; \mu s$ is

$$I = I_0e^{-t/\tau} = (100 \text{ mA})e^{-(5.0 \mu s)/(10 \mu s)} = 61 \text{ mA}$$

b. To find the time at which a particular current is reached we need to go back to Equation 33.57 and solve for *t*:

$$t = -\frac{L}{R}\ln\left(\frac{I}{I_0}\right) = -\tau\ln\left(\frac{I}{I_0}\right)$$

The time at which the current has decayed to 1 mA (1% of I_0) is

$$t = -(10 \; \mu s)\ln\left(\frac{1 \text{ mA}}{100 \text{ mA}}\right) = 46 \; \mu s$$

ASSESS For all practical purposes, the current has decayed away in $\approx 50 \; \mu s$. The inductance in this circuit is not large, so a short decay time is not surprising.

STOP TO THINK 33.7   Rank in order, from largest to smallest, the time constants τ_a, τ_b, and τ_c of these three circuits.

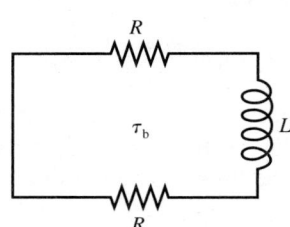

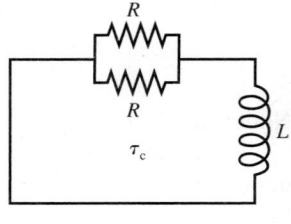

R
τ_a
L

R
τ_b
L
R

R
R
τ_c
L

Induction heating

Induction heating uses induced currents to heat metal objects to high temperatures for applications such as surface hardening, brazing, or even melting. To illustrate the idea, consider a copper wire formed into a 4.0 cm × 4.0 cm square loop and placed in a magnetic field—perpendicular to the plane of the loop—that oscillates with 0.010 T amplitude at a frequency of 1000 Hz. What is the wire's initial temperature rise, in °C/min?

MODEL The changing magnetic flux through the loop will induce a current that, because of the wire's resistance, will heat the wire. Eventually, when the wire gets hot, heat loss through radiation and/or convection will limit the temperature rise, but initially we can consider the temperature change due only to the heating by the current. Assume that the wire's diameter is much less than the 4.0 cm width of the loop.

VISUALIZE **FIGURE 33.48** shows the copper loop in the magnetic field. The wire's cross-section area A is unknown, but our assumption of a thin wire means that the loop has a well-defined area L^2. Values of copper's resistivity, density, and specific heat were taken from tables inside the back cover of the book. We've used subscripts to distinguish between mass density ρ_{mass} and resistivity ρ_{elec}, a potentially confusing duplication of symbols.

FIGURE 33.48 A copper wire being heated by induction.

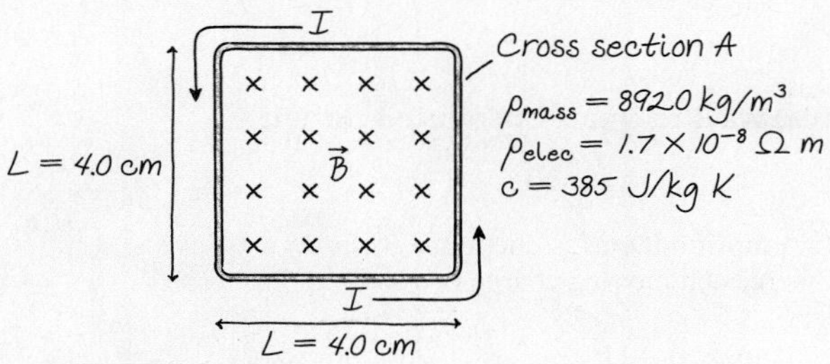

$\rho_{mass} = 8920 \text{ kg/m}^3$
$\rho_{elec} = 1.7 \times 10^{-8} \, \Omega \, \text{m}$
$c = 385 \text{ J/kg K}$
$L = 4.0 \text{ cm}$

SOLVE Power dissipation by a current, $P = I^2R$, heats the wire. As long as heat losses are negligible, we can use the heating rate and the wire's specific heat c to calculate the rate of temperature change. Our first task is to find the induced current. According to Faraday's law,

$$I = \frac{\mathcal{E}}{R} = -\frac{1}{R}\frac{d\Phi_m}{dt} = -\frac{L^2}{R}\frac{dB}{dt}$$

where R is the loop's resistance and $\Phi_m = L^2B$ is the magnetic flux through a loop of area L^2. The oscillating magnetic field can be written $B = B_0\cos\omega t$, with $B_0 = 0.010$ T and $\omega = 2\pi \times 1000$ Hz $= 6280$ rad/s. Thus

$$\frac{dB}{dt} = -\omega B_0\sin\omega t$$

from which we find that the induced current oscillates as

$$I = \frac{\omega B_0 L^2}{R}\sin\omega t$$

As the current oscillates, the power dissipation in the wire is

$$P = I^2R = \frac{\omega^2 B_0^2 L^4}{R}\sin^2\omega t$$

The power dissipation also oscillates, but very rapidly in comparison to a temperature rise that we expect to occur over seconds or minutes. Consequently, we are justified in replacing the oscillating P with its *average* value P_{avg}. Recall that the time average of the function $\sin^2\omega t$ is $\frac{1}{2}$, a result that can be proven by integration or justified by noticing that a graph of $\sin^2\omega t$ oscillates symmetrically between 0 and 1. Thus the average power dissipation in the wire is

$$P_{avg} = \frac{\omega^2 B_0^2 L^4}{2R}$$

Recall that power is the *rate* of energy transfer. In this case, the power dissipated in the wire is the wire's heating rate: $dQ/dt = P_{avg}$, where here Q is heat, not charge. Using $Q = mc\Delta T$, from thermodynamics, we can write

$$\frac{dQ}{dt} = mc\frac{dT}{dt} = P_{avg} = \frac{\omega^2 B_0^2 L^4}{2R}$$

To complete the calculation, we need the mass and resistance of the wire. The wire's total length is $4L$, and its cross-section area is A. Thus

$$m = \rho_{mass}V = 4\rho_{mass}LA$$

$$R = \frac{\rho_{elec}(4L)}{A} = \frac{4\rho_{elec}L}{A}$$

Substituting these into the heating equation, we have

$$4\rho_{mass}LAc\frac{dT}{dt} = \frac{\omega^2 B_0^2 L^3 A}{8\rho_{elec}}$$

Interestingly, the wire's cross-section area cancels. The wire's temperature initially increases at the rate

$$\frac{dT}{dt} = \frac{\omega^2 B_0^2 L^2}{32\rho_{elec}\rho_{mass}c}$$

All the terms on the right-hand side are known. Evaluating, we find

$$\frac{dT}{dt} = 3.3 \text{ K/s} = 200°\text{C/min}$$

ASSESS This is a rapid but realistic temperature rise for a small object, although the rate of increase will slow as the object begins losing heat to the environment through radiation and/or convection. Induction heating can increase an object's temperature by several hundred degrees in a few minutes.

SUMMARY

The goal of Chapter 33 has been to understand and apply electromagnetic induction.

General Principles

Faraday's Law

MODEL Make simplifying assumptions.

VISUALIZE Use Lenz's law to determine the direction of the **induced current.**

SOLVE The **induced emf** is

$$\mathcal{E} = \left| \frac{d\Phi_m}{dt} \right|$$

Multiply by N for an N-turn coil.
The size of the induced current is $I = \mathcal{E}/R$.

ASSESS Is the result reasonable?

Lenz's Law

There is an induced current in a closed conducting loop if and only if the magnetic flux through the loop is changing. The direction of the induced current is such that the induced magnetic field opposes the *change* in the flux.

Magnetic flux

Magnetic flux measures the amount of magnetic field passing through a surface.

$$\Phi_m = \vec{A} \cdot \vec{B} = AB\cos\theta$$

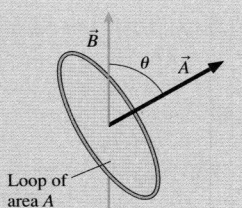

Loop of area A

Important Concepts

Three ways to change the flux

1. A loop moves into or out of a magnetic field.

2. The loop changes area or rotates.

3. The magnetic field through the loop increases or decreases.

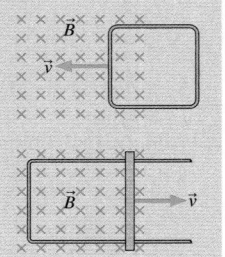

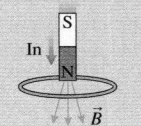

Two ways to create an induced current

1. A **motional emf** is due to magnetic forces on moving charge carriers.

2. An induced electric field is due to a changing magnetic field.

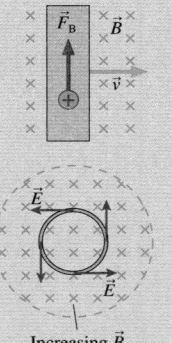

Increasing $\vec{B}$

Applications

Inductors

Solenoid inductance $L_{\text{solenoid}} = \dfrac{\mu_0 N^2 A}{l}$

Potential difference $\Delta V_L = -L\dfrac{dI}{dt}$

Energy stored $U_L = \frac{1}{2}LI^2$

Magnetic energy density $u_B = \dfrac{1}{2\mu_0}B^2$

LC circuit

Oscillates at $\omega = \sqrt{\dfrac{1}{LC}}$

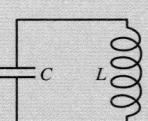

LR circuit

Exponential change with $\tau = \dfrac{L}{R}$

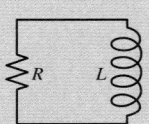

Terms and Notation

electromagnetic induction	area vector, $\vec{A}$	induced magnetic field	inductor
induced current	Lenz's law	electromagnetic wave	*LC* circuit
motional emf	induced emf, $\mathcal{E}$	primary coil	*LR* circuit
generator	Faraday's law	secondary coil	time constant, τ
eddy current	induced electric field	transformer	
magnetic flux, Φ_m	Coulomb electric field	inductance, L	
weber, Wb	non-Coulomb electric field	henry, H	

CONCEPTUAL QUESTIONS

1. What is the direction of the induced current in FIGURE Q33.1?

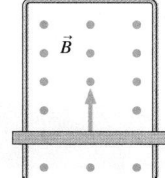

FIGURE Q33.1

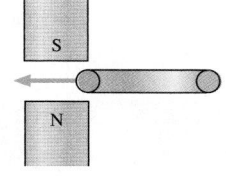

FIGURE Q33.2

2. You want to insert a loop of copper wire between the two permanent magnets in FIGURE Q33.2. Is there an attractive magnetic force that tends to *pull* the loop in, like a magnet pulls on a paper clip? Or do you need to *push* the loop in against a repulsive force? Explain.

3. A vertical, rectangular loop of copper wire is half in and half out of the horizontal magnetic field in FIGURE Q33.3. (The field is zero beneath the dashed line.) The loop is released and starts to fall. Is there a net magnetic force on the loop? If so, in which direction? Explain.

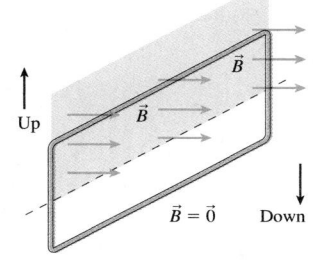

FIGURE Q33.3

4. Does the loop of wire in FIGURE Q33.4 have a clockwise current, a counterclockwise current, or no current under the following circumstances? Explain.
   a. The magnetic field points out of the page and is increasing.
   b. The magnetic field points out of the page and is constant.
   c. The magnetic field points out of the page and is decreasing.

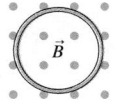

FIGURE Q33.4

FIGURE Q33.5

5. The two loops of wire in FIGURE Q33.5 are stacked one above the other. Does the upper loop have a clockwise current, a counterclockwise current, or no current at the following times? Explain.
   a. Before the switch is closed.
   b. Immediately after the switch is closed.
   c. Long after the switch is closed.
   d. Immediately after the switch is reopened.

6. FIGURE Q33.6 shows a bar magnet being pushed toward a conducting loop from below, along the axis of the loop.
   a. What is the current direction in the loop? Explain.
   b. Is there a magnetic force on the loop? If so, in which direction? Explain.
   Hint: A current loop is a magnetic dipole.
   c. Is there a force on the magnet? If so, in which direction?

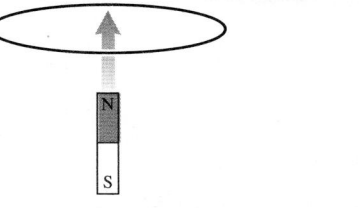

FIGURE Q33.6

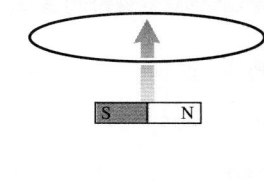

FIGURE Q33.7

7. A bar magnet is pushed toward a loop of wire as shown in FIGURE Q33.7. Is there a current in the loop? If so, in which direction? If not, why not?

8. FIGURE Q33.8 shows a bar magnet, a coil of wire, and a current meter. Is the current through the meter right to left, left to right, or zero for the following circumstances? Explain.
   a. The magnet is inserted into the coil.
   b. The magnet is held at rest inside the coil.
   c. The magnet is withdrawn from the left side of the coil.

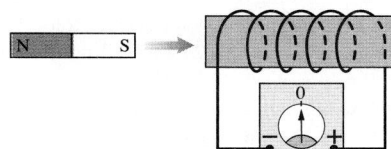

FIGURE Q33.8

9. Is the magnetic field strength in FIGURE Q33.9 increasing, decreasing, or steady? Explain.

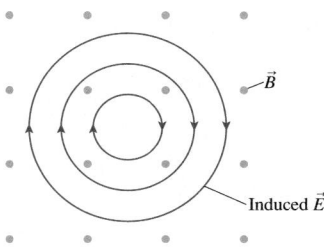

FIGURE Q33.9

10. An inductor with a 2.0 A current stores energy. At what current will the stored energy be twice as large?

11. a. Can you tell which of the inductors in FIGURE Q33.11 has the larger current through it? If so, which one? Explain.
    b. Can you tell through which inductor the current is changing more rapidly? If so, which one? Explain.
    c. If the current enters the inductor from the bottom, can you tell if the current is increasing, decreasing, or staying the same? If so, which? Explain.

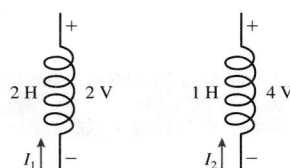

FIGURE Q33.11

12. An LC circuit oscillates at a frequency of 2000 Hz. What will the frequency be if the inductance is quadrupled?

13. Rank in order, from largest to smallest, the three time constants τ_a to τ_c for the three circuits in FIGURE Q33.13. Explain.

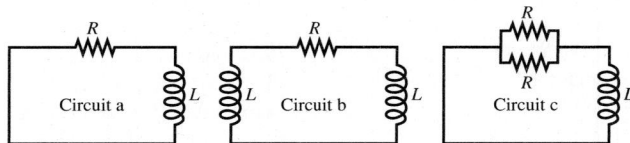

FIGURE Q33.13

14. For the circuit of FIGURE Q33.14:
    a. What is the battery current immediately after the switch closes? Explain.
    b. What is the battery current after the switch has been closed a long time? Explain.

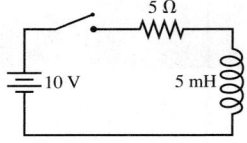

FIGURE Q33.14

EXERCISES AND PROBLEMS

Problems labled ▨ integrate material from earlier chapters.

Exercises

Section 33.2 Motional emf

1. | The earth's magnetic field strength is 5.0×10^{-5} T. How fast would you have to drive your car to create a 1.0 V motional emf along your 1.0-m-long radio antenna? Assume that the motion of the antenna is perpendicular to $\vec{B}$.

2. | A potential difference of 0.050 V is developed across the 10-cm-long wire of FIGURE EX33.2 as it moves through a magnetic field perpendicular to the page. What are the strength and direction (in or out) of the magnetic field?

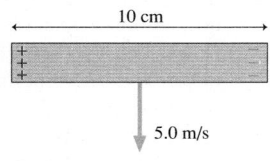

FIGURE EX33.2

3. ‖ A 10-cm-long wire is pulled along a U-shaped conducting rail in a perpendicular magnetic field. The total resistance of the wire and rail is 0.20 Ω. Pulling the wire at a steady speed of 4.0 m/s causes 4.0 W of power to be dissipated in the circuit.
   a. How big is the pulling force?
   b. What is the strength of the magnetic field?

Section 33.3 Magnetic Flux

4. | What is the magnetic flux through the loop shown in FIGURE EX33.4?

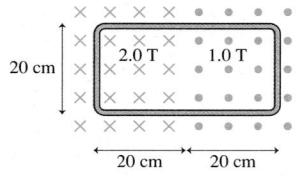

FIGURE EX33.4

5. ‖ FIGURE EX33.5 shows a 2.0-cm-diameter solenoid passing through the center of a 6.0-cm-diameter loop. The magnetic field inside the solenoid is 0.20 T. What is the magnetic flux through the loop when it is perpendicular to the solenoid and when it is tilted at a 60° angle?

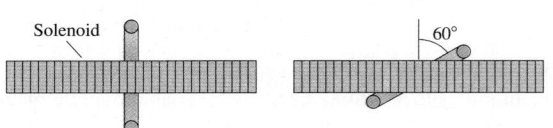

FIGURE EX33.5

6. ‖ What is the magnetic flux through the loop shown in FIGURE EX33.6?

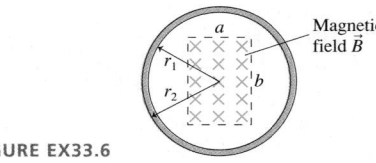

FIGURE EX33.6

Section 33.4 Lenz's Law

7. | There is a cw induced current in the conducting loop shown in FIGURE EX33.7. Is the magnetic field inside the loop increasing in strength, decreasing in strength, or steady?

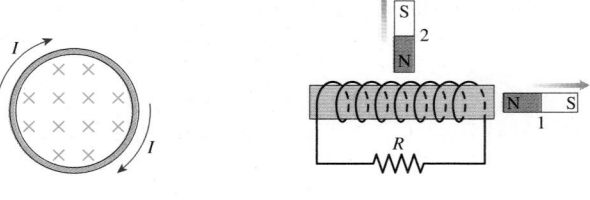

FIGURE EX33.7   FIGURE EX33.8

8. | A solenoid is wound as shown in FIGURE EX33.8.
   a. Is there an induced current as magnet 1 is moved away from the solenoid? If so, what is the current direction through resistor R?
   b. Is there an induced current as magnet 2 is moved away from the solenoid? If so, what is the current direction through resistor R?

9. ‖ The current in the solenoid of FIGURE EX33.9 is increasing. The solenoid is surrounded by a conducting loop. Is there a current in the loop? If so, is the loop current cw or ccw?

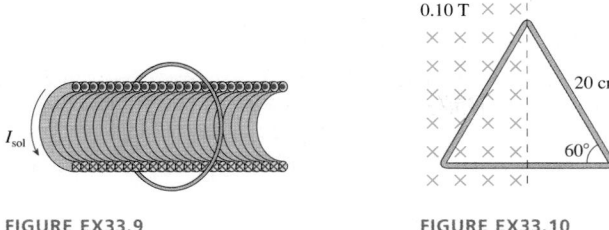

FIGURE EX33.9

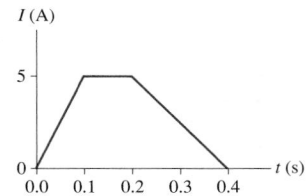

FIGURE EX33.10

10. | The metal equilateral triangle in FIGURE EX33.10, 20 cm on each side, is halfway into a 0.10 T magnetic field.
   a. What is the magnetic flux through the triangle?
   b. If the magnetic field strength decreases, what is the direction of the induced current in the triangle?

Section 33.5 Faraday's Law

11. | FIGURE EX33.11 shows a 10-cm-diameter loop in three different magnetic fields. The loop's resistance is 0.20 Ω. For each, what are the size and direction of the induced current?

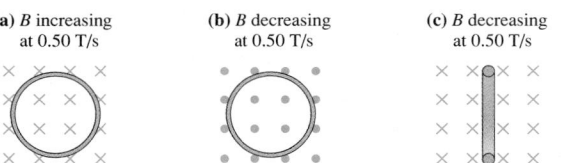

FIGURE EX33.11

12. | The loop in FIGURE EX33.12 is being pushed into the 0.20 T magnetic field at 50 m/s. The resistance of the loop is 0.10 Ω. What are the direction and the magnitude of the current in the loop?

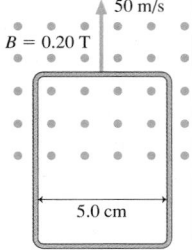

FIGURE EX33.12

13. ‖ A 1000-turn coil of wire 1.0 cm in diameter is in a magnetic field that increases from 0.10 T to 0.30 T in 10 ms. The axis of the coil is parallel to the field. What is the emf of the coil?

14. | The resistance of the loop in FIGURE EX33.14 is 0.20 Ω. Is the magnetic field strength increasing or decreasing? At what rate (T/s)?

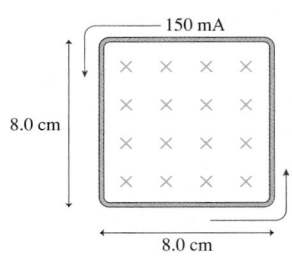

FIGURE EX33.14

Section 33.6 Induced Fields

15. ‖ FIGURE EX33.15 shows the current as a function of time through a 20-cm-long, 4.0-cm-diameter solenoid with 400 turns. Draw a graph of the induced electric field strength as a function of time at a point 1.0 cm from the axis of the solenoid.

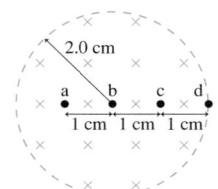

FIGURE EX33.15

16. ‖ The magnetic field inside a 5.0-cm-diameter solenoid is 2.0 T and decreasing at 4.0 T/s. What is the electric field strength inside the solenoid at a point (a) on the axis and (b) 2.0 cm from the axis?

17. ‖ The magnetic field in FIGURE EX33.17 is decreasing at the rate 0.10 T/s. What is the acceleration (magnitude and direction) of a proton initially at rest at points a to d?

FIGURE EX33.17

Section 33.8 Inductors

18. | What is the potential difference across a 10 mH inductor if the current through the inductor drops from 150 mA to 50 mA in 10 μs? What is the direction of this potential difference? That is, does the potential increase or decrease along the direction of the current?

19. | The maximum allowable potential difference across a 200 mH inductor is 400 V. You need to raise the current through the inductor from 1.0 A to 3.0 A. What is the minimum time you should allow for changing the current?

20. | A 100 mH inductor whose windings have a resistance of 4.0 Ω is connected across a 12 V battery having an internal resistance of 2.0 Ω. How much energy is stored in the inductor?

21. ‖ How much energy is stored in a 3.0-cm-diameter, 12-cm-long solenoid that has 200 turns of wire and carries a current of 0.80 A?

Section 33.9 *LC* Circuits

22. ‖ An FM radio station broadcasts at a frequency of 100 MHz. What inductance should be paired with a 10 pF capacitor to build a receiver circuit for this station?

23. | A 2.0 mH inductor is connected in parallel with a variable capacitor. The capacitor can be varied from 100 pF to 200 pF. What is the range of oscillation frequencies for this circuit?

24. | An MRI machine needs to detect signals that oscillate at BIO very high frequencies. It does so with an *LC* circuit containing a 15 mH coil. To what value should the capacitance be set to detect a 450 MHz signal?

Section 33.10 *LR* Circuits

25. ‖ What value of resistor R gives the circuit in FIGURE EX33.25 a time constant of 25 μs?

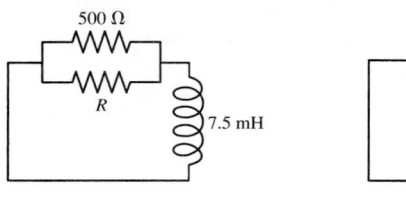

FIGURE EX33.25

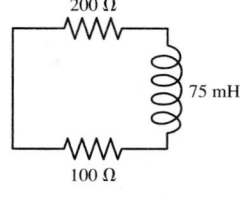

FIGURE EX33.26

26. ‖ At $t = 0$ s, the current in the circuit in FIGURE EX33.26 is I_0. At what time is the current $\frac{1}{2}I_0$?

Problems

27. ‖ FIGURE P33.27 shows a 10 cm × 10 cm square bent at a 90° angle. A uniform 0.050 T magnetic field points downward at a 45° angle. What is the magnetic flux through the loop?

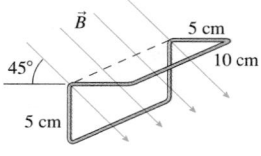

FIGURE P33.27

28. ‖ A 5.0-cm-diameter coil has 20 turns and a resistance of 0.50 Ω. A magnetic field perpendicular to the coil is $B = 0.020t + 0.010t^2$, where B is in tesla and t is in seconds.
    a. Find an expression for the induced current $I(t)$ as a function of time.
    b. Evaluate I at $t = 5$ s and $t = 10$ s.

29. ‖ A 20 cm × 20 cm square loop has a resistance of 0.10 Ω. A magnetic field perpendicular to the loop is $B = 4t - 2t^2$, where B is in tesla and t is in seconds. What is the current in the loop at $t = 0.0$ s, $t = 1.0$ s, and $t = 2.0$ s?

30. ‖ A 100-turn, 2.0-cm-diameter coil is at rest in a horizontal plane. A uniform magnetic field 60° away from vertical increases from 0.50 T to 1.50 T in 0.60 s. What is the induced emf in the coil?

31. ‖ A 100-turn, 8.0-cm-diameter coil is made of 0.50-mm-diameter copper wire. A magnetic field is parallel to the axis of the coil. At what rate must B increase to induce a 2.0 A current in the coil?

32. ‖ A circular loop made from a flexible, conducting wire is shrinking. Its radius as a function of time is $r = r_0 e^{-\beta t}$. The loop is perpendicular to a steady, uniform magnetic field B. Find an expression for the induced emf in the loop at time t.

33. ‖ A 10 cm × 10 cm square loop lies in the xy-plane. The magnetic field in this region of space is $B = (0.30t\,\hat{\imath} + 0.50t^2\,\hat{k})$ T, where t is in s. What is the emf induced in the loop at (a) $t = 0.5$ s and (b) $t = 1.0$ s?

34. ‖ A 20 cm × 20 cm square loop of wire lies in the xy-plane with its bottom edge on the x-axis. The resistance of the loop is 0.50 Ω. A magnetic field parallel to the z-axis is given by $B = 0.80y^2t$, where B is in tesla, y in meters, and t in seconds. What is the size of the induced current in the loop at $t = 0.50$ s?

35. ‖ A 2.0 cm × 2.0 cm square loop of wire with resistance 0.010 Ω has one edge parallel to a long straight wire. The near edge of the loop is 1.0 cm from the wire. The current in the wire is increasing at the rate of 100 A/s. What is the current in the loop?

36. ‖ The rectangular loop in FIGURE P33.36 has 0.020 Ω resistance. What is the induced current in the loop at this instant?

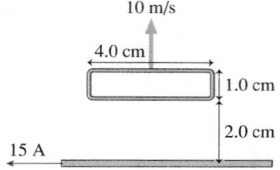

FIGURE P33.36

37. ‖ FIGURE P33.37 shows a 4.0-cm-diameter loop with resistance 0.10 Ω around a 2.0-cm-diameter solenoid. The solenoid is 10 cm long, has 100 turns, and carries the current shown in the graph. A positive current is cw when seen from the left. Find the current in the loop at (a) $t = 0.5$ s, (b) $t = 1.5$ s, and (c) $t = 2.5$ s.

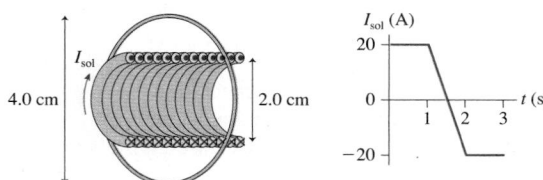

FIGURE P33.37

38. ‖ FIGURE P33.38 shows a 1.0-cm-diameter loop with $R = 0.50$ Ω inside a 2.0-cm-diameter solenoid. The solenoid is 8.0 cm long, has 120 turns, and carries the current shown in the graph. A positive current is cw when seen from the left. Determine the current in the loop at $t = 0.010$ s.

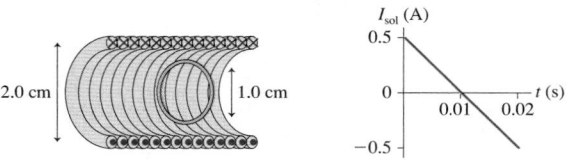

FIGURE P33.38

39. ‖ FIGURE P33.39 shows two 20-turn coils tightly wrapped on the same 2.0-cm-diameter cylinder with 1.0-mm-diameter wire. The current through coil 1 is shown in the graph. Determine the current in coil 2 at (a) $t = 0.05$ s and (b) $t = 0.25$ s. A positive current is into the page at the top of a loop. Assume that the magnetic field of coil 1 passes entirely through coil 2.

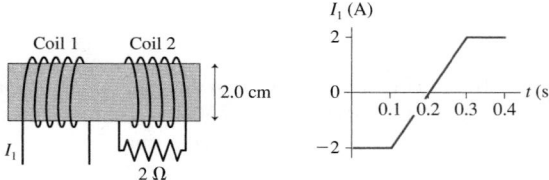

FIGURE P33.39

40. ‖ A 50-turn, 4.0-cm-diameter coil with $R = 0.50$ Ω surrounds a 2.0-cm-diameter solenoid. The solenoid is 20 cm long and has 200 turns. The 60 Hz current through the solenoid is $I_{sol} = (0.50$ A$)\sin(2\pi ft)$. Find an expression for I_{coil}, the induced current in the coil as a function of time.

41. ‖ A loop antenna, such as is used on older televisions to pick up UHF broadcasts, is 25 cm in diameter. The plane of the loop is perpendicular to the oscillating magnetic field of a 150 MHz electromagnetic wave. The magnetic field through the loop is $B = (20$ nT$)\sin \omega t$.

a. What is the maximum emf induced in the antenna?

b. What is the maximum emf if the loop is turned 90° to be perpendicular to the oscillating electric field?

42. ‖ A 40-turn, 4.0-cm-diameter coil with $R = 0.40\ \Omega$ surrounds a 3.0-cm-diameter solenoid. The solenoid is 20 cm long and has 200 turns. The 60 Hz current through the solenoid is $I = I_0 \sin(2\pi ft)$. What is I_0 if the maximum induced current in the coil is 0.20 A?

43. | Electricity is distributed from electrical substations to neighborhoods at 15,000 V. This is a 60 Hz oscillating (AC) voltage. Neighborhood transformers, seen on utility poles, step this voltage down to the 120 V that is delivered to your house.

a. How many turns does the primary coil on the transformer have if the secondary coil has 100 turns?

b. No energy is lost in an ideal transformer, so the output power P_{out} from the secondary coil equals the input power P_{in} to the primary coil. Suppose a neighborhood transformer delivers 250 A at 120 V. What is the current in the 15,000 V line from the substation?

44. ‖ A small, 2.0-mm-diameter circular loop with $R = 0.020\ \Omega$ is at the center of a large 100-mm-diameter circular loop. Both loops lie in the same plane. The current in the outer loop changes from +1.0 A to −1.0 A in 0.10 s. What is the induced current in the inner loop?

45. ‖ The square loop shown in FIGURE P33.45 moves into a 0.80 T magnetic field at a constant speed of 10 m/s. The loop has a resistance of 0.10 Ω, and it enters the field at $t = 0$ s.

a. Find the induced current in the loop as a function of time. Give your answer as a graph of I versus t from $t = 0$ s to $t = 0.020$ s.

b. What is the maximum current? What is the position of the loop when the current is maximum?

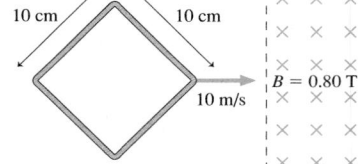

FIGURE P33.45

46. ‖ The L-shaped conductor in FIGURE P33.46 moves at 10 m/s across a stationary L-shaped conductor in a 0.10 T magnetic field. The two vertices overlap, so that the enclosed area is zero, at $t = 0$ s. The conductor has a resistance of 0.010 ohms *per meter.*

a. What is the direction of the induced current?

b. Find expressions for the induced emf and the induced current as functions of time.

c. Evaluate $\mathcal{E}$ and I at $t = 0.10$ s.

FIGURE P33.46

47. ‖ A 4.0-cm-long slide wire moves outward with a speed of 100 m/s in a 1.0 T magnetic field. (See Figure 33.26.) At the instant the circuit forms a 4.0 cm × 4.0 cm square, with $R = 0.010\ \Omega$ on each side, what are

a. The induced emf?

b. The induced current?

c. The potential difference between the two ends of the moving wire?

48. ‖ A 20-cm-long, zero-resistance slide wire moves outward, on zero-resistance rails, at a steady speed of 10 m/s in a 0.10 T magnetic field. (See Figure 33.26.) On the opposite side, a 1.0 Ω carbon resistor completes the circuit by connecting the two rails. The mass of the resistor is 50 mg.

a. What is the induced current in the circuit?

b. How much force is needed to pull the wire at this speed?

c. If the wire is pulled for 10 s, what is the temperature increase of the carbon? The specific heat of carbon is 710 J/kg K.

49. ‖ Your camping buddy has an idea for a light to go inside your tent. He happens to have a powerful (and heavy!) horseshoe magnet that he bought at a surplus store. This magnet creates a 0.20 T field between two pole tips 10 cm apart. His idea is to build the hand-cranked generator shown in FIGURE P33.49. He thinks you can make enough current to fully light a 1.0 Ω lightbulb rated at 4.0 W. That's not super bright, but it should be plenty of light for routine activities in the tent.

a. Find an expression for the induced current as a function of time if you turn the crank at frequency f. Assume that the semicircle is at its highest point at $t = 0$ s.

b. With what frequency will you have to turn the crank for the maximum current to fully light the bulb? Is this feasible?

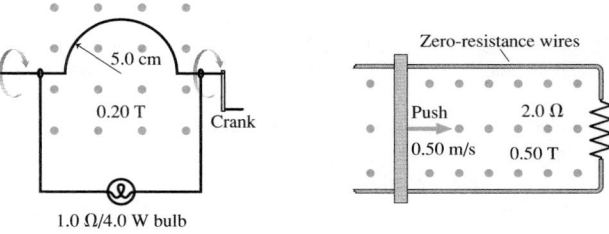

FIGURE P33.49   FIGURE P33.50

50. | The 10-cm-wide, zero-resistance slide wire shown in FIGURE P33.50 is pushed toward the 2.0 Ω resistor at a steady speed of 0.50 m/s. The magnetic field strength is 0.50 T.

a. How big is the pushing force?

b. How much power does the pushing force supply to the wire?

c. What are the direction and magnitude of the induced current?

d. How much power is dissipated in the resistor?

51. ‖ One way to determine a magnetic field strength is to measure the emf induced in a rotating coil. To calibrate a large magnet in your laboratory, you attach a 2.0-cm-diameter, 100-turn coil to the end of a motor-driven shaft, place the coil between the pole tips of the magnet, and rotate it at different frequencies. The emf oscillates, so you use a voltmeter that measures its amplitude. The table shows your data:

Frequency (Hz)	Voltage (mV)
10	380
15	610
20	780
25	1020
30	1160

Use an appropriate graph of the data to determine the magnetic field strength.

52. ‖ You've decided to make the magnetic projectile launcher shown in FIGURE P33.52 for your science project. An aluminum

bar of length l slides along metal rails through a magnetic field B. The switch closes at $t = 0$ s, while the bar is at rest, and a battery of emf $\mathcal{E}_{bat}$ starts a current flowing around the loop. The battery has internal resistance r. The resistance of the rails and the bar are effectively zero.

a. Show that the bar reaches a terminal speed v_{term}, and find an expression for v_{term}.

b. Evaluate v_{term} for $\mathcal{E}_{bat} = 1.0$ V, $r = 0.10\ \Omega$, $l = 6.0$ cm, and $B = 0.50$ T.

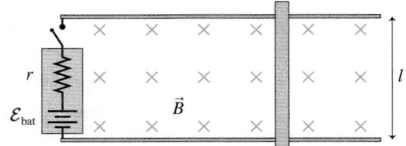

FIGURE P33.52

53. ||| A slide wire of length l, mass m, and resistance R slides down a U-shaped metal track that is tilted upward at angle θ. The track has zero resistance and no friction. A vertical magnetic field B fills the loop formed by the track and the slide wire.

a. Find an expression for the induced current I when the slide wire moves at speed v.

b. Show that the slide wire reaches a terminal speed v_{term}, and find an expression for v_{term}.

54. || **FIGURE P33.54** shows a U-shaped conducting rail that is oriented vertically in a horizontal magnetic field. The rail has no electric resistance and does not move. A slide wire with mass m and resistance R can slide up and down without friction while maintaining electrical contact with the rail. The slide wire is released from rest.

a. Show that the slide wire reaches a terminal speed v_{term}, and find an expression for v_{term}.

b. Determine the value of v_{term} if $l = 20$ cm, $m = 10$ g, $R = 0.10\ \Omega$, and $B = 0.50$ T.

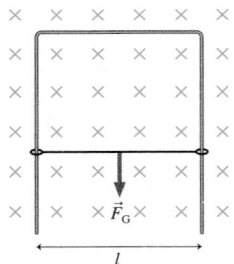

FIGURE P33.54   **FIGURE P33.55**

55. || Experiments to study vision often need to track the movements
BIO of a subject's eye. One way of doing so is to have the subject sit in a magnetic field while wearing special contact lenses with a coil of very fine wire circling the edge. A current is induced in the coil each time the subject rotates his eye. Consider the experiment of **FIGURE P33.55** in which a 20-turn, 6.0-mm-diameter coil of wire circles the subject's cornea while a 1.0 T magnetic field is directed as shown. The subject begins by looking straight ahead. What emf is induced in the coil if the subject shifts his gaze by 5° in 0.20 s?

56. || A 10-turn coil of wire having a diameter of 1.0 cm and a resistance of 0.20 Ω is in a 1.0 mT magnetic field, with the coil oriented for maximum flux. The coil is connected to an uncharged 1.0 μF capacitor rather than to a current meter. The coil is quickly pulled out of the magnetic field. Afterward, what is the voltage across the capacitor?

Hint: Use $I = dq/dt$ to relate the *net* change of flux to the amount of charge that flows to the capacitor.

57. ||| The magnetic field at one place on the earth's surface is 55 μT in strength and tilted 60° down from horizontal. A 200-turn coil having a diameter of 4.0 cm and a resistance of 2.0 Ω is connected to a 1.0 μF capacitor rather than to a current meter. The coil is held in a horizontal plane and the capacitor is discharged. Then the coil is quickly rotated 180° so that the side that had been facing up is now facing down. Afterward, what is the voltage across the capacitor? See the Hint in Problem 56.

58. || The magnetic field inside a 4.0-cm-diameter superconducting solenoid varies sinusoidally between 8.0 T and 12.0 T at a frequency of 10 Hz.

a. What is the maximum electric field strength at a point 1.5 cm from the solenoid axis?

b. What is the value of B at the instant E reaches its maximum value?

59. || Equation 33.26 is an expression for the induced electric field inside a solenoid ($r < R$). Find an expression for the induced electric field outside a solenoid ($r > R$) in which the magnetic field is changing at the rate dB/dt.

60. || A solenoid inductor has an emf of 0.20 V when the current through it changes at the rate 10.0 A/s. A steady current of 0.10 A produces a flux of 5.0 μWb per turn. How many turns does the inductor have?

61. || a. What is the magnetic energy density at the center of a 4.0-cm-diameter loop carrying a current of 1.0 A?

b. What current in a straight wire gives the magnetic energy density you found in part a at a point 2.0 cm from the wire?

62. | MRI (magnetic resonance imaging) is a medical technique
BIO that produces detailed "pictures" of the interior of the body. The patient is placed into a solenoid that is 40 cm in diameter and 1.0 m long. A 100 A current creates a 5.0 T magnetic field inside the solenoid. To carry such a large current, the solenoid wires are cooled with liquid helium until they become superconducting (no electric resistance).

a. How much magnetic energy is stored in the solenoid? Assume that the magnetic field is uniform within the solenoid and quickly drops to zero outside the solenoid.

b. How many turns of wire does the solenoid have?

63. | One possible concern with MRI (see Problem 62) is turning
BIO the magnetic field on or off too quickly. Bodily fluids are conductors, and a changing magnetic field could cause electric currents to flow through the patient. Suppose a typical patient has a maximum cross-section area of 0.060 m². What is the smallest time interval in which a 5.0 T magnetic field can be turned on or off if the induced emf around the patient's body must be kept to less than 0.10 V?

64. || **FIGURE P33.64** shows the current through a 10 mH inductor. Draw a graph showing the potential difference ΔV_L across the inductor for these 6 ms.

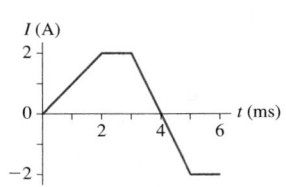   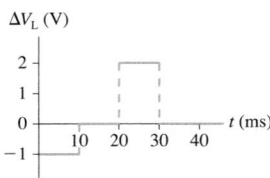

FIGURE P33.64   **FIGURE P33.65**

65. || **FIGURE P33.65** shows the potential difference across a 50 mH inductor. The current through the inductor at $t = 0$ s is 0.20 A.

Draw a graph showing the current through the inductor from $t = 0$ s to $t = 40$ ms.

66. ‖ The current through inductance L is given by $I = I_0 \sin \omega t$.
   a. Find an expression for the potential difference ΔV_L across the inductor.
   b. The maximum voltage across the inductor is 0.20 V when $L = 50 \ \mu H$ and $f = 500$ kHz. What is I_0?

67. ‖ The current through inductance L is given by $I = I_0 e^{-t/\tau}$.
   a. Find an expression for the potential difference ΔV_L across the inductor.
   b. Evaluate ΔV_L at $t = 0$, 1, 2, and 3 ms if $L = 20$ mH, $I_0 = 50$ mA, and $\tau = 1.0$ ms.

68. ‖ An LC circuit is built with a 20 mH inductor and an 8.0 pF capacitor. The capacitor voltage has its maximum value of 25 V at $t = 0$ s.
   a. How long is it until the capacitor is first fully discharged?
   b. What is the inductor current at that time?

69. ‖ An LC circuit has a 10 mH inductor. The current has its maximum value of 0.60 A at $t = 0$ s. A short time later the capacitor reaches its maximum potential difference of 60 V. What is the value of the capacitance?

70. ‖ An electric oscillator is made with a 0.10 μF capacitor and a 1.0 mH inductor. The capacitor is initially charged to 5.0 V. What is the maximum current through the inductor as the circuit oscillates?

71. ‖‖ In recent years it has been possible to buy a 1.0 F capacitor. This is an enormously large amount of capacitance. Suppose you want to build a 1.0 Hz oscillator with a 1.0 F capacitor. You have a spool of 0.25-mm-diameter wire and a 4.0-cm-diameter plastic cylinder. How long must your inductor be if you wrap it with 2 layers of closely spaced turns?

72. ‖ For your final exam in electronics, you're asked to build an LC circuit that oscillates at 10 kHz. In addition, the maximum current must be 0.10 A and the maximum energy stored in the capacitor must be 1.0×10^{-5} J. What values of inductance and capacitance must you use?

73. ‖ The switch in FIGURE P33.73 has been in position 1 for a long time. It is changed to position 2 at $t = 0$ s.
   a. What is the maximum current through the inductor?
   b. What is the first time at which the current is maximum?

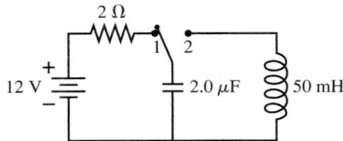

FIGURE P33.73

74. ‖ The 300 μF capacitor in FIGURE P33.74 is initially charged to 100 V, the 1200 μF capacitor is uncharged, and the switches are both open.
   a. What is the maximum voltage to which you can charge the 1200 μF capacitor by the proper closing and opening of the two switches?
   b. How would you do it? Describe the sequence in which you would close and open switches and the times at which you would do so. The first switch is closed at $t = 0$ s.

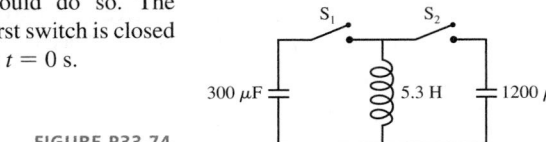

FIGURE P33.74

75. | The switch in FIGURE P33.75 has been open for a long time. It is closed at $t = 0$ s.
   a. What is the current through the battery immediately after the switch is closed?
   b. What is the current through the battery after the switch has been closed a long time?

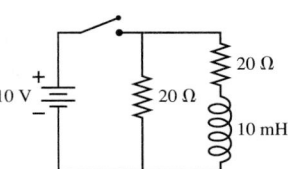

FIGURE P33.75

76. ‖ The switch in FIGURE P33.76 has been open for a long time. It is closed at $t = 0$ s. What is the current through the 20 Ω resistor
   a. immediately after the switch is closed?
   b. after the switch has been closed a long time?
   c. immediately after the switch is reopened?

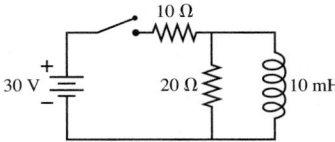

FIGURE P33.76

77. ‖ The switch in FIGURE P33.77 has been open for a long time. It is closed at $t = 0$ s.
   a. After the switch has been closed for a long time, what is the current in the circuit? Call this current I_0.
   b. Find an expression for the current I as a function of time. Write your expression in terms of I_0, R, and L.
   c. Sketch a current-versus-time graph from $t = 0$ s until the current is no longer changing.

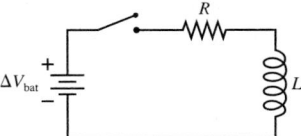

FIGURE P33.77

78. ‖ To determine the inductance of an unmarked inductor, you set up the circuit shown in FIGURE P33.78. After moving the switch from a to b at $t = 0$ s, you monitor the resistor voltage with an oscilloscope. Your data are as follows:

Time (μs)	Voltage (V)
0	9.0
10	6.7
20	4.6
30	3.2
40	2.5

Use an appropriate graph of the data to determine the inductance.

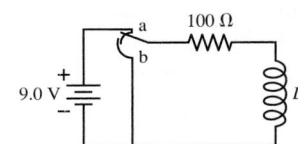

FIGURE P33.78

Challenge Problems

79. The metal wire in FIGURE CP33.79 moves with speed v parallel to a straight wire that is carrying current I. The distance between the two wires is d. Find an expression for the potential difference between the two ends of the moving wire.

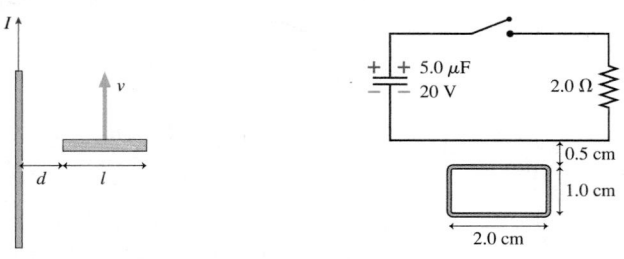

FIGURE CP33.79   FIGURE CP33.80

80. A rectangular metal loop with 0.050 Ω resistance is placed next to one wire of the RC circuit shown in FIGURE CP33.80. The capacitor is charged to 20 V with the polarity shown, then the switch is closed at $t = 0$ s.
    a. What is the direction of current in the loop for $t > 0$ s?
    b. What is the current in the loop at $t = 5.0$ μs? Assume that only the circuit wire next to the loop is close enough to produce a significant magnetic field.

81. A closed, square loop is formed with 40 cm of wire having $R = 0.10$ Ω, as shown in FIGURE CP33.81. A 0.50 T magnetic field is perpendicular to the loop. At $t = 0$ s, two diagonally opposite corners of the loop begin to move apart at 0.293 m/s.
    a. How long does it take the loop to collapse to a straight line?
    b. Find an expression for the induced current I as a function of time while the loop is collapsing. Assume that the sides remain straight lines during the collapse.
    c. Evaluate I at four or five times during the collapse, then draw a graph of I versus t.

FIGURE CP33.81

82. Let's look at the details of eddy-current braking. A square loop, length l on each side, is shot with velocity v_0 into a uniform magnetic field B. The field is perpendicular to the plane of the loop. The loop has mass m and resistance R, and it enters the field at $t = 0$ s. Assume that the loop is moving to the right along the x-axis and that the field begins at $x = 0$ m.
    a. Find an expression for the loop's velocity as a function of time as it enters the magnetic field. You can ignore gravity, and you can assume that the back edge of the loop has not entered the field.
    b. Calculate and draw a graph of v over the interval 0 s $\leq t \leq 0.04$ s for the case that $v_0 = 10$ m/s, $l = 10$ cm, $m = 1.0$ g, $R = 0.0010$ Ω, and $B = 0.10$ T. The back edge of the loop does not reach the field during this time interval.

83. An 8.0 cm × 8.0 cm square loop is halfway into a magnetic field perpendicular to the plane of the loop. The loop's mass is 10 g and its resistance is 0.010 Ω. A switch is closed at $t = 0$ s, causing the magnetic field to increase from 0 to 1.0 T in 0.010 s.
    a. What is the induced current in the square loop?
    b. With what speed is the loop "kicked" away from the magnetic field?
    Hint: What is the impulse on the loop?

84. A 2.0-cm-diameter solenoid is wrapped with 1000 turns per meter. 0.50 cm from the axis, the strength of an induced electric field is 5.0×10^{-4} V/m. What is the rate dI/dt with which the current through the solenoid is changing?

85. High-frequency signals are often transmitted along a *coaxial cable*, such as the one shown in FIGURE CP33.85. For example, the cable TV hookup coming into your home is a coaxial cable. The signal is carried on a wire of radius r_1 while the outer conductor of radius r_2 is grounded. A soft, flexible insulating material fills the space between them, and an insulating plastic coating goes around the outside.
    a. Find an expression for the inductance per meter of a coaxial cable. To do so, consider the flux through a rectangle of length l that spans the gap between the inner and outer conductors.
    b. Evaluate the inductance per meter of a cable having $r_1 = 0.50$ mm and $r_2 = 3.0$ mm.

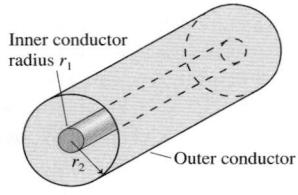

FIGURE CP33.85

STOP TO THINK ANSWERS

Stop to Think 33.1: d. According to the right-hand rule, the magnetic force on a positive charge carrier is to the right.

Stop to Think 33.2: No. The charge carriers in the wire move parallel to $\vec{B}$. There's no magnetic force on a charge moving parallel to a magnetic field.

Stop to Think 33.3: $F_b = F_d > F_a = F_c$. $\vec{F}_a$ is zero because there's no field. $\vec{F}_c$ is also zero because there's no current around the loop. The charge carriers in both the right and left edges are pushed to the bottom of the loop, creating a motional emf but no current. The currents at b and d are in opposite directions, but the forces on the segments in the field are both to the left and of equal magnitude.

Stop to Think 33.4: Clockwise. The wire's magnetic field as it passes through the loop is into the page. The flux through the loop decreases

into the page as the wire moves away. To oppose this decrease, the induced magnetic field needs to point into the page.

Stop to Think 33.5: d. The flux is increasing into the loop. To oppose this increase, the induced magnetic field needs to point out of the page. This requires a ccw induced current. Using the right-hand rule, the magnetic force on the current in the left edge of the loop is to the right, away from the field. The magnetic forces on the top and bottom segments of the loop are in opposite directions and cancel each other.

Stop to Think 33.6: b or f. The potential decreases in the direction of increasing current and increases in the direction of decreasing current.

Stop to Think 33.7: $\tau_c > \tau_a > \tau_b$. $\tau = L/R$, so smaller total resistance gives a larger time constant. The parallel resistors have total resistance $R/2$. The series resistors have total resistance $2R$.

34 Electromagnetic Fields and Waves

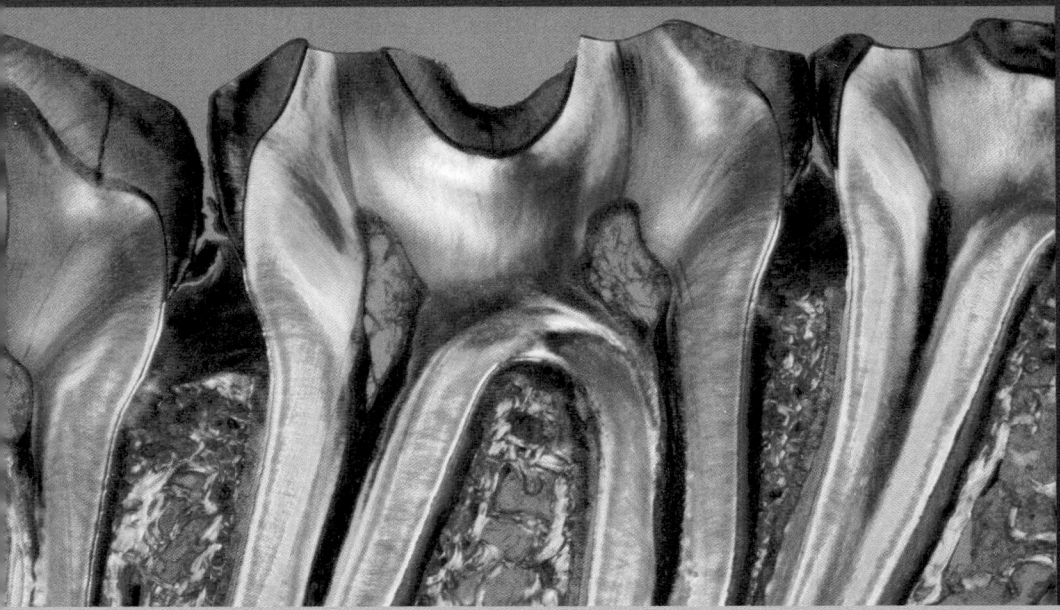

A thin section of molar teeth seen in polarized light. The rainbow of colors arises because different biological materials have different effects on the light's polarization.

▶ **Looking Ahead** The goal of Chapter 34 is to study the properties of electromagnetic fields and waves.

Maxwell's Theory of Electromagnetism

All of electricity and magnetism is based on four equations for the fields, called **Maxwell's equations,** and one equation that tells us how charges respond to fields.

Gauss's law: Charged particles create electric fields.
Faraday's law: Electric fields can also be created by changing magnetic fields.
Gauss's law for magnetism: There are no isolated magnetic poles.
Ampère-Maxwell law: Magnetic fields can be created by currents or by changing electric fields.

◀ **Looking Back**
Section 27.4 Gauss's law
Section 32.6 Ampère's law
Section 33.5 Faraday's law

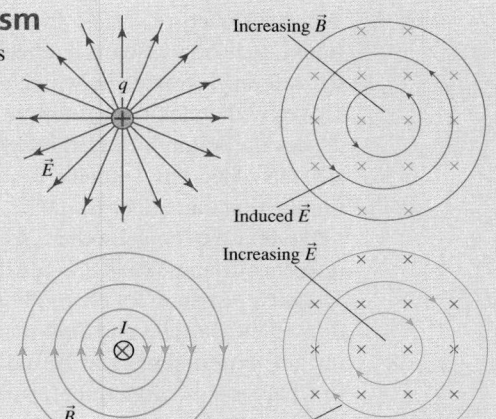

Electric and magnetic fields exert forces on charged particles. When combined, the net force on charge q is called the **Lorentz force law:**

$$\vec{F} = q\,(\vec{E} + \vec{v} \times \vec{B})$$

Electromagnetic Waves

You'll learn that Maxwell's equations predict the existence of self-sustaining **electromagnetic waves** that travel through space without the presence of charges or currents.

- $\vec{E}$ and $\vec{B}$ are perpendicular to each other and to the direction of travel.
- In vacuum $v_{em} = 1/\sqrt{\epsilon_0 \mu_0} = c$, the speed of light.

Electromagnetic waves are often **polarized,** meaning that the electric field always oscillates in the same plane. You'll learn to calculate the intensity of light transmitted through a polarizing filter.

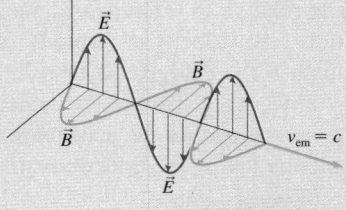

No light is transmitted through *crossed polarizers* whose axes are perpendicular to each other.

Field Transformations

Electric and magnetic fields turn out not to be separate, independent entities. Whether the field at a point is electric or magnetic depends on your motion relative to the charges and currents.

You'll learn how to transform the fields measured in reference frame A to a second frame B moving relative to A.

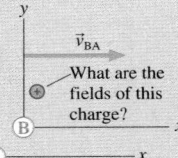

◀ **Looking Back**
Section 4.4 Relative motion

34.1 *E* or *B*? It Depends on Your Perspective

FIGURE 34.1 Brittney carries a charge past Alec.

(a)

Charge *q* moves with velocity $\vec{v}$ relative to Alec.

(b)

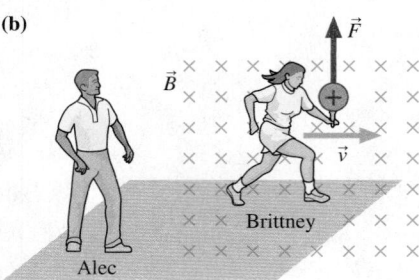

Charge *q* moves through a magnetic field established by Alec.

It seems clear, after the last nine chapters, that charges create electric fields and that moving charges, or currents, create magnetic fields. But consider **FIGURE 34.1a**, where Brittney, carrying charge *q*, runs past Alec with velocity $\vec{v}$. Alec sees a moving charge, and he knows that this charge creates a magnetic field. But from Brittney's perspective, the charge is at rest. Stationary charges don't create magnetic fields, so Brittney claims that the magnetic field is zero. Is there, or is there not, a magnetic field?

Or what about the situation in **FIGURE 34.1b**? Now Brittney is carrying the charge through a magnetic field that Alec has created. Alec sees a charge moving in a magnetic field, so he knows there's a force $\vec{F} = q\vec{v} \times \vec{B}$ on the charge. But for Brittney the charge is still at rest. Stationary charges don't experience magnetic forces, so Brittney claims that $\vec{F} = \vec{0}$.

Now, we may be a bit uncertain about magnetic fields, but surely there can be no disagreement over forces. After all, forces cause observable and measurable effects, so Alec and Brittney should be able to agree on whether or not the charge experiences a force. Further, if Brittney runs with constant velocity, then both Alec and Brittney are in *inertial reference frames*. You learned in Chapter 4 that these are the reference frames in which Newton's laws are valid, so we can't say that there's anything abnormal or unusual about Alec's and Brittney's observations. We have a paradox.

This paradox has arisen because magnetic fields and forces depend on velocity, but we haven't looked at the issue of velocity *with respect to what* or velocity *as measured by whom*. The resolution of this paradox will lead us to the conclusion that $\vec{E}$ and $\vec{B}$ are not, as we've been assuming, separate and independent entities. They are closely intertwined.

Reference Frames

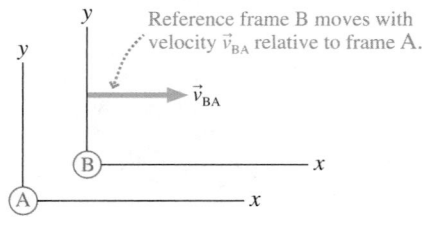

FIGURE 34.2 Reference frames A and B.

Reference frame B moves with velocity $\vec{v}_{BA}$ relative to frame A.

We introduced reference frames and relative motion in Chapter 4. To remind you, **FIGURE 34.2** shows two reference frames labeled A and B. You can think of these as the reference frames in which Alec and Brittney, respectively, are at rest. Frame B moves with velocity $\vec{v}_{BA}$ with respect to frame A. That is, an observer (Alec) at rest in A sees the origin of B (Brittney) go past with velocity $\vec{v}_{BA}$. Of course, Brittney would say that Alec has velocity $\vec{v}_{AB} = -\vec{v}_{BA}$ relative to her reference frame. There's no implication that either frame is "at rest." All we know is that the two reference frames move relative to each other. We will stipulate that both reference frames are inertial reference frames, so $\vec{v}_{BA}$ is constant.

FIGURE 34.3 shows a charged particle C. Experimenters in frame A measure the motion of the particle and find that its velocity *relative to frame A* is $\vec{v}_{CA}$. At the same instant, experimenters in B find that the particle's velocity *relative to frame B* is $\vec{v}_{CB}$. In Chapter 4, we found that $\vec{v}_{CA}$ and $\vec{v}_{CB}$ are related by

$$\vec{v}_{CA} = \vec{v}_{CB} + \vec{v}_{BA} \tag{34.1}$$

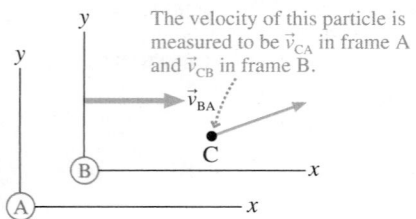

FIGURE 34.3 The particle's velocity is measured in both frame A and frame B.

The velocity of this particle is measured to be $\vec{v}_{CA}$ in frame A and $\vec{v}_{CB}$ in frame B.

Equation 34.1, the *Galilean transformation of velocity*, tells us that the velocity of the particle relative to reference frame A is its velocity relative to frame B plus (vector addition!) the velocity of frame B relative to frame A.

Suppose the charged particle in Figure 34.3 is accelerating, as it would if acted on by a net force. How does its acceleration $\vec{a}_{CA}$, as measured by experimenters in frame A, compare to the acceleration $\vec{a}_{CB}$ measured in frame B? We can answer this question by taking the time derivative of Equation 34.1:

$$\frac{d\vec{v}_{CA}}{dt} = \frac{d\vec{v}_{CB}}{dt} + \frac{d\vec{v}_{BA}}{dt}$$

The derivatives of $\vec{v}_{CA}$ and $\vec{v}_{CB}$ are the particle's accelerations $\vec{a}_{CA}$ and $\vec{a}_{CB}$ in frames A and B, respectively. But $\vec{v}_{BA}$ is a *constant* velocity, so $d\vec{v}_{BA}/dt = \vec{0}$. Thus the Galilean transformation of acceleration is simply

$$\vec{a}_{CA} = \vec{a}_{CB} \tag{34.2}$$

Brittney and Alec may measure different positions and velocities for a particle, but they *agree* on its acceleration. And if they agree on its acceleration, they must, by using Newton's second law, agree on the force acting on the particle. That is, **experimenters in all inertial reference frames agree about the force acting on a particle.** This conclusion is the key to understanding how different experimenters see electric and magnetic fields.

The Transformation of Electric and Magnetic Fields

Imagine that Alec has measured the electric field $\vec{E}_A$ and the magnetic field $\vec{B}_A$ in reference frame A. Our investigations thus far give us no reason to think that Brittney's measurements of the fields will differ from Alec's. After all, it seems like the fields are just "there," waiting to be measured. Thus our expectation is that Brittney, in frame B, will measure $\vec{E}_B = \vec{E}_A$ and $\vec{B}_B = \vec{B}_A$.

To find out if this is true, Alec establishes a region of space with a uniform magnetic field $\vec{B}_A$ but no electric field ($\vec{E}_A = \vec{0}$). Then, as shown in **FIGURE 34.4**, he shoots a positive charge q through the magnetic field. At an instant when q is moving horizontally with velocity $\vec{v}_{CA}$, Alec observes that the particle experiences force $\vec{F}_A = q(\vec{E}_A + \vec{v}_{CA} \times \vec{B}_A) = q\vec{v}_{CA} \times \vec{B}_A$. The direction of the force, given by the right-hand rule, is straight up.

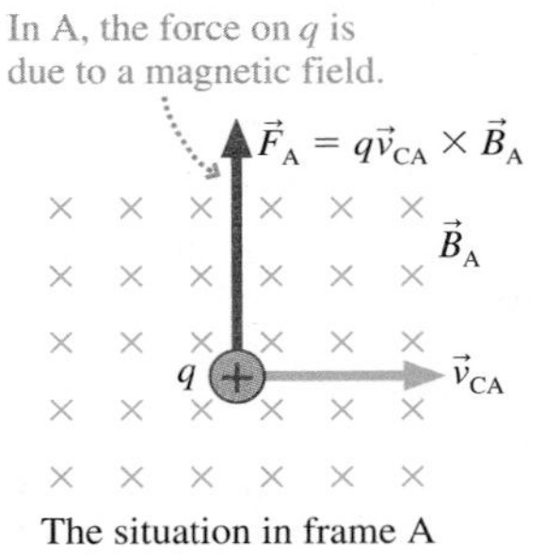

Suppose that Brittney, in frame B, runs alongside the charge with the same velocity: $\vec{v}_{BA} = \vec{v}_{CA}$. To her, in frame B, the charge is at rest. Nonetheless, because both experimenters must agree about forces, Brittney *must* observe the same upward force on the charge that Alec observed. But there is *no* magnetic force on a stationary charge, so how can this be?

Because Brittney sees a stationary charge being acted on by an upward force, her only possible conclusion is that there is an upward-pointing *electric field*. After all, the electric field was initially defined in terms of the force experienced by a stationary charge. If the electric field in frame B is $\vec{E}_B$, then the force on the charge is $\vec{F}_B = q\vec{E}_B$. But we know that $\vec{F}_B = \vec{F}_A$, and Alec has already measured $\vec{F}_A = q\vec{v}_{CA} \times \vec{B}_A = q\vec{v}_{BA} \times \vec{B}_A$. Thus we're led to the conclusion that

$$\vec{E}_B = \vec{v}_{BA} \times \vec{B}_A \tag{34.3}$$

As Brittney runs past Alec, she finds that at least part of Alec's magnetic field has become an electric field! **Whether a field is seen as "electric" or "magnetic" depends on the motion of the reference frame relative to the sources of the field.**

FIGURE 34.5 shows the situation from Brittney's perspective. There is a force on charge q, the same force that Alec measured in Figure 34.4, but Brittney attributes this force to an electric field rather than a magnetic field. (Brittney needs a moving charge to measure magnetic forces, so we'll need a different experiment to see whether or not there's a magnetic field in frame B.)

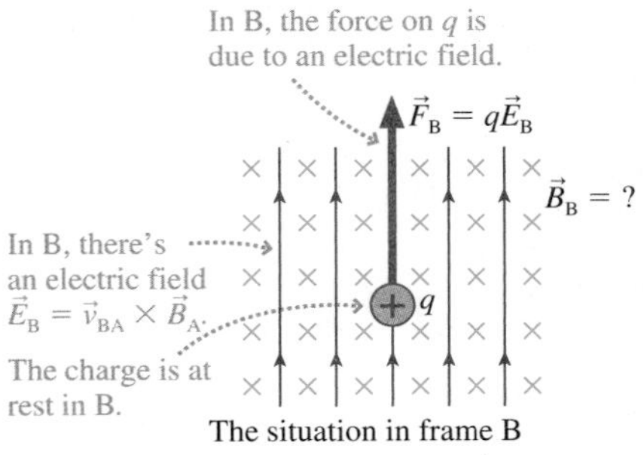

More generally, suppose that an experimenter in reference frame A creates both an electric field $\vec{E}_A$, and a magnetic field $\vec{B}_A$. A charge moving in A with velocity $\vec{v}_{CA}$ experiences the force $\vec{F}_A = q(\vec{E}_A + \vec{v}_{CA} \times \vec{B}_A)$ shown in **FIGURE 34.6a** on the next page. The charge is at rest in a reference frame B that moves with velocity $\vec{v}_{BA} = \vec{v}_{CA}$ so the force in B can be due only to an electric field: $\vec{F}_B = q\vec{E}_B$. Equating the forces, because experimenters in all inertial reference frames agree about forces, we find that

$$\vec{E}_B = \vec{E}_A + \vec{v}_{BA} \times \vec{B}_A \tag{34.4}$$

FIGURE 34.4 A charged particle moves through a magnetic field in reference frame A and experiences a magnetic force.

In A, the force on q is due to a magnetic field.

$\vec{F}_A = q\vec{v}_{CA} \times \vec{B}_A$

$\vec{B}_A$

q $\vec{v}_{CA}$

The situation in frame A

FIGURE 34.5 In frame B, the charge experiences an electric force.

In B, the force on q is due to an electric field.

$\vec{F}_B = q\vec{E}_B$

$\vec{B}_B = ?$

In B, there's an electric field
$\vec{E}_B = \vec{v}_{BA} \times \vec{B}_A$

The charge is at rest in B.

The situation in frame B

Equation 34.4 transforms the electric and magnetic fields measured in reference frame A into the electric field measured in a frame B that moves relative to A with velocity $\vec{v}_{BA}$. FIGURE 34.6b shows the outcome. Although we used a charge as a probe to find Equation 34.4, the equation is strictly about fields in different reference frames; it makes no mention of charges.

FIGURE 34.6 A charge in reference frame A experiences electric and magnetic forces. The charge experiences the same force in frame B, but it is due only to an electric field.

(a) The electric and magnetic fields in frame A

(b) The electric field in frame B, where the charged particle is at rest

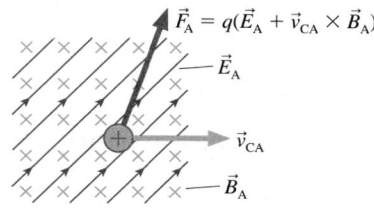
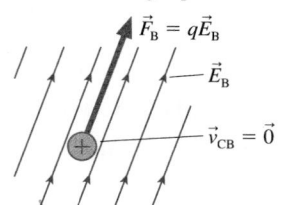

EXAMPLE 34.1 **Transforming the electric field**

A laboratory experimenter has created the parallel electric and magnetic fields $\vec{E} = 10{,}000\,\hat{\imath}$ V/m and $\vec{B} = 0.10\,\hat{\imath}$ T. A proton is shot into these fields with velocity $\vec{v} = 1.0 \times 10^5\,\hat{\jmath}$ m/s. What is the electric field in the proton's reference frame?

MODEL Let the laboratory be reference frame A and a frame moving with the proton be reference frame B. The relative velocity is $\vec{v}_{BA} = 1.0 \times 10^5\,\hat{\jmath}$ m/s.

VISUALIZE FIGURE 34.7 shows the geometry. The laboratory fields, now labeled A, are parallel to the x-axis while $\vec{v}_{BA}$ is in the y-direction. Thus $\vec{v}_{BA} \times \vec{B}_A$ points in the negative z-direction.

SOLVE $\vec{v}_{BA}$ and $\vec{B}_A$ are perpendicular, so the magnitude of $\vec{v}_{BA} \times \vec{B}_A$ is $(1.0 \times 10^5\,\text{m/s})(0.10\,\text{T})(\sin 90°) = 10{,}000$ V/m. Thus the electric field in frame B, the proton's frame, is

$$\vec{E}_B = \vec{E}_A + \vec{v}_{BA} \times \vec{B}_A = (10{,}000\,\hat{\imath} - 10{,}000\,\hat{k})\text{ V/m}$$

$$= (14{,}000\text{ V/m, } 45°\text{ below the } x\text{-axis})$$

FIGURE 34.7 Finding electric field $\vec{E}_B$.

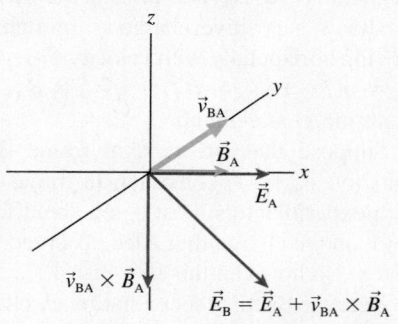

ASSESS The force on the proton is the same in both reference frames. But in the proton's reference frame that force is due entirely to an electric field tilted 45° below the x-axis.

To find a transformation equation for the magnetic field, FIGURE 34.8a shows charge q at rest in reference frame A. Alec measures the fields of a stationary point charge, which we know are

$$\vec{E}_A = \frac{1}{4\pi\epsilon_0} \frac{q}{r^2}\hat{r} \qquad \vec{B}_A = \vec{0}$$

What are the fields at this point in space as measured by Brittney in frame B? We can use Equation 34.4 to find $\vec{E}_B$. Because $\vec{B}_A = \vec{0}$, the electric field in frame B is

$$\vec{E}_B = \vec{E}_A = \frac{1}{4\pi\epsilon_0} \frac{q}{r^2}\hat{r} \tag{34.5}$$

In other words, Coulomb's law is still valid in a frame in which the point charge is moving.

But Brittney also measures a magnetic field $\vec{B}_B$, because, as seen in FIGURE 34.8b, charge q is moving in reference frame B. The magnetic field of a moving point charge is given by the Biot-Savart law:

$$\vec{B}_B = \frac{\mu_0}{4\pi} \frac{q}{r^2}\vec{v}_{CB} \times \hat{r} = -\frac{\mu_0}{4\pi} \frac{q}{r^2}\vec{v}_{BA} \times \hat{r} \tag{34.6}$$

where we used the fact that the charge's velocity in frame B is $\vec{v}_{CB} = -\vec{v}_{BA}$.

It will be useful to rewrite Equation 34.6 as

$$\vec{B}_B = -\frac{\mu_0}{4\pi}\frac{q}{r^2}\vec{v}_{BA} \times \hat{r} = -\epsilon_0\mu_0\vec{v}_{BA} \times \left(\frac{1}{4\pi\epsilon_0}\frac{q}{r^2}\hat{r}\right)$$

The expression in parentheses is simply $\vec{E}_A$, the electric field in frame A, so we have

$$\vec{B}_B = -\epsilon_0\mu_0\vec{v}_{BA} \times \vec{E}_A \qquad (34.7)$$

Equation 34.7 expresses the remarkable idea that **the Biot-Savart law for the magnetic field of a moving point charge is nothing other than the Coulomb electric field of a stationary point charge transformed into a moving reference frame.**

We will assert without proof that if the experimenters in frame A create a magnetic field $\vec{B}_A$ in addition to the electric field $\vec{E}_A$, then the magnetic field $\vec{B}_B$ measured in frame B is

$$\vec{B}_B = \vec{B}_A - \epsilon_0\mu_0\vec{v}_{BA} \times \vec{E}_A \qquad (34.8)$$

This is a general transformation matching Equation 34.4 for the electric field $\vec{E}_B$.

Notice something interesting. The constant μ_0 has units of T m/A; those of ϵ_0 are $C^2/N\,m^2$. By definition, 1 T = 1 N/A m and 1 A = 1 C/s. Consequently, the units of $\epsilon_0\mu_0$ turn out to be s^2/m^2. In other words, the quantity $1/\sqrt{\epsilon_0\mu_0}$, with units of m/s, is a speed. But what speed? The constants are well known from measurements of static electric and magnetic fields, so we can compute

$$\frac{1}{\sqrt{\epsilon_0\mu_0}} = \frac{1}{\sqrt{(8.85 \times 10^{-12}\,C^2/N\,m^2)(1.26 \times 10^{-6}\,T\,m/A)}} = 3.00 \times 10^8\,m/s$$

Of all the possible values you might get from evaluating $1/\sqrt{\epsilon_0\mu_0}$, what are the chances it equals c, the speed of light? It is not a random coincidence. In Section 34.5 we'll show that electric and magnetic fields can exist as a *traveling wave*, and that the wave speed is predicted by the theory to be none other than

$$v_{em} = c = \frac{1}{\sqrt{\epsilon_0\mu_0}} \qquad (34.9)$$

For now, we'll go ahead and write $\epsilon_0\mu_0 = 1/c^2$. With this, our **Galilean field transformation equations** are

$$\vec{E}_B = \vec{E}_A + \vec{v}_{BA} \times \vec{B}_A$$
$$\vec{B}_B = \vec{B}_A - \frac{1}{c^2}\vec{v}_{BA} \times \vec{E}_A \qquad (34.10)$$

where $\vec{v}_{BA}$ is the velocity of reference frame B relative to frame A and where, to reiterate, the fields are measured *at the same point in space* by experimenters *at rest* in each reference frame.

NOTE ▶ We'll see shortly that these equations are valid only if $v_{BA} \ll c$. ◀

We can no longer believe that electric and magnetic fields have a separate, independent existence. Changing from one reference frame to another mixes and rearranges the fields. Different experimenters watching an event will agree on the outcome, such as the deflection of a charged particle, but they will ascribe it to different combinations of fields. Our conclusion is that **there is a single electromagnetic field that presents different faces, in terms of $\vec{E}$ and $\vec{B}$, to different viewers.**

FIGURE 34.8 A charge at rest in frame A is moving in frame B and creates a magnetic field $\vec{B}_B$.

(a) In frame A, the static charge creates an electric field but no magnetic field.

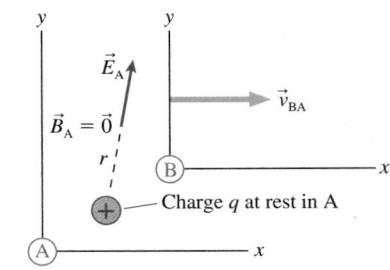

(b) In frame B, the moving charge creates both an electric and a magnetic field.

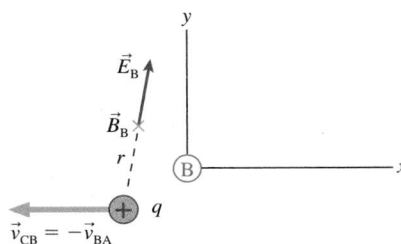

EXAMPLE 34.2 **Two views of a magnet**

The 1.0 T field of a large laboratory magnet points straight up. A rocket flies past the laboratory, parallel to the ground, at 1000 m/s. What are the fields between the magnet's pole tips as measured—very quickly!—by scientists on the rocket?

MODEL Let the laboratory be reference frame A and a frame moving with the rocket be reference frame B.

VISUALIZE **FIGURE 34.9** shows the magnet and establishes the coordinate systems. The relative velocity is $\vec{v}_{BA} = 1000\,\hat{\imath}$ m/s.

FIGURE 34.9 The rocket and the magnet.

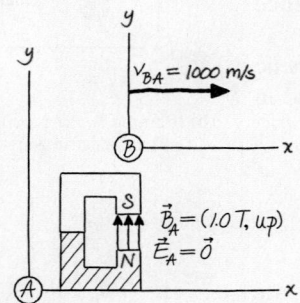

SOLVE The fields in the laboratory reference frame are $\vec{E}_A = \vec{0}$ and $\vec{B}_A = 1.0\hat{\jmath}$ T. Transforming the fields to the rocket's reference frame gives first, for the electric field,

$$\vec{E}_B = \vec{E}_A + \vec{v}_{BA} \times \vec{B}_A = \vec{v}_{BA} \times \vec{B}_A$$

From the right-hand rule, $\vec{v}_{BA} \times \vec{B}_A$ is out of the page, in the z-direction. $\vec{v}_{BA}$ and $\vec{B}_A$ are perpendicular, so

$$\vec{E}_B = v_{BA}B_A\hat{k} = 1000\,\hat{k} \text{ V/m}$$

Similarly, for the magnetic field,

$$\vec{B}_B = \vec{B}_A - \frac{1}{c^2}\vec{v}_{BA} \times \vec{E}_A = \vec{B}_A = 1.0\hat{\jmath} \text{ T}$$

Thus the rocket scientists measure

$$\vec{E}_B = 1000\,\hat{k} \text{ V/m} \quad \text{and} \quad \vec{B}_B = 1.0\hat{\jmath} \text{ T}$$

Almost Relativity

FIGURE 34.10 Two charges moving parallel to each other.

(a)

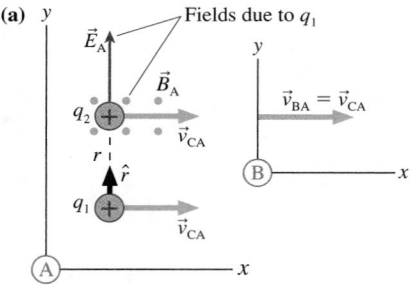

Fields seen in frame A

(b)

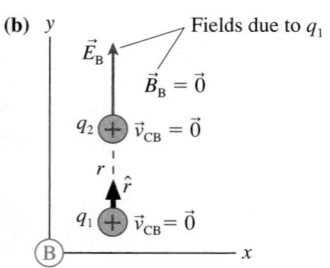

Fields seen in frame B

FIGURE 34.10a shows two positive charges moving side by side through frame A with velocity $\vec{v}_{CA}$. Charge q_1 creates an electric field and a magnetic field at the position of charge q_2. These are

$$\vec{E}_A = \frac{1}{4\pi\epsilon_0}\frac{q_1}{r^2}\hat{\jmath} \quad \text{and} \quad \vec{B}_A = \frac{\mu_0}{4\pi}\frac{q_1 v_{CA}}{r^2}\hat{k}$$

where r is the distance between the charges, and we've used $\hat{r} = \hat{\jmath}$ and $\vec{v} \times \hat{r} = v\hat{k}$.

How are the fields seen in frame B, which moves with $\vec{v}_{BA} = \vec{v}_{CA}$ and in which the charges are at rest? From the field transformation equations,

$$\vec{B}_B = \vec{B}_A - \frac{1}{c^2}\vec{v}_{BA} \times \vec{E}_A = \frac{\mu_0}{4\pi}\frac{q_1 v_{CA}}{r^2}\hat{k} - \frac{1}{c^2}\left(v_{CA}\hat{\imath} \times \frac{1}{4\pi\epsilon_0}\frac{q_1}{r^2}\hat{\jmath}\right)$$

(34.11)

$$= \frac{\mu_0}{4\pi}\frac{q_1 v_{CA}}{r^2}\left(1 - \frac{1}{\epsilon_0\mu_0 c^2}\right)\hat{k}$$

where we used $\hat{\imath} \times \hat{\jmath} = \hat{k}$. But $\epsilon_0\mu_0 = 1/c^2$, so the term in parentheses is zero and thus $\vec{B}_B = \vec{0}$. This result was expected because q_1 is at rest in frame B and shouldn't create a magnetic field.

The transformation of the electric field is similar:

$$\vec{E}_B = \vec{E}_A + \vec{v}_{BA} \times \vec{B}_A = \frac{1}{4\pi\epsilon_0}\frac{q_1}{r^2}\hat{\jmath} + v_{BA}\hat{\imath} \times \frac{\mu_0}{4\pi}\frac{q_1 v_{CA}}{r^2}\hat{k}$$

(34.12)

$$= \frac{1}{4\pi\epsilon_0}\frac{q_1}{r^2}(1 - \epsilon_0\mu_0 v_{BA}^2)\hat{\jmath} = \frac{1}{4\pi\epsilon_0}\frac{q_1}{r^2}\left(1 - \frac{v_{BA}^2}{c^2}\right)\hat{\jmath}$$

where we used $\hat{\imath} \times \hat{k} = -\hat{\jmath}$, $\vec{v}_{CA} = \vec{v}_{BA}$, and $\epsilon_0\mu_0 = 1/c^2$. **FIGURE 34.10b** shows the charges and fields in frame B.

But now we have a problem. In frame B where the two charges are at rest and separated by distance r, the electric field due to charge q_1 should be simply

$$\vec{E}_B = \frac{1}{4\pi\epsilon_0}\frac{q_1}{r^2}\hat{j}$$

The field transformation equations have given a "wrong" result for the electric field $\vec{E}_B$.

It turns out that the field transformations of Equations 34.10, which are based on Galilean relativity, aren't quite right. We would need Einstein's relativity—a topic that we'll take up in Chapter 36—to give the correct transformations. However, the Galilean field transformations in Equations 34.10 are equivalent to the relativistically correct transformations when $v \ll c$, in which case $v^2/c^2 \ll 1$. You can see that the two expressions for $\vec{E}_B$ do, in fact, agree if v_{BA}^2/c^2 can be neglected.

Thus our use of the field transformation equations has an additional rule: Set v^2/c^2 to zero. This is an acceptable rule for speeds $v < 10^7$ m/s. Even with this limitation, our investigation has provided us with a deeper understanding of electric and magnetic fields.

STOP TO THINK 34.1   The first diagram shows electric and magnetic fields in reference frame A. Which diagram shows the fields in frame B?

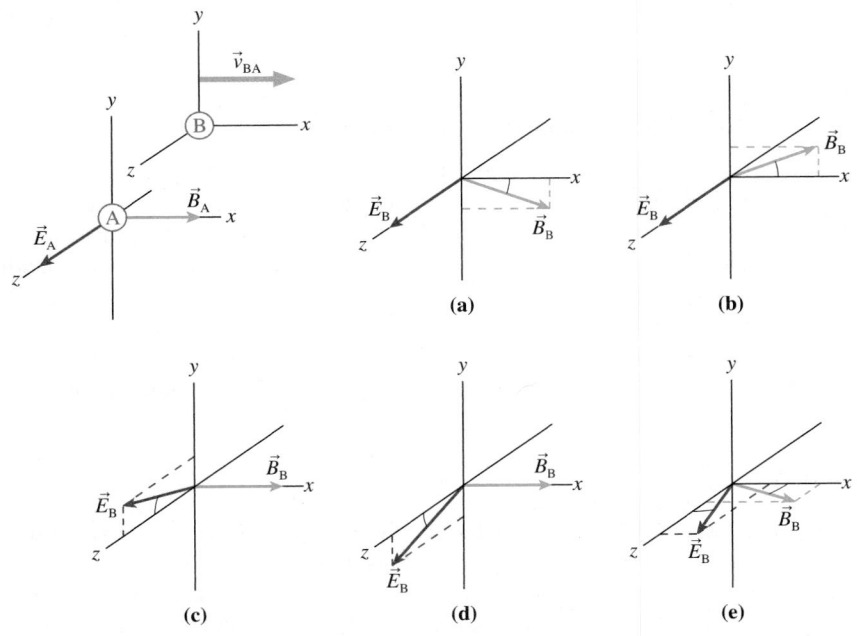

(a)   (b)   (c)   (d)   (e)

Faraday's Law Revisited

The transformation of electric and magnetic fields gives us new insight into Faraday's law. **FIGURE 34.11a** on the next page shows a reference frame A in which a conducting loop is moving with velocity $\vec{v}$ into a magnetic field. You learned in Chapter 33 that the magnetic field exerts a magnetic force $\vec{F}_B = q\vec{v} \times \vec{B} = (qvB,\text{ upward})$ on the charges in the leading edge of the wire, creating an emf $\mathcal{E} = vLB$ and an induced current in the loop. We called this a *motional emf*.

How do things appear to an experimenter who is in frame B that moves with the loop at velocity $\vec{v}_{BA} = \vec{v}$ and for whom the loop is at rest? We have learned the important lesson that experimenters in different inertial reference frames agree about the outcome of any experiment; hence an experimenter in frame B agrees that there is an

FIGURE 34.11 A motional emf as seen in two different reference frames.

FIGURE 34.11 A motional emf as seen in two different reference frames.

(a) Laboratory frame A

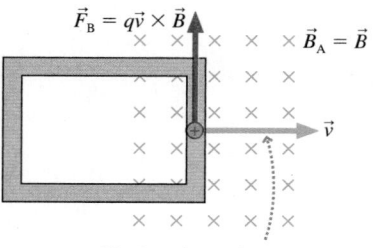

The loop is moving to the right.

(b) Loop frame B

The induced electric field points up.

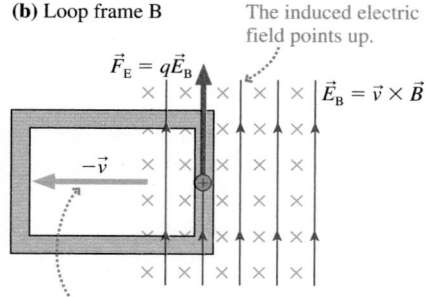

The magnetic field is moving to the left.

FIGURE 34.12 A Gaussian surface enclosing a charge.

Gaussian surface

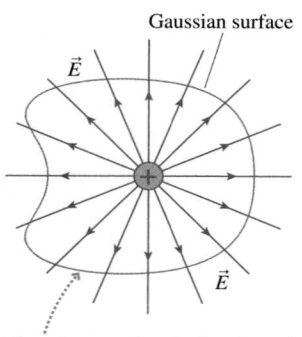

There is a net electric flux through this surface that encloses a charge.

FIGURE 34.13 There is no net flux through a Gaussian surface around a magnetic dipole.

Gaussian surface

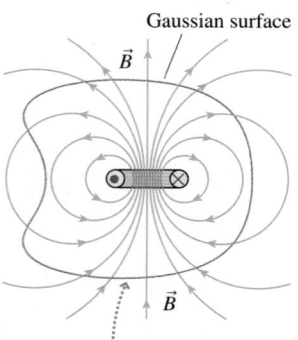

There is no net magnetic flux through this closed surface.

induced current in the loop. But the charges are at rest in frame B so there cannot be any magnetic force on them. How is the emf established in frame B?

We can use the field transformations to determine that the fields in frame B are

$$\vec{E}_B = \vec{E}_A + \vec{v} \times \vec{B}_A = \vec{v} \times \vec{B}$$
$$\vec{B}_B = \vec{B}_A - \frac{1}{c^2}\vec{v} \times \vec{E}_A = \vec{B} \qquad (34.13)$$

where we used the fact that $\vec{E}_A = \vec{0}$ in frame A.

An experimenter in the loop's frame sees not only a magnetic field but also the electric field $\vec{E}_B$ shown in **FIGURE 34.11b**. The magnetic field exerts no force on the charges, because they're at rest in this frame, but the electric field does. The force on charge q is $\vec{F}_E = q\vec{E}_B = q\vec{v} \times \vec{B} = (qvB, \text{ upward})$. This is the same force as was measured in the laboratory frame, so it will cause the same emf and the same current. The outcome is identical, as we knew it had to be, but the experimenter in B attributes the emf to an electric field whereas the experimenter in A attributes it to a magnetic field.

Field $\vec{E}_B$ is, in fact, the *induced electric field* of Faraday's law. Faraday's law, fundamentally, is a statement that **a changing magnetic field creates an electric field.** But only in frame B, the frame of the loop, is the magnetic field changing. Thus the induced electric field is seen in the loop's frame but not in the laboratory frame.

34.2 The Field Laws Thus Far

Let's remind ourselves where we are in terms of discovering laws about the electromagnetic field. Gauss's law, which you studied in Chapter 27, states a very general property of the electric field. It says that charges create electric fields in such a way that the electric flux of the field is the same through *any* closed surface surrounding the charges. **FIGURE 34.12** illustrates this idea by showing the field lines passing through a Gaussian surface enclosing a charge.

The mathematical statement of Gauss's law for the electric field says that for any *closed* surface enclosing total charge Q_{in}, the net electric flux through the surface is

$$(\Phi_e)_{\text{closed surface}} = \oint \vec{E} \cdot d\vec{A} = \frac{Q_{in}}{\epsilon_0} \qquad (34.14)$$

The circle on the integral sign indicates that the integration is over a closed surface. Gauss's law is the first of what will turn out to be four *field equations*.

There's an analogous equation for magnetic fields, an equation we implied in Chapter 32—where we noted that isolated north or south poles do not exist—but didn't explicitly write it down. **FIGURE 34.13** shows a Gaussian surface around a magnetic dipole. Magnetic field lines form continuous curves, without starting or stopping, so every field line leaving the surface at some point must reenter it at another. Consequently, the net magnetic flux over a *closed* surface is zero.

We've shown only one surface and one magnetic field, but this conclusion turns out to be a general property of magnetic fields. Because every north pole is accompanied by a south pole, we can't enclose a "net pole" within a surface. Thus Gauss's law for magnetic fields is

$$(\Phi_m)_{\text{closed surface}} = \oint \vec{B} \cdot d\vec{A} = 0 \qquad (34.15)$$

Equation 34.14 is the mathematical statement that Coulomb electric field lines start and stop on charges. Equation 34.15 is the mathematical statement that magnetic field lines form closed loops; they don't start or stop (i.e., there are no isolated magnetic poles).

These two versions of Gauss's law are important statements about what types of fields can and cannot exist. They will become two of Maxwell's equations.

The third field law we've established is Faraday's law:

$$\mathcal{E} = \oint \vec{E} \cdot d\vec{s} = -\frac{d\Phi_m}{dt} \qquad (34.16)$$

where the line integral of $\vec{E}$ is around the closed curve that bounds the surface through which the magnetic flux Φ_m is calculated. Equation 34.16 is the mathematical statement that an electric field (and thus an emf $\mathcal{E}$) can also be created by a changing magnetic field. The correct use of Faraday's law requires a convention for determining when fluxes are positive and negative. The sign convention will be given in the next section, where we discuss the fourth and last field equation—an analogous equation for magnetic fields.

34.3 The Displacement Current

We introduced Ampère's law in Chapter 32 as an alternative to the Biot-Savart law for calculating the magnetic field of a current. Whenever total current $I_{through}$ passes through an area bounded by a closed curve, the line integral of the magnetic field around the curve is

$$\oint \vec{B} \cdot d\vec{s} = \mu_0 I_{through} \qquad (34.17)$$

FIGURE 34.14 illustrates the geometry of Ampère's law. The sign of each current can be determined by using Tactics Box 34.1. In this case, $I_{through} = I_1 - I_2$.

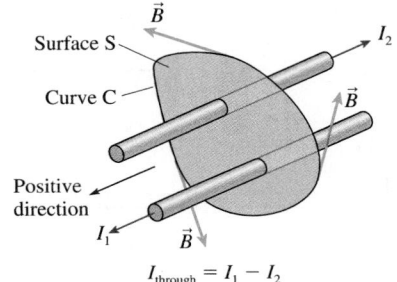

FIGURE 34.14 Ampère's law relates the line integral of $\vec{B}$ around curve C to the current passing through surface S.

$$I_{through} = I_1 - I_2$$

TACTICS
BOX 34.1 **Determining the signs of flux and current**   (MP)

❶ For a surface S bounded by a closed curve C, choose either the clockwise (cw) or counterclockwise (ccw) direction around C.

❷ Curl the fingers of your *right* hand around the curve in the chosen direction, with your thumb perpendicular to the surface. Your thumb defines the positive direction.

- A flux Φ through the surface is positive if the field is in the same direction as your thumb, negative if the field is in the opposite direction.

- A current through the surface in the direction of your thumb is positive, in the direction opposite your thumb is negative.

Exercises 4–6 ▨

Ampère's law is the formal statement that **currents create magnetic fields.** Although Ampère's law can be used to calculate magnetic fields in situations with a high degree of symmetry, it is more important as a statement about what types of magnetic field can and cannot exist.

Something Is Missing

Nothing restricts the bounded surface of Ampère's law to being flat. It's not hard to see that any current passing through surface S_1 in FIGURE 34.15 on the next page must also pass through the curved surface S_2. To interpret Ampère's law properly, we have to say that the current $I_{through}$ is the net current passing through *any* surface S that is bounded by curve C.

FIGURE 34.15 The *net* current passing through the flat surface S₁ also passes through the curved surface S₂.

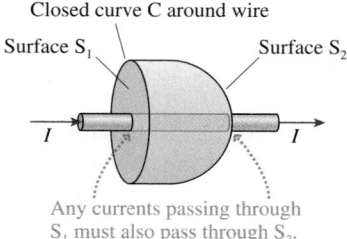

Closed curve C around wire

Surface S₁   Surface S₂

I I

Any currents passing through S₁ must also pass through S₂.

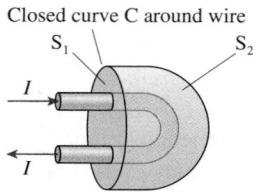

Closed curve C around wire

S₁   S₂

I

I

Even in this case, the *net* current through S₁, namely zero, matches the net current through S₂.

FIGURE 34.16 There is no current through surface S₂ as the capacitor charges, but there is a changing electric flux.

(a)   Cross section through a closed curve C around the wire

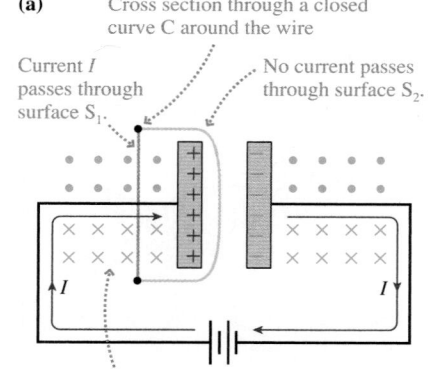

Current I passes through surface S₁.

No current passes through surface S₂.

I I

This is the magnetic field of the current I that is charging the capacitor.

(b)

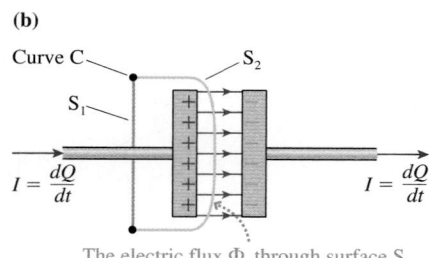

Curve C   S₂

S₁

$I = \dfrac{dQ}{dt}$ $I = \dfrac{dQ}{dt}$

The electric flux Φ_e through surface S₂ increases as the capacitor charges.

But this leads to an interesting puzzle. **FIGURE 34.16a** shows a capacitor being charged. Current I, from the left, brings positive charge to the left capacitor plate. The same current carries charges away from the right capacitor plate, leaving the right plate negatively charged. This is a perfectly ordinary current in a conducting wire, and you can use the right-hand rule to verify that its magnetic field is as shown.

Curve C is a closed curve encircling the wire on the left. The current passes through surface S₁, a flat surface across C, and we could use Ampère's law to find that the magnetic field is that of a straight wire. But what happens if we try to use surface S₂ to determine I_{through}? Ampère's law says that we can consider *any* surface bounded by curve C, and surface S₂ certainly qualifies. But *no* current passes through S₂. Charges are brought to the left plate of the capacitor and charges are removed from the right plate, but *no* charge moves across the gap between the plates. Surface S₁ has $I_{\text{through}} = I$, but surface S₂ has $I_{\text{through}} = 0$. Another dilemma!

It would appear that Ampère's law is either wrong or incomplete. Maxwell was the first to recognize the seriousness of this problem. He noted that there may be no current passing through S₂, but, as **FIGURE 34.16b** shows, there is an electric flux Φ_e through S₂ due to the electric field inside the capacitor. Furthermore, this flux is *changing* with time as the capacitor charges and the electric field strength grows. Faraday had discovered the significance of a changing magnetic flux, but no one had considered a changing electric flux.

The current I passes through S₁, so Ampère's law applied to S₁ gives

$$\oint \vec{B} \cdot d\vec{s} = \mu_0 I_{\text{through}} = \mu_0 I$$

We believe this result because it gives the correct magnetic field for a current-carrying wire. Now the line integral depends only on the magnetic field at points on curve C, so its value won't change if we choose a different surface S to evaluate the current. The problem is with the right side of Ampère's law, which would incorrectly give zero if applied to surface S₂. We need to modify the right side of Ampère's law to recognize that an electric flux rather than a current passes through S₂.

The electric flux between two capacitor plates of surface area A is

$$\Phi_e = EA$$

The capacitor's electric field is $E = Q/\epsilon_0 A$; hence the flux is actually independent of the plate size:

$$\Phi_e = \frac{Q}{\epsilon_0 A} A = \frac{Q}{\epsilon_0} \tag{34.18}$$

The *rate* at which the electric flux is changing is

$$\frac{d\Phi_e}{dt} = \frac{1}{\epsilon_0} \frac{dQ}{dt} = \frac{I}{\epsilon_0} \tag{34.19}$$

where we used $I = dQ/dt$. The flux is changing with time at a rate directly proportional to the charging current I.

Equation 34.19 suggests that the quantity $\epsilon_0(d\Phi_e/dt)$ is in some sense "equivalent" to current I. Maxwell called the quantity

$$I_{\text{disp}} = \epsilon_0 \frac{d\Phi_e}{dt} \tag{34.20}$$

the **displacement current.** He had started with a fluid-like model of electric and magnetic fields, and the displacement current was analogous to the displacement of a fluid. The fluid model has since been abandoned, but the name lives on despite the fact that nothing is actually being displaced.

Maxwell hypothesized that the displacement current was the "missing" piece of Ampère's law, so he modified Ampère's law to read

$$\oint \vec{B} \cdot d\vec{s} = \mu_0(I_{\text{through}} + I_{\text{disp}}) = \mu_0 \left(I_{\text{through}} + \epsilon_0 \frac{d\Phi_e}{dt} \right) \tag{34.21}$$

Equation 34.21 is now known as the Ampère-Maxwell law. When applied to Figure 34.16b, the Ampère-Maxwell law gives

$$S_1: \quad \oint \vec{B} \cdot d\vec{s} = \mu_0 \left(I_{\text{through}} + \epsilon_0 \frac{d\Phi_e}{dt} \right) = \mu_0(I + 0) = \mu_0 I$$

$$S_2: \quad \oint \vec{B} \cdot d\vec{s} = \mu_0 \left(I_{\text{through}} + \epsilon_0 \frac{d\Phi_e}{dt} \right) = \mu_0(0 + I) = \mu_0 I$$

where, for surface S_2, we used Equation 34.19 for $d\Phi_e/dt$. Surfaces S_1 and S_2 now both give the same result for the line integral of $\vec{B} \cdot d\vec{s}$ around the closed curve C.

NOTE ▶ The displacement current I_{disp} between the capacitor plates is numerically equal to the current I in the wires to and from the capacitor, so in some sense it allows "current" to be conserved all the way through the capacitor. Nonetheless, the displacement current is *not* a flow of charge. The displacement current is equivalent to a real current in that it creates the same magnetic field, but it does so with a changing electric flux rather than a flow of charge. ◀

The Induced Magnetic Field

Ordinary Coulomb electric fields are created by charges, but a second way to create an electric field is by having a changing magnetic field. That's Faraday's law. Ordinary magnetic fields are created by currents, but now we see that a second way to create a magnetic field is by having a changing electric field. Just as the electric field created by a changing $\vec{B}$ is called an induced electric field, the magnetic field created by a changing $\vec{E}$ is called an *induced magnetic field.*

FIGURE 34.17 shows the close analogy between induced electric fields, governed by Faraday's law, and induced magnetic fields, governed by the second term in the Ampère-Maxwell law. An increasing solenoid current causes an increasing magnetic field. The changing magnetic field, in turn, induces a circular electric field. The negative sign in Faraday's law dictates that the induced electric field direction is ccw when seen looking along the magnetic field direction.

An increasing capacitor charge causes an increasing electric field. The changing electric field, in turn, induces a circular magnetic field. But the sign of the Ampère-Maxwell law is positive, the opposite of the sign of Faraday's law, so the induced magnetic field direction is cw when you're looking along the electric field direction.

FIGURE 34.17 The close analogy between an induced electric field and an induced magnetic field.

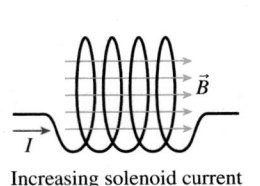

Increasing solenoid current   Increasing $\vec{B}$

Faraday's law describes an induced electric field.

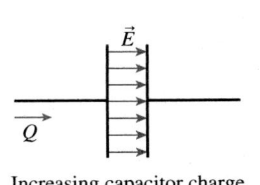

Increasing capacitor charge   Increasing $\vec{E}$

The Ampère-Maxwell law describes an induced magnetic field.

EXAMPLE 34.3 **The fields inside a charging capacitor**

A 2.0-cm-diameter parallel-plate capacitor with a 1.0 mm spacing is being charged at the rate 0.50 C/s. What is the magnetic field strength inside the capacitor at a point 0.50 cm from the axis?

MODEL The electric field inside a parallel-plate capacitor is uniform. As the capacitor is charged, the changing electric field induces a magnetic field.

VISUALIZE **FIGURE 34.18** shows the fields. The induced magnetic field lines are circles concentric with the capacitor.

FIGURE 34.18 The magnetic field strength is found by integrating around a closed curve of radius r.

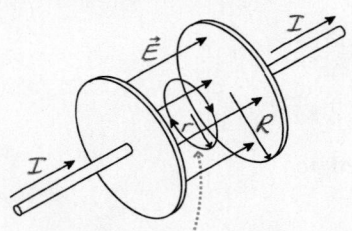

The magnetic field line is a circle concentric with the capacitor. The electric flux through this circle is $\pi r^2 E$.

SOLVE The electric field of a parallel-plate capacitor is $E = Q/\epsilon_0 A = Q/\epsilon_0 \pi R^2$. The electric flux through the circle of radius r (not the full flux of the capacitor) is

$$\Phi_e = \pi r^2 E = \pi r^2 \frac{Q}{\epsilon_0 \pi R^2} = \frac{r^2}{R^2} \frac{Q}{\epsilon_0}$$

Thus the Ampère-Maxwell law is

$$\oint \vec{B} \cdot d\vec{s} = \epsilon_0 \mu_0 \frac{d\Phi_e}{dt} = \epsilon_0 \mu_0 \frac{d}{dt}\left(\frac{r^2}{R^2}\frac{Q}{\epsilon_0}\right) = \mu_0 \frac{r^2}{R^2}\frac{dQ}{dt}$$

The magnetic field is everywhere tangent to the circle of radius r, so the integral of $\vec{B} \cdot d\vec{s}$ around the circle is simply $BL = 2\pi r B$. With this value for the line integral, the Ampère-Maxwell law becomes

$$2\pi r B = \mu_0 \frac{r^2}{R^2}\frac{dQ}{dt}$$

and thus

$$B = \frac{\mu_0}{2\pi}\frac{r}{R^2}\frac{dQ}{dt} = (2.0 \times 10^{-7}\,\text{T m/A})\frac{0.0050\,\text{m}}{(0.010\,\text{m})^2}(0.50\,\text{C/s})$$

$$= 5.0 \times 10^{-6}\,\text{T}$$

If a changing magnetic field can induce an electric field and a changing electric field can induce a magnetic field, what happens when both fields change simultaneously? That is the question that Maxwell was finally able to answer after he modified Ampère's law to include the displacement current, and it is the subject to which we turn next.

STOP TO THINK 34.2 The electric field in four identical capacitors is shown as a function of time. Rank in order, from largest to smallest, the magnetic field strength at the outer edge of the capacitor at time T.

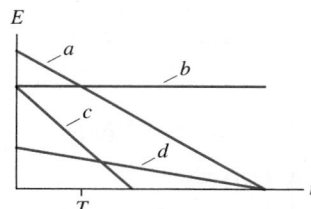

34.4 Maxwell's Equations

James Clerk Maxwell was a young, mathematically brilliant Scottish physicist. In 1855, barely 24 years old, he presented a paper to the Cambridge Philosophical Society entitled "On Faraday's Lines of Force." It had been 30 years and more since the major discoveries of Oersted, Ampère, Faraday, and others, but electromagnetism remained a loose collection of facts and "rules of thumb" without a consistent theory to link these ideas together.

Maxwell's goal was to synthesize this body of knowledge and to form a *theory* of electromagnetic fields. The critical step along the way was his recognition of the need to include a displacement-current term in Ampère's law.

Maxwell's theory of electromagnetism is embodied in four equations that we today call **Maxwell's equations.** These are

$$\oint \vec{E} \cdot d\vec{A} = \frac{Q_{in}}{\epsilon_0} \qquad \text{Gauss's law}$$

$$\oint \vec{B} \cdot d\vec{A} = 0 \qquad \text{Gauss's law for magnetism}$$

$$\oint \vec{E} \cdot d\vec{s} = -\frac{d\Phi_m}{dt} \qquad \text{Faraday's law}$$

$$\oint \vec{B} \cdot d\vec{s} = \mu_0 I_{through} + \epsilon_0 \mu_0 \frac{d\Phi_e}{dt} \qquad \text{Ampère-Maxwell law}$$

Maxwell's claim is that these four equations are a *complete* description of electric and magnetic fields. They tell us how fields are created by charges and currents, and also how fields can be induced by the changing of other fields. We need one more equation for completeness, an equation that tells us how matter responds to electromagnetic fields. The general force equation

$$\vec{F} = q(\vec{E} + \vec{v} \times \vec{B}) \qquad \text{(Lorentz force law)}$$

is known as the *Lorentz force law.* **Maxwell's equations for the fields, together with the Lorentz force law to tell us how matter responds to the fields, form the complete theory of electromagnetism.**

Maxwell's equations bring us to the pinnacle of classical physics. When combined with Newton's three laws of motion, his law of gravity, and the first and second laws of thermodynamics, we have all of classical physics—a total of just 11 equations.

While some physicists might quibble over whether all 11 are truly fundamental, the important point is not the exact number but how few equations we need to describe the overwhelming majority of our experience of the physical world. It seems as if we could have written them all on page 1 of this book and been finished, but it doesn't work that way. Each of these equations is the synthesis of a tremendous number of physical phenomena and conceptual developments. To know physics isn't just to know the equations, but to know what the equations *mean* and how they're used. That's why it's taken us so many chapters and so much effort to get to this point. Each equation is a shorthand way to summarize a book's worth of information!

Let's summarize the physical meaning of the five electromagnetic equations:

Classical physics

Newton's first law
Newton's second law
Newton's third law
Newton's law of gravity
Gauss's law
Gauss's law for magnetism
Faraday's law
Ampère-Maxwell law
Lorentz force law
First law of thermodynamics
Second law of thermodynamics

- **Gauss's law:** Charged particles create an electric field.
- **Faraday's law:** An electric field can also be created by a changing magnetic field.
- **Gauss's law for magnetism:** There are no isolated magnetic poles.
- **Ampère-Maxwell law, first half:** Currents create a magnetic field.
- **Ampère-Maxwell law, second half:** A magnetic field can also be created by a changing electric field.
- **Lorentz force law, first half:** An electric force is exerted on a charged particle in an electric field.
- **Lorentz force law, second half:** A magnetic force is exerted on a charge moving in a magnetic field.

These are the *fundamental ideas* of electromagnetism. Other important ideas, such as Ohm's law, Kirchhoff's laws, and Lenz's law, despite their practical importance, are not fundamental ideas. They can be derived from Maxwell's equations, sometimes with the addition of empirically based concepts such as resistance.

It's true that Maxwell's equations are mathematically more complex than Newton's laws and that their solution, for many problems of practical interest, requires advanced mathematics. Fortunately, we have the mathematical tools to get just far enough into Maxwell's equations to discover their most startling and revolutionary implication—the prediction of electromagnetic waves.

34.5 Electromagnetic Waves

It had been known since the early 19th century, from experiments on interference and diffraction, that light is a wave. We studied the wave properties of light in Part V, but at that time we were not able to determine just what is "waving."

Faraday speculated that light was somehow connected with electricity and magnetism, but Maxwell, using his equations of the electromagnetic field, was the first to understand that light is an oscillation of the electromagnetic field. Maxwell was able to predict that

■ Electromagnetic waves can exist at any frequency, not just at the frequencies of visible light. This prediction was the harbinger of radio waves.
■ All electromagnetic waves travel in a vacuum with the same speed, a speed that we now call the *speed of light*.

A general wave equation can be derived from Maxwell's equations, but the necessary mathematical techniques are beyond the level of this textbook. We'll adopt a simpler approach in which we *assume* an electromagnetic wave of a certain form and then show that it's consistent with Maxwell's equations. After all, the wave can't exist *unless* it's consistent with Maxwell's equations.

To begin, we're going to assume that electric and magnetic fields can exist independently of charges and currents in a *source-free* region of space. This is a very important assumption because it makes the statement that **fields are real entities.** They're not just cute pictures that tell us about charges and currents, but real things that can exist all by themselves. Our assertion is that the fields can exist in a self-sustaining mode in which a changing magnetic field creates an electric field (Faraday's law) that in turn changes in just the right way to re-create the original magnetic field (the Ampère-Maxwell law).

The source-free Maxwell's equations, with no charges or currents, are

$$\oint \vec{E} \cdot d\vec{A} = 0 \qquad \oint \vec{E} \cdot d\vec{s} = -\frac{d\Phi_m}{dt}$$

$$\oint \vec{B} \cdot d\vec{A} = 0 \qquad \oint \vec{B} \cdot d\vec{s} = \epsilon_0 \mu_0 \frac{d\Phi_e}{dt} \tag{34.22}$$

Any electromagnetic wave traveling in empty space must be consistent with these equations.

Let's postulate that an electromagnetic plane wave traveling with speed v_{em} has the characteristics shown in FIGURE 34.19. It's a useful picture, and one that you'll see in any textbook, but a picture that can be very misleading if you don't think about it carefully. $\vec{E}$ and $\vec{B}$ are *not* spatial vectors. That is, they don't stretch spatially in the y- or z-direction for a certain distance. Instead, these vectors are showing the values of the electric and magnetic fields along a single line, the x-axis. An $\vec{E}$ vector pointing in the y-direction says that *at this position* on the x-axis, where the vector's tail is, the electric field points in the y-direction and has a certain strength. Nothing is "reaching" to a point in space above the x-axis. In fact, this picture contains no information about the fields anywhere other than right on the x-axis.

However, we are assuming that this is a *plane wave,* which, you'll recall from Chapter 20, is a wave for which the fields are the same *everywhere* in any yz-plane, perpendicular to the x-axis. FIGURE 34.20a shows a small section of the xy-plane where, at this instant of time,

Large radar installations like this one are used to track rockets and missiles.

FIGURE 34.19 A sinusoidal electromagnetic wave.

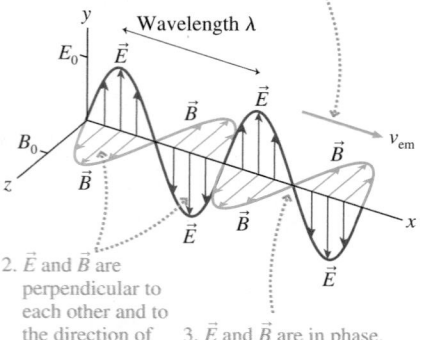

1. A sinusoidal wave with frequency f and wavelength λ travels with wave speed v_{em}.

2. $\vec{E}$ and $\vec{B}$ are perpendicular to each other and to the direction of travel. The fields have amplitudes E_0 and B_0.

3. $\vec{E}$ and $\vec{B}$ are in phase. That is, they have matching crests, troughs, and zeros.

$\vec{E}$ is pointing up and $\vec{B}$ is pointing toward you. The field strengths vary with x, the direction of travel, but not with y. As the wave moves forward, the fields that are now in the x_1-plane will soon arrive in the x_2-plane, and those now in the x_2-plane will move to x_3.

FIGURE 34.20b shows a section of the yz-plane that slices the x-axis at x_2. These fields are moving out of the page, coming toward you. The fields are the same everywhere in this plane, which is what we mean by a plane wave. If you watched a movie of the event, you would see the $\vec{E}$ and $\vec{B}$ fields at each point in this plane *oscillating* in time, but always synchronized with all the other points in the plane.

Gauss's Laws

Now that we understand the shape of the electromagnetic field, we can check its consistency with Maxwell's equations. This field is a sinusoidal wave, so the components of the fields are

$$E_x = 0 \quad E_y = E_0 \sin\left(2\pi(x/\lambda - ft)\right) \quad E_z = 0$$
$$B_x = 0 \quad B_y = 0 \quad\quad\quad\quad B_z = B_0 \sin\left(2\pi(x/\lambda - ft)\right) \tag{34.23}$$

where E_0 and B_0 are the amplitudes of the oscillating electric and magnetic fields.

FIGURE 34.21 shows an imaginary box—a Gaussian surface—centered on the x-axis. Both electric and magnetic field vectors exist at each point in space, but the figure shows them separately for clarity. $\vec{E}$ oscillates along the y-axis, so all electric field lines enter and leave the box through the top and bottom surfaces; no electric field lines pass through the sides of the box.

FIGURE 34.21 A closed surface can be used to check Gauss's law for the electric and magnetic fields.

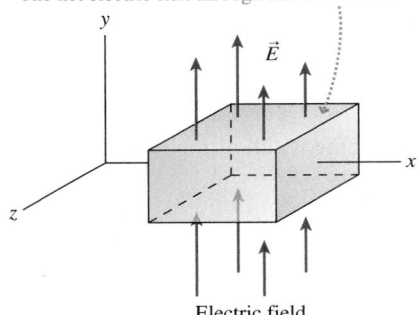

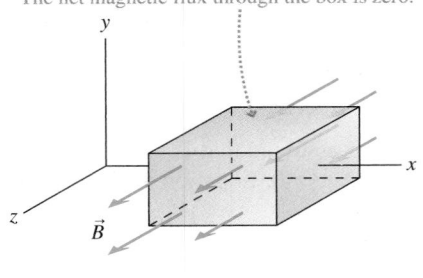

Because this is a plane wave, the magnitude of each electric field vector entering the bottom of the box is exactly matched by the electric field vector leaving the top. The electric flux through the top of the box is equal in magnitude but opposite in sign to the flux through the bottom, and the flux through the sides is zero. Thus the *net* electric flux is $\Phi_e = 0$. There is no charge inside the box because there are no sources in this region of space, so we also have $Q_{in} = 0$. Hence the electric field of a plane wave is consistent with the first of the source-free Maxwell's equations, Gauss's law.

The exact same argument applies to the magnetic field. The net magnetic flux is $\Phi_m = 0$; thus the magnetic field is consistent with the second of Maxwell's equations.

Faraday's Law

Faraday's law is concerned with the changing magnetic flux through a closed curve. We'll apply Faraday's law to a narrow rectangle in the xy-plane, shown in FIGURE 34.22, with height h and width Δx. We'll assume Δx to be so small that $\vec{B}$ is essentially constant over the width of the rectangle.

FIGURE 34.20 Interpreting the electromagnetic wave of Figure 34.19.

(a) Wave traveling to the right

(b) Wave coming toward you

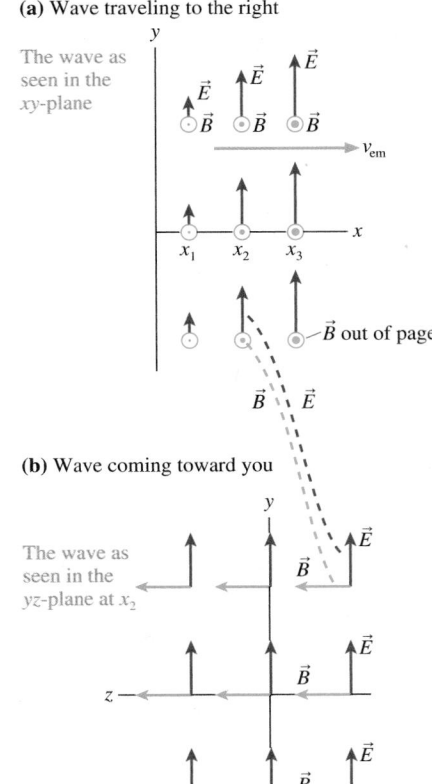

FIGURE 34.22 Faraday's law can be applied to a narrow rectangle in the xy-plane.

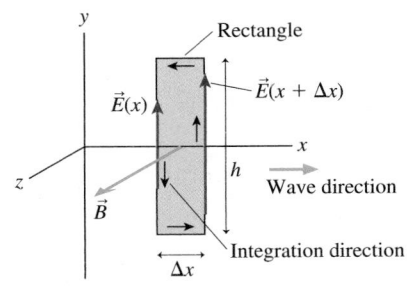

The magnetic field $\vec{B}$ points in the z-direction, perpendicular to the rectangle. The magnetic flux through the rectangle is $\Phi_m = B_z A_{rectangle} = B_z h \Delta x$, hence the flux *changes* at the rate

$$\frac{d\Phi_m}{dt} = \frac{d}{dt}(B_z h \Delta x) = \frac{\partial B_z}{\partial t} h \Delta x \qquad (34.24)$$

The ordinary derivative dB_z/dt, which is the full rate of change of B from all possible causes, becomes a partial derivative $\partial B_z/\partial t$ in this situation because the change in magnetic flux is due entirely to the change of B with time and not at all to the spatial variation of B.

According to our sign convention, we have to go around the rectangle in a ccw direction to make the flux positive. Thus we must also use a ccw direction to evaluate the line integral

$$\oint \vec{E} \cdot d\vec{s} = \int_{right} \vec{E} \cdot d\vec{s} + \int_{top} \vec{E} \cdot d\vec{s} + \int_{left} \vec{E} \cdot d\vec{s} + \int_{bottom} \vec{E} \cdot d\vec{s} \qquad (34.25)$$

The electric field $\vec{E}$ points in the y-direction, hence $\vec{E} \cdot d\vec{s} = 0$ at all points on the top and bottom edges, and these two integrals are zero.

Along the left edge of the loop, at position x, $\vec{E}$ has the same value at every point. Figure 34.22 shows that the direction of $\vec{E}$ is *opposite* to $d\vec{s}$, thus $\vec{E} \cdot d\vec{s} = -E_y(x) ds$. On the right edge of the loop, at position $x + \Delta x$, $\vec{E}$ is *parallel* to $d\vec{s}$ and $\vec{E} \cdot d\vec{s} = E_y(x + \Delta x) ds$. Thus the line integral of $\vec{E} \cdot d\vec{s}$ around the rectangle is

$$\oint \vec{E} \cdot d\vec{s} = -E_y(x)h + E_y(x + \Delta x)h = [E_y(x + \Delta x) - E_y(x)]h \qquad (34.26)$$

NOTE ▶ $E_y(x)$ indicates that E_y is a function of the position x. It is *not* E_y multiplied by x. ◀

You learned in calculus that the derivative of the function $f(x)$ is

$$\frac{df}{dx} = \lim_{\Delta x \to 0} \left[\frac{f(x + \Delta x) - f(x)}{\Delta x} \right]$$

We've assumed that Δx is very small. If we now let the width of the rectangle go to zero, $\Delta x \to 0$, Equation 34.26 becomes

$$\oint \vec{E} \cdot d\vec{s} = \frac{\partial E_y}{\partial x} h \Delta x \qquad (34.27)$$

We've used a partial derivative because E_y is a function of both position x and time t.

Now, using Equations 34.24 and 34.27, we can write Faraday's law as

$$\oint \vec{E} \cdot d\vec{s} = \frac{\partial E_y}{\partial x} h \Delta x = -\frac{d\Phi_m}{dt} = -\frac{\partial B_z}{\partial t} h \Delta x$$

The area $h \Delta x$ of the rectangle cancels, and we're left with

$$\frac{\partial E_y}{\partial x} = -\frac{\partial B_z}{\partial t} \qquad (34.28)$$

Equation 34.28, which compares the rate at which E_y varies with position to the rate at which B_z varies with time, is a *required condition* that an electromagnetic wave must satisfy to be consistent with Maxwell's equations. We can use Equations 34.23 for E_y and B_z to evaluate the partial derivatives:

$$\frac{\partial E_y}{\partial x} = \frac{2\pi E_0}{\lambda} \cos\left(2\pi(x/\lambda - ft)\right)$$

$$\frac{\partial B_z}{\partial t} = -2\pi f B_0 \cos\left(2\pi(x/\lambda - ft)\right)$$

Thus the required condition of Equation 34.28 is

$$\frac{\partial E_y}{\partial x} = \frac{2\pi E_0}{\lambda}\cos\left(2\pi(x/\lambda - ft)\right) = -\frac{\partial B_z}{\partial t} = 2\pi f B_0 \cos\left(2\pi(x/\lambda - ft)\right)$$

Canceling the many common factors, and multiplying by λ, we're left with

$$E_0 = (\lambda f)B_0 = v_{em}B_0 \qquad (34.29)$$

where we used the fact that $\lambda f = v$ for any sinusoidal wave.

Equation 34.29, which came from applying Faraday's law, tells us that the field amplitudes E_0 and B_0 of an electromagnetic wave are not arbitrary. **Once the amplitude B_0 of the magnetic field wave is specified, the electric field amplitude E_0 must be $E_0 = v_{em}B_0$.** Otherwise the fields won't satisfy Maxwell's equations.

The Ampère-Maxwell Law

We have one equation to go, but this one will now be easier. The Ampère-Maxwell law is concerned with the changing electric flux through a closed curve. FIGURE 34.23 shows a very narrow rectangle of width Δx and length l in the xz-plane. The electric field is perpendicular to this rectangle; hence the electric flux through it is $\Phi_e = E_y A_{rectangle} = E_y l \Delta x$. This flux is changing at the rate

$$\frac{d\Phi_e}{dt} = \frac{d}{dt}(E_y l \Delta x) = \frac{\partial E_y}{\partial t} l \Delta x \qquad (34.30)$$

The line integral of $\vec{B} \cdot d\vec{s}$ around this closed rectangle is calculated just like the line integral of $\vec{E} \cdot d\vec{s}$ in Figure 34.22. $\vec{B}$ is perpendicular to $d\vec{s}$ on the narrow ends, so $\vec{B} \cdot d\vec{s} = 0$. The field at *all* points on the left edge, at position x, is $\vec{B}(x)$, and this field is parallel to $d\vec{s}$ to make $\vec{B} \cdot d\vec{s} = B_z(x)\,ds$. Similarly, $\vec{B} \cdot d\vec{s} = -B_z(x + \Delta x)\,ds$ at all points on the right edge, where $\vec{B}$ is opposite to $d\vec{s}$.

Thus, if we let $\Delta x \rightarrow 0$,

$$\oint \vec{B} \cdot d\vec{s} = B_z(x)l - B_z(x + \Delta x)l = -[B_z(x + \Delta x) - B_z(x)]l$$
$$= -\frac{\partial B_z}{\partial x} l \Delta x \qquad (34.31)$$

Equations 34.30 and 34.31 can now be used in the Ampère-Maxwell law:

$$\oint \vec{B} \cdot d\vec{s} = -\frac{\partial B_z}{\partial x}l\Delta x = \epsilon_0\mu_0\frac{d\Phi_e}{dt} = \epsilon_0\mu_0\frac{\partial E_y}{\partial t}l\Delta x$$

The area of the rectangle cancels, and we're left with

$$\frac{\partial B_z}{\partial x} = -\epsilon_0\mu_0\frac{\partial E_y}{\partial t} \qquad (34.32)$$

Equation 34.32 is a second required condition that the fields must satisfy. If we again evaluate the partial derivatives, using Equations 34.23 for E_y and B_z, we find

$$\frac{\partial E_y}{\partial t} = -2\pi f E_0 \cos\left(2\pi(x/\lambda - ft)\right)$$

$$\frac{\partial B_z}{\partial x} = \frac{2\pi B_0}{\lambda}\cos\left(2\pi(x/\lambda - ft)\right)$$

With these, Equation 34.32 becomes

$$\frac{\partial B_z}{\partial x} = \frac{2\pi B_0}{\lambda}\cos\left(2\pi(x/\lambda - ft)\right) = -\epsilon_0\mu_0\frac{\partial E_y}{\partial t} = 2\pi\epsilon_0\mu_0 f E_0 \cos\left(2\pi(x/\lambda - ft)\right)$$

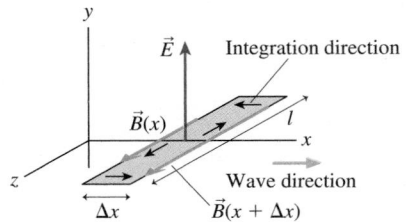

FIGURE 34.23 The Ampère-Maxwell law can be applied to a narrow rectangle in the xz-plane.

A final round of cancellations and another use of $\lambda f = v_{em}$ leave us with

$$E_0 = \frac{B_0}{\epsilon_0 \mu_0 \lambda f} = \frac{B_0}{\epsilon_0 \mu_0 v_{em}} \tag{34.33}$$

The last of Maxwell's equations gives us another constraint between E_0 and B_0.

The Speed of Light

But how can Equation 34.29, which required $E_0 = v_{em} B_0$, and Equation 34.33 both be true at the same time? The one and only way is if

$$\frac{1}{\epsilon_0 \mu_0 v_{em}} = v_{em}$$

from which we find

$$v_{em} = \frac{1}{\sqrt{\epsilon_0 \mu_0}} = 3.00 \times 10^8 \text{ m/s} = c \tag{34.34}$$

This is a remarkable conclusion. The constants ϵ_0 and μ_0 are from electrostatics and magnetostatics, where they determine the size of $\vec{E}$ and $\vec{B}$ due to point charges. Coulomb's law and the Biot-Savart law, where ϵ_0 and μ_0 first appeared, have nothing to do with waves. Yet Maxwell's theory of electromagnetism ends up predicting that electric and magnetic fields can form a self-sustaining electromagnetic wave *if* that wave travels at the specific speed $v_{em} = 1/\sqrt{\epsilon_0 \mu_0}$. No other speed will satisfy Maxwell's equations.

We've made no assumption about the frequency of the wave, so apparently all electromagnetic waves, regardless of their frequency, travel (in vacuum) at the same speed $v_{em} = 1/\sqrt{\epsilon_0 \mu_0}$. We call this speed c, the "speed of light," but it applies equally well from low-frequency radio waves to ultrahigh-frequency x rays.

STOP TO THINK 34.3 An electromagnetic wave is propagating in the positive x-direction. At this instant of time, what is the direction of $\vec{E}$ at the center of the rectangle?

a. In the positive x-direction
c. In the positive y-direction
e. In the positive z-direction
b. In the negative x-direction
d. In the negative y-direction
f. In the negative z-direction

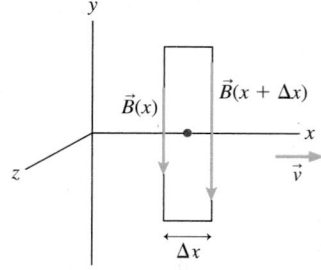

34.6 Properties of Electromagnetic Waves

We've demonstrated that one very specific sinusoidal wave is consistent with Maxwell's equations. It's possible to show that *any* electromagnetic wave, whether it's sinusoidal or not, must satisfy four basic conditions:

1. The fields $\vec{E}$ and $\vec{B}$ are perpendicular to the direction of propagation $\vec{v}_{em}$. Thus an electromagnetic wave is a transverse wave.
2. $\vec{E}$ and $\vec{B}$ are perpendicular to each other in a manner such that $\vec{E} \times \vec{B}$ is in the direction of $\vec{v}_{em}$.
3. The wave travels in vacuum at speed $v_{em} = 1/\sqrt{\epsilon_0 \mu_0} = c$.
4. $E = cB$ at any point on the wave.

In this section, we'll look at some other properties of electromagnetic waves.

Energy and Intensity

Waves transfer energy. Ocean waves erode beaches, sound waves set your eardrums vibrating, and light from the sun warms the earth. The energy flow of an electromagnetic wave is described by the **Poynting vector** $\vec{S}$, defined as

$$\vec{S} \equiv \frac{1}{\mu_0} \vec{E} \times \vec{B} \qquad (34.35)$$

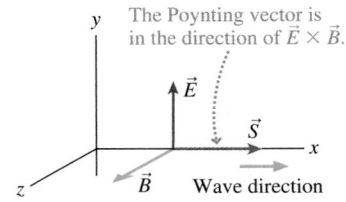

FIGURE 34.24 The Poynting vector.

The Poynting vector is in the direction of $\vec{E} \times \vec{B}$.

The Poynting vector, shown in FIGURE 34.24, has two important properties:

1. The Poynting vector points in the direction in which an electromagnetic wave is traveling. You can see this by looking back at Figure 34.19.
2. It is straightforward to show that the units of S are W/m^2, or power (joules per second) per unit area. Thus the magnitude S of the Poynting vector measures the rate of energy transfer per unit area of the wave.

Because $\vec{E}$ and $\vec{B}$ of an electromagnetic wave are perpendicular to each other, and $E = cB$, the magnitude of the Poynting vector is

$$S = \frac{EB}{\mu_0} = \frac{E^2}{c\mu_0} = c\epsilon_0 E^2$$

The Poynting vector is a function of time, oscillating from zero to $S_{max} = E_0^2/c\mu_0$ and back to zero twice during each period of the wave's oscillation. That is, the energy flow in an electromagnetic wave is not smooth. It "pulses" as the electric and magnetic fields oscillate in intensity. We're unaware of this pulsing because the electromagnetic waves that we can sense—light waves—have such high frequencies.

Of more interest is the *average* energy transfer, averaged over one cycle of oscillation, which is the wave's **intensity** I. In our earlier study of waves, we defined the intensity of a wave to be $I = P/A$, where P is the power (energy transferred per second) of a wave that impinges on area A. Because $E = E_0 \sin\left(2\pi(x/\lambda - ft)\right)$, and the average over one period of $\sin^2\left(2\pi(x/\lambda - ft)\right)$ is $\frac{1}{2}$, the intensity of an electromagnetic wave is

$$I = \frac{P}{A} = S_{avg} = \frac{1}{2c\mu_0} E_0^2 = \frac{c\epsilon_0}{2} E_0^2 \qquad (34.36)$$

Equation 34.36 relates the intensity of an electromagnetic wave, a quantity that is easily measured, to the amplitude of the wave's electric field.

The intensity of a plane wave, with constant electric field amplitude E_0, would not change with distance. But a plane wave is an idealization; there are no true plane waves in nature. You learned in Chapter 20 that, to conserve energy, the intensity of a wave far from its source decreases with the inverse square of the distance. If a source with power P_{source} emits electromagnetic waves *uniformly* in all directions, the electromagnetic wave intensity at distance r from the source is

$$I = \frac{P_{source}}{4\pi r^2} \qquad (34.37)$$

Equation 34.37 simply expresses the recognition that the energy of the wave is spread over a sphere of surface area $4\pi r^2$.

EXAMPLE 34.4 **Fields of a cell phone**

A digital cell phone broadcasts a 0.60 W signal at a frequency of 1.9 GHz. What are the amplitudes of the electric and magnetic fields at a distance of 10 cm, about the distance to the center of the user's brain?

MODEL Treat the cell phone as a point source of electromagnetic waves.

Continued

SOLVE The intensity of a 0.60 W point source at a distance of 10 cm is

$$I = \frac{P_{source}}{4\pi r^2} = \frac{0.60 \text{ W}}{4\pi (0.10 \text{ m})^2} = 4.78 \text{ W/m}^2$$

We can find the electric field amplitude from the intensity:

$$E_0 = \sqrt{\frac{2I}{c\epsilon_0}} = \sqrt{\frac{2(4.78 \text{ W/m}^2)}{(3.00 \times 10^8 \text{ m/s})(8.85 \times 10^{-12} \text{ C}^2/\text{N m}^2)}}$$

$$= 60 \text{ V/m}$$

The amplitudes of the electric and magnetic fields are related by the speed of light. This allows us to compute

$$B_0 = \frac{E_0}{c} = 2.0 \times 10^{-7} \text{ T}$$

ASSESS The electric field amplitude is modest; the magnetic field amplitude is very small. This implies that the interaction of electromagnetic waves with matter is mostly due to the electric field.

STOP TO THINK 34.4 An electromagnetic wave is traveling in the positive y-direction. The electric field at one instant of time is shown at one position. The magnetic field at this position points

a. In the positive x-direction.
b. In the negative x-direction.
c. In the positive y-direction.
d. In the negative y-direction.
e. Toward the origin.
f. Away from the origin.

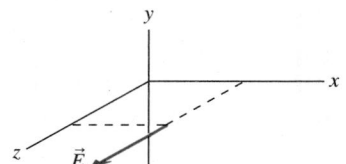

Radiation Pressure

Electromagnetic waves transfer not only energy but also momentum. An object gains momentum when it absorbs electromagnetic waves, much as a ball at rest gains momentum when struck by a ball in motion.

Suppose we shine a beam of light on an object that completely absorbs the light energy. If the object absorbs energy during a time interval Δt, its momentum changes by

$$\Delta p = \frac{\text{energy absorbed}}{c}$$

This is a consequence of Maxwell's theory, which we'll state without proof.

The momentum change implies that the light is exerting a force on the object. Newton's second law, in terms of momentum, is $F = \Delta p/\Delta t$. The radiation force due to the beam of light is

$$F = \frac{\Delta p}{\Delta t} = \frac{(\text{energy absorbed})/\Delta t}{c} = \frac{P}{c}$$

where P is the power (joules per second) of the light.

It's more interesting to consider the force exerted on an object per unit area, which is called the **radiation pressure** p_{rad}. The radiation pressure on an object that absorbs all the light is

$$p_{rad} = \frac{F}{A} = \frac{P/A}{c} = \frac{I}{c} \tag{34.38}$$

where I is the intensity of the light wave. The subscript on p_{rad} is important in this context to distinguish the radiation pressure from the momentum p.

Artist's conception of a future spacecraft powered by radiation pressure from the sun.

EXAMPLE 34.5 **Solar sailing**

A low-cost way of sending spacecraft to other planets would be to use the radiation pressure on a solar sail. The intensity of the sun's electromagnetic radiation at distances near the earth's orbit is about 1300 W/m². What size sail would be needed to accelerate a 10,000 kg spacecraft toward Mars at 0.010 m/s²?

MODEL Assume that the solar sail is perfectly absorbing.

SOLVE The force that will create a 0.010 m/s² acceleration is $F = ma = 100$ N. We can use Equation 34.38 to find the sail

area that, by absorbing light, will receive a 100 N force from the sun:

$$A = \frac{cF}{I} = \frac{(3.00 \times 10^8 \text{ m/s})(100 \text{ N})}{1300 \text{ W/m}^2} = 2.3 \times 10^7 \text{ m}^2$$

ASSESS If the sail is a square, it would need to be 4.8 km × 4.8 km, or roughly 3 mi × 3 mi. This is large, but not entirely out of the question with thin films that can be unrolled in space. But how will the crew return from Mars?

Antennas

We've seen that an electromagnetic wave is self-sustaining, independent of charges or currents. However, charges and currents are needed at the *source* of an electromagnetic wave. We'll take a brief look at how an electromagnetic wave is generated by an antenna.

FIGURE 34.25 is the electric field of an electric dipole. If the dipole is vertical, the electric field $\vec{E}$ at points along a horizontal line is also vertical. Reversing the dipole, by switching the charges, reverses $\vec{E}$. If the charges were to oscillate back and forth, switching position at frequency f, then $\vec{E}$ would oscillate in a vertical plane. The changing $\vec{E}$ would then create an induced magnetic field $\vec{B}$, which could then create an $\vec{E}$, which could then create a $\vec{B}, \ldots$, and an electromagnetic wave at frequency f would radiate out into space.

FIGURE 34.25 An electric dipole creates an electric field that reverses direction if the dipole charges are switched.

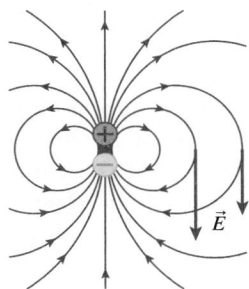
Positive charge on top

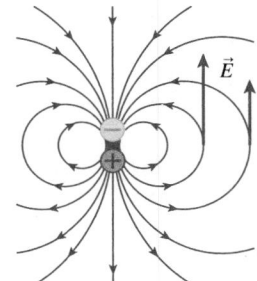
Negative charge on top

This is exactly what an **antenna** does. FIGURE 34.26 shows two metal wires attached to the terminals of an oscillating voltage source. The figure shows an instant when the top wire is negative and the bottom is positive, but these will reverse in half a cycle. The wire is basically an oscillating dipole, and it creates an oscillating electric field. The oscillating $\vec{E}$ induces an oscillating $\vec{B}$, and they take off as an electromagnetic wave at speed $v_{em} = c$. The wave does need oscillating charges as a *wave source*, but once created it is self-sustaining and independent of the source. The antenna might be destroyed, but the wave could travel billions of light years across the universe, bearing the legacy of James Clerk Maxwell.

FIGURE 34.26 An antenna generates a self-sustaining electromagnetic wave.

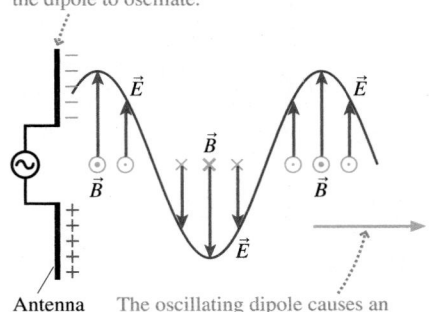

An oscillating voltage causes the dipole to oscillate.

Antenna wire

The oscillating dipole causes an electromagnetic wave to move away from the antenna at speed $v_{em} = c$.

STOP TO THINK 34.5 The amplitude of the oscillating electric field at your cell phone is 4.0 μV/m when you are 10 km east of the broadcast antenna. What is the electric field amplitude when you are 20 km east of the antenna?

a. 1.0 μV/m   b. 2.0 μV/m
c. 4.0 μV/m   d. There's not enough information to tell.

34.7 Polarization

The plane of the electric field vector $\vec{E}$ and the Poynting vector $\vec{S}$ (the direction of propagation) is called the **plane of polarization** of an electromagnetic wave. Figure 34.27 shows two electromagnetic waves moving along the *x*-axis. The electric field in FIGURE 34.27a oscillates vertically, so we would say that this wave is *vertically polarized*. Similarly the wave in FIGURE 34.27b is *horizontally polarized*. Other polarizations are possible, such as a wave polarized 30° away from horizontal.

FIGURE 34.27 The plane of polarization is the plane in which the electric field vector oscillates.

(a) Vertical polarization

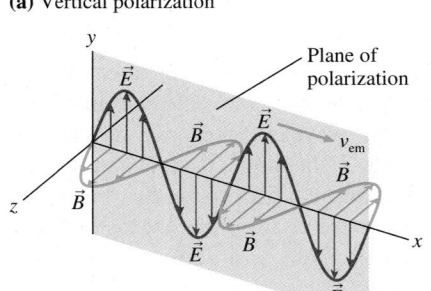

(b) Horizontal polarization

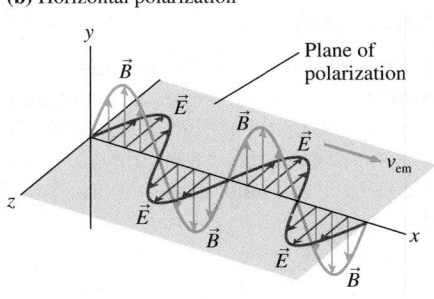

NOTE ▶ This use of the term "polarization" is completely independent of the idea of *charge polarization* that you learned about in Chapter 25. ◀

Some wave sources, such as lasers and radio antennas, emit *polarized* electromagnetic waves with a well-defined plane of polarization. By contrast, most natural sources of electromagnetic radiation are unpolarized, emitting waves whose electric fields oscillate randomly with all possible orientations.

A few natural sources are *partially polarized,* meaning that one direction of polarization is more prominent than others. The light of the sky at right angles to the sun is partially polarized because of how the sun's light scatters from air molecules to create skylight. Bees and other insects make use of this partial polarization to navigate. Light reflected from a flat, horizontal surface, such as a road or the surface of a lake, has a predominantly horizontal polarization. This is the rationale for using polarizing sunglasses.

The most common way of artificially generating polarized visible light is to send unpolarized light through a *polarizing filter*. The first widely used polarizing filter was invented by Edwin Land in 1928, while he was still an undergraduate student. He developed an improved version, called Polaroid, in 1938. Polaroid, as shown in FIGURE 34.28, is a plastic sheet containing very long organic molecules known as polymers. The sheets are formed in such a way that the polymers are all aligned to form a grid, rather like the metal bars in a barbecue grill. The sheet is then chemically treated to make the polymer molecules somewhat conducting.

As a light wave travels through Polaroid, the component of the electric field oscillating parallel to the polymer grid drives the conduction electrons up and down the molecules. The electrons absorb energy from the light wave, so the parallel component of $\vec{E}$ is absorbed in the filter. But the conduction electrons can't oscillate perpendicular to the molecules, so the component of $\vec{E}$ perpendicular to the polymer grid passes through without absorption. Thus the light wave emerging from a polarizing filter is polarized perpendicular to the polymer grid.

FIGURE 34.28 A polarizing filter.

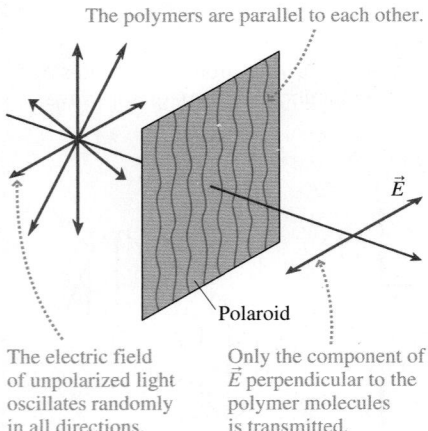

The polymers are parallel to each other.

Polaroid

The electric field of unpolarized light oscillates randomly in all directions.

Only the component of $\vec{E}$ perpendicular to the polymer molecules is transmitted.

Malus's Law

Suppose a *polarized* light wave of intensity I_0 approaches a polarizing filter. What is the intensity of the light that passes through the filter? FIGURE 34.29 shows that an oscillating electric field can be decomposed into components parallel and perpendicular to the polarizer's axis (i.e., the polarization direction transmitted by the polarizer). If we call the polarizer axis the y-axis, then the incident electric field is

$$\vec{E}_{\text{incident}} = E_\perp \hat{\imath} + E_\parallel \hat{\jmath} = E_0 \sin\theta\, \hat{\imath} + E_0 \cos\theta\, \hat{\jmath} \qquad (34.39)$$

where θ is the angle between the incident plane of polarization and the polarizer axis.

If the polarizer is ideal, meaning that light polarized parallel to the axis is 100% transmitted and light perpendicular to the axis is 100% blocked, then the electric field of the light transmitted by the filter is

$$\vec{E}_{\text{transmitted}} = E_\parallel \hat{\jmath} = E_0 \cos\theta\, \hat{\jmath} \qquad (34.40)$$

Because the intensity depends on the square of the electric field amplitude, you can see that the transmitted intensity is related to the incident intensity by

$$I_{\text{transmitted}} = I_0 \cos^2\theta \qquad \text{(incident light polarized)} \qquad (34.41)$$

This result, which was discovered experimentally in 1809, is called **Malus's law.**

FIGURE 34.30a shows that Malus's law can be demonstrated with two polarizing filters. The first, called the *polarizer,* is used to produce polarized light of intensity I_0. The second, called the *analyzer,* is rotated by angle θ relative to the polarizer. As the photographs of **FIGURE 34.30b** show, the transmission of the analyzer is (ideally) 100% when $\theta = 0°$ and steadily decreases to zero when $\theta = 90°$. Two polarizing filters with perpendicular axes, called *crossed polarizers,* block all the light.

FIGURE 34.29 An incident electric field can be decomposed into components parallel and perpendicular to a polarizer's axis.

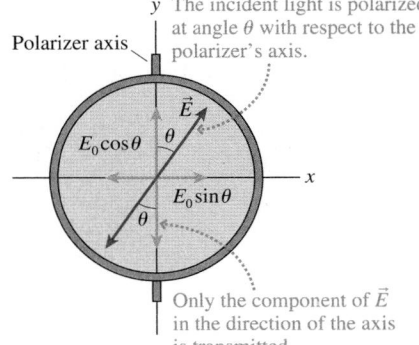

Polarizer axis

y The incident light is polarized at angle θ with respect to the polarizer's axis.

$E_0 \cos\theta$

$\vec{E}$

θ

$E_0 \sin\theta$

θ

x

Only the component of $\vec{E}$ in the direction of the axis is transmitted.

FIGURE 34.30 The intensity of the transmitted light depends on the angle between the polarizing filters.

(a)

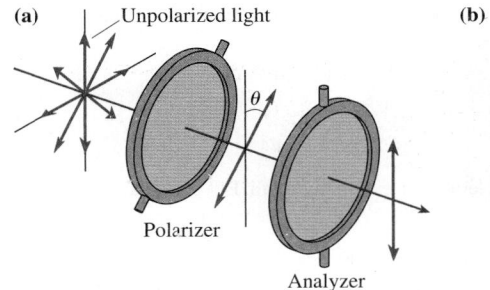

Unpolarized light

θ

Polarizer

Analyzer

(b)

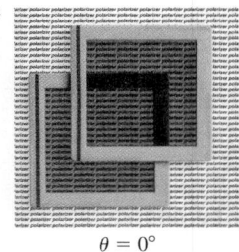

$\theta = 0°$

$\theta = 45°$

$\theta = 90°$

Suppose the light incident on a polarizing filter is *unpolarized,* as is the light incident from the left on the polarizer in Figure 34.30a. The electric field of unpolarized light varies randomly through all possible values of θ. Because the *average* value of $\cos^2\theta$ is $\frac{1}{2}$, the intensity transmitted by a polarizing filter is

$$I_{\text{transmitted}} = \frac{1}{2} I_0 \qquad \text{(incident light unpolarized)} \qquad (34.42)$$

In other words, a polarizing filter passes 50% of unpolarized light and blocks 50%.

In polarizing sunglasses, the polymer grid is aligned horizontally (when the glasses are in the normal orientation) so that the glasses transmit vertically polarized light. Most natural light is unpolarized, so the glasses reduce the light intensity by 50%. But *glare*—the reflection of the sun and the skylight from roads and other horizontal surfaces—has a strong horizontal polarization. This light is almost completely blocked by the Polaroid, so the sunglasses "cut glare" without affecting the main scene you wish to see.

You can test whether your sunglasses are polarized by holding them in front of you and rotating them as you look at the glare reflecting from a horizontal surface. Polarizing sunglasses substantially reduce the glare when the glasses are "normal" but not when the glasses are 90° from normal. (You can also test them against a pair of sunglasses known to be polarizing by seeing if all light is blocked when the lenses of the two pairs are crossed.)

The vertical polarizer blocks the horizontally polarized glare from the surface of the water.

STOP TO THINK 34.6 Unpolarized light of equal intensity is incident on four pairs of polarizing filters. Rank in order, from largest to smallest, the intensities I_a to I_d transmitted through the second polarizer of each pair.

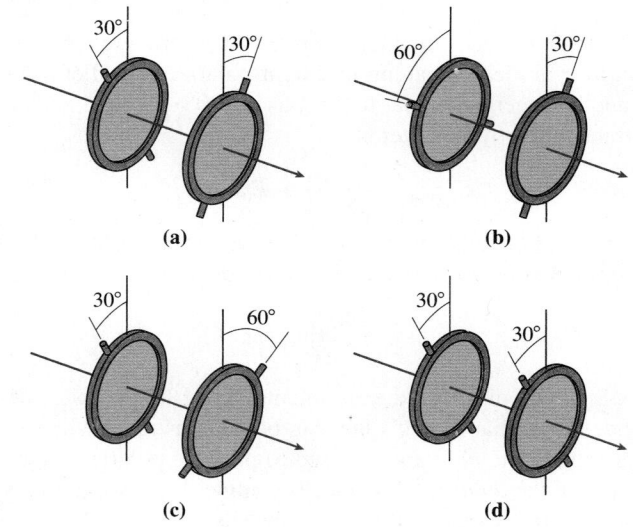

(a)   (b)

(c)   (d)

CHALLENGE EXAMPLE 34.6 **Light propulsion**

Future space rockets might propel themselves by firing laser beams, rather than exhaust gases, out the back. The acceleration would be small, but it could continue for months or years in the vacuum of space. Consider a 1200 kg unmanned space probe powered by a 15 MW laser. After one year, how far will it have traveled and how fast will it be going?

MODEL Assume the laser efficiency is so high that it can be powered for a year with a negligible mass of fuel.

SOLVE Light waves transfer not only energy but also momentum, which is how they exert a radiation-pressure force. We found that the radiation force of a light beam of power P is

$$F = \frac{P}{c}$$

From Newton's third law, the emitted light waves must exert an equal-but-opposite reaction force on the source of the light. In this case, the emitted light exerts a force of this magnitude on the space probe to which the laser is attached. This reaction force causes the probe to accelerate at

$$a = \frac{F}{m} = \frac{P}{mc} = \frac{15 \times 10^6 \text{ W}}{(1200 \text{ kg})(3.0 \times 10^8 \text{ m/s})}$$
$$= 4.2 \times 10^{-5} \text{ m/s}^2$$

As expected, the acceleration is extremely small. But one year is a large amount of time: $\Delta t = 3.15 \times 10^7$ s. After one year of acceleration,

$$v = a\Delta t = 1300 \text{ m/s}$$

$$d = \tfrac{1}{2}a(\Delta t)^2 = 2.1 \times 10^{10} \text{ m}$$

The space probe will have traveled 2.1×10^{10} m and will be going 1300 m/s.

ASSESS Even after a year, the speed is not exceptionally fast—only about 2900 mph. But the probe will have traveled a substantial distance, about 25% of the distance to Mars.

SUMMARY

The goal of Chapter 34 has been to study the properties of electromagnetic fields and waves.

General Principles

Maxwell's Equations

These equations govern electromagnetic fields:

$$\oint \vec{E} \cdot d\vec{A} = \frac{Q_{\text{in}}}{\epsilon_0}$$ Gauss's law

$$\oint \vec{B} \cdot d\vec{A} = 0$$ Gauss's law for magnetism

$$\oint \vec{E} \cdot d\vec{s} = -\frac{d\Phi_m}{dt}$$ Faraday's law

$$\oint \vec{B} \cdot d\vec{s} = \mu_0 I_{\text{through}} + \epsilon_0 \mu_0 \frac{d\Phi_e}{dt}$$ Ampère-Maxwell law

Maxwell's equations tell us that:

An electric field can be created by
- Charged particles
- A changing magnetic field

A magnetic field can be created by
- A current
- A changing electric field

Lorentz Force

This force law governs the interaction of charged particles with electromagnetic fields:

$$\vec{F} = q(\vec{E} + \vec{v} \times \vec{B})$$

- An electric field exerts a force on any charged particle.
- A magnetic field exerts a force on a moving charged particle.

Field Transformations

Fields measured in reference frame A to be $\vec{E}_A$ and $\vec{B}_A$ are found in frame B to be

$$\vec{E}_B = \vec{E}_A + \vec{v}_{BA} \times \vec{B}_A$$

$$\vec{B}_B = \vec{B}_A - \frac{1}{c^2} \vec{v}_{BA} \times \vec{E}_A$$

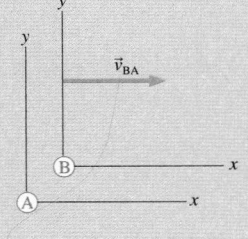

Important Concepts

Induced fields

An induced electric field is created by a changing magnetic field.

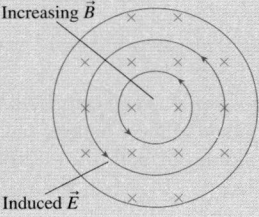

An induced magnetic field is created by a changing electric field.

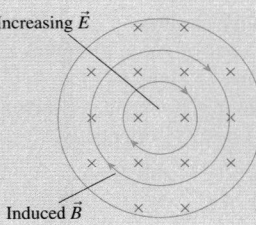

These fields can exist independently of charges and currents.

An electromagnetic wave is a self-sustaining electromagnetic field.
- An em wave is a transverse wave with $\vec{E}$, $\vec{B}$, and $\vec{v}_{\text{em}}$ mutually perpendicular.
- An em wave propagates with speed $v_{\text{em}} = c = 1/\sqrt{\epsilon_0 \mu_0}$.
- The electric and magnetic field strengths are related by $E = cB$.
- The **Poynting vector** $\vec{S} = (\vec{E} \times \vec{B})/\mu_0$ is the energy transfer in the direction of travel.
- The wave **intensity** is $I = P/A = (1/2c\mu_0)E_0^2 = (c\epsilon_0/2)E_0^2$.

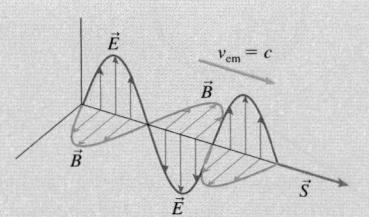

Applications

Polarization

The electric field and the Poynting vector define the **plane of polarization.** The intensity of polarized light transmitted through a polarizing filter is given by Malus's law:

$$I = I_0 \cos^2 \theta$$

where θ is the angle between the electric field and the polarizer axis.

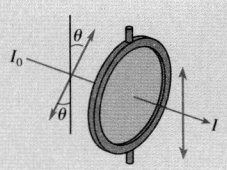

Terms and Notation

Galilean field transformation equations	Poynting vector, $\vec{S}$	antenna
displacement current	intensity, I	plane of polarization
Maxwell's equations	radiation pressure, p_{rad}	Malus's law

CONCEPTUAL QUESTIONS

1. Andre is flying his spaceship to the left through the laboratory magnetic field of **FIGURE Q34.1**.
   a. Does Andre see a magnetic field? If so, in which direction does it point?
   b. Does Andre see an electric field? If so, in which direction does it point?

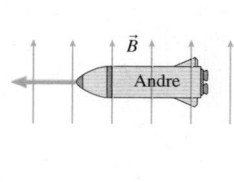

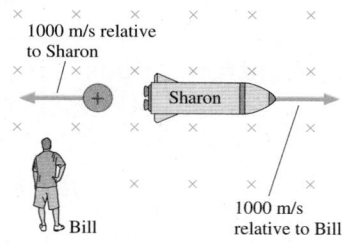

FIGURE Q34.1   **FIGURE Q34.2**

2. Sharon drives her rocket through the magnetic field of **FIGURE Q34.2** traveling to the right at a speed of 1000 m/s as measured by Bill. As she passes Bill, she shoots a positive charge backward at a speed of 1000 m/s relative to her.
   a. According to Sharon, what kind of force or forces act on the charge? In which directions? Explain.
   b. According to Bill, what kind of force or forces act on the charge? In which directions? Explain.

3. If you curl the fingers of your right hand as shown, are the electric fluxes in **FIGURE Q34.3** positive or negative?

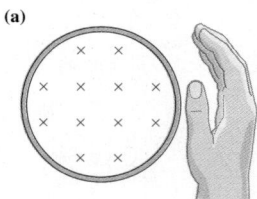   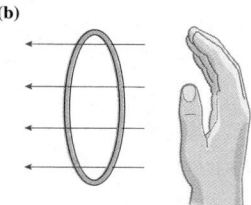

FIGURE Q34.3

4. What is the current through surface S in **FIGURE Q34.4** if you curl your right fingers in the direction of the arrow?

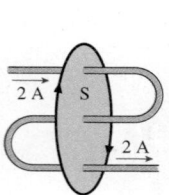

FIGURE Q34.4   **FIGURE Q34.5**

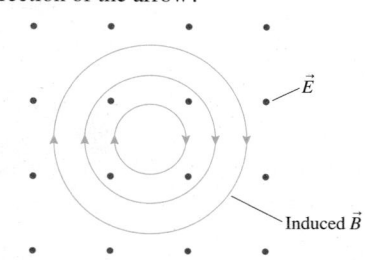

5. Is the electric field strength in **FIGURE Q34.5** increasing, decreasing, or not changing? Explain.

6. Do the situations in **FIGURE Q34.6** represent possible electromagnetic waves? If not, why not?

(a) 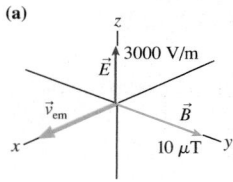   (b)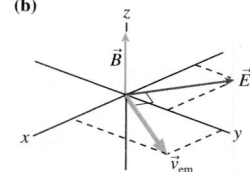

FIGURE Q34.6

7. In what directions are the electromagnetic waves traveling in **FIGURE Q34.7**?

(a) 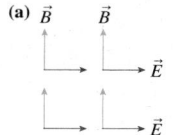   (b)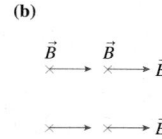

FIGURE Q34.7

8. The intensity of an electromagnetic wave is 10 W/m². What will the intensity be if:
   a. The amplitude of the electric field is doubled?
   b. The amplitude of the magnetic field is doubled?
   c. The amplitudes of both the electric and the magnetic fields are doubled?
   d. The frequency is doubled?

9. Older televisions used a *loop antenna* like the one in **FIGURE Q34.9**. How does this antenna work?

FIGURE Q34.9

10. A vertically polarized electromagnetic wave passes through the five polarizers in **FIGURE Q34.10**. Rank in order, from largest to smallest, the transmitted intensities I_a to I_e.

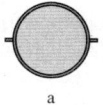    

a   b   c   d   e

FIGURE Q34.10

EXERCISES AND PROBLEMS

Problems labeled ▦ integrate material from earlier chapters.

Exercises

Section 34.1 *E* or *B*? It Depends on Your Perspective

1. | A rocket cruises past a laboratory at 1.00×10^6 m/s in the positive x-direction just as a proton is launched with velocity (in the laboratory frame) $\vec{v} = (1.41 \times 10^6 \,\hat{\imath} + 1.41 \times 10^6 \,\hat{\jmath})$ m/s. What are the proton's speed and its angle from the y-axis in (a) the laboratory frame and (b) the rocket frame?

2. | FIGURE EX34.2 shows the electric and magnetic field in frame A. A rocket in frame B travels parallel to one of the axes of the A coordinate system. Along which axis must the rocket travel, and in which direction, in order for the rocket scientists to measure (a) $B_B > B_A$, (b) $B_B = B_A$, and (c) $B_B < B_A$?

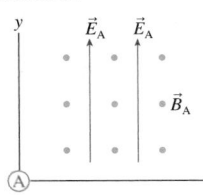

FIGURE EX34.2

3. ‖ Scientists in the laboratory create a uniform electric field $\vec{E} = 1.0 \times 10^6 \,\hat{k}$ V/m in a region of space where $\vec{B} = \vec{0}$. What are the fields in the reference frame of a rocket traveling in the positive x-direction at 1.0×10^6 m/s?

4. | Laboratory scientists have created the electric and magnetic fields shown in FIGURE EX34.4. These fields are also seen by scientists that zoom past in a rocket traveling in the x-direction at 1.0×10^6 m/s. According to the rocket scientists, what angle does the electric field make with the axis of the rocket?

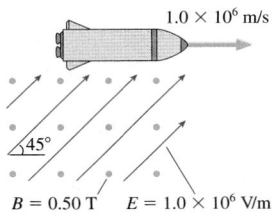

FIGURE EX34.4

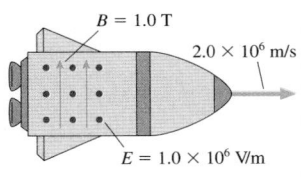

FIGURE EX34.5

5. | A rocket zooms past the earth at $v = 2.0 \times 10^6$ m/s. Scientists on the rocket have created the electric and magnetic fields shown in FIGURE EX34.5. What are the fields measured by an earthbound scientist?

Section 34.2 The Field Laws Thus Far

Section 34.3 The Displacement Current

6. ‖ The magnetic field is uniform over each face of the box shown in FIGURE EX34.6. What are the magnetic field strength and direction on the front surface?

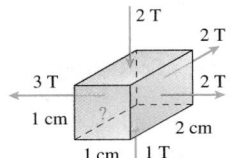

FIGURE EX34.6

7. | Show that the quantity $\epsilon_0 (d\Phi_e/dt)$ has units of current.

8. ‖ Show that the displacement current inside a parallel-plate capacitor can be written $C(dV_C/dt)$.

9. | What capacitance, in μF, has its potential difference increasing at 1.0×10^6 V/s when the displacement current in the capacitor is 1.0 A?

10. ‖ A 10-cm-diameter parallel-plate capacitor has a 1.0 mm spacing. The electric field between the plates is increasing at the rate 1.0×10^6 V/m s. What is the magnetic field strength (a) on the axis, (b) 3.0 cm from the axis, and (c) 7.0 cm from the axis?

11. ‖ A 5.0-cm-diameter parallel-plate capacitor has a 0.50 mm gap. What is the displacement current in the capacitor if the potential difference across the capacitor is increasing at 500,000 V/s?

Section 34.5 Electromagnetic Waves

12. | What is the electric field amplitude of an electromagnetic wave whose magnetic field amplitude is 2.0 mT?

13. | What is the magnetic field amplitude of an electromagnetic wave whose electric field amplitude is 10 V/m?

14. | The magnetic field of an electromagnetic wave in a vacuum is $B_z = (3.00 \,\mu\text{T}) \sin((1.00 \times 10^7)x - \omega t)$, where x is in m and t is in s. What are the wave's (a) wavelength, (b) frequency, and (c) electric field amplitude?

15. ‖ The electric field of an electromagnetic wave in a vacuum is $E_y = (20.0 \text{ V/m}) \cos((6.28 \times 10^8)x - \omega t)$, where x is in m and t is in s. What are the wave's (a) wavelength, (b) frequency, and (c) magnetic field amplitude?

Section 34.6 Properties of Electromagnetic Waves

16. | A radio wave is traveling in the negative y-direction. What is the direction of $\vec{E}$ at a point where $\vec{B}$ is in the positive x-direction?

17. | a. What is the magnetic field amplitude of an electromagnetic wave whose electric field amplitude is 100 V/m?
   b. What is the intensity of the wave?

18. | A radio receiver can detect signals with electric field amplitudes as small as 300 μV/m. What is the intensity of the smallest detectable signal?

19. ‖ A helium-neon laser emits a 1.0-mm-diameter laser beam with a power of 1.0 mW. What are the amplitudes of the electric and magnetic fields of the light wave?

20. ‖ A 200 MW laser pulse is focused with a lens to a diameter of 2.0 μm.
   a. What is the laser beam's electric field amplitude at the focal point?
   b. What is the ratio of the laser beam's electric field to the electric field that keeps the electron bound to the proton of a hydrogen atom? The radius of the electron orbit is 0.053 nm.

21. ‖ A radio antenna broadcasts a 1.0 MHz radio wave with 25 kW of power. Assume that the radiation is emitted uniformly in all directions.
   a. What is the wave's intensity 30 km from the antenna?
   b. What is the electric field amplitude at this distance?

22. ‖ At what distance from a 10 W point source of electromagnetic waves is the magnetic field amplitude 1.0 μT?

23. | A 1000 W carbon-dioxide laser emits light with a wavelength of 10 μm into a 3.0-mm-diameter laser beam. What force does the laser beam exert on a completely absorbing target?

Section 34.7 Polarization

24. | FIGURE EX34.24 shows a vertically polarized radio wave of frequency 1.0×10^6 Hz traveling into the page. The maximum electric field strength is 1000 V/m. What are
    a. The maximum magnetic field strength?
    b. The magnetic field strength and direction at a point where $\vec{E} = $ (500 V/m, down)?

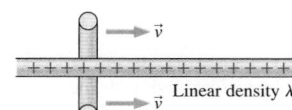

FIGURE EX34.24

25. || Only 25% of the intensity of a polarized light wave passes through a polarizing filter. What is the angle between the electric field and the axis of the filter?
26. || A 200 mW vertically polarized laser beam passes through a polarizing filter whose axis is 35° from horizontal. What is the power of the laser beam as it emerges from the filter?
27. || Unpolarized light with intensity 350 W/m² passes first through a polarizing filter with its axis vertical, then through a second polarizing filter. It emerges from the second filter with intensity 131 W/m². What is the angle from vertical of the axis of the second polarizing filter?

Problems

28. || What is the force (magnitude and direction) on the proton in FIGURE P34.28? Give the direction as an angle cw or ccw from vertical.

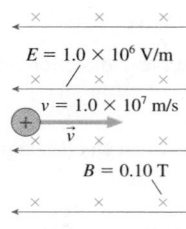

FIGURE P34.28

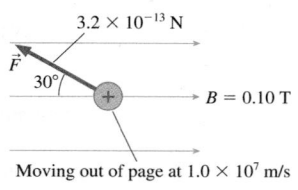

FIGURE P34.29

29. || What are the electric field strength and direction at the position of the proton in FIGURE P34.29?
30. | What electric field strength and direction will allow the electron in FIGURE P34.30 to pass through this region of space without being deflected?

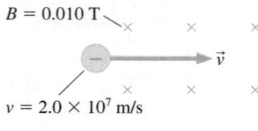

FIGURE P34.30

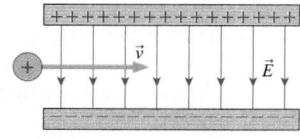

FIGURE P34.31

31. | A proton is fired with a speed of 1.0×10^6 m/s through the parallel-plate capacitor shown in FIGURE P34.31. The capacitor's electric field is $\vec{E} = (1.0 \times 10^5$ V/m, down).
    a. What magnetic field $\vec{B}$, both strength and direction, must be applied to allow the proton to pass through the capacitor with no change in speed or direction?

b. Find the electric and magnetic fields in the proton's reference frame.
    c. How does an experimenter in the proton's frame explain that the proton experiences no force as the charged plates fly by?
32. ||| An electron travels with $\vec{v} = 5.0 \times 10^6\,\hat{i}$ m/s through a point in space where $\vec{E} = (2.0 \times 10^5\,\hat{i} - 2.0 \times 10^5\,\hat{j})$ V/m and $\vec{B} = -0.10\,\hat{k}$ T. What is the force on the electron?
33. || A very long, 1.0-mm-diameter wire carries a 2.5 A current from left to right. Thin plastic insulation on the wire is positively charged with linear charge density 2.5 nC/cm. A mosquito 1.0 cm from the center of the wire would like to move in such a way as to experience an electric field but no magnetic field. How fast and which direction should she fly?
34. || In FIGURE P34.34, a circular loop of radius r travels with speed v along a charged wire having linear charge density λ. The wire is at rest in the laboratory frame, and it passes through the center of the loop.
    a. What are $\vec{E}$ and $\vec{B}$ at a point on the loop as measured by a scientist in the laboratory? Include both strength and direction.
    b. What are the fields $\vec{E}$ and $\vec{B}$ at a point on the loop as measured by a scientist in the frame of the loop?
    c. Show that an experimenter in the loop's frame sees a current $I = \lambda v$ passing through the center of the loop.
    d. What electric and magnetic fields would an experimenter in the loop's frame calculate at distance r from the current of part c?
    e. Show that your fields of parts b and d are the same.

FIGURE P34.34

35. || The magnetic field inside a 4.0-cm-diameter superconducting solenoid varies sinusoidally between 8.0 T and 12.0 T at a frequency of 10 Hz.
    a. What is the maximum electric field strength at a point 1.5 cm from the solenoid axis?
    b. What is the value of B at the instant E reaches its maximum value?
36. || A simple series circuit consists of a 150 Ω resistor, a 25 V battery, a switch, and a 2.5 pF parallel-plate capacitor (initially uncharged) with plates 5.0 mm apart. The switch is closed at $t = 0$ s.
    a. After the switch is closed, find the maximum electric flux and the maximum displacement current through the capacitor.
    b. Find the electric flux and the displacement current at $t = 0.50$ ns.
37. || A wire with conductivity σ carries current I. The current is increasing at the rate dI/dt.
    a. Show that there is a displacement current in the wire equal to $(\epsilon_0/\sigma)(dI/dt)$.
    b. Evaluate the displacement current for a copper wire in which the current is increasing at 1.0×10^6 A/s.
38. || A 10 A current is charging a 1.0-cm-diameter parallel-plate capacitor.
    a. What is the magnetic field strength at a point 2.0 mm radially from the center of the wire leading to the capacitor?
    b. What is the magnetic field strength at a point 2.0 mm radially from the center of the capacitor?

39. ‖ FIGURE P34.39 shows the voltage across a 0.10 μF capacitor. Draw a graph showing the displacement current through the capacitor as a function of time.

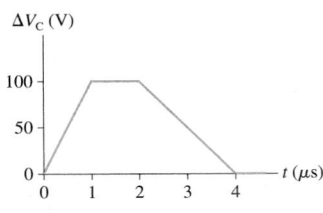

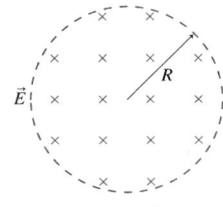

FIGURE P34.39   FIGURE P34.40

40. ‖ FIGURE P34.40 shows the electric field inside a cylinder of radius $R = 3.0$ mm. The field strength is increasing with time as $E = 1.0 \times 10^8 t^2$ V/m, where t is in s. The electric field outside the cylinder is always zero, and the field inside the cylinder was zero for $t < 0$.
   a. Find an expression for the electric flux Φ_e through the entire cylinder as a function of time.
   b. Draw a picture showing the magnetic field lines inside and outside the cylinder. Be sure to include arrowheads showing the field's direction.
   c. Find an expression for the magnetic field strength as a function of time at a distance $r < R$ from the center. Evaluate the magnetic field strength at $r = 2.0$ mm, $t = 2.0$ s.
   d. Find an expression for the magnetic field strength as a function of time at a distance $r > R$ from the center. Evaluate the magnetic field strength at $r = 4.0$ mm, $t = 2.0$ s.

41. ‖ A 1.0 μF capacitor is discharged, starting at $t = 0$ s. The displacement current through the plates is $I_{disp} = (10 \text{ A})\exp(-t/2.0 \text{ }\mu s)$. What was the capacitor's initial voltage $(\Delta V_C)_0$?

42. ‖ At one instant, the electric and magnetic fields at one point of an electromagnetic wave are $\vec{E} = (200\,\hat{\imath} + 300\,\hat{\jmath} - 50\,\hat{k})$ V/m and $\vec{B} = B_0(7.3\,\hat{\imath} - 7.3\,\hat{\jmath} + a\,\hat{k})$ μT.
   a. What are the values of a and B_0?
   b. What is the Poynting vector at this time and position?

43. ‖ a. Show that u_E and u_B, the energy densities of the electric and magnetic fields, are equal to each other in an electromagnetic wave. In other words, show that the wave's energy is divided equally between the electric field and the magnetic field.
   b. What is the total energy density in an electromagnetic wave of intensity 1000 W/m²?

44. ‖ Assume that a 7.0-cm-diameter, 100 W lightbulb radiates all its energy as a single wavelength of visible light. Estimate the electric and magnetic field strengths at the surface of the bulb.

45. ‖ The intensity of sunlight reaching the earth is 1360 W/m².
   a. What is the power output of the sun?
   b. What is the intensity of sunlight on Mars?

46. ‖ A cube of water 10 cm on a side is placed in a microwave beam having $E_0 = 11$ kV/m. The microwaves illuminate one face of the cube, and the water absorbs 80% of the incident energy. How long will it take to raise the water temperature by 50°C? Assume that the water has no heat loss during this time.

47. ‖ A laser beam passes through a converging lens with a focal length f. At what distance past the lens has the laser beam's (a) intensity and (b) electric field strength increased by a factor of 4?

48. ‖ When the Voyager 2 spacecraft passed Neptune in 1989, it was 4.5×10^9 km from the earth. Its radio transmitter, with which it sent back data and images, broadcast with a mere 21 W of power. Assuming that the transmitter broadcast equally in all directions,
   a. What signal intensity was received on the earth?
   b. What electric field amplitude was detected?
   The received signal was somewhat stronger than your result because the spacecraft used a directional antenna, but not by much.

49. ‖ In reading the instruction manual that came with your garage-door opener, you see that the transmitter unit in your car produces a 250 mW signal and that the receiver unit is supposed to respond to a radio wave of the correct frequency if the electric field amplitude exceeds 0.10 V/m. You wonder if this is really true. To find out, you put fresh batteries in the transmitter and start walking away from your garage while opening and closing the door. Your garage door finally fails to respond when you're 42 m away. Are the manufacturer's claims true?

50. ‖ The maximum electric field strength in air is 3.0 MV/m. Stronger electric fields ionize the air and create a spark. What is the maximum power that can be delivered by a 1.0-cm-diameter laser beam propagating through air?

51. ‖ A LASIK vision-correction system uses a laser that emits
BIO 10-ns-long pulses of light, each with 2.5 mJ of energy. The laser beam is focused to a 0.85-mm-diameter circle on the cornea. What is the electric field amplitude of the light wave at the cornea?

52. ‖ The intensity of sunlight reaching the earth is 1360 W/m². Assuming all the sunlight is absorbed, what is the radiation-pressure force on the earth? Give your answer (a) in newtons and (b) as a fraction of the sun's gravitational force on the earth.

53. ‖ For radio and microwaves, the depth of penetration into the
BIO human body is proportional to $\lambda^{1/2}$. If 27 MHz radio waves penetrate to a depth of 14 cm, how far do 2.4 GHz microwaves penetrate?

54. ‖ A laser beam shines straight up onto a flat, black foil of mass m.
   a. Find an expression for the laser power P needed to levitate the foil.
   b. Evaluate P for a foil with a mass of 25 μg.

55. ‖ For a science project, you would like to horizontally suspend an 8.5 by 11 inch sheet of black paper in a vertical beam of light whose dimensions exactly match the paper. If the mass of the sheet is 1.0 g, what light intensity will you need?

56. ‖ You've recently read about a chemical laser that generates a 20-cm-diameter, 25 MW laser beam. One day, after physics class, you start to wonder if you could use the radiation pressure from this laser beam to launch small payloads into orbit. To see if this might be feasible, you do a quick calculation of the acceleration of a 20-cm-diameter, 100 kg, perfectly absorbing block. What speed would such a block have if pushed *horizontally* 100 m along a frictionless track by such a laser?

57. ‖ An 80 kg astronaut has gone outside his space capsule to do some repair work. Unfortunately, he forgot to lock his safety tether in place, and he has drifted 5.0 m away from the capsule. Fortunately, he has a 1000 W portable laser with fresh batteries that will operate it for 1.0 h. His only chance is to accelerate himself toward the space capsule by firing the laser in the opposite direction. He has a 10-h supply of oxygen. How long will it take him to reach safety?

58. ‖ Unpolarized light of intensity I_0 is incident on three polarizing filters. The axis of the first is vertical, that of the second is 45° from vertical, and that of the third is horizontal. What light intensity emerges from the third filter?

Challenge Problems

59. An electron travels with $\vec{v} = 5.0 \times 10^6 \hat{\imath}$ m/s through a point in space where $\vec{B} = 0.10 \hat{\jmath}$ T. The force on the electron at this point is $\vec{F} = (9.6 \times 10^{-14} \hat{\imath} - 9.6 \times 10^{-14} \hat{k})$ N. What is the electric field?

60. A 4.0-cm-diameter parallel-plate capacitor with a 1.0 mm spacing is charged to 1000 V. A switch closes at $t = 0$ s, and the capacitor is discharged through a wire with 0.20 Ω resistance.
    a. Find an expression for the magnetic field strength inside the capacitor at $r = 1.0$ cm as a function of time.
    b. Draw a graph of B versus t.

61. The radar system at an airport broadcasts 11 GHz microwaves with 150 kW of power. An approaching airplane with a 31 m² cross section is 30 km away. Assume that the radar broadcasts uniformly in all directions and that the airplane scatters microwaves uniformly in all directions. What is the electric field strength of the microwave signal received back at the airport 200 μs later?

62. Large quantities of dust should have been left behind after the creation of the solar system. Larger dust particles, comparable in size to soot and sand grains, are common. They create shooting stars when they collide with the earth's atmosphere. But very small dust particles are conspicuously absent. Astronomers believe that the very small dust particles have been blown out of the solar system by the sun. By comparing the forces on dust particles, determine the diameter of the smallest dust particles that can remain in the solar system over long periods of time. Assume that the dust particles are spherical, black, and have a density of 2000 kg/m³. The sun emits electromagnetic radiation with power 3.9×10^{26} W.

63. Consider current I passing through a resistor of radius r, length L, and resistance R.
    a. Determine the electric and magnetic fields at the surface of the resistor. Assume that the electric field is uniform throughout, including at the surface.
    b. Determine the strength and direction of the Poynting vector at the surface of the resistor.
    c. Show that the flux of the Poynting vector (i.e., the integral of $\vec{S} \cdot d\vec{A}$) over the surface of the resistor is $I^2 R$. Then give an interpretation of this result.

64. Unpolarized light of intensity I_0 is incident on a stack of 7 polarizing filters, each with its axis rotated 15° cw with respect to the previous filter. What light intensity emerges from the last filter?

STOP TO THINK ANSWERS

Stop to Think 34.1: b. $\vec{v}_{AB}$ is parallel to $\vec{B}_A$ hence $\vec{v}_{AB} \times \vec{B}_A$ is zero. Thus $\vec{E}_B = \vec{E}_A$ and points in the positive z-direction. $\vec{v}_{AB} \times \vec{E}_A$ points down, in the negative y-direction, so $-\vec{v}_{AB} \times \vec{E}_A/c^2$ points in the positive y-direction and causes $\vec{B}_B$ to be angled upward.

Stop to Think 34.2: $B_c > B_a > B_d > B_b$. The induced magnetic field strength depends on the *rate dE/dt* at which the electric field is changing. Steeper slopes on the graph correspond to larger magnetic fields.

Stop to Think 34.3: e. $\vec{E}$ is perpendicular to $\vec{B}$ and to $\vec{v}$, so it can only be along the z-axis. According to the Ampère-Maxwell law, $d\Phi_e/dt$ has the same sign as the line integral of $\vec{B} \cdot d\vec{s}$ around the closed curve. The integral is positive for a cw integration. Thus, from the right-hand rule, $\vec{E}$ is either into the page (negative z-direction) and increasing, or out of the page (positive z-direction) and decreasing. We can see from the figure that B is decreasing in strength as the wave moves from left to right, so E must also be decreasing. Thus $\vec{E}$ points along the positive z-axis.

Stop to Think 34.4: a. The Poynting vector $\vec{S} = (\vec{E} \times \vec{B})/\mu_0$ points in the direction of travel, which is the positive y-direction. $\vec{B}$ must point in the positive x-direction in order for $\vec{E} \times \vec{B}$ to point upward.

Stop to Think 34.5: b. The intensity along a line from the antenna decreases inversely with the square of the distance, so the intensity at 20 km is $\frac{1}{4}$ that at 10 km. But the intensity depends on the square of the electric field amplitude, or, conversely, E_0 is proportional to $I^{1/2}$. Thus E_0 at 20 km is $\frac{1}{2}$ that at 10 km.

Stop to Think 34.6: $I_d > I_a > I_b = I_c$. The intensity depends on $\cos^2 \theta$, where θ is the angle *between* the axes of the two filters. The filters in d have $\theta = 0°$. The two filters in both b and c are crossed ($\theta = 90°$) and transmit no light at all.

35 AC Circuits

Transmission lines carry alternating current at voltages as high as 500,000 V.

▶ **Looking Ahead** The goal of Chapter 35 is to understand and apply basic techniques of AC circuit analysis.

AC Electricity

The wires that transport electricity across the country—the *grid*—use alternating current, called **AC.**

Transformers allow an oscillating voltage to be "stepped up" to a higher voltage so that power can be delivered using lower currents that don't overheat the wires. Smaller transformers bring the voltage down to 120 V.

Capacitors and Inductors

You'll learn that capacitors and inductors are much more useful in AC circuits than they were in DC circuits.

The peak current and peak voltage of a capacitor or an inductor are related by a resistance-like quantity called **reactance,** also measured in ohms. An inductor's reactance increases with frequency; that of a capacitor decreases.

◀ **Looking Back**
Section 29.5 Capacitors
Section 33.8 Inductors

RLC Circuits

A circuit that is especially important in communication electronics is the series *RLC* **circuit,** consisting of a resistor, capacitor, and inductor.

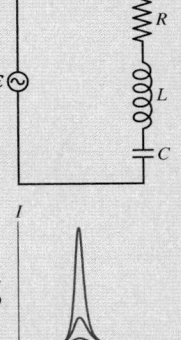

You'll learn that an *RLC* circuit exhibits *resonance*, allowing it to be tuned to a specific frequency.

Phasors

Voltages and currents oscillate, so the mathematics of AC circuits is similar to that of simple harmonic motion.

You'll learn a new way to represent oscillating quantities with rotating vectors called **phasors.** The instantaneous value of a phasor is its horizontal projection.

◀ **Looking Back**
Chapter 14 Simple harmonic motion and resonance

Filter Circuits

Simple circuits consisting of resistors and capacitors can act as *filters.*

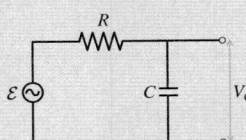

You'll see how this circuit transmits low frequencies to the output—the capacitor voltage— but blocks high frequencies. It is called a **low-pass filter.**

◀ **Looking Back**
Chapter 31 Circuit analysis

Phase and Power

The emf and the current of an AC circuit oscillate with the same frequency but usually not in phase with each other. You'll find that the *phase difference* limits an emf's ability to deliver power because the current and voltage aren't pushing and pulling together.

The power delivered to, say, a motor is reduced by a quantity called the **power factor.**

FIGURE 35.1 An oscillating emf can be represented as a graph or as a phasor diagram.

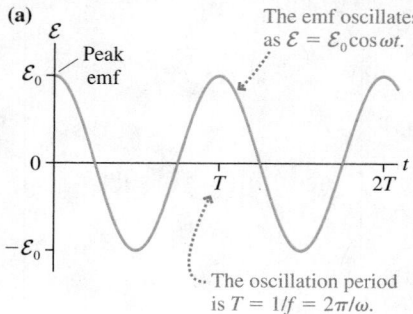

(a)

(b)

The length of the phasor is $\mathcal{E}_0$

The phasor rotates ccw at angular frequency ω.

The *phase angle* is ωt.

The tip of the phasor goes once around the circle in time T.

The instantaneous emf value $\mathcal{E}_0 \cos \omega t$ is the projection of the phasor onto the horizontal axis.

FIGURE 35.2 The correspondence between a phasor and points on a graph.

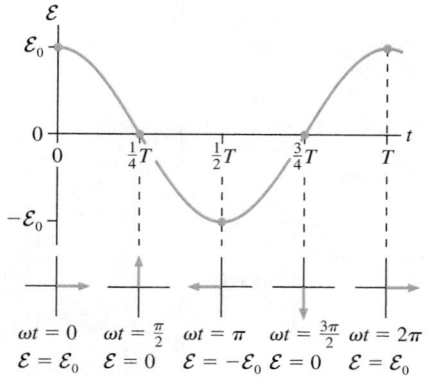

$\omega t = 0$ $\omega t = \frac{\pi}{2}$ $\omega t = \pi$ $\omega t = \frac{3\pi}{2}$ $\omega t = 2\pi$
$\mathcal{E} = \mathcal{E}_0$ $\mathcal{E} = 0$ $\mathcal{E} = -\mathcal{E}_0$ $\mathcal{E} = 0$ $\mathcal{E} = \mathcal{E}_0$

35.1 AC Sources and Phasors

One of the examples of Faraday's law cited in Chapter 33 was an electric generator. A turbine, which might be powered by expanding steam or falling water, causes a coil of wire to rotate in a magnetic field. As the coil spins, the emf and the induced current oscillate sinusoidally. The emf is alternately positive and negative, causing the charges to flow in one direction and then, a half cycle later, in the other. The oscillation frequency of the *grid* in North and South America is $f = 60$ Hz, whereas most of the rest of the world uses a 50 Hz oscillation.

The generator's peak emf—the peak voltage—is a fixed, unvarying quantity, so it might seem logical to call a generator an *alternating voltage source*. Nonetheless, circuits powered by a sinusoidal emf are called **AC circuits,** where AC stands for *alternating current.* By contrast, the steady-current circuits you studied in Chapter 31 are called **DC circuits,** for *direct current.*

AC circuits are not limited to the use of 50 Hz or 60 Hz power-line voltages. Audio, radio, television, and telecommunication equipment all make extensive use of AC circuits, with frequencies ranging from approximately 10^2 Hz in audio circuits to approximately 10^9 Hz in cell phones. These devices use *electrical oscillators* rather than generators to produce a sinusoidal emf, but the basic principles of circuit analysis are the same.

You can think of an AC generator or oscillator as a battery whose output voltage undergoes sinusoidal oscillations. The instantaneous emf of an AC generator or oscillator, shown graphically in **FIGURE 35.1a,** can be written

$$\mathcal{E} = \mathcal{E}_0 \cos \omega t \qquad (35.1)$$

where $\mathcal{E}_0$ is the peak or maximum emf and $\omega = 2\pi f$ is the angular frequency in radians per second. Recall that the units of emf are volts. As you can imagine, the mathematics of AC circuit analysis are going to be very similar to the mathematics of simple harmonic motion.

An alternative way to represent the emf and other oscillatory quantities is with the *phasor diagram* of **FIGURE 35.1b.** A **phasor** is a vector that rotates *counterclockwise* (ccw) around the origin at angular frequency ω. The length or magnitude of the phasor is the maximum value of the quantity. For example, the length of an emf phasor is $\mathcal{E}_0$. The angle ωt is the *phase angle,* an idea you learned about in Chapter 14, where we made a connection between circular motion and simple harmonic motion.

The quantity's instantaneous value, the value you would measure at time t, is the projection of the phasor onto the horizontal axis. This is also analogous to the connection between circular motion and simple harmonic motion. **FIGURE 35.2** helps you visualize the phasor rotation by showing how the phasor corresponds to the more familiar graph at several specific points in the cycle.

> **STOP TO THINK 35.1** The magnitude of the instantaneous value of the emf represented by this phasor is
>
> a. Increasing.
> b. Decreasing.
> c. Constant.
> d. It's not possible to tell without knowing t.

Resistor Circuits

In Chapter 31 you learned to analyze a circuit in terms of the current I, voltage V, and potential difference ΔV. Now, because the current and voltage are oscillating, we will use lowercase i to represent the *instantaneous* current through a circuit element and v for the circuit element's *instantaneous* voltage.

FIGURE 35.3 shows the instantaneous current i_R through a resistor R. The potential difference across the resistor, which we call the *resistor voltage* v_R, is given by Ohm's law:

$$v_R = i_R R \qquad (35.2)$$

FIGURE 35.4 shows a resistor R connected across an AC emf $\mathcal{E}$. Notice that the circuit symbol for an AC generator is ———. We can analyze this circuit in exactly the same way we analyzed a DC resistor circuit. Kirchhoff's loop law says that the sum of all the potential differences around a closed path is zero:

$$\sum \Delta V = \Delta V_{\text{source}} + \Delta V_{\text{res}} = \mathcal{E} - v_R = 0 \qquad (35.3)$$

The minus sign appears, just as it did in the equation for a DC circuit, because the potential *decreases* when we travel through a resistor in the direction of the current. We find from the loop law that $v_R = \mathcal{E} = \mathcal{E}_0 \cos \omega t$. This isn't surprising because the resistor is connected directly across the terminals of the emf.

The resistor voltage in an AC circuit can be written

$$v_R = V_R \cos \omega t \qquad (35.4)$$

where V_R is the peak or maximum voltage. You can see that $V_R = \mathcal{E}_0$ in the single-resistor circuit of Figure 35.4. Thus the current through the resistor is

$$i_R = \frac{v_R}{R} = \frac{V_R \cos \omega t}{R} = I_R \cos \omega t \qquad (35.5)$$

where $I_R = V_R/R$ is the peak current.

NOTE ▶ Ohm's law applies to both the instantaneous *and* peak currents and voltages. ◀

The resistor's instantaneous current and voltage are in phase, both oscillating as $\cos \omega t$. FIGURE 35.5 shows the voltage and the current simultaneously on a graph and as a phasor diagram. The fact that the current phasor is shorter than the voltage phasor has no significance. Current and voltage are measured in different units, so you can't compare the length of one to the length of the other. Showing the two different quantities on a single graph—a tactic that can be misleading if you're not careful—illustrates that they oscillate in phase and that their phasors rotate together at the same angle and frequency.

FIGURE 35.3 Instantaneous current i_R through a resistor.

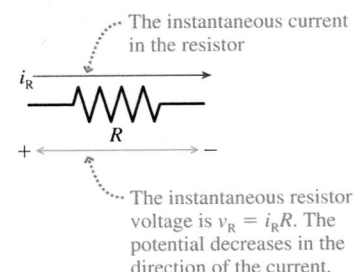

The instantaneous current in the resistor

The instantaneous resistor voltage is $v_R = i_R R$. The potential decreases in the direction of the current.

FIGURE 35.4 An AC resistor circuit.

This is the current direction when $\mathcal{E} > 0$. A half cycle later it will be in the opposite direction.

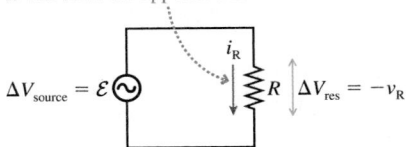

FIGURE 35.5 Graph and phasor diagram of the resistor current and voltage.

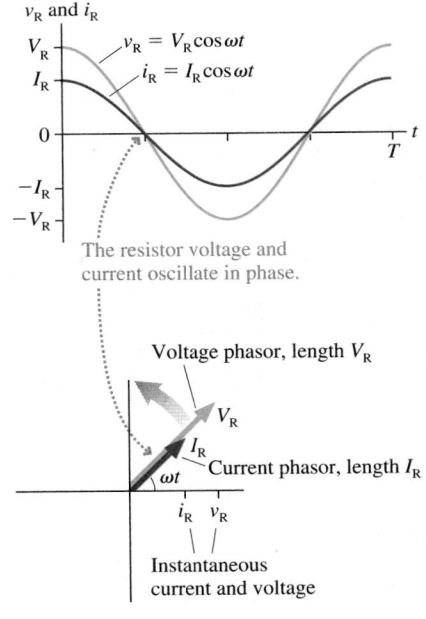

The resistor voltage and current oscillate in phase.

Voltage phasor, length V_R

Current phasor, length I_R

Instantaneous current and voltage

EXAMPLE 35.1 Finding resistor voltages

In the circuit of FIGURE 35.6, what are (a) the peak voltage across each resistor and (b) the instantaneous resistor voltages at $t = 20$ ms?

VISUALIZE Figure 35.6 shows the circuit diagram. The two resistors are in series.

SOLVE a. The equivalent resistance of the two series resistors is $R_{\text{eq}} = 5\ \Omega +$ 15 $\Omega = 20\ \Omega$. The instantaneous current through the equivalent resistance is

FIGURE 35.6 An AC resistor circuit.

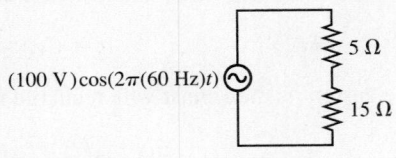

$(100\ \text{V})\cos(2\pi(60\ \text{Hz})t)$ 5 Ω 15 Ω

$$i_R = \frac{v_R}{R_{\text{eq}}} = \frac{\mathcal{E}_0 \cos \omega t}{R_{\text{eq}}} = \frac{(100\ \text{V})\cos(2\pi(60\ \text{Hz})t)}{20\ \Omega}$$

$$= (5.0\ \text{A}) \cos(2\pi(60\ \text{Hz})t)$$

Continued

The peak current is $I_R = 5.0$ A, and this is also the peak current through the two resistors that form the 20 Ω equivalent resistance. Hence the peak voltage across each resistor is

$$V_R = I_R R = \begin{cases} 25 \text{ V} & 5 \text{ Ω resistor} \\ 75 \text{ V} & 15 \text{ Ω resistor} \end{cases}$$

b. The instantaneous current at $t = 0.020$ s is

$$i_R = (5.0 \text{ A}) \cos \left(2\pi(60 \text{ Hz})(0.020 \text{ s})\right) = 1.55 \text{ A}$$

The resistor voltages at this time are

$$v_R = i_R R = \begin{cases} 7.7 \text{ V} & 5 \text{ Ω resistor} \\ 23.2 \text{ V} & 15 \text{ Ω resistor} \end{cases}$$

ASSESS The sum of the instantaneous voltages, 30.9 V, is what you would find by calculating $\mathcal{E}$ at $t = 20$ ms. This self-consistency gives us confidence in the answer.

STOP TO THINK 35.2 The resistor whose voltage and current phasors are shown here has resistance R

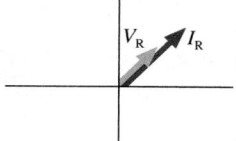

a. > 1 Ω
b. < 1 Ω
c. It's not possible to tell.

35.2 Capacitor Circuits

FIGURE 35.7 An AC capacitor circuit.

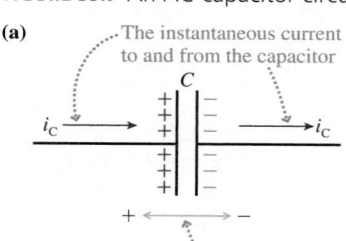

(a) The instantaneous current to and from the capacitor

The instantaneous capacitor voltage is $v_C = q/C$. The potential decreases from $+$ to $-$.

(b)

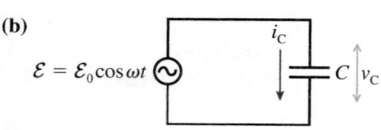

FIGURE 35.7a shows current i_C charging a capacitor with capacitance C. The instantaneous capacitor voltage is $v_C = q/C$, where $\pm q$ is the charge on the two capacitor plates at this instant. It is useful to compare Figure 35.7a to Figure 35.3 for a resistor.

FIGURE 35.7b, where capacitance C is connected across an AC source of emf $\mathcal{E}$, is the most basic capacitor circuit. The capacitor is in parallel with the source, so the capacitor voltage equals the emf: $v_C = \mathcal{E} = \mathcal{E}_0 \cos \omega t$. It will be useful to write

$$v_C = V_C \cos \omega t \tag{35.6}$$

where V_C is the peak or maximum voltage across the capacitor. You can see that $V_C = \mathcal{E}_0$ in this single-capacitor circuit.

To find the current to and from the capacitor, we first write the charge

$$q = C v_C = C V_C \cos \omega t \tag{35.7}$$

The current is the *rate* at which charge flows through the wires, $i_C = dq/dt$, thus

$$i_C = \frac{dq}{dt} = \frac{d}{dt}(C V_C \cos \omega t) = -\omega C V_C \sin \omega t \tag{35.8}$$

We can most easily see the relationship between the capacitor voltage and current if we use the trigonometric identity $-\sin(x) = \cos(x + \pi/2)$ to write

$$i_C = \omega C V_C \cos \left(\omega t + \frac{\pi}{2}\right) \tag{35.9}$$

In contrast to a resistor, a capacitor's current and voltage are *not* in phase. In FIGURE 35.8a, a graph of the instantaneous voltage v_C and current i_C, you can see that the current peaks one-quarter of a period *before* the voltage peaks. The phase angle of the current phasor on the phasor diagram of FIGURE 35.8b is $\pi/2$ rad—a quarter of a circle—larger than the phase angle of the voltage phasor.

We can summarize this finding:

The AC current of a capacitor *leads* the capacitor voltage by $\pi/2$ rad, or 90°.

The current reaches its peak value I_C at the instant the capacitor is fully discharged and $v_C = 0$. The current is zero at the instant the capacitor is fully charged.

A simple harmonic oscillator provides a mechanical analogy of the 90° phase difference between current and voltage. You learned in Chapter 14 that the position and velocity of a simple harmonic oscillator are

$$x = A \cos \omega t$$

$$v = \frac{dx}{dt} = -\omega A \sin \omega t = -v_{max} \sin \omega t = v_{max} \cos\left(\omega t + \frac{\pi}{2}\right)$$

You can see that the velocity of an oscillator leads the position by 90° in the same way that the capacitor current leads the voltage.

Capacitive Reactance

We can use Equation 35.9 to see that the peak current to and from a capacitor is $I_C = \omega C V_C$. This relationship between the peak voltage and peak current looks much like Ohm's law for a resistor if we define the **capacitive reactance** X_C to be

$$X_C \equiv \frac{1}{\omega C} \tag{35.10}$$

With this definition,

$$I_C = \frac{V_C}{X_C} \quad \text{or} \quad V_C = I_C X_C \tag{35.11}$$

The units of reactance, like those of resistance, are ohms.

NOTE ▶ Reactance relates the *peak* voltage V_C and current I_C. But reactance differs from resistance in that it does *not* relate the instantaneous capacitor voltage and current because they are out of phase. That is, $v_C \neq i_C X_C$. ◀

A resistor's resistance R is independent of the emf frequency. In contrast, as FIGURE 35.9 shows, a capacitor's reactance X_C depends inversely on the frequency. The reactance becomes very large at low frequencies (i.e., the capacitor is a large impediment to current). This makes sense because $\omega = 0$ would be a nonoscillating DC circuit, and we know that a steady DC current cannot pass through a capacitor. The reactance decreases as the frequency increases until, at very high frequencies, $X_C \approx 0$ and the capacitor begins to act like an ideal wire. This result has important consequences for how capacitors are used in many circuits.

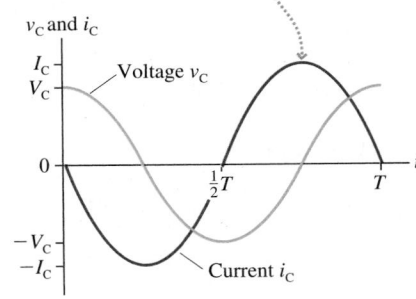

FIGURE 35.8 Graph and phasor diagrams of the capacitor current and voltage.

(a) i_C peaks $\frac{1}{4}T$ before v_C peaks. We say that the current *leads* the voltage by 90°.

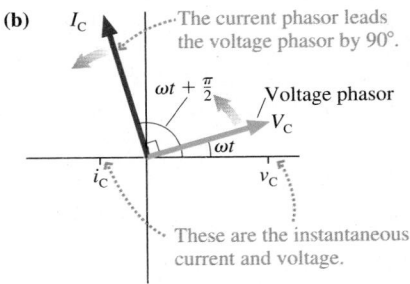

(b) The current phasor leads the voltage phasor by 90°.

These are the instantaneous current and voltage.

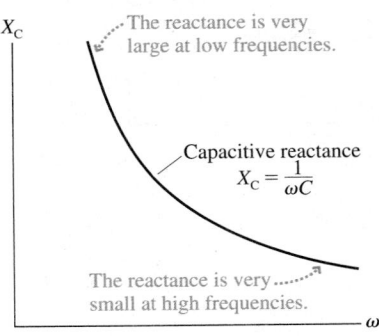

FIGURE 35.9 The capacitive reactance as a function of frequency.

The reactance is very large at low frequencies.

Capacitive reactance $X_C = \frac{1}{\omega C}$

The reactance is very small at high frequencies.

> **EXAMPLE 35.2** | **Capacitive reactance**
>
> What is the capacitive reactance of a 0.10 μF capacitor at a 100 Hz audio frequency and at a 100 MHz FM-radio frequency?
>
> **SOLVE** At 100 Hz,
>
> $$X_C(\text{at 100 Hz}) = \frac{1}{\omega C} = \frac{1}{2\pi(100 \text{ Hz})(1.0 \times 10^{-7} \text{ F})} = 16{,}000 \ \Omega$$
>
> Increasing the frequency by a factor of 10^6 decreases X_C by a factor of 10^6, giving
>
> $$X_C(\text{at 100 MHz}) = 0.016 \ \Omega$$
>
> **ASSESS** A capacitor with a substantial reactance at audio frequencies has virtually no reactance at FM-radio frequencies.

EXAMPLE 35.3   **Capacitor current**

A 10 μF capacitor is connected to a 1000 Hz oscillator with a peak emf of 5.0 V. What is the peak current to the capacitor?

VISUALIZE Figure 35.7b showed the circuit diagram. It is a simple one-capacitor circuit.

SOLVE The capacitive reactance at $\omega = 2\pi f = 6280$ rad/s is

$$X_C = \frac{1}{\omega C} = \frac{1}{(6280 \text{ rad/s})(10 \times 10^{-6} \text{ F})} = 16 \ \Omega$$

The peak voltage across the capacitor is $V_C = \mathcal{E}_0 = 5.0$ V; hence the peak current is

$$I_C = \frac{V_C}{X_C} = \frac{5.0 \text{ V}}{16 \ \Omega} = 0.31 \text{ A}$$

ASSESS Using reactance is just like using Ohm's law, but don't forget it applies to only the *peak* current and voltage, not the instantaneous values.

STOP TO THINK 35.3   What is the capacitive reactance of "no capacitor," just a continuous wire?

a. 0   b. ∞ c. Undefined

35.3 *RC* Filter Circuits

FIGURE 35.10 An *RC* circuit driven by an AC source.

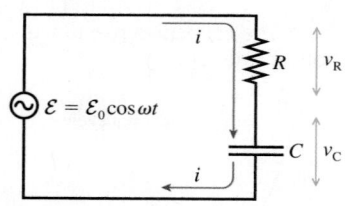

You learned in Chapter 31 that a resistance R causes a capacitor to be charged or discharged with time constant $\tau = RC$. We called this an *RC* circuit. Now that we've looked at resistors and capacitors individually, let's explore what happens if an *RC* circuit is driven continuously by an alternating current source.

FIGURE 35.10 shows a circuit in which a resistor R and capacitor C are in series with an emf $\mathcal{E}$ oscillating at angular frequency ω. Before launching into a formal analysis, let's try to understand qualitatively how this circuit will respond as the frequency is varied. If the frequency is very low, the capacitive reactance will be very large, and thus the peak current I_C will be very small. The peak current through the resistor is the same as the peak current to and from the capacitor (conservation of current requires $I_R = I_C$); hence we expect the resistor's peak voltage $V_R = I_R R$ to be very small at very low frequencies.

On the other hand, suppose the frequency is very high. Then the capacitive reactance approaches zero and the peak current, determined by the resistance alone, will be $I_R = \mathcal{E}_0/R$. The resistor's peak voltage $V_R = IR$ will approach the peak source voltage $\mathcal{E}_0$ at very high frequencies.

This reasoning leads us to expect that V_R will *increase* steadily from 0 to $\mathcal{E}_0$ as ω is increased from 0 to very high frequencies. Kirchhoff's loop law has to be obeyed, so the capacitor voltage V_C will *decrease* from $\mathcal{E}_0$ to 0 during the same change of frequency. A quantitative analysis will show us how this behavior can be used as a *filter*.

The goal of a quantitative analysis is to determine the peak current I and the two peak voltages V_R and V_C as functions of the emf amplitude $\mathcal{E}_0$ and frequency ω. Our analytic procedure is based on the fact that the instantaneous current i is the same for two circuit elements in series.

Using phasors to analyze an *RC* circuit

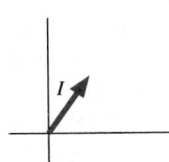

Begin by drawing a current phasor of length *I*. This is the starting point because the series circuit elements have the same current *i*. The angle at which the phasor is drawn is not relevant.

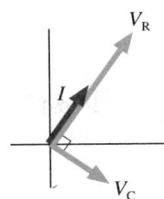

The current and voltage of a resistor are in phase, so draw a resistor voltage phasor of length V_R parallel to the current phasor *I*. The capacitor current leads the capacitor voltage by 90°, so draw a capacitor voltage phasor of length V_C that is 90° behind [i.e., clockwise (cw) from] the current phasor.

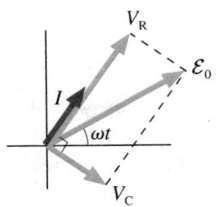

The series resistor and capacitor are in parallel with the emf, so their *instantaneous* voltages satisfy $v_R + v_C = \mathcal{E}$. This is a *vector* addition of phasors, so draw the emf phasor as the vector sum of the two voltage phasors. The emf is $\mathcal{E} = \mathcal{E}_0 \cos \omega t$, hence the emf phasor is at angle ωt.

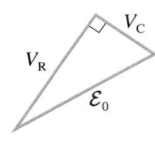

The length of the emf phasor, $\mathcal{E}_0$, is the hypotenuse of a right triangle formed by the resistor and capacitor phasors. Thus $\mathcal{E}_0^2 = V_R^2 + V_C^2$.

The relationship $\mathcal{E}_0^2 = V_R^2 + V_C^2$ is based on the peak values, not the instantaneous values, because the peak values are the lengths of the sides of the right triangle. The peak voltages are related to the peak current *I* via $V_R = IR$ and $V_C = IX_C$, thus

$$\mathcal{E}_0^2 = V_R^2 + V_C^2 = (IR)^2 + (IX_C)^2 = (R^2 + X_C^2)I^2$$
$$= (R^2 + 1/\omega^2 C^2)I^2 \tag{35.12}$$

Consequently, the peak current in the *RC* circuit is

$$I = \frac{\mathcal{E}_0}{\sqrt{R^2 + X_C^2}} = \frac{\mathcal{E}_0}{\sqrt{R^2 + 1/\omega^2 C^2}} \tag{35.13}$$

Knowing *I* gives us the two peak voltages:

$$V_R = IR = \frac{\mathcal{E}_0 R}{\sqrt{R^2 + X_C^2}} = \frac{\mathcal{E}_0 R}{\sqrt{R^2 + 1/\omega^2 C^2}}$$

$$V_C = IX_C = \frac{\mathcal{E}_0 X_C}{\sqrt{R^2 + X_C^2}} = \frac{\mathcal{E}_0/\omega C}{\sqrt{R^2 + 1/\omega^2 C^2}} \tag{35.14}$$

Frequency Dependence

Our goal was to see how the peak current and voltages vary as functions of the frequency ω. Equations 35.13 and 35.14 are rather complex and best interpreted by looking at graphs. FIGURE 35.11 is a graph of V_R and V_C versus ω.

You can see that our qualitative predictions have been borne out. That is, V_R increases from 0 to $\mathcal{E}_0$ as ω is increased, while V_C decreases from $\mathcal{E}_0$ to 0. The explanation for this behavior is that the capacitive reactance X_C decreases as ω increases. For low frequencies, where $X_C \gg R$, the circuit is primarily capacitive. For high frequencies, where $X_C \ll R$, the circuit is primarily resistive.

The frequency at which $V_R = V_C$ is called the **crossover frequency** ω_c. The *crossover* frequency is easily found by setting the two expressions in Equations 35.14 equal to each other. The denominators are the same and cancel, as does $\mathcal{E}_0$, leading to

$$\omega_c = \frac{1}{RC} \tag{35.15}$$

In practice, $f_c = \omega_c/2\pi$ is also called the crossover frequency.

FIGURE 35.11 Graph of the resistor and capacitor peak voltages as functions of the emf angular frequency ω.

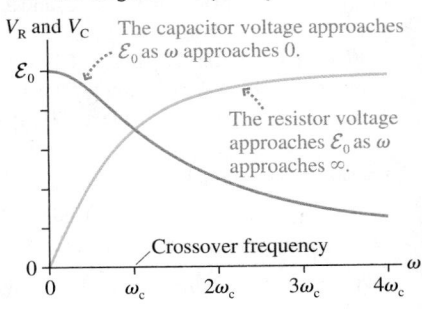

The capacitor voltage approaches $\mathcal{E}_0$ as ω approaches 0.

The resistor voltage approaches $\mathcal{E}_0$ as ω approaches ∞.

Crossover frequency

We'll leave it as a homework problem to show that $V_R = V_C = \mathcal{E}_0/\sqrt{2}$ when $\omega = \omega_c$. This may seem surprising. After all, shouldn't V_R and V_C add up to $\mathcal{E}_0$?

No! V_R and V_C are the *peak values* of oscillating voltages, not the instantaneous values. The instantaneous values do, indeed, satisfy $v_R + v_C = \mathcal{E}$ at all instants of time. But the resistor and capacitor voltages are out of phase with each other, as the phasor diagram shows, so the two circuit elements don't reach their peak values at the same time. The peak values are related by $\mathcal{E}_0^2 = V_R^2 + V_C^2$, and you can see that $V_R = V_C = \mathcal{E}_0/\sqrt{2}$ satisfies this equation.

NOTE ▶ It's very important in AC circuit analysis to make a clear distinction between instantaneous values and peak values of voltages and currents. Relationships that are true for one set of values may not be true for the other. ◀

Filters

FIGURE 35.12 Low-pass and high-pass filter circuits.

(a) Low-pass filter

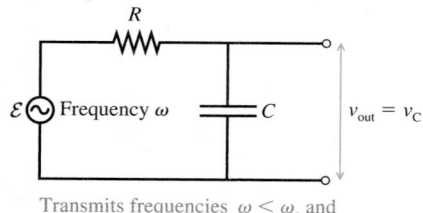

Transmits frequencies $\omega < \omega_c$ and blocks frequencies $\omega > \omega_c$

(b) High-pass filter

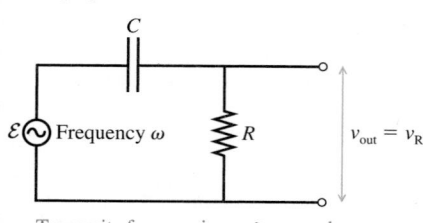

Transmits frequencies $\omega > \omega_c$ and blocks frequencies $\omega < \omega_c$

FIGURE 35.12a is the circuit we've just analyzed; the only difference is that the capacitor voltage v_C is now identified as the *output voltage* v_{out}. This is a voltage you might measure or, perhaps, send to an amplifier for use elsewhere in an electronic instrument. You can see from the capacitor voltage graph in Figure 35.11 that the peak output voltage is $V_{out} \approx \mathcal{E}_0$ if $\omega \ll \omega_c$, but $V_{out} \approx 0$ if $\omega \gg \omega_c$. In other words,

■ If the frequency of an input signal is well below the crossover frequency, the input signal is transmitted with little loss to the output.

■ If the frequency of an input signal is well above the crossover frequency, the input signal is strongly attenuated and the output is very nearly zero.

This circuit is called a **low-pass filter**.

The circuit of **FIGURE 35.12b**, which instead uses the resistor voltage v_R for the output v_{out}, is a **high-pass filter**. The output is $V_{out} \approx 0$ if $\omega \ll \omega_c$, but $V_{out} \approx \mathcal{E}_0$ if $\omega \gg \omega_c$. That is, an input signal whose frequency is well above the crossover frequency is transmitted without loss to the output.

Filter circuits are widely used in electronics. For example, a high-pass filter designed to have $f_c = 100$ Hz would pass the audio frequencies associated with speech ($f > 200$ Hz) while blocking 60 Hz "noise" that can be picked up from power lines. Similarly, the high-frequency hiss from old vinyl records can be attenuated with a low-pass filter, allowing the lower-frequency audio signal to pass.

A simple RC filter suffers from the fact that the crossover region where $V_R \approx V_C$ is fairly broad. More sophisticated filters have a sharper transition from off ($V_{out} \approx 0$) to on ($V_{out} \approx \mathcal{E}_0$), but they're based on the same principles as the RC filter analyzed here.

EXAMPLE 35.4 **Designing a filter**

For a science project, you've built a radio to listen to AM radio broadcasts at frequencies near 1 MHz. The basic circuit is an antenna, which produces a very small oscillating voltage when it absorbs the energy of an electromagnetic wave, and an amplifier. Unfortunately, your neighbor's short-wave radio broadcast at 10 MHz interferes with your reception. Having just finished physics, you decide to solve this problem by placing a filter between the antenna and the amplifier. You happen to have a 500 pF capacitor. What frequency should you select as the filter's crossover frequency? What value of resistance will you need to build this filter?

MODEL You need a low-pass filter to block signals at 10 MHz while passing the lower-frequency AM signal at 1 MHz.

VISUALIZE The circuit will look like the low-pass filter in Figure 35.12a. The oscillating voltage generated by the antenna will be the emf, and v_{out} will be sent to the amplifier.

SOLVE You might think that a crossover frequency near 5 MHz, about halfway between 1 MHz and 10 MHz, would work best. But 5 MHz is a factor of 5 higher than 1 MHz while only a factor of 2 less than 10 MHz. A crossover frequency the same factor above 1 MHz as it is below 10 MHz will give the best results. In practice, choosing $f_c = 3$ MHz would be sufficient. You can then use Equation 35.15 to select the proper resistor value:

$$R = \frac{1}{\omega_c C} = \frac{1}{2\pi(3 \times 10^6 \text{ Hz})(500 \times 10^{-12} \text{ F})}$$

$$= 106 \ \Omega \approx 100 \ \Omega$$

ASSESS Rounding to 100 Ω is appropriate because the crossover frequency was determined to only one significant figure. Such "sloppy design" is adequate when the two frequencies you need to distinguish are well separated.

STOP TO THINK 35.4 Rank in order, from largest to smallest, the crossover frequencies $(\omega_c)_a$ to $(\omega_c)_d$ of these four circuits.

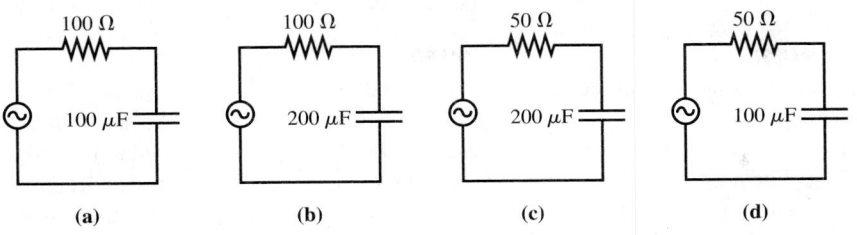

(a)   (b)   (c)   (d)

35.4 Inductor Circuits

FIGURE 35.13a shows the instantaneous current i_L through an inductor. If the current is changing, the instantaneous inductor voltage is

$$v_L = L\frac{di_L}{dt} \tag{35.16}$$

You learned in Chapter 33 that the potential decreases in the direction of the current if the current is increasing ($di_L/dt > 0$) and increases if the current is decreasing ($di_L/dt < 0$).

FIGURE 35.13b is the simplest inductor circuit. The inductor L is connected across the AC source, so the inductor voltage equals the emf: $v_L = \mathcal{E} = \mathcal{E}_0 \cos \omega t$. We can write

$$v_L = V_L \cos \omega t \tag{35.17}$$

where V_L is the peak or maximum voltage across the inductor. You can see that $V_L = \mathcal{E}_0$ in this single-inductor circuit.

We can find the inductor current i_L by integrating Equation 35.17. First, we use Equation 35.17 to write Equation 35.16 as

$$di_L = \frac{v_L}{L}dt = \frac{V_L}{L}\cos \omega t\, dt \tag{35.18}$$

Integrating gives

$$i_L = \frac{V_L}{L}\int \cos \omega t\, dt = \frac{V_L}{\omega L}\sin \omega t = \frac{V_L}{\omega L}\cos\left(\omega t - \frac{\pi}{2}\right)$$
$$= I_L \cos\left(\omega t - \frac{\pi}{2}\right) \tag{35.19}$$

where $I_L = V_L/\omega L$ is the peak or maximum inductor current.

NOTE ▶ Mathematically, Equation 35.19 could have an integration constant i_0. An integration constant would represent a constant DC current through the inductor, but there is no DC source of potential in an AC circuit. Hence, on physical grounds, we set $i_0 = 0$ for an AC circuit. ◀

We define the **inductive reactance,** analogous to the capacitive reactance, to be

$$X_L \equiv \omega L \tag{35.20}$$

FIGURE 35.13 Using an inductor in an AC circuit.

(a) The instantaneous current through the inductor

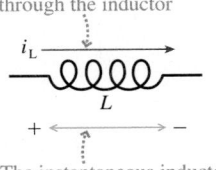

i_L

L

The instantaneous inductor voltage is $v_L = L(di_L/dt)$.

(b)

$\mathcal{E} = \mathcal{E}_0 \cos \omega t$ i_L L v_L

Then the peak current $I_L = V_L/\omega L$ and the peak voltage are related by

$$I_L = \frac{V_L}{X_L} \quad \text{or} \quad V_L = I_L X_L \tag{35.21}$$

FIGURE 35.14 The inductive reactance as a function of frequency.

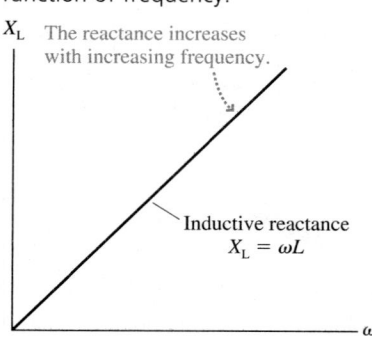

X_L The reactance increases with increasing frequency.

Inductive reactance
$X_L = \omega L$

ω

FIGURE 35.14 shows that the inductive reactance increases as the frequency increases. This makes sense. Faraday's law tells us that the induced voltage across a coil increases as the time rate of change of $\vec{B}$ increases, and $\vec{B}$ is directly proportional to the inductor current. For a given peak current I_L, $\vec{B}$ changes more rapidly at higher frequencies than at lower frequencies, and thus V_L is larger at higher frequencies than at lower frequencies.

FIGURE 35.15a is a graph of the inductor voltage and current. You can see that the current peaks one-quarter of a period *after* the voltage peaks. The angle of the current phasor on the phasor diagram of **FIGURE 35.15b** is $\pi/2$ rad less than the angle of the voltage phasor. We can summarize this finding:

The AC current through an inductor *lags* the inductor voltage by $\pi/2$ rad, or 90°.

FIGURE 35.15 Graphs and phasor diagrams of the inductor current and voltage.

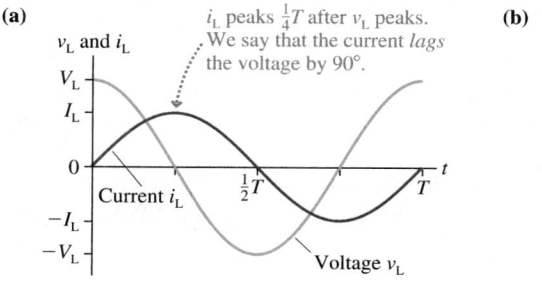

(a)

v_L and i_L

i_L peaks $\frac{1}{4}T$ after v_L peaks. We say that the current *lags* the voltage by 90°.

V_L
I_L

0

Current i_L

$\frac{1}{2}T$

T

t

$-I_L$
$-V_L$

Voltage v_L

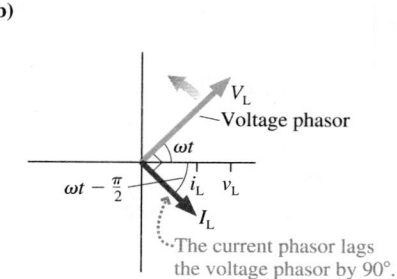

(b)

V_L

Voltage phasor

ωt

$\omega t - \frac{\pi}{2}$

i_L v_L

I_L

The current phasor lags the voltage phasor by 90°.

Current and voltage of an inductor

A 25 μH inductor is used in a circuit that oscillates at 100 kHz. The current through the inductor reaches a peak value of 20 mA at $t = 5.0$ μs. What is the peak inductor voltage, and when, closest to $t = 5.0$ μs, does it occur?

MODEL The inductor current lags the voltage by 90°, or, equivalently, the voltage reaches its peak value one-quarter period *before* the current.

VISUALIZE The circuit looks like Figure 35.13b.

SOLVE The inductive reactance at $f = 100$ kHz is

$$X_L = \omega L = 2\pi(1.0 \times 10^5 \text{ Hz})(25 \times 10^{-6} \text{ H}) = 16 \text{ }\Omega$$

Thus the peak voltage is $V_L = I_L X_L = (20 \text{ mA})(16 \text{ }\Omega) = 320 \text{ mV}$. The voltage peak occurs one-quarter period before the current peaks, and we know that the current peaks at $t = 5.0$ μs. The period of a 100 kHz oscillation is 10.0 μs, so the voltage peaks at

$$t = 5.0 \text{ }\mu\text{s} - \frac{10.0 \text{ }\mu\text{s}}{4} = 2.5 \text{ }\mu\text{s}$$

35.5 The Series *RLC* Circuit

FIGURE 35.16 A series *RLC* circuit.

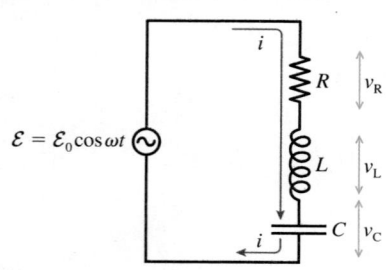

$\mathcal{E} = \mathcal{E}_0 \cos \omega t$

i

R v_R

L v_L

C v_C

i

The circuit of **FIGURE 35.16**, where a resistor, inductor, and capacitor are in series, is called a **series *RLC* circuit**. The series *RLC* circuit has many important applications because, as you will see, it exhibits resonance behavior.

The analysis, which is very similar to our analysis of the *RC* circuit in Section 35.3, will be based on a phasor diagram. Notice that the three circuit elements are in series with each other and, together, are in parallel with the emf. We can draw two conclusions that form the basis of our analysis:

1. The instantaneous current of all three elements is the same: $i = i_R = i_L = i_C$.
2. The sum of the instantaneous voltages matches the emf: $\mathcal{E} = v_R + v_L + v_C$.

Using phasors to analyze an *RLC* circuit

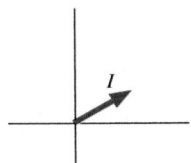 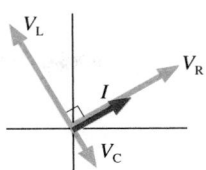 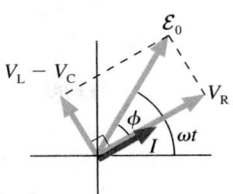 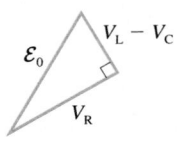

Begin by drawing a current phasor of length *I*. This is the starting point because the series circuit elements have the same current *i*.

The current and voltage of a resistor are in phase, so draw a resistor voltage phasor parallel to the current phasor *I*. The capacitor current leads the capacitor voltage by 90°, so draw a capacitor voltage phasor that is 90° behind the current phasor. The inductor current *lags* the voltage by 90°, so draw an inductor voltage phasor 90° ahead of the current phasor.

The instantaneous voltages satisfy $\mathcal{E} = v_R + v_L + v_C$. In terms of phasors, this is a *vector* addition. We can do the addition in two steps. Because the capacitor and inductor phasors are in opposite directions, their vector sum has length $V_L - V_C$. Adding the resistor phasor, at right angles, then gives the emf phasor $\mathcal{E}$ at angle ωt.

The length $\mathcal{E}_0$ of the emf phasor is the hypotenuse of a right triangle. Thus
$$\mathcal{E}_0^2 = V_R^2 + (V_L - V_C)^2$$

If $V_L > V_C$, which we've assumed, then the instantaneous current *i* lags the emf by a phase angle ϕ. We can write the current, in terms of ϕ, as

$$i = I\cos(\omega t - \phi) \tag{35.22}$$

Of course, there's no guarantee that V_L will be larger than V_C. If the opposite is true, $V_L < V_C$, the emf phasor is on the other side of the current phasor. Our analysis is still valid if we consider ϕ to be negative when *i* is ccw from $\mathcal{E}$. Thus ϕ can be anywhere between $-90°$ and $+90°$.

Now we can continue much as we did with the *RC* circuit. Based on the right triangle, $\mathcal{E}_0^2$ is

$$\mathcal{E}_0^2 = V_R^2 + (V_L - V_C)^2 = [R^2 + (X_L - X_C)^2]I^2 \tag{35.23}$$

where we wrote each of the peak voltages in terms of the peak current *I* and a resistance or a reactance. Consequently, the peak current in the *RLC* circuit is

$$I = \frac{\mathcal{E}_0}{\sqrt{R^2 + (X_L - X_C)^2}} = \frac{\mathcal{E}_0}{\sqrt{R^2 + (\omega L - 1/\omega C)^2}} \tag{35.24}$$

The three peak voltages, if you need them, are then found from $V_R = IR$, $V_L = IX_L$, and $V_C = IX_C$.

Impedance

The denominator of Equation 35.24 is called the **impedance** *Z* of the circuit:

$$Z = \sqrt{R^2 + (X_L - X_C)^2} \tag{35.25}$$

Impedance, like resistance and reactance, is measured in ohms. The circuit's peak current can be written in terms of the source emf and the circuit impedance as

$$I = \frac{\mathcal{E}_0}{Z} \tag{35.26}$$

Equation 35.26 is a compact way to write *I*, but it doesn't add anything new to Equation 35.24.

Phase Angle

FIGURE 35.17 The current is not in phase with the emf.

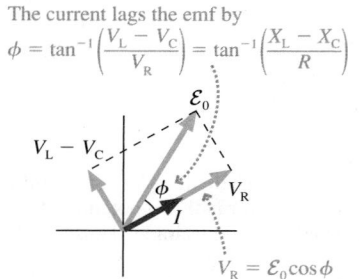

The current lags the emf by
$$\phi = \tan^{-1}\left(\frac{V_L - V_C}{V_R}\right) = \tan^{-1}\left(\frac{X_L - X_C}{R}\right)$$

$V_R = \mathcal{E}_0 \cos\phi$

It is often useful to know the phase angle ϕ between the emf and the current. You can see from **FIGURE 35.17** that

$$\tan\phi = \frac{V_L - V_C}{V_R} = \frac{(X_L - X_C)I}{RI}$$

The current I cancels, and we're left with

$$\phi = \tan^{-1}\left(\frac{X_L - X_C}{R}\right) \tag{35.27}$$

We can check that Equation 35.27 agrees with our analyses of single-element circuits. A resistor-only circuit has $X_L = X_C = 0$ and thus $\phi = \tan^{-1}(0) = 0$ rad. In other words, as we discovered previously, the emf and current are in phase. An AC inductor circuit has $R = X_C = 0$ and thus $\phi = \tan^{-1}(\infty) = \pi/2$ rad, agreeing with our earlier finding that the inductor current lags the voltage by 90°.

Other relationships can be found from the phasor diagram and written in terms of the phase angle. For example, it is useful to write the peak resistor voltage as

$$V_R = \mathcal{E}_0 \cos\phi \tag{35.28}$$

Notice that the resistor voltage oscillates in phase with the emf only if $\phi = 0$ rad.

Resonance

Suppose we vary the emf frequency ω while keeping everything else constant. There is very little current at very low frequencies because the capacitive reactance $X_C = 1/\omega C$ is very large. Similarly, there is very little current at very high frequencies because the inductive reactance $X_L = \omega L$ becomes very large.

If I approaches zero at very low and very high frequencies, there should be some intermediate frequency where I is a maximum. Indeed, you can see from Equation 35.24 that the denominator will be a minimum, making I a maximum, when $X_L = X_C$, or

$$\omega L = \frac{1}{\omega C} \tag{35.29}$$

The frequency ω_0 that satisfies Equation 35.29 is called the **resonance frequency**:

$$\omega_0 = \frac{1}{\sqrt{LC}} \tag{35.30}$$

This is the frequency for *maximum current* in the series *RLC* circuit. The maximum current

$$I_{max} = \frac{\mathcal{E}_0}{R} \tag{35.31}$$

is that of a purely resistive circuit because the impedance is $Z = R$ at resonance.

You'll recognize ω_0 as the oscillation frequency of the *LC* circuit we analyzed in Chapter 33. The current in an ideal *LC* circuit oscillates forever as energy is transferred back and forth between the capacitor and the inductor. This is analogous to an ideal, frictionless simple harmonic oscillator in which the energy is transformed back and forth between kinetic and potential.

Adding a resistor to the circuit is like adding damping to a mechanical oscillator. The emf is then a sinusoidal driving force, and the series *RLC* circuit is directly analogous to the driven, damped oscillator that you studied in Chapter 14. A mechanical oscillator exhibits *resonance* by having a large-amplitude response when the driving frequency matches the system's natural frequency. Equation 35.30 is the natural frequency of the series *RLC* circuit, the frequency at which the current would like to oscillate. Consequently, the circuit has a large current response when the oscillating emf matches this frequency.

FIGURE 35.18 shows the peak current I of a series *RLC* circuit as the emf frequency ω is varied. Notice how the current increases until reaching a maximum at frequency ω_0, then decreases. This is the hallmark of a resonance.

As R decreases, causing the damping to decrease, the maximum current becomes larger and the curve in Figure 35.18 becomes narrower. You saw exactly the same behavior for a driven mechanical oscillator. The emf frequency must be very close to ω_0 in order for a lightly damped system to respond, but the response at resonance is very large.

For a different perspective, FIGURE 35.19 graphs the instantaneous emf $\mathcal{E} = \mathcal{E}_0 \cos \omega t$ and current $i = I \cos(\omega t - \phi)$ for frequencies below, at, and above ω_0. The current and the emf are in phase at resonance ($\phi = 0$ rad) because the capacitor and inductor essentially cancel each other to give a purely resistive circuit. Away from resonance, the current decreases *and* begins to get out of phase with the emf. You can see, from Equation 35.27, that the phase angle ϕ is negative when $X_L < X_C$ (i.e., the frequency is below resonance) and positive when $X_L > X_C$ (the frequency is above resonance).

Resonance circuits are widely used in radio, television, and communication equipment because of their ability to respond to one particular frequency (or very narrow range of frequencies) while suppressing others. The selectivity of a resonance circuit improves as the resistance decreases, but the inherent resistance of the wires and the inductor coil keeps R from being 0 Ω.

FIGURE 35.18 A graph of the current I versus emf frequency for a series *RLC* circuit.

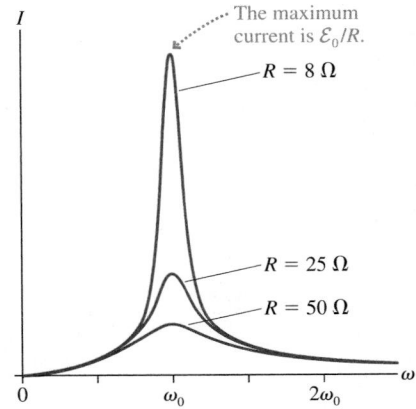

FIGURE 35.19 Graphs of the emf $\mathcal{E}$ and the current i at frequencies below, at, and above the resonance frequency ω_0.

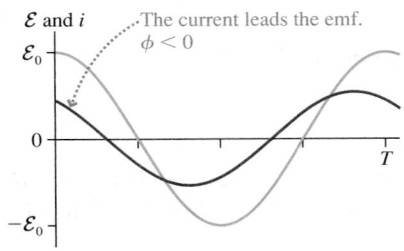

Below resonance: $\omega < \omega_0$

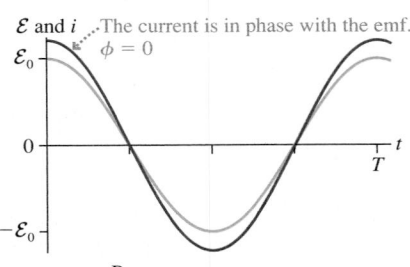

Resonance: $\omega = \omega_0$
Maximum current

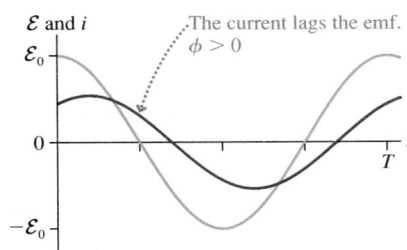

Above resonance: $\omega > \omega_0$

EXAMPLE 35.6 **Designing a radio receiver**

An AM radio antenna picks up a 1000 kHz signal with a peak voltage of 5.0 mV. The tuning circuit consists of a 60 μH inductor in series with a variable capacitor. The inductor coil has a resistance of 0.25 Ω, and the resistance of the rest of the circuit is negligible.

a. To what value should the capacitor be tuned to listen to this radio station?
b. What is the peak current through the circuit at resonance?
c. A stronger station at 1050 kHz produces a 10 mV antenna signal. What is the current at this frequency when the radio is tuned to 1000 kHz?

MODEL The inductor's 0.25 Ω resistance can be modeled as a resistance in series with the inductance, hence we have a series *RLC* circuit. The antenna signal at $\omega = 2\pi \times 1000$ kHz is the emf.

VISUALIZE The circuit looks like Figure 35.16.

SOLVE a. The capacitor needs to be tuned to where it and the inductor are resonant at $\omega_0 = 2\pi \times 1000$ kHz. The appropriate value is

$$C = \frac{1}{L\omega_0^2} = \frac{1}{(60 \times 10^{-6}\text{ H})(6.28 \times 10^6\text{ rad/s})^2}$$

$$= 4.2 \times 10^{-10}\text{ F} = 420\text{ pF}$$

b. $X_L = X_C$ at resonance, so the peak current is

$$I = \frac{\mathcal{E}_0}{R} = \frac{5.0 \times 10^{-3}\text{ V}}{0.25\ \Omega} = 0.020\text{ A} = 20\text{ mA}$$

c. The 1050 kHz signal is "off resonance," so we need to compute $X_L = \omega L = 396\ \Omega$ and $X_C = 1/\omega C = 361\ \Omega$ at $\omega = 2\pi \times 1050$ kHz. The peak voltage of this signal is $\mathcal{E}_0 = 10$ mV. With these values, Equation 35.24 for the peak current is

$$I = \frac{\mathcal{E}_0}{\sqrt{R^2 + (X_L - X_C)^2}} = 0.28\text{ mA}$$

ASSESS These are realistic values for the input stage of an AM radio. You can see that the signal from the 1050 kHz station is strongly suppressed when the radio is tuned to 1000 kHz.

A series *RLC* circuit has $V_C = 5.0$ V, $V_R = 7.0$ V, and $V_L = 9.0$ V. Is the frequency above, below, or equal to the resonance frequency?

35.6 Power in AC Circuits

A primary role of the emf is to supply energy. Some circuit devices, such as motors and lightbulbs, use the energy to perform useful tasks. Other circuit devices dissipate the energy as an increased thermal energy in the components and the surrounding air. Chapter 31 examined the topic of power in DC circuits. Now we can perform a similar analysis for AC circuits.

The emf supplies energy to a circuit at the rate

$$p_{\text{source}} = i\mathcal{E} \tag{35.32}$$

where i and $\mathcal{E}$ are the instantaneous current from and potential difference across the emf. We've used a lowercase p to indicate that this is the instantaneous power. We need to look at the power losses in individual circuit elements.

Resistors

A resistor dissipates energy at the rate

$$p_R = i_R v_R = i_R^2 R \tag{35.33}$$

We can use $i_R = I_R \cos \omega t$ to write the resistor's instantaneous power loss as

$$p_R = i_R^2 R = I_R^2 R \cos^2 \omega t \tag{35.34}$$

FIGURE 35.20 shows the instantaneous power graphically. You can see that, because the cosine is squared, the power oscillates twice during every cycle of the emf. The energy dissipation peaks both when $i_R = I_R$ and when $i_R = -I_R$.

In practice, we're more interested in the *average power* than in the instantaneous power. The **average power** P is the total energy dissipated per second. We can find P_R for a resistor by using the identity $\cos^2(x) = \frac{1}{2}(1 + \cos 2x)$ to write

$$P_R = I_R^2 R \cos^2 \omega t = I_R^2 R \left[\frac{1}{2}(1 + \cos 2\omega t) \right] = \frac{1}{2}I_R^2 R + \frac{1}{2}I_R^2 R \cos 2\omega t$$

The $\cos 2\omega t$ term oscillates positive and negative twice during each cycle of the emf. Its average, over one cycle, is zero. Thus the average power loss in a resistor is

$$P_R = \frac{1}{2}I_R^2 R \qquad \text{(average power loss in a resistor)} \tag{35.35}$$

It is useful to write Equation 35.35 as

$$P_R = \left(\frac{I_R}{\sqrt{2}} \right)^2 R = (I_{\text{rms}})^2 R \tag{35.36}$$

where the quantity

$$I_{\text{rms}} = \frac{I_R}{\sqrt{2}} \tag{35.37}$$

is called the **root-mean-square current,** or rms current, I_{rms}. Technically, an rms quantity is the square root of the average, or mean, of the quantity squared. For a sinusoidal oscillation, the rms value turns out to be the peak value divided by $\sqrt{2}$.

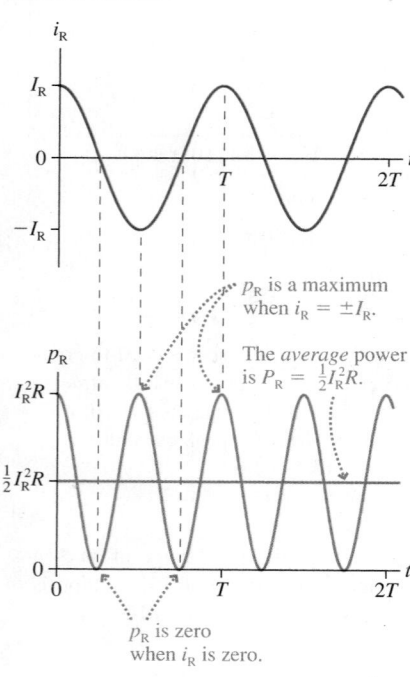

FIGURE 35.20 The instantaneous power loss in a resistor.

p_R is a maximum when $i_R = \pm I_R$.

The *average* power is $P_R = \frac{1}{2}I_R^2 R$.

p_R is zero when i_R is zero.

The rms current allows us to compare Equation 35.36 directly to the energy dissipated by a resistor in a DC circuit: $P = I^2R$. You can see that the average power loss of a resistor in an AC circuit with $I_{\text{rms}} = 1$ A is the same as in a DC circuit with $I = 1$ A. **As far as power is concerned, an rms current is equivalent to an equal DC current.**

Similarly, we can define the root-mean-square voltage and emf:

$$V_{\text{rms}} = \frac{V_{\text{R}}}{\sqrt{2}} \qquad \mathcal{E}_{\text{rms}} = \frac{\mathcal{E}_0}{\sqrt{2}} \qquad\qquad (35.38)$$

The resistor's average power loss in terms of the rms quantities is

$$P_{\text{R}} = (I_{\text{rms}})^2 R = \frac{(V_{\text{rms}})^2}{R} = I_{\text{rms}} V_{\text{rms}} \qquad\qquad (35.39)$$

and the average power supplied by the emf is

$$P_{\text{source}} = I_{\text{rms}} \mathcal{E}_{\text{rms}} \qquad\qquad (35.40)$$

The power rating on a lightbulb is its average power at $V_{\text{rms}} = 120$ V.

The single-resistor circuit that we analyzed in Section 35.1 had $V_{\text{R}} = \mathcal{E}$ or, equivalently, $V_{\text{rms}} = \mathcal{E}_{\text{rms}}$. You can see from Equations 35.39 and 35.40 that the power loss in the resistor exactly matches the power supplied by the emf. This must be the case in order to conserve energy.

NOTE ▶ Voltmeters, ammeters, and other AC measuring instruments are calibrated to give the rms value. An AC voltmeter would show that the "line voltage" of an electrical outlet in the United States is 120 V. This is $\mathcal{E}_{\text{rms}}$. The peak voltage $\mathcal{E}_0$ is larger by a factor of $\sqrt{2}$, or $\mathcal{E}_0 = 170$ V. The power-line voltage is sometimes specified as "120 V/60 Hz," showing the rms voltage and the frequency. ◀

EXAMPLE 35.7 **Lighting a bulb**

A 100 W incandescent lightbulb is plugged into a 120 V/60 Hz outlet. What is the resistance of the bulb's filament? What is the peak current through the bulb?

MODEL The filament in a lightbulb acts as a resistor.

VISUALIZE FIGURE 35.21 is a simple one-resistor circuit.

FIGURE 35.21 An AC circuit with a lightbulb as a resistor.

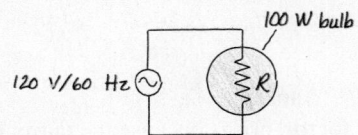

SOLVE A bulb labeled 100 W is designed to dissipate an average 100 W at $V_{\text{rms}} = 120$ V. We can use Equation 35.39 to find

$$R = \frac{(V_{\text{rms}})^2}{P_{\text{R}}} = \frac{(120 \text{ V})^2}{100 \text{ W}} = 144 \ \Omega$$

The rms current is then found from

$$I_{\text{rms}} = \frac{P_{\text{R}}}{V_{\text{rms}}} = \frac{100 \text{ W}}{120 \text{ V}} = 0.833 \text{ A}$$

The peak current is $I_{\text{R}} = \sqrt{2} I_{\text{rms}} = 1.18$ A.

ASSESS Calculations with rms values are just like the calculations for DC circuits.

Capacitors and Inductors

In Section 35.2, we found that the instantaneous current to a capacitor is $i_{\text{C}} = -\omega C V_{\text{C}} \sin \omega t$. Thus the instantaneous energy dissipation in a capacitor is

$$p_{\text{C}} = v_{\text{C}} i_{\text{C}} = (V_{\text{C}} \cos \omega t)(-\omega C V_{\text{C}} \sin \omega t) = -\frac{1}{2} \omega C V_{\text{C}}^2 \sin 2\omega t \qquad (35.41)$$

where we used $\sin(2x) = 2\sin(x)\cos(x)$.

FIGURE 35.22 on the next page shows Equation 35.41 graphically. Energy is transferred into the capacitor (positive power) as it is charged, but, instead of being dissipated, as it would be by a resistor, the energy is stored as potential energy in the capacitor's electric field. Then, as the capacitor discharges, this energy is given back to the circuit. Power is the rate at which energy is *removed* from the circuit, hence p is negative as the capacitor transfers energy back into the circuit.

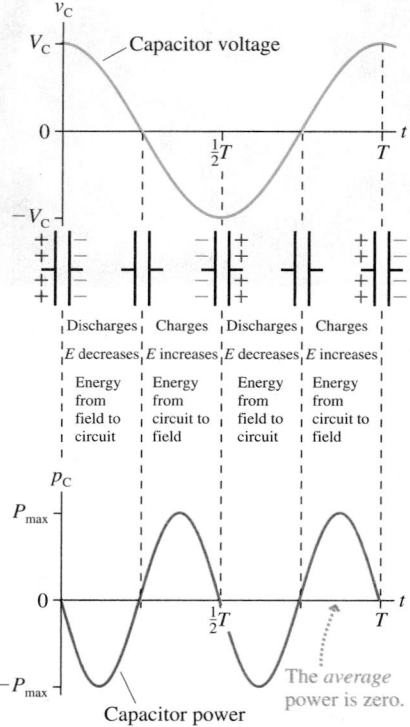

Using a mechanical analogy, a capacitor is like an ideal, frictionless simple harmonic oscillator. Kinetic and potential energy are constantly being exchanged, but there is no dissipation because none of the energy is transformed into thermal energy. The important conclusion is that **a capacitor's average power loss is zero: $P_C = 0$.**

The same is true of an inductor. An inductor alternately stores energy in the magnetic field, as the current is increasing, then transfers energy back to the circuit as the current decreases. The instantaneous power oscillates between positive and negative, but **an inductor's average power loss is zero: $P_L = 0$.**

NOTE ▶ We're assuming ideal capacitors and inductors. Real capacitors and inductors inevitably have a small amount of resistance and dissipate a small amount of energy. However, their energy dissipation is negligible compared to that of the resistors in most practical circuits. ◀

The Power Factor

In an *RLC* circuit, energy is supplied by the emf and dissipated by the resistor. But an *RLC* circuit is unlike a purely resistive circuit in that the current is not in phase with the potential difference of the emf.

We found in Equation 35.22 that the instantaneous current in an *RLC* circuit is $i = I\cos(\omega t - \phi)$, where ϕ is the angle by which the current lags the emf. Thus the instantaneous power supplied by the emf is

$$p_{\text{source}} = i\mathcal{E} = (I\cos(\omega t - \phi))(\mathcal{E}_0 \cos \omega t) = I\mathcal{E}_0 \cos \omega t \cos(\omega t - \phi) \qquad (35.42)$$

We can use the expression $\cos(x - y) = \cos(x)\cos(y) + \sin(x)\sin(y)$ to write the power as

$$p_{\text{source}} = (I\mathcal{E}_0 \cos \phi)\cos^2 \omega t + (I\mathcal{E}_0 \sin \phi)\sin \omega t \cos \omega t \qquad (35.43)$$

In our analysis of the power loss in a resistor and a capacitor, we found that the average of $\cos^2 \omega t$ is $\frac{1}{2}$ and the average of $\sin \omega t \cos \omega t$ is zero. Thus we can immediately write that the *average* power supplied by the emf is

$$P_{\text{source}} = \frac{1}{2}I\mathcal{E}_0 \cos \phi = I_{\text{rms}}\mathcal{E}_{\text{rms}} \cos \phi \qquad (35.44)$$

The rms values, you will recall, are $I/\sqrt{2}$ and $\mathcal{E}_0/\sqrt{2}$.

The term $\cos \phi$, called the **power factor,** arises because the current and the emf in a series *RLC* circuit are not in phase. Because the current and the emf aren't pushing and pulling together, the source delivers less energy to the circuit.

We'll leave it as a homework problem for you to show that the peak current in an *RLC* circuit can be written $I = I_{\text{max}} \cos \phi$, where $I_{\text{max}} = \mathcal{E}_0/R$ was given in Equation 35.31. In other words, the current term in Equation 35.44 is a function of the power factor. Consequently, the average power is

$$P_{\text{source}} = P_{\text{max}} \cos^2 \phi \qquad (35.45)$$

where $P_{\text{max}} = \frac{1}{2}I_{\text{max}}\mathcal{E}_0$ is the *maximum* power the source can deliver to the circuit.

The source delivers maximum power only when $\cos \phi = 1$. This is the case when $X_L - X_C = 0$, requiring either a purely resistive circuit or an *RLC* circuit operating at the resonance frequency ω_0. The average power loss is zero for a purely capacitive or purely inductive load with, respectively, $\phi = -90°$ or $\phi = +90°$, as found above.

Motors of various types, especially large industrial motors, use a significant fraction of the electric energy generated in industrialized nations. Motors operate most efficiently, doing the maximum work per second, when the power factor is as close to

Industrial motors use a significant fraction of the electric energy generated in the United States.

1 as possible. But motors are inductive devices, due to their electromagnet coils, and if too many motors are attached to the electric grid, the power factor is pulled away from 1. To compensate, the electric company places large capacitors throughout the transmission system. The capacitors dissipate no energy, but they allow the electric system to deliver energy more efficiently by keeping the power factor close to 1.

Finally, we found in Equation 35.28 that the resistor's peak voltage in an *RLC* circuit is related to the emf peak voltage by $V_R = \mathcal{E}_0 \cos\phi$ or, dividing both sides by $\sqrt{2}$, $V_{rms} = \mathcal{E}_{rms} \cos\phi$. We can use this result to write the energy loss in the resistor as

$$P_R = I_{rms} V_{rms} = I_{rms} \mathcal{E}_{rms} \cos\phi \qquad (35.46)$$

But this expression is P_{source}, as we found in Equation 35.44. Thus we see that the energy supplied to an *RLC* circuit by the emf is ultimately dissipated by the resistor.

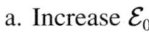 **STOP TO THINK 35.6** The emf and the current in a series *RLC* circuit oscillate as shown. Which of the following (perhaps more than one) would increase the rate at which energy is supplied to the circuit?

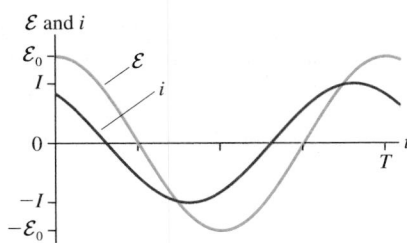

a. Increase $\mathcal{E}_0$

b. Increase L

c. Increase C

d. Decrease $\mathcal{E}_0$

e. Decrease L

f. Decrease C

CHALLENGE EXAMPLE 35.8   **Power in an *RLC* circuit**

An audio amplifier drives a series *RLC* circuit consisting of an 8.0 Ω loudspeaker, a 160 μF capacitor, and a 1.5 mH inductor. The amplifier output is 15.0 V rms at 500 Hz.

a. What power is delivered to the speaker?

b. What maximum power could the amplifier deliver, and how would the capacitor have to be changed for this to happen?

MODEL The emf and voltage of an *RLC* circuit are not in phase, and that affects the power delivered to the circuit. All the power is dissipated by the circuit's resistance, which in this case is the loudspeaker.

VISUALIZE The circuit looks like Figure 35.16.

SOLVE a. The power delivered by the emf is $P_{source} = I_{rms}\mathcal{E}_{rms} \cos\phi$, where ϕ is the phase angle between the emf and the current. In an AC circuit, the current is $I = \mathcal{E}/Z$, where Z is the impedance. To calculate Z, we need the reactances of the capacitor and inductor, and these, in turn, depend on the frequency. At 500 Hz, the angular frequency is $\omega = 2\pi(500 \text{ Hz}) = 3140$ rad/s. With this, we can find

$$X_C = \frac{1}{\omega C} = \frac{1}{(3140 \text{ rad/s})(160 \times 10^{-6} \text{ F})} = 1.99 \ \Omega$$

$$X_L = \omega L = (3140 \text{ rad/s})(0.0015 \text{ H}) = 4.71 \ \Omega$$

Now we can calculate the impedance:

$$Z = \sqrt{R^2 + (X_L - X_C)^2} = 8.45 \ \Omega$$

and thus

$$I_{rms} = \frac{\mathcal{E}_{rms}}{Z} = \frac{15.0 \text{ V}}{8.45 \ \Omega} = 1.78 \text{ A}$$

Lastly, we need the phase angle between the emf and the current:

$$\phi = \tan^{-1}\left(\frac{X_L - X_C}{R}\right) = 18.8°$$

The power factor is $\cos(18.8°) = 0.947$, and thus the power delivered by the emf is

$$P_{source} = I_{rms}\mathcal{E}_{rms} \cos\phi = (1.78 \text{ A})(15.0 \text{ V})(0.947) = 25 \text{ W}$$

b. Maximum power is delivered when the current is in phase with the emf, making the power factor 1.00. This occurs when $X_C = X_L$, making the impedance $Z = R = 8.0 \ \Omega$ and the current $I_{rms} = \mathcal{E}_{rms}/R = 1.88$ A. Then

$$P_{source} = I_{rms}\mathcal{E}_{rms} \cos\phi = (1.88 \text{ A})(15.0 \text{ V})(1.00) = 28 \text{ W}$$

To deliver maximum power, we need to change the capacitance to make $X_C = X_L = 4.71 \ \Omega$. The required capacitance is

$$C = \frac{1}{(3140 \text{ rad/s})(4.71 \ \Omega)} = 68 \ \mu\text{F}$$

So delivering maximum power requires lowering the capacitance from 160 μF to 68 μF.

ASSESS Changing the capacitor not only increases the power factor, it also increases the current. Both contribute to the higher power.

SUMMARY

The goal of Chapter 35 has been to understand and apply basic techniques of AC circuit analysis.

Important Concepts

AC circuits are driven by an emf

$$\mathcal{E} = \mathcal{E}_0 \cos \omega t$$

that oscillates with angular frequency $\omega = 2\pi f$.

Phasors can be used to represent the oscillating emf, current, and voltage.

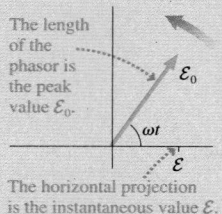

The length of the phasor is the peak value $\mathcal{E}_0$.

$\mathcal{E}_0$

ωt

$\mathcal{E}$

The horizontal projection is the instantaneous value $\mathcal{E}$.

Basic circuit elements

Element	i and v	Resistance/ reactance	I and V	Power
Resistor	In phase	R is fixed	$V = IR$	$I_{rms}V_{rms}$
Capacitor	i leads v by 90°	$X_C = 1/\omega C$	$V = IX_C$	0
Inductor	i lags v by 90°	$X_L = \omega L$	$V = IX_L$	0

For many purposes, especially calculating power, the **root-mean-square** (rms) quantities

$$V_{rms} = V/\sqrt{2} \qquad I_{rms} = I/\sqrt{2} \qquad \mathcal{E}_{rms} = \mathcal{E}_0/\sqrt{2}$$

are equivalent to the corresponding DC quantities.

Key Skills

Using phasor diagrams

- Start with a phasor (v or i) common to two or more circuit elements.

- The sum of instantaneous quantities is vector addition.

- Use the Pythagorean theorem to relate peak quantities.

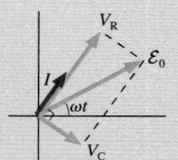

V_R

$\mathcal{E}_0$

I

ωt

V_C

For an RC circuit, shown here,

$$v_R + v_C = \mathcal{E}$$

$$V_R^2 + V_C^2 = \mathcal{E}_0^2$$

Kirchhoff's laws

Loop law The sum of the potential differences around a loop is zero.

Junction law The sum of currents entering a junction equals the sum leaving the junction.

Instantaneous and peak quantities

Instantaneous quantities v and i generally obey different relationships than peak quantities V and I.

Applications

RC filter circuits

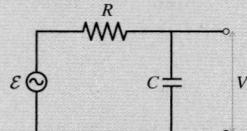

$$V_C = \frac{\mathcal{E}_0 X_C}{\sqrt{R^2 + X_C^2}}$$

$$V_C \rightarrow \mathcal{E}_0 \text{ as } \omega \rightarrow 0$$

A **low-pass filter** transmits low frequencies and blocks high frequencies.

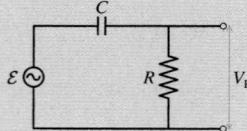

$$V_R = \frac{\mathcal{E}_0 R}{\sqrt{R^2 + X_C^2}}$$

$$V_R \rightarrow \mathcal{E}_0 \text{ as } \omega \rightarrow \infty$$

A **high-pass filter** transmits high frequencies and blocks low frequencies.

Series RLC circuits

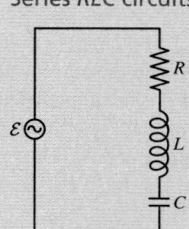

$I = \mathcal{E}_0/Z$ where Z is the **impedance**

$$Z = \sqrt{R^2 + (X_L - X_C)^2}$$

$$V_R = IR \qquad V_L = IX_L \qquad V_C = IX_C$$

When $\omega = \omega_0 = 1/\sqrt{LC}$ (the **resonance frequency**), the current in the circuit is a maximum $I_{max} = \mathcal{E}_0/R$.

In general, the current i lags behind $\mathcal{E}$ by the **phase angle** $\phi = \tan^{-1}\left((X_L - X_C)/R\right)$.

The power supplied by the emf is $P_{source} = I_{rms}\mathcal{E}_{rms}\cos\phi$, where $\cos\phi$ is called the **power factor**.

The power lost in the resistor is $P_R = I_{rms}V_{rms} = (I_{rms})^2 R$.

Terms and Notation

AC circuit	crossover frequency, ω_c	series *RLC* circuit	root-mean-square current, I_{rms}
DC circuit	low-pass filter	impedance, Z	power factor, $\cos\phi$
phasor	high-pass filter	resonance frequency, ω_0	
capacitive reactance, X_C	inductive reactance, X_L	average power, P	

CONCEPTUAL QUESTIONS

1. FIGURE Q35.1 shows emf phasors a, b, and c.
   a. For each, what is the instantaneous value of the emf?
   b. At this instant, is the magnitude of each emf increasing, decreasing, or holding constant?

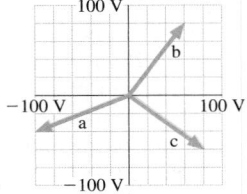

FIGURE Q35.1

2. A resistor is connected across an oscillating emf. The peak current through the resistor is 2.0 A. What is the peak current if:
   a. The resistance R is doubled?
   b. The peak emf $\mathcal{E}_0$ is doubled?
   c. The frequency ω is doubled?

3. A capacitor is connected across an oscillating emf. The peak current through the capacitor is 2.0 A. What is the peak current if:
   a. The capacitance C is doubled?
   b. The peak emf $\mathcal{E}_0$ is doubled?
   c. The frequency ω is doubled?

4. A low-pass *RC* filter has a crossover frequency $f_c = 200$ Hz. What is f_c if:
   a. The resistance R is doubled?
   b. The capacitance C is doubled?
   c. The peak emf $\mathcal{E}_0$ is doubled?

5. An inductor is connected across an oscillating emf. The peak current through the inductor is 2.0 A. What is the peak current if:
   a. The inductance L is doubled?
   b. The peak emf $\mathcal{E}_0$ is doubled?
   c. The frequency ω is doubled?

6. The resonance frequency of a series *RLC* circuit is 1000 Hz. What is the resonance frequency if:
   a. The resistance R is doubled?
   b. The inductance L is doubled?
   c. The capacitance C is doubled?
   d. The peak emf $\mathcal{E}_0$ is doubled?

7. In the series *RLC* circuit represented by the phasors of FIGURE Q35.7, is the emf frequency less than, equal to, or greater than the resonance frequency ω_0? Explain.

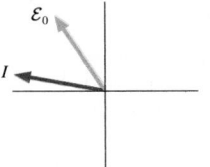

FIGURE Q35.7

8. The resonance frequency of a series *RLC* circuit is less than the emf frequency. Does the current lead or lag the emf? Explain.

9. The current in a series *RLC* circuit lags the emf by 20°. You cannot change the emf. What two different things could you do to the circuit that would increase the power delivered to the circuit by the emf?

10. The average power dissipated by a resistor is 4.0 W. What is P_R if:
    a. The resistance R is doubled while $\mathcal{E}_0$ is held fixed?
    b. The peak emf $\mathcal{E}_0$ is doubled while R is held fixed?
    c. Both are doubled simultaneously?

EXERCISES AND PROBLEMS

Problems labeled [ ] integrate material from earlier chapters.

Exercises

Section 35.1 AC Sources and Phasors

1. | The emf phasor in FIGURE EX35.1 is shown at $t = 15$ ms.
   a. What is the angular frequency ω? Assume this is the first rotation.
   b. What is the instantaneous value of the emf?

$\mathcal{E}_0 = 12$ V   Phasor at $t = 15$ ms

30°

FIGURE EX35.1

2. ‖ The emf phasor in FIGURE EX35.2 is shown at $t = 2.0$ ms.
   a. What is the angular frequency ω? Assume this is the first rotation.
   b. What is the peak value of the emf?

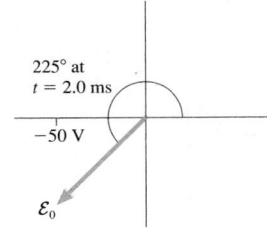

225° at $t = 2.0$ ms

-50 V

$\mathcal{E}_0$

FIGURE EX35.2

3. | A 110 Hz source of emf has a peak voltage of 50 V. Draw the emf phasor at $t = 3.0$ ms.

4. ‖ Draw the phasor for the emf $\mathcal{E} = (170 \text{ V})\cos\big((2\pi \times 60 \text{ Hz})t\big)$ at $t = 60$ ms.

5. | A 200 Ω resistor is connected to an AC source with $\mathcal{E}_0 = 10$ V. What is the peak current through the resistor if the emf frequency is (a) 100 Hz? (b) 100 kHz?

6. | FIGURE EX35.6 shows voltage and current graphs for a resistor.
   a. What is the emf frequency f?
   b. What is the value of the resistance R?
   c. Draw the resistor's voltage and current phasors at $t = 15$ ms.

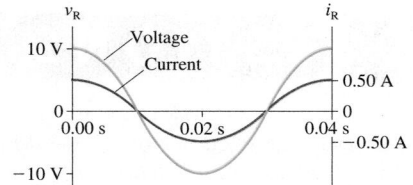

FIGURE EX35.6

Section 35.2 Capacitor Circuits

7. | A 0.30 μF capacitor is connected across an AC generator that produces a peak voltage of 10 V. What is the peak current to and from the capacitor if the emf frequency is (a) 100 Hz? (b) 100 kHz?

8. | The peak current to and from a capacitor is 10 mA. What is the peak current if
   a. The emf frequency is doubled?
   b. The emf peak voltage is doubled (at the original frequency)?

9. | A 20 nF capacitor is connected across an AC generator that produces a peak voltage of 5.0 V.
   a. At what frequency f is the peak current 50 mA?
   b. What is the instantaneous value of the emf at the instant when $i_C = I_C$?

10. || A capacitor is connected to a 15 kHz oscillator. The peak current is 65 mA when the rms voltage is 6.0 V. What is the value of the capacitance C?

11. | A capacitor has a peak current of 330 μA when the peak voltage at 250 kHz is 2.2 V.
   a. What is the capacitance?
   b. If the peak voltage is held constant, what is the peak current at 500 kHz?

Section 35.3 RC Filter Circuits

12. | A high-pass RC filter is connected to an AC source with a peak voltage of 10.0 V. The peak capacitor voltage is 6.0 V. What is the resistor voltage?

13. | A high-pass RC filter with a crossover frequency of 1000 Hz uses a 100 Ω resistor. What is the value of the capacitor?

14. | A low-pass RC filter with a crossover frequency of 1000 Hz uses a 100 Ω resistor. What is the value of the capacitor?

15. || What are V_R and V_C if the emf frequency in FIGURE EX35.15 is 10 kHz?

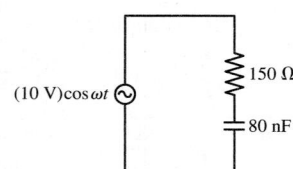

FIGURE EX35.15

16. | A low-pass filter consists of a 100 μF capacitor in series with a 159 Ω resistor. The circuit is driven by an AC source with a peak voltage of 5.00 V.

   a. What is the crossover frequency f_c?
   b. What is V_C when $f = \frac{1}{2}f_c$, f_c, and $2f_c$?

17. | A high-pass filter consists of a 1.59 μF capacitor in series with a 100 Ω resistor. The circuit is driven by an AC source with a peak voltage of 5.00 V.
   a. What is the crossover frequency f_c?
   b. What is V_R when $f = \frac{1}{2}f_c$, f_c, and $2f_c$?

Section 35.4 Inductor Circuits

18. | The peak current through an inductor is 10 mA. What is the peak current if
   a. The emf frequency is doubled?
   b. The emf peak voltage is doubled (at the original frequency)?

19. | A 20 mH inductor is connected across an AC generator that produces a peak voltage of 10 V. What is the peak current through the inductor if the emf frequency is (a) 100 Hz? (b) 100 kHz?

20. | An inductor is connected to a 15 kHz oscillator. The peak current is 65 mA when the rms voltage is 6.0 V. What is the value of the inductance L?

21. | A 500 μH inductor is connected across an AC generator that produces a peak voltage of 5.0 V.
   a. At what frequency f is the peak current 50 mA?
   b. What is the instantaneous value of the emf at the instant when $i_L = I_L$?

22. || An inductor has a peak current of 330 μA when the peak voltage at 45 MHz is 2.2 V.
   a. What is the inductance?
   b. If the peak voltage is held constant, what is the peak current at 90 MHz?

Section 35.5 The Series RLC Circuit

23. | A series RLC circuit has a 200 kHz resonance frequency. What is the resonance frequency if
   a. The resistor value is doubled?
   b. The capacitor value is doubled?

24. | A series RLC circuit has a 200 kHz resonance frequency. What is the resonance frequency if the capacitor value is doubled and, at the same time, the inductor value is halved?

25. || What capacitor in series with a 100 Ω resistor and a 20 mH inductor will give a resonance frequency of 1000 Hz?

26. | What inductor in series with a 100 Ω resistor and a 2.5 μF capacitor will give a resonance frequency of 1000 Hz?

27. | A series RLC circuit consists of a 50 Ω resistor, a 3.3 mH inductor, and a 480 nF capacitor. It is connected to an oscillator with a peak voltage of 5.0 V. Determine the impedance, the peak current, and the phase angle at frequencies (a) 3000 Hz, (b) 4000 Hz, and (c) 5000 Hz.

28. || At what frequency f do a 1.0 μF capacitor and a 1.0 μH inductor have the same reactance? What is the value of the reactance at this frequency?

Section 35.6 Power in AC Circuits

29. | The heating element of a hair drier dissipates 1500 W when connected to a 120 V/60 Hz power line. What is its resistance?

30. || A resistor dissipates 2.0 W when the rms voltage of the emf is 10.0 V. At what rms voltage will the resistor dissipate 10.0 W?

31. || For what absolute value of the phase angle does a source deliver 75% of the maximum possible power to an RLC circuit?

32. ‖ The motor of an electric drill draws a 3.5 A rms current at the power-line voltage of 120 V rms. What is the motor's power if the current lags the voltage by 20°?

33. ‖ A series *RLC* circuit attached to a 120 V/60 Hz power line draws a 2.4 A rms current with a power factor of 0.87. What is the value of the resistor?

34. ‖ A series *RLC* circuit with a 100 Ω resistor dissipates 80 W when attached to a 120 V/60 Hz power line. What is the power factor?

Problems

35. ‖ a. For an *RC* circuit, find an expression for the angular frequency at which $V_R = \frac{1}{2}\mathcal{E}_0$.
    b. What is V_C at this frequency?

36. ‖ a. For an *RC* circuit, find an expression for the angular frequency at which $V_C = \frac{1}{2}\mathcal{E}_0$.
    b. What is V_R at this frequency?

37. ‖ a. Evaluate V_C in **FIGURE P35.37** at emf frequencies 1, 3, 10, 30, and 100 kHz.
    b. Graph V_C versus frequency. Draw a smooth curve through your five points.

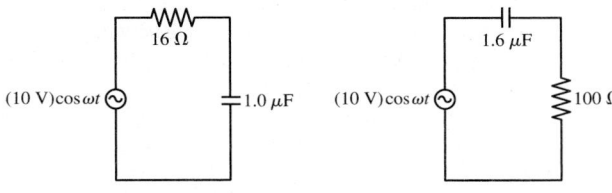

FIGURE P35.37   **FIGURE P35.38**

38. ‖ a. Evaluate V_R in **FIGURE P35.38** at emf frequencies 100, 300, 1000, 3000, and 10,000 Hz.
    b. Graph V_R versus frequency. Draw a smooth curve through your five points.

39. ‖ For an *RC* filter circuit, show that $V_R = V_C = \mathcal{E}_0/\sqrt{2}$ at $\omega = \omega_c$.

40. ‖‖ When two capacitors are connected in parallel across a 10.0 V rms, 1.00 kHz oscillator, the oscillator supplies a total rms current of 545 mA. When the same two capacitors are connected to the oscillator in series, the oscillator supplies an rms current of 126 mA. What are the values of the two capacitors?

41. ‖ Show that Equation 35.27 for the phase angle ϕ of a series *RLC* circuit gives the correct result for a capacitor-only circuit.

42. ‖ a. What is the peak current supplied by the emf in **FIGURE P35.42**?
    b. What is the peak voltage across the 3.0 μF capacitor?

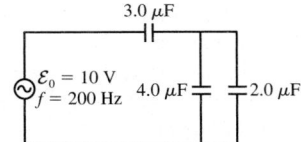

FIGURE P35.42

43. ‖ You have a resistor and a capacitor of unknown values. First, you charge the capacitor and discharge it through the resistor. By monitoring the capacitor voltage on an oscilloscope, you see that the voltage decays to half its initial value in 2.5 ms. You then use the resistor and capacitor to make a low-pass filter. What is the crossover frequency f_c?

44. ‖ **FIGURE P35.44** shows a parallel *RC* circuit.
    a. Use a phasor-diagram analysis to find expressions for the peak currents I_R and I_C.
    Hint: What do the resistor and capacitor have in common? Use that as the initial phasor.
    b. Complete the phasor analysis by finding an expression for the peak emf current I.

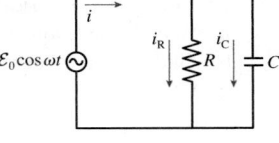

FIGURE P35.44

45. ‖ **FIGURE P35.45** shows voltage and current graphs for a capacitor.
    a. What is the emf frequency f?
    b. What is the value of the capacitance C?

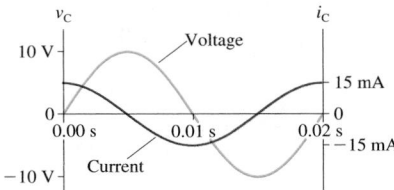

FIGURE P35.45

46. ‖ **FIGURE P35.46** shows voltage and current graphs for an inductor.
    a. What is the emf frequency f?
    b. What is the value of the inductance L?

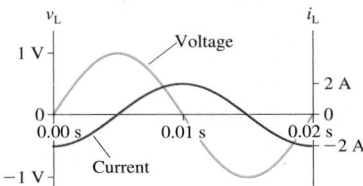

FIGURE P35.46

47. ‖ Use a phasor diagram to analyze the *RL* circuit of **FIGURE P35.47**. In particular,
    a. Find expressions for I, V_R, and V_L.
    b. What is V_R in the limits $\omega \to 0$ and $\omega \to \infty$?
    c. If the output is taken from the resistor, is this a low-pass or a high-pass filter? Explain.
    d. Find an expression for the crossover frequency ω_c.

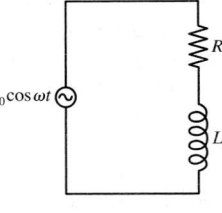

FIGURE P35.47

48. ‖ A series *RLC* circuit consists of a 100 Ω resistor, a 0.15 H inductor, and a 30 μF capacitor. It is attached to a 120 V/60 Hz power line. What are (a) the emf $\mathcal{E}_{rms}$, (b) the phase angle ϕ, and (c) the average power loss?

49. ‖ A series *RLC* circuit consists of a 25 Ω resistor, a 0.10 H inductor, and a 100 μF capacitor. It draws a 2.5 A rms current when attached to a 60 Hz source. What are (a) the emf $\mathcal{E}_{rms}$, (b) the phase angle ϕ, and (c) the average power loss?

50. ‖ For the circuit of **FIGURE P35.50**,
    a. What is the resonance frequency, in both rad/s and Hz?
    b. Find V_R and V_L at resonance.
    c. How can V_L be larger than $\mathcal{E}_0$? Explain.

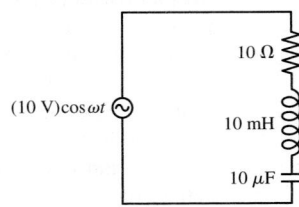

FIGURE P35.50

51. | For the circuit of FIGURE P35.51,
    a. What is the resonance frequency, in both rad/s and Hz?
    b. Find V_R and V_C at resonance.
    c. How can V_C be larger than $\mathcal{E}_0$? Explain.

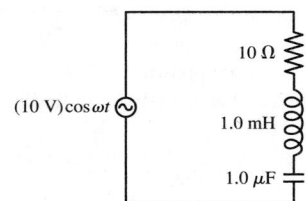

FIGURE P35.51

52. ‖ In FIGURE P35.52, what is the current supplied by the emf when (a) the frequency is very small and (b) the frequency is very large?

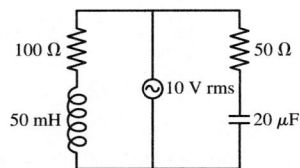

FIGURE P35.52

53. ‖ The current lags the emf by 30° in a series *RLC* circuit with $\mathcal{E}_0 = 10$ V and $R = 50\ \Omega$. What is the peak current through the circuit?

54. ‖ A series *RLC* circuit consists of a 50 Ω resistor, a 3.3 mH inductor, and a 480 nF capacitor. It is connected to a 5.0 kHz oscillator with a peak voltage of 5.0 V. What is the instantaneous current *i* when
    a. $\mathcal{E} = \mathcal{E}_0$?
    b. $\mathcal{E} = 0$ V and is decreasing?

55. ‖ A series *RLC* circuit consists of a 50 Ω resistor, a 3.3 mH inductor, and a 480 nF capacitor. It is connected to a 3.0 kHz oscillator with a peak voltage of 5.0 V. What is the instantaneous emf $\mathcal{E}$ when
    a. $i = I$?
    b. $i = 0$ A and is decreasing?
    c. $i = -I$?

56. ‖ A series *RLC* circuit consists of a 100 Ω resistor, a 10 mH inductor, and a 1.0 nF capacitor. It is connected to an oscillator with an rms voltage of 10 V. What is the power supplied to the circuit if (a) $\omega = \frac{1}{2}\omega_0$? (b) $\omega = \omega_0$? (c) $\omega = 2\omega_0$?

57. ‖ Show that the power factor of a series *RLC* circuit is $\cos\phi = R/Z$.

58. ‖ For a series *RLC* circuit, show that
    a. The peak current can be written $I = I_{max}\cos\phi$.
    b. The average power can be written $P_{source} = P_{max}\cos^2\phi$.

59. ‖ The tuning circuit in an FM radio receiver is a series *RLC* circuit with a 0.200 μH inductor.
    a. The receiver is tuned to a station at 104.3 MHz. What is the value of the capacitor in the tuning circuit?
    b. FM radio stations are assigned frequencies every 0.2 MHz, but two nearby stations cannot use adjacent frequencies. What is the maximum resistance the tuning circuit can have if the peak current at a frequency of 103.9 MHz, the closest frequency that can be used by a nearby station, is to be no more than 0.10% of the peak current at 104.3 MHz? The radio is still tuned to 104.3 MHz, and you can assume the two stations have equal strength.

60. ‖ A television channel is assigned the frequency range from 54 MHz to 60 MHz. A series *RLC* tuning circuit in a TV receiver resonates in the middle of this frequency range. The circuit uses a 16 pF capacitor.
    a. What is the value of the inductor?
    b. In order to function properly, the current throughout the frequency range must be at least 50% of the current at the resonance frequency. What is the minimum possible value of the circuit's resistance?

61. ‖ Lightbulbs labeled 40 W, 60 W, and 100 W are connected to a 120 V/60 Hz power line as shown in FIGURE P35.61. What is the rate at which energy is dissipated in each bulb?

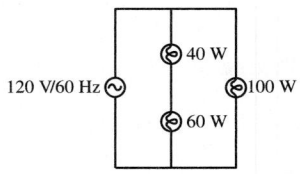

FIGURE P35.61

62. ‖ Commercial electricity is generated and transmitted as *three-phase electricity*. Instead of a single emf, three separate wires carry currents for the emfs $\mathcal{E}_1 = \mathcal{E}_0\cos\omega t$, $\mathcal{E}_2 = \mathcal{E}_0\cos(\omega t + 120°)$, and $\mathcal{E}_3 = \mathcal{E}_0\cos(\omega t - 120°)$ over three parallel wires, each of which supplies one-third of the power. This is why the long-distance transmission lines you see in the countryside have three wires. Suppose the transmission lines into a city supply a total of 450 MW of electric power, a realistic value.
    a. What would be the current in each wire if the transmission voltage were $\mathcal{E}_0 = 120$ V rms?
    b. In fact, transformers are used to step the transmission-line voltage up to 500 kV rms. What is the current in each wire?
    c. Big transformers are expensive. Why does the electric company use step-up transformers?

63. ‖ Commercial electricity is generated and transmitted as *three-phase electricity*. Instead of a single emf $\mathcal{E} = \mathcal{E}_0\cos\omega t$, three separate wires carry currents for the emfs $\mathcal{E}_1 = \mathcal{E}_0\cos\omega t$, $\mathcal{E}_2 = \mathcal{E}_0\cos(\omega t + 120°)$, and $\mathcal{E}_3 = \mathcal{E}_0\cos(\omega t - 120°)$. This is why the long-distance transmission lines you see in the countryside have three parallel wires, as do many distribution lines within a city.
    a. Draw a phasor diagram showing phasors for all three phases of a three-phase emf.
    b. Show that the sum of the three phases is zero, producing what is referred to as *neutral*. In *single-phase* electricity, provided by the familiar 120 V/60 Hz electric outlets in your home, one side of the outlet is neutral, as established at a nearby electrical substation. The other, called the *hot side,* is one of the three phases. (The round opening is connected to ground.)
    c. Show that the potential difference between any two of the phases has the rms value $\sqrt{3}\,\mathcal{E}_{rms}$, where $\mathcal{E}_{rms}$ is the familiar single-phase rms voltage. Evaluate this potential difference for $\mathcal{E}_{rms} = 120$ V. Some high-power home appliances, especially electric clothes dryers and hot-water heaters, are designed to operate between two of the phases rather than between one phase and neutral. Heavy-duty industrial motors are designed to operate from all three phases, but full three-phase power is rare in residential or office use.

64. ‖ A motor attached to a 120 V/60 Hz power line draws an 8.0 A current. Its average energy dissipation is 800 W.
    a. What is the power factor?
    b. What is the rms resistor voltage?
    c. What is the motor's resistance?
    d. How much series capacitance needs to be added to increase the power factor to 1.0?

Challenge Problems

65. The small transformers that power many consumer products produce a 12.0 V rms, 60 Hz emf. Design a circuit using resistors and capacitors that uses the transformer voltage as an input and produces a 6.0 V rms output that leads the input voltage by 45°.

66. FIGURE CP35.66 shows voltage and current graphs for a series *RLC* circuit.
    a. What is the resistance *R*?
    b. If $L = 200 \ \mu H$, what is the resonance frequency?

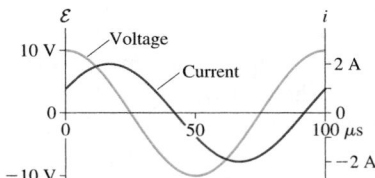

FIGURE CP35.66

67. You're the operator of a 15,000 V rms, 60 Hz electrical substation. When you get to work one day, you see that the station is delivering 6.0 MW of power with a power factor of 0.90.
    a. What is the rms current leaving the station?
    b. How much series capacitance should you add to bring the power factor up to 1.0?
    c. How much power will the station then be delivering?

68. a. Show that the average power loss in a series *RLC* circuit is

    $$P_{\text{avg}} = \frac{\omega^2 \mathcal{E}_{\text{rms}}^2 R}{\omega^2 R^2 + L^2(\omega^2 - \omega_0^2)^2}$$

    b. Prove that the energy dissipation is a maximum at $\omega = \omega_0$.

69. a. Show that the peak inductor voltage in a series *RLC* circuit is maximum at frequency

    $$\omega_{\text{L}} = \left(\frac{1}{\omega_0^2} - \frac{1}{2} R^2 C^2 \right)^{-1/2}$$

    b. A series *RLC* circuit with $\mathcal{E}_0 = 10.0$ V consists of a 1.0 Ω resistor, a 1.0 μH inductor, and a 1.0 μF capacitor. What is V_{L} at $\omega = \omega_0$ and at $\omega = \omega_{\text{L}}$?

70. The telecommunication circuit shown in FIGURE CP35.70 has a parallel inductor and capacitor in series with a resistor.
    a. Use a phasor diagram to show that the peak current through the resistor is

    $$I = \frac{\mathcal{E}_0}{\sqrt{R^2 + \left(\dfrac{1}{X_{\text{L}}} - \dfrac{1}{X_{\text{C}}} \right)^{-2}}}$$

    Hint: Start with the inductor phasor v_{L}.
    b. What is *I* in the limits $\omega \to 0$ and $\omega \to \infty$?
    c. What is the resonance frequency ω_0? What is *I* at this frequency?

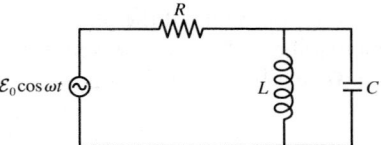

FIGURE CP35.70

71. Consider the parallel *RLC* circuit shown in FIGURE CP35.71.
    a. Show that the current drawn from the emf is

    $$I = \mathcal{E}_0 \sqrt{\frac{1}{R^2} + \left(\frac{1}{\omega L} - \omega C \right)^2}$$

    Hint: Start with a phasor that is common to all three circuit elements.
    b. What is *I* in the limits $\omega \to 0$ and $\omega \to \infty$?
    c. Find the frequency for which *I* is a minimum.
    d. Sketch a graph of *I* versus ω.

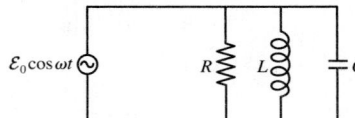

FIGURE CP35.71

STOP TO THINK ANSWERS

Stop to Think 35.1: a. The instantaneous emf value is the projection down onto the horizontal axis. The emf is negative but increasing in magnitude as the phasor, which rotates ccw, approaches the horizontal axis.

Stop to Think 35.2: c. Voltage and current are measured using different scales and units. You can't compare the length of a voltage phasor to the length of a current phasor.

Stop to Think 35.3: a. There is "no capacitor" when the separation between the two capacitor plates becomes zero and the plates touch. Capacitance *C* is inversely proportional to the plate spacing *d*, hence $C \to \infty$ as $d \to 0$. The capacitive reactance is inversely proportional to *C*, so $X_{\text{C}} \to 0$ as $C \to \infty$.

Stop to Think 35.4: $(\omega_{\text{c}})_{\text{d}} > (\omega_{\text{c}})_{\text{c}} = (\omega_{\text{c}})_{\text{a}} > (\omega_{\text{c}})_{\text{b}}$. The crossover frequency is $1/RC$.

Stop to Think 35.5: Above. $V_{\text{L}} > V_{\text{C}}$ tells us that $X_{\text{L}} > X_{\text{C}}$. This is the condition above resonance, where X_{L} is increasing with ω while X_{C} is decreasing.

Stop to Think 35.6: a, b, and **f.** You can always increase power by turning up the voltage. The current leads the emf, telling us that the circuit is primarily capacitive. The current can be brought into phase with the emf, thus maximizing the power, by decreasing *C* or increasing *L*.

SUMMARY

Electricity and Magnetism

Mass and charge are the two most fundamental properties of matter. The first five parts of this text were investigations of the properties and interactions of masses. Part VI has been a study of the physics of charge—what charge is and how charges interact.

Electric and magnetic fields were introduced to enable us to understand the long-range forces of electricity and magnetism. The field concept is subtle, but it is an essential part of our modern understanding of the physical universe. One charge— the source charge—alters the space around it by creating an electric field and, if the charge is moving, a magnetic field. Other charges experience forces exerted *by the fields*. Thus the

electric and magnetic fields are the agents by which charges interact.

Faraday's discovery of electromagnetic induction led scientists to recognize that the fields are *real* and can exist independently of charges. The most vivid confirmation of this reality was Maxwell's discovery of electromagnetic waves—the quintessential electromagnetic phenomenon.

Part VI has introduced many new phenomena, concepts, and laws. The knowledge structure table draws together the major ideas about charges and fields, and it briefly summarizes some of the most important applications of electricity and magnetism.

KNOWLEDGE STRUCTURE VI   **Electricity and Magnetism**

ESSENTIAL CONCEPTS   Charge, dipole, field, potential, emf
BASIC GOALS   How do charged particles interact?
What are the properties and characteristics of electromagnetic fields?

GENERAL PRINCIPLES

Coulomb's law
$$\vec{E}_{\text{point charge}} = \frac{1}{4\pi\epsilon_0}\frac{q}{r^2}\hat{r} = \left(\frac{1}{4\pi\epsilon_0}\frac{q}{r^2}, \text{away from } q\right)$$

Biot-Savart law
$$\vec{B}_{\text{point charge}} = \frac{\mu_0}{4\pi}\frac{q\vec{v}\times\hat{r}}{r^2} = \left(\frac{\mu_0}{4\pi}\frac{qv\sin\theta}{r^2}, \text{direction of right-hand rule}\right)$$

Faraday's law $\mathcal{E} = |d\Phi_{\text{m}}/dt|$ $I_{\text{induced}} = \mathcal{E}/R$ in the direction of Lenz's law

Lenz's law   An induced current flows around a conducting loop in the direction such that the induced magnetic field opposes the *change* in the magnetic flux.

Lorentz force law $\vec{F}_{\text{on } q} = q(\vec{E} + \vec{v}\times\vec{B})$

Superposition   The electric or magnetic field due to multiple charges is the vector sum of the field of each charge. This principle was used to derive the fields of many special charge distributions, such as wires, planes, and loops.

FIELD AND POTENTIAL The electric field of charges can also be described in terms of an electric potential V:

$$V_{\text{point charge}} = \frac{q}{4\pi\epsilon_0 r}$$

- The electric field is perpendicular to equipotential surfaces and in the direction of decreasing potential.

- The potential energy of charge q is $U = qV$. The total energy $K + qV$ of a group of charges is conserved.

ELECTROMAGNETIC WAVES All the properties of electromagnetic fields are summarized mathematically in four equations called *Maxwell's equations*. From Maxwell's equations we learn that electromagnetic fields can exist independently of charges as an *electromagnetic wave*.

- An em wave travels at speed $c = 1/\sqrt{\epsilon_0\mu_0}$.

- $\vec{E}$ and $\vec{B}$ are perpendicular to each other and to the direction of travel, with $E = cB$.

Electric and magnetic properties of materials

- Charges move through conductors but not through insulators.

- Conductors and insulators are *polarized* in an electric field.

- A magnetic moment in a magnetic field experiences a torque.

Model of current and conductivity

- The charge carriers in metals are electrons.

- emf $\rightarrow$ electric field $\rightarrow$ current density $J = \sigma E \rightarrow I = JA$

Applications to circuits

- Circuits obey Kirchhoff's loop law (conservation of energy) and junction law (conservation of current).

- Resistors control the current: $I = \Delta V/R$ (Ohm's law).

- Capacitors store charge $Q = C\Delta V$ and energy $V_{\text{C}} = \frac{1}{2}C(\Delta V_{\text{C}})^2$.

The Telecommunications Revolution

In 1800, the year that Alessandro Volta invented the battery and Thomas Jefferson was elected president, the fastest a message could travel was the speed of a man or woman on horseback. News took three days to travel from New York to Boston, and well over a month to reach the frontier outpost of Cincinnati.

But Hans Oersted's 1820 discovery that a current creates a magnetic field introduced revolutionary changes to communications. The American scientist Joseph Henry, who shares with Faraday credit for the discovery of electromagnetic induction, saw a simple electromagnet in 1825. Inspired, he set about improving the device. By 1830, Henry was able to send current through more than a mile of wire to activate an electromagnet and strike a bell.

In 1835, Henry met an entrepreneur interested in the commercial development of electric technology—Samuel F. B. Morse. Morse was one of the most prominent American artists of the early 19th century, but he also had an abiding interest in technology. In the 1830s, he invented the famous code that bears his name—Morse code—and began to experiment with electromagnets.

With advice and encouragement from Henry, Morse developed the first practical telegraph. The first telegraph line, between Washington, D.C., and Baltimore, began operating in 1844; the first message sent was "What hath God wrought?" For the first time, long-distance communications could take place essentially instantaneously.

Telegraph communication advanced as quickly as wire could be strung, and a worldwide network had been established by 1875. But the telegraph didn't hold its monopoly for long, as other inventors began to think about using electromagnetic devices to transmit speech. The first to succeed was Alexander Graham Bell, who invented the telephone in 1876.

The telegraph and telephone provided electromagnetic communication over wires, but the discovery of electromagnetic waves opened up another possibility—wireless communication at the speed of light. Radio technology developed rapidly in the late 19th century, and in 1901 the Italian inventor Guglielmo Marconi sent and received the first transatlantic radio message. World War I prompted further development of radio, because of the need to communicate with military units as they moved about, and by 1925 more than 1000 radio stations were operating in the United States.

Radio and, later, television spanned the globe by 1960, but radio stations reached a few hundred miles at best, and television transmission was limited to each city. National broadcasts within the United States required the signal to be transmitted via microwave relays to local stations for rebroadcast. Network television shows were possible, but not live-from-the-scene broadcasts. Journalists had to film events, then return the film to the studio for broadcast. Television images from overseas could only be seen the next day, after film was flown back to the United States.

The first communications satellite was launched by NASA in 1960, followed two years later by a more practical satellite, Telstar, that used solar power to amplify signals received from earth and beam them back down. The first live transatlantic television transmission was made on July 11, 1962, and was broadcast throughout the United States.

Plans were made for a system of roughly 100 satellites, so that one would always be overhead, but another idea soon proved more practical. In 1945, 12 years before space flight began, the science-fiction writer Arthur C. Clarke proposed placing satellites in orbits 22,300 miles above the earth. A satellite at this altitude orbits with a 24-hour period, so from the ground it appears to hang stationary in space. We now call this a *geosynchronous orbit*. One such satellite would allow microwave communication between two points one-third of a world apart, so just three geosynchronous satellites would span the entire earth.

Much more energy is required to reach geosynchronous orbit than to reach low-earth orbit, but rocket technology was advancing faster than NASA could build Telstar satellites. The first commercial communications satellite was placed in geosynchronous orbit in 1965, and, for the first time, television images could be broadcast live to anywhere in the world. Today all of the world's intercontinental television and much of the intercontinental telephone traffic travel via microwaves to and from a cluster of these artificial stars floating high above the earth.

Today, in the 21st century, information and images span the world as quickly as or more quickly than they once moved through a small village. You can talk to friends or relatives anywhere around the globe, and each day's news brings live images from remote places. Telecommunication unites our world, and the technologies of telecommunications are direct descendants of Coulomb, Ampère, Oersted, Henry, and—most of all—Michael Faraday.

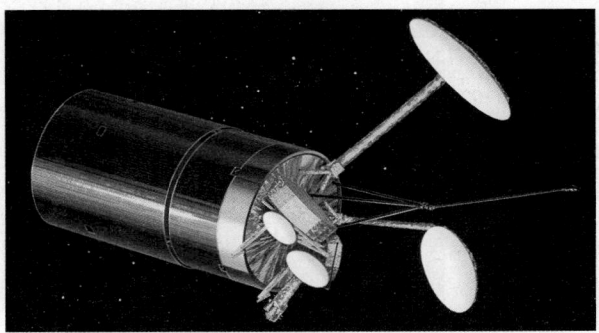

This INTELSAT telecommunications satellite is 12 m (40 ft) long.

Relativity and Quantum Physics

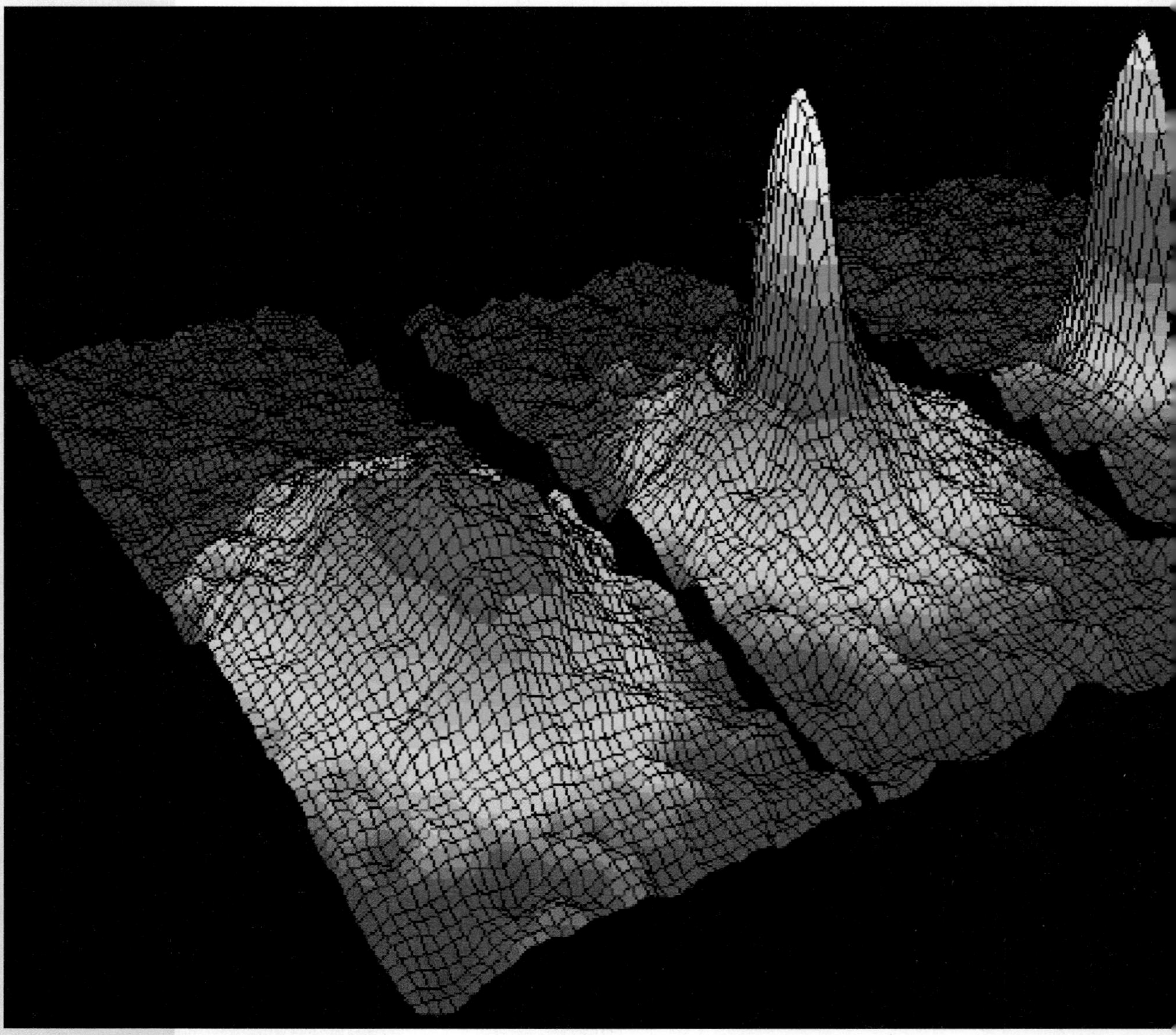

This three-frame sequence shows a gas of a few thousand rubidium atoms condensing into a single quantum state known as a Bose-Einstein condensate. This phenomenon was predicted by Einstein in 1925 but not observed until 1995, when physicists learned how to use lasers to cool the atoms to temperatures below 200 nanokelvin.

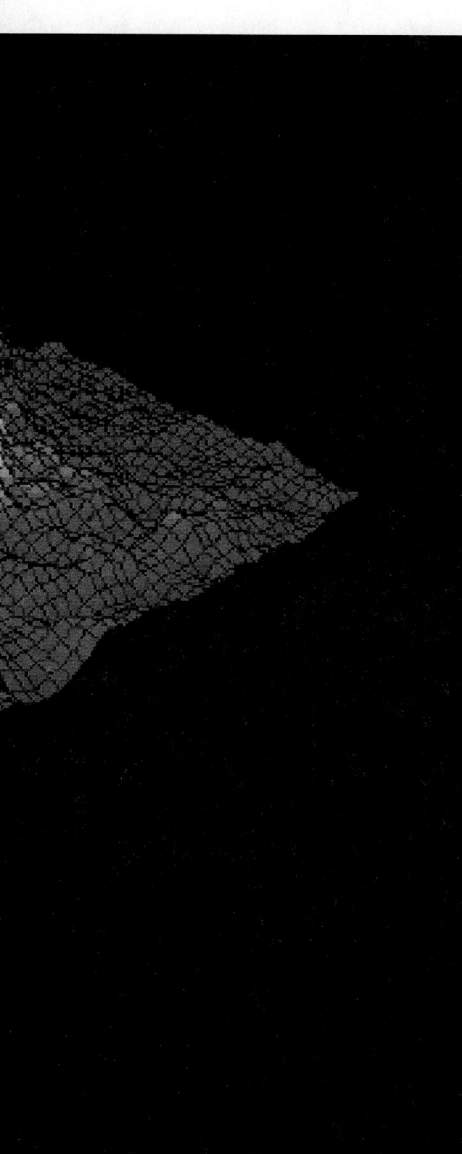

OVERVIEW

Contemporary Physics

Our journey into physics is nearing its end. We began roughly 350 years ago with Newton's discovery of the laws of motion. Part VI brought us to the end of the 19th century, just over 100 years ago. Along the way you've learned about the motion of particles, the conservation of energy, the physics of waves, and the electromagnetic interactions that hold atoms together and generate light waves. We begin the last phase of our journey with confidence.

Newton's mechanics and Maxwell's electromagnetism were the twin pillars of science at the end of the 19th century and the basis for much of engineering and applied science in the 20th century. Despite the successes of these theories, a series of discoveries starting around 1900 and continuing into the first few decades of the 20th century profoundly altered our understanding of the universe at the most fundamental level.

- Einstein's theory of relativity forced scientists to completely revise their concepts of space and time. Our exploration of these fascinating ideas will end with perhaps the most famous equation in physics: Einstein's $E = mc^2$.
- Experimenters found that the classical distinction between *particles* and *waves* breaks down at the atomic level. Light sometimes acts like a particle, while electrons and even entire atoms sometimes act like waves. We will need a new theory of light and matter—quantum physics—to explain these phenomena.

These two theories form the basis for physics as it is practiced today, and they are now having a significant impact on 21st-century engineering.

The complete theory of quantum physics, as it was developed in the 1920s, describes atomic particles in terms of an entirely new concept called a *wave function*. One of our most important tasks in Part VII will be to learn what a wave function is, what laws govern its behavior, and how to relate wave functions to experimental measurements. We will concentrate on one-dimensional models that, while not perfect, will be adequate for understanding the essential features of scanning tunneling microscopes, various semiconductor devices, radioactive decay, and other applications.

We'll complete our study of quantum physics with an introduction to atomic and nuclear physics. You will learn where the electron-shell model of chemistry comes from, how atoms emit and absorb light, what's inside the nucleus, and why some nuclei undergo radioactive decay.

The quantum world with its wave functions and probabilities can seem strange and mysterious, yet quantum physics gives the most definitive and accurate predictions of any physical theory ever devised. The contemporary perspective of quantum physics will be a fitting end to our journey.

36 Relativity

The Large Hadron Collider, the world's highest-energy particle accelerator, is built in a 27-km-circumference tunnel near Geneva, Switzerland. It accelerates protons to 99.999999% of the speed of light.

▶ **Looking Ahead** The goal of Chapter 36 is to understand how Einstein's theory of relativity changes our concepts of space and time.

Principle of Relativity

Einstein's theory of relativity is based on a simple-sounding principle: The laws of physics are the same in every inertial reference frame. This seemingly innocuous statement will force us to completely rethink our ideas of space and time.

The most well-known consequence of this principle is that light travels at the same speed c in all inertial reference frames.

You'll learn why it is that no object or information can travel faster than the speed of light.

Space

The physical length of an object is *less* when the object is moving in a reference frame than when it is at rest in that reference frame. This is **length contraction.**

To us, the Fermilab Accelerator is 3.9 miles in circumference. To protons in the accelerator, moving at 0.999999c, the circumference is only 30 feet.

Mass and Energy

You'll learn the significance of relativity's famous equation, $E = mc^2$. Mass can be transformed into energy, and energy into mass, as long as the total energy is conserved.

The sun is powered by the conversion of 4 billion kilograms of matter into energy every second. Even so, the sun will continue to shine for billions of years.

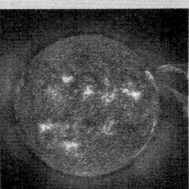

Reference Frames

You'll learn to work with **events** whose position in space and time of occurrence are measured by experimenters in different **inertial reference frames.**

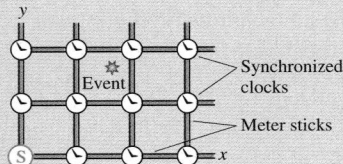

Synchronized clocks

Meter sticks

◀ **Looking Back**
Section 4.4 Reference frames and relative velocity

Time

The time interval between ticks of a clock is *longer* when the clock is moving in a reference frame than when it is at rest in that reference frame. This is **time dilation.**

You'll learn about the **twin paradox.** If an astronaut travels to a distance star and back at a speed close to that of light, she'll be younger than her identical twin when she returns.

Applications of Relativity

Abstract though it may seem, relativity is important for modern technologies such as PET scans (positron-electron tomography) in medicine and nuclear energy. Relativity also underlies our understanding of the physics of stars and galaxies.

Your GPS device receives signals from precision clocks in orbiting satellites. The clocks must be corrected for relativistic effects in order for the GPS system to work.

36.1 Relativity: What's It All About?

What do you think of when you hear the phrase "theory of relativity"? A white-haired Einstein? $E = mc^2$? Black holes? Time travel? Perhaps you've heard that the theory of relativity is so complicated and abstract that only a handful of people in the whole world really understand it.

There is, without doubt, a certain mystique associated with relativity, an aura of the strange and exotic. The good news is that understanding the ideas of relativity is well within your grasp. Einstein's *special theory of relativity,* the portion of relativity we'll study, is not mathematically difficult at all. The challenge is conceptual because relativity questions deeply held assumptions about the nature of space and time. In fact, that's what relativity is all about—space and time.

What's Special About Special Relativity?

Einstein's first paper on relativity, in 1905, dealt exclusively with inertial reference frames, reference frames that move relative to each other with constant velocity. Ten years later, Einstein published a more encompassing theory of relativity that considered accelerated motion and its connection to gravity. The second theory, because it's more general in scope, is called *general relativity.* General relativity is the theory that describes black holes, curved spacetime, and the evolution of the universe. It is a fascinating theory but, alas, very mathematical and outside the scope of this textbook.

Motion at constant velocity is a "special case" of motion—namely, motion for which the acceleration is zero. Hence Einstein's first theory of relativity has come to be known as **special relativity.** It is special in the sense of being a restricted, special case of his more general theory, not special in the everyday sense meaning distinctive or exceptional. Special relativity, with its conclusions about time dilation and length contraction, is what we will study.

Albert Einstein (1879–1955) was one of the most influential thinkers in history.

36.2 Galilean Relativity

Relativity is the process of relating measurements in one reference frame to those in a different reference frame moving *relative to* the first. To appreciate and understand what is new in Einstein's theory, we need a firm grasp of the ideas of relativity that are embodied in Newtonian mechanics. Thus we begin with *Galilean relativity.*

Reference Frames

Suppose you're passing me as we both drive in the same direction along a freeway. My car's speedometer reads 55 mph while your speedometer shows 60 mph. Is 60 mph your "true" speed? That is certainly your speed relative to someone standing beside the road, but your speed relative to me is only 5 mph. Your speed is 120 mph relative to a driver approaching from the other direction at 60 mph.

An object does not have a "true" speed or velocity. The very definition of velocity, $v = \Delta x/\Delta t$, assumes the existence of a coordinate system in which, during some time interval Δt, the displacement Δx is measured. The best we can manage is to specify an object's velocity relative to, or with respect to, the coordinate system in which it is measured.

Let's define a **reference frame** to be a coordinate system in which experimenters equipped with meter sticks, stopwatches, and any other needed equipment make position and time measurements on moving objects. Three ideas are implicit in our definition of a reference frame:

- A reference frame extends infinitely far in all directions.
- The experimenters are at rest in the reference frame.
- The number of experimenters and the quality of their equipment are sufficient to measure positions and velocities to any level of accuracy needed.

The first two ideas are especially important. It is often convenient to say "the laboratory reference frame" or "the reference frame of the rocket." These are shorthand expressions for "a reference frame, infinite in all directions, in which the laboratory (or the rocket) and a set of experimenters happen to be at rest."

NOTE ▶ A reference frame is not the same thing as a "point of view." That is, each person or each experimenter does not have his or her own private reference frame. **All experimenters at rest relative to each other share the same reference frame.** ◀

FIGURE 36.1 The standard reference frames S and S'.

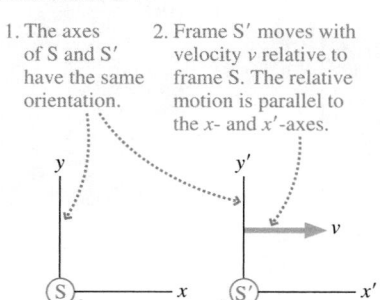

1. The axes of S and S' have the same orientation.

2. Frame S' moves with velocity v relative to frame S. The relative motion is parallel to the x- and x'-axes.

3. The origins of S and S' coincide at $t = 0$. This is our definition of $t = 0$.

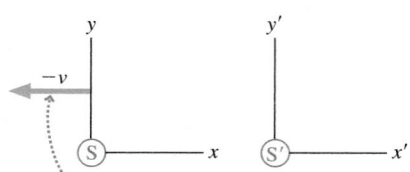

4. Alternatively, frame S moves with velocity $-v$ relative to frame S'.

FIGURE 36.1 shows two reference frames called S and S'. The coordinate axes in S are x, y, z and those in S' are x', y', z'. Reference frame S' moves with velocity v relative to S or, equivalently, S moves with velocity $-v$ relative to S'. There's no implication that either reference frame is "at rest." Notice that the zero of time, when experimenters start their stopwatches, is the instant that the origins of S and S' coincide.

We will restrict our attention to *inertial reference frames,* implying that the relative velocity v is constant. You should recall from Chapter 5 that an **inertial reference frame** is a reference frame in which Newton's first law, the law of inertia, is valid. In particular, an inertial reference frame is one in which an isolated particle, one on which there are no forces, either remains at rest or moves in a straight line at constant speed.

Any reference frame moving at constant velocity with respect to an inertial reference frame is itself an inertial reference frame. Conversely, a reference frame accelerating with respect to an inertial reference frame is *not* an inertial reference frame. Our restriction to reference frames moving with respect to each other at constant velocity—with no acceleration—is the "special" part of special relativity.

NOTE ▶ An inertial reference frame is an idealization. A true inertial reference frame would need to be floating in deep space, far from any gravitational influence. In practice, an earthbound laboratory is a good approximation of an inertial reference frame because the accelerations associated with the earth's rotation and motion around the sun are too small to influence most experiments. ◀

STOP TO THINK 36.1 Which of these is an inertial reference frame (or a very good approximation)?

a. Your bedroom
b. A car rolling down a steep hill
c. A train coasting along a level track
d. A rocket being launched
e. A roller coaster going over the top of a hill
f. A sky diver falling at terminal speed

FIGURE 36.2 The position of an exploding firecracker is measured in reference frames S and S'.

At time t, the origin of S' has moved distance vt to the right. Thus $x = x' + vt$.

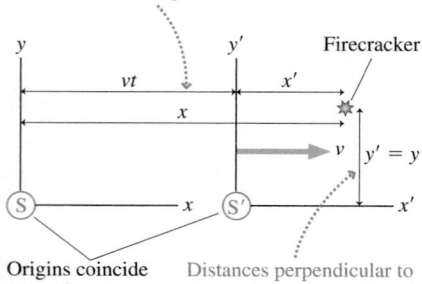

Origins coincide at $t = 0$.

Distances perpendicular to the motion are not affected. Thus $y' = y$ and $z' = z$.

The Galilean Transformations

Suppose a firecracker explodes at time t. The experimenters in reference frame S determine that the explosion happened at position x. Similarly, the experimenters in S' find that the firecracker exploded at x' in their reference frame. What is the relationship between x and x'?

FIGURE 36.2 shows the explosion and the two reference frames. You can see from the figure that $x = x' + vt$, thus

$$
\begin{aligned}
x &= x' + vt & x' &= x - vt \\
y &= y' & \text{or} \quad y' &= y \\
z &= z' & z' &= z
\end{aligned}
\tag{36.1}
$$

These are the *Galilean transformations of position*. If you know a position measured by the experimenters in one inertial reference frame, you can calculate the position that would be measured by experimenters in any other inertial reference frame.

Suppose the experimenters in both reference frames now track the motion of the object in FIGURE 36.3 by measuring its position at many instants of time. The experimenters in S find that the object's velocity is $\vec{u}$. During the *same time interval* Δt, the experimenters in S′ measure the velocity to be $\vec{u}'$.

NOTE ▶ In this chapter, we will use v to represent the velocity of one reference frame relative to another. We will use $\vec{u}$ and $\vec{u}'$ to represent the velocities of objects with respect to reference frames S and S′. ◀

We can find the relationship between $\vec{u}$ and $\vec{u}'$ by taking the time derivatives of Equations 36.1 and using the definition $u_x = dx/dt$:

$$u_x = \frac{dx}{dt} = \frac{dx'}{dt} + v = u'_x + v$$

$$u_y = \frac{dy}{dt} = \frac{dy'}{dt} = u'_y$$

The equation for u_z is similar. The net result is

$$
\begin{array}{ccc}
u_x = u'_x + v & & u'_x = u_x - v \\
u_y = u'_y & \text{or} & u'_y = u_y \\
u_z = u'_z & & u'_z = u_z
\end{array}
\qquad (36.2)
$$

Equations 36.2 are the *Galilean transformations of velocity*. If you know the velocity of a particle in one inertial reference frame, you can find the velocity that would be measured by experimenters in any other inertial reference frame.

NOTE ▶ In Section 4.4 you learned the Galilean transformation of velocity as $\vec{v}_{CB} = \vec{v}_{CA} + \vec{v}_{AB}$, where $\vec{v}_{AB}$ means "the velocity of A relative to B." Equations 36.2 are equivalent for relative motion parallel to the *x*-axis but are written in a more formal notation that will be useful for relativity. ◀

FIGURE 36.3 The velocity of a moving object is measured in reference frames S and S′.

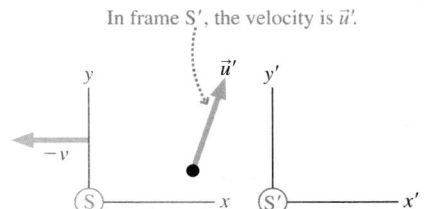

The object's velocity in frame S is $\vec{u}$.

In frame S′, the velocity is $\vec{u}'$.

EXAMPLE 36.1 **The speed of sound**

An airplane is flying at speed 200 m/s with respect to the ground. Sound wave 1 is approaching the plane from the front, sound wave 2 is catching up from behind. Both waves travel at 340 m/s relative to the ground. What is the speed of each wave relative to the plane?

MODEL Assume that the earth (frame S) and the airplane (frame S′) are inertial reference frames. Frame S′, in which the airplane is at rest, moves at $v = 200$ m/s relative to frame S.

VISUALIZE FIGURE 36.4 shows the airplane and the sound waves.

FIGURE 36.4 Experimenters in the plane measure different speeds for the waves than do experimenters on the ground.

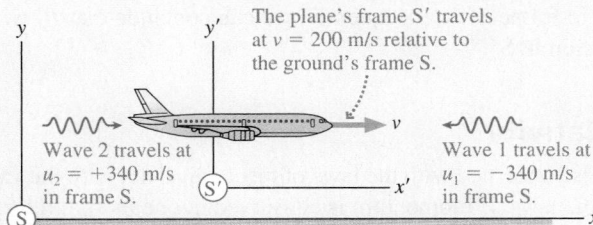

The plane's frame S′ travels at $v = 200$ m/s relative to the ground's frame S.

Wave 2 travels at $u_2 = +340$ m/s in frame S.

Wave 1 travels at $u_1 = -340$ m/s in frame S.

SOLVE The speed of a mechanical wave, such as a sound wave or a wave on a string, is its speed *relative to its medium*. Thus the *speed of sound* is the speed of a sound wave through a reference frame in which the air is at rest. This is reference frame S, where wave 1 travels with velocity $u_1 = -340$ m/s and wave 2 travels with velocity $u_2 = +340$ m/s. Notice that the Galilean transformations use *velocities*, with appropriate signs, not just speeds.

The airplane travels to the right with reference frame S′ at velocity v. We can use the Galilean transformations of velocity to find the velocities of the two sound waves in frame S′:

$$u'_1 = u_1 - v = -340 \text{ m/s} - 200 \text{ m/s} = -540 \text{ m/s}$$

$$u'_2 = u_2 - v = 340 \text{ m/s} - 200 \text{ m/s} = 140 \text{ m/s}$$

ASSESS This isn't surprising. If you're driving at 50 mph, a car coming the other way at 55 mph is approaching you at 105 mph. A car coming up behind you at 55 mph is gaining on you at the rate of only 5 mph. Wave speeds behave the same. Notice that a mechanical wave appears to be stationary to a person moving at the wave speed. To a surfer, the crest of the ocean wave remains at rest under his or her feet.

Ocean waves are approaching the beach at 10 m/s. A boat heading out to sea travels at 6 m/s. How fast are the waves moving in the boat's reference frame?

a. 16 m/s   b. 10 m/s   c. 6 m/s   d. 4 m/s

The Galilean Principle of Relativity

FIGURE 36.5 Experimenters in both reference frames test Newton's second law by measuring the force on a particle and its acceleration.

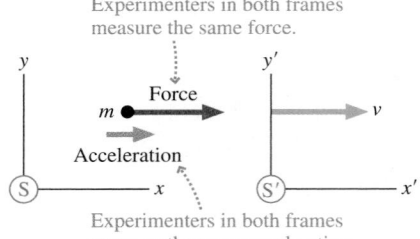

Experimenters in reference frames S and S′ measure different values for position and velocity. What about the force on and the acceleration of the particle in FIGURE 36.5? The strength of a force can be measured with a spring scale. The experimenters in reference frames S and S′ both see the *same reading* on the scale (assume the scale has a bright digital display easily seen by all experimenters), so both conclude that the force is the same. That is, $F' = F$.

We can compare the accelerations measured in the two reference frames by taking the time derivative of the velocity transformation equation $u' = u - v$. (We'll assume, for simplicity, that the velocities and accelerations are all in the x-direction.) The relative velocity v between the two reference frames is *constant*, with $dv/dt = 0$, thus

$$a' = \frac{du'}{dt} = \frac{du}{dt} = a \qquad (36.3)$$

Experimenters in reference frames S and S′ measure different values for an object's position and velocity, but they *agree* on its acceleration.

If $F = ma$ in reference frame S, then $F' = ma'$ in reference frame S′. Stated another way, if Newton's second law is valid in one inertial reference frame, then it is valid in all inertial reference frames. Because other laws of mechanics, such as the conservation laws, follow from Newton's laws of motion, we can state this conclusion as the *Galilean principle of relativity:*

> **Galilean principle of relativity** The laws of mechanics are the same in all inertial reference frames.

The Galilean principle of relativity is easy to state, but to understand it we must understand what is and is not "the same." To take a specific example, consider the law of conservation of momentum. FIGURE 36.6a shows two particles about to collide. Their total momentum in frame S, where particle 2 is at rest, is $P_i = 9.0$ kg m/s. This is an isolated system, hence the law of conservation of momentum tells us that the momentum after the collision will be $P_f = 9.0$ kg m/s.

FIGURE 36.6b has used the velocity transformation to look at the same particles in frame S′ in which particle 1 is at rest. The initial momentum in S′ is $P_i' = -18$ kg m/s. Thus it is not the *value* of the momentum that is the same in all inertial reference frames. Instead, the Galilean principle of relativity tells us that the *law* of momentum conservation is the same in all inertial reference frames. If $P_f = P_i$ in frame S, then it must be true that $P_f' = P_i'$ in frame S′. Consequently, we can conclude that P_f' will be -18 kg m/s after the collision in S′.

FIGURE 36.6 Total momentum measured in two reference frames.

(a) Collision seen in frame S

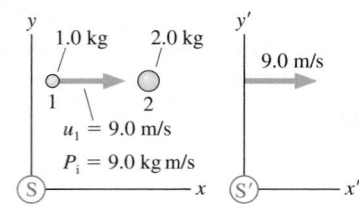

(b) Collision seen in frame S′

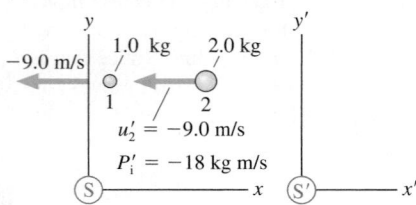

Using Galilean Relativity

The principle of relativity is concerned with the laws of mechanics, not with the values that are needed to satisfy the laws. If momentum is conserved in one inertial reference frame, it is conserved in all inertial reference frames. Even so, a problem may be easier to solve in one reference frame than in others.

Elastic collisions provide a good example of using reference frames. You learned in Chapter 10 how to calculate the outcome of a perfectly elastic collision between two particles in the reference frame in which particle 2 is initially at rest. We can use that information together with the Galilean transformations to solve elastic-collision problems in any inertial reference frame.

TACTICS
BOX 36.1 **Analyzing elastic collisions**

❶ Transform the initial velocities of particles 1 and 2 from frame S to reference frame S' in which particle 2 is at rest.

❷ The outcome of the collision in S' is given by

$$u'_{1f} = \frac{m_1 - m_2}{m_1 + m_2} u'_{1i}$$

$$u'_{2f} = \frac{2m_1}{m_1 + m_2} u'_{1i}$$

❸ Transform the two final velocities from frame S' back to frame S.

Exercises 4–5

EXAMPLE 36.2 | **An elastic collision**

A 300 g ball moving to the right at 2.0 m/s has a perfectly elastic collision with a 100 g ball moving to the left at 4.0 m/s. What are the direction and speed of each ball after the collision?

MODEL The velocities are measured in the laboratory frame, which we call frame S.

VISUALIZE **FIGURE 36.7a** shows both the balls and a reference frame S' in which ball 2 is at rest.

SOLVE The three steps of Tactics Box 36.1 are illustrated in **FIGURE 36.7b**. We're given u_{1i} and u_{2i}. The Galilean transformations of these velocities to frame S', using $v = -4.0$ m/s, are

$$u'_{1i} = u_{1i} - v = (2.0 \text{ m/s}) - (-4.0 \text{ m/s}) = 6.0 \text{ m/s}$$

$$u'_{2i} = u_{2i} - v = (-4.0 \text{ m/s}) - (-4.0 \text{ m/s}) = 0 \text{ m/s}$$

The 100 g ball is at rest in frame S', which is what we wanted. The velocities after the collision are

$$u'_{1f} = \frac{m_1 - m_2}{m_1 + m_2} u'_{1i} = 3.0 \text{ m/s}$$

$$u'_{2f} = \frac{2m_1}{m_1 + m_2} u'_{1i} = 9.0 \text{ m/s}$$

We've finished the collision analysis, but we're not done because these are the post-collision velocities in frame S'. Another application of the Galilean transformations tells us that the post-collision velocities in frame S are

$$u_{1f} = u'_{1f} + v = (3.0 \text{ m/s}) + (-4.0 \text{ m/s}) = -1.0 \text{ m/s}$$

$$u_{2f} = u'_{2f} + v = (9.0 \text{ m/s}) + (-4.0 \text{ m/s}) = 5.0 \text{ m/s}$$

Thus the 300 g ball rebounds to the left at a speed of 1.0 m/s and the 100 g ball is knocked to the right at a speed of 5.0 m/s.

FIGURE 36.7 Using reference frames to solve an elastic-collision problem.

(a)

$u_{1i} = 2.0$ m/s   −4.0 m/s   Frame S' moves with particle 2.

$u_{2i} = -4.0$ m/s

The collision takes place in frame S.

(b)

❶ Transform the velocities to frame S' in which particle 2 is at rest.

$u'_{1i} = 6.0$ m/s

$u'_{2f} = 9.0$ m/s

$u'_{1f} = 3.0$ m/s

❷ Analyze the collision in frame S'.

❸ Transform the post-collision velocities back to frame S.

$u_{1f} = -1.0$ m/s

$u_{2f} = 5.0$ m/s

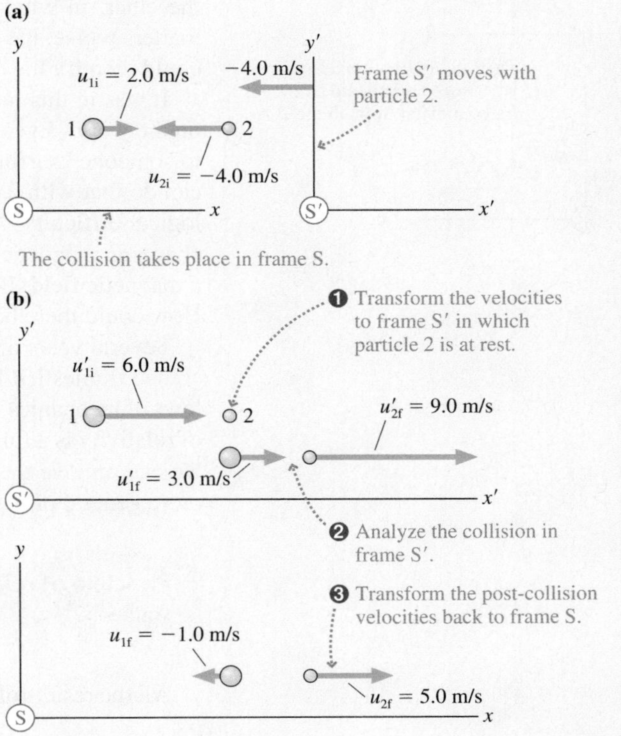

ASSESS You can easily verify that momentum is conserved: $P_f = P_i = 0.20$ kg m/s. The calculations in this example were easy. The important point of this example, and one worth careful thought, is the *logic* of what we did and why we did it.

36.3 Einstein's Principle of Relativity

The 19th century was an era of optics and electromagnetism. Thomas Young demonstrated in 1801 that light is a wave, and by midcentury scientists had devised techniques for measuring the speed of light. Faraday discovered electromagnetic induction in 1831, setting in motion a series of events leading to Maxwell's conclusion, in 1864, that light is an electromagnetic wave.

If light is a wave, what is the medium in which it travels? This was perhaps *the* most important scientific question of the second half of the 19th century. The medium in which light waves were assumed to travel was called the **ether.** Experiments to measure the speed of light were assumed to be measuring its speed through the ether. But just what *is* the ether? What are its properties? Can we collect a jar full of ether to study? Despite the significance of these questions, efforts to detect the ether or measure its properties kept coming up empty handed.

Maxwell's theory of electromagnetism didn't help the situation. The crowning success of Maxwell's theory was his prediction that light waves travel with speed

$$c = \frac{1}{\sqrt{\epsilon_0 \mu_0}} = 3.00 \times 10^8 \text{ m/s}$$

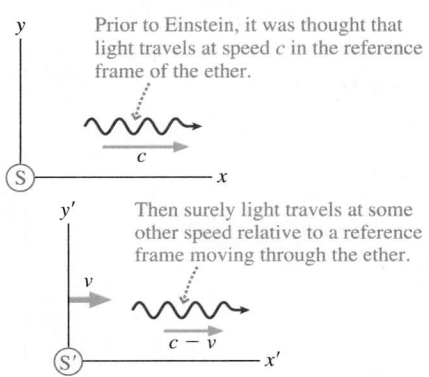

FIGURE 36.8 It seems as if the speed of light should differ from *c* in a reference frame moving through the ether.

Prior to Einstein, it was thought that light travels at speed *c* in the reference frame of the ether.

Then surely light travels at some other speed relative to a reference frame moving through the ether.

This is a very specific prediction with no wiggle room. The difficulty with such a specific prediction was the implication that Maxwell's laws of electromagnetism are valid *only* in the reference frame of the ether. After all, as FIGURE 36.8 shows, the light speed should certainly be larger or smaller than *c* in a reference frame moving through the ether, just as the sound speed is different to someone moving through the air.

As the 19th century closed, it appeared that Maxwell's theory did not obey the classical principle of relativity. There was just one reference frame, the reference frame of the ether, in which the laws of electromagnetism seemed to be true. And to make matters worse, the fact that no one had been able to detect the ether meant that no one could identify the one reference frame in which Maxwell's equations "worked."

It was in this muddled state of affairs that a young Albert Einstein made his mark on the world. Even as a teenager, Einstein had wondered how a light wave would look to someone "surfing" the wave, traveling alongside the wave at the wave speed. You can do that with a water wave or a sound wave, but light waves seemed to present a logical difficulty. An electromagnetic wave sustains itself by virtue of the fact that a changing magnetic field induces an electric field and a changing electric field induces a magnetic field. But to someone moving with the wave, *the fields would not change.* How could there be an electromagnetic wave under these circumstances?

Several years of thinking about the connection between electromagnetism and reference frames led Einstein to the conclusion that *all* the laws of physics, not just the laws of mechanics, should obey the principle of relativity. In other words, the principle of relativity is a fundamental statement about the nature of the physical universe. Thus we can remove the restriction in the Galilean principle of relativity and state a much more general principle:

> **Principle of relativity** All the laws of physics are the same in all inertial reference frames.

All the results of Einstein's theory of relativity flow from this one simple statement.

The Constancy of the Speed of Light

If Maxwell's equations of electromagnetism are laws of physics, and there's every reason to think they are, then, according to the principle of relativity, Maxwell's equations must be true in *every* inertial reference frame. On the surface this seems to be an innocuous statement, equivalent to saying that the law of conservation of momentum is true in every inertial reference frame. But follow the logic:

1. Maxwell's equations are true in all inertial reference frames.
2. Maxwell's equations predict that electromagnetic waves, including light, travel at speed $c = 3.00 \times 10^8$ m/s.
3. Therefore, **light travels at speed c in all inertial reference frames.**

FIGURE 36.9 shows the implications of this conclusion. *All* experimenters, regardless of how they move with respect to each other, find that *all* light waves, regardless of the source, travel in their reference frame with the *same* speed c. If Cathy's velocity toward Bill and away from Amy is $v = 0.9c$, Cathy finds, by making measurements in her reference frame, that the light from Bill approaches her at speed c, not at $c + v = 1.9c$. And the light from Amy, which left Amy at speed c, catches up from behind at speed c *relative to Cathy,* not the $c - v = 0.1c$ you would have expected.

Although this prediction goes against all shreds of common sense, the experimental evidence for it is strong. Laboratory experiments are difficult because even the highest laboratory speed is insignificant in comparison to c. In the 1930s, however, physicists R. J. Kennedy and E. M. Thorndike realized that they could use the earth itself as a laboratory. The earth's speed as it circles the sun is about 30,000 m/s. The *relative* velocity of the earth in January differs by 60,000 m/s from its velocity in July, when the earth is moving in the opposite direction. Kennedy and Thorndike were able to use a very sensitive and stable interferometer to show that the numerical values of the speed of light in January and July differ by less than 2 m/s.

More recent experiments have used unstable elementary particles, called π mesons, that decay into high-energy photons of light. The π mesons, created in a particle accelerator, move through the laboratory at 99.975% the speed of light, or $v = 0.99975c$, as they emit photons at speed c in the π meson's reference frame. As **FIGURE 36.10** shows, you would expect the photons to travel through the laboratory with speed $c + v = 1.99975c$. Instead, the measured speed of the photons in the laboratory was, within experimental error, 3.00×10^8 m/s.

In summary, *every* experiment designed to compare the speed of light in different reference frames has found that light travels at 3.00×10^8 m/s in every inertial reference frame, regardless of how the reference frames are moving with respect to each other.

How Can This Be?

You're in good company if you find this impossible to believe. Suppose I shot a ball forward at 50 m/s while driving past you at 30 m/s. You would certainly see the ball traveling at 80 m/s relative to you and the ground. What we're saying with regard to light is equivalent to saying that the ball travels at 50 m/s relative to my car and *at the same time* travels at 50 m/s relative to the ground, even though the car is moving across the ground at 30 m/s. It seems logically impossible.

You might think that this is merely a matter of semantics. If we can just get our definitions and use of words straight, then the mystery and confusion will disappear. Or perhaps the difficulty is a confusion between what we "see" versus what "really happens." In other words, a better analysis, one that focuses on what really happens, would find that light "really" travels at different speeds in different reference frames.

Alas, what "really happens" is that light travels at 3.00×10^8 m/s in every inertial reference frame, regardless of how the reference frames are moving with respect to each other. It's not a trick. There remains only one way to escape the logical contradictions.

The definition of velocity is $u = \Delta x/\Delta t$, the ratio of a distance traveled to the time interval in which the travel occurs. Suppose you and I both make measurements on an object as it moves, but you happen to be moving relative to me. Perhaps I'm standing on the corner, you're driving past in your car, and we're both trying to measure the velocity of a bicycle. Further, suppose we have agreed in advance to measure the position of the bicycle

FIGURE 36.9 Light travels at speed c in all inertial reference frames, regardless of how the reference frames are moving with respect to the light source.

This light wave leaves Amy at speed c relative to Amy. It approaches Cathy at speed c relative to Cathy.

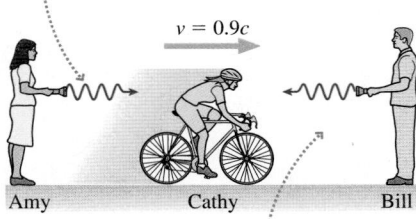

This light wave leaves Bill at speed c relative to Bill. It approaches Cathy at speed c relative to Cathy.

FIGURE 36.10 Experiments find that the photons travel through the laboratory with speed c, not the speed $1.99975c$ that you might expect.

A photon is emitted at speed c relative to the π meson. Measurements find that the photon's speed in the laboratory reference frame is also c.

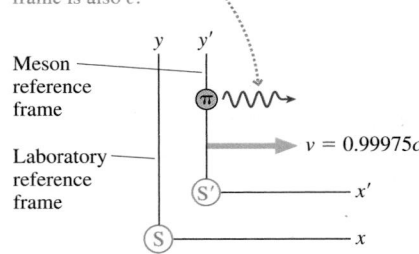

first as it passes the tree in FIGURE 36.11, then later as it passes the lamppost. Your $\Delta x'$, the bicycle's displacement, differs from my Δx because of your motion relative to me, causing you to calculate a bicycle velocity u' in your reference frame that differs from its velocity u in my reference frame. This is just the Galilean transformations showing up again.

FIGURE 36.11 Measuring the velocity of an object by appealing to the basic definition $u = \Delta x/\Delta t$.

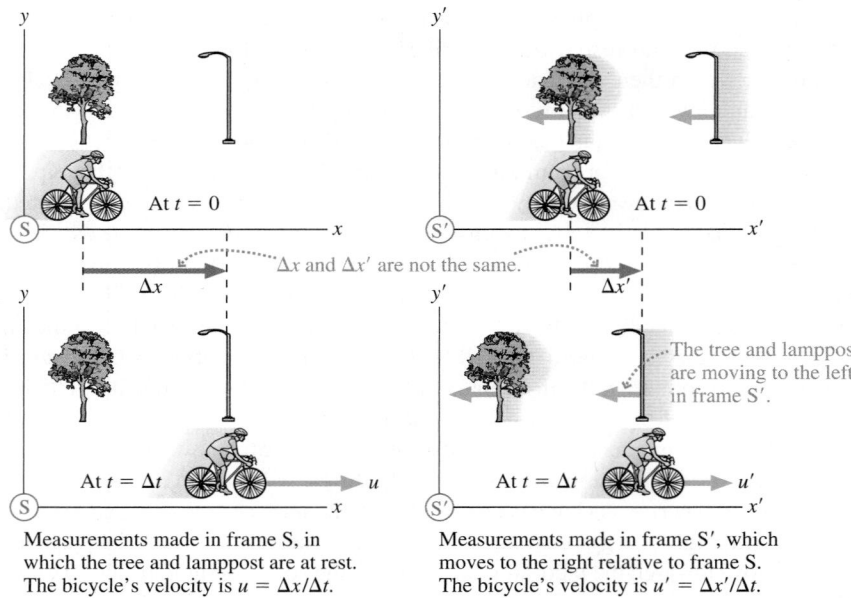

Measurements made in frame S, in which the tree and lamppost are at rest. The bicycle's velocity is $u = \Delta x/\Delta t$.

Measurements made in frame S', which moves to the right relative to frame S. The bicycle's velocity is $u' = \Delta x'/\Delta t$.

Now let's repeat the measurements, but this time let's measure the velocity of a light wave as it travels from the tree to the lamppost. Once again, your $\Delta x'$ differs from my Δx, and the obvious conclusion is that your light speed u' differs from my light speed u. The difference will be very small if you're driving past in your car, very large if you're flying past in a rocket traveling at nearly the speed of light. Although this conclusion seems obvious, it is wrong. Experiments show that, for a light wave, we'll get the *same* values: $u' = u$.

The only way this can be true is if your Δt is not the same as my Δt. If the time it takes the light to move from the tree to the lamppost in your reference frame, a time we'll now call $\Delta t'$, differs from the time Δt it takes the light to move from the tree to the lamppost in my reference frame, then we might find that $\Delta x'/\Delta t' = \Delta x/\Delta t$. That is, $u' = u$ even though you are moving with respect to me.

We've assumed, since the beginning of this textbook, that time is simply time. It flows along like a river, and all experimenters in all reference frames simply use it. For example, suppose the tree and the lamppost both have big clocks that we both can see. Shouldn't we be able to agree on the time interval Δt the light needs to move from the tree to the lamppost?

Perhaps not. It's demonstrably true that $\Delta x' \neq \Delta x$. It's experimentally verified that $u' = u$ for light waves. Something must be wrong with *assumptions* that we've made about the nature of time. The principle of relativity has painted us into a corner, and our only way out is to reexamine our understanding of time.

36.4 Events and Measurements

To question some of our most basic assumptions about space and time requires extreme care. We need to be certain that no assumptions slip into our analysis unnoticed. Our goal is to describe the motion of a particle in a clear and precise way, making the barest minimum of assumptions.

Events

The fundamental element of relativity is called an **event.** An event is a physical activity that takes place at a definite point in space and at a definite instant of time. An exploding firecracker is an event. A collision between two particles is an event. A light wave hitting a detector is an event.

Events can be observed and measured by experimenters in different reference frames. An exploding firecracker is as clear to you as you drive by in your car as it is to me standing on the street corner. We can quantify where and when an event occurs with four numbers: the coordinates (x, y, z) and the instant of time t. These four numbers, illustrated in FIGURE 36.12, are called the **spacetime coordinates** of the event.

The spatial coordinates of an event measured in reference frames S and S′ may differ. It now appears that the instant of time recorded in S and S′ may also differ. Thus the spacetime coordinates of an event measured by experimenters in frame S are (x, y, z, t) and the spacetime coordinates of the *same event* measured by experimenters in frame S′ are (x', y', z', t').

The motion of a particle can be described as a sequence of two or more events. We introduced this idea in the preceding section when we agreed to measure the velocity of a bicycle and then of a light wave by making measurements when the object passed the tree (first event) and when the object passed the lamppost (second event).

FIGURE 36.12 The location and time of an event are described by its spacetime coordinates.

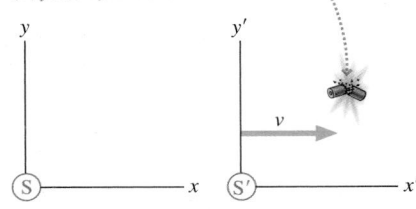

An event has spacetime coordinates (x, y, z, t) in frame S and different spacetime coordinates (x', y', z', t') in frame S′.

Measurements

Events are what "really happen," but how do we learn about an event? That is, how do the experimenters in a reference frame determine the spacetime coordinates of an event? This is a problem of *measurement.*

We defined a reference frame to be a coordinate system in which experimenters can make position and time measurements. That's a good start, but now we need to be more precise as to *how* the measurements are made. Imagine that a reference frame is filled with a cubic lattice of meter sticks, as shown in FIGURE 36.13. At every intersection is a clock, and all the clocks in a reference frame are *synchronized.* We'll return in a moment to consider how to synchronize the clocks, but assume for the moment it can be done.

Now, with our meter sticks and clocks in place, we can use a two-part measurement scheme:

- The (x, y, z) coordinates of an event are determined by the intersection of the meter sticks closest to the event.
- The event's time t is the time displayed on the clock nearest the event.

You can imagine, if you wish, that each event is accompanied by a flash of light to illuminate the face of the nearest clock and make its reading known.

Several important issues need to be noted:

1. The clocks and meter sticks in each reference frame are imaginary, so they have no difficulty passing through each other.
2. Measurements of position and time made in one reference frame must use only the clocks and meter sticks in that reference frame.
3. There's nothing special about the sticks being 1 m long and the clocks 1 m apart. The lattice spacing can be altered to achieve whatever level of measurement accuracy is desired.
4. We'll assume that the experimenters in each reference frame have assistants sitting beside every clock to record the position and time of nearby events.
5. Perhaps most important, t is the time at which the event *actually happens,* not the time at which an experimenter sees the event or at which information about the event reaches an experimenter.
6. All experimenters in one reference frame agree on the spacetime coordinates of an event. In other words, **an event has a unique set of spacetime coordinates in each reference frame.**

FIGURE 36.13 The spacetime coordinates of an event are measured by a lattice of meter sticks and clocks.

The spacetime coordinates of this event are measured by the nearest meter stick intersection and the nearest clock.

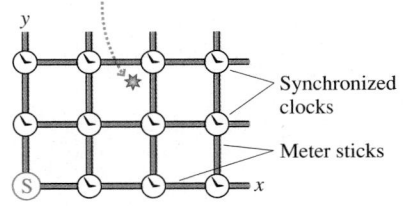

Synchronized clocks

Meter sticks

Reference frame S

Reference frame S′ has its own meter sticks and its own clocks.

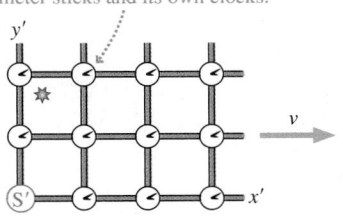

Reference frame S′

A carpenter is working on a house two blocks away. You notice a slight delay between seeing the carpenter's hammer hit the nail and hearing the blow. At what time does the event "hammer hits nail" occur?

a. At the instant you hear the blow
b. At the instant you see the hammer hit
c. Very slightly before you see the hammer hit
d. Very slightly after you see the hammer hit

Clock Synchronization

It's important that all the clocks in a reference frame be **synchronized,** meaning that all clocks in the reference frame have the same reading at any one instant of time. Thus we need a method of synchronization. One idea that comes to mind is to designate the clock at the origin as the *master clock.* We could then carry this clock around to every clock in the lattice, adjust that clock to match the master clock, and finally return the master clock to the origin.

This would be a perfectly good method of clock synchronization in Newtonian mechanics, where time flows along smoothly, the same for everyone. But we've been driven to reexamine the nature of time by the possibility that time is different in reference frames moving relative to each other. Because the master clock would *move,* we cannot assume that the moving master clock would keep time in the same way as the stationary clocks.

We need a synchronization method that does not require moving the clocks. Fortunately, such a method is easy to devise. Each clock is resting at the intersection of meter sticks, so by looking at the meter sticks, the assistant knows, or can calculate, exactly how far each clock is from the origin. Once the distance is known, the assistant can calculate exactly how long a light wave will take to travel from the origin to each clock. For example, light will take 1.00 μs to travel to a clock 300 m from the origin.

NOTE ▶ It's handy for many relativity problems to know that the speed of light is $c = 300$ m/μs. ◀

To synchronize the clocks, the assistants begin by setting each clock to display the light travel time from the origin, but they don't start the clocks. Next, as FIGURE 36.14 shows, a light flashes at the origin and, simultaneously, the clock at the origin starts running from $t = 0$ s. The light wave spreads out in all directions at speed c. A photodetector on each clock recognizes the arrival of the light wave and, without delay, starts the clock. The clock had been preset with the light travel time, so each clock as it starts reads exactly the same as the clock at the origin. Thus all the clocks will be synchronized after the light wave has passed by.

Events and Observations

We noted above that t is the time the event *actually happens.* This is an important point, one that bears further discussion. Light waves take time to travel. Messages, whether they're transmitted by light pulses, telephone, or courier on horseback, take time to be delivered. An experimenter *observes* an event, such as an exploding firecracker, only *at a later time* when light waves reach his or her eyes. But our interest is in the event itself, not the experimenter's observation of the event. The time at which the experimenter sees the event or receives information about the event is not when the event actually occurred.

Suppose at $t = 0$ s a firecracker explodes at $x = 300$ m. The flash of light from the firecracker will reach an experimenter at the origin at $t_1 = 1.0$ μs. The sound of the explosion will reach a sightless experimenter at the origin at $t_2 = 0.88$ s. Neither of these is the time t_{event} of the explosion, although the experimenter can work backward from these times, using known wave speeds, to determine t_{event}. In this example, the spacetime coordinates of the event—the explosion—are (300 m, 0 m, 0 m, 0 s).

FIGURE 36.14 Synchronizing clocks.

1. This clock is preset to 1.00 μs, the time it takes light to travel 300 m.

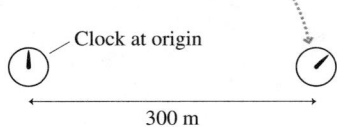

Clock at origin

300 m

2. At $t = 0$ s, a light flashes at the origin and the origin clock starts running. A very short time later, seen here, a light wave has begun to move outward.

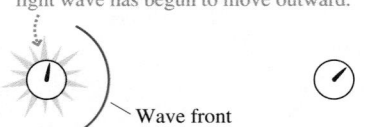

Wave front

3. The clock starts when the light wave reaches it. It is now synchronized with the origin clock.

EXAMPLE 36.3   **Finding the time of an event**

Experimenter A in reference frame S stands at the origin looking in the positive x-direction. Experimenter B stands at $x = 900$ m looking in the negative x-direction. A firecracker explodes somewhere between them. Experimenter B sees the light flash at $t = 3.0$ μs. Experimenter A sees the light flash at $t = 4.0$ μs. What are the spacetime coordinates of the explosion?

MODEL   Experimenters A and B are in the same reference frame and have synchronized clocks.

VISUALIZE   FIGURE 36.15 shows the two experimenters and the explosion at unknown position x.

SOLVE   The two experimenters observe light flashes at two different instants, but there's only one event. Light travels at 300 m/μs, so the additional 1.0 μs needed for the light to reach experimenter A implies that distance $(x - 0$ m$)$ is 300 m longer than distance $(900$ m $- x)$. That is,

$$(x - 0 \text{ m}) = (900 \text{ m} - x) + 300 \text{ m}$$

This is easily solved to give $x = 600$ m as the position coordinate of the explosion. The light takes 1.0 μs to travel 300 m to experimenter B, 2.0 μs to travel 600 m to experimenter A. The light is

FIGURE 36.15   The light wave reaches the experimenters at different times. Neither of these is the time at which the event actually happened.

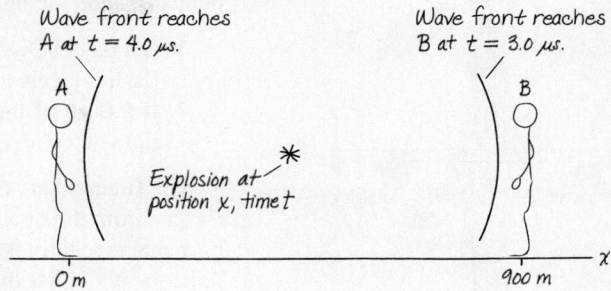

received at 3.0 μs and 4.0 μs, respectively; hence it was emitted by the explosion at $t = 2.0$ μs. The spacetime coordinates of the explosion are (600 m, 0 m, 0 m, 2.0 μs).

ASSESS   Although the experimenters *see* the explosion at different times, they agree that the explosion actually *happened* at $t = 2.0$ μs.

Simultaneity

Two events 1 and 2 that take place at different positions x_1 and x_2 but at the *same time* $t_1 = t_2$, as measured in some reference frame, are said to be **simultaneous** in that reference frame. Simultaneity is determined by when the events actually happen, not when they are seen or observed. In general, simultaneous events are *not* seen at the same time because of the difference in light travel times from the events to an experimenter.

EXAMPLE 36.4   **Are the explosions simultaneous?**

An experimenter in reference frame S stands at the origin looking in the positive x-direction. At $t = 3.0$ μs she sees firecracker 1 explode at $x = 600$ m. A short time later, at $t = 5.0$ μs, she sees firecracker 2 explode at $x = 1200$ m. Are the two explosions simultaneous? If not, which firecracker exploded first?

MODEL   Light from both explosions travels toward the experimenter at 300 m/μs.

SOLVE   The experimenter *sees* two different explosions, but perceptions of the events are not the events themselves. When did the explosions *actually* occur? Using the fact that light travels at 300 m/μs, we can see that firecracker 1 exploded at $t_1 = 1.0$ μs and firecracker 2 also exploded at $t_2 = 1.0$ μs. The events *are* simultaneous.

STOP TO THINK 36.4   A tree and a pole are 3000 m apart. Each is suddenly hit by a bolt of lightning. Mark, who is standing at rest midway between the two, sees the two lightning bolts at the same instant of time. Nancy is at rest under the tree. Define event 1 to be "lightning strikes tree" and event 2 to be "lightning strikes pole." For Nancy, does event 1 occur before, after, or at the same time as event 2?

36.5 The Relativity of Simultaneity

We've now established a means for measuring the time of an event in a reference frame, so let's begin to investigate the nature of time. The following "thought experiment" is very similar to one suggested by Einstein.

FIGURE 36.16 on the next page shows a long railroad car traveling to the right with a velocity v that may be an appreciable fraction of the speed of light. A firecracker is tied

FIGURE 36.16 A railroad car traveling to the right with velocity v.

The firecrackers will make burn marks on the ground at the positions where they explode.

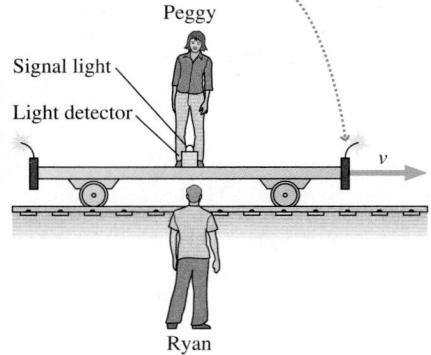

to each end of the car, just above the ground. Each firecracker is powerful enough so that, when it explodes, it will make a burn mark on the ground at the position of the explosion.

Ryan is standing on the ground, watching the railroad car go by. Peggy is standing in the exact center of the car with a special box at her feet. This box has two light detectors, one facing each way, and a signal light on top. The box works as follows:

1. If a flash of light is received at the detector facing right, as seen by Ryan, before a flash is received at the left detector, then the light on top of the box will turn green.
2. If a flash of light is received at the left detector before a flash is received at the right detector, or if two flashes arrive simultaneously, the light on top will turn red.

The firecrackers explode as the railroad car passes Ryan, and he sees the two light flashes from the explosions simultaneously. He then measures the distances to the two burn marks and finds that he was standing exactly halfway between the marks. Because light travels equal distances in equal times, Ryan concludes that the two explosions were simultaneous in his reference frame, the reference frame of the ground. Further, because he was midway between the two ends of the car, he was directly opposite Peggy when the explosions occurred.

FIGURE 36.17a shows the sequence of events in Ryan's reference frame. Light travels at speed c in all inertial reference frames, so, although the firecrackers were moving, the light waves are spheres centered on the burn marks. Ryan determines that the light wave coming from the right reaches Peggy and the box before the light wave coming from the left. Thus, according to Ryan, the signal light on top of the box turns green.

FIGURE 36.17 Exploding firecrackers seen in two different reference frames.

(a) The events in Ryan's frame

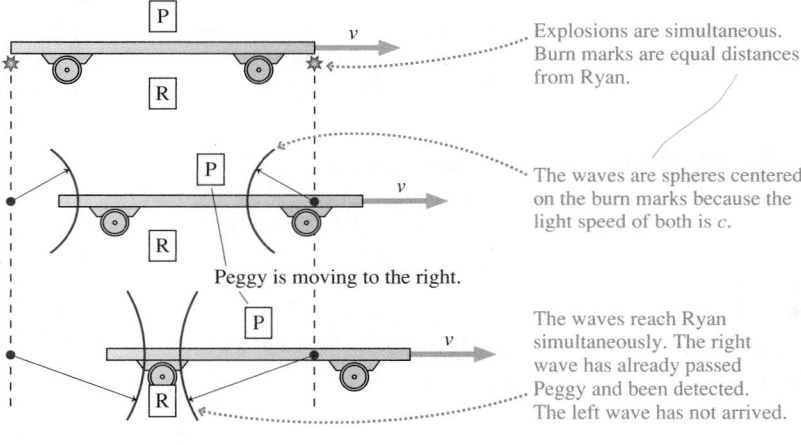

Explosions are simultaneous. Burn marks are equal distances from Ryan.

The waves are spheres centered on the burn marks because the light speed of both is c.

Peggy is moving to the right.

The waves reach Ryan simultaneously. The right wave has already passed Peggy and been detected. The left wave has not arrived.

(b) The events in Peggy's frame

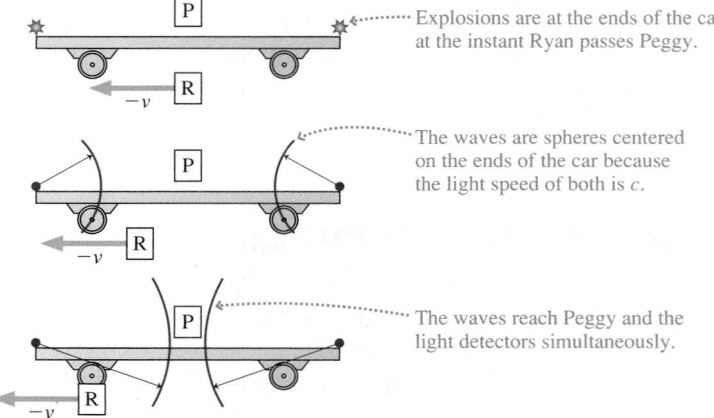

Explosions are at the ends of the car at the instant Ryan passes Peggy.

The waves are spheres centered on the ends of the car because the light speed of both is c.

The waves reach Peggy and the light detectors simultaneously.

How do things look in Peggy's reference frame, a reference frame moving to the right at velocity v relative to the ground? As FIGURE 36.17b shows, Peggy sees Ryan moving to the left with speed v. Light travels at speed c in all inertial reference frames, so the light waves are spheres centered on the ends of the car. If the explosions are simultaneous, as Ryan has determined, the two light waves reach her and the box simultaneously. Thus, according to Peggy, the signal light on top of the box turns red!

Now the light on top must be either green or red. *It can't be both!* Later, after the railroad car has stopped, Ryan and Peggy can place the box in front of them. Either it has a red light or a green light. Ryan can't see one color while Peggy sees the other. Hence we have a paradox. It's impossible for Peggy and Ryan both to be right. But who is wrong, and why?

What do we know with absolute certainty?

1. Ryan detected the flashes simultaneously.
2. Ryan was halfway between the firecrackers when they exploded.
3. The light from the two explosions traveled toward Ryan at equal speeds.

The conclusion that the explosions were simultaneous in Ryan's reference frame is unassailable. The light is green.

Peggy, however, made an assumption. It's a perfectly ordinary assumption, one that seems sufficiently obvious that you probably didn't notice, but an assumption nonetheless. Peggy assumed that the explosions were simultaneous.

Didn't Ryan find them to be simultaneous? Indeed, he did. Suppose we call Ryan's reference frame S, the explosion on the right event R, and the explosion on the left event L. Ryan found that $t_R = t_L$. But Peggy has to use a different set of clocks, the clocks in her reference frame S′, to measure the times t'_R and t'_L at which the explosions occurred. The fact that $t_R = t_L$ in frame S does *not* allow us to conclude that $t'_R = t'_L$ in frame S′.

In fact, in frame S′ the right firecracker must explode *before* the left firecracker. Figure 36.17b, with its assumption about simultaneity, was incorrect. FIGURE 36.18 shows the situation in Peggy's reference frame, with the right firecracker exploding first. Now the wave from the right reaches Peggy and the box first, as Ryan had concluded, and the light on top turns green.

One of the most disconcerting conclusions of relativity is that **two events occurring simultaneously in reference frame S are *not* simultaneous in any reference frame S′ moving relative to S.** This is called the **relativity of simultaneity**.

The two firecrackers *really* explode at the same instant of time in Ryan's reference frame. And the right firecracker *really* explodes first in Peggy's reference frame. It's not a matter of when they see the flashes. Our conclusion refers to the times at which the explosions actually occur.

The paradox of Peggy and Ryan contains the essence of relativity, and it's worth careful thought. First, review the logic until you're certain that there *is* a paradox, a logical impossibility. Then convince yourself that the only way to resolve the paradox is to abandon the assumption that the explosions are simultaneous in Peggy's reference frame. If you understand the paradox and its resolution, you've made a big step toward understanding what relativity is all about.

FIGURE 36.18 The real sequence of events in Peggy's reference frame.

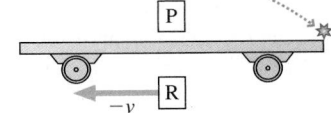

The right firecracker explodes first.

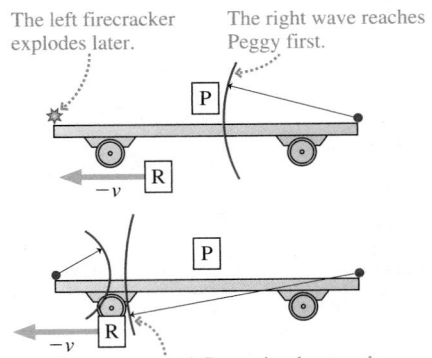

The left firecracker explodes later.   The right wave reaches Peggy first.

The waves reach Ryan simultaneously. The left wave has not reached Peggy.

STOP TO THINK 36.5 A tree and a pole are 3000 m apart. Each is hit by a bolt of lightning. Mark, who is standing at rest midway between the two, sees the two lightning bolts at the same instant of time. Nancy is flying her rocket at $v = 0.5c$ in the direction from the tree toward the pole. The lightning hits the tree just as she passes by it. Define event 1 to be "lightning strikes tree" and event 2 to be "lightning strikes pole." For Nancy, does event 1 occur before, after, or at the same time as event 2?

36.6 Time Dilation

The principle of relativity has driven us to the logical conclusion that time is not the same for two reference frames moving relative to each other. Our analysis thus far has been mostly qualitative. It's time to start developing some quantitative tools that will allow us to compare measurements in one reference frame to measurements in another reference frame.

FIGURE 36.19a shows a special clock called a *light clock*. The light clock is a box with a light source at the bottom and a mirror at the top, separated by distance h. The light source emits a very short pulse of light that travels to the mirror and reflects back to a light detector beside the source. The clock advances one "tick" each time the detector receives a light pulse, and it immediately, with no delay, causes the light source to emit the next light pulse.

Our goal is to compare two measurements of the interval between two ticks of the clock: one taken by an experimenter standing next to the clock and the other by an experimenter moving with respect to the clock. To be specific, **FIGURE 36.19b** shows the clock at rest in reference frame S′. We call this the **rest frame** of the clock. Reference frame S′ moves to the right with velocity v relative to reference frame S.

Relativity requires us to measure *events,* so let's define event 1 to be the emission of a light pulse and event 2 to be the detection of that light pulse. Experimenters in both reference frames are able to measure where and when these events occur *in their frame*. In frame S, the time interval $\Delta t = t_2 - t_1$ is one tick of the clock. Similarly, one tick in frame S′ is $\Delta t' = t_2' - t_1'$.

To be sure we have a clear understanding of the relativity result, let's first do a classical analysis. In frame S′, the clock's rest frame, the light travels straight up and down, a total distance $2h$, at speed c. The time interval is $\Delta t' = 2h/c$.

FIGURE 36.20a shows the operation of the light clock as seen in frame S. The clock is moving to the right at speed v in S, thus the mirror moves distance $\frac{1}{2}v(\Delta t)$ during the time $\frac{1}{2}(\Delta t)$ in which the light pulse moves from the source to the mirror. The distance traveled by the light during this interval is $\frac{1}{2}u_{\text{light}}(\Delta t)$, where u_{light} is the speed of light in frame S. You can see from the vector addition in **FIGURE 36.20b** that the speed of light in frame S is $u_{\text{light}} = (c^2 + v^2)^{1/2}$. (Remember, this is a classical analysis in which the speed of light *does* depend on the motion of the reference frame relative to the light source.)

The Pythagorean theorem applied to the right triangle in Figure 36.20a is

$$h^2 + \left(\frac{1}{2}v\,\Delta t\right)^2 = \left(\frac{1}{2}u_{\text{light}}\Delta t\right)^2 = \left(\frac{1}{2}\sqrt{c^2 + v^2}\Delta t\right)^2$$

$$= \left(\frac{1}{2}c\,\Delta t\right)^2 + \left(\frac{1}{2}v\,\Delta t\right)^2 \tag{36.4}$$

The term $\left(\frac{1}{2}v\,\Delta t\right)^2$ is common to both sides and cancels. Solving for Δt gives $\Delta t = 2h/c$, identical to $\Delta t'$. In other words, a classical analysis finds that the clock ticks at exactly the same rate in both frame S and frame S′. This shouldn't be surprising. There's only one kind of time in classical physics, measured the same by all experimenters independent of their motion.

The principle of relativity changes only one thing, but that change has profound consequences. According to the principle of relativity, light travels at the same speed in *all* inertial reference frames. In frame S′, the rest frame of the clock, the light simply goes straight up and back. The time of one tick,

$$\Delta t' = \frac{2h}{c} \tag{36.5}$$

is unchanged from the classical analysis.

FIGURE 36.19 The ticking of a light clock can be measured by experimenters in two different reference frames.

(a) A light clock

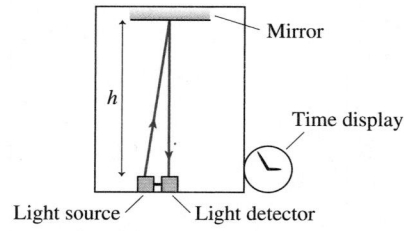

(b) The clock is at rest in frame S′.

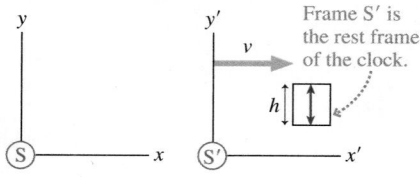

FIGURE 36.20 A classical analysis of the light clock.

(a)

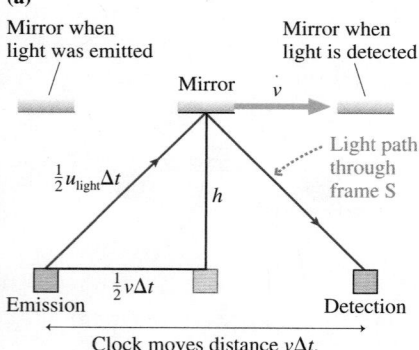

(b)

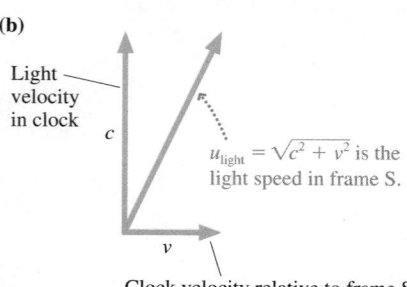

FIGURE 36.21 shows the light clock as seen in frame S. The difference from Figure 36.20a is that the light now travels along the hypotenuse at speed c. We can again use the Pythagorean theorem to write

$$h^2 + \left(\frac{1}{2}v\,\Delta t\right)^2 = \left(\frac{1}{2}c\,\Delta t\right)^2 \tag{36.6}$$

Solving for Δt gives

$$\Delta t = \frac{2h/c}{\sqrt{1 - v^2/c^2}} = \frac{\Delta t'}{\sqrt{1 - v^2/c^2}} \tag{36.7}$$

The time interval between two ticks in frame S is *not* the same as in frame S′.

It's useful to define $\beta = v/c$, the velocity as a fraction of the speed of light. For example, a reference frame moving with $v = 2.4 \times 10^8$ m/s has $\beta = 0.80$. In terms of β, Equation 36.7 is

$$\Delta t = \frac{\Delta t'}{\sqrt{1 - \beta^2}} \tag{36.8}$$

NOTE ▶ The expression $(1 - v^2/c^2)^{1/2} = (1 - \beta^2)^{1/2}$ occurs frequently in relativity. The value of the expression is 1 when $v = 0$, and it steadily decreases to 0 as $v \to c$ (or $\beta \to 1$). The square root is an imaginary number if $v > c$, which would make Δt imaginary in Equation 36.8. Time intervals certainly have to be real numbers, suggesting that $v > c$ is not physically possible. One of the predictions of the theory of relativity, as you've undoubtedly heard, is that nothing can travel faster than the speed of light. Now you can begin to see why. We'll examine this topic more closely in Section 36.9. In the meantime, we'll require v to be less than c. ◀

Proper Time

Frame S′ has one important distinction. It is the *one and only* inertial reference frame in which the light clock is at rest. Consequently, it is the one and only inertial reference frame in which the times of both events—the emission of the light and the detection of the light—are measured by the *same* reference-frame clock. You can see that the light pulse in Figure 36.19a starts and ends at the same position. In Figure 36.21, the emission and detection take place at different positions in frame S and must be measured by different reference-frame clocks, one at each position.

The time interval between two events that occur at the *same position* is called the **proper time** $\Delta\tau$. Only one inertial reference frame measures the proper time, and it does so with a single clock that is present at both events. An inertial reference frame moving with velocity $v = \beta c$ relative to the proper-time frame must use two clocks to measure the time interval: one at the position of the first event, the other at the position of the second event. The time interval between the two events in this frame is

$$\Delta t = \frac{\Delta\tau}{\sqrt{1 - \beta^2}} \geq \Delta\tau \qquad \text{(time dilation)} \tag{36.9}$$

The "stretching out" of the time interval implied by Equation 36.9 is called **time dilation.** Time dilation is sometimes described by saying that "moving clocks run slow." This is not an accurate statement because it implies that some reference frames are "really" moving while others are "really" at rest. The whole point of relativity is that all inertial reference frames are equally valid, that all we know about reference frames is how they move relative to each other. A better description of time dilation is the statement that **the time interval between two ticks is the shortest in the reference frame in which the clock is at rest.** The time interval between two ticks is longer (i.e., the clock "runs slower") when it is measured in any reference frame in which the clock is moving.

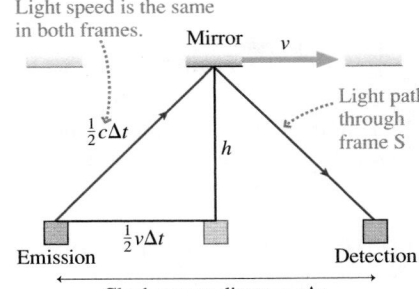

FIGURE 36.21 A light clock analysis in which the speed of light is the same in all reference frames.

Light speed is the same in both frames.

Mirror v

$\frac{1}{2}c\Delta t$ h Light path through frame S

Emission $\frac{1}{2}v\Delta t$ Detection

Clock moves distance $v\Delta t$.

NOTE ▶ Equation 36.9 was derived using a light clock because the operation of a light clock is clear and easy to analyze. But the conclusion is really about time itself. *Any* clock, regardless of how it operates, behaves the same. ◀

EXAMPLE 36.5 From the sun to Saturn

Saturn is 1.43×10^{12} m from the sun. A rocket travels along a line from the sun to Saturn at a constant speed of $0.9c$ relative to the solar system. How long does the journey take as measured by an experimenter on earth? As measured by an astronaut on the rocket?

MODEL Let the solar system be in reference frame S and the rocket be in reference frame S′ that travels with velocity $v = 0.9c$ relative to S. Relativity problems must be stated in terms of *events*. Let event 1 be "the rocket and the sun coincide" (the experimenter on earth says that the rocket passes the sun; the astronaut on the rocket says that the sun passes the rocket) and event 2 be "the rocket and Saturn coincide."

FIGURE 36.22 Pictorial representation of the trip as seen in frames S and S′.

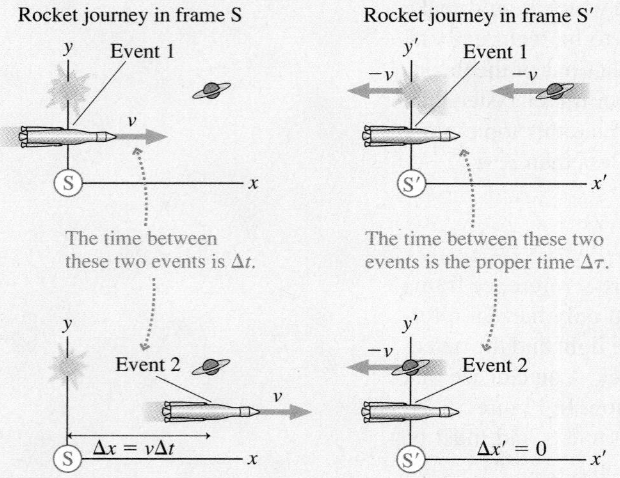

Rocket journey in frame S

Rocket journey in frame S′

The time between these two events is Δt.

The time between these two events is the proper time $\Delta \tau$.

$\Delta x = v\Delta t$

$\Delta x' = 0$

VISUALIZE FIGURE 36.22 shows the two events as seen from the two reference frames. Notice that the two events occur at the *same position* in S′, the position of the rocket, and consequently can be measured by *one* clock.

SOLVE The time interval measured in the solar system reference frame, which includes the earth, is simply

$$\Delta t = \frac{\Delta x}{v} = \frac{1.43 \times 10^{12} \text{ m}}{0.9 \times (3.00 \times 10^8 \text{ m/s})} = 5300 \text{ s}$$

Relativity hasn't abandoned the basic definition $v = \Delta x/\Delta t$, although we do have to be sure that Δx and Δt are measured in just one reference frame and refer to the same two events.

How are things in the rocket's reference frame? The two events occur at the *same position* in S′ and can be measured by *one* clock, the clock at the origin. Thus the time measured by the astronauts is the *proper time* $\Delta \tau$ between the two events. We can use Equation 36.9 with $\beta = 0.9$ to find

$$\Delta \tau = \sqrt{1 - \beta^2}\, \Delta t = \sqrt{1 - 0.9^2}\,(5300 \text{ s}) = 2310 \text{ s}$$

ASSESS The time interval measured between these two events by the astronauts is less than half the time interval measured by experimenters on earth. The difference has nothing to do with when earthbound astronomers *see* the rocket pass the sun and Saturn. Δt is the time interval from when the rocket actually passes the sun, as measured by a clock at the sun, until it actually passes Saturn, as measured by a synchronized clock at Saturn. The interval between *seeing* the events from earth, which would have to allow for light travel times, would be something other than 5300 s. Δt and $\Delta \tau$ are different because *time is different* in two reference frames moving relative to each other.

STOP TO THINK 36.6 Molly flies her rocket past Nick at constant velocity v. Molly and Nick both measure the time it takes the rocket, from nose to tail, to pass Nick. Which of the following is true?

    a. Both Molly and Nick measure the same amount of time.
    b. Molly measures a shorter time interval than Nick.
    c. Nick measures a shorter time interval than Molly.

Experimental Evidence

Is there any evidence for the crazy idea that clocks moving relative to each other tell time differently? Indeed, there's plenty. An experiment in 1971 sent an atomic clock around the world on a jet plane while an identical clock remained in the laboratory. This was a difficult experiment because the traveling clock's speed was so small compared to c, but measuring the small differences between the time intervals was just barely within the capabilities of atomic clocks. It was also a more complex experiment than we've analyzed because the clock accelerated as it moved around a circle. The scientists found that, upon its return, the eastbound clock, traveling faster than the laboratory on a rotating earth, was 60 ns behind the stay-at-home clock, which was exactly as predicted by relativity.

Very detailed studies have been done on unstable particles called *muons* that are created at the top of the atmosphere, at a height of about 60 km, when high-energy cosmic rays collide with air molecules. It is well known, from laboratory studies, that stationary muons decay with a *half-life* of 1.5 μs. That is, half the muons decay within 1.5 μs, half of those remaining decay in the next 1.5 μs, and so on. The decays can be used as a clock.

The muons travel down through the atmosphere at very nearly the speed of light. The time needed to reach the ground, assuming $v \approx c$, is $\Delta t \approx (60,000 \text{ m})/ (3 \times 10^8 \text{ m/s}) = 200 \ \mu$s. This is 133 half-lives, so the fraction of muons reaching the ground should be $\approx \left(\frac{1}{2}\right)^{133} = 10^{-40}$. That is, only 1 out of every 10^{40} muons should reach the ground. In fact, experiments find that about 1 in 10 muons reach the ground, an experimental result that differs by a factor of 10^{39} from our prediction!

The discrepancy is due to time dilation. In FIGURE 36.23, the two events "muon is created" and "muon hits ground" take place at two different places in the earth's reference frame. However, these two events occur at the *same position* in the muon's reference frame. (The muon is like the rocket in Example 36.5.) Thus the muon's internal clock measures the proper time. The time-dilated interval $\Delta t = 200 \ \mu$s in the earth's reference frame corresponds to a proper time $\Delta \tau \approx 5 \ \mu$s in the muon's reference frame. That is, in the muon's reference frame it takes only 5 μs from creation at the top of the atmosphere until the ground runs into it. This is 3.3 half-lives, so the fraction of muons reaching the ground is $\left(\frac{1}{2}\right)^{3.3} = 0.1$, or 1 out of 10. We wouldn't detect muons at the ground at all if not for time dilation.

The details are beyond the scope of this textbook, but dozens of high-energy particle accelerators around the world that study quarks and other elementary particles have been designed and built on the basis of Einstein's theory of relativity. The fact that they work exactly as planned is strong testimony to the reality of time dilation.

FIGURE 36.23 We wouldn't detect muons at the ground if not for time dilation.

A muon travels $\approx$450 m in 1.5 μs. We would not detect muons at ground level if the half-life of a moving muon were 1.5 μs.

Muon is created.

Because of time dilation, the half-life of a high-speed muon is long enough in the earth's reference frame for about 1 in 10 muons to reach the ground.

Muon hits ground.

The Twin Paradox

The most well-known relativity paradox is the twin paradox. George and Helen are twins. On their 25th birthday, Helen departs on a starship voyage to a distant star. Let's imagine, to be specific, that her starship accelerates almost instantly to a speed of 0.95c and that she travels to a star that is 9.5 light years (9.5 ly) from earth. Upon arriving, she discovers that the planets circling the star are inhabited by fierce aliens, so she immediately turns around and heads home at 0.95c.

A **light year,** abbreviated ly, is the distance that light travels in one year. A light year is vastly larger than the diameter of the solar system. The distance between two neighboring stars is typically a few light years. For our purpose, we can write the speed of light as $c = 1$ ly/year. That is, light travels 1 light year per year.

This value for c allows us to determine how long, according to George and his fellow earthlings, it takes Helen to travel out and back. Her total distance is 19 ly and, due to her rapid acceleration and rapid turn-around, she travels essentially the entire distance at speed $v = 0.95c = 0.95$ ly/year. Thus the time she's away, as measured by George, is

$$\Delta t_{\text{G}} = \frac{19 \text{ ly}}{0.95 \text{ ly/year}} = 20 \text{ years} \tag{36.10}$$

George will be 45 years old when his sister Helen returns with tales of adventure.

While she's away, George takes a physics class and studies Einstein's theory of relativity. He realizes that time dilation will make Helen's clocks run more slowly than his clocks, which are at rest relative to him. Her heart—a clock—will beat fewer

The global positioning system (GPS), which allows you to pinpoint your location anywhere in the world to within a few meters, uses a set of orbiting satellites. Because of their motion, the atomic clocks on these satellites keep time differently from clocks on the ground. To determine an accurate position, the software in your GPS receiver must carefully correct for time-dilation effects.

times and the minute hand on her watch will go around fewer times. In other words, she's aging more slowly than he is. Although she is his twin, she will be younger than he is when she returns.

Calculating Helen's age is not hard. We simply have to identify Helen's clock, because it's always with Helen as she travels, as the clock that measures proper time $\Delta\tau$. From Equation 36.9,

$$\Delta t_{\text{H}} = \Delta\tau = \sqrt{1 - \beta^2}\,\Delta t_{\text{G}} = \sqrt{1 - 0.95^2}\,(20 \text{ years}) = 6.25 \text{ years} \quad (36.11)$$

George will have just celebrated his 45th birthday as he welcomes home his 31-year-and-3-month-old twin sister.

This may be unsettling because it violates our commonsense notion of time, but it's not a paradox. There's no logical inconsistency in this outcome. So why is it called "the twin paradox"?

Helen, knowing that she had quite of bit of time to kill on her journey, brought along several physics books to read. As she learns about relativity, she begins to think about George and her friends back on earth. Relative to her, they are all moving away at 0.95c. Later they'll come rushing toward her at 0.95c. Time dilation will cause their clocks to run more slowly than her clocks, which are at rest relative to her. In other words, as FIGURE 36.24 shows, Helen concludes that people on earth are aging more slowly than she is. Alas, she will be much older than they when she returns.

Finally, the big day arrives. Helen lands back on earth and steps out of the starship. George is expecting Helen to be younger than he is. Helen is expecting George to be younger than she is.

Here's the paradox! It's logically impossible for each to be younger than the other at the time they are reunited. Where, then, is the flaw in our reasoning? It seems to be a symmetrical situation—Helen moves relative to George and George moves relative to Helen—but symmetrical reasoning has led to a conundrum.

But are the situations really symmetrical? George goes about his business day after day without noticing anything unusual. Helen, on the other hand, experiences three distinct periods during which the starship engines fire, she's crushed into her seat, and free dust particles that had been floating inside the starship are no longer, in the starship's reference frame, at rest or traveling in a straight line at constant speed. In other words, George spends the entire time in an inertial reference frame, *but Helen does not.* The situation is *not* symmetrical.

The principle of relativity applies *only* to inertial reference frames. Our discussion of time dilation was for inertial reference frames. Thus George's analysis and calculations are correct. Helen's analysis and calculations are *not* correct because she was trying to apply an inertial reference frame result while traveling in a noninertial reference frame.

Helen is younger than George when she returns. This is strange, but not a paradox. It is a consequence of the fact that time flows differently in two reference frames moving relative to each other.

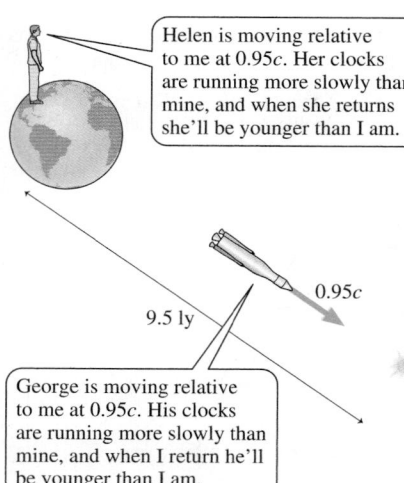

FIGURE 36.24 The twin paradox.

Helen is moving relative to me at 0.95c. Her clocks are running more slowly than mine, and when she returns she'll be younger than I am.

0.95c

9.5 ly

George is moving relative to me at 0.95c. His clocks are running more slowly than mine, and when I return he'll be younger than I am.

36.7 Length Contraction

We've seen that relativity requires us to rethink our idea of time. Now let's turn our attention to the concepts of space and distance. Consider the rocket that traveled from the sun to Saturn in Example 36.5. FIGURE 36.25a shows the rocket moving with velocity v through the solar system reference frame S. We define $L = \Delta x = x_{\text{Saturn}} - x_{\text{sun}}$ as the distance between the sun and Saturn in frame S or, more generally, the *length* of the spatial interval between two points. The rocket's speed is $v = L/\Delta t$, where Δt is the time measured in frame S for the journey from the sun to Saturn.

FIGURE 36.25 L and L' are the distances between the sun and Saturn in frames S and S'.

(a) Reference frame S: The solar system is stationary.

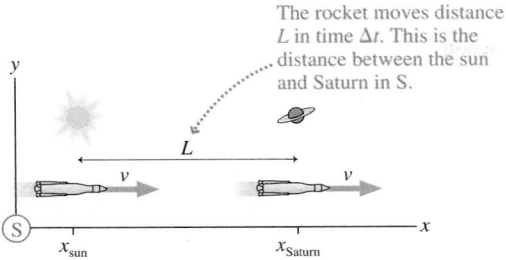

The rocket moves distance L in time Δt. This is the distance between the sun and Saturn in S.

(b) Reference frame S': The rocket is stationary.

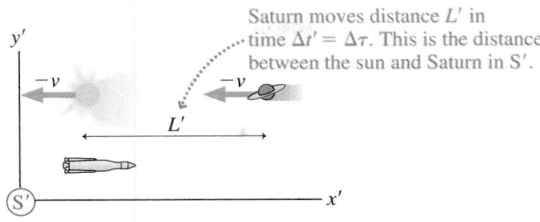

Saturn moves distance L' in time $\Delta t' = \Delta \tau$. This is the distance between the sun and Saturn in S'.

FIGURE 36.25b shows the situation in reference frame S', where the rocket is at rest. The sun and Saturn move to the left at speed $v = L'/\Delta t'$, where $\Delta t'$ is the time measured in frame S' for Saturn to travel distance L'.

Speed v is the relative speed between S and S' and is the same for experimenters in both reference frames. That is,

$$v = \frac{L}{\Delta t} = \frac{L'}{\Delta t'} \tag{36.12}$$

The time interval $\Delta t'$ measured in frame S' is the proper time $\Delta \tau$ because both events occur at the same position in frame S' and can be measured by one clock. We can use the time-dilation result, Equation 36.9, to relate $\Delta \tau$ measured by the astronauts to Δt measured by the earthbound scientists. Then Equation 36.12 becomes

$$\frac{L}{\Delta t} = \frac{L'}{\Delta \tau} = \frac{L'}{\sqrt{1 - \beta^2}\,\Delta t} \tag{36.13}$$

The Δt cancels, and the distance L' in frame S' is

$$L' = \sqrt{1 - \beta^2}\,L \tag{36.14}$$

Surprisingly, we find that **the distance between two objects in reference frame S' is *not the same* as the distance between the same two objects in reference frame S.**

Frame S, in which the distance is L, has one important distinction. It is the *one and only* inertial reference frame in which the objects are at rest. Experimenters in frame S can take all the time they need to measure L because the two objects aren't going anywhere. The distance L between two objects, or two points on one object, measured in the reference frame in which the objects are at rest is called the **proper length** ℓ. Only one inertial reference frame can measure the proper length.

We can use the proper length ℓ to write Equation 36.14 as

$$L' = \sqrt{1 - \beta^2}\,\ell \le \ell \tag{36.15}$$

This "shrinking" of the distance between two objects, as measured by an experiment moving with respect to the objects, is called **length contraction.** Although we derived length contraction for the distance between two distinct objects, it applies equally well to the length of any physical object that stretches between two points along the x- and x'-axes. The length of an object is greatest in the reference frame in which the object is at rest. The object's length is less (i.e., the length is contracted) when it is measured in any reference frame in which the object is moving.

The Stanford Linear Accelerator (SLAC) is a 2-mi-long electron accelerator. The accelerator's length is less than 1 m in the reference frame of the electrons.

The distance from the sun to Saturn

In Example 36.5 a rocket traveled along a line from the sun to Saturn at a constant speed of $0.9c$ relative to the solar system. The Saturn-to-sun distance was given as 1.43×10^{12} m. What is the distance between the sun and Saturn in the rocket's reference frame?

MODEL Saturn and the sun are, at least approximately, at rest in the solar system reference frame S. Thus the given distance is the proper length ℓ.

SOLVE We can use Equation 36.15 to find the distance in the rocket's frame S':

$$L' = \sqrt{1 - \beta^2}\,\ell = \sqrt{1 - 0.9^2}\,(1.43 \times 10^{12}\ \text{m})$$
$$= 0.62 \times 10^{12}\ \text{m}$$

ASSESS The sun-to-Saturn distance measured by the astronauts is less than half the distance measured by experimenters on earth. L' and ℓ are different because *space is different* in two reference frames moving relative to each other.

FIGURE 36.26 Carmen and Dan each measure the length of the other's meter stick as they move relative to each other.

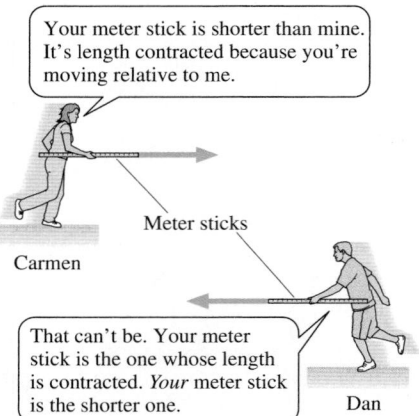

Your meter stick is shorter than mine. It's length contracted because you're moving relative to me.

Meter sticks

Carmen

That can't be. Your meter stick is the one whose length is contracted. *Your* meter stick is the shorter one.

Dan

FIGURE 36.27 Distance d is the same in both coordinate systems.

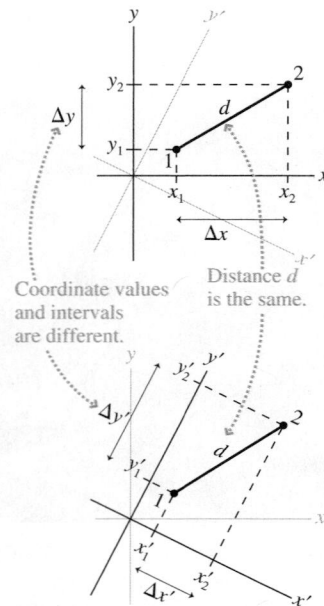

Measurements in the xy-system

Coordinate values and intervals are different.

Distance d is the same.

Measurements in the $x'y'$-system

The conclusion that space is different in reference frames moving relative to each other is a direct consequence of the fact that time is different. Experimenters in both reference frames agree on the relative velocity v, leading to Equation 36.12: $v = L/\Delta t = L'/\Delta t'$. We had already learned that $\Delta t' < \Delta t$ because of time dilation. Thus L' *has* to be less than L. That is the only way experimenters in the two reference frames can reconcile their measurements.

To be specific, the earthly experimenters in Examples 36.5 and 36.6 find that the rocket takes 5300 s to travel the 1.43×10^{12} m between the sun and Saturn. The rocket's speed is $v = L/\Delta t = 2.7 \times 10^8$ m/s $= 0.9c$. The astronauts in the rocket find that it takes only 2310 s for Saturn to reach them after the sun has passed by. But there's no conflict, because they also find that the distance is only 0.62×10^{12} m. Thus Saturn's speed toward them is $v = L'/\Delta t' = (0.62 \times 10^{12}\ \text{m})/(2310\ \text{s}) = 2.7 \times 10^8$ m/s $= 0.9c$.

Another Paradox?

Carmen and Dan are in their physics lab room. They each select a meter stick, lay the two side by side, and agree that the meter sticks are exactly the same length. Then, for an extra-credit project, they go outside and run past each other, in opposite directions, at a relative speed $v = 0.9c$. FIGURE 36.26 shows their experiment and a portion of their conversation.

Now, Dan's meter stick can't be both longer and shorter than Carmen's meter stick. Is this another paradox? No! Relativity allows us to compare the *same* events as they're measured in two different reference frames. This did lead to a real paradox when Peggy rolled past Ryan on the train. There the signal light on the box turns green (a single event) or it doesn't, and Peggy and Ryan have to agree about it. But the events by which Dan measures the length (in Dan's frame) of Carmen's meter stick are *not the same events* as those by which Carmen measures the length (in Carmen's frame) of Dan's meter stick.

There's no conflict between their measurements. In Dan's reference frame, Carmen's meter stick has been length contracted and is less than 1 m in length. In Carmen's reference frame, Dan's meter stick has been length contracted and is less than 1m in length. If this weren't the case, if both agreed that one of the meter sticks was shorter than the other, then we could tell which reference frame was "really" moving and which was "really" at rest. But the principle of relativity doesn't allow us to make that distinction. Each is moving relative to the other, so each should make the same measurement for the length of the other's meter stick.

The Spacetime Interval

Forget relativity for a minute and think about ordinary geometry. FIGURE 36.27 shows two ordinary coordinate systems. They are identical except for the fact that one has been rotated relative to the other. A student using the xy-system would measure coordinates (x_1, y_1) for point 1 and (x_2, y_2) for point 2. A second student, using the $x'y'$-system, would measure (x'_1, y'_1) and (x'_2, y'_2).

The students soon find that none of their measurements agree. That is, $x_1 \neq x_1'$ and so on. Even the intervals are different: $\Delta x \neq \Delta x'$ and $\Delta y \neq \Delta y'$. Each is a perfectly valid coordinate system, giving no reason to prefer one over the other, but each yields different measurements.

Is there *anything* on which the two students can agree? Yes, there is. The distance d between points 1 and 2 is independent of the coordinates. We can state this mathematically as

$$d^2 = (\Delta x)^2 + (\Delta y)^2 = (\Delta x')^2 + (\Delta y')^2 \qquad (36.16)$$

The quantity $(\Delta x)^2 + (\Delta y)^2$ is called an **invariant** in geometry because it has the same value in any Cartesian coordinate system.

Returning to relativity, is there an invariant in the spacetime coordinates, some quantity that has the *same value* in all inertial reference frames? There is, and to find it let's return to the light clock of Figure 36.21. FIGURE 36.28 shows the light clock as seen in reference frames S′ and S″. The speed of light is the same in both frames, even though both are moving with respect to each other and with respect to the clock.

Notice that the clock's height h is common to both reference frames. Thus

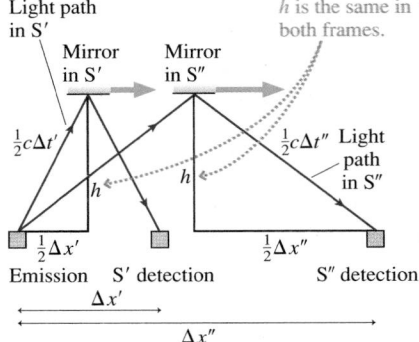

FIGURE 36.28 The light clock seen by experimenters in reference frames S′ and S″.

$$h^2 = \left(\frac{1}{2}c\,\Delta t'\right)^2 - \left(\frac{1}{2}\Delta x'\right)^2 = \left(\frac{1}{2}c\,\Delta t''\right)^2 - \left(\frac{1}{2}\Delta x''\right)^2 \qquad (36.17)$$

The factor $\frac{1}{2}$ cancels, allowing us to write

$$c^2(\Delta t')^2 - (\Delta x')^2 = c^2(\Delta t'')^2 - (\Delta x'')^2 \qquad (36.18)$$

Let us define the **spacetime interval** s between two events to be

$$s^2 = c^2(\Delta t)^2 - (\Delta x)^2 \qquad (36.19)$$

What we've shown in Equation 36.18 is that **the spacetime interval s has the same value in all inertial reference frames.** That is, the spacetime interval between two events is an invariant. It is a value that all experimenters, in all reference frames, can agree upon.

EXAMPLE 36.7 **Using the spacetime interval**

A firecracker explodes at the origin of an inertial reference frame. Then, 2.0 μs later, a second firecracker explodes 300 m away. Astronauts in a passing rocket measure the distance between the explosions to be 200 m. According to the astronauts, how much time elapses between the two explosions?

MODEL The spacetime coordinates of two events are measured in two different inertial reference frames. Call the reference frame of the ground S and the reference frame of the rocket S′. The spacetime interval between these two events is the same in both reference frames.

SOLVE The spacetime interval (or, rather, its square) in frame S is

$$s^2 = c^2(\Delta t)^2 - (\Delta x)^2 = (600\text{ m})^2 - (300\text{ m})^2 = 270{,}000\text{ m}^2$$

where we used $c = 300$ m/μs to determine that $c\,\Delta t = 600$ m. The spacetime interval has the same value in frame S′. Thus

$$s^2 = 270{,}000\text{ m}^2 = c^2(\Delta t')^2 - (\Delta x')^2$$
$$= c^2(\Delta t')^2 - (200\text{ m})^2$$

This is easily solved to give $\Delta t' = 1.85$ μs.

ASSESS The two events are closer together in both space and time in the rocket's reference frame than in the reference frame of the ground.

Einstein's legacy, according to popular culture, was the discovery that "everything is relative." But it's not so. Time intervals and space intervals may be relative, as were the intervals Δx and Δy in the purely geometric analogy with which we opened this section, but some things are *not* relative. In particular, the spacetime interval s between two events is not relative. It is a well-defined number, agreed on by experimenters in each and every inertial reference frame.

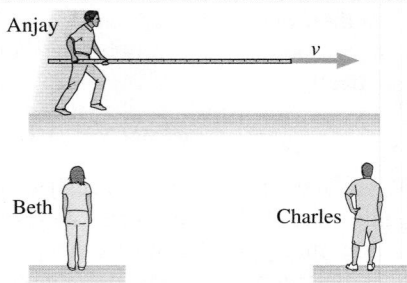

STOP TO THINK 36.7 Beth and Charles are at rest relative to each other. Anjay runs past at velocity v while holding a long pole parallel to his motion. Anjay, Beth, and Charles each measure the length of the pole at the instant Anjay passes Beth. Rank in order, from largest to smallest, the three lengths L_A, L_B, and L_C.

Anjay

v

Beth

Charles

36.8 The Lorentz Transformations

The Galilean transformation $x' = x - vt$ of classical relativity lets us calculate the position x' of an event in frame S′ if we know its position x in frame S. Classical relativity, of course, assumes that $t' = t$. Is there a similar transformation in relativity that would allow us to calculate an event's spacetime coordinates (x', t') in frame S′ if we know their values (x, t) in frame S? Such a transformation would need to satisfy three conditions:

1. Agree with the Galilean transformations in the low-speed limit $v \ll c$.
2. Transform not only spatial coordinates but also time coordinates.
3. Ensure that the speed of light is the same in all reference frames.

FIGURE 36.29 The spacetime coordinates of an event are measured in inertial reference frames S and S′.

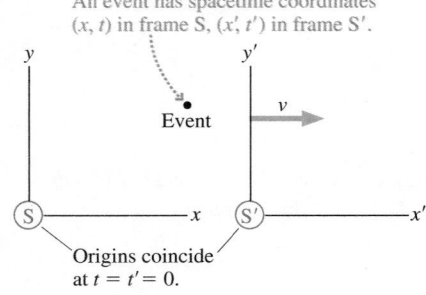

An event has spacetime coordinates (x, t) in frame S, (x', t') in frame S′.

Event

v

Origins coincide at $t = t' = 0$.

We'll continue to use reference frames in the standard orientation of **FIGURE 36.29**. The motion is parallel to the x- and x'-axes, and we *define* $t = 0$ and $t' = 0$ as the instant when the origins of S and S′ coincide.

The requirement that a new transformation agree with the Galilean transformation when $v \ll c$ suggests that we look for a transformation of the form

$$x' = \gamma(x - vt) \quad \text{and} \quad x = \gamma(x' + vt') \tag{36.20}$$

where γ is a dimensionless function of velocity that satisfies $\gamma \to 1$ as $v \to 0$.

To determine γ, we consider the following two events:

Event 1: A flash of light is emitted from the origin of both reference frames $(x = x' = 0)$ at the instant they coincide $(t = t' = 0)$.

Event 2: The light strikes a light detector. The spacetime coordinates of this event are (x, t) in frame S and (x', t') in frame S′.

Light travels at speed c in both reference frames, so the positions of event 2 are $x = ct$ in S and $x' = ct'$ in S′. Substituting these expressions for x and x' into Equation 36.20 gives

$$ct' = \gamma(ct - vt) = \gamma(c - v)t$$
$$ct = \gamma(ct' + vt') = \gamma(c + v)t' \tag{36.21}$$

We solve the first equation for t', by dividing by c, then substitute this result for t' into the second:

$$ct = \gamma(c + v)\frac{\gamma(c - v)t}{c} = \gamma^2(c^2 - v^2)\frac{t}{c}$$

The t cancels, leading to

$$\gamma^2 = \frac{c^2}{c^2 - v^2} = \frac{1}{1 - v^2/c^2}$$

Thus the γ that "works" in the proposed transformation of Equation 36.20 is

$$\gamma = \frac{1}{\sqrt{1 - v^2/c^2}} = \frac{1}{\sqrt{1 - \beta^2}} \qquad (36.22)$$

You can see that $\gamma \rightarrow 1$ as $v \rightarrow 0$, as expected.

The transformation between t and t' is found by requiring that $x = x$ if you use Equation 36.20 to transform a position from S to S′ and then back to S. The details will be left for a homework problem. Another homework problem will let you demonstrate that the y and z measurements made perpendicular to the relative motion are not affected by the motion. We tacitly assumed this condition in our analysis of the light clock.

The full set of equations are called the **Lorentz transformations.** They are

$$
\begin{aligned}
x' &= \gamma(x - vt) & x &= \gamma(x' + vt') \\
y' &= y & y &= y' \\
z' &= z & z &= z' \\
t' &= \gamma(t - vx/c^2) & t &= \gamma(t' + vx'/c^2)
\end{aligned}
\qquad (36.23)
$$

The Lorentz transformations transform the spacetime coordinates of *one* event. Compare these to the Galilean transformation equations in Equations 36.1.

NOTE ▶ These transformations are named after the Dutch physicist H. A. Lorentz, who derived them prior to Einstein. Lorentz was close to discovering special relativity, but he didn't recognize that our concepts of space and time have to be changed before these equations can be properly interpreted. ◀

Using Relativity

Relativity is phrased in terms of *events;* hence relativity problems are solved by interpreting the problem statement in terms of specific events.

PROBLEM-SOLVING
STRATEGY 36.1   **Relativity**

MODEL Frame the problem in terms of events, things that happen at a specific place and time.

VISUALIZE A pictorial representation defines the reference frames.

- Sketch the reference frames, showing their motion relative to each other.
- Show events. Identify objects that are moving with respect to the reference frames.
- Identify any proper time intervals and proper lengths. These are measured in an object's rest frame.

SOLVE The mathematical representation is based on the Lorentz transformations, but not every problem requires the full transformation equations.

- Problems about time intervals can often be solved using time dilation: $\Delta t = \gamma \Delta \tau$.
- Problems about distances can often be solved using length contraction: $L = \ell/\gamma$.

ASSESS Are the results consistent with Galilean relativity when $v \ll c$?

EXAMPLE 36.8 Ryan and Peggy revisited

Peggy is standing in the center of a long, flat railroad car that has firecrackers tied to both ends. The car moves past Ryan, who is standing on the ground, with velocity $v = 0.8c$. Flashes from the exploding firecrackers reach him simultaneously $1.0\ \mu s$ after the instant that Peggy passes him, and he later finds burn marks on the track 300 m to either side of where he had been standing.

a. According to Ryan, what is the distance between the two explosions, and at what times do the explosions occur relative to the time that Peggy passes him?

b. According to Peggy, what is the distance between the two explosions, and at what times do the explosions occur relative to the time that Ryan passes her?

MODEL Let the explosion on Ryan's right, the direction in which Peggy is moving, be event R. The explosion on his left is event L.

VISUALIZE Peggy and Ryan are in inertial reference frames. As FIGURE 36.30 shows, Peggy's frame S' is moving with $v = 0.8c$ relative to Ryan's frame S. We've defined the reference frames such that Peggy and Ryan are at the origins. The instant they pass, by definition, is $t = t' = 0$ s. The two events are shown in Ryan's reference frame.

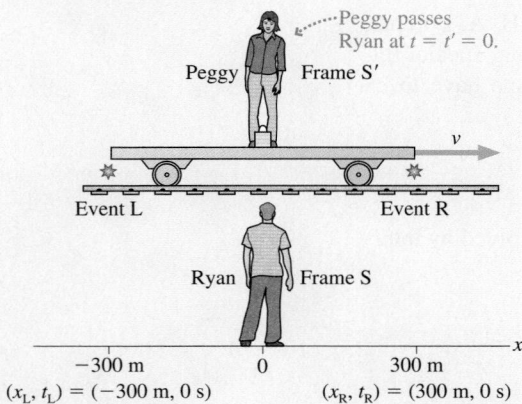

FIGURE 36.30 A pictorial representation of the reference frames and events.

SOLVE a. The two burn marks tell Ryan that the distance between the explosions was $L = 600$ m. Light travels at $c = 300$ m/μs, and the burn marks are 300 m on either side of him, so Ryan can determine that each explosion took place $1.0\ \mu s$ before he saw the flash. But this was the instant of time that Peggy passed him, so Ryan concludes that the explosions were simultaneous with each other and with Peggy's passing him. The spacetime coordinates of the two events in frame S are $(x_R, t_R) = (300\ \text{m}, 0\ \mu s)$ and $(x_L, t_L) = (-300\ \text{m}, 0\ \mu s)$.

b. We already know, from our qualitative analysis in Section 36.5, that the explosions are *not* simultaneous in Peggy's reference frame. Event R happens before event L in S', but we don't know how they compare to the time at which Ryan passes Peggy. We can now use the Lorentz transformations to relate the spacetime coordinates of these events as measured by Ryan to the spacetime coordinates as measured by Peggy. Using $v = 0.8c$, we find that γ is

$$\gamma = \frac{1}{\sqrt{1 - v^2/c^2}} = \frac{1}{\sqrt{1 - 0.8^2}} = 1.667$$

For event L, the Lorentz transformations are

$$x'_L = 1.667((-300\ \text{m}) - (0.8c)(0\ \mu s)) = -500\ \text{m}$$

$$t'_L = 1.667((0\ \mu s) - (0.8c)(-300\ \text{m})/c^2) = 1.33\ \mu s$$

And for event R,

$$x'_R = 1.667((300\ \text{m}) - (0.8c)(0\ \mu s)) = 500\ \text{m}$$

$$t'_R = 1.667((0\ \mu s) - (0.8c)(300\ \text{m})/c^2) = -1.33\ \mu s$$

According to Peggy, the two explosions occur 1000 m apart. Furthermore, the first explosion, on the right, occurs $1.33\ \mu s$ before Ryan passes her at $t' = 0$ s. The second, on the left, occurs $1.33\ \mu s$ after Ryan goes by.

ASSESS Events that are simultaneous in frame S are *not* simultaneous in frame S'. The results of the Lorentz transformations agree with our earlier qualitative analysis.

A follow-up discussion of Example 36.8 is worthwhile. Because Ryan moves at speed $v = 0.8c = 240$ m/μs relative to Peggy, he moves 320 m during the $1.33\ \mu s$ between the first explosion and the instant he passes Peggy, then another 320 m before the second explosion. Gathering this information together, FIGURE 36.31 shows the sequence of events in Peggy's reference frame.

The firecrackers define the ends of the railroad car, so the 1000 m distance between the explosions in Peggy's frame is the car's length L' in frame S'. The car is at rest in frame S', hence length L' is the proper length: $\ell = 1000$ m. Ryan is measuring the length of a moving object, so he should see the car length contracted to

$$L = \sqrt{1 - \beta^2}\,\ell = \frac{\ell}{\gamma} = \frac{1000\ \text{m}}{1.667} = 600\ \text{m}$$

And, indeed, that is exactly the distance Ryan measured between the burn marks.

Finally, we can calculate the spacetime interval s between the two events. According to Ryan,

$$s^2 = c^2(\Delta t^2) - (\Delta x)^2 = c^2(0 \ \mu s)^2 - (600 \ \text{m})^2 = -(600 \ \text{m})^2$$

Peggy computes the spacetime interval to be

$$s^2 = c^2 (\Delta t')^2 - (\Delta x')^2 = c^2 (2.67 \ \mu s)^2 - (1000 \ \text{m})^2 = -(600 \ \text{m})^2$$

Their calculations of the spacetime interval agree, showing that s really is an invariant, but notice that s itself is an imaginary number.

Length

We've already introduced the idea of length contraction, but we didn't precisely define just what we mean by the *length* of a moving object. The length of an object at rest is clear because we can take all the time we need to measure it with meter sticks, surveying tools, or whatever we need. But how can we give clear meaning to the length of a moving object?

A reasonable definition of an object's length is the distance $L = \Delta x = x_R - x_L$ between the right and left ends when the positions x_R and x_L are measured *at the same time t*. In other words, length is the distance spanned by the object at *one instant* of time. Measuring an object's length requires *simultaneous* measurements of two positions (i.e., two events are required); hence the result won't be known until the information from two spatially separated measurements can be brought together.

FIGURE 36.32 shows an object traveling through reference frame S with velocity v. The object is at rest in reference frame S' that travels with the object at velocity v; hence the length in frame S' is the proper length ℓ. That is, $\Delta x' = x_R' - x_L' = \ell$ in frame S'.

At time t, an experimenter (and his or her assistants) in frame S makes simultaneous measurements of the positions x_R and x_L of the ends of the object. The difference $\Delta x = x_R - x_L = L$ is the length in frame S. The Lorentz transformations of x_R and x_L are

$$x_R' = \gamma(x_R - vt)$$
$$x_L' = \gamma(x_L - vt) \tag{36.24}$$

where, it is important to note, t is the *same* for both because the measurements are simultaneous.

Subtracting the second equation from the first, we find

$$x_R' - x_L' = \ell = \gamma(x_R - x_L) = \gamma L = \frac{L}{\sqrt{1 - \beta^2}}$$

Solving for L, we find, in agreement with Equation 36.15, that

$$L = \sqrt{1 - \beta^2}\, \ell \tag{36.25}$$

This analysis has accomplished two things. First, by giving a precise definition of length, we've put our length-contraction result on a firmer footing. Second, we've had good practice at relativistic reasoning using the Lorentz transformation.

NOTE ▶ Length contraction does not tell us how an object would *look*. The visual appearance of an object is determined by light waves that arrive simultaneously at the eye. These waves left points on the object at different times (i.e., *not* simultaneously) because they had to travel different distances to the eye. The analysis needed to determine an object's visual appearance is considerably more complex. Length and length contraction are concerned only with the *actual* length of the object at one instant of time. ◀

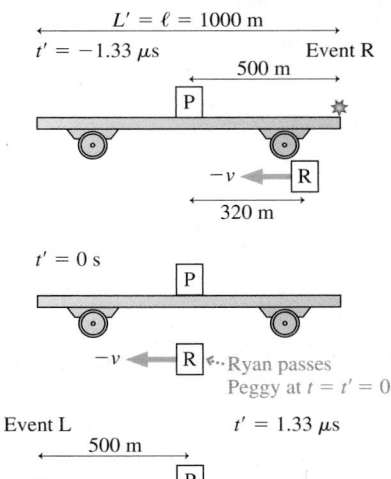

FIGURE 36.31 The sequence of events as seen in Peggy's reference frame.

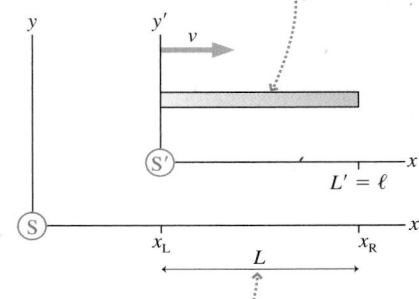

FIGURE 36.32 The length of an object is the distance between *simultaneous* measurements of the positions of the end points.

The Binomial Approximation

The binomial approximation

If $x \ll 1$, then $(1 + x)^n \approx 1 + nx$.

You've met the binomial approximation earlier in this text and in your calculus class. The binomial approximation is useful when we need to calculate a relativistic expression for a nonrelativistic velocity $v \ll c$. Because $v^2/c^2 \ll 1$ in these cases, we can write

$$\text{If } v \ll c: \begin{cases} \sqrt{1 - \beta^2} = (1 - v^2/c^2)^{1/2} \approx 1 - \dfrac{1}{2}\dfrac{v^2}{c^2} \\[2mm] \gamma = \dfrac{1}{\sqrt{1 - \beta^2}} = (1 - v^2/c^2)^{-1/2} \approx 1 + \dfrac{1}{2}\dfrac{v^2}{c^2} \end{cases} \tag{36.26}$$

The following example illustrates the use of the binomial approximation.

EXAMPLE 36.9 **The shrinking school bus**

An 8.0-m-long school bus drives past at 30 m/s. By how much is its length contracted?

MODEL The school bus is at rest in an inertial reference frame S' moving at velocity $v = 30$ m/s relative to the ground frame S. The given length, 8.0 m, is the proper length ℓ in frame S'.

SOLVE In frame S, the school bus is length contracted to

$$L = \sqrt{1 - \beta^2}\,\ell$$

The bus's velocity v is much less than c, so we can use the binomial approximation to write

$$L \approx \left(1 - \frac{1}{2}\frac{v^2}{c^2}\right)\ell = \ell - \frac{1}{2}\frac{v^2}{c^2}\ell$$

The *amount* of the length contraction is

$$\ell - L = \frac{1}{2}\frac{v^2}{c^2}\ell = \frac{1}{2}\left(\frac{30 \text{ m/s}}{3.0 \times 10^8 \text{ m/s}}\right)^2 (8.0 \text{ m})$$

$$= 4.0 \times 10^{-14} \text{ m} = 40 \text{ fm}$$

where 1 fm = 1 femtometer = 10^{-15} m.

ASSESS The bus "shrinks" by only slightly more than the diameter of the nucleus of an atom. It's no wonder that we're not aware of length contraction in our everyday lives. If you had tried to calculate this number exactly, your calculator would have shown $\ell - L = 0$ because the difference between ℓ and L shows up only in the 14th decimal place. A scientific calculator determines numbers to 10 or 12 decimal places, but that isn't sufficient to show the difference. The binomial approximation provides an invaluable tool for finding the very tiny difference between two numbers that are nearly identical.

The Lorentz Velocity Transformations

FIGURE 36.33 The velocity of a moving object is measured to be u in frame S and u' in frame S'.

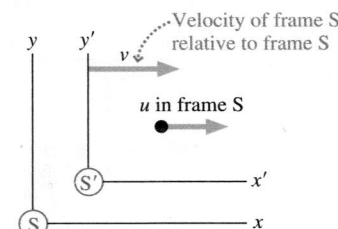

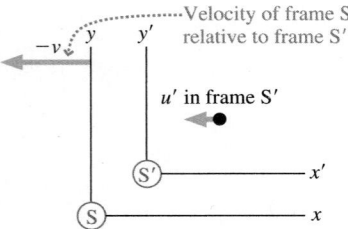

FIGURE 36.33 shows an object that is moving in both reference frame S and reference frame S'. Experimenters in frame S determine that the object's velocity is u, while experimenters in frame S' find it to be u'. For simplicity, we'll assume that the object moves parallel to the x- and x'-axes.

The Galilean velocity transformation $u' = u - v$ was found by taking the time derivative of the position transformation. We can do the same with the Lorentz transformation if we take the derivative with respect to the time in each frame. Velocity u' in frame S' is

$$u' = \frac{dx'}{dt'} = \frac{d(\gamma(x - vt))}{d(\gamma(t - vx/c^2))} \tag{36.27}$$

where we've used the Lorentz transformations for position x' and time t'.

Carrying out the differentiation gives

$$u' = \frac{\gamma(dx - v\,dt)}{\gamma(dt - v\,dx/c^2)} = \frac{dx/dt - v}{1 - v(dx/dt)/c^2} \tag{36.28}$$

But dx/dt is u, the object's velocity in frame S, leading to

$$u' = \frac{u - v}{1 - uv/c^2} \tag{36.29}$$

You can see that Equation 36.29 reduces to the Galilean transformation $u' = u - v$ when $v \ll c$, as expected.

The transformation from S′ to S is found by reversing the sign of v. Altogether,

$$u' = \frac{u - v}{1 - uv/c^2} \quad \text{and} \quad u = \frac{u' + v}{1 + u'v/c^2} \qquad (36.30)$$

Equations 36.30 are the Lorentz velocity transformation equations.

NOTE ▶ It is important to distinguish carefully between v, which is the relative velocity between two reference frames, and u and u', which are the velocities of an *object* as measured in the two different reference frames. ◀

EXAMPLE 36.10 **A really fast bullet**

A rocket flies past the earth at $0.90c$. As it goes by, the rocket fires a bullet in the forward direction at $0.95c$ with respect to the rocket. What is the bullet's speed with respect to the earth?

MODEL The rocket and the earth are inertial reference frames. Let the earth be frame S and the rocket be frame S′. The velocity of frame S′ relative to frame S is $v = 0.90c$. The bullet's velocity in frame S′ is $u' = 0.95c$.

SOLVE We can use the Lorentz velocity transformation to find

$$u = \frac{u' + v}{1 + u'v/c^2} = \frac{0.95c + 0.90c}{1 + (0.95c)(0.90c)/c^2} = 0.997c$$

The bullet's speed with respect to the earth is 99.7% of the speed of light.

NOTE ▶ Many relativistic calculations are much easier when velocities are specified as a fraction of c. ◀

ASSESS In Newtonian mechanics, the Galilean transformation of velocity would give $u = 1.85c$. Now, despite the very high speed of the rocket and of the bullet with respect to the rocket, the bullet's speed with respect to the earth remains less than c. This is yet more evidence that objects cannot travel faster than the speed of light.

Suppose the rocket in Example 36.10 fired a laser beam in the forward direction as it traveled past the earth at velocity v. The laser beam would travel away from the rocket at speed $u' = c$ in the rocket's reference frame S′. What is the laser beam's speed in the earth's frame S? According to the Lorentz velocity transformation, it must be

$$u = \frac{u' + v}{1 + u'v/c^2} = \frac{c + v}{1 + cv/c^2} = \frac{c + v}{1 + v/c} = \frac{c + v}{(c + v)/c} = c \qquad (36.31)$$

Light travels at speed c in both frame S and frame S′. This important consequence of the principle of relativity is "built into" the Lorentz transformations.

36.9 Relativistic Momentum

In Newtonian mechanics, the total momentum of a system is a conserved quantity. Further, as we've seen, the law of conservation of momentum, $P_f = P_i$, is true in all inertial reference frames *if* the particle velocities in different reference frames are related by the Galilean velocity transformations.

The difficulty, of course, is that the Galilean transformations are not consistent with the principle of relativity. It is a reasonable approximation when all velocities are very much less than c, but the Galilean transformations fail dramatically as velocities approach c. It's not hard to show that $P'_f \neq P'_i$ if the particle velocities in frame S′ are related to the particle velocities in frame S by the Lorentz transformations.

There are two possibilities:

1. The so-called law of conservation of momentum is not really a law of physics. It is approximately true at low velocities but fails as velocities approach the speed of light.
2. The law of conservation of momentum really is a law of physics, but the expression $p = mu$ is not the correct way to calculate momentum when the particle velocity u becomes a significant fraction of c.

Momentum conservation is such a central and important feature of mechanics that it seems unlikely to fail in relativity.

The classical momentum, for one-dimensional motion, is $p = mu = m(\Delta x/\Delta t)$. Δt is the time to move distance Δx. That seemed clear enough within a Newtonian framework, but now we've learned that experimenters in different reference frames disagree about the amount of time needed. So whose Δt should we use?

One possibility is to use the time measured *by the particle*. This is the proper time $\Delta \tau$ because the particle is at rest in its own reference frame and needs only one clock. With this in mind, let's redefine the momentum of a particle of mass m moving with velocity $u = \Delta x/\Delta t$ to be

$$p = m\frac{\Delta x}{\Delta \tau} \tag{36.32}$$

We can relate this new expression for p to the familiar Newtonian expression by using the time-dilation result $\Delta \tau = (1 - u^2/c^2)^{1/2}\Delta t$ to relate the proper time interval measured by the particle to the more practical time interval Δt measured by experimenters in frame S. With this substitution, Equation 36.32 becomes

$$p = m\frac{\Delta x}{\Delta \tau} = m\frac{\Delta x}{\sqrt{1 - u^2/c^2}\,\Delta t} = \frac{mu}{\sqrt{1 - u^2/c^2}} \tag{36.33}$$

You can see that Equation 36.33 reduces to the classical expression $p = mu$ when the particle's speed $u \ll c$. That is an important requirement, but whether this is the "correct" expression for p depends on whether the total momentum P is conserved when the velocities of a system of particles are transformed with the Lorentz velocity transformation equations. The proof is rather long and tedious, so we will assert, without actual proof, that the momentum defined in Equation 36.33 does, indeed, transform correctly. **The law of conservation of momentum is still valid in all inertial reference frames *if* the momentum of each particle is calculated with Equation 36.33.**

The factor that multiplies mu in Equation 36.33 looks much like the factor γ in the Lorentz transformation equations for x and t, but there's one very important difference. The v in the Lorentz transformation equations is the velocity of a *reference frame*. The u in Equation 36.33 is the velocity of a particle moving *in* a reference frame.

With this distinction in mind, let's define the quantity

$$\gamma_p = \frac{1}{\sqrt{1 - u^2/c^2}} \tag{36.34}$$

where the subscript p indicates that this is γ for a particle, not for a reference frame. In frame S′, where the particle moves with velocity u', the corresponding expression would be called γ_p'. With this definition of γ_p, the momentum of a particle is

$$p = \gamma_p mu \tag{36.35}$$

EXAMPLE 36.11 **Momentum of a subatomic particle**

Electrons in a particle accelerator reach a speed of $0.999c$ relative to the laboratory. One collision of an electron with a target produces a muon that moves forward with a speed of $0.95c$ relative to the laboratory. The muon mass is 1.90×10^{-28} kg. What is the muon's momentum in the laboratory frame and in the frame of the electron beam?

MODEL Let the laboratory be reference frame S. The reference frame S′ of the electron beam (i.e., a reference frame in which the electrons are at rest) moves in the direction of the electrons at $v = 0.999c$. The muon velocity in frame S is $u = 0.95c$.

SOLVE γ_p for the muon in the laboratory reference frame is

$$\gamma_p = \frac{1}{\sqrt{1 - u^2/c^2}} = \frac{1}{\sqrt{1 - 0.95^2}} = 3.20$$

Thus the muon's momentum in the laboratory is

$$p = \gamma_p mu = (3.20)(1.90 \times 10^{-28}\text{ kg})(0.95 \times 3.00 \times 10^8\text{ m/s})$$
$$= 1.73 \times 10^{-19}\text{ kg m/s}$$

The momentum is a factor of 3.2 larger than the Newtonian momentum mu. To find the momentum in the electron-beam reference frame, we must first use the velocity transformation equation to find the muon's velocity in frame S':

$$u' = \frac{u - v}{1 - uv/c^2} = \frac{0.95c - 0.999c}{1 - (0.95c)(0.999c)/c^2} = -0.962c$$

In the laboratory frame, the faster electrons are overtaking the slower muon. Hence the muon's velocity in the electron-beam frame is negative. γ'_p for the muon in frame S' is

$$\gamma'_p = \frac{1}{\sqrt{1 - u'^2/c^2}} = \frac{1}{\sqrt{1 - 0.962^2}} = 3.66$$

The muon's momentum in the electron-beam reference frame is

$$p' = \gamma'_p \, mu'$$
$$= (3.66)(1.90 \times 10^{-28} \text{ kg})(-0.962 \times 3.00 \times 10^8 \text{ m/s})$$
$$= -2.01 \times 10^{-19} \text{ kg m/s}$$

ASSESS From the laboratory perspective, the muon moves only slightly slower than the electron beam. But it turns out that the muon moves faster with respect to the electrons, although in the opposite direction, than it does with respect to the laboratory.

The Cosmic Speed Limit

FIGURE 36.34a is a graph of momentum versus velocity. For a Newtonian particle, with $p = mu$, the momentum is directly proportional to the velocity. The relativistic expression for momentum agrees with the Newtonian value if $u \ll c$, but p approaches ∞ as $u \rightarrow c$.

The implications of this graph become clear when we relate momentum to force. Consider a particle subjected to a constant force, such as a rocket that never runs out of fuel. If F is constant, we can see from $F = dp/dt$ that the momentum is $p = Ft$. If Newtonian physics were correct, a particle would go faster and faster as its velocity $u = p/m = (F/m)t$ increased without limit. But the relativistic result, shown in **FIGURE 36.34b**, is that the particle's velocity asymptotically approaches the speed of light ($u \rightarrow c$) as p approaches ∞. Relativity gives a very different outcome than Newtonian mechanics.

The speed c is a "cosmic speed limit" for material particles. A force cannot accelerate a particle to a speed higher than c because the particle's momentum becomes infinitely large as the speed approaches c. The amount of effort required for each additional increment of velocity becomes larger and larger until no amount of effort can raise the velocity any higher.

Actually, at a more fundamental level, c is a speed limit for *any* kind of **causal influence**. If I throw a rock and break a window, my throw is the *cause* of the breaking window and the rock is the *causal influence*. If I shoot a laser beam at a light detector that is wired to a firecracker, the light wave is the *causal influence* that leads to the explosion. A causal influence can be any kind of particle, wave, or information that travels from A to B and allows A to be the cause of B.

For two unrelated events—a firecracker explodes in Tokyo and a balloon bursts in Paris—the relativity of simultaneity tells us that they may be simultaneous in one reference frame but not in others. Or in one reference frame the firecracker may explode before the balloon bursts but in some other reference frame the balloon may burst first. These possibilities violate our commonsense view of time, but they're not in conflict with the principle of relativity.

For two causally related events—A *causes* B—it would be nonsense for an experimenter in any reference frame to find that B occurs before A. No experimenter in any reference frame, no matter how it is moving, will find that you are born before your mother is born. If A causes B, then it must be the case that $t_A < t_B$ in *all* reference frames.

Suppose there exists some kind of causal influence that *can* travel at speed $u > c$. **FIGURE 36.35** shows a reference frame S in which event A occurs at position $x_A = 0$. The faster-than-light causal influence—perhaps some yet-to-be-discovered "z ray"— leaves A at $t_A = 0$ and travels to the point at which it will cause event B. It arrives at x_B at time $t_B = x_B/u$.

How do events A and B appear in a reference frame S' that travels at an ordinary speed $v < c$ relative to frame S? We can use the Lorentz transformations to find out.

FIGURE 36.34 The speed of a particle cannot reach the speed of light.

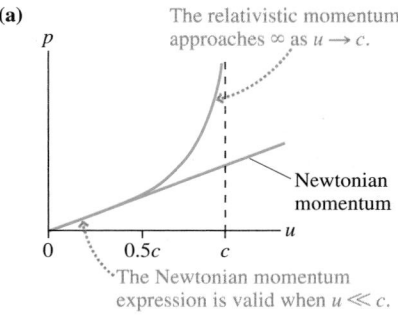

(a)

The relativistic momentum approaches ∞ as $u \rightarrow c$.

Newtonian momentum

The Newtonian momentum expression is valid when $u \ll c$.

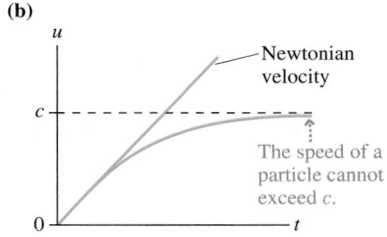

(b)

Newtonian velocity

The speed of a particle cannot exceed c.

FIGURE 36.35 Assume that a causal influence can travel from A to B at a speed $u > c$.

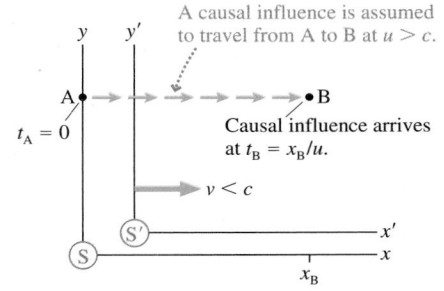

A causal influence is assumed to travel from A to B at $u > c$.

Causal influence arrives at $t_B = x_B/u$.

Because $x_A = 0$ and $t_A = 0$, it's easy to see that $x_A' = 0$ and $t_A' = 0$. That is, the origins of S and S′ overlap at the instant the causal influence leaves event A. More interesting is the time at which this influence reaches B in frame S′. The Lorentz time transformation for event B is

$$t_B' = \gamma\left(t_B - \frac{vx_B}{c^2}\right) = \gamma t_B\left(1 - \frac{v(x_B/t_B)}{c^2}\right) = \gamma t_B\left(1 - \frac{vu}{c^2}\right) \qquad (36.36)$$

where we first factored out t_B, then made use of the fact that $u = x_B/t_B$ in frame S.

We're assuming $u > c$, so let $u = \alpha c$ where $\alpha > 1$ is a constant. Then $vu/c^2 = \alpha v/c$. Now follow the logic:

1. If $v > c/\alpha$, which is possible because $\alpha > 1$, then $vu/c^2 > 1$.
2. If $vu/c^2 > 1$, then the term $(1 - vu/c^2)$ is negative and $t_B' < 0$.
3. If $t_B' < 0$, then event B happens *before* event A in reference frame S′.

In other words, if a causal influence can travel faster than c, then there exist reference frames in which the effect happens before the cause. We know this can't happen, so our assumption $u > c$ must be wrong. **No causal influence of any kind—particle, wave, or yet-to-be-discovered z rays—can travel faster than c.**

The existence of a cosmic speed limit is one of the most interesting consequences of the theory of relativity. "Warp drive," in which a spaceship suddenly leaps to faster-than-light velocities, is simply incompatible with the theory of relativity. Rapid travel to the stars will remain in the realm of science fiction unless future scientific discoveries find flaws in Einstein's theory and open the doors to yet-undreamed-of theories. While we can't say with certainty that a scientific theory will never be overturned, there is currently not even a hint of evidence that disagrees with the special theory of relativity.

36.10 Relativistic Energy

Energy is our final topic in this chapter on relativity. Space, time, velocity, and momentum are changed by relativity, so it seems inevitable that we'll need a new view of energy.

In Newtonian mechanics, a particle's kinetic energy $K = \frac{1}{2}mu^2$ can be written in terms of its momentum $p = mu$ as $K = p^2/2m$. This suggests that a relativistic expression for energy will likely involve both the square of p and the particle's mass. We also hope that energy will be conserved in relativity, so a reasonable starting point is with the one quantity we've found that is the same in all inertial reference frames: the spacetime interval s.

Let a particle of mass m move through distance Δx during a time interval Δt, as measured in reference frame S. The spacetime interval is

$$s^2 = c^2(\Delta t)^2 - (\Delta x)^2 = \text{invariant}$$

We can turn this into an expression involving momentum if we multiply by $(m/\Delta\tau)^2$, where $\Delta\tau$ is the proper time (i.e., the time measured by the particle). Doing so gives

$$(mc)^2\left(\frac{\Delta t}{\Delta\tau}\right)^2 - \left(\frac{m\Delta x}{\Delta\tau}\right)^2 = (mc)^2\left(\frac{\Delta t}{\Delta\tau}\right)^2 - p^2 = \text{invariant} \qquad (36.37)$$

where we used $p = m(\Delta x/\Delta\tau)$ from Equation 36.32.

Now Δt, the time interval in frame S, is related to the proper time by the time-dilation result $\Delta t = \gamma_p \Delta\tau$. With this change, Equation 36.37 becomes

$$(\gamma_p mc)^2 - p^2 = \text{invariant}$$

Finally, for reasons that will be clear in a minute, we multiply by c^2, to get

$$(\gamma_p mc^2)^2 - (pc)^2 = \text{invariant} \qquad (36.38)$$

To say that the right side is an *invariant* means it has the same value in all inertial reference frames. We can easily determine the constant by evaluating it in the reference frame in which the particle is at rest. In that frame, where $p = 0$ and $\gamma_p = 1$, we find that

$$(\gamma_p mc^2)^2 - (pc)^2 = (mc^2)^2 \qquad (36.39)$$

Let's reflect on what this means before taking the next step. The spacetime interval s has the same value in all inertial reference frames. In other words, $c^2 (\Delta t)^2 - (\Delta x)^2 = c^2(\Delta t')^2 - (\Delta x')^2$. Equation 36.39 was derived from the definition of the spacetime interval; hence the quantity mc^2 is also an invariant having the same value in all inertial reference frames. In other words, if experimenters in frames S and S' both make measurements on this particle of mass m, they will find that

$$(\gamma_p mc^2)^2 - (pc)^2 = (\gamma_p' mc^2)^2 - (p'c)^2 \qquad (36.40)$$

Experimenters in different reference frames measure different values for the momentum, but experimenters in all reference frames agree that momentum is a conserved quantity. Equations 36.39 and 36.40 suggest that the quantity $\gamma_p mc^2$ is also an important property of the particle, a property that changes along with p in just the right way to satisfy Equation 36.39. But what is this property?

The first clue comes from checking the units. γ_p is dimensionless and c is a velocity, so $\gamma_p mc^2$ has the same units as the classical expression $\frac{1}{2} mv^2$—namely, units of energy. For a second clue, let's examine how $\gamma_p mc^2$ behaves in the low-velocity limit $u \ll c$. We can use the binomial approximation expression for γ_p to find

$$\gamma_p mc^2 = \frac{mc^2}{\sqrt{1 - u^2/c^2}} \approx \left(1 + \frac{1}{2}\frac{u^2}{c^2}\right)mc^2 = mc^2 + \frac{1}{2}mu^2 \qquad (36.41)$$

The second term, $\frac{1}{2}mu^2$, is the low-velocity expression for the kinetic energy K. This is an energy associated with motion. But the first term suggests that the concept of energy is more complex than we originally thought. It appears that **there is an inherent energy associated with mass itself.**

With that as a possibility, subject to experimental verification, let's define the **total energy** E of a particle to be

$$E = \gamma_p mc^2 = E_0 + K = \text{rest energy} + \text{kinetic energy} \qquad (36.42)$$

This total energy consists of a **rest energy**

$$E_0 = mc^2 \qquad (36.43)$$

and a relativistic expression for the *kinetic energy*

$$K = (\gamma_p - 1)mc^2 = (\gamma_p - 1)E_0 \qquad (36.44)$$

This expression for the kinetic energy is very nearly $\frac{1}{2}mu^2$ when $u \ll c$ but, as FIGURE 36.36 shows, differs significantly from the classical value for very high velocities.

Equation 36.43 is, of course, Einstein's famous $E = mc^2$, perhaps the most famous equation in all of physics. Before discussing its significance, we need to tie up some loose ends. First, notice that the right-hand side of Equation 36.39 is the square of the rest energy E_0. Thus we can write a final version of that equation:

$$E^2 - (pc)^2 = E_0^2 \qquad (36.45)$$

The quantity E_0 is an *invariant* with the same value mc^2 in *all* inertial reference frames.

Second, notice that we can write

$$pc = (\gamma_p mu)c = \frac{u}{c}(\gamma_p mc^2)$$

FIGURE 36.36 The relativistic kinetic energy.

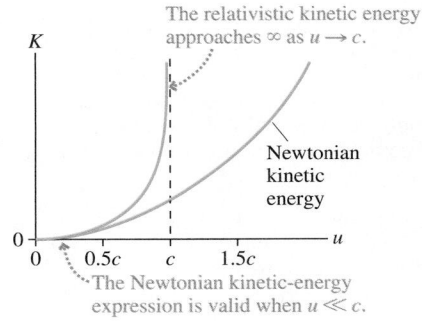

The relativistic kinetic energy approaches ∞ as $u \to c$.

Newtonian kinetic energy

The Newtonian kinetic-energy expression is valid when $u \ll c$.

But $\gamma_p mc^2$ is the total energy E and $u/c = \beta_p$, where the subscript p, as on γ_p, indicates that we're referring to the motion of a particle within a reference frame, not the motion of two reference frames relative to each other. Thus

$$pc = \beta_p E \qquad (36.46)$$

FIGURE 36.37 shows the "velocity-energy-momentum triangle," a convenient way to remember the relationships among the three quantities.

FIGURE 36.37 The velocity-energy-momentum triangle.

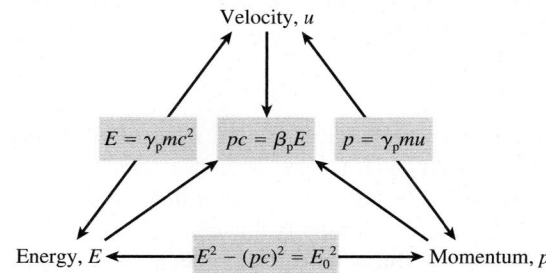

EXAMPLE 36.12 | Kinetic energy and total energy

Calculate the rest energy and the kinetic energy of (a) a 100 g ball moving with a speed of 100 m/s and (b) an electron with a speed of 0.999c.

MODEL The ball, with $u \ll c$, is a classical particle. We don't need to use the relativistic expression for its kinetic energy. The electron is highly relativistic.

SOLVE a. For the ball, with $m = 0.10$ kg,

$$E_0 = mc^2 = 9.0 \times 10^{15} \text{ J}$$

$$K = \frac{1}{2}mu^2 = 500 \text{ J}$$

b. For the electron, we start by calculating

$$\gamma_p = \frac{1}{(1 - u^2/c^2)^{1/2}} = 22.4$$

Then, using $m_e = 9.11 \times 10^{-31}$ kg, we find

$$E_0 = mc^2 = 8.2 \times 10^{-14} \text{ J}$$

$$K = (\gamma_p - 1)E_0 = 170 \times 10^{-14} \text{ J}$$

ASSESS The ball's kinetic energy is a typical kinetic energy. Its rest energy, by contrast, is a staggeringly large number. For a relativistic electron, on the other hand, the kinetic energy is more important than the rest energy.

STOP TO THINK 36.8 An electron moves through the lab at 99% the speed of light. The lab reference frame is S and the electron's reference frame is S′. In which reference frame is the electron's rest mass larger?

a. In frame S, the lab frame
b. In frame S′, the electron's frame
c. It is the same in both frames.

FIGURE 36.38 An inelastic collision between two balls of clay does not seem to conserve the total energy E.

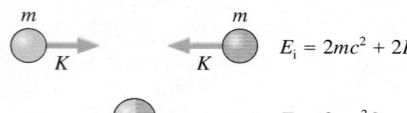

Mass-Energy Equivalence

Now we're ready to explore the significance of Einstein's famous equation $E = mc^2$. FIGURE 36.38 shows two balls of clay approaching each other. They have equal masses and equal kinetic energies, and they slam together in a perfectly inelastic collision to form one large ball of clay at rest. In Newtonian mechanics, we would say that the initial energy $2K$ is dissipated by being transformed into an equal amount of thermal energy, raising the temperature of the coalesced ball of clay. But Equation 36.42, $E = E_0 + K$, doesn't say anything about thermal energy. The total energy before the collision is $E_i = 2mc^2 + 2K$, with the factor of 2 appearing because there are two masses. It seems like the total energy after the collision, when the clay is at rest, should be $2mc^2$, but this value doesn't conserve total energy.

There's ample experimental evidence that energy is conserved, so there must be a flaw in our reasoning. The statement of energy conservation is

$$E_f = Mc^2 = E_i = 2mc^2 + 2K \qquad (36.47)$$

where M is the mass of clay after the collision. But, remarkably, this requires

$$M = 2m + \frac{2K}{c^2} \qquad (36.48)$$

In other words, **mass is not conserved.** The mass of clay after the collision is larger than the mass of clay before the collision. Total energy can be conserved only if kinetic energy is transformed into an "equivalent" amount of mass.

The mass increase in a collision between two balls of clay is incredibly small, far beyond any scientist's ability to detect. So how do we know if such a crazy idea is true?

FIGURE 36.39 shows an experiment that has been done countless times in the last 50 years at particle accelerators around the world. An electron that has been accelerated to $u \approx c$ is aimed at a target material. When a high-energy electron collides with an atom in the target, it can easily knock one of the electrons out of the atom. Thus we would expect to see two electrons leaving the target: the incident electron and the ejected electron. Instead, *four* particles emerge from the target: three electrons and a positron. A *positron,* or positive electron, is the antimatter version of an electron, identical to an electron in all respects other than having charge $q = +e$.

In chemical-reaction notation, the collision is

$$e^- \text{ (fast)} + e^- \text{ (at rest)} \rightarrow e^- + e^- + e^- + e^+$$

An electron and a positron have been *created,* apparently out of nothing. Mass $2m_e$ before the collision has become mass $4m_e$ after the collision. (Notice that charge has been conserved in this collision.)

Although the mass has increased, it wasn't created "out of nothing." This is an inelastic collision, just like the collision of the balls of clay, because the kinetic energy after the collision is less than before. In fact, if you measured the energies before and after the collision, you would find that the decrease in kinetic energy is exactly equal to the energy equivalent of the two particles that have been created: $\Delta K = 2m_e c^2$. The new particles have been created *out of energy!*

Particles can be created from energy, and particles can return to energy. FIGURE 36.40 shows an electron colliding with a positron, its antimatter partner. When a particle and its antiparticle meet, they *annihilate* each other. The mass disappears, and the energy equivalent of the mass is transformed into light. In Chapter 38, you'll learn that light is *quantized,* meaning that light is emitted and absorbed in discrete chunks of energy called *photons.* For light with wavelength λ, the energy of a photon is $E_{\text{photon}} = hc/\lambda$, where $h = 6.63 \times 10^{-34}$ J s is called *Planck's constant.* Photons carry momentum as well as energy. Conserving both energy and momentum in the annihilation of an electron and a positron requires the emission in opposite directions of two photons of equal energy.

If the electron and positron are fairly slow, so that $K \ll mc^2$, then $E_i \approx E_0 = mc^2$. In that case, energy conservation requires

$$E_f = 2E_{\text{photon}} = E_i \approx 2m_e c^2 \qquad (36.49)$$

Hence the wavelength of the emitted photons is

$$\lambda = \frac{hc}{m_e c^2} \approx 0.0024 \text{ nm} \qquad (36.50)$$

The tracks of elementary particles in a bubble chamber show the creation of an electron-positron pair. The negative electron and positive positron spiral in opposite directions in the magnetic field.

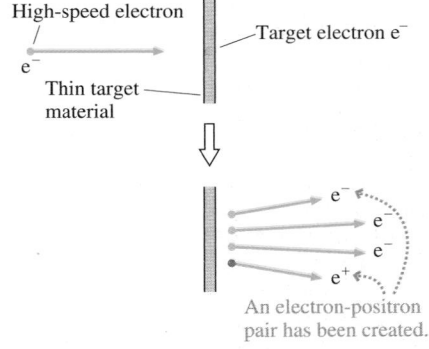

FIGURE 36.39 An inelastic collision between electrons can create an electron-positron pair.

An electron-positron pair has been created.

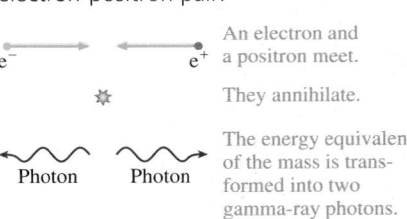

FIGURE 36.40 The annihilation of an electron-positron pair.

An electron and a positron meet.

They annihilate.

The energy equivalent of the mass is transformed into two gamma-ray photons.

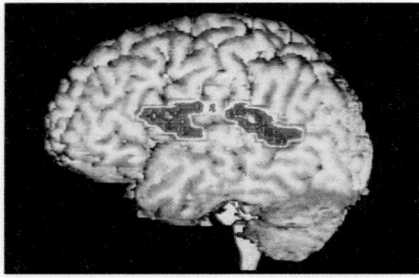

Positron-electron annihilation (a PET scan) provides a noninvasive look into the brain.

This is an extremely short wavelength, even shorter than the wavelengths of x rays. Photons in this wavelength range are called *gamma rays*. And, indeed, the emission of 0.0024 nm gamma rays is observed in many laboratory experiments in which positrons are able to collide with electrons and thus annihilate. In recent years, with the advent of gamma-ray telescopes on satellites, astronomers have found 0.0024 nm photons coming from many places in the universe, especially galactic centers—evidence that positrons are abundant throughout the universe.

Positron-electron annihilation is also the basis of the medical procedure known as a positron-emission tomography, or PET scans. A patient ingests a very small amount of a radioactive substance that decays by the emission of positrons. This substance is taken up by certain tissues in the body, especially those tissues with a high metabolic rate. As the substance decays, the positrons immediately collide with electrons, annihilate, and create two gamma-ray photons that are emitted back to back. The gamma rays, which easily leave the body, are detected, and their trajectories are traced backward into the body. The overlap of many such trajectories shows quite clearly the tissue in which the positron emission is occurring. The results are usually shown as false-color photographs, with redder areas indicating regions of higher positron emission.

Conservation of Energy

The creation and annihilation of particles with mass, processes strictly forbidden in Newtonian mechanics, are vivid proof that neither mass nor the Newtonian definition of energy is conserved. Even so, the *total* energy—the kinetic energy *and* the energy equivalent of mass—remains a conserved quantity.

> **Law of conservation of total energy** The energy $E = \sum E_i$ of an isolated system is conserved, where $E_i = (\gamma_p)_i m_i c^2$ is the total energy of particle i.

Mass and energy are not the same thing, but, as the last few examples have shown, they are *equivalent* in the sense that mass can be transformed into energy and energy can be transformed into mass as long as the total energy is conserved.

Probably the most well-known application of the conservation of total energy is nuclear fission. The uranium isotope ^{236}U, containing 236 protons and neutrons, does not exist in nature. It can be created when a ^{235}U nucleus absorbs a neutron, increasing its atomic mass from 235 to 236. The ^{236}U nucleus quickly fragments into two smaller nuclei and several extra neutrons, a process known as **nuclear fission**. The nucleus can fragment in several ways, but one is

$$\text{n} + {}^{235}\text{U} \rightarrow {}^{236}\text{U} \rightarrow {}^{144}\text{Ba} + {}^{89}\text{Kr} + 3\text{n}$$

Ba and Kr are the atomic symbols for barium and krypton.

This reaction seems like an ordinary chemical reaction—until you check the masses. The masses of atomic isotopes are known with great precision from many decades of measurement in instruments called mass spectrometers. If you add up the masses on both sides, you find that the mass of the products is 0.185 u smaller than the mass of the initial neutron and ^{235}U, where, you will recall, 1 u = 1.66×10^{-27} kg is the atomic mass unit. In kilograms the mass loss is 3.07×10^{-28} kg.

Mass has been lost, but the energy equivalent of the mass has not. As **FIGURE 36.41** shows, the mass has been converted to kinetic energy, causing the two product nuclei and three neutrons to be ejected at very high speeds. The kinetic energy is easily calculated: $\Delta K = m_{\text{lost}} c^2 = 2.8 \times 10^{-11}$ J.

This is a very tiny amount of energy, but it is the energy released from *one* fission. The number of nuclei in a macroscopic sample of uranium is on the order of N_A, Avogadro's number. Hence the energy available if *all* the nuclei fission is enormous. This energy, of course, is the basis for both nuclear power reactors and nuclear weapons.

FIGURE 36.41 In nuclear fission, the energy equivalent of lost mass is converted into kinetic energy.

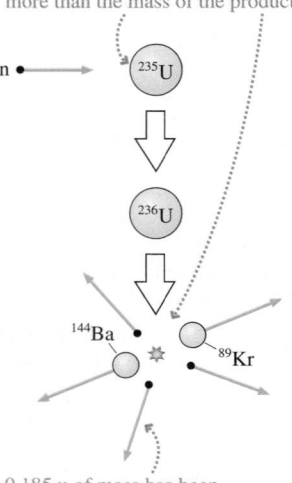

The mass of the reactants is 0.185 u more than the mass of the products.

n

^{235}U

^{236}U

^{144}Ba

^{89}Kr

0.185 u of mass has been converted into kinetic energy.

We started this chapter with an expectation that relativity would challenge our basic notions of space and time. We end by finding that relativity changes our understanding of mass and energy. Most remarkable of all is that each and every one of these new ideas flows from one simple statement: The laws of physics are the same in all inertial reference frames.

CHALLENGE EXAMPLE 36.13 **Goths and Huns**

The rockets of the Goths and the Huns are each 1000 m long in their rest frame. The rockets pass each other, virtually touching, at a relative speed of 0.8c. The Huns have a laser cannon at the rear of their rocket that fires a deadly laser beam perpendicular to the rocket's motion. The captain of the Huns wants to send a threatening message to the Goths by "firing a shot across their bow." He tells his first mate, "The Goths' rocket is length contracted to 600 m. Fire the laser cannon at the instant the tail of their rocket passes the nose of ours. The laser beam will cross 400 m in front of them."

But things are different in the Goths' reference frame. The Goth captain muses, "The Huns' rocket is length contracted to 600 m, 400 m shorter than our rocket. If they fire as the nose of their ship passes the tail of ours, the lethal laser beam will pass right through our side."

The first mate on the Huns' rocket fires as ordered. Does the laser beam blast the Goths or not?

MODEL Both rockets are inertial reference frames. Let the Huns' rocket be frame S and the Goths' rocket be frame S'. S' moves with velocity $v = 0.8c$ relative to S. We need to describe the situation in terms of events.

VISUALIZE Begin by considering the situation from the Huns' reference frame, as shown in **FIGURE 36.42**.

FIGURE 36.42 The situation seen by the Huns.

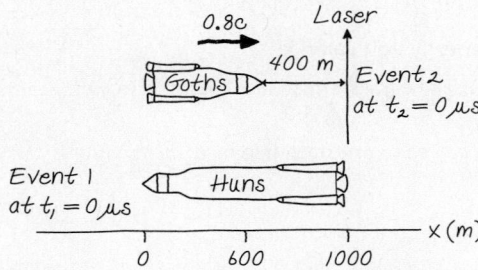

SOLVE The key to resolving the paradox is that two events simultaneous in one reference frame are not simultaneous in a different reference frame. The Huns do, indeed, see the Goths' rocket length contracted to $L_{Goths} = (1 - (0.8)^2)^{1/2} (1000 \text{ m}) = 600 \text{ m}$. Let event 1 be the tail of the Goths' rocket passing the nose of the Huns' rocket. Since we're free to define the origin of our coordinate system, we define this event to be at time $t_1 = 0 \text{ } \mu\text{s}$ and at position $x_1 = 0 \text{ m}$. Then, in the Huns' reference frame, the spacetime coordinates of event 2, the firing of the laser cannon,

are $(x_2, t_2) = (1000 \text{ m}, 0 \text{ } \mu\text{s})$. The nose of the Goths' rocket is at $x = 600 \text{ m}$ at $t = 0 \text{ } \mu\text{s}$; thus the laser cannon misses the Goths by 400 m.

Now we can use the Lorentz transformations to find the spacetime coordinates of the events in the Goths' reference frame. The nose of the Huns' rocket passes the tail of the Goths' rocket at $(x'_1, t'_1) = (0 \text{ m}, 0 \text{ } \mu\text{s})$. The Huns fire their laser cannon at

$$x'_2 = \gamma(x_2 - vt_2) = \frac{5}{3}(1000 \text{ m} - 0 \text{ m}) = 1667 \text{ m}$$

$$t'_2 = \gamma\left(t_2 - \frac{vx_2}{c^2}\right) = \frac{5}{3}\left(0 \text{ } \mu\text{s} - (0.8)\frac{1000 \text{ m}}{300 \text{ m}/\mu\text{s}}\right) = -4.444 \text{ } \mu\text{s}$$

where we calculated $\gamma = 5/3$ for $v = 0.8c$. Events 1 and 2 are *not* simultaneous in S'. The Huns fire the laser cannon 4.444 μs *before* the nose of their rocket reaches the tail of the Goths' rocket. The laser is fired at $x'_2 = 1667 \text{ m}$, missing the nose of the Goths' rocket by 667 m. **FIGURE 36.43** shows how the Goths see things.

FIGURE 36.43 The situation seen by the Goths.

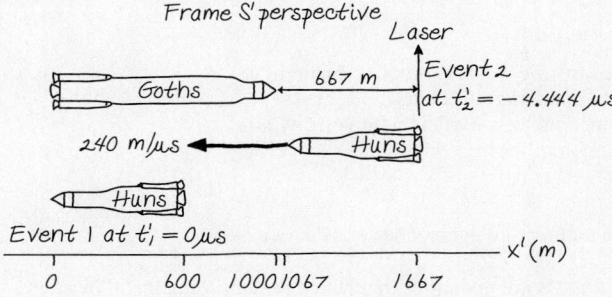

In fact, since the Huns' rocket is length contracted to 600 m, the nose of the Huns' rocket is at $x' = 1667 \text{ m} - 600 \text{ m} = 1067 \text{ m}$ at the instant they fire the laser cannon. At a speed of $v = 0.8c = 240 \text{ m}/\mu\text{s}$, in 4.444 μs the nose of the Huns' rocket travels $\Delta x' = (240 \text{ m}/\mu\text{s})(4.444 \text{ } \mu\text{s}) = 1067 \text{ m}$—exactly the right distance to be at the tail of the Goths' rocket at $t'_1 = 0 \text{ } \mu\text{s}$. We could also note that the 667 m "miss distance" in the Goths' frame is length contracted to $(1 - (0.8)^2)^{1/2} (667 \text{ m}) = 400 \text{ m}$ in the Huns' frame—exactly the amount by which the Huns think they miss the Goths' rocket.

ASSESS Thus we end up with a consistent explanation. The Huns miss the Goths' rocket because, to them, the Goths' rocket is length contracted. The Goths find that the Huns miss because event 2 (the firing of the laser cannon) occurs before event 1 (the nose of one rocket passing the tail of the other). The 400 m distance of the miss in the Huns' reference frame is the length-contracted miss distance of 667 m in the Goths' reference frame.

SUMMARY

The goal of Chapter 36 has been to understand how Einstein's theory of relativity changes our concepts of space and time.

General Principles

Principle of Relativity All the laws of physics are the same in all inertial reference frames.

- The speed of light c is the same in all inertial reference frames.
- No particle or causal influence can travel at a speed greater than c.

Important Concepts

Space

Spatial measurements depend on the motion of the experimenter relative to the events. An object's length is the difference between *simultaneous* measurements of the positions of both ends.

Proper length ℓ is the length of an object measured in a reference frame in which the object is at rest. The object's length in a frame in which the object moves with velocity v is

$$L = \sqrt{1 - \beta^2}\,\ell \le \ell$$

This is called **length contraction.**

Time

Time measurements depend on the motion of the experimenter relative to the events. Events that are simultaneous in reference frame S are not simultaneous in frame S′ moving relative to S.

Proper time $\Delta\tau$ is the time interval between two events measured in a reference frame in which the events occur at the same position. The time interval between the events in a frame moving with relative velocity v is

$$\Delta t = \Delta\tau/\sqrt{1 - \beta^2} \ge \Delta\tau$$

This is called **time dilation.**

Momentum

The law of conservation of momentum is valid in all inertial reference frames if the momentum of a particle with velocity u is $p = \gamma_\text{p} m u$, where

$$\gamma_\text{p} = 1/\sqrt{1 - u^2/c^2}$$

The momentum approaches ∞ as $u \to c$.

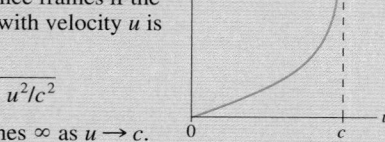

Energy

The law of conservation of energy is valid in all inertial reference frames if the energy of a particle with velocity u is $E = \gamma_\text{p} m c^2 = E_0 + K$.

Rest energy $E_0 = mc^2$

Kinetic energy $K = (\gamma_\text{p} - 1)mc^2$

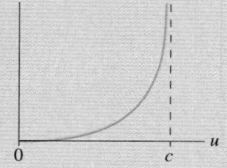

Invariants

Invariants are quantities that have the same value in all inertial reference frames.

Spacetime interval: $s^2 = (c\Delta t)^2 - (\Delta x)^2$

Particle rest energy: $E_0^2 = (mc^2)^2 = E^2 - (pc)^2$

Mass-energy equivalence

Mass m can be transformed into energy $E = mc^2$.

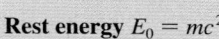

Energy can be transformed into mass $m = \Delta E/c^2$.

Applications

An event happens at a specific place in space and time. Spacetime coordinates are (x, t) in frame S and (x', t') in frame S′.

A reference frame is a coordinate system with meter sticks and clocks for measuring events. Experimenters at rest relative to each other share the same reference frame.

The Lorentz transformations transform spacetime coordinates and velocities between reference frames S and S′.

$$x' = \gamma(x - vt) \qquad x = \gamma(x' + vt')$$
$$y' = y \qquad\qquad y = y'$$
$$z' = z \qquad\qquad z = z'$$
$$t' = \gamma(t - vx/c^2) \qquad t = \gamma(t' + vx'/c^2)$$
$$u' = \frac{u - v}{1 - uv/c^2} \qquad u = \frac{u' + v}{1 + u'v/c^2}$$

where u and u' are the x- and x'-components of an object's velocity.

$$\beta = \frac{v}{c} \quad \text{and} \quad \gamma = 1/\sqrt{1 - v^2/c^2} = 1/\sqrt{1 - \beta^2}$$

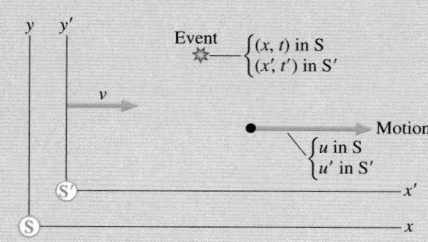

Terms and Notation

special relativity	spacetime coordinates,	time dilation	causal influence
reference frame	(x, y, z, t)	light year, ly	total energy, E
inertial reference frame	synchronized	proper length, ℓ	rest energy, E_0
Galilean principle of relativity	simultaneous	length contraction	law of conservation of total
ether	relativity of simultaneity	invariant	energy
principle of relativity	rest frame	spacetime interval, s	nuclear fission
event	proper time, $\Delta\tau$	Lorentz transformations	

CONCEPTUAL QUESTIONS

1. **FIGURE Q36.1** shows two balls. What are the speed and direction of each (a) in a reference frame that moves with ball 1 and (b) in a reference frame that moves with ball 2?

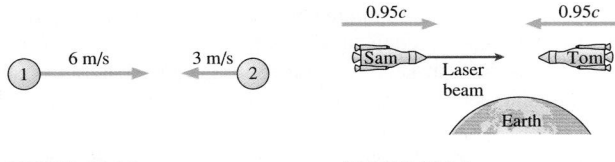

FIGURE Q36.1   **FIGURE Q36.2**

2. Teenagers Sam and Tom are playing chicken in their rockets. As **FIGURE Q36.2** shows, an experimenter on earth sees that each is traveling at $0.95c$ as he approaches the other. Sam fires a laser beam toward Tom.
   a. What is the speed of the laser beam relative to Sam?
   b. What is the speed of the laser beam relative to Tom?
3. Firecracker A is 300 m from you. Firecracker B is 600 m from you in the same direction. You see both explode at the same time. Define event 1 to be "firecracker A explodes" and event 2 to be "firecracker B explodes." Does event 1 occur before, after, or at the same time as event 2? Explain.
4. Firecrackers A and B are 600 m apart. You are standing exactly halfway between them. Your lab partner is 300 m on the other side of firecracker A. You see two flashes of light, from the two explosions, at exactly the same instant of time. Define event 1 to be "firecracker A explodes" and event 2 to be "firecracker B explodes." According to your lab partner, based on measurements he or she makes, does event 1 occur before, after, or at the same time as event 2? Explain.
5. **FIGURE Q36.5** shows Peggy standing at the center of her railroad car as it passes Ryan on the ground. Firecrackers attached to the ends of the car explode. A short time later, the flashes from the two explosions arrive at Peggy at the same time.

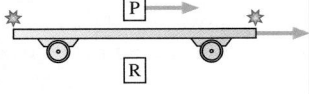

FIGURE Q36.5

   a. Were the explosions simultaneous in Peggy's reference frame? If not, which exploded first? Explain.
   b. Were the explosions simultaneous in Ryan's reference frame? If not, which exploded first? Explain.
6. **FIGURE Q36.6** shows a rocket traveling from left to right. At the instant it is halfway between two trees, lightning simultaneously (in the rocket's frame) hits both trees.

   a. Do the light flashes reach the rocket pilot simultaneously? If not, which reaches her first? Explain.
   b. A student was sitting on the ground halfway between the trees as the rocket passed overhead. According to the student, were the lightning strikes simultaneous? If not, which tree was hit first? Explain.

FIGURE Q36.6

7. Your friend flies from Los Angeles to New York. She carries an accurate stopwatch with her to measure the flight time. You and your assistants on the ground also measure the flight time.
   a. Identify the two events associated with this measurement.
   b. Who, if anyone, measures the proper time?
   c. Who, if anyone, measures the shorter flight time?
8. As the meter stick in **FIGURE Q36.8** flies past you, you simultaneously measure the positions of both ends and determine that $L < 1$ m.

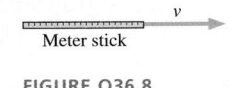

FIGURE Q36.8

   a. To an experimenter in frame S', the meter stick's frame, did you make your two measurements simultaneously? If not, which end did you measure first? Explain.
   b. Can experimenters in frame S' give an explanation for why your measurement is less than 1 m?
9. A 100-m-long train is heading for an 80-m-long tunnel. If the train moves sufficiently fast, is it possible, according to experimenters on the ground, for the entire train to be inside the tunnel at one instant of time? Explain.
10. Particle A has half the mass and twice the speed of particle B. Is the momentum p_A less than, greater than, or equal to p_B? Explain.
11. Event A occurs at spacetime coordinates (300 m, 2 μs).
    a. Event B occurs at spacetime coordinates (1200 m, 6 μs). Could A possibly be the cause of B? Explain.
    b. Event C occurs at spacetime coordinates (2400 m, 8 μs). Could A possibly be the cause of C? Explain.

EXERCISES AND PROBLEMS

Problems labeled [ ] integrate material from earlier chapters.

Exercises

Section 36.2 Galilean Relativity

1. ‖ At $t = 1.0$ s, a firecracker explodes at $x = 10$ m in reference frame S. Four seconds later, a second firecracker explodes at $x = 20$ m. Reference frame S′ moves in the x-direction at a speed of 5.0 m/s. What are the positions and times of these two events in frame S′?

2. ‖ A firecracker explodes in reference frame S at $t = 1.0$ s. A second firecracker explodes at the same position at $t = 3.0$ s. In reference frame S′, which moves in the x-direction at speed v, the first explosion is detected at $x′ = 4.0$ m and the second at $x′ = -4.0$ m.
   a. What is the speed of frame S′ relative to frame S?
   b. What is the position of the two explosions in frame S?

3. ‖ A sprinter crosses the finish line of a race. The roar of the crowd in front approaches her at a speed of 360 m/s. The roar from the crowd behind her approaches at 330 m/s. What are the speed of sound and the speed of the sprinter?

4. ‖ A baseball pitcher can throw a ball with a speed of 40 m/s. He is in the back of a pickup truck that is driving away from you. He throws the ball in your direction, and it floats toward you at a lazy 10 m/s. What is the speed of the truck?

5. ‖ A newspaper delivery boy is riding his bicycle down the street at 5.0 m/s. He can throw a paper at a speed of 8.0 m/s. What is the paper's speed relative to the ground if he throws the paper (a) forward, (b) backward, and (c) to the side?

Section 36.3 Einstein's Principle of Relativity

6. ‖ An out-of-control alien spacecraft is diving into a star at a speed of 1.0×10^8 m/s. At what speed, relative to the spacecraft, is the starlight approaching?

7. ‖ A starship blasts past the earth at 2.0×10^8 m/s. Just after passing the earth, it fires a laser beam out the back of the starship. With what speed does the laser beam approach the earth?

8. ‖ A positron moving in the positive x-direction at 2.0×10^8 m/s collides with an electron at rest. The positron and electron annihilate, producing two gamma-ray photons. Photon 1 travels in the positive x-direction and photon 2 travels in the negative x-direction. What is the speed of each photon?

Section 36.4 Events and Measurements

Section 36.5 The Relativity of Simultaneity

9. ‖ Your job is to synchronize the clocks in a reference frame. You are going to do so by flashing a light at the origin at $t = 0$ s. To what time should the clock at $(x, y, z) = (30$ m, 40 m, 0 m$)$ be preset?

10. ‖ Bjorn is standing at $x = 600$ m. Firecracker 1 explodes at the origin and firecracker 2 explodes at $x = 900$ m. The flashes from both explosions reach Bjorn's eye at $t = 3.0$ μs. At what time did each firecracker explode?

11. ‖ Bianca is standing at $x = 600$ m. Firecracker 1, at the origin, and firecracker 2, at $x = 900$ m, explode simultaneously. The flash from firecracker 1 reaches Bianca's eye at $t = 3.0$ μs. At what time does she see the flash from firecracker 2?

12. ‖ You are standing at $x = 9.0$ km. Lightning bolt 1 strikes at $x = 0$ km and lightning bolt 2 strikes at $x = 12.0$ km. Both flashes reach your eye at the same time. Your assistant is standing at $x = 3.0$ km. Does your assistant see the flashes at the same time? If not, which does she see first, and what is the time difference between the two?

13. ‖ You are standing at $x = 9.0$ km and your assistant is standing at $x = 3.0$ km. Lightning bolt 1 strikes at $x = 0$ km and lightning bolt 2 strikes at $x = 12.0$ km. You see the flash from bolt 2 at $t = 10$ μs and the flash from bolt 1 at $t = 50$ μs. According to your assistant, were the lightning strikes simultaneous? If not, which occurred first, and what was the time difference between the two?

14. ‖ Jose is looking to the east. Lightning bolt 1 strikes a tree 300 m from him. Lightning bolt 2 strikes a barn 900 m from him in the same direction. Jose sees the tree strike 1.0 μs before he sees the barn strike. According to Jose, were the lightning strikes simultaneous? If not, which occurred first, and what was the time difference between the two?

15. ‖ You are flying your personal rocketcraft at 0.9c from Star A toward Star B. The distance between the stars, in the stars' reference frame, is 1.0 ly. Both stars happen to explode simultaneously in your reference frame at the instant you are exactly halfway between them. Do you see the flashes simultaneously? If not, which do you see first, and what is the time difference between the two?

Section 36.6 Time Dilation

16. ‖ A cosmic ray travels 60 km through the earth's atmosphere in 400 μs, as measured by experimenters on the ground. How long does the journey take according to the cosmic ray?

17. ‖ At what speed, as a fraction of c, does a moving clock tick at half the rate of an identical clock at rest?

18. ‖ An astronaut travels to a star system 4.5 ly away at a speed of 0.9c. Assume that the time needed to accelerate and decelerate is negligible.
   a. How long does the journey take according to Mission Control on earth?
   b. How long does the journey take according to the astronaut?
   c. How much time elapses between the launch and the arrival of the first radio message from the astronaut saying that she has arrived?

19. ‖ a. How fast must a rocket travel on a journey to and from a distant star so that the astronauts age 10 years while the Mission Control workers on earth age 120 years?
   b. As measured by Mission Control, how far away is the distant star?

20. ‖ You fly 5000 km across the United States on an airliner at 250 m/s. You return two days later at the same speed.
   a. Have you aged more or less than your friends at home?
   b. By how much?
   Hint: Use the binomial approximation.

21. ‖ At what speed, in m/s, would a moving clock lose 1.0 ns in 1.0 day according to experimenters on the ground?
   Hint: Use the binomial approximation.

Section 36.7 Length Contraction

22. | At what speed, as a fraction of c, will a moving rod have a length 60% that of an identical rod at rest?

23. | Jill claims that her new rocket is 100 m long. As she flies past your house, you measure the rocket's length and find that it is only 80 m. Should Jill be cited for exceeding the $0.5c$ speed limit?

24. ‖ A muon travels 60 km through the atmosphere at a speed of $0.9997c$. According to the muon, how thick is the atmosphere?

25. ‖ A cube has a density of 2000 kg/m^3 while at rest in the laboratory. What is the cube's density as measured by an experimenter in the laboratory as the cube moves through the laboratory at 90% of the speed of light in a direction perpendicular to one of its faces?

26. | Our Milky Way galaxy is 100,000 ly in diameter. A spaceship crossing the galaxy measures the galaxy's diameter to be a mere 1.0 ly.
    a. What is the speed of the spaceship relative to the galaxy?
    b. How long is the crossing time as measured in the galaxy's reference frame?

27. ‖ A human hair is about 50 μm in diameter. At what speed, in m/s, would a meter stick "shrink by a hair"?
    Hint: Use the binomial approximation.

Section 36.8 The Lorentz Transformations

28. | An event has spacetime coordinates $(x, t) = (1200 \text{ m}, 2.0 \mu\text{s})$ in reference frame S. What are the event's spacetime coordinates (a) in reference frame S$'$ that moves in the positive x-direction at $0.8c$ and (b) in reference frame S$''$ that moves in the negative x-direction at $0.8c$?

29. ‖ A rocket travels in the x-direction at speed $0.6c$ with respect to the earth. An experimenter on the rocket observes a collision between two comets and determines that the spacetime coordinates of the collision are $(x', t') = (3.0 \times 10^{10} \text{ m}, 200 \text{ s})$. What are the spacetime coordinates of the collision in earth's reference frame?

30. ‖ In the earth's reference frame, a tree is at the origin and a pole is at $x = 30$ km. Lightning strikes both the tree and the pole at $t = 10 \mu$s. The lightning strikes are observed by a rocket traveling in the x-direction at $0.5c$.
    a. What are the spacetime coordinates for these two events in the rocket's reference frame?
    b. Are the events simultaneous in the rocket's frame? If not, which occurs first?

31. ‖ A rocket cruising past earth at $0.8c$ shoots a bullet out the back door, opposite the rocket's motion, at $0.9c$ relative to the rocket. What is the bullet's speed relative to the earth?

32. ‖ A laboratory experiment shoots an electron to the left at $0.9c$. What is the electron's speed relative to a proton moving to the right at $0.9c$?

33. ‖ A distant quasar is found to be moving away from the earth at $0.8c$. A galaxy closer to the earth and along the same line of sight is moving away from us at $0.2c$. What is the recessional speed of the quasar as measured by astronomers in the other galaxy?

Section 36.9 Relativistic Momentum

34. | A proton is accelerated to $0.999c$.
    a. What is the proton's momentum?
    b. By what factor does the proton's momentum exceed its Newtonian momentum?

35. ‖ At what speed is a particle's momentum twice its Newtonian value?

36. ‖‖ A 1.0 g particle has momentum 400,000 kg m/s. What is the particle's speed?

37. ‖ What is the speed of a particle whose momentum is mc?

Section 36.10 Relativistic Energy

38. | What are the kinetic energy, the rest energy, and the total energy of a 1.0 g particle with a speed of $0.8c$?

39. | A quarter-pound hamburger with all the fixings has a mass of 200 g. The food energy of the hamburger (480 food calories) is 2 MJ.
    a. What is the energy equivalent of the mass of the hamburger?
    b. By what factor does the energy equivalent exceed the food energy?

40. | How fast must an electron move so that its total energy is 10% more than its rest mass energy?

41. | At what speed is a particle's kinetic energy twice its rest energy?

42. ‖ At what speed is a particle's total energy twice its rest energy?

Problems

43. | A 50 g ball moving to the right at 4.0 m/s overtakes and collides with a 100 g ball moving to the right at 2.0 m/s. The collision is perfectly elastic. Use reference frames and the Chapter 10 result for perfectly elastic collisions to find the speed and direction of each ball after the collision.

44. | A billiard ball has a perfectly elastic collision with a second billiard ball of equal mass. Afterward, the first ball moves to the left at 2.0 m/s and the second to the right at 4.0 m/s. Use reference frames and the Chapter 10 result for perfectly elastic collisions to find the speed and direction of each ball before the collision.

45. ‖ The diameter of the solar system is 10 light hours. A spaceship crosses the solar system in 15 hours, as measured on earth. How long, in hours, does the passage take according to passengers on the spaceship?
    Hint: $c = 1$ light hour per hour.

46. | A 30-m-long rocket train car is traveling from Los Angeles to New York at $0.5c$ when a light at the center of the car flashes. When the light reaches the front of the car, it immediately rings a bell. Light reaching the back of the car immediately sounds a siren.
    a. Are the bell and siren simultaneous events for a passenger seated in the car? If not, which occurs first and by how much time?
    b. Are the bell and siren simultaneous events for a bicyclist waiting to cross the tracks? If not, which occurs first and by how much time?

47. ‖‖ The star Alpha goes supernova. Ten years later and 100 ly away, as measured by astronomers in the galaxy, star Beta explodes.
    a. Is it possible that the explosion of Alpha is in any way responsible for the explosion of Beta? Explain.
    b. An alien spacecraft passing through the galaxy finds that the distance between the two explosions is 120 ly. According to the aliens, what is the time between the explosions?

48. ‖ Two events in reference frame S occur 10 μs apart at the same point in space. The distance between the two events is 2400 m in reference frame S$'$.
    a. What is the time interval between the events in reference frame S$'$?
    b. What is the velocity of S$'$ relative to S?

49. ⫼ A starship voyages to a distant planet 10 ly away. The explorers stay 1 yr, return at the same speed, and arrive back on earth 26 yr after they left. Assume that the time needed to accelerate and decelerate is negligible.
    a. What is the speed of the starship?
    b. How much time has elapsed on the astronauts' chronometers?

50. ⫼ In Section 36.6 we saw that muons can reach the ground because of time dilation. But how do things appear in the muon's reference frame, where the muon's half-life is only 1.5 μs? How can a muon travel the 60 km to reach the earth's surface before decaying? Resolve this apparent paradox. Be as quantitative as you can in your answer.

51. ⫼ The Stanford Linear Accelerator (SLAC) accelerates electrons to $c = 0.99999997c$ in a 3.2-km-long tube. If they travel the length of the tube at full speed (they don't, because they are accelerating), how long is the tube in the electrons' reference frame?

52. ⫼ In an attempt to reduce the extraordinarily long travel times for voyaging to distant stars, some people have suggested traveling at close to the speed of light. Suppose you wish to visit the red giant star Betelgeuse, which is 430 ly away, and that you want your 20,000 kg rocket to move so fast that you age only 20 years during the round trip.
    a. How fast must the rocket travel relative to earth?
    b. How much energy is needed to accelerate the rocket to this speed?
    c. Compare this amount of energy to the total energy used by the United States in the year 2010, which was roughly 1.0×10^{20} J.

53. ⎮ A rocket traveling at $0.5c$ sets out for the nearest star, Alpha Centauri, which is 4.25 ly away from earth. It will return to earth immediately after reaching Alpha Centauri. What distance will the rocket travel and how long will the journey last according to (a) stay-at-home earthlings and (b) the rocket crew? (c) Which answers are the correct ones, those in part a or those in part b?

54. ⫼ The star Delta goes supernova. One year later and 2 ly away, as measured by astronomers in the galaxy, star Epsilon explodes. Let the explosion of Delta be at $x_D = 0$ and $t_D = 0$. The explosions are observed by three spaceships cruising through the galaxy in the direction from Delta to Epsilon at velocities $v_1 = 0.3c$, $v_2 = 0.5c$, and $v_3 = 0.7c$.
    a. What are the times of the two explosions as measured by scientists on each of the three spaceships?
    b. Does one spaceship find that the explosions are simultaneous? If so, which one?
    c. Does one spaceship find that Epsilon explodes before Delta? If so, which one?
    d. Do your answers to parts b and c violate the idea of causality? Explain.

55. ⫼ Two rockets approach each other. Each is traveling at $0.75c$ in the earth's reference frame. What is the speed of one rocket relative to the other?

56. ⫼ A rocket fires a projectile at a speed of $0.95c$ while traveling past the earth. An earthbound scientist measures the projectile's speed to be $0.90c$. What was the rocket's speed?

57. ⫼ Through what potential difference must an electron be accelerated, starting from rest, to acquire a speed of $0.99c$?

58. ⫼ What is the speed of a proton after being accelerated from rest through a 50×10^6 V potential difference?

59. ⫼ The half-life of a muon at rest is 1.5 μs. Muons that have been accelerated to a very high speed and are then held in a circular storage ring have a half-life of 7.5 μs.
    a. What is the speed of the muons in the storage ring?
    b. What is the total energy of a muon in the storage ring? The mass of a muon is 207 times the mass of an electron.

60. ⫼ A solar flare blowing out from the sun at $0.9c$ is overtaking a rocket as it flies away from the sun at $0.8c$. According to the crew on board, with what speed is the flare gaining on the rocket?

61. ⫼ This chapter has assumed that lengths perpendicular to the direction of motion are not affected by the motion. That is, motion in the x-direction does not cause length contraction along the y- or z-axes. To find out if this is really true, consider two spray-paint nozzles attached to rods perpendicular to the x-axis. It has been confirmed that, when both rods are at rest, both nozzles are exactly 1 m above the base of the rod. One rod is placed in the S reference frame with its base on the x-axis; the other is placed in the S' reference frame with its base on the x'-axis. The rods then swoop past each other and, as FIGURE P36.61 shows, each paints a stripe across the other rod.

We will use proof by contradiction. Assume that objects perpendicular to the motion *are* contracted. An experimenter in frame S finds that the S' nozzle, as it goes past, is less than 1 m above the x-axis. The principle of relativity says that an experiment carried out in two different inertial reference frames will have the same outcome in both.

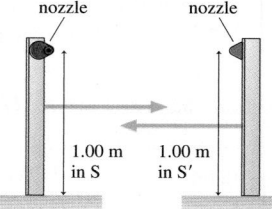

a. Pursue this line of reasoning and show that you end up with a logical contradiction, two mutually incompatible situations.
b. What can you conclude from this contradiction?

FIGURE P36.61

62. ⫼ Derive the Lorentz transformations for t and t'.
    Hint: See the comment following Equation 36.22.

63. ⫼ a. Derive a velocity transformation equation for u_y and u'_y. Assume that the reference frames are in the standard orientation with motion parallel to the x- and x'-axes.
    b. A rocket passes the earth at $0.8c$. As it goes by, it launches a projectile at $0.6c$ perpendicular to the direction of motion. What is the projectile's speed in the earth's reference frame?

64. ⫼ What is the momentum of a particle whose total energy is four times its rest energy? Give your answer as a multiple of mc.

65. ⫼ a. What are the momentum and total energy of a proton with speed $0.99c$?
    b. What is the proton's momentum in a different reference frame in which $E' = 5.0 \times 10^{-10}$ J?

66. ⫼ At what speed is the kinetic energy of a particle twice its Newtonian value?

67. ⎮ A typical nuclear power plant generates electricity at the rate of 1000 MW. The efficiency of transforming thermal energy into electrical energy is $\frac{1}{3}$ and the plant runs at full capacity for 80% of the year. (Nuclear power plants are down about 20% of the time for maintenance and refueling.)
    a. How much thermal energy does the plant generate in one year?
    b. What mass of uranium is transformed into energy in one year?

68. ⎮ The sun radiates energy at the rate 3.8×10^{26} W. The source of this energy is fusion, a nuclear reaction in which mass is transformed into energy. The mass of the sun is 2.0×10^{30} kg.
    a. How much mass does the sun lose each year?
    b. What percent is this of the sun's total mass?
    c. Estimate the lifetime of the sun.

69. ‖ The radioactive element radium (Ra) decays by a process known as *alpha decay,* in which the nucleus emits a helium nucleus. (These high-speed helium nuclei were named alpha particles when radioactivity was first discovered, long before the identity of the particles was established.) The reaction is $^{226}\text{Ra} \rightarrow ^{222}\text{Rn} + ^{4}\text{He}$, where Rn is the element radon. The accurately measured atomic masses of the three atoms are 226.025, 222.017, and 4.003. How much energy is released in each decay? (The energy released in radioactive decay is what makes nuclear waste "hot.")

70. ‖ The nuclear reaction that powers the sun is the fusion of four protons into a helium nucleus. The process involves several steps, but the net reaction is simply $4p \rightarrow ^{4}\text{He} + \text{energy}$. The mass of a helium nucleus is known to be 6.64×10^{-27} kg.
    a. How much energy is released in each fusion?
    b. What fraction of the initial rest mass energy is this energy?

71. ‖‖ An electron moving to the right at $0.9c$ collides with a positron moving to the left at $0.9c$. The two particles annihilate and produce two gamma-ray photons. What is the wavelength of the photons?

72. ‖ Consider the inelastic collision $e^- + e^- \rightarrow e^- + e^- + e^- + e^+$ in which an electron-positron pair is produced in a head-on collision between two electrons moving in opposite directions at the same speed. This is similar to Figure 36.39, but both of the initial electrons are moving.
    a. What is the threshold kinetic energy? That is, what minimum kinetic energy must each electron have to allow this process to occur?
    b. What is the speed of an electron with this kinetic energy?

Challenge Problems

73. Two rockets, A and B, approach the earth from opposite directions at speed $0.8c$. The length of each rocket measured in its rest frame is 100 m. What is the length of rocket A as measured by the crew of rocket B?

74. Two rockets are each 1000 m long in their rest frame. Rocket Orion, traveling at $0.8c$ relative to the earth, is overtaking rocket Sirius, which is poking along at a mere $0.6c$. According to the crew on Sirius, how long does Orion take to completely pass? That is, how long is it from the instant the nose of Orion is at the tail of Sirius until the tail of Orion is at the nose of Sirius?

75. Some particle accelerators allow protons (p^+) and antiprotons (p^-) to circulate at equal speeds in opposite directions in a device called a *storage ring.* The particle beams cross each other at various points to cause $p^+ + p^-$ collisions. In one collision, the outcome is $p^+ + p^- \rightarrow e^+ + e^- + \gamma + \gamma$, where γ represents a high-energy gamma-ray photon. The electron and positron are ejected from the collision at $0.9999995c$ and the gamma-ray photon wavelengths are found to be 1.0×10^{-6} nm. What were the proton and antiproton speeds prior to the collision?

76. A very fast pole vaulter lives in the country. One day, while practicing, he notices a 10.0-m-long barn with the doors open at both ends. He decides to run through the barn at $0.866c$ while carrying his 16.0-m-long pole. The farmer, who sees him coming, says, "Aha! This guy's pole is length contracted to 8.0 m. There will be a short interval of time when the pole is entirely inside the barn. If I'm quick, I can simultaneously close both barn doors while the pole vaulter and his pole are inside." The pole vaulter, who sees the farmer beside the barn, thinks to himself, "That farmer is crazy. The barn is length contracted and is only 5.0 m long. My 16.0-m-long pole cannot fit into a 5.0-m-long barn. If the farmer closes the doors just as the tip of my pole reaches the back door, the front door will break off the last 11.0 m of my pole."

Can the farmer close the doors without breaking the pole? Show that, when properly analyzed, the farmer and the pole vaulter agree on the outcome. Your analysis should contain both quantitative calculations and written explanation.

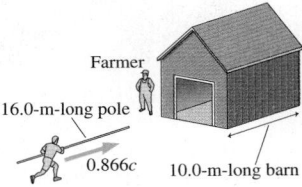

FIGURE CP36.76

Stop to Think 36.1: a, c, and f. These move at constant velocity, or very nearly so. The others are accelerating.

Stop to Think 36.2: a. $u' = u - v = -10$ m/s $- 6$ m/s $= -16$ m/s. The *speed* is 16 m/s.

Stop to Think 36.3: c. Even the light has a slight travel time. The event is the hammer hitting the nail, not your seeing the hammer hit the nail.

Stop to Think 36.4: At the same time. Mark is halfway between the tree and the pole, so the fact that he *sees* the lightning bolts at the same time means they *happened* at the same time. It's true that Nancy *sees* event 1 before event 2, but the events actually occurred before she sees them. Mark and Nancy share a reference frame, because they are at rest relative to each other, and all experimenters in a reference frame, after correcting for any signal delays, *agree* on the spacetime coordinates of an event.

Stop to Think 36.5: After. This is the same as the case of Peggy and Ryan. In Mark's reference frame, as in Ryan's, the events are simultaneous. Nancy *sees* event 1 first, but the time when an event is seen is not when the event actually happens. Because all experimenters in a reference frame agree on the spacetime coordinates of an event, Nancy's position in her reference frame cannot affect the order of the events. If Nancy had been passing Mark at the instant the lightning strikes occur in Mark's frame, then Nancy would be equivalent to Peggy. Event 2, like the firecracker at the front of Peggy's railroad car, occurs first in Nancy's reference frame.

Stop to Think 36.6: c. Nick measures proper time because Nick's clock is present at both the "nose passes Nick" event and the "tail passes Nick" event. Proper time is the smallest measured time interval between two events.

Stop to Think 36.7: $L_A > L_B = L_C$. Anjay measures the pole's proper length because it is at rest in his reference frame. Proper length is the longest measured length. Beth and Charles may *see* the pole differently, but they share the same reference frame and their *measurements* of the length agree.

Stop to Think 36.8: c. The rest energy E_0 is an invariant, the same in all inertial reference frames. Thus $m = E_0/c^2$ is independent of speed.

37 The Foundations of Modern Physics

Studies of the light emitted by gas discharge tubes helped lay the foundations of modern physics.

▶ **Looking Ahead** The goal of Chapter 37 is to learn about the structure and properties of atoms.

Matter and Light

Except for relativity, everything you've studied until now was known by 1900. But within the span of just a few years, right around 1900, investigations into the structure of matter and the properties of light led to many discoveries at odds with classical physics.

Our goal is to establish the experimental basis for new theories of matter and light that arose in the first decades of the 20th century. We cannot see atoms, so what is the *evidence* for our current understanding of the atomic structure of matter?

Electrons

Experiments to study electrical conduction in gases found that unknown "rays" travel outward from the cathode—the negative electrode. You'll learn how J. J. Thomson discovered that these **cathode rays** are a stream of subatomic particles—electrons.

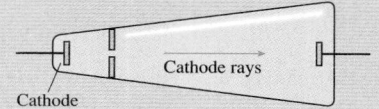

A cathode-ray tube, or CRT, was, until very recently, the "picture tube" of televisions and computer monitors.

X Rays

At very high voltages, the cathode-ray tube itself is a source of rays, highly penetrating rays called **x rays.** You'll learn that x rays are electromagnetic waves with wavelengths much shorter than those of visible light.

The penetrating power of x rays led to their use in medicine almost immediately. Today, x rays have applications ranging from inspecting machine parts to deciphering the structure of biological molecules.

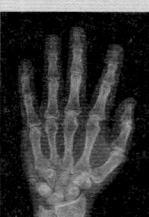

Emission and Absorption

You'll learn how **spectroscopy** is based on the interference of light and how scientists used spectroscopy to study the emission and absorption of light. This provided many new clues about the structure of matter.

Atoms emit a **discrete spectrum** consisting of many discrete wavelengths. Each element has a unique spectrum, an *electromagnetic fingerprint* that can identify that element.

The Nucleus

How are atoms built? You'll learn how Ernest Rutherford used the particles emitted in radioactive decay to discover that atoms have an incredibly tiny **nucleus.**

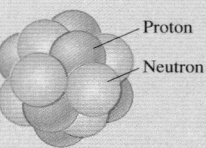

The nucleus, which is unbelievably dense, consists of positive protons and neutral neutrons.

◀ **Looking Back**
Section 28.1–28.2 Electric potential energy

Rutherford's Model of the Atom

The discovery of the electron and the nucleus led Rutherford to propose a solar-system model of the atom in which negative electrons orbit a tiny, dense, positive nucleus.

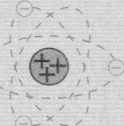

An element's atomic number is the number of protons in the nucleus. An atom with three protons is lithium.

◀ **Looking Back**
Section 16.2 Atomic masses

37.1 **Matter and Light**

The idea that matter consists of small, indivisible particles can be traced to the Greek philosophers Leucippus and Democritus in the 5th century BCE. They called these particles *atoms*, Greek for "not divisible." Atomism was not widely accepted, due in no small part to the complete lack of evidence, but atomic ideas never died.

Things began to change in the early years of the 19th century. The English chemist John Dalton argued that chemical reactions could be understood if each chemical element consisted of identical atoms. An important feature of Dalton's work, which made it more science than speculation, was his experimental effort to determine the relative masses of the atoms of different elements.

The evidence for atoms grew stronger as thermodynamics and the kinetic theory of gases developed in the mid-19th century. Slight deviations from the ideal-gas law at high pressures, which could be understood if the atoms were beginning to come into close proximity to one another, led to a rough estimate of atomic sizes. By 1890, it was widely accepted that atoms exist and have diameters of approximately 10^{-10} m.

Other scientists of the 19th century were trying to understand what light is. Newton, as we have noted, favored a *corpuscular* theory whereby small particles of light travel in straight lines. However, the situation changed when, in 1801, Thomas Young demonstrated the interference of light with his celebrated double-slit experiment. But if light is a wave, what is waving? Studies of electricity and magnetism by Ampère, Faraday, and others led Maxwell to the realization that light is an electromagnetic wave.

This was the situation at the end of the 19th century, when a series of discoveries began to reveal that the theories of Newton and Maxwell were not sufficient to explain the properties of atoms. New theories of matter and light at the atomic level, collectively called *modern physics*, arose in the early decades of the 20th century. Modern physics, including relativity and quantum physics, is our topic for the final part of this textbook.

A difficulty, however, is that we cannot directly see, feel, or manipulate atoms. To know what the theories of modern physics are attempting to explain, and whether they are successful, we must start with *experimental evidence* about the behavior of atoms and light. That is the primary purpose of this chapter and the next.

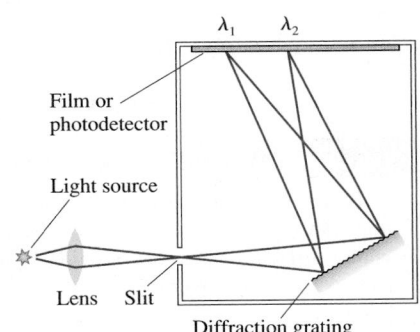

FIGURE 37.1 A grating spectrometer is used to study the emission of light.

37.2 **The Emission and Absorption of Light**

The interference and diffraction of light, discovered early in the 19th century, soon led to practical tools for measuring the wavelengths of light. The most important tool, still widely used today, is the **spectrometer,** such as the one shown in FIGURE 37.1. The heart of a spectrometer is a diffraction grating that diffracts different wavelengths of light at different angles. Making the grating slightly curved, like a spherical mirror, focuses the interference fringes onto a *photographic plate* or (more likely today) an electronic array detector. Each wavelength in the light is focused to a different position on the detector, producing a distinctive pattern called the **spectrum** of the light. Spectroscopists discovered very early that there are two distinct types of spectra.

Continuous Spectra and Blackbody Radiation

Cool lava is black, but lava heated to a high temperature glows red and, if hot enough, yellow. A tungsten wire, dark gray at room temperature, emits bright white light when heated by a current—thus becoming the bright filament in an incandescent lightbulb. Hot, self-luminous objects, such as the lava or the lightbulb, form a rainbow-like **continuous spectrum** in which light is emitted at every possible wavelength. FIGURE 37.2 is a continuous spectrum.

FIGURE 37.2 The continuous spectrum of an incandescent lightbulb.

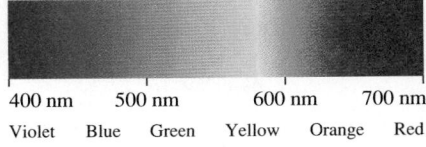

400 nm   500 nm   600 nm   700 nm
Violet   Blue   Green   Yellow   Orange   Red

Black lava glows brightly when hot.

This temperature-dependent emission of electromagnetic waves was called *thermal radiation* when we studied it as the mechanism of heat transfer in Chapter 17. Recall that the heat energy Q radiated in a time interval Δt by an object with surface area A and absolute temperature T is given by

$$\frac{Q}{\Delta t} = e\sigma A T^4 \tag{37.1}$$

where $\sigma = 5.67 \times 10^{-8}$ W/m^{2}K^4 is the Stefan-Boltzmann constant. Notice the very strong fourth-power dependence on temperature.

The parameter e in Equation 37.1 is the *emissivity* of the surface, a measure of how effectively it radiates. A perfectly absorbing—and thus perfectly emitting—object with $e = 1$ is called a *blackbody*, and the thermal radiation emitted by a blackbody is called **blackbody radiation.** Charcoal is an excellent approximation of a blackbody.

Our interest in Chapter 17 was the amount of energy radiated. Now we want to examine the spectrum of that radiation. If we measure the spectrum of a blackbody at three temperatures, 3500 K, 4500 K, and 5500 K, the data appear as in FIGURE 37.3. These continuous curves are called *blackbody spectra*. There are four important features of the spectra:

FIGURE 37.3 Blackbody radiation spectra.

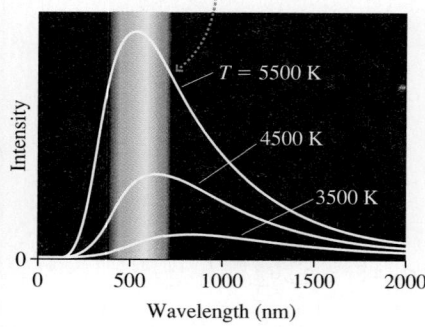

- All blackbodies at the same temperature emit exactly the same spectrum. **The spectrum depends on only an object's temperature, not the material of which it is made.**
- Increasing the temperature increases the radiated intensity at *all* wavelengths. **Making the object hotter causes it to emit more radiation across the entire spectrum.**
- Increasing the temperature causes the peak intensity to shift toward shorter wavelengths. **The higher the temperature, the shorter the wavelength of the peak of the spectrum.**
- The visible rainbow that we see is only a small portion of the continuous blackbody spectrum. Much of the emission is infrared. Extremely hot objects, such as stars, emit a significant fraction of their radiation at ultraviolet wavelengths.

The wavelength corresponding to the peak of the intensity graph is given by

$$\lambda_{\text{peak}}(\text{in nm}) = \frac{2.90 \times 10^6 \text{ nm K}}{T} \tag{37.2}$$

where T must be in kelvin. Equation 37.2 is known as **Wien's law.**

EXAMPLE 37.1 **Finding peak wavelengths**

What are the peak wavelengths and the corresponding spectral regions for thermal radiation from the sun, a glowing ball of gas with a surface temperature of 5800 K, and from the earth, whose average surface temperature is 15°C?

MODEL The sun and the earth are well approximated as blackbodies.

SOLVE The sun's wavelength of peak intensity is given by Wien's law:

$$\lambda_{\text{peak}} = \frac{2.90 \times 10^6 \text{ nm K}}{5800 \text{ K}} = 500 \text{ nm}$$

This is right in the middle of the visible spectrum. The earth's wavelength of peak intensity is

$$\lambda_{\text{peak}} = \frac{2.90 \times 10^6 \text{ nm K}}{288 \text{ K}} = 10,000 \text{ nm}$$

where we converted the surface temperature to kelvin before computing. This is rather far into the infrared portion of the spectrum, which is not surprising because we don't "see" the earth glowing.

ASSESS The difference between these two wavelengths is quite important for understanding the earth's greenhouse effect. Most of the energy from the sun—its spectrum is much like the highest curve in Figure 37.3—arrives as visible light. The earth's atmosphere is transparent to visible wavelengths, so this energy reaches the ground and is absorbed. The earth must radiate an equal amount of energy back to space, but it does so with long-wavelength infrared radiation. These wavelengths are strongly absorbed by some gases in the atmosphere, so the atmosphere acts as a blanket to keep the earth's surface warmer than it would be otherwise.

That all blackbodies at the same temperature emit the same spectrum was an unexpected discovery. Why should this be? It seemed that a combination of thermodynamics and Maxwell's new theory of electromagnetic waves ought to provide a convincing explanation, but scientists of the late 19th century failed to come up with a theoretical justification for the curves seen in Figure 37.3.

Discrete Spectra

Michael Faraday, who discovered electromagnetic induction, wanted to know whether an electric current could pass through a gas. To find out, he sealed metal electrodes into a glass tube, lowered the pressure with a primitive vacuum pump, and then attached an electrostatic generator. When he started the generator, the gas inside the tube began to glow with a bright purple color! Faraday's device, called a **gas discharge tube,** is shown in FIGURE 37.4.

The purple color Faraday saw is characteristic of nitrogen, the primary component of air. You are more likely familiar with the reddish-orange color of a neon discharge tube, but neon was not discovered until long after Faraday's time. If light from a neon discharge tube is passed through a spectrometer, it produces the spectrum seen in FIGURE 37.5. This is called a **discrete spectrum** because it contains only discrete, individual wavelengths. Further, each kind of gas emits a unique spectrum—a *spectral fingerprint*—that distinguishes it from every other gas.

The discrete emission spectrum of a hot, low-density gas stands in sharp contrast to the continuous blackbody spectrum of a glowing solid. Not only do gases emit discrete wavelengths, but it was soon discovered that they also absorb discrete wavelengths. FIGURE 37.6a shows an absorption experiment in which white light passes through a sample of gas. Without the gas, the white light would expose the film with a continuous rainbow spectrum. Any wavelengths absorbed by the gas are missing, and the film is dark at that wavelength. FIGURE 37.6b shows, for sodium vapor, that only certain discrete wavelengths are absorbed.

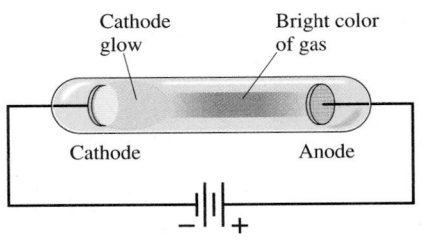
FIGURE 37.4 Faraday's gas discharge tube.

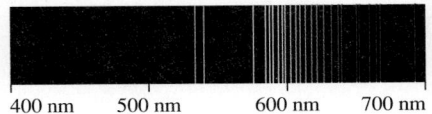
FIGURE 37.5 The discrete spectrum of a neon discharge tube.

FIGURE 37.6 Measuring an absorption spectrum.

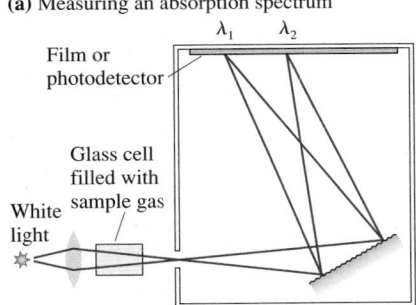
(a) Measuring an absorption spectrum

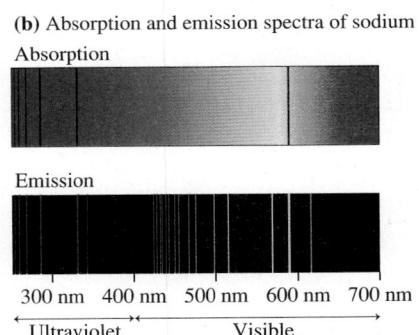
(b) Absorption and emission spectra of sodium

Although the emission and absorption spectra of a gas are both discrete, there is an important difference: **Every wavelength absorbed by the gas is also emitted, but *not every emitted wavelength is absorbed.*** Figure 37.6b shows that the wavelengths in the absorption spectrum are a subset of those in the emission spectrum. All the absorption wavelengths are prominent in the emission spectrum, but there are many emission wavelengths for which no absorption occurs.

What causes atoms to emit or absorb light? Why a discrete spectrum? Why are some wavelengths emitted but not absorbed? Why is each element different? Nineteenth-century physicists struggled with these questions but could not answer them. Ultimately, their inability to understand the emission and absorption of light forced scientists to the unwelcome realization that classical physics was simply incapable of providing an understanding of atoms.

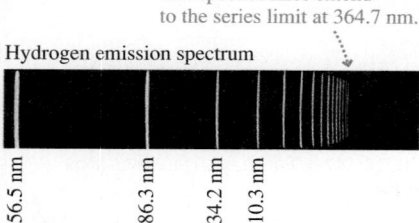

The spectral lines extend to the series limit at 364.7 nm.

Hydrogen emission spectrum

656.5 nm
486.3 nm
434.2 nm
410.3 nm

The only encouraging sign came from an unlikely source. While the spectra of other atoms have dozens or even hundreds of wavelengths, the emission spectrum of hydrogen, seen in **FIGURE 37.7**, is very simple and regular. If any spectrum could be understood, it should be that of the first element in the periodic table. The breakthrough came in 1885, not by an established and recognized scientist but by a Swiss schoolteacher, Johann Balmer. Balmer showed that the wavelengths in the hydrogen spectrum could be represented by the simple formula

$$\lambda = \frac{91.18 \text{ nm}}{\left(\dfrac{1}{2^2} - \dfrac{1}{n^2}\right)} \qquad n = 3, 4, 5, \ldots \qquad (37.3)$$

This formula predicts a series of spectral lines of gradually decreasing wavelength, converging to the series limit wavelength of 364.7 nm as $n \rightarrow \infty$. This series of spectral lines is now called the **Balmer series.**

Later experimental evidence, as ultraviolet and infrared spectroscopy developed, showed that Balmer's result could be generalized to

$$\lambda = \frac{91.18 \text{ nm}}{\left(\dfrac{1}{m^2} - \dfrac{1}{n^2}\right)} \qquad m = 1, 2, 3, \ldots \qquad n = m + 1, m + 2, \ldots \qquad (37.4)$$

We now refer to Equation 37.4 as the **Balmer formula,** although Balmer himself suggested only the original version of Equation 37.3 in which $m = 2$. Other than at the highest levels of resolution, where new details appear that need not concern us in this text, the Balmer formula accurately describes *every* wavelength in the emission spectrum of hydrogen.

The Balmer formula is what we call *empirical knowledge*. It is an accurate mathematical representation found empirically—that is, through experimental evidence—but it does not rest on any physical principles or physical laws. Yet the formula was so simple that it must, everyone agreed, have a simple explanation. It would take 30 years to find it.

STOP TO THINK 37.1 These spectra are due to the same element. Which one is an emission spectrum and which is an absorption spectrum?

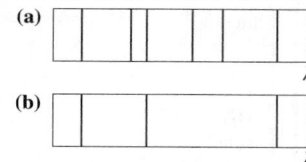

(a)

λ

(b)

λ

37.3 Cathode Rays and X Rays

Faraday's invention of the gas discharge tube had two major repercussions. One set of investigations, as we've seen, led to the development of spectroscopy. Another set led to the discovery of the electron.

In addition to the bright color of the gas in a discharge tube, Figure 37.4 shows a separate, constant glow around the negative electrode (i.e., the cathode) called the **cathode glow.** As vacuum technology improved, scientists made two discoveries:

1. At lower pressures, the cathode glow became more extended.
2. If the cathode glow extended to the wall of the glass tube, the glass itself emitted a greenish glow—*fluorescence*—at that point.

In fact, a solid object sealed inside a low-pressure tube casts a *shadow* on the glass wall, as shown in **FIGURE 37.8**. This suggests that the cathode emits *rays* of some form that

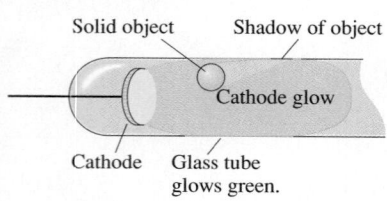

Solid object   Shadow of object

Cathode glow

Cathode   Glass tube glows green.

travel in straight lines but are easily blocked. These rays, which are invisible but cause the glass to glow where they strike it, were quickly dubbed **cathode rays.** This name lives on today in the *cathode-ray tube* that forms the picture tube in older televisions and computer-display terminals. But naming the rays did nothing to explain them. What were they?

Crookes Tubes

The most systematic studies on the new cathode rays were carried out during the 1870s by the English scientist Sir William Crookes. Crookes devised a set of glass tubes, such as the one shown in **FIGURE 37.9**, that could be used to make careful studies of cathode rays. These tubes, today called **Crookes tubes,** generate a small glowing spot where the cathode rays strike the face of the tube.

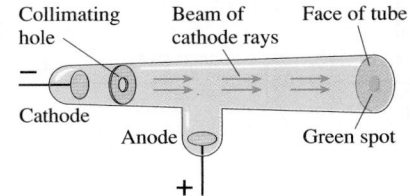

FIGURE 37.9 A Crookes tube.

The work of Crookes and others demonstrated that

1. There is an electric current in a tube in which cathode rays are emitted.
2. The rays are deflected by a magnetic field *as if* they are negative charges.
3. Cathodes made of any metal produce cathode rays. Furthermore, the ray properties are independent of the cathode material.

Crookes's experiments led to more questions than they answered. Were the cathode rays some sort of particles? Or a wave? Were the rays themselves the carriers of the electric current, or were they something else that happened to be emitted whenever there was a current? Item 3 is worthy of note because it suggests that the cathode rays are a *fundamental* entity, not a part of the element from which they are emitted.

It is important to realize how difficult these questions were at the time and how experimental evidence was used to answer them. Crookes suggested that molecules in the gas collided with the cathode, somehow acquired a negative charge (i.e., became negative ions), and then "rebounded" with great speed as they were repelled by the negative cathode. These "charged molecules" would travel in a straight line, be deflected by a magnetic field, and cause the tube to glow where they struck the glass. Crookes's theory predicted, of course, that the negative ions should also be deflected by an electric field, but his experimental efforts were inconclusive. Otherwise, Crookes's model seemed to explain the observations.

However, Crookes's theory was immediately attacked. Critics noted that the cathode rays could travel the length of a 90-cm-long tube with no discernible deviation from a straight line. But the mean free path for molecules, due to collisions with other molecules, is only about 6 mm at the pressure in Crookes's tubes. There was no chance at all that molecules could travel in a straight line for 150 times their mean free path! Crookes's theory, seemingly adequate when it was proposed, was wildly inconsistent with subsequent observations.

But if cathode rays were not particles, what were they? An alternative theory was that the cathode rays were electromagnetic waves. After all, light travels in straight lines, casts shadows, and can, under the right circumstances, cause materials to fluoresce. It was known that hot metals emit light—incandescence—so it seemed plausible that the cathode could be emitting waves. The major obstacle for the wave theory was the deflection of cathode rays by a magnetic field. But the theory of electromagnetic waves was quite new at the time, and many characteristics of these waves were still unknown. Visible light was not deflected by a magnetic field, but it was easy to think that some other form of electromagnetic waves might be so influenced.

The controversy over particles versus waves was intense. British scientists generally favored particles, but their continental counterparts preferred waves. Such controversies are an integral part of science, for they stimulate the best minds to come forward with new ideas and new experiments.

X Rays

FIGURE 37.10 Röntgen's x-ray tube.

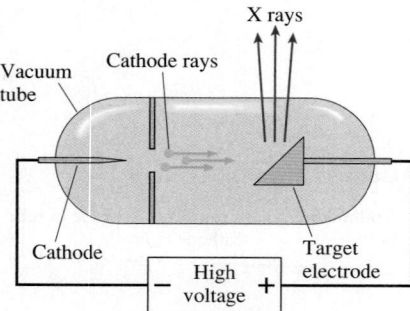

FIGURE 37.10 Röntgen's x-ray tube.

The German physicist Wilhelm Röntgen, also studying cathode rays, made a remarkable discovery in 1895. He had sealed a cathode and a metal target electrode into a vacuum tube, as shown in FIGURE 37.10, and then applied a much higher voltage than normally used to produce cathode rays. He happened, by chance, to leave a sealed envelope containing photographic film near the vacuum tube, and was later surprised to discover that the film had been exposed. This serendipitous discovery was the beginning of the study of x rays.

Röntgen quickly found that the vacuum tube was the source of whatever was exposing the film. Not having any idea what was coming from the tube, he called them **x rays,** using the algebraic symbol x as meaning "unknown." X rays were unlike anything, particle or wave, ever discovered. Röntgen was not successful at reflecting the rays or at focusing the rays with a lens. He showed that they travel in straight lines, like particles, but they also pass right through most solid materials, something no known particle could do.

Scientists soon began to suspect that x rays were an electromagnetic wave with a wavelength much shorter than that of visible light. However, it wasn't until 20 years after their discovery that this was verified by the diffraction of x rays, showing that they have wavelengths in the range 0.01 nm to 10 nm. The production and the properties of x rays seemed far outside the scope of Maxwell's theory of electromagnetic waves.

37.4 The Discovery of the Electron

Shortly after Röntgen's discovery of x rays, the young English physicist J. J. Thomson began using them to study electrical conduction in gases. He found that x rays could discharge an electroscope and concluded that they must be ionizing the air molecules, thereby making the air conductive.

This simple observation was of profound significance. Until then, the only form of ionization known was the creation of positive and negative ions in solutions where, for example, a molecule such as NaCl splits into two smaller charged pieces. Although the underlying process was not yet understood, the fact that two atoms could acquire charge as a molecule splits apart did not jeopardize the idea that the atoms themselves were indivisible. But after observing that even monatomic gases, such as helium, could be ionized by x rays, Thomson realized that **the atom itself must have charged constituents that could be separated!** This was the first direct evidence that the atom is a complex structure, not a fundamental, indivisible unit of matter.

Thomson was also investigating the nature of cathode rays. Other scientists, using a Crookes tube like the one shown in FIGURE 37.11a, had measured an electric current in a cathode-ray beam. Although its presence seemed to demonstrate that the rays are charged particles, proponents of the wave model argued that the current might be a separate, independent event that just happened to be following the same straight line as the cathode rays.

Thomson realized that he could use magnetic deflection of the cathode rays to settle the issue. He built a modified tube, shown in FIGURE 37.11b, in which the collecting electrode was off to the side. With no magnetic field, the cathode rays struck the center of the tube face and created a greenish spot on the glass. No current was measured by the electrode under these circumstances. Thomson then placed the tube in a magnetic field to deflect the cathode rays to the side. He could determine their trajectory by the location of the green spot as it moved across the face of the tube. Just at the point when the field was strong enough to deflect the cathode rays onto the electrode, a current was detected! At an even stronger field, when the cathode rays were deflected completely to the other side of the electrode, the current ceased.

FIGURE 37.11 Experiments to measure the electric current in a cathode-ray tube.

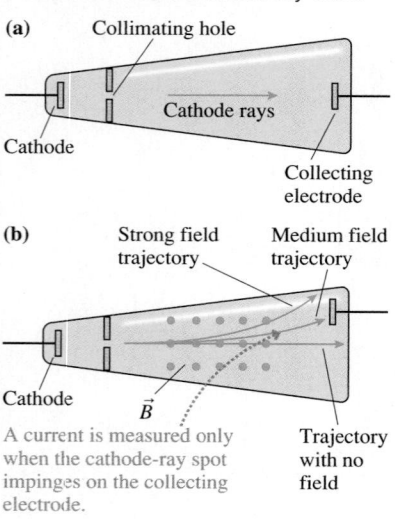

A current is measured only when the cathode-ray spot impinges on the collecting electrode.

This was the first conclusive demonstration that cathode rays really are negatively charged particles. But why were they not deflected by an electric field? Thomson's experience with the x-ray ionization of gases soon led him to recognize that the rapidly moving cathode-ray particles must be colliding with the few remaining gas molecules in the tube with sufficient energy to *ionize* them by splitting them into charged pieces. The electric field created by these charges neutralized the field of the electrodes, hence there was no deflection.

Fortunately, vacuum technology was getting ever better. By using the most sophisticated techniques of his day, Thomson was able to lower the pressure to a point where ionization of the gas was not a problem. Then, just as he had expected, the cathode rays *were* deflected by an electric field!

Thomson's experiment was a decisive victory for the charged-particle model, but it still did not indicate anything about the nature of the particles. What were they?

J. J. Thomson.

Thomson's Crossed-Field Experiment

Thomson could measure the deflection of cathode-ray particles for various strengths of the magnetic field, but magnetic deflection depends both on the particle's charge-to-mass ratio q/m *and* on its speed. Measuring the charge-to-mass ratio, and thus learning something about the particles themselves, requires some means of determining their speed. To do so, Thomson devised the experiment for which he is most remembered.

Thomson built a tube containing the parallel-plate electrodes visible in the photo in FIGURE 37.12a. He then placed the tube in a magnetic field. FIGURE 37.12b shows that the electric and magnetic fields were perpendicular to each other, thus creating what came to be known as a **crossed-field experiment.**

The magnetic field, which is perpendicular to the particle's velocity $\vec{v}$, exerts a magnetic force on the charged particle of magnitude

$$F_B = qvB \tag{37.5}$$

The magnetic field alone would cause a negatively charged particle to move along an *upward* circular arc. The particle doesn't move in a complete circle because the velocity is large and because the magnetic field is limited in extent. As you learned in Chapter 32, the radius of the arc is

$$r = \frac{mv}{qB} \tag{37.6}$$

The net result is to *deflect* the beam of particles upward. It is a straightforward geometry problem to determine the radius of curvature r from the measured deflection.

Thomson's new idea was to create an electric field between the parallel-plate electrodes that would exert a *downward* force on the negative charges, pushing them back toward the center of the tube. The magnitude of the electric force on each particle is

$$F_E = qE \tag{37.7}$$

Thomson adjusted the electric field strength until the cathode-ray beam, in the presence of both electric and magnetic fields, had no deflection and was seen exactly in the center of the tube.

Zero deflection occurs when the magnetic and electric forces exactly balance each other, as FIGURE 37.12c shows. The force vectors point in opposite directions, and their magnitudes are equal when

$$F_B = qvB = F_E = qE$$

Notice that the charge q cancels. Once E and B are set, a charged particle can pass undeflected through the crossed fields only if its speed is

$$v = \frac{E}{B} \tag{37.8}$$

FIGURE 37.12 Thomson's crossed-field experiment to measure the velocity of cathode rays. The photograph shows his original tube and the coils he used to produce the magnetic field.

(a)

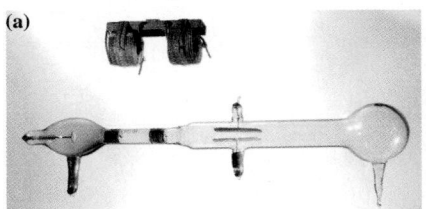

(b)

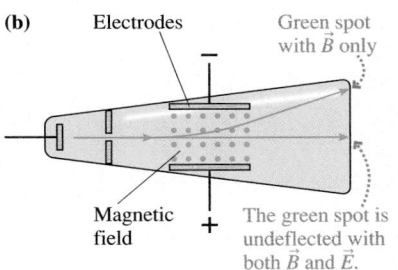

(c)

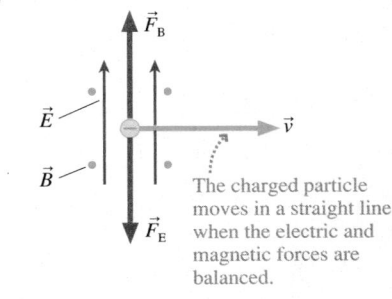

The charged particle moves in a straight line when the electric and magnetic forces are balanced.

By balancing the magnetic force against the electric force, Thomson could determine the speed of the charged-particle beam. Once he knew v, he could use Equation 37.6 to find the charge-to-mass ratio:

$$\frac{q}{m} = \frac{v}{rB} \tag{37.9}$$

Thomson found that the charge-to-mass ratio of cathode rays is $q/m \approx 1 \times 10^{11}$ C/kg. This seems not terribly accurate in comparison to the modern value of 1.76×10^{11} C/kg, but keep in mind both the experimental limitations of his day and the fact that, prior to his work, no one had *any* idea of the charge-to-mass ratio.

EXAMPLE 37.2 A crossed-field experiment

An electron is fired between two parallel-plate electrodes that are 5.0 mm apart and 3.0 cm long. A potential difference ΔV between the electrodes establishes an electric field between them. A 3.0-cm-wide, 1.0 mT magnetic field overlaps the electrodes and is perpendicular to the electric field. When $\Delta V = 0$ V, the electron is deflected by 2.0 mm as it passes between the plates. What value of ΔV will allow the electron to pass through the plates without deflection?

MODEL Assume that the fields between the electrodes are uniform and that they are zero outside the electrodes.

VISUALIZE FIGURE 37.13 shows an electron passing through the magnetic field between the plates when $\Delta V = 0$ V. The curvature has been exaggerated to make the geometry clear.

FIGURE 37.13 The electron's trajectory in Example 37.2.

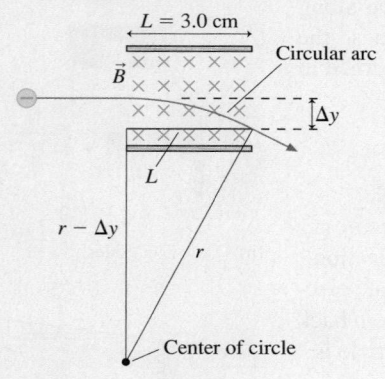

SOLVE We can find the needed electric field, and thus ΔV, if we know the electron's speed. We can find the electron's speed from the radius of curvature of its circular arc in a magnetic field. Figure 37.13 shows a right triangle with hypotenuse r and width L. We can use the Pythagorean theorem to write

$$(r - \Delta y)^2 + L^2 = r^2$$

where Δy is the electron's deflection in the magnetic field. This is easily solved to find the radius of the arc:

$$r = \frac{(\Delta y)^2 + L^2}{2\,\Delta y} = \frac{(0.0020 \text{ m})^2 + (0.030 \text{ m})^2}{2(0.0020 \text{ m})} = 0.226 \text{ m}$$

The speed of an electron traveling along an arc with this radius is found from Equation 37.6:

$$v = \frac{erB}{m} = 4.0 \times 10^7 \text{ m/s}$$

Thus the electric field allowing the electron to pass through without deflection is

$$E = vB = 40,000 \text{ V/m}$$

The electric field of a parallel-plate capacitor of spacing d is related to the potential difference by $E = \Delta V/d$, so the necessary potential difference is

$$\Delta V = Ed = (40,000 \text{ V/m})(0.0050 \text{ m}) = 200 \text{ V}$$

ASSESS A fairly small potential difference is sufficient to counteract the magnetic deflection.

The Electron

Thomson next measured q/m for different cathode materials. Finding them all to be the same, he concluded that all metals emit the *same* cathode rays. Thomson then compared his result to the charge-to-mass ratio of the hydrogen ion, known from electrolysis to have a value of $\approx 1 \times 10^8$ C/kg. This value was roughly 1000 times smaller than for the cathode-ray particles, which could imply that a cathode-ray particle has a much larger charge than a hydrogen ion, or a much smaller mass, or some combination of these.

Electrolysis experiments suggested the existence of a basic unit of charge, so it was tempting to assume that the cathode-ray charge was the same as the charge of a hydrogen ion. However, cathode rays were so different from the hydrogen ion that such an assumption could not be justified without some other evidence. To provide that evidence, Thomson called attention to previous experiments showing that cathode rays can penetrate thin metal foils but atoms cannot. This can be true, Thomson argued, only if cathode-ray particles are vastly smaller and thus much less massive than atoms.

In a paper published in 1897, Thomson assembled all of the evidence to announce the discovery that cathode rays are negatively charged particles, that they are much less massive ($\approx 0.1\%$) than atoms, and that they are identical when generated by different elements. In other words, Thomson had discovered a **subatomic particle,** one of the constituents of which atoms themselves are constructed. In recognition of the role this particle plays in electricity, it was later named the **electron.** By 1900 it was clear to all that electrons were a fundamental building block of atoms. Thomson was awarded the Nobel Prize in 1906.

STOP TO THINK 37.2 Thomson's conclusion that cathode-ray particles are *fundamental* constituents of atoms was based primarily on which observation?

 a. They have a negative charge.
 b. They are the same from all cathode materials.
 c. Their mass is much less than that of hydrogen.
 d. They penetrate very thin metal foils.

37.5 The Fundamental Unit of Charge

Thomson measured the electron's charge-to-mass ratio and *surmised* that the mass must be much smaller than that of an atom, but clearly it was desirable to measure the charge q directly. This was done in 1906 by the American scientist Robert Millikan.

The **Millikan oil-drop experiment,** as we call it today, is illustrated in FIGURE 37.14. A squeeze-bulb atomizer sprayed out a very fine mist of oil droplets. Millikan found that some of these droplets were charged from friction in the sprayer. The charged droplets slowly settled toward a horizontal pair of parallel-plate electrodes, where a few droplets passed through a small hole in the top plate. Millikan observed the drops by shining a bright light between the plates and using an eyepiece to see the droplets' reflections. He then established an electric field by applying a voltage to the plates.

A drop will remain suspended between the plates, moving neither up nor down, if the electric field exerts an upward force on a charged drop that exactly balances the downward gravitational force. The forces balance when

$$m_{\text{drop}}g = q_{\text{drop}}E \tag{37.10}$$

and thus the charge on the drop is measured to be

$$q_{\text{drop}} = \frac{m_{\text{drop}}g}{E} \tag{37.11}$$

Notice that m and q are the mass and charge of the oil droplet, not of an electron. But because the droplet is charged by acquiring (or losing) electrons, the charge of the droplet should be related to the electron's charge.

The field strength E could be determined accurately from the voltage applied to the plates, so the limiting factor in measuring q_{drop} was Millikan's ability to determine the mass of these small drops. Ideally, the mass could be found by measuring a drop's diameter and using the known density of the oil. However, the drops were too small ($\approx 1\ \mu\text{m}$) to measure accurately by viewing through the eyepiece.

FIGURE 37.14 Millikan's oil-drop apparatus to measure the fundamental unit of charge.

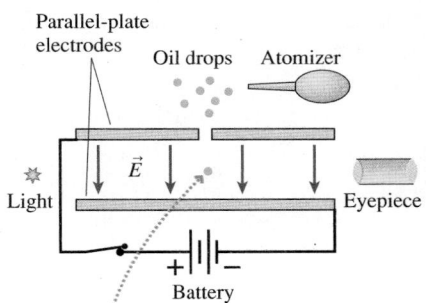

The upward electric force on a negatively charged droplet balances the downward gravitational force.

Instead, Millikan devised an ingenious method to find the size of the droplets. Objects this small are *not* in free fall. The air resistance forces are so large that the drops fall with a very small but constant speed. The motion of a sphere through a viscous medium is a problem that had been solved in the 19th century, and it was known that the sphere's terminal speed depends on its radius and on the viscosity of air. By timing the droplets' fall with a stopwatch, then using the known viscosity of air, Millikan could calculate their radii, compute their masses, and, finally, arrive at a value for their charge. Although it was a somewhat roundabout procedure, Millikan was able to measure the charge on a droplet with an accuracy of ±0.1%.

Millikan measured many hundreds of droplets, some for hours at a time, under a wide variety of conditions. He found that some of his droplets were positively charged and some negatively charged, but **all had charges that were integer multiples of a certain minimum charge value.** Millikan concluded that "the electric charges found on ions all have either exactly the same value or else some small exact multiple of that value." That value, the *fundamental unit of charge* that we now call *e*, is measured to be

$$e = 1.60 \times 10^{-19} \text{ C}$$

We can then combine the measured *e* with the measured charge-to-mass ratio *e/m* to find that the mass of the electron is

$$m_{\text{elec}} = 9.11 \times 10^{-31} \text{ kg}$$

Taken together, the experiments of Thomson, Millikan, and others provided overwhelming evidence that electric charge comes in discrete units and that *all* charges found in nature are multiples of a fundamental unit of charge we call *e*.

EXAMPLE 37.3 **Suspending an oil drop**

Oil has a density of 860 kg/m³. A 1.0-μm-diameter oil droplet acquires 10 extra electrons as it is sprayed. What potential difference between two parallel plates 1.0 cm apart will cause the droplet to be suspended in air?

MODEL Assume a uniform electric field $E = \Delta V/d$ between the plates.

SOLVE The magnitude of the charge on the drop is $q_{\text{drop}} = 10e$. The mass of the charge is related to its density ρ and volume V by

$$m_{\text{drop}} = \rho V = \frac{4}{3}\pi R^3 \rho = 4.50 \times 10^{-16} \text{ kg}$$

where the droplet's radius is $R = 5.0 \times 10^{-7}$ m. The electric field that will suspend this droplet against the force of gravity is

$$E = \frac{m_{\text{drop}} g}{q_{\text{drop}}} = 2760 \text{ V/m}$$

Establishing this electric field between two plates spaced by $d = 0.010$ m requires a potential difference

$$\Delta V = Ed = 27.6 \text{ V}$$

ASSESS Experimentally, this is a very convenient voltage.

37.6 The Discovery of the Nucleus

By 1900, it was clear that atoms are not indivisible but, instead, are constructed of charged particles. Atomic sizes were known to be $\approx 10^{-10}$ m, but the electrons common to all atoms are much smaller and much less massive than the smallest atom. How do they "fit" into the larger atom? What is the positive charge of the atom? Where are the charges located inside the atoms?

Thomson proposed the first model of an atom. Because the electrons are very small and light compared to the whole atom, it seemed reasonable to think that the positively charged part would take up most of the space. Thomson suggested that the atom consists of a spherical "cloud" of positive charge, roughly 10^{-10} m in diameter, in which the smaller negative electrons are embedded. The positive charge exactly balances the negative, so the atom as a whole has no net charge. This model of the atom has often been called the "plum-pudding model" or the "raisin-cake model" for reasons that should be clear from FIGURE 37.15.

Thomson was never able to make any predictions that would enable his model to be tested, and the Thomson atom did not stand the test of time. His model is of interest today primarily to remind us that our current models of the atom are by no means obvious. Science has many sidesteps and dead ends as it progresses.

One of Thomson's students was a New Zealander named Ernest Rutherford. While Rutherford and Thomson were studying the ionizing effects of x rays, in 1896, the French physicist Antoine Henri Becquerel announced the discovery that some new form of "rays" were emitted by crystals of uranium. These rays, like x rays, could expose film, pass through objects, and ionize the air. Yet they were emitted continuously from the uranium without having to "do" anything to it. This was the discovery of **radioactivity,** a topic we'll study in Chapter 42.

With x rays only a year old and cathode rays not yet completely understood, it was not known whether all these various kinds of rays were truly different or merely variations of a single type. Rutherford immediately began a study of these new rays. He quickly discovered that at least two *different* rays are emitted by a uranium crystal. The first, which he called **alpha rays,** were easily absorbed by a piece of paper. The second, **beta rays,** could penetrate through at least 0.1 inch of metal.

Thomson soon found that beta rays have the same charge-to-mass ratio as cathode rays. The beta rays turned out to be high-speed electrons emitted by the uranium crystal. Rutherford, using similar techniques, showed that alpha rays are *positively* charged particles. By 1906 he had measured their charge-to-mass ratio to be

$$\frac{q}{m} = \frac{1}{2}\frac{e}{m_{\text{H}}}$$

where m_{H} is the mass of a hydrogen atom. This value could indicate either a singly ionized hydrogen molecule H_2^+ ($q = e$, $m = 2m_{\text{H}}$) *or* a doubly ionized helium atom He^{++} ($q = 2e$, $m = 4m_{\text{H}}$).

In an ingenious experiment, Rutherford sealed a sample of radium—an emitter of alpha radiation—into a glass tube. Alpha rays could not penetrate the glass, so the particles were contained within the tube. Several days later, Rutherford used electrodes in the tube to create a discharge and observed the spectrum of the emitted light. He found the characteristic wavelengths of helium, but not those of hydrogen. Alpha rays (or alpha particles, as we now call them) consist of doubly ionized helium atoms (bare helium nuclei) emitted at high speed ($\approx 3 \times 10^7$ m/s) from the sample.

The First Nuclear Physics Experiment

Rutherford soon realized that he could use these high-speed particles to probe inside other atoms. In 1909, Rutherford and his students set up the experiment shown in FIGURE 37.16 to shoot alpha particles at very thin metal foils. Some of the alpha particles penetrated the foil, but the beam of alpha particles that did so became somewhat spread out. This was not surprising. According to Thomson's raisin-cake model of the atom, the electric forces exerted on the positive alpha particle by the positive atomic charges should roughly cancel the forces from the negative electrons, causing the alpha particles to be deflected only slightly.

At Rutherford's suggestion, his students then set up the apparatus to see if any alpha particles were deflected at *large* angles. It took only a few days to find the answer. Not only were alpha particles deflected at large angles, but a very few were reflected almost straight backward toward the source!

How can we understand this result? FIGURE 37.17a on the next page shows that only a small deflection is expected for an alpha particle passing through a Thomson atom. But if an atom has a small, positive core, such as the one in FIGURE 37.17b, a few of the

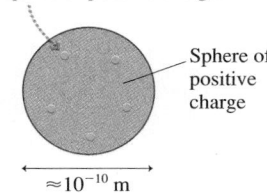

FIGURE 37.15 Thomson's raisin-cake model of the atom.

Thomson proposed that small, negative electrons are embedded in a sphere of positive charge.

Sphere of positive charge

$\approx 10^{-10}$ m

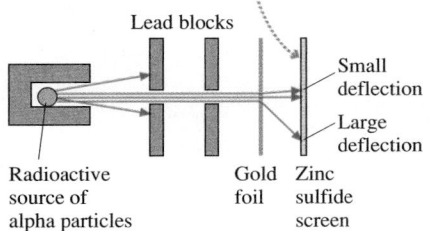

FIGURE 37.16 Rutherford's experiment to shoot high-speed alpha particles through a thin gold foil.

The alpha particles make little flashes of light where they hit the screen.

Lead blocks

Small deflection

Large deflection

Radioactive source of alpha particles

Gold foil

Zinc sulfide screen

FIGURE 37.17 Alpha particles interact differently with a concentrated positive nucleus than they would with the spread-out charge in Thomson's model.

(a)

Alpha

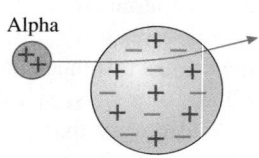

The alpha particle is only slightly deflected by a Thomson atom because forces from the spread-out positive and negative charges nearly cancel.

(b)

Alpha

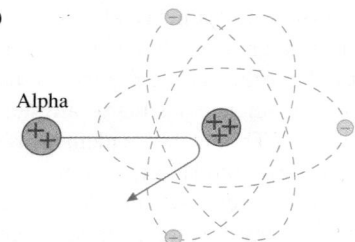

If the atom has a concentrated positive nucleus, some alpha particles will be able to come very close to the nucleus and thus feel a very strong repulsive force.

alpha particles can come very close to the core. Because the electric force varies with the inverse square of the distance, the very large force of this very close approach can cause a large-angle scattering or a backward deflection of the alpha particle.

Thus the discovery of large-angle scattering of alpha particles led Rutherford to envision an atom in which negative electrons orbit an unbelievably small, massive, positive **nucleus,** rather like a miniature solar system. This is the **nuclear model of the atom.** Notice that nearly all of the atom is empty space—the void!

EXAMPLE 37.4 **A nuclear physics experiment**

An alpha particle is shot with a speed of 2.0×10^7 m/s directly toward the nucleus of a gold atom. What is the distance of closest approach to the nucleus?

MODEL Energy is conserved in electric interactions. Assume that the gold nucleus, which is much more massive than the alpha particle, does not move. Also recall that the exterior electric field and potential of a sphere of charge can be found by treating the total charge as a point charge at the center.

VISUALIZE **FIGURE 37.18** is a pictorial representation. The motion is in and out along a straight line.

FIGURE 37.18 A before-and-after pictorial representation of an alpha particle colliding with a nucleus.

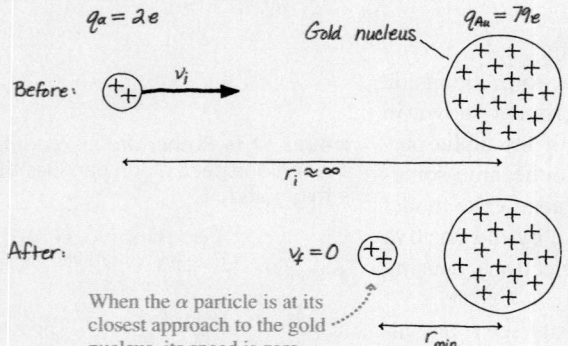

When the α particle is at its closest approach to the gold nucleus, its speed is zero.

SOLVE We are not interested in how long the collision takes or any of the details of the trajectory, so using conservation of energy rather than Newton's laws is appropriate. Initially, when the alpha particle is very far away, the system has only kinetic energy. At the moment of closest approach, just before the alpha particle is reflected, the charges are at rest and the system has only potential energy. The conservation of energy statement $K_f + U_f = K_i + U_i$ is

$$0 + \frac{1}{4\pi\epsilon_0} \frac{q_\alpha q_{Au}}{r_{min}} = \frac{1}{2}mv_i^2 + 0$$

where q_α is the alpha-particle charge and we've treated the gold nucleus as a point charge q_{Au}. The mass m is that of the alpha particle. The solution for r_{min} is

$$r_{min} = \frac{1}{4\pi\epsilon_0} \frac{2q_\alpha q_{Au}}{mv_i^2}$$

The alpha particle is a helium nucleus, so $m = 4$ u $= 6.64 \times 10^{-27}$ kg and $q_\alpha = 2e = 3.20 \times 10^{-19}$ C. Gold has atomic number 79, so $q_{Au} = 79e = 1.26 \times 10^{-17}$ C. We can then calculate

$$r_{min} = 2.7 \times 10^{-14} \text{ m}$$

This is only about 1/10,000 the size of the atom itself!

ASSESS We ignored the atom's electrons in this example. In fact, they make almost no contribution to the alpha particle's trajectory. The alpha particle is exceedingly massive compared to the electrons, and the electrons are spread out over a distance very large compared to the size of the nucleus. Hence the alpha particle easily pushes them aside without any noticeable change in its velocity.

Rutherford went on to make careful experiments of how the alpha particles scattered at different angles. From these experiments he deduced that the diameter of the atomic nucleus is $\approx 1 \times 10^{-14}$ m $= 10$ fm (1 fm $= 1$ femtometer $= 10^{-15}$ m), increasing a little for elements of higher atomic number and atomic mass.

It may seem surprising to you that the Rutherford model of the atom, with its solar-system analogy, was not Thomson's original choice. However, scientists at the time could not imagine matter having the extraordinarily high density implied by a small nucleus. Neither could they understand what holds the nucleus together, why the positive charges do not push each other apart. Thomson's model, in which the positive charge was spread out and balanced by the negative electrons, actually made more sense. It would be several decades before the forces holding the nucleus together began to be understood, but Rutherford's evidence for a very small nucleus was indisputable.

STOP TO THINK 37.3 If the alpha particle has a positive charge, which way will it be deflected in the magnetic field?

   a. Up   b. Down   c. Into the page   d. Out of the page

The Electron Volt

The joule is a unit of appropriate size in mechanics and thermodynamics, where we dealt with macroscopic objects, but it is poorly matched to the needs of atomic physics. It will be very useful to have an energy unit appropriate to atomic and nuclear events.

FIGURE 37.19 shows an electron accelerating (in a vacuum) from rest across a parallel-plate capacitor with a 1.0 V potential difference. What is the electron's kinetic energy when it reaches the positive plate? We know from energy conservation that $K_f + qV_f = K_i + qV_i$, where $U = qV$ is the electric potential energy. $K_i = 0$ because the electron starts from rest, and the electron's charge is $q = -e$. Thus

$$K_f = -q(V_f - V_i) = -q\,\Delta V = e\,\Delta V = (1.60 \times 10^{-19}\ \text{C})(1.0\ \text{V})$$

$$= 1.60 \times 10^{-19}\ \text{J}$$

Let us define a new unit of energy, called the **electron volt,** as

$$1\ \text{electron volt} = 1\ \text{eV} \equiv 1.60 \times 10^{-19}\ \text{J}$$

With this definition, the kinetic energy gained by the electron in our example is

$$K_f = 1\ \text{eV}$$

In other words, **1 electron volt is the kinetic energy gained by an electron (or proton) if it accelerates through a potential difference of 1 volt.**

> **NOTE** ▶ The abbreviation eV uses a lowercase e but an uppercase V. Units of keV (10^3 eV), MeV (10^6 eV), and GeV (10^9 eV) are common. ◀

The electron volt can be a troublesome unit. One difficulty is its unusual name, which looks less like a unit than, say, "meter" or "second." A more significant difficulty is that the name suggests a relationship to volts. But *volts* are units of electric potential, whereas this new unit is a unit of energy! It is crucial to distinguish between the *potential V*, measured in volts, and an *energy* that can be measured either in joules or in electron volts. You can now use electron volts anywhere that you would previously have used joules.

> **NOTE** ▶ To reiterate, the electron volt is a unit of *energy,* convertible to joules, and not a unit of potential. Potential is always measured in volts. However, the joule remains the SI unit of energy. It will be useful to express energies in eV, but you *must* convert this energy to joules before doing most calculations. ◀

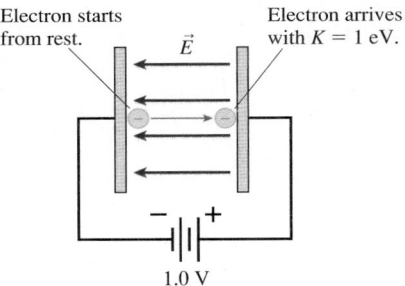

FIGURE 37.19 An electron accelerating across a 1 V potential difference gains 1 eV of kinetic energy.

EXAMPLE 37.5 **The speed of an alpha particle**

Alpha particles are usually characterized by their kinetic energy in MeV. What is the speed of an 8.3 MeV alpha particle?

SOLVE Alpha particles are helium nuclei, having $m = 4\,\text{u} = 6.64 \times 10^{-27}$ kg. The kinetic energy of this alpha particle is 8.3×10^6 eV. First, we convert the energy to joules:

$$K = 8.3 \times 10^6 \,\text{eV} \times \frac{1.60 \times 10^{-19}\,\text{J}}{1.00\,\text{eV}} = 1.33 \times 10^{-12}\,\text{J}$$

Now we can find the speed:

$$K = \frac{1}{2}mv^2 = 1.33 \times 10^{-12}\,\text{J}$$

$$v = \sqrt{\frac{2K}{m}} = 2.0 \times 10^7 \,\text{m/s}$$

This was the speed of the alpha particle in Example 37.4.

EXAMPLE 37.6 **Energy of an electron**

In a simple model of the hydrogen atom, the electron orbits the proton at 2.19×10^6 m/s in a circle with radius 5.29×10^{-11} m. What is the atom's energy in eV?

MODEL The electron has a kinetic energy of motion, and the electron + proton system has an electric potential energy.

SOLVE The potential energy is that of two point charges, with $q_{\text{proton}} = +e$ and $q_{\text{elec}} = -e$. Thus

$$E = K + U = \frac{1}{2}m_{\text{elec}}v^2 + \frac{1}{4\pi\epsilon_0}\frac{(e)(-e)}{r} = -2.17 \times 10^{-18}\,\text{J}$$

Conversion to eV gives

$$E = -2.17 \times 10^{-18}\,\text{J} \times \frac{1\,\text{eV}}{1.60 \times 10^{-19}\,\text{J}} = -13.6\,\text{eV}$$

ASSESS The negative energy reflects the fact that the electron is *bound* to the proton. You would need to *add* energy to remove the electron.

Using the Nuclear Model

The nuclear model of the atom makes it easy to understand and picture such processes as ionization. Because electrons orbit a positive nucleus, an x-ray photon or a rapidly moving particle, such as another electron, can knock one of the orbiting electrons away, creating a positive ion. Removing one electron makes a singly charged ion, with $q = +e$. Removing two electrons creates a doubly charged ion, with $q = +2e$. This is shown for lithium (atomic number 3) in FIGURE 37.20.

FIGURE 37.20 Different ionization stages of the lithium atom ($Z = 3$).

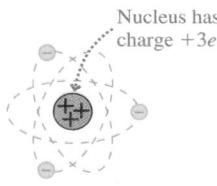
Nucleus has charge $+3e$.

Neutral Li

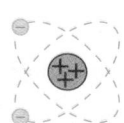
Singly charged Li$^+$

Doubly charged Li^{++}

The nuclear model also allows us to understand why, during chemical reactions and when an object is charged by rubbing, electrons are easily transferred but protons are not. The protons are tightly bound in the nucleus, shielded by all the electrons, but outer electrons are easily stripped away. Rutherford's nuclear model has explanatory power that was lacking in Thomson's model.

EXAMPLE 37.7 **The ionization energy of hydrogen**

What is the minimum energy required to ionize a hydrogen atom?

SOLVE In Example 37.6 we found that the atom's energy is $E_i = -13.6$ eV. Ionizing the atom means removing the electron and taking it very far away. As $r \rightarrow \infty$, the potential energy becomes zero. Further, using the least possible energy to ionize the atom will leave the electron, when it is very far away, very nearly at rest. Thus the atom's energy after ionization is

$E_f = K_f + U_f = 0 + 0 = 0$ eV. This is *larger* than E_i by 13.6 eV, so the minimum energy that is required to ionize a hydrogen atom is 13.6 eV. This is called the atom's *ionization energy*. If the electron receives ≥ 13.6 eV (2.17×10^{-18} J) of energy from a photon, or in a collision with another electron, or by any other means, it will be knocked out of the atom and leave a H$^+$ ion behind.

Carbon is the sixth element in the periodic table. How many electrons are in a C^{++} ion?

37.7 Into the Nucleus

Chapter 42 will discuss nuclear physics in more detail, but it will be helpful to give a brief overview of the nucleus. The relative masses of many of the elements were known from chemistry experiments by the mid-19th century. By arranging the elements in order of ascending mass, and noting recurring regularities in their chemical properties, the Russian chemist Dmitri Mendeleev first proposed the periodic table of the elements in 1872. But what did it mean to say that hydrogen was atomic number 1, helium number 2, lithium number 3, and so on?

It soon became known that hydrogen atoms can be only singly ionized, producing H^+. A doubly ionized H^{++} is never observed. Helium, by contrast, can be both singly and doubly ionized, creating He^+ and He^{++}, but He^{+++} is not observed. Once Thomson discovered the electron and Millikan established the fundamental unit of charge, it seemed fairly clear that a hydrogen atom contains only one electron and one unit of positive charge, helium has two electrons and two units of positive charge, and so on. Thus the **atomic number** of an element, which is always an integer, describes the number of electrons (of a neutral atom) and the number of units of positive charge in the nucleus. The atomic number is represented by Z, so hydrogen is $Z = 1$, helium $Z = 2$, and lithium $Z = 3$. Elements are listed in the periodic table by their atomic number.

Rutherford's discovery of the nucleus soon led to the recognition that the positive charge is associated with a positive subatomic particle called the **proton.** The proton's charge is $+e$, equal in magnitude but opposite in sign to the electron's charge. Further, because nearly all the atomic mass is associated with the nucleus, the proton is much more massive than the electron. According to Rutherford's nuclear model, atoms with atomic number Z consist of Z negative electrons, with net charge $-Ze$, orbiting a massive nucleus that contains protons and has net charge $+Ze$. The Rutherford atom went a long way toward explaining the periodic table.

But there was a problem. Helium, with atomic number 2, has twice as many electrons as hydrogen. Lithium, $Z = 3$, has three electrons. But it was known from chemistry measurements that helium is *four times* as massive as hydrogen and lithium is *seven times* as massive. If a nucleus contains Z protons to balance the Z orbiting electrons, and if nearly all the atomic mass is contained in the nucleus, then helium should be simply twice as massive as hydrogen and lithium three times as massive.

The Neutron

About 1910, Thomson and his student Francis Aston developed a device called a **mass spectrometer** for measuring the charge-to-mass ratios of atomic ions. As Aston and others began collecting data, they soon found that many elements consist of atoms of *differing* mass! Neon, for example, had been assigned an atomic mass of 20. But Aston found, as the data of FIGURE 37.21 show, that while 91% of neon atoms have mass $m = 20$ u, 9% have $m = 22$ u and a very small percentage have $m = 21$ u. Chlorine was found to be a mixture of 75% chlorine atoms with $m = 35$ u and 25% chlorine atoms with $m = 37$ u, both having atomic number $Z = 17$.

These difficulties were not resolved until the discovery, in 1932, of a third subatomic particle. This particle has essentially the same mass as a proton but *no* electric charge. It is called the **neutron.** Neutrons reside in the nucleus, with the protons, where they contribute to the mass of the atom but not to its charge. As you'll see in Chapter 42, neutrons help provide the "glue" that holds the nucleus together.

FIGURE 37.21 The mass spectrum of neon.

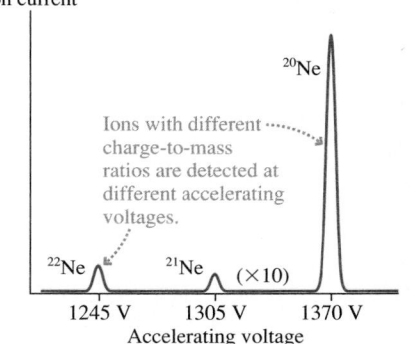

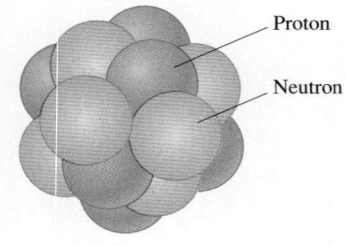

FIGURE 37.22 The nucleus of an atom contains protons and neutrons.

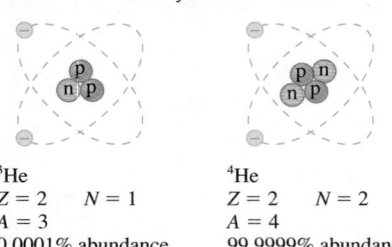

FIGURE 37.23 The two isotopes of helium. ^{3}He is only 0.0001% abundant.

^{3}He
$Z = 2$ $N = 1$
$A = 3$
0.0001% abundance

^{4}He
$Z = 2$ $N = 2$
$A = 4$
99.9999% abundance

The neutron was the missing link needed to explain why atoms of the same element can have different masses. We now know that every atom with atomic number Z has a nucleus containing Z protons with charge $+Ze$. In addition, as shown in **FIGURE 37.22**, the nucleus contains N neutrons. There are a *range* of neutron numbers that happily form a nucleus with Z protons, creating a series of nuclei having the same Z-value (i.e., they are all the same chemical element) but different masses. Such a series of nuclei are called **isotopes.**

Chemical behavior is determined by the orbiting electrons. All isotopes of one element have the same number Z of orbiting electrons and have the same chemical properties. But different isotopes of the same element can have quite different nuclear properties.

An atom's **mass number** A is defined to be $A = Z + N$. It is the total number of protons and neutrons in a nucleus. The mass number, which is dimensionless, is *not* the same thing as the atomic mass m. By definition, A is an integer. But because the proton and neutron masses are both ≈ 1 u, the mass number A is *approximately* the mass in atomic mass units.

The notation used to label isotopes is ^AZ, where the mass number A is given as a *leading* superscript. The proton number Z is not specified by an actual number but, equivalently, by the chemical symbol for that element. The most common isotope of neon has $Z = 10$ protons and $N = 10$ neutrons. Thus it has mass number $A = 20$ and it is labeled ^{20}Ne. The neon isotope ^{22}Ne has $Z = 10$ protons (that's what makes it neon) and $N = 12$ neutrons. Helium has the two isotopes shown in **FIGURE 37.23**. The rare ^{3}He is only 0.0001% abundant, but it can be isolated and has important uses in scientific research.

> **STOP TO THINK 37.5**   Carbon is the sixth element in the periodic table. How many protons and how many neutrons are there in a nucleus of the isotope ^{14}C?

37.8 Classical Physics at the Limit

At the start of the 19th century, only a few scientists thought that matter consists of atoms. By century's end, there was substantial evidence not only for atoms but also for the existence of charged subatomic particles. The explorations into atomic structure culminated with Rutherford's nuclear model.

Rutherford's nuclear model of the atom matched the experimental evidence about the *structure* of atoms, but it had one serious shortcoming. According to Maxwell's theory of electricity and magnetism, the orbiting electrons in a Rutherford atom should act as small antennas and radiate electromagnetic waves. That sounds encouraging, because we know that atoms can emit light, but the radiation of electromagnetic waves means the atoms would continuously lose energy. As **FIGURE 37.24** shows, this would cause the electrons to spiral into the nucleus! Calculations showed that a Rutherford atom can last no more than about a microsecond. In other words, classical Newtonian mechanics and electromagnetism predict that an atom in which electrons orbit a nucleus would be highly unstable and would immediately self-destruct. This clearly does not happen.

The experimental efforts of the late 19th and early 20th centuries had been impressive, and there could be no doubt about the existence of electrons, about the small positive nucleus, and about the unique discrete spectrum emitted by each atom. But the theoretical framework for understanding such observations had lagged behind. As the new century dawned, physicists could not explain the structure of atoms, could not explain the stability of matter, could not explain discrete spectra or blackbody radiation, and could not explain the origin of x rays or radioactivity.

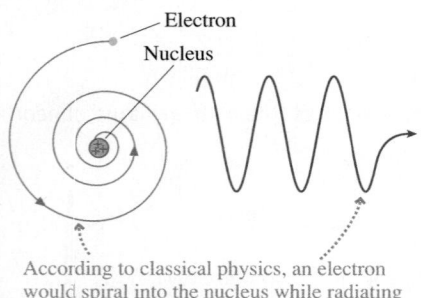

FIGURE 37.24 The fate of a Rutherford atom.

Electron

Nucleus

According to classical physics, an electron would spiral into the nucleus while radiating energy as an electromagnetic wave.

Yet few physicists were willing to abandon the successful and long-cherished theories of classical physics. Most considered these "problems" with atoms to be minor discrepancies that would soon be resolved. But classical physics had, indeed, reached its limit, and a whole new generation of brilliant young physicists, with new ideas, was about to take the stage. Among the first was an unassuming young man in Bern, Switzerland. His scholastic record had been mediocre, and the best job he could find upon graduation was as a clerk in the patent office, examining patent applications. His name was Albert Einstein.

CHALLENGE EXAMPLE 37.8 **Radioactive decay**

The cesium isotope ^{137}Cs, with $Z = 55$, is radioactive and decays by beta decay. A beta particle is observed in the laboratory with a kinetic energy of 300 keV. With what kinetic energy was the beta particle ejected from the 12.4-fm-diameter nucleus?

MODEL A beta particle is an electron that was ejected from the nucleus of an atom during a radioactive decay. Energy is conserved in the decay.

VISUALIZE **FIGURE 37.25** shows a before-and-after pictorial representation. The electron starts by being ejected from the nucleus with kinetic energy K_i. It has electric potential energy U_i due to its interaction with the nucleus. The potential energy due to the atom's orbiting electrons is negligible because they are so far away

FIGURE 37.25 A before-and-after pictorial representation of the beta decay.

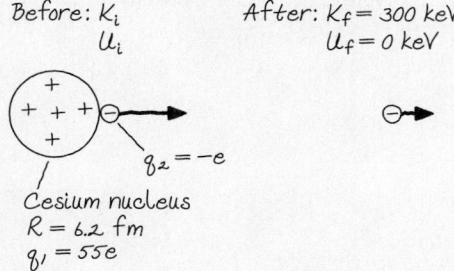

in comparison to a nuclear radius. The detected electron is very far from the nucleus, so $U_f = 0$.

SOLVE The conservation of energy statement is $K_f + U_f = K_i + U_i$. The electron starts outside the nucleus, even though at the surface, so the spherical nucleus can be treated as a point charge with $q_1 = 55e$. The electron has $q_2 = -e$, so the initial electron-nucleus potential energy is

$$U_i = \frac{Kq_1q_2}{r_i}$$

$$= \frac{(9.0 \times 10^9 \, \text{N}\,\text{m}^2/\text{C}^2)(55 \times 1.60 \times 10^{-19}\,\text{C})(-1.60 \times 10^{-19}\,\text{C})}{6.20 \times 10^{-15}\,\text{m}}$$

$$= -2.04 \times 10^{-12}\,\text{J} \times \frac{1\,\text{eV}}{1.60 \times 10^{-19}\,\text{J}} = -12.8\,\text{MeV}$$

To be detected in the laboratory with $K_f = 300$ keV $= 0.3$ MeV, the electron had to be ejected from the nucleus with

$$K_i = K_f + U_f - U_i = 0.3\,\text{MeV} + 0\,\text{MeV} + 12.8\,\text{MeV}$$

$$= 13.1\,\text{MeV}$$

ASSESS A negative electron is very strongly attracted to the nucleus. It's not surprising that it has to be ejected from the nucleus with an enormous amount of kinetic energy to be able to escape at all.

SUMMARY

The goal of Chapter 37 has been to learn about the structure and properties of atoms.

Important Concepts/Experiments

Nineteenth-century scientists focused on understanding matter and light. Faraday's invention of the gas discharge tube launched two important avenues of inquiry:

- Atomic structure.
- Atomic spectra.

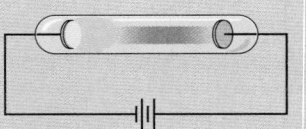

Cathode Rays and Atomic Structure

Thomson found that cathode rays are negative, subatomic particles. These were soon named **electrons**. Electrons are

- Constituents of atoms.
- The fundamental units of negative charge.

Rutherford discovered the atomic **nucleus**. His nuclear model of the atom proposes

- A very small, dense positive nucleus.
- Orbiting negative electrons.

Later, different **isotopes** were recognized to contain different numbers of **neutrons** in a nucleus with the same number of **protons.**

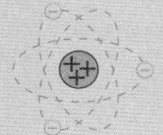

Atomic Spectra and the Nature of Light

The spectra emitted by the gas in a discharge tube consist of discrete wavelengths.

- Every element has a unique spectrum.
- Every spectral line in an element's absorption spectrum is present in its emission spectrum, but not all emission lines are seen in the absorption spectrum.

The wavelengths of the hydrogen emission spectrum are

$$\lambda = \frac{91.18 \text{ nm}}{\left(\dfrac{1}{m^2} - \dfrac{1}{n^2}\right)} \qquad m = 1, 2, 3, \ldots \qquad n = m + 1, m + 2, \ldots$$

The end of classical physics. . .
Atomic spectra had to be related to atomic structure, but no one could understand how. Classical physics could not explain

- The stability of matter.
- Discrete atomic spectra.
- Continuous blackbody spectra.

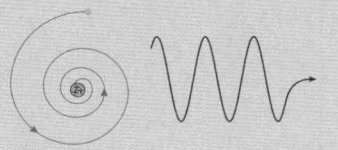

Applications

Millikan's oil-drop experiment measured the fundamental unit of charge:

$$e = 1.60 \times 10^{-19} \text{ C}$$

One electron volt (1 eV) is the energy an electron or proton (charge $\pm e$) gains by accelerating through a potential difference of 1 V:

$$1 \text{ eV} = 1.60 \times 10^{-19} \text{ J}$$

Terms and Notation

spectrometer	Balmer formula	Millikan oil-drop experiment	proton
spectrum	cathode glow	radioactivity	mass spectrometer
continuous spectrum	cathode rays	alpha rays	neutron
blackbody radiation	Crookes tube	beta rays	isotope
Wien's law	x rays	nucleus	mass number, A
gas discharge tube	crossed-field experiment	nuclear model of the atom	
discrete spectrum	subatomic particle	electron volt, eV	
Balmer series	electron	atomic number, Z	

CONCEPTUAL QUESTIONS

1. a. Summarize the experimental evidence *prior* to the research of Thomson by which you might conclude that cathode rays are some kind of particle.

   b. Summarize the experimental evidence *prior* to the research of Thomson by which you might conclude that cathode rays are some kind of wave.

2. Thomson observed deflection of the cathode-ray particles due to magnetic and electric fields, but there was no observed deflection due to gravity. Why not?

3. What was the significance of Thomson's experiment in which an off-center electrode was used to collect charge deflected by a magnetic field?

4. What is the evidence by which we know that an electron from an iron atom is identical to an electron from a copper atom?

5. a. Describe the experimental evidence by which we know that the nucleus is made up not just of protons.

   b. The neutron is not easy to isolate or control because it has no charge that would allow scientists to manipulate it. What evidence allowed scientists to determine that the mass of the neutron is almost the same as the mass of a proton?

6. Rutherford studied alpha particles using the crossed-field technique Thomson had invented to study cathode rays. Assuming

that $v_{alpha} \approx v_{cathode ray}$ (which turns out to be true), would the deflection of an alpha particle by a magnetic field be larger, smaller, or the same as the deflection of a cathode-ray particle by the same field? Explain.

7. Once Thomson showed that atoms consist of very light negative electrons and a much more massive positive charge, why didn't physicists immediately consider a solar-system model of electrons orbiting a positive nucleus? Why would physicists in 1900 object to such a model?

8. Explain why the observation of alpha particles scattered at very large angles led Rutherford to reject Thomson's model of the atom and to propose a nuclear model.

9. Identify the element, the isotope, and the charge state of each atom in **FIGURE Q37.9**. Give your answer in symbolic form, such as $^4\text{He}^+$ or $^8\text{Be}^-$.

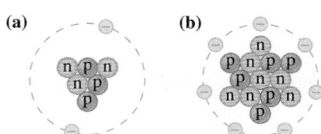

FIGURE Q37.9

EXERCISES AND PROBLEMS

Problems labeled [ ] integrate material from earlier chapters.

Exercises

Section 37.2 The Emission and Absorption of Light

1. | Figure 37.7 identified the wavelengths of four lines in the Balmer series of hydrogen.
   a. Determine the Balmer formula n and m values for these wavelengths.
   b. Predict the wavelength of the fifth line in the spectrum.

2. | What are the wavelengths of spectral lines in the Balmer series with $n = 6, 8$, and 10?

3. ‖ The wavelengths in the hydrogen spectrum with $m = 1$ form a series of spectral lines called the Lyman series. Calculate the wavelengths of the first four members of the Lyman series.

4. | Two of the wavelengths emitted by a hydrogen atom are 102.6 nm and 1876 nm.
   a. What are the m and n values for each of these wavelengths?
   b. For each of these wavelengths, is the light infrared, visible, or ultraviolet?

5. | What temperature, in °C, is a blackbody whose emission spectrum peaks at (a) 300 nm and (b) 3.00 μm?

6. ‖ A 2.0-cm-diameter metal sphere is glowing red, but a spectrum shows that its emission spectrum peaks at an infrared wavelength of 2.0 μm. How much power does the sphere radiate?

7. ‖ A ceramic cube 3.0 cm on each side radiates heat at 630 W. At what wavelength, in μm, does its emission spectrum peak?

Section 37.3 Cathode Rays and X Rays
Section 37.4 The Discovery of the Electron

8. | The current in a Crookes tube is 10 nA. How many electrons strike the face of the glass tube each second?

9. | An electron in a cathode-ray beam passes between 2.5-cm-long parallel-plate electrodes that are 5.0 mm apart. A 2.0 mT, 2.5-cm-wide magnetic field is perpendicular to the electric field between the plates. The electron passes through the electrodes without being deflected if the potential difference between the plates is 600 V.
   a. What is the electron's speed?
   b. If the potential difference between the plates is set to zero, what is the electron's radius of curvature in the magnetic field?

10. | Electrons pass through the parallel electrodes shown in **FIGURE EX37.10** with a speed of 5.0×10^6 m/s. What magnetic field strength and direction will allow the electrons to pass through without being deflected? Assume that the magnetic field is confined to the region between the electrodes.

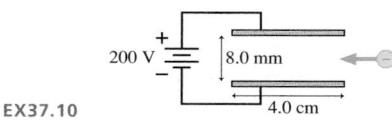

FIGURE EX37.10

Section 37.5 The Fundamental Unit of Charge

11. | A 0.80-μm-diameter oil droplet is observed between two parallel electrodes spaced 11 mm apart. The droplet hangs motionless if the upper electrode is 20 V more positive than the lower electrode. The density of the oil is 885 kg/m^3.
    a. What is the droplet's mass?
    b. What is the droplet's charge?
    c. Does the droplet have a surplus or a deficit of electrons? How many?

12. || An oil droplet with 15 excess electrons is observed between two parallel electrodes spaced 12 mm apart. The droplet hangs motionless if the upper electrode is 25 V more positive than the lower electrode. The density of the oil is 860 kg/m^3. What is the radius of the droplet?

13. || Suppose that in a hypothetical oil-drop experiment you measure the following values for the charges on the drops: 3.99×10^{-19} C, 6.65×10^{-19} C, 2.66×10^{-19} C, 10.64×10^{-19} C, and 9.31×10^{-19} C. What is the largest value of the fundamental unit of charge that is consistent with your measurements?

Section 37.6 The Discovery of the Nucleus

Section 37.7 Into the Nucleus

14. | Determine:
    a. The speed of a 300 eV electron.
    b. The speed of a 3.5 MeV H$^+$ ion.
    c. The specific type of particle that has 2.09 MeV of kinetic energy when moving with a speed of 1.0×10^7 m/s.

15. | Determine:
    a. The speed of a 7.0 MeV neutron.
    b. The speed of a 15 MeV helium atom.
    c. The specific type of particle that has 1.14 keV of kinetic energy when moving with a speed of 2.0×10^7 m/s.

16. || Express in eV (or keV or MeV if more appropriate):
    a. The kinetic energy of an electron moving with a speed of 5.0×10^6 m/s.
    b. The potential energy of an electron and a proton 0.10 nm apart.
    c. The kinetic energy of a proton that has accelerated from rest through a potential difference of 5000 V.

17. || Express in eV (or keV or MeV if more appropriate):
    a. The kinetic energy of a Li^{++} ion that has accelerated from rest through a potential difference of 5000 V.
    b. The potential energy of two protons 10 fm apart.
    c. The kinetic energy, just before impact, of a 200 g ball dropped from a height of 1.0 m.

18. | A parallel-plate capacitor with a 1.0 mm plate separation is charged to 75 V. With what kinetic energy, in eV, must a proton be launched from the negative plate if it is just barely able to reach the positive plate?

19. | How many electrons, protons, and neutrons are contained in the following atoms or ions: (a) ^{6}Li, (b) ^{16}O$^+$, and (c) ^{13}N^{++}?

20. | How many electrons, protons, and neutrons are contained in the following atoms or ions: (a) ^{10}B, (b) ^{13}N$^+$, and (c) ^{17}O^{+++}?

21. | Write the symbol for an atom or ion with:
    a. five electrons, five protons, and six neutrons.
    b. five electrons, six protons, and eight neutrons.

22. | Write the symbol for an atom or ion with:
    a. one electron, one proton, and two neutrons.
    b. seven electrons, eight protons, and ten neutrons.

23. | Consider the gold isotope ^{197}Au.
    a. How many electrons, protons, and neutrons are in a neutral ^{197}Au atom?
    b. The gold nucleus has a diameter of 14.0 fm. What is the density of matter in a gold nucleus?
    c. The density of lead is 11,400 kg/m^3. How many times the density of lead is your answer to part b?

24. || Consider the lead isotope ^{207}Pb.
    a. How many electrons, protons, and neutrons are in a neutral ^{207}Pb atom?
    b. The lead nucleus has a diameter of 14.2 fm. What is the electric field strength at the surface of a lead nucleus?

Problems

25. || What is the total energy, in MeV, of
    a. A proton traveling at 99% of the speed of light?
    b. An electron traveling at 99% of the speed of light?
    Hint: This problem uses relativity.

26. | What is the velocity, as a fraction of c, of
    a. A proton with 500 GeV total energy?
    b. An electron with 2.0 GeV total energy?
    Hint: This problem uses relativity.

27. | You learned in Chapter 36 that mass has an equivalent amount of energy. What are the energy equivalents in MeV of the rest masses of an electron and a proton?

28. || The factor γ appears in many relativistic expressions. A value $\gamma = 1.01$ implies that relativity changes the Newtonian values by approximately 1% and that relativistic effects can no longer be ignored. At what kinetic energy, in MeV, is $\gamma = 1.01$ for (a) an electron, (b) a proton, and (c) an alpha particle?

29. | The fission process $n + {}^{235}U \rightarrow {}^{236}U \rightarrow {}^{144}Ba + {}^{89}Kr + 3n$ converts 0.185 u of mass into the kinetic energy of the fission products. What is the total kinetic energy in MeV?

30. || An electron in a cathode-ray beam passes between 2.5-cm-long parallel-plate electrodes that are 5.0 mm apart. A 1.0 mT, 2.5-cm-wide magnetic field is perpendicular to the electric field between the plates. If the potential difference between the plates is 150 V, the electron passes through the electrodes without being deflected. If the potential difference across the plates is set to zero, through what angle is the electron deflected as it passes through the magnetic field?

31. || The two 5.0-cm-long parallel electrodes in FIGURE P37.31 are spaced 1.0 cm apart. A proton enters the plates from one end, an equal distance from both electrodes. A potential difference $\Delta V = 500$ V across the electrodes deflects the proton so that it strikes the outer end of the lower electrode. What magnetic field strength and direction will allow the proton to pass through undeflected while the 500 V potential difference is applied? Assume that both the electric and magnetic fields are confined to the space between the electrodes.

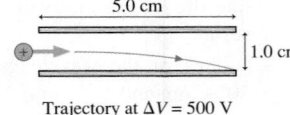

Trajectory at $\Delta V = 500$ V

FIGURE P37.31

32. ‖ An unknown charged particle passes without deflection through crossed electric and magnetic fields of strengths 187,500 V/m and 0.1250 T, respectively. The particle passes out of the electric field, but the magnetic field continues, and the particle makes a semicircle of diameter 25.05 cm. What is the particle's charge-to-mass ratio? Can you identify the particle?

33. ‖ In one of Thomson's experiments he placed a thin metal foil in the electron beam and measured its temperature rise. Consider a cathode-ray tube in which electrons are accelerated through a 2000 V potential difference, then strike a 10 mg copper foil. What is the electron-beam current if the foil temperature rises 6.0°C in 10 s? Assume no loss of energy by radiation or other means. The specific heat of copper is 385 J/kg K.

34. ‖ A neutral lithium atom has three electrons. As you will discover in Chapter 41, two of these electrons form an "inner core," but the third—the valence electron—orbits at much larger radius. From the valence electron's perspective, it is orbiting a spherical ball of charge having net charge $+1e$ (i.e., the three protons in the nucleus and the two inner-core electrons). The energy required to ionize a lithium atom is 5.14 eV. According to Rutherford's nuclear model of the atom, what are the orbital radius and speed of the valence electron?
    Hint: Consider the energy needed to remove the electron *and* the force needed to give the electron a circular orbit.

35. ‖ The diameter of an atom is 1.2×10^{-10} m and the diameter of its nucleus is 1.0×10^{-14} m. What percent of the atom's volume is occupied by mass and what percent is empty space?

36. ‖ A ^{222}Rn atom (radon) in a 0.75 T magnetic field undergoes radioactive decay, emitting an alpha particle in a direction perpendicular to $\vec{B}$. The alpha particle begins cyclotron motion with a radius of 45 cm. With what energy, in MeV, was the alpha particle emitted?

37. ‖‖ The diameter of an aluminum atom of mass 27 u is approximately 1.2×10^{-10} m. The diameter of the nucleus of an aluminum atom is approximately 8×10^{-15} m. The density of solid aluminum is 2700 kg/m^3.
    a. What is the average density of an aluminum atom?
    b. Your answer to part a was larger than the density of solid aluminum. This suggests that the atoms in solid aluminum have spaces between them rather than being tightly packed together. What is the average volume per atom in solid aluminum? If this volume is a sphere, what is the radius?
       Hint: The volume *per* atom is not the same as the volume *of* an atom.
    c. What is the density of the aluminum nucleus? By what factor is the nuclear density larger than the density of solid aluminum?

38. ‖ The charge-to-mass ratio of a nucleus, in units of e/u, is $q/m = Z/A$. For example, a hydrogen nucleus has $q/m = 1/1 = 1$.
    a. Make a graph of charge-to-mass ratio versus proton number Z for nuclei with $Z = 5, 10, 15, 20, \ldots, 90$. For A, use the average atomic mass shown on the periodic table of elements in Appendix B. Show each of these 18 nuclei as a dot, but don't connect the dots together as a curve.
    b. Describe any trend that you notice in your graph.
    c. What's happening in the nuclei that is responsible for this trend?

39. ‖ If the nucleus is a few fm in diameter, the distance between the centers of two protons must be ≈2 fm.

a. Calculate the repulsive electric force between two protons that are 2.0 fm apart.
b. Calculate the attractive gravitational force between two protons that are 2.0 fm apart. Could gravity be the force that holds the nucleus together?
c. Your answers to parts a and b imply that there must be some other force that binds the nucleus together and prevents the protons from pushing each other out. What characteristics of this force can you deduce from the discussion of the atom and the nucleus in this chapter?

40. ‖‖ A proton is shot straight outward from the surface of a 1.0-mm-diameter glass bead that has been charged to 0.20 nC. If the proton is launched with 520 eV of kinetic energy, what is its kinetic energy, in eV, when it is 2.0 mm from the surface?

41. ‖ In a head-on collision, the closest approach of a 6.24 MeV alpha particle to the center of a nucleus is 6.00 fm. The nucleus is in an atom of what element? Assume the nucleus remains at rest.

42. ‖ Through what potential difference would you need to accelerate an alpha particle, starting from rest, so that it will just reach the surface of a 15-fm-diameter ^{238}U nucleus?

43. ‖ The oxygen nucleus ^{16}O has a radius of 3.0 fm.
    a. With what speed must a proton be fired toward an oxygen nucleus to have a turning point 1.0 fm from the surface? Assume the nucleus remains at rest.
    b. What is the proton's kinetic energy in MeV?

44. ‖ To initiate a nuclear reaction, an experimental nuclear physicist wants to shoot a proton *into* a 5.50-fm-diameter ^{12}C nucleus. The proton must impact the nucleus with a kinetic energy of 3.00 MeV. Assume the nucleus remains at rest.
    a. With what speed must the proton be fired toward the target?
    b. Through what potential difference must the proton be accelerated from rest to acquire this speed?

Challenge Problems

45. An alpha particle approaches a ^{197}Au nucleus with a speed of 1.50×10^7 m/s. As **FIGURE CP37.45** shows, the alpha particle is scattered at a 49° angle at the slower speed of 1.49×10^7 m/s. In what direction does the ^{197}Au nucleus recoil, and with what speed?

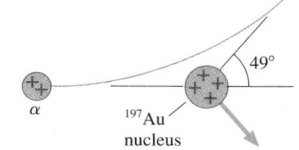

FIGURE CP37.45

46. Physicists first attempted to understand the hydrogen atom by applying the laws of classical physics. Consider an electron of mass m and charge $-e$ in a circular orbit of radius r around a proton of charge $+e$.
    a. Use Newtonian physics to show that the total energy of the atom is $E = -e^2/8\pi\epsilon_0 r$.
    b. Show that the potential energy is -2 times the electron's kinetic energy. This result is called the *virial theorem*.
    c. The minimum energy needed to ionize a hydrogen atom (i.e., to remove the electron) is found experimentally to be 13.6 eV. From this information, what are the electron's speed and the radius of its orbit?

47. Consider an oil droplet of mass m and charge q. We want to determine the charge on the droplet in a Millikan-type experiment. We will do this in several steps. Assume, for simplicity, that the charge is positive and that the electric field between the plates points upward.

    a. An electric field is established by applying a potential difference to the plates. It is found that a field of strength E_0 will cause the droplet to be suspended motionless. Write an expression for the droplet's charge in terms of the suspending field E_0 and the droplet's weight mg.

    b. The field E_0 is easily determined by knowing the plate spacing and measuring the potential difference applied to them. The larger problem is to determine the mass of a microscopic droplet. Consider a mass m falling through a viscous medium in which there is a retarding or drag force. For very small particles, the retarding force is given by $F_{\text{drag}} = -bv$ where b is a constant and v the droplet's velocity. The sign recognizes that the drag force vector points upward when the droplet is falling (negative v). A falling droplet quickly reaches a constant speed, called the *terminal speed*. Write an expression for the terminal speed v_{term} in terms of m, g, and b.

    c. A spherical object of radius r moving slowly through the air is known to experience a retarding force $F_{\text{drag}} = -6\pi\eta rv$

    where η is the *viscosity* of the air. Use this and your answer to part b to show that a spherical droplet of density ρ falling with a terminal velocity v_{term} has a radius

    $$r = \sqrt{\frac{9\eta v_{\text{term}}}{2\rho g}}$$

    d. Oil has a density 860 kg/m³. An oil droplet is suspended between two plates 1.0 cm apart by adjusting the potential difference between them to 1177 V. When the voltage is removed, the droplet falls and quickly reaches constant speed. It is timed with a stopwatch, and falls 3.00 mm in 7.33 s. The viscosity of air is 1.83×10^{-5} kg/m s. What is the droplet's charge?

    e. How many units of the fundamental electric charge does this droplet possess?

48. A classical atom orbiting at frequency f would emit electromagnetic waves of frequency f because the electron's orbit, seen edge-on, looks like an oscillating electric dipole.

    a. At what radius, in nm, would the electron orbiting the proton in a hydrogen atom emit light with a wavelength of 600 nm?

    b. What is the total mechanical energy of this atom?

STOP TO THINK ANSWERS

Stop to Think 37.1: a is emission, b is absorption. All wavelengths in the absorption spectrum are seen in the emission spectrum, but not all wavelengths in the emission spectrum are seen in the absorption spectrum.

Stop to Think 37.2: b. This observation says that all electrons are the same.

Stop to Think 37.3: b. From the right-hand rule with $\vec{v}$ to the right and $\vec{B}$ out of the page.

Stop to Think 37.4: 4. Neutral carbon would have six electrons. C⁺⁺ is missing two.

Stop to Think 37.5: 6 protons and 8 neutrons. The number of protons is the atomic number, which is 6. That leaves $14 - 6 = 8$ neutrons.

38 Quantization

A scanning tunneling microscope image shows an electron standing wave in a "quantum corral" made from 60 iron atoms.

▶ **Looking Ahead** The goal of Chapter 38 is to understand the quantization of energy for light and matter.

Waves and Particles

You've learned that light is an electromagnetic *wave*.

But matters are more complex.

An electron is a basic, subatomic *particle*.

Or maybe not.

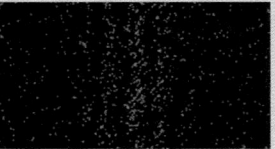

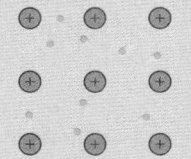

As we found in Chapter 22, light exhibits interference and diffraction, the hallmarks of waviness.

At very low intensity, light hits the screen in "chunks." Sometimes light acts like a particle.

Our model of conduction in metals was based on the motion of particle-like electrons moving through a lattice of fixed ions.

Shooting electrons through two closely spaced slits produces an interference pattern. Sometimes matter acts like a wave.

◀ **Looking Back**
Chapter 22 Interference and diffraction

You'll learn that light and matter have characteristics of both particles *and* waves.

Photons

The **photon model** of light says that

- Light consists of discrete, massless "chunks" called **photons.**
- The energy of a photon of frequency f is *quantized*: $E_{\text{photon}} = hf$, where h is a constant.

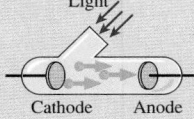

Light

Cathode   Anode

In the *photoelectric effect*, short-wavelength light ejects electrons from a metal surface but long-wavelength light does not. You'll learn how this is evidence for photons.

Matter Waves

You'll learn that the wave-like properties of matter are described by the **de Broglie wavelength** $\lambda = h/mv$. Wave properties are not noticeable for macroscopic matter but are essential to understand matter at the atomic level.

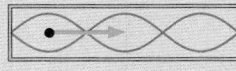

A classical particle confined to a box would bounce back and forth. But reflected matter waves set up a standing wave. You'll see how this leads to quantized energy levels.

Bohr's Atomic Model

By adding quantum ideas to Rutherford's solar-system model of the atom, Bohr created an atomic model in which the electrons can orbit only with certain discrete energies. These are the **energy levels** of the atom.

You'll learn to use *energy-level diagrams* to understand the discrete emission and absorption spectra of gases.

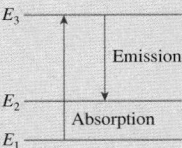

◀ **Looking Back**
Section 37.6 Rutherford's atom

38.1 The Photoelectric Effect

FIGURE 38.1 Lenard's experimental device to study the photoelectric effect.

Ultraviolet light causes the metal cathode to emit electrons. This is the photoelectric effect.

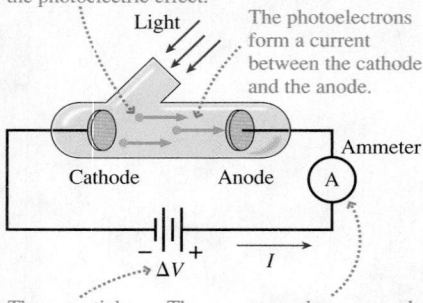

The photoelectrons form a current between the cathode and the anode.

The potential difference can be changed or reversed.

The current can be measured while the potential difference, the light frequency, and the light intensity are varied.

In 1886, Heinrich Hertz, who was the first to demonstrate that electromagnetic waves can be artificially generated, noticed that a negatively charged electroscope could be discharged by shining ultraviolet light on it. Hertz's observation caught the attention of J. J. Thomson, who inferred that the ultraviolet light was causing the electrode to emit electrons, thus restoring itself to electric neutrality. The emission of electrons from a substance due to light striking its surface came to be called the **photoelectric effect.** The emitted electrons are often called *photoelectrons* to indicate their origin, but they are identical in every respect to all other electrons.

Although this discovery might seem to be a minor footnote in the history of science, it soon became a, or maybe *the,* pivotal event that opened the door to new ideas.

Characteristics of the Photoelectric Effect

It was not the discovery itself that dealt the fatal blow to classical physics, but the specific characteristics of the photoelectric effect found around 1900 by one of Hertz's students, Phillip Lenard. Lenard built a glass tube, shown in FIGURE 38.1, with two facing electrodes and a window. After removing the air from the tube, he allowed light to shine on the cathode.

Lenard found a counterclockwise current (clockwise flow of electrons) through the ammeter whenever ultraviolet light was shining on the cathode. There are no junctions in this circuit, so the current must be the same all the way around the loop. The current in the space between the cathode and the anode consists of electrons moving freely through the evacuated space between the electrodes (i.e., not inside a wire) at the *same rate* (same number of electrons per second) as the current in the wire. There is no current if the electrodes are in the dark, so electrons don't spontaneously leap off the cathode. Instead, the light causes electrons to be ejected from the cathode at a steady rate.

Lenard used a battery to establish an adjustable potential difference ΔV between the two electrodes. He then studied how the current I varied as the potential difference and the light's frequency and intensity were changed. Lenard made the following observations:

FIGURE 38.2 The photoelectric current as a function of the light frequency f for light of constant intensity.

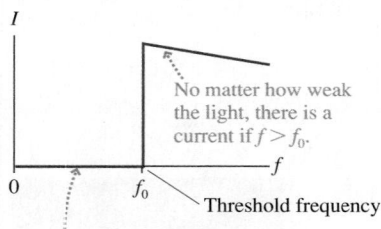

No matter how weak the light, there is a current if $f > f_0$.

Threshold frequency

No matter how intense the light, there is no current if $f < f_0$.

1. The current I is directly proportional to the light intensity. If the light intensity is doubled, the current also doubles.
2. The current appears without delay when the light is applied. To Lenard, this meant within the ≈ 0.1 s with which his equipment could respond. Later experiments showed that the current begins in less than 1 ns.
3. Photoelectrons are emitted *only* if the light frequency f exceeds a **threshold frequency** f_0. This is shown in the graph of FIGURE 38.2.
4. The value of the threshold frequency f_0 depends on the type of metal from which the cathode is made.
5. If the potential difference ΔV is more than about 1 V positive (anode positive with respect to the cathode), the current does not change as ΔV is increased. If ΔV is made negative (anode negative with respect to the cathode), by reversing the battery, the current decreases until, at some voltage $\Delta V = -V_{stop}$ the current reaches zero. The value of V_{stop} is called the **stopping potential.** This behavior is shown in FIGURE 38.3.
6. The value of V_{stop} is the same for both weak light and intense light. A more intense light causes a larger current, as Figure 38.3 shows, but in both cases the current ceases when $\Delta V = -V_{stop}$.

FIGURE 38.3 The photoelectric current as a function of the battery potential.

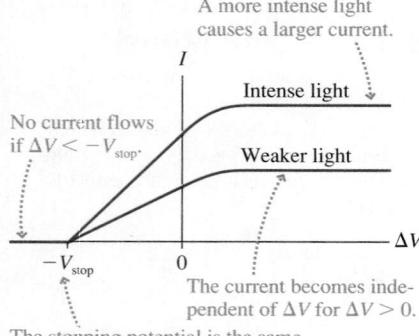

A more intense light causes a larger current.

Intense light

No current flows if $\Delta V < -V_{stop}$.

Weaker light

The current becomes independent of ΔV for $\Delta V > 0$.

The stopping potential is the same for intense light and weak light.

NOTE ▶ We're defining V_{stop} to be a *positive* number. The potential difference that stops the electrons is $\Delta V = -V_{stop}$, with an explicit minus sign. ◀

Classical Interpretation of the Photoelectric Effect

The mere existence of the photoelectric effect is not, as is sometimes assumed, a difficulty for classical physics. You learned in Chapter 25 that electrons are the charge carriers in a metal. The electrons move freely but are bound inside the metal and do not spontaneously spill out of an electrode at room temperature. But a piece of metal heated to a sufficiently high temperature *does* emit electrons in a process called **thermal emission.** The electron gun in an older television or computer display terminal starts with the thermal emission of electrons from a hot tungsten filament.

A useful analogy, shown in FIGURE 38.4, is the water in a swimming pool. Water molecules do not spontaneously leap out of the pool if the water is calm. To remove a water molecule, you must do *work* on it to lift it upward, against the force of gravity. A minimum energy is needed to extract a water molecule, namely the energy needed to lift a molecule that is right at the surface. Removing a water molecule that is deeper requires more than the minimum energy. People playing in the pool add energy to the water, causing waves. If sufficient energy is added, a few water molecules will gain enough energy to splash over the edge and leave the pool.

Similarly, a *minimum* energy is needed to free an electron from a metal. To extract an electron, you would need to exert a force on it and pull it (i.e., do *work* on it) until its speed is large enough to escape. The minimum energy E_0 needed to free an electron is called the **work function** of the metal. Some electrons, like the deeper water molecules, may require more energy than E_0 to escape, but all will require *at least* E_0. Different metals have different work functions; Table 38.1 provides a short list. Notice that work functions are given in electron volts.

Heating a metal, like splashing in the pool, increases the thermal energy of the electrons. At a sufficiently high temperature, the kinetic energy of a small percentage of the electrons may exceed the work function. These electrons can "make it out of the pool" and leave the metal. In practice, there are only a few elements, such as tungsten, for which thermal emission can become significant before the metal melts!

Suppose we could raise the temperature of only the electrons, not the crystal lattice. One possible way to do this is to shine a light wave on the surface. Because electromagnetic waves are absorbed by the conduction electrons, not by the positive ions, the light wave heats only the electrons. Eventually the electrons' energy is transferred to the crystal lattice, via collisions, but if the light is sufficiently intense, the *electron temperature* may be significantly higher than the temperature of the metal. In 1900, it was plausible to think that an intense light source could cause the thermal emission of electrons without melting the metal.

The Stopping Potential

Photoelectrons leave the cathode with kinetic energy. An electron with energy E_{elec} inside the metal loses energy ΔE as it escapes, so it emerges as a photoelectron with $K = E_{\text{elec}} - \Delta E$. The work function energy E_0 is the *minimum* energy needed to remove an electron, so the *maximum* possible kinetic energy of a photoelectron is

$$K_{\text{max}} = E_{\text{elec}} - E_0 \qquad (38.1)$$

Some photoelectrons reach the anode, creating a measurable current, but many do not. However, as FIGURE 38.5 shows:

- A positive anode attracts the photoelectrons. Once all electrons reach the anode, which happens for ΔV greater than about 1 V, a further increase in ΔV does not cause any further increase in the current I. That is why the graph lines become horizontal on the right side of Figure 38.3.
- A negative anode repels the electrons. However, photoelectrons leaving the cathode with sufficient kinetic energy can still reach the anode. The current steadily decreases as the anode voltage becomes increasingly negative until, at the stopping potential, *all* electrons are turned back and the current ceases. This was the behavior observed on the left side of Figure 38.3.

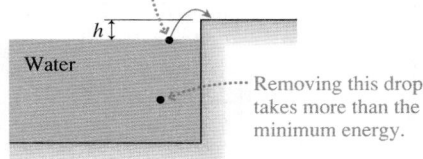

FIGURE 38.4 A swimming pool analogy of electrons in a metal.

The *minimum* energy to remove a drop of water from the pool is *mgh*.

Removing this drop takes more than the minimum energy.

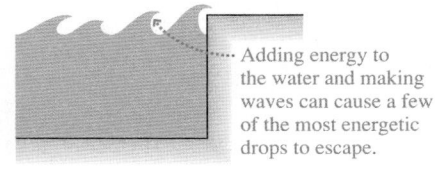

Adding energy to the water and making waves can cause a few of the most energetic drops to escape.

TABLE 38.1 The work function for some of the elements

Element	E_0 (eV)
Potassium	2.30
Sodium	2.75
Aluminum	4.28
Tungsten	4.55
Copper	4.65
Iron	4.70
Gold	5.10

FIGURE 38.5 The photoelectron current depends on the anode potential.

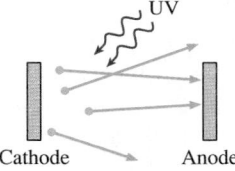

$\Delta V = 0$: The photoelectrons leave the cathode in all directions. Only a few reach the anode.

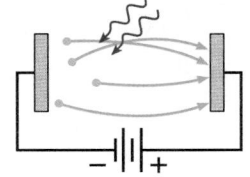

$\Delta V > 0$: A positive anode attracts the photoelectrons to the anode.

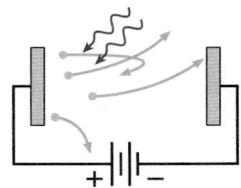

$\Delta V < 0$: A negative anode repels the electrons. Only the very fastest make it to the anode.

FIGURE 38.6 Energy is conserved.

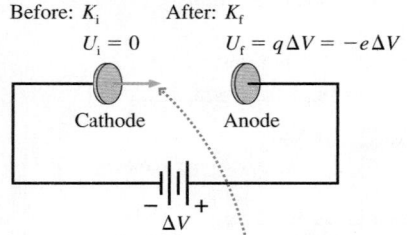

Energy is transformed from kinetic to potential as an electron moves from cathode to anode.

Let the cathode be the point of zero potential energy, as shown in FIGURE 38.6. An electron emitted from the cathode with kinetic energy K_i has initial total energy

$$E_i = K_i + U_i = K_i + 0 = K_i$$

When the electron reaches the anode, which is at potential ΔV relative to the cathode, it has potential energy $U = q\,\Delta V = -e\,\Delta V$ and final total energy

$$E_f = K_f + U_f = K_f - e\,\Delta V$$

From conservation of energy, $E_f = E_i$, the electron's final kinetic energy is

$$K_f = K_i + e\,\Delta V \qquad (38.2)$$

The electron speeds up ($K_f > K_i$) if ΔV is positive. The electron slows down if ΔV is negative, but it still reaches the anode ($K_f > 0$) if K_i is large enough.

An electron with initial kinetic energy K_i will stop just as it reaches the anode if the potential difference is $\Delta V = -K_i/e$. The potential difference that turns back the very fastest electrons, those with $K = K_{max}$, and thus stops the current is

$$\Delta V_{\text{stop fastest electrons}} = -\frac{K_{max}}{e}$$

By definition, the potential difference that causes the electron current to cease is $\Delta V = -V_{stop}$, where V_{stop} is the stopping potential. The stopping potential is

$$V_{stop} = \frac{K_{max}}{e} \qquad (38.3)$$

Thus the stopping potential tells us the maximum kinetic energy of the photoelectrons.

EXAMPLE 38.1 | **The classical photoelectric effect**

A photoelectric-effect experiment is performed with an aluminum cathode. An electron inside the cathode has a speed of 1.5×10^6 m/s. If the potential difference between the anode and cathode is -2.00 V, what is the highest possible speed with which this electron could reach the anode?

MODEL Energy is conserved.

SOLVE If the electron escapes with the maximum possible kinetic energy, its kinetic energy at the anode will be given by Equation 38.2 with $\Delta V = -2.00$ V. The electron's initial kinetic energy is

$$E_{elec} = \frac{1}{2}mv^2 = \frac{1}{2}(9.11 \times 10^{-31}\ \text{kg})(1.5 \times 10^6\ \text{m/s})^2$$

$$= 1.025 \times 10^{-18}\ \text{J} = 6.41\ \text{eV}$$

Its maximum possible kinetic energy as it leaves the cathode is

$$K_i = K_{max} = E_{elec} - E_0 = 2.13\ \text{eV}$$

where $E_0 = 4.28$ eV is the work function of aluminum. Thus the kinetic energy at the anode, given by Equation 38.2, is

$$K_f = K_i + e\,\Delta V = 2.13\ \text{eV} - (e)(2.00\ \text{V}) = 0.13\ \text{eV}$$

Notice that the electron loses 2.00 eV of *energy* as it moves through the *potential* difference of -2.00 V, so we can compute the final kinetic energy in eV without having to convert to joules. However, we must convert K_f to joules to find the final speed:

$$K_f = \frac{1}{2}mv_f^2 = 0.13\ \text{eV} = 2.1 \times 10^{-20}\ \text{J}$$

$$v_f = \sqrt{\frac{2K_f}{m}} = 2.1 \times 10^5\ \text{m/s}$$

Limits of the Classical Interpretation

A classical analysis has provided a possible explanation of observations 1 and 5 above. But nothing in this explanation suggests that there should be a threshold frequency, as Lenard found. If a weak intensity at a frequency just slightly above f_0 can generate a current, why can't a strong intensity at a frequency just slightly below f_0 do so?

What about Lenard's observation that the current starts instantly? If the photo-electrons are due to thermal emission, it should take some time for the light to raise the electron temperature sufficiently high for some to escape. The experimental evidence was in sharp disagreement. And more intense light would be expected to heat the electrons to a higher temperature. Doing so should increase the maximum kinetic energy of the photoelectrons and thus should increase the stopping potential V_{stop}. But as Lenard found, the stopping potential is the same for strong light as it is for weak light.

Although the mere presence of photoelectrons did not seem surprising, classical physics was unable to explain the observed behavior of the photoelectrons. The threshold frequency and the instant current seemed particularly anomalous.

38.2 Einstein's Explanation

Albert Einstein, seen in FIGURE 38.7, was a little-known young man of 26 in 1905. He had recently graduated from the Polytechnic Institute in Zurich, Switzerland, with the Swiss equivalent of a Ph.D. in physics. Although his mathematical brilliance was recognized, his overall academic record was mediocre. Rather than pursue an academic career, Einstein took a job with the Swiss Patent Office in Bern. This was a fortuitous choice because it provided him with plenty of spare time to think about physics.

FIGURE 38.7 A young Einstein.

In 1905, Einstein published his initial paper on the theory of relativity, the subject for which he is most well known to the general public. He also published another paper, on the nature of light. In it Einstein offered an exceedingly simple but amazingly bold idea to explain Lenard's photoelectric-effect data.

A few years earlier, in 1900, the German physicist Max Planck had been trying to understand the details of the rainbow-like blackbody spectrum of light emitted by a glowing hot object. As we noted in the preceding chapter, this problem didn't yield to a classical physics analysis, but Planck found that he could calculate the spectrum perfectly if he made an unusual assumption. The atoms in a solid vibrate back and forth around their equilibrium positions with frequency f. You learned in Chapter 14 that the energy of a simple harmonic oscillator depends on its amplitude and can have *any* possible value. But to predict the spectrum correctly, Planck had to assume that the oscillating atoms are *not* free to have any possible energy. Instead, the energy of an atom vibrating with frequency f has to be one of the specific energies $E = 0, hf, 2hf, 3hf, \ldots$, where h is a constant. That is, the vibration energies are *quantized*.

Planck was able to determine the value of the constant h by comparing his calculations of the spectrum to experimental measurements. The constant that he introduced into physics is now called **Planck's constant.** Its contemporary value is

$$h = 6.63 \times 10^{-34} \, \text{J s} = 4.14 \times 10^{-15} \, \text{eV s}$$

The first value, with SI units, is the proper one for most calculations, but you will find the second to be useful when energies are expressed in eV.

Einstein was the first to take Planck's quantization idea seriously. He went even further and suggested that **electromagnetic radiation itself is quantized!** That is, light is not really a continuous wave but, instead, arrives in small packets or bundles of energy. Einstein called each packet of energy a **light quantum,** and he postulated that the energy of one light quantum is directly proportional to the frequency of the light. That is, each quantum of light has energy

$$E = hf \tag{38.4}$$

where h is Planck's constant and f is the frequency of the light.

The idea of light quanta is subtle, so let's look at an analogy with raindrops. A downpour has a torrent of raindrops, but in a light shower the drops are few. The difference between "intense" rain and "weak" rain is the *rate* at which the drops arrive. An intense rain makes a continuous noise on the roof, so you are not aware of the individual drops, but the individual drops become apparent during a light rain.

Similarly, intense light has so many quanta arriving per second that the light seems continuous, but very weak light consists of only a few quanta per second. And just as raindrops come in different sizes, with larger-mass drops having larger kinetic energy, higher-frequency light quanta have a larger amount of energy. Although this analogy is not perfect, it does provide a useful mental picture of light quanta arriving at a surface.

EXAMPLE 38.2 | **Light quanta**

The retina of your eye has three types of color photoreceptors, called *cones*, with maximum sensitivities at 437 nm, 533 nm, and 575 nm. For each, what is the energy of one quantum of light having that wavelength?

MODEL The energy of light is quantized.

SOLVE Light with wavelength λ has frequency $f = c/\lambda$. The energy of one quantum of light at this wavelength is

$$E = hf = \frac{hc}{\lambda}$$

The calculation requires λ to be in m, but it is useful to have Planck's constant in eV s. At 437 nm, we have

$$E = \frac{(4.14 \times 10^{-15} \text{ eV s})(3.00 \times 10^8 \text{ m/s})}{437 \times 10^{-9} \text{ m}} = 2.84 \text{ eV}$$

Carrying out the same calculation for the other two wavelengths gives $E = 2.33$ eV at 533 nm and $E = 2.16$ eV at 575 nm.

ASSESS The electron volt turns out to be more convenient than the joule for describing the energy of light quanta. Because these wavelengths span a good fraction of the visible spectrum of 400–700 nm, you can see that visible light corresponds to light quanta having energy of roughly 2–3 eV.

Einstein's Postulates

Einstein framed three postulates about light quanta and their interaction with matter:

1. Light of frequency f consists of discrete quanta, each of energy $E = hf$. Each photon travels at the speed of light c.
2. Light quanta are emitted or absorbed on an all-or-nothing basis. A substance can emit 1 or 2 or 3 quanta, but not 1.5. Similarly, an electron in a metal cannot absorb half a quantum but, instead, only an integer number.
3. A light quantum, when absorbed by a metal, delivers its entire energy to *one* electron.

NOTE ▶ These three postulates—that light comes in chunks, that the chunks cannot be divided, and that the energy of one chunk is delivered to one electron—are crucial for understanding the new ideas that will lead to quantum physics. They are completely at odds with the concepts of classical physics, where energy can be continuously divided and shared, so they deserve careful thought. ◀

Let's look at how Einstein's postulates apply to the photoelectric effect. If Einstein is correct, the light of frequency f shining on the metal is a flow of light quanta, each of energy hf. Each quantum is absorbed by *one* electron, giving that electron an energy $E_{\text{elec}} = hf$. This leads us to several interesting conclusions:

1. An electron that has just absorbed a quantum of light energy has $E_{\text{elec}} = hf$. (The electron's thermal energy at room temperature is so much less than hf that we can neglect it.) **FIGURE 38.8** shows that this electron can escape from the metal, becoming a photoelectron, if

$$E_{\text{elec}} = hf \geq E_0 \qquad (38.5)$$

FIGURE 38.8 The creation of a photoelectron.

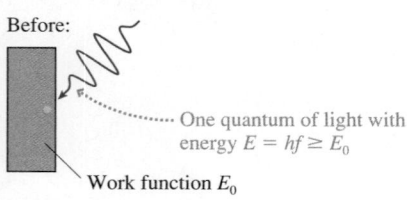

Before:

One quantum of light with energy $E = hf \geq E_0$

Work function E_0

After:

A single electron has absorbed the entire energy of the light quantum and has escaped.

where, you will recall, the work function E_0 is the minimum energy needed to free an electron from the metal. As a result, there is a *threshold frequency*

$$f_0 = \frac{E_0}{h} \tag{38.6}$$

for the ejection of photoelectrons. If f is less than f_0, even by just a small amount, none of the electrons will have sufficient energy to escape no matter how intense the light. But even very weak light with $f \geq f_0$ will give a few electrons sufficient energy to escape **because each light quantum delivers all of its energy to one electron.** This threshold behavior is exactly what Lenard observed.

NOTE ▶ The threshold frequency is directly proportional to the work function. Metals with large work functions, such as iron, copper, and gold, exhibit the photoelectric effect only when illuminated by high-frequency ultraviolet light. Photoemission occurs with lower-frequency visible light for metals with smaller values of E_0, such as sodium and potassium. ◀

2. A more intense light means *more quanta* of the same energy, not more energetic quanta. These quanta eject a larger number of photoelectrons and cause a larger current, exactly as observed.
3. There is a distribution of kinetic energies, because different photoelectrons require different amounts of energy to escape, but the *maximum* kinetic energy is

$$K_{\max} = E_{\text{elec}} - E_0 = hf - E_0 \tag{38.7}$$

As we noted in Equation 38.3, the stopping potential V_{stop} is directly proportional to $K_{\max}$. Einstein's theory predicts that the stopping potential is related to the light frequency by

$$V_{\text{stop}} = \frac{K_{\max}}{e} = \frac{hf - E_0}{e} \tag{38.8}$$

The stopping potential does *not* depend on the intensity of the light. Both weak light and intense light will have the same stopping potential, which Lenard had observed but which could not previously be explained.

4. If each light quantum transfers its energy hf to just one electron, that electron *immediately* has enough energy to escape. The current should begin instantly, with no delay, exactly as Lenard had observed.

Using the swimming pool analogy again, FIGURE 38.9 shows a pebble being thrown into the pool. The pebble increases the energy of the water, but the increase is shared among all the molecules in the pool. The increase in the water's energy is barely enough to make ripples, not nearly enough to splash water out of the pool. But suppose *all* the pebble's energy could go to *one drop* of water that didn't have to share it. That one drop of water could easily have enough energy to leap out of the pool. Einstein's hypothesis that a light quantum transfers all its energy to one electron is equivalent to the pebble transferring all its energy to one drop of water.

A Prediction

Not only do Einstein's hypotheses explain all of Lenard's observations, they also make a new prediction. According to Equation 38.8, the stopping potential should be a linearly increasing function of the light's frequency f. We can rewrite Equation 38.8 in terms of the threshold frequency $f_0 = E_0/h$ as

$$V_{\text{stop}} = \frac{h}{e}(f - f_0) \tag{38.9}$$

FIGURE 38.9 A pebble transfers energy to the water.

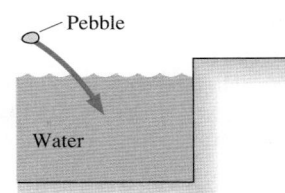

Classically, the energy of the pebble is shared by all the water molecules. One pebble causes only very small waves.

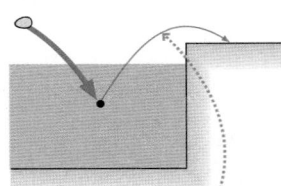

If the pebble could give *all* its energy to one drop, that drop could easily splash out of the pool.

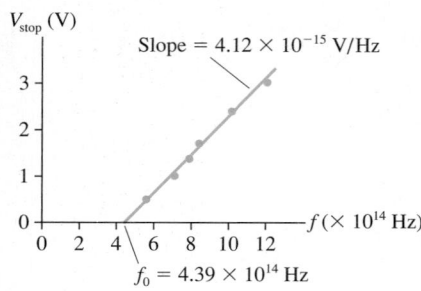

A graph of the stopping potential V_{stop} versus the light frequency f should start from zero at $f = f_0$, then rise linearly with a slope of h/e. In fact, the slope of the graph provides a way to measure Planck's constant h.

Lenard had not measured the stopping potential for different frequencies, so Einstein offered this as an untested prediction of his postulates. Robert Millikan, known for his oil-drop experiment to measure e, took up the challenge. Some of Millikan's data for a cesium cathode are shown in **FIGURE 38.10**. As you can see, Einstein's prediction of a linear relationship between f and V_{stop} was confirmed.

Millikan measured the slope of his graph and multiplied it by the value of e (which he had measured a few years earlier in the oil-drop experiment) to find h. His value agreed with the value that Planck had determined in 1900 from an entirely different experiment. Light quanta, whether physicists liked the idea or not, were real.

EXAMPLE 38.3 **The photoelectric threshold frequency**

What are the threshold frequencies and wavelengths for photoemission from sodium and from aluminum?

SOLVE Table 38.1 gives the sodium work function as $E_0 = 2.75$ eV. Aluminum has $E_0 = 4.28$ eV. We can use Equation 38.6, with h in units of eV s, to calculate

$$f_0 = \frac{E_0}{h} = \begin{cases} 6.64 \times 10^{14} \text{ Hz} & \text{sodium} \\ 10.34 \times 10^{14} \text{ Hz} & \text{aluminum} \end{cases}$$

These frequencies are converted to wavelengths with $\lambda = c/f$, giving

$$\lambda = \begin{cases} 452 \text{ nm} & \text{sodium} \\ 290 \text{ nm} & \text{aluminum} \end{cases}$$

ASSESS The photoelectric effect can be observed with sodium for $\lambda < 452$ nm. This includes blue and violet visible light but not red, orange, yellow, or green. Aluminum, with a larger work function, needs ultraviolet wavelengths $\lambda < 290$ nm.

EXAMPLE 38.4 **Maximum photoelectron speed**

What is the maximum photoelectron speed if sodium is illuminated with light of 300 nm?

SOLVE The light frequency is $f = c/\lambda = 1.00 \times 10^{15}$ Hz, so each light quantum has energy $hf = 4.14$ eV. The maximum kinetic energy of a photoelectron is

$$K_{max} = hf - E_0 = 4.14 \text{ eV} - 2.75 \text{ eV} = 1.39 \text{ eV}$$
$$= 2.22 \times 10^{-19} \text{ J}$$

Because $K = \frac{1}{2}mv^2$, where m is the electron's mass, not the mass of the sodium atom, the maximum speed of a photoelectron leaving the cathode is

$$v_{max} = \sqrt{\frac{2K_{max}}{m}} = 6.99 \times 10^5 \text{ m/s}$$

Note that we had to convert K_{max} to SI units of J before calculating a speed in m/s.

STOP TO THINK 38.1 The work function of metal A is 3.0 eV. Metals B and C have work functions of 4.0 eV and 5.0 eV, respectively. Ultraviolet light shines on all three metals, creating photoelectrons. Rank in order, from largest to smallest, the stopping potentials for A, B, and C.

38.3 Photons

Einstein was awarded the Nobel Prize in 1921 not for his theory of relativity, as many suppose, but for his explanation of the photoelectric effect. Although Planck had made the first suggestion, it was Einstein who showed convincingly that energy is quantized. Quanta of light energy were later given the name **photons.**

But just what are photons? To begin our explanation, let's return to the experiment that showed most dramatically the wave nature of light—Young's double-slit interference experiment. We will make a change, though: We will dramatically lower the light intensity by inserting filters between the light source and the slits. The fringes will be too dim to see by eye, so we will replace the viewing screen with a detector that can build up an image over time.

What would we predict for the outcome of this experiment? If light is a wave, there is no reason to think that the nature of the interference fringes will change. The detector should continue to show alternating light and dark bands.

FIGURE 38.11 shows the actual outcome at four different times. At early times, contrary to our prediction, the detector shows not dim interference fringes but discrete, bright dots. If we didn't know that light is a wave, we would interpret the dots as evidence that light is a stream of some type of particle-like objects. They arrive one by one, seemingly randomly, and each is localized at a specific point on the detector. (Waves, you will recall, are not localized at a specific point in space.)

As the detector builds up the image for a longer period of time, we see that these dots are not entirely random. They are grouped into bands at *exactly* the positions where we expected to see bright constructive-interference fringes. No dot ever appears at points of destructive interference. After a long time, the individual dots overlap and the image looks like the photographs of interference fringes in Chapter 22.

We're detecting individual photons! Most light sources—even very dim sources—emit such vast numbers of photons that you are aware of only their wave-like superposition, just as you notice only the roar of a heavy rain on your roof and not the individual raindrops. But at extremely low intensities the light begins to appear as a stream of individual photons, like the random patter of raindrops when it is barely sprinkling. Each dot on the detector in Figure 38.11 signifies a point where one particle-like photon delivered its energy and caused a measurable signal.

But photons are certainly not classical particles. Classical particles, such as Newton's corpuscles of light, would travel in straight lines through the two slits of a double-slit experiment and make just two bright areas on the detector. Instead, as Figure 38.11 shows, the *particle*-like photons seem to be landing at places where a *wave* undergoes constructive interference, thus forming the bands of dots.

Today, it is quite feasible to do this experiment with a light intensity so low that only one photon at a time is passing through the double-slit apparatus. But if one photon at a time can build up a wave-like interference pattern, what is the photon interfering with? The only possible answer is that **the photon is interfering *with itself.*** Nothing else is present. But if each photon interferes with itself, rather than with other photons, then each photon, despite the fact that it is a particle-like object, must somehow go through *both* slits! Photons seem to be both wave-like *and* particle-like at the same time.

This all seems pretty crazy, but it's the way light actually behaves. **Sometimes light exhibits particle-like behavior and sometimes it exhibits wave-like behavior.** The thing we call *light* is stranger and more complex than it first appeared, and there is no way to reconcile these seemingly contradictory behaviors. We have to accept nature as it is rather than hoping that nature will conform to our expectations. Furthermore, as we will see, this half-wave/half-particle behavior is not restricted to light.

The Photon Model of Light

The **photon model** of light consists of three basic postulates:

1. Light consists of discrete, massless units called *photons*. A photon travels in vacuum at the speed of light.
2. Each photon has energy

$$E_{\text{photon}} = hf \qquad (38.10)$$

where f is the frequency of the light and $h = 6.63 \times 10^{-34}$ Js is Planck's constant. In other words, the light comes in discrete "chunks" of energy hf.
3. The superposition of a sufficiently large number of photons has the characteristics of a classical light wave.

FIGURE 38.11 A double-slit experiment performed with light of very low intensity.

(a) Image after a very short time

(b) Image after a slightly longer time

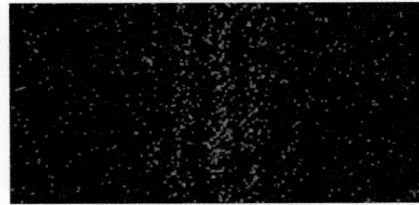

(c) Continuing to build up the image

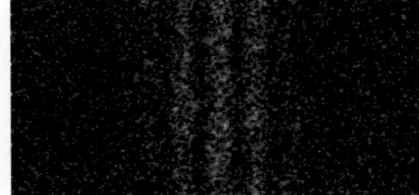

(d) Image after a very long time

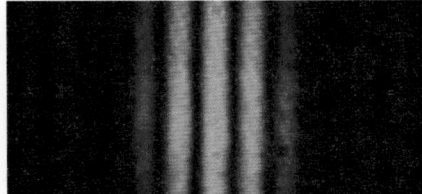

FIGURE 38.12 A wave packet has wave-like and particle-like properties.

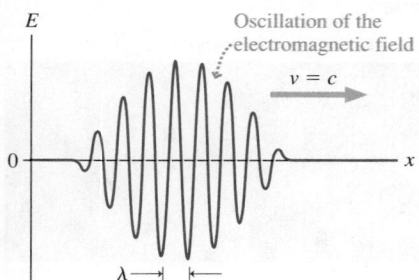

Photons are sometimes visualized as **wave packets.** The electromagnetic wave shown FIGURE 38.12 has a wavelength and a frequency, yet it is also discrete and fairly localized. But this cannot be exactly what a photon is because a wave packet would take a finite amount of time to be emitted or absorbed. This is contrary to much evidence that the entire photon is emitted or absorbed in a single instant; there is no point in time at which the photon is "half absorbed." The wave packet idea, although useful, is still too classical to represent a photon.

In fact, there simply is no "true" mental representation of a photon. Analogies such as raindrops or wave packets can be useful, but none is perfectly accurate. We can detect photons, measure the properties of photons, and put photons to practical use, but the ultimate nature of the photon remains a mystery. To paraphrase Gertrude Stein, "A photon is a photon is a photon."

The Photon Rate

Light, in the raindrop analogy, consists of a stream of photons. For monochromatic light of frequency f, N photons have a total energy $E_{light} = Nhf$. We are usually more interested in the *power* of the light, or the rate (in joules per second, or watts) at which the light energy is delivered. The power is

$$P = \frac{dE_{light}}{dt} = \frac{dN}{dt}hf = Rhf \qquad (38.11)$$

where $R = dN/dt$ is the *rate* at which photons arrive or, equivalently, the number of photons per second.

EXAMPLE 38.5 **The photon rate in a laser beam**

The 1.0 mW light beam of a helium-neon laser ($\lambda = 633$ nm) shines on a screen. How many photons strike the screen each second?

SOLVE The light-beam power, or energy delivered per second, is $P = 1.0$ mW $= 0.0010$ J/s. The frequency of the light is $f = c/\lambda = 4.74 \times 10^{14}$ Hz. The number of photons striking the screen per second, which is the *rate* of arrival of photons, is

$$R = \frac{P}{hf} = 3.2 \times 10^{15} \text{ photons per second}$$

ASSESS That is a lot of photons per second. No wonder we are not aware of individual photons!

STOP TO THINK 38.2 The intensity of a beam of light is increased but the light's frequency is unchanged. Which one (or perhaps more than one) of the following is true?

a. The photons travel faster.
b. Each photon has more energy.
c. The photons are larger.
d. There are more photons per second.

38.4 Matter Waves and Energy Quantization

Prince Louis-Victor de Broglie was a French graduate student in 1924. It had been 19 years since Einstein had shaken the world of physics by introducing photons and blurring the distinction between a particle and a wave. As de Broglie thought about these issues, it seemed that nature should have some kind of symmetry. If light waves could have a particle-like nature, why shouldn't material particles have some kind of wave-like nature? In other words, could **matter waves** exist?

With no experimental evidence to go on, de Broglie reasoned by analogy with Einstein's equation $E = hf$ for the photon and with some of the ideas of his theory of relativity. The details need not concern us, but they led de Broglie to postulate that *if* a material particle of momentum $p = mv$ has a wave-like nature, then its wavelength must be given by

$$\lambda = \frac{h}{p} = \frac{h}{mv} \qquad (38.12)$$

where h is Planck's constant. This is called the **de Broglie wavelength.**

EXAMPLE 38.6 **The de Broglie wavelength of an electron**

What is the de Broglie wavelength of a 1.0 eV electron?

SOLVE An electron with $1.0\,\text{eV} = 1.6 \times 10^{-19}\,\text{J}$ of kinetic energy has speed

$$v = \sqrt{\frac{2K}{m}} = 5.9 \times 10^5 \text{ m/s}$$

Although fast by macroscopic standards, this is a slow electron because it gains this speed by accelerating through a potential difference of a mere 1 V. Its de Broglie wavelength is

$$\lambda = \frac{h}{mv} = 1.2 \times 10^{-9}\,\text{m} = 1.2\text{ nm}$$

ASSESS The electron's wavelength is small, but it is similar to the wavelengths of x rays and larger than the approximately 10^{-10} m spacing of atoms in a crystal.

What would it mean for matter—an electron or a proton or a baseball—to have a wavelength? Would it obey the principle of superposition? Would it exhibit interference and diffraction? The classic test of "waviness" is Young's double-slit experiment. FIGURE 38.13 shows the intensity pattern recorded after 50 keV electrons passed through two slits separated by 1.0 μm. The pattern is clearly a double-slit interference pattern, and the spacing of the fringes is exactly as predicted for a wavelength given by de Broglie's formula. And because the electron beam was weak, with one electron at a time passing through the apparatus, it would appear that each electron—like photons—somehow went through both slits, then interfered with itself before striking the detector!

Surprisingly, electrons—also neutrons—exhibit all the behavior we associate with waves. But electrons and neutrons are subatomic particles. What about entire atoms, aggregates of many fundamental particles? Amazing as it seems, research during the 1980s demonstrated that whole atoms, and even molecules, can produce interference patterns.

FIGURE 38.14 shows an *atom interferometer.* You learned in Chapter 22 that an interferometer, such as the Michelson interferometer, works by dividing a wave front

FIGURE 38.13 A double-slit interference pattern created with electrons.

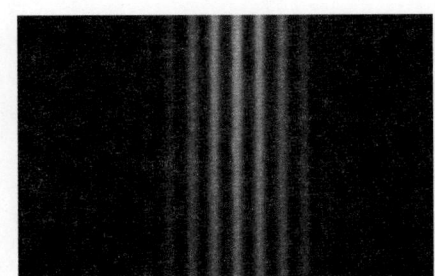

FIGURE 38.14 An atom interferometer.

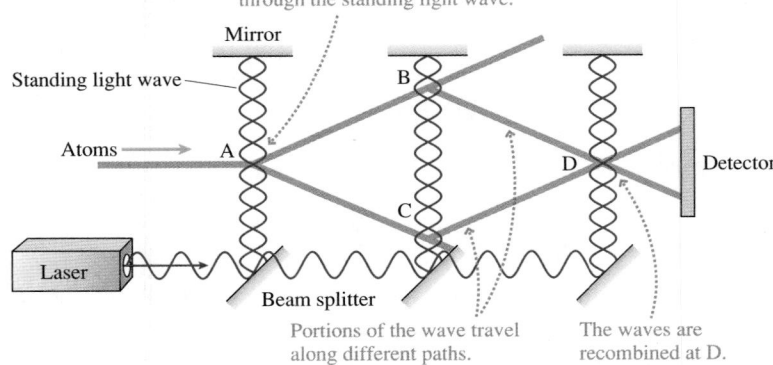

The atom wave is divided at A by diffracting through the standing light wave.

Mirror

Standing light wave

B

Atoms A

C

D Detector

Laser

Beam splitter

Portions of the wave travel along different paths.

The waves are recombined at D.

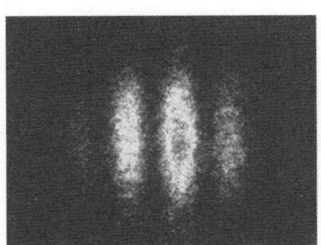

Detector output

into two waves, sending the two waves along separate paths, then recombining them. For light waves, wave division can be accomplished by sending light through the *periodic* slits in a diffraction grating. In an atom interferometer, the atom's matter wave is divided by sending atoms through the *periodic* intensity of a standing light wave.

You can see in the figure that a laser creates three parallel *standing waves* of light, each with nodes spaced a distance $\lambda/2$ apart. The wavelength is chosen so that the light waves exert small forces on an atom in the laser beam. Because the intensity along a standing wave alternates between maximum at the antinodes and zero intensity at the nodes, an atom crossing the laser beam experiences a *periodic* force field. A particle-like atom would be deflected by this periodic force, but a wave is *diffracted*. After being diffracted by the first standing wave at A, an atom is, in some sense, traveling toward both point B *and* point C.

The second standing wave diffracts the atom waves again at points B and C, directing some of them toward D where, with a third diffraction, they are recombined after having traveled along different paths. The detector image shows interference fringes, exactly as would be expected for a wave but completely at odds with the expectation for particles.

The atom interferometer is fascinating because it completely inverts everything we previously learned about interference and diffraction. The scientists who studied the wave nature of light during the 19th century aimed light (a wave) at a diffraction grating (a periodic structure of matter) and found that it diffracted. Now we aim atoms (matter) at a standing wave (a periodic structure of light) and find that the atoms diffract. The roles of light and matter have been reversed!

Quantization of Energy

The fact that matter has wave-like properties is not merely a laboratory curiosity; the implications are profound. Foremost among them is that the energy of matter, like that of light, is quantized.

We'll illustrate quantization with a simple system that physicists call "a particle in a box." FIGURE 38.15a shows a particle of mass m moving in one dimension as it bounces back and forth with speed v between the ends of a box of length L. The width of the box is irrelevant, so we'll call this a *one-dimensional box*. We'll assume that the collisions at the ends are perfectly elastic, so the particle's energy—entirely kinetic—never changes. According to classical physics, there are no restrictions on the particle's speed or energy.

But if matter has wave-like properties, perhaps we should consider the particle in a box to be a *wave* reflecting back and forth between the ends of the box, as shown in FIGURE 38.15b. These are the conditions that create standing waves. You learned in Chapter 21 that a standing wave of length L *must* have one of the wavelengths given by

$$\lambda_n = \frac{2L}{n} \qquad n = 1, 2, 3, 4, \ldots \qquad (38.13)$$

If the confined particle has wave-like properties, it should satisfy both Equation 38.13 *and* the de Broglie relationship $\lambda = h/mv$. That is, a particle in a box should obey the relationship

$$\lambda_n = \frac{h}{mv} = \frac{2L}{n}$$

Thus the particle's speed must be

$$v_n = n\left(\frac{h}{2Lm}\right) \qquad n = 1, 2, 3, \ldots \qquad (38.14)$$

In other words, the particle cannot bounce back and forth with just any speed. Rather, it can have *only* those specific speeds v_n, given by Equation 38.14, for which the de Broglie wavelength creates a standing wave in the box.

FIGURE 38.15 A particle confined in a box of length L.

(a) A classical particle bounces back and forth.

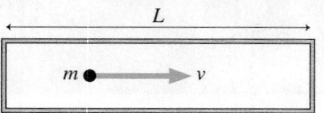

(b) A reflected wave creates a standing wave.

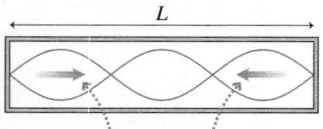

Matter waves travel in both directions.

Thus the particle's energy, which is purely kinetic energy, is

$$E_n = \frac{1}{2}mv_n^2 = n^2\frac{h^2}{8mL^2} \qquad n = 1, 2, 3, \ldots \qquad (38.15)$$

De Broglie's hypothesis about the wave-like properties of matter leads us to the remarkable conclusion that **a particle confined in a box can have only certain energies.** We say that its energy is **quantized.** The energy of the particle in the box can be $1(h^2/8mL^2)$, or $4(h^2/8mL^2)$, or $9(h^2/8mL^2)$, but it *cannot* have an energy between these values.

The possible values of the particle's energy are called **energy levels,** and the integer n that characterizes the energy levels is called the **quantum number.** The quantum number can be found by counting the antinodes, just as you learned to do for standing waves on a string. The standing wave shown in Figure 38.15 is $n = 3$, thus its energy is E_3.

We can rewrite Equation 38.15 in the useful form

$$E_n = n^2 E_1 \qquad (38.16)$$

where

$$E_1 = \frac{h^2}{8mL^2} \qquad (38.17)$$

is the **fundamental quantum of energy** for a particle in a one-dimensional box. It is analogous to the fundamental frequency f_1 of a standing wave on a string.

EXAMPLE 38.7 **The energy levels of a virus**

A 30-nm-diameter virus is about the smallest imaginable macroscopic particle. What is the fundamental quantum of energy for this virus if confined in a one-dimensional cell of length 1.0 μm? The density of a virus is very close to that of water.

MODEL Model the virus as a particle in a box.

SOLVE The mass of a virus is $m = \rho V$, where the volume is $\frac{4}{3}\pi r^3$. A quick calculation shows that a 30-nm-diameter virus has mass $m = 1.4 \times 10^{-20}$ kg. The confinement length is $L = 1.0 \times 10^{-6}$ m. From Equation 38.17, the fundamental quantum of energy is

$$E_1 = \frac{h^2}{8mL^2} = \frac{(6.63 \times 10^{-34}\,\text{J s})^2}{8(1.4 \times 10^{-20}\,\text{kg})(1.0 \times 10^{-6}\,\text{m})^2}$$

$$= 3.9 \times 10^{-36}\,\text{J} = 2.5 \times 10^{-17}\,\text{eV}$$

ASSESS This is such an incredibly small amount of energy that there is no hope of distinguishing between energies of E_1 or $4E_1$ or $9E_1$. For any macroscopic particle, even one this tiny, the allowed energies will *seem* to be perfectly continuous. We will not observe the quantization.

EXAMPLE 38.8 **The energy levels of an electron**

As a very simple model of a hydrogen atom, consider an electron confined in a one-dimensional box of length 0.10 nm, about the size of an atom. What are the first three allowed energy levels?

SOLVE We can use Equation 38.17, with $m_{\text{elec}} = 9.11 \times 10^{-31}$ kg and $L = 1.0 \times 10^{-10}$ m, to find that the fundamental quantum of energy is $E_1 = 6.0 \times 10^{-18}$ J = 38 eV. Thus the first three allowed energies of an electron in a 0.10 nm box are

$$E_1 = 38\,\text{eV}$$

$$E_2 = 4E_1 = 152\,\text{eV}$$

$$E_3 = 9E_1 = 342\,\text{eV}$$

ASSESS You'll soon see that the results are way off. This model of a hydrogen atom is *too* simple to capture essential details.

It is the *confinement* of the particle in a box that leads to standing matter waves and thus energy quantization. Our goal is to extend this idea to atoms. An atom is certainly more complicated than a one-dimensional box, but an electron is "confined" within an atom. Thus an electron in an atom must be some kind of three-dimensional standing wave and, like the particle in a box, must have quantized energies. De Broglie's idea is steering us toward a new theory of matter.

What is the quantum number of this particle confined in a box?

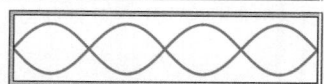

38.5 Bohr's Model of Atomic Quantization

Thomson's electron and Rutherford's nucleus made it clear that the atom has a *structure* of some sort. The challenge at the beginning of the 20th century was to deduce, from experimental evidence, the correct structure. The difficulty of this task cannot be exaggerated. The evidence about atoms, such as observations of atomic spectra, was very indirect, and experiments were carried out with only the simplest measuring devices.

Rutherford's nuclear model was the most successful of various proposals, but Rutherford's model failed to explain why atoms are stable or why their spectra are discrete. A missing piece of the puzzle, although not recognized as such for a few years, was Einstein's 1905 introduction of light quanta. If light comes in discrete packets of energy, which we now call photons, and if atoms emit and absorb light, what does that imply about the structure of the atoms?

This was the question posed by the Danish physicist Niels Bohr, shown as a young man in FIGURE 38.16. After receiving his doctoral degree in physics in 1911, Bohr went to England to work in Rutherford's laboratory. Rutherford had just, within the previous year, completed his development of the nuclear model of the atom. Rutherford's model certainly contained a kernel of truth, but Bohr wanted to understand how a solar-system-like atom could be stable and not radiate away all its energy. He soon recognized that Einstein's light quanta had profound implications for the structure of atoms. In 1913, Bohr proposed a new model of the atom in which he added quantization to Rutherford's nuclear atom.

The basic assumptions of the **Bohr model of the atom** are as follows:

FIGURE 38.16 Niels Bohr.

Understanding Bohr's model

Electrons can exist in only certain allowed orbits.

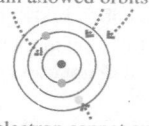

An electron cannot exist here, where there is no allowed orbit.

This is one stationary state. This is another stationary state.

Stationary states

These other states are **excited states**.

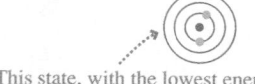

This state, with the lowest energy E_1, is the **ground state**. It is stable and can persist indefinitely.

Photon emission

Excited-state electron

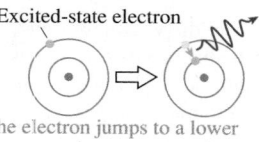

The electron jumps to a lower energy stationary state and emits a photon.

Photon absorption

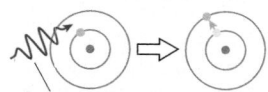

Approaching photon
The electron absorbs the photon and jumps to a higher energy stationary state.

Collisional excitation

Approaching particle   Particle loses energy.

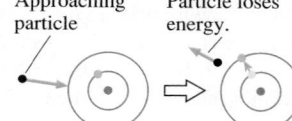

The particle transfers energy to the atom in the collision and excites the atom.

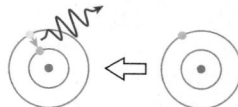

An atom in an excited state jumps to lower states, emitting a photon at each jump.

1. The electrons in an atom can exist in only certain *allowed orbits*. A particular arrangement of electrons in these orbits is called a **stationary state.**

2. Each stationary state has a discrete, well-defined energy E_n. That is, atomic energies are *quantized*. The stationary states are labeled by the *quantum number n* in order of increasing energy: $E_1 < E_2 < E_3 < \cdots$.

3. An atom can undergo a **transition** or **quantum jump** from one stationary state to another by emitting or absorbing a photon whose energy is exactly equal to the energy difference between the two stationary states.

4. Atoms can also move from a lower energy state to a higher energy state by absorbing energy in a collision with an electron or other atom in a process called **collisional excitation.**

The excited atoms soon jump down to lower states, eventually ending in the stable ground state.

Bohr's model builds upon Rutherford's model, but it adds two new ideas that are derived from Einstein's ideas of quanta. The first, expressed in assumption 1, is that only certain electron orbits are "allowed" or can exist. The second, expressed in assumption 3, is that **the atom can jump from one state to another by emitting or absorbing a photon of just the right frequency to conserve energy.**

According to Einstein, a photon of frequency f has energy $E_{\text{photon}} = hf$. If an atom jumps from an initial state with energy E_i to a final state with energy E_f, energy will be conserved if the atom emits or absorbs a photon with $E_{\text{photon}} = \Delta E_{\text{atom}} = |E_f - E_i|$. This photon must have frequency

$$f_{\text{photon}} = \frac{\Delta E_{\text{atom}}}{h} \qquad (38.18)$$

if it is to add or carry away exactly the right amount of energy. The total energy of the atom-plus-light system is conserved.

NOTE ▶ When an atom is excited to a higher energy level by absorbing a photon, the photon vanishes. Thus energy conservation requires $E_{\text{photon}} = \Delta E_{\text{atom}}$. When an atom is excited to a higher energy level in a collision with a particle, such as an electron or another atom, the particle still exists after the collision and still has energy. Thus energy conservation requires the less stringent condition $E_{\text{particle}} \geq \Delta E_{\text{atom}}$. ◀

The implications of Bohr's model are profound. In particular:

1. **Matter is stable.** An atom in its ground state has no states of any lower energy to which it can jump. It can remain in the ground state forever.

2. **Atoms emit and absorb a *discrete spectrum*.** Only those photons whose frequencies match the energy *intervals* between the stationary states can be emitted or absorbed. Photons of other frequencies cannot be emitted or absorbed without violating energy conservation.

3. **Emission spectra can be produced by collisions.** In a gas discharge tube, the current-carrying electrons moving through the tube occasionally collide with the atoms. A collision transfers energy to an atom and can kick the atom to an excited state. Once the atom is in an excited state, it can emit photons of light—a discrete emission spectrum—as it jumps back down to lower-energy states.

4. **Absorption wavelengths are a subset of the wavelengths in the emission spectrum.** Recall that all the lines seen in an absorption spectrum are also seen in emission, but many emission lines are *not* seen in absorption. According to Bohr's model, most atoms, most of the time, are in their lowest energy state, the $n = 1$ ground state. Thus the absorption spectrum consists of *only* those transitions such as $1 \rightarrow 2$, $1 \rightarrow 3$, ... in which the atom jumps from $n = 1$ to a higher value of n by absorbing a photon. Transitions such as $2 \rightarrow 3$ are *not* observed because there are essentially no atoms in $n = 2$ at any instant of time. On the other hand, atoms that have been excited to the $n = 3$ state by collisions can emit photons corresponding to transitions $3 \rightarrow 1$ *and* $3 \rightarrow 2$. Thus the wavelength corresponding to $\Delta E_{\text{atom}} = E_3 - E_1$ is seen in both emission and absorption, but transitions with $\Delta E_{\text{atom}} = E_3 - E_2$ occur in emission only.

5. **Each element in the periodic table has a unique spectrum.** The energies of the stationary states are the energies of the orbiting electrons. Different elements, with different numbers of electrons, have different stable orbits and thus different stationary states. States with different energies emit and absorb photons of different wavelengths.

EXAMPLE 38.9 **The wavelength of an emitted photon**

An atom has stationary states with energies $E_j = 4.00$ eV and $E_k = 6.00$ eV. What is the wavelength of a photon emitted in a quantum jump from state k to state j?

MODEL To conserve energy, the emitted photon must have exactly the energy lost by the atom in the quantum jump.

SOLVE The atom can jump from the higher energy state k to the lower energy state j by emitting a photon. The atom's change in energy is $\Delta E_{atom} = |E_j - E_k| = 2.00$ eV, so the photon energy must be $E_{photon} = 2.00$ eV.

The photon frequency is

$$f = \frac{E_{photon}}{h} = \frac{2.00 \text{ eV}}{4.14 \times 10^{-15} \text{ eV s}} = 4.83 \times 10^{14} \text{ Hz}$$

The wavelength of this photon is

$$\lambda = \frac{c}{f} = 621 \text{ nm}$$

ASSESS 621 nm is a visible-light wavelength. Notice that the wavelength depends on the *difference* between the atom's energy levels, not the *values* of the energies.

Energy-Level Diagrams

An **energy-level diagram,** such as the one shown in FIGURE 38.17, is a useful pictorial representation of the stationary-state energies. An energy-level diagram is less a graph than it is a picture. The vertical axis represents energy, but the horizontal axis is not a scale. Think of this as a picture of a ladder in which the energies are the rungs of the ladder. The lowest rung, with energy E_1, is the ground state. Higher rungs are labeled by their quantum numbers, $n = 2, 3, 4, \ldots$.

FIGURE 38.17 An energy-level diagram.

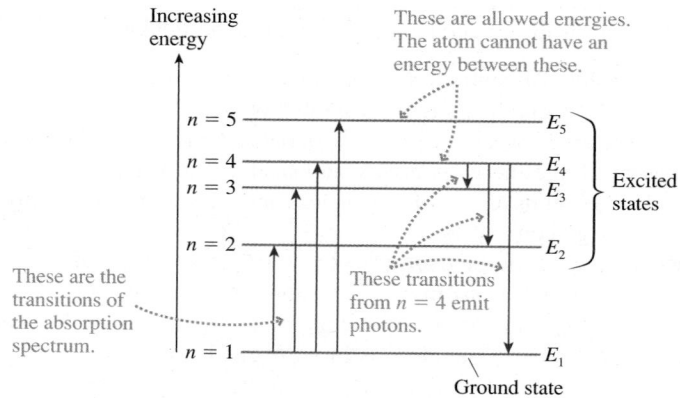

Energy-level diagrams are especially useful for showing transitions, or quantum jumps, in which a photon of light is emitted or absorbed. As examples, Figure 38.17 shows upward transitions in which a photon is absorbed by a ground-state atom ($n = 1$) and downward transitions in which a photon is emitted from an $n = 4$ excited state.

EXAMPLE 38.10 **Emission and absorption**

An atom has stationary states $E_1 = 0.00$ eV, $E_2 = 3.00$ eV, and $E_3 = 5.00$ eV. What wavelengths are observed in the absorption spectrum and in the emission spectrum of this atom?

MODEL Photons are emitted when an atom undergoes a quantum jump from a higher energy level to a lower energy level. Photons are absorbed in a quantum jump from a lower energy level to a higher energy level. But most of the atoms are in the $n = 1$ ground state, so the only quantum jumps seen in the absorption spectrum start from the $n = 1$ state.

VISUALIZE FIGURE 38.18 shows an energy-level diagram for the atom.

FIGURE 38.18 The atom's energy-level diagram.

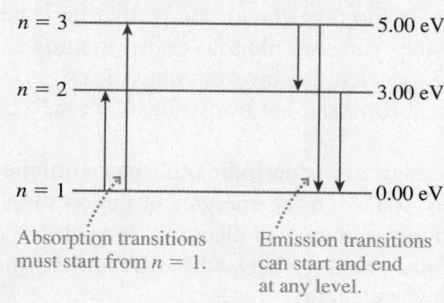

SOLVE This atom will absorb photons on the $1 \rightarrow 2$ and $1 \rightarrow 3$ transitions, with $\Delta E_{1 \rightarrow 2} = 3.00$ eV and $\Delta E_{1 \rightarrow 3} = 5.00$ eV. From $f = \Delta E_{\text{atom}}/h$ and $\lambda = c/f$, we find that the wavelengths in the absorption spectrum are

$$1 \rightarrow 2 \qquad f = 3.00 \text{ eV}/h = 7.25 \times 10^{14} \text{ Hz}$$

$$\lambda = 414 \text{ nm (blue)}$$

$$1 \rightarrow 3 \qquad f = 5.00 \text{ eV}/h = 1.21 \times 10^{15} \text{ Hz}$$

$$\lambda = 248 \text{ nm (ultraviolet)}$$

The emission spectrum will also have the 414 nm and 248 nm wavelengths due to the $2 \rightarrow 1$ and $3 \rightarrow 1$ quantum jumps from excited states 2 and 3 to the ground state. In addition, the emission spectrum will contain the $3 \rightarrow 2$ quantum jump with $\Delta E_{3 \rightarrow 2} = -2.00$ eV that is *not* seen in absorption because there are too few atoms in the $n = 2$ state to absorb. We found in Example 38.9 that a 2.00 eV transition corresponds to a wavelength of 621 nm. Thus the emission wavelengths are

$$2 \rightarrow 1 \qquad \lambda = 414 \text{ nm (blue)}$$

$$3 \rightarrow 1 \qquad \lambda = 248 \text{ nm (ultraviolet)}$$

$$3 \rightarrow 2 \qquad \lambda = 621 \text{ nm (orange)}$$

STOP TO THINK 38.4 A photon with a wavelength of 414 nm has energy $E_{\text{photon}} = 3.00$ eV. Do you expect to see a spectral line with $\lambda = 414$ nm in the emission spectrum of the atom represented by this energy-level diagram? If so, what transition or transitions will emit it? Do you expect to see a spectral line with $\lambda = 414$ nm in the absorption spectrum? If so, what transition or transitions will absorb it?

$n = 4$ ———————	6.00 eV
$n = 3$ ———————	5.00 eV
$n = 2$ ———————	2.00 eV
$n = 1$ ———————	0.00 eV

38.6 The Bohr Hydrogen Atom

Bohr's hypothesis was a bold new idea, yet there was still one enormous stumbling block: What *are* the stationary states of an atom? Everything in Bohr's model hinges on the existence of these stationary states, of there being only certain electron orbits that are allowed. But nothing in classical physics provides any basis for such orbits. And Bohr's model describes only the *consequences* of having stationary states, not how to find them. If such states really exist, we will have to go beyond classical physics to find them.

To address this problem, Bohr did an explicit analysis of the hydrogen atom. The hydrogen atom, with only a single electron, was known to be the simplest atom. Furthermore, as we discussed in Chapter 37, Balmer had discovered a fairly simple formula that characterized the wavelengths in the hydrogen emission spectrum. Anyone with a successful model of an atom was going to have to *predict,* from theory, Balmer's formula for the hydrogen atom.

Bohr's paper followed a rather circuitous line of reasoning. That is not surprising because he had little to go on at the time. But our goal is a clear explanation of the ideas, not a historical study of Bohr's methods, so we are going to follow a different analysis using de Broglie's matter waves. De Broglie did not propose matter waves until 1924, 11 years after Bohr's paper, but with the clarity of hindsight we can see that treating the electron as a wave provides a more straightforward analysis of the hydrogen atom. Although our route will be different from Bohr's, we will arrive at the same point, and, in addition, we will be in a much better position to understand the work that came after Bohr.

NOTE ▶ Bohr's analysis of the hydrogen atom is sometimes called the *Bohr atom.* It's important not to confuse this analysis, which applies only to hydrogen, with the more general postulates of the *Bohr model of the atom.* Those postulates, which we looked at in Section 38.5, apply to any atom. To make the distinction clear, we'll call Bohr's analysis of hydrogen the *Bohr hydrogen atom.* ◀

The Stationary States of the Hydrogen Atom

FIGURE 38.19 shows a Rutherford hydrogen atom, with a single electron orbiting a nucleus that consists of a single proton. We will assume a circular orbit of radius r and speed v. We will also assume, to keep the analysis manageable, that the proton remains stationary while the electron revolves around it. This is a reasonable assumption because the proton is roughly 1800 times as massive as the electron. With these assumptions, the atom's energy is the kinetic energy of the electron plus the potential energy of the electron-proton interaction. This is

$$E = K + U = \frac{1}{2}mv^2 + \frac{1}{4\pi\epsilon_0}\frac{q_{\text{elec}}q_{\text{proton}}}{r} = \frac{1}{2}mv^2 - \frac{e^2}{4\pi\epsilon_0 r} \qquad (38.19)$$

where we used $q_{\text{elec}} = -e$ and $q_{\text{proton}} = +e$.

NOTE ▶ m is the mass of the electron, *not* the mass of the entire atom. ◀

Now, the electron, as we are coming to understand it, has both particle-like and wave-like properties. First, let us treat the electron as a charged particle. The proton exerts a Coulomb electric force on the electron:

$$\vec{F}_{\text{elec}} = \left(\frac{1}{4\pi\epsilon_0}\frac{e^2}{r^2}, \text{ toward center} \right) \qquad (38.20)$$

This force gives the electron an acceleration $\vec{a}_{\text{elec}} = \vec{F}_{\text{elec}}/m$ that also points to the center. This is a centripetal acceleration, causing the particle to move in its circular orbit. The centripetal acceleration of a particle moving in a circle of radius r at speed v *must* be v^2/r, thus

$$a_{\text{elec}} = \frac{F_{\text{elec}}}{m} = \frac{e^2}{4\pi\epsilon_0 m r^2} = \frac{v^2}{r} \qquad (38.21)$$

Rearranging, we find

$$v^2 = \frac{e^2}{4\pi\epsilon_0 m r} \qquad (38.22)$$

Equation 38.22 is a *constraint* on the motion. The speed v and radius r must satisfy Equation 38.22 if the electron is to move in a circular orbit. This constraint is not unique to atoms; we earlier found a similar relationship between v and r for orbiting satellites.

Now let's treat the electron as a de Broglie wave. In Section 38.4 we found that a particle confined to a one-dimensional box sets up a standing wave as it reflects back and forth. A standing wave, you will recall, consists of two traveling waves moving in opposite directions. When the round-trip distance in the box is equal to an integer number of wavelengths ($2L = n\lambda$), the two oppositely traveling waves interfere constructively to set up the standing wave.

Suppose that, instead of traveling back and forth along a line, our wave-like particle travels around the circumference of a circle. The particle will set up a standing wave, just like the particle in the box, if there are waves traveling in both directions and if the round-trip distance is an integer number of wavelengths. This is the idea we want to carry over from the particle in a box. As an example, **FIGURE 38.20** shows a standing wave around a circle with $n = 10$ wavelengths.

The mathematical condition for a circular standing wave is found by replacing the round-trip distance $2L$ in a box with the round-trip distance $2\pi r$ on a circle. Thus a circular standing wave will occur when

$$2\pi r = n\lambda \qquad n = 1, 2, 3, \ldots \qquad (38.23)$$

But the de Broglie wavelength for a particle *has* to be $\lambda = h/p = h/mv$. Thus the standing-wave condition for a de Broglie wave is

$$2\pi r = n\frac{h}{mv}$$

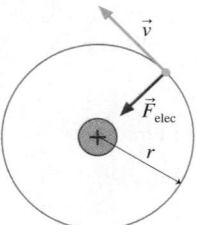

FIGURE 38.19 A Rutherford hydrogen atom. The size of the nucleus is greatly exaggerated.

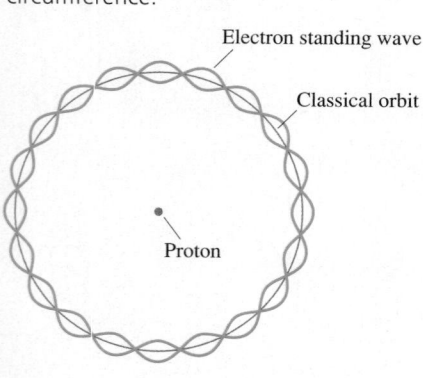

FIGURE 38.20 An $n = 10$ electron standing wave around the orbit's circumference.

Electron standing wave

Classical orbit

Proton

Binding Energy and Ionization Energy

It is important to understand why the energies of the stationary states are negative. Because the potential energy of two charged particles is $U = q_1 q_2 / 4 \pi \epsilon_0 r$, the zero of potential energy occurs at $r = \infty$ where the particles are infinitely far apart. The state of zero total energy corresponds to having the electron at rest ($K = 0$) and infinitely far from the proton ($U = 0$). This situation, which is the case of two "free particles," occurs in the limit $n \to \infty$, for which $r_n \to \infty$ and $v_n \to 0$.

An electron and a proton bound into an atom have *less* energy than two free particles. We know this because we would have to do work (i.e., add energy) to pull the electron and proton apart. If the bound atom's energy is lower than that of two free particles, and if the total energy of two free particles is zero, then it must be the case that the atom has a *negative* amount of energy.

Thus $|E_n|$ is the **binding energy** of the electron in stationary state n. In the ground state, where $E_1 = -13.60$ eV, we would have to add 13.60 eV to the electron to free it from the proton and reach the zero energy state of two free particles. We can say that the electron in the ground state is "bound by 13.60 eV." An electron in an $n = 3$ orbit, where it is farther from the proton and moving more slowly, is bound by only 1.51 eV. That is the amount of energy you would have to supply to remove the electron from an $n = 3$ orbit.

Removing the electron entirely leaves behind a positive ion, H^+ in the case of a hydrogen atom. (The fact that H^+ happens to be a proton does not alter the fact that it is also an atomic ion.) Because nearly all atoms are in their ground state, the binding energy $|E_1|$ of the ground state is called the **ionization energy** of an atom. Bohr's analysis predicts that the ionization energy of hydrogen is 13.60 eV. FIGURE 38.22 illustrates the ideas of binding energy and ionization energy.

We can test this prediction by shooting a beam of electrons at hydrogen atoms. A projectile electron can knock out an atomic electron if its kinetic energy K is greater than the atom's ionization energy, leaving an ion behind. But a projectile electron will be unable to cause ionization if its kinetic energy is less than the atom's ionization energy. This is a fairly straightforward experiment to carry out, and the evidence shows that the ionization energy of hydrogen is, indeed, 13.60 eV.

FIGURE 38.22 Binding energy and ionization energy.

The *binding energy* is the energy needed to remove an electron from its orbit.

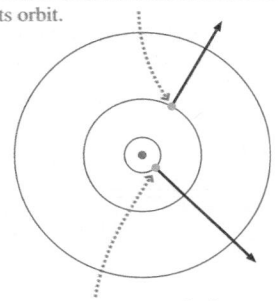

The *ionization energy* is the energy needed to create an ion by removing a ground-state electron.

Quantization of Angular Momentum

The angular momentum of a particle in circular motion, whether it is a planet or an electron, is

$$L = mvr$$

You will recall that angular momentum is conserved in orbital motion because a force directed toward a central point exerts no torque on the particle. Bohr used conservation of energy explicitly in his analysis of the hydrogen atom, but what role does conservation of angular momentum play?

The condition that a de Broglie wave for the electron set up a standing wave around the circumference was given, in Equation 38.23, as

$$2 \pi r = n \lambda = n \frac{h}{mv}$$

Multiplying by mv and dividing by 2π, we can rewrite this equation as

$$mvr = n \frac{h}{2\pi} = n\hbar \qquad (38.32)$$

But mvr is the angular momentum L for a particle in a circular orbit. It appears that the angular momentum of an orbiting electron cannot have just any value. Instead, it must satisfy

$$L = n\hbar \qquad n = 1, 2, 3, \ldots \qquad (38.33)$$

Thus angular momentum also is quantized! The electron's angular momentum must be an integer multiple of Planck's constant $\hbar$.

The quantization of angular momentum is a direct consequence of this wave-like nature of the electron. We will find that the quantization of angular momentum plays a major role in the behavior of more complex atoms, leading to the idea of electron shells that you likely have studied in chemistry.

STOP TO THINK 38.5   What is the quantum number of this hydrogen atom?

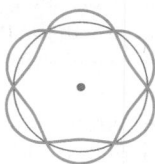

38.7 The Hydrogen Spectrum

Our analysis of the hydrogen atom has revealed stationary states, but how do we know whether the results make any sense? The most important experimental evidence that we have about the hydrogen atom is its spectrum, so the primary test of the Bohr hydrogen atom is whether it correctly predicts the spectrum.

The Hydrogen Energy-Level Diagram

FIGURE 38.23 The energy-level diagram of the hydrogen atom.

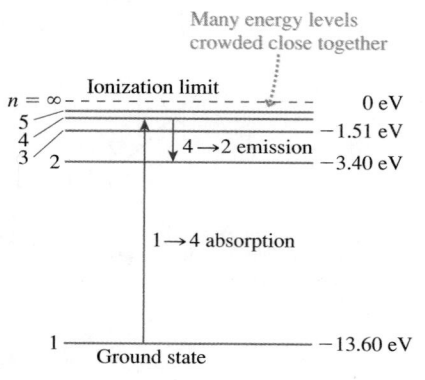

FIGURE 38.23 is an energy-level diagram for the hydrogen atom. As we noted earlier, the energies are like the rungs of a ladder. The lowest rung is the ground state, with $E_1 = -13.60$ eV. The top rung, with $E = 0$ eV, corresponds to a hydrogen ion in the limit $n \rightarrow \infty$. This top rung is called the **ionization limit**. In principle there are an infinite number of rungs, but only the lowest few are shown. The higher values of n are all crowded together just below the ionization limit at $n = \infty$.

The figure shows a $1 \rightarrow 4$ transition in which a photon is absorbed and a $4 \rightarrow 2$ transition in which a photon is emitted. For two quantum states m and n, where $n > m$ and E_n is the higher energy state, an atom can *emit* a photon in an $n \rightarrow m$ transition or *absorb* a photon in an $m \rightarrow n$ transition.

The Emission Spectrum

According to the third assumption of Bohr's model of atomic quantization, the frequency of the photon emitted in an $n \rightarrow m$ transition is

$$f = \frac{\Delta E_{\text{atom}}}{h} = \frac{E_n - E_m}{h} \tag{38.34}$$

We can use Equation 38.30 for the energies E_n and E_m to predict that the emitted photon has frequency

$$f = \frac{1}{h}\left\{\left[-\frac{1}{n^2}\left(\frac{1}{4\pi\epsilon_0}\frac{e^2}{2a_{\text{B}}}\right)\right] - \left[-\frac{1}{m^2}\left(\frac{1}{4\pi\epsilon_0}\frac{e^2}{2a_{\text{B}}}\right)\right]\right\}$$

$$= \frac{1}{4\pi\epsilon_0}\frac{e^2}{2ha_{\text{B}}}\left(\frac{1}{m^2} - \frac{1}{n^2}\right) \tag{38.35}$$

The frequency is a positive number because $m < n$ and thus $1/m^2 > 1/n^2$.

We are more interested in wavelength than frequency, because wavelengths are the quantity measured by experiment. The wavelength of the photon emitted in an $n \rightarrow m$ quantum jump is

$$\lambda_{n \rightarrow m} = \frac{c}{f} = \frac{8\pi\epsilon_0 hca_{\text{B}}/e^2}{\left(\dfrac{1}{m^2} - \dfrac{1}{n^2}\right)} \tag{38.36}$$

This looks rather gruesome, but notice that the numerator is simply a collection of various constants. The value of the numerator, which we can call λ_0, is

$$\lambda_0 = \frac{8\pi\epsilon_0 h c a_B}{e^2} = 9.112 \times 10^{-8} \text{ m} = 91.12 \text{ nm}$$

With this definition, our prediction for the wavelengths in the hydrogen emission spectrum is

$$\lambda_{n\to m} = \frac{\lambda_0}{\left(\dfrac{1}{m^2} - \dfrac{1}{n^2}\right)} \quad m = 1, 2, 3, \ldots \quad n = m+1, m+2, \ldots \quad (38.37)$$

This should look familiar. It is the Balmer formula from Chapter 37! However, there is one *slight* difference: Bohr's analysis of the hydrogen atom has predicted $\lambda_0 = 91.12$ nm, whereas Balmer found, from experiment, that $\lambda_0 = 91.18$ nm. Could Bohr have come this close but then fail to predict the Balmer formula correctly?

The problem, it turns out, is in our assumption that the proton remains at rest while the electron orbits it. In fact, *both* particles rotate about their common center of mass, rather like a dumbbell with a big end and a small end. The center of mass is very close to the proton, which is far more massive than the electron, but the proton is not entirely motionless. The good news is that a more advanced analysis can account for the proton's motion. It changes the energies of the stationary states ever so slightly—about 1 part in 2000—but that is precisely what is needed to give a revised value:

$$\lambda_0 = 91.18 \text{ nm when corrected for the nuclear motion}$$

It works! Unlike all previous atomic models, **the Bohr hydrogen atom correctly predicts the discrete spectrum of the hydrogen atom.** FIGURE 38.24 shows the *Balmer series* and the *Lyman series* transitions on an energy-level diagram. Only the Balmer series, consisting of transitions ending on the $m = 2$ state, gives visible wavelengths, and this is the series that Balmer initially analyzed. The Lyman series, ending on the $m = 1$ ground state, is in the ultraviolet region of the spectrum and was not measured until later. These series, as well as others in the infrared, are observed in a discharge tube where collisions with electrons excite the atoms upward from the ground state to state n. They then decay downward by emitting photons. Only the Lyman series is observed in the absorption spectrum because, as noted previously, essentially all the atoms in a quiescent gas are in the ground state.

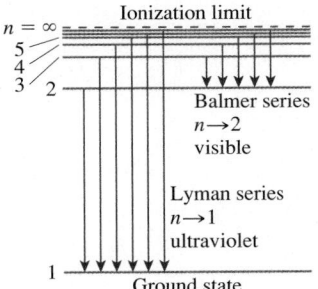

FIGURE 38.24 Transitions producing the Lyman series and the Balmer series of lines in the hydrogen spectrum.

EXAMPLE 38.12 **Hydrogen absorption**

Whenever astronomers look at distant galaxies, they find that the light has been strongly absorbed at the wavelength of the $1 \to 2$ transition in the Lyman series of hydrogen. This absorption tells us that interstellar space is filled with vast clouds of hydrogen left over from the Big Bang. What is the wavelength of the $1 \to 2$ absorption in hydrogen?

SOLVE Equation 38.37 predicts the *absorption* spectrum of hydrogen if we let $m = 1$. The absorption seen by astronomers is from the ground state of hydrogen ($m = 1$) to its first excited state ($n = 2$). The wavelength is

$$\lambda_{1\to 2} = \frac{91.18 \text{ nm}}{\left(\dfrac{1}{1^2} - \dfrac{1}{2^2}\right)} = 121.6 \text{ nm}$$

ASSESS This wavelength is far into the ultraviolet. Ground-based astronomy cannot observe this region of the spectrum because the wavelengths are strongly absorbed by the atmosphere, but with space-based telescopes, first widely used in the 1970s, astronomers see 121.6 nm absorption in nearly every direction they look.

Hydrogen-Like Ions

An ion with a *single* electron orbiting Z protons in the nucleus is called a **hydrogen-like ion.** Z is the atomic number and describes the number of protons in the nucleus. He^+, with one electron circling a $Z = 2$ nucleus, and Li^{++}, with one electron and a

$Z = 3$ nucleus, are hydrogen-like ions. So is U^{+91}, with one lonely electron orbiting a $Z = 92$ uranium nucleus.

Any hydrogen-like ion is simply a variation on the Bohr hydrogen atom. The only difference between a hydrogen-like ion and neutral hydrogen is that the potential energy $-e^2/4\pi\epsilon_0 r$ becomes, instead, $-Ze^2/4\pi\epsilon_0 r$. Hydrogen itself is the $Z = 1$ case. If we repeat the analysis of the previous sections with this one change, we find:

$$r_n = \frac{n^2 a_B}{Z} \qquad E_n = -\frac{13.60 Z^2 \text{ eV}}{n^2}$$

$$v_n = Z\frac{v_1}{n} \qquad \lambda_0 = \frac{91.18 \text{ nm}}{Z^2} \tag{38.38}$$

As the nuclear charge increases, the electron moves into a smaller-diameter, higher-speed orbit. Its ionization energy $|E_1|$ increases significantly, and its spectrum shifts to shorter wavelengths. Table 38.3 compares the ground-state atomic diameter $2r_1$, the ionization energy $|E_1|$, and the first wavelength $3 \rightarrow 2$ in the Balmer series for hydrogen and the first two hydrogen-like ions.

TABLE 38.3 Comparison of hydrogen-like ions with $Z = 1$, 2, and 3

Ion	Diameter $2r_1$	Ionization energy $|E_1|$	Wavelength of $3 \rightarrow 2$
H ($Z = 1$)	0.106 nm	13.6 eV	656 nm
He$^+$ ($Z = 2$)	0.053 nm	54.4 eV	164 nm
Li^{++} ($Z = 3$)	0.035 nm	122.4 eV	73 nm

Success and Failure

Bohr's analysis of the hydrogen atom seemed to be a resounding success. By introducing Einstein's ideas about light quanta, Bohr was able to provide the first understanding of discrete spectra and to predict the Balmer formula for the wavelengths in the hydrogen spectrum. And the Bohr hydrogen atom, unlike Rutherford's model, was stable. There was clearly some validity to the idea of stationary states.

But Bohr was completely unsuccessful at explaining the spectra of any other neutral atom. His method did not work even for helium, the second element in the periodic table with a mere two electrons. Something inherent in Bohr's assumptions seemed to work correctly for a single electron but not in situations with two or more electrons.

It is important to make a distinction between the Bohr model of atomic quantization, described in Section 38.5, and the Bohr hydrogen atom. The Bohr model assumes that stationary states exist, but it does not say how to find them. We found the stationary states of a hydrogen atom by requiring that an integer number of de Broglie waves fit around the circumference of the orbit, setting up standing waves. The difficulty with more complex atoms is not the Bohr model but the method of finding the stationary states. Bohr's model of the atomic quantization remains valid, and we will continue to use it, but the procedure of fitting standing waves to a circle is just too simple to find the stationary states of complex atoms. We need to find a better procedure.

Einstein, de Broglie, and Bohr carried physics into uncharted waters. Their successes made it clear that the microscopic realm of light and atoms is governed by quantization, discreteness, and a blurring of the distinction between particles and waves. Although Bohr was clearly on the right track, his inability to extend the Bohr hydrogen atom to more complex atoms made it equally clear that the complete and correct theory remained to be discovered. Bohr's theory was what we now call "semi-classical," a hybrid of classical Newtonian mechanics with the new ideas of quanta. Still missing was a complete theory of motion and dynamics in a quantized universe—a *quantum* mechanics.

Hydrogen fluorescence

Fluorescence is the absorption of light at one wavelength followed by emission at a longer wavelength. Suppose a hydrogen atom in its ground state absorbs an ultraviolet photon with a wavelength of 95.10 nm. Immediately after the absorption, the atom undergoes a quantum jump with $\Delta n = 3$. What is the wavelength of the photon emitted in this quantum jump?

MODEL Photons are emitted and absorbed as an atom undergoes quantum jumps from one energy level to another. The Bohr model gives the energy levels of the hydrogen atom.

VISUALIZE **FIGURE 38.25** shows the process. To be absorbed, the photon energy has to match exactly the energy *difference* between the ground state of hydrogen and an excited state with quantum number n. After excitation, the atom emits a photon as it jumps downward in a $n \rightarrow n - 3$ transition.

FIGURE 38.25 The process of fluorescence in hydrogen. Energy levels are not drawn to scale.

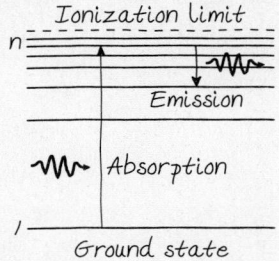

SOLVE The energy of the absorbed photon is

$$E = hf = \frac{hc}{\lambda} = \frac{(4.14 \times 10^{-15} \text{ eV s})(3.00 \times 10^8 \text{ m/s})}{95.10 \times 10^{-9} \text{ m}} = 13.06 \text{ eV}$$

The atom's initial energy is $E_1 = -13.60 \text{ eV}$, the energy of the ground state of hydrogen. Absorbing a 13.06 eV photon raises the atom's energy to $E_n = E_1 + 13.06 \text{ eV} = -0.54 \text{ eV}$. The energy levels of hydrogen are given by

$$E_n = -\frac{13.60 \text{ eV}}{n^2}$$

The quantum number of the energy level with -0.54 eV is

$$n = \sqrt{-\frac{13.60 \text{ eV}}{(-0.54 \text{ eV})}} = 5$$

We see that the absorption is a $1 \rightarrow 5$ transition; thus the emission, with $\Delta n = 3$, must be a $5 \rightarrow 2$ transition. The energy of the $n = 2$ state is

$$E_2 = -\frac{13.60 \text{ eV}}{2^2} = -3.40 \text{ eV}$$

Consequently, the energy of the emitted photon is

$$E_{\text{photon}} = \Delta E_{\text{atom}} = (-0.54 \text{ eV}) - (-3.40 \text{ eV}) = 2.86 \text{ eV}$$

Inverting the energy-wavelength relationship that we started with, we find

$$\lambda = \frac{hc}{E_{\text{photon}}} = \frac{(4.14 \times 10^{-15} \text{ eV s})(3.00 \times 10^8 \text{ m/s})}{2.86 \text{ eV}} = 434 \text{ nm}$$

When atomic hydrogen gas is irradiated with ultraviolet light having a wavelength of 95.10 nm, it fluoresces at the visible wavelength of 434 nm. (It also fluoresces at infrared and ultraviolet wavelengths in downward transitions with other values of Δn.)

ASSESS The $5 \rightarrow 2$ transition is a member of the Balmer series, and a 434 nm spectral line was shown in the hydrogen spectrum of Figure 37.7. It is important to notice that the 13.06 eV photon energy does not match any energy level of the hydrogen atom. Instead, it matches the *difference* between two levels because that conserves energy in a quantum jump between those two levels. Photons with nearby wavelengths, such as 94 nm or 96 nm, would not be absorbed at all because their energy does not match the difference of any two energy levels in hydrogen.

SUMMARY

The goal of Chapter 38 has been to understand the quantization of energy for light and matter.

General Principles

Light has particle-like properties

- The energy of a light wave comes in discrete packets called light quanta or photons.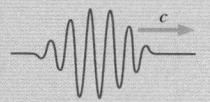

- For light of frequency f, the energy of each photon is $E = hf$, where h is **Planck's constant.**

- For a light wave that delivers power P, photons arrive at rate R such that $P = Rhf$.

- Photons are "particle-like" but are not classical particles.

Matter has wave-like properties

- The **de Broglie wavelength** of a "particle" of mass m is $\lambda = h/mv$.

- The wave-like nature of matter is seen in the interference patterns of electrons, neutrons, and entire atoms.

- When a particle is confined, it sets up a de Broglie standing wave. The fact that standing waves have only certain allowed wavelengths leads to the conclusion that a confined particle has only certain allowed energies. That is, energy is quantized.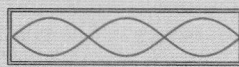

Important Concepts

Einstein's Model of Light

- Light consists of quanta of energy $E = hf$.

- Quanta are emitted and absorbed on an all-or-nothing basis.

- When a light quantum is absorbed, it delivers all its energy to *one* electron.

Bohr's Model of the Atom

- An atom can exist in only certain stationary states. The allowed energies are quantized. State n has energy E_n.

- An atom can jump from one stationary state to another by emitting or absorbing a photon with $E_{photon} = hf = \Delta E_{atom}$.

- Atoms can be excited in inelastic collisions.

- Atoms seek the $n = 1$ **ground state.** Most atoms, most of the time, are in the ground state.

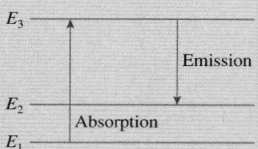

Applications

Photoelectric effect

Light can eject electrons from a metal only if $f \geq f_0 = E_0/h$, where E_0 is the metal's **work function.**

The **stopping potential** that stops even the fastest electrons is

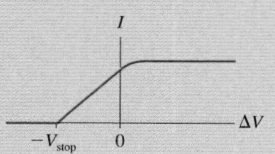

$$V_{stop} = \frac{h}{e}(f - f_0)$$

Particle in a box

A particle confined to a one-dimensional box of length L sets up de Broglie standing waves. The allowed energies are

$$E_n = \frac{1}{2}mv_n^2 = n^2\frac{h^2}{8mL^2} \qquad n = 1, 2, 3, \ldots$$

The Bohr hydrogen atom

The stationary states are found by requiring an integer number of de Broglie wavelengths to fit around the circumference of the electron's orbit: $2\pi r = n\lambda$.

This leads to energy quantization with

$$r_n = n^2 a_B \qquad v_n = \frac{v_1}{n} \qquad E_n = -\frac{13.60 \text{ eV}}{n^2}$$

where $a_B = 0.0529$ nm is the **Bohr radius.** The Bohr hydrogen atom successfully predicts the Balmer formula for the hydrogen spectrum. Angular momentum is also quantized, with $L = n\hbar$.

Terms and Notation

photoelectric effect	wave packet	stationary state	ionization energy
threshold frequency, f_0	matter wave	excited state	ionization limit
stopping potential, V_{stop}	de Broglie wavelength	ground state	hydrogen-like ion
thermal emission	quantized	transition	
work function, E_0	energy level	quantum jump	
Planck's constant, h or $\hbar$	quantum number, n	collisional excitation	
light quantum	fundamental quantum of	energy-level diagram	
photon	energy, E_1	Bohr radius, a_B	
photon model	Bohr model of the atom	binding energy	

CONCEPTUAL QUESTIONS

1. a. A negatively charged electroscope can be discharged by shining an ultraviolet light on it. How does this happen?
   b. You might think that an ultraviolet light shining on an initially uncharged electroscope would cause the electroscope to become positively charged as photoelectrons are emitted. In fact, ultraviolet light has no noticeable effect on an uncharged electroscope. Why not?

2. a. Explain why the graphs of Figure 38.3 are mostly horizontal for $\Delta V > 0$.
   b. Explain why photoelectrons are ejected from the cathode with a range of kinetic energies, rather than all electrons having the same kinetic energy.
   c. Explain the reasoning by which we claim that the stopping potential V_{stop} indicates the maximum kinetic energy of the electrons.

3. How would the graph of Figure 38.2 look *if* classical physics provided the correct description of the photoelectric effect? Draw the graph and explain your reasoning. Assume that the light intensity remains constant as its frequency and wavelength are varied.

4. How would the graphs of Figure 38.3 look *if* classical physics provided the correct description of the photoelectric effect? Draw the graph and explain your reasoning. Include curves for both weak light and intense light.

5. FIGURE Q38.5 is the current-versus-potential-difference graph for a photoelectric-effect experiment with an unknown metal. *If* classical physics provided the correct description of the photoelectric effect, how would the graph look if:
   a. The light was replaced by an equally intense light with a shorter wavelength? Draw it.
   b. The metal was replaced by a different metal with a smaller work function? Draw it.

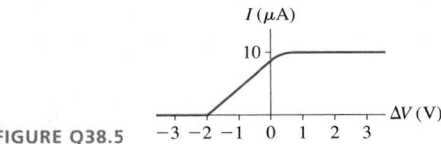

FIGURE Q38.5

6. Metal 1 has a larger work function than metal 2. Both are illuminated with the same short-wavelength ultraviolet light. Do photoelectrons from metal 1 have a higher speed, a lower speed, or the same speed as photoelectrons from metal 2? Explain.

7. Electron 1 is accelerated from rest through a potential difference of 100 V. Electron 2 is accelerated from rest through a potential difference of 200 V. Afterward, which electron has the larger de Broglie wavelength? Explain.

8. An electron and a proton are each accelerated from rest through a potential difference of 100 V. Afterward, which particle has the larger de Broglie wavelength? Explain.

9. FIGURE Q38.9 is a simulation of the electrons detected behind two closely spaced slits. Each bright dot represents one electron. How will this pattern change if
   a. The electron-beam intensity is increased?
   b. The electron speed is reduced?
   c. The electrons are replaced by neutrons?
   d. The left slit is closed?
   Your answers should consider the number of dots on the screen and the spacing, width, and positions of the fringes.

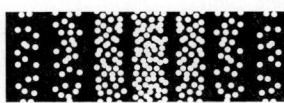

FIGURE Q38.9

10. Imagine that the horizontal box of Figure 38.15 is instead oriented vertically. Also imagine the box to be on a neutron star where the gravitational field is so strong that the particle in the box slows significantly, nearly stopping, before it hits the top of the box. Make a *qualitative* sketch of the $n = 3$ de Broglie standing wave of a particle in this box.
    Hint: The nodes are *not* uniformly spaced.

11. If an electron is in a *stationary state* of an atom, is the electron at rest? If not, what does the term mean?

12. FIGURE Q38.12 shows the energy-level diagram of Element X.
    a. What is the ionization energy of Element X?
    b. An atom in the ground state absorbs a photon, then emits a photon with a wavelength of 1240 nm. What conclusion can you draw about the energy of the photon that was absorbed?
    c. An atom in the ground state has a collision with an electron, then emits a photon with a wavelength of 1240 nm. What conclusion can you draw about the initial kinetic energy of the electron?

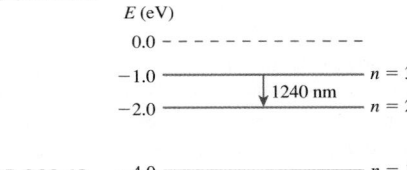

FIGURE Q38.12

EXERCISES AND PROBLEMS

Problems labeled ▨ integrate material from earlier chapters.

Exercises

Section 38.1 The Photoelectric Effect

Section 38.2 Einstein's Explanation

1. ‖ How many photoelectrons are ejected per second in the experiment represented by the graph of FIGURE EX38.1?

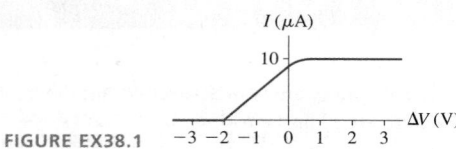

FIGURE EX38.1

2. | Which metals in Table 38.1 exhibit the photoelectric effect for (a) light with $\lambda = 400$ nm and (b) light with $\lambda = 250$ nm?

3. | Photoelectrons are observed when a metal is illuminated by light with a wavelength less than 388 nm. What is the metal's work function?

4. ‖ Electrons in a photoelectric-effect experiment emerge from an aluminum surface with a maximum kinetic energy of 1.30 eV. What is the wavelength of the light?

5. | You need to design a photodetector that can respond to the entire range of visible light. What is the maximum possible work function of the cathode?

6. ‖ A photoelectric-effect experiment finds a stopping potential of 1.56 V when light of 200 nm is used to illuminate the cathode.
   a. From what metal is the cathode made?
   b. What is the stopping potential if the intensity of the light is doubled?

Section 38.3 Photons

7. | a. Determine the energy, in eV, of a photon with a 550 nm wavelength.
   b. Determine the wavelength of a 7.5 keV x-ray photon.

8. | What is the wavelength, in nm, of a photon with energy (a) 0.30 eV, (b) 3.0 eV, and (c) 30 eV? For each, is this wavelength visible, ultraviolet, or infrared light?

9. | What is the energy, in eV, of (a) a 450 MHz radio-frequency photon, (b) a visible-light photon with a wavelength of 450 nm, and (c) an x-ray photon with a wavelength of 0.045 nm?

10. | An FM radio station broadcasts with a power of 10 kW at a frequency of 101 MHz.
    a. How many photons does the antenna emit each second?
    b. Should the broadcast be treated as an electromagnetic wave or discrete photons? Explain.

11. | For what wavelength of light does a 100 mW laser deliver 2.50×10^{17} photons per second?

12. | A red laser with a wavelength of 650 nm and a blue laser with a wavelength of 450 nm emit laser beams with the same light power. How do their rates of photon emission compare? Answer this by computing R_{red}/R_{blue}.

13. | A 100 W incandescent lightbulb emits about 5 W of visible light. (The other 95 W are emitted as infrared radiation or lost as heat to the surroundings.) The average wavelength of the visible light is about 600 nm, so make the simplifying assumption that all the light has this wavelength. How many visible-light photons does the bulb emit per second?

Section 38.4 Matter Waves and Energy Quantization

14. ‖ At what speed is an electron's de Broglie wavelength (a) 1.0 pm, (b) 1.0 nm, (c) 1.0 μm, and (d) 1.0 mm?

15. ‖ Through what potential difference must an electron be accelerated from rest to have a de Broglie wavelength of 500 nm?

16. ‖ What is the de Broglie wavelength of a neutron that has fallen 1.0 m in a vacuum chamber, starting from rest?

17. | a. What is the de Broglie wavelength of a 200 g baseball with a speed of 30 m/s?
    b. What is the speed of a 200 g baseball with a de Broglie wavelength of 0.20 nm?

18. | The diameter of the nucleus is about 10 fm. What is the kinetic energy, in MeV, of a proton with a de Broglie wavelength of 10 fm?

19. | What is the quantum number of an electron confined in a 3.0-nm-long one-dimensional box if the electron's de Broglie wavelength is 1.0 nm?

20. | The diameter of the nucleus is about 10 fm. A simple model of the nucleus is that protons and neutrons are confined within a one-dimensional box of length 10 fm. What are the first three energy levels, in MeV, for a proton in such a box?

21. ‖ What is the length of a one-dimensional box in which an electron in the $n = 1$ state has the same energy as a photon with a wavelength of 600 nm?

Section 38.5 Bohr's Model of Atomic Quantization

22. | FIGURE EX38.22 is an energy-level diagram for a simple atom. What wavelengths appear in the atom's (a) emission spectrum and (b) absorption spectrum?

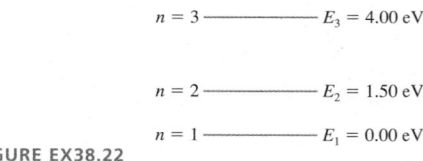

FIGURE EX38.22

23. ‖ An electron with 2.00 eV of kinetic energy collides with the atom shown in FIGURE EX38.22.
    a. Is the electron able to excite the atom? Why or why not?
    b. If your answer to part a was yes, what is the electron's kinetic energy after the collision?

24. ‖ The allowed energies of a simple atom are 0.00 eV, 4.00 eV, and 6.00 eV.
    a. Draw the atom's energy-level diagram. Label each level with the energy and the quantum number.
    b. What wavelengths appear in the atom's emission spectrum?
    c. What wavelengths appear in the atom's absorption spectrum?

25. ‖ The allowed energies of a simple atom are 0.00 eV, 4.00 eV, and 6.00 eV. An electron traveling with a speed of 1.30×10^6 m/s collides with the atom. Can the electron excite the atom to the $n = 2$ stationary state? The $n = 3$ stationary state? Explain.

Section 38.6 The Bohr Hydrogen Atom

26. ‖ What is the radius of a hydrogen atom whose electron moves at 7.3×10^5 m/s?

27. ‖ What is the radius of a hydrogen atom whose electron is bound by 0.378 eV?

28. ‖ a. What quantum number of the hydrogen atom comes closest to giving a 100-nm-diameter electron orbit?
    b. What are the electron's speed and energy in this state?

29. ‖ a. Calculate the de Broglie wavelength of the electron in the $n = 1$, 2, and 3 states of the hydrogen atom. Use the information in Table 38.2.
    b. Show numerically that the circumference of the orbit for each of these stationary states is exactly equal to n de Broglie wavelengths.
    c. Sketch the de Broglie standing wave for the $n = 3$ orbit.

30. ‖ How much energy does it take to ionize a hydrogen atom that is in its first excited state?

31. ‖ Show, by calculation, that the first three states of the hydrogen atom have angular momenta $\hbar$, $2\hbar$, and $3\hbar$, respectively.

Section 38.7 The Hydrogen Spectrum

32. ‖ Determine the wavelengths of all the possible photons that can be emitted from the $n = 4$ state of a hydrogen atom.

33. ‖ What is the third-longest wavelength in the absorption spectrum of hydrogen?

34. ‖ Is a spectral line with wavelength 656.5 nm seen in the absorption spectrum of hydrogen atoms? Why or why not?

35. ‖ Find the radius of the electron's orbit, the electron's speed, and the energy of the atom for the first three stationary states of He^+.

Problems

36. ‖ A ruby laser emits an intense pulse of light that lasts a mere 10 ns. The light has a wavelength of 690 nm, and each pulse has an energy of 500 mJ.
    a. How many photons are emitted in each pulse?
    b. What is the *rate* of photon emission, in photons per second, during the 10 ns that the laser is "on"?

37. ‖ In a photoelectric-effect experiment, the wavelength of light shining on an aluminum cathode is decreased from 250 nm to 200 nm. What is the change in the stopping potential?

38. ‖ The wavelengths of light emitted by a firefly span the visible
BIO spectrum but have maximum intensity near 550 nm. A typical flash lasts for 100 ms and has a power output of 1.2 mW. How many photons does a firefly emit in one flash if we assume that all light is emitted at the peak intensity wavelength of 550 nm?

39. ‖ *Dinoflagellates* are single-cell organisms that float in the
BIO world's oceans. Many types are bioluminescent. When disturbed, a typical bioluminescent dinoflagellate emits 10^8 photons in a 0.10-s-long flash of wavelength 460 nm. What is the power of the flash?

40. ‖ Potassium and gold cathodes are used in a photoelectric-effect experiment. For each cathode, find:
    a. The threshold frequency.
    b. The threshold wavelength.
    c. The maximum photoelectron ejection speed if the light has a wavelength of 220 nm.
    d. The stopping potential if the wavelength is 220 nm.

41. ‖ The maximum kinetic energy of photoelectrons is 2.8 eV. When the wavelength of the light is increased by 50%, the maximum energy decreases to 1.1 eV. What are (a) the work function of the cathode and (b) the initial wavelength of the light?

42. ‖ In a photoelectric-effect experiment, the stopping potential at a wavelength of 400 nm is 25.7% of the stopping potential at a wavelength of 300 nm. Of what metal is the cathode made?

43. ‖ The graph in FIGURE P38.43 was measured in a photoelectric-effect experiment.
    a. What is the work function (in eV) of the cathode?
    b. What experimental value of Planck's constant is obtained from these data?

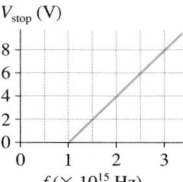

FIGURE P38.43

44. ‖ A metal cathode whose work function is 3.3 eV is illuminated with 15 μW of light having a wavelength of 300 nm. The efficiency of converting photons to photoelectrons is 12%. What current is measured in the experiment?

45. ‖ In a photoelectric-effect experiment, the stopping potential was measured for several different wavelengths of incident light. The data are as follows:

Wavelength (nm)	Stopping potential (V)
500	0.19
450	0.48
400	0.83
350	1.28
300	1.89
250	2.74

Use an appropriate graph of the data to determine (a) the metal used for the cathode and (b) an experimental value for Planck's constant.

46. ‖ The relationship between momentum and energy from Einstein's theory of relativity is $E^2 - (pc)^2 = E_0^2$, where, in this context, $E_0 = mc^2$ is the rest energy rather than the work function.
    a. A photon is a massless particle. What is a photon's momentum p in terms of its energy E?
    b. Einstein also claimed that the energy of a photon is related to its frequency by $E = hf$. Use this and your result from part a to write an expression for the wavelength λ of a photon in terms of its momentum p.
    c. Your result for part b is for a "particle-like wave." Suppose you thought this expression should also apply to a "wave-like particle." What is your expression for λ if you replace p with the classical-mechanics expression for the momentum of a particle of mass m? Is this a familiar-looking expression?

47. ‖ A red blood cell is a 7.0-μm-diameter, 2.0-μm-thick disk with
BIO a density of 1100 kg/m³. What is the de Broglie wavelength of
a red blood cell moving through a capillary at 4.0 mm/s? Do we
need to be concerned with the wave nature of blood cells when
describing the flow of blood?

48. ‖ The electron interference pattern of Figure 38.13 was made
by shooting electrons with 50 keV of kinetic energy through two
slits spaced 1.0 μm apart. The fringes were recorded on a detector 1.0 m behind the slits.
   a. What was the speed of the electrons? (The speed is large
      enough to justify using relativity, but for simplicity do this as
      a nonrelativistic calculation.)
   b. Figure 38.13 is greatly magnified. What was the actual spac-
      ing on the detector between adjacent bright fringes?

49. ‖ An experiment was performed in which neutrons were shot
through two slits spaced 0.10 nm apart and detected 3.5 m behind the slits. FIGURE P38.49 shows the detector output. Notice
the 100 μm scale on the figure. To one significant figure, what
was the speed of the neutrons?

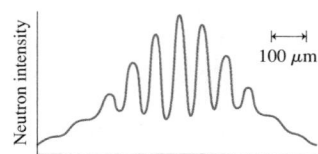

FIGURE P38.49

50. ‖ The electrons in a cathode-ray tube are accelerated through
a 250 V potential difference and then shot through a 33-nm-
diameter circular aperture. What is the diameter of the bright
spot on an electron detector 1.5 m behind the aperture?

51. ‖ An electron confined in a one-dimensional box is observed,
at different times, to have energies of 12 eV, 27 eV, and 48 eV.
What is the length of the box?

52. ‖ An electron confined in a one-dimensional box emits a
200 nm photon in a quantum jump from $n = 2$ to $n = 1$. What is
the length of the box?

53. ‖‖ A proton confined in a one-dimensional box emits a 2.0 MeV
gamma-ray photon in a quantum jump from $n = 2$ to $n = 1$.
What is the length of the box?

54. ‖ Consider a small virus having a diameter of 10 nm. The atoms
BIO of the intracellular fluid are confined within the virus. Suppose
we model the virus as a 10-nm-long "box." What is the ground-
state energy (in eV) of a sodium ion confined in this box?

55. ‖ The absorption spectrum of an atom consists of the wavelengths 200 nm, 300 nm, and 500 nm.
   a. Draw the atom's energy-level diagram.
   b. What wavelengths are seen in the atom's emission spectrum?

56. ‖ The first three energy levels of
the fictitious element X are shown in
FIGURE P38.56.
   a. What is the ionization energy of
      element X?
   b. What wavelengths are observed in
      the absorption spectrum of element
      X? Express your answers in nm.
   c. State whether each of your wave-
      lengths in part b corresponds to ultraviolet, visible, or infra-
      red light.

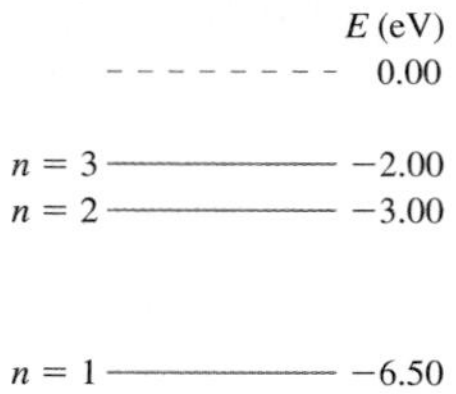

E (eV)
-------- 0.00

$n = 3$ ———— -2.00
$n = 2$ ———— -3.00

$n = 1$ ———— -6.50

FIGURE P38.56

57. ‖ The first three energy levels of the fictitious element X were
shown in FIGURE P38.56. An electron with a speed of 1.4×10^6 m/s

collides with an atom of element X. Shortly afterward, the atom
emits a photon with a wavelength of 1240 nm. What was the
electron's speed after the collision? Assume that, because the
atom is much more massive than the electron, the recoil of the
atom is negligible.
Hint: The energy of the photon is *not* the energy transferred to
the atom in the collision.

58. ‖ Starting from Equation 38.29, derive Equation 38.30.

59. ‖ Calculate *all* the wavelengths of *visible* light in the emission
spectrum of the hydrogen atom.
Hint: There are infinitely many wavelengths in the spectrum,
so you'll need to develop a strategy for this problem rather than
using trial and error.

60. ‖ An electron with a speed of 2.1×10^6 m/s collides with a
hydrogen atom, exciting the atom to the highest possible energy
level. The atom then undergoes a quantum jump with $\Delta n = 1$.
What is the wavelength of the photon emitted in the quantum
jump?

61. ‖ a. What wavelength photon does a hydrogen atom emit in a
      $200 \rightarrow 199$ transition?
   b. What is the *difference* in the wavelengths emitted in a
      $199 \rightarrow 2$ transition and a $200 \rightarrow 2$ transition?

62. ‖ Draw an energy-level diagram, similar to Figure 38.23, for the
He⁺ ion. On your diagram:
   a. Show the first five energy levels. Label each with the values
      of n and E_n.
   b. Show the ionization limit.
   c. Show all possible emission transitions from the $n = 4$ energy
      level.
   d. Calculate the wavelengths (in nm) for each of the transitions
      in part c and show them alongside the appropriate arrow.

63. ‖ What are the wavelengths of the transitions $3 \rightarrow 2$, $4 \rightarrow 2$,
and $5 \rightarrow 2$ in the hydrogen-like ion O⁺⁷? In what spectral range
do these lie?

64. ‖ Two hydrogen atoms collide head-on. The collision brings
both atoms to a halt. Immediately after the collision, both atoms
emit a 121.6 nm photon. What was the speed of each atom just
before the collision?

65. ‖ A beam of electrons is incident upon a gas of hydrogen atoms.
   a. What minimum speed must the electrons have to cause
      the emission of 656 nm light from the $3 \rightarrow 2$ transition of
      hydrogen?
   b. Through what potential difference must the electrons be ac-
      celerated to have this speed?

Challenge Problems

66. Ultraviolet light with a wavelength of 70.0 nm shines on a gas
of hydrogen atoms in their ground states. Some of the atoms are
ionized by the light. What is the kinetic energy of the electrons
that are freed in this process?

67. In the atom interferometer experiment of Figure 38.14, laser-
cooling techniques were used to cool a dilute vapor of sodium
atoms to a temperature of 0.0010 K = 1.0 mK. The ultracold atoms passed through a series of collimating apertures to form the
atomic beam you see entering the figure from the left. The standing light waves were created from a laser beam with a wavelength of 590 nm.
   a. What is the rms speed v_{rms} of a sodium atom ($A = 23$) in a
      gas at temperature 1.0 mK?

b. By treating the laser beam as if it were a diffraction grating, calculate the first-order diffraction angle of a sodium atom traveling with the rms speed of part a.

c. How far apart are points B and C if the second standing wave is 10 cm from the first?

d. Because interference is observed between the two paths, each individual atom is apparently present at both point B *and* point C. Describe, in your own words, what this experiment tells you about the nature of matter.

68. Consider a hydrogen atom in stationary state n.

a. Show that the orbital period of an electron in quantum state n is $T = n^3 T_1$, and find a numerical value for T_1.

b. On average, an atom stays in the $n = 2$ state for 1.6 ns before undergoing a quantum jump to the $n = 1$ state. On average, how many revolutions does the electron make before the quantum jump?

69. Consider an electron undergoing cyclotron motion in a magnetic field. According to Bohr, the electron's angular momentum must be quantized in units of $\hbar$.

a. Show that allowed radii for the electron's orbit are given by $r_n = (n\hbar/eB)^{1/2}$, where $n = 1, 2, 3,\dots$.

b. Compute the first four allowed radii in a 1.0 T magnetic field.

c. Find an expression for the allowed energy levels E_n in terms of $\hbar$ and the cyclotron frequency f_{cyc}.

70. The *muon* is a subatomic particle with the same charge as an electron but with a mass that is 207 times greater: $m_\mu = 207 m_e$. Physicists think of muons as "heavy electrons." However, the muon is not a stable particle; it decays with a half-life of 1.5 μs into an electron plus two neutrinos. Muons from cosmic rays are sometimes "captured" by the nuclei of the atoms in a solid. A captured muon orbits this nucleus, like an electron, until it decays. Because the muon is often captured into an excited orbit ($n > 1$), its presence can be detected by observing the photons emitted in transitions such as $2 \rightarrow 1$ and $3 \rightarrow 1$.

Consider a muon captured by a carbon nucleus ($Z = 6$). Because of its large mass, the muon orbits well *inside* the electron cloud and is not affected by the electrons. Thus the muon "sees" the full nuclear charge Ze and acts like the electron in a hydrogen-like ion.

a. What are the orbital radius and speed of a muon in the $n = 1$ ground state? Note that the mass of a muon differs from the mass of an electron.

b. What is the wavelength of the $2 \rightarrow 1$ muon transition?

c. Is the photon emitted in the $2 \rightarrow 1$ transition infrared, visible, ultraviolet, or x ray?

d. How many orbits will the muon complete during 1.5 μs? Is this a sufficiently large number that the Bohr model "makes sense," even though the muon is not stable?

STOP TO THINK ANSWERS

Stop to Think 38.1: $V_A > V_B > V_C$. For a given wavelength of light, electrons are ejected with more kinetic energy from metals with smaller work functions because it takes less energy to remove an electron. Faster electrons need a larger negative voltage to stop them.

Stop to Think 38.2: d. Photons always travel at c, and a photon's energy depends only on the light's frequency, not its intensity.

Stop to Think 38.3: $n = 4$. There are four antinodes.

Stop to Think 38.4: Not in absorption. In emission from the $n = 3$ to $n = 2$ transition. The photon energy has to match the energy *difference* between two energy levels. Absorption is from the ground state, at $E_1 = 0.00$ eV. There's no energy level at 3.00 eV to which the atom could jump.

Stop to Think 38.5: $n = 3$. Each antinode is half a wavelength, so this standing wave has three full wavelengths in one circumference.

39 Wave Functions and Uncertainty

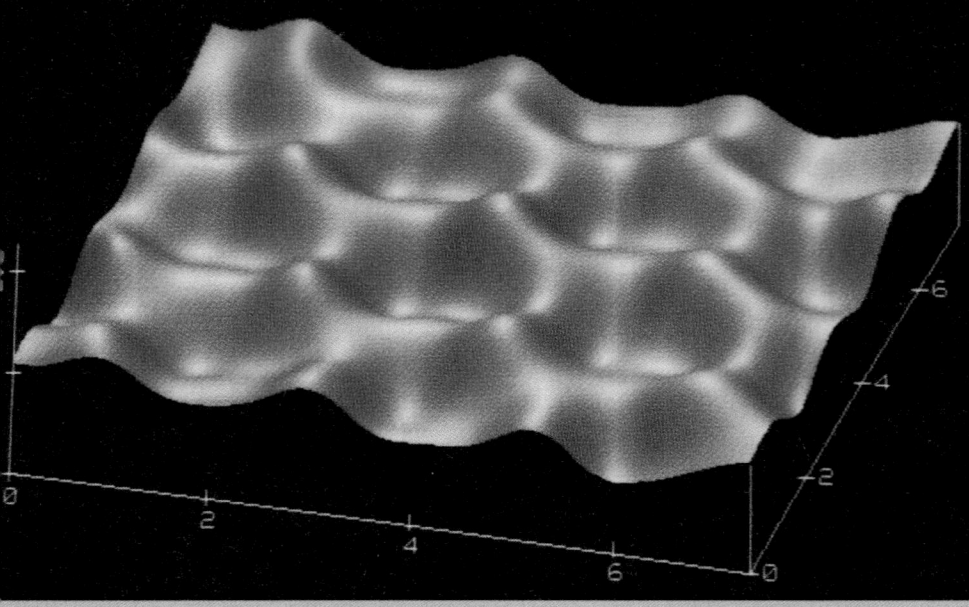

The surface of graphite, imaged with atomic resolution by a scanning tunneling microscope. The hexagonal ridges show the most probable locations of the electrons.

▶ **Looking Ahead** The goal of Chapter 39 is to introduce and learn to use the wave-function description of matter.

Quantum Mechanics

This chapter and the next will introduce the essential ideas of **quantum mechanics** in one dimension. The full theory is beyond the scope of this textbook, but we can delve far enough into quantum mechanics to learn, in the final two chapters, how it solves the problems of atomic and nuclear structure.

Despite the strange and unfamiliar aspects of quantum mechanics, its predictions are verified with amazing precision. It is the most successful physical theory ever devised.

Wave Functions

You'll learn that the probability of finding a particle is determined by the particle's **wave function** $\psi(x)$.

The wave function is an oscillatory function.

The square of the wave function is the **probability density**.

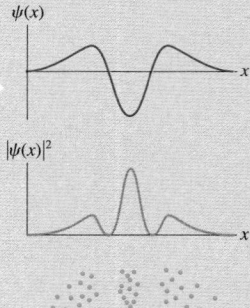

The particle is most likely found where the probability density peaks.

The wave function $\psi(x)$ is a *wave-like function* that can be used to make probabilistic predictions, but nothing is actually waving.

You'll learn how to interpret the wave function in different situations.

◀ **Looking Back**
Section 38.3 Photons
Section 38.4 de Broglie wavelength

Waves and Particles

The wave function reconciles the experimental evidence that matter has both particle-like and wave-like properties. The probability of detecting a *particle* is governed by a *wave-like* function that can exhibit interference.

In the double-slit experiment, interference fringes in the wave function indicate that particle-like electrons are most likely to be detected where a wave would exhibit bright fringes.

Probability

You'll learn that quantum mechanics deals with *probabilities*. We cannot say exactly where an electron is or how it's moving, but we can make accurate statements about the probability of locating the electron in a region of space.

If an experiment detects N particles, the *expected number* in a given region of space is N times the probability of being in that region.

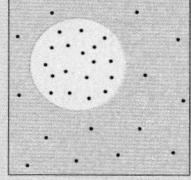

Uncertainty

You'll learn that the **Heisenberg uncertainty principle** places a fundamental limit on how well you can know a particle's position and speed.

Because matter has wave-like properties, a particle simply does not have a precise position or a precise speed. Our knowledge of a particle is inherently uncertain.

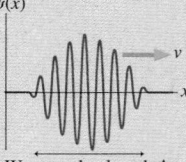

Wave packet length Δx

◀ **Looking Back**
Section 21.8 Beats

39.1 Waves, Particles, and the Double-Slit Experiment

You may feel surprise at how slowly we have been building up to quantum mechanics. Why not just write it down and start using it? There are two reasons. First, quantum mechanics explains microscopic phenomena that we cannot directly sense or experience. It was important to begin by learning how light and atoms behave. Otherwise, how would you know if quantum mechanics explains anything? Second, the concepts we'll need in quantum mechanics are rather abstract. Before launching into the mathematics, we need to establish a connection between theory and experiment.

We will make the connection by returning to the double-slit interference experiment, an experiment that goes right to the heart of wave–particle duality. The significance of the double-slit experiment arises from the fact that both light and matter exhibit the same interference pattern. Regardless of whether photons, electrons, or neutrons pass through the slits, their arrival at a detector is a particle-like event. That is, they make a collection of discrete dots on a detector. Yet our understanding of how interference "works" is based on the properties of waves. Our goal is to find the connection between the wave description and the particle description of interference.

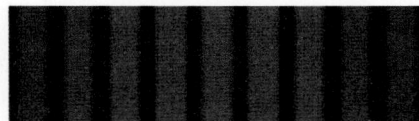

Interference fringes in an optical double-slit interference experiment.

A Wave Analysis of Interference

The interference of light can be analyzed from either a wave perspective or a photon perspective. Let's start with a wave analysis. FIGURE 39.1 shows light waves passing through a double slit with slit separation d. You should recall that the lines in a wavefront diagram represent wave crests, spaced one wavelength apart. The bright fringes of constructive interference occur where two crests or two troughs overlap. The graphs and the picture of the detection screen (notice that they're aligned vertically) show the outcome of the experiment.

You studied interference and the double-slit experiment in Chapters 21 and 22. The two waves traveling from the slits to the viewing screen are traveling waves with displacements

$$D_1 = a\sin(kr_1 - \omega t)$$

$$D_2 = a\sin(kr_2 - \omega t)$$

where a is the amplitude of each wave, $k = 2\pi/\lambda$ is the wave number, and r_1 and r_2 are the distances from the two slits. The "displacement" of a light wave is not a physical displacement, as in a water wave, but a change in the electromagnetic field.

According to the principle of superposition, these two waves add together where they meet at a point on the screen to give a wave with net displacement $D = D_1 + D_2$. Previously (see Equation 22.12) we found that the amplitude of their superposition is

$$A(x) = 2a\cos\left(\frac{\pi dx}{\lambda L}\right) \tag{39.1}$$

where x is the horizontal coordinate on the screen, measured from $x = 0$ in the center.

The function $A(x)$, the top graph in Figure 39.1, is called the *amplitude function*. It describes the amplitude A of the light wave as a function of the position x on the viewing screen. The amplitude function has maxima where two crests from individual waves overlap and add constructively to make a larger wave with amplitude $2a$. $A(x)$ is zero at points where the two individual waves are out of phase and interfere destructively.

If you carry out a double-slit experiment in the lab, what you observe on the screen is the light's *intensity*, not its amplitude. A wave's intensity I is proportional to the *square* of the amplitude. That is, $I \propto A^2$, where $\propto$ is the "is proportional to" symbol.

FIGURE 39.1 The double-slit experiment with light.

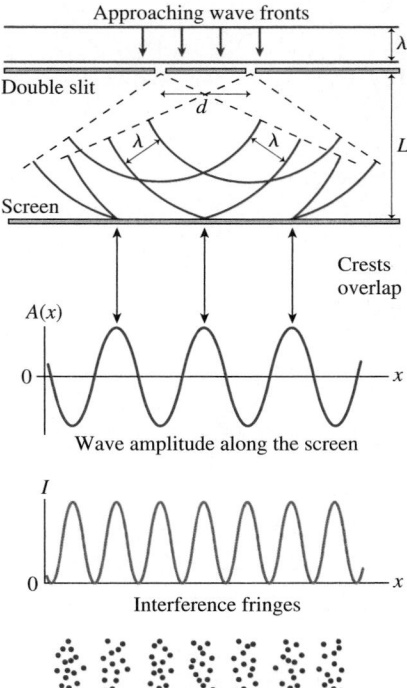

Using Equation 39.1 for the amplitude at each point, we find the intensity $I(x)$ as a function of position x on the screen is

$$I(x) = C\cos^2\left(\frac{\pi dx}{\lambda L}\right) \tag{39.2}$$

where C is a proportionality constant.

The lower graph in Figure 39.1 shows the intensity as a function of position along the screen. This graph shows the alternating bright and dark interference fringes that you see in the laboratory. In other words, the intensity of the wave is the *experimental reality* that you observe and measure.

Probability

Before discussing photons, we need to introduce some ideas about probability. Imagine throwing darts at a dart board while blindfolded. FIGURE 39.2 shows how the board might look after your first 100 throws. From this information, can you predict where your 101st throw is going to land? We'll assume that all darts hit the board.

No. The position of any individual dart is *unpredictable*. No matter how hard you try to reproduce the previous throw, a second dart will not land at the same place. Yet there is clearly an overall *pattern* to where the darts strike the board. Even blindfolded, you had a general sense of where the center of the board was, so each dart was *more likely* to land near the center than at the edge.

Although we can't predict where any individual dart will land, we can use the information in Figure 39.2 to determine the *probability* that your next throw will land in region A or region B or region C. Because 45 out of 100 throws landed in region A, we could say that the *odds* of hitting region A are 45 out of 100, or 45%.

Now, 100 throws isn't all that many. If you throw another 100 darts, perhaps only 43 will land in region A. Then maybe 48 of the next 100 throws. Imagine that the total number of throws N_{tot} becomes extremely large. Then the **probability** that any particular throw lands in region A is defined to be

$$P_A = \lim_{N_{tot} \to \infty} \frac{N_A}{N_{tot}} \tag{39.3}$$

In other words, the probability that the outcome will be A is the fraction of outcomes that are A in an enormously large number of trials. Similarly, $P_B = N_B/N_{tot}$ and $P_C = N_C/N_{tot}$ as $N_{tot} \to \infty$. We can give probabilities as either a decimal fraction or a percentage. In this example, $P_A \approx 45\%$, $P_B \approx 35\%$, and $P_C \approx 20\%$. We've used $\approx$ rather than $=$ because 100 throws isn't enough to determine the probabilities with great precision.

What is the probability that a dart lands in either region A *or* region B? The number of darts landing in either A *or* B is $N_{A\ or\ B} = N_A + N_B$, so we can use the definition of probability to learn that

$$\begin{aligned} P_{A\ or\ B} &= \lim_{N_{tot} \to \infty} \frac{N_{A\ or\ B}}{N_{tot}} = \lim_{N_{tot} \to \infty} \frac{N_A + N_B}{N_{tot}} \\ &= \lim_{N_{tot} \to \infty} \frac{N_A}{N_{tot}} + \lim_{N_{tot} \to \infty} \frac{N_B}{N_{tot}} = P_A + P_B \end{aligned} \tag{39.4}$$

That is, **the probability that the outcome will be A *or* B is the sum of P_A and P_B.** This important conclusion is a general property of probabilities.

Each dart lands *somewhere on* the board. Consequently, the probability that a dart lands in A *or* B *or* C must be 100%. And, in fact,

$$P_{somewhere} = P_{A\ or\ B\ or\ C} = P_A + P_B + P_C = 0.45 + 0.35 + 0.20 = 1.00$$

Thus another important property of probabilities is that **the sum of the probabilities of all possible outcomes must equal 1.**

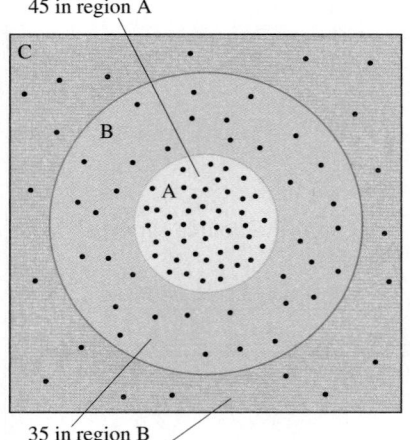

FIGURE 39.2 One hundred throws at a dart board.

45 in region A

C

B

A

35 in region B
20 in region C

Suppose exhaustive trials have established that the probability of a dart landing in region A is P_A. If you throw N darts, how many do you *expect* to land in A? This value, called the **expected value,** is

$$N_{A \text{ expected}} = NP_A \qquad (39.5)$$

The expected value is your best prediction of the outcome of an experiment.

If $P_A = 0.45$, your *best prediction* is that 27 of 60 throws (45% of 60) will land in A. Of course, predicting 27 and actually getting 27 aren't the same thing. You would predict 30 heads in 60 flips of a coin, but you wouldn't be surprised if the actual number were 28 or 31. Similarly, the number of darts landing in region A might be 24 or 29 instead of 27. In general, the agreement between actual values and expected values improves as you throw more darts.

STOP TO THINK 39.1 Suppose you roll a die 30 times. What is the expected number of 1's *and* 6's?

A Photon Analysis of Interference

Now let's look at the double-slit results from a photon perspective. We know, from experimental evidence, that the interference pattern is built up photon by photon. The bottom portion of Figure 39.1 shows the pattern made on a detector after the arrival of the first few dozen photons. It is clearly a double-slit interference pattern, but it's made, rather like a newspaper photograph, by piling up dots in some places but not others.

The arrival position of any particular photon is *unpredictable.* That is, nothing about how the experiment is set up or conducted allows us to predict exactly where the dot of an individual photon will appear on the detector. Yet there is clearly an overall pattern. There are some positions at which a photon is *more likely* to be detected, other positions at which it is *less likely* to be found.

If we record the arrival positions of many thousands of photons, we will be able to determine the *probability* that a photon will be detected at any given location. If 50 out of 50,000 photons land in one small area of the screen, then each photon has a probability of $50/50{,}000 = 0.001 = 0.1\%$ of being detected there. The probability will be zero at the interference minima because no photons at all arrive at those points. Similarly, the probability will be a maximum at the interference maxima. The probability will have some in-between value on the sides of the interference fringes.

FIGURE 39.3a shows a narrow strip with width δx and height H. (We will assume that δx is very small in comparison with the fringe spacing, so the light's intensity over δx is very nearly constant.) Think of this strip as a very narrow detector that can detect and count the photons landing on it. Suppose we place the narrow strip at position x. We'll use the notation $N(\text{in } \delta x \text{ at } x)$ to indicate the number of photons that hit the detector at this position. The value of $N(\text{in } \delta x \text{ at } x)$ varies from point to point. $N(\text{in } \delta x \text{ at } x)$ is large if x happens to be near the center of a bright fringe; $N(\text{in } \delta x \text{ at } x)$ is small if x is in a dark fringe.

Suppose N_{tot} photons are fired at the slits. The *probability* that any one photon ends up in the strip at position x is

$$\text{Prob(in } \delta x \text{ at } x) = \lim_{N_{\text{tot}} \to \infty} \frac{N(\text{in } \delta x \text{ at } x)}{N_{\text{tot}}} \qquad (39.6)$$

As FIGURE 39.3b shows, Equation 39.6 is an empirical method for determining the probability of the photons hitting a particular spot on the detector.

FIGURE 39.3 A strip of width δx at position x.

(a) The number of photons in this narrow strip when it is at position x is $N(\text{in } \delta x \text{ at } x)$.

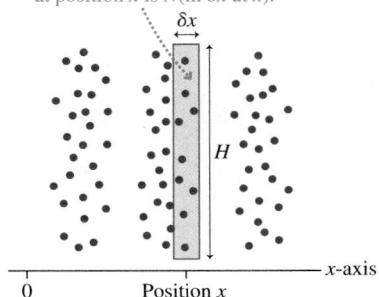

(b) $N(\text{in } \delta x \text{ at } x_1) = 12$ $N(\text{in } \delta x \text{ at } x_2) = 3$

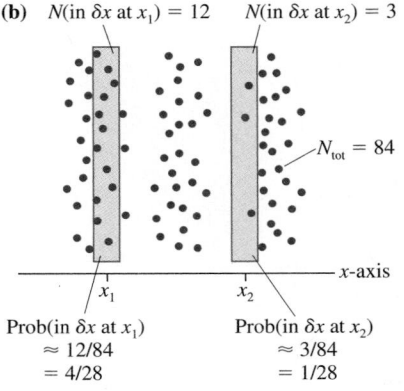

$N_{\text{tot}} = 84$

Prob(in δx at x_1)
$\approx 12/84$
$= 4/28$

Prob(in δx at x_2)
$\approx 3/84$
$= 1/28$

Alternatively, suppose we can calculate the probabilities from a theory. In that case, the *expected value* for the number of photons landing in the narrow strip when it is at position x is

$$N(\text{in } \delta x \text{ at } x) = N \times \text{Prob}(\text{in } \delta x \text{ at } x) \qquad (39.7)$$

We cannot predict what any individual photon will do, but we can predict the fraction of the photons that should land in this little region of space. $\text{Prob}(\text{in } \delta x \text{ at } x)$ is the probability that it will happen.

39.2 Connecting the Wave and Photon Views

The wave model of light describes the interference pattern in terms of the wave's intensity $I(x)$, a continuous-valued function. The photon model describes the interference pattern in terms of the probability $\text{Prob}(\text{in } \delta x \text{ at } x)$ of detecting a photon. These two models are very different, yet Figure 39.1 shows a clear correlation between the *intensity of the wave* and the *probability of detecting photons*. That is, photons are more likely to be detected at those points where the wave intensity is high and less likely to be detected at those points where the wave intensity is low.

The intensity of a wave is $I = P/A$, the ratio of light power P (joules per second) to the area A on which the light falls. The narrow strip in Figure 39.3a has area $A = H \delta x$. If the light intensity at position x is $I(x)$, the amount of light energy E falling onto this narrow strip during each second is

$$E(\text{in } \delta x \text{ at } x) = I(x)A = I(x)H \delta x = H I(x) \delta x \qquad (39.8)$$

The notation $E(\text{in } \delta x \text{ at } x)$ refers to the energy landing on this narrow strip if you place it at position x.

From the photon perspective, energy E is due to the arrival of N photons, each of which has energy hf. The number of photons that arrive in the strip each second is

$$N(\text{in } \delta x \text{ at } x) = \frac{E(\text{in } \delta x \text{ at } x)}{hf} = \frac{H}{hf}I(x) \delta x \qquad (39.9)$$

We can then use Equation 39.6, the definition of probability, to write the *probability* that a photon lands in the narrow strip δx at position x as

$$\text{Prob}(\text{in } \delta x \text{ at } x) = \frac{N(\text{in } \delta x \text{ at } x)}{N_{\text{tot}}} = \frac{H}{hfN_{\text{tot}}}I(x) \delta x \qquad (39.10)$$

Equation 39.10 is a critical link between the wave model and the photon model. It tells us that the probability of detecting a photon is proportional to the intensity of the light at that point and to the width of the detector.

As a final step, recall that the light intensity $I(x)$ is proportional to $|A(x)|^2$, the square of the amplitude function. Consequently,

$$\text{Prob}(\text{in } \delta x \text{ at } x) \propto |A(x)|^2 \delta x \qquad (39.11)$$

where the various constants in Equation 39.10 have all been incorporated into the unspecified proportionality constant of Equation 39.11.

In other words, **the probability of detecting a photon at a particular point is directly proportional to the square of the light-wave amplitude function at that point.** If the wave amplitude at point A is twice that at point B, then a photon is four times as likely to land in a narrow strip at A as it is to land in an equal-width strip at B.

NOTE ▶ Equation 39.11 is the connection between the particle perspective and the wave perspective. It relates the probability of observing a particle-like event—the arrival of a photon—to the amplitude of a continuous, classical wave. This connection will become the basis of how we interpret the results of quantum-physics calculations. ◀

Probability Density

We need one last definition. Recall that the mass of a wire or string of a length L can be expressed in terms of the linear mass density μ as $m = \mu L$. Similarly, the charge along a length L of wire can be expressed in terms of the linear charge density λ as $Q = \lambda L$. If the length had been very short—in which case we might have denoted it as δx—and if the density varied from point to point, we could have written

$$\text{mass(in length } \delta x \text{ at } x) = \mu(x)\,\delta x$$

$$\text{charge(in length } \delta x \text{ at } x) = \lambda(x)\,\delta x$$

where $\mu(x)$ and $\lambda(x)$ are the linear densities at position x. Writing the mass and charge this way separates the role of the density from the role of the small length δx.

Equation 39.11 looks similar. Using the mass and charge densities as analogies, as shown in FIGURE 39.4, we can define the **probability density** $P(x)$ such that

$$\text{Prob(in } \delta x \text{ at } x) = P(x)\,\delta x \qquad (39.12)$$

In one dimension, probability density has SI units of m^{-1}. Thus the probability density multiplied by a length, as in Equation 39.12, yields a dimensionless probability.

NOTE ▶ $P(x)$ itself is *not* a probability, just as the linear mass density λ is not, by itself, a mass. You must multiply the probability density by a length, as shown in Equation 39.12, to find an actual probability. ◀

By comparing Equation 39.12 to Equation 39.11, you can see that the photon probability density is directly proportional to the square of the light-wave amplitude:

$$P(x) \propto |A(x)|^2 \qquad (39.13)$$

The probability density, unlike the probability itself, is independent of the width δx and depends on only the amplitude function.

Although we were inspired by the double-slit experiment, nothing in our analysis actually depends on the double-slit geometry. Consequently, Equation 39.13 is quite general. It says that for *any* experiment in which we detect photons, **the probability density for detecting a photon is directly proportional to the square of the amplitude function of the corresponding electromagnetic wave.** We now have an explicit connection between the wave-like and the particle-like properties of the light.

FIGURE 39.4 The probability density is analogous to the linear mass density.

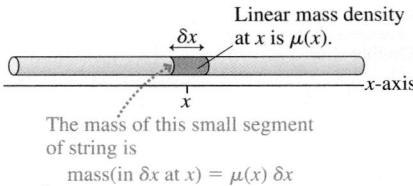

The mass of this small segment of string is
$$\text{mass(in } \delta x \text{ at } x) = \mu(x)\,\delta x$$

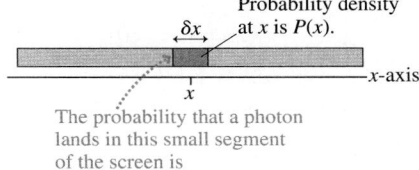

The probability that a photon lands in this small segment of the screen is
$$\text{Prob(in } \delta x \text{ at } x) = P(x)\,\delta x$$

EXAMPLE 39.1 **Calculating the probability density**

In an experiment, 6000 out of 600,000 photons are detected in a 1.0-mm-wide strip located at position $x = 50$ cm. What is the probability density at $x = 50$ cm?

SOLVE The probability that a photon arrives at this particular strip is

$$\text{Prob(in 1.0 mm at } x = 50 \text{ cm)} = \frac{6000}{600,000} = 0.010$$

Thus the probability density $P(x) = \text{Prob(in } \delta x \text{ at } x)/\delta x$ at this position is

$$P(50 \text{ cm}) = \frac{\text{Prob(in 1.0 mm at } x = 50 \text{ cm)}}{0.0010 \text{ m}} = \frac{0.010}{0.0010 \text{ m}}$$

$$= 10 \text{ m}^{-1}$$

STOP TO THINK 39.2 The figure shows the detection of photons in an optical experiment. Rank in order, from largest to smallest, the square of the amplitude function of the electromagnetic wave at positions A, B, C, and D.

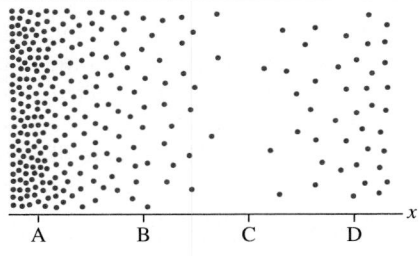

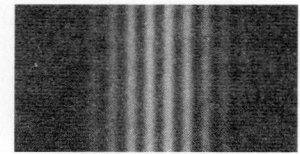

Electrons create interference fringes.

FIGURE 39.5 The double-slit experiment with electrons.

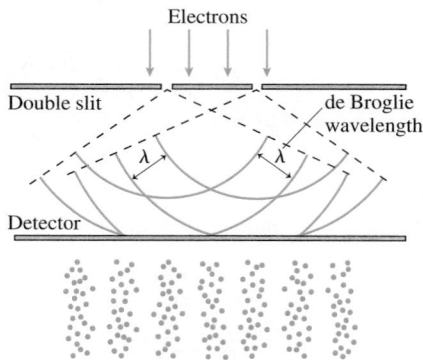

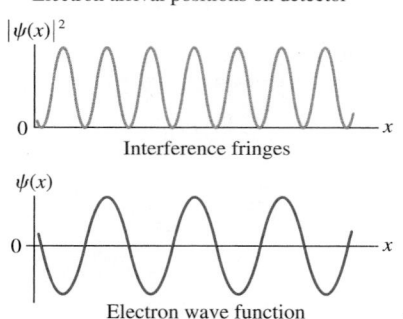

FIGURE 39.6 The square of the wave function is the probability density for detecting the electron at position x.

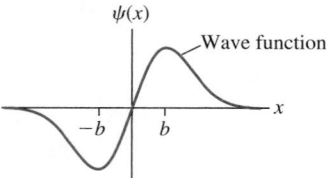

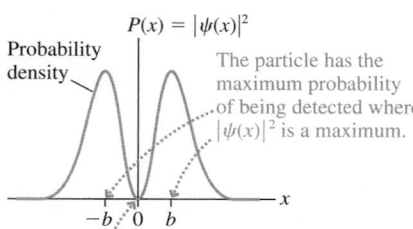

The particle has zero probability of being detected where $|\psi(x)|^2 = 0$.

39.3 The Wave Function

Now let's look at the interference of matter. Electrons passing through a double-slit apparatus create the same interference patterns as photons. The pattern is built up electron by electron, but there is no way to predict where any particular electron will be detected. Even so, we can establish the *probability* of an electron landing in a narrow strip of width δx by measuring the positions of many individual electrons.

For light, we were able to relate the photon probability density $P(x)$ to the amplitude of an electromagnetic wave. But there is no wave for electrons like electromagnetic waves for light. So how do we find the probability density for electrons? We have reached the point where we must make an inspired leap beyond classical physics. Let us *assume* that there is some kind of continuous, wave-like function for matter that plays a role analogous to the electromagnetic amplitude function $A(x)$ for light. We will call this function the **wave function** $\psi(x)$, where ψ is a lowercase Greek psi. The wave function is a function of position, which is why we write it as $\psi(x)$.

To connect the wave function to the real world of experimental measurements, we will interpret $\psi(x)$ in terms of the *probability* of detecting a particle at position x. If a matter particle, such as an electron, is described by the wave function $\psi(x)$, then the probability Prob(in δx at x) of finding the particle within a narrow region of width δx at position x is

$$\text{Prob(in } \delta x \text{ at } x) = |\psi(x)|^2 \delta x = P(x)\,\delta x \qquad (39.14)$$

That is, the probability density $P(x)$ for finding the particle is

$$P(x) = |\psi(x)|^2 \qquad (39.15)$$

With Equations 39.14 and 39.15, we are *defining* the wave function $\psi(x)$ to play the same role for material particles that the amplitude function $A(x)$ does for photons. The only difference is that $P(x) = |\psi(x)|^2$ is for particles, whereas Equation 39.13 for photons is $P(x) \propto |A(x)|^2$. The difference is that the electromagnetic field amplitude $A(x)$ had previously been defined through the laws of electricity and magnetism. $|A(x)|^2$ is *proportional* to the probability density for finding a photon, but it is not directly *the* probability density. In contrast, we do not have any preexisting definition for the wave function $\psi(x)$. Thus we are free to define $\psi(x)$ so that $|\psi(x)|^2$ is *exactly* the probability density. That is why we used $=$ rather than $\propto$ in Equation 39.15.

FIGURE 39.5 shows the double-slit experiment with electrons. This time we will work backward. From the observed distribution of electrons, which represents the probabilities of their landing in any particular location, we can deduce that $|\psi(x)|^2$ has alternating maxima and zeros. The oscillatory wave function $\psi(x)$ is the square root *at each point* of $|\psi(x)|^2$. Notice the very close analogy with the amplitude function $A(x)$ in Figure 39.1.

NOTE ▶ $|\psi(x)|^2$ is uniquely determined by the data, but the wave function $\psi(x)$ is *not* unique. The alternative wave function $\psi'(x) = -\psi(x)$—an upside-down version of the graph in Figure 39.5—would be equally acceptable. ◀

FIGURE 39.6 is a different example of a wave function. After squaring it *at each point*, as shown in the bottom half of the figure, we see that this wave function represents a particle most likely to be detected very near $x = -b$ or $x = +b$. These are the points where $|\psi(x)|^2$ is a maximum. There is zero likelihood of finding the particle right in the center. The particle is more likely to be detected at some positions than at others, but we cannot predict what its exact location will be at any given time.

NOTE ▶ One of the difficulties in learning to use the concept of a wave function is coming to grips with the fact that there is no "thing" that is waving. There is no disturbance associated with a physical medium. The wave function $\psi(x)$ is simply a *wave-like function* (i.e., it oscillates between positive and negative values) that can be used to make probabilistic predictions about atomic particles. ◄

A Little Science Methodology

Equation 39.14 defines the wave function $\psi(x)$ for a particle in terms of the probability of finding the particle at different positions x. But our interests go beyond merely characterizing experimental data. We would like to develop a new *theory* of matter. But just what is a theory? Although this is not a book on scientific methodology, we can loosely say that a physical theory needs two basic ingredients:

1. A *descriptor,* a mathematical quantity used to describe our knowledge of a physical object.
2. One or more *laws* that govern the behavior of the descriptor.

For example, Newtonian mechanics is a theory of motion. The primary descriptor in Newtonian mechanics is a particle's *position* $x(t)$ as a function of time. This describes our knowledge of the particle at all times. The position is governed by *Newton's laws.* These laws, especially the second law, are mathematical statements of how the descriptor changes in response to forces. If we predict $x(t)$ for a known set of forces, we feel confident that an experiment carried out at time t will find the particle right where predicted.

Newton's theory of motion *assumes* that a particle's position is well defined at every instant of time. The difficulty facing physicists early in the 20th century was the astounding discovery that **the position of an atomic-size particle is** *not* **well defined.** An electron in a double-slit experiment must, in some sense, go through *both* slits to produce an electron interference pattern. It simply does not have a well-defined position as it interacts with the slits. But if the position function $x(t)$ is not a valid descriptor for matter at the atomic level, what is?

We will assert that the wave function $\psi(x)$ is the *descriptor* of a particle in quantum mechanics. In other words, the wave function tells us everything we can know about the particle. The wave function $\psi(x)$ plays the same leading role in quantum mechanics that the position function $x(t)$ plays in classical mechanics.

Whether this hypothesis has any merit will not be known until we see if it leads to predictions that can be verified. And before we can do that, we need to learn the "law of psi." What new law of physics determines the wave function $\psi(x)$ in a given situation? We will answer this question in the next chapter.

It may seem to you, as we go along, that we are simply "making up" ideas. Indeed, that is at least partially true. The inventors of entirely new theories use their existing knowledge as a guide, but ultimately they have to make an inspired guess as to what a new theory should look like. Newton and Einstein both made such leaps, and the inventors of quantum mechanics had to make such a leap. We can attempt to make the new ideas *plausible,* but ultimately a new theory is simply a bold assertion that must be tested against reality via controlled experiments. The wave-function theory of quantum mechanics passed the only test that really matters in science—it works!

STOP TO THINK 39.3   This is the wave function of a neutron. At what value of x is the neutron most likely to be found?

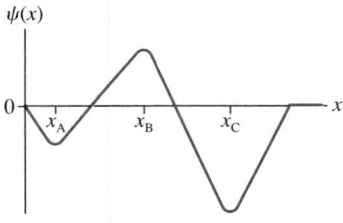

39.4 Normalization

In our discussion of probability we noted that the dart has to hit the wall *somewhere*. The mathematical statement of this idea is the requirement that $P_A + P_B + P_C = 1$. That is, the probabilities of all the mutually exclusive outcomes *must* add up to 1.

Similarly, a photon or electron has to land *somewhere* on the detector after passing through an experimental apparatus. Consequently, the probability that it will be detected at *some* position is 100%. To make use of this requirement, consider an experiment in which an electron is detected on the x-axis. As FIGURE 39.7 shows, we can divide the region between positions x_L and x_R into N adjacent narrow strips of width δx.

The probability that any particular electron lands in the narrow strip i at position x_i is

$$\text{Prob(in } \delta x \text{ at } x_i) = P(x_i)\,\delta x$$

where $P(x_i) = |\psi(x_i)|^2$ is the probability density at x_i. The probability that the electron lands in the strip at x_1 *or* x_2 *or* x_3 *or* ... is the sum

$$\text{Prob(between } x_L \text{ and } x_R) = \text{Prob(in } \delta x \text{ at } x_1)$$

$$+ \text{Prob(in } \delta x \text{ at } x_2) + \cdots \quad (39.16)$$

$$= \sum_{i=1}^{N} P(x_i)\,\delta x = \sum_{i=1}^{N} |\psi(x_i)|^2\,\delta x$$

That is, **the probability that the electron lands *somewhere* between x_L and x_R is the sum of the probabilities of landing in each narrow strip.**

If we let the strips become narrower and narrower, then $\delta x \to dx$ and the sum becomes an integral. Thus the probability of finding the particles in the range $x_L \leq x \leq x_R$ is

$$\text{Prob(in range } x_L \leq x \leq x_R) = \int_{x_L}^{x_R} P(x)\,dx = \int_{x_L}^{x_R} |\psi(x)|^2\,dx \quad (39.17)$$

As FIGURE 39.8a shows, we can interpret Prob(in range $x_L \leq x \leq x_R$) as the area under the probability density curve between x_L and x_R.

> **NOTE** ▶ The integral of Equation 39.17 is needed when the probability density changes over the range x_L to x_R. For sufficiently narrow intervals, over which $P(x)$ remains essentially constant, the expression Prob(in δx at x) = $P(x)\,\delta x$ is still valid and is easier to use. ◀

Now let the detector become infinitely wide, so that the probability that the electron will arrive *somewhere* on the detector becomes 100%. The statement that the electron has to land *somewhere* on the x-axis is expressed mathematically as

$$\int_{-\infty}^{\infty} P(x)\,dx = \int_{-\infty}^{\infty} |\psi(x)|^2\,dx = 1 \quad (39.18)$$

Equation 39.18 is called the **normalization condition**. Any wave function $\psi(x)$ must satisfy this condition; otherwise we would not be able to interpret $|\psi(x)|^2$ as a probability density. As FIGURE 39.8b shows, Equation 39.18 tells us that the total area under the probability density curve must be 1.

> **NOTE** ▶ The normalization condition integrates the *square* of the wave function. We don't have any information about what the integral of $\psi(x)$ might be. ◀

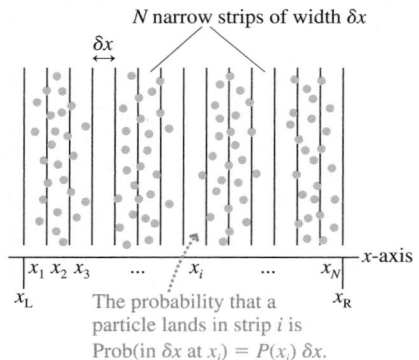

FIGURE 39.7 Dividing the entire detector into many small strips of width δx.

N narrow strips of width δx

x_L The probability that a particle lands in strip i is Prob(in δx at x_i) = $P(x_i)\,\delta x$.

FIGURE 39.8 The area under the probability density curve is a probability.

(a)

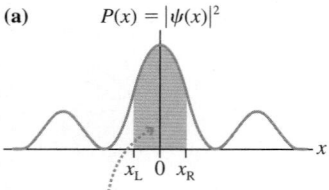

$P(x) = |\psi(x)|^2$

The area under the curve between x_L and x_R is the probability of finding the particle between x_L and x_R.

(b)

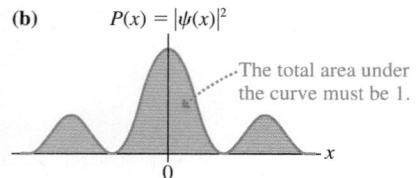

$P(x) = |\psi(x)|^2$

The total area under the curve must be 1.

EXAMPLE 39.2   **Normalizing and interpreting a wave function**

FIGURE 39.9 shows the wave function of a particle confined within the region between $x = 0$ nm and $x = L = 1.0$ nm. The wave function is zero outside this region.

a. Determine the value of the constant c that makes this a normalized wave function.
b. Draw a graph of the probability density $P(x)$.
c. Draw a dot picture showing where the first 40 or 50 particles might be found.
d. Calculate the probability of finding the particle in a region of width $\delta x = 0.01$ nm at positions $x_1 = 0.05$ nm, $x_2 = 0.50$ nm, and $x_3 = 0.95$ nm.

FIGURE 39.9 The wave function of Example 39.2.

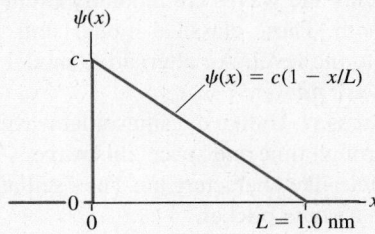

MODEL The probability of finding the particle is determined by the probability density $P(x)$.

VISUALIZE The wave function is shown in Figure 39.9.

SOLVE a. The wave function is $\psi(x) = c(1 - x/L)$ between 0 and L, 0 otherwise. This is a function that decreases linearly from $\psi = c$ at $x = 0$ to $\psi = 0$ at $x = L$. The constant c is the height of this wave function. The particle *has* to be in the region $0 \le x \le L$ with probability 1, and only one value of c will make it so. We can determine c by using Equation 39.18, the normalization condition. Because the wave function is zero outside the interval from 0 to L, the integration limits are 0 to L. Thus

$$1 = \int_0^L |\psi(x)|^2 \, dx = c^2 \int_0^L \left(1 - \frac{x}{L}\right)^2 dx$$

$$= c^2 \int_0^L \left(1 - \frac{2x}{L} + \frac{x^2}{L^2}\right) dx$$

$$= c^2 \left[x - \frac{x^2}{L} + \frac{x^3}{3L^2}\right]_0^L = \frac{1}{3}c^2 L$$

The solution for c is

$$c = \sqrt{\frac{3}{L}} = \sqrt{\frac{3}{1.0 \text{ nm}}} = 1.732 \text{ nm}^{-1/2}$$

Note the unusual units for c. Although these are not SI units, we can correctly compute probabilities as long as δx has units of nm. A multiplicative constant such as c is often called a *normalization constant*.

b. The wave function is

$$\psi(x) = (1.732 \text{ nm}^{-1/2})\left(1 - \frac{x}{1.0 \text{ nm}}\right)$$

Thus the probability density is

$$P(x) = |\psi(x)|^2 = (3.0 \text{ nm}^{-1})\left(1 - \frac{x}{1.0 \text{ nm}}\right)^2$$

This probability density is graphed in FIGURE 39.10a.

FIGURE 39.10 The probability density $P(x)$ and the detected positions of particles.

(a) $P(x)$ (nm^{-1})

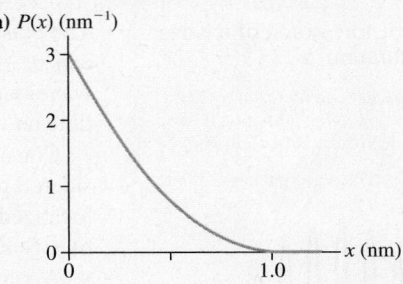

(b)

c. Particles are most likely to be detected at the left edge of the interval, where the probability density $P(x)$ is maximum. The probability steadily decreases across the interval, becoming zero at $x = 1.0$ nm. FIGURE 39.10b shows how a group of particles described by this wave function might appear on a detection screen.

d. $P(x)$ is essentially constant over the small interval $\delta x = 0.01$ nm, so we can use

$$\text{Prob(in } \delta x \text{ at } x) = P(x)\,\delta x = |\psi(x)|^2 \, \delta x$$

for the probability of finding the particle in a region of width δx at the position x. We need to evaluate $|\psi(x)|^2$ at the three positions $x_1 = 0.05$ nm, $x_2 = 0.50$ nm, and $x_3 = 0.95$ nm. Doing so gives

$$\text{Prob(in 0.01 nm at } x_1 = 0.05 \text{ nm}) = c^2(1 - x_1/L)^2 \, \delta x$$
$$= 0.0270 = 2.70\%$$

$$\text{Prob(in 0.01 nm at } x_2 = 0.50 \text{ nm}) = c^2(1 - x_2/L)^2 \, \delta x$$
$$= 0.0075 = 0.75\%$$

$$\text{Prob(in 0.01 nm at } x_3 = 0.95 \text{ nm}) = c^2(1 - x_3/L)^2 \, \delta x$$
$$= 0.00008 = 0.008\%$$

The value of the constant a is

a. $a = 2.0 \text{ mm}^{-1}$
b. $a = 1.0 \text{ mm}^{-1}$
c. $a = 0.5 \text{ mm}^{-1}$
d. $a = 2.0 \text{ mm}^{-1/2}$
e. $a = 1.0 \text{ mm}^{-1/2}$
f. $a = 0.5 \text{ mm}^{-1/2}$

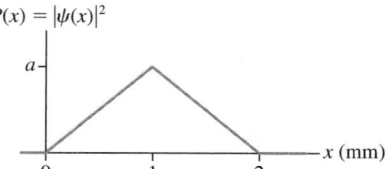

39.5 Wave Packets

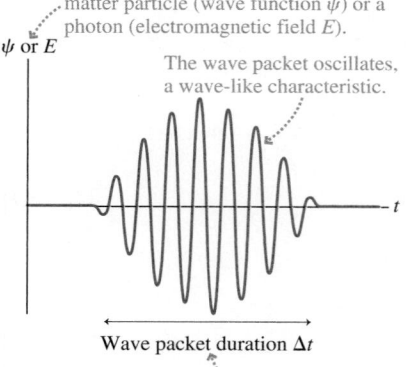

FIGURE 39.11 History graph of a wave packet with duration Δt.

A wave packet can represent either a matter particle (wave function ψ) or a photon (electromagnetic field E).

The wave packet oscillates, a wave-like characteristic.

Wave packet duration Δt

The wave packet is localized, a particle-like characteristic.

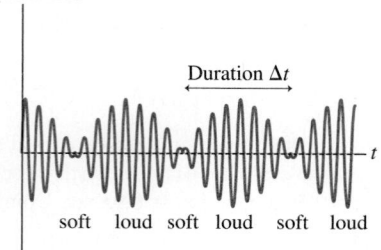

FIGURE 39.12 Beats are a series of wave packets.

Displacement

Duration Δt

soft loud soft loud soft loud

The classical physics ideas of particles and waves are mutually exclusive. An object can be one or the other, but not both. These classical models fail to describe the wave–particle duality seen at the atomic level. An alternative model with both particle and wave characteristics is a *wave packet*.

Consider the wave shown in FIGURE 39.11. Unlike the sinusoidal waves we have considered previously, which stretch through time and space, this wave is bunched up, or localized. The localization is a particle-like characteristic. The oscillations are wave-like. Such a localized wave is called a **wave packet.**

A wave packet travels through space with constant speed v, just like a photon in a light wave or an electron in a force-free region. A wave packet has a wavelength, hence it will undergo interference and diffraction. But because it is also localized, a wave packet has the possibility of making a "dot" when it strikes a detector. We can visualize a light wave as consisting of a very large number of these wave packets moving along together. Similarly, we can think of a beam of electrons as a series of wave packets spread out along a line.

Wave packets are not a perfect model of photons or electrons (we need the full treatment of quantum physics to get a more accurate description), but they do provide a useful way of thinking about photons and electrons in many circumstances.

You might have noticed that the wave packet in Figure 39.11 looks very much like one cycle of a beat pattern. You will recall that beats occur if we superimpose two waves of frequencies f_1 and f_2 where the two frequencies are very similar: $f_1 \approx f_2$. FIGURE 39.12, which is copied from Chapter 21 where we studied beats, shows that the loud, soft, loud, soft, ... pattern of beats corresponds to a series of wave packets.

In Chapter 21, the beat frequency (number of pulses per second) was found to be

$$f_{\text{beat}} = f_1 - f_2 = \Delta f \tag{39.19}$$

where Δf is the *range* of frequencies that are superimposed to form the wave packet. Figure 39.12 defines Δt as the duration of each beat or each wave packet. This interval of time is equivalent to the *period* T_{beat} of the beat. Because period and frequency are inverses of each other, the duration Δt is

$$\Delta t = T_{\text{beat}} = \frac{1}{f_{\text{beat}}} = \frac{1}{\Delta f}$$

We can rewrite this as

$$\Delta f \Delta t = 1 \tag{39.20}$$

Equation 39.20 is nothing new; we are simply writing what we already knew in a different form. Equation 39.20 is a combination of three things: the relationship $f = 1/T$ between period and frequency, writing T_{beat} as Δt, and the specific knowledge that the beat frequency f_{beat} is the difference Δf of the two frequencies contributing to the wave packet. As the frequency separation gets smaller, the duration of each beat gets longer.

When we superimpose two frequencies to create beats, the wave packet repeats over and over. A more advanced treatment of waves, called Fourier analysis, reveals that a single, *nonrepeating* wave packet can be created through the superposition of *many* waves of very similar frequency. FIGURE 39.13 illustrates this idea. At one instant of time, all the waves interfere constructively to produce the maximum amplitude of the wave packet. At other times, the individual waves get out of phase and their superposition tends toward zero.

FIGURE 39.13 A single wave packet is the superposition of many component waves of similar wavelength and frequency.

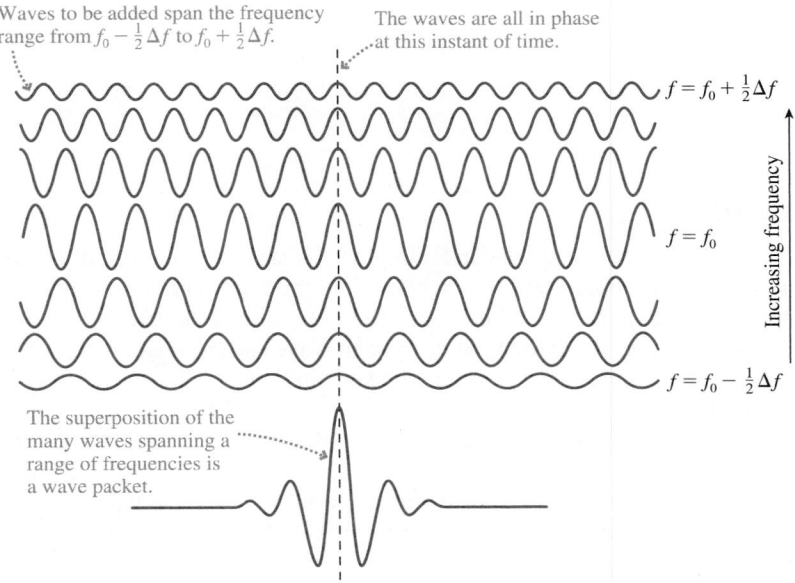

Waves to be added span the frequency range from $f_0 - \frac{1}{2}\Delta f$ to $f_0 + \frac{1}{2}\Delta f$.

The waves are all in phase at this instant of time.

$f = f_0 + \frac{1}{2}\Delta f$

Increasing frequency

$f = f_0$

$f = f_0 - \frac{1}{2}\Delta f$

The superposition of the many waves spanning a range of frequencies is a wave packet.

Suppose a single nonrepeating wave packet of duration Δt is created by the superposition of *many* waves that span a range of frequencies Δf. We'll not prove it, but Fourier analysis shows that for *any* wave packet

$$\Delta f \Delta t \approx 1 \qquad (39.21)$$

The relationship between Δf and Δt for a general wave packet is not as precise as Equation 39.20 for beats. There are two reasons for this:

1. Wave packets come in a variety of shapes. The exact relationship between Δf and Δt depends somewhat on the shape of the wave packet.
2. We have not given a precise definition of Δt and Δf for a general wave packet. The quantity Δt is "about how long the wave packet lasts," while Δf is "about the range of frequencies needing to be superimposed to produce this wave packet." For our purposes, we will not need to be any more precise than this.

Equation 39.21 is a purely classical result that applies to waves of any kind. It tells you the range of frequencies you need to superimpose to construct a wave packet of duration Δt. Alternatively, Equation 39.21 tells you that a wave packet created as a superposition of various frequencies cannot be arbitrarily short but *must* last for a time interval $\Delta t \approx 1/\Delta f$.

EXAMPLE 39.3 **Creating radio-frequency pulses**

A short-wave radio station broadcasts at a frequency of 10.000 MHz. What is the range of frequencies of the waves that must be superimposed to broadcast a radio-wave pulse lasting 0.800 μs?

MODEL A pulse of radio waves is an electromagnetic wave packet, hence it must satisfy the relationship $\Delta f \Delta t \approx 1$.

VISUALIZE FIGURE 39.14 shows the pulse.

FIGURE 39.14 A pulse of radio waves.

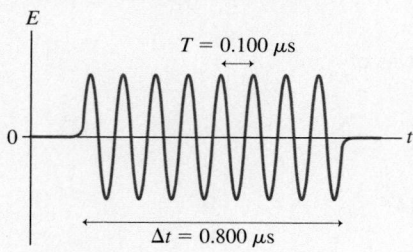

SOLVE The period of a 10.000 MHz oscillation is 0.100 μs. A pulse 0.800 μs in duration is 8 oscillations of the wave. Although the station broadcasts at a nominal frequency of 10.000 MHz, this pulse is not a pure 10.000 MHz oscillation. Instead, the pulse has been created by the superposition of many waves whose frequencies span

$$\Delta f \approx \frac{1}{\Delta t} = \frac{1}{0.800 \times 10^{-6}\ \text{s}} = 1.250 \times 10^6\ \text{Hz} = 1.250\ \text{MHz}$$

This range of frequencies will be centered at the 10.000 MHz broadcast frequency, so the waves that must be superimposed to create this pulse span the frequency range

$$9.375\ \text{MHz} \le f \le 10.625\ \text{MHz}$$

Bandwidth

Short-duration pulses, like the one in Example 39.3, are used to transmit digital information. Digital signals are sent over a phone line by brief tone pulses, over satellite links by brief radio pulses like the one in the example, and through optical fibers by brief laser-light pulses. Regardless of the type of wave and the medium through which it travels, any wave pulse must obey the fundamental relationship $\Delta f \Delta t \approx 1$.

Sending data at a higher rate (i.e., more pulses per second) requires that the pulse duration Δt be shorter. But a shorter-duration pulse must be created by the superposition of a *larger* range of frequencies. Thus the medium through which a shorter-duration pulse travels must be physically able to transmit the full range of frequencies.

The range of frequencies that can be transmitted through a medium is called the **bandwidth** Δf_B of the medium. The shortest possible pulse that can be transmitted through a medium is

$$\Delta t_{\text{min}} \approx \frac{1}{\Delta f_B} \tag{39.22}$$

A pulse shorter than this would require a larger range of frequencies than the medium can support.

The concept of bandwidth is extremely important in digital communications. A higher bandwidth permits the transmission of shorter pulses and allows a higher data rate. A standard telephone line does not have a very high bandwidth, and that is why a modem is limited to sending data at the rate of roughly 50,000 pulses per second. A 0.80 μs pulse can't be sent over a phone line simply because the phone line won't transmit the range of frequencies that would be needed.

An optical fiber is a high-bandwidth medium. A fiber has a bandwidth $\Delta f_B > 1$ GHz and thus can transmit laser-light pulses with duration $\Delta t < 1$ ns. As a result, more than 10^9 pulses per second can be sent along an optical fiber, which is why optical-fiber networks now form the backbone of the Internet.

Uncertainty

There is another way of thinking about the time-frequency relationship $\Delta f \Delta t \approx 1$. Suppose you want to determine *when* a wave packet arrives at a specific point in space, such as at a detector of some sort. At what instant of time can you say that the wave packet is detected? When the front edge arrives? When the maximum amplitude arrives? When the back edge arrives? Because a wave packet is spread out in time, there is not a unique and well-defined time t at which the packet arrives. All we can say is that it arrives within some interval of time Δt. We are *uncertain* about the exact arrival time.

Similarly, suppose you would like to know the oscillation frequency of a wave packet. There is no precise value for f because the wave packet is constructed from many waves within a range of frequencies Δf. All we can say is that the frequency is within this range. We are *uncertain* about the exact frequency.

The time-frequency relationship $\Delta f \Delta t \approx 1$ tells us that the uncertainty in our knowledge about the arrival time of the wave packet is related to our uncertainty about the packet's frequency. The more precisely and accurately we know one quantity, the less precisely we will be able to know the other.

FIGURE 39.15 shows two different wave packets. The wave packet of FIGURE 39.15a is very narrow and thus very localized in time. As it travels, our knowledge of when it will arrive at a specified point is fairly precise. But a very wide range of frequencies Δf is required to create a wave packet with a very small Δt. The price we pay for being fairly certain about the time is a very large uncertainty Δf about the frequency of this wave packet.

FIGURE 39.15b shows the opposite situation: The wave packet oscillates many times and the frequency of these oscillations is pretty clear. Our knowledge of the frequency is good, with minimal uncertainty Δf. But such a wave packet is so spread out that there is a very large uncertainty Δt as to its time of arrival.

In practice, $\Delta f \Delta t \approx 1$ is really a lower limit. Technical limitations may cause the uncertainties in our knowledge of f and t to be even larger than this relationship implies. Consequently, a better statement about our knowledge of a wave packet is

$$\Delta f \Delta t \geq 1 \qquad (39.23)$$

The fact that waves are spread out makes it meaningless to specify an exact frequency and an exact arrival time simultaneously. This is an inherent feature of waviness that applies to all waves.

STOP TO THINK 39.5   What minimum bandwidth must a medium have to transmit a 100-ns-long pulse?

a. 1 MHz   b. 10 MHz   c. 100 MHz   d. 1000 MHz

FIGURE 39.15 Two wave packets with different Δt.

(a)

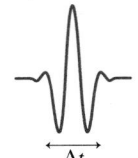

This wave packet has a large frequency uncertainty Δf.

(b)

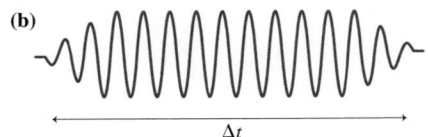

This wave packet has a small frequency uncertainty Δf.

39.6 The Heisenberg Uncertainty Principle

If matter has wave-like aspects and a de Broglie wavelength, then the expression $\Delta f \Delta t \geq 1$ must somehow apply to matter. How? And what are the implications?

Consider a particle with velocity v_x as it travels along the x-axis with deBroglie wavelength $\lambda = h/p_x$. Figure 39.11 showed a *history graph* (ψ versus t) of a wave packet that might represent the particle as it passes a point on the x-axis. It will be more useful to have a *snapshot graph* (ψ versus x) of the wave packet traveling along the x-axis.

The time interval Δt is the duration of the wave packet as the particle passes a point in space. During this interval, the packet moves forward

$$\Delta x = v_x \Delta t = \frac{p_x}{m} \Delta t \qquad (39.24)$$

where $p_x = mv_x$ is the x-component of the particle's momentum. The quantity Δx, shown in FIGURE 39.16, is the length or spatial extent of the wave packet. Conversely, we can write the wave packet's duration in terms of its length as

$$\Delta t = \frac{m}{p_x} \Delta x \qquad (39.25)$$

FIGURE 39.16 A snapshot graph of a wave packet.

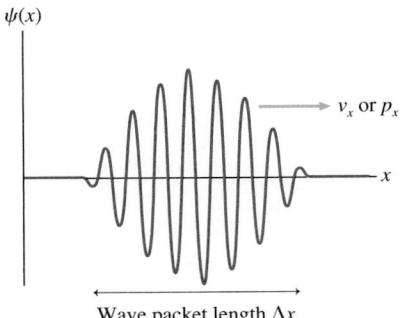

Wave packet length Δx

You will recall that any wave with sinusoidal oscillations must satisfy the wave condition $\lambda f = v$. For a material particle, where λ is the de Broglie wavelength, the frequency f is

$$f = \frac{v}{\lambda} = \frac{p_x/m}{h/p_x} = \frac{p_x^2}{hm}$$

If the momentum p_x should vary by the small amount Δp_x, the frequency will vary by the small amount Δf. Assuming that $\Delta f \ll f$ and $\Delta p_x \ll p_x$ (reasonable assumptions), we can treat Δf and Δp_x as if they were differentials df and dp_x. Taking the derivative, we find

$$\Delta f = \frac{2p_x \Delta p_x}{hm} \tag{39.26}$$

Multiplying together these expressions for Δt and Δf, we find that

$$\Delta f \Delta t = \frac{2p_x \Delta p_x}{hm} \frac{m \Delta x}{p_x} = \frac{2}{h} \Delta x \Delta p_x \tag{39.27}$$

Because $\Delta f \Delta t \geq 1$ for any wave, one last rearrangement of Equation 39.27 shows that a matter wave must obey the condition

$$\Delta x \Delta p_x \geq \frac{h}{2} \qquad \text{(Heisenberg uncertainty principle)} \tag{39.28}$$

This statement about the relationship between the position and momentum of a particle was proposed by Werner Heisenberg, creator of one of the first successful quantum theories. Physicists often just call it the **uncertainty principle**.

NOTE ▶ In statements of the uncertainty principle, the right side is sometimes $h/2$, as we have it, but other times it is just h or contains various factors of π. The specific number is not especially important because it depends on exactly how Δx and Δp are defined. The important idea is that the product of Δx and Δp_x for a particle cannot be significantly less than Planck's constant h. A similar relationship for $\Delta y \Delta p_y$ applies along the y-axis. ◀

What Does It Mean?

Heisenberg's uncertainty principle is a statement about our *knowledge* of the properties of a particle. If we want to know *where* a particle is located, we measure its position x. That measurement is not absolutely perfect but has some uncertainty Δx. Likewise, if we want to know *how fast* the particle is going, we need to measure its velocity v_x or, equivalently, its momentum p_x. This measurement also has some uncertainty Δp_x.

Uncertainties are associated with all experimental measurements, but better procedures and techniques can reduce those uncertainties. Newtonian physics places no limits on how small the uncertainties can be. A Newtonian particle at any instant of time has an exact position x and an exact momentum p_x, and with sufficient care we can measure both x and p_x with such precision that the product $\Delta x \Delta p_x \rightarrow 0$. There are no inherent limits to our knowledge about a classical, or Newtonian, particle.

Heisenberg, however, made the bold and original statement that our knowledge has real limitations. No matter how clever you are, and no matter how good your experiment, you *cannot* measure both x and p_x simultaneously with arbitrarily good precision. Any measurements you make are limited by the condition that $\Delta x \Delta p_x \geq h/2$. **Our knowledge about a particle is *inherently* uncertain.**

Why? Because of the wave-like nature of matter. The "particle" is spread out in space, so there simply is not a precise value of its position x. Similarly, the de Broglie relationship between momentum and wavelength implies that we cannot know the momentum of a wave packet any more exactly than we can know its wavelength or frequency. Our belief that position and momentum have precise values is tied to our classical concept of a particle. As we revise our ideas of what atomic particles are like, we will also have to revise our old ideas about position and momentum.

EXAMPLE 39.4   **The uncertainty of a dust particle**

A 1.0-μm-diameter dust particle ($m \approx 10^{-15}$ kg) is confined within a 10-μm-long box. Can we know with certainty if the particle is at rest? If not, within what range is its velocity likely to be found?

MODEL All matter is subject to the Heisenberg uncertainty principle.

SOLVE If we know *for sure* that the particle is at rest, then $p_x = 0$ with no uncertainty. That is, $\Delta p_x = 0$. But then, according to the uncertainty principle, the uncertainty in our knowledge of the particle's position would have to be $\Delta x \to \infty$. In other words, we would have no knowledge at all about the particle's position—it could be anywhere! But that is not the case. We know the particle is *somewhere* in the box, so the uncertainty in our knowledge of its position is at most $\Delta x = L = 10\ \mu$m. With a finite Δx, the uncertainty Δp_x *cannot* be zero. We cannot know with certainty if the particle is at rest inside the box. No matter how hard we try to bring the particle to rest, the uncertainty in our knowledge of

the particle's momentum will be $\Delta p_x \approx h/(2\,\Delta x) = h/2L$. We've assumed the most accurate measurements possible so that the $\geq$ in Heisenberg's uncertainty principle becomes $\approx$. Consequently, the range of possible velocities is

$$\Delta v_x = \frac{\Delta p_x}{m} \approx \frac{h}{2mL} \approx 3.0 \times 10^{-14}\ \text{m/s}$$

This range of possible velocities will be centered on $v_x = 0$ m/s if we have done our best to have the particle be at rest. Thus all we can know with certainty is that the particle's velocity is somewhere within the interval -1.5×10^{-14} m/s $\leq v \leq 1.5 \times 10^{-14}$ m/s.

ASSESS For practical purposes you might consider this to be a satisfactory definition of "at rest." After all, a particle moving with a speed of 1.5×10^{-14} m/s would need 6×10^{10} s to move a mere 1 mm. That is about 2000 years! Nonetheless, we can't know if the particle is "really" at rest.

EXAMPLE 39.5   **The uncertainty of an electron**

What range of velocities might an electron have if confined to a 0.10-nm-wide region, about the size of an atom?

MODEL Electrons are subject to the Heisenberg uncertainty principle.

SOLVE The analysis is the same as in Example 39.4. If we know that the electron's position is located within an interval $\Delta x \approx 0.1$ nm, then the best we can know is that its velocity is within the range

$$\Delta v_x = \frac{\Delta p_x}{m} \approx \frac{h}{2mL} \approx 4 \times 10^6\ \text{m/s}$$

Because the *average* velocity is zero, the best we can say is that the electron's velocity is somewhere in the interval -2×10^6 m/s $\leq v \leq 2 \times 10^6$ m/s. It is simply not possible to know the electron's velocity any more precisely than this.

ASSESS Unlike the situation in Example 39.4, where Δv was so small as to be of no practical consequence, our uncertainty about the electron's velocity is enormous—about 1% of the speed of light!

Once again, we see that even the smallest of macroscopic objects behaves very much like a classical Newtonian particle. Perhaps a 1-μm-diameter particle is slightly fuzzy and has a slightly uncertain velocity, but it is far beyond the measuring capabilities of even the very best instruments to detect this wave-like behavior. In contrast, the effects of the uncertainty principle at the atomic scale are stupendous. We are unable to determine the velocity of an electron in an atom-size container to any better accuracy than about 1% of the speed of light.

STOP TO THINK 39.6   Which of these particles, A or B, can you locate more precisely?

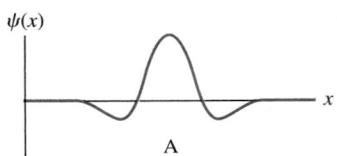

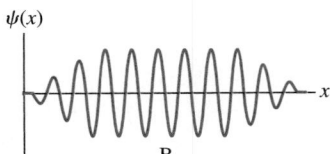

CHALLENGE EXAMPLE 39.6 **The probability of finding a particle**

A particle is described by the wave function

$$\psi(x) = \begin{cases} 0 & x < 0 \\ ce^{-x/L} & x \geq 0 \end{cases}$$

where $L = 1$ nm.

a. Determine the value of the constant c.
b. Draw graphs of $\psi(x)$ and the probability density $P(x)$.
c. If 10^6 particles are detected, how many are expected to be found in the region $x \geq 1$ nm?

MODEL The probability of finding a particle is determined by the probability density $P(x)$.

SOLVE a. The wave function is an exponential $\psi(x) = ce^{-x/L}$ that extends from $x = 0$ to $x = +\infty$. Equation 39.18, the normalization condition, is

$$1 = \int_{-\infty}^{\infty} |\psi(x)|^2\,dx = c^2 \int_0^{\infty} e^{-2x/L}\,dx = -\frac{c^2 L}{2} e^{-2x/L}\Big|_0^{\infty} = \frac{c^2 L}{2}$$

We can solve this for the normalization constant c:

$$c = \sqrt{\frac{2}{L}} = \sqrt{\frac{2}{1\ \text{nm}}} = 1.414\ \text{nm}^{-1/2}$$

b. The probability density is

$$P(x) = |\psi(x)|^2 = (2.0\ \text{nm}^{-1})e^{-2x/(1.0\ \text{nm})}$$

The wave function and the probability density are graphed in **FIGURE 39.17**.

c. The probability of finding a particle in the region $x \geq 1$ nm is the shaded area under the probability density curve in Figure 39.17. We must use Equation 39.17 and integrate to find a numerical value. The probability is

FIGURE 39.17 The wave function and probability density of Example 39.6.

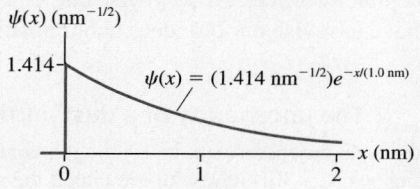

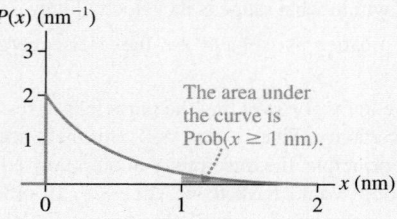

$$\text{Prob}(x \geq 1\ \text{nm}) = \int_{1\ \text{nm}}^{\infty} |\psi(x)|^2\,dx$$

$$= (2.0\ \text{nm}^{-1})\int_{1\ \text{nm}}^{\infty} e^{-2x/(1.0\ \text{nm})}\,dx$$

$$= (2.0\ \text{nm}^{-1})\left(-\frac{1.0\ \text{nm}}{2}\right)e^{-2x/(1.0\ \text{nm})}\Big|_{1\ \text{nm}}^{\infty}$$

$$= e^{-2} = 0.135 = 13.5\%$$

The number of particles expected to be found at $x \geq 1$ nm is

$$N_{\text{detected}} = N \cdot \text{Prob}(x \geq 1\ \text{nm}) = (10^6)(0.135) = 135{,}000$$

ASSESS There is a 13.5% chance of detecting a particle beyond 1 nm and thus an 86.5% chance of finding it within the interval $0 \leq x \leq 1$ nm. Unlike classical physics, we cannot make an exact prediction of a particle's position.

SUMMARY

The goal of Chapter 39 has been to introduce and learn to use the wave-function description of matter.

General Principles

Wave Functions and the Probability Density

We cannot predict the exact trajectory of an atomic-level particle such as an electron. The best we can do is to predict the **probability** that a particle will be found in some region of space. The probability is determined by the particle's wave function $\psi(x)$.

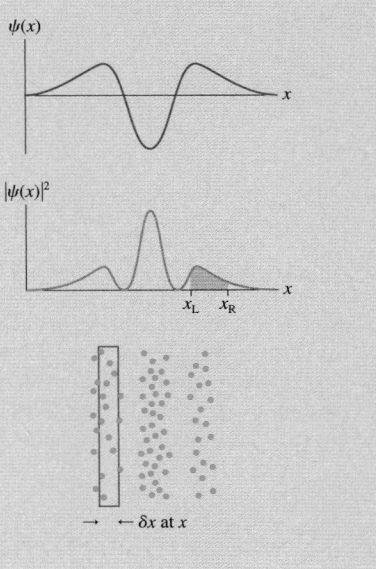

- $\psi(x)$ is a continuous, wave-like (i.e., oscillatory) function.

- The probability that a particle will be found in the narrow interval δx at position x is

$$\text{Prob(in } \delta x \text{ at } x) = |\psi(x)|^2 \, \delta x$$

- $|\psi(x)|^2$ is the probability density $P(x)$.

- For the probability interpretation of $\psi(x)$ to make sense, the wave function must satisfy the normalization condition:

$$\int_{-\infty}^{\infty} P(x) \, dx = \int_{-\infty}^{\infty} |\psi(x)|^2 \, dx = 1$$

That is, it is certain that the particle is *somewhere* on the x-axis.

- For an extended interval

$$\text{Prob}(x_L \leq x \leq x_R) = \int_{x_L}^{x_R} |\psi(x)|^2 \, dx = \text{area under the curve}$$

$\rightarrow \leftarrow \delta x$ at x

Heisenberg Uncertainty Principle

A particle with wave-like characteristics does not have a precise value of position x or a precise value of momentum p_x. Both are uncertain. The position uncertainty Δx and momentum uncertainty Δp_x are related by $\Delta x \, \Delta p_x \geq h/2$. The more you try to pin down the value of one, the less precisely the other can be known.

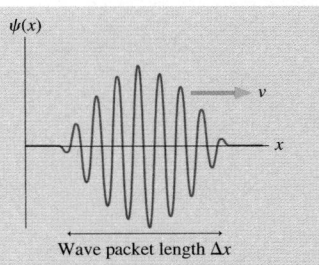

Wave packet length Δx

Important Concepts

The probability that a particle is found in region A is

$$P_A = \lim_{N_{\text{tot}} \to \infty} \frac{N_A}{N_{\text{tot}}}$$

If the probability is known, the expected number of A outcomes in N trials is $N_A = NP_A$.

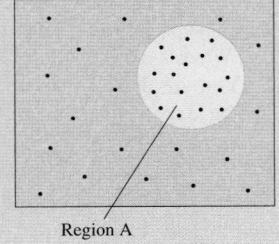

Region A

A wave packet of duration Δt can be created by the superposition of many waves spanning the frequency range Δf. These are related by

$$\Delta f \, \Delta t \approx 1$$

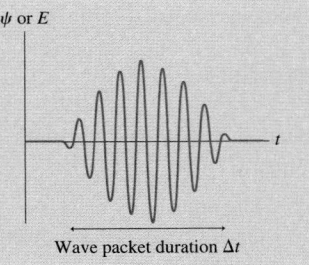

Wave packet duration Δt

Terms and Notation

probability	wave function, $\psi(x)$	bandwidth, Δf_B
expected value	normalization condition	uncertainty principle
probability density, $P(x)$	wave packet	

CONCEPTUAL QUESTIONS

1. **FIGURE Q39.1** shows the probability density for photons to be detected on the *x*-axis.
   a. Is a photon more likely to be detected at $x = 0$ m or at $x = 1$ m? Explain.
   b. One million photons are detected. What is the expected number of photons in a 1-mm-wide interval at $x = 0.50$ m?

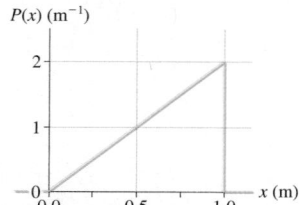

$P(x)$ (m^{-1})

FIGURE Q39.1

2. What is the difference between the probability and the probability density?
3. For the electron wave function shown in **FIGURE Q39.3**, at what position or positions is the electron most likely to be found? Least likely to be found? Explain.

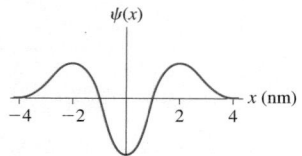

$\psi(x)$

FIGURE Q39.3

4. **FIGURE Q39.4** shows the dot pattern of electrons landing on a screen.
   a. At what value or values of *x* is the electron probability density at maximum? Explain.
   b. Can you tell at what value or values of *x* the electron wave function $\psi(x)$ is most positive? If so, where? If not, why not?

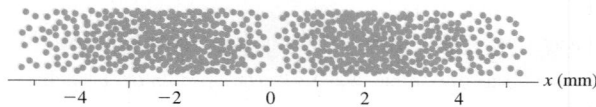

FIGURE Q39.4

5. What is the value of the constant *a* in **FIGURE Q39.5**?

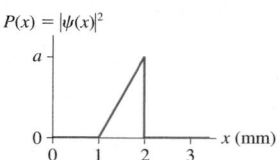

$P(x) = |\psi(x)|^2$

FIGURE Q39.5

6. **FIGURE Q39.6** shows wave packets for particles 1, 2, and 3. Which particle can have its velocity known most precisely? Explain.

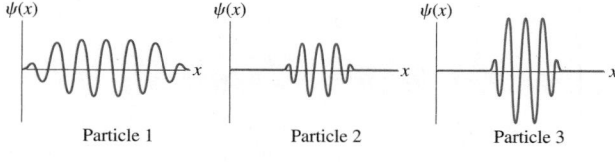

Particle 1   Particle 2   Particle 3

FIGURE Q39.6

EXERCISES AND PROBLEMS

Problems labeled ▓ integrate material from earlier chapters.

Exercises

Section 39.1 Waves, Particles, and the Double-Slit Experiment

1. ‖ An experiment has four possible outcomes, labeled A to D. The probability of A is $P_A = 40\%$ and of B is $P_B = 30\%$. Outcome C is twice as probable as outcome D. What are the probabilities P_C and P_D?
2. ‖ Suppose you toss three coins into the air and let them fall on the floor. Each coin shows either a head or a tail.
   a. Make a table in which you list all the possible outcomes of this experiment. Call the coins A, B, and C.
   b. What is the probability of getting two heads and one tail? Explain.
   c. What is the probability of getting *at least* two heads?

3. | Suppose you draw a card from a regular deck of 52 cards.
   a. What is the probability that you draw an ace?
   b. What is the probability that you draw a spade?
4. | You are dealt 1 card each from 1000 decks of cards. What is the expected number of picture cards (jacks, queens, and kings)?
5. | Make a table in which you list all possible outcomes of rolling two dice. Call the dice A and B. What is the probability of rolling (a) a 7, (b) any double, and (c) a 6 or an 8? You can give the probabilities as fractions, such as 3/36.

Section 39.2 Connecting the Wave and Photon Views

6. | In one experiment, 2000 photons are detected in a 0.10-mm-wide strip where the amplitude of the electromagnetic wave is 10 V/m. How many photons are detected in a nearby 0.10-mm-wide strip where the amplitude is 30 V/m?

7. ‖ In one experiment, 6000 photons are detected in a 0.10-mm-wide strip where the amplitude of the electromagnetic wave is 200 V/m. What is the wave amplitude at a nearby 0.20-mm-wide strip where 3000 photons are detected?

8. ‖ 1.0×10^{10} photons pass through an experimental apparatus. How many of them land in a 0.10-mm-wide strip where the probability density is 20 m^{-1}?

9. | When 5×10^{12} photons pass through an experimental apparatus, 2.0×10^9 land in a 0.10-mm-wide strip. What is the probability density at this point?

Section 39.3 The Wave Function

10. | What are the units of ψ? Explain.

11. | FIGURE EX39.11 shows the probability density for an electron that has passed through an experimental apparatus. If 1.0×10^6 electrons are used, what is the expected number that will land in a 0.010-mm-wide strip at (a) $x = 0.000$ mm and (b) 2.000 mm?

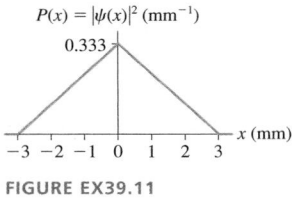

FIGURE EX39.11

12. ‖ In an interference experiment with electrons, you find the most intense fringe is at $x = 7.0$ cm. There are slightly weaker fringes at $x = 6.0$ and 8.0 cm, still weaker fringes at $x = 4.0$ and 10.0 cm, and two very weak fringes at $x = 1.0$ and 13.0 cm. No electrons are detected at $x < 0$ cm or $x > 14$ cm.
   a. Sketch a graph of $|\psi(x)|^2$ for these electrons.
   b. Sketch a possible graph of $\psi(x)$.
   c. Are there other possible graphs for $\psi(x)$? If so, draw one.

13. | FIGURE EX39.13 shows the probability density for an electron that has passed through an experimental apparatus. What is the probability that the electron will land in a 0.010-mm-wide strip at (a) $x = 0.000$ mm, (b) $x = 0.500$ mm, (c) $x = 1.000$ mm, and (d) $x = 2.000$ mm?

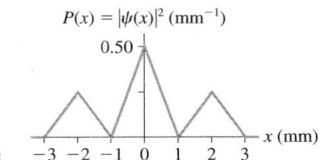

FIGURE EX39.13

Section 39.4 Normalization

14. ‖ FIGURE EX39.14 is a graph of $|\psi(x)|^2$ for an electron.
   a. What is the value of a?
   b. Draw a graph of the wave function $\psi(x)$. (There is more than one acceptable answer.)
   c. What is the probability that the electron is located between $x = 1.0$ nm and $x = 2.0$ nm?

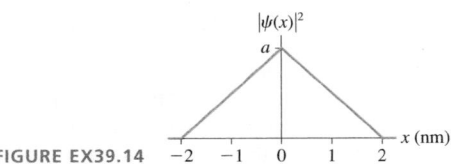

FIGURE EX39.14

15. ‖ FIGURE EX39.15 is a graph of $|\psi(x)|^2$ for a neutron.
   a. What is the value of a?
   b. Draw a graph of the wave function $\psi(x)$. (There is more than one acceptable answer.)
   c. What is the probability that the neutron is located at a position with $|x| \geq 2$ fm?

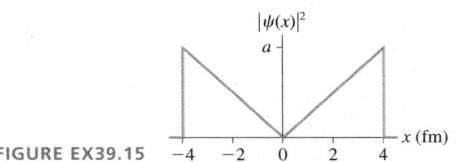

FIGURE EX39.15

16. ‖ FIGURE EX39.16 shows the wave function of an electron.
   a. What is the value of c?
   b. Draw a graph of $|\psi(x)|^2$.
   c. What is the probability that the electron is located between $x = -1.0$ nm and $x = 1.0$ nm?

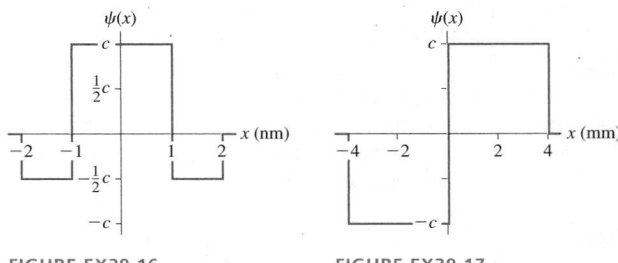

FIGURE EX39.16   FIGURE EX39.17

17. ‖ FIGURE EX39.17 shows the wave function of a neutron.
   a. What is the value of c?
   b. Draw a graph of $|\psi(x)|^2$.
   c. What is the probability that the neutron is located between $x = -1.0$ mm and $x = 1.0$ mm?

Section 39.5 Wave Packets

18. | What minimum bandwidth is needed to transmit a pulse that consists of 100 cycles of a 1.00 MHz oscillation?

19. ‖ A radio-frequency amplifier is designed to amplify signals in the frequency range 80 MHz to 120 MHz. What is the shortest-duration radio-frequency pulse that can be amplified without distortion?

20. ‖ Sound waves of 498 Hz and 502 Hz are superimposed at a temperature where the speed of sound in air is 340 m/s. What is the length Δx of one wave packet?

21. ‖ A 1.5-μm-wavelength laser pulse is transmitted through a 2.0-GHz-bandwidth optical fiber. How many oscillations are in the shortest-duration laser pulse that can travel through the fiber?

Section 39.6 The Heisenberg Uncertainty Principle

22. ‖ A thin solid barrier in the xy-plane has a 10-μm-diameter circular hole. An electron traveling in the z-direction with $v_x = 0$ m/s passes through the hole. Afterward, is it certain that v_x is still zero? If not, within what range is v_x likely to be?

23. ‖ Andrea, whose mass is 50 kg, thinks she's sitting at rest in her 5.0-m-long dorm room as she does her physics homework. Can Andrea be sure she's at rest? If not, within what range is her velocity likely to be?

24. ‖ What is the minimum uncertainty in position, in nm, of an electron whose velocity is known to be between 3.48×10^5 m/s and 3.58×10^5 m/s?

25. ‖ A proton is confined within an atomic nucleus of diameter 4.0 m. Use a one-dimensional model to estimate the smallest range of speeds you might find for a proton in the nucleus.

Problems

26. ⏐ A 1.0-mm-diameter sphere bounces back and forth between two walls at $x = 0$ mm and $x = 100$ mm. The collisions are perfectly elastic, and the sphere repeats this motion over and over with no loss of speed. At a random instant of time, what is the probability that the center of the sphere is
   a. At exactly $x = 50.0$ mm?
   b. Between $x = 49.0$ mm and $x = 51.0$ mm?
   c. At $x \geq 75$ mm?

27. ⏐ A radar antenna broadcasts electromagnetic waves with a period of 0.100 ns. What range of frequencies would need to be superimposed to create a 1.0-ns-long radar pulse?

28. ‖ Ultrasound pulses with a frequency of 1.000 MHz are transmitted into water, where the speed of sound is 1500 m/s. The spatial length of each pulse is 12 mm.
   a. How many complete cycles are contained in one pulse?
   b. What range of frequencies must be superimposed to create each pulse?

29. ‖ FIGURE P39.29 shows a *pulse train*. The period of the pulse train is $T = 2\Delta t$, where Δt is the duration of each pulse. What is the maximum pulse-transmission rate (pulses per second) through an electronics system with a 200 kHz bandwidth? (This is the bandwidth allotted to each FM radio station.)

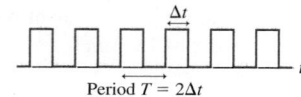

FIGURE P39.29   Period $T = 2\Delta t$

30. ‖ Consider a single-slit diffraction experiment using electrons. (Single-slit diffraction was described in Section 22.4.) Using Figure 39.5 as a model, draw
   a. A dot picture showing the arrival positions of the first 40 or 50 electrons.
   b. A graph of $|\psi(x)|^2$ for the electrons on the detection screen.
   c. A graph of $\psi(x)$ for the electrons. Keep in mind that ψ, as a wave-like function, oscillates between positive and negative.

31. ⏐ An experiment finds electrons to be uniformly distributed over the interval 0 cm $\leq x \leq 2$ cm, with no electrons falling outside this interval.
   a. Draw a graph of $|\psi(x)|^2$ for these electrons.
   b. What is the probability that an electron will land within the interval 0.79 to 0.81 cm?
   c. If 10^6 electrons are detected, how many will be detected in the interval 0.79 to 0.81 cm?
   d. What is the probability density at $x = 0.80$ cm?

32. ‖ In an experiment with 10,000 electrons, which land symmetrically on both sides of $x = 0$ cm, 5000 are detected in the range -1.0 cm $\leq x \leq +1.0$ cm, 7500 are detected in the range -2.0 cm $\leq x \leq +2.0$ cm, and all 10,000 are detected in the range -3.0 cm $\leq x \leq +3.0$ cm. Draw a graph of a probability density that is consistent with these data. (There may be more than one acceptable answer.)

33. ‖ FIGURE P39.33 shows $|\psi(x)|^2$ for the electrons in an experiment.
   a. Is the electron wave function normalized? Explain.
   b. Draw a graph of $\psi(x)$ over this same interval. Provide a numerical scale on both axes. (There may be more than one acceptable answer.)
   c. What is the probability that an electron will be detected in a 0.0010-cm-wide region at $x = 0.00$ cm? At $x = 0.50$ cm? At $x = 0.999$ cm?
   d. If 10^4 electrons are detected, how many are expected to land in the interval -0.30 cm $\leq x \leq 0.30$ cm?

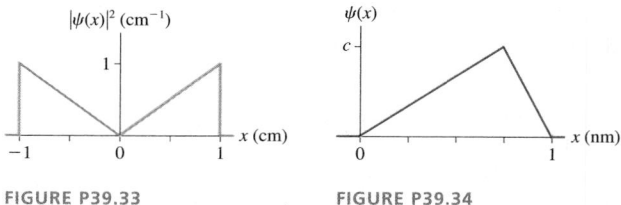

FIGURE P39.33   FIGURE P39.34

34. ‖ FIGURE P39.34 shows the wave function of a particle confined between $x = 0$ nm and $x = 1.0$ nm. The wave function is zero outside this region.
   a. Determine the value of the constant c, as defined in the figure.
   b. Draw a graph of the probability density $P(x) = |\psi(x)|^2$.
   c. Draw a dot picture showing where the first 40 or 50 particles might be found.
   d. Calculate the probability of finding the particle in the interval 0 nm $\leq x \leq 0.25$ nm.

35. ‖‖ FIGURE P39.35 shows the wave function of a particle confined between $x = -4.0$ mm and $x = 4.0$ mm. The wave function is zero outside this region.
   a. Determine the value of the constant c, as defined in the figure.
   b. Draw a graph of the probability density $P(x) = |\psi(x)|^2$.
   c. Draw a dot picture showing where the first 40 or 50 particles might be found.
   d. Calculate the probability of finding the particle in the interval -2.0 mm $\leq x \leq 2.0$ mm.

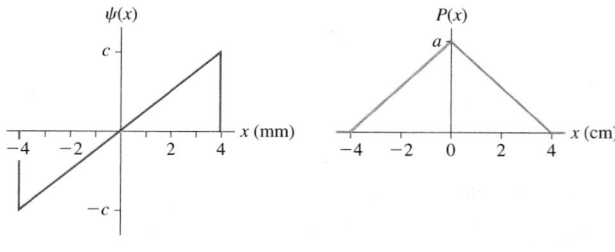

FIGURE P39.35   FIGURE P39.36

36. ‖ FIGURE P39.36 shows the probability density for finding a particle at position x.
   a. Determine the value of the constant a, as defined in the figure.
   b. At what value of x are you most likely to find the particle? Explain.
   c. Within what range of positions centered on your answer to part b are you 75% certain of finding the particle?
   d. Interpret your answer to part c by drawing the probability density graph and shading the appropriate region.

37. ‖ An electron that is confined to $x \geq 0$ nm has the normalized wave function

$$\psi(x) = \begin{cases} 0 & x < 0 \text{ nm} \\ (1.414 \text{ nm}^{-1/2})e^{-x/(1.0 \text{ nm})} & x \geq 0 \text{ nm} \end{cases}$$

where x is in nm.
  a. What is the probability of finding the electron in a 0.010-nm-wide region at $x = 1.0$ nm?
  b. What is the probability of finding the electron in the interval $0.50 \text{ nm} \leq x \leq 1.50 \text{ nm}$?

38. ‖ A particle is described by the wave function

$$\psi(x) = \begin{cases} ce^{x/L} & x \leq 0 \text{ mm} \\ ce^{-x/L} & x \geq 0 \text{ mm} \end{cases}$$

where $L = 2.0$ mm.
  a. Sketch graphs of both the wave function and the probability density as functions of x.
  b. Determine the normalization constant c.
  c. Calculate the probability of finding the particle within 1.0 mm of the origin.
  d. Interpret your answer to part b by shading the region representing this probability on the appropriate graph in part a.

39. ‖ Consider the electron wave function

$$\psi(x) = \begin{cases} c\sqrt{1 - x^2} & |x| \leq 1 \text{ cm} \\ 0 & |x| \geq 1 \text{ cm} \end{cases}$$

where x is in cm.
  a. Determine the normalization constant c.
  b. Draw a graph of $\psi(x)$ over the interval $-2 \text{ cm} \leq x \leq 2 \text{ cm}$. Provide numerical scales on both axes.
  c. Draw a graph of $|\psi(x)|^2$ over the interval $-2 \text{ cm} \leq x \leq 2 \text{ cm}$. Provide numerical scales.
  d. If 10^4 electrons are detected, how many will be in the interval $0.00 \text{ cm} \leq x \leq 0.50 \text{ cm}$?

40. ‖ Consider the electron wave function

$$\psi(x) = \begin{cases} c\sin\left(\dfrac{2\pi x}{L}\right) & 0 \leq x \leq L \\ 0 & x < 0 \text{ or } x > L \end{cases}$$

  a. Determine the normalization constant c. Your answer will be in terms of L.
  b. Draw a graph of $\psi(x)$ over the interval $-L \leq x \leq 2L$.
  c. Draw a graph of $|\psi(x)|^2$ over the interval $-L \leq x \leq 2L$.
  d. What is the probability that an electron is in the interval $0 \leq x \leq L/3$?

41. ‖ The probability density for finding a particle at position x is

$$P(x) = \begin{cases} \dfrac{a}{(1 - x)} & -1 \text{ mm} \leq x < 0 \text{ mm} \\ b(1 - x) & 0 \text{ mm} \leq x \leq 1 \text{ mm} \end{cases}$$

and zero elsewhere.
  a. You will learn in Chapter 40 that the wave function must be a *continuous* function. Assuming that to be the case, what can you conclude about the relationship between a and b?
  b. Determine values for a and b.
  c. Draw a graph of the probability density over the interval $-2 \text{ mm} \leq x \leq 2 \text{ mm}$.
  d. What is the probability that the particle will be found to the left of the origin?

42. ‖ A pulse of light is created by the superposition of many waves that span the frequency range $f_0 - \frac{1}{2}\Delta f \leq f \leq f_0 + \frac{1}{2}\Delta f$, where $f_0 = c/\lambda$ is called the *center frequency* of the pulse. Laser technology can generate a pulse of light that has a wavelength of 600 nm and lasts a mere 6.0 fs (1 fs = 1 femtosecond $= 10^{-15}$ s).
  a. What is the center frequency of this pulse of light?
  b. How many cycles, or oscillations, of the light wave are completed during the 6.0 fs pulse?
  c. What range of frequencies must be superimposed to create this pulse?
  d. What is the spatial length of the laser pulse as it travels through space?
  e. Draw a snapshot graph of this wave packet.

43. ‖‖ What is the smallest one-dimensional box in which you can confine an electron if you want to know for certain that the electron's speed is no more than 10 m/s?

44. ‖ A small speck of dust with mass 1.0×10^{-13} g has fallen into the hole shown in FIGURE P39.44 and appears to be at rest. According to the uncertainty principle, could this particle have enough energy to get out of the hole? If not, what is the deepest hole of this width from which it would have a good chance to escape?

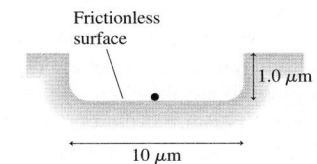

Frictionless surface

1.0 μm

10 μm

FIGURE P39.44

45. ‖ You learned in Chapter 37 that, except for hydrogen, the mass of a nucleus with atomic number Z is larger than the mass of the Z protons. The additional mass was ultimately discovered to be due to neutrons, but prior to the discovery of the neutron it was suggested that a nucleus with mass number A might contain A protons and $(A - Z)$ electrons. Such a nucleus would have the mass of A protons, but its net charge would be only Ze.
  a. We know that the diameter of a nucleus is approximately 10 fm. Model the nucleus as a one-dimensional box and find the minimum range of speeds that an electron would have in such a box.
  b. What does your answer imply about the possibility that the nucleus contains electrons? Explain.

46. ‖ a. Starting with the expression $\Delta f \Delta t \approx 1$ for a wave packet, find an expression for the product $\Delta E \Delta t$ for a photon.
  b. Interpret your expression. What does it tell you?
  c. The Bohr model of atomic quantization says that an atom in an excited state can jump to a lower-energy state by emitting a photon. The Bohr model says nothing about how long this process takes. You'll learn in Chapter 41 that the time any particular atom spends in the excited state before emitting a photon is unpredictable, but the *average lifetime* Δt of many atoms can be determined. You can think of Δt as being the uncertainty in your knowledge of how long the atom spends in the excited state. A typical value is $\Delta t \approx 10$ ns. Consider an atom that emits a photon with a 500 nm wavelength as it jumps down from an excited state. What is the uncertainty in the energy of the photon? Give your answer in eV.
  d. What is the *fractional uncertainty* $\Delta E/E$ in the photon's energy?

Challenge Problems

47. FIGURE CP39.47 shows 1.0-μm-diameter dust particles ($m = 1.0 \times 10^{-15}$ kg) in a vacuum chamber. The dust particles are released from rest above a 1.0-μm-diameter hole, fall through the hole (there's just barely room for the particles to go through), and land on a detector at distance d below.

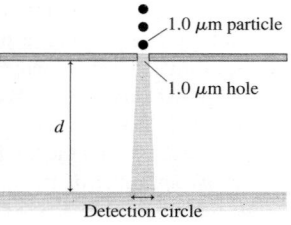

1.0 μm particle

1.0 μm hole

Detection circle

FIGURE CP39.47

    a. If the particles were purely classical, they would all land in the same 1.0-μm-diameter circle. But quantum effects don't allow this. If $d = 1.0$ m, by how much does the diameter of the circle in which most dust particles land exceed 1.0 μm? Is this increase in diameter likely to be detectable?

    b. Quantum effects would be noticeable if the detection-circle diameter increased by 10% to 1.1 μm. At what distance d would the detector need to be placed to observe this increase in the diameter?

48. Physicists use laser beams to create an *atom trap* in which atoms are confined within a spherical region of space with a diameter of about 1 mm. The scientists have been able to cool the atoms in an atom trap to a temperature of approximately 1 nK, which is extremely close to absolute zero, but it would be interesting to know if this temperature is close to any limit set by quantum physics. We can explore this issue with a one-dimensional model of a sodium atom in a 1.0-mm-long box.

    a. Estimate the *smallest* range of speeds you might find for a sodium atom in this box.

    b. Even if we do our best to bring a group of sodium atoms to rest, individual atoms will have speeds within the range you found in part a. Because there's a distribution of speeds, suppose we estimate that the root-mean-square speed v_{rms} of the atoms in the trap is half the value you found in part a. Use this v_{rms} to estimate the temperature of the atoms when they've been cooled to the limit set by the uncertainty principle.

49. The wave function of a particle is

$$\psi(x) = \sqrt{\frac{b}{\pi(x^2 + b^2)}}$$

where b is a positive constant. Find the probability that the particle is located in the interval $-b \le x \le b$.

50. The wave function of a particle is

$$\psi(x) = \begin{cases} \dfrac{b}{(1 + x^2)} & -1 \text{ mm} \le x < 0 \text{ mm} \\ c(1 + x)^2 & 0 \text{ mm} \le x \le 1 \text{ mm} \end{cases}$$

and zero elsewhere.

    a. You will learn in Chapter 40 that the wave function must be a *continuous* function. Assuming that to be the case, what can you conclude about the relationship between b and c?

    b. Draw graphs of the wave function and the probability density over the interval -2 mm $\le x \le 2$ mm.

    c. What is the probability that the particle will be found to the right of the origin?

51. Consider the electron wave function

$$\psi(x) = \begin{cases} cx & |x| \le 1 \text{ nm} \\ \dfrac{c}{x} & |x| \ge 1 \text{ nm} \end{cases}$$

where x is in nm.

    a. Determine the normalization constant c.

    b. Draw a graph of $\psi(x)$ over the interval -5 nm $\le x \le 5$ nm. Provide numerical scales on both axes.

    c. Draw a graph of $|\psi(x)|^2$ over the interval -5 nm $\le x \le 5$ nm. Provide numerical scales.

    d. If 10^6 electrons are detected, how many will be in the interval -1.0 nm $\le x \le 1.0$ nm?

STOP TO THINK ANSWERS

Stop to Think 39.1: 10. The probability of a 1 is $P_1 = \frac{1}{6}$. Similarly, $P_6 = \frac{1}{6}$. The probability of a 1 *or* a 6 is $P_{1 \text{ or } 6} = \frac{1}{6} + \frac{1}{6} = \frac{1}{3}$. Thus the expected number is $30(\frac{1}{3}) = 10$.

Stop to Think 39.2: A > B = D > C. $|A(x)|^2$ is proportional to the density of dots.

Stop to Think 39.3: x_C. The probability is largest at the point where the *square* of $\psi(x)$ is largest.

Stop to Think 39.4: b. The area $\frac{1}{2}a(2 \text{ mm})$ must equal 1.

Stop to Think 39.5: b. $\Delta t = 1.0 \times 10^{-7}$ s. The bandwidth is $\Delta f_B = 1/\Delta t = 1.0 \times 10^7$ Hz = 10 MHz.

Stop to Think 39.6: A. Wave packet A has a smaller spatial extent Δx. The wavelength isn't relevant.

40 One-Dimensional Quantum Mechanics

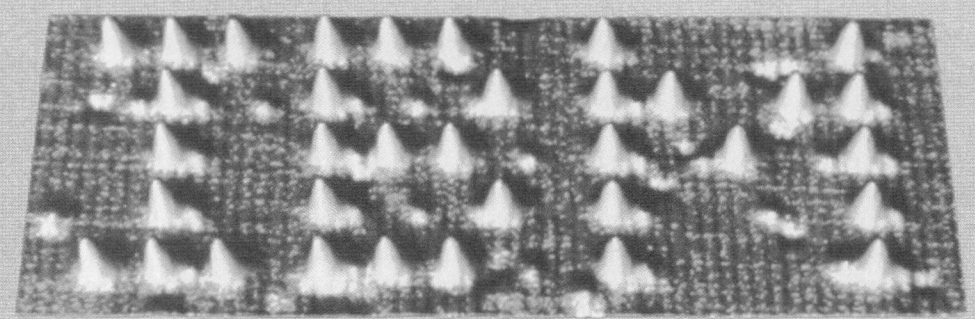

▶ **Looking Ahead** The goal of Chapter 40 is to understand and apply the essential ideas of quantum mechanics.

The Law of Psi

The basic law of quantum mechanics is the **Schrödinger equation.** It plays a role analogous to Newton's second law in classical mechanics.

We will limit our study to quantum mechanics in one dimension so as to focus on the physics without becoming bogged down by mathematical details.

You'll learn how the solutions to the Schrödinger equation predict the energy levels of a quantum system.

◀ **Looking Back**
Sections 39.3–39.4 Wave functions and normalization

Quantum Models

Classical systems are described in terms of forces. In contrast, a quantum system is described by a potential-energy function $U(x)$.

You'll learn to use **potential wells** to model different physical situations. A region in which $E < U_0$ is forbidden to a classical particle but not always to a quantum particle.

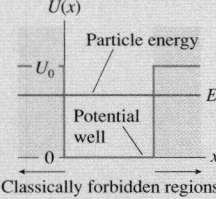

◀ **Looking Back**
Section 10.6 Energy diagrams

Tunneling

A surprising conclusion of quantum mechanics is that a wave function can penetrate some distance into a classically forbidden region.

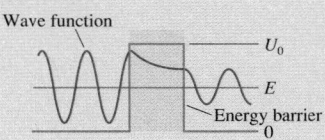

You'll learn that a particle can **tunnel** through a potential-energy barrier, emerging on the other side with no loss of energy. This totally nonclassical behavior is the basis of the scanning tunneling microscope.

Quantum Mechanics

Quantum mechanics is not just for physicists anymore. A knowledge of quantum mechanics is needed to understand the properties of materials and to design semiconductor devices. Quantum mechanics will be even more important in the near future for atomic level engineering of nanostructures and the development of quantum computing.

◀ **Looking Back**
Sections 38.4–38.5 Matter waves and the Bohr model of quantization

Wave Functions

You'll learn to understand why wave functions have the shapes they do.

■ The wave function oscillates between the classical turning points.
■ The wave function decays exponentially in a classically forbidden region.

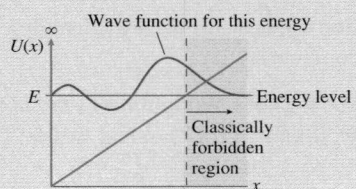

Quantum Applications

You'll study practical applications of quantum mechanics:

■ Quantum-well lasers
■ Molecular bonds
■ Nuclear energy levels

A quantum-well laser is made with a layer of gallium arsenide only about 1 nm thick. Electrons confined within this layer have discrete, quantized energy levels.

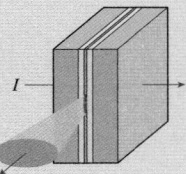

40.1 Schrödinger's Equation: The Law of Psi

Erwin Schrödinger.

In the winter of 1925, just before Christmas, the Austrian physicist Erwin Schrödinger gathered together a few books and headed off to a villa in the Swiss Alps. He had recently learned of de Broglie's 1924 suggestion that matter has wave-like properties, and he wanted some time free from distractions to think about it. Before the trip was over, Schrödinger had discovered the law of quantum mechanics.

Schrödinger's goal was to predict the outcome of atomic experiments, a goal that had eluded classical physics. The mathematical equation that he developed is now called the **Schrödinger equation.** It is the law of quantum mechanics in the same way that Newton's laws are the laws of classical mechanics. It would make sense to call it Schrödinger's law, but by tradition it is called simply the Schrödinger equation.

You learned in Chapter 39 that a matter particle is characterized in quantum physics by its wave function $\psi(x)$. If you know a particle's wave function, you can predict the probability of detecting it in some region of space. That's all well and good, but Chapter 39 didn't provide any method for determining wave functions. The Schrödinger equation is the missing piece of the puzzle. It is an equation for finding a particle's wave function $\psi(x)$.

Consider an atomic particle with mass m and mechanical energy E whose interactions with the environment can be characterized by a one-dimensional potential-energy function $U(x)$. The Schrödinger equation for the particle's wave function is

$$\frac{d^2\psi}{dx^2} = -\frac{2m}{\hbar^2}\left[E - U(x)\right]\psi(x) \qquad \text{(the Schrödinger equation)} \qquad (40.1)$$

This is a differential equation whose solution is the wave function $\psi(x)$ that we seek. Our first goal is to learn what this equation means and how it is used.

Justifying the Schrödinger Equation

The Schrödinger equation can be neither derived nor proved. It is not an outgrowth of any previous theory. Its success depended on its ability to explain the various phenomena that had refused to yield to a classical-physics analysis and to make new predictions that were subsequently verified.

Although the Schrödinger equation cannot be derived, the reasoning behind it can at least be made *plausible*. De Broglie had postulated a wave-like nature for matter in which a particle of mass m, velocity v, and momentum $p = mv$ has a wavelength

$$\lambda = \frac{h}{p} = \frac{h}{mv} \qquad (40.2)$$

Schrödinger's goal was to find a *wave equation* for which the solution would be a wave function having the de Broglie wavelength.

An oscillatory wave-like function with wavelength λ is

$$\psi(x) = \psi_0 \sin\left(\frac{2\pi x}{\lambda}\right) \qquad (40.3)$$

where ψ_0 is the amplitude of the wave function. Suppose we take a second derivative of $\psi(x)$:

$$\frac{d^2\psi}{dx^2} = \frac{d}{dx}\frac{d\psi}{dx} = \frac{d}{dx}\left[\frac{2\pi}{\lambda}\psi_0\cos\left(\frac{2\pi x}{\lambda}\right)\right] = -\frac{(2\pi)^2}{\lambda^2}\psi_0\sin\left(\frac{2\pi x}{\lambda}\right)$$

We can use Equation 40.3 to write this as

$$\frac{d^2\psi}{dx^2} = -\frac{(2\pi)^2}{\lambda^2}\psi(x) \qquad (40.4)$$

Equation 40.4 relates the wavelength λ to a combination of the wave function $\psi(x)$ and its second derivative.

NOTE ▶ These manipulations are not specific to quantum mechanics. Equation 40.4 is well known for classical waves, such as sound waves and waves on a string. ◀

Schrödinger's insight was to identify λ with the de Broglie wavelength of a particle. We can write the de Broglie wavelength in terms of the particle's kinetic energy K as

$$\lambda = \frac{h}{mv} = \frac{h}{\sqrt{2m\left(\frac{1}{2}mv^2\right)}} = \frac{h}{\sqrt{2mK}} \tag{40.5}$$

Notice that **the de Broglie wavelength increases as the particle's kinetic energy decreases.** This observation will play a key role.

If we square this expression for λ and substitute it into Equation 40.4, we find

$$\frac{d^2\psi}{dx^2} = -\frac{(2\pi)^2 2mK}{h^2}\psi(x) = -\frac{2m}{\hbar^2}K\psi(x) \tag{40.6}$$

where $\hbar = h/2\pi$. Equation 40.6 is a differential equation for the function $\psi(x)$. The solution to this equation is the sinusoidal wave function of Equation 40.3, where λ is the de Broglie wavelength for a particle with kinetic energy K.

Our derivation of Equation 40.6 assumed that the particle's kinetic energy K is constant. The energy diagram of FIGURE 40.1a reminds you that a particle's kinetic energy remains constant as it moves along the x-axis only if its potential energy U is constant. In this case, the de Broglie wavelength is the same at all positions.

FIGURE 40.1 The de Broglie wavelength changes as a particle's kinetic energy changes.

(a)

The kinetic energy $K = E - U$ is constant.

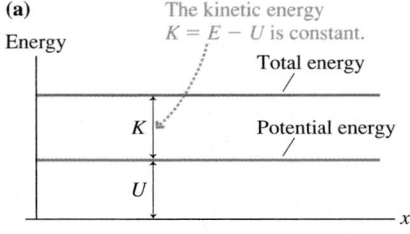

(b)

The kinetic energy decreases as x increases.

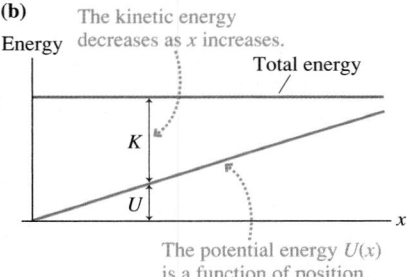

The potential energy $U(x)$ is a function of position.

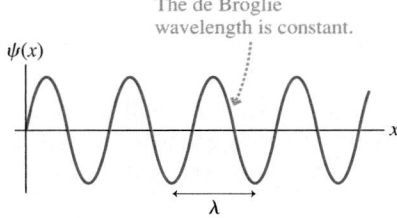

The de Broglie wavelength is constant.

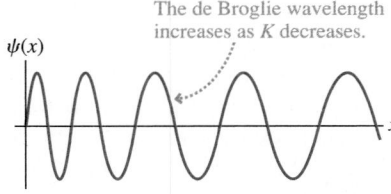

The de Broglie wavelength increases as K decreases.

In contrast, FIGURE 40.1b shows the energy diagram for a particle whose kinetic energy is *not* constant. This particle speeds up or slows down as it moves along the x-axis, transforming potential energy to kinetic energy or vice versa. Consequently, its de Broglie wavelength changes with position.

Suppose a particle's potential energy—gravitational or electric or any other kind of potential energy—is described by the function $U(x)$. That is, the potential energy is a *function of position* along the axis of motion. For example, the potential energy of a spring is $\frac{1}{2}kx^2$.

If E is the particle's total mechanical energy, its kinetic energy at position x is

$$K = E - U(x) \tag{40.7}$$

If we use this expression for K in Equation 40.6, that equation becomes

$$\frac{d^2\psi}{dx^2} = -\frac{2m}{\hbar^2}\left[E - U(x)\right]\psi(x)$$

This is Equation 40.1, the Schrödinger equation for the particle's wave function $\psi(x)$.

NOTE ▶ This has not been a derivation of the Schrödinger equation. We've made a *plausibility argument,* based on de Broglie's hypothesis about matter waves, but only experimental evidence will show if this equation has merit. ◀

STOP TO THINK 40.1 Three de Broglie waves are shown for particles of equal mass. Rank in order, from fastest to slowest, the speeds of particles a, b, and c.

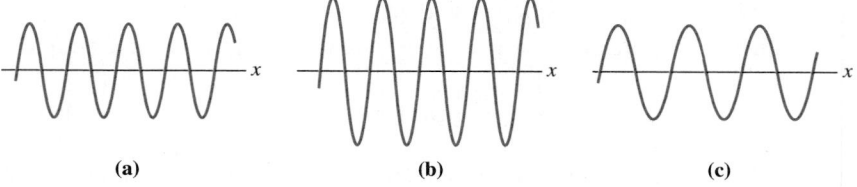

(a)   **(b)**   **(c)**

Quantum-Mechanical Models

Long ago, in your study of Newtonian mechanics, you learned the importance of *models.* To understand the motion of an object, we made simplifying assumptions: that the object could be represented by a particle, that friction could be described in a simple way, that air resistance could be neglected, and so on. Models allowed us to understand the primary features of an object's motion without getting lost in the details.

The same holds true in quantum mechanics. The exact description of a microscopic atom or a solid is extremely complicated. Our only hope for using quantum mechanics effectively is to make a number of simplifying assumptions—that is, to make a **quantum-mechanical model** of the situation. Much of this chapter will be about building and using quantum-mechanical models.

The test of a model's success is its agreement with experimental measurement. Laboratory experiments cannot measure $\psi(x)$, and they rarely make direct measurements of probabilities. Thus it will be important to tie our models to measurable quantities such as wavelengths, charges, currents, times, and temperatures.

There's one very important difference between models in classical mechanics and quantum mechanics. Classical models are described in terms of *forces,* and Newton's laws are a connection between force and motion. The Schrödinger equation for the wave function is written in terms of *energies.* Consequently, quantum-mechanical modeling involves finding a potential-energy function $U(x)$ that describes a particle's interactions with its environment.

FIGURE 40.2 reminds you how to interpret an energy diagram. We will use energy diagrams extensively in this and the remaining chapters to portray quantum-mechanical models. **A review of Section 10.6, where energy diagrams were introduced, is highly recommended.**

FIGURE 40.2 Interpreting an energy diagram.

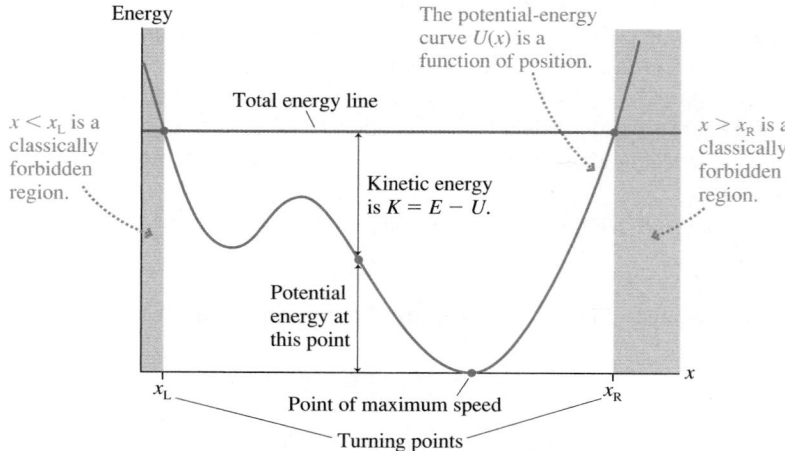

40.2 Solving the Schrödinger Equation

The Schrödinger equation is a second-order differential equation, meaning that it is a differential equation for $\psi(x)$ involving its second derivative. However, this textbook does not assume that you know how to solve differential equations. As we did with Newton's laws, we will restrict ourselves to situations for which you already have the mathematical skills from your calculus class.

The solution to an algebraic equation is simply a number. For example, $x = 3$ is the solution to the equation $2x = 6$. In contrast, the solution to a differential equation is a *function*. You saw this idea in the preceding section, where Equation 40.6 was constructed so that the function $\psi(x) = \psi_0 \sin(2\pi x/\lambda)$ was a solution.

The Schrödinger equation can't be solved until the potential-energy function $U(x)$ has been specified. Different potential-energy functions result in different wave functions, just as different forces lead to different trajectories in classical mechanics. Once $U(x)$ has been specified, the solution of the differential equation is a *function* $\psi(x)$. We will usually display the solution as a graph of $\psi(x)$ versus x.

Restrictions and Boundary Conditions

Not all functions $\psi(x)$ make *acceptable* solutions to the Schrödinger equation. That is, some functions may satisfy the Schrödinger equation but not be physically meaningful. We have previously encountered restrictions in our solutions of algebraic equations. We insist, for physical reasons, that masses be positive rather than negative numbers, that positions be real rather than imaginary numbers, and so on. Mathematical solutions not meeting these restrictions are rejected as being unphysical.

Because we want to interpret $|\psi(x)|^2$ as a probability density, we have to insist that the function $\psi(x)$ be one for which this interpretation is possible. The conditions or restrictions on acceptable solutions are called the **boundary conditions.** You will see, in later examples, how the boundary conditions help us choose the correct solution for $\psi(x)$. The primary conditions the wave function must obey are:

1. $\psi(x)$ is a continuous function.
2. $\psi(x) = 0$ if x is in a region where it is physically impossible for the particle to be.
3. $\psi(x) \rightarrow 0$ as $x \rightarrow +\infty$ and $x \rightarrow -\infty$.
4. $\psi(x)$ is a normalized function.

The last is not, strictly speaking, a boundary condition but is an auxiliary condition we require for the wave function to have a useful interpretation. Boundary condition 3 is needed to enable the normalization integral $\int |\psi(x)|^2 \, dx$ to converge.

Once boundary conditions have been established, there are general approaches to solving the Schrödinger equation: Use general mathematical techniques for solving second-order differential equations, solve the equation numerically on a computer, or make a physically informed guess.

More advanced courses make extensive use of the first and second approaches. However, we are not assuming a knowledge of differential equations, so you will not be asked to use these methods. The third, although it sounds almost like cheating, is widely used in simple situations where we can use physical arguments to infer the form of the wave function. The upcoming examples will illustrate this third approach.

A quadratic algebraic equation has two different solutions. Similarly, a second-order differential equation has two independent solutions $\psi_1(x)$ and $\psi_2(x)$. By "independent solutions" we mean that $\psi_2(x)$ is not just a constant multiple of $\psi_1(x)$, such as $3\psi_1(x)$, but that $\psi_1(x)$ and $\psi_2(x)$ are totally different functions.

Suppose that $\psi_1(x)$ and $\psi_2(x)$ are known to be two independent solutions of the Schrödinger equation. A theorem you will learn in differential equations states that a *general solution* of the equation can be written as

$$\psi(x) = A\psi_1(x) + B\psi_2(x) \qquad (40.8)$$

where A and B are constants whose values are determined by the boundary conditions. Equation 40.8 is a powerful statement, although one that will make more sense after you see it applied in upcoming examples. The main point is that **if we can find two independent solutions $\psi_1(x)$ and $\psi_2(x)$ by guessing, then Equation 40.8 is the general solution to the Schrödinger equation.**

Quantization

We've asserted that the Schrödinger equation is the law of quantum mechanics, but thus far we've not said anything about quantization. Although the particle's total energy E appears in the Schrödinger equation, it is treated in the equation as an unspecified constant. However, it will turn out that there are *no* acceptable solutions for most values of E. That is, there are no functions $\psi(x)$ that satisfy both the Schrödinger equation *and* the boundary conditions. Acceptable solutions exist only for *discrete* values of E. The energies for which solutions exist are the quantized energies of the system. Thus, as you'll see, the Schrödinger equation has quantization built in.

Problem Solving in Quantum Mechanics

Our problem-solving strategy for classical mechanics focused on identifying and using forces. In quantum mechanics we're interested in *energy* rather than forces. The critical step in solving a problem in quantum mechanics is to determine the particle's potential-energy function $U(x)$. Identifying the interactions that cause a potential energy is the *physics* of the problem. Once the potential-energy function is known, it is "just mathematics" to solve for the wave function.

PROBLEM-SOLVING STRATEGY 40.1 Quantum-mechanics problems

MODEL Determine a potential-energy function that describes the particle's interactions. Make simplifying assumptions.

VISUALIZE The potential-energy curve is the pictorial representation.

- Draw the potential-energy curve.
- Identify known information.
- Establish the boundary conditions that the wave function must satisfy.

SOLVE The Schrödinger equation is the mathematical representation.

- Utilize the boundary conditions.
- Normalize the wave functions.
- Draw graphs of $\psi(x)$ and $|\psi(x)|^2$.
- Determine the allowed energy levels.
- Calculate probabilities, wavelengths, or other specific quantities.

ASSESS Check that your result has the correct units, is reasonable, and answers the question.

The solutions to the Schrödinger equation are the *stationary states* of the system. Bohr had postulated the existence of stationary states, but he didn't know how to find them. Now we have a strategy for finding them.

Bohr's idea of transitions, or quantum jumps, between stationary states remains very important in Schrödinger's quantum mechanics. The system can jump from one stationary state, characterized by wave function $\psi_i(x)$ and energy E_i, to another state, characterized by $\psi_f(x)$ and E_f, by emitting or absorbing a photon of frequency

$$f = \frac{\Delta E}{h} = \frac{|E_f - E_i|}{h}$$

Thus the solutions to the Schrödinger equation will allow us to predict the emission and absorption spectra of a quantum system. These predictions will test the validity of Schrödinger's theory.

40.3 A Particle in a Rigid Box: Energies and Wave Functions

FIGURE 40.3 shows a particle of mass m confined in a rigid, one-dimensional box of length L. The walls of the box are assumed to be perfectly rigid, and the particle undergoes perfectly elastic reflections from the ends. This situation, which we looked at in Chapter 38, is known as a "particle in a box."

A classical particle bounces back and forth between the walls of the box. There are no restrictions on the speed or kinetic energy of a classical particle. In contrast, a wave-like particle characterized by a de Broglie wavelength sets up a standing wave as it reflects back and forth. In Chapter 38, we found that a standing de Broglie wave automatically leads to energy quantization. That is, only certain discrete energies are allowed. However, our hypothesis of a de Broglie standing wave was just a guess, with no real justification, because we had no theory about how a wave-like particle ought to behave.

We will now revisit this problem from the new perspective of quantum mechanics. The basic questions we want to answer in this, and any quantum-mechanics problem, are:

- What are the allowed energies of the particle?
- What is the wave function associated with each energy?
- In which part of the box is the particle most likely to be found?

We can use Problem-Solving Strategy 40.1 to answer these questions.

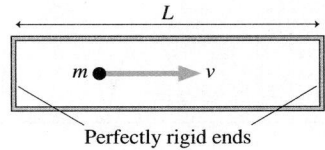

FIGURE 40.3 A particle in a rigid box of length L.

Model: Identify a Potential-Energy Function

By a *rigid box* we mean a box whose walls are so sturdy that they can confine a particle no matter how fast the particle moves. Furthermore, the walls are so stiff that they do not flex or give as the particle bounces. No real container has these attributes, so the rigid box is a *model* of a situation in which a particle is extremely well confined. Our first task is to characterize the rigid box in terms of a potential-energy function.

Let's establish a coordinate axis with the boundaries of the box at $x = 0$ and $x = L$. The rigid box has three important characteristics:

1. The particle can move freely between 0 and L at constant speed and thus with constant kinetic energy.
2. No matter how much kinetic energy the particle has, its turning points are at $x = 0$ and $x = L$.
3. The regions $x < 0$ and $x > L$ are forbidden. The particle cannot leave the box.

A potential-energy function that describes the particle in this situation is

$$U_{\text{rigid box}}(x) = \begin{cases} 0 & 0 \leq x \leq L \\ \infty & x < 0 \quad \text{or} \quad x > L \end{cases} \tag{40.9}$$

Inside the box, the particle has only kinetic energy. The infinitely high potential-energy barriers prevent the particle from ever having $x < 0$ or $x > L$ no matter how much kinetic energy it may have. It is this potential energy for which we want to solve the Schrödinger equation.

Visualize: Establish Boundary Conditions

FIGURE 40.4 is the energy diagram of a particle in the rigid box. You can see that $U = 0$ and $E = K$ inside the box. The upward arrows labeled ∞ indicate that the potential energy becomes infinitely large at the walls of the box ($x = 0$ and $x = L$).

NOTE ▶ Figure 40.4 is not a picture of the box. It is a graphical representation of the particle's total, kinetic, and potential energy. ◀

FIGURE 40.4 The energy diagram of a particle in a rigid box of length L.

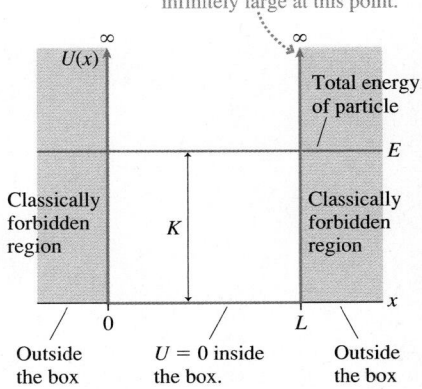

FIGURE 40.5 Applying boundary conditions to the wave function of a particle in a box.

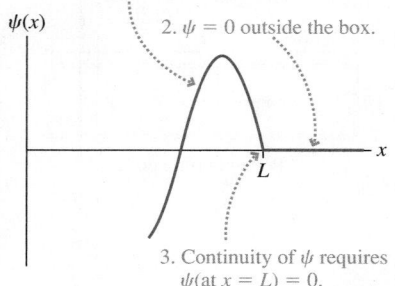

1. Inside the box, ψ is oscillating in some way still to be determined.

$\psi(x)$

2. $\psi = 0$ outside the box.

L

3. Continuity of ψ requires $\psi(\text{at } x = L) = 0$.

Next, we need to establish the boundary conditions that the solution must satisfy. Because it is physically impossible for the particle to be outside the box, we require

$$\psi(x) = 0 \quad \text{for } x < 0 \quad \text{or} \quad x > L \tag{40.10}$$

That is, there is zero probability (i.e., $|\psi(x)|^2 = 0$) of finding the particle outside the box.

Furthermore, the wave function must be a *continuous* function. That is, there can be no break in the wave function at any point. Because the solution is zero everywhere outside the box, continuity requires that the wave function inside the box obey the two conditions

$$\psi(\text{at } x = 0) = 0 \quad \text{and} \quad \psi(\text{at } x = L) = 0 \tag{40.11}$$

In other words, as FIGURE 40.5 shows, the oscillating wave function inside the box must go to zero at the boundaries to be continuous with the wave function outside the box. This requirement of the wave function is equivalent to saying that a standing wave on a string must have a node at the ends.

Solve I: Find the Wave Functions

At all points *inside* the box the potential energy is $U(x) = 0$. Thus the Schrödinger equation inside the box is

$$\frac{d^2\psi}{dx^2} = -\frac{2mE}{\hbar^2}\psi(x) \tag{40.12}$$

There are two aspects to solving this equation:

1. For what values of E does Equation 40.12 have physically meaningful solutions?
2. What are the solutions $\psi(x)$ for those values of E?

To begin, let's simplify the notation by defining $\beta^2 = 2mE/\hbar^2$. Equation 40.12 is then

$$\frac{d^2\psi}{dx^2} = -\beta^2\psi(x) \tag{40.13}$$

We're going to solve this differential equation by guessing! Can you think of any functions whose second derivative is a *negative* constant times the function itself? Two such functions are

$$\psi_1(x) = \sin\beta x \quad \text{and} \quad \psi_2(x) = \cos\beta x \tag{40.14}$$

Both are solutions to Equation 40.13 because

$$\frac{d^2\psi_1}{dx^2} = \frac{d^2}{dx^2}(\sin\beta x) = -\beta^2\sin\beta x = -\beta^2\psi_1(x)$$

$$\frac{d^2\psi_2}{dx^2} = \frac{d^2}{dx^2}(\cos\beta x) = -\beta^2\cos\beta x = -\beta^2\psi_2(x)$$

Furthermore, these are *independent* solutions because $\psi_2(x)$ is not a multiple or a rearrangement of $\psi_1(x)$. Consequently, according to Equation 40.8, the general solution to the Schrödinger equation for the particle in a rigid box is

$$\psi(x) = A\sin\beta x + B\cos\beta x \tag{40.15}$$

where

$$\beta = \frac{\sqrt{2mE}}{\hbar} \tag{40.16}$$

The constants A and B must be determined by using the boundary conditions of Equation 40.11. First, the wave function must go to zero at $x = 0$. That is,

$$\psi(\text{at } x = 0) = A \cdot 0 + B \cdot 1 = 0 \tag{40.17}$$

This boundary condition can be satisfied only if $B = 0$. The $\cos \beta x$ term may satisfy the differential equation in a mathematical sense, but it is not a physically meaningful solution for this problem because it does not satisfy the boundary conditions. Thus the physically meaningful solution is

$$\psi(x) = A \sin \beta x$$

The wave function must also go to zero at $x = L$. That is,

$$\psi(\text{at } x = L) = A \sin \beta L = 0 \qquad (40.18)$$

This condition could be satisfied by $A = 0$, but then we wouldn't have a wave function at all! Fortunately, that isn't necessary. This boundary condition is also satisfied if $\sin \beta L = 0$, which requires

$$\beta L = n\pi \quad \text{or} \quad \beta_n = \frac{n\pi}{L} \qquad n = 1, 2, 3, \dots \qquad (40.19)$$

Notice that n starts with 1, not 0. The value $n = 0$ would give $\beta = 0$ and make $\psi = 0$ at all points, a physically meaningless solution.

Thus the solutions to the Schrödinger equation for a particle in a rigid box are

$$\psi_n(x) = A \sin \beta_n x = A \sin\left(\frac{n\pi x}{L}\right) \qquad n = 1, 2, 3, \dots \qquad (40.20)$$

We've found a whole *family* of solutions, each corresponding to a different value of the integer n. These wave functions represent the stationary states of the particle in the box. The constant A remains to be determined.

Solve II: Find the Allowed Energies

Equation 40.16 defined β. The boundary condition of Equation 40.19 then placed restrictions on the possible values of β:

$$\beta_n = \frac{\sqrt{2mE_n}}{\hbar} = \frac{n\pi}{L} \qquad n = 1, 2, 3, \dots \qquad (40.21)$$

where the value of β and the energy associated with the integer n have been labeled β_n and E_n. We can solve for E_n by squaring both sides:

$$E_n = n^2 \frac{\pi^2 \hbar^2}{2mL^2} = n^2 \frac{h^2}{8mL^2} \qquad n = 1, 2, 3, \dots \qquad (40.22)$$

where, in the last step, we used the definition $\hbar = h/2\pi$. For a particle in a box, **these energies are the only values of E for which there are physically meaningful solutions to the Schrödinger equation.** That is, the particle's energy is quantized! It is worth emphasizing that quantization is not inherent in the wave function itself but arises because the boundary conditions—the physics of the situation—are satisfied by only a small subset of the mathematical solutions to the Schrödinger equation.

It is useful to write the energies of the stationary states as

$$E_n = n^2 E_1 \qquad (40.23)$$

where E_n is the energy of the stationary state with *quantum number n*. The smallest possible energy $E_1 = h^2/8mL^2$ is the energy of the $n = 1$ *ground state*. These allowed energies are shown in the *energy-level diagram* of FIGURE 40.6. Recall, from Chapter 38, that an energy-level diagram is not a graph (the horizontal axis doesn't represent anything) but a "ladder" of allowed energies.

Equation 40.22 is identical to the energies we found in Chapter 38 by requiring the de Broglie wave of a particle in a box to form a standing wave. Only now we have a theory that tells not only the energies but also the wave functions.

FIGURE 40.6 The energy-level diagram for a particle in a box.

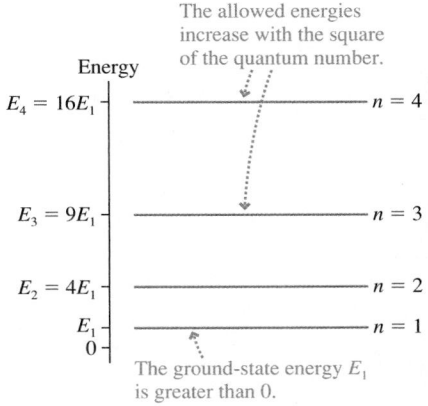

An electron in a box

An electron is confined to a rigid box. What is the length of the box if the energy difference between the first and second states is 3.0 eV?

MODEL Model the electron as a particle in a rigid one-dimensional box of length L.

SOLVE The first two quantum states, with $n = 1$ and $n = 2$, have energies E_1 and $E_2 = 4E_1$. Thus the energy difference between the states is

$$\Delta E = 3E_1 = \frac{3h^2}{8mL^2} = 3.0 \text{ eV} = 4.8 \times 10^{-19} \text{ J}$$

The length of the box for which $\Delta E = 3.0$ eV is

$$L = \sqrt{\frac{3h^2}{8m\,\Delta E}} = 6.14 \times 10^{-10} \text{ m} = 0.614 \text{ nm}$$

ASSESS The expression for E_1 is in SI units, so energies must be in J, not eV.

Solve III: Normalize the Wave Functions

We can determine the constant A by requiring the wave functions to be normalized. The normalization condition, which we found in Chapter 39, is

$$\int_{-\infty}^{\infty} |\psi(x)|^2 \, dx = 1$$

This is the mathematical statement that the particle must be *somewhere* on the x-axis. The integration limits extend to $\pm\infty$, but here we need to integrate only from 0 to L because the wave function is zero outside the box. Thus

$$\int_0^L |\psi_n(x)|^2 \, dx = A_n^2 \int_0^L \sin^2\left(\frac{n\pi x}{L}\right) dx = 1 \qquad (40.24)$$

or

$$A_n = \left[\int_0^L \sin^2\left(\frac{n\pi x}{L}\right) dx \right]^{-1/2} \qquad (40.25)$$

We placed a subscript n on A_n because it is possible that the normalization constant is different for each wave function in the family. This is a standard integral. We will leave it as a homework problem for you to show that its value, for any n, is

$$A_n = \sqrt{\frac{2}{L}} \qquad n = 1, 2, 3, \ldots \qquad (40.26)$$

We now have a complete solution to the problem. The normalized wave function for the particle in quantum state n is

$$\psi_n(x) = \begin{cases} \sqrt{\dfrac{2}{L}} \sin\left(\dfrac{n\pi x}{L}\right) & 0 \le x \le L \\ 0 & x < 0 \text{ and } x > L \end{cases} \qquad (40.27)$$

40.4 A Particle in a Rigid Box: Interpreting the Solution

Our solution to the quantum-mechanical problem of a particle in a box tells us that:

1. The particle must have energy $E_n = n^2 E_1$, where $n = 1, 2, 3, \ldots$ is the quantum number. $E_1 = h^2/8mL^2$ is the energy of the $n = 1$ ground state.
2. The wave function for a particle in quantum state n is

$$\psi_n(x) = \begin{cases} \sqrt{\dfrac{2}{L}} \sin\left(\dfrac{n\pi x}{L}\right) & 0 \le x \le L \\ 0 & x < 0 \text{ and } x > L \end{cases}$$

These are the stationary states of the system.

3. The probability density for finding the particle at position x inside the box is

$$P_n(x) = |\psi_n(x)|^2 = \frac{2}{L} \sin^2\left(\frac{n\pi x}{L}\right) \tag{40.28}$$

FIGURE 40.7 Wave functions and probability densities for a particle in a rigid box of length L.

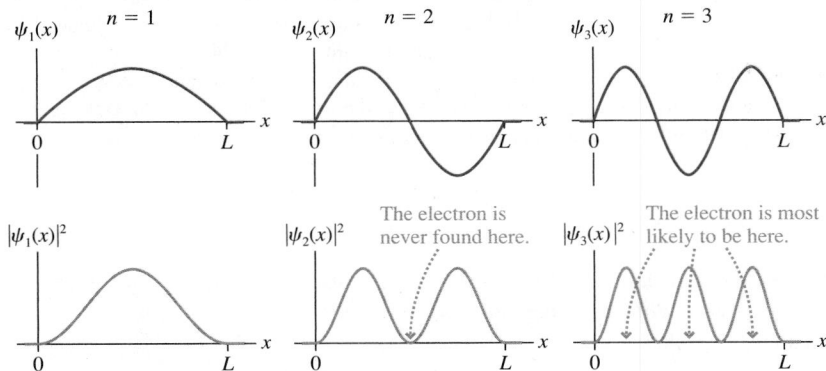

A graphical presentation will make these results more meaningful. FIGURE 40.7 shows the wave functions $\psi(x)$ and the probability densities $P(x) = |\psi(x)|^2$ for quantum states $n = 1$ to 3. Notice that the wave functions go to zero at the boundaries and thus are continuous with $\psi = 0$ outside the box.

The wave functions $\psi(x)$ for a particle in a rigid box are analogous to standing waves on a string that is tied at both ends. You can see that $\psi_n(x)$ has $(n - 1)$ nodes (zeros), excluding the ends, and n antinodes (maxima and minima). This is a general result for any wave function, not just for a particle in a rigid box.

FIGURE 40.8 shows another way in which energies and wave functions are shown graphically in quantum mechanics. First, the graph shows the potential-energy function $U(x)$ of the particle. Second, the allowed energies are shown as horizontal lines (total energy lines) across the potential-energy graph. These are labeled with the quantum number n and the energy E_n. Third—and this is a bit tricky—the wave function for each n is drawn *as if* the energy line were the zero of the y-axis. That is, the graph of $\psi_n(x)$ is drawn on top of the E_n energy line. This allows energies and wave functions to be displayed simultaneously, but it does *not* imply that ψ_2 is in any sense "above" ψ_1. Both oscillate sinusoidally about zero, as Figure 40.7 shows.

FIGURE 40.8 An alternative way to show the potential-energy diagram, the energies, and the wave functions.

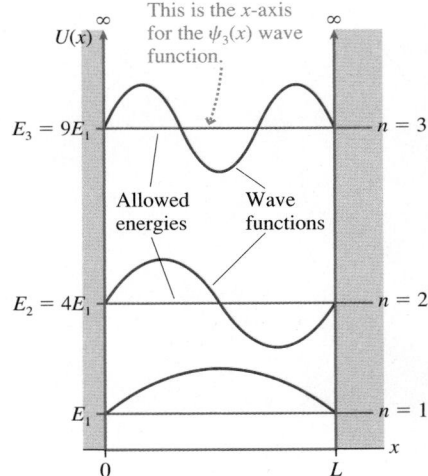

EXAMPLE 40.2 **Energy levels and quantum jumps**

A semiconductor device known as a *quantum-well device* is designed to "trap" electrons in a 1.0-nm-wide region. Treat this as a one-dimensional problem.

a. What are the energies of the first three quantum states?
b. What wavelengths of light can these electrons absorb?

MODEL Model an electron in a quantum-well device as a particle confined in a rigid box of length $L = 1.0$ nm.

VISUALIZE FIGURE 40.9 shows the first three energy levels and the transitions by which an electron in the ground state can absorb a photon.

FIGURE 40.9 Energy levels and quantum jumps for an electron in a quantum-well device.

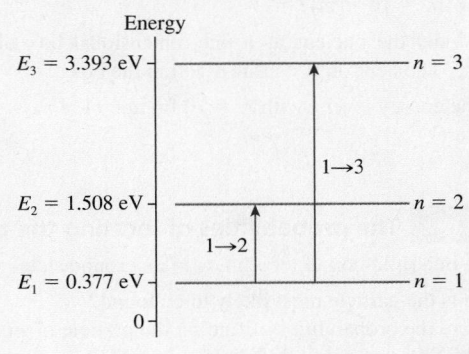

Continued

SOLVE a. The particle's mass is $m = m_e = 9.11 \times 10^{-31}$ kg. The allowed energies, in both J and eV, are

$$E_1 = \frac{h^2}{8mL^2} = 6.03 \times 10^{-20} \text{ J} = 0.377 \text{ eV}$$

$$E_2 = 4E_1 = 1.508 \text{ eV}$$

$$E_3 = 9E_1 = 3.393 \text{ eV}$$

b. An electron spends most of its time in the $n = 1$ ground state. According to Bohr's model of stationary states, the electron can absorb a photon of light and undergo a transition, or quantum jump, to $n = 2$ or $n = 3$ if the light has frequency $f = \Delta E/h$. The wavelengths, given by $\lambda = c/f = hc/\Delta E$, are

$$\lambda_{1 \to 2} = \frac{hc}{E_2 - E_1} = 1098 \text{ nm}$$

$$\lambda_{1 \to 3} = \frac{hc}{E_3 - E_1} = 411 \text{ nm}$$

ASSESS In practice, various complications usually make the $1 \to 3$ transition unobservable. But quantum-well devices do indeed exhibit strong absorption and emission at the $\lambda_{1 \to 2}$ wavelength. In this example, which is typical of quantum-well devices, the wavelength is in the near-infrared portion of the spectrum. Devices such as these are used to construct the semiconductor lasers used in DVD players and laser printers.

NOTE ▶ The wavelengths of light emitted or absorbed by a quantum system are determined by the *difference* between two allowed energies. Quantum jumps involve two stationary states. ◀

Zero-Point Motion

The lowest energy state in Example 40.2, the ground state, has $E_1 = 0.38$ eV. There is no stationary state having $E = 0$. Unlike a classical particle, **a quantum particle in a box cannot be at rest!** No matter how much its energy is reduced, such as by cooling it toward absolute zero, it cannot have energy less than E_1.

The particle motion associated with energy E_1, called the **zero-point motion,** is a consequence of Heisenberg's uncertainty principle. Because the particle is somewhere in the box, its position uncertainty is $\Delta x = L$. If the particle were at rest in the box, we would know that its velocity and momentum are exactly zero with *no* uncertainty: $\Delta p_x = 0$. But then $\Delta x \Delta p_x = 0$ would violate the Heisenberg uncertainty principle. One of the conclusions that follow from the uncertainty principle is that **a confined particle cannot be at rest.**

Although the particle's position and velocity are uncertain, the particle's energy in each state can be calculated with a high degree of precision. This distinction between a precise energy and uncertain position and velocity seems strange, but it is just our old friend the standing wave. In order to *have* a stationary state at all, the de Broglie waves have to form standing waves. Only for very precise frequencies, and thus precise energies, can the standing-wave pattern appear.

EXAMPLE 40.3 | **Nuclear energies**

Protons and neutrons are tightly bound within the nucleus of an atom. If we use a one-dimensional model of a nucleus, what are the first three energy levels of a neutron in a 10-fm-diameter nucleus (1 fm = 10^{-15} m)?

MODEL Model the nucleus as a one-dimensional box of length $L = 10$ fm. The neutron is confined within the box.

SOLVE The energy levels, with $L = 10$ fm and $m = m_n = 1.67 \times 10^{-27}$ kg, are

$$E_1 = \frac{h^2}{8mL^2} = 3.29 \times 10^{-13} \text{ J} = 2.06 \text{ MeV}$$

$$E_2 = 4E_1 = 8.24 \text{ MeV}$$

$$E_3 = 9E_1 = 18.54 \text{ MeV}$$

ASSESS You've seen that an electron confined in an atom-size space has energies of a few eV. A neutron confined in a nucleus-size space has energies of a few *million* eV.

EXAMPLE 40.4 | **The probabilities of locating the particle**

A particle in a rigid box of length L is in its ground state.

a. Where is the particle most likely to be found?
b. What are the probabilities of finding the particle in an interval of width $0.01L$ at $x = 0.00L, 0.25L,$ and $0.50L$?

c. What is the probability of finding the particle between $L/4$ and $3L/4$?

MODEL The wave functions for a particle in a rigid box have been determined.

VISUALIZE FIGURE 40.10 shows the probability density $P_1(x) = |\psi_1(x)|^2$ in the ground state.

FIGURE 40.10 Probability density for a particle in the ground state.

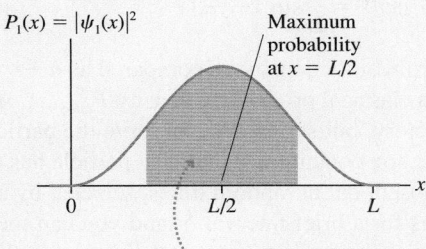

$P_1(x) = |\psi_1(x)|^2$

Maximum probability at $x = L/2$

The probability of being in the interval from $L/4$ to $3L/4$ is the area under the curve.

SOLVE a. The particle is most likely to be found at the point where the probability density $P(x)$ is a maximum. You can see from Figure 40.10 that the point of maximum probability for $n = 1$ is $x = L/2$.

b. For a *small* width δx, the probability of finding the particle in δx at position x is

$$\text{Prob(in } \delta x \text{ at } x) = P_1(x)\,\delta x = |\psi_1(x)|^2\,\delta x = \frac{2}{L}\sin^2\left(\frac{\pi x}{L}\right)\delta x$$

The interval $\delta x = 0.01L$ is sufficiently small for this to be valid. The probabilities of finding the particle are

Prob(in $0.01L$ at $x = 0.00L$) $= 0.000 = 0.0\%$

Prob(in $0.01L$ at $x = 0.25L$) $= 0.010 = 1.0\%$

Prob(in $0.01L$ at $x = 0.50L$) $= 0.020 = 2.0\%$

c. You learned in Chapter 39 that the probability of being in an interval is the area under the probability-density curve. We calculate this by integrating:

$$\text{Prob}\left(\text{in interval } \frac{1}{4}L \text{ to } \frac{3}{4}L\right) = \int_{L/4}^{3L/4} P_1(x)\,dx$$

$$= \frac{2}{L}\int_{L/4}^{3L/4} \sin^2\left(\frac{\pi x}{L}\right) dx$$

$$= \left[\frac{x}{L} - \frac{1}{\pi}\sin\left(\frac{\pi x}{L}\right)\cos\left(\frac{\pi x}{L}\right)\right]_{L/4}^{3L/4}$$

$$= \frac{1}{2} + \frac{1}{\pi} = 0.818$$

The integral of $\sin^2$ was taken from the table of integrals in Appendix A.

ASSESS If a particle in a box is in the $n = 1$ ground state, there is an 81.8% chance of finding it in the center half of the box. The probability is greater than 50% because, as you can see in Figure 40.10, the probability density $P_1(x)$ is larger near the center of the box than near the boundaries.

This has been a lengthy presentation of the particle-in-a-box problem. However, it was important that we explore the method of solution completely. Future examples will now go more quickly because many of the issues discussed here will not need to be repeated.

STOP TO THINK 40.2   A particle in a rigid box in the $n = 2$ stationary state is most likely to be found

a. In the center of the box.
b. One-third of the way from either end.
c. One-quarter of the way from either end.
d. It is equally likely to be found at any point in the box.

40.5 The Correspondence Principle

Suppose we confine an electron in a microscopic box, then allow the box to get bigger and bigger. What started out as a quantum-mechanical situation should, when the box becomes macroscopic, eventually look like a classical-physics situation. Similarly, a classical situation such as two charged particles revolving about each other should begin to exhibit quantum behavior as the orbit size becomes smaller and smaller.

These examples suggest that there should be some in-between size, or energy, for which the quantum-mechanical solution corresponds in some way to the solution of classical mechanics. Niels Bohr put forward the idea that the *average* behavior of a quantum system should begin to look like the classical solution in the limit that the quantum number becomes very large—that is, as $n \rightarrow \infty$. Because the radius of the Bohr hydrogen atom is $r = n^2 a_B$, the atom becomes a macroscopic object as n

becomes very large. Bohr's idea, that the quantum world should blend smoothly into the classical world for high quantum numbers, is today known as the **correspondence principle.**

Our quantum knowledge of a particle in a box is given by its probability density

$$P_{quant}(x) = |\psi_n(x)|^2 = \frac{2}{L} \sin^2\left(\frac{n\pi x}{L}\right) \tag{40.29}$$

To what classical quantity can the probability density be compared as $n \to \infty$?

Interestingly, we can also define a classical probability density $P_{class}(x)$. A classical particle follows a well-defined trajectory, but suppose we observe the particle at random times. For example, suppose the box containing a classical particle has a viewing window. The window is normally closed, but at random times, selected by a random-number generator, the window opens for a brief interval δt and you can measure the particle's position. When the window opens, what is the probability that the particle will be in a narrow interval δx at position x?

The probability of finding a classical particle within a small interval δx is equal to the *fraction of its time* that it spends passing through δx. That is, you're more likely to find the particle in those intervals δx where it spends lots of time, less likely to find it in a δx where it spends very little time.

Consider a classical particle oscillating back and forth between two turning points with period T. The time it spends moving from one turning point to the other is $\frac{1}{2}T$. As it moves between the turning points, it passes once through the interval δx at position x, taking time δt to do so. Consequently, the probability of finding the particle within this interval is

$$\text{Prob}_{class}(\text{in } \delta x \text{ at } x) = \text{fraction of time spent in } \delta x = \frac{\delta t}{\frac{1}{2}T} \tag{40.30}$$

The amount of time needed to pass through δx is $\delta t = \delta x / v(x)$, where $v(x)$ is the particle's velocity at position x. Thus the probability of finding the particle in the interval δx at position x is

$$\text{Prob}_{class}(\text{in } \delta x \text{ at } x) = \frac{\delta x / v(x)}{\frac{1}{2}T} = \frac{2}{Tv(x)} \delta x \tag{40.31}$$

You learned in Chapter 39 that the probability is related to the probability density by

$$\text{Prob}_{class}(\text{in } \delta x \text{ at } x) = P_{class}(x)\,\delta x$$

Thus the classical probability density for finding a particle at position x is

$$P_{class}(x) = \frac{2}{Tv(x)} \tag{40.32}$$

where the velocity $v(x)$ is expressed as a function of x. **Classically, a particle is more likely to be found where it is moving slowly, less likely to be found where it is moving quickly.**

NOTE ▶ Our derivation of Equation 40.32 made no assumptions about the particle's motion other than the requirement that it be periodic. This is the classical probability density for any oscillatory motion. ◀

FIGURE 40.11a is the motion diagram of a classical particle in a rigid box of length L. The particle's speed is a *constant* $v(x) = v_0$ as it bounces back and forth between the walls. The particle travels distance $2L$ during one round trip, so the period is $T = 2L/v_0$. Consequently, the classical probability density for a particle in a box is

$$P_{class}(x) = \frac{2}{(2L/v_0)v_0} = \frac{1}{L} \tag{40.33}$$

$P_{class}(x)$ is independent of x, telling us that the particle is equally likely to be found *anywhere* in the box.

FIGURE 40.11 The classical probability density is indicated by the density of dots in a motion diagram.

(a) Uniform speed

Particle in an empty box

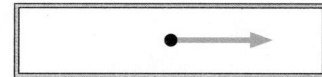

Motion diagram

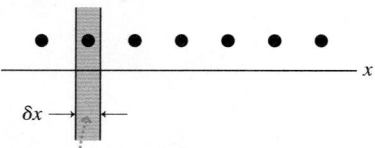

The probability of finding the particle in δx is the fraction of time the particle spends in δx.

(b) Nonuniform speed

Particle on a spring

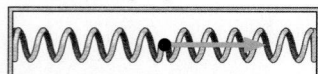

Motion diagram

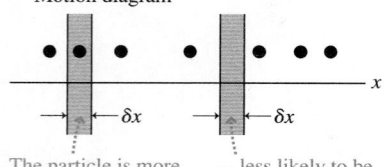

The particle is more likely to be found where it's moving slowly,...

... less likely to be found where it's moving quickly.

In contrast, FIGURE 40.11b shows a particle with nonuniform speed. A mass on a spring slows down near the turning points, so it spends more time near the ends of the box than in the middle. Consequently the classical probability density for this particle is a maximum at the edges and a minimum at the center. We'll look at this classical probability density again later in the chapter.

EXAMPLE 40.5 **The classical probability of locating the particle**

A classical particle is in a rigid 10-cm-long box. What is the probability that, at a random instant of time, the particle is in a 1.0-mm-wide interval at the center of the box?

SOLVE The particle's probability density is

$$P_{class}(x) = \frac{1}{L} = \frac{1}{10 \text{ cm}} = 0.10 \text{ cm}^{-1}$$

The probability that the particle is in an interval of width $\delta x = 1.0 \text{ mm} = 0.10 \text{ cm}$ is

$$\text{Prob(in } \delta x \text{ at } x = 5 \text{ cm}) = P(x)\delta x = (0.10 \text{ cm}^{-1})(0.10 \text{ cm})$$
$$= 0.010 = 1.0\%$$

ASSESS The classical probability is 1.0% because 1.0 mm is 1% of the 10 cm length.

FIGURE 40.12 shows the quantum and the classical probability densities for the $n = 1$ and $n = 20$ quantum states of a particle in a rigid box. Notice that:

- The quantum probability density oscillates between a minimum of 0 and a maximum of $2/L$, so it oscillates around the classical probability density $1/L$.
- For $n = 1$, the quantum and classical probability densities are quite different. The ground state of the quantum system will be very nonclassical.
- For $n = 20$, *on average* the quantum particle's behavior looks very much like that of the classical particle.

FIGURE 40.12 The quantum and classical probability densities for a particle in a box.

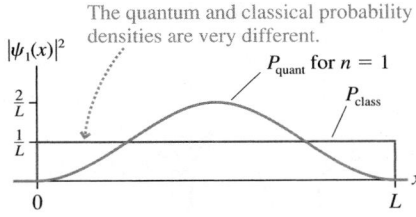
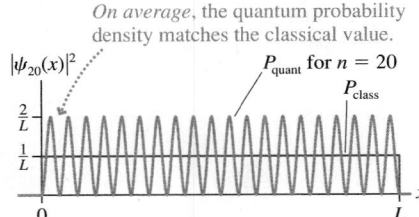

As n gets even bigger and the number of oscillations increases, the probability of finding the particle in an interval δx will be the same for both the quantum and the classical particles as long as δx is large enough to include several oscillations of the wave function. As Bohr predicted, the quantum-mechanical solution "corresponds" to the classical solution in the limit $n \rightarrow \infty$.

40.6 Finite Potential Wells

Figure 40.4, the potential-energy diagram for a particle in a rigid box, is an example of a **potential well,** so named because the graph of the potential-energy "hole" looks like a well from which you might draw water. The rigid box was an *infinite* potential well. There was no chance that a particle inside could escape the infinitely high walls.

A more realistic model of a confined particle is the *finite* potential well shown in FIGURE 40.13a on the next page. A particle with total energy $E < U_0$ is confined within the well, bouncing back and forth between turning points at $x = 0$ and $x = L$. The regions $x < 0$ and $x > L$ are **classically forbidden regions** for a particle with $E < U_0$. However, the particle will escape the well if it manages to acquire energy $E > U_0$.

For example, consider an electron confined within a metal or semiconductor. An electron with energy less than the work function moves freely until it reaches the edge, where it reflects to stay within the solid. But the electron *can* escape if it somehow—such

FIGURE 40.13 A finite potential well of width L and depth U_0.

(a) $U = 0$ inside the well.

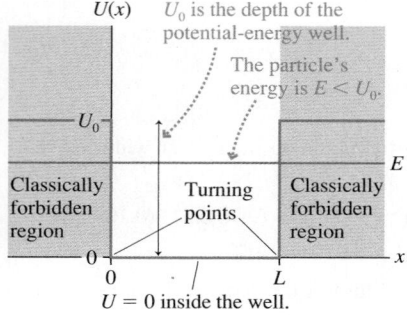

(b) $U = 0$ outside the well.

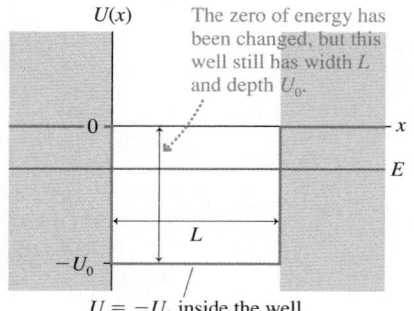

$U = -U_0$ inside the well.

as absorbing energy from a photon—acquires an energy larger than the work function. Similarly, a neutron is confined within the nucleus by the nuclear force, but it *can* escape the nucleus if it has enough energy. The electron, the neutron, and many other particles that are confined can be modeled as a particle in a finite potential well, so it is one of the most important models in quantum mechanics. The Schrödinger equation depends on the *shape* of the potential-energy function, not the cause.

FIGURE 40.13b is the same potential well, simply redrawn to place the zero of potential energy—which, you will recall, is arbitrary—at the level of the "energy plateau." Both have width L and depth U_0 so both have the same wave functions and the same energy levels relative to the bottom of the well. Which one we use is a matter of convenience.

Although it is possible to solve the Schrödinger equation exactly for the finite potential well, the result is cumbersome and not especially illuminating. Instead, we'll present the results of numerical calculations. The derivation of the wave functions and energy levels is not as important as understanding and interpreting the results.

As a first example, consider an electron in a 2.0-nm-wide potential well of depth $U_0 = 1.0$ eV. These are reasonable parameters for an electron in a semiconductor device. **FIGURE 40.14a** is a graphical presentation of the allowed energies and wave functions. For comparison, **FIGURE 40.14b** shows the first three energy levels and wave functions for a rigid box ($U_0 \rightarrow \infty$) with the same 2.0 nm width.

FIGURE 40.14 Energy levels and wave functions for a finite potential well. For comparison, the energies and wave functions are shown for a rigid box of equal width.

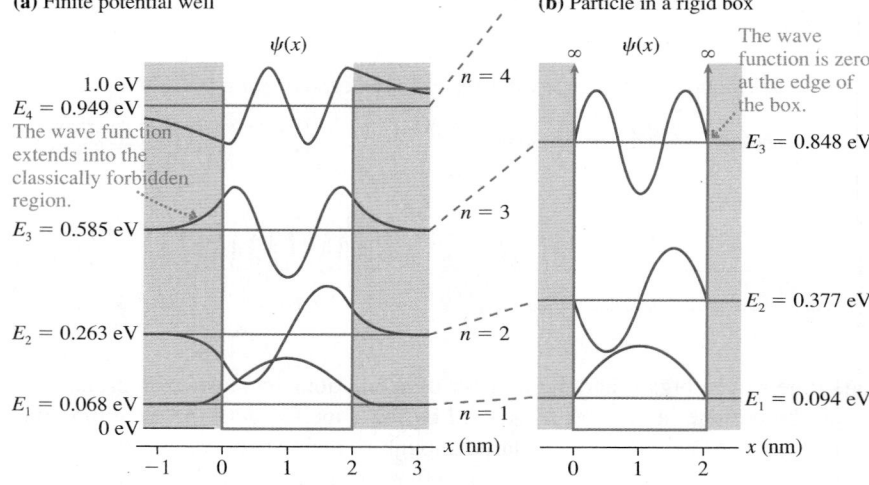

The quantum-mechanical solution for a particle in a finite potential well has some important properties:

- The particle's energy is quantized. A particle in the potential well *must* be in one of the stationary states with quantum numbers $n = 1, 2, 3, \ldots$.
- There are only a finite number of **bound states**—four in this example, although the number will be different in other examples. These wave functions represent electrons confined to, or bound in, the potential well. There are no stationary states with $E > U_0$ because such a particle would not remain in the well.
- The wave functions are qualitatively similar to those of a particle in a rigid box, but the energies are somewhat lower. This is because the wave functions are slightly more spread out horizontally. A slightly longer de Broglie wavelength corresponds to a lower velocity and thus a lower energy.
- Most interesting, perhaps, is that the wave functions of Figure 40.14a extend into the classically forbidden regions. It is as though a tennis ball penetrated partly *through* the racket's strings before bouncing back, but without breaking the strings.

EXAMPLE 40.6 | **Absorption spectrum of an electron**

What wavelengths of light are absorbed by a semiconductor device in which electrons are confined in a 2.0-nm-wide region with a potential-energy depth of 1.0 eV?

MODEL The electron is in the finite potential well whose energies and wave functions were shown in Figure 40.14a.

SOLVE Photons can be absorbed if their energy $E_{photon} = hf$ exactly matches the energy difference ΔE between two energy levels. Because most electrons are in the $n = 1$ ground state, the absorption transitions are $1 \to 2$, $1 \to 3$, and $1 \to 4$.

The absorption wavelengths $\lambda = c/f$ are

$$\lambda_{n \to m} = \frac{hc}{\Delta E} = \frac{hc}{|E_n - E_m|}$$

For this example, we find

$\Delta E_{1-2} = 0.195$ eV	$\lambda_{1 \to 2} = 6.37\ \mu m$
$\Delta E_{1-3} = 0.517$ eV	$\lambda_{1 \to 3} = 2.40\ \mu m$
$\Delta E_{1-4} = 0.881$ eV	$\lambda_{1 \to 4} = 1.41\ \mu m$

ASSESS These transitions are all infrared wavelengths.

STOP TO THINK 40.3 This is a wave function for a particle in a finite quantum well. What is the particle's quantum number?

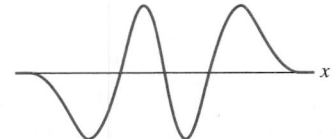

The Classically Forbidden Region

The extension of a particle's wave functions into the classically forbidden region is an important difference between classical and quantum physics. Let's take a closer look at the wave function in the region $x \geq L$ of Figure 40.13a. The potential energy in the classically forbidden region is U_0; thus the Schrödinger equation for $x \geq L$ is

$$\frac{d^2\psi}{dx^2} = -\frac{2m}{\hbar^2}(E - U_0)\psi(x)$$

We're assuming a confined particle, with E less than U_0, so $E - U_0$ is negative. It will be useful to reverse the order of these and write

$$\frac{d^2\psi}{dx^2} = \frac{2m}{\hbar^2}(U_0 - E)\psi(x) = \frac{1}{\eta^2}\psi(x) \tag{40.34}$$

where

$$\eta^2 = \frac{\hbar^2}{2m(U_0 - E)} \tag{40.35}$$

is a *positive* constant. As a homework problem, you can show that the units of η are meters.

The Schrödinger equation of Equation 40.34 is one we can solve by guessing. We simply need to think of two functions whose second derivatives are a positive constant times the functions themselves. Two such functions, as you can quickly confirm, are $e^{x/\eta}$ and $e^{-x/\eta}$. Thus, according to Equation 40.8, the general solution of the Schrödinger equation for $x \geq L$ is

$$\psi(x) = Ae^{x/\eta} + Be^{-x/\eta} \quad \text{for } x \geq L \tag{40.36}$$

One requirement of the wave function is that $\psi \to 0$ as $x \to \infty$. The function $e^{x/\eta}$ diverges as $x \to \infty$, so the only way to satisfy this requirement is to set $A = 0$. Thus

$$\psi(x) = Be^{-x/\eta} \quad \text{for } x \geq L \tag{40.37}$$

This is an exponentially decaying function. Notice that all the wave functions in Figure 40.14a look like exponential decays for $x > L$.

The wave function must also be continuous. Suppose the oscillating wave function within the potential well ($x \leq L$) has the value ψ_{edge} when it reaches the classical boundary at $x = L$. To be continuous, the wave function of Equation 40.37 has to match this value at $x = L$. That is,

$$\psi(\text{at } x = L) = Be^{-L/\eta} = \psi_{edge} \tag{40.38}$$

This boundary condition at $x = L$ is sufficient to determine that the constant B is

$$B = \psi_{\text{edge}} e^{L/\eta} \qquad (40.39)$$

If we use the Equation 40.39 result for B in Equation 40.37, we find that the wave function in the classically forbidden region of a finite potential well is

$$\psi(x) = \psi_{\text{edge}} e^{-(x-L)/\eta} \quad \text{for } x \geq L \qquad (40.40)$$

In other words, **the wave function oscillates until it reaches the classical turning point at $x = L$, then it decays exponentially within the classically forbidden region.** A similar analysis could be done for $x \leq 0$.

FIGURE 40.15 shows the wave function in the classically forbidden region. You can see that the wave function at $x = L + \eta$ has decreased to

$$\psi(\text{at } x = L + \eta) = e^{-1}\psi_{\text{edge}} = 0.37\psi_{\text{edge}}$$

Although an exponential decay does not have a sharp ending point, the parameter η measures "about how far" the wave function extends past the classical turning point before the probability of finding the particle has decreased nearly to zero. This distance is called the **penetration distance:**

$$\text{penetration distance } \eta = \frac{\hbar}{\sqrt{2m(U_0 - E)}} \qquad (40.41)$$

A classical particle reverses direction at the $x = L$ turning point. But atomic particles are not classical. Because of wave–particle duality, an atomic particle is "fuzzy" with no well-defined edge. Thus an atomic particle can spread a distance of roughly η into the classically forbidden region.

The penetration distance is unimaginably small for any macroscopic mass, but it can be significant for atomic particles. Notice that the penetration distance depends inversely on the quantity $U_0 - E$, the distance of the energy level below the top of the potential well. You can see in Figure 40.14a that η is much larger for the $n = 4$ state, near the top of the potential well, than for the $n = 1$ state.

NOTE ▶ In making use of Equation 40.41, you *must* use SI units of J s for $\hbar$ and J for the energies. The penetration distance η is then in meters. ◀

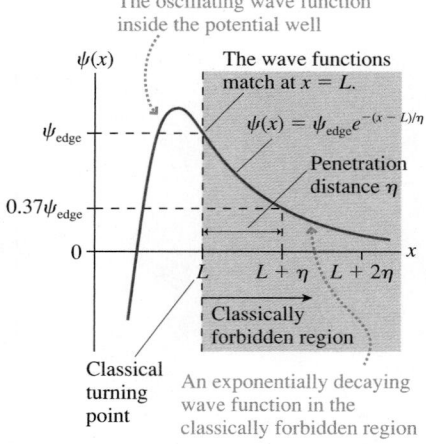

FIGURE 40.15 The wave function in the classically forbidden region.

EXAMPLE 40.7 **Penetration distance of an electron**

An electron is confined in a 2.0-nm-wide region with a potential-energy depth of 1.00 eV. What are the penetration distances into the classically forbidden region for an electron in the $n = 1$ and $n = 4$ states?

MODEL The electron is in the finite potential well whose energies and wave functions were shown in Figure 40.14a.

SOLVE The ground state has $U_0 - E_1 = 1.000 \text{ eV} - 0.068 \text{ eV} = 0.932 \text{ eV}$. Similarly, $U_0 - E_4 = 0.051 \text{ eV}$ in the $n = 4$ state. We can use Equation 40.41 to calculate

$$\eta = \frac{\hbar}{\sqrt{2m(U_0 - E)}} = \begin{cases} 0.20 \text{ nm} & n = 1 \\ 0.86 \text{ nm} & n = 4 \end{cases}$$

ASSESS These values are consistent with Figure 40.14a.

Quantum-Well Devices

In Part VI we developed a model of electrical conductivity in which the valence electrons of a metal form a loosely bound "sea of electrons." The typical speed of an electron is the rms speed:

$$v_{\text{rms}} = \sqrt{\frac{3k_B T}{m}}$$

where k_B is Boltzmann's constant. Hence at room temperature, where $v_{rms} \approx 1 \times 10^5$ m/s, the de Broglie wavelength of a typical conduction electron is

$$\lambda \approx \frac{h}{mv_{rms}} \approx 7 \text{ nm}$$

There is a range of wavelengths because the electrons have a range of speeds, but this is a typical value.

You've now seen many times that wave effects are significant only when the sizes of physical structures are comparable to or smaller than the wavelength. Because the de Broglie wavelength of conduction electrons is only a few nm, quantum effects are insignificant in electronic devices whose features are larger than about 100 nm. The electrons in macroscopic devices can be treated as classical particles, which is how we analyzed electric current in Chapter 30.

However, devices smaller than about 100 nm do exhibit quantum effects. Some semiconductor devices, such as the semiconductor lasers used in fiber-optic communications, now incorporate features only a few nm in size. Quantum effects play an important role in these devices.

FIGURE 40.16a shows a *semiconductor diode laser* through which a current travels from left to right. In the center is a very thin layer of the semiconductor gallium arsenide (GaAs). It is surrounded on either side by layers of gallium aluminum arsenide (GaAlAs), and these in turn are embedded within the larger structure of the diode. The electrons within the central GaAs layer begin to emit laser light when the current through the diode exceeds a *threshold current*. The laser beam diverges because of diffraction through the "slit" of the GaAs layer, with the wider axis of the laser beam corresponding to the narrower portion of the lasing region.

You can learn in a solid-state physics or materials engineering course that the electric potential energy of an electron is slightly lower in GaAs than in GaAlAs. This makes the GaAs layer a potential well for electrons, with higher-potential-energy GaAlAs "walls" on either side. As a result, the electrons become trapped within the thin GaAs layer. Such a device is called a **quantum-well laser.**

As an example, FIGURE 40.16b shows a quantum-well device with a 1.0-nm-thick GaAs layer in which the electron's potential energy is 0.300 eV lower than in the surrounding GaAlAs layers. A numerical solution of the Schrödinger equation finds that this potential well has only a *single* quantum state, $n = 1$ with $E_1 = 0.125$ eV. Every electron trapped in this quantum well has the *same* energy—a very nonclassical result! The fact that the electron energies are so well defined, in contrast to the range of electron energies in bulk material, is what makes this a useful device. You can also see from the probability density $|\psi|^2$ that the electrons are more likely to be found in the center of the layer than at the edges. This concentration of electrons makes it easier for the device to begin laser action.

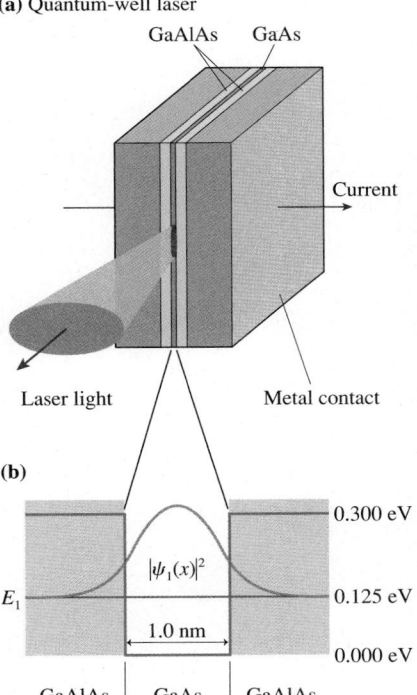

FIGURE 40.16 A semiconductor diode laser with a single quantum well.

(a) Quantum-well laser

GaAlAs GaAs

Current

Laser light Metal contact

(b)

E_1

$|\psi_1(x)|^2$

0.300 eV

0.125 eV

1.0 nm

0.000 eV

GaAlAs GaAs GaAlAs

Nuclear Physics

The nucleus of an atom consists of an incredibly dense assembly of protons and neutrons. The positively charged protons exert extremely strong electric repulsive forces on each other, so you might wonder how the nucleus keeps from exploding. During the 1930s, physicists found that protons and neutrons also exert an *attractive* force on each other. This force, one of the fundamental forces of nature, is called the *strong force*. It is the force that holds the nucleus together.

The primary characteristic of the strong force, other than its strength, is that it is a *short-range* force. The attractive strong force between two *nucleons* (a nucleon is either a proton or a neutron; the strong force does not distinguish between them) rapidly decreases to zero if they are separated by more than about 2 fm. This is in sharp contrast to the long-range nature of the electric force.

A reasonable model of the nucleus is to think of the protons and neutrons as particles in a nuclear potential well that is created by the strong force. The diameter of the

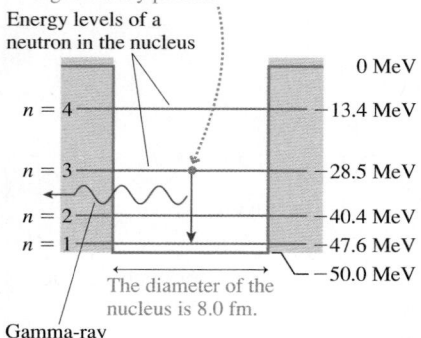

FIGURE 40.17 There are four allowed energy levels for a neutron in this nuclear potential well.

A radioactive decay has left the neutron in the $n = 3$ excited state. The neutron jumps to the $n = 1$ ground state, emitting a gamma-ray photon.

Energy levels of a neutron in the nucleus

0 MeV

$n = 4$ — 13.4 MeV

$n = 3$ — 28.5 MeV

$n = 2$ — 40.4 MeV
$n = 1$ — 47.6 MeV
— 50.0 MeV

The diameter of the nucleus is 8.0 fm.

Gamma-ray emission

potential well is equal to the diameter of the nucleus (this varies with atomic mass), and nuclear physics experiments have found that the depth of the potential well is ≈ 50 MeV.

The real potential well is three-dimensional, but let's make a simplified model of the nucleus as a one-dimensional potential well. FIGURE 40.17 shows the potential energy of a neutron along an x-axis passing through the center of the nucleus. Notice that the zero of energy has been chosen such that a "free" neutron, one outside the nucleus, has $E = 0$. Thus the potential energy inside the nucleus is -50 MeV. The 8.0 fm diameter shown is appropriate for a nucleus having atomic mass number $A \approx 40$, such as argon or potassium. Lighter nuclei will be a little smaller, heavier nuclei somewhat larger. (The potential-energy diagram for a proton is similar, but is complicated a bit by the electric potential energy.)

A numerical solution of the Schrödinger equation finds the four stationary states shown in Figure 40.17. The wave functions have been omitted, but they look essentially identical to the wave functions in Figure 40.14a. The major point to note is that the allowed energies differ by several *million* electron volts! These are enormous energies compared to those of an electron in an atom or a semiconductor. But recall that the energies of a particle in a rigid box, $E_n = n^2h^2/8mL^2$, are proportional to $1/L^2$. Our previous examples, with nanometer-size boxes, found energies in the eV range. When the box size is reduced to femtometers, the energies jump up into the MeV range.

It often happens that the nuclear decay of a radioactive atom leaves a neutron in an excited state. For example, Figure 40.17 shows a neutron that has been left in the $n = 3$ state by a previous radioactive decay. This neutron can now undergo a quantum jump to the $n = 1$ ground state by emitting a photon with energy

$$E_{\text{photon}} = E_3 - E_1 = 19.1 \text{ MeV}$$

and wavelength

$$\lambda_{\text{photon}} = \frac{c}{f} = \frac{hc}{E_{\text{photon}}} = 6.50 \times 10^{-5} \text{ nm}$$

This photon is $\approx 10^7$ times more energetic, and its wavelength $\approx 10^7$ times smaller, than the photons of visible light! These extremely high-energy photons are called **gamma rays**. Gamma-ray emission is, indeed, one of the primary processes in the decay of radioactive elements.

Our one-dimensional model cannot be expected to give accurate results for the energy levels or gamma-ray energies of any specific nucleus. Nonetheless, this model does provide a reasonable understanding of the energy-level structure in nuclei and correctly predicts that nuclei can emit photons having energies of several million electron volts. This model, when extended to three dimensions, becomes the basis for the *shell model* of the nucleus in which the protons and neutrons are grouped in various shells analogous to the electron shells around an atom that you remember from chemistry. You can learn more about nuclear physics and the shell model in Chapter 42.

40.7 Wave-Function Shapes

Bound-state wave functions are standing de Broglie waves. In addition to boundary conditions, two other factors govern the shapes of wave functions:

1. The de Broglie wavelength is inversely dependent on the particle's speed. Consequently, the node spacing is smaller (shorter wavelength) where the kinetic energy is larger, and the spacing is larger (longer wavelength) where the kinetic energy is smaller.

2. A classical particle is more likely to be found where it is moving more slowly. In quantum mechanics, the probability of finding the particle increases as the wave-function amplitude increases. Consequently, the wave-function amplitude is larger where the kinetic energy is smaller, and it is smaller where the kinetic energy is larger.

We can use this information to draw reasonably accurate wave functions for the different allowed energies in a potential-energy well.

TACTICS
BOX 40.1 **Drawing wave functions**

❶ Draw a graph of the potential energy $U(x)$. Show the allowed energy E as a horizontal line. Locate the classical turning points.

❷ Draw the wave function as a continuous, oscillatory function between the turning points. The wave function for quantum state n has n antinodes and $(n - 1)$ nodes (excluding the ends).

❸ Make the wavelength longer (larger node spacing) and the amplitude higher in regions where the kinetic energy is smaller. Make the wavelength shorter and the amplitude lower in regions where the kinetic energy is larger.

❹ Bring the wave function to zero at the edge of an infinitely high potential-energy "wall."

❺ Let the wave function decay exponentially inside a classically forbidden region where $E < U$. The penetration distance η increases as E gets closer to the top of the potential-energy well.

Exercises 10–13

EXAMPLE 40.8 **Sketching wave functions**

FIGURE 40.18 shows a potential-energy well and the allowed energies for the $n = 1$ and $n = 4$ quantum states. Sketch the $n = 1$ and $n = 4$ wave functions.

VISUALIZE The steps of Tactics Box 40.1 have been followed to sketch the wave functions shown in FIGURE 40.19.

FIGURE 40.18 A potential-energy well.

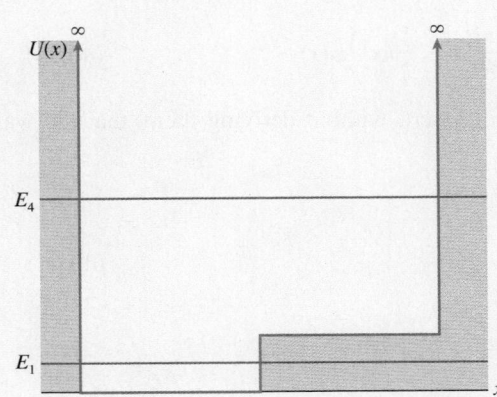

FIGURE 40.19 The $n = 1$ and $n = 4$ wave functions.

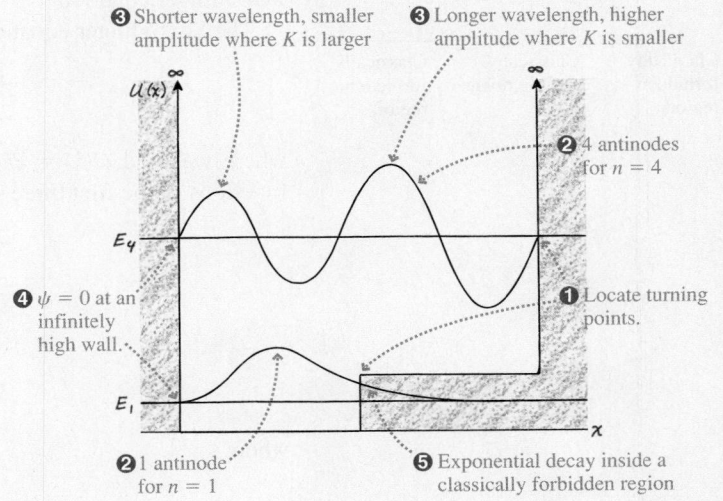

STOP TO THINK 40.4 For which potential energy $U(x)$ is this an appropriate $n = 4$ wave function?

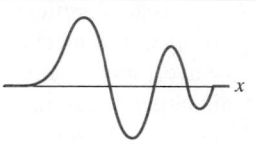

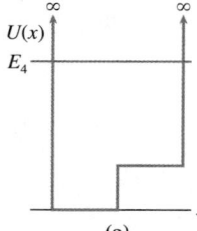

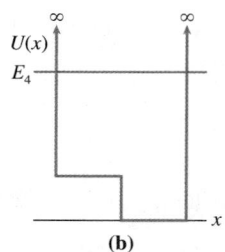

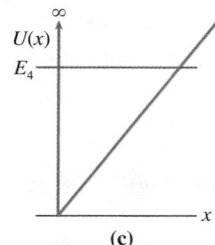

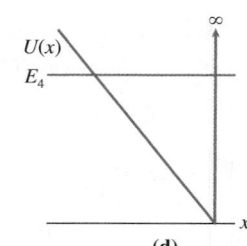

(a)　　　　　　(b)　　　　　　(c)　　　　　　(d)

40.8 The Quantum Harmonic Oscillator

Simple harmonic motion is exceptionally important in classical physics, where it serves as a prototype for more complex oscillations. As you might expect, a microscopic oscillator—the **quantum harmonic oscillator**—is equally important as a model of oscillations at the atomic level.

The defining characteristic of simple harmonic motion is a linear restoring force: $F = -kx$, where k is the spring constant. The corresponding potential-energy function, as you learned in Chapter 10, is

$$U(x) = \frac{1}{2}kx^2 \tag{40.42}$$

where we'll assume that the equilibrium position is $x_e = 0$. The potential energy of a harmonic oscillator is shown in FIGURE 40.20. It is a potential-energy well with curved sides.

A classical particle of mass m oscillates with angular frequency

$$\omega = \sqrt{\frac{k}{m}} \tag{40.43}$$

between the two turning points where the energy line crosses the parabolic potential-energy curve. As you've learned, this classical description fails if m represents an atomic particle, such as an electron or an atom. In that case, we need to solve the Schrödinger equation to find the wave functions.

The Schrödinger equation for a quantum harmonic oscillator is

$$\frac{d^2\psi}{dx^2} = -\frac{2m}{\hbar^2}\left(E - \frac{1}{2}kx^2\right)\psi(x) \tag{40.44}$$

where we used $U(x) = \frac{1}{2}kx^2$. We will assert, without deriving them, that the wave functions of the first three states are

$$\psi_1(x) = A_1 e^{-x^2/2b^2}$$

$$\psi_2(x) = A_2\frac{x}{b}e^{-x^2/2b^2} \tag{40.45}$$

$$\psi_3(x) = A_3\left(1 - \frac{2x^2}{b^2}\right)e^{-x^2/2b^2}$$

where

$$b = \sqrt{\frac{\hbar}{m\omega}} \tag{40.46}$$

FIGURE 40.20 The potential energy of a harmonic oscillator.

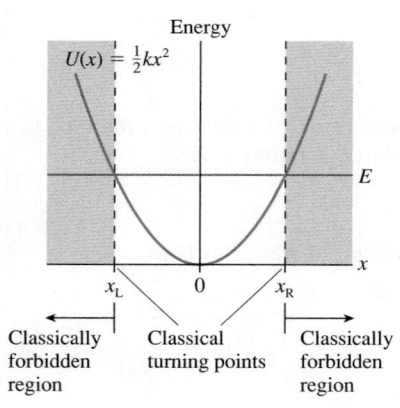

Classically forbidden region · Classical turning points · Classically forbidden region

The constant b has dimensions of length. We will leave it as a homework problem for you to show that b is the classical turning point of an oscillator in the $n = 1$ ground state. The constants A_1, A_2, and A_3 are normalization constants. For example, A_1 can be found by requiring

$$\int_{-\infty}^{\infty} |\psi_1(x)|^2 \, dx = A_1^2 \int_{-\infty}^{\infty} e^{-x^2/b^2} \, dx = 1 \qquad (40.47)$$

The completion of this calculation also will be left as a homework problem.

As expected, stationary states of a quantum harmonic oscillator exist only for certain discrete energy levels, the quantum states of the oscillator. The allowed energies are given by the simple equation

$$E_n = \left(n - \frac{1}{2}\right)\hbar\omega \qquad n = 1, 2, 3, \ldots \qquad (40.48)$$

where ω is the classical angular frequency, Equation 40.43, and n is the quantum number.

NOTE ▶ The ground-state energy of the quantum harmonic oscillator is $E_1 = \frac{1}{2}\hbar\omega$. An atomic mass on a spring can *not* be brought to rest. This is a consequence of the uncertainty principle. ◀

FIGURE 40.21 shows the first three energy levels and wave functions of a quantum harmonic oscillator. Notice that the energy levels are equally spaced by $\Delta E = \hbar\omega$. This result differs from the particle in a box, where the energy levels get increasingly farther apart. Also notice that the wave functions, like those of the finite potential well, extend beyond the turning points into the classically forbidden region.

FIGURE 40.21 The first three energy levels and wave functions of a quantum harmonic oscillator.

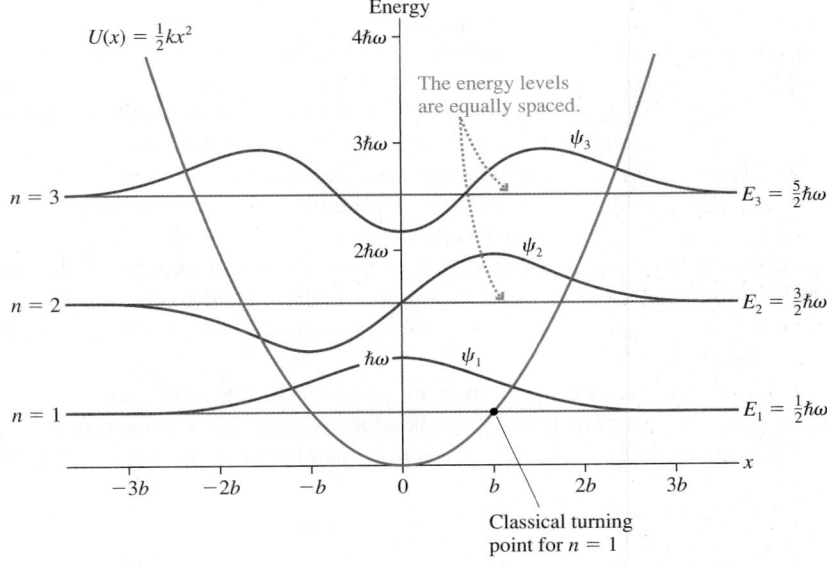

FIGURE 40.22 shows the probability density $|\psi(x)|^2$ for the $n = 11$ state of a quantum harmonic oscillator. Notice how the node spacing and the amplitude both increase as the particle moves away from the equilibrium position at $x = 0$. This is consistent with item 3 of Tactics Box 40.1. The particle slows down as it moves away from the origin, causing its de Broglie wavelength *and* the probability of finding it to increase.

Section 40.5 introduced the classical probability density $P_{class}(x)$ and noted that a classical particle is most likely to be found where it is moving the slowest. Figure 40.22

FIGURE 40.22 The quantum and classical probability densities for the $n = 11$ state of a quantum harmonic oscillator.

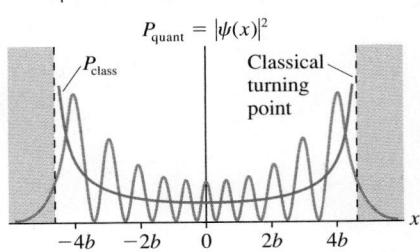

shows $P_{class}(x)$ for a classical particle with the same total energy as the $n = 11$ quantum state. You can see that *on average* the quantum probability density $|\psi(x)|^2$ mimics the classical probability density. This is just what the correspondence principle leads us to expect.

EXAMPLE 40.9 | **Light emission by an oscillating electron**

An electron in a harmonic-oscillator potential well emits light of wavelength 600 nm as it jumps from one level to the next lowest level. What is the "spring constant" of the restoring force?

MODEL The electron is a quantum harmonic oscillator.

SOLVE A photon is emitted as the electron undergoes the quantum jump $n \rightarrow n - 1$. We can use Equation 40.48 for the energy levels to find that the electron loses energy

$$\Delta E = E_n - E_{n-1} = \left(n - \frac{1}{2}\right)\hbar\omega_e - \left(n - 1 - \frac{1}{2}\right)\hbar\omega_e = \hbar\omega_e$$

$\Delta E = \hbar\omega_e$ for *all* transitions, independent of n, because the energy levels of the quantum harmonic oscillator are equally spaced. We need to distinguish the harmonic oscillations of the electron from the oscillations of the light wave, hence the subscript e on ω_e.

The emitted photon has energy $E_{photon} = hf_{ph} = \Delta E$. Thus

$$\hbar\omega_e = \frac{h}{2\pi}\omega_e = hf_{ph} = \frac{hc}{\lambda}$$

The wavelength of the light is $\lambda = 600$ nm, so the classical angular frequency of the oscillating electron is

$$\omega_e = 2\pi\frac{c}{\lambda} = 3.14 \times 10^{15} \text{ rad/s}$$

The electron's angular frequency is related to the spring constant of the restoring force by

$$\omega_e = \sqrt{\frac{k}{m}}$$

Thus $k = m\omega_e^2 = 9.0$ N/m.

Molecular Vibrations

We've made many uses of the idea that atoms are held together by spring-like molecular bonds. We've always assumed that the bonds could be modeled as classical springs. The classical model is acceptable for some purposes, but it fails to explain some important features of molecular vibrations. Not surprisingly, the quantum harmonic oscillator is a better model of a molecular bond.

FIGURE 40.23 shows the potential energy of two atoms connected by a molecular bond. Nearby atoms attract each other through a polarization force, much as a charged rod picks up small pieces of paper. If the atoms get too close, a *repulsive* force between the negative electrons pushes them apart. The equilibrium separation at which the attractive and repulsive forces are balanced is r_0, and two classical atoms would be at rest at this separation. But quantum particles, even in their lowest energy state, have $E > 0$. Consequently, the molecule *vibrates* as the two atoms oscillate back and forth along the bond.

U_{dissoc} is the energy at which the molecule will *dissociate* and the two atoms will fly apart. Dissociation can occur at very high temperatures or after the molecule has absorbed a high-energy (ultraviolet) photon, but under typical conditions a molecule has energy $E \ll U_{dissoc}$. In other words, the molecule is in an energy level near the bottom of the potential well.

You can see that the lower portion of the potential well is very nearly a parabola. Consequently, we can model a molecular bond as a quantum harmonic oscillator. The energy associated with the molecular vibration is quantized and can have *only* the values

$$E_{vib} \approx \left(n - \frac{1}{2}\right)\hbar\omega \qquad n = 1, 2, 3, \ldots \tag{40.49}$$

where ω is the angular frequency with which the atoms would vibrate if the bond were a classical spring. The molecular potential-energy curve is not exactly that of a harmonic oscillator, hence the $\approx$ sign, but the model is very good for low values of the quantum number n. The energy levels calculated by Equation 40.49 are called the **vibrational energy levels** of the molecule. The first few vibrational energy levels are shown in Figure 40.23.

At room temperature, most molecules are in the $n = 1$ vibrational ground state. Their vibrational motion can be excited by absorbing photons of frequency $f = \Delta E/h$.

FIGURE 40.23 The potential energy of a molecular bond and a few of the allowed energies.

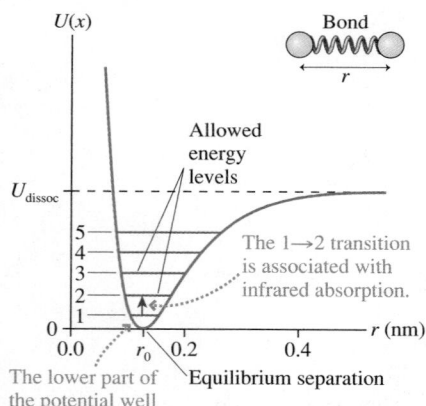

The lower part of the potential well is nearly a parabola.

This frequency is usually in the infrared region of the spectrum, and these *vibrational transitions* give each molecule a unique and distinctive infrared absorption spectrum.

As an example, FIGURE 40.24 shows the infrared absorption spectrum of acetone. The vertical axis is the percentage of the light intensity passing all the way through the sample. The sample is essentially transparent at most wavelengths (transmission ≈ 100%), but there are two prominent absorption features. The transmission drops to ≈75% at $\lambda = 3.3\ \mu$m and to a mere 7% at $\lambda = 5.8\ \mu$m. The 3.3 μm absorption is due to the $n = 1$ to $n = 2$ transition in the vibration of a C–CH₃ carbon-methyl bond. The 5.8 μm absorption is the $1 \rightarrow 2$ transition of a vibrating C=O carbon-oxygen double bond.

Absorption spectra are known for thousands of molecules, and chemists routinely use absorption spectroscopy to identify the chemicals in a sample. A specific bond has the same absorption wavelength regardless of the larger molecule in which it is embedded; thus the presence of that absorption wavelength is a "signature" that the bond is present within a molecule.

FIGURE 40.24 The absorption spectrum of acetone.

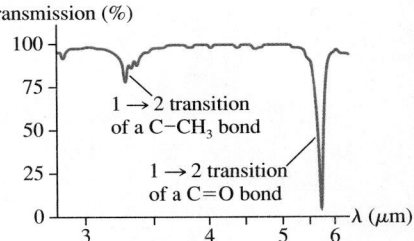

STOP TO THINK 40.5 Which probability density represents a quantum harmonic oscillator with $E = \frac{5}{2}\hbar\omega$?

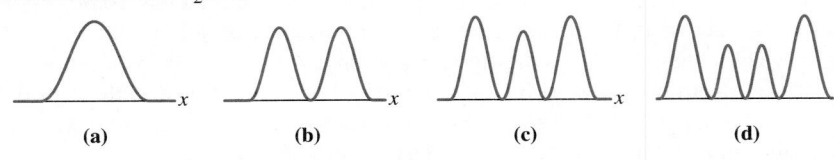

(a)   (b)   (c)   (d)

40.9 More Quantum Models

In this section we'll look at two more examples of quantum-mechanical models.

A Particle in a Capacitor

Many semiconductor devices are designed to confine electrons within a layer only a few nanometers thick. If a potential difference is applied across the layer, the electrons act very much as if they are trapped within a microscopic capacitor.

FIGURE 40.25a shows two capacitor plates separated by distance L. The left plate is positive, so the electric field points to the right with strength $E = \Delta V_0/L$. Because of its negative charge, an electron launched from the left plate is slowed by a *retarding* force. The electron makes it across to the right plate if it starts with sufficient kinetic energy; otherwise, it reaches a turning point and then is pushed back toward the positive plate.

This classical analysis is a valid model of a macroscopic capacitor. But if L becomes sufficiently small, comparable to the de Broglie wavelength of an electron, then the wave-like properties of the electron cannot be ignored. We need a quantum-mechanical model.

Let's establish a coordinate system with $x = 0$ at the left plate and $x = L$ at the right plate. We define the electric potential to be zero at the positive plate. The potential *decreases* in the direction of the field, so the potential inside the capacitor (see Section 28.5) is

$$V(x) = -Ex = -\frac{\Delta V_0}{L}x$$

The electron, with charge $q = -e$, has potential energy

$$U(x) = qV(x) = +\frac{e\,\Delta V_0}{L}x \qquad 0 < x < L \qquad (40.50)$$

This potential energy increases linearly for $0 < x < L$. If we assume that the capacitor plates act like the walls of a rigid box, then $U(x) \rightarrow \infty$ at $x = 0$ and $x = L$.

FIGURE 40.25 An electron in a capacitor.

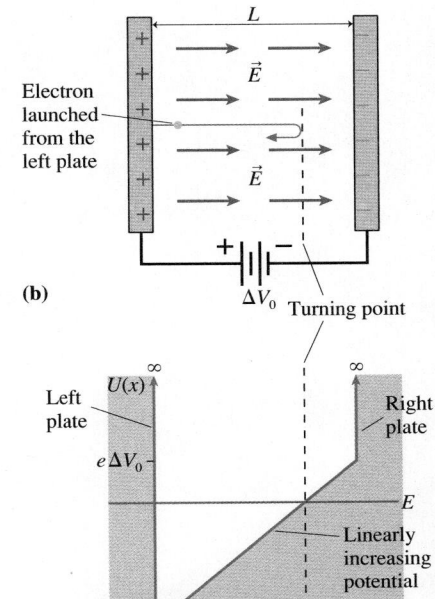

FIGURE 40.25b shows the electron's potential-energy function. It is the particle-in-a-rigid-box potential with a sloping "floor" due to the electric field. The figure also shows the total energy line E of an electron in the capacitor. The energy is purely kinetic at $x = 0$, where $K = E$, but it is converted to potential energy as the electron moves to the right. The right turning point occurs where the energy line E crosses the potential-energy curve $U(x)$. If the electron is a classical particle, it must reverse direction at this point.

NOTE ▶ This is also the shape of the potential energy for a microscopic bouncing ball that is trapped between a floor at $y = 0$ and a ceiling at $y = L$. ◀

It is physically impossible for the electron to be outside the capacitor, so the wave function must be zero for $x < 0$ and $x > L$. The continuity of ψ requires the same boundary conditions as for a particle in a rigid box: $\psi = 0$ at $x = 0$ and at $x = L$. The wave functions inside the capacitor are too complicated to find by guessing, so we have solved the Schrödinger equation numerically and will present the results graphically.

FIGURE 40.26 shows the wave functions and probability densities for the first five quantum states of an electron confined in a 5.0-nm-thick layer that has a 0.80 V potential difference across it. Each allowed energy is represented as a horizontal line, with the numerical values shown on the right. They range from $E_1 = 0.23$ eV up to $E_5 = 0.81$ eV. An electron *must* have one of the allowed energies shown in the figure. An electron cannot have $E = 0.30$ eV in this capacitor because no de Broglie wave with that energy can match the necessary boundary conditions.

FIGURE 40.26 Energy levels, wave functions, and probability densities for an electron in a 5.0-nm-wide capacitor with a 0.80 V potential difference.

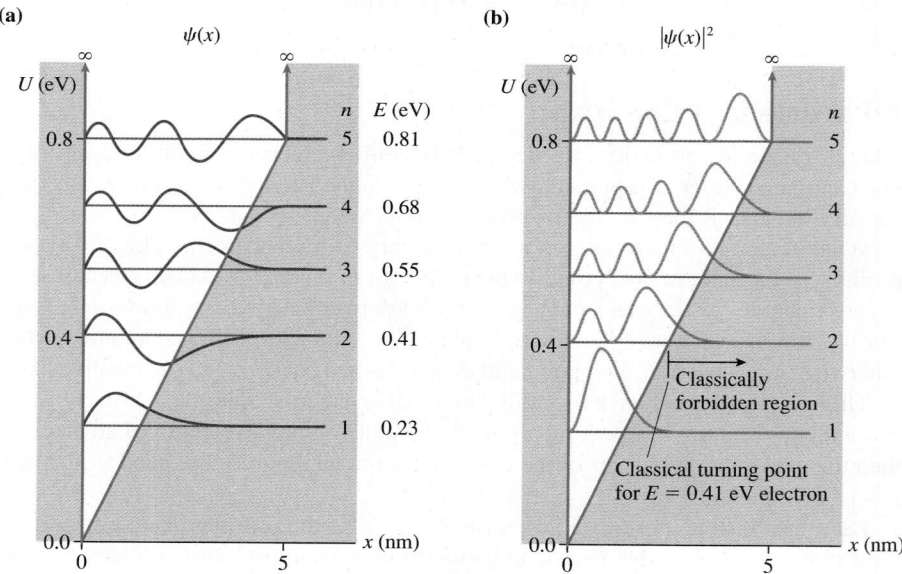

NOTE ▶ Remember that each wave function and probability density is graphed as if its energy line is the zero of the y-axis. ◀

We can make some observations about the Schrödinger equation solutions:

1. The energies E_n become more closely spaced as n increases, at least to $n = 5$. This contrasts with the particle in a box, for which E_n became more widely spaced.
2. The spacing between the nodes of a wave function is not constant but increases toward the right. This is because an electron on the right side of the capacitor has less kinetic energy and thus a slower speed and a larger de Broglie wavelength.

3. The height of the probability density $|\psi|^2$ increases toward the right. That is, we are more likely to find the electron on the right side of the capacitor than on the left. But this also makes sense if, classically, the electron is moving more slowly when on the right side and thus spending more time there than on the left side.

4. The electron penetrates *beyond* the classical turning point into the classically forbidden region.

EXAMPLE 40.10 **The emission spectrum of an electron in a capacitor**

What are the wavelengths of photons emitted by electrons in the $n = 4$ state of Figure 40.26?

SOLVE Photon emission occurs as the electrons make $4 \rightarrow 3$, $4 \rightarrow 2$, and $4 \rightarrow 1$ quantum jumps. In each case, the photon frequency is $f = \Delta E/h$ and the wavelength is

$$\lambda = \frac{c}{f} = \frac{hc}{\Delta E}$$

The energies of the quantum jumps, which can be read from Figure 40.26a, are $\Delta E_{4 \rightarrow 3} = 0.13$ eV, $\Delta E_{4 \rightarrow 2} = 0.27$ eV, and

$\Delta E_{4 \rightarrow 1} = 0.45$ eV. Thus

$$\lambda_{4 \rightarrow 3} = 9500 \text{ nm} = 9.5 \text{ } \mu\text{m}$$

$$\lambda_{4 \rightarrow 2} = 4600 \text{ nm} = 4.6 \text{ } \mu\text{m}$$

$$\lambda_{4 \rightarrow 1} = 2800 \text{ nm} = 2.8 \text{ } \mu\text{m}$$

ASSESS The $n = 4$ electrons in this device emit three distinct infrared wavelengths.

The Covalent Bond

You probably recall from chemistry that a **covalent molecular bond,** such as the bond between the two atoms in molecules such as H_2 and O_2, is a bond in which the electrons are shared between the atoms. The basic idea of covalent bonding can be understood with a one-dimensional quantum-mechanical model.

The simplest molecule, the hydrogen molecular ion H_2^+, consists of two protons and one electron. Although it seems surprising that such a system could be stable, the two protons form a molecular bond with one electron. This is the simplest covalent bond.

How can we model the H_2^+ ion? To begin, **FIGURE 40.27a** shows a one-dimensional model of a hydrogen *atom* in which the electron's Coulomb potential energy, with its $1/r$ dependence, has been approximated by a finite potential well of width 0.10 nm ($\approx 2a_B$) and depth 24.2 eV. You learned in Chapter 38 that an electron in the ground state of the Bohr hydrogen atom orbits the proton with radius $r_1 = a_B$ (the Bohr radius) and energy $E_1 = -13.6$ eV. A numerical solution of the Schrödinger equation finds that the ground-state energy of this finite potential well is $E_1 = -13.6$ eV. This model of a hydrogen atom is very oversimplified, but it does have the correct size and ground-state energy.

We can model H_2^+ by bringing two of these potential wells close together. The molecular bond length of H_2^+ is known to be ≈ 0.12 nm, so **FIGURE 40.27b** shows potential wells with 0.12 nm between their centers. This is a model of H_2^+, not a complete H_2 molecule, because this is the potential energy of a single electron. (Modeling H_2 is more complex because we would need to consider the repulsion between the two electrons.)

FIGURE 40.28 on the next page shows the allowed energies, wave functions, and probability densities for an electron with this potential energy. The $n = 1$ wave function has a high probability of being found within the classically forbidden region *between* the two protons. In other words, an electron in this quantum state really is "shared" by the protons and spends most of its time between them.

In contrast, an electron in the $n = 2$ energy level has zero probability of being found between the two protons because the $n = 2$ wave function has a node at the center. The probability density shows that an $n = 2$ electron is "owned" by one proton or the other rather than being shared.

To learn the consequences of these wave functions we need to calculate the total energy of the molecule: $E_{\text{mol}} = E_{\text{p-p}} + E_{\text{elec}}$. The $n = 1$ and $n = 2$ energies shown in Figure 40.28 are the energies E_{elec} of the electron. At the same time, the protons

FIGURE 40.27 A molecule can be modeled as two closely spaced potential wells, one representing each atom.

(a) Simple one-dimensional model of a hydrogen atom

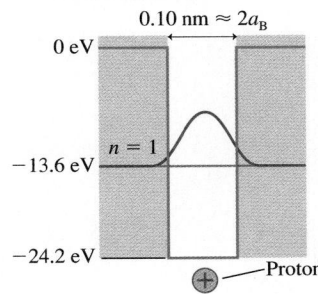

(b) An H_2^+ molecule modeled as an electron with two protons separated by 0.12 nm

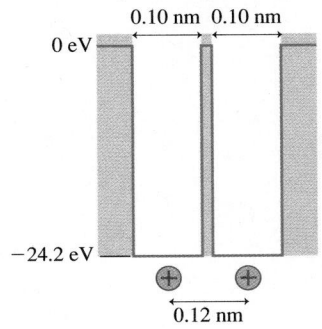

FIGURE 40.28 The wave functions and probability densities of the electron in H_2^+.

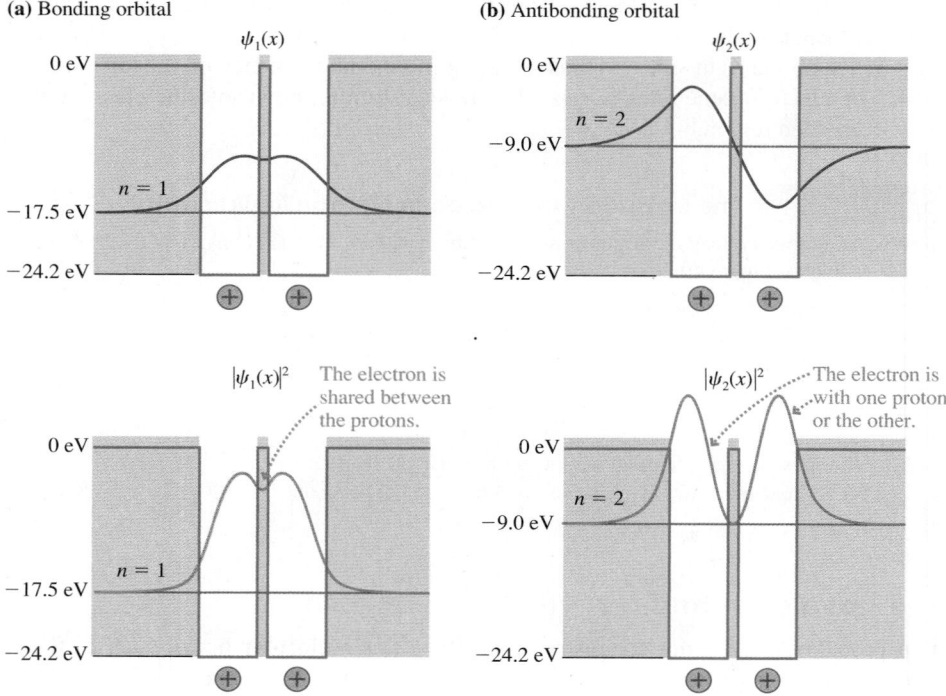

repel each other and have electric potential energy E_{p-p}. It's not hard to calculate that $E_{p-p} = 12.0$ eV for two protons separated by 0.12 nm. Thus

$$E_{mol} = E_{p-p} + E_{elec} = \begin{cases} 12.0 \text{ eV} - 17.5 \text{ eV} = -5.5 \text{ eV} & n = 1 \\ 12.0 \text{ eV} - 9.0 \text{ eV} = +3.0 \text{ eV} & n = 2 \end{cases}$$

The $n = 1$ molecular energy is less than zero, showing that this is a *bound state*. The $n = 1$ wave function is called a **bonding molecular orbital.** Although the protons repel each other, the shared electron provides sufficient "glue" to hold the system together. The $n = 2$ molecular energy is positive, so this is *not* a bound state. The system would be more stable as a hydrogen atom and a distant proton. The $n = 2$ wave function is called an **antibonding molecular orbital.**

Both E_{elec} and E_{p-p} depend on the separation between the protons, which we assumed to be 0.12 nm in this calculation. If we were to calculate and graph E_{mol} for many different values of the proton separation, the graph would look like the molecular-bond energy curve shown in Figure 40.23. In other words, a molecular bond has an equilibrium length where the bond energy is a minimum *because* of the interplay between E_{p-p} and E_{elec}.

Although real molecular wave functions are more complex than this one-dimensional model, the $n = 1$ wave function captures the essential idea of a covalent bond. Notice that a "classical" molecule cannot have a covalent bond because the electron would not be able to exist in the classically forbidden region. Covalent bonds can be understood only within the context of quantum mechanics. In fact, the explanation of molecular bonds was one of the earliest successes of quantum mechanics.

40.10 Quantum-Mechanical Tunneling

FIGURE 40.29a shows a ball rolling toward a hill. A ball with sufficient kinetic energy can go over the top of the hill, slowing down as it ascends and speeding up as it rolls down the other side. A ball with insufficient energy rolls partway up the hill, then reverses direction and rolls back down.

FIGURE 40.29 A hill is an energy barrier to a rolling ball.

(a)

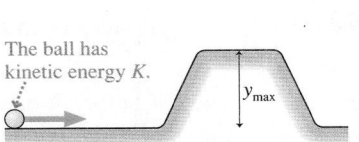

The ball has kinetic energy K.

y_{max}

(b)

$U(x)$

A ball with this energy slows down while going over the hill, but it makes it over.

$E > U_0$

U_0

$E < U_0$

0

A ball with this energy reverses direction at the turning point.

$U_0 = mgy_{max}$

Turning point

x

We can think of the hill as an "energy barrier" of height $U_0 = mgy_{max}$. As **FIGURE 40.29b** shows, a ball incident from the left with energy $E > U_0$ can go over the barrier (i.e., roll over the hill), but a ball with $E < U_0$ will reflect from the energy barrier at the turning point. According to the laws of classical physics, a ball that is incident on the energy barrier from the left with $E < U_0$ will never be found on the right side of the barrier.

> NOTE ▶ Figure 40.29b is not a "picture" of the energy barrier. And when we say that a ball with energy $E > U_0$ can go "over" the barrier, we don't mean that the ball is thrown from a higher elevation in order to go over the top of the hill. The ball rolls *on the ground* the entire time, as Figure 40.29a shows, and Figure 40.29b describes the kinetic and potential energy of the ball as it rolls. A higher total energy line means a larger initial kinetic energy, not a higher elevation. ◀

FIGURE 40.30 shows the situation from the perspective of quantum mechanics. As you've learned, quantum particles can penetrate with an exponentially decreasing wave function into the classically forbidden region of an energy barrier. Suppose that the barrier is very narrow. Although the wave function decreases within the barrier, starting at the classical turning point, it hasn't vanished when it reaches the other side. In other words, there is some probability that a quantum particle will pass *through* the barrier and emerge on the other side!

It is very much as if the ball of Figure 40.29a gets to the turning point and then, instead of reversing direction and rolling back down, tunnels its way *through* the hill and emerges on the other side. Although this feat is strictly forbidden in classical mechanics, it is apparently acceptable behavior for quantum particles. The process is called **quantum-mechanical tunneling.**

The process of tunneling through a potential-energy barrier is one of the strangest and most unexpected predictions of quantum mechanics. Yet it does happen, and you will see that it even has many practical applications.

> NOTE ▶ The word "tunneling" is used as a metaphor. If a classical particle really did tunnel, it would expend energy doing so and emerge on the other side with less energy. Quantum-mechanical tunneling requires no expenditure of energy. The total energy line is at the same height on both sides of the barrier. A particle that tunnels through a barrier emerges with *no* loss of energy. That is why the de Broglie wavelength is the same on both sides of the potential barrier in Figure 40.30. ◀

To simplify our analysis of tunneling, **FIGURE 40.31** shows an idealized energy barrier of height U_0 and width w. We've superimposed the wave function on top of the energy diagram so that you can see how it aligns with the potential energy. The wave function to the left of the barrier is a sinusoidal oscillation with amplitude A_L. The wave function *within* the barrier is the decaying exponential we found in Equation 40.40:

$$\psi_{in}(0 \leq x \leq w) = \psi_{edge}e^{-x/\eta} = A_L e^{-x/\eta} \tag{40.51}$$

FIGURE 40.30 A quantum particle can penetrate through the energy barrier.

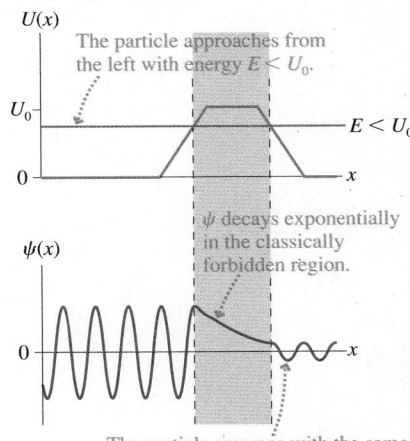

$U(x)$

The particle approaches from the left with energy $E < U_0$.

U_0

$E < U_0$

0

x

ψ decays exponentially in the classically forbidden region.

$\psi(x)$

0

x

The particle emerges with the same de Broglie wavelength after tunneling through the energy barrier.

FIGURE 40.31 Tunneling through an idealized energy barrier.

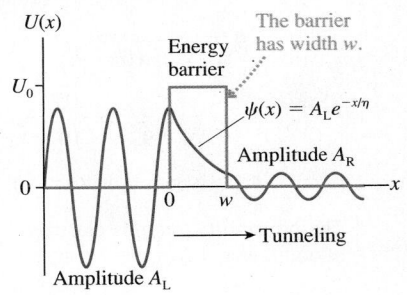

$U(x)$

Energy barrier

The barrier has width w.

U_0

$\psi(x) = A_L e^{-x/\eta}$

Amplitude A_R

0

w

x

Tunneling

Amplitude A_L

where we've assumed $\psi_{edge} = A_L$. The penetration distance η was given in Equation 40.41 as

$$\eta = \frac{\hbar}{\sqrt{2m(U_0 - E)}}$$

NOTE ▶ You *must* use SI units when calculating values of η. Energies must be in J and $\hbar$ in J s. The penetration distance η has units of meters. ◀

The wave function decreases exponentially within the barrier, but before it can decay to zero, it emerges again on the right side ($x > w$) as an oscillation with amplitude

$$A_R = \psi_{in}(\text{at } x = w) = A_L e^{-w/\eta} \tag{40.52}$$

The probability that the particle is to the left of the barrier is proportional to $|A_L|^2$, and the probability of finding it to the right of the barrier is proportional to $|A_R|^2$. Thus the probability that a particle striking the barrier from the left will emerge on the right is

$$P_{tunnel} = \frac{|A_R|^2}{|A_L|^2} = (e^{-w/\eta})^2 = e^{-2w/\eta} \tag{40.53}$$

This is the probability that a particle will tunnel through the energy barrier.

Now, our analysis, we have to say, has not been terribly rigorous. For example, we assumed that the oscillatory wave functions on the left and the right were exactly at a maximum where they reached the barrier at $x = 0$ and $x = w$. There is no reason this has to be the case. We have taken other liberties, which experts will spot, but—fortunately—it really makes no difference. Our result, Equation 40.53, turns out to be perfectly adequate for most applications of tunneling.

Because the tunneling probability is an exponential function, it is *very* sensitive to the values of w and η. The tunneling probability can be substantially reduced by even a small increase in the thickness of the barrier. The parameter η, which measures how far the particle can penetrate into the barrier, depends both on the particle's mass and on $U_0 - E$. A particle with E only slightly less than U_0 will have a larger value of η and thus a larger tunneling probability than will an identical particle with less energy.

EXAMPLE 40.11 **Electron tunneling**

a. Find the probability that an electron will tunnel through a 1.0-nm-wide energy barrier if the electron's energy is 0.10 eV less than the height of the barrier.

b. Find the tunneling probability if the barrier in part a is widened to 3.0 nm.

c. Find the tunneling probability if the electron in part a is replaced by a proton with the same energy.

SOLVE a. An electron with energy 0.10 eV less than the height of the barrier has $U_0 - E = 0.10$ eV $= 1.60 \times 10^{-20}$ J. Thus its penetration distance is

$$\eta = \frac{\hbar}{\sqrt{2m(U_0 - E)}}$$

$$= \frac{1.05 \times 10^{-34} \text{ J s}}{\sqrt{2(9.11 \times 10^{-31} \text{ kg})(1.60 \times 10^{-20} \text{ J})}}$$

$$= 6.18 \times 10^{-10} \text{ m} = 0.618 \text{ nm}$$

The probability that this electron will tunnel through a barrier of width $w = 1.0$ nm is

$$P_{tunnel} = e^{-2w/\eta} = e^{-2(1.0 \text{ nm})/(0.618 \text{ nm})} = 0.039 = 3.9\%$$

b. Changing the width to $w = 3.0$ nm has no effect on η. The new tunneling probability is

$$P_{tunnel} = e^{-2w/\eta} = e^{-2(3.0 \text{ nm})/(0.618 \text{ nm})} = 6.0 \times 10^{-5}$$

$$= 0.006\%$$

Increasing the width by a factor of 3 decreases the tunneling probability by a factor of 660!

c. A proton is more massive than an electron. Thus a proton with $U_0 - E = 0.10$ eV has $\eta = 0.014$ nm. Its probability of tunneling through a 1.0-nm-wide barrier is

$$P_{tunnel} = e^{-2w/\eta} = e^{-2(1.0 \text{ nm})/(0.014 \text{ nm})} \approx 1 \times 10^{-64}$$

For practical purposes, the probability that a proton will tunnel through this barrier is zero.

ASSESS If the probability of a proton tunneling through a mere 1 nm is only 10^{-64}, you can see that a macroscopic object will "never" tunnel through a macroscopic distance!

Quantum-mechanical tunneling seems so obscure that it is hard to imagine practical applications. Surprisingly, it is the physics behind one of today's most important technical tools, as we describe in the next section.

The Scanning Tunneling Microscope

Diffraction limits the resolution of an optical microscope to objects no smaller than about a wavelength of light—roughly 500 nm. This is more than 1000 times the size of an atom, so there is no hope of resolving atoms or molecules via optical microscopy. Electron microscopes are similarly limited by the de Broglie wavelength of the electrons. Their resolution is much better than that of an optical microscope, but still not quite at the level of resolving individual atoms.

This situation changed dramatically in 1981 with the invention of the **scanning tunneling microscope,** or STM as it is usually called. The STM allowed scientists, for the first time, to "see" surfaces literally atom by atom. The atomic-resolution photos at the beginning of Chapter 39 and this chapter demonstrate the power of an STM. These pictures and many others you have likely seen (but may not have known where they came from) are stupendous, but how are they made?

FIGURE 40.32a shows how the scanning tunneling microscope works. A conducting probe with a *very* sharp tip, just a few atoms wide, is brought to within a few tenths of a nanometer of a surface. Preparing the tips and controlling the spacing are both difficult technical challenges, but scientists have learned how to do both. Once positioned, the probe can mechanically scan back and forth across the surface.

When we analyzed the photoelectric effect, you learned that electrons are bound inside metals by an amount of energy called the *work function* E_0. A typical work function is 4 or 5 eV. This is the energy that must be supplied—by a photon or otherwise—to remove an electron from the metal. In other words, the electron's energy in the metal is E_0 less than its energy outside the metal.

This fact is the basis for the potential-energy diagram of FIGURE 40.32b. The small air gap between the sample and the probe tip is a potential-energy barrier. The energy of an electron in the metal of the sample or the probe tip is lower than the energy of an electron in the air by ≈ 4 eV, the work function. The absorption of a photon with $E_{photon} > 4$ eV would lift the electron *over* the barrier, from the sample to the probe. This is just the photoelectric effect. Alternatively, electrons can tunnel *through* the barrier if it is sufficiently narrow. This creates a *tunneling current* from the sample into the probe.

In operation, the tunneling current is recorded as the probe tip scans across the surface. You saw above that the tunneling current is extremely sensitive to the barrier thickness. As the tip scans over the position of an atom, the gap decreases by ≈ 0.1 nm and the current increases. The gap is larger when the tip is between atoms, so the current drops. Today's STMs can sense changes in the gap of as little as 0.001 nm, or about 1% of an atomic diameter! The images you see are computer-generated from the current measurements at each position.

The STM has revolutionized the science and engineering of microscopic objects. STMs are now used to study everything from how surfaces corrode and oxidize, a topic of great practical importance in engineering, to how biological molecules are structured. Another example of quantum mechanics working for you!

FIGURE 40.32 A scanning tunneling microscope.

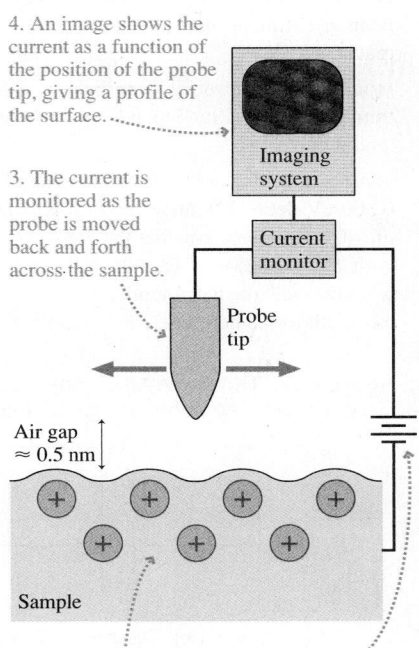

(a)

4. An image shows the current as a function of the position of the probe tip, giving a profile of the surface.

Imaging system

3. The current is monitored as the probe is moved back and forth across the sample.

Current monitor

Probe tip

Air gap ≈ 0.5 nm

Sample

1. The sample can be modeled as positive ion cores in an electron "sea."

2. The small positive voltage causes electrons to tunnel across the narrow air gap between the probe tip and the sample.

(b)

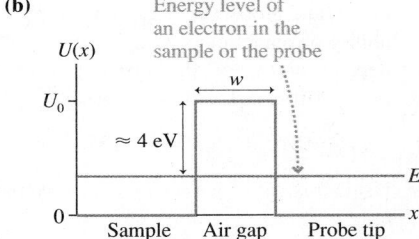

Energy level of an electron in the sample or the probe

$U(x)$

U_0

w

≈ 4 eV

E

0

Sample   Air gap   Probe tip

x

STOP TO THINK 40.6   A particle with energy E approaches an energy barrier with height $U_0 > E$. If U_0 is slowly decreased, the probability that the particle reflects from the barrier

a. Increases.
b. Decreases.
c. Does not change.

Tunneling in semiconductors

Quantum-mechanical tunneling can be important in semiconductors. Consider a 1.0-nm-thick layer of GaAs sandwiched between 4.0-nm-thick layers of GaAlAs. This is the situation explored in Figure 40.16, where we learned that the electron's potential energy is 0.300 eV lower in GaAs than in GaAlAs. An electron in the GaAs layer can tunnel through the GaAlAs to escape, but this doesn't happen instantly. In quantum mechanics, we can't predict exactly when tunneling will occur, only the probability of it happening. Estimate the time at which the probability of escape has reached 50%.

MODEL The electron is a particle in a finite potential well. The tunneling probability depends on the height and thickness of a potential barrier.

VISUALIZE **FIGURE 40.33** shows the potential well. An electron in a 0.300-eV-deep, 1.0-nm-wide well is exactly the situation of Figure 40.16, so we know that the electron has a single quantum state with $E_1 = 0.125$ eV. The wave function decreases exponentially with distance into the potential barriers, but a very tiny amplitude—too small to see here—still exists at the far edge of the barrier.

FIGURE 40.33 The potential energy of an electron in a layer of GaAs sandwiched between layers of GaAlAs.

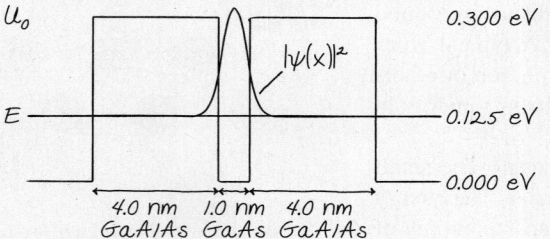

SOLVE We can approach this problem by thinking of the electron as a particle bouncing back and forth between the walls of the potential well. Each time it hits a wall, it has probability P_{tunnel} of tunneling and probability $P_{reflect} = 1 - P_{tunnel}$ of reflecting. The probability of tunneling is $P_{tunnel} = e^{-2w/\eta}$. The penetration distance η depends on $U_0 - E$, the "distance" from the energy level to the top of the barrier, which in this case is

$$U_0 - E = 0.300 \text{ eV} - 0.125 \text{ eV} = 0.175 \text{ eV} = 2.8 \times 10^{-20} \text{ J}$$

Using this value, we can calculate the penetration distance to be

$$\eta = \frac{\hbar}{\sqrt{2m(U_0 - E)}} = 0.465 \text{ nm}$$

We then find the probability of tunneling through a 4.0-nm-wide barrier to be

$$P_{tunnel} = e^{-2w/\eta} = 3.4 \times 10^{-8}$$

It's a very small probability, as expected. The probability of *not* tunneling, of reflecting back into the well, is then

$$P_{reflect} = 1 - P_{tunnel} = 0.999999966$$

You've seen that the probability of A *or* B happening is $P_A + P_B$. Similarly, the probability of A *and* B happening, assuming they are independent events, is $P_A \times P_B$. The probability of a head in a coin toss is $\frac{1}{2}$. If you toss two coins, the probability that A is a head *and* B is a head is $\frac{1}{2} \times \frac{1}{2} = \frac{1}{4}$. If you toss three coins, the probability that all three are heads is $\left(\frac{1}{2}\right)^3 = \frac{1}{8}$. If the electron is still in the potential well after N bounces, it had to reflect N times. The probability of this happening is

$$P_{in \text{ well}} = P_{reflect} \times P_{reflect} \times P_{reflect} \times \cdots \times P_{reflect} = (P_{reflect})^N$$

Because $P_{reflect} < 1$, the probability of still being in the potential well decreases as N increases.

We've focused not on P_{escape} but on $P_{in \text{ well}} = 1 - P_{escape}$ because staying in the well requires N specific events to happen. Escape, on the other hand, could have occurred on any of N attempts, so a direct calculation of P_{escape} is much more complicated. If the probability of escape is 50%, then it's also 50% probable that the electron is still in the potential well. We can find the number of reflections needed to get to the 50% probability by taking the logarithm of both sides of the equation:

$$\log(P_{in \text{ well}}) = \log\left((P_{reflect})^N\right) = N \log(P_{reflect})$$

$$N = \frac{\log(P_{in \text{ well}})}{\log(P_{reflect})} = \frac{\log(0.50)}{\log(0.999999966)} = 2.0 \times 10^7$$

After 20 million reflections, the electron is 50% likely to have escaped. Although that's a large number of reflections, it doesn't take long because the electron is moving only a very small distance between reflections at a fairly high speed. The electron's energy inside the potential well is entirely kinetic, $K = E = 0.125 \text{ eV} = 2.0 \times 10^{-20}$ J, so its speed is

$$v = \sqrt{\frac{2K}{m}} = 2.1 \times 10^5 \text{ m/s}$$

The time between reflections is the time needed to travel across the 1.0-nm-wide GaAs layer:

$$\Delta t = \frac{1.0 \times 10^{-9} \text{ m}}{2.1 \times 10^5 \text{ m/s}} = 4.8 \times 10^{-15} \text{ s}$$

Thus the time needed for 2.0×10^7 reflections is

$$t_{50\%} = N\Delta t = 9.6 \times 10^{-8} \text{ s} = 96 \text{ ns}$$

Because we're making only an estimate, we can say that an electron has a 50% probability of tunneling out of the GaAs layer within about 100 ns.

ASSESS Even though the tunneling probability is very tiny, tunneling takes place very rapidly on a human time scale. An increasing number of semiconductor devices make practical use of this *tunneling current*. Note that no energy is lost in the tunneling process; "tunneling" is a metaphor, not a process that requires work. The electron emerges with 0.125 eV of kinetic energy.

SUMMARY

The goal of Chapter 40 has been to understand and apply the essential ideas of quantum mechanics.

General Principles

The Schrödinger Equation (the law of psi)

$$\frac{d^2\psi}{dx^2} = -\frac{2m}{\hbar^2}\big[E - U(x)\big]\psi(x)$$

This equation determines the wave function $\psi(x)$ and, through $\psi(x)$, the probabilities of finding a particle of mass m with potential energy $U(x)$.

Boundary conditions

- $\psi(x)$ is a continuous function.
- $\psi(x) \to 0$ as $x \to \pm\infty$.
- $\psi(x) = 0$ in a region where it is physically impossible for the particle to be.
- $\psi(x)$ is normalized.

Shapes of wave functions

- The wave function oscillates in the region between the classical turning points.
- State n has n antinodes.
- Node spacing and amplitude increase as kinetic energy K decreases.
- $\psi(x)$ decays exponentially in a classically forbidden region.

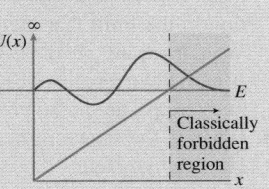

Quantum-mechanical models are characterized by the particle's potential-energy function $U(x)$.

- Wave-function solutions exist for only certain values of E. Thus energy is quantized.
- Photons are emitted or absorbed in quantum jumps.

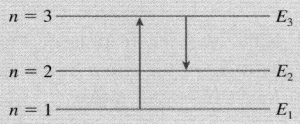

Important Concepts

Quantum-mechanical tunneling

A wave function can penetrate into a classically forbidden region with

$$\psi(x) = \psi_{\text{edge}}e^{-(x-L)/\eta}$$

where the **penetration distance** is

$$\eta = \frac{\hbar}{\sqrt{2m(U_0 - E)}}$$

The probability of tunneling through a barrier of width w is

$$P_{\text{tunnel}} = e^{-2w/\eta}$$

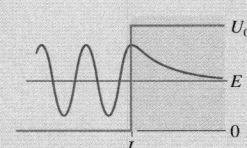

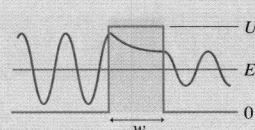

The **correspondence principle** says that the quantum world blends smoothly into the classical world for high quantum numbers. This is seen by comparing $|\psi(x)|^2$ to the classical probability density

$$P_{\text{class}} = \frac{2}{Tv(x)}$$

P_{class} expresses the idea that a classical particle is more likely to be found where it is moving slowly.

Applications

Particle in a rigid box: $\quad E_n = n^2\dfrac{h^2}{8mL^2} \qquad n = 1, 2, 3, \ldots$

Quantum harmonic oscillator: $E_n = (n - \tfrac{1}{2})\hbar\omega \qquad n = 1, 2, 3, \ldots$

Other applications were studied through numerical solution of the Schrödinger equation.

Terms and Notation

Schrödinger equation	bound state	bonding molecular orbital
quantum-mechanical model	penetration distance, η	antibonding molecular orbital
boundary conditions	quantum-well laser	quantum-mechanical tunneling
zero-point motion	gamma rays	scanning tunneling microscope (STM)
correspondence principle	quantum harmonic oscillator	
potential well	vibrational energy levels	
classically forbidden regions	covalent molecular bond	

CONCEPTUAL QUESTIONS

1. The correspondence principle says that the *average* behavior of a quantum system should begin to look like the Newtonian solution in the limit that the quantum number becomes very large. What is meant by "the *average* behavior" of a quantum system?
2. A particle in a potential well is in the $n = 5$ quantum state. How many peaks are in the probability density $P(x) = |\psi(x)|^2$?
3. What is the quantum number of the particle in FIGURE Q40.3? How can you tell?

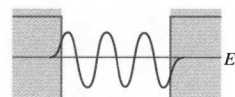

FIGURE Q40.3

4. Rank in order, from largest to smallest, the penetration distances η_a to η_c of the wave functions corresponding to the three energy levels in FIGURE Q40.4.

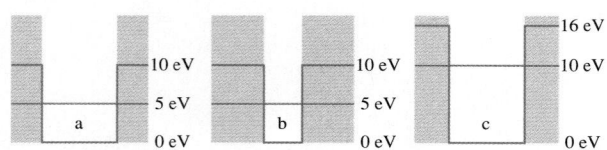

FIGURE Q40.4

5. Consider a quantum harmonic oscillator.
   a. What happens to the spacing between the nodes of the wave function as $|x|$ increases? Why?

b. What happens to the heights of the antinodes of the wave function as $|x|$ increases? Why?
   c. Sketch a reasonably accurate graph of the $n = 8$ wave function of a quantum harmonic oscillator.
6. FIGURE Q40.6 shows two possible wave functions for an electron in a linear triatomic molecule. Which of these is a bonding orbital and which is an antibonding orbital? Explain how you can distinguish them.

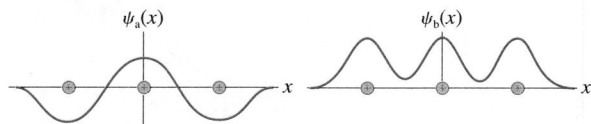

FIGURE Q40.6

7. Four quantum particles, each with energy E, approach the potential-energy barriers seen in FIGURE Q40.7 from the left. Rank in order, from largest to smallest, the tunneling probabilities $(P_{tunnel})_a$ to $(P_{tunnel})_d$.

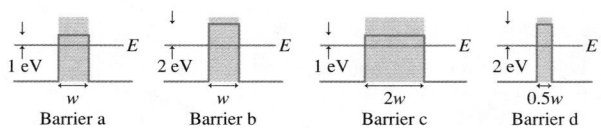

FIGURE Q40.7

EXERCISES AND PROBLEMS

Problems labeled ▨ integrate material from earlier chapters.

Exercises

Sections 40.3–40.4 A Particle in a Rigid Box

1. ‖ An electron in a rigid box absorbs light. The longest wavelength in the absorption spectrum is 600 nm. How long is the box?
2. | The electrons in a rigid box emit photons of wavelength 1484 nm during the $3 \rightarrow 2$ transition.
   a. What kind of photons are they—infrared, visible, or ultraviolet?
   b. How long is the box in which the electrons are confined?

3. ‖ FIGURE EX40.3 shows the wave function of an electron in a rigid box. The electron energy is 6.0 eV. How long is the box?

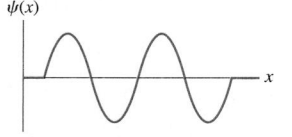

FIGURE EX40.3

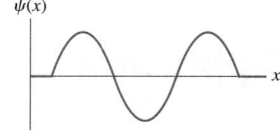

FIGURE EX40.4

4. | FIGURE EX40.4 shows the wave function of an electron in a rigid box. The electron energy is 12.0 eV. What is the energy of the electron's ground state?

Section 40.6 Finite Potential Wells

5. | Show that the penetration distance η has units of m.

6. | a. Sketch graphs of the probability density $|\psi(x)|^2$ for the four states in the finite potential well of Figure 40.14a. Stack them vertically, similar to the Figure 40.14a graphs of $\psi(x)$.

   b. What is the probability that a particle in the $n = 2$ state of the finite potential well will be found at the center of the well? Explain.

   c. Is your answer to part b consistent with what you know about waves? Explain.

7. | A finite potential well has depth $U_0 = 2.00$ eV. What is the penetration distance for an electron with energy (a) 0.50 eV, (b) 1.00 eV, and (c) 1.50 eV?

8. ‖ An electron in a finite potential well has a 1.0 nm penetration distance into the classically forbidden region. How far below U_0 is the electron's energy?

9. ‖ The energy of an electron in a 2.00-eV-deep potential well is 1.50 eV. At what distance into the classically forbidden region has the amplitude of the wave function decreased to 25% of its value at the edge of the potential well?

10. | A helium atom is in a finite potential well. The atom's energy is 1.0 eV below U_0. What is the atom's penetration distance into the classically forbidden region?

Section 40.7 Wave-Function Shapes

11. | Sketch the $n = 4$ wave function for the potential energy shown in FIGURE EX40.11.

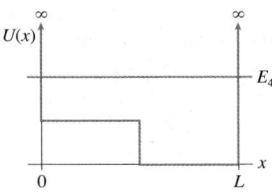

FIGURE EX40.11   FIGURE EX40.12

12. | Sketch the $n = 8$ wave function for the potential energy shown in FIGURE EX40.12.

13. | The graph in FIGURE EX40.13 shows the potential-energy function $U(x)$ of a particle. Solution of the Schrödinger equation finds that the $n = 3$ level has $E_3 = 0.5$ eV and that the $n = 6$ level has $E_6 = 2.0$ eV.

   a. Redraw this figure and add to it the energy lines for the $n = 3$ and $n = 6$ states.

   b. Sketch the $n = 3$ and $n = 6$ wave functions. Show them as oscillating about the appropriate energy line.

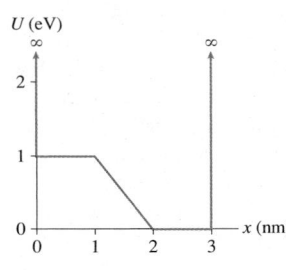

FIGURE EX40.13   FIGURE EX40.14

14. | Sketch the $n = 1$ and $n = 7$ wave functions for the potential energy shown in FIGURE EX40.14.

Section 40.8 The Quantum Harmonic Oscillator

15. | An electron in a harmonic potential well absorbs a photon with a wavelength of 400 nm as it undergoes a $1 \rightarrow 2$ quantum jump. What wavelength is absorbed in a $1 \rightarrow 3$ quantum jump?

16. | An electron is confined in a harmonic potential well that has a spring constant of 2.0 N/m.

   a. What are the first three energy levels of the electron?

   b. What wavelength photon is emitted if the electron undergoes a $3 \rightarrow 1$ quantum jump?

17. | An electron is confined in a harmonic potential well that has a spring constant of 12.0 N/m. What is the longest wavelength of light that the electron can absorb?

18. | An electron confined in a harmonic potential well emits a 1200 nm photon as it undergoes a $3 \rightarrow 2$ quantum jump. What is the spring constant of the potential well?

19. | Two adjacent energy levels of an electron in a harmonic potential well are known to be 2.0 eV and 2.8 eV. What is the spring constant of the potential well?

Section 40.10 Quantum-Mechanical Tunneling

20. ‖ What is the probability that an electron will tunnel through a 0.45 nm gap from a metal to a STM probe if the work function is 4.0 eV?

21. ‖ An electron approaches a 1.0-nm-wide potential-energy barrier of height 5.0 eV. What energy electron has a tunneling probability of (a) 10%, (b) 1.0%, and (c) 0.10%?

Problems

22. ‖ A 2.0-μm-diameter water droplet is moving with a speed of 1.0 μm/s in a 20-μm-long box.

   a. Estimate the particle's quantum number.

   b. Use the correspondence principle to determine whether quantum mechanics is needed to understand the particle's motion or if it is "safe" to use classical physics.

23. ‖ Suppose that $\psi_1(x)$ and $\psi_2(x)$ are both solutions to the Schrödinger equation for the same potential energy $U(x)$. Prove that the superposition $\psi(x) = A\psi_1(x) + B\psi_2(x)$ is also a solution to the Schrödinger equation.

24. ‖ Figure 40.27a modeled a hydrogen atom as a finite potential well with rectangular edges. A more realistic model of a hydrogen atom, although still a one-dimensional model, would be the electron + proton electrostatic potential energy in one dimension:

$$U(x) = -\frac{e^2}{4\pi\epsilon_0 |x|}$$

   a. Draw a graph of $U(x)$ versus x. Center your graph at $x = 0$.

   b. Despite the divergence at $x = 0$, the Schrödinger equation can be solved to find energy levels and wave functions for the electron in this potential. Draw a horizontal line across your graph of part a about one-third of the way from the bottom to the top. Label this line E_2, then, on this line, sketch a plausible graph of the $n = 2$ wave function.

   c. Redraw your graph of part a and add a horizontal line about two-thirds of the way from the bottom to the top. Label this line E_3, then, on this line, sketch a plausible graph of the $n = 3$ wave function.

25. ‖ a. Derive an expression for $\lambda_{2\rightarrow1}$, the wavelength of light emitted by a particle in a rigid box during a quantum jump from $n = 2$ to $n = 1$.
   b. In what length rigid box will an electron undergoing a $2 \rightarrow 1$ transition emit light with a wavelength of 694 nm? This is the wavelength of a ruby laser.

26. ‖ Model an atom as an electron in a rigid box of length 0.100 nm, roughly twice the Bohr radius.
   a. What are the four lowest energy levels of the electron?
   b. Calculate all the wavelengths that would be seen in the emission spectrum of this atom due to quantum jumps between these four energy levels. Give each wavelength a label $\lambda_{n\rightarrow m}$ to indicate the transition.
   c. Are these wavelengths in the infrared, visible, or ultraviolet portion of the spectrum?
   d. The stationary states of the Bohr hydrogen atom have negative energies. The stationary states of this model of the atom have positive energies. Is this a physically significant difference? Explain.
   e. Compare this model of an atom to the Bohr hydrogen atom. In what ways are the two models similar? Other than the signs of the energy levels, in what ways are they different?

27. ‖ Show that the normalization constant A_n for the wave functions of a particle in a rigid box has the value given in Equation 40.26.

28. ‖ A particle confined in a rigid one-dimensional box of length 10 fm has an energy level $E_n = 32.9$ MeV and an adjacent energy level $E_{n+1} = 51.4$ MeV.
   a. Determine the values of n and $n + 1$.
   b. Draw an energy-level diagram showing all energy levels from 1 through $n + 1$. Label each level and write the energy beside it.
   c. Sketch the $n + 1$ wave function on the $n + 1$ energy level.
   d. What is the wavelength of a photon emitted in the $n + 1 \rightarrow n$ transition? Compare this to a typical visible-light wavelength.
   e. What is the mass of the particle? Can you identify it?

29. ‖ Consider a particle in a rigid box of length L. For each of the states $n = 1$, $n = 2$, and $n = 3$:
   a. Sketch graphs of $|\psi(x)|^2$. Label the points $x = 0$ and $x = L$.
   b. Where, in terms of L, are the positions at which the particle is *most* likely to be found?
   c. Where, in terms of L, are the positions at which the particle is *least* likely to be found?
   d. Determine, by examining your $|\psi(x)|^2$ graphs, if the probability of finding the particle in the left one-third of the box is less than, equal to, or greater than $\frac{1}{3}$. Explain your reasoning.
   e. *Calculate* the probability that the particle will be found in the left one-third of the box.

30. ‖‖ For a particle in a finite potential well of width L and depth U_0, what is the ratio of the probability Prob(in δx at $x = L + \eta$) to the probability Prob(in δx at $x = L$)?

31. ‖ For the quantum-well laser of Figure 40.16, *estimate* the probability that an electron will be found within one of the GaAlAs layers rather than in the GaAs layer. Explain your reasoning.

32. ‖ Use the data from Figure 40.24 to calculate the first three vibrational energy levels of a C=O carbon-oxygen double bond.

33. ‖ Verify that the $n = 1$ wave function $\psi_1(x)$ of the quantum harmonic oscillator really is a solution of the Schrödinger equation. That is, show that the right and left sides of the Schrödinger equation are equal if you use the $\psi_1(x)$ wave function.

34. ‖ Show that the constant b used in the quantum-harmonic-oscillator wave functions (a) has units of length and (b) is the classical turning point of an oscillator in the $n = 1$ ground state.

35. ‖ a. Determine the normalization constant A_1 for the $n = 1$ ground-state wave function of the quantum harmonic oscillator. Your answer will be in terms of b.
   b. Write an expression for the probability that a quantum harmonic oscillator in its $n = 1$ ground state will be found in the classically forbidden region.
   c. (Optional) Use a numerical integration program to evaluate your probability expression of part b.
   Hint: It helps to simplify the integral by making a change of variables to $u = x/b$.

36. ‖ a. Derive an expression for the classical probability density $P_{class}(x)$ for a simple harmonic oscillator with amplitude A.
   b. Graph your expression between $x = -A$ and $x = +A$.
   c. Interpret your graph. Why is it shaped as it is?

37. ‖ a. Derive an expression for the classical probability density $P_{class}(y)$ for a ball that bounces between the ground and height h. The collisions with the ground are perfectly elastic.
   b. Graph your expression between $y = 0$ and $y = h$.
   c. Interpret your graph. Why is it shaped as it is?

38. ‖ Figure 40.17 showed that a typical nuclear radius is 4.0 fm. As you'll learn in Chapter 42, a typical energy of a neutron bound inside the nuclear potential well is $E_n = -20$ MeV. To find out how "fuzzy" the edge of the nucleus is, what is the neutron's penetration distance into the classically forbidden region as a fraction of the nuclear radius?

39. ‖ Even the smoothest mirror finishes are "rough" when viewed at a scale of 100 nm. When two very smooth metals are placed in contact with each other, the actual distance between the surfaces varies from 0 nm at a few points of real contact to ≈ 100 nm. The average distance between the surfaces is ≈ 50 nm. The work function of aluminum is 4.3 eV. What is the probability that an electron will tunnel between two pieces of aluminum that are 50 nm apart? Give your answer as a power of 10 rather than a power of e.

40. ‖ A proton's energy is 1.0 MeV below the top of a 10-fm-wide energy barrier. What is the probability that the proton will tunnel through the barrier?

Challenge Problems

41. A typical electron in a piece of metallic sodium has energy $-E_0$ compared to a free electron, where E_0 is the 2.7 eV work function of sodium.
   a. At what distance *beyond* the surface of the metal is the electron's probability density 10% of its value *at* the surface?
   b. How does this distance compare to the size of an atom?

42. In a nuclear physics experiment, a proton is fired toward a $Z = 13$ nucleus with the diameter and neutron energy levels shown in Figure 40.17. The nucleus, which was initially in its ground state, subsequently emits a gamma ray with wavelength 1.73×10^{-4} nm. What was the *minimum* initial speed of the proton?
   Hint: Don't neglect the proton-nucleus collision.

43. A particle of mass m has the wave function $\psi(x) = A x \exp(-x^2/a^2)$ when it is in an allowed energy level with $E = 0$.
   a. Draw a graph of $\psi(x)$ versus x.
   b. At what value or values of x is the particle most likely to be found?
   c. Find and graph the potential-energy function $U(x)$.

44. In most metals, the atomic ions form a regular arrangement called a *crystal lattice*. The conduction electrons in the sea of electrons move through this lattice. FIGURE CP40.44 is a one-dimensional model of a crystal lattice. The ions have mass m, charge e, and an equilibrium separation b.

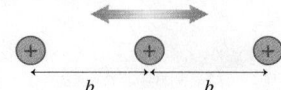

FIGURE CP40.44

a. Suppose the middle charge is displaced a very small distance ($x \ll b$) from its equilibrium position while the outer charges remain fixed. Show that the net electric force on the middle charge is given approximately by

$$F = -\frac{e^2}{b^3\pi\epsilon_0}x$$

In other words, the charge experiences a linear restoring force.

b. Suppose this crystal consists of aluminum ions with an equilibrium spacing of 0.30 nm. What are the energies of the four lowest vibrational states of these ions?

c. What wavelength photons are emitted during quantum jumps between *adjacent* energy levels? Is this wavelength in the infrared, visible, or ultraviolet portion of the spectrum?

45. a. What is the probability that an electron will tunnel through a 0.50 nm air gap from a metal to a STM probe if the work function is 4.0 eV?

b. The probe passes over an atom that is 0.050 nm "tall." By what factor does the tunneling current increase?

c. If a 10% current change is reliably detectable, what is the smallest height change the STM can detect?

46. Tennis balls traveling faster than 100 mph routinely bounce off tennis rackets. At some sufficiently high speed, however, the ball will break through the strings and keep going. The racket is a potential-energy barrier whose height is the energy of the slowest string-breaking ball. Suppose that a 100 g tennis ball traveling at 200 mph is just sufficient to break the 2.0-mm-thick strings. Estimate the probability that a 120 mph ball will tunnel through the racket without breaking the strings. Give your answer as a power of 10 rather than a power of e.

STOP TO THINK ANSWERS

Stop to Think 40.1: $v_a = v_b > v_c$. The de Broglie wavelength is $\lambda = h/mv$, so slower particles have longer wavelengths. The wave amplitude is not relevant.

Stop to Think 40.2: c. The $n = 2$ state has a node in the middle of the box. The antinodes are centered in the left and right halves of the box.

Stop to Think 40.3: $n = 4$. There are four antinodes and three nodes (excluding the ends).

Stop to Think 40.4: d. The wave function reaches zero abruptly on the right, indicating an infinitely high potential-energy wall. The exponential decay on the left shows that the left wall of the potential energy is *not* infinitely high. The node spacing and the amplitude increase steadily in going from right to left, indicating a *steadily* decreasing kinetic energy and thus a steadily increasing potential energy.

Stop to Think 40.5: c. $E = (n - \frac{1}{2})\hbar\omega$, so $\frac{5}{2}\hbar\omega$ is the energy of the $n = 3$ state. An $n = 3$ state has 3 antinodes.

Stop to Think 40.6: b. The probability of tunneling through the barrier increases as the difference between E and U_0 decreases. If the tunneling probability increases, the reflection probability must decrease.

41 Atomic Physics

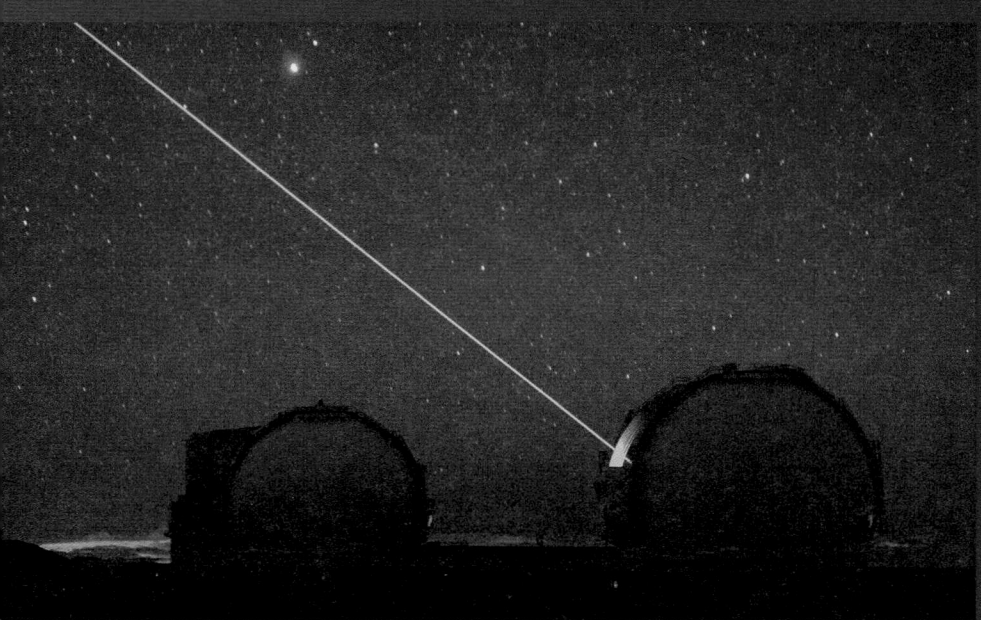

Lasers are one of the most important applications of the quantum-mechanical properties of atoms and light.

▶ **Looking Ahead** The goal of Chapter 41 is to understand the structure and properties of atoms.

The Hydrogen Atom

You'll learn to interpret the hydrogen atom as a three-dimensional wave function giving the probability of locating the electron in a region of space.

We'll need three quantum numbers to describe the state of the electron. The wave function is often pictured graphically as an **electron cloud.**

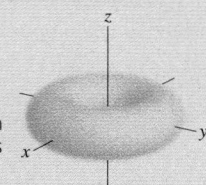

◀ **Looking Back**
Sections 39.3–39.4 Wave functions

Multielectron Atoms

The quantum-mechanical model of the atom is able to explain the properties of multielectron atoms, including their energy levels, ionization energies, and spectra.

You'll use atomic energy-level diagrams like this to understand which states are occupied and how spectra are produced.

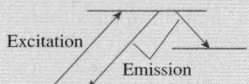

Atomic Spectra

You'll learn to interpret spectra in terms of *excitation* followed by *emission*. You'll also learn that emission doesn't occur instantly; instead, the excited state has a **lifetime** of typically a few nanoseconds.

- Excitation is by the absorption of a photon or by collision with another particle.
- Emission obeys **selection rules** that allow some quantum jumps but not others.

Atomic Models

Our understanding of the atom has evolved as the experimental evidence has grown. You should review:

- The atom as an indivisible object.
- Thomson's plum-pudding model.
- Rutherford's solar-system model.
- Bohr's semi-classical model.

You will learn how Schrödinger's quantum-mechanical model of the atom finally succeeds at explaining *all* the experimental evidence about atoms.

◀ **Looking Back**
Sections 38.5–38.7 The Bohr model

Electron Spin

In addition to having an inherent mass and an inherent charge, the electron has an inherent magnetic moment called the **electron spin.** As a result, a fourth quantum number is needed to specify a quantum state completely

You'll learn how the **Pauli exclusion principle,** which says that only one electron can occupy each quantum state, is the key to understanding the periodic table of the elements.

1 H			
3 Li	4 Be		
11 Na	12 Mg		
19	20	21	22

Lasers

You'll learn that lasers work because of the **stimulated emission** of light, a process in which an incoming photon causes an excited atom to emit an identical photon.

In some lasers, an intense burst of light from a flashlamp creates a **population inversion** with more atoms in an excited state than in the ground state.

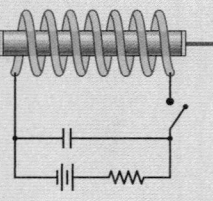

41.1 The Hydrogen Atom: Angular Momentum and Energy

Bohr's concept of stationary states provided a means of understanding both the stability of atoms and the quantum jumps that lead to discrete spectra. Yet, as we have seen, the Bohr model was not successful for any neutral atom other than hydrogen. Is Schrödinger's quantum mechanics better at explaining atomic structure than other models? The answer, as you can probably anticipate, is a decisive yes. This chapter is an overview of how quantum mechanics finally provides us with an understanding of atomic structure and atomic properties.

Let's begin with a quantum-mechanical model of the hydrogen atom. FIGURE 41.1 shows an electron at distance r from a proton. The proton is much more massive than the electron, so we will assume that the proton remains at rest at the origin.

As you learned in Chapter 40, the problem-solving procedure in quantum mechanics consists of two basic steps:

1. Specify a potential-energy function.
2. Solve the Schrödinger equation to find the wave functions, allowed energy levels, and other quantum properties.

The first step is easy. The proton and electron are charged particles with $q = \pm e$, so the potential energy of a hydrogen atom as a function of the electron distance r is

$$U(r) = -\frac{1}{4\pi\epsilon_0}\frac{e^2}{r} \tag{41.1}$$

The difficulty arises with the second step. The Schrödinger equation of Chapter 40 was for one-dimensional problems. Atoms are three-dimensional, and the three-dimensional Schrödinger equation turns out to be a partial differential equation whose solution is outside the scope of this textbook. Consequently, we'll present results without derivation or proof. The good news is that you have learned enough quantum mechanics to interpret and use the results.

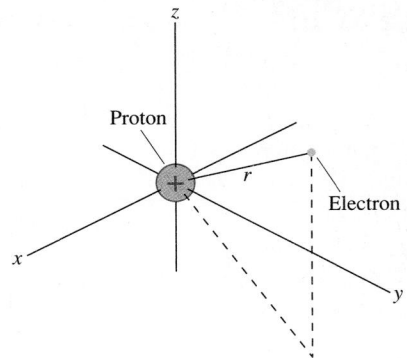

FIGURE 41.1 The electron in a hydrogen atom is distance r from the proton.

Stationary States of Hydrogen

In one dimension, energy quantization appeared as a consequence of *boundary conditions* on the wave function. That is, only for certain discrete energies, characterized by the quantum number n, did solutions to the Schrödinger equation satisfy the boundary conditions. In three dimensions, the wave function must satisfy *three* different boundary conditions. Consequently, solutions to the three-dimensional Schrödinger equation have *three* quantum numbers and *three* quantized parameters.

Solutions to the Schrödinger equation for the hydrogen atom potential energy exist only if three conditions are satisfied:

1. The atom's energy must be one of the values

$$E_n = -\frac{1}{n^2}\left(\frac{1}{4\pi\epsilon_0}\frac{e^2}{2a_B}\right) = -\frac{13.60\ \text{eV}}{n^2} \qquad n = 1, 2, 3, \ldots \tag{41.2}$$

where $a_B = 4\pi\epsilon_0\hbar^2/me^2 = 0.0529$ nm is the Bohr radius. The integer n is called the **principal quantum number**. These energies are the same as those predicted by the Bohr model of the hydrogen atom.

2. The orbital angular momentum L of the electron's orbit must be one of the values

$$L = \sqrt{l(l+1)}\,\hbar \qquad l = 0, 1, 2, 3, \ldots, n-1 \tag{41.3}$$

The integer l is called the **orbital quantum number**.

3. The z-component of the angular momentum L_z must be one of the values

$$L_z = m\hbar \qquad m = -l, -l+1, \ldots, 0, \ldots, l-1, l \qquad (41.4)$$

The integer m is called the **magnetic quantum number.**

In other words, each stationary state of the hydrogen atom is identified by a triplet of quantum numbers (n, l, m). Each quantum number is associated with a physical property of the atom.

NOTE ▶ The energy of the stationary state depends only on the principal quantum number n, not on l or m. ◀

EXAMPLE 41.1 **Listing quantum numbers**

List all possible states of a hydrogen atom that have energy $E = -3.40$ eV.

SOLVE Energy depends only on the principal quantum number n. States with $E = -3.40$ eV have

$$n = \sqrt{\frac{-13.60 \text{ eV}}{-3.40 \text{ eV}}} = 2$$

An atom with principal quantum number $n = 2$ could have either $l = 0$ or $l = 1$, but $l \geq 2$ is excluded by the rule $l \leq n - 1$. If $l = 0$, the only possible value for the magnetic quantum number m is $m = 0$. If $l = 1$, then the atom could have $m = -1$, $m = 0$, or $m = +1$. Thus the possible quantum numbers are

n	l	m
2	0	0
2	1	1
2	1	0
2	1	-1

These four states all have the same energy.

TABLE 41.1 Symbols used to represent quantum number l

l	Symbol
0	s
1	p
2	d
3	f

Hydrogen turns out to be unique. For all other elements, the allowed energies depend on both n and l (but not m). Consequently, it is useful to label the stationary states by their values of n and l. The lowercase letters shown in Table 41.1 are customarily used to represent the various values of quantum number l. These symbols come from spectroscopic notation used in prequantum-mechanics days, when some spectral lines were classified as **s**harp, others as **p**rincipal, and so on.

Using these symbols, we call the ground state of the hydrogen atom, with $n = 1$ and $l = 0$, the $1s$ state. The $3d$ state has $n = 3$, $l = 2$. In Example 41.1, we found one $2s$ state (with $l = 0$) and three $2p$ states (with $l = 1$), all with the same energy.

Angular Momentum Is Quantized

A planet orbiting the sun has two different angular momenta: *orbital angular momentum* due to its orbit around the sun (a 365-day period for the earth) and *rotational angular momentum* as it rotates on its axis (a 24-hour period for the earth). We introduced angular momentum in Chapter 12, and a brief review of Section 12.11 is highly recommended.

A classical model of the hydrogen atom would be similar. Although a circular orbit is possible, it's more likely that the electron would follow an elliptical orbit with the proton at one focus of the ellipse. Further, the orbit need not lie in the xy-plane. **FIGURE 41.2** shows a classical orbit tilted at angle θ below the xy-plane. The electron, like a planet, has orbital angular momentum, and Figure 41.2 reminds you that the orbital angular momentum vector $\vec{L}$ is perpendicular to the plane of the orbit. (The electron also has a quantum version of rotational angular momentum, called *spin*, that we'll introduce in Section 41.3.) The orbital angular momentum vector has component $L_z = L\cos\theta$ along the z-axis.

Classically, L and L_z can have any values. Not so in quantum mechanics. Quantum conditions 2 and 3 tell us that **the electron's orbital angular momentum is quantized.** The magnitude of the orbital angular momentum must be one of the discrete values

$$L = \sqrt{l(l+1)}\,\hbar = 0, \sqrt{2}\hbar, \sqrt{6}\hbar, \sqrt{12}\hbar, \ldots$$

where l is an integer. Simultaneously, the z-component L_z must have one of the values $L_z = m\hbar$, where m is an integer between $-l$ and l. No other values of L or L_z allow the wave function to satisfy the boundary conditions.

FIGURE 41.2 The angular momentum of an elliptical orbit.

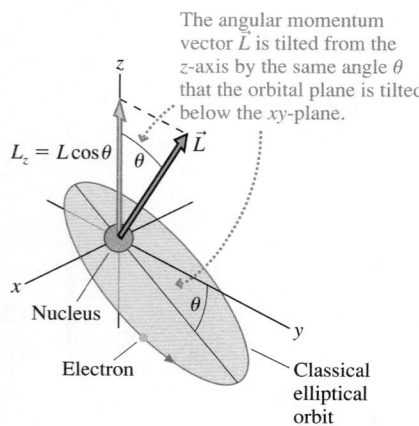

The angular momentum vector $\vec{L}$ is tilted from the z-axis by the same angle θ that the orbital plane is tilted below the xy-plane.

$L_z = L\cos\theta$

Nucleus

Electron

Classical elliptical orbit

The quantization of angular momentum places restrictions on the shape and orientation of the electron's orbit. To see this, consider a hydrogen atom with orbital quantum number $l = 2$. In this state, the *magnitude* of the electron's angular momentum must be $L = \sqrt{6}\hbar = 2.45\hbar$. Furthermore, the angular momentum vector must point in a *direction* such that $L_z = m\hbar$, where m is one of only five integers in the range $-2 \leq m \leq 2$.

The combination of these two requirements allows $\vec{L}$ to point only in certain directions in space, as shown in FIGURE 41.3. This is a rather unusual figure that requires a little thought to understand. Suppose $m = 0$ and thus $L_z = 0$. With no z-component, the angular momentum vector $\vec{L}$ must lie somewhere in the xy-plane. Furthermore, because the length of $\vec{L}$ is constrained to be $2.45\hbar$, the tip of $\vec{L}$ must lie somewhere on the circle labeled $m = 0$. These values of $\vec{L}$ correspond to classical orbits tipped into a vertical plane.

Similarly, $m = 2$ requires $\vec{L}$ to lie along the surface of the cone whose height is $2\hbar$ and whose side has length $2.45\hbar$. These values of $\vec{L}$ correspond to classical orbits tilted slightly out of the xy-plane. Notice that $\vec{L}$ **cannot point directly along the z-axis.** The maximum possible value of L_z, when $m = l$, is $(L_z)_{max} = l\hbar$. But $l < \sqrt{l(l+1)}$, so $(L_z)_{max} < L$. The angular momentum vector *must* have either an x- or a y-component (or both). In other words, the corresponding classical orbit cannot lie in the xy-plane.

An angular momentum vector $\vec{L}$ tilted at angle θ from the z-axis corresponds to an orbit tilted at angle θ out of the xy-plane. The quantization of angular momentum restricts the orbital planes to only a few discrete angles. For quantum state (n, l, m), the angle of the angular momentum vector is

$$\theta_{lm} = \cos^{-1}\left(\frac{L_z}{L}\right) = \cos^{-1}\left(\frac{m\hbar}{\sqrt{l(l+1)}\hbar}\right) = \cos^{-1}\left(\frac{m}{\sqrt{l(l+1)}}\right) \quad (41.5)$$

Angles θ_{22}, θ_{21}, and θ_{20} are labeled in Figure 41.3. Orbital planes at other angles are not allowed because they don't satisfy the quantization conditions for angular momentum.

FIGURE 41.3 The five possible orientations of the angular momentum vector for $l = 2$. The angular momentum vectors all have length $L = \sqrt{6}\hbar = 2.45\hbar$.

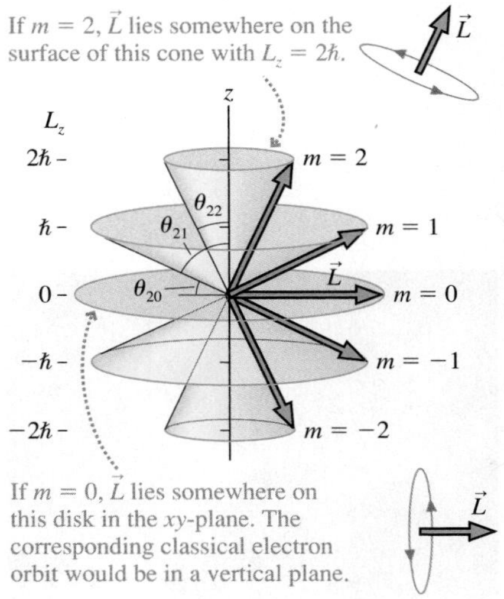

If $m = 2$, $\vec{L}$ lies somewhere on the surface of this cone with $L_z = 2\hbar$.

If $m = 0$, $\vec{L}$ lies somewhere on this disk in the xy-plane. The corresponding classical electron orbit would be in a vertical plane.

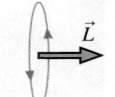

EXAMPLE 41.2 | **The angle of the angular momentum vector**

What is the angle between $\vec{L}$ and the z-axis for a hydrogen atom in the stationary state $(n, l, m) = (4, 2, 1)$?

SOLVE The angle θ_{21} is labeled in Figure 41.3. The state $(4, 2, 1)$ has $l = 2$ and $m = 1$, thus

$$\theta_{21} = \cos^{-1}\left(\frac{1}{\sqrt{6}}\right) = 65.9°$$

ASSESS This quantum state corresponds to a classical orbit tilted 65.9° away from the xy-plane.

NOTE ▶ The ground state of hydrogen, with $l = 0$, has *zero* angular momentum. A classical particle cannot orbit unless it has angular momentum, but apparently a quantum particle does not have this requirement. ◀

Energy Levels of the Hydrogen Atom

The energy of the hydrogen atom is quantized. Only those energies given by Equation 41.2 allow the wave function to satisfy the boundary conditions. The allowed energies of hydrogen depend only on the principal quantum number n, but for other atoms the energies will depend on both n and l. In anticipation of using both quantum numbers, FIGURE 41.4 is an *energy-level diagram* for the hydrogen atom in which the rows are labeled by n and the columns by l. The left column contains all of the $l = 0$ s states, the next column is the $l = 1$ p states, and so on.

Because the quantum condition of Equation 41.3 requires $n > l$, the s states begin with $n = 1$, the p states begin with $n = 2$, and the d states with $n = 3$. That is, the lowest-energy d state is $3d$ because states with $n = 1$ or $n = 2$ cannot have $l = 2$. For hydrogen, where the energy levels do not depend on l, the energy-level diagram shows that the $3s$, $3p$, and $3d$ states have equal energy. Figure 41.4 shows only the first few energy levels for each value of l, but there really are an infinite number of levels, as $n \to \infty$, crowding together beneath $E = 0$. The dashed line at $E = 0$ is the atom's *ionization limit,* the energy of a hydrogen atom in which the electron has been moved infinitely far away to form an H^+ ion.

FIGURE 41.4 Energy-level diagram for the hydrogen atom.

Quantum number l	0	1	2	3
Symbol	s	p	d	f

n	$E = 0$ eV		Ionization limit	
4	−0.85 eV	$4s$	$4p$	$4d$	$4f$
3	−1.51 eV	$3s$	$3p$	$3d$	
2	−3.40 eV	$2s$	$2p$		
1	−13.60 eV	$1s$	Ground state		

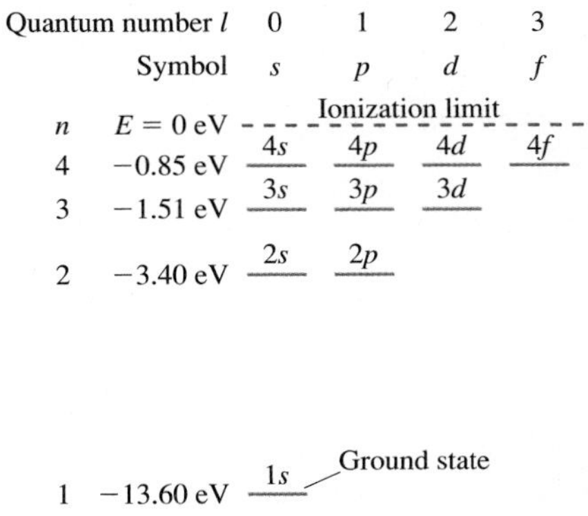

The lowest energy state, the $1s$ state with $E_1 = -13.60$ eV, is the *ground state* of hydrogen. The value $|E_1| = 13.60$ eV is the **ionization energy,** the *minimum* energy that would be needed to form a hydrogen ion by removing the electron from the ground state. All of the states with $n > 1$ (i.e., the states with energy higher than the ground state) are *excited states.*

STOP TO THINK 41.1   What are the quantum numbers n and l for a hydrogen atom with $E = -(13.60/9)$ eV and $L = \sqrt{2}\hbar$?

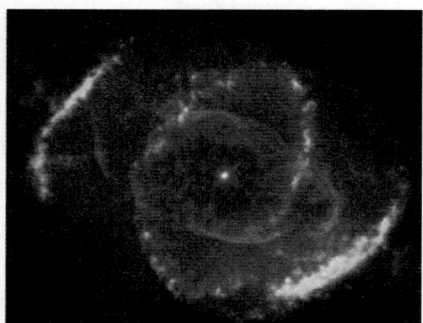

The red color of this nebula is due to the emission of light from hydrogen atoms. The atoms are excited by intense ultraviolet light from the star in the center. They then emit red light ($\lambda = 656$ nm) in a $3 \rightarrow 2$ transition, part of the Balmer series of spectral lines emitted by hydrogen.

41.2 The Hydrogen Atom: Wave Functions and Probabilities

You learned in Chapter 40 that the probability of finding a particle in a small interval of width δx at the position x is given by

$$\text{Prob(in } \delta x \text{ at } x) = |\psi(x)|^2 \delta x = P(x) \delta x$$

where $P(x) = |\psi(x)|^2$ is the probability density. This interpretation of $|\psi(x)|^2$ as a probability density lies at the heart of quantum mechanics. However, $P(x)$ was for a one-dimensional wave function.

For a three-dimensional atom, the wave function is $\psi(x, y, z)$, a function of three variables. We now want to consider the probability of finding a particle in a small *volume* of space δV at the position described by the three coordinates (x, y, z). This probability is

$$\text{Prob(in } \delta V \text{ at } x, y, z) = |\psi(x, y, z)|^2 \delta V \qquad (41.6)$$

We can still interpret the square of the wave function as a probability density.

In one-dimensional quantum mechanics we could simply graph $P(x)$ versus x. Portraying the probability density of a three-dimensional wave function is more of a challenge. One way to do so, shown in FIGURE 41.5, is to use denser shading to indicate regions of larger probability density. That is, the amplitude of ψ is larger, and the electron is more likely to be found in regions where the shading is darker. These figures show the probability densities of the $1s$, $2s$, and $2p$ states of hydrogen. As you can see, the probability density in three dimensions creates what is often called an **electron cloud** around the nucleus.

FIGURE 41.5 The probability densities of the electron in the $1s$, $2s$, and $2p$ states of hydrogen.

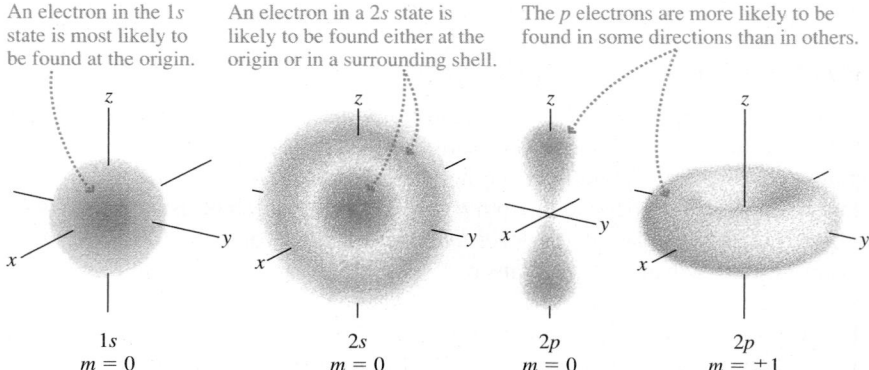

These figures contain a lot of information. For example, notice how the p electrons have directional properties. These directional properties allow p electrons to "reach out" toward nearby atoms, forming molecular bonds. The quantum mechanics of bonding goes beyond what we can study in this text, but the electron-cloud pictures of the p electrons begin to suggest how bonds could form.

Radial Wave Functions

In practice, the probability of finding the electron at a certain point in space is often less useful than the probability of finding the electron at a certain *distance* from the nucleus. That is, what is the probability that the electron is to be found within the small range of distances δr at the distance r?

It turns out that the solutions to the three-dimensional Schrödinger equation, the wave functions $\psi(x, y, z)$, can be written in a form that focuses on the electron's radial distance r from the proton. The portion of the wave function that depends only on r is called the **radial wave function.** These functions, which depend on the quantum numbers n and l, are designated $R_{nl}(r)$. The first three radial wave functions are

$$R_{1s}(r) = \frac{1}{\sqrt{\pi a_B^3}} e^{-r/a_B}$$

$$R_{2s}(r) = \frac{1}{\sqrt{8\pi a_B^3}} \left(1 - \frac{r}{2a_B}\right) e^{-r/2a_B} \qquad (41.7)$$

$$R_{2p}(r) = \frac{1}{\sqrt{24\pi a_B^3}} \left(\frac{r}{2a_B}\right) e^{-r/2a_B}$$

where a_B is the Bohr radius.

The radial wave functions may seem mysterious, because we haven't shown where they come from, but they are essentially the same as the one-dimensional wave functions $\psi(x)$ you learned to work with in Chapter 40. In fact, these radial wave functions are mathematically similar to the one-dimensional wave functions of the simple harmonic oscillator. One important difference, however, is that r ranges from 0 to ∞. For one-dimensional wave functions, x ranged from $-\infty$ to ∞.

NOTE ▶ Don't be confused by the notation. R is not a radius but, like ψ, is the symbol for a wave function, the *radial* wave function. It is a function of the distance r from the proton. ◀

FIGURE 41.6 shows the radial wave functions for the $1s$ and $2s$ states. Notice that the radial wave function is nonzero at $r = 0$, the position of the nucleus. This is surprising, but it is consistent with our observation in Figure 41.5 that the $1s$ and $2s$ electrons have a strong probability of being found at the origin.

Our purpose for introducing the radial wave functions was to determine the probability of finding the electron a certain *distance* from the nucleus. **FIGURE 41.7** shows a shell of radius r and thickness δr centered on the nucleus. The probability of finding the electron at distance r from the nucleus is equivalent to the probability that the electron is located somewhere within this shell. The volume of a thin shell is its surface area multiplied by its thickness δr. The surface area of a sphere is $4\pi r^2$, so the volume of this thin shell is

$$\delta V = 4\pi r^2 \delta r$$

Just as $|\psi(x)|^2$ is the probability in one dimension of finding a particle within an interval δx, the probability of locating the electron within this spherical shell can be written in terms of the *radial* wave function $R_{nl}(r)$ as

$$\text{Prob(in } \delta r \text{ at } r) = |R_{nl}(r)|^2 \delta V = 4\pi r^2 |R_{nl}(r)|^2 \delta r \qquad (41.8)$$

If we define the **radial probability density** $P_r(r)$ for state nl as

$$P_r(r) = 4\pi r^2 |R_{nl}(r)|^2 \qquad (41.9)$$

then, exactly analogous to the one-dimensional quantum mechanics of Chapter 40, we can write the probability of finding the electron within a small interval δr at distance r as

$$\text{Prob(in } \delta r \text{ at } r) = P_r(r)\delta r \qquad (41.10)$$

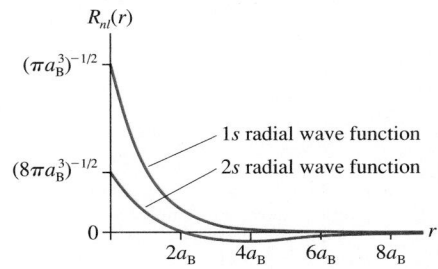

FIGURE 41.6 The $1s$ and $2s$ radial wave functions of hydrogen.

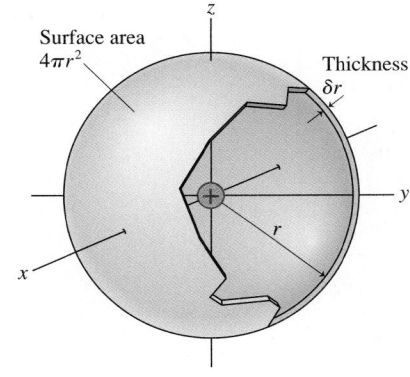

FIGURE 41.7 The radial probability density gives the probability of finding the electron in a spherical shell of thickness δr at radius r.

FIGURE 41.8 The radial probability densities for $n = 1$, 2, and 3.

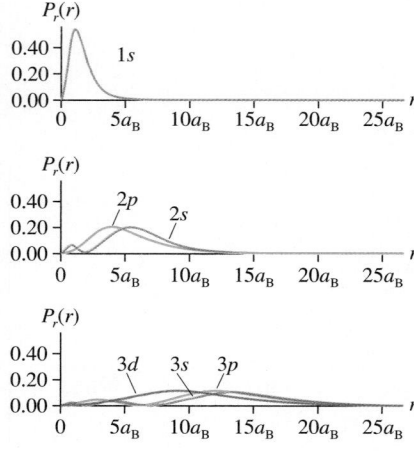

FIGURE 41.9 More circular orbits have larger angular momenta.

The circular orbit has the largest angular momentum. The electron stays at a constant distance from the nucleus.

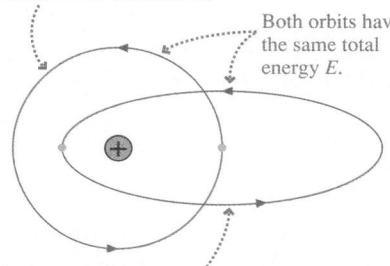

Both orbits have the same total energy E.

The elliptical orbit has a smaller angular momentum. Compared to the circular orbit, the electron gets both closer to and farther from the nucleus.

The radial probability density tells us the relative likelihood of finding the electron at distance r from the nucleus. The volume factor $4\pi r^2$ reflects the fact that more space is available in a shell of larger r, and this additional space increases the probability of finding the electron at that distance.

The probability of finding the electron between r_{min} and r_{max} is

$$\text{Prob}(r_{min} \leq r \leq r_{max}) = \int_{r_{min}}^{r_{max}} P_r(r)\, dr = 4\pi \int_{r_{min}}^{r_{max}} r^2 |R_{nl}(r)|^2\, dr \quad (41.11)$$

The electron must be *somewhere* between $r = 0$ and $r = \infty$, so the integral of $P_r(r)$ between 0 and ∞ must equal 1. This normalization condition was used to determine the constants in front of the radial wave functions of Equations 41.7.

FIGURE 41.8 shows the radial probability densities for the $n = 1$, 2, and 3 states of the hydrogen atom. You can see that the $1s$, $2p$, and $3d$ states, with maxima at a_B, $4a_B$, and $9a_B$, respectively, are following the pattern $r_{peak} = n^2 a_B$ of the radii of the orbits in the Bohr hydrogen atom. There we simply bent a one-dimensional de Broglie wave into a circle of that radius. Now we have a three-dimensional wave function for which the electron is *most likely* to be this distance from the nucleus, although it *could* be found at other values of r. The physical situation is very different in quantum mechanics, but it is good to see that various aspects of the Bohr model of the hydrogen atom can be reproduced.

But why is it the $3d$ state that agrees with the Bohr atom rather than $3s$ or $3p$? All states with the same value of n form a collection of "orbits" having the same energy. In **FIGURE 41.9**, the state with $l = n - 1$ has the largest angular momentum of the group. Consequently, the maximum-l state corresponds to a circular classical orbit and matches the circular orbits of the Bohr atom. Notice that the radial probability densities for the $2p$ and $3d$ states have a single peak, corresponding to a classical orbit at a constant distance.

States with smaller l correspond to elliptical orbits. You can see in Figure 41.8 that the radial probability density of a $3s$ electron has a peak close to the nucleus. The $3s$ electron also has a good chance of being found *farther* from the nucleus than a $3d$ electron, suggesting an orbit that alternately swings in near the nucleus, then moves out past the circular orbit with the same energy. This distinction between circular and elliptical orbits will be important when we discuss the energy levels in multielectron atoms.

NOTE ▶ In quantum mechanics, nothing is really orbiting. However, the probability densities for the electron to be, or not to be, any given distance from the nucleus mimic certain aspects of classical orbits and provide a useful analogy. ◀

You can see in Figure 41.8 that the most likely distance from the nucleus of an $n = 1$ electron is approximately a_B. The distance of an $n = 2$ electron is most likely to be between about $3a_B$ and $7a_B$. An $n = 3$ electron is most likely to be found between about $8a_B$ and $15a_B$. In other words, the radial probability densities give the clear impression that each value of n has a fairly well-defined range of radii where the electron is most likely to be found. This is the basis of the **shell model** of the atom that is used in chemistry.

However, there's one significant puzzle. In Figure 41.5, the fuzzy sphere representing the $1s$ ground state is densest at the center, where the electron is most likely to be found. This maximum density at $r = 0$ agrees with the $1s$ radial wave function of Figure 41.6, which is a maximum at $r = 0$, but it seems to be in sharp disagreement with the $1s$ graph of Figure 41.8, which is *zero* at the nucleus and peaks at $r = a_B$.

To resolve this puzzle, we must distinguish between the probability density $|\psi(x, y, z)|^2$ and the *radial* probability density $P_r(r)$. The $1s$ wave function, and thus the $1s$ probability density, really does peak at the nucleus. But $|\psi(x, y, z)|^2$ is the probability of being in a small volume δV, such as a small box with sides δx, δy, and δz, whereas $P_r(r)$ is the probability of being in a spherical shell of thickness δr. Compared to $r = 0$, the probability density $|\psi(x, y, z)|^2$ is smaller at any *one* point having $r = a_B$. But the volume of *all* points with $r \approx a_B$ (i.e., the volume of the spherical shell at $r = a_B$) is so large that the radial probability density P_r peaks at this distance.

To use a mass analogy, consider a fuzzy ball that is densest at the center. Even though the density away from the center has decreased, a spherical shell of modest radius r can have *more total mass* than a small-radius spherical shell of the same thickness simply because it has so much more volume.

EXAMPLE 41.3   **Maximum probability**

Show that an electron in the $2p$ state is most likely to be found at $r = 4a_B$.

SOLVE We can use the $2p$ radial wave function from Equations 41.7 to write the radial probability density

$$P_r(r) = 4\pi r^2 |R_{2p}(r)|^2 = 4\pi r^2 \left[\frac{1}{\sqrt{24\pi a_B^3}} \left(\frac{r}{2a_B} \right) e^{-r/2a_B} \right]^2$$

$$= Cr^4 e^{-r/a_B}$$

where $C = (24a_B^5)^{-1}$ is a constant. This expression for $P_r(r)$ was graphed in Figure 41.8.

Maximum probability occurs at the point where the first derivative of $P_r(r)$ is zero:

$$\frac{dP_r}{dr} = C(4r^3)(e^{-r/a_B}) + C(r^4)\left(\frac{-1}{a_B} e^{-r/a_B} \right)$$

$$= Cr^3 \left(4 - \frac{r}{a_B} \right) e^{-r/a_B} = 0$$

This expression is zero only if $r = 4a_B$, so $P_r(r)$ is maximum at $r = 4a_B$. An electron in the $2p$ state is most likely to be found at this distance from the nucleus.

STOP TO THINK 41.2   How many maxima will there be in a graph of the radial probability density for the $4s$ state of hydrogen?

41.3 The Electron's Spin

Recall, from Chapter 32, that an orbiting electron generates a microscopic *magnetic moment* $\vec{\mu}$. **FIGURE 41.10** reminds you that a magnetic moment, like a compass needle, has north and south poles. Consequently, a magnetic moment in an external magnetic field experiences forces and torques. In the early 1920s, the German physicists Otto Stern and Walter Gerlach developed a technique to measure the magnetic moments of atoms. Their apparatus, shown in **FIGURE 41.11**, prepares an *atomic beam* by evaporating atoms out of a hole in an "oven." These atoms, traveling in a vacuum, pass through a *nonuniform* magnetic field. Reducing the size of the upper pole tip makes the field stronger toward the top of the magnet, weaker toward the bottom.

FIGURE 41.10 An orbiting electron generates a magnetic moment.

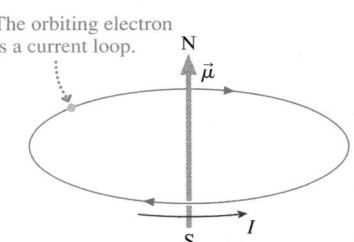

The orbiting electron is a current loop.

A current loop generates a magnetic moment with north and south magnetic poles.

FIGURE 41.11 The Stern-Gerlach experiment.

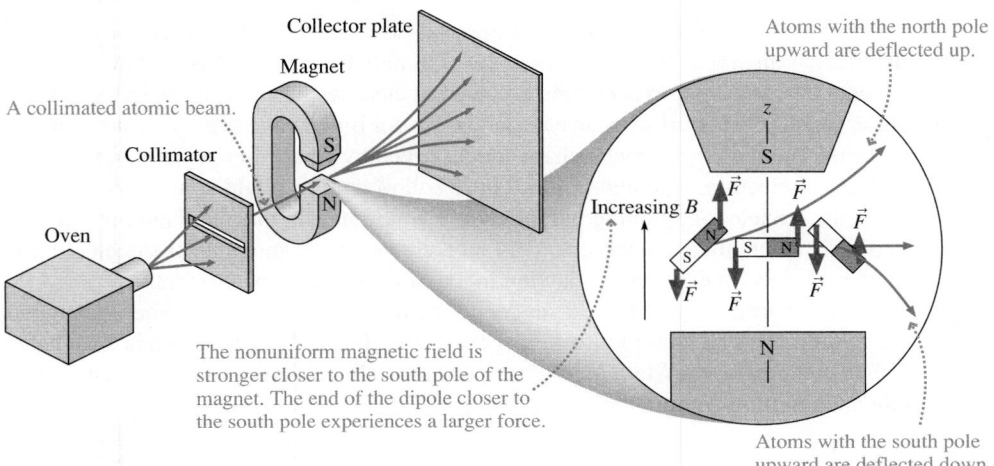

Collector plate

Magnet

A collimated atomic beam.

Collimator

Oven

The nonuniform magnetic field is stronger closer to the south pole of the magnet. The end of the dipole closer to the south pole experiences a larger force.

Atoms with the north pole upward are deflected up.

Increasing B

Atoms with the south pole upward are deflected down.

A magnetic moment experiences a *net force* in the nonuniform magnetic field because the field exerts forces of different strengths on the moment's north and south poles. If we define a z-axis to point upward, then an atom whose magnetic moment

FIGURE 41.12 Distribution of the atoms on the collector plate.

(a)

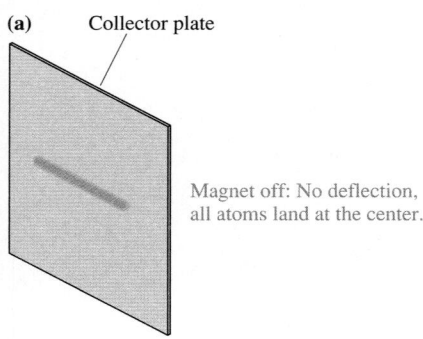

Collector plate

Magnet off: No deflection, all atoms land at the center.

(b)

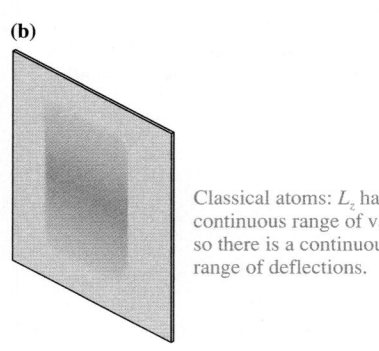

Classical atoms: L_z has a continuous range of values, so there is a continuous range of deflections.

(c)

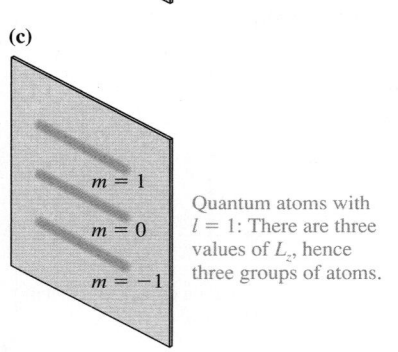

$m = 1$
$m = 0$
$m = -1$

Quantum atoms with $l = 1$: There are three values of L_z, hence three groups of atoms.

FIGURE 41.13 The outcome of the Stern-Gerlach experiment for hydrogen atoms.

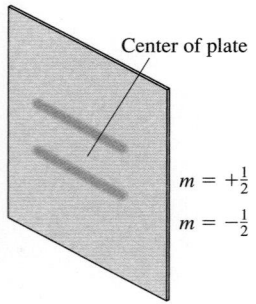

Center of plate

$m = +\frac{1}{2}$
$m = -\frac{1}{2}$

vector $\vec{\mu}$ is tilted upward ($\mu_z > 0$) has an upward force on its north pole that is larger than the downward force on its south pole. As the figure shows, this atom is deflected upward as it passes through the magnet. A downward-tilted magnetic moment ($\mu_z < 0$) experiences a net downward force and is deflected downward. A magnetic moment perpendicular to the field ($\mu_z = 0$) feels no net force and passes through the magnet without deflection. In other words, an atom's deflection as it passes through the magnet is proportional to μ_z, the z-component of its magnetic moment.

It's not hard to show, although we will omit the proof, that an atom's magnetic moment is proportional to the electron's orbital angular momentum: $\vec{\mu} \propto \vec{L}$. Because the deflection of an atom depends on μ_z, measuring the deflections in a nonuniform field provides information about the L_z values of the atoms in the atomic beam. The measurements are made by allowing the atoms to stick on a collector plate at the end of the apparatus. After the experiment has been run for several hours, the collector plate is removed and examined to learn how the atoms were deflected.

With the magnet off, the atoms pass through without deflection and land along a narrow line at the center, as shown in **FIGURE 41.12a**. If the orbiting electrons are classical particles, they should have a continuous range of angular momenta. Turning on the magnet should produce a continuous range of vertical deflections, and the distribution of atoms collected on the plate should look like **FIGURE 41.12b**. But if angular momentum is *quantized,* as Bohr had suggested several years earlier, the atoms should be deflected to discrete positions on the collector plate.

For example, an atom with $l = 1$ has three distinct values of L_z corresponding to quantum numbers $m = -1$, 0, and 1. This leads to a prediction of the three distinct groups of atoms shown in **FIGURE 41.12c**. There should always be an *odd* number of groups because there are $2l + 1$ values of L_z.

In 1927, with Schrödinger's quantum theory brand new, the Stern-Gerlach technique was used to measure the magnetic moment of hydrogen atoms. The ground state of hydrogen is $1s$, with $l = 0$, so the atoms should have *no* magnetic moment and there should be *no* deflection at all. Instead, the experiment produced the two-peaked distribution shown in **FIGURE 41.13**.

Because the hydrogen atoms were deflected, they *must* have a magnetic moment. But where does it come from if $L = 0$? Even stranger was the deflection into two groupings, rather than an odd number. The deflection is proportional to L_z, and $L_z = m\hbar$ where m ranges in integer steps from $-l$ to $+l$. The experimental results would make sense only if $l = \frac{1}{2}$, allowing m to take the two possible values $-\frac{1}{2}$ and $+\frac{1}{2}$. But according to Schrödinger's theory, the quantum numbers l and m must be integers.

An explanation for these observations was soon suggested, then confirmed: The electron has an *inherent* magnetic moment. After all, the electron has an inherent gravitational character, its mass m_e, and an inherent electric character, its charge $q_e = -e$. These are simply part of what an electron is. Thus it is plausible that an electron should also have an inherent magnetic character described by a built-in magnetic moment $\vec{\mu}_e$. A classical electron, if thought of as a little ball of charge, could spin on its axis as it orbits the nucleus. A spinning ball of charge would have a magnetic moment associated with its angular momentum. This inherent magnetic moment of the electron is what caused the unexpected deflection in the Stern-Gerlach experiment.

If the electron has an inherent magnetic moment, it must have an inherent angular momentum. This angular momentum is called the electron's **spin**, which is designated $\vec{S}$. The outcome of the Stern-Gerlach experiment tells us that the z-component of this spin angular momentum is

$$S_z = m_s\hbar \quad \text{where } m_s = +\tfrac{1}{2} \text{ or } -\tfrac{1}{2} \tag{41.12}$$

The quantity m_s is called the **spin quantum number.**

The z-component of the spin angular momentum vector is determined by the electron's orientation. The $m_s = +\frac{1}{2}$ state, with $S_z = +\frac{1}{2}\hbar$, is called the **spin-up** state, and the $m_s = -\frac{1}{2}$ state is called the **spin-down** state. It is convenient to picture a little angular momentum vector that can be drawn ↑ for an $m_s = +\frac{1}{2}$ state and ↓ for an $m_s = -\frac{1}{2}$ state. We will use this notation in the next section. Because the electron must be either spin-up or spin-down, a hydrogen atom in the Stern-Gerlach experiment will be deflected either up or down. This causes the two groups of atoms seen in Figure 41.13. No atoms have $S_z = 0$, so there are no undeflected atoms in the center.

NOTE ▶ The atom has spin angular momentum *in addition* to any orbital angular momentum that the electrons may have. Only in *s* states, for which $L = 0$, can we see the effects of "pure spin." ◀

The spin angular momentum *S* is analogous to Equation 41.3 for *L*:

$$S = \sqrt{s(s+1)}\hbar = \frac{\sqrt{3}}{2}\hbar \qquad (41.13)$$

where *s* is a quantum number with the single value $s = \frac{1}{2}$. *S* is the *inherent* angular momentum of the electron. Because of the single value of *s*, physicists usually say that the electron has "spin one-half." **FIGURE 41.14**, which should be compared to Figure 41.3, shows that the terms "spin up" and "spin down" refer to S_z, not the full spin angular momentum. As was the case with $\vec{L}$, it's not possible for $\vec{S}$ to point along the *z*-axis.

NOTE ▶ The term "spin" must be used with caution. Although a classical charged particle could generate a magnetic moment by spinning, the electron most assuredly is *not* a classical particle. It is not spinning in any literal sense. It simply has an inherent magnetic moment, just as it has an inherent mass and charge, and that magnetic moment makes it look *as if* the electron is spinning. It is a convenient figure of speech, not a factual statement. **The electron has a spin, but it is not a spinning electron!** ◀

The electron's spin has significant implications for atomic structure. The solutions to the Schrödinger equation could be described by the three quantum numbers *n*, *l*, and *m*, but the Stern-Gerlach experiment implies that this is not a complete description of an atom. Knowing that a ground-state atom has quantum numbers $n = 1$, $l = 0$, and $m = 0$ is not sufficient to predict whether the atom will be deflected up or down in a nonuniform magnetic field. We need to add the spin quantum number m_s to make our description complete. (Strictly speaking, we also need to add the quantum number *s*, but it provides no additional information because its value never changes.) So we really need *four* quantum numbers (n, l, m, m_s) to characterize the stationary states of the atom. The spin orientation does not affect the atom's energy, so a ground-state electron in hydrogen could be in either the $(1, 0, 0, +\frac{1}{2})$ spin-up state or the $(1, 0, 0, -\frac{1}{2})$ spin-down state.

The fact that *s* has the single value $s = \frac{1}{2}$ has other interesting implications. The correspondence principle tells us that a quantum particle begins to "act classical" in the limit of large quantum numbers. But *s* cannot become large! **The electron's spin is an intrinsic quantum property of the electron that has *no* classical counterpart.**

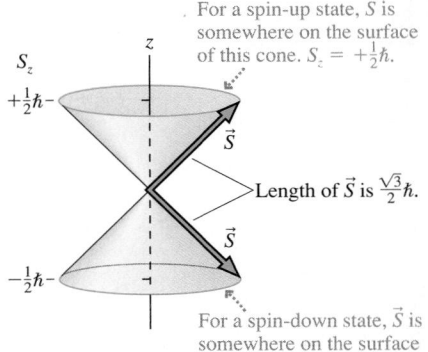

FIGURE 41.14 The spin angular momentum has two possible orientations.

For a spin-up state, $\vec{S}$ is somewhere on the surface of this cone. $S_z = +\frac{1}{2}\hbar$.

Length of $\vec{S}$ is $\frac{\sqrt{3}}{2}\hbar$.

For a spin-down state, $\vec{S}$ is somewhere on the surface of this cone. $S_z = -\frac{1}{2}\hbar$.

STOP TO THINK 41.3 Can the spin angular momentum vector lie in the *xy*-plane? Why or why not?

41.4 Multielectron Atoms

The Schrödinger-equation solution for the hydrogen atom matches the experimental evidence, but so did the Bohr hydrogen atom. The real test of Schrödinger's theory is how well it works for multielectron atoms. A neutral multielectron atom consists of *Z* electrons surrounding a nucleus with *Z* protons and charge $+Ze$. *Z*, the *atomic number,* is the order in which elements are listed in the periodic table. Hydrogen is $Z = 1$, helium $Z = 2$, lithium $Z = 3$, and so on.

The potential-energy function of a multielectron atom is that of Z electrons interacting with the nucleus *and* Z electrons interacting *with each other*. The electron-electron interaction makes the atomic-structure problem more difficult than the solar-system problem, and it proved to be the downfall of the simple Bohr model. The planets in the solar system do exert attractive gravitational forces on each other, but their masses are so much less than that of the sun that these planet-planet forces are insignificant for all but the most precise calculations. Not so in an atom. The electron charge is the same as the proton charge, so the electron-electron repulsion is just as important to atomic structure as is the electron-nucleus attraction.

The potential energy due to electron-electron interactions fluctuates rapidly in value as the electrons move and the distances between them change. Rather than treat this interaction in detail, we can reasonably consider each electron to be moving in an *average* potential due to all the other electrons. That is, electron i has potential energy

$$U(r_i) = -\frac{Ze^2}{4\pi\epsilon_0 r_i} + U_{\text{elec}}(r_i) \tag{41.14}$$

where the first term is the electron's interaction with the Z protons in the nucleus and U_{elec} is the average potential energy due to all the other electrons. Because each electron is treated independently of the other electrons, this approach is called the **independent particle approximation,** or IPA. This approximation allows the Schrödinger equation for the atom to be broken into Z separate equations, one for each electron.

A major consequence of the IPA is that **each electron can be described by a wave function having the same four quantum numbers n, l, m, and m_s used to describe the single electron of hydrogen.** Because m and m_s do not affect the energy, we can still refer to electrons by their n and l quantum numbers, using the same labeling scheme that we used for hydrogen.

A major difference, however, is that the energy of an electron in a multielectron atom depends on both n *and* l. Whereas the $2s$ and $2p$ states in hydrogen had the same energy, their energies are different in a multielectron atom. The difference arises from the electron-electron interactions that do not exist in a single-electron hydrogen atom.

FIGURE 41.15 shows an energy-level diagram for the electrons in a multielectron atom. For comparison, the hydrogen-atom energies are shown on the right edge of the figure. The comparison is quite interesting. States in a multielectron atom that have small values of l are significantly lower in energy than the corresponding state in hydrogen. For each n, the energy increases as l increases until the maximum-l state has an energy very nearly that of the same n in hydrogen. Can we understand this pattern?

Indeed we can. Recall that states of lower l correspond to elliptical classical orbits and the highest-l state corresponds to a circular orbit. Except for the smallest values of n, an electron in a circular orbit spends most of its time *outside* the electron cloud of the remaining electrons. This is illustrated in **FIGURE 41.16**. The outer electron is orbiting a ball of charge consisting of Z protons and $(Z-1)$ electrons. This ball of charge has *net* charge $q_{\text{net}} = +e$, so the outer electron "thinks" it is orbiting a proton. An electron in a maximum-l state is nearly indistinguishable from an electron in the hydrogen atom; thus its energy is very nearly that of hydrogen.

The low-l states correspond to elliptical orbits. A low-l electron penetrates in very close to the nucleus, which is no longer shielded by the other electrons. The electron's interaction with the Z protons in the nucleus is much stronger than the interaction it would have with the single proton in a hydrogen nucleus. This strong interaction *lowers* its energy in comparison to the same state in hydrogen.

As we noted earlier, a quantum electron does not really orbit. Even so, the probability density of a $3s$ electron has in-close peaks that are missing in the probability density of a $3d$ electron, as you should confirm by looking back at Figure 41.8. Thus a low-l electron really does have a likelihood of being at small r, where its interaction with the Z protons is strong, whereas a high-l electron is most likely to be farther from the nucleus.

FIGURE 41.15 An energy-level diagram for electrons in a multielectron atom.

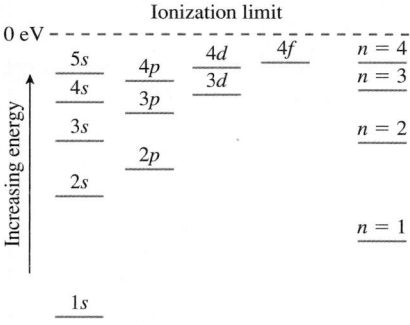

FIGURE 41.16 High-l and low-l orbitals in a multielectron atom.

A high-l electron corresponds to a circular orbit. It stays outside the core of inner electrons and sees a net charge of $+e$, so it behaves like an electron in a hydrogen atom.

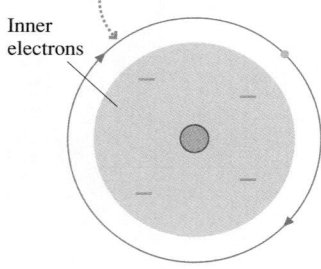

High-l state

A low-l electron corresponds to an elliptical orbit. It penetrates into the core and interacts strongly with the nucleus. The electron-nucleus force is attractive, so this interaction lowers the electron's energy.

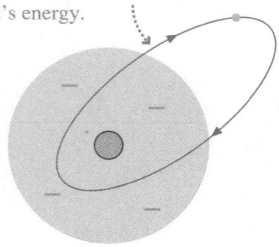

Low-l state

The Pauli Exclusion Principle

By definition, the ground state of a quantum system is the state of lowest energy. What is the ground state of an atom having Z electrons and Z protons? Because the $1s$ state is the lowest energy state in the independent particle approximation, it seems that the ground state should be one in which all Z electrons are in the $1s$ state. However, this idea is not consistent with the experimental evidence.

In 1925, the young Austrian physicist Wolfgang Pauli hypothesized that no two electrons in a quantum system can be in the same quantum state. That is, **no two electrons can have exactly the same set of quantum numbers** (n, l, n, m_s). If one electron is present in a state, it *excludes* all others. This statement, which is called the **Pauli exclusion principle,** turns out to be an extremely profound statement about the nature of matter.

The exclusion principle is not applicable to the hydrogen atom, which has only a single electron. But in helium, with $Z = 2$ electrons, we must make sure that the two electrons are in different quantum states. This is not difficult. For a $1s$ state, with $l = 0$, the only possible value of the magnetic quantum number is $m = 0$. But there are *two* possible values of m_s, namely $+\frac{1}{2}$ and $-\frac{1}{2}$. If a first electron is in the spin-up $1s$ state $(1, 0, 0, +\frac{1}{2})$, a second $1s$ electron can still be added to the atom as long as it is in the spin-down state $(1, 0, 0, -\frac{1}{2})$. This is shown schematically in FIGURE 41.17a, where the dots represent electrons on the rungs of the "energy ladder" and the arrows represent spin-up or spin-down.

The Pauli exclusion principle does not prevent both electrons of helium from being in the $1s$ state as long as they have opposite values of m_s, so we predict this to be the ground state. A list of an atom's occupied energy levels is called its **electron configuration.** The electron configuration of the helium ground state is written $1s^2$, where the superscript 2 indicates two electrons in the $1s$ energy level. An excited state of the helium atom might be the electron configuration $1s2s$. This state is shown in FIGURE 41.17b. Here, because the two electrons have different values of n, there is no restriction on their values of m_s.

The states $(1, 0, 0, +\frac{1}{2})$ and $(1, 0, 0, -\frac{1}{2})$ are the only two states with $n = 1$. The ground state of helium has one electron in each of these states, so all the possible $n = 1$ states are filled. Consequently, the electron configuration $1s^2$ is called a **closed shell.** Because the two electron magnetic moments point in opposite directions, we can predict that helium has *no* net magnetic moment and will be undeflected in a Stern-Gerlach apparatus. This prediction is confirmed by experiment.

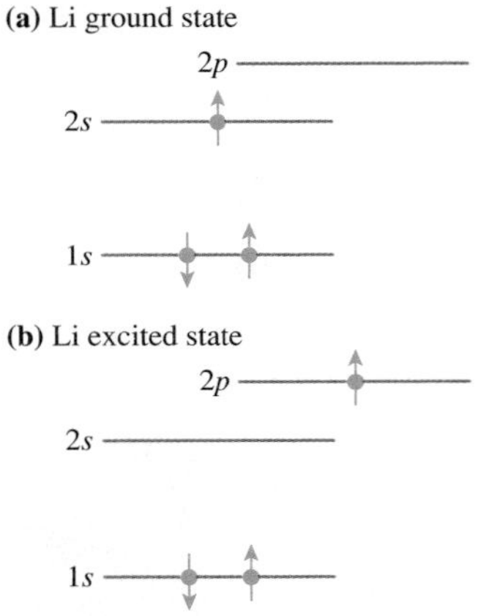

The next element, lithium, has $Z = 3$ electrons. The first two electrons can go into $1s$ states, with opposite values of m_s, but what about the third electron? The $1s^2$ shell is closed, and there are no additional quantum states having $n = 1$. The only option for the third electron is the next energy state, $n = 2$. The $2s$ and $2p$ states had equal energies in the hydrogen atom, but they do *not* in a multi-electron atom. As Figure 41.15 showed, a lower-l state has lower energy than a higher-l state with the same n. The $2s$ state of lithium is lower in energy than $2p$, so lithium's third ground-state electron will be $2s$. This requires $l = 0$ and $m = 0$ for the third electron, but the value of m_s is not relevant because there is only a single electron in $2s$. FIGURE 41.18a shows the electron configuration with the $2s$ electron being spin-up, but it could equally well be spin-down. The electron configuration for the lithium ground state is written $1s^2 2s$. This indicates two $1s$ electrons and a single $2s$ electron.

FIGURE 41.19a shows the probability density of electrons in the $1s^2 2s$ ground state of lithium. You can see the $2s$ electron shell surrounding the inner $1s^2$ core. For comparison, FIGURE 41.19b shows the *first excited state* of lithium, in which the $2s$ electron has been excited to the $2p$ energy level. This forms the $1s^2 2p$ configuration, also shown in FIGURE 41.18b.

FIGURE 41.17 The ground state and the first excited state of helium.

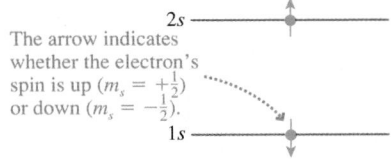

(a) He ground state

(b) He excited state

FIGURE 41.18 The ground state and the first excited state of lithium.

(a) Li ground state

(b) Li excited state

FIGURE 41.19 Electron clouds for the lithium electron configurations $1s^2 2s$ and $1s^2 2p$.

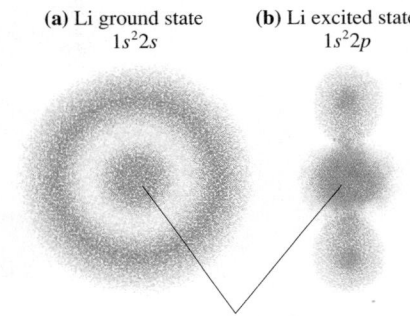

(a) Li ground state
$1s^2 2s$

(b) Li excited state
$1s^2 2p$

Inner core of two $1s$ electrons

The Schrödinger equation accurately predicts the energies of the $1s^2 2s$ and the $1s^2 2p$ configurations of lithium, but the Schrödinger equation does not tell us which states the electrons actually occupy. The electron spin and the Pauli exclusion principle were the final pieces of the puzzle. Once these were added to Schrödinger's theory, the initial phase of quantum mechanics was complete. Physicists finally had a successful theory for understanding the structure of atoms.

41.5 The Periodic Table of the Elements

The 19th century was a time when scientists were discovering new elements and studying their chemical properties. Several chemists in the 1860s began to point out the regular recurrence of chemical properties. For example, there are obvious similarities among the alkali metals lithium, sodium, potassium, and cesium. But attempts at organization were hampered by the fact that many elements had yet to be discovered.

The Russian chemist Dmitri Mendeléev was the first to propose, in 1867, a *periodic* arrangement of the elements. He did so by explicitly pointing out "gaps" where, according to his hypothesis, undiscovered elements should exist. He could then predict the expected properties of the missing elements. The subsequent discovery of these elements verified Mendeléev's organizational scheme, which came to be known as the *periodic table of the elements*.

FIGURE 41.20 shows a modern periodic table. (A larger version can be found in Appendix B.) The significance of the periodic table to a physicist is the implication that there is a basic regularity or periodicity to the *structure* of atoms. Any successful theory of the atom needs to explain *why* the periodic table looks the way it does.

FIGURE 41.20 The modern periodic table of the elements, showing the atomic number Z of each.

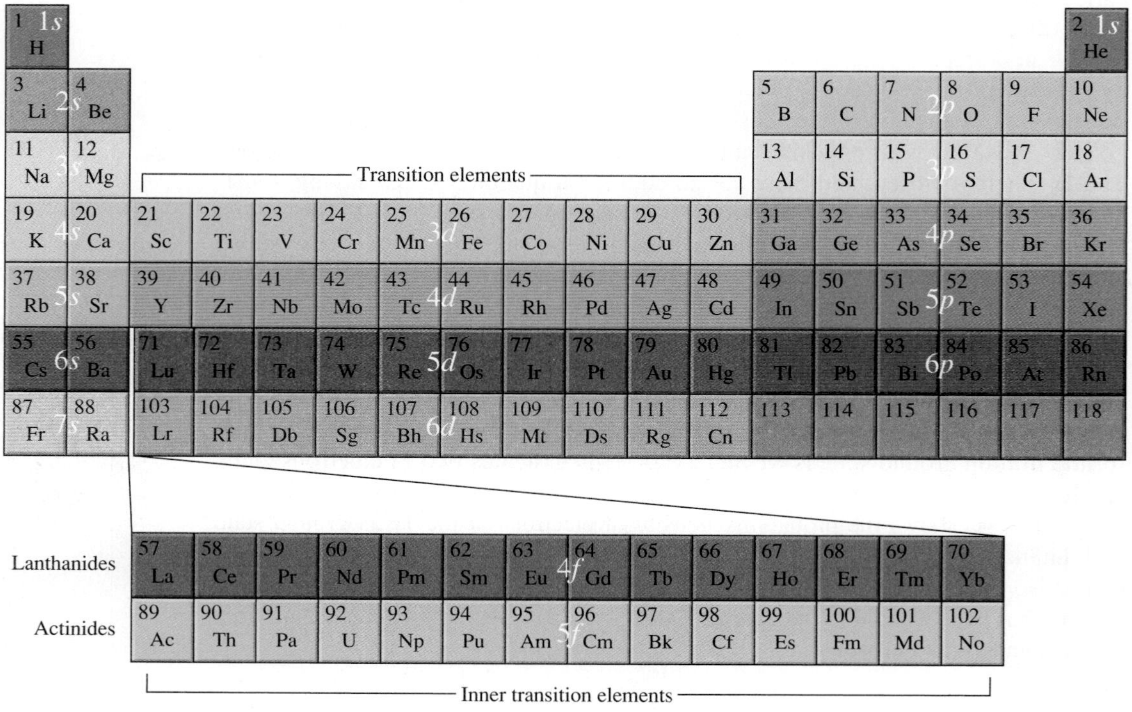

The First Two Rows

Quantum mechanics successfully explains the structure of the periodic table. We need three basic ideas to see how this works:

1. The energy levels of an atom are found by solving the Schrödinger equation for multielectron atoms. Figure 41.15, a very important figure for understanding the periodic table, showed that the energy depends on the quantum numbers n and l.

2. For each value l of the orbital quantum number, there are $2l + 1$ possible values of the magnetic quantum number m and, for each of these, two possible values of the spin quantum number m_s. Consequently, each energy *level* in Figure 41.15 is actually $2(2l + 1)$ different *states*. Each of these states has the same energy.

3. The ground state of the atom is the lowest-energy electron configuration that is consistent with the Pauli exclusion principle.

We used these ideas in the last section to look at the elements helium ($Z = 2$) and lithium ($Z = 3$). Four-electron beryllium ($Z = 4$) comes next. The first two electrons go into $1s$ states, forming a closed shell, and the third goes into $2s$. There is room in the $2s$ level for a second electron as long as its spin is opposite that of the first $2s$ electron. Thus the third and fourth electrons occupy states $(2, 0, 0, +\frac{1}{2})$ and $(2, 0, 0, -\frac{1}{2})$. These are the only two possible $2s$ states. All the states with the same values of n and l are called a **subshell,** so the fourth electron closes the $2s$ subshell. (The outer two electrons are called a subshell, rather than a shell, because they complete only the $2s$ possibilities. There are still spaces for $2p$ electrons.) The ground state of beryllium, shown in FIGURE 41.21, is $1s^2 2s^2$.

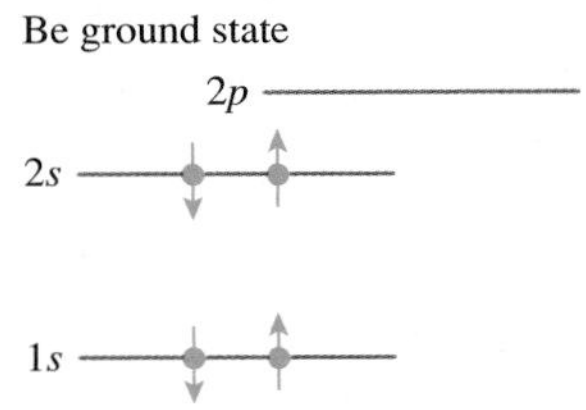

These principles can continue to be applied as we work our way through the elements. There are $2l + 1$ values of m associated with each value of l, and each of these can have $m_s = \pm\frac{1}{2}$. This gives, altogether, $2(2l + 1)$ distinct quantum states in each nl subshell. Table 41.2 lists the number of states in each subshell.

Boron ($1s^2 2s^2 2p$) opens the $2p$ subshell. The remaining possible $2p$ states are filled as we continue across the second row of the periodic table. These elements are shown in FIGURE 41.22. With neon ($1s^2 2s^2 2p^6$), which has six $2p$ electrons, the $n = 2$ shell is complete, and we have another closed shell. The second row of the periodic table is eight elements wide because of the two $2s$ electrons *plus* the six $2p$ electrons needed to fill the $n = 2$ shell.

FIGURE 41.21 The ground state of beryllium ($Z = 4$).

TABLE 41.2 Number of states in each subshell of an atom

Subshell	l	Number of states
s	0	2
p	1	6
d	2	10
f	3	14

FIGURE 41.22 Filling the $2p$ subshell with the elements boron through neon.

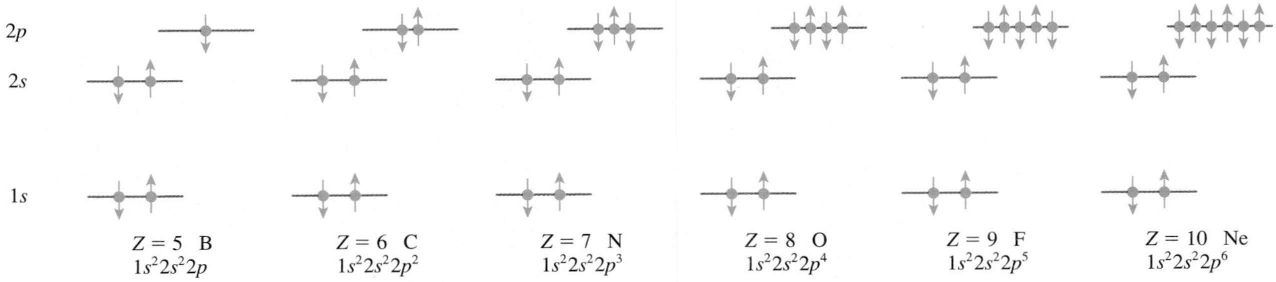

| $Z = 5$ B | $Z = 6$ C | $Z = 7$ N | $Z = 8$ O | $Z = 9$ F | $Z = 10$ Ne |
| $1s^2 2s^2 2p$ | $1s^2 2s^2 2p^2$ | $1s^2 2s^2 2p^3$ | $1s^2 2s^2 2p^4$ | $1s^2 2s^2 2p^5$ | $1s^2 2s^2 2p^6$ |

Elements with $Z > 10$

The third row of the periodic table is similar to the second. The two $3s$ states are filled in sodium and magnesium. The two columns on the left of the periodic table represent the two electrons that can go into an s subshell. Then the six $3p$ states are filled, one by one, in aluminum through argon. The six columns on the right represent the six electrons of the p subshell. Argon ($Z = 18$, $1s^2 2s^2 2p^6 3s^2 3p^6$) is another inert gas, although this may seem surprising because the $3d$ subshell is still open.

The fourth row is where the periodic table begins to get complicated. You might expect the closure of the 3p subshell in argon to be followed, starting with potassium ($Z = 19$), by filling the 3d subshell. But if you look back at Figure 41.15, where the energies of the different nl states are shown, you will see that the 3d state is slightly *higher* in energy than the 4s state. Because the ground state is the *lowest energy state* consistent with the Pauli exclusion principle, potassium finds it more favorable to fill a 4s state than to fill a 3d state. Thus the ground-state configuration of potassium is $1s^2 2s^2 2p^6 3s^2 3p^6 4s$ rather than the expected $1s^2 2s^2 2p^6 3s^2 3p^6 3d$.

At this point, we begin to see a competition between increasing n and decreasing l. The highly elliptical characteristic of the 4s state brings part of its orbit in so close to the nucleus that its energy is less than that of the more circular 3d state. The 4p state, though, reverts to the "expected" pattern. We find that

$$E_{4s} < E_{3d} < E_{4p}$$

so the states across the fourth row are filled in the order 4s, then 3d, and finally 4p.

Because there had been no previous d states, the 3d subshell "splits open" the periodic table to form the 10-element-wide group of *transition elements*. Most commonly occurring metals are transition elements, and their metallic properties are determined by their partially filled d subshell. The 3d subshell closes with zinc, at $Z = 30$, then the next six elements fill the 4p subshell up to krypton, at $Z = 36$.

Things get even more complex starting in the sixth row, but the ideas are familiar. The $l = 3$ subshell (f electrons) becomes a possibility with $n = 4$, but it turns out that the 5s, 5p, and 6s states are all lower in energy than 4f. Not until barium ($Z = 56$) fills the 6s subshell (and lanthanum ($Z = 57$) adds a 5d electron) is it energetically favorable to add a 4f electron. Immediately after barium you have to switch down to the *lanthanides* at the bottom of the table. The lanthanides fill in the 4f states.

The 4f subshell is complete with $Z = 70$ ytterbium. Then $Z = 71$ lutetium through $Z = 80$ mercury complete the transition-element 5d subshell, followed by the 6p subshell in the six elements thallium through radon at the end of the sixth row. Radon, the last inert gas, has $Z = 86$ electrons and the ground-state configuration

$$1s^2 2s^2 2p^6 3s^2 3p^6 4s^2 3d^{10} 4p^6 5s^2 4d^{10} 5p^6 6s^2 4f^{14} 5d^{10} 6p^6$$

This is frightening to behold, but we can now understand it!

EXAMPLE 41.4 **The ground state of arsenic**

Predict the ground-state electron configuration of arsenic.

SOLVE The periodic table shows that arsenic (As) has $Z = 33$, so we must identify the states of 33 electrons. Arsenic is in the fourth row, following the first group of transition elements. Argon ($Z = 18$) filled the 3p subshell, then calcium ($Z = 20$) filled the 4s subshell. The next 10 elements, through zinc ($Z = 30$), filled the 3d subshell. The 4p subshell starts filling with gallium ($Z = 31$), and arsenic is the third element in this group, so it will have three 4p electrons. Thus the ground-state configuration of arsenic is

$$1s^2 2s^2 2p^6 3s^2 3p^6 4s^2 3d^{10} 4p^3$$

The white lettering on the periodic table of Figure 41.20 summarizes the results, showing the subshells as they are filled. It is especially important to note how the electron's spin is absolutely essential for understanding the periodic table. Explaining the periodic table of the elements is a remarkable success of the quantum model of the atom.

Ionization Energies

Ionization energy is the minimum energy needed to remove a ground-state electron from an atom and leave a positive ion behind. The ionization energy of hydrogen is 13.60 eV because the ground-state energy is $E_1 = -13.60$ eV. FIGURE 41.23 shows the ionization energies of the first 60 elements in the periodic table.

FIGURE 41.23 Ionization energies of the elements up to $Z = 60$.

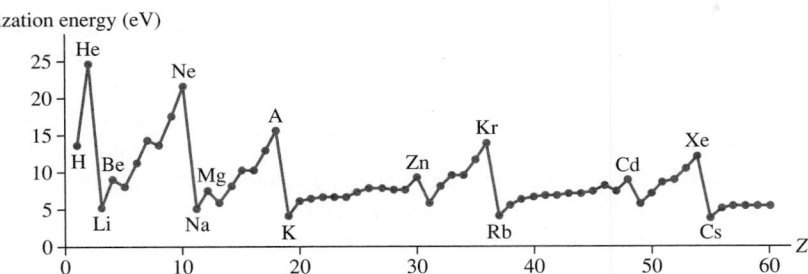

The ionization energy is different for each element, but there's a clear pattern to the values. Ionization energies are ≈ 5 eV for the alkali metals, on the left edge of the periodic table, then increase steadily to ≥ 15 eV for the inert gases before plunging back to ≈ 5 eV. Can the quantum theory of atoms explain this recurring pattern in the ionization energies?

Indeed it can. The inert-gas elements (helium, neon, argon, . . .) in the right column of the periodic table have *closed shells*. A closed shell is a very stable structure, and that is why these elements are chemically nonreactive (i.e., inert). It takes a large amount of energy to pull an electron out of a stable closed shell; thus the inert gases have the largest ionization energies.

The alkali metals, in the left column of the periodic table, have a single *s*-electron outside a closed shell. This electron is easily disrupted, which is why these elements are highly reactive and have the lowest ionization energies. Between the edges of the periodic table are elements such as beryllium ($1s^2 2s^2$) with a closed $2s$ subshell. You can see in Figure 41.23 that the closed subshell gives beryllium a larger ionization energy than its neighbors lithium ($1s^2 2s$) or boron ($1s^2 2s^2 2p$). However, a closed subshell is not nearly as tightly bound as a closed shell, so the ionization energy of beryllium is much less than that of helium or neon.

All in all, you can see that the basic idea of shells and subshells, which follows from the Schrödinger-equation energy levels and the Pauli principle, provides a good understanding of the recurring features in the ionization energies.

STOP TO THINK 41.4   Is the electron configuration $1s^2 2s^2 2p^4 3s$ a ground-state configuration or an excited-state configuration?

a. Ground-state   b. Excited-state
c. It's not possible to tell without knowing which element it is.

41.6 Excited States and Spectra

The periodic table organizes information about the *ground states* of the elements. These states are chemically most important because most atoms spend most of the time in their ground states. All the chemical ideas of valence, bonding, reactivity, and so on are consequences of these ground-state atomic structures. But the periodic table does not tell us anything about the excited states of atoms. It is the excited states that hold the key to understanding atomic spectra, and that is the topic to which we turn next.

FIGURE 41.24 The [Ne]3s ground state of the sodium atom and some of the excited states.

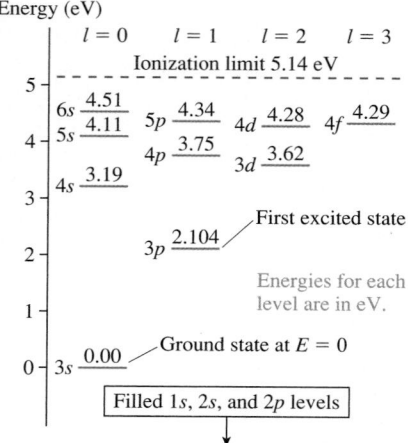

Sodium ($Z = 11$) is a multielectron atom that we will use as a prototypical atom. The ground-state electron configuration of sodium is $1s^2 2s^2 2p^6 3s$. The first 10 electrons completely fill the $n = 1$ and $n = 2$ shells, creating a *neon core*, while the $3s$ electron is a valence electron. It is customary to represent this configuration as [Ne]3s or, more simply, as just $3s$.

The excited states of sodium are produced by raising the valence electron to a higher energy level. The electrons in the neon core are unchanged. Thus the excited states can be labeled [Ne]*nl* or, more simply, just *nl*. FIGURE 41.24 is an energy-level diagram showing the ground state and some of the excited states of sodium. Notice that the $1s$, $2s$, and $2p$ states of the neon core are not shown on the diagram. These states are filled and unchanging, so only the states available to the valence electron are shown.

Figure 41.24 has a new feature: The zero of energy has been shifted to the ground state. As we have discovered many times, the zero of energy can be located where it is most convenient. For analyzing spectra it is convenient to let the ground state have $E = 0$. With this choice, the excited-state energies tell us how far each state is above the ground state. The ionization limit now occurs at the value of the atom's ionization energy, which is 5.14 eV for sodium.

The first energy level above $3s$ is $3p$, so the *first excited state* of sodium is $1s^2 2s^2 2p^6 3p$, written as [Ne]3p or, more simply, $3p$. The valence electron is excited, while the core electrons are unchanged. This state is followed, in order of increasing energy, by [Ne]4s, [Ne]3d, and [Ne]4p. Notice that the order of excited states is exactly the same order ($3p$–$4s$–$3d$–$4p$) that explained the fourth row of the periodic table.

Other atoms with a single valence electron have energy-level diagrams similar to that of sodium. Things get more complicated when there is more than one valence electron, so we'll defer those details to more advanced courses. Nevertheless, you already can *utilize* the information shown on an energy-level diagram without having to understand precisely *why* each level is where it is.

Excitation by Absorption

Left to itself, an atom will be in its lowest-energy ground state. How does an atom get into an excited state? The process of getting an atom into an excited state is called **excitation,** and there are two basic mechanisms: absorption and collision. We'll begin by looking at excitation by absorption.

One postulate of the Bohr model is that an atom can jump from one stationary state, of energy E_1, to a higher-energy state E_2 by absorbing a photon of frequency

$$f = \frac{\Delta E_{\text{atom}}}{h} = \frac{E_2 - E_1}{h} \qquad (41.15)$$

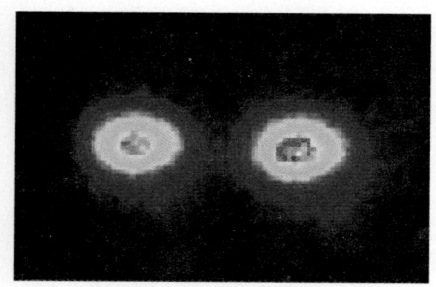

The dots of light are being emitted by two beryllium ions held in a device called an ion trap. Each ion, which is excited by an invisible ultraviolet laser, emits about 10^6 visible-light photons per second.

Because we are interested in spectra, it is more useful to write Equation 41.15 in terms of the wavelength:

$$\lambda = \frac{c}{f} = \frac{hc}{\Delta E_{\text{atom}}} = \frac{1240 \text{ eV nm}}{\Delta E \text{ (in eV)}} \qquad (41.16)$$

The final expression, which uses the value $hc = 1240$ eV nm, gives the wavelength in nanometers *if* ΔE_{atom} is in electron volts.

Bohr's idea of quantum jumps remains an integral part of our interpretation of the results of quantum mechanics. By absorbing a photon, an atom jumps from its ground state to one of its excited states. However, a careful analysis of how the electrons in an atom interact with a light wave shows that not every conceivable transition can occur. The **allowed transitions** must satisfy a **selection rule:** A transition (either absorption or emission) from a state in which the valence electron has orbital quantum number l_1 to another with orbital quantum number l_2 is allowed only if

$$\Delta l = |l_2 - l_1| = 1 \quad \text{(selection rule for emission and absorption)} \qquad (41.17)$$

That is, the electron's orbital quantum number must change by exactly 1. Thus an atom in an *s* state ($l = 0$) can absorb a photon and be excited to a *p* state ($l = 1$) but *not* to another *s* state or to a *d* state. An atom in a *p* state ($l = 1$) can emit a photon by dropping to a lower-energy *s* state *or* to a lower-energy *d* state but not to another *p* state.

EXAMPLE 41.5 | **Absorption in hydrogen**

What is the longest wavelength in the absorption spectrum of hydrogen? What is the transition?

SOLVE The longest wavelength corresponds to the smallest energy change ΔE_{atom}. Because the atom starts from the $1s$ ground state, the smallest energy change occurs for absorption to the first $n = 2$ excited state. The energy change is

$$\Delta E_{\text{atom}} = E_2 - E_1 = \frac{-13.6 \text{ eV}}{2^2} - \frac{-13.6 \text{ eV}}{1^2} = 10.2 \text{ eV}$$

The wavelength of this transition is

$$\lambda = \frac{1240 \text{ eV nm}}{10.2 \text{ eV}} = 122 \text{ nm}$$

This is an ultraviolet wavelength. Because of the selection rule, the transition is $1s \rightarrow 2p$, not $1s \rightarrow 2s$.

EXAMPLE 41.6 | **Absorption in sodium**

What is the longest wavelength in the absorption spectrum of sodium? What is the transition?

SOLVE The sodium ground state is [Ne]$3s$. The lowest excited state is the $3p$ state. $3s \rightarrow 3p$ is an allowed transition ($\Delta l = 1$), so this is the longest wavelength. You can see from the data in Figure 41.24 that $\Delta E_{\text{atom}} = 2.104$ eV for this transition.

The corresponding wavelength is

$$\lambda = \frac{1240 \text{ eV nm}}{2.104 \text{ eV}} = 589 \text{ nm}$$

ASSESS This wavelength (yellow color) is a prominent feature in the spectrum of sodium. Because the ground state has $l = 0$, absorption *must* be to a *p* state. The *s* states and *d* states of sodium cannot be excited by absorption.

Collisional Excitation

A particle traveling with a speed of 1.0×10^6 m/s has a kinetic energy of 2.85 eV. If this particle collides with a ground-state sodium atom, a portion of its energy can be used to excite the atom to its $3p$ state. This process is called **collisional excitation** of the atom.

Collisional excitation differs from excitation by absorption in one very fundamental way. In absorption, the photon disappears. Consequently, *all* of the photon's energy must be transferred to the atom. Conservation of energy requires $E_{\text{photon}} = \Delta E_{\text{atom}}$. In contrast, the particle is still present after collisional excitation and can carry away some kinetic energy. That is, the particle does *not* have to transfer its entire energy to the atom. If the particle has an incident kinetic energy of 2.85 eV, it could transfer 2.10 eV to the sodium atom, thereby exciting it to the $3p$ state, and still depart the collision with an energy of 0.75 eV.

To excite the atom, the incident energy of the particle merely has to *exceed* ΔE_{atom}. That is $E_{\text{particle}} \geq \Delta E_{\text{atom}}$. There's a threshold energy for exciting the atom, but no upper limit. It is all a matter of energy conservation. **FIGURE 41.25** shows the idea graphically.

Collisional excitation by electrons is the predominant method of excitation in electrical discharges such as fluorescent lights, street lights, and neon signs. A gas is placed in a tube at reduced pressure (≈ 1 mm of Hg), then a fairly high voltage (≈ 1000 V) between electrodes at the ends of the tube causes the gas to ionize, creating a current in which both ions and electrons are charge carriers. The mean free path of electrons between collisions is large enough for the electrons to gain several eV of kinetic energy as they accelerate in the electric field. This energy is then transferred to the gas atoms upon collision. The process does not work at atmospheric pressure because the mean free path between collisions is too short for the electrons to gain enough kinetic energy to excite the atoms.

NOTE ▶ There are no selection rules for collisional excitation. Any state can be excited if the colliding particle has sufficient energy. ◀

FIGURE 41.25 Excitation by photon absorption and electron collision.

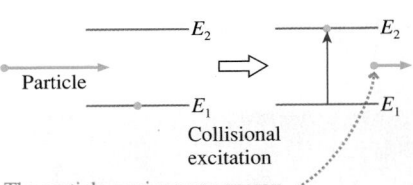

The photon disappears. Energy conservation requires $E_{\text{photon}} = E_2 - E_1$.

Absorption

Collisional excitation

The particle carries away energy. Energy conservation requires $E_{\text{particle}} \geq E_2 - E_1$.

EXAMPLE 41.7 **Excitation of hydrogen**

Can an electron traveling at 2.0×10^6 m/s cause a hydrogen atom to emit the prominent red spectral line ($\lambda = 656$ nm) in the Balmer series?

MODEL The electron must have sufficient energy to excite the upper state of the transition.

SOLVE The electron's energy is $E_{elec} = \frac{1}{2}mv^2 = 11.4$ eV. This is significantly larger than the 1.89 eV energy of a photon with wavelength 656 nm, but don't confuse the energy of the photon with the energy of the excitation. The red spectral line in the

Balmer series is emitted by an $n = 3$ to $n = 2$ quantum jump with $\Delta E_{atom} = 1.89$ eV. But to cause this emission, the electron must excite an atom from its *ground state*, with $n = 1$, up to the $n = 3$ level. The necessary excitation energy is

$$\Delta E_{atom} = E_3 - E_1 = (-1.51 \text{ eV}) - (-13.60 \text{ eV})$$

$$= 12.09 \text{ eV}$$

The electron does *not* have sufficient energy to excite the atom to the state from which the emission would occur.

FIGURE 41.26 Generation of an emission spectrum.

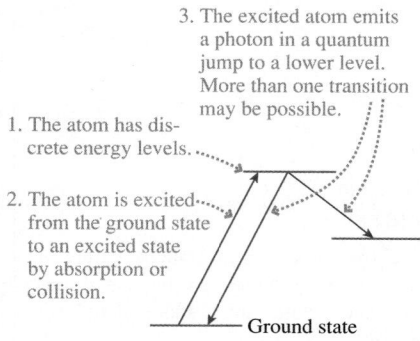

3. The excited atom emits a photon in a quantum jump to a lower level. More than one transition may be possible.

1. The atom has discrete energy levels.

2. The atom is excited from the ground state to an excited state by absorption or collision.

Ground state

FIGURE 41.27 The emission spectrum of sodium.

(a)

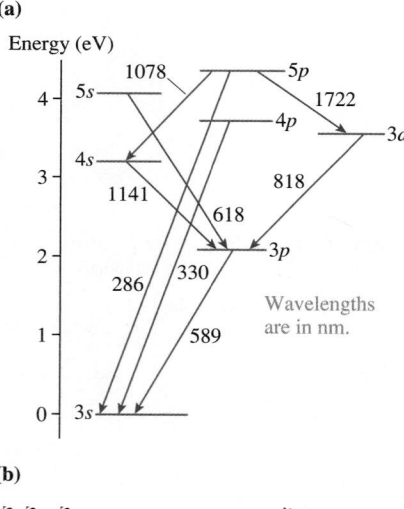

(b)

Emission Spectra

The absorption of light is an important process, but it is the emission of light that really gets our attention. The overwhelming bulk of sensory information that we perceive comes to us in the form of light. With the small exception of cosmic rays, all of our knowledge about the cosmos comes to us in the form of light and other electromagnetic waves emitted in various processes.

Understanding emission hinges on the three ideas shown in **FIGURE 41.26**. Once we have determined the energy levels of an atom, by solving the Schrödinger equation, we can immediately predict its emission spectrum. Conversely, we can use the measured emission spectrum to determine an atom's energy levels.

As an example, **FIGURE 41.27a** shows some of the transitions and wavelengths observed in the emission spectrum of sodium. This diagram makes the point that each wavelength represents a quantum jump between two well-defined energy levels. Notice that the selection rule $\Delta l = 1$ is being obeyed in the sodium spectrum. The $5p$ levels can undergo quantum jumps to $3s$, $4s$, or $3d$ but *not* to $3p$ or $4p$.

FIGURE 41.27b shows the emission spectrum of sodium as it would be recorded in a spectrometer. (Many of the lines seen in this spectrum start from higher excited states that are not seen in the rather limited energy-level diagram of Figure 41.27a.) By comparing the spectrum to the energy-level diagram, you can recognize that the spectral lines at 589 nm, 330 nm, 286 nm, and 268 nm form a *series* of lines due to all the possible $np \rightarrow 3s$ transitions. They are the dominant features in the sodium spectrum.

The most obvious visual feature of sodium emission is its bright yellow color, produced by the emission wavelength of 589 nm. This is the basis of the *flame test* used in chemistry to test for sodium: A sample is held in a Bunsen burner, and a bright yellow glow indicates the presence of sodium. The 589 nm emission is also prominent in the pinkish-yellow glow of the common sodium-vapor street lights. These operate by creating an electrical discharge in sodium vapor. Most sodium-vapor lights use high-pressure lamps to increase their light output. The high pressure, however, causes the formation of Na_2 molecules, and these molecules, which have a different spectral fingerprint, emit the pinkish portion of the light.

Some cities close to astronomical observatories use low-pressure sodium lights, and these emit the distinctive yellow 589 nm light of sodium. The glow of city lights is a severe problem for astronomers, but the very specific 589 nm emission from sodium is easily removed with a *sodium filter*. The light from the telescope is passed through a container of sodium vapor, and the sodium atoms *absorb* only the unwanted 589 nm photons without disturbing any other wavelengths! However, this cute trick does not work for the other wavelengths emitted by high-pressure sodium lamps or light from other sources.

Color in Solids

It is worth concluding this section with a few remarks about color in solids. Whether it is the intense multihued colors of a stained glass window, the bright colors of flowers or paint, or the deep luminescent red of a ruby, most of the colors we perceive in our lives come from solids rather than free atoms. The basic principles are the same, but the details are different for solids.

An excited atom in a gas has little choice but to give up its energy by emitting a photon. Its only other option, which is rare for gas atoms, is to collide with another atom and transfer its energy into the kinetic energy of recoil. But the atoms in a solid are in intimate contact with each other at all times. Although an excited atom in a solid has the option of emitting a photon, it is often more likely that the energy will be converted, via interactions with neighboring atoms, to the thermal energy of the solid. A process in which an atom is de-excited without radiating is called a **nonradiative transition.**

This is what happens in pigments, such as those in paints, plants, and dyes. Pigment molecules absorb certain wavelengths of light but not other wavelengths. The energy-level structure of a molecule is complex, so the absorption consists of "bands" of wavelengths rather than discrete spectral lines. But instead of re-radiating the energy by photon emission, as a free atom would, the pigment molecules undergo nonradiative transitions and convert the energy into increased thermal energy. That is why darker objects get hotter in the sun than lighter objects.

When light falls on an object, it can be either absorbed or reflected. If *all* wavelengths are reflected, the object is perceived as white. Any wavelengths absorbed by the pigments are removed from the reflected light. A pigment with blue-absorbing properties converts the energy of blue-wavelength photons into thermal energy, but photons of other wavelengths are reflected without change. A blue-absorbing pigment reflects the red and yellow wavelengths, causing the object to be perceived as the color orange!

Some solids, though, are a little different. The color of many minerals and crystals is due to so-called *impurity atoms* embedded in them. For example, the gemstone ruby is a very simple and common crystal of aluminum oxide, called corundum, that happens to have chromium atoms present at the concentration of about one part in a thousand. Pure corundum is transparent, so all of a ruby's color comes from these chromium impurity atoms.

FIGURE 41.28 shows what happens when ruby is illuminated by white light. The chromium atoms have a group of excited states that absorb all wavelengths shorter than about 600 nm—that is, everything except orange and red. Unlike the pigments in red glass, which convert all the absorbed energy into thermal energy, the chromium atoms dissipate only a small amount of heat as they undergo a nonradiative transition to another excited state. From there they emit a photon with $\lambda = hc/(E_2 - E_1) \approx 690$ nm (dark red color) as they jump back to the ground state.

The net effect is that short-wavelength photons, rather than being completely absorbed, are *re-radiated* as longer-wavelength photons. This is why rubies sparkle and have such intense color, whereas red glass is a dull red color. The color of other minerals and gems is due to different impurity atoms, but the principle is the same.

The colors in a stained-glass window are due to the selective absorption of light.

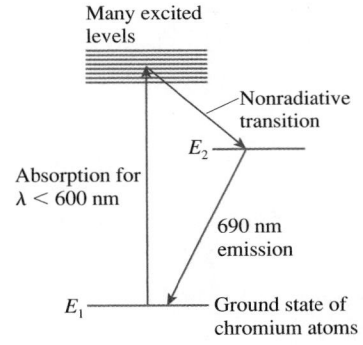

FIGURE 41.28 Absorption and emission in a crystal of ruby.

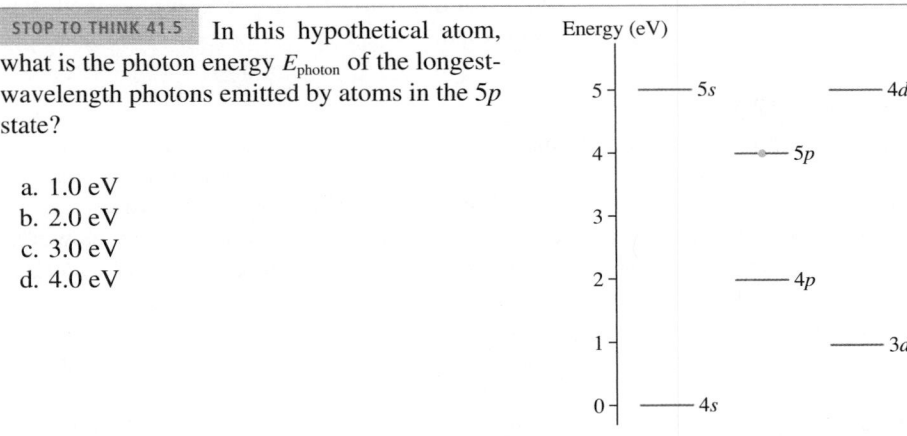

STOP TO THINK 41.5   In this hypothetical atom, what is the photon energy E_{photon} of the longest-wavelength photons emitted by atoms in the $5p$ state?

a. 1.0 eV
b. 2.0 eV
c. 3.0 eV
d. 4.0 eV

41.7 Lifetimes of Excited States

Excitation of an atom, by either absorption or collision, leaves it in an excited state. From there it jumps back to a lower energy level by emitting a photon. How long does this process take? There are actually two questions here. First, how long does an atom remain in an excited state before undergoing a quantum jump to a lower state? Second, how long does the transition last as the quantum jump is occurring?

Our best understanding of the quantum physics of atoms is that quantum jumps are instantaneous. The absorption or emission of a photon is an all-or-nothing event, so there is not a time when a photon is "half emitted." The prediction that quantum jumps are instantaneous has troubled many physicists, but careful experimental tests have never revealed any evidence that the jump itself takes a measurable amount of time.

The time spent in the excited state, waiting to make a quantum jump, is another story. **FIGURE 41.29** shows experimental data for the length of time that doubly charged xenon ions Xe^{++} spend in a certain excited state. In this experiment, a pulse of electrons was used to excite the atoms to the excited state. The number of excited-state atoms was then monitored by detecting the photons emitted—one by one!—as the excited atoms jumped back to the ground state. The number of photons emitted at time t is directly proportional to the number of excited-state atoms present at time t. As the figure shows, the number of atoms in the excited state decreases *exponentially* with time, and virtually all have decayed within 25 ms of their creation.

FIGURE 41.29 Experimental data for the photon emission rate from an excited state in Xe^{++}.

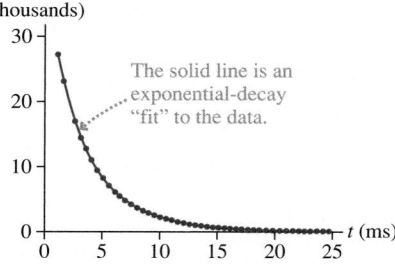

Photon counts (thousands)

The solid line is an exponential-decay "fit" to the data.

Figure 41.29 has two important implications. First, atoms spend time in the excited state before undergoing a quantum jump back to a lower state. Second, the length of time spent in the excited state is not a constant value but varies from atom to atom. If every excited xenon ion lived for 5 ms in the excited state, then we would detect *no* photons for 5 ms, a big burst right at 5 ms as they all decay, then no photons after that. Instead, the data tell us that there is a *range* of times spent in the excited state. Some undergo a quantum jump and emit a photon after 1 ms, others after 5 ms or 10 ms, and a few wait as long as 20 or 25 ms.

Consider an experiment in which N_0 excited atoms are created at time $t = 0$. As the curve in Figure 41.29 shows, the number of excited atoms remaining at time t is well described by the exponential function

$$N_{exc} = N_0 e^{-t/\tau} \tag{41.18}$$

where τ is the point in time at which $e^{-1} = 0.368 = 36.8\%$ of the original atoms remain in the excited state. Thus 63.2% of the atoms, nearly two-thirds, have emitted a photon and jumped to the lower state by time $t = \tau$. The interval of time τ is called the **lifetime** of the excited state. From Figure 41.29 we can deduce that the lifetime of this state in Xe^{++} is ≈ 4 ms because that is the point in time at which the curve has decayed to 36.8% of its initial value.

This lifetime in Xe^{++} is abnormally long, which is why the state was studied. More typical excited-state lifetimes are a few nanoseconds. Table 41.3 gives some measured values of excited-state lifetimes. Whatever the value of τ, the number of excited-state atoms decreases exponentially. Why is this?

TABLE 41.3 Some excited-state lifetimes

Atom	State	Lifetime (ns)
Hydrogen	$2p$	1.6
Sodium	$3p$	17
Neon	$3p$	20
Potassium	$4p$	26

The Decay Equation

Quantum mechanics is about probabilities. We cannot say exactly where the electron is located, but we can use quantum mechanics to calculate the *probability* that the electron is located in a small interval Δx at position x. Similarly, we cannot say exactly when an excited electron will undergo a quantum jump and emit a photon. However, we can use quantum mechanics to find the *probability* that the electron will undergo a quantum jump during a small time interval Δt at time t.

Let us assume that the probability of an excited atom emitting a photon during time interval Δt is *independent* of how long the atom has been waiting in the excited state. For example, a newly excited atom may have a 10% probability of emitting a photon within the 1 ns interval from 0 ns to 1 ns. If it survives until $t = 7$ ns, our assumption is that it still has a 10% probability of emitting a photon during the 1 ns interval from 7 ns to 8 ns.

This assumption, which can be justified with a detailed analysis, is similar to flipping coins. The probability of a head on your first flip is 50%. If you flip seven heads in a row, the probability of a head on your eighth flip is still 50%. It is *unlikely* that you will flip seven heads in a row, but doing so does not influence the eighth flip. Likewise, it may be *unlikely* for an excited atom to live for 7 ns, but doing so does not affect its probability of emitting a photon during the next 1 ns.

If Δt is small, the probability of photon emission during time interval Δt is directly proportional to Δt. That is, if the emission probability in 1 ns is 1%, it will be 2% in 2 ns and 0.5% in 0.5 ns. (This logic fails if Δt gets too big. If the probability is 70% in 20 ns, we can *not* say that the probability would be 140% in 40 ns because a probability > 1 is meaningless.) We will be interested in the limit $\Delta t \rightarrow dt$, so the concept is valid and we can write

$$\text{Prob(emission in } \Delta t \text{ at time } t) = r\,\Delta t \qquad (41.19)$$

where r is called the **decay rate** because the number of excited atoms decays with time. It is a probability *per second*, with units of s^{-1}, and thus is a rate. For example, if an atom has a 5% probability of emitting a photon during a 2 ns interval, its decay rate is

$$r = \frac{P}{\Delta t} = \frac{0.05}{2 \text{ ns}} = 0.025 \text{ ns}^{-1} = 2.5 \times 10^7 \text{ s}^{-1}$$

NOTE ▶ Equation 41.19 is directly analogous to Prob(found in Δx at x) $= P\,\Delta x$, where P, which had units of m^{-1}, was the probability density. ◀

FIGURE 41.30 shows N_{exc} atoms in an excited state. During a small time interval Δt, the number of these atoms that we expect to undergo a quantum jump and emit a photon is N_{exc} multiplied by the probability of decay. That is,

$$\text{number of photons in } \Delta t \text{ at time } t = N_{\text{exc}} \times \text{Prob(emission in } \Delta t \text{ at } t)$$
$$= rN_{\text{exc}}\,\Delta t \qquad (41.20)$$

Now the *change* in N_{exc} is the *negative* of Equation 41.20. For example, suppose 1000 excited atoms are present at time t and each has a 5% probability of emitting a photon in the next 1 ns. On average, the number of photons emitted during the next 1 ns will be $1000 \times 0.05 = 50$. Consequently, the number of excited atoms changes by $\Delta N_{\text{exc}} = -50$, with the minus sign indicating a decrease.

Thus the *change* in the number of atoms in the excited state is

$$\Delta N_{\text{exc}}(\text{in } \Delta t \text{ at } t) = -N_{\text{exc}} \times \text{Prob(decay in } \Delta t \text{ at } t) = -rN_{\text{exc}}\,\Delta t \quad (41.21)$$

Now let $\Delta t \rightarrow dt$. Then $\Delta N_{\text{exc}} \rightarrow dN_{\text{exc}}$ and Equation 41.21 becomes

$$\frac{dN_{\text{exc}}}{dt} = -rN_{\text{exc}} \qquad (41.22)$$

Equation 41.22 is a *rate equation* because it describes the *rate* at which the excited-state population changes. If r is large, the population will decay at a rapid rate and will have a short lifetime. Conversely, a small value of r implies that the population will decay slowly and will live a long time.

The rate equation is a differential equation, but we solved a similar equation for *RC* circuits in Chapter 31. First, we rewrite Equation 41.22 as

$$\frac{dN_{\text{exc}}}{N_{\text{exc}}} = -r\,dt$$

Then we integrate both sides from $t = 0$, when the initial excited-state population is N_0, to an arbitrary time t when the population is N_{exc}. That is,

$$\int_{N_0}^{N_{\text{exc}}} \frac{dN_{\text{exc}}}{N_{\text{exc}}} = -r \int_0^t dt \qquad (41.23)$$

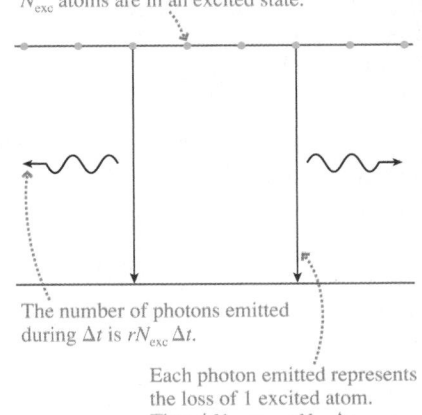

FIGURE 41.30 The number of atoms that emit photons during Δt is directly proportional to the number of excited atoms.

N_{exc} atoms are in an excited state.

The number of photons emitted during Δt is $rN_{\text{exc}}\,\Delta t$.

Each photon emitted represents the loss of 1 excited atom. Thus $\Delta N_{\text{exc}} = -rN_{\text{exc}}\,\Delta t$.

Both are well-known integrals, giving

$$\ln N_{exc} \Big|_{N_0}^{N_{exc}} = \ln N_{exc} - \ln N_0 = \ln\left(\frac{N_{exc}}{N_0}\right) = -rt$$

We can solve for the number of excited atoms at time t by taking the exponential of both sides, then multiplying by N_0. Doing so gives

$$N_{exc} = N_0 e^{-rt} \qquad (41.24)$$

Notice that $N_{exc} = N_0$ at $t = 0$, as expected. Equation 41.24, the *decay equation*, shows that the excited-state population decays exponentially with time, as we saw in the experimental data of Figure 41.29.

It will be more convenient to write Equation 41.24 as

$$N_{exc} = N_0 e^{-t/\tau} \qquad (41.25)$$

where

$$\tau = \frac{1}{r} = \text{the } lifetime \text{ of the excited state} \qquad (41.26)$$

This is the definition of the lifetime we used in Equation 41.18 to describe the experimental results. The lifetime is the inverse of the decay rate r.

EXAMPLE 41.8 **The lifetime of an excited state in mercury**

The mercury atom has two valence electrons. One is always in the $6s$ state, the other is in a state with quantum numbers n and l. One of the excited states in mercury is the state designated $6s6p$. The decay rate of this state is $7.7 \times 10^8 \text{ s}^{-1}$.

a. What is the lifetime of this state?

b. If 1.0×10^{10} mercury atoms are created in the $6s6p$ state at $t = 0$, how many photons will be emitted during the first 1.0 ns?

SOLVE a. The lifetime is

$$\tau = \frac{1}{r} = \frac{1}{7.7 \times 10^8 \text{ s}^{-1}} = 1.3 \times 10^{-9} \text{ s} = 1.3 \text{ ns}$$

b. If there are $N_0 = 1.0 \times 10^{10}$ excited atoms at $t = 0$, the number still remaining at $t = 1.0$ ns is

$$N_{exc} = N_0 e^{-t/\tau} = (1.0 \times 10^{10}) e^{-(1.0 \text{ ns})/(1.3 \text{ ns})} = 4.63 \times 10^9$$

This result implies that 5.37×10^9 atoms undergo quantum jumps during the first 1.0 ns. Each of these atoms emits one photon, so the number of photons emitted during the first 1.0 ns is 5.37×10^9.

STOP TO THINK 41.6 An equal number of excited A atoms and excited B atoms are created at $t = 0$. The decay rate of B atoms is twice that of A atoms: $r_B = 2r_A$. At $t = \tau_A$ (i.e., after one lifetime of A atoms has elapsed), the ratio N_B/N_A of the number of excited B atoms to the number of excited A atoms is

a. >2 b. 2 c. 1 d. $\frac{1}{2}$ e. $<\frac{1}{2}$

41.8 Stimulated Emission and Lasers

We have seen that an atom can jump from a lower-energy level E_1 to a higher-energy level E_2 by absorbing a photon. FIGURE 41.31a illustrates the basic absorption process, with a photon of frequency $f = \Delta E_{atom}/h$ disappearing as the atom jumps from level 1 to level 2. Once in level 2, as shown in FIGURE 41.31b, the atom can emit a photon of the same frequency as it jumps back to level 1. This transition is called **spontaneous emission**.

In 1917, four years after Bohr's proposal of stationary states in atoms but still prior to de Broglie and Schrödinger, Einstein was puzzled by how quantized atoms reach thermodynamic equilibrium in the presence of electromagnetic radiation. Einstein found that absorption and spontaneous emission were not sufficient to allow a collection of atoms to reach thermodynamic equilibrium. To resolve this difficulty, Einstein proposed a third mechanism for the interaction of atoms with light.

The left half of **FIGURE 41.31c** shows a photon with frequency $f = \Delta E_{atom}/h$ approaching an *excited* atom. If a photon can induce the $1 \rightarrow 2$ transition of absorption, then Einstein proposed that it should also be able to induce a $2 \rightarrow 1$ transition. In a sense, this transition is a *reverse absorption*. But to undergo a reverse absorption, the atom must *emit* a photon of frequency $f = \Delta E_{atom}/h$. The end result, as seen in the right half of Figure 41.31c, is an atom in level 1 plus *two* photons! Because the first photon induced the atom to emit the second photon, this process is called **stimulated emission.**

Stimulated emission occurs only if the first photon's frequency exactly matches the $E_2 - E_1$ energy difference of the atom. This is precisely the same condition that absorption has to satisfy. More interesting, the emitted photon is *identical* to the incident photon. This means that as the two photons leave the atom they have exactly the same frequency and wavelength, are traveling in exactly the same direction, and are exactly in phase with each other. In other words, **stimulated emission produces a second photon that is an exact clone of the first.**

Stimulated emission is of no importance in most practical situations. Atoms typically spend only a few nanoseconds in an excited state before undergoing spontaneous emission, so the atom would need to be in an extremely intense light wave for stimulated emission to occur prior to spontaneous emission. Ordinary light sources are not nearly intense enough for stimulated emission to be more than a minor effect; hence it was many years before Einstein's prediction was confirmed. No one had doubted Einstein because he had clearly demonstrated that stimulated emission was necessary to make the energy equations balance, but it seemed no more important than would pennies to a millionaire balancing her checkbook. At least, that is, until 1960, when a revolutionary invention appeared that made explicit use of stimulated emission: the laser.

Lasers

The word **laser** is an acronym for **l**ight **a**mplification by the **s**timulated **e**mission of **r**adiation. The first laser, a ruby laser, was demonstrated in 1960, and several other kinds of lasers appeared within a few months. The driving force behind much of the research was the American physicist Charles Townes. Townes was awarded the Nobel Prize in 1964 for the invention of the maser, an earlier device using microwaves, and his theoretical work leading to the laser.

Today, lasers do everything from being the light source in fiber-optic communications to measuring the distance to the moon and from playing your DVD to performing delicate eye surgery. But what is a laser? Basically it is a device that produces a beam of highly *coherent* and essentially monochromatic (single-color) light as a result of stimulated emission. **Coherent** light is light in which all the electromagnetic waves have the same phase, direction, and amplitude. It is the coherence of a laser beam that allows it to be very tightly focused or to be rapidly modulated for communications.

Let's take a brief look at how a laser works. **FIGURE 41.32** represents a system of atoms that have a lower energy level E_1 and a higher energy level E_2. Suppose there are N_1 atoms in level 1 and N_2 atoms in level 2. Left to themselves, all the atoms would soon end up in level 1 because of the spontaneous emission $2 \rightarrow 1$. To prevent this, we can imagine that some type of excitation mechanism, perhaps an electrical discharge, is continuing to produce new excited atoms in level 2.

Let a photon of frequency $f = (E_2 - E_1)/h$ be incident on this group of atoms. Because it has the correct frequency, it could be absorbed by one of the atoms in

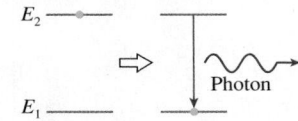

FIGURE 41.31 Three types of radiative transitions.

(a) Absorption

(b) Spontaneous emission

(c) Stimulated emission

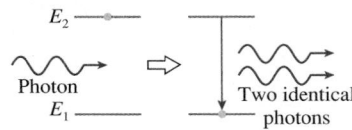

FIGURE 41.32 Energy levels 1 and 2, with populations N_1 and N_2.

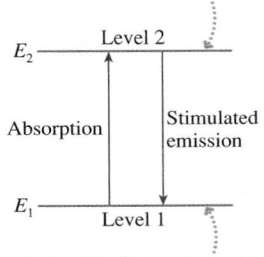

N_2 atoms in level 2. Photons of energy $E_{photon} = E_2 - E_1$ can cause these atoms to undergo stimulated emission.

N_1 atoms in level 1. These atoms can absorb photons of energy $E_{photon} = E_2 - E_1$.

Charles Townes.

level 1. Another possibility is that it could cause stimulated emission from one of the level 2 atoms. Ordinarily $N_2 \ll N_1$, so absorption events far outnumber stimulated emission events. Even if a few photons were generated by stimulated emission, they would quickly be absorbed by the vastly larger group of atoms in level 1.

But what if we could somehow arrange to place *every* atom in level 2, making $N_1 = 0$? Then the incident photon, upon encountering its first atom, will cause stimulated emission. Where there was initially one photon of frequency f, now there are two. These will strike two additional excited-state atoms, again causing stimulated emission. Then there will be four photons. As FIGURE 41.33 shows, there will be a *chain reaction* of stimulated emission until all N_2 atoms emit a photon of frequency f.

FIGURE 41.33 Stimulated emission creates a chain reaction of photon production in a population of excited atoms.

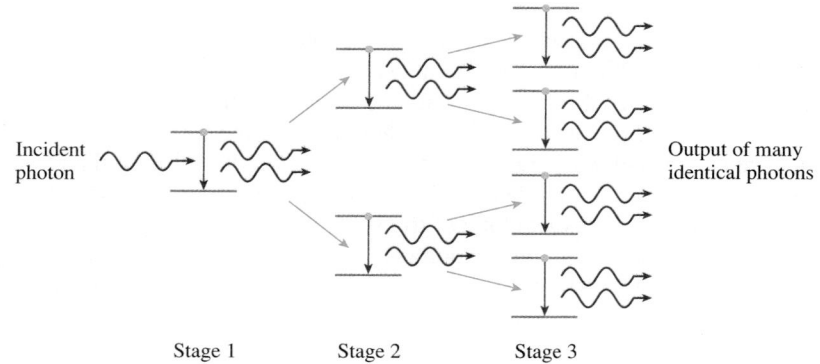

Incident photon

Output of many identical photons

Stage 1   Stage 2   Stage 3

In stimulated emission, each emitted photon is *identical* to the incident photon. The chain reaction of Figure 41.33 will lead not just to N_2 photons of frequency f, but to N_2 identical photons, all traveling together in the same direction with the same phase. If N_2 is a large number, as would be the case in any practical device, the one initial photon will have been amplified into a gigantic coherent pulse of light! A collection of excited-state atoms is called an *optical amplifier*.

As FIGURE 41.34 shows, the stimulated emission is sustained by placing the *lasing medium*—the sample of atoms that emits the light—in an **optical cavity** consisting of two facing mirrors. One of the mirrors will be partially transmitting so that some of the light emerges as the *laser beam*.

Although the chain reaction of Figure 41.33 illustrates the idea most clearly, it is not necessary for every atom to be in level 2 for amplification to occur. All that is needed is to have $N_2 > N_1$ so that stimulated emission exceeds absorption. Such a situation is called a **population inversion**. The process of obtaining a population inversion is called **pumping**, and we will look at two specific examples. Pumping is the technically difficult part of designing and building a laser because normal excitation mechanisms do not create population inversions. In fact, lasers would likely have been discovered accidentally long before 1960 if population inversions were easy to create.

FIGURE 41.34 Lasing takes place in an optical cavity.

The counterpropagating waves interact repeatedly with the atoms, allowing the light intensity to build up to a high level.

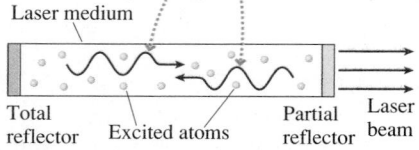

Laser medium

Total reflector   Excited atoms   Partial reflector   Laser beam

The Ruby Laser

The first laser to be developed was a ruby laser. FIGURE 41.35a shows the energy-level structure of the chromium atoms that gives ruby its optical properties. Normally, the number of atoms in the ground-state level E_1 far exceeds the number of excited-state atoms with energy E_2. That is, $N_2 \ll N_1$. Under these circumstances 690 nm light is absorbed rather than amplified. But suppose that we could *rapidly* excite more than half the chromium atoms to level E_2. Then we would have a population inversion ($N_2 > N_1$) between levels E_1 and E_2.

This can be accomplished by *optically pumping* the ruby with a very intense pulse of white light from a *flashlamp*. A flashlamp is like a camera flash, only vastly more intense. In the basic arrangement of FIGURE 41.35b, a helical flashlamp is coiled around a ruby rod that has mirrors bonded to its end faces. The lamp is fired by discharging a high-voltage capacitor through it, creating a very intense light pulse lasting just a few microseconds. This intense light excites nearly all the chromium atoms from the ground state to the upper energy levels. From there, they quickly ($\approx 10^{-8}$ s) decay nonradiatively to level 2. With $N_2 > N_1$, a population inversion has been created.

Once a photon initiates the laser pulse, the light intensity builds quickly into a brief but incredibly intense burst of light. A typical output pulse lasts 10 ns and has an energy of 1 J. This gives a *peak power* of

$$P = \frac{\Delta E}{\Delta t} = \frac{1\ \text{J}}{10^{-8}\ \text{s}} = 10^8\ \text{W} = 100\ \text{MW}$$

One hundred megawatts of light power! That is more than the electrical power used by a small city. The difference, of course, is that a city consumes that power continuously but the laser pulse lasts a mere 10 ns. The laser cannot fire again until the capacitor is recharged and the laser rod cooled. A typical firing rate is a few pulses per second, so the laser is "on" only a few billionths of a second out of each second.

Ruby lasers have been replaced by other pulsed lasers that, for various practical reasons, are easier to operate. However, they all operate with the same basic idea of rapid optical pumping to upper states, rapid nonradiative decay to level 2 where the population inversion is formed, then rapid buildup of an intense optical pulse.

The Helium-Neon Laser

The familiar red laser used in lecture demonstrations, laboratories, and supermarket checkout scanners is the helium-neon laser, often called a HeNe laser. Its output is a *continuous,* rather than pulsed, wavelength of 632.8 nm. The medium of a HeNe laser is a mixture of $\approx 90\%$ helium and $\approx 10\%$ neon gases. As FIGURE 41.36a shows, the gases are sealed in a glass tube, then an electrical discharge is established along the bore of the tube. Two mirrors are bonded to the ends of the discharge tube, one a total reflector and the other having $\approx 2\%$ transmission so that the laser beam can be extracted.

The atoms that lase are the neon atoms, but the pumping method involves the helium atoms. The electrons in the discharge collisionally excite the $1s2s$ state of helium. This state has a very low spontaneous decay rate (i.e., a very long lifetime) because a decay back to the $1s^2$ state would violate the Δl selection rule, so it is possible to build up a fairly large population (but not an inversion) of excited helium atoms in the $1s2s$ state. The energy of the $1s2s$ state is 20.6 eV.

Interestingly, an excited state of neon, the $5s$ state, also has an energy of 20.6 eV. If a $1s2s$ excited helium atom collides with a ground-state neon atom, as frequently happens, the excitation energy can be transferred from one atom to the other! Written as a chemical reaction, the process is

$$\text{He}^* + \text{Ne} \rightarrow \text{He} + \text{Ne}^*$$

where the asterisk indicates the atom is in an excited state. This process, called **excitation transfer,** is very efficient for the $5s$ state because the process is *resonant*—a perfect energy match. Thus the two-step process of collisional excitation of helium, followed by excitation transfer between helium and neon, pumps the neon atoms into the excited $5s$ state. This is shown in FIGURE 41.36b.

The $5s$ energy level in neon is ≈ 1.95 eV above the $3p$ state. The $3p$ state is very nearly empty of population, both because it is not efficiently populated in the discharge and because it undergoes very rapid spontaneous emission to the $3s$ states. Thus the large number of atoms pumped into the $5s$ state creates a population inversion with respect to the lower $3p$ state. These are the necessary conditions for laser action.

Because the lower level of the laser transition is normally empty of population, placing only a small fraction of the neon atoms in the $5s$ state creates a population inversion. Thus a fairly modest pumping action is sufficient to create the inversion and start the laser. Furthermore, a HeNe laser can maintain a *continuous* inversion and thus sustain

FIGURE 41.35 A flashlamp-pumped ruby laser.

(a)

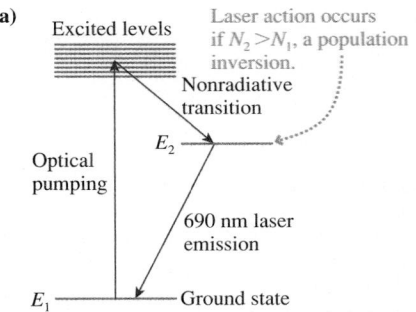

(b)

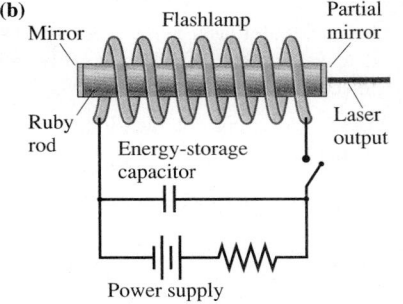

FIGURE 41.36 A HeNe laser.

(a) He/Ne gas mixture

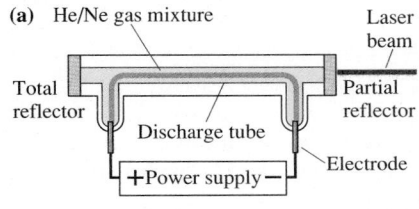

(b)

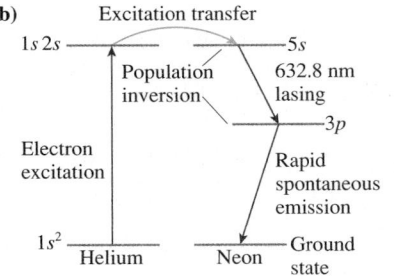

continuous lasing. The electrical discharge continuously creates 5s excited atoms in the upper level, via excitation transfer, and the rapid spontaneous decay of the 3p atoms from the lower level keeps its population low enough to sustain the inversion.

A typical helium-neon laser has a power output of 1 mW = 10^{-3} J/s at 632.8 nm in a 1-mm-diameter laser beam. As you can show in a homework problem, this output corresponds to the emission of 3.2×10^{15} photons per second. Other continuous lasers operate by similar principles, but can produce much more power. The argon laser, which is widely used in scientific research, can produce up to 20 W of power at green and blue wavelengths. The carbon dioxide laser produces output power in excess of 1000 W at the infrared wavelength of 10.6 μm. It is used in industrial applications for cutting and welding.

EXAMPLE 41.9 An ultraviolet laser

An ultraviolet laser generates a 10 MW, 5.0-ns-long light pulse at a wavelength of 355 nm. How many photons are in each pulse?

SOLVE The energy of each light pulse is the power multiplied by the duration:

$$E_{pulse} = P\,\Delta t = (1.0 \times 10^7 \text{ W})(5.0 \times 10^{-9} \text{ s}) = 0.050 \text{ J}$$

Each photon in the pulse has energy

$$E_{photon} = hf = \frac{hc}{\lambda} = 3.50 \text{ eV} = 5.60 \times 10^{-19} \text{ J}$$

Because $E_{pulse} = NE_{photon}$, the number of photons is

$$N = E_{pulse}/E_{photon} = 8.9 \times 10^{16} \text{ photons}$$

CHALLENGE EXAMPLE 41.10 Electron probability in hydrogen

What is the probability that a 1s hydrogen electron is found at a distance from the proton that is less than half the Bohr radius?

MODEL The Schrödinger model of the hydrogen atom represents the electron as a wave function. We can't say exactly where the electron is, but we can calculate the probability of finding it in a specified region of space.

SOLVE We're interested in finding the electron not at a certain *point* in space but within a certain *distance* from the nucleus. For this we use the radial probability density

$$P_r(r) = 4\pi r^2 |R_{nl}(r)|^2$$

where $R_{nl}(r)$ is the radial wave function, rather than the square of the wave function $\psi(x, y, z)$. The probability of finding the electron at a distance between r_{min} and r_{max} is

$$\text{Prob}(r_{min} \le r \le r_{max}) = \int_{r_{min}}^{r_{max}} P_r(r)\,dr$$

$$= 4\pi \int_{r_{min}}^{r_{max}} r^2 |R_{nl}(r)|^2\,dr$$

The 1s radial wave function was given in Equations 41.7:

$$R_{1s}(r) = \frac{1}{\sqrt{\pi a_B^3}} e^{-r/a_B}$$

where a_B is the Bohr radius. We specify that the electron is less than half the Bohr radius from the proton by setting $r_{min} = 0$ and $r_{max} = \frac{1}{2}a_B$. Thus the probability we seek is

$$\text{Prob}(r \le \tfrac{1}{2}a_B) = 4\pi \int_0^{a_B/2} r^2 |R_{1s}(r)|^2\,dr$$

$$= \frac{4\pi}{\pi a_B^3} \int_0^{a_B/2} r^2 e^{-2r/a_B}\,dr$$

To evaluate this integral, it will be useful to change variables. Let $u = 2r/a_B$, so that the exponential can be written more simply as e^{-u}. Turning this around, we have $r = \frac{1}{2}a_B u$ and thus

$$r^2\,dr = \left(\tfrac{1}{2}a_B u\right)^2 \left(\tfrac{1}{2}a_B\,du\right) = \tfrac{1}{8}a_B^3 u^2\,du$$

A change of variables requires a corresponding change of limits: When $r = 0$, $u = 0$ also; when $r = \frac{1}{2}a_B$, $u = 1$. With these substitutions, the probability calculation becomes

$$\text{Prob}(r \le \tfrac{1}{2}a_B) = \frac{1}{2}\int_0^1 u^2 e^{-u}\,du$$

This looks much nicer! Notice that all the a_B have disappeared, so our answer will be a numerical value.

This is not an easy integral, but it is a common one. It can be found in integral tables, such as in Appendix A, or evaluated with mathematical software. The result is

$$\text{Prob}\left(r \le \tfrac{1}{2}a_B\right) = \frac{1}{2}\left[-(u^2 + 2u + 2)e^{-u}\right]_0^1$$

$$= \frac{1}{2}\left[2 - 5e^{-1}\right] = 0.080$$

The probability that a 1s hydrogen electron is less than half the Bohr radius from the proton is 0.080, or 8.0%.

ASSESS The probability is small, but that is not unexpected. The graph of the radial probability density in Figure 41.8 shows that the probability peaks at $r = a_B$ and then decreases rather slowly. We can see that the area under that curve from $r = 0$ to $r = \frac{1}{2}a_B$ is not large. The electron can be found much closer to the proton than one Bohr radius, but not with a large probability.

SUMMARY

The goal of Chapter 41 has been to understand the structure and properties of atoms.

Important Concepts

Hydrogen Atom

The three-dimensional Schrödinger equation has stationary-state solutions for the hydrogen atom potential energy only if three conditions are satisfied:

- Energy $E_n = -13.60 \text{ eV}/n^2$ $n = 1, 2, 3, \ldots$
- Angular momentum $L = \sqrt{l(l+1)}\hbar$ $l = 0, 1, 2, 3, \ldots, n-1$
- z-component of angular momentum
  $L_z = m\hbar$ $m = -l, -l+1, \ldots, 0, \ldots, l-1, l$

Each state is characterized by **quantum numbers** (n, l, m), but the energy depends only on n.

The probability of finding the electron within a small distance interval δr at distance r is

$$\text{Prob(in } \delta r \text{ at } r) = P_r(r)\delta r$$

where $P_r(r) = 4\pi r^2 |R_{nl}(r)|^2$ is the **radial probability density.**

Graphs of $P_r(r)$ suggest that the electrons are arranged in shells.

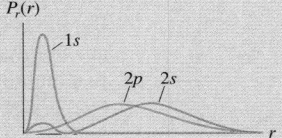

Multielectron Atoms

The potential energy is electron-nucleus plus electron-electron. In the **independent particle approximation,** each electron is described by the same quantum numbers (n, l, m, m_s) used for the hydrogen atom. The energy of a state depends on n and l. For each n, energy increases as l increases.

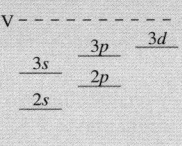

- High-l states correspond to circular orbits. These stay outside the core.

- Low-l states correspond to elliptical orbits. These penetrate the core to interact more strongly with the nucleus. This interaction lowers their energy.

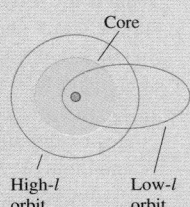

Electron spin

The electron has an inherent angular momentum $\vec{S}$ and magnetic moment $\vec{\mu}$ *as if* it were spinning. The spin angular momentum has a fixed magnitude $S = \sqrt{s(s+1)}\hbar$, where $s = \frac{1}{2}$. The z-component is $S_z = m_s\hbar$, where $m_s = \pm\frac{1}{2}$. These two states are called **spin-up** and **spin-down.** Each atomic state is fully characterized by the four quantum numbers (n, l, m, m_s).

The Pauli exclusion principle says that no more than one electron can occupy each quantum state. The periodic table of the elements is based on the fact that the ground state is the lowest-energy electron configuration compatible with the Pauli principle.

Applications

Atomic spectra are generated by excitation followed by a photon-emitting quantum jump.

- Excitation by absorption or collision
- Quantum-jump selection rule $\Delta l = \pm 1$

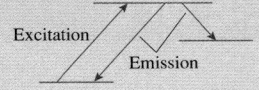

Lifetimes of excited states

The excited-state population decreases exponentially as

$$N_{\text{exc}} = N_0 e^{-t/\tau}$$

where $\tau = 1/r$ is the **lifetime** and r is the **decay rate.** It's not possible to predict when a particular atom will decay, but the *probability* is

$$\text{Prob(in } \delta t \text{ at } t) = r\delta t$$

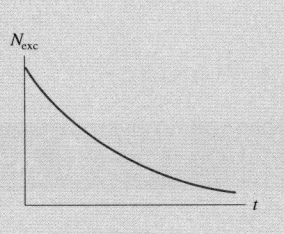

Stimulated emission of an excited state can be caused by a photon with $E_{\text{photon}} = E_2 - E_1$. Laser action can occur if $N_2 > N_1$, a condition called a population inversion.

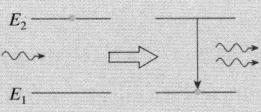

Terms and Notation

principal quantum number, n	spin quantum number, m_s	subshell	spontaneous emission
orbital quantum number, l	spin-up	excitation	stimulated emission
magnetic quantum number, m	spin-down	allowed transition	laser
ionization energy	independent particle	selection rule	coherent
electron cloud	approximation (IPA)	collisional excitation	optical cavity
radial wave function, $R_{nl}(r)$	Pauli exclusion principle	nonradiative transition	population inversion
radial probability density, $P_r(r)$	electron configuration	lifetime, τ	pumping
shell model	closed shell	decay rate, r	excitation transfer
spin			

CONCEPTUAL QUESTIONS

1. Consider the two hydrogen-atom states $5d$ and $4f$. Which has the higher energy? Explain.
2. What is the difference between the *probability density* and the *radial probability density?*
3. What is the difference between l and L?
4. What is the difference between s and S?
5. **FIGURE Q41.5** shows the outcome of a Stern-Gerlach experiment with atoms of element X.
   a. Do the peaks represent different values of the atom's total angular momentum or different values of the z-component of its angular momentum? Explain.
   b. What angular momentum quantum numbers characterize these four peaks?
6. Does each of the configurations in **FIGURE Q41.6** represent a possible electron configuration of an element? If so, (i) identify the element and (ii) determine whether this is the ground state or an excited state. If not, why not?

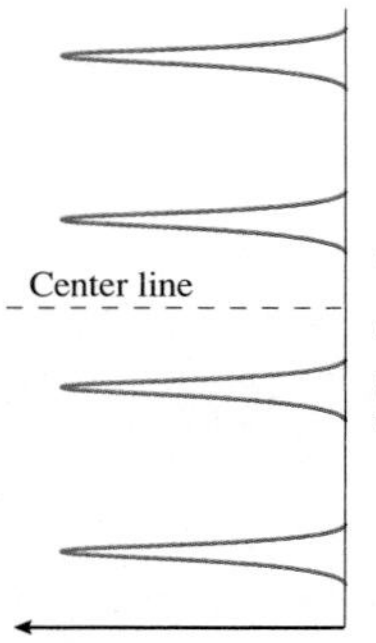

Center line - - - -

Number of atoms

Collection plate

FIGURE Q41.5

7. What *is* an atom's ionization energy? In other words, if you know the ionization energy of an atom, what is it that you know about the atom?
8. Figure 41.23 shows that the ionization energy of cadmium ($Z = 48$) is larger than that of its neighbors. Why is this?
9. A neon discharge tube emits a bright reddish-orange spectrum, but a glass tube filled with neon is completely transparent. Why doesn't the neon in the tube absorb orange and red wavelengths?
10. The hydrogen atom $1s$ wave function is a maximum at $r = 0$. But the $1s$ radial probability density, shown in Figure 41.8, peaks at $r = a_B$ and is zero at $r = 0$. Explain this paradox.
11. In a multielectron atom, the lowest-l state for each n ($2s$, $3s$, $4s$, etc.) is significantly lower in energy than the hydrogen state having the same n. But the highest-l state for each n ($2p$, $3d$, $4f$, etc.) is very nearly equal in energy to the hydrogen state with the same n. Explain.
12. In **FIGURE Q41.12**, a photon with energy 2.0 eV is incident on an atom in the p state. Does the atom undergo an absorption transition, a stimulated emission transition, or neither? Explain.

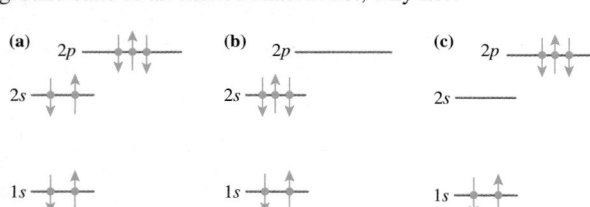

FIGURE Q41.6

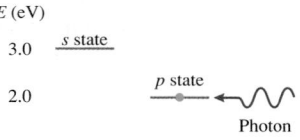

E (eV)

3.0 ___s state___

2.0 ___p state___ ◄~~~

Photon

FIGURE Q41.12 0.0 ___s state___

EXERCISES AND PROBLEMS

Problems labeled ▨ integrate material from earlier chapters.

Exercises

Sections 41.1–41.2 The Hydrogen Atom

1. | What is the angular momentum of a hydrogen atom in (a) a $6s$ state and (b) a $4f$ state? Give your answers as a multiple of $\hbar$.

2. | List the quantum numbers, excluding spin, of (a) all possible $3p$ states and (b) all possible $3d$ states.
3. | A hydrogen atom has orbital angular momentum 3.65×10^{-34} J s.
   a. What letter ($s, p, d,$ or f) describes the electron?
   b. What is the atom's minimum possible energy? Explain.
4. | What is the maximum possible angular momentum L (as a multiple of $\hbar$) of a hydrogen atom with energy -0.544 eV?

5. | What are E and L (as a multiple of $\hbar$) of a hydrogen atom in the $6f$ state?

Section 41.3 The Electron's Spin

6. ‖ When all quantum numbers are considered, how many different quantum states are there for a hydrogen atom with $n = 1$? With $n = 2$? With $n = 3$? List the quantum numbers of each state.

7. | How many lines of atoms would you expect to see on the collector plate of a Stern-Gerlach apparatus if the experiment is done with (a) lithium and (b) beryllium? Explain.

Section 41.4 Multielectron Atoms

Section 41.5 The Periodic Table of the Elements

8. | Predict the ground-state electron configurations of Mg, Sr, and Ba.

9. | Predict the ground-state electron configurations of Al, Ga, and In.

10. | Identify the element for each of these electron configurations. Then determine whether this configuration is the ground state or an excited state.
   a. $1s^2 2s^2 2p^5$
   b. $1s^2 2s^2 2p^6 3s^2 3p^6 4s^2 3d^{10} 4p$

11. | Identify the element for each of these electron configurations. Then determine whether this configuration is the ground state or an excited state.
   a. $1s^2 2s^2 2p^5 3s$
   b. $1s^2 2s^2 2p^6 3s^2 3p^6 4s^2 3d^2$

Section 41.6 Excited States and Spectra

12. | Show that $hc = 1240\ \text{eV}\,\text{nm}$.

13. ‖ What is the electron configuration of the second excited state of lithium?

14. ‖ An electron accelerates through a 12.5 V potential difference, starting from rest, and then collides with a hydrogen atom, exciting the atom to the highest energy level allowed. List all the possible quantum-jump transitions by which the excited atom could emit a photon and the wavelength (in nm) of each.

15. | a. Is a $4p \rightarrow 4s$ transition allowed in sodium? If so, what is its wavelength (in nm)? If not, why not?
   b. Is a $3d \rightarrow 4s$ transition allowed in sodium? If so, what is its wavelength (in nm)? If not, why not?

Section 41.7 Lifetimes of Excited States

16. | An excited state of an atom has a 25 ns lifetime. What is the probability that an excited atom will emit a photon during a 0.50 ns interval?

17. | 1.0×10^6 sodium atoms are excited to the $3p$ state at $t = 0$ s. How many of these atoms remain in the $3p$ state at (a) $t = 10$ ns, (b) $t = 30$ ns, and (c) $t = 100$ ns?

18. | A hydrogen atom is in the $2p$ state. How much time must elapse for there to be a 1% chance that this atom will undergo a quantum jump to the ground state?

19. ‖ 1.0×10^6 atoms are excited to an upper energy level at $t = 0$ s. At the end of 20 ns, 90% of these atoms have undergone a quantum jump to the ground state.
   a. How many photons have been emitted?
   b. What is the lifetime of the excited state?

20. ‖ 1.00×10^6 sodium atoms are excited to the $3p$ state at $t = 0$ s. At what time have 8.0×10^5 photons been emitted?

Section 41.8 Stimulated Emission and Lasers

21. | A 1.0 mW helium-neon laser emits a visible laser beam with a wavelength of 633 nm. How many photons are emitted per second?

22. ‖ In LASIK surgery, a laser is used to reshape the cornea of the
   BIO eye to improve vision. The laser produces extremely short pulses of light, each containing 1.0 mJ of energy.
   a. There are 9.7×10^{14} photons in each pulse. What is the wavelength of the laser?
   b. Each pulse lasts a mere 20 ns. What is the average power delivered to the cornea during a pulse?

23. | A laser emits 1.0×10^{19} photons per second from an excited state with energy $E_2 = 1.17$ eV. The lower energy level is $E_1 = 0$ eV.
   a. What is the wavelength of this laser?
   b. What is the power output of this laser?

Problems

24. ‖ a. Draw a diagram similar to Figure 41.3 to show all the possible orientations of the angular momentum vector $\vec{L}$ for the case $l = 3$. Label each $\vec{L}$ with the appropriate value of m.
   b. What is the minimum angle between $\vec{L}$ and the z-axis?

25. ‖ There exist subatomic particles whose spin is characterized by $s = 1$, rather than the $s = \frac{1}{2}$ of electrons. These particles are said to have a spin of one.
   a. What is the magnitude (as a multiple of $\hbar$) of the spin angular momentum S for a particle with a spin of one?
   b. What are the possible values of the spin quantum number?
   c. Draw a vector diagram similar to Figure 41.14 to show the possible orientations of $\vec{S}$.

26. ‖ A hydrogen atom in its fourth excited state emits a photon with a wavelength of 1282 nm. What is the atom's maximum possible orbital angular momentum (as a multiple of $\hbar$) after the emission?

27. ‖ A hydrogen atom has $l = 2$. What are the (a) minimum (as a multiple of $\hbar$) and (b) maximum values of the quantity $(L_x^2 + L_y^2)^{1/2}$?

28. | Calculate (a) the radial wave function and (b) the radial probability density at $r = \frac{1}{2} a_B$ for an electron in the $1s$ state of hydrogen. Give your answers in terms of a_B.

29. ‖ For an electron in the $1s$ state of hydrogen, what is the probability of being in a spherical shell of thickness $0.010 a_B$ at distance (a) $\frac{1}{2} a_B$, (b) a_B, and (c) $2a_B$ from the proton?

30. ‖ Prove that the normalization constant of the $1s$ radial wave function of the hydrogen atom is $(\pi a_B^3)^{-1/2}$, as given in Equations 41.7.
   Hint: A useful definite integral is

$$\int_0^\infty x^n e^{-\alpha x}\,dx = \frac{n}{\alpha^{n+1}}$$

31. ‖ Prove that the normalization constant of the $2p$ radial wave function of the hydrogen atom is $(24\pi a_B^3)^{-1/2}$, as shown in Equations 41.7.
   Hint: See the hint in Problem 30.

32. ‖ Prove that the radial probability density peaks at $r = a_B$ for the $1s$ state of hydrogen.

33. ‖ a. Calculate and graph the hydrogen radial wave function $R_{2p}(r)$ over the interval $0 \leq r \leq 8a_B$.
    b. Determine the value of r (in terms of a_B) for which $R_{2p}(r)$ is a maximum.
    c. Example 41.3 and Figure 41.8 showed that the radial probability density for the $2p$ state is a maximum at $r = 4a_B$. Explain why this differs from your answer to part b.

34. ‖ In general, an atom can have both orbital angular momentum and spin angular momentum. The *total* angular momentum is defined to be $\vec{J} = \vec{L} + \vec{S}$. The total angular momentum is quantized in the same way as $\vec{L}$ and $\vec{S}$. That is, $J = \sqrt{j(j+1)}\hbar$, where j is the total angular momentum quantum number. The z-component of $\vec{J}$ is $J_z = L_z + S_z = m_j\hbar$, where m_j goes in integer steps from $-j$ to $+j$. Consider a hydrogen atom in a p state, with $l = 1$.
    a. L_z has three possible values and S_z has two. List all possible combinations of L_z and S_z. For each, compute J_z and determine the quantum number m_j. Put your results in a table.
    b. The number of values of J_z that you found in part a is too many to go with a single value of j. But you should be able to divide the values of J_z into two groups that correspond to two values of j. What are the allowed values of j? Explain. In a classical atom, there would be no restrictions on how the two angular momenta $\vec{L}$ and $\vec{S}$ can combine. Quantum mechanics is different. You've now shown that there are only two allowed ways to add these two angular momenta.

35. ‖ Draw a series of pictures, similar to Figure 41.22, for the ground states of K, Sc, Co, and Ge.

36. ‖ Draw a series of pictures, similar to Figure 41.22, for the ground states of Ca, Ni, As, and Kr.

37. ‖ a. What downward transitions are possible for a sodium atom in the $6s$ state? (See Figure 41.24.)
    b. What are the wavelengths of the photons emitted in each of these transitions?

38. ‖ The $5d \rightarrow 3p$ transition in the emission spectrum of sodium has a wavelength of 499 nm. What is the energy of the $5d$ state?

39. ‖ A sodium atom emits a photon with wavelength 818 nm shortly after being struck by an electron. What minimum speed did the electron have before the collision?

40. ‖ The ionization energy of an atom is known to be 5.5 eV. The emission spectrum of this atom contains only the four wavelengths 310.0 nm, 354.3 nm, 826.7 nm, and 1240.0 nm. Draw an energy-level diagram with the fewest possible energy levels that agrees with these experimental data. Label each level with an appropriate l quantum number.
    Hint: Don't forget about the Δl selection rule.

41. ‖ **FIGURE P41.41** shows the first few energy levels of the lithium atom. Make a table showing all the allowed transitions in the emission spectrum. For each transition, indicate
    a. The wavelength, in nm.
    b. Whether the transition is in the infrared, the visible, or the ultraviolet spectral region.
    c. Whether or not the transition would be observed in the lithium absorption spectrum.

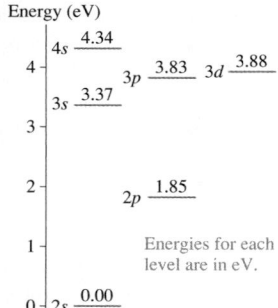

FIGURE P41.41

42. ‖ **FIGURE P41.42** shows a few energy levels of the mercury atom.
    a. Make a table showing all the allowed transitions in the emission spectrum. For each transition, indicate the photon wavelength, in nm.
    b. What minimum speed must an electron have to excite the 492-nm-wavelength blue emission line in the Hg spectrum?

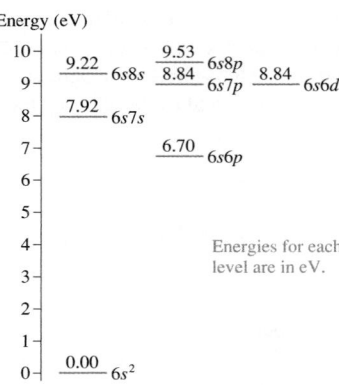

FIGURE P41.42

43. ‖‖ Suppose you put five electrons into a 0.50-nm-wide one-dimensional rigid box (i.e., an infinite potential well).
    a. Use an energy-level diagram to show the electron configuration of the ground state.
    b. What is the ground-state energy of this configuration?

44. ‖ Three electrons are in a one-dimensional rigid box (i.e., an infinite potential well) of length 0.50 nm. Two are in the $n = 1$ state and one is in the $n = 6$ state. The selection rule for the rigid box allows only those transitions for which Δn is odd.
    a. Draw an energy-level diagram. On it, show the filled levels and show all transitions that could emit a photon.
    b. What are all the possible wavelengths that could be emitted by this system?

45. ‖ a. What is the decay rate for the $2p$ state of hydrogen?
    b. During what interval of time will 10% of a sample of $2p$ hydrogen atoms decay?

46. ‖ An atom in an excited state has a 1.0% chance of emitting a photon in 0.10 ns. What is the lifetime of the excited state?

47. ‖ a. Find an expression in terms of τ for the *half-life* $t_{1/2}$ of a sample of excited atoms. The half-life is the time at which half of the excited atoms have undergone a quantum jump and emitted a photon.
    b. What is the half-life of the $3p$ state of sodium?

48. ‖ In fluorescence microscopy, an important tool in biology, a laser beam is absorbed by target molecules in a sample. These molecules are then imaged by a microscope as they emit longer-wavelength photons in quantum jumps back to lower energy levels, a process known as *fluorescence*. A variation on this technique is *two-photon excitation*. If two photons are absorbed simultaneously, their energies add. Consequently, a molecule that is normally excited by a photon of energy E_{photon} can be excited by the simultaneous absorption of two photons having half as much energy. For this process to be useful, the sample must be irradiated at the very high intensity of at least 10^{32} photons/m² s. This is achieved by concentrating the laser power into very short pulses (100 fs pulse length) and then focusing the laser beam to a small spot. The laser is fired at the rate of 10^8 pulses each second. Suppose a biologist wants to use two-photon excitation

to excite a molecule that in normal fluorescence microscopy would be excited by a laser with a wavelength of 420 nm. If she focuses the laser beam to a 2.0-μm-diameter spot, what minimum energy must each pulse have?

49. ‖ An electrical discharge in a neon-filled tube maintains a *steady* population of 1.0×10^9 atoms in an excited state with $\tau = 20$ ns. How many photons are emitted per second from atoms in this state?

50. ‖ A ruby laser emits a 100 MW, 10-ns-long pulse of light with a wavelength of 690 nm. How many chromium atoms undergo stimulated emission to generate this pulse?

Challenge Problems

51. Two excited energy levels are separated by the very small energy difference ΔE. As atoms in these levels undergo quantum jumps to the ground state, the photons they emit have nearly identical wavelengths λ.
    a. Show that the wavelengths differ by

    $$\Delta \lambda = \frac{\lambda^2}{hc} \Delta E$$

    b. In the Lyman series of hydrogen, what is the wavelength difference between photons emitted in the $n = 20$ to $n = 1$ transition and photons emitted in the $n = 21$ to $n = 1$ transition?

52. What is the probability of finding a $1s$ hydrogen electron at distance $r > a_B$ from the proton?

53. Prove that the most probable distance from the proton of an electron in the $2s$ state of hydrogen is $5.236a_B$.

54. Find the distance, in terms of a_B, between the two peaks in the radial probability density of the $2s$ state of hydrogen.
    Hint: This problem requires a numerical solution.

55. Suppose you have a machine that gives you pieces of candy when you push a button. Eighty percent of the time, pushing the button gets you two pieces of candy. Twenty percent of the time, pushing the button yields 10 pieces. The *average* number of pieces per push is $N_{avg} = 2 \times 0.80 + 10 \times 0.20 = 3.6$. That is, 10 pushes should get you, on average, 36 pieces. Mathematically, the average value when the probabilities differ is $N_{avg} = \sum(N_i \times \text{Probability of } i)$. We can do the same thing in quantum mechanics, with the difference that the sum becomes an integral. If you measured the distance of the electron from the proton in many hydrogen atoms, you would get many values, as indicated

by the radial probability density. But the *average* value of r would be

$$r_{avg} = \int_0^\infty r P_r(r)\, dr$$

Calculate the average value of r in terms of a_B for the electron in the $1s$ and the $2p$ states of hydrogen.

56. An atom in an excited state has a 1.0% chance of emitting a photon in 0.20 ns. How long will it take for 25% of a sample of excited atoms to decay?

57. The 1997 Nobel Prize in physics went to Steven Chu, Claude Cohen-Tannoudji, and William Phillips for their development of techniques to slow, stop, and "trap" atoms with laser light. To see how this works, consider a beam of rubidium atoms (mass 1.4×10^{-25} kg) traveling at 500 m/s after being evaporated out of an oven. A laser beam with a wavelength of 780 nm is directed against the atoms. This is the wavelength of the $5s \rightarrow 5p$ transition in rubidium, with $5s$ being the ground state, so the photons in the laser beam are easily absorbed by the atoms. After an average time of 15 ns, an excited atom spontaneously emits a 780-nm-wavelength photon and returns to the ground state.
    a. The energy-momentum-mass relationship of Einstein's theory of relativity is $E^2 = p^2c^2 + m^2c^4$. A photon is massless, so the momentum of a photon is $p = E_{photon}/c$. Assume that the atoms are traveling in the positive x-direction and the laser beam in the negative x-direction. What is the initial momentum of an atom leaving the oven? What is the momentum of a photon of light?
    b. The total momentum of the atom and the photon must be conserved in the absorption processes. As a consequence, how many photons must be absorbed to bring the atom to a halt?

    NOTE ▶ Momentum is also conserved in the emission processes. However, spontaneously emitted photons are emitted in random directions. Averaged over many absorption/emission cycles, the net recoil of the atom due to emission is zero and can be ignored. ◀

    c. Assume that the laser beam is so intense that a ground-state atom absorbs a photon instantly. How much time is required to stop the atoms?
    d. Use Newton's second law in the form $F = \Delta p/\Delta t$ to calculate the force exerted on the atoms by the photons. From this, calculate the atoms' acceleration as they slow.
    e. Over what distance is the beam of atoms brought to a halt?

STOP TO THINK ANSWERS

Stop to Think 41.1: $n = 3, l = 1$, or a $3p$ **state.**

Stop to Think 41.2: 4. You can see in Figure 41.8 that the ns state has n maxima.

Stop to Think 41.3: No. $m_s = \pm\frac{1}{2}$, so the z-component S_z cannot be zero.

Stop to Think 41.4: b. The atom would have less energy if the $3s$ electron were in a $2p$ state.

Stop to Think 41.5: c. Emission is a quantum jump to a lower-energy state. The $5p \rightarrow 4p$ transition is not allowed because $\Delta l = 0$ violates the selection rule. The lowest-energy allowed transition is $5p \rightarrow 3d$, with $E_{photon} = \Delta E_{atom} = 3.0$ eV.

Stop to Think 41.6: e. Because $r_B = 2r_A$, the ratio is $e^{-2}/e^{-1} = e^{-1} < \frac{1}{2}$.

42 Nuclear Physics

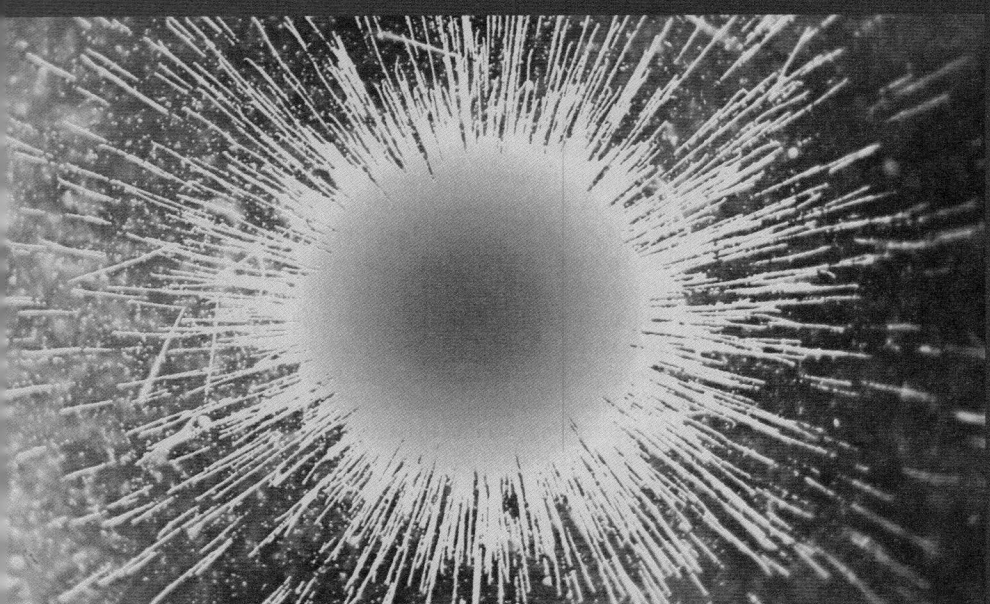

A photographic emulsion records the tracks of alpha particles emitted by a speck of radium.

▶ **Looking Ahead** The goal of Chapter 42 is to understand the physics of the nucleus and some applications of nuclear physics.

Nuclear Structure

You'll learn how the nucleus is constructed, what holds it together, and why some nuclei are more stable than others.

The nucleus consists of positively charged *protons* and electrically neutral *neutrons*. Together, these are called **nucleons.** The nuclear diameter is only a few femtometers.

◀ **Looking Back**
Sections 37.6–37.7 The nucleus

Nuclear Stability

More than 3000 isotopes are known, but only 266 have a stable nucleus. In a graph of neutron number against proton number, the stable nuclei all cluster near a well-defined **line of stability.**

As nuclei grow, the neutron number increases faster than the proton number. You'll learn how this is necessary to hold the nucleus together.

Nuclear Decay

You'll learn the three basic ways in which an unstable nucleus can decay.

- Alpha decay: Emission of a ^{4}He nucleus (alpha particle).
- Beta decay: Emission of an electron or positron (beta particle).
- Gamma decay: Emission of a high-energy photon (gamma ray).

Nuclear decay releases large amounts of energy. This plutonium sphere—used to power spacecraft—will glow for decades from the energy of alpha decay.

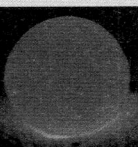

The Shell Model

The force holding the nucleus together is a fundamental force of nature called the **strong force.** It is a *short-range force* whose influence extends over only a few femtometers.

You'll learn to use a **shell model** of nuclear structure, analogous to the electron shells of atoms, to explain nuclear properties.

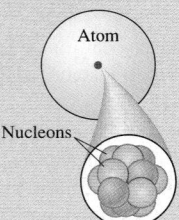

◀ **Looking Back**
Section 40.6 Finite potential wells

Half-Lives

The number of unstable nuclei in a sample decreases exponentially with time. We describe nuclear decay by its **half-life,** the time for half the atoms to decay. Half-lives range from microseconds to billions of years.

Nuclei that decay don't vanish—they become some other kind of nucleus called the **daughter nucleus.** You'll learn how to identify the daughter nucleus of a decay.

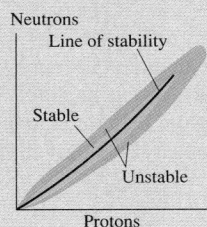

Nuclear Applications

You'll learn about some of the applications of nuclear physics, which range from measuring ages to curing diseases. You'll also learn how **radiation dose** is measured and what it means.

This image of the brain of a stroke patient was made with nuclei that decay by emitting gamma-ray photons. The damaged area, with reduced activity, is clearly visible.

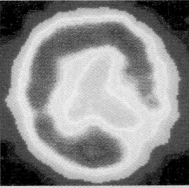

42.1 Nuclear Structure

The 1890s was a decade of mysterious rays. Cathode rays were being studied in several laboratories, and, in 1895, Röntgen discovered x rays. In 1896, after hearing of Röntgen's discovery, the French scientist A. H. Becquerel wondered if mineral crystals that fluoresce after exposure to sunlight were emitting x rays. He put a piece of film in an opaque envelope, then placed a crystal on top and left it in the sun. To his delight, the film in the envelope was exposed.

Becquerel thought he had discovered x rays coming from crystals, but his joy was short lived. He soon found that the film could be exposed equally well simply by being stored in a closed drawer with the crystals. Further investigation showed that the crystal, which happened to be a mineral containing uranium, was spontaneously emitting some new kind of ray. Rather than finding x rays, as he had hoped, Becquerel had discovered what became known as *radioactivity*.

Ernest Rutherford soon took up the investigation and found not one but three distinct kinds of rays emitted from crystals containing uranium. Not knowing what they were, he named them for their ability to penetrate matter and ionize air. The first, which caused the most ionization and penetrated the least, he called *alpha rays*. The second, with intermediate penetration and ionization, were *beta rays,* and the third, with the least ionization but the largest penetration, became *gamma rays.*

Within a few years, Rutherford was able to show that alpha rays are helium nuclei emitted from the crystal at very high velocities. These became the projectiles that he used in 1909 to probe the structure of the atom. The outcome of that experiment, as you learned in Chapter 37, was Rutherford's discovery that atoms have a very small, dense nucleus at the center.

Rutherford's discovery of the nucleus may have settled the question of atomic structure, but it raised many new issues for scientific research. Foremost among them were:

- What is nuclear matter? What are its properties?
- What holds the nucleus together? Why doesn't the repulsive electrostatic force blow it apart?
- What is the connection between the nucleus and radioactivity?

These questions marked the beginnings of **nuclear physics,** the study of the properties of the atomic nucleus.

Nucleons

The nucleus is a tiny speck in the center of a vastly larger atom. As **FIGURE 42.1** shows, the nuclear diameter of roughly 10^{-14} m is only about 1/10,000 the diameter of the atom. Even so, the nucleus is more than 99.9% of the atom's mass. What we call *matter* is overwhelmingly empty space!

The nucleus is composed of two types of particles: *protons* and *neutrons*, which together are referred to as **nucleons.** The role of the neutrons, which have nothing to do with keeping electrons in orbit, is an important issue that we'll address in this chapter. Table 42.1 summarizes the basic properties of protons and neutrons.

As you can see, protons and neutrons are virtually identical other than that the proton has one unit of the fundamental charge e whereas the neutron is electrically neutral. The neutron is slightly more massive than the proton, but the difference is very small, only about 0.1%. Notice that the proton and neutron, like the electron, have an *inherent angular momentum* and magnetic moment with spin quantum number $s = \frac{1}{2}$. As a consequence, protons and neutrons obey the Pauli exclusion principle.

The number of protons Z is the element's **atomic number.** In fact, an element is identified by the number of protons in the nucleus, not by the number of orbiting electrons. Electrons are easily added and removed, forming negative and positive ions, but doing so doesn't change the element. The **mass number** A is defined to be $A = Z + N$,

FIGURE 42.1 The nucleus is a tiny speck within an atom.

This picture of an atom would need to be 10 m in diameter if it were drawn to the same scale as the dot representing the nucleus.

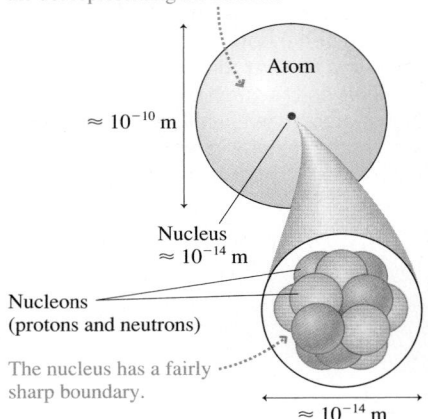

Atom

$\approx 10^{-10}$ m

Nucleus
$\approx 10^{-14}$ m

Nucleons
(protons and neutrons)

The nucleus has a fairly sharp boundary.

$\approx 10^{-14}$ m

TABLE 42.1 Protons and neutrons

	Proton	**Neutron**
Number	Z	N
Charge q	$+e$	0
Spin s	$\frac{1}{2}$	$\frac{1}{2}$
Mass, in u	1.00728	1.00866

where N is the **neutron number.** The mass number is the total number of nucleons in a nucleus.

NOTE ▶ The mass number, which is dimensionless, is *not* the same thing as the atomic mass m. We'll look at actual atomic masses later. ◀

Isotopes and Isobars

It was discovered early in the 20th century that not all atoms of the same element (same Z) have the same mass. There are a *range* of neutron numbers that happily form a nucleus with Z protons, creating a group of nuclei having the same Z-value (i.e., they are all the same chemical element) but different A-values. The atoms of an element with different values of A are called **isotopes** of that element.

Chemical behavior is determined by the orbiting electrons. All isotopes of one element have the same number of orbiting electrons (if the atoms are electrically neutral) and thus have the same chemical properties, but different isotopes of the same element can have quite different nuclear properties.

The notation used to label isotopes is $^A Z$, where the mass number A is given as a *leading* superscript. The proton number Z is not specified by an actual number but, equivalently, by the chemical symbol for that element. Hence ordinary carbon, which has six protons and six neutrons in the nucleus, is written ^{12}C and pronounced "carbon twelve." The radioactive form of carbon used in carbon dating is ^{14}C. It has six protons, making it carbon, and eight neutrons.

More than 3000 isotopes are known. The majority of these are **radioactive,** meaning that the nucleus is not stable but, after some period of time, will either fragment or emit some kind of subatomic particle in an effort to reach a more stable state. Many of these radioactive isotopes are created by nuclear reactions in the laboratory and have only a fleeting existence. Only 266 isotopes are **stable** (i.e., nonradioactive) and occur in nature. We'll begin to look at the issue of nuclear stability in the next section.

The *naturally occurring* nuclei include the 266 stable isotopes and a handful of radioactive isotopes with such long half-lives, measured in billions of years, that they also occur naturally. The most well-known example of a naturally occurring radioactive isotope is the uranium isotope ^{238}U. For each element, the fraction of naturally occurring nuclei represented by one particular isotope is called the **natural abundance** of that isotope.

Although there are many radioactive isotopes of the element iodine, iodine occurs *naturally* only as ^{127}I. Consequently, we say that the natural abundance of ^{127}I is 100%. Most elements have multiple naturally occurring isotopes. The natural abundance of ^{14}N is 99.6%, meaning that 996 out of every 1000 naturally occurring nitrogen atoms are the isotope ^{14}N. The remaining 0.4% of naturally occurring nitrogen is the isotope ^{15}N, with one extra neutron.

A series of nuclei having the same A-value (the same mass number) but different values of Z and N are called **isobars.** For example, the three nuclei ^{14}C, ^{14}N, and ^{14}O are isobars with $A = 14$. Only ^{14}N is stable; the other two are radioactive.

Atomic Mass

You learned in Chapter 16 that atomic masses are specified in terms of the *atomic mass unit* u, defined such that the atomic mass of the isotope ^{12}C is exactly 12 u. The conversion to SI units is

$$1 \text{ u} = 1.6605 \times 10^{-27} \text{ kg}$$

Alternatively, we can use Einstein's $E_0 = mc^2$ to express masses in terms of their energy equivalent. The energy equivalent of 1 u of mass is

$$E_0 = (1.6605 \times 10^{-27} \text{ kg})(2.9979 \times 10^8 \text{ m/s})^2$$
$$= 1.4924 \times 10^{-10} \text{ J} = 931.49 \text{ MeV}$$

(42.1)

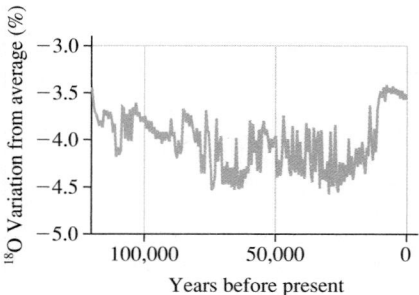

When water freezes to make snow crystals, the fraction of molecules containing ^{18}O is greater for snow that forms at higher atmospheric temperatures. Snow accumulating over tens of thousands of years has built up a thick ice sheet in Greenland. A core sample of this ice gives a record of the isotopic composition of the snow that fell over this time period. Higher numbers on the graph correspond to higher average temperatures. Broad trends, such as the increase in temperature at the end of the last ice age, are clearly seen.

Thus the atomic mass unit can be written

$$1 \text{ u} = 931.49 \text{ MeV}/c^2$$

It may seem unusual, but the units MeV/c^2 are units of mass.

NOTE ▶ We're using more significant figures than usual. Many nuclear calculations look for the small difference between two masses that are almost the same. Those two masses must be calculated or specified to four or five significant figures if their difference is to be meaningful. ◀

Table 42.2 shows the atomic masses of the electron, the nucleons, and three important light elements. Appendix C contains a more complete list. Notice that the mass of a hydrogen atom is the sum of the masses of a proton and an electron. But a quick calculation shows that the mass of a helium atom (2 protons, 2 neutrons, and 2 electrons) is 0.03038 u less than the sum of the masses of its constituents. The difference is due to the binding energy of the nucleus, a topic we'll look at in Section 42.2.

The isotope ^{2}H is a hydrogen atom in which the nucleus is not simply a proton but a proton and a neutron. Although the isotope is a form of hydrogen, it is called **deuterium.** The natural abundance of deuterium is 0.015%, or about 1 out of every 6700 hydrogen atoms. Water made with deuterium (sometimes written D_2O rather than H_2O) is called *heavy water.*

NOTE ▶ Don't let the name *deuterium* cause you to think this is a different element. Deuterium is an isotope of hydrogen. Chemically, it behaves just like ordinary hydrogen. ◀

The *chemical* atomic mass shown on the periodic table of the elements is the *weighted average* of the atomic masses of all naturally occurring isotopes. For example, chlorine has two stable isotopes: ^{35}Cl, with $m = 34.97$ u, is 75.8% abundant and ^{37}Cl, at 36.97 u, is 24.2% abundant. The average, weighted by abundance, is $0.758 \times 34.97 + 0.242 \times 36.97 = 35.45$. This is the value shown on the periodic table and is the correct value for most chemical calculations, but it is not the mass of any particular isotope of chlorine.

NOTE ▶ The atomic masses of the proton and the neutron are both ≈ 1 u. Consequently, the value of the mass number A is *approximately* the atomic mass in u. The approximation $m \approx A$ u is sufficient in many contexts, such as when we're calculating the masses of atoms in the kinetic theory of gases, but in nuclear physics calculations, we almost always need the more accurate mass values that you find in Table 42.2 or Appendix C. ◀

TABLE 42.2 Some atomic masses

Particle	Symbol	Mass (u)	Mass (MeV/c^2)
Electron	e	0.00055	0.51
Proton	p	1.00728	938.28
Neutron	n	1.00866	939.57
Hydrogen	^{1}H	1.00783	938.79
Deuterium	^{2}H	2.01410	1876.12
Helium	^{4}He	4.00260	3728.40

Nuclear Size and Density

Unlike the atom's electron cloud, which is quite diffuse, the nucleus has a fairly sharp boundary. Experimentally, the radius of a nucleus with mass number A is found to be

$$R = r_0 A^{1/3} \tag{42.2}$$

where $r_0 = 1.2$ fm. Recall that 1 fm = 1 femtometer = 10^{-15} m.

As FIGURE 42.2 shows, the radius is proportional to $A^{1/3}$. Consequently, the volume of the nucleus (proportional to R^3) is directly proportional to A, the number of nucleons. A nucleus with twice as many nucleons will occupy twice as much volume. This finding has three implications:

- Nucleons are incompressible. Adding more nucleons doesn't squeeze the inner nucleons into a smaller volume.
- The nucleons are tightly packed, looking much like the drawing in Figure 42.1.
- Nuclear matter has a constant density.

FIGURE 42.2 The nuclear radius and volume as a function of A.

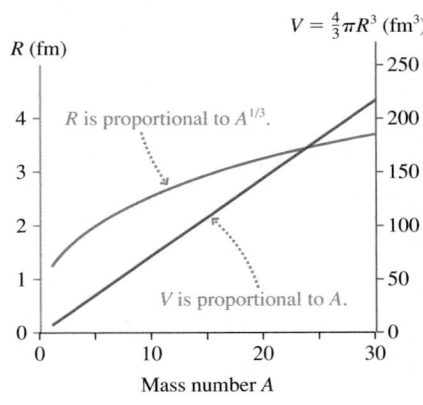

FIGURE 42.3 Density profiles of three nuclei.

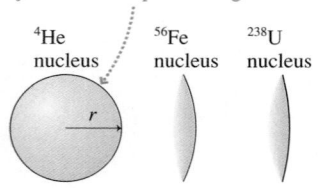

Imagine the nucleus is a drop of liquid. Its density is the same up to the edge of the drop.

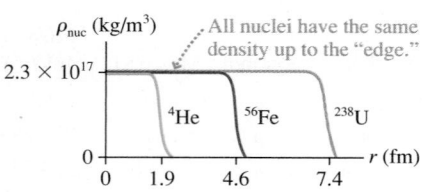

In fact, we can use Equation 42.2 to estimate the density of nuclear matter. Consider a nucleus with mass number A. Its mass, within 1%, is A atomic mass units. Thus

$$\rho_{nuc} \approx \frac{A \text{ u}}{\frac{4}{3}\pi R^3} = \frac{A \text{ u}}{\frac{4}{3}\pi r_0^3 A} = \frac{1 \text{ u}}{\frac{4}{3}\pi r_0^3} = \frac{1.66 \times 10^{-27} \text{ kg}}{\frac{4}{3}\pi (1.2 \times 10^{-15} \text{ m})^3} \quad (42.3)$$

$$= 2.3 \times 10^{17} \text{ kg/m}^3$$

The fact that A cancels means that **all nuclei have this density.** It is a staggeringly large density, roughly 10^{14} times larger than the density of familiar liquids and solids. One early objection to Rutherford's model of a nuclear atom was that matter simply couldn't have a density this high. Although we have no direct experience with such matter, nuclear matter really is this dense.

FIGURE 42.3 shows the density profiles of three nuclei. The constant density right to the edge is analogous to that of a drop of incompressible liquid, and, indeed, one successful model of many nuclear properties is called the **liquid-drop model.** Notice that the range of nuclear radii, from small helium to large uranium, is not quite a factor of 4. The fact that ^{56}Fe is a fairly typical atom in the middle of the periodic table is the basis for our earlier assertion that the nuclear diameter is roughly 10^{-14} m, or 10 fm.

> **STOP TO THINK 42.1** Three electrons orbit a neutral ^{6}Li atom. How many electrons orbit a neutral ^{7}Li atom?

42.2 Nuclear Stability

We've noted that fewer than 10% of the known nuclei are stable (i.e., not radioactive). Because nuclei are characterized by two independent numbers, N and Z, it is useful to show the known nuclei on a plot of neutron number N versus proton number Z. **FIGURE 42.4** shows such a plot. Stable nuclei are represented by blue diamonds and unstable, radioactive nuclei by red dots.

FIGURE 42.4 Stable and unstable nuclei shown on a plot of neutron number N versus proton number Z.

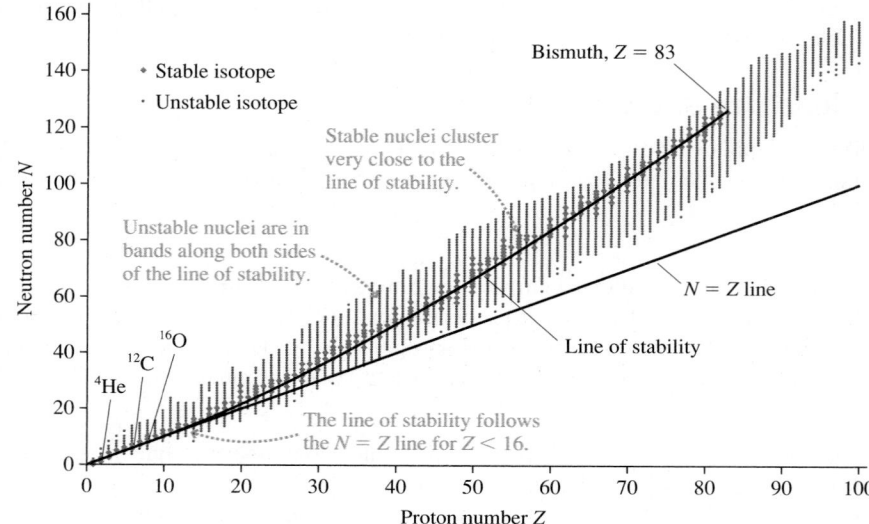

We can make several observations from this graph:

- The stable nuclei cluster very close to the curve called the **line of stability.**
- There are no stable nuclei with $Z > 83$ (bismuth).
- Unstable nuclei are in bands along both sides of the line of stability.
- The lightest elements, with $Z < 16$, are stable when $N \approx Z$. The familiar elements ^{4}He, ^{12}C, and ^{16}O all have equal numbers of protons and neutrons.
- As Z increases, the number of neutrons needed for stability grows increasingly larger than the number of protons. The N/Z ratio is ≈ 1.2 at $Z = 40$ but has grown to ≈ 1.5 at $Z = 80$.

STOP TO THINK 42.2 The isobars corresponding to one specific value of A are found on the plot of Figure 42.4 along

a. A vertical line.

b. A horizontal line.

c. A diagonal line that goes up and to the left.

d. A diagonal line that goes up and to the right.

Binding Energy

A nucleus is a *bound system.* That is, you would need to supply energy to disperse the nucleons by breaking the nuclear bonds between them. FIGURE 42.5 shows this idea schematically.

You learned a similar idea in atomic physics. The energy levels of the hydrogen atom are negative numbers because the bound system has less energy than a free proton and electron. The energy you must supply to an atom to remove an electron is called the *ionization energy.*

In much the same way, the energy you would need to supply to a nucleus to disassemble it into individual protons and neutrons is called the **binding energy.** Whereas ionization energies of atoms are only a few eV, the binding energies of nuclei are tens or hundreds of MeV, energies large enough that their mass equivalent is not negligible.

Consider a nucleus with mass m_{nuc}. It is found experimentally that m_{nuc} is *less* than the total mass $Zm_p + Nm_n$ of the Z protons and N neutrons that form the nucleus, where m_p and m_n are the masses of the proton and neutron. That is, the energy equivalent $m_{nuc}c^2$ of the nucleus is less than the energy equivalent $(Zm_p + Nm_n)c^2$ of the individual nucleons. The binding energy B of the nucleus (not the entire atom) is defined as

$$B = (Zm_p + Nm_n - m_{nuc})c^2 \qquad (42.4)$$

This is the energy you would need to supply to disassemble the nucleus into its pieces.

The practical difficulty is that laboratory scientists use mass spectroscopy to measure *atomic* masses, not nuclear masses. The atomic mass m_{atom} is m_{nuc} plus the mass Zm_e of Z orbiting electrons. (Strictly speaking, we should allow for the binding energy of the electrons, but these binding energies are roughly a factor of 10^6 smaller than the nuclear binding energies and can be neglected in all but the most precise measurements and calculations.)

Fortunately, we can switch from the nuclear mass to the atomic mass by the simple trick of both adding and subtracting Z electron masses. We begin by writing Equation 42.4 in the equivalent form

$$B = (Zm_p + Zm_e + Nm_n - m_{nuc} - Zm_e)c^2 \qquad (42.5)$$

FIGURE 42.5 The nuclear binding energy.

The binding energy is the energy that would be needed to disassemble a nucleus into individual nucleons.

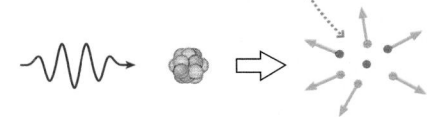

Energy Nucleus Disassembled nucleus

$$B \quad + \quad m_{nuc}c^2 \quad = \quad (Zm_p + Nm_n)c^2$$

Now $m_{nuc} + Zm_e = m_{atom}$, the atomic mass, and $Zm_p + Zm_e = Z(m_p + m_e) = Zm_H$, where m_H is the mass of a hydrogen *atom*. Finally, we use the conversion factor $1\ u = 931.49\ MeV/c^2$ to write $c^2 = 931.49\ MeV/u$. The binding energy is then

$$B = (Zm_H + Nm_n - m_{atom}) \times (931.49\ MeV/u) \qquad (42.6)$$
$$\text{(binding energy)}$$

where all three masses are in atomic mass units.

EXAMPLE 42.1 **The binding energy of iron**

What is the binding energy of the ^{56}Fe nucleus?

SOLVE The isotope ^{56}Fe has $Z = 26$ and $N = 30$. The atomic mass of ^{56}Fe, found in Appendix C, is 55.9349 u. Thus the mass difference between the ^{56}Fe nucleus and its constituents is

$$B = 26(1.0078\ u) + 30(1.0087\ u) - 55.9349\ u = 0.529\ u$$

where, from Table 42.2, 1.0078 u is the mass of the hydrogen atom. Thus the binding energy of ^{56}Fe is

$$B = (0.529\ u) \times (931.49\ MeV/u) = 493\ MeV$$

ASSESS The binding energy is extremely large, the energy equivalent of more than half the mass of a proton or a neutron.

The nuclear binding energy increases as A increases simply because there are more nuclear bonds. A more useful measure for comparing one nucleus to another is the quantity B/A, called the *binding energy per nucleon*. Iron, with $B = 493$ MeV and $A = 56$, has 8.80 MeV per nucleon. This is the amount of energy, on average, you would need to supply in order to remove *one* nucleon from the nucleus. Nuclei with larger values of B/A are more tightly held together than nuclei with smaller values of B/A.

FIGURE 42.6 The curve of binding energy.

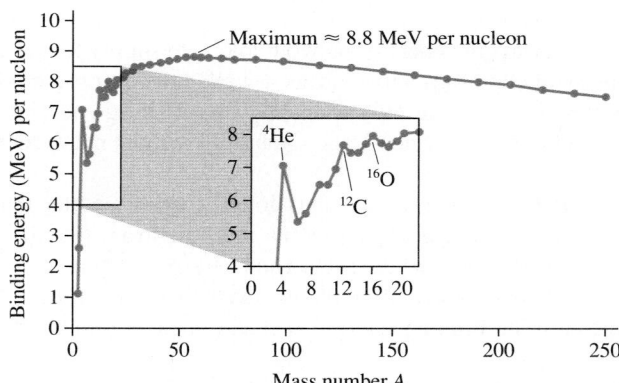

FIGURE 42.6 is a graph of the binding energy per nucleon versus mass number A. The line connecting the points is often called the **curve of binding energy.** This curve has three important features:

■ There are peaks in the binding energy curve at $A = 4$, 12, and 16. The one at $A = 4$, corresponding to ^{4}He, is especially pronounced. As you'll see, these peaks, which represent nuclei more tightly bound than their neighbors, are due to *closed shells* in much the same way that the graph of atomic ionization energies (see Figure 41.23) peaked for closed electron shells.

■ The binding energy per nucleon is *roughly* constant at ≈ 8 MeV per nucleon for $A > 20$. This suggests that, as a nucleus grows, there comes a point where the nuclear bonds are *saturated*. Each nucleon interacts only with its nearest neighbors, the ones it's actually touching. This, in turn, implies that the nuclear force is a *short-range* force.

- The curve has a broad maximum at $A \approx 60$. This will be important for our understanding of radioactivity. In principle, heavier nuclei could become *more* stable (more binding energy per nucleon) by breaking into smaller pieces. Lighter nuclei could become *more* stable by fusing together into larger nuclei. There may not always be a mechanism for such nuclear transformations to take place, but *if* there is a mechanism, it is energetically favorable for it to occur.

42.3 The Strong Force

Rutherford's discovery of the atomic nucleus was not immediately accepted by all scientists. Their primary objection was that the protons would blow themselves apart at tremendously high speeds due to the extremely large electrostatic forces between them at a separation of a few femtometers. No known force could hold the nucleus together.

It soon became clear that a previously unknown force of nature operates within the nucleus to hold the nucleons together. This new force had to be stronger than the repulsive electrostatic force; hence it was named the **strong force**. It is also called the *nuclear force*.

The strong force has four important properties:

1. It is an *attractive* force between any two nucleons.
2. It does not act on electrons.
3. It is a *short-range* force, acting only over nuclear distances.
4. Over the range where it acts, it is *stronger* than the electrostatic force that tries to push two protons apart.

The fact that the strong force is short-range, in contrast to the long-range $1/r^2$ electric, magnetic, and gravitational forces, is apparent from the fact that we see no evidence for nuclear forces outside the nucleus.

FIGURE 42.7 summarizes the three interactions that take place within the nucleus. Whether the strong force between two protons is the same strength as the force between two neutrons or between a proton and a neutron is an important question that can be answered experimentally. The primary means of investigating the strong force is to accelerate a proton to very high speed, using a cyclotron or some other particle accelerator, then to study how the proton is scattered by various target materials.

The conclusion of many decades of research is that the strong force between two nucleons is independent of whether they are protons or neutrons. Charge is the basis for electromagnetic interactions, but it is of no relevance to the strong force. Protons and neutrons are identical as far as nuclear forces are concerned.

Potential Energy

Unfortunately, there's no simple formula to calculate the strong force or the potential energy of two nucleons interacting via the strong force. FIGURE 42.8 is an experimentally determined potential-energy diagram for two interacting nucleons, with r the distance between their centers. The potential-energy minimum at $r \approx 1$ fm is a point of stable equilibrium.

Recall that the force is the negative of the slope of a potential-energy diagram. The steeply rising potential for $r < 1$ fm represents a strongly repulsive force. That is, the nucleon "cores" strongly repel each other if they get too close together. The force is attractive for $r > 1$ fm, where the slope is positive, and it is strongest where the slope is steepest, at $r \approx 1.5$ fm. The strength of the force quickly decreases for $r > 1.5$ fm and is zero for $r > 3$ fm. That is, the strong force represented by this potential energy is effective only over a very short range of distances.

Notice how small the electrostatic energy of two protons is in comparison to the potential energy of the strong force. At $r \approx 1.0$ fm, the point of stable equilibrium, the magnitude of the nuclear potential energy is ≈ 100 times larger than the electrostatic potential energy.

FIGURE 42.7 The strong force is the same between any two nucleons.

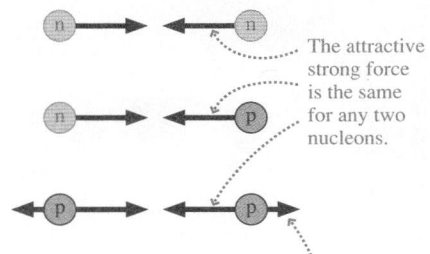

The attractive strong force is the same for any two nucleons.

Two protons also experience a smaller electrostatic repulsive force.

FIGURE 42.8 The potential-energy diagram for two nucleons interacting via the strong force.

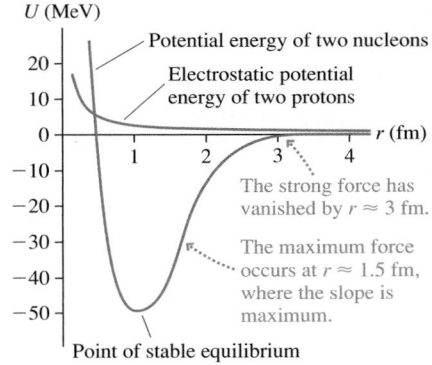

Potential energy of two nucleons

Electrostatic potential energy of two protons

The strong force has vanished by $r \approx 3$ fm.

The maximum force occurs at $r \approx 1.5$ fm, where the slope is maximum.

Point of stable equilibrium

But why does the nucleus have neutrons at all? The answer is related to the short range of the strong force. Protons throughout the nucleus exert repulsive electrostatic forces on each other, but, because of the short range of the strong force, a proton feels an attractive force only from the very few other protons with which it is in close contact. Even though the strong force at its maximum is much larger than the electrostatic force, there wouldn't be enough attractive nuclear bonds for an all-proton nucleus to be stable. Because neutrons participate in the strong force but exert no repulsive forces, **the neutrons provide the extra "glue" that holds the nucleus together.** In small nuclei, where most nucleons are in contact, one neutron per proton is sufficient for stability. Hence small nuclei have $N \approx Z$. But as the nucleus grows, the repulsive force increases faster than the binding energy. More neutrons are needed for stability, causing heavy nuclei to have $N > Z$.

42.4 The Shell Model

Figure 42.8 shows the potential energy of *two* interacting nucleons. To solve Schrödinger's equation for the nucleus, we would need to know the total potential energy of *all* interacting nucleon pairs within the nucleus, including both the strong force and the electrostatic force. This is far too complex to be a tractable problem.

We faced a similar situation with multielectron atoms. Calculating an atom's exact potential energy is exceedingly complicated. To simplify the problem, we made a *model* of the atom in which each electron moves independently with an *average* potential energy due to the nucleus and all other electrons. That model, although not perfect, correctly predicted electron shells and explained the periodic table of the elements.

The **shell model** of the nucleus, using multielectron atoms as an analogy, was proposed in 1949 by Maria Goeppert-Mayer. The shell model considers each nucleon to move independently with an *average* potential energy due to the strong force of all the other nucleons. For the protons, we also have to include the electrostatic potential energy due to the other protons.

FIGURE 42.9 shows the average potential energy of a neutron and a proton. Here r is the distance from the center of the nucleus, not the nucleon–nucleon distance as it was in Figure 42.8. On average, a nucleon's interactions with neighboring nucleons are independent of the nucleon's position inside the nucleus; hence the constant potential energy inside the nucleus. You can see that, to a good approximation, a nucleon appears to be a particle in a finite potential well, a quantum-mechanics problem you studied in Chapter 40.

Maria Goeppert-Mayer received the 1963 Nobel Prize in Physics for her work in nuclear physics.

FIGURE 42.9 The average potential energy of a neutron and a proton.

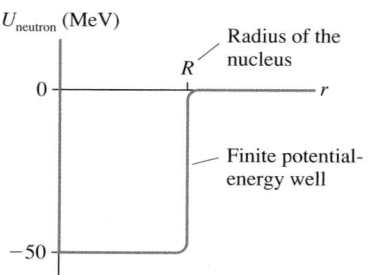

The average neutron potential energy is due to the strong force.

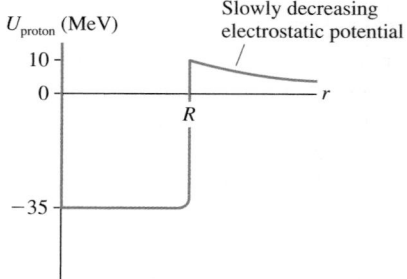

The average proton potential energy is due to the strong force and the electric force. This potential-well depth is for $Z \approx 30$.

Three observations are worthwhile:

1. The depth of the neutron's potential-energy well is ≈ 50 MeV for all nuclei. The radius of the potential-energy well is the nuclear radius $R = r_0 A^{1/3}$.

2. For protons, the positive electrostatic potential energy "lifts" the potential-energy well. The lift varies from essentially none for very light elements to a significant fraction of the well depth for very heavy elements. The potential energy shown in the figure would be appropriate for a nucleus with $Z \approx 30$.

3. Outside the nucleus, where the strong force has vanished, a proton's potential energy is $U = (Z-1)e^2/4\pi\epsilon_0 r$ due to its electrostatic interaction with the $(Z-1)$ other protons within the nucleus. This positive potential energy decreases slowly with increasing distance.

The task of quantum mechanics is to solve for the energy levels and wave functions of the nucleons in these potential-energy wells. Once the energy levels are found, we build up the nuclear state, just as we did with atoms, by placing all the nucleons in the lowest energy levels consistent with the Pauli principle. The Pauli principle affects nucleons, just as it did electrons, because they are spin-$\frac{1}{2}$ particles. Each energy level can hold only a certain number of spin-up particles and spin-down particles, depending on the quantum numbers. Additional nucleons have to go into higher energy levels.

Low-Z Nuclei

As an example, we'll consider the energy levels of low-Z nuclei ($Z < 8$). Because these nuclei have so few protons, we can use a reasonable approximation that neglects the electrostatic potential energy due to proton-proton repulsion and considers only the much larger nuclear potential energy. In that case, the proton and neutron potential-energy wells and energy levels are the same.

FIGURE 42.10 shows the three lowest energy levels and the maximum number of nucleons that the Pauli principle allows in each. Energy values vary from nucleus to nucleus, but the spacing between these levels is several MeV. It's customary to draw the proton and neutron potential-energy diagrams and energy levels back to back. Notice that the radial axis for the proton potential-energy well points to the right, while the radial axis for the neutron potential-energy well points to the left.

Let's apply this model to the $A = 12$ isobar. Recall that an isobar is a series of nuclei with the same total number of neutrons and protons. **FIGURE 42.11** shows the energy-level diagrams of ^{12}B, ^{12}C, and ^{12}N. Look first at ^{12}C, a nucleus with six protons and six neutrons. You can see that exactly six protons are allowed in the $n = 1$ and $n = 2$ energy levels. Likewise for the six neutrons. Thus ^{12}C has a closed $n = 2$ proton shell and a closed $n = 2$ neutron shell.

NOTE ▶ Protons and neutrons are different particles, so the Pauli principle is not violated if a proton and a neutron have the same quantum numbers. ◀

FIGURE 42.10 The three lowest energy levels of a low-Z nucleus. The neutron energy levels are on the left, the proton energy levels on the right.

The neutron radial distance is measured to the left.

The proton potential energy is nearly identical to the neutron potential energy when Z is small.

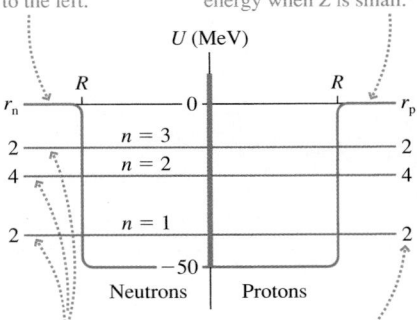

These are the first three allowed energy levels. They are spaced several MeV apart.

These are the maximum numbers of nucleons allowed by the Pauli principle.

FIGURE 42.11 The $A = 12$ isobar has to place 12 nucleons in the lowest available energy levels.

A ^{12}B nucleus could lower its energy if a neutron could turn into a proton.

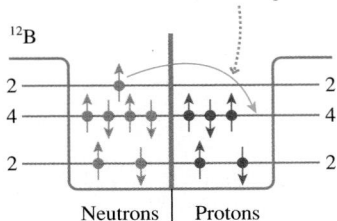

^{12}B

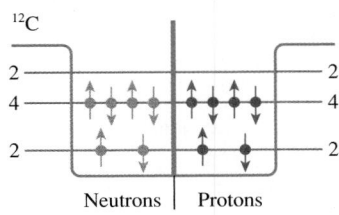

^{12}C

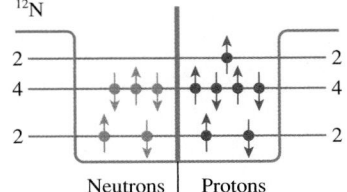

^{12}N

^{12}N has seven protons and five neutrons. The sixth proton fills the $n = 2$ proton shell, so the seventh proton has to go into the $n = 3$ energy level. The $n = 2$ neutron shell has one vacancy because there are only five neutrons. ^{12}B is just the opposite, with the seventh neutron in the $n = 3$ energy level. You can see from the diagrams that the ^{12}B and ^{12}N nuclei have significantly more energy—by several MeV—than ^{12}C.

In atoms, electrons in higher energy levels decay to lower energy levels by emitting a photon as the electron undergoes a quantum jump. That can't happen here because the higher-energy nucleon in ^{12}B is a neutron whereas the vacant lower energy level is that of a proton. But an analogous process could occur *if* a neutron could somehow turn into a proton. And that's exactly what happens! We'll explore the details in Section 42.6, but both ^{12}B and ^{12}N decay into ^{12}C in the process known as *beta decay*.

^{12}C is just one of three low-Z nuclei in which both the proton and neutron shells are full. The other two are ^{4}He (filling both $n = 1$ shells with $Z = 2$, $N = 2$) and ^{16}O (filling both $n = 3$ shells with $Z = 8$, $N = 8$). If the analogy with closed electron shells is valid, these nuclei should be more tightly bound than nuclei with neighboring values of A. And indeed, we've already noted that the curve of binding energy (Figure 42.6) has peaks at $A = 4$, 12, and 16. The shell model of the nucleus satisfactorily explains these peaks. Unfortunately, the shell model quickly becomes much more complex as we go beyond $n = 3$. Heavier nuclei do have closed shells, but there's no evidence for them in the curve of binding energy.

High-Z Nuclei

We can use the shell model to give a qualitative explanation for one more observation, although the details are beyond the scope of this text. FIGURE 42.12 shows the neutron and proton potential-energy wells of a high-Z nucleus. In a nucleus with many protons, the electrostatic potential energy lifts the proton potential-energy well higher than the neutron potential-energy well. Protons and neutrons now have a different set of energy levels.

As a nucleus is "built," by the addition of protons and neutrons, the proton energy well and the neutron energy well must fill to just about the same height. If there were neutrons in energy levels above vacant proton levels, the nucleus would lower its energy by using beta decay to change the neutron into a proton. Similarly, beta decay would change a proton into a neutron if there were a vacant neutron energy level beneath a filled proton level. **The net result of beta decay is to keep the levels on both sides filled to just about the same energy.**

Because the neutron potential-energy well starts at a lower energy, *more neutron states* are available than proton states. Consequently, a high-Z nucleus will have more neutrons than protons. This conclusion is consistent with our observation in Figure 42.4 that $N > Z$ for heavy nuclei.

FIGURE 42.12 The proton energy levels are displaced upward in a high-Z nucleus.

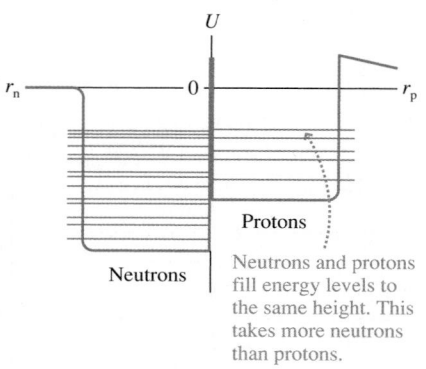

Neutrons

Protons

Neutrons and protons fill energy levels to the same height. This takes more neutrons than protons.

42.5 Radiation and Radioactivity

Becquerel's 1896 discovery of "rays" from crystals of uranium prompted a burst of activity. In England, J. J. Thomson and, especially, his student and protégé Ernest Rutherford worked to identify the unknown rays. Using combinations of electric and magnetic fields, much as Thomson had done in his investigations of cathode rays, they found three distinct types of radiation. FIGURE 42.13 shows the basic experimental procedure, and Table 42.3 on the next page summarizes the results.

FIGURE 42.13 Identifying radiation by its deflection in a magnetic field.

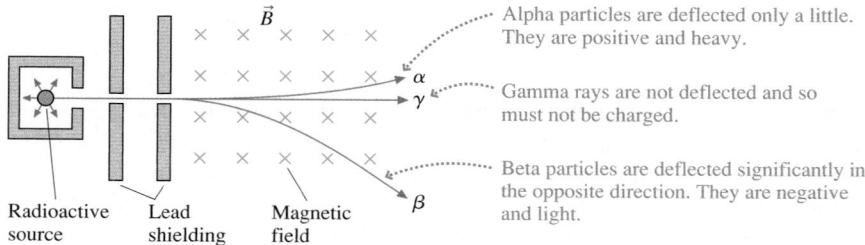

Alpha particles are deflected only a little. They are positive and heavy.

Gamma rays are not deflected and so must not be charged.

Beta particles are deflected significantly in the opposite direction. They are negative and light.

Radioactive source   Lead shielding   Magnetic field

TABLE 42.3 Three types of radiation

Radiation	Identification	Charge	Stopped by
Alpha, α	^{4}He nucleus	$+2e$	Sheet of paper
Beta, β	Electron	$-e$	Few mm of aluminum
Gamma, γ	High-energy photon	0	Many cm of lead

Within a few years, as Rutherford and others deduced the basic structure of the atom, it became clear that these emissions of radiation were coming from the atomic nucleus. We now define *radioactivity* or *radioactive decay* to be the spontaneous emission of particles or high-energy photons from unstable nuclei as they decay from higher-energy to lower-energy states. Radioactivity has nothing to do with the orbiting valence electrons.

NOTE ▶ The term "radiation" merely means something that is *radiated outward,* similar to the word "radial." Electromagnetic waves are often called "electromagnetic radiation." Infrared waves from a hot object are referred to as "thermal radiation." Thus it was no surprise that these new "rays" were also called radiation. Unfortunately, the general public has come to associate the word "radiation" with *nuclear radiation,* something to be feared. It is important, when you use the term, to be sure you're not conveying a wrong impression to a listener or a reader. ◀

Ionizing Radiation

Electromagnetic waves, from microwaves through ultraviolet radiation, are absorbed by matter. The absorbed energy increases an object's thermal energy and its temperature, which is why objects sitting in the sun get warm.

In contrast to visible-light photon energies of a few eV, the energies of the alpha and beta particles and the gamma-ray photons of nuclear decay are typically in the range 0.1–10 MeV, a factor of roughly 10^6 larger. These energies are much larger than the ionization energies of atoms and molecules. Rather than simply being absorbed and increasing an object's thermal energy, nuclear radiation *ionizes* matter and *breaks* molecular bonds. Nuclear radiation (and also x rays, which behave much the same in matter) is called **ionizing radiation.**

An alpha or beta particle traveling through matter creates a trail of ionization, as shown in FIGURE 42.14. Because the ionization energy of an atom is ≈ 10 eV, a particle with 1 MeV of kinetic energy can ionize $\approx 100,000$ atoms or molecules before finally stopping. The low-mass electrons are kicked sideways, but the much more massive positive ions barely move and form the trail. This ionization is the basis for the **Geiger counter,** one of the most well-known detectors of nuclear radiation. FIGURE 42.15 shows how a Geiger counter works. The important thing to remember is that a Geiger counter detects only *ionizing radiation.*

Ionizing radiation damages materials. Ions drive chemical reactions that wouldn't otherwise occur. Broken molecular bonds alter the workings of molecular machinery, especially in large biological molecules. It is through these mechanisms—ionization and bond breaking—that nuclear radiation can cause mutations or tumors. We'll look at the biological issues in Section 42.7.

NOTE ▶ Ionizing radiation causes structural damage to materials, but **irradiated objects do not become radioactive.** Ionization drives chemical processes involving the electrons. An object could become radioactive only if its nuclei were somehow changed, and that does not happen. ◀

STOP TO THINK 42.3   A very bright spotlight shines on a Geiger counter. Does it click?

FIGURE 42.14 Alpha and beta particles create a trail of ionization as they pass through matter.

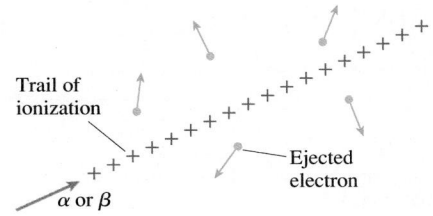

Trail of ionization

Ejected electron

α or β

FIGURE 42.15 A Geiger counter.

1. Ejected electrons cause a chain reaction of ionization of the gas as they accelerate toward the positive wire.

2. Thousands of electrons reach the wire, causing a surge of current.

3. The negative current pulse in the wire causes the "click" of the Geiger counter.

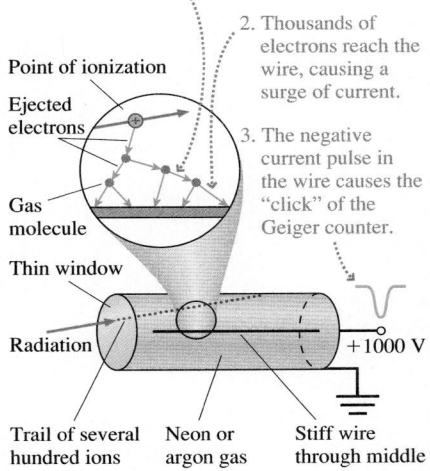

Point of ionization

Ejected electrons

Gas molecule

Thin window

Radiation

+1000 V

Trail of several hundred ions

Neon or argon gas

Stiff wire through middle

Nuclear Decay and Half-Lives

Rutherford discovered experimentally that the number of radioactive atoms in a sample decreases exponentially with time. The reason is that radioactive decay is a *random process*. That is, we can predict only the *probability* that a nucleus will decay, not the exact moment. We encountered exactly this situation when we investigated the lifetimes of excited states of atoms in Section 41.7.

As we did with atoms, let *r* be the *decay rate*, the probability of decay within the next second by the emission of an alpha or beta particle or a gamma-ray photon. Then the probability of decay within a small time interval Δt is

$$\text{Prob(decay in time interval } \Delta t) = r \, \Delta t \tag{42.7}$$

This equation was also the starting point in our analysis of the spontaneous emission of photons by atoms. The mathematical analysis is exactly the same as in Section 41.7, to which you should refer, leading to the exponential-decay equation

$$N = N_0 e^{-t/\tau} \tag{42.8}$$

where $\tau = 1/r$ is the *lifetime* of the nucleus.

FIGURE 42.16 shows the decrease of *N* with time. The number of radioactive nuclei decreases from N_0 at $t = 0$ to $e^{-1}N_0 = 0.368N_0$ at time $t = \tau$. In practical terms, the number decreases by roughly two-thirds during one lifetime.

FIGURE 42.16 The number of radioactive atoms decreases exponentially with time.

Number of nuclei remaining

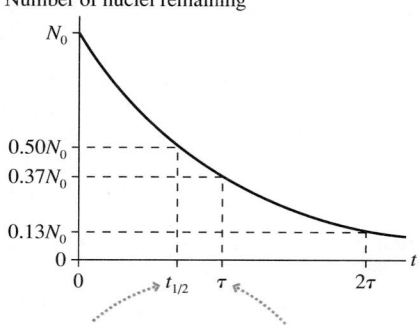

The half-life is the time in which half the nuclei decay.

The lifetime is the time at which the number of nuclei is e^{-1}, or 37%, of the initial number.

NOTE ▶ An important aspect of exponential decay is that you can choose any instant you wish to be $t = 0$. The number of radioactive nuclei present at that instant is N_0. If at one instant you have 10,000 radioactive nuclei whose lifetime is $\tau = 10$ min, you'll have roughly 3680 nuclei 10 min later. The fact that you may have had more than 10,000 nuclei earlier isn't relevant. ◀

In practice, it's much easier to measure the time at which half of a sample has decayed than the time at which 36.8% has decayed. Let's define the **half-life** $t_{1/2}$ as the time interval in which half of a sample of radioactive atoms decays. The half-life is shown in Figure 42.16.

The half-life is easily related to the lifetime τ because we know, by definition, that $N = \frac{1}{2}N_0$ at $t = t_{1/2}$. Thus, according to Equation 42.8,

$$\frac{N_0}{2} = N_0 e^{-t_{1/2}/\tau} \tag{42.9}$$

The N_0 cancels, and we can then take the natural logarithm of both sides to find

$$\ln\left(\frac{1}{2}\right) = -\ln 2 = -\frac{t_{1/2}}{\tau} \tag{42.10}$$

With one final rearrangement we have

$$t_{1/2} = \tau \ln 2 = 0.693\tau \tag{42.11}$$

Equation 42.8 can be written in terms of the half-life as

$$N = N_0\left(\frac{1}{2}\right)^{t/t_{1/2}} \tag{42.12}$$

Thus $N = N_0/2$ at $t = t_{1/2}$, $N = N_0/4$ at $t = 2t_{1/2}$, $N = N_0/8$ at $t = 3t_{1/2}$, and so on. **No matter how many nuclei there are, the number decays by half during the next half-life.**

NOTE ▶ Half the nuclei decay during one half-life, but don't fall into the trap of thinking that all will have decayed after two half-lives. ◀

FIGURE 42.17 shows the half-life graphically. This figure also conveys two other important ideas:

1. Nuclei don't vanish when they decay. The decayed nuclei have merely become some other kind of nuclei, called the *daughter nuclei.*
2. The decay process is random. We can predict that half the nuclei will decay in one half-life, but we can't predict which ones.

Each radioactive isotope, such as ^{14}C, has its own half-life. That half-life doesn't change with time as a sample decays. If you've flipped a coin 10 times and, against all odds, seen 10 heads, you may feel that a tail is overdue. Nonetheless, the probability that the next flip will be a head is still 50%. After 10 half-lives have gone by, $(1/2)^{10} = 1/1024$ of a radioactive sample is still there. There was nothing special or distinctive about these nuclei, and, despite their longevity, each remaining nucleus has exactly a 50% chance of decay during the next half-life.

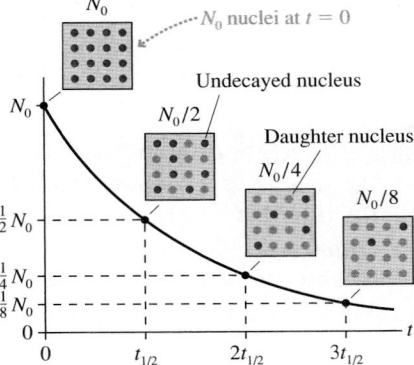

FIGURE 42.17 Half the nuclei decay during each half-life.

EXAMPLE 42.2   The decay of iodine

The iodine isotope ^{131}I, which has an eight-day half-life, is used in nuclear medicine. A sample of ^{131}I containing 2.00×10^{12} atoms is created in a nuclear reactor.

a. How many ^{131}I atoms remain 36 hours later when the sample is delivered to a hospital?
b. The sample is constantly getting weaker, but it remains usable as long as there are at least 5.0×10^{11} ^{131}I atoms. What is the maximum delay before the sample is no longer usable?

MODEL The number of ^{131}I atoms decays exponentially.

SOLVE a. The half-life is $t_{1/2} = 8$ days $= 192$ h. After 36 h have elapsed,

$$N = (2.00 \times 10^{12})\left(\frac{1}{2}\right)^{36/192} = 1.76 \times 10^{12} \text{ nuclei}$$

b. The time after creation at which 5.0×10^{11} ^{131}I atoms remain is given by

$$5.0 \times 10^{11} = 0.50 \times 10^{12} = (2.0 \times 10^{12})\left(\frac{1}{2}\right)^{t/8 \text{ days}}$$

To solve for t, we first write this as

$$\frac{0.50}{2.00} = 0.25 = \left(\frac{1}{2}\right)^{t/8 \text{ days}}$$

Now we take the logarithm of both sides. Either natural logarithms or base–10 logarithms can be used, but we'll use natural logarithms:

$$\ln(0.25) = -1.39 = \frac{t}{t_{1/2}}\ln(0.5) = -0.693\frac{t}{t_{1/2}}$$

Solving for t gives

$$t = 2.00 t_{1/2} = 16 \text{ days}$$

ASSESS The weakest usable sample is one-quarter of the initial sample. You saw in Figure 42.17 that a radioactive sample decays to one-quarter of its initial number in 2 half-lives.

Activity

The **activity** R of a radioactive sample is the number of decays per second. This is simply the absolute value of dN/dt, or

$$R = \left|\frac{dN}{dt}\right| = rN = rN_0 e^{-t/\tau} = R_0 e^{-t/\tau} = R_0\left(\frac{1}{2}\right)^{t/t_{1/2}} \tag{42.13}$$

where $R_0 = rN_0$ is the activity at $t = 0$. The activity of a sample decreases exponentially along with the number of remaining nuclei.

The SI unit of activity is the **becquerel,** defined as

$$1 \text{ becquerel} = 1 \text{ Bq} \equiv 1 \text{ decay/s or } 1 \text{ s}^{-1}$$

An older unit of activity, but one that continues in widespread use, is the **curie.** The curie was originally defined as the activity of 1 g of radium. Today, the conversion factor is

$$1 \text{ curie} = 1 \text{ Ci} \equiv 3.7 \times 10^{10} \text{ Bq}$$

One curie is a substantial activity. The radioactive samples used in laboratory experiments are typically $\approx 1 \ \mu\text{Ci}$ or, equivalently, $\approx 40{,}000 \ \text{Bq}$. These samples can be handled with only minor precautions. Larger sources of radioactivity require lead shielding and special precautions to prevent exposure to high levels of radiation.

EXAMPLE 42.3 | **A laboratory source**

The isotope ^{137}Cs is a standard laboratory source of gamma rays. The half-life of ^{137}Cs is 30 years.

a. How many ^{137}Cs atoms are in a 5.0 μCi source?
b. What is the activity of the source 10 years later?

MODEL The number of ^{137}Cs atoms decays exponentially.

SOLVE a. The number of atoms can be found from $N_0 = R_0/r$. The activity in SI units is

$$R = 5.0 \times 10^{-6} \ \text{Ci} \times \frac{3.7 \times 10^{10} \ \text{Bq}}{1 \ \text{Ci}} = 1.85 \times 10^5 \ \text{Bq}$$

To find the decay rate, first convert the half-life to seconds:

$$t_{1/2} = 30 \ \text{years} \times \frac{3.15 \times 10^7 \ \text{s}}{1 \ \text{year}} = 9.45 \times 10^8 \ \text{s}$$

Then

$$r = \frac{1}{\tau} = \frac{\ln 2}{t_{1/2}} = 7.33 \times 10^{-10} \ \text{s}^{-1}$$

Thus the number of ^{137}Cs atoms is

$$N_0 = \frac{R_0}{r} = \frac{1.85 \times 10^5 \ \text{Bq}}{7.33 \times 10^{-10} \ \text{s}^{-1}} = 2.5 \times 10^{14} \ \text{atoms}$$

b. The activity decreases exponentially, just like the number of nuclei. After 10 years,

$$R = R_0 \left(\frac{1}{2} \right)^{t/t_{1/2}} = (5.0 \ \mu\text{Ci}) \left(\frac{1}{2} \right)^{10/30} = 4.0 \ \mu\text{Ci}$$

ASSESS Although N_0 is a very large number, it is a very small fraction ($\approx 10^{-10}$) of a mole. The sample is about 60 ng (nanograms) of ^{137}Cs.

Radioactive Dating

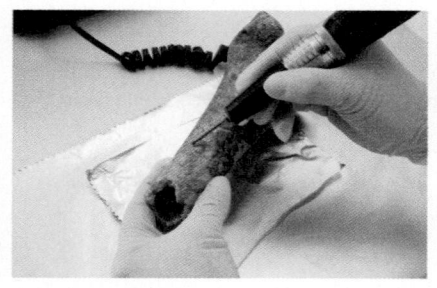

A researcher is extracting a small sample of an ancient bone. She will determine the age of the bone by measuring the ratio of ^{14}C to ^{12}C.

Many geological and archeological samples can be dated by measuring the decays of naturally occurring radioactive isotopes. Because we have no way to know N_0, the initial number of radioactive nuclei, radioactive dating depends on the use of ratios.

The most well-known dating technique is carbon dating. The carbon isotope ^{14}C has a half-life of 5730 years, so any ^{14}C present when the earth formed 4.5 billion years ago would long since have decayed away. Nonetheless, ^{14}C is present in atmospheric carbon dioxide because high-energy cosmic rays collide with gas molecules high in the atmosphere. These cosmic rays are energetic enough to create ^{14}C nuclei from nuclear reactions with nitrogen and oxygen nuclei. The creation and decay of ^{14}C have reached a steady state in which the $^{14}\text{C}/^{12}\text{C}$ ratio is 1.3×10^{-12}. That is, atmospheric carbon dioxide has ^{14}C at the concentration of 1.3 parts per trillion. As small as this is, it's easily measured by modern chemical techniques.

All living organisms constantly exchange carbon dioxide with the atmosphere, so the $^{14}\text{C}/^{12}\text{C}$ ratio in living organisms is also 1.3×10^{-12}. When an organism dies, the ^{14}C in its tissue begins to decay and no new ^{14}C is added. Objects are dated by comparing the measured $^{14}\text{C}/^{12}\text{C}$ ratio to the 1.3×10^{-12} value of living material.

Carbon dating is used to date skeletons, wood, paper, fur, food material, and anything else made of organic matter. It is quite accurate for ages to about 15,000 years, roughly three half-lives of ^{14}C. Beyond that, the difficulty of measuring such a small ratio and some uncertainties about the cosmic ray flux in the past combine to decrease the accuracy. Even so, items are dated to about 50,000 years with a fair degree of reliability.

Other isotopes with longer half-lives are used to date geological samples. Potassium-argon dating, using ^{40}K with a half-life of 1.25 billion years, is especially useful for dating rocks of volcanic origin.

EXAMPLE 42.4 **Carbon dating**

Archeologists excavating an ancient hunters' camp have recovered a 5.0 g piece of charcoal from a fireplace. Measurements on the sample find that the ^{14}C activity is 0.35 Bq. What is the approximate age of the camp?

MODEL Charcoal, from burning wood, is almost pure carbon. The number of ^{14}C atoms in the wood has decayed exponentially since the branch fell off a tree. Because wood rots, it is reasonable to assume that there was no significant delay between when the branch fell off the tree and the hunters burned it.

SOLVE The ^{14}C/^{12}C ratio was 1.3×10^{-12} when the branch fell from the tree. We first need to determine the present ratio, then use the known ^{14}C half-life $t_{1/2} = 5730$ years to calculate the time needed to reach the present ratio. The number of ordinary ^{12}C nuclei in the sample is

$$N(^{12}\text{C}) = \left(\frac{5.0 \text{ g}}{12 \text{ g/mol}}\right)(6.02 \times 10^{23} \text{ atoms/mol})$$

$$= 2.5 \times 10^{23} \text{ nuclei}$$

The number of ^{14}C nuclei can be found from the activity to be $N(^{14}\text{C}) = R/r$, but we need to determine the ^{14}C decay rate r. After converting the half-life to seconds, $t_{1/2} = 5730$ years $= 1.807 \times 10^{11}$ s, we can compute

$$r = \frac{1}{\tau} = \frac{1}{t_{1/2}/\ln 2} = 3.84 \times 10^{-12} \text{ s}^{-1}$$

Thus

$$N(^{14}\text{C}) = \frac{R}{r} = \frac{0.35 \text{ Bq}}{3.84 \times 10^{-12} \text{ s}^{-1}} = 9.1 \times 10^{10} \text{ nuclei}$$

and the present ^{14}C/^{12}C ratio is $N(^{14}\text{C})/N(^{12}\text{C}) = 0.36 \times 10^{-12}$. Because this ratio has been decaying with a half-life of 5730 years, the time needed to reach the present ratio is found from

$$0.36 \times 10^{-12} = (1.3 \times 10^{-12})\left(\frac{1}{2}\right)^{t/t_{1/2}}$$

To solve for t, we first write this as

$$\frac{0.36}{1.3} = 0.277 = \left(\frac{1}{2}\right)^{t/t_{1/2}}$$

Now we take the logarithm of both sides:

$$\ln(0.277) = -1.28 = \frac{t}{t_{1/2}}\ln(0.5) = -0.693\frac{t}{t_{1/2}}$$

Thus the age of the hunters' camp is

$$t = 1.85t_{1/2} = 10,600 \text{ years}$$

ASSESS This is a realistic example of how radioactive dating is done.

STOP TO THINK 42.4 A sample starts with 1000 radioactive atoms. How many half-lives have elapsed when 750 atoms have decayed?

a. 0.25
b. 1.5
c. 2.0
d. 2.5

42.6 Nuclear Decay Mechanisms

This section will look in more detail at the mechanisms of the three types of radioactive decay.

Alpha Decay

An alpha particle, symbolized as α, is a ^{4}He nucleus, a strongly bound system of two protons and two neutrons. An unstable nucleus that ejects an alpha particle will lose two protons and two neutrons, so we can write the decay as

$$^A\text{X}_Z \rightarrow {}^{A-4}\text{Y}_{Z-2} + \alpha + \text{energy} \tag{42.14}$$

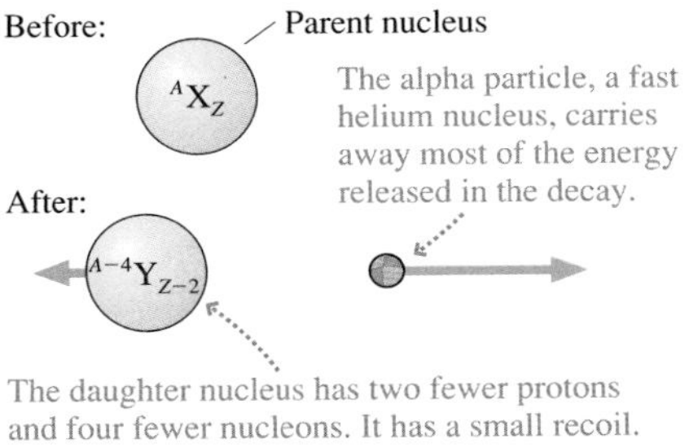

FIGURE 42.18 shows the alpha-decay process. The original nucleus X is called the **parent nucleus,** and the decay-product nucleus Y is the **daughter nucleus.** This reaction can occur only when the mass of the parent nucleus is greater than the mass of the daughter nucleus plus the mass of an alpha particle. This requirement is met for heavy,

FIGURE 42.18 Alpha decay.

Before: Parent nucleus

$^A\text{X}_Z$

The alpha particle, a fast helium nucleus, carries away most of the energy released in the decay.

After:

$^{A-4}\text{Y}_{Z-2}$

The daughter nucleus has two fewer protons and four fewer nucleons. It has a small recoil.

---

end as the ^{207}Pb isotope of lead, a stable nucleus.

Notice that some nuclei can decay by either alpha *or* beta decay. Thus there are a variety of paths that a decay can follow, but they all end at the same point.

STOP TO THINK 42.5 The cobalt isotope ^{60}Co ($Z = 27$) decays to the nickel isotope ^{60}Ni ($Z = 28$). The decay process is

a. Alpha decay.
b. Beta-minus decay.
c. Beta-plus decay.
d. Electron capture.
e. Gamma decay.

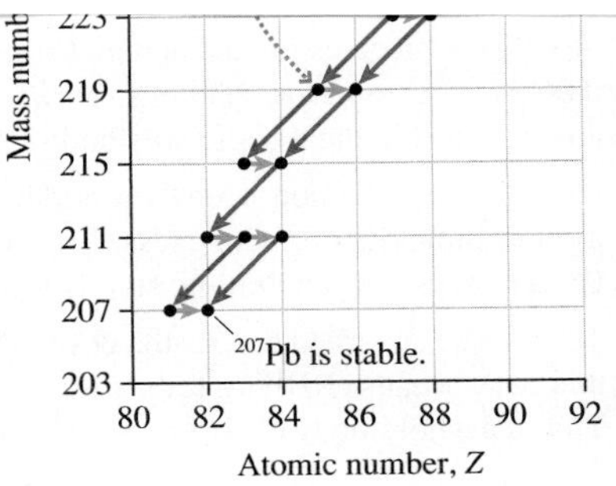

^{207}Pb is stable.

Atomic number, Z

Once you stop to think of it, the process n → p⁺ + e⁻ seems ludicrous, not because it violates mass-energy conservation but because we have no idea *how* a neutron could turn into a proton. Alpha decay may be a strange process because tunneling in general goes against our commonsense notions, but it is a perfectly ordinary quantum-mechanical process. Now we're suggesting that one of the basic building blocks of matter can somehow morph into a different basic building block.

To make matters more confusing, measurements in the 1930s found that beta decay didn't seem to conserve either energy or momentum. Faced with these difficulties, the Italian physicist Enrico Fermi made two bold suggestions:

1. A previously unknown fundamental force of nature is responsible for beta de-

42.7 Biological Applications of Nuclear Physics

Nuclear physics has brought both peril and promise to society. Radiation can cause tumors, but it also can be used to cure some cancers. This section is a brief survey of medical and biological applications of nuclear physics.

Radiation Dose

Nuclear radiation, which is ionizing radiation, disrupts a cell's machinery by altering and damaging the biological molecules. The consequences of this disruption vary from genetic mutations to uncontrolled cell multiplication (i.e., tumors) to cell death.

Beta and gamma radiation can penetrate the entire body and damage internal organs. Alpha radiation has less penetrating ability, but it deposits all its energy in a very small, localized volume. Internal organs are usually safe from alpha radiation, but the skin is very susceptible, as are the lungs if radioactive dust is inhaled.

Biological effects of radiation depend on two factors. The first is the physical factor of how much energy is absorbed by the body. The second is the biological factor of how tissue reacts to different forms of radiation.

The **absorbed dose** of radiation is the energy of ionizing radiation absorbed per kilogram of tissue. The SI unit of absorbed dose is the **gray,** abbreviated Gy. It is defined as

$$1 \text{ gray} = 1 \text{ Gy} \equiv 1.00 \text{ J/kg of absorbed energy}$$

The absorbed dose depends only on the energy absorbed, not at all on the type of radiation or on what the absorbing material is.

Biologists and biophysicists have found that a 1 Gy dose of gamma rays and a 1 Gy dose of alpha particles have different biological consequences. To account for such differences, the **relative biological effectiveness** (RBE) is defined as the biological effect of a given dose relative to the biological effect of an equal dose of x rays.

Table 42.4 shows the relative biological effectiveness of different forms of radiation. Larger values correspond to larger biological effects. Beta radiation and neutrons have a range of values because the biological effect varies with the energy of the particle. Alpha radiation has the largest RBE because the energy is deposited in the smallest volume.

The product of the absorbed dose with the RBE is called the **dose equivalent.** Dose equivalent is measured in **sieverts,** abbreviated Sv. To be precise,

$$\text{dose equivalent in Sv} = \text{absorbed dose in Gy} \times \text{RBE}$$

1 Sv of radiation produces the same biological damage regardless of the type of radiation. An older but still widely used unit for dose equivalent is the **rem,** defined as 1 rem = 0.010 Sv. Small radiation doses are measured in millisievert (mSv) or millirem (mrem).

TABLE 42.4 Relative biological effectiveness of radiation

Radiation type	RBE
X rays	1
Gamma rays	1
Beta particles	1–2
Neutrons	5–20
Alpha particles	20

EXAMPLE 42.7 **Radiation exposure**

A 75 kg laboratory technician working with the radioactive isotope ¹³⁷Cs receives an accidental 1.0 mSv exposure. ¹³⁷Cs emits 0.66 MeV gamma-ray photons. How many gamma-ray photons are absorbed in the technician's body?

MODEL The radiation dose is a combination of deposited energy and biological effectiveness. The RBE for gamma rays is 1. Gamma rays are penetrating, so this is a whole-body exposure.

SOLVE The absorbed dose is the dose in Sv divided by the RBE. In this case, because RBE = 1, the dose is 0.0010 Gy = 0.0010 J/kg. This is a whole-body exposure, so the total energy deposited in the

technician's body is 0.075 J. The energy of each absorbed photon is 0.66 MeV, but this value must be converted into joules. The number of photons in 0.075 J is

$$N = \frac{0.075 \text{ J}}{(6.6 \times 10^5 \text{ eV/photon})(1.60 \times 10^{-19} \text{ J/eV})}$$

$$= 7.1 \times 10^{11} \text{ photons}$$

ASSESS The energy deposited, 0.075 J, is very small. Radiation does its damage not by thermal effects, which would require substantially more energy, but by ionization.

Table 42.5 gives some basic information about radiation exposure. We are all exposed to a continuous natural background of radiation from cosmic rays and from naturally occurring radioactive atoms (uranium and other atoms in the uranium decay series) in the ground, the atmosphere, and even the food we eat. This background averages about 3 mSv per year, although there are wide regional variations depending on the soil type and the elevation. (Higher elevations have a larger exposure to cosmic rays.)

Medical x rays vary significantly. The average person in the United States receives approximately 0.6 mSv per year from all medical sources. All other sources, such as fallout from atmospheric nuclear tests many decades ago, nuclear power plants, and industrial uses of radioactivity, add up to less than 0.1 mSv per year.

The question inevitably arises: What is a safe dose? This remains a controversial topic and the subject of ongoing research. The effects of large doses of radiation are easily observed. The effects of small doses are hard to distinguish from other natural and environmental causes. Thus there's no simple or clear definition of a safe dose. A prudent policy is to avoid unnecessary exposure to radiation but not to worry over exposures less than the natural background. It's worth noting that the μCi radioactive sources used in laboratory experiments provide exposures *much* less than the natural background, even if used on a regular basis.

TABLE 42.5 Radiation exposure

Radiation source	Typical exposure (mSv)
CT scan	10
Natural background (1 year)	3
Mammogram x ray	0.8
Chest x ray	0.3
Dental x ray	0.03

Medical Uses of Radiation

Radiation can be put to good use killing cancer cells. This area of medicine is called *radiation therapy*. Gamma rays are the most common form of radiation, often from the isotope ^{60}Co. As FIGURE 42.25 shows, the gamma rays are directed along many different lines, all of which intersect the tumor. The goal is to provide a lethal dose to the cancer cells without overexposing nearby tissue. The patient and the radiation source are rotated around each other under careful computer control to deliver the proper dose.

Other tumors are treated by surgically implanting radioactive "seeds" within or next to the tumor. Alpha particles, which are very damaging locally but don't penetrate far, can be used in this fashion.

Radioactive isotopes are also used as *tracers* in diagnostic procedures. This technique is based on the fact that all isotopes of an element have identical chemical behavior. As an example, a radioactive isotope of iodine is used in the diagnosis of certain thyroid conditions. Iodine is an essential element in the body, and it concentrates in the thyroid gland. A doctor who suspects a malfunctioning thyroid gland gives the patient a small dose of sodium iodide in which some of the normal ^{127}I atoms have been replaced with ^{131}I. (Sodium iodide, which is harmless, dissolves in water and can simply be drunk.) The ^{131}I isotope, with a half-life of eight days, undergoes beta decay and subsequently emits a gamma-ray photon that can be detected.

The radioactive iodine concentrates inside the thyroid gland within a few hours. The doctor then monitors the gamma-ray photon emissions over the next few days to see how the iodine is being processed within the thyroid and how quickly it is eliminated from the body.

Other important radioactive tracers include the chromium isotope ^{51}Cr, which is taken up by red blood cells and can be used to monitor blood flow, and the xenon isotope ^{133}Xe, which is inhaled to reveal lung functioning. Radioactive tracers are *noninvasive,* meaning that the doctor can monitor the inside of the body without surgery.

Magnetic Resonance Imaging

The proton, like the electron, has an inherent angular momentum (spin) and an inherent magnetic moment. You can think of the proton as being like a little compass needle that can be in one of two positions, the positions we call spin up and spin down.

A compass needle aligns itself with an external magnetic field. This is the needle's lowest-energy position. Turning a compass needle by hand is like rolling a ball uphill; you're giving it energy, but, like the ball rolling downhill, it will realign itself with the

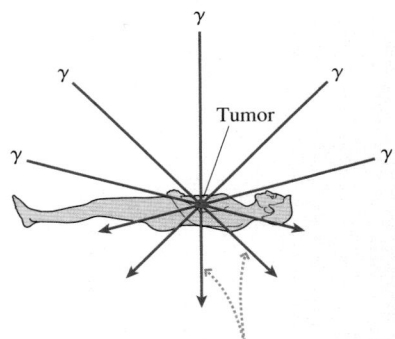

FIGURE 42.25 Radiation therapy is designed to deliver a lethal dose to the tumor without damaging nearby tissue.

Gamma radiation is incident along many lines, all of which intersect the tumor.

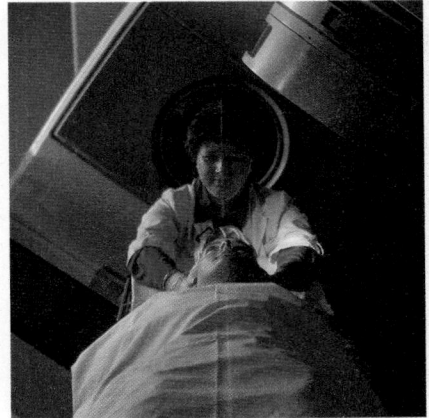

Radiation therapy is a beneficial use of nuclear physics.

lowest-energy position when you remove your finger. There is, however, an *unstable equilibrium* position, like a ball at the top of a hill, in which the needle is anti-aligned with the field. The slightest jostle will cause it to flip around, but the needle will be steady in its upside-down configuration if you can balance it perfectly.

A proton in a magnetic field behaves similarly, but with a major difference: Because the proton's energy is quantized, the proton cannot assume an intermediate position. It's either aligned with the magnetic field (the spin-up orientation) or anti-aligned (spin-down). FIGURE 42.26a shows these two quantum states. Turning on a magnetic field lowers the energy of a spin-up proton and increases the energy of an anti-aligned, spin-down proton. In other words, the magnetic field creates an *energy difference* between these states.

FIGURE 42.26 Nuclear magnetic resonance is possible because spin-up and spin-down protons have slightly different energies in a magnetic field.

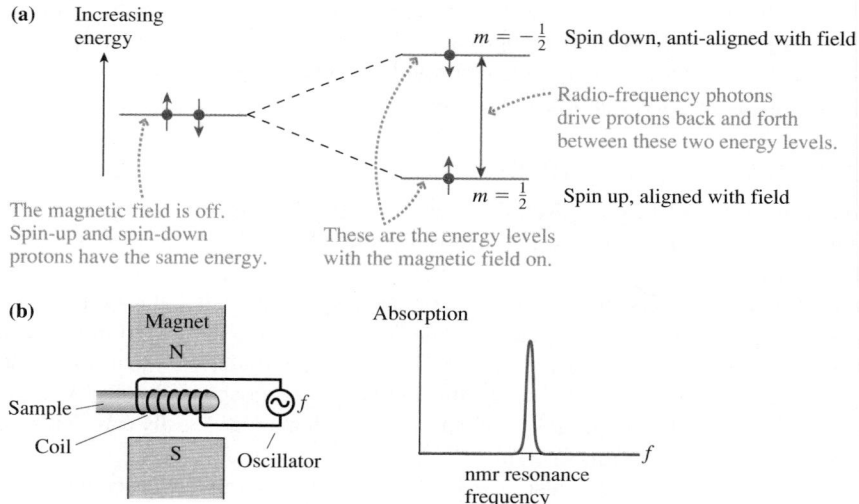

(a) Increasing energy

$m = -\frac{1}{2}$ Spin down, anti-aligned with field

Radio-frequency photons drive protons back and forth between these two energy levels.

$m = \frac{1}{2}$ Spin up, aligned with field

The magnetic field is off. Spin-up and spin-down protons have the same energy.

These are the energy levels with the magnetic field on.

(b) Magnet N

Sample

Coil

S   Oscillator f

Absorption

nmr resonance frequency f

The energy difference is very tiny, only about 10^{-7} eV. Nonetheless, photons whose energy matches the energy difference cause the protons to move back and forth between these two energy levels as the photons are absorbed and emitted. In effect, the photons are causing the proton's spin to flip back and forth rapidly. The photon frequency, which depends on the magnetic field strength, is typically about 100 MHz, similar to FM radio frequencies.

FIGURE 42.26b shows how this behavior is put to use. A sample containing protons is placed in a magnetic field. A coil is wrapped around the sample, and a variable frequency AC source drives a current through this coil. The protons absorb power from the coil when its frequency is just right to flip the spin back and forth; otherwise, no power is absorbed. A *resonance* is seen by scanning the coil through a small range of frequencies.

This technique of observing the spin flip of nuclei (the technique also works for nuclei other than hydrogen) in a magnetic field is called **nuclear magnetic resonance,** or nmr. It has many applications in physics, chemistry, and materials science. Its medical use exploits the fact that tissue is mostly water, and two out of the three nuclei in a water molecule are protons. Thus the human body is basically a sample of protons, with the proton density varying as the tissue density varies.

The medical procedure known as **magnetic resonance imaging,** or MRI, places the patient in a spatially varying magnetic field. The variations in the field cause the proton absorption frequency to vary from point to point. From the known shape of the field and measurements of the frequencies that are absorbed, and how strongly, sophisticated computer software can transform the raw data into detailed images such as the one shown in FIGURE 42.27.

FIGURE 42.27 Magnetic resonance imaging shows internal organs in exquisite detail.

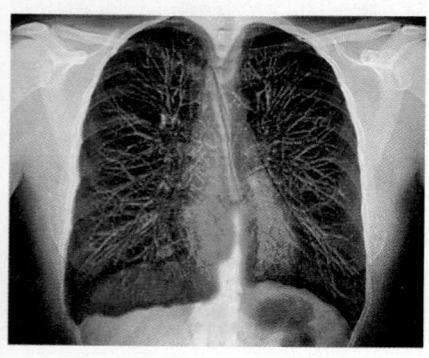

As an interesting footnote, the technique was still being called *nuclear magnetic resonance* when it was first introduced into medicine. Unfortunately, doctors soon found that many patients were afraid of it because of the word "nuclear." Hence the alternative term "magnetic resonance imaging" was coined. It is true that the public perception of nuclear technology is not always positive, but equally true that nuclear physics has made many important and beneficial contributions to society.

CHALLENGE EXAMPLE 42.8 | **A radioactive tracer**

An 85 kg patient swallows a 30 μCi beta emitter that is to be used as a tracer. The isotope's half-life is 5.0 days. The average energy of the beta particles is 0.35 MeV, and they have an RBE (relative biological effectiveness) of 1.5. Ninety percent of the beta particles are stopped inside the patient's body and 10% escape. What total dose equivalent does this patient receive?

MODEL Beta radiation penetrates the body—enough that 10% of the particles escape—so this is a whole-body exposure. Even the escaping particles probably deposit some energy in the body, but we'll assume that the dose is from only those particles that stop inside the body.

SOLVE The dose equivalent is the absorbed dose in Gy multiplied by the RBE of 1.5. The absorbed dose is the energy absorbed per kilogram of tissue, so we need to find the total energy absorbed from the time the patient swallows the emitter until it has all decayed. The sample's initial activity R_0 is related to the nuclear decay rate r and the initial number of radioactive atoms N_0 by $R_0 = rN_0$. Thus the number of radioactive atoms in the sample, all of which are going to decay and emit a beta particle, is

$$N_0 = \frac{R_0}{r} = \tau R_0 = \frac{t_{1/2}}{\ln 2} R_0$$

In developing this relationship, we used first the fact that the lifetime τ is the inverse of the decay rate, then the connection between the lifetime and the half-life.

The initial activity is given in microcuries. Converting to becquerels, we have

$$R_0 = (30 \times 10^{-6} \text{ Ci}) \times \frac{3.7 \times 10^{10} \text{ Bq}}{1 \text{ Ci}}$$

$$= 1.1 \times 10^6 \text{ Bq} = 1.1 \times 10^6 \text{ decays/s}$$

The half-life in seconds is

$$t_{1/2} = 5.0 \text{ days} \times \frac{86,400 \text{ s}}{1 \text{ day}} = 4.3 \times 10^5 \text{ s}$$

Thus the total number of beta decays over the course of several weeks, as the sample completely decays, is

$$N_0 = \frac{t_{1/2}}{\ln 2} R_0 = \frac{(4.3 \times 10^5 \text{ s})(1.1 \times 10^6 \text{ decays/s})}{\ln 2} = 6.8 \times 10^{11}$$

Ninety percent of these decays deposit, on average, 0.35 MeV in the body, so the absorbed energy is

$$E_{\text{abs}} = (0.90)(6.8 \times 10^{11}) \left((3.5 \times 10^5 \text{ eV}) \times \frac{1.60 \times 10^{-19} \text{ J}}{1 \text{ eV}} \right)$$

$$= 0.034 \text{ J}$$

This is not a lot of energy in an absolute sense, but it is all damaging, ionizing radiation. The absorbed dose is

$$\text{absorbed dose} = \frac{0.034 \text{ J}}{85 \text{ kg}} = 4.0 \times 10^{-4} \text{ Gy}$$

and thus the dose equivalent is

$$\text{dose equivalent} = 1.5 \times (4.0 \times 10^{-4} \text{ Gy}) = 0.60 \text{ mSv}$$

ASSESS This dose, typical of many medical uses of radiation, is about 20% of the yearly radiation dose from the natural background. Although one should always avoid unnecessary radiation, this dose would not cause concern if there were a medical reason for it.

SUMMARY

The goal of Chapter 42 has been to understand the physics of the nucleus and some applications of nuclear physics.

General Principles

The Nucleus

The nucleus is a small, dense, positive core at the center of an atom.

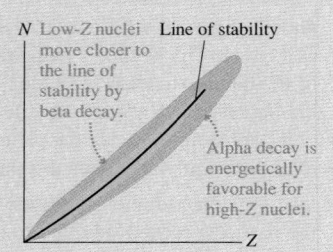

Z protons: charge $+e$, spin $\frac{1}{2}$

N neutrons: charge 0, spin $\frac{1}{2}$

The mass number is $A = Z + N$.

The nuclear radius is $R = r_0 A^{1/3}$, where $r_0 = 1.2$ fm. Typical radii are a few fm.

Nuclear forces

Attractive strong force
- Acts between any two nucleons
- Is short range, < 3 fm
- Is felt between nearest neighbors

Repulsive electric force
- Acts between two protons
- Is long range
- Is felt across the nucleus

Nuclear Stability

Most nuclei are not stable. Unstable nuclei undergo radioactive decay. Stable nuclei cluster along the **line of stability** in a plot of the isotopes.

Three mechanisms by which unstable nuclei decay:

Decay	Particle	Mechanism	Energy	Penetration
α	^{4}He nucleus	tunneling	few MeV	low
β	e^-	$n \rightarrow p^+ + e^-$	≈ 1 MeV	medium
	e^+	$p^+ \rightarrow n + e^+$	≈ 1 MeV	medium
γ	photon	quantum jump	≈ 1 MeV	high

Important Concepts

Shell model

Each nucleon moves with an average potential energy due to all other nucleons.

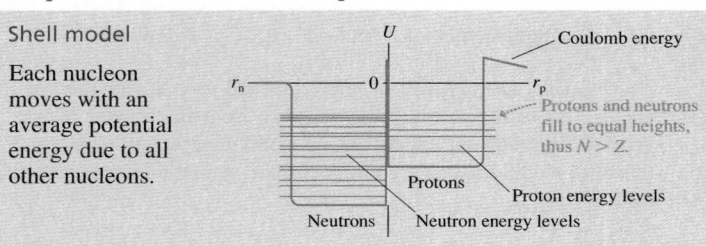

Curve of binding energy

The average binding energy per nucleon has a broad maximum at $A \approx 60$.

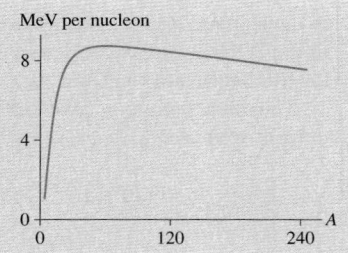

Applications

Radioactive decay

The number of undecayed nuclei decreases exponentially with time t:

$$N = N_0 \exp(-t/\tau)$$
$$= N_0 (1/2)^{t/t_{1/2}}$$

The lifetime τ is $1/r$, where r is the decay rate. The **half-life**

$$t_{1/2} = \tau \ln 2 = 0.693\tau$$

is the time in which half of any sample decays.

Measuring radiation

The **activity** $R = rN$ of a radioactive sample, measured in becquerels or curies, is the number of decays per second.

The **absorbed dose** is measured in gray, where

$$1 \text{ Gy} \equiv 1.00 \text{ J/kg of absorbed energy}$$

The **relative biological effectiveness** (RBE) is the biological effect of a dose relative to the biological effects of x rays.

The **dose equivalent** is measured in Sv, where Sv = Gy × RBE. One Sv of radiation produces the same biological effect regardless of the type of radiation. Dose equivalent is also measured in rem, where 1 rem = 0.010 Sv.

Terms and Notation

nuclear physics	liquid-drop model	half-life, $t_{1/2}$	gray, Gy
nucleon	line of stability	activity, R	relative biological effectiveness
atomic number, Z	binding energy, B	becquerel, Bq	(RBE)
mass number, A	curve of binding energy	curie, Ci	dose equivalent
neutron number, N	strong force	parent nucleus	sievert, Sv
isotope	shell model	daughter nucleus	rem
radioactive	alpha decay	electron capture	nuclear magnetic resonance (nmr)
stable	beta decay	weak interaction	magnetic resonance imaging (MRI)
natural abundance	gamma decay	neutrino	
isobar	ionizing radiation	decay series	
deuterium	Geiger counter	absorbed dose	

CONCEPTUAL QUESTIONS

1. Consider the atoms ^{16}O, ^{18}O, ^{18}F, ^{18}Ne, and ^{20}Ne. Some of the following questions may have more than one answer. Give all answers that apply.
   a. Which are isotopes?
   b. Which are isobars?
   c. Which have the same chemical properties?
   d. Which have the same number of neutrons?

2. a. Is the binding energy of a nucleus with $A = 200$ more than, less than, or equal to the binding energy of a nucleus with $A = 60$? Explain.
   b. Is a nucleus with $A = 200$ more tightly bound, less tightly bound, or bound equally tightly as a nucleus with $A = 60$? Explain.

3. a. How do we know the strong force exists?
   b. How do we know the strong force is short range?

4. Does each nuclear energy-level diagram in FIGURE Q42.4 represent a nuclear ground state, an excited nuclear state, or an impossible nucleus? Explain.

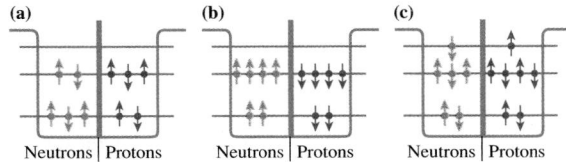

FIGURE Q42.4

5. Are the following decays possible? If not, why not?
   a. $^{232}Th\,(Z = 90) \rightarrow\ ^{236}U\,(Z = 92) + \alpha$
   b. $^{238}Pu\,(Z = 94) \rightarrow\ ^{236}U\,(Z = 92) + \alpha$
   c. $^{11}B\,(Z = 5) \rightarrow\ ^{11}B\,(Z = 5) + \gamma$
   d. $^{33}P\,(Z = 15) \rightarrow\ ^{32}S\,(Z = 16) + e^-$

6. Nucleus A decays into nucleus B with a half-life of 10 s. At $t = 0$ s, there are 1000 A nuclei and no B nuclei. At what time will there be 750 B nuclei?

7. What kind of decay, if any, can occur for the nuclei in FIGURE Q42.7?

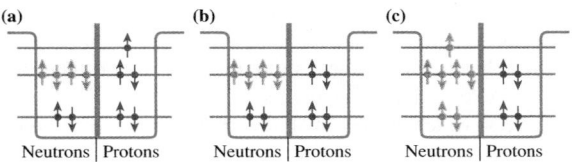

FIGURE Q42.7

8. Apple A in FIGURE Q42.8 is strongly irradiated by nuclear radiation for 1 hour. Apple B is not irradiated. Afterward, in what ways are apples A and B different?

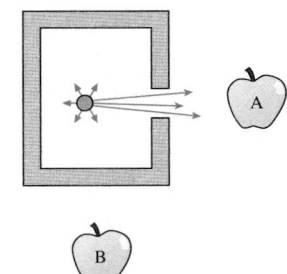

FIGURE Q42.8

9. The three isotopes ^{212}Po, ^{137}Cs, and ^{90}Sr decay as $^{212}Po \rightarrow\ ^{208}Pb + \alpha$, $^{137}Cs \rightarrow\ ^{137}Ba + e^- + \gamma$, and $^{90}Sr \rightarrow\ ^{90}Y + e^-$. Which of these isotopes would be most useful as a biological tracer? Why?

EXERCISES AND PROBLEMS

See Appendix C for data on atomic masses, isotopic abundance, radioactive decay modes, and half-lives.

Problems labeled ▨ integrate material from earlier chapters.

Exercises

Section 42.1 Nuclear Structure

1. | How many protons and how many neutrons are in (a) ^{3}He, (b) ^{32}P, (c) ^{32}S, and (d) ^{238}U?
2. | How many protons and how many neutrons are in (a) ^{6}Li, (b) ^{54}Cr, (c) ^{54}Fe, and (d) ^{220}Rn?
3. | Calculate the nuclear diameters of (a) ^{4}He, (b) ^{40}Ar, and (c) ^{220}Rn.
4. | Which stable nuclei have a diameter of 7.46 fm?
5. | Calculate the mass, radius, and density of the nucleus of (a) ^{7}Li and (b) ^{207}Pb. Give all answers in SI units.

Section 42.2 Nuclear Stability

6. | Use data in Appendix C to make your own chart of stable and unstable nuclei, similar to Figure 42.4, for all nuclei with $Z \leq 8$. Use a blue or black dot to represent stable isotopes, a red dot to represent isotopes that undergo beta-minus decay, and a green dot to represent isotopes that undergo beta-plus decay or electron-capture decay.
7. | a. What is the smallest value of A for which there are two stable nuclei? What are they?
   b. For which values of A less than this are there *no* stable nuclei?
8. | Calculate (in MeV) the total binding energy and the binding energy per nucleon for ^{3}H and for ^{3}He.
9. | Calculate (in MeV) the total binding energy and the binding energy per nucleon for ^{54}Cr and for ^{54}Fe.
10. ‖ Calculate (in MeV) the binding energy per nucleon for ^{3}He and ^{4}He. Which is more tightly bound?
11. ‖ Calculate (in MeV) the binding energy per nucleon for ^{14}O and ^{16}O. Which is more tightly bound?
12. | Calculate the chemical atomic mass of neon.

Section 42.3 The Strong Force

13. ‖ Use the potential-energy diagram in Figure 42.8 to estimate the strength of the strong force between two nucleons separated by 1.5 fm.
14. ‖ Use the potential-energy diagram in Figure 42.8 to sketch an approximate graph of the strong force between two nucleons versus the distance r between their centers.
15. ‖ Use the potential-energy diagram in Figure 42.8 to estimate the ratio of the gravitational potential energy to the nuclear potential energy for two neutrons separated by 1.0 fm.

Section 42.4 The Shell Model

16. | a. Draw energy-level diagrams, similar to Figure 42.11, for all $A = 10$ nuclei listed in Appendix C. Show all the occupied neutron and proton levels.

b. Which of these nuclei are stable? What is the decay mode of any that are radioactive?

17. | a. Draw energy-level diagrams, similar to Figure 42.11, for all $A = 14$ nuclei listed in Appendix C. Show all the occupied neutron and proton levels.
   b. Which of these nuclei are stable? What is the decay mode of any that are radioactive?

Section 42.5 Radiation and Radioactivity

18. | The radium isotope ^{226}Ra has a half-life of 1600 years. A sample begins with 1.00×10^{10} ^{226}Ra atoms. How many are left after (a) 200 years, (b) 2000 years, and (c) 20,000 years?
19. | The barium isotope ^{131}Ba has a half-life of 12 days. A 250 μg sample of ^{131}Ba is prepared. What is the mass of ^{131}Ba after (a) 1 day, (b) 10 days, and (c) 100 days?
20. | The radioactive hydrogen isotope ^{3}H, called *tritium*, has a half-life of 12 years.
   a. What are the decay mode and the daughter nucleus of tritium?
   b. What are the lifetime and the decay rate of tritium?
21. | A sample of 1.0×10^{10} atoms that decay by alpha emission has a half-life of 100 min. How many alpha particles are emitted between $t = 50$ min and $t = 200$ min?
22. ‖ The half-life of ^{60}Co is 5.27 years. The activity of a ^{60}Co sample is 3.50×10^9 Bq. What is the mass of the sample?
23. ‖ What is the half-life in days of a radioactive sample with 5.0×10^{15} atoms and an activity of 5.0×10^8 Bq?

Section 42.6 Nuclear Decay Mechanisms

24. | Identify the unknown isotope X in the following decays.
   a. ^{230}Th $\rightarrow$ X $+ \alpha$
   b. ^{35}S $\rightarrow$ X $+ e^- + \bar{\nu}$
   c. X $\rightarrow$ ^{40}K $+ e^+ + \nu$
   d. ^{24}Na $\rightarrow$ ^{24}Mg $+ e^- + \bar{\nu} \rightarrow$ X $+ \gamma$
25. | Identify the unknown isotope X in the following decays.
   a. X $\rightarrow$ ^{224}Ra $+ \alpha$
   b. X $\rightarrow$ ^{207}Pb $+ e^- + \bar{\nu}$
   c. ^{7}Be $+ e^- \rightarrow$ X $+ \nu$
   d. X $\rightarrow$ ^{60}Ni $+ \gamma$
26. | a. What are the isotopic symbols of all $A = 17$ isobars?
   b. Which of these are stable nuclei?
   c. For those that are not stable, identify both the decay mode and the daughter nucleus.
27. | a. What are the isotopic symbols of all $A = 19$ isobars?
   b. Which of these are stable nuclei?
   c. For those that are not stable, identify both the decay mode and the daughter nucleus.
28. ‖ What is the energy (in MeV) released in the alpha decay of ^{239}Pu?
29. ‖ An unstable nucleus undergoes alpha decay with the release of 5.52 MeV of energy. The combined mass of the parent and daughter nuclei is 452 u. What was the parent nucleus?
30. ‖ What is the total energy (in MeV) released in the beta-minus decay of ^{3}H?

31. ‖ What is the total energy (in MeV) released in the beta-minus decay of ^{24}Na?

32. ‖ What is the total energy (in MeV) released in the beta decay of a neutron?

Section 42.7 Biological Applications of Nuclear Physics

33. | 1.5 Gy of gamma radiation are directed into a 150 g tumor during radiation therapy. How much energy does the tumor absorb?
BIO

34. | The doctors planning a radiation therapy treatment have determined that a 100 g tumor needs to receive 0.20 J of gamma radiation. What is the dose in gray?
BIO

35. ‖ A 50 kg laboratory worker is exposed to 20 mJ of beta radiation with RBE = 1.5. What is the dose equivalent in mrem?
BIO

36. | How many gray of gamma-ray photons cause the same biological damage as 30 Gy of alpha radiation?
BIO

Problems

37. ‖‖ a. What initial speed must an alpha particle have to just touch the surface of a ^{197}Au gold nucleus before being turned back? Assume the nucleus stays at rest.
   b. What is the initial energy (in MeV) of the alpha particle?
   Hint: The alpha particle is not a point particle.

38. ‖‖ Particle accelerators fire protons at target nuclei for investigators to study the nuclear reactions that occur. In one experiment, the proton needs to have 20 MeV of kinetic energy as it impacts a ^{207}Pb nucleus. With what initial kinetic energy (in MeV) must the proton be fired toward the lead target? Assume the nucleus stays at rest.
   Hint: The proton is not a point particle.

39. ‖ Stars are powered by nuclear reactions that fuse hydrogen into helium. The fate of many stars, once most of the hydrogen is used up, is to collapse, under gravitational pull, into a *neutron star*. The force of gravity becomes so large that protons and electrons are fused into neutrons in the reaction $p^+ + e^- \rightarrow n + \nu$. The entire star is then a tightly packed ball of neutrons with the density of nuclear matter.
   a. Suppose the sun collapses into a neutron star. What will its radius be? Give your answer in km.
   b. The sun's rotation period is now 27 days. What will its rotation period be after it collapses?
   Rapidly rotating neutron stars emit pulses of radio waves at the rotation frequency and are known as *pulsars*.

40. ‖ The element gallium has two stable isotopes: ^{69}Ga with an atomic mass of 68.92 u and ^{71}Ga with an atomic mass of 70.92 u. A periodic table shows that the chemical atomic mass of gallium is 69.72 u. What is the percent abundance of ^{69}Ga?

41. ‖ You learned in Chapter 41 that the binding energy of the electron in a hydrogen atom is 13.6 eV.
   a. By how much does the mass decrease when a hydrogen atom is formed from a proton and an electron? Give your answer both in atomic mass units and as a percentage of the mass of the hydrogen atom.
   b. By how much does the mass decrease when a helium nucleus is formed from two protons and two neutrons? Give your

answer both in atomic mass units and as a percentage of the mass of the helium nucleus.
   c. Compare your answers to parts a and b. Why do you hear it said that mass is "lost" in nuclear reactions but not in chemical reactions?

42. ‖‖ Use the graph of binding energy to estimate the total energy released if a nucleus with mass number 240 fissions into two nuclei with mass number 120.

43. ‖‖ Use the graph of binding energy to estimate the total energy released if three ^{4}He nuclei fuse together to form a ^{12}C nucleus.

44. ‖ Could a ^{56}Fe nucleus fission into two ^{28}Al nuclei? Your answer, which should include some calculations, should be based on the curve of binding energy.

45. ‖‖ What energy (in MeV) alpha particle has a de Broglie wavelength equal to the diameter of a ^{238}U nucleus?

46. ‖ What is the age in years of a bone in which the ^{14}C/^{12}C ratio is measured to be 1.65×10^{-13}?

47. ‖ The activity of a sample of the cesium isotope ^{137}Cs, with a half-life of 30 years, is 2.0×10^8 Bq. Many years later, after the sample has fully decayed, how many beta particles will have been emitted?

48. ‖ A 115 mCi radioactive tracer is made in a nuclear reactor. When it is delivered to a hospital 16 hours later its activity is 95 mCi. The lowest usable level of activity is 10 mCi.
BIO
   a. What is the tracer's half-life?
   b. For how long after delivery is the sample usable?

49. ‖ The radium isotope ^{223}Ra, an alpha emitter, has a half-life of 11.43 days. You happen to have a 1.0 g cube of ^{223}Ra, so you decide to use it to boil water for tea. You fill a well-insulated container with 100 mL of water at 18°C and drop in the cube of radium.
   a. How long will it take the water to boil?
   b. Will the water have been altered in any way by this method of boiling? If so, how?

50. ‖ How many half-lives must elapse until (a) 90% and (b) 99% of a radioactive sample of atoms has decayed?

51. ‖ A sample contains radioactive atoms of two types, A and B. Initially there are five times as many A atoms as there are B atoms. Two hours later, the numbers of the two atoms are equal. The half-life of A is 0.50 hour. What is the half-life of B?

52. ‖ Radioactive isotopes often occur together in mixtures. Suppose a 100 g sample contains ^{131}Ba, with a half-life of 12 days, and ^{47}Ca, with a half-life of 4.5 days. If there are initially twice as many calcium atoms as there are barium atoms, what will be the ratio of calcium atoms to barium atoms 2.5 weeks later?

53. ‖ The technique known as potassium-argon dating is used to date old lava flows. The potassium isotope ^{40}K has a 1.28 billion year half-life and is naturally present at very low levels. ^{40}K decays by two routes: 89% undergo beta-minus decay into ^{40}Ca while 11% undergo electron capture to become ^{40}Ar. Argon is a gas, and there is no argon in flowing lava because the gas escapes. Once the lava solidifies, any argon produced in the decay of ^{40}K is trapped inside and cannot escape. A geologist brings you a piece of solidified lava in which you find the ^{40}Ar/^{40}K ratio to be 0.013. What is the age of the rock?

54. ‖ The half-life of the uranium isotope ^{235}U is 700 million years. The earth is approximately 4.5 billion years old. How much more ^{235}U was there when the earth formed than there is today? Give your answer as the then-to-now ratio.

55. ‖ A chest x ray uses 10 keV photons with an RBE of 0.85. A BIO 60 kg person receives a 0.30 mSv dose from one chest x ray that exposes 25% of the patient's body. How many x ray photons are absorbed in the patient's body?

56. ‖ The rate at which a radioactive tracer is lost from a patient's BIO body is the rate at which the isotope decays *plus* the rate at which the element is excreted from the body. Medical experiments have shown that stable isotopes of a particular element are excreted with a 6.0 day half-life. A radioactive isotope of the same element has a half-life of 9.0 days. What is the effective half-life of the isotope in a patient's body?

57. ‖ The plutonium isotope ^{239}Pu has a half-life of 24,000 years BIO and decays by the emission of a 5.2 MeV alpha particle. Plutonium is not especially dangerous if handled because the activity is low and the alpha radiation doesn't penetrate the skin. However, there are serious health concerns if even the tiniest particles of plutonium are inhaled and lodge deep in the lungs. This could happen following any kind of fire or explosion that disperses plutonium as dust. Let's determine the level of danger.

   a. Soot particles are roughly 1 μm in diameter, and it is known that these particles can go deep into the lungs. How many atoms are in a 1.0-μm-diameter particle of ^{239}Pu? The density of plutonium is 19,800 kg/m^3.

   b. What is the activity, in Bq, of a 1.0-μm-diameter particle?

   c. The activity of the particle is very small, but the penetrating power of alpha particles is also very small. The alpha particles are all stopped, and each deposits its energy in a 50-μm-diameter sphere around the particle. What is the dose, in mSv/year, to this small sphere of tissue in the lungs? Assume that the tissue density is that of water.

   d. Is this exposure likely to be significant? How does it compare to the natural background of radiation exposure?

Challenge Problems

58. The uranium isotope ^{238}U is naturally present at low levels in BIO many soils. One of the nuclei in the decay series of ^{238}U is the radon isotope ^{222}Rn, which decays by emitting a 5.50 MeV alpha particle with $t_{1/2} = 3.82$ days. Radon is a gas, and it tends to seep from soil into basements. The Environmental Protection Agency recommends that homeowners take steps to remove radon, by pumping in fresh air, if the radon activity exceeds 4 pCi per liter of air.

   a. How many ^{222}Rn atoms are there in 1 m^3 of air if the activity is 4 pCi/L?

   b. The range of alpha particles in air is ≈3 cm. Suppose we model a person as a 180-cm-tall, 25-cm-diameter cylinder with a mass of 65 kg. Only decays within 3 cm of the cylinder can cause exposure, and only ≈50% of the decays direct the alpha particle toward the person. Determine the dose in mSv per year for a person who spends the entire year in a room where the activity is 4 pCi/L.

   c. Does the EPA recommendation seem appropriate? Why?

59. Estimate the stopping distance in air of a 5.0 MeV alpha particle. Assume that the particle loses on average 30 eV per collision.

60. Beta-plus decay is $^{A}X_Z \rightarrow {^{A}Y_{Z-1}} + e^+ + \nu$.

   a. Determine the mass threshold for beta-plus decay. That is, what is the minimum atomic mass m_X for which this decay is energetically possible? Your answer will be in terms of the atomic mass m_Y and the electron mass m_e.

   b. Can ^{13}N undergo beta-plus decay into ^{13}C? If so, how much energy is released in the decay?

61. All the very heavy atoms found in the earth were created long ago by nuclear fusion reactions in a supernova, an exploding star. The debris spewed out by the supernova later coalesced into the gases from which the sun and the planets of our solar system were formed. Nuclear physics suggests that the uranium isotopes ^{235}U and ^{238}U should have been created in roughly equal numbers. Today, 99.28% of uranium is ^{238}U and only 0.72% is ^{235}U. How long ago did the supernova occur?

62. It might seem strange that in beta decay the positive proton, which is repelled by the positive nucleus, remains in the nucleus while the negative electron, which is attracted to the nucleus, is ejected. To understand beta decay, let's analyze the decay of a free neutron that is at rest in the laboratory. We'll ignore the antineutrino and consider the decay $n \rightarrow p^+ + e^-$. The analysis requires the use of relativistic energy and momentum, from Chapter 36.

   a. What is the total kinetic energy, in MeV, of the proton and electron?

   b. Write the equation that expresses the conservation of relativistic energy for this decay. Your equation will be in terms of the three masses m_n, m_p, and m_e and the relativistic factors γ_p and γ_e.

   c. Write the equation that expresses the conservation of relativistic momentum for this decay. Let v represent speed, rather than velocity, then write any minus signs explicitly.

   d. You have two simultaneous equations in the two unknowns v_p and v_e. To help in solving these, first prove that $\gamma v = (\gamma^2 - 1)^{1/2}c$.

   e. Solve for v_p and v_e. (It's easiest to solve for γ_p and γ_e, then find v from γ.) First get an algebraic expression for each, in terms of the masses. Then evaluate each, giving v as a fraction of c.

   f. Calculate the kinetic energy in MeV of the proton and the electron. Verify that their sum matches your answer to part a.

   g. Now explain why the electron is ejected in beta decay while the proton remains in the nucleus.

63. Alpha decay occurs when an alpha particle tunnels through the Coulomb barrier. **FIGURE CP42.63** shows a simple one-dimensional model of the potential-energy well of an alpha particle in a nucleus with $A \approx 235$. The 15 fm width of this one-dimensional potential-energy well is the *diameter* of the nucleus. Further, to keep the model simple, the Coulomb barrier has been modeled as a 20-fm-wide, 30-MeV-high rectangular potential-energy barrier. The goal of this problem is to calculate the half-life of an alpha particle in the energy level $E = 5.0$ MeV.

   a. What is the kinetic energy of the alpha particle while inside the nucleus? What is its kinetic energy after it escapes from the nucleus?

   b. Consider the alpha particle within the nucleus to be a point particle bouncing back and forth with the kinetic energy you stated in part a. What is the particle's *collision rate,* the number of times per second it collides with a wall of the potential?

c. What is the tunneling probability P_{tunnel}?

d. P_{tunnel} is the probability that on any one collision with a wall the alpha particle tunnels through instead of reflecting. The probability of *not* tunneling is $1 - P_{tunnel}$. Hence the probability that the alpha particle is still inside the nucleus after N collisions is $(1 - P_{tunnel})^N \approx 1 - NP_{tunnel}$, where we've used the binomial approximation because $P_{tunnel} \ll 1$. The half-life is the *time* at which half the nuclei have not yet decayed. Use this to determine (in years) the half-life of the nucleus.

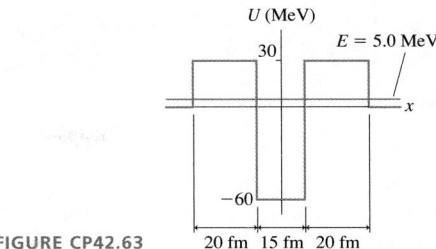

FIGURE CP42.63

STOP TO THINK ANSWERS

Stop to Think 42.1: 3. Different isotopes of an element have different numbers of neutrons but the same number of protons. The number of electrons in a neutral atom matches the number of protons.

Stop to Think 42.2: c. To keep A constant, increasing N by 1 (going up) requires decreasing Z by 1 (going left).

Stop to Think 42.3: No. A Geiger counter responds only to ionizing radiation. Visible light is not ionizing radiation.

Stop to Think 42.4: c. One-quarter of the atoms are left. This is one-half of one-half, or $(1/2)^2$.

Stop to Think 42.5: b. An increase of Z with no change in A occurs when a neutron changes to a proton and an electron, ejecting the electron.

VII Relativity and Quantum Physics

Niels Bohr was right on target with his remark, "Anyone who is not shocked by quantum theory has not understood it." Quantum mechanics *is* shocking. The predictability of Newtonian physics has been replaced by a mysterious world in which physical entities that by all rights should be waves sometimes act like particles. Electrons and neutrons somehow produce wave-like interference with themselves. These discoveries stood common sense on its head.

According to quantum mechanics, the wave function and its associated probabilities are *all we can know* about an atomic particle. This idea is so unsettling that many great scientists were reluctant to accept it. Einstein famously said, "God does not play dice with the universe." But Einstein was wrong. As strange as it seems, this is the way that nature really is.

As we conclude our journey into physics, the knowledge structure for Part VII summarizes the important ideas of relativity and quantum physics. Whether you're shocked or not, these are the scientific theories behind the emerging technologies of the 21st century.

KNOWLEDGE STRUCTURE VII **Relativity and Quantum Physics**

ESSENTIAL CONCEPTS Reference frame, event, atom, photon, quantization, wave function, probability density
BASIC GOALS What are the properties and characteristics of space and time?
How do we know about light and atoms?
How are atomic and nuclear phenomena explained by energy levels, wave functions, and photons?

GENERAL PRINCIPLES   **Principle of relativity**   All the laws of physics are the same in all inertial reference frames.

Schrödinger's equation $$\frac{d^2\psi}{dx^2} = -\frac{2m}{\hbar^2}[E - U(x)]\psi(x)$$

Pauli exclusion principle   No more than one electron or nucleon can occupy the same quantum state.

Uncertainty principle $\Delta x\,\Delta p \geq h/2$

RELATIVITY It follows from the principle of relativity that:

- The speed of light c is the same in all inertial reference frames. No particle or causal influence can travel faster than c.

- Length contraction: The length of an object in a reference frame in which the object moves with speed v is
$$L = \sqrt{1 - \beta^2}\,\ell \leq \ell$$
where ℓ is the proper length and $\beta = v/c$.

- Time dilation: The proper time interval $\Delta\tau$ between two events is measured in a reference frame in which the two events occur at the same position. The time interval Δt in a frame moving with relative speed v is
$$\Delta t = \Delta\tau/\sqrt{1 - \beta^2} \geq \Delta\tau$$

- $E = mc^2$ is the energy equivalent of mass. Mass can be transformed into energy and energy into mass.

QUANTUM PHYSICS Quantum systems are described by a wave function $\psi(x)$.

- The probability that a particle will be found in the narrow interval δx at position x is Prob(in δx at x) $= P(x)\,\delta x$. The probability density is $P(x) = |\psi(x)|^2$.

- The wave function must be normalized
$$\int_{-\infty}^{\infty} |\psi(x)|^2\,dx = 1$$

- The wave function can penetrate into a classically forbidden region with penetration distance
$$\eta = \frac{\hbar}{\sqrt{2m(U_0 - E)}}$$

- A particle can tunnel through an energy barrier of height U_0 and width w with probability $P_{\text{tunnel}} = e^{-2w/\eta}$.

Properties of light

- A photon of light of frequency f has energy $E_{\text{photon}} = hf$.

- Photons are emitted and absorbed on an all-or-nothing basis.

Properties of atoms

- Quantized energy levels, found by solving the Schrödinger equation, depend on quantum numbers n and l.

- An atom can jump from one state to another by emitting or absorbing a photon of energy $E_{\text{photon}} = \Delta E_{\text{atom}}$.

- The ground-state electron configuration is the lowest-energy configuration consistent with the Pauli principle.

Properties of nuclei

- The nucleus is held together by the strong force, an attractive short-range force between any two nucleons.

- Nuclei are stable only if the proton and neutron numbers fall along the line of stability.

- Unstable nuclei decay by alpha, beta, or gamma decay. The number of nuclei decreases exponentially with time.

Quantum Computers

All the systems we studied in Part VII were in a single, well-defined quantum state. For example, a hydrogen atom was in the $1s$ state or, perhaps, the $2p$ state. But there's another possibility. Some quantum systems can exist in a *superposition* of two or more quantum states.

We hinted at the possibility of superposition when we re-examined the double-slit interference experiment in the light of quantum physics. We noted that a photon or electron must, in some sense, go through both slits and then interfere with itself to produce the dot-by-dot buildup of an interference pattern on the screen. Suppose we say that an electron that has passed through the top slit in the figure is in quantum state ψ_a. An electron that has passed through the bottom slit is in state ψ_b.

FIGURE PSVII.1 The electron emerging from the double slit is in a superposition state.

To say that the electron goes through both slits is to say that the electron emerges from the double slit in the *superposition state* $\psi = \alpha\psi_a + \beta\psi_b$, where the coefficients α and β must satisfy $\alpha^2 + \beta^2 = 1$. (Notice that this is like finding the magnitude of a vector from its components.) If we were to detect the electron, α^2 and β^2 are the probabilities that we would find it to be in state ψ_a or state ψ_b, respectively. But until we detect it, the electron exists in the superposition of *both* states ψ_a and ψ_b. It is this superposition that allows the electron to interfere with itself to produce the interference pattern.

But what does this have to do with computers? As you know, everything a modern digital computer does, from surfing the Internet to crunching numbers, is accomplished by manipulating binary strings of 0s and 1s. The *concept* of computing with binary bits goes back to Charles Babbage in the mid-19th century, but it wasn't until the mid-20th century that scientists and engineers developed the technology that gives this concept a physical representation.

A binary bit is always a 1 or a 0; there's no in-between state. These are represented in a modern microprocessor by small capacitors that are either charged or uncharged. Suppose we wanted to represent information not with capacitors but with a quantum system that has two states. We could say that the system represents a 0 when it is in state ψ_a and a 1 when it is in ψ_b. Such a quantum system is an ordinary binary bit as long as the system is in one state or the other.

But the quantum system, unlike a classical bit, has the possibility of being in a superposition state. Using 0 and 1, rather than ψ_a and ψ_b, we could say that the system can be in the state $\psi = \alpha \cdot 0 + \beta \cdot 1$. This basic unit of quantum computing is called a *qubit*. It may seem at first that we could do the same thing with a classical system by allowing the capacitor charge to vary, but a partially charged capacitor is still a single, well-defined state. In contrast, the qubit—like the electron that goes through both slits—is simultaneously in both state 0 *and* state 1.

To illustrate the possibilities, suppose you have three classical bits and three qubits. The three bits can represent eight different numbers (000 to 111), but only one at a time. The three qubits represent all eight numbers *simultaneously*. To perform a mathematical operation, you must do it eight times on the three bits to learn all the possible outcomes. But you would learn all eight outcomes simultaneously from *one* operation on the three qubits. In general, computing with n qubits provides a theoretical improvement of 2^n over computing with n bits.

We say "theoretical" because quantum computing is still mostly in the concept stage, much as digital computers were 150 years ago. What kind of quantum systems can actually be placed in an appropriate superposition state? How do you manipulate qubits? How do you read information in and out? What kinds of computations would be improved by quantum computing?

These are all questions that are being actively researched today. Quantum computing is in its infancy, and the technology for making a real quantum computer is largely unknown. Just as Charles Babbage couldn't possibly have imagined today's computers, the uses of tomorrow's quantum computers are still unforeseen. But, quite possibly, there are uses that some of you may help to invent.

FIGURE PSVII.2 This string of beryllium ions held in an ion trap is being studied as a possible quantum computer. The quantum states of the ions are manipulated with laser beams.

Mathematics Review

Algebra

Using exponents:
$$a^{-x} = \frac{1}{a^x} \qquad a^x a^y = a^{(x+y)} \qquad \frac{a^x}{a^y} = a^{(x-y)} \qquad (a^x)^y = a^{xy}$$

$$a^0 = 1 \qquad a^1 = a \qquad a^{1/n} = \sqrt[n]{a}$$

Fractions:
$$\left(\frac{a}{b}\right)\left(\frac{c}{d}\right) = \frac{ac}{bd} \qquad \frac{a/b}{c/d} = \frac{ad}{bc} \qquad \frac{1}{1/a} = a$$

Logarithms:
If $a = e^x$, then $\ln(a) = x$ $\qquad \ln(e^x) = x \qquad\qquad e^{\ln(x)} = x$

$$\ln(ab) = \ln(a) + \ln(b) \qquad\qquad \ln\left(\frac{a}{b}\right) = \ln(a) - \ln(b) \qquad \ln(a^n) = n\ln(a)$$

The expression $\ln(a + b)$ cannot be simplified.

Linear equations: The graph of the equation $y = ax + b$ is a straight line. a is the slope of the graph. b is the y-intercept.

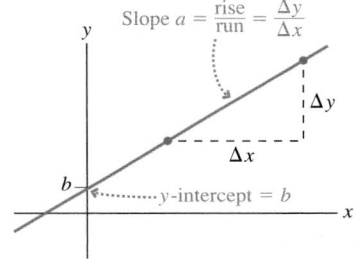

Proportionality: To say that y is proportional to x, written $y \propto x$, means that $y = ax$, where a is a constant. Proportionality is a special case of linearity. A graph of a proportional relationship is a straight line that passes through the origin. If $y \propto x$, then

$$\frac{y_1}{y_2} = \frac{x_1}{x_2}$$

Quadratic equation: The quadratic equation $ax^2 + bx + c = 0$ has the two solutions $x = \dfrac{-b \pm \sqrt{b^2 - 4ac}}{2a}$.

Geometry and Trigonometry

Area and volume:

Rectangle

$A = ab$

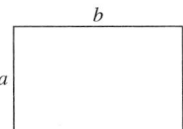

Rectangular box

$V = abc$

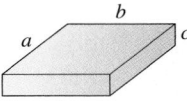

Triangle

$A = \frac{1}{2}ab$

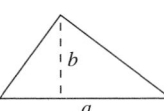

Right circular cylinder

$V = \pi r^2 l$

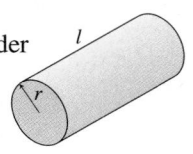

Circle

$C = 2\pi r$
$A = \pi r^2$

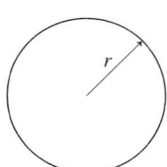

Sphere

$A = 4\pi r^2$
$V = \frac{4}{3}\pi r^3$

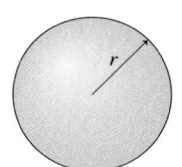

Arc length and angle: The angle θ in radians is defined as $\theta = s/r$.

The arc length that spans angle θ is $s = r\theta$.

2π rad $= 360°$

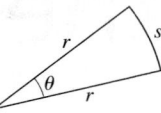

Right triangle: Pythagorean theorem $c = \sqrt{a^2 + b^2}$ or $a^2 + b^2 = c^2$

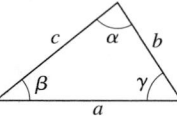

$$\sin\theta = \frac{b}{c} = \frac{\text{far side}}{\text{hypotenuse}} \qquad \theta = \sin^{-1}\left(\frac{b}{c}\right)$$

$$\cos\theta = \frac{a}{c} = \frac{\text{adjacent side}}{\text{hypotenuse}} \qquad \theta = \cos^{-1}\left(\frac{a}{c}\right)$$

$$\tan\theta = \frac{b}{a} = \frac{\text{far side}}{\text{adjacent side}} \qquad \theta = \tan^{-1}\left(\frac{b}{a}\right)$$

General triangle: $\alpha + \beta + \gamma = 180° = \pi$ rad

Law of cosines $c^2 = a^2 + b^2 - 2ab\cos\gamma$

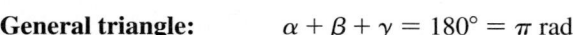

Identities:

$$\tan\alpha = \frac{\sin\alpha}{\cos\alpha} \qquad\qquad \sin^2\alpha + \cos^2\alpha = 1$$

$$\sin(-\alpha) = -\sin\alpha \qquad\qquad \cos(-\alpha) = \cos\alpha$$

$$\sin(\alpha \pm \beta) = \sin\alpha\cos\beta \pm \cos\alpha\sin\beta \qquad \cos(\alpha \pm \beta) = \cos\alpha\cos\beta \mp \sin\alpha\sin\beta$$

$$\sin(2\alpha) = 2\sin\alpha\cos\alpha \qquad\qquad \cos(2\alpha) = \cos^2\alpha - \sin^2\alpha$$

$$\sin(\alpha \pm \pi/2) = \pm\cos\alpha \qquad\qquad \cos(\alpha \pm \pi/2) = \mp\sin\alpha$$

$$\sin(\alpha \pm \pi) = -\sin\alpha \qquad\qquad \cos(\alpha \pm \pi) = -\cos\alpha$$

Expansions and Approximations

Binomial expansion: $(1 + x)^n = 1 + nx + \dfrac{n(n-1)}{2}x^2 + \cdots$

Binomial approximation: $(1 + x)^n \approx 1 + nx$ if $x \ll 1$

Trigonometric expansions: $\sin\alpha = \alpha - \dfrac{\alpha^3}{3!} + \dfrac{\alpha^5}{5!} - \dfrac{\alpha^7}{7!} + \cdots$ for α in rad

$\cos\alpha = 1 - \dfrac{\alpha^2}{2!} + \dfrac{\alpha^4}{4!} - \dfrac{\alpha^6}{6!} + \cdots$ for α in rad

Small-angle approximation: If $\alpha \ll 1$ rad, then $\sin\alpha \approx \tan\alpha \approx \alpha$ and $\cos\alpha \approx 1$.

The small-angle approximation is excellent for $\alpha < 5°$ (≈ 0.1 rad) and generally acceptable up to $\alpha \approx 10°$.

Calculus

The letters a and n represent constants in the following derivatives and integrals.

Derivatives

$$\frac{d}{dx}(a) = 0$$

$$\frac{d}{dx}(ax) = a$$

$$\frac{d}{dx}\left(\frac{a}{x}\right) = -\frac{a}{x^2}$$

$$\frac{d}{dx}(ax^n) = anx^{n-1}$$

$$\frac{d}{dx}\left(\ln(ax)\right) = \frac{1}{x}$$

$$\frac{d}{dx}(e^{ax}) = ae^{ax}$$

$$\frac{d}{dx}\left(\sin(ax)\right) = a\cos(ax)$$

$$\frac{d}{dx}\left(\cos(ax)\right) = -a\sin(ax)$$

Integrals

$$\int x\,dx = \frac{1}{2}x^2$$

$$\int x^2\,dx = \frac{1}{3}x^3$$

$$\int \frac{1}{x^2}\,dx = -\frac{1}{x}$$

$$\int x^n\,dx = \frac{x^{n+1}}{n+1} \qquad n \neq -1$$

$$\int \frac{dx}{x} = \ln x$$

$$\int \frac{dx}{a+x} = \ln(a+x)$$

$$\int \frac{x\,dx}{a+x} = x - a\ln(a+x)$$

$$\int \frac{dx}{\sqrt{x^2 \pm a^2}} = \ln\left(x + \sqrt{x^2 \pm a^2}\right)$$

$$\int \frac{x\,dx}{\sqrt{x^2 \pm a^2}} = \sqrt{x^2 \pm a^2}$$

$$\int \frac{dx}{x^2 + a^2} = \frac{1}{a}\tan^{-1}\left(\frac{x}{a}\right)$$

$$\int \frac{dx}{(x^2 + a^2)^2} = \frac{1}{2a^3}\tan^{-1}\left(\frac{x}{a}\right) + \frac{x}{2a^2(x^2 + a^2)}$$

$$\int \frac{dx}{(x^2 \pm a^2)^{3/2}} = \frac{\pm x}{a^2\sqrt{x^2 \pm a^2}}$$

$$\int \frac{x\,dx}{(x^2 \pm a^2)^{3/2}} = -\frac{1}{\sqrt{x^2 \pm a^2}}$$

$$\int e^{ax}\,dx = \frac{1}{a}e^{ax}$$

$$\int xe^{-x}\,dx = -(x+1)e^{-x}$$

$$\int x^2 e^{-x}\,dx = -(x^2 + 2x + 2)e^{-x}$$

$$\int \sin(ax)\,dx = -\frac{1}{a}\cos(ax)$$

$$\int \cos(ax)\,dx = \frac{1}{a}\sin(ax)$$

$$\int \sin^2(ax)\,dx = \frac{x}{2} - \frac{\sin(2ax)}{4a}$$

$$\int \cos^2(ax)\,dx = \frac{x}{2} + \frac{\sin(2ax)}{4a}$$

$$\int_0^\infty x^n e^{-ax}\,dx = \frac{n!}{a^{n+1}}$$

$$\int_0^\infty e^{-ax^2}\,dx = \frac{1}{2}\sqrt{\frac{\pi}{a}}$$

Periodic Table of Elements

Atomic number — 27
Symbol — Co
Atomic mass — 58.9

Transition elements

Inner transition elements

An atomic mass in brackets is that of the longest-lived isotope of an element with no stable isotopes.

Period	Group 1	2																	18
1	1 H 1.0																	2 He 4.0
2	3 Li 6.9	4 Be 9.0										5 B 10.8	6 C 12.0	7 N 14.0	8 O 16.0	9 F 19.0	10 Ne 20.2
3	11 Na 23.0	12 Mg 24.3										13 Al 27.0	14 Si 28.1	15 P 31.0	16 S 32.1	17 Cl 35.5	18 Ar 39.9
4	19 K 39.1	20 Ca 40.1	21 Sc 45.0	22 Ti 47.9	23 V 50.9	24 Cr 52.0	25 Mn 54.9	26 Fe 55.8	27 Co 58.9	28 Ni 58.7	29 Cu 63.5	30 Zn 65.4	31 Ga 69.7	32 Ge 72.6	33 As 74.9	34 Se 79.0	35 Br 79.9	36 Kr 83.8
5	37 Rb 85.5	38 Sr 87.6	39 Y 88.9	40 Zr 91.2	41 Nb 92.9	42 Mo 95.9	43 Tc [98]	44 Ru 101.1	45 Rh 102.9	46 Pd 106.4	47 Ag 107.9	48 Cd 112.4	49 In 114.8	50 Sn 118.7	51 Sb 121.8	52 Te 127.6	53 I 126.9	54 Xe 131.3
6	55 Cs 132.9	56 Ba 137.3	71 Lu 175.0	72 Hf 178.5	73 Ta 180.9	74 W 183.9	75 Re 186.2	76 Os 190.2	77 Ir 192.2	78 Pt 195.1	79 Au 197.0	80 Hg 200.6	81 Tl 204.4	82 Pb 207.2	83 Bi 209.0	84 Po [209]	85 At [210]	86 Rn [222]
7	87 Fr [223]	88 Ra [226]	103 Lr [262]	104 Rf [265]	105 Db [268]	106 Sg [271]	107 Bh [272]	108 Hs [270]	109 Mt [276]	110 Ds [281]	111 Rg [280]	112 Cn [285]	113	114	115	116	117	118

Lanthanides 6

| 57 La 138.9 | 58 Ce 140.1 | 59 Pr 140.9 | 60 Nd 144.2 | 61 Pm 144.9 | 62 Sm 150.4 | 63 Eu 152.0 | 64 Gd 157.3 | 65 Tb 158.9 | 66 Dy 162.5 | 67 Ho 164.9 | 68 Er 167.3 | 69 Tm 168.9 | 70 Yb 173.0 |

Actinides 7

| 89 Ac [227] | 90 Th 232.0 | 91 Pa 231.0 | 92 U 238.0 | 93 Np [237] | 94 Pu [244] | 95 Am [243] | 96 Cm [247] | 97 Bk [247] | 98 Cf [251] | 99 Es [252] | 100 Fm [257] | 101 Md [258] | 102 No [259] |

Atomic and Nuclear Data

Atomic Number (Z)	Element	Symbol	Mass Number (A)	Atomic Mass (u)	Percent Abundance	Decay Mode	Half-Life $t_{1/2}$
0	(Neutron)	n	1	1.008 665		β^-	10.4 min
1	Hydrogen	H	1	1.007 825	99.985	stable	
	Deuterium	D	2	2.014 102	0.015	stable	
	Tritium	T	3	3.016 049		β^-	12.33 yr
2	Helium	He	3	3.016 029	0.000 1	stable	
			4	4.002 602	99.999 9	stable	
			6	6.018 886		β^-	0.81 s
3	Lithium	Li	6	6.015 121	7.50	stable	
			7	7.016 003	92.50	stable	
			8	8.022 486		β^-	0.84 s
4	Beryllium	Be	7	7.016 928		EC	53.3 days
			9	9.012 174	100	stable	
			10	10.013 534		β^-	1.5×10^6 yr
5	Boron	B	10	10.012 936	19.90	stable	
			11	11.009 305	80.10	stable	
			12	12.014 352		β^-	0.020 2 s
6	Carbon	C	10	10.016 854		β^+	19.3 s
			11	11.011 433		β^+	20.4 min
			12	12.000 000	98.90	stable	
			13	13.003 355	1.10	stable	
			14	14.003 242		β^-	5 730 yr
			15	15.010 599		β^-	2.45 s
7	Nitrogen	N	12	12.018 613		β^+	0.011 0 s
			13	13.005 738		β^+	9.96 min
			14	14.003 074	99.63	stable	
			15	15.000 108	0.37	stable	
			16	16.006 100		β^-	7.13 s
			17	17.008 450		β^-	4.17 s
8	Oxygen	O	14	14.008 595		EC	70.6 s
			15	15.003 065		β^+	122 s
			16	15.994 915	99.76	stable	
			17	16.999 132	0.04	stable	
			18	17.999 160	0.20	stable	
			19	19.003 577		β^-	26.9 s
9	Fluorine	F	17	17.002 094		EC	64.5 s
			18	18.000 937		β^+	109.8 min
			19	18.998 404	100	stable	
			20	19.999 982		β^-	11.0 s
10	Neon	Ne	19	19.001 880		β^+	17.2 s
			20	19.992 435	90.48	stable	
			21	20.993 841	0.27	stable	
			22	21.991 383	9.25	stable	

Atomic Number (Z)	Element	Symbol	Mass Number (A)	Atomic Mass (u)	Percent Abundance	Decay Mode	Half-Life $t_{1/2}$
11	Sodium	Na	22	21.994 434		β^+	2.61 yr
			23	22.989 770	100	stable	
			24	23.990 961		β^-	14.96 hr
12	Magnesium	Mg	24	23.985 042	78.99	stable	
			25	24.985 838	10.00	stable	
			26	25.982 594	11.01	stable	
13	Aluminum	Al	27	26.981 538	100	stable	
			28	27.981 910		β^-	2.24 min
14	Silicon	Si	28	27.976 927	92.23	stable	
			29	28.976 495	4.67	stable	
			30	29.973 770	3.10	stable	
			31	30.975 362		β^-	2.62 hr
15	Phosphorus	P	30	29.978 307		β^+	2.50 min
			31	30.973 762	100	stable	
			32	31.973 908		β^-	14.26 days
16	Sulfur	S	32	31.972 071	95.02	stable	
			33	32.971 459	0.75	stable	
			34	33.967 867	4.21	stable	
			35	34.969 033		β^-	87.5 days
			36	35.967 081	0.02	stable	
17	Chlorine	Cl	35	34.968 853	75.77	stable	
			36	35.968 307		β^-	3.0×10^5 yr
			37	36.965 903	24.23	stable	
18	Argon	Ar	36	35.967 547	0.34	stable	
			38	37.962 732	0.06	stable	
			39	38.964 314		β^-	269 yr
			40	39.962 384	99.60	stable	
			42	41.963 049		β^-	33 yr
19	Potassium	K	39	38.963 708	93.26	stable	
			40	39.964 000	0.01	β^+	1.28×10^9 yr
			41	40.961 827	6.73	stable	
20	Calcium	Ca	40	39.962 591	96.94	stable	
			42	41.958 618	0.64	stable	
			43	42.958 767	0.13	stable	
			44	43.955 481	2.08	stable	
			47	46.954 547		β^-	4.5 days
			48	47.952 534	0.18	stable	
24	Chromium	Cr	50	49.946 047	4.34	stable	
			52	51.940 511	83.79	stable	
			53	52.940 652	9.50	stable	
			54	53.938 883	2.36	stable	
26	Iron	Fe	54	53.939 613	5.9	stable	
			55	54.938 297		EC	2.7 yr
			56	55.934 940	91.72	stable	
			57	56.935 396	2.1	stable	
			58	57.933 278	0.28	stable	

Atomic Number (Z)	Element	Symbol	Mass Number (A)	Atomic Mass (u)	Percent Abundance	Decay Mode	Half-Life $t_{1/2}$
27	Cobalt	Co	59	58.933 198	100	stable	
			60	59.933 820		β^-	5.27 yr
28	Nickel	Ni	58	57.935 346	68.08	stable	
			60	59.930 789	26.22	stable	
			61	60.931 058	1.14	stable	
			62	61.928 346	3.63	stable	
			64	63.927 967	0.92	stable	
29	Copper	Cu	63	62.929 599	69.17	stable	
			65	64.927 791	30.83	stable	
47	Silver	Ag	107	106.905 091	51.84	stable	
			109	108.904 754	48.16	stable	
48	Cadmium	Cd	106	105.906 457	1.25	stable	
			109	108.904 984		EC	462 days
			110	109.903 004	12.49	stable	
			111	110.904 182	12.80	stable	
			112	111.902 760	24.13	stable	
			113	112.904 401	12.22	stable	
			114	113.903 359	28.73	stable	
			116	115.904 755	7.49	stable	
53	Iodine	I	127	126.904 474	100	stable	
			129	128.904 984		β^-	1.6×10^7 yr
			131	130.906 124		β^-	8 days
54	Xenon	Xe	128	127.903 531	1.9	stable	
			129	128.904 779	26.4	stable	
			130	129.903 509	4.1	stable	
			131	130.905 069	21.2	stable	
			132	131.904 141	26.9	stable	
			133	132.905 906		β^-	5.4 days
			134	133.905 394	10.4	stable	
			136	135.907 215	8.9	stable	
55	Cesium	Cs	133	132.905 436	100	stable	
			137	136.907 078		β^-	30 yr
56	Barium	Ba	131	130.906 931		EC	12 days
			133	132.905 990		EC	10.5 yr
			134	133.904 492	2.42	stable	
			135	134.905 671	6.59	stable	
			136	135.904 559	7.85	stable	
			137	136.905 816	11.23	stable	
			138	137.905 236	71.70	stable	
79	Gold	Au	197	196.966 543	100	stable	
81	Thallium	Tl	203	202.972 320	29.524	stable	
			205	204.974 400	70.476	stable	
			207	206.977 403		β^-	4.77 min
82	Lead	Pb	204	203.973 020	1.4	stable	
			205	204.974 457		EC	1.5×10^7 yr

Atomic Number (Z)	Element	Symbol	Mass Number (A)	Atomic Mass (u)	Percent Abundance	Decay Mode	Half-Life $t_{1/2}$
			206	205.974 440	24.1	stable	
			207	206.975 871	22.1	stable	
			208	207.976 627	52.4	stable	
			210	209.984 163		α, β^-	22.3 yr
			211	210.988 734		β^-	36.1 min
83	Bismuth	Bi	208	207.979 717		EC	3.7×10^5 yr
			209	208.980 374	100	stable	
			211	210.987 254		α	2.14 min
			215	215.001 836		β^-	7.4 min
84	Polonium	Po	209	208.982 405		α	102 yr
			210	209.982 848		α	138.38 days
			215	214.999 418		α	0.001 8 s
			218	218.008 965		α, β^-	3.10 min
85	Astatine	At	218	218.008 685		α, β^-	1.6 s
			219	219.011 294		α, β^-	0.9 min
86	Radon	Rn	219	219.009 477		α	3.96 s
			220	220.011 369		α	55.6 s
			222	222.017 571		α, β^-	3.823 days
87	Francium	Fr	223	223.019 733		α, β^-	22 min
88	Radium	Ra	223	223.018 499		α	11.43 days
			224	224.020 187		α	3.66 days
			226	226.025 402		α	1 600 yr
			228	228.031 064		β^-	5.75 yr
89	Actinium	Ac	227	227.027 749		α, β^-	21.77 yr
			228	228.031 015		β^-	6.15 hr
90	Thorium	Th	227	227.027 701		α	18.72 days
			228	228.028 716		α	1.913 yr
			229	229.031 757		α	7 300 yr
			230	230.033 127		α	75.000 yr
			231	231.036 299		α, β^-	25.52 hr
			232	232.038 051	100	α	1.40×10^{10} yr
			234	234.043 593		β^-	24.1 days
91	Protactinium	Pa	231	231.035 880		α	32.760 yr
			234	234.043 300		β^-	6.7 hr
92	Uranium	U	233	233.039 630		α	1.59×10^5 yr
			234	234.040 946		α	2.45×10^5 yr
			235	235.043 924	0.72	α	7.04×10^8 yr
			236	236.045 562		α	2.34×10^7 yr
			238	238.050 784	99.28	α	4.47×10^9 yr
93	Neptunium	Np	236	236.046 560		EC	1.15×10^5 yr
			237	237.048 168		α	2.14×10^6 yr
94	Plutonium	Pu	238	238.049 555		α	87.7 yr
			239	239.052 157		α	2.412×10^4 yr
			240	240.053 808		α	6 560 yr
			242	242.058 737		α	3.73×10^6 yr

ActivPhysics OnLine™ Activities (MP) www.masteringphysics.com

The following list gives the activity numbers and titles of the ActivPhysics activities available in the Pearson eText and the Study Area of MasteringPhysics, followed by the corresponding textbook page references.

PhET Simulations (MP) www.masteringphysics.com

The following list gives the titles of the PhET simulations available in the Pearson eText and in the Study Area of MasteringPhysics, with the corresponding textbook section and page references.

*Indicates an associated tutorial is available in the MasteringPhysics Item Library.

APPENDIX

D

Answers

Answers to Odd-Numbered Exercises and Problems

Chapter 1

1.

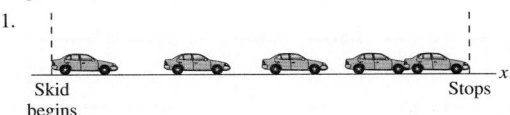

3.

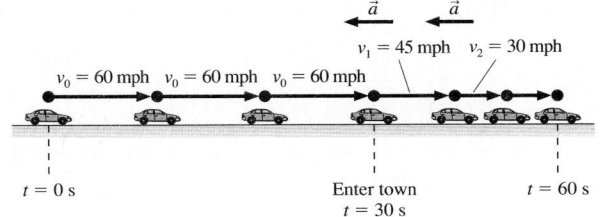

5.

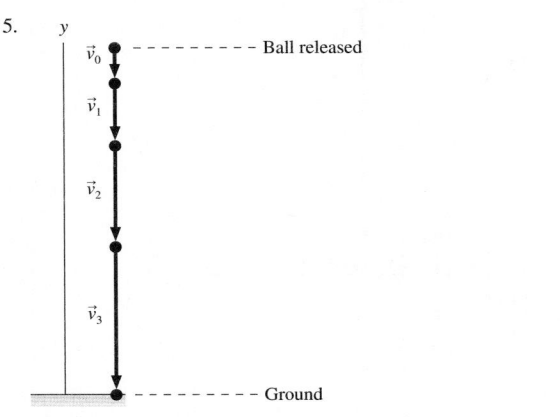

7.

9. a. Greater
   b.

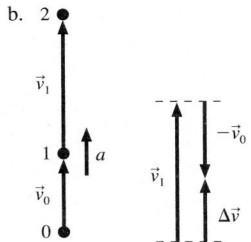

11. a.

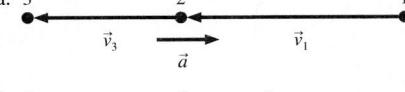

13.

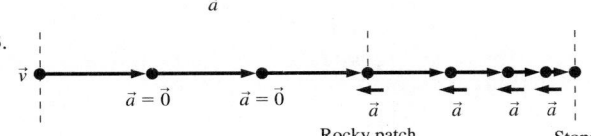

15.

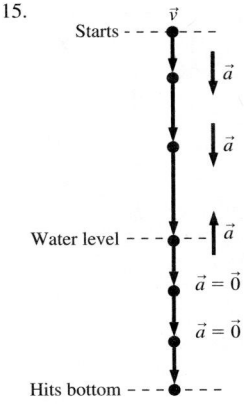

17.

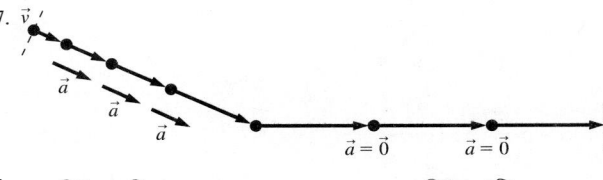

21.

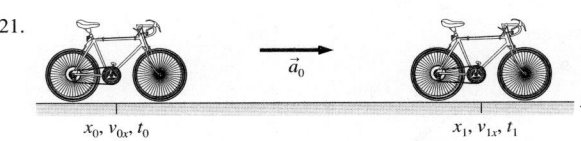

Known
$v_{0x} = 0$ m/s $t_0 = 0$ s $x_0 = 0$ m
$a_0 = 1.5$ m/s^2
$v_1 = 7.5$ m/s

Find
x_1

23. a. 6.15×10^{-3} s   b. 27.2×10^3 m   c. 31.1 m/s
    d. 7.2×10^{-2} m/s
25. a. 4×10^3 s   b. 9×10^4 s   c. 3×10^7 s   d. 65.5 m/s
27. a. 12 in   b. 50 mph   c. 3 mi   d. 0.3 in
29. a. 1.95×10^3 b. 65.1   c. 2.3   d. 0.0800
31. 50 ft or 15 m, about 8 times my height
33. 9.8×10^{-9} m/s $= 40$ μm/h
35. **Pictorial representation**

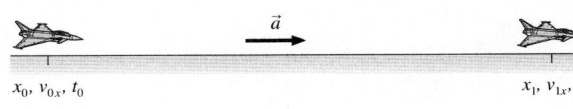

Motion diagram

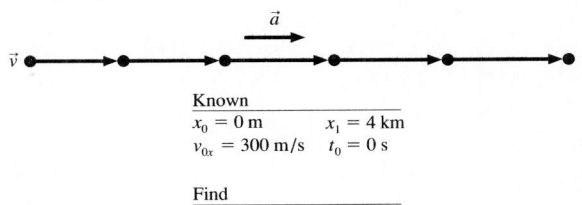

Known
$x_0 = 0$ m $x_1 = 4$ km
$v_{0x} = 300$ m/s $t_0 = 0$ s

Find
a_x

37.

Pictorial representation   **Motion diagram**

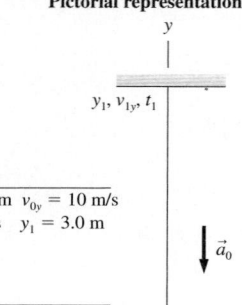
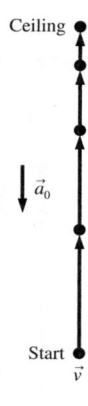

Known
$y_0 = 0$ m $v_{0y} = 10$ m/s
$t_0 = 0$ s $y_1 = 3.0$ m
$a_{0y} < 0$

Find
t_1

39. **Pictorial representation**

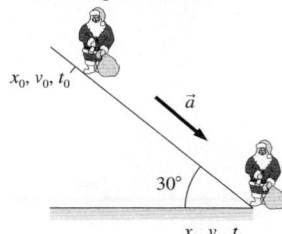

30°

Known
$x_0 = 0$ m $x_1 = 10$ m/s
$t_0 = 0$ s $a_x = (9.8$ m/s$^2) \sin (30°)$
$v_{0x} = 0$ m/s

Motion diagram

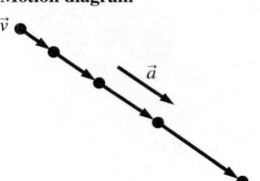

Find
v_{1x}

41. **Pictorial representation**

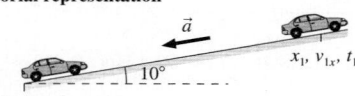

Motion diagram

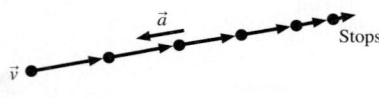

Known
$x_0 = 0$ m $v_{1x} = 0$ m/s
$t_0 = 0$ s $a_x = -(9.8$ m/s$^2) \sin (10°)$
$v_{0x} = 30$ m/s

Find
x_1

43.   **Pictorial representation**

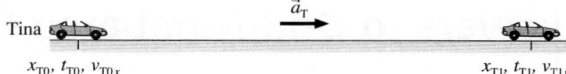

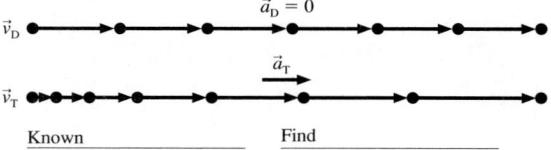

Motion diagram

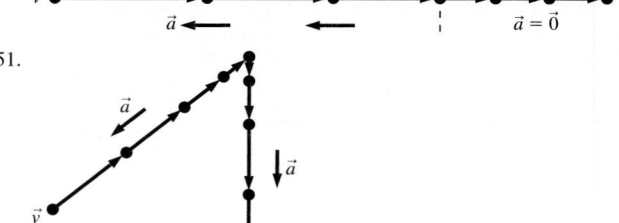

Known   Find
$x_{D0} = 0$ m $x_{T0} = 0$ m a_{Tx}
$t_{D0} = 0$ s $t_{T0} = 0$ s
$v_{D0x} = 30$ m/s $v_{T0x} = 0$ m/s
$a_{D0x} = 0$ m/s^2

49.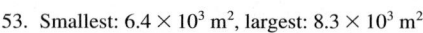

51.

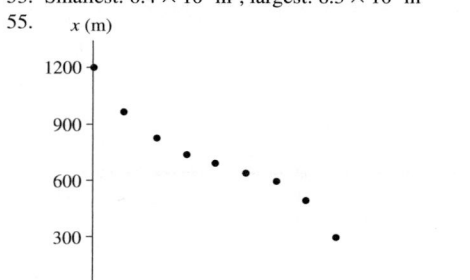

53. Smallest: 6.4×10^3 m^2, largest: 8.3×10^3 m^2

55.

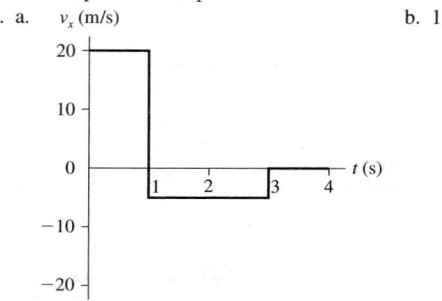

Chapter 2

1. a. Beth   b. 20 min
3. a. 48 mph   b. 50 mph
5. a. v_x (m/s)   b. 1 s

7. 8.0 cm

9.

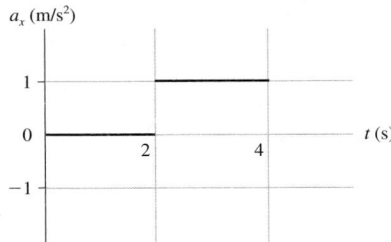

11. a. 6.0 m   b. 4.0 m/s   c. 2.0 m/s^2
13. 8.8 m/s^2; this is reasonable for a jet
15. -2.8 m/s^2
17. a. 78.4 m   b. -39.2 m/s
19. 3.2 s
21. a. 64 m   b. 7.1 s
23. a. 7 m   b. 7 m/s   c. 4 m/s^2
25. 16 m/s
27. -10 m/s, -20 m/s, 75 m/s
29. a. Zero at $t = 0$ s, 1 s, 2 s, 3 s, . . . ; most positive at $t = 0.5$ s, 2.5 s, most negative at $t = 1.5$ s, 3.5 s, . . .
   b.

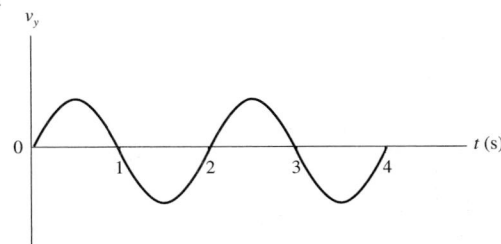

31. a. 0 s and 3 s   b. 12 m and -18 m/s^2; -15 m and 18 m/s^2
33. 2.0 m/s^3
35.

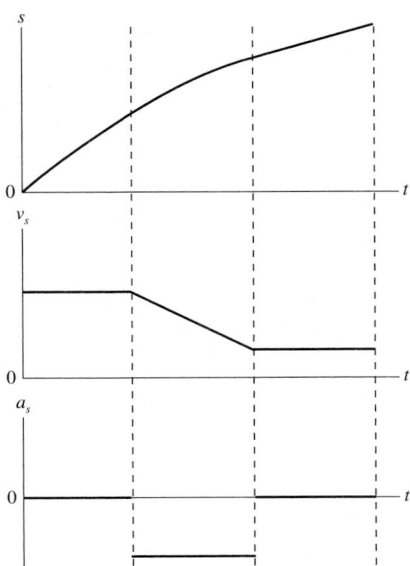

37.

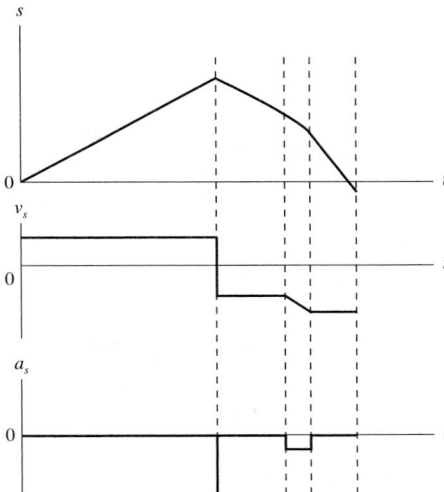

39.

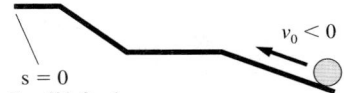

$s = 0$
Ball rolls off left edge

41. a. 2.7 m/s^2 b. 28%   c. 1.3×10^2 m $= 4.3 \times 10^2$ ft
43. a. 100 m   b. -2 m/s^2 c. 11.5 s
45. a. 5 m   b. 22 m/s
47. 5.2 cm
49. a. 54.8 km   b. 228 s
   c.

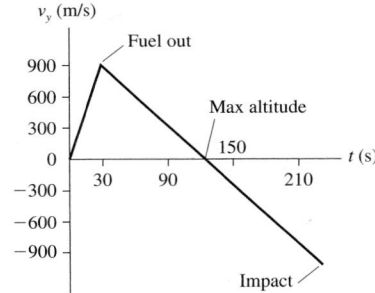

51. 19.7 m
53. 216 m
55. 9.9 m/s
57. a. $v_f = \sqrt{2gh}$ b. 2.4 m/s
59. Yes
61. a. gh/d b. 70 m/s^2
63. 5.7 m/s
65. a. 900 m   b. 60 m/s
67. 17.2 m
69. 4.4 m/s^2
71. a. Yes   b. 7.9 m/s^2
73. c. 17.2 m/s
75. c. 750 m
77. 5.5 m/s^2
79. a. 10 s   b. 3.8 m/s^2 c. 5.6%
81. 12.5 m/s
83. 4500 m/s^2

Chapter 3

1. a.

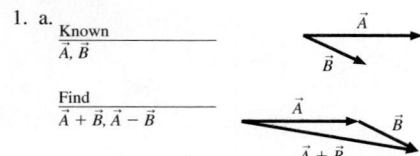

b.

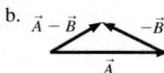

3. a. $-E\cos\theta$, $E\sin\theta$ b. $E_x = -E\sin\phi$, $E_y = E\cos\phi$
5. 6 m
7. a. 0 m/s, -10 m/s   b. 17 m/s², -10 m/s²   c. -60 N, 80 N
9. 2.2 T, 27° below the $+x$-axis
11. a. 7.2, 56° below the $+x$-axis
    b. 94 m, 58° above the $+x$-axis
    c. 45 m/s, 63° above the $-x$-axis
    d. 6.3 m/s², 18° right of the $-y$-axis
13. a. $1\hat{\imath} + 3\hat{\jmath}$
    b.

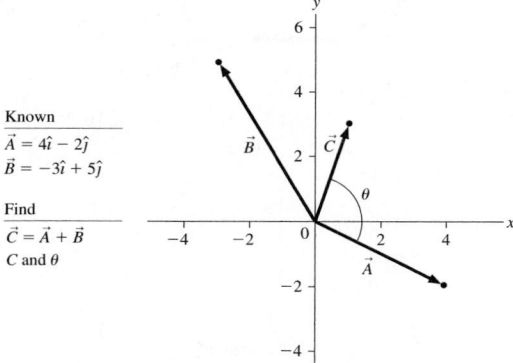

    c. 3.2, 72° above the $+x$-axis
15. a. $10\hat{\imath} + 2\hat{\jmath}$
    b.

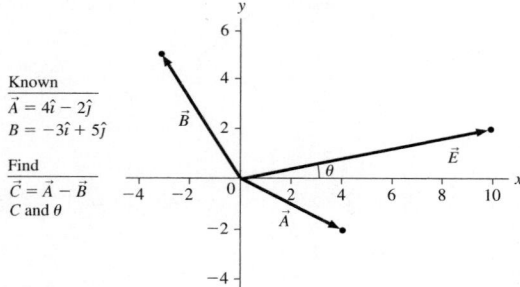

    c. 10, 11° above the $+x$-axis
17. a. $B_x = -4.3$ m, $B_y = 2.5$ m   b. $B_x = -2.5$ m, $B_y = 4.3$ m
19. a.

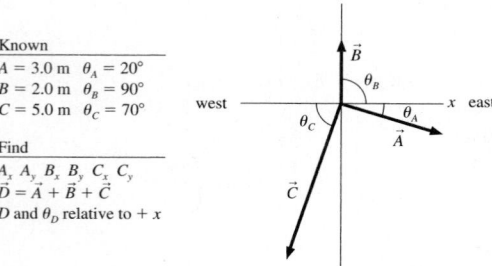

b. $\vec{A} = (2.8\hat{\imath} + 1.0\hat{\jmath})$ m, $\vec{B} = (2.0\hat{\jmath})$ m, $\vec{C} = (-1.7\hat{\imath} - 4.7\hat{\jmath})$ m
c. 3.9 m, 74° below the $+x$-axis
21. a. 0 m, 26 m, 160 m   b. $(10\hat{\imath} + 8.0\hat{\jmath})t$ m/s
    c. 0 m/s, 26 m/s, 64 m/s
23. a. $-4\hat{\imath} + 3\hat{\jmath}$,   b. 5.0, 37° above the $-x$-axis
25. $-1\hat{\imath} - 3\hat{\jmath}$
27. $\vec{B} = \dfrac{1}{\sqrt{2}}\hat{\imath} + \dfrac{1}{\sqrt{2}}\hat{\jmath}$
29. a. 100 m lower   b. 5.0 km
31. a.

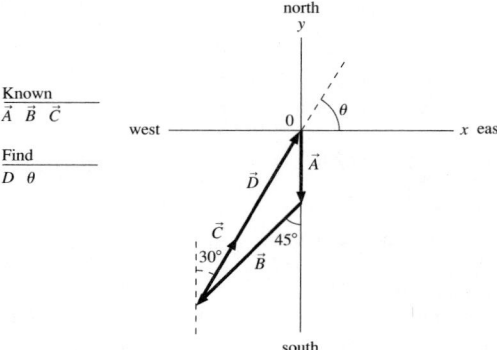

b. 360 m, 59° north of east   c. Yes
33. 7.5 m
35. a. 34°   b. 1.7 m/s
37. a. 50 m
    b.

39. -15 m/s
41. 49° west of south
43. a. 0.50 N up-slope   b. 1.7 N
    c. 1.7 N, 73° clockwise from the $-x$-axis

Chapter 4

1. a.

3. H
5. E
7. 111 m
9. a. 62° above the $+x$-axis   b. 180 cm
11. a. $\vec{v}(0) = (2.0\hat{\imath} + 4.0\hat{\jmath})$ m/s, $\vec{v}(2) = (2.0\hat{\imath} + 0.0\hat{\jmath})$ m/s,
    $\vec{v}(3) = (2.0\hat{\imath} - 2.0\hat{\jmath})$ m/s   b. 2.0 m/s²   c. 63° above the $+x$-axis
13. a. 0.064 s   b. 780 m/s
15. 2.0 km/h
17. a. 300 m   b. 3.2 m/s
19. a. 0 rad/s   b. $-\dfrac{\pi}{2}$ rad/s   c. 3π rad/s

21.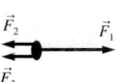

23. 8.0°/h
25. 1680 km/h, 1040 mph
27. a. 3.0×10^4 m/s   b. 2.0×10^{-7} rad/s   c. 6.0×10^{-3} m/s^2
29. a. -100 rad/s^2 b. 0 rad/s^2 c. 50 rad/s^2
31. 9.5 rev
33. 47 rad/s^2
35. a. -2.6 m/s^2 b. 31 rev
37. $\vec{r} = (710\hat{i} - 400\hat{j} + 160\hat{k}) \times 10^3$ km
39. a. $\dfrac{v_0^2 \sin^2 \theta}{2g}$

    b. $h = 14.4$ m, 28.8 m, 43.2 m; $d = 99.8$ m, 115.2 m, 99.8 m
41. a. 81 m higher   b. 34 m
43. a. 6 times farther, or 276 m farther
    b. 6 times longer, or 12.8 s longer
45. Clears by 1.0 m
47. No
49. 3.7 m
51. 470 m/s^2
53. a. 42° west of north   b. 45 s
55. 69 m/s at 21° from the vertical
57. a. 40 m/s^2 b. 80 m/s^2
59. a. 0.97 m/s^2 b. $14g$
61. a. 420 m/s   b. 200 m/s
63. a. 12 m/s   b. 36 rev
65. a. -100 rad/s^2 b. 50 rev
67. 0.75 rad/s^2
69. a. $v = \sqrt{2\alpha \Delta \theta R}$ b. $a = 2\alpha \Delta \theta R$
71. 26 rad/s^2
73. 5500 rpm
75. b. 30 m west
77. 6.6×10^{12} m/s^2
79. 110 m/s
81. 4.8 m/s
83. 59 m away, 11 m lower
85. 10°

Chapter 5

1. Tension, gravity
3. Thrust, normal force, gravity, air resistance
5. Gravity, air resistance
7. a. 2.4 m/s^2 b. 0.60 m/s^2
9. 12/25
11. 3.7 s
13. a. 1 N   b. 2 N
15. 3.6 N

17. a. ≈ 0.05 N   b. ≈ 100 N
19.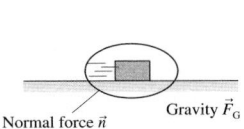

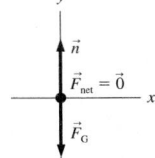

25.   **Force identification**   **Free-body diagram**

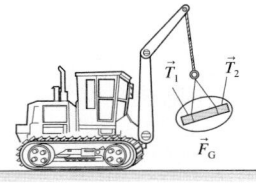

27.   **Force identification**   **Free-body diagram**

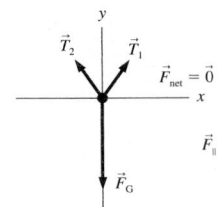

29.   **Motion diagrams**

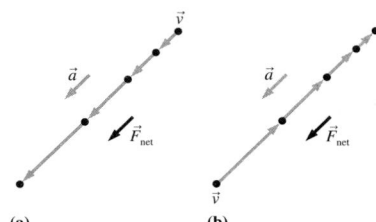

31.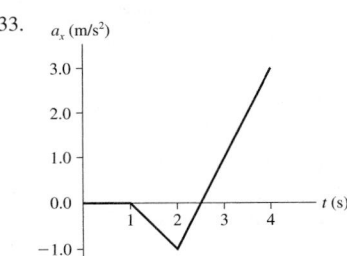

33.

35. a. 5.0 m/s^2 b. 30 m/s^2
    c. 10 m/s^2 d. 2.5 m/s^2

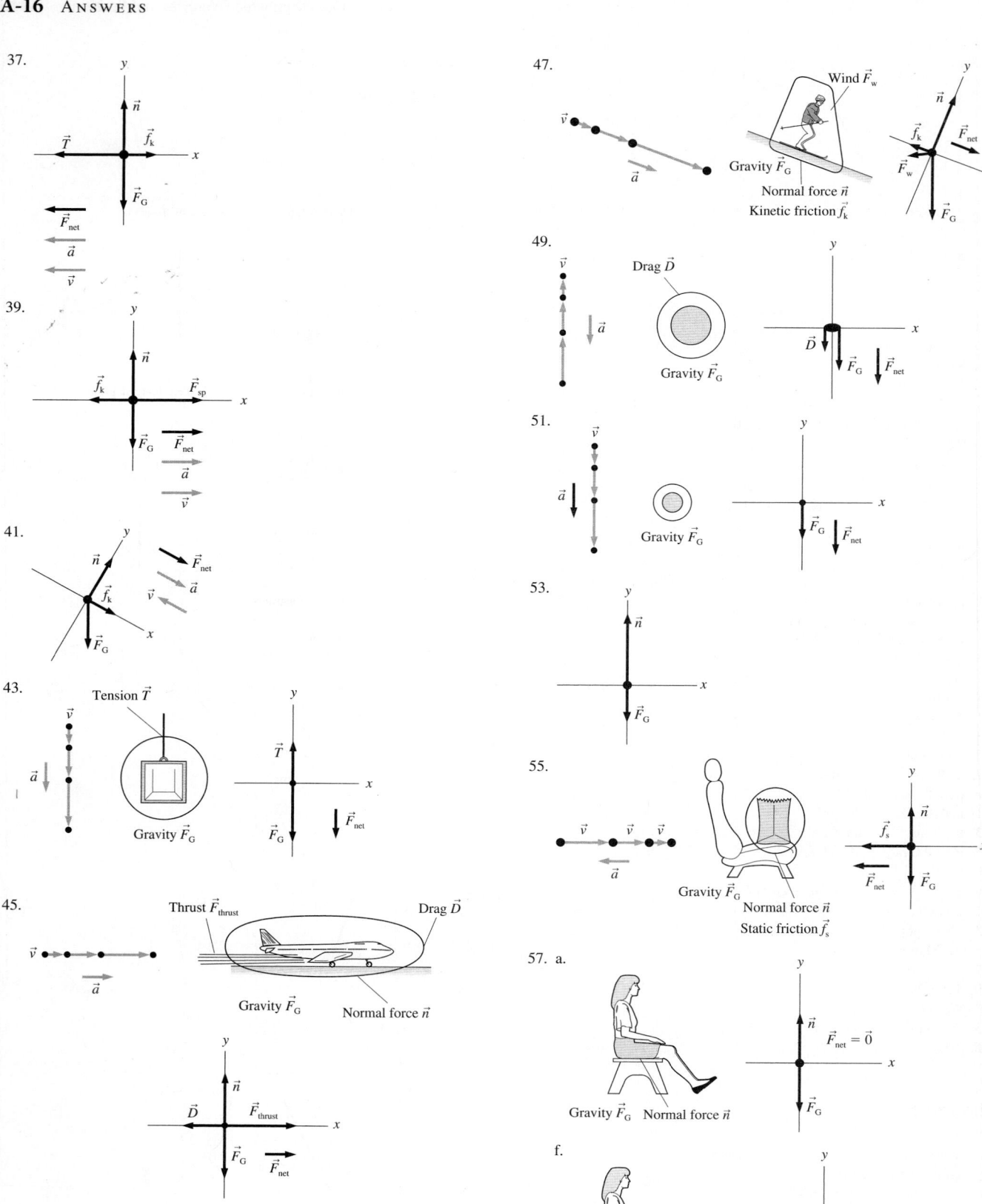

Chapter 6

1. $T_1 = 86.7$ N, $T_2 = 50.0$ N
3. 147 N
5. 800 N
7. a. $a_x = 1.5$ m/s^2, $a_y = 0$ m/s^2 b. $a_x = -0.28$ m/s^2, $a_y = 0$ m/s^2
9. 0 m/s^2, 4 m/s
11. a. 490 N   b. 490 N   c. 740 N   d. 240 N
13. a. 540 N   b. $m = 55$ kg; $mg = 210$ N
15. 1035 N, 740 N, 590 N
17. 0.250
19. a.–b.

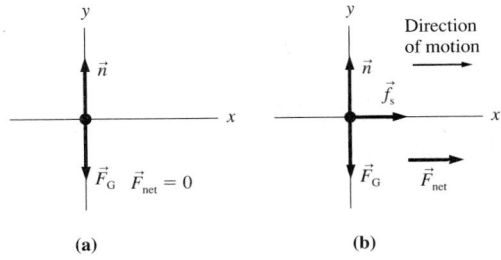

   c. 4.9 m/s^2
21. 10,000 N
23. 2.6 km
25. 69 g
27. 9 m/s
29. a. 0.0036 N   b. 0.010 N
31. a. Down   b. 77 kg   c. 39 m
33. a. 59 N   b. 67.8°   c. 79 N
35. a. −12 N   b. 12 N
37. 3.1 m
39. a. 6700 N   b. 600 μs
   c.

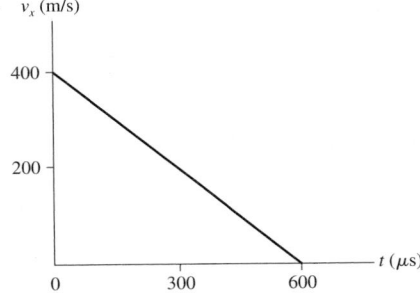

41. a. $\sqrt{2g(h - \mu_k \sqrt{L^2 - h^2})}$ b. 9.9 m/s
43. 0.165
45. 2.7 m/s
47. 190 kN
49. 0.12
51. a. $\dfrac{v_0^2}{2\mu_s g}$ b. 14 m
53. a. Yes   b. Yes
55. Stay at rest
57. 37 m/s
59. 3/8 in
61. a. 0 N   b. 220 N
63. a. $\dfrac{F_0}{m}\dfrac{T}{2}$ b. $\dfrac{F_0}{m}\dfrac{T^2}{3}$
65. a. 9.4×10^{-10} N, 5.7×10^{-13} N   b. 1.8 m/s^2, 130 m/s^2
67. b. 0.36 s
69. c. 102 m
71. c. 2.8 m/s^2

73. 13 m/s^2
75. a. $v_0 e^{-\frac{6\pi \eta R t}{m}}$ b. 61 s
77. b. 160 s, 480 s   b. No

Chapter 7

1. a.   **Interaction diagram**

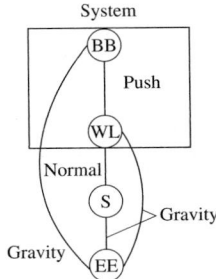

   BB = Barbells
   WL = Weight lifter
   S = Surface   EE = Entire Earth

b. The system is the weightlifter and barbell.

c.   **Free-body diagrams**

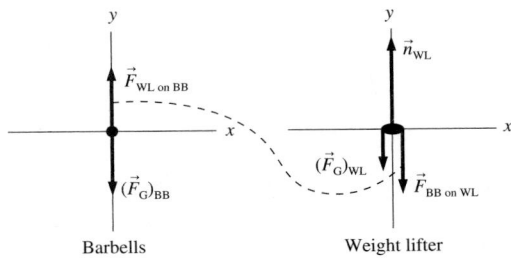

3. a.   **Interaction diagram**

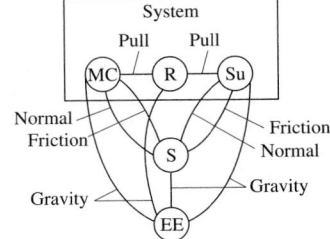

   MC = Mountain climber
   R = Rope   Su = Supply bag
   S = Surface
   EE = Entire Earth

b. The system consists of the mountain climber, rope, and bag of supplies.

c. **Free-body diagrams**

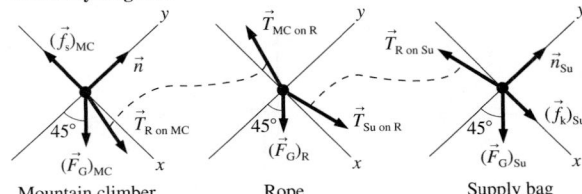

5. a. **Interaction diagram**

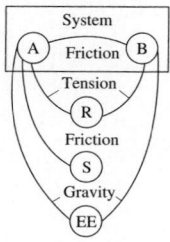

b. The system consists of the two blocks.

c. **Free-body diagrams**

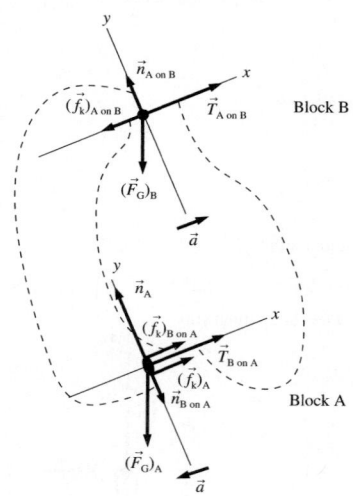

7. a. 7.8×10^2 N   b. 1.6×10^3 N
9. a. 3000 N   b. 3000 N
11. 5.0 kg
13. a. 32 N   b. 19 N   c. 16 N   d. 3.2 N
15. a. 20 N   b. 21 N
17. a. ~ 0.05 N   b. ~ 100 N
19. a. **Interaction diagram**

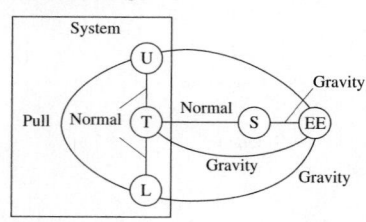

U = Upper magnet
L = Lower magnet
T = Table
S = Surface
EE = Entire Earth

Known

$(F_G)_T = 20.0$ N
$(F_G)_U = 2.0$ N
$(F_G)_L = 2.0$ N
$F_{U \text{ on } L} = 3(F_G)_L$

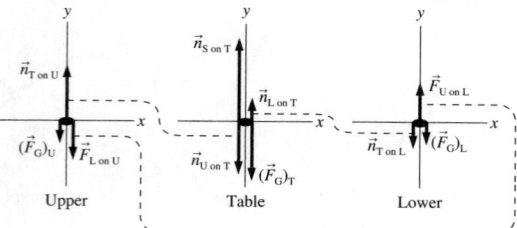

b. $(F_G)_U = 2.0$ N, $n_{T \text{ on } U} = 8.0$ N, $F_{L \text{ on } U} = 6.0$ N, $(F_G)_T = 20$ N, $n_{U \text{ on } T} = 8.0$ N, $n_{L \text{ on } T} = 4.0$ N, $n_{S \text{ on } T} = 24$ N, $n_{S \text{ on } T} = 24$ N, $n_{T \text{ on } L} = 4.0$ N, $F_{U \text{ on } L} = 6.0$ N
21. 60 N
23. 2.7×10^2 N
25. No
27. 99 m
29. a. 2.3×10^2 N   b. 0.20 m/s
31. 1.5 s
33. a. 3.9 N   b. 2.2 m/s^2
35. 1.8 m/s^2
37. 160 N
39. $T_1 = 100$ N, $T_2 = T_3 = T_5 = F = 50$ N, $T_4 = 150$ N
41. a. 1.8 kg   b. 1.3 m/s^2
43. a. 0.67 m   b. Slides back down
45. a. 8.2×10^3 N   b. 4.8×10^2 N
47. $a = 2T/m - \mu_r g$
49. $F = (m_1 + m_2)g \tan \theta$
53. a. 1.0 m/s   b. 90 N   c. 90 N
55. 3.3 m/s^2
57. a. $2a_{3y} + a_{2y} + a_{1y} = 0$ m/s^2
   c. $a_{1y} - 2.2$ m/s^2, $a_{2y} = 2.9$ m/s^2, $a_{3y} = -0.32$ m/s^2

Chapter 8

1. 39 m
3. a. 56 h   b. 0.092°   c. Yes
5. 6.8 kN
7. 6.6×10^{15} rev/s
9. 2.01×10^{20} N
11. 1.58 m/s^2
13. 22 m/s
15. 3
17. 30 rpm
19. 8.2 s
21. 4.5 m/s
23. 105 m
25. a. $y = \frac{1}{2}x$ b. Straight line   c. 1090 m
27. a. 165 m   b. Straight line
29. a. 24.0 h   b. 0.223 m/s^2 c. 0 N
31. 5.5 m/s
33. a. $\sqrt{\dfrac{g}{L \sin \theta}}$ b. 72 rpm
35. 4 m/s
37. a. 2.9 m/s   b. 14 N
39. Horizontal circle
41. $a_A = -9.8$ m/s^2, $a_B = -12.9$ m/s^2, $a_C = -6.7$ m/s^2
43. a. 320 N, 1400 N   b. 5.7 s
45. a. $\sqrt{\dfrac{g}{L}}$ b. 30 rpm
47. 0.38 N
49. a. $\sqrt{gL}$ b. 2.6 m/s, 5.9 mph
51. 1.4 m
53. 13 N
55. a. 6.6 rad/s   b. 44 N
57. b. 20 rad/s = 190 rpm
59. a. $\theta = \frac{1}{2}\tan^{-1}(mg/F)$ b. 11.5%
61. $\sqrt{gy}$
63. $T_1 = 14.2$ N, $T_2 = 8.3$ N
65. 37 km
67. $\dfrac{rv_0}{r + v_0 \mu_k t}$

Chapter 9

1. a. 4.5×10^4 kg m/s   b. 8.0 kg m/s
3. 4 N s
5. 1.5×10^3 N
7. 800 N in the $-x$-direction
9. 1.0 m/s to the left
11. a. 1.5×10^4 N s   b. 30 s, 110 m/s
13. 0.2 s
15. 5.0×10^2 kg
17. 4.8 m/s
19. 3.0×10^2 m/s
21. 3.6 m/s
23. $(2\hat{i} + 4\hat{j})$ kg m/s
25. 45° north of east at 1.7 m/s
27. a. 6.4 m/s   b. $F_{avg} = 3.6 \times 10^2$ N $= 612(F_G)_B$
29. 0.497 m
31. 7.5×10^{-10} kg m/s
33. 8.0×10^2 N
35. $v_f = \dfrac{m_1}{(m_1 + m_2)}\sqrt{\dfrac{2dF}{m_1}}$
37. 2.2×10^{-10}%
39. 32 m/s
41. $v_0/\sqrt{2}$, 45° east of north
43. a. $v_{bullet} = \dfrac{m + M}{m}\sqrt{2\mu_k g d}$ b. 4.4×10^2 m/s
45. 28 m/s
47. 20 m/s downward
49. 2.0 m/s
51. 8.6 m/s
53. $3v_0$
55. 16 m/s
57. 4.5 km
59. 14.0 u
61. a. -1.4×10^{-22} kg m/s
    b. and c. 1.4×10^{-22} kg m/s in the direction of the electron
63. 0.85 m/s, 72° below the $+x$-axis
65. 2.0×10^3 m/s
67. b. $(v_{ix})_2 = 6.0$ m/s
69. c. $(v_{fx})_1 = -12$ m/s
71. 1.5×10^7 m/s in the forward direction
73. 1.2 km/s

Chapter 10

1. The bullet
3. 112 km/h
5. a. 25.1 m   b. 10 m/s   c. 22 m/s
7. 2
9. 4.5 m/s
11. 4.2 m/s
13. 98 N/m
15. 31 cm
17. a. 49 N   b. 1450 N/m   c. 3.4 cm
19. 18 J
21. 4.0 cm
23. Go hungry
25. a. Right   b. 17.3 m/s   c. $x = 1.0$ m and 6.0 m
27. a. 7.7 m/s   b. 10 m/s
29. 0.86 m/s and 2.9 m/s
31. a. -5.0 m/s and 5.0 m/s   b. 2.5 m/s
33. a. 0.34 m
35. 81 m/s

37. a. 2.3 m/s   b. $v(\theta) = \sqrt{9 - 1.96(1 - \cos\theta)}$ m/s
39. 9.7 J
41. $v_0/\sqrt{2}$
43. a. 2.6 m/s   b. 33 cm
45. 2.0×10^5 N/m
47. a. 2.2×10^4 N/m   b. 18 m/s
49. a. 3.3 m/s   b. 11.8 cm   c. 0.83 m/s, 6.5 cm
51. a. $\sqrt{\left[2\left(\dfrac{m+M}{m}\right)^2 gd - \left(\dfrac{m+M}{m^2}\right)\dfrac{M^2 g^2}{k} + k\left(\dfrac{m+M}{m^2}\right)(d - Mg/k)^2\right]}$
    b. 450 m/s
53. $\frac{5}{2}R$
55. 7.9 m/s
57. 100 g ball: 0.80 m/s to the left; 400 g ball: 2.2 m/s to the right
59. a. $x_1 = \dfrac{\pi}{3}$ and $x_2 = \dfrac{2\pi}{3}$ b. $\dfrac{\pi}{3}$: unstable; $\dfrac{2\pi}{3}$: stable
61. a. Yes   b. 1.0 mJ/m
63. c. 15 m/s
65. c. 32 cm
67. 93 cm
69. 20 N/m
71. a. 1.5 m   b. 20 cm
73. a. 4.6 cm   b. $v_{2 kg} = 1.33$ m/s and $v_{1 kg} = 5.3$ m/s
75. 100 g ball rebounds to 79°; 200 g ball rebounds to 14.7°

Chapter 11

1. a. -18 b. 10
3. a. 125°   b. 67°
5. a. 11   b. -4.6 c. 0
7. a. -6.0 J   b. 12 J
9. 0 J
11. 1.25×10^4 J, -7.92×10^3 J, -4.58×10^3 J
13. AB $= 0$ J, BC $= 0$ J, CD $= -4$ J, DE $= 4$ J
15. 3.7 m/s, 6.6 m/s, 9.7 m/s
17. 8 N
19. -60 N at $x = 1$ m, 15 N at $x = 4$ m
21. 2.5 N, 0.40 N, 0.16 N
23. 1360 m/s
25.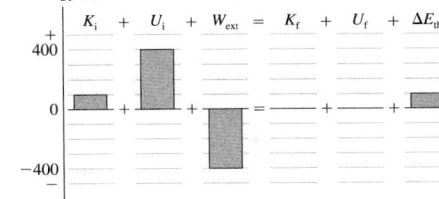
27. -1 J of work is done to the environment
29. a. 9.80×10^5 J   b. 1.96×10^4 W
31. 42 m²
33. 5.5×10^4 liters
37. a. 50 J   b. 50 J   c. 50 J, yes
39. a.   b. -2 J   c. 22 m/s

41. a.
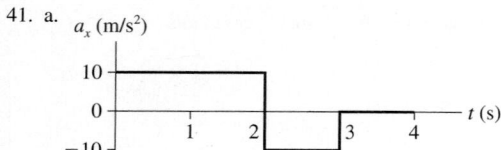

b. x (m)
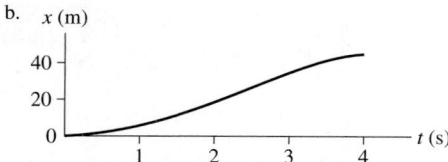

c. K (J)
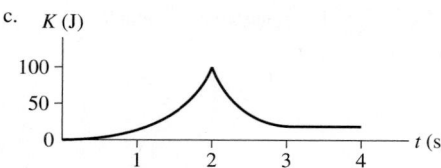

d. F (N)
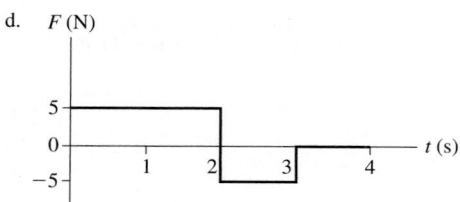

e. 10 N s, -5 N s   f. 20 m/s, 10 m/s

g. F_x (N)
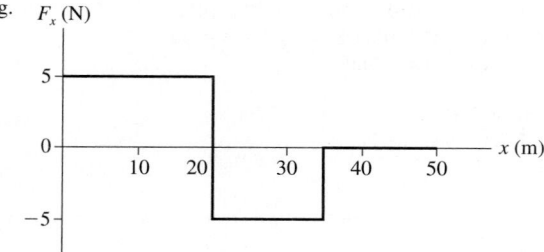

h. 100 J, -75 J   i. 20 m/s, 10 m/s
43. a. 2.3×10^2 J   b. 2.3×10^2 N   c. 6.8 kW
45. 16 m/s
47. 0.54 m
49. 0.12 km
51. a. $v_f = \sqrt{2gh(m - \mu_k M)/(M + m)}$ b. $v_f = \sqrt{2gmh/(M + m)}$
53. 10 m/s
55. a. 0.51 m   b. 0.38 m
57. a. 14 m/s   b. 32 m
61. a. N   b. m^{-1} c. $\pi/(2c)$

d. F_x
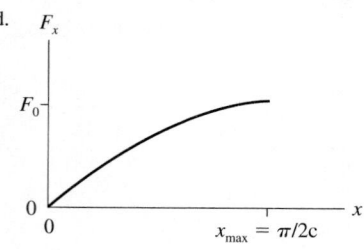

e. $\sqrt{v_0^2 + 2F_0/(mc)}$
63. a. 78 J   b. 1.3 W

65. a. -0.25 kJ   b. 2.6×10^5 kg
67. a. 6.53 m/s^2 b. 11/7 m/s   c. 1.79 s
69. c. 6.3 m/s
71. c. 3.2 kW
73. 6.7 m
75. 24 W

Chapter 12

1. 13.2 m/s
3. a. 1.5 m/s   b. 13 rev
5. 4.7×10^6 m
7. $x_{cm} = 6.7$, $y_{cm} = 5.0$
9. 2.57×10^{29} J
11. a. 0.032 kg m^2 b. 16 J
13. a. (5.7 cm, 4.6 cm)   b. 0.0066 kg m^2
15. a. (0.060 m, 0.040 m)   b. 0.0020 kg m^2 c. 0.0013 kg m^2
17. a. 3.8×10^{-5} kg m^2 b. 1.14×10^{-4} kg m^2
19. 4.3 N m
21. 12.5 kN m
23. 8.0 N m
25. 0.28 N m in the ccw direction
27. 8.0 rad/s
29. No
31. 1.5 m
33. 0.38 J
35. 43 cm
37. a. (21, into the page)   b. (24, out of the page)
39. a. $-\hat{j}$ b. $\vec{0}$
41. a. $n\hat{i}$ b. $2\hat{j}$ c. $1\hat{k}$
43. $1.20\hat{k}$ kg m^2/s
45. (2.1 kg m^2/s, out of the page)
47. 91 rpm
49. 7.5 cm
51. 28 m/s
53. a. 0.010 kg m^2 b. 0.030 kg m^2
55. $\dfrac{M}{3L}[(L - d)^3 + d^3]$
57. $\dfrac{1}{6}ML^2$
59. 0.91 m
61. $F_1 = 750$ N, $F_2 = 1000$ N
63. 1.0 m
65. 31 kg
67. a. 39 mN   b. 38 rpm
69. 1.1 s
71. 1.6 N
73. 4.3 m
75. a. $\sqrt{2g/R}$ b. $\sqrt{8gR}$
77. $\dfrac{20Tr}{13MR^2}$
79. 1.2 rad/s
81. 22 rpm
83. 4.0 rpm
85. Emily
87. 67°

Chapter 13

1. 6.00×10^{-4}
3. 2.18
5. 2.3×10^{-7} N

7. a. 274 m/s^2 b. 5.90 × 10^{-3} m/s^2
9. 2.4 km
11. a. 3.0 × 10^{24} kg   b. 0.89 m/s^2
13. 60.2 km/s
15. 4.21 × 10^4 m/s
17. 4.4 × 10^{11} m, 1.7 × 10^4 m/s
19. 1600 earth days
21. a. $T_2 = 250$ min, $T_3 = 459$ min   b. $F_2 = 20{,}000$ N, $F_3 = 4{,}440$ N
    c. 1.50
23. 4.2 h
25. 46 kg and 104 kg
27. 3.0 × 10$^{-7}\hat{\jmath}$ N
29. −1.96 × 10^{-7} J
31. 12 cm
33. a. 3.02 km/s   b. 3.13 km/s   c. 3.6%
35. 4.2 × 10^5 m
37. 33 km/s
39. 2.78 km/s
41. 3.0 × 10^4 m/s
43. 3.7 × 10^5 m/s
45. 6.7 × 10^8 J
47. a. 7.0 m/s   b. 12 m/s
49. 6.71 × 10^7 m
51. a. $y = \left(\dfrac{q}{p}\right)x + \dfrac{\log C}{p}$ b. Straight line
    c. q/p d. 1.996 × 10^{30} kg
53. a. 2.1 × 10^8 y   b. 24   c. 1.9 × 10^{41} kg   d. 9.4 × 10^{10}
55. 3.71 km/s
57. 4.49 km/s
59. c. 6.21 × 10^7 m
61. c. 1680 m/s
63. 1.50 × 10^9 m
65. Crash
67. 11.8%
69. a. $-\dfrac{GMm}{L}\ln\left(\dfrac{x + L/2}{x - L/2}\right)$ b. $-GMm\left(\dfrac{4}{4x^2 - L^2}\right), x \geq \dfrac{L}{2}$

Chapter 14

1. 2.27 ms
3. a. 13 cm   b. 9.0 cm
5. a. 10 cm   b. 0.50 Hz   c. +120°
9. $x(t) = (8.0\text{ cm})\cos[(\pi\text{ rad/s})t - \pi\text{ rad}]$
11. a. 2.8 s b. 1.4 s   c. 2.0 s   d. 1.4 s
13. a. 0.50 s   b. 4π rad/s   c. 5.5 cm   d. 0.45 rad
    e. 70 cm/s   f. 8.8 m/s^2 g. 0.049 J   h. 3.8 cm
15. a. 10 cm   b. 35 cm/s
17. a. 0.17 kg   b. 0.57 m/s
19. a. 4.0 s   b. 5.7 s   c. 2.8 s   d. 4.0 s
21. a. 2.0 s   b. 2.1 s
23. 3.67 m/s^2
25. 0.079 N/m
27.

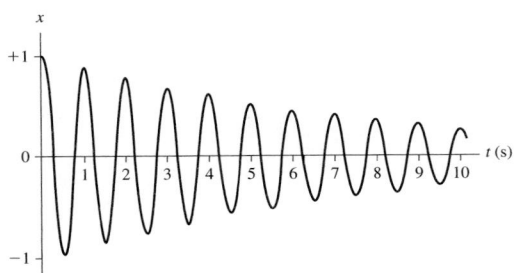

29. 250 N/m
31. a. $\frac{2}{3}\pi$ rad   b. −13.6 cm/s   c. 15.7 cm/s
33. 0.41 s
35. a. 55 kg   b. 0.73 m/s
37. a. $\dfrac{3}{4}$ b. $\dfrac{A}{\sqrt{2}}$
39. a. 6.4 cm   b. 160 cm/s^2 c. −6.4 cm   d. 28 cm/s
41. 1.02 m/s
43. a. 3.2 Hz   b. 7.1 cm 5.0 J
45. a. 1.1 Hz   b. 23 cm   c. 4.1 cm below equilibrium point
47. 1.7 Hz
49. 0.72
51. a. 7.5 m   b. 0.45 m/s
53. 0.65 m/s
55. 0.66 s
57. $\dfrac{1}{2\pi}\sqrt{\dfrac{5\,g}{7\,R}}$
59. 8.7 × 10^{-2} kg m^2
61. a. 2.0 Hz   b. 1.2 cm
63. 7.9 × 10^{13} Hz
65. a. Highest point   b. 2.5 Hz
67. a. 9.5 N/m   b. 0.010 kg/s
69. 25 s
71. 236 oscillations
75. 1.6 Hz
77. 1.8 Hz
79. a. $\Delta T = \dfrac{T}{2}\dfrac{\Delta m}{m}$ b. 2.001s

Chapter 15

1. 50 mL
3. 1.4 × 10^5 kg
5. 1.1 × 10^3 atmospheres
7. a. 6.3 m^3 b. 1.2 × 10^5 Pa
9. 3.2 km
11. 10.3 m
13. 3.5 cm
15. 6.7 × 10^2 kg/m^3
17. 44 N
19. 8.4 cm
21. 56 kg
23. a. 1.0 m/s, 16 m/s   b. 3.1 × 10^{-4} m^3/s
25. 110 kPa
27. 5.5 × 10^9 N/m^2
29. 1 mm
31. 0.20%
33. a. 5.8 kN   b. 6.0 kN
35. 27 psi
37. 5.27 × 10^{18} kg
39. a. 106 kPa   b. 4.4 kPa 4.4 kPa
41. 55 cm
43. 7.5 cm
45. a. $F = \rho gDWL$ b. $F = \frac{1}{2}\rho gD^2L$ c. 0.78 kN, 1.4 kN
47. 8.01%
49. a. $\rho_{\text{liq}}Agx$ b. 0.62 J
51. 8.9 × 10^2 kg/m^3
53. 18 cm
55. 5.2 cm
57. 3.5 m/s
59. 187 nm/s
61. 28 cm
63. 4.4 cm

65. a. $v = \sqrt{2g(h-y)}$ b. $x = v\sqrt{2y/g}$ c. $y = h/2, x_{max} = h/2,$
67. 1 mm
69. 1 L
71. $\dfrac{h}{l} = \left(1 - \dfrac{\rho_0}{\rho_f}\right)^{1/3}$
73. b. $(F_{net})_y = -\rho_f A g y$ c. $\rho_f A g$ e. 18.9 s

Chapter 16

1. 1900 cm^3
3. 8.0 cm
5. 4.8×10^{23} atoms
7. a. 6.02×10^{28} atoms/m^3 b. 3.28×10^{28} atoms/m^3
9. 6.8 cm^3
11. $-127°F = 88°C = 185$ K; $136°F = 58°C = 331$ K
13. a. 171°Z   b. 671°C = 944 K
15. a. 2   b. Unchanged
17. a. $1.27V_0$ b. $2V_0$
19. 2.4×10^{22} molecules
21. 7.4 kg/m^3
23. a. $V_2 = V_1$ b. $T_2 = \dfrac{T_1}{3}$
25. 2.6 atm
27. a. 9500 kPa
    b.
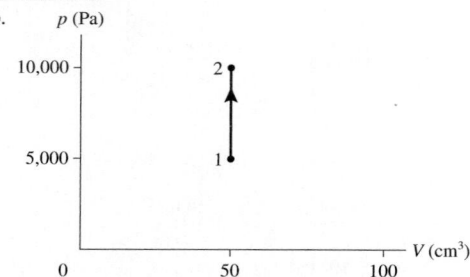
29. a. 48 atm
31. a. Isobaric   b. 118°C   c. 9.35×10^{-3} mol
33. 0.228 nm
35. 3.3×10^{26} protons
37. 1.1×10^{15} particles/m^3
39. 380 K = 107°C
41. 1.8 g
43. $\dfrac{3}{2}T_0$
45. 2.4 m
47. 35 psi
49. 155 cm^3
51. 24 cm
53. No
55.
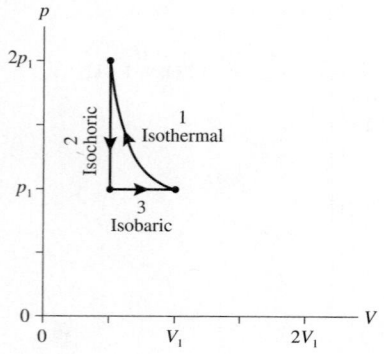

57. a. 880 kPa   b. $T_2 = 323°C, T_3 = -49°C, T_4 = 398°C$
59. a. $T_1 = 122$ K, $T_2 = 366$ K   b. Isobaric   c. 3 atm
61. 2364°C
63. a. 4.0 atm, $-73°C$
    b.
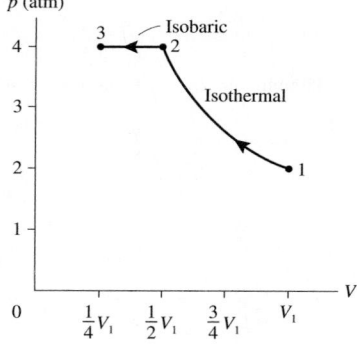
65.   b.
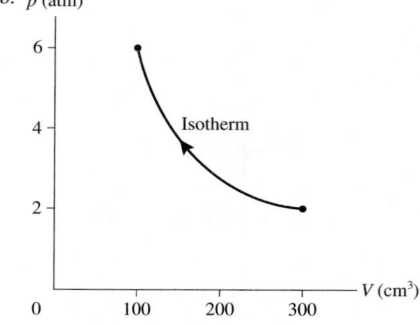
    c. 6 atm
67.   b.
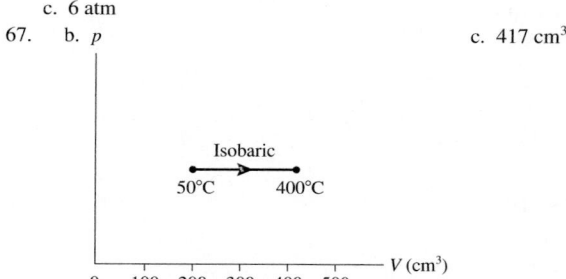
    c. 417 cm^3
69. a. 23 cm   b. 7.5 cm
71. 93 cm^3
73. a. 4.0×10^5 Pa   b. Irreversible

Chapter 17

1. 60 J
3. 200 cm^3
5.
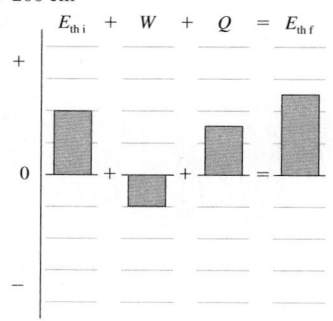

7.

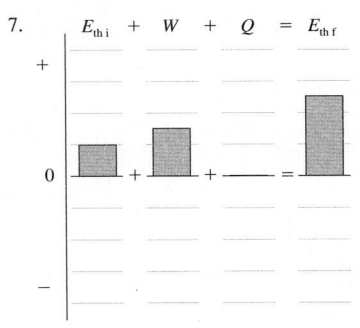

9. 60 J
11. 52 kJ
13. a. 36°C   b. 3000 J
15. 0.98 g
17. 272°C, 522°F
19. Iron
21. a. 31 J   b. 60°C
23. 2.5 kJ
25. a. 1.9×10^{-3} m³   b. 74°C
27. 16 kW
29. 230 W
31. 16 kJ
33. 15 m
35. 6.6 h
37. 12 J/s
39. −56°C
41. Aluminum
43. 650 J/kg K
45. a. 2.0 kJ/kg K   b. 2.7 kJ/kg K   c. −20°C, 40°C
    d. 4.0×10^4 J/kg, 1.2×10^5 J/kg
47. 2.4×10^6 L
49. a. 5.5 kJ   b. 3.4 kJ
    c. *p* (atm)

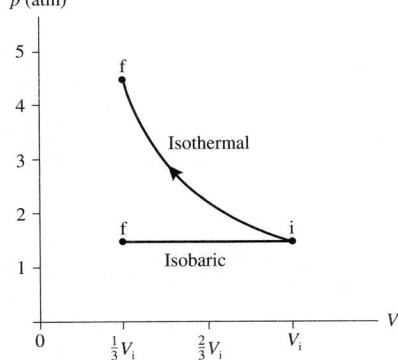

51. a. 350 Pa   b. 4.9×10^{20} c. 110°C   d. 26 cm   e. −0.57 J
53. a. 3.1 atm   b. 9.7 L
    c. *p* (atm)

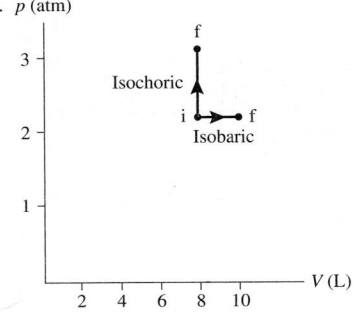

55. a. T_1 b. $-nRT_1 \ln 2$ c. $nRT_1 \ln 2$
59. −330 J, 0 J
61. $\gamma = 1.29$
63. a. 0.15 kJ   b. −91 J
65. a.

Point	*p* (atm)	*T* (°C)	*V* (cm³)
1	1.0	133	1000
2	5.0	1757	1000
3	1.0	1757	5000

    b. $W_{1\to2} = 0$, $W_{2\to3} = -815$ J, $W_{3\to1} = 405$ J   c. $Q_{1\to2} = 609$ J,
    $Q_{2\to3} = 815$ J, $Q_{3\to1} = -1.01$ kJ
67. 28°C
69. a. 5.5 kK   b. 0 J   c. 54 kJ   d. 20
    e. *p* (atm)

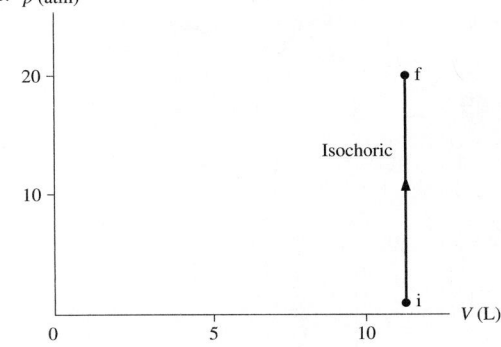

71. 110°C
73. −18°C
75. b. 217°C
77. a.

Point	*p* (atm)	*T* (°C)	*V* (cm³)
1	3.0	946	1000
2	1.0	946	3000
3	0.48	310	3000

    b. −334 J, 0 J, 239 J   c. 334 J, −239 J, 0 J
79. 15 atm
81. 150 J

Chapter 18

1. 5.5×10^{24}
3. 0.023 Pa
5. a. 300 nm   b. 600 nm
7. 13 cm
9. a. $(0\hat{i} + 0\hat{j})$ b. 57 m/s   c. 60 m/s
11. a. 289 K   b. 200 kPa
13. 6.5×10^{25} s⁻¹
15. 283 m/s
17. −246°C
19. 300 m/s
21. 0.43 cm/s
23. a. Doubles   b. $\sqrt{2}$ c. Same
25. a. 4.1×10^{-16} J   b. 7.0×10^5 m/s
27. 580 m/s
29. 3.6×10^7 J
31. 93 kJ
33. a. 3.80×10^5 J   b. 2.25×10^{-9} m   c. 0 J
35. 5000 J
37. 61
39. a. Helium   b. 1370 m/s   c. 1.86 μm

41. a. 4×10^{-22} atm   b. 270 m/s   c. 2.5×10^5 m
43. 1.004
45. 1.9×10^4 Pa
47. 29 J/mol K
49. a. $(E_{He})_i = 1900$ J, $(E_O)_i = 3100$ J
    b. $(E_{He})_f = 2700$ J, $(E_O)_f = 2300$ J
    c. 850 J from oxygen to helium   d. 436 K
51. 7
55. a. Increase factor of 2   b. Increase by factor of 4
    c. Increase by factor of 4   d. Same
57. a. 4   b. 1   c. 16
59. a. 141,000 K   b. 10,100 K
61. a. 2.0×10^6 J   b. 4.8×10^{-6} c. 0.0013 K
63. $\dfrac{15n + 3}{2} p_i V_i$
65. c. 436 K; 850 J is transferred from oxygen to helium

Chapter 19

1. a. 250 J   b. 150 J
3. a. 0.27   b. 15 kJ
5. a. 200 J   b. 250 J
7. 96,000
9.

	ΔE_{th}	W_s	Q
A	+	0	+
B	−	+	0
C	0	+	+
D	−	−	−

11. 40 J
13. a. 30 J, 0.15 kJ   b. 0.21
15. 285 J
17. 0.24
19. a. (b)   b. (a)
21. 7°C
23. a. 25%   b. 232°C
25. 135°C
27. 1.7
29. a. 60 J   b. −23°C
31. 1.7 MJ
37. 8.3%
39. 47°C
41. 218
43. 8.57 J
45. No
47. a. 48 m   b. 32%
49. 37%
51. a. 5.0 kW   b. 1.7
53. a.

	W_s (J)	Q (J)	ΔE_{th}
$1 \to 2$	3.04	16.97	13.93
$2 \to 3$	0	−10.13	−10.13
$3 \to 1$	−1.52	−5.32	−3.80
Net	1.52	1.52	0

b. 9.0%   c. 13 W

55. a.

	W_s (kJ)	Q (kJ)	ΔE_{th} (kJ)
$1 \to 2$	0.991	2.476	1.486
$2 \to 3$	0	−1.693	−1.693
$3 \to 1$	−0.207	0	0.207
Net	0.783	0.783	0

b. 0.32
57. a. $p_1 = 100$ kPa $V_1 = 2690$ cm^3 $T_1 = 269$ K
    b.

	ΔE_{th} (J)	W_s (J)	Q (J)
$1 \to 2$	327	−327	0
$2 \to 3$	0	553	553
$3 \to 1$	−327	−131	−458
Net	0	95	95

c. 17%
59. a.

	p (atm)	T (K)	V (cm^3)
1	1.0	406	1000
2	5.0	2030	1000
3	1.0	2030	5000

b. 29%   c. 80%
61. a. $T_1 = 1620$ K $T_2 = 2407$ K $T_3 = 6479$ K
    b.

	ΔE_{th} (J)	W_s (J)	Q (J)
$1 \to 2$	327	−327	0
$2 \to 3$	1692	677	2369
$3 \to 1$	−2019	0	−2019
Net	0	350	350

c. 15%
63. a. $W_{net} = 350$ J   b. $\eta = 0.24$
65. b. 1.1×10^3°C
67. b. $Q_C = 80$ J
69. b. 10 J   c. 0.13

Chapter 20

1. 110 N
3. 2.0 m
5.

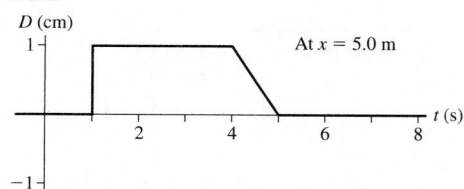

7.

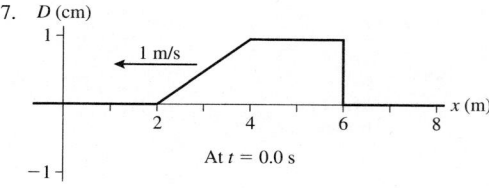

9. Equilibrium
t = 0 s

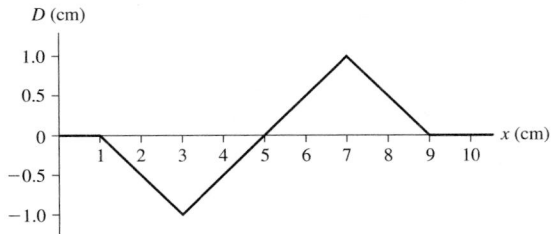

11. a. 3.1 rad/m   b. 9.5 m/s
13. a. 11 Hz   b. 1.1 m   c. 13 m/s
15. $\frac{\pi}{2}$ rad, $\frac{3}{2}\pi$ rad
17. 2.5 m
19. 1500 m/s
21. a. 1.5 GHz   b. 990 nm
23. a. 2.96 m   b. 116 Hz
25. a. 1.5×10^{-11} s   b. 3.4 mm
27. a. 1.88×10^8 m/s   b. 4.48×10^{14} Hz
29. 6.0×10^5 J
31. 110 dB
33. a. 65 dB   b. 105 dB
35. 5.0 W
37. a. 650 Hz   b. 560 Hz
39. 38.1 m/s
41. a. 0.80 m   b. $\frac{1}{2}\pi$ rad
   c. $D(x, t) = (2.0 \text{ mm})\sin(2.5\pi x - 10\pi t + \frac{1}{2}\pi)$
43. $\frac{v_0}{2}$

45.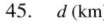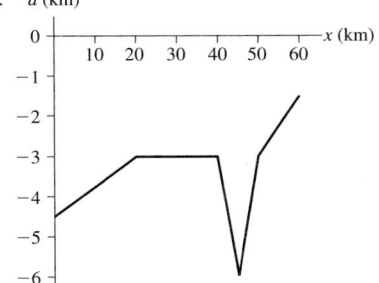

47. 410 ms
49. a. 440 Hz   b. 3.4 m
51. a. $-y$-direction   b. y-axis   c. 0.701 m, 350 m/s, 2.00 ms
53. a. 12.6 N   b. 2.00 cm   c. 12.8 m/s
55. $D(x, t) = (0.010 \text{ mm})\sin[(\pi \text{ rad/m})x - (400\pi \text{ rad/s})t + \frac{1}{2}\pi \text{ rad}]$
57. -19 m/s, 0 m/s, 19 m/s
59. 8
61. 9.4 m/s
63. a. 0.095 W/m^2 b. 1.6 MW/m^2
65. a. 6.67×10^4 W   b. $8.5 \times 10^{10} \text{ W/m}^2$
67. 50 m
69. 1.3
71. 21 min
75. Receding at 1.5×10^6 m/s
77. 0.07°C
81. 29 s

Chapter 21

1.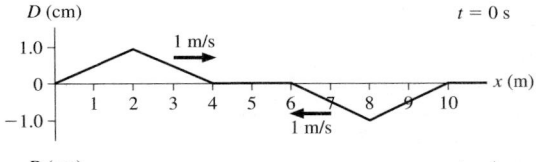
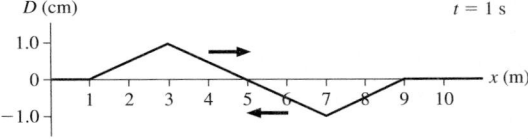
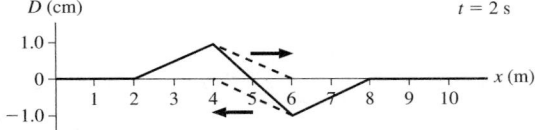
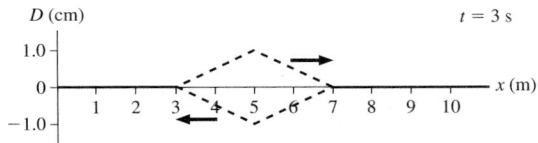
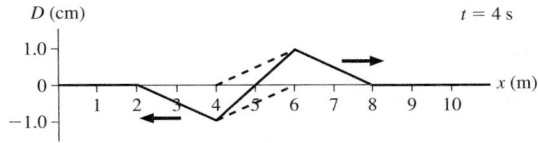
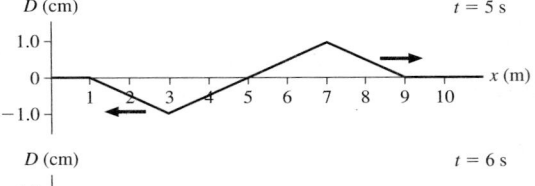
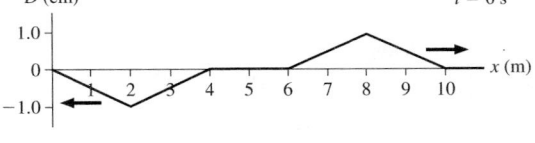

3.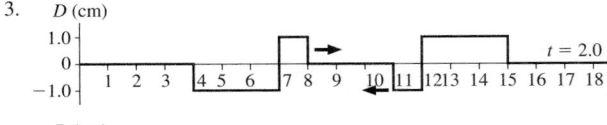
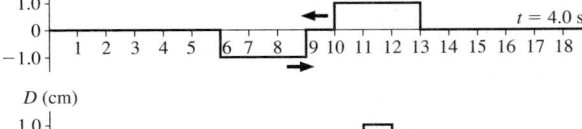
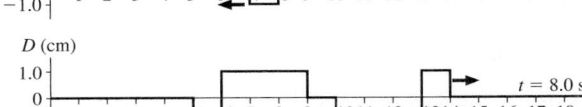

5. *D* (cm)

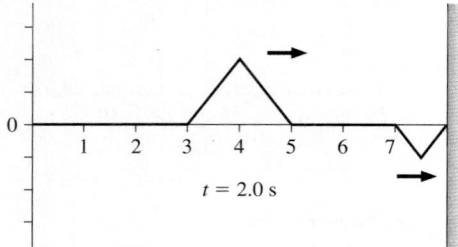

$t = 2.0$ s

D (cm)

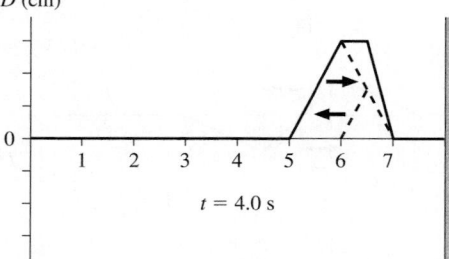

$t = 4.0$ s

D (cm)

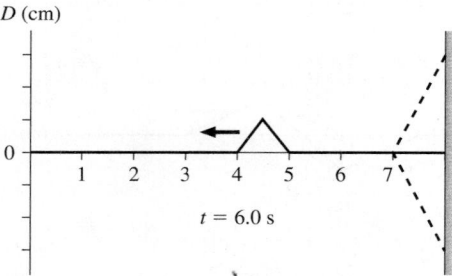

$t = 6.0$ s

D (cm)

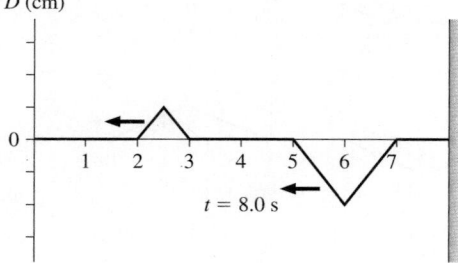

$t = 8.0$ s

7. 50 Hz
9. a. 4.8 m, 2.4 m, 1.6 m   b. 75 Hz
11. 12 kg
13. a. 2.42 m, 1.21 m, 0.807 m   b. 4.84 m, 1.61 m, 0.968 m
15. 512 Hz
17. 2180 N
19. a. 80 cm   b. 100 cm
21. 216 nm
23. a. In phase
    b.

	r_1	r_2	Δr	C/D
P	3λ	4λ	λ	C
Q	$\frac{7}{2}\lambda$	2λ	$\frac{3}{2}\lambda$	D
R	$\frac{5}{2}\lambda$	$\frac{7}{2}\lambda$	λ	C

25. Perfect destructive
27. 203 Hz
29. 1.26 cm

31. $A(x = 10$ cm$) = 0.62$ cm, $A(x = 20$ cm$) = 1.18$ cm,
    $A(x = 30$ cm$) = 1.62$ cm, $A(x = 40$ cm$) = 1.90$ cm,
    $A(x = 50$ cm$) = 2.00$ cm
33. 1.4 cm
35. 180 Hz
37. 28.4 cm
39. 18 cm
41. 140 N/m
43. 6.1 cm
45. $9\mu_0/4$
47. 13.0 cm
49. 580 Hz, 4.9 kHz
51. 12.1 kHz
53. 450 N
55. 93 m
57. 7.9 cm
59. a. 850 Hz   b. $-\pi/2$ rad
61. 7.2 cm
63. 20
65. 170 Hz
67. 1/3
69. a. a b. 1.0 m   c. 9
71. a. 5   b. 4.6 mm
73. 7.0 m/s
75. 4.0 cm, 35 cm, 65 cm
77. 2.0 kg
79. a. $\lambda_1 = 20.0$ m, $\lambda_2 = 10.0$ m, $\lambda_3 = 6.67$ m
    b. $v_1 = 5.59$ m/s, $v_2 = 3.95$ m/s, $v_3 = 3.22$ m/s
    d. $T_1 = 3.58$ s, $T_2 = 2.53$ s, $T_3 = 2.07$ s

Chapter 22

1. 0.023 rad = 1.3°
3. 1000 nm
5. 0.36 mm
7. 0.286°
9. 1.6°, 3.2°
11. 530
13. 7.9 μm
15. 0.20 mm
17. 0.50 mm
19. 4.0 mm
21. 7.6 m
23. 0.015 rad = 0.87°
25. 0.25 mm
27. 400 nm
29. 0.2895 mm
31. a. Single slit   b. 0.15 mm
33. 1.67 m
35. 3 mW/m²
37. 12.0 μm
39. 667.8 nm
41. 25 cm
43. 3
45. a. 1230 lines/mm   b. 46.5°
47. 670 lines/mm
49. 16°
51. 800 lines/mm
53. a. 2   b. 1.15   c. 1
55. 670 nm
57. 0.12 mm
59. a. 550 nm   b. 0.40 mm
61. 50 cm

63. a. 22.3°   b. 16.6°
65. 19
67. a. Dark   b. 1.597
69. a. No   b. 0.044°   c. 4.6 mm   d. 1.5 m
71. b. 0.022°, 0.058°
73. b. −11.5°, −53.1°
75. a. 0.52 mm   b. 0.074°   c. 1.3 m

Chapter 23

1. a. 3.3 ns   b. 75 cm, 67 cm, 46 cm
3. 0.40 ns
5. 30°
7. 6.1 m
9. 433 cm
11. 16°
13. 1.39
15. 76.7°
17. 3.2 cm
19. 1.52
21. 1.48
23. 1600 nm
25. 6.0 cm behind the lens, inverted
27. 7.5 cm in front of the lens, upright
29. 68 cm
31. 200 cm
33. 36 cm
35. 40 cm in front of mirror, inverted
37. 12 cm behind mirror, upright
39. a. 3   b. B(+1.0 m, −2.0 m), C(−1.0 m, +2.0 m), D(+1.0 m, +2.0 m)
41. 10 m
43. 1.7
45. a. 87 cm   b. 65 cm   c. 43 cm
47. 4.0 m
49. a. Total internal reflection   b. Refraction at 72°   c. 18 cm
51. 1.58
53. 1.0°
55. 2.00
57. b. −15 cm, 1.5 cm, agree
59. b. 50 cm, 0.67 cm, agree
61. b. −20 cm, 0.33 cm, agree
63. 15.1 cm
65. −15 cm, 0.75 cm, behind, upright
67. Concave, 3.6 cm
69. 67 cm, 1.0 m
71. a. 5.9 cm   b. 6.0 cm
73. 16 cm
75. 13 cm
79. a. $t = \dfrac{n_1}{c}\sqrt{x^2 + a^2} + \dfrac{n_2}{c}\sqrt{(w-x)^2 + b^2}$

   b. $0 = \dfrac{n_1 x}{c\sqrt{x^2 + a^2}} - \dfrac{n_2(w-x)}{c\sqrt{(w-x)^2 + b^2}}$
81. b. 1.574

Chapter 24

1. b. $s_2' = 49$ cm, $h_2' = 4.6$ cm
3. b. $s_2' = 30$ cm, $h_2' = 6.0$ cm
5. b. $s_2' = -3.33$ cm, $h_2' = 0.66$ cm
7. 5.0
9. 3.0 mm
11. 6.0 mm
13. a. Myopia   b. 100 cm

15. 6.3 cm
17. 5.0 cm
19. 6.0 mm
21. a. 8.0 cm   b. 1.2 cm
23. Upright image, 1.0 cm tall, 6.4 cm to left of the second lens
25. a. Both images 2.0 cm tall; one upright 10 cm left of lens, the other inverted 20 cm to right of lens.
27. a. $f_2 + f_1$ b. $\dfrac{f_2}{|f_1|} w_1$
29. 16 cm placed 80 cm from screen
31. 23 cm
33. 5.0 cm
35. a. +3.0 D as objective   b. −1.5   c. 0.56 m
37. 4.6 mm
39. 15 km
41. a. 3.8 cm   b. Sun is too bright
43. 3.5 m
45. b. $\Delta n_2 = \dfrac{1}{2}\Delta n_1$ c. Crown converging, flint diverging   d. 4.18 cm

Chapter 25

1. a. Electrons added   b. 7.5×10^{10}
3. 2.5×10^{10}
5. 1.9×10^5
9. Right negatively charged, left positively charged
13. a. 0.056 N   b. 2.9
15. a. 58 N   b. 4.7×10^{-35} N   c. 1.2×10^{36}
17. $-(4.1 \times 10^{-4}\,\text{N})\hat{\jmath}$
19. a. 1.3×10^{14} m/s² toward bead   b. 2.4×10^{17} m/s² away from bead
21. a. $(6.4\hat{\imath} + 1.6\hat{\jmath}) \times 10^{-17}$ N
   b. $-(6.4\hat{\imath} + 1.6\hat{\jmath}) \times 10^{-17}$ N   c. 4.0×10^{10} m/s²   d. 7.3×10^{13} m/s²
23. $-4.5 \times 10^4\,\hat{r}$ N/C (i.e., toward the bead)
25. 3.3×10^6 N/C, downward
27. $-6.8 \times 10^4 \hat{\imath}$ N/C, $3.0 \times 10^4 \hat{\imath}$ N/C, $(8.1 \times 10^3\,\hat{\imath} - 3.9 \times 10^4 \hat{\jmath})$ N/C
29. a. 0.36 m/s² toward glass bead   b. 0.18 m/s² toward plastic bead
31. 82 nC
33. 3.1×10^{-4} N, upward
35. 4.3×10^{-3} N, 253° ccw
37. 2.0×10^{-4} N, 45° cw
39. $-1.0 \times 10^{-3}\hat{\imath}$ N
41. $(1.02 \times 10^{-5}\hat{\imath} + 2.2 \times 10^{-5}\hat{\jmath})$ N
43. 0.68 nC
45. $(F_{\text{net}})_x = \dfrac{-2KQqa}{(a^2 + y^2)^{3/2}}$
47. $(2 - \sqrt{2})\dfrac{KQq}{L^2}$
49. $-\dfrac{4}{9}q$, $x = \dfrac{1}{3}L$
51. 6.6×10^{15} rev/s
53. a. 2.3×10^{-6} b. 4.3×10^7 N/C, upward
55. 33 nC
57. a. 1.1×10^{18} m/s²   b. 1.0×10^{-12} N   c. 6.3×10^6 N/C   d. 69 nC
59. 0.75 μC
61. 1.8×10^5 N/C, 60° ccw from the +x-axis; 1.8×10^5 N/C, 60° cw from the −x-axis
63. a. (4.0 cm, 1.0 cm)   b. (0.0 cm, 2.0 cm)   c. (−2.0 cm, −2.0 cm)
65. a. $\vec{E}_1 = (8.5\hat{\imath} - 2.8\hat{\jmath})$ kN/C, $\vec{E}_2 = 10\,\hat{\imath}$ kN/C,
   $\vec{E}_3 = (8.5\hat{\imath} + 2.8\hat{\jmath})$ kN/C   c. $27\hat{\imath}$ kN/C
67. 14°
69. b. 22 nC
71. b. 5.1 nC

73. 0.11 μC
75. 1.7×10^{-4} N

Chapter 26

1. 7.6×10^3 N/C along the $+x$-axis
3. 1.0×10^4 N/C at 11° below the $+x$-axis
5. a. 36 N/C b. 18 N/C
7. 4000 N/C
9. 1.3×10^5 N/C, 0.0 N/C, 1.3×10^5 N/C
11. a. 2.6×10^4 N/C, left b. 2.6×10^{-5} N, right
13. a. 7.6×10^4 N/C, left b. 7.6×10^{-5} N, right
15. 27 nC
17. 1.9 cm
19. 2.7×10^{11}
21. a. 3.6×10^6 N/C. b. 8.3×10^5 m/s
23. 18 cm
25. 3.1×10^{-21} N m
27. 9.0×10^{-13} N$\vec{p}$
29. a. $(-9.7 \times 10^4 \hat{i} + 9.2 \times 10^4 \hat{j})$ N/C
    b. 1.34×10^5 N/C, 136°ccw from the $+x$-axis
31. $\dfrac{1}{4\pi\epsilon_0 L^2} Q (\sqrt{2} - 1)(\hat{i} + \hat{j})$
33. a. $\dfrac{2qx}{4\pi\epsilon_0(x^2 + s^2/4)^{3/2}}$
    b. 0 N/C, 768,000 N/C, 576,000 N/C, 358,000 N/C, 158,000 N/C
35. a. $\dfrac{2q}{4\pi\epsilon_0}\left[\dfrac{1}{x^2} - \dfrac{x}{(x^2 + d^2)^{3/2}}\right]\hat{i}$
37. $\dfrac{1}{4\pi\epsilon_0} \dfrac{8\lambda d}{4y^2 + d^2}$
39. -0.056 nC
41. $\dfrac{Q}{4\pi\epsilon_0} \dfrac{1}{x\sqrt{x^2 + L^2}}\hat{i} - \dfrac{Q}{4\pi\epsilon_0 Lx}\left(1 - \dfrac{x}{\sqrt{x^2 + L^2}}\right)\hat{j}$
43. a. $\dfrac{R}{\sqrt{2}}$ b. $\dfrac{2}{3\sqrt{3}} \dfrac{Q}{4\pi\epsilon_0 R^2}$
45. c. $\dfrac{1}{4\pi\epsilon_0} \dfrac{2Q}{\pi R^2}(\hat{i} + \hat{j})$
47. 1.41×10^5 N/C
49. 2.2 mm
51. 1.19×10^7 m/s
53. a. $\dfrac{\frac{4}{3}\pi r^3 \rho g + qE}{6\pi\eta r}$ b. 0.067 mm/s c. 0.049 mm/s
55. 6.56×10^{15} Hz
57. a. $\dfrac{\text{C}^2\,\text{s}^2}{\text{kg}}$ b. $\left(\dfrac{1}{4\pi\epsilon_0}\right)^2 \dfrac{2q^2\alpha}{r^5}$, toward ion
59. b. 1.0 mm
61. b. $\dfrac{R}{\sqrt{3}}$
63. 4.2×10^{-4} N
65. a.

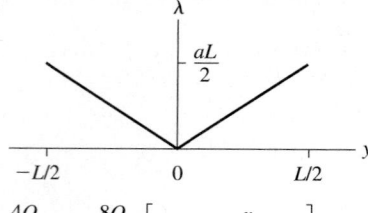

b. $\dfrac{4Q}{L^2}$ c. $\dfrac{8Q}{4\pi\epsilon_0 L^2}\left[1 - \dfrac{x}{\sqrt{x^2 + L^2/4}}\right]$

67. a. $\dfrac{2\eta}{4\pi\epsilon_0} \ln\left(\dfrac{2x + L}{2x - L}\right)\hat{i}$

c.

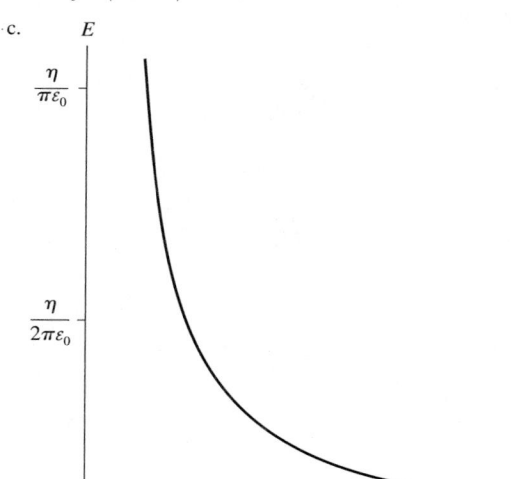

69. -2.3 nC/m
71. a. $k = \dfrac{qQ}{4\pi\epsilon_0 R^3}$ c. 2.0×10^{12} Hz

Chapter 27

1.

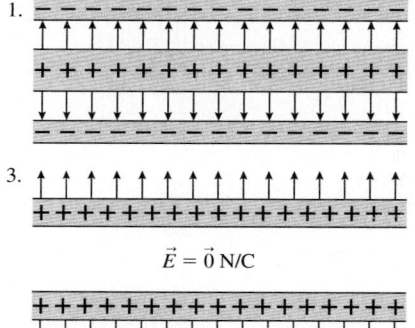

3.

$\vec{E} = \vec{0}$ N/C

5. No charge
7. Into the front face of the cube; field strength must exceed 5 N/C
9. 1.0 N m^2/C
11. 1.4×10^3 N/C
13. a. 0.0 N m^2/C b. 3.0×10^{-2} N m^2/C
15. 3.5×10^{-4} N m^2/C
19. $+2q$, $+q$, $-3q$
21. 0.11 kN m^2/C
23. -1.00 N m^2/C
25. 2.7×10^{-5} C/m^2
27. a. $\vec{E} = (25\hat{k})$ kN/C, upward from the plate
    b. 0.0 N/C c. 2.5 kN/C, downward from the plate
29. a. -0.39 N m^2/C, 0.23 N m^2/C, 0.39 N m^2/C, -0.23 N m^2/C b. 0 N m^2/C
31. a. -3.5 N m^2/C b. 1.2 N m^2/C
33. 0.19 kN m^2/C
35. a. 2.0 kN/C b. 0.25 kN m^2/C c. 2.2 nC
37. a. -100 nC b. $+50$ nC

39. a. 2.4×10^{-6} C/m^3
    b. 1 nC, 10 nC, 80 nC   c. 5 kN/C, 9.0 kN/C, 1.8×10^4 N/C
41. -4.51×10^5 C
43. 2.5×10^4 N/C, outward; 0 N/C; 7.9×10^3 N/C, outward
45. $\vec{0}$ N/C, $\dfrac{1}{4\pi\epsilon_0}\dfrac{Q}{r^2}\hat{r}$
47. $\vec{0}$ N/C, $(\eta/2\epsilon_0)\hat{j}$, $-(\eta/2\epsilon_0)\hat{j}$, $\vec{0}$ N/C
49. $(\eta/2\epsilon_0)\hat{j}$, $\vec{0}$ N/C, $(\eta/2\epsilon_0)\hat{j}$, $-(\eta/2\epsilon_0)\hat{j}$
51. a. $\dfrac{\lambda}{2\pi\epsilon_0 r}\dfrac{\vec{r}}{r}$ b. $\dfrac{3\lambda}{2\pi\epsilon_0}\dfrac{\vec{r}}{r}$
53. a. $\dfrac{1}{4\pi\epsilon_0}\dfrac{Q}{r^2}\hat{r}$ b. $\vec{E} = \vec{0}$ c. $\dfrac{1}{4\pi\epsilon_0}\dfrac{Q}{r^2}\left(\dfrac{r^3 - R_{in}^{\ 3}}{R_{out}^{\ 3} - R_{in}^{\ 3}}\right)\hat{r}$
55. a. $\dfrac{\lambda L^2 dy}{4\pi\epsilon_0\left[y^2 + (L/2^2)\right]}$ b. $\lambda L/(4\epsilon_0)Q_{in}/\epsilon_0$
57. a. $C = \dfrac{Q}{4\pi R}$ b. $\dfrac{1}{4\pi\epsilon_0}\dfrac{Q}{Rr}\hat{r}$ c. Yes
59. a. $\dfrac{Q}{4\pi\epsilon_0 R^2}$ b. $\dfrac{3Qr^3}{2\pi R^6}$

Chapter 28

1. 1.4×10^5 m/s
3. 2.1×10^6 m/s
5. -2.2×10^{-19} J
7. 4.8×10^{-6} J
9. a. $-1.0\ \mu$J   b. $1.0\ \mu$J
11. 1.87×10^7 m/s
13. -8.4×10^4 V
15. a. Lower   b. -0.712 V
17. a. 1.5 V   b. 2.1×10^{-11} C
19. a. 200 V   b. 6.3×10^{-9} C
21. a. 1800 V, 1800 V, 900 V   b. 0 V, -900 V
23. a. 27 V   b. 4.3×10^{-18} J
25. -1600 V
27. a. $\pm\infty$ b. 0, $\pm\infty$
    c.

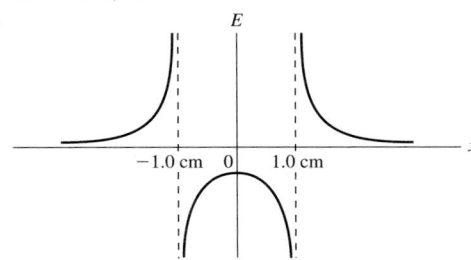

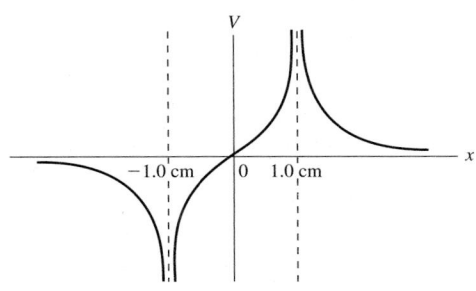

29. a. Positive, positive   b. 1
    c.

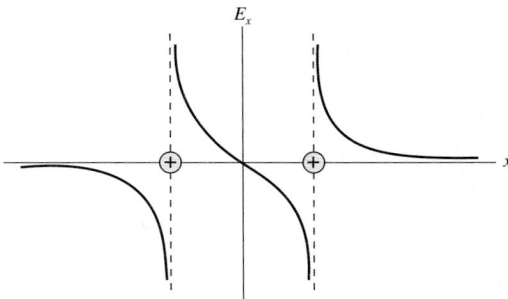

31. 1.4×10^{-3} N
33. a. $+103$ V   b. 5.40×10^4 V/m
35. ± 12 cm
37. 0.49 m/s
39. a. 1.1×10^{-20} J   b. 2×10^{21} ions
41. 54 kHz
43. a. 2.1×10^6 V/m   b. 9.4×10^7 m/s
45. a. 0.85 m   b. 2.6 m
47. 8.0×10^7 m/s
49. -5.1×10^{-19} J
51. 310 nC
53. 6.8 fm
55. a. Yes   c. 8.21×10^8 m/s
57. a. 2.1×10^{-10} C, 3.0 kV/m, 15 V   b. 2.1×10^{-10} C, 3.0 kV/m, 30 V
    c. 2.1×10^{-10} C, 0.75 kV/m, 3.8 V
59. a. $\dfrac{V_0}{R}$ b. 100 kV/m
61. a. $8.3\ \mu$C   b. 3.3×10^6 V/m
63. 2.1 kV, b is higher
65. a. $\dfrac{2q}{4\pi\epsilon_0 x}\dfrac{1}{\sqrt{1 + s^2/4x^2}}$
    b.

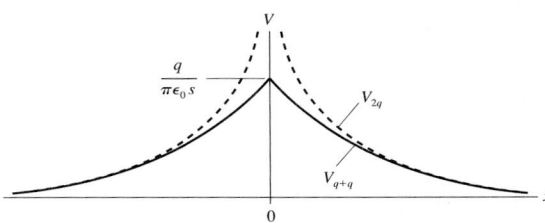

67. $(Q/4\pi\epsilon_0 L)\ln\left[(x + L/2)/(x - L/2)\right]$
69. $Q/4\pi\epsilon_0 R$
71. b. q_1 and q_2 are 10 nC and 30 nC
73. b. 6.0 cm
75. $v_A = 0.018$ m/s, $v_B = 0.011$ m/s
79. a. $\dfrac{1}{4\pi\epsilon_0}\dfrac{q}{R}dq$ b. $\dfrac{1}{4\pi\epsilon_0}\dfrac{Q^2}{2R}$ c. 2.3×10^{-13} J
81. $\dfrac{3Q}{8\pi\epsilon_0 R^3}\left(R\sqrt{R^2 + z^2} + \ln\left(\dfrac{|z|}{R + \sqrt{R^2 + z^2}}\right)\right)$

Chapter 29

1. -200 V
3. -0.30 kV
5. 1.5×10^{-6} J
7. 3.0 C
9. $-(20\hat{j})$ kV/m

11.

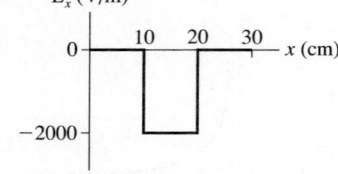

13. -1.0 kV/m
15. a. 27 V/m   b. 3.7 V/m
17. a. 13 pF   b. 1.3 nC
19. 3.0 V
21. 32 μF
23. 150 μF, in series
25. 1.4 kV
27. 1/2
29. a. 1.1×10^{-7} J   b. 0.71 J/m^3
31. a. 62 pC, 9.0 V, 29 kV/m   b. 20 pC, 9.0 V, 90 kV/m
33. a. A   b. -70 V
35. a.

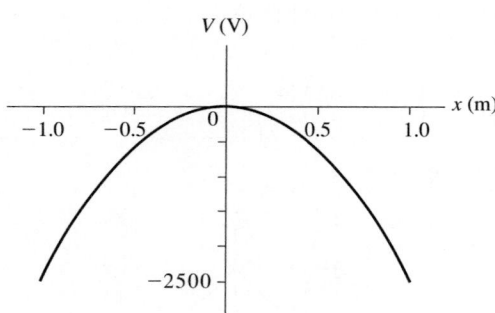

b. $V(x) = -(2500\,x^2)$ V
37. a. $\vec{E} = -(1.4 \times 10^7\,\hat{\imath})$ V/m, $V = 7 \times 10^4$ V
    b. $E = 0.0$ V/m, $V = 1.4 \times 10^5$ V
    c. $\vec{E} = 1.4 \times 10^7\,\hat{\imath}$ V/m, $V = 7 \times 10^4$ V
39. $\vec{E}_{\text{disk}}(z) = \dfrac{Q}{2\pi\epsilon_0 R^2}\left[1 - \dfrac{z}{\sqrt{R^2 + z^2}}\right]\hat{k}$
41. Point 1: 3750 V/m, downward; point 2: 7500 V/m, upward
43. 1000 V/m, 127° ccw from the +x-axis
45. $Q_{1f} = 2$ nC, $Q_{2f} = 4$ nC
47. 1.1 nC
49. a. ± 32pC, 9.0 V   b. ± 16pC, 9.0 V
51. 7.5 μF
53. 5.0 V, 15 V, 10 V
55. $Q_1 = 45\ \mu$C, $V_1 = 9$ V; $Q_2 = 22\ \mu$C $V_2 = 5.4$ V; and $Q_3 = 22\ \mu$C,
    $V_3 = 3.6$ V
57. a. $\dfrac{3}{2}C$ b. 0 V
59. $Q_1 = 0.83$ mC, $Q_2 = Q_3 = 0.67$ mC,
    $\Delta V_1 = 55$ V, $\Delta V_2 = 34$ V, $\Delta V_3 = 22$ V
61. $Q_1 = 33\ \mu$C, $Q_2 = 67\ \mu$C, $\Delta V_1' = \Delta V_2' = 3.3$ V
63. a. 5.7×10^{-7} J   b. 11.4×10^{-7} J
    c. Work was done on the capacitor.
65. 0.85 kV

67. 0.13 F
69. 2.4×10^{-14} J
73. b. $(10 - z^2)$ V, with z in meters
75. b. 2 μF
77. a. $V = \dfrac{q}{4\pi\epsilon_0}\left[\dfrac{1}{\sqrt{x^2 + (y - s/2)^2}} - \dfrac{1}{\sqrt{x^2 + (y + s/2)^2}}\right]$

   b. $V = \dfrac{qsy}{4\pi\epsilon_0(x^2 + y^2)^{3/2}}$

   c. $E_x = \dfrac{qs(3xy)}{4\pi\epsilon_0(x^2 + y^2)^{5/2}}$, $E_y = \dfrac{qs(2y^2 - x^2)}{4\pi\epsilon_0(x^2 + y^2)^{5/2}}$

   d. $\vec{E}_{\text{on-axis}} = \dfrac{2p}{4\pi\epsilon_0 r^3}\hat{\jmath}$, yes

   e. $E_{\text{bisecting axis}} = -\dfrac{p}{4\pi\epsilon_0 r^3}\hat{\jmath}$, yes

79. a. $V_r = \dfrac{1}{4\pi\epsilon_0}\dfrac{Q}{R}\left[\dfrac{3}{2} - \dfrac{r^2}{2R^2}\right]$ b. 3/2
    c.

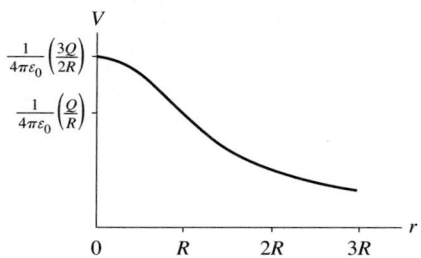

81. a. $\dfrac{2\pi\epsilon_0}{\ln(R_2/R_1)}$ b. 31 pF/m

Chapter 30

1. 3.0 d
3. 7.6×10^{26} electrons
5. 0.023 V/m
7. 1.0×10^{19} s^{-1}
9. a. 0.80 A   b. 7.0×10^7 A/m^2
11. 130 C
13. 1.88×10^{22}
15. 2.6 mA
17. a. 6.3×10^5 A/m^2 b. 6.5×10^{-5} m/s
19. 1.68 A
21. 5.0×10^{-8} Ω m
23. a. 1.64×10^{-3} V/m   b. 1.10×10^{-5} m/s
25. Tungsten
27. $\dfrac{1}{2}$
29. Tungsten
31. a. 30 m   b. 1.0 A
33. 4100 Ω
35. 380
37. 0.64 mm
39. Yes, 2.2×10^5 Ω^{-1}m^{-1}
41. a. 75 nA   b. 130 s
43. a. 6.6×10^{15} Hz   b. 1.05×10^{-3} A
45. a. 120 C   b. 0.45 mm
47. 1.4 Ω m
49. 0.50 mm

51. a. $E = \dfrac{I}{4\pi\sigma r^2}$ b. $E_{inner} = 3.3 \times 10^{-4}$ V/m, $E_{outer} = 5.3 \times 10^{-5}$ N/C

53. a. $I(t) = (10 \text{ A})e^{-t/2.0\,s}$ b. 10 A

   c.   *I* (A)

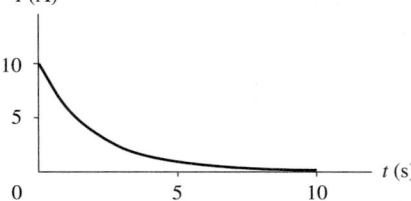

55. 2.0 A, 5.0×10^{-5} m/s

57. 7.2 mm

59. 0.16 V/m

61. 2R

63. a. 4.2×10^5 A   b. Decrease   c. 1.1×10^{-5} J

65. 1.8×10^8 A/m²

67. a. 2.5 C   b. 1.8 cm

69. 1.01×10^{23}

71. a. 9.4×10^{15} b. 115 A/m²

73. a. $\eta = \dfrac{\epsilon_0 I}{A}\left(\dfrac{1}{\sigma_2} - \dfrac{1}{\sigma_1}\right)$ b. 3.7×10^{-18} C

Chapter 31

1.

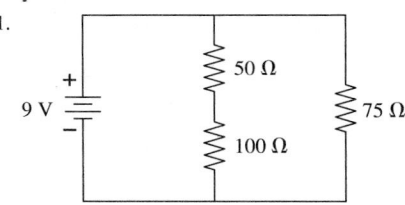

3. 1 A to left

5. a. 0.9 A ccw

   b.   *V* (V)

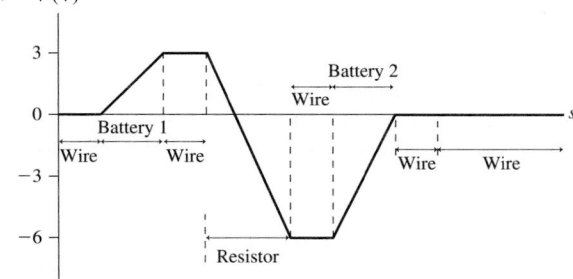

7. 9.60 Ω, 12.5 A

9. 60 W bulb is brighter

11. a. 11.6 A   b. 10.4 Ω

13. 75 Ω

15. a. 0.65 Ω   b. 3.5 W

17. 3.2%

19. 240 Ω

21. 40 Ω

23. 183 Ω

25. 9 V, 1 V

29. 2 ms

31. a. 36 μC, 0.36 A   b. 22 μC, 0.22 A   c. 4.9 μC, 49 mA

33. 18 μF

35. D

37. 93 W

39.

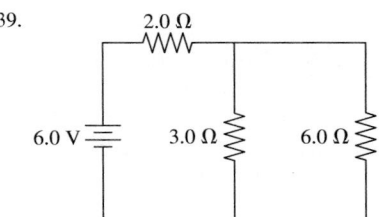

41. 7 Ω

43. 60 V, 10 Ω

45. 9.0 V, 0.50 Ω

47. 1.8 V

49. 1.0 A, 2.0 A, 15 V

51. \$65 for the incandescent bulb, \$20 for the fluorescent tube

53. a. 0.231 A   b. 0.214 S   c. 7.4%

55. 900 Ω

57. a. 0.505 Ω   b. 0.500 Ω

59.

Resistor	Potential difference (V)	Current (A)
3 Ω	6.0	2.0
4 Ω	6.0	1.5
48 Ω	6.0	0.125
16 Ω	6.0	0.375

61.

Resistor	Potential difference (V)	Current (A)
24 Ω	6.00	0.25
3 Ω	3.00	1.00
5 Ω	3.75	0.75
4 Ω	2.25	0.56
12 Ω	2.25	0.19

63. 9/25 A, left to right

65. 150 V, bottom

67. 0.41 A, left to right

69. a. 65 kΩ   b. 87 V

71. 73 Ω

73. a. $\mathcal{E}$ b. $C\mathcal{E}$ c. $+dQ/dt$

   d. $\left(\dfrac{\mathcal{E}}{R}\right)e^{-t/\tau}$

   $I/(\mathcal{E}/R)$

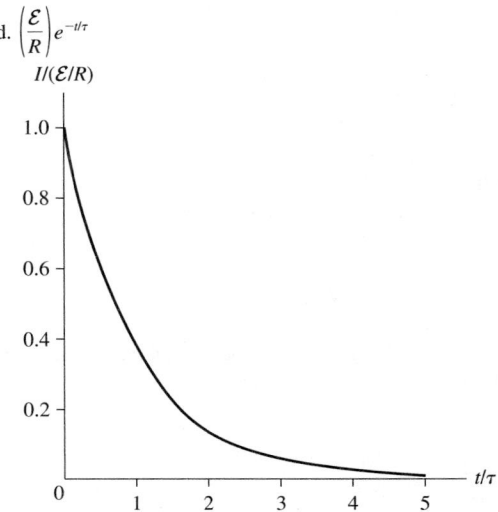

75. 2.0 m, 0.49 mm

77. 20 V

79. a. $\mathcal{E}^2 C$ b. $\mathcal{E}^2 C/2$ c. $\mathcal{E}^2 C/2$ d. Yes

81. 0.60 A

Chapter 32

1. $\vec{B}_1 = (2.0 \text{ mT, into the page})$, $\vec{B}_2 = (4.0 \text{ mT, into the page})$
3. a. 0 T   b. $1.60 \times 10^{-15} \hat{k}$ T   c. $-4.0 \times 10^{-16} \hat{k}$ T
5. $-1.13 \times 10^{-15} \hat{k}$ T
7. 6.3×10^6 m/s in the $+z$-direction
9. 4.0 cm, 0.40 mm, 20 μm to 2.0 μm, 0.20 μm
11. a. 20 A   b. 1.6×10^{-3} m
13. $2.0 \times 10^{-4} \hat{\imath}$ T, $4.0 \times 10^{-4} \hat{\imath}$ T, $2.0 \times 10^{-4} \hat{\imath}$ T
15. a. 0.025 A m^2 b. 1.5 μT
17. 1.4 cm
19. 0.071 T m
21. 7.00 A
23. 1.26×10^{-6} T m
25. 1.0 mm
27. a. $8.0 \times 10^{-13} \hat{\jmath}$ N   b. $5.7 \times 10^{-13} (-\hat{\jmath} - \hat{k})$ N
29. 1.6×10^{-3} T
31. 81 mT
33. 0.131 T, out of page
35. 3.0 Ω
37. 7.5×10^{-4} N m
39. a. 1.26×10^{-11} N m   b. Rotated by $\pm 90°$
41. 0.040 μA
43. $(5.2 \times 10^{-5}$ T, out of page), $\vec{0}$ T
45. 0.77R
47. $(7.9 \times 10^{-5}$ T, into page)
49. #18, 4.1 A
51. a. 1.13×10^{10} A   b. 0.014 A/m^2 c. 1.3×10^6 A/m^2
53. a. 5.7×10^{-6} A   b. 2.9×10^{-8} A m^2
55. $\dfrac{\mu_0 I}{4R}$
57. 0; $\dfrac{\mu_0 I}{2\pi r}\left(\dfrac{r^2 - R_1^2}{R_2^2 - R_1^2}\right)$; $\dfrac{\mu_0 I}{2\pi r}$
59. 1.50 mT, 30° ccw from the $+x$-axis
61. 2.9×10^{-3} T
63. 2.4×10^{10} m/s^2, up
65.

	Ion	Accelerating voltage (V)
a.	O_2^+	96.793
b.	N_2^+	110.25
c.	CO^+	110.29

67. 0.12 T
69. 87 mT
71. a. $\dfrac{\mu g \tan\theta}{I}$, down   b. 11 mT, down
73. 13 T
75. a. $2\pi RIB \sin\theta$ b. 4.3×10^{-3} N
77. a. $\dfrac{\mu_0 IL}{4\pi d \sqrt{(L/2)^2 + d^2}}$ b. $\dfrac{\sqrt{2}\mu_0 I}{\pi R}$ c. 0.900
79. a. $\dfrac{3I}{2\pi R^2}$ b. $\dfrac{\mu_0}{2\pi}\dfrac{Ir^2}{R^3}$ c. Yes
81. a. Horizontal and to the left above the sheet; horizontal and to the right below the sheet   b. $\frac{1}{2}\mu_0 J_s$

Chapter 33

1. 2.0×10^4 m/s
3. a. 1.0 N   b. 2.2 T
5. 6.3×10^{-5} Wb in both cases
7. Decreasing
9. Clockwise current
11. a. 3.9 mV, 20 mA, ccw   b. 3.9 mV, 20 mA, ccw   c. No current

13. 1.6 V
15. $E (\times 10^{-4} \text{ V/m})$

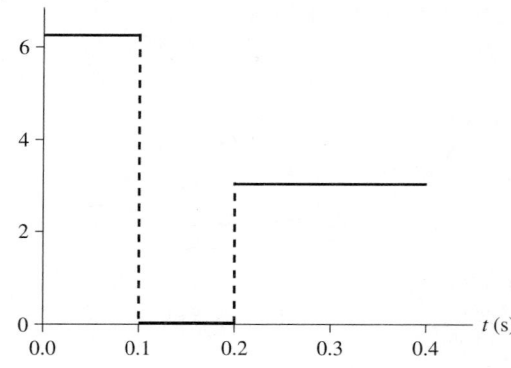

17. a. 4.8×10^4 m/s^2, up   b. 0   c. 4.8×10^4 m/s^2, down
    d. 9.6×10^4 m/s^2, down
19. 1.0 ms
21. 9.5×10^{-5} J
23. 250 kHz to 360 kHz
25. 750 Ω
27. 3.5×10^{-4} Wb
29. 1.6 A, 0.0 A, -1.6 A
31. 8.7 T/s
33. a. -0.0050 V   b. 0.0100 V
35. 44 μA
37. a. 0 μA   b. 160 μA   c. 0 μA
39. a. 0.0 A   b. 79 μA
41. a. 0.93 V   b. 0 V
43. a. 12 500   b. 2.0 A
45. a. I (A)

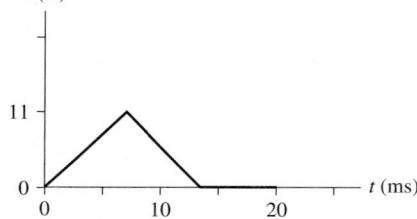

    b. 11 A when halfway in
47. a. 4.0 V   b. 100 A   c. 3.0 V
49. a. $(4.9 \times 10^{-3}) f \sin(2\pi f t)$ A   b. 4.1×10^2 Hz, not feasible
51. 0.28 T
53. a. $(vlB\cos\theta)/R$ b. $(mgR\tan\theta)/l^2 B^2 \cos\theta$
55. 2.5×10^{-4} V
57. 12 V
59. $(R^2/2r)/(dB/dt)$
61. a. 3.9×10^{-4} J/m^3 b. 3.1 A
63. 3.0 s
65. I (A)

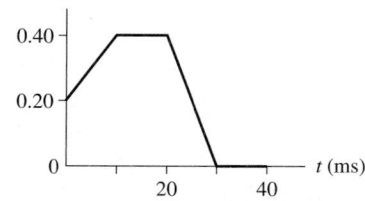

67. a. $\Delta V_L = \left(\dfrac{LI_0}{\tau}\right) e^{-t/\tau}$ b. 0.37 V
69. 1.0 μF
71. 0.50 m
73. a. 76 mA   b. 0.50 ms

75. a. 0.50 A   b. 1.0 A

77. a. $\Delta V_{bat}/R$ b. $I = I_0(1 - e^{-t/(L/R)})$

79. $(\mu v_0 I/2\pi) \ln[(d + l)/d]$

81. a. 0.10 s   b. $2.93\left(\dfrac{(0.10)^2 - 2[0.0707 + (0.293)t]^2}{\sqrt{(0.10)^2 - [0.0707 + (0.293)t]^2}}\right)$ A

c.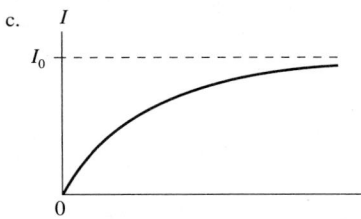

83. a. 32 A   b. 1.3 m/s

85. a. $(\mu_0/2\pi) \ln(r_2/r_1)$ b. 0.36 μH/m

Chapter 34

1. a. $(2.0 \times 10^6$ m/s, 45° from the y-axis) 45°
   b. $(1.47 \times 10^6$ m/s, 16.2° from the y'-axis) 16.2°

3. $-1.0 \times 10^6 \hat{k}$ V/m, $-1.11 \times 10^{-5} \hat{j}$ T

5. 16.3° above the $+x$-axis

9. 1.0 μF

11. 17 μA

13. 3.3×10^{-8} T

15. a. 10.0 nm   b. 3.00×10^{16} Hz   c. 6.67×10^{-8} T

17. a. 3.33×10^{-7} T   b. 13.3 W/m^2

19. 980 V/m, 3.3 μT

21. a. 2.2×10^{-6} W/m^2 b. 0.041 V/m

23. 3.3×10^{-6} N

25. 60°

27. 30°

29. $(1.73 \times 10^6$ V/m, left)

31. a. (0.10 T, into page)   b. 0 V/m, (0.10 T, into page)

33. 1.0×10^7 m/s parallel to the current

35. a. 0.94 V/m   b. 10 T

37. b. 1.5×10^{-13} A

39. I_{disp} (A)

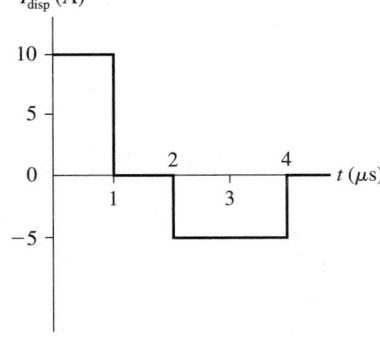

41. 20 V

43. b. 6.67×10^{-6} J/m^3

45. a. 3.85×10^{26} W   b. 589 W/m^2

47. a. $(1/2)f$ b. $(3/4)f$

49. Yes

51. 1.8×10^7 V/m

53. 1.3 m

55. 4.9×10^7 W/m^2

57. 8.8 h

59. $(-6.0 \times 10^5 \hat{\imath} + 1.0 \times 10^5 \hat{\jmath})$ V/m

61. 5.2 μV/m

63. a. $E = IR/L$, $B = \dfrac{\mu_0 I}{2\pi r} IR/L$ b. $(I^2 R/2\pi rL$, radially inward)

Chapter 35

1. a. 22×10^2 rad/s   b. -10 V

3.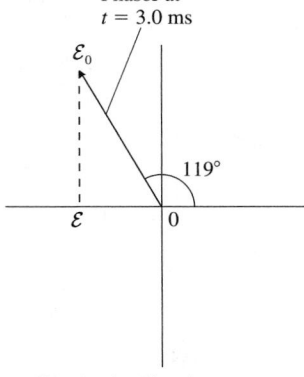

5. a. 50 mA   b. 50 mA

7. a. 1.9 mA   b. 1.9 A

9. a. 80 Hz   b. 0 V

11. a. 95 pF   b. 660 μA

13. 1.6 μF

15. $V_R = 6.0$ V, $V_C = 8.0$ V

17. a. 1000 Hz   b. 2.24 V, 3.53 V, 4.47 V

19. a. 0.80 A   b. 0.80 mA

21. a. 3.2×10^4 Hz   b. 0 V

23. a. 200 kHz   b. 141 kHz

25. 1.3 μF

27. a. 70 Ω, 72 mA, $-44°$ b. 50 Ω, 0.10 A, 0°   c. 62 Ω, 80 mA, 37°

29. 9.6 Ω

31. 30°

33. 44 Ω

35. a. $(\sqrt{3}RC)^{-1}$ b. $\sqrt{3}\mathcal{E}_0/2$

37. a. 9.95 V, 9.57 V, 7.05 V, 3.15 V, 0.990 V

b. V_C (V)

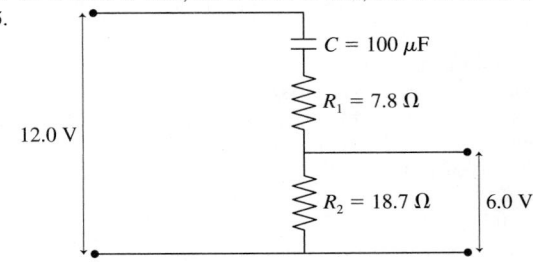

43. 44 Hz

45. a. 50 Hz   b. 4.8 μF

47. a. $\mathcal{E}_0/\sqrt{R^2 + \omega^2 L^2}$, $\mathcal{E}_0 R/\sqrt{R^2 + \omega^2 L^2}$, $\mathcal{E}_0 \omega L/\sqrt{R^2 + \omega^2 L^2}$

b. $V_R \rightarrow \mathcal{E}_0$, $V_R \rightarrow 0$ c. Low pass   d. R/L

49. a. 69 V   b. 24°   c. 0.17 kW

51. a. 5.0×10^3 Hz   b. 10 V, 32 V

53. 0.17 A

55. a. 3.6 V   b. 3.5 V   c. -3.6 V

59. a. 11.6 pF   b. 1.49×10^{-3} Ω

61. 14 W in 40 W bulb, 9.6 W in 60 W bulb, 100 W in 100 W bulb

65.

67. a. 0.44 kA b. 1.8×10^{-4} F c. 7.4 MW
69. b. 10 V, 12 V
71. b. ∞, ∞ c. $1/\sqrt{LC}$
  d.

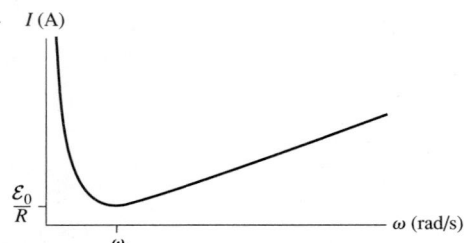

Chapter 36

1. $x'_1 = 5.0$ m at $t = 1.0$ s, $x'_2 = -5.0$ m at $t = 5.0$ s
3. $v_{sound} = 345$ m/s, $v_{sprinter} = 15$ m/s
5. a. 13 m/s b. 3.0 m/s c. 9.4 m/s
7. 3.0×10^8 m/s
9. 167 ns
11. 2.0 μs
13. No, bolt 2 hits 20 μs before bolt 1.
15. Yes
17. $0.866c$
19. a. $0.9965c$ b. 59.8 ly
21. 46 m/s
23. Yes
25. 4600 kg/m^3
27. 3.0×10^6 m/s
29. $x = 8.3 \times 10^{10}$ m, $t = 330$ s
31. $0.36c$
33. $0.71c$
35. $0.80c$
37. $0.707c$
39. a. 1.8×10^{16} J b. 9.0×10^9
41. $0.943c$
43. $u_{50 \text{ final}} = 1.33$ m/s to the right, $u_{100 \text{ final}} = 3.33$ m/s to the right
45. 11.2 h
47. a. No b. 67.1 y
49. a. $0.80c$ b. 16 y
51. 0.78 m
53. a. 17 y b. 15 y c. Both
55. $0.96c$
57. 3.1×10^6 V
59. a. $0.98c$ b. 8.5×10^{-11} J
61. b. Lengths perpendicular to the motion are not affected.
63. a. $u'_y = u_y/\gamma\left(1 - u_x v/c^2\right)$ b. 0.877c
65. a. 3.5×10^{-18} kg m/s, 1.1×10^{-9} J b. 1.6×10^{-18} kg m/s
67. a. 7.6×10^{16} J b. 0.84 kg
69. 7.5×10^{13} J
71. 1 pm
73. 22 m
75. $0.85c$

Chapter 37

1. a. $(m, n) = (2, 3), (2, 4), (2, 5), (2, 6)$ b. 397.1 nm
3. 121.6 nm, 102.6 nm, 97.3 nm, 95.0 nm
5. a. 9.39×10^4°C b. 694°C
7. 2.4 μm
9. a. 6.0×10^7 m/s b. 17 cm
11. a. 2.4×10^{-16} kg b. 1.3×10^{-18} C c. 8
13. 1.33×10^{19} C
15. a. 3.7×10^7 m/s b. 2.7×10^7 m/s c. Electron

17. a. 10 keV b. 0.14 MeV c. 1.2×10^{19} eV
19. a. 3 electrons, 3 protons, 3 neutrons
  b. 7 electrons, 8 protons, 8 neutrons
  c. 5 electrons, 7 protons, 6 neutrons
21. a. ^{11}B b. ^{14}C$^+$
23. a. 79 electrons, 79 protons, 118 neutrons
  b. 2.29×10^{17} kg/m^3
  c. 2.01×10^{13}
25. a. 6660 MeV b. 3.6 MeV
27. a. 0.512 MeV b. 939 MeV
29. 173 MeV
31. 46 mT, into the page
33. 1.2 μA
35. 0.000000000058% contains mass, 99.999999999942% empty space
37. a. 5.0×10^4 kg/m^3
  b. 1.7×10^{-29} m^3, 1.6×10^{-10} m
  c. 1.7×10^{17} kg/m^3, 6.2×10^{13}
39. a. 58 N b. 4.7×10^{-35} N
41. Aluminum
43. a. 2.3×10^7 m/s b. 2.9 MeV
45. 2.52×10^5 m/s, 65.1° below $+x$-axis
47. a. mg/E_0 b. mg/b d. 2.4×10^{18} C e. 15

Chapter 38

1. 6.25×10^{13} electrons/s
3. 3.20 eV
5. 1.78 eV
7. a. 2.26 eV b. 0.166 nm
9. a. 1.86×10^{-6} eV b. 2.76 eV c. 27.6 keV
11. 497 nm
13. 1×10^{19} photons/s
15. 6.0×10^{-6} V
17. a. 1.1×10^{-34} m b. 1.7×10^{-23} m/s
19. 6
21. 0.427 nm
23. a. Yes b. 0.50 eV
25. $n = 2$: yes; $n = 3$: no
27. 1.90 nm
29. a. 0.332 nm, 0.665 nm, 0.997 nm
  c.

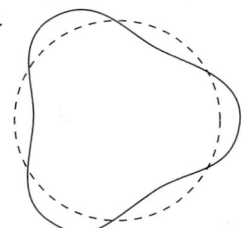

33. 97.26 nm
35.

n	r_n (nm)	v_n (m/s)	E_n (eV)
1	0.026	4.38×10^6	-54.4
2	0.106	2.19×10^6	-13.6
3	0.238	1.46×10^6	-6.0

37. 1.24 V
39. 4.3×10^{-10} W
41. a. 2.3 eV b. 244 nm
43. a. 4.14 eV b. 6.4×10^{-34} J s
45. a. Potassium b. 4.24×10^{-15} eV s
47. 2.0×10^{-18} m, no
49. 200 m/s
51. 0.35 nm

53. 18 fm

55. a.

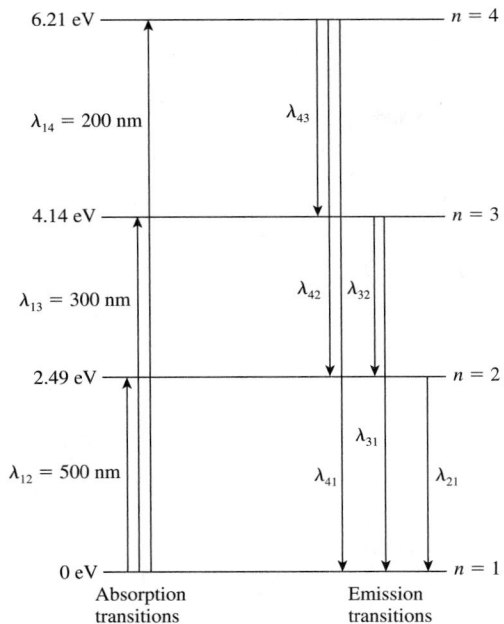

b. 200 nm, 300 nm, 334 nm, 500 nm, 601 nm, 753 nm

57. 6.2×10^5 m/s

59. 410.3 nm, 434.2 nm, 486.3 nm, 656.5 nm

61. a. 0.362 m   b. 0.000368 nm

63. $3 \rightarrow 2$:10.28 nm, $4 \rightarrow 2$:7.62 nm, $5 \rightarrow 2$: 6.80 nm; all ultraviolet

65. a. 2.06×10^6 m/s   b. 12.09 V

67. a. 1.0 m/s   b. 3.2°   c. 1.1 cm

69. b. 25.7 nm, 36.3 nm, 44.5 nm, 51.4 nm   c. $n(\pi \hbar f_{cyc})$

Chapter 39

1. $P_C = 0.20$, $P_D = 0.10$

3. a. 7.7%   b. 25%

5. a. 1/6   b. 1/6 c. 5/18

7. 100 V/m

9. 4.0 m^{-1}

11. a. 3300   b. 1100

13. a. 5.0×10^{-3} b. 2.5×10^{-3} c. 0   d. 2.5×10^{-3}

15. a. 0.25 fm^{-1} b.   c. 0.75

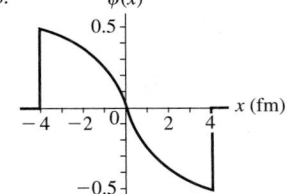

17. a. 0.354 mm$^{-1/2}$

b.

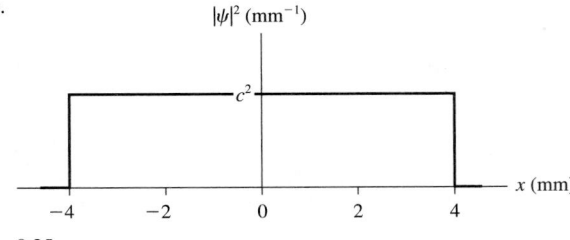

c. 0.25

19. 25 ns

21. 1.0×10^5

23. -0.65×10^{-36}m/s $\leq v_x \leq 0.65 \times 10^{-36}$ m/s

25. 0.0 m/s $\leq v_x \leq 2.5 \times 10^7$ m/s

27. 9.5 GHz $\leq f \leq$ 10.5 GHz

29. 1.0×10^5 pulses/s

31. a.

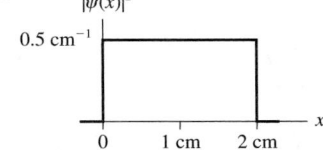

b. 1%   c. 10^4 d. 0.5 cm^{-1}

33. a. Yes   b.

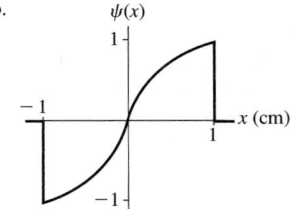

c. 0.000, 0.00050, 0.0010   d. 900

35. a. $\sqrt{3/8}$ mm$^{-1/2}$

b.

c.

d. 0.13

37. a. 0.27%   b. 32%

39. a. 0.87 cm$^{-1/2}$

b.

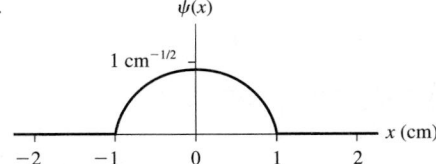

c.

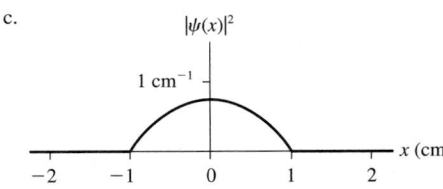

d. 3.4×10^3

41. a. $a = b$ b. $a = b = 0.84$

c.   d. 58.1%

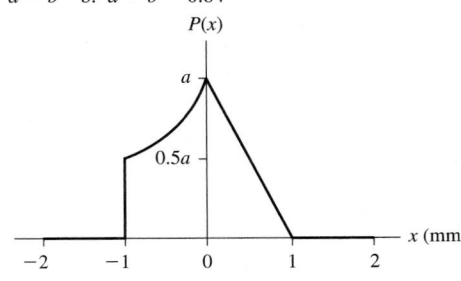

43. 18 μm
45. a. 0 m/s to 1.8×10^{10} m/s
  b. The speed in part a exceeds the speed of light, so it is impossible.
47. a. 1.5×10^{-13} m b. 4.4×10^{11} m
49. 50%
51. a. $c = \sqrt{3/8}$
  b.

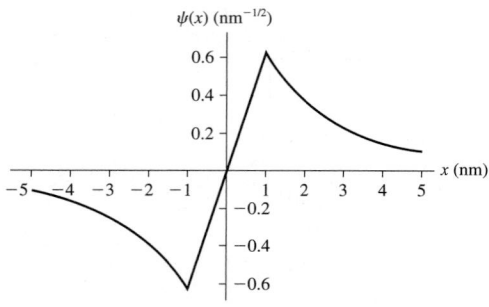

  c.

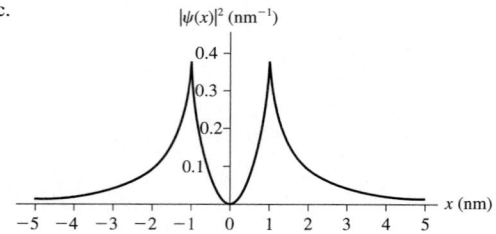

  d. 2.5×10^5

Chapter 40

1. 0.739 nm
3. 1.0 nm
7. a. 0.159 nm b. 0.159 nm c. 0.275 nm
9. 0.38 nm
11.

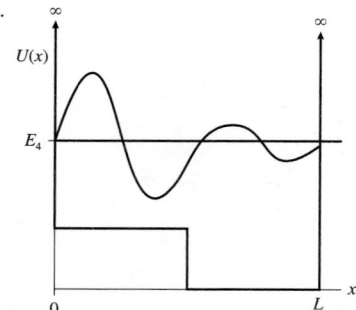

13.

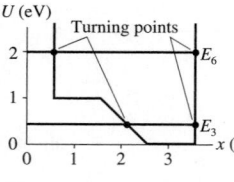
(a)

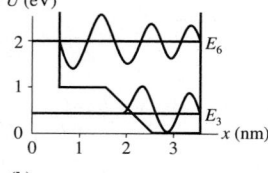
(b)

15. 200 nm
17. 519 nm
19. 1.4 N/m

21. a. 4.95 eV b. 4.80 eV c. 4.55 eV
25. a. $\lambda_{2 \rightarrow 1} = 8mcL^2/3h$ b. 0.795 nm
29. a.

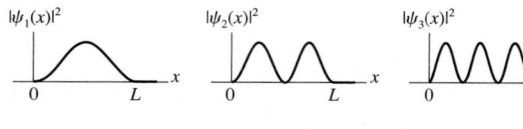

$n =$	1	2	3
b. Most likely	$\frac{1}{2}L$	$\frac{1}{4}L, \frac{3}{4}L$	$\frac{1}{6}L, \frac{3}{6}L, \frac{5}{6}L$
c. Least likely	$0, L$	$0, \frac{1}{2}L, L$	$0, \frac{1}{3}L, \frac{2}{3}L, L$
d. Prob in left $\frac{1}{3}$ from graph	$> \frac{1}{3}$	$> \frac{1}{3}$	$\frac{1}{3}$
e. Prob in left $\frac{1}{3}$ calculated	0.195	0.402	0.333

31. 10%
35. a. $A_1 = \dfrac{1}{(\pi b^2)^{1/4}}$ b. $\text{Prob}(x < -b \text{ or } x > b) = \dfrac{2}{\sqrt{\pi b^2}} \displaystyle\int_b^\infty e^{-x^2/b^2} dx$
  c. 15.7%

37. a. $P_{\text{class}}(x) = \dfrac{1}{2h\sqrt{1 - (y/h)}}$
  b. $P_{\text{class}}(y)$

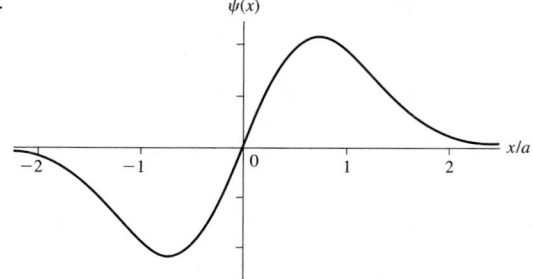

39. 10^{-463}
41. a. 0.136 nm b. One atomic diameter
43. a.

b. $\pm a/\sqrt{2}$ c. $U(x) = \dfrac{2\hbar^2}{ma^2}\left(\left(\dfrac{x}{a}\right)^2 - \dfrac{3}{2}\right)$

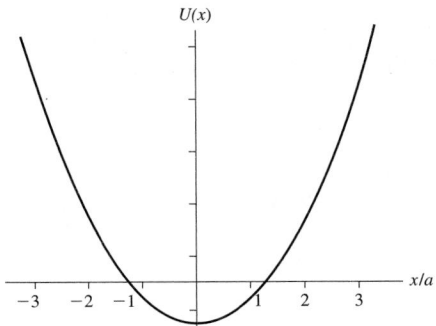

45. a. 3.4×10^{-5} b. 2.8   c. 0.005 nm

Chapter 41

1. a. 0   b. $\sqrt{12}\hbar$
3. a. f b. -0.85 eV
5. -0.378 eV; $\sqrt{12}\hbar$
7. a. 2   b. 1
9. $1s^2 2s^2 2p^6 3s^2 3p$, $1s^2 2s^2 2p^6 3s^2 3p^6 4s^2 3d^{10} 4p$,
   $1s^2 2s^2 2p^6 3s^2 3p^6 4s^2 3d^{10} 4p^6 5s^2 4d^{10} 5p$
11. a. Excited state of Ne   b. Ground state of Ti
13. $1s^2 3s$
15. a. Yes, 2.21 μm   b. No
17. a. 5.6×10^5 b. 1.7×10^5 c. 3.0×10^3
19. a. 9.0×10^5 b. 8.7 ns
21. 3.2×10^{15} s^{-1}
23. a. 1.06 μm   b. 1.9 W
25. a. $\sqrt{2}\hbar$ b. $-1, 0, 1$

c.

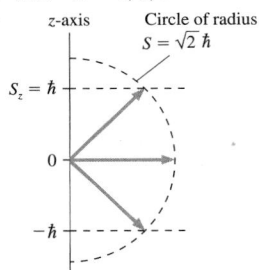

27. a. $\sqrt{2}\hbar$ b. $\sqrt{6}\hbar$
29. a. 3.7×10^{-3} b. 5.4×10^{-3} c. 2.9×10^{-3}
33. a. $R_{2p}(r) = \dfrac{A_{2p}}{2a_B} r e^{-r/2a_B}$

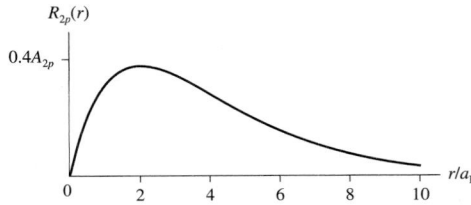

b. $2a_B$

35.

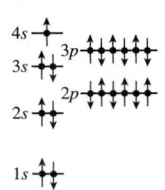

K
$(1s^2\, 2s^2\, 2p^6\, 3s^2\, 3p^6\, 4s^1)$

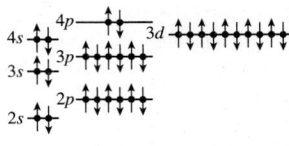

Ge
$(1s^2\, 2s^2\, 2p^6\, 3s^2\, 3p^6\, 4s^2\, 3d^{10}\, 4p^2)$

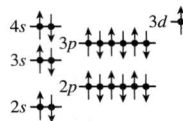

Sc
$(1s^2\, 2s^2\, 2p^6\, 3s^2\, 3p^6\, 4s^2\, 3d)$

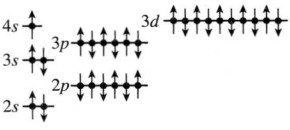

Cu
$(1s^2\, 2s^2\, 2p^6\, 3s^2\, 3p^6\, 4s\, 3d^{10})$

37. a.

Transition	$6s \to 5p$	$6s \to 4p$	$6s \to 3p$
b. λ(nm)	7290	1630	515

39. 1.13×10^6 m/s

41.

Transition	(a) Wavelength	(b) Type	(c) Absorption
$2p \to 2s$	670 nm	VIS	Yes
$3s \to 2p$	816 nm	IR	No
$3p \to 2s$	324 nm	UV	Yes
$3p \to 3s$	2696 nm	IR	No
$3d \to 2p$	611 nm	VIS	No
$3d \to 3p$	24800 nm	IR	No
$4s \to 2p$	498 nm	VIS	No
$4s \to 3p$	2430 nm	IR	No

43. a.   Energy   b. 28.7 eV

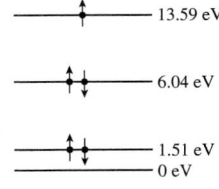

45. a. 6.3×10^8 s^{-1} b. 0.17 ns
47. a. $\tau \ln 2$ b. 12 ns
49. 5.0×10^{16}
51. b. 0.021 nm
55. $1.5a_B$, $5.0a_B$

57. a. $p_{atom} = 7.0 \times 10^{-23}$ kg m/s; $p_{photon} = -8.5 \times 10^{-28}$ kg m/s
    b. 82000 photons   c. 1.2 ms   d. -5.7×10^{-20} N, -4.0×10^5 m/s^2
    e. 31 cm

Chapter 42

1.

		Protons	Neutrons
a.	^{3}He	2	1
b.	^{32}P	15	17
c.	^{32}S	16	16
d.	^{238}U	92	146

3. a. 3.8 fm   b. 8.2 fm   c. 14.5 fm
5. a. $m = 9.988 \times 10^{-27}$ kg; $r = 2.2 \times 10^{-15}$ m; $\rho = 2.3 \times 10^{17}$ kg/m^3
   b. $m = 3.437 \times 10^{-25}$ kg; $r = 7.1 \times 10^{-15}$ m; $\rho = 2.3 \times 10^{17}$ kg/m^3
7. a. ^{36}S and ^{36}Ar   b. 5, 8
9. ^{54}Cr: 474 MeV, 8.78 MeV; ^{54}Fe: 472 MeV, 8.74 MeV
11. ^{14}O = 7.05 MeV; ^{16}O = 7.98 MeV; ^{16}O is more tightly bound
13. 8000 N
15. 2.3×10^{-38}
17. a. ^{14}C

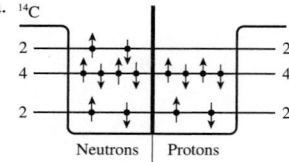

^{14}N

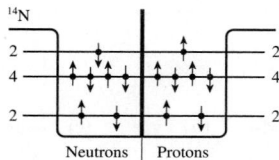

^{14}O

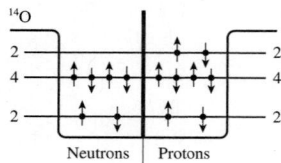

    b. ^{14}N is stable; ^{14}C undergoes beta-minus decay and ^{14}O undergoes
    beta-plus decay

19. a. 236 μg   b. 140 μg   c. 0.775 μg
21. 4.6×10^9
23. 80 d
25. a. ^{228}Th   b. ^{207}Tl   c. ^{7}Li   d. ^{60}Ni
27. a. ^{19}O, ^{19}F, ^{19}Ne   b. ^{19}F
    c. ^{19}O decays by β^- to ^{19}F; ^{19}Ne decays by β^+ to ^{19}F
29. ^{228}Th
31. 5.51 MeV
33. 0.225 J
35. 60 mrem
37. a. 3.5×10^7 m/s   b. 25 MeV
39. a. 12.7 km   b. 780 μs
41. a. 1.46×10^{-8} u, 1.45×10^{-6}%   b. 0.0304 u, 0.76%
43. 6.0 MeV
45. 0.93 MeV
47. 2.7×10^{17}
49. a. 19 s   b. No
51. 1.2 h
53. 210 million years
55. 3.3×10^{12}
57. a. 2.6×10^7 b. 0.024 Bq   c. 1.9×10^5 mSv
    d. Yes, many times more than the background radiation
59. 15 cm
61. $\approx$ 6 billion years ago
63. a. $K_{in} = 65.0$ MeV; $K_{out} = 5.0$ MeV   b. 3.7×10^{21} collisions/s
    c. 6.6×10^{-39} d. 650 million years

Credits

Page **xvi**: Educational Development Center, Inc.
Page **xxix**: IBM Research, Almaden Research Center.

PART I OPENER
Page **1**: Sharon Green/Pacific Stock/Photolibrary.

CHAPTER 1
Page **2** Top: iStockphoto. Page **2** Bottom: Bureau International des Poids et Mesures. Page **3** Top left: Jason Edwards/National Geographic/Alamy. Page **3** Top right: Shutterstock. Page **3** Bottom left: Richard Megna/Fundamental Photographs. Page **3** Bottom right: Brand X Pictures/Alamy. Page **10** Top: Shutterstock. Page **10** Bottom: Shutterstock. Page **13**: Transtock/Alamy. Page **21**: Shutterstock. Page **23**: U.S. Department of Commerce. Page **24**: Bureau International des Poids et Mesures.

CHAPTER 2
Page **33** Top: Steve Allen/Photo Researchers, Inc. Page **33** Bottom: Dominic Ebenbichler/Reuters. Page **34**: Nicosan/Alamy. Page **40**: NASA. Page **52**: Jim Sugar/Corbis. Page **53**: Michel Bureau/BIOSphoto/Photolibrary.

CHAPTER 3
Page **69**: Simon Hathaway/Alamy. Page **69** Bottom: Paul Chesley/National Geographic Stock. Page **74**: Patrick LaCroix/Alamy.

CHAPTER 4
Page **85** Top: Danita Delimont/Alamy. Page **85** Middle: Richard Bartz. Page **85** Bottom: Shutterstock. Page **91**: Tony Freeman/PhotoEdit, Inc. Page **93**: Richard Megna/Fundamental Photographs. Page **98**: Photolibrary RF. Page **102**: Guy Croft SciTech/Alamy.

CHAPTER 5
Page **116** Top: p17/ZUMA Press/Newscom. Page **116** 2nd from Top: Tony Freeman/PhotoEdit, Inc. Page **116** Middle left: Pearson Science/Creative Digital Vision. Page **116** Middle 2nd from left: Dave Pattison/Alamy. Page **116** Middle 3rd from left: Randy Allbritton/Getty Images - PhotoDisc. Page **116** Middle right: 81a/Alamy. Page **116** Bottom: Shutterstock. Page **117** Top: Tony Freeman/PhotoEdit, Inc. Page **117** 2nd from Top: Brian Drake/Photolibrary. Page **117** 3rd from Top: Getty Images/Comstock Images. Page **117** 4th from Top: Dorling Kindersley. Page **117** 5th from Top: Shutterstock. Page **117** Bottom: Pearson Science/Creative Digital Vision. Page **129**: fStop/Alamy. Page **130**: Photodisc/AGE Fotostock.

CHAPTER 6
Page **138** Top: Tom Tracy Photography/Alamy. Page **138** Bottom left: NASA. Page **138** Bottom right: NASA. Page **139**: Akira Kaede/Photodisc/Getty Images. Page **147**: NASA. Page **152** Top: Pearson Science/Eric Schrader. Page **152** Bottom: Patrick Behar/Agence Vandystadt/Photo Researchers, Inc. Page **164**: Richard Megna/Fundamental Photographs.

CHAPTER 7
Page **167** Top: Shutterstock. Page **167** Bottom: iStockphoto. Page **169**: Shutterstock. Page **171**: Koji Aoki/Alfo Foto Agency RF/AGE Fotostock. Page **189**: Peter Langer/DanitaDelimont.com/Newscom.

CHAPTER 8
Page **191** Top: imagebroker.net/SuperStock. Page **191** Left: Antony Nettle/Alamy. Page **191** Right: Tony Freeman/PhotoEdit, Inc. Page **193**: Getty Images/Digital Vision. Page **195**: Shutterstock. Page **200**: Getty Images/Digital Vision. Page **201**: NASA.

PART II OPENER
Page **218**: Hank Morgan/Photo Researchers, Inc.

CHAPTER 9
Page **220** Top: Alexey Stiop/Alamy. Page **220** Middle left: Shutterstock. Page **220** Middle right: iStockphoto. Page **220** Bottom: Shutterstock. Page **221**: Russ Kinne/Jupiter Images RF. Page **228**: bl1/ZUMA Press/Newscom. Page **236**: Richard Megna/Fundamental Photographs.

CHAPTER 10
Page **245** Top: Oleksiy Maksymenko Photography/Alamy. Page **245** Bottom: Dorling Kindersley. Page **246** Left: Kyodo/Newscom. Page **246** Middle: Thinkstock/Stockbyte/John Foxx. Page **246** Right: Shutterstock. Page **257**: Paul Harris/Paul Harris BWP Media USA/Newscom. Page **258**: iStockphoto. Page **265**: Dorling Kindersley.

CHAPTER 11
Page **278** Top: REUTERS/Bobby Yip. Page **278** Middle: x99/ZUMA Press/Newscom. Page **278** Bottom: Shutterstock. Page **281**: Max Faulkner/MCT/Newscom. Page **289**: PETER COSGROVE/AFP/Newscom. Page **298**: JOEL SAGET/AFP/Getty Images/Newscom.

PART III OPENER
Page **310**: NASA-GSFC, data from NOAA GOES.

CHAPTER 12
Page **312** Top: Gouhier-Hahn-Nebinger/MCT/Newscom. Page **312** Middle: Fotolia. Page **312** Bottom: Visions of America, LLC/Alamy. Page **322**: iStockphoto. Page **330**: Shutterstock. Page **333**: Chris Nash/Getty Images. Page **334**: Richard Megna/Fundamental Photographs. Page **341**: PCN Photography/Alamy.

CHAPTER 13
Page **354** Top: NASA. Page **354** Left: NASA. Page **354** Right: PNT.GOV. Page **356**: American Institute of Physics/Emilio Segre Visual Archives. Page **358**: Corbis RF. Page **361**: NASA/Ralph White. Page **366**: NASA.

CHAPTER 14
Page **377** Top: iStockphoto. Page **377** Right: iStockphoto. Page **377** Left: Superstock/Art Life Images. Page **382**: Serge Kozak/AGE Fotostock. Page **388**: Thomas D. Rossing. Page **392**: Richard Megna/Fundamental Photographs. Page **395**: Hornbil Images/Alamy. Page **399**: Martin Bough/Fundamental Photographs. Page **404**: Harold G. Craighead.

CHAPTER 15
Page **407** Top: Chris A Crumley/Alamy. Page **407** Middle left: Richard Megna/Fundamental Photographs. Page **407** Middle right: TechArt.de. Page **407** Bottom: Shutterstock. Page **415**: Joseph Sinnot/Fundamental Photographs. Page **422**: Louise Murray/Alamy. Page **423**: Alix/Photo Researchers, Inc. Page **424**: TechArt.de. Page **425**: Don Farrall/Getty Images - Photodisc. Page **431**: Shutterstock.

PART IV OPENER
Page **442:** ImageState/Alamy.

CHAPTER 16
Page **444** Top: Frans Lanting Studio/Alamy. Page **444** Middle: NASA. Page **444** Bottom: Fotolia. Page **447:** Richard Megna/Fundamental Photographs. Page **449** Both: David Young-Wolff/PhotoEdit, Inc. Page **452:** Vibe Images/Alamy.

CHAPTER 17
Page **469** Top: Ted Kinsman/Peter Arnold Inc./Photolibrary. Page **469** Bottom left: Shutterstock. Page **469** Bottom middle: iStockphoto. Page **469** Bottom right: Clouds Hill Imaging Ltd./Photo Researchers, Inc. Page **471:** Alexey Dudoladov/Getty Images RF. Page **477:** Gordon Saunders/Shutterstock. Page **482:** Getty Images/Stocktrek Images. Page **491** Left: Cn Boon/Alamy. Page **491** Middle left: Andrew Davidhazy. Page **491** Middle right: Pascal Goetgheluck/Photo Researchers, Inc. Page **491** Right: Corbis RF. Page **492:** Science Photos/Alamy. Page **493:** Johns Hopkins University Applied Physics Laboratory.

CHAPTER 18
Page **502** Top: Jose Mendes/Fotolia. Page **502** Bottom: Rick Lord/Shutterstock. Page **519:** Getty Images – Photodisc.

CHAPTER 19
Page **526** Top: imagebroker.net/SuperStock. Page **526** Bottom left: Peter Bowater/Alamy. Page **526** Bottom right: Shutterstock. Page **529:** Peter Bowater/Alamy. Page **532:** Shutterstock. Page **536:** Malcolm Fife/Getty Images - Photodisc.

PART IV SUMMARY
Page **557:** AGE Fotostock/SuperStock.

PART V OPENER
Page **558:** AguaSonic Acoustics/Photo Researchers, Inc.

CHAPTER 20
Page **560** Top: EpicStockMedia/Shutterstock. Page **560** Middle left: iStockphoto. Page **560** Middle: B. Benoit/Photo Researchers, Inc. Page **560** Middle right: David Parker/Photo Researchers, Inc. Page **560** Bottom left: Oleksiy Maksymenko/Alamy. Page **560** Bottom right: Aaron Kohr/Shutterstock. Page **561:** Dudarev Mikhail/Shutterstock. Page **562:** Uri Haber-Schaim - From PSSC Physics 7th Edition by Haber-Schaim, Dodge, Gardner, Shore. Kendall/Hunt Publishing Company, Dubuque, Iowa (C)1991. Page **566:** Aflo Foto Agency/Alamy. Page **575:** B. Benoit/Photo Researchers, Inc. Page **576:** David Parker/Photo Researchers, Inc. Page **581:** NOAA/AP Images. Page **583:** Space Telescope Science Institute.

CHAPTER 21
Page **591** Top: Peter Aprahamian/SPL/Photo Researchers, Inc. Page **591** Middle left: Pearson Science/Creative Digital Vision. Page **591** Middle: Richard Megna/Fundamental Photographs. Page **591** Middle right: Randy Knight. Page **591** Bottom left: Olga Miltsova/Shutterstock. Page **591** Bottom right: Pearson Science/Eric Schrader. Page **593:** Richard Megna/Fundamental Photographs. Page **594:** Prelinger Archives. Page **596:** Education Development Center, Inc. Page **597:** Pearson Science/Creative Digital Vision. Page **603:** Brian Atkinson/Alamy. Page **609:** Pearson Science/Eric Schrader. Page **611:** Richard Megna/Fundamental Photographs. Page **614** Both: Randy Knight.

CHAPTER 22
Page **627** Top: Pxlxl/Dreamstime. Page **627** Middle left: Irina Pusepp/Alamy. Page **627** Middle: Michael W. Davidson. Page **627** Middle right: Dieter Zawischa. Page **627** Middle left: Jane Pang/iStockphoto. Page **627** Bottom left: Shutterstock. Page **627** Bottom middle: langdu/Shutterstock. Page **627** Bottom right: CENCO Physics/Fundamental Photographs. Page **628** Top: Richard Megna/Fundamental Photographs. Page **628** Bottom: Todd Gipstein/National Geographic Stock. Page **635:** Holographix, LLC. Page **636:** Andrew Bargery/Alamy. Page **636:** inset: Jian Zi. Page **639:** Ken Kay/Fundamental Photographs. Page **644:** CENCO Physics/Fundamental Photographs. Page **645** Top: Philippe Plailly/Photo Researchers, Inc. Page **645** Bottom: Rod Nave. Page **651:** Pyma/Shutterstock.

CHAPTER 23
Page **655** Top: 68images.com - Axel Schmies/Alamy. Page **655** Middle left: Ivinst/iStockphoto. Page **655** Middle: Sciencephotos/Alamy. Page **655** Middle right: GIPhotoStock/Photo Researchers, Inc. Page **655** Bottom left: Shutterstock. Page **655** Bottom 2nd from left: Leslie Garland/Alamy. Page **655** Bottom 3rd from left: piluhinAlamy. Page **655** Bottom right: Shutterstock. Page **655** Bottom right: iStockphoto. Page **661:** Pearson Science/Creative Digital Vision. Page **664:** Richard Megna/Fundamental Photographs. Page **669:** Pearson Science/Eric Schrader. Page **670:** Shutterstock. Page **671** Both: Richard Megna/Fundamental Photographs. Page **675:** Getty Images - Stockbyte. Page **683:** Shutterstock. Page **685:** Bajande/Photo Researchers, Inc.

CHAPTER 24
Page **694** Top: Don Hammond/Alamy. Page **694** Middle left: iStockphoto. Page **694** Middle: Shutterstock. Page **694** Middle right: Radius Images/Alamy. Page **694** Bottom left: Shutterstock. Page **694** Bottom middle: Shutterstock. Page **698** Both: Richard Megna/Fundamental Photographs. Page **699** Top: Canon U.S.A., Inc. Page **699** Middle: NASA. Page **699** Bottom: Garden World/Alamy. Page **701:** Tetra Images/Alamy. Page **705** Top: Tetra Images/Alamy. Page **705** Bottom: Biophoto Associates/Photo Researchers, Inc. Page **710:** Dr. Jeremy Burgess/Photo Researchers, Inc.

PART V SUMMARY
Page **717:** WL Delft Hydraulics.

PART VI OPENER
Page **718:** Lucidio Studio/Getty Images.

CHAPTER 25
Page **720** Top: A. T. Willett/Alamy. Page **720** Left: George Resch/Fundamental Photographs. Page **720** Bottom and Right: Shutterstock. Page **729:** Charles D. Winters/Photo Researchers, Inc. Page **737** Both: John Eisele.

CHAPTER 26
Page **750** Top: Shutterstock. Page **750** Bottom: Fotolia. Page **767:** Pacific Northwest National Laboratory.

CHAPTER 27
Page **780** Top: Alfred Benjamin/Photo Researchers, Inc. Page **780** Bottom: Pearson Science/Eric Schrader.

CHAPTER 28
Page **810** Top: NASA. Page **810** Middle left: iStockphoto. Page **810** Middle: iStockphoto. Page **810** Bottom left: NASA. Page **810** Bottom right: Dorling Kindersley. Page **818:** Dorling Kindersley. Page **827:** Tony Freeman/Photo Edit, Inc. Page **828:** Deborah Zemek.

Index

Astronomical Data

Planetary body	Mean distance from sun (m)	Period (years)	Mass (kg)	Mean radius (m)
Sun	—	—	1.99×10^{30}	6.96×10^{8}
Moon	$3.84 \times 10^{8}*$	27.3 days	7.36×10^{22}	1.74×10^{6}
Mercury	5.79×10^{10}	0.241	3.18×10^{23}	2.43×10^{6}
Venus	1.08×10^{11}	0.615	4.88×10^{24}	6.06×10^{6}
Earth	1.50×10^{11}	1.00	5.98×10^{24}	6.37×10^{6}
Mars	2.28×10^{11}	1.88	6.42×10^{23}	3.37×10^{6}
Jupiter	7.78×10^{11}	11.9	1.90×10^{27}	6.99×10^{7}
Saturn	1.43×10^{12}	29.5	5.68×10^{26}	5.85×10^{7}
Uranus	2.87×10^{12}	84.0	8.68×10^{25}	2.33×10^{7}
Neptune	4.50×10^{12}	165	1.03×10^{26}	2.21×10^{7}

*Distance from earth

Typical Coefficients of Friction

Material	Static μ_s	Kinetic μ_k	Rolling μ_r
Rubber on concrete	1.00	0.80	0.02
Steel on steel (dry)	0.80	0.60	0.002
Steel on steel (lubricated)	0.10	0.05	
Wood on wood	0.50	0.20	
Wood on snow	0.12	0.06	
Ice on ice	0.10	0.03	

Melting/Boiling Temperatures and Heats of Transformation

Substance	T_m (°C)	L_f (J/kg)	T_b (°C)	L_v (J/kg)
Water	0	3.33×10^{5}	100	22.6×10^{5}
Nitrogen (N_2)	−210	0.26×10^{5}	−196	1.99×10^{5}
Ethyl alcohol	−114	1.09×10^{5}	78	8.79×10^{5}
Mercury	−39	0.11×10^{5}	357	2.96×10^{5}
Lead	328	0.25×10^{5}	1750	8.58×10^{5}

Properties of Materials

Substance	ρ (kg/m³)	c (J/kg K)
Air at STP*	1.28	
Ethyl alcohol	790	2400
Gasoline	680	
Glycerin	1260	
Mercury	13,600	140
Oil (typical)	900	
Seawater	1030	
Water	1000	4190
Aluminum	2700	900
Copper	8920	385
Gold	19,300	129
Ice	920	2090
Iron	7870	449
Lead	11,300	128
Silicon	2330	703

*Standard temperature (0°C) and pressure (1 atm)

Molar Specific Heats of Gases

Gas	C_P (J/mol K)	C_V (J/mol K)
Monatomic Gases		
He	20.8	12.5
Ne	20.8	12.5
Ar	20.8	12.5
Diatomic Gases		
H_2	28.7	20.4
N_2	29.1	20.8
O_2	29.2	20.9

Indices of Refraction

Material	Index of refraction
Vacuum	1 exactly
Air	1.0003
Water	1.33
Glass	1.50
Diamond	2.42

Resistivity and Conductivity of Conductors

Metals	Resistivity (Ω m)	Conductivity ($\Omega^{-1}m^{-1}$)
Aluminum	2.8×10^{-8}	3.5×10^{7}
Copper	1.7×10^{-8}	6.0×10^{7}
Gold	2.4×10^{-8}	4.1×10^{7}
Iron	9.7×10^{-8}	1.0×10^{7}
Silver	1.6×10^{-8}	6.2×10^{7}
Tungsten	5.6×10^{-8}	1.8×10^{7}
Nichrome	1.5×10^{-6}	6.7×10^{5}
Carbon	3.5×10^{-5}	2.9×10^{4}

Atomic and Nuclear Data

Atom	Z	Mass (u)	Mass (MeV/c^2)
Electron	—	0.00055	0.51
Proton	—	1.00728	938.28
Neutron	—	1.00866	939.57
^{1}H	1	1.00783	938.79
^{2}H	1	2.01410	
^{4}He	2	4.00260	
^{12}C	6	12.00000	
^{14}C	6	14.00324	
^{14}N	7	14.00307	
^{16}O	8	15.99492	
^{20}Ne	10	19.99244	
^{27}Al	13	26.98154	
^{40}Ar	18	39.96238	
^{207}Pb	82	206.97444	
^{238}U	92	238.05078	

Hydrogen Atom Energies and Radii

n	E_n (eV)	r_n (nm)
1	−13.60	0.053
2	−3.40	0.212
3	−1.51	0.476
4	−0.85	0.848
5	−0.54	1.322

Work Functions of Metals

Metal	E_0 (eV)
Potassium	2.30
Sodium	2.75
Aluminum	4.28
Tungsten	4.55
Iron	4.65
Copper	4.70
Gold	5.10